Handbook of Metal-Microbe Interactions and Bioremediation

Handbook of Metal-Microbe Interactions and Bioremediation

Edited by
Surajit Das and Hirak Ranjan Dash

CRC Press
Taylor & Francis Group
Boca Raton London New York

CRC Press is an imprint of the
Taylor & Francis Group, an **informa** business

CRC Press
Taylor & Francis Group
6000 Broken Sound Parkway NW, Suite 300
Boca Raton, FL 33487-2742

First issued in paperback 2020

ISBN-13: 978-1-4987-6242-7 (hbk)
ISBN-13: 978-0-367-65809-0 (pbk)

Library of Congress Cataloging-in-Publication Data

Names: Das, Surajit, editor. | Dash, Hirak Ranjan, editor.
Title: Handbook of metal-microbe interactions and bioremediation /
[edited by] Surajit Das and Hirak Ranjan Dash.
Description: Boca Raton : Taylor & Francis, 2017. | Includes
bibliographical references.
Identifiers: LCCN 2016028316| ISBN 9781498762427 (hardback : alk. paper) |
ISBN 9781498762434 (ebook)
Subjects: | MESH: Metals--toxicity | Metals--metabolism | Environmental
Pollutants--toxicity | Biodegradation, Environmental | Environmental
Pollution | Microbiological Processes
Classification: LCC RA1231.M52 | NLM QV 290 | DDC 615.9/253--dc23
LC record available at https://lccn.loc.gov/2016028316

Visit the Taylor & Francis Web site at
http://www.taylorandfrancis.com

and the CRC Press Web site at
http://www.crcpress.com

Contents

SECTION I Introduction to Metal Contamination and Microbial Bioremediation

SECTION II Metal–Microbe Interactions

SECTION III Molecular Approaches in Metal Bioremediation

SECTION IV Interdisciplinary Approaches for Enhanced Metal Bioremediation

SECTION V Advances in Metal Bioremediation Research

Preface

Both natural and anthropogenic sources contribute to the global pollutants budget every year, which includes recalcitrant, xenobiotics, polycyclic aromatic hydrocarbons, plastics, and toxic metals. However, toxic metals in the environment go unnoticed compared to other toxic elements, and they pose potential hazards on the ecosystems due to their high toxicity and biomagnification at each trophic level. Though metals cannot be removed from the environment completely, they can be transformed or sequestered from one form to another to achieve a lower toxic effect. In this process, microorganisms play an important role due to their small size, low-complexity genetic system, and huge adaptability for the application in the bioremediation of metals. Many studies have been conducted around the world so far for the utilization of microbial entities, that is, bacteria, fungi, algae, and actinomycetes for metal bioremediation practices. However, a complete understanding of the genetic mechanism of resistance to toxic metals in the tiny microorganisms is necessary prior to their suitable application in bioremediation.

To combat toxic metal pollution in the environment, many conventional and advanced techniques have been employed. However, none of them have reached to deliver the optimum output. Popular conventional practices of metal treatment include excavation, physical separation, thermal treatment, hydrometallurgical treatments, leaching and extraction, and electrokinetic separation. However, each of these techniques has its loopholes. Thus, as an alternative, bioremediation practices have been widely exercised, but most of the research on bioremediation of toxic metals is limited to the laboratory scale of treatment. *In situ* and real-time application of bioremediation practice is far beyond the reach of the current research. Thus, a proper understanding of the nature and mode of genetic mechanism of resistance, the propagation of resistant genotype, and the improvement of strains at the genetic level to implement a concrete bioremediation strategy for *in situ* application are needed.

This book brings together most of the current understanding on both *in situ* and *ex situ* metal bioremediation strategies and their utilization in contaminated environments. The chapters have been written by experts who have contributed substantially to answer the fundamentals questions on the basic and practical issues in metal bioremediation process and are currently actively engaged in resolving these issues. Section I of this book introduces the readers to toxic metals, their sources in the environment, toxic effects, current remediation strategies, and the advantage of microbial bioremediation over current bioremediation practices. Section II deals with metal–microbe interaction, genetic machinery, and enzyme-catalyzed metal transformations. Additionally, as an alternative to microbial–metal transformation, biosorption by live cells and metal-binding proteins and microbial extracellular polymer-mediated biosorption are also discussed in further detail. The improvement of the metabolic potential of microbes at the genetic level is discussed in Section III using various processes of genetic manipulation, discovery of novel metabolic pathways, and construction of GMOs. Section IV deals with the exploitation of interdisciplinary approaches of metal bioremediation that includes combinatorial approach of microbial and phytoremediation, immobilization technique, bioreactor-based approach, in silico studies, and bacterial nanoparticle-based bioremediation of toxic metals. Section V describes the recent advances on metal bioremediation practices with respect to the most common toxic metals, that is, arsenic, cadmium, copper, mercury, nickel, lead, zinc, chromium, and radionuclide wastes.

We have tried our best to share the available knowledge around the globe regarding the advancements in bioremediation practices of metals as a key point of reference to everyone involved in the research and development of metal bioremediation practices. The concept and practical aspects described in this book regarding toxic metal bioremediation serve as an important and rationalized resource material. Additionally, a broad perspective toward researchers, students, professionals, policy-makers, and practitioners in metal bioremediation practices with maximum output is highlighted.

Editors

Dr. Surajit Das has been an Assistant Professor at the Department of Life Science, National Institute of Technology, Rourkela, Odisha, India, since 2009. He earlier served the Amity Institute of Biotechnology, Amity University Uttar Pradesh, Noida, India. He received his Ph.D. in Marine Biology (Microbiology) from Annamalai University (Centre of Advanced Study in Marine Biology), Tamil Nadu, India. He has been an awardee of the Endeavour Research Fellowship of the Australian government for carrying out postdoctoral research on marine microbial technology at the University of Tasmania. He has multiple research interests with core research programs on marine microbiology. As the group leader of a research in the Laboratory of Environmental Microbiology and Ecology, he is currently conducting a study on a biofilm-based bioremediation of PAHs and heavy metals by marine bacteria, nanoparticle-based drug delivery, and nanobioremediation and the metagenomic approach for exploring the diversity of catabolic gene and immunoglobulins in Indian major carps, with the help of research grants from the Department of Biotechnology, Ministry of Science and Technology; the Indian Council of Agricultural Research; and the Ministry of Environment, Forests and Climate Change of the government of India. Recognizing his work, the National Environmental Science Academy in New Delhi had conferred to him the 2007 Junior Scientist of the Year award for marine microbial diversity. He is the recipient of the Young Scientist Award in Environmental Microbiology from the Association of Microbiologists of India in 2009. Dr. Das is also the recipient of the Ramasamy Padayatchiar Endowment Merit Award given by government of Tamil Nadu for the year 2002–2003 from Annamalai University. He is a member of the International Union for Conservation of Nature (IUCN) Commission on Ecosystem Management, South Asia, and a life member of the Association of Microbiologists of India, the Indian Science Congress Association, the National Academy of Biological Sciences, and the National Environmental Science Academy, New Delhi. He is the IUCN-nominated expert (observer) on "synthetic biology" of the Biosafety Clearing-House set up by the Cartagena Protocol on Biosafety. He is also a member of the International Association for Ecology. He is an academic editor of *PLOS ONE*, an associate editor (for subjects on ecological and evolutionary microbiology) of *BMC Microbiology*, and an editorial board member (for environmental biotechnology) of the journals *QScience Connect* and *Nusantara Bioscience*. He has written 4 books and authored more than 50 research publications in leading national and international journals.

Dr. Hirak Ranjan Dash received his Ph.D. from the National Institute of Technology (Department of Life Science), Rourkela, India, and is currently working as a Scientific Officer (DNA Fingerprinting) at the Forensic Science Laboratory in Sagar, Madhya Pradesh, India. He also earned his M.Sc. in Microbiology from Orissa University of Agriculture and Technology, Bhubaneswar, Odisha, India. During his research at the Laboratory of Environmental Microbiology and Ecology, he worked on the mercury bioremediation potential of marine bacteria isolated from the Bay of Bengal, Odisha, India. His research interests include molecular microbiology, microbial bioremediation, marine microbiology, microbial phylogeny, genetic manipulation of bacterial systems, and microbial diversity. He has developed a number of microbial techniques for the assessment of mercury pollution in marine environments. The discovery of a novel approach for mercury resistance, that is, intracellular biosorption in marine bacteria, was reported by him. He has successfully constructed a transgenic marine bacterium for enhanced utilization in mercury removal by simultaneous mercury volatilization and sequestration. He has also worked in the field of genotyping and antibiotic resistance mechanism of pathogenic *Vibrio* and *Staphylococcus* spp. He has written 1 book, has published 18 research papers and 7 book chapters, and has 10 conference proceedings in his credit.

Contributors

Fahmi A. Abu Al-Rub
Chemical Engineering Department
Faculty of Engineering, Jordan University of Science &
    Technology
Irbid, Jordan

Ananya Acharya
Department of Biological Sciences
Indian Institute of Science Education and Research (IISER)
    Kolkata
Mohanpur, Nadia, West Bengal, India

Ismael Acosta-Rodríguez
Research Center and Post-grade Studies
Autonomous University of San Luis Potosí
San Luis Potosí, México

Azhar Alhasawi
Faculty of Science & Engineering
Laurentian University
Sudbury, Ontariao, Canada

María Julia Amoroso
Pilot Plant of Industrial and Microbiological Process
    (PROIMI)-CONICET
and
Faculty of Biochemistry
Chemicals & Pharmaceuticals
National University of Tucuman
Tucuman, Argentina

Sharmila Anishetty
Centre for Biotechnology
Anna University
Chennai, Tamil Nadu, India

Varun P. Appanna
Faculty of Science & Engineering
Laurentian University
Sudbury, Ontariao, Canada

Vasu D. Appanna
Faculty of Science & Engineering
Laurentian University
Sudbury, Ontariao, Canada

Bhavna Arora
Earth & Environmental Sciences Area
Lawrence Berkeley National Laboratory
Berkeley, California

Madhurima Bakshi
Department of Environmental Science
University of Calcutta
Kolkata, West Bengal, India

Eva Baldikova
Global Change Research Institute
Academy of Sciences of the Czech Republic
and
Department of Applied Chemistry
University of South Bohemia
Ceske Budejovice, Czech Republic

Mainak Banerjee
Department of Chemistry
Shiv Nadar University
Dadri, Uttar Pradesh, India

Matías R. Barrionuevo
Environmental Chemistry Area
Institute of Sciences
National University of General Sarmiento-CONICET

Irene C. Lazzarini Behrmann
Environmental Chemistry Area
Institute of Sciences
National University of General Sarmiento-CONICET

Kiron Bhakat
Department of Microbiology
University of Kalyani
Kalyani, West Bengal, India

Ram Naresh Bharagava
Department of Environmental Microbiology
Babasaheb Bhimrao Ambedkar University
Lucknow, Uttar Pradesh, India

Chiranjib Bhattacharjee
Department of Chemical Engineering
Jadvapur University
Kolkata, West Bengal, India

Nanthi S. Bolan
Global Centre for Environmental Remediation
The University of Newcastle Callaghan
New South Wales, Australia
and
Cooperative Research Centre for Contamination Assessment
    and Remediation of the Environment
The University of Newcastle Callaghan
New South Wales, Australia

Sutapa Bose
Department of Earth Sciences
Indian Institute of Science Education and Research (IISER)
   Kolkata
Mohanpur, Nadia, West Bengal, India

Nicholas Bouskill
Earth & Environmental Sciences Area
Lawrence Berkeley National Laboratory
Berkeley, California

Eoin Brodie
Earth & Environmental Sciences Area
Lawrence Berkeley National Laboratory
Berkeley, California

Odile Bruneel
HydroSciences Montpellier
UMR HSM 5569 (UM, CNRS, IRD)
University of Montpellier
Montpellier, France

Rosa Olivia Cañizares-Villanueva
Department of Biotechnology and Bioengineering
Cinvestav
Mexico City, México

Juan F. Cárdenas-González
Research Center and Post-grade Studies
Autonomous University of San Luis Potosí
San Luis Potosí, México

Corinne Casiot
HydroSciences Montpellier
UMR HSM 5569 (UM, CNRS, IRD)
University of Montpellier
Montpellier, France

Domenic Castignetti
Biology Department
Loyola University Chicago
1032 West Sheridan Road
Chicago, Illinois

Lucia Cavalca
Dipartimento di Scienze per gli Alimenti la Nutrizione e
   l'Ambiente
Università degli Studi di Milano
Milano, Italy

Ashish Chalana
Department of Chemistry
Shiv Nadar University
Dadri, Uttar Pradesh, India

Jaya Chakraborty
Department of Life Science
National Institute of Technology
Rourkela, Odisha, India

N. Chandrasekaran
Centre for Nanobiotechnology
VIT University
Vellore, Tamil Nadu, India

Gouri Chaudhuri
Centre for Nanobiotechnology
VIT University
Vellore, Tamil Nadu, India

Punarbasu Chaudhuri
Department of Environmental Science
University of Calcutta
Kolkata, West Bengal, India

Nitin Chauhan
Applied Microbiology Laboratory
Centre for Rural Development and Technology Indian
   Institute of Technology Delhi
New Delhi, India

Yiwei Cheng
Earth & Environmental Sciences Area
Lawrence Berkeley National Laboratory
Berkeley, California

Katarzyna Chojnacka
Department of Advanced Material
Wroclaw University of Technology
Wroclaw, Poland

Girish Choppala
Southern Cross GeoScience
Southern Cross University
Lismore, New South Wales, Australia

Pankaj Chowdhary
Department of Environmental Microbiology
Laboratory for Bioremediation and Metagenomics Research
Babasaheb Bhimrao Ambedkar University
Lucknow, Uttar Pradesh, India

Verónica Leticia Colin
Pilot Plant of Industrial and Microbiological Process
   (PROIMI)-CONICET
Tucuman, Argentina

Anna Corsini
Dipartimento di Scienze per gli Alimenti la Nutrizione e
   l'Ambiente
Università degli Studi di Milano
Milano, Italy

F. Costa
Centre of Biological Engineering
University of Minho
Braga, Portugal

Dhilna Damodharan
Department of Chemical Engineering
National Institute of Technology Karnataka
Surathkal, Karnataka, India

María Alejandra Daniel
Environmental Chemistry Area
Institute of Sciences
National University of General Sarmiento-CONICET
Buenos Aires, Argentina

Ikram Dahmani
Laboratory of Microbiology and Molecular Biology
    (LMBM)
FLS, University Mohammed V
Rabat, Maroc

Surajit Das
Department of Life Science
National Institute of Technology
Rourkela, Odisha, India

Tapan Kumar Das
Department of Biochemistry and Biophysics
University of Kalyani
Kalyani, West Bengal, India

Ayusman Dash
Department of Biological Sciences
Indian Institute of Science Education and Research (IISER)
    Kolkata
Mohanpur, Nadia, West Bengal, India

Hirak Ranjan Dash
Department of Life Science
National Institute of Technology
Rourkela, Odisha, India

Siddhartha Datta
Department of Chemical Engineering
Jadavpur University
Kolkata, West Bengal, India

Anna Dawiec
Department of Chemical Engineering
Wroclaw University of Technology
Wroclaw, Poland

Chirayu Desai
P. D. Patel Institute of Applied Science
Charotar University of Science and Technology
Changa, Gujarat, India

Mohammad M. Fares
Chemistry Department
Faculty of Science and Arts, Jordan University of Science &
    Technology
Irbid, Jordan

Lidia Fernandez-Rojo
HydroSciences Montpellier
UMR HSM 5569 (UM, CNRS, IRD)
University of Montpellier
Montpellier, France

María Laura Ferreira
Environmental Chemistry Area
Institute of Sciences
National University of General Sarmiento-CONICET
Buenos Aires, Argentina

Soma Ghosh
Department of Biotechnology
Indian Institute of Technology Kharagpur
Kharagpur, West Bengal, India

Deepak Gola
Applied Microbiology Laboratory
Centre for Rural Development and Technology Indian
    Institute of Technology Delhi
New Delhi, India

Elzbieta Gumienna-Kontecka
Faculty of Chemistry
University of Wrocław
Wrocław, Poland

Anirban Das Gupta
Department of Microbiology
University of Calcutta
Kolkata, West Bengal, India

Ozge Hanay
Department of Environmental Engineering
Firat University
Elazığ, Turkey

Marina Héry
HydroSciences Montpellier
UMR HSM 5569 (UM, CNRS, IRD)
University of Montpellier
Montpellier, France

Maziya Ibrahim
Centre for Biotechnology
Anna University
Chennai, Tamil Nadu, India

Ekramul Islam
Department of Microbiology
University of Kalyani
Kalyani, West Bengal, India

Jaya Mary Jacob
Department of Chemical Engineering
National Institute of Technology Karnataka
Surathkal, Karnataka, India

Kunal Jain
Post Graduate Department of Biosciences
Sardar Patel University
Bakrol, Gujarat, India

Jana Jass
The Life Science Center
The School of Science and Technology
Örebro University
Örebro, Sweden

Gaurav Kaithwas
Department of Pharmaceutical Sciences
Babasaheb Bhimrao Ambedkar University
Lucknow, Uttar Pradesh, India

Shankar Prasad Kanaujia
Department of Biosciences and Bioengineering
Indian Institute of Technology Guwahati
Guwahati, Assam, India

Munther Kandah
Chemical Engineering Department
Faculty of Engineering, Jordan University of Science &
    Technology
Irbid, Jordan

Tiyasha Kanjilal
Department of Chemical Engineering
Jadvapur University
Kolkata, West Bengal, India

Dugeshwar Karley
Water & Steam Chemistry Division
BARC Facilities
Kalpakkam, Tamil Nadu, India

Ramesh Karri
Department of Chemistry
Shiv Nadar University
Dadri, Uttar Pradesh, India

Sufia K. Kazy
Department of Biotechnology
National Institute of Technology Durgapur
West Bengal, India

Eric King
Earth & Environmental Sciences Area
Lawrence Berkeley National Laboratory
Berkeley, California

Erika Kothe
Microbial Communication
Friedrich Schiller University
Jena, Germany

Rajesh Kumar
Water & Steam Chemistry Division
BARC Facilities
Kalpakkam, Tamil Nadu, India

Anitha Kunhikrishnan
Department of Agro-Food Safety
National Academy of Agricultural Science
Wanju-gun, Jeollabuk, Republic of Korea

Elia Laroche
HydroSciences Montpellier
UMR HSM 5569 (UM, CNRS, IRD)
University of Montpellier
Montpellier, France

Joe Lemire
Department of Biological Sciences
University of Calgary
Calgary, Alberta, Canada

Yiwen Liu
Centre for Technology in Water and Wastewater
University of Technology Sydney
Sydney, New South Wales, Australia

Drewniak Lukasz
Laboratory of Environmental Pollution Analysis
Faculty of Biology
University of Warsaw
Warsaw, Poland

Datta Madamwar
Post Graduate Department of Biosciences
Sardar Patel University
Bakrol, Gujarat, India

Shouvik Mahanty
Department of Environmental Science
University of Calcutta
Kolkata, West Bengal, India

Anushree Malik
Applied Microbiology Laboratory
Centre for Rural Development and Technology Indian
    Institute of Technology Delhi
New Delhi, India

Abul Mandal
School of Bioscience
Systems Biology Research Center
University of Skövde
Skövde, Sweden

Neelam Mangwani
National Innovation Foundation-India
Ahmedabad, Gujarat, India

Alfredo de Jesús Martínez Roldán
CONACYT-Tecnológico Nacional de México/Instituto
 Tecnológico de Durango
Dirección de Posgrado e Investigación
Maestría en Sistemas Ambientales
Durango, México

Víctor M. Martínez-Juárez
Autonomous University of Hidalgo State
Agricultural Sciences Institute
Tulancingo de Bravo Hidalgo, México

Kenneth Mbene
Department of Chemistry
Higher Teacher Training College of the University of
 Yaoundé
Yaounde, Cameroon

Orlando Melchy-Antonio
Department of Biotechnology and Bioengineering
Cinvestav
Mexico City, México

María de Guadalupe Moctezuma-Zárate
Research Institute and Post-grade Studies
Autonomous University of San Luis Potosí
San Luis Potosí, México

A.A. Mohamed Hatha
Department of Marine Biology, Microbiology and
 Biochemistry
Cochin University of Science and Technology
Cochin, Kerala, India

Raj Mohan B.
Department of Chemical Engineering
National Institute of Technology Karnataka Surathkal
Karnataka, India

Balaram Mohapatra
Department of Biotechnology
Indian Institute of Technology Kharagpur
Kharagpur, West Bengal, India

John W. Moreau
School of Earth Sciences
University of Melbourne
Parkville, Victoria, Australia

Amitava Mukherjee
Centre for Nanobiotechnology
VIT University
Vellore, India

Goutam Mukherjee
Department of Microbiology
University of Calcutta
Kolkata, West Bengal, India

Samir Kumar Mukherjee
Department of Microbiology
University of Kalyani
Kalyani, West Bengal, India

Aniruddha Mukhopadhyay
Department of Environmental Science
University of Kolkata
Kolkatta, West Bengal, India

Catherine N. Mulligan
Department of Building, Civil and Environmental
 Engineering
Concordia University
Montreal, Quebec, Canada

Karthick Muthuvel
Department of Chemistry
Shiv Nadar University
Dadri, Uttar Pradesh, India

Noor Nahar
Systems Biology Research Center
University of Skövde
Skövde, Sweden

Niloofar Nasirpour
Department of Chemical Engineering
Tarbiat Modares University
Tehran, Iran

Arijit Nath
Department of Chemical Engineering
Jadvapur University
Kolkata, West Bengal, India

Neelu N. Nawani
Microbial Diversity Research Centre
Dr. D. Y. Patil Biotechnology and Bioinformatics Institute,
 Dr. D. Y. Patil Vidyapeeth
Pune, Maharashtra, India

Bing-Jie Ni
State Key Laboratory of Pollution Control and Resources
    Reuse
Tongji University
Yangpu, Shanghai, People's Republic of China
and
Advanced Water Management Centre
The University of Queensland
St. Lucia, Queensland, Australia

Sanaz Orandi
School of Chemical Engineering
University of Adelaide
Adelaide, South Australia, Australia

Malgorzata Ostrowska
Faculty of Chemistry
University of Wrocław
Wrocław, Poland

Jin Hee Park
Geologic Environment Division
Korea Institute of Geoscience and Mineral Resources
Daejeon, Republic of Korea

Dhiraj Paul
Department of Biotechnology
Indian Institute of Technology Kharagpur
Kharagpur, West Bengal, India

Lai Peng
Laboratory of Microbial Ecology and Technology
Ghent University
Ghent, Belgium

Hugo Virgilio Perales-Vela
Dr. Hugo V. Perales Vela Profesor TitularUniversidad
    Nacional Autónoma de México Facultad de Estudios
    Superiores Iztacala Unidad de Morfología y Función
Lab. de Bioquímica

Adriana Rodríguez Pérez
Research Institute and Post-grade Studies
Autonomous University of San Luis Potosí
San Luis Potosí, México

Daria Podstawczyk
Department of Chemical Engineering
Wroclaw University of Technology
Wroclaw, Poland

Marta Alejandra Polti
Pilot Plant of Industrial and Microbiological Process
    (PROIMI)-CONICET
and
Faculty of Natural Sciences and Miguel Lillo Institute
National University of Tucuman
Tucuman, Argentina

Kristyna Pospiskova
Regional Centre of Advanced Technologies and Materials
Palacky University
Olomouc, Czech Republic

Dharmar Prabaharan
National Facility for Marine Cyanobacteria, (Sponsored by
    DBT, Govt. of India)
Department of Marine Biotechnology
Bharathidasan University
Tiruchirappalli, Tamil Nadu, India

Neha Pradhan
Department of Earth Sciences
Indian Institute of Science Education and Research (IISER)
    Kolkata
Mohanpur, Nadia, West Bengal, India

Aminur Rahman
Systems Biology Research Center
University of Skövde
Skövde, Sweden

Silvana A. Ramírez
Environmental Chemistry Area
Institute of Sciences
National University of General Sarmiento-CONICET
Buenos Aires, Argentina

Vijayaraghavan Rashmi
PG and Research
Department of Microbiology
Srimad Andavan Arts and Science College (Autonomous)
Tiruchirappalli, Tamil Nadu, India

T. Subba Rao
Water & Steam Chemistry Division
BARC Facilities
Kalpakkam, Tamil Nadu, India

Martin Reinicke
Microbial Communication
Friedrich Schiller University
Jena, Germany

Gouriprasanna Roy
Department of Chemistry
Shiv Nadar University
Dadri, Uttar Pradesh, India

Rajdeep Roy
Centre for Nanobiotechnology
VIT University
Vellore, Tamil Nadu, India

Rusha Roy
Department of Microbiology
University of Calcutta
Kolkata, West Bengal, India

Jibin Sadasivan
Department of Biological Sciences
Indian Institute of Science Education and Research (IISER)
    Kolkata
Mohanpur, Nadia, West Bengal, India

Ivo Safarik
Department of Nanobiotechnology
Biology Centre
Academy of Sciences of the Czech Republic
Ceske Budejovice, Czech Republic
and
Regional Centre of Advanced Technologies and Materials
Palacky University
Olomouc, Czech Republic

Mirka Safarikova
Biology Centre
Academy of Sciences of the Czech Republic
Ceske Budejovice, Czech Republic

Papita Das
Department of Chemical Engineering
Jadavpur University
Kolkata, West Bengal, India

Pinaki Sar
Department of Biotechnology
Indian Institute of Technology Kharagpur
Kharagpur, West Bengal, India

Angana Sarkar
Department of Biotechnology
Indian Institute of Technology Kharagpur
Kharagpur, West Bengal, India

Keka Sarkar
Department of Microbiology
University of Kalyani
Kalyani, West Bengal, India

Anandakumar Saha
Department of Zoology
University of Rajshahi
Rajshahi, Bangladesh

Pramita Sen
Department of Chemical Engineering
Jadvapur University
Kolkata, West Bengal, India

Samya Sen
Department of Microbiology
University of Kalyani
Kalyani, West Bengal, India

Gülşad Uslu Senel
Department of Environmental Engineering
Firat University
Elazığ, Turkey

Shubhalakshmi Sengupta
Department of Environmental Science
University of Kolkata
Kolkata, West Bengal, India

Balaji Seshadri
Global Centre for Environmental Remediation
The University of Newcastle
Callaghan, New South Wales, Australia
and
Cooperative Research Centre for Contamination Assessment
    and Remediation of the Environment
The University of Newcastle
Callaghan, New South Wales Australia

H. Seshadri
Safety Research Institute
Atomic Energy Regulatory Board
Kalpakkam, Tamil Nadu, India

Ziauddin A. Shaikh
Department of Biochemical Engineering & Biotechnology
Indian Institute of Technology Delhi
New Delhi, India

Seyed Abbas Shojaosadati
Department of Chemical Engineering
Tarbiat Modares University
Tehran, Iran

Sudhir K. Shukla
Water & Steam Chemistry Division
BARC Facilities
Kalpakkam, Tamil Nadu, India

Mohandass ShylajaNaciyar
National Facility for Marine Cyanobacteria, (Sponsored by
    DBT, Govt. of India)
Department of Marine Biotechnology
Bharathidasan University
Tiruchirappalli, Tamil Nadu, India

Alok Kumar Sil
Department of Microbiology
University of Calcutta
Kolkata, West Bengal, India

B. Silva
Centre of Biological Engineering
University of Minho
Braga, Portugal

Sangram Sinha
Department of Botany
Vivekananda Mahavidyalaya
Haripal, Hooghly, India

Griselda Sosa
Environmental Chemistry Area
Institute of Sciences
National University of General Sarmiento-CONICET
Buenos Aires, Argentina

M. Sowmya
School of Environmental Sciences
Mahatma Gandhi University
Kottayam, Kerala, India

Kurtis Stefan
Biology Department
Loyola University Chicago
1032 West Sheridan Road
Chicago, Illinois

Ahalyaa Subramanian
Centre for Biotechnology
Anna University
Chennai, Tamil Nadu, India

Jing Sun
State Key Laboratory of Pollution Control and Resources
    Reuse
Tongji University
Yangpu, Shanghai, People's Republic of China

Agnieszka Szebesczyk
Institute of Cosmetology
Public Higher Professional Medical School in Opole
Opole, Poland

T. Tavares
Centre of Biological Engineering
University of Minho
Braga, Portugal

Sujeenthar Tharmalingam
Faculty of Science & Engineering
Laurentian University
Sudbury, Ontariao, Canada

Raymond J. Turner
Department of Biological Sciences
University of Calgary
Calgary, Alberta, Canada

Lakshmanan Uma
National Facility for Marine Cyanobacteria, (Sponsored by
    DBT, Govt. of India)
Department of Marine Biotechnology
Bharathidasan University
Tiruchirappalli, Tamil Nadu, India

Maribel Vázquez-Hernández
Department of Biotechnology and Bioengineering
Cinvestav
Mexico City, México

Francis Vincent
Water & Steam Chemistry Division
BARC Facilities
Kalpakkam, Tamil Nadu, India

Diana L. Vullo
Environmental Chemistry Area
Institute of Sciences
National University of General Sarmiento-CONICET
Buenos Aires, Argentina

Anna Witek-Krowiak
Department of Chemical Engineering
Wroclaw University of Technology
Wroclaw, Poland

Uhrynowski Witold
Laboratory of Environmental Pollution Analysis
Faculty of Biology
University of Warsaw
Warsaw, Poland

Ashutosh Yadav
Department of Environmental Microbiology
Babasaheb Bhimrao Ambedkar University
Lucknow, Uttar Pradesh, India

Yubo Yan
Global Centre for Environmental Remediation
The University of Newcastle
Callaghan, New South Wales, Australia

and

Jiangsu Key Laboratory of Chemical Pollution Control and
    Resources Reuse
Nanjing University of Science and Technology
Nanjing, Jiangsu, People's Republic of China

Priti Prabhakar Yewale
Microbial Diversity Research Centre
Dr. D. Y. Patil Biotechnology and Bioinformatics Institute,
    Dr. D. Y. Patil Vidyapeeth
Pune, Maharashtra, India

Anita Zalts
Environmental Chemistry Area
Institute of Sciences
National University of General Sarmiento-CONICET
Buenos Aires, Argentina

Seyed Morteza Zamir
Department of Chemical Engineering
Tarbiat Modares University
Tehran, Iran

Karolina Zdyb
Faculty of Chemistry
University of Wrocław
Wrocław, Poland

Section I

Introduction to Metal Contamination and Microbial Bioremediation

1 Metals and Their Toxic Effects
An Introduction to Noxious Elements

Jaya Chakraborty, Hirak Ranjan Dash, and Surajit Das

CONTENTS

ABSTRACT

An escalating rise in urbanization and population dynamics has established a large number of industries releasing toxic metals into the environment. These metals have a wide range of applications in agricultural, domestic, industrial, medical, pharmaceutical, and technological aspects, resulting in their extensive circulation in the environment. Primary threats to the flora, fauna, and human health are associated with the most toxic metals in the environment, that is, arsenic (As), cadmium (Cd), chromium (Cr), lead (Pb), mercury (Hg), and silver (Ag) as reviewed by the World Health Organization (WHO). Biological factors such as species characteristics, trophic interactions, and physiological adaptation play an important role in deducing the presence of metal in the environment. Metal toxicity is dependent on diverse factors such as dosage, route of exposure, presence of chemical species, age, gender, genetics, and nutritional status of the exposed individuals. On exposure, they cause various organ deformities and restrict normal physiological functions of the body. Many conventional strategies have been practiced for reducing their toxicities; however, several challenges have led to limited field implementation of these techniques. The use of microbes is a cheaper alternative for achieving remediation by converting highly toxic metals to less-toxic/nontoxic forms and can also be utilized for large-scale metal removal. Therefore, this chapter summarizes the environmental occurrence, global toxicity, and potential human adversities caused by the most toxic metals in the environment and their strategic conventional and bioremediation treatment approaches.

1.1　INTRODUCTION

With increasing population dynamics and rapid mechanization, there has been a considerable rise in environmental pollution. The liberation of industrial effluents leads to the accumulation of toxic metals and persistent organic compounds in the environment. These contaminants have direct and indirect effects on the flora and fauna, including the aquatic, terrestrial, and atmospheric habitats. Soil, water, and air contaminated with organic compounds and metals have drawn global concern for their eradication, as many of them are mutagenic and carcinogenic. As toxic metals discharged into the environment cannot be degraded completely under natural conditions, they impart many adverse effects on the health of the ecosystem. Thus, metals, their relevant properties, and their relation to environment, agriculture, and human health have been studied with great importance throughout the globe.

A group of elements exhibiting metallic properties with a specific gravity >5 including the transition metals, lanthanides, and actinides are termed as heavy metals (Gadd et al., 1992). Their presence in the environment is dependent on many physical factors like temperature, phase association, adsorption, and sequestration. Chemical factors influencing their presence involve their speciation at thermodynamic equilibrium, complexation kinetics, lipid solubility, and octanol/water partition coefficients (Hamelink et al., 1994). Biological factors such as species characteristics, trophic interactions, and biological/physiological adaptation play an important role in determining the presence of metal in the environment (Verkleji, 1993). Prolonged exposure to heavy metals such as arsenic (As), cadmium (Cd), chromium (Cr), copper, lead (Pb), nickel, mercury (Hg), silver (Ag), and zinc can cause deleterious health effects in humans. On increased metal concentration, they accumulate in the body thereby disrupting the functions of vital organs and glands such as heart, brain, kidneys, bone, and liver. They function by displacing the vital nutritional minerals from specific active sites of certain enzymes hindering their biological function. However, these toxic metals are introduced into the body by consumption of foods, beverages, skin exposure, and the inhaled air. Thus, in order to make the environment healthier and safer, contaminated land and water bodies containing toxic metals need to be remediated by the application of various advanced techniques.

The most commonly used techniques for toxic metal removal from the environment include chemical precipitation, oxidation or reduction, filtration, ion exchange, reverse osmosis, membrane technology, evaporation, and electrochemical treatment (Dixit et al., 2015). However, most of the physicochemical techniques are less efficient at low metal concentrations. Instead, biological methods such as biosorption, bioaccumulation, and biotransformation can be implemented as the striking alternatives to the conventional physicochemical techniques. This chapter summarizes the biomagnification attributes of the toxic metals and their noxious effects on human health with strategic treatment of conventional and bioremediation practices.

1.2 METALS AND ENVIRONMENT

Metal is a solid material with hard, shiny, malleable, fusible, and ductile properties with good electrical and thermal conductivity (Housecroft and Sharpe, 2008). It includes arsenic, gold, silver, aluminum, copper, chromium, cobalt, nickel, mercury, cadmium, lead, and beryllium. Among them, naturally occurring metals with a high atomic weight and density five times greater than that of water are termed as heavy metals. But the International Union of Pure and Applied Chemistry technical report has replaced the term *heavy metal* with *toxic metal* due to the absence of any coherent scientific basis (Duffus, 2002). Based on their toxicity, As, Cd, Cr, Pb, Hg, and Ag are ranked on the top priority for floral, faunal, and public health implications. Toxic manifestations are dependent on dose, route of exposure, chemical species, genetics, and nutritional status of exposed individuals (Tchounwou et al., 2012). Some of them are classified as human carcinogens according to the U.S. Environmental Protection Agency (USEPA) and the International Agency for Research on Cancer (IARC). They are widely used in industries, medical equipment, agriculture, laboratories, metallurgy, microelectronics, etc., accelerating their concentration in the environment. They are important components in automobiles, appliances, tools, computers, and other electronic devices and are also crucial for highways, bridges, railroads, airports infrastructure, electrical utilities, and food production and distribution (Hudson et al., 1999). Incremental increase in their concentration over a specific limit is a potential threat to environment and human health.

The entry of toxic metals into plant, animal, and human tissue occurs via air inhalation, food, and pollution of the environment. As, Cd, Co, Ni, Pb, Sb, V, Zn, Pt, Pd, and Rh are the major airborne metal contaminants. These metals can pollute the groundwater, lakes, streams, and rivers by leaching from industrial and consumer waste. Acid rain aggravates this process by releasing toxic metals trapped in soils. Similarly, plants are also affected by the toxic metals through water uptake (Singh et al., 2011). Animals feeding on these plants get infected, and humans feeding on the plants and animals are again affected in the food chain. Toxic metal contamination also occurs by absorption through skin contact with soil. With a persistent nature, these metals remain accumulated in the organism resisting their removal. Anthropogenic sources like agricultural activities and metallurgical activities of mining, smelting, and metal finishing also release toxic metals into the environment. Industrial waste disposal, energy production, and energy transportation also introduce toxic metals into the environment (Bradl, 2005). They also enter the environment through metal corrosion, deposition, soil erosion, metal leaching, sediment resuspension, and metal evaporation from water resources to soil and groundwater (Nriagu, 1989). Weathering and volcanic eruptions significantly contribute to metal pollution (Shallari et al., 1998). Paper processing, petroleum combustion, burning of coal, plastic textile, and wood preservation also lead to metal accumulation in the environment (Pacyna, 1996; Arruti et al., 2010). Biogeochemical cycle of metals occurs with their transport and mobilization between different biotic and abiotic systems. These cycles interplay between the metal concentration and degradation. Atmospheric residence time of a particular metal is defined as the ratio of the total mass of a chemical in a particular system to either the total emission rate or the total removal rate, under steady-state conditions. Among all toxic metals, Hg and Pb have a greater residence time leading to increased toxicity.

Many remediation approaches like excavation, land fill, thermal treatment, electroreclamation, and soil capping have been developed. These chemical methods depend on extension, depth, and kind of contamination but are an expensive and nonenvironment friendly approach. Therefore, updated methods are being established for the complete reduction of these toxic elements from the environment.

1.3 BIOMAGNIFICATION: ROLE IN INCREASED TOXICITY OF METALS

With increasing industrialization, our ecosystem is threatened by emerging pollutants with adverse effects on vegetation, wildlife, and humans. In a polluted ecosystem, bioaccumulation of toxic metals in organisms is a major integrative indicator of metal exposure (Phillips and Rainbow, 2013). It is defined as the increase in concentration of a substance in an organism with comparison to its food in a single species or in respective trophic levels (Taylor, 1983). It is referred to as secondary poisoning because mass of the pollutant is conserved along the food chain with increasing concentrations. Metal and metalloid bioaccumulation acts as an exposure indicator due to its lack of metabolism in organisms and environment. This is a complex process dependent on multiple exposure routes through various diets. Geochemical effects determine not only variable patterns of metal bioavailability but also various species that affect biomagnification (Chapman et al., 1996; Cain et al., 2004). The bioaccumulation of various toxic metals and their

biomagnification are interlinked, depicted as an example of aquatic ecosystem (Figure 1.1). Metal toxicity is species specific in nature and is dependent on the rate of uptake of metal internally by the organisms (Roesijadi and Robinson, 1994; Wallace et al., 2003).

As, Cd, Pb, and Hg are in the top 10 priority list of the hazardous substances (ATSDR, 1999). In a food chain, from lower to higher trophic levels, organisms accumulate a higher concentration of metal. In a study of roadside samples of arthropods, it was observed that lead was concentrated within the arthropod classified as predators (Price et al., 1974). Another report very well stated that a sample of European mantids concentrated higher amount of metals indicating the insects as bioaccumulator of Pb (Giles et al., 1973). Cd is bioaccumulated within birds, animals, and humans with increasing concentration from primary to secondary and tertiary consumers (Hughes et al., 1980). Cd is highly bioaccumulated by the pelagic seabirds feeding on fish (Nicholson and Osborn, 1983). In addition, snails feeding on detritus also accumulated high levels of Cd (Williamson, 1979). In marine crustaceans, metals are accumulated directly from food, building up metal concentration in the food web (Hook and Fisher, 2001). However, in a food chain, trophic transfer potential (TTP) can be determined from the biodynamic parameters of weight-specific ingestion rate, assimilation efficiency (AE), and rate constant of loss. Among all metals, Hg as methylmercury efficiently biomagnifies and eliminates very slowly in proportion to biomass (Reinfelder et al., 1998). The ultimate sink of toxic metals is noted in the aquatic ecosystems.

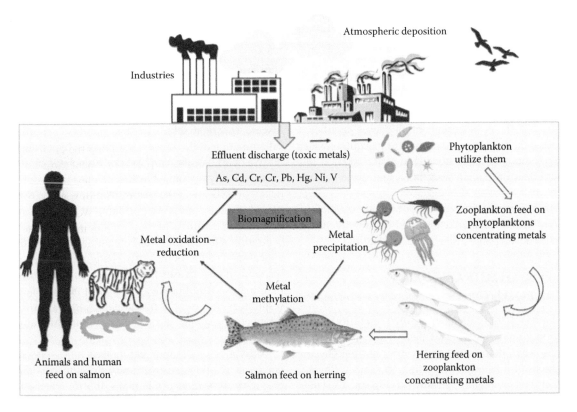

FIGURE 1.1 Bioaccumulation and biomagnification of toxic metals in the aquatic ecosystem.

Fish situated on the top of aquatic food chain accumulates these metals from food, water, and sediments (Yilmaz et al., 2007; Zhao et al., 2012). On consumption of affected fish, its beneficial role is nullified by the effect of toxic metal affecting humans adversely. The major threats include renal failure, damage to liver, cardiovascular diseases, and even death. Therefore, monitoring of the aquatic ecosystem and quality of fish is required before human consumption (Meche et al., 2010; Rahman et al., 2012).

Heavy metal intrusion into body through metal-contaminated food crops poses serious health hazards in animals and human beings (Tripathi et al., 2001). Bioaccessibility rate is dependent on the level of metals present in the body and acts as an index of metal exposure (Khan et al., 2008). For normal human body functioning, permissible levels of metals are required, above which harmful manifestations can occur. Children are more prone to heavy metal stress than adults. Some metals like Cu, Zn, and Fe are essential for the body. But their concentration above the permissible limit induces growth proliferation and cancer with many other defects. Higher concentration of copper restricts oxidation and reduction of copper ions, generating high reactive oxygen species (ROS) adversely modifying proteins, lipids, and nucleic acids (Birben et al., 2012). Similarly, Ni ions render DNA damage including DNA–protein cross-links, DNA strand breaks, and chromosomal aberrations (Armendariz and Vulpe, 2003). Increased dietary intake of Zn, Cr, and Cd leads to breast cancer mortalities (Pasha et al., 2010). Fe, a vital part of proteins, helps in maintaining normal physiological function, but its increase in levels as free iron molecules hampers normal body functioning by disrupting metabolic activities. Bioaccumulation causes the risk of cancer by damaging the tissues by acting as a catalyst for free radical ion formation attacking cellular membranes, DNA strand breaks, enzyme inactivation, lipid peroxidation, and polysaccharide depolymerization (Freeman, 2004). Pb intake in excessive conditions causes cancer of stomach, small intestine, large intestine, ovary, kidney, and lungs; myeloma; lymphoma; and leukemia (Reddy et al., 2004). As bioaccumulation causes significant health hazards in humans by its exposure from drinking water and consumption of As-rich seafood (Liu et al., 2004). It has carcinogenic effect on human skin, lung, and urinary bladder and also causes various skin diseases (Chiou et al., 1995; IPCS, 2001). This overall illustration of biomagnification of several metals with its side effects should be monitored at a global scale for developing various control measures.

1.4 ADVERSE EFFECTS OF TOXIC METAL POLLUTANTS

Bioaccumulation of toxic metals and their inability to get metabolized in human body makes them highly toxic. They cause various organ deformities and restrict normal physiological functions of the body. Metals like As, Cd, Cr, Pb, Hg, and Ag affect various organs as shown in Figure 1.2. Their presence in higher concentrations causes various abnormalities in several organs, whereas their deficiency causes certain imbalances in the human body (Table 1.1). Therefore, a stable balance needs to be maintained for proper metabolism and functioning of all organs and organ systems. The mechanisms of adverse effects by the toxic metals have been described further.

1.4.1 ARSENIC

Arsenic (As) is a well-known carcinogen that affects about 150 million people. This element has an ionic character forming cationic and anionic compounds and is stable in oxidation states −III, 0, +III, and +V. Among these, +III and +V oxidation states are the most common and are extremely toxic. The inorganic forms of As, that is, trivalent arsenite, As(III), and pentavalent arsenite, As(V), are more toxic as well as persistent in nature than the organic forms (Shankar and Shanker, 2014). Generally, high amounts of As contamination are found in groundwater due to natural and anthropogenic sources. Natural sources of As in the environment are primarily due to the result of natural geological local bedrock (Garelick et al., 2008). Additionally, wood preservatives, electronics as semiconductors, medicines and botanicals agriculture products (e.g., fungicides, herbicides, pesticides, and silvicides), nonferrous alloys, desiccants, animal feed additives, glass, and ceramics are the major source of anthropogenic As in the environment. Major affected countries facing As contamination include China, Ghana, Vietnam, Canada, Laos, Mexico, Taiwan, United States, Japan, Bangladesh, India, Chile, and Argentina (Mandal and Suzuki, 2002). The maximum residual limit (MDL) for As in water is 10 µg/L as demarcated by the WHO and the USEPA.

Most of the detrimental effects of As include disruption of general protein metabolism with a higher rate of toxicity by reacting with sulfhydryl groups of existing cysteine residues (Rai et al., 2011). It causes numerous skin diseases and has carcinogenic effects (Das et al., 2014). As causes cancer in lung, urinary bladder, skin, prostate, liver, and nose. It also causes stillbirths, postneonatal mortality, ischemic heart disease (heart attack), diabetes mellitus, nephritis (chronic inflammation of the kidneys), nephrosis, hypertension, hypertensive heart disease, emphysema, bronchitis, chronic airway, obstruction, lymphoma, blackfoot disease, and developmental deficits (USEPA, 2001). As in drinking water over a longer period of time causes As poisoning or arsenicosis. The only preventive measure is to supply drinking water low in As concentration. On drinking As-contaminated water, the physical appearance of the body starts changing such as the color of the skin. It also causes hard patches on the palms and soles of the feet. Additionally, symptoms of weakness, chronic respiratory disease, peripheral neuropathy, liver fibrosis, and peripheral vascular disease also occur (Mazumder et al., 1998; NRC, 1999).

As is a class I human carcinogen and is associated with depletion of global 5-hydroxymethylcytosine (5-hmC) that originates skin cancer in humans. It also induces sex-specific changes in 5-hmC that is an epigenetic marker associated with cancer (Niedzwiecki et al., 2015). Many reports have

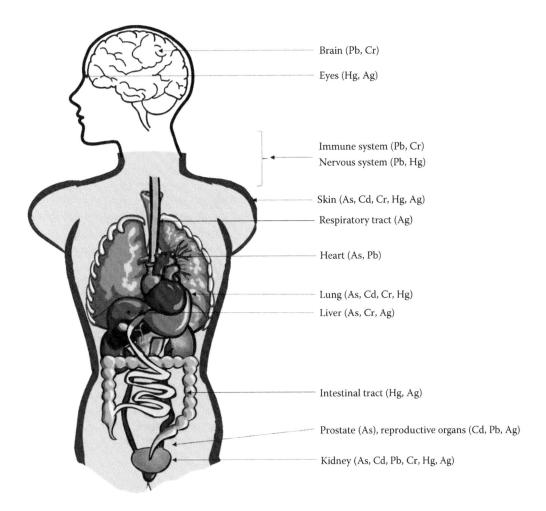

Brain (Pb, Cr)

Eyes (Hg, Ag)

Immune system (Pb, Cr)

Nervous system (Pb, Hg)

Skin (As, Cd, Cr, Hg, Ag)

Respiratory tract (Ag)

Heart (As, Pb)

Lung (As, Cd, Cr, Hg)

Liver (As, Cr, Ag)

Intestinal tract (Hg, Ag)

Prostate (As), reproductive organs (Cd, Pb, Ag)

Kidney (As, Cd, Pb, Cr, Hg, Ag)

FIGURE 1.2 Various organs and organ system affected by toxic metals.

emphasized global DNA methylation in blood on Ar exposure using surrogate markers of DNA methylation, such as long interspersed nucleotide element-1 (LINE-1) (Argos, 2015). As exposure also affects activator protein-1 (AP-1) expression by disturbing the expression of *c-fos* gene and plays an indirect role in fluoride-induced bone toxicity (Zeng et al., 2014). The AP-1 is a transcription factor belonging to the c-Fos, c-Jun, ATF, and JDP families, and its gene expression is regulated in the presence of cytokines, growth factors, stress, and microbial infections (Hess et al., 2004). It regulates many cellular processes including differentiation, proliferation, and apoptosis. As induces epigenetic modifications like DNA methylation in gene promoter regions, regulating gene expression and histone tail modifications. Double capacity of As inducing mutations and epimutations is the primary cause of As-induced carcinogenesis (Bustaffa et al., 2014).

In order to prevent As contamination and counteract its deleterious effect, the use of antioxidant compounds is very useful. Phytochemicals with antioxidant properties help in developing anti-As strategies derived from plants (Ramos and Soria, 2014). Chronic As infection can be treated using chelators that can form a ring structure with the metal or metalloid. In toxic metal poisoning, the administration of the chelating agent forms a chelate structure with the metal with greater

water solubility and is excreted by the kidneys (Mazumder, 2000). Since earlier times, dimercaptosuccinic acid and 2,3-dimercapto-1-propanesulfonic acid are developed as water-soluble analogs of dimercaprol that functions as heavy metal chelators (Petrunkin, 1956; Liang et al., 1957).

1.4.2 Cadmium

Cd in the environment is obtained in the form of a by-product during refining of zinc and other metals as well as agricultural runoffs and mining or battery industries. Chronic effect of Cd exposure first came into recognition as an epidemic of bone disease *itai-itai* in Japan during the 1960s (Nordberg, 1974). The toxicology of Cd is associated with uptake by crops, accumulation in animal tissues, and enrichment of soil and water from anthropogenic sources. The mechanism of Cd toxicity is relatively complex that impairs cell's metal sensing machinery causing it to malfunction and pump out the wrong metal ions including the essential elements (Begg et al., 2015). Cd enters into the environment by volcanic eruption, weathering, erosion, river transports, and human activities like mining, tobacco smoking, smelting, combustion and incineration of municipal waste, batteries of nickel and Cd, phosphate fertilizers, and recycling of electric and electronic

TABLE 1.1

Sources and Adverse Health Effects of Toxic Metals

Element	Sources	Effects	Deficiency	Excess	References
As	Industrial dust and polluted water	Lungs, kidneys, skin, and liver	Not reported	Loss of hair, kidney, liver damage, depigmentation, nephropathy, respiratory and skin cancer, perforation of nasal septum, and dermatoses	USEPA (2001); Kakehashi et al. (2013)
Cd	Industrial dusts and fumes, polluted water and food, Ni–Cd batteries, and fertilizers	Lungs, kidney, bones, blood, and skin	Not reported	Inhibits acetylcholine, hypertension, *itai-itai*, lung impairment, renal and kidney damage, dopaminergic systems in children, prostate dysfunction, anemia, atherosclerosis, aminoaciduria, bone demineralization glucosuria, osteomalacia, and emphysema; affects blood pressure	Bernard (2008); Caciari et al. (2012); Pizent et al. (2012); Evrenoglou et al. (2013)
Cr	Airborne dust emissions, contaminated landfills, tobacco smoke, and chemical plant effluent	Skin, lungs, kidneys, liver, circulatory system, and nervous system of brain	Impaired glucose tolerance and weight loss	Acute tubular necrosis, skin disease, cancer, ulcer, respiratory cancer, hemorrhagic diathesis, toxic nephritis, circulatory collapse, liver damage, acute multisystem organ failure, and coma	Wedeen and Qian (1991); ATSDR (2012)
Pb	Industrial fumes, air, water emissions, and Pb paints	Affects nervous and immune system, kidneys, reproductive and developmental systems, cardiovascular system, and brain	Not reported	Causes high blood pressure, headache neuropathy, encephalopathy, memory loss CNS damage, anemia, peripheral neuropathy, learning with decreased intelligence, attention deficit disorder, and behavior issue	NRDC (2000); USEPA (2015)
Hg	Industrial fumes, vapor, polluted water, coal-fired and chlor-alkali power plants, dental fillings, and polluted fish consumption	Lungs, skin, eyes, gingiva, kidneys, gastrointestinal tract, central nervous system, and thyroid gland	Not reported	Chronic rhinitis and sinusitis, respiratory cancer, dermatitis, neurological and behavioral disorders, tremors, emotional instability, insomnia, memory loss, neuromuscular changes, and headache	Dash and Das (2012); MPCA (2013); WHO (2016)
Ag	Smelting by-products, photographic and electrical supplies, coal combustion, mine tailings, and electroplating	Kidney, skin, gonads, eye, liver, respiratory and intestinal tract	Impaired immunity	Argyria, argyrosis, breathing problems, damage reproductive tissues, respiratory irritation, gastrointestinal disturbances, and renal failure	ATSDR (1990); Eisler (1997); Drake and Hazelwood (2005)

waste (WHO, 2000; UNEP, 2008). Maximum Cd enters the environment from paint, from pigments, from electroplating, and in making polyvinyl chloride plastics. Cd then enters the aquatic organisms by certain crops and agricultural runoffs and accumulates in the food chain (WHO, 2007). The permissible limit of Cd in water is 3 µg/L and in air is 5 ng/m^3 (WHO, 2008). The entry of Cd to the aquatic ecosystems poses threat to the benthic biota and can also enter the water column. Salinity, dissolved organic matter, hydrogen ion concentration, and chemical forms of Cd govern the bioavailability of Cd in aquatic biota. Freshwater fish and some invertebrate fish are mostly affected by Cd. However, aquatic animals respond to Cd by producing metal-binding proteins known as "metallothioneins."

Cd is highly persistent in humans, with a biological half-life of 20–30 years, and is classified as a category I (human) carcinogen. Skeletal deformities, bone loss, kidney damage,

and pain can result if humans are contaminated with Cd. Cd blocks Vitamin D synthesis preventing required calcium deposition in bones making it soft and brittle. It also inhibits the enzymes required for resorption by kidney and excretes glucose, protein, and red blood cells in the urine causing anemia (Wright and Welbourn, 2002; Landis and Yu, 2003). It also causes nephrotoxicity, carcinogenicity, teratogenicity, and endocrine and reproductive toxicities. At the cellular level, Cd affects cell proliferation, differentiation, apoptosis, and other cellular activities. Cd also induces genomic instability by interacting with DNA repair mechanism generating ROS and induces apoptosis (Rani et al., 2014). Cd causes epigenetic changes in DNA expression and regulates transport pathways of kidney tubule. Cd mimics the physiologic action of Zn or Mg, inhibiting heme synthesis, and impairs the function of mitochondria and induces apoptosis. In addition, depletion of glutathione occurs due to binding of Cd to the sulfhydryl group (Valko et al., 2005; Cannino et al., 2009; Moulis, 2010; Schauder et al., 2010; Vesey, 2010; Wang et al., 2012). Cd binds with cysteine, glutamate, histidine, and aspartate ligands leading to the deficiency of iron (Castagnetto et al., 2002). Having same oxidation states, Cd can replace zinc present in metallothionein, thereby inhibiting it from acting as a free radical scavenger within the cell.

1.4.3 CHROMIUM

Cr, an inorganic toxic metal, is a major environmental pollutant and exists in two toxic forms: Cr(III) and Cr(VI). Due to high solubility, mobility dominance, stability, and rapid permeability, Cr(VI) is considered to be more toxic. It is steel-gray, lustrous, hard-crystalline, odorless, and tasteless metal that takes a high polish. It exists naturally as chromite ($FeOCr_2O_3$) ores extensively distributed in air, water, soil, and food. It enters soil and water by erosion and volcanic eruptions. Its concentration in soil ranges between 1 and 300 mg/kg, in seawater 5–800 µg/L, and in rivers and lakes 26 µg/L to 5.2 mg/L (Kotaś and Stasicka, 2000). In India, 2000–3000 tons of Cr is introduced into the environment annually from tannery industries containing 2000–5000 mg/L Cr, which is very high as compared to the recommended permissible limit of 2 mg/L (Belay, 2010).

Cr(VI) compounds have toxic, genotoxic, mutagenic, and carcinogenic effects on humans, animals, plants, and microbes (Mount and Hockett, 2000; Feng et al., 2003; Ackerley et al., 2004). It causes skin irritation, ulceration, eardrum perforation, and lung carcinoma (Madhavi et al., 2013). Cr(VI) has been identified by the USEPA as one of the 17 chemicals posing the greatest threats to humans (Grevatt, 1998). The Institute of Medicine (IOM) of the National Research Council has determined Cr(III) as an essential nutrient for normal energy metabolism. Intake of 20–45 µg Cr(III)/day is adequate for adolescents and adults. In fact, an average plasma Cr concentration of 2–3 nmol/L and an average urinary Cr excretion of 0.22 µg/L are safe as reported by IOM. Chromodulin, referred to as glucose tolerance factor, is a biological target for Cr, although details are still unknown. In rat liver cells, Cr acts as substrate for divalent cation uptake transporter DNMT1 and substitutes Zn on metallothionein or Fe on ferritin. This leads to the substitution of wrong metal cofactor into metalloproteins and disrupts normal cellular functions (Hartwig, 2001; Casalino et al., 2002).

Cr(VI) causes DNA strand breaks, base modifications, and lipid peroxidation, thereby disrupting cellular integrity inducing toxic, as well as mutagenic effects (Mattia et al., 2004). Cr(VI) is used worldwide in a variety of applications, including pigment and textile production; leather tanneries; wood processing; chrome plating; metallurgical and chemical industries; stainless steel factories; welding; cement manufacturing factories; ceramic, glass, and photographic industries; and catalytic converter production for automobiles, heat resistance, and as an antirust agent in cooling plants (Costa and Klein, 2006). Cr(VI) exposure causes higher risk of asthma, skin ulcers, lesions in nasal septum, and respiratory system

cancer. It causes cytotoxicity, genotoxicity, immunotoxicity, reproductive toxicity, allergic dermatitis, and carcinogenic effects in human and laboratory animals (Von Burg and Liu, 1993; Salnikow and Zhitkovich, 2008; Li et al., 2011). Cr(VI) exposure increases the risk of poor semen quality and sperm abnormalities that lead to infertility originating developmental problems in children (Bonde, 1993).

Cr is reduced to form Cr–ascorbate, Cr–glutathione, and Cr–cysteine cross-links depleting cellular antioxidants and disrupting the redox balance in the cell. Cr(VI) induces oxidative stress leading to cytotoxicity and carcinogenicity. Due to high ROS production level, lipid and DNA are directly affected causing lipid peroxidation, DNA damage, and many cellular injuries causing apoptosis and necrosis (Thompson et al., 2011). Metabolic intermediates of Cr(VI), Cr(V), Cr(IV), and Cr(III) readily form Cr–DNA adducts with ascorbate-, glutathione-, histidine-, and cysteine-containing proteins. However, DNA repair plays an important role in antagonizing Cr(VI)-induced DNA damage (Zhitkovich, 2005). Ternary Cr–DNA adducts are strong DNA inhibitors of DNA replication and transcription. In addition, replication inhibition of ternary Cr–DNA adduct requires mismatch repair (MMR) proteins. Therefore, MMR-null mice and human cells became resistant to Cr(VI)-induced toxicity (Reynolds et al., 2009). In addition, Cr(VI) reduction also generates DNA or chromosome lesions with abasic sites, single- and double-stranded breaks, protein–Cr–DNA cross-links, and DNA inter-/intrastrand cross-links (Nickens et al., 2010; Zhitkovich, 2011). It has been reported that p53 point mutations occur at higher frequencies in lung tumors from chromate workers (Kondo et al., 1997). However, orally exposed Cr(VI) animals have very few DNA mutations. Cr(VI) exposure induces genomic changes causing microsatellite instability and chromosome aberrations *in vitro* and *in vivo* (Wise and Wise, 2012).

Cr(VI) induces epigenetic modifications leading to carcinogenicity (Arita and Costa, 2009). The first evidence of abnormally induced DNA methylation by Cr(VI) was observed by Klein et al (2002), which silenced *gpt* transgene. In addition, Cr(VI) inhalation increased DNA methylation in the promoter of tumor suppressor gene P16 that leads to P16 gene silencing. Cr(VI) exposure also targets posttranscriptional histone tail modification and microRNA expression. It results in decreased histone acetylation and increased histone biotinylation and altered histone methylation in both global and gene-specific manner (Zhou et al., 2009; Xia et al., 2014).

1.4.4 LEAD

Pb is naturally found in combination with different elements forming various minerals. It is hazardous, highly toxic, persistent, and a cumulative poison. It is classified as a Group B2 carcinogen (possible human carcinogen) by the IARC. Major sources of Pb exposure include Pb in paints, gasoline, water distribution systems, and food. Air emission from automobile also produces Pb. It also enters soil by flaking, chipping, and weathering of paint. Pb contamination in water occurs in the Pb pipes or connectors, Pb solder connecting pipes and fumes, brass fixtures, and Pb-lined tanks in water coolers. Acidic water also poses serious problems increasing the amount of Pb that will leech from Pb plumbing. Pb in air is emitted by smelters, refineries, incinerators, power plants, manufacturing operations, and many recycling processes.

Prolonged Pb exposure causes anemia, reproductive impairment, renal failure, and neurodegenerative damage (Fowler, 1998; Tong et al., 2000; Lam et al., 2007). It accumulates in bones and other organs. The Centers for Disease Control has defined blood Pb levels of 10 mcg/dL as toxic. The toxic impacts on public health include brain and nervous system damage that causes mental impairment in children. Second, reproductive system interference with premature infants and low births is also a cause of Pb contamination. It induces circulatory disorders by decreasing O_2 absorption with an increase in blood pressure. It also causes behavioral abnormalities, learning impairment, decreased hearing, and impaired cognitive functions in human and experimental animals (Bressler et al., 1999). Recent studies have shown that apoptosis might be associated with Pb-induced oxidative stress and DNA damage in cancer cells (Yedjou et al., 2006, 2010). Bone function is affected directly and indirectly by Pb. Its indirect effect is the modification of bone cell function by altering the circulation levels of hormones, for example, 1, 25-dihydroxyvitamin D3. The bone cell function is directly affected by disturbing them to respond to hormonal regulation. Also, the components of the bone matrix such as collagen or bone sialoproteins are impaired by Pb (Pounds et al., 1991). Pb precedes Hg in toxicity to plants and humans as well as in its occurrence and distribution on globe from among all the heavy metals (ATSDR, 2003; Grover et al., 2010; Shahid et al., 2011). Pb blocks L-type calcium channel and impairs dendritic spine outgrowth in hippocampal neurons in rats. It significantly decreases the mRNA levels of NDR2 and their substrate Rabin3 in developmental rat brain affecting NDR1/2 signal expression (Du et al., 2015).

Pb is highly toxic and affects the nervous system by changing the neurotransmitter levels. It impairs vitamin D metabolism and affects human reproductive system by decreasing the sexual hormonal levels. It causes impaired lymphocyte function and impaired antibody formation. High-dose poisoning in gastrointestinal tract by severe cramping and nausea is a side effect. Pb exposure can produce encephalopathy and signs of ataxia, coma, convulsions, death, hyperirritability, and stupor. Pb inhibits body's ability to make hemoglobin by inhibiting D-aminolevulinic acid dehydratase and ferrochelatase activity. It causes two types of anemia, hemolytic and frank anemia, accompanied by basophilic stripping of the erythrocytes. It interferes the heme synthesis and diminishes red blood cell survival. Pb-causing anemia is hypochromic and normo- or microcytic with associated reticulocytosis. Pb induces osteoporosis in animals (Puzas et al., 1992). The USEPA has classified elemental Pb and inorganic Pb compounds as Group 2B probable human carcinogens (ATSDR, 1999). National toxicology program classifies Pb and its compounds as reasonably anticipated to be carcinogen (NTP, 2004).

1.4.5 Mercury

Hg is a naturally occurring shiny-white metal and odorless liquid and accumulates very fast with various chronic toxicities. It occurs naturally as the insoluble HgS ore (cinnabar), as soluble inorganic complexes of Hg^{+2}, Hg^{+1}, or $(Hg_2)^{2+}$ with counterions such as acetate, nitrate, and halides, and as organomercurials generated by microbial and anthropogenic processes. Hg^0 is stable at specific temperature and pressure in liquid state with high volatile monoatomic vapor (Barkay et al., 2003). Hg exists mainly in three forms: metallic elements, inorganic salts, and organic compounds, each of which possesses different toxicity and bioavailability. Hg is used in batteries, switches, thermostats, electrodes, and medical devices, also including the Hg–Ag amalgam of dental restorations (Crinnion, 2000; Barkay et al., 2003; Richardson et al., 2011). Hg affects marine environment and the primary sources include anthropogenic activities such as agriculture, municipal wastewater discharges, mining, incineration, and discharges of industrial wastewater (Chen et al., 2012). These Hg compounds are consumed by the microorganisms forming methyl Hg, thus undergoing biomagnification disturbing aquatic life. Aquatic animal contaminated with Hg consumed by human is the primary route for human exposure to methyl Hg that is a neurotoxic compound (Trasande et al., 2005).

The U.S. Government Agency for Toxic Substances and Disease Registry has ranked Hg as the third most toxic element on the planet after As and Pb (Rice et al., 2014). Human activities have nearly tripled the amount of Hg in atmosphere, and its burden is increasing 1.5% per year. The target organ for Hg toxicity is the brain, but it also impairs nerves, kidneys, and muscle function. It disrupts the membrane potential and interrupts with intracellular calcium homeostasis. Due to high stability constant, Hg binds to the freely available thiols present in the enzymes (Patrick, 2002). It damages the tertiary and quaternary protein structure and alters the cellular function by attaching to the selenohydryl and sulfhydryl groups undergoing reaction with methyl Hg. Hg interferes with transcription and translation of protein synthesis, resulting in disappearance of ribosomes, eliminating endoplasmic reticulum, and interferes with the activity of natural killer cells. Hg vapors cause bronchitis, asthma, and temporary respiratory problems. Hg induces free radical formation and affects cellular integrity. Due to strong affinity for sulfur ligands, Hg easily binds low-molecular-weight thiols involved in intracellular redox homeostasis (Oram et al., 1996; Valko et al., 2005). Enzymes having cysteine in the active site are hampered in the presence of Hg (Carvalho et al., 2008). It also displaces metal ion cofactors, disturbs structural stability, and forms stable cross-link between intra- and interprotein cysteine residues (O'Connor et al., 1993; Soskine et al., 2002). It also reacts more strongly with selenol of selenocysteine across all domains of life (Khan and Wang, 2009).

Methylmercury is the most toxic and hazardous form of Hg. It causes neurological alterations in humans and experimental animals because it increases the ROS. It causes neurodegenerative diseases like lateral sclerosis, Parkinson's disease, and Alzheimer's disease. It is believed that the mechanisms are related to the toxic increase in ROS (Roulet et al., 1998; Bridges and Zalups, 2010). Hg with a high affinity toward sulfhydryl (−SH) group inhibits microtubular organization in CNS development (Ponce et al., 1994). Binding of −SH group adversely interferes with the intracellular signaling of multiple receptors (muscarinic, nicotinic, dopaminergic) blocking Ca^{++} channels in neurons (Castoldi et al., 2000). Inorganic Hg also increases the permeability of chloride channels of GABA A receptors in the dorsal root ganglion associated with neuronal hyperpolarization (Mottet et al., 1997). Hg exposure has an increased risk of hypertension, myocardial infarction, coronary dysfunction, and atherosclerosis with the risk of developing cardiovascular diseases (Guallar et al., 2002; Fillion et al., 2006). Hg inactivates the enzyme "paraoxonase," therefore increasing LDL oxidation process that has an important anti-atherosclerotic action (Hulthe and Fagerberg, 2002). Methyl Hg induces phospholipase D activation through oxidative stress and thiol-redox alterations (Sherwani et al., 2013). Hg vapors produce harmful effects on digestive, nervous, and immune systems and fatal devising effects on lungs and kidneys.

Inorganic Hg is corrosive to the skin, eyes, and gastrointestinal tract and on ingestion induces kidney toxicity. Inhalation of Hg causes neurological and behavioral disorders causing tremors, insomnia, memory loss, neuromuscular effects, headaches, and cognitive–motor dysfunction. It also causes many reproductive disorders and cardiac effects like hypertension and atherosclerosis in humans (Sørensen et al., 2004; Virtanen et al., 2005; Scheuhammer et al., 2007). In a report by Silva et al. (2005), it was demonstrated that $HgCl_2$ has a sex-specific immunotoxic effect on cytokine production by thymocytes, lymph node cells, and splenocytes in BALB/c mice. It also affects fertilization by limiting the propagation of individual species decreasing their reproductive potential (Gerhard et al., 1998).

1.4.6 SILVER

Ag is a rare, naturally occurring, ductile, and malleable element slightly harder than gold. It has the highest electrical and thermal conductivity and lowest contact resistance having a wide range of applications (Nordberg and Gerhardsson, 1988). Ag occurs commonly as elemental Ag (0 oxidation state) and the monovalent Ag ion (+1 oxidation state). They are utilized for surgical prosthesis, splints, fungicides, coinage, and in the treatment of mental illness, epilepsy, nicotine addition, gastroenteritis, syphilis, and gonorrhea (Gulbranson et al., 2000). Overexposure to Ag nitrate causes acute symptoms like decreased blood pressure, diarrhea, stomach irritation, and decreased respiration. However, chronic symptoms from long-term exposure include fatty acid degeneration of liver and kidneys, blood cell alteration, and argyria

and/or argyrosis (irreversible skin pigmentation) (Nordberg and Gerhardsson, 1988; Gulbranson et al., 2000). Soluble Ag compounds accumulate in small amounts in the brain and muscles (Fung and Bowen, 1996). The target tissues for Ag deposition are the skin, eye, brain, liver, kidney spleen, and bone marrow; however, it is not absorbed into the brain or central or peripheral nervous systems and does not pass the blood–brain barrier (Zheng et al., 2003).

In vitro studies demonstrate the interaction of Ag^+ with the hydroxyapatite complex and can displace calcium and magnesium ions. In addition to that, Ag^+ induces calcium release from the sarcoplasmic reticulum in skeletal muscle by acting on the calcium-release channels and calcium-pump mechanisms through oxidizing sulfhydryl groups. Therefore, bone and cartilage are susceptible to sustainable Ag^+ release as antibiotic in bone cements, orthopedic pins, and dental devices (Tupling and Green, 2002; Lansdown, 2009). Dust laden with Ag nitrate and Ag oxide can cause breathing problems, lung and throat irritation, and stomach pain. Skin contact causes mild allergic conditions, rashes, swelling, and inflammation.

Drinking water containing 2.6 g/L Ag is life-threatening. To monitor the concentration of Ag, the USEPA suggested that the level of Ag in drinking water should not be more than 0.05 mg/L. Later, the USEPA suggested that drinking water levels of Ag should be within 1.142 mg/L (South African Bureau of Standards Commercial, 2005).

1.5 GLOBAL STATUS OF METAL TOXICITY

Toxic metals like As, Cd, Cr, Hg, Pb, and Ag occur naturally in the environment. In addition, anthropogenic activities significantly increase their concentration contaminating the environment with adverse effects on humans such as cardiovascular diseases, developmental abnormalities, neurological and neurobehavioral disorders, diabetes, hearing loss, hematologic and immunologic disorders, and various types of cancer. These metals are systemic toxicants, which address adverse health effects in humans. Eradication of these toxic metals is the primary enterprise for reducing their concentrations; therefore, it is essential to know the global toxicity status of these metals.

1.5.1 ARSENIC

As is ubiquitously present in the environment and is one of the most toxic metals present in the environment. It is also classified as Class I carcinogen by the IARC (IARC, 2004). As inhalation and ingestion exposure causes internal cancers. It is usually present in the drinking water posing threat to human health and ecosystem. Among surface water, rain water, and groundwater, the maximum As contamination is present in the groundwater. As per WHO guidelines, the permissible limit of As in drinking water should not exceed 10 mg/L, but 30 million people from Bangladesh and India have an intake of 50 mg/L. Most human beings relying on drinking water are at higher risk of As exposure than those using surface water (Smedley and Kinniburgh, 2001). Globally, As pollution in groundwater is mostly present in Bangladesh; India (West Bengal); China (Inner Mongolia); California, Nevada, Alaska, and Utah in the United States; Afghanistan; Argentina; Australia; Chile; Korea; Mexico; Nepal; Pakistan; Cambodia; Myanmar; Vietnam; and Taiwan (Garelick and Jones, 2008). In Afghanistan, in Ghazni metropolitan area, 97 of 171 wells tested (total number of wells—3000) did not show As, 74 wells contained As, and around 56 of these had As ranging from 10 to 500 µg/L. The estimated number of people who are potentially at risk is around 500,000. In Argentina, Cordoba contained 100–2000 µg/L As in drinking water. As was present in surface water, shallow wells, and thermal springs in Salta and Jujuy provinces in northwestern Argentina with As levels ranging from 52 to 1,045 µg/L, whereas three thermal springs had As levels between 126 and 10,650 µg/L (De Sastre et al., 1992). These contaminations were natural and related to tertiary–quaternary volcanic deposits, postvolcanic geysers, and thermal springs. Nearly about 100 million people are at high risk of drinking As-contaminated water around India and Bangladesh, which needs to be addressed at the earliest (Chowdhury et al., 2006).

1.5.2 CADMIUM

Cd is a nonessential toxic metal with toxic manifestations on plants, animals, and human health. The global Cd production as well as toxicity has increased four times from 1950 to 1990. The estimated world resource of cadmium is about 6 million tons. Cd pollution is mostly present in China, Japan, United States, Belgium, Canada, Mexico, Kazakhstan, Germany, Russia, and Australia. With a short residence time, in air, it has long-range transport in atmosphere. Atmospheric pollutant Cd occurs mostly in northern European countries like Norway, Denmark, Sweden, Netherlands, Belgium, and Germany. In Norway, $1–10$ ng/m^3 of Cd is present in the urban areas, whereas $1–20$ ng/m^3 in the industrial areas. However, emission up to 100 ng/m^3 also occurs. Mean annual deposition rates of Cd are $0.02–0.08$ mg/m^2 in remote areas, $0.04–0.4$ mg/m^2 in rural areas, $0.2–3.3$ mg/m^2 in urban areas, and $0.8–3.3$ mg/m^2 in industrial areas. In the mid-1970s and 1980s, there was a downward trend in Cd concentration in northern Europe. At the global level, the smelting of nonferrous metal ores has been estimated to be the largest human source of Cd release in the aquatic environment (Nriagu and Pacyna, 1988). Another important source of global Cd pollution represents liquid effluent discharge into fresh and coastal waters produced by air pollution control (gas scrubbing) agents mixed with site drainage waters (Nriagu and Pacyna, 1988). A GESAMP study of the Mediterranean Sea indicated that this source was comparable in magnitude to the total river inputs of Cd to the region (GESAMP, 1985). Similarly, large Cd inputs to the North Sea (110–430 tons/year) have also been estimated, based on the extrapolation from measurements of Cd deposition along the coast (Van Aalst et al., 1983). However, another approach based on

model simulation yielded a modest annual Cd input of 14 tons (Krell and Roeckner, 1988).

The average Cd content of seawater has been given as about 0.1 µg/L or less (Korte, 1983). WHO (1992) reported that current measurement of dissolved Cd in surface waters of the open oceans is <5 ng/L. The vertical distribution of dissolved Cd in ocean waters is characterized by surface depletion and deep water enrichment, which corresponds to the pattern of nutrient concentration in these areas (Boyle et al., 1976). This distribution is considered to be the result of the absorption of Cd by phytoplankton in surface waters and its transport to the depths, incorporation to biological debris, and subsequent release. In contrast, Cd is enriched in the surface water of areas of upwelling, and this also leads to elevated levels in plankton (Martin and Broenkow, 1975; Boyle et al., 1976). Oceanic sediments underlying these areas of high productivity can contain markedly elevated Cd levels as a result of inputs associated with biological debris (Simpson, 1981). Cd levels of up to 5 mg/kg have been reported in river and lake sediments and from 0.03 to 1 mg/kg in marine sediments (Korte, 1983).

1.5.3 CHROMIUM

Tanning industry releases large amount of Cr mostly operating in low- and middle-income countries. The percentage of these countries contributing to light and heavy leather materials increased from 35% to 56% and 26% to 56%, respectively, between 1970 and 1995 (Jenkins et al., 2004). In Hazaribagh, a large tanning region of Bangladesh having 200 separate tanneries is estimated to dispose 7.7 million liters of wastewater and 88 million tons of solid waste annually. This contaminated nearly all the surface and groundwater systems with high levels of Cr. The regions mostly impacted by Cr pollution are South Asia, Central America, South America, and Africa (Bhuiyan et al., 2011). Pakistan showed a Cr contamination ranging from <0.001 to 9.8 mg/L with the highest pollution of well water from Kasur and Punjab province (Tariq et al., 2008). The surface water contamination was found in the range of 0.16–0.29 mg/L in Bara River, Nowshera, KPK (Nazif et al., 2006). A high content of Cr was observed in leaf and edible portion of vegetables, that is, 3.74 and 7.56 mg/kg in the area of Peshawar, Pakistan (Perveen et al., 2012). A significantly higher concentration of Cr (3.93 mg/kg) was reported in spinach irrigated with wastewater containing higher Cr content in Hassanabdal area, Pakistan, while irrigation with clean water results in 0.004 mg/kg Cr concentration in same vegetables (Lone et al., 2003).

1.5.4 LEAD

Pb is highly toxic and persistent in the environment, significantly inducing motor dysfunction and cognitive impairment in children. Its poisoning has been reported 5000 years ago in Iran, and still its preventive measures are being invented (Karrari et al., 2012). Mining industries release

Pb contaminating water. Pb has its higher incidence in the United States (Landrigan and Todd, 1994). The WHO expresses the limit for blood Pb levels (BLL) as 1.9 µmol/L (40 µg/dL) for men and 1.4 µmol/L (30 µg/dL) for women of child-bearing age (Abdollahi et al., 1996). Pb contamination has also been observed in Pearl Valley of Jammu and Kashmir that ranged between 1.8 and 4.7 mg/L (Javaid et al., 2008). According to the WHO guidelines, Pb concentrations were seen to be 466 times higher in South Jammu and Kashmir (WHO, 2011). Punjab was also reported to be contaminated with Pb (Ullah et al., 2009). The highest Pb contamination of 2.34 mg/L was reported in Hattar Industrial State representing the water as most hazardous for soil, plants, and human beings (Manzoor et al., 2006). The highest level of Pb (121 mg/kg) was observed in various regions of Pakistan in the coastal sediments of the Arabian sea along the urban Karachi (Siddique et al., 2009) followed by 49.5 mg/kg from surficial sediments of Lyari River (Mashiatullah et al., 2013), with Islamabad facing a serious problem of Pb toxicity. It has become an important issue of public health for children in China affecting one-quarter of the population in China. Pb pollution and its effects on children have become an important issue of public health in China (Hua et al., 2005).

1.5.5 MERCURY

Hg is a toxic metal naturally present in the environment. It is a global pollutant because of its ability to be transported over long distances. Hg, the only metal that is liquid at room temperature, is a highly potent neurotoxin. The metal does not degrade and builds up in the food chain, as it is absorbed by flora and fauna. China is top among all the Asian countries emitting 28% of global Hg. Others include India, Japan, Kazakhstan, and North Korea (Li et al., 2009). It has been observed that China, Japan, India, and Kazakhstan release 604.7, 143.5, 149.9, and 43.9 mg of Hg, respectively (Streets et al., 2005). The WHO classifies Hg as one of the top 10 chemicals of major public health concern that has been daunting the international community for years. In 2010, 1960 metric tons of Hg was emitted globally. It has also been notified by the USEPA that U.S. sources are responsible for about 3% of global Hg emissions, of which 60% of Hg deposits are from domestic and man-made sources and 30% are generated by power plants. It also reported that the U.S.-based sources account for 37% of total Hg deposition in Georgia, 58% in North Carolina, 62% in South Carolina, and 68% in Florida (Dufault et al., 2009). Mercury contaminated hotspots include Minnesota and Northern Wisconsin, with a level of Hg deposition of 12.5 and 10 µg for 2 years, respectively (Engstrom et al., 1994). UNEP (2008) reported that the overall annual release of Hg in Europe was 145.2 mg/L, out of which the highest contribution, 52%, is from stationary combustion sources. The rest of the 38% is by the industrial processes such as chlor-alkali, ferrous and nonferrous metals, and cement production. Remaining 10% is from waste incinerators and emissions from various Hg uses (Pirrone et al., 2010). Due to

incessant rise in human population dynamics, increased economic growth, and demand in energy, Asian countries are heading high toward global Hg burden than Europe and North America (Pirrone et al., 2010). Hg is distributed globally and attaining regions of Greenland roughly with 13 metric tons of Hg. To protect global population and environment, 97 countries have signed the Minamata Convention on Hg, to protect human health and the environment from anthropogenic emissions and releases of Hg and Hg compounds.

1.5.6 SILVER

The toxicity, chemistry, and bioavailability of Ag make it a concerned environmental pollutant (Luoma et al., 1995). It ranks second after Hg pollution (Ratte, 1999). It largely affects marine habitats as it primarily enters the marine environment from mining, photographic industry, combustion of wastes, cloud seeding, and electronic applications (ATSDR, 1990; Lam and Wang, 2006). Ag exceeds in concentration in drinking water. The average Ag concentration in natural water is 0.2–0.3 µg/L. In the United States, Ag levels in drinking water untreated with disinfection process varied between "nondetectable" and 5 µg/L. It was also reported in a survey of Canadian tap water, in which only 0.1% of the samples contained more than 1–5 ng of Ag/L (USEPA, 1980). Ag contamination is present in North Pacific waters, probably from industrial emissions in Asia. The University of California, Santa Cruz researchers have measured Ag concentrations 50 times greater than the natural background level. It has also been observed that Ag concentrations in the North Pacific trace the atmospheric depositions of industrial aerosols from Asia, with the highest concentrations in those waters closest to the Asian mainland. With rising industrialization, atmospheric pollution from Asia is becoming a serious problem for the western United States, and this may be a valuable tool for tracing those emissions, both in the atmosphere and in water. In San Francisco Bay, Ag and copper from industrial discharges reached high levels in the 1970s and early 1980s that stopped reproducing of some important primary marine organisms and some species of invertebrates disappeared from mudflats (Stephens, 2005).

1.6 STRATEGIES FOR METAL TREATMENT AND BIOREMEDIATION

Improving technologies and progress in industrial arena is leading to toxic metal contamination in water, soil, and sediments by industrial and natural processes. This is a global problem leading to acute and chronic toxicities in plants, animals, and human beings causing negative impacts on the ecosystem (Rathnayak et al., 2009). Toxic metals lacking the property of degradation are threat to the environment, as they pollute the ecosystem by accumulating in various niches of food chain (Ashraf and Ali, 2007). Therefore, immediate action is required to combat their deleterious effect on the ecosystem.

1.6.1 STRATEGIES FOR METAL TREATMENT

The most promising conventional techniques for remediation involve chemical precipitation, conventional coagulation, and adsorption by activated carbons, adsorption by natural materials, ion exchange, and reverse osmosis (USEPA, 2007). Other techniques used are containment, extraction, oxidation–reduction, stabilization/solidification, surfactant-enhanced aquifer remediation, immobilization, soil washing, air stripping, precipitation, vitrification, thermal desorption, membrane separation, solid vapor extraction, electrochemical treatment, and *in situ* oxidation (Farhadian et al., 2008). Electrokinetic remediation is another effective measure for the extraction of contaminants. However, several challenges have led to limited field implementation of these techniques. The reasons include lower risk of exposure to adsorbed contaminated area, cost of treatment, technical challenges, complex environmental parameters, and the need for acidification to induce desorption (Alshawabkeh, 2009). The treatment of maximum contaminant levels in water and soil by the present techniques is unrealistic and cost prohibitive at

many sites, and more realistic goals are essential to make progress. Therefore, the use of microbes is a cheaper alternative for remediation that converts toxic metals to less-toxic/nontoxic forms and also can be used for *in situ* large-scale metal removal. This *in situ* technique is safe and effective as compared to other physicochemical techniques (Eccles and Wigfield, 1995).

Microbes have unique characteristic features making them potential candidates for bioremediation. Microbes, namely, algae, bacteria, and fungi, can thrive in diverse terrestrial, aquatic, and marine environments with various genetic constitutions providing them durability in adverse conditions (Das et al., 2016). Bacteria with the distinctive characteristics of small size, high surface area to volume ratio, efficient transfer of genetic traits, and adaptability make them the effective aspirants for metal bioremediation (Figure 1.3).

The primary attribute for metal removal is the use of bacteria that are highly resistant to these toxic metals. Some bacteria have the specific resistance and efflux genes helping them to survive the metal stress conditions. The various

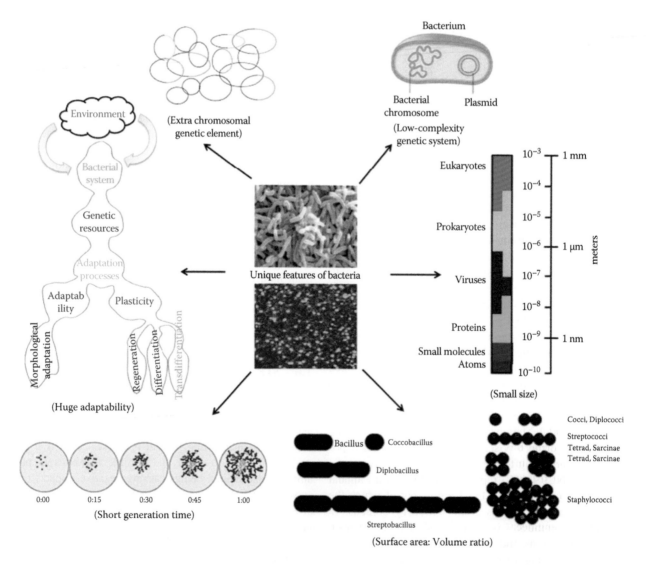

FIGURE 1.3 Certain unique features of bacteria as versatile and potential candidates for use in bioremediation technology.

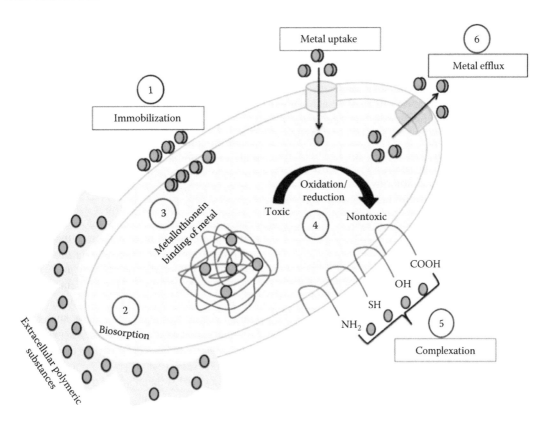

FIGURE 1.4   Characteristic features for metal–microbe interaction in bacteria required for toxic metal bioremediation.

mechanisms include biosorption and sequestration of metals on bacterial cell walls via complexation, coordination, chelation mechanisms, ion-exchange adsorption, and inorganic microprecipitation (biocrystallization), intracellular assimilation, precipitation, immobilization, and change in metal speciation levels by oxidation–reduction reactions (Mullen et al., 1989; Stelting et al., 2010) (Figure 1.4). For large-scale environmental cleanup, suitable assessment and full-scale ecosystem interactions should be studied for their implementation for effective remediation without any adverse environmental impact. Factors like magnitude, toxicity, mobility of contaminants, proximity of human and environmental receptors, planned site use, and ability to properly monitor are important and should be taken into consideration. *In situ* techniques of biosparging, bioventing, and bioaugmentation and *ex situ* techniques of land farming, composting, biopiling, use of bioreactors, flocculation, microfiltration, and electrodialysis are commonly used for toxic metal bioremediation.

The synergistic association of microbes in a microbial consortium instead of pure cultures is used for improved and speedy toxic metal bioremediation. No matter whatever enzymes are involved, metal remediation is very challenging due to its persistent character. Therefore, engineered microorganisms have come to the forefront that have modulated genetic makeup for enhanced toxic metal bioremediation. Genetic engineering of metal sequestration gene into metal-resistant bacteria is a strategy used for enhanced bioremediation potential. Reports of insertion of the *bmtA* gene encoding metallothionein into a suitable vector and its

transformation into marine bacteria have been conducted and successfully employed in highly metal-contaminated environments (Naik and Dubey, 2011). Kivela et al. (2008) reported *Pseudoalteromonas haloplanktis* possessing a shuttle plasmid encoding suppressor for amber mutation that has been used for genetic manipulation for bioremediation. In another study, *merA* gene from *Bacillus thuringiensis* was transformed into another Hg accumulating *Bacillus cereus* for simultaneous volatilization and accumulation of Hg (Dash et al., 2014; Dash and Das, 2015).

Biofilm- and biosurfactant-mediated remediation of toxic metals is also a strategy for enhanced removal of toxic metals. Biofilms are multilayered, 3D sessile structures encapsulated in hydrated extracellular polymeric substances (EPSs) on a substratum (Priester et al., 2007). It is a rich matrix of polysaccharides, proteins, and nucleic acids (Jain et al., 2013). This rich exopolymer content of the biofilms can entrap and biosorb positively charged metals leading to metal precipitation and removal. The presence of EPSs has active sites for metal binding (Volesky, 1990). Likewise, biosurfactants also aid in metal remediation by ion exchange, precipitation–dissolution, and counter binding. They are extensive group of structurally diverse surface-active compounds with a hydrophilic part, consisting of amino acid or peptide anions or cations, mono- or polysaccharides, and a hydrophobic part consisting of saturated or unsaturated fatty acids produced by a variety of microorganisms (Rosenberg and Ron, 1999). Biosurfactant-assisted removal of heavy metal ions by complex formation and succeeding mobilization has received much interest.

1.6.2 Bioremediation

Microorganisms play vital role in environmental restoration by removing toxic metals with multiple mechanisms like transformation between toxic and nontoxic forms. Many reports have suggested the utilization of genetic machineries of As-resistant bacteria for application in bioremediation of As-contaminated environments. *Marinomonas communis* is the first nongenetically engineered potent As-accumulating bacterium accumulating 2290 μg As/g (dry weight) that is the highest value ever reported for any strain (Takeuchi et al., 2007). Other bacterial strains belonging to the genera *Aeromonas*, *Bacillus*, and *Pseudomonas* indigenous to polluted sediments of the Orbetello Lagoon have also been reported showing tremendous potential of As resistance as well as bioremediation (Pepi et al., 2007). Additionally, a recent study on bioremediation of As by bacteria concluded that As-resistant bacteria are widespread in the polluted soil environments that are valuable candidates for bioremediation of As-contaminated ecosystems (Ghodsi et al., 2011). Many γ-proteobacterium and *Firmicutes* isolated from As-affected area showed the potential to oxidize arsenite to arsenate as well as to reduce arsenate to arsenite (Banerjee et al., 2011).

Owing to the high level of toxicity and increasing level of Cd pollution around the globe, many Cd-resistant bacteria have been discovered to act as rescue to fight with the toxic metal pollution. Indigenous Cd-resistant *Pseudomonas aeruginosa* has been mutated to increase its Cd removal efficiency from 38.68% to 53.58% (Kermani et al., 2010). Additionally, a marine *Bacillus safensis* has also been reported for the reduction of Cd at a level of 98.10% (Priyalaxmi et al., 2014). Another strain of *P. aeruginosa* KUCd1 exhibits high Cd accumulation under *in vitro* aerobic conditions by removing >89% of soluble Cd suggesting its utility in Cd bioremediation (Sinha and Mukherjee, 2009). A recent study revealed the practical utility of Cd-resistant bacteria in application of reduced uptake of Cd in rice plants by 61% (Siripornadulsil and Siripornadulsil, 2013). Both biomass and EPSs of a biofilm-forming marine bacterium *P. aeruginosa* JP-11 have been reported to remove 29.5%–58.7% Cd by expressing *czcABC* genes (Chakraborty and Das, 2014). Thus, Cd-resistant bacteria from the unexplored environments may be exploited further for their application in the removal of Cd from the contaminated environments.

The presence of chromate-reducing enzyme and fast reduction of Cr(VI) in Cr-resistant *Lactobacillus* strains have proven to be useful for Cr detoxification from gastrointestinal tract as well as bioremediation of Cr(VI) from contaminated environment (Mishra et al., 2012). Many bacterial strains such as *Enterobacter* sp. and *Pseudomonas* sp. have also been isolated and utilized in Cr(VI) reduction under anaerobic conditions (Kamaludeen et al., 2003). Cell-free extracts of *Bacillus* sp. JDM-2-1 and *Staphylococcus capitis* have also shown the reduction of 83% and 70% of Cr(VI), respectively, with the active participation of a 25 kDa inducible protein suggesting its utility in bioremediation of toxic Cr (Zahoor and Rehman, 2009). Other bacterial strains such as *B. pumilus*, *A. faecalis*, and *Staphylococcus* sp. have also

shown promising result for Cr(VI) reduction at a level of 95%, 97%, and 91%, respectively, from Cr-containing wastewater (Shakoori et al., 2010).

Metallothioneins in Pb-resistant bacteria have been exploited extensively to study the bioaccumulation of Pb in bacterial cells. *P. aeruginosa* WI-1 resisting 0.6 mM Pb nitrate has been reported to accumulate 26.5 mg Pb/g dry weight of cells. Additionally, a significant intracellular accumulation of Pb was found at a level of 19 mg and 22 mg Pb per gram dry weight in *Salmonella choleraesuis* strain 4A and *Proteus penneri* strain GM10, respectively (Naik et al., 2012a). Other bacterial strains such as *B. megaterium*, *P. aeruginosa*, *S. choleraesuis*, and *P. penneri* bioaccumulating high amount of Pb have also been exploited for application in bioremediation of Pb-contaminated environments (Naik et al., 2012b). Reclamation of Pb by using microbial precipitation method has also been found to be effective, eco-friendly, and affordable (Naik and Dubey, 2013). In another approach, many Pb-precipitating bacteria have been studied for their Pb bioremediation potential. *Vibrio harveyi* has been reported to precipitate as $Pb_9 (PO_4)_6$ (Mire et al., 2004). *Providencia alcalifaciens* 2EA was also found to precipitate Pb(II) (Naik and Dubey, 2013) that showed the similar mechanism as that of *Klebsiella* sp. (Aiking et al., 1985).

Many bacterial species have been isolated and characterized throughout the globe for their Hg resistance genotype and Hg removal efficiency to be suitably applied in bioremediation. A strain of *Pseudomonas putida* isolated from Hg-contaminated soil has been reported to possess the potential of Hg removal at a magnitude of 85.2% with optimum conditions of pH 7.0 and 30°C (Xu et al., 2012). Further, a simple, robust, and effective biotechnology for Hg remediation in polluted wastewater has been demonstrated by Wagner-Döbler et al. (2000). When pure culture of seven Hg-resistant *Pseudomonas* strains was immobilized in a carrier material of packed bed bioreactor, Hg retention efficiency of 97% was obtained within 10 h of inoculation. This suggests the utility and efficacy of Hg-resistant bacteria for bioremediation application *in situ*. Both *P. aeruginosa* and *Klebsiella pneumoniae* have been demonstrated for their Hg removal potential from synthesized water polluted by Hg in fixed and fluidized bed bioreactor to obtain 100% Hg cleanup rate (Dzairi et al., 2004). The *mer* operon–mediated volatilization has been demonstrated to remove >90% of inorganic Hg in a marine bacterium *B. thuringiensis* PW-05 (Dash et al., 2014). Additionally, Hg biosorption has also been successfully employed to sequester Hg from the contaminated environment at a level of 40.06% by *B. cereus* BW-03 (De et al., 2014). The same bacteria were constructed transgenic *B. cereus* BW-03(pPW-05) with Hg biosorption capability, by transforming a plasmid harboring *mer* operon of a marine bacterium *B. thuringiensis* PW-05. This was able to remove >99% of Hg supplement *in vitro* by simultaneous volatilization (>53%) and biosorption (~40%) (Dash and Das, 2015). Thus, Hg-resistant bacteria and bacterial *mer* operon should be exploited further to obtain a consensus strategy of Hg removal from the contaminated environments.

Ag uptake and removal was reported by a *Pseudomonas diminuta* in a chloride-free medium containing Ag ions (50 mM) in batch culture. It showed higher accumulation of Ag inside the cell during early exponential phase as compared to the amount bound at the cell surface suggesting possible metal uptake during bacterial growth (Ibrahim et al., 2001). Ag remediation was reported by four EPSs producing marine bacteria that showed the accumulation of 333 mg/g EPS (Deschatre et al., 2013).

1.7 CONCLUSION AND FUTURE PROSPECTS

Environmental pollution by toxic metals is expanding in each century and demands a global solution to this problem. As, Cd, Cr, Pb, Hg, and Ag are systemic toxicants causing adverse diseases of cardiovascular abnormalities, neurologic and neurobehavioral disorders, developmental and reproductive abnormalities, immunologic and hematologic disorders, and tumorigenesis. The presence of metal in various speciations plays a significant role in metal toxicokinetics. To elucidate advancing ways of toxic metal control, determination of metal interaction with human is prerequisite to trace out the interaction and possibilities of metal removal from the environment. The use of conventional techniques for the removal of metals is widely practiced, but they are not productive, cheap, and environmental friendly. Thus, microbes are used as potential candidates of bioremediation that can adapt quickly to the changing noxious environment and be utilized for toxic metal remediation. A large number of microbial biomass is yet to be explored from the uncultured plethora with more effective functions in bioremediation. These genes can be cloned and expressed in culturable fractions for achieving increased bioremediation of toxic metals. Thus, an in-depth understanding of community structure and metabolic functioning of indigenous microbial communities is required for improved biological treatment strategies. Therefore, understanding on molecular aspects of diverse microbes is essential for engineering them to optimize their enzyme production, metabolic pathways, and growth conditions to meet the demand for toxic metal removal. Understanding of metal chemistry with human and metal remediation by biological means by characterizing new microbial communities will be a massive technological leap toward sustainable development. With modern scientific upliftment, the emergence of specialized techniques for bioremediation and cleanup of contaminated sites and effluents can be designed further by monitoring the genome, transcriptome, and proteome modification approach of the *in situ* culturable as well as unculturable microbes.

REFERENCES

Abdollahi, M., Sadeghi Mojarad, A.L.I., Jalali, N. 1996. Lead toxicity in employees of a paint factory. *Medical Journal of the Islamic Republic of Iran* 10(3): 203–206.

Ackerley, D.F., Gonzalez, C.F., Keyhan, M., Blake, R., Matin, A. 2004. Mechanism of chromate reduction by the *Escherichia coli* protein, NfsA, and the role of different chromate reductases in minimizing oxidative stress during chromate reduction. *Environmental Microbiology* 6(8): 851–860.

Agency for Toxic Substances and Disease Registry (ATSDR). 1990. Toxicological profile for Ag. U.S. Department of Health and Human Services, Public Health Service, Atlanta, GA.

Agency for Toxic Substances and Disease Registry (ATSDR). 1999, 2003. Priority list of hazardous substances. U.S. Department of Health and Human Services. Public Health Service. Atlanta. GA.

Agency for Toxic Substances and Disease Registry (ATSDR). 2012. Toxicological profile for chromium.

Aiking, H., Govers, H., Van't Riet, J. 1985. Detoxification of mercury, cadmium, and lead in *Klebsiella aerogenes* NCTC 418 growing in continuous culture. *Applied and Environmental Microbiology* 50(5): 1262–1267.

Alshawabkeh, A.N. 2009. Electrokinetic soil remediation: Challenges and opportunities. *Separation Science and Technology* 44(10): 2171–2187.

Argos, M. 2015. Arsenic exposure and epigenetic alterations: Recent findings based on the illumina 450 K DNA methylation array. *Current Environmental Health Reports* 2(2): 137–144.

Arita, A., Costa, M. 2009. Epigenetics in metal carcinogenesis: Nickel, arsenic, chromium and cadmium. *Metallomics* 1(3): 222–228.

Armendariz, A.D., Vulpe, C.D. 2003. Nutritional and genetic copper overload in a mouse fibroblast cell line. *Nutrition* 133: 203–282.

Arruti, A., Fernández-Olmo, I., Irabien, A. 2010. Evaluation of the contribution of local sources to trace metals levels in urban PM2.5 and PM10 in the Cantabria region (Northern Spain). *Journal of Environmental Monitoring* 12(7): 1451–1458.

Ashraf, R., Ali, T.A. 2007. Effect of heavy metals on soil microbial community and mung beans seed germination. *Pakistan Journal of Botany* 39(2): 629.

Banerjee, S., Datta, S., Chattyopadhyay, D., Sarkar, P. 2011. Arsenic accumulating and transforming bacteria isolated from contaminated soil for potential use in bioremediation. *Journal of Environmental Science and Health, Part A* 46(14): 1736–1747.

Barkay, T., Miller, S.M., Summers, A.O. 2003. Bacterial mercury resistance from atoms to ecosystems. *FEMS Microbiology Reviews* 27(2–3): 355–384.

Begg, S.L., Eijkelkamp, B.A., Luo, Z., Couñago, R.M., Morey, J.R., Maher, M.J., Paton, J.C. 2015. Dysregulation of transition metal ion homeostasis is the molecular basis for cadmium toxicity in *Streptococcus pneumoniae*. *Nature Communications* 6. doi: 10.1038/ncomms7418.

Belay, A.A. 2010. Impacts of chromium from tannery effluent and evaluation of alternative treatment options. *Journal of Environmental Protection* 1(01): 53.

Bernard, A. 2008. Cadmium and its adverse effects on human health. *Indian Journal of Medical Research* 128(4): 557–564.

Bhuiyan, M.A.H., Suruvi, N.I., Dampare, S.B., Islam, M.A., Quraishi, S.B., Ganyaglo, S., Suzuki, S. 2011. Investigation of the possible sources of heavy metal contamination in lagoon and canal water in the tannery industrial area in Dhaka, Bangladesh. *Environmental Monitoring and Assessment* 175(1–4): 633–649.

Birben, E., Sahiner, U.M., Sackesen, C., Erzurum, S., Kalayci, O. 2012. Oxidative stress and antioxidant defense. *World Allergy Organization Journal* 5(1): 9–19.

Bonde, J.P. 1993. The risk of male sub fecundity attributable to welding of metals: Studies of semen quality, infertility, adverse pregnancy outcome and childhood malignancy. *International Journal of Andrology* 16 (Suppl. 1): 1–29.

Boyle, E.A., Sclater, F., Edmond, J.M. 1976. On the marine geochemistry of Cd. *Nature* 263(5572): 42–44.

Bradl, H. (ed.). 2005. *Heavy Metals in the Environment: Origin, Interaction and Remediation*, Vol. 6. Academic Press, Orlando, FL.

Bridges, C.C., Zalups, R.K. 2010. Transport of inorganic Hg and methylmercury in target tissues and organs. *Journal of Toxicology and Environmental Health Part B* 13(5): 385–410.

Bustaffa, E., Stoccoro, A., Bianchi, F., Migliore, L. 2014. Genotoxic and epigenetic mechanisms in arsenic carcinogenicity. *Archives of Toxicology* 88(5): 1043–1067.

Caciari, T., Sancini, A., Tomei, F., Antetomaso, L., Tomei, G., Scala, B., Schifano, M.P. 2012. Cadmium blood/urine levels and blood pressure in workers occupationally exposed to urban stressor. *Annali di Lgiene: Medicina Preventiva e di Comunita* 24(5): 417–428.

Cain, D.J., Luoma, S.N., Wallace, W.G. 2004. Linking metal bioaccumulation of aquatic insects to their distribution patterns in a mining-impacted river. *Environmental Toxicology and Chemistry* 23(6): 1463–1473.

Cannino, G., Ferruggia, E., Luparello, C., Rinaldi, A.M. 2009. Cadmium and mitochondria. *Mitochondrion* 9: 377–384.

Carvalho, C.M., Chew, E.H., Hashemy, S.I., Lu, J., Holmgren, A. 2008. Inhibition of the human thioredoxin system: A molecular mechanism of mercury toxicity. *Journal of Biological Chemistry* 283(18): 11913–11923.

Casalino, E., Calzaretti, G., Sblano, C., Landriscina, C. 2002. Molecular inhibitory mechanisms of antioxidant enzymes in rat liver and kidney by cadmium. *Toxicology* 179(1): 37–50.

Castagnetto, J.M., Hennessy, S.W., Roberts, V.A., Getzoff, E.D., Tainer, J.A., Pique, M.E. 2002. MDB: The metalloprotein database and browser at the Scripps Research Institute. *Nucleic Acids Research* 30(1): 379–382.

Castoldi, A.F., Barni, S., Turin, I., Gandini, C., Manzo, L. 2000. Early acute necrosis, delayed apoptosis and cytoskeletal breakdown in cultured cerebellar granule neurons exposed to methylmercury. *Journal of Neuroscience Research* 59(6): 775–787.

Chakraborty, J., Das, S. 2014. Characterization and cadmium-resistant gene expression of biofilm-forming marine bacterium *Pseudomonas aeruginosa* JP-11. *Environmental Science and Pollution Research* 21(24): 14188–14201.

Chapman, P.M., Allen, H.E., Godtfredsen, K., Z'Graggen, M.N. 1996. Policy analysis, peer reviewed: Evaluation of bioaccumulation factors in regulating metals. *Environmental Science and Technology* 30(10): 448A–452A.

Chen, C.W., Chen, C.F., Dong, C.D. 2012. Distribution and accumulation of mercury in sediments of Kaohsiung River Mouth, Taiwan. *APCBEE Procedia* 1: 153–158.

Chiou, H.Y., Hsueh, Y.M., Liaw, K.F., Horng, S.F., Chiang, M.H., Pu, Y.S., Chen, C.J. 1995. Incidence of internal cancers and ingested inorganic Arsenic: A seven-year follow-up study in Taiwan. *Cancer Research* 55(6): 1296–1300.

Chowdhury, M.A.I., Uddin, M.T., Ahmed, M.F., Ali, M.A., Uddin, S.M. 2006. How Arsenic contamination of groundwater does causes severity and health hazard in Bangladesh. *Journal of Applied Sciences* 6: 1275–1286.

Costa, M., Klein, C.B. 2006. Toxicity and carcinogenicity of chromium compounds in humans. *Critical Reviews in Toxicology* 36(2): 155–163.

Crinnion, W.J. 2000. Environmental medicine, part three: Long-term effects of chronic low-dose mercury exposure. *Alternative Medicine Review: A Journal of Clinical Therapeutic* 5(3): 209–223.

Das, S., Dash, H.R., Chakraborty, J. 2016. Genetic basis and importance of metal resistant genes in bacteria for bioremediation of contaminated environments with toxic metal pollutants. *Applied Microbiology and Biotechnology* 100: 2967–2984.

Das, S., Raj., R., Mangwani, N., Dash, H.R., Chakraborty, J. 2014. Heavy metals and hydrocarbons: Adverse effects and mechanism of toxicity. In Das, S. (ed.) *Microbial Biodegradation and Bioremediation*. Elsevier Inc, USA, ISBN-13: 978-0128000212, pp. 23–54.

Dash, H.R., Das, S. 2012. Bioremediation of mercury and the importance of bacterial *mer* genes. *International Biodeterioration & Biodegradation* 75: 207–213.

Dash, H.R., Das, S. 2015. Bioremediation of inorganic Hg through volatilization and biosorption by transgenic *Bacillus cereus* BW-03 (pPW-05). *International Biodeterioration & Biodegradation* 103: 179–185.

Dash, H.R., Das, S. 2015. Bioremediation of inorganic mercury through volatilization and biosorption by transgenic *Bacillus cereus* BW-03(pPW-05). *International Biodeterioration and Biodegradation* 103: 179–185.

Dash, H.R., Mangwani, N., Das, S. 2014. Characterization and potential application in Hg bioremediation of highly Hg-resistant marine bacterium *Bacillus thuringiensis* PW-05. *Environmental Science and Pollution Research* 21(4): 2642–2653.

De, J., Dash H.R., Das., S. 2014. Mercury pollution and bioremediation-A case study on biosorption by a mercury-resistant marine bacterium. In Das, S. (ed.) *Microbial Biodegradation and Bioremediation*. Elsevier Inc, USA, ISBN-13:978-0128000212, pp. 137–166.

De Sastre, M.S.R., Varillas, A., Kirschbaum, P. May 1992. Arsenic content in water in the northwest area of Argentina. In *International Seminar Proceedings: Arsenic in the Environment and Its Incidence on Health*. Santiago, Chile, pp. 25–29.

Deschatre, M., Ghillebaert, F., Guezennec, J., Colin, C.S. 2013. Sorption of copper (II) and Ag (I) by four bacterial exopolysaccharides. *Applied Biochemistry and Biotechnology* 171(6): 1313–1327.

Dixit, R., Malaviya, D., Pandiyan, K., Singh, U.B., Sahu, A., Shukla, R., Paul, D. 2015. Bioremediation of heavy metals from soil and aquatic environment: An overview of principles and criteria of fundamental processes. *Sustainability* 7(2): 2189–2212.

Drake, P.L., Hazelwood, K.J. 2005. Exposure-related health effects of silver and silver compounds: A review. *Annals of Occupational Hygiene* 49(7): 575–585.

Du, Y., Ge, M.M., Xue, W., Yang, Q.Q., Wang, S., Xu, Y., Wang, H.L. 2015. Chronic lead exposure and mixed factors of Gender × Age × Brain regions interactions on dendrite growth, Spine Maturity and NDR Kinase. *PLoS ONE* 10(9): e0138112.

Dufault, R., LeBlanc, B., Schnoll, R., Cornett, C., Schweitzer, L., Wallinga, D., Lukiw, W.J. 2009. Hg from chlor-alkali plants: Measured concentrations in food product sugar. *Environmental Health* 8(1): 2.

Duffus, J.H. 2002. "Heavy metals"-a meaningless term? *Pure and Applied Chemistry* 74: 793–807.

Dzairi, F.Z., Zeroual, Y., Moutaouakkil, A., Taoufik, J., Talbi, M., Loutfi, M., Blaghen, M. 2004. Bacterial volatilization of Hg by immobilized bacteria in fixed and fluidized bed bioreactors. *Annals of Microbiology* 54: 353–364.

Eccles, J.S., Wigfield, A. 1995. In the mind of the actor: The structure of adolescents' achievement task values and expectancy-related beliefs. *Personality and Social Psychology Bulletin* 21: 215–225.

Eisler, R. 1997. Ag hazards to fish, wildlife and invertebrates: A synoptic review. U.S. Department of the Interior, National Biological Service, Washington, DC, 44pp. Biological Report 32 and Contaminant Hazard Reviews Report 32.

Engstrom, D.R., Swain, E.B., Henning, T.A., Brigham, M.E., Brezonik, P.L. 1994. Atmospheric Hg deposition to lakes and watersheds. A quantitative reconstruction from multiple sediment cores. *Advances in Chemistry* 237: 33–66.

Evrenoglou, L., Partsinevelou, S.A., Stamatis, P., Lazaris, A., Patsouris, E., Kotampasi, C., Nicolopoulou-Stamati, P. 2013. Children exposure to trace levels of heavy metals at the north zone of Kifissos River. *Science of the Total Environment* 443: 650–661.

Farhadian, M., Vachelard, C., Duchez, D., Larroche, C. 2008. In situ bioremediation of monoaromatic pollutants in groundwater: A review. *Bioresource Technology* 99(13): 5296–5308.

Feng, Z., Hu, W., Rom, W.N., Costa, M., Tang, M.S. 2003. Cr (VI) exposure enhances polycyclic aromatic hydrocarbon–DNA binding at the p53 gene in human lung cells. *Carcinogenesis* 24(4): 771–778.

Fillion, M., Mergler, D., Passos, C.J., Larribe, F., Lemire, M., Guimarães, J.R. 2006. A preliminary study of Hg exposure and blood pressure in the Brazilian Amazon. *Environmental Health* 5(1): 29.

Fowler, B.A. 1998. Roles of Pb-binding proteins in mediating Pb bioavailability. *Environmental Health Perspectives* 106(Suppl. 6): 1585.

Freeman, H.J. 2004. Risk of gastrointestinal malignancies and mechanisms of cancer development with obesity and its treatment. *Best Practice & Research Clinical Gastroenterology* 18(6): 1167–1175.

Fung, M.C., Bowen, D.L. 1996. Ag products for medical indications: Risk–benefit assessment. *Clinical Toxicology* 34(1): 119–126.

Gadd, G.M. 1992. Microbial control of heavy metal pollution. In Fry, J.C., Gadd, G.M., Herbert, R.A., Jones, C.W., Watson-Craik, I. (eds.) *Microbial Control of Pollution*. Cambridge University Press, Cambridge, pp. 59–88.

Garelick, H., Jones, H. 2008. Mitigating arsenic pollution. *Chemistry International*. 30. https://www.iupac.org/publications/ci/2008/3004/2_garelick.html.

Garelick, H., Jones, H., Dybowska, A., Valsami-Jones, E. 2008. *Arsenic Pollution Sources*. Springer, New York, pp. 17–60.

Gerhard, I., Monga, B., Waldbrenner, A., Runnebaum, B. 1998. Heavy metals and fertility. *Journal of Toxicology and Environmental Health Part A* 54(8): 593–611.

GESAMP. 1985. IMO/FAO/UNESCO/WHO/IAEA/UN/UNEP Joint Group of Experts on the Scientific Aspects of Marine Pollution: Atmospheric transport of contaminants into the Mediterranean region. World Meteorological Association, Geneva, Switzerland. Reports and Studies No. 26.

Ghodsi, H., MehranHoodaji, A., Gheisari, M.M. 2011. Investigation of bioremediation of Arsenic by bacteria isolated from contaminated soil. *African Journal of Microbiology Research* 5(32): 5889–5895.

Giles, F.E., Middleton, S.G., Grau, J.G. 1973. Evidence for the accumulation of atmospheric lead by insects in areas of high traffic density. *Environmental Entomology* 2(2): 299–300.

Grevatt, P.C. 1998. Toxicological review of hexavalent chromium. Support of summary information on the Integrated Risk Information System (IRIS), U.S. Environmental Protection Agency, Washington, DC.

Grover, P., Rekhadevi, P., Danadevi, K., Vuyyuri, S., Mahboob, M., Rahman, M. 2010. Genotoxicity evaluation in workers occupationally exposed to lead. *International Journal of Hygiene and Environmental Health* 213(2): 99–106.

Guallar, E., Sanz-Gallardo, M.I., Veer, P.V.T., Bode, P., Aro, A., Gómez-Aracena, J., Kok, F.J. 2002. Mercury, fish oils, and the risk of myocardial infarction. *New England Journal of Medicine* 347(22): 1747–1754.

Gulbranson, S.H., Hud, J.A., Hansen, R.C. 2000. Argyria following the use of dietary supplements containing colloidal Ag protein. *Cutis* 66(5): 373–378.

Hamelink, J., Landrum, P.F., Bergman, H., Benson, W.H. 1994. *Bioavailability: Physical, Chemical, and Biological Interactions*. CRC Press, Boca Raton, FL.

Hartwig, A. 2001. Zinc finger proteins as potential targets for toxic metal ions: Differential effects on structure and function. *Antioxidants and Redox Signaling* 3(4): 625–634.

Hess, J., Angel, P., Schorpp-Kistner, M. 2004. AP-1 subunits: Quarrel and harmony among siblings. *Journal of Cell Science* 117(25): 5965–5973.

Hook, S.E., Fisher, N.S. 2001. Reproductive toxicity of metals in calanoid copepods. *Marine Biology* 138(6): 1131–1140.

Housecroft, C.E., Sharpe, A.G. 2008. *Inorganic Chemistry*, 3rd edn. Pearson Education, Harlow, U.K.

Hua, L., Yang, W.H., Liu, X.Z., Huang, H.K., Xie, G.X. 2005. A survey on blood lead and cadmium levels of 1490 children. *Chinese Journal of Child Health Care* 14: 304–305.

Hudson, T.L., Fox, F.D., Plumlee, G.S. 1999. *Metal Mining and the Environment*. Environmental Awareness Series 3, American Geological Institute, Alexandra, VA.

Hughes, M.K., Lepp, N.W., Phipps, D.A., MacFadyen, A. 1980. *Aerial Heavy Metal Pollution and Terrestrial Ecosystems*. Academic Press Inc. Ltd., London, U.K., pp. 217–327.

Hulthe, J., Fagerberg, B. 2002. Circulating oxidized LDL is associated with subclinical atherosclerosis development and inflammatory cytokines (AIR Study). *Arteriosclerosis, Thrombosis, and Vascular Biology* 22(7): 1162–1167.

IARC. 2004. *Monographs on the Evaluation of Carcinogenic Risks to Humans*. International Agency for Research on Cancer, Lyon, France.

Ibrahim, Z., Ahmad, W.A., Baba, A.B. 2001. Bioaccumulation of Ag and the isolation of metal-binding protein from *P. diminuta*. *Brazilian Archives of Biology and Technology* 44(3): 223–225.

IPCS. 2001. Environmental health criteria on arsenic and arsenic compounds. In Gomez-Caminero, A., Howe, P., Hughes, M., Kenyon, E., Lewis, D.R., Moore, M., Ng, J., Aitio, A., Becking, G. (eds.) *Arsenic and Arsenic Compounds*. Environmental Health Criteria Series, No. 224, WHO, Geneva, Switzerland, p. 521.

Jain, K., Parida, S., Mangwani, N., Dash, H.R., Das, S. 2013. Isolation and characterization of biofilm-forming bacteria and associated extracellular polymeric substances from oral cavity. *Annals of Microbiology* 63(4): 1553–1562.

Javaid, S., Shah, S.G.S., Chaudhary, A.J., Khan, M.H. 2008. Assessment of trace metal contamination of drinking water in the Pearl Valley, Azad Jammu and Kashmir. *Clean–Soil, Air, Water* 36(2): 216–221.

Jenkins, R., Barton, J., Hesselberg, J. 2002. Chapter 7: The global tanning industry: A commodity chain approach. In *Environmental Regulation in the New Global Economy: The Impact on Industry and Competitiveness*. Edward Elgar Publishing, p. 368. doi: 10.4337/9781781950418.

Kakehashi, A., Wei, M., Fukushima, S., Wanibuchi, H. 2013. Oxidative stress in the carcinogenicity of chemical carcinogens. *Cancers* 5(4): 1332–1354.

Kamaludeen, S.P., Megharaj, M., Juhasz, A.L., Sethunathan, N., Naidu, R. 2003. Cr-microorganism interactions in soils: remediation implications. *Reviews of Environmental Contamination and Toxicology*. 178: 93–164.

Karrari, P., Mehrpour, O., Abdollahi, M. 2012. A systematic review on status of lead pollution and toxicity in Iran; Guidance for preventive measures. *DARU Journal of Pharmaceutical Sciences* 20(1): 1.

Kermani, A.J.N., Ghasemi, M.F., Khosravan, A., Farahmand, A., Shakibaie, M.R. 2010. Cd bioremediation by metal-resistant mutated bacteria isolated from active sludge of industrial effluent. *Iranian Journal of Environmental Health Science & Engineering* 7(4): 279–286.

Khan, M.A., Wang, F. 2009. Hg-selenium compounds and their toxicological significance: Toward a molecular understanding of the Hg-selenium antagonism. *Environmental Toxicology and Chemistry* 28(8): 1567–1577.

Khan, S., Cao, Q., Lin, A.J., Zhu, Y.G. 2008. Concentrations and bioaccessibility of polycyclic aromatic hydrocarbons in wastewater-irrigated soil using in vitro gastrointestinal test. *Environmental Science and Pollution Research-International* 15(4): 344–353.

Kivela, H.M., Madonna, S., Krupovic, M., Tutino, M.L., Bamford, J.K.H. 2008. Genetics for *Pseudoalteromonas* provides tools to manipulate marine bacterial virus PM2. *Journal of Bacteriology* 190: 1298–1307.

Klein, C.B., Su, L., Bowser, D., Leszczynska, J. 2002. Chromate-induced epimutations in mammalian cells. *Environmental Health Perspectives* 110(Suppl. 5): 739.

Kondo, K., Hino, N., Sasa, M., Kamamura, Y., Sakiyama, S., Tsuyuguchi, M., Monden, Y. 1997. Mutations of the p53 gene in human lung cancer from chromate-exposed workers. *Biochemical and Biophysical Research Communications* 239(1): 95–100.

Korte, F. 1983. Ecotoxicology of cadmium: General overview. *Ecotoxicology and Environmental Safety* 7(1): 3–8.

Kotaś, J., Stasicka, Z. 2000. Cr occurrence in the environment and methods of its speciation. *Environmental Pollution* 107(3): 263–283.

Krell, U., Roeckner, E. 1988. Model simulation of the atmospheric input of lead and cadmium into the North Sea. *Atmospheric Environment (1967)* 22(2): 375–381.

Lam, I.K., Wang, W.X. 2006. Accumulation and elimination of aqueous and dietary Ag in *Daphnia magna*. *Chemosphere* 64(1): 26–35.

Lam, T.V., Agovino, P., Niu, X., Roché, L. 2007. Linkage study of cancer risk among lead-exposed workers in New Jersey. *Science of the Total Environment* 372(2): 455–462.

Landis, W.G., Yu, M.-H. 2003. *Introduction to Environmental Toxicology: Impacts of Chemicals Upon Ecological Systems.* CRC Press, Boca Raton, FL.

Landrigan, P.J., Todd, A.C. 1994. Lead poisoning. *Western Journal of Medicine* 161(2): 153.

Lansdown, A.B.G. 2009. Cartilage and bone as target tissues for toxic materials. *General, Applied and Systems Toxicology.* doi: 10.1002/9780470744307.gat071.

Li, P., Feng, X.B., Qiu, G.L., Shang, L.H., Li, Z.G. 2009. Hg pollution in Asia: A review of the contaminated sites. *Journal of Hazardous Materials* 168(2): 591–601.

Li, Z.H., Li, P., Randak, T. 2011. Evaluating the toxicity of environmental concentrations of waterborne Cr (VI) to a model teleost, *Oncorhynchus mykiss*: A comparative study of in vivo and in vitro. *Comparative Biochemistry and Physiology Part C: Toxicology & Pharmacology* 153(4): 402–407.

Liang, Y.I., Chu, C.C., Tsen, Y.L., Ting, K.S. 1957. Studies on anti-bilharzial drugs. VI. The antidotal effects of sodium dimercaptosuccinate and BAL-glucoside against tartar emetic. *Acta Physiologica Sinica* 21(1): 24–32.

Liu, C.W., Jang, C.S., Liao, C.M. 2004. Evaluation of arsenic contamination potential using indicator kriging in the Yun–Lin aquifer (Taiwan). *Science of the Total Environment* 321(1): 173–188.

Lone, M.I., Saleem, S., Mahmood, T., Saifullah, K., Hussain, G. 2003. Heavy metal contents of vegetables irrigated by sewage/tubewell water. *International Journal of Agriculture and Biology* 5(4): 533–535.

Luoma, S.N., Ho, Y.B., Bryan, G.W. 1995. Fate, bioavailability and toxicity of silver in estuarine environments. *Marine Pollution Bulletin* 31(1): 44–54.

Madhavi, V., Reddy, A.V.B., Reddy, K.G., Madhavi, G., Prasad, T.N.K.V. 2013. An overview on research trends in remediation of Cr. *Research Journal of Recent Sciences* 2: 71–83.

Mandal, B.K., Suzuki, K.T. 2002. Arsenic round the world: A review. *Talanta* 58(1): 201–235.

Manzoor, S., Shah, M.H., Shaheen, N., Khalique, A., Jaffar, M. 2006. Multivariate analysis of trace metals in textile effluents in relation to soil and groundwater. *Journal of Hazardous Materials* 137(1): 31–37.

Martin, J.H., Broenkow, W.W. 1975. Cd in plankton: Elevated concentrations off Baja California. *Science* 190: 884–885.

Mashiatullah, A., Chaudhary, M.Z., Ahmad, N., Javed, T., Ghaffar, A. 2013. Metal pollution and ecological risk assessment in marine sediments of Karachi Coast, Pakistan. *Environmental Monitoring and Assessment* 185(2): 1555–1565.

Mattia, G.D., Bravi, M.C., Laurenti, O., Luca, O.D., Palmeri, A., Sabatucci, A., Ghiselli, A. 2004. Impairment of cell and plasma redox state in subjects professionally exposed to chromium. *American Journal of Industrial Medicine* 46(2): 120–125.

Mazumder, D.G. 2000. Diagnosis and treatment of chronic arsenic poisoning. United Nations Synthesis Report on Arsenic in Drinking Water. http://www.who.int/water_sanitation_health/dwq/arsenicun4.pdf.

Mazumder, D.N., Steinmaus, C., Bhattacharya, P., von Ehrenstein, O.S., Ghosh, N., Gotway, M., Sil, A., Balmes, J.R., Haque, R., Hira-Smith, M.M., Smith, A.H. 2005. Bronchiectasis in persons with skin lesions resulting from arsenic in drinking water. *Epidemiology* 16: 760–765.

Meche, A., Martins, M.C., Lofrano, B.E., Hardaway, C.J., Merchant, M., Verdade, L. 2010. Determination of heavy metals by inductively coupled plasma-optical emission spectrometry in fish from the Piracicaba River in Southern Brazil. *Microchemical Journal* 94(2): 171–174.

Minnesota Pollution Control Agency (MPCA). 2013. https://www.pca.state.mn.us/sites/default/files/c-s3-13b.pdf.

Mire, C.E., Tourjee, J.A., O'Brien, W.F., Ramanujachary, K.V., Hecht, G.B. 2004. Lead precipitation by *Vibrio harveyi*: Evidence for novel quorum-sensing interactions. *Applied and Environmental Microbiology* 70(2): 855–864.

Mishra, R., Sinha, V., Kannan, A., Upreti, R.K. 2012. Reduction of Cr-VI by Cr resistant *Lactobacilli*: A prospective bacterium for bioremediation. *Toxicology International* 19(1): 25.

Mottet, N.K., Vahter, M.E., Charleston, J.S., Friberg, L.T. 1997. Metabolism of methylmercury in the brain and its toxicological significance. *Metal Ions in Biological Systems* 34: 371.

Moulis, J.M. 2010. Cellular mechanisms of cadmium toxicity related to the homeostasis of essential metals. *Biometals* 23: 877–896.

Mount, D.R., Hockett, J.R. 2000. Use of toxicity identification evaluation methods to characterize, identify, and confirm hexavalent Cr toxicity in an industrial effluent. *Water Research* 34(4): 1379–1385.

Mullen, M.D., Wolf, D.C., Ferris, F.G., Beveridge, T.J., Flemming, C.A., Bailey, G.W. 1989. Bacterial sorption of heavy metals. *Applied and Environmental Microbiology* 55(12): 3143–3149.

Naik, M.M., Dubey, S.K. 2011. Lead-enhanced siderophore production and alteration in cell morphology in a Pb-resistant *Pseudomonas aeruginosa* strain 4EA. *Current Microbiology* 62(2): 409–414.

Naik, M.M., Dubey, S.K. 2013. Lead resistant bacteria: Lead resistance mechanisms, their applications in Pb bioremediation and biomonitoring. *Ecotoxicology and Environmental Safety* 98: 1–7.

Naik, M.M., Pandey, A., Dubey, S.K. 2012a. *Pseudomonas aeruginosa* strain WI-1 from Mandovi estuary possesses metallothionein to alleviate lead toxicity and promotes plant growth. *Ecotoxicology and Environmental Safety* 79: 129–133.

Naik, M.M., Shamim, K., Dubey, S.K. 2012b. Biological characterization of Pb resistant bacteria to explore role of bacterial metallothionein in lead resistance. *Current Science* 103(4): 00113891.

National Research Council (NRC). 1999. *Ar in Drinking Water*. National Academic Press, Washington, DC.

National Resources Defense Council (NRDC). 2000. https://www.nrdc.org/resources/get-lead-out-lead-your-area.

Nazif, W., Perveen, S., Shah, S.A. 2006. Evaluation of irrigation water for heavy metals of Akbarpura area. *Journal of Agricultural and Biological science* 1(1): 51–54.

Nicholson, J.K., Osborn, D. 1983. Kidney lesions in pelagic seabirds with high tissue levels of Cd and Hg. *Journal of Zoology* 200(1): 99–118.

Nickens, K.P., Patierno, S.R., Ceryak, S. 2010. Chromium genotoxicity: A double-edged sword. *Chemico-Biological Interactions* 188(2): 276–288.

Niedzwiecki, M.M., Liu, X., Hall, M.N., Thomas, T., Slavkovich, V., Ilievski, V., Graziano, J.H. 2015. Sex-specific associations of arsenic exposure with global DNA methylation and hydroxymethylation in leukocytes: Results from two studies in Bangladesh. *Cancer Epidemiology Biomarkers and Prevention* 24(11): 1748–1757.

Nordberg, G.F. 1974. Health hazards of environmental cadmium pollution. *Ambio* 3:55.

Nordberg, G., Gerhardsson, L. 1988. Silver. In *Handbook on Toxicity of Inorganic Compounds*, Seiler, H.G., Sigel, H., Sigel, A. (eds.). Marcel Dekker, New York, pp. 619–624.

Nriagu, J.O. 1989. A global assessment of natural sources of atmospheric trace metals. *Nature* 338: 47–49.

Nriagu, J.O., Pacyna, J.M. 1988. Quantitative assessment of worldwide contamination of air, water and soils by trace metals. *Nature* 333(6169): 134–139.

NTP (National Toxicology Program). 2004. Report on carcinogens, 11th edn. U.S. Department of Health and Human Service, Public Health Service, National Toxicology Program, Research Triangle Park, NC.

O'Connor, T.R., Graves, R.J., de Murcia, G., Castaing, B., Laval, J. 1993. Fpg protein of *Escherichia coli* is a zinc finger protein whose cysteine residues have a structural and/or functional role. *Journal of Biological Chemistry* 268(12): 9063–9070.

Oram, P.D., Fang, X., Fernando, Q., Letkeman, P., Letkeman, D. 1996. The formation constants of Hg (II)-glutathione complexes. *Chemical Research in Toxicology* 9(4): 709–712.

Pacyna, J.M. 1996. Monitoring and assessment of metal contaminants in the air. In Chang, L.W. (ed.) *Toxicology of Metals*. CRC Press, Boca Raton, FL, pp. 9–28.

Pasha, Q., Malik, S.A., Shaheen, N., Shah, M.H. 2010. Investigation of trace metals in the blood plasma and scalp hair of gastrointestinal cancer patients in comparison with controls. *Clinica Chimica Acta* 411(7): 531–539.

Patrick, L. 2002. Hg toxicity and antioxidants: Part 1: Role of glutathione and alpha-lipoic acid in the treatment of mercury toxicity. *Alternative Medicine Review* 7(6): 456–471.

Pepi, M., Volterrani, M., Renzi, M., Marvasi, M., Gasperini, S., Franchi, E., Focardi, S.E. 2007. Arsenic-resistant bacteria isolated from contaminated sediments of the Orbetello Lagoon, Italy, and their characterization. *Journal of Applied Microbiology* 103(6): 2299–2308.

Perveen, S., Samad, A.B.D.U.S., Nazif, W., Shah, S. 2012. Impact of sewage water on vegetables quality with respect to heavy metals in Peshawar, Pakistan. *Pakistan Journal of Botany* 44(6): 1923–1931.

Petrunkin, V.E. 1956. Synthesis and properties of dimercapto derivatives of alkylsulfonic acids. *Ukr Khem Zh* 22: 603–607.

Phillips, D.J., Rainbow, P.S. 2013. *Biomonitoring of Trace Aquatic Contaminants*, Vol. 37. Springer Science and Business Media New York, USA.

Pirrone, N., Cinnirella, S., Feng, X., Finkelman, R.B., Friedli, H.R., Leaner, J., Telmer, K. 2010. Global mercury emissions to the atmosphere from anthropogenic and natural sources. *Atmospheric Chemistry and Physics* 10(13): 5951–5964.

Pizent, A., Tariba, B., Živković, T. 2012. Reproductive toxicity of metals in men. *Archives of Industrial Hygiene and Toxicology* 63(Suppl. 1): 35–45.

Ponce, R.A., Kavanagh, T.J., Mottet, N.K., Whittaker, S.G., Faustman, E.M. 1994. Effects of methylmercury on the cell cycle of primary rat CNS cells in vitro. *Toxicology and Applied Pharmacology* 127(1): 83–90.

Pounds, J.G., Long, G.J., Rosen, J.F. 1991. Cellular and molecular toxicity of lead in bone. *Environmental Health Perspectives* 91: 17.

Price, P.W., Rathcke, B.J., Gentry, D.A. 1974. Lead in terrestrial arthropods: Evidence for biological concentration. *Environmental Entomology* 3(3): 370–372.

Priester, J.H., Horst, A.M., Van De Werfhorst, L.C., Saleta, J.L., Mertes, L.A., Holden, P.A. 2007. Enhanced visualization of microbial biofilms by staining and environmental scanning electron microscopy. *Journal of Microbiological Methods* 68(3): 577–587.

Priyalaxmi, R., Murugan, A., Raja, P., Raj, K.D. 2014. Bioremediation of cadmium by *Bacillus safensis* (JX126862), a marine bacterium isolated from mangrove sediments. *International Journal of Current Microbiology and Applied Science* 3(12): 326–335.

Puzas, J.E., Sickel, M.J., Felter, M.E. 1992. Osteoblasts and chondrocytes are important target cells for the toxic effects of lead. *Neurotoxicology* 13: 800–806.

Rahman, M.S., Molla, A.H., Saha, N., Rahman, A. 2012. Study on heavy metals levels and its risk assessment in some edible fishes from Bangshi River, Savar, Dhaka, Bangladesh. *Food Chemistry* 134(4): 1847–1854.

Rai, A., Tripathi, P., Dwivedi, S., Dubey, S., Shri, M., Kumar, S., Tripathi, P.K. et al. 2011. Arsenic tolerances in rice (*Oryza sativa*) have a predominant role in transcriptional regulation of a set of genes including sulphur assimilation pathway and antioxidant system. *Chemosphere* 82(7): 986–995.

Ramos, E.S.I., Soria, E.A. 2014. Ar immunotoxicity and immunomodulation by phytochemicals: Potential relations to develop chemopreventive approaches. *Recent Patents on Inflammation and Allergy Drug Discovery* 8(2): 92–103.

Rani, A., Kumar, A., Lal, A., Pant, M. 2014. Cellular mechanisms of Cd-induced toxicity: A review. *International Journal of Environmental Health Research* 24(4): 378–399.

Rathnayake, I., Megharaj, M., Bolan, N., Naidu, R. 2009. Tolerance of heavy metals by gram positive soil bacteria. Doctoral dissertation, World Academy of Science Engineering and Technology.

Ratte, H.T. 1999. Bioaccumulation and toxicity of silver compounds: A review. *Environmental Toxicology and Chemistry* 18(1): 89–108.

Reddy, S.B., Charles, M.J., Raju, G.N., Deddy, B.S., Reddy, T.S., Lakshmi, P.R., Vijayan, V. 2004. Trace elemental analysis of cancer-afflicted intestine by PIXE technique. *Biological Trace Element Research* 102(1–3): 265–281.

Reinfelder, J.R., Fisher, N.S., Luoma, S.N., Nichols, J.W., Wang, W.X. 1998. Trace element trophic transfer in aquatic organisms: A critique of the kinetic model approach. *Science of the Total Environment* 219(2): 117–135.

Reynolds, M.F., Peterson-Roth, E.C., Bespalov, I.A., Johnston, T., Gurel, V.M., Menard, H.L., Zhitkovich, A. 2009. Rapid DNA double-strand breaks resulting from processing of Cr-DNA crosslinks by both MutS dimers. *Cancer Research* 69(3): 1071–1079.

Rice, K.M., Walker, E.M., Wu, M., Gillette, C., Blough, E.R. 2014. Environmental mercury and its toxic effects. *Journal of Preventive Medicine and Public Health* 47(2): 74–83.

Richardson, G.M., Wilson, R., Allard, D., Purtill, C., Douma, S., Graviere, J. 2011. Mercury exposure and risks from dental amalgam in the US population, post-2000. *Science of the Total Environment* 409(20): 4257–4268.

Roesijadi, G., Robinson, W.E. 1994. Metal regulation in aquatic animals: Mechanisms of uptake, accumulation and release. *Aquatic Toxicology* 102: 125–133.

Rosenberg, E., Ron, E.Z. 1999. High- and low-molecular-mass microbial surfactants. *Applied Microbiology and Biotechnology* 52(2): 154–162.

Roulet, M., Lucotte, M., Canuel, R., Rheault, I., Tran, S., Gog, Y.D.F., Mergler, D. 1998. Distribution and partition of total Hg in waters of the Tapajós River Basin, Brazilian Amazon. *Science of the Total Environment* 213(1): 203–211.

Salnikow, K., Zhitkovich, A. 2008. Genetic and epigenetic mechanisms in metal carcinogenesis and cocarcinogenesis: Nickel, arsenic, and chromium. *Chemical Research in Toxicology* 21(1): 28–44.

Schauder, A., Avital, A., Malik, Z. 2010. Regulation and gene expression of heme synthesis under heavy metal exposure— Review. *Journal of Environmental Pathology, Toxicology and Oncology* 29(2): 137–158.

Scheuhammer, A.M., Meyer, M.W., Sandheinrich, M.B., Murray, M.W. 2007. Effects of environmental methylmercury on the health of wild birds, mammals, and fish. *AMBIO: A Journal of the Human Environment* 36(1): 12–19.

Shahid, M., Pinelli, E., Pourrut, B., Silvestre, J., Dumat, C. 2011. Lead-induced genotoxicity to *Vicia faba* L. roots in relation with metal cell uptake and initial speciation. *Ecotoxicology and Environmental Safety* 74(1): 78–84.

Shakoori, F.R., Tabassum, S., Rehman, A., Shakoori, A.R. 2010. Isolation and characterization of Cr^{6+} reducing bacteria and their potential use in bioremediation of Cr containing wastewater. *Pakistan Journal of Zoology* 42(6): 651–658.

Shallari, S., Schwartz, C., Hasko, A., Morel, J.L. 1998. Heavy metals in soils and plants of serpentine and industrial sites of Albania. *Science of the Total Environment* 209(2): 133–142.

Shankar, S., Shanker, U. 2014. Arsenic contamination of groundwater: A review of sources, prevalence, health risks, and strategies for mitigation. *The Scientific World Journal*. doi: org/10.1155/2014/304524.

Sherwani, S.I., Pabon, S., Patel, R.B., Sayyid, M.M., Hagele, T., Kotha, S.R., Parinandi, N.L. 2013. Eicosanoid signaling and vascular dysfunction: Methylmercury-induced phospholipase D activation in vascular endothelial cells. *Cell Biochemistry and Biophysics* 67(2): 317–329.

Siddique, A., Mumtaz, M., Zaigham, N.A., Mallick, K.A., Saied, S., Zahir, E., Khwaja, H.A. 2009. Heavy metal toxicity levels in the coastal sediments of the Arabian Sea along the urban Karachi (Pakistan) region. *Marine Pollution Bulletin* 58(9): 1406–1414.

Silva, I.A., El Nabawi, M., Hoover, D., Silbergeld, E.K. 2005. Prenatal $HgCl_2$ exposure in BALB/c mice: Gender-specific effects on the ontogeny of the immune system. *Developmental and Comparative Immunology* 29(2): 171–183.

Simpson, W.R. 1981. A critical review of cadmium in the marine environment. *Progress in Oceanography* 10(1): 1–70.

Singh, R., Gautam, N., Mishra, A., Gupta, R. 2011. Heavy metals and living systems: An overview. *Indian Journal of Pharmacology* 43(3): 246.

Sinha, S., Mukherjee, S.K. 2009. *Pseudomonas aeruginosa* KUCd1, a possible candidate for cadmium bioremediation. *Brazilian Journal of Microbiology* 40(3): 655–662.

Siripornadulsil, S., Siripornadulsil, W. 2013. Cadmium-tolerant bacteria reduce the uptake of Cd in rice: Potential for microbial bioremediation. *Ecotoxicology and Environmental Safety* 94: 94–103.

Smedley, P.L., Kinniburgh, D.G. 2001. Chapter 1: Sources and behavior of arsenic in natural water. In *United Nations Synthesis Report on Arsenic in Drinking Water*. World Health Organization, Geneva, Switzerland. http://www.bvsde.ops-oms.org/bvsacd/who/arsin.pdf.

Sørensen, K., Kristensen, K.S., Bang, L.E., Svendsen, T.L., Wiinberg, N., Buttenschön, L., Talleruphuus, U. 2004. Increased systolic ambulatory blood pressure and microalbuminuria in treated and non-treated hypertensive smokers. *Blood Pressure* 13(6): 362–368.

Soskine, M., Steiner-Mordoch, S., Schuldiner, S. 2002. Crosslinking of membrane-embedded cysteines reveals contact points in the EmrE oligomer. *Proceedings of the National Academy of Sciences of the United States of America* 99(19): 12043–12048.

South African Bureau of Standards Commercial. 2005. Sterility of ionic colloidal silver 19 ppm.

Stelting, S., Burns, R.G., Sunna, A., Visnovsky, G., Bunt, C. 2010. Immobilization of *Pseudomonas* sp. strain ADP: A stable inoculant for the bioremediation of atrazine. In *Nineteenth World Congress of Soil Science, Soil Solutions for a Changing World*, Brisbane, Queensland, Australia.

Stephens, T. 2005. Survey finds Ag contamination in North Pacific waters, probably from industrial emissions in Asia. *Currents Online* 9(28): 14–20.

Streets, D.G., Hao, J., Wu, Y., Jiang, J., Chan, M., Tian, H., Feng, X. 2005. Anthropogenic mercury emissions in China. *Atmospheric Environment* 39(40): 7789–7806.

Takeuchi, M., Kawahata, H., Gupta, L.P., Kita, N., Morishita, Y., Ono, Y., Komai, T. 2007. Arsenic resistance and removal by marine and non-marine bacteria. *Journal of Biotechnology* 127(3): 434–442.

Tariq, S.R., Shah, M.H., Shaheen, N., Jaffar, M., Khalique, A. 2008. Statistical source identification of metals in groundwater exposed to industrial contamination. *Environmental Monitoring and Assessment* 138(1–3): 159–165.

Taylor, D. 1983. The significance of the accumulation of Cd by aquatic organisms. *Ecotoxicology and Environmental Safety* 7(1): 33–42.

Tchounwou, P.B., Yedjou, C.G., Patlolla, A.K., Sutton, D.J. 2012. Heavy metal toxicity and the environment. In Andreas, L. (ed.) *Molecular, Clinical and Environmental Toxicology*. Springer, Basel, Switzerland, pp. 133–164.

Thompson, C.M., Proctor, D.M., Haws, L.C., Hébert, C.D., Grimes, S.D., Shertzer, H.G., Harris, M.A. 2011. Investigation of the mode of action underlying the tumorigenic response induced in B6C3F1 mice exposed orally to hexavalent Cr. *Toxicological Sciences* 123(1): 58–70.

Tong, S., Schirnding, Y.E.V., Prapamontol, T. 2000. Environmental Pb exposure: A public health problem of global dimensions. *Bulletin of the World Health Organization* 78(9): 1068–1077.

Trasande, L., Landrigan, P.J., Schechter, C. 2005. Public health and economic consequences of methyl mercury toxicity to the developing brain. *Environmental Health Perspectives* 113(5): 590–596.

Tripathi, R.M., Raghunath, R., Mahapatra, S., Sadasivan, S. 2001. Blood lead and its effect on Cd, Cu, Zn, Fe and hemoglobin levels of children. *Science of the Total Environment* 277(1): 161–168.

Tupling, R., Green, H. 2002. Ag ions induce Ca^{2+} release from the SR in vitro by acting on the Ca^{2+} release channel and the Ca^{2+} pump. *Journal of Applied Physiology* 92(4): 1603–1610.

Ullah, R., Malik, R.N., Qadir, A. 2009. Assessment of groundwater contamination in an industrial city, Sialkot, Pakistan. *African Journal of Environmental Science and Technology* 3(12).

United Nations Environment Program (UNEP). 2008. Draft final review of scientific information on cadmium. United Nations Environment Programme, Chemicals Branch. http://www.unep.org/hazardoussubstances/Portals/9/Lead_Cadmium/docs/Interim_reviews/Final_UNEP_Cadmium_review_Nov_2008.pdf.

United States Environmental Protection Agency (USEPA). 2001. Arsenic compounds. http://www3.epa.gov./airtoxics/hlthef/Ar.html. Accessed on Oct 21, 2016.

United States Environmental Protection Agency (USEPA). 2007. *Treatment Technologies for Hg in Soil, Waste, and Water*. Office of Superfund Remediation and Technology Innovation, Washington, DC.

United States Environmental Protection Agency (USEPA). 2015. Lead poisoning: A historical perspective. Washington, DC.

U.S. Environmental Protection Agency (USEPA). 1980. *Ambient Water Quality Criteria for Silver*. U.S. Environmental Protection Agency, Washington, DC, EPA 440/5-80-071.

Valko, M.M.H.C.M., Morris, H., Cronin, M.T.D. 2005. Metals, toxicity and oxidative stress. *Current Medicinal Chemistry* 12(10): 1161–1208.

Van Aalst, R.M., Van Ardenne, R.A.M., De Kruk, J.F., Lems, T. 1983. Pollution of the North Sea from the atmosphere. Organization for Applied Scientific Research (TNO), Apeldoorn, the Netherlands, Report No. C182/152.

Verkleji Housecroft, C.E., Sharpe, A.G. 2008. *Inorganic Chemistry*. Prentice Hall, Harlow, U.K.

Verkleji, J.A.S. 1993. The effects of heavy metals stress on higher plants and their use as bio monitors. In Markert, B. (ed.) *Plant as Bioindicators: Indicators of Heavy Metals in the Terrestrial Environment*. VCH, New York, pp. 415–424.

Vesey, D.A. 2010. Transport pathways for cadmium in the intestine and kidney proximal tubule: Focus on the interaction with essential metals. *Toxicology Letters* 198: 13–19.

Virtanen, J.K., Voutilainen, S., Rissanen, T.H., Mursu, J., Tuomainen, T.P., Korhonen, M.J., Salonen, J.T. 2005. Mercury, fish oils, and risk of acute coronary events and cardiovascular disease, coronary heart disease, and all-cause mortality in men in eastern Finland. *Arteriosclerosis, Thrombosis and Vascular Biology* 25(2): 228–233.

Volesky, B. 1990. *Biosorption of Heavy Metals*. CRC Press, Boca Raton, FL.

Von Burg, R., Liu, D. 1993. Chromium and hexavalent chromium. *Journal of Applied Toxicology* 13(3): 225–230.

Wagner-Döbler, I., Von Canstein, H., Li, Y., Timmis, K.N., Deckwer, W.D. 2000. Removal of Hg from chemical wastewater by microorganisms in technical scale. *Environmental Science and Technology* 34(21): 4628–4634.

Wallace, W.G., Lee, B.G., Luoma, S.N. 2003. Subcellular compartmentalization of cadmium and Zn in two bivalves. I. Significance of metal-sensitive fractions (MSF) and biologically detoxified metal (BDM). *Marine Ecology Progress Series* 249: 183–197.

Wang, B., Li, Y., Shao, C., Tan, Y., Cai, L. 2012. Cadmium and its epigenetic effects. *Current Medicinal Chemistry* 19(16): 2611–2620.

Wedeen, R.P., Qian, L.F. 1991. Chromium-induced kidney disease. *Environmental Health Perspectives* 92: 71.

WHO (World Health Organization). 1992. Cd. Environmental Health Criteria 134. World Health Organization, International Programme on Chemical Safety (IPCS), Geneva, Switzerland.

WHO (World Health Organization). 2000. Cd. In *Air Quality Guidelines for Europe*, 2nd edn. World Health Organization Regional Office for Europe, Copenhagen, Denmark. ISBN: 92 890 1358 3. http://www.euro.who.int/__data/assets/pdf_file/0005/74732/E71922.pdf.

WHO (World Health Organization). 2007. *Health Risks of Heavy Metals from Long-Range Trans-Boundary Air Pollution*. World Health Organization Regional Office for Europe, Copenhagen, Denmark.

WHO (World Health Organization). 2008. Cd. In *Guidelines for Drinking-Water Quality*, 3rd edn. Incorporating 1st and 2nd addenda. Vol. 1. Recommendations. World Health Organization, Geneva, Switzerland, pp. 317–319. ISBN:978 92 4 154761 1. http://www.who.int/water_sanitation_health/dwq/fulltext.pdf.

WHO (World Health Organization). 2011. *Guidelines for Drinking-Water Quality*, 4th edn. World Health Organization, Geneva, Switzerland.

WHO (World Health Organization). 2016. Mercury and health. http://www.who.int/mediacentre/factsheets/fs361/en/. Accessed on Oct 21, 2016.

Williamson, P. 1979. Comparison of metal levels in invertebrate detritivores and their natural diets: Concentration factors reassessed. *Oecologia* 44(1): 75–79.

Wise, S.S., Wise, J.P. 2012. Chromium and genomic stability. *Mutation Research/Fundamental and Molecular Mechanisms of Mutagenesis* 733(1): 78–82.

Wright, D.A., Welbourn, P. 2002. *Environmental Toxicology*. Cambridge University Press, Cambridge, U.K.

Xia, B., Ren, X.H., Zhuang, Z.X., Yang, L.Q., Huang, H.Y., Pang, L., Zou, F. 2014. Effect of hexavalent Cr on histone biotinylation in human bronchial epithelial cells. *Toxicology Letters* 228(3): 241–247.

Xu, H., Cao, D.J., Tian, Z.F. 2012. Isolation and identification of a mercury resistant strain. *Environment Protection Engineering* 38(4): 67–75.

Yedjou, C., Steverson, M., Tchounwou, P. 2006. Lead nitrate-induced oxidative stress in human liver carcinoma (HepG2) cells. *Metal Ions in Biology and Medicine* 9: 293–297.

Yedjou, C.G., Tchounwou, C.K., Haile, S., Edwards, F., Tchounwou, P.B. 2010. N-acetyl-cysteine protects against DNA damage associated with lead toxicity in HepG2 cells. *Ethnicity and Disease* 20(1 Suppl. 1): S1.

Yilmaz, F., Özdemir, N., Demirak, A., Tuna, A.L. 2007. Heavy metal levels in two fish species *Leuciscus cephalus* and *Lepomis gibbosus*. *Food Chemistry* 100(2): 830–835.

Zahoor, A., Rehman, A. 2009. Isolation of Cr(VI) reducing bacteria from industrial effluents and their potential use in bioremediation of chromium containing wastewater. *Journal of Environmental Sciences* 21(6): 814–820.

Zeng, Q.B., Xu, Y.Y., Yu, X., Yang, J., Hong, F., Zhang, A.H. 2014. Arsenic may be involved in fluoride-induced bone toxicity through PTH/PKA/AP1 signaling pathway. *Environmental Toxicology and Pharmacology* 37(1): 228–233.

Zhao, S., Feng, C., Quan, W., Chen, X., Niu, J., Shen, Z. 2012. Role of living environments in the accumulation characteristics of heavy metals in fishes and crabs in the Yangtze River Estuary, China. *Marine Pollution Bulletin* 64(6): 1163–1171.

Zheng, W., Aschner, M., Ghersi-Egea, J.F. 2003. Brain barrier systems: A new frontier in metal neurotoxicological research. *Toxicology and Applied Pharmacology* 192(1): 1–11.

Zhitkovich, A. 2005. Importance of Cr-DNA adducts in mutagenicity and toxicity of Cr (VI). *Chemical Research in Toxicology* 18(1): 3–11.

Zhitkovich, A. 2011. Cr in drinking water: Sources, metabolism, and cancer risks. *Chemical Research in Toxicology* 24(10): 1617–1629.

Zhou, X., Li, Q., Arita, A., Sun., H., Costa, M. 2009. Effects of nickel, chromate, and arsenite on histone 3 lysine methylation. *Toxicology and Applied Pharmacology* 236(1): 78–84.

2 Sources of Metal Pollution, Global Status, and Conventional Bioremediation Practices

Priti Prabhakar Yewale, Aminur Rahman, Noor Nahar,
Anandakumar Saha, Jana Jass, Abul Mandal, and Neelu N. Nawani

CONTENTS

ABSTRACT

Pollution control has become a priority task for global regulatory authorities. The framing of regulations, guidelines, and implementation of pollution awareness and control programs has begun at a massive scale. Heavy metals that are one of the most challenging pollutants that affect humans, animals, plants, and the ecosystem health. The sources of different metals and their toxicities are described. Current approaches in bioremediation are addressed along with the challenges posed by them. Furthermore, recent developments in biotechnology that offer novel ways to recover metals from contaminated sites are discussed.

2.1   INTRODUCTION

The world is under a rapid revolution of inventions made through science and technology. In the twentieth century, rapid urbanization and industrialization have been established in developed and developing countries. These have resulted in increased anthropogenic activities. Many of the human anthropogenic activities are interfering with the ecosystem, and this has created an imbalance in the geological cycles. One of the major problems is heavy metal pollution and its impact. All anthropogenic activities like industrial processes release toxic elements in the atmosphere and ultimately into the environment that causes major ecological problems.

Environmental pollution is the presence of a pollutant (substance present beyond the level of tolerance) in the ecosystem, like air, water, and soil. The pollutant may be toxic and may harm living beings in the environment.

A metal is a material (element, compound, or alloy) that has good thermal and electrical conductivity. Heavy metals are generally defined as metals with high density (greater than 5 g/cm^3) and are poisonous at very low concentrations. Generally, metals are not decomposed naturally. They settle down, persist in the environment for a long period of time, accumulate into the living organisms, and magnify in the food chain (Figure 2.1).

In nature, metals are present in the core of the earth. They are present in different forms of ores that include sulfides and oxides. Among these metals, lead, cobalt, iron, arsenic, lead-zinc, nickel, gold, and silver exist in the form of sulfides while aluminum, selenium, antimony, and gold are present in the form of oxides. Some of these metals such as iron, copper, and cobalt are present as oxides and sulfides. These sulfides and oxides of metals are called minerals. The ore minerals are found together, for example, sulfides of cadmium, mercury, and lead are present along with sulfides of iron (pyrite, FeS_2) and copper (chalcopyrite, $CuFeS_2$). Some of the metals present in minor quantities are obtained as by-products during the recovery of major metals. Naturally, all the elements are present at environment-friendly concentrations. These elements are required by living organisms in very small quantities. If the quantity/concentration exceeds the desired or required level, it interferes with the normal metabolic function of the organism. Due to human activities and industrialization, these elements (especially heavy metals) are deposited at a faster rate and in higher concentrations. Natural processes for the release of these metals include mineral rock processing, volcanic eruption, and microbial activities, whereas anthropogenic sources include fossil fuel burning, mining, agricultural activities, waste incineration, and industrial processes.

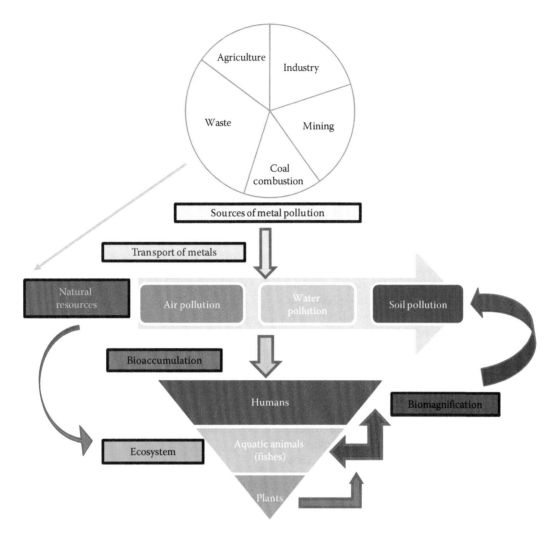

FIGURE 2.1 Fate of heavy metals in the ecosystem.

2.2 SOURCES OF HEAVY METAL POLLUTION

The following are the major sources from which metals are generally liberated into natural resources.

2.2.1 AGRICULTURE AND FORESTRY

Agriculture and forestry are directed toward faster and massive growth of specific crop plants and plants of economic value. As the essential elements are required in trace amounts for the growth of plants and animals, they should be provided externally for their desired growth.

2.2.1.1 Fertilizers

There are varying soil profiles gifted by nature, and these natural profiles have changed due to anthropogenic activities. Some soils are deficient in certain nutrient elements required by plants (Co, Cu, Fe, Mn, Mo, Ni, and Zn); therefore, these elements have to be provided to the crops in the form of fertilizers.

- *Natural fertilizers*: Natural fertilizers include compost and manure; compost is usually produced by the landfilling of domestic waste generated by household activities (Okedeyi et al., 2014). It may contain some metallic materials that contaminate it. In poultry and pig farms, the diet given to animals for higher mass production consists of zinc, copper, arsenic, and selenium as its major constituents. These metals leach out through the poultry droppings and pig excreta and enter the ecosystem through soil or water.

- *Chemical fertilizers*: Chemical fertilizers are also called synthetic fertilizers. Chemical fertilizers have essential macronutrients and micronutrients in a definite proportion. Excessive and repeated application of these chemical fertilizers causes bioaccumulation and biomagnification of toxic metals in the ecosystem. For example, phosphate fertilizers are a major source of cadmium because the concentration of cadmium varies greatly according to the substrate phosphate rock used during the production of fertilizers. Some fertilizers are produced from industrial waste that is rich in trace elements like zinc. Similarly, it also contains a high concentration of mercury, cadmium, zinc, arsenic, lead, copper, and chromium. The use of these fertilizers results in the uptake of metals by plants and ultimately enters the food chain. Superphosphate and urea fertilizers are widely used. Superphosphate fertilizers are reported as the major source of trace elements like copper, vanadium, and zinc, and urea is the source for zinc, cadmium, lead, and nickel (WHO, 2007; Benson et al., 2014).

2.2.1.2 Pesticides

Some metals and their compounds are required in small quantities for the growth of living beings. An excess of any of these compounds results in the disturbance of biochemical and metabolic pathways. This mechanism forms the basic principle for the manufacture of pesticides so that the compound can destroy the pest by interfering in the basic metabolic or regulatory functions of the cells. Several pesticide formulations contain metals, and their application leads to the liberation of metals like copper, mercury, manganese, cadmium, and lead that eventually contaminate the soil, groundwater, freshwater, and living ecosystems (Wuana and Okieimen, 2011).

Some pesticides, although banned in developed countries, are still being used in the third-world nations and are a cause of major concern. Some of these include dichlorodiphenyltrichloroethane (DDT) that is persistent due to its resistance toward microbial enzymes present in the rhizosphere. Aldrin and dieldrin both accumulate into the food chain and result in biomagnification. Hexachlorobenzene (HCB) is produced commercially for use as fungicide. Heptachlor is a broad-spectrum insecticide that persists in the ecosystem for long durations. Lindane has wide applications, and it is used to treat plants, animals, soil, and seeds as well as in water treatment and repair of buildings. Being widely used, its exposure to all natural resources is drastic.

2.2.2 Mining and Smelting

Mining is the extraction of a few hundred million metric tons of metals from its ore. Along with the metal, an equal volume of waste is produced in mining. Smelting, one of the processes in mining, is used for the extraction of metal in a free state. In mining, the waste generated is disposed directly into a nearby site; this waste is known as tailing. Tailings may contain processing effluents and high concentration of metals. Tailings are a very rich source of mercury, copper, arsenic, zinc, and cadmium. It is well known that mercury gets exposed in very high concentrations during gold mining and cadmium is released as a by-product in the refining of zinc.

2.2.2.1 Acid Mine Drainage

When mine tailings are exposed to oxygen and water, they form acids, especially when sulfides are adequately present. The toxic metals present in the surroundings dissolve in the acid and form a solution. This solution is a major contaminating source for soil and water as it is the richest source of toxic heavy metals. The acid mine drains are said to be disastrous, especially because there is no solution to neutralize acid drains. Due to very high acidity, soil profiles are also damaged, which ultimately destroys the vegetation.

2.2.3 Fossil Fuel Combustion and Coal

Combustion of fossil fuels and residual oils is one of the main sources of air pollution. As the era of industrialization has developed, excessive use of fossil fuels in power generation, vehicles, industrial furnaces, and heaters has occurred. Combustion of fossil fuels during all these activities releases very harmful effluents in the air and water (solid and liquid pollutants). Almost all the toxic metals (Pb, Zn, Cu, Cr, Hg, Cd, As) are released in the atmosphere. These metals are oxidized, and their oxides contaminate soil and water. Lead is emitted in the combustion of petrochemicals where gasoline is the major source for the emission of lead in its toxic form (tetraethyl lead). The electric power sector is the largest source of toxic pollutants in the United States due to coal ash and coal waste, which contains toxic heavy metals (Benson et al., 2014).

2.2.4 Industrial Processes

Metal processing factories are one of the most inevitable sources of metal deposition. During processing, metals are released into the environment either through gas emission, liquid waste effluents, or solid matter. The gaseous emissions released by industries, or during mining, and from exhausts of vehicles contain metals in gaseous form resulting in the formation of metal oxides and particulate matter in the atmosphere (Okedeyi et al., 2014). Metallurgical processes like electroplating, leather tanning, production of ferrous alloys, and nonferrous materials generate metal wastes. In electroplating, very high concentrations of chromium and lead are generated. In leather and tanning process, chromium is generated and, if discharged in the environment, these metals accumulate in wastewaters.

2.2.5 Waste Disposal

2.2.5.1 Municipal Waste

Municipal solid waste contains domestic waste generated by society (it includes organic waste, plastic waste, medical waste, chemical waste, electronic waste [e-waste]). Metals are introduced into the municipal solid waste by a variety of waste streams. The input streams of these metals include used batteries, dust, discarded paints, leaked motor oils, broken or unused glass, and wrappers made from coatings of heavy metals. Solid waste is disposed by landfilling and/or incineration where landfilling leachate is rich in toxic metals like chromium, cadmium, lead, mercury, nickel, copper, and zinc. According to the review article proposed by Smith (2009),

Zn and Pb are present in higher concentrations in municipal waste. Leachate reaches to the soil and groundwater and becomes the source of metal deposition (Smith, 2009).

2.2.5.1.1 E-Waste

The whole globe is running fast on the track of electronics and electronic gadgets that is said to be the digital revolution. This digital revolution is also responsible for the accumulation of metals in the environment at faster rate. Handling e-waste is one of the priority issues in the list of global problems. Nearly 30–50 million tons of e-waste is generated every year worldwide (CPCB, 2011). Some hazardous metals that leach out through e-waste include mercury present in fluorescent tubes, flat screen monitors, and mechanical doorbells and cadmium found in corrosive-resistant alloys used in marine and aviation environment, nickel–cadmium batteries, and light-sensitive resistors. If the batteries are not recycled properly, cadmium leaches out in the soil and hampers the soil health and microflora. Lead occurs mainly in CRT monitors, lead–acid batteries, and PVCs (Sitaramaiah and Kumari, 2014).

2.3 HEAVY METALS OF MAJOR CONCERN

Some of the toxic heavy metals that are of major concern for public health and environment are as follows.

2.3.1 ARSENIC

Arsenic is said to be a slow death mineral because it slowly accumulates into the body and disrupts the body functions one by one. The chemical symbol of arsenic is As; its atomic number is 30 and atomic mass is 75. Its occurrence in the earth's crust is around 0.00005%. Arsenic is found in both inorganic and organic forms. Trivalent and pentavalent arsenates are the most predominant inorganic forms, whereas monomethylarsinic acid, dimethylarsenic acid, and trimethylarsine oxide are the organic forms of arsenic.

2.3.1.1 Uses and Possible Sources of Release in the Ecosystem

Volcanic eruption and weathering of rocks release arsenic in different forms.

- *Mining*: Arsenics are found in the sulfides of Co, Cu, Pb, Ag, Sb, and Fe ores. During mining and smelting, this metal is released into tailings.
- *Pesticides*: Toxic As^{3+} occurs in herbicides, fungicides, and insecticides. Lead arsenate, which is an insecticide, contains pentavalent arsenate.
- *Poultry diet*: Diet given to poultry is rich in arsenic and it is used to enhance their growth. Banned compounds that have been used in poultry are roxarsone, arsanilic acid, nitarsone, and carbarsone (WHO, 2011).
- *Fuel combustion*: Burning of low-grade brown coal releases arsenic in the atmosphere. Industrial

processes of dyes, wine yards, ceramics, etc., include arsenic. Orpiment is the most toxic arsenic compound used in the tanning industry.

The Environmental Protection Agency (EPA) has documented the laws to limit the release of arsenic in the environment by all industries. The global survey of the EPA assessed that pesticides are the major arsenic contaminating source of groundwater. Therefore, many pesticides like DDT are banned by the EPA. According to the Occupational Safety and Health Administration, the permissible exposure limit for workers is 10 µg/m (Ziemacki et al., 1989; Aftab et al., 2013). The WHO has declared inorganic arsenic as human carcinogen, and the EPA has classified it as a toxic compound.

2.3.1.2 Potential Route of Exposure and Health Effects

The most inevitable source of arsenic intake is by drinking of contaminated water and diet. Inhalation of dust and fumes also leads to arsenic toxicity.

2.3.1.2.1 Mechanism of Toxicity

Inorganic trivalent arsenate (As^{3+}) is the most toxic form of arsenic. As^{3+} binds the sulfhydryl (SH) group of proteins and inactivates hundreds of enzymes involved in metabolic pathways. Along with this, it also phosphorylates many enzymes and inactivates them. Arsenic hampers the body functions by interrupting cellular respiration by the inactivation of mitochondrial enzymes. Arsenic affects the normal bodily functions by coagulating proteins, forms complexes with coenzymes, and inhibits the production of biomolecules subsequently inducing apoptosis. It creates an imbalance in the level of antioxidants (WHO, 2011; Tchounwou et al., 2012).

2.3.1.2.2 Diseases

Depending on the oxidation state and concentration, arsenic affects the cardiovascular system, the peripheral nervous system, and the gastrointestinal tract. Inhalation of arsine gas at 150 ppm causes immediate death; 70–180 mg of arsenic can be fatal. Arsenic toxicity is responsible for a disorder (anti-immune disorder) that is similar to the Guillain–Barré syndrome. It results into nerve inflammation and muscular weakness. In all its oxidation states, it acts as a carcinogen causing liver, skin, and bladder cancer. Chronic toxicity causes changes in the skin pigmentation and peripheral vascular disease (WHO, 2011; Tchounwou et al., 2012).

2.3.2 CADMIUM

Cadmium is one of the most critical heavy metal pollutants. It is said to be macro mineral that destroys the human body. Cadmium is a soft, light-colored metal with Cd as its chemical symbol. Its atomic number is 48 and atomic mass is 112.4. It is widely distributed in the earth's crust. Its concentration in the earth's crust is around 0.1 ppm (mg/kg). Its main occurrence is in sedimentary rocks and marine phosphates. It is chemically similar to zinc.

2.3.2.1 Uses and Possible Sources of Release in the Ecosystem

Naturally, cadmium is found in Zn ores and is released during volcanic eruptions.

- *Mining*: Cadmium is released as a by-product in the refining of zinc.
- *Fertilizers*: The use of phosphate fertilizers also contributes to the exposure of cadmium in the soil and groundwater.
- *Industrial processes/chemicals*: It is used in the production of paints, stabilizers, pigments, and detergents. So, waste generated from these factories is the major source for the release of cadmium. Nickel–cadmium batteries, wine bottle wraps, and mirror coatings are also cadmium emitters.
- *Fossil fuels*: Refined petroleum products are a reserve for cadmium.
- *Sewage sludge*: Sewage sludge is rich in cadmium and its derivatives. In the United States, the annual loading of Cd by sewage sludge is 190 kg/km² (Chibuike and Obiora, 2014).
- *Vegetables and food*: The organic form of cadmium is accumulated in leafy vegetables, potatoes, grains, liver, mushroom, dried seaweed, and cocoa powder and in the shells of crustaceans.

The EPA has announced cadmium as a highly toxic heavy metal.

2.3.2.2 Potential Route of Exposure and Health Effects

Cadmium enters the biological cycle by ingestion of contaminated food and water. Inhalation of cigarette smoke is the most prominent source of cadmium poisoning. Around 1–2 μg Cd is present in one cigarette and 10% of this is inhaled during smoking (WHO, 2007).

2.3.2.2.1 Mechanism of Toxicity

Cadmium is chemically similar to zinc. Zinc is the essential micronutrient required for plant and animal growth. Due to this property, cadmium affects cellular metabolism by replacing zinc. In the blood, it binds to ferritin and decreases hemoglobin. It also decreases the absorption of copper in the liver and plasma (WHO, 2011).

2.3.2.2.2 Health Effects

Cadmium is toxic at extremely low levels. Exposure to cadmium affects the circulatory system (blood pressure and myocardial dysfunctions). In humans, long-term exposure to cadmium results in kidney dysfunction. If cadmium is inhaled via dust and fumes in higher concentration, it severely damages the respiratory system. Cadmium causes bone defects, ouch-ouch disease, osteomalacia, osteoporosis, and spontaneous fractures because it interrupts calcium signaling. It causes the itai-itai disease characterized by severe bone pain. Cadmium is classified as a carcinogen by several regulatory authorities. It is a possible carcinogen involved in lung cancer (Tchounwou et al., 2012).

2.3.3 Chromium

Chromium is a lustrous, hard, steel-gray metallic element resistant to tarnish and corrosion. The name chromium was derived from the Greek word *chroma*, which means color. Its chemical symbol is Cr, atomic number is 24, and atomic mass is 52. It exists in divalent to hexavalent states. In nature, chromium is abundant in its trivalent state. Natural sources of chromium are chromite ($FeOCr_2O_3$) and crocoisite ($PbCrO_4$) ores. Chromium does not exist in its elemental form in nature, and its occurrence in the earth's crust is around 140 ppm.

2.3.3.1 Uses and Possible Sources of Release in the Ecosystem

- *Mining*: Mine tailing of the chromite ore is a very strong source of chromium release.
- *Fossil fuel combustion*: Fuel burning is the major source of pollution by all heavy metals.
- *Industrial processes*: These include the tanning industry; stainless steel production; chrome pigment production; production of sodium chromate, potassium chromate, ferrochromium, and dichromate; and electroplating. Potassium chromate is used in very huge quantities in tanning and leather industry as well as in dying.
- *Fertilizers*: Urea fertilizers release chromium into the rhizosphere.

2.3.3.2 Potential Route of Exposure and Health Effects

Chromium can enter the body system due to ingestion of chromium-contaminated food and water as well as by inhalation.

2.3.3.2.1 Mechanism of Toxicity

It is an essential dietary element. Cr^{6+} is the most toxic form of chromium because it readily passes through the cell membrane, and it is a strong oxidizing agent. It disrupts the proteins and lipids and results in the destruction of cell membrane integrity (Tchounwou et al., 2012).

2.3.3.2.2 Health Effects

Toxicity of chromium depends on its form, and its first target organ is the lung. Chromium damages the respiratory tract and causes nose inflammation and asthma. It also disrupts the reproductive function as its exposure results in decreased sperm count. Chromium acts as allergen and chronic exposure of chromium affects the nervous system, the cardiovascular system, and the gastrointestinal tract. It leads to stomach ulcers and inflammation of the intestine (Ziemacki et al., 1989).

2.3.4 Mercury

Mercury is the quixotic bad boy of the periodic table (exquisitely beautiful but deadly). The chemical symbol of mercury is Hg and its Latin name is hydragyrum (water silver) since it is liquid like water and shiny like silver. The atomic number of mercury is 80 and its atomic weight is 200.6. It exists in three forms, namely, elemental form (Hg), inorganic form (Hg^{2+}, Hg_2^{2+}), and organic form (CH_3Hg, CH_2CH_3Hg). Mercury is a rare element in the earth's crust as it is found only 0.08 ppm (mg/kg) by mass. Its most common ore found on earth is cinnabar (HgS).

2.3.4.1 Uses and Possible Sources of Release in the Ecosystem

Naturally, mercury is exposed during volcanic eruptions and mineral rocking. Nearly one-third of mercury in the environment is released due to natural processes.

- *Mining*: Mercury has an uncanny ability to bind precious metals like gold and silver. Therefore, mercury is released in gold mining, where burning releases mercury in the air and pure gold is isolated. As an example, if the content of mercury in a gold ore is 100 mg/kg, then for every hundred tons of ore, 1 kg of mercury is potentially released in the environment. This is a major source of mercury release and should be controlled.
- *Pesticides*: Several fungicides contain mercury that is toxic to the target fungus. But it also gets accumulated in the soil and living organisms.
- *Medical waste:* Mercury is used in the production of antiseptics, antisyphilitics, and antidepressants. It is also used in the manufacturing of thermometers, fluorescent tubes, barometers, and thermistors. When these instruments are broken, then they directly release mercury in the surrounding. Dental amalgams are a major source of mercury, and 10% of mercury emissions are contributed by medical waste.
- *Industrial processes*: Mercury is used in the production of paints and sodium hydroxide (NaOH).
- *Combustion of fossil fuels and coal*: Burning of fossil fuels in power plants, industries, and vehicles releases mercury in the air.
- *Pollution*: Mercury is a disaster causing heavy metal because it volatizes rapidly. It enters the environment through leaching of soil, acid rain, burning of coal, industrial processes, medical waste and mining. After its release, mercury exists in organic (methyl/ethylmercury), inorganic (mercurous (Hg_2^{2+}), and mercuric (Hg^{2+})) forms in the environment. Microorganisms convert the inorganic form into organic form by biotransformation. The most toxic form among these is methylmercury, which is soluble in water and volatizes in the air. The regulatory authority, EPA, has declared mercury is very toxic and is a serious health hazard (WHO, 2007).

2.3.4.2 Potential Route of Exposure and Health Effects

Inhalation of mercury vapors and ingestion of seafood or food contaminated with mercury are the prominent routes of exposure. Mercury is a neurotoxin and has a strong affinity for SH groups in proteins, enzymes, hemoglobin, and serum albumin. The central nervous system is affected by the damage to the blood–brain barrier. Transfer of metabolites such as amino acids to/from the brain is not properly regulated. In the stomach, at low pH, mercury is converted into its highly stable oxidizing states (Hg^{2+}). This binds with biomolecules, specially enzymes, and replaces the hydrogen atoms present in the molecule. Conformational changes in the proteins disrupt their function, and there is a generation of oxidative stress. Mercury in its organic and inorganic form alters the calcium homeostasis as a result of oxidative stress (Tchounwou et al., 2012).

2.3.4.2.1 Health Effects

In urine, mercury is found in its inorganic form. Mercury acts as an allergen in its metallic form. Inorganic forms of mercury cause anxiety, restlessness, and kidney problems. All are reversible symptoms, meaning they disappear after the removal of exposure to mercury. Acute mercury exposure damages the lungs, brain, and kidneys and causes gastrointestinal disorders. Long-term mercury exposure results in hypertension (Chibuike and Obiora, 2014). In the reproductive stage of females, acute mercury exposure results in spontaneous abortions or the fetus is born with disorders. In the blood, mercury is associated with its organic form. Mercury (monomethyl mercury, dimethyl mercury) poisoning leads to skin problems characterized by rashes on the hands and feet and leads to sensitivity of different tissues and organs (Ziemacki et al., 1989; WHO, 2007; Tchounwou et al., 2012).

2.3.5 LEAD

Lead is a horror mineral because of its violent action. Lead is a grayish soft metal with the chemical symbol Pb. Its atomic number is 82 and its atomic weight is 207.2. Lead concentration in the earth's crust is around 10–30 ppm. It is present as sulfides (PbS, $PbSO_4$) and oxides ($PbCO_3$). Its inorganic form is Pb^{2+}, and its most polluting organic form is tetramethyl lead and tetraethyl lead. Lead oxides and lead hydroxides contaminate the soil and groundwater.

2.3.5.1 Uses and Possible Sources of Release in the Ecosystem

- *Mining*: Lead is released during mining and smelting of sulfide ores.
- *Industrial processes*: Lead is used in the production of paints, metal products (pipes), and ammunitions, and lead azide or lead styphnate is used in firearms.
- *Pesticides*: Lead arsenate is used as a pesticide that contaminates the soil and groundwater due to leaching of lead.
- *Fertilizers*: Urea fertilizers are a potential source for release of lead in the environment.
- *E-waste*: Lead is used in the production of lead–acid chargeable batteries and is also a component of CRT monitors.

2.3.5.2 Potential Route of Exposure and Health Effects

Inhalation of contaminated soil and dust is the most prominent source for exposure to lead. Lead can also get absorbed through ingestion of contaminated food and contaminated drinking water and paints. It is a toxic cumulative poison.

2.3.5.2.1 Mechanism of Action

Lead has the ability to mimic the functions of calcium. Therefore, lead inhibits calcium binding in calcium-dependent signaling. It alters the enzyme activity by binding to SH and amino groups of enzymes and binds to biomolecules by replacing calcium and also accumulates in the bones. Cellular damage is caused by the formation of reactive oxygen species (Tchounwou et al., 2012).

2.3.5.2.2 Health Effects

Lead is not required in biological processes by organisms. Exposure of lead is very dangerous as it acts as a neurotoxin. The inorganic form of lead is accumulated in the kidney and bones followed by the liver, lungs, and heart. It affects the central nervous system, cardiovascular system, endocrine system, reproductive system (spontaneous abortion, decreased sperm count), and hematopoietic system (Tchounwou et al., 2012). The inorganic form of lead damages the gastrointestinal tract and urinary tract and causes bloody urine. Lead also inhibits the synthesis of hemoglobin. In children, lead poisoning affects the intelligent quotient (Ziemacki et al., 1989; WHO, 2007; Tchounwou et al., 2012).

2.3.6 Manganese

Manganese is a gray-white metal with the chemical symbol Mn. Its atomic number is 25 and its atomic mass is 54.93. It is found abundantly in the core of the earth. Its average concentration in the earth's crust is 0.11% (1100 ppm). It is not classified as a heavy metal by the EPA (EPA, 2001). It does not exist in its elemental state but occurs as oxides, carbonates, and silicates. Its most abundant ore is pyrolusite (MnO_2) and rhodochrosite ($MnCO_3$).

2.3.6.1 Uses and Possible Sources of Release in the Ecosystem

- *Mining*: Manganese occurs in the ore of iron.
- *Industrial processes*: It is used in the manufacturing of dry cell batteries and in leather and textile industries and iron–steel processing industries. Combustion of coal and fossil fuels releases manganese in the air.
- *Fertilizers*: Manganese is an essential micronutrient of plants and animals. Therefore, fertilizers (organic and chemical) constitute manganese in trace amounts.
- *Pesticides*: Maneb and mancozeb pesticides are rich in manganese content.

2.3.6.2 Potential Route of Exposure and Health Effects

The uptake of manganese is via foods like spinach, tea, herbs, rice, soya beans, and oysters. It also enters the body by inhalation of fumes and dust produced during welding of iron rods. It is not considered a toxic element (Meyers et al., 2008).

2.3.6.2.1 Mechanism of Action

Manganese acts as a cofactor in many enzymatic reactions. So its deficiency as well as higher concentration inhibits metabolic pathways.

2.3.6.2.2 Health Effects

It is an essential micronutrient required for growth. It is generally nontoxic, but at higher concentrations, it disturbs metabolic pathways. It is one of the elements responsible for obesity, defects in glucose metabolism, impairment of blood clotting, skin problems, changes in cholesterol levels, skeleton problems, birth defects, and changes in hair color. Its chronic exposure causes Parkinson's disease, respiratory failure, and pneumonia (Ziemacki et al., 1989).

2.3.7 Nickel

Nickel is a light-colored hard metal with Ni as its chemical symbol. Its atomic number is 28 and its atomic mass is 58.69. It is found in the earth's crust at a concentration of 58–94 ppm (mg/kg). Its most common ore is the pentlandite ore (sulfide ore).

2.3.7.1 Uses and Possible Sources of Release in the Ecosystem

- *Mining*: Nickel is released during the smelting of its ore.
- *Combustion of fossil fuel*: The major source of nickel emission is combustion of fossil fuel and residual oil burning.
- *Fertilizers*: It is present in urea and superphosphate fertilizers.
- *Industry*: It is used in the manufacturing of nickel–cadmium batteries and ceramics, in the electroplating industry, and in steel production (Sood et al., 2012).

2.3.7.2 Potential Route of Exposure and Health Effects

Breathing of polluted air, drinking of contaminated water, and ingestion of food are the sources for potential exposure. Smoking of cigarette is the major source for nickel exposure. It is considered as a moderate concern metallic element because of human carcinogenicity and toxicity to plants and aquatic organisms (CONC Report 2014).

2.3.7.2.1 Health Effects

Accumulation of higher concentrations of nickel in the body results in lung cancer, larynx cancer, lung embolism, respiratory failure, birth defects, asthma, chronic bronchitis, and allergic reactions.

2.3.8 Zinc

Its chemical symbol is Zn, atomic number is 30, and atomic mass is 65.4. It is one of the essential micronutrients required for the growth of plants and animals.

2.3.8.1 Uses and Possible Sources of Release in the Ecosystem

- *Mining*: Zinc ore mining releases the metal into the soil and water.
- *Fertilizers*: Zinc is one of the constituents of urea fertilizers. It is reported that zinc is present in excessive amounts in urea and superphosphate fertilizers.
- *Industry*: Zinc is used in the production of chemicals, paints, petrochemicals, dyes, textile, leather, and tanning. Waste produced during all these processes contains heavy metals.

2.3.8.2 Health Effects

As zinc is an essential micronutrient required for growth, it is considered to be relatively nontoxic, but it is toxic to aquatic life (Garbarino, 1995). Zinc deficiency causes birth defects. If taken in excessive amounts, it can cause hampered growth and problems in reproduction. Patients suffering from zinc toxicity suffer from vomiting, bloody urine, and diarrhea. In severe poisoning, patients are at the risk of kidney and liver failure.

2.3.9 Copper

Copper is a ductile, reddish-brown metallic element. Its name is derived from the Latin word "cuprum," and it is symbolized as Cu. Its atomic number is 29 and its atomic mass is 63.54. It is found in the earth's crust in the form of sulfides. Its different ores include cuprite (Cu_2O), chalcocite (Cu_2S), chalcopyrite ($CuFeS_2$), bornite (Cu_2FeS_3), and malachite ($CuCO_3.Cu(OH)_2$).

2.3.9.1 Uses and Possible Sources of Release in the Ecosystem

- *Fertilizers*: Copper is present in phosphate fertilizers.
- *Pesticides*: It is a very important ingredient in insecticides and fungicides.
- *Industry*: Copper-derived materials are present in the brewery equipment, copper wires, heaters, cooking pots, and ornaments.
- *Municipal and industrial wastes*: It is one of the sources for copper discharge. Waste containing discarded copper wires, copper utensils, etc., contribute in the discharge of copper in the environment (WHO, 2011).

2.3.9.2 Potential Route of Exposure and Health Effects

Copper enters the body by inhalation and ingestion of copper-containing food. Generally, it is not toxic to humans (EPA, 2001).

TABLE 2.1

United States Environmental Protection Agency Maximum Contamination Levels for Heavy Metals in Air, Soil, and Water

	Recommended Maximum Limits of Heavy Metals		
Metal	In Air/Particulate Matter ($\mu g/m^3$)	In Soil (mg/kg or ppm)	In Drinking Water (mg/L)
As	ND	50–300	0.01
Cd	0.005	1–3	0.005
Cr	ND	100–150	0.05
Hg	ND	ND	0.001
Pb	0.1–0.3	420	0.01
Mn	0.15	ND	0.05
Ni	1	30–75	0.02
Zn	ND	150–300	ND
Cu	ND	50–140	2
Reference	WHO (2000)	Okedeyi et al. (2014)	EPA (2001)

ND, not defined.

2.3.9.2.1 Health Effects

It is an essential micronutrient required by plants and animals and is involved in many metabolic reactions. Copper is a part of oxidation–reduction processes and acts as a stimulant for the activity of hemoglobin, in the process of hardening of collagen, hair keratinization, melanin synthesis, and lipid metabolism, and affects the properties of myelin sheath of nerve fibers. The body requires copper in acceptable concentrations, but its low dose and high dose cause various health effects. The estimated adult dietary intake for copper is between 2 and 4 mg/day.

Ingestion of high amount of copper leads to mental disorders, anemia, arthritis/rheumatoid arthritis, hypertension, nausea/vomiting, hyperactivity, schizophrenia, insomnia, autism, stuttering, postpartum psychosis, inflammation and enlargement of the liver, heart problems, and cystic fibrosis (EPA, 2001). The maximum contamination level of the heavy metals in the environment has been listed in Table 2.1.

2.3.10 Effect of Heavy Metal Pollution on Plants

As plants are the primary producers in the food chain, they are the base for a balanced geological cycle. Due to their immobilization, they are exposed to all types of attacks. One of the major evading attacks is heavy metal exposure. Due to heavy metal exposure, the growth of the plant is hampered, photosynthetic ability is reduced because of pigment destruction, and physiological characters get affected, such as height of the plant decreases. Seed germination decreases, morphology of leaves changes, and rate of transpiration reduces due to the accumulation of specific heavy metals. Beneficial microflora present in the rhizosphere is affected, which leads to the reduced growth of plant, and oxidative stress is also induced in the plants (Reale et al., 2015).

2.4 GLOBAL STATUS OF HEAVY METAL POLLUTION

Heavy metals are released in the air, water, and soil. The pollution of these natural resources by heavy metals in various countries is discussed in the following.

2.4.1 Europe

The European Monitoring and Evaluation Programme (EMEP) in 2015 conducted a survey where several countries were reviewed for the status of air pollution. As per the report, global lead emission has reduced up to 90% due to ban on the gasoline leaded petroleum and emission of mercury and cadmium is reduced by 60%. According to the assessment, the southwestern part of Germany, the Benelux region, the southern part of Poland, and the northern part of Italy and the Balkans are the most polluting zones for Hg, Pb (10–20 ng/m^3), and Cd (0.2–0.6 ng/m^3). In Europe, mercury emission was less in 2010 as compared to the emission levels in 1990. Hungary and the Benelux countries were declared as hot spots for lead and cadmium emission in the air. Central and South America, sub-Saharan Africa, and Southeast Asia are polluted by mercury deposition due to small-scale gold mining (Ilyin et al., 2015). According to the assessment report of EU, distribution of Cd in European countries varies from 83% (Republic of Moldova) to 7% (Spain). Cd deposition in soil is in the range of 30–70 g/km^2. The profile of lead deposition was 10–30 ng/m^3 in 1990 that decreased to 5–15 ng/m^3 in 2003. The United Kingdom showed maximum pollution due to lead explosion. Lead distribution in soil varied in the range of 10–70 mg/kg in European countries (WHO, 2007).

2.4.2 Africa

In African countries, the major source of soil pollution is waste released during mining of gold. In Zambia, 1000 mg/kg of Pb was found in soil around mining areas. High levels of lead were reported in the Natalspruit stream in Johannesburg, South Africa. One major accident of lead poisoning had occurred in 2010 in Nigeria during gold mining. In this accident, 400 children died and 2000 children suffered from permanent disabilities. In the Dandora region, soil pollution due to lead was 10 times higher than acceptable limits. The source of pollution was waste generated from dry cell battery processes. High concentration of cadmium was reported in Ghana, Municipal Lake in Cameroon, and Nigeria from industrial waste disposal (Chibuike and Obiora, 2014). According to the regional assessment report (WHO, 2014) on chemicals of public health concern and their management in the African Region, the major source for mercury and lead emissions is gold mining. The soil and water pollution by lead is maximum due to mining in Zambia. Arsenic concentration in drinking water is beyond the maximum acceptable limit in many countries including Botswana, Burkina Faso, Cameroon, Ethiopia, Ghana, and Nigeria (WHO, 2014).

In Nigeria, the study conducted by Fagbote and Olanipekun (2010) on heavy metal pollution of soil in the region of Agbabu bitumen deposit area concludes that the presence of heavy metals (Fe, Cu, Mn, Cr, Zn, Pb, Cd, Ni, V) in the dry soil is due to the natural occurrence of bitumen (Fagbote and Olanipekun, 2010). Okedeyi et al. (2014) have reported the total concentration of metals in the soil samples near the three coal-fired power plants in South Africa. The total metal concentration in soil ranged from 0.05 ± 0.02 to 1836 ± 70 μg/g, 0.08 ± 0.05 to 1744 ± 29 μg/g, and 0.07 ± 0.04 to 1735 ± 91 μg/g in Matla, Lethabo, and Rooiwal power plants, respectively (Okedeyi et al., 2014).

2.4.3 America

Old reports about the pollution of the Mississippi River have indicated that heavy metal pollution of water was due to the disposal of mine tailings. The tailings release most of the inorganic Pb and Cd in water as sediments while Cr and Cu are released in residual organic form. Concentration of Pb was in the range of 0.35–0.45 mg/L in the area of 600–1900 km from mining spots in the river water. According to this survey, the metal loading rate was Cd, 2.2 kg/day; Cr, 29 kg/day; Cu, 17 kg/day; Pb, 4.3 kg/day; and Hg, 0.17 kg/day from the wastewater discharge into the Mississippi River (Garbarino et al., 1995). Recent reports indicate improvement in the levels of metal loading that have decreased due to the strict implementation of regulations.

Gray et al. (2015) tested the concentration of Hg in drinking water samples in the United States; the reported values were below 2000 ng/L, but the concentration in mine wastewater leachate was 0.001–760 μg/L. The concentration of Hg in surface soils near mining regions in Texas (United States) was 3.8–11 μg/g of soil (Gray et al., 2015). Hurtado-Jiménezi and Gardea-Torresdey (2006) have studied well water contamination due to arsenic in Mexico. The mean concentration of arsenic in 129 well samples was found in the range of 14.7 to 101.9 μg/L. Of the 129 well samples, 14% wells (17 cities) had arsenic beyond the maximum permissible limits governed by the WHO (Hurtado-Jiménez and Gardea-Torresdey, 2006).

2.4.4 Asia

East Asia is a highly polluted region as indicated by several studies. Chaudhari et al. (2012) have reported heavy metal concentration in the air in Nagpur City in India. The concentration of Zn and Fe was higher than the other heavy metals (Pb, Cd, Ni, Cr). Mohanraj et al. (2004) have reported the concentration of Zn, Cu, Pb, Ni, Cr, and Cd in the air in Coimbatore, India, where among all the metals, the highest deposition of Pb (2147 ng/m^3) was observed. According to the report submitted by the Ministry of Water Resources, Central Water Commission, Government of India in the year 2014, the concentration of As in the river water samples was 0.00–9.47 μg/L. Cd concentration in the Yamuna River sample was the highest at 4 μg/L. Eleven major Indian rivers exceeded the concentration of Cr (>50 μg/L). Hg level was within the acceptable limits of BIS. From the total water samples tested,

9%–10% samples had excess levels of Ni (CWC report 2014). Das et al. (2013) describe the heavy accumulation (16–417 mg/kg) of arsenic in the soils of central India. Climate change and human pollution are reported to enhance the reducing conditions in deep cretaceous and tertiary aquifers, thereby increasing the availability of marcasite that can be the reason of possible arsenic contamination in groundwater (Keesari et al., 2015); this shows that natural processes may be accelerated by human pollution leading to faster rate of deterioration of water quality.

Waseem et al. (2014) have reviewed the heavy metal pollution status in Pakistan. According to the discussion, almost all the water resources are contaminated with arsenic three to five times greater than the regulatory limits recommended by the WHO. Heavy metals like Cd, Cr, Zn, Mn, Ni, Pb, and As were detected above the maximum acceptable limits of the WHO. In Lahore, the soil is highly contaminated with Cd, Cr, Pb, and Zn due to mining activities. The values exceeded the regulatory limits recommended by global agencies. Heavy metal loads of the soils in the study area were 137.5 mg/kg for Cu, 305.2 mg/kg for Cr, 51.3 mg/kg for Co, 79.0 mg/kg for Ni, and 139.0 mg/kg for Zn (Waseem et al., 2014).

Bangladesh is the most polluted country in terms of arsenic contamination; the concentration of As in wells is around 0–1660 ppb. There are 35–77 million people in Bangladesh who are at the risk of arsenic poisoning (WHO, 2011).

The heavy metal concentration in the Xiawangang River in China was reported by Jiang et al. (2013) as follows: Cu, 0.31–0.82 mg/L; Cd, 0.03–0.32 mg/L; Pb, 0.64–1.21 mg/L; and Zn, 2.79–5.69 mg/L. The estimated soil pollutants were 13.8–512.1 µg/g for Cd, 213.9–920.5 µg/g for Cu, 308.2–5,146.3 µg/g for Pb, and 1,898.1–14,105.5 µg/g for Zn.

2.4.5 AUSTRALIA

The world's largest lead smelter is located at Port Pirie, South Australia. Mining of ores has drastically polluted water in most of the rivers. Contaminants detected in river water samples were Cd, Zn, Cu, Pb, and Hg. The King River in Tasmania, Captains Flat and Molonglo River in New South Wales (mainly Zn, also Cu, Cd, Pb, and As), and Rum Jungle and Finniss River in the Northern Territory (Cu, Zn) show signs of metal pollution. Also, two rivers in Victoria (Lerderderg River and Goulburn River) that contain elevated mercury levels due to deposition of waste from old gold mining areas in the river water have been well studied (Gray et al., 2015). The heavy metal study reports in Rosebery town in Tasmania, Australia, show the presence of lead (4590 mg/kg) and arsenic (646 mg/kg). The water sample analysis showed very high levels of Mn (15,100 µg/L) (LEAD Action News, 2010).

According to a national survey, soil in 4 million hectares of arable land is severely contaminated due to increased industrialization. The main contaminants detected in the soil are Cd, Ni, Cu, As, Hg, Pb, DDT, and PAHs. The average concentration of cadmium in the soil was around 0.07–1.2 mg/kg all over the world (Waseem et al., 2014), indicating the major emission of heavy metals is due to metal

processing and industrial operations. All continents are heavily polluted with As (drinking water) that has exceeded the maximum acceptable limits. Therefore, arsenic pollution is a major problem in the globe. The Cd and Cu accumulation in the soil and water samples is also comparatively higher in the industrial zones. Smelters and mining are the major sources for the release of Pb, Cr, and Hg in the atmosphere as well as in water and soil.

2.5 CONVENTIONAL BIOREMEDIATION PRACTICES FOR HEAVY METAL REMOVAL

The development of eco-friendly practices and processes for a healthy future is the primary goal of all regulatory authorities. But heavy metal removal from any contaminated areas is tricky due to the inherent properties of heavy metals. Many processes, strategies, and technologies are studied and under development for the recovery and recycling of heavy metals. Some conventional practices for heavy metal removal or recovery are discussed in the following.

2.5.1 PHYSIOCHEMICAL PROCESSES

2.5.1.1 Chemical Precipitation

Chemical precipitation is the most commonly used method for the removal of soluble heavy metals in wastewater by the addition of an agent that converts the soluble metal into insoluble form that is recovered later by filtration/skimming.

- *Hydroxide precipitation*: In this method, lime is used to precipitate metals in the hydroxide compounds. It is used for the recovery of copper, iron, lead, and zinc.
- *Carbonate precipitation*: Carbonate precipitation uses sodium/potassium carbonate as a precipitating agent. It removes lead and cadmium.
- *Sulfide precipitation*: It is used when the concentration of heavy metals is high because most of the metals are stable as sulfides (Chaudhari et al., 2012).

2.5.1.2 Coagulation

The coagulation method is used after the step of chemical precipitation. Chemical precipitation forms very small precipitates that take time to settle down due to which coagulation is required. In coagulation, alum is used widely because it is cheaper and is an alkaline pH controller. Coagulation involves the formation of flocs that are easily removed by skimming. Pang et al. (2009) have studied the efficiency of aluminum sulfate, polyaluminum chloride, and magnesium chloride for the removal of Zn, Cu, Pb, and Fe. They have efficiently (99%) removed Pb from wastewater.

2.5.1.3 Adsorption

It is the most preferred method for the removal of toxic substances from waste. Adsorption is based on the phenomenon of capturing various substances present in the solution (wastewater) on porous or active material (adsorbent). Activated charcoal (carbon) is used in many industrial waste treatment plants due to its highly porous structure with high surface area. But due to its global and wide applications, it has become costly. Extensive studies have been carried out to use different varieties of low-cost adsorbents. Saif et al. (2015) have reported the use of activated carbon prepared from the pine cones of *Pinus roxburghii*. The maximum adsorption capacity of activated carbon (prepared from pine cones) was 14.2 mg/g for Cu (II), 31.4 Ni(II) and 29.6 mg/g Cr(VI) respectively.

2.5.1.4 Reverse Osmosis

It is a membrane-based technology and pressure-derived phenomenon. It uses a semipermeable membrane that acts as barrier and prevents the passage of pollutant/heavy metals through it. The removal efficiency of heavy metals (Cu^{2+}, Ni^{2+}, As^{5+}) by this process is around 90%–98% (Holan and Volesky, 1993).

2.5.2 Electrochemical Processes

2.5.2.1 Ion Exchange

Ion exchange is the process of exchange of ions between two electrolytes. In wastewater treatments, ion exchange is one of the most widely recommended methods for heavy metal removal. In the ion exchange process, hydrogen ion present on the ion exchanger is replaced by a metal ion present in the surrounding medium. Strong acid cation exchangers are used for the extraction of Zn^{2+}, Cu^{2+}, Co^{2+}, Ni^{2+}, Cd^{2+}, and Pb^{2+} (Chaudhari et al., 2012).

2.5.2.2 Electrocoagulation

Electrocoagulation is an advanced technique used in industrial wastewater treatment plants. This technique includes the application of electric charge to the water and neutralization of surface charge on the metal so as to coagulate the metal compounds selectively (Fenglian and Wang, 2011).

2.5.3 Disadvantages of Conventional Methods of Remediation

1. All the methods mentioned earlier are costly.
2. Methods like precipitation and adsorption produce a huge amount of metallic waste; recycling of such waste is again a major problem.
3. Electrochemical processes are dependent on electricity. Power consumption is tremendous in these processes.
4. Efficiency of metal removal is comparatively low, 1–100 mg/L (Wuana and Okieimen, 2011).

2.5.4 Biological Processes

2.5.4.1 Bioremediation

All these conventional processes have some disadvantages, for example, precipitation and coagulation produce very huge amount of metallic sludge that is again an issue for disposal. Bioremediation is a very convenient and efficient way for the removal of heavy metals. Bioremediation may be defined as the use of biological material for the removal of pollutants from the environment. In a bioremediation process, microorganisms use the pollutant as a source of energy and convert them into useful energy compounds through biochemical redox reactions.

Bioremediation is classified into different types on the basis of mechanism involved in the heavy metal removal.

- *Biomineralization*: It refers to the complete oxidation of organic pollutants.
- *Biotransformation*: The process involves conversion of toxic compounds into less toxic or harmless components.
- *Biosorption*: It may be defined as the adsorption of metals on biological surfaces or microbes.
- *Bioaccumulation*: Bioaccumulation may be defined as the accumulation of metals in the microbial cell and prevention of its mobility (Rajendran et al., 2003).

Biosorption is one of the most promising bioremediation techniques. The process of biosorption uses biological material as adsorbent while any activated surface is utilized in adsorption. There are different opinions about the mechanism of biosorption. Some reports explain the process as a passive uptake of metals by dead biomass through physiological processes (Rajendran et al., 2003). The intracellular active transport of metals by living cell is said to be bioaccumulation. Some biological materials accumulate heavy metals from wastewater through metabolically mediated processes or physicochemical interactions. Wide research is carried out globally to study how different low-cost and strong biosorbents can be developed and used in bioremediation. Agricultural waste (crop residues), waste sludge, algal biomass, fungal mycelia, bacterial biofilms, fly ash, tanning-containing biomass, chitosan, clay, peat moss, etc., are being developed as biosorbents.

Depending on the location where the metal compound is adsorbed, biosorption is classified as follows:

1. *Extracellular biosorption*: The metal compound is found extracellularly. The interactions involved in such biosorption include electrostatic interactions and chemical reactions between metal ions and compounds released in response to the metals. Electrostatic interactions also occur with dead biomass and heavy metals.
2. *Cell surface biosorption*: In this type of biosorption, metals form complexes with the cell wall components. Covalent bonds are formed between metal compound and surface reacting groups of the cell membrane. The mechanism involved is ion exchange, and amino groups, sulfide groups, and carboxyl groups are involved in the interaction with metal ions.

3. *Intracellular biosorption*: In this biosorption, intracellular transport of metal ions takes place. Metal ions are accumulated inside the cell. In this process, living cells are required.

Biosorption can be done by using agricultural waste like peanut husk charcoal, natural zeolite, and fly ash as low-cost adsorbent for the removal of Cu^{2+} and Zn^{2+} from wastewater. Some indicate the use of *Cajanus cajan* husk and Bengal gram (*Cicer arietinum*) husk for the extraction of Cr^{2+}, Fe^{3+}, Ni^{2+}, and Hg^{2+} from wastewater. Peanut shells have biosorption capacity for Cu(II) and Cr(III) ions, with monolayer sorption capacities of 25.39 mg Cu^{2+} and 27.86 mg Cr^{3+} per gram biomass, respectively (Wuana and Okieimen, 2011).

Biosorption of heavy metals can also be done using algal biomass. Algae are a diverse group of eukaryotes with an outstanding capacity of photosynthesis, faster reproduction cycles, and low nutritional requirement. Algal biomass is an efficient biosorbent because of its structure and composition. It has a high metal-binding capacity due to the functional groups (amino, carboxyl, hydroxyl, and sulfate) of proteins, lipids, glycolipids, and glycoproteins present on the cell wall. Many researchers have reported the application of algal biomass in heavy metal removal. Saunders et al. (2012) have reported the use of algal biomass for biosorption of heavy metals with *Hydrodictyon* sp., *Oedogonium* sp., and *Rhizoclonium* sp. from wastewater generated through coal-fired power generation plant. Metal adsorption rate was 137 mg/kg dry weight (Saunders et al., 2012). Holan et al. (1993) have reported the efficiency of marine algal biomass for the uptake of Ni and Pb. Abdel et al. (2012) have studied the factors (pH, contact time, and concentration of adsorbent) in the effective removal of Cd^{2+} and Pb^{2+} by *Anabaena* sp. where the biosorption capacity was 111.1 mg/g for Cd^{2+} and 121.95 mg/g for Pb^{2+}.

Biosorption of heavy metals is also done using fungal mycelia. Fungi are widely used in pharmaceutical, brewing/food, and biofuel industries for the production of a wide variety of products. A large amount of fungal mycelia are produced as waste during industrial processes. These waste mycelia can be used as adsorbents for metal recovery or metal extraction from contaminated waste. Joshi et al. (2012) have isolated 76 fungal isolates from industrial wastewater and studied for the biosorption of Pb, CD, Cr, and Ni. *Aspergillus terreus* was the highest accumulator than other isolates. Park et al. (2005) have studied the application of dead fungal biomass (*Aspergillus niger*) for the biotransformation of Cr^{6+} (toxic form) into Cr^{3+} (less toxic form). Aftab et al. (2013) describe the ability of *Aspergillus flavus* NA9 for the effective removal of Zn from industrial effluents. The initial concentration of Zn was 600 mg/L that was reduced to 287.8 ± 11.1 mg/L due to biosorption potential of the fungus.

Biosorption of heavy metals is also done using bacterial biomass. The chemical composition of the cell wall of bacteria includes functional groups like phosphate, amine, carboxyl, and sulfates that are responsible for selective heavy metal removal. Carboxyl groups have a high metal-binding capacity (Pang et al., 2009). Kumar et al. (2012) have studied the biosorption capacity of different bacterial cultures (*Staphylococcus*, *Pseudomonas*, *Bacillus*) to reduce the metals Ni, Cu, Cr, Pb, Zn, and Cd from solid waste (landfill), sludge, and industrial wastewater. Live and dead biomass of bacterial cells can both be used for the adsorption of metals. *Lysinibacillus* sp. BA2, a nickel-tolerant strain from bauxite mine, could adsorb 238.04 mg of nickel on 1 g of dead biomass and 196.32 mg on 1 g of live biomass (Desale et al., 2014a). *Pseudomonas* sp. has been reported to be very efficient in heavy metal removal among other species while *Staphylococcus* sp. reduced 93% of lead from all the contaminated samples. Oves et al. (2012) have reported the potential of *Bacillus thuringiensis* OSM 29 for the removal of Cd, Cr, Cu, Pb, and Ni from soil irrigated with metal-contaminated water. The strain had removal efficiency ranging from 87% to 94% for all metals. Kamika and Momba (2013) have studied the heavy metal removal capacity of *Pseudomonas putida*, *Bacillus licheniformis*, and *Peranema* sp. *P. putida* had the maximum biosorption capacity compared to the other strains, Co, 71%; Ni, 51%; Mn, 45%; V, 83%; Pb, 96%; Ti, 100%; and Cu, 49%. Rahman et al. (2014, 2015a,b) reported the accumulation of arsenics by *Lysinibacillus* sp. B1-CDA where genomic studies revealed the presence of nearly 123 proteins involved in metal binding and transport of metal ions. Rahman et al. (2015c) also reported the ability of *Enterobacter cloacae* B2-DHA in the remediation of chromium. One gram dry biomass of these cells accumulates 320 mg chromium. Although some microbes have unusual properties like tolerance to heavy metals, additional properties of the ability to degrade toxic pollutants or the ability to enhance plant growth can be useful in the reclamation of polluted habitats. For example, Desale et al. (2014b) reported the plant growth promoting abilities of heavy metal and salt-tolerant *Halobacillus* sp. and *Halomonas* sp. in the reclamation of salt marshes that have accumulated heavy metal deposits due to the discharge of heavy metal wastes in the sea.

Bioremediation of heavy metals using plants is called phytoremediation. This is an emerging technology that is cost-effective and sustainable. Extensively studied plants for bioremediation and genetic transformation include *Brassica* sp., *Thlaspi* sp., and *Nicotiana* sp. for the extraction of Pb, Hg, Cr, Zn, and As. There are different mechanisms of phytoremediation by which plants can reduce heavy metal concentration from the contaminated sites.

The mechanisms of phytoremediation are given as follows:

1. *Phytoextraction/phytoaccumulation*: It is the process of extraction of metals by plant cells from contaminated site into their parts (roots, stems, or leaves). This process includes the use of hyperaccumulators (the plants are able to accumulate metals 10–500 times higher than other plants). The mechanism followed by hyperaccumulators is the storage of heavy metals in the vacuole. The species of Brassicaceae have been reported to be hyperaccumulators. *Thlaspi* sp. and *Arabidopsis* sp. are hyperaccumulators for

toxic heavy metals. Another way of phytoextraction is the use of chelators. Chelators enhance the uptake of heavy metals. Some heavy metals form insoluble precipitates that cannot be absorbed by plants; addition of chelators prevents the precipitation and enhances the easy uptake by plants. Chelators also mediate the transport of metal compounds inside the plant. Ethylenediaminetetraacetic acid, citric acid, ammonium sulfate, elemental sulfur, etc., are used as chelating agents (Maiti et al., 2003). *Arabidopsis thaliana* PCS1 protein, which is involved in the binding of heavy metals in plants, is essential for the formation of chelating complex with cadmium or arsenite (Nahar et al., 2014). Chayapan et al. (2015) have reported that more than 10,000 mg of zinc/kg of plant parts could be accumulated by *Colocasia esculenta* L. Sood et al. (2012) describe the ability of *Azolla* for the removal of heavy metals from sewage sludge and industrial effluents.

2. *Phytotransformation*: The process of phytoremediation involves the uptake of metal ions/compounds from the surrounding and conversion of these compounds into nontoxic form. For example, Gardea-Torresdey et al. (2000) have reported the conversion of Cr^{4+} (toxic form) into Cr^{3+} (less toxic form) by *Avena monida* (oat) biomass (Gardea-Torresdey et al., 2000).

3. *Phytostabilization*: It refers to the immobilization of metals at the interface of soil and plant parts (roots) thereby preventing their discharge into the ecosystem. Heavy metals are mainly stabilized by the formation of precipitates and reduction of metal ions (Ziemacki et al., 1989).

4. *Phytovolatalization*: Phytovolatalization refers to the uptake of metal compounds, their conversion into volatile compounds, and subsequent release into the atmosphere. This technique is used in the removal of mercury. Mercuric ion is volatized into elemental mercury (Wuana and Okieimen, 2011).

5. *Rhizofiltration*: Rhizofiltration refers to the extraction of heavy metals from polluted water by plant roots and accumulation into the plant biomass. Various aquatic plant species are used for rhizofiltration. These include sharp dock (*Polygonum amphibium* L.), duck weed (*Lemna minor* L.), water hyacinth (*Eichhornia crassipes*), water lettuce (*Pistia stratiotes*), water dropwort (*Oenanthe javanica*), calamus (*Lepironia articulate*), and pennywort (*Hydrocotyle umbellate* L.). Brunet et al. (2008) described the ability of grass pea (*Lathyrus sativus* L.) to accumulate lead into roots. Meyers et al. (2008) have used *Brassica juncea* for the uptake of lead and studied its localization into root tissue.

Interestingly, a combined approach of microbial and phytoremediation can be more useful in the restoration of disturbed habitats.

2.5.4.2 Genetic Engineering

Recent biotechnological developments have documented the genetic modification of microorganisms and plants for efficient and specific heavy metal removal. In this approach, the genes responsible for heavy metal resistance, heavy metal tolerance, and detoxification of metals are integrated and expressed in living organisms. These genetic transformants are used for bioremediation. The plants studied extensively for genetic transformation to extract the heavy metals are *A. thaliana*, *Nicotiana tabacum*, and *Brassica* sp. *A. thaliana* has been modified by inserting bacterial genes to convert arsenate into a nontoxic form. Transgenic poplar trees were produced by the biotransformation of ScYCF1 (yeast cadmium factor 1) gene that encodes a transporter that sequesters toxic metal(loid)s into the vacuoles of budding yeast and tested for the bioaccumulation of Cd, Zn, and Pb (Shim et al., 2013).

2.5.4.3 Advantages of Bioremediation over Conventional Remediation Practices

1. The substrate for bioremediation is recycled waste in the case of biosorption. Therefore, the bioremediation process is low cost.
2. No additional waste is generated as in the case of conventional practices.
3. The methods are selective, so specific metal recovery is possible.
4. The biosorbents can be reused. After completion of one cycle of biosorption, biosorbents are treated to remove the metals and reused for subsequent cycles.
5. Site disruption is minimal.

2.5.4.4 Factors Limiting Application of Bioremediation Technology

For the success of bioremediation, optimum conditions (environmental and physiological) are required. It includes pH, temperature, nutrients, availability of and contact between polluted samples and bioremediators, type of soil, and population of microbes (Das et al., 2008).

In bioremediation, the pH preferred is neutral as, if the pH increases, it may affect the binding capacity of anionic groups present on the bacterial cell and inhibit specific metal-binding interactions between microbes and contaminating compounds. The optimum temperature required for bioremediation usually is 20°C–40°C (mesophilic conditions). Soil texture and its organic content also affect the ionic processes in the rhizosphere. The organic content of soil enhances the cation exchange phenomenon between microbes and soil, whereas sandy soil limits the cation exchange process. Because metals are positively charged, humic substances in organic soil form negatively charged particles and enhance the soil–metal-binding interactions. All these factors influence the bioremediation process and because metals are present in the various states/forms, the unavailability of adequate conditions limits the remedial process. Also, to achieve success in a bioremediation process, high concentration of desired microorganisms is required. Despite the challenges, bioremedial practices seem to offer an environment-friendly solution to metal pollution.

2.6 FUTURE PROSPECTS

Recent developments in biotechnology give novel ways to recover or handle harmful pollutants. Genetically engineered microbes or plants offer effective ways to resist the pollutant or remove the pollutants by several of the mechanisms described in this chapter. Although pollutants like metals are harmful, they are essential for various industrial processes, and their use can be minimized but cannot be completely banned. Thus, the recovery of metals from originating point sources and their effective management is the best possible way to control heavy metal depositions in natural resources. Besides this, strict regulations and their appropriate implementation are also essential to prevent unwarranted discharge of harmful wastes in natural water bodies or soils.

ACKNOWLEDGMENTS

This chapter is inspired from a research grant on bioremediation, and the authors are thankful to the Swedish International Development Cooperation Agency (SIDA, Grant no. AKT-2010-018) for the support. The authors are also thankful to Dr. D. Y. Patil Vidyapeeth, Pune, and University of Skövde, Sweden, for infrastructure support to implement the project on bioremediation.

REFERENCES

Abdel Aty, A., Ammar, N., Ghafar, H. et al. 2012. Biosorption of cadmium and lead from aqueous solution by fresh water alga *Anabaena sphaerica* biomass. *Journal of Advanced Research* 4: 367–374.

Aftab, K., Akhtar, K., Jabbar, A. 2013. Physico-chemical study for zinc removal and recovery onto native/chemically modified *Aspergillus flavus* NA9 from industrial effluent. *Water Research* 47: 4238–4246.

Benson, N., Anake, W., Etesin, U. 2014. Trace metals levels in inorganic fertilizers commercially available in Nigeria. *Journal of Scientific Research & Reports* 3: 610–620.

Brunet, J., Repellin, A., Varrault, G. 2008. Lead accumulation in the roots of grass pea (*Lathyrus sativus* L.): A novel plant for phytoremediation systems? *Comptes Rendus Biology* 331: 859–864.

Central Pollution Control Board. 2011. Implementation of E-Waste Rules 2011. Central Pollution Control Board, Delhi, India.

Central Water Commission Report. 2014. Status of trace and toxic metals in Indian rivers. Government of India, Ministry of Water Resources, Central Water Commission. http://www.cwc.nic.in/main/downloads/Trace%20&%20Toxic%20Report%2025%20June%202014.pdf. Accessed on December 20, 2015.

Chaudhari, P., Gupta, R., Gajghate, D. 2012. Heavy metal pollution of ambient air in Nagpur City. *Environmental Monitoring and Assessment* 184: 2487–2496.

Chayapan, P., Kruatrachue, M., Meetam, M. 2015. Phytoremediation potential of Cd and Zn by wetland plants, *Colocasia esculenta* L. Schott., *Cyperus malaccensis* Lam. and *Typha angustifolia* L. grown in hydroponics. *Journal of Environmental Biology* 36: 1179–1183.

Chibuike, G., Obiora, S. 2014. Heavy metal polluted soils: Effect on plants and bioremediation methods. *Applied and Environmental Soil Science* 2014: 1–12.

Das, N., Vimala, R., Karthika, P. 2008. Biosorption of heavy metals: An overview. *Indian Journal of Biotechnology* 7: 159–169.

Das, S., Jean, J., Kar, S. 2013. Bioaccessibility and health risk assessment of arsenic in arsenic-enriched soils, Central India. *Ecotoxicology and Environmental Safety* 92: 252–257.

Desale, P., Kashyap, D., Nawani, N., Nahar, N., Rahman, A., Kapadnis, B., Mandal, A. 2014a. Biosorption of nickel by *Lysinibacillus* sp. BA2 native to bauxite mine. *Ecotoxicology and Environmental Safety* 107: 260–268.

Desale, P., Patel, B., Singh, S., Malhotra, A., Nawani, N. 2014b. Plant growth promoting properties of *Halobacillus* sp. and *Halomonas* sp. in presence of salinity and heavy metals. *Journal of Basic Microbiology* 54: 781–791.

EPA. 2001. Parameters of water quality: Interpretation and standards. https://www.epa.ie/pubs/advice/water/quality/Water_Quality.pdf. Accessed on December 12, 2015.

Fagbote, E., Olanipekun, E. 2010. Evaluation of the status of heavy metal pollution of soil and plant (chromolaenaodorata) of agbabu bitumen deposit area, Nigeria. *American-Eurasian Journal of Scientific Research* 5: 241–248.

Fenglian, F., Wang, Q. 2011. Removal of metal ions from Wastewaters: A review. *Journal of Environmental Management* 92: 407–418.

Garbarino, J., Hayes, H., Roth, D. 1995. Heavy metals in the Mississippi River. In: Robert H. Meade *Contaminants in the Mississippi River*. U.S. Geological Survey Circular 1133, Reston, VA.

Gardea-Torresdey, J., Tiemann, K., Armendariz, V. 2000. Characterization of Cr (VI) binding and reduction to Cr (III) by the agricultural byproducts of *Avena monida* (oat) biomass. *Journal of Hazardous Materials* 80: 175–188.

Gray, J., Theodorakos, P., Fey, D. 2015. Mercury concentrations and distribution in soil, water, mine waste leachates, and air in and around mercury mine in the Big Bend region, Texas, USA. *Environmental Geochemistry and Health* 37: 35–48.

Holan, Z., Volesky, B. 1993. Biosorption of lead and nickel by biomass of marine algae. *Biotechnology and Bioengineering* 43: 1001–1009.

Hurtado-Jiménez, R., Gardea-Torresdey, J. 2006. Arsenic in drinking water in the Los Altos de Jalisco region of Mexico. *Pan American Journal of Public Health* 20: 236–247.

Ilyin, I., Rozovskaya, O., Travnikov, O. 2015. Heavy metals: Analysis of long-term trends, country-specific research and progress in mercury regional and global modeling. EMEP Status Report 2/2015.

Jiang, M., Zeng, G., Zhang, C. 2013. Assessment of heavy metal contamination in the surrounding soils and surface sediments in Xiawangang River, Qingshuitang District. *PLoS One* 8: e71176.

Joshi, P., Swarup, A., Maheshwari, S. 2012. Bioremediation of heavy metals in liquid media through fungi isolated from contaminated sources. *Indian Journal of Microbiology* 51: 482–487.

Kamika, L., Momba, M. 2013. Assessing the resistance and bioremediation ability of selected bacterial and protozoan species to heavy metals in metal-rich industrial wastewater. *BMC Microbiology* 13: 28.

Keesari, T., Ramakumar, K., BalaKrishnaPrasad, M., Chidambaram, S., Perumal, P., Prakash, D., Nawani, N. 2015. Microbial evaluation of groundwater and its implications on redox condition of a multi-layer sedimentary aquifer system. *Environmental Processes* 2: 331–346.

Kumar, A., Singh-Bisht, B., Joshi, V. 2012. Biosorption of heavy metals by four acclimated microbial species, *Bacillus* spp., *Pseudomonas* spp., *Staphylococcus* spp. and *Aspergillus niger*. *Journal of Biology and Environmental Science* 4: 97–108.

LEAD Action News 2010. Heavy metal poisoning in an Australian Lead Mining Town: The view from the Ramparts. *LEAD Action News* 11(2): 1–52.

Maiti, R., Pinero, J., Oreja, J. 2003. Plant based bioremediation and mechanisms of heavy metal tolerance of plants: A review. *Indian National Science Academy* B70: 1–12.

Meyers, D., Auchterlonie, G., Webb, R. 2008. Uptake and localisation of lead in the root system of *Brassica juncea*. *Environmental Pollution* 153: 323–332.

Mohanraj, R., Azeez, P., Priscilla, T. 2004. Heavy metals in airborne particulate matter of urban Coimbatore. *Archives of Environmental Contamination and Toxicology* 47: 162–167.

Nahar, N., Rahman, A., Moś, M., Warzecha, T., Ghosh, S., Hossain, K., Nawani, N., Mandal, A. 2014. In silico and in vivo studies of molecular structures and mechanisms of AtPCS1 protein involved in binding arsenite and/or cadmium in plant cells. *Journal of Molecular Modeling* 20: 2104.

Okedeyi, O., Dube, S., Awofolu, O. 2014. Assessing the enrichment of heavy metals in surface soil and plant (*Digitaria eriantha*) around coal-fired power plants in South Africa. *Environmental Science and Pollution Research* 21: 4686–4696.

Oves, M., Khan., Zaidi, A. 2012. Biosorption of heavy metals by *Bacillus thuringiensis* strain OSM29 originating from industrial effluent contaminated north Indian soil. *Saudi Journal of Biological Sciences* 20: 121–129.

Pang, F., Teng, S., Teng, T. 2009. Heavy metals removal by hydroxide precipitation and coagulation-flocculation methods from aqueous solutions. *Water Quality Research Journal of Canada* 44: 2.

Park, D., Yun, Y., Jo, J. 2005. Mechanism of hexavalent chromium removal by dead fungal biomass of *Aspergillus niger*. *Water Research* 39: 533–540.

Rahman, A., Nahar, N., Nawani, N. et al. 2014. Isolation of a *Lysinibacillus* strain B1-CDA showing potentials for arsenic bioremediation. *Journal of Environmental Science and Health, Part A* 49: 1349–1360.

Rahman, A., Nahar, N., Nawani, N., Jass, J., Ghosh, S., Olsson, B., Mandal, A. 2015a. Data in support of the comparative genome analysis of *Lysinibacillus* B1-CDA, a bacterium that accumulates arsenics. *Data in Brief* 5: 579–585.

Rahman, A., Nahar, N., Nawani, N., Jass, J., Ghosh, S., Olsson, B., Mandal, A. 2015b. Comparative genome analysis of *Lysinibacillus* B1-CDA, a bacterium that accumulates arsenics. *Genomics* 106: 384–392.

Rahman, A., Nahar, N., Nawani, N., Jass, J., Hossain, K., Saud, Z., Saha, A., Ghosh, S., Olsson, B., Mandal, A. 2015c. Bioremediation of hexavalent chromium (VI) by a soil borne bacterium, *Enterobacter cloacae* B2-DHA. *Journal of Environmental Science and Health, Part A* 50: 1136–1147.

Rajendran, P., Muthukrishnan, J., Gunasekaran, P. 2003. Microbes in heavy metal remediation. *Indian Journal of Experimental Biology* 41: 935–944.

Reale, L., Ferranti, F., Mantilacci, S. 2015. Cyto-histological and morpho-physiological responses of common duckweed (*Lemna minor* L.) to chromium. *Chemosphere* 145: 98–105.

Saif, M., Zia, K., Fazal-ur-Rehman. 2015. Removal of heavy metals by adsorption onto activated carbon derived from pine cones of *Pinus roxburghii*. *Water Environment Research* 87: 291–297.

Saunders, R., Paul, N., Hu, Y. 2012. Sustainable sources of biomass for bioremediation of heavy metals in waste water derived from coal-fired power generation. *PLoS One* 7(5): e36470.

Shim, D., Kim, S., Song, W. 2013. Transgenic poplar trees expressing yeast cadmium factor 1 exhibit the characteristics necessary for the phytoremediation of mine tailing soil. *Chemosphere* 90: 1478–1486.

Sitaramaiah, Y., Kumari, M. 2014. Impact of electronic waste leading to environmental pollution. *Journal of Chemical and Pharmaceutical Sciences*. Special issue 3: 39–42.

Smith, S. 2009. Critical review of the bioavailability and impacts of heavy metals in municipal solid waste composts compared to sewage sludge. *Environment International* 35: 142–156.

Sood, A., Uniyal, P., Prasanna, R. 2012. Phytoremediation potential of aquatic macrophyte, *Azolla*. *Ambio* 41: 122–137.

Tchounwou, P., Yedjou, C., Patlolla, A. 2012. Heavy metals toxicity and the environment. *EXS* 101: 133–164.

Waseem, A., Arshad, J., Iqbal, F. 2014. Pollution status of Pakistan: A retrospective review on heavy metal contamination of water, soil, and vegetables. *BioMed Research International* 2014. Article ID: 813206.

WHO. 2000. Air quality guidelines for Europe. http://www.euro.who.int/__data/ assets/pdf_file/0005/74732/E71922.pdf. Accessed on December 16, 2015.

WHO. 2007. Health risks of heavy metals from long-range transboundary air pollution. http://www.euro.who.int/__data/assets/pdf_file/0007/78649/E91044.pdf. Accessed on December 22, 2015.

WHO. 2011. Adverse health effects of heavy metals in children. http://www.who.int/ceh/capacity/heavy_metals.pdf. Accessed on December 10, 2015.

WHO. 2014. Chemicals of public health concern and their management in the African Region. http://apps.who.int/iris/bitstream/10665/178166/5/9789290232810 pdf. Accessed on December 20, 2015.

Wuana, R., Okieimen, F. 2011. Heavy metals in contaminated soils: A review of sources, chemistry, risks and best available strategies for remediation. *International Scholarly Research Network (ISRN) Ecology* 2011: 1–20.

Ziemacki, G., Viviano, G., Merly, F. 1989. Heavy metals: Resources and environmental presence. *Annalidel l'Istituto Superiore di Sanità* 25: 521–526.

3 Microbes
Natural Scavengers of Toxic Metals and Their Role in Bioremediation

Drewniak Lukasz and Uhrynowski Witold

CONTENTS

ABSTRACT

In this chapter, the mechanisms of microbially mediated mobilization and immobilization of heavy metals are reviewed in the context of bioremediation of contaminated environments. Microbial oxidative and reductive dissolution and organotrophic leaching of toxic elements from minerals, tailings, contaminated soil, and sediments are described as the main strategies for the treatment of solid-phase environments. In turn, biosorption and bioprecipitation are presented as the main driving forces for scavenging metal and metalloid ions from water and wastewater. All of these heavy metal transformations have been considered in terms of their self-purification potential as well as their applicability in various engineering remediation technologies.

3.1   INTRODUCTION

The ability to acquire nutrients from the environment is one of the fundamental physiological features of all living organisms. As the complexity of the organism increases, the cells undergo specialization, which may narrow the range of assimilable substrates and result in limited adaptability. This is not the case for microorganisms that have developed highly efficient and versatile mechanisms for the uptake of various nutrients, enabling them to survive in even the most extreme environments, where the typically preferred substrates are unavailable. Therefore, among the molecules used by microorganisms are not only the biogenic elements such as C, H, O, N, P, and S and their compounds but also metals and metalloids, including those of a potentially toxic nature, known as heavy metals (e.g., As, Cd, Cr, Cu, Ni, U).

Microorganisms may obtain heavy metal ions either directly from aqueous solutions or from the surface of minerals and other solid matter (e.g., soils, sediments, dumps, industrial waste) (Gadd, 2010a), following their dissolution caused by redox reactions with cell exudates. These ions can be used as sources of energy and terminal electron acceptors in respiration; they also constitute components of enzymes involved in other cellular processes. Scavenging and transformation of heavy metals are often connected with detoxification and resistance mechanisms, but they may also be carried out as side processes, unrelated to the stimulation of growth.

A vast majority of microorganisms possess and utilize mechanisms for the detoxification of heavy metals based on biosorption on the cell wall, entrapment in extracellular polymeric substances (EPS), precipitation, and complexation. All these microbial processes play an important role in biogeochemical cycles of metals and metalloids found in various environments and may also be applicable in bioremediation of both aqueous and solid-phase environments. Bioremediation of heavy metal–contaminated waters (such as groundwaters, leachates) is often carried out with the use of biosorption or biomineralization processes, in which metals are transformed into insoluble, chemically inert forms. In turn, the treatment of contaminated soil, sediments, and industrial waste may involve microbial dissolution of metal contaminants, chelation and complexation of metals by strains producing siderophores, and bioleaching caused by growing cells.

In this chapter, we describe microbial scavengers of metals and metalloids, with particular emphasis on the mechanisms of their transformation by both live and dead bacterial cells, leading to mobilization or immobilization of heavy metals. We also present examples of microbial bioremediation of contaminated water, wastewater, and soils by mobilization and immobilization strategies.

3.2 MOBILIZATION OF METALS AND METALLOIDS FROM SOLID PHASE

Metal and metalloid ions may be toxic to microbial cells even at low concentrations. However, some microorganisms developed efficient resistance mechanisms (Nies, 2003), and some of them even integrated heavy metals into the metabolism. For example, heavy metal ions can be used by numerous chemolithoautotrophic bacteria as sources of energy in electron-donating reactions or as terminal electron acceptors. Metal ions released in redox processes may also constitute structural elements of enzymes. Therefore, under certain conditions, microorganisms may intensify the natural physicochemical processes of dissolution of solid material, leading to the release of the required substrates.

To obtain metal and metalloid ions from minerals and other solid phases, chemolithotrophic microorganisms may carry out oxidation or reduction processes. In turn, heterotrophic bacteria can produce complexing agents, mainly carboxylic acids. Such microbial activity may prove beneficial for the mining industry, as it may cause the release of, for example,

precious metals from off-balance ore material in the process known as bioleaching. On the other hand, the same microbial processes occurring spontaneously in the environment may lead to dissemination of various elements in an uncontrolled manner. This is particularly worrisome as many microorganisms are capable of utilizing potentially toxic compounds, whose increased bioavailability may cause pollution.

However, in certain cases, dissolution of metals is desirable, and the involvement of microorganisms in the release of pollutants from contaminated soil, sediments, and industrial waste is favored (Gadd, 2010b), as the treatment of such material still remains one of the most costly environmental challenges (Sobecky and Coombs, 2009). In bioremediation, the controlled release of pollutants from solid material is often used in order to avoid an uncontrolled spread.

Given the earlier discussion, microorganisms are treated as key agents in heavy metal transformation in various environments, since they may be responsible for the release and dissemination of many metals and metalloids or may contribute to the (self) purification of contaminated areas. The paths and mechanisms of microbial transformation of heavy metals and the consequences of this interaction are closely linked to both the physiological requirements of the microorganisms and their response to environmental conditions.

Understanding of the mechanisms of solubilization and the methods of controlling the process are important for successful bioremediation. For this reason, the role of microorganisms in the biotransformation of minerals and other solid matter is still intensively studied (de Menezes et al., 2012). In this

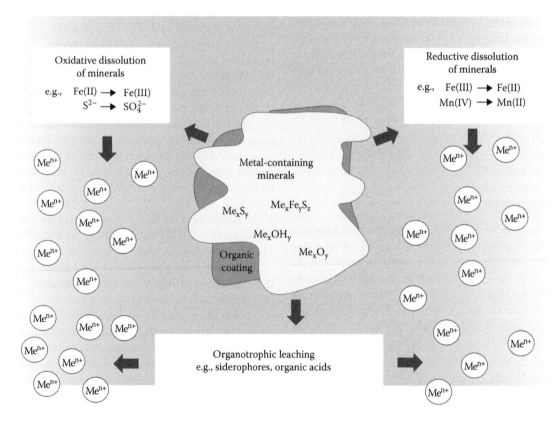

FIGURE 3.1 Schematic representation of the major microbial mechanisms leading to the mobilization of heavy metals.

chapter, we will present the latest findings on (1) oxidative and (2) reductive dissolution, as well as (3) protonolysis by organic acids and complexation by organic ligands (Figure 3.1). All these processes play a crucial role in the mobilization of heavy metals in different ecological niches and have found application in the bioremediation of solid-phase environments.

3.2.1 Oxidative Dissolution of Minerals

Oxidative dissolution of solid matter carried out by chemolithotrophic microorganisms is usually connected with metabolic processes, in which both metal and nonmetal elements, constituting the material, may be used as energy sources in electron-donating reactions. In these processes, oxidation of inorganic compounds is coupled with the reduction of molecular oxygen or, in the case of anaerobic conditions, oxyanions (e.g., nitrate or chlorate ions), which act as terminal electron acceptors (Rhine et al., 2006; Sun et al., 2009). The released energy is used for CO_2 fixation and cell growth.

In most cases, mineral dissolution processes are closely related to the oxidation of sulfides, leading to the formation of elemental sulfur or sulfuric acid, under anaerobic and aerobic conditions, respectively. The presence of the latter compound leads to the decrease of the pH and promotes further dissolution of minerals, causing the release of various metal ions (including heavy metals) from otherwise insoluble compounds (Smith and Melville, 2004).

Although oxidation of sulfur can occur spontaneously under the influence of oxygen, the activity of thiobacteria (e.g., *Acidithiobacillus*, *Thiothrix*, or *Thiovolum*) or archaeons (e.g., *Sulfolobus*) can greatly increase the efficiency of the process (Appelo and Postma, 2005). For example, in the presence of oxygen, *Acidithiobacillus thiooxidans*, an acidophilic representative of sulfur bacteria, produces sulfuric acid. In turn, under anaerobic conditions, metal sulfides can be oxidized by green and purple sulfur bacteria (e.g., *Allochromatium*, *Chlorobium*) causing the formation of elemental sulfur deposits (Frigaard and Dahl, 2009). Interestingly, some species of bacteria, including *Thiobacillus denitrificans*, can oxidize sulfides with simultaneous reduction of nitrates, causing denitrification (Beller et al., 2006).

Sulfur-oxidizing bacteria are capable of oxidizing sulfides of various metals, which are insoluble in water, including As_2S_3, As_4S_4, CdS, CoS, Cu_2S, NiS, PbS, and ZnS. The reaction proceeds in acidic pH, and polysulfides and elemental sulfur are formed as intermediate products. In turn, disulfides, such as FeS_2, MoS_2, and WS_2, are first oxidized by Fe(III), leading to the formation of thiosulfates, which are then degraded to sulfates (Sand et al., 2001).

Apart from the sulfur-oxidizing bacteria, autotrophic leaching of metal ions from solid matter may be carried out by iron-oxidizing bacteria, mainly the representatives of the genera *Acidithiobacillus* and *Leptospirillum* (Johnson and Hallberg, 2003). The activity of these microorganisms leads to the production of ferric iron, a strong oxidizing agent that contributes to the mobilization of various heavy metals co-occurring with pyrite and other Fe–S minerals.

Oxidation of pyrite may be carried out via an aerobic pathway, where Fe(II) is released to the surrounding along with two protons, one of which is then used to oxidize ferrous iron to Fe(III). Subsequently, in aqueous environment, ferric iron rapidly forms iron(III) hydroxide, insoluble in pH > 3.5, and the resulting protons cause pH to decrease. In strongly acidic pH, $Fe(OH)_3$ is not formed, and further oxidation of pyrite occurs in the reaction with Fe(III), which is faster than that with molecular oxygen. Under anaerobic conditions, ferric ions can also initiate pyrite oxidation, which is carried out as long as Fe(III) is present in the environment (Evangelou, 2001). Since Fe(III) formation is the limiting step in pyrite dissolution, sulfur and ferrous iron oxidation catalyzed by microorganisms may greatly speed up the process. The oxidation reactions may be carried out by bacteria attached directly to the mineral surface or in the exopolysaccharide excreted by some species. Exopolymers form a reaction zone, where the generated ferric ions are complexed to glucuronic acid residues and promote effective bioleaching (Welch and Ullman, 1999; Sand et al., 2001).

Iron oxidation may also be carried out by iron- and sulfur-oxidizing bacteria, for example, *A. ferrooxidans* (Osorio et al., 2013). Interestingly, some bacterial species can carry out iron and sulfur oxidation processes interchangeably, depending on the oxygen concentration in the surrounding. For example, *Sulfobacillus thermosulfidooxidans*, a gram positive facultative chemotrophic strain, isolated from mining wastes and mineral deposits, typically uses Fe(II) as electron donor. However, under microaerophilic conditions, this strain can also perform the reverse process. It reduces ferric iron by drawing energy from the oxidation of sulfur compounds. In such case, oxidation of pyrite (FeS) may generate Fe(III) and H_2SO_4 (Norris et al., 1996). This may prove to be beneficial for bioleaching reactions.

As it was explained earlier, oxidation of sulfide minerals may result in the acidification of waters and their contamination with sulfate anions and heavy metal ions (Johnson and Hallberg, 2003). This is a common problem in various mining areas, where metals, such as copper, zinc, or lead (Aykol et al., 2003; Lee et al., 2004), are extracted. Iron minerals can also be accompanied by other metals and metalloids, which together may form composite sulfides such as FeAsS or $CuFeS_2$ (Sand et al., 2001; Drewniak and Sklodowska, 2013). Oxidative dissolution of these minerals by sulfur- or iron-oxidizing bacteria typically leads to the release of iron along with arsenic and copper, respectively.

Dissemination of arsenic into the environment may also result from the activity of arsenite-oxidizing bacteria. These chemolithoautotrophic microorganisms promote the dissolution of arsenic-bearing minerals, increasing its concentration in the aqueous phase (Drewniak and Sklodowska, 2013). Some microorganisms utilize the solubilized arsenite ions as substrates for cellular respiration. Arsenates produced in the electron chain reactions are less mobile and thus less toxic form of arsenic. In contrast to As(III), arsenate ions can be effectively removed from water using different natural sorbents (Bochkarev et al., 2010). For this reason, As-oxidizing bacteria are considered important agents in bioremediation (Zouboulis and Katsoyianni, 2005).

The earlier discussion is just one example showing that it is possible to take advantage of the naturally occurring processes. Although solid matter dissolution may be harmful to the environment, microorganisms may be effectively used to increase the efficiency of metal extraction from ores (bioleaching) or to prevent the uncontrolled dissemination of toxicants (bioremediation).

Bioleaching of copper sulfide on a commercial scale was first carried out in the Rio Tinto mines (Amils et al., 2004). The leaching of other metals, including iron and zinc, occurred simultaneously. Nowadays, autotrophic leaching of metal sulfides by *Acidithiobacillus* species and other acidophilic bacteria is well established for use in industrial-scale biomining processes (Feng et al., 2015). For example, controlled dissolution of off-balance ores and industrial waste lead by iron-oxidizing bacteria found application in bioremediation of polluted waters (Johnson and Hallberg, 2005) and soil (Zagury et al., 1994). Autotrophic leaching has been used to remediate metal-contaminated solid materials including soil and red mud (sludge), the main waste product of aluminum extraction from bauxite (Gadd, 2001). Recently, acidophilic chemolithotrophic bacteria were used to remove heavy metals from activated sludge (Nicolova et al., 2015) and gray forest soil (Georgiev et al., 2015).

In the bioleaching and bioremediation processes, it is important to consider the entire microbial community, not only the chemolithoautotrophic organisms that are directly involved in dissolution reactions. For example, heterotrophic bacteria may also be important for effective detoxification of the environment, as they utilize organic compounds, which may inhibit autotrophic growth. Also other autochthonous bacteria, colonizing the same ecological niche, may constitute important components of many bioremediation systems. They can have a direct impact on the efficiency of the mobilization process and may contribute to the colonization of solid matter, for example, by biofilm production.

3.2.2 Reductive Dissolution of Minerals

The reaction pathways by which metal and metalloids can be mobilized from solid matter by microorganisms greatly depend on the physicochemical conditions of a given environment. Unlike oxidation processes, reductive dissolution cannot take place in the presence of oxygen that, for energetic reasons, is the preferred terminal electron acceptor for microorganisms capable of aerobic respiration. Only under anaerobic (reductive) conditions, other elements, including oxidized forms of metals (e.g., ferric iron, manganese(IV)) and metalloids (e.g., arsenate, selenate), can be used to generate energy in electron-accepting reactions, in which either organic or inorganic molecules may serve as electron donors. Microbial reduction processes are still being intensively studied, as they play a crucial role in the biogeochemical cycles of various elements and, especially, their mobilization from solid matter. In this section, the mechanisms behind these processes and the environmental impact of such reactions, as well as their potential applicability in bioremediation technologies, will be discussed.

In literature, there is much geochemical and microbiological evidence suggesting that Fe(III) reduction was one of the crucial forms of respiration on the early Earth (Vargas et al., 1998). Despite its very low solubility in pH > 3.5, iron (hydr) oxide constitutes one of the most widespread components of soils and sediments (Lloyd, 2003). It is natural, therefore, that numerous archaeons and bacteria inhabiting subsurface environments are capable of producing energy using ferric iron as the terminal electron acceptor (Vargas et al., 1998). In such communities, members of the Geobacteraceae family are considered the key players (Lloyd, 2003). In turn, Shewanellaceae representatives, for example, *Shewanella putrefaciens* (Lovley, 1993) or *S. oneidensis* MR-1, are important iron reducers found in aquatic sediments (Bennett et al., 2015).

Geobacteraceae are capable of Fe(III) reduction using a wide range of organic compounds as electron donors, mainly acetate and lactate (Call and Logan, 2011), and also toluene, phenol, benzoate, and other aromatic compounds (Lovley, 1993). In turn, the range of potential electron donors utilized by Shewanellaceae is restricted to small organic acids and hydrogen (Lloyd, 2003). Several other bacterial and archaeal species can oxidize H_2 during iron respiration, including *Wolinella succinogenes* (Lovley and Phillips, 1988) and *Archaeoglobus fulgidus* (Vargas et al., 1998), respectively.

Ferric iron can be biologically reduced only when efficient means of electron transfer between the cell and the mineral are provided. Preferably, microorganisms attach to the surface of the material (Coker et al., 2012), allowing for *in situ* activity of the dissimilatory ferric iron reductase (Schroder et al., 2003). Studies on the mechanisms of Fe(III) reduction of *S. oneidensis* and *Geobacter sulfurreducens* indicated the role of outer membrane-bound c-type cytochromes in the electron transport chain (Lloyd, 2003). If direct contact is impossible, bacteria can export electrons from the cell to Fe(III) using pili or nanowires as conductors (Reguera et al., 2005). Alternatively, ferric iron reduction may be carried out via extracellular electron shuttles, such as humic acids or quinolones (Lovley et al., 1996; Lloyd, 2003). The oxidized form of the shuttle molecule can act as an electron acceptor. Microbial reduction allows for the recovery of the shuttle so that the cycle may continue (Nevin and Lovley, 2002). Simultaneous occurrence of both direct and indirect mechanisms may further increase the rate of reductive dissolution.

Microorganisms carrying out dissimilatory iron reduction can also influence the dissolution of metals and metalloids co-occurring with Fe minerals (Lloyd, 2003). It must be noted that biogenic ferrous ions released in the process can act as reducing agents, contributing to the mobilization of a variety of contaminants, including Cr(VI), Tc(VII), and U(VI) (Gadd, 2010b). Interestingly, most of the ferric iron-reducing bacteria can also reduce Mn(IV) (Vandieken et al., 2012). Bacteria capable of manganese reduction are usually described together with iron reducers (see reviews by

Lovely, 2013), as in many cases one species can carry out transformation of both metals, for example, *Shewanella oneidensis* and *Geobacter metallireducens* (Lovley et al., 1989). One of the few examples of microorganisms that can reduce Fe(III), but not Mn(IV), is *Ferribacterium limneticum* (Lovley and Phillips, 1988; Cummings et al., 1999).

Although microbial reduction of Mn(IV) is regarded as similar in many ways to Fe(III) reduction, the mechanisms of both transformations differ in several aspects. First, in contrast to Fe(III) reduction, in which both MtrC and OmcA reductases are involved, only one terminal c-type cytochrome is needed for manganese(IV) transformation (Lin et al., 2012). Second, it was demonstrated that manganese(IV) reduction is a two-electron process, while the transformation of Fe(III) to Fe(II) requires only one electron. Moreover, it was shown that manganese(IV) reduction proceeds by two successive one-electron transfer steps, and soluble Mn(III) is produced as the transient intermediate (Luther, 2005).

Studies suggest that the first electron transfer reaction is only a reductive solubilization step leading to the increase of manganese bioavailability. Subsequently, Mn(III) species undergo further reduction process, which takes place outside the cell. For this reason, manganese compounds may be considered as versatile reagents as, depending on the environmental conditions, they can act either as electron acceptors or as electron donors, forming soluble Mn(II) or insoluble Mn(IV), respectively (Luther, 2005).

Manganese(IV) oxides can strongly absorb heavy metals, preventing their extraction from contaminated soils. As in the case of iron ores, solubilization of Mn(IV) compounds can also be increased with the addition of humic acids or similar molecules, which may function as electron shuttles. However, uncontrolled microbial manganese respiration may also contribute to the mobilization of co-occurring trace metals, which may contaminate waters (Lee et al., 2011).

As indicated earlier, iron- and manganese-reducing bacteria may simultaneously facilitate the release of other elements from solid matter into the aqueous phase. However, the efficiency of the cosolubilization process depends on the crystalline nature of the mineral. In some cases, microorganisms may also directly reduce metalloid oxyanions (e.g., arsenates or selenates) in anoxic environments.

Reductive dissolution of arsenic minerals may be carried out either for detoxification or energy production reasons. Paradoxically, As(V) reduction results in the formation of more mobile and toxic As(III) (Stolz and Oremland, 1999; Oremland and Stolz, 2003). However, as the entering of As(V) oxyanions into the cell is unavoidable due to their structural similarity to phosphate ions, many microbial arsenic resistance mechanisms involve specialized efflux proteins for the active transport of As(III) outside the cell (Mukhopadhyay et al., 2002).

In dissimilatory (i.e., respiratory) arsenic reduction, microorganisms utilize arsenic compounds as the terminal electron acceptors and organic (e.g., acetate, butyrate, lactate) (Newman et al., 1998) or inorganic compounds, such as sulfides (Hoeft and Bruggen, 2004), as electron donors.

The resulting arsenite is released into the surrounding (e.g., Blum et al., 1998). Thus, dissimilatory reduction is regarded as the main process leading to the mobilization of arsenic from secondary minerals (Drewniak and Sklodowska, 2013).

The first two strains capable of dissimilatory arsenate reduction, which have been described in the literature, were *Sulfurospirillum arsenophilum* MIT13 and *Sulfurospirillum barnesii* SES-3T (Stolz and Oremland, 1999). Since then, dissimilatory arsenate-reducing bacteria (DARBs) have been isolated from a variety of environments contaminated with arsenic. For example, *Citrobacter* sp. NC-1 and *Bacillus* sp. SF-1 were extracted from arsenic-polluted sediments (Frigaard and Bryant, 2008), and *Chrysiogenes arsenatis* was isolated from gold mine wastewaters (Langdona et al., 2003). Interestingly, bacteria with similar characteristics are also being found in nonarsenic-polluted geothermal waters and waters contaminated with other metalloids (Copeland et al., 2007).

The use of As(V) as electron acceptor may only be favorable when no other respiratory substrates, allowing for the production of more energy, are available. For example, the presence of such compounds as nitrate ions may hinder the process (Lloyd, 2003). Therefore, many DARBs are capable of using other molecules, including sulfates, thiosulfates, or nitrates as electron acceptors (Paez-Espino et al., 2009). Several dissimilatory Fe(III)-reducing *Shewanellaceae* representatives, including *Shewanella* ANA-3 (Tufano et al., 2008), *S. oneidensis* (Bennett et al., 2015), or *Shewanella* sp. O23S (Drewniak et al., 2015), are also capable of dissimilatory arsenate reduction (Jiang et al., 2013a). Some strains, for example, *Thermus strain* HR13, have been found to reduce As(V) to As(III) under anaerobic conditions and carry out the opposite reaction under aerobic conditions (Gihring and Banfield, 2001).

3.2.3 Complexation by Organic Acids and Ligands

Metal and metalloids present in rocks, secondary minerals, sediments, and waste materials become available only when they are extracted from the crystal lattice or other matrix that binds these elements. One of the main mechanisms of microbial release of metals from solid-phase materials is dissolution mediated by organic agents produced by the cells (Drever and Stillings, 1997). These organic agents include organic acids and organic ligands that are synthesized by the cells and excreted when growth is limited by the absence of an essential nutrient or as metabolic by-products.

Organic acids may directly affect mineral dissolution by (1) protonation of metal–oxygen bonds, what results in their destabilization, or (2) chelation of the metal by anions formed during acids, dissociation (Golab and Orlowska, 1988; Gadd, 2010b). Solubilization of minerals by organic acids may also proceed indirectly—as a result of the lowering of the solution saturation state due to complexation of metals in the solution (Bennett et al., 1988). Microorganisms that are involved in metal leaching by organic acids, production require an organic carbon source for growth and energy supply, as they do not take any advantage from solids, degradation.

An interesting example of microbial solubilization of solid metal by organic acid is dissolution of uranium and other toxic metals from industry waste by *Clostridium* sp. (Francis et al., 1994; Francis and Dodge, 2008). The activity of the tested strain leads to the solubilization of the waste material by the production of organic acids (acetic, butyric, propionic, formic, pyruvic, lactic, isobutyric, valeric, and isocaproic acids) from glucose fermentation and subsequent immobilization of uranium and other metals (Cd, Cr, Cu, Ni, Pb) by reductive precipitation and biosorption. Another example is bioweathering of arsenic-containing apatite $(Ca_2(PO_4)_3(F,Cl,OH))$, which is solubilized by the production of gluconic acid by *Burkholderia fungorum* (ATCC BAA-463) growing in a phosphorus-limited medium (Mailloux et al., 2009). In turn, Brantley et al. (2001) showed that low-molecular-weight organic acids (e.g., acetic, formic, oxalic, and citric) produced by *Arthrobacter* sp. yield protons to solution and decrease pH, but the effect plays only a supportive role in the chelation of metals by organic ligands.

Regardless of the produced organic acids and the mechanisms of solubilization, heterotrophic bacteria that are capable of fermenting glucose and other simple organic compounds play a significant role in many bioremediation systems and microbial communities involved in the cleanup of the environments contaminated with heavy metals. An example is the use of a mixed culture of heterotrophic bacteria producing peroxides and organic acids for the remediation of an alkaline soil, heavily polluted with radionuclides (mainly U and Ra) and heavy metals (Cu, Zn, Cd, Pb) (Groudev et al., 2015). The main sources of organic substrates for heterotrophic bacteria were plant compost, hay, and straw. The products (organic acids) obtained from their degradation were responsible for solubilization and complexation of heavy metals.

Organic ligands operate mainly by chelation of metal ions in aqueous phase, but they may also bind to atoms on the mineral surface and promote detachment of surface metal centers (Huang et al., 2011). Similarly to organic acids, metal complexing organic ligands may also indirectly promote mineral dissolution by lowering the relative saturation of the solution phase with respect to the mineral constituent (Huang et al., 2011). The largest class of organic ligands involved in complexation and chelation of metals are siderophores. These low-molecular-weight complexing agents are produced and secreted by fungi and bacteria only under conditions of iron deficiency (Raymond et al., 1984; Gadd, 2010a). Although siderophores production is directed only to Fe assimilation, these organic compounds are also able to effectively bind many other metal and metalloids, including As, Cd, Co, Cu, Cr, Mg, Mn, Ni, Pb, Sn, and Zn (Aiken et al., 2003; Malik, 2004; Schalk et al., 2011; Johnstone and Nolan, 2015). For example, the pyochelin siderophore produced by *Pseudomonas aeruginosa* chelates a wide range of metal ions, that is, Ag^+, Al^{3+}, Cd^{2+}, Co^{2+}, Cr^{2+}, Cu^{2+}, Eu^{3+}, Ga^{3+}, Hg^{2+}, Mn^{2+}, Ni^{2+}, Pb^{2+}, Sn^{2+}, Tb^{3+}, Tl^+, and Zn^{2+} (Braud et al., 2009). A very important feature of siderophores produced by bacteria is their ability to bind not only to solubilized metal ions but also to metals/metalloids on the mineral surface.

An interesting example are iron-binding ligands produced by *Pseudomonas* spp. that help them to acquire iron from insoluble arsenopyrite and other minerals found in gold mine ores (Drewniak et al., 2010). Mobilization of several metals (i.e., Fe, Ni, and Co) from waste material in the presence of siderophores produced by *P. fluorescens* was demonstrated for acid-leached ore of a former uranium mine (Edberg et al., 2010). Arsenic removal from metal-contaminated soil with the involvement of siderophores was presented for *Agrobacterium radiobacter* (Wang et al., 2011). The metal-binding abilities of microorganisms producing siderophores may be applied in the bioremediation of metal-contaminated soil and mine waste material.

3.3 IMMOBILIZATION OF HEAVY METALS

Immobilization of metals and metalloids by microbial biomass is an integral part of biogeochemical cycles of these and other elements in the environment. It depends on several factors, including physicochemical conditions, the presence of certain groups of microorganisms and their relationships with other biotic components of the ecosystems, as well as on the anthropogenic activities. Processes leading to the scavenging of metals from the environment by microbial biomass can be metabolically independent (dead biomass) or they may depend on the presence of active and growing cells (living biomass). Immobilization may occur by (1) adsorption to cell walls (biosorption) and entrapment in EPS, (2) intracellular accumulation and deposition, localization, and sequestration, or (3) precipitation in and around the cells as a result of metabolite release or reduction (Gadd, 2010b) (Figure 3.2). These biotransformations rarely occur separately and usually are interconnected, cumulatively driving or supporting self-cleaning processes in many heavy metal–contaminated environments. Microbial attenuation of heavy metals by sorption, precipitation, reduction, or other immobilization processes has been described for many natural aquatic and solid-phase (soils and sediments) ecosystems, as well as for areas with anthropogenic pollution (e.g., acid mine drainage and other industrial leachates) (Gadd, 2010b; Nicolova et al., 2014; Groudev et al., 2015). Applicability of most of the processes mentioned earlier is still investigated in laboratory scale. Some of them, however, have already been implemented.

3.3.1 BIOSORPTION

Biosorption is one of the most important microbial mechanisms of scavenging heavy metals from the environment, as practically every bacterial, fungal, and algal cell is able to bind heavy metals to the cell wall or other outer-membrane structural elements (Fomina and Gadd, 2014). It seems that the binding potential is determined by the type of biomass, as well as the specificity and chemical character of heavy metals (Volesky and Holan, 1995). It is worth emphasizing that biosorption has been documented for most metals, metalloids, and various radionuclides (Gadd, 2009; Wang and

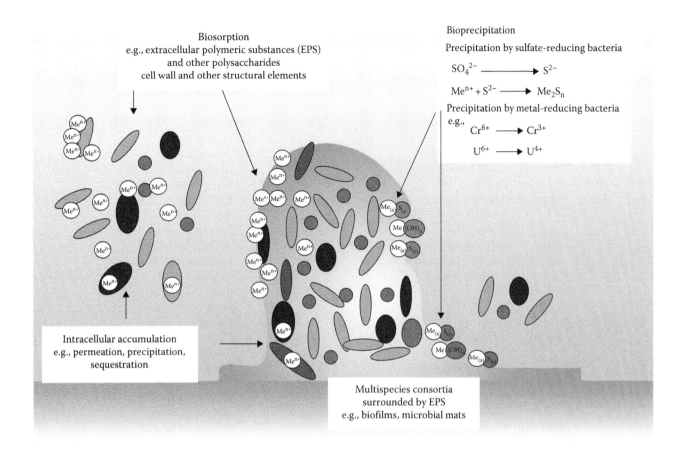

FIGURE 3.2 Microbiological processes involved in the immobilization of metals and metalloids.

Chen, 2009, 2014; Dhankhar and Hooda, 2011; Fomina and Gadd, 2014; Won et al., 2014). Among them, the most intensively studied were the elements that are known environmental pollutants, for example, As, Cd, Co, Cu, Cr, Pb, and U (Gadd, 2009; Gadd and Fomina, 2011).

Biosorption of heavy metals involves both passive and active processes. Passive biosorption is a metabolically independent process, while in the active processes, living biomass accumulates metals and metalloids. Passive biosorption may occur on the surface of living cells, for example, complexation of adsorbates by the cell wall components and/or EPS, or it can take place on dead biomass or cell fragments, which may also have potential for binding metals (Malik, 2004). In the external parts of the microbial cells, there are various structural components containing functional groups, which may interact with metal species, for example, carboxyl, phosphate, hydroxyl, amino, and thiol (Fomina and Gadd, 2014). Peptidoglycan carboxyl groups and phosphate groups found in cell walls are the main binding sites for metal cations in bacterial cells (Gadd, 2009). Pseudomurein, sulfonated polysaccharides, and glycoproteins forming the archaeal cell walls contain carboxyl and sulfate groups that are also characterized by a high metal affinity. Functional groups in polysaccharides such as cellulose, chitin, and alginate, which are components of fungal or algal cell walls, have also been found to play a very important role in the binding of metals (Wang and Chen, 2009). EPS found in capsules, slimes, and sheaths produced by many microorganism cells also play an important role in the binding of metal ions, sulfides, and oxides (Flemming, 1995; Comte et al., 2008; Gadd, 2009). EPS may consist of the following molecules: polysaccharides, proteins, nucleic acids, lipids, humic-like substances, and glycoproteins (Sheng et al., 2010). These EPS biomolecules may contain different functional groups, such as carboxyl, phosphoric, amine, and hydroxyl groups (Liu and Fang, 2002).

Many of the functional groups mentioned earlier as well as the metal-binding metabolites excreted by microbial cells may interact with metal ions by ion exchange, complexation, coordination, adsorption, electrostatic interaction, chelation, and precipitation (Veglio and Beolchini, 1997; Gutnick and Bach, 2000; Vijayaraghavan and Yun, 2008). Interaction of microbial cells with metal ions may occur by one of the mechanisms mentioned earlier or by a combination thereof.

Biosorption processes have been described and compared for thousands of environmental isolates and pure cultures of bacteria and fungi, in many reviews (Ahluwalia and Goyal, 2007; Wang and Chen, 2009; Park et al., 2010; Fomina and Gadd, 2014). Selected examples of bacterial and fungal species, whose biomass was found to have high biosorption

TABLE 3.1

Sorption Capacity of Microbial Biomass and Natural Mineral Sorbents

Metal	Microbes	Sorbent	Sorption Capacity (mg·g⁻¹)	Reference
Cd	Bacteria	*Staphylococcus xylosus*	250.0	Ziagova et al. (2007)
		Aeromonas caviae	155.3	Loukidou et al. (2004)
		Pseudomonas sp.	278.0	Ziagova et al. (2007)
	Fungi	*Saccharomyces cerevisiae*	79.2	Al-Saraj et al. (1999)
		Penicillium canescens	102.7	Say et al. (2003)
		Trametes versicolor	120.6	Arica et al. (2001)
	Minerals	Natural Jordanian sorbent		Al-Degs et al. (2006)
		Sepiolite	29.1	Hojati and Khademi (2013)
Cu	Bacteria	*Pseudomonas putida*	96.9	Uslu and Tanyol (2006)
		Streptomyces coelicolor	66.7	Ozturk et al. (2004)
		Bacillus firmus	381	Salehizadeh and Shojaosadati (2003)
	Fungi	*Saccharomyces cerevisiae*	6.4	Al-Saraj et al. (1999)
		Penicillium chrysogenum	108.3	Deng and Ting (2005)
		Aspergillus terreus	160	Gulati et al. (2002)
	Minerals	Bog iron ore	25.2	Rzepa et al. (2009)
		Peat	14.3	Ho and McKay (2000)
		Zeolite	11.0	Bajda et al. (2004)
Pb	Bacteria	*Corynebacterium glutamicum*	567.7	Choi and Yun (2004)
		Pseudomonas putida	270.4	Uslu and Tanyol (2006)
		Streptomyces rimosus	135.0	Selatnia et al. (2004)
	Fungi	*Saccharomyces cerevisiae*	79.2	Al-Saraj et al. (1999)
		Penicillium canescens	213.2	Say et al. (2003)
		Aspergillus niger	28.9	Dursun et al. (2003)
	Minerals	Bog iron ore	97.0	Rzepa et al. (2009)
		Peat	24.7	Ho and McKay (2000)
		Zeolite	28.3	Bajda et al. (2004)
Zn	Bacteria	*Aphanothece halophytica*	133	Incharoensakdi and Kitjaharn (2002)
		Acidithiobacillus ferrooxidans	172.4	Liu et al. (2004)
		Streptomyces rimosus	80.0	Mameri et al. (1999)
	Fungi	*Saccharomyces cerevisiae*	23.4	Al-Saraj et al. (1999)
		Penicillium notatum	23	Kapoor and Viraraghavan (1995)

capabilities, are presented in Table 3.1. Sorption capacities of several (commonly used) natural mineral sorbents are shown as a reference.

As can be seen in Table 3.1, metal sorption capacity of the microbial biomass may be competitive with the commonly used mineral sorbents. Despite this, most biosorption processes have not been implemented and are still investigated at the laboratory scale (Fomina and Gadd, 2014). One of the reasons is the fact that biosorption capacity of most of the isolates described in the literature is not as high as in the examples given in Table 3.1 and their biomass does not have a sufficiently high metal-binding capacity and selectivity for heavy metals, which is required for use in full-scale biosorption processes. Among the other problems associated with the implementation of bacterial and fungal sorbents, the following should be mentioned: (1) an insufficient understanding of the mechanisms, kinetics, and thermodynamics of the process, (2) the process of regeneration and reuse in online processes is complex and expensive, and (3) the high cost of immobilization of sorbents.

Biosorption capabilities greatly differ between the biomass made of live and dead cells. Most studies reported that dead biomass has a greater sorption capacity and numerous advantages compared to the active and growing cells as, for example, it is not limited by the presence of high concentrations of toxicants nor by the absence of the required growth media and nutrients in the feed solution (Dhankhar and Hooda, 2011; Prithviraj et al., 2014). However, the applicability of dead biomass may be subject to the same restrictions as the mineral sorbents, which can be used efficiently only when the concentration of pollutants is high, owing to the formation of an equilibrium state between the sorbent and the sorbate, described by Langmuir isotherms. As the concentrations of metals in the environment are usually much lower than the ones tested in the laboratory experiments, sorption processes at environmental conditions do not reach the expected efficiency.

The situation is much different in the case of application of active and growing cells, which can effectively remove heavy metals even at relatively low concentrations of pollutants. This is due to the fact that physical sorption on the cell surfaces is supported by the continuous metabolic uptake of metals and precipitation by metabolic exudates from the organism (Malik, 2004). Scavenging abilities of growing cells are also assisted by irreversible processes, which take place during detoxification (e.g., the metals diffuse into the cells and bind to intracellular proteins or chelations and are incorporated into intracellular sites) and/or by such accompanying processes as bioprecipitation. For several environmental isolates, the potential application of live microbial cells for metal scavenging has been reported (Prithviraj et al., 2014), but biosorption processes enhanced by accumulation and precipitation are generally assigned to multispecies consortia and biofilms. These consortia contain a mixture of specialized strains (in sorption and accumulation), and they additionally offer a microenvironment, which could be very beneficial for metal precipitation (e.g., alkaline pH, high concentrations of CO_2) and may provide an entrapment barrier for dispersed solids due to the high exopolymer content (Malik, 2004). Such multispecies consortia naturally form biofilms and microbial mats in many heavy metal–contaminated environments including hot springs, freshwater lake sediments (Gough and Stahl, 2011), acid streams, and mine drainage (Sanchez-Andrea et al., 2011; Hogsden and Harding, 2012; Zirnstein et al., 2012; Streten-Joyce et al., 2013; Mendez-Garcia et al., 2014). In many cases, biofilms and microbial mats play a crucial role in self-purification of such environments, as they can effectively retain heavy metals in the matrix. An example may be the natural biofilm found in Piquette Pb-Zn deposit, which is dominated by sulfate-reducing bacteria (SRB) of the family *Desulfobacteraceae* (Tennyson, Wisconsin) (Labrenz et al., 2000). This biofilm entraps zinc and other metal and nonmetal ions and is involved in sphalerite (ZnS) precipitation. The phenomenon of biosorption assisted by precipitation with the use of active and growing cells is used in practice (in pilot and industrial scale) in many engineering treatment approaches, such as permeable reactive barriers (PRB) or other passive systems (Ledin and Pedersen, 1996; Groudev et al., 2008a).

In PRB systems, various reactive (inorganic and organic) materials are used for treating heavy metal–contaminated waters. Activated sludge is one of the examples of microbial biomass materials used in PRB, as it conducts the removal of heavy metal from the contaminated groundwater. Sorption of heavy metals by activated sludge depends on the presence of microorganisms, whose cell walls and other structural elements contain functional groups, for example, carboxyl, hydroxyl, or phenolic, and which may additionally be surrounded by EPS (Yuncu et al., 2006). Biosorption processes carried out by microbial consortia may also be used in active systems, in which rotating biological contactor supports the immobilized growing biofilms. Orandi and Lewis (2013) showed that an indigenous algal-microbial biofilm from acid mine drainage has a high potential for the immobilization of a variety of elements from AMD in a photo-rotating biological contactor.

3.3.2 Biomineralization

Microbial metabolic processes may contribute to mineral precipitation and thus may lead to metal immobilization in many environments. Biomineralization processes may result from (1) microbial oxidation/reduction of metal species, (2) interaction with metabolite exudates (e.g., sulfide, oxalate), or (3) nucleation and formation of mineral precipitates on the surface of cells (Beveridge, 1989; Fortin and Beveridge, 1997; Gadd, 2009). Microbial mineralization processes are very often accompanied by sorption and accumulation. Therefore, in most cases, bioprecipitation of heavy metals is not directly controlled by the organism and occurs as a response to the interactions between metal (or metalloid) and metabolic exudates from the organism.

Microbial-mediated oxidation of metals may be the cause of biomineralization by chemical precipitation on the surface of cells or by extracellular precipitation. A direct removal of heavy metal ions from aqueous environments is carried out, for example, by manganese oxidizing bacteria. Several studies have reported that the bacterial oxidation of Mn(III) leads to the formation of Mn(II) and Mn(IV) oxides, which are then deposited on the surface of cells (Tebo et al., 2005). This process can be carried out nonenzymatically, as a result of nonspecific interactions with cellular or extracellular products, or may be catalyzed by multicopper oxidases (Tebo et al., 2005).

Immobilization of metals in and/or around the cells is also attributed to microbial reduction of metals. A number of bacteria are able to anaerobically reduce U(VI) to U(IV) and Cr(VI) to Cr(III) using a variety of organic electron donors (Gadd, 2010b). Microbial reduction of hexavalent U(VI) to tetravalent U(IV) may lead to the formation of uranium ores such as uraninite (UO_2) and, thus, to the effective uranium removal from contaminated environments (Landa, 2005; Renshaw et al., 2005). Reductive precipitation of metals may also proceed by an indirect mechanism. Iron-reducing bacteria promote metal precipitation by the production of (bio)genic Fe, which can catalyze the reduction of metals or may remove metals by sorption (Gadd, 2010b; Williamson et al., 2014).

Scavenging of heavy metals by biomineralization is commonly carried out by SRB. Under reducing conditions, SRB can generate hydrogen sulfide, which reacts with metal ions (e.g., Cd, Co, Cu, Fe, Ni, Pb, or Zn) and leads to the precipitation of secondary sulfide minerals (Ludwig et al., 2002), according to the following reactions (Kovach et al. 1995):

$$SO_4^{2-} + 2CH_2O \rightarrow H_2S + 2HCO_3^- \tag{3.1}$$

$$H_2S + Me^{2+} \rightarrow MeS_{(s)} + 2H^+ \tag{3.2}$$

As can be seen from the reaction equation (3.1), dissimilatory sulfate reduction is coupled with organic compound oxidation and leads to the production of carbonate, which may directly contribute to the precipitation of metals. Carbonate production may also increase the pH and indirectly lead to the precipitation of metals (mostly in the form of hydroxides)

(Warren et al., 2001; Ludwig et al., 2002; Perez-Gonzalez et al., 2010). A number of bacterial strains and consortia have been shown to mediate this reaction and be involved in metal attenuation by precipitation (Baker et al., 2003; Church et al., 2007; Barbosa et al., 2014; Janyasuthiwong et al., 2015). For this reason, SRB constitute the most common group of microorganisms used in the bioremediation of heavy metal–contaminated environments.

Usually, the SRB are used in passive systems such as wetlands or PRB or upward flow bioreactors (Da Silva et al., 2007; Groudev et al., 2010; Uster et al., 2015). In the passive systems, the source of SRB and organic components are locally available low-cost materials such as mushroom compost, manure of cow, chicken, horse and sheep, municipal compost, sawdust, peat, straw, and leaf compost (Costa et al., 2008; Groudev et al., 2008b; Zhang and Wang, 2014; Uster et al., 2015). These organic substrates constitute well-known sources of not only SRB consortia but also other groups of microorganisms (e.g., cellulolytic, denitrifying), which may also increase SRB growth and enhance heavy metal removal through the production of intermediate metabolites. Furthermore, the structure of the organic matter added to the system may function as a sorption barrier (Zhang and Wang, 2014). Applications of such organic carbon–based sulfate-reducing systems require controlled pH conditions, as most SRB prefer pH > 5. This is especially important in the purification of acid mine drainage (Siegrist et al., 2008) and other environments with very low pH (1.5–3.0), where neutralization of the effluent is necessary before it reaches the reactor. Naturally, some of the SRB are also capable of low-pH sulfidogenesis and metal attenuation, and they may find application in the remediation of acidic environments (Kolmert and Johnson, 2001; Kimura et al., 2006; Sierra-Alvarez et al., 2006), but in most cases limestone neutralization is performed.

3.4 SUMMARY

In this chapter, the mechanisms of microbially mediated mobilization and immobilization of heavy metals have been reviewed in the context of bioremediation of contaminated environments. Microbial oxidative and reductive dissolution and organotrophic leaching of toxic elements from minerals, tailings, contaminated soil, and sediments have been described as the main strategies for the treatment of solid-phase environments. In turn, biosorption and bioprecipitation have been presented as the main driving force for scavenging of metal and metalloid ions from water and wastewater. All of the heavy metal transformations mentioned earlier have been considered in terms of self-purification potential as well as their applicability in various engineering remediation technologies.

REFERENCES

Ahluwalia, S. S. and Goyal, D. 2007. Microbial and plant derived biomass for removal of heavy metals from wastewater. *Bioresource Technology*, 98, 2243–2257.

Aiken, A. M., Peyton, B. M., Apel, W. A., and Petersen, J. N. 2003. Heavy metal induced inhibition of *Aspergillus niger* reductase: Applications for rapid contaminant detection in aqueous samples. *Analytica Chimica Acta*, 480, 131–142.

Al-Degs, Y. S., El-Barghouthi, M. I., Issa, A. A., Khraisheh, M. A., and Walker, G. M. 2006. Sorption of Zn(II), Pb(II), and Co(II) using natural sorbents: Equilibrium and kinetic studies. *Water Research*, 40, 2645–2658.

Al-Saraj, M., Abdel-Latif, M. S., El-Nahal, I., and Baraka, R. 1999. Bioaccumulation of some hazardous metals by sol-gel entrapped microorganisms. *Journal of Non-Crystalline Solids*, 248, 137–140.

Amils, R. G. T., Gómez, F., Fernández-Remolar, D., Rodríguez, N., Malki, M., Zuluaga, J., Aguilera, A., and Amaral-Zettler, L. A. 2004. Importance of chemolithotrophy for early life on earth: The tinto river (iberian pyritic belt) case. In Seckbach, J. (ed.) *Origins*. Amsterdam, the Netherlands: Springer.

Appelo, C. A. J. and Postma, D. 2005. *Geochemistry, Groundwater and Pollution*, 2nd edn., Leiden, the Netherlands: A.A. Balkema Publishers.

Arica, M. Y., Kacar, Y., and Genc, O. 2001. Entrapment of white-rot fungus trametes versicolor in ca-alginate beads: Preparation and biosorption kinetic analysis for cadmium removal from an aqueous solution. *Bioresource Technology*, 80, 121–129.

Aykol, A., Budakoglu, M., Kumral, M., Gultekin, A. H., Turhan, M., Esenli, V., Yavuz, F., and Orgun, Y. 2003. Heavy metal pollution and acid drainage from the abandoned balya pb-zn sulfide mine, NW Anatolia, Turkey. *Environmental Geology*, 45, 198–208.

Bajda, T., Franus, W., Manecki, M., Mozgawa, W., and Sikora, M. 2004. Sorption of heavy metals on natural zeolite and smectite-zeolite shale from the polish flysch carpathians. *Polish Journal of Environmental Studies*, 13, 4.

Baker, B. J., Moser, D. P., Macgregor, B. J., Fishbain, S., Wagner, M., Fry, N. K., Jackson, B. et al. 2003. Related assemblages of sulphate-reducing bacteria associated with ultradeep gold mines of South Africa and deep basalt aquifers of Washington State. *Environmental Microbiology*, 5, 267–277.

Barbosa, L. P., Costa, P. F., Bertolino, S. M., Silva, J. C. C., Guerra-Sa, R., Leao, V. A., and Teixeira, M. C. 2014. Nickel, manganese and copper removal by a mixed consortium of sulfate reducing bacteria at a high COD/sulfate ratio. *World Journal of Microbiology & Biotechnology*, 30, 2171–2180.

Beller, H. R., Chain, P. S. G., Letain, T. E., Chakicherla, A., Larimer, F. W., Richardson, P. M., Coleman, M. A., Wood, A. P., and Kelly, D. P. 2006. The genome sequence of the obligately chemolithoautotrophic, facultatively anaerobic bacterium *Thiobacillus denitrificans*. *Journal of Bacteriology*, 188, 1473–1488.

Bennett, B. D., Brutinel, E. D., and Gralnick, J. A. 2015. A ferrous iron exporter mediates iron resistance in *Shewanella oneidensis* Mr-1. *Applied and Environmental Microbiology*, 81, 7938–7944.

Bennett, P. C., Melcer, M. E., Siegel, D. I., and Hassett, J. P. 1988. The dissolution of quartz in dilute aqueous-solutions of organic-acids at 25-degrees-C. *Geochimica et Cosmochimica Acta*, 52, 1521–1530.

Beveridge, T. J. 1989. Role of cellular design in bacterial metal accumulation and mineralization. *Annual Review of Microbiology*, 43, 147–171.

Blum, J. S., Bindi, A. B., Buzzelli, J., Stolz, J. F., and Oremland, R. S. 1998. *Bacillus arsenicoselenatis*, sp. nov., and *Bacillus selenitireducens*, sp. nov.: Two haloalkaliphiles from Mono Lake, California that respire oxyanions of selenium and arsenic. *Archives of Microbiology*, 171, 19–30.

Bochkarev, G. R., Pushkareva, G. I., and Kovalenko, K. A. 2010. Natural sorbent and catalyst to remove arsenic from natural and waste waters. *Journal of Mining Science*, 46, 197–202.

Brantley, S. L., Liermann, L., Bau, M., and Wu, S. 2001. Uptake of trace metals and rare earth elements from hornblende by a soil bacterium. *Geomicrobiology Journal*, 18, 37–61.

Braud, A., Hannauer, M., Mislin, G. L. A., and Schalk, I. J. 2009. The *Pseudomonas aeruginosa* pyochelin-iron uptake pathway and its metal specificity. *Journal of Bacteriology*, 191, 3517–3525.

Call, D. F. and Logan, B. E. 2011. Lactate oxidation coupled to iron or electrode reduction by *Geobacter sulfurreducens* PCA. *Applied and Environmental Microbiology*, 77, 8791–8794.

Chang, Y. C., Nawata, A., Junk, K., and Kikuchi, S. 2012. Isolation and characterization of an arsenate-reducing bacterium and its application for arsenic extraction from contaminated soil. *Journal of Industrial Microbiology and Biotechnology* 39(1), 37–44.

Choi, S. B. and Yun, Y. S. 2004. Lead biosorption by waste biomass of *Corynebacterium glutamicum* generated from lysine fermentation process. *Biotechnology Letters*, 26, 331–336.

Church, C. D., Wilkin, R. T., Alpers, C. N., Rye, R. O., and Mccleskey, R. B. 2007. Microbial sulfate reduction and metal attenuation in pH 4 acid mine water. *Geochemical Transactions*, 8, 10.

Coker, V. S., Byrne, J. M., Telling, N. D., Van Der Laan, G., Lloyd, J. R., Hitchcock, A. P., Wang, J., and Pattrick, R. A. D. 2012. Characterisation of the dissimilatory reduction of Fe(III)-oxyhydroxide at the microbe–mineral interface: The application of STXM-XMCD. *Geobiology*, 10, 347–354.

Comte, S., Guibaud, G., and Baudu, M. 2008. Biosorption properties of extracellular polymeric substances (Eps) towards Cd, Cu and Pb for different pH values. *Journal of Hazardous Materials*, 151, 185–193.

Copeland, R. C., Lytle, D. A., and Dionysiou, D. D. 2007. Desorption of arsenic from drinking water distribution system solids. *Environmental Monitoring and Assessment*, 127, 523–535.

Costa, M. C., Martins, M., Jesus, C., and Duarte, J. C. 2008. Treatment of acid mine drainage by sulphate-reducing bacteria using low cost matrices. *Water Air and Soil Pollution*, 189, 149–162.

Cummings, D. E., Caccavo, F., Spring, S., and Rosenzweig, R. F. 1999. *Ferribacterium limneticum*, gen. nov., sp. nov., an Fe(III)-reducing microorganism isolated from mining-impacted freshwater lake sediments. *Archives of Microbiology*, 171, 183–188.

Da Silva, M. L. B., Johnson, R. L., and Alvarez, P. J. J. 2007. Microbial characterization of groundwater undergoing treatment with a permeable reactive iron barrier. *Environmental Engineering Science*, 24, 1122–1127.

De Menezes, A., Clipson, N., and Doyle, E. 2012. Comparative metatranscriptomics reveals widespread community responses during phenanthrene degradation in soil. *Environmental Microbiology*, 14, 2577–2588.

Deng, S. B. and Ting, Y. P. 2005. Characterization of PEI-modified biomass and biosorption of Cu(II), Pb(II) and Ni(II). *Water Research*, 39, 2167–2177.

Dhankhar, R. and Hooda, A. 2011. Fungal biosorption: An alternative to meet the challenges of heavy metal pollution in aqueous solutions. *Environmental Technology*, 32, 467–491.

Drever, J. I. and Stillings, L. L. 1997. The role of organic acids in mineral weathering. *Colloids and Surfaces A: Physicochemical and Engineering Aspects*, 120, 167–181.

Drewniak, L., Matlakowska, R., Rewerski, B., and Sklodowska, A. 2010. Arsenic release from gold mine rocks mediated by the activity of indigenous bacteria. *Hydrometallurgy*, 104, 437–442.

Drewniak, L. and Sklodowska, A. 2013. Arsenic-transforming microbes and their role in biomining processes. *Environmental Science and Pollution Research*, 20, 7728–7739.

Drewniak, L., Stasiuk, R., Uhrynowski, W., and Sklodowska, A. 2015. *Shewanella* sp. O23s as a driving agent of a system utilizing dissimilatory arsenate-reducing bacteria responsible for self-cleaning of water contaminated with arsenic. *International Journal of Molecular Sciences*, 16, 14409–14427.

Dursun, A. Y., Uslu, G., Cuci, Y., and Aksu, Z. 2003. Bioaccumulation of copper(II), lead(II) and chromium(VI) by growing *Aspergillus niger*. *Process Biochemistry*, 38, 1647–1651.

Edberg, F., Kalinowski, B. E., Holmstrom, S. J. M., and Holm, K. 2010. Mobilization of metals from uranium mine waste: The role of pyoverdines produced by *Pseudomonas fluorescens*. *Geobiology*, 8, 278–292.

Evangelou, V. P. 2001. Pyrite microencapsulation technologies: Principles and potential field application. *Ecological Engineering*, 17, 165–178.

Feng, S. S., Yang, H. L., and Wang, W. 2015. Improved chalcopyrite bioleaching by *Acidithiobacillus* sp. via direct step-wise regulation of microbial community structure. *Bioresource Technology*, 192, 75–82.

Flemming, H. C. 1995. Sorption sites in biofilms. *Water Science and Technology*, 32, 27–33.

Fomina, M. and Gadd, G. M. 2014. Biosorption: Current perspectives on concept, definition and application. *Bioresource Technology*, 160, 3–14.

Fortin, D. and Beveridge, T. J. 1997. Role of the bacterium *Thiobacillus* in the formation of silicates in acidic mine tailings. *Chemical Geology*, 141, 235–250.

Francis, A. J. and Dodge, C. J. 2008. Bioreduction of uranium(VI) complexed with citric acid by Clostridia affects its structure and solubility. *Environmental Science & Technology*, 42, 8277–8282.

Francis, A. J., Dodge, C. J., Lu, F., Halada, G. P., and Clayton, C. R. 1994. XPS and XANES studies of uranium reduction by *Clostridium* sp. *Environmental Science & Technology*, 28, 636–639.

Frigaard, N. U. and Bryant, D. A. 2008. Genomic and evolutionary perspectives on sulfur metabolism in green sulfur bacteria. *Microbial Sulfur Metabolism*, 60–76.

Frigaard, N. U. and Dahl, C. 2009. Sulfur metabolism in phototrophic sulfur bacteria. *Advances in Microbial Physiology*, 54, 103–200.

Gadd, G. M. 2001. Microbial metal transformations. *Journal of Microbiology*, 39, 83–88.

Gadd, G. M. 2009. Biosorption: Critical review of scientific rationale, environmental importance and significance for pollution treatment. *Journal of Chemical Technology and Biotechnology*, 84, 13–28.

Gadd, G. M. 2010a. Metals, minerals and microbes: Geomicrobiology and bioremediation. *Microbiology*, 156, 609–643.

Gadd, G. M. 2010b. Metals, minerals and microbes: Geomicrobiology and bioremediation. *Microbiology*, 156, 609–643.

Gadd, G. M. and Fomina, M. 2011. Uranium and fungi. *Geomicrobiology Journal*, 28, 471–482.

Georgiev, P. G., Groudev, S., Spasova, I., and Nicolova, M. 2015. Remediation of a grey forest soil contaminated with heavy metals by means of leaching at acidic pH. *Journal of Soils and Sediments*, 16(4), 1288–1299.

Gihring, T. M. and Banfield, J. F. 2001. Arsenite oxidation and arsenate respiration by a new *Thermus* isolate. *FEMS Microbiology Letters*, 204, 335–340.

Golab, Z. and Orlowska, B. 1988. The effect of amino and organic-acids produced by the selected microorganisms on metal leaching. *Acta Microbiologica Polonica*, 37, 83–93.

Gough, H. L. and Stahl, D. A. 2011. Microbial community structures in anoxic freshwater lake sediment along a metal contamination gradient. *ISME Journal*, 5, 543–558.

Groudev, S., Georgiev, P., Spasova, I., and Nicolova, M. 2008a. Bioremediation of acid mine drainage in a uranium deposit. *Hydrometallurgy*, 94, 93–99.

Groudev, S. N., Spasova, I. I., Nicolova, V. N., and Georgiev, P. S. 2008. Bioremediation in situ of polluted soil in a uranium deposit. In Annable, M. D., Teodorescu, M., Hlavinek, P., Diels, L. (eds.), *Methods and Techniques for Clean-up Contaminated Sites, NATO Science for Peace and Security Series* – C: Environmental Security. Springer, Dordrecht, pp. 25–34.

Groudev, S., Spasova, I., Nicolova, M., and Georgiev, P. 2010. In situ bioremediation of contaminated soils in uranium deposits. *Hydrometallurgy*, 104, 518–523.

Groudev, S. N., Georgiev, P. S., Spasova, I. I., and Nicolova. M. V. 2015. Bioremediation of an alkaline soil heavily polluted with radionuclides and heavy metals. *XVI Balkan Mineral Processing Congress*, Belgrade, Serbia, pp. 1003–1006.

Gulati, R., Saxena, R. K., and Gupta, R. 2002. Fermentation waste of *Aspergillus terreus*: A potential copper biosorbent. *World Journal of Microbiology & Biotechnology*, 18, 397–401.

Gutnick, D. L. and Bach, H. 2000. Engineering bacterial biopolymers for the biosorption of heavy metals; new products and novel formulations. *Applied Microbiology and Biotechnology*, 54, 451–460.

Ho, Y. S. and Mckay, G. 2000. The kinetics of sorption of divalent metal ions onto *Sphagnum* moss peat. *Water Research*, 34, 735–742.

Hoeft, M. and Bruggen, M. 2004. Feedback in active galactic nucleus heating of galaxy clusters. *Astrophysical Journal*, 617, 896–902.

Hogsden, K. L. and Harding, J. S. 2012. Consequences of acid mine drainage for the structure and function of benthic stream communities: A review. *Freshwater Science*, 31, 108–120.

Hojati, S. and Khademi, H. 2013. Cadmium sorption from aqueous solutions onto iranian sepiolite: Kinetics and isotherms. *Journal of Central South University*, 20, 3627–3632.

Huang, M., Li, Y., and Sumner, M. E. 2011. *Handbook of Soil Sciences: Resource Management and Environmental Impacts*, Boca Raton, FL: CRC Press.

Incharoensakdi, A. and Kitjaharn, P. 2002. Zinc biosorption from aqueous solution by a halotolerant cyanobacterium *Aphanothece halophytica*. *Current Microbiology*, 45, 261–264.

Janyasuthiwong, S., Rene, E. R., Esposito, G., and Lens, P. N. L. 2015. Effect of pH on Cu, Ni and Zn removal by biogenic sulfide precipitation in an inversed fluidized bed bioreactor. *Hydrometallurgy*, 158, 94–100.

Jiang, S., Lee, J. H., Kim, D., Kanaly, R. A., Kim, M. G., and Hur, H. G. 2013a. Differential arsenic mobilization from as-bearing ferrihydrite by iron-respiring *Shewanella* strains with different arsenic-reducing activities. *Environmental Science & Technology*, 47, 8616–8623.

Johnson, D. B. and Hallberg, K. B. 2003. The microbiology of acidic mine waters. *Research in Microbiology*, 154, 466–473.

Johnson, D. B. and Hallberg, K. B. 2005. Acid mine drainage remediation options: A review. *Science of the Total Environment*, 338, 3–14.

Johnstone, T. C. and Nolan, E. M. 2015. Beyond iron: Non-classical biological functions of bacterial siderophores. *Dalton Transactions*, 44, 6320–6339.

Kapoor, A. and T. Viraraghavan. 1995. Fungal biosorption – an alternative treatment option for heavy metal bearing wastewaters: A review. *Bioresource Technology* 53(3), 195–206.

Kimura, S., Hallberg, K. B., and Johnson, D. B. 2006. Sulfidogenesis in low pH (3.8–4.2) media by a mixed population of acidophilic bacteria. *Biodegradation*, 17, 57–65.

Kolmert, A. and Johnson, D. B. 2001. Remediation of acidic waste waters using immobilised, acidophilic sulfate-reducing bacteria. *Journal of Chemical Technology and Biotechnology*, 76, 836–843.

Kovach, M. E., Elzer, P. H., Hill, D. S., Robertson, G. T., Farris, M. A., Roop, R. M. 2nd, and Peterson, K. M. 1995. Four new derivatives of the broad-host-range cloning vector pBBR1MCS, carrying different antibiotic-resistance cassettes. *Gene*, 166, 175–176.

Labrenz, M., Druschel, G. K., Thomsen-Ebert, T., Gilbert, B., Welch, S. A., Kemner, K. M., Logan, G. A. et al. 2000. Formation of sphalerite (Zns) deposits in natural biofilms of sulfate-reducing bacteria. *Science*, 290, 1744–1747.

Landa, E. R. 2005. Microbial biogeochemistry of uranium mill tailings. *Advances in Applied Microbiology*, 57, 113–130.

Langdona, C. J., Piearce, T. G., Mehargd, A. A., and Semplec, K. T. 2003. Interactions between earthworms and arsenic in the soil environment: A review. *Environmental Pollution*, 124, 13.

Ledin, M. and Pedersen, K. 1996. The environmental impact of mine wastes: Roles of microorganisms and their significance in treatment of mine wastes. *Earth-Science Reviews*, 41, 67–108.

Lee, J. H., Kennedy, D. W., Dohnalkova, A., Moore, D. A., Nachimuthu, P., Reed, S. B., and Fredrickson, J. K. 2011. Manganese sulfide formation via concomitant microbial manganese oxide and thiosulfate reduction. *Environmental Microbiology*, 13, 3275–3288.

Lee, J. Y., Choi, J. C., Yi, M. J., Kim, J. W., Cheon, J. Y., and Lee, K. K. 2004. Evaluation of groundwater chemistry affected by an abandoned metal mine within a dam construction site, South Korea. *Quarterly Journal of Engineering Geology and Hydrogeology*, 37, 241–256.

Lin, H., Szeinbaum, N. H., Dichristina, T. J., and Taillefert, M. 2012. Microbial Mn(IV) reduction requires an initial one-electron reductive solubilization step. *Geochimica et Cosmochimica Acta*, 99, 179–192.

Liu, H. and Fang, H. H. P. 2002. Characterization of electrostatic binding sites of extracellular polymers by linear programming analysis of titration data. *Biotechnology and Bioengineering*, 80, 806–811.

Liu, H. L., Chen, B. Y., Lan, Y. W., and Cheng, Y. C. 2004. Improved cultivation of the indigenous *Acidithiobacillus thiooxidans* BC1 by a fed-batch process. *Journal of the Chinese Institute of Chemical Engineers*, 35, 195–201.

Lloyd, J. R. 2003. Microbial reduction of metals and radionuclides. *FEMS Microbiology Reviews*, 27, 411–425.

Loukidou, M. X., Karapantsios, T. D., Zouboulis, A. I., and Matis, K. A. 2004. Diffusion kinetic study of cadmiurn(II) biosorption by *Aeromonas caviae*. *Journal of Chemical Technology and Biotechnology*, 79, 711–719.

Lovley, D. (2013). Dissimilatory Fe(III)- and Mn(IV)-reducing prokaryotes. In Rosenberg, E., DeLong, E. F., Lory, S., Stackebrandt, E., and Thompson, F. (eds.) *The Prokaryotes: Prokaryotic Physiology and Biochemistry*. Springer, Berlin, Heidelberg, pp. 287–308.

Lovley, D. R. 1993. Dissimilatory metal reduction. *Annual Review of Microbiology*, 47, 263–290.

Lovley, D. R., Coates, J. D., Bluntharris, E. L., Phillips, E. J. P., and Woodward, J. C. 1996. Humic substances as electron acceptors for microbial respiration. *Nature*, 382, 445–448.

Lovley, D. R. and Phillips, E. J. P. 1988. Novel mode of microbial energy-metabolism: Organic-carbon oxidation coupled to dissimilatory reduction of iron or manganese. *Applied and Environmental Microbiology*, 54, 1472–1480.

Lovley, D. R., Phillips, E. J. P., and Lonergan, D. J. 1989. Hydrogen and formate oxidation coupled to dissimilatory reduction of iron or manganese by *Alteromonas putrefaciens*. *Applied and Environmental Microbiology*, 55, 700–706.

Ludwig, R. D., Mcgregor, R. G., Blowes, D. W., Benner, S. G., and Mountjoy, K. 2002. A permeable reactive barrier for treatment of heavy metals. *Ground Water*, 40, 59–66.

Luther, G. W. 2005. Manganese(II) oxidation and Mn(IV) reduction in the environment: Two one-electron transfer steps versus a single two-electron step. *Geomicrobiology Journal*, 22, 195–203.

Mailloux, B. J., Alexandrova, E., Keimowitz, A. R., Wovkulich, K., Freyer, G. A., Herron, M., Stolz, J. F. et al. 2009. Microbial mineral weathering for nutrient acquisition releases arsenic. *Applied and Environmental Microbiology*, 75, 2558–2565.

Malik, A. 2004. Metal bioremediation through growing cells. *Environment International*, 30, 261–278.

Mameri, N., Boudries, N., Addour, L., Belhocine, D., Lounici, H., Grib, H., and Pauss, A. 1999. Batch zinc biosorption by a bacterial nonliving *Streptomyces rimosus* biomass. *Water Research*, 33, 1347–1354.

Mendez-Garcia, C., Mesa, V., Sprenger, R. R., Richter, M., Diez, M. S., Solano, J., Bargiela, R. et al. 2014. Microbial stratification in low pH oxic and suboxic macroscopic growths along an acid mine drainage. *Isme Journal*, 8, 1259–1274.

Mukhopadhyay, R., Rosen, B. P., Pung, L. T., and Silver, S. 2002. Microbial arsenic: From geocycles to genes and enzymes. *FEMS Microbiology Reviews*, 26, 311–325.

Nevin, K. P. and Lovley, D. R. 2002. Mechanisms for accessing insoluble Fe(III) oxide during dissimilatory Fe(III) reduction by *Geothrix fermentans*. *Applied and Environmental Microbiology*, 68, 2294–2299.

Newman, D. K., Ahmann, D., and Morel, F. M. M. 1998. A brief review of microbial arsenate respiration. *Geomicrobiology Journal*, 15, 255–268.

Nicolova, M., Spasova, I., Georgiev, P., and Groudev, S. 2014. Bioremediation of polluted waters in a uranium deposit by means of passive system. *Annual of the University of Mining and Geology St Ivan Rilski*, 57, 4.

Nicolova, M., Spasova, I., Georgiev, P., and Groudev, S. N. 2015. Microbial removal of heavy metals from activated sludge for producing a high-quality compost. *Annual of the University of Mining and Geology*, 58, 5.

Nies, D. H. 2003. Efflux-mediated heavy metal resistance in prokaryotes. *FEMS Microbiology Reviews*, 27, 313–339.

Norris, P. R., Clark, D. A., Owen, J. P., and Waterhouse, S. 1996. Characteristics of *Sulfobacillus acidophilus* sp. nov. and other moderately thermophilic mineral-sulphide-oxidizing bacteria. *Microbiology*, 142, 775–783.

Orandi, S. and Lewis, D. M. 2013. Biosorption of heavy metals in a photo-rotating biological contactor: A batch process study. *Applied Microbiology and Biotechnology*, 97, 5113–5123.

Oremland, R. S. and Stolz, J. F. 2003. The ecology of arsenic. *Science*, 300, 939–944.

Osorio, H., Mangold, S., Denis, Y., Nancucheo, I., Esparza, M., Johnson, D. B., Bonnefoy, V., Dopson, M., and Holmes, D. S. 2013. Anaerobic sulfur metabolism coupled to dissimilatory iron reduction in the extremophile *Acidithiobacillus ferrooxidans*. *Applied and Environmental Microbiology*, 79, 2172–2181.

Ozturk, A., Artan, T., and Ayar, A. 2004. Biosorption of nickel(II) and copper(II) ions from aqueous solution by *Streptomyces coelicolor* A3(2). *Colloids and Surfaces B: Biointerfaces*, 34, 105–111.

Paez-Espino, D., Tamames, J., De Lorenzo, V., and Canovas, D. 2009. Microbial responses to environmental arsenic. *Biometals*, 22, 117–130.

Park, D., Yun, Y. S., and Park, J. M. 2010. The past, present, and future trends of biosorption. *Biotechnology and Bioprocess Engineering*, 15, 86–102.

Perez-Gonzalez, T., Jimenez-Lopez, C., Neal, A. L., Rull-Perez, F., Rodriguez-Navarro, A., Fernandez-Vivas, A., and Ianez-Pareja, E. 2010. Magnetite biomineralization induced by *Shewanella oneidensis*. *Geochimica et Cosmochimica Acta*, 74, 967–979.

Prithviraj, D., Deboleena, K., Neelu, N., Noor, N., Aminur, R., Balasaheb, K., and Abul, M. 2014. Biosorption of nickel by *Lysinibacillus* sp. BA2 native to bauxite mine. *Ecotoxicology and Environmental Safety*, 107, 260–268.

Raymond, K. N., Muller, G., and Matzanke, B. F. 1984. Complexation of iron by siderophores: A review of their solution and structural chemistry and biological function. *Topics in Current Chemistry*, 123, 49–102.

Reguera, G., Mccarthy, K. D., Mehta, T., Nicoll, J. S., Tuominen, M. T., and Lovley, D. R. 2005. Extracellular electron transfer via microbial nanowires. *Nature*, 435, 1098–1101.

Renshaw, J. C., Butchins, L. J. C., Livens, F. R., May, I., Charnock, J. M., and Lloyd, J. R. 2005. Bioreduction of uranium: Environmental implications of a pentavalent intermediate. *Environmental Science & Technology*, 39, 5657–5660.

Rhine, E. D., Phelps, C. D., and Young, L. Y. 2006. Anaerobic arsenite oxidation by novel denitrifying isolates. *Environmental Microbiology*, 8, 899–908.

Rzepa, G., Bajda, T., and Ratajczak, T. 2009. Utilization of bog iron ores as sorbents of heavy metals. *Journal of Hazardous Materials*, 162, 1007–1013.

Salehizadeh, H. and Shojaosadati, S. A. 2003. Removal of metal ions from aqueous solution by polysaccharide produced from *Bacillus firmus*. *Water Research*, 37, 4231–4235.

Sanchez-Andrea, I., Rodriguez, N., Amils, R., and Sanz, J. L. 2011. Microbial diversity in anaerobic sediments at Rio Tinto, a naturally acidic environment with a high heavy metal content. *Applied and Environmental Microbiology*, 77, 6085–6093.

Sand, W., Gehrke, T., Jozsa, P. G., and Schippers, A. 2001. (Bio) chemistry of bacterial leaching: Direct vs. indirect bioleaching. *Hydrometallurgy*, 59, 159–175.

Say, R., Yimaz, N., and Denizli, A. 2003. Removal of heavy metal ions using the fungus *Penicillium canescens*. *Adsorption Science & Technology*, 21, 643–650.

Schalk, I. J., Hannauer, M., and Braud, A. 2011. New roles for bacterial siderophores in metal transport and tolerance. *Environmental Microbiology*, 13, 2844–2854.

Schroder, I., Johnson, E., and De Vries, S. 2003. Microbial ferric iron reductases. *FEMS Microbiology Reviews*, 27, 427–447.

Selatnia, A., Boukazoula, A., Kechid, N., Bakhti, M. Z., Chergui, A., and Kerchich, Y. 2004. Biosorption of lead (II) from aqueous solution by a bacterial dead *Streptomyces rimosus* biomass. *Biochemical Engineering Journal*, 19, 127–135.

Sheng, G. P., Yu, H. Q., and Li, X. Y. 2010. Extracellular polymeric substances (Eps) of microbial aggregates in biological wastewater treatment systems: A review. *Biotechnology Advances*, 28, 882–894.

Siegrist, P. T., Comte, N., Holzmeister, J., Sutsch, G., Koepfli, P., Namdar, M., Duru, F., Brunckhorst, C., Scharf, C., and Kaufmann, P. A. 2008. Effects of Av delay programming on ventricular resynchronisation: Role of radionuclide ventriculography. *European Journal of Nuclear Medicine and Molecular Imaging*, 35, 1516–1522.

Sierra-Alvarez, R., Karri, S., Freeman, S., and Field, J. A. 2006. Biological treatment of heavy metals in acid mine drainage using sulfate reducing bioreactors. *Water Science and Technology*, 54, 179–185.

Smith, J. and Melville, M. D. 2004. Iron monosulfide formation and oxidation in drain-bottom sediments of an acid sulfate soil environment. *Applied Geochemistry*, 19, 1837–1853.

Sobecky, P. A. and Coombs, J. M. 2009. Horizontal gene transfer in metal and radionuclide contaminated soils. *Methods in Molecular Biology*, 532, 455–472.

Stolz, J. F. and Oremland, R. S. 1999. Bacterial respiration of arsenic and selenium. *FEMS Microbiology Reviews*, 23, 615–627.

Streten-Joyce, C., Manning, J., Gibb, K. S., Neilan, B. A., and Parry, D. L. 2013. The chemical composition and bacteria communities in acid and metalliferous drainage from the wet-dry tropics are dependent on season. *Science of the Total Environment*, 443, 65–79.

Sun, W., Sierra-Alvarez, R., Fernandez, N., Sanz, J. L., Amils, R., Legatzki, A., Maier, R. M., and Field, J. A. 2009. Molecular characterization and in situ quantification of anoxic arsenite-oxidizing denitrifying enrichment cultures. *FEMS Microbiology Ecology*, 68, 72–85.

Tebo, B. M., Johnson, H. A., Mccarthy, J. K., and Templeton, A. S. 2005. Geomicrobiology of manganese(II) oxidation. *Trends in Microbiology*, 13, 421–428.

Tufano, K. J., Reyes, C., Saltikov, C. W., and Fendorf, S. 2008. Reductive processes controlling arsenic retention: Revealing the relative importance of iron and arsenic reduction. *Environmental Science & Technology*, 42, 8283–8289.

Uslu, G. and Tanyol, M. 2006. Equilibrium and thermodynamic parameters of single and binary mixture biosorption of lead(II) and copper(II) ions onto *Pseudomonas putida*: Effect of temperature. *Journal of Hazardous Materials*, 135, 87–93.

Uster, B., O'sullivan, A. D., Ko, S. Y., Evans, A., Pope, J., Trumm, D., and Caruso, B. 2015. The use of mussel shells in upward-flow sulfate-reducing bioreactors treating acid mine drainage. *Mine Water and the Environment*, 34, 442–454.

Vandieken, V., Pester, M., Finke, N., Hyun, J. H., Friedrich, M. W., Loy, A., and Thamdrup, B. 2012. Three manganese oxide-rich marine sediments harbor similar communities of acetate-oxidizing manganese-reducing bacteria. *Isme Journal*, 6, 2078–2090.

Vargas, M., Kashefi, K., Blunt-Harris, E. L., and Lovley, D. R. 1998. Microbiological evidence for Fe(III) reduction on early earth. *Nature*, 395, 65–67.

Veglio, F. and Beolchini, F. 1997. Removal of metals by biosorption: A review. *Hydrometallurgy*, 44, 301–316.

Vijayaraghavan, K. and Yun, Y. S. 2008. Bacterial biosorbents and biosorption. *Biotechnology Advances*, 26, 266–291.

Volesky, B. and Holan, Z. R. 1995. Biosorption of heavy-metals. *Biotechnology Progress*, 11, 235–250.

Wang, J. L. and Chen, C. 2009. Biosorbents for heavy metals removal and their future. *Biotechnology Advances*, 27, 195–226.

Wang, J. L. and Chen, C. 2014. Chitosan-based biosorbents: Modification and application for biosorption of heavy metals and radionuclides. *Bioresource Technology*, 160, 129–141.

Wang, Q., Xiong, D., Zhao, P., Yu, X., Tu, B., and Wang, G. 2011. Effect of applying an arsenic-resistant and plant growth-promoting rhizobacterium to enhance soil arsenic phytoremediation by *Populus deltoides* Lh05-17. *Journal of Applied Microbiology*, 111, 1065–1074.

Warren, L. A., Maurice, P. A., Parmar, N., and Ferris, F. G. 2001. Microbially mediated calcium carbonate precipitation: Implications for interpreting calcite precipitation and for solid-phase capture of inorganic contaminants. *Geomicrobiology Journal*, 18, 93–115.

Welch, S. A. and Ullman, W. J. 1999. The effect of microbial glucose metabolism on bytownite feldspar dissolution rates between 5 degrees and 35 degrees C. *Geochimica et Cosmochimica Acta*, 63, 3247–3259.

Williamson, A. J., Morris, K., Law, G. T. W., Rizoulis, A., Charnock, J. M., and Lloyd, J. R. 2014. Microbial reduction of U(VI) under alkaline conditions: Implications for radioactive waste geodisposal. *Environmental Science & Technology*, 48, 13549–13556.

Won, S. W., Kotte, P., Wei, W., Lim, A., and Yun, Y. S. 2014. Biosorbents for recovery of precious metals. *Bioresource Technology*, 160, 203–212.

Yuncu, B., Sanin, F. D., and Yetis, U. 2006. An investigation of heavy metal biosorption in relation to C/N ratio of activated sludge. *Journal of Hazardous Materials*, 137, 990–997.

Zagury, G. J., Narasiah, K. S., and Tyagi, R. D. 1994. Adaptation of indigenous iron-oxidizing bacteria for bioleaching of heavy-metals in contaminated soils. *Environmental Technology*, 15, 517–530.

Zhang, M. L. and Wang, H. X. 2014. Organic wastes as carbon sources to promote sulfate reducing bacterial activity for biological remediation of acid mine drainage. *Minerals Engineering*, 69, 81–90.

Ziagova, M., Dimitriadis, G., Aslanidou, D., Papaioannou, X., Tzannetaki, E. L., and Liakopoulou-Kyriakides, M. 2007. Comparative study of Cd(II) and Cr(VI) biosorption on *Staphylococcus xylosus* and *Pseudomonas* sp. in single and binary mixtures. *Bioresource Technology*, 98, 2859–2865.

Zirnstein, I., Arnold, T., Krawczyk-Barsch, E., Jenk, U., Bernhard, G., and Roske, I. 2012. Eukaryotic life in biofilms formed in a uranium mine. *Microbiologyopen*, 1, 83–94.

Zouboulis, Al. and Katsoyianni, I. A. 2005. Recent advances in the bioremediation of arsenic-contaminated groundwater. *Environment International*, 31, 7.

4 Insight into Microbe-Assisted Bioremediation Technologies for the Amendment of Toxic Metal–Contaminated Sites

Sangram Sinha, Ekramul Islam, Kiron Bhakat, and Samir Kumar Mukherjee

CONTENTS

ABSTRACT

Essentially, the awareness of heavy metal pollution has gained the focus of public interest since the evolution of delicate analytical techniques, making it possible to detect them even in very small traces. With the advent of modern analytical detection procedures, now it is possible to measure a thousandth of an mg/kg for certain matrixes. This in turn helped toxicologists to pursue biological experiments to follow up the effects of individual substances down to the smallest concentrations. Depending on the kind and depth of contamination, different remediation techniques were developed to decontaminate the sites. The available methods include physical, biological, and chemical procedures. But biological procedures exploiting microorganisms are advantageous in many respects over the conventional chemical methods. The biological method has been emerging as the most promising method because it is safe and cost-effective over the other existing procedures. Biological treatment has been broadly termed as bioremediation, which could be defined as a process that exploits the genetic diversity and metabolic versatility of living organisms to enhance the rate or extent of pollution destruction. The selection and exploitation of inherent physiological or metabolic property of heavy metal–resistant microbial strains could thus be a valuable tool for decontaminating affected sites. Artificially, bacterial strain also could be made to ameliorate heavy metal contamination from soil or water by introducing foreign genes from other sources. Metal removal therefore could be achieved either through indigenous microorganisms or genetically modified microorganisms or by introducing metal accumulator plants. The use of rhizobacteria in phytoremediation technologies is now being considered to play an important role on enhanced detoxification of soil by using complex plant–microbe–metal–soil interactions under suitable conditions. The properties of plants used for phytoremediation, for example, biomass production, contaminant uptake, plant nutrition, and health, are improved by rhizobacteria, but it is important to select rhizobacteria that can survive and succeed when used in phytoremediation practices. Considering this background, an assessment on the current status of technology deployment and suggestions for future bioremediation research relating metal decontamination is discussed. The roles of plant-associated microbes in metal mobilization/immobilization and in the application of these processes in heavy metal phytoremediation are reviewed, which might give insight to develop better future strategies to decontaminate metal-contaminated sites, crop fields in particular.

4.1 INTRODUCTION

The progress of modern civilization demanding huge industrialization requires rapid exploration of natural resources leading to a huge release of industrial effluents containing heavy metals. The increase in heavy metal concentration in the environment poses severe threat to human health and other organisms (Jarup, 2003; Zhuang et al., 2009; Islam and Sar, 2011; Liu et al., 2013). Furthermore, it is slowly accumulating in the food chain and exerting permanent harmful effects (Liu et al., 2013). Therefore, it has been a serious concern to mitigate all environmental issues related to elimination or cutback of hazards due to heavy metal contaminations. Unlike organic pollutants, heavy metals cannot be degraded and hence persist in the soil; therefore, decontaminating the metal-contaminated/metal-polluted sites is a challenging task (Rajkumar et al., 2010; Ma et al., 2011). Increased public awareness and strict legal constraints on the environmental release of heavy metals necessitate an effective and affordable technology for the removal, containment, or neutralization of their toxic effects in soil. In order to clean up contaminated sites, heavy metals should be extracted and concentrated by suitable techniques for disposal in appropriate sites. Remedial goals for heavy metal–rich wastes or contaminated sites could be achieved through the following ways: (1) complete removal of metals from the system, (2) precipitation and immobilization of metals, (3) concentration and thus reduction in volume of contaminated matrices, and (4) compartmentalization of metals to a part of the environment in which their harmful effect is reduced. Several traditional remediation technologies such as thermal processes, physical separation, electrochemical methods, washing, burial, and stabilization/solidification are available (Kumpiene et al., 2008; Bolan et al., 2014; Tsang and Yip, 2014). Conventional remediation methods lack specificity, are prohibitory expensive, and more importantly destruct landscape and biodiversity (Kumpiene et al., 2008; Tsang and Yip, 2014). The ultimate goal of any soil remediation process must be not only to remove the metals from the polluted soil but, most importantly, to restore the continued capacity of the soil to exert its function as per its potential (i.e., to recover soil quality). Keeping these views in mind, microorganism- and plant-based bioremediation strategies are gaining considerable importance. Exploitation of microorganism and plants or both of them for removing heavy metals and other toxic compounds from the environment has been termed as bioremediation (Rajkumar et al., 2010; Ali et al., 2013; Shutcha et al., 2015). Bioremediation is a multidisciplinary approach, but its main component exploits microbial catabolic abilities to reduce, eliminate, accumulate, immobilize, or transform different hazardous materials to nontoxic products (Gadd, 2010). Owing to their long-standing evolutionary history, microbes have evolved several mechanisms to survive and flourish in metal-rich environments. They often execute a number of strategies to tolerate the presence of heavy metals or to use them as terminal electron acceptors and to interact with these elements affecting their geochemistry, fate, and environmental toxicity (Gadd, 2010). The multiplicity of mechanisms by which microbes interact with metals includes passive sequestration onto the cells (biosorption), active intracellular accumulation (bioaccumulation), redox transformation, and complexation/precipitation with microbial products like polysaccharide, siderophore, and metallothionein. Some of these natural microbial processes may be harnessed as the basis of potential biotechnological strategies for the abatement of toxic metal pollution. Nevertheless, using microorganisms alone, we could reduce the toxicity of metals or trapping it; however, it will be in the soil in the long run.

Phytoextraction is one of the widely accepted key processes of phytoremediation that exploits the trait of the natural ability of some plants to accumulate (hyperaccumulate) heavy metals in concentrated form in their harvestable parts and thereby remove them from the soil. This process depends on (1) the growth of plants at the contaminated site, (2) the capacity of the plant root system to access the pollutants in the soil, and (3) the metal accumulation capacity of plant. Several chemical amendments, such as chelators and limestone, have been used to boost the phytoextraction or phytostabilization process. Even though these amendments increase the efficiency of phytoextraction/phytostabilization, some chemical chelators are not only phytotoxic but also toxic to soil microorganisms that affect soil fertility (Park et al., 2011).

Microbial-assisted phytoextraction (combining plant–microbial remediation strategies) is another promising alternative technology that employs biodiversity to remove metal pollutants from the soil to overcome such constrains (Sessitsch et al., 2013; Langella et al., 2014). This process depends mostly on the establishment of complex plant–microbe interactions in the vicinity of the rhizosphere that affect plant growth and increase metal uptake efficiency by altering the mobility and bioavailability. Since the primary objective of remediating metal-contaminated sites is to limit the leachability of metals, *in situ* immobilization might offer a promising resolution (Bolan et al., 2014).

This chapter focuses on the issues for possible exploitation of microbial activities either by itself or through assisting hyperaccumulators for detoxifying or removing heavy metals from contaminated sites as a part of sustainable development. Additionally, newer approaches such as using designer plants and rhizospheric manipulation targeting bioremediation of metal-contaminated sites in cheaper and safer ways have also been discussed.

4.2 SOURCES OF HEAVY METAL POLLUTION

Two main categories of sources of heavy metal pollution are known so far: (1) natural sources and (2) anthropogenic sources. Heavy metals are present in the Earth's crust since its formation. But the pedogenic processes, natural weathering of rocks, soil erosion, volcanic eruption, etc., are the major causes of bioavailability of heavy metals in the environment. However, anthropogenic activities and rapid urbanization are adding further load of heavy metal pollution in the environment. Anthropogenic sources of metal pollution can be divided into five main groups: (1) metalliferous mining and smelting, (2) industry, (3) natural deposition, (4) agriculture, and (5) waste disposal. Worldwide, mining activities have been intensified during the last few decades due to the increasing market demand of raw materials. This disturbs natural habitat, hampers the soil's bacterial flora, and interferes with the geochemical cycling, thus creating an accumulation of toxic heavy metals high above the critical level and affecting human health. The balance of heavy metal load in the soil environment can be calculated using the following mathematical formula (Lombi and Gerzabek, 1998):

$$M_{total} = (M_p + M_a + M_f + M_{ag} + M_{ow} + M_{ip}) - (M_{cr} + M_l)$$

where

M is the heavy metal
p is the parent materials
a is the atmospheric deposition
f is the fertilizer source
ag is the agrochemical source
ow is the organic waste source
ip is the inorganic pollutants
cr is the crop removal
l is the losses by leaching, complexion, volatilization, and others

The contribution of anthropogenic activities in heavy metal pollution has been found to be threefold higher than the natural process (Sposito and Page, 1984). Most of the heavy metals occur naturally with soil particles in an immobilized form, which is not readily bioavailable for uptake. Unlike geogenic inputs, toxic metals added through anthropogenic activities usually have a high bioavailability for uptake (Bolan et al., 2014). The tolerance limits of certain heavy metals are summarized in Table 4.1.

4.3 SURVIVAL STRATEGIES OF BACTERIA IN METAL-POLLUTED ENVIRONMENTS

Although some metals play an important role acting as trace elements in the life process, at higher concentrations, they exhibit a wide array of toxic effects in different organisms. Toxic effects of metals can arise due to leaching from natural geogenic events, more commonly associated with the anthropogenic release of toxic metals in the environment. However, physicochemical nature of the environment and the chemical behavior of the metal species greatly affect the metal toxicity (Gadd, 2010). Despite apparent toxicity, several microorganisms could grow and even flourish in metal-enriched/metal-polluted environment and exhibit a variety of mechanisms, both active and passive, which contribute to resistance (Avery, 2001; Holden and Adams, 2003; Fomina et al., 2005; Gadd, 2010). The mechanism of microbial resistance to metals is ubiquitous, ranging from a few percent to nearly 100% in pristine environments and heavily polluted environments, respectively (Silver and Phung, 2009).

All the survival mechanisms of microbes may lead to decreased or increased mobility of metals. These include changes in redox state, organic and inorganic precipitation, production of metal-chelating peptides and proteins (e.g., metallothioneins, phytochelatins), active transport, efflux and intracellular compartmentalization, metal binding to cell walls and other structural components (White and Gadd, 1998; Gadd, 2004, 2005), or even solubilization of metal from organic and inorganic sources (Gadd, 2007).

Resistance mechanisms toward various metals include efflux or enzymatic detoxification and release or volatilization from cells, for example, Hg^{2+} reduction to Hg^0 (Nies, 1992, 1999, 2003; Nies and Silver, 1995; Silver and Phung, 1996, 2009; Osman and Cavet, 2008). Resistance mechanism might be conferred by the genes carried by plasmids or chromosomes or both. Plasmid-borne resistance genes are observed for many toxic metals and metalloids, for example, AsO_2^-, AsO_4^{3-}, Cd^{2+}, Co^{2+}, CrO_4^{2-}, Cu^{2+}, Hg^{2+}, Ni^{2+}, and Zn^{2+}. Chromosomal systems are also frequently observed, for example, Hg^{2+} resistance and Cd^{2+} efflux in *Bacillus* and efflux of arsenic in *Escherichia coli* (Silver and Phung, 1996; Rosen, 2002). Several researches have been carried out in detail for many metals including Ag, As, Cd, Co, Cu, Hg, Ni, Pb, and Zn, and in many cases, genes responsible for resistance have been explored and mechanisms were proposed (Osman and Cavet, 2008; Silver and Phung, 2009). Microbially synthesized metal-binding peptides and proteins, for example, metallothioneins and phytochelatins, also regulate metal ion homeostasis and affect toxic responses (Eide, 2000; Avery, 2001). So the wide ranges of biochemical protection strategy of microorganisms from potentially toxic metals can be used for cost-effective and eco-friendly bioremediation applications (Figure 4.1).

TABLE 4.1
Maximum Contamination Levels for Heavy Metal Concentration in a Safe Environment

Heavy Metal	Air (mg/m³)	Sludge/ Soil (ppm)	Drinking Water (ppm)	Water Supporting Aquatic Life (ppm)
			Maximum Concentration	
Cd	0.1–0.2	85	0.005	0.0008
Pb	NA[a]	420	0.01(0.0)	0.0058
Zn	1.5	7500	5.00	0.0766
Hg	NA	<1	0.002	0.05
Ca	5	Tolerable	50	Tolerable > 50
Ag	0.01	NA	0.0	0.1
As	NA	NA	0.01	NA

Source: Duruibe, J.O. et al., *Int. J. Phys. Sci.*, 2, 112, 2007.
[a] Not available.

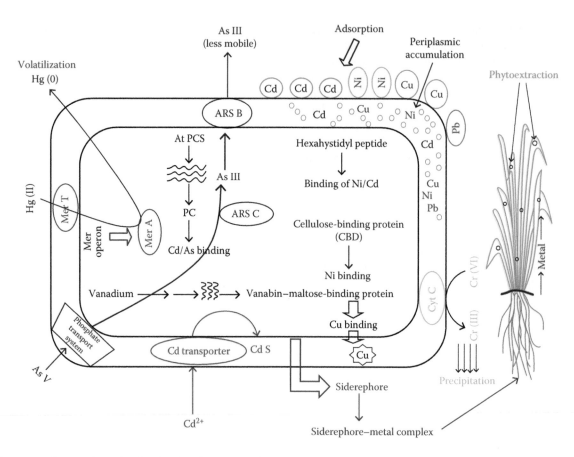

FIGURE 4.1 A glimpse of metal–microbe–plant interactions for possible decontamination strategy.

4.4 REACTIONS OF HEAVY METAL(LOID)S IN SOILS: AN INSIGHT OF MICROBE–METAL INTERACTION

Metals directly and/or indirectly affect the biology of microbes by regulating their growth, metabolism, and differentiation. Interaction between microbes and metals and metallic compounds depends on the redox state of metals, type of organism, interacting environment, metabolic activity, and structural components of the organism. The fate of interaction leads to influence metal speciation and therefore solubility, mobility, bioavailability, and toxicity (Gadd, 2010). Microbes regulate the biogeochemistry of the metals leading to their mobilization and immobilization that depend on the microniche.

4.4.1 Biosorption

Microorganisms are numerous and ubiquitous in soil. With small sizes, high numbers, and a high surface-to-volume ratio, microorganisms provide a large reactive area with the surrounding environment. The negative net charge on the large cell envelope makes these organisms prone to attach metal cations from the exterior to their cell surface. Metal biosorption is generally considered as a rapid physiochemical process by which metal ions get associated with as a result of immediate cell contact (Volesky, 1990a,b). The cell biomass of both living and nonliving organisms can take part in the biosorption

process. But in the case of living cells, metabolic pathways may partly contribute to the biosorption process (Gadd and White, 1993) as well. Microbially produced polysaccharide takes up an active role in immobilizing heavy metals and therefore its bioavailability (Iyer et al., 2005). For example, *Klebsiella aerogenes* produce exopolysaccharides that help them survive under Cd stress conditions by removing Cd from the growth medium (Scott and Palmer, 1990). Besides, the cell wall microbial biomass contains structural molecules like protein, lipid, and carbohydrate that retain the heavy metal by complexing it. The binding involves various bonds like ionic interaction, van der Waals force, and electrostatic interaction.

4.4.2 Intracellular Accumulation

Microbial interaction with metal elements is a frequent event that often leads to cellular accumulation of metals from the environment (Gadd and White, 1993). Once the metals enter into the cell, they may be compartmentalized and/or converted to innocuous forms by binding or precipitation in the form of phosphide, sulfide, carbide, or hydroxide. The metals find their entry into the bacterial cell through a metal transporter (Summers and Silver, 1978), which actually transports important physiological cations. *Pseudomonas aeruginosa* was found to accumulate Ni in the form of phosphide salts (Sar et al., 2001). The accumulated metal was mainly partitioned into the periplasm and membrane (Naz et al., 2005; Sinha and

Mukherjee, 2009). Intracellular accumulation of the metal was also noticed in pseudomonads where the accommodated metal was restricted to periplasm mainly in the form of metal sulfide (Aiking et al., 1982; Higham et al., 1984, 1986; Kazy et al., 1999; Sinha and Mukherjee, 2009).

4.4.3 Volatilization

Microorganisms often detoxify a metal ion by converting it into a less soluble and therefore into a less toxic compound through the volatilization process. This is achieved by oxidation, methylation, reduction, and demethylation of the compounds (Thayer, 1989). Microbes like methanoarchaea, obligately or facultatively fermentative bacteria, or sulfate reducers proved to be involved in metal volatilization like Hg, As, Se, and Sb (Meyer et al., 2007). Hg volatilization is the well-known example of this process, which occurs during geochemical cycling of Hg by certain microbes. Hg methylation and demethylation along with oxidoreduction causes volatilization of Hg (Barkay et al., 1989). The plasmids of many anaerobic and aerobic bacteria mediate cleavage of mercury and cause reduction of the latter into volatile elemental form (Hg^0) thereby effectively detoxifying Hg from the cell (Wood and Wang, 1983; Brim et al., 2003). Similarly, biovolatilization of As has been proposed to play a significant role in As biogeochemical cycling and is also explorable as a potential tool for arsenic decontamination (Wang et al., 2014; Mallick et al., 2015). Although biovolatilization processes seem to take part primarily in the detoxification of the microhabitats and thus is of local importance, the transformations of toxic metal(loid)s to volatile forms are creating a global concern due to the increasing mobility of those elements and raising higher toxicity of the derivatives as compared to their other inorganic counterparts (Meyer et al., 2007).

4.4.4 Extracellular Precipitation

Microorganisms release a diverse range of specific and nonspecific metal-binding compounds into the environment, which can act to ameliorate the presence of toxic metals. Phosphates, oxalates, and sulfides produced by the microbes facilitate extracellular immobilization of available heavy metals. Nonspecific metal-binding microbial metabolites, for example, organic acids, can complex metals and affect their mobility and thus their toxicity (Gadd, 1990). Microbially produced undefined macromolecules can also bind significant amounts of potentially toxic metals. These include various organic acids that arise during humification and extracellular polymeric substances (Mukherjee and Asanuma, 1997; Spark et al., 1997). Microorganisms detoxify heavy metals externally by converting them into less soluble sulfide salts. For example, sulfate-reducing bacteria (SRB) can produce millimolar concentrations of sulfide, and in environments such as anaerobic sediments, SRB can also generate significant amounts of sulfide. The high environmental sulfide concentrations produced by SRB lead to the precipitation of any metal ions present as metal sulfides. Metal sulfides, other than those of the alkali metals, have very low solubility products, meaning that metal sulfides are essentially insoluble under neutral, anaerobic conditions. As a result of the combination of neutral pH, low E_h, and high sulfide conditions, soluble metal ion concentrations in the microenvironment of SRB are enormously low. These conditions permit SRB to grow in environments containing high levels of toxic metals, and SRB are being explored for the purpose of bioremediation of metal-contaminated land and water by precipitating contaminated metal species, such as metal sulfide (White et al., 1998). Microorganisms, in response to low levels of metals, also produce specific extracellular metal-binding compounds to take up essential metals from the ambient environment. The common example in this regard is the production of siderophores in response to low iron content in the environment. Siderophores are Fe(III) scavenging compounds (500–1000 Da) of low molecular mass produced by many microorganisms and act by complexing and solubilizing insoluble Fe(III) in a form that can be transported into the cell using specific transport mechanisms (Neilands, 1981). Although siderophores are iron(III)-binding compounds, they are also hypothesized to bind other metals such as magnesium, manganese, chromium(III), gallium(III), cadmium, and certain radionuclides (Birch and Bachofen, 1990; Sinha and Mukherjee, 2009).

4.5 EXPLOITATION OF MICROBIAL ACTIVITIES IN BIOREMEDIATION

4.5.1 Heavy Metal Bioremediation by Whole-Cell Bacteria

4.5.1.1 Active Sequestration

Living cells could accumulate heavy metals within the cell employing its metabolic activity. There are many heavy metal accumulating indigenous bacteria that have been reported by several workers (Higham et al., 1986; Wang et al., 1997; Sar et al., 2001). But the efficacy of bacteria-mediated heavy metal accumulation has been reported to be magnified by genetic manipulation. A potential approach of upgrading bioremediation processes is to genetically engineer bacterial strains to confer increased abilities to accumulate toxic metals. Initiatives to improve the metal bioaccumulation have been made by overexpressing metal-binding peptides, such as polyhistidines (Sousa et al., 1996) or metallothioneins (Sousa et al., 1998; Mejare and Bulow, 2001), in microbes. Phytochelatin in plants helps to accumulate metal; therefore, scientists are trying to clone and overexpress phytochelatin synthase genes into bacteria or plants for the metal removal purpose (Sauge-Merle et al., 2003). Genetically engineered *E. coli*, expressing *vanabin* genes from the vanadium-rich ascidian *Ascidia sydneiensis samea*, has been exploited for effective copper accumulation (Ueki et al., 2003). The other biotechnological approach relies on the expression of bacterial surface metal-binding proteins to develop whole cell tools for better sequestration of toxic metals. Short metal-binding peptides, such as hexahistidyl peptides, have been introduced into bacterial surface proteins to facilitate improved metal-binding capacity. Using this strategy, both *E. coli* (Sousa et al., 1996) and *Staphylococcus carnosus* (Samuelson et al., 2000) strains with an enhanced ability to bind Ni^{2+} and Cd^{2+} have been developed. Attempts have been made in order to surface display of metal-binding proteins, which could serve as potential bioremediation agents in the contaminated site. For example, surface display of combinatorially engineered cellulose-binding domains from *Trichoderma reesei* cellulase Cel7A on staphylococci was generated for effective nickel accumulation purpose (Wernérus et al., 2001). Cloning and expression of *mer* operon to *Deinococcus geothermalis* from *E. coli* has been reported, and huge Hg reduction potentiality has been achieved in the strain (Brim et al., 2003).

4.5.1.2 Passive Sequestration

Microbial biomass can passively bind a large amount of metal(s), a phenomenon commonly referred to as biosorption (Macaskie and Dean, 1985; McHale and McHale, 1994; Li and Yu, 2014), thus providing a cost-effective way out for industrial wastewater management. Biosorption is possible by both living and nonliving biomass. It mainly involves cell surface binding, ion exchange, and microprecipitation (Gadd, 1990; Muraleedharan et al., 1991). Binding of metal to bacterial surface goes after two major processes: the first one is the stoichiometric interaction between cell surface molecules and metal and the second one is the inorganic deposition of increased amounts of metal(s) (Beveridge and Murry, 1980). Microbial metal-binding capacity differs according to the degree of affinity toward metal. Particular microbial biomass preferably binds to certain metal(s), while a broad range of bacteria do not exhibit specific binding to any metal (Hosea et al., 1986). Gram-positive bacteria contain glycoproteins on the outer surface of their cell wall that facilitated the binding of Cd^{2+} compared to other molecules. The phosphoryl groups of the LPS and phospholipids act as the most feasible binding sites for metal cations in the *E. coli* outer membrane. Most of the metal deposition occurred at the polar head regions of constituent membranes or along the peptidoglycan in the purified *E. coli* K12 cell envelopes. In *Streptomyces longwoodensis*, phosphate residues were suggested to be the primary constituents responsible for uranium binding (Crist et al., 1981). Among various studied bacteria, *Bacillus* sp. has been frequently identified as potential metal sequestrating microorganism and has been considered as commercial biosorbent preparation. In addition, there are reports on the biosorption of metal(s) using *Pseudomonas* sp., *Zoogloea ramigera*, and *Streptomyces* sp. (Noberg and Persson, 1984; Mullen et al., 1989).

4.5.2 Microbial Assisted Phytoremediation

During the absorption of water, plants not only take up nutrients but also uptake metals/metalloids from the soil. Phytoremediation refers to the utilization of higher plants to take out, transform, or stabilize contaminants, including toxic metals in water, sediments, or soils (Dary et al., 2010). Metal hyper accumulating plants are generally applied for efficient phytoextraction as they could accumulate large amounts of metal(loid)s in shoots without any visible symptoms such as wilting or necrosis of leaves, root discoloration, and plant growth. More than 400 plants across various families are known that could perform this kind of job (Prasad and Freitas, 2003). Certain plants are used to immobilize metals or reduce the bioavailability of metals in the soil; the process is known as phytostabilization (Shutcha et al., 2015). This process prefers to stabilize metals by accumulation in plant roots or precipitation within the rhizosphere rather than removing them from soil. Phytovolatilization is another process of phytoremediation that uses plants to extract metals from the soil and transforms them into a volatile form and then releases them into the atmosphere (Pulford and Watson, 2003; Mench et al., 2006). Volatile heavy metals, such as mercury (Hg), arsenic (As), and selenium (Se), could be removed from soil by this approach. Attempts have been made to boost the efficiency of phytoremediation by manipulating the mechanism of metal metabolism in plants through genetic modification, which provides a powerful tool to improve enhanced accumulation and transformation of heavy metals in plants (Kotrba, 2013).

Despite its positive aspects, the industrial application of phytoextraction is impaired due to the following constraints: (1) lack or slow growth of plants at contaminated sites due to nutrient limitation, metal toxicity, and lack of soil structure, (2) reduced capacity of plants to penetrate in the soil through their root system and access the pollutants, (3) the time required for effective phytoremediation is often long, and (4) potential for contaminants to get entry in the food chain through the consumption of plants accumulating toxic metals. To remove such constrains, several chemical amendments, such as EDTA and limestone, have been used to enhance either phytoextraction or phytostabilization process. Even though these amendments increase the efficiency of phytoextraction/phytostabilization, some chemical amendments (e.g., EDTA) are not only phytotoxic but also toxic to soil microorganisms that decrease soil fertility.

Microbial-assisted phytoextraction is another promising alternative technology that exploits microbial metabolic activities in the rhizosphere to boost the phytoremediation. Soil-dwelling microbes actively take part in the plant growth, enhance solubility and availability of metals to plants, and enhance metal translocation from soil to plant. Therefore, the success of this process depends on the establishment of robust plant–microbe interactions.

Rhizobacteria, which have the ability to colonize the root environment, actively take part in the phytoextraction process (Kloepper et al., 1991; Kloepper, 1994). Helpful rhizospheric bacteria, the plant-growth-promoting rhizobacteria (PGPR), are defined by three intrinsic characteristics: (1) they must have the ability to colonize the root, (2) they must survive and multiply in microhabitats associated with the root surface in combination with other organisms, at least for the time needed to express their plant promotion/protection traits, and (3) they must enhance the growth of the host plant. The rhizosphere is an area encircling the plant root system known to have enhanced biomass productivity. Rhizosphere bacteria obtain nutrients excreted from roots, such as organic acids, enzymes, amino acids, and complex carbohydrates (Anderson et al., 1993; Yee et al., 1998; Shim et al., 2000). In return, the plants are benefited by PGPR through various ways. The PGPR enhance plant growth by atmospheric nitrogen fixation, phytohormone production, expressing specific enzymatic system and protecting plants from diseases and pathogen-depressing substances and chelating agents (Kamnev and van der Lelie, 2000; Sinha and Mukherjee, 2008). Free-living bacteria as well as PGPR can promote plant growth directly by increasing bioavailable phosphorus, fixing nitrogen, sequestering trace elements, producing plant hormones, and lowering of plant ethylene levels (Glick et al., 1999). Furthermore, the root tips make available for a steady-state redox condition and a structural surface for bacterial colonization. The plant root system is responsible for soil aeration and distribution of bacteria through soil and penetrates otherwise impermeable soil layers while fetching soluble forms of hazardous substances in the soil water toward the plant and the microbes. Phytoremediation technologies using PGPR are now being considered to play an important agronomic role as PGPR can promote plant growth well in the metal-contaminated fields (Burd et al., 2000; Ali et al., 2013) and enhance detoxification of soil (Mayak et al., 2004; Sinha and Mukherjee, 2008; Mallick and Mukherjee, 2015; Mallick et al., 2014, 2015). The properties of plants used for phytoremediation, for example, biomass production, low-level contaminant uptake (Meers et al., 2010; Huang et al., 2011), plant nutrition, and health are enhanced by PGPR, but it is important to select appropriate PGPR candidates that can persist and beneficially interact when used in phytoremediation practices. There are many reports regarding bioremediation by PGPR. Kuiper et al. (2004) have reviewed beneficial plant–microbe interaction for effective rhizoremediation purposes. Abou-Shanab et al. (2003) reported that the addition of *Microbacterium liquefaciens*, *M. arabinogalactanolyticum*, and *Sphingomonas macrogoltabidus* to *Alyssum murale* grown in serpentine soil significantly augmented Ni uptake when compared with the uninoculated controls as a result of soil pH reduction. Burd et al. (1998, 2000) reported that some PGPR can significantly increase the growth of plants in the presence of heavy metals and thus allow plants to develop longer roots and shoots during the initial stages of growth (Glick et al., 1998). Once the seedling is established, the bacteria extend help to acquire sufficient iron by the plant for growth. Reduced chromium uptake stimulated seed germination and growth of wheat in the presence of potassium dichromate (Hasnain and Sabri, 1996) by chromium-resistant pseudomonads, isolated from paint industry effluents. Pairing PGPR with arbuscular mycorrhizal fungi (AMF) have been reported to increase the efficiency of phytoremediation. The bacterial strain *Brevibacillus*, isolated from Zn-contaminated soil, showed that it consistently enhanced plant growth, N and P accumulation, as well as the nodule number and mycorrhizal infection that demonstrated its plant-growth-promoting features. The amount of Zn acquired per root weight unit was reduced by the bacterial strain or AMF and particularly by the mixed bacterium-AMF inocula (Vivas et al., 2006). Soil rhizobacteria also directly affect metal bioavailability in the

vicinity of the roots by changing their chemical properties, such as pH, organic matter content, and redox state. These can aid in the leaching of contaminants from soils. The bioavailability of heavy metals in soils is a function of its solubility with pH with organic matter content being the main controlling factors. Rhizospheric bacteria have been documented for rhizoremediation of various toxic substances including volatile organic carbon contaminants, parathion (Anderson et al., 1993), atrazine (Anderson and Coats, 1995), trichloroethylene (Yee et al., 1998; Shim et al., 2000), and polychlorinated biphenyls (Brazil et al., 1995; Villacieros et al., 2005). Although the role of PGPR is significant in the phytoremediation strategies, research in this area is very limited and requires field studies to support laboratory-based observations (Lucy et al., 2004).

4.5.3 USE OF DESIGNER PLANTS

Metal-polluted soil is often cocontaminated with organic pollutants. Both metal and organic pollutants enter plant tissues and not only affect plant growth but also rule out the use of the plant biomass for other purposes due to associated toxicity of these contaminants (Sayler and Ripp, 2000). This scenario is more problematic for sites contaminated with a cocktail of pollutants, such as waste from refinery industries containing different heavy metals, surfactants, emulsifiers, and other pollutants and wastes produced by mining industries (Abhilash et al., 2012). A novel remediation technology using "designer" plants has been projected as a possible solution to these particularly challenging remediation issues. This technology relies on the combined ability of plants (removal of inorganic pollutant) and soil-dwelling microbes. Following this approach, customized plant systems could be produced, whose rhizosphere/rhizoplane is colonized with microbes able to degrade complex organic contaminants, while the same plants already harbor pollutant-degrading endophytic microbes. In this approach, transgenic plant inducted with microbial genes for the detoxification and accumulation of metals (Maestri and Marmiroli, 2011) would be of importance for enhancing uptake, transport, and volatilization of various heavy metals (Ruis and Daniell, 2009; Maestri and Marmiroli, 2011; Bitther et al., 2012).

4.6 APPLICATION OF MOLECULAR TOOLS IN METAL BIOREMEDIATION

In the last few decades, a multitude of research in various dimensions has been carried out on basic and applied aspects of microbial interaction with metals and plants with the bioremediation perspective. These studies mainly include the isolation and identification of superior microorganisms that can efficiently interact with toxic metals that could resist and/or accumulate or transform metal, their biochemical and genetic characterization, elucidation of microbe–metal–plant interactions mechanisms, and all related studies relevant for bioremediation. In spite of considerable research interests and its few commercialization, the success of bioremediation remains confined and yet to be appropriately implemented.

There are various reasons of failures including, but not limited to, (1) lack of information of bacteria inhabiting the contaminated sites (since most of the microbes are not readily culturable), (2) narrow knowledge of their metabolic capabilities in contaminated environment, and (3) how the indigenous communities respond to environmental changes within such habitats. It has been suggested that to develop successful phytoextraction strategies, understanding the geochemistry of the contaminated site along with a detailed profile of the microbial communities and plant involved in key physiological processes is extremely essential. More precisely, microbial community profiling must be done in terms of structure (to know who is there), phenotypic potential (to know their metabolic capabilities), function (to know the realized metabolic potential), and their interaction with the physical environments (Rittmann et al., 2006). Effective phytoextraction strategies should encompass an understanding of the fundamental molecular and physiological processes of remediating populations. Ideally, all studies should be carried out in the most ecologically relevant context, but the trade-off is that *in situ*, it is hard to manage enough variables to make the results interpretable (Thompson et al., 2010; Sar and Islam, 2012). Currently, the emergence of "genomics era" using high-throughput DNA sequencing and methods to analyze gene expression has shown potential to revolutionize the field as it provides a global insight into the microbial metabolic activity within contaminated environment, irrespective of their culturability. Currently, application of technologies, so-called "omics technologies," related to functional analyses of microbial communities at various levels of "transcriptomes, proteomes, and metabolomes" for studying microbial activities and their roles in environmental hazard monitoring and decontamination, are being emphasized (Desai et al., 2010).

Significant advances in understanding the roles of plant-associated microbes in metal mobilization/immobilization have been made; however, additional advances are required in the application of these processes in heavy metal phytoremediation (Sessitsch et al., 2013). For example, complete genome sequences for several environmentally relevant microorganisms, uptake mechanism of microbial chelator–metal complex by the plants, factors influencing the solubility and plant availability of nutrients/heavy metals, and cell signaling that occurs between plant roots and microbes; these types of analysis will surely demonstrate the usefulness of exploring the mechanism of metal–microbe–plant interactions (Desai et al., 2010; Sessitsch et al., 2013). Moreover, such knowledge may enable us to manipulate the rhizosphere processes, for example, increasing rhizosphere microorganisms, inoculating the microbial strains with various plant-growth-promoting features, and coinoculating microbial consortia would yield better results for effective phytoremediation. An integrated approach for rhizosphere management is important here, since soil environments are highly complex and heterogeneous. Also, an improved fundamental knowledge of the role of microbes, both culturable and nonculturable, in metal ion speciation in rhizosphere soils would be crucial for the successful formulation of microbial inoculum in the metal-contaminated fields.

4.7 UNREVEALED AREAS, FUTURE PERSPECTIVES, AND CONCLUSION

Since bioremediation is based on natural attenuation, general people consider it more suitable than other technologies. However, in spite of huge prospects, it has many limitations as well. The cleanup of soil contaminated with toxic metals is one of the most difficult tasks for environmental engineering, presenting a different set of problems, due to the dynamic nature of metals in soils. Therefore, not all strategies developed so far for metal decontamination would be explorable in every field condition. Moreover, applicability of certain strategies depends on the demographic sets of the regions. Additionally, a bioremediation process is dependent on the growth and survival of the microorganisms that vary with seasonal changes and mostly relies on the nutrient requirement to support their persistence in complex soil environment. Furthermore, microbial activity, due to direct exposure to changes in environmental factors, cannot be controlled and can cause problematic application of treatment additives. However, considering the significant merits of bioremediation, future studies should be directed to elucidate factors involved in improving *in situ* bioremediation strategies and to understand the factors affecting complex microbe–metal–plant interactions for its applicability and adaptability in all the possible metal stress conditions as well. There is every possibility of applying microorganisms or an integrated management for the decontamination of metals from the environment for successful bioremediation, replacing the existing physicochemical methods after a proper trial in actual field conditions.

REFERENCES

Abhilash, P.C., Powell, J.R., Singh, H.B., Singh, B.K. 2012. Plant-microbe interactions: Novel applications for exploitation in multipurpose remediation technologies. *Trends in Biotechnology* 30: 416–420.

Abou-Shanab, R.A., Angle, J.S., Delorme, T.A., Chaney, R.L., van Berkum, P., Moawad, H., Ghanem, K., Ghozlan, H.A. 2003. Rhizobacterial effects on nickel extraction from soil and uptake by *Alyssum murale*. *New Phytology* 158: 219–224.

Aiking, H., Kok, K., Heerikhuizen, H.V., Riet, J.V.T. 1982. Adaptation to cadmium by *Klebsiella aerogenes* growing in continuous culture proceeds mainly via formation of cadmium sulfide. *Applied and Environmental Microbiology* 44: 938–944.

Ali, H., Khan, E., Sajad, M.A. 2013. Phytoremediation of heavy metals—Concepts and applications. *Chemosphere* 91: 869–881.

Anderson, T.A., Coats, J.R. 1995. An overview of microbial degradation in the rhizosphere and its implication for bioremediation. In: Skipper, H.D., Turco, R.F. (Eds.) *Bioremediation: Science and Applications*, Vol. 43. Soil Science Society of America, Madison, WI, pp. 135–143.

Anderson, T.A., Guthrie, E.A., Walton, B.T. 1993. Bioremediation in the rhizosphere. *Environmental Science & Technology* 27: 2630–2636.

Avery, S.V. 2001. Metal toxicity in yeast and the role of oxidative stress. *Advances in Applied Microbiology* 49: 111–142.

Barkay, T., Liebert, C., Gillman, M. 1989. Environmental significance of the potential for *mer*(Tn21)-mediated reduction of Hg^{2+} to Hg^0 in natural waters. *Applied and Environmental Microbiology* 55: 1196–1202.

Beveridge, T.J., Murry, R.G.E. 1980. Sites of metal deposition in the cell wall of *Bacillus subtilis*. *Journal of Bacteriology* 141: 876–887.

Birch, L., Bachofen, R. 1990. Complexing agents from microorganisms. *Experientia* 46: 827–834.

Bitther, O.P., Pilon-Smits, E.A.H., Meagher, R.B., Doty, S. 2012. Biotechnological approaches for phytoremediation. In: Arie Altman, A., Hasegawa, P.M. (Eds.) *Plant Biotechnology and Agriculture*. Academic Press, Oxford, U.K., pp. 309–328.

Bolan, N., Kunhikrishnan, A., Thangarajan, R., Kumpiene, J., Park, J., Makino, T., Kirkham, M.B., Scheckel, K. 2014. Remediation of heavy metal(loid)s contaminated soils—To mobilize or to immobilize? *Journal of Hazardous Materials* 266: 141–166.

Brazil, G.M., Kenefick, L., Callanan, M., Haro, A., de Lorenzo, V., Dowling, D.N., O'Gara, F. 1995. Construction of a rhizosphere pseudomonad with potential to degrade polychlorinated biphenyls and detection of *bph* gene expression in the rhizosphere. *Applied and Environmental Microbiology* 61: 1946–1952.

Brim, H., Venkateshwaran, A., Kostandarithes, H.M., Fredrickson, J.K., Daly, M.J. 2003. Engineering *Deinococcus geothermalis* for bioremediation of high temperature radioactive waste environments. *Applied and Environmental Microbiology* 69: 4575–4582.

Burd, G.I., Dixon, D.G., Glick, B.R. 1998. A plant growth-promoting bacterium that decreases nickel toxicity in plant seedlings. *Applied and Environmental Microbiology* 64: 3663–3668.

Burd, G.I., Dixon, G.D., Glick, B.R. 2000. Plant growth promoting bacteria that decrease heavy metal toxicity in plants. *Canadian Journal of Microbiology* 46: 237–245.

Crist, R.H., Oberholser, K., Shank, N., Nguyen, M. 1981. Nature of bonding between metallic ions and algal cell walls. *Environmental Science & Technology* 15: 1212–1217.

Dary, M., Chamber-Pérez, M., Palomares, A., Pajuelo, E. 2010. "In situ" phytostabilisation of heavy metal polluted soils using *Lupinus luteus* inoculated with metal resistant plant-growth promoting rhizobacteria. *Journal of Hazardous Materials* 177: 323–330.

Desai, C., Pathak, H., Madamwar, D. 2010. Advances in molecular and "-omics" technologies to gauge microbial communities and bioremediation at xenobiotic/anthropogen contaminated sites. *Bioresource Technology* 101: 1558–1569.

Duruibe, J.O., Ogwuegbu, M.O.C., Egwurugwu, J.N. 2007. Heavy metal pollution and human biotoxic effects. *International Journal of Physical Sciences* 2(5): 112–118.

Eide, D.J. 2000. Metal ion transport in eukaryotic microorganisms: Insights from *Saccharomyces cerevisiae*. *Advances in Microbial Physiology* 43: 1–38.

Fomina, M., Burford, E.P., Gadd, G.M. 2005. Toxic metals and fungal communities. In: Dighton, J., White, J.F., Oudemans, P. (Eds.) *The Fungal Community. Its Organization and Role in the Ecosystem*. CRC Press, Boca Raton, FL, pp. 733–758.

Gadd, G.M. 1990. Heavy metal accumulation by bacteria and other microorganisms. *Experientia* 46: 834–840.

Gadd, G.M. 2004. Microbial influence on metal mobility and application for bioremediation. *Geoderma* 122: 109–119.

Gadd, G.M. 2005. Microorganisms in toxic metal polluted soils. In: Buscot, F., Varma, A. (Eds.) *Microorganisms in Soils: Roles in Genesis and Functions*. Springer-Verlag, Berlin, Germany, pp. 325–356.

Gadd, G.M. 2007. Geomycology: Biogeochemical transformations of rocks, minerals, metals and radionuclides by fungi, bioweathering and bioremediation. *Mycology Research* 111: 3–49.

Gadd, G.M. 2010. Metals, minerals and microbes: Geomicrobiology and bioremediation. *Microbiology* 156: 609–643.

Gadd, G.M., White, C. 1993. Microbial treatment of heavy metal pollution—A working biotechnology? *Trends in Biotechnology* 11: 353–358.

Glick, B.R., Patten, C.L., Holguin, G., Penrose, D.M. 1999. *Biochemical and Genetic Mechanisms Used by Plant Growth Promoting Bacteria*. Imperial College Press, London, U.K.

Glick, B.R., Penrose, D.M., Li, J.P. 1998. A model for the lowering of plant ethylene concentrations by plant growth-promoting bacteria. *Journal of Theoretical Biology* 190: 63–68.

Hasnain, S., Sabri, A.N. 1996. Growth stimulation of *Triticum Aestivum* seedlings under Cr-stresses by non rhizospheric pseudomonad strains. *Proceedings of the Seventh International Symposium on Biological Nitrogen Fixation with Non-Legumes*. Kluwer Academic Publishers, Dordrecht, the Netherlands, p. 36.

Higham, D.P., Sadler, P.J., Scawen, M.D. 1984. Cadmium-resistant *Pseudomonas putida* synthesizes novel cadmium proteins. *Science* 225: 1043–1046.

Higham, D.P., Sadler, P.J., Scawen, M.D. 1986. Cadmium-binding proteins in *Pseudomonas putida*: Pseudothioneins. *Environmental Health Perspectives* 65: 5–11.

Holden, J.F., Adams, M.W.W. 2003. Microbe-metal interactions in marine hydrothermal vents. *Current Opinion in Chemical Biology* 7: 160–165.

Hosea, M., Greene, B., McPherson, R., Henzl, M., Alexander, M.D., Darnall, D.W. 1986. Accumulation of elemental gold on the alga *Chlorella vulgaris*. *Inorganica Chimica Acta* 123: 161–165.

Huang, H., Yu, N., Wang, L., Gupta, D.K., He, Z., Wang, K., Zhu, Z., Yan, X., Li, T., Yang, X.E. 2011. The phytoremediation potential of bioenergy crop *Ricinus communis* for DDTs and cadmium co-contaminated soil. *Bioresource Technology* 102: 11034–11038.

Islam, E., Sar, P. 2011. Molecular assessment on impact of uranium ore contamination in soil bacterial diversity. *International Biodeterioration and Biodegradation* 65: 1043–1051.

Iyer, A., Mody, K., Jha, B. 2005. Biosorption of heavy metals by a marine bacterium. *Marine Pollution Bulletin* 50: 340–343.

Jarup, L. 2003. Hazards of heavy metal contamination. *British Medical Bulletin* 68: 167–182.

Kamnev, A.A., van der Lelie, D. 2000. Chemical and biological parameters as tools to evaluate and improve heavy metal phytoremediation. *Bioscience Reports* 20: 239–258.

Kazy, S.K., Sar, P., Asthana, R.K., Singh, S.P. 1999. Copper uptake and its compartmentalization in *Pseudomonas aeruginosa* strains: Chemical nature of cellular metal. *World Journal of Microbiology and Biotechnology* 15: 599–605.

Kloepper, J.W. 1994. Plant growth-promoting rhizobacteria (other systems). In: Okon, Y.Y. (Ed.) *Azospirillum/Plant Associations*. CRC Press, Boca Raton, FL, pp. 111–118.

Kloepper, J.W., Zablotowick, R.M., Tipping, E.M., Lifshitz, R. 1991. Plant growth promotion mediated by bacterial rhizosphere colonizers. In: Keister, D.L., Cregan, P.B. (Eds.) *The Rhizosphere and Plant Growth*. Kluwer Academic Publishers, Dordrecht, the Netherlands, pp. 315–326.

Kotrba, P. 2013. Transgenic approaches to enhance phytoremediation of heavy metal-polluted soils. In: Gupta, D.K. (Ed.) *Plant-Based Remediation Processes*. Springer, Heidelberg, Germany, pp. 239–271.

Kuiper, I., Lagendijk, E.L., Bloemberg, G.V., Lugtenberg, B.J.J. 2004. Rhizoremediation: A beneficial plant-microbe interaction. *Molecular Plant-Microbe Interaction* 17: 6–15.

Kumpiene, J., Lagerkvist, A., Maurice, C. 2008. Stabilization of As, Cr, Cu, Pb and Zn in soil using amendments—A review. *Waste Management* 28: 215–225.

Langella, F., Grawunder, A., Stark, R., Weist, A., Merten, D., Hafenburg, G., Büchel, G., Kothe, E. 2014. Microbially assisted phytoremediation approaches for two multi-element contaminated sites. *Environmental Science and Pollution Research* 21: 6845–6858.

Li, W., Yu, H. 2014. Insight into the roles of microbial extracellular polymer substances in metal biosorption. *Bioresource Technology* 160: 15–23.

Liu, X., Song, Q., Tang, Y., Li, W., Xu, J., Wu, J., Wang, F., Brookes, P.C. 2013. Human health risk assessment of heavy metals in soil–vegetable system: A multi-medium analysis. *Science of the Total Environment* 463–464: 530–540.

Lombi, E., Gerzabek, M.H. 1998. Determination of mobile heavy metal fraction in soil: Results of a pot experiment with sewage sludge. *Communications in Soil Science and Plant Analysis* 29: 2545–2556.

Lucy, M., Reed, E., Glick, B.R. 2004. Application of free living plant growth promoting rhizobacteria. *Antonie van Leeuwenhoek* 86: 1–25.

Ma, Y., Prasad, M.N.V., Rajkumar, M., Freitas, H. 2011. Plant growth promoting rhizobacteria and endophytes accelerate phytoremediation of metalliferous soils. *Biotechnology Advances* 29: 248–258.

Macaskie, L.E., Dean, A.C.R. 1985. Strontium accumulation by immobilized cells of a *Citrobacter* sp. *Biotechnology Letters* 7: 457–462.

Maestri, E., Marmiroli, M. 2011. Genetic and molecular aspects of metal tolerance and hyperaccumulation. In: Gupta, D.K., Sandalio, L.M. (Eds.) *Metal Toxicity in Plants: Perception, Signalling and Remediation*. Springer, Berlin, Germany, pp. 41–61.

Mallick, I., Hossain, S.T., Sinha, S., Mukherjee, S.K. 2014. *Brevibacillus* sp. KUMAs2, a bacterial isolate for possible bioremediation of arsenic in rhizosphere. *Ecotoxicology and Environmental Safety* 107: 236–244.

Mallick, I., Islam, E., Mukherjee, S.K. 2015. Fundamentals and application potential of arsenic resistant bacteria for bioremediation in rhizosphere: A review. *Soil and Sediment Contamination* 24: 704–718.

Mallick, I., Mukherjee, S.K. 2015. Bioremediation potential of an arsenic immobilizing strain *Brevibacillus* sp. KUMAs1 in the rhizosphere of chilli plant. *Environmental Earth Sciences* 74: 6757–6765.

Mayak, S., Tirosh, T., Glick, B.R. 2004. Plant growth-promoting bacteria that confer resistance to water stress in tomatoes and peppers. *Plant Science* 166: 525–530.

McHale, A.P., McHale, S. 1994. Microbial biosorption of metals: Potential in the treatment of metal pollution. *Biotechnology Advances* 12: 647–652.

Meers, E., van Slycken, S., Adriaensen, K., Ruttens, A., Vangronsveld, J., Du Laing, G., Witters, N., Thewys, T., Tack, F.M. 2010. The use of bio-energy crops (*Zea mays*) for "phytoattenuation" of heavy metals on moderately contaminated soils: A field experiment. *Chemosphere* 78: 35–41.

Mejare, M., Bulow, L. 2001. Metal-binding proteins and peptides in bioremediation and phytoremediation of heavy metals. *Trends in Biotechnology* 19: 67–73.

Mench, M., Vangronsveld, J., Bleeker, P., Ruttens, A., Geebelen, W., Lepp, N. 2006. Phytostabilisation of metal-contaminated sites. In: Echevarria, G., Goncharova, N., Morel, J.-L. (Eds.) *Phytoremediation of Metal-Contaminated Soils*. Springer, Dordrecht, the Netherlands, pp. 109–190.

Meyer, J., Schmidt, A., Michalke, K., Hensel, R. 2007. Volatilisation of metals and metalloids by the microbial population of an alluvial soil. *Systematic and Applied Microbiology* 30: 229–238.

Mukherjee, S.K., Asanuma, S. 1997. Al binding to the EPS and DNA of the bradyrhizobial cells exposed to Al stress. *Kyushu Agricultural Research* 60: 71.

Mullen, M.D., Wolf, D.C., Ferris, F.G., Beveridge, T.J., Fleming, C.A., Bailey, G.W. 1989. Bacterial sorption of heavy metals. *Applied and Environmental Microbiology* 55: 3143–3149.

Muraleedharan, T.R., Iyengar, L., Venkobachar, C. 1991. Biosorption: An attractive alternative for metal removal and recovery. *Current Science* 61: 379–385.

Naz, N., Young, H.K., Ahmed, N., Gadd, G.M. 2005. Cadmium accumulation and DNA homology with metal resistance genes in sulfate-reducing bacteria. *Applied and Environmental Microbiology* 71: 4610–4618.

Neilands, J.B. 1981. Microbial iron compounds. *Annual Review in Biochemistry* 50: 715–731.

Nies, D.H. 1992. Resistance to cadmium, cobalt, zinc, and nickel in microbes. *Plasmid* 27: 17–28.

Nies, D.H. 1999. Microbial heavy metal resistance. *Applied Microbiology and Biotechnology* 51: 730–750.

Nies, D.H. 2003. Efflux-mediated heavy metal resistance in prokaryotes. *FEMS Microbiology Review* 27: 313–339.

Nies, D.H., Silver, S. 1995. Ion efflux systems involved in bacterial metal resistances. *Journal of Industrial Microbiology* 14: 186–199.

Noberg, A.B., Persson, H. 1984. Accumulation of heavy-metal ions by *Zoogloea ramigera*. *Biotechnology and Bioengineering* 26: 239–246.

Osman, D., Cavet, J.S. 2008. Copper homeostasis in bacteria. *Advances in Applied Microbiology* 65: 217–247.

Park, J.H., Lamb, D., Paneerselvam, P., Choppala, G., Bolan, N., Chung, J.W. 2011. Role of organic amendments on enhanced bioremediation of heavy metal(loid) contaminated soils. *Journal of Hazardous Materials* 185: 549–574.

Prasad, M.N.V., Freitas, H. 2003. Metal hyperaccumulation in plants: Biodiversity prospecting for phytoremediation technology. *Journal of Biotechnology* 6: 285–321.

Pulford, I.D., Watson, C. 2003. Phytoremediation of heavy metal contaminated land by trees—A review. *Environmental International* 29: 529–540.

Rajkumar, M., Ae, N., Prasad, M.N.V., Freitas, H. 2010. Potential of siderophore-producing bacteria for improving heavy metal phytoextraction. *Trends in Biotechnology* 28: 142–149.

Rittmann, B.E., Hausner, M., Loffler, F., Love, N.G., Muyzer, G., Okabe, S., Oerther, D.B., Peccia, J., Raskin, L., Wagner, M. 2006. A vista for microbial ecology and environmental biotechnology. *Environmental Science & Technology* 40: 1096–1103.

Rosen, B.P. 2002. Transport and detoxification systems for transition metals, heavy metals and metalloids in eukaryotic and prokaryotic microbes. *Comparative Biochemistry and Physiology Part A* 133: 689–693.

Ruis, O.N., Daniell, H. 2009. Genetic engineering to enhance mercury phytoremediation. *Current Opinion in Biotechnology* 20: 213–219.

Samuelson, P., Wernérus, H., Svedberg, M., Ståhl, S. 2000. Staphylococcal surface display of metal-binding polyhistidyl peptides. *Applied and Environmental Microbiology* 66: 1243–1248.

Sar, P., Islam, E. 2012. Metagenomics approaches in microbial bioremediation of metals and radionuclides In: Satyanarayana, T. et al. (Eds.) *Microorganisms in Environmental Management: Microbes and Environment.* Springer Science + Business Media B.V, Dordrecht, the Netherlands, pp. 525–546.

Sar, P., Kazy, S.K., Singh, S.P. 2001. Intracellular nickel accumulation by *Pseudomonas aeruginosa* and its chemical nature. *Letter in Applied Microbiology* 32: 257–261.

Sauge-Merle, S., Cuiné, S., Carrier, P., Lecomte-Pradines, C., Luu, D.T., Peltier, G. 2003. Enhanced toxic metal accumulation in engineered bacterial cells expressing *Arabidopsis thaliana* phytochelatin synthase. *Applied and Environmental Microbiology* 69: 490–494.

Sayler, G.S., Ripp, S. 2000. Field applications of genetically engineered microorganisms for bioremediation process. *Current Opinion in Biotechnology* 11: 286–289.

Scott, J.A., Palmer, S.J. 1990. Site of cadmium uptake in bacteria used for biosorption. *Applied Microbiology and Biotechnology* 33: 221–225.

Sessitsch, A., Kuffner, M., Kidd, P., Vangronsveld, J., Wenzel, W.W., Fallmann, K., Puschenreiter, M. 2013. The role of plant-associated bacteria in the mobilization and phytoextraction of trace elements in contaminated soils. *Soil Biology and Biochemistry* 60: 182–194.

Shim, H., Chauhan, S., Ryoo, D., Bowers, K., Thomas, S.M., Canada, K.A., Burken, J.G., Wood, T.K. 2000. Rhizosphere competitiveness of trichloroethylene-degrading poplar-colonizing recombinant bacteria. *Applied and Environmental Microbiology* 66: 4673–4678.

Shutcha, M.N., Faucon, M.P., Kissi, C.K., Colinet, G., Mahy, G., Luhembwe, M.N., Visser, M., Meerts, P. 2015. Three years of phytostabilisation experiment of bare acidic soil extremely contaminated by copper smelting using plant biodiversity of metal-rich soils in tropical Africa (Katanga, DR Congo). *Ecological Engineering* 82: 81–90.

Silver, S., Phung, L.T. 1996. Bacterial heavy metal resistance: New surprises. *Annual Review in Microbiology* 50: 753–789.

Silver, S., Phung, L.T. 2009. Heavy metals, bacterial resistance. In: Schaechter, M. (Ed.) *Encyclopedia of Microbiology.* Elsevier, Oxford, U.K., pp. 220–227.

Sinha, S., Mukherjee, S.K. 2008. Cadmium-induced siderophore production by a high Cd resistant bacterial strain relieved Cd-toxicity in plants through root colonization. *Current Microbiology* 56: 55–60.

Sinha, S., Mukherjee, S.K. 2009. *Pseudomonas aeruginosa* KUCd1, a possible candidate for cadmium bioremediation. *Brazilian Journal of Microbiology* 40: 655–662.

Sousa, C., Cebolla, A., de Lorenzo, V. 1996. Enhanced metalloadsorption of bacterial cells displaying poly-His peptides. *Nature Biotechnology* 14: 1017–1020.

Sousa, C., Kotrba, P., Ruml, T., Cebolla, A., de Lorenzo, V. 1998. Metalloadsorption by *Escherichia coli* cells displaying yeast and mammalian metallothioneins anchored to the outer membrane protein LamB. *Journal of Bacteriology* 180: 2280–2284.

Spark, K.M., Wells, J.D., Johnson, B.B. 1997. The interaction of a humic acid with heavy metals. *Australian Journal of Soil Research* 35: 89–101.

Sposito, G., Page, A.L. 1984. Cycling of metal ions in the soil environment. In: Sigel, H. (Ed.) *Metal Ions in Biological Systems, Circulation of Metals in the Environment.* Marcel Dekker, Inc., New York, pp. 287–332.

Summers, A.O., Silver, S. 1978. Microbial transformation of metals. *Annual Review in Microbiology* 32: 637–672.

Thayer, J.S. 1989. Methylation; its role in environmental mobility in heavy elements. *Applied Organometallic Chemistry* 3: 123–128.

Thompson, D., Chourey, K., Wickham, G., Thieman, S., VerBerkmoes, N., Zhang, B., McCarthy, A., Rudisill, M., Shah, M., Hettich, R. 2010. Proteomics reveals a core molecular response of *Pseudomonas putida* F1 to acute chromate challenge. *BMC Genomics* 11: 311.

Tsang, D.C.W., Yip, A.C.K. 2014. Comparing chemical-enhanced washing and waste based stabilisation approach for soil remediation. *Journal of Soil Sediment* 14: 936–947.

Ueki, T., Yasuhisa, S., Nobuo, Y., Hitoshi, M. 2003. Bioaccumulation of copper ions by *Escherichia coli* expressing vanabin genes from the vanadium-rich ascidian *Ascidia sydneiensis samea.* *Applied and Environmental Microbiology* 69: 6442–6446.

Villacieros, M., Whelan, C., Mackova, M., Molgaard, J., Sanchez-Contreras, M., Lloret, J., de Carcer, D.A. et al. 2005. Polychlorinated biphenyl rhizoremediation by *Pseudomonas fluorescens* F113 derivatives, using a *Sinorhizobium meliloti* nod system to drive *bph* gene expression. *Applied and Environmental Microbiology* 71: 2687–2694.

Vivas, A., Biró, B., Ruíz-Lozano, J.M., Barea, J.M., Azcón, R. 2006. Two bacterial strains isolated from a Zn-polluted soil enhance plant growth and mycorrhizal efficiency under Zn-toxicity. *Chemosphere* 62: 1523–1533.

Volesky, B. 1990a. Biosorption and biosorbents. In: Volesky, B. (Ed.) *Biosorption of Heavy Metals.* CRC Press, Boston, MA, pp. 3–5.

Volesky, B. 1990b. Removal and recovery of heavy metals by biosorption. In: Volesky, B. (Ed.) *Biosorption of Heavy Metals.* CRC Press, Boston, MA, pp. 7–43.

Wang, C.L., Michels, P.C., Dawson, S.C., Kitisakkul, S., Baross, J.A., Keasling, J.D., Clark, D.S. 1997. Cadmium removal by a new strain of *Pseudomonas aeruginosa* in aerobic culture. *Applied and Environmental Microbiology* 63: 4075–4078.

Wang, P., Sun, G., Jia, Y., Meharg, A.A., Zhu, Y. 2014. A review on completing arsenic biogeochemical cycle: Microbial volatilization of arsines in environment. *Journal of Environmental Science* 26: 371–381.

Wernérus, H., Lehtiö, J., Teeri, T., Nygren, P.A., Ståhl, S. 2001. Generation of metal-binding *Staphylococci* through surface display of combinatorially engineered cellulose-binding domains. *Applied and Environmental Microbiology* 67: 4678–4684.

White, C., Gadd, G.M. 1998. Accumulation and effects of cadmium on sulphate-reducing bacterial biofilms. *Microbiology* 144: 1407–1415.

White, C., Sharman, A.K., Gadd, G.M. 1998. An integrated microbial process for the bioremediation of soil contaminated with toxic metals. *Nature Biotechnology* 16: 570–575.

Wood, J.M., Wang, H.W. 1983. Microbial resistance to heavy metals. *Environmental Science & Technology* 17: 583–590.

Yee, D.C., Maynard, J.A., Wood, T.K. 1998. Rhizoremediation of trichloroethylene by a recombinant, root-colonizing *Pseudomonas fluorescens* strain expressing toluene ortho-monooxygenase constitutively. *Applied and Environmental Microbiology* 64: 112–118.

Zhuang, P., Mcbridge, B.B., Xia, H.P., Li, N.Y., Li, Z.A. 2009. Health risk from heavy metals via consumption of food crops in the vicinity of Dabaoshan mine, South China. *Science of the Total Environment* 407: 1551–1561.

5 Biotransformation of Heavy Metal(loid)s in Relation to the Remediation of Contaminated Soils*

Anitha Kunhikrishnan, Girish Choppala, Balaji Seshadri, Jin Hee Park, Kenneth Mbene, Yubo Yan, and Nanthi S. Bolan

CONTENTS

ABSTRACT

The dynamics of trace elements in soils is dependent on both their physicochemical interactions with inorganic and organic soil constituents and their biological interactions linked to the microbial activities of soil–plant systems. Microorganisms control the transformation (microbial or biotransformation) of trace elements by several mechanisms that include oxidation, reduction, methylation, demethylation, complex formation, and biosorption. Microbial transformation plays a major role in the behavior and fate of toxic elements, especially arsenic (As), chromium (Cr), mercury (Hg), and selenium (Se) in soils and sediments. Biotransformation processes can alter the speciation and redox state of these elements and hence control their solubility and subsequent mobility. These processes play an important role in the bioavailability, mobility, ecotoxicity, and environmental health of these trace elements. A greater understanding of biotransformation processes is necessary to efficiently manage and utilize them for contaminant removal and to develop *in situ* bioremediation technologies. In this chapter, the key microbial transformation processes, including biosorption, redox reactions, and methylation/demethylation reactions controlling the fate and behavior of As, Cr, Hg, and Se, are addressed. The factors affecting these processes in relation to the bioavailability and remediation of trace elements in the environment are also examined, and possible future research directions are recommended.

5.1 INTRODUCTION

Heavy metal(loid)s include both biologically essential (e.g., cobalt [Co], copper [Cu], chromium [Cr], manganese [Mn], and zinc [Zn]) and nonessential (e.g., cadmium [Cd], lead [Pb], and mercury [Hg]) elements. The biologically essential elements are required in low concentrations and so are termed as "micronutrients." The nonessential elements are toxic and hence are known as "toxic elements" (Adriano 2001). Both groups are toxic to plants, animals, and/or humans at exorbitant concentrations (Alloway 1990; Adriano 2001).

Soil represents the major sink for trace elements released into the biosphere through both geogenic (i.e., weathering or pedogenic) and anthropogenic (i.e., human activities) processes. The dynamics of trace elements in soils is dependent both on their physicochemical reactions with inorganic and organic soil components and on biological interactions associated with the microbial activities of soil–plant systems (Adriano 2001). Two approaches have been used to examine the interaction between microbes and trace elements in soils (Alexander 1999): (1) the influence of trace elements on microbial populations and functions and (2) the influence

* This chapter is a condensed version of Bolan et al. (2013a).

67

and role of microbes on the transformation of trace elements. Microorganisms control the transformation (microbial or biotransformation) of trace elements by various mechanisms that include oxidation, reduction, methylation, demethylation, complex formation, and biosorption (Alexander 1999).

Microbial transformation plays a key role in the behavior and fate of toxic trace elements, especially arsenic (As), Cr, Hg, and selenium (Se) in soils and sediments. Biotransformation processes alter the speciation and oxidation/reduction state of these trace elements in soils, thereby controlling the metal solubility and its subsequent mobility (Gadd 2010). These processes play a major role in the bioavailability, mobility, ecotoxicology, and environmental health of these trace elements. For example, microbial reduction/methylation of trace elements and its consequences to human health has received attention primarily from a series of widespread poisoning incidents including "Gasio-gas" poisoning resulted from converting arsenic trioxide in wallpaper glue into volatile poisonous trimethylarsine or "Gasio-gas" (Adriano et al. 2004) and As contamination of surface and groundwaters mediated through redox reactions of geogenic As (Mahimairaja et al. 2005).

Microbial reduction and methylation reactions have also been identified as important mechanisms for detoxifying toxic elements (Zhang and Frakenberger 2003). These processes are particularly important for those elements (e.g., As, Hg, and Se) that are able to form methyl or metal(loid)-hydride compounds. Thus, a greater understanding of biotransformation processes will help to monitor the environmental fate of the trace elements, particularly through the food web, and will help to develop *in situ* bioremediation technologies that are environmentally compatible. By understanding the biochemistry of biotransformation processes, these processes can be readily managed and efficiently utilized for the removal of contaminants. This chapter examines the influence of microbial processes on the accumulation and transformation of heavy metal(loid)s (with emphasis on As, Cr, Hg, and Se).

5.2 SOURCES AND SPECIATION OF TRACE ELEMENTS

Arsenic is a naturally occurring element, the major source of which is weathering of igneous and sedimentary rocks, including coal. Significant anthropogenic sources of As include fossil fuel combustion, leaching from mining wastes and landfills, mineral processing, and application of a range of agricultural by-products (e.g., poultry manure) (Christen 2001). Although the anthropogenic As source is becoming more important, in Bangladesh and West Bengal, recent incidents of extensive As contamination of groundwaters transported by rivers from sedimentary rocks in the Himalayas over thousands of years are of geological origin (Mahimairaja et al. 2005).

In soils, As is present as arsenite [As(III)], arsenate [As(V)], and organic As (monomethylarsonic acid (MMA) and dimethylarsinic acid (DMA) or cacodylic acid) (Sadiq 1997; Smith et al. 1998; Mahimairaja et al. 2005). Arsenic species are adsorbed onto iron (Fe), Mn, and aluminum (Al) compounds (Smith et al. 1998). In aquatic systems, As concentration in suspended solids and sediments is many times higher than that in water, indicating that the suspended solids are a good scavenging agent and sediments a sink for As (Mahimairaja et al. 2005).

Chromium enters the soil environment mainly through industrial waste disposal from coal-fired power plants, electroplating activities, leather tanning, timber treatment, and mineral ore and petroleum refining (Bolan et al. 2003a; Choppala et al. 2012). Chromium exists as hexavalent [Cr(VI)] and trivalent [Cr(III)] forms. While Cr(VI) is toxic and highly soluble in water, Cr(III) is less toxic, insoluble in water, and hence less mobile in soils (Barnhart 1997; Kosolapov et al. 2004). In soils, Cr exists mainly as Cr(III) unless oxidizing agents such as manganous oxide [Mn(IV)] are present (Gong and Donahoe 1997).

The burning of fossil fuels and gold recovery in mining are the major sources of Hg (Pacyna et al. 2001; de Lacerda 2003). Mercury forms salts in two ionic states, mercurous [Hg(I)] and mercuric [Hg(II)], with the latter much more common in the environment than the former (Schroeder and Munthe 1998). Mercury also forms organometallic compounds, many of which have industrial and agricultural uses. Elemental Hg gives rise to a vapor that is only slightly soluble in water but is problematic because of easy transport in the atmosphere (Boening 2000). The environmental Hg cycle has four strongly interconnected compartments: atmospheric, terrestrial, aquatic, and biotic. The atmospheric compartment is dominated by gaseous Hg(0), although Hg(II) dominates the fluxes in the aquatic and terrestrial compartments. The terrestrial compartment is dominated by Hg(II) sorbed to organic matter in soils. While the biotic compartment is dominated by methyl Hg, the aquatic compartment is dominated by Hg(II)–ligand pairs in water and Hg(II) in sediments (Wiener et al. 2003).

The principal sources of Se for commercial applications are Cu-bearing ore and sulfur (S) deposits. Selenium is used in xerography, as a semiconductor in photocell, and also used in the manufacture of batteries, glass, electronic equipments, antidandruff products, veterinary therapeutic agents, feed additives, and fertilizers. Selenium can be found in four different oxidation states: selenate [Se(VI); SeO_4^{2-}], selenite [Se(IV); SeO_3^{2-}], elemental selenium [Se(0); Se^0], and selenide [Se(-II); Se^{2-}]. Selenate and Se(IV) are common ions in natural waters and soils. Selenides and Se(0) are the common Se species in acidic soils that are under reducing conditions and are rich in organic matter. At moderate redox potential (Eh) either $HSeO_3^-$ or SeO_3^{2-} is the predominant form. At high Eh in well-aerated alkaline soils, the highly soluble SeO_4^{2-} is the predominant form.

Reduced Se compounds include volatile methylated species such as dimethyl selenide [DMSe, $Se(CH_3)_2$], dimethyl diselenide [DMDSe, $Se_2(CH_3)_2$], and dimethyl selenone [$(CH_3)_2 SeO_2$] and S-containing amino acids including selenomethionine, selenocysteine, and selenocystine. Inorganic reduced Se forms include mineral selenides and hydrogen selenide (H_2Se).

5.3 BIOTRANSFORMATION PROCESSES

The microbial processes involved in transforming trace elements in soils and sediments are grouped into three categories that include bioaccumulation, oxidation/reduction, and methylation/demethylation (Figure 5.1). Microorganisms can accumulate organometal(loid)s, a phenomenon relevant to toxicant transfer to higher organisms (i.e., biomagnification). In addition, many microorganisms are also capable of degrading and detoxifying organometal(loid)s through demethylation and dealkylation processes. Several organometal(loid) transformations are potentially useful for environmental bioremediation (Geoffrey and Gadd 2007; Gadd 2010).

5.3.1 BIOACCUMULATION

The physicochemical mechanisms by which trace elements are removed are encompassed by the general term *biosorption*. Biosorption includes adsorption, ion exchange, entrapment, and metabolic uptake, which are features of both living and dead biomass and their derived products (Ahalya et al. 2003). In living cells, biosorption is directly and indirectly influenced by metabolism. Metabolism-dependent mechanisms of trace

element removal that occur in living microorganisms include precipitation as sulfides, complexation by siderophores and other metabolites, sequestration by metal(loid)-binding proteins and peptides, transport, and intracellular compartmentation (White et al. 1995). Microorganisms exhibit a strong ability to accumulate (bioaccumulation) trace elements from substrates containing very low concentrations (Robinson et al. 2006). Both bacteria and fungi bioaccumulate trace elements, and the bioaccumulation process is activated by two processes (Schiewer and Volesky 2000; Adriano et al. 2004): (1) sorption (i.e., biosorption) of trace elements by microbial biomass and its by-products and (2) physiological uptake of trace elements by microorganisms through metabolically active and passive processes (Table 5.1).

Metal(loid) biosorption onto bacterial and fungal biomass is rapid and temperature dependent (Ledin et al. 1999; Dursun 2006). A wide range of binding groups, including carboxyl, amine, hydroxyl, phosphate (P), and sulfhydryl, have been shown to contribute to biosorption of trace elements. Both living and dead biomass act as biosorptive agents, and the magnitude of the phenomenon is directly related to these functional groups. Bacteria capable of producing large quantities of extracellular polymers are of a polysaccharide nature,

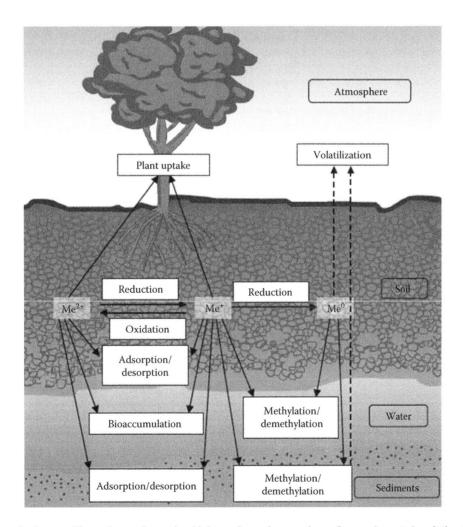

FIGURE 5.1 Schematic diagram illustrating various microbial transformation reactions of trace elements in relation to remediation.

TABLE 5.1

Selected References on the Biosorption of Arsenic, Chromium, Mercury, and Selenium in Soil and Aquatic Environments

Trace Element	Medium	Biomass	Observation	Reference
Arsenic	Water	*Penicillium chrysogenum*	Amine-based surfactants or with a cationic polyelectrolyte increased biosorption of As(V).	Loukidou et al. (2003)
	Water	*Lessonia nigrescens*	As(V) achieved better sorption at low pH.	Hansen et al. (2006)
	Water	Acid-washed crab shells	As(V) bound to amide groups present in crab shells.	Niu et al. (2007)
	Water	FeCl$_3$ pretreated tea fungal mat	Fe in fungal mat formed As(V)–Fe oxide bonds.	Murugesan et al. (2006)
	Water	*Staphylococcus xylosus*	As(V) and As(III) were adsorbed on the Fe(II) pretreated biomass surface through interaction with FeOH and FeOH$_2^+$ groups.	Aryal et al. (2010)
	Soil	*Rhodococcus* sp. WB.12	Maximum sorption capacity of As(III) was 77.3 mg g^{-1} at 30°C, pH 7.0, and indicated the involvement of several functional groups.	Prasad et al. (2011)
Chromium	Tannery effluent	*Bacillus megaterium* and *B. coagulans*	Dead biomass sorbed more Cr(VI) than living cells.	Srinath et al. (2002)
	Water	*Pinus sylvestris* cone biomass	pH of the aqueous phase strongly affected the sorption capacity of Cr(VI).	Ucun et al. (2002)
	Water	*Ocimum basilicum* seeds	Porous swollen outer layer of seeds increased the sorption of Cr(VI).	Melo and D'Souza (2004)
	Water	*Sargassum* sp.	Maximum adsorption of Cr(III) was observed at 30°C, pH 3.5, and was 1.31 mmol g^{-1}.	Cossich et al. (2004)
	Water	*Ecklonia* brown alga	Cr(III) adsorbs on –COOH functional groups.	Yun et al. (2001)
	Water	*Ceramium virgatum*—red algae	Biosorption of Cr(III) and Cr(VI) was taken place by chemisorption.	Sari and Tuzen (2008)
Mercury	Water	*Ulva lactuca*—algae	Sorption depended on the pH of solution and concentration of Hg.	Zeroual et al. (2003)
	Water	*Ricinus communis* leaves powder	Biosorption efficiencies increased with increasing contact time and initial metal(loid) concentration.	Al Rmalli et al. (2008)
	Water	*Penicillium oxalicum* var. *Armeniaca, Tolypocladium* sp.	The greater efficiency may be due to deacetylation treatment used for cleaning the waste biomass.	Svecova et al. (2006)
	Water	*Lessonia nigrescens* and *L. trabeculata*	Presence of Cl$^-$ and competing ions such as Cd(II), Zn(II), and Ni(II) decreased Hg sorption.	Reategui et al. (2010)
	Water	Magnetically modified *Saccharomyces cerevisiae* ssp. *uvarum*	Biosorption of Hg^{2+} increased with an increase in pH and reached plateau at pH 5.5.	Yavuz et al. (2006)
	Water	Estuarine *Bacillus* sp.	Changes in pH of solution had great effect on sorption of Hg on biomass.	Green-Ruiz (2006)
Selenium	Water	Thiolated egg shell membranes	Thiol groups formed by modification increased the biosorption of Se(IV) and Se(VI) species.	Yang et al. (2011)
	Water	Wheat bran	Se(IV) and Se(VI) species were sorbed on biomass by ion-exchange process.	Hasan et al. (2010)
	Water	Green algae—*Cladophora hutchinsiae*	Sorption of Se(IV) by biomass was through chemical ion-exchange mechanism.	Tuzen and Sari (2010)

have anionic properties, and are involved in removing soluble metal(loid) ions from solution by an ion-exchange process (Kodukula et al. 1994; Iyer et al. 2005). Macrofungi, such as *Agaricus*, can bioaccumulate Cd and Hg from soils/compost containing low concentrations of these elements (Tüezen et al. 1998). Many fungal products, such as glucans, mannans, melanins, chitins, and chitosans, have been shown to act as efficient biosorption agents (Gadd 1990).

Several trace elements are essential for many metabolic functions of microorganisms, and their uptake involves both active and passive processes (Zouboulis et al. 2004). In the case of metabolically active process, microorganisms exhibit specific mechanisms for the uptake of trace elements, which involve carrier systems associated with active ionic influxes across the cell membrane. Generally, the metabolically active process is slower than passive absorption, requiring the presence of suitable energy source and ambient conditions. Within the cell, microorganisms may convert metal(loid) ions into innocuous forms by precipitation or binding. For example, sulfate-reducing bacteria such as *Desulfovibrio* release hydrogen sulfide (H$_2$S), thus forming insoluble metal(loid) sulfide precipitates (Sass et al. 2009). Some bacteria produce siderophores in the form of phenols, catechols, or hydroxamates as part of their overall Fe uptake strategy (Renshaw et al. 2002; Hider and Kong 2010). Bacteria also produce metal(loid)-binding proteins, such as metallothionein that could bind metal(loid)s, thereby acting as detoxicants. For example, both *Pseudomonas putida* and *Escherichia coli* have been shown to produce low-molecular-weight proteins that bind Cd (Mejáre and Bülow 2001).

5.3.2 REDOX REACTIONS

Trace elements, including As, Cr, Hg, and Se, are most commonly subjected to microbial oxidation/reduction reactions (Table 5.2). Redox reactions influence the speciation and mobility of trace elements. While metals are generally less soluble in their higher oxidation state, the solubility of metalloids depends on both the oxidation state and the ionic form (Ross 1994). The oxidation/reduction reactions for various metal(loid)s and the optimum redox values for these reactions are given in Table 5.3. The redox reactions are grouped into two categories, assimilatory and dissimilatory (Brock and Madigan 1991). In assimilatory reactions, the metal(loid) substrate will serve a role in the physiology and metabolic functioning of the organism by acting as terminal electron acceptor. In contrast, for dissimilatory reactions the metal(loid) substrate has no known role in the physiology of the species responsible for the reaction and indirectly initiates redox reactions.

Arsenic in soils and sediments can be oxidized to As(V) by bacteria (Battaglia-Brunet et al. 2002; Bachate et al. 2012; Table 5.2). Since As(V) is strongly retained by inorganic soil components, microbial oxidation results in the immobilization of As. Under well-drained conditions, As would present as $H_2AsO_4^-$ in acidic soil and as $HAsO_4^{2-}$ in alkaline soils. Under reduced conditions, As(III) dominates in soils, but elemental arsenic [As(0)] and arsine (H_2As) can also be present. The distribution and mobilization of As species in the soil and sediments are controlled both by microbially mediated transformation of the As species and by adsorption (Adriano et al. 2004; Mahimairaja et al. 2005). In sediments, the reduction and methylation reactions of As are generally controlled by bacterial degradation of organic matter followed by the reduction and use of sulfate as the terminal electron acceptor (Adriano et al. 2004). Ferrous iron [Fe(II)] can also serve as an electron acceptor in bacterial oxidization of organic matter, resulting in the decomposition of ferric [Fe(III)] oxides and hydroxides.

Although Cr(III) is strongly retained on soil particles, Cr(VI) is very weakly adsorbed in soils that are net negatively charged and is readily available for plant uptake and leaching to groundwater (Leita et al. 2011). Oxidation of Cr(III) to Cr(VI) is primarily mediated abiotically through oxidizing agents such as Mn(IV) and to a lesser extent by Fe(III), whereas reduction of Cr(VI) to Cr(III) is mediated through both abiotic and biotic processes (Choppala et al. 2012). Oxidation of Cr(III) to Cr(VI) can enhance the mobilization and bioavailability of Cr. Chromate can be reduced to Cr(III)

in environments where a ready source of electrons [Fe(II)] is available (Hsu et al. 2009; Chen et al. 2010).

Dissimilatory Se(IV) reduction to Se(0) is the major biological transformation for the remediation of Se oxyanions in anoxic sediments (Lens et al. 2006). Selenite is readily reduced to the elemental state by chemical reductants such as sulfide or hydroxylamine or biochemically by systems such as glutathione reductase. Hence, precipitation of Se in its elemental form, which has been associated with bacterial dissimilatory Se(VI) reduction, has great environmental significance (Oremland et al. 1989, 2004).

Mercury undergoes abiological reduction process in soils and sediments. Microorganisms play a major role in reducing reactive Hg(II) to nonreactive Hg(0), which may be subjected to volatilization losses. Bacteria play a major role in the reduction of Hg(II) than eukaryotic phytoplankton. Mercury-resistant bacteria can transform ionic mercury [Hg(II)] to metallic mercury [Hg(0)] by enzymatic reduction (Von Canstein et al. 2002). The dissimilatory metal(loid)-reducing bacterium *Shewanella oneidensis* has been shown to reduce Hg(II) to Hg(0), which requires the presence of electron donors (Wiatrowski et al. 2006).

5.3.3 METHYLATION/DEMETHYLATION

Methylated derivatives of As, Hg, and Se can be derived through chemical and biological mechanisms and cause a transformation in volatility, solubility, toxicity, and mobility. The major microbial methylating agents are methylcobalamin (CH_3CoB_{12}), involved in the methylation of Hg, and S-adenosylmethionine (SAM), involved in the methylation of As and Se. Biological methylation (biomethylation) may result in metal(loid) detoxification, since methylated derivatives may be excreted readily from cells and are often volatile and may be less toxic, for example, organoarsenicals. However, for Hg, methylation may not play a major role in detoxification, because of the existence of more efficient resistance mechanisms, for example, reduction of Hg(II) to Hg(0) (Gadd 1993).

Although methylation of metal(loid)s occurs through both chemical (abiotic) and biological processes, biomethylation is considered to be the dominant process in soils and aquatic environments (Table 5.4). Thayer and Brinckman (1982) grouped methylation into two categories: transmethylation and fission-methylation. Transmethylation refers to the transfer of an intact methyl group from one compound (methyl donor) to another compound (methyl acceptor). Fission-methylation refers to the fission of a compound (methyl source), not necessarily containing a methyl group, so as to eliminate a molecule such as formic acid. The fission molecule is subsequently captured by another compound that is reduced to a methyl group.

Arsenic undergoes a series of biological transformation in aquatic systems, yielding a large number of compounds, especially organoarsenicals (Maher and Butler 1988). Benthic microbes are capable of methylating As under both aerobic and anaerobic conditions to produce methylarsines and methyl arsenic compounds with a generic

TABLE 5.2

Selected References on the Redox Reactions of Arsenic, Chromium, Mercury, and Selenium in Soil and Aquatic Environments

Trace Element	Medium	Observation	Reference
Arsenic(V)	Soil	Dissimilatory As(V)–reducing bacterium, *Bacillus selenatarsenatis* increased the removal of As from contaminated soils.	Yamamura et al. (2008)
	Soil	Both biotic and abiotic (S^{2-}, Fe^{2+}, and $H_2(g)$) factors were responsible for the reduction of As(V).	Jones et al. (2000)
	Soil	Attachment of *Shewanella putrefaciens* cells to oxide mineral surfaces promoted As(V) desorption, thereby facilitating its reduction.	Huang et al. (2011)
Arsenic(III)	Soil	Fe(III) and Mn(III) oxides oxidized As(III) to As(V) through electron transfer reaction.	Mahimairaja et al. (2005)
	Water	The oxidation rate of As(III) to As(V) and its subsequent adsorption was high in the presence of synthetic birnessite (MnO_2).	Manning et al. (2002)
	Water	In the presence of dissolved Fe(III) and near ultraviolet light, As(III) oxidized to As(V) in water.	Emett and Khoe (2001)
	Water	As(III) oxidation rate increased in the presence of *Euglena mutabilis*, a detoxification pathway.	Casiot et al. (2004)
	Water	As(III) was oxidized to As(V) within hours in the presence of magnetite but reaction was quenched by the addition of ascorbic acid.	Chiu and Hering (2000)
	Water	Bacteria attached to submerged macrophytes mediated the rapid As(III) oxidation reaction.	Wilkie and Hering (1998)
	Water	Ferrihydrite oxidized As(III) to As(V) in the presence of light (photooxidation).	Bhandari et al. (2011)
Chromium(VI)	Soil	Application of Fe(II) under flow conditions increased the reduction of Cr(VI).	Franco et al. (2009)
	Soil	Root exudates of *Typha latifolia* and *Carex lurida* increased sulfide species, which facilitated Cr(VI) reduction in sediment pore water.	Zazo et al. (2008)
	Soil	Organic amendments increased DOC, which reduced Cr(VI) to Cr(III) in soils.	Bolan et al. (2003a)
	Soil	Addition of glucose promoted both biotic and abiotic Cr(VI) reduction in soils.	Leita et al. (2011)
	Water	Microbial fuel cells reduced Cr(VI) with the help of mixed consortium in autotrophic conditions.	Tandukar et al. (2009)
	Water	Chitosan–Fe^0 nanoparticles reduced Cr(VI) and pH of the solution played an important role.	Geng et al. (2009)
Chromium(III)	Soil	Cr(III) was oxidized by atmospheric oxygen at high temperature to Cr(VI) in tannery sludge contaminated sites.	Apte et al. (2006)
	Soil	Hydrous Mn(IV) oxides reacted with Cr(III) hydroxides and influenced the rate of Cr(III) oxidation.	Landrot et al. (2009)
	Soil	In the alkaline soil with moderate organic C at the ore processing site, most of the Cr(VI) remained dissolved after H_2O_2 had decayed, indicating the mobilization of Cr(VI). As oxidation of organic C promotes disintegration of soil structure, it increases the access of solution to Cr(VI) mineral phases.	Rock et al. (2001)
	Soil	The Mn oxide salts, birnessite and todorokite, had the same capacity to oxidize Cr(III) to Cr(VI) in soils.	Kim et al. (2002)
	Water	Natural oxidation of Cr(III) to Cr(VI) by Mn (hydr)oxides formed by microbial activity.	Ndung'u et al. (2010)
Mercury(II)	Water	Carboxylic groups in the humic acids reduced Hg(II) to Hg(0).	Allard and Arsenie (1991)
	Water	Photochemical and humic-mediated reduction of Hg(II) were two important components of the Hg(0) flux in the marine environment.	Costa and Liss (1999)
	Water	Sulfite reduced Hg(II) to insoluble $HgSO_3$ and finally converted it to Hg(0).	Munthe et al. (1991)
	Water	Humic substances, especially sulfur-containing ligands in terrestrial environment, reduced Hg(II) to Hg(0).	Rocha et al. (2000)
Selenium(VI)	Soil and sediments	In suboxic conditions, green rust [Fe(II,III)] reduced Se(VI) to Se(0).	Myneni et al. (1997)
	Water	Microorganisms reduced Se(VI) to Se(IV) and reduction increased in the presence of lactate.	Maiers et al. (1988)
	Soil	Organic amendments and low oxygen level in the soil reduced Se(VI) to Se(IV).	Guo et al. (1999)
	Soil and water	*Moraxella bovis* and bacterial consortia reduced Se(VI) to Se(IV) and Se(0).	Biswas et al. (2011)
	Water	*Bacillus pumilus* CrK08 had the highest capacity to reduce Se(VI) to Se(IV).	Ikram and Faisal (2010)
	Water	Nearly half of the total sorbed Se(IV) was reduced to Se(0) by Fe(II) sorbed on calcite within 24 h.	Chakraborty et al. (2010)

(Continued)

TABLE 5.2 (*Continued*)

Selected References on the Redox Reactions of Arsenic, Chromium, Mercury, and Selenium in Soil and Aquatic Environments

Trace Element	Medium	Observation	Reference
Selenium(VI) and (IV)	Water	Rice straw, a good source of carbon and energy, helped several bacteria in reducing Se(VI) and Se(IV) to Se(0).	Frankenberger et al. (2005)
	Water	Formic acid and methanol/ethanol promoted photoreduction of Se(VI) and Se(IV) to Se(0) in the presence of TiO_2.	Tan et al. (2003)
Se(0)	Sediments	*Bacillus selenitireducens* and some other bacteria were capable to reduce Se(0) to Se(II) in anoxic sediments.	Herbel et al. (2003)

TABLE 5.3

Oxidation–Reduction Reactions of Selected Elements and the Optimum Redox Potentials

Trace Element	Transformation	Reaction	E_o (mV)
Arsenic(0)	$As(0) + 3H_2O \rightarrow H_3AsO_3 + 3H^+ + 3e^-$	Oxidation	+250
Arsenic(V)	$H_3AsO_4 + 2H^+ + 2e^- \rightarrow H_3AsO_3 + H_2O$	Reduction	+560
Cadmium(0)	$Cd \rightarrow Cd^{2+} + 2e^-$	Oxidation	+402
Cadmium(II)	$Cd^{2+} + 2e^- \rightarrow Cd$	Reduction	−400
Chromium(III)	$2Cr^{3+} + 3H_2O + 2MnO_4^- \rightarrow 2Cr_2O_7^{2-} + 6H^+ + 2MnO_2$	Oxidation	+350
Chromium(VI)	$Cr_2O_7^{2-} + 14H^+ + 6e^- \rightarrow 2Cr^{3+} + 7H_2O$	Reduction	+1360
Iron(II)	$Fe^{2+} + 2e^- \rightarrow Fe(s)$	Oxidation	−440
Iron(III)	$Fe^{3+} + e^- \rightarrow Fe^{2+}$	Reduction	+770
Manganese(II)	$Mn^{2+} + 2e^- \rightarrow Mn$	Reduction	−1180
Manganese(IV)	$Mn^{4+} + 2e^- \rightarrow Mn^{2+}$	Reduction	+1210
Manganese(VII)	$MnO_4^- + 8H^+ + 5e^- \rightarrow Mn^{2+} + 4H_2O$	Reduction	+1510
Manganese(VII)	$MnO_4^- + e^- \rightarrow MnO_4^{2-}$	Reduction	+564
Mercury	$Hg_2^{2+} + 2e^- \rightarrow 2Hg$	Reduction	+790
Nitrogen(V)	$2NO_3^- + 4H^+ + 2e^- \rightarrow 2NO_2 + 2H_2O$	Reduction	+803
Selenium(0)	$Se(0)/H_2Se$	Reduction	−730
Selenium(VI)	SeO_4^{2-}/SeO_3^{2-}	Reduction	+440
Selenium(VI)	$SeO_3^{2-}/Se(0)$	Reduction	+180
Sulfur(II)	$S^{2-} + 2H^+ \rightarrow H_2S$	Reduction	−220
Sulfur(VI)	$SO_4^{2-} + H_2O + 2e^- \rightarrow SO_3^{2-} + 2OH^-$	Reduction	−520

formula $(CH_3)_nAs(O)(OH)_{3-n}$, where i may be 1, 2, or 3. Monomethylarsonic acid and DMA are common organoarsenicals in river water. Methylation may play a significant role in the mobilization of As by releasing it from the sediments to aqueous environment (Anderson and Bruland 1991; Wang and Mulligan 2006). In the sediments–aquatic system, methylation occurs only in the sediments because the thermodynamics of water or aquatic environments are not favorable for methylation (Duester et al. 2008).

Mercury is methylated through biotic and abiotic pathways, although microbial methylation mediated mainly through dissimilatory sulfate- and iron-reducing bacteria is generally regarded as the dominant environmental pathway (Musante 2008; Graham et al. 2012). Methylation of Hg occurs under both aerobic and anaerobic conditions (Rodríguez Martín-Doimeadios et al. 2004). The Hg(II) ions can be biologically methylated to form either monomethyl or dimethyl Hg under

anaerobic conditions, and they are highly toxic and relatively more biologically mobile compared to other forms (Adriano et al. 2004). The main methylation mechanism for Hg involves nonenzymatic transfer of methyl groups of methylcobalamin (a vitamin B_{12} derivative, produced by many microorganisms) to Hg(II) ions (Ullrich et al. 2001; Equations 5.1 and 5.2). Methylation occurs both enzymatically and nonenzymatically; ionic species Hg(II) are required for proceeding biological methylation to produce methyl Hg:

$$Hg(II) + 2R\text{--}CH_3 \rightarrow CH_3HgCH_3 \rightarrow CH_3Hg^+ \quad (5.1)$$

$$Hg(II) + 2R\text{--}CH_3 \rightarrow CH_3Hg^+ + R\text{--}CH_3 \rightarrow CH_3HgCH_3 \quad (5.2)$$

Selenium biomethylation is of interest because it represents a potential mechanism for removing Se from contaminated

TABLE 5.4

Selected References on the Methylation/Demethylation of Arsenic, Mercury, and Selenium in Soils

Trace Element	Organisms	Observations	Reference
Arsenic	*Cyanobacterium* sp. (*Synechocystis* sp. and *Nostoc* sp.)	Methylated As(III) to less toxic trimethylarsine as an end product in water and soil.	Yin et al. (2011)
	Cyanidioschyzon sp. (Eukaryotic alga)	Methylated As(III) to form trimethylarsine oxide and dimethylarsenate.	Qin et al. (2009)
	Streptomyces sp.	Demethylated methylarsine acid to As(III) species.	Yoshinaga et al. (2011)
	Ecklonia radiata	Arsenosugars were degraded to As(V) with dimethylarsinoylethanol and dimethylarsinate as intermediate species and observed that nonextractable, recalcitrant As increased in decaying algae.	Navratilova et al. (2011)
	Methanogenic archaea	*Methanobacterium* sp. cultures were most effective in producing volatile As derivatives.	Michalke et al. (2000)
Mercury	*Polygonum densiflorum*	The *Polygonum* sp. found in periphyton associated to macrophyte roots may be the dominating Hg methylating family among sulfate-reducing bacteria.	Achá et al. (2011)
	Desulfomicrobium escambiense	*Desulfomicrobium* sp. had the highest methylating capability of Hg (45%) and methylation was strain specific but not species or genus dependent.	Bridou et al. (2011)
	Desulfovibrio desulfuricans	*D. desulfuricans* ND132 strain has the ability to produce methyl Hg and its further degradation. However, the presence of sulfide decreased the methylation process.	Gilmour et al. (2011)
	Geobacter sp.	*Geobacter* sp. strain CLFeRB, an iron-reducing bacterium, methylated Hg at rates comparable to sulfate-reducing bacteria.	Fleming et al. (2006)
Selenium	*Escherichia coli*	*Escherichia coli* cells encoding the bacterial thiopurine methyltransferase have methylated Se(IV) and methylselenocysteine into DMSe and DMDSe.	Ranjard et al. (2002)
	Enterobacter cloacae	Reduced toxic oxyanions of Se to insoluble Se(0) and Se biomethylation were protein-/peptide-limited process.	Frankenberger Jr and Arshad (2001)
	Acremonium falciforme	Methylated inorganic Se more rapidly than organic forms.	Thompson-Eagle et al. (1989)
	Alternaria alternate (black microfungi)	Volatilized Se to DMSe and DMDSe even after the immobilization by goethite.	Peitzsch et al. (2010)

environments, and it is believed that methylated compounds, such as dimethyl selenenyl sulfide (CH_3SeSCH_3), are less toxic than dissolved Se oxyanions. Fungi are more active in the methylation of Se in soils although some Se-methylating bacterial isolates have also been identified (Adriano et al. 2004). Hydrogen-oxidizing methanogens such as *Methanobacterium omelianskii* are involved in the reductive methylation, while methylotrophic bacteria carry out demethylation. Dimethyl selenide can be demethylated in anoxic sediments as well as anaerobically by an obligate methylotroph similar to *Methanococcoides methylutens* in pure culture.

5.4 FACTORS AFFECTING BIOTRANSFORMATION PROCESSES

Biotransformation of As, Cr, Hg, and Se in soils, sediments, and aquatic systems is affected by biological functioning of the system, as measured by microbial activity, the bioavailability of the metal(loid)s ions as measured by speciation, and the physicochemical characteristics of the media such as pH, moisture content, and temperature. The biotransformation processes can also be manipulated through the addition of inorganic and organic amendments (Park et al. 2011a).

Although the oxidation/reduction and methylation/demethylation reactions occur through both chemical and biological processes, the biological process is considered to predominate in soils, sediments, and aquatic systems, especially in the absence of adequate levels of inorganic sources of electron donor/acceptor such as Fe^{2+} and Mn^{2+} ions. For example, Losi et al. (1994), Bolan et al. (2003a), and Choppala et al. (2012) have shown that the addition of organic manure caused a greater increase in the biological reduction than the chemical reduction of Cr(VI), which suggests that the supply of microorganisms is more important than the supply of organic carbon (C) in the manure-induced Cr(VI) reduction. Addition of organic manure has often been shown to increase the microbial activity of soils through increased supply of both C and nutrient sources (Wardle 1992; Marinari et al. 2000).

The rate of biotransformation of trace elements depends on the bioavailability of the metal(loid) concerned, as measured by its concentration and speciation. For example, reactive and free Hg ions are required to initiate biotransformation reactions involving microorganisms, and complexation of Hg with dissolved organic carbon (DOC) decreases methylation (Miskimmin et al. 1992; Ravichandran 2004). Similarly,

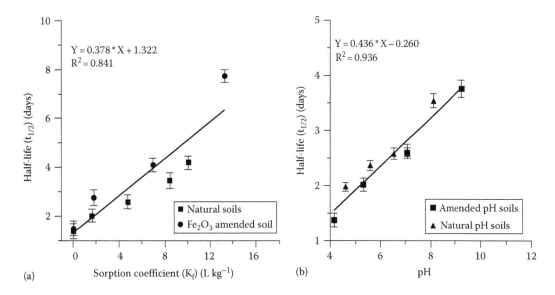

FIGURE 5.2 Relationship between (a) chromate (Cr(VI)) adsorption as measured by K_f values and Cr(VI) reduction as measured by half-life, (b) pH and half-life for Cr(VI) reduction. (From Choppala, G. et al., *J. Hazard. Mater.*, 261, 718, 2013. With permission.)

bacterial-induced Se reduction has often been shown to be influenced by the initial Se concentration. For example, Lortie et al. (1992) observed that Se reduction rate by *Pseudomonas stutzeri* increased with increasing Se(VI) and Se(IV) concentrations up to 19 mM. At Se concentration more than 19 mM, Se(IV) reduction decreased, while Se(VI) reduction remained constant, which might be attributed to the higher toxicity of the Se(IV) than Se(VI). Choppala et al. (2013) observed that the addition of Fe(III) oxide to Cr(VI)-contaminated soils resulted in a decrease in the rate of reduction of Cr(VI) as measured by half-life values (Figure 5.2a). This phenomenon may be due to the increased retention of Cr(VI) by Fe(III) oxide, thereby decreasing the bioavailability of Cr(VI) for microorganisms.

Soil pH is an important property that affects biotransformation processes through its effects on the microorganisms, supply of protons, and adsorption and speciation of metal(loid)s. For example, Kelly et al. (2003) and Roy et al. (2009) observed that the extent of methylation decreased as soil pH increased. This has been attributed to the unavailability of Hg(II) at high pH due to its stronger adsorption and to the reduced supply of methylating organic matter at high pH. pH can also influence the bioaccumulation of trace elements by microorganisms. An increase in pH increases surface charge, thereby promoting the surface binding of trace elements (Yin et al. 2002). Furthermore, energy-dependent trace element uptake is frequently pH dependent (Öztürk et al. 2004), and maximum rates are observed between 5 and 7 (Sağlam et al. 1999; Bishnoi et al. 2007). pH has been shown to influence the methylation rate by controlling organic C compounds with functional groups that would otherwise bind Hg (Adriano 2001). Protons are required for reducing Cr(VI) to Cr(III) (Equation 5.3). It has often been observed that Cr(VI) reduction, being a proton consumption (or hydroxyl release) reaction, increases as

soil pH decreases (Eary and Rai 1991; Choppala et al. 2012; Figure 5.2b):

$$2Cr_2O_7 + 3C^0 + 16H^+ \rightarrow 4Cr(III) + 3CO_2 + 8H_2O \quad (5.3)$$

Mercury methylation decreased with decreasing pH of sediment, and methylation was not detected at a pH value <5.0, which may be related to the unavailability of inorganic Hg (Ramial et al. 1985). Baker et al. (1983) reported that the methylation of Hg from inorganic mercuric chloride occurred in a narrow pH range of 5.5–6.5, perhaps because the microbial population had been adapted to a pH of 5.8 of the tested sediment. Kelly et al. (2003) investigated the influence of increasing protons on the uptake of Hg by an aquatic bacterium. A small decrease in pH (7.3–6.3) significantly increased Hg uptake by bacteria.

Fulladosa et al. (2004) studied the effect of pH on As(V) or As(III) speciation, and the resulting toxicity was investigated using the Microtox bioassay. Within a 5.0–8.0 pH range, EC_{50} values for As(V) were found to decrease as pH became basic, reflecting an increase in toxicity; for As(III), the EC_{50} values were almost unchanged within a 6.0–8.0 pH range and were lowered only at pH 9.0. They observed that the highest toxicity to *Vibrio fischeri* occurred at basic pH values when $HAsO_4^{2-}$ and H_2AsO^{3-} species representing As(V) and As(III) were predominant.

Thompson-Eagle et al. (1989) showed that DMSe production by *Alternaria alternata* was affected by the reaction mixture pH. The optimum pH for methylation of Se was 6.5 (Frankenberger and Arshad 2001). Dissolved Se(VI) constituted 95% of the total soluble Se at pH 9 and decreased to 75% at pH 6.5. pH greatly influenced Se speciation, solubility, and volatilization, therefore indicating that pH is an important factor in the Se biogeochemistry (Masscheleyn et al. 1990).

Soil moisture influences redox reactions by controlling the activity of microorganisms and also redox conditions of the microenvironment (Alexander 2000). An increase in moisture content and the amount of available C tend to increase the net loss of methyl Hg (Schlüter 2000; Oiffer and Siciliano 2009). Alternate wetting and drying enhances the release of volatile Se compounds, which is attributed to the release of nutrients through organic matter mineralization (Hechun et al. 1996). Most studies showed that irrespective of the matrix type, greater quantities of volatile Se are released under aerobic than anaerobic conditions (Thompson-Eagle and Frankenberger 1990). When the soil moisture content is high, the diffusion of O_2 is limited and local anaerobic environment is created. In aerobic soils the predominant As species is As(V) in soil pore water, while As(III) comprised up to 80% of the total As in anaerobic soils. Chemical reducing conditions increase As(III) in anaerobic soils. Chemical conversion of As(V) to As(III) may reduce the microbial activity and production of MMA (Haswell et al. 1985).

The methylation rate for Hg is likely to be higher under anaerobic than aerobic conditions because the former conditions hinder the conversion of methyl Hg. More than 90% of the methyl Hg was formed by biochemical processes under anaerobic conditions (Berman and Bartha 1986; Akagi et al. 1995). The prevalence of methylcobalamin and a natural methylating agent that transfers methyl groups under anaerobic conditions may facilitate the methylation of inorganic Hg (Ullrich et al. 2001).

Temperature influences biotransformation reactions of trace elements, mainly by controlling microbial activity and functions (Alexander 1999). Schwesig and Matzner (2001) and Heyes et al. (2006) observed that both the production and the volatilization loss of methyl Hg were directly proportional to temperature. Kocman and Horvat (2010) noticed a strong positive correlation between the soil surface temperature and Hg emission flux. They suggested that this thermally controlled emission of Hg from soils depended on the equilibrium of Hg(0) between the soil matrix and the soil gas.

Temperature is one of the major factors that affect Se volatilization (Frankenberger and Karlson 1994a). For every 10°C increase in the temperature, the vapor pressure of volatile Se raises three- to fourfold (Karlson et al. 1994). Dungan and Frankenberger (2000) observed that the optimum temperature for Se volatilization was 35°C, with the rate of Se volatilization increasing as the temperature increased from 12°C to 35°C. At 40°C, the rate of Se volatilization was slightly less than 35°C but greater than at 30°C. Camargo et al. (2003) found that Cr(VI) reduction by a Cr-resistant bacteria (*Bacillus* sp.) increased with an increase in soil temperature with maximum reduction occurring at 30°C.

Rhizosphere influences the biotransformation of trace elements through its effect on microbial activity, pH, and the release of organic compounds. For example, plant roots enhance the reduction of metal(loid)s such as As and Cr externally, by releasing root exudates, or internally through endogenous metal(loid) reductase enzyme activity in the root

mainly from an increase in microbial activity (Dhankher et al. 2006; Xu et al. 2007). For example, Bolan et al. (2013b) observed higher rates of As(V) and Cr(VI) reduction in rhizosphere than nonrhizosphere soils (Figure 5.3), which they attributed to several reasons that included increased microbial activity, DOC, and organic acid production and to decreased pH and Eh in the rhizosphere soil (Hinsinger et al. 2009). Xu et al. (2007) demonstrated for the first time that tomato and rice roots rapidly reduced As(V) to As(III) internally, some of which was actively effluxed to the growing medium, resulting in an increase in As(III) in the medium.

It is well known that As speciation depends on soil pH. Acidification of the rhizosphere immobilizes As(V) in soil under oxic conditions. Normally, the number of microorganisms in the rhizosphere is one order of magnitude greater than in bulk soil, because of higher nutrients and C availability in the rhizosphere (Marschner 1995; Fitz and Wenzel 2002). Since various bacteria, fungi, yeasts, and algae transform As compounds by oxidation, reduction, demethylation, and methylation, the rhizosphere is an important factor affecting the biotransformation of As (Frankenberger and Arshad 2002).

The rhizosphere environment can reduce Cr(VI) to Cr(III), because the pH-dependent redox reaction is related to Fe^{2+}, organic matter, and S contents in the rhizosphere (Zeng et al. 2008). Chen et al. (2000) reported that Cr(VI) reduction in a fresh wheat rhizosphere was induced by the decrease in pH. Microbial metabolism in the rhizosphere also caused the reduction of Cr(VI) to Cr(III). Low-molecular-weight organic acids, such as formic and acetic acids in rhizosphere, can contribute to Cr(VI) reduction and to Cr(III) chelation (Bluskov et al. 2005).

Biotransformation of trace elements in soils and sediments is affected by inorganic and organic amendments (Park et al. 2011a) (Table 5.5). Most biotransformation reactions require an energy source, which is often organic, but can be inorganic too. For example, nitrate (NO_3^-) addition has often been shown to affect the microbial reduction of Se. In general, NO_3^- is a potential inhibitor of Se(VI) reduction under anaerobic condition because it can act as an electron acceptor. Losi and Frankenberger (1997a) observed a correlation between potential dissimilarity Se reduction (DSeR) and potential denitrification, indicating possible involvement of NO_3^--respiring microbes in DSeR.

It has often been shown that addition of organic matter–rich soil amendments enhances the reduction of metal(loid)s such as Cr and Se (Frankenberger and Karlson 1994b; Park et al. 2011a; Figure 5.4). For example, several studies showed that the addition of cattle manure enhanced the reduction of Cr(VI) to Cr(III) (Losi et al. 1994; Cifuentes et al. 1996; Higgins et al. 1998). Similarly, various organic materials, such as powdered leaves, *Pinus sylvestris* bark, black carbon prepared from the dry biomass of rice straw, and plant weed *Solanum elaeagnifolium*, are highly effective in reducing Cr(VI) toxicity (Alves et al. 1993; Hsu et al. 2009). An increase in Cr(VI) reduction in the presence of organic manure composts could be due to the supply of C and protons and the stimulation of microorganisms (Losi et al. 1994).

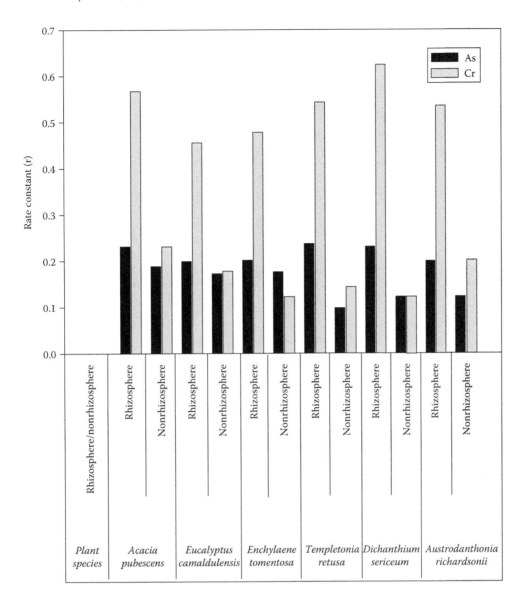

FIGURE 5.3 Rate of reduction of arsenate (As(V)) and chromate(Cr(VI)) in rhizosphere and nonrhizosphere soils of different plant species. (From Bolan, N.S. et al., *Rev. Environ. Contam. Toxicol.*, 225, 1, 2013a; Bolan, N.S. et al., *Plant Soil*, 367, 615, 2013b. With permission.)

For example, Choppala et al. (2013) observed a decrease in Cr(VI) toxicity in soils treated with black carbon, which was attributed to the supply of electrons for the reduction of toxic Cr(VI) to nontoxic Cr (III) species.

Organic amendments such as biosolids and manure significantly reduce the potential environmental risks of As contamination under highly anoxic conditions (Carbonell-Barrachina et al. 1999). Similarly, Yadav et al. (2009) observed that dairy sludge and biofertilizer reduced As and Cr bioavailability and increased plant growth. The addition of manure increased the loss of As from contaminated soil through methylation to volatile As by microbes. The rates of As loss were closely related to the microbial respiration because of nutrient supplementation to microbes. Bioaugmentation by As-methylating fungi increased H_2As evolution rates in field contaminated soils (Edvantoro et al. 2004).

5.5 IMPLICATIONS FOR REMEDIATION

Bioaccumulation and biotransformation processes play a major role in the decontamination and remediation of trace element contaminated soil, sediment, and water (Figure 5.1). Microorganisms play a vital role in transforming trace elements, thereby influencing their bioavailability and remediation (Figure 5.5). From toxicological or environmental viewpoints, these processes are important for three reasons. They may alter (1) the toxicity, (2) the water solubility, and/or (3) the mobility of the element (Alexander 1999). An increase in solubility can be exploited to bioremediate insoluble forms of elements in soil, because the biotransformed product in the solution phase becomes readily mobile and bioavailable. On the contrary, a decrease in element solubility can be utilized to remove the element from surface or groundwater through

TABLE 5.5

Selected References on the Remediation of Arsenic, Chromium, Mercury, and Selenium Toxicity by Organic Amendments

Trace Element	Amendment	Plant Used	Observation	Reference
Arsenic	Municipal solid waste, biosolids compost	*Pteris vittata*	Increased soil water-soluble As and reduced As(V) to As(III) but decreased leaching in the presence of fern.	Cao et al. (2003)
	Biosolids compost	*Daucus carota* L. and *Lactuca sativa* L.	Biosolids compost reduced plant As uptake by 79%–86%, which might be due to adsorption of As by biosolids organic matter.	Cao and Ma (2004)
	Green waste compost and biochar	*Miscanthus* species	Green waste compost substantially increased the plant yield; however, it increased water-soluble and surface-adsorbed fractions of As.	Hartley et al. (2009)
	Dairy sludge	*Jatropha curcas* L.	Application of dairy sludge decreased DTPA-extractable As in soil.	Yadav et al. (2009)
Chromium	Cow manure	*Festuca arundinacea*	Reduced Cr in roots and no change in Cr concentration in shoots.	Banks et al. (2006)
	Biosolids compost	*Brassica juncea*	Reduced Cr concentration in plant tissue.	Bolan et al. (2003a)
	Cattle compost and straw	*Lactuca sativa*	Cr content in aerial biomass decreased with the addition of amendments, which may be due to decreased association of Cr with carbonates and amorphous oxides and increased association with humic substances.	Rendina et al. (2006)
	Hog manure and cattle dung compost	*Triticum vulgare*	Hog manure decreased soil available Cr(VI), attributed to its low C/N ratio and thus increased microbial reduction of Cr(VI).	Lee et al. (2006)
	Bark of *Pinus radiata*	*Helianthus annuus*	Reduced availability of Cr for plant uptake.	Bolan and Thiagarajan (2001)
	Biosolids compost	*Sesbania punicea* and *Sesbania virgata*	Decreased Cr in plant extract.	Branzini and Zubillaga (2010)
	Dairy sludge	*Jatropha curcas* L.	Application of dairy sludge decreased DTPA-extractable Cr in soil.	Yadav et al. (2009)
	Farm yard manure	*Spinacia oleracea*	Increased root and shoot growth by decreasing Cr(VI) toxicity.	Singh et al. (2007)
Mercury	Humic acid	*Lactuca sativa* and *Brassica chinensis*	Humic acids decreased the amount of Hg in soil and translocation of Hg into plants.	Wang et al. (1997)
	Green waste compost	*Vulpia myuros* L.	Addition of compost showed negative relationship with soluble Hg and Hg tissue concentration, which may be due to the adsorption by compost.	Heeraman et al. (2001)
	Reactivated carbon	—	Powder reactivated carbon (PAC) increased the stabilization/solidification of Hg in the solid wastes and pretreating the PAC with carbon disulfide (CS_2) increased adsorption efficiency.	Zhang and Bishop (2002)
	Fulvic acid	—	Presence of fulvic acid increased the adsorption of Hg on goethite, which might be due to the strong affinity between sulfur groups within the fulvic acid and Hg.	Bäckström et al. (2003)
	Humic acid	*Brassica juncea*	Mercury translocation to aerial tissues of plant was restricted in the presence of humic acid.	Moreno et al. (2005a)
	Thiosulfates	*Brassica juncea*	Thiosulfates increased Hg accumulation in the plant, and Hg could be removed by phytoextraction from contaminated soils.	Moreno et al. (2005b)
Selenium	Insoluble (casein) and soluble (casamino acids) organic amendments	—	Organic amendments enhanced Se removal by providing an energy source and methyl donor to the methylating microorganisms, which increased Se volatilization from soil.	Zhang and Frankenberger (1999)
	Orange peel, cattle manure, gluten, and casein	—	The addition of organic amendments promoted the volatilization of Se; gluten was more effective, increased volatilization by 1.2- to 3.2-fold over the control.	Calderone et al. (1990)
	Compost manure and gluten	—	Reduction of Se(VI) to Se(IV) increased in the presence of organic amendments under low oxygen concentration, thereby retarding Se mobility.	Guo et al. (1999)
	Press mud and poultry manure	*Triticum aestivum* L. and *Brassica napus*	Application of amendments reduced Se accumulation by enhancing volatilization, thereby reducing the transfer of Se from soil to plant.	Dhillon et al. (2010)
	Poultry manure, sugarcane press mud, and farmyard manure	*Triticum aestivum* L. and *Brassica napus*	Addition of organic amendments decreased Se accumulation and increased grain quality; however, the extent of reduction depended on the type of organic amendment applied.	Sharma et al. (2011)

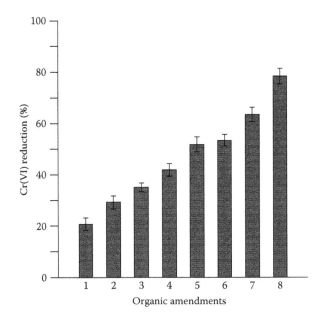

FIGURE 5.4 Effect of organic amendments on the reduction of chromate(Cr(VI)) in soil. 1, soil; 2, soil + horse manure; 3, soil + farmyard manure; 4, soil + fish manure; 5, soil + spent manure; 6, soil + piggery manure; 7, soil + poultry manure; 8, soil + biosolids compost. (From Bolan, N.S. et al., *J. Environ. Qual.*, 32, 120, 2003b. With permission.)

precipitation. In some cases, gaseous metal(loid) products can be removed through volatilization.

Arsenic can be reduced to As(0), which is subsequently precipitated as As_2S_3 as a result of microbial sulfate reduction.

Because As(III) is more soluble than As(V), the latter can be reduced using bacteria in soil and subsequently leached. Conversely, As(III) is oxidized to As(V) using microbes, which is subsequently precipitated using ferric ions (Williams and Silver 1984). *Desulfotomaculum auripigmentum* reduces both As(V) to As(III) and sulfate to H_2S and leads to As_2S_3 precipitation (Newman et al. 1997). Since Cr(III) is less soluble than Cr(VI), the reduction reaction will eventually result in the immobilization of Cr, thereby lowering its mobility and transport. Reduction of Cr(VI) to Cr(III), and subsequent hydroxide precipitation of Cr(III) ion, is the most common method for treating Cr(VI)-contaminated industrial effluents (Blowes et al. 1997; James 2001). Similarly, Choppala et al. (2013) have noticed that reducing Cr(VI) to Cr(III) in variable charge soils is likely to result in the adsorption of Cr(III) (Equation 5.4) from an increase in pH-induced negative charges and precipitation of $Cr(OH)_3$ (Equation 5.5), resulting from the reduction-induced release of OH$^-$ ions. Microbial reduction can be accomplished by direct reduction of Cr(VI) to Cr(III), using C as an energy source. Indirect reduction is achieved by sulfate addition under reduced conditions resulting in the release of H_2S that subsequently reduces Cr(VI) to Cr(III) and causes the precipitation of Cr as Cr_2S_3 (Smith et al. 2002; Vainshtein et al. 2003):

$$Cr(III) + Soil \rightarrow Cr^{3+} + H_2O + H^+$$

(Adsorption–proton release reaction)   (5.4)

$$Cr(III) + H_2O \rightarrow Cr(OH)_3 + H^+$$

(Precipitation–proton release reaction)   (5.5)

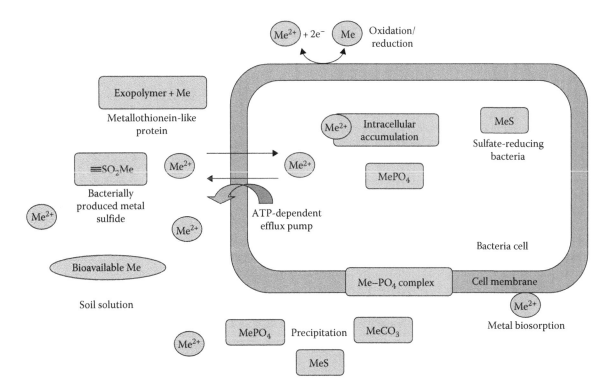

FIGURE 5.5 Microbial aspects of trace element immobilization. (Modified from Park, J.H. et al., *Pedologist*, 54, 162, 2011b. With permission.)

Selenium is prone to various forms of microbial transformations, some of which may have applicability in bioremediation (Ehrlich 1996; Losi and Frankenberger 1997b). Biological immobilization of Se(VI) by dissimilatory reduction to Se(0) is a practical approach for remediation. For example, anaerobic bacteria can be introduced into the contaminated medium supplied with a C source as an electron donor and Se(VI) an electron acceptor (Equations 5.6 and 5.7). The extent of Se(VI) reduction depends upon the availability of the C source, and the reduced Se in the elemental form, which is insoluble, can be physically separated from contaminated water. Similarly, methylation reaction can be used to form gaseous metal(loid) species, which can easily be removed through volatilization (Thompson-Eagle and Frankenberger 1992; Bañuelos and Lin 2007):

$$4CH_3COO^- + 3SeO_4^{2-} \rightarrow 3Se^0 + 8CO_2 + 4H_2O + 4H^+ \quad (5.6)$$

$$CH_3COO^- + 4SeO_4^{2-} \rightarrow 4SeO_3^{2-} + 2HCO_3^- + H^+ \quad (5.7)$$

Bioremediation of metal(loid)-contaminated sites can also be facilitated indirectly through the transformation of associated compounds. For example, bioleaching of metal(loid)s using acidification produced during the biological oxidation of pyrite has practical application for removing metal(loid)s in tailings and coal mine spoils (Sand et al. 2001). Similarly, microorganisms produce H_2S from sulfate and inorganic P from organic P; both H_2S and inorganic P form insoluble precipitates with a number of metal(loid)s (Douglas and Beveridge 1998; Bolan et al. 2003b; Kosolapov et al. 2004). Park et al. (2011b) noted that the inoculation of P-solubilizing bacteria enhanced the immobilization of Pb, thereby reducing its subsequent mobility and bioavailability.

5.6 SUMMARY AND CONCLUSIONS

Biotransformation processes can influence the solubility and subsequent mobility of these trace elements, especially As, Cr, Hg, and Se in soils and sediments. Biotransformation processes can be readily managed and enhanced for efficient removal of contaminants, and thus, a greater understanding of biotransformation processes will help to monitor the environmental fate of the trace elements, particularly through the food web, and will help to develop *in situ* bioremediation technologies.

The microbial transformation processes are considered as protective mechanisms to increase microbial resistance to toxic trace elements. These mechanisms are biochemical in nature and generally render the metal(loid) ion ineffective in disturbing the normal biochemical processes of the cell. The mechanisms include efflux pumps that remove the ions from the cell, enzymatic reduction of metal(loid)s to less toxic elemental forms, chelation by enzymatic polymers (i.e., metallothionein), binding the metal(loid) to cell wall surfaces, precipitating insoluble inorganic complexes (usually sulfides and oxides) at the cell surface, and biomethylation with

subsequent transport through the cell membrane by diffusion. Soil amendments can be used to manipulate biotransformation processes, thereby the managing the bioavailability and remediation of trace elements.

Since most of the environments that receive wastes containing trace elements are anoxic (e.g., subsurface saturated soils or organic-rich marsh sediments), biochemically mediated transformations of metal(loid)s play a vital role in their mobilization and bioavailability. Desorption and remobilization of metal(loid)s, such as Cr and As from sediments, are controlled by pH, Eh, and metal(loid) concentration in the sediment interstitial water, as well as by contents in total Fe, Mn, and mineral hydrous oxides. Physical disturbances of the sediments by storm or flooding may move the underlying sediments to oxidizing environments where the sulfides undergo oxidation resulting in the release of large quantities of metal(loid)s into the water. Chemolithotrophic bacteria play a major role in the oxidation process, thereby enhancing the mobilization of metal(loid)s. The use of fungi species offers a way to leach As from industrial waste sites. Organic acid production by heterotrophic organisms and sulfuric acid generation by microorganisms, such as *Thiobacillus*, also offer some promising approach for As extraction. Similarly, depending on the nature of metal(loid)s present in soil, the rhizosphere-induced redox reactions have implications to both their bioavailability to higher plants and microorganisms and remediation of contaminated soils. For example, while rhizo-reduction decreases Cr bioavailability, it increases that of As.

While methylation of inorganic Hg and Se in aquatic systems is the norm, it can also occur in the terrestrial environment. As Hg(0) and Se(0) formation removes reactive Hg and Se from soil, sediments, and water where these could otherwise be methylated, this process plays an important role in their eventual removal from these systems. Although there is huge potential, field application of dissimilatory metal(loid) reduction as a bioremediation tool is still very limited. Most bioremediation technologies are based on biotransformation processes that are designed to remove metal(loid)s mainly from aquatic systems. The viability and metabolic activity of microorganisms are the major limiting factors on efficiency of biotransforming metal(loid)s in soils. Even though genetically engineered microorganisms showed higher biotransformation of metal(loid)s, the application of genetically modified organisms into the environment is a matter of concern for many. Therefore, it is important to manipulate these biotransformation reactions by controlling the factors affecting them and also by using appropriate soil amendments. This will enable the sustainable management of trace element contamination to mitigate their environmental and health impacts.

REFERENCES

Achá, D., Hintelmann, H., Yee, J. 2011. Importance of sulfate reducing bacteria in mercury methylation and demethylation in periphyton from Bolivian Amazon region. *Chemosphere* 82: 911–916.

Adriano, D.C. 2001. *Trace Elements in the Terrestrial Environments: Biogeochemistry, Bioavailability, and Risks of Heavy Metals.* Springer-Verlag, New York.

Adriano, D.C., Wenzel, W.W., Vangronsveld, J., Bolan, N.S. 2004. Role of assisted natural remediation in environmental cleanup. *Geoderma* 122: 121–142.

Ahalya, N., Ramachandra, T.V., Kanamadi, R.D. 2003. Biosorption of heavy metals. *Research Journal of Chemistry and Environment* 7: 71–79.

Akagi, H., Malm, O., Branches, F., Kinjo, Y., Kashima, Y., Guimaraes, J.R.D., Oliveira, R.B. et al. 1995. Human exposure to mercury due to goldmining in the Tapajos river basin, Amazon, Brazil: Speciation of mercury in human hair, blood and urine. *Water, Air, and Soil Pollution* 80: 85–94.

Alexander, M. 1999. *Biodegradation and Bioremediation*, 2nd edn. Academic Press, San Diego, CA.

Alexander, M. 2000. Aging, bioavailability, and overestimation of risk from environmental pollutants. *Environmental Science & Technology* 34: 4259–4265.

Allard, B., Arsenie, I. 1991. Abiotic reduction of mercury by humic substances in aquatic system—An important process for the mercury cycle. *Water Air & Soil Pollution* 56(1): 457–464.

Alloway, B.J. 1990. Soil processes and the behaviour of metals. In: Alloway, B.J. (ed.) *Heavy Metals in Soils.* Blackie & Son Ltd., Glasgow, Scotland, pp. 7–28.

Al Rmalli, S.W., Dahmani, A.A., Abuein, M.M., Gleza, A.A. 2008. Biosorption of mercury from aqueous solutions by powdered leaves of castor tree (*Ricinus communis* L.). *Journal of Hazardous Materials* 152: 955–959.

Alves, M.M., Beça, C.G., De Carvalho, R.G., Castanheira, J.M., Pereira, M.S., Vasconcelos, L.A.T. 1993. Chromium removal in tannery wastewaters "polishing" by *Pinus sylvestris* bark. *Water Research* 27: 1333–1338.

Anderson, L.C., Bruland, K.W. 1991. Biogeochemistry of arsenic in natural waters: The importance of methylated species. *Environmental Science & Technology* 25: 420–427.

Apte, A.D., Tare, V., Bose, P. 2006. Extent of oxidation of Cr (III) to Cr (VI) under various conditions pertaining to natural environment. *Journal of Hazardous Materials* 128: 164–174.

Aryal, M., Ziagova, M., Liakopoulou-Kyriakides, M. 2010. Study on arsenic biosorption using Fe (III)-treated biomass of *Staphylococcus xylosus Chemical Engineering Journal* 162: 178–185.

Bachate, S.P., Khapare, R.M., Kodam, K.M. 2012. Oxidation of arsenite by two β-proteobacteria isolated from soil. *Applied Microbiology and Biotechnology* 93: 2135–2145.

Bäckström, M., Dario, M., Karlsson, S., Allard, B. 2003. Effects of a fulvic acid on the adsorption of mercury and cadmium on goethite. *Science of the Total Environment* 304: 257–268.

Baker, M.D., Inniss, W.E., Mayfield, C.I., Wong, P.T.S., Chau, Y.K. 1983. Effect of pH on the methylation of mercury and arsenic by sediment microorganisms. *Environmental Technology Letters* 4: 89–100.

Banks, M.K., Schwab, A.P., Henderson, C. 2006. Leaching and reduction of chromium in soil as affected by soil organic content and plants. *Chemosphere* 62: 255–264.

Banuelos, G.S., Li, Z.Q. 2007. Acceleration of selenium volatilization in seleniferous agricultural drainage sediments amended with methionine and casein. *Environmental Pollution* 150: 306–312.

Barnhart, J. 1997. Chromium chemistry and implications for environmental fate and toxicity. *Soil and Sediment Contamination* 6: 561–568.

Battaglia-Brunet, F., Dictor, M.C., Garrido, F., Crouzet, C., Morin, D., Dekeyser, K., Clarens, M., Baranger, P. 2002. An arsenic (III)-oxidizing bacterial population: Selection, characterization, and performance in reactors. *Journal of Applied Microbiology* 93: 656–667.

Berman, M., Bartha, R. 1986. Levels of chemical versus biological methylation of mercury in sediments. *Bulletin of Environmental Contamination and Toxicology* 36: 401–404.

Bhandari, N., Reeder, R.J., Strongin, D.R. 2011. Photoinduced oxidation of arsenite to arsenate on ferrihydrite. *Environmental Science & Technology* 45: 2783–2789.

Bishnoi, N.R., Kumar, R., Kumar, S., Rani, S. 2007. Biosorption of Cr (III) from aqueous solution using algal biomass *Spirogyra* spp. *Journal of Hazardous Materials* 145: 142–147.

Biswas, K.C., Barton, L.L., Tsui, W.L., Shuman, K., Gillespie, J., Eze, C.S. 2011. A novel method for the measurement of elemental selenium produced by bacterial reduction of selenite. *Journal of Microbiological Methods* 86: 140–144.

Blowes, D.W., Ptacek, C.J., Jambor, J.L. 1997. In-situ remediation of Cr (VI)-contaminated groundwater using permeable reactive walls: Laboratory studies. *Environmental Science & Technology* 31: 3348–3357.

Bluskov, S., Arocena, J., Omotoso, O., Young, J. 2005. Uptake, distribution, and speciation of chromium in *Brassica juncea*. *International Journal of Phytoremediation* 7: 153–165.

Boening, D.W. 2000. Ecological effects, transport, and fate of mercury: A general review. *Chemosphere* 40: 1335–1351.

Bolan, N.S., Adriano, D.C., Naidu, R. 2003a. Role of phosphorus in immobilization and bioavailability of heavy metals in the soil–plant system. *Reviews of Environmental Contamination and Toxicology* 177: 1–44.

Bolan, N.S., Adriano, D.C., Natesan, R., Koo, B.J. 2003b. Effects of organic amendments on the reduction and phytoavailability of chromate in mineral soil. *Journal of Environmental Quality* 32: 120–128.

Bolan, N.S., Choppala, G., Kunhikrishnan, A., Park, J.H., Naidu, R. 2013a. Microbial transformation of trace elements in soils in relation to bioavailability and remediation. *Reviews of Environmental Contamination and Toxicology* 225: 1–56.

Bolan, N.S., Kunhikrishnan, A., Gibbs, J. 2013b. Rhizoreduction of chromate and arsenate in Australian native grass, shrub and tree vegetation. *Plant and Soil* 367: 615–625.

Bolan, N.S., Thiagarajan, S. 2001. Retention and plant availability of chromium in soils as affected by lime and organic matter amendments. *Soil Research* 39: 1091–1104.

Branzini, A., Zubillaga, M. 2010. Assessing phytotoxicity of heavy metals in remediated soil. *International Journal of Phytoremediation* 12: 335–342.

Bridou, R., Monperrus, M., Gonzalez, P.R., Guyoneaud, R., Amouroux, D. 2011. Simultaneous determination of mercury methylation and demethylation capacities of various sulfate-reducing bacteria using species-specific isotopic tracers. *Environmental Toxicology and Chemistry* 30: 337–344.

Brock, T.D., Madigan, T. 1991. *Biology of Microorganisms*. Prentice Hall, Englewood Cliffs, NJ.

Calderone, S.J., Frankenberger, W.T., Parker, D.R., Karlson, U. 1990. Influence of temperature and organic amendments on the mobilization of selenium in sediments. *Soil Biology and Biochemistry* 22: 615–620.

Camargo, F.A.O., Okeke, B.C., Bento, F.M., Frankenberger, W.T. 2003. In vitro reduction of hexavalent chromium by a cell-free extract of *Bacillus* sp. ES 29 stimulated by Cu²⁺. *Applied Microbiology and Biotechnology* 62: 569–573.

Cao, X., Ma, L.Q. 2004. Effects of compost and phosphate on plant arsenic accumulation from soils near pressure-treated wood. *Environmental Pollution* 132: 435–442.

Cao, X., Ma, L.Q., Shiralipour, A. 2003. Effects of compost and phosphate amendments on arsenic mobility in soils and arsenic uptake by the hyperaccumulator, *Pteris vittata* L. *Environmental Pollution* 126: 157–167.

Carbonell-Barrachina, A., Jugsujinda, A., DeLaune, R.D., Patrick, W.H., Burló, F., Sirisukhodom, S., Anurakpongsatorn, P. 1999. The influence of redox chemistry and pH on chemically active forms of arsenic in sewage sludge-amended soil. *Environment International* 25: 613–618.

Casiot, C., Bruneel, O., Personné, J.C., Leblanc, M., Elbaz-Poulichet, F. 2004. Arsenic oxidation and bioaccumulation by the acidophilic protozoan, *Euglena mutabilis*, in acid mine drainage (Carnoules, France). *Science of the Total Environment* 320: 259–267.

Chakraborty, S., Bardelli, F., Charlet, L. 2010. Reactivities of Fe(II) on calcite: Selenium reduction. *Environmental Science & Technology* 44: 1288–1294.

Chen, C.P., Juang, K.W., Lin, T.H., Lee, D.Y. 2010. Assessing the phytotoxicity of chromium in Cr (VI)-spiked soils by Cr speciation using XANES and resin extractable Cr (III) and Cr (VI). *Plant and Soil* 334: 299–309.

Chen, N.C., Kanazawa, S., Horiguchi, T. 2000. Chromium (VI) reduction in wheat rhizosphere. *Pedosphere* 10: 31–36.

Chiu, V.Q., Hering, J.G. 2000. Arsenic adsorption and oxidation at manganite surfaces. 1. Method for simultaneous determination of adsorbed and dissolved arsenic species. *Environmental Science & Technology* 34: 2029–2034.

Choppala, G., Bolan, N., Seshadri, B. 2013. Chemodynamics of chromium reduction in soils: Implications to bioavailability. *Journal of Hazardous Materials* 261: 718–724.

Choppala, G.K., Bolan, N.S., Megharaj, M., Chen, Z., Naidu, R. 2012. The influence of biochar and black carbon on reduction and bioavailability of chromate in soils. *Journal of Environmental Quality* 41: 1175–1184.

Christen, K. 2001. Chickens, manure, and arsenic. *Environmental Science & Technology* 35: 184A–185A.

Cifuentes, F., Lindemann, W., Barton, L. 1996. Chromium sorption and reduction in soil with implications to bioremediation. *Soil Science* 161: 233–241.

Cossich, E.S., Da Silva, E.A., Tavares, C.R.G., Cardozo Filho, L., Ravagnani, T.M.K. 2004. Biosorption of chromium (III) by biomass of seaweed *Sargassum* sp. in a fixed-bed column. *Adsorption* 10: 129–138.

Costa, M., Liss, P.S. 1999. Photoreduction of mercury in sea water and its possible implications for Hg⁰ air–sea fluxes. *Marine Chemistry* 68: 87–95.

de Lacerda, L. 2003. Updating global Hg emissions from small-scale gold mining and assessing its environmental impacts. *Environmental Geology* 43: 308–314.

Dhankher, O.P., Rosen, B.P., McKinney, E.C., Meagher, R.B. 2006. Hyperaccumulation of arsenic in the shoots of *Arabidopsis* silenced for arsenate reductase (ACR2). *Proceedings of the National Academy of Sciences of the United States of America* 103: 5413–5418.

Dhillon, K., Dhillon, S., Dogra, R. 2010. Selenium accumulation by forage and grain crops and volatilization from seleniferous soils amended with different organic materials. *Chemosphere* 78: 548–556.

Douglas, S., Beveridge, T.J. 1998. Mineral formation by bacteria in natural microbial communities. *FEMS Microbiology Ecology* 26: 79–88.

Duester, L., Vink, J.M., Hirner, A.V. 2008. Methylantimony and -arsenic species in sediment pore water tested with the sediment or fauna incubation experiment. *Environmental Science & Technology* 42: 5866–5871.

Dungan, R.S., Frankenberger, Jr., W.T. 2000. Factors affecting the volatilization of dimethylselenide by *Enterobacter cloacae* SLD1a-1. *Soil Biology and Biochemistry* 32: 1353–1358.

Dursun, A.Y. 2006. A comparative study on determination of the equilibrium, kinetic and thermodynamic parameters of biosorption of copper (II) and lead (II) ions onto pretreated *Aspergillus niger*. *Biochemical Engineering Journal* 28: 187–195.

Eary, L.E., Rai, D. 1991. Chromate reduction by subsurface soils under acidic conditions. *Soil Science Society of America Journal* 55: 676–683.

Edvantoro, B.B., Naidu, R., Megharaj, M., Merrington, G., Singleton, I. 2004. Microbial formation of volatile arsenic in cattle dip site soils contaminated with arsenic and DDT. *Applied Soil Ecology* 25: 207–217.

Ehrlich, H.L. 1996. *Geomicrobiology*, 3rd edn. Dekker, New York.

Emett, M.T., Khoe, G.H. 2001. Photochemical oxidation of arsenic by oxygen and iron in acidic solutions. *Water Research* 35: 649–656.

Fitz, W.J., Wenzel, W.W. 2002. Arsenic transformations in the soil–rhizosphere–plant system: Fundamentals and potential application to phytoremediation. *Journal of Biotechnology* 99: 259–278.

Fleming, E.J., Mack, E.E., Green, P.G., Nelson, D.C. 2006. Mercury methylation from unexpected sources: Molybdate-inhibited freshwater sediments and an iron-reducing bacterium. *Applied Environmental Microbiology* 72: 457–464.

Franco, D.V., Da Silva, L.M., Jardim, W.F. 2009. Chemical reduction of hexavalent chromium present in contaminated soil using a packed-bed column reactor. *CLEAN* 37: 858–865.

Frankenberger, Jr., W.T., Arshad, M. 2001. Bioremediation of selenium-contaminated sediments and water. *Biofactors* 14: 241–254.

Frankenberger, Jr., W.T., Arshad, M. 2002. Volatilization of arsenic. In: Frankenberger, W. (ed.) *Environmental Chemistry of Arsenic*. Marcel Dekker, New York, pp. 363–380.

Frankenberger, Jr., W.T., Arshad, M., Siddique, T., Han, S.K., Okeke, B.C., Zhang, Z. 2005. Bacterial diversity in selenium reduction of agricultural drainage water amended with rice straw. *Journal of Environmental Quality* 34: 217–226.

Frankenberger, Jr., W.T., Karlson, U. 1994a. Soil management factors affecting volatilization of selenium from dewatered sediments. *Geomicrobiological Journal* 12: 265–278.

Frankenberger, Jr., W.T., Karlson, U. 1994b. Microbial volatilization of selenium from soils and sediments. In: Frankenberger, Jr., W.T., Benson, S. (eds.) *Selenium in the Environment*. Marcel Dekker, New York, pp. 369–387.

Fulladosa, E., Murat, J.C., Martinez, M., Villaescusal, I. 2004. Effect of pH on arsenate and arsenite toxicity to luminescent bacteria (*Vibrio fischeri*). *Archives of Environmental Contamination and Toxicology* 46: 176–182.

Gadd, G.M. 1990. Heavy metal accumulation by bacteria and other microorganisms. *Cellular and Molecular Life Sciences* 46: 834–840.

Gadd, G.M. 1993. Microbial formation and transformation of organometallic and organometalloid compounds. *FEMS Microbiological Reviews* 11: 297–316.

Gadd, G.M. 2010. Metals, minerals and microbes: Geomicrobiology and bioremediation. *Microbiology* 156: 609–643.

Geng, B., Jin, Z., Li, T., Qi, X. 2009. Kinetics of hexavalent chromium removal from water by chitosan-Fe^0 nanoparticles. *Chemosphere* 75: 825–830.

Geoffrey, M., Gadd, G. 2007. Geomycology: Biogeochemical transformations of rocks, minerals, metals and radionuclides by fungi, bioweathering and bioremediation. *Mycological Research* 111: 3–49.

Gilmour, C.C., Elias, D.A., Kucken, A.M., Brown, S.D., Palumbo, A.V., Schadt, C.W., Wall, J.D. 2011. Sulfate-reducing bacterium *Desulfovibrio desulfuricans* ND132 as a model for understanding bacterial mercury methylation. *Applied and Environmental Microbiology* 77: 3938–3951.

Gong, C., Donahoe, R.J. 1997. An experimental study of heavy metal attenuation and mobility in sandy loam soils. *Applied Geochemistry* 12: 243–254.

Graham, A.M., Aiken, G.R., Gilmour, C.C. 2012. Dissolved organic matter enhances microbial mercury methylation under sulfidic conditions. *Environmental Science & Technology* 46: 2715–2723.

Green-Ruiz, C. 2006. Mercury(II) removal from aqueous solutions by nonviable *Bacillus* sp. from a tropical estuary. *Bioresource Technology* 97: 1907–1911.

Guo, L., Frankenberger, Jr., W.T., Jury, W.A. 1999. Evaluation of simultaneous reduction and transport of selenium in saturated soil columns. *Water Resources Research* 35: 663–669.

Hansen, H.K., Ribeiro, A., Mateus, E. 2006. Biosorption of arsenic(V) with *Lessonia nigrescens*. *Minerals Engineering* 19: 486–490.

Hartley, W., Dickinson, N.M., Riby, P., Lepp, N.W. 2009. Arsenic mobility in brownfield soils amended with green waste compost or biochar and planted with *Miscanthus*. *Environmental Pollution* 157: 2654–2662.

Hasan, S., Ranjan, D., Talat, M. 2010. Agro-industrial waste "wheat bran" for the biosorptive remediation of selenium through continuous up-flow fixed-bed column. *Journal of Hazardous Materials* 181: 1134–1142.

Haswell, S.J., O'Neill, P., Bancroft, K.C. 1985. Arsenic speciation in soil–pore waters from mineralized and unmineralized areas of south–west England. *Talanta* 32: 69–72.

Hechun, P., Guangshen, L., Zhiyun, Y., Yetang, H. 1996. Acceleration of selenate reduction by alternative drying and wetting of soils. *Chinese Journal of Geochemistry* 15: 278–284.

Heeraman, D., Claassen, V., Zasoski, R. 2001. Interaction of lime, organic matter and fertilizer on growth and uptake of arsenic and mercury by Zorro fescue (*Vulpia myuros* L.). *Plant and Soil* 234: 215–231.

Herbel, M.J., Blum, J.S., Oremland, R.S., Borglin, S.E. 2003. Reduction of elemental selenium to selenide: Experiments with anoxic sediments and bacteria that respire Se-oxyanions. *Geomicrobiology Journal* 20: 587–602.

Heyes, A., Mason, R.P., Kim, E.H., Sunderland, E. 2006. Mercury methylation in estuaries: Insights from using measuring rates using stable mercury isotopes. *Marine Chemistry* 102: 134–147.

Hider, R.C., Kong, X. 2010. Chemistry and biology of siderophores. *Natural Product Reports* 27: 637–657.

Higgins, T.E., Halloran, A., Dobbins, M., Pittignano, A. 1998. In situ reduction of hexavalent chromium in alkaline soils enriched with chromite ore processing residue. *Journal of the Air & Waste Management Association* 48: 1100–1106.

Hinsinger, P., Bengough, G., Vetterlein, D., Young, I.M. 2009. Rhizosphere: Biophysics, biochemistry and ecological relevance. *Plant and Soil* 321: 117–152.

Hsu, N.H., Wang, S.L., Lin, Y.C., Sheng, G.D., Lee, J.F. 2009. Reduction of Cr (VI) by crop-residue-derived black carbon. *Environmental Science & Technology* 43: 8801–8806.

Huang, J.H., Voegelin, A., Pombo, S.A., Lazzaro, A., Zeyer, J., Kretzschmar, R. 2011. Influence of arsenate adsorption to ferrihydrite, goethite, and boehmite on the kinetics of arsenate reduction by *Shewanella putrefaciens* strain CN-32. *Environmental Science & Technology* 45: 7701–7709.

Ikram, M., Faisal, M. 2010. Comparative assessment of selenite (SeIV) detoxification to elemental selenium (Se0) by *Bacillus* sp. *Biotechnology Letters* 32: 1255–1259.

Iyer, A., Mody, K., Jha, B. 2005. Biosorption of heavy metals by a marine bacterium. *Marine Pollution Bulletin* 50: 340–343.

James, B.R. 2001. Remediation-by-reduction strategies for chromate-contaminated soils. *Environmental Geochemistry and Health* 23: 175–179.

Jones, C.A., Langner, H.W., Anderson, K., McDermott, T.R., Inskeep, W.P. 2000. Rates of microbially mediated arsenate reduction and solubilization. *Soil Science Society of America Journal* 64: 600–608.

Karlson, U., Frankenberger, Jr., W.T., Spencer, W.F. 1994. Physicochemical properties of dimethyl selenide. *Journal of Chemical and Engineering Data* 39: 608–610.

Kelly, C., Rudd, J.W.M., Holoka, M. 2003. Effect of pH on mercury uptake by an aquatic bacterium: Implications for Hg cycling. *Environmental Science & Technology* 37: 2941–2946.

Kim, J.G.D., Chusuei, J.B., Deng, C.C. 2002. Oxidation of chromium(III) to (VI) by manganese oxides. *Soil Science Society of America Journal* 66: 306–315.

Kocman, D., Horvat, M. 2010. A laboratory based experimental study of mercury emission from contaminated soils in the River Idrijca catchment. *Atmospheric Chemistry and Physics* 10: 1417–1426.

Kodukula, P.S., Patterson, J.W., Surampalli, R.Y. 1994. Sorption and precipitation of metals in activated-sludge. *Biotechnology and Bioengineering* 43: 874–880.

Kosolapov, D., Kuschk, P., Vainshtein, M., Vatsourina, A., Wiessner, A., Kästner, M., Müller, R. 2004. Microbial processes of heavy metal removal from carbon-deficient effluents in constructed wetlands. *Engineering in Life Sciences* 4: 403–411.

Landrot, G., Ginder-Vogel, M., Sparks, D.L. 2009. Kinetics of chromium (III) oxidation by manganese (IV) oxides using quick scanning X-ray absorption fine structure spectroscopy (Q-XAFS). *Environmental Science & Technology* 44: 143–149.

Ledin, M., Krantz-Rulcker, C., Allard, B. 1999. Microorganisms as metal sorbents: Comparison with other soil constituents in multi-compartment systems. *Soil Biology and Biochemistry* 31: 1639–1648.

Lee, D.Y., Shih, Y.N., Zheng, H.C., Chen, C.P., Juang, K.W., Lee, J.F., Tsui, L. 2006. Using the selective ion exchange resin extraction and XANES methods to evaluate the effect of compost amendments on soil chromium (VI) phytotoxicity. *Plant and Soil* 281: 87–96.

Leita, L., Margon, A., Sinicco, T., Mondini, C. 2011. Glucose promotes the reduction of hexavalent chromium in soil. *Geoderma* 164: 122–127.

Lens, P., Van Hullebusch, E., Astratinei, V. 2006. Bioconversion of selenate in methanogenic anaerobic granular sludge. *Journal of Environmental Quality* 35: 1873–1883.

Lortie, L., Gould, W.D., Rajan, S., McCready, R.G.L., Cheng, K.J. 1992. Reduction of selenate and selenite to elemental selenium by a *Pseudomonas stutzeri* isolate. *Applied and Environmental Microbiology* 58: 4042–4044.

Losi, M., Amrhein, C., Frankenberger, Jr., W. 1994. Factors affecting chemical and biological reduction of hexavalent chromium in soil. *Environmental Toxicology and Chemistry* 13: 1727–1735.

Losi, M.E., Frankenberger, Jr., W.T. 1997a. Reduction of selenium oxyanions by *Enterobacter cloacae* strain SLD1a-1: Reduction of selenate to selenite. *Environmental Toxicology and Chemistry* 16: 1851–1858.

Losi, M.E., Frankenberger, W.T. 1997b. Bioremediation of selenium in soil and water. *Soil Science* 162: 692–702.

Loukidou, M.X., Matis, K.A., Zouboulis, A.I., Liakopoulou-Kyriakidou, M. 2003. Removal of As(V) from wastewaters by chemically modified fungal biomass. *Water Research* 37: 4544–4552.

Maher, W., Butler, E. 1988. Arsenic in the marine environment. *Applied Organometallic Chemistry* 2: 191–214.

Mahimairaja, S., Bolan, N.S., Adriano, D., Robinson, B. 2005. Arsenic contamination and its risk management in complex environmental settings. *Advances in Agronomy* 86: 1–82.

Maiers, D., Wichlacz, P., Thompson, D., Bruhn, D. 1988. Selenate reduction by bacteria from a selenium-rich environment. *Applied Environmental Microbiology* 54: 2591–2593.

Manning, B.A., Fendorf, S.E., Bostick, B., Suarez, D.L. 2002. Arsenic (III) oxidation and arsenic (V) adsorption reactions on synthetic birnessite. *Environmental Science & Technology* 36: 976–981.

Marinari, S., Masciandaro, G., Ceccanti, B., Grego, S. 2000. Influence of organic and mineral fertilisers on soil biological and physical properties. *Bioresource Technology* 72: 9–17.

Marschner, H. 1995. *Mineral Nutrition of Higher Plants*, 2nd edn. Academic Press, London, U.K.

Masscheleyn, P.H., Delaune, R.D., Patrick, Jr., W.H. 1990. Transformations of selenium as affected by sediment oxidation-reduction potential and pH. *Environmental Science & Technology* 24: 91–96.

Mejáre, M., Bülow, L. 2001. Metal-binding proteins and peptides in bioremediation and phytoremediation of heavy metals. *Trends in Biotechnology* 19: 67–73.

Melo, J., D'Souza, S. 2004. Removal of chromium by mucilaginous seeds of *Ocimum basilicum*. *Bioresource Technology* 92: 151–155.

Michalke, K., Wickenheiser, E.B., Mehring, M., Hirner, A.V., Hensel, R. 2000. Production of volatile derivatives of metal (loid) s by microflora involved in anaerobic digestion of sewage sludge. *Applied and Environmental Microbiology* 66: 2791–2796.

Miskimmin, B.M., Rudd, J.W.M., Kelly, C.A. 1992. Influences of DOC, pH, and microbial respiration rates of mercury methylation and demethylation in lake water. *Canadian Journal of Fisheries and Aquatic Sciences* 49: 17–22.

Moreno, F.N., Anderson, C.W., Stewart, R.B., Robinson, B.H., Ghomshei, M., Meech, J.A. 2005a. Induced plant uptake and transport of mercury in the presence of sulphur-containing ligands and humic acid. *New Phytologist* 166: 445–454.

Moreno, F.N., Anderson, C.W., Stewart, R.B., Robinson, B.H., Nomura, R., Ghomshei, M., Meech, J.A. 2005b. Effect of thioligands on plant-Hg accumulation and volatilisation from mercury-contaminated mine tailings. *Plant and Soil* 275: 233–246.

Munthe, J., Xiao, Z., Lindqvist, O. 1991. The aqueous reduction of divalent mercury by sulfite. *Water Air and Soil Pollution* 56: 621–630.

Murugesan, G., Sathishkumar, M., Swaminathan, K. 2006. Arsenic removal from groundwater by pretreated waste tea fungal biomass. *Bioresource Technology* 97: 483–487.

Musante, A. 2008. The role of mercury speciation in its methylation by methylcobalamin (vitamin-B12). Bachelor thesis, Wheaton College, Norton, MA.

Myneni, S., Tokunaga, T.K., Brown, Jr., G.E. 1997. Abiotic selenium redox transformations in the presence of Fe(II, III) oxides. *Science* 278: 1106–1109.

Navratilova, J., Raber, G., Fisher, S.J., Francesconi, K.A. 2011. Arsenic cycling in marine systems: Degradation of arsenosugars to arsenate in decomposing algae, and preliminary evidence for the formation of recalcitrant arsenic. *Environmental Chemistry* 8: 44–51.

Ndung'u, K., Friedrich, S., Gonzalez, A.R., Flegal, A.R. 2010. Chromium oxidation by manganese (hydr) oxides in a California aquifer. *Applied Geochemistry* 25: 377–381.

Newman, D.K., Beveridge, T.J., Morel, F. 1997. Precipitation of arsenic trisulfide by *Desulfotomaculum auripigmentum*. *Applied and Environmental Microbiology* 63: 2022–2028.

Niu, C.H., Volesky, B., Cleiman, D. 2007. Biosorption of arsenic(V) with acid-washed crab shells. *Water Research* 41: 2473–2478.

Oiffer, L., Siciliano, S.D. 2009. Methyl mercury production and loss in Arctic soil. *Science of the Total Environment* 407: 1691–1700.

Oremland, R.S., Herbel, M.J., Blum, J.S., Langley, S., Beveridge, T.J., Ajayan, P.M., Curran, S. 2004. Structural and spectral features of selenium nanospheres produced by Se-respiring bacteria. *Applied and Environmental Microbiology* 70: 52–60.

Oremland, R.S., Hollibaugh, J.T., Maest, A.S., Presser, T.S., Miller, L.G., Culbertson, C.W. 1989. Selenate reduction to elemental selenium by anaerobic bacteria in sediments and culture: Biogeochemical significance of a novel, sulfate-independent respiration. *Applied and Environmental Microbiology* 55: 2333–2343.

Öztürk, A., Artan, T., Ayar, A. 2004. Biosorption of nickel(II) and copper(II) ions from aqueous solution by *Streptomyces coelicolor* A3(2). *Colloids and Surfaces B: Biointerfaces* 34: 105–111.

Pacyna, E., Pacyna, J., Pirrone, N. 2001. European emissions of atmospheric mercury from anthropogenic sources in 1995. *Atmospheric Environment* 35: 2987–2996.

Park, J.H., Bolan, N., Megharaj, M., Naidu, R. 2011a. Isolation of phosphate solubilizing bacteria and their potential for lead immobilization in soil. *Journal of Hazardous Materials* 185: 829–836.

Park, J.H., Bolan, N., Megharaj, M., Naidu, R., Chung, J.W. 2011b. Bacterial-assisted immobilization of lead in soils: Implications for remediation. *Pedologist* 54: 162–174.

Park, J.H., Lamb, D., Paneerselvam, P., Choppala, G., Bolan, N., Chung, J.W. 2011c. Role of organic amendments on enhanced bioremediation of heavy metal (loid) contaminated soils. *Journal of Hazardous Materials* 185: 549–574.

Peitzsch, M., Kremer, D., Kersten, M. 2010. Microfungal alkylation and volatilization of selenium adsorbed by goethite. *Environmental Science & Technology* 44: 129–135.

Prasad, K.S., Srivastava, P., Subramanian, V., Paul, J. 2011. Biosorption of As(III) ion on *Rhodococcus* sp. WB-12: Biomass characterization and kinetic studies. *Separation Science and Technology* 46: 2517–2525.

Qin, J., Lehr, C.R., Yuan, C., Le, X.C., McDermott, T.R., Rosen, B.P. 2009. Biotransformation of arsenic by a Yellowstone thermoacidophilic eukaryotic alga. *Proceedings of the National Academy of Sciences of the United States of America* 106: 5213–5217.

Ramial, P.S., Rudd, J.W.M., Furutam, A., Xun, L. 1985. The effect of pH on methyl mercury production and decomposition in lake sediments. *Canadian Journal of Fisheries and Aquatic Sciences* 42: 685–692.

Ranjard, L., Prigent-Combaret, C., Nazaret, S., Cournoyer, B. 2002. Methylation of inorganic and organic selenium by the bacterial thiopurine methyltransferase. *Journal of Bacteriology* 184: 3146–3149.

Ravichandran, M., 2004. Interactions between mercury and dissolved organic matter—A review. *Chemosphere* 55: 319–331.

Reategui, M., Maldonado, H., Ly, M., Guibal, E. 2010. Mercury(II) biosorption using *Lessonia* sp. kelp. *Applied Biochemistry and Biotechnology* 162: 805–822.

Rendina, A., Barros, M., de Iorio, A. 2006. Phytoavailability and solid-phase distribution of chromium in a soil amended with organic matter. *Bulletin in Environmental Contamination and Toxicology* 76: 1031–1037.

Renshaw, J.C., Robson, G.D., Trinci, A.P., Wiebe, M.G., Livens, F.R., Collison, D., Taylor, R.J. 2002. Fungal siderophores: Structures, functions and applications. *Mycological Research* 106: 1123–1142.

Robinson, B., Bolan, N.S., Mahimairaja, S., Clothier, B. 2006. Solubility, mobility and bioaccumulation of trace elements: Abiotic processes in the rhizosphere. In: Prasad, M., Sajwan, K., Naidu, R. (eds.) *Trace Elements in the Environment: Biogeochemistry, Biotechnology and Bioremediation.* CRC Press, London, U.K., pp. 97–110.

Rocha, J.C., Junior, É.S., Zara, L.F., Rosa, A.H., dos Santos, A., Burba, P. 2000. Reduction of mercury (II) by tropical river humic substances (Rio Negro)—A possible process of the mercury cycle in Brazil. *Talanta* 53: 551–559.

Rock, M.L., James, B.R., Helz, G.R. 2001. Hydrogen peroxide effects on chromium oxidation state and solubility in four diverse, chromium-enriched soils. *Environmental Science & Technology* 35: 4054–4059.

Rodríguez Martín-Doimeadios, R., Tessier, E., Amouroux, D., Guyoneaud, R., Duran, R., Caumette, P., Donard, O. 2004. Mercury methylation/demethylation and volatilization pathways in estuarine sediment slurries using species-specific enriched stable isotopes. *Marine Chemistry* 90: 107–123.

Ross, S.M. 1994. Retention, transformation and mobility of toxic metals in soils. In: Ross, S.M. (ed.) *Toxic Metals in Soil–Plant Systems.* Wiley, New York, pp. 63–152.

Roy, V., Amyot, M., Carignan, R. 2009. Beaver ponds increase methylmercury concentrations in Canadian shield streams along vegetation and pond-age gradients. *Environmental Science & Technology* 43: 5605–5611.

Sadiq, M. 1997. Arsenic chemistry in soils: An overview of thermodynamic predictions and field observations. *Water Air and Soil Pollution* 93: 117–136.

Sağlam, N., Say, R., Denizli, A., Patır, S., Arıca, M.Y. 1999. Biosorption of inorganic mercury and alkylmercury species on to *Phanerochaete chrysosporium* mycelium. *Process Biochemistry* 34: 725–730.

Sand, W., Gehrke, T., Jozsa, P.G., Schippers, A. 2001. Biochemistry of bacterial leaching-direct vs. indirect bioleaching. *Hydrometallurgy* 59: 159–175.

Sari, A., Tuzen, M. 2008. Biosorption of total chromium from aqueous solution by red algae (*Ceramium virgatum*): Equilibrium, kinetic and thermodynamic studies. *Journal of Hazardous Materials* 160: 349–355.

Sass, H., Ramamoorthy, S., Yarwood, C., Langner, H., Schumann, P., Kroppenstedt, R.M., Rosenzweig, R.F. 2009. *Desulfovibrio idahonensis* sp. nov. sulfate-reducing bacteria isolated from a metal (loid)-contaminated freshwater sediment. *International Journal of Systematic and Evolutionary Microbiology* 59: 2208–2214.

Schiewer, S., Volesky, B. 2000. Biosorption processes for heavy metal removal. In: Lovley, D.R. (ed.) *Environmental Microbe-Metal Interactions.* ASM Press, Washington, DC, pp. 329–362.

Schlüter, K. 2000. Review: Evaporation of mercury from soils. An integration and synthesis of current knowledge. *Environmental Geology* 39: 249–271.

Schroeder, W.H., Munthe, J. 1998. Atmospheric mercury—An overview. *Atmospheric Environment* 32: 809–822.

Schwesig, D., Matzner, E. 2001. Dynamics of mercury and methylmercury in forest floor and runoff of a forested watershed in Central Europe. *Biogeochemistry* 53: 181–200.

Sharma, S., Bansal, A., Dogra, R., Dhillon, S.K., Dhillon, K.S. 2011. Effect of organic amendments on uptake of selenium and biochemical grain composition of wheat and rape grown on seleniferous soils in northwestern India. *Journal of Plant Nutrition and Soil Science* 174: 269–275.

Singh, G., Brar, M., Malhi, S. 2007. Decontamination of chromium by farm yard manure application in spinach grown in two texturally different Cr-contaminated soils. *Journal of Plant Nutrition* 30: 289–308.

Smith, E., Naidu, R., Alston, A.M. 1998. Arsenic in the soil environment: A review. *Advances in Agronomy* 66: 149–195.

Smith, W.A., Apel, W.A., Petersen, J.N., Peyton, B.M. 2002. Effect of carbon and energy source on bacterial chromate reduction. *Bioremediation Journal* 6: 205–215.

Srinath, T., Verma, T., Ramteke, P., Garg, S. 2002. Chromium(VI) biosorption and bioaccumulation by chromate resistant bacteria. *Chemosphere* 48: 427–435.

Svecova, L., Spanelova, M., Kubal, M., Guibal, E. 2006. Cadmium, lead and mercury biosorption on waste fungal biomass issued from fermentation industry. I. Equilibrium studies. *Separation and Purification Technology* 52: 142–153.

Tan, T., Beydoun, D., Amal, R. 2003. Effects of organic whole scavengers on the photocatalytic reduction of selenium anions. *Journal of Photochemistry and Photobiology A* 159: 273–280.

Tandukar, M., Huber, S.J., Onodera, T., Pavlostathis, S.G. 2009. Biological chromium (VI) reduction in the cathode of a microbial fuel cell. *Environmental Science & Technology* 43: 8159–8165.

Thayer, J.S., Brinckman, F.E. 1982. The biological methylation of metals. *Advances in Organometallic Chemistry* 20: 313–356.

Thompson-Eagle, E., Frankenberger, Jr., W.T. 1990. Site volatilization of selenium with alternative sources of protein for microbial deselenification at evaporation ponds. *Journal of Environmental Quality* 19: 125–129.

Thompson-Eagle, E., Frankenberger, Jr., W.T., Karlson, U. 1989. Volatilization of selenium by *Alternaria alternata.* *Applied Environmental Microbiology* 55: 1406–1413.

Thompson-Eagle, E.T., Frankenberger, Jr., W.T. 1992. Bioremediation of soils contaminated with selenium. In: Lal, R., Stewart, B.A. (eds.) *Advances in Soil Science.* Springer, New York, pp. 261–310.

Tüzen, M., Özdemir, M., Demirbaş, A. 1998. Heavy metal bioaccumulation by cultivated *Agaricusbisporus* from artificially enriched substrates. *Zeitschrift für Lebensmittel-untersuchung und -Forschung A* 206: 417–419.

Tuzen, M., Sari, A. 2010. Biosorption of selenium from aqueous solution by green algae (*Cladophora hutchinsiae*) biomass: equilibrium, thermodynamic and kinetic studies. *Chemical Engineering Journal* 158: 200–206.

Ucun, H., Bayhan, Y.K., Kaya, Y., Cakici, A., Algur, O. 2002. Biosorption of chromium (VI) from aqueous solution by cone biomass of *Pinus sylvestris*. *Bioresource Technology* 85: 155–158.

Ullrich, S.M., Tanton, T.W., Abdrashitova, S.A. 2001. Mercury in the aquatic environment: A review of factors affecting methylation. *Critical Reviews in Environmental Science & Technology* 31: 241–293.

Vainshtein, M., Kuschk, P., Mattusch, J., Vatsourina, A., Wiessner, A. 2003. Model experiments on the microbial removal of chromium from contaminated groundwater. *Water Research* 37: 1401–1405.

Von Canstein, H., Kelly, S., Li, Y., Wagner-Dobler, I. 2002. Species diversity improves the efficiency of mercury-reducing biofilms under changing environmental conditions. *Applied and Environmental Microbiology* 68: 2829–2837.

Wang, D., Qing, C., Guo, T., Guo, Y. 1997. Effects of humic acid on transport and transformation of mercury in soil-plant systems. *Water Air and Soil Pollution* 95: 35–43.

Wang, S., Mulligan, C.N. 2006. Natural attenuation processes for remediation of arsenic contaminated soils and groundwater. *Journal of Hazardous Materials* 138: 459–470.

Wardle, D.A. 1992. A comparative assessment of factors which influence microbial biomass carbon and nitrogen levels in soil. *Biological Reviews* 67: 321–358.

White, C., Wilkinson, S.C., Gadd, G.M. 1995. The role of microorganisms in biosorption of toxic metals and radionuclides. *International Biodeterioration and Biodegradation* 35: 17–40.

Wiatrowski, H.A., Ward, P.M., Barkay, T. 2006. Novel reduction of mercury (II) by mercury-sensitive dissimilatory metal reducing bacteria. *Environmental Science & Technology* 40: 6690–6696.

Wiener, J.G., Gilmour, C.C., Krabbenhoft, D.P. 2003. Mercury strategy for the bay-delta ecosystem: A unifying framework for science, adaptive management, and ecological restoration. Report to the California Bay Delta authority, Sacramento, CA.

Wilkie, J.A., Hering, J.G. 1998. Rapid oxidation of geothermal arsenic(III) in streamwaters of the eastern Sierra Nevada. *Environmental Science & Technology* 32: 657–662.

Williams, J.W., Silver, S. 1984. Bacterial resistance and detoxification of heavy metals. *Enzyme and Microbial Technology* 12: 530–537.

Xu, X., McGrath, S., Zhao, F. 2007. Rapid reduction of arsenate in the medium mediated by plant roots. *New Phytologist* 176: 590–599.

Yadav, S.K., Juwarkar, A.A., Kumar, G.P., Thawale, P.R., Singh, S.K., Chakrabarti, T. 2009. Bioaccumulation and phytotranslocation of arsenic, chromium and zinc by *Jatropha curcas* L.: Impact of dairy sludge and biofertilizer. *Bioresource Technology* 100: 4616–4622.

Yamamura, S., Watanabe, M., Kanzaki, M., Soda, S., Ike, M. 2008. Removal of arsenic from contaminated soils by microbial reduction of arsenate and quinone. *Environmental Science & Technology* 42: 6154–6159.

Yang, T., Chen, M.L., Hu, X.W., Wang, Z.W., Wang, J.H., Dasgupta, P.K. 2011. Thiolated eggshell membranes sorb and speciate inorganic selenium. *Analyst* 136: 83–89.

Yavuz, H., Denizli, A., Güngüneş, H., Safarikova, M., Safarik, I. 2006. Biosorption of mercury on magnetically modified yeast cells. *Separation and Purification Technology* 52: 253–260.

Yin, X., Chen, J., Qin, J., Sun, G., Rosen, B., Zhu, Y. 2011. Biotransformation and volatilization of arsenic by three photosynthetic cyanobacteria. *Plant Physiology* 156: 1631–1638.

Yin, Y., Impellitteri, C.A., You, S.J., Allen, H.E. 2002. The importance of organic matter distribution and extract soil: Solution ratio on the desorption of heavy metals from soils. *Science of the Total Environment* 287: 107–119.

Yoshinaga, M., Cai, Y., Rosen, B.P. 2011. Demethylation of methylarsonic acid by a microbial community. *Environmental Microbiology* 13: 1205–1215.

Yun, Y.S., Park, D., Park, J.M., Volesky, B. 2001. Biosorption of trivalent chromium on the brown seaweed biomass. *Environmental Science & Technology* 35: 4353–4358.

Zazo, J.A., Paull, J.S., Jaffe, P.R. 2008. Influence of plants on the reduction of hexavalent chromium in wetland sediments. *Environmental Pollution* 156: 29–35.

Zeng, F., Chen, S., Miao, Y., Wu, F., Zhang, G. 2008. Changes of organic acid exudation and rhizosphere pH in rice plants under chromium stress. *Environmental Pollution* 155: 284–289.

Zeroual, Y., Moutaouakkil, A., Dzairi, F.Z., Talbi, M., Chung, P.U., Lee, K., Blaghen, M. 2003. Biosorption of mercury from aqueous solution by Ulva lactuca biomass. *Bioresource Technology* 90: 349–351.

Zhang, J., Bishop, P.L. 2002. Stabilization/solidification (S/S) of mercury-containing wastes using reactivated carbon and Portland cement. *Journal of Hazardous Materials* 92: 199–212.

Zhang, Y., Frankenberger, W.T. 2003. Factors affecting removal of selenate in agricultural drainage water utilizing rice straw. *Science of the Total Environment* 305: 207–216.

Zhang, Y.Q., Frankenberger, W.T. 1999. Effects of soil moisture, depth, and organic amendments on selenium volatilization. *Journal of Environmental Quality* 28: 1321–1326.

Zouboulis, A., Loukidou, M., Matis, K. 2004. Biosorption of toxic metals from aqueous solutions by bacteria strains isolated from metal-polluted soils. *Process Biochemistry* 39: 909–916.

6 Microbes
Potential Arsenal to Combat Metal Toxicity

Anirban Das Gupta, Goutam Mukherjee, Rusha Roy, and Alok Kumar Sil

CONTENTS

ABSTRACT

Toxic heavy metals are often considered to be a group of metals and metalloids that have a relatively high density, occur in multiple oxidation states, and exert toxic effects on living organisms even upon exposure in low concentrations. Most of these heavy metals have widespread industrial applications. As a result of their industrial applications and irresponsible anthropogenic activities, the natural environment has been contaminated with these toxic metals. Therefore, scientists and engineers must pay utmost attention to remediate and mitigate the harmful effect of toxic metal pollution. In this context, biological remediation, specifically microbe-mediated remediation, could potentially be an effective solution to this problem since microbe-mediated remediation is a cost-effective, non-invasive, and efficient method to mitigate the harmful effects of toxic heavy metals. Microorganisms are the most ubiquitous, metabolically diverse, and abundant group of organisms in the world. They are capable of remediating heavy metals by a wide range of pathways that may include oxidation or reduction of heavy metals, immobilization of heavy metals in the environment, or entrapment of heavy metals within the microbial cell. Therefore, compared to conventional physicochemical remediation, microbe-mediated metal remediation may be a more suitable strategy for the remediation of metals from the contaminated environment.

6.1   INTRODUCTION

Heavy metals are loosely defined as a group of metals and metalloids that have relatively high density, occur in multiple oxidation states, and exert extreme toxic effects on living organisms (Brathwaite and Rabone, 1985; Srivastava and Goyal, 2010). Heavy metals, often referred to as toxic heavy metals, occur naturally in the earth's crust in high concentration as mineral deposits and can also become concentrated at a particular region due to anthropogenic activities (Srivastava and Goyal, 2010). The toxic heavy metals typically bind to cellular components and can interfere with vital functioning of living cells. These toxic metals can enter plants through accumulation via roots and animals and human beings through ingestion of metal-contaminated plants, inhalation, direct contact, etc. The toxic heavy metals that are of grave concern due to toxic effects on living systems include arsenic, lead, cadmium, chromium, and mercury (World Health Organization, 2015). Some of these toxic heavy metals have widespread industrial uses, and so these have accumulated in the environment as a result of extensive use and often irresponsible disposal. Owing to its toxicity, Arsenic (As) had many agricultural uses particularly as insecticide and fungicide. It was also used as medicine in the eighteenth and nineteenth centuries (Gibaud and Jaouen, 2010). As was also used in chemical weapons (Agent Blue during the Vietnam War) for its high toxicity (Westing, 1971). Since the detection of its toxic effects, the industrial application of this metal has been significantly reduced. Lead (Pb) has been used extensively as an antiknocking agent in automobile engines (Seyferth, 2003). The compound tetraethyl lead (TEL) was mixed in gasoline as an antiknocking agent, and it resulted in widespread contamination of the natural environment. Progressively, the use of lead as an antiknocking agent was banned in industrialized countries. However, some underdeveloped countries and growing economies still use leaded gasoline, and TEL is still used in aviation gasoline (Seyferth, 2003). Moreover, the use of lead is prevalent in other industries that include paints, weapons, toys, and other manufacturing sectors. Cadmium (Cd) also has extensive industrial applications, particularly in batteries. The most common use of cadmium is the nickel–cadmium in rechargeable batteries. Moreover, cadmium is required for large-scale electroplating processes for its ability to resist corrosion of steel surfaces (Scoullos et al., 2012). Chromium (Cr) is another toxic heavy metal that has many industrial applications. The most common use of this metal, particularly in developing nations, is in the tannery industry (Covington, 1997). Chrome alum and chromium sulfates are well-known, extremely effective, and cheap tanning agents (Covington, 1997). Cr is also used in metallurgy and dyeing sectors. Nowadays, Cr is also being used in concrete and reinforcements as Cr-containing salts are effective binding agents (Giergiczny and Król, 2008). Historically, mercury (Hg) had many industrial and medicinal uses. However, as the toxic effects of Hg became increasingly clearer, the industrial application of Hg also diminished (Scoullos et al., 2012). However, Hg is still commonly used in thermometers, barometers, and other measuring instruments. Mercury also has some pharmaceutical uses as it is often used as an antiseptic due to its toxicity. After the discovery of its toxic effects, the industrial application of this metal has greatly diminished.

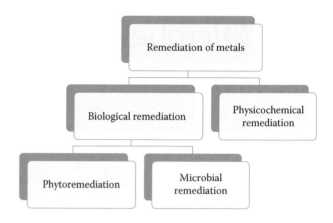

FIGURE 6.1 Metal bioremediation scheme.

Heavy metals exert their toxicity on humans and other living organisms in a variety of ways (Duruibe et al., 2007). They have detrimental toxic effects on vital organ systems that include the nervous system, digestive system, reproductive system, digestive system, and skeletal system of human beings even upon very low exposure (Graeme and Pollack, 1998). Some toxic heavy metals are also well-known mutagens and carcinogens for humans and animals. Owing to their widespread distribution in the biosphere and high toxic effects on living systems in very low exposure concentrations (Chowdhury and Chandra, 1986), these heavy metals are considered to be extremely hazardous contaminants of the natural environment. Therefore, environmental engineers and scientists must pay utmost attention toward the remediation and mitigation of toxic effects of these heavy metals. To this end, scientist and engineers have come up with several remediation strategies. These can be broadly classified into two categories: physicochemical remediation and biological remediation or bioremediation (Figure 6.1). Physicochemical remediation strategies broadly include the use of physical or chemical agents to remove and/or remediate from an environmental contaminant source. For example, ceramic filters and nanofilters are used to remove heavy metals like As from water. This is an extensively documented practice (Sato et al., 2002; Košutić et al., 2005). High-temperature volatilization is also used in some cases to remove heavy metals like Hg and As from contaminated soils. Chelating agents are often used to immobilize heavy metals to a particular localized region in heavy metal–contaminated soils (Leštan et al., 2008). In the past 15 years, bioremediation of heavy metals has received immense attention from the scientific community as bioremediation often offers an *in situ*, cost-effective, noninvasive, and efficient strategy for the remediation and mitigation of toxic effects of heavy metals (Barker et al., 2002). Bioremediation commonly involves two distinct remediation strategies: phytoremediation and microbial remediation (Mejáre and Bülow, 2001; Guo et al., 2010). Both of these remediation strategies involve the use of biomass (live or dead). However, the type of biomass that will be used in a process depends on the goal of the process. Live biomass is often used to accumulate and subsequently quarantine heavy metals within living systems (Guo et al., 2010). Alternatively, living biomass converts them to less toxic oxidation states. On the other hand, dead biomass is largely used as effective filtering agents to remove heavy metals from contaminated environmental sources. Microorganisms are the most ubiquitous group of living organisms in the world and are capable of driving major ecosystem processes like nutrient cycling (van der Heijden et al., 2008) and maintenance of ecosystem health (Lugtenberg and Kamilova, 2009). These microbial processes are catalyzed and supported by several enzymes and other chemicals that may be intracellular or may be secreted by the microorganisms into the environment. Microorganisms have evolved different resistance mechanisms to avoid metal toxicity. They have the ability to transform metals in the environment, and this transformation serves different functions for the microorganisms. Microbe-mediated metal transformations can be classified into two broad groups: redox conversions of inorganic forms and conversions from inorganic to organic form and vice versa, most commonly through methylation and demethylation. Oxidation of iron, sulfur, manganese, and arsenic by microorganisms is related to energy production (Santini et al., 2000). On the other hand, reduction of metals typically occurs through dissimilatory reduction wherein microorganisms utilize metals as a terminal electron acceptor during anaerobic respiration. For example, oxyanions of As and Cr are used in microbial anaerobic respiration as terminal electron acceptors (Stolz and Oremland, 1999; Niggemyer et al., 2001). Additionally, many microorganisms possess reduction mechanisms that are not coupled to respiration but instead impart metal resistance, for example, reduction of Hg(II) to Hg(0) (Brim et al., 2000; Wagner-Döbler et al., 2000).

Microbial methylation plays an important role in the biogeochemical cycle of metals, since methylated compounds are oftentimes volatile. For example, mercury [Hg(II)] can be biomethylated by various bacterial species (e.g., *Pseudomonas* sp., *Escherichia* sp., *Bacillus* sp., and *Clostridium* sp.) to methylmercury, which is gaseous in nature (Pan-Hou and Imura, 1982; Compeau and Bartha, 1985; Pongratz and Heumann, 1999). This is the most poisonous and most easily accumulated form of mercury. Also, biomethylation of arsenic to gaseous arsines (Gao and Burau, 1997) and biomethylation of lead to dimethyl lead (Pongratz and Heumann, 1999) have been observed in various soil environments. Apart from redox conversions and methylation reactions, acidophilic iron- and sulfur-oxidizing bacteria are able to extract elevated concentrations of arsenic, copper, cadmium, cobalt, nickel, and zinc from contaminated soils (White et al., 1997; Seidel et al., 2000; Groudev et al., 2001; Löser et al., 2001). Metals can also be precipitated as insoluble sulfides by the metabolic action of sulfate-reducing bacteria (White et al., 1997; Lloyd et al., 2001). Sulfate-reducing bacteria are anaerobic heterotrophs utilizing a range of organic substrates with SO_4^{2-} as the terminal electron acceptor. In summary, microbiological processes can either solubilize metals, thereby increasing their bioavailability and potential toxicity, or immobilize them, and thereby reducing the bioavailability of metals. These biotransformations are important components of biogeochemical cycles of metals and may be exploited in the bioremediation of metal-contaminated soils (Lovley and Coates, 1997; Gadd, 2000; Barkay and Schaefer, 2001; Lloyd and Lovley, 2001).

6.2 ARSENIC

6.2.1 SOURCE OF ARSENIC IN THE ENVIRONMENT

Arsenic is a hazardous toxic heavy metal that is widely found in the lithosphere as a naturally occurring mineral. The most important natural source of distribution of arsenic in the biosphere is weathering of minerals and erosion. However, the most significant reason for the distribution and contamination of arsenic in the environment is anthropogenic activity (Järup, 2003). Overutilization of groundwater in areas with a high deposit of arsenic minerals has resulted in extreme arsenic pollution of the environment and contamination of the food chain (Ng et al., 2003). Water-soluble arsenic minerals that contaminate groundwater upon drinking enter the living system and become bioavailable. Moreover, As was used as a food preservative and fungicide, which led to arsenic contamination of environment.

6.2.2 MODE OF ARSENIC TOXICITY

Within the environment, As occurs in inorganic and organic forms and it exists largely in a trivalent arsenate (Arsenic III) and pentavalent arsenite (Arsenic V) oxidation states. Inorganic arsenic compounds exert much higher toxicity than organic arsenic compounds, and the toxicity is dependent on its oxidation state. Unlike other major toxic heavy metals such as Pb and Cr, which exert higher toxicity at higher oxidation state, trivalent arsenic is more toxic than pentavalent arsenic. Arsenic(III) is known to be approximately 100 times more toxic than the Arsenic(V) (Cervantes et al., 1994). Toxicity of Arsenic(III) stems from its ability to bind to protein sulfhydryl groups (Gebel, 2000). Thus, Arsenic(III) inhibits enzyme reactions requiring free sulfhydryl groups, leading to cell death. Moreover, arsenate is an analogue of phosphate, and therefore, the arsenate enters living cells through phosphate uptake pathways. Arsenate interferes with key cellular processes that include oxidative phosphorylation and ATP synthesis. Excess As within the cell also causes oxidative stress as a result of production of reactive oxygen species (ROS) (Vigo et al., 2006).

6.2.3 EFFECTS OF ARSENIC EXPOSURE

The first symptoms of sustained exposure to high levels of inorganic arsenic (e.g., through drinking water and food) are usually observable in the skin and include pigmentation changes, lesions, and hard patches on the palms and soles (hyperkeratosis). These occur as a precursor to skin cancer and these effects are manifested within 5 years of exposure. Additionally, long-term exposure may also cause bladder and lung cancers (Ratnaike, 2003). On the other hand, immediate effects of acute arsenic exposure include nausea and vomiting, severe abdominal pain, and violent diarrhea. These are followed by numbness, muscle cramps, and ultimately death, in cases of extreme poisoning (Ratnaike, 2003).

6.2.4 REMEDIATION STRATEGY

An effective strategy for As remediation from the environments involves microbial metabolic activities wherein As(III) is oxidized by arsenite-oxidizing bacteria (AsOB) through respiratory processes that utilize As(III) as an electron donor (Hu et al., 2015). Different groups of AsOB are capable of surviving in aerobic as well as anaerobic environments and As oxidation is facilitated by nitrate availability in anaerobic environments (Hu et al., 2015). Microorganisms thus transform As(III) to a much less toxic As(V), which is also less mobile in soil and aquatic environments (Stolz et al., 2006; Hu et al., 2015). Therefore, microorganism-mediated As oxidation contributes toward the attenuation of As by converting As(III) to a less toxic As(V) and by decreasing its bioavailability through immobilization. AsOB uses a cluster of genes known as *aox* gene for the oxidation of As(III) in soil and aquatic environments (Hu et al., 2015). Examples of AsOB include *Acidovorax* sp., *Albidiferax ferrireducens*, and many other microorganisms belonging to *Acidovorax* genera. However, apart from their independent action, they also exert their remediation potential in combination with plants.

An important area of concern with regard to As contamination of food chain is rice, since rice is the world's single most important food crop and among agricultural crops rice is the single most important route of exposure for the uptake of inorganic As (Su et al., 2010). In this context, recent studies have shown that root iron plaques of paddy plants act as effective barrier to As uptake because of their ability to accumulate metals (Hu et al., 2015). Root iron plaques are formed as a result of oxidation of Fe(II) to Fe(III) by the active participation of rhizosphere microorganisms in physiologically active roots. Thus, root iron plaques of rice plants growing in As-contaminated areas immobilize As in the rhizosphere region preventing its accumulation in the aboveground paddy biomass (Su et al., 2010; Hu et al., 2015). Microbe-mediated remediation of As through oxidation and immobilization provides an effective way to control As pollution in a wide variety of soil and aquatic environments.

6.3 LEAD

6.3.1 SOURCE OF LEAD IN THE ENVIRONMENT

Lead (Pb) poisoning is now recognized as most alarming in developing countries, like India and Brazil. Though there are many preventive measures, the cases of lead poisoning are reported continually. Due to some unique properties of

lead like softness, low melting point, resistant to corrosion, ductility, and high malleability, it is widely used in automobile, painting, battery, chemical, and plastic industries. Thus, lead contaminates our environment and accumulates in the biological system. Besides, it persists in biological system for prolonged time. The exposure to human occurs mainly through ingestion, inhalation, or occasionally through skin contact. Tetraethyl lead, used as an activator of gasoline, is adsorbed by the skin and also enters into our body by inhalation. Free Pb from painting and battery industries also enters into our body by ingestion (Levitt et al., 1984; Jones et al., 1999). It was estimated that about 30%–40% of inhaled Pb dust is deposited into the lungs and about 90% of that goes into blood circulation. But in case of ingested Pb, only 15%–20% is absorbed.

6.3.2 MODE OF LEAD TOXICITY

The current reference level of concentration of Pb is 5 µg/dL for healthy children and 25 µg/dL for healthy adults. If the Pb concentration in blood increases beyond the reference level, it exerts many physiological problems. A Pb concentration of 50 µg/dL in blood circulation affects the nervous system, redox regulatory system, reproductive system, hematopoietic system, and cardiovascular system severely (Pearce, 2007). Its most detrimental effect is on the nervous system, where it blocks the *N*-methyl-D-aspartate. This is an important receptor for the maturation of brain plasticity. It also affects the communication between astrocytes (a star-shaped glial cell of the central nervous system) and endothelial cells. In the peripheral nervous system, Pb causes the demyelinization of the axon (Brent, 2006). In a mother's womb, Pb passes through placenta and accumulates into the fetus that causes many developmental abnormalities like learning disability of the child. Besides, hormonal imbalance is also observed as Pb replaces the Ca^{2+} in Ca–calmodulin-dependent signaling pathway (Needleman, 2004). Pb also enhances the production of ROS by inhibiting the activity of glutathione peroxidase and causes cellular death by damaging DNA, proteins, and lipids (Ahamed and Siddiqui, 2007).

6.3.3 REMEDIATION STRATEGY

For the reduction of Pb burden from the environment, remediation is considered to be the most efficient process over the other physiochemical methods like precipitation, ion exchange, reverse osmosis, evaporation, and sorption as the method is very easy, efficient, and cost-effective (Gadd and White, 1993; Chatterjee et al., 2012). It was observed that some yeast cells like *Rhodotorula mucilaginosa* efficiently adsorb Pb from contaminated sites (Chatterjee et al., 2012). Also, many ureolytic bacteria mineralize the Pb as $PbCO_3$ while they enzymatically hydrolyze urea to ammonia and carbonate ion:

$$2(NH_2)_2CO + 2H_2O \xrightarrow{\text{Urease}} CO_3^{2-} + 2NH_3$$

$$CO_3{}^{2-} + Pb^{2+} \rightarrow PbCO_3$$

The urease activity is stimulated by urea supplementation. Thus, microorganisms belonging to this group produce ammonia and carbonate ion. The carbonate ion in turn gets precipitated as Pb carbonate while reacting with lead salt (Kang et al., 2015). Besides this, bioabsorption is also exploited for the bioremediation of Pb. In this case, bacteria produce extracellular polymeric substance, which can absorb the metal form the environment. *Pseudomonas aeruginosa* is a good example of this kind. It has been observed that when sulfate-reducing bacteria *Desulfovibrio desulfuricans* form a biofilm on a quartz surface anaerobically, it immobilized Pb efficiently with the production of hydrogen sulfide (Beyenal and Lewandowski, 2004). The hydrogen sulfide causes precipitation of the redox-sensitive minerals of Pb. In nature, there are many acidotolerant thermophilic bacteria (*Desulfovibrio salexigens* DSM 2538, *Desulfovibrio africanus* DSM 2603, and *Desulfovibrio carbinolicus* DSM 3852) that are able to adsorb Pb from industrial effluent (Umrania, 2006). Another method of Pb remediation is executed by *Alcaligenes eutrophus* that involves siderophore-mediated metal solubilization (Diels et al., 1999). The solubilized lead is precipitated with biomass and separated from soil by flocculation process.

6.4 CADMIUM

6.4.1 Source of Cadmium in the Environment

The major source of Cd in the environment are anthropogenic activities, but the volcanic activities are also another source of widespread distribution of Cd in the environment. Cd is mainly used as an anticorrosive agent in the steel industry and is an important constituent for the battery industry (Scoullos et al., 2001). Cd sulfide is also used as a pigment in the plastic industry. Apart from these, Cd is also an important by-product in metal mining and extraction and fertilizer industries. Therefore, industrial effluents are the major source of Cd contamination in water (Galal-Gorchev, 1991).

6.4.2 Effects of Cadmium Exposure

Accumulation of a high amount of Cd in human leads to osteomalacia, renal disturbance, and lung insufficiencies. Also, it has the potential to cause cancer in humans.

6.4.3 Mode of Cadmium Toxicity

High amount of Cd absorption into the body leads to the inhibition of calcium-dependent cell signaling. Due to the same oxidation state, it can replace Ca^{2+} and Zn^{2+} and causes an imbalance in cellular metabolism. It also induces the generation of ROS and lipid peroxidation by inhibiting glutathione peroxidase activity (Congiu, 2000). By inducing the generation of free radicals, it also damages DNA and thus acts as a mutagen (Sutoo et al., 1990).

6.4.4 Remediation Strategy

Immobilization constitutes an effective strategy compared to conventional methods like chemical precipitation and filtration through ion-exchange resin to decontaminate the cadmium-contaminated soil. Researchers have identified many Cd-resistant organisms, which can tolerate high Cd concentration during their growth (Sinha and Mukherjee, 2009). They are efficient in accumulating Cd from the environment. An example of this kind of bacteria is *P. aeruginosa* KUCd1 (Sinha and Mukherjee, 2009).

6.5 CHROMIUM

6.5.1 Source of Chromium in the Environment

The metal chromium (Cr) is an abundant element that occurs in the environment in the form of minerals in multiple oxidation states (François, 1992). The most commonly occurring Cr mineral is chromite ($FeCr_2O_4$). Cr is used in many industries for multiple purposes, and as a result of industrial discharge, Cr contamination of rivers, wetlands, and solid waste disposal sites are fairly common phenomenon. Although Cr occurs in multiple oxidation states, ranging from zerovalent to hexavalent state, the commonly occurring forms of Cr are the trivalent (III) and hexavalent (VI) states (Losi et al., 1994). Cr in both of these oxidation states is stable, and Cr(VI) has strong oxidizing ability.

Chromium compounds, in either the chromium(III) or chromium(VI) forms, are used for chrome plating, the manufacture of dyes and pigments, tanning leather, and preservation of wood (Convington et al., 1997). Smaller amounts are used in drilling muds, textiles, and toners for copying machines (Convington et al., 1997).

6.5.2 Effects of Chromium Exposure

Immediate symptoms of Cr exposure include perforation of the nasal septum, spasm, asthmatic reactions, and bronchial carcinomas. Other effects from acute inhalation exposure to Cr include gastrointestinal and neurological effects. The major effects of Cr(VI) ingestion include nausea, abdominal pain, and even hemorrhage in cases of extreme exposure. Dermal exposure causes skin irritation and skin burns in humans. Cr(VI) is a well-known mutagen, and several epidemiological studies (conducted by the USEPA) of industrial workers exposed to Cr(VI) have conclusively established that inhaled chromium is a human carcinogen, resulting in an increased risk of lung cancer.

6.5.3 Mode of Cr Toxicity

Cr(VI) is much more toxic than Cr(III) for both acute and chronic exposures. The respiratory tract is the primary target for Cr(VI) following inhalation exposure in humans, particularly workers associated with Cr-based industries. Cr(VI) ions enter living cells through phosphate and sulfate transport pathways. Within the cells, Cr(VI) ions produce oxidative stress owing to its acutely high oxidation potential. This leads to disruption of cell signaling, causing DNA damage and cell death (Katz and Salem, 1993).

6.5.4 Remediation Strategy

Since Cr(VI) is more harmful, highly soluble, and more readily bioavailable in nature than Cr(III), the most effective remediation strategy encompasses reduction of Cr(VI) to Cr(III). This reduction can be mediated by the use of chemical nonenzymatic reducing agents or biotic enzyme-mediated reduction of Cr(VI) (Lovley and Phillips, 1994; Krishna and Philip, 2005). The ability of certain microorganisms to reduce highly toxic and highly soluble Cr(VI) to less toxic and less soluble Cr(III) has led to the utilization of Cr(VI)-reducing microorganisms to effectively remediate Cr(VI) pollution in soil as well as aquatic environment (Lovley and Phillips, 1994). For example, various species of *Desulfovibrio* are capable of performing cytochrome c3 mediated Cr (VI) reduction (Lovely et al., 1994; Viti et al., 2003; Krishna, 2005). *Ochrobactrum anthropi* are also capable of reducing Cr(VI) to Cr(III) (Cheng et al., 2010). Aside from microbial remediation, phytoremediation is also considered to be an effective strategy for treating Cr contamination (Mangkoedihardjo et al., 2008).

6.6 MERCURY

6.6.1 Source of Mercury in the Environment

Mercury (Hg) poisoning, known as hydrargyria or mercurialism, is a threat to our environment. In normal atmospheric pressure and temperature, mercury is a liquid metal that generally exists in three different valance states, that is, Hg^0 (metallic mercury), Hg^{2+} (mercuric mercury), and Hg_2^{2+} (mercurous mercury), and they maintain an equilibrium among them (Wang et al., 2007). The predominant forms of Hg are $Hg(OH)_2$ and $HgCl_2$. Apart from these, organomercurials such as aryl or short- or long-chain alkyl forms are also a major source of Hg and mainly found in industrial effluents. A large portion of Hg is emitted by coal-burning power plants and smelting industry. There are other sources too, such as battery, fluorescent lamp, fertilizer, and cement industries. Biofuel burning is another major source of Hg contamination in the environment. An individual may get exposure to Hg through a number of different ways that include dental amalgam fillings, household products, fluorescent light bulbs, broken thermometers, and industrial settings (Wang et al., 2004; Zhang and Wong, 2007).

6.6.2 Effects of Mercury Exposure

Mercury salts (both organic and inorganic) are potentially more toxic than that of the free mercury. The poisoning of mercury causes several diseases and the toxicity of mercury depends upon the form of mercury as well as the dosage of exposure. Exposure to metallic mercury (Hg^0) vapor causes bronchitis and bronchiolitis (Zhang and Wong, 2007). Mercurous mercury induces the toxicity of mercuric mercury (Wang et al., 2004). Exposure to mercuric mercury (Hg^{2+}) causes extensive precipitation of erythrocytic proteins, vomiting, necrosis of gut mucosal layer, and bloody diarrhea with abdominal pain. Nephrotoxicity, acute dermatitis, asthma, and suppression of development and activity of natural killer cells are also reported. Deposition in testicles reduces the rate of spermatogenesis (Kasuya, 1972; Wang et al., 2004). Organic mercury causes chronic fatigue syndrome. It damages many parts of the brain and peripheral nervous systems. It also inhibits the activity of NK cells and deregulates the ratio between Th1 and Th2.

6.6.3 Mode of Mercury Toxicity

Both Hg^{2+} and organic Hg are capable of binding with sulfhydryl and selenohydryl group of amino acids and thus alter the quaternary and tertiary structure of proteins. This in turn affects basic cellular metabolic processes like transcription land translation. Organic Hg also causes mitochondrial dysfunction that leads to the generation of ROS and oxidative stress.

6.6.4 Remediation Strategy

Existing literature shows that a wide range of Gram-positive and Gram-negative bacteria exhibit resistance against the toxic effects of mercury and it can mainly be attributed to their ability to transform the toxic form of mercury (Hg^{2+}) to a less toxic and/or nontoxic form (or Hg^+/Hg^0) (Sinha et al., 2012). The resistant bacteria synthesize thiols that reduce the mercuric ion or there may be a presence of permeability barrier that resist the entry of mercury into the cells. Extensive studies on mercury resistance revealed that many of these bacteria, if not all, harbor *mer* operon, which is present as an extrachromosomal gene cluster (i.e., present in a plasmid) or as a chromosomal component or as a transposon-bound component (Huang et al., 2010). The presence of this operon confers Hg resistance to these bacteria. Figure 6.2 presents an outline of *mer* operon, and the functions of different genes are described in Table 6.1.

6.7 CONCLUSION

In conclusion, our environment is contaminated with various metals mainly through anthropogenic activities. Although a few of these metals may be important for some biological function, all of them are toxic at even at low or moderate concentration. Therefore, the current situation needs tools and technologies to remediate contaminated metals from the environment. It is evident from the existing literature that microbes have the potential to remediate metals and thus, if microbes

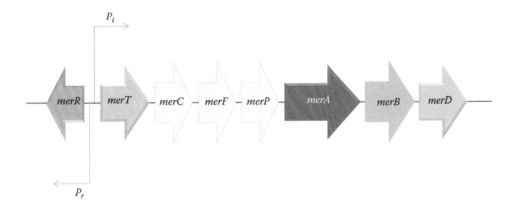

FIGURE 6.2 Schematic representation of *mer* operon.

TABLE 6.1
The List of Genes in *mer* Operon and Their Functions

Genes	Gene Product	Location	Function
merA	Mercuric ion reductase	Cytoplasm	Conversion of Hg^{2+} to Hg^0
merB	Organomercurial lyase	Cytoplasm	Lysis of $C–Hg^+$ bond
merP	Periplasmic mercuric ion-binding protein	Periplasm	Transfer of Hg^{2+} ion to integral membrane proteins
merT	Mercuric ion transport protein	Inner membrane	Transport of mercuric ion
merD	Regulatory protein	Cytoplasm	Negatively regulate mer operon
merR	Regulatory protein (repressor)	Cytoplasm	Positively regulate mer operon
merC	Mercuric ion transport protein	Inner membrane	Transport of mercuric ion
merF	Mercuric ion transport protein	Inner membrane	Transport of mercuric ion

can properly be selected, they can provide cost-effective eco-friendly method to get rid of the hazardous effects originated from contaminating metals.

REFERENCES

Ahamed, M., Siddiqui, M.K.J. 2007. Low level lead exposure and oxidative stress: Current opinions. *Clinica Chimica Acta* 383: 57–64.

Barkay, T., Schaefer, J. 2001. Metal and radionuclide bioremediation: Issues, considerations and potentials. *Current Opinion in Microbiology* 4: 318–323.

Barker, A.V., Bryson, G.M. 2002. Bioremediation of heavy metals and organic toxicants by composting. *The Scientific World Journal* 2: 407–420.

Beyenal, H., Lewandowski, Z. 2004. Dynamics of lead immobilization in sulfate reducing biofilms. *Water Research* 38: 2726–2736.

Brathwaite, R.L., Rabone, S.D.C. 1985. Heavy metal sulphide deposits and geochemical surveys for heavy metals in New Zealand. *Journal of the Royal Society of New Zealand* 15: 363–370.

Brent, J.A. 2006. Review of medical toxicology. *Clinical Toxicology* 44: 355.

Brim, H., McFarlan, S.C., Fredrickson, J.K., Minton, K.W., Zhai, M., Wackett, L.P., Daly, M.J. 2000. Engineering Deinococcus radiodurans for metal remediation in radioactive mixed waste environments. *Nature Biotechnology* 18: 85–90.

Cervantes, C., Ramirez, J., Silver, S. 1994. Resistance to arsenic compounds in microorganisms. *FEMS Microbiology Reviews* 15: 355–367.

Chatterjee, S., Mukherjee, A., Sarkar, A., Roy, P. 2012. Bioremediation of lead by lead resistant microorganisms, isolated from industrial sample. *Advances in Bioscience and Biotechnology* 3: 290–295.

Cheng, Y., Yan, F., Huang, F., Chu, W., Pan, D., Chen, Z., Wu, Z. 2010. Bioremediation of Cr(VI) and immobilization as Cr(III) by *Ochrobactrum anthropi*. *Environmental Science & Technology* 44: 6357–6363.

Chowdhury, B.A., Chandra, R.K. 1986. Biological and health implications of toxic heavy metal and essential trace element interactions. *Progress in Food & Nutrition Science* 11: 55–113.

Compeau, G.C., Bartha, R. 1985. Sulfate-reducing bacteria: Principal methylators of mercury in anoxic estuarine sediment. *Applied and Environmental Microbiology* 50: 498–502.

Congiu, L., Chicca, M., Pilastro, A., Turchetto, M., Tallandini, L. 2000. Effects of chronic dietary cadmium on hepatic glutathione levels and glutathione peroxidase activity in starlings (*Sturnus vulgaris*). *Archives of Environmental Contamination and Toxicology* 38: 357–361.

Covington, A.D. 1997. Modern tanning chemistry. *Chemical Society Review* 26: 111–126.

Diels, L., De Smet, M., Hooyberghs, L., Corbisier, P. 1999. Heavy metals bioremediation of soil. *Molecular Biotechnology* 12: 149–158.

Duruibe, J.O., Ogwuegbu, M.O.C., Egwurugwu, J.N. 2007. Heavy metal pollution and human biotoxic effects. *International Journal of Physical Sciences* 2: 112–118.

Gadd, G.M. 2000. Bioremedial potential of microbial mechanisms of metal mobilization and immobilization. *Current Opinion in Biotechnology* 11: 271–279.

Gadd, G.M., White, C. 1993. Microbial treatment of metal pollution—A working biotechnology? *Trends in Biotechnology* 11: 353–359.

Galal-Gorchev, H. 1991. Dietary intake of pesticide residues: Cadmium, mercury, and lead. *Food Additives & Contaminants* 8: 793–806.

Gao, S., Burau, R.G. 1997. Environmental factors affecting rates of arsine evolution from and mineralization of arsenicals in soil. *Journal of Environmental Quality* 26: 753–763.

Gebel, T. 2000. Confounding variables in the environmental toxicology of arsenic. *Toxicology* 144: 155–162.

Gibaud, S., Jaouen, G. 2010. Arsenic-based drugs: From fowler's solution to modern anticancer chemotherapy. In: *Medicinal Organometallic Chemistry*, Jaouen, G., Metzler-Nolte, N. (eds.), Springer, Berlin, Germany, pp. 1–20.

Giergiczny, Z., Król, A. 2008. Immobilization of heavy metals (Pb, Cu, Cr, Zn, Cd, Mn) in the mineral additions containing concrete composites. *Journal of Hazardous Materials* 160: 247–255.

Graeme, K.A., Pollack, C.V. 1998. Heavy metal toxicity, part I: Arsenic and mercury. *The Journal of Emergency Medicine* 16: 45–56.

Groudev, S.N. 2001. Biobeneficiation of mineral raw materials. In: *Mineral Biotechnology*, Kawatra S.K., Natarajan, K.A. (eds.), SME, Littleton, pp. 37–54.

Guo, H., Luo, S., Chen, L., Xiao, X., Xi, Q., Wei, W., He, Y. 2010. Bioremediation of heavy metals by growing hyperaccumulaor endophytic bacterium *Bacillus* sp. L14. *Bioresource Technology* 101: 8599–8605.

Hu, M., Li, F., Liu, C., Wu, W. 2015. The diversity and abundance of As(III) oxidizers on root iron plaque is critical for arsenic bioavailability to rice. *Scientific Reports* 5: 13611. doi:10.1038/srep13611.

Huang, C.C., Chien, M.F., Lin, K.H. 2010. Bacterial mercury resistance of TnMERI1 and its application in bioremediation. In: *Interdisciplinary Studies on Environmental Chemistry—Biological Responses to Contaminants*, Hamamura, N., Suzuki, S., Mendo, S., Barroso, C.M., Iwata, H., Tanabe, S. (eds.), TERRAPUB, Tokyo, Japan, Vol. 3, pp. 21–29.

Järup, L. 2003. Hazards of heavy metal contamination. *British Medical Bulletin* 68: 167–182.

Jones, T.F., Moore, W.L., Craig, A.S., Reasons, R.L., Schaffner, W. 1999. Hidden threats: Lead poisoning from unusual sources. *Pediatrics* 104 (Suppl. 6): 1223–1225.

Kang, C.H., Oh, S.J., Shin, Y., Han, S.H., Nam, I.H., So, J.S. 2015. Bioremediation of lead by ureolytic bacteria isolated from soil at abandoned metal mines in South Korea. *Ecological Engineering* 74: 402–407.

Kasuya, M. 1972. Effects of inorganic, aryl, alkyl and other mercury compounds on the outgrowth of cells and fibers from dorsal root ganglia in tissue culture. *Toxicology and Applied Pharmacology* 23: 136–146.

Katz, S.A., Salem, H. 1993. The toxicology of chromium with respect to its chemical speciation: A review. *Journal of Applied Toxicology* 13: 217–224.

Košutić, K., Furač, L., Sipos, L., Kunst, B. 2005. Removal of arsenic and pesticides from drinking water by nanofiltration membranes. *Separation and Purification Technology* 42: 137–144.

Krishna, K.R., Philip, L. 2005. Bioremediation of Cr(VI) in contaminated soils. *Journal of Hazardous Materials* 121: 109–117.

Leštan, D., Luo, C.L., Li, X.D. 2008. The use of chelating agents in the remediation of metal-contaminated soils: A review. *Environmental Pollution* 153: 3–13.

Levitt, C., Godes, J., Eberhardt, M., Ing, R., Simpson, J.M. 1984. Sources of lead poisoning. *JAMA* 252: 3127–3128.

Lloyd, J.R., Lovley, D.R. 2001. Microbial detoxification of metals and radionuclides. *Current Opinion in Biotechnology* 12: 248–253.

Lloyd, J.R., Mabbett, A.N., Williams, D.R., Macaskie, L.E. 2001. Metal reduction by sulphate-reducing bacteria: Physiological diversity and metal specificity. *Hydrometallurgy* 59: 327–337.

Losi, M.E., Amrhein, C., Frankenberger, Jr., W.T. 1994. Environmental biochemistry of chromium. In: *Reviews of Environmental Contamination and Toxicology*. Springer, New York, Vol. 136, pp. 91–121.

Löser, C., Seidel, H., Hoffmann, P., Zehnsdorf, A. 2001. Remediation of heavy metal-contaminated sediments by solid-bed bioleaching. *Environmental Geology* 40: 643–650.

Lovley, D.R., Coates, J.D. 1997. Bioremediation of metal contamination. *Current Opinion in Biotechnology* 8: 285–289.

Lovley, D.R., Phillips, E.J., 1994. Reduction of chromate by *Desulfovibrio vulgaris* and its c3 cytochrome. *Applied and Environmental Microbiology* 60: 726–728.

Lugtenberg, B., Kamilova, F. 2009. Plant-growth-promoting rhizobacteria. *Annual Review of Microbiology* 63: 541–556.

Mangkoedihardjo, S., Ratnawati, R., Alfianti, N., 2008. Phytoremediation of hexavalent chromium polluted soil using *Pterocarpus indicus* and *Jatropha curcas* L. *World Applied Sciences Journal* 4: 338–342.

Mejáre, M., Bülow, L. 2001. Metal-binding proteins and peptides in bioremediation and phytoremediation of heavy metals. *Trends in Biotechnology* 19: 67–73.

Needleman, H. 2004. Lead poisoning. *Annual Review of Medicine* 55: 209–222.

Ng, J.C., Wang, J., Shraim, A. 2003. A global health problem caused by arsenic from natural sources. *Chemosphere* 52: 1353–1359.

Niggemyer, A., Spring, S., Stackebrandt, E., Rosenzweig, R.F. 2001. Isolation and characterization of a novel As (V)-reducing bacterium: implications for arsenic mobilization and the genus Desulfitobacterium. *Applied and Environmental Microbiology* 67: 5568–5580.

Pan-Hou, H.S., Imura, N. 1982. Involvement of mercury methylation in microbial mercury detoxication. *Archives of Microbiology* 131: 176–177.

Pearce, J.M.S. 2007. Burton's line in lead poisoning. *European Neurology* 57: 118–119.

Pongratz, R., Heumann, K.G. 1999. Production of methylated mercury, lead, and cadmium by marine bacteria as a significant natural source for atmospheric heavy metals in polar regions. *Chemosphere* 39: 89–102.

Ratnaike, R.N. 2003. Acute and chronic arsenic toxicity. *Postgraduate Medical Journal* 79: 391–396.

Santini, J.M., Sly, L.I., Schnagl, R.D., Macy, J.M. 2000. A new chemolithoautotrophic arsenite-oxidizing bacterium isolated from a gold mine: phylogenetic, physiological, and preliminary biochemical studies. *Applied and Environmental Microbiology* 66: 92–97.

Sato, Y., Kang, M., Kamei, T., Magara, Y. 2002. Performance of nanofiltration for arsenic removal. *Water Research* 36: 3371–3377.

Scoullos, M., Vonkeman, G.H., Thornton, I., Makuch, Z. 2001. *Mercury-Cadmium-Lead Handbook for Sustainable Heavy Metals Policy and Regulation*, Scoullos, M. (ed.) Springer Science & Business Media. Springer, The Netherlands.

Scoullos, M., Vonkeman, G.H., Thornton, I., Makuch, Z. 2001. *Mercury–Cadmium–Lead Handbook for Sustainable Heavy Metals Policy and Regulation*. Springer Science & Business Media, Springer Publishing.

Seidel, H., Ondruschka, J., Morgenstern, P., Wennrich, R., Hoffmann, P. 2000. Bioleaching of heavy metal-contaminated sediments by indigenous *Thiobacillus* spp.: Metal solubilization and sulfur oxidation in the presence of surfactants. *Applied Microbiology and Biotechnology* 54: 854–857.

Seyferth, D. 2003. The rise and fall of tetraethyl lead 2. *Organometallics* 22: 5154–5178.

Sinha, A., Pant, K.K., Khare, S.K. 2012. Studies on mercury bioremediation by alginate immobilized mercury tolerant Bacillus cereus cells. *International Biodeterioration & Biodegradation* 71: 1–8.

Sinha, S., Mukherjee, S.K. 2009. Pseudomonas aeruginosa KUCd1, a possible candidate for cadmium bioremediation. *Brazilian Journal of Microbiology* 40: 655–662.

Srivastava, S., Goyal, P. 2010. *Novel Biomaterials: Decontamination of Toxic Metals from Wastewater*. Springer Science & Business Media. Springer, Berlin Heidelberg.

Stolz, J.F., Basu, P., Santini, J.M., Oremland, R.S. 2006. Arsenic and selenium in microbial metabolism. *Annual Review of Microbiology* 60: 107–130.

Stolz, J.F., Oremland, R.S. 1999. Bacterial respiration of arsenic and selenium. *FEMS Microbiology Reviews* 23: 615–627.

Su, Y.H., McGrath, S.P., Zhao, F.J. 2010. Rice is more efficient in arsenite uptake and translocation than wheat and barley. *Plant and Soil* 328: 27–34.

Sutoo, D.E., Akiyama, K., Imamiya, S. 1990. A mechanism of cadmium poisoning: The cross effect of calcium and cadmium in the calmodulin-dependent system. *Archives of Toxicology* 64: 161–164.

Umrania, V.V. 2006. Bioremediation of toxic heavy metals using acidothermophilic autotrophes. *Bioresource Technology* 97: 1237–1242.

Van Der Heijden, M.G., Bardgett, R.D., Van Straalen, N.M. 2008. The unseen majority: Soil microbes as drivers of plant diversity and productivity in terrestrial ecosystems. *Ecology Letters* 11: 296–310.

Vigo, J.B., Ellzey, J.T. 2006. Effects of arsenic toxicity at the cellular level: a review. *Texas Journal of Microscopy* 37: 45–49.

Viti, C., Pace, A., Giovannetti, L. 2003. Characterization of Cr(VI)-resistant bacteria isolated from chromium-contaminated soil by tannery activity. *Current Microbiology* 46: 0001–0005.

Wagner-Döbler, I., Lünsdorf, H., Lübbehüsen, T., Von Canstein, H.F., Li, Y. 2000. Structure and species composition of mercury-reducing biofilms. *Applied and Environmental Microbiology* 66: 4559–4563.

Wang, Q., Kim, D., Dionysiou, D.D., Sorial, G.A., Timberlake, D. 2004. Sources and remediation for mercury contamination in aquatic systems—A literature review. *Environmental Pollution* 131: 323–336.

Wang, X., Andrews, L., Riedel, S., Kaupp, M. 2007. Mercury is a transition metal: The first experimental evidence for HgF_4. *Angewandte Chemie International Edition* 46: 8371–8375.

Westing, A.H. 1971. Forestry and the war in South Vietnam. *Journal of Forestry* 69: 777–783.

White, C., Sayer, J.A., Gadd, G.M. 1997. Microbial solubilization and immobilization of toxic metals: Key biogeochemical processes for treatment of contamination. *FEMS Microbiology Reviews* 20: 503–516.

World Health Organisation. 2015. Ten chemicals of major public health concern. http://www.who.int/ipcs/assessment/public_health/chemicals_phc/en/.

Zhang, L., Wong, M.H. 2007. Environmental mercury contamination in China: Sources and impacts. *Environment International* 33: 108–121.

Section II

Metal–Microbe Interactions

7 Metal–Microbe Interactions and Microbial Bioremediation of Toxic Metals

Gülşad Uslu Senel and Ozge Hanay

CONTENTS

ABSTRACT

Heavy metals are natural components of the earth's crust and are widely distributed in the environment, but indiscriminate use in industries for various human purposes has distorted their geochemical cycles and biochemical balance. Industries often release excess amount of heavy metals such as cadmium, copper, lead, nickel, and zinc into natural soil and aquatic environments. Exposure to such heavy metals for a prolonged period may have deleterious effects on human life and aquatic biota. Microorganisms play a crucial role in the remediation of heavy metals by biotransforming them into nontoxic forms. Understanding the biochemical and molecular mechanism of metal accumulation has several biotechnological implications for the restoration of metal-contaminated sites. In view of this, this chapter explores the abilities of microorganisms to resist, accumulate, and transform heavy metals into innocuous forms. This chapter summarizes fundamental insights regarding genetic and molecular basis of metal tolerance in microbes, with special reference to the metal-binding genes involved in tolerance and detoxification. These strategies can be further utilized to overcome the bottlenecks associated with the heavy metal remediation process.

7.1 INTRODUCTION

Heavy metal pollution is one of the utmost important environmental concerns these days. The occurrence of heavy metals in industrial effluents is a major environmental and ecological hazard. Heavy metals such as copper, cadmium, and lead have been extensively used in the industry. They are considered as the potentially toxic metals with cumulative toxicity to humans and various aquatic organisms (Uslu et al., 2003; Malik, 2004).

Metal contaminants are commonly found in soils, sediments, and water matrix. Metal pollutants can be produced through industrial processes such as mining, ore processing, smelting, refining, and electroplating. A crucial factor in the remediation of metals is that metals are nonbiodegradable but can be made less toxic by changing their valence state configuration and transforming them through sorption, methylation, and complexation. These conversions alter the movement and affect the bioavailability of metals. At low concentrations, metals can serve as important components in life processes. They play a crucial role in enzyme functioning by acting as their cofactors. However, above certain threshold concentrations, metals can become toxic to many species.

Unfortunately, microorganisms can affect the reactivity and mobility of metals. Microorganisms affecting metal reactivity and mobility can be used to detoxify some metals and prevent further metal contamination (Adeniji, 2004).

Several methods are available for removing heavy metals from waste streams, including chemical precipitation, membrane filtration, ion exchange, and carbon adsorption (Uslu et al., 2003, 2011; Uslu and Tanyol, 2006). These traditional methods are often ineffective and/or very expensive when used for the removal of heavy metal ions at very low concentrations, using of microorganisms (for removal of heavy metals) offers a potential alternative (Dursun et al., 2003).

Bioremediation is the most effective management process and is an attractive and successful cleaning technique for a polluted environment. It offers a cost-effective remediation technique, compared to other remediation methods. Bioremediation is a natural process and does not produce toxic by-products. Bioremediation results in complete mineralization of the contaminants in the environment, thus providing a permanent solution for restoring metal-contaminated sites (Nandini et al., 2015). Almost all metal–microbe interactions have been studied as a means for eliminating, detoxifying, and recovering inorganic and organic metal or radionuclide pollutant. Bioremediation technology uses microorganisms to reduce, eliminate, and transform contaminants present in soils, sediments, water, and air to benign products (Tabak et al., 2005). The microorganisms involved in this process may belong to bacteria, fungi, yeast, and algae (Kulshreshtha et al., 2014). Various genes present in the microorganisms and the biochemical reactions catalyzed by them result in activity, growth, and reproduction of that organism.

Mechanisms governed by microorganisms by which they remediate heavy metals include biosorption (metal sorption to cell surface by physiochemical mechanisms), bioleaching (heavy metal mobilization through the excretion of organic acids or methylation reactions), biomineralization (heavy metal immobilization through the formation of insoluble sulfides or polymeric complexes), intracellular accumulation, and enzyme-catalyzed transformation (redox reactions) (Lloyd, 2002). Generally, the sites contaminated with heavy metals are the sources of metal-tolerant microorganisms (Gadd, 1993).

Bioremediation techniques are divided into three categories: *in situ*, *ex situ* solid, and *ex situ* slurry. With *in situ* techniques, the soil and associated ground water is treated in place without excavation, while it is excavated prior to treatment with *ex situ* applications.

This chapter is aimed to describe metal–microbe interactions and bioremediation and also its principles and applications for toxic metals in detail.

7.2 BIOREMEDIATION

Bioremediation technology utilizes microorganisms to reduce, eliminate, contain, or transform to benign product contaminants present in soils, sediments, water, or air. According to

EPA, bioremediation is a "treatment that uses naturally occurring organisms for disrupting hazardous substances into less toxic or nontoxic substances." Bioremediation is not a new technology. For example, both composting of agricultural material and sewage treatment of household waste are based on the use of microorganisms to catalyze chemical transformation. Such environmental technologies have been practiced by humankind since the beginning of chronicled history.

Bioremediation is an alternative to traditional remediation technologies such as landfilling or incineration. Bioremediation is often caused by the indigenous microorganisms that exist at the polluted site. However, these practices may consume very long time, and some processes cannot be carried out by the indigenous population. Bioremediation depends on the presence of the appropriate microorganisms in the correct amounts and combinations and on the appropriate environmental conditions. Although prokaryotes—Bacteria and Archaea—are usually the agents accountable for most bioremediation strategies, eukaryotes such as fungi and algae can also transform and degrade contaminants. Microorganisms already living in contaminated environments are often well adapted to survival in the presence of existing contaminants and to the temperature, pH, and oxidation–reduction potential of the site. Various bioremediation techniques for the removal of toxic metals have been studied, and they are mainly confined to the terrestrial and freshwater environments (Das et al., 2008). However, the characteristic feature of a marine environment is unique in nature, and the microorganisms dwelling in those environmental conditions have enormous potential to overcome the incessant changing pattern of pH, temperature, salinity, sea temperature, and other variable parameters (Giller et al., 1998; Dash et al., 2013). The phytoremediation ability depends on the ion uptake mechanism of each species depending on their genetic, physiological, anatomical, and morphological characteristics (Rahman and Hasegawa, 2011).

According to EPA (2001, 2002), on the basis of removal and transportation of wastes for treatment, there are basically two methods: *in situ* bioremediation and *ex situ* bioremediation.

7.2.1 IN SITU BIOREMEDIATION

In situ bioremediation is no need to excavate or remove soils or water in order to accomplish remediation. The pollution is eliminated directly at the place where it occurs or at the site of contamination so it may be less expensive, creates less dust, and is possible to treat a large volume of soil and cause less release of contaminants. *In situ* biodegradation comprises providing oxygen and nutrients by circulating aqueous solutions through contaminated soils to stimulate naturally occurring bacteria to degrade organic contaminants. It can be used for soil and groundwater (Girma, 2015).

Most often, *in situ* bioremediation is a practical approach for the degradation of contaminants in saturated soils and groundwater. It is a superior method to cleaning contaminated environments since it is cheaper and uses harmless microbial organisms to degrade the chemicals and also a safer method in degrading harmful compounds.

In situ bioremediation can be two types. These are intrinsic bioremediation and engineered *in situ* bioremediation. *In situ* bioremediation deals with stimulation of indigenous or naturally occurring microbial inhabitants by feeding them nutrients and oxygen to increase their metabolic activity, whereas engineered *in situ* bioremediation method involves the introduction of certain microorganisms to the site of contamination. When site conditions are not suitable, engineered systems have to be introduced to that particular site. Engineered *in situ* bioremediation quickens the degradation process by enhancing the physicochemical conditions to encourage the growth of microorganisms. Oxygen, electron acceptors, and nutrients (nitrogen and phosphorus) promote microbial growth (Evans and Furlong, 2003). For *in situ* remediation, it is important that the remediation process be as noninvasive and environmentally benign as possible if the end product is intended to be a healthy productive ecosystem.

In situ bioremediation is a technique that can be used to decrease the distribution of metal contaminants by applying biological treatment to harmful chemicals present in soil and groundwater (Adeniji, 2004). *In situ* bioremediation has the ability to transform contaminants to less toxic compounds, making this a promising environmental cleanup technique (Adeniji, 2004). It accelerates contaminant desorption and dissolution by treating pollutants close to their source. Methods such as pump and treat only remove or destroy contaminants in groundwater, but not those contaminants sorbed in soil or solids in the aquifer (Adeniji, 2004). Although the use of bioremediation is rapidly growing, *in situ* bioremediation of metals is not widely understood (Adeniji, 2004). The many intricacies of microbial metabolism related to metals have not been grasped completely because effective usage of microbes to clean up metals is a complex matter. Furthermore, the process of verifying whether or not complete bioremediation has actually occurred (meaning the metal contaminant changed forms through abiotic chemical reactions) is an issue researchers are still struggling with in the labs. Demonstrating that lab-grown microbes have the potential to degrade a pollutant is not enough to sufficiently prove that microorganisms can completely clean up a site (Adeniji, 2004). Moreover, there is still a need for researchers to verify that the bioremediation of metal pollutants is a permanent process.

7.2.1.1 Types of *In Situ* Bioremediation

7.2.1.1.1 Intrinsic Bioremediation

Intrinsic bioremediation manages the distinctive abilities of naturally occurring microbial communities to degrade environmental pollutants without taking any engineering steps to enhance the process. This approach deals with stimulation of indigenous or naturally occurring microbial inhabitants by feeding them nutrients and oxygen to increase their metabolic activity.

7.2.1.1.2 Engineered In Situ Bioremediation

The second approach involves the introduction of certain microorganism to the site of contamination. Engineered *in situ* bioremediation accelerates the degradation process by enhancing the physicochemical conditions to encourage the growth of microorganism.

7.2.2 Ex Situ Bioremediation

This process requires excavation of contaminated soil or pumping of groundwater to facilitate microbial degradation. This technique has more disadvantages than advantages. *Ex situ* bioremediation practices involve the excavation or removal of contaminated soil from ground. Depending on the state of the pollutant to be removed, *ex situ* bioremediation is classified as solid-phase and slurry-phase systems.

7.2.2.1 *Ex Situ* Bioremediation Techniques

Ex situ bioremediation techniques are usually aerobic and involve treatment of contaminated soils or sediments using slurry- or solid-phase systems.

7.2.2.2 Slurry-Phase Bioremediation

Slurry-phase bioremediation is a batch treatment technique in which excavated soils/sediments are mixed with water and treated in reactor vessels or in contained ponds or lagoons (Blackburn and Hafker, 1993; Troy, 1994). Effective bioremediation has been obtained with slurry-phase systems for soils and sediments contaminated with a wide range of organic compounds, including pesticides, petroleum hydrocarbons, pentachlorophenol, polychlorinated biphenyls, creosote coal tars, and wood-preserving wastes (Yare, 1991; Troy, 1994). Slurry-phase bioremediation can take place on-site or the soil can be removed and transported to a remote location for treatment (USEPA, 1990). The process generally takes place in a tank or vessel (a "bioreactor") but can also take place in a lagoon. During treatment, the oxygen and nutrient content, pH, and temperature of the slurry are adjusted and maintained at levels suitable for aerobic microbial growth. When the desired level of treatment has been achieved, the unit is emptied and a second volume of soil is treated.

7.2.2.3 Solid-Phase Bioremediation

Solid-phase bioremediation includes land farming (soil treatment units), compost heaps, and engineered biopiles (Litchfield, 1991; Field et al., 1995). In this system, soil is treated in aboveground treatment areas equipped with collection systems to prevent any contaminant from escaping the treatment. Moisture, heat, nutrients, and oxygen are controlled to enhance bioremediation for the application of this treatment. These systems are relatively simple to operate and maintain, require a large amount of space, and cleanups require more time to complete than slurry-phase processes. Solid-phase bioremediation can be implemented in any of the following ways.

7.2.2.4 Contained Solid-Phase Bioremediation

Where the excavated soils are not slurred with water; the contaminated soils are simply blended to achieve a homogeneous texture. Occasionally, textural or bulk amendments, nutrients, moisture, pH adjustment, and microbes are added. The soil is then placed in an enclosed building, vault, tank, or vessel. In addition, since the soil mass is enclosed, rainfall and runoff are eliminated, and volatile organic carbon emissions can be controlled.

7.2.2.5 Composting

This method is similar to contained solid-phase bioremediation, but it does not employ added microorganisms if carried out in an enclosed vessel. It is usually conducted outdoors rather than in an enclosed space. The two basic types of unenclosed composting are open and static windrow systems. In open windrow systems, the compost is stacked in elongated piles, whereas in static windrow systems, the piles are aerated by a forced air system. The waste is not protected from variation in natural environmental conditions, such as rainfall and temperature fluctuations.

Composting is a technique that involves uniting contaminated soil with nonhazardous organic amendments such as manure or agricultural wastes. The presence of these organic resources supports the development of a rich microbial population and raised temperature characteristic of composting (Cunningham and Philip, 2000; Girma, 2015).

7.2.2.6 Land Farming

Land farming is a simple technique in which contaminated soil is mined and spread over a prepared bed and periodically tilled until pollutants are degraded. Since land farming has the potential to reduce monitoring and maintenance costs, as well as cleanup liabilities, it has received much attention as a disposal alternative (EPA, 2003). The soil is spread in thin lifts up to half inch thick. The soil is tilled periodically thereby providing oxygen. Microorganisms, nutrients, and moisture may also be added. Clay or plastic liners may be installed in the field prior to the placement of the contaminated soil, which prevents the leaching of the contaminants into groundwater. Treatment is achieved through biodegradation, in combination with aeration and possibly photooxidation in sunlight. These processes are most active in warm, moist, sunny conditions and are completely diminished or arrested during winter months when temperature is cold and snow covers the ground.

7.2.2.7 Biopiles

Biopiles are a hybrid of land farming and composting. Essentially, engineered cells are constructed as aerated composted piles. Characteristically used for the treatment of surface contamination with petroleum hydrocarbons, they are a refined form of land farming that incline to control physical losses of the contaminants by leaching and volatilization. Biopiles provide a favorable environment for indigenous aerobic and anaerobic microorganisms (EPA, 2003). Slurry-phase bioremediation is a relatively more rapid process compared to the other treatment processes. Contaminated soil is combined with water and other additives in a large tank called a bioreactor and mixed to keep the microorganisms, which are already present in the soil, in contact with the contaminants in the soil. Nutrients and oxygen are added and conditions in the bioreactor are controlled to create the optimum environment for the microorganisms to reduce the contaminants. When the treatment is accomplished, the water is removed from the solids, which are disposed of or treated further if they still contain pollutants (Cunningham and Philip, 2000; Pal et al., 2010; Girma, 2015).

7.2.2.8 Bioreactor

Bioreactor is a containment vessel and apparatus used to create a three-phase (solid, liquid, and gas) mixing condition to increase the bioremediation rate of soil-bound and water-soluble pollutants as water slurry of the contaminated soil and biomass capable of degrading target contaminants. In general, the rate and extent of biodegradation are greater in a bioreactor system than in *in situ* or solid-phase systems because the contained environment is more manageable and hence more controllable and predictable. Despite the advantages of reactor systems, there are some disadvantages. The contaminated soil requires pretreatment or alternatively the contaminant can be stripped from the soil via soil washing or physical extraction before being placed in a bioreactor (EPA, 2003).

7.2.3 Advantages and Disadvantages of Bioremediation

Developed economies made higher use of low-cost *in situ* bioremediation technologies such as monitored natural attenuation, while their emerging counterparts appeared to focus on occasionally more expensive *ex situ* technologies. Bioremediation techniques are typically more economical than traditional approaches such as incineration, and some pollutants can be treated on-site, thus decreasing exposure risks for cleanup personnel or potentially wider exposure as a result of transportation accidents. Since bioremediation is based on natural attenuation, the public considers it more acceptable than other technologies. Table 7.1 shows the advantages and disadvantages of bioremediation.

7.3 MICROORGANISMS AND HEAVY METALS

Microorganisms have evolved various mechanisms of metal resistance, and scientists have tried to exploit genetic/metabolic basis of all such mechanisms for the production of superior strains. Silver et al. (2001) reported that although chromosomal genes may be involved, bacterial resistance to heavy metals is often conferred by-products of genes simulated on plasmids rendering genetic manipulation for strain improvement easy and feasible.

A possible way to confer resistance to heavy metals on microorganisms is to progressively increase the concentration of the toxicants microbes are exposed to and perceive microbial metabolism and proliferation. Scientists have explored this attribute to ascertain if tolerance of higher concentrations can consequently aid in biodegradation of such toxicants. Donmez and Aksu (2001b) tried to adapt strains of

TABLE 7.1

Advantages and Disadvantages of Microbial Bioremediation

Advantages	Disadvantages
Bioremediation is considered to be a natural process, and therefore, it is accepted as a waste treatment process for a contaminated environment.	Bioremediation can be applied to only those compounds that can be biodegraded.
Bioremediation is proven to be less expensive than other technologies that are used for remediation purposes.	All type of compounds cannot undergo bioremediation process.
Bioremediation can be carried out on-site without causing any problem of normal activities.	There are some chances that the products of biodegradation may be more toxic than the parent compound.
The end products of the treatment are usually nontoxic and include water, bacterial biomass, and carbon dioxide.	Biological processes are highly specific. Important factors are required for the complete result of the process, which includes the microbial populations capable of degrading the contaminants, site condition, suitable growth conditions for microbes, and appropriate levels of nutrients and contaminants.
Since there is a complete destruction of contaminants, this eliminates the chance of future liability associated with treatment and disposal of contaminated material.	It is difficult to extrapolate from bench- and pilot-scale studies to full-scale field operations.
The complete destruction of target pollutants is possible.	Contaminants may be present as solids, liquids, and gases.
As compared with other treatment processes, this technique relatively uses less equipment.	Bioremediation consumes more time compared to other treatment options.

Candida species (isolated from sewage) to copper and nickel by serial subcultures in copper- and nickel-supplemented growth medium. Adapted cells could grow well in the presence of higher metal concentrations, while nonadapted ones perished. Moreover, the specific metal uptake capacity and metal removal (%) by adapted cells were higher than the nonadapted cells at all the concentrations tested. Although the authors could not explicate the mechanism of adaptation, they inferred constitutive synthesis of metallothionein (other copper-binding proteins or changes in the genetic makeup) as one of the aspects in adaptation. Baillet et al. (1997) adapted *Thiobacillus ferrooxidans* strain via successive exposure to higher concentrations of cadmium. The biomass (living but not growing) of adapted strain had cadmium uptake capacity of 0.31 g/g dry weight as compared to 0.21 g/g dry weight in the case of nonadapted cells.

7.3.1 METAL–MICROORGANISM INTERACTIONS

The way microbes interact with metals depends in part on whether the organisms are prokaryotic or eukaryotic. Both types of microbes have the capability to bind metal ions present in the external environment at the cell surface or to transport them into the cell for various intracellular functions. In some cases, these processes involve highly specific biochemical pathways that have evolved to protect the microbial cell from toxic heavy metals.

Bioremediation is a complex interaction of biological, chemical, and physical processes; a detailed understanding of different mechanisms is required in order to utilize the most suitable mechanism in different conditions in different contaminated sites. Although there are many ways for microbes to interact with toxic metals, due to such complexity, many of these mechanisms are still under investigation (Tabak et al., 2005).

Microorganisms can interact with metals via many mechanisms some of which may be used as the basis of potential bioremediation approaches. The major types of interaction are summarized in Figure 7.1. In addition to the mechanisms outlined, accumulation of metals by plants (phytoremediation) warrants attention as an additional conventional route for the bioremediation of metal contamination. In some cases, these processes involve extremely specific biochemical pathways that have evolved to protect the microbial cell from toxic heavy metals. Other mechanisms of potential commercial importance depend on the production of biogenic ligands that can complex metals, resulting in their mobilization from contaminated soils. The mobilized metals can then be pumped out of the soil or sediment and trapped in a bioreactor on the surface (Lloyd, 2002; Tabak, 2005).

7.3.2 METAL TOXICITY TO MICROORGANISMS

Microorganisms have developed numerous mechanisms of metal resistance, and scientists have tried to exploit genetic/metabolic basis of all such mechanisms for the production of superior strains (Adams et al., 2014). Metals play an important role in the life processes of microbes. Some metals such as chromium (Cr), calcium (Ca), magnesium (Mg), manganese (Mn), copper (Cu), sodium (Na), nickel (Ni), and zinc (Zn) are essential as micronutrients for various metabolic functions and for redox functions. Other metals have no biological role, for example, cadmium (Cd), lead (Pb), mercury (Hg), aluminum (Al), gold (Au), and silver (Ag). They are nonessential and potentially toxic to soil microbes. Some of them, for example, Cd^{2+}, Ag^{2+}, and Hg^{2+}, tend to bind the SH groups of enzymes and inhibit their activity (Turpeinen, 2002).

Soil contamination by heavy metals may suppress or even kill parts of the microbial community in soil. Interactions of metals

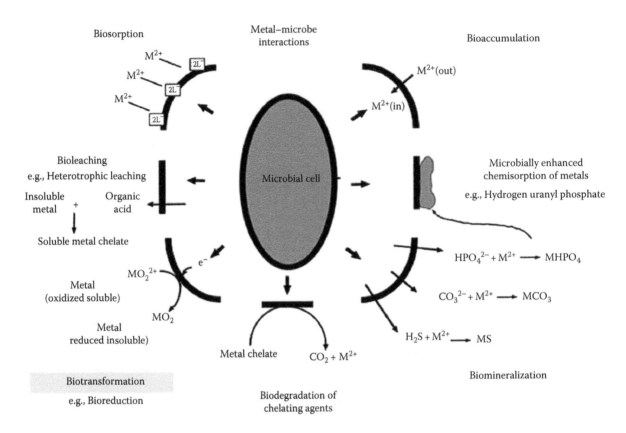

FIGURE 7.1 Major mechanisms of microremediation. (From Tabak, H.H. et al., *Rev. Environ. Sci. Biotechnol.*, 4, 115, 2005. With permission.)

with cellular proteins/enzymes are more commonly involved in causing toxicity than interaction with membranes. Binding affects the structure and function of proteins and enzymes.

Bioremediation is the naturally occurring process in which microorganisms or plants either restrain or transform environmental contaminants to innocuous state end products (Mueller, 1996). During bioremediation, microbes utilize chemical contaminants in the soil as an energy source, and through redox potential, they can metabolize the target pollutant into usable energy for microbes. Although multitudes of reactions are adopted by microbes to degrade and transform pollutants but all the energy yielding reactions are oxidation-reduction reactions and the typical electron acceptors are oxygen, nitrates, sulfate and carbon dioxide. For bioremediation, it is important that effective microorganisms and plants may reduce the pollutants into harmless products by various enzymatic actions. Microbes cannot degrade heavy metals directly, but they can alter the valence states of metals, which may convert them into immobile or less toxic forms.

At high concentrations, either metal ions can completely inhibit the microbial population by inhibiting their various metabolic activities or organisms can develop resistance or tolerance to the elevated levels of metals. Unlike many other pollutants, metals cannot undergo biodegradation and produce less toxic, less mobile, and/or less bioavailable products; heavy metals are difficult to be removed from contaminated environment. These metals cannot be degraded biologically

and are ultimately interminable, though the speciation and bioavailability of metals may change with variation in the environmental factors. Some metals such as zinc, copper, nickel, and chromium are vital or beneficial micronutrients for plants, animals, and microorganisms (Olson et al., 2001), while others (e.g., cadmium, mercury, and lead) have no known biological and/or physiological functions (Gadd, 1992). However, the higher concentration of these metals has great effects on the microbial communities in soils in several ways: (1) it may lead to a reduction of total microbial biomass (Chaudri et al., 1993), (2) it decreases numbers of specific populations, or (3) it may alter microbial community structure (Gray and Smith, 2005). Thus, at high concentrations, either metal ions can completely inhibit the microbial population by inhibiting their various metabolic activities like protein denaturation, inhibition of cell division, and cell membrane disruption or organisms can develop resistance or tolerance to the elevated levels of metals. Thus, at high concentrations, either metal ions can completely inhibit the microbial population by inhibiting their numerous metabolic activities (Figure 7.2) or organisms can develop resistance or tolerance to the elevated levels of metals.

In contrast, resistance is the ability of microbes to survive in higher concentrations of toxic substances by detoxification mechanisms, activated in direct response to the presence of the same pollutant (Ahemad, 2012). Toxic heavy metals, therefore, need to be either completely removed from the contaminated soil, transformed, or immobilized, producing

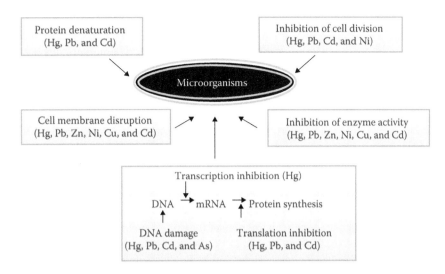

FIGURE 7.2 Heavy metal toxicity mechanism to microorganism. (Modified from Ahemad, M., *IIOABJ*, 3, 39, 2012.)

much less or nontoxic species. However, in order to endure and proliferate in metal-contaminated soils, tolerance has to be present both in microbes and their associative hosts (Ahemad et al., 2009).

7.3.3 MICROORGANISMS USED IN BIOREMEDIATION

The bioremediation practices may be conducted by the autochthonous microorganisms, which naturally inhabit the soil/water environment undergoing purification, or by other microorganisms, which derive from different environments. There are a number of microorganisms that can be used to remove metal from the environment, such as bacteria, fungi, yeast, and algae (White et al., 1997). Microorganisms can be isolated from almost any environmental conditions. Microbes can adapt and grow at subzero temperatures, as well as extreme heat, desert environments, in water, with an excess of oxygen and in anaerobic conditions, with the presence of precarious compounds, or on any waste stream. Because of the adaptability of microbes and other biological systems, these can be used to degrade or remediate environmental hazards. The main requirements are an energy source and a carbon source (Vidali, 2001). Because of the adaptableness of microbes and other biological systems, these can be used to degrade or remediate environmental hazards. Natural organisms, either indigenous or extraneous (introduced), are the prime agents used for bioremediation (Prescott et al., 2002). The organisms that are utilized vary, depending on the chemical nature of the polluting agents, and are to be selected carefully as they only survive within a limited range of chemical contaminants (Prescott et al., 2002; Dubey, 2004). Since numerous types of pollutants are to be encountered in a contaminated site, diverse types of microorganisms are likely to be required for effective mediation (Watanabe et al., 2001). The first patent for a biological remediation agent was registered in 1974, being a strain of *Pseudomonas putida* that was able to degrade petroleum (Prescott et al., 2002; Glazer and Nikaido, 2007).

7.4 BIOREMEDIATION APPLICATION

Microbial bioremediation is defined as the process by which microorganisms are stimulated to rapidly degrade the hazardous contaminants to environmentally safe levels in soil, subsurface materials, water, sludge, and residues. Microbial activity is proved to play an important role in remediating metals in soil residues. Most reports on this technique are restricted to laboratory studies. *Deinococcus geothermalis* was examined for the microbial bioremediation of mercury in radioactive wastes (Brim et al., 2003), synergistic use of combined bacterial plant processes has also been reported for the cleaning of the environment from mercury (Glick, 2003), a metal-resistant *Cupriavidus metallidurans* strain was examined for its potential to bioremediate mercury (Rojas et al., 2011), and hydrocarbon-utilizing haloarchaea was reported to remediate mercury under hypersaline conditions (Al-Mailem et al., 2011). Microbial surfactant was also reported to be beneficial in the bioremediation of mercury (Singh and Cameotra, 2004). Geiger et al. (1993) studied the influence of different remediation methods on the heavy metal uptake by *Lactuca sativa* at Dornach (NW, Switzerland), where the soil is calcareous and heavy metal polluted. As the different methods used were not satisfactory regarding the normal development of the plant and its heavy metal content, they suggested to use heavy metal–tolerant plants or physiologically adapted plants to high tissue concentrations to stabilize or remove the heavy metals, respectively. Adhikari et al. (2004) also defined bioremediation as a process of cleaning up hazardous wastes with microorganisms or plants and as the safest method of clearing soil of pollutants.

Nanda and Abraham (2013) have evaluated the effect of As, Cr, Mg, and Cu on some essential soil bacteria such as *Azotobacter*, *Pseudomonas*, and *Rhizobium*. They found As to be the most toxic of all followed by Cr, Mg, and Cu. The interaction between microorganisms, plant roots, and amendment might have a greater impact on both the increase of nutrient uptake and migration of metal uptake (Smith, 1994).

As a result of plant root microbial interaction, the migrations of contaminants to groundwater are reduced by immobilization. Establishing a vegetative cover on contaminated sites can retain contaminants in place, thus reducing their loss via erosion and percolation into the soil profile (Pulford and Watson, 2003). When revegetation of contaminated soil is combined with soil amendments, such as organic matter, the mobility of contaminants in the soil can be further reduced (Mench et al., 2000).

Iwahori et al. (2000) found that normal bacterial community was sensitive to 0.7 mM of Hg_2^+ concentration and the most resistant *T. ferrooxidans*[SUG 2.2] could grow in an autotrophic medium supplemented with Fe^{2+} at pH 2.5 containing 6 mM Hg^{2+} concentration, where the amount of mercury volatilized by resistant cells increased to 62% when Fe^{2+} was added. Recent wastewater bioremediation techniques for heavy metal removal have focused on biosorptive mechanisms using nonviable microbial biomass rather than bioaccumulation by living microbial cells (Donmez and Aksu, 1999).

Umrania (2006) investigations were carried out to isolate microbial strains from soil, mud, and water samples from a metallurgically contaminated environment for bioremediation of toxic heavy metals. As a result of primary and secondary screening, various 72 acidothermophilic autotrophic microbes were isolated and adapted for metal tolerance and biosorption potentiality. The multimetal tolerance was developed with higher gradient of concentrations of Ag, As, Bi, Cd, Cr, Co, Cu, Hg, Li, Mo, Pb, Sn, and Zn. The isolates were checked for their biosolubilization ability with copper containing metal sulfide ores. In case of chalcopyrite 85.82% and in coverlet as high as 97.5%, copper solubilization occurred in the presence of 10^{-3} M multi–heavy metals on the fifth day at 55°C and pH 2.5. Chemical analyses were carried out by inductively coupled plasma spectroscopy for metal absorption. The selected highly potential isolate (ATh-14) displayed a maximum adsorption of Ag 73%, followed by Pb 35%, Zn 34%, As 19%, Ni 15%, and Cr 9% in chalcopyrite.

Silver et al. (2001) reported that although chromosomal genes may be involved, bacterial resistance to heavy metals is often conferred by-products of genes simulated on plasmids rendering genetic manipulation for strain improvement easy and feasible. Donmez and Aksu (2001) tried to adapt strains of *Candida* species (isolated from sewage) to copper and nickel by serial subcultures in copper- and nickel-supplemented growth medium. Adapted cells could grow well in the presence of higher metal concentrations, while nonadapted ones perished. Moreover, the specific metal uptake capacity and metal removal (%) by adapted cells were higher than the nonadapted cells at all the concentrations tested. Although the authors could not explain the mechanism of adaptation, they implied constitutive synthesis of metallothionein (other copper-binding proteins or changes in the genetic makeup) as one of the factors in adaptation. Other nonconventional techniques reported by Soares et al. (2002) are the comparison of flocculent strain and nonflocculent strain of *Saccharomyces cerevisiae* for metal bioremediation. They inferred that flocculent strain S646-1b accumulated more copper than nonflocculent S646-8D strain in the first 10 minutes of contact with the metal. The authors attributed this to the presence of additional metal binding on the cell surface of flocculent strains. Katarina et al. (2004) reported the removal of arsenic from contaminated soil using bioremediation. This research corroborates the fact that arsenic can be reduced using a biological treatment scheme. The reduction in arsenic might be as a result of the presence of *Bacillus* species in the soil samples that have high adsorptive ability due to high peptidoglycan and teichoic acid content in their cell walls, which enables an ion-exchange reaction.

Studies on the remediation of cadmium conducted by Ike et al. (2007) report a technique of successfully remediating cadmium-contaminated soil using symbiosis between leguminous plants and recombinant rhizobia. Bagot et al. (2005) selected bacteria (*Bacillus*), fungus, and actinomycetes for bioremediation of cadmium-contaminated agricultural soils. Their results indicate that *Bacillus* and actinomycetes effectively reduced cadmium. The slight reduction of cadmium concentration in this study suggests a necessity for optimization of the biochemical process of cellular sequestration and growth of *Bacillus* sp. (Adams et al., 2014).

Cheung et al. (2003) reported that the microbial reduction of chromium is commonly catalyzed by soluble enzymes with *Pseudomonas* spp. and *Escherichia coli* capable of secreting chromium reductases. They also reported that through the utilization of membrane-associated reductases, *Pseudomonas maltophilia* and *Bacillus megaterium* could detoxify chromium (Adams et al., 2014). The presence of *Pseudomonas* sp., *Bacillus* sp., and *E. coli* and subsequent decrease in the concentration of chromium during the experiment is in agreement with the findings of these researchers. Santhosh (2008) studied the biosorption of heavy metals by *Paenibacillus polymyxa* and reported that the heavy metal cobalt was reduced to up to 90% primarily due to the binding of the metal to the cell walls through the extracellular polymeric substances and precipitation of sulfides. This study identified the presence of *E. coli* and the possibility of biosorption to the cell surfaces and following reduction in the concentration of cobalt.

Ghodsi et al. (2011) studied bioremediation by isolating arsenate-resistant bacteria from arsenic-contaminated soil and the investigation of arsenite bioremediation efficiency by the most resistant isolates. The maximum percentage of arsenite removal potential (92%) and arsenite bioaccumulation (36%) by *Bacillus macerans* was found. Narayanan et al. (2011) studied the bioremediation on effluents from magnesite and bauxite mines using *Thiobacillus* spp. and *Pseudomonas* spp. The results of biosorption process showed the *T. ferrooxidans* reduced/absorbed some heavy metals from mines (Cd, Ca, Zn, Cr, Mn, and Pb) and *Pseudomonas aeruginosa* absorbed most of the metals than *T. ferrooxidans*. Both species effectively absorbed Cd, Ca, and Zn followed by Pb. Iqbal and Edyvean (2007) conducted a study on the ability of loaf a sponge-immobilized fungal biomass to remove lead ions from aqueous solution. A new biosorbent was established by immobilizing a white rot basidiomycete *Phanerochaete chrysosporium* within a low-cost and easily available matrix of loofa sponge. There are some nonconventional approaches to strain selection. Although *S. cerevisiae* is often used for metal bioremediation, recently reported comparison of a flocculent and nonflocculent strain for Cu^{2+} removal appears to be a novel and interesting approach (Soares et al., 2002).

Boricha and Fulekar (2009) found that *Pseudomonas plecoglossicida* is a novel organism for the bioremediation of cypermethrin while *P. aeruginosa*, *Bacillus* sp., *Streptomyces* sp., and *Pseudomonas fluorescens* were effective against chromium (Akhtar et al., 2013). Similarly, fungal biosorbents include *Aspergillus*, and Pan et al. (2009) observed similar results using *Penicillium* and *Fusarium* as bioremediation agents, as well as Ting and Choong (2009) in their comparison between the ability of a *Trichoderma* isolate to bioaccumulate and biosorb. The identification of microorganisms that can catalyze the reduction of Cr(VI) to Cr(III) has led to the suggestions that bioremediation of Cr(VI)-contaminated effluents would be a low-cost, low-technology alternative for presently used methods (Ohtake and Silver, 1994).

Microbial bioremediation is defined as the process by which microorganisms are stimulated to rapidly degrade the hazardous contaminants to environmentally safe levels in soil, subsurface materials, water, sludge, and residues. Microbial activity is proved to play an important role in remediating metals in soil residues. Studies on interaction of microorganisms with heavy metals have an increasing interest in the past 20 years. Microbial metal uptake can either occur actively (bioaccumulation) or passively (biosorption). A study carried out by Irma et al. (2013) revealed that the *Aspergillus fumigatus* fungal isolated from a contaminated site has good biosorption capacity toward selected heavy metals. Vargas et al. (2009) showed efficient detoxification of multipolluted heavy metals by fungi isolated from compost.

Collins and Stotzky (1989) stated that several metals are essential for biological systems and present in a range of certain concentration. The metals have been always associated with metalloproteins and enzymes as cofactors. Its low concentrations lead to a decrease in metabolic activity, while high concentrations could act in a deleterious way by blocking essential functional groups, displacing other metal ions, or modifying the active conformation of biological molecules. Besides, they are toxic for both higher organisms and microorganisms.

White et al. (2006) reported that bacterial remediation is the process of using metal-reducing bacteria to break down the contaminants. The metal-reducing bacteria are able to reduce very toxic soluble forms into less toxic forms. Macaskie et al. (2005) found that sulfate-reducing bacteria successfully treat the metal leachates generated by sulfuric-acid producing *Thiobacillus* sp. However, Meysami and Baheri (2003) reported that white rot fungi have the ability to transform the pollutants from the contaminants in soil through ligninolytic enzymes. Moreover, Lang et al. (1998) have successfully used the production and activity of ligninolytic enzymes in the

bioremediation of contaminated soil from under field conditions. According to Bindler et al. (1999), lead contamination is mainly restricted to surface soil in boreal forests, which are rich in humus contents. However, sometimes it is also possible that lead forms a complex with dissolved organic matter and migrates from the surface soil layer to mineral soil and possibly contaminates the underground water.

Boonchan et al. (2000) and Joanne et al. (2008) reported that optimum pH for bioremediation is between 6.0 and 8.9. They opined that changes from initial levels of pH could be as a result of the release of acidic and alkaline intermediates and final products during biodegradation of hydrocarbons, which has an effect on the pH. The findings of this research indicate that pH level fell within a suitable range to support microbial growth in all the treatment categories and control. Katarina et al. (2004) reported the removal of arsenic from contaminated soil using bioremediation. This research corroborates the fact that arsenic can be reduced using a biological treatment scheme. The decrease of arsenic might be as a result of the occurrence of *Bacillus* species in the soil samples that have high adsorptive capacity due to high peptidoglycan and teichoic acid content in their cell walls, which facilitates an ion-exchange reaction.

REFERENCES

Adams, G.O., Tawari-Fufeyin, P., Igelenyah, E., and Odukoya, E. 2014. Assessment of heavy metals bioremediation potential of microbial consortia from poultry litter and spent oil contaminated site. *International Journal of Environmental Bioremediation & Biodegradation* 2: 84–92.

Adeniji, A. 2004. Bioremediation of arsenic, chromium, lead, and mercury. U.S. Environmental Protection Agency, Office of Solid Waste and Emergency Response Technology Innovation Office, Washington, DC.

Adhikari, T., Manna, M.C., Singh, M.V., and Wanjari, R.H. 2004. Bioremediation measure to minimize heavy metals accumulation in soils and crops irrigated with city effluent. *Food, Agriculture and Environment* 2(1): 266–270.

Ahemad, M. 2012. Implications of bacterial resistance against heavy metals in bioremediation: A review. *IIOABJ* 3: 39–46.

Ahemad, M., Khan, M.S., Zaidi, A., and Wani, P.A. 2009. Remediation of herbicides contaminated soil using microbes. In: *Microbes in Sustainable Agriculture*, Khan, M.S., Zaidi, A., and Musarrat, J. (eds.). Nova Science Publishers, New York, pp. 261–284.

Akhtar, M.S., Chali, B., and Azam, T. 2013. Bioremediation of arsenic and lead by plants and microbes from contaminated soil. *Research in Plant Sciences* 1: 68–73.

Al-Mailem, D.M., Al-Awadhi, H., Sorkhoh, N.A., Eliyas, M., and Radwan, S.S. 2011. Mercury resistance and volatilization by oil utilizing haloarchaea under hypersaline conditions. *Extremophiles* 15: 39–44.

Bagot, D., Lebeau, T., Jezequel, K., Fabre, B., 2005. Selection of microorganisms for bioremediation of agricultural soils contaminated by cadmium. In: Lichtfouse, E., Schwarzbauer, J., Robert, D. (eds.) Environmental chemistry. Springer, pp. 215–222.

Baillet, F., Magnin, J.P., Cheruy, A., and Ozil, P. 1997. Cadmium tolerance and uptake by a *Thiobacillus ferrooxidans* biomass. *Environmental Technology* 18: 631–638.

Bindler, R., Branvall, M.L., and Renberg, I. 1999. Natural lead concentrations in pristine boreal forest soils and past pollution trends: A reference for critical load models. *Environmental Science and Technology* 33: 3362–3367.

Blackburn, J.W. and Hafker, W.R. 1993. The impact of biochemistry, bioavailability and bioactivity on the selection of bioremediation techniques. *Trends in Biotechnology* 11: 328–333.

Boonchan, S., Britz, M.L., and Stanley, G.A. 2000. Degradation and mineralisation of high-molecular weight polycyclic aromatic hydrocarbons by defined fungal-bacterial co cultures. *Applied Environmental Microbiology* 66: 1007–1019.

Boricha, H. and Fulekar, M.F. 2009. *Pseudomonas plecoglossicida* as a novel organism for the bioremediation of cypermethrin. *Biology and Medicine* 1: 1–10.

Brim, H., McFarlan, S.C., Fredrickson, J.K., Minton, K., Zhai, M., Wackett, L.P., and Daly, M.J. 2003. Engineering *Deinococcus radiodurans* for metal remediation in radioactive mixed waste environments. *Nature Biotechnology* 18: 85–90.

Chaudri, A.M., McGrath, S.P., Giller, K.E., Rietz, E., and Sauerbeck, D.R. 1993. Enumeration of indigenous *Rhizobium leguminosarum* biovar *Trifolii* in soils previously treated with metal-contaminated sewage sludge. *Soil Biology & Biochemistry* 25: 301–309.

Cheung, K.C., Poon, B.H.T., Lan, C.Y., and Wong, M.H. 2003. Assessment of metal and nutrient concentrations in river water and sediment collected from the cities in the Pearl River Delta, South China. *Chemosphere* 52: 1431–1440.

Collins, Y.E. and Stotzky, G. 1989. Factors affecting the toxicity of heavy metals to microbes. In: *Metal Ions and Bacteria*, Beveridge, T.J. and Doyle, R.J. (eds.). Wiley, Toronto, Ontario, Canada, pp. 31–90.

Cunningham, C.J. and Philip, J.C. 2000. Comparison of bioaugmentation and biostimulation in ex situ treatment of diesel contaminated soil. *Land Contamination and Reclamation* 8(4) (University of Edinburgh, Scotland).

Das, N., Vimala, R., and Karthika, P. 2008. Biosorption of heavy metals—An overview. *Indian Journal of Biotechnology* 7: 159–169.

Dash, H.R., Mangwani, N., Chakraborty, J., Kumari, S., and Das, S. 2013. Marine bacteria: Potential candidates for enhanced bioremediation. *Applied Microbiology and Biotechnology* 97: 561–571.

Donmez, G. and Aksu, Z. 1999. The effect of copper(II) ions on growth and bioaccumulation properties of some yeasts. *Process Biochemistry* 35: 135–142.

Donmez, G. and Aksu, Z. 2001a. Bioaccumulation of copper(II) and nickel(II) by the non-adapted and adapted growing *Candida* spp. *Water Research* 35: 1425–1434.

Donmez, G. and Aksu, Z. 2001b. The effect of copper (II) ions on growth and bioaccumulation properties of some yeasts. *Process Biochemistry* 35: 135–142.

Dubey, R.C. 2004. *A Text Book of Biotechnology*, 3rd edn. Chand & Company Ltd., New Delhi, India, pp. 365–375.

Dursun, A.Y., Ulsu, G., Cuci, Y., and Aksu, Z. 2003. Bioaccumulation of copper(II), lead(II) and chromium(VI) by growing *Aspergillus niger. Process Biochemistry* 38: 1647–1651.

EPA. 2001. Remediation case studies. Federal Remediation Technology Roundtable. Report No. 542-F-01-032.

EPA. 2002. *Handbook on In Situ Treatment of Hazardous Waste Contaminated Soils.*

Evans, G.M. and Furlong, J.C. 2003. *Environmental Biotechnology Theory and Application.* John Wiley & Sons, West Sussex, U.K.

Field, J.A., Stams, A.J.M., Kato, M., and Schraa, G. 1995. Enhanced biodegradation of aromatic pollutants in cocultures of anaerobic and aerobic bacterial consortia. *Antonie van Laeuwenhoek* 67: 47–77.

Gadd, G.M. 1992. Metals and microorganisms: A problem of definition. *FEMS Microbiology Letters* 100: 197–204.

Gadd, G.M. 1993. Microbial formation and transformation of organometallic and organometalloid compounds. *FEMS Microbiology Reviews* 11: 297–316.

Geiger, G., Federer, P., and Sticher, H. 1993. Reclamation of heavy metal-contaminated soils: Field studies and germination experiments. *Journal of Environmental Quality* 22: 201–207.

Ghodsi, H., Hoodaji, M., Tahmourespour, A., and Gheisari, M.M. 2011. Investigation of bioremediation of arsenic by bacteria isolated from contaminated soil. *African Journal of Microbiology Research* 5: 5889–5895.

Giller, K.E., Witter, E., and McGrath, S.P. 1998. Toxicity of heavy metals to microorganisms and microbial process in agricultural soils: A review. *Soil Biology & Biochemistry* 30: 1389–1414.

Girma, G. 2015. Microbial bioremediation of some heavy metals in soils: An updated review. *Egyptian Academic Journal of Biological Sciences* 7: 29–45.

Glazer, A.N. and Nikaido, H. 2007. *Microbial Biotechnology: Fundamentals of Applied Microbiology*, 2nd edn. Cambridge University Press, Cambridge, New York, pp. 510–528.

Glick, B.R. 2003. Phytoremediation: Synergistic use of plants and bacteria to clean up the environment. *Biotechnology Advances* 21: 383–393.

Gray, E.J. and Smith, D.L. 2005. Intracellular and extracellular PGPR: Commonalities and distinctions in the plant-bacterium signaling processes. *Soil Biology & Biochemistry* 37: 395–412.

Ike, A., Sriprang, R., Ono, H., Murooka, Y., and Yamashita, M. 2007. Bioremediation of cadmium contaminated soil using symbiosis between leguminous plant and recombinant rhizobia with the MTL4 and the PCS genes. *Chemosphere* 66: 1670–1676.

Iqbal M. and Edyvean R.G.J. 2007. Ability of loofa sponge—Immobilized fungal biomass to remote lead ions from aqueous solution. *Pakistan Journal of Botany* 39: 231–238.

Iwahori, K., Takeuchi, F., Kamimura, K., and Sugio, T. 2000. Ferrous iron-dependent volatilization of mercury by the plasma membrane of *Thiobacillus ferrooxidans. Applied and Environmental Microbiology* 66: 3823–3827.

Joanne, W.M., Linda, S.M., and Christopher, W.J. 2008. *Prescott, Harley and Kleins Microbiology*, 7th edn. McGraw-Hill Publishers, New York, pp. 101–504.

Katarina, C., Darina, S., Vladimir, K., Miloslava, P., Jana, H., Andrea Puškrov, B., and Ferianc, P. 2004. Identification and characterization of eight cadmium resistant bacterial isolates from a cadmium-contaminated sewage sludge. *Biologia* (*Bratislava*) 59: 817–827.

Khan, M.W.A. and Ahmad, M. 2006. Detoxification and bioremediation potential of a *Pseudomonas fluorescens* isolate against the major Indian water pollutants. *Journal of Environmental Science* 41: 659–674.

Kulshreshtha, A., Agrawal, R., Barar, M., Saxena, S. 2014. A review on bioremediation of heavy metals in contaminated water. *IOSR Journal of Environmental Science, Toxicology and Food Technology* 8: 2319–2402.

Lang, E., Nerud, F., and Zadrazil, F. 1998. Production of ligninolytic enzymes by *Pleurotus* sp., and *Dichomitus squalens* in soil and lignocellulose substrate as influence by soil microorganisms. *Microbiology* 167: 239–244.

Litchfield, C.D. 1991. Practices, potential and pitfalls in the application of biotechnology to environmental problems. In: *Environmental Biotechnology for Waste Treatment*, Sayler, G.S., Fox, R.D., and Blackburn, J.W. (eds.). Plenum Press, New York, pp. 147–157.

Lloyd, J.R. 2002. Bioremediation of metals: The application of microorganisms that make and break minerals. *Microbiology Today* 29: 67–69.

Macaskie, L.E., Empson, R.M., Cheetham, A.K., Grey, C.P., and Skarnulis, A.J. 2005. Uranium bioaccumulation by a *Citrobacter* sp. as a result of enzymically mediated growth of polycrystalline. *Science* 257: 782–784.

Malik, A. 2004. Metal bioremediation through growing cells. *Environment International* 30: 261–278.

Mench, M., Vangronsveld, H., Clisters, N., Lepp, W., and Edwards, R. 2000. In situ metal immobilization and phytostabilization of contaminated soils. In: *Phytoremediation of Contaminated Soil and Water*, Terry, N. and Banuelos, G. (eds.). Lewis Publishers, Boca Raton, FL, pp. 323–358.

Meysami, P. and Baheri, H. 2003. Pre-screening of fungi and bulking agents for contaminated soil bioremediation. *Environmental Research* 7: 881–887.

Mueller, E.G. 1996. A glutathione reductase mutant of yeast accumulates high levels of oxidized glutathione and requires thioredoxin for growth. *Molecular Biology* 7: 1805–1813.

Nanda, S. and Abraham, J. 2013. Remediation of heavy metal contaminated soil. *African Journal of Biotechnology* 12: 3099–3109.

Nandini, M.R., Udayashankara, T.H., and Madhukar, M. 2015. A review on the bioremediation process for the removal of heavy metals. *International Journal for Scientific Research & Development* 3(5): 2321–0613.

Narayanan, M. and Natarajan, D. 2011. Bioremediation on effluents from magnesite and bauxite mines using *Thiobacillus* spp. and *Pseudomonas* spp. *Journal of Bioremediation & Biodegradation* 2: 115.

Ohtake, H. and Silver, S. 1994. Bacterial detoxification of toxic chromate. In: *Biological Degradation and Bioremediation of Toxic Chemicals*, Chaudhary, G.R. (ed.) Chapman and Hall, London, pp. 403–415.

Olson, J.W., Mehta, N.S., and Maier, RJ. 2001. Requirement of nickel metabolism protein HypA and HypB for full activity of both hydrogenase and urease in *Helicobacter pylori*. *Molecular Microbiology* 39: 176–182.

Pan X., Meng X., Zhang D., and Wang J. 2009. Biosorption of strontium ion by immobilized *Aspergillus niger*. *International Journal of Environment and Pollution* 37: 276–288.

Prescott, L.M., Harley, J.P., and Klein, D.A. 2002. *Microbiology*, 5th edn. McGraw-Hill, New York, pp. 10–14.

Pulford, I.D. and Watson, C. 2003. Phytoremediation of heavy metal contaminated land by trees—A review. *Environment International* 29: 529–540.

Rahman, M.A. and Hasegawa, H. 2011. Aquatic arsenic: Phytoremediation using floating macrophytes. *Chemosphere* 83: 633–646.

Rojas, L.A., Yanez, C., Gonzalez, M., Lobos, S., Smalla, K., and Seeger, M. 2011. Characterization of the metabolically modified heavy metal resistant *Cupriavidus metallidurans* strain MSR33 gene rated for mercury bioremediation. *PLoS ONE* 6: 10.

Shazia, I., Uzma, G., Sadia R., and Ara, T. 2013. Bioremediation of Heavy Metals using Isolates of Filamentous Fungus *Aspergillus fumigatus* Collected from Polluted Soil of Kasur, Pakistan. *International Research Journal of Biological Sciences*, ISSN 2278-3202, 2(12): 66–73.

Silver, S., Phung, L.T., Lo, J.-F., and Gupta, A. 2001. Toxic metal resistances: Molecular biology and the potential for bioremediation. In: *Industrial and Environmental Biotechnology*, Nuzhat, A., Qureshi, F.M., Khan, O., and Khan, Y. (eds.). Horizon Scientific Press, Wymondham, U.K., pp. 33–41.

Singh, P. and Cameotra, S.S. 2004. Enhancement of metal bioremediation by use of microbial surfactants. *Biochemical and Biophysical Research Communications* 319: 291–297.

Smith, S.R. 1994. Effect of soil pH on availability to crop of metals in sewage sludge treated soils. I. Nickel, copper and zinc uptake and toxicity to ryegrass. *Environmental Pollution* 85: 321–327.

Soares, E.V., Geoffrey, D.C., Duarte, F., and Soares, H.M.V.M. 2002. Use of *Saccharomyces cerevisiae* for Cu^{2+} removal from solution: The advantages of using a flocculent strain. *Biotechnology Letters* 24: 663–666.

Tabak, H.H., Lens, P., van Hullebusch, E.D., and Dejonghe, W. 2005. Developments in bioremediation of soils and sediments polluted with metals and radionuclides. Microbial processes and mechanisms affecting bioremediation of metal contamination and influencing metal toxicity and transport. *Reviews in Environmental Science and Biotechnology* 4: 115–156.

Ting, A.S.Y. and Choong, C.C. 2009. Bioaccumulation and biosorption efficacy of *Trichoderma* isolate SP2F1 in removing copper Cu(II) from aqueous solutions. World *Journal of Microbiology and Biotechnology* 25: 1431–1437.

Troy, M.A. 1994. Bioengineering of soils and groundwaters. In: *Bioremediation*, Baker, K.H. and Herson, D.S. (eds.). McGraw-Hill, New York, pp. 173–201.

Turpeinen, R. 2002. Interaction between metals, microbes and plants—Bioremediation of arsenic and lead contaminated soils. PhD, University of Helsinki, Helsinki, Finland.

Umrania, V.V. 2006. Bioremediation of toxic heavy metals using acidothermophilic autotrophes. *Bioresource Technology* 97: 1237–1242.

USEPA. 1990. Slurry bioremediation engineering bulletin. EPA 68-C8-0062. Office of Emergency and Remedial Response, Washington, DC.

Uslu, G., Dursun, A.Y., Ekiz, H.I., and Aksu, Z., 2003. The effect of Cd (II), Pb (II) and Cu (II) ions on the growth and bioaccumulation properties of *Rhizopus arrhizus*. *Process Biochemistry* 39: 105–110.

Uslu, G., Gültekin, G., and Tanyol M. 2011. Bioaccumulation of Copper(II) and Cadmium (II) from aqueous solution by *Pseudomonas putida*. *Fresenius Environmental Bulletin* 20(7a): 1812–1820.

Uslu, G. and Tanyol, M. 2006. Equilibrium and thermodynamic parameters of single and binary mixture biosorption of lead (II) and copper (II) ions onto *Pseudomonas putida*: Effect of temperature. *Journal of Hazardous Materials* 135: 87–93.

Uslu, G., Dursun, A.Y., Ekiz, H.İ., and Aksu, Z. 2003. The effect of Cd (II), Pb (II) and Cu (II) ions on the growth and bioaccumulation properties of *Rhizopus arrhizus*. *Process Biochemistry* 39: 105–110.

Vidali, M. 2001. Bioremediation: An overview. *Pure and Applied Chemistry* 73: 1163–1172.

White, C., Sayer, J.A., and Gadd, G.M. 1997. Microbial solubilization and immobilization of toxic metals: key biochemical processes for treatment of contamination. *FEMS Microbiology Reviews*. 20: 503–516.

White, C., Shaman, A.K., and Gadd, G.M. 2006. An integrated microbial process for the bioremediation of soil contaminated with toxic metals. *Nature Biotechnology* 16: 572–575.

Yare, B.S. 1991. A comparison of soil-phase and slurry-phase bioremediation of PNA-containing soils. In: *On Site Bioreclamation: Processes for Xenobiotics and Hydrocarbon Treatment*, Hinchee, R.E. and Olfenbuttel, R.F. (eds.). Butterworth-Heinemann, Stoneham, MA, pp. 173–187.

8 Interaction between Plants, Metals, and Microbes

Jibin Sadasivan, Neha Pradhan, Ayusman Dash,
Ananya Acharya, and Sutapa Bose

CONTENTS

ABSTRACT

Rigorous study has been carried out to understand the interactions between plants-microbes-metals. Our understanding of these interactions is incomplete due to the difficulty of studying these complex processes under controlled yet natural conditions. Thus, developing novel methodologies to study these interactions under natural conditions is crucial. Joint efforts of scientists working in areas of plant biology, microbial ecology, and pedology need to develop systems where careful analysis could be performed on-site to understand the minute details of interactions between plants and microbes. It is now well understood that roots are rhizospheric mediators facilitating communication between the plant and microorganisms in the soil. Ecological knowledge supports 'above-ground' interactions that could potentially be used to understand 'below-ground' interactions between plants and microbes. An understanding of the molecular process involved in the actual secretion of molecules by roots is de rigueur in order to develop molecular markers for this process. Finally, improvement of the knowledge of the key factors affecting the interactions between microbes, plants and metals, from the molecular to the ecosystem level can systematically lead to the scientific development of better plants capable of absorbing more nutrients, detoxifying soils with better efficiency, or eliminating invasive weeds and pathogens effectively.

8.1 INTRODUCTION

With the advent of urbanization and industrialization, there has been a substantial hike in the discharge of wastes into the environment that has led to the accumulation of heavy metals in the niche shared by plants, animals, and microbes. These metals have been classically found to be posing alarming threats to the growth and health of not only the organisms but also the ecosystem as a whole even though slow depletion of heavy metals also takes place through leaching, plant uptake, erosion, and deflation. The net indiscriminate release of heavy metals into the soil and water has become a chief concern worldwide, as they cannot be broken down to nontoxic forms and therefore have long-lasting effects. Many of them are toxic even at very low concentrations.

There have been several techniques to remove these heavy metals. Some of these include precipitation reactions, redox reactions, filtration, ion-exchange and size exclusion chromatography, osmosis, and electrochemical treatment. But most of these techniques are not very effective when the concentrations of heavy metals are less than a critical value (Ahluwalia et al., 2007). Most heavy metal salts being water soluble cannot be separated by physical separation methods (Hussein et al., 2004). Additionally, these methods dealing with physical and chemical properties are not worth it in terms of price as well as quality when the concentration of heavy metals is very low. Alternately, biological methods like biosorption and/or bioaccumulation for the removal of heavy metals may be an alluring alternative to physicochemical methods (Kapoor et al., 1995). The use of microorganisms and plants for remediation purposes is, thus, a possible solution for heavy metal pollution since it includes sustainable remediation technologies to rectify and reestablish the natural conditions of soil.

However, the response to heavy metals depends on the concentration and availability of heavy metals and is a complex process that is controlled by multiple factors, such as type of metal, the nature of the medium, and species of microbes and plants. Thus, it is highly essential to understand the details of the interactions between the plants, metals, and microbes and utilize the idea to deal with this disastrous problem. In addition, an in-depth knowledge about the interactions can provide greater awareness of the ecological and health effects of the toxic metals.

Thus, understanding the interactions between plants, metals, and microbes opens up a plethora of amazing and innovative promises for the removal of heavy metals and recovery of heavy metals in polluted water and lands. Since microorganisms and plants have evolved to acquire strategies for their survival in heavy metal–polluted habitats, it is possible that they are adopting quite different detoxifying mechanisms such as biosorption, bioaccumulation, biotransformation, and biomineralization, which can be exploited for bioremediation either *ex situ* or *in situ* (Dixit et al., 2015).

8.2 INTERACTIONS BETWEEN PLANTS AND METALS

8.2.1 HYPERSENSITIVITY, HYPERTOLERANCE, AND HYPERACCUMULATION

Metals play a wide variety of roles in many key physiological processes in plants. Some metals are essential for plant survival, whereas some metals are nonessential. Metals such as zinc and copper are essential to plant physiology, but an elevated level has serious deleterious effects. Zinc acts as a protein-stabilizing element and plays a life-sustaining role as a cofactor in a plethora of enzymes (Clarke and Berg, 1998; Clemens, 2001). Electron transfer chain, which is essential for ATP production (the energy currency of a cell), contains copper-containing cytochrome c oxidase and plastocyanin. But copper is known to produce reactive oxygen species that can oxidize various cellular components and destroy the cell. On the other hand, nonessential elements such as lead and cadmium pose serious threat to protein integrity due to their affinity toward nitrogen and sulfur atoms in amino acid side chains. To counter such harmful effects, plants have devised an array of homeostatic measures that intend to check and regulate metal uptake, transport, chelation, and sequestration with an aim to keep a track of the metal load and neutralize the underlying deleterious effects. Inability to maintain any of these processes creates hypersensitivity toward the respective metals (Clemens, 2001).

Hypertolerance, which means higher level of tolerance, toward a metal is seen in some plant species growing in dangerously elevated concentrations of that metal such as in soils contaminated with zinc and nickel due to mining activities (Ernst, 1974; Chaney et al., 1997; Clemens, 2001). Another homeostatic mechanism is hyperaccumulation with around 400 identified plant species practicing it. This applies to plants that can accumulate a dry weight of cadmium, nickel, lead, and zinc higher than 0.01%, 0.1%, 0.1%, and 1%, respectively. This technique is avidly exploited by employing hyperaccumulating plants to alleviate the higher level of metals in metal-contaminated soils, a practice widely known as phytoremediation (Chaney et al., 1997; Clemens, 2001). Of all the hyperaccumulators identified till date, about 3/4 of them are nickel hyperaccumulators (Baker and Brooks, 1989; Clemens, 2001).

8.2.2 NANOMATERIAL–PLANT INTERACTION

Nanomaterials are commonly used in a wide variety of commercial products. Their diverse usage accounts to its many beneficial aspects, but a look at its harmful effects beg serious investigation. Nanoparticle contamination in soils may occur by accidental spillage or sewage disposal during industrial manufacture and transport of nanoparticle-derived products (Fabrega et al., 2011; Andreotti et al., 2015). This may then be manifested in the food chain affecting a vast number of organisms (Pradhan et al., 2014; Andreotti et al., 2015).

Past literature cite the positive and negative effects of metallic nanoparticles on plant growth and survival and its cumulative uptake by these plants (Lin and Xing, 2007; Zhu et al., 2008; Shah and Belozerova, 2009). For example, antimicrobial properties of CuO nanoparticles are exploited in manufacturing plastics and textiles, and its electrical properties are exploited in manufacturing films, polymers, and ceramics (Kim et al., 2012; Llorens et al., 2012; Abramova et al., 2013). Metals such as copper are readily taken up by salt marsh plants such as *Halimione portulacoides* and *Phragmites australis*, thus providing an alternative for the phytoremediation of metal-contaminated soils (Almeida et al., 2004; Weis and Weis, 2004; Andreotti et al., 2015). But metals, in nanoparticle form, lower these plants' phytoremediation potential as a consequence of lesser bioavailability in this form, leading to a lesser metal uptake by these plants. Experiments, where *H. portulacoides* and *P. australis* were exposed to copper nanoparticle, showed a relative decrease in copper accumulation in their roots than when exposed to ionic copper (Andreotti et al., 2015).

8.2.3 Aluminum–Plant Interaction

Aluminum is a metal that is found in two ionic states +2 and +3 and shows high affinity toward negatively charged ions (Mossor-Pietraszewska, 2001; Imadi et al., 2015). Natural source of aluminum is aluminum silicate. It is essential to living beings in negligibly small quantities such that even a micromolar concentration of its +3 state is deemed to be toxic and has deleterious effects on plants (Ma et al., 2001; Silva 2012; Gupta et al., 2013; Imadi et al. 2015). Some plants form aluminum acid complexes to neutralize its toxic effects (Ma et al., 2001). Aluminum mostly inhibits the growth of roots in plants and attacks both the structural (cell wall, cytoskeleton, plasma membrane) and functional (calcium signalling and downstream transduction pathways) integrity of root cells (Kochian et al., 2004; Ma, 2007; Poschenrieder et al., 2008; Imadi et al., 2015). Aluminum causes iron deficiency in rice and wheat (Rout et al., 2001). Aluminum creates oxidative stress in plant cells by generating reactive oxygen species that disrupts cell organelles and depletes ATP (Yamamoto et al., 2002). Loss of mitochondrial transmembrane potential results in a lack of ATP supply, thereby leading to apoptosis and programmed cell death (Panda and Matsumoto, 2007; Imadi et al., 2015).

Globally, aluminum stands as one of the key players in decreasing the productivity of crops deriving nutrition from acidic soil. Aluminum causes thickening of leaf epidermis and reduces leaf size with marginal curling and chlorosis beneath the leaf margins (Matsumoto et al., 1976; Pavan and Bingham, 1982; Imadi et al., 2015). It inhibits shoot growth in coffee plants and has been shown to inhibit lateral branching in the shoot of soya bean plants. Cytokinins generated from shoot meristem promote lateral branching of shoot. Aluminum targets this shoot meristem thereby curtailing the formation of lateral branches (Pavan and Bingham, 1982; Pan et al., 1988; Imadi et al., 2015). But the most deleterious effect of aluminum is observed on root growth in plants growing in acidic soil, where it curtails root elongation within few hours of its intake in apical root tissues (Sujatha and Mehar, 2015). Aluminum in monomeric form is the culprit of such a toxicity-induced root growth inhibition (He et al., 2012). This fact is corroborated by a study on sunflowers and soya beans, where higher phosphorus-to-aluminum ratio promoted root elongation and reduced monomeric aluminum content available for root uptake (Alva et al., 1986; Ali et al., 2011; Imadi et al., 2015; Sujatha and Mehar, 2015).

Aluminum uptake starts from the root cap and the mucilage covered epidermal cells, wherefrom it takes an apoplastic route to enter cortex or directly enter stele via plasmalemma (Roy et al., 1988; Imadi et al., 2015).

Aluminum interferes with plasma membrane ATPases and lipids, Golgi apparatus, and plasmalemma. It also reduces the photosynthetic efficiency of plants by reducing the chlorophyll concentration in leaves (Roy et al., 1988). In an alga named *Chara coralline*, nearly 99% of total aluminum taken up is trafficked to its cell wall and cytosol (Rengel and Reid, 1997). Lanthanum is a well-known competitive inhibitor of aluminum for the binding site of aluminum on plasma membrane that thereby reduces aluminum uptake by plants (Ma and Hiradate, 2000; Imadi et al., 2015).

Aluminum affects plant genes such as cell wall–responsive genes in mostly aluminum-sensitive varieties of plants (Maron et al., 2008). For example, in *Medicago truncatula*, approximately twofold upregulation and downregulation was observed in 324 and 267 genes, respectively, in the presence of aluminum. The genes that were upregulated functioned in cell wall and stress response, while the genes that were downregulated functioned in protein metabolism and cell division (Chandran et al., 2008; Imadi et al., 2015). Similar lines of investigation on *Arabidopsis thaliana* demonstrated higher activity of oxidative stress genes upon aluminum treatment suggesting that these genes protect plants from aluminum toxicity (Richards et al., 1998; Ezaki et al., 2000; Imadi et al., 2015).

Aluminum in +3 state competitively binds to calcium binding site of calmodulin (calcium-binding messenger protein). This leads to loss of function of calcium-dependent enzymes such as phospholipase-C and disruption of calcium-responsive metabolic pathways (Jones and Kochian, 1997; Sivaguru et al., 2005; Imadi et al., 2015).

Enhanced malic acid formation protects wheat seedlings from the toxic effects of aluminum. This enhancement is the result of an aluminum-specific response (Delhaize et al., 1993). Glutathione-S transferase, superoxide dismutase, and catalase are among the few compounds synthesized in aluminum-tolerant plants to neutralize reactive oxygen species generated from aluminum exposure (Darkó et al., 2004; Imadi et al., 2015).

8.2.4 Copper–Plant Interaction

Copper is a ubiquitous element found in rocks and minerals or in enzymes such as hemocyanin and cytochrome oxidase (Ademar Avelar Ferreira et al., 2015; Saglam et al., 2015).

Copper is a heavy metal with three valence states, namely, 0, +1, and +2. It is an element essential in limited quantities for many plant physiological processes including respiration and protein kinetics, whereas an excess of it has deleterious effects on plants such as leaf chlorosis, reduced root development, and branching (Guzel and Terzi, 2013; Nair and Chung, 2015; Saglam et al., 2015). Copper mainly deposits in the roots and is transported in small quantities to shoots (Marschner, 2012). As water and other nutrients gain entry to plants via roots, so a defective root growth indirectly affects the growth of other plant parts and the plant as a whole. A report stated that copper entry led to root cell membrane disruption in maize root tissues, resulting in the efflux of organic and inorganic substances (Liu et al., 2014; Saglam et al., 2015). Prolonged copper stress in roots leads to decreased stomatal pore sizes and decreased intercellular mesophilic spaces and increased stomatal count, causing reduced transpiration in leaves (Panou-Filotheou et al., 2001). It has also been seen that copper stress leads to accumulation of abscisic acid in sunflower seedlings that reduces transpiration and thereby prevents copper translocation from root to shoot (Zengin and Kirbag, 2007). Copper is also responsible for the inhibition of primary roots favoring lateral root formation mediated via regional auxin distribution (Yuan et al., 2013; Saglam et al., 2015). One major symptom of copper stress is abnormal photosynthesis demarcated by reduced rate of photosynthesis, reduced RUBISCO efficiency, and reduced activity of Electron Transport System(ETS) and Photosystem II (PSII) systems (Saglam et al., 2015).

To tackle copper stress, plants have learnt to store proline in tissues (Ku et al., 2012). Copper stress tends to generate reactive oxygen species and proline acts as a reactive oxygen species scavenger. Other scavengers include superoxide dismutase and peroxidase whose enhanced antioxidant potential upon copper stress has been studied in *Lolium perenne* and *Festuca arundinacea* plants (Zhao et al., 2010; Saglam et al., 2015). But prolonged copper exposure blocks antioxidant activity as observed in *Solanum lycopersicon* (Chamseddine et al., 2009). Apart from creating oxidative stress, reactive oxygen species also acts as a messenger to stimulate preventive mechanisms against copper toxicity. Melatonin is another hormone known to rescue plants against copper stress, whereby external administration in pea plants rescues it from death (Tan et al., 2007; Saglam et al., 2015). A major signalling pathway activated by copper toxicity is mitogen-activated protein kinase as seen in rice plants (Yeh et al., 2007) and *Medicago sativa* (Jonak et al., 2004). It plays key role in the upregulation of metal transporter and chelator production genes to suppress copper levels in plant tissues (Saglam et al., 2015).

8.2.5 Manganese–Plant Interaction

Manganese toxicity is demarcated by brown spots on leaves and stems indicating necrosis (Horst and Marschner, 1978; Wu, 1994). The spots spreading a time gradient starting from lower leaves to upper leaves incrementing in size and number to end up becoming brown and succumbing to death

(Elamin and Wilcox, 1986). Decrease in internodal length is observed in *Cucumis sativus* upon manganese toxicity (Crawford Jr et al., 1989). "Crinkle leaf" is a condition that causes chlorosis and browning of newly formed leaf, petiole, and stem upon manganese toxicity (Horst and Marschner, 1978; Wu, 1994; Bachman and Miller, 1995). In few species of plants, manganese causes chlorosis of older leaves first that gradually spreads toward younger leaves (Gupta, 1972; Elamin and Wilcox, 1986; Bachman and Miller, 1995). Manganese toxicity on a tree named *Eucalyptus gummifera*, located in Australia, caused apoptosis of terminal bud tissues and disruption of leaf morphology (Winterhalder, 1963; Reichman, 2002).

Manganese stress causes iron accumulation in roots minimizing the transport of iron to shoot tissues. Thus, iron deficiency is one of the outcomes of manganese toxicity that is displayed in terms of leaf chlorosis (Sideris, 1950). A study was done where iron complexed with EDTA was externally applied to manganese-stressed *E. gummifera*. As expected, this caused loss of chlorosis in the chlorotic leaves (Winterhalder, 1963). The effects of "crinkle leaf" condition mentioned earlier mimic those of calcium deficiency such as reduced calcium transport to leaves. Manganese is also observed to cause magnesium deficiency in needles of *Pinus radiata* by preventing the normal magnesium transport across plant tissues (Safford, 1975; Reichman, 2002).

8.2.6 Zinc–Plant Interaction

Zinc toxicity is demarcated by chlorosis and deposition of anthocyanin, thereby giving a reddish outlook to young leaves (Harmens et al., 1993; Fontes and Cox, 1995; Lee et al., 1996). Zinc causes reduction in leaf size and vertical alignment of leaves, while in some species it causes browning (Ren et al., 1993; Fontes and Cox, 1995). Zinc has the potential to cause necrosis and apoptosis in leaves, while primary root growth inhibition and yellow colorization of roots mark its other symptoms (Harmens et al., 1993; Ren et al., 1993; Reichman, 2002).

When it comes to chlorosis caused by zinc toxicity, it is simultaneously marked with iron deficiency (Wallace and Abou-Zamzam, 1989; Ren et al., 1993). Retrieval of leaves from chlorosis was achieved by the treatment of leaves with ammonium ferric sulfate (Chaney, 1993; O'Sullivan et al., 1997). Surprisingly, zinc toxicity in grasses was marked by symptoms of phosphorus reduction although phosphorus content remained unchanged (Plenderleith, 1984). The only plausible explanation could be that phosphorus reduction occurs by changes in phosphorus synthesis within tissues while keeping the phosphorus uptake constant (Reichman, 2002) (Figure 8.1).

8.3 INTERACTIONS BETWEEN METALS AND MICROBES

Metals are essential components of the ecosystem, which are required for the proliferation of all kingdom of life and

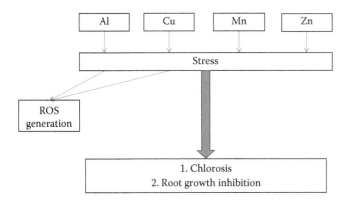

FIGURE 8.1 Schematic representation of metal stress and its effect on plants.

perform a variety of structural and chemical functions. The biologically available concentration of metals depends on various biological and geological processes. Trace amounts of certain metal types are required by various organisms for their life processes. However, most of these organisms cannot tolerate high or low concentration of these metals. Hence, the presence of an optimum concentration of certain metals is essential for microbial growth.

8.3.1 Soil Microbes

Along with certain macroorganisms such as earthworm, insects, and small vertebrates, soil contains complex microbial community including bacteria, viruses, protozoa, and fungi. The abundance and diversity of these groups in the soil depend on various physical, chemical, and biological conditions. Although soil does not provide all the nutrients required for microbial growth, it is one of the richest reservoirs of microorganisms. One gram of agricultural soil may contain several billions of colony-forming units of microorganism belonging to thousands of different species (Rosselló-Mora and Amann, 2001). Apart from their presence, these tiny organisms may also contribute to certain soil structure and fertility.

Prokaryotic organisms such as bacteria and archaea are the most abundant microorganisms in soil that serve much important purpose including nitrogen fixation. A teaspoon of soil contains 100 million to 1 billion bacterial cells (Lenart-boroń and Boroń, 2014). *Acinetobacter, Agrobacterium, Alcaligenes, Arthrobacter, Bacillus, Brevibacterium, Caulobacter, Cellulomonas, Clostridium, Corynebacterium, Flavobacterium, Hyphomicrobium, Metallogenium, Micrococcus, Mycobacterium, Pseudomonas, Sarcina, Streptococcus,* and *Xanthomonas* are some of the most commonly encountered bacterial genera in soil (Tate, 2000). The diversity of soil microorganism comprises different levels of biological organizations that include species variability, species richness, and species evenness and functional groups within communities (Ranjard and Richaume, 2001).

8.3.2 Effect of Industrialization in Soil Microflora

Industrialization results in rapid release of pollutants in the atmosphere. Improper management of hazardous waste materials has led to the scarcity of clean water and affects the soil thereby limiting crop production. Heavy metal contamination due to natural and anthropogenic activities is a global concern. This also affects the variety of microorganisms present in soil and water. Chemical compounds enter the ecosystem and accumulate in soil and water systems, mainly due to anthropological activities. Inappropriate and careless disposal of hazardous chemicals often results in polluting the environment that has harmful effects in microbes, plants, animals, and humans. Heavy metals, which are one of the major pollutants, cannot be destroyed and remains in the soil for long time. They can only be transformed from one state to another (Gisbert et al., 2003). Trace amounts of heavy metals occur naturally due to pedogenetic processes of weathering parent materials and are rarely toxic (Pierzyński et al., 2000). However, the disturbance and acceleration of the naturally slow geochemical cycles by man result in rapid accumulation of one or more heavy metals above the defined background level.

Heavy metals, such as lead, zinc, chromium, cadmium, nickel, and platinum, emitted by vehicles enter into the environment (Indeka and Karczun, 1999). Extensive use of fertilizers, pesticides, and biosolids such as composts, livestock manures, and municipal sewage sludge are also some of the significant sources of heavy metals in soil (Atafar et al., 2010). Other sources of heavy metals in the environment include stack emission such as dust emission from storage areas and heavy industry such as mining and metallurgy.

8.3.3 Effect of Metals on Microbes

Tolerance to a particular metal depends on the function and concentration. Organisms can tolerate metals with biological function in high concentration, but metals without biological function are tolerated in only very minute concentrations (Haferburg and Kothe, 2007). Metals such as Fe, Ni, Zn, Co, and Cu are very important for microbial activities when they occur at lower concentrations (Bruins, Kapil, & Oehme, 2000). These metals are often involved in metabolic functions, structural demands, or redox reactions. For instance, *Alcaligenes faecalis* can oxidize AsO_2^{2-} to AsO_2^{4-}, and *Pseudomonas fluorescens LB 300* or *Enterobacter cloacae* are involved in redox reactions. However, at higher concentrations, these metals have inhibitory roles, which will affect the activity of various microbial enzymes, decomposition processes, and the microbial diversity (Tyler, 1974). Often, the toxic concentration and high ionic nature of heavy metals result in enzyme inactivation that happens due to the displacement of metals with toxic metals having similar structure. Toxic metals also affect the protein and nucleic acid structure and render it inactive that ultimately results in the destruction of the entire cell.

8.3.4 Role of Microbes in Nutrient Extraction

Microbial depletion in the soil has an adverse effect on the environment. Plants require many elements from the soil for their life processes and largely depend on microbes to extract these elements and incorporate them into the organic molecules. When the microbes break the nutrients, the nutrient dissolves in soil water, where they can be assessed by the plant roots. Amount of elements such as nitrogen in the soil depends on nitrifying bacteria. Nutrients such as iron, calcium, and magnesium are present in relatively large amount in soil. However, microbes play a minor role in extraction of these elements. Sulfides are common soil mineral, but plants cannot absorb sulfur in this form. *Thiobacillus* bacteria can convert sulfides to sulfates, which plants can use for their life processes. Similarly, phosphorous in the soil is tightly bound to the soil particles and is not readily available for plants. Certain fungi such as *Penicillium radicum* and *Penicillium bilaiae* can assist plants to extract phosphorous from the soil. Plants cannot utilize atmospheric nitrogen and instead require it as nitrate or ammonia. Though a small amount on nitrogen is converted by lightening, plants depend on some free living soil microbes such as *Rhizobium, Azotobacter chroococcum, Azotobacter brasilense, Agrobacterium radiobacter*, and *Bacillus polymyxa* to fix nitrogen into usable form. Hence, any reduction in microbial population will have adverse effects in soil fertility that will also affect organisms in the upper trophic levels.

8.3.5 Microbes and Metal Toxicity

The toxicity of metals to microbes opens a new field of drug discovery. Copper and silver have adverse effects in heterotrophic bacteria. This effect is used in antiseptic preparations. Increased concentration of lead on soil surface layers affects microflora. High concentration of heavy metals affects the metabolic processes in microbes. Heavy metal contamination results in reduction in microbial biomass and their diversity and ultimately affects the community structure (Doelman, 1986). Certain metals such as iron is essential for bacterial growth. Iron acts as an essential cofactor in diverse physiological processes including respiration, deoxynucleotide biosynthesis, and DNA replication (Andreini et al., 2008). Iron deprivation strategies to kill microbes are a potential therapeutic strategy that can be adopted. However, the challenge lies in the need of the same metal by host organism in their survival. Hence, targeting the iron metabolizing pathways in pathogens that do not affect the host will have more therapeutic applications.

The challenge lies in the emergence of metal-resistant bacteria in clinical and laboratory conditions. Continuous metal exposure may lead to the establishment of tolerant species or strain. Metal resistance is acquired by expressing efflux transporters that help in maintaining lower cytosolic metal concentrations as in the case of *Cupriavidus metallidurans* (Nies 2003) or by releasing metal-binding compounds into extracellular surroundings or by cytosolic sequestration mechanisms.

Siderophores are a diverse group of specialized Fe^{3+} biding metabolites that supply iron to microbes (Neilands, 1995).

Microbes play important role in the environment by helping nutrient extraction. However, various microbes raise health concerns in humans also. Microbes depend on various metals for their life processes. Optimum concentration of metals is required for microbes. Therefore, metals are useful both in promoting and in controlling microbes. Reducing high level of metals from the polluted site is important to promote microbial growth and improve soil fertility. But remediation of metal-polluted site is very difficult and expensive. The best way to minimize the contamination is to prevent it. Phytoremediation is one of the best techniques for the treatment of metal-polluted sites. It is based on the use of special kind of plants to decontaminate soil by inactivating metal in rhizosphere or translocating them to aerial parts. Using of microorganisms to detoxify metals could be another approach in which the microorganism detoxifies metals by valence transformation, extracellular chemical precipitation, or volatilization (Lone, He, Stoffella, & Yang, 2008).

8.4 INTERACTIONS BETWEEN PLANTS AND MICROBES

One of the distinctive characteristics of organisms that make them living is their instinctive ability to interact with one another be it ones of the same kind or of another. Therefore, it is possible to have such interactions between the plants and microbes as they coexist in several different niches. Any kind of interactions can be broadly classified into beneficial (nonnegative) and harmful. Thus, it also applies for plants and microbial interaction. All of these interactions lead to the production of a particular kind of signal that elicits a specific response in its corresponding interacting partner. This consequently leads to a series of signaling pathways giving rise to the transduction of information that ultimately results in beneficial or harmful interactions (Clarke et al., 1992). Soil is traditionally considered to be an excellent medium for the growth and development of both plants and microbes, and thus, it can be expected that most of the existing plant–microbe interactions occur below the ground in the soil.

8.4.1 Belowground Interactions: Key Factors Affecting the Interactions

In the association of two complex living organisms, it is obvious that several abiotic factors play a key role. There are several players that can potentially alter the numbers and variety of organisms involved in a particular kind of interaction.

Root exudates, decaying root material, differences in concentrations of respiratory gases, and changes in pH, water, and mineral nutrients tend to affect the growth, proliferation, and interaction of microbes in the rhizosphere of the plants (Rovira, 1965). There are little quantitative data on the relative amounts of exudates and cell debris, but it is now well established that root exudates play a key role in the stimulation of selective microorganisms around roots, while moribund root cells and root hairs will make significant contributions in the older portions of the roots (Rovira, 1965).

It is often stated that even the type of the plant is crucial. For example, legumes exert greater rhizosphere stimulation than nonlegumes. It is found that when rhizosphere counts were expressed on a soil weight basis, the rhizosphere effects in decreasing order were red clover > flax > oats > wheat > maize > barley, but on a root weight basis, the order of rhizoplane counts were maize > wheat > red clover > oats > barley > flax (Rouatt and Katznelson, 1961). This factor is so important that difference between two soybean lines (nodulating and nonnodulating) by just a single gene leads to a completely different set of rhizospheric microbiota (Elkan, 1961). The number of organisms is also estimated on either a soil weight or a root weight basis, which increases with age of the plant (Louw and Webley, 1959). The dominance of particular fungal species on the root surface is influenced by the soil. There is also evidence that the bacterial flora associated with roots is affected by soil pH (Welte and Trolldenier, 1961). There is ample evidence that the addition of fertilizer increases the rhizosphere population of most plants (Macura, 1958). Environmental factors are also known to affect the interactions between plants and microbes. For example, when wheat was grown at two light intensities, the total numbers of associated bacteria and fungi varied. This variation in terms of quality and quantity of associated microbes can potentially affect the interactions between plants and microbes like extent of symbiosis, extent of pathogenicity, nodulating capacity, and exudation ability (Mugabo, 2014; Figure 8.2).

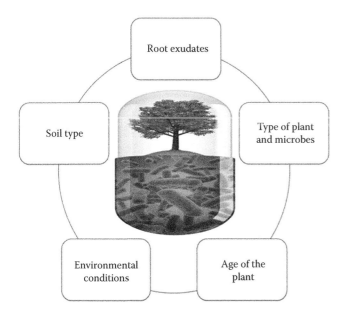

FIGURE 8.2 A diagrammatic representation of the key factors affecting the interactions between plants and microbes.

The release of carbon compounds from plants increases the microbial biomass and activity in the rhizosphere relative to the bulk soil (Bending, 2003). Plants can have specific associations that satiate their desire. The rhizosphere-to-bulk (total)-soil ratios for bacteria, actinomycetes, and fungi mostly have values around 2–20, 5–10, and 10–20, respectively (Morgan et al., 2005). In relatively lower-aged plants, the rhizospheric zone surrounding the roots is r-strategists dominated, which are species with generally faster growth rates and capacities to utilize simpler substrates than their counterparts. However, the roots that are mature are more populated by bacterial communities with relatively slow growth rates and the capacities to degrade complex substrates (k-strategists) dominate (Andrews and Harris, 1986).

8.4.2 TYPES OF INTERACTIONS BETWEEN PLANTS AND MICROBES

8.4.2.1 Beneficial Interactions between Plants and Microbes

The beneficial or nonnegative–plant-microbe interactions are caused by symbiotic or nonsymbiotic bacteria and a highly specialized group of fungi (mycorrhizal fungi). Symbiotic bacteria mostly include nitrogen-fixing bacteria that associate themselves with the roots of the plants in the process of nodulation. Within the nodules formed, the host plant provides the bacteria with the necessary nutrients (e.g., carbohydrates) that they need for their metabolism and survival. In return, the rhizobial bacteria fix the atmospheric form of inert nitrogen gas into ammonium ions, which are further converted into usable forms like amides or ureides, which are then passed to the plant xylem for further usage (Morgan et al., 2005). Mycorrhizal fungi are mostly symbionts that remain associated with a wide range of plant communities all across the globe. The establishment of the mycorrhizal network offers several advantages that help in the acquisition of mineral nutrients by the plants. Fungal hyphae extend beyond the area of nutrient depletion surrounding the root and enhance the surface area for the absorption of nutrients relative to nonmycorrhizal roots. Hyphae are able to extend into soil pores that are too small for roots to enter in some mycorrhizal fungi that can access forms of N and P that are unavailable to nonmycorrhizal plants, particularly organic forms (Morgan et al., 2005). In addition to this, mycorrhizal fungi provide protection to the host plant against root and shoot pathogens (Whipps, 2004).

Beneficial plant-associated microbes are also known to stimulate the plant growth and enhance their resistance to degenerative diseases and abiotic stresses. *Azospirillum*, *Bacillus*, *Pseudomonas*, *Rhizobium*, *Streptomyces*, etc., are a few bacterial genera that belong to this category. These are commonly known as plant growth–promoting rhizobacteria (PGPR). Growth-promoting substances are produced by these soil microorganisms that can influence indirectly on the overall growth and development of the plants (Tyagi et al., 2014).

8.4.2.2 Harmful Interactions between Plants and Microbes

It is obvious that not all interactions are going to affect the plant in a positive or a neutral fashion. There could definitely be ones that negatively affect the plants. There have been numerous instances cited in the literature that support this statement. Economically relevant pathogens such as the fungus-like oomycete *Phytophthora infestans* responsible for Irish potato famine cause dramatic yield losses in crops such as potato and tomato (Fry, 2008).

In addition, interactions are not just one to one or specific. There could even be a complex interrelationship. One of the brilliant examples of the interplay is plant pathogenic fungi belonging to the genus *Rhizopus*, infamous for causing rice seedling blight. This disease is usually initiated in plants by an abnormal swelling of the seedling roots without any sign of infection by the pathogen. Interestingly, rhizoxin is not biosynthesized by the fungus itself, but by endosymbiotic, that is, intracellular living, bacteria of the genus *Burkholderia* (Laila, 2005). This is a remarkable complex symbiotic pathogenic relationship that extends the fungus–plant interaction to a third, bacterial, key player.

However, in the end, there is extensive evidence for commonalities between pathogenic and symbiotic lifestyles (Rey and Schornack, 2013). Both these types of interactions follow similar processes of identification based on development, cell penetration in plants, and redifferentiation and coevolution of the host cells to establish intracellular interfaces for nutrient and signal exchange (Parniske, 2000). Undecorated chitin oligomers of microbial origin, known to be potential inducers of plant immunity (Petutschnig et al., 2010), were found recently to also activate symbiosis-related signalling (Genre et al., 2013). Furthermore, effector proteins were described recently in mycorrhizal fungi and have been elaborated as hallmarks of animal and plant pathogens and help suppress, defense, and reprogram the host based on the demand of the situation (Plett et al., 2011).

8.4.3 Advantages of Interactions between Plants and Microbes

There could be several reasons as to why plants would like to interact with microbes or vice versa. Such interactions are responsible for a number of intrinsic processes such as carbon sequestration, ecosystem functioning, and nutrient cycling. The composition and quantity of microbes in the soil influence the ability of a plant to obtain nutrients (Singh et al., 2004). Plants can influence the net effect of their interactions with the microbes through the deposition of small metabolites, amino acids, secreted enzymes, mucilage, and cell lysates into the rhizosphere that attract or inhibit the growth of specific microorganisms (Grayston et al., 1998). Thus, it is a two-way interrelationship that maintains the ecological balance. In addition to the fact that the secretion of allelopathic compounds inhibits pathogenic associations, it is also demonstrated that they

could change the microbial population dynamics altogether (Fons et al., 2003).

The microbes contribute to the growth of plants and affect its interactions in both positive and negative ways. It is shown that fungicide treatments affected the interactions between the plant *Centaurea maculosa* and its neighboring plant species (Callaway et al., 2004). The biomass of *C. maculosa* was increased in untreated soils when grown with two particular native grass species. However, this effect was not seen when *C. maculosa* was grown alone or with these two grasses in benomyl-treated soils. This indirectly suggests that mycorrhizal fungi associated with these grasses favor the growth of *C. maculosa*. However, when the same experiment was conducted using *C. maculosa* and the forb *Gaillardia aristata*, the opposite effect was observed, with *G. aristata*–associated fungi apparently having detrimental effects on *C. maculosa* growth. No effects were seen when *C. maculosa* was grown in the presence of different soil microbial communities when competing plants were absent, indicating that these effects are not direct or one to one, but a part of the far more complicated system (Bais et al., 2006).

8.4.4 Conclusion

This section depicts several aspects of plant–microbe interaction. In spite of that, one cannot completely argue that we have a complete picture of the interactions that occur between the plants and the microbes and the role of the either parties in mediating some of these processes. Our understanding of these interactions is incomplete due to the difficulty of studying these complex processes under controlled yet natural conditions. Thus, developing novel methodologies to study these interactions under natural conditions is needed. Joint efforts of scientists working in areas of plant biology, microbial ecology, and pedology need to develop systems where studies could be performed on-site to understand the minute details of interactions between plants and microbes. It is now well understood that roots are rhizosphere mediators facilitating communication between the plant and microorganisms in the soil. Ecological knowledge supports that aboveground interactions could potentially be used to understand belowground interactions between plants and microbes. An understanding of the molecular process involved in the actual secretion of molecules by roots is also needed in order to develop molecular markers for this process. Finally, improvement of the knowledge of the key factors affecting the interactions between microbes and plants from the molecular to the ecosystem level can systematically lead to the scientific development of better plants capable of absorbing more nutrients, detoxifying soils with better efficiency or eliminating invasive weeds and pathogens effectively.

In the end, a clear understanding of the interactions between plants and microbes will enable us to generate crop plants with appropriate nutrient quality and pathogen resistivity without compromising either with the benefits obtained in terms of symbiosis or disrupting the ecological balance.

8.5 INTERACTION BETWEEN PLANTS, METALS, AND MICROBES

In the earlier parts of this chapter, it has been elaborately discussed regarding the plant–metal, metal–microbe, and the plant–microbe interactions. In this section, the center stage is offered to the idea of the interaction between plants, metals, and microbes as a whole. A general idea is formulated based on the prediscussed sections that the symbiotic relationship between microbes and plants helps in the process of remediation of heavy metals in the soil and aids in the growth of the plant. This section is dedicated to determine the relevance of this "general idea."

8.5.1 HIGHLIGHTS D'INTERACTION

Symbiosis is a broader heading for the participants of the action. As long as both the parties are benefited, it is appropriate to pool them under the same heading as "symbionts." Plants and microbes also exist in a symbiotic relationship, and this useful clue has been widely studied upon and experimented with. Metals, however, in this particular scenario play a despicable role of sorts. Metals, specifically heavy metals (As, Cd, Fe, Pb, Cr, Cu, etc.), present in the soil, interfere with the natural proceedings of the plants as well as microbes inhabiting in it.

Heavy metals find their way into the soil system as a result of unethical disposal of industrial waste. The metals are usually found in the bioavailable or nonbioavailable forms (Sposito, 2000). Contamination of the rhizospheric soil by heavy metals affects the symbiotic relationship between plants and microbes (Kamnev, 2008). Nevertheless, there are exceptions. Several metals in small concentrations prove to be a source of nutrient both for plants and for microbes such as iron (Lasat, 2000). Low concentration of metals facilitates numerous physiological processes of plants (Kamnev, 2008).

8.5.2 DETRIMENTAL EFFECT OF METALS ON PLANT AND MICROBE INTERACTION

The presence of heavy metals leads to damages to the cell membrane and the DNA structure, brings about alteration to the enzyme specificity, and interrupts the cellular functions of the microbial cell (Visioli et al., 2015). The presence of heavy metals in the soil also affects the PGPR in the rhizosphere (Chaudhary and Khan, 2015). The growth of plants is affected when the surrounding soil is laden with a high concentration of toxic heavy metals. Heavy metal contamination in the soil also interrupts with the physiological process of the plant cells, namely, photosynthesis (Wani et al., 2007). Respiration, protein synthesis, and carbohydrate metabolism in plant cells are also inactivated under the influence of heavy metals in the soil (Shakolnik, 1984).

However, there are reports of plants that are accumulators of metals where they avoid a similar fate. These plants accumulate the metals mostly in the aerial parts (Khan et al., 2008). These members of the plant kingdom generally thrive in the areas restricted to challenging growth conditions. In this case, the plant grows in soil contaminated with toxic heavy metals. Plants resort to several mechanisms in order to deal with the high concentration of toxic metals in the soil. The mechanisms are phytoextraction, phytodegradation, rhizodegradation, phytostabilization, and phytovolatilization (Khan et al., 2008). All these mechanisms are subsections of the broader heading of phytoremediation. Phytoextraction deals with the mechanisms of dissolution, root absorption, root-to-shoot transport, and storage in order to remediate soil from heavy metal contaminants (Schnoor, 1997). Phytodegradation, also termed as phytotransformation, works based on the factors such as uptake efficiency, transpiration rate of the plant, and concentration of metal in the soil (Miller, 1996). Phytovolatilization helps in the release of the contaminants from the soil in their volatile form into the atmosphere by means of the process of transpiration by plants (Newman et al., 1997). Phytostabilization helps in the immobilization of the contaminants in the roots of the plants and thereby reduces the mobility of the contaminants and their extent of migration into groundwater or the deeper layers of soil (Burken and Schnoor, 1997). Rhizodegradation, also termed as phytostimulation, helps in the remediation process by disintegrating the heavy metal contaminants in the soil with the help of the bacteria present in the rhizosphere (Han et al., 2006).

Just like plants, several organisms exist in the microbial kingdom that are the accumulators of heavy metals and thereby help in the remediation of the metals present in the soil (Rajendran et al., 2003). *Escherichia coli* is one of the examples of heavy metal accumulator. It finds its use in remediating soil from Cd contamination (Khan et al., 2015). *Bacillus arsenics* help in remediating soil from As contamination (Podder and Majumder, 2015).

8.5.3 ADVANTAGE OF PLANT–METAL–MICROBE INTERACTION

The greatest advantage of this typical interaction of plant–metal–microbe is its "remediatory" effect on soil. Soil is often contaminated with pollutants that are not easy to degrade. Hyperaccumulators resolve the problem easily and effectively by accumulating the heavy metals in their aerial parts (Cervantes et al., 2001). This process is aided by the microorganisms that inhabit the rhizospheric region around the roots of the hyperaccumulators and also by those microbes that reside within the endospheric region of the plants (Kabata-Pendias, 2011).

Hyperaccumulators mainly inhabit in the tropical and temperate zones around the globe (Wu et al., 2006a). These hyperaccumulators are also identified by the name "metallophytes" (Baker and Brooks, 1989). A small number of genes are believed to be responsible for the hypertolerance observed in the metallophytes (Clemens, 2001). These metallophytes characteristically produce little biomass, and their growth rate is slow compared to nonaccumulators (Chaudhary and Khan, 2015). However, this field has not been well traversed and provides a substantial window for exploration.

The growth of these plants is primarily aided by the microorganisms that dwell in the immediate soil region, closer to the roots, namely, the rhizospheric region and also within the plant system, namely, the endospheric region. The rhizosphere is a multifaceted milieu comprising several life forms such as plant roots, invertebrates, fungi, bacteria, virus, and both organic and inorganic fractions (Manara, 2012). The microbes in the rhizospheric region are metal tolerants and facilitate the growth of the plants by several mechanisms, explicitly, by proliferation of root hairs thereby increasing the surface area, facilitating the metal uptake and channelling them to the aerial parts of the plants (Wu et al., 2009). The rhizosphere bacteria adsorb nutrients secreted by the roots in the form of organic acids, carbohydrates, enzymes, and amino acids (Kamnev, 2008). In return, the microbes help in the uptake and bioavailability of the metals in the soil. There exist several transporters, chelators, and chaperones for heavy metals in plants as well as microbes that have been identified over the years that help in the uptake of heavy metals (Lemanceau et al., 2009). The transporters have been cloned and thoroughly explored in order to study about the phenomenon of metal trafficking (Wu et al., 2006). By channelling the toxic heavy metals in the aerial parts of the plant, the microbes help in immobilizing the metals and hence remediate the contaminated soil and also prevent leaching of the these metals into deeper strata of the soil.

The immobilization process is carried out by microbes with the help of these methods: biosorption and intracellular accumulation; metal-binding peptides, protein, polysaccharides, and other biomolecules; metal precipitation by sulfate-reducing bacteria; bacterial and fungal oxidation; and phosphatase-mediated metal precipitation (Gadd, 2004). Biosorption is carried out by microbes by metabolically or physicochemically mediating the metals in the soil for uptake

(Wu et al., 2006). Metal-binding peptides such as metallothioneins and phytochelatins get expressed that help in the accumulation of metals (Wu et al., 2006). The mobilization process by microbes is carried out with the help of several methods: heterotrophic leaching, autotrophic leaching, siderophores, and biomethylation (Gadd, 2004). Leaching of metal by microbes takes place with the formation of organic and inorganic acids, oxidation and reduction reactions, and the elimination of by-products (Manara, 2012). They also alleviate the bioavailability of several metals by participating in redox transformations (Gadd, 2004).

Examples of typical microorganisms associated with the hyperaccumulators are *Arthrobacter, Microbacterium, Bacillus,* and *Curtobacterium,* belonging to the house of Gram-positive bacteria (Han et al., 2006). *Pseudomonas, Sphingomonas,* and *Variovorax* are some of the examples of Gram-negative bacteria widely studied in association with the hyperaccumulators (Han et al., 2006).

Studies have also revealed that the incorporation of the typical microorganisms helping plants with their survival in heavy metal–contaminated soil can also help nonhyperaccumulators to also accrue toxic metals in the aerial parts of the plants (Rout and Das, 2003). The microbes protect the plants from the toxic effects of the heavy metals by the incorporation of two mechanisms—the microbes adsorb the heavy metals and they form various bacterial populations on the surface of the roots (Pischik et al., 2009).

Several examples of hyperaccumulators and microbes that help in accumulating metals are discussed as follows, and they are further enlisted in Table 8.1.

Alyssum bertolonii accumulates Ni with the help of *Leifsonia* sp., *Curtobacterium* sp., etc. The microbe belongs to the endospheric region and aids in the root colonizing ability of the plants (Barzanti et al., 2007). *Alyssum murale* also accumulates Ni, and it helps the plant in Ni uptake from the soil (Abou-Shanab et al., 2003, 2006). *Alyssum pintodasylvae* accumulates Ni and helps in an increased rate of plant

TABLE 8.1
List of Certain Examples of Hyperaccumulators, the Metals They Remediate, and the Microbes Aiding in the Process of Remediation

Hyperaccumulator	Metal Remediated	Aiding Microbe(s)	Reference
Alyssum bertolonii	Ni	*Leifsonia* sp., *Curtobacterium* sp.	Barzanti et al. (2007)
Alyssum murale	Ni	*Leifsonia* sp., *Curtobacterium* sp.	Abou-Shanab et al. (2003,2007)
Alyssum pintodasylvae	Ni	*Leifsonia* sp., *Curtobacterium* sp.	Cabello-Conejo et al. (2014)
Alyssum serpyllifolium	Ni	*Leifsonia* sp., *Curtobacterium* sp.	Becerra-Castro et al. (2011,2013)
Arabidopsis halleri	Cd, Zn	*Chryseobacterium* sp., *Tsukamurella* sp.	Farinati et al. (2011)
Brassica napus	Pb	*Chryseobacterium* sp., *Tsukamurella* sp.	Sheng et al. (2008)
Brassica juncea	Cd, Ni	*Rhodococcus* sp., *Flavobacterium* sp., *Psychrobacter* sp.,	Ma et al. (2011)
Chenopodium ambrosioides	Pb, Zn	*Exiguobacterium* sp., *Paenibacillus* sp., *Comamonas* sp.,	Zhang et al. (2012)
Noccaea caerulescens	Ni	*Kocuria rhizophila*	Visioli et al. (2014)
Pteris multifida	As	*Massilia* sp., *Brevundimonas* sp., *Paracoccus* sp., *Roseomonas* sp.	Zhu et al. (2014)
Pteris vittata	As	*Naxibacter* sp., *Acinetobacter* sp., *Caryophanon* sp.	Zhu et al. (2014)

biomass accumulation (Cabello-Conejo et al., 2014). *Alyssum serpyllifolium* also accumulates Ni and helps in the solubilization of the metal and its increased accumulation in the shoots of the plants (Becerra-Castro et al., 2011, 2013).

Arabidopsis halleri accumulates Cd–Zn with the help of *Chryseobacterium* sp., *Tsukamurella* sp., etc. The microbes reside in the rhizospheric region and aid in increasing Cd–Zn shoot content and chlorophyll content, and they also help in increasing the photosynthesis and stress-related abiotic proteins (Farinati et al., 2011). *Brassica napus* accumulates Pb and aids in boosting the process of root elongation and uptake of Pb from the soil (Sheng et al., 2008). *Brassica juncea* accumulates Cd and Ni with the help of *Rhodococcus* sp., *Flavobacterium* sp., *Psychrobacter* sp., etc. The microbes dwell in the endospheric region and help in Ni uptake and also boost root and shoot elongation (Ma et al., 2011).

Chenopodium ambrosioides accumulates Pb–Zn with the help of *Exiguobacterium* sp., *Paenibacillus* sp., *Comamonas* sp., etc. The microbes reside in the rhizospheric region, and they help in Pb uptake and increase plant biomass accumulation (Zhang et al., 2012). *Noccaea caerulescens* accumulates Ni with the help of *Kocuria rhizophila*. The microbe inhabits the endospheric region and helps in root elongation (Visioli et al., 2014). *Pteris multifida* accumulates As with the help of *Massilia* sp., *Brevundimonas* sp., *Paracoccus* sp., *Roseomonas* sp., etc. The microbe increases the As tolerance for the plant and resides in the endospheric region (Zhu et al., 2014).

Pteris vittata also accumulates As with the help of *Naxibacter* sp., *Acinetobacter* sp., *Caryophanon* sp., etc. The microbes reside in both the endospheric and rhizospheric region, and they help in As uptake, solubilization, and tolerance (Zhu et al., 2014).

8.6 FUTURE TRENDS

Microbes can significantly affect the distribution of metals in the environment with or without an association with plants and vice versa. Understanding these interactions is indirectly a promise for effective, economical, and eco-friendly metal bioremediation technology for industrial exploitation and pollution-free environment. A good and efficient metal biosorbent could probably in the near future replace the conventional use of commercial ion-exchange resins. Interdisciplinary approach between biotechnologists and metallurgists to bring lab-scale bioremediation process to land-scale technology will be acceptable for industrialists (Rajendran et al., 2003). Thus, it is indeed very crucial that we try to understand and appreciate the interactions between plants, metals, and microbes so as to accelerate the process of remediation and make this planet a better place to live in.

8.7 CONCLUSION

The plant–microbe–metal interaction is a broad field that requires extensive study in several fields such as the study of the characteristics of the microorganisms associated with the hyperaccumulators. Application of next-generation sequence analysis can provide with a better understanding of the rhizome and endospheric population of the hyperaccumulators. There lies a huge scope for exploration to study about interaction between rhizospheric bacteria and hyperaccumulators as well as nonhyperaccumulators. This particular interaction can be widely applied to remediate soil with heavy metals that proved to be toxic and are not particularly degradation friendly.

REFERENCES

Abou-Shanab, R.A., J.S. Angle, T.A. Delorme, R.L. Chaney, P. van Berkum, H. Moawad, K. Ghanem, and H.A. Ghozlan. 2003. Rhizobacterial effects on nickel extraction from soil and uptake by *Alyssum murale*. *New Phytologist* 158: 219–224.

Abou-Shanab, R.A., P. van Berkum, and J.S. Angle. 2007. Heavy metal resistance and genotypic analysis of metal resistance genes in Gram-positive and Gram-negative bacteria present in Ni-rich serpentine soil and in the rhizosphere of *Alyssum murale*. *Chemosphere* 68: 360–336.

Abramova, A., A.Gedanken, V. Popov, E.H. Ooi, T.J. Mason, E.M. Joyce, and V. Bayazitov. 2013. A sonochemical technology for coating of textiles with antibacterial nanoparticles and equipment for its implementation. *Materials Letters* 96: 121–124.

Ademar Avelar Ferreira, P., C.A. Ceretta, H.T. Hildebrand Soriani, Luiz Tiecher, C.R. Fonsêca Sousa Soares, L.V. Rossato, and P. Cornejo. 2015. *Rhizophagus clarus* and phosphate alter the physiological responses of *Crotalaria juncea* cultivated in soil with a high Cu level. *Applied Soil Ecology* 91: 37–47.

Ahluwalia, S.S. and D. Goyal. 2007. Microbial and plant derived biomass for removal of heavy metals from wastewater. *Bioresource Technology* 98: 2243–2257.

Ali, S., F. Zeng, L. Qiu, and G. Zhang. 2011. The effect of chromium and aluminum on growth, root morphology, photosynthetic parameters and transpiration of the two barley cultivars. *Biologia Plantarum* 55: 291–296.

Almeida, C.M.R., A.P. Mucha, and M.T.S.D. Vasconcelos. 2004. Influence of the sea rush *Juncus maritimus* on metal concentration and speciation in estuarine sediment colonized by the plant. *Environmental Science and Technology* 38: 3112–3118.

Alva, A.K., D.G. Edwards, C.J. Asher, and F.P.C. Blamey. 1986. Effects of phosphorus/aluminum molar ratio and calcium concentration on plant response to aluminum toxicity. *Soil Science Society of America Journal* 50: 133–137.

Andreini, C., I. Bertini, G. Cavallaro, G.L. Holliday, and J.M. Thornton. 2008. Metal ions in biological catalysis: From enzyme databases to general principles. *Journal of Biological Inorganic Chemistry* 13: 1205–1218.

Andreotti, F., A.P. Mucha, C. Caetano, P. Rodrigues, C. Rocha Gomes, and C.M.R. Almeida. 2015. Interactions between salt marsh plants and Cu nanoparticles—Effects on metal uptake and phytoremediation processes. *Ecotoxicology and Environmental Safety* 120: 303–309.

Andrews, J.H. and R.F. Harris. 1986. r-selection and k-selection and microbial ecology. *Advances in Microbial Ecology* 9: 9–147.

Atafar, Z., A. Mesdaghinia, J. Nouri, M. Homaee, M. Yunesian, M. Ahmadimoghaddam, and A.H. Mahvi. 2010. Effect of fertilizer application on soil heavy metal concentration. *Environmental Monitoring and Assessment* 160(1–4): 83–89.

Bachman, G.R. and W.B. Miller. 1995. Iron chelate inducible iron/manganese toxicity in zonal geranium. *Journal of Plant Nutrition* 18: 1917–1929.

Bais, H.P., T.L. Weir, L.G. Perry, S. Gilroy, and J.M. Vivanco. 2006. The role of root exudates in rhizosphere interactions with plants and other organisms. *Annual Review of Plant Biology* 57: 233–266.

Baker, A.J.M. and R.R. Brooks. 1989. Terrestrial higher plants which hyperaccumulate metallic elements—A review of their distribution, ecology and phytochemistry. *Biorecovery* 1: 81–126.

Barzanti, R., F. Ozino, M. Bazzicalupo, R. Gabbrielli, F. Galardi, C. Gonnelli and A. Mengoni. 2007. Isolation and characterization of endophytic bacteria from the nickel hyperaccumulator plant *Alyssum bertolonii*. *Microbial Ecology* 53(2): 306–316.

Becerra-Castro, C., P. Kidd, M. Kuffner, Á Prieto-Fernández, S. Hann, C. Monterroso, A. Sessitsch, W. Wenzel, and M. Puschenreiter. 2013. Bacterially induced weathering of ultramafic rock and its implications for phytoextraction. *Applied and environmental microbiology* 79(17): 5094–5103.

Becerra-Castro, C., Á. Prieto-Fernández, V. Álvarez-Lopez, C. Monterroso, M.I. Cabello-Conejo, M.J. Acea, P.S. Kidd. 2011. Nickel solubilizing capacity and characterization of rhizobacteria isolated from hyperaccumulating and non-hyperaccumulating subspecies of *Alyssum serpyllifolium*. *International Journal of Phytoremediation* 13(suppl 1): 229–244.

Bending G.D. 2003. The rhizosphere and its microorganisms. *Encyclopaedia of Applied Plant Sciences* 3: 1123–1129.

Bruins, M.R., S. Kapil, and F.W. Oehme. 2000. Microbial resistance to metals in the environment. *Ecotoxicology and Environmental Safety* 45(3): 198–207.

Burken, J.G. and J.L. Schnoor. 1997. Uptake and metabolism of atrazine by poplar trees. *Environmental Science and Technology* 31: 1399–1406.

Cabello-Conejo, M.I., C. Becerra-Castro, A. Prieto-Fernández, C. Monterroso, A. Saavedra-Ferro, M. Mench, and P.S. Kidd. 2014. Rhizobacterial inoculants can improve nickel phytoextraction by the hyperaccumulator *Alyssum pintodasilvae*. *Plant and Soil* 379(1–2): 35–50.

Callaway, R.M., G.C. Thelen, S. Barth, P.W. Ramsey, and J.E. Gannon. 2004. Soil fungi alter interactions between the invader *Centaurea maculosa* and North American natives. *Ecology* 85: 1062–1071.

Cervantes, C., J. Campos-Garcia, S. Devars, F. Gutierrez-Corona, H. Loza-Tavera, J. Torres-Guzman, and R. Moreno-Sanchez. 2001. Interactions of chromium with micro-organisms and plants. *FEMS Microbiology Reviews* 25: 335–347.

Chamseddine, M., B.A. Wided, H. Guy, C. Marie-Edith, and J. Fatma. 2009. Cadmium and copper induction of oxidative stress and antioxidative response in tomato (*Solanum lycopersicon*) leaves. *Plant Growth Regulation* 57: 89–99.

Chandran, D., N. Sharopova, S. Ivashuta, J.S. Gantt, K.A. VandenBosch, and D.A. Samac. 2008. Transcriptome profiling identified novel genes associated with aluminum toxicity, resistance and tolerance in *Medicago truncatula*. *Planta* 228: 151–166.

Chaney, R.L. 1993. Zinc phytotoxicity. *Developments in Plant and Soil Sciences* 55: 135–150.

Chaney, R.L., M. Malik, Y.M. Li, S.L. Brown, E.P. Brewer, J.S. Angle, and A.J.M. Baker. 1997. Phytoremediation of soil metals. *Current Opinion in Biotechnology* 8: 279–284.

Chaudhary, K. and S. Khan. 2015. Plant microbe-interaction in heavy metal contaminated soil. *Indian Research Journal of Genetics and Biotechnology* 7: 235–240.

Clarke, H.R., J.A. Leigh, and C.J. Douglas. 1992. Molecular signals in the interactions between plants and microbes. *Cell* 71: 191–199.

Clarke, N.D. and J.M. Berg. 1998. Zinc fingers in *Caenorhabditis elegans*: Finding families and probing pathways. *Science* 282: 2018–2022.

Clemens, S. 2001. Molecular mechanisms of plant metal tolerance and homeostasis. *Planta* 212: 475–486.

Crawford, Jr, T.W., J.L. Stroehlein, and R.O. Kuehl. 1989. Manganese and rates of growth and mineral accumulation in cucumber. *Journal of the American Society for Horticultural Science* (*USA*) 114: 300–306.

Darkó, É., H. Ambrus, É. Stefanovits-Bányai, J. Fodor, F. Bakos, and B. Barnabás. 2004. Aluminium toxicity, Al tolerance and oxidative stress in an Al-sensitive wheat genotype and in Al-tolerant lines developed by in vitro microspore selection. *Plant Science* 166: 583–591.

Delhaize, E., P.R. Ryan, and P.J. Randall. 1993. Aluminum tolerance in wheat (*Triticum aestivum* 1). *Plant Physiology* 103: 695–702.

Dixit, R., D. Malaviya, K. Pandiyan, U.B. Singh, A. Sahu, R. Shukla, B.P. Singh, J.P. Rai, P.K. Sharma, H. Lade, and D. Paul, D. 2015. Bioremediation of heavy metals from soil and aquatic environment: An overview of principles and criteria of fundamental processes. *Sustainability* 7: 2189–2212.

Doelman, P. 1986. Resistance of soil microbial communities to heavy metals. In: Jensen, V., Kioller, A., and Sorensen, C.H. (eds.) *Microbial Communities in Soil*. London, U.K.: Elsevier Applied Science Publishers, pp. 369–384.

Elamin, O.M. and G.E. Wilcox. 1986. Effect of magnesium and manganese nutrition on muskmelon growth and manganese toxicity. *Journal of the American Society for Horticultural Science* 111: 582–587.

Elkan, G.H. 1961. A nodulation-inhibiting root excretion from a non-nodulating soybean strain. *Canadian Journal of Microbiology* 7: 851–856.

Ernst, W. 1974. *Schwermetallvegetation der erde*. Stuttgart, Germany: Gustav Fischer Verlag.

Ezaki, B., R.C. Gardner, Y. Ezaki, and H. Matsumoto. 2000. Expression of aluminum-induced genes in transgenic arabidopsis plants can ameliorate aluminum stress and/or oxidative stress. *Plant Physiology* 122: 657–665.

Fabrega, J., S.N. Luoma, C.R. Tyler, T.S. Galloway, and J.R. Lead. 2011. Silver nanoparticles: Behaviour and effects in the aquatic environment. *Environment International* 37: 517–531.

Farinati, S., G. DalCorso, M. Panigati, and A. Furini. 2011. Interaction between selected bacterial strains and *Arabidopsis halleri* modulates shoot proteome and cadmium and zinc accumulation. *Journal of Experimental Botany* 62: 3433–3447.

Fons, F., N. Amellal, C. Leyval, N. Saint-Martin, and M. Henry. 2003. Effects of *Gypsophila saponins* on bacterial growth kinetics and on selection of subterranean clover rhizosphere bacteria. *Canadian Journal of Microbiology* 49: 367–373.

Fontes, R.L.F. and F.R. Cox. 1995. Effects of sulfur supply on soybean plants exposed to zinc toxicity. *Journal of Plant Nutrition* 18: 1893–1906.

Fry, W. 2008. Phytophthora infestans: The plant (and R gene) destroyer. *Molecular and Plant Pathology* 9: 385–402.

Gadd, G.M. 2004. Microbial influence on metal mobility and application for bioremediation. *Geoderma* 122: 109–119.

Genre, A., M. Chabaud, C. Balzergue, V. Puech-Pages, M. Novero, D.G. Barker et al. 2013. Short-chain chitin oligomers from arbuscular mycorrhizal fungi trigger nuclear Ca(2+) spiking in *Medicago truncatula* roots and their production is enhanced by strigolactone. *New Phytologist* 198: 190–202.

Gisbert, C., R. Ros, A. De Haro, D.J. Walker, M.P. Bernal, R. Serrano, and J. Navarro-Aviñó. 2003. A plant genetically modified that accumulates Pb is especially promising for phytoremediation. *Biochemical and Biophysical Research Communications* 303(2): 440–445.

Grayston, S.J., S.Q. Wang, C.D. Campbell, and A.C. Edwards. 1998. Selective influence of plant species on microbial diversity in the rhizosphere. *Soil Biology and Biochemistry* 30: 369–378.

Gupta, N., S.S. Gaurav, and A. Kumar. 2013. Molecular basis of aluminium toxicity in plants: A review. *American Journal of Plant Sciences* 4: 21–37.

Gupta, U.C. 1972. Effects of manganese and lime on yield and on the concentrations of manganese, molybdenum, boron, copper and iron in the boot stage tissue of barley. *Soil Science* 114: 131–136.

Guzel, S. and R. Terzi. 2013. Exogenous hydrogen peroxide increases dry matter production, mineral content and level of osmotic solutes in young maize leaves and alleviates deleterious effects of copper stress. *Botanical Studies* 54: 2–10.

Haferburg, G. and E. Kothe. 2007. Microbes and metals: Interactions in the environment. *Journal of Basic Microbiology* 47(6): 453–467.

Han, F., X.Q. Shan, S.Z. Zhang, B. Wen, and G. Owens. 2006. Enhanced cadmium accumulation in maize roots—The impact of organic acids. *Plant and Soil* 289: 355–368.

Harmens, H., N.G.C.P.B. Gusmao, P.R. Denhartog, J.A.C.Verkleij, and W.H.O. Ernst. 1993. Uptake and transport of zinc in zinc-sensitive and zinc-tolerant silene vulgaris. *Journal of Plant Physiology* 141: 309–315.

He, H.Y., L.F. He, M.H. Gu, and X.F. Li. 2012. Nitric oxide improves aluminum tolerance by regulating hormonal equilibrium in the root apices of rye and wheat. *Plant Science* 183: 123–130.

Horst, W.J. and H. Marschner. 1978. Effect of excessive manganese supply on uptake and translocation of calcium in bean plants (*Phaseolus vulgaris* L.). *Zeitschrift Für Pflanzenphysiologie* 87: 137–148.

Hussein, H., S. Farag, and H. Moawad. 2004. Isolation and characterization of Pseudomonas resistant to heavy metals contaminants. *Arab Journal of Biotechnology* 7: 13–22.

Imadi, S.R., S. Waseem, A.G. Kazi, M.M. Azooz, and P. Ahmad. 2015. Chapter 1. Aluminum toxicity in plants: An overview. In: Ahmad, P. (ed.) *Plant Metal Interaction, Emerging Remediation Techniques.* Amsterdam, the Netherlands: Elsevier.

Indeka L. and Z. Karczun. 1999. Accumulation of selected heavy metals in soils along busy traffic routes. *Ecology and Technology* 6: 174–180.

Jonak, C., H. Nakagami, and H. Hirt. 2004. Heavy metal stress. Activation of distinct mitogen-activated protein kinase pathways by copper and cadmium. *Plant Physiology* 136: 3276–3283.

Jones, D.L. and L.V. Kochian. 1997. Aluminum interaction with plasma membrane lipids and enzyme metal binding sites and its potential role in Al cytotoxicity. *FEBS Letters* 400: 51–57.

Kabata-Pendias, A. 2011. *Trace Elements in Soils and Plants,* 4th edn. Boca Raton, FL: CRC Press, Taylor & Francis.

Kamnev, A.A. 2008. Metals in soil versus plant–microbe interactions: Biotic and chemical interferences. *Plant-Microbe Interactions* 291–318.

Kapoor, A. and T. Viraraghvan. 1995. Fungal biosorption—An alternative treatment option for heavy metal bearing wastewater: A review. *Bioresource Technology* 53, 195–206.

Khan, M.S., A. Zaidi, P.A. Wani, and M. Oves. 2008. Role of plant growth promoting rhizobacteria in the remediation of the metal contaminated soils. *Environmental Chemistry Letters* 7: 1–19

Khan, Z., M.A. Nisar, Z.F. Hussain, M.N. Arshad, and A. Rehman. 2015. Cadmium resistance mechanism in *Escherichia coli* P4 and its potential use to bioremediate environmental cadmium. *Applied Microbiology and Biotechnology* 99: 10745–10757.

Kim, S., S. Lee, and I. Lee. 2012. Alteration of phytotoxicity and oxidant stress potential by metal oxide nanoparticles in *Cucumis sativus. Water, Air, & Soil Pollution* 223: 2799–2806.

Kochian, L.V., O.A. Hoekenga, and M.A. Pineros. 2004. How do crop plants tolerate acid soils? Mechanisms of aluminum tolerance and phosphorous efficiency. *Annual Review of Plant Biology* 55: 459–493.

Ku, G., M. Zhou, S. Song, Q. Huang, J. Hazle, and C. Li. 2012. Copper sulfide nanoparticles as a new class of photoacoustic contrast agent for deep tissue imaging at 1064 nm. *ACS Nano* 6: 7489–7496.

Laila, P., P. Martinez, and C. Hertweck. 2005. Pathogenic fungus harbours endosymbiotic bacteria for toxin production. *Nature* 437: 884–888.

Lasat, M.M. 2000. Phytoextraction of metals from contaminated soil: A review of plant/soil/metal interaction and assessment of pertaining agronomic issues. *Journal of Hazardous Substance Research* 2: 1–25.

Lee, C.W., M.B. Jackson, M.E. Duysen, T.P. Freeman, and J.R. Self. 1996. Induced micronutrient toxicity in "Touchdown" Kentucky bluegrass. *Crop Science* 36: 705–712.

Lemanceau, P., P. Bauer, S. Kraemer, and J.F. Briat. 2009. Iron dynamics in the rhizosphere as a case study for analyzing interactions between soil, plants and microbes. *Plant Soil* 321: 513–535.

Lenart-boroń, A. and P. Boroń. 2014. The effect of industrial heavy metal pollution on microbial abundance and diversity in soils—A review. Environmental risk assessment of soil contamination. Edited by Maria C. Hernandez-Soriano. ISBN 978-953-51-1235-8 Publisher: InTech.

Lin, D. and B. Xing. 2007. Phytotoxicity of nanoparticles: Inhibition of seed germination and root growth. *Environmental Pollution* 150: 243–250.

Liu, J.J., Z. Wei, and J.H. Li. 2014. Effects of copper on leaf membrane structure and root activity of maize seedling. *Botanical Studies* 55: 1–6.

Llorens, A., E. Lloret, P.A. Picouet, R. Trbojevich, and A. Fernandez. 2012. Metallic-based micro and nanocomposites in food contact materials and active food packaging. *Trends in Food Science and Technology* 24: 19–29.

Lone, M. I., Z. He, P.J. Stoffella, and X. Yang. 2008. Phytoremediation of heavy metal polluted soils and water: progresses and perspectives. *Journal of Zhejiang University—Science B* 9(3): 210–220.

Louw, H.A. and D.M. Webley. 1959. The bacteriology of the root region of the oat plant grown under controlled pot culture conditions. *Journal of Applied Bacteriology* 22: 216–226.

Ma, J.F. 2007. Syndrome of aluminum toxicity and diversity of aluminum resistance in higher plants. *International Review of Cytology* 264: 225–252.

Ma, J.F. and S. Hiradate. 2000. Form of aluminium for uptake and translocation in buckwheat (*Fagopyrum esculentum* Moench). *Planta* 211: 355–360.

Ma, J.F., P.R. Ryan, and E. Delhaize. 2001. Aluminium tolerance in plants and the complexing role of organic acids. *Trends in Plant Science* 6: 273–278.

Ma, Y., M.N.V. Prasad, M. Rajkumar, and H. Freitas. 2011. Plant growth promoting rhizobacteria and endophytes accelerate phytoremediation of metalliferous soil. *Biotechnology Advances* 29: 248–258.

Macura, J. 1958. Continuous-flow method for the study of microbiological processes in soil samples. *Nature* 182: 1796–1797.

Manara, A. 2012. Plant responses to heavy metal toxicity, In: Furini, A. (ed.) *Plants and Heavy Metals*, Springer Briefs in Biometals. Springer, Netherlands, pp. 27–53.

Maron, L.G., M. Kirst, C. Mao, M.J. Milner, M. Menossi, and L.V. Kochian. 2008. Transcriptional profiling of aluminum toxicity and tolerance responses in maize roots. *New Phytologist* 179: 116–128.

Marschner, P. 2012. *Marschner's Mineral Nutrition of Higher Plants*. Academic press.

Matsumoto, H., E. Hirasawa, S. Morimura, and E. Takahashi. 1976. Localization of aluminium in tea leaves. *Plant & Cell Physiology* 631: 627–631.

Miller, R. 1996. Phytoremediation, technology overview report, Series O, Vol. 3. Ground-water Remediation Technologies Analysis Center, Pittsburgh, PA.

Morgan, J.A.W., G.D. Bending, and P.J. White. 2005. Biological costs and benefits to plant–microbe interactions in the rhizosphere. *Journal of Experimental Botany* 56: 1729–1739.

Mossor-Pietraszewska, T. 2001. Effect of aluminium on plant growth and metabolism. *Acta Biochimica Polonica* 48: 673–686.

Mugabo, J.P., B.S. Bhople, A. Kumar, H. Erneste, B. Emmanuel, and Y.N. Singh. 2014. Contribution of arbuscular mycorrhizal fungi (am fungi) and rhizobium inoculation on crop growth and chemical properties of rhizospheric soils in high plants. *IOSR Journal of Agriculture and Veterinary Science* 7: 45–55.

Nair, P.M.G. and I.M. Chung. 2015. Study on the correlation between copper oxide nanoparticles induced growth suppression and enhanced lignification in Indian mustard (*Brassica juncea* L.). *Ecotoxicology and Environmental Safety* 113: 302–313.

Neilands, J.B. 1952. A crystalline organo-iron pigment from a rust fungus (Ustilago sphaerogena). *Journal of the American Chemical Society* 74(19): 4846–4847.

Newman, L.A., S.E. Strand, N. Choe, J. Duffy, G. Ekuan, M. Ruszaj, J. Shurtleff, J. Wilmoth, P. Heilman, and M.P. Gordon. 1997. Uptake and bio transformation of trichloroethylene by hybrid poplars. *Environmental Science and Technology* 31: 1062–1067.

Nies, D.H. 2003. Efflux-mediated heavy metal resistance in prokaryotes. *FEMS Microbiology Reviews* 27(2–3): 313–339.

O'Sullivan, J.N., C.J. Asher, and F.P.C. Blamey. 1997. *Nutrient Disorders of Sweet Potato*, Vol. 48. Canberra, Australian Capital Territory, Australia: Australian Centre for International Agricultural Research Canberra.

Pan, W.L., A.G. Hopkins, and W.A. Jackson. 1988. Aluminum-inhibited shoot development in soybean: A possible consequence of impaired cytokinin supply. *Communications in Soil Science & Plant Analysis* 19: 1143–1153.

Panda, S.K. and H. Matsumoto. 2007. Molecular physiology of aluminum toxicity and tolerance in plants. *Botanical Review* 73: 326–347.

Panou-Filotheou, H., A.M. Bosabalidis, and S. Karataglis. 2001. Effects of copper toxicity on leaves of oregano (*Origanum vulgare* subsp. *hirtum*). *Annals of Botany* 88: 207–214.

Parniske, M. 2000. Intracellular accommodation of microbes by plants: A common developmental program for symbiosis and disease. *Current Opinion in Plant Biology* 3: 320–328.

Pavan, M.A. and F.T. Bingham. 1982. Toxicity of aluminum to coffee seedlings grown in nutrient solution. *Soil Science Society of America Journal* 46: 993–997.

Petutschnig, E.K., A.M. Jones, L. Serazetdinova, U. Lipka, and V. Lipka. 2010. The lysin motif receptor-like kinase (LysM-RLK) CERK1 is a major chitin-binding protein in *Arabidopsis thaliana* and subject to chitin-induced phosphorylation. *Journal of Chemical Biology* 285: 28902–28911.

Pierzyński G.M., J.T. Sims, and G.F. Vance. 2000. *Soils and Environmental Quality*. London, U.K.: CRC Press.

Pischik, V.N., N.A. Provorov, N.I. Vorobyov, E.P. Chizevskaya, V.I. Safronova, N.A. Tuev, and A.P. Kozhemyakov. 2009. Interactions between plants and associated bacteria in soils contaminated with heavy metals. *Microbiology* 78: 785–793.

Plenderleith, R.W. 1984. An evaluation of the tolerance of a range of tropical grasses to excessive soil levels of copper and zinc. MSc thesis. University of Queensland, Brisbane, Queensland, Australia.

Plett, J.M., M. Kemppainen, S.D. Kale, A. Kohler, V. Legue, and A. Brun. 2011. A secreted effector protein of *Laccaria bicolor* is required for symbiosis development. *Current Biology* 21: 1197–1203.

Podder, M.S. and C.B. Majumder. 2015. Modelling of optimum conditions for bioaccumulation of As (III) and As (V) by response surface methodology (RSM). *Journal of Environmental of Chemical Engineering* 3: 1986–2001.

Poschenrieder, C., B. Gunsé, I. Corrales, and J. Barceló. 2008. A glance into aluminum toxicity and resistance in plants. *Science of the Total Environment* 400: 356–368.

Pradhan, A., S. Seena, D. Dobritzsch, S. Helm, K. Gerth, M. Dobritzsch, and F. Cássio. 2014. Physiological responses to nanoCuO in fungi from non-polluted and metal-polluted streams. *Science of the Total Environment* 466–467: 556–563.

Ranjard, L. and A. Richaume. 2001. Quantitative and qualitative microscale distribution of bacteria in soil. *Research in Microbiology*, 15(8): 707–716.

Rajendran, P., J. Muthukrishnan, and P. Gunasekaran. 2003. Microbes in heavy metal remediation. *Indian Journal of Experimental Biology* 41: 935–944.

Reichman, S.M. 2002. *The Responses of Plants to Metal Toxicity: A Review Focusing on Copper, Manganese and Zinc*. Melbourne, Victoria, Australia: Australian Minerals and Energy Environment Foundation, pp. 1–54.

Ren, F., T. Liu, H. Liu, and B. Hu. 1993. Influence of zinc on the growth, distribution of elements, and metabolism of one-year old American ginseng plants. *Journal of Plant Nutrition* 16: 393–405.

Rengel, Z. and R.J. Reid. 1997. Uptake of Al across the plasma membrane of plant cells. *Plant and Soil* 192: 31–35.

Rey, T. and S. Schornack. 2013. Interactions of beneficial and detrimental root colonizing filamentous microbes with plant hosts. *Genome Biology* 14: 121–125.

Richards, K.D., E.J. Schott, Y.K. Sharma, K.R. Davis, and R.C. Gardner. 1998. Aluminum induces oxidative stress genes in *Arabidopsis thaliana*. *Plant Physiology* 116: 409–418.

Rosselló-Mora, R and Amann, R. 2001. The species concept for prokaryotes. *FEMS Microbiology Reviews* 25(1): 39–67.

Rouatt, J.W. and H. Katznelson. 1961. A study of the bacteria on the root surface and in the rhizosphere soil of crop plants. *Journal of Applied Bacteriology* 24: 164–171.

Rout, G.R. and P. Das. 2003. Effect of metal toxicity on plant growth and metabolism: I. Zinc. *Agronomie* 23: 3–11.

Rout, G.R., S. Samantaray, and P. Das. 2001. Aluminium toxicity in plants: A review. *Agronomie* 21: 3–21.

Rovira A.D. 1965. Interactions between plant roots and soil microorganisms. *Annual Review of Microbiology* 19: 241–266.

Roy, A.K., A. Sharma, and G. Talukder. 1988. Some aspects of aluminum toxicity in plants. *The Botanical Review* 54: 145–178.

Safford, L.O. 1975. Effect of manganese level in nutrient solution on growth and magnesium content of *Pinus radiata* seedlings. *Plant and Soil* 42: 293–297.

Saglam, A., F. Yetissin, M. Demiralay, and R. Terzi. 2015. Chapter 2. Copper stress and responses in plants. In: Ahmad, P. (ed.) *Plant Metal Interaction, Emerging Remediation Techniques*. Amsterdam, the Netherlands: Elsevier.

Schnoor, J.L. 1997. Phytoremediation, technology overview report, Series E, Vol. 1. Ground-water Remediation Technologies Analysis Center, Pittsburgh, PA.

Shah, V. and I. Belozerova. 2009. Influence of metal nanoparticles on the soil microbial community and germination of lettuce seeds. *Water, Air, and Soil Pollution* 197: 143–148.

Shakolnik, N.Y. 1984. *Trace Elements in Plants*. New York: Elsevier, pp. 140–171.

Sheng, X.F., J.J. Xia, C.Y. Jiang, L.Y. He, and M. Qian. 2008. Characterization of heavy metal-resistant endophytic bacteria from rape (*Brassica napus*) roots and their potential in promoting the growth and lead accumulation of rape. *Environmental Pollution* 156: 1164–1170.

Sideris, C.P. 1950. Manganese interference in the absorption and translocation of radioactive iron in *Ananas comosus* (L.) Merr. *Plant Physiology* 25: 307–321.

Silva, S. 2012. Aluminium toxicity targets in plants. *Journal of Botany*. 2012: 8 pp. Article ID 219462, doi:10.1155/2012/219462.

Singh, B.K., P. Millard, A.S. Whiteley, and J.C. Murrell. 2004. Unravelling rhizosphere-microbial interactions: Opportunities and limitations. *Trends in Microbiology* 12: 386–393.

Sivaguru, M., Y. Yamamoto, Z. Rengel, S.J. Ahn, and H. Matsumoto. 2005. Early events responsible for aluminum toxicity symptoms in suspension-cultured tobacco cells. *The New Phytologist* 165: 99–109.

Sposito, F.G. 2000. The chemistry of soils. In: Maier, R.M., Pepper, I.L., and Gerba, C.B. (eds.) *Environmental Microbiology*. London, U.K.: Academic Press, pp. 406.

Sujatha, K. and S.K. Mehar. 2015. Toxic effects of aluminium in plants. *Indian Journal of Plant Sciences* 4: 1–4.

Tan, D.X., L.C. Manchester, P. Helton, and R.J. Reiter. 2007. Phytoremediative capacity of plants enriched with melatonin. *Plant Signaling & Behavior* 2: 514–516.

Tate, R.L. 2000. *Soil Microbiology*, 2nd edn. Hoboken, NJ: John Wiley & Sons Inc.

Tyagi, S., R. Singh, and S. Javeria. 2014. Effect of climate change on plant-microbe interaction: An overview. *European Journal of Molecular Biotechnology* 5: 149–156.

Tyler, G. 1974. Heavy metal pollution and soil enzymatic activity. *Plant and Soil* 41: 303–311.

Visioli, G., S. D'Egidio, and A.M. Sanangelantoni. 2015. The bacterial rhizobiome of hyperaccumulators: Future perspectives based on omics analysis and advanced microscopy. *Frontiers in Plant Science* 5: 752.

Visioli, G., S. D'Egidio, S. Vamerali, M. Mattarozzi, and A.M. Sanangelantoni. 2014. Culturable endophytic bacteria enhance Ni translocation in the hyperaccumulator *Noccaea caerulescens*. *Chemosphere* 117: 538–544.

Wallace, A. and A.M. Abou-Zamzam. 1989. Calcium-zinc interactions and growth of bush beans in solution culture. *Soil Science* 147: 442–443.

Wani, P.A., M.S. Khan, and A. Zaidi. 2007. Impact of heavy metal toxicity on plant growth, symbiosis, seed yield and nitrogen and metal uptake in chickpea. *Australian Journal of Experimental Agriculture* 47: 712–720.

Weis, J.S. and P. Weis. 2004. Metal uptake, transport and release by wetland plants: Implications for phytoremediation and restoration. *Environment International* 30: 685–700.

Welte, E. and G. Trolldenier. 1961. The influence of the hydrogen ion concentration of the soil in the rhizosphere. *Naturwissenschaften* 48: 509.

Whipps, J.M. 2004. Prospects and limitations for mycorrhizas in biocontrol of root pathogens. *Canadian Journal of Botany* 82: 1198–1227.

Winterhalder, E.K. 1963. Differential resistance of two species of *Eucalyptus* to toxic soil manganese levels. *Australian Journal of Science* 25: 363–364.

Wu, C.H., S.M. Bernard, G.L. Andersen, and W. Chen. 2009. Developing microbe-plant interactions for applications in plant-growth promotion and disease control, production of useful compounds, remediation and carbon sequestration. *Microbial Biotechnology* 2: 428–440.

Wu, C.H., T.K. Wood, A. Mulchandani, and W. Chen. 2006a. Engineering plant-microbe symbiosis for rhizoremediation of heavy metals. *Applied and Environmental Microbiology* 72(2): 1129–1134.

Wu, S. 1994. Effect of manganese excess on the soybean plant cultivated under various growth conditions. *Journal of Plant Nutrition* 17: 991–1003.

Wu, S.C., K.C. Cheung, Y.M. Luo, and M.H. Wong. 2006b. Effects of inoculation of plant growth-promoting rhizobacteria on metal uptake by *Brassica juncea*. *Environmental Pollution* 140: 124–135.

Yamamoto, Y., Y. Kobayashi, S. Devi, S. Rikiishi, and H. Matsumoto. 2002. Aluminium toxicity is associated with mitochondrial dysfunction and the production of reactive oxygen species in plant cells. *Plant Physiology* 128: 63–72.

Yeh, C.M., P.S. Chien, and H.J. Huang. 2007. Distinct signalling pathways for induction of MAP kinase activities by cadmium and copper in rice roots. *Journal of Experimental Botany* 58: 659–671.

Yuan, H.M., H.H. Xu, W.C. Liu, and Y.T. Lu. 2013. Copper regulates primary root elongation through PIN1-mediated auxin redistribution. *Plant and Cell Physiology* 54: 766–778.

Zengin, F.K. and S. Kirbag. 2007. Effects of copper on chlorophyll, proline, protein and abscisic acid level of sunflower (*Helianthus annuus* L.) seedlings. *Journal of Environmental Biology/Academy of Environmental Biology, India* 28: 561–566.

Zhang, W.H., Z. Huang, L.Y. He, and X.F. Sheng. 2012. Assessment of bacterial communities and characterization of lead-resistant bacteria in the rhizosphere soils of metal-tolerant *Chenopodium ambrosioides* grown on lead-zinc mine tailings. *Chemosphere* 87: 1171–1178.

Zhao, H., H. Fu, and R. Qiao. 2010. Copper-catalyzed direct amination of ortho-functionalized haloarenes with sodium azide as the amino source. *Journal of Organic Chemistry* 75: 3311–3316.

Zhu, H., J. Han, J.Q. Xiao, and Y. Jin. 2008. Uptake, translocation, and accumulation of manufactured iron oxide nanoparticles by pumpkin plants. *Journal of Environmental Monitoring* 10: 713–717.

Zhu, L.J., D.X. Guan, J. Luo, B. Rathinasabapathi, and L.Q. Ma. 2014. Characterization of arsenic-resistant endophytic bacteria from hyperaccumulators *Pteris vittata* and *Pteris multifida*. *Chemosphere* 113: 9–16.

9 Toxic Metals in the Environment
Threats on Ecosystem and Bioremediation Approaches

Ashutosh Yadav, Pankaj Chowdhary, Gaurav Kaithwas, and
Ram Naresh Bharagava

CONTENTS

ABSTRACT

Heavy metals are ubiquitous environmental contaminants in an industrialized society, and thus, the concern over the possible health hazards and ecosystem effects of heavy metals has increased. However, like organic pollutants, metals are not degraded and are accumulated in environments as well as in living organisms where they cause toxic, genotoxic, mutagenic, and carcinogenic effects. Contamination of soil and water with toxic metals represents a serious threat for the ecosystem and human health and, thus, requires the proper implementation of appropriate remedial measures. Although the application of many bioremediation and phytoremediation cleanup technologies is rapidly expanding, these approaches also have many limitations that should be addressed carefully for the proper implementation of cleanup technologies so that the contaminated environments can be restored.

9.1 INTRODUCTION

Rapid industrialization and modernization around the world have produced the unfortunate consequences of releasing various types of wastes containing different types of toxic metals into the environment. Heavy metals are elements with an atomic weight between 63.5 and 200.6 and specific gravity greater than 5.0 (Srivastava and Majumder, 2008). Although metal ions are essential as trace elements, at higher concentrations, these become major sources of environmental pollution and cause severe health hazards. Due to a wide range of industrial and agricultural activities, large quantities of chemical contaminants are released into the environment and cause serious environmental problems and health hazards directly or indirectly.

The contamination of environments (aquatic and terrestrial) with various toxic metals is common in India and elsewhere due to a number of industrial as well as anthropogenic activities. Different types of industrial wastes (solid and liquid) containing a number of toxic metals such as copper (Cu), cadmium (Cd), chromium (Cr), zinc (Zn), nickel (Ni), mercury (Hg), arsenic (As), and lead (Pb) are directly or indirectly discharged into the environment without adequate treatment in developing countries (Nagajyoti et al., 2010; Barakat, 2011; Dixit et al., 2015). Heavy metals are ubiquitous environmental contaminants in an industrialized society. Concern over possible health hazards and ecosystem effects of heavy metals has increased. Like organic contaminants, heavy metals are not degradable and tend to accumulate in living organisms where they cause toxic, genotoxic, mutagenic, and carcinogenic effects (Rajendran et al., 2003).

Since food chain contamination is one of the major routes for the entry of toxic metals into the animal and human body (Chandra et al., 2011). Contamination of soil and water with toxic metals represents a serious threat for the ecosystem and human health and requires the implementation of appropriate remedial measures. The application of bioremediation and phytoremediation cleanup technologies is rapidly expanding, and according to an estimate, the worldwide demand for these biological technologies is thought to be valued up to $1.5 billion per annum.

Although the microbial activity in soil accounts for the degradation of most of the organic contaminants, chemical and physical mechanisms can also provide significant transformation pathways for these compounds. The specific remediation processes that have been applied to clean up the contaminated sites include natural attenuation, land farming, biopiling or composting, use of bioreactor, bioventing, soil vapor extraction, thermal desorption, incineration, soil washing, and landfilling (Abhilash et al., 2012; Dixit et al., 2015). Biological remediation using microbes and plants is generally considered as an environment-friendly, safe, and less expensive method for the removal of toxic metals from the contaminated site.

Microorganisms have a primary catalytic role in the degradation and/or mineralization of various contaminants or converting them into the nontoxic by-products during the bioremediation of contaminants. Plants have an inherent ability to detoxify soil by the direct/indirect uptake of the contaminants followed by their subsequent transport and accumulation into different plant parts (Singh et al., 2015). In this way, plants play an important role in the bioremediation of pollutants from the contaminated environments.

9.2 TOXIC METALS: SOURCES AND CONTAMINATION IN ENVIRONMENT

The rapid industrialization and urbanization has given rise to the problem of heavy metal contamination into the environment. The environments (soil, water, and air) have been severely contaminated with organic and inorganic pollutants because different types of pollutants are released into the environment through different industrial as well as natural activities. A number of toxic metals such as copper (Cu), cadmium (Cd), chromium (Cr), zinc (Zn), nickel (Ni), mercury (Hg), arsenic (As), lead (Pb), cobalt (Co), and selenium (Se) are known as serious environmental pollutants and act as potentially toxic or carcinogenic agents in nature at very low concentrations and cause serious health hazards in humans and animals, if they enter into the food chain (Hutchinson and Meema, 1987; Jarup, 2003; Khan et al., 2008).

Nowadays, the contamination of soil and water ecosystems with toxic metals is a major environmental problem. There are a number of natural and anthropogenic activities that act as major sources of metal contamination in the environment. However, most significant natural sources of metal contamination in the environment are weathering of minerals, erosion, and volcanic activity, while anthropogenic sources largely depend upon human activities such as mining, smelting, electroplating, use of pesticides, and discharge of phosphate fertilizer as well as application of biosolids (e.g., livestock manures, composts, and municipal sewage sludge) and atmospheric deposition.

A mass balance of heavy metals in the soil environment can be expressed by using the following formula (Khan et al., 2008):

$$M_{total} = (M_p + M_a + M_f + M_{ag} + M_{ow} + M_{ip}) - (M_{cr} + M_l)$$

where

M is the heavy metal
p is the parent material
a is the atmospheric deposition
f is the fertilizer source
ag is the agrochemical source
ow is the organic waste source
ip is the inorganic pollutant
cr is the crop removal
l is losses by leaching, volatilization, and other processes

It is estimated that the emission of several heavy metals in the atmosphere from anthropogenic sources is one to three times higher than that of natural sources.

In the environment, heavy metals enter naturally from pedogenic processes of weathering of parent materials and also through various anthropogenic sources. There are different sources of heavy metal contamination in the environment as shown in Figure 9.1.

9.2.1 NATURAL SOURCES

The most important natural source of heavy metal contamination in the environment is the geological parent material or rock outcroppings. The composition and concentration of heavy metals largely depend on the rock type and environmental conditions, activating the weathering process. The geological plant materials generally have high concentrations of Cr, Cd, Cu, Zn, Ni, Pb, Hg, Co, Mn, Sn, etc.

However, class-wise heavy metal concentrations vary within rocks. Sedimentary rocks are mainly responsible for soil formation and contribute only a minute quantity of toxic metals to the environment, since these are not generally or easily weathered (Nagajyoti et al., 2010). Within the class of sedimentary rocks, shale has the highest concentrations of Cr, Mn, Co, Ni, Cu, Zn, Cd, Sn, Hg, and Pb followed by limestone and sandstone (Nagajyoti et al., 2010). But many igneous rocks such as olivine, augite, and hornblende are reported to have contributed a considerable amount of Mn, Co, Ni, Cu, and Zn into the soil (Nagajyoti et al., 2010).

9.2.2 ANTHROPOGENIC SOURCES

The anthropogenic input of various toxic metals into the environment has increased drastically since the Industrial Revolution because different metals are widely used in various industrial processes. Industries such as metallurgical, chemical, refractory brick, leather, wood preservation, pigments, and dyes are the major sources of toxic metal contamination in the environment. In this way, millions of people worldwide working in various industries such as pigment production, chrome plating, stainless steel welding, and leather tanning are exposed to various toxic metals leading to the development of various types of serious environmental problems and health hazards.

9.2.2.1 Agricultural Sources of Heavy Metal

The organic and inorganic fertilizers, including those used in liming, sewage sludge, irrigation waters, and pesticides

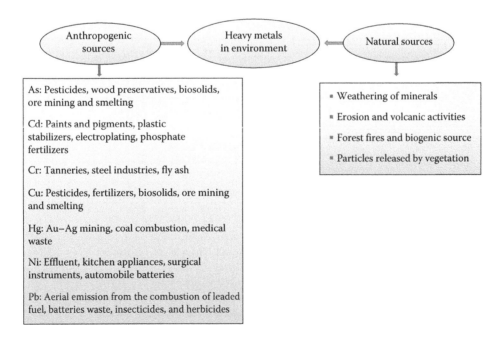

FIGURE 9.1 Various natural and anthropogenic sources of metal contamination in environment.

used in agriculture practice to enhance crop yield, are also one of the major sources of metal contamination in agricultural soil and aquatic ecosystem. Fungicides, inorganic fertilizers, and phosphate fertilizers have significant amount of Cd, Cr, Ni, Pb, and Zn depending on their sources. In crop plants, Cd gets accumulated in very high concentration in leaves, which may be consumed by animals or human beings. In addition, cadmium enrichment may also occur due to the application of sewage sludge, manure, and limes (Yanqun et al., 2005). Although the quantity of heavy metals in agricultural soil is very small, repeated use of phosphate fertilizer and long persistence time for metals can cause the levels of metals in agricultural soil to increase continuously. Animal manure enriches agricultural soil by adding Mn, Zn, Cu, and Co and sewage sludge by adding Zn, Cr, Pb, Ni, Cd, and Cu (Yanqun et al., 2005; Nagajyoti et al., 2010). The increase in heavy metal contamination in agricultural soil largely depends on the rate of the application of fertilizers, pesticides, fungicides, and animal/green manure as well as on the physicochemical characteristics of agricultural soils. In soil, the liming process contributes more heavy metal than nitrate fertilizers and compost refuse. Sewage sludge used to enhance the fertility of agricultural soil is also one of the major sources of heavy metal contamination. Several heavy metal–based pesticides used to control the diseases of grains, fruit crops, and vegetables add a good amount of heavy metals in agricultural soil (Ross, 1994).

9.2.2.2 Industrial Sources of Heavy Metals

Industries are the major sources of soil, water, and air pollution contributing a variety of pollutants into the environment. However, the major industrial processes contributing various toxic metals into the environment include mining, refinement, such as spoil heaps and tailings, transport of ores, smelting and metal finishing and recycling of metals, and tanning. Mining operations emit different types of toxic metals depending on the type of mining process. For example, coal mines are the major sources of As, Cd, Fe, etc., which enrich the soil mainly around the coalfield directly or indirectly. The use of Hg in gold mining and its transport from old mines to various industries have become a major source of Hg contamination into the environment. Leather industries using a large amount of chromium salts during the tanning process add a significant amount of chromium metal into soil and water ecosystem (Mishra and Bharagava, 2015). High-temperature processing of metals such as smelting and casting emits metals in particulate and vapor forms. This vapor form of metals such as As, Cd, Cu, Pb, Sn, and Zn combines with water in the atmosphere and forms aerosols, which may be either dispersed by wind (dry deposition) or precipitated in rainfall (wet deposition) causing contamination of soil and water ecosystems. The contamination of soil and water ecosystems with toxic metals mainly occurs through the runoff from erosion of mine wastes, dusts produced during the transport of crude ores, corrosion of metals, and leaching of heavy metals to soil and groundwater, and metal-containing wastewaters discharged from various industries as shown in Table 9.1.

TABLE 9.1

Various Types of Metals or Their Compounds in Wastewaters Discharged from Different Industries

Sr. No.	Industry	Heavy Metals
1.	Mining operations and ore processing	Al, As, Cd, Hg, Mn, Mo, Pb, Pb
2.	Metallurgy and electroplating	Ag, As, Be, Bi, Cd, Cr, Cu, Hg, In, Pb, Ni, Zn
3.	Chemical industries	Al, As, Ba, Cd, Cr, Cu, Fe, Ga, Hg, Os, Pb, Sn Ta Ti, Zn
4.	Dyes and pigments	Al, As, Cd, Cu, Fe, Pb, Sb, TI, Ti
5.	Ink manufacturing	Co, Cu, Fe, Hg, Ni
6.	Pottery and porcelain	As, Cr, Sb
7.	Alloys	Be, Ga, In, Os, Pd, Ta
8.	Print	Ba, Cr, Os, Pb, Ti, Zn
9.	Photography	Ag, Au, Cd, Cr, Mo, Pb
10.	Glass	As, Ba, Co, Ni, Ti
11.	Paper mill	Al, Cr, Cu, Hg, Pd, Sb, Ta, Ti
12.	Leather industry	Al, As, Ba, Cr, Cu, Fe, Hg, Zn
13.	Pharmaceuticals	Al, Cu, Fe, Ga, Hg, Os, Ta
14.	Textiles	Al, Ag, Ba, Cd, Cu, Fe, Hg, Os, Ni, Sb
15.	Nuclear technology	Ba, Cd, In
16.	Fertilizers	Al, As, Cd, Cr, Cu, Fe, Hg, Mn, Pd, Ni, Zn
17.	Chlor-alkali production	Al, As, Cd, Cr, Fe, Hg, Mn, Pd, Sn, Zn
18.	Petroleum refining	Al, As, Cd, Cr, Fe, Ga, Hg, Pb, Ni, Zn

Sources: Modified from Dixit, R. et al., *Sustainability*, 7, 2189, 2015; Mishra, S. and Bharagava, R.N., *J. Environ. Sci. Health C*, 2015, 34(1):1–32; Nagajyoti, P.C. et al., *Environ. Chem. Lett.*, 8, 199, 2010.

9.2.2.3 Heavy Metals from Domestic Waste

Domestic wastes constitute the largest single source of metal contamination in the environment. Domestic wastes may consist of untreated or partially treated wastewaters, which have been passed through the filters of biological treatment plants and discharged into the receiving water bodies and end up into the sea from coastal residential areas. The use of detergents creates a serious environmental problem, since common household detergent products can affect the water quality up to a significant level. Angino et al. (1970) have reported that most of the enzyme in detergents contained trace amounts of many metals such as Fe, Mn, Cr, Co, and Zn. With regard to the pollution resulting from urbanized areas, there is an increasing awareness that the urban runoff presents serious health hazards of heavy metal contamination.

9.2.2.4 Heavy Metals from Airborne Sources

Metals from airborne sources are generally released as particulates contained in the gas stream. Some metals such as Cd, Pb, and As can also be volatilized during high-temperature processing and are converted into oxides and condensed as fine particulates unless a reducing atmosphere is maintained. Stack or duct emissions of air, gas, or vapor streams and fugitive emissions, such as dust from storage areas or waste piles, act as major sources of toxic metal contamination in environments. Stack emissions can be distributed over a wide area by natural air currents until dry and wet precipitation mechanisms remove them from the gas stream. Fugitive emissions are distributed over a much smaller area because emissions are made near the ground. The concentration of metals in fugitive emissions is lower as compared to the stack emissions. The solid particles present in smoke from fires and in other emissions from factory chimneys get eventually deposited on land or sea. Most of fossil fuels contain different types of heavy metals and act as a source of contamination, which is continuing on a large scale, since the industrial revolution began. For example, plants growing in soil and adjacent to smelting works showed very high concentration of Cd, Pb, Zn, etc. In addition, motor vehicles are also one of the major sources of Cu, Cd, Pb, and Hg for air contamination, whereas tyres and lubricant oils act as major sources of Zn, Ni, and Cd for contamination in the environment (USEPA, 1997).

9.2.2.5 Other Sources of Heavy Metals

The other sources of heavy metals include refuse incineration, landfills, and transportation (automobiles, diesel-powered vehicles, and aircrafts). The fly ash produced from coal burning and of commercial waste products are the two main anthropogenic sources that add Cu, Cd, Cr, Fe, Ni, Hg, Mn, Al, and Ti and galvanized metals (primarily Zn), whereas oil burning contributes V, Fe, Pb, and Ni into the environment (Al-Hiyaly et al., 1988).

9.3 ENVIRONMENTAL POLLUTION AND HEALTH HAZARDS FROM TOXIC METALS

Industrialization and extraction of natural resources have resulted in large-scale environmental contamination and severe health hazards to animals and human beings. The contamination of soils, groundwater, sediments, surface water, and air with toxic metals and chemicals is one of the major problems the world is facing today. The need to remediate these natural resources (soil, water, and air) has led to the development of new technologies that emphasizes mainly on the destruction of pollutants rather than the conventional approaches of their safe disposal into the environment because of their potential to enter the food chain.

Anthropogenic activities like metalliferous mining and smelting, agriculture, waste disposal, or industry discharge add a significant amount of different toxic metals such as Cu, Cd, Cr, Zn, Ni, Hg, Pb, Co, Pd, and As into the environment, and these metals, if entered in the food chain, cause severe health hazards in animals and human beings. However, some of these metals like arsenic, copper, iron, and nickel are also required in the body in trace amounts for various metabolic activities, but their high concentrations cause various cytotoxic, carcinogenic, mutagenic, and metallic disorders in living organisms as shown in Table 9.2. Plants, for their proper growth and optimum performance, also require some metals in very small amounts, but the increasing concentration of several metals in soil and water ecosystem due to various anthropogenic activities has created an alarming situation for all life-forms.

9.3.1 CADMIUM (CD)

Cadmium occurs naturally in ores together with zinc, lead, and copper. Cadmium compounds are used as stabilizers in PVC products, in color pigments, in several alloys, and most commonly in rechargeable nickel and cadmium batteries. Metallic cadmium has been mostly used as an anticorrosing agent (cadmiation). Cadmium is also present as a pollutant in phosphate fertilizers. Cadmium-containing products are rarely recycled, but frequently dumped together with household wastes, thereby contaminating the environment, especially if the waste is incinerated.

9.3.1.1 Health Hazards

Inhalation of cadmium fumes or particles can be life-threatening. Although acute pulmonary effects and deaths are uncommon, sporadic cases are still occurring (Seidal et al., 1993; Barbee and Prince, 1999). Cadmium exposure may also cause kidney damage. The first sign of renal lesion is tubular dysfunction, which is evidenced by an increased excretion of low-molecular-weight proteins (such as β2-microglobulin and α1-microglobulin [protein HC]) or enzymes (such as N-acetyl-β-D-glucosaminidase) (WHO, 1992; Jarup et al., 1998). It has been suggested that tubular damage is reversible, but there is overwhelming evidence that the cadmium-induced tubular damage is indeed irreversible (Jarup et al., 1998).

According to WHO, a urinary excretion of 10 nmol/mmol creatinine (corresponding to *circa* 200 mg Cd/kg kidney cortex) would constitute a "critical limit" below which kidney damage would not occur (WHO, 1992). Several reports have shown that a lower cadmium level may also cause kidney damage and/or bone effects (Jarup, 2003).

TABLE 9.2

Applications of Various Toxic Metals in Different Industries and Their Health Hazards in Living Organisms

Metals	Applications	Health Hazards	References
Cr	Tanning, paints pigment, fungicide	Cancer, nephritis, ulceration, and hair loss	Salem et al. (2000), Mishra and Bharagava (2015)
Hg	Coal vinyl chlorides, electrical batteries, thermometers	Autoimmune disease, depression, drowsiness, fatigue, hair loss, insomnia, loss of memory, restlessness, disturbance of vision, tremors, temper outbursts, brain damage, lung and kidney failure	Neustadt and Pieczenik (2007), Ainza et al. (2010), Gulati et al. (2010)
Pb	Plastic, paint, pipe, batteries, gasoline, autoexhaust	Neurotoxic and risk of cardiovascular disease	Salem et al. (2000), Padmavathiamma and Li (2007)
Cd	Fertilizer, plastic, pigments	Carcinogenic, mutagenic, endocrine disrupter, kidney damage, lung damage, and fragile bones, affect calcium regulation in biological systems	Salem et al. (2000), Degraeve (1991)
Zn	Fertilizer	Dizziness, fatigue, vomiting, renal damage, and cramps	Hess and Schmid (2002)
Co	Vitamin B-12, wood preservative	Diarrhea, low blood pressure, and paralysis	Wang (2006), Barceloux and Barceloux (1999)
Se	Coal, sulfur	Affects endocrine function with dietary exposure of around 300 μg/day; impairment of natural killer cells activity; hepatotoxicity, gastrointestinal disturbances; damage of liver, kidney, and spleen; nervousness	Vinceti et al. (2001)
Ni	Electroplating	Cancer of lungs, allergic disease such as itching, immunotoxic, neurotoxic, teratogenic, carcinogenic, genotoxic, and mutagenic, affects fertility and hair loss	Salem et al. (2000), Khan et al. (2007), Das et al. (2008), Duda-Chodak and Baszczyk (2008)
Be	Coal, rocket fuel	Carcinogenic, acute and chronic poison	IPCS (1990)
Cu	Electronics, wood preservative, architecture	Brain and kidney damage, liver cirrhosis that results from elevated levels, chronic anemia, stomach and intestine irritation	Bonham et al. (2002)
As	Pesticides, treated wood products, herbicides	Affects essential cellular processes such as oxidative phosphorylation and ATP synthesis, arsenicosis, carcinogen and cancer	Ng et al. (2003), Tripathi et al. (2007), Meng et al. (2015), WHO (2001)
Ba	Appropriate dust control equipment and industrial controls	Cause cardiac arrhythmias, respiratory failure, gastrointestinal dysfunction, muscle twitching, and elevated blood pressure	Acobs et al. (2002)

9.3.2 Mercury (Hg)

Mercury is a chemical element with the symbol Hg, and its atomic number is 80. Metallic mercury is used in thermometers, barometers, and instruments used to measure blood pressure. Mercury is largely used in the electrochemical process of chlorine manufacturing, where mercury is used as an electrode in the chlor-alkali industry.

9.3.2.1 Health Hazards

Acute mercury exposure may give rise to lung damage. Chronic poisoning is characterized by neurological and psychological symptoms such as tremor, changes in personality, restlessness, anxiety, sleep disturbance, and depression (Jarup, 2003). Metallic mercury may cause kidney damage, which is reversible if exposure is stopped. It has been also possible to detect proteinuria at relatively low levels of occupational exposure.

9.3.3 Pb

Lead is a chemical element in the carbon group with the symbol Pb, atomic number 82, atomic mass of 207.2, density of

11.4 g/cm^3, melting point of 327.4°C, and boiling point of 1750°C. It is a naturally occurring, bluish-gray metal usually found in the form of minerals combined with other elements such as sulfur (i.e., PbS, $PbSO_4$) or oxygen ($PbCO_3$), and its concentration ranges from 10 to 30 mg/kg in the Earth's crust. Lead is a soft, malleable, and heavy post-transition metal. Metallic lead has a bluish-white color after being freshly cut, but it soon turns into a dull grayish color when exposed to air. The general population gets exposed to lead from air and food in roughly equal proportions. However, occupational exposure to inorganic lead mainly occurs in mines and smelters as well as welding of lead painted metal, and in battery plants; moreover, low or moderate exposure may take place in the glass industry. High levels of air emissions may pollute areas near the lead mines and smelters. Airborne lead can be deposited in soil and water and, thus, finally reaches the human's or animal's body.

9.3.3.1 Health Hazards

The symptoms of acute lead poisoning are headache, irritability, abdominal pain, and various disorders related to the nervous system (Steenland and Boffetta, 2000). Lead enceph-alopathy is characterized by sleeplessness and restlessness. Children may be affected with behavioral disturbances and learning and concentration difficulties. In severe cases of lead encephalopathy, the affected person may suffer from acute psychosis, confusion, and reduced consciousness. People who have been exposed to lead for a long period of time may suffer from memory deterioration, prolonged reaction time, and reduced ability to understand. Individuals with average blood lead levels under 3 µmol/L may show signs of peripheral nervous symptoms with reduced nerve conduction velocity and reduced dermal sensibility.

9.3.4 As

Arsenic is a widely distributed metalloid, which occurs in rock, soil, water, and air. Inorganic arsenic is a metalloid present in the group VA and period 4 of the periodic table that is found in a wide variety of mineral ores such as As_2O_3 and can be recovered by processing of ores containing mostly Cu, Pb, Zn, Ag, Au, etc. It is also present in ashes from coal combustion. Arsenic has the following properties: it has an atomic number of 33, atomic mass of 75, density of 5.72 g/cm^3, melting point of 817°C, and boiling point of 613°C, exhibits fairly complex chemistry, and can also be present in several oxidation states (−III, 0, III, V). In aerobic environments, As (V) is dominant, usually in arsenate (AsO_4^{3-}) forms in various protonation states: H_3AsO_4, $H_2AsO_4^-$, $HAsO_4^{2-}$, and AsO_4^{3-}. Arsenate and other anionic forms of arsenic behave as a chelator and can precipitate in the presence of metal cations.

Since arsenic is often present in anionic form, it does not form complexes with simple anions such as Cl$^-$ and SO_4^{2-}. Arsenic speciation also includes organ metallic forms such as methyl arsenic acid ($CH_3)AsO_2H_2$ and dimethyl arsenic acid ($CH_3)_2AsO_2H$. Many As compounds adsorb strongly to soils and are therefore transported only over short distances in groundwater and surface water. Arsenic is reported to cause skin damage, increased risk of cancer, and problems with circulatory system.

9.3.4.1 Health Hazards

Inorganic arsenic is acutely toxic, and its intake in large quantities may lead to gastrointestinal symptoms, severe disturbances in cardiovascular and central nervous systems, and even death also. Arsenic exposure through drinking water is reported to cause skin, lung, kidney, and bladder cancer (WHO, 2001). In an affected person, skin cancer is preceded by directly observable precancerous lesions. Uncertainties in the estimation of past exposures are important while assessing the exposure–response relationships, but arsenic contamination in drinking water at a level of 100 µg/L leads to the development of cancer, whereas the concentration of arsenic from 50 to 100 µg/L is found to be associated with the precursors of skin cancer. The relationship between arsenic exposure and health effects are not so clear. Though there is strong evidence for the cause hypertension and cardiovascular diseases, the evidence is only suggestive for diabetes and reproductive effects and weak for cerebrovascular disease, long-term neurological effects, and cancer at the sites other than the lung, bladder, kidney, skin, etc.

9.3.5 Cr

Chromium is a d-block transition metal placed in group VIB in the periodic table and has the atomic number of 24, atomic mass of 52, density of 7.19 g/cm^3, melting point of 1875°C, and boiling point of 2665°C. It is one of the less common elements and does not occur naturally in elemental form but only in compounds. Chromium is mined as a primary ore product in the form of mineral chromate ($FeCr_2O_4$). The major sources of Cr contamination in the environment include the discharge from electroplating industries and Cr-containing wastes from other sources. In nature, chromium mainly exits in trivalent and hexavalent forms. Chromium is a naturally occurring heavy metal that exists in air, water, soil, and food. It is now considered as one of the major environmental pollutants due to its toxicity for ecological, nutritional, and environmental reasons.

9.3.5.1 Health Hazards

Cr(VI) can enter the body when people breathe air, eat food, or drink water contaminated with it. Cr(VI) is also found in house dust and soil, which can be ingested or inhaled out of the various forms of chromium. Cr(VI) is the most toxic and common form of metal in nature. Many Cr(VI) compounds have been found to be carcinogenic in nature, but the evidence to date indicates that the carcinogenicity is site specific and limited to the lungs and the sinonasal cavity and dependent on the intensity of exposure (Salem et al., 2000). Inhaling relatively high concentrations of Cr(VI) can cause a runny nose, sneezing, itching, nosebleeds, ulcers, and holes in the nasal septum. Short-term high-level inhalational exposure can cause adverse effects at the contact site, including ulcers, irritation of the nasal mucosa, and holes in the nasal septum. Ingestion of very high Cr(VI) doses can cause kidney and liver damage, nausea, irritation of the gastrointestinal tract, stomach ulcers, convulsions, and death, while dermal exposures may cause skin ulcers or allergic reactions. Cr(VI) is one of the most highly allergenic metals, second to nickel, and studies on mice given high doses of Cr(VI) have shown reproductive abnormalities including reduced litter size and decreased fetal weight (ATSDR, 2000; Mishra and Bharagava, 2015).

9.4 APPROACHES FOR THE BIOREMEDIATION OF TOXIC METALS FROM CONTAMINATED SITES

Nowadays, with the growth of industrialization and exploitation of natural resources, there has been a considerable increase in the discharge of industrial wastes into the environment mainly in soil and aquatic resources, which has led to the accumulation of various toxic metals into the environment. The contamination of soil, groundwater sediments, surface water, and air with toxic metals and chemicals is one of the major threats the world is facing today, as these cannot be broken down into nontoxic forms and, therefore, has long-lasting effects on the ecosystem. According to a recent study, the need to remediate these natural resources has led to the development of new technologies that emphasized on the destruction of pollutants rather on the conventional approaches to prevent their entry into the food chain for the safe disposal of pollutants into the environment (Fulekar, 2010; Dixit et al., 2015).

Bioremediation is an innovative and promising technology available for the removal and recovery of metals from the contaminated sites. Since microorganisms have developed various strategies for their survival in metal-contaminated habitats, these organisms are well known to develop and adopt different detoxifying mechanisms such as biosorption, bioaccumulation, biotransformation, and biomineralization (Fulekar, 2010).

The quality of life on Earth is linked to the overall quality of the environment. The environmental problems and health hazards associated with the contaminated sites are now increasing day by day in many countries. Large quantities of organic (chemicals) and inorganic (toxic metals) pollutants are released into the environment each year as a result of various natural and anthropogenic activities, which is a major threat for environment, human, and animal health.

Many microorganisms living in soil and water naturally use organic and inorganic pollutants that are harmful to living beings and the environment. These microorganisms are capable to transform these chemical pollutants into nontoxic products. In addition to microbes, plants can also be used to clean up the contaminated soil, water, or air: this is known as phytoremediation.

9.4.1 Bioremediation of Metal-Contaminated Sites by Using Microbes Used in Processes

Microorganisms play an important role in nutritional chains, which are important parts of the geochemical cycles of carbon, nitrogen, sulfur, and phosphorous in the ecosystem.

Bioremediation involves the removal/degradation/transformation of contaminants converting them into nontoxic forms. Some examples of bioremediation technologies are phytoremediation, bioventing, bioleaching, land farming, bioreactor, composting, and biostimulation.

Microorganisms that carry out the biodegradation process in contaminated environments are identified as active members of microbial consortiums. Microbes get adapted and grow at subzero temperatures as well as in extreme heat, in desert conditions, in water, with an excess of oxygen, and in anaerobic conditions in the presence of hazardous compounds or on any waste stream (Singh et al., 2014). Many studies have demonstrated that microbes have the ability to remove/transform the toxic metals into nontoxic form in the contaminated soil and water. Many microorganisms such as *Pseudomonas* spp., *Alcaligenes* spp., *Arthrobacter* spp., *Bacillus* spp., *Corynebacterium* spp., *Flavobacterium* spp., *Azotobacter* spp., *Rhodococcus* spp., *Mycobacterium* spp., *Nocardia* spp., *Methosinus* spp., methanogens, *Aspergillus niger*, *Pleurotus ostreatus*, *Rhizopus arrhizus*, *Stereum hirsutum*, and *Ganoderma applanatum* are also (Rajendran et al., 2003) reported to have used different types of metals in their biochemical activities as shown in Table 9.3.

9.4.1.1 Types of Bioremediation

On the basis of removal and transportation of wastes for the treatment, bioremediation is mainly of two types.

9.4.1.1.1 In Situ Bioremediation

In situ bioremediation means that there is no need to excavate or remove the contaminated soils or water in order to accomplish the remediation process. It involves the supply of oxygen and nutrients by circulating the aqueous solutions through contaminated soil and water to stimulate the naturally occurring microbes for the degradation of organic contaminants (Vidali, 2001). Most often, *in situ* bioremediation is mainly applied for the degradation of contaminants in saturated soils and groundwater. It is a superior method to clean the contaminated

---

TABLE 9.3

Microbes Utilizing Different Heavy Metals in Various Biochemical Activities

Microorganism	Metals	References
Zooglea spp.	U, Cu, Ni	Sar and D'Souza (2001)
Citrobacter spp.	Co, Ni, Cd	Sar and D'Souza (2001)
Bacillus spp.	Cu, Zn	Philip et al. (2000), Rajendran et al. (2003)
Citrobacter spp.	Cd, U, Pb	Rajendran et al. (2003)
Chlorella vulgaris	Au, Cu, Ni, U, Pb, Hg, Zn	Rajendran et al. (2003)
Pleurotus ostreatus	Cd, Cu, Zn	Rajendran et al. (2003)
Aspergillus niger	Cd, Zn, Ag, Th, U	Rajendran et al. (2003)
Rhizopus arrhizus	Ag, Hg, P, Cd, Pb, Ca	Favero et al. (1991), Rajendran et al. (2003)
Stereum hirsutum	Cd, Co, Cu, Ni	Gabriel et al. (1994, 1996)
Phormidium valderium	Cd, Pb	Gabriel et al. (1994, 1996)
Ganoderma applanatum	Cu, Hg, Pb	Gabriel et al. (1994, 1996)

environments since it is cheaper and uses microbial organisms for the degradation of chemical pollutants. Chemotaxis is an important mechanism used to study *in situ* bioremediation process because microorganisms with chemotactic abilities can move into an area containing organic/inorganic contaminants. Thus, by enhancing the cells' chemotactic abilities, *in situ* bioremediation can be used for the effective degradation/metabolization/mineralization of organic/inorganic pollutants. However, *in situ* bioremediation may be of the following types:

9.4.1.1.1.1 Bioventing Bioventing is the most common *in situ* treatment process, which involves the supply of air and nutrients through wells to contaminated soils to stimulate the indigenous bacteria. Bioventing employs low airflow rates and provides only a limited amount of oxygen necessary for the biodegradation of pollutants, while minimizing the volatilization and release of contaminants into the atmosphere. It works for simple hydrocarbons and can be used where the contamination is deep under the surface.

9.4.1.1.1.2 Biosparging Biosparging involves the injection of air under pressure below the water table to increase the groundwater oxygen content and enhance the rate of biological degradation of contaminants by naturally occurring bacteria. Biosparging increases the mixing of contaminants and microbes in the saturated zone and thereby increases the contact time between soil and groundwater. Due to its low installation cost, small-diameter air injection points allow the considerable flexibility in designing and construction of the biosparging system.

9.4.1.1.1.3 Bioaugmentation It is the addition of potential indigenous or exogenous microorganisms to the contaminated sites to enhance the degradation/remediation of metal pollutants. There are two factors that limit the use of added microbial cultures in a land treatment unit: (1) It is difficult for the nonindigenous cultures to compete with the indigenous population to develop and sustain the useful population level for the effective bioremediation. (2) Most soils with a long-term exposure to biodegradable waste have indigenous microorganisms that are effective in degrading chemical pollutant, if the land treatment unit is well managed.

9.4.1.1.2 Ex Situ Bioremediation
This technique involves the excavation or removal of contaminated soil from the original sites to treatment site such as waste treatment plants.

9.4.1.1.2.1 Land Farming Land farming is a simple technique in which the contaminated soil is excavated and spread over a prepared bed and periodically tilled until the metal pollutants are degraded. The ultimate goal is to stimulate the indigenous microorganisms and finally facilitate the aerobic degradation of metal contaminants. Land farming has the potential to reduce the monitoring and maintenance costs as well as cleanup liabilities, but treatment of the contaminated sites is only limited up to a depth of 10–35 cm.

9.4.1.1.2.2 Composting Composting is a technique that involves combining of contaminated soil organic amendments such as manure or agricultural wastes to promote the development of a rich microbial population and elevated temperature, which is the main characteristic feature of the composting process.

9.4.1.1.2.3 Biopiles It is a hybrid of land farming and composting in which the engineered cells are constructed as aerated composted piles and typically used for the treatment of surface contamination with petroleum hydrocarbons. It is a refined version of land farming that tends to control the physical losses of contaminants by leaching and volatilization. Biopiles provide a favorable environment for the indigenous aerobic and anaerobic microorganisms to perform the bioremediation process at an optimum level.

9.4.1.1.2.4 Bioreactors Bioreactors (slurry or aqueous reactors) are used for the *ex situ* treatment of contaminated soil and water pumped up from a contaminated plume. Bioremediation in reactors involves the processing of contaminated solid material (soil, sediment, sludge) or water through an engineered system. A slurry bioreactor may be defined as a containment vessel apparatus used to create a three-phase (solid, liquid, and gas) mixing condition to increase the bioremediation of soil-bound and water-soluble pollutants as water slurry of the contaminated soil and biomass (usually indigenous microorganisms) capable of degrading target contaminants. In general, the rate and extent of biodegradation are greater in a bioreactor system than in *in situ* or in solid-phase systems because the contained environment is more manageable and, hence, more controllable and predictable. The contaminated soil requires pretreatment (e.g., excavation), or alternatively the contaminant can be stripped from the soil via soil washing or physical extraction (e.g., vacuum extraction) before being placed in a bioreactor.

9.4.2 PLANT/PHYTOREMEDIATION

Phytoremediation refers to the use of plants and associated microorganisms to partially or completely remove the selected contaminants from soil, sludge, sediments, wastewater, groundwater, etc. It can also be used for the removal of radionuclides, organic pollutants, and toxic metals from contaminated sites (Dixit et al., 2015). Phytoremediation utilizes a variety of processes and of plant species (Table 9.4), with the help of the plants' physical characteristics, to aid in the remediation of contaminated sites. Over the recent years, a special emphasis has been given to phytoremediation since this technique can be exploited for the remediation of toxic metal–polluted soils, as it is a cost-effective, eco-friendly, and efficient *in situ* remediation technology driven by solar energy. The phytoremediation technique includes a number of different processes such as phytoextraction, phytofiltration, phytostabilization, phytovolatilization, and phytodegradation as shown in Table 9.5 and Figure 9.2.

TABLE 9.4

Various Plant Species Reported Capable for the Phytoremediation of Different Heavy Metals

Heavy Metal	Plants Species	References
Cd	Castor (*Ricinus communis*)	Huang et al. (2011), Pandey (2013)
Cd, Pb, Zn	Corn (*Zea mays*)	Meers et al. (2010)
Cd, Cu, Pb, Zn	*Populus* spp. (*Populus deltoides, Populus nigra, Populus trichocarpa*)	Ruttens et al. (2011), Abhilash et al. (2012)
Cd, Cu, Pb, Zn	*Salix* spp. (*Salix viminalis, Salix fragilis*)	Pulford and Watson (2003), Volk et al. (2006), Ruttens et al. (2011)
Cd, Cu, Ni, Pb	Jatropha (*Jatropha curcas* L.)	Abhilash et al. (2009), Jamil et al. (2009)
Hg	*Populus deltoides*	Che et al. (2003)
Se	*Brassica juncea, Astragalus bisulcatus*	Bitther et al. (2012)
Zn	*Populus canescens*	Bittsanszkya et al. (2005)

TABLE 9.5

Different Phytoremediation Approaches and Their Applications to Various Contaminated Sites

Technique	Plant Mechanism	Contaminated Sites
Phytoextraction	Uptake and concentration of metals via direct uptake into the plant tissue with subsequent removal of the plants (hyperaccumulation).	Soil
Phytostabilization	Root exudates cause metal to precipitate and become less available (complexation).	Soil, groundwater, mine tailing
Phytotransformation	Plant uptake and degradation of organic and inorganic compounds.	Surface water, groundwater
Rhizofiltration	Uptake of metals into plant roots, rhizosphere accumulation.	Organic and inorganic contaminated soil
Phytovolatilization	Plants evaporate and transpire selenium, mercury, and volatile hydrocarbons.	Soil and groundwater
Phytodegradation	Enhances microbial degradation in rhizosphere.	Soil, groundwater within rhizosphere

Sources: Modified from Dixit, R. et al., *Sustainability*, 7, 2189, 2015; Ghosh, M. and Singh, S.P., *Asian J. Energy Environ.*, 6(04), 214, 2005; Vidali, M., *Pure Appl. Chem.*, 73, 1163, 2001.

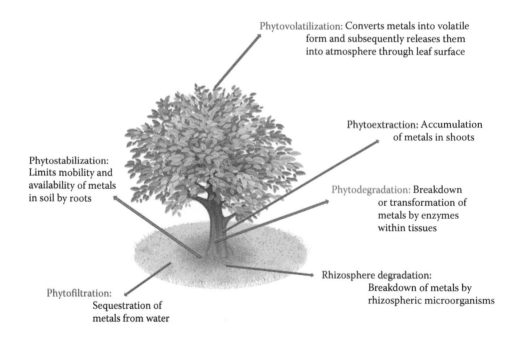

Phytovolatilization: Converts metals into volatile form and subsequently releases them into atmosphere through leaf surface

Phytoextraction: Accumulation of metals in shoots

Phytostabilization: Limits mobility and availability of metals in soil by roots

Phytodegradation: Breakdown or transformation of metals by enzymes within tissues

Phytofiltration: Sequestration of metals from water

Rhizosphere degradation: Breakdown of metals by rhizospheric microorganisms

FIGURE 9.2   Various phytoremediation processes used by plants to remove toxic metals from contaminated sites.

9.4.2.1 Phytoextraction

Phytoextraction is also known as phytoaccumulation in which metals are removed from the contaminated sites by taking advantage of the plants' ability to (hyper-) absorb and accumulate or translocate metals and/or metalloids and by concentrating them within the biomass. The purpose of this type of remediation is to reduce the metal concentration in contaminated soils so that they can be used profitably for agriculture, forestry, horticulture, grazing, etc.

9.4.2.2 Phytostabilization

Phytostabilization is also known as phytoimmobilization, in which plants in combination with soil additives mechanically stabilize the sites and reduce the transfer of metals to the other compartments of ecosystem and finally into the food chain. The "stabilized" organic or inorganic compound normally gets incorporated into the plant lining or into the soil humus. The basis for phytostabilization is that metals do not degrade, so capturing them *in situ* is often the best alternative. This approach is particularly applicable when low-concentration, diffused, and vast areas of contamination are to be treated. Plants restrict the metal pollutants by creating a zone around the roots where the pollutant is precipitated and stabilized. When phytostabilization is undertaken, the plants used do not absorb the targeted pollutant(s) into plant tissue.

9.4.2.3 Phytostimulation

In phytostimulation plant, roots promote the development of rhizospheric microorganisms that are capable for the degradation of metal contaminants and microbes utilize plant root exudates as a carbon and energy source.

9.4.2.4 Phytovolatilization/Rhizovolatilization

Phytovolatilization/rhizovolatilization employs the metabolic capabilities of plants and associated rhizospheric microorganisms to transform metal pollutants into volatile compounds that are released into the atmosphere. Some ions, such as elements of subgroups II, V, and VI of the periodic table like mercury, selenium, and arsenic, are absorbed by roots, get converted into less toxic forms, and are released into the surrounding environment.

9.4.2.5 Phytodegradation

In phytodegradation, the metal contaminants get degraded or mineralized by the specific plant enzymes or exudates, and then the organic component is utilized by the rhizospheric microbes as carbon and energy sources, while metal component is taken up by the plant roots and used in various metabolic activities. However, phytoremediation being as cost-effective and eco-friendly technology also has many limitations as shown in Table 9.6, due to which this technology is not very effective for a wide range of metal-contaminated sites.

9.4.2.6 Rhizofiltration

Rhizofiltration is the use of plants to absorb, concentrate, and/or precipitate metal contaminants in the aqueous system. Rhizofiltration is also used to partially treat the industrial and agricultural runoff.

Plants that can accumulate large quantities of metals by natural methods have been identified and are used to

TABLE 9.6
Advantages and Limitations of Various Phytoremediation Techniques

Sr. No.	Advantages	Limitations
1.	Low cost and aesthetically pleasing (no excavation require).	The plant must be able to grow in the polluted media.
2.	Easy to implement and maintain. Plants are a cheap and renewable resource, easily available	Introduction of inappropriate or invasive plant species should be avoided (nonnative species may affect biodiversity)
3.	Does not require expensive equipment or highly specialized personnel.	Restricted to sites with low contaminant concentration in environment.
4.	Reduces the amount of waste going to landfills.	Although faster than natural attenuation, it requires long time periods (several years)
5.	Applicable to a wide variety of inorganic and organic contaminants	Limited by depth (roots) and solubility and availability of the contaminant.
6.	Removal of secondary air or water-borne wastes.	Climatic conditions are the limiting factor.
7.	Less noisy than other remediation methods. Actually, trees may reduce noise from industrial activities.	Amendments and cultivation practices may have negative consequences on contaminant mobility.
8.	Enhanced regulatory and public acceptance.	Introduction of exotic plant species may affect biodiversity.
9.	Environmentally friendly, aesthetically pleasing, socially accepted, low-tech alternative	Contaminants may be transferred to another medium, the environment, and/or the food chain.

Sources: Modified from Chandra, R. et al., Phytoremediation of environmental pollutants: An eco-sustainable green technology to environmental management, in: Chandra, R., ed., *Advances in Biodegradation and Bioremediation of Industrial Waste*, CRC Press, Taylor & Francis Group, Boca Raton, FL, 2015, pp. 1–30; Alkorta, I. et al., *Rev. Environ. Sci. Bio/Technol.*, 3, 71, 2004.

remediate the metal-contaminated sites. These plants are known as hyperaccumulators and are often found growing in areas having elevated metal concentrations in soil. But, unfortunately at high metal content, even hyperaccumulating plants grow slow and attain only a small height. Thus, high metal content inhibits the plant growth even if they are capable for the hyperaccumulation of metals. However, depending upon the amount of metals at a particular site and the type of soil, even hyperaccumulating plants may require 15–20 years to remediate the contaminated sites. This time frame is usually too slow for practical application. Hence, there is a need to search the plant species, which can grow fast and accumulate the greater amounts of biomass in addition to being tolerant to one or more metals.

9.4.3 Role of Plant-Growth-Promoting Rhizobacteria in Bioremediation of Metal-Contaminated Sites

The plant-growth-promoting rhizobacteria (PGPR) enhance plant growth by fixing the atmospheric nitrogen, producing phytohormones, enhancing specific enzymatic activity, and producing antibiotics and other pathogen-depressing substances such as siderophores and chelating agents; all these factors protect plants from various diseases.

Microbial cells produce and sense signal molecules allowing the whole microbial population to spread as a biofilm over the root surface and initiating a concerted action when a particular population density is achieved. This phenomenon

TABLE 9.7

Plant Growth–Promoting Rhizobacteria Used in Bioremediation for Various Toxic Metals

Bacteria	Plant	Heavy Metals	Conditions of Experiments	Role of PGPR
Brevundimonas	None	Cd	Culture media	Sequestered cadmium directly from solution.
Variovorax paradoxus, Rhodococcus spp., *Flavobacterium*	*Brassica juncea*	Cd	Experiment in Petri dishes	Stimulating root elongation.
Azotobacter chroococcum	*Brassica juncea*	Pb, Zn	Experiments in greenhouse	Stimulated plant growth.
Burkholderia sp. J62	Maize and tomato	Pb, Cd	Experiments in greenhouse	Role of PGPR, increased the biomass of maize and tomato plant significantly; the increased Pb and Cd content in tissue varied from 38% to 192% and from 5% to 191%, respectively.
Methylobacterium oryzae, Burkholderia spp.	*Lycopersicon esculentum*	Ni, Cd	Gnotobiotic and pot culture experiments	None.
Xanthomonas spp., *Azomonas* spp., *Pseudomonas* spp., *Bacillus* spp.	*Brassica napus*	Cd	Experiment in pots	Stimulated plant growth and increased cadmium accumulation.
Pseudomonas sp., *Bacillus* spp.	Mustard	Cr(VI)	Experiment in pots	Stimulated plant growth and decreased Cr(VI) content.
Ochrobactrum, Bacillus cereus	Mungbean	Cr(VI)	Experiment in pots	Lowers the toxicity of chromium to seedlings by reducing Cr(VI) to Cr(III).
Kluyvera ascorbata	Indian mustard, canola, tomato	Ni, Pb, Zn	Experiment in pots	Both strains decreased some plant growth inhibition by heavy metals, no increase of metal uptake with either strain over noninoculated plants.
Brevibacillus	*Trifolium repens*	Zn	Experiment in pots	Enhanced plant growth and nutrition of plants and decreased zinc concentration in plant tissues.
Pseudomonas spp.	Soybean, mungbean, wheat	Ni, Cd, Cr	Experiment in pots	Promotes growth of plants.
Pseudomonas fluorescens	Soybean	Hg	Experiment in greenhouse	Increased plant growth.
Ochrobactrum intermedium	Sunflower	Cr(VI)	Experiment in pots	Increased plant growth and decreased Cr(VI) uptake.

Sources: Modified from Khan, M.S. et al., *Environ. Chem. Lett.*, 7, 1, 2009; Jiang, C.Y. et al., *Chemosphere*, 72, 157, 2008; Sheng, X. and Xia, J., *Chemosphere*, 64, 1036, 2006.

is known as "quorum sensing," which in combination with other regulatory systems expands the range of environmental signals targeting gene expression beyond the population density. The nitrogen-fixing rhizobial bacteria get chemotactically attracted toward the legume roots by the root exudates, adhere to and colonize on the root surface, and activate rhizobial nodulation genes/nod factors. Many quorum-sensing signal molecules, such as *N*-acyl-homoserine lactones (AHLs), are produced, which regulate the expression and repression of many symbiotic genes. However, free-living as well as symbiotic PGPR enhance the plant growth directly by providing bioavailable phosphorus for plant uptake; fixing nitrogen for plant use; sequestering trace elements like iron for plants by siderophores; producing plant hormones like auxins, cytokinins, and gibberellins; and lowering of plant ethylene levels (Table 9.7). The use of PGPR in phytoremediation technologies is now being considered to play an important role as adding PGPR can aid plant growth on contaminated sites and enhance the remediation of metal-contaminated sites.

9.5 CHALLENGES AND FUTURE PROSPECTS

Rapid industrialization and technology development have resulted in many adverse side effects like severe soil and water contamination with various industrial pollutants leading to the degradation of soil health, although a number of *in situ* and *ex situ* bioremediation approaches have been developed to combat the heavy metal contamination in the environment. However, due to certain limitations of conventional methods, the use of microbes has arisen as a time-saver for bioremediation. Bioremediation is a fast-developing field, in last 10 years, many field applications have been made all over the world. This clearly holds promise for effective, economical, and eco-friendly metal bioremediation technology of industrial exploitation and a pollution-free environment. However, bioremediation technology has also certain limitations: several microorganisms cannot transform toxic metals into nontoxic forms, and these have inhibitory effects on the microbial activity. Thus, future studies should be focused on the various factors involved in the *in situ* or *ex situ* bioremediation of metal-contaminated sites. To date, commercial phytoextraction has been constrained by the expectation that site remediation should be achieved in a time comparable to other cleanup technologies. So far, most of the phytoremediation experiments have taken place in the lab scale, where plants are grown in a hydroponic setting and fed with heavy metal diets.

9.6 SUMMARY

Metal contamination issues are becoming increasingly common throughout the world with many studies on metal toxicity, particularly in areas with high anthropogenic activities. Heavy metals are ubiquitous environmental contaminants in an industrialized society, and thus, the concern over the possible health hazards and ecosystem effects of heavy metals has been increased. However, like organic pollutants, metals are not degraded and are accumulated in environments as well as in living organisms where they cause toxic, genotoxic, mutagenic, and carcinogenic effects. Contamination of soil and water ecosystems with toxic metals represents a serious threat for the ecosystem and human health and, thus, requires proper implementation of appropriate remedial measures. Although the application of many bioremediation and phytoremediation cleanup technologies is rapidly expanding, these approaches also have many limitations that should be addressed carefully for the proper implementation of cleanup technologies so that the contaminated environments can be restored.

ACKNOWLEDGMENT

The financial support received as the "Major Research Project" (Grant no.: SB/EMEQ-357/2013) from the "Science and Engineering Research Board" (SERB), Department of Science & Technology (DST), Government of India (GOI), New Delhi, is duly acknowledged.

REFERENCES

Abhilash, P.C., S. Jamil, and N. Singh. 2009. Transgenic plants for enhanced biodegradation and phytoremediation of organic xenobiotics. *Biotechnology Advances* 27:474–488.

Abhilash, P.C., J.R. Powell, H.B. Singh, and B.K. Singh. 2012. Plant-microbe interactions: Novel applications for exploitation in multipurpose remediation technologies. *Trends in Biotechnology* 30:416–420.

Acobs, I.A., J. Taddeo, K. Kelly, and C. Valenziano. 2002. Poisoning as a result of barium styphnate explosion. *American Journal of Industrial Medicine* 41:285–288.

Agency for Toxic Substances and Disease Registry (ATSDR). 2000. Toxicological profile for chromium. U.S. Department of Health and Human Services, Public Health Service, Atlanta, GA.

Ainza, C., J. Trevors, and M. Saier. 2010. Environmental mercury rising. *Water, Air, & Soil Pollution* 205:47–48.

Al-Hiyaly, S.A., T. McNeilly, and A.D. Bradshaw. 1988. The effect of zinc concentration from electricity pylons—Evolution in replicated situation. *New Physiologist* 110:571–580.

Alkorta, I., J. Hernandez-Allica, J.M. Becerril, I. Amezaga, I. Albizu, and C. Garbisu. 2004. Recent findings on the phytoremediation of soils contaminated with environmentally toxic heavy metals and metalloids such as zinc, cadmium, lead, and arsenic. *Reviews in Environmental Science and Bio/Technology* 3:71–90.

Angino, E.E., L.M. Magnuson, T.C. Waugh, O.K. Galle, and J. Bredfeldt. 1970. Arsenic in detergents—Possible danger and pollution hazard. *Science* 168:389–392.

Barakat, M.A. 2011. New trends in removing heavy metals from industrial wastewater. *Arabian Journal of Chemistry* 4:361–377.

Barbee, J.J. and T.S. Prince. 1999. Acute respiratory distress syndrome in a welder exposed to metal fumes. *Southern Medical Journal* 92:510–512.

Barceloux, D.G. and D. Barceloux. 1999. Cobalt. *Journal of Toxicology: Clinical Toxicology* 37:201–216.

Bitther, O.P., E.A.H. Pilon-Smits, R.B. Meagher, and S. Doty. 2012. Biotechnological approaches for phytoremediation. In *Plant Biotechnology and Agriculture*, A. Altman and P.M. Hasegawa (Eds.), pp. 309–328. Oxford, U.K.: Academic Press.

Bittsanszkya, A., T. Kömives, G. Gullner, G. Gyulai, J. Kiss, L. Heszky, L. Radimszky, and H. Rennenberg. 2005. Ability of transgenic poplars with elevated glutathione content to tolerate zinc (2^+) stress. *Environmental International* 31:251–254.

Bonham, M., J.M. O'Connor, M.B. Hannigan, and J.J. Strain. 2002. The immune system as a physiological indicator of marginal copper status? *British Journal of Nutrition* 87:393–403.

Chandra, R., R.N. Bharagava, A. Kapley, and H.J. Purohit. 2011. Bacterial diversity, organic pollutants and their metabolites in two aeration lagoons of common effluent treatment plant (CETP) during the degradation and detoxification of tannery wastewater. *Bioresource Technology* 102:2333–2341.

Chandra, R., G. Saxena, and V. Kumar. 2015. Phytoremediation of environmental pollutants: An eco-sustainable green technology to environmental management. In *Advances in Biodegradation and Bioremediation of Industrial Waste*, R. Chandra (Ed.), pp. 1–30. Boca Raton, FL: CRC Press, Taylor & Francis Group.

Che, D., R.B. Meagher, A.C. Heaton, A. Lima, C.L. Rugh, and S.A. Merkle. 2003. Expression of mercuric ion reductase in Eastern cottonwood (*Populus deltoides*) confers mercuric ion reduction and resistance. *Plant Biotechnology Journal* 1:311–319.

Das, N., R. Vimala, and P. Karthika. 2008. Biosorption of heavy metals—An overview. *Indian Journal Biotechnology* 7:159–169.

Degraeve, N. 1991. Carcinogenic, teratogenic and mutagenic effects of cadmium. *Mutation Research/Reviews in Genetic Toxicology* 86:115–135.

Dixit, R., Wasiullah, D. Malaviya, K. Pandiyan, U.B. Singh, A. Sahu et al. 2015. Bioremediation of heavy metals from soil and aquatic environment: An overview of principles and criteria of fundamental processes. *Sustainability* 7:2189–2212.

Duda-Chodak, A. and U. Baszczyk. 2008. The impact of nickel on human health. *Journal of Elementology* 13:685–696.

Favero, N., P. Costa, and M.L. Massimino. 1991. In vitro uptake of cadmium by basidiomycete *Pleurotus ostreatus*. *Biotechnology Letters* 10:701–704.

Fulekar, M.H. 2010. *Bioremediation Technology: Recent Advances*. Dordrecht, the Netherlands: Springer.

Gabriel, J., O. Kofronova, P. Rychlovsky, and M. Krenzelok. 1996. Accumulation and effect of cadmium in the wood rotting basidiomycete, *Daedalea quercina*. *Bulletin of Environmental Contamination and Toxicology* 57:383–390.

Gabriel, J., M. Mokrejs, J. Bily, and P. Rychlovsky. 1994. Accumulation of heavy metal by some Woodrooting fungi. *Folia Microbiologica* 39:115–118.

Ghosh, M. and S.P. Singh. 2005. A review on phytoremediation of heavy metals and utilization of it's by products. *Asian Journal on Energy and Environment* 6(04):214–231.

Gulati, K., B. Banerjee, S. Lall Bala, and A. Ray. 2010. Effects of diesel exhaust heavy metals and pesticides on various organ systems: Possible mechanisms and strategies for prevention and treatment. *Indian Journal of Experimental Biology* 48:710–721.

Hess, R. and B. Schmid. 2002. Zinc supplement overdose can have toxic effects. *Journal of Pediatric Hematology/Oncology* 24:582–584.

Huang, H., N. Yu, L. Wang, D.K. Gupta, Z. He, K. Wang, Z. Zhu, X. Yan, T. Li, and X. Yang. 2011. The phytoremediation potential of bioenergy crop *Ricinus communis* for DDTs and cadmium co-contaminated soil. *Bioresource Technology* 102:10034–11038.

Hutchinson, T.C. and K.M. Meema (Eds.). 1987. *Lead, Mercury, Cadmium and Arsenic in the Environment*, pp. 279–303. Chichester, U.K.: John Wiley & Sons.

International Programme on Chemical Safety (IPCS). 1990. Beryllium: Environmental health criteria 106. World Health Organization. Geneva, Switzerland. Retrieved April 10, 2011. http://www.inchem.org/documents/ehc/ehc/ehc106.htm.

Jarup, L. 2003. Hazards of heavy metal contamination. *British Medical Bulletin* 68:167–182.

Jarup, L., Berglund, M., Elinder, C.G., Nordberg, G., and Vahter, M. 1998. Health effects of cadmium exposure—a review of the literature and a risk estimate. HYPERLINK "http://www.sjweh.fi/" *Scandinavian Journal of Work, Environment & Health* 24:1–51.

Jamil, S., P.C. Abhilash, N. Singh, and P.N. Sharma. 2009. *Jatropha curcas*: A potential crop for phytoremediation of coal fly ash. *Journal of Hazards Material* 172:269–275.

Jiang, C.Y., X.F. Sheng, M. Qian, and Q.Y. Wang. 2008. Isolation and characterization of a heavy metal-resistant *Burkholderia* sp. from heavy metal-contaminated paddy field soil and its potential in promoting plant growth and heavy metal accumulation in metal polluted soil. *Chemosphere* 72:157–164.

Khan, M.A., I. Ahmad, and I. Rahman. 2007. Effect of environmental pollution on heavy metals content of *Withania somnifera*. *Journal of the Chinese Chemical Society* 54:339–343.

Khan, M.S., A. Zaidi, P.A. Wani, and M. Oves. 2009. Role of plant growth promoting rhizobacteria in the remediation of metal contaminated soils. *Environmental Chemistry Letter* 7:1–19.

Khan, S., Q. Cao, Y.M. Zheng, Y.Z. Huang, and Y.G. Zhu. 2008. Health risks of heavy metals in contaminated soils and food crops irrigated with wastewater in Beijing, China. *Environmental Pollution* 152:686–692.

Meers, E., S.S. Van, K. Adriaensen, A. Ruttens, J. Vangronsveld, G.L. Du, N. Witters, T. Thewys, and F.M. Tack. 2010. The use of bio-energy crops (*Zea mays*) for "phytoattenuation" of heavy metals on moderately contaminated soils: A field experiment. *Chemosphere* 78:35–41.

Meng, D., D. Wei, Z. Tan, A. Lin, and Y. Du. 2015. The potential risk assessment for different arsenic species in the aquatic environment. *Journal of Environmental Science* 27:1–8.

Mishra, S. and R.N. Bharagava. 2015. Toxic and genotoxic effects of hexavalent chromium in environment and its bioremediation strategies. *Journal of Environmental Science and Health*, 34(1):1–32.

Nagajyoti, P.C., K.D. Lee, and T.V.M. Sreekanth. 2010. Heavy metals, occurrence and toxicity for plants: A review. *Environmental Chemistry Letters* 8:199–216.

Neustadt, J. and S. Pieczenik. 2007. Toxic-metal contamination: Mercury. *Integrative Medicine* 6:36–37.

Ng, J.C., J. Wang, and A. Shraim. 2003. A global health problem caused by arsenic from natural sources. *Chemosphere* 5:1353–1359.

Padmavathiamma, P.K. and L.Y. Li. 2007. Phytoremediation technology: Hyperaccumulation metals in plants. *Water Air and Soil Pollution* 184:105–126.

Pandey, V.C. 2013. Suitability of *Ricinus communis* L. cultivation for phytoremediation of fly ash disposal sites. *Ecological Engineering* 57:336–341.

Philip, L., L. Iyengar, and L. Venkobacher. 2000. Site of interaction of copper on *Bacillus polymyxa*. *Water Air and Soil Pollution* 119:11–21.

Pulford, I.D. and C. Watson. 2003. Phytoremediation of heavy metal-contaminated land by trees—A review. *Environmental International* 29:529–540.

Rajendran, P., J. Muthukrishnan, and P. Gunasekaran. 2003. Microbes in heavy metal remediation. *Indian Journal of Experimental Biology* 41:935–944.

Ross, S.M. 1994. *Toxic Metals in Soil–Plant Systems*, p. 469. Chichester, U.K.: Wiley.

Ruttens, A., J. Boulet, N. Weyens, K. Smeets, K. Adriaensen, E. Meers et al. 2011. Short rotation coppice culture of willows and poplars as energy crops on metal contaminated agricultural soils. *International Journal of Phytoremediation* 13:194–207.

Salem, H.M., E.A. Eweida, and A. Farag. 2000. *Heavy Metals in Drinking Water and Their Environmental Impact on Human Health*, pp. 542–556. Giza, Egypt: ICEHM 2000 Cairo University.

Sar, P. and S.F. D'Souza. 2001. Biosorptive Uranium uptake by *Pseudomonas* strain: Characterization and equilibrium studies. *Journal of Chemical Technology and Biotechnology* 76:1286–1294.

Seidal, K., N. Jorgensen, C.G. Linder, B. Jogren, and M. Vahter. 1993. Fatal cadmium-induced pneumonitis. *Scandinavian Journal of Work, Environment & Health* 19:429–431.

Sheng, X. and J. Xia. 2006. Improvement of rape (*Brassica napus*) plant growth and cadmium uptake by cadmium–resistant bacteria. *Chemosphere* 64:1036–1042.

Singh, R., P. Singh, and R. Sharma. 2014. Microorganism as a tool of bioremediation technology for cleaning environment: A review. *Proceedings of the International Academy of Ecology and Environmental Sciences* 4:1–6.

Singh, S.N., S.K. Goyal, and S.R. Singh. 2015. Bioremediation of heavy metals polluted soils and their effect on plants. *Agriways* 3:19–24.

Srivastava, N.K. and C.B. Majumder. 2008. Novel biofiltration methods for the treatment of heavy metals from industrial wastewater. *Journal of Hazards Materials* 51:1–8.

Steenland, K. and P. Boffetta. 2000. Lead and cancer in humans: Where are we now? *American Journal of Industrial Medicine* 38:295–299.

Tripathi, R.D., S. Srivastava, S. Mishra, N. Singh, R. Tuli, D.K. Gupta, and F.J.M. Maathuis. 2007. Arsenic hazards: Strategies for tolerance and remediation by plants. *Trends in Biotechnology* 25:158–165.

USEPA. 1997. Recent developments for in-situ treatment of metals contaminated soils. Report. U.S. Environmental Protection Agency, Office of Solid Waste and Emergency Response. Washington D.C.

Vidali, M. 2001. Bioremediation: An overview. *Pure Applied Chemistry* 73:1163–1172.

Vinceti, M., E.T. Wei, C. Malagoli, M. Bergomi, and G. Vivoli. 2001. Adverse health effects of selenium in humans. *Reviews of Environmental Health* 16:233–251.

Volk, T.A., L.P. Abrahamson, C.A. Nowak, L.B. Smart, P.J. Tharakan, and E.H. White. 2006. The development of short-rotation willow in the northeastern United States for bioenergy and bioproducts, agro forestry and phytoremediation. *Biomass Bioenergy* 30:715–727.

Wang, S. 2006. Cobalt—Its recovery, recycling, and application. *Journal of the Minerals, Metals and Materials Society* 58:47–50.

WHO. 1992. *Cadmium*. Environmental Health Criteria, Vol. 134. Geneva, Switzerland: World Health Organization.

WHO. 2001. *Arsenic and Arsenic Compounds*. Environmental Health Criteria, Vol. 224. Geneva, Switzerland: World Health Organization.

Yanqun, Z., L. Yuan, C. Jianjun, C. Haiyan, Q. Li, and C. Schratz. 2005. Hyper accumulation of Pb, Zn and Cd in herbaceous grown on lead-zinc mining area in Yunnan, China. *Environmental International* 31:755–762.

10 Biochemical Pathways in Bacteria to Survive Metal-Contaminated Environments

Joe Lemire and Raymond J. Turner

CONTENTS

ABSTRACT

This chapter reviews the biochemistries employed by microorganisms to survive metal-contaminated environments. Certainly, it does not exhaustively discuss all known mechanisms but highlights those that are important for practitioners of bioremediation. We first focus on mechanisms that microorganisms use to immobilize metals either intracellularly or extracellularly. Second, we focus on mechanisms of metal resistance that bacteria use to survive in the presence of toxic metals, but not necessarily remove them or change their speciation. Unfortunately, little data is available regarding the capacity of prokaryotes to survive exposure to multiple metal toxins. The information here provides a bacterial toolkit list that can aid a practitioner of bioremediation in developing an effective remediation strategy.

10.1 INTRODUCTION

The anthropogenic use of metals is ever increasing. Mining activities, although economically powerful, result in the generation of tailings waste that can be rife in residual metals—some of them toxic to human health and the environment. Further, mining operations in the energy industry also produce wastes that are high in a variety of metals, particularly bitumen mining and coal burning (Linak and Wendt, 1994; Mahdavi et al., 2012). Aside from traditional mining activities, metals and metal nanoparticles (NPs) are gaining a

heightened presence in the environment due to their allure for medical and commercial purposes as antimicrobials, antiodorants, and antifouling agents. Notably, silver (Ag) is seeing extensive use as an antimicrobial agent in endotracheal tubes, wound dressings, and catheters (Srinivasan et al., 2006; Kollef et al., 2008; Lemire et al., 2015), in addition to numerous commercial products (Jung et al., 2008; Lemire et al., 2013). In the clinic, copper (Cu) and titanium (Ti) are gaining notoriety as efficacious antimicrobial surface coatings (Grass et al., 2011; Hasan et al., 2013; Lemire et al., 2013). Also on the rise, metals as NP formulations—many of which have unique chemistries that remain unexplored—are a potentially latent environmental hazard. Indeed, the *Nano Age* has the capacity to revolutionize materials and medicine. But questions remain regarding the toxicity of NPs to human health and the environment. For example, though the toxicity of many metals in their elemental or ionic state has been established, NPs have size and surface properties that may impart different mechanisms of toxicity in their own right (Donaldson et al., 2004; Moore, 2006; Balbus et al., 2007). Needless to say, the necessity for novel technologies to decontaminate metal-poisoned environments remains.

Apart from anthropogenic introduction of metals into the environment, metals are a natural moiety in soil, marine, and waterways. Thus, microorganisms have evolved elegant molecular mechanisms to survive in the presence of metals. Indeed, the physiology and biochemistry that microorganisms exploit to survive in metal-rich environments have been explored quite extensively (Silver, 1996; Silver and Phung, 1996; Nies,1999; Cheung and Gu, 2007; Lemire et al., 2013). Truly, bacteria have even evolved molecular mechanisms to combat less environmentally ubiquitous metals such as Ag—reducing our capacity to it for medical purposes (Gupta and Silver 1998; Gupta et al., 1999). In fact, Ag-resistant isolates have already been observed in the clinic (Randall et al., 2015). Nevertheless, the same molecular modules employed by microorganisms to survive in metal-rich environments can provide us with tools to exploit for remediating metal-contaminated environments. Here, we will explore the arsenal of biochemical processes that allow bacteria to resist metal intoxication. Namely, we will elaborate on what is known about extracellular/intracellular sequestration of metals, chemical modification of metals, and biomineralization of metals. Since the focus of this study is on bioremediation, this chapter will only discuss other pathways of metal resistance—metal efflux, reduction in metal uptake, enzyme repair mechanisms, and use of alternate biochemical pathways—as survival strategies in an environment that

is cocontaminated, where the goal is to remediate an organic pollutant, but metals remain present.

10.2 STRATEGIES TO TOLERATE METALS IN A METAL-CONTAMINATED ENVIRONMENT

Bacteria employ a variety of strategies to resist metal stress. However, there is no single strategy that allows for complete resistance to all toxic metals due to the unique physicochemical properties of metal atoms and the variety of speciation states. Thus, each metal-contaminated environment will require a distinctive strategy for the metal(s) and species present. Currently, very few studies address the molecular and biochemical mechanisms employed by microorganisms to survive multimetal challenges. Though they are fundamentally different, the mechanisms that bacteria use to resist and adapt to metals can be categorized by function (Figure 10.1). In this section, we describe the established mechanisms of metal resistance that can potentially be used to sequester, accumulate, and ultimately recover metals from a metal-rich environment.

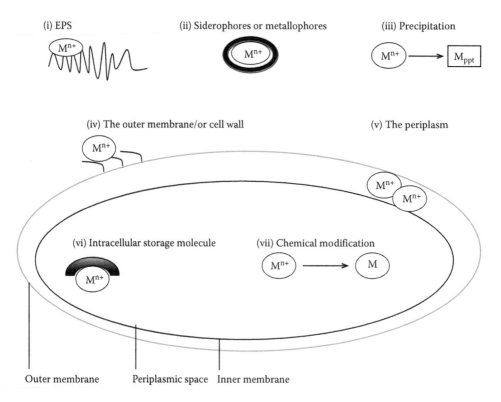

FIGURE 10.1 Strategies used by microorganisms to immobilize metals in the extracellular environment. (1) Microorganisms produce EPSs composed of DNA, protein, and carbohydrates. (2) Microorganisms can produce organic siderophores or metallophores that bind to the metal and subsequently immobilize it or allow for its uptake. (3) Microorganisms can produce anions that react with metals. This chemical reaction leads to the precipitation of the metal salt in the extracellular environment. (4) The outer membrane of some Gram-negative bacteria or polymeric substances (e.g., LPSs) present on the outer membrane can immobilize metals in the extracellular environment. Gram-positive bacteria, which does not have the outer membrane and LPS, have teichoic acids in their cell wall, which can function to immobilize metals. (5) The periplasmic space offers a unique physicochemical environment that is conducive to the immobilization of some metals within the space. (6) Some intracellular metal storage molecules can bind to select metals preventing them from reacting with cellular constituents in the cytoplasm. (7) Metals can undergo specific redox or covalent reactions inside the cell. These reactions change the chemical reactivity of the metal and may alter the bioavailability of the metal or reduce its toxicity. This process can change the oxidation state of the metal, create metal salt precipitates, or generate organometallic small-molecule compounds.

10.2.1 EXTRACELLULAR METAL SEQUESTRATION AND PRECIPITATION

When challenged by a metal stressor, microorganisms can upregulate the expression of extracellular polymeric substances (EPSs), siderophores, and low-molecular-weight ligands (metallophores), containing chemical functional groups that can coordinate with metal ions (Teitzel et al., 2006; Harrison et al., 2007; Dimkpa et al., 2008; More et al., 2014).

The EPSs generated by microbial biofilms are rife with DNA, various carbohydrate polymers, and proteins with the capacity to sequester metals (More et al., 2014). Indeed, the production of EPS, with polyanionic moieties, makes the biofilm mode of life ideal for microbial survival in environments with elevated metal ion concentrations. Still, additional active processes offer microorganisms extracellular defense against metal intoxication. For example, siderophores, which are deployed by many bacteria and fungi to acquire essential metals from their extracellular environment, have also been demonstrated to protect against toxic metal stress (Schalk et al., 2011). Experimentally, it has been demonstrated that siderophores, pyochelin, produced by *Pseudomonas aeruginosa* to recover iron from their extracellular environment, can chelate a variety of metals including Ag, aluminum (Al), cadmium (Cd), Cu, cobalt (Co), and gallium (Ga) (to name a few)—albeit at much lower affinities than Fe (Braud et al., 2009).

Low-molecular-weight ligands (metallophores), like siderophores, are also produced and excreted by some microorganisms into their extracellular environment to bind metals. They are ideal for metal detoxification because they can be readily transported out of the cell and are generally soluble. Solubility is not a requirement; however, *Pseudomonas fluorescens* has been experimentally demonstrated to produce a gelatinous, hydrophobic, phospholipid (phosphatidylethanolamine) moiety that sequesters Al (Appanna et al., 1994; Appanna and Pierre, 1996). Other phosphate-based metallophores are ideal metallophores to bind hard acid metals, for example, in 2000, Macaskie et al. reported on the capacity of *Citrobacter* sp. to enzymatically bioprecipitate uranium using extracellular phosphatases (Macaskie et al., 2000). Another metallophore, reduced sulfur—produced as a metabolic byproduct by sulfate-reducing bacteria (SRB)—efficiently leads to metal sulfides that will precipitate out the metal in a biologically inert state. SRB are a nuisance in the oil industry due to their capacity to oxidize the Fe^0 to Fe^{2+} in pipelines and cause corrosion (Enning and Garrelfs, 2014). Nonetheless, this biological process can be exploited for bioremediation, as these organisms are part of an environmental community that lead to a synergistic protection of other organisms. Additionally, this biochemistry leading to abiotic reduction of metals can be useful for bioremediation purposes. SRB have a demonstrated capacity to generate ZnS (Labrenz et al., 2000), and although SRB have been experimentally observed to remove other metals such as arsenic (As), Cu, and nickel (Ni) (Jong and Parry, 2003), it is not clear whether or not they are being directly reduced to metal sulfides. For example, Moreau et al. recently demonstrated that As, Cu, and lead (Pb) from acid mine drainage mineralized with biogenic FeS_2 aggregates, and not as their individual sulfide species (Moreau et al., 2013).

Calcite formation is another extracellular metal precipitation strategy. Microbial-induced calcium (Ca) carbonate precipitation is coupled to the hydrolysis of urea within the cell, which gives the products of ammonia and bicarbonate. The latter of these molecules then lead to the biomineralization of metals extracellularly [(1) $COO(HH_2)_2 + H_2O \rightarrow NH_2COOH + NH_3$, (2) $NH_2COOH + H_2O \rightarrow NH_3 + H_2CO_3$, (3) $H_2CO_3 \rightarrow 2H^+ + 2CO_3^{2-}$, (4) $CO_3^{2-} + Ca^{2+} \rightarrow CaCO_3$] (Dhami et al., 2013). The production of calcite through this process of flushing soil with Ca as the metal, followed by urea, is gaining popularity for fixing cracks in cement and stabilizing soil from erosion and has been a demonstrated capacity to precipitate out strontium and arsenic (Achal et al., 2012).

Another extracellular sink for metals is the cell wall and the outer membrane (of some Gram-negative cells). Metals may bind to or precipitate at the cell surface through interactions with proteins, carbohydrates, and lipopolysaccharides (LPSs). For example, Ag and lanthanum (La) were found to mostly precipitate at the cell surface (Mullen et al., 1989). Further studies on the metal-binding capacity of LPS specifically demonstrated the capacity of B-band LPS to be involved in the precipitation of Cu, La, and Fe (Langley and Beveridge, 1999). Additionally, a study by Macaskie et al. (2000) suggests the involvement of the phosphate groups within LPS to be nucleation points for uranium biomineralization.

For Gram negatives, the peptidoglycan component is a thin layer mesh between the inner membrane and the outer membrane. Yet for Gram positives, there is no outer membrane and the peptidoglycan layer is much thicker. This peptidoglycan has hydroxyl, carbonyl, and amino functional groups and thus can chelate a variety of metals. In addition to the peptidoglycan, there are teichoic acids and lipoteichoic acids that serve as chelating agents (Wang and Chen, 2009; Brown et al., 2013). In addition to the cell wall of both Gram-negative and Gram-positive bacteria, there is also a capsule of polysaccharides as well as protein extrusions (pilin and flagella) or protein coat referred to the S-layer that are available for metal binding. These polysaccharides and proteins, with their functional grouped amino acids, can also act as metal binding sites. Although they typically have weak affinities for metals, due to the high concentration of peptidoglycan, polysaccharides, and proteins on the surface of a cell, the metal absorptivity can be in the micromolar range.

A relatively novel discovery is the production of nanowires by *Shewanella oneidensis*, *Synechocystis* PCC6803, and *Pelotomaculum thermopropionicum* (Gorby et al., 2006). To solubilize Fe and manganese (Mn) oxyhydroxide minerals in the soil, *S. oneidensis* use cytochromes (MtrC and OmcA) to supply electrons for the reduction of insoluble Mn(IV) and Fe(III) minerals to soluble Mn(III) and Fe(II) via a nanowire structure (Gorby et al., 2006). The potential for using this process for bioremediation is yet to be explored.

Extracellular metal defenses deployed by microorganisms offer nonspecific mechanisms of resistance toward metal contaminants, some of which have been explored for remediation approaches. The advantages offered by extracellular strategies for metal resistance for bioremediation are that (1) the metal is neutralized outside the cell, thus the bacteria will likely survive and may be retrained or continue to produce-increasing biosorption capacity, and (2) there is the potential to collect the metal from the extracellular medium.

10.2.2 Intracellular Metal Sequestration

As companion to extracellular biochemical and chemical interactions with metals, other metal resistance processes involve intracellular sequestration of the metals. Within Gram-negative cells, there are two compartments where metal sequestration may occur: (1) the periplasmic space and (2) within the cytoplasm. Although Gram-positive bacteria do not have a periplasmic space *per se*, there are biochemical reactions similar to what is found in the periplasm, in a space between the cell membrane and the thick cell wall, as this cell wall separates this space and allows for unique physicochemical state not unlike the periplasm.

In many cases, toxic metals have been observed to concentrate in the periplasmic space of some Gram-negative prokaryotes, but the mechanisms governing the accumulation and sequestration of some metals, within that compartment, remain to be described. For example, palladium (Pd) has been observed to accumulate in the periplasm of *S. oneidensis* following the reduction from Pd(II) to Pd(0), yet no mechanism was described (De Windt et al., 2005). This is in contrast to Ag, where a mechanism was described for Ag-NP formation in the periplasm of a Ag-resistant *Escherichia coli* strain via a periplasmic nitrate reductase (Lin et al., 2014). The chalcogen tellurium (Te) also accumulates in the periplasm of *E. coli* in a reduction-mediated manner (Taylor et al., 1988; Turner, 2013; Diaz-Vasquez et al., 2015). Indeed, other periplasmic-based metal resistance mechanisms exist, as the periplasm offers a unique redox environment compared to the extracellular milieu or the cytoplasm. However, in many cases after the metal ion is either reduced or oxidized, it is either pumped out of the cell or delivered to the cytoplasm for further biochemistry.

Typically, some form of a storage molecule governs metal accumulation in the cytoplasm, whether it is a low-molecular-weight compound or a protein. Indeed, some metals are regulated within submillimolar concentrations; thus, it is rare to observe an accumulation of free metals in the cytoplasm. Typically, Fe is bound in bacterioferritin as an Fe oxide, providing a sink and a storage for this vital biological metal (Carrondo, 2003). To date, there is relatively little evidence to suggest that bacterioferritin can accumulate any other metal in a meaningful manner. Another potential molecular metal sink in the bacterial cytoplasm is metallothioneins (MTs). MTs are thought to be the main storage molecule of the essential metal, Zn (Blindauer, 2011). The presence of multiple cysteine residues in the metal binding site of MTs led researchers to postulate that it may bind other thiophilic metals. Indeed,

MTs have been observed to bind other soft metals including Cu and Cd, making MTs a potential ideal molecule to remediate Cu- and Cd-contaminated sites (Blindauer, 2011). A key thiophilic moiety that was found to be important in protecting many bacteria from a variety of toxic metals is glutathione (in Gram-negative and mycothiol in Gram-positive prokaryotes). Glutathione can form a variety of metal chelates, RS_x–M_y, as with *E. coli* and Cd toxicity, for example (Taylor et al., 1988). These molecules also participate in redox chemistry such as Painter-type reactions with oxyanions ($2RSH + MO_y^{n-} \rightarrow RS$–$M$–$SR \rightarrow RSSR + M^0$). Typically, mutants of key components in the glutathione biosynthetic pathway are found to be more sensitive to metal toxicity (Helbig et al., 2008; Zannoni et al., 2008).

There exist select cases where a storage molecule was not responsible for metal accumulation within the cytoplasm of microorganisms. For example, Pb has been observed binding to polyphosphates in phototrophic microbes (Burgos et al., 2013), magnetotactic bacteria accumulate magnetite (Lefevre and Bazylinski, 2013), and *Rhodanobacter* sp. forms uranyl phosphates (Sousa et al., 2013). Of course, these examples are specific and extreme and should be considered for very specific biochemical adaptation to the organism's environmental niche.

10.2.3 Chemical Modification

Here, we discuss some of the chemistries employed by microorganisms to detoxify metals. Though the reactions often take place in discreet locations within or outside the cell, they often do not remain in that compartment. Thus, bioremediation strategies need to include a strategy for the collection of the metal following chemical modification by the microorganism.

Oxidation/reduction reactions are common occurrences within a microbial cell. Both As and Cr are detoxified by the bacterial cell via reduction. In As-rich environments, the genes for As reduction from As(V) to As(III)—*arsC*— are found in multiple species of microorganisms (Escudero et al., 2013). Following reduction, As is then exported from the cell using As(III) exporters. Meanwhile, Cr(VI), the more soluble and toxic oxidation state, is reduced to Cr(III)—the less soluble species—exemplified by bacterial isolates from tannery sites (Batool et al., 2012). Another strategy used by microorganisms to detoxify As is methylation (Messens and Silver, 2006), though it remains unclear whether or not this enzyme is specific for As (Escudero et al., 2013). Mercury (Hg) is also methylated by *Desulfovibrio desulfuricans* (Choi et al., 1994); however, methyl mercury is far more toxic to most other organisms.

Another interesting decontamination strategy deployed by some prokaryotes is the biomineralization of metals into NPs—nanoparticlization. Though the mechanism has not been detailed, recent work by Zonaro et al. (2015) demonstrated the capacity of *Stenotrophomonas maltophilia* and *Ochrobactrum* sp. to generate NPs of Te and selenium (Se), respectively. However, NP formation can lead to a metal species with unique physical and chemical properties that may lead to enhanced toxicity to other bacteria.

Here, we described several approaches that prokaryotes use to protect themselves from the toxicity of metals that could also be coupled for the bioremediation of metals in a fashion that provides the possibility to recover the metals. Whether sequestering or immobilizing the metals in the extracellular environment or within the cell, these methods can be used to recover metals for removal from the environment that is being bioremediated. The approaches mentioned earlier do not encompass all of the methods microorganisms use to resist heavy metals, but those that offer the potential for removing metals for recovery purposes.

10.3 OTHER STRATEGIES TO SURVIVE METAL-CONTAMINATED ENVIRONMENTS FOR REMEDIATION OF COCONTAMINATION

The mechanisms of resistance microorganisms use to survive metal poisoning have been described in detail for over 50 years. Many of these resistance mechanisms are valuable to a bioremediation practitioner, not for direct cleanup of a metal toxicant *per se*, but for the survival of the microorganism in a metal-rich environment—such as a cocontaminated site. These

mechanisms are discreet from those previously described in that they bestow metal resistance to the microorganism but do not necessarily sequester or accumulate the metal for retrieval (Figure 10.2).

10.3.1 Metal Efflux

Several resistance determinants on chromosome or mobile genetic elements encode for efflux pumps or transporters that carry toxic metals out of the cell (Nies, 1999). Signaling modules that respond to nanomolar to zeptomolar concentrations of metals regulate many of these transporters (Changela et al., 2003). These transporters are active and are fueled by ATP hydrolysis or a chemiosmotic potential (Lemire et al., 2013). For example, As, Cu, Ag, Zn, and Cd are all effluxed from the microbial cell via a P-type ATPase or $M^{2+}/H^{+}(K^{+})$ antiporters (Arguello et al., 2007; Besaury et al., 2013; Kolaj-Robin et al., 2015).

10.3.2 Reduction in Metal Uptake

While microbial cells enhance their export of toxic metals, they also concert an effort to reduce the uptake of these same toxic metals by downregulating key metal importers. The importers may be specific to the metal because it is

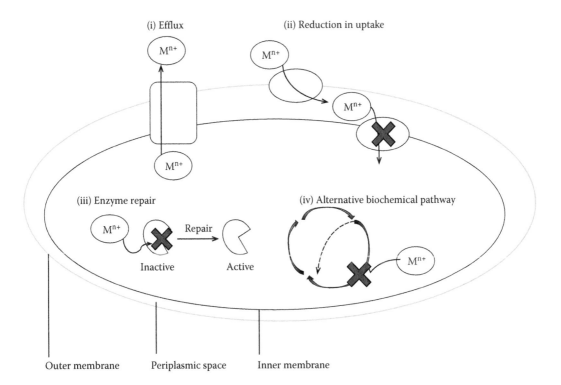

FIGURE 10.2 Strategies deployed by microorganisms to survive in metal-rich environments. (1) Microorganisms encode for metal efflux transporters in their chromosomal DNA or on mobile genetic elements. Using the hydrolysis of ATP or a chemiosmotic gradient, the metals can be effluxed out of the microbial cell. (2) To reduce the quantity of toxic metals that are being transported inside the cell, microorganisms can downregulate or inhibit metal importers. (3) Microorganisms carry a robust repertoire of biochemical systems that can repair enzymes that have been inactivated by a toxic metal or the reaction by-products of toxic metals—including ROS. (4) If repair is not possible, a toxic metal can inhibit a biochemical pathway or process. In such cases, some microorganisms have evolved alternate pathways for essential biochemical pathways that are not inhibited by toxic metals.

required, yet excessive concentrations can be lethal. Or a metal may get into the cell via ionic/molecular mimicry for a transporter of required ions. Typically, reduction in metal uptake and enhanced metal efflux are complement systems required for the survival of the cell (Ma et al., 2009). In some circumstances, specific mutations in transporters have been demonstrated to protect the bacterial cell from toxicity. For example, in Cr-rich environments, mutations within sulfate importers (a pathway with established capacity to import Cr (VI)) bestow resistance to bacteria (Ramirez-Diaz et al., 2008).

10.3.3 ENZYME REPAIR MECHANISMS

Some enzymes—especially those dependent upon metals or those with redox-sensitive functional groups—tend to be sensitive to metal poisoning. For example, enzymes that contain [Fe-S] cluster are particularly sensitive to soft acid metals such as Ag, Zn, Cd, and Hg (Xu and Imlay, 2012). Often, the catalytic by-products of toxic metals such as reactive oxygen species (ROS) produced via Fenton reactions will also impose a toxic effect on enzymes. Cellular chaperones, antioxidants, and enzymes involved in antioxidant homeostasis and repair enzymes can all be invoked to overcome the stress imposed by the toxic metal or its by-products (Harrison et al., 2009; Mailloux et al., 2011).

10.3.4 ALTERNATE BIOCHEMICAL PATHWAYS

When a biological pathway is hindered by the presence of a toxic metal, bacterial cells can bypass these pathways by using alternative enzymes that are not affected by that metal. For example, in the presence of Al and Ga, *P. fluorescens* utilizes alternate means to generate ATP through a modified tricarboxylic acid cycle (Mailloux et al., 2007; Chenier et al., 2008; Lemire et al., 2010). In this manner, the essential functioning of the cell can proceed regardless of the presence of the toxicant.

Here, we discussed the mechanisms employed by prokaryotes to survive a toxic metal ion–rich environment. These are important considerations for species that will use complex organic pollutants for energy and/or carbon source and that environment is also in the presence of metal toxins.

10.4 CONCLUSIONS AND PERSPECTIVE

This chapter reviewed the biochemistries employed by microorganisms to survive metal-contaminated environments. Certainly, it does not exhaustively discuss all known mechanisms but highlights those that are important for practitioners of bioremediation. Here, we focused first on mechanisms that microorganisms use to immobilize metals either intracellularly or extracellularly. Second, we focused on mechanisms of metal resistance that bacteria use to survive in the presence of toxic metals, but not necessarily remove them or change their species. Unfortunately, little is known about how multimetal

exposures may affect the capacity of prokaryotes to survive. The information here provides a list of bacterial toolkit list that can aid a practitioner of bioremediation in developing an effective remediation strategy.

REFERENCES

Achal, V., Pan, X., Fu, Q., Zhang, D. 2012. Biomineralization based remediation of As(III) contaminated soil by *Sporosarcina gingensisoli*. *Journal of Hazardous Materials* 201–202:178–184.

Appanna, V.D., Kepes, M., Rochon, P. 1994. Aluminum tolerance in *Pseudomonas fluorescens* ATCC 13525: Involvement of a gelatinous lipid-rich residue. *FEMS Microbiology Letters* 119:295–301.

Appanna, V.D., Pierre, M.S. 1996. Aluminum elicits exocellular phosphatidylethanolamine production in *Pseudomonas fluorescens*. *Applied and Environmental Microbiology* 62:2778–2782.

Arguello, J.M., Eren, E., Gonzalez-Guerrero, M. 2007. The structure and function of heavy metal transport P1B-ATPases. *Biometals* 20:233–248.

Balbus, J.M., Maynard, A.D., Colvin, V.L., Castranova, V., Daston, G.P., Denison, R.A., Dreher, K.L. et al. 2007. Meeting report: Hazard assessment for nanoparticles: Report from an interdisciplinary workshop. *Environmental Health Perspectives* 115:1654–1659.

Batool, R., Yrjala, K., Hasnain, S. 2012. Hexavalent chromium reduction by bacteria from tannery effluent. *Journal of Microbiology and Biotechnology* 22:547–554.

Besaury, L., Bodilis, J., Delgas, F., Andrade, S., De la Iglesia, R., Ouddane, B., Quillet, L. 2013. Abundance and diversity of copper resistance genes *cusA* and *copA* in microbial communities in relation to the impact of copper on Chilean marine sediments. *Marine Pollution Bulletin* 67:16–25.

Blindauer, C.A. 2011. Bacterial metallothioneins: Past, present, and questions for the future. *Journal of Biology and Inorganic Chemistry* 16:1011–1024.

Braud, A., Hannauer, M., Mislin, G.L., Schalk, I.J. 2009. The *Pseudomonas aeruginosa* pyochelin-iron uptake pathway and its metal specificity. *Journal of Bacteriology* 191:3517–3525.

Brown, S., Santa Maria, J.P. Jr., Walker, S. 2013. Wall teichoic acids of gram-positive bacteria. *Annual Review of Microbiology* 67:313–336.

Burgos, A., Maldonado, J., De Los Rios, A., Sole, A., Esteve, I. 2013. Effect of copper and lead on two consortia of phototrophic microorganisms and their capacity to sequester metals. *Aquatic Toxicology* 140–141:324–336.

Carrondo, M.A. 2003. Ferritins, iron uptake and storage from the bacterioferritin viewpoint. *EMBO Journal* 22:1959–1968.

Changela, A., Chen, K., Xue, Y., Holschen, J., Outten, C.E., O'Halloran, T.V., Mondragon, A. 2003. Molecular basis of metal-ion selectivity and zeptomolar sensitivity by CueR. *Science* 301:1383–1387.

Chenier, D., Beriault, R., Mailloux, R., Baquie, M., Abramia, G., Lemire, J., Appanna, V. 2008. Involvement of fumarase C and NADH oxidase in metabolic adaptation of *Pseudomonas fluorescens* cells evoked by aluminum and gallium toxicity. *Applied and Environmental Microbiology* 74:3977–3984.

Cheung, K.H., Gu, J.D. 2007. Mechanism of hexavalent chromium detoxification by microorganisms and bioremediation application potential: A review. *International Biodeterioration and Biodegradation* 59:8–15.

Choi, S.C., Chase, T., Bartha, R. 1994. Metabolic pathways leading to mercury methylation in *Desulfovibrio desulfuricans* LS. *Applied and Environmental Microbiology* 60:4072–4077.

De Windt, W., Aelterman, P., Verstraete, W. 2005. Bioreductive deposition of palladium (0) nanoparticles on *Shewanella oneidensis* with catalytic activity towards reductive dechlorination of polychlorinated biphenyls. *Environmental Microbiology* 7:314–325.

Dhami, N.K., Reddy, M.S., Mukherjee, A. 2013. Biomineralization of calcium carbonates and their engineered applications: A review. *Frontiers in Microbiology* 4:314.

Diaz-Vasquez, W.A., Abarca-Lagunas, M.J., Cornejo, F.A., Pinto, C.A., Arenas, F.A., Vasquez, C.C. 2015. Tellurite-mediated damage to the *Escherichia coli* NDH-dehydrogenases and terminal oxidases in aerobic conditions. *Archieves in Biochemistry and Biophysics* 566:67–75.

Dimkpa, C.O., Svatos, A., Dabrowska, P., Schmidt, A., Boland, W., Kothe, E. 2008. Involvement of siderophores in the reduction of metal-induced inhibition of auxin synthesis in *Streptomyces* spp. *Chemosphere* 74:19–25.

Donaldson, K., Stone, V., Tran, C.L., Kreyling, W., Borm, P.J.A. 2004. Nanotoxicology. *Occupational and Environmental Medicine* 61:727–728.

Enning, D., Garrelfs, J. 2014. Corrosion of iron by sulfate-reducing bacteria: New views of an old problem. *Applied and Environmental Microbiology* 80:1226–1236.

Escudero, L.V., Casamayor, E.O., Chong, G., Pedros-Alio, C., Demergasso, C. 2013. Distribution of microbial arsenic reduction, oxidation and extrusion genes along a wide range of environmental arsenic concentrations. *PLoS ONE* 8:e78890.

Gorby, Y.A., Yanina, S., McLean, J.S., Rosso, K.M., Moyles, D., Dohnalkova, A., Beveridge, T.J. et al. 2006. Electrically conductive bacterial nanowires produced by *Shewanella oneidensis* strain MR-1 and other microorganisms. *Proceedings of the National Academy of Sciences of the United States of America* 103:11358–11363.

Grass, G., Rensing, C., Solioz, M. 2011. Metallic copper as an antimicrobial surface. *Applied and Environmental Microbiology* 77:1541–1547.

Gupta, A., Matsui, K., Lo, J.F., Silver, S. 1999. Molecular basis for resistance to silver cations in *Salmonella*. *Nature Medicine* 5:183–188.

Gupta, A., Silver, S. 1998. Molecular genetics: Silver as a biocide: Will resistance become a problem? *Nature Biotechnology* 16:888.

Harrison, J.J., Ceri, H., Turner, R.J. 2007. Multimetal resistance and tolerance in microbial biofilms. *Nature Reviews Microbiology* 5:928–938.

Harrison, J.J., Tremaroli, V., Stan, M.A., Chan, C.S., Vacchi-Suzzi, C., Heyne, B.J., Parsek, M.R., Ceri, H., Turner, R.J. 2009. Chromosomal antioxidant genes have metal ion-specific roles as determinants of bacterial metal tolerance. *Environmental Microbiology* 11:2491–2509.

Hasan, J., Crawford, R.J., Ivanova, E.P. 2013. Antibacterial surfaces: The quest for a new generation of biomaterials. *Trends in Biotechnology* 31:295–304.

Helbig, K., Grosse, C., Nies, D.H. 2008. Cadmium toxicity in glutathione mutants of *Escherichia coli*. *Journal of Bacteriology* 190:5439–5454.

Jong, T., Parry, D.L. 2003. Removal of sulfate and heavy metals by sulfate reducing bacteria in short-term bench scale upflow anaerobic packed bed reactor runs. *Water Research* 37:3379–3389.

Jung, W.K., Koo, H.C., Kim, K.W., Shin, S., Kim, S.H., Park, Y.H. 2008. Antibacterial activity and mechanism of action of the silver ion in *Staphylococcus aureus* and *Escherichia coli*. *Applied and Environmental Microbiology* 74:2171–2178.

Kolaj-Robin, O., Russell, D., Hayes, K.A., Pembroke, J.T., Soulimane, T. 2015. Cation diffusion facilitator family: Structure and function. *FEBS Letters* 589:1283–1295.

Kollef, M.H., Afessa, B., Anzueto, A., Veremakis, C., Kerr, K.M., Margolis, B.D., Craven, D.E. et al. 2008. Silver-coated endotracheal tubes and incidence of ventilator-associated pneumonia: The NASCENT randomized trial. *JAMA* 300:805–813.

Labrenz, M., Druschel, G.K., Thomsen-Ebert, T., Gilbert, B., Welch, S.A., Kemner, K.M., Logan, G.A. et al. 2000. Formation of sphalerite (ZnS) deposits in natural biofilms of sulfate-reducing bacteria. *Science* 290:1744–1747.

Langley, S., Beveridge, T.J. 1999. Effect of O-side-chain-lipopolysaccharide chemistry on metal binding. *Applied and Environmental Microbiology* 65:489–498.

Lefevre, C.T., Bazylinski, D.A. 2013. Ecology, diversity, and evolution of magnetotactic bacteria. *Microbiology and Molecular Biology Review* 77:497–526.

Lemire, J., Mailloux, R., Auger, C., Whalen, D., Appanna, V.D. 2010. *Pseudomonas fluorescens* orchestrates a fine metabolic-balancing act to counter aluminium toxicity. *Environmental Microbiology* 12:1384–1390.

Lemire, J.A., Harrison, J.J., Turner, R.J. 2013. Antimicrobial activity of metals: Mechanisms, molecular targets and applications. *Nature Reviews Microbiology* 11:371–384.

Lemire, J.A., Kalan, L., Bradu, A., Turner, R.J. 2015. Silver oxynitrate, an unexplored silver compound with antimicrobial and antibiofilm activity. *Antimicrobial Agents and Chemotherapy* 59:4031–4039.

Lin, I.W.S., Lok, C.N., Che, C.M. 2014. Biosynthesis of silver nanoparticles from silver(I) reduction by the periplasmic nitrate reductase c-type cytochrome subunit NapC in a silver-resistant *E. coli*. *Chemical Science* 5:3144–3150.

Linak, W.P., Wendt, J.O.L. 1994. Trace metal transformation mechanisms during coal combustion. *Fuel Processing Technology* 39:173–198.

Ma, Z., Jacobsen, F.E., Giedroc, D.P. 2009. Coordination chemistry of bacterial metal transport and sensing. *Chemical Review* 109:4644–4681.

Macaskie, L.E., Bonthrone, K.M., Yong, P., Goddard, D.T. 2000. Enzymically mediated bioprecipitation of uranium by a *Citrobacter* sp.: A concerted role for exocellular lipopolysaccharide and associated phosphatase in biomineral formation. *Microbiology* 146:1855–1867.

Mahdavi, H., Ulrich, A.C., Liu, Y. 2012. Metal removal from oil sands tailings pond water by indigenous micro-alga. *Chemosphere* 89:350–354.

Mailloux, R.J., Beriault, R., Lemire, J., Singh, R., Chenier, D.R., Hamel, R.D., Appanna, V.D. 2007. The tricarboxylic acid cycle, an ancient metabolic network with a novel twist. *PLoS ONE* 2:e690.

Mailloux, R.J., Lemire, J., Appanna, V.D. 2011. Metabolic networks to combat oxidative stress in *Pseudomonas fluorescens*. *Antonie Van Leeuwenhoek* 99:433–442.

Messens, J., Silver, S. 2006. Arsenate reduction: Thiol cascade chemistry with convergent evolution. *Journal of Molecular Biology* 362:1–17.

Moore, M.N. 2006. Do nanoparticles present ecotoxicological risks for the health of the aquatic environment? *Environment International* 32:967–976.

More, T.T., Yadav, J.S.S., Yan, S., Tyagi, R.D., Surampalli, R.Y. 2014. Extracellular polymeric substances of bacteria and their potential environmental applications. *Journal of Environmental Management* 144:1–25.

Moreau, J.W., Fournelle, J.H., Banfield, J.F. 2013. Quantifying heavy metals sequestration by sulfate-reducing bacteria in an acid mine drainage-contaminated natural wetland. *Frontiers in Microbiology* 4:43.

Mullen, M.D., Wolf, D.C., Ferris, F.G., Beveridge, T.J., Flemming, C.A., Bailey, G.W. 1989. Bacterial sorption of heavy metals. *Applied and Environmental Microbiology* 55:3143–3149.

Nies, D.H. 1999. Microbial heavy-metal resistance. *Applied Microbiology and Biotechnology* 51:730–750.

Ramirez-Diaz, M.I., Diaz-Perez, C., Vargas, E., Riveros-Rosas, H., Campos-Garcia, J., Cervantes, C. 2008. Mechanisms of bacterial resistance to chromium compounds. *Biometals* 21:321–332.

Randall, C.P., Gupta, A., Jackson, N., Busse, D., O'Neill, A.J. 2015. Silver resistance in Gram-negative bacteria: A dissection of endogenous and exogenous mechanisms. *Journal of Antimicrobial Chemotherapy* 70:1037–1046.

Schalk, I.J., Hannauer, M., Braud, A. 2011. New roles for bacterial siderophores in metal transport and tolerance. *Environmental Microbiology* 13:2844–2854.

Silver, S. 1996. Bacterial resistances to toxic metal ions: A review. *Gene* 179:9–19.

Silver, S., Phung, L.T. 1996. Bacterial heavy metal resistance: New surprises. *Annual Review of Microbiology* 50:753–789.

Sousa, T., Chung, A.P., Pereira, A., Piedade, A.P., Morais, P.V. 2013. Aerobic uranium immobilization by *Rhodanobacter* A2–61 through formation of intracellular uranium-phosphate complexes. *Metallomics* 5:390–397.

Srinivasan, A., Karchmer, T., Richards, A., Song, X., Perl, T.M. 2006. A prospective trial of a novel, silicone-based, silver-coated foley catheter for the prevention of nosocomial urinary tract infections. *Infection Control & Hospital Epidemiology* 27:38–43.

Taylor, D.E., Walter, E.G., Sherburne, R., Bazett-Jones, D.P. 1988. Structure and location of tellurium deposited in *Escherichia coli* cells harbouring tellurite resistance plasmids. *Journal of Ultrastructure and Molecular Structure Research* 99:18–26.

Teitzel, G.M., Geddie, A., De Long, S.K., Kirisits, M.J., Whiteley, M., Parsek, M.R. 2006. Survival and growth in the presence of elevated copper: Transcriptional profiling of copper-stressed *Pseudomonas aeruginosa*. *Journal of Bacteriology* 188:7242–7256.

Turner, R. 2013. Bacterial tellurite resistance. In Kretsinger, R., Uversky, V., Permyakov, E. (eds.), *Encyclopedia of Metalloproteins*. Springer, New York, pp. 219–223.

Wang, J., Chen, C. 2009. Biosorbents for heavy metals removal and their future. *Biotechnology Advances* 27:195–226.

Xu, F.F., Imlay, J.A. 2012. Silver(I), mercury(II), cadmium(II), and zinc(II) target exposed enzymic iron-sulfur clusters when they toxify *Escherichia coli*. *Applied Environmental Microbiology* 78:3614–3621.

Zannoni, D., Borsetti, F., Harrison, J.J., Turner, R.J. 2008. The bacterial response to the chalcogen metalloids Se and Te. *Advances in Microbial Physiology* 53:1–72.

Zonaro, E., Lampis, S., Turner, R.J., Qazi, S.J.S., Vallini, G. 2015. Biogenic selenium and tellurium nanoparticles synthesized by environmental microbial isolates efficaciously inhibit bacterial planktonic cultures and biofilms. *Frontiers in Microbiology* 6:584. doi: 10.3389/fmicb.2015.00584.

11 Decontamination of Multiple-Metal Pollution by Microbial Systems
The Metabolic Twist

*Sujeenthar Tharmalingam, Azhar Alhasawi,
Varun P. Appanna, and Vasu D. Appanna*

CONTENTS

ABSTRACT

Metal pollution is an ongoing global problem that affects all living organisms. Bioremediation provides an environmentally friendly tool to manage metal pollutants. This process does not only render the toxic metals biologically inactive but also helps confine these pollutants to a very limited area. However, to be effective and operate in a natural environment, the metabolic networks that enable microbes to process these metals need to be properly fine-tuned. In this review, the pivotal role metabolism plays in ensuring the decontamination of a multiple-metal environment is elaborated. The enzymes, transport systems, metabolites, and cofactors that participate in the uptake, transformation, binding, and immobilization of the metals are discussed, and the biological processes that orchestrate a constant supply of these moieties are unraveled. For instance, the reprogramming of the tricarboxylic acid cycle, glyoxylate shunt, glycolysis, and pentose phosphate pathway aimed at generating oxalate, NADPH, ATP, and lipids that are essential in the elimination and sequestration of Al, Ga, Zn, Fe, and Ca is delineated. Hence, metabolic engineering is a crucial component of bioremediation technology and its success.

11.1 INTRODUCTION

Environmental metal contamination due to anthropogenic activities is of great public concern due to its negative impact on biological functions of all living organisms (Lemire et al., 2010b). Industrial, domestic, agricultural, medical, and technological applications have contributed to unnatural toxic metal distribution. For example, environmental pollution produced from mining, foundries and smelters, metal-based industrial operations, electroplating, pesticides, fertilizer discharge, and biosolids have been immense sources of environmental metal contamination (He et al., 2005; Tchounwou et al., 2012). Environmental metal toxicity can also occur through natural processes, including metal corrosion, atmospheric deposition, soil erosion/leaching of metal ions, sediment resuspension, and metal evaporation from water sources to soil or groundwater (Singh et al., 2011). Even natural events such as weathering and volcanic eruptions contribute significantly to metal pollution (Nriagu and Pacyna, 1988). While weathering of the earth's minerals is a slow geochemical process, metals released from anthropogenic activities have resulted in the accumulation of metals in soils and waters causing major health concerns to humans, plants, animals, and aquatic species (D'Amore et al., 2005).

Metals are recovered from ores as minerals in numerous forms. These metals are naturally found as sulfides (iron, arsenic, lead, lead–zinc, cobalt, gold–silver, and nickel) or oxides (aluminum, manganese, gold, selenium, and antimony). For example, lead, cadmium, arsenic, and mercury are usually found together with sulfides of iron or copper (Duruibe et al., 2007). Some metals are present in either sulfide or oxide form as in iron, copper, and cobalt. Many metals are nevertheless essential nutrients required for proper biochemical and physiological functions. Of all the metals, cadmium, chromium, mercury, and lead are among the greatest public health concerns and are known to produce multiple organ damage (Tchounwou et al., 2012). The U.S. Environmental Protection Agency has also labeled these metals as carcinogenic. Mining activities coupled with the growth of industrial metal usage have resulted in considerable increase in toxic discharge of metal pollutants into the soil and water system (Dixit et al., 2015). Metals such as mercury (Hg), lead (Pb), cadmium (Cd), aluminum (Al), and chromium (Cr) and metalloids like arsenic (As) are major disruptors of the ecosystem balance (Godt et al., 2006; Jomova et al., 2011; Bernhoft, 2012). Furthermore, numerous studies demonstrate that increasing industrial demand has resulted in concurrent increase in metal pollution (Figure 11.1) (He et al., 2015). Although physical and chemical methods have been

developed to detoxify metal-contaminated regions, these techniques are ineffective and expensive. Bioremediation is a microbial-based process of removing or immobilizing metals or other pollutants (Gaur et al., 2014). This chapter will focus on the use of microorganisms for the management of multiple-metal pollution and how metabolism is critical for the effectiveness of this process.

11.2 GLOBAL CONCERNS REGARDING METAL POLLUTION

Mining contributes to the majority of environmental metal pollution since metals are readily leached out and released to the surrounding. During mining, remnant metals left behind as tailings in partially covered pits result in substantial amounts of metals being mobilized and transported in the environment (Rodriguez et al., 2009). Peplow (1999) reported that even though hard-rock mines usually operate for about 5–15 years until minerals are depleted, environmental metal contamination persists hundreds of years after the cessation of mining activities (Peplow, 1999). In addition, sloppy mine sites further increase metal-leaching process by downstream acid water currents that run off to the sea beds. Furthermore, mine ores dumped on ground surface for manual dressing expose metals to air and rain, further contributing to the generation of acid mine drainage (Johnson and Hallberg, 2005). Mining activities also generate acid mine drainage as a result of oxidizing bacteria that convent pyrite and other sulfide minerals in the aquifer into methylated organic forms (Casiot et al., 2003). Examples of these include monomethylmercury and dimethylcadmium. From here, metals are transported through rivers and streams either dissolved in water or as part of suspended sediments. These metals are stored in the riverbed as

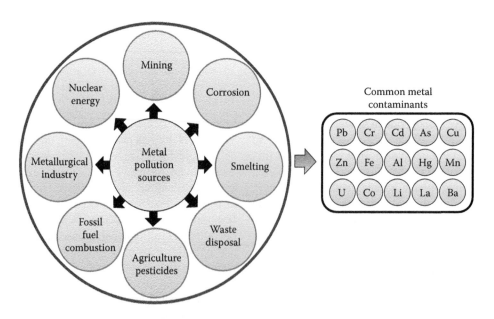

FIGURE 11.1 The origin of common metal pollutants.

sediments and sometimes seep into the groundwater contaminating numerous water sources (Vezzaro et al., 2015). Indeed, several studies have shown elevated metal contamination in wells located close to the mining area (Garbarino et al., 1995; Peplow, 1999).

When metal contamination occurs in agriculture soils, the crops from these areas are known to accumulate these metals (Conesa et al., 2006). Similarly, marine organisms that live in metal-polluted waters also help in the concentration of these toxicants (Cardwell et al., 2002). Animals and humans that consume these metal-contaminated crops and aquatic species become exposed to metals resulting in numerous health problems. Hence, metal contamination perturbs the entire ecosystem through the food chain ultimately contaminating humans (Dudka and Miller, 1999). An example of major metal contamination is evident in Brazil, a worldwide leader in mining and agricultural activities. Iron and copper mining activities in southern Brazil have resulted in elevated levels of Mn and As in surrounding water bodies (Machado et al., 2011). More importantly, agricultural fertilizer use in Brazil had been under extreme scrutiny in 2015 when studies demonstrated that in order to reduce costs fertilizers were produced using industrial and mine-waste products (Nascimento and Chartone-Souza, 2003; Mirlean and Roisenberg, 2006). These raw materials contained high levels of Cd, Pb, and Cr, which directly resulted in metal contamination in the agricultural crops and underground water beds (Mirlean et al., 2007). A similar problem was seen in leafy vegetables of Guyana where mine-waste fertilizer use resulted with increased levels of Cd, Pb, Cu, Zn, Co, Ni, Mn, and Fe (Nankishore, 2014). Recently, Brazil's Vale–BHP dam burst, which housed mine tailings, poses an even more ecological concern regarding metal contamination to surrounding water sources. Likewise, water and agricultural metal contamination is prevalent in many countries and poses immense threat to the global ecological balance and ultimately human health via the food chain and impure water sources for drinking. Thus, effective legislation and metal detoxification systems need to be implemented to reduce global metal pollution.

11.3 METAL TOXICITY IN HUMANS

Environmental toxins in air, water, and soil that are beyond the tolerance limit of the ecosystem produce numerous health risks to the biosphere. Metal toxins enter the body via food, air, and water and accumulate in tissues (Oliver and Gregory, 2015). Biotoxicity results when metals are consumed above the recommended limits. When ingested, the acidic stomach oxidizes the metals in stable states that then enter the body to form conjugates to proteins and enzymes creating strong and stable chemical bonds (Bridges and Zalups, 2005). Binding of metals to biomolecules prevents its function thereby disrupting cellular and physiological homeostasis (Tapiero and Tew, 2003). Metals damage cellular organelles and components such as cell membrane, mitochondria, lysosome,

endoplasmic reticulum, nuclei, and enzymes involved in metabolism, detoxification, and damage repair (Sharma and Dietz, 2009). Metal ions interact with cell components such as DNA and nuclear proteins resulting in damaged DNA and protein conformational changes that result in cell cycle modulation, carcinogenesis, and apoptosis. In addition, free metals induce reactive oxygen species (ROS) production, which contributes to toxicity and carcinogenicity of metals (Valko et al., 2005).

The essential metals are crucial for enzymes involved in oxidation–reduction reactions. For example, copper is an important cofactor for oxidative stress–related enzymes including catalase, superoxide dismutase, peroxidase, cytochrome c oxidase, ferroxidases, and monoamine oxidase (Tapiero et al., 2003). However, excessive exposure to copper is linked to cellular damage and Wilson disease. Similarly, numerous metals are required for biological activity; however, excess amounts can produce cellular and tissue damage leading to various adverse effects. Metal toxicity–induced cellular changes translate into a wide variety of physiology abnormalities including gastrointestinal disorders, diarrhea, stomatitis, tremor, hemoglobinuria, ataxia, paralysis, vomiting, convulsion, depression, and pneumonia (Fraga, 2005; Duruibe et al., 2007). Cellular and biochemical toxicity of some metals is described.

11.3.1 LEAD, ALUMINUM, CADMIUM, ARSENIC, GALLIUM, AND RADIONUCLIDES: TOXICITY

High concentrations of lead (Pb) are mined for the production of lead-acid batteries, ammunitions, and a range of metal products and devices to shield x-rays. Pb has been also used in paints and ceramic products. Pb-contaminated paint found on the interior surfaces contributes greatly to the source of elevated Pb in households and human exposure (Papanikolaou et al., 2005). Pb exposure occurs via inhalation of lead-contaminated dust particles and ingestion of lead-containing food, water, and paints (Gidlow, 2004). Once ingested, Pb is taken into the kidney, followed by the liver and soft tissues including the heart and brain, and also the bones. In blood, Pb is bound to erythrocytes and eliminated slowly via urine (Gidlow, 2004). Pb that accumulates in the skeleton is released very slowly with a half-life of over 20 years (Rabinowitz, 1991). The brain is most susceptible to Pb toxicity leading to headache, loss of memory, irritability, and various other symptoms related to the nervous system such as behavioral changes and concentration difficulties (Goyer et al., 2004). Long-term low-level Pb exposure in children is associated with diminished intellectual capacity (Rummo et al., 1979). Acute Pb exposure also leads to proximal renal tubular damage and long-term exposure promotes kidney damage (Loghman-Adham, 1997). At the cellular level, Pb disrupts biological functions by binding to sulfhydryl groups present on numerous enzymes thus negatively impacting their function. Pb is also able to mimic and inhibit the actions of calcium, iron, and zinc by displacing these essential metals from active sites of various enzymes resulting in improper function (Fullmer et al., 1985; Lemire et al., 2008b). One of the main groups of enzymes disrupted with Pb toxicity includes those involved in the production of heme biosynthesis such as delta-aminolevulinic acid dehydratase and ferrochelatase. The latter enzyme catalyzes the joining of protoporphyrin with iron to form heme. Thus, Pb toxicity routinely leads to development of anemia (Smith et al., 1995). Furthermore, the buildup of heme precursors directly and indirectly leads to neuronal damage, thus explaining many of the negative effects mediated by lead toxicity on the nervous system.

Aluminum (Al) is a trivalent metal that is most abundant metallic element of the earth's crust (Auger et al., 2013a). However, aluminum is toxic to biological systems. Al is present as complexes that are largely inaccessible; however, anthropogenic activity has increased its bioavailability (Verstraeten et al., 2008). Industrialization has brought out increased acid rain, which results in lowered soil pH, thus allowing for increased leaching of Al into groundwater. This is highly undesirable as now the Al can bioaccumulate in vegetation and enter the human system through the food chain (Pina and Cervantes, 1996). Al bioaccumulates in the kidney, liver, and brain (Mailloux et al., 2011a). Al has prooxidant properties at concentrations greater than 0.1 mg/mL and is acutely toxic to aquatic fauna (Exley et al., 1991). Al also interferes with iron transport systems, disrupts plant nodulation, and inhibits photosynthesis and nitrogen fixation (Appanna, 1989). Al interferes with the cellular functions of biologically important metals such as Ca, Mg, and Fe primarily by increasing intracellular Ca concentrations, has increased affinity for ATP thus inhibiting Mg interaction with ATP, and competes for binding sites with biomolecules that utilize Fe (Pina and Cervantes, 1996; Lemire et al., 2010a; Auger et al., 2013a). In addition to these, Al interferes with phosphate-rich membrane lipids due to the negative charges produced by phosphate groups. When Al binds to phospholipids, the cell membrane becomes rigid and loses its function. However, the high propensity for Al to bind to phospholipids also allows it to be sequestered and detoxified. Similarly, the dense negative charges present on the phosphate-rich DNA backbone attract Al, thereby interfering with DNA replication and transcriptional processes (Johnson and Wood, 1990). Furthermore, Al binds much stronger than Mg, thereby disrupting the crucial effects of Mg on DNA replication (Lemire et al., 2010b). In addition, Al disrupts Fe-dependent enzymes leading to production of ROS by producing free Fe in cells and by disrupting enzymes involved with oxidation–reduction reactions, further abrogating cellular oxidative stress (Middaugh et al., 2005). Reduction in Fe activity also negatively affects ATP production via inhibition of citric acid cycle and electron transport chain that cannot function without Fe (Lemire et al., 2010b). Specifically, enzymes affected include aconitase, succinate dehydrogenase, fumarase, complex I, and complex IV. Since ATP is the main energy source of cells, organisms under Al toxicity must adapt and develop alternate energy sources to

survive. Al toxicity can lead to obesity and neurological disorders (Lemire and Appanna, 2011; Han et al., 2013).

Cadmium (Cd) is produced as a by-product from extracting zinc, lead, and copper ores (Safarzadeha et al., 2007). Cd is primarily used as an anticorrosion agent and for phosphate fertilizer production and serves as a stabilizer for PVC products, color pigments, numerous alloys, and now popular nickel–cadmium rechargeable batteries (Jarup, 2003). Cd-containing fertilizers and contaminated soils lead to increased Cd uptake in crops and subsequent human exposure through consumption. Cigarette smoking also significantly contributes to increased Cd levels in the blood (Satarug and Moore, 2004). Nevertheless, food is the primary source of Cd intake in humans. Cd exposure leads to kidney damage initiated by tubular dysfunction evident by secretion of low-molecular-weight proteins (Jarup, 2002). Prolonged Cd exposure further damages the kidney by producing drastically reduced glomerular filtration rate. Studies have shown that Cd-mediated kidney damage is associated with increased cardiovascular damage. Long-term Cd intake also contributes to skeletal damage evident by low bone mineral density and fractures (Blumenthal et al., 1995). Cd is also a carcinogen, with increased exposure increasing the chance of prostate and renal cell carcinomas (Goyer et al., 2004). At the cellular level, Cd increases the formation of ROS and promotes lipid peroxidation. Cd also depletes antioxidants such as glutathione and disrupts protein functions by binding to sulfhydryl groups (Jarup, 2003). Furthermore, Cd exposure promotes the production of inflammatory cytokines (Yucesoy et al., 1997).

Arsenic (As) is a widely distributed metalloid, and anthropogenic activities such as smelting of nonferrous metals and production of energy from fossil fuels have led to As contamination in soil, air, and water (Han et al., 2003). Human exposure results predominantly from As-contaminated food products and drinking water (Duxbury et al., 2003). Ingestion of high amounts of As leads to death, and exposure to low levels results in nausea, vomiting, reduced red and white blood cell count, irregular heart rhythm, damaged blood vessels, and pain sensation in the extremities (Hughes, 2002). Inorganic As also contributes to increased risk of liver, bladder, and lung cancer (Mushak and Crocetti, 1995). There is also strong evidence linking As exposure with hypertension, cardiovascular disease, cerebrovascular disease, and long-term neurological effects (Chen et al., 2009). At the cellular level, As exposure leads to cellular apoptosis by disrupting the cell's energy system via allosteric inhibition of pyruvate dehydrogenase, the enzyme responsible for the oxidation of pyruvate to acetyl-CoA by NAD^+ (Kumagai and Sumi, 2007). As also interferes with thiamine function resulting in symptoms of thiamine deficiency (Nandi et al., 2005). In addition, As poisoning elevates lactate levels leading to lactic acidosis and stimulates the production of ROS. Furthermore, As mediates neurological defects by interfering with voltage-gated potassium channels, which contributes to disruption of cellular electrolytic balance (Jarup, 2003).

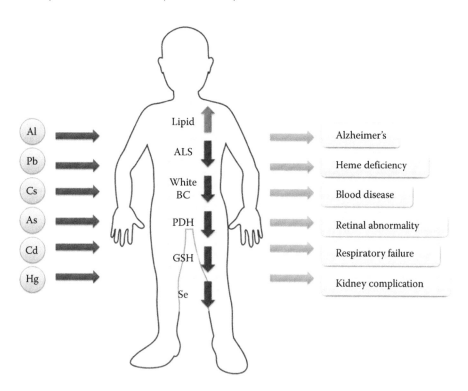

FIGURE 11.2 Impact of metal pollutants on human health. ALS, aminolevulinate synthase; white BC, white blood cells; PDH, pyruvate dehydrogenase; GSH, reduced glutathione.

Gallium (Ga) is found as a trivalent salt in zinc and bauxite ores and readily processed through smelting (Moskalyk, 2003). This is a vital component of semiconductors made from gallium arsenide or gallium nitride (Beriault et al., 2007). Ga is an Fe mimetic due to their close physical resemblances; thus, Ga mediates its toxicity by interfering with Fe metabolism (Al-Aoukaty et al., 1992). Substitution of Ga for Fe allows incorporation of Ga into numerous metalloproteins of the tricarboxylic acid (TCA) cycle, thereby arresting cell growth (Auger et al., 2013a). Unbound Fe is a prooxidant resulting in ROS formation. Ga causes increased unbound Fe since Ga occupies ferritin, the storage protein for Fe, thus causing increased unbound Fe ultimately leading to cellular oxidative stress (Lemire et al., 2010a; Bignucolo et al., 2013).

Naturally occurring radionuclides include uranium, radium, and radon. Numerous studies have demonstrated that acute radionuclide inhalation results in inflammatory responses in the nasal passages and kidney (Kao et al., 1994; Van der Meeren et al., 2008). In humans, chronic uranium exposure has been attributed to chronic lung disease, while radon exposure is linked to acute leucopenia, anemia, necrosis of jaw, and various other effects (Brugge and Buchner, 2012). Radionuclide exposure predominantly results in cancer formation. Oral exposure of radium in humans causes bone, head, and nasal passage tumors, while nasal exposure causes lung cancers (Harrison and Muirhead, 2003). Similarly, uranium intake in humans triggers lung cancer and lymphatic and hematopoietic tumors (Roscoe et al., 1989). Radionuclides mediate cellular

toxicity by damaging DNA and degrading important molecular proteins required for cell survival (Figure 11.2).

11.4 METAL TOXICITY IN PLANTS

The detrimental effects of metal accumulation in plants are of global concern due to its adverse effects on food safety, crop growth, and environmental health of soil organisms (Gonzalez and Gonzalez-Chavez, 2006). Long-term exposure of phosphatic fertilizers, sewage sludge usage, dust particles from smelters, industrial waste products, and bad watering practices have immensely contributed to metal toxicity in agricultural soils (Yadav, 2010). Metals of concern to agricultural corps include Cd, Cu, Zn, Ni, Co, Cr, Pb, and As. Cd toxicity is of particular importance in crops since it can bioaccumulate at high levels in leaves, which can then result in human exposure via the food chain (de Vries et al., 2007).

Metal toxicity in plants can be classified into redox active (Fe, Cu, Cr, Co) or redox inactive (Cd, Zn, Ni, Al) interactions (Hossain et al., 2012). The redox active metals contribute to cellular oxidative stress via redox reactions whereby hydrogen peroxide and super oxide radicals are produced (Schutzendubel et al., 2001). Redox inactive metals promote oxidative stress indirectly by disrupting a variety of cellular functions including antioxidant defense systems and electron transport chain and induction of lipid peroxidation by promoting lipoxygenase activity (Hossain et al., 2012). Another mechanism for metal toxicity in plants includes binding of

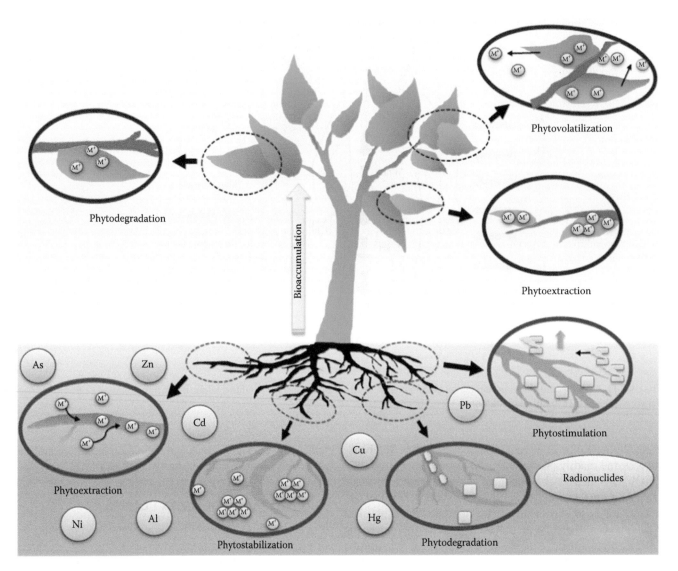

FIGURE 11.3 Interaction of metal pollutants with plants. (Adapted from Favas, P.J.C. et al., Phytoremediation of soils contaminated with metals and metalloids at mining areas: Potential of native flora, in: Hernandez-Soriano, M.C., ed., *Environmental Risk Assessment of Soil Contamination*, InTech, Rijeka, Croatia, 2014, pp. 485–517. With permission.)

metals to oxygen, nitrogen, and sulfur atoms. This results in the binding of metals to cysteine residues present in enzymatic and protein structures, thus preventing proper molecular functions (Battin and Brumaghim, 2009). For example, Cd interaction with sulfhydryl groups of enzymes and structural proteins results in misfolded and inactive cellular functions (Palma et al., 2002). Furthermore, toxic metals further inhibit protein function by displacing essential cofactor metal ions that catalyze biochemical reactions. For instance, Co, Ni, and, Zn displace Mg in ribulose-1,5-bisphosphate-carboxylase/oxygenase resulting in reduced enzyme activity (Hossain et al., 2012). Similarly, Cd in place of Ca in calmodulin resulted in the reduction of calmodulin-dependent phosphodiesterase activity in radish (Rivetta et al., 1997). Finally, toxic metals mediate plant cell membrane damage by oxidizing and cross-linking protein thiols and by inhibiting ion gradient

maintenance pumps such as sodium potassium ATPase. These effects cause changes in composition and fluidity of plant lipid membranes (Cooke et al., 1991). Thus, metals mediate toxicity in plants by stimulating ROS, inhibiting protein functions by binding to sulfhydryl groups, and displacing essential metal ions from specific regions of proteins.

The molecular effects mediated by toxic metals directly result in detrimental cellular effects such as reduced plant growth, cell necrosis, turgor loss, leaf chlorosis, decreased germination, and reduced photosynthesis, ultimately leading to plant death (Peralta et al., 2001). In addition, metals negatively impact water uptake, nutrient transport, transpiration, and metabolism. The crippled photosynthesis pathways result from metals disrupting essential interaction of proteins and lipids with thylakoid membranes, thus damaging light-harvesting complexes and photosystem II (Hossain et al., 2012).

Therefore, plants require a balanced environment and healthy soil devoid of toxic metals for proper growth. Thus, metal toxicity poses immense threat to the agricultural industry and ultimately human health (Figure 11.3).

11.5 PHYSICOCHEMICAL METAL POLLUTION MANAGEMENT TECHNIQUES

Specialized methods have been developed for the removal of environmental metal pollution from waste discharges and for remediation and restoration of contaminated soils and groundwaters. Techniques for the removal of metal toxins from solution include chemical precipitation, coagulation, ion exchange, solvent extraction, reverse osmosis, cementation, complexation, oxidation–reduction, membrane technology, evaporation, and electrochemical treatment (Gupta et al., 2012; Dixit et al., 2015). Most importantly, metals cannot be destroyed or modified by chemical or thermal methods; thus, methods for metal remediation attempt to change the form or phase of the metals. Predominantly, metals are transformed into insoluble form hence preventing reentry of the metals into the environment in a process known as solidification or vitrification (Mulligan et al., 2001). In solidification, metal toxins are extracted from a contaminated environment and immobilized using lime, cement kiln dust, calcified clays, soluble silicates, and various other additives to form an insoluble solid mass (Tai and Jou, 1999). During vitrification, the metal toxins are added into a glass matrix by applying current thus immobilizing the toxic metals into the matrix and preventing its reentry into the environment.

Chemical precipitation is the most common method for metal remediation and is based on the low solubility of metal hydroxides. This lengthy procedure involves pretreatment, pH adjustment, clarification, sludge thickening, sludge dewatering, and effluent polishing (Matlock et al., 2002). During this procedure, addition of alkaline substances results in precipitation of heavy voluminous colloidal suspensions, which are then recovered by mechanical entrapment (Dixit et al., 2015). Chemical precipitation is very effective in removing

high amounts of metal toxins from wastewater solutions and mining effluents. However, the efficiency of the chemical precipitation technique is dependent on the rates of formation and settling of metal hydroxide precipitates. Since precipitation formation is slow with low metal concentrations, there is concurrent increased consumption of lime and caustic soda. Another problem is that this technique generates toxic sludge that needs to be dewatered, stabilized, and disposed to prevent reentry into the environment. In fact, most physicochemical remediation methods are ineffective when metals are less than 100 mg/L (Mulligan et al., 2001). Furthermore, most metals are water soluble, and once dissolved with wastewater the physical separation methods are futile, are expensive, and have limits in removal capacity (Figure 11.4).

11.6 MICROBIAL METAL DETOXIFICATION STRATEGIES

The environment contains natural detoxifying microorganisms that are able to break down hazardous material into less toxic or nontoxic substances. Studies indicate that naturally occurring microbes can be utilized as biological adsorbents to remove heavy metals from water at a low-cost and eco-friendly manner (Davis et al., 2003; Juwarkar et al., 2010). The use of microorganisms and plants for the remediation of metal pollution is the preferred method over existing physical and chemical metal decontamination techniques since microbes allow for sustainable technology and help the environment to reestablish the natural condition of the soil (Dixit et al., 2015). The ability of the microbes to remove metals depends on the type of metal, nature of medium, and microbial species involved. Furthermore, microorganisms provide the possibility to recover metals from mineral sources not accessible via conventional mining methods. For example, microbes such as bacteria and fungi have natural abilities to convert metals into water-soluble forms and serve as natural catalysts for metal leaching (White et al., 1998). This technology can also be used to recover useful metals from industrial wastes. For instance, bioleaching had been employed in the Rio Tinto mines of Spain more than 100 years ago whereby heaps of low-grade copper ores up to 200,000 tons were bioleached. Here, *Thiobacillus ferrooxidans* were identified as the leachates, which were used to solubilize the metal (Lopez-Archilla et al., 2001).

Microbial systems are known to develop intricate strategies to cope with metal-polluted habitats (Dixit et al., 2015). These metal detoxification methods include biosorption, bioaccumulation, biotransformation, and biomineralization resulting in an economical and biologically innocuous material (Lloyd and Lovley, 2001). More specifically, microbial-based bioremediation techniques include autotrophic and heterotrophic leaching mechanisms, reductive precipitation, sulfate reduction, and metal sulfide precipitation. Examples of microbial remediation include bioleaching of heavy metals, biooxidation of gold ores, desulfurization of coal and oil, tertiary recovery

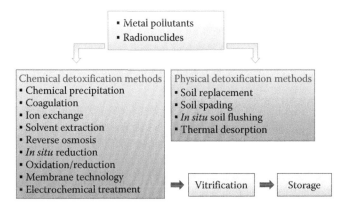

FIGURE 11.4 Chemical and physical metal decontamination technologies.

of oil, and biosorption of metal ions (Yu et al., 2006; Shukla et al., 2010). Biotransformed and concentrated forms of metals can then be stored in backfills or via vitrification (Auger et al., 2013a). Metal-resistant microbes are naturally present in contaminated sites and their capacity to remove metals greatly varies. Numerous studies have reported the isolation and selection of natural microbial strains that have high tolerance for metals (Congeevaram et al., 2007; Joshi et al., 2011). For example, microbes were isolated from metal-polluted environment to identify fungi strains that could mediate the bioremediation of toxic metals (Zafar et al., 2007).

Proper microbial growth and development require small amounts of certain metals. However, the total adsorption capacity of the organisms depends on total biomass and metabolic enzymatic state (Dixit et al., 2015). Thus, the ability of the microbe to adapt its metabolic state upon metal toxicity and subsequent sequestration of the metal is the basis of bioremediation. Microbes are naturally able to adapt and detoxify metal; therefore, the use of microbes possesses inherent and undiscovered potential for metal remediation (Figure 11.5) (Nicolaou et al., 2010). Bacteria, fungi, and plants have been successfully employed in numerous environments to detoxify metal pollution. Methods utilized by microorganisms to survive toxic metal conditions include redox transformations, production of metal-binding proteins, organic/inorganic precipitation, active transport, metal efflux pumps, intracellular compartmentalization, and metal solubilization using organic and inorganic biomolecules (Gadd, 2010). Since microbes are unable to convert inorganic metals into harmless compounds, bioremediation strategies predominantly depend on the adaptive metabolizing capabilities of these microbes. Therefore, microbes build resistance in metal environment primarily by adapting its metabolic state via modulating the expression and function of transport systems and enzymes, which resist metal-mediated toxic effects (Booth et al., 2011, 2015).

Microbial-mediated bioremediation processes depend on its ability to modulate mobilization or immobilization characteristics of toxic metals. Microbes mediate metal mobilization by protonation, chelation, and chemical transformation, while metal immobilization is achieved via precipitation or crystallization of insoluble compounds, uptake, and intracellular sequestration (Gadd, 2000; Valls and de Lorenzo, 2002). However, redox reactions can mobilize or immobilize metals depending on the metal species generated. Toxic metals sequestered in soils or sediments can be removed via biosolubilization, whereas metals in aqueous environments (lakes and water bodies) lend themselves for ready immobilization with the aid of microbial systems. The following sections will discuss in detail the detoxification techniques mediated by microbial systems.

11.7 MICROBIAL-MEDIATED METAL MOBILIZATION PROCESSES

Microorganisms utilize H^+/ATPase pumps and carbon dioxide respiration to create an acidic environment, which aids in metal leaching (del Dacera and Babel, 2006). Increased protons due to the acid environment compete with metals in the metal–anion complex, which frees the metal cations. Similarly, microbes mediate the efflux of organic acids and siderophores, which can promote metal leaching by providing protons and metal-complexing anions (Gadd, 2004). For example, citrate and oxalic acid form stable complexes with numerous metals and have been shown to successfully leach toxic metals from a variety of wastewaters and contaminated soils. Similarly, siderophores are specific Fe ligands that are able to bind to other metals such as Mg, Mn, Cr, and Ga and radionuclides like plutonium, thus removing these free cations from the polluted environment (Konyi et al., 1997). Metal biotransformation via redox reactions can also promote metal leaching. Microbes biotransform metals predominantly by oxidation of metal sulfides (Suzuki, 2001). These basic redox detoxification reactions covalently add oxygen atoms to metals by enzymes such as monooxygenases, dioxygenases, hydroxylases, oxidative dehydrogenases, or ROS generated by peroxidases (Torres Pazmino et al., 2010; Dixit et al., 2015). Essentially, these microbes serve as oxidizing agents for metals, which results in metal electron loss. These electrons are accepted by electron acceptors such as nitrate, sulfate, and ferric oxides. Microbes can also oxidize metals via the use of ferric iron. Microbes oxidize ferrous iron into ferric iron, and ferric iron oxidizes metal sulfides (Weber et al., 2006). Metal oxidation in turn reproduces the ferrous iron again, which can then be used by the microbes to continue the cycle. Unlike the aerobic degradation–mediated loss of metal electrons, anaerobic degradation of metal toxins involves anoxic electron acceptors (Lovley and Coates, 1997). Both aerobic- and anaerobic-mediated losses of metal electrons lead to changes in the physical properties of the toxic metals, resulting in increased metal water solubility (Dixit et al., 2015). Another method of metal mobilization includes microbial-mediated metal methylation, which results in altered metal solubility, volatility, and toxicity (Ranjard et al., 2002; Gadd, 2004). Volatilization of methylated metals such as selenium results in removal from soil. Similarly, biomethylation of arsenic compounds leads to volatile forms that are evaporated from a contaminated environment (Bentley and Chasteen, 2002). However, volatilization is not an effective bioremediation method as the toxic metals can contaminate surrounding regions.

Microbial metal uptake pathways play an important role in bioremediation techniques, which utilize metal immobilization methods for detoxification. Microorganisms take up metals actively through bioaccumulation or passively through

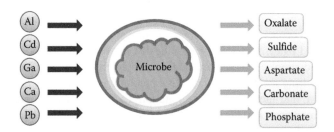

FIGURE 11.5 Microbial metal detoxification processes.

adsorption (Velasquez and Dussan, 2009). Bioremediation using biosorption refers to the ability of the microorganism to adsorb metal species from the contaminated environment to cellular surfaces. The microbial cell walls composed of polysaccharides, lipids, and proteins provide functional groups such as carboxylate, hydroxyl, amino, and phosphate groups that can bind metal ions (Gadd, 2009). In addition, biochemical moieties such as peptidoglycan carboxyl groups, phosphate groups, and structural component chitin are effective biosorbent molecules found on bacterial and fungal cell walls (Moriyon and Lopez-Goni, 1998). Similarly, microbes can utilize active membrane transport systems to uptake and bioaccumulate metal species from the surrounding (Barkay and Schaefer, 2001). However, passive biosorption seems to be more feasible than active bioaccumulation for large-scale applications since microbes require additional nutrients to maintain active uptake of heavy metals. Once metals enter the microbe, various methods are employed to sequester the metals such as precipitation, localization within intracellular structures, or translocalization into specific structures (Gadd, 2004). Microbes also secrete biofilms that can sequester metals externally making it easier to subsequently remove the metals from the environment while leaving the microbes intact (Singh et al., 2006). These secreted biosurfactants bind metals with stronger affinity and form complexes before being desorbed into the soil matrix. Here, the biosurfactants serve as a biosorbent for metal ions.

Microorganisms have developed numerous nonspecific metal-binding proteins that aid in detoxifying metals once inside the cell. Some examples include organic acids, alcohols, and macromolecules such as polysaccharides and humic and fulvic acids (Gadd, 2004). During toxic stress, microbes increase the expression of low-molecular-weight metal-binding proteins such as metallothioneins and various other molecules that contain glutamic acids and cysteine at amino terminals resulting in increased affinity for metal cations (Valls et al., 2000). Redox reactions that result in lower metal oxidative state aid in metal precipitation and therefore reduced toxicity. In addition, indirect metal reduction reactions are routinely undertaken by sulfate-reducing bacterial systems (Utgikar et al., 2002). Thus, the various metal immobilization methods employed by microorganisms transform the physical and chemical state of the metal leading to the formation of insoluble gelatinous precipitates for easy excretion while also preventing reabsorption (Figure 11.6) (Mailloux et al., 2011b).

11.7.1 Metabolic Adaptation to Metal Pollutants by Microbial Systems

Free metal ions initiate oxidative stress pathways resulting in ROS production and cellular damage. Therefore, microbes must increase the production of factors that can negate the increase in ROS generated by metal toxicity for proper cellular survival (Auger et al., 2013a). Thus, classical mechanisms to counteract ROS such as catalase, superoxide dismutase, and glutathione peroxidase/reductase are increased in microbes exposed to toxic metals (Sinha et al., 2005). In addition to these, microbes upregulate enzymes that increase the production of NADPH (crucial for regeneration of ROS scavenging molecules) such as isocitrate dehydrogenase, malic enzyme, glucose-6-phosphate dehydrogenase, NAD phosphorylating enzyme, and NAD kinase (Beriault et al., 2005; Middaugh et al., 2005; Singh et al., 2008). Similarly, the production of NADH is decreased to prevent the formation of ROS. This is accomplished by downregulating expression levels of ICDH-NAD, α-ketoglutarate dehydrogenase, and H_2O-generating NADH oxidase (Chenier et al., 2008). The overall increase in NADPH production allows to promote the production of ROS scavengers and also serves as an important factor for lipogenesis, which is important in the production of phosphatidylethanolamine (PE) (Appanna and Pierre, 1996; Hamel and Appanna, 2003; Auger et al., 2013a).

PE and other organic acids consist of negatively charged carboxylic acid functional groups, which allow for chelation

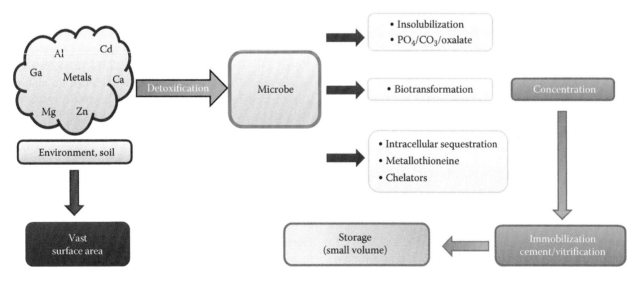

FIGURE 11.6 Overview of metal bioremediation strategies.

and subsequent removal of toxic metals (Appanna et al., 2003a; Hamel and Appanna, 2003; Lemire et al., 2008a). Therefore, in response to metal toxicity, microbes increased the production of oxalate-CoA transferase (OCT), the enzyme responsible for the production of oxalyl-CoA from succinyl-CoA (Singh et al., 2009). Succinyl-CoA is used for ATP production from substrate-level phosphorylation (SLP), whereas oxalyl-CoA is a precursor for oxalate, a dicarboxylic acid and potent chelator of Al and other ions (Hamel and Appanna, 2001; Auger et al., 2013a). Another metabolic adjustment made by microbes to adapt for metal toxicity includes the production of oxaloacetate, another dicarboxylic acid similar to oxalate (Appanna et al., 2003b; Lemire et al., 2008a). Oxaloacetate is an important precursor to a polycarboxylic aluminophore, and the production of this metabolite is controlled by numerous enzymes including malate dehydrogenase and pyruvate carboxylase, both of which are upregulated during metal toxicity (Reinoso et al., 2013; Alhasawi et al., 2015a). Similarly, enzymes that lower oxaloacetate production such as phosphoenolpyruvate carboxykinase and nucleoside diphosphate kinase have reduced expression during metal toxicity. Here, the oxaloacetate and its derivatives serve as a metal chelator to sequester free metal ions. Therefore, the overall increase in oxalate and oxaloacetate contributes to metal precipitate formation and subsequent secretion. Individual metal detoxification mechanisms employed by microorganisms are described in detail in the following.

Organisms adapt to Al toxicity in several ways. For example, plants release organic acids and phenolic compounds to chelate Al and prevent its toxicity (Krill et al., 2010). Fungi uses negatively charged metabolites such as citrate and malate to chelate and decrease free Al. General mechanism for Al removal is by first chelating Al with organic acids, then compartmentalization thus effectively limiting available Al (Auger et al., 2013a). Clay and MAM-4 strain of *Providencia rettgeri* have been used to remove Al, Cu, and Co from watershed (Abo-Amer et al., 2013). Since *P. rettgeri* has high tolerance for metal uptake and compartmentalization, in addition to the high negatively charged nature of clay, 87% removal of trivalent metals was possible with this technique (Abo-Amer, 2012; Auger et al., 2013a). *P. rettgeri* was able to increase the biosorption of metals by mediating the addition of hydroxyl, carboxyl, and phosphate groups to metals. Other defense mechanisms developed by microbes to prevent Al toxicity include production of exopolysaccharide, reduction of cell membrane negative charge, and production of aluminum–phosphate complexes that are insoluble (Appanna and Preston, 1987; Appanna, 1989; Avelar Ferreira et al., 2012).

Al also interferes with biomolecules that utilize Fe; therefore, iron–sulfur clusters used by the TCA cycle and the electron transport chain are heavily disrupted in the presence of Al (Hamel and Appanna, 2001; Yamamoto et al., 2002; Mailloux et al., 2007). The overall effect of Al results in decreased aerobic production of ATP (Grose et al., 2006). To counteract this loss, microbes deploy SLP during glycolysis and a modified TCA cycle (Singh et al., 2009). The modified TCA cycle includes increased production of isocitrate

dehydrogenase, which cleaves isocitrate to produce succinate and glyoxylate thus bypassing Al-mediated inhibition of aconitase (Middaugh et al., 2005). These types of metabolic modifications allow microbes to survive Al toxicity (Figure 11.7a) (Auger et al., 2013a).

Ga interferes with Fe metabolism; therefore, microbes exposed to Ga demonstrated reduced energy production due to crippled aerobic machinery and increased ROS production (Olakanmi et al., 2000). Thus, microbes involved in detoxification of Ga increase the production of antioxidant enzymes such as SOD, catalase, glutathione peroxidase, and NADPH-producing enzymes (Beriault et al., 2007; Lau et al., 2008). In addition, Ga stress results in the upregulation of an H_2O-dependent NADH oxidase NAD^+ for production from NADH and NAD kinase that produces NADK (Beriault et al., 2007; Lemire et al., 2010a). NADK ensures a steady supply of $NADP^+$ for NADPH production by using ATP molecules to phosphorylate NAD^+. The reverse reaction is catalyzed by NADPase; thus, Ga-stressed microbes downregulate the production of this enzyme. Hence, microorganisms detoxify Ga by increasing NADPH production while decreasing NADH synthesis in order to neutralize the oxidative environment mediated by Ga (Beriault et al., 2007; Chenier et al., 2008; Alhasawi et al., 2015b). In addition, Ga stress reduced energy production by inhibiting Fe-dependent ACN and fumarase (Chenier et al., 2008). Microbes adapt to ACN inhibition by upregulating NADP-ICDH and isocitrate lyase (ICL), which pushes citrate metabolism without the use of ACN. Similarly, Fe-independent isoform of fumarase is upregulated during Ga stress, which allows for the formation of malate from fumarate. These numerous metabolic adaptive changes are required for microorganisms to survive and detoxify Ga stress (Figure 11.7b).

Microorganisms deposit exocellular calcite, a crystalline $CaCO_3$, in order to survive extracellular Ca^{2+}-stressed environments (Anderson et al., 1992). The calcite secreted by the microbes sequesters the Ca^{2+} forming an exocellular white crystalline precipitate (Konyi et al., 1997). This adaptive mechanism is beneficial for the organism as the intracellular Ca^{2+} levels can be maintained at low levels required for proper cellular function while the extracellular Ca^{2+} is sequestered and unable to reenter the organisms (Lemire et al., 2010a). Interestingly, the production of the insoluble precipitate increases once the microbes have reached the stationary growth phase since during exponential growth phase intracellular Ca^{2+} levels are tightly maintained by Ca^{2+}/ATPase pumps and Ca^{2+}/proton antiporters. This phenomenon may be the reason why microorganisms favor biomineralization over the use of ATP-dependent pumps (Lemire et al., 2010a). Ca^{2+} biomineralization requires macromolecules that are acidic, aspartate/glutamate rich, and contain polysaccharides. The negative charges produced by aspartate/glutamate aid in Ca^{2+} binding and control crystal growth (Appanna and St Pierre, 1996). Calcite produced by microbes under Ca^{2+} stress requires the maintenance of an environment supersaturated in carbonate and Ca^{2+} ions. Thus, the production of carbonic anhydrase is increased in Ca^{2+}-stressed microorganisms, an

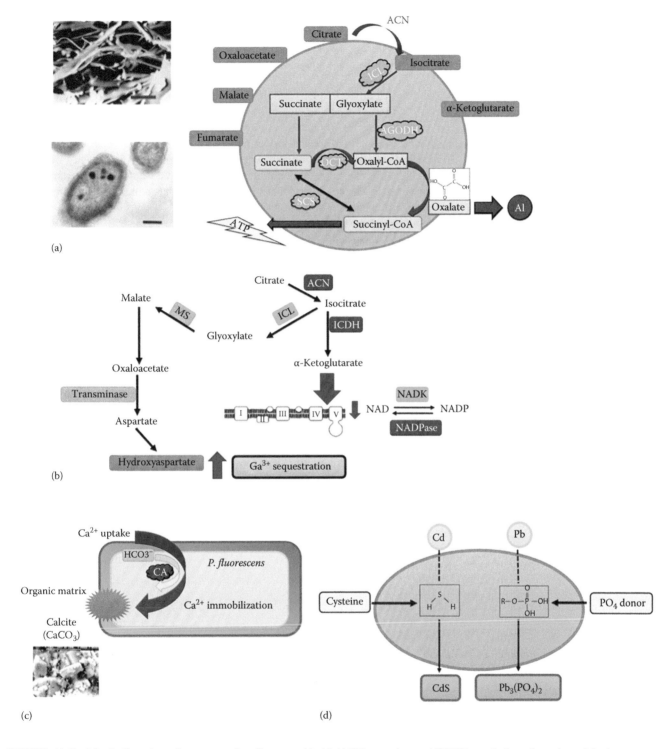

FIGURE 11.7 Metabolic adaptation to metal pollutants. (a) Al-(ACN, aconitase; AGODH, acylating glyoxylate dehydrogenase; OCT, oxalyl-CoA transferase; SCS, succinyl-CoA synthetase); (b) Ga-(ACN, aconitase; ICDH, isocitrate dehydrogenase; ICL, isocitrate lyase; MS, malate synthase; NADK, NAD kinase; NADPase, NADP phosphatase); (c) Ca-(CA, carbonic anhydrase); (d) Cd, Pb. (a: Transmission electron micrograph Al deposit in lipids [Modified from Appanna, V. D. and St Pierre, M., *J. Biotechnol.*, 48, 129, 1996] and intracellularly [Modified from Auger, C. et al., *Encycl. Metalloprot.*, 2013, 800, 2013b]; c: Ca immobilization as calcite [Modified from Anderson, S. et al., *FEBS Lett.*, 308, 94, 1992].)

enzyme that produces bicarbonate. The carbonate formed by carbonic anhydrase reacts with the mobilized Ca^{2+} to form $CaCO_3$ crystals in a very organized manner, a feature critical to the formation of calcite (Figure 11.7c) (Lemire et al., 2010a).

Calcite precipitation has been utilized in strontium (Sr^{2+})-contaminated regions for bioremediation purposes (Anderson and Appanna, 1994; Achal et al., 2012). During bioremediation, there was an initial decrease in surrounding pH attributed to microbial oxidation of organic acids, followed by pH increase due to NH_3^+ generation. The decrease in extracellular Sr^{2+} concentration was associated with production of strontianite that was identified as 20–70 nm sized (Kang et al., 2015). Therefore, formation of sparingly soluble Sr^{2+} precipitates sequestered Sr^{2+} and carbon dioxide into a more stable and less toxic form such as strontianite. These results demonstrate that detoxification of metal-contaminated environment through biomineralization of carbonate minerals is a feasible bioremediation technique.

Mercury (Hg) detoxification involves the reduction of Hg^{2+} to the volatile Hg^0 form catalyzed by the inducible enzyme mercuric ion reductase (Dash et al., 2014). Mercuric ion reductase is a flavoprotein coded by the *merA* gene and is inducible by subinhibitory concentrations of mercuric ions and a variety of organomercurial molecules (Nascimento and Chartone-Souza, 2003). In addition to *merA*, microbes also upregulate the expression *merB*, which codes for organomercurial lyase. Mercuric ion reductase and organomercurial lyase are responsible for reducing highly toxic organomercurial compounds such as methylmercury and phenylmercuric acetate into almost nontoxic volatile elemental Hg (Misra, 1992). For example, Zhang et al. (2012) demonstrated that *Pseudomonas putida* SP-1 was able to volatize 89% of Hg, thus validating the use of Hg-resistant bacteria in Hg decontamination (Zhang et al., 2012). In this instance, the pollutant is not concentrated but scattered over a large area, thereby diminishing its toxicity.

Cd bioremediation was predominantly performed via microbial bioprecipitation as cadmium sulfide (Sharma et al., 2000). Bioprecipitation was achieved by upregulation of cysteine desulfhydrase, the enzyme responsible for producing sulfide, ammonia, and pyruvate from D-cysteine (Bai et al., 2008). Subcellular fractionation analysis demonstrated that Cd was mostly removed from intracellular compartments and transformed by precipitation on the cell wall. Usually, sulfate-reducing bacteria and aerobic bacteria can produce hydrogen sulfide and therefore precipitate Cd; however, the sulfide needed for this is achieved under strict anaerobic or strict aerobic conditions (Bai et al., 2008). Thus, the use of phototrophic bacteria that are anaerobically photoautotrophic and photoheterotrophic during light phase and aerobically chemoheterotrophic during dark phase is able to produce sulfide under aerobic conditions and precipitate metal sulfide complexes on the cell wall. Microbes also detoxify Cd via mycelia biosorption. Here, Cd toxicity increased the production of oxalic acid, intracellular proline contents, and antioxidative enzymes, which resulted in detoxification of toxic Cd levels (Chakraborty and Das, 2014).

Microbes remove Pb from contaminated soils by adding phosphates to Pb forming complexes called pyromorphites (Al-Aoukaty et al., 1991; Rhee et al., 2014). This complex is insoluble and, when ingested, cannot be absorbed in the digestive tract, thus making it far less toxic than freely available Pb. However, addition of phosphate can have negative impact on the soil as it would reduce pH and cause further leaching of other unwanted toxic meals. However, similar to pyromorphites, certain fungal microbes are able to convert lead into chloropyromorphites (Rhee et al., 2014). This nontoxic mineral produces a much more stable form of phosphate–lead interaction than pyromorphites, thus preventing its breakdown and subsequent lead availability or soil pH modulation. Furthermore, some studies demonstrated that Pb is also detoxified by microbial-induced calcite precipitation (Figure 11.7d).

11.7.2 Phytoremediation

Phytoremediation refers to the use of plants to partially or completely remediate metal contaminants (Mani et al., 2015). Plants have been effectively used to remove contamination from radionucleotides, organic pollutants, and metals. This is a cost-effective method that relies on solar energy. Phytoremediation techniques include phytoextraction, phytofiltration, phytostabilization, phytovolatilization, and phytodegradation (Figure 11.3) (Lee, 2013). Phytoremediation begins with phytoextraction whereby there is uptake of metals by roots and subsequent translocation to the shoots where the metals bioaccumulate (Singh et al., 2003). Next, the metals undergo phytofiltration where metals are absorbed thus minimizing their movement back into the environment. Examples of phytofiltration include rhizofiltration (roots), blastofiltration (seeds), or caulofiltration (shoots). Rhizofiltration is the use of plant roots to absorb, precipitate, and concentrate pollutants. In addition, metals sometimes undergo further immobilization changes, which include phytostabilization and phytoimmobilization (Dixit et al., 2015). Unlike phytoextraction, phytostabilization focuses mainly on sequestering pollutants in soil near the root, but not in plant tissues. Here, the plants are used to immobilize or stabilize toxic metals in the soil thus decreasing the bioavailability of toxins in the surrounding environment and reducing further environmental degradation by leaching (Lee, 2013). Some plants also perform phytovolatilization. For example, toxic metals such as Se, As, and Hg are biomethylated into volatile biomolecules (Chaney et al., 1997). In particular, Se volatilization as methyl selenate was reported as a major mechanism of Se removal by plants. Once the metals are volatilized, it is further degraded in the atmosphere or remains in the air as pollutants (Lee, 2013).

Phytoremediation has been successfully utilized for the removal of uranium from contaminated soils using sunflower plants (Chang et al., 2005). Sunflowers effectively uptake uranium via rhizofiltration and demonstrate high capacity for uranium accumulation. In order for effective uptake, uranium needs to be in solution. Thus, uranium desorption from soil to solution is increased by the addition of organic acids such as citric acid (Huang et al., 1998). Once uranium is concentrated in sunflower plants, the crops are harvested, burned, and further remediated by vitrification.

11.8 MICROBIAL MANAGEMENT OF MULTIPLE METALS

Single metal bioremediation detoxification mechanisms have been well characterized; however, cellular and metabolic strategies for the removal of multiple metals from contaminated environments need to be elucidated (Appanna et al., 1996b). In the natural setting, the toxic environment consists of numerous metals. For example, Migliorini et al. (2004) demonstrated that anthropogenic-mediated contaminated sites consisted of numerous metal toxins including Pb, Sb, Ni, Zn, Mn, As, and Cu (Migliorini et al., 2004). Similarly, Joshi et al. (2011) demonstrated the removal of Pb, Cd, Cr, and Ni using fungi collected from sewage, sludge, and industrial effluents containing elevated metal levels (Joshi et al., 2011). Therefore, microbial survival in the natural environment is challenged with multiple metals rather than single metal contamination; thus, microorganisms must develop global bioremediation mechanisms to combat multiple-metal stresses for proper survival (Figure 11.8).

11.8.1 DECONTAMINATION OF MULTIPLE-METAL POLLUTION BY POLYPHOSPHATE PRECIPITATION

An effective and widely employed microbial method for detoxifying multiple-metal pollution involves metal precipitation with microbial-generated phosphate species (Al-Aoukaty et al., 1991). Phosphate unselectively precipitates many metals and reduces their bioavailability via the formation of insoluble metal phosphate species (Olaniran et al., 2013). Certain metals are more readily able to precipitate in the presence of phosphate than others. For example, cobalt demonstrates minimal precipitation when exposed to phosphates, whereas ionic nickel is rapidly precipitated in the presence of phosphate (Olaniran et al., 2013). In addition, pH levels modulate phosphate-metal precipitation where elevated pH results in increased precipitation while acidic pH promotes the release of free ionic metal species from phosphates (Korkeala and Pekkanen, 1978; Franklin et al., 2000). At reduced pH, increased proton availability saturates phosphate metal-binding sites; therefore, metals are unable to form insoluble precipitates since much of the phosphate has been protonated (Olaniran et al., 2011).

Microbial-mediated phosphate precipitation has been employed for detoxification of radioactive uranium particles and heavy metal–contaminated subsurface sediments in contaminated soils of nuclear weapons research regions (Beazley et al., 2007; Martinez et al., 2007; Newsome et al., 2015). Here, microbes mediate the phosphate precipitation of toxic uranium(VI) into a reduced crystalline uranium(IV) phosphate mineral (Newsome et al., 2015). The resulting mineral was drastically more resistant to oxidative remobilization than the initial products of microbial uranium(VI) reduction. Furthermore, microbes promoted the development of reducing conditions upon metal toxicity by increasing the production of organophosphates. Thus, microbes stimulated with glycerol phosphate under anaerobic conditions are able to increase organophosphate production and subsequent

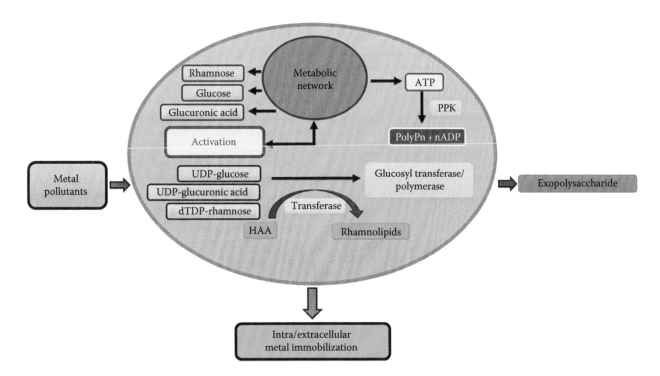

FIGURE 11.8 Multiple-metal sequestration by biopolymers. HAA, 3-(3-hydroxyalkanoyloxy) alkanoic acid; PPK, polyphosphate kinase.

uranium-phosphate precipitation (Beazley et al., 2007). In addition, overexpression of polyphosphate kinase and exo-polyphosphatases has been shown to further increase microbial-mediated radionuclide detoxification via metal phosphate precipitation (Singh et al., 2007). Similarly, radiation-resistant bacteria overexpressing nonspecific phosphatases led to increased uranium bioprecipitation from nuclear wastes (Misra et al., 2012). Likewise, increased enzymatic efficiency of chromate and uranyl reductases resulted in improved enzymatic kinetics and better radionuclide and heavy metal bioremediation effects (Barak et al., 2006). Therefore, microbes capable of hydrolyzing organophosphate substrates promote phosphate-based mineralization of multiple-metal contaminants (Martinez et al., 2007).

Similarly, fungi are able to release inorganic phosphate (P_i) and have been shown to remediate Pb bioprecipitation when grown in the presence of organic phosphate substrates. For example, in the presence of glycerol 2-phosphate or phytic acid, microbes expelled liberated P_i into the medium resulting in complete removal of Pb from solution and extensive precipitation of Pb-containing pyromorphite or oxalate minerals around the biomass (Liang et al., 2016). Here, microbial-mediated organic phosphate hydrolysis led to pyromorphite formation, while oxalate excretion produced lead oxalate conjugation (Debela et al., 2010).

Similar to bacterial and fungal examples provided earlier, the basidiomycetous yeast *Cryptococcus humicola* was shown to be adapt to multiple metals including Mn, Co, Ni, Zn, lanthanum, and Cd cations at concentrations toxic for many types of yeast (Andreeva et al., 2013, 2014). *C. humicola* achieved this feat by increasing the production of acid-soluble inorganic polyphosphates. In the presence of heavy metals, *C. humicola* revealed polyphosphate accumulation in the cell wall and cytoplasmic inclusions. In normal yeasts, heavy metal polyphosphates accumulate in vacuoles resulting in organelle morphological disturbances and cellular dysfunction. Therefore, increased polyphosphate accumulation and cellular localization in cell wall provided the mechanism for *C. humicola*–mediated removal of numerous heavy metals from contaminated regions (Andreeva et al., 2014). Thus, polyphosphate-mediated metal bioremediation is employed by all types of microorganisms and is an effective method to detoxify multiple methods in a nonspecific manner.

11.8.2 Management of Multiple-Metal Pollution by Biosurfactants

Metals are usually tightly bound to colloidal particles and organic matter in the environment; thus, treatment of polluted soils with surfactants aids in separating metals into soluble species. Rhamnolipid biosurfactants produced from *Pseudomonas aeruginosa* are extremely potent in mobilizing multiple metals and decontaminating contaminated soils (Juwarkar et al., 2007). These biosurfactants are nontoxic to soil structure and easily biodegradable compared to synthetic surfactants, which are currently being used to mobilize metal contaminants. These biosurfactants also increase the

availability of recalcitrant or sorbed contaminants from soil (Juwarkar et al., 2007). The rhamnolipid biosurfactant was shown to mobilize multiple metals including Cd, Cr, Pb, Cu, and Ni (Rahman et al., 2002). Therefore, microbial-produced biosurfactants can be used to effectively bioremediate multiple metals from polluted soils.

11.8.3 Management of Multiple-Metal Pollution by Genetically Modified Microbial Systems

In order to produce microbes with increased bioremediation potential, genetically engineered microorganisms have been generated through recombinant DNA technology to provide organisms with specific and tailored characteristics for more efficient metal detoxification capabilities (Sayler and Ripp, 2000). Bioremediation with genetic engineering can also modify proteins involved in the control of cell growth and metabolic adaptability, thereby further promoting bioremediation efficiency. Furthermore, heterologous expression of metal-binding proteins such as metallothionein or phytochelatins has been shown to increase resistance for metal toxicity and promote the bioremediation of multiple metals (Sauge-Merle et al., 2003). Similarly, genes that modify the metal-binding sites on the outer membrane proteins are being genetically engineered into microbes to allow for better bioadsorption of toxic metals. For example, bacteria that express phytochelatin 20 on cell surface result in 25 times more ability to take up Cd or Hg compared to wild-type strains (Bae et al., 2001, 2003; Dixit et al., 2015). Furthermore, new approaches whereby bacterial designer organisms are created by utilizing bacteria genes responsible for metal degradation genetically engineered into plants. Since plants are excellent bioaccumulators, the expression of the bacterial enzymes can degrade the bioaccumulated metals (Eapen and D'Souza, 2005; Dixit et al., 2015). Conversely, others have proposed the use of biofilter where genetically modified microbes are fixed to porous medium to detoxify organic and inorganic contaminants from wastewater (Srivastava and Majumder, 2008). Indeed, the major concern with the use of genetically modified microbes is the unknown consequence of introducing microbes with foreign genes into the environment and the potential detrimental effect these modified organisms may have on the balance of the biota. This may be circumvented by the use of genetically modified organisms in strictly regulated *ex situ* bioremediation setup whereby the genetically modified microbes may be easily regulated and controlled from the natural environment.

11.8.4 Management of Multiple-Metal Pollution by *Pseudomonas fluorescens*

Pseudomonas fluorescens has been extensively employed in order to study adaptive mechanisms under multiple-metal stress (Paulsen et al., 2005). *P. fluorescens* is a rod-shaped gram-negative bacterium that provides excellent bioremediation potential due to its ability to adapt its metabolic state when exposed to extreme environmental conditions.

P. fluorescens has great metabolic flexibility and is considered nonpathogenic to mammals (Anderson and Appanna, 1993; Appanna et al., 1996a; Auger et al., 2013a). In addition, *P. fluorescens* is extremely resistant to metals since it chelates metals and translocates this complex across bacterial membrane where it is degraded intracellularly. Furthermore, it excludes the metals from its cell as an insoluble lipid precipitate using PE (Appanna et al., 1996b). Studies have shown that *P. fluorescens* grown in millimolar amounts of Al, Fe, Ga, Ca, and Zn adapted to toxic environments by decreasing enzymes involved in the production of NADH, increasing enzymes important for NADPH formation, and an alternate route of ATP production (Appanna et al., 1995; Lemire et al., 2010a). Here, ATP was produced using a modified TCA cycle whereby the organism employed alternated glycoxyl shunt, which produced ATP concomitantly with the dicarboxylic acid oxalate. Oxalate further sequestered and detoxified the metals as lipid complexes (Alhasawi et al., 2015b).

Organisms must adapt in order to survive and flourish in highly toxic multiple-metal environments. *P. fluorescens* undergoes SLP in order to produce energy using nonoxidative and noniron-dependent manner (Singh et al., 2009). For example, *P. fluorescens* utilizes succinate and oxalate produced by glyoxylate shunt, which produced substrates for ATP-generating enzymes such as OCT and succinyl-CoA transferase (Mailloux et al., 2006). Another SLP pathway utilized by *P. fluorescens* includes glycolysis (Singh et al., 2005). NADH is also affected due to disruption of ETC chain causing NADH accumulation. To prevent this, *P. fluorescens* reduced the function of NADH-producing enzymes in the TCA cycle including NAD-dependent isocitrate dehydrogenase, α-ketoglutarate dehydrogenase, and malate dehydrogenase (Mailloux et al., 2009; Lemire et al., 2010a). Apart from adapting its metabolic pathways, organisms must also develop methods to sequester and remove metals such as Al. Indeed, organisms use charged anionic organic acids such as citrate, malate, and oxalate to mobilize toxic metals. Here, *P. fluorescens* uses oxalate to sequester Al. Oxalate is a dicarboxylic acid capable of efficiently binding metals and, when it binds to Ca, forms Ca-oxalate precipitates (Gadd, 2004). A similar technique of Al sequestering occurs in plants. Thus, *P. fluorescens* modifies its metabolic profile to promote the production of oxalate. This is accomplished by upregulation of ICL and downregulation of NAD-ICDH, which shunts citric acid toward glyoxylate production (Hamel et al., 2004). These changes are accompanied by increased acetylation of glyoxylate dehydrogenase which promotes oxidation of glyoxylate to form oxalate (Hamel et al., 2004). *P. fluorescens* also utilizes PE and other lipids to sequester Al within vesicles thereby effectively excluding the Al from toxic effects to cellular components. In addition, phospholipids are very effective in entrapping metals due to their highly dense negative charge that allows it to interact with metals, which characteristically contain high, charges (Lemire et al., 2010a). The end result is immobilization of metals as an insoluble gelatinous precipitate that not only permits its removal but

also prevents reabsorption. Similar techniques are utilized to remove a variety of other metals.

Another method in which *P. fluorescens* adapts to metal toxicity is by upregulating enzymes involved in the production of NADPH such as G6PDH and ICDH-NADP$^+$ (Auger et al., 2013a). Increased NADPH creates a reducing environment and protects the organism from metal-induced oxidative stress. Indeed, the cell uses its own ROS defense system to counteract small amounts of ROS, which is produced by body's own energy production system during oxidative phosphorylation for ATP synthesis. These include superoxide dismutase, which catalyzes conversion of superoxide into oxygen and hydrogen peroxide; catalase, which converts hydrogen peroxide into oxygen and water; and peroxiredoxins, α-ketoacids, and glutathione peroxidases, which all reduce intracellular H_2O_2 production (Mates et al., 1999). The thiol group of the reduced glutathione tripeptide composed of L-cysteine, L-glutamic acid, and glycine reduces ROS. GSH also keeps the antioxidants vitamin E and C in active forms. Furthermore, GSH is a cofactor for glutathione peroxidase. These proteins collectively reduce H_2O_2 to water while creating glutathione disulfide (GSSG) as a by-product. Glutathione reductase then cleaves GSSG to give two GSH molecules. Glutathione reductase is upregulated during oxidative stress such as in Ga-stressed organisms, and its activity is dependent on NADPH (Beriault et al., 2007).

NADH-producing enzymes such as KGDH are drastically diminished when *P. fluorescens* is stressed with metal toxicity (Singh et al., 2009). NADH is a prooxidant; thus, there needs to be appropriate balance between NADPH and NADH for proper cell survival. Not only does reduced KGDH contribute to reduced NADH production but also allows the accumulation of α-ketoglutarate that is a prooxidant. ACN is severely limited during Ga toxicity since Fe sites are occupied with Ga (Singh et al., 2007). To overcome this, *P. fluorescens* upregulates NADP-ICDH and ICL; this allows for processing of citrate without ACN. *P. fluorescens* also upregulates fumarase, which catalyzes fumarate to malate (Hamel and Appanna, 2001; Lemire et al., 2008a). Since isoforms A and B of fumarate are Fe dependent, *P. fluorescens* upregulates Fe-independent isoform of fumarate, isoform C.

Hg, Pb, As, and Ga are toxic and even trace amounts need to be removed for proper cell survival. Metal chelators have been used in the past to form water-soluble compounds that are later removed through excretion such as in the case with BAL (Vilensky, 2003). Bacteria use endogenous Fe chelators such as siderophores for metabolic processes. These siderophores contain a β-hydroxyaspartate that binds Fe but can also bind Ga and a variety of other metals (Nair et al., 2007). The binding of metals to β-hydroxyaspartate allows the complex to be sequestered and removed. During metal toxicity, *P. fluorescens* increased the production of β-hydroxyaspartate by upregulating malate dehydrogenase thereby increasing metal sequestration (Lemire et al., 2010a). Therefore, *P. fluorescens* demonstrates that numerous metabolic changes are required for survival in multiple-metal stressed environments (Figure 11.9).

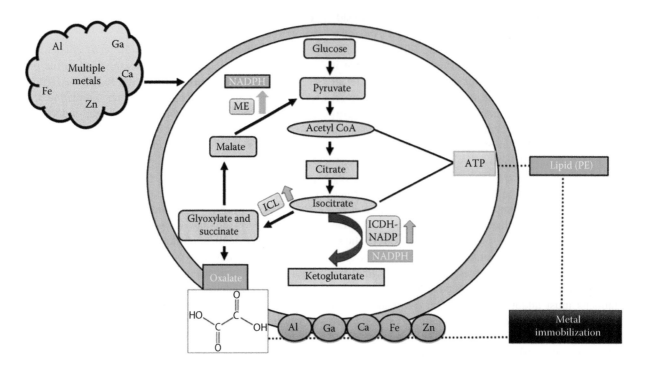

FIGURE 11.9 Metabolic networks in multiple-metal decontamination in *Pseudomonas fluorescens*. ICL, isocitrate lyase; ICDH-NADP, isocitrate dehydrogenase-NADP dependent; ME, malic enzyme.

11.9 CONCLUSION

Microorganisms have developed numerous strategies to adapt to metal toxicity. These adaptive survival processes utilized by microorganisms can be tailored to decontaminate metal-polluted environments. Compared to physicochemical remediation techniques, the use of microbes for detoxifying polluted areas permits the reestablishment of the natural condition of the environment while providing sustainable technology that is cheap, effective, and renewable. Moreover, organisms with

adaptive metabolic networks aimed at numerous pollutants can detoxify multiple metals, a situation common in nature. Hence, understanding the metabolic mechanisms responsible for the transformation and sequestration of multiple metals is crucial in order to improve the efficacy and versatility of microbial-mediated bioremediation technologies. Natural adaptation, accelerated acclimatization, and metabolic reprogramming are the key ingredients that will drive the effectiveness, the acceptability, and profitability of bioremediation technologies propelled by microbes (Figure 11.10).

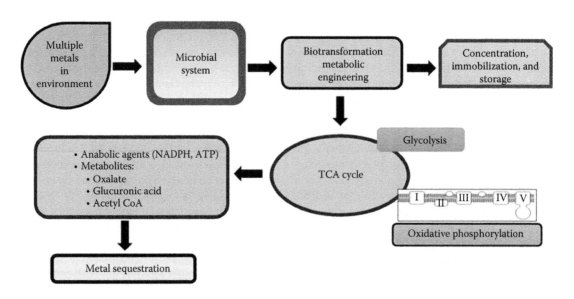

FIGURE 11.10 The pivotal role of microbial metabolic engineering in metal bioremediation.

ACKNOWLEDGMENTS

This work is supported by Laurentian University, Northern Ontario Heritage Fund Corporation, Industry Canada and NATO collaborative linkage grant. A. Alhasawi is the recipient of a doctoral fellowship from the Ministry of Higher Education of Saudi Arabia.

REFERENCES

Abo-Amer, A. E. 2012. Characterization of a strain of *Pseudomonas putida* isolated from agricultural soil that degrades cadusafos (an organophosphorus pesticide). *World Journal of Microbiology and Biotechnology* 28: 805–814.

Abo-Amer, A. E., Ramadan, A. B., Abo-State, M., Abu-Gharbia, M. A., and Ahmed, H. E. 2013. Biosorption of aluminum, cobalt, and copper ions by *Providencia rettgeri* isolated from wastewater. *Journal of Basic Microbiology* 53: 477–488.

Achal, V., Pan, X., and Zhang, D. 2012. Bioremediation of strontium (Sr) contaminated aquifer quartz sand based on carbonate precipitation induced by Sr resistant *Halomonas* sp. *Chemosphere* 89: 764–768.

Al-Aoukaty, A., Appanna, V. D., and Falter, H. 1992. Gallium toxicity and adaptation in *Pseudomonas fluorescens*. *FEMS Microbiology Letters* 71: 265–272.

Al-Aoukaty, A., Appanna, V. D., and Huang, J. 1991. Exocellular and intracellular accumulation of lead in *Pseudomonas fluorescens* ATCC 13525 is mediated by the phosphate content of the growth medium. *FEMS Microbiology Letters* 67: 283–290.

Alhasawi, A., Castonguay, Z., Appanna, N. D., Auger, C., and Appanna, V. D. 2015a. Glycine metabolism and antioxidative defence mechanisms in *Pseudomonas fluorescens*. *Microbiological Research* 171: 26–31.

Alhasawi, A., Costanzi, J., Auger, C., Appanna, N. D., and Appanna, V. D. 2015b. Metabolic reconfigurations aimed at the detoxification of a multi-metal stress in *Pseudomonas fluorescens*: Implications for the bioremediation of metal pollutants. *Journal of Biotechnology* 200: 38–43.

Anderson, S. and Appanna, V. D. 1993. Indium detoxification in *Pseudomonas fluorescens*. *Environmental Pollution* 82: 33–37.

Anderson, S. and Appanna, V. D. 1994. Deposition of crystalline strontium carbonate by *Pseudomonas fluorescens*. *FEMS Microbiology Letters* 116: 6.

Anderson, S., Appanna, V. D., Huang, J., and Viswanatha, T. 1992. A novel role for calcite in calcium homeostasis. *FEBS Letters* 308: 94–96.

Andreeva, N., Ryazanova, L., Dmitriev, V., Kulakovskaya, T., and Kulaev, I. 2013. Adaptation of *Saccharomyces cerevisiae* to toxic manganese concentration triggers changes in inorganic polyphosphates. *FEMS Yeast Research* 13: 463–470.

Andreeva, N., Ryazanova, L., Dmitriev, V., Kulakovskaya, T., and Kulaev, I. 2014. Cytoplasmic inorganic polyphosphate participates in the heavy metal tolerance of *Cryptococcus humicola*. *Folia Microbiologica* (*Praha*) 59: 381–389.

Appanna, V. D. 1989. Galactosyl residue in exopolysaccharide from rhizobium meliloti JJ-1 exposed to manganese in furanoid. *Antonie van Leeuwenhoek* 56: 357–360.

Appanna, V. D., Gazsó, L. G., and Pierre, M. S. 1996a. Influence of chromium(III) on the ability of *Pseudomonas fluorescens* to adapt to a multiple-metal stress. *Microbiological Research* 151: 4.

Appanna, V. D., Gazsó, L. G., and Pierre, M. S. 1996b. Multiple-metal tolerance in *Pseudomonas fluorescens* and its biotechnological significance. *Journal of Biotechnology* 52: 6.

Appanna, V. D., Hamel, R., Mackenzie, C., Kumar, P., and Kalyuzhnyi, S. V. 2003a. Adaptation of *Pseudomonas fluorescens* to Al-citrate: Involvement of tricarboxylic acid and glyoxylate cycle enzymes and the influence of phosphate. *Current Microbiology* 47: 521–527.

Appanna, V. D., Hamel, R. D., and Levasseur, R. 2003b. The metabolism of aluminum citrate and biosynthesis of oxalic acid in *Pseudomonas fluorescens*. *Current Microbiology* 47: 32–39.

Appanna, V. D., Mayer, R. E., and St-Pierre, M. 1995. Aluminum tolerance of *Pseudomonas fluorescens* in a phosphate-deficient medium. *Bulletin in Environmental Contamination and Toxicology* 55: 404–411.

Appanna, V. D. and Pierre, M. S. 1996. Aluminum elicits exocellular phosphatidylethanolamine production in *Pseudomonas fluorescens*. *Applied Environmental Microbiology* 62: 2778–2782.

Appanna, V. D. and Preston, C. M. 1987. Manganese elicits the synthesis of a novel exopolysaccharide in an arctic *Rhizobium*. *FEBS Letters* 215: 4.

Appanna, V. D. and St Pierre, M. 1996. Cellular response to a multiple-metal stress in *Pseudomonas fluorescens*. *Journal of Biotechnology* 48: 129–136.

Auger, C., Han, S., Appanna, V. P., Thomas, S. C., Ulibarri, G., and Appanna, V. D. 2013a. Metabolic reengineering invoked by microbial systems to decontaminate aluminum: Implications for bioremediation technologies. *Biotechnology Advances* 31: 266–273.

Auger, C., Lemire, J., Appanna, V. P., and Appanna, V. D. 2013b. Gallium in bacteria: Metabolic and medical implications. *Encyclopedia of Metalloproteins* 2013: 800–807.

Avelar Ferreira, P. A., Bomfeti, C. A., Lima Soares, B., and De Souza Moreira, F. M. 2012. Efficient nitrogen-fixing rhizobium strains isolated from amazonian soils are highly tolerant to acidity and aluminium. *World Journal of Microbiology and Biotechnology* 2012: 13.

Bae, W., Mehra, R. K., Mulchandani, A., and Chen, W. 2001. Genetic engineering of *Escherichia coli* for enhanced uptake and bioaccumulation of mercury. *Applied Environmental Microbiology* 67: 5335–5338.

Bae, W., Wu, C. H., Kostal, J., Mulchandani, A., and Chen, W. 2003. Enhanced mercury biosorption by bacterial cells with surface-displayed merR. *Applied Environmental Microbiology* 69: 3176–3180.

Bai, H. J., Zhang, Z. M., Yang, G. E., and Li, B. Z. 2008. Bioremediation of cadmium by growing *Rhodobacter sphaeroides*: Kinetic characteristic and mechanism studies. *Bioresource Technology* 99: 7716–7722.

Barak, Y., Ackerley, D. F., Dodge, C. J., Banwari, L., Alex, C., Francis, A. J., and Matin, A. 2006. Analysis of novel soluble chromate and uranyl reductases and generation of an improved enzyme by directed evolution. *Applied Environmental Microbiology* 72: 7074–7082.

Barkay, T. and Schaefer, J. 2001. Metal and radionuclide bioremediation: Issues, considerations and potentials. *Current Opinion in Microbiology* 4: 318–323.

Battin, E. E. and Brumaghim, J. L. 2009. Antioxidant activity of sulfur and selenium: A review of reactive oxygen species scavenging, glutathione peroxidase, and metal-binding antioxidant mechanisms. *Cell Biochemistry and Biophysics* 55: 1–23.

Beazley, M. J., Martinez, R. J., Sobecky, P. A., Webb, S. M., and Taillefert, M. 2007. Uranium biomineralization as a result of bacterial phosphatase activity: Insights from bacterial isolates from a contaminated subsurface. *Environmental Science & Technology* 41: 5701–5707.

Bentley, R. and Chasteen, T. G. 2002. Microbial methylation of metalloids: Arsenic, antimony, and bismuth. *Microbiology and Molecular Biology Review* 66: 250–271.

Beriault, R., Chenier, D., Singh, R., Middaugh, J., Mailloux, R., and Appanna, V. 2005. Detection and purification of glucose 6-phosphate dehydrogenase, malic enzyme, and NADP-dependent isocitrate dehydrogenase by blue native polyacrylamide gel electrophoresis. *Electrophoresis* 26: 2892–2897.

Beriault, R., Hamel, R., Chenier, D., Mailloux, R. J., Joly, H., and Appanna, V. D. 2007. The overexpression of NADPH-producing enzymes counters the oxidative stress evoked by gallium, an iron mimetic. *Biometals* 20: 165–176.

Bernhoft, R. A. 2012. Mercury toxicity and treatment: A review of the literature. *Journal of Environmental Public Health* 2012: 460508.

Bignucolo, A., Appanna, V. P., Thomas, S. C., Auger, C., Han, S., Omri, A., and Appanna, V. D. 2013. Hydrogen peroxide stress provokes a metabolic reprogramming in *Pseudomonas fluorescens*: Enhanced production of pyruvate. *Journal of Biotechnology* 167: 309–315.

Blumenthal, N. C., Cosma, V., Skyler, D., Legeros, J., and Walters, M. 1995. The effect of cadmium on the formation and properties of hydroxyapatite in vitro and its relation to cadmium toxicity in the skeletal system. *Calcified Tissue International* 56: 316–322.

Booth, S. C., Weljie, A. M., and Turner, R. J. 2015. Metabolomics reveals differences of metal toxicity in cultures of *Pseudomonas pseudoalcaligenes* KF707 grown on different carbon sources. *Frontiers in Microbiology* 6: 827.

Booth, S. C., Workentine, M. L., Weljie, A. M., and Turner, R. J. 2011. Metabolomics and its application to studying metal toxicity. *Metallomics* 3: 1142–1152.

Bridges, C. C. and Zalups, R. K. 2005. Molecular and ionic mimicry and the transport of toxic metals. *Toxicology and Applied Pharmacology* 204: 274–308.

Brugge, D. and Buchner, V. 2012. Radium in the environment: Exposure pathways and health effects. *Review of Environmental Health* 27: 1–17.

Cardwell, A. J., Hawker, D. W., and Greenway, M. 2002. Metal accumulation in aquatic macrophytes from southeast Queensland, Australia. *Chemosphere* 48: 653–663.

Casiot, C., Morin, G., Juillot, F., Bruneel, O., Personne, J. C., Leblanc, M., Duquesne, K., Bonnefoy, V., and Elbaz-Poulichet, F. 2003. Bacterial immobilization and oxidation of arsenic in acid mine drainage (Carnoulès creek, France). *Water Research* 37: 2929–2936.

Chakraborty, J. and Das, S. 2014. Characterization and cadmium-resistant gene expression of biofilm-forming marine bacterium *Pseudomonas aeruginosa* JP-11. *Environmental Science and Pollution Research International* 21: 14188–14201.

Chaney, R. L., Malik, M., Li, Y. M., Brown, S. L., Brewer, E. P., Angle, J. S., and Baker, A. J. 1997. Phytoremediation of soil metals. *Current Opinion in Biotechnology* 8: 279–284.

Chang, P., Kim, K. W., Yoshida, S., and Kim, S. Y. 2005. Uranium accumulation of crop plants enhanced by citric acid. *Environmental Geochemistry and Health* 27: 529–538.

Chen, Y., Parvez, F., Gamble, M., Islam, T., Ahmed, A., Argos, M., Graziano, J. H., and Ahsan, H. 2009. Arsenic exposure at low-to-moderate levels and skin lesions, arsenic metabolism, neurological functions, and biomarkers for respiratory and cardiovascular diseases: Review of recent findings from the health effects of arsenic longitudinal study (heals) in Bangladesh. *Toxicology and Applied Pharmacology* 239: 184–192.

Chenier, D., Beriault, R., Mailloux, R., Baquie, M., Abramia, G., Lemire, J., and Appanna, V. 2008. Involvement of fumarase C and NADH oxidase in metabolic adaptation of *Pseudomonas fluorescens* cells evoked by aluminum and gallium toxicity. *Applied and Environmental Microbiology* 74: 3977–3984.

Conesa, H. M., Faz, A., and Arnaldos, R. 2006. Heavy metal accumulation and tolerance in plants from mine tailings of the semiarid Cartagena-La union mining district (SE Spain). *Science of the Total Environment* 366: 1–11.

Congeevaram, S., Dhanarani, S., Park, J., Dexilin, M., and Thamaraiselvi, K. 2007. Biosorption of chromium and nickel by heavy metal resistant fungal and bacterial isolates. *Journal of Hazardous Materials* 146: 270–277.

Cooke, D. T., Munkonge, F. M., Burden, R. S., and James, C. S. 1991. Fluidity and lipid composition of oat and rye shoot plasma membrane: Effect of sterol perturbation by xenobiotics. *Biochimica et Biophysica Acta* 1061: 156–162.

D'amore, J. J., Al-Abed, S. R., Scheckel, K. G., and Ryan, J. A. 2005. Methods for speciation of metals in soils: A review. *Journal of Environmental Quality* 34: 1707–1745.

Dash, H. R., Mangwani, N., and Das, S. 2014. Characterization and potential application in mercury bioremediation of highly mercury-resistant marine bacterium *Bacillus thuringiensis* PW-05. *Environmental Science and Pollution Research International* 21: 2642–2653.

Davis, T. A., Volesky, B., and Mucci, A. 2003. A review of the biochemistry of heavy metal biosorption by brown algae. *Water Research* 37: 4311–4330.

Debela, F., Arocena, J. M., Thring, R. W., and Whitcombe, T. 2010. Organic acid-induced release of lead from pyromorphite and its relevance to reclamation of Pb-contaminated soils. *Chemosphere* 80: 450–456.

Del Dacera, D. M. and Babel, S. 2006. Use of citric acid for heavy metals extraction from contaminated sewage sludge for land application. *Water Science and Technology* 54: 129–135.

De Vries, W., Lofts, S., Tipping, E., Meili, M., Groenenberg, J. E., and Schutze, G. 2007. Impact of soil properties on critical concentrations of cadmium, lead, copper, zinc, and mercury in soil and soil solution in view of ecotoxicological effects. *Review on Environmental Contamination and Toxicology* 191: 47–89.

Dixit, R., Wasiullah, D., M., Pandiyan, K., Singh, U. B., Sahu, A., Shukla, R., Singh, B. P. et al. 2015. Bioremediation of heavy metals from soil and aquatic environment: An overview of principles and criteria of fundamental processes. *Sustainability* 7: 23.

Dudka, S. and Miller, W. P. 1999. Accumulation of potentially toxic elements in plants and their transfer to human food chain. *Journal of Environmental Science and Health Part B* 34: 681–708.

Duruibe, J. O., Ogweugbu, M. O., and Egwurugwu, J. N. 2007. Heavy metal pollution and human biotoxic effects. *International Journal of Physical Sciences* 2: 7.

Duxbury, J. M., Mayer, A. B., Lauren, J. G., and Hassan, N. 2003. Food chain aspects of arsenic contamination in Bangladesh: Effects on quality and productivity of rice. *Journal of Environmental Science and Health A Toxic/Hazardous Substances & Environmental Engineering* 38: 61–69.

Eapen, S. and D'souza, S. F. 2005. Prospects of genetic engineering of plants for phytoremediation of toxic metals. *Biotechnology Advances* 23: 97–114.

Exley, C., Chappell, J. S., and Birchall, J. D. 1991. A mechanism for acute aluminium toxicity in fish. *Journal of Theoretical Biology* 151: 417–428.

Favas, P. J. C., Pratas, J., Varun, M., D'souza, R., and Paul, M. S. 2014. Phytoremediation of soils contaminated with metals and metalloids at mining areas: Potential of native flora. In: *Environmental Risk Assessment of Soil Contamination*, Ed. M. C. Hernandez-Soriano, InTech, Rijeka, Croatia. pp. 485–517.

Fraga, C. G. 2005. Relevance, essentiality and toxicity of trace elements in human health. *Molecular Aspects of Medicine* 26: 235–244.

Franklin, N. M., Stauber, J. L., Markich, S. J., and Lim, R. P. 2000. pH-dependent toxicity of copper and uranium to a tropical freshwater alga (*Chlorella* sp.). *Aquatic Toxicology* 48: 275–289.

Fullmer, C. S., Edelstein, S., and Wasserman, R. H. 1985. Lead-binding properties of intestinal calcium-binding proteins. *Journal of Biological Chemistry* 260: 6816–6819.

Gadd, G. M. 2000. Bioremedial potential of microbial mechanisms of metal mobilization and immobilization. *Current Opinion in Biotechnology* 11: 271–279.

Gadd, G. M. 2004. Microbial influence on metal mobility and application for bioremediation. *Geoderma* 122: 11.

Gadd, G. M. 2009. Biosorption: Critical review of scientific rationale, environmental importance and significance for pollution treatment. *Journal of Chemical Technology and Biotechnology* 84: 16.

Gadd, G. M. 2010. Metals, minerals and microbes: Geomicrobiology and bioremediation. *Microbiology* 156: 609–643.

Garbarino, J. R., Hayes, H., Roth, D., Antweider, R., Brinton, T. I., and Taylor, H. 1995. Contaminants in the Mississippi river, U.S. Geological Survey Circular 1133, Denver, CO.

Gaur, N., Flora, G., Yadav, M., and Tiwari, A. 2014. A review with recent advancements on bioremediation-based abolition of heavy metals. *Environmental Science: Process & Impacts* 16: 180–193.

Gidlow, D. A. 2004. Lead toxicity. *Occupational Medicine (London)* 54: 76–81.

Godt, J., Scheidig, F., Grosse-Siestrup, C., Esche, V., Brandenburg, P., Reich, A., and Groneberg, D. A. 2006. The toxicity of cadmium and resulting hazards for human health. *Journal of Occupational Medicine and Toxicology* 1: 22.

Gonzalez, R. C. and Gonzalez-Chavez, M. C. 2006. Metal accumulation in wild plants surrounding mining wastes. *Environmental Pollution* 144: 84–92.

Goyer, R. A., Liu, J., and Waalkes, M. P. 2004. Cadmium and cancer of prostate and testis. *Biometals* 17: 555–558.

Grose, J. H., Joss, L., Velick, S. F., and Roth, J. R. 2006. Evidence that feedback inhibition of NAD kinase controls responses to oxidative stress. *Proceedings of the National Academy of Sciences of the United States of America* 103: 7601–7606.

Gupta, V. K., Ali, I., Saleh, T. A., Nayak, A., and Agarwal, S. 2012. Chemical treatment technologies for waste-water recycling—An overview. *RSC Advances* 2: 9.

Hamel, R. and Appanna, V. D. 2003. Aluminum detoxification in *Pseudomonas fluorescens* is mediated by oxalate and phosphatidylethanolamine. *Biochimica et Biophysica Acta* 1619: 70–76.

Hamel, R., Appanna, V. D., Viswanatha, T., and Puiseux-Dao, S. 2004. Overexpression of isocitrate lyase is an important strategy in the survival of *Pseudomonas fluorescens* exposed to aluminum. *Biochemical and Biophysical Research Communication* 317: 1189–1194.

Hamel, R. D. and Appanna, V. D. 2001. Modulation of TCA cycle enzymes and aluminum stress in *Pseudomonas fluorescens*. *Journal of Inorganic Biochemistry* 87: 1–8.

Han, F. X., Su, Y., Monts, D. L., Plodinec, M. J., Banin, A., and Triplett, G. E. 2003. Assessment of global industrial-age anthropogenic arsenic contamination. *Naturwissenschaften* 90: 395–401.

Han, S., Lemire, J., Appanna, V. P., Auger, C., Castonguay, Z., and Appanna, V. D. 2013. How aluminum, an intracellular ROS generator promotes hepatic and neurological diseases: The metabolic tale. *Cell Biology and Toxicology* 29: 75–84.

Harrison, J. D. and Muirhead, C. R. 2003. Quantitative comparisons of cancer induction in humans by internally deposited radionuclides and external radiation. *International Journal of Radiation Biology* 79: 1–13.

He, Z., Shentu, J., Yang, X., Baligar, V. C., Zhang, T., and Stofella, P. J. 2015. Heavy metal contamination of soils: Sources, indicators, and assessment. *Journal of Environmental Indicators* 9: 2.

He, Z. L., Yang, X. E., and Stoffella, P. J. 2005. Trace elements in agroecosystems and impacts on the environment. *Journal of Trace Elements and Medicine Biology* 19: 125–140.

Hossain, M. A., Piyatida, P., Teixeira Da Silva, J. A., and Fujita, M. 2012. Molecular mechanism of heavy metal toxicity and tolerance in plants: Central role of glutathione in detoxification of reactive oxygen species and methylglyoxal and in heavy metal chelation. *Journal of Botany* 2012: 37.

Huang, J. W., Blaylock, M. J., Kapulnik, Y., and Ensley, B. D. 1998. Phytoremediation of uranium-contaminated soils: Role of organic acids in triggering uranium hyperaccumulation in plants. *Environmental Science & Technology* 32: 5.

Hughes, M. F. 2002. Arsenic toxicity and potential mechanisms of action. *Toxicology Letters* 133: 1–16.

Jarup, L. 2002. Cadmium overload and toxicity. *Nephrology and Dialysis Transplantation* 17 (Suppl. 2): 35–39.

Jarup, L. 2003. Hazards of heavy metal contamination. *British Medical Bulletin* 68: 16.

Johnson, A. C. and Wood, M. 1990. DNA, a possible site of action of aluminum in *Rhizobium* spp. *Applied and Environmental Microbiology* 56: 3629–3633.

Johnson, D. B. and Hallberg, K. B. 2005. Acid mine drainage remediation options: A review. *Science of the Total Environment* 338: 3–14.

Jomova, K., Jenisova, Z., Feszterova, M., Baros, S., Liska, J., Hudecova, D., Rhodes, C. J., and Valko, M. 2011. Arsenic: Toxicity, oxidative stress and human disease. *Journal of Applied Toxicology* 31: 95–107.

Joshi, P. K., Swarup, A., Maheshwari, S., Kumar, R., and Singh, N. 2011. Bioremediation of heavy metals in liquid media through fungi isolated from contaminated sources. *Indian Journal of Microbiology* 51: 482–487.

Juwarkar, A. A., Nair, A., Dubey, K. V., Singh, S. K., and Devotta, S. 2007. Biosurfactant technology for remediation of cadmium and lead contaminated soils. *Chemosphere* 68: 1996–2002.

Juwarkar, A. A., Singh, S. K., and Mudhoo, A. 2010. A comprehensive overview of elements in bioremediation. *Reviews in Environmental Science and Biotechnology* 9: 74.

Kang, S., Yumi, K., Lee, Y. J., and Roh, Y. 2015. Microbially induced precipitation of strontianite nanoparticles. *Journal of Nanoscience and Nanotechnology* 15: 5362–5365.

Kao, C. H., Lin, H. T., Yu, S. L., Wang, S. J., and Yeh, S. H. 1994. Relationship of alveolar permeability and lung inflammation in patients with active diffuse infiltrative lung disease detected by 99tcm-dtpa radioaerosol inhalation lung scintigraphy and quantitative 67ga lung scans. *Nuclear Medicine Communications* 15: 850–854.

Konyi, J., Koska, P., Berzsenyi, G., Gazso, L. G., and Appanna, V. D. 1997. The role of microorganisms in the mobility of radionuclides in soil. I. Examination of resistance to strontium, cesium, cobalt and zinc. *Journal of Radioecology* 5: 8.

Korkeala, H. and Pekkanen, T. J. 1978. The effect of pH and potassium phosphate buffer on the toxicity of cadmium for bacteria. *Acta Veterinaria Scandinavica* 19: 93–101.

Krill, A. M., Kirst, M., Kochian, L. V., Buckler, E. S., and Hoekenga, O. A. 2010. Association and linkage analysis of aluminum tolerance genes in maize. *PLoS ONE* 5: E9958.

Kumagai, Y. and Sumi, D. 2007. Arsenic: Signal transduction, transcription factor, and biotransformation involved in cellular response and toxicity. *Annual Review in Pharmacology and Toxicology* 47: 243–262.

Lau, A. T., Wang, Y., and Chiu, J. F. 2008. Reactive oxygen species: Current knowledge and applications in cancer research and therapeutic. *Journal of Cellular Biochemistry* 104: 657–667.

Lee, J. H. 2013. An overview of phytoremediation as a potentially promising technology for environmental pollution control. *Biotechnology and Bioprocess Engineering* 18: 9.

Lemire, J. and Appanna, V. D. 2011. Aluminum toxicity and astrocyte dysfunction: A metabolic link to neurological disorders. *Journal of Inorganic Biochemistry* 105: 1513–1517.

Lemire, J., Auger, C., Bignucolo, A., Appanna, V. P., and Appanna, V. D. 2010a. Metabolic strategies deployed by *Pseudomonas fluorescens* to combat metal pollutants: Biotechnological prospects. *Current Research, Technology and Education Topics in Applied Microbiology and Microbial Biotechnology* 1: 11.

Lemire, J., Kumar, P., Mailloux, R., Cossar, K., and Appanna, V. D. 2008a. Metabolic adaptation and oxaloacetate homeostasis in *P. fluorescens* exposed to aluminum toxicity. *Journal of Basic Microbiology* 48: 252–259.

Lemire, J., Mailloux, R., and Appanna, V. D. 2008b. Zinc toxicity alters mitochondrial metabolism and leads to decreased ATP production in hepatocytes. *Journal of Applied Toxicology* 28: 175–182.

Lemire, J., Mailloux, R., Auger, C., Whalen, D., and Appanna, V. D. 2010b. *Pseudomonas fluorescens* orchestrates a fine metabolic-balancing act to counter aluminium toxicity. *Environmental Microbiology* 12: 1384–1390.

Liang, X., Kierans, M., Ceci, A., Hillier, S., and Gadd, G. M. 2016. Phosphatase-mediated bioprecipitation of lead by soil fungi. *Environmental Microbiology*, 18: 219–231.

Lloyd, J. R. and Lovley, D. R. 2001. Microbial detoxification of metals and radionuclides. *Current Opinion in Biotechnology* 12: 248–253.

Loghman-Adham, M. 1997. Renal effects of environmental and occupational lead exposure. *Environmental Health Perspectives* 105: 928–938.

Lopez-Archilla, A. I., Marin, I., and Amils, R. 2001. Microbial community composition and ecology of an acidic aquatic environment: The Tinto river, Spain. *Microbial Ecology* 41: 20–35.

Lovley, D. R. and Coates, J. D. 1997. Bioremediation of metal contamination. *Current Opinion in Biotechnology* 8: 285–289.

Machado, W., Rodrigues, A. P., Bidone, E. D., Sella, S. M., and Santelli, R. E. 2011. Evaluation of Cu potential bioavailability changes upon coastal sediment resuspension: An example on how to improve the assessment of sediment dredging environmental risks. *Environmental Science and Pollution Research International* 18: 1033–1036.

Mailloux, R. J., Beriault, R., Lemire, J., Singh, R., Chenier, D. R., Hamel, R. D., and Appanna, V. D. 2007. The tricarboxylic acid cycle, an ancient metabolic network with a novel twist. *PLoS ONE* 2: E690.

Mailloux, R. J., Hamel, R., and Appanna, V. D. 2006. Aluminum toxicity elicits a dysfunctional TCA cycle and succinate accumulation in hepatocytes. *Journal of Biochemical and Molecular Toxicology* 20: 198–208.

Mailloux, R. J., Lemire, J., and Appanna, V. D. 2011a. Hepatic response to aluminum toxicity: Dyslipidemia and liver diseases. *Experimental Cell Research* 317: 2231–2238.

Mailloux, R. J., Lemire, J., and Appanna, V. D. 2011b. Metabolic networks to combat oxidative stress in *Pseudomonas fluorescens*. *Antonie van Leeuwenhoek* 99: 433–442.

Mailloux, R. J., Puiseux-Dao, S., and Appanna, V. D. 2009. Alpha-ketoglutarate abrogates the nuclear localization of HIF-1alpha in aluminum-exposed hepatocytes. *Biochimie* 91: 408–415.

Mani, D., Kumar, C., and Patel, N. K. 2015. Integrated micro-biochemical approach for phytoremediation of cadmium and zinc contaminated soils. *Ecotoxicology and Environmental Safety* 111: 86–95.

Martinez, R. J., Beazley, M. J., Taillefert, M., Arakaki, A. K., Skolnick, J., and Sobecky, P. A. 2007. Aerobic uranium (VI) bioprecipitation by metal-resistant bacteria isolated from radionuclide- and metal-contaminated subsurface soils. *Environmental Microbiology* 9: 3122–3133.

Mates, J. M., Perez-Gomez, C., and Nunez De Castro, I. 1999. Antioxidant enzymes and human diseases. *Clinical Biochemistry* 32: 595–603.

Matlock, M. M., Howerton, B. S., and Atwood, D. A. 2002. Chemical precipitation of heavy metals from acid mine drainage. *Water Research* 36: 4757–4764.

Middaugh, J., Hamel, R., Jean-Baptiste, G., Beriault, R., Chenier, D., and Appanna, V. D. 2005. Aluminum triggers decreased aconitase activity via Fe-S cluster disruption and the overexpression of isocitrate dehydrogenase and isocitrate lyase: A metabolic network mediating cellular survival. *Journal of Biological Chemistry* 280: 3159–3165.

Migliorini, M., Pigino, G., Bianchi, N., Bernini, F., and Leonzio, C. 2004. The effects of heavy metal contamination on the soil arthropod community of a shooting range. *Environmental Pollution* 129: 331–340.

Mirlean, N. and Roisenberg, A. 2006. The effect of emissions of fertilizer production on the environment contamination by cadmium and arsenic in southern Brazil. *Environmental Pollution* 143: 335–340.

Mirlean, N., Roisenberg, A., and Chies, J. O. 2007. Metal contamination of vineyard soils in wet subtropics (southern Brazil). *Environmental Pollution* 149: 10–17.

Misra, C. S., Appukuttan, D., Kantamreddi, V. S., Rao, A. S., and Apte, S. K. 2012. Recombinant *D. radiodurans* cells for bioremediation of heavy metals from acidic/neutral aqueous wastes. *Bioengineering Bugs* 3: 44–48.

Misra, T. K. 1992. Bacterial resistances to inorganic mercury salts and organomercurials. *Plasmid* 27: 4–16.

Moriyon, I. and Lopez-Goni, I. 1998. Structure and properties of the outer membranes of *Brucella abortus* and *Brucella melitensis*. *International Microbiology* 1: 19–26.

Moskalyk, R. R. 2003. Gallium: The backbone of the electronics industry. *Minerals Engineering* 16: 9.

Mulligan, C. N., Yong, R. N., and Gibbs, B. F. 2001. Remediation technologies for metal-contaminated soils and groundwater: An evaluation. *Engineering Geology* 60: 11.

Mushak, P. and Crocetti, A. F. 1995. Risk and revisionism in arsenic cancer risk assessment. *Environmental Health Perspectives* 103: 684–689.

Nair, A., Juwarkar, A. A., and Singh, S. K. 2007. Production and characterization of siderophores and its application in arsenic removal from contaminated soil. *Water, Air, and Soil Pollution* 180: 14.

Nandi, D., Patra, R. C., and Swarup, D. 2005. Effect of cysteine, methionine, ascorbic acid and thiamine on arsenic-induced oxidative stress and biochemical alterations in rats. *Toxicology* 211: 26–35.

Nankishore, A. 2014. Heavy metal levels in leafy vegetables from selected markets in Guyana. *Journal of Agricultural Technology* 10: 13.

Nascimento, A. M. and Chartone-Souza, E. 2003. Operon *mer*: Bacterial resistance to mercury and potential for bioremediation of contaminated environments. *Genetics and Molecular Research* 2: 92–101.

Newsome, L., Morris, K., Trivedi, D., Bewsher, A., and Lloyd, J. R. 2015. Biostimulation by glycerol phosphate to precipitate recalcitrant Uranium(IV) phosphate. *Environmental Science & Technology* 49: 9.

Nicolaou, S. A., Gaida, S. M., and Papoutsakis, E. T. 2010. A comparative view of metabolite and substrate stress and tolerance in microbial bioprocessing: From biofuels and chemicals, to biocatalysis and bioremediation. *Metabolic Engineering* 12: 307–331.

Nriagu, J. O. and Pacyna, J. M. 1988. Quantitative assessment of worldwide contamination of air, water and soils by trace metals. *Nature* 333: 134–139.

Olakanmi, O., Britigan, B. E., and Schlesinger, L. S. 2000. Gallium disrupts iron metabolism of mycobacteria residing within human macrophages. *Infection and Immunity* 68: 5619–5627.

Olaniran, A. O., Balgobind, A., and Pillay, B. 2011. Quantitative assessment of the toxic effects of heavy metals on 1,2-dichloroethane biodegradation in co-contaminated soil under aerobic condition. *Chemosphere* 85: 839–847.

Olaniran, A. O., Balgobind, A., and Pillay, B. 2013. Bioavailability of heavy metals in soil: Impact on microbial biodegradation of organic compounds and possible improvement strategies. *International Journal of Molecular Science* 14: 10197–10228.

Oliver, M. A. and Gregory, P. J. 2015. Soil, food security and human health: A review. *European Journal of Soil Science* 66: 20.

Palma, J. M., Sandalio, L. M., Corpas, F., Romero-Puertas, M. C., McCarthy, I., and Del Río, L. A. 2002. Plant proteases, protein degradation, and oxidative stress: Role of peroxisomes. *Plant Physiology and Biochemistry* 40: 10.

Papanikolaou, N. C., Hatzidaki, E. G., Belivanis, S., Tzanakakis, G. N., and Tsatsakis, A. M. 2005. Lead toxicity update. a brief review. *Medical Science Monitoring* 11: Ra329–Ra336.

Paulsen, I. T., Press, C. M., Ravel, J., Kobayashi, D. Y., Myers, G. S., Mavrodi, D. V., Deboy, R. T. et al. 2005. Complete genome sequence of the plant commensal *Pseudomonas fluorescens* Pf-5. *Nature Biotechnology* 23: 873–878.

Peplow, D. 1999. Environmental impacts of mining in eastern Washington, Center for Water and Watershed Studies Fact Sheet. University of Washington, Seattle, WA.

Peralta, J. R., Gardea-Torresdey, J. L., Tiemann, K. J., Gomez, E., Arteaga, S., Rascon, E., and Parsons, J. G. 2001. Uptake and effects of five heavy metals on seed germination and plant growth in alfalfa (*Medicago sativa* L.). *Bulletin of Environmental Contamination and Toxicology* 66: 727–734.

Pina, R. G. and Cervantes, C. 1996. Microbial interactions with aluminium. *Biometals* 9: 311–316.

Rabinowitz, M. B. 1991. Toxicokinetics of bone lead. *Environmental Health Perspectives* 91: 33–37.

Rahman, K. S., Rahman, T. J., Mcclean, S., Marchant, R., and Banat, I. M. 2002. Rhamnolipid biosurfactant production by strains of *Pseudomonas aeruginosa* using low-cost raw materials. *Biotechnology Progress* 18: 1277–1281.

Ranjard, L., Prigent-Combaret, C., Nazaret, S., and Cournoyer, B. 2002. Methylation of inorganic and organic selenium by the bacterial thiopurine methyltransferase. *Journal of Bacteriology* 184: 3146–3149.

Reinoso, C. A., Appanna, V. D., and Vasquez, C. C. 2013. Alpha-ketoglutarate accumulation is not dependent on isocitrate dehydrogenase activity during tellurite detoxification in *Escherichia coli*. *Biomedical Research International* 2013: 784190.

Rhee, Y. J., Hillier, S., Pendlowski, H., and Gadd, G. M. 2014. Pyromorphite formation in a fungal biofilm community growing on lead metal. *Environmental Microbiology* 16: 1441–1451.

Rivetta, A., Negrini, N., and Cocucci, M. 1997. Involvement of Ca^{2+}-calmodulin in Cd^{2+} toxicity during the early phases of radish (*Raphanus sativus* L.) seed germination. *Plant, Cell and Environment* 5: 9.

Rodriguez, L., Ruiz, E., Alonso-Azcarate, J., and Rincon, J. 2009. Heavy metal distribution and chemical speciation in tailings and soils around a Pb-Zn mine in Spain. *Journal of Environmental Management* 90: 1106–1116.

Roscoe, R. J., Steenland, K., Halperin, W. E., Beaumont, J. J., and Waxweiler, R. J. 1989. Lung cancer mortality among non-smoking uranium miners exposed to radon daughters. *JAMA* 262: 629–633.

Rummo, J. H., Routh, D. K., Rummo, N. J., and Brown, J. F. 1979. Behavioral and neurological effects of symptomatic and asymptomatic lead exposure in children. *Archives of Environmental Health* 34: 120–124.

Safarzadeha, M. S., Bafghia, M. S., Moradkhanib, D., and Ilkhchid, M. O. 2007. A review on hydrometallurgical extraction and recovery of cadmium from various resources. *Minerals Engineering* 20: 10.

Satarug, S. and Moore, M. R. 2004. Adverse health effects of chronic exposure to low-level cadmium in foodstuffs and cigarette smoke. *Environmental Health Perspectives* 112: 1099–1103.

Sauge-Merle, S., Cuine, S., Carrier, P., Lecomte-Pradines, C., Luu, D. T., and Peltier, G. 2003. Enhanced toxic metal accumulation in engineered bacterial cells expressing *Arabidopsis thaliana* phytochelatin synthase. *Applied and Environmental Microbiology* 69: 490–494.

Sayler, G. S. and Ripp, S. 2000. Field applications of genetically engineered microorganisms for bioremediation processes. *Current Opinion in Biotechnology* 11: 286–289.

Schutzendubel, A., Schwanz, P., Teichmann, T., Gross, K., Langenfeld-Heyser, R., Godbold, D. L., and Polle, A. 2001. Cadmium-induced changes in antioxidative systems, hydrogen peroxide content, and differentiation in scots pine roots. *Plant Physiology* 127: 887–898.

Sharma, P. K., Balkwill, D. L., Frenkel, A., and Vairavamurthy, M. A. 2000. A new *Klebsiella planticola* strain (Cd-1) grows anaerobically at high cadmium concentrations and precipitates cadmium sulfide. *Applied and Environmental Microbiology* 66: 3083–3087.

Sharma, S. S. and Dietz, K. J. 2009. The relationship between metal toxicity and cellular redox imbalance. *Trends in Plant Science* 14: 43–50.

Shukla, K. P., Singh, N. K., and Sharma, S. 2010. Bioremediation: Developments, current practices and perspectives. *Genetic Engineering and Biotechnology Journal* 2010: 1.

Singh, O. V., Labana, S., Pandey, G., Budhiraja, R., and Jain, R. K. 2003. Phytoremediation: An overview of metallic ion decontamination from soil. *Applied Microbiology and Biotechnology* 61: 405–412.

Singh, R., Beriault, R., Middaugh, J., Hamel, R., Chenier, D., Appanna, V. D., and Kalyuzhnyi, S. 2005. Aluminum-tolerant *Pseudomonas fluorescens*: ROS toxicity and enhanced NADPH production. *Extremophiles* 9: 367–373.

Singh, R., Gautam, N., Mishra, A., and Gupta, R. 2011. Heavy metals and living systems: An overview. *Indian Journal of Pharmacology* 43: 246–253.

Singh, R., Lemire, J., Mailloux, R. J., and Appanna, V. D. 2008. A novel strategy involved in [corrected] anti-oxidative defense: The conversion of NADH into NADPH by a metabolic network. *PLoS ONE* 3: E2682.

Singh, R., Lemire, J., Mailloux, R. J., Chenier, D., Hamel, R., and Appanna, V. D. 2009. An ATP and oxalate generating variant tricarboxylic acid cycle counters aluminum toxicity in *Pseudomonas fluorescens*. *PLoS ONE* 4: E7344.

Singh, R., Mailloux, R. J., Puiseux-Dao, S., and Appanna, V. D. 2007. Oxidative stress evokes a metabolic adaptation that favors increased NADPH synthesis and decreased NADH production in *Pseudomonas fluorescens*. *Journal of Bacteriology* 189: 6665–6675.

Singh, R., Paul, D., and Jain, R. K. 2006. Biofilms: Implications in bioremediation. *Trends Microbiology* 14: 389–397.

Sinha, S., Saxena, R., and Singh, S. 2005. Chromium induced lipid peroxidation in the plants of *Pistia stratiotes* l: Role of antioxidants and antioxidant enzymes. *Chemosphere* 58: 595–604.

Smith, C. M., Wang, X., Hu, H., and Kelsey, K. T. 1995. A polymorphism in the delta-aminolevulinic acid dehydratase gene may modify the pharmacokinetics and toxicity of lead. *Environmental Health Perspectives* 103: 248–253.

Srivastava, N. K. and Majumder, C. B. 2008. Novel biofiltration methods for the treatment of heavy metals from industrial wastewater. *Journal of Hazardous Materials* 151: 1–8.

Suzuki, I. 2001. Microbial leaching of metals from sulfide minerals. *Biotechnology Advances* 19: 119–132.

Tai, H. S. and Jou, C. J. 1999. Immobilization of chromium-contaminated soil by means of microwave energy. *Journal of Hazardous Materials* 65: 267–275.

Tapiero, H. and Tew, K. D. 2003. Trace elements in human physiology and pathology: Zinc and metallothioneins. *Biomedicals and Pharmacotherapeutics* 57: 399–411.

Tapiero, H., Townsend, D. M., and Tew, K. D. 2003. Trace elements in human physiology and pathology: Copper. *Biomedicals and Pharmacotherapeutics* 57: 386–398.

Tchounwou, P. B., Yedjou, C. G., Patlolla, A. K., and Sutton, D. J. 2012. Heavy metal toxicity and the environment. *EXS* 101: 133–164.

Torres Pazmino, D. E., Winkler, M., Glieder, A., and Fraaije, M. W. 2010. Monooxygenases as biocatalysts: Classification, mechanistic aspects and biotechnological applications. *Journal of Biotechnology* 146: 9–24.

Utgikar, V. P., Harmon, S. M., Chaudhary, N., Tabak, H. H., Govind, R., and Haines, J. R. 2002. Inhibition of sulfate-reducing bacteria by metal sulfide formation in bioremediation of acid mine drainage. *Environmental Toxicology* 17: 40–48.

Valko, M., Morris, H., and Cronin, M. T. 2005. Metals, toxicity and oxidative stress. *Current Medicinal Chemistry* 12: 1161–1208.

Valls, M. and De Lorenzo, V. 2002. Exploiting the genetic and biochemical capacities of bacteria for the remediation of heavy metal pollution. *FEMS Microbiology Review* 26: 327–338.

Valls, M., De Lorenzo, V., Gonzalez-Duarte, R., and Atrian, S. 2000. Engineering outer-membrane proteins in *Pseudomonas putida* for enhanced heavy-metal bioadsorption. *Journal of Inorganic Biochemistry* 79: 219–223.

Van Der Meeren, A., Tourdes, F., Gremy, O., Grillon, G., Abram, M. C., Poncy, J. L., and Griffiths, N. 2008. Activation of alveolar macrophages after plutonium oxide inhalation in rats: Involvement in the early inflammatory response. *Radiation Research* 170: 591–603.

Velasquez, L. and Dussan, J. 2009. Biosorption and bioaccumulation of heavy metals on dead and living biomass of *Bacillus sphaericus*. *Journal of Hazardous Materials* 167: 713–716.

Verstraeten, S. V., Aimo, L., and Oteiza, P. I. 2008. Aluminium and lead: Molecular mechanisms of brain toxicity. *Archives of Toxicology* 82: 789–802.

Vezzaro, L., Sharma, A. K., Ledin, A., and Mikkelsen, P. S. 2015. Evaluation of stormwater micropollutant source control and end-of-pipe control strategies using an uncertainty-calibrated integrated dynamic simulation model. *Journal of Environmental Management* 151: 56–64.

Vilensky, J. A. and Redman, K. 2003. British anti-Lewisite (dimercaprol): an amazing history. *Annals of Emergency Medicine* 41: 378–383.

Weber, K. A., Achenbach, L. A., and Coates, J. D. 2006. Microorganisms pumping iron: Anaerobic microbial iron oxidation and reduction. *Nature Review Microbiology* 4: 752–764.

White, C., Sharman, A. K., and Gadd, G. M. 1998. An integrated microbial process for the bioremediation of soil contaminated with toxic metals. *Nature Biotechnology* 16: 572–575.

Yadav, S. K. 2010. Heavy metals toxicity in plants: An overview on the role of glutathione and phytochelatins in heavy metal stress tolerance of plants. *South African Journal of Botany* 76: 13.

Yamamoto, Y., Kobayashi, Y., Devi, S. R., Rikiishi, S., and Matsumoto, H. 2002. Aluminum toxicity is associated with mitochondrial dysfunction and the production of reactive oxygen species in plant cells. *Plant Physiology* 128: 63–72.

Yu, B., Xu, P., Shi, Q., and Ma, C. 2006. Deep desulfurization of diesel oil and crude oils by a newly isolated *Rhodococcus erythropolis* strain. *Applied and Environmental Microbiology* 72: 54–58.

Yucesoy, B., Turhan, A., Ure, M., Imir, T., and Karakaya, A. 1997. Effects of occupational lead and cadmium exposure on some immunoregulatory cytokine levels in man. *Toxicology* 123: 143–147.

Zafar, S., Aqil, F., and Ahmad, I. 2007. Metal tolerance and biosorption potential of filamentous fungi isolated from metal contaminated agricultural soil. *Bioresource Technology* 98: 2557–2561.

Zhang, W., Chen, L., and Liu, D. 2012. Characterization of a marine-isolated mercury-resistant *Pseudomonas putida* strain sp1 and its potential application in marine mercury reduction. *Applied Microbiology and Biotechnology* 93: 1305–1314.

12 New Trends in Microbial Biosorption Modeling and Optimization

Daria Podstawczyk, Anna Dawiec,
Anna Witek-Krowiak, and Katarzyna Chojnacka

CONTENTS

ABSTRACT

This chapter describes the latest trends in microorganism-induced heavy metal ions biosorption. Since biosorption is a similar process to adsorption, conventional equilibrium and kinetic models can be successfully applied. However, most of those models, like the Langmuir isotherm, which originally assumes monolayer adsorption by gases on activated carbon, have been developed in order to simulate simple cases of adsorption. The conventional models have been further adjusted to securely fit other adsorption-based processes such as microbial biosorption. Biosorption is a very complex process depending on the surface chemistry taking place in the microorganism's cell wall and the solution of physico-chemical conditions; hence, it requires detailed research on the modeling. Several models presented in this chapter have been used for a long time; however, they have been only recently applied for modeling heavy metal ions biosorption by microorganisms.

This chapter has been divided into two parts. The first section presents conventional models such as the Langmuir and Freundlich isotherms and chemical reaction–based kinetic models applied for modeling heavy metal ions' removal by microorganisms. The second part deals with rather rarely used models such as competitive, mechanistic, and numerical simulations as well as novel tools like artificial neural networks and response surface methodology. The application of the new techniques in mathematical modeling of microbial biosorption may lead to the development of successful methods for the designing and scaling up of biosorption processes for the treatment of industrial metal-bearing effluents.

12.1 INTRODUCTION

12.1.1 Biosorption and Microbial Biosorbents

It is the high efficiency, cost-effectiveness, and simplicity of the biological methods of pollutants removal that attracted the interest of scientists. They include biosorption as a passive and physicochemical binding of chemical species to active sites on the surface of the biosorbent. Biosorption is a technique that can be applied to remove barely degradable toxins, especially heavy metal ions, from aqueous solutions. Biosorbent microorganisms encompass bacteria, fungi, and algae, all of which are characterized by a high sorption capacity.

Metal ions biosorption by microbial biomass mainly depends on the components of the cell surface. The bacterial surface contains peptidoglycan and teichoic and lipoteichoic acids, whereas algal and fungal cell wall is rich in cellulose, chitin, alginate, or glycan. Due to the type of cellular components, several functional groups are present on the bacterial cell wall, including the most popular carboxyl, amine, and hydroxyl groups, being responsible for heavy metal uptake by microorganisms in biosorption process.

12.1.2 Biosorption Mechanisms

In biosorption, the binding mechanism of the sorbate onto the biomass surface is based on physicochemical interactions. The interactions include electrostatic interactions, covalent binding, ions exchange, microprecipitation, chelation, and complexation. The mechanism of biosorption can be investigated by several analytical techniques but also by the observation of the kinetics progress, isotherm shape, results of thermodynamic studies, and system response to various process parameters.

FTIR spectroscopy analysis of microorganism surface before and after biosorption gives information about functional groups on the surface of the biomass involved in heavy metal ion binding. Similarly, the surface groups present on biosorbents surface can be identified by the potentiometric titration, giving information about their cation exchange capacity (Markai et al., 2003). Blocking of selected functional groups allows the qualitative and quantitative estimation of the contribution of particular groups to the binding of sorbates on the biosorbent surface (Chojnacka et al., 2005). Desorption studies with various eluents also reveal the bonding type and strength, physical adsorption can be found using water as an eluent, and stronger bonds require other eluents like salts, chelating agents, acids, and bases (Bai et al., 2014). Physical adsorption dominates when the surface area contains micropores. In the case of microorganisms, external area is

rather smooth, thus contribution of physisorption in biosorption is small (Chojnacka et al., 2005). The interpretation of the surface analysis results with SEM-EDX before and after biosorption makes possible the understanding of the structural changes of surface morphology and the distribution of various elements on the biomass surface (Dmytryk et al., 2014). Assuming ion exchange to be a predominant mechanism, the release of protons (decrease in pH as observed during the experiments) and ions of alkaline and earth metals might be expected to take place while heavy metal ions are being bound (Chakravarty and Banerjee, 2008). Electrostatic interactions can also be a part of biosorption (Chakravarty and Banerjee, 2012). At higher pH values, precipitation of insoluble metal oxides, hydroxides, and nanocrystals on the biomass surface can occur (Chen et al., 2015).

12.1.3 Factors Affecting Microbial Biosorption

Biosorption strongly depends on operational parameters of which the most important include pH, temperature, biosorbent and sorbate concentrations, ionic strength, and the presence of other substances in the solution. For dynamic column systems, we should additionally control the flow velocity of the sorbate solution and the height of the fixed bed. The pH of the solution is one of the parameters that has a colossal influence on biosorption. The pH of the environment influences both the chemistry of the compounds that are present in the solution and the degree of dissociation of the groups present at the surface of the microbial biosorbent. The effect of pH also depends on the physiological state of biomass, the type of microorganism (Gram-positive or Gram-negative bacteria, algae, or fungi), and their modification or immobilization technique (Saravanan et al., 2013). Each biosorbent has its own pH at zero point charge (pHzpc) value when the net surface charge equals zero. At pH lower than the pHzpc, the overall charge of the biomass surface is positive; at pH higher than the pHzpc, the overall surface charge is negative. At lower pH heavy metals are in hydrated form in the solution. If biosorption mainly takes the form of ion exchange, protons compete with metal cations for the binding sites on the biomass surface (Ahmad et al., 2011). The higher pH means the higher availability of ligands for metal ion binding. Alkaline pH favors the precipitation of insoluble metal hydroxides of selected heavy metal ions, such as $Cu(II)$, $Cd(II)$, $Cr(III)$, $Mn(II)$, and $Pb(II)$ (Haq et al., 2015; Hou et al., 2015). For anions (e.g., $Cr(VI)$) uptake is favored at low pH (1–3).

The amount of the sorbent used in the process is a very important parameter: the higher the concentration of the sorbent, the greater the availability of active sites that can effectively bind ions, which enhances the efficiency of the process. For a given initial concentration of the sorbate in a solution, there is an optimum value of the concentration of the biosorbent above which the efficiency of the process cannot be increased. It is very important to determine this value from the economic point of view. An increase in the concentration of the sorbate enhances the driving force of biosorption, which makes sorption more efficient (Ahmady-Asbchin et al., 2015).

The effect of temperature is not so significant within the range of 10°C–30°C. The usual uptake of heavy metal ions rises with the increase in the temperature (Calero et al., 2009). At temperatures higher than 40°C, the structure of the outer surface of cells could be irreversibly changed. Typically, sorption capacity decreases, probably by reducing the quantity of actively bond heavy metal ions (Hou et al., 2015).

The intensity of the flow of the feed solution in the columns is an important parameter that affects biosorption because it directly determines the contact time between the sorbate and the biomass. The lower the intensity of the flow of the feed solution, the higher the adsorption degree at the surface of the biomass due to a longer contact of the solution containing heavy metal ions with the microbial biomass (Samuel et al., 2013). The greater the amount of biomass (higher bed), the more efficiently the process develops; when a larger amount of biosorbent is available, a larger area for mass transfer is also available, which means there are more active sites that are capable of adsorbing heavy metal ions (Podder and Majumder, 2016a).

12.2 CONVENTIONAL MATHEMATICAL MODELS OF HEAVY METAL IONS BIOSORPTION BY MICROORGANISMS

12.2.1 EQUILIBRIUM MODELING

The correlation between the sorbed and the aqueous metal ion concentrations at equilibrium can be described by several isotherm models. Isotherms are prepared for constant environmental parameters, such as pH, temperature, ionic strength, or agitation speed. Uptake of heavy metal ions increases with the rise in heavy metal concentrations in the solution. Depending on the mechanism the system could reach saturation at high concentrations. Equilibrium models help to describe the experimental data and predict how the system will behave in a broad range of heavy metal ion concentrations. Mathematical modeling of the isotherms is an important tool for the scaling up and optimization of biosorption. A well-selected model can hypothesize the mechanism of ion binding on the surface. Conventional models of adsorption, adapted to the needs of the mathematical description of biosorption, are empirical multiparameter models fitting well given experimental points. Adsorption model assumptions are usually not fulfilled in the case of such a complex process as biosorption. For single-metal solutions the most popular adsorption isotherms are the two-parameter models of Langmuir, Freundlich, Temkin, and Dubinin–Radushkevich and the three-parameter ones of Sips and Redlich–Peterson, with those of Langmuir and Freundlich being the most widely used (Table 12.1).

12.2.1.1 Langmuir Isotherm

The main assumption of the Langmuir model (Langmuir, 1916) is the existence of a specified number of homogenous adsorption centers on the surface of the adsorbent. Each of them can adsorb only one molecule of adsorbate (localized adsorption) by chemical or physical interactions. When all the adsorption centers on the surface of the biosorbent are filled

and a monomolecular layer is created, adsorption reaches the maximum uptake q_{max} (Equation 12.1):

$$q = q_{max} \frac{K_L c_e}{1 + K_L c_e} \tag{12.1}$$

where

q_{max} is the maximum amount of the compound bound by 1 g of biosorbent
c_e is the concentration of the adsorbate in equilibrium
K_L is the Langmuir constant

The Langmuir constant presents the adsorbent–adsorbate affinity, describing the efficiency of the whole process. This value enables determining dimensionless constant R_L (Equation 12.2) with the evaluation of favorable or unfavorable process for Langmuir biosorption:

$$R_L = \frac{1}{1 + K c_0} \tag{12.2}$$

The value of constant R_L indicates that biosorption is favorable ($0 < R_L < 1$), unfavorable ($R_L > 1$), linear ($R_L = 1$), or irreversible ($R_L = 0$).

The Langmuir isotherm is also helpful in determining the specific surface monolayer of the adsorbed metal ion for a given adsorbent (Equation 12.3):

$$S = \frac{q_{max} N A}{M} \tag{12.3}$$

where

N is the Avogadro number (6.022×10^{23})
A is the cross-sectional area of the ion (Å²)
M is the molar mass of the ion

The Langmuir isotherm is the most popular model for biosorption (Table 12.1). Biosorption isotherm curves are mostly L-shape, reaching saturation at higher concentrations (Chowdhury and Mulligan, 2011). On the basis of this model the biosorption capacity of microbial biomass can be predicted, which is one of the key characteristics of a biosorbent.

12.2.1.2 Freundlich Isotherm

The Freundlich isotherm (Freundlich, 1906) is an empirical equation for adsorption that occurs on the heterogeneous surface. The model assumes mobility of a sorbate at the binding site of an adsorbent, which promotes the formation of a multilayer. The equation of the isotherm has the following form (Equation 12.4):

$$q = K_F C^{\frac{1}{n_F}} \tag{12.4}$$

where

K_F is the Freundlich constant
n_F is an index related to intensity of adsorption considering heterogeneity of the surface of the biosorbent

The value of $1/n_F$ gives information about heterogeneity: the smaller the $1/n_F$, the greater the heterogeneity. A value higher

TABLE 12.1

Biosorption of Heavy Metal Ions by Microorganism—Isotherm Models

Biosorbent	Heavy Metal	pH	Temperature	Isotherm Model	Regression Method	Reference
Rhodococcus opacus	Al(III)	5	25	L, F, T, DR, D, RP	nl	Cayllahua and Torem (2010)
Streptomyces rimosus	Al(III)	4	25	L, F, DR	l	Tassist et al. (2010)
Rhodococcus sp.	As(III)	7	30	L, F, DR	n	Prasad et al. (2011)
Kocuria rhizophila	Cd(II)	8	35	L, F	l	Haq et al. (2015)
Klebsiella sp.	Cd(II)	5.5	30	L, F	Fuzzy l	Hou et al. (2015)
Microcystis aeruginosa	Cd(II)	6	15, 25, 35	L, F, T	—	Wang et al. (2014)
P. aeruginosa	Cd(II)	5	12	L, F	n	Kőnig-Péter et al. (2014)
Streptomyces sp.	Cd(II)	6	28	L, F, DR	l	Yuan et al. (2009)
Sphaerotilus natans	Cd(II)	3–6	—	L, F, RP	nl	Pagnanelli et al. (2002)
K. rhizophila	Cr(III)	4	35	L, F	l	Haq et al. (2015)
S. rimosus	Cr(III)	4.8	20	L, F, T	l	Sahmoune et al. (2009)
S. rimosus	Cr(III)	4.8	20	L, F, T	l	Sahmoune et al. (2009)
Bacillus salmalaya	Cr(VI)	3	25	L, F	l	Dadrasnia et al. (2015)
B. salmalaya	Cr(VI)	3	25	L, F	l	Dadrasnia et al. (2015)
Enteromorpha sp.	Cr(VI)	2	30	L, F, DR, T	l	Rangabhashiyam et al. (2015)
Pseudomonas putida	Cr(VI)	3	35	L, F, DR	l	Garg et al. (2015)
Pseudomonas sp.	Cu(II)	4.5	28	L, F	l	Zhang et al. (2014)
Streptomyces lunalinharesii	Cu(II)	5	25	L, F	l	Veneu et al. (2013)
S. natans	Cu(II)	3–6	—	L, F, RP	nl	Pagnanelli et al. (2002)
Klebsiella sp.	Mn(II)	5	30	L, F		Hou et al. (2015)
Pseudomonas sp.	Mn(II)	6	25	L, F	nl	Gialamouidis et al. (2010)
Staphylococcus xylosus	Mn(II)	6	25	L, F	nl	Gialamouidis et al. (2010)
Blakeslea trispora	Mn(II)	6	25	L, F	nl	Gialamouidis et al. (2010)
Pseudomonas sp.	Ni(II)	4.5	28	L, F	l	Zhang et al. (2014)
Psychrotrophic Bacillus sp.	Pb(II)	5	15	L, F	l	Ren et al. (2015)
Stenotrophomonas maltophilia	Pb(II)	5.5	30	L, F	l	Wierzba (2015)
B. subtilis	Pb(II)	5.5	30	L, F	l	Wierzba (2015)
Bacillus gibsonii	Pb(II)	4	40	L, F	l	Zhang et al. (2013)
P. aeruginosa	Pb(II)	5	12	L, F	n	Kőnig-Péter et al. (2014b)
R. opacus	Pb(II)	5	15, 25, 35	L, F	l	Bueno et al. (2011)
Streptomyces fradiae	Pb(II)	5	25	L, F	l	Kirova et al. (2015)
Fusarium sp.	Th(IV)	5	25	L, F, T	l	Yang et al. (2015)
Chlamydomonas reinhardtii	U(VI)	4.5	25	L, F, DR, RP	l	Erkaya et al. (2014)
P. aeruginosa	Zn(II)	6	25	L, F	l	Ahmady-Asbchin et al. (2015)
Escherichia coli	Zn(II)	5	22	L, DR	n	Kőnig-Peter et al. (2014a)
Chlorella vulgaris	Zn(II)	5	22	L, DR	n	Kőnig-Peter et al. (2014a)
S. lunalinharesii	Zn(II)	6	25	L, F	l	Veneu et al. (2013)
Aspergillus awamori	Ni(II)	6	25	L, F, DR, RP	l, nl	Shahverdi et al. (2015)

Note: L, Langmuir; F, Freundlich; DR, Dubinin–Radushkevich; RP, Redlich–Peterson; T, Temkin; n, nonlinear; l, linear.

than 1 indicates a strong adsorption of the solvent, penetration of the solute into the adsorbent, strong intermolecular attraction within the adsorbent layers, and monofunctional nature of the adsorbate (Malkoc et al., 2015).

12.2.1.3 Temkin Isotherm

The Temkin isotherm (Temkin, 1940) describes the process of adsorption on heterogeneous solids (nonuniform surface energy). The isotherm corresponds to continuous decay energy of adsorption. Additionally, it assumes the reduction of the heat of adsorption of sorbate molecules because of the adsorbent–adsorbate interactions:

$$q_e = \frac{RT}{b_T} \ln(k_T c_e) \qquad (12.5)$$

where

b_T is the adsorption energy

k_T is the Temkin equilibrium constant

12.2.1.4 Dubinin–Radushkevich Model

This model (Dubinin and Radushkevich, 1947) assumes a heterogeneous surface with a Gaussian energy distribution of binding sites. It has often been applied to distinguish the physical and chemical mechanisms of adsorption based on the magnitude of the mean energy per molecule of adsorbate; a value between 1 and 8 kJ/mol indicates physical adsorption, between 8 and 16 kJ/mol, ion-exchange mechanism, whereas a value within the range of 20–40 kJ/mol indicates chemical adsorption (Sadaf et al., 2014):

$$q_e = q_D \exp\left[-k_D\left(RT\ln\left\{1+\frac{1}{c_e}\right\}\right)^2\right] \quad (12.6)$$

where

q_D is the Dubinin–Radushkevich model uptake capacity
k_D is the model constant

This model is more general than that of Langmuir; it is not based on ideal assumptions (equipotential binding sites, an absence of steric hindrance between adsorbed particles and solutes, surface homogeneity) (Hasan et al., 2012).

12.2.1.5 Sips Isotherm

This is a three-parameter model and it is a combination of the Langmuir and Freundlich isotherms (Sips, 1948). At low sorbate concentrations, the model approaches the Freundlich isotherms; at high sorbate concentrations, the model approaches the monolayer Langmuir isotherm (Equation 12.7):

$$q = q_{max}\frac{K_S c_e^{ns}}{1+K_S c_e^{ns}} \quad (12.7)$$

where K_S and n_S are the Sips constants.

12.2.1.6 Redlich–Peterson Isotherm

At low concentrations of sorbate the Redlich–Peterson isotherm (Redlich and Peterson, 1959) approximates the Henry law and at high concentrations to the Freundlich isotherm. This three-parameter model is suitable to describe the biosorption equilibrium within a wide range of adsorbate concentrations (Equation 12.8):

$$q_e = \frac{K_{RP} c_e}{1+a_{RP} c_e^{b_{RP}}} \quad (12.8)$$

where K_{RP}(L/g), a_{RP} (L/mg), and b_{RP} are the RP parameters. The value of b_{RP} is between 0 and 1.

12.2.2 KINETICS MODELING

Generally, metal ion sorption on a microorganism (adsorbent material) may be performed by the following four stages:

1. *Bulk diffusion*: Migration of metal ions from the bulk to the adsorbent's surface
2. *Film diffusion*: Diffusion of metal ions through the boundary layer to the adsorbent's surface
3. *Pore diffusion/intraparticle diffusion*: Transport of metal ions from the surface into the pores of the molecule
4. *Chemical reaction*: Adsorption of metal ions at an active site on the adsorbent material via complexation, ion exchange, or chelation

Kinetic modeling provides detailed information on adsorption efficiency, uptake rates, and rate-controlling steps of the metal ion sorption. Moreover, kinetic modeling is essential to depict the response of the adsorption system to different experimental conditions, the various properties of biosorbents, and/or the model parametric susceptibility to process parameters (Park et al., 2010). In most instances, the adsorbent's size and morphology, the initial metal ion concentration, the uptake capacity of biosorbent, the mass transfer coefficient, and the metal ion diffusion influence the model outcome. Kinetic modeling can be carried out using chemical reaction–based or particle diffusion–based models (Ho, 2006; Kumar and Gaur, 2011a). It is the pseudo-first-order model and the pseudo-second-order model that are frequently used when chemical reaction–based modeling is analyzed, whereas when particle diffusion–based modeling is concerned the intraparticle diffusion model (Webber and Morris model) is widely employed (Crini and Badot, 2008).

12.2.2.1 Reaction-Sorption Control Models

The most frequently used reaction-sorption control adsorption kinetics are presented in Figure 12.1. The values of pseudo-first-order and pseudo-second-order kinetic model parameters are predominantly determined by the linearization method. A plot of $\ln(q_e - q_t)$ and t/q_t in the function of time t should yield a linear relationship for first-order and second-order reactions, respectively. The kinetic model parameters can be calculated from the slope and intercept of these plots. However, frequently, the nonlinear regression modeling provides a more accurate estimation of kinetic model parameters compared to linear methods.

The reaction-sorption control adsorption kinetics used for biosorption description of metal ions on selected microorganisms are summarized in Table 12.2. As can be observed, the pseudo-first-order model and pseudo-second-order model are mostly applied to estimate the adsorption kinetics. In most cases, the pseudo-second-order model was selected to describe the biosorption kinetic due to a higher model fit accuracy (R^2 value) as compared to coefficients of the determination obtained in the pseudo-first-order modeling. This fact indicated that the second-order mechanism can be predominant and that chemisorption may be the rate-limiting step controlling the biosorption process of metal ions on selected microorganisms. Still, several researchers observed that both models significantly explain the binding of metal ions on the adsorbent surface. The suitability of both models for the description of kinetic adsorption indicated that both physical adsorption and chemisorption are involved (Mishra et al., 2013). As can be seen in Table 12.2, only several researchers investigated additional reaction-sorption kinetics models. Engin et al. (2012) performed kinetic studies of Cu(II) biosorption on *Spirulina platensis* and dried activated sludge. Investigators applied three kinetic models to reach the experimental result: pseudo-first order, pseudo-second order, and the Elovich model (Elovich, 1959). The Elovich model can be expressed as follows:

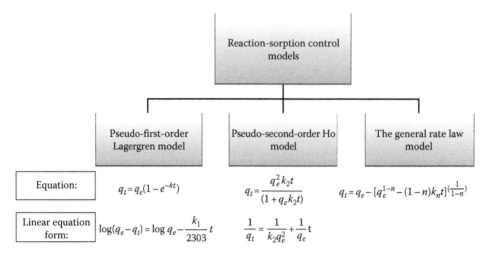

q_e, q_t: the amounts of metal ion retained on the weight unit of biosorbent at equilibrium and at time t;

k_1, k_2, k_n: the rate constant of pseudo-first order, pseudo-second order, and n-rate kinetics equation

FIGURE 12.1 Reaction-sorption control adsorption kinetics and their linear forms. (Modified from Ho, Y.S., *J. Hazard. Mater.*, B136, 681, 2006; Crini, G. and Badot, P.-M., *Prog. Polym. Sci.*, 33, 399, 2008.)

$$\frac{dq_t}{dt} = \propto \exp(-\beta q_t) \qquad (12.9)$$

or in integrated form as (for $\alpha\beta t > t$, and boundary conditions $t = 0$ to $t = t$, $q_t = 0$ to $q_t = q_e$)

$$q_t = \frac{1}{\beta}\ln(\alpha\beta) + \frac{1}{\beta}\ln(t) \qquad (12.10)$$

where

q_t is the initial adsorption rate
α and β are the desorption constants

A plot of q_t in function of $\ln(t)$ should be a straight line, and model parameters can be estimated from a slope $(1/\beta)$ and an intercept $(1/\beta\ln(\alpha\beta))$ (Engin et al., 2012). In general the Elovich kinetic model is applicable to chemisorption kinetics. This approach is valid for systems in which the adsorbent surface is heterogeneous.

Kumar and Gaur (2011) explored the suitability of chemical reaction–based kinetic models (pseudo-first-order, pseudo-second-order, and the general rate law model) for defining the biosorption of Cu(II), Pb(II), and Cd(II) by *Phormidium* sp., which dominates in cyanobacterial mats. All three reaction-sorption-based kinetic models fitted significantly the biosorption experimental results. Yet, researchers concluded that these models do not accurately depict the kinetics of metal ion adsorption by microorganisms, owing to the fact that an unjustifiable change in the rate constant or reaction order with different concentrations of metal ions and biomass in the solution can occur/can have occurred.

12.2.2.2 Diffusion-Sorption Control Models

As it was mentioned, the sorption of any adsorbate molecule on the adsorbent particle involves four main stages, from which the external mass transfer and intraparticle diffusion play a predominant role in the overall mass transport. Intraparticle diffusion has often been applied to investigate the internal diffusion mechanism of the biosorption of metal ions onto microorganism. As it can be seen in Table 12.3, the most frequently used is the Weber–Morris kinetic model (Weber and Morris, 1963).

The Weber–Morris model is derived from the simplification of Crank's solutions (Crank, 1970) and can be represented as follows:

$$q_t = k_i t^{0.5} + I \qquad (12.11)$$

where

k_i is the intraparticle diffusion rate constant
I is the intercept

The k_i can be determined from the initial linear slope of the plotted function q_t versus $(t)^{0.5}$. According to the Weber–Morris model if the intraparticle diffusion is a predominant step, a linear plot q_t versus $(t)^{0.5}$ should pass through the origin, or else some additional mechanism can occur, in parallel to intraparticle diffusion (Li et al., 2011).

In most analyzed literature reports, Kumar and Gaur (2011a), Li et al. (2011), and Mishra et al. (2013), the graph of q_t versus $(t)^{0.5}$ had three linear regions, which indicates that more than one mechanism affects the sorption phenomena onto cell. The first part of the curve represents the diffusion of metal ions from the bulk up to the boundary layer—external mass transfer. The second region is the intraparticle diffusion, and the third part corresponds to the saturation of the sorption process, which is the final equilibrium step, where intraparticle diffusion slowly decreases. The first stage is mostly the most rapid. The second part of the slope is the intraparticle diffusion rate constant (k_i), whereas intercept (I) evaluates the

TABLE 12.2

Kinetic Models of Metal Ions Biosorption by Selected Microorganisms in Reaction-Sorption Control Models

Microorganism	Metal Ion	Metal Ion Concentration (mg/L)	Kinetic Model	R^2	Reference
Rhizopus oligosporus	Ni(II)	20–100	Pseudo-second order	0.9909–0.9998	Ozsoy and van Leeuwen (2011)
Bacillus, Pseudomonas, Klebsiella, and *Escherichia*	Ni(II)	65.39 g/mol	Pseudo-first order	0.2410 0.4794	Arjomandzadegan et al. (2014)
	Zn(II)	58.69 g/mol	Pseudo-second order	0.9648 0.9992	
Chlorella vulgaris	Pb(II)	100	Pseudo-second order	1	El-Naas et al. (2007)
Spirulina platensis	Cu(II)	40,80,160	Pseudo-first order	0.934–0.967	Engin et al. (2012)
			Pseudo-second order	0.997–0.999	
			Elovich model	0.954–0.967	
Dried activated sludge	Cu(II)	40,80,160	Pseudo-first order	0.916–0.933	Engin et al. (2012)
			Pseudo-second order	0.998–0.999	
			Elovich model	0.95–0.983	
Bacillaceae bacteria	Co(II)		Pseudo-second order	0.997	Fosso-Kankeu et al. (2011)
	Ni(II)			0.9992	
	Mg(II)			0.9853	
	Ca(II)			0.9955	
Brevundimonas spp.	Co(II)			0.3508	
	Ni(II)			0.8734	
	Mg(II)			0.8108	
	Ca(II)			0.8434	
P. aeruginosa	Co(II)			0.9991	
	Ni(II)			0.9998	
	Mg(II)			0.9937	
	Ca(II)			0.9136	
Bacillus cereus	Cd(II)	50	Pseudo-second order	0.976	Huang et al. (2013)
Bacillus laterosporus	Cd(II)		Pseudo-first order	0.9275 0.9469	Kulkarni et al. (2014)
	Ni(II)		Pseudo-second order	0.9999 0.9989	
Immobilized microorganisms (B350)- alginate entrapment	Pb(II)	100	Pseudo-first order	0.623	Li et al. (2011)
			Pseudo-second order	0.999	
NaOH-treated immobilized microorganisms (B350)- alginate entrapment	Pb(II)	100	Pseudo-first order	0.830	Li et al. (2011)
			Pseudo-second order	0.999	
Pseudomonas taiwanensis	Cr(VI)	20–200	Zero-order kinetic	0.934–0.999	Majumder et al. (2014)
			Three-half-order kinetic model	0.959–0.999	
Zinc sequestering bacterium VMSDCM	Zn(II)		Pseudo-first order	0.982–0.990	Mishra (2015)
			Pseudo-second order	0.979–0.996	
Rhizoplane bacterial isolates (*P. putida* and *B. cereus*)	Cu(II)	500	Pseudo-first order	0.965–0.949	Padmapriya and Murugesan (2015)
			Pseudo-second order	1–0.999	
Lysinibacillus sp. BA2 (dead)	Ni(II)	300	Pseudo-first order	0.972	Prithviraj et al. (2014)
			Pseudo-second order	0.982	
Lysinibacillus sp. BA2 (live)	Ni(II)	300	Pseudo-first order	0.99	Prithviraj et al. (2014)
			Pseudo-second order	0.966	
Living B350 biomass	Pb(II) Zn(II)		Pseudo-first order	0.811 0.802	Qu et al. (2014)
			Pseudo-second order	0.998 0.999	

(Continued)

TABLE 12.2 (*Continued*)

Kinetic Models of Metal Ions Biosorption by Selected Microorganisms in Reaction-Sorption Control Models

Microorganism	Metal Ion	Metal Ion Concentration (mg/L)	Kinetic Model	R^2	Reference
Chlorella pyrenoidosa	Zn(II)	10,20,30,100	Pseudo-second order	0.957–0.979	Rezaei et al. (2012)
Ceratocystis paradoxa MSR2	Cr(VI)	25–125	Pseudo-first order	0.97–0.998	Samuel et al. (2015)
			Pseudo-second order	0.967–0.991	
			Elovich model	0.963–0.990	
Dry Baker's Yeast	Cd(II)	25–150	Pseudo-second order	1	Stanescu et al. (2015)
Acinetobacter haemolyticus	Cr(III)	50–100	Pseudo-first order	0.007–0.058	Yahya et al. (2012)
			Pseudo-second order	0.981–0.996	
Phormidium sp.	Pb(II)	0.1–1 mM	Pseudo-first order	0.961–0.994	Kumar and Gaur (2011a)
	Cu(II)				
	Cd(II)				
				0.932–0.997	
				0.977–0.995	
			Pseudo-second order	0.981–0.993	
				0.953–0.999	
				0.991–0.998	
			General order rate law	0.986–0.994	
				0.965–0.999	
				0.995–0.999	

TABLE 12.3

Kinetic Models of Metal Ions Biosorption by Selected Microorganisms in Diffusion-Sorption Control Models

Microorganism	Metal Ion	Kinetic Model	Reference
Chlorella vulgaris	Pb(II)	Weber–Morris	El-Naas et al. (2007)
Immobilized microorganisms (B350)	Pb(II)	Weber–Morris	Li et al. (2011)
Aeromonas caviae	Cd(II)	Intraparticle diffusion model; external mass transfer model	Loukidou et al. (2004)
Zinc sequestering bacterium VMSDCM	Zn(II)	Intraparticle diffusion model; Bengham's film diffusion	Mishra (2015)
Ceratocystis paradoxa MSR2	Cr(VI)	Intraparticle diffusion model	Samuel et al. (2015)
Phormidium sp.	Pb(II)	Weber–Morris	Kumar and Gaur (2011a)
	Cu(II)	External mass transfer model	
	Cd(II)	Boyd model	
Penicillium camemberti	Mn(II)	Weber–Morris	Khalilnezhad et al. (2014)

thickness of the boundary layer. When intercept (*I*) is larger than 0, the intraparticle diffusion is not at the rate-limiting stage (Gupta et al., 2009).

Kumar and Gaur (2011a) proposed an external mass transfer model, which is represented by (Equation 12.12)

$$\left[\frac{d\left(C_t/C_0\right)}{dt} \right]_{t=0} = -k_{ES} \qquad (12.12)$$

where k_{ES} is the rate constant of the external mass transfer model. The external mass transfer constant can be calculated by the slope of the plot of C_t/C_0 versus t. Researchers compared the external mass transfer constant values, determined for the sorption of Pb(II), Cu(II), and Cd(II) metal ions onto *Phormidium* sp. The k_{ES} of Pb(II) ions decreased with rising Pb(II) and biomass concentration in the system, which implies that high Pb(II) concentrations diminish the external mass transfer resistance to the biosorption of Pb(II) ions. The external mass transfer rate constant of Cu(II) and Cd(II) ions did not demonstrate specific trends when the metal or biomass concentration in the solution increased. Kumar and Gaur (2011a) suggested that this phenomena might be associated with the smaller ionic size of Cu(II) (73 pm) and Cd(II) (109 pm) compared to the Pb(II) (120 pm) ionic size.

12.2.3 THERMODYNAMIC MODELING

As it was presented in the previous section, adsorption isotherms and kinetic studies are the crucial features involved in the modeling of adsorption processes. Thermochemistry is yet another feature made use of in the investigation of metal ions adsorption phenomena. Temperature is an essential parameter that affects not only the diffusion rate of an adsorbate like a metal ion to the adsorbent but also the solubility of the adsorbate. An increase in temperature is also followed by a rise in the diffusivity of the metal ions, and, therefore, by extension, in the adsorption rate if diffusion is the rate-limiting step (Crini and Badot, 2008). The fundamental thermodynamic parameters such as the Gibbs free energy change (ΔG), the enthalpy change (ΔH), and the entropy change (ΔS) express the adsorption characteristics of different materials and can be evaluated by employing the thermodynamic equilibrium coefficient gained at different adsorption conditions (temperature and concentrations). These parameters provide information about possibilities of adsorption mechanisms. The equations of basic thermodynamic expressions are presented in Figure 12.2.

Thermodynamics postulates that the entropy change is the driving force in an isolated system. To consider the spontaneity of the process, the Gibbs free energy change (ΔG) value must be evaluated. The ΔG parameter is the fundamental criterion for spontaneity, and its negative value signifying the spontaneity of the process reaction. The Gibbs free energy change can be estimated according to the Gibbs equation. In the biosorption process ΔG can be calculated when the adsorption is reversible and the equilibrium adsorption data are determined. The enthalpy change (ΔH) and the entropy

change (ΔS) can be estimated by van't Hoff plot (as a function of $1/T$) from the slope and intercept of the plot, respectively. Supposing the ΔS and ΔG have negative values, the enthalpy change value needs to be negative too. The negative value of ΔH indicates that adsorption is exothermic. Ordinarily, when ΔH range from -4 to -40 kJ/mol, the physical adsorption on solids occurs, whereas in the range from -40 to -800 kJ/mol, the chemical adsorption does (Crini and Badot, 2008).

Table 12.4 presents the results of the thermodynamic modeling of the metal ions biosorption by selected microorganisms. Engin et al. (2012), Kulkarni et al. (2014), Li et al. (2011), and Rezaei et al. (2012) examined the thermodynamics of metal ions biosorption by *S. platensis*, *Bacillus laterosporus*, immobilized strains (B350), and *Chlorella pyrenoidosa*, respectively. They reported that the value of ΔG was negative, indicating a spontaneous nature of metal ion adsorption. Engin et al. (2012) observed that when process temperatures increase, the feasibility of biosorption decreases, which is discrepant with the results obtained of Kulkarni et al. (2014), Li et al. (2011), and Rezaei et al. (2012). Researchers observed that a rise in the process temperatures resulted in a decrease in ΔG values. Engin et al. (2012) estimated the negative value of ΔH and ΔS, which implies the exothermic nature of and a decrease in the randomness at the solid/solution interface

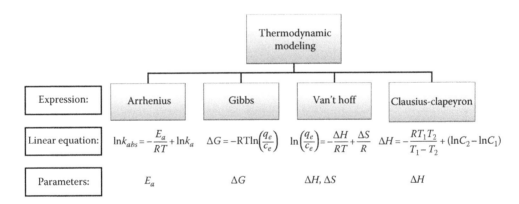

FIGURE 12.2 Basic thermodynamic expressions. (Modified from Crini, G. and Badot, P.-M., *Prog. Polym. Sci.*, 33, 399, 2008.)

TABLE 12.4

Thermodynamic Modeling of Metal Ions Biosorption by Selected Microorganisms

Microorganism	Metal Ion	T (°C)	ΔG (kJ/mol)	ΔH (kJ/mol)	ΔS (J/kmol)	Reference
Spirulina platensis	Cu(II)	25, 30, 35, 40	−3.07, −2.41, −1.75, −1.09	−42.42	−0.13	Engin et al. (2012)
Dried activated sludge	Cu(II)	25, 30, 35, 40	−4.48, −4.24, −3.99, −3.74	−19.44	−0.05	Engin et al. (2012)
Bacillus laterosporus	Cd(II)	20, 30, 40	−17.193, −17.956, −18.739	5.45	77.27	Kulkarni et al. (2014)
	Ni(II)		−10.343, −11.177, −12.719	24.33	117.9	
Immobilized microorganisms (B350)-alginate entrapment	Pb(II)	25, 30, 35, 40, 45	−3.22, −3.56, −4.42, −4.91, −5.73	34.64	126.54	Li et al. (2011)
NaOH-treated immobilized microorganisms (B350)-alginate entrapment	Pb(II)	25, 30, 35, 40, 45	−4.47, −4.75, −5.05, −6.25, −6.72	31.25	119.16	Li et al. (2011)
Chlorella pyrenoidosa	Zn(II)	10, 20, 25, 30, 35, 40	−13.61, −37.39, −51.84, −68.01, −84.74, −98.31	3.34	13.24	Rezaei et al. (2012)

during biosorption, respectively. The positive values of ΔH and ΔS demonstrated both the endothermic nature of biosorption and the increased randomness at the biosorbent metal solution interface during the adsorption of heavy metal ions.

12.2.4 Fixed-Bed Column Modeling

The modeling of a continuous biosorption process is crucial for estimating the biosorptive performance of a semitechnical or industrial process, based on laboratory-scale results. Mathematical modeling gives a potentiality of analyzing the experimental data and predicting the system response to initial process conditions. In laboratory-scale experiments the fixed-bed column is widely used for the continuous modeling of biosorption (Park et al., 2010). The efficiency of fixed-bed column operations is defined by the breakthrough concentration curve of a metal ion, which is generally *S shaped*. Many parameters influence the biosorption performance of the column, which include the initial metal ion concentration, flow rate, biomass dosage, fixed-bed height, length of sorption zone, or operating time. An appropriate biosorbent is characterized by a high uptake and removal performance, steep breakthrough concentration curve, earlier exhaustion, and brief mass transfer zone (Vijayaraghavan and Yun, 2008).

The breakthrough concentration curves of the fixed-bed column can be modeled by mathematical as well as statistical approaches (artificial neural network [ANN], response surface methodology [RSM]); however, the mathematical modeling is more often applied. Diverse mathematical models have been expanded to depict the metal ions' biosorption in fixed-bed column systems. Among these, the Bohart–Adams, Thomson, Yoon–Nelson, or Yan models are the most extensively adapted (Witek-Krowiak et al., 2013).

Table 12.5 presents mathematical models used in modeling continuous metal ions biosorption onto microorganisms. As can be seen, Thomas and Bohart–Adams models are among the most popular.

The Thomas model assumes the Langmuir kinetics of adsorption–desorption as well as no axial dispersion (Saadi et al., 2013). The equation of the Thomas model for the adsorption column has the following form (Thomas, 1944):

$$\frac{C_t}{C_0} = \frac{1}{1 + \exp\left[((K_{Th}q_e x)/Q) - K_{Th}C_0 t\right]} \quad (12.13)$$

where

K_{Th} is the Thomas model constant
q_e is the determined adsorption uptake
x is the mass of adsorbent
Q is the initial flow rate
C_0 is the initial metal ion concentration
C_t is the effluent solution concentration

The Thomas model can also be expressed in the linear form as follows:

$$\ln\left(\frac{C_0}{C_t} - 1\right) = \frac{K_{Th}q_e x}{Q} - K_{Th}C_0 t \quad (12.14)$$

TABLE 12.5
Modeling of Continuous Biosorption Process (Fixed-Bed Column Modeling)

Microorganism	Metal Ion	Column Parameters	R^2	Model	Reference
Paenibacillus polymyxa	Cu(II)	Flow rate: 0.5–1.5 mL/min	0.989–0.989	Thomas model	Çolak et al. (2013)
	Ni(II)		0.982–0.987		
P. polymyxa	Cu(II)	C_0: 100–200 mg/L	0.989–0.990	Thomas model	Çolak et al. (2013)
	Ni(II)		0.982–0.992		
Saccharomyces cerevisiae	Cd(II)	Flow rate: 0.5–4 mL/min	0.9524–0.996	Thomas model	Galedar and Younesi (2013)
	Ni(II)		0.9472–0.9976	Yan model	
	Co(II)			Breakthrough analysis	
Lyngbya putealis HH-15	Cr(VI)	Flow rate: 1 mL/min	0.9494–0.9979	Adam-Bohart model	Kiran and Kaushik (2008)
		C_0: 10 mg/L			
(PAA/HCl)-modified *Escherichia coli*	Pt(II)	Flow rate: 2 mL/min	0.998	Thomas model	Mao et al. (2015)
		C_0: 100 mg/L			
Aspergillus niger	Cu(II)	Flow rate: 1.6–9.8 mL/min		The fixed-bed model	Mukhopadhyay et al. (2011)
				Reduced lumped diffusion model	
Brevundimonas vesicularis	Pb(II)	Flow rate: 1.5 mL/min	0.9893	Thomas model	Resmi et al. (2010)
		C_0: 100 mg/L			
Cystoseira indica	Th(II)	C_0: 10–90 mg/L	0.98–0.99	Mass transfer model	Riazi et al. (2014)
			0.98–0.99	Clark model	
Penicillium	Cu(II)	C_0, bed height, operating time, flow rate, initial pH		Plackett–Burman design	Xiao et al. (2013)
				Box–Behnken design	

The Bohart–Adams model is commonly used for describing the initial stage of the breakthrough curve. This model assumes that the rate of sorption is proportional to the concentration of the sorbate and the residual capacity of the sorbent (Saadi et al., 2013). The expression of the Bohart–Adams model is as follows:

$$\frac{C_t}{C_0} = \exp\left(K_{AB}C_0t - K_{AB}N_0 \frac{h}{U_0} \right) \quad (12.15)$$

where

K_{AB} (L/mg min) is the biosorption rate constant
N_0 (mg/L) is the max ion adsorption capacity per unit volume of the column
U_0 (cm/min) is the linear velocity of influent solution
h (cm) is the bed depth

The parameters k_{AB} and N_0 can be determined from the intercept and slope of the linear plot of $\ln(C_t/C_0)$ versus t.

In the fixed-bed column system, the actual sorption capacity (q_0) of the adsorbent can be calculated by Equation 12.16 from the experimental data of the curve (Witek-Krowiak et al., 2013):

$$q_0 = \frac{C_0Q}{1000x} \int_0^t \left(1 - \frac{C_{out}}{C_0} \right) dt \quad (12.16)$$

where C_{out} is solution concentration in the effluent (mg/L).

The maximum adsorption capacity in the column can be expressed as follows:

$$q_e = \frac{Q}{1000x} \int_0^t (C_0 - C_t) dt \quad (12.17)$$

$$q_{e\max} = \frac{q_{total}}{x} \quad (12.18)$$

where $q_{e\max}$ is the maximum adsorption capacity of the sorbent (mg/g), $t = t_{total}$.

12.3 RECENT TRENDS IN MICROBIAL BIOSORPTION MODELING

12.3.1 EQUILIBRIUM MODELING

12.3.1.1 Multicomponent Modeling

The optimal design of the processes based on biosorption phenomena requires accurate adsorption equilibrium data for mono- and multicomponent systems. These data can be obtained either by conducting a series of experiments or by modeling the adsorption equilibrium. So far, the biosorption of metal ions has been usually studied in single-metal systems. Typically, various metal ions coexist in actual metal wastewater with different concentrations and forms. Thus, in order to consider biosorption as a useful process to treat polluted industrial waters, the mechanism of competitive adsorption types in a multimetal system/in multimetal systems should be investigated. The knowledge about how a single-metal ion competes with that of another to occupy binding sites may help design the biosorption-based process on an industrial scale.

In a multicomponent system, not only do the biomass surface and physicochemical properties of the solution (pH, temperature, initial ion concentration, biomass dosage, ionic strength) have an influence on the biosorption of the particular metal but also the number of metal ions that compete for the active sites on the biosorbent surface (Fagundes-Klen et al., 2007). Therefore, it is important to determine the ions selectivity in the solution by the biosorbent material. According to the literature review (Table 12.6), the multicomponent Langmuir model was the most frequently applied model fitted to the multicomponent biosorption data. The mathematical equation of the Langmuir isotherm for a

TABLE 12.6
The Literature Review on the Biosorption of Multimetal Systems by Microorganisms

Biosorbent Type	Metal Ions	Modeling	Models	Reference
Arthrospira (Spirulina) platensis	Ni (II), Zn(II), Pb(II)	Equilibrium	Multicomponent Langmuir and Freundlich isotherms	Rodrigues et al. (2012)
Wine-processing waste sludge	Cr(III), Ni(II), Pb(II)	Kinetics	Competitive kinetic sorption	Liu et al. (2009)
Rhizopus arrhizus	Cu(II), Pb(II)	Equilibrium	Multicomponent Langmuir and Freundlich isotherms	Alimohamadi et al. (2005)
Staphylococcus xylosus and *Pseudomonas* sp.	Cr(VI), Cd(II)	Equilibrium	Multicompetitive isotherms	Ziagova et al. (2007)
Aspergillus niger	Cd(II), Zn(II), Co(II)	Equilibrium	Langmuir, Freundlich, Temkin and Sips multicomponent competitive models	Hajahmadi et al. (2015)
Oscillatoria angustissima	Cu(II), Zn(II), Co(II)	Effect of pH	Response surface methodology	Mohapatra and Gupta (2005)
Activated sludge	Cd(II), Zn(II)	Effect of parameters	Response surface methodology	Remenárová et al. (2012)

multimetal system can be expressed as follows (Rodrigues et al., 2012):

$$q_{L,i} = \frac{q_{\max,L,i} K_{L,i} c_{e,i}}{1 + \sum_{j=1}^{n} K_{L,j} c_{e,j}} \tag{12.19}$$

where

n is the component number ($n = 2$ for binary systems)
$q_{L,i}$ is the sorption capacity at equilibrium for sorbate "i"
$c_{e,i}$ and $c_{e,j}$ are the equilibrium concentrations of sorbates "i" and "j"

The theoretical maximum sorption capacity for sorbate "i" is $q_{\max,i}$, while $K_{L,i}$ and $K_{L,j}$ (L/mg) are the Langmuir equilibrium constants for sorbates "i" and "j." The model is developed on the basis of single-component biosorption data, and it assumes the inhibitory effect on the biosorption of multimetal ions.

The second most widely used model is the Freundlich binary system equation proposed by Sheindorf et al. (1981), which assumes that each component individually executes the Freundlich isotherm, and can be given as (Rodrigues et al., 2012)

$$q_{eq,i} = K_{F,i} c_{e,i} \left(\sum_{j=1}^{n} a_{ij} c_{e,j} \right)^{\frac{1}{n_i - 1}} \tag{12.20}$$

where

the Freundlich constants $K_{F,i}$ and n_i are obtained from single-metal equilibrium experiments
a_{ij} is the competition coefficient in the multicomponent system

Luna et al. (2010) investigated the competitive biosorption of Cd(II) and Zn(II) ions onto *Sargassum filipendula* from single-component and binary systems. They tested seven isotherm models for modeling the binary system including not only the most popular competitive Langmuir and Freundlich models but also uncompetitive and partial-competitive Langmuir isotherms as well as the extended Freundlich isotherm and the Langmuir–Freundlich isotherm for a binary system. The uncompetitive model assumes that the two sorbates could be simultaneously attached to the same binding site. The model for the binary system can be expressed as follows (Apiratikul and Pavasant, 2006):

$$q_{L,i} = q_{\max,L} \frac{K_{L,i} c_{e,i} + K_{L,ij} c_{e,i} c_{e,j}}{1 + K_{L,i} c_{e,i} + K_{L,j} c_{e,j} + K_{L,ij} c_{e,i} c_{e,j}} \tag{12.21}$$

where all parameters are obtained from multicomponent sorption data.

Although *S. filipendula* is macroalgae, the results prove that the model can be applied for any biosorption system.

The partial competitive isotherm model assumes that one metal could be taken up by only one binding site and also by the already occupied binding sites (Equation 12.22):

$$q_{L,i} = q_{\max,L} \frac{K_{L,i} c_{e,i} + K_{L,i} K_{L,ij} c_{e,i} c_{e,j}}{1 + K_{L,i} c_{e,i} + K_{L,j} c_{e,j} + (K_{L,ij} K_{L,i} + K_{L,ji} K_{L,j}) c_{e,i} c_{e,j}} \tag{12.22}$$

where all parameters are obtained from multicomponent sorption data.

Apart from the Langmuir and Freundlich multimetal models, Hajahmadi et al. (2015) fitted also the Temkin and Sips isotherm to the equilibrium data of the multicompetitive biosorption of Zn(II), Co(II), and Cd(II) from ternary mixture onto pretreated dried *Aspergillus niger* biomass. The Temkin and Sips models for multimetal systems are similar to those for single-component ones and are expressed as follows:

$$q_{eq,i} = b_{T,i} \sum_{i=1}^{n} \ln(K_{T,i} c_{e,i}) \tag{12.23}$$

$$q_{S,i} = \frac{K_s c_e^{n_s}}{1 + \sum_{i=1}^{n} a_{s,i} c_{e,i}^{n_s}} \tag{12.24}$$

where

$b_{T,i}$ is the Temkin constant
$K_{T,i}$ is the Temkin isotherm constant at equilibrium for sorbate "i," while $a_{s,i}$, $K_{s,i}$, and $n_{s,i}$ are the Sips parameters at equilibrium for sorbate "i."

Table 12.6 presents the literature review on the microbial biosorption of multimetal systems. As mentioned earlier, the most frequently used isotherm models are those of Langmuir and Freundlich with their extensions and modifications. The growing interest in investigating multimetal systems is associated with the developing research on biosorption with microorganisms and their application in actual industrial systems on a large scale. As shown in Table 12.6, researchers also investigated the effect of operating parameters such as pH and temperature on the removal of metal ions from their multicomponent solutions. Determining the influence of various conditions on the biosorption of multicomponent systems is crucial due to the fact that metal ions can exist in various forms and concentrations depending on the physicochemical properties of the solution, while the quantity and quality of binding sites on the microbial biomass can change with changing external conditions. Remenárová et al. (2012) applied the response surface methodology for the investigation of interactions and competitive effects in a binary metal system Cd(II)–Zn(II) sorbed by activated sludge. The researchers planned their experiments on the basis of the Box–Behnken experiment design, while the behavior of the binary sorption system was explained empirically by the second-order polynomial model. The results of their investigation revealed strong interactions between the initial concentration of coion, solution pH, and sorption capacity of the primary ion, indicating that the microbial biosorption is highly dependent on the concentration of competitive metal ions and pH as one of the operating process parameters. A similar conclusion was reached by Mohapatra and Gupta (2005), who reported the biosorption of Zn(II), Cu(II), and Co(II) onto *Oscillatoria angustissima* biomass from binary and ternary metal solutions, as a function of pH and metal concentrations via RSM using the central composite design (CCD). They revealed that lower concentrations of the competing metal ions did not affect Cu(II), which proves that in some cases the presence of cocations can be neglected.

The examination of the effects of multimetal systems in various combinations is more representative of the actual environmental problems, since real wastewaters contain more than one heavy metal ion. Further research on the microbial biosorption in multicomponent systems may lead to the design and the development of highly selective biosorbents for the treatment of actual industrial wastewaters polluted with toxic metal ions.

12.3.1.2 Mechanistic Models

The complexity of the biosorption by microorganisms highlights the importance of developing mechanistic models that are worked out on the basis of microbial sorption mechanism. The current progress in mechanistic modeling can be explained by the developing biosorption characterization techniques (scanning electron microscopy, Fourier transform infrared spectroscopy) that help in assessing the biosorption behavior in a mechanistic way. Metal sequestration by biomass is often attained by one or more mechanisms that involve a passive metal ion binding (biosorption) on the cell surface or in the pores or an active intracellular uptake (bioaccumulation) (Paul et al., 2012). Additionally, biosorption itself can be divided into several submechanisms including ion exchange, microprecipitation, and complexation. A better understanding of the interactions between active sites and ionic species in the solution may be achieved by mechanistic modeling. Mechanistic models are advanced tools for biosorption study that can be used to screen the physical and chemical aspects involved in metal ions biosorption (Pagnanelli et al., 2004). These models are not usually simple mathematical expressions derived from experimental data, but they present a series of equations representing physicochemical reactions between biosorbent surface active sites and metallic ions. Biosorption

is influenced by a few operating parameters. One of the most important factors affecting biosorption is pH due to its high impact on the form and acid–base properties of metal species in the solution.

Many microorganisms have been shown to possess unique properties on their cell walls that allow them to adsorb and retain metal ions. Reactive functional groups on the microbial surface have been found to be responsible for metal ions uptake. Adsorption of metallic ions onto the cell wall may be due to electrostatic (related to the charge of the adsorbent and metal species) or nonelectrostatic forces (Ngwenya and Chirwa, 2015).

The modeling of microbial biosorption by mechanism-based models can be divided into three main stages consisting of several substeps (Figure 12.3). At first, the biosorbent before and after adsorption should be properly characterized in terms of functional groups (by means of FT-IR) and microelemental content on the surface (SEM-EDX), the surface morphology (SEM), and the point of zero charge. The detailed biosorbent characterization mentioned earlier provides a knowledge about the physical, chemical, and electrical properties of the microbial cell wall, and it is the first step toward the biosorption mechanism evaluation. The biggest challenge is the development of a model of adequate fit to capture all the mechanisms that control microbial biosorption, which is implemented in the third stage of the modeling. Due to the fact that mechanistic models may contain a large number of parameters, it is vital to construct the model in such a way that it remains computationally viable and solves all the issues. In this step, a number of experimental tests are also performed in order to obtain values of regression parameters of the mechanistic model (Pagnanelli, 2011). The last step involves parameter estimation and the obtained model validation.

The mechanistic model is usually presented by a series of reactions occurring between the biosorbent surface and metallic ions in the solution depending on the mechanism of the metallic species uptake by biomass. Esposito et al. (2002) developed a pH-related noncompetitive model that was used to simulate the biosorption behavior of Cu(II) and Cd(II) on *Sphaerotilus natans* as a pH function. The model assumes (1) a noncompetitive mechanism of interaction for heavy metal and hydrogen ions in solution; (2) one type of binding site on

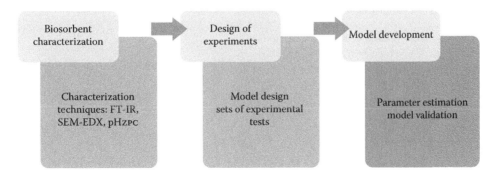

FIGURE 12.3 The mechanistic modeling procedure for microbial biosorption.

the cell wall; and neglects (3) the ion charge on the microbial surface. The series of the reactions representing the model are as follows:

$$S + H \rightarrow SH \tag{12.25}$$

$$S + M \rightarrow SM \tag{12.26}$$

$$SH + M \rightarrow SHM \tag{12.27}$$

where

S is the biosorption site with weakly acidic properties
M is the heavy metal in solution

$$\frac{[SM]}{[S]} = \frac{[SHM]}{[SH]} \tag{12.28}$$

The following mechanistic model can be derived from the relative equilibrium constant obtained on the basis of each reaction equation when combined with the noncompetitive hypothesis (Equation 12.28):

$$q = \frac{S_{TOT}c_e}{((1 + 10^{-pH})/K_1)(K_M + c_e)} \tag{12.29}$$

where

S_{TOT} is the total of active sites
K_M is the metal ion-active site equilibrium constant
K_1 is the inverse of proton-active site equilibrium constant

Recently, increasing attention has been paid to the modeling of the effect of pH on the heavy metal ions biosorption equilibrium (Pagnanelli et al., 2004, 2013; Chojnacka et al., 2005; Kumar and Gaur, 2011b).

Also, the ion-exchange effect on biosorption mechanistic modeling has been investigated. Schiewer (1999) proposed a model that takes into account both the effect of proton binding and ion exchange. It assumes two kinds of binding sites, carboxyl (C) and sulfate (S) groups, to be present on the biomass cell wall. The reactions for the binding of protons and metal ions to these sites are

$$-C^- + H^+ \leftrightarrow -CH \tag{12.30}$$

$$-S^- + H^+ \rightarrow -SH \tag{12.31}$$

$$2-C^- + M^{2+} \leftrightarrow 2-CM_{0.5} \tag{12.32}$$

$$-2S^- + M^{2+} \leftrightarrow 2-SM_{0.5} \tag{12.33}$$

From the equilibrium equation obtained for the reactions (Equations 12.30 through 12.33), Schiewer and Volesky (1995) derived an isotherm equation predicting the binding of protons and metal ions' as a function of metal ion concentration and pH:

$$q_H = CH + SH = C_t \frac{K_{CH}[H]}{1 + K_{CH}\left[H\right] + (K_{CM}[M])^{0.5}}$$

$$+ S_t \frac{K_{SH}[H]}{1 + K_{SH}\left[H\right] + (K_{SM}[M])^{0.5}} \tag{12.34}$$

$$q_M = CM_{0.5} + SM_{0.5} = C_t \frac{K_{CM}[M]^{0.5}}{1 + K_{CH}\left[H\right] + (K_{CM}[M])^{0.5}}$$

$$+ S_t \frac{K_{SM}[M]^{0.5}}{1 + K_{SH}\left[H\right] + (K_{SM}[M])^{0.5}} \tag{12.35}$$

where

C_t and S_t are the total binding sites (free [ionized] or occupied by protons or metal ions)
K_{CM} and K_{SM} are the metal ion-active site equilibrium constants
K_{CH} and K_{SH} are the proton-active site equilibrium constants

Plazinski and Rudzinski (2010) presented a pH-dependent model successfully applied for heavy metal ions biosorption by algal biomass. It includes not only the effect of pH on the adsorption equilibrium but also the reaction of protonation/deprotonation of acidic binding sites, the hydrolysis of metal ions M^{2+} to obtain $M(OH)^+$, the binding of metallic ions, and the formation of the surface complex with the hydrolyzed metal ion.

Microbial cell walls are mostly composed of polysaccharides, proteins, and lipids that consist of surface organic functional groups (active sites) such as amino, carboxylic, phosphate, and hydroxyl. The surface complexation theory postulates that the sorption be described in terms of chemical reactions between these surface functional groups and metallic species in the solution. So far, surface complexation models have effectively been applied to bind protons and metal ions onto various bacterial surfaces, including *Pseudomonas pseudoalcaligenes* (Liu et al., 2013) and *Bacillus subtilis* (Markai et al., 2003; Mishra et al., 2010; Moon and Peacock, 2011).

Markai et al. (2003) described the acid–base properties of carboxylic, phosphate, and hydroxyl surface functional groups (Equations 12.36 through 12.41) and the corresponding equilibrium equations as well as the interaction reactions between the 3+ charged metal ion and the deprotonated surface sites on the bacterial cell wall (Equations 12.42 through 12.47):

$$RCOOH \leftrightarrow RCOO^- + H^+ \tag{12.36}$$

$$K_1 = \frac{\left[RCOO^-\right]a_{H^+}}{[RCOOH]} \tag{12.37}$$

$$RPO_4H \leftrightarrow RPO_4^- + H^+ \tag{12.38}$$

$$K_2 = \frac{\left[RPO_4^-\right]a_{H^+}}{[RPO_4H]} \tag{12.39}$$

$$ROH \leftrightarrow RO^- + H^+ \tag{12.40}$$

$$K_3 = \frac{\left[RO^-\right]a_{H^+}}{[ROH]} \tag{12.41}$$

$$M^{3+} + RCOO^- \leftrightarrow RCOOM^{2+} \qquad (12.42)$$

$$K_4 = \frac{\left[RCOOM^{2+}\right]}{\left[RCOO^-\right]a_{M^{3+}}} \qquad (12.43)$$

$$M^{3+} + 2RCOO^- \leftrightarrow (RCOO)_2M^+ \qquad (12.44)$$

$$K_5 = \frac{\left[(RCOO)_2M^+\right]}{\left[R-COO^-\right]^2 a_{M^{3+}}} \qquad (12.45)$$

$$M^{3+} + R-PO_4^- \leftrightarrow R-PO_4M^{2+} \qquad (12.46)$$

$$K_6 = \frac{\left[R-PO_4^-\right]}{[R-PO_4H]a_{M^{3+}}} \qquad (12.47)$$

The prediction of biosorption performance by mechanistic models can help to determine the optimal operating conditions, including pH, for selective binding of heavy metals.

12.3.2 KINETICS MODELING

12.3.2.1 Numerical Simulations

Although analytical methods may help to understand the mechanism through the model, the numerical tools can be used to solve complex issues like geometry and the flow in the column and reactors as well as diffusion through the particle of the sorbent.

The sorption rate is very important for practical applications, design, and scaling up of biosorption. Due to the fact that various biosorbents have different porosities and surface morphologies, two mechanisms may control intraparticle diffusion, which are pore or surface diffusion (Viegas et al., 2014). It is obvious that more than one step can contribute to process performance. In order to simplify the overall kinetic model, it is assumed that one of the cocurrent processes dominates over the others (the rate-controlling step). The most popular differential equations applied for modeling adsorption are the homogeneous surface, pore volume, and homogeneous intraparticle diffusion models.

The diffusional models are based on a few assumptions:

- The surface and the structure of the biosorbent/adsorbent are homogeneous or quasi-homogeneous.
- The intraparticle diffusion is the predominant mass transfer mechanism and may be either surface or pore volume diffusion.
- Adsorption of sorbate on an active site is immediate.
- The process is isothermal, and the equilibrium is described by the isotherm equation.
- The diffusion coefficient of counterions released as a result of ion exchange with metal ions is several times higher than that of metal ions adsorbed; thus, the overall sorption rate is just controlled by the movement of heavy metal ions (Yang and Volesky, 1999).

The differential equations of the homogenous intraparticle diffusion model are as follows (Yang and Volesky, 1996):

$$\varepsilon \frac{\partial c_r}{\partial t} + \rho \frac{\partial q}{\partial t} = D_e \left[\frac{1}{r^2} \frac{\partial}{\partial r} \left(r^2 \frac{\partial c_r}{\partial r} \right) \right] \qquad (12.48)$$

$$V \frac{dc_b}{dt} = -D_e S_t \left(\frac{\partial c_r}{\partial t} \right)_{r=R} \qquad (12.49)$$

with the initial and boundary conditions (Equations 12.50 through 12.53):

$$t = 0, \quad c_r = 0 \qquad (12.50)$$

$$t = 0, \quad c_b = c_0 \qquad (12.51)$$

$$r = 0, \quad \frac{\partial c_r}{\partial r} = 0 \qquad (12.52)$$

$$r = R, \quad (c_r)_{r=R} = c_b \qquad (12.53)$$

where

c is the metal concentrations in the bulk solution
c_r is the intraparticle metal concentration in the pore solution
q is the metal concentration in the bulk solution
D_e is the intraparticle diffusivity
S_t is the total area of the particles
V is the volume of the batch reactor
R is the average radius of the particles

The homogeneous surface diffusion model (HSDM) is similar to the homogeneous intraparticle diffusion model. The difference is in the expression of differential equations and the coefficient, which is a surface diffusion coefficient in HSDM and effective diffusion coefficient for the intraparticle model. The differential mass equation can be given as follows:

$$\frac{\partial q}{\partial r} = \frac{D_s}{r^2} \frac{\partial}{\partial r} \left(r^2 \frac{\partial q}{\partial r} \right) \qquad (12.54)$$

with the initial and boundary conditions (Equations 12.55 through 12.57):

$$t = 0, \quad q = 0 \qquad (12.55)$$

$$r = 0, \quad \frac{\partial q}{\partial r} = 0 \qquad (12.56)$$

$$r = R, \quad \rho_p D_s (1-\varepsilon) \frac{\partial q}{\partial r} = k_f (c_b - c_s) \qquad (12.57)$$

where
r is the radial direction
q is the mass of sorbate adsorbed at r
R is the radius of the particle

D_s is the surface diffusion coefficient
k_f is the film mass transfer coefficient
ε is the porosity of the composite
ρ_p is the adsorbent particle density
c_s is the sorbate concentration at the biosorbent's surface
c_b is the dye concentration in the bulk phase

Although biosorbents are not uniformly composed and should be considered as a heterogeneous material and require high knowledge about surface morphology and particle porosity and other properties, microorganisms due to their gel-like nature could be considered as a quasi-homogeneous phase for mathematical modeling (Yang and Volesky, 1999).

Only a few attempts have been made at modeling batch microbial biosorption dynamics by numerical simulations (Figure 12.4). Vilar et al. (2007) investigated the batch dynamics in the biosorption of several heavy metal ions (Pb(II), Cd(II), Zn(II), Cu(II), Cr(III), Ni(II)) on algae *Gelidium sesquipedale*, which is actually macroalgae, but the researchers proved that the model can be successfully applied for biosorption systems. They assumed that the rate was controlled by the homogeneous diffusion inside the particle in the direction normal to the surface of the particles. The concentration profile inside the particle was predicted by the HDM, which was solved numerically. The obtained average diffusion coefficients had a magnitude two times lower than that of the metal ion diffusion coefficient in water at 20°C due to the resistance inside the biosorbent particles. The satisfactory fit obtained in their study indicated that microbial biosorption dynamic mechanism can be evaluated on the basis of numeric models.

Loukidou and co-workers (2004) examined several numerical diffusional (intraparticle and external) rate models in order to identify a suitable rate expression for the biosorption of Cd(II) on *Aeromonas caviae*. They have proved that biosorption can be a complex process that may be described more correctly by more than one rate model.

Categorized according to the operation mode, adsorption can be generally divided into static (batch) and dynamic adsorption. In batch adsorption, a given amount of adsorbent contacts with a certain volume of a solute, while in column mode, adsorbent solution flows through a column packed with adsorbent (Xu et al., 2013). Continuous biosorption onto microbial cells can be carried out in traditional fixed-bed columns only after a prior immobilization of biomass in a polymer matrix to obtain particles with appropriate dimensions and enhanced mechanical stability (Pagnanelli et al., 2004). Due to the complexity of the adsorption column and reactor systems, their mathematical modeling is much more difficult to attain than batch processes. This is primarily because the models used currently in their analytical form are simplistic and do not account for all aspects of adsorption. To choose or develop a suitable model, accuracy and convenience should be considered simultaneously (Xu et al., 2013).

Conventional reactors include fixed-bed columns with immobilized biosorbent. A membrane reactor, in which cells are separated from the treated solution on a membrane, can provide an alternative to them. There are only very few reports on dynamic simulations and models developed to represent a continuous biosorption process of heavy metal ions on microorganisms in membrane reactors (Figure 12.4). Gaining

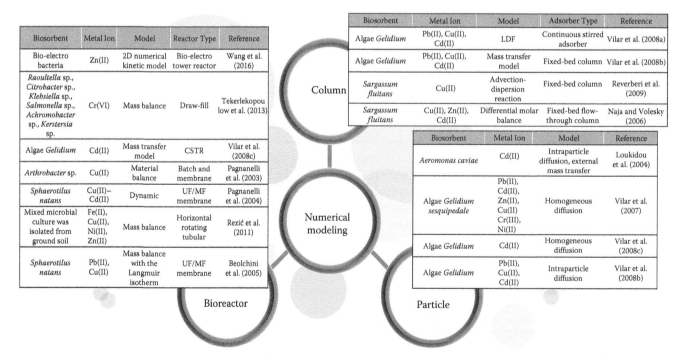

FIGURE 12.4 Literature review on the numeric models applied for modeling of the microbial biosorption.

insight into the biosorption mechanism is extremely important for the analysis and design of the reactor. According to Pagnanelli et al. (2004), the dynamic model for biosorption of metal systems in a membrane reactor combines the metal mass balances in the system with the equilibrium equations. The model assumes

- The constant reactor volume, biomass concentration, pH, and temperature in the reactor during the process
- No metal retention by the membrane
- Perfect mixing conditions in the reactor
- No bulk diffusion effect around the microbial cells
- Immediate adsorption of sorbate on an active site

The dynamic model differential equations for binary systems are as follows:

$$\frac{dc_i}{dt} = \frac{(c_{0i} - c_i)}{\tau} - x \frac{dq_i(c_i, c_j)}{dt} \qquad (12.58)$$

where
$q_i(c_i, c_j)$ is the equilibrium equation giving the specific metal uptake
x is the biomass concentration
τ is the residence time
t is the process time

The model derived from Pagnanelli et al. (2003) can be given as

$$V \frac{dc_i}{dt} = F(c_{0i} - c_i) - x \cdot V \frac{dq_{i,j}(c_i, c_j)}{dt} \qquad (12.59)$$

where
V is the constant reaction volume
F is the inlet flow rate

The model was tested for biosorption on *S. natans* in the Cd(II)–Cu(II) binary system (Pagnanelli et al., 2004) and for Cu(II) adsorption on *Arthrobacter* sp. (Pagnanelli et al., 2003) as well as for the biosorption of Cr(VI) and Fe(III) on *Rhizopus arrhizus* (Sağ et al., 2001).

Wang et al. (2016) invented the bio-electro tower reactor combining the electrochemistry technique with microbial biosorption. They developed an innovative numeric kinetic model that uses mass transfer, which contains the isotherm equation—Langmuir equation. This model can be an alternative to the pseudo-first-order and the pseudo-second-order kinetic model. It works on the assumption that (1) the supporting material on which the biofilm is disposed is porous, (2) the equilibrium is described by the Langmuir model, (3) the wall effect of the tower can be neglected, and (4) the effect of acceleration on the ions' movement due to the electric field only has impact on mass transfer coefficient K (Equation 12.60). The numeric model is based on the momentum and continuity equations (Equations 12.61 and 12.62) and the Blake–Kozeny equation (Equation 12.63):

$$\frac{\partial c_L}{\partial t} + u \cdot \nabla c_L = D \nabla^2 c_L - K\alpha(c_L - c_e) \qquad (12.60)$$

$$\frac{\partial u}{\partial t} + u \cdot \nabla u = -\frac{1}{\rho} \nabla p + \nu \nabla^2 u - \Delta p \qquad (12.61)$$

$$\nabla u = 0 \qquad (12.62)$$

$$\frac{|\Delta p|}{L} = \frac{150\mu}{D_p^2} \frac{(1-\varepsilon)^2}{\varepsilon^3} u \qquad (12.63)$$

where
u is the liquid velocity
D_p is the particle diameter
K is the mass transfer coefficient
α is the surface area of the unit volume
c_e is the equilibrium ion concentration, whose concentration inside the biofilm is c_s

The researchers state that the pseudo-first and pseudo-second equations that are widely used to describe the biosorption rate have limitations and cannot be used to explain the complex process conducted in the reactor. Thus, they developed a novel model that explains equilibrium sorption capacity as a time-dependent variable. The model was successfully applied for the biosorption of Zn(II) by bacteria in the bio-electro tower reactor.

Vilar et al. (2008b) stated that the highest performance and sorbent capacity can be achieved for biosorption conducted in the packed bed column as compared to completely mixed systems. The modeling of the dynamic sorption can be difficult, though, due to the fact that the flow properties, mass transfer, and equilibrium of the fluid have to be simulated simultaneously both in time and in space of the sorption column (Volesky, 1990). Computer-aided simulation may help to overcome such difficulties. Particularly, numerical methods may be useful to solve complex problems involving both mass transfer and equilibrium problems as well as stimulating the fluid flow in the column. The application of the numeric models to describe biosorption in the fixed-bed column has not been frequently reported (Figure 12.4). Riazi et al. (2014) applied the mass transfer model developed on the basis of the component mass balance of the bulk phase:

$$\frac{\partial c_i(z,t)}{\partial t} = D_z \frac{\partial^2 c_i(z,t)}{\partial z^2} - u_i \frac{\partial c_i(z,t)}{\partial z} - \frac{(1-\varepsilon)}{\varepsilon} \rho_p \frac{\partial q_i(z,t)}{\partial t} \qquad (12.64)$$

with the initial and boundary conditions (Equations 12.24 through 12.26):

$$t = 0, \quad c_i = 0 \qquad (12.65)$$

$$z = 0, \quad \frac{\partial c_i}{\partial z} = \frac{u}{D_z}(c - c_{in}) \qquad (12.66)$$

$$z = L, \quad \frac{\partial c_i}{\partial z} = 0 \qquad (12.67)$$

where
  D_z is the axial dispersion coefficient
  L is the length of the column

The model has been developed on the following assumption:

- Biosorbent particles have a spherical shape, and they are packed uniformly in the fixed bed.
- Isothermal biosorption.
- Constant velocity in the direction of the axis.
- The mass and velocity gradients are neglected in the radial direction of the bed.
- There occurs intraparticle mass transfer.
- The axial dispersion occurs in the bulk phase.
- The equilibrium of the biosorption is represented by conventional isotherm equation.

Although the original mass transfer model was developed on the basis of the modeling of adsorption on the activated carbon, an ion-exchange model rather than a model of activated carbon sorption should be used due to the ion exchange that occurs between the biosorbent and sorbate solution. Kratochvil et al. (1997) described the dynamics of multicomponent ion exchange in the fixed bed by the following differential equation:

$$0 = \frac{\partial c_i(z,t)}{\partial z} - \frac{1}{Pe_c} \frac{\partial^2 c_i(z,t)}{\partial z^2} + \frac{\partial c_i(z,t)}{\partial t} + D_{gi} \frac{\partial q_i(z,t)}{\partial t} \qquad (12.68)$$

The initial and boundary conditions are specified as follows:

$$t = 0, \quad c_i = 0 \qquad (12.69)$$

$$z = 0, \quad c_i = c_{0i} + \frac{1}{Pe_c} \frac{\partial c_i}{\partial z} \qquad (12.70)$$

$$z = L, \quad \frac{\partial c_i}{\partial z} = 0 \qquad (12.71)$$

where D_{gi} is the dimensionless solute distribution parameter, while Pe_c is the dimensionless column Peclet number. Similar assumptions were made as for the mass transfer balance: biosorption is considered as an isobaric and isothermal process, physical properties of the solution are constant, and radial dispersion is negligible (Cossich et al., 2004).

12.3.3 Metamodels

Recently, urgent attention has been paid to the modeling of bioprocesses by sophisticated techniques such as RSM based on the design of experiments (DoEs) and artificial neural networks (ANNs). These tools are particularly useful for describing problematic complex biological systems such as biosorption by microorganisms, which cannot be easily solved by available conventional methods. The modeling research using RSM and ANN can help in improving the design of optimal biosorption-based processes, employing feasible approaches toward the efficient heavy metal–containing wastewater treatment. The use of metamodels can help in the reduction of unnecessary experimental tests by indicating the most relevant experiments and providing tools to optimize and scale up of the process, which in turn may prove more cost and time efficient (Figure 12.5).

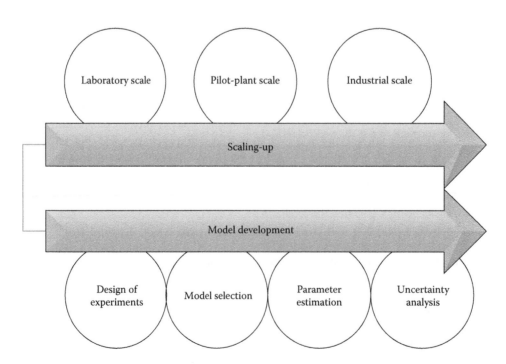

FIGURE 12.5 Biosorption process modeling and scaling-up procedure.

12.3.3.1 Design of Experiments

The DoEs serve the purpose of achieving the highest accuracy of model fitting with the minimal number of experimental tests.

Several plan methods have been applied for microbial biosorption modeling (Figure 12.7), including the CCD, Box–Behnken design, Doehlert matrix, as well as full or fractional factorial (FFD) design.

The choice of a particular design depends on the specific issue to be solved and the equation to be fitted. For example, the Box–Behnken plan consists of a fewer number of experiments (15 for three factors) than other designs; however, it was created to estimate only the second-order models. The Box and Behnken plan is a three-level incomplete factorial design with several replications of the central point, which improves precision.

In the full factorial design all input parameters are set only at two levels, −1 and 1, representing the highest and the lowest coded values within the range of an appropriate variable. FFD consists of all possible combinations of variables with multiple levels collected in the 27-run design. Whereas the full factorial design allows for determining the main and low-order interaction effects with great efficiency, fitting of the second- and higher-order equations may be problematic (Witek-Krowiak et al., 2014).

The CCD contains as much information as the FFD does; however, it requires fewer number of experimental points than the full factorial design. Moreover, both linear and quadratic equations can be easily estimated on the basis of CCD due to the high precision in predicting coefficients. The CCD comprises the full factorial or fractional factorial design at two levels ($2n$), center points (cp), which correspond to the middle level of the factors, and axial points ($2n$), which in turn depend on specific properties desired for the design and the number of parameters related (Myers and Montgomery, 2002; Witek-Krowiak et al., 2014).

By appropriate DoEs, the goal is to optimize a response of the system (dependent variable), which is affected by several independent factors (input variables).

12.3.3.2 Response Surface Methodology

Microbial biosorption is influenced by parameters as pH, temperature, specific and coexisting metal ion concentrations, biomass dosage, time, etc. The effect of each factor can be easily determined by applying the OVAT method (one variable at time), while the relationship among more than one variables and their collective impact on the biosorption performance cannot be evaluated in an easy way. One of the methodologies for modeling the complex nonlinear issue is the response surface technique.

RSM is a set of mathematical techniques that describe the relationship between a few independent variables and one or more dependent variables (Witek-Krowiak et al., 2014). The technique is used to optimize the response surface that is influenced by various process parameters. The result of the RSM modeling is an equation that quantifies the relationship between the input parameters and the obtained response surfaces (Raissi and Farsani, 2009).

Numerous variables may affect the response of the system under study, and it is practically impossible to identify and control all of them. While considering biosorption modeling by RSM, it is vital to select those of the parameters that have a relevant contribution to the overall performance. Statistical tools such as the Plackett–Burman design allow for selecting the most important variables. This design is simple and enables the study of the effect of a large number of factors in a relatively low number of experimental runs, which in turn leads to time reduction.

Selection of the most important variables is a preliminary DoEs, while in the second stage the experimental runs are conducted in accordance with the selected appropriate DoE. The role of the DoEs is to choose the minimal number of runs that best simulate a given response surface. The most popular DoEs have been summarized in the previous section.

The next step requires a suitable approximation for the actual functional relationship between independent variables and the response surface. Two important models are commonly used in RSM. These include the first-degree model

$$Y = \beta_0 + \sum_{i=1}^{k} \beta_i X_i + \epsilon \qquad (12.72)$$

and the second-degree model

$$Y = \beta_0 + \sum_{i=1}^{k} \beta_i X_i + \sum_{j=2}^{k} \sum_{i=1}^{k} \beta_{ij} X_i X_j + \sum_{i=1}^{k} \beta_{ii} X_i^2 + \epsilon \qquad (12.73)$$

where

ϵ is a random error
X_i are coded input values

The independent variables have to be coded a priori to +1, 0, and −1 representing the highest, medium, and lowest value within the range of a particular variable. The β_i are coefficients that are obtained by the least square method.

Expression (12.73) involves main effects, interaction effects, and quadratic effects.

A very important step in the modeling by RSM is a verification of the model. The accuracy of the model fitting is usually evaluated by the analysis of variance (ANOVA). ANOVA explains every variation in the statistically obtained model, and the importance of each model parameters. The significance of the obtained equation is evaluated by the F-test for an appropriate confidence level as well as the lack-of-fit test, while the coefficients and their importance in the model are assessed adapting the t-test and the p-value. Usually, the larger the t-value and F-value and the lower probability of p-value for a given confidence level, the model parameter is considered as significant (Sudamalla et al., 2012). It is also important to calculate a coefficient of determination (R^2), which is a quantitative measure of the statistical model fitness.

One of the disadvantages of RSM, while considering bio-sorption modeling, is an inability to extrapolate the function outside the range of the collected data. This is due to the complex biological nature of the microorganism's surface whose properties may vary depending on the operating conditions in the solution. Even a small change of the parameters may affect biosorption performance significantly. Removing insignificant factors from the resulting model may allow for the extrapolation (Podstawczyk et al., 2015).

The visualization of the predicted model equation can be obtained by the 3D and contour plot of response surface. The response surface plot is the theoretical 3D plot showing the relationship between the response and two independent variables. These plots give useful information about the model fitted, but they may not represent the real behavior of the system. The contour plots allow for visualization of the shape of the response surface. When the contour plot displays ellipses or circles, the center of the system corresponds to the maximum or minimum response (Baş and Boyac, 2007). Thus, the plots may help in finding the optimal values of variables that provide the highest value of biosorption performance, for example, the maximum biosorption capacity.

The objective of the last stage is to optimize the process by calculating the optimal values of influencing parameters. The optimization is implemented by calculating the first derivatives of the objective function, which equals zero. Then, the existence of the local extreme calculated by the maximization of the equation should be compared to the model graphical presentation.

Summing up, modeling microbial biosorption by RSM can be implemented in accordance with the procedure presented in Figure 12.6.

Recently, the application of RSM has been widely investigated in the case of Cd(II) biosorption on *A. niger* (Amini and Younesi, 2009), Cu(II) adsorption on *Aspergillus terreus* (Cerino-Córdova et al., 2012), and Ni(II) binding by *Leucobacter* sp. (Qu et al., 2011). The latest relevant research on microbial biosorption modeling using RSM is summarized in Figure 12.7.

12.3.3.3 Artificial Neural Network

The ANN is a parallel model made up of signal processing elements called neurons. Neurons are interconnected by means of weighted linkages that are synaptic connections (synapses) holding information. Adapting the weights under the training algorithm, ANN allows for learning from experimental data sets (Falamarzi et al., 2014; Podstawczyk et al., 2015).

Microbial biosorption used for the treatment of metal ion–containing wastewater is a complex process due to a large number of parameters that affect the adsorption performance and mechanism. ANNs have the ability to recognize the relationships between parameters and their effect on the response by creating connections between many different processing elements in the input–output system.

The accuracy of the ANN model may be affected by changing the structure of the network. Although a number of ANN architectures have been proposed, the feedforward neural networks (FNN—"multilayer perceptrons") have been so far the most frequently used (Table 12.7).

A simple three-layer feedforward neural network consists of an input layer, a hidden layer, and an output layer interconnected by links with weights. The number of neurons in input and output layers depends on the respective number of inputs and outputs being considered (Di Massimo et al., 1991). The selection of the number of hidden layers and number of

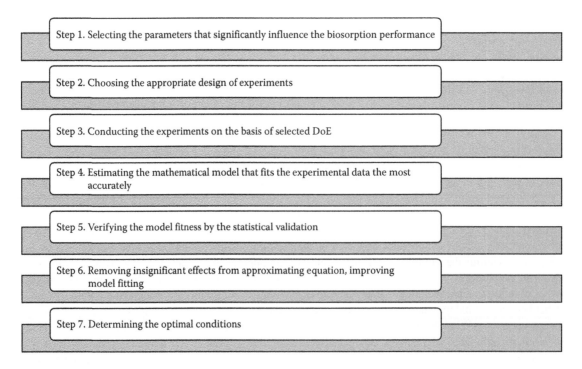

Step 1. Selecting the parameters that significantly influence the biosorption performance

Step 2. Choosing the appropriate design of experiments

Step 3. Conducting the experiments on the basis of selected DoE

Step 4. Estimating the mathematical model that fits the experimental data the most accurately

Step 5. Verifying the model fitness by the statistical validation

Step 6. Removing insignificant effects from approximating equation, improving model fitting

Step 7. Determining the optimal conditions

FIGURE 12.6 RSM modeling procedure.

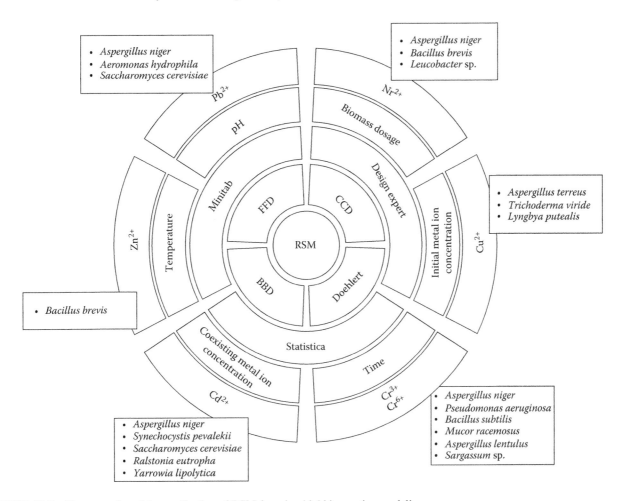

FIGURE 12.7 The examples of the application of RSM for microbial biosorption modeling.

TABLE 12.7

Literature Review on the ANN Applied for Modeling of the Microbial Biosorption

Input Variables	Output Variables	Architecture/Algorithm	Microorganisms	Metal	Software	Reference
c_0, X, T, t	%S	Backpropagation with three-layer architecture	*Bacillus cereus* biomass	As(III)	MATLAB® 7.0	Giri et al. (2011)
Q, h, c_0	c/c_0	Orthogonal least square learning algorithm	*Pseudomonas aeruginosa*	Cd(II)	MATLAB	Saha et al. (2010)
pH, t, c_0	q_e	Feedforward backpropagation	*Bacillus* sp.	Total Cr	MATLAB 7.1	Masood et al. (2012)
pH, X, t, c_0	%R	Feedforward backpropagation (BP) with Levenberg–Marquardt (LM) training algorithm	*Botryococcus braunii*	As(III), As(V)	MATLAB	Podder and Majumder (2016)
pH, t, X, T, c_0	q_e	Feedforward BP with LM training algorithm	*B. subtilis*	Cd(II)	MATLAB	Ahmad et al. (2014)

Note: c_0, initial metal ion concentration; t, time; T, temperature; X, biomass dosage or inoculum size; h, bed height; Q, feed flow; %S, percent sorption; %R, percent remediation.

neurons in each hidden layer is a challenging issue and should be specified during the network training. In fact, the hidden layers determine the topology of the network.

Once the ANN topology is defined, the set of inputs and outputs is used to properly train the model. The inputs are experimental points that have to be collected a priori. The biggest disadvantage of ANN modeling is that it requires a large number of experiments, which can be troublesome in the case of biosorption. Since biosorption is affected by many factors including pH, ionic strength, temperature, metal ion concentration, and biomass dosage, the other issue is to select those of the parameters that significantly influence the biosorption performance. Unfortunately, so far there are no unique solution to these problems and no method for determining which set of inputs is useful for the effective determination of the required output value (Witek-Krowiak et al., 2014). As can be seen in Table 12.7, the percentage removal or equilibrium biosorption capacity is usually employed as the response (output). Subsequently, input data should be divided into training and validation sets for accurately simulating the network and avoiding its overtraining (the network error is verified instantly).

In a multilayer perceptron, the signal enters the network through the input layer, and it is processed layer by layer by adjusting the weights, eventually reaching the output layer and generating the error (neural network performance). The mathematical expression of the network is as follows (Podstawczyk et al., 2015):

$$y = f(\mathbf{w}'\mathbf{x} + b) = f\left(\sum_{i=1}^{n} w_i x_i + b\right) = f(net) \quad (12.74)$$

$$\mathbf{w} = (w_1, w_2, \ldots, w_n) \quad (12.75)$$

$$\mathbf{x} = (x_1, x_2, \ldots, x_n) \quad (12.76)$$

where $f(\mathbf{w}'\mathbf{x} + b)$ is an activation function defined as a scalar product of weights ($\mathbf{w}$) and inputs ($\mathbf{x}$) vectors plus a bias (b).

This error is minimized by adapting the weights through the learning algorithm. The most popular neural network learning algorithm is the backpropagation (BP) method. In the BP, the error is adjusted with an increasing direction of the error function, while weights are updated countercurrently.

In the last stage of an ANN simulation, the model is verified and validated on the basis of the error function, while weights are updated to reach the minimal error. Generally, the modeling with ANNs consists of several main steps (Figure 12.8).

Although ANNs allow for predicting the output on the basis of the input data without the need to explicitly define the relationship between them, which is especially important in the case of complex and highly nonlinear issues such as biosorption (Witek-Krowiak et al., 2014), there have been only a few reports related to the modeling of microbial biosorption with this tool (Table 12.7). This can be due to the fact that experiments on biosorption by microorganisms require considerable amount of time, while ANN modeling can be implemented by computer-aided software packages alone. Still, the research in this field has been increasing for the last decade. For instance, Ahmad et al. (2014) investigated the effect of pH, contact time, biosorbent dosage, temperature, and initial cadmium ion concentration on the biosorption capacity of immobilized *B. subtilis* beads by applying ANN model based on feedforward backpropagation and Levenberg–Marquardt algorithm.

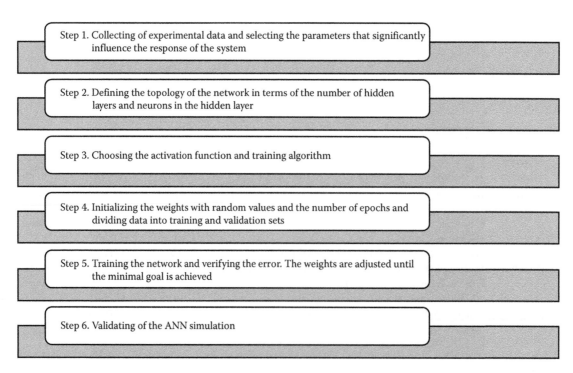

Step 1. Collecting of experimental data and selecting the parameters that significantly influence the response of the system

Step 2. Defining the topology of the network in terms of the number of hidden layers and neurons in the hidden layer

Step 3. Choosing the activation function and training algorithm

Step 4. Initializing the weights with random values and the number of epochs and dividing data into training and validation sets

Step 5. Training the network and verifying the error. The weights are adjusted until the minimal goal is achieved

Step 6. Validating of the ANN simulation

FIGURE 12.8 ANN modeling procedure.

12.4 CONCLUSION

The overview presented in this chapter indicates that various models can be used in the modeling of biosorption processes. The models developed so far take into account all the mechanisms that control the behavior of the biosorption like macroscopic and microscopic mass transfer, fluid flow, equilibrium, and physical and electrical phenomena as well as the effect of process parameters on the biosorption performance. Since biosorption is a complex process, its modeling is very challenging. Some of the models are complicated; thus, they require a solid computer-aided mathematical and numerical background. The improvement of experimental techniques and computational power provide the possibility for developing complex models such as numerical simulation of the dynamics of biosorption.

It is also important to select only significantly relevant parameters from all of the parameters that affect biosorption in order to reduce the complexity of the simulation. The best way to determine relevant factors from the range of operating conditions can be the application of the DoEs and statistical analysis that help in experimental data processing and isolation. The models that determine the relationships between input and output variables can be effective tools to describe the phenomena and the impact of great importance for control, optimization, and gaining a better theoretical knowledge.

RSM and ANNs allow in a simple way for modeling the effect of selected process parameters on the biosorption efficiency. The simplicity of the fitting means that an appropriate experimental data set is sufficient to simulate the relationship between each input variable and their impact on the process performance by means of computer software. Both RSM and ANN are powerful tools for biosorption identification, control, and optimization. On a small scale, the process analysis done by means of these models may help to effectively scale it up for industrial application.

The current progress in biosorbent characterization techniques and computer software has been used to unravel the underlying mechanisms that regulate biosorption. The ability to explore the mechanism of the adsorption by microbial cells may allow for the replacement of conventional models with mechanistic ones and therefore for promoting the development of efficient bioprocess models for industrial applications. It will be possible to update the simplified models that describe microbial biosorption with more detailed models based on the sorption mechanisms that capture the ion exchange between a sorbate solution and biosorbent active sites as well as competitive effects in multimetal systems.

REFERENCES

Ahmad, I., Ahmad, F., Pichtel, J. 2011. *Bacterial Biosorption: Microbes and Microbial Technology Agricultural and Environmental Applications*. Springer, New York.

Ahmad, M. F., Haydar, S., Bhatti, A. A. et al. 2014. Application of artificial neural network for the prediction of biosorption capacity of immobilized *Bacillus subtilis* for the removal of cadmium ions from aqueous solution. *Biochemical Engineering Journal* 84: 83–90.

Ahmady-Asbchin, S., Safari, M., Tabaraki, R. 2015. Biosorption of Zn (II) by *Pseudomonas aeruginosa* isolated from a site contaminated with petroleum. *Desalination and Water Treatment* 54: 3372–3379.

Alimohamadi, M., Abolhamd, G., Keshtkar, A. 2005. Pb(II) and Cu(II) biosorption on *Rhizopus arrhizus* modeling mono- and multi-component systems. *Minerals Engineering* 18: 1325–1330.

Amini, M., Younesi, H. 2009. Biosorption of Cd(II), Ni(II) and Pb(II) from aqueous solution by dried biomass of *Aspergillus niger*: Application of response surface methodology to the optimization of process parameters. *Clean: Soil, Air, Water* 37: 776–786.

Apiratikul, R., Pavasant, P. 2006. Sorption isotherm model for binary component sorption of copper, cadmium, and lead ions using dried green macroalga, *Caulerpa lentillifera*. *Chemical Engineering Journal* 119: 135–145.

Arjomandzadegan, M., Rafiee, P., Moraveji, M. K. et al. 2014. Efficacy evaluation and kinetic study of biosorption of nickel and zinc by bacteria isolated from stressed conditions in a bubble column. *Asian Pacific Journal of Tropical Medicine* 7: 194–198.

Bai, J., Yang, X., Du, R. et al. 2014. Biosorption mechanisms involved in immobilization of soil Pb by *Bacillus subtilis* DBM in a multi-metal-contaminated soil. *Journal of Environmental Sciences* 26: 2056–2064.

Baş, D., Boyacı, I. H. 2007. Modeling and optimization I: Usability of response surface methodology. *Journal of Food Engineering* 78: 836–845.

Beolchini, F., Pagnanelli, F., Toro, L. et al. 2005. Continuous biosorption of copper and lead in single and binary systems using *Sphaerotilus natans* cells confined by a membrane: Experimental validation of dynamic models. *Hydrometallurgy* 76: 73–85.

Bueno, B. Y. M., Torem, M. L., de Carvalho, R. J. et al. 2011. Fundamental aspects of biosorption of lead (II) ions onto a *Rhodococcus opacus* strain for environmental applications. *Minerals Engineering* 24: 1619–1624.

Calero, M., Hernáinz, F., Blázquez, G., Martín-Lara, M. A., & Tenorio, G. 2009. Biosorption kinetics of Cd (II), Cr (III) and Pb (II) in aqueous solutions by olive stone. *Brazilian Journal of Chemical Engineering*, 26(2): 265–273.

Cayllahua, J. E. B., Torem, M. L. 2010. Biosorption of aluminum ions onto *Rhodococcus opacus* from wastewaters. *Chemical Engineering Journal* 161: 1–8.

Cerino-Córdova, F. J., García-León, A. M., Soto-Regalado, E. et al. 2012. Experimental design for the optimization of copper biosorption from aqueous solution by *Aspergillus terreus*. *Journal of Environmental Management* 95: S77–S82.

Chakravarty, R., Banerjee P. C. 2008. Morphological changes in an acidophilic bacterium induced by heavy metals. *Extremophiles* 12: 279–284.

Chakravarty, R., Banerjee P. C. 2012. Mechanism of cadmium binding on the cell wall of an acidophilic bacterium. *Bioresource Technology* 108: 176–183.

Chen, Z., Pan, X., Chen, H. et al. 2015. Investigation of lead(II) uptake by *Bacillus thuringiensis* 016. *World Journal of Microbiology and Biotechnology* 31: 1729–1736.

Chojnacka, K., Chojnacki A., Górecka, H. 2005. Biosorption of Cr^{3+}, Cd^{2+} and Cu^{2+} ions by blue–green algae *Spirulina* sp.: Kinetics, equilibrium and the mechanism of the process. *Chemosphere* 59: 75–84.

Chowdhury, M. R. I., Mulligan, C. N. 2011. Biosorption of arsenic from contaminated water by anaerobic biomass. *Journal of Hazardous Materials* 190: 486–492.

Çolak, F., Olgun, A., Atar, N., Yazıcıoglu, D. 2013. Heavy metal resistances and biosorptive behaviors of *Paenibacillus polymyxa*: Batch and column studies. *Journal of Industrial and Engineering Chemistry* 19: 863–869.

Cossich, E. S., Da Silva, E. A., Tavares, C. R. G. et al. 2004. Biosorption of chromium(III) by biomass of seaweed *Sargassum* sp. in a fixed-bed column. *Adsorption* 10: 129–138.

Crank, J. 1970. *Mathematics of Diffusion*. Clarendon Press, Oxford, U.K.

Crini, G., Badot, P.-M. 2008. Application of chitosan, a natural aminopolysaccharide, for dye removal by aqueous solutions by adsorption processes using batch studies: A review of recent literature. *Progress in Polymer Science* 33: 399–447.

Dadrasnia, A., Chuan Wei, K. S., Shahsavari, N. et al. 2015. Biosorption potential of *Bacillus salmalaya* strain 139SI for removal of Cr(VI) from aqueous solution. *International Journal of Environmental Research of Public Health* 12: 15321–15338.

Di Massimo, C., Willis, M. J., Montague, G. A. et al. 1991. Bioprocess model building using artificial neural networks. *Bioprocess Engineering* 7: 77–82.

Dmytryk, A., Saeid, A., Chojnacka, K. 2014. Biosorption of microelements by spirulina: Towards technology of mineral feed supplements. *The Scientific World Journal* 2014: ID 356328.

Dubinin, M. M., Radushkevich, L. V. 1947. The equation of the characteristic curve of the activated charcoal. *Proceedings of the USSR Academy of Sciences USSR Physical Chemistry Section* 55: 331–337.

El-Naas, M. H., Abu Al-Rub, F., Ashour, I., Al Marzouqi, M. 2007. Effect of competitive interference on the biosorption of lead(II) by *Chlorella vulgaris*. *Chemical Engineering and Processing* 46: 1391–1399.

Elovich, S. J. 1959. *Proceeding of the Second International Congress on Surface Activity*, Vol. 11, Schulman, J. H. (ed.). Academic Press, Inc., New York, p. 253.

Engin, G. O., Muftuoglu, B., Senturk, E. 2012. Dynamic biosorption characteristics and mechanisms of dried activated sludge and *Spirulina platensis* for the removal of Cu^{2+} ions from aqueous solutions. *Desalination and Water Treatment* 47: 310–321.

Erkaya, I. A., Arica, M. Y., Akbulut, A. et al. 2014. Biosorption of uranium(VI) by free and entrapped *Chlamydomonas reinhardtii*: Kinetic, equilibrium and thermodynamic studies. *Journal of Radioanalytical Nuclear Chemistry* 299: 1993–2003.

Esposito, A., Pagnanelli, F., Veglio, F. 2002. pH-related equilibria models for biosorption in single metal systems. *Chemical Engineering Science* 57: 307–313.

Fagundes-Klen, M. R., Ferri, P., Martins, T. D. et al. 2007. Equilibrium study of the binary mixture of cadmium–zinc ions biosorption by the *Sargassum filipendula* species using adsorption isotherms models and neural network. *Biochemical Engineering Journal* 34: 136–146.

Falamarzi, Y., Palizdan, N., Huang, Y. F. et al. 2014. Estimating evapotranspiration from temperature and wind speed data using artificial and wavelet neural networks (WNNs). *Agricultural Water Management* 140: 26–36.

Fosso-Kankeu, E., Mulaba-Bafubiandi, A. F., Mamba, B. B. et al. 2011. Prediction of metal-adsorption behaviour in the remediation of water contamination using indigenous microorganisms. *Journal of Environmental Management* 92: 2786–2793.

Freundlich, H. M. F. 1906. Uber die adsorption in losungen. *Zeitschrift für Physikalische Chemie* 57(A): 385–470.

Galedar, M., Younesi, H. 2013. Biosorption of ternary cadmium, nickel and cobalt ions from aqueous solution onto *Saccharomyces cerevisiae* cells: Batch and column studies. *American Journal of Biochemistry and Biotechnology* 9: 47–60.

Garg, S. K., Singh, K., Tripathi, M. 2015. Optimization of process variables for hexavalent chromium biosorption by psychrotrophic *Pseudomonas putida* SKG-1 isolate. *Desalination and Water Treatment* 57: 19865–19876..

Gialamouidis, D., Mitrakas, M., Liakopoulou-Kyriakides, M. 2010. Equilibrium, thermodynamic and kinetic studies on biosorption of Mn(II) from aqueous solution by *Pseudomonas* sp., *Staphylococcus xylosus* and *Blakeslea trispora* cells. *Journal of Hazardous Materials* 182: 672–680.

Giri, A. K., Patel, R. K., Mahapatra, S. S. 2011. Artificial neural network (ANN) approach for modelling of arsenic (III) biosorption from aqueous solution by living cells of *Bacillus cereus* biomass. *Chemical Engineering Journal* 178: 15–25.

Gupta, S., Kumar, D., Gaur, J. P. 2009. Kinetic and isotherm modeling of lead(II) sorption onto some waste plant materials. *Chemical Engineering Journal* 148: 226–233.

Hajahmadi, Z., Younesi, H., Bahramifar, N. et al. 2015. Multicomponent isotherm for biosorption of Zn(II), Co(II) and Cd(II) from ternary mixture onto pretreated dried *Aspergillus niger* biomass. *Water Resources and Industry* 11: 71–80.

Haq, F., Butt, M., Ali, H. et al. 2015. Biosorption of cadmium and chromium from water by endophytic *Kocuria rhizophila*: Equilibrium and kinetic studies. *Desalination and Water Treatment* 57: 19946–19958.

Hasan, H. A., Abdullah, S. R. S., Kofli, N. T. et al. 2012. Isotherm equilibria of Mn^{2+} biosorption in drinking water treatment by locally isolated *Bacillus* species and sewage activated sludge. *Journal of Environmental Management* 111: 34–43.

Ho, Y. S., 2006. Review of second-order models for adsorption systems. *Journal of Hazardous Materials* B136: 681–689.

Hou, Y., Cheng, K., Li, Z. et al. 2015. Biosorption of cadmium and manganese using free cells of *Klebsiella* sp. isolated from waste water. *PLoS ONE* 10(10): e0140962.

Huang, F., Dang, Z., Guo, C.-L. et al. 2013. Biosorption of Cd(II) by live and dead cells of *Bacillus cereus* RC-1 isolated from cadmium-contaminated soil. *Colloids and Surfaces B: Biointerfaces* 107: 11–18.

Khalilnezhad, R., Olya, M. E., Khosravi, M., Marandi, R. 2014. Manganese biosorption from aqueous solution by *Penicillium camemberti* biomass in the batch and fix bed reactors: A kinetic study. *Applied Biochemistry and Biotechnology* 174: 1919–1934.

Kiran, B., Kaushik, A. 2008. Cyanobacterial biosorption of Cr(VI): Application of two parameter and Bohart Adams models for batch and column studies. *Chemical Engineering Journal* 144: 391–399.

Kirova, G., Velkova, Z., Stoytcheva, M. et al. 2015. Biosorption of Pb(II) ions from aqueous solutions by waste biomass of *Streptomyces fradiae* pretreated with NaOH. *Biotechnology & Biotechnological Equipment* 29: 689–695.

Kőnig-Péter, A., Csudai, C., Felinger, A. et al. 2014a. Potential of various biosorbents for Zn(II) removal. *Water, Air, & Soil Pollution* 225: 2089.

Kőnig-Péter, A., Kocsis, B., Kilár, F. et al. 2014. Bioadsorption characteristics of *Pseudomonas aeruginosa* PAOI. *Journal of the Serbian Chemical Society* 79: 495–508.

Kratochvil, D., Volesky, B., Demopoulos, G. 1997. Optimizing Cu removal/recovery in a biosorption column. *Water Research* 31: 2327–2339.

Kulkarni, R. M., Vidya Shetty, K., Srinikethan, G. 2014. Cadmium (II) and nickel (II) biosorption by *Bacillus laterosporus* (MTCC 1628). *Journal of the Taiwan Institute of Chemical Engineers* 45: 1628–1635.

Kumar, D., Gaur, J. P. 2011a. Chemical reaction- and particle diffusion-based kinetic modeling of metal biosorption by a *Phormidium* sp.-dominated cyanobacterial mat. *Bioresource Technology* 102: 633–640.

Kumar, D., Gaur, J. P. 2011b. Metal biosorption by two cyanobacterial mats in relation to pH, biomass concentration, pretreatment and reuse. *Bioresource Technology* 102: 2529–2535.

Langmuir, I. 1916. Constitution and fundamental properties of solids and liquids. *Journal of American Chemical Society* 38: 2221–2295.

Li, X., Wang, Y., Li, Y. et al. 2011. Biosorption behaviors of biosorbents based on microorganisms immobilized by Ca-alginate for removing lead (II) from aqueous solution. *Biotechnology and Bioprocess Engineering* 16: 808–820.

Liu, C.-C., Wang, M.-K., Chiou, C.-S. 2009. Biosorption of chromium, copper and zinc by wine-processing waste sludge: Single and multi-component system study. *Journal of Hazardous Materials* 171: 386–392.

Liu, R., Song, Y., Tang, H. 2013. Application of the surface complexation model to the biosorption of Cu(II) and Pb(II) ions onto *Pseudomonas pseudoalcaligenes* biomass. *Adsorption Science & Technology* 31: 1–16.

Loukidou, M. X., Karapantsios, T. D., Zouboulis, A. I. et al. 2004. Diffusion kinetic study of cadmium(II) biosorption by *Aeromonas caviae*. *Journal of Chemical Technology and Biotechnology* 79: 711–719.

Luna, A. S., Costa, A. L. H., da Costa, A. C. A. et al. 2010. Competitive biosorption of cadmium(II) and zinc(II) ions from binary systems by *Sargassum filipendula*. *Bioresource Technology* 101: 5104–5111.

Majumder, S., Raghuvanshi, S., Gupta, S. 2014. Estimation of kinetic parameters for bioremediation of Cr(VI) from wastewater using *Pseudomonas taiwanensis*, an isolated strain from enriched mixed culture. *Bioremediation Journal* 18: 236–247.

Malkoc, S., Kaynak, E., Guven, K. 2015. Biosorption of zinc(II) on dead and living biomass of *Variovorax paradoxus* and *Arthrobacter viscosus*. *Desalination and Water Treatment* 57: 15445–15454.

Mao, J., Kim, S., Wu, X. H. 2015. A sustainable cationic chitosan/*E. coli* fiber biosorbent for Pt(IV) removal and recovery in batch and column systems. *Separation and Purification Technology* 143: 32–39.

Markai, S., Andrès, Y., Montavon, G. et al. 2003. Study of the interaction between europium (III) and *Bacillus subtilis*: Fixation sites, biosorption modeling and reversibility. *Journal of Colloid and Interface Science* 262: 351–361.

Masood, F., Ahmad, M., Ansari, M. A. et al. 2012. Prediction of biosorption of total chromium by *Bacillus* sp. using artificial neural network. *Bulletin of Environmental Contamination and Toxicology* 88: 563–570.

Mishra, V. 2015. Modelling of the batch biosorption system: Study on exchange of protons with cell wall-bound mineral ions. *Environmental Technology* 24: 3194–3200.

Mishra, V., Balomajumder, C., Agarwal, V. K. 2010. Zn(II) ion biosorption onto surface of eucalyptus leaf biomass: Isotherm, kinetic, and mechanistic modelling. *Clean: Soil, Air, Water* 38: 1062–1073.

Mishra, V., Balomajumder, C., Agarwal, V. K. 2013. Dynamic, mechanistic, and thermodynamic modeling of Zn(II) ion biosorption onto zinc sequestering bacterium VMSDCM. *Clean: Soil, Air, Water* 41: 883–889.

Mohapatra, H., Gupta, R. 2005. Concurrent sorption of Zn(II), Cu(II) and Co(II) by *Oscillatoria angustissima* as a function of pH in binary and ternary metal solutions. *Bioresource Technology* 96: 1387–1398.

Moon, E. M., Peacock, C. L. 2011. Adsorption of Cu(II) to *Bacillus subtilis*: A pH-dependent EXAFS and thermodynamic modelling study. *Geochimica et Cosmochimica Acta* 75: 6705–6719.

Mukhopadhyay, M., Kaur, T., Khanna, R. 2011. Fixed bed and reduced lumped diffusion model parameter estimation of copper biosorption using *Aspergillus niger* biomass. *The Canadian Journal of Chemical Engineering* 90: 1011–1016.

Myers, R. H., Montgomery, D. C. 2002. Response surface methodology. In *Bloomfield*, Cressie, P., Fisher, N. A. C., Johnstone, N. I. et al. (eds.). John Wiley & Sons Inc, Canada.

Naja, G., Volesky, B. 2006. Multi-metal biosorption in a fixed-bed flow-through column. *Colloids and Surfaces A: Physicochemical and Engineering Aspects* 281: 194–201.

Ngwenya, N., Chirwa, E. M. N. 2015. Characterisation of surface uptake and biosorption of cationic nuclear fission products by sulphate-reducing bacteria. *Water SA* 41: 314–324.

Ozsoy, H. D., van Leeuwen, J. (Hans). 2011. Fungal biosorption of Ni (II) ions. In *Sustainable Bioenergy and Bioproducts Value Added Engineering Applications*, Gopalakrishnan, K., (Hans) van Leeuwen, J., Brown, R. C. (eds.). Springer-Verlag, London, U.K., pp. 45–59.

Padmapriya, G., Murugesan A. G. 2015. Biosorption of copper ions using rhizoplane bacterial isolates isolated from *Eicchornia crassipes* ((Mart.) solms) with kinetic studies. *Desalination and Water Treatment* 53: 3513–3520.

Pagnanelli, F. 2011. Equilibrium, kinetic and dynamic modelling of biosorption processes. In *Microbial Biosorption of Metals*, Kotrba, P., Mackova, M., Macek, I. (eds.). Springer, Dordrecht, the Netherlands, pp. 59–118.

Pagnanelli, F., Beolchini, F., Di Biase, A. 2004. Biosorption of binary heavy metal systems onto *Sphaerotilus natans* cells confined in an UF/MF membrane reactor: Dynamic simulations by different Langmuir-type competitive models. *Water Research* 38: 1055–1061.

Pagnanelli, F., Beolchini, F., Esposito, A. 2003. Mechanistic modeling of heavy metal biosorption in batch and membrane reactor systems. *Hydrometallurgy* 71: 201–208.

Pagnanelli, F., Esposito, A., Toro, L. et al. 2002. Copper and cadmium biosorption onto *Sphaerotilus natans*: Application and discrimination of commonly used adsorption models. *Separation Science and Technology* 37: 677–699.

Pagnanelli, F., Jbari, N., Trabucco, F. et al. 2013. Biosorption-mediated reduction of Cr(VI) using heterotrophically-grown *Chlorella vulgaris*: Active sites and ionic strength effect. *Chemical Engineering Journal* 231: 94–102.

Park, D., Yun, Y.-S., Park, J. M. 2010. The past, present, and future trends of biosorption. *Biotechnology and Bioprocess Engineering* 15: 86–102.

Paul, M. L., Samuel, J., Chandrasekaran, N. et al. 2012. Comparative kinetics, equilibrium, thermodynamic and mechanistic studies on biosorption of hexavalent chromium by live and heat killed biomass of *Acinetobacter junii* VITSUKMW2, an indigenous chromite mine isolate. *Chemical Engineering Journal* 187: 104–113.

Plazinski, W., Rudzinski, W. 2010. Heavy metals binding to biosorbents. Insights into Non-Competitive Models from a simple pH-dependent model. *Colloids and Surfaces B: Biointerfaces* 80: 133–137.

Podder, M. S., Majumder, C. B. 2016a. Fixed-bed column study for As(III) and As(V) removal and recovery by bacterial cells immobilized on sawdust/$MnFe_2O_4$ composite. *Biochemical Engineering Journal* 105: 114–135.

Podder, M.S., Majumder, C.B. 2016. The use of artificial neural network for modelling of phycoremediation of toxic elements As(III) and As(V) from wastewater using *Botryococcus braunii*. *Spectrochimica Acta Part A: Molecular and Biomolecular Spectroscopy* 155: 130–145.

Podstawczyk, D., Witek-Krowiak, A., Dawiec, A., Bhatnagar, A. 2015. Biosorption of copper(II) ions by flax meal: Empirical modeling and process optimization by response surface methodology (RSM) and artificial neural network (ANN) simulation. *Ecological Engineering* 83: 364–379.

Prasad, K. S., Srivastava, P., Subramanian, V. et al. 2011. Biosorption of As(III) Ion on *Rhodococcus* sp. WB-12: Biomass characterization and kinetic studies. *Separation Science and Technology* 46: 2517–2525.

Prithviraj, D., Deboleena, K., Neelu, N. et al. 2014. Biosorption of nickel by *Lysinibacillus* sp. BA2 native to bauxite mine. *Ecotoxicology and Environmental Safety* 107: 260–268.

Qu, Y., Li, H., Li, A. et al. 2011. Identification and characterization of *Leucobacter* sp. N-4 for Ni (II) biosorption by response surface methodology. *Journal of Hazardous Materials* 190: 869–875.

Qu, Y., Meng, Q., Zhao, Q., Ye, Z. 2014. Biosorption of Pb(II) and Zn(II) from aqueous solutions by living B350 biomass. *Desalination and Water Treatment* 55: 1832–1839.

Raissi, S., Farsani, R.-E. 2009. Statistical process optimization through multi-response surface methodology. *World Academy of Science, Engineering and Technology* 51: 267–271.

Rangabhashiyam, S., Suganya, E., Alen, V. L. 2015. Equilibrium and kinetics studies of hexavalent chromium biosorption on a novel green macroalgae *Enteromorpha sp. Research on Chemical Intermediates* 42: 1275–1294.

Redlich, O., Peterson, D. L. 1959. A useful adsorption isotherm. *Journal of Physical Chemistry* 63: 1024.

Remenárová, L., Pipíška, M., Horník, M. et al. 2012. Biosorption of cadmium and zinc by activated sludge from single and binary solutions: Mechanism, equilibrium and experimental design study. *Journal of the Taiwan Institute of Chemical Engineers* 43: 433–443.

Ren, G., Jin, Y., Zhang, Ch. et al. 2015. Characteristics of *Bacillus* sp. PZ-1 and its biosorption to Pb(II). *Ecotoxicology and Environmental Safety* 117: 141–148.

Resmi, G., Thampi, S. G., Chandrakaran, S., Elias, P. 2010. Biosorption of lead by immobilized biomass of *Brevundimonas vesicularis*: Batch and column studies. *Separation Science and Technology* 45: 2356–2362.

Reverberi, A. P., Dovì, V. G., Fabiano, B. et al. 2009. Using non-equilibrium biosorption column modelling for improved process efficiency. *Journal of Cleaner Production* 17: 963–968.

Rezaei, H., Kulkarni, S. D., Saptarshi, P. G. 2012. Study of physical chemistry on biosorption of zinc by using *Chlorella pyrenoidosa*. *Russian Journal of Physical Chemistry A* 86: 1332–1339.

Rezić, T., Zeiner, M., Šantek, B. et al. 2011. Mathematical modeling of Fe(II), Cu(II), Ni(II) and Zn(II) removal in a horizontal rotating tubular bioreactor. *Bioprocess and Biosystems Engineering* 34: 1067–1080.

Riazi, M., Keshtkar, A. R., Moosavian, M. A. 2014. Batch and continuous fixed-bed column biosorption of thorium(IV) from aqueous solutions: Equilibrium and dynamic modeling. *Journal of Radioanalytical and Nuclear Chemistry* 301: 493–503.

Rodrigues, M. L., Ferreira, L. S., Monteiro de Carvalho, J. C. et al. 2012. Metal biosorption onto dry biomass of *Arthrospira* (*Spirulina*) *platensis* and *Chlorella vulgaris*: Multi-metal systems. *Journal of Hazardous Materials* 217–218: 246–255.

Saadi, Z., Saadi, R., Fazaeli, R. 2013. Fixed-bed adsorption dynamics of Pb (II) adsorption from aqueous solution using nanostructured γ-alumina. *Journal of Nanostructure in Chemistry* 3: 48.

Sadaf, S., Bhatti, H. N., Nausheen, S., Noreen, S. 2014. Potential use of low-cost lignocellulosic waste for the removal of direct violet 51 from aqueous solution: Equilibrium and breakthrough studies. *Archives of Environmental Contamination and Toxicology* 66: 557–571.

Sağ, Y., Yalçuk, A., Kutsal, T. 2001. Use of a mathematical model for prediction of the performance of the simultaneous biosorption of Cr(VI) and Fe(III) on *Rhizopus arrhizus* in a semi-batch reactor. *Hydrometallurgy* 59: 77–87.

Saha, D., Bhowal, A., Datta, S. 2010. Artificial neural network modeling of fixed bed biosorption using radial basis approach. *Heat Mass Transfer* 46: 431–436.

Sahmoune, M. N., Louhab, K., Boukhiar, A. et al. 2009. Kinetic and equilibrium models for the biosorption of Cr(III) on *Streptomyces rimosus*. *Toxicological & Environmental Chemistry* 91: 1291–1303.

Samuel, J., Pulimi, M., Paul, M. L. et al. 2013. Batch and continuous flow studies of adsorptive removal of Cr(VI) by adapted bacterial consortia immobilized in alginate beads. *Bioresource Technology* 128: 423–430.

Samuel, M. S., Abigail, M. E. A., Ramalingam, C. 2015. Biosorption of Cr(VI) by *Ceratocystis paradoxa* MSR2 using isotherm modelling, kinetic study and optimization of batch parameters using response surface methodology. *PLoS ONE* 10:e0118999.

Saravanan, N., Kannadasan, T., Basha, C. A. et al. 2013. Biosorption of textile dye using immobilized bacterial (*Pseudomonas aeruginosa*) and fungal (*Phanerochate chrysosporium*) cells. *American Journal of Environmental Science* 9: 377–387.

Schiewer, S. 1999. Modelling complexation and electrostatic attraction in heavy metal biosorption by *Sargassum* biomass. *Journal of Applied Phycology* 11: 79–87.

Schiewer, S., Volesky, B. 1995. Modeling of the proton-metal ion exchange in biosorption. *Environmental Science & Technology* 29: 3049–3058.

Shahverdi, F., Ahmadi, M., Avazmoghadam, S. 2015. Isotherm models for the nickel(II) biosorption using dead fungal biomass of *Aspergillus awamori*: Comparison of various error functions. *Desalination and Water Treatment* 57: 19846–19856.

Sheindorf, C. H., Rebhun, M., Sheintuch, M. 1981. A Freundlich-type multicomponent isotherm. *Journal of Colloid and Interface Science* 79: 136–142.

Sips, R. 1948. Combined form of Langmuir and Freundlich equations. *The Journal of Chemical Physics* 16: 490–495.

Stanescu, A. M., Stoica, L., Constantin, C., Bacioiu, G. 2015. Modelling and kinetics of Cd(II) biosorption onto inactive instant dry baker's yeast. *Revista de Chimie (Bucharest)* 66: 173–177.

Sudamalla, P., Saravanan, P., Matheswaran, M. 2012. Optimization of operating parameters using response surface methodology for adsorption of crystal violet by activated carbon prepared from mango kernel. *Sustainable Environment Research* 22: 1–7.

Tassist, A., Lounici, H., Abdi, N. et al. 2010. Equilibrium, kinetic and thermodynamic studies on aluminum biosorption by a mycelial biomass (*Streptomyces rimosus*). *Journal of Hazardous Materials* 183: 35–43.

Temkin, M.J., Pyzhev, V. 1940. Kinetics of Ammonia Synthesis on Promoted Iron Catalysts. *Acta Physicochimica URSS*. 12, 217–222.

Thomas, H. C. 1944. Heterogeneous ion exchange in a flowing system. *Journal of the American Chemical Society* 66: 1466–1664.

Veneu, D. M., Torem, M. L., Pino, G. A. H. 2013. Fundamental aspects of copper and zinc removal from aqueous solutions using a *Streptomyces lunalinharesii* strain. *Minerals Engineering* 48: 44–50.

Viegas, R. M. C., Campinas, M., Rosa, H. C. M. J. 2014. How do the HSDM and Boyd's model compare for estimating intraparticle diffusion coefficients in adsorption processes. *Adsorption* 20: 737–746.

Vijayaraghavan, K., Yun Y.-S. 2008. Bacterial biosorbents and biosorption. *Biotechnology Advances* 26: 266–291.

Vilar, V. J. P., Botelho, C. M. S., Boaventura, R. A. R. 2007. Modeling equilibrium and kinetics of metal uptake by algal biomass in continuous stirred and packed bed adsorbers. *Adsorption* 13: 587–601.

Vilar, V. J. P., Botelho, C. M. S., Boaventura, R. A. R. 2008a. Metal biosorption by algae *Gelidium* derived materials from binary solutions in a continuous stirred adsorber. *Chemical Engineering Journal* 141: 42–50.

Vilar, V. J. P., Loureiro, J. M., Botelho, C. M. S. et al. 2008b. Continuous biosorption of Pb/Cu and Pb/Cd in fixed-bed column using algae *Gelidium* and granulated agar extraction algal waste. *Journal of Hazardous Materials* 154: 1173–1182.

Vilar, V. J. P., Santos, S. C. R., Martins, R. J. E. et al. 2008c. Cadmium uptake by algal biomass in batch and continuous (CSTR and packed bed column) adsorbers. *Biochemical Engineering Journal* 42: 276–289.

Volesky, B. 1990. Detoxification of metal-bearing effluents: Biosorption for the next century. *Hydrometallurgy* 59: 203–216.

Wang, H., Liu, Y.-G., Hu, X.-J. et al. 2014. Removal of cadmium from aqueous solution by immobilized *Microcystis aeruginosa*: Isotherms, kinetics and thermodynamics. *Journal of Central South University* 21: 2810–2818.

Wang, T., Wang, H., Li, C. et al. 2016. The numerical model of biosorption of Zn^{2+} and its application to the bio-electro tower reactor (BETR). *Chemosphere* 146: 233–237.

Weber, W. J., Morris J. C. 1963. Kinetics of adsorption on carbon solution. *Journal of the Sanitary Engineering Division ASCE* 89: 31–60.

Wierzba, S. 2015. Biosorption of lead(II), zinc(II) and nickel(II) from industrial wastewater by *Stenotrophomonas maltophilia* and *Bacillus subtilis*. *Polish Journal of Chemical Technology* 17: 79–87.

Witek-Krowiak, A., Chojnacka, K., Podstawczyk, D. 2013. Enrichment of soybean meal with microelements during the process of biosorption in a fixed-bed column. *Journal of Agriculture and Food Chemistry* 61: 8436–8443.

Witek-Krowiak, A., Chojnacka, K., Podstawczyk, D. et al. 2014. Application of response surface methodology and artificial neural network methods in modelling and optimization of biosorption process. *Bioresource Technology* 160: 150–160.

Xiao, G., Zhang, X., Su, H., Tan, T. 2013. Plate column biosorption of Cu(II) on membrane-type biosorbent (MBS) of *Penicillium* biomass: Optimization using statistical design methods. *Bioresource Technology* 143: 490–498.

Xu, Z., Cai, J., Pan, B. 2013. Mathematically modeling fixed-bed adsorption in aqueous systems. *Journal of Zhejiang University-SCIENCE A (Applied Physics & Engineering)* 4: 155–176.

Yahya, S. K., Zakariab, Z. A., Samin, J. et al. 2012. Isotherm kinetics of Cr(III) removal by non-viable cells of *Acinetobacter haemolyticus*. *Colloids and Surfaces B: Biointerfaces* 94: 362–368.

Yang, J., Volesky, B. 1996. Intraparticle diffusivity of Cd ions in a new biosorbent material. *Journal of Chemical Technology and Biotechnology* 66: 355–364.

Yang, J., Volesky, B. 1999. Cadmium biosorption rate in protonated *Sargassum* biomass. *Environmental Science & Technology* 33: 751–757.

Yang, S. K., Tan, N., Wu, W. L. et al. 2015. Biosorption of thorium(IV) from aqueous solution by living biomass of marine-derived fungus *Fusarium* sp. #ZZF51. *Journal of Radioanalytical and Nuclear Chemistry* 306: 99–105.

Yuan, H.-P., Zhang, J.-H., Lu, Z.-H. et al. 2009. Studies on biosorption equilibrium and kinetics of Cd^{2+} by *Streptomyces* sp. K33 and HL-12. *Journal of Hazardous Materials* 164: 423–431.

Zhang, B., Fan, R., Bai, Z. et al. 2013. Biosorption characteristics of *Bacillus gibsonii* S-2 waste biomass for removal of lead (II) from aqueous solution. *Environmental Science and Pollution Research* 20: 1367–1373.

Zhang, J., Yang, K., Wang, H. et al. 2014. Biosorption of copper and nickel ions using *Pseudomonas* sp. in single and binary metal systems. *Desalination and Water Treatment* 57: 2799–2808.

Ziagova, M., Dimitriadis, G., Aslanidou, D. et al. 2007. Comparative study of Cd(II) and Cr(VI) biosorption on *Staphylococcus xylosus* and *Pseudomonas* sp. in single and binary mixtures. *Bioresource Technology* 98: 2859–2865.

13 Understanding Toxic Metal–Binding Proteins and Peptides

Shankar Prasad Kanaujia

CONTENTS

ABSTRACT

It is estimated that ~30% of a proteome binds to metal ion(s) for accomplishing its catalytic activities and structural complexities. Although metals are essential for the survival of a cell, they show cytotoxicity below or above a threshold concentration. In addition, toxic heavy metals, such as mercury and arsenic, present in the environment pose a higher level of toxicity to cells once inhaled, especially in humans. Thus, cells persistently produce a large number of metal-binding proteins to regulate the concentration and to detoxify the effect of toxic heavy metal ions inside the cell. Hence, for several years, metal-binding proteins, mostly from bacteria and plants, have been exploited for the bioremediation of water and land sites polluted by toxic heavy metals generally produced as by-products from various types of industries. Thus, the identification, characterization, and understanding of the molecular interaction(s) of metalloproteins with their cognate metal ions become necessary to design novel proteins/peptides to specifically chelate toxic heavy metals that could be used for bioremediation purposes. In this chapter, the aim is to describe and explore the existing knowledge of metal-binding proteins and peptides from different domains of life, which can be used for the bioremediation of heavy metal ions from the environment. It summarizes the structural view of the existing knowledge about various metal-binding proteins from different organisms in the prospect of bioremediation of toxic metals.

13.1 INTRODUCTION

About one-third of the proteome of an organism is estimated to require metal ions for its biological function(s) (Barondeau and Getzoff, 2004; Waldron and Robinson, 2009). Based on their biological functions and effects, metal ions have been classified into three major classes: (1) essential metals (e.g., Na, K, Mg, Ca, Mn, Fe, Co, Ni, Cu, Zn, Mo, and W) with known biological functions, (2) toxic metals (e.g., Ag, Al, As, Au, Cd, Hg, Pb, Sb, Se, Sn, Ti), and (3) nonessential, nontoxic metals (e.g., Cs, Rb, Sr, and T) with no known biological effects (Roane and Pepper, 2000). Essential metal ions are known to play a key role in various biological processes such as metabolism, electron transport, and survival of every organism, including microbial virulence (Waldron and Robinson, 2009; Diaz-Ochoa et al., 2014). However, at higher concentrations, these metal ions show cytotoxicity effect, thus necessitating a control of intracellular metal concentration in each organism.

In contrast, heavy metals such as Ag, As, Au, Cd, Cu, Hg, Ni, Pb, and Se are universally toxic to biological organisms. These can also have carcinogenic and mutagenic effects on an organism in addition to being cytotoxic (Salem et al., 2000). A vast amount of heavy metals are present in the environment and are a source of health and environmental problems. In humans, these heavy metals, once absorbed or ingested, cause acute toxicity, antibiotic resistance, mental retardation, many other pathological complications, and even death (Table 13.1) (Wuana and Okieimen, 2011).

In order to make the environment healthier for human beings, contaminated water and land sites need to be decontaminated and made free from heavy metals. In the past, several techniques such as chemical precipitation, oxidoreduction,

filtration, reverse osmosis, membrane technology, evaporation, and electrochemical treatment have been employed to get rid of heavy metals (Dixit et al., 2015). However, as most heavy metal salts are water soluble, it becomes difficult to separate them by physical separation methods (Peters et al., 1985). In addition, physical techniques become ineffective when the concentration of heavy metals decreases below 100 mg/L (Ahluwalia and Goyal, 2007).

In bacteria, there are two major defense mechanisms: (1) production of degradative enzymes for the target pollutants (biotransformation and biomineralization) (Silver, 1996; Schelert et al., 2004; Orell et al., 2012) and (2) resistance to relevant heavy metals (biosorption, bioaccumulation) (Iram et al., 2015). These mechanisms can alternatively be classified into

TABLE 13.1
Toxic Effects of Some Heavy Metals on Human Health (EPA, USA)

Heavy Metal	EPA Regulatory Limit (ppm)	Use	Toxic Effects	References
Arsenic (As)	0.01	Insecticide, organoarsenic as poultry feed, and glass and wood preservation	Affects oxidative phosphorylation and ATP synthesis and carcinogen	Tripathi et al. (2007)
Barium (Ba)	2.0	Guttering, bearing alloys, and spark plugs	Causes cardiac arrhythmias, respiratory failure, gastrointestinal dysfunction, and elevated blood pressure	Jacobs et al. (2002)
Beryllium (Be)	0.005	Coal, rocket fuel	Carcinogen, acute, and chronic poison	Kuschner (1981)
Cadmium (Cd)	5.0	Fertilizer, plastic, and pigment	Carcinogenic, mutagenic, lung damage, fragile bones, affects calcium regulation, kidney damage, injury in CNS, and mental retardation	Degraeve (1981); Salem et al. (2000)
Cobalt (Co)	0.04	Vitamin B12	Diarrhea, low blood pressure, and paralysis	Basketter et al. (2003)
Chromium (Cr)	0.1	Tanning, paints, pigment, and fungicide	Hair loss, nephrite, cancer, and ulceration	Salem et al. (2000)
Copper (Cu)	1.3	Wire and cable, electric motors, antibiofouling, and antimicrobial	Brain and kidney damage, chronic anemia, stomach and intestine irritation	Salem et al. (2000); Wuana and Okieimen (2011)
Lead (Pb)	15	Plastic, paint, pipe, batteries, gasoline, and auto exhaust	Impaired development in children, reduced intelligence, short-term memory loss, cardiovascular disease, and neurotoxicity	Salem et al. (2000); Padmavathiamma and Li (2007); Wuana and Okieimen (2011)
Mercury (Hg)	2.0	Coal vinyl chlorides and electrical batteries	Autoimmune disease, depression, drowsiness, fatigue, hair loss, insomnia, memory loss, restlessness, tremors, brain damage, and lung and kidney failure	Neustadt and Pieczenik (2007); Ainza et al. (2010); Gulati et al. (2010)
Nickel (Ni)	0.2	Electroplating	Itching; cancer of the lungs, nose, sinuses, throat; immunotoxic, neurotoxic, genotoxic, affects fertility, hair loss, teratogenic, carcinogenic, genotoxic, and mutagenic effects	Salem et al. (2000)
Silver (Ag)	0.10	Antibacterial cream, dressings, and urinary catheters	Causes the skin and other tissues to turn gray if exposed, breathing problems, lung and throat irritation, and stomach pain	Chopra (2007)
Selenium (Se)	50	Coal and sulfur	Affects endocrine function, natural killer cells activity and gastrointestinal, damage of liver, kidney, spleen, and nervousness	Vinceti et al. (2001)
Zinc (Zn)	5	Fertilizer, batteries, and alloys	Vomiting, renal damage, and cramps	Johnson et al. (2007)

Note: EPA, Environmental Protection Agency.

active (Sikkema et al., 1995; Roane and Pepper, 2000) and passive mechanisms of metal resistance. Examples of active metal resistance systems include efflux pumps (Baker-Austin et al., 2005) or the proton gradient (Mangold et al., 2013). The passive mechanisms of metal resistance involve (1) detoxification of free metal ions by forming metal complexes (Di Toro et al., 2001; Dopson et al., 2014), (2) generating chemiosmotic positive internal membrane potential (Baker-Austin and Dopson, 2007; Slonczewski et al., 2009), and (3) creating competition between proton and metal to adhere to the cell surface (Mangold, 2012). In addition, some microorganisms such as acidophiles form biofilm that can absorb heavy metal ions on the cell surface (Teitzel and Parsek, 2003; Harrison et al., 2007).

In eukaryotes and plants, detoxification of heavy metals is generally obtained by (1) the formation of metal complexes, (2) metal binding to the cell wall, (3) reduced uptake across the cell membrane, (4) active efflux, (5) compartmentalization, and (6) chelation (Prasad, 1995; Rensing et al., 1999).

Common proteins involved in metal ion trafficking include transmembrane (TM) transporters, sequestration/storage proteins, and metallochaperones that transport specific metal ions through the cellular milieu and assist in their incorporation into specific metalloproteins and, occasionally, reductases that catalyze the reduction of the metal ions to redox states that are easier to transport or eliminate (O'Halloran and Culotta, 2000; Mukhopadhyay and Rosen, 2002); these proteins can be exploited for bioremediation (Malik, 2004). Hence, the use of microorganisms and plants for the decontamination of water and land sites has attracted growing attention (Dixit et al., 2015).

It is thus of great interest to better characterize the interactions between proteins and metals and to analyze structural factors governing metal binding and thermodynamic stabilization of proteins. Studies in this direction will benefit the understanding of the molecular factors governing heavy metal toxicity and speciation in cells and will also aid in developing new molecules for selective binding of a metal that could be used for biodetection or bioremediation purposes as well as for developing protein-based biosensors. In this chapter, the focus is to describe and explore the existing knowledge of metal-binding proteins and peptides that can be used for the bioremediation of heavy metal ions from the environment.

13.2 METALLOTHIONEINS

Metallothioneins (MTs), a group of low-molecular-weight (0.5–14 kDa) proteins consisting of high cysteine amino acid residues, are present in vast classes of life, including prokaryotes, protozoa, plants, yeast, invertebrates (insects, mollusks, echinoderms), and vertebrates. In eukaryotes, they are localized in the membrane of the Golgi apparatus (Felizola et al., 2014). MTs have the capacity to bind both essential (such as Zn, Cu, Se) and toxic and nonessential (such as Ag, As, Cd, Hg) heavy metals through the thiol group of their cysteine residues, which represent ~30% of

their constituent amino acid residues (Huang et al., 2004; Singh et al., 2008; Duncan, 2009; Freisinger and Vasak, 2013). Although the exact biological function of MTs is not clear, experimental studies indicate that MTs are involved in providing protection from metal toxicity and oxidative stress and in the regulation of physiological metals (Ruttkay-Nedecky et al., 2013). Thus, MTs are suggested to play a primary role in metal storage, transport, and detoxification (Manso et al., 2011).

There are several inducers of MTs, such as hormones, cytotoxic agents, and metals, which are regulated at the transcriptional level by specific promoter sequences (Hijova, 2004). Nonessential metals, such as Cd, Hg, Pb, Bi, Ag, Au, and Pt, are sequestered by MTs (Miles et al., 2000). In humans, ~17 genes located on chromosome 16q13 encode MTs (Takahashi, 2012). They consist of 61–68 amino acid residues having ~25%–30% cysteines and no aromatic amino acids (Sutherland et al., 2012). MTs from humans are classified into four major isoforms: MT1 (subtypes A, B, E, F, G, H, L, M, X), MT2, MT3, and MT4 that are expressed mainly in the liver and kidneys (Raudenska et al., 2014). The differences between the MT isoforms arise mainly from changes in the amino acid sequences (Thirumoorthy et al., 2011). MT1 and MT2 are present in most tissues and are highly inducible by metals (such as Ag^+, Bi^{3+}, Cd^{2+}, Co^{2+}, Hg^{2+}, Ni^{2+}, and Zn^{2+}), alkylating and oxidizing agents, glucocorticoids, cytokines, and inflammatory signals (Sutherland and Stillman, 2011). MT3 is detected mainly in the central nervous system (CNS), but it is also observed in the heart, kidneys, and reproductive organs, whereas MT4 can be found in the cells of squamous epithelium of the mouth, upper gastrointestinal tract, and skin. The presence of MTs in all types of mammalian cells suggests that they play an important role as intracellular (cytoplasm, mitochondria, nucleus, and lysosome) as well as extracellular (plasma, urine, pancreatic and amniotic fluid, milk) active molecules (Milnerowicz, 1997; Sliwinska-Mosson et al., 2009; Sabolic et al., 2010).

At the primary structure level, MT1 and MT2 are similar; MT3 contains an additional threonine element in the N-terminal part and an acidic hexapeptide in the C-terminal region (Figure 13.1). In addition, MT3 contains the Cys_6-Pro-Cys-Pro_9 motif, which does not exist in other MTs. The interaction of metal ions and cysteine residues dominates the secondary structure of the protein. A total of 20 cysteine residues are found in primary sequence with repetitions such as Cys-X-Cys, Cys-X-Cys-Cys, and Cys-Cys-X-Cys-Cys where X denotes any amino acid except cysteine. The N-terminal region (β-domain) consists of three divalent metal (Cd^{2+} or Zn^{2+})-binding sites or six sites for monovalents (Cu^+) with nine cysteinyl sulfurs (Figure 13.1). The C-terminal region (or α-domain) of the protein is capable of binding four divalent (Zn^{2+} or Cd^{2+}) or six monovalent (Cu^+) metal ions by 11 cysteine residues (Figure 13.1) (Braun et al., 1992; Luber and Reiher, 2010; Sutherland and Stillman, 2011). In summary, the sulfhydryl groups of the cysteine residues can bind 7 and 12 mol of divalent and monovalent

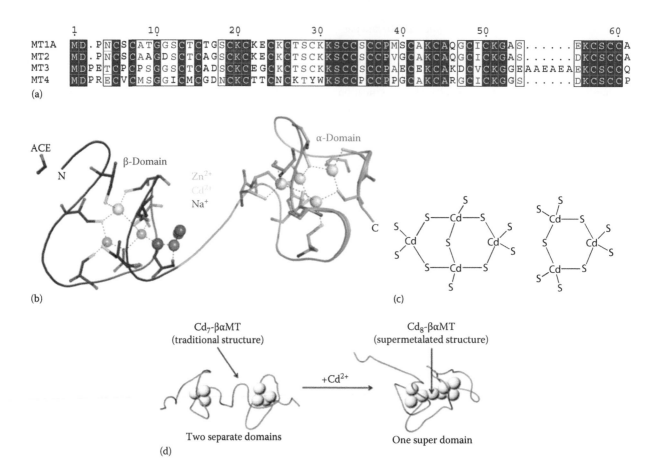

FIGURE 13.1 (a) Comparison of four isoforms of human MT sequences. Sequence comparisons were made using the Clustal Omega program (Sievers and Higgins, 2014). The sequences were derived from the Swiss-Prot database (MT1A: P04731, MT2: P02795, MT3: P25713, MT4: P47944). Identical positions are enclosed with a red box. (b) Tertiary structure of rat MT2 (PDB id: 4MT2) with α- and β-domains colored in pink and blue, respectively. The metal ions (Cd^{2+}, yellow; Zn^{2+}, gray; Na$^+$, magenta) are shown as spheres. The amino acid residues involved in the metal interaction are shown as sticks. Water molecules are shown as red spheres. (c) Schematic representation of the metal coordination to the cysteine residues of the protein. The cluster structure of two metal thiolates formed in vertebrate MTs: M4IICys11 cluster of the α-domain (left) and M3IICys9 cluster of the β-domain (right). Only the coordinating sulfur atoms of the Cys residues are shown. (d) A proposed model of the possible structure for the supermetalated Cd$_8$-βα-rhMT1A. (Modified from Sutherland, D.E. et al., *J. Am. Chem. Soc.*, 134, 3290, 2012. With permission.)

metal ions per mol of MT, respectively (Figure 13.1) (Rigby Duncan and Stillman, 2006).

MTs were initially classified into three classes (I–III) based on the position of cysteine residues in the primary structure (Fowler et al., 1987; Kojima, 1991). However, this classification system does not allow to clearly differentiate the patterns of structural similarities. Thus, MTs were later classified into different families based on their phylogeny (UniProtKB) (Table 13.2).

MTs bind their substrates in a sequential, noncooperative manner (Ngu and Stillman, 2009). After filling four places in the α-domain, metals bind to the β-domain. Although naturally occurring protein binds Zn^{2+} ion, other metal ions such as Ag$^+$, Cd^{2+}, Cu$^+$, Hg^{2+}, Pb^{2+}, Pd^{2+}, and Pt$^{2+/4+}$ having higher affinity for thiolate may substitute the Zn^{2+} ion (Krezel and Maret, 2007; Ngu and Stillman, 2009). Recently, it was demonstrated that human MT1A binds seven Zn^{2+} or Cd^{2+} ions and subsequently binds the eighth structurally significant

Cd^{2+} ion, which leads to the creation of the supermetalated form Cd$_8$-βα-rhMT1A (where hMT1 is human MT1), forming a single domain structure (Figure 13.1) (Sutherland et al., 2010, 2012).

Prokaryotic MTs have also been identified in a few cyanobacterial strains of the genus *Synechococcus*. These MTs, encoded by the *smtA* gene, contain fewer cysteine residues than mammalian MTs (9 compared to 20) (Blindauer, 2011). In 2002, Blindauer and colleagues identified numerous eubacterial MTs of the SmtA/BmtA family, including those in *Pseudomonas aeruginosa*, *P. putida* KT 2440, *Magnetospirillum*, and *Staphylococcus epidermidis* (Blindauer et al., 2001, 2002). SmtA proteins have been shown to have a physiological role, that is, metal binding (Figure 13.2) (Blindauer, 2013). GatA, a putative MT of *Escherichia coli*, binds a single zinc ion and was ultimately reclassified as a zinc-finger protein (Rensing and Grass, 2003).

TABLE 13.2

Families of Metallothioneins

Family	Organism	Pattern	PDB ids
1	Vertebrate	K-x(1,2)-C-C-x-C-C-P-x(2)-C	Mouse Cd7-MT-1 (1DFS, 1DFT), humanCd7-MT-2 (1HMU, 2HMU), rabbit Cd7-MT-2 (1MRB, 2MRB), ratCd7-MT-2 (1MRT, 2MRT, 4MT2)
2	Mollusk	C-x-C-x(3)-C-T-G-x(3)-C-x-C-x(3)-C-x-C-K	
3	Crustacean	P-[GD]-P-C-C-x(3,4)-C-x-C	Blue crab Cd6-MT-I (1DMC, 1DMD, 1DME, 1DMF)
4	Echinodermata	P-D-x-K-C-V-C-C-x(5)-C-x-C-x(4)-C-C-x(4)-C-C-x(4,6)-C-C	Sea urchin Cd7-MTA (1QJK, 1QJL)
5	Dipteral	C-G-x(2)-C-x-C-x(2)-Q-x(5)-C-x-C-x(2)-D-C-x-C	
6	Nematode	K-C-C-x(3)-C-C	
7	Ciliate		
8	Fungi-I	C-G-C-S-x(4)-C-x-C-x(3,4)-C-x-C-S-x-C	*N. crassa* Cu6-MT (1T2Y)
9	Fungi-II		
10	Fungi-III		
11	Fungi-IV	C-X-K-C-x-C-x(2)-C-K-C	
12	Fungi-V		1AOO, 1AQQ, 1AQR, 1AQS
13	Fungi-VI		
14	Prokaryota	K-C-A-C-x(2)-C-L-C	
15	Plant	[YFH]-x(5,25)-C-[SKD]-C-[GA]-[SDPAT]-x(0,1)-C-x-[CYF]	
99	Phytochelatins		

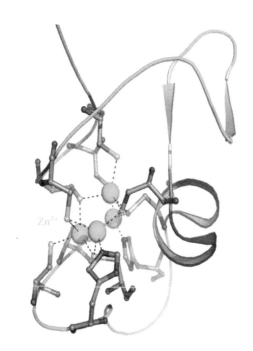

FIGURE 13.2 Tertiary structure of prokaryotic metallothionein (also known as SmtA) (PDB id: 1JJD). The four Zn^{2+} ions bound to the protein are shown as green spheres. The amino acid residues coordinating the metal ions are shown as ball-and-stick model.

13.3 PHYTOCHELATINS

Phytochelatins (PCs) are the members of class III MTs and are responsible for the formation of metal complexes and are important for heavy metal detoxification (Zenk, 1996; Pal and Rai, 2010). PCs are oligomers of glutathione (GSH) produced by the enzyme PC synthase (PCS) (Rauser, 1995; Cobbett,

2000; Vatamaniuk et al., 2001). A list of genes found in different organisms and the functions of their products is given in Table 13.3.

PCs are induced in all autotrophic plants (including maize, rapeseed, rice, soybean, tobacco, and wheat) analyzed so far, as well as in fungi, nematodes, and algae, including cyanobacteria (Gupta et al., 2013; Bundy and Kille, 2014). The biosynthesis of PCs is induced by many metals, including Ag, Au, Cd, Cu, Hg, Ni, Pb, and Zn, with Cd being the strongest inducer (Rauser, 1995; Hirata et al., 2005). PCs have not been reported in animal species, supporting the notion that MTs may well perform most of the functions normally performed by PCs in plants. The identification of PCS, however, in some animal species suggests that the mechanism of both MTs and PCs contribute to metal detoxification and/or metabolism (Rea et al., 2004; Bundy and Kille, 2014).

PCs are structurally related to glutathione (GSH) having the general structure (γ-Glu-Cys)$_n$-Gly where n has been reported to be as high as 11 but is generally in the range of 2–5. In addition, a number of structural variants, for example, (γ-Glu-Cys)$_n$-β-Ala, (γ-Glu-Cys)$_n$-Ser, and (γ-Glu-Cys)$_n$-Glu, have been identified in some plant species (Rauser, 1995, 1999; Zenk, 1996). A list of PC-like peptides is provided in Table 13.4. The N-terminal sequence of PC-synthase shows high conservation, whereas the C-terminal part is highly variable (Figure 13.3).

13.4 METAL-BINDING PEPTIDES

Naturally occurring metal-binding proteins and peptides, such as MTs and PCs, are very rich in cysteine residues (Mejare and Bulow, 2001). In addition, histidines are known to have high affinity for transition metal ions such as Zn^{2+},

TABLE 13.3

Genes Involved in Phytochelatin Biosynthesis and Function

Organism	Gene/Locus	Activity/Function	References
PC biosynthesis			
Schizosaccharomyces	Gsh1	GCS/GSH biosynthesis	
pombe	Gsh2	GS/GSH biosynthesis	
	PCS1	PC synthase/PC biosynthesis	Clemens et al. (1999); Ha et al. (1999)
Arabidopsis thaliana	CAD1	PC synthase/PC biosynthesis	Howden et al. (1995); Ha et al. (1999)
	CAD2	GCS/GSH biosynthesis	Cobbett et al. (1998)
PC functions			
Schizosaccharomyces	Hmt1	PC-Cd vacuolar membrane ABC-type transporter	Ortiz et al. (1995)
pombe	Ade2,6,7,8	Metabolism of cysteine sulfinate to products involved in sulfide biosynthesis; also required for adenine biosynthesis	Juang et al. (1993)
	Hmt2	Mitochondrial sulfide; quinine oxidoreductase/ detoxification of sulfide	Vande Weghe and Ow (1999)
Candida albicans	Hem2	Porphobilinogen synthase/siroheme biosynthesis (cofactor for sulfite reductase)	Hunter and Mehra (1998)
Other Cd-detoxification mechanisms			
Saccharomyces cerevisiae	YCF1	GSH-Cd vacuolar membrane ABC-type transporter	Li et al. (1997)

Source: Cobbett, C.S., *Curr. Opin. Plant Biol.*, 3, 211, 2000.

TABLE 13.4

Phytochelatin-Like Peptides

Type	Structure	Where They Can Be Found	Precursor
Phytochelatin	$(\gamma\text{-Glu-Cys})_n\text{-Gly}$	Many organisms	Glutathione
Homophytochelatin	$(\gamma\text{-Glu-Cys})_n\text{-Ala}$	Legumes	Homoglutathione
Desglycine phytochelatin	$(\gamma\text{-Glu-Cys})_n$	Maize, yeasts	
Hydroxymethyl-phytochelatin	$(\gamma\text{-Glu-Cys})_n\text{-Ser}$	Grasses	Hydroxymethyl-glutathione
iso-Phytochelatin (Glu)	$(\gamma\text{-Glu-Cys})_n\text{-Glu}$	Maize	Glutamylcysteinyl-glutamate
iso-Phytochelatin (Gln)	$(\gamma\text{-Glu-Cys})_n\text{-Gln}$	Horseradish	

Co^{2+}, Ni^{2+}, and Cu^{2+}. Therefore, various peptides comprising different sequences of cysteine or histidine residues have been exploited for chelating metal ions. Recently, metal-binding peptides containing either histidine $(GHHPHG)_2$ or cysteine (GCGCPCGCG) residues were fused to LamB and expressed on the *E. coli* cell surface (Kotrba et al., 1999b). In an another study, an expression of a cysteine-rich metal-binding motif $(CGCCG)_3$ fused with maltose-binding protein in *E. coli* exhibited a 10-fold increase in the binding of Cd^{2+} and Hg^{2+} (Pazirandeh et al., 1998). Moreover, a His-6-peptide has been overexpressed on the cell surface of *E. coli* as a fusion to the outer membrane (OM) protein LamB (Sousa et al., 1996). Another hexapeptide (HSQKVF) was overexpressed as a fusion to the surface exposed area

of the OmpA in *E. coli* that exhibited a higher tolerance to toxic concentrations of Cd^{2+} to the cells compared to control cells (Mejare et al., 1998). Another hexapeptide (SYHHHH) containing four histidine residues was selected for binding to zinc and was expressed on the surface of *E. coli* (Mejare et al., 1998). A list of some metal-binding peptides and proteins expressed on *E. coli* cell surface for Cd^{2+} absorption is given in Table 13.5.

13.5 MERCURY-BINDING PROTEINS

Mercury (Hg) in its oxidized form (Hg^{2+}) is a highly toxic element that has a pronounced and adverse effect on the ecosystem (Magos and Clarkson, 2006; Boyd et al., 2009) due to its high affinity for protein sulfhydryl groups, which upon binding destabilize protein structure and decrease enzyme activity (Tamas et al., 2014). However, it has been found that some aerobic bacterial and archaeal species have evolved mechanisms to resist the metal by degrading organomercuric compounds by the reduction of inorganic Hg^{2+} to gaseous Hg^0 (a nontoxic form) (Barkay et al., 2003; Boyd and Barkay, 2012).

Genetic analysis reveals that the mercury resistance (mer) system is encoded by the *mer* operon. It is estimated that ~1%–10% of cultured heterotrophic, aerobic microbes present in diverse environments possess *mer* systems (Barkay, 1987; Barkay et al., 2010). Thus, the *mer* system has been the basis for the development of biological tools for Hg^{2+} bioremediation and the management of Hg^{2+}-contaminated environments. It consists of the homodimeric flavin-dependent disulfide oxidoreductase mercuric reductase (MerA) enzyme. It also encodes organomercury lyase (MerB), a periplasmic Hg^{2+}-scavenging protein (MerP), one or more inner membrane (IM) spanning proteins (MerC, MerE, MerF, MerG,

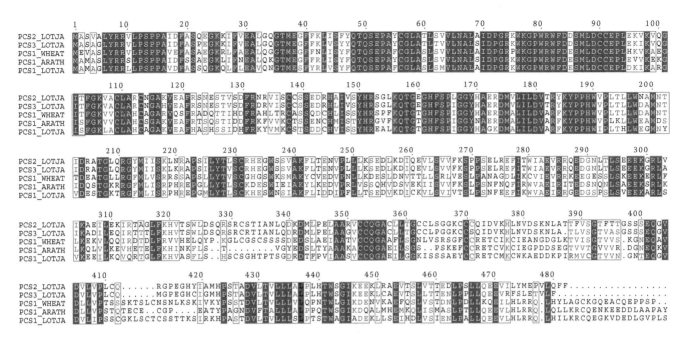

FIGURE 13.3 Comparison of amino acid sequence of PCS isoforms from different organisms (PCS1, *Arabidopsis thaliana* (Q9S7Z3); PCS1, *Lotus japonicus* (Q2TSC7); PCS1, *Triticum aestivum* (Q9SWW5); PCS2, *L. japonicus* (Q2TE74); and PCS3, *L. japonicus* (Q2QKL5)). The multiple sequence alignment was performed with the program Clustal Omega (Sievers and Higgins, 2014) and shows that the amino terminal domains exhibit high sequence conservation compared to the carboxyl terminal of the protein.

TABLE 13.5

Metal-Binding Peptides and Proteins Expressed on the Surface of *Escherichia coli* and Their Effect on the Cells upon Cadmium Exposure

Peptide/Protein	Expression Site	Increase in Cd Accumulations	Reference
His6, single or tandem expressed	OM, LamB	5- and 11-fold	Sousa et al. (1996)
Human MT	OM, Lpp	66-fold	Jacobs et al. (1989)
Mammalian MT	OM, LamB	15–20-fold	Sousa et al. (1998)
CdBP (HSQKVF)	OM, OmpA	Increased Cd tolerance	Mejare et al. (1998)
CP (GCGCPCGCG)	OM, LamB	Fourfold	Kotrba et al. (1999a)
HP (GHHPHG)2	OM, LamB	Threefold	Kotrba et al. (1999a)
MT α-domain	OM, LamB	17-fold	Kotrba et al. (1999b)
1–12 tandem repeats of *Neurospora crassa* MT	Periplasm, MBP	10–65-fold	Mauro and Pazirandeh (2000)
(CGCCG)3	OM, LamB	10-fold	Pazirandeh et al. (1998)

Notes: OM, outer membrane; LamB, calcium binding; Lpp, protein; OmpA, cysteine-containing peptide; HP, histidine-containing peptide.

MerH, and MerT), and regulatory proteins (MerD and/or MerR) (Figure 13.4). The overall expression of *mer* is regulated by MerR, which acts as a transcriptional repressor or activator in the absence or presence of Hg^{2+}, respectively (Figure 13.4) (Mathema et al., 2011).

Structurally, MerA proteins are composed of two main components: a multidomain catalytic core (structurally homologous to pyridine nucleotide disulfide oxidoreductase proteins) and certain thioredoxin reductases (Figure 13.5) (Schiering et al., 1991; Williams et al., 2000). The N-terminal contains one or two repeats of a sequence homologous with other "soft" metal ion trafficking proteins/domains (Hong et al., 2014; Lian et al., 2014). Studies on mercuric ion reductase (MerA) have shown that the homodimeric catalytic core has two interfacial active sites (Figure 13.5) (Distefano et al., 1990). In each site, the cofactor flavin adenine dinucleotide mediates electron transfer between pyridine nucleotide and Hg^{2+} bound to an inner pair of cysteine thiols in one subunit (Engst and Miller, 1999). A second pair of cysteine thiols from the C-terminal region of the other subunit serves as a ligand exchange conduit for the transfer of Hg^{2+} from thiols in solution to the inner thiols at the site of reduction (Hong et al., 2014; Lian et al., 2014). The N-terminal domain of the protein (NmerA) also contains a pair of cysteine thiols in a GMTCXXC sequence motif that is conserved in the soft metal ion trafficking proteins that share a common βαββαβ structural fold (Ledwidge et al., 2010). The presence of NmerA enhances the survival rate of organisms with high oxidative stress (Figure 13.6) (Ledwidge et al., 2005).

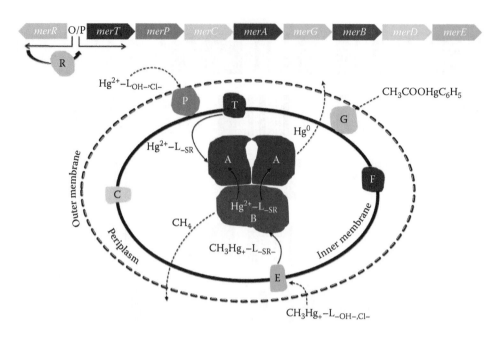

FIGURE 13.4 The *mer* operon system. A generic *mer* operon with genes depicting those that are present in some but not the majority of operons (top). The outer cell wall is shown as a broken line assuming that not all microbes have an OM. The solid line arrows depict transport or transformations. The diffusion of molecules is represented as broken line arrows. The ligand types are denoted as subscripts to L (ligand). The colors for the *mer* genes and proteins have been kept same for the clarity. (Redrawn from Boyd, E.S. and Barkay, T., *Front. Microbiol.*, 3, 349, 2012.)

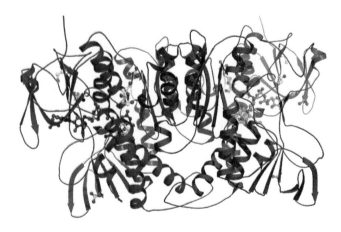

FIGURE 13.5 Homodimeric catalytic core of MerA. Tertiary structure of MerA from *P. aeruginosa* (PDB id: 1ZK7). Functional groups of conserved residues in binding pathway and active site are shown in ball-and-stick model.

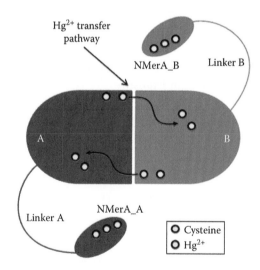

FIGURE 13.6 Schematic representation of the MerA (full-length) homodimer and the scheme for the transfer of Hg^{2+} from the solvent into the catalytic sites of the MerA core. Hg^{2+} is first bound by the NmerA cysteines and is then delivered to the C-terminal cysteines of the other monomer. Finally, Hg^{2+} is transferred to a pair of cysteines in the catalytic site of the core to be reduced to Hg^0. (Redrawn from Hong, L. et al., *Biophys. J.*, 107, 393, 2014. With permission.)

Naturally, mercury can exist in three different forms: elemental mercury (Hg^0), methylmercury (CH_3Hg^+), and ionic mercury (Hg^{2+}). Even though the exact composition of the *mer* operon varies among bacterial strains, those resistant to high levels of CH_3Hg^+ encode two enzymes (MerA and MerB) that transform CH_3Hg^+ to Hg^0 (Wahba et al., 2016). The organomercurial lyase MerB forms a dimer (Figure 13.7) that cleaves the carbon–mercury bond of CH_3Hg^+, releases methane (Figure 13.4) (Lafrance-Vanasse et al., 2009; Wahba

et al., 2016), and directly transfers the Hg^{2+} to the mercurial reductase, MerA. MerA reduces the Hg^{2+} to Hg^0, which is volatile and diffuses out of the bacteria (Figure 13.4) (Summers and Sugarman, 1974).

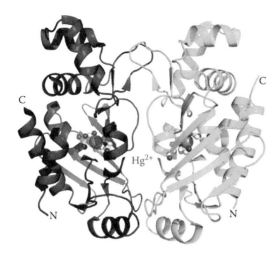

FIGURE 13.7 Crystal structure of MerB from *E. coli*. The dimeric form of MerB (PDB id: 3F0P). The colors of the two subunits are green and blue. The metal ion (Hg^{2+}) bound to MerB is shown as orange spheres.

The MerC, MerE, MerF, MerH, and MerT proteins are embedded in the IM and are thought to function by equivalent mechanisms, with mercury-binding cysteines within the TM helix facing toward the periplasmic side and functional cysteines on the cytoplasmic face of the IM (Figure 13.8) (Wilson et al., 2000; Sone et al., 2013a; Tian et al., 2014). These membrane proteins are responsible for transporting

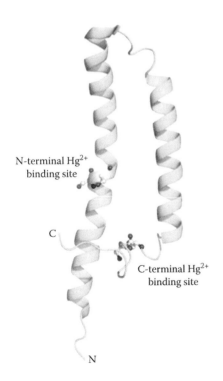

FIGURE 13.8 The 3D structure of MerF (PDB id: 2MOZ) is shown as a cartoon in cyan with the two mercury-binding sites labeled. The amino acids required for binding to the metal ion are shown in ball-and-stick model.

Hg^{2+} across the cell membrane and delivering it to MerA, a multidomain enzyme that reduces Hg^{2+} to Hg^0. The MerE protein is a broad-spectrum mercury transporter that anchors the transport of $CH_3Hg(I)$ and Hg^{2+} across bacterial IM (Kiyono et al., 2009; Sone et al., 2010, 2013b). Each protein has a cysteine pair located in the first TM region: Cys–Cys in MerT and MerF, Cys–Pro–Cys in MerE, and Cys–Ala–Ala–Cys in MerC, which have been shown to be essential for Hg^{2+} transport into the cell (Sahlman et al., 1999; Lu et al., 2013; Sone et al., 2013a). The MerT protein shows a limited homology to HisM/HisQ (histidine transport proteins) and MalF (maltose transport system) proteins, which require periplasmic binding proteins (PBPs or SBPs) for function. The MerH protein is mostly found in archaea and is suggested to be a mercury metallochaperone that plays a critical role in mediating mercuric chloride resistance across the IM via a pair of cysteine residues located but only when coexpressed with MerA (Schue et al., 2009; Schelert et al., 2013). The metal ion trafficking by MerH is thought to have two distinct roles: (1) by analogy to the actions of bacterial MerP and the N-terminal bacterial MerA motif, archaeal MerH could facilitate metal transfer to MerA for the reduction and efflux of its volatile state, or (2) MerH can also control the derepression of transcription at the promoter of merH indicating its role in trafficking Hg^{2+} to the MerR transcription factor.

The MerG protein is involved in phenylmercury resistance, and it helps in reducing cell permeability toward phenylmercury (Kiyono and Pan-Hou, 1999). The MerP protein binds Hg^{2+} ions in the periplasm and delivers them to the membrane transport proteins (MerC/MerE/MerF/MerH/MerT). After the Hg^{2+} is transported through the membrane, MerA reduces Hg^{2+} to Hg^0. The MerP protein contains the $\beta\alpha\beta\beta\alpha\beta$ fold characteristic of "ferredoxin-like" proteins (Figure 13.9) (Steele and Opella, 1997). The metal-binding site is located at the N-terminal of the protein and consists of the highly conserved sequence XMXCXXC (GMTCXXC in MerP; X denotes any amino acid) housing the two reactive cysteine residues; a motif found mostly in metallochaperones and metal-transporting ATPases (Figure 13.9) (Powlowski and Sahlman, 1999; Serre et al., 2004).

The MerR protein is a dual-function transcriptional regulator that coordinates the expression of the *mer* operon (Brown et al., 2003). MerR belongs to the CmtR/CadC/ArsR/SmtB family of proteins (Busenlehner et al., 2003; Osman and Cavet, 2010). The apo-MerR (metal free) binds to the *mer* operator and/or promoter region as a repressor to block the transcription initiation, whereas Hg^{2+}-bound MerR acts as an activator (Figure 13.10) (Chang et al., 2015). Thus, MerR tightly regulates the expression of *mer* operon and can switch from a transcriptional repressor to a transcriptional activator in an Hg^{2+}-dependent manner (Guo et al., 2010). The MerD protein is suggested to be a coregulator that downregulates the expression of *mer* operon (Champier et al., 2004). It is proposed that MerD plays a dynamic role in destabilizing the Hg-bound MerR–DNA complex to

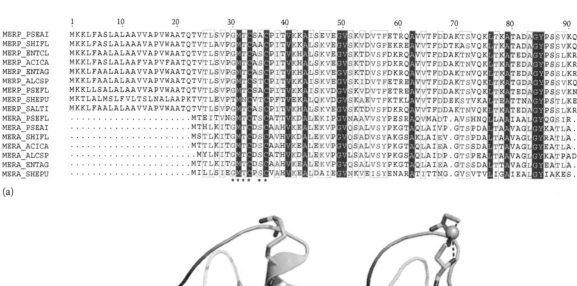

(a)

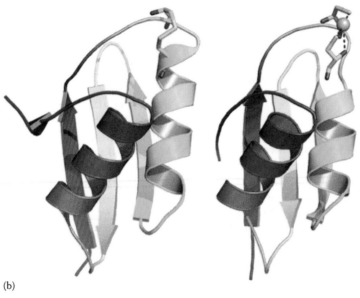

(b)

FIGURE 13.9   (a) Sequence alignment of MerP and N-terminal extensions of MerA from different organisms (MerP: *P. aeruginosa* (P04131), *Shigella flexneri* (P04129), *Enterobacter cloacae* (P0A218), *Acinetobacter calcoaceticus* (Q52107), *Enterobacter agglomerans* (P0A217), *Alcaligenes* sp. (P94186), *Pseudomonas fluorescens* (Q51770), *Shewanella putrefaciens* (Q54463), and *Salmonella typhi* (P0A216); MerA: *P. fluorescens* (Q51772), *P. aeruginosa* (P00392), *S. flexneri* (P08332), *A. calcoaceticus* (Q52109), *Alcaligenes* sp. (P94188), *E. agglomerans* (P94702), and *S. putrefaciens* (Q54465)). The identical and similar positions in the alignment are enclosed in red box. The highly conserved sequence GMTCXXC (X denotes any amino acid) in the metal-binding domain is marked with an asterisk (*). (b) The 3D structure of MerP (left) oxidized form (PDB id: 1OSD) and (right) reduced form (PDB id: 1AFJ). The metal ion (sphere in gray) and the coordinating amino acid residues (ball-and-stick model) are also shown for clarity.

promote the synthesis of MerR, which then will regain *mer* OP (operator), thus switching on or off the expression of the *mer* operon depending on the presence or absence of Hg^{2+} (Champier et al., 2004).

13.6   ARSENIC-BINDING PROTEINS

Arsenic (As) is one of the most ubiquitous environmental toxins and a human carcinogen. Enzymes that are shown to be inhibited by As include glutathione reductase (Styblo et al., 1997), glutathione S-transferase, glutathione peroxidase (Chouchane and Snow, 2001), thioredoxin reductase (Lu et al., 2007), thioredoxin peroxidase (Chang et al., 2003), DNA ligases, Arg–tRNA protein transferase (Li and Pickart, 1995), trypanothione reductase (Cunningham et al., 1994), IκB kinase β (IKKβ) (Kapahi et al., 2000),

pyruvate kinase galectin (Lin et al., 2006), protein tyrosine phosphatase (Rehman et al., 2012), JNK phosphatase, Wip1 phosphatase (Yoda et al., 2008), and E3 ligases c-CBL and SIAH1 (Mao et al., 2010). Resistance to arsenite (As^{3+}) and antimonite (Sb^{3+}) in many prokaryotes is conferred by an *ars* operon, which often contains three genes, *arsR*, *arsB*, and *arsC* (Dopson et al., 2001; Dhuldhaj et al., 2013). In some organisms, two additional genes, namely, *arsA* and *arsD* (first identified in *E. coli*), are found in *ars* operon (Figure 13.11) (Rosen, 2002; Li et al., 2010). Cells expressing the *arsRDABC* operon are more resistant toward As^{5+} and As^{3+} than those expressing only *arsRBC* genes. The *arsA* gene encodes the transport ATPase (Figures 13.11 and 13.12) (Zhou et al., 2001), and the *arsB* gene encodes the membrane translocase subunit of an arsenic pump (Figures 13.11 and 13.13). The last gene of the operon, *arsC*, encodes

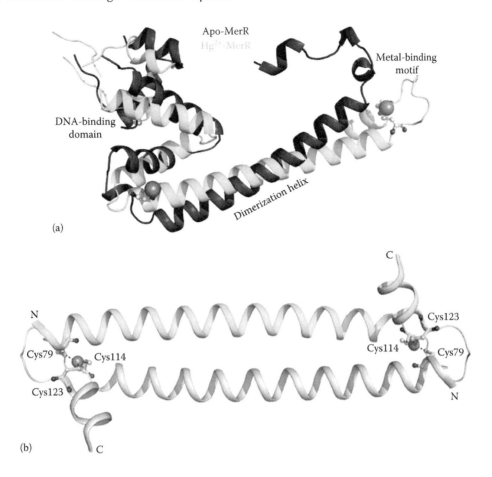

FIGURE 13.10 Tertiary structure MerR. (a) Cartoon representations of the MerR in apo (blue, PDB id: 4UA2) and Hg²⁺-bound (cyan, PDB id: 4UA1) states. For both structures, the DNA-binding domain, dimerization helix, and metal-binding motif are labeled. (b) The metal-binding sites at the dimeric interface of MerR. Three cysteine residues are involved in the interaction with the metal (Hg²⁺) ion (shown in orange).

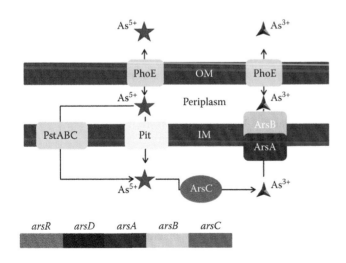

FIGURE 13.11 Transport and resistance mechanisms of As⁵⁺ in *E. coli*. PhoE: phosphate uptake ABC transporter permease; PstABC, phosphate transporter; Pit, inorganic phosphate transporter family; OM, outer membrane; IM, inner membrane.

an arsenate reductase that catalyzes the reduction of As⁵⁺ to As³⁺ prior to extrusion (Martin et al., 2001). The *arsR* and *arsD* genes encode regulatory proteins; both are homodimeric and show high affinity for the *arsRDABC* promoter (Figure 13.14) (Itou et al., 2008). The expression of ArsR protein is alleviated by the presence of As³⁺, Sb³⁺, and Bi³⁺ (Xu et al., 1996).

Another As-sensing regulator known in the literature is the AioXRS system, which was proposed to induce the arsenite oxidase-encoding *aioBA* operon in the presence of As³⁺ in *Agrobacterium tumefaciens* (Kashyap et al., 2006), *Herminiimonas arsenicoxydans* (Koechler et al., 2010), and *Rhizobium* sp. strain NT-26 (Sardiwal et al., 2010). The *aioXSR* genes have been detected upstream of the *aioAB* (formerly known as AroA and AoxB, homologous to ArrA and ArrB) operon in several bacteria, including *Thiomonas arsenitoxydans* (Stolz et al., 2010; Lett et al., 2011; Slyemi et al., 2013; Jiang et al., 2014; Moinier et al., 2014). In some bacteria, another operon *arx* (containing *arxRSXABCD*) is required for chemoautotrophic growth with As³⁺ coupled to NO₃⁻ (nitrate)

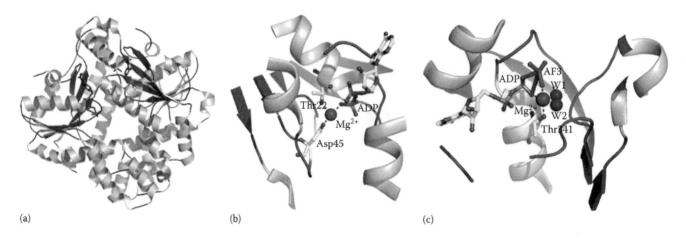

(a)   (b)   (c)

FIGURE 13.12 3D structure of ArsA from *E. coli*. (a) The overall tertiary structure of ArsA (PDB id: 1IHU), (b) and (c) metal-binding sites of ArsA. The amino acid residues and ligands coordinating metal ion(s) are shown in ball-and-stick model. The water molecules are shown as purple spheres.

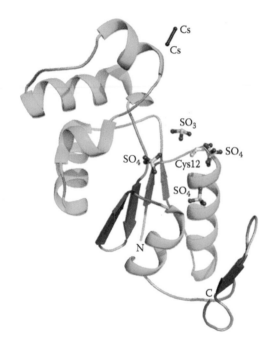

FIGURE 13.13 Tertiary structure of ArsC from *E. coli* (PDB id: 1J9B). The metal binds at the interface of two domains of ArsC. The active site cysteine residue is shown in ball-and-stick model.

respiration (Zargar et al., 2010, 2012). Arsenite oxidase (AioAB) catalyzes the conversion of As³⁺ to As⁵⁺ to detoxify and gain energy (Osborne et al., 2013; Wang et al., 2015a). The enzyme arsenite oxidase converts the highly toxic arsenite (As³⁺) to the relatively more innocuous arsenate (As⁵⁺). The 3D structure of arsenite oxidase is shown in Figure 13.15 (Ellis et al., 2001; Warelow et al., 2013).

13.7 IRON-BINDING PROTEINS

Bacteria have evolved several strategies to acquire and manage iron (Fe) that include (1) deployment of molecules having high affinity for iron or heme scavenging from the surroundings

(Genco and Dixon, 2001), (2) storage of intracellular iron as nutrient source in limiting external supplies (Smith, 2004), (3) employment of redox stress resistance systems to minimize the damage caused by iron-induced reactive oxygen species (Ma et al., 1999), and (4) transcriptional regulation of iron-binding proteins to acquire and store iron (Escolar et al., 1999). The iron stimulon consists of a large number of genes under the control of many different transcriptional regulators, Fur being the most studied regulator (Fillat, 2014). More than 90 genes in *E. coli* (McHugh et al., 2003), 87 in *P. aeruginosa* (Ochsner et al., 2002), and 46 in *Bacillus subtilis* (Baichoo et al., 2002) are known to be regulated by Fur. Figure 13.16 shows the prediction of genes to be regulated by Fur in an extreme acidophile, *Acidithiobacillus ferrooxidans* (Quatrini et al., 2007; Osorio et al., 2008). Iron acquisition includes FeoPABC (Fe²⁺) and MntH (Fe²⁺/Mn²⁺, an ortholog of NRAMP family) for the uptake of ferric iron by means of TonB-dependent OM ferrisiderophore receptors (Tdr) (Figure 13.16) (Andrews et al., 2003; Quatrini et al., 2007; Osorio et al., 2008; Lau et al., 2016). The structures of FeoABC are shown in Figure 13.17 (Koster et al., 2009; Hung et al., 2012; Lau et al., 2013).

Most organisms acquire iron either through siderophore-mediated or siderophore-independent ferric ion transport (Clarke et al., 2001; Boukhalfa and Crumbliss, 2002; Winkelmann, 2002; Raymond et al., 2003; Harrington and Crumbliss, 2009). The majority of pathogenic bacteria employ siderophore-dependent iron acquisition systems while competing for host iron (Miethke and Marahiel, 2007). In contrast, *Haemophilus influenzae* and pathogenic *Neisseria* spp. utilize a highly preserved siderophore-independent high-affinity iron acquisition system (Morton and Williams, 1990; Strange et al., 2011). This system employs specific cell surface receptors that can directly bind host iron-binding proteins such as transferrin (Tf) or lactoferrin (Lf) (Bartnikas, 2012). Intracellular pathogens capture iron from host proteins and import it into their periplasm through an energy-dependent TonB-mediated transport (Strange et al., 2011). The uptake

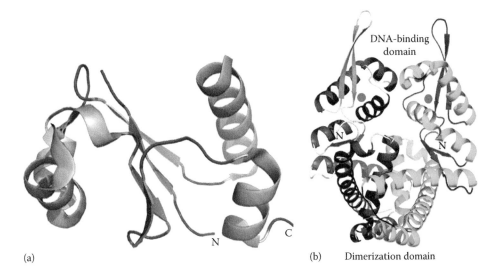

FIGURE 13.14 Tertiary structure of *ars* operon transcriptional regulator(s). (a) 3D structure of ArsD from *E. coli* (PDB id: 3KGK). (b) Overall tertiary and dimeric structure of ArsR from *Pyrococcus horikoshii* (PDB id: 1ULY). The DNA-binding and oligomerization domains are labeled. Two metal-binding sites are indicated by two orange spots.

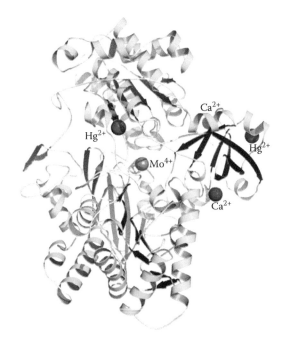

FIGURE 13.15 Tertiary structure of arsenic oxidase (AioAB) from *Alcaligenes faecalis* (PDB id: 1G8K). The molybdenum (Mo^{4+}, orange), mercury (Hg^{2+}, magenta), and calcium (Ca^{2+}, red) ions are shown as spheres. The Fe_3S_4 and Fe_2S_2 iron–sulfur clusters are shown in ball-and-stick model.

of free iron from periplasm to cytosol is mediated through FbpABC transporter, composed of a ferric ion-binding protein (FbpA) and an IM ABC transporter comprising a membrane permease (FbpB) and an ATP-binding protein (FbpC) regulated by Fur (Strange et al., 2011). Much like mammalian transferrins, bacterial transferrin FbpA binds to a single (Fe^{3+}) ion using a common set of amino acid residues with a very high affinity (2.4×10^{18} M^{-1}) (Parker Siburt et al., 2012).

Another set of proteins known as TbpAB binds to transferrin for the uptake of iron (Biville et al., 2014). Once iron crosses the OM, it is deposited into FbpA that further docks to FbpBC to release the iron into cytoplasm.

Another OM transporter FptA helps assimilating iron through the siderophore pyochelin (Pch) that has been shown to chelate a wide range of metal ions such as Ag^+, Al^{3+}, Cd^{2+}, Co^{2+}, Cr^{2+}, Cu^{2+}, Hg^{2+}, Mn^{2+}, Ni^{2+}, Pb^{2+}, Tl^+, and Zn^{2+}, though with lower affinities, in addition to Fe^{3+} (Cobessi et al., 2005; Braud et al., 2009; Cunrath et al., 2015). In some organisms, the siderophore pyoverdine (PVD) is used to chelate iron. The metal complexes with PVD (ferriPVD) are transported back inside the cell across a specific OM receptor, FpvAB (Hartney et al., 2013). The energy required is provided by the H^+-motive force of the IM via the TonB–ExbB–ExbD complex, with which FpvA interacts in the periplasm (Adams et al., 2006; Nader et al., 2011). In *Yersinia* HPI, two IM proteins YbtPQ (members of ABC transporters), which are regulated by AraC-like transcription factor YbtA, help in acquiring iron (Brem et al., 2001; Koh et al., 2016). Another ABC transporter, YfeABCDE, transports Fe and Mn and is regulated by Fur (Perry et al., 2012). The Yfe/Sit systems of A-1 cluster of solute-binding proteins (SBPs) are known to acquire Mn and Fe in several bacteria (Fetherston et al., 2012). Another Fur-regulated ferric reductase involved in iron homeostasis is YqjH, which requires NADPH for its activity (Wang et al., 2011). The expression of *yqjH* is also regulated by another transcription factor encoded by *yqjI* gene, which is a Ni^{2+}-binding repressor (Wang et al., 2014).

Intracellular iron is also detoxified by the storage of excess iron in ferritin-like molecules found in both prokaryotes and eukaryotes (Smith, 2004; Theil and Goss, 2009; Yao et al., 2011). These molecules use O_2 and H_2O_2 as electron acceptors to oxidize Fe^{2+} to Fe^{3+} that is further internalized as a mineral (Osorio et al., 2008). Eukaryotic ferritins are made up of

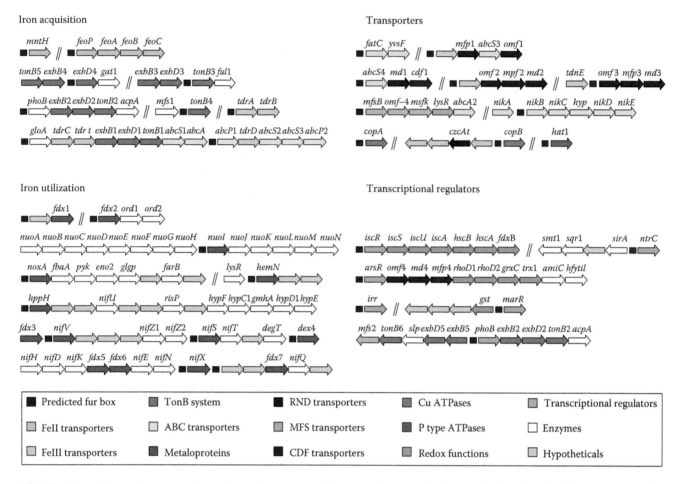

Iron acquisition

mntH feoP feoA feoB feoC

tonB5 exbB4 exbD4 gat1 exbB3 exbD3 tonB3 ful1

phoB exbB2 exbD2 tonB2 acpA mfs1 tonB4 tdrA tdrB

gloA tdrC tdr t exbB1 exbD1 tonB1 abcS1 abcA abcP1 tdrD abcS2 abcS3 abcP2

Transporters

fatC yvsF mfp1 abcS3 omf1

abcS4 md1 cdf1 omf2 mpf2 md2 tdnE omf3 mfp3 md3

mfsB omf−4 msfk lysR abcA2 nikA nikB nikC hyp nikD nikE

copA czcAt copB hat1

Iron utilization

fdx1 fdx2 ord1 ord2

nuoA nuoB nuoC nuoD nuoE nuoF nuoG nuoH nuoI nuoJ nuoK nuoL nuoM nuoN

noxA fbaA pyk eno2 glgp farB lysR hemN

hppH nifU risP hypF hypC1 gmhA hypD1 hypE

fdx3 nifV nifZ1 nifZ2 nifS nifT degT dex4

nifH nifD nifK fdx5 fdx6 nifE nifN nifX fdx7 nifQ

Transcriptional regulators

iscR iscS iscU iscA hscB hscA fdxB smt1 sqr1 sirA ntrC

arsR omf4 md4 mfp4 rhoD1 rhoD2 grxC trx1 amiC hfytil

irr gst marR

mfs2 tonB6 slp exbD5 exbB5 phoB exbB2 exbD2 tonB2 acpA

■ Predicted fur box	■ TonB system	■ RND transporters	■ Cu ATPases	■ Transcriptional regulators
■ FeII transporters	■ ABC transporters	■ MFS transporters	■ P type ATPases	▢ Enzymes
■ FeIII transporters	■ Metaloproteins	■ CDF transporters	■ Redox functions	▢ Hypotheticals

FIGURE 13.16 Schematic representation of predicted Fur-regulated gene clusters and their associated predicted Fur boxes grouped into four main functional categories: (1) iron acquisition, (2) iron utilization, (3) transporters, and (4) transcriptional regulators. Arrows represent the direction of transcription of each gene and are not up to scale. The double-hashed line separating independent gene clusters indicates their noncontiguous nature in the genome. (From Quatrini, R. et al., *Nucleic Acids Res.*, 35, 2153, 2007. With permission.)

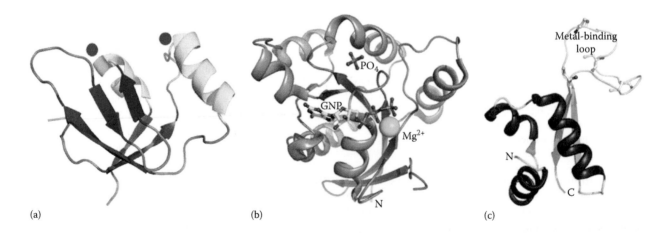

(a) (b) (c)

FIGURE 13.17 Tertiary structures of FeoABC. The overall structure of (a) FeoA from *E. coli* (PDB id: 2LX9); the metal-binding sites are shown as orange dots. (b) FeoB from *Methanocaldococcus jannaschii* (PDB id: 2WJI); the metal ion (Mg^{2+}) bound in the crystal structure is shown as sphere in green, and ligands are shown in ball-and-stick model. (c) FeoC from *Klebsiella pneumoniae* (PDB id: 2K02); the metal-binding loop is labeled.

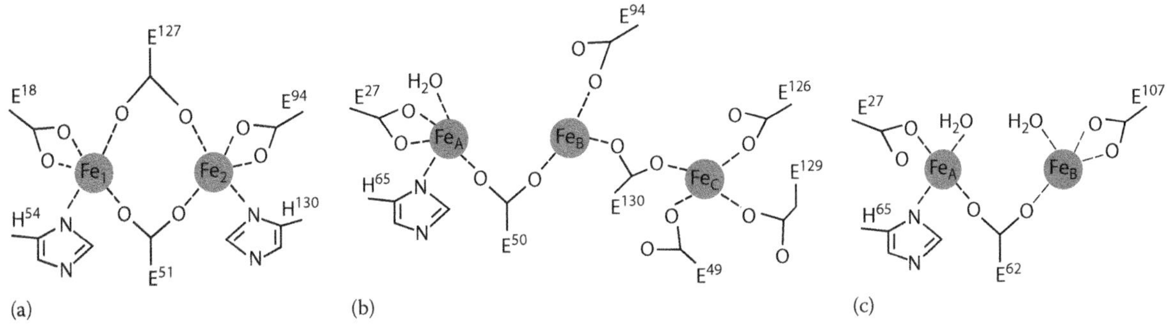

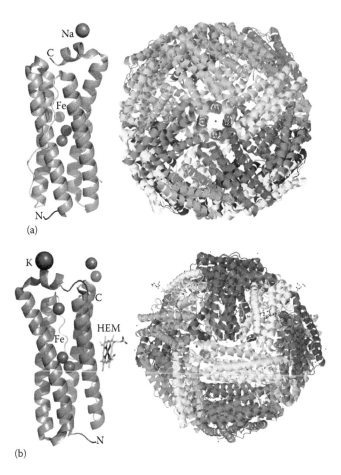

FIGURE 13.18 A schematic view of (a) a symmetrical ferroxidase center typical of bacterioferritin (Bfr) where Fe_1 and Fe_2 are coordinated by two glutamate residues (the numbering scheme of residues is as per the sequence of *P. aeruginosa* BfrB (PaBfrB)), (b) *E. coli* ferroxidase center and archaeal Ftn ("site C" [Fec] is included in addition to ferroxidase iron Fe_A and Fe_B; the numbering scheme of residues is as per the sequence of *E. coli* FtnA [EcFtnA]), and (c) ferroxidase center of human ferritin adapted from the crystal structure of its Tb^{3+} derivative (Lawson et al., 1991). (From Yao, H. et al., *Biochemistry*, 50, 5236, 2011. With permission.)

FIGURE 13.19 Tertiary and quaternary structures of FtnA and BfrB. (a) Left: tertiary structure of FtnA from *P. aeruginosa* (PDB id: 3R2R). Fe^{2+} and Na^+ ions are shown as orange and purple spheres, respectively. Right: the overall quaternary (24-mer) structure of FtnA. The iron metal ions are stored in the void space generated by the quaternary structure. (b) Left: tertiary structure of BfrB from *P. aeruginosa* (PDB id: 5D8P). Fe^{2+} and Na^+ ions are shown as orange and violet spheres, respectively. The HEM group bound to the protein is shown in ball-and-stick model. Right: the quaternary structure of BfrB (24 mer). The iron metals are stored in the void space generated by the overall structure.

two isostructures (H and L) assembled in a 24-mer quaternary structure (Liu and Theil, 2005). In contrast, bacteria contain two different ferritin-like molecules, that is, ferritins (Ftn) and bacterioferritins (Bfr) (Smith, 2004; Wang et al., 2015b). Like eukaryotic ferritins, bacterial Ftns are composed of 24 subunits assembled into a spherical protein shell wherein the iron can be stored (Figures 13.18 and 13.19). On the other hand, bacterioferritins contain heme groups (12 per biological molecule) that chelate iron (Andrews, 2010; Le Brun et al., 2010).

13.8 COPPER-BINDING PROTEINS

The presence of excess copper (Cu) affects the activity of various enzymes such as sulfite oxidase and ammonium persulfate reductase (Xia et al., 2010). It also affects lipid oxidation (Hong et al., 2012), replaces several other metal ions bound to proteins (Sellin et al., 1987), forms spurious disulfide bonds (Hiniker et al., 2005), and oxidizes and degrades iron–sulfur clusters in proteins (Macomber and Imlay, 2009). Copper can induce the generation of free radicals through the Haber–Weiss and Fenton reactions, which are capable of damaging cell membranes, proteins, and DNA; thus, copper shows high cytotoxicity, in general (Halliwell and Gutteridge, 1984; Hordyjewska et al., 2014). Several organisms, acidophiles in particular, show tolerance to copper even at higher concentrations (>100 mM) (Navarro et al., 2013). Genome analyses of these organisms show that they contain a high number of genes encoding copper-binding proteins. In bacteria, one of the most studied systems that control copper content is encoded by the *cop* operon (Shafeeq et al., 2011). One such set of genes is CopABCD involved in copper import or export regulated by CopRS proteins (Figure 13.20) (Adaikkalam and Swarup, 2005; Zhang et al., 2006; Hu et al., 2009; Jayakanthan et al., 2012; Fu et al., 2013).

In addition to *cop* operon, a plasmid-encoded copper resistance system *pco* encoding PcoABCDRSE proteins has been described in bacteria (Munson et al., 2000). In *E. coli*, four structural genes *pcoABCD* are involved in copper resistance. PcoA is a multicopper oxidase, which is able

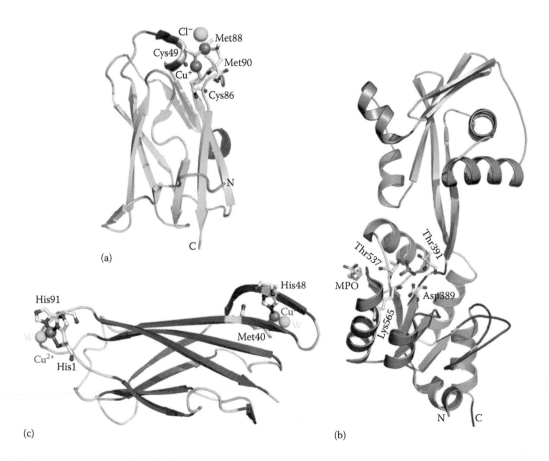

FIGURE 13.20 Tertiary structure of CopABC. (a) The overall structure and metal-binding site of CopA (PDB id: 4F2F). The metal ions (Cu+, orange and Cl−, cyan) are shown as sphere. (b) The tertiary structure and metal-binding site of CopB (PDB id: 3SKY). (c) 3D structure of CopC (PDB id: 2C9Q). The metal ions (Cu2+, orange) and water molecules (W, cyan) are shown as sphere. The active site residues known to be involved in metal binding are shown in ball-and-stick model.

to oxidize PcoC-bound Cu^+ to its less toxic form Cu^{2+} (Djoko et al., 2008). Both the proteins, PcoC and CopC, are highly homologous, containing β-barrels that bind both Cu^+ and Cu^{2+} at sites separated by ~30 Å (Djoko et al., 2007). These sites are tailored to bind their individual ions with high affinity (K_D ~ 10^{-13} M):Cu^+(His)(Met)$_{2\ or\ 3}$ (trigonal or tetrahedral), Cu^{2+}(His)$_2$A(Nterm)(OH$_2$) (Djoko et al., 2007). In addition, *pcoR* and *pcoS* genes encode their regulatory factors much like *copR* and *copS* in *cop*-encoded proteins. The *pcoR* (receiver) gene products are transcriptional regulators that receive regulatory information from the *pcoS* (sensor) gene products (Munson et al., 2000).

In Gram-positive bacterium *B. subtilis*, two proteins YcnJK (a homolog of CopCD) play an important role in copper metabolism (Chillappagari et al., 2009). The *ycnK* gene located upstream to *ycnJ* was shown to encode a transcriptional regulator that acts, in addition to regulator CsoR, as a copper-specific repressor for *ycnJ* (Chillappagari et al., 2009). In *E. coli*, copper homeostasis is mainly controlled by CueR, a MerR-like transcriptional regulator. CueR is responsive to Cu^+ but also shows affinities toward Ag^+ and Au^+ (Rademacher and Masepohl, 2012). Multicopper oxidases (MCOs) are enzymes that are critically involved in copper detoxification and homeostasis; genes encoding

MCO have been identified in animals, plants, insects, fungi, and bacteria (Wherland et al., 2014). MCOs have also been shown to be required for the virulence of some pathogens such as *Salmonella* (Achard et al., 2010), *Xanthomonas campestris* (Hsiao et al., 2011), and *Mycobacterium tuberculosis* (Rowland and Niederweis, 2013). In addition, an MCO, namely, CueO, oxidizes Cu^+ to Cu^{2+}, the latter ion having only limited potential to enter the cytoplasm (Cortes et al., 2015). A large subfamily of MCOs is called laccases, which are mostly found in plants, fungi, and microorganisms (Moin and Omar, 2014). Another substrate-binding protein, namely, CeuP, a periplasmic copper-binding protein, is involved in the transfer of copper ions to SodCII. Biophysical analyses suggest that CeuP can bind both Cu^{2+} and Zn^{2+} (Yoon et al., 2013).

A second group of proteins that has a crucial role in maintaining copper levels within the cell is the *cut* family. In *E. coli*, *cutABCDEF* genes have been shown to be involved in copper uptake, intracellular storage, and efflux (Rouch et al., 1989). One of the most studied members of this family is the CutC protein, which is suggested to be directly or indirectly involved in protection from excess amount of copper (Latorre et al., 2011). The tertiary structure of human CutC adopts a typical TIM-barrel fold present in cytoplasmic

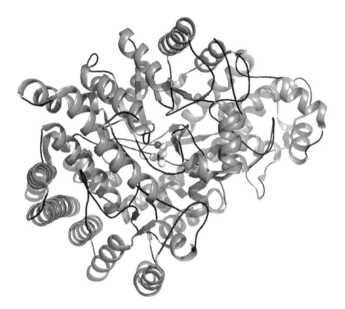

FIGURE 13.21 A cartoon representation of the tertiary structure of CutC (PDB id: 5A0U) showing the location of the potential Cu⁺-binding site. The choline ion (CHT) bound in the active site of the CutC protein is shown in ball-and-stick model.

proteins (Figure 13.21) (Kalnins et al., 2015), with two conserved cysteine residues for potential participation in copper coordination. A laccase-like MCO, namely, CutO, was described recently from *Rhodobacter capsulatus*, purple nonsulfur photosynthetic bacteria (Wiethaus et al., 2006). In *M. tuberculosis*, an MT, namely, MymT, is involved in copper homeostasis (Gold et al., 2008). MymT is most likely the first MT from a Gram-positive bacterium for which a function, that is, binding Cu⁺ and protecting the cell from copper toxicity, is known. Although the Cys–X–His–X–X–Cys–X–Cys motif of MymT is mirrored in SmtA, the function of the second motif (Cys–His–Cys–(X)₂–G–(X)₂–Tyr–Arg–Cys–Thr–Cys) of MymT is unknown (Festa et al., 2011; Rowland and Niederweis, 2012).

13.9 NICKEL- AND COBALT-BINDING PROTEINS

In general, nickel (Ni) uptake systems are classified into two broad groups: (1) single-component Ni^{2+} permeases and (2) ABC transporters (Eitinger and Mandrand-Berthelot, 2000; Rodionov et al., 2006). The single-component permeases include three different families, namely, NiCoT, HupE/UreJ, and UreH, whereas the multicomponent ABC transporters include NikMQO/CbiMQO and NikABCDE families. One of the most studied nickel uptake systems of bacteria is *nik* operon, encoding several proteins such as NikABCDER, which is required under anaerobic growth conditions, and Ni-(L-His)₂ complex, which acts as a substrate (Table 13.6) (Lebrette et al., 2013). The NikA protein is a periplasmic nickel-binding protein, NikB and NikC proteins are integral IM proteins, and NikD and NikE are membrane-associated proteins with intrinsic ATPase activity (Figure 13.22) (Cherrier et al., 2008; Lebrette et al., 2015).

TABLE 13.6
List of Genes, Proteins Encoded by *nik* Operon, and Their Function

Gene	Protein Encoded	Functional Role
nikA	ABC transporter Ni²⁺-binding component	Binds Ni²⁺ in the periplasm and transfers it to the permease and may also play a role in chemotaxis away from toxic concentrations of nickel
nikB and *nikC*	ABC transporter permease components	Transport of Ni²⁺ across the inner membrane
nikD and *nikE*	ABC transporter ATP-binding components	Hydrolyzes ATP to provide energy for the transport process
nikR	Transcriptional regulator of the *nikABCDE* operon	Downregulates the expression of the *nik* operon in the presence of excess Ni²⁺ ions

In *E. coli*, the expression of *nikABCDE* is regulated by the NikR transcription factor (Chivers and Tahirov, 2005; Chivers et al., 2012). The *nikR* expression is regulated by two promoters: (1) FNR regulon and (2) a 51 bp upstream region of the start site of the *nikR* gene. The FNR-controlled regulation of *nikABCDE–nikR* takes place at a NikR-binding site located upstream of the *nikA* gene (Figure 13.23) (West et al., 2012).

The protein HupE shows homology to UreJ, encoded in several urease gene clusters as found in *Bordetella bronchiseptica*. The association of HupE and UreJ proteins with Ni-dependent enzymes led to the assignment of their role as a nickel permease. In *Rhizobium leguminosarum*, synthesis of the hydrogen uptake (Hup) hydrogenase requires a total of 18 genes (*hupSLCDEFGHIJK-hypABFCDEX*) (Eitinger et al., 2005). HupN is a TM protein initially found in *Bradyrhizobium japonicum* that has a high homology with HoxN (Ni²⁺-binding protein) of *Alcaligenes eutrophus* (Fu et al., 1994). In addition, another periplasmic SBP, namely, CeuE, present both in Gram-negative and Gram-positive bacteria can capture (Ni²⁺/Co²⁺) ions or complexes. Once in the periplasm, the metal ions are transferred to IM translocators FecDE (Shaik et al., 2014). In *H. pylori*, another set of proteins, ExbB/EsbD/TonB, are proposed to import nickel captured from the environment by an unknown nickelophore (Schauer et al., 2007).

13.10 URANIUM-BINDING PROTEINS

In nature, uranium (U) is predominantly found in its hexavalent oxidation state U^{6+} as the linear dioxouranyl form (UO_2^{2+}). The coordination of uranyl to proteins is similar to that of Ca²⁺, which form electrostatic interactions, preferentially with hard donor oxygen ligands arranged in distorted octahedral or pentagonal bipyramidal structures. Currently, several plants and microorganisms are used for the removal of heavy metals like mercury, lead, and cadmium

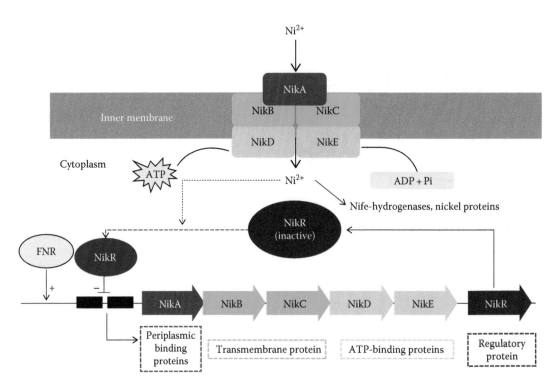

FIGURE 13.22 Regulation of nickel (Ni^{2+}) ion uptake by components of *nik* operon.

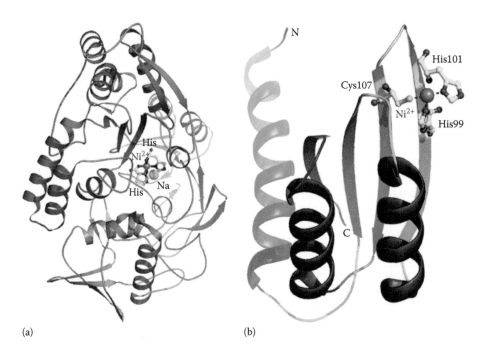

(a)   (b)

FIGURE 13.23 Tertiary structure of NikA and NikR. (a) The overall tertiary structure of NikA from *S. aureus* (PDB id: 4XKR). The cognate metal (Ni^{2+}) ion bound to the active site of the molecule is shown as an orange sphere. The Ni-(L-His)(2-methyl-thiazolidine dicarboxylate) molecule, to which Ni^{2+} is bound, is shown in ball-and-stick model. (b) Cartoon representation of the 3D structure of NikR from *Helicobacter pylori* (PDB id: 3PHT). The active site residues (ball-and-stick model) bound to the metal (sphere) ion are also shown.

(Gaur et al., 2014). In addition, magnetotactic bacteria biomineralize magnetite (Fe_3O_4) or greigite (Fe_3S_4) within their magnetosomes (Posfai and Dunin-Borkowski, 2009). In nature, certain bacteria, fungi, and plants reduce U^{6+} to U^{4+}, thus making them a promising tool for bioremediation by decreasing the bioavailability of U (Williams et al., 2013). In addition, a number of serum proteins, including transferrin, serum albumin, designed peptides, and DNA, have been found to interact with uranyl mainly through carboxylic groups like aspartate and glutamate, of which ceruloplasmin, hemopexin, and complement proteins show a high capacity to bind U with a stoichiometry of >1 (Vidaud et al., 2005; Pible et al., 2006).

Among several metal sensors reported for the detection and treatment of U poisoning, a catalytic beacon sensor for U-based DNAzyme was established recently (Liu et al., 2007). However, these sensors show low selectivity in detecting U in contaminated soils. Recently, a chimeric spider silk protein was tagged with U-recognition motif to treat U exposures to humans (Gorden et al., 2003) and to conduct environmental recovery operations, including monitoring (Liu et al., 2007) nuclear waste management (Armstrong et al., 2012), developmental biology (Hillson et al., 2007), and clinical toxicology (Vicente-Vicente et al., 2010). Such remarkable properties of these proteins prompt interest in their functionalization for further enhancement of their properties, such as heavy metal sequestration and recovery (Spiess et al., 2010). In another study, a chimeric protein that binds U along with 33-amino acid residues from calmodulin protein was designed (Krishnaji and Kaplan, 2013). However, the mechanism of uranium interaction with proteins at the molecular level is at its infancy (Averseng et al., 2010; Michon et al., 2010; Pible et al., 2010).

13.11 ZINC-BINDING PROTEINS

Although zinc (Zn) is essential for enzyme-mediated catalysis and its structure, Zn from zinc sulfide (ZnS) causes toxicity by forming metal complexes with cellular components. Zn import is mediated by high affinity ZnuABC, MntH, and zupT IRT-like protein (ZIP) transporters (Kambe et al., 2014). Znu, first identified in *E. coli*, belongs to the ABC transporter family (2005). ZnuABC includes three components: the ZnuA periplasmic Zn^{2+}-binding protein (Figure 13.24) (Yatsunyk et al., 2008), the ZnuB integral membrane transporter responsible for Zn^{2+} transport across the IM, and the ZnuC ATPase that couples ATP hydrolysis to the transport process.

The ZevAB uptake system in bacteria mostly functions in zinc-limiting conditions, in contrast to ZnuABC that requires a rich culture medium for optimal growth (Rosadini et al., 2011). In bacteria, there are two major types of zinc import systems for high- and low-affinity zinc uptake (Hantke, 2005). In *E. coli*, high-affinity zinc uptake is mediated via the cooperation of ZnuABC and ZinT (formerly known as YodA) (Colaco et al., 2016). On the other hand, ZupT, a low-affinity zinc uptake transporter, is a cytoplasmic membrane protein and a member of the ZIP family of divalent metal ion transporters (ZRT, IRT-like protein) regulated by Zur transcriptional regulator. ZipB is another protein that can selectively bind Zn^{2+} and Cd^{2+}, whereas it rejects transition metal ions (Lin et al., 2010). ZraP is a periplasmic zinc-binding protein and protects the periplasmic enzymes from high zinc concentrations to a certain extent. In *E. coli*, two proteins ZraS (HydH) and ZraR (HydG) play a role in sensing and regulating the expression of ZraP in a Zn^{2+}- and Pb^{2+}-ion dependent manner. The protein ZraS senses the

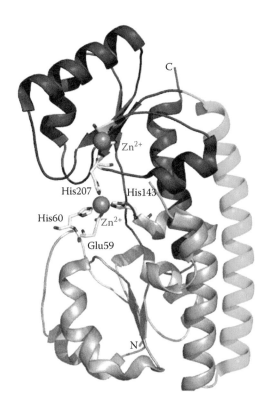

FIGURE 13.24 Structure of ZnuA. Cartoon diagram of the tertiary structure of ZnuA from *E. coli* (PDB id: 2PRS) showing N-terminal domain (pink), the C-terminal domain (blue), and the connecting α-helix (green). The zinc (Zn^{2+}) ions are represented as orange spheres. The active site residues coordinating the metal ion(s) are shown in ball-and-stick model.

presence of extracellular Zn^{2+} or Pb^{2+} and activates ZraR via phosphorylation that further induces the expression of ZraP (Jaroslawiecka and Piotrowska-Seget, 2014). The C-terminal ZraP domain facilitates the modulation of another zinc transporter ZntA (Noll et al., 1998). The *tro* operon, first identified in the spirochaete Gram-negative bacterium *Treponema pallidum*, encodes five proteins (an ABC transporter *troABCD* and a DtxR-like transcriptional regulator *troR*) (Lee et al., 1999; Brett et al., 2008). The DtxR family of transcriptional regulators is also shown to be involved in Mn and Fe homeostasis (Brett et al., 2008).

13.12 CHROMIUM-BINDING PROTEINS

Chromate (CrO_4^{2-}) is produced as a by-product of numerous industrial processes, including chrome-plating, pigment production, thermonuclear weapons manufacture, and welding. As a consequence, chromate is found to be the second most common heavy metal contaminant in waste sites with a concentration as high as 170 mM (Riley et al., 1992). Chromate bears structural similarity to SO_4^{2-} and is readily taken up by bacterial and mammalian cells through the SO_4^{2-} transport system (Cervantes et al., 2001). In some bacteria, chromium resistance has been observed by involving precipitation and enzyme-mediated conversion of Cr^{6+}

(Cummings et al., 2006). Two potential reduction systems involving C-type cytochromes (ApaC) and a predicted NADPH-dependent Cr^{6+} reductase have been identified in some acidophiles (Magnuson et al., 2010). Using sequence analyses, two classes of novel bacterial chromate reductases were identified (Park et al., 2002). Two enzymes from Class I, namely, ChrR of *P. putida* and YieF of *E. coli*, reduce chromate to Cr^{3+} (Park et al., 2002; Ackerley et al., 2004). Class II chromate reductases, which also show nitroreductase activity, bear no homology to Class I enzymes but show homology with a chromate reductase from *Pseudomonas ambigua* (Suzuki et al., 1992). Two proteins, namely, NfsA (oxygen-insensitive nitroreductase) from *E. coli* and ChfN from *B. subtilis*, members of the Class II family, also possess chromate reductase activity (Ackerley et al., 2004).

13.13 MOLYBDENUM-BINDING/ TUNGSTATE-BINDING PROTEINS

Molybdate (MoO_4^{2-}) and tungstate (WO_4^{2-}) anions are the primary sources of essential metals molybdenum (Mo) and tungsten (W) in bacterial cells (Hille, 2002). These soluble oxyanions (MoO_4^{2-} and WO_4^{2-}) enter cells through specific ATP-binding cassette (ABC) transport systems, which, in prokaryotes, are divided into three different families: Mod,

Wtp, and Tup (Figure 13.25) (Schwarz et al., 2007). The genes encoding the three components are organized in an operon (*mod/wtpABC*) or (*tupABC*) regulated by ModE in the case of the *modABC* operon (Schwarz et al., 2009; Hagen, 2011). In *E. coli*, MoO_4^{2-} may be taken up through three transport systems: (1) the high-affinity ModABC system, (2) the low-affinity CysPTWA (SulT) sulfate–thiosulfate permease, and (3) a nonspecific low-efficiency anion transporter that transports molybdate, sulfate, selenate, and selenite (Self et al., 2001; Pau, 2004; Aguilar-Barajas et al., 2011). The transporter ModABC, a member of the MolT family, is also known to import WO_4^{2-} and SO_4^{2-} (Figure 13.25) (Self et al., 2001). The periplasmic molybdate-binding protein ModA (Figure 13.26) specifically binds MoO_4^{2-} or WO_4^{2-} with a K_d of ~20 nM (Chan et al., 2010). The integral membrane protein ModB builds the membrane channel of the ModABC transporter. The protein ModC possesses the ATPase activity that energizes MoO_4^{2-} transport. The expression of ModABC transporting proteins is regulated by the transcriptional factor ModE in the presence of MoO_4^{2-}. The active form of ModE regulates the expression of ModA by binding to its operator (Anderson et al., 1997; Self et al., 2001). The transcriptional factor belonging to the TunR family is proposed to control WO_4^{2-} and MoO_4^{2-} homeostasis in sulfate-reducing deltaproteobacteria (Kazakov et al., 2013).

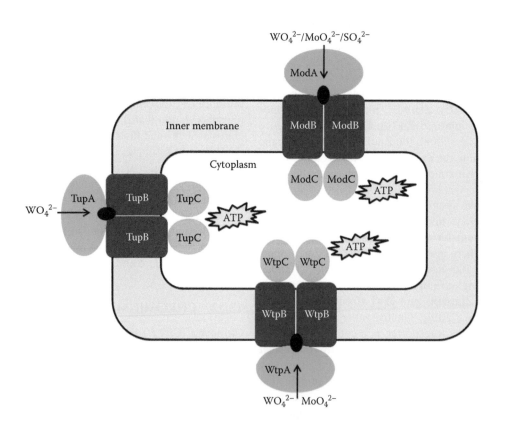

FIGURE 13.25 Diagrammatic view of bacterial oxyanion importers. All three importer systems (1) molybdate transporters (ModABC) and (2) tungstate transporters (TupABC and WtpABC) are shown schematically. The ModABC importer can transport all three oxyanions (molybdate (MoO_4^{2-}), tungstate (WO_4^{2-}), and sulfate (SO_4^{2-})). The WtpABC importer can allow the uptake of tungstate (WO_4^{2-}) as well as molybdate (MoO_4^{2-}). The TupABC importer specifically allows the transport of tungstate (WO_4^{2-}).

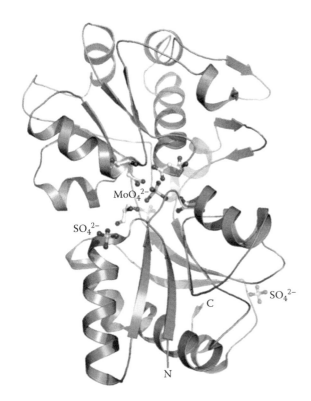

FIGURE 13.26 Structure of the ModA. The overall 3D structure of ModA from *Methanosarcina acetivorans* (PDB id: 3K6W). The two domains of ModA are shown in pink and blue. The molybdate (MoO_4^{2-}) ion, sulfate (SO_4^{2-}) ions, and active site residues are shown in ball-and-stick model.

13.14 TELLURIUM-BINDING PROTEINS

Both organic and inorganic tellurite (TeO_3^{2-}) soluble salts have been found to be toxic for most forms of life even at very low concentrations (Chasteen et al., 2009). The oxyanion form of tellurium (tellurite TeO_3^{2-}) is highly toxic for most microorganisms as it oxidizes cellular thiols like cysteines and generates oxidative stress (Calderon et al., 2009; Chasteen et al., 2009; Rigobello et al., 2011). Tellurite-resistant bacteria have been isolated that often reduce TeO_3^{2-} to its elemental form tellurium ($Te°$), which usually accumulates as black deposits inside the cell (Ottosson et al., 2010; Chien et al., 2011). Conversion of TeO_3^{2-} to $Te°$ involves unspecific reductases, such as nitrate reductase or catalase, but also leads to the production of superoxide radicals (Calderon et al., 2006; Castro et al., 2008; Chasteen et al., 2009). In *E. coli*, the *ter* operon encodes TerZABCDE involved in tellurite resistance (Ponnusamy et al., 2011). Interestingly, the expression of *terXYW* genes is required to express either *terZABCDEF* or *terZA* in *E. coli* cells to avoid lethality associated with the expression of the latter gene clusters. In *Bacillus anthracis*, two genes *yceGH* have been found to provide tellurite resistance (Franks et al., 2014). Another tellurite resistance protein TrgB, first identified in *Rhodobacter sphaeroides*, consists of two domains exhibiting similarity to γ-glutamyl cyclotransferases and the

ADP-ribose pyrophosphatase subfamily of nudix (nucleoside diphosphate) hydrolases (Marchler-Bauer et al., 2011).

13.15 CADMIUM-BINDING PROTEINS

Cadmium (Cd^{2+}) is a highly stable and toxic heavy metal ion, which once released into the environment remains nondegradable in nature. Cd^{2+} shows affinity toward many essential respiratory enzymes and may cause oxidative stress and cancer (Nies, 2003; Banjerdkji et al., 2005). Cadmium (Cd^{2+}) is mostly taken up inside the cell by Zn^{2+} and/or Mn^{2+} transport systems of Gram-negative and Gram-positive bacteria (Tynecka et al., 1981; Archibald and Duong, 1984; Laddaga and Silver, 1985; Laddaga et al., 1985; Burke and Pfister, 1986). However, Gram-negative bacteria such as *P. aeruginosa* and *E. coli* show less sensitivity toward Cd^{2+} than Gram-positive bacteria such as *Staphylococcus aureus*, *B. subtilis*, and *Streptococcus faecium* (Morozzi et al., 1986). In *Lactobacillus plantarum*, another protein MntA, a Mn^{2+}-binding protein, is reported to bind Cd^{2+} as well (Hao et al., 1999).

13.16 LEAD-BINDING PROTEINS

Of all the heavy metals, lead (Pb^{2+}) is the most abundant metal used since prehistoric times; thus, uptake of Pb^{2+} by the environment and exposure to humans have consequently increased (Smith, 1984). Pb^{2+} can affect heme synthesis and impair psychological and neurobehavioral functions (Tong et al., 2000). Pb^{2+} resistance has been reported in both Gram-negative and Gram-positive bacteria isolated from Pb^{2+}-contaminated soils (Roane, 1999). Bacterial strains having the ability to accumulate in the form of Pb–phosphate complex have been isolated (Levinson and Mahler, 1998). Furthermore, bacterial strains have also been identified that encode proteins for Pb resistance, *pbr* operon being one among them (Borremans et al., 2001). Recently, in *P. aeruginosa* WI-1 strain, an MT BmtA was suggested to confer the resistance against Pb^{2+} ions (Naik et al., 2012a; Naik and Dubey, 2013). Thus, the bioremediation of lead present in adulterated environmental sites is possible by employing microorganisms such as *Bacillus megaterium*, *P. aeruginosa* WI-1, *Salmonella choleraesuis* 4A, and *Proteus penneri* GM10 (Naik et al., 2012a,b).

REFERENCES

Achard, M.E., Tree, J.J., Holden, J.A., Simpfendorfer, K.R., Wijburg, O.L., Strugnell, R.A., McEwan, A.G. 2010. The multi-copper-ion oxidase CueO of *Salmonella enterica* serovar Typhimurium is required for systemic virulence. *Infection and Immunity* 78: 2312–231.

Ackerley, D.F., Gonzalez, C.F., Keyhan, M., Blake, R., Matin, A. 2004. Mechanism of chromate reduction by the *Escherichia coli* protein, NfsA, and the role of different chromate reductases in minimizing oxidative stress during chromate reduction. *Environmental Microbiology* 6: 851–860.

Adaikkalam, V., Swarup, S. 2005. Characterization of copABCD operon from a copper-sensitive *Pseudomonas putida* strain. *Canadian Journal of* Microbiology 51: 209–216.

Adams, H., Zeder-Lutz, G., Greenwald, J., Schalk, I.J., Celia, H., Pattus, F. 2006. Interaction of TonB with outer membrane receptor FpvA of *Pseudomonas aeruginosa*. *Journal of Bacteriology* 188: 5752–5761.

Aguilar-Barajas, E, Daaz-Perez, C., Ramirez-Diaz, M.I., Riveros-Rosas, H., Cervantes, C. 2011. Bacterial transport of sulfate, molybdate, and related oxyanions. *Biometals* 24: 687–707.

Ahluwalia, S.S., Goyal, D. 2007. Microbial and plant derived biomass for removal of heavy metals from wastewater. *Bioresource Technology* 98: 2243–2257.

Ainza, C., Trevors, J., Saier, M. 2010. Environmental mercury rising. *Water, Air and Soil Pollution* 205: 47–48.

Anderson, L.A., Palmer, T., Price, N.C., Bornemann, S., Boxer, D.H., Pau, R.N. 1997. Characterisation of the molybdenum responsive ModE regulatory protein and its binding to the promoter region of the modABCD (molybdenum transport) operon of *Escherichia coli*. *European Journal of Biochemistry* 246: 119–126.

Andrews, S.C. 2010. The Ferritin-like superfamily: Evolution of the biological iron storeman from a rubrerythrin-like ancestor. *Biochimica et Biophysica Acta* 1800: 691–705.

Andrews, S.C., Robinson, A.K., Rodriguez-Quinones, F. 2003. Bacterial iron homeostasis. *FEMS Microbiology Reviews* 27: 215–237.

Archibald, F.S., Duong, M. 1984. Manganese acquisition by *Lactobacillus plantarum*. *Journal of Bacteriology* 158: 1–8.

Armstrong, C.R., Nyman, M., Shvareva, T., Sigmon, G.E., Burns, P.C., Navrotsky, A. 2012. Uranyl peroxide enhanced nuclear fuel corrosion in seawater. *Proceedings of the National Academy and Sciences of the United States of America* 109: 1874–1877.

Averseng, O., Hagege, A., Taran, F., Vidaud, C. 2010. Surface plasmon resonance for rapid screening of uranyl affine proteins. *Analytical Chemistry* 82: 9797–9802.

Baichoo, N., Wang, T., Ye, R., Helmann, J.D. 2002. Global analysis of the *Bacillus subtilis* Fur regulon and the iron starvation stimulon. *Molecular Microbiology* 45: 1613–1629.

Baker-Austin, C., Dopson, M. 2007. Life in acid: pH homeostasis in acidophiles. *Trends in Microbiology* 15: 165–171.

Baker-Austin, C., Dopson, M., Wexler, M., Sawers, G., Bond, P.L. 2005. Molecular insight into extreme copper resistance in the extremophilic archaeon "*Ferroplasma acidarmanus*" FerI. *Microbiology* 151: 2637–2646.

Banjerdkji, P., Vattanaviboon, P., Mongkolsuk, S. 2005. Exposure to cadmium elevates expression of genes in the oxyR and OhrR regulons and induces cross–resistance to peroxide killing treatment in *Xanthomonas campestris*. *Applied and Environmental Microbiology* 71: 1843–1849.

Barkay, T. 1987. Adaptation of aquatic microbial communities to hg stress. *Applied and Environmental Microbiology* 53: 2725–2732.

Barkay, T., Kritee, K., Boyd, E., Geesey, G. 2010. A thermophilic bacterial origin and subsequent constraints by redox, light and salinity on the evolution of the microbial mercuric reductase. *Environmental Microbiology* 12: 2904–2917.

Barkay, T., Miller, S.M., Summers, A.O. 2003. Bacterial mercury resistance from atoms to ecosystems. *FEMS Microbiology Reviews* 27: 355–384.

Barondeau, D.P., Getzoff, E.D. 2004. Structural insights into protein–metal ion partnerships. *Current Opinion in Structural Biology* 14: 765–774.

Bartnikas, T.B. 2012. Known and potential roles of transferrin in iron biology. *Biometals* 25: 677–686.

Basketter, D.A., Angelini, G., Ingber, A., Kern, P.S., Menne, T. 2003. Nickel, chromium and cobalt in consumer products: Revisiting safe levels in the new millennium. *Contact Dermatitis* 49: 1–7.

Biville, F., Brezillon, C., Giorgini, D., Taha, M.K. 2014. Pyrophosphate-mediated iron acquisition from transferrin in *Neisseria meningitidis* does not require TonB activity. *PLoS ONE* 9: e107612.

Blindauer, C.A. 2011. Bacterial metallothioneins: Past, present, and questions for the future. *Journal of Biological and Inorganic Chemistry* 16: 1011–1024.

Blindauer, C.A. 2013. Lessons on the critical interplay between zinc binding and protein structure and dynamics. *Journal of Inorganic Biochemistry* 121: 145–155.

Blindauer, C.A., Harrison, M.D., Parkinson, J.A., Robinson, A.K., Cavet, J.S., Robinson, N.J., Sadler, P.J. 2001. A metallothionein containing a zinc finger within a four-metal cluster protects a bacterium from zinc toxicity. *Proceedings of the National Academy of Science of the United States of America* 98: 9593–9598.

Blindauer, C.A., Harrison, M.D., Robinson, A.K., Parkinson, J.A., Bowness, P.W., Sadler, P.J., Robinson, N.J. 2002. Multiple bacteria encode metallothioneins and SmtA-like zinc fingers. *Molecular Microbiology* 45: 1421–1432.

Borremans, B., Hobman, J.L., Provoost, A., Brown, N.L., van Der Lelie, D. 2001. Cloning and functional analysis of the *pbr* lead resistance determinant of *Ralstonia metallidurans* CH34. *Journal of Bacteriology* 183: 5651–5658.

Boukhalfa, H., Crumbliss, A.L. 2002. Chemical aspects of siderophore mediated iron transport. *Biometals* 15: 325–339.

Boyd, E.S., Barkay, T. 2012. The mercury resistance operon: From an origin in a geothermal environment to an efficient detoxification machine. *Frontiers in Microbiology* 3: 349.

Boyd, E.S., King, S., Tomberlin, J.K., Nordstrom, D.K., Krabbenhoft, D.P., Barkay, T., Geesey, G.G. 2009. Methylmercury enters an aquatic food web through acidophilic microbial mats in Yellowstone National Park, Wyoming. *Environmental Microbiology* 11: 950–959.

Braud, A., Hannauer, M., Mislin, G.L., Schalk, I.J. 2009. The *Pseudomonas aeruginosa* pyochelin-iron uptake pathway and its metal specificity. *Journal of Bacteriology* 191: 3517–3525.

Braun, W., Vasak, M., Robbins, A.H., Stout, C.D, Wagner, G., Kagi, J.H., Wuthrich, K. 1992. Comparison of the NMR solution structure and the x-ray crystal structure of rat metallothionein-2. *Proceedings of the National Academy of Science of the United States of America* 89: 10124–10128.

Brem, D., Pelludat, C., Rakin, A., Jacobi, C.A., Heesemann, J. 2001. Functional analysis of yersiniabactin transport genes of *Yersinia enterocolitica*. *Microbiology* 147: 1115–1127.

Brett, P.J., Burtnick, M.N., Fenno, J.C., Gherardini, F.C. 2008. *Treponema denticola* TroR is a manganese- and iron-dependent transcriptional repressor. *Molecular Microbiology* 70: 396–409.

Brown, N.L., Stoyanov, J.V., Kidd, S.P., Hobman, J.L. 2003. The MerR family of transcriptional regulators. *FEMS Microbiology Reviews* 27: 145–163.

Bundy, J.G., Kille, P. 2014. Metabolites and metals in Metazoa—What role do phytochelatins play in animals? *Metallomics* 6: 1576–1582.

Burke, B.E., Pfister, R.M. 1986. Cadmium transport by a Cd^{2+}-sensitive and a Cd^{2+}-resistant strain of *Bacillus subtilis*. *Canadian Journal of Microbiology* 32: 539–542.

Busenlehner, L.S., Pennella, M.A., Giedroc, D.P. 2003. The SmtB/ArsR family of metalloregulatory transcriptional repressors: Structural insights into prokaryotic metal resistance. *FEMS Microbiology Reviews* 27: 131–143.

Calderon, I.L., Arenas, F.A., Perez, J.M., Fuentes, D.E., Araya, M.A., Saavedra, C.P., Tantaleán, J.C., Pichuantes, S.E., Youderian, P.A., Vasquez, C.C. 2006. Catalases are NAD(P)H-dependent tellurite reductases. *PLoS ONE* 1: e70.

Calderon, I.L., Elias, A.O., Fuentes, E.L., Pradenas, G.A., Castro, M.E., Arenas, F.A., Perez, J.M., Vasquez, C.C. 2009. Tellurite-mediated disabling of [4Fe–4S] clusters of *Escherichia coli* dehydratases. *Microbiology* 155: 1840–1846.

Castro, M.E., Molina, R., Diaz, W., Pichuantes, S.E., Vasquez C.C. 2008. The dihydrolipoamide dehydrogenase of *Aeromonas caviae* ST exhibits NADH-dependent tellurite reductase activity. *Biochemical and Biophysica Research Communications* 375: 91–94.

Cervantes, C., Campos-Garcia, J., Devars, S., Gutiérrez-Corona, F., Loza-Tavera, H., Torres-Guzmán, J.C., Moreno-Sánchez, R. 2001. Interactions of chromium with microorganisms and plants. *FEMS Microbiology Reviews* 25: 335–347.

Champier, L., Duarte, V., Michaud-Soret, I., Coves, J. 2004. Characterization of the MerD protein from *Ralstonia metallidurans* CH34: A possible role in bacterial mercury resistance by switching off the induction of the mer operon. *Molecular Microbiology* 52: 1475–1485.

Chan, S., Giuroiu, I., Chernishof, I., Sawaya, M.R., Chiang, J., Gunsalus, R.P., Arbing, M.A., Perry, L.J. 2010. Apo and ligand-bound structures of ModA from the archaeon *Methanosarcina acetivorans*. *Acta Crystallographica Sect F Structural Biology and Crystallization Communications* 66: 242–250.

Chang, C.C., Lin, L.Y., Zou, X.W., Huang, C.C., Chan, N.L. 2015. Structural basis of the mercury(II)-mediated conformational switching of the dual-function transcriptional regulator MerR. *Nucleic Acids Research* 43: 7612–7623.

Chang, K.N., Lee, T.C., Tam, M.F., Chen, Y.C., Lee, L.W., Lee, S.Y., Lin, P.J., Huang, R.N. 2003. Identification of galectin I and thioredoxin peroxidase II as two arsenic-binding proteins in Chinese hamster ovary cells. *Biochemical Journal* 371: 495.

Chasteen, T.G., Fuentes, D.E., Tantalean, J.C., Vasquez, C.C. 2009. Tellurite: History, oxidative stress, and molecular mechanisms of resistance. *FEMS Microbiology Reviews* 33: 820–832.

Cherrier, M.V., Cavazza, C., Bochot, C., Lemaire, D., Fontecilla-Camps, J.C. 2008. Structural characterization of a putative endogenous metal chelator in the periplasmic nickel transporter NikA. *Biochemistry* 47: 9937–9943.

Chien, C.C., Jiang, M.H., Tsai, M.R., Chien, C.C. 2011. Isolation and characterization of an environmental cadmium- and tellurite-resistant *Pseudomonas* strain. *Environmental Toxicology and Chemistry* 30: 2202–2207.

Chillappagari, S., Miethke, M., Trip, H., Kuipers, O.P., Marahiel, M.A. 2009. Copper acquisition is mediated by YcnJ and regulated by YcnK and CsoR in *Bacillus subtilis*. *Journal of Bacteriology* 191: 2362–2370.

Chivers, P.T., Benanti, E.L., Heil-Chapdelaine, V., Iwig, J.S., Rowe, J.L. 2012. Identification of Ni-(L-His) 2 as a substrate for NikABCDE-dependent nickel uptake in *Escherichia coli*. *Metallomics* 4: 1043–1050.

Chivers, P.T., Tahirov, T.H. 2005. Structure of *Pyrococcus horikoshii* NikR: Nickel sensing and implications for the regulation of DNA recognition. *Journal of Molecular Biology* 348: 597–607.

Chopra, I. 2007. The increasing use of silver-based products as antimicrobial agents: A useful development or a cause for concern? *Journal of Antimicrobial Chemotherapy* 59: 587–590.

Chouchane, S., Snow, E.T. 2001. In vitro effect of arsenical compounds on glutathione-related enzymes. *Chemical Research in Toxicology* 14: 517–522.

Clarke, T.E., Tari, L.W., Vogel, H.J. 2001. Structural biology of bacterial iron uptake systems. *Current Topics in Medicinal Chemistry* 1: 7–30.

Clemens, S., Kim, E.J., Neumann, D., Schroeder, J.I. 1999. Tolerance to toxic metals by a gene family of phytochelatin synthases from plants and yeast. *The EMBO Journal* 18: 3325–3333.

Cobbett, C.S. 2000. Phytochelatin biosynthesis and function in heavy-metal detoxification. *Current Opinion in Plant Biology* 3: 211–216.

Cobbett, C.S., May, M.J., Howden, R., Rolls, B. 1998. The glutathione-deficient, cadmium-sensitive mutant, cad2-1, of *Arabidopsis thaliana* is deficient in glutamylcysteine synthetase. *The Plant Journal* 16: 73–78.

Cobessi, D., Celia, H., Pattus, F. 2005. Crystal structure at high resolution of ferric-pyochelin and its membrane receptor FptA from *Pseudomonas aeruginosa*. *Journal of Molecular Biology* 352: 893–904.

Colaco, H.G., Santo, P.E., Matias, P.M., Bandeiras, T.M., Vicente, J.B. 2016. Roles of *Escherichia coli* ZinT in cobalt, mercury and cadmium resistance and structural insights into the metal binding mechanism. *Metallomics* 8: 327–336.

Cortes, L., Wedd, A.G., Xiao, Z. 2015. The functional roles of the three copper sites associated with the methionine-rich insert in the multicopper oxidase CueO from *E. coli*. *Metallomics* 7: 776–785.

Cummings, D.E., Fendorf, S., Singh, N.K., Sani, R.K., Peyton, B.M., Magnuson, T.S. 2006. Reduction of Cr(VI) under acidic conditions by the facultative Fe(III)-reducing bacterium *Acidiphilium cryptum*. *Environmental Science and Technology* 41: 146–152.

Cunningham, M.L., Zvelebil, M.J.J.M., Fairlamb, A.H. 1994. Mechanism of inhibition of trypanothione reductase and glutathione reductase by trivalent organic arsenicals. *European Journal of Biochemistry* 221: 285.

Cunrath, O., Gasser, V., Hoegy, F., Reimmann, C., Guillon, L., Schalk, I.J. 2015. A cell biological view of the siderophore pyochelin iron uptake pathway in *Pseudomonas aeruginosa*. *Environmental Microbiology* 17: 171–185.

Degraeve, N. 1981. Carcinogenic, teratogenic and mutagenic effects of cadmium. *Mutation Research* 86: 115–135.

Dhuldhaj, U.P., Yadav, I.C., Singh, S., Sharma, N.K. 2013. Microbial interactions in the arsenic cycle: Adoptive strategies and applications in environmental management. *Reviews of Environmental Contamination Toxicology* 224: 1–38.

Di Toro, D.M., Allen, H.E., Bergman, H.L., Meyer, J.S., Paquin, P.R., Santore, R.C. 2001. Biotic ligand model of the acute toxicity of metals. 1. Technical basis. *Environmental Toxicology and Chemistry* 20: 2383–2396.

Diaz-Ochoa, V.E., Jellbauer, S., Klaus, S., Raffatellu, M. 2014. Transition metal ions at the crossroads of mucosal immunity and microbial pathogenesis. *Frontiers in Cellular and Infection Microbiology* 4: 1–10.

Distefano, M.D., Moore, M.J., Walsh, C.T. 1990. Active site of mercuric reductase resides at the subunit interface and requires Cys135 and Cys140 from one subunit and Cys558 and Cys559 from the adjacent subunit: Evidence from in vivo and in vitro heterodimer formation. *Biochemistry* 29: 2703–2713.

Dixit, R., Wasiullah, D., Malaviya, D., Pandiyan, K., Singh, U.B., Sahu, A., Shukla, R. et al. 2015. Bioremediation of heavy metals from soil and aquatic environment: An overview of principles and criteria of fundamental processes. *Sustainability* 7: 2189–2212.

Djoko, K.Y., Xiao, Z., Huffman, D.L., Wedd, A.G. 2007. Conserved mechanism of copper binding and transfer. A comparison of the copper-resistance proteins PcoC from *Escherichia coli* and CopC from *Pseudomonas syringae*. *Inorganic Chemistry* 46: 4560–4508.

Djoko, K.Y., Xiao, Z., Wedd, A.G. 2008. Copper resistance in *E. coli*: The multicopper oxidase PcoA catalyzes oxidation of copper(I) in Cu(I) Cu(II)-PcoC. *Chembiochem* 9: 1579–1582.

Dopson, M., Lindström, E.B., Hallberg, K.B. 2001. Chromosomally encoded arsenical resistance of the moderately thermophilic acidophile *Acidithiobacillus caldus*. *Extremophiles* 5: 247–255.

Dopson, M., Ossandon, F., Lövgren, L., Holmes, D.S. 2014. Metal resistance or tolerance? Acidophiles confront high metal loads via both abiotic and biotic mechanisms. *Frontiers in Microbiology* 5: 157.

Duncan, K. 2009. Metallothioneins and related chelators. In: Sigel, A., Sigel, H., Sigel, R.K.O. (eds). *Metal ions in life sciences*, Vol. 5. Angewandte Chemie International Edition.

Eitinger, T., Mandrand-Berthelot, M.A. 2000. Nickel transport systems in microorganisms. *Archives of Microbiology* 173: 1–9.

Eitinger, T., Suhr, J., Moore, L., Smith, J.A. 2005. Secondary transporters for nickel and cobalt ions: Theme and variations. *Biometals* 18: 399–405.

Ellis, P.J., Conrads, T., Hille, R., Kuhn, P. 2001. Crystal structure of the 100 kDa arsenite oxidase from *Alcaligenes faecalis* in two crystal forms at 1.64 A and 2.03 A. *Structure* 9: 125–132.

Engst, S., Miller, S.M. 1999. Rapid reduction of Hg(II) by mercuric ion reductase does not require the conserved C-terminal cysteine pair using HgBr2 as the substrate. *Biochemistry* 38: 853–854.

Escolar, L., Perez-Martin, J., De Lorenzo, V. 1999. Opening the iron box: Transcriptional metalloregulation by the Fur protein. *Journal of Bacteriology* 181: 6223–6229.

Felizola, S.J., Nakamura, Y., Arata, Y., Ise, K., Satoh, F., Rainey, W.E., Midorikawa, S., Suzuki, S., Sasano, H. 2014. Metallothionein-3 (MT-3) in the human adrenal cortex and its disorders. *Endocrine Pathology* 25: 229–235.

Festa, R.A., Jones, M.B., Butler-Wu, S., Sinsimer, D., Gerads, R., Bishai, W.R., Peterson, S.N., Darwin, K.H. 2011. A novel copper-responsive regulon in *Mycobacterium tuberculosis*. *Molecular Microbiology* 79: 133–148.

Fetherston, J.D., Mier, I., Truszczynska, H., Perry, R.D. 2012. The Yfe and Feo transporters are involved in microaerobic growth and virulence of *Yersinia pestis* in bubonic plague. *Infection Immunity* 80: 3880–3891.

Fillat, M.F. 2014. The FUR (ferric uptake regulator) superfamily: Diversity and versatility of key transcriptional regulators. *Archives of Biochemistry and Biophysics* 546: 41–52.

Fowler, B.A., Hildebrand, C.E., Kojima, Y., Webb, M. 1987. Nomenclature of metallothionein. *Experientia Supplementum* 52: 19–22.

Franks, S.E., Ebrahimi, C., Hollands, A., Okumura, C.Y., Aroian, R.V., Nizet, V., McGillivray, S.M. 2014. Novel role for the yceGH tellurite resistance genes in the pathogenesis of *Bacillus anthracis*. *Infection Immunity* 82: 1132–1140.

Freisinger, E., Vasak, M. 2013. Cadmium in metallothioneins. *Metal Ions in Life Science* 11: 339–371.

Fu, C., Javedan, S., Moshiri, F., Maier, R.J. 1994. Bacterial genes involved in incorporation of nickel into a hydrogenase enzyme. *Proceedings of the National Academy of Science of the United States of America* 91: 5099–5103.

Fu, Y., Tsui, H.C., Bruce, K.E., Sham, L.T., Higgins, K.A., Lisher, J.P., Kazmierczak, K.M. et al. 2013. A new structural paradigm in copper resistance in *Streptococcus pneumoniae*. *Nature Chemical Biology* 9: 177–183.

Gaur, N., Flora, G., Yadav, M., Tiwari, A. 2014. A review with recent advancements on bioremediation-based abolition of heavy metals. *Environmental Science: Process and Impacts* 16: 180–193.

Genco, C.A., Dixon, D.W. 2001. Emerging strategies in microbial haem capture. *Molecular Microbiology* 39: 1–11.

Gold, B., Deng, H., Bryk, R., Vargas, D., Eliezer, D., Roberts, J., Jiang, X., Nathan, C. 2008. Identification of a copper-binding metallothionein in pathogenic mycobacteria. *Nature Chemical Biology* 4: 609–616.

Gorden, A.E., Xu, J., Raymond, K.N., Durbin, P. 2003. Rational design of sequestering agents for plutonium and other actinides. *Chemical Reviews* 103: 4207–4282.

Gulati, K., Banerjee, B., Bala Lall, S., Ray, A. 2010. Effects of diesel exhaust, heavy metals and pesticides on various organ systems: Possible mechanisms and strategies for prevention and treatment. *Indian Journal of Experimental Biology* 48: 710–721.

Guo, H.B., Johs, A., Parks, J.M., Olliff, L., Miller, S.M., Summers, A.O., Liang, L., Smith, J.C. 2010. Structure and conformational dynamics of the metalloregulator MerR upon binding of Hg(II). *Journal of Molecular Biology* 398: 555–568.

Gupta, D.K., Huang, H.G., Corpas, F.J. 2013. Lead tolerance in plants: Strategies for phytoremediation. *Environmental Science and Pollution Research* 20: 2150–2161.

Ha, S.B., Smith, A.P., Howden, R., Dietrich, W.M., Bugg, S., O'Connell, M.J., Goldsbrough, P.B., Cobbett, C.S. 1999. Phytochelatin synthase genes from *Arabidopsis* and the yeast, *Schizosaccharomyces pombe*. *Plant Cell* 11: 1153–1164.

Hagen, W.R. 2011. Cellular uptake of molybdenum and tungsten. Coord. *Chemical Reviews* 255: 1117–1128.

Halliwell, B., Gutteridge, J.M. 1984. Free radicals, lipid peroxidation, and cell damage. *The Lancet* 324: 1095.

Hantke, K. 2005. Bacterial zinc uptake and regulators. *Current Opinion in Microbiology* 8: 196–202.

Hao, Z., Chen, S., Wilson, D.B. 1999. Cloning, expression, and characterization of cadmium and manganese uptake genes from *Lactobacillus plantarum*. *Applied and Environmental Microbiology* 65: 4746–4752.

Harrington, J.M., Crumbliss, A.L. 2009. The redox hypothesis in siderophore-mediated iron uptake. *Biometals* 22: 679–689.

Harrison, J.J., Ceri, H., Turner, R.J. 2007. Multimetal resistance and tolerance in microbial biofilms. *Nature Reviews Microbiology* 5: 928–938.

Hartney, S.L., Mazurier, S., Girard, M.K., Mehnaz, S., Davis, E.W. 2nd, Gross, H., Lemanceau, P., Loper, J.E. 2013. Ferric-pyoverdine recognition by Fpv outer membrane proteins of *Pseudomonas protegens* Pf-5. *Journal of Bacteriology* 195: 765–776.

Hijova, E. 2004. Metallothioneins and zinc: Their functions and interactions. *Bratislavske Lekarske Listy* 105: 230–234.

Hille, R. 2002. Molybdenum and tungsten in biology. *Trends in Biochemical Sciences* 27: 360–367.

Hillson, N.J., Hu, P., Andersen, G.L., Shapiro, L. 2007. *Caulobacter crescentus* as a whole-cell uranium biosensor. *Applied and Environmental Microbiology* 73: 7615–7621.

Hiniker, A., Collet, J.F., Bardwell, J.C. 2005. Copper stress causes an in vivo requirement for the *Escherichia coli* disulfide isomerase DsbC. *Journal of Biological Chemistry* 280: 33785–33791.

Hirata, K., Tsuji, N., Miyamoto, K. 2005. Biosynthetic regulation of phytochelatins, heavy metal-binding peptides. *Journal of Bioscience and Bioengineering* 100: 593–599.

Hong, L., Sharp, M.A., Poblete, S., Biehl, R., Zamponi, M., Szekely, N., Appavou, M.S. et al. 2014. Structure and dynamics of a compact state of a multidomain protein, the mercuric ion reductase. *Biophysical Journal* 107: 393–400.

Hong, R., Kang, T.Y., Michels, C.A., Gadura, N. 2012. Membrane lipid peroxidation in copper alloy-mediated contact killing of *Escherichia coli*. *Applied and Environmental Microbiology* 78: 1776–1784.

Hordyjewska, A., Popiolek, L., Kocot, J. 2014. The many "faces" of copper in medicine and treatment. *Biometals* 27: 611–621.

Howden, R., Goldsbrough, P.B., Andersen, C.R., Cobbett, C.S. 1995. Cadmium sensitive, cad1, mutants of *Arabidopsis thaliana* are phytochelatin deficient. *Plant Physiology* 107: 1059–1066.

Hsiao, Y.M., Liu, Y.F., Lee, P.Y., Hsu, P.C., Tseng, S.Y. Pan, Y.C. 2011. Functional characterization of copA gene encoding multicopper oxidase in *Xanthomonas campestris* pv. campestris. *Journal of Agricultural and Food Chemistry* 59: 9290–9302.

Hu, Y.H., Wang, H.L., Zhang, M., Sun, L. 2009. Molecular analysis of the copper-responsive CopRSCD of a pathogenic *Pseudomonas fluorescens* strain. *Journal of Microbiology* 47: 277–286.

Huang, M., Shaw, C.F., Petering, D. 2004. Interprotein metal exchange between transcription factor IIIa and apo-metallothionein. *Journal of Inorganic Biochemistry* 98: 639–648.

Hung, K.W., Juan, T.H., Hsu, Y.L., Huang, T.H. 2012. NMR structure note: The ferrous iron transport protein C (FeoC) from *Klebsiella pneumoniae*. *Journal of Biomolecular NMR* 53: 161–165.

Hunter, T.C., Mehra, R.K. 1998. A role for HEM2 in cadmium tolerance. *Journal of Inorganic Biochemistry* 69: 293–303.

Iram, S., Shabbir, R., Zafar, H., Javaid, M. 2015. Biosorption and bioaccumulation of copper and lead by heavy metal-resistant fungal isolates. *Arabian Journal for Science and Engineering* 40: 1867–1873.

Itou, H., Yao, M., Watanabe, N., Tanaka, I. 2008. Crystal structure of the PH1932 protein, a unique archaeal ArsR type winged-HTH transcription factor from *Pyrococcus horikoshii* OT3. *Proteins* 70: 1631–1634.

Jacobs, F.A., Romeyer, F.M., Beauchemin, M., Brousseau, R. 1989. Human metallothionein-II is synthesized as a stable membrane-localized fusion protein in *Escherichia coli*. *Gene* 83: 95–103.

Jacobs, I.A., Taddeo, J., Kelly, K., Valenziano, C. 2002. Poisoning as a result of barium styphnate explosion. *American Journal of Industrial Medicine* 41: 285–288.

Jaroslawiecka, A., Piotrowska-Seget, Z. 2014. Lead resistance in micro-organisms. *Microbiology* 160: 12–25.

Jayakanthan, S., Roberts, S.A., Weichsel, A., Arguello, J.M., McEvoy, M.M. 2012. Conformations of the apo-, substrate-bound and phosphate-bound ATP-binding domain of the Cu(II) ATPase CopB illustrate coupling of domain movement to the catalytic cycle. *Bioscience Reports* 32: 443–453.

Jiang, Z., Li, P., Jiang, D., Wu, G., Dong, H., Wang, Y., Li, B., Wang, Y., Guo, Q. 2014. Diversity and abundance of the arsenite oxidase gene aioA in geothermal areas of Tengchong, Yunnan, China. *Extremophiles* 18: 161–170.

Johnson, A.R., Munoz, A., Gottlieb, J.L., Jarrard, D.F. 2007. High dose zinc increases hospital admissions due to genitourinary complications. *Journal of Urology* 177: 639–643.

Juang, R.H., MacCue, K.F., Ow, D.W. 1993. Two purine biosynthetic enzymes that are required for cadmium tolerance in *Schizosaccharomyces pombe* utilize cysteine sulfinate in vitro. *Arch Biochem Biophys* 304: 392–401.

Kalnins, G., Kuka, J., Grinberga, S., Makrecka-Kuka, M., Liepinsh, E., Dambrova, M., Tars, K. 2015. Structure and function of CutC choline lyase from human microbiota bacterium *Klebsiella pneumoniae*. *The Journal of Biological Chemistry* 290: 21732–21740.

Kambe, T., Hashimoto, A., Fujimoto, S. 2014. Current understanding of ZIP and ZnT zinc transporters in human health and diseases. *Cellular and Molecular Life Science* 71: 3281–3295.

Kapahi, P., Takahashi, T., Natoli, G., Adams, S.R., Chen, Y., Tsien, R.Y., Karin, M. 2000. Inhibition of NF-κB activation by arsenite through reaction with a critical cysteine in the activation loop of IκB kinase. *The Journal of Biological Chemistry* 275: 36062–360626.

Kashyap, D.R., Botero, L.M., Franck, W.L., Hassett, D.J., McDermott, T.R. 2006. Complex regulation of arsenite oxidation in *Agrobacterium tumefaciens*. *Journal of Bacteriology* 188: 1081–1088.

Kazakov, A.E., Rajeev, L., Luning, E.G., Zane, G.M., Siddartha, K., Rodionov, D.A., Dubchak, I. et al. 2013. New family of tungstate-responsive transcriptional regulators in sulfate-reducing bacteria. *Journal of Bacteriology* 195: 4466–4475.

Kiyono, M., Pan-Hou, H. 1999. The merG gene product is involved in phenylmercury resistance in *Pseudomonas* strain K-62. *Journal of Bacteriology* 181: 726–730.

Kiyono, M., Sone, Y., Nakamura, R., Pan-Hou, H., Sakabe, K., 2009. The MerE protein encoded by transposon Tn21 is a broad mercury transporter in *Escherichia coli*. *FEBS Letters* 583: 1127–1131.

Koechler, S., Cleiss-Arnold, J., Proux, C., Sismeiro, O., Dillies, M.A., Goulhen-Chollet, F., Hommais, F. et al. 2010. Multiple controls affect arsenite oxidase gene expression in *Herminiimonas arsenicoxydans*. *BMC Microbiology* 10: 1.

Koh, E.I., Hung, C.S., Henderson, J.P. 2016. The Yersinia bactin-associated ATP binding cassette proteins YbtP and YbtQ enhance *E. coli* fitness during high titer cystitis. *Infection Immunity* 84: 1312–1319.

Kojima, Y. 1991. Definitions and nomenclature of metallothioneins. *Methods in Enzymology* 205: 8–10.

Koster, S., Wehner, M., Herrmann, C., Kuhlbrandt, W., Yildiz, O. 2009. Structure and function of the FeoB G-domain from *Methanococcus jannaschii*. *Journal Molecular Biology* 39: 405–419.

Kotrba, P., Doleckova, L., de Lorenzo, V., Ruml, T. 1999a. Enhanced bioaccumulation of heavy metal ions by bacterial cells due to surface display of short metal binding peptides. *Applied and Environmental Microbiology* 65: 1092–1098.

Kotrba, P., Pospisil, P., de Lorenzo, V., Ruml, T. 1999b. Enhanced metallosorption of *Escherichia coli* cells due to surface display of beta- and alpha-domains of mammalian metallothionein as a fusion to LamB protein. *Journal of Receptors and Signal Transduction Research* 19: 703–715.

Krezel, A., Maret, W. 2007. Dual nanomolar and picomolar Zn(II) binding properties of metallothionein. *Journal of American Chemical Society* 129: 10911–10921.

Krishnaji, S.T., Kaplan, D.L. 2013. Bioengineered chimeric spider silk-uranium binding proteins. *Macromolecular Bioscience* 13: 256–264.

Kuschner, M. 1981. The carcinogenicity of beryllium. *Environmental Health Perspectives* 40: 101–105.

Laddaga, R.A., Bessen, R., Silver, S. 1985. Cadmium-resistant mutant of *Bacillus subtilis* 168 with reduced cadmium transport. *Journal of Bacteriology* 162: 1106–1110.

Laddaga, R.A., Silver, S. 1985. Cadmium uptake in *Escherichia coli* K-12. *Journal of Bacteriology* 162: 1100–1105.

Lafrance-Vanasse, J., Lefebvre, M., Di Lello, P., Sygusch, J., Omichinski, J.G. 2009. Crystal structures of the organomercurial lyase MerB in its free and mercury-bound forms: Insights into the mechanism of methylmercury degradation. *Journal of Biological Chemistry* 284: 938–944.

Latorre, M., Olivares, F., Reyes-Jara, A., Lopez, G., González, M. 2011. CutC is induced late during copper exposure and can modify intracellular copper content in *Enterococcus faecalis*. *Biochemical and Biophysical Research Communication* 406: 633–637.

Lau, C.K., Ishida, H., Liu, Z., Vogel, H.J. 2013. Solution structure of *Escherichia coli* FeoA and its potential role in bacterial ferrous iron transport. *Journal of Bacteriology* 195: 46–55.

Lau, C.K., Krewulak, K.D., Vogel, H.J. 2016. Bacterial ferrous iron transport: The Feo system. *FEMS Microbiol Reviews* 40: 273–298.

Lawson, D.M., Artymiuk, P.J., Yewdall, S.J., Smith, J.A., Livingstone, J.C., Treffry, A., Luzzago, A., Levi, S., Arosio, P., Cesareni, G., Thomas, C.D., Shaw, W.V., Harrison, P.M. 1991. Solving the Structure of Human H Ferritin by Genetically Engineering Intermolecular Crystal Contacts. *Nature* 349: 541–544.

Le Brun, N.E., Crow, A., Murphy, M.E.P., Mauk, A.G., Moore, G.R. 2010. Iron core mineralization in prokaryotic ferritins. *Biochimica et Biophysica Acta* 1800: 732–744.

Lebrette, H., Borezée-Durant, E., Martin, L., Richaud, P., Boeri Erba, E., Cavazza, C. 2015. Novel insights into nickel import in *Staphylococcus aureus*: The positive role of free histidine and structural characterization of a new thiazolidine-type nickel chelator. *Metallomics* 7: 613–621.

Lebrette, H., Iannello, M., Fontecilla-Camps, J.C., Cavazza, C. 2013. The binding mode of Ni-(L-His) 2 in NikA revealed by X-ray crystallography. *Journal of Inorganic Biochemistry* 121: 16–18.

Ledwidge, R., Hong, B., Miller, S.M. 2010. NmerA of Tn501 mercuric ion reductase: Structural modulation of the pKa values of the metal binding cysteine thiols. *Biochemistry* 49: 8988–8998.

Ledwidge, R., Patel, B., Dong, A., Fiedler, D., Falkowski, M., Zelikova, J., Summers, A.O., Pai, E.F., Miller, S.M. 2005. NmerA, the metal binding domain of mercuric ion reductase, removes Hg^{2+} from proteins, delivers it to the catalytic core, and protects cells under glutathione-depleted conditions. *Biochemistry* 44: 11402–11416.

Lee, Y.H., Deka, R.K., Norgard, M.V., Radolf, J.D., Hasemann, C.A. 1999. *Treponema pallidum* TroA is a periplasmic zinc-binding protein with a helical backbone. *Nature Structural and Molecular Biology* 6: 628–633.

Lett, M.C., Muller, D., Lievremont, D., Silver, S., Santini, J. 2011. Unified nomenclature for genes involved in prokaryotic aerobic arsenite oxidation. *Journal of Bacteriology* 194: 207–208.

Levinson, H., Mahler, I. 1998. Phosphatase activity and lead resistance in *Citrobacter freundii* and *Staphylococcus aureus*. *FEMS Microbiology Letters* 161: 135–138.

Li, B., Lin, J., Mi, S., Lin, J. 2010. Arsenic resistance operon structure in *Leptospirillum ferriphilum* and proteomic response to arsenic stress. *Bioresource Technology* 101: 9811–9814.

Li, J., Pickart, C.M. 1995. Binding of phenylarsenoxide to Arg-tRNA protein transferase is independent of vicinal thiols. *Biochemistry* 34: 15829–15837.

Li, Z.S., Lu, Y.P., Zhen, R.G., Szczypka, M., Thiele, D.J., Rea, P.A. 1997. A new pathway for vacuolar cadmium sequestration in *Saccharomyces cerevisiae*: YCF1-catalyzed transport of bis (glutathionato)-cadmium. *Proceedings of the National Academy of Science of the United States of America* 94: 42–47.

Lian, P., Guo, H.B., Riccardi, D., Dong, A., Parks, J.M., Xu, Q., Pai, E.F. et al. 2014. X-ray structure of a Hg^{2+} complex of mercuric reductase (MerA) and quantum mechanical/molecular mechanical study of Hg^{2+} transfer between the C-terminal and buried catalytic site cysteine pairs. *Biochemistry* 53: 7211–7222.

Lin, C.H., Huang, C.F., Chen, W.Y., Chang, Y.Y., Ding, W.H., Lin, M.S., Wu, S.H., Huang, R.N. 2006. Characterization of the interaction of galectin-1 with sodium arsenite. *Chemical Research in Toxicology* 19: 469–474.

Lin, W., Chai, J., Love, J., Fu, D. 2010. Selective electrodiffusion of zinc ions in a Zrt-, Irt-like protein, ZIPB. *The Journal of Biological Chemistry* 285: 39013–39020.

Liu, J., Brown, A.K., Meng, X., Cropek, D.M., Istok, J.D., Watson, D.B., Lu, Y. 2007. A catalytic beacon sensor for uranium with parts-per-trillion sensitivity and million fold selectivity. *Proceedings of the National Academy Science of the United States of America* 104: 2056–2061.

Liu, X., Theil, E.C. 2005. Ferritins: Dynamic management of biological iron and oxygen chemistry. *Accounts of Chemical Research* 38: 167–175.

Lu, G.J., Tian, Y., Vora, N., Marassi, F.M., Opella, S.J. 2013. The structure of the mercury transporter MerF in phospholipid bilayers: A large conformational rearrangement results from N-terminal truncation. *Journal of the American Chemical Society* 135: 9299–9302.

Lu, J., Chew, E.H., Holmgren, A. 2007. Targeting thioredoxin reductase is a basis for cancer therapy by arsenic trioxide. *Proceedings of the National Academy of Science of the United States of America* 104: 12288–12293.

Luber, S., Reiher, M. 2010. Theoretical Raman optical activity study of the beta domain of rat metallothionein. *The Journal of Physical Chemistry B* 114: 1057–1063.

Ma, J.F., Ochsner, U.A., Klotz, M.G., Nanayakkara, V.K., Howell, M.L., Johnson, Z., Posey, J.E., Vasil, M.L., Monaco, J.J., Hassett, D.J. 1999. Bacterioferritin A modulates catalase A (KatA) activity and resistance to hydrogen peroxide in *Pseudomonas aeruginosa*. *Journal of Bacteriology* 181: 3730–3742.

Macomber, L., Imlay, J.A. 2009. The iron-sulfur clusters of dehydratases are primary intracellular targets of copper toxicity. *Proceedings of the National Academy of Science of the United States of America* 106: 8344–8349.

Magnuson, T.S., Swenson, M.W., Paszczynski, A.J., Deobald, L.A., Kerk, D., Cummings, D.E. 2010. Proteogenomic and functional analysis of chromate reduction in *Acidiphilium cryptum* JF-5, an Fe(III)-respiring acidophile. *Biometals* 23: 1129–1138.

Magos, L., Clarkson, T.W. 2006. Overview of the clinical toxicity of mercury. *Annals of Clinical Biochemistry* 43: 257–268.

Malik, A. 2004. Metal bioremediation through growing cells. *Environment International* 30: 261–278.

Mangold, S. 2012. Growth and survival of *Acidithiobacilli* in acidic, metal rich environments. Doctoral thesis, Umeå Universitet.

Mangold, S., Potrykus, J., Björn, E., Lövgren, L., Dopson, M. 2013. Extreme zinc tolerance in acidophilic microorganisms from the bacterial and archaeal domains. *Extremophiles* 17: 75–85.

Manso, Y., Adlard, P.A., Carrasco, J., Vasak, M., Hidalgo, J. 2011. Metallothionein and brain inflammation. *Journal of Biological Inorganic Chemistry* 16: 1103–1113.

Mao, J.H., Sun, X.Y., Liu, J.X., Zhang, Q.Y., Liu, P., Huang, Q.H., Li, K.K., Chen, Q., Chen, Z., Chen, S.J. 2010. As4S4 targets RING-type E3 ligase c-CBL to induce degradation of BCR-ABL in chronic myelogenous leukemia. *Proceedings of the National Academy of Science of the United States of America* 107: 21683.

Marchler-Bauer, A., Lu, S., Anderson, J.B., Chitsaz, F., Derbyshire, M.K., DeWeese-Scott, C., Fong, J.H. et al. 2011. CDD: A conserved domain database for the functional annotation of proteins. *Nucleic Acids Research* 39: 225–229.

Martin, P., DeMel, S., Shi, J., Gladysheva, T., Gatti, D.L., Rosen, B.P., Edwards, B.F. 2001. Insights into the structure, solvation, and mechanism of ArsC arsenate reductase, a novel arsenic detoxification enzyme. *Structure* 9: 1071–1081.

Mathema, V.B., Thakuri, B.C., Sillanpaa, M. 2011. Bacterial mer operon-mediated detoxification of mercurial compounds: A short review. *Archives of Microbiology* 193: 837–844.

Mauro, J.M., Pazirandeh, M. 2000. Construction and expression of functional multidomain polypeptides in *Escherichia coli*: Expression of the *Neurospora crassa* metallothionein gene. *Letters in Applied Microbiology* 30: 161–166.

McHugh, J.P., Rodríguez-Quinones, F., Abdul-Tehrani, H., Svistunenko, D.A., Poole, R.K., Cooper, C.E., Andrews, S.C. 2003. Global iron-dependent gene regulation in *Escherichia coli*. A new mechanism for iron homeostasis. *The Journal of Biological Chemistry* 278: 29478–29486.

Mejare, M., Bulow, L. 2001. Metal-binding proteins and peptides in bioremediation and phytoremediation of heavy metals. *Trends in Biotechnology* 19: 67–73.

Mejare, M., Ljung, S., Bülow, L. 1998. Selection of cadmium specific hexapeptides and their expression as OmpA fusion proteins in *Escherichia coli*. *Protein Engineering* 11: 489–494.

Michon, J., Frelon, S., Garnier, C., Coppin, F. 2010. Determinations of uranium (VI) binding properties with some metalloproteins (transferrin, albumin, metallothionein and ferritin) by fluorescence quenching. *Journal of Fluorescence* 20: 581–590.

Miethke, M., Marahiel, M.A. 2007. Siderophore-based iron acquisition and pathogen control. *Microbiology and Molecular Biology Reviews* 71: 413–451.

Miles, A.T., Hawksworth, G.M., Beattie, J.H., Rodilla, V. 2000. Induction, regulation, degradation, and biological significance of mammalian metallothioneins. *Critical Reviews in Biochemistry Molecular Biology* 35: 35–70.

Milnerowicz, H. 1997. Influence of tobacco smoking on metallothioneins isoforms contents in human placenta, amniotic fluid and milk. *International Journal of Occupational Medicine and Environmental Health* 10: 395–403.

Moin, S.F., Omar, M.N. 2014. Laccase enzymes: Purification, structure to catalysis and tailoring. *Protein and Peptide Letters* 21: 707–713.

Moinier, D., Slyemi, D., Byrne, D., Lignon, S., Lebrun, R., Talla, E., Bonnefoy, V. 2014. An ArsR/SmtB family member is involved in the regulation by arsenic of the arsenite oxidase operon in *Thiomonas arsenitoxydans*. *Applied and Environmental Microbiology* 80: 6413–6426.

Morozzi, G., Cenci, G., Scardazza, F., Pitzurra, M. 1986. Cadmium uptake by growing cells of gram-positive and gram-negative bacteria. *Microbios* 48: 27–35.

Morton, D.J., Williams, P. 1990. Siderophore-independent acquisition of transferrin-bound iron by *Haemophilus influenzae* type b. *Microbiology* 136: 927–933.

Mukhopadhyay, R., Rosen, B.P. 2002. Arsenate reductases in prokaryotes and eukaryotes. *Environmental Health Perspectives* 110: 745–748.

Munson, G.P., Lam, D.L., Outten, F.W., O'Halloran, T.V. 2000. Identification of a copper-responsive two-component system on the chromosome of *Escherichia coli* K-12. *Journal of Bacteriology* 182: 5864–5871.

Nader, M., Journet, L., Meksem, A., Guillon, L., Schalk, I.J. 2011. Mechanism of ferrisiderophore uptake by *Pseudomonas aeruginosa* outer membrane transporter FpvA: No diffusion channel formed at any time during ferrisiderophore uptake. *Biochemistry* 50: 2530–2540.

Naik, M.M., Dubey, S.K. 2013. Lead resistant bacteria: Lead resistance mechanisms, their applications in lead bioremediation and biomonitoring. *Ecotoxicology and Environmental Safety* 98: 1–7.

Naik, M.M., Pandey, A., Dubey, S.K. 2012a. *Pseudomonas aeruginosa* strain WI-1 from Mandovi estuary possesses metallothionein to alleviate lead toxicity and promotes plant growth. *Ecotoxicology and Environmental Safety* 79: 129–133.

Naik, M.M., Shamim, K., Dubey, S.K. 2012b. Biological characterization of lead resistant bacteria to explore role of bacterial metallothionein in lead resistance. *Current Science* 103: 1–3.

Navarro, C.A., von Bernath, D., Jerez, C.A. 2013. Heavy metal resistance strategies of acidophilic bacteria and their acquisition: Importance for biomining and bioremediation. *Biological Research* 46: 363–371.

Neustadt, J., Pieczenik, S. 2007. Toxic-metal contamination: Mercury. *Integrative Medicine-Innovision Communications* 6: 36–37.

Ngu, T.T., Stillman, M.J. 2009. Metalation of metallothioneins. *IUBMB Life* 61: 438–446.

Nies, D.H. 2003. Efflux mediated heavy metal resistance in prokaryotes. *FEMS Microbiology Reviews* 27: 313–339.

Noll, M., Petrukhin, K., Lutsenko, S. 1998. Identification of a novel transcription regulator from *Proteus mirabilis*, PMTR, revealed a possible role of YJAI protein in balancing zinc in *Escherichia coli*. *The Journal of Biological Chemistry* 273: 21393–21401.

O'Halloran, T.V., Culotta, V.C. 2000. Metallochaperones, an intracellular shuttle service for metal ions. *The Journal of Biological Chemistry* 275: 25057–25060.

Ochsner, U.A., Wilderman, P.J., Vasil, A.I., Vasil, M.L. 2002. GeneChipR expression analysis of the iron starvation response in *Pseudomonas aeruginosa*: Identification of novel pyoverdine biosynthesis genes. *Molecular Microbiology* 45: 1277–1287.

Orell, A., Navarro, C.A., Rivero, M., Aguilar, J.S., Jerez, C.A. 2012. Inorganic polyphosphates in extremophiles and their possible functions. *Extremophiles* 16: 573–583.

Ortiz, D.F., Ruscitti, T., McCue, K.F., Ow, D.W. 1995. Transport of metal-binding peptides by HMT1, a fission yeast ABC-type vacuolar membrane protein. *The Journal of Biological Chemistry* 270: 4721–4728.

Osborne, T.H., Heath, M.D., Martin, A.C.R., Pankowski, J.A., Hudson-Edwards, K.A., Santini, J.M. 2013. Cold-adapted arsenite oxidase from a psychrotolerant Polaromonas species. *Metallomics* 5: 318–324.

Osman, D., Cavet, J.S. 2010. Bacterial metal-sensing proteins exemplified by ArsR-SmtB family repressors. *Nature Product Reports* 27: 668–680.

Osorio, H., Martínez, V., Nieto, P.A., Holmes, D.S., Quatrini, R. 2008. Microbial iron management mechanisms in extremely acidic environments: Comparative genomics evidence for diversity and versatility. *BMC Microbiology* 8: 203.

Ottosson, L.G., Logg, K., Ibstedt, S., Sunnerhagen, P., Kall, M., Blomberg, A. Warringer. J. 2010. Sulfate assimilation mediates tellurite reduction and toxicity in *Saccharomyces cerevisiae*. *Eukaryotic Cell* 9: 1635–1647.

Padmavathiamma, P.K., Li, L.Y. 2007. Phytoremediation technology: Hyperaccumulation metals in plants. *Water, Air and Soil Pollution* 184: 105–126.

Pal, R., Rai, J.P. 2010. Phytochelatins: Peptides involved in heavy metal detoxification. *Applied Biochemistry and Biotechnology* 160: 945–963.

Park, C.H., Gonzalez, C.F., Ackerley, D.F., Keyhan, M., Matin, A. 2002. Molecular engineering of soluble bacterial proteins with chromate reductase activity. In *Proceedings of the First International Conference on Remediation of Contaminated Sediments*, Venice, Italy, Vol. III. Hinchee, R.E., Porta, A., Pellei, M. (eds.). Columbus, OH: Batelle Press, pp. 103–111.

Parker Siburt, C.J., Mietzner, T.A., Crumbliss, A.L. 2012. FbpA—A bacterial transferrin with more to offer. *Biochimica et Biophysica Acta* 1820: 379–392.

Pau, R.N. 2004. Molybdenum uptake and homeostasis. In: Klipp, W., Masepohl, B., Gallon, J.R., Newton, W.E. (eds). *Genetics and Regulation of Nitrogen Fixation in Free-Living Bacteria*. Kluwer Academic Publishers, Netherlands, pp. 225–256.

Pazirandeh, M., Wells, B.M., Ryan, R.L. 1998. Development of bacterium-based heavy metal biosorbents: Enhanced uptake of cadmium and mercury by *Escherichia coli* expressing a metal binding motif. *Applied and Environmental Microbiology* 64: 4068–4072.

Perry, R.D., Craig, S.K., Abney, J., Bobrov, A.G., Kirillina, O., Mier, IJr., Truszczynska, H., Fetherston, J.D. 2012. Manganese transporters Yfe and MntH are Fur-regulated and important for the virulence of *Yersinia pestis*. *Microbiology* 158: 804–815.

Peters, R.W., Young, K., Bhattacharayan, D. 1985. Evaluation of recent treatment technique for removal of heavy metals from industrial wastewater. *AICHE Symposium Series* 81: 1695–1703.

Pible, O., Guilbaud, P., Pellequerm J.L., Vidaud, C., Quemeneur, E. 2006. Structural insights into protein-uranyl interaction: Towards an in silico detection method. *Biochimie* 88: 1631–1638.

Pible, O., Vidaud, C., Plantevin, S., Pellequer, J.L., Quemeneur, E. 2010. Predicting the disruption by UO(2)(2+) of a protein-ligand interaction. *Protein Science* 19: 2219–2230.

Ponnusamy, D., Hartson, S.D., Clinkenbeard, K.D. 2011. Intracellular *Yersinia pestis* expresses general stress response and tellurite resistance proteins in mouse macrophages. *Veterinary Microbiology* 150: 146–151.

Posfai, M., Dunin-Borkowski, R.E. 2009. Magnetic nanocrystals in organisms. *Elements* 5: 235–240.

Powlowski, J., Sahlman, L. 1999. Reactivity of the two essential cysteine residues of the periplasmic mercuric ion-binding protein, MerP. *The Journal of Biological Chemistry* 274: 33320–33326.

Prasad, M.N.V. 1995. Cadmium toxicity and tolerance in vascular plants. *Environmental and Experimental Botany* 35: 525–545.

Quatrini, R., Lefimil, C., Veloso, F.A., Pedroso, I., Holmes, D.S., Jedlicki, E. 2007. Bioinformatic prediction and experimental verification of Fur-regulated genes in the extreme acidophile *Acidithiobacillus ferrooxidans*. *Nucleic Acids Research* 35: 2153–2166.

Rademacher, C., Masepohl, B. 2012. Copper-responsive gene regulation in bacteria. *Microbiology* 158: 2451–2464.

Raudenska, M., Gumulec, J., Podlaha, O., Sztalmachova, M., Babula, P., Eckschlager, T., Adam, V., Kizek, R., Masarik, M. 2014. Metallothionein polymorphisms in pathological processes. *Metallomics* 6: 55–68.

Rauser, W.E. 1995. Phytochelatins and related peptides. Structure, biosynthesis, and function. *Plant Physiology* 109: 1141–1149.

Rauser, W.E. 1999. Structure and function of metal chelators produced by plants: The case for organic acids, amino acids, phytin, and metallothioneins. *Cell Biochemistry and Biophysics* 31: 19–48.

Raymond, K.N., Dertz, E.A., Kim, S.S. 2003. Enterobactin: An archetype for microbial iron transport. *Proceedings of the National Academy of Science of the United States of America* 100: 3584–3588.

Rea, P.A., Vatamaniuk, O.K., Rigden, D.J. 2004. Weeds, worms, and more. Papain's long-lost cousin, phytochelatin synthase. *Plant Physiology* 136: 2463–2474.

Rehman, K., Chen, Z., Wang, W.W., Wang, Y.W., Sakamoto, A., Zhang, Y.F., Naranmandura, H., Suzuki, N. 2012. Mechanisms underlying the inhibitory effects of arsenic compounds on protein tyrosine phosphatase (PTP). *Toxicology and Applied Pharmacology* 263: 273–280.

Rensing, C., Ghosh, M., Rosen, B.P. 1999. Families of soft-metal-ion-transporting ATPases. *Journal of Bacteriology* 181: 5891–5897.

Rensing, C., Grass, G. 2003. *Escherichia coli* mechanisms of copper homeostasis in a changing environment. *FEMS Microbiology Reviews* 27: 197–213.

Rigby Duncan, K.E., Stillman, M.J. 2006. Metal-dependent protein folding: Metallation of metallothionein. *Journal of Inorganic Biochemistry* 100: 2101–2107.

Rigobello, M.P., Folda, A., Citta, A., Scutari, G., Gandin, V., Fernandes, A.P., Rundlo, A.K., Marzano, C., Bjornstedt, M., Bindoli, A. 2011. Interaction of selenite and tellurite with thiol-dependent redox enzymes: Kinetics and mitochondrial implications. *Free Radical Biology and Medicine* 50: 1620–1629.

Riley, R.G., Zachara, J.M., Wobber, F.J. 1992. Chemical contaminants on DOE lands and selection of contaminants mixtures for subsurface science research. Report DOE/ER-0547T. Pacific Northwest Lab, Richland, WA.

Roane, T.M. 1999. Lead resistance in two isolates from heavy metalcontaminated soils. *Microbial Ecology* 37: 218–224.

Roane, T.M., Pepper, I.L. 2000. Microorganisms and metal pollution. In: Maier, R.M., Pepper, I.L., Gerba, C.P. (eds). *Environmental Microbiology*. Academic Press, California, USA, pp. 403–423.

Rodionov, D.A., Hebbeln, P., Gelfand, M.S., Eitinger, T. 2006. Comparative and functional genomic analysis of prokaryotic nickel and cobalt uptake transporters: Evidence for a novel group of ATP-binding cassette transporters. *Journal of Bacteriology* 188: 317–327.

Rosadini, C.V., Gawronski, J.D., Raimunda, D., Argüello, J.M., Akerley, B.J. 2011. A novel zinc binding system, ZevAB, is critical for survival of nontypeable *Haemophilus influenzae* in a murine lung infection model. *Infection Immunity* 79: 3366–3376.

Rosen, B.P. 2002. Biochemistry of arsenic detoxification. *FEBS Letters* 529: 86–92.

Rouch, D., Camakaris, J., Lee, B.T.O. 1989. Copper transport in *E. coli*. In: Hamer, D.H., Winge, D.R. (eds). *Metal Ion Homeostasis, Molecular Biology and Chemistry*. New York, N.Y: Alan R. Liss, Inc., pp. 469–477.

Rowland, J.L., Niederweis, M. 2012. Resistance mechanisms of *Mycobacterium tuberculosis* against phagosomal copper overload. *Tuberculosis* 92: 202–210.

Rowland, J.L., Niederweis, M. 2013. A multicopper oxidase is required for copper resistance in *Mycobacterium tuberculosis*. *Journal of Bacteriology* 195: 3724–3733.

Ruttkay-Nedecky, B., Nejdl, L., Gumulec, J., Zitka, O., Masarik, M., Eckschlager, T., Stiborova, M., Adam, V., Kizek, R. 2013. The role of metallothionein in oxidative stress. *International Journal of Molecular Science* 14: 6044–6066.

Sabolic, I., Breljak, D., Skarica, M., Herak-Kramberger, C.M. 2010. Role of metallothionein in cadmium traffic and toxicity in kidneys and other mammalian organs. *Biometals* 23: 897–926.

Sahlman, L., Hagglof, E.M., Powlowski, J. 1999. Roles of the four cysteine residues in the function of the integral inner membrane Hg^{2+}-binding protein, MerC. *Biochemical and Biophysical Research Communications* 255: 307–311.

Salem, H.M., Eweida, E.A., Farag, A. 2000. Heavy metals in drinking water and their environmental impact on human health. In *ICEHM 2000*. Giza, Egypt: Cairo University, pp. 542–556.

Sardiwal, S., Santini, J.M., Osborne, T.H., Djordjevic, S. 2010. Characterization of a two-component signal transduction system that controls arsenite oxidation in the chemolithoautotroph NT-26. *FEMS Microbiology Letters* 313: 20–28.

Schauer, K., Gouget, B., Carriere, M., Labigne, A., de Reuse, H. 2007. Novel nickel transport mechanism across the bacterial outer membrane energized by the TonB/ExbB/ExbD machinery. *Molecular Microbiology* 63: 1054–1068.

Schelert, J., Dixit, V., Hoang, V., Simbahan, J., Drozda, M., Blum, P. 2004. Occurrence and characterization of mercury resistance in the hyperthermophilic archaeon *Sulfolobus solfataricus* by use of gene disruption. *Journal of Bacteriology* 186: 427–437.

Schelert, J., Rudrappa, D., Johnson, T., Blum, P. 2013. Role of MerH in mercury resistance in the archaeon *Sulfolobus solfataricus*. *Microbiology* 59: 1198–208.

Schiering, N., Kabsch, W., Moore, M.J., Distefano, M.D., Walsh, C.T., Pai, E.F. 1991. Structure of the detoxification catalyst mercuric ion reductase from *Bacillus* sp. strain RC607. *Nature* 352: 168–172.

Schue, M., Dover, L.G., Besra, G.S., Parkhill, J., Brown, N.L. 2009. Sequence and analysis of a plasmid-encoded mercury resistance operon from *Mycobacterium marinum* identifies MerH, a new mercuric ion transporter. *Journal of Bacteriology* 191: 439–444.

Schwarz, G., Hagedoorn, P., Fischer, K. 2007. Molybdate and tungstate: uptake, homeostasis, cofactors, and enzymes. In: Nies, D.H., Silver, S. (eds). *Molecular microbiology of heavy metals*, Vol. 6. Springer-Verlag, Berlin, pp. 421–451.

Schwarz, G., Mendel, R.R., Ribbe, M.W. 2009. Molybdenum cofactors, enzymes and pathways. *Nature* 460: 839–847.

Self, W.T., Grunden, A.M., Hasona, A., Shanmugam, K.T. 2001. Molybdate transport. *Research in Microbiology* 152: 311–321.

Sellin, S., Eriksson, L.E., Mannervik, B. 1987. Electron paramagnetic resonance study of the active site of copper-substituted human glyoxalase I. *Biochemistry* 26: 6779–6784.

Serre, L., Rossy, E., Pebay-Peyroula, E., Cohen-Addad, C., Coves, J. 2004. Crystal structure of the oxidized form of the periplasmic mercury-binding protein MerP from *Ralstonia metallidurans* CH34. *Journal of Molecular Biology* 339: 161–171.

Shafeeq, S., Yesilkaya, H., Kloosterman, T.G., Narayanan, G., Wandel, M., Andrew, P.W., Kuipers, O.P., Morrissey, J.A. 2011. The cop operon is required for copper homeostasis and contributes to virulence in *Streptococcus pneumoniae*. *Molecular Microbiology* 81: 1255–1270.

Shaik, M.M., Cendron, L., Salamina, M., Ruzzene, M., Zanotti, G. 2014. Helicobacter pylori periplasmic receptor CeuE (HP1561) modulates its nickel affinity via organic metallophores. *Molecular Microbiology* 91: 724–735.

Sievers, F. Higgins, D.G. 2014. Clustal Omega, accurate alignment of very large numbers of sequences. *Methods in Molecular Biology* 1079: 105–116.

Sikkema, J., de Bont, J.A.M., Poolman, B. 1995. Mechanisms of membrane toxicity of hydrocarbons. *Microbiological Reviews* 59: 201–222.

Silver, S. 1996. Bacterial resistances to toxic metal ions—A review. *Gene* 179: 9–19.

Singh, S., Mulchandani, A., Chen, W. 2008. Highly selective and rapid arsenic removal by metabolically engineered *Escherichia coli* cells expressing *Fucus vesiculosus* metallothionein. *Applied Environmental Microbiology* 74: 2924–2927.

Sliwinska-Mosson, M., Milnerowicz, H., Rabczynski, J., Milnerowicz, S. 2009. Immunohistochemical localization of metallothioneins and p53 protein in pancreatic serous cystadenomas. *Archivum Immunologiae et Therapiae Experimentalis* 57: 295–301.

Slonczewski, J.L., Fujisawa, M., Dopson, M., Krulwich, T.A. 2009. Cytoplasmic pH measurement and homeostasis in bacteria and archaea. *Advances in Microbial Physiology* 55: 1–79.

Slyemi, D., Moinier, D., Talla, E., Bonnefoy, V. 2013. Organization and regulation of the arsenite oxidase operon of the moderately acidophilic and facultative chemoautotrophic *Thiomonas arsenitoxydans*. *Extremophiles* 17: 911–920.

Smith, J.L. 2004. The physiological role of ferritin-like compounds in bacteria. *Critical Reviews in Microbiology* 30: 173–185.

Smith, M.A. 1984. Lead in history. In: Lansdown, R., Yule, W. (eds). *The lead debate: the environmental toxicology and child health*. London, Croom Helm, pp. 7–24.

Sone, Y., Nakamura, R., Pan-Hou, H., Itoh, T., Kiyono, M. 2013a. Role of MerC, MerE, MerF, MerT, and/or MerP in resistance to mercurials and the transport of mercurials in *Escherichia coli*. *Biological and Pharmaceutical Bulletin* 6: 1835–1841.

Sone, Y., Nakamura, R., Pan-Hou, H., Sato, M.H., Itoh, T., Kiyono, M. 2013b. Increase methylmercury accumulation in *Arabidopsis thaliana* expressing bacterial broad-spectrum mercury transporter MerE. *AMB Express* 3: 52.

Sone, Y., Pan-Hou, H., Nakamura, R., Sakabe, K., Kiyono, M. 2010. Roles played by MerE and MerT in the transport of inorganic and organic mercury compounds in Gram-negative bacteria. *Journal of Health Science* 56: 123–127.

Sousa, C., Cebolla, A., de Lorenzo, V. 1996. Enhanced metalloadsorption of bacterial cells displaying poly-His peptides. *Nature Biotechnology* 14: 1017–1020.

Sousa, C., Kotrba, P., Ruml, T., Cebolla, A., De Lorenzo, V. 1998. Metalloadsorption by *Escherichia coli* cells displaying yeast and mammalian metallothioneins anchored to the outer membrane protein LamB. *Journal of Bacteriology* 180: 2280–2284.

Spiess, K., Lammel, A., Scheibel, T. 2010. Recombinant spider silk proteins for applications in biomaterials. *Macromolecular Bioscience* 10: 998–1007.

Steele, R.A., Opella, S.J. 1997. Structures of the reduced and mercury-bound forms of MerP, the periplasmic protein from the bacterial mercury detoxification system. *Biochemistry* 36: 6885–6895.

Stolz, J.F., Basu, P., Oremland, R.S. 2010. Microbial arsenic metabolism: New twists on an old poison. *Microbe* 5: 53–59.

Strange, H.R., Zola, T.A., Cornelissen, C.N. 2011. The fbpABC operon is required for Ton-independent utilization of xenosiderophores by *Neisseria gonorrhoeae* strain FA19. *Infection Immunity* 79: 267–278.

Styblo, M., Serves, S.V., Cullen, W.R., Thomas, D.J. 1997. Comparative inhibition of yeast glutathione reductase by arsenicals and arsenothiols. *Chemical Research in Toxicology* 10: 27–33.

Summers, A.O., Sugarman, L.I., 1974. Cell-free mercury(II)-reducing activity in a plasmid-bearing strain of *Escherichia coli*. *Journal of Bacteriology* 119: 242–249

Sutherland, D.E., Willans, M.J., Stillman, M.J. 2010. Supermetalation of the beta domain of human metallothionein 1a. *Biochemistry* 49: 3593–3601.

Sutherland, D.E.K., Stillman, M.J. 2011. The "magic numbers" of metallothionein. *Metallomics* 3: 444–463.

Sutherland, D.E.K., Willans, M.J., Stillman, M.J. 2012. Single domain metallothioneins: Supermetalation of human MT 1a. *Journal of American Chemical Society* 134: 3290–3299.

Suzuki, T., Miyata, N., Horitsu, H., Kawai, K., Takamizawa, K., Tai, Y., Okazaki, M. 1992. NAD(P)H-dependent chromium(VI) reductase of *Pseudomonas ambigua* G-1: A Cr(V) intermediate is formed during the reduction of Cr(VI) to Cr(III). *Journal of Bacteriology* 174: 5340–5534.

Takahashi, S. 2012. Molecular functions of metallothionein and its role in the hematological malignancies. *Journal of Hematology and Oncology* 5: 1–8.

Tamas, M.J., Sharma, S.K., Ibstedt, S., Jacobson, T., Christen, P. 2014. Heavy metals and metalloids as a cause for protein misfolding and aggregation. *Biomolecules* 4: 252–267.

Teitzel, G.M., Parsek, M.R. 2003. Heavy metal resistance of biofilm and planktonic *Pseudomonas aeruginosa*. *Applied and Environmental Microbiology* 69: 2313–2320.

Theil, E.C., Goss, D.J. 2009. Living with iron (and oxygen): Questions and answers about iron homeostasis. *Chemical Reviews* 109: 4568–4579.

Thirumoorthy, N., Sunder, A.S., Kumar K.M., Ganesh G.N.K., Chatterjee, M. 2011. A review of metallothionein isoforms and their role in pathophysiology. *World Journal of Surgical Oncology* 9: 54.

Tian, Y., Lu, G.J., Marassi, F.M., Opella, S.J. 2014. Structure of the membrane protein MerF, a bacterial mercury transporter, improved by the inclusion of chemical shift anisotropy constraints. *Journal of Biomolecular NMR* 60: 67–71.

Tong, S., Schirnding, Y.E.V., Prapamonto, T. 2000. Environmental lead exposure: A public health problem of global dimensions. *Bulletin of WHO* 78: 1068–1077.

Tripathi, R.D., Srivastava, S., Mishra, S., Singh, N., Tuli, R., Gupta, D.K., Maathuis, F.J.M. 2007. Arsenic hazards: Strategies for tolerance and remediation by plants. *Trends in Biotechnology* 25: 158–165.

Tynecka, Z., Gos, Z., Zajac, J. 1981. Reduced cadmium transport determined by a resistance plasmid in *Staphylococcus aureus*. *Journal of Bacteriology* 147: 305–312.

Vande Weghe, J.G., Ow, D.W. 1999. A fission yeast gene for mitochondrial sulfide oxidation. *Journal of Biological Chemistry* 274: 13250–13257.

Vatamaniuk, O.K., Bucher, E.A., Ward, J.T., Rea, P.A. 2001. A new pathway for heavy metal detoxification in animals. Phytochelatin synthase is required for cadmium tolerance in *Caenorhabditis elegans*. *Journal of Biological Chemistry* 276: 20817–20820.

Vicente-Vicente, L., Quiros, Y., Pérez-Barriocanal, F., López-Novoa, J.M., López-Hernández, F.J., Morales, A.I. 2010. Nephrotoxicity of uranium: Pathophysiological, diagnostic and therapeutic perspectives. *Toxicological Sciences* 118: 324–347.

Vidaud, C., Dedieu, A., Basset, C., Plantevin, S., Dany, I., Pible, O., Quéméneur, E. 2005. Screening of human serum proteins for uranium binding. *Chemical Research in Toxicology* 18: 946–953.

Vinceti, M., Wei, E.T., Malagoli, C., Bergomi, M., Vivoli, G. 2001. Adverse health effects of selenium in humans. *Reviews on Environmental Health* 16: 233–251.

Wahba, H.M., Lecoq, L., Stevenson, M., Mansour, A., Cappadocia, L., Lafrance-Vanasse, J., Wilkinson, K.J., Sygusch, J., Wilcox, D.E., Omichinski, J.G. 2016. Structural and biochemical characterization of a copper-binding mutant of the organomercurial lyase MerB: Insight into the key role of the active site aspartic acid in Hg-carbon bond cleavage and metal binding specificity. *Biochemistry* 55: 1070–1081.

Waldron, K.J., Robinson, N.J. 2009. How do bacterial cells ensure that metalloproteins get the correct metal? *Nature Reviews Microbiology* 7: 25–35.

Wang, Q., Warelow, T.P., Kang, Y.S., Romano, C., Osborne, T.H., Lehr, C.R., Bothner, B., McDermott, T.R., Santini, J.M., Wang, G. 2015a. Arsenite oxidase also functions as an antimonite oxidase. *Applied and Environmental Microbiology* 81: 1959–1965.

Wang, S., Blahut, M., Wu, Y., Philipkosky, K.E., Outten, F.W. 2014. Communication between binding sites is required for YqjI regulation of target promoters within the yqjH-yqjI intergenic region. *Journal of Bacteriology* 196: 3199–3207.

Wang, S., Wu, Y., Outten, F.W. 2011. Fur and the novel regulator YqjI control transcription of the ferric reductase gene yqjH in *Escherichia coli*. *Journal of Bacteriology* 193: 563–574.

Wang, Y., Yao, H., Cheng, Y., Lovell, S., Battaile, K.P., Midaugh, C.R., Rivera, M. 2015b. Characterization of the bacterioferritin/bacterioferritin associated ferredoxin protein-protein interaction in solution and determination of binding energy hot spots. *Biochemistry* 54: 6162–6175.

Warelow, T.P., Oke, M., Schoepp-Cothenet, B., Dahl, J.U., Bruselat, N., Sivalingam, G.N., Leimkühler, S., Thalassinos, K., Kappler, U., Naismith, J.H., Santini, J.M. 2013. The respiratory arsenite oxidase: Structure and the role of residues surrounding the Rieske Cluster. *PLoS One* 8: e72535.

West, A.L., Evans, S.E., González, J.M., Carter, L.G., Tsuruta, H., Pozharski, E., Michel, S.L. 2012. Ni(II) coordination to mixed sites modulates DNA binding of HpNikR via a long-range effect. *Proceedings of the National Academy of Science of the United States of America* 109: 5633–5638.

Wherland, S., Farver, O., Pecht, I. 2014. Multicopper oxidases: Intramolecular electron transfer and O$_2$ reduction. *Journal of Biological Inorganic Chemistry* 19: 541–554.

Wiethaus, J., Wildner, G.F., Masepohl, B. 2006. The multicopper oxidase CutO confers copper tolerance to *Rhodobacter capsulatus*. *FEMS Microbiology Letters* 256: 67–74.

Williams, C.H., Arscott, L.D., Muller, S., Lennon, B.W., Ludwig, M.L., Wang, P.F., Veine, D.M., Becker, K., Schirmer, R.H. 2000. Thioredoxin reductase two modes of catalysis have evolved. *European Journal of Biochemistry* 267: 6110–6117.

Williams, K.H., Bargar, J.R., Lloyd, J.R., Lovley, D.R. 2013. Bioremediation of uranium-contaminated groundwater: A systems approach to subsurface biogeochemistry. *Current Opinion in Biotechnology* 24: 489–497.

Wilson, J.R., Leang, C., Morby, A.P., Hobman, J.L., Brown, N.L. 2000. MerF is a mercury transport protein: Different structures but a common mechanism for mercuric ion transporters? *FEBS Letters* 472: 78–82.

Winkelmann, G. 2002. Microbial siderophore-mediated transport. *Biochemical Society Transactions* 30: 691–696.

Wuana, R.A., Okieimen, F.E. 2011. Heavy metals in contaminated soils: A review of sources, chemistry, risks and best available strategies for remediation. *ISRN Ecology* 2011: 1–20.

Xia, L., Yin, C., Cai, L., Qiu, G., Qin, W., Peng, B., Liu, J. 2010. Metabolic changes of *Acidithiobacillus caldus* under Cu^{2+} stress. *Journal of Basic Microbiology* 50: 591–598.

Xu, C., Shi, W., Rosen, B.P. 1996. The chromosomal arsR gene of *Escherichia coli* encodes a trans-acting metalloregulatory protein. *Journal of Biological Chemistry* 271: 2427–2432.

Yao, H., Jepkorir, G., Lovell, S., Nama, P.V., Weeratunga, S., Battaile, K.P., Rivera, M. 2011. Two distinct ferritin-like molecules in *Pseudomonas aeruginosa*: The product of the bfrA gene is a bacterial ferritin (FtnA) and not a bacterioferritin (Bfr). *Biochemistry* 50: 5236–5248.

Yatsunyk, L.A., Easton, J.A., Kim, L.R., Sugarbaker, S.A., Bennett, B., Breece, R.M., Vorontsov, I.I., Tierney, D.L., Crowder, M.W. Rosenzweig, A.C. 2008. Structure and metal binding properties of ZnuA, a periplasmic zinc transporter from *Escherichia coli*. *Journal of Biological Inorganic Chemistry* 13: 271–288.

Yoda, A., Toyoshima, K., Watanabe, Y., Onishi, N., Hazaka, Y., Tsukuda, Y., Tsukada, J., Kondo, T., Tanaka, Y., Minami, Y.J. 2008. Arsenic trioxide augments Chk2/p53-mediated apoptosis by inhibiting oncogenic Wip1 phosphatase. *Journal of Biological Chemistry* 283: 18969–18979.

Yoon, B.Y., Kim, Y.H., Kim, N., Yun, B.Y., Kim, J.S., Lee, J.H., Cho, H.S., Lee, K., Ha, N.C. 2013. Structure of the periplasmic copper-binding protein CueP from *Salmonella enterica* serovar Typhimurium. *Acta Crystallographica Section D: Biological Crystallography* 69: 1867–1875.

Zargar, K., Conrad, A., Bernick, D.L., Lowe, T.M., Stolc, V., Hoeft, S., Oremland, R.S., Stolz, J., Saltikov, C.W. 2012. ArxA, a new clade of arsenite oxidase within the DMSO reductase family of molybdenum oxidoreductases. *Environmental Microbiology* 14: 1635–1645.

Zargar, K., Hoeft, S, Oremland, R.S., Saltikov, C.W. 2010. Identification of a novel arsenite oxidase gene, arxA, in the haloalkaliphilic, arsenite-oxidizing bacterium *Alkalilimnicola ehrlichii* strain MLHE-1. *Journal of Bacteriology* 192: 3755–3762.

Zenk, M.H. 1996. Heavy metal detoxification in higher plants a review. *Gene* 179: 21–30.

Zhang, L., Koay, M., Maher, M.J., Xiao, Z., Wedd, A.G. 2006. Intermolecular transfer of copper ions from the CopC protein of *Pseudomonas syringae*. Crystal structures of fully loaded Cu(I)Cu(II) forms. *Journal of American Chemical Society* 128: 5834–50.

Zhou, T., Radaev, S., Rosen, B.P., Gatti, D.L. 2001. Conformational changes in four regions of the *Escherichia coli* ArsA ATPase link ATP hydrolysis to ion translocation. *The Journal of Biological Chemistry* 276: 30414–30422.

14 Microbial Communities in Metal-Contaminated Environments
Adaptation and Function in Soil

Erika Kothe and Martin Reinicke

CONTENTS

ABSTRACT

Metal contamination is increasingly influencing soil habitats. In order to develop strategies for future land use and restoration of plant growth in such areas, a deeper understanding of the microbiology and the processes occurring in metal-contaminated soil is essential. Here, we review microbial communities in soil with special emphasis on heavy metal contamination through acid mine drainage that is associated with metal mining operations world over. Interactions with plant roots are one specific niche in soil, and hence, interactions between soil microbes and plant roots are covered. The changes that occur in soil when heavy metals are present and the mechanisms by which the microorganisms in the rhizosphere are responding to metal stress are discussed and methods compared that can be used to study microbial rhizosphere communities. An outlook on the application of the gained insight for bioremediation approaches, which are microbially aided or governed, is given at the end of this chapter.

14.1 INTRODUCTION

Soil is one of the most complex terrestrial ecosystems, and its fertility and function in global biogeochemical cycles are influencing humankind, as soil forms the highly important interface between the geosphere, hydrosphere, and atmosphere (van Camp et al., 2004). Soil quality is defined as the ability to sustain life between ecosystem function and land use for the production of nutrients as well as impacting air quality and water resources (Doran and Safley, 1997). Not only human, animal, and plant health are controlled by soil function, but also erosion control, water quality and accessibility, and biodiversity and regulation of (micro) climate are dependent on soil functions. These valid ecosystem functions result from processes in soil (Brussaard, 2012), like decomposition and transformation of organic materials. At the same time, contaminant persistence and ecotoxicity are governed by soil processes of microbial biodegradation and mineralization (McCarthy and Williams, 1992; De Boer et al., 1999). Approximately 90% of soil processes are dependent on microbial activities, with microbes constituting 80% of soil living biomass (Lynch, 1988). Both spatial and temporal variations depend not only on multiple parameters, including matrix structure, soil oxygen contents, water and nutrient accessibility, temperature, land use and management, and vegetation but also on contaminations (Ward et al., 1998; Nusslein and Tiedje, 1999; Sandaa et al., 1999; Marschner et al., 2001; Horner-Devine et al., 2004).

With this chapter, we address the specific communities present in metal-contaminated land and their role in soil formation. Since this is the prerequisite of future land use, the importance of microorganisms in establishing a soil cover cannot be overestimated (Joshi et al., 2014). Still, many plans for remediation action or use of metal-contaminated sites do not include the management of soil microorganisms. Here, we want to show what should be expected from a successful plan for remediation actions on sites influenced by heavy metals through anthropogenic use, for example, with former mining activities. We discuss microbe–plant interactions, metal resistance mechanisms within the microbes, specific communities, and their adaptation to metal contamination and, in a concluding paragraph, develop a scheme for future microbial-mediated remediation actions.

14.2 MICROBE–PLANT INTERACTIONS IN THE RHIZOSPHERE

The vegetation cover exerts a high influence on belowground biodiversity and microbial population functions. The plant root exudates allow for enhanced microbial growth in the rhizosphere, which may be highly specific for a given plant species and even cultivar (Smalla et al., 2001; Nannipieri et al., 2008). The symbioses of plants and microbes cover the entire range from mutually beneficial to pathogenic. Both plant growth–promoting bacteria support plant nutrition, including nitrogen fixation by plant-associated rhizobia or free living *Azotobacter*, *Bacillus*, *Clostridium*, or *Klebsiella* (Narula et al., 2012; Wagner, 2012), act through phytohormone production (Costacurta and Vanderleyden, 1995), or control pathogens (Folman et al., 2001). Additional activities may prevent metal uptake while allowing for better iron nutrition through the formation of secondary metabolites like siderophores (Dimkpa et al., 2008a, 2009a; Albarracín et al., 2010).

Not to the least, microbes, like plants, are involved in weathering by excreting organic acids, enzymes, and other metabolites (Ehrlich, 1996; Wengel et al., 2006; Hopf et al., 2009; Gadd, 2010; Seifert et al., 2011). Together, all these microbial activities lead to genesis of soil. The fertility of the resulting substrate is by far the most prominent microbial function used by man (Figure 14.1) and hence should be understood when contaminated land is reclaimed in order to be able to offer successful remediation strategies.

14.3 SOURCES AND EFFECTS OF HEAVY METAL CONTAMINATION ON SOIL AND RHIZOSPHERE

14.3.1 ANTHROPOGENIC SOURCES

Mining operations are prominently involved in metal pollution in the environment (Johnson and Hallberg, 2005), with increasing demands for metals, exerting increasing strain on ecosystems as well as human welfare (Sterritt and Lester, 1980; Nriagu, 1990). Sulfidic ores specifically are prone to lead to environmental dangers since acid mine drainage (AMD) increases metal mobility and makes metal input hardly measurable, since it will persist for a long time and is self-sustaining through microbial activity (Adam et al., 1997; Bruins et al., 2000). The acidic habitat can be inhabited by acidophilic, iron-oxidizing bacteria like *Gallionella ferruginea* (Stumm and Morgan, 1981), *Acidithiobacillus ferrooxidans* (Kelly and Wood, 2000), or other species of the genera *Thiobacillus*, *Ferroplasma*, *Acidimicrobium*, or *Acidobacterium* (Hallberg and Johnson, 2001). The microbially catalyzed, biogenic oxidation of Fe(II) to Fe(III) enhances the abiotic pyrite oxidation, which can be used with microbial leaching of low-grade ores (Krebs et al., 1997; Rawlings and Johnson, 2007; Johnson et al., 2013; Schippers et al., 2014).

14.3.2 SOIL PROCESSES

The translocation of metals in soil is determined by soil parameters like pH, redox potential, content, and composition of the organic fraction or sorption capacity by, for example, clay minerals (Brümmer et al., 1986; Clemente et al., 2006; Merdy et al., 2009; Violante et al., 2010). Aside from geochemical

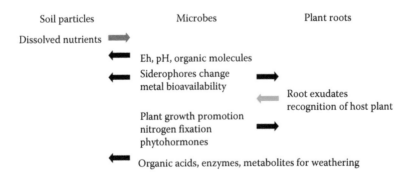

FIGURE 14.1 Microbial impact on soil. The interactions within soil between the geosphere, plants, and microbial community consist of a multitude of specific reactions, some of which are summarized here.

processes, microbial activity is a determinant for metal mobility and bioavailability in soil. Degradation of organic molecules may decrease sorption and increase water-soluble, metal-organic complexes (Kabata-Pendias and Pendias, 2001), while reduction of iron and manganese oxides releases metals to the soil and water (Francis and Dodge, 1990).

Other microbial activities in soil lead to reduction of metal loads. Biogenically formed Fe(II) or sulfides can act as reductants (Burdige and Nealson, 1986; Nealson and Myers, 1992), while enzymatic change of reduction state may lead to reduced mobility (Machemer and Wildeman, 1992; Lovley, 1993). The changed metal reduction state, at the same time, leads to a change in toxicity and bioavailability (Cottenie et al., 1980; Michalke, 2003).

14.3.3 METAL UPTAKE AND ECOTOXICOLOGY

Many metals are essential micronutrients, for example, copper, iron, magnesium, or nickel (Burgess and Lowe, 1996; Watt and Ludden, 1999; McCall et al., 2000; Sauer and Yachandra, 2004); others like cadmium, silver, or mercury are not known to exert biological functions (Doelman et al., 1994). Different complexes or oxidation states may reduce toxicity, like protection from metabolization in, for example, arsenobetaine (Florence, 1989) or Cr(III) (Florence, 1989; Guerrero-Romero and Rodríguez-Morán, 2005). Metal toxicity may exert a negative effect on enzymatic activities in soil (Doelman and Haanstra, 1986; Kunito et al., 2001), while, at the same time, mutagenicity contributes to changes in microbial community composition (Hengstler et al., 2003; Zhou et al., 2008). The mutagenic activity mainly is due to the formation of reactive oxygen species through the chemical Fenton reaction (Keyer and Imlay, 1996; Lemire et al., 2013). In addition, copper, silver, and cadmium are expected to lead to structural changes in membrane lipids leading to cell lysis (Slawson et al., 1992; Dibrov et al., 2002; Hong et al., 2012). These processes pertain, necessarily, also to the rhizosphere soil. Therefore, the changes in microbial predominance due to metals in the soil will be specifically important.

14.4 HEAVY METAL CAUSED CHANGES OF SOIL MICROFLORA

In addition to microbial biomass, microbial diversity is an indicator of soil quality (Schloter et al., 2003; Sharma et al., 2011). High metal loads generally reduce microbial biomass (Valsecchi et al., 1995; Wang et al., 2007), enzymatic and microbial activity (Lee et al., 2002; Frey et al., 2006; Oliveira and Pampulha, 2006; Wang et al., 2007; Sprocati et al., 2014), and community structure and compositions (Baath et al., 1998; Kandeler et al., 2000; Bamborough and Cummings, 2009; Desai et al., 2009). Adaptation to metal contamination leads to higher prevalence of metal tolerant or resistant taxa (Babich and Stotzky, 1977; Doelman et al., 1994; Pennanen et al., 1996; Bruins et al., 2000). The selection process is controlled by organic contents, pH, metal concentration and species, exposition time, and other habitat factors

(Sandaa et al., 1999; Ellis et al., 2003; Mirete et al., 2007; Dhal et al., 2011; Pereira et al., 2014). The change in community structure often is associated with functional differences to microbial communities in pristine soils (Giller et al., 1998; Frey et al., 2006), and the functional differences may directly influence ecosystem stability (Seybold et al., 1999).

While reduction in biomass is a well-documented effect of metal contamination, the adaptation of microbial communities to contaminants in soil is associated with an initial diversification. Only at very high contamination levels, a reduction in biodiversity is expected. A stable ecosystem generally is dominated by few taxa, and their repression at intermediate contamination levels paves the way for new species to invade resulting in a higher diversity (Connell, 1978; Giller et al., 1998). At very high contamination, biodiversity is lost and only highly resistant microorganisms can survive. An alternative hypothesis of linear species loss is less likely (Giller et al., 1998; Wohl et al., 2004).

In order to survive the harsh conditions in metal-contaminated soil and in the resulting rhizosphere, metal resistance mechanisms are needed with the microbes present in the soil. These may allow not only for survival but also for the modification of the bioavailability of metal species in soil and in the rhizosphere.

14.5 BACTERIAL METAL RESISTANCE MECHANISMS

14.5.1 METAL RESISTANCE AND TOLERANCE

With respect to adaptation to metal contamination, bacteria that can cope with high environmental metal loads are positively selected. Thus, from metal-contaminated soil, (highly) resistant bacterial strains may be isolated (Doelman et al., 1994; Margesin and Schinner, 1996; Hassen et al., 1998; Schmidt et al., 2005; Jiang et al., 2008; Bafana et al., 2010). Schmidt et al. (2009) isolated two strains of *Streptomyces mirabilis* from an AMD-influenced site that were able to grow at nickel or zinc concentrations of more than 100 mM. Other studies identified strains with metal resistance against diverse heavy metals, for example, *Cupriavidus metallidurans* (Nies, 2003), *Arthrobacter ramosus* (Bafana et al., 2010), or *Pseudomonas aeruginosa* (Chen et al., 2006), as well as species of the genera *Acidocella* (Ghosh et al., 1997), *Acidiphilium* (San Martin-Uriz et al., 2014), and *Streptomyces* (Amoroso et al., 2002). A differentiation between resistance and tolerance toward metals refers to active cellular mechanisms for resistance, while for tolerance, passive adsorption may be sufficient to sequester metals and thus lower free metal ion concentrations.

14.5.2 INTRA- AND EXTRACELLULAR MECHANISMS

Intracellular metal homeostasis depends on resistance mechanisms, which cluster into five general groups: intra- and extracellular detoxification, efflux through suitable metal transport systems, intracellular sequestration, extracellular sequestration, and sorption to the cell surface (Haferburg and Kothe, 2007).

Mostly, resistance factors are encoded on plasmids or transposons, which enable horizontal gene transfer (Silver and Phung, 1996; Osborn et al., 1997; Abou-Shanab et al., 2007).

Efflux will allow survival, albeit without changing the bioavailability of metals. The substrate spectra may allow detoxification of more than one metal (Silver and Phung, 1996). Most efflux transporters are powered by proton antiport, for example, with the copper, zinc, and cadmium accepting Czc system in Gram-negative bacteria like *Cupriavidus necator* (formerly *Ralstonia eutropha*; Nies, 1995). Other transporters are fueled by ATP hydrolysis, for example, the cadmium and zinc accepting P-type ATPase CadA in Gram-positive *Staphylococcus aureus* (Nies, 1999).

More likely to lead to stable conditions with reduced metal motilities' in the habitat are mechanisms based on extracellular or intracellular sequestration/binding. In *Escherichia coli*, *pco* codes for copper-binding proteins in the periplasm (Brown et al., 1995), and nickel has been shown to be bound to polyphosphate granula (volutin) in *Staphylococcus aureus* (Gonzalez and Jensen, 1998).

Extracellular mechanisms include the excretion of chelators or sorption to the cell surface. Siderophores can act both in iron acquisition limiting in most soils (Andrews et al., 2003) while binding other metals like nickel or cadmium and preventing their entry into the cell, for example, the desferroxamines in streptomycetes (Dimkpa et al., 2008a). A third extracellular mechanism could be the formation of biominerals, which is widely distributed among bacteria (Lowenstam and Weiner, 1989; Skinner, 2005). Locally dominating minerals indeed have been found to be formed under laboratory conditions by isolates from AMD-influenced areas (Haferburg et al., 2008).

14.5.3   IMPACT ON THE ENVIRONMENT

Resistance of microorganisms is a prerequisite for the restoration of microbial activity in metal-contaminated soils (Gitay et al., 1996). The resistant microbes provide activities leading not only to soil fertility but also to reduction of metal bioavailability and hence to support less tolerant taxa (Stephen et al., 1999). The measurement of microbial diversity as a result of this selection process thus is an important tool to address potential environmental changes through microbial activities (Figure 14.2).

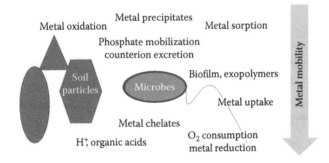

FIGURE 14.2   Microbial influence on metal mobility in soil. Major physiological processes are indicated depending on the effect on metal mobility within the soil water phase.

14.6   MEASURING MICROBIAL DIVERSITY

Uncontaminated soils carry high cell densities with a high genetic diversity of up to 13,000 genomes per gram of soil dry weight (Torsvik et al., 1994; Euzeby, 1997). Estimates for the number of cultivable bacteria range between 0.1% and 10% of species, which rises doubt on the representative nature of cultivation approaches (Amann and Schleifer, 1995; Rondon et al., 2000; Hall, 2007). In addition, soil water content and environmental factors influence numbers of isolated bacteria per gram soil (Table 14.1). Molecular approaches have been developed to fill this gap. These approaches, however, may be biased by other factors like differences in cell wall stability, PCR primer efficiency, or still unknown processes. Generally, DNA-dependent approaches not only show higher prevalences of proteobacteria but also yield taxa like Acidobacteria, Planctomycetes, Verrucomicrobia, or Armatimonadetes (Liesack and Stackebrandt, 1992; Jones et al., 2009). Amplification of 16S rDNA, which covers the variable regions V1 through V9 of the small subunit ribosome RNA, allows for species identification based on support by databank entries (Olsen et al., 1986; Woese, 1987). Fingerprinting methods like GC content of metagenomic DNA, nucleic acid reassociation kinetics (Torsvik et al., 1990), denaturing gradient gel electrophoresis (Muyzer et al., 1993), terminal restriction fragment length polymorphism (Tiedje et al., 1999), single-strand conformation polymorphism (Orita et al., 1989), or ribosomal intergenic spacer analysis (Fisher and Triplett, 1999) may be more cost-effective, but these methods often are error prone and specific care must be taken to identify the correct control to be added with many technical and biological replicates (Kuske et al., 1997; Head et al., 1998; Rappé and Giovannoni, 2003). Without additional sequencing, genera mainly may be identified while species often cannot be distinguished (Hartmann and Widmer, 2006; Nocker et al., 2007).

Cloning amplified 16S rDNA of metagenomic DNA and subsequent sequencing allows for higher resolution of bacterial/archaeal community composition (Hugenholtz et al., 1998). However, again a bias is exerted and specifically taxon sampling or ensuring sufficient sequences to be recovered is essential (Dunbar et al., 2002).

The development of metagenomic tools has enabled more detailed studies (Margulies et al., 2005; Ansorge, 2009). An increase in sequences made the development of more databases and software applications necessary, like ribosomal database project (Larsen et al., 1993; Maidak et al., 1994; Roh et al., 2010). Like all DNA-based methods, metagenome sequencing is dependent on DNA extraction (Thakuria et al., 2008; Lombard et al., 2011), primers used (Engelbrektson et al., 2010; Gonzalez et al., 2012), and biological variability including copy number or sequence divergence (Coenye and Vandamme, 2003). However, the numbers of sequences of a specific species allow for quantification (Nocker et al., 2007; Gonzalez et al., 2012). However, it so far seems advisable to have the synergistic information gain with both cultivation and DNA-based methods. This will allow a holistic picture of community structures and the related microbial abundances and functions.

TABLE 14.1

Colony-Forming Units Are Dependent on Time of Sampling

	Plot 1	Standard Deviation	Plot 2	Standard Deviation	Plot 3	Standard Deviation
Spring	2.03×10^4	3.77×10^3	1.43×10^6	1.70×10^4	5.47×10^7	4.11×10^6
Summer	3.93×10^6	1.47×10^5	4.63×10^6	3.47×10^4	1.88×10^7	2.35×10^6
Fall	2.00×10^4	1.00×10^4	9.00×10^4	1.00×10^4	3.09×10^6	7.40×10^5
Next spring	1.10×10^5	1.00×10^3	1.40×10^5	1.00×10^4	2.47×10^6	4.10×10^5
Next fall	6.18×10^5	9.17×10^4	5.90×10^6	6.93×10^5	1.66×10^7	1.96×10^6

The examples are taken from the test field site (see Section 14.6).

Biomass determination (Tabacchioni et al., 2000), assays for metabolic diversity (Garland, 1996), and soil respiration additionally allow for insight into community functions, especially when contaminated soil is analyzed (Islam et al., 2011; Liu et al., 2012; Moreno et al., 2012).

14.7 CONCEPTS OF MICROBIALLY AIDED REMEDIATION OF HEAVY METAL–CONTAMINATED SOIL

14.7.1 Ecotoxicological Implications

The potential risk of contaminant dispersal via wind or water and uptake into food chains necessitates approaches for remediation. Depending on area size, content of sulfidic minerals that enhance AMD formation, concentration and bioavailability of metals, distribution and proximity of inhabited or ecologically important habitats, and the feasibility of a biological concept for remediation have to be assessed (Adam et al., 1997; Downing and Madeisky, 1997; Mulligan et al., 2001; Akcil and Koldas, 2006). To achieve a long-term protection for the environment and the population, methods to lower toxicity and/or bioavailability may be optimized with the addition of microbiological measures (Martin and Ruby, 2004; Kothe and Büchel, 2014; Langella et al., 2014; Nicoara et al., 2014; Phieler et al., 2014; Sprocati et al., 2014).

14.7.2 On-Site Approaches

While *ex situ* methods are cost intensive and destroy the soil properties and functioning, *in situ* approaches may allow for lower risk to the environment and protect the already established habitats with starting succession. However, highly contaminated areas, often bare ground or brownfield sites, may not be permissible for bioremediation. In heterogeneous, large-sized, medium to low contaminated areas, however, microbial activities might be governed by, for example, establishing oxidizing/reducing conditions, establishing pH conditions more circumneutral, or inoculating microbes with desired functions like metal immobilization, metal chelation, or enhancement of plant growth for phytoextraction or phytostabilization. Amendments of inorganic (clay minerals,

phosphate, calcite, fly ash, etc.) or organic substrates (compost, straw, acetate, or inoculation with microorganisms) may be used to lower mobility and toxicity of prevalent metals through sorption, complex formation, and precipitation (Hashimoto et al., 2009), which aids recultivation and plant growth. An example for the role of amendments in establishing changed functions in microbial populations can be seen with Figure 14.3.

14.7.3 Microbially Aided Phytoremediation

The use of metal-tolerant plants can be used for phytostabilization, where the metal is converted into less mobile species, or phytoextraction into harvestable biomass (Salt et al., 1995). Since microorganisms change bioavailability and may enhance plant growth, they may contribute to either process. The amendment of microbes capable to survive within the autochthonous microbial population, best achieved with strains isolated from the specific site, is mandatory for success. Additional amendments like compost or topsoil may aid the processes of soil formation on the generally nutrient-limited, poor substrates and may aid the transfer of metals into plant biomass (Li et al., 2007). Microbial additives need to be selected for the reduction of metal stress (Dimkpa et al., 2009b; Albarracín et al., 2010) or phytohormone production (Dimkpa et al., 2008b; Langella et al., 2014).

14.8 CONCLUSION

The identification of soil microbes has gained interest over the last years, and only with the advent of new methodologies and their combination have we been able to support bioremediation approaches in an educated way. The feasibility of a phytoremediation, for example, can only be evaluated when the basic principles governing metal resistance and physiological abilities of microorganisms present can be assessed and—in the best case—be combined to achieve the desired results. The knowledge of microbial communities and their adaptive strategies to cope with high metal loads will change the rhizosphere. This, in turn, impacts plant performance. Hence, the deeper understanding of microbial communities and their adaptation is relevant for microbially aided phytoremediation

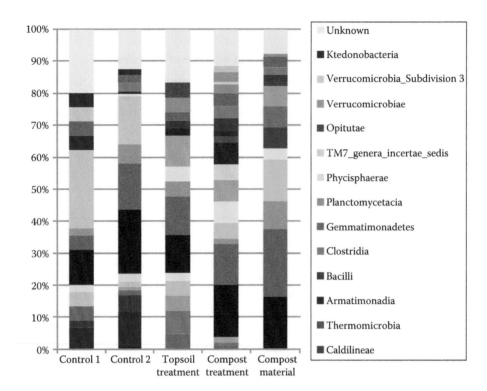

FIGURE 14.3 Restoration of microbial community structure in heavy metal–polluted soils. The initial community (control) with low biodiversity was amended by addition of 5–10 cm topsoil mixed with the upper 30 cm of substrate. These one-time additions led to a more diverse community visible for both, with better effects visible from compost addition (topsoil or compost treatments).

approaches that may be helpful for revegetation and potential future land use of former mining site.

ACKNOWLEDGMENTS

The work was funded by the excellence graduate school JSMC and DFG GRK1257 "Alteration and element mobility at microbe-mineral interphases" within the "Jena School for Microbial Communication" and by the EU-FP 7 project UMBRELLA. The authors particularly thank the FSU Jena Applied Geology staff, especially G. Büchel, D. Merten, U. Buhler, I. Kamp, and G. Weinzierl.

REFERENCES

Abou-Shanab, R.A.I., van Berkum, P., and Angle, J.S. 2007. Heavy metal resistance and genotypic analysis of metal resistance genes in gram-positive and gram-negative bacteria present in Ni-rich serpentine soil and in the rhizosphere of *Alyssum murale*. *Chemosphere* 68: 360–367.

Adam, K., Kourtis, A., Gazea, B., and Kontopoulos, A. 1997. Evaluation of static tests used to predict the potential for acid drainage generation at sulphide mine sites. *Trends in Mineral and Metallurgy A* 106: A1–A8.

Akcil, A. and Koldas, S. 2006. Acid mine drainage (AMD): Causes, treatment and case studies. *Journal of Cleaner Production* 14: 1139–1145.

Albarracín, V.H., Amoroso, M.J., and Abate, C.M. 2010. Bioaugmentation of copper polluted soil microcosms with *Amycolatopsis tucumanensis* to diminish phytoavailable copper for *Zea mays* plants. *Chemosphere* 79: 131–137.

Amann, L. and Schleifer, K.H. 1995. Phylogenetic identification and *in situ* detection of individual microbial cells without cultivation. *Microbiology Reviews* 59: 143–169.

Amoroso, M.J., Guillermo, O., and Guillermo, R.C. 2002. Estimation of growth inhibition by copper and cadmium in heavy metal tolerant actinomycetes. *Journal of Basic Microbiology* 42: 231–237.

Andrews, S.C., Robinson, A.K., and Rodríguez-Quiñones, F. 2003. Bacterial iron homeostasis. *FEMS Microbiology Reviews* 27: 215–237.

Ansorge, W.J. 2009. Next-generation DNA sequencing techniques. *New Biotechnology* 25: 195–203.

Baath, E., Diaz-Ravina, M., Frostegard, A., and Campbell, C.D. 1998. Effect of metal-rich sludge amendments on the soil microbial community. *Applied and Environmental Microbiology* 64: 238–245.

Babich, H. and Stotzky, G. 1977. Sensitivity of various bacteria, including actinomycetes, and fungi to cadmium and influence of pH on sensitivity. *Applied and Environmental Microbiology* 33: 681–695.

Bafana, A., Krishnamurthi, K., Patil, M., and Chakrabarti, T. 2010. Heavy metal resistance in *Arthrobacter ramosus* strain G2 isolated from mercuric salt-contaminated soil. *Journal of Hazardous Materials* 177: 481–486.

Bamborough, L. and Cummings, S.P. 2009. The impact of increasing heavy metal stress on the diversity and structure of the bacterial and actinobacterial communities of metallophytic grassland soil. *Biology and Fertility of Soils* 45: 273–280.

Brown, N.L., Barrett, S.R., Camakaris, J., Lee, B.T.O., and Rouch, D.A. 1995. Molecular genetics and transport analysis of the copper-resistance determinant (*pco*) from *Escherichia coli* plasmid pRJ1004. *Molecular Microbiology* 17: 1153–1166.

Bruins, M.R., Kapil, S., and Oehme, F.W. 2000. Microbial resistance to metals in the environment. *Ecotoxicology and Environmental Safety* 45: 198–207.

Brümmer, G.W., Gerth, J., and Herms, U. 1986. Heavy-metal species, mobility and availability in soils. *Zeitschr Pflanz Bodenk* 149: 382–398.

Brussaard, L. 2012. Ecosystem services provided by the soil biota. In *Soil Ecology and Ecosystem Services*, Wall, D.H. (ed.). Oxford, U.K.: Oxford University Press, pp. 45–58.

Burdige, D.J. and Nealson, K.H. 1986. Chemical and microbiological studies of sulfide-mediated manganese reduction. *Geomicrobiology Journal* 4: 361–387.

Burgess, B.K. and Lowe, D.J. 1996. Mechanism of molybdenum nitrogenase. *Chemical Review* 96: 2983–3012.

Chen, B.Y., Wu, C.H., and Chang, J.S. 2006. An assessment of the toxicity of metals to *Pseudomonas aeruginosa* PU21 (Rip64). *Bioresource Technology* 97: 1880–1886.

Clemente, R., Escolar, Á., and Bernal, M.P. 2006. Heavy metals fractionation and organic matter mineralisation in contaminated calcareous soil amended with organic materials. *Bioresource Technology* 97: 1894–1901.

Coenye, T. and Vandamme, P. 2003. Intragenomic heterogeneity between multiple 16S ribosomal RNA operons in sequenced bacterial genomes. *FEMS Microbiology Letters* 228: 45–49.

Connell, J.H. 1978. Diversity in tropical rain forests and coral reefs—High diversity of trees and corals is maintained only in a non-equilibrium state. *Science* 199: 1302–1310.

Costacurta, A. and Vanderleyden, J. 1995. Synthesis of phytohormones by plant-associated bacteria. *Critical Reviews in Microbiology* 21: 1–18.

Cottenie, A., Camerlynck, R., Verloo, M., and Dhaese, A. 1980. Fractionation and determination of trace-elements in plants, soils and sediments. *Pure and Applied Chemistry* 52: 45–53.

De Boer, W., Gerards, S., Gunnewiek, P.J.A., and Modderman, R. 1999. Response of the chitinolytic microbial community to chitin amendments of dune soils. *Biology and Fertility of Soils* 29: 170–177.

Desai, C., Parikh, R.Y., Vaishnav, T., Shouche, Y.S., and Madamwar, D. 2009. Tracking the influence of long-term chromium pollution on soil bacterial community structures by comparative analyses of 16S rRNA gene phylotypes. *Research in Microbiology* 160: 1–9.

Dhal, P., Islam, E., Kazy, S., and Sar, P. 2011. Culture-independent molecular analysis of bacterial diversity in uranium-ore-/-mine waste-contaminated and non-contaminated sites from uranium mines. *Biotechnology* 1: 261–272.

Dibrov, P., Dzioba, J., Gosink, K.K., and Hase, C.C. 2002. Chemiosmotic mechanism of antimicrobial activity of Ag^+ in *Vibrio cholerae*. *Antimicrobial Agents and Chemotherapy* 46: 2668–2670.

Dimkpa, C., Svatos, A., Merten, D., Büchel, G., and Kothe, E. 2008a. Hydroxamate siderophores produced by *Streptomyces acidiscabies* E13 bind nickel and promote growth in cowpea (*Vigna unguiculata* L.) under nickel stress. *Canadian Journal of Microbiology* 54: 163–172.

Dimkpa, C., Weinand, T., and Asch, F. 2009a. Plant-rhizobacteria interactions alleviate abiotic stress conditions. *Plant, Cell & Environment* 32: 1682–1694.

Dimkpa, C.O., Merten, D., Svatos, A., Büchel, G., and Kothe, E. 2009b. Metal-induced oxidative stress impacting plant growth in contaminated soil is alleviated by microbial siderophores. *Soil Biology and Biochemistry* 41: 154–162.

Dimkpa, C.O., Svatos, A., Dabrowska, P., Schmidt, A., Boland, W., and Kothe, E. 2008b. Involvement of siderophores in the reduction of metal-induced inhibition of auxin synthesis in *Streptomyces* spp. *Chemosphere* 74: 19–25.

Doelman, P. and Haanstra, L. 1986. Short-term and long-term effects of heavy-metals on urease activity in soils. *Biology and Fertility of Soils* 2: 213–218.

Doelman, P., Jansen, E., Michels, M., and van Til, M. 1994. Effects of heavy-metals in soil on microbial diversity and activity as shown by the sensitivity-resistance index, an ecologically relevant parameter. *Biology and Fertility of Soils* 17: 177–184.

Doran, J.W. and Safley, M. 1997. Defining and assessing soil health and sustainable productivity. In *Biological Indicators of Soil Health*, Pankhurst, C., Doube, B.M., and Gupta, V.V.S.R. (eds.). Wallingford, CT: CAB International, pp. 1–28.

Downing, B.W. and Madeisky, H.E. 1997. Lithogeochemical methods for acid rock drainage studies and prediction. *Exploration and Mining Geology* 6: 367–379.

Dunbar, J., Barns, S.M., Ticknor, L.O., and Kuske, C.R. 2002. Empirical and theoretical bacterial diversity in four Arizona soils. *Applied and Environmental Microbiology* 68: 3035–3045.

Ehrlich, H.L. 1996. How microbes influence mineral growth and dissolution. *Chemical Geology* 132: 5–9.

Ellis, R.J., Morgan, P., Weightman, A.J., and Fry, J.C. 2003. Cultivation-dependent and -independent approaches for determining bacterial diversity in heavy-metal-contaminated soil. *Applied and Environmental Microbiology* 69: 3223–3230.

Engelbrektson, A., Kunin, V., Wrighton, K.C., Zvenigorodsky, N., Chen, F., Ochman, H., and Hugenholtz, P. 2010. Experimental factors affecting PCR-based estimates of microbial species richness and evenness. *ISME Journal* 4: 642–647.

Euzeby, J.P. 1997. List of bacterial names with standing in nomenclature: A folder available on the internet. *International Journal of Systematic Bacteriology* 47: 590–592.

Fisher, M.M. and Triplett, E.W. 1999. Automated approach for ribosomal intergenic spacer analysis of microbial diversity and its application to freshwater bacterial communities. *Applied and Environmental Microbiology* 65: 4630–4636.

Florence, T.M. 1989. Trace element speciation in biological systems. In *Trace Element Speciation: Analytical Methods and Problems*, Batley, G.E. (ed.). Boca Raton, FL: CRC Press, pp. 319–343.

Folman, L.B., Postma, J., and van Veen, J.A. 2001. Ecophysiological characterization of rhizosphere bacterial communities at different root locations and plant developmental stages of cucumber grown on rockwool. *Microbial Ecology* 42: 586–597.

Francis, A.J. and Dodge, C.J. 1990. Anaerobic microbial remobilization of toxic metals coprecipitated with iron oxide. *Environmental Science & Technology* 24: 373–378.

Frey, B., Stemmer, M., Widmer, F., Luster, J., and Sperisen, C. 2006. Microbial activity and community structure of a soil after heavy metal contamination in a model forest ecosystem. *Soil Biology and Biochemistry* 38: 1745–1756.

Gadd, G.M. 2010. Metals, minerals and microbes: Geomicrobiology and bioremediation. *Microbiology* 156: 609–643.

Garland, J.L. 1996. Patterns of potential C source utilization by rhizosphere communities. *Soil Biology and Biochemistry* 28: 223–230.

Ghosh, S., Mahapatra, N.R., and Banerjee, P.C. 1997. Metal resistance in *Acidocella* strains and plasmid-mediated transfer of this characteristic to *Acidiphilium multivorum* and *Escherichia coli*. *Applied and Environmental Microbiology* 63: 4523–4527.

Giller, K.E., Witter, E., and McGrath, S.P. 1998. Toxicity of heavy metals to microorganisms and microbial processes in agricultural soils: A review. *Soil Biology and Biochemistry* 30: 1389–1414.

Gitay, H., Wilson, J.B., and Lee, W.G. 1996. Species redundancy: A redundant concept? *Journal of Ecology* 84: 121–124.

Gonzalez, H. and Jensen, T.E. 1998. Nickel sequestering by polyphosphate bodies in *Staphylococcus aureus*. *Microbios* 93: 179–185.

Gonzalez, J.M., Portillo, M.C., Belda-Ferre, P., and Mira, A. 2012. Amplification by PCR artificially reduces the proportion of the rare biosphere in microbial communities. *PLoS ONE* 7: e29973.

Guerrero-Romero, F. and Rodríguez-Morán, M. 2005. Complementary therapies for diabetes: The case for chromium, magnesium, and antioxidants. *Archieves in Medical Research* 36: 250–257.

Haferburg, G., Kloess, G., Schmitz, W., and Kothe, E. 2008. "Ni-struvite"—A new biomineral formed by a nickel resistant *Streptomyces acidiscabies*. *Chemosphere* 72: 517–523.

Haferburg, G. and Kothe, E. 2007. Microbes and metals: Interactions in the environment. *Journal of Basic Microbiology* 47: 453–467.

Hall, N. 2007. Advanced sequencing technologies and their wider impact in microbiology. *Journal of Experimental Biology* 210: 1518–1525.

Hallberg, K.B. and Johnson, D.B. 2001. Biodiversity of acidophilic prokaryotes. *Advances in Applied Microbiology* 49: 37–84.

Hartmann, M. and Widmer, F. 2006. Community structure analyses are more sensitive to differences in soil bacterial communities than anonymous diversity indices. *Applied and Environmental Microbiology* 72: 7804–7812.

Hashimoto, Y., Matsufuru, H., Takaoka, M., Tanida, H., and Sato, T. 2009. Impacts of chemical amendment and plant growth on lead speciation and enzyme activities in a shooting range soil: An X-ray absorption fine structure investigation. *Journal of Environmental Quality* 38: 1420–1428.

Hassen, A., Saidi, N., Cherif, M., and Boudabous, A. 1998. Resistance of environmental bacteria to heavy metals. *Bioresource Technology* 64: 7–15.

Head, I.M., Saunders, J.R., and Pickup, R.W. 1998. Microbial evolution, diversity, and ecology: A decade of ribosomal RNA analysis of uncultivated microorganisms. *Microbial Ecology* 35: 1–21.

Hengstler, J.G., Bolm-Audorff, U., Faldum, A., Janssen, K., Reifenrath, M., Götte, W., Jung, D., Mayer-Popken, O., Fuchs, J., Gebhard, S., Bienfait, H.G., Schlink, K., Dietrich, C., Faust, D., Epe, B. and Oesch, F. 2003. Occupational exposure to heavy metals: DNA damage induction and DNA repair inhibition prove co-exposures to cadmium, cobalt and lead as more dangerous than hitherto expected. *Carcinogenesis* 24: 63–73.

Hong, R., Kang, T.Y., Michels, C.A., and Gadura, N. 2012. Membrane lipid peroxidation in copper alloy-mediated contact killing of *Escherichia coli*. *Applied and Environmental Microbiology* 78: 1776–1784.

Hopf, J., Langenhorst, F., Pollok, K., Merten, D., and Kothe, E. 2009. Influence of microorganisms on biotite dissolution: An experimental approach. *Chemical Erde-Geochemistry* 69: 45–56.

Horner-Devine, M.C., Carney, K.M., and Bohannan, B.J.M. 2004. An ecological perspective on bacterial biodiversity. *Proceedings of Royal Society of Britain, Biological Sciences* 271: 113–122.

Hugenholtz, P., Goebel, B.M., and Pace, N.R. 1998. Impact of culture-independent studies on the emerging phylogenetic view of bacterial diversity. *Journal of Bacteriology* 180: 4765–4774.

Islam, E., Dhal, P.K., Kazy, S.K., and Sar, P. 2011. Molecular analysis of bacterial communities in uranium ores and surrounding soils from Banduhurang open cast uranium mine, India: A comparative study. *Journal of Environmental Science and Health A* 46: 271–280.

Jiang, C.-Y., Sheng, X.-F., Qian, M., and Wang, Q.-Y. 2008. Isolation and characterization of a heavy metal-resistant *Burkholderia* sp. from heavy metal-contaminated paddy field soil and its potential in promoting plant growth and heavy metal accumulation in metal-polluted soil. *Chemosphere* 72: 157–164.

Johnson, D.B., Grail, B., and Hallberg, K. 2013. A new direction for biomining: Extraction of metals by reductive dissolution of oxidized ores. *Minerals* 3: 49–58.

Johnson, D.B. and Hallberg, K.B. 2005. Acid mine drainage remediation options: A review. *Science of the Total Environment* 338: 3–14.

Jones, R.T., Robeson, M.S., Lauber, C.L., Hamady, M., Knight, R., and Fierer, N. 2009. A comprehensive survey of soil acidobacterial diversity using pyrosequencing and clone library analyses. *ISME Journal* 3: 442–453.

Joshi, S.R., Kalita, D., Kumar, R., Nongkhlaw, M., and Swar, P.K. 2014. Metal-microbe interaction and bioremediation. In *Radionuclide Contamination and Remediation through Plants*, Gupta, D.K. and Walther, C. (eds.). Heidelberg, Germany: Springer, pp. 235–251.

Kabata-Pendias, A. and Pendias, H. 2001. *Trace Elements in Soils and Plants*. Boca Raton, FL: CRC Press.

Kandeler, E., Tscherko, D., Bruce, K.D., Stemmer, M., Hobbs, P.J., Bardgett, R.D., and Amelung, W. 2000. Structure and function of the soil microbial community in microhabitats of a heavy metal polluted soil. *Biology and Fertility of Soils* 32: 390–400.

Kelly, D.P. and Wood, A.P. 2000. Reclassification of some species of *Thiobacillus* to the newly designated genera *Acidithiobacillus* gen. nov., *Halothiobacillus* gen. nov. and *Thermithiobacillus* gen. nov. *International Journal of Systematic and Evolutionary Microbiology* 50: 511–516.

Keyer, K. and Imlay, J.A. 1996. Superoxide accelerates DNA damage by elevating free-iron levels. *Proceedings of the National Academy of Science of the United States of America* 93: 13635–13640.

Kothe, E. and Büchel, G. 2014. UMBRELLA: Using MicroBes for the REgulation of heavy metaL mobiLity at ecosystem and landscape scAle. *Environmental Science and Pollution Research* 21: 6761–6764.

Krebs, W., Brombacher, C., Bosshard, P.P., Bachofen, R., and Brandl, H. 1997. Microbial recovery of metals from solids. *FEMS Microbiology Reviews* 20: 605–617.

Kunito, T., Saeki, K., Goto, S., Hayashi, H., Oyaizu, H., and Matsumoto, S. 2001. Copper and zinc fractions affecting microorganisms in long-term sludge-amended soils. *Bioresource Technology* 79: 135–146.

Kuske, C.R., Barns, S.M., and Busch, J.D. 1997. Diverse uncultivated bacterial groups from soils of the arid southwestern United States that are present in many geographic regions. *Applied and Environmental Microbiology* 63: 3614–3621.

Langella, F., Grawunder, A., Stark, R. et al. 2014. Microbially assisted phytoremediation approaches for two multi-element contaminated sites. *Environmental Science and Pollution Research* 21: 6845–6858.

Larsen, N., Olsen G.J., Maidak, B.L. et al. 1993. The ribosomal database project. *Nucleic Acids Research* 21: 3021–3023.

Lee, I.-S., Kim, O.K., Chang, Y.-Y., Bae, B., Kim, H.H., and Baek, K.H. 2002. Heavy metal concentrations and enzyme activities in soil from a contaminated Korean shooting range. *Journal of Bioscience and Bioengineering* 94: 406–411.

Lemire, J.A., Harrison, J.J., and Turner, R.J. 2013. Antimicrobial activity of metals: Mechanisms, molecular targets and applications. *Nature Reviews Microbiology* 11: 371–384.

Li, W.C., Ye, Z.H., and Wong, M.H. 2007. Effects of bacteria on enhanced metal uptake of the Cd/Zn-hyperaccumulating plant, *Sedum alfredii. Journal of Experimental Botany* 58: 4173–4182.

Liesack, W. and Stackebrandt, E. 1992. Occurrence of novel groups of the domain Bacteria as revealed by analysis of genetic material isolated from an australian terrestrial environment. *Journal of Bacteriology* 174: 5072–5078.

Liu, Y., Zhou, T., Crowley, D. et al. 2012. Decline in topsoil microbial quotient, fungal abundance and C utilization efficiency of rice paddies under heavy metal pollution across South China. *PLoS ONE* 7: e38858.

Lombard, N., Prestat, E., van Elsas, J.D., and Simonet, P. 2011. Soil-specific limitations for access and analysis of soil microbial communities by metagenomics. *FEMS Microbiology and Ecology* 78: 31–49.

Lovley, D.R. 1993. Dissimilatory metal reduction. *Annual Review of Microbiology* 47: 263–290.

Lowenstam, H.A. and Weiner, S. 1989. *On Biomineralization.* Oxford, U.K.: Oxford University Press.

Lynch, J.M. 1988. The terrestrial environment. In *Microorganisms in Action: Concepts and Applications in Microbial Ecology,* Lynch, J.M. and Hobbie, J.E. (eds.). Oxford, U.K.: Blackwell, pp. 103–131.

Machemer, S.D. and Wildeman, T.R. 1992. Adsorption compared with sulfide precipitation as metal removal processes from acid-mine drainage in a constructed wetland. *Journal of Contamination and Hydrology* 9: 115–131.

Maidak, B.L., Cole, J.R., Lilburn, T.G. et al. 1994. The ribosomal database project. *Nucleic Acids Research* 22: 3485–3487.

Margesin, R. and Schinner, F. 1996. Bacterial heavy metal tolerance—Extreme resistance to nickel in *Arthrobacter* spp. strains. *Journal of Basic Microbiology* 36: 269–282.

Margulies, M., Engholm, M., Altman, W.E. et al. 2005. Genome sequencing in microfabricated high-density picolitre reactors. *Nature* 437: 376–380.

Marschner, P., Yang, C.H., Lieberei, R., and Crowley, D.E. 2001. Soil and plant specific effects on bacterial community composition in the rhizosphere. *Soil Biology and Biochemistry* 33: 1437–1445.

Martin, T.A. and Ruby, M.V. 2004. Review of in situ remediation technologies for lead, zinc, and cadmium in soil. *Remediation Journal* 14: 35–53.

McCall, K.A., Huang, C.-C., and Fierke, C.A. 2000. Function and mechanism of zinc metalloenzymes. *Journal of Nutrition* 130: 1437S–1446S.

McCarthy, A.J. and Williams, S.T. 1992. Actinomycetes as agents of biodegradation in the environment—A review. *Gene* 115: 189–192.

Merdy, P., Gharbi, L.T., and Lucas, Y. 2009. Pb, Cu and Cr interactions with soil: Sorption experiments and modelling. *Colloids and Surfaces A: Physicochemical and Engineering Aspects* 347: 192–199.

Michalke, B. 2003. Element speciation definitions, analytical methodology, and some examples. *Ecotoxicology and Environmental Safety* 56: 122–139.

Mirete, S., de Figueras, C.G., and Gonzalez-Pastor, J.E. 2007. Novel nickel resistance genes from the rhizosphere metagenome of plants adapted to acid mine drainage. *Applied and Environmental Microbiology* 73: 6001–6011.

Moreno, M.L., Piubeli, F., Bonfá, M.R.L., García, M.T., Durrant, L.R., and Mellado, E. 2012. Analysis and characterization of cultivable extremophilic hydrolytic bacterial community in heavy-metal-contaminated soils from the Atacama Desert and their biotechnological potentials. *Journal of Applied Microbiology* 113: 550–559.

Mulligan, C.N., Yong R.N., and Gibbs, B.F. 2001. Remediation technologies for metal-contaminated soils and groundwater: An evaluation. *Engineering Geology* 60: 193–207.

Muyzer, G., Dewaal, E.C., and Uitterlinden, A.G. 1993. Profiling of complex microbial populations by denaturing gradient gel electrophoresis analysis of polymerase chain reaction-amplified genes coding for 16S ribosomal RNA. *Applied and Environmental Microbiology* 59: 695–700.

Nannipieri, P., Ascher, J., Ceccherini, M.T. et al. 2008. Effects of root exudates in microbial diversity and activity in rhizosphere soils. In *Molecular Mechanisms of Plant and Microbe Eoexistence,* Nautiyal, C. and Dion, P. (eds.). Heidelberg, Germany: Springer, pp. 339–365.

Narula, N., Reinicke, M., Haferburg, G., Kothe, E., and Behl, R.K. 2012. Plant-microbe interaction in heavy metal-contaminated soils. In *Bio-Geo Interactions in Metal-Contaminated Soils,* Kothe, E. and Varma, A. (eds.). Heidelberg, Germany: Springer, pp. 143–162.

Nealson, K.H. and Myers, C.R. 1992. Microbial reduction of manganese and iron—New approaches to carbon cycling. *Applied and Environmental Microbiology* 58: 439–443.

Nicoara, A., Neagoe, A., Stancu, P. et al. 2014. Coupled pot and lysimeter experiments assessing plant performance in microbially assisted phytoremediation. *Environmental Science and Pollution Research* 21: 6905–6920.

Nies, D.H. 1995. The cobalt, zinc and cadmium efflux system CzcABC from *Alcaligenes eutrophus*—Functions as a cation-proton antiporter in *Escherichia coli. Journal of Bacteriology* 177: 2707–2712.

Nies, D.H. 1999. Microbial heavy-metal resistance. *Applied Microbiology and Biotechnology* 51: 730–750.

Nies, D.H. 2003. Efflux-mediated heavy metal resistance in prokaryotes. *FEMS Microbiology Reviews* 27: 313–339.

Nocker, A., Burr, M., and Camper, A.K. 2007. Genotypic microbial community profiling: A critical technical review. *Microbial Ecology* 54: 276–289.

Nriagu, J.O. 1990. Global metal pollution—Poisoning the biosphere. *Environment* 32: 6–8.

Nusslein, K. and Tiedje, J.M. 1999. Soil bacterial community shift correlated with change from forest to pasture vegetation in a tropical soil. *Applied and Environmental Microbiology* 65: 3622–3626.

Oliveira, A. and Pampulha, M.E. 2006. Effects of long-term heavy metal contamination on soil microbial characteristics. *Journal of Bioscience and Bioengineering* 102: 157–161.

Olsen, G.J., Lane, D.J., Giovannoni, S.J., Pace, N.R., and Stahl, D.A. 1986. Microbial ecology and evolution—A ribosomal-RNA approach. *Annual Review of Microbiology* 40: 337–365.

Orita, M., Suzuki, Y., Sekiya, T., and Hayashi, K. 1989. Rapid and sensitive detection of point mutations and DNA polymorphisms using the polymerase chain reaction. *Genomics* 5: 874–879.

Osborn, A.M., Bruce, K.D., Strike, P., and Ritchie, D.A. 1997. Distribution, diversity and evolution of the bacterial mercury resistance (*mer*) operon. *FEMS Microbiology Reviews* 19: 239–262.

Pennanen, T., Frostegard, A., Fritze, H., and Baath E. 1996. Phospholipid fatty acid composition and heavy metal tolerance of soil microbial communities along two heavy metal-polluted gradients in coniferous forests. *Applied and Environmental Microbiology* 62: 420–428.

Pereira, L.B., Vicentini, R., and Ottoboni, L.M.M. 2014. Changes in the bacterial community of soil from a neutral mine drainage channel. *PLoS ONE* 9: e96605.

Phieler, R., Voit, A., and Kothe, E. 2014. Microbially supported phytoremediation of heavy metal contaminated soils: Strategies and applications. *Advances in Biochemical Engineering and Biotechnology* 141: 211–235.

Rappé, M.S. and Giovannoni, S.J. 2003. The uncultured microbial majority. *Annual Review of Microbiology* 57: 369–394.

Rawlings, D.E. and Johnson, D.B. 2007. *Biomining.* Heidelberg, Germany: Springer.

Roh, S.W., Kim, K.H., Nam, Y.D., Chang, H.W., Park, E.J., and Bae, J.W. 2010. Investigation of archaeal and bacterial diversity in fermented seafood using barcoded pyrosequencing. *ISME Journal* 4: 1–16.

Rondon, M.R., August, P.R., Bettermann, A.D. et al. 2000. Cloning the soil metagenome: A strategy for accessing the genetic and functional diversity of uncultured microorganisms. *Applied and Environmental Microbiology* 66: 2541–2547.

Salt, D.E., Blaylock, M., Kumar, N.P.B.A., Dushenkov, V., Ensley, B.D., Chet, I., and Raskin, I. 1995. Phytoremediation—A novel strategy for the removal of toxic metals from the environment using plants. *BioTechnology* 13: 468–474.

Sandaa, R.A., Torsvik, V., Enger, O., Daae, F.L., Castberg, T., and Hahn, D. 1999. Analysis of bacterial communities in heavy metal-contaminated soils at different levels of resolution. *FEMS Microbiology Ecology* 30: 237–251.

San Martin-Uriz, P., Mirete, S., Alcolea, P.J., Gomez, M.J., Amils, R., and Gonzalez-Pastor, J.E. 2014. Nickel resistance determinants in *Acidiphilium* sp. PM identified by genome-wide functional screening. *PLoS ONE* 9: e95041.

Sauer, K. and Yachandra, V.K. 2004. The water-oxidation complex in photosynthesis. *BBA-Bioenergetics* 1655: 140–148.

Schippers, A., Hedrich, S., Vasters, J., Drobe, M., Sand, W., and Willscher, S. 2014. Biomining: Metal recovery from ores with microorganisms. In *Geobiotechnology*, Schippers, A., Glombitza, F., and Sand, W. (eds.). Heidelberg, Germany: Springer, pp. 1–47.

Schloter, M., Dilly, O., and Munch, J.C. 2003. Indicators for evaluating soil quality. *Agricultural Ecosystem and Environment* 98: 255–262.

Schmidt, A., Haferburg, G., Schmidt, A., Lischke, U., Merten, D., Gherghel, G., Büchel, G., and Kothe, E. 2009. Heavy metal resistance to the extreme: *Streptomyces* strains from a former uranium mining area. *Chemical Erde-Geochemistry* 69: 35–44.

Schmidt, A., Haferburg, G., Sineriz, M., Merten, D., Büchel, G., and Kothe, E. 2005. Heavy metal resistance mechanisms in actinobacteria for survival in AMD contaminated soils. *Chemical Erde-Geochemistry* 65S1: 131–144.

Seifert, A.G., Trumbore, S., Xu, X.M., Zhang, D.C., Kothe, E., and Gleixner, G. 2011. Variable effects of labile carbon on the carbon use of different microbial groups in black slate degradation. *Geochim Cosmochim Acta* 75: 2557–2570.

Seybold, C.A., Herrick, J.E., and Brejda, J.J. 1999. Soil resilience: A fundamental component of soil quality. *Soil Science* 164: 224–234.

Sharma, S., Ramesh, A., Sharma, M., Joshi, O., Govaerts, B., Steenwerth, K., and Karlen, D. 2011. Microbial community structure and diversity as indicators for evaluating soil quality. In *Biodiversity, Biofuels, Agroforestry and Conservation Agriculture*, Lichtfouse, E. (ed.). Amsterdam, the Netherlands: Springer, pp. 317–358.

Silver, S. and Phung, L.T. 1996. Bacterial heavy metal resistance: New surprises. *Annual Review of Microbiology* 50: 753–789.

Skinner, H.C.W. 2005. Biominerals. *Mineralogical Magazine* 69: 621–641.

Slawson, R.M., van Dyke, M.I., Lee, H., and Trevors, J.T. 1992. Germanium and silver resistance, accumulation, and toxicity in microorganisms. *Plasmid* 27: 72–79.

Smalla, K., Wieland, G., Buchner, A. et al. 2001. Bulk and rhizosphere soil bacterial communities studied by denaturing gradient gel electrophoresis: Plant-dependent enrichment and seasonal shifts revealed. *Applied and Environmental Microbiology* 67: 4742–4751.

Sprocati, A.R., Alisi, C., Tasso, F. et al. 2014. Bioprospecting at former mining sites across Europe: Microbial and functional diversity in soils. *Environmental Science and Pollution Research* 21: 6824–6835.

Stephen, J.R., Chang, Y.J., Macnaughton, S.J., Kowalchuk, G.A., Leung, K.T., Flemming, C.A., and White, D.C. 1999. Effect of toxic metals on indigenous soil β-subgroup proteobacterium ammonia oxidizer community structure and protection against toxicity by inoculated metal-resistant bacteria. *Applied and Environmental Microbiology* 65: 95–101.

Sterritt, R.M. and Lester, J.N. 1980. Interactions of heavy metals with bacteria. *Science of the Total Environment* 14: 5–17.

Stumm, W. and Morgan, J.J. 1981. *Aquatic Chemistry: An Introduction Emphasizing Chemical Equilibria in Natural Waters.* New York: Wiley.

Tabacchioni, S., Chiarini, L., Bevivino, A., Cantale, C., and Dalmastri, C. 2000. Bias caused by using different isolation media for assessing the genetic diversity of a natural microbial population. *Microbial Ecology* 40: 169–176.

Thakuria, D., Schmidt, O., MacSiurtain, M., Egan, D., and Doohan, F.M. 2008. Importance of DNA quality in comparative soil microbial community structure analyses. *Soil Biology and Biochemistry* 40: 1390–1403.

Tiedje, J.M., Asuming-Brempong, S., Nusslein, K., Marsh, T.L., and Flynn, S.J. 1999. Opening the black box of soil microbial diversity. *Applied Soil Ecology* 13: 109–122.

Torsvik, V., Goksoyr, J., and Daae, F.L. 1990. High diversity in DNA of soil bacteria. *Applied and Environmental Microbiology* 56: 782–787.

Torsvik, V., Goksøyr, J., Daae, F.L., Sørheim, R., Michalsen, J., and Salte, K. 1994. Use of DNA analysis to determine the diversity of microbial communities. In *Beyond the Biomass: Compositional and Functional Analysis of Soil Microbial Communities*, Ritz, K., Dighton, J., and Giller, K.E. (eds.). Chichester, U.K: Wiley, pp. 39–48.

Valsecchi, G., Gigliotti, C., and Farini, A. 1995. Microbial biomass, activity and organic-matter accumulation in soils contaminated with heavy metals. *Biology and Fertility of Soils* 20: 253–259.

van Camp, L., Bujarrabal, B., Gentile, A.-R., Jones, R.J.A., Montanarella, L., Olazabal, C., and Selvaradjou, S.-K. 2004. Reports of the technical working groups established under the thematic strategy for soil protection. Luxembourg, Europe: European Community.

Violante, A., Cozzolino, V., Perelomov, L., Caporale, A.G., and Pigna, M. 2010. Mobility and bioavailability of heavy metals and metalloids in soil environments. *Journal of Soil Science and Plant Nutrition* 10: 268–292.

Wagner, S.C. 2012. Biological nitrogen fixation. *Nature Education and Knowledge* 3: 15.

Wang, Y.P., Shi, J.Y., Wang, H., Lin, Q., Chen, X.C., and Chen, Y.X. 2007. The influence of soil heavy metals pollution on soil microbial biomass, enzyme activity, and community composition near a copper smelter. *Ecotoxicology and Environmental Safety* 67: 75–81.

Ward, D.M., Ferris, M.J., Nold, S.C., and Bateson, M.M. 1998. A natural view of microbial biodiversity within hot spring cyanobacterial mat communities. *Microbiology and Molecular Biology Review* 62: 1353–1370.

Watt, R.K. and Ludden, P.W. 1999. Nickel-binding proteins. *Cellular and Molecular Life Sciences* 56: 604–625.

Wengel, M., Kothe, E., Schmidt, C.M., Heide, K., and Gleixner, G. 2006. Degradation of organic matter from black shales and charcoal by the wood-rotting fungus *Schizophyllum commune* and release of DOC and heavy metals in the aqueous phase. *Science of the Total Environment* 367: 383–393.

Woese, C.R. 1987. Bacterial evolution. *Microbiology Review* 51: 221–271.

Wohl, D.L., Arora, S., and Gladstone, J.R. 2004. Functional redundancy supports biodiversity and ecosystem function in a closed and constant environment. *Ecology* 85: 1534–1540.

Zhou, S., Wei, C.H., Liao, C.D., and Wu, H.Z. 2008. Damage to DNA of effective microorganisms by heavy metals: Impact on wastewater treatment. *Journal of Environmental Science* 20: 1514–1518.

Section III

Molecular Approaches in Metal Bioremediation

15 Siderophores
Microbial Tools for Iron Uptake and Resistance to Other Metals

Karolina Zdyb, Malgorzata Ostrowska, Agnieszka Szebesczyk, and Elzbieta Gumienna-Kontecka

CONTENTS

ABSTRACT

Metal ions can have a twofold effect on all living organisms, both essential, allowing to perform vital functions, and toxic. To receive the indispensable and to deal with toxic metals, microorganisms have developed mechanisms to control the level of the absorption, storage, and elimination of metal ions. One of the known systems is siderophore-mediated uptake, mainly applied to the iron assimilation, but it has been proven that microbial pathogens utilize structurally diverse siderophores to scavenge and control other metals. This ability not only allows bacteria and fungi to gain the necessary nutrients like iron but also provides protection against the toxic metal ions. Here, the interactions of bacterial and fungal siderophores with essential and heavy metal ions and actinides are reported.

15.1 INTRODUCTION

All living organisms identified to date, and among them microorganisms, need metal ions to perform vital functions. For example, the essential transition metals, such as iron, copper, manganese, molybdenum, nickel, vanadium, and tungsten, are used by microorganisms as cofactors for enzymes, particularly those responsible for redox or hydrolytic reactions. Zinc, also essential, is found in proteolytic enzymes and in recognition motif of transcription factors. Although metals are fairly widespread in Earth's crust, they are not directly available to be involved in microbial cell processes. In the environment, many of them form insoluble precipitates, especially under aerobic conditions, or are components of mineral rocks. When in a host, they are often withheld by proteins as a defensive strategy but can be also excreted in excess as an antimicrobial weapon. Besides essential metal ions, microorganisms, during their life, meet a lot of other metals such as lead, arsenic, chromium, cadmium, or actinides; most of these do not absolutely have biological function and are toxic to bacteria or fungi.

To deal with both the vital and toxic metals, microorganisms have evolved a range of mechanisms, which allow them to control the level of metal ions absorbed, used, stored, and excreted. One of such systems is siderophore-mediated uptake. Despite that it is dedicated to iron assimilation, there is a growing evidence that microbial pathogens capitalize on siderophore's structural diversity to specialize their function in scavenging and accurately controlling also other metals in order to defend against their toxicity.

15.2 SIDEROPHORES AS IRON CARRIERS

Almost every organism requires iron for carrying out the processes necessary for survival (Crichton 2001). Among others, iron ions are found in proteins responsible for oxygen transport, ribonucleotide reduction, activation and decomposition of peroxides, and transport of electrons via a wide range of electron carriers (Pierre et al. 2002). Therefore, iron deficiency can be a life-threatening condition, but equally dangerous is iron excess. Biological processes in aerobic living systems are the source of superoxide and hydrogen peroxide, which may enter redox cycles of Fe(II) and Fe(III) iron forms. Free Fe(II) ions can undergo reaction of Fenton cycle, with reactive oxygen species (ROS) production (Pierre and Fontecave1999; Pierre et al. 2002):

$$Fe(II) + H_2O_2 \rightleftharpoons Fe(III) + OH^- + OH^- \qquad (15.1)$$

ROS, such as hydroxyl radicals, react very fast with cell components, among other lipids, proteins, or nucleic acids, destroying vital structures and disrupting essential processes. When the cell cannot cope with the amount of ROS, oxidative stress occurs, resulting even in cell death. Therefore, iron homeostasis, that is, the tight balance between iron uptake, storage, and excretion, is crucial for organism survival.

Contrary to iron excess, iron deficiency appears more frequently. Anemia is the most common result of iron shortage in humans, especially women and children (Denic and Agarwal 2007). As iron is present not only in hemoglobin but also in enzymes such as dehydrogenases and cytochromes, its deficiency reduces aerobic capacity and influences both body and brain condition (Coad and Pedley 2014). In plants with iron deficiency, chlorosis is observed, as a result of increased protein participation in carbon assimilation during photosynthesis (Solti et al. 2008; López-Millán et al. 2013). Yeasts living on iron-depleted media present dysfunctions of hemoproteins and Fe–S clusters, which leads to changes in glucose metabolism and biosynthesis of amino acids and lipids (Shakoury-Elizeh et al. 2010; Philpott et al. 2012). An insufficient amount of iron in bacteria results in growth retardation, decreased level of DNA and RNA synthesis, and changes in cell morphology (Messenger and Barclay 1983).

Iron uptake in humans is tightly regulated on the level of absorption in gastrointestinal tract, as we lack effective mechanisms of removing iron excess (Gumienna-Kontecka et al. 2014).

Plants have evolved two main strategies of iron assimilation. Strategy I involves the reduction of soluble forms of Fe(III) (e.g., chelates) to Fe(II) by (1) reductases and (2) lowering the pH, locally near the roots, as a result of ATPase's activity (Romheld 1987). This strategy, depending on the availability of soluble forms of chelated Fe(III), is not applicable when plants are growing on alkaline (e.g., calcareous) soil. In such cases, graminaceous plants (i.e., grasses, cereals, and rice) are able to secrete Fe(III)-binding molecules (Strategy II). Fe(III) chelate complexes are transported by specific carriers to the roots (Schmidt et al. 2014).

For microorganisms, the evolution of photosynthesis and appearance of dioxygen molecules was a turning point in terms of iron acquisition. First, oxygen was readily dissolved in the oceans, oxidizing all easily available Fe(II) ions to Fe(III). Ferric ions form insoluble hydroxides; therefore, the concentration of available iron ions decreased to 10^{-10} M, including all soluble complexes of Fe(III) hydroxides (Braun and Killmann 1999; Boukhalfa and Crumbliss 2002). Microorganisms' iron requirements are approximately 10^{-6} M; thus, they were forced to create mechanisms to compete with precipitation. Additionally, despite a large quantity of iron being found in vertebrates, it is not easily available,

because it is tightly bound in hemoproteins (e.g., hemoglobin), storage proteins (e.g., ferritin), transporters (e.g., transferrin), or other proteins and organic molecules. Due to its utility, iron is a subject to competition between host and pathogen. Therefore, microorganisms have developed a number of different mechanisms of iron uptake and assimilation, especially relevant during infection. To consume iron from the most abundant source, heme, microorganisms use direct heme uptake systems or highly specific proteins, hemophores (Caza and Kronstad 2013). Hemolysins, secreted, for example, by *Candida albicans*, are able to cause the lysis and destruction of erythrocytes (Crichton 2001; Drago-Serrano et al. 2006). Abundant iron sources in host are also Fe transporters, such as transferrin. In the cell membrane of microorganism, specific receptors are localized and able to bind Fe–transferrin complexes before they can be recognized by receptors of host cells (Crichton 2001; Caza and Kronstad 2013). Additionally, some virulent bacteria and fungi, such as *C. albicans* or *Neisseria meningitidis*, can utilize iron from ferritin, the iron storage protein.

To bind and deliver iron ions to the cell, microorganisms also secrete low-molecular-weight compounds called siderophores (Greek term for "iron carriers"). They present high affinity for Fe(III), even if other ions are present in high concentrations. Such specificity results from the ligands' appropriate binding groups and the molecules' architecture, which increase the Fe(III)–siderophore complex stability (Crumbliss and Harrington 2009). Iron is a hard Lewis acid; thus, it forms the most stable complexes with hard donor atoms, such as oxygen; it prefers octahedral geometry. High affinity for iron allows siderophores to compete with iron precipitation as hydroxides and with other chelating agents present in environment. Ferric iron binding by siderophores increases its solubility in water as well as enables the control of redox activity. Siderophores' ferric complexes are characterized by reduction potentials in the range of −82 mV for rhizoferrin to −750 mV for enterobactin (Hider and Kong 2010). For Fe(III)/Fe(II) redox couple in water, the reduction potential is +770 mV. The more negative reduction potential, the more difficult reduction of Fe(III) complex. For most siderophores, such potential is not only negative enough to prevent massive Fe(II) formation and ROS production but also high enough to provide hydrolytic mechanism of iron release inside the cell.

Interestingly, siderophores could be used not only for direct ferric iron binding but also for reoxidation of Fe(II). Microorganisms living on the mineral rocks use reductases to reduce insoluble Fe(III) contained in minerals, such as goethite and hematite, to more soluble Fe(II). Through redox reaction resulted from interaction with siderophore, ferrous ion is again oxidized to Fe(III) and imported into the cell (Farkas et al. 2001, 2003).

The most common groups involved in siderophores' iron binding are catechol, hydroxamic acid, and α-hydroxycarboxylic acid (Figure 15.1).

All these groups are able to form with Fe(III) five-membered chelating ring, which is an effective ring size for Fe(III) ion in view of metal ion size and bite angles (McDougall et al. 1978).

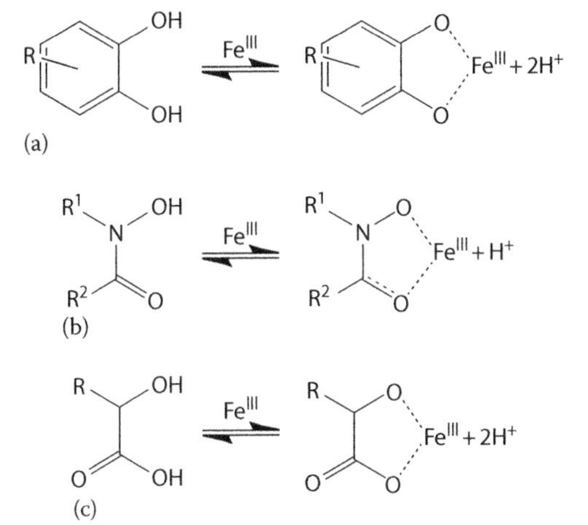

(a)

(b)

(c)

FIGURE 15.1 The most common siderophore-binding groups and their scheme of coordination: (a) catechol, (b) hydroxamate, and (c) α-hydroxycarboxylate.

Other less common donor groups are carboxylic acid or amine moiety and heterocyclic structures, such as hydroxypyridinone, oxazole, or thiazoline (Scarrow et al. 1985). To fulfil iron coordination sphere and form octahedral complexes, siderophores usually possess three bidentate donor groups (of one type or various types) arranged in a linear, tripodal, exocyclic, or endocyclic structure (Figure 15.2).

Siderophores form a family of at least 500 different compounds (Hider and Kong 2010); structures of siderophores discussed in this chapter are given in Figure 15.3.

Desferrioxamine B (DFOB) is a linear hydroxamate siderophore, synthesized by *Streptomyces pilosus*

(Bickel et al. 1960). The ferrioxamine family of siderophores is not only produced by Gram-positive and Gram-negative bacteria, but ferrioxamines can also be utilized as ferric complexes by nonproducers, for example, pathogenic fungal species, such as *C. albicans, Saccharomyces cerevisiae, Cryptococcus neoformans,* or *Rhizopus oryzae* (Boelaert et al. 1993; Howard 1999). The phenomenon of ability to recognize and utilize ferric complexes of exogenous siderophores (or xenosiderophores) is thought to be one of the virulence factors of pathogenic microorganisms and an adaptive advantage to different environmental conditions.

In DFOB, three hydroxamates are involved in Fe(III) binding and formation of octahedral complex with a stability constant of $\log\beta_{110} = 30.6$. Ferrioxamines in general contain three hydroxamate groups, arranged in linear or endocyclic structures (Figure 15.3). DFOB is used in clinical removal of iron excess in patients suffering from, for example, hemochromatosis, genetic disease manifested by excessive accumulation of iron in tissues (Gumienna-Kontecka et al. 2014). Being commercially available, DFOB has been the most extensively studied.

Another hydroxamate siderophore is ferrichrome, produced by fungi *Ustilago sphaerogena*. Ferrichrome family is one of the largest siderophores' class and includes molecules with exocyclic structure (Howard 1999), where three identical arms are attached nonsymmetrically to a chiral hexapeptide ring. The ring is composed of various amino acids, such as glycine, alanine, or serine, and the backbone of arms is formed by N^δ-acyl-N^δ-hydroxyl-L-ornithine units. Three hydroxamates, located at the arms, chelate ferric ion and form the complex characterized by a stability constant of

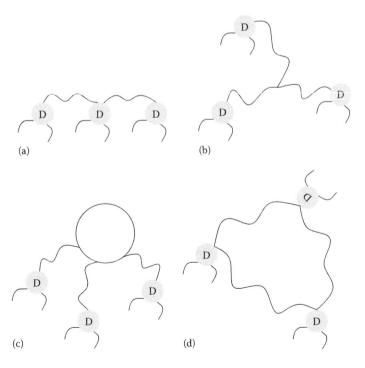

FIGURE 15.2 Schematic structures of siderophores: (a) linear, (b) tripodal, (c) exocyclic, (d) endocyclic; "D" stands for donor groups. (Adapted from Crumbliss, A.L. and Harrington, J.M., *Adv. Inorg. Chem.*, 61, 179, 2009. With permission.)

$\log\beta_{110} = 29.07$ (Anderegg et al. 1963). Although ferrichrome is synthesized by fungi, its Fe(III) complexes are recognized by some pathogenic bacteria (Budde and Leong 1989).

Coprogen is a linear, chiral, hydroxamate siderophore produced by *Penicillium* spp., *Neurospora crassa*, and a number of human pathogens (Höfte 1992; Howard 1999). The stability constant of ferric complex of coprogen is $\log\beta_{110} = 29.35$ (Wong et al. 1983).

Dihydroxamate siderophores, linear (rhodotorulic acid [RA]) and cyclic (alcaligin), due to their tetradentate character, are unable to form with Fe(III) ions stable 1:1 metal-to-ligand complex. Therefore, to complete the coordination sphere of ferric ion, they form complexes of Fe_2L_3 stoichiometry (Boukhalfa et al. 2000; Spasojevic et al. 2001). The stability constants of binuclear complexes were determined to be $\log\beta_{230} = 62.2$ and 64.7, and estimated stability

constants for FeL complexes were $\log\beta_{110} = 21.9$ and 23.5 for RA and alcaligin, respectively (Carrano et al. 1979; Hou et al. 1996).

Enterobactin is a tris-catecholate, exocyclic siderophore, produced by bacteria from *Enterobacteriaceae* and *Streptomycetaceae* families. Three catechol arms are situated on trilactone ring. Catechol siderophores form the most stable complexes with ferric ions; for Fe(III)–enterobactin complex, the stability constant is $\log\beta_{110} \sim 49$ (Loomis and Raymond 1991). Such high affinity results from the presence of two hard oxygen donor atoms, the ability to form five-membered chelating ring, and delocalized electron density of catechol donor groups, what stabilizes ferric state (Crumbliss and Harrington 2009). Enterobactin is also able to bind iron from transferrin (Carrano et al. 1979; Raymond and Telford 1995).

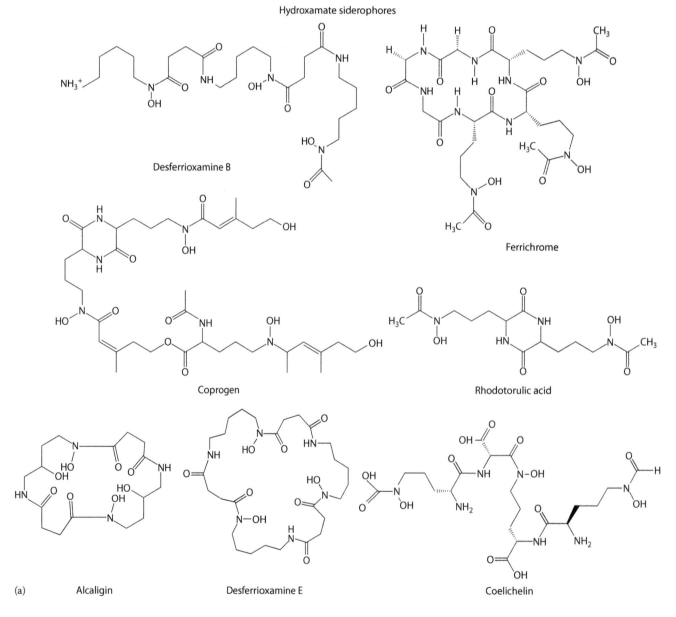

Hydroxamate siderophores

Desferrioxamine B

Ferrichrome

Coprogen

Rhodotorulic acid

(a)   Alcaligin   Desferrioxamine E   Coelichelin

FIGURE 15.3 (a) Chemical structures of the siderophores discussed in this chapter. *(Continued)*

Aerobactin is a mixed-type marine siderophore, containing two hydroxamates and α-hydroxycarboxylate as binding groups on citric acid skeleton. Ferric complex of aerobactin is characterized by $\log\beta_{110} = 27.6$. In Fe(III)–aerobactin complex, iron is bound by two hydroxamates and two hydroxyls from α-hydroxycarboxylic group. Such complex can undergo photoreduction, but the stability constant of Fe(III)–photoproduct complex was estimated to be $\log\beta_{110} = 28.6$, and this complex was effective iron source for *Vibrio* sp. in bacterial growth test (Küpper et al. 2006). Of importance, aerobactin is more efficient in iron scavenging under lower pH conditions, where the effectiveness of catecholate siderophores is decreased due to the protonation of their binding groups (Abergel et al. 2006).

Pyoverdines (PVDs) are siderophores with unique structural features and fluorescent properties (Boukhalfa et al. 2006).

The general structure of these chromopeptides comprises an octapeptide chain, dicarboxylic acid (or its amide), and a dihydroxyquinoline chromophore (Fuchs and Budzikiewicz 2001). Depending on the strain and growth medium, the chromophore and peptide chain may vary.

Pyoverdine PaA (PVD_{PaA}) is produced by *Pseudomonas aeruginosa*, a common pathogenic bacteria. PVD_{PaA} is a fluorescent Fe(III) biological ligand with exocyclic structure. It binds ferric ion by two hydroxamates and one catechol group from dihydroxyquinoline-type moiety attached to peptide chain and forms a complex with stability similar to DFOB, coprogen, or ferrichrome, determined as $\log\beta_{110} = 30.8$ (Albrecht-Gary et al. 1994; Wasielewski et al. 2002).

PVD Pp is excreted by *Pseudomonas putida*, the model environmental bacteria. It binds ferric ion by

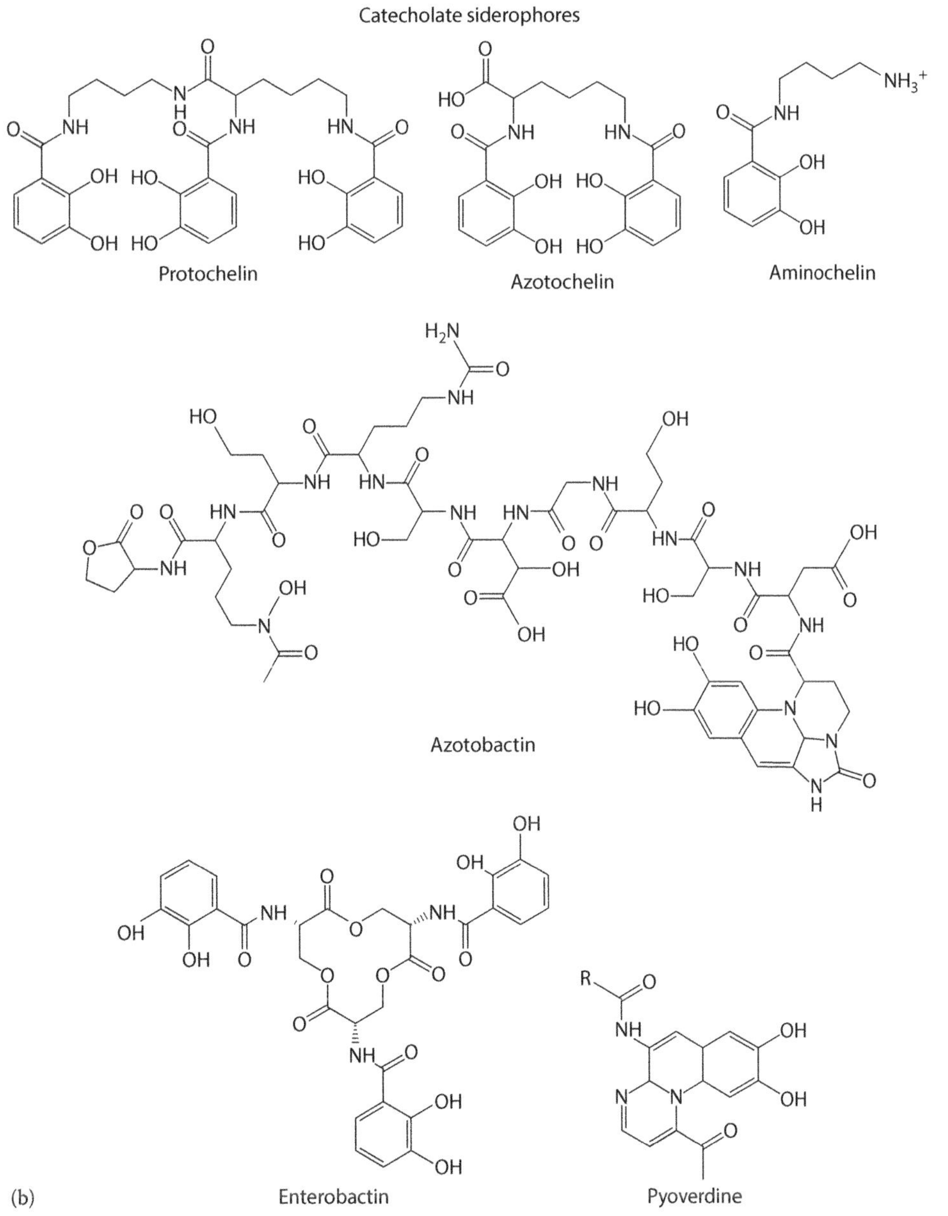

FIGURE 15.3 (*Continued*)   (b) Chemical structures of the siderophores discussed in this chapter.   (*Continued*)

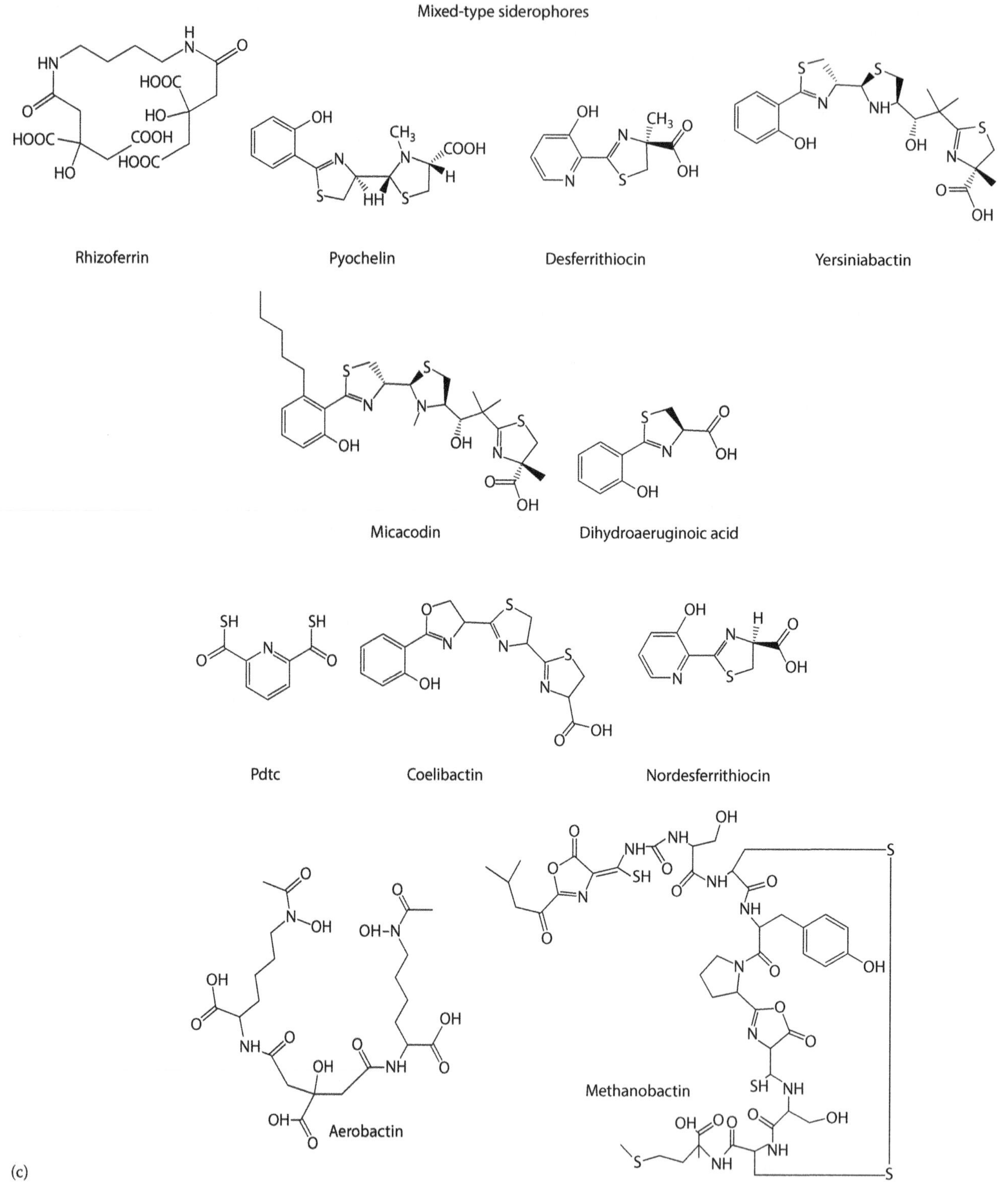

Mixed-type siderophores

Rhizoferrin   Pyochelin   Desferrithiocin   Yersiniabactin

Micacodin   Dihydroaeruginoic acid

Pdtc   Coelibactin   Nordesferrithiocin

Methanobactin

Aerobactin

(c)

FIGURE 15.3 (Continued)   (c) Chemical structures of the siderophores discussed in this chapter.

oxygen donor atoms from catechol, hydroxamate, and α-hydroxycarboxylic groups, with $\log\beta_{110} = 31.06$, very similar to PVD_{PaA}.

PVD-like fluorescent siderophore, azotobactin, is excreted by *Azotobacter vinelandii*, a nitrogen-fixing soil bacterium (Palanché et al. 2004). Chromophore, derived from 2,3-diamino-6,7 dihydroxyquinoline, is bound to decapeptide chain. Two of three iron-binding sites are localized on the chain, that is, hydroxamate and α-hydroxycarboxyl, while the last catechol donor remains on chromophore

(Palanché et al. 2004). Stability constant determined for hexadentate complex was $\log\beta_{110} = 28.1$.

Rhizoferrin is a polycarboxylate siderophore produced by, for example, zygomycetes such as fungi *Rhizopus microsporus* or *R. arrhizus* and bacterium *Ralstonia picketti*. It binds iron by carboxylates and α-hydroxycarboxylates with $\log\beta_{110} = 25.3$ (Carrano et al. 1996).

Pyochelin (PCH) is synthesized and secreted by *P. aeruginosa* and *Burkholderia cepacia*, opportunistic bacteria causing lethal lung infections in patients with defects in immune system (Rivault et al. 2006). It is a low-molecular-weight thiazoline derivative. Due to its poor water solubility, stability constants for Fe(III) complexes were determined in alcoholic solution and were $\log\beta_{110} = 17.3$ and $\log\beta_{120} = 28.8$ for monochelate (FeL) and bischelate (FeL$_2$), respectively (Brandel et al. 2012).

Yersiniabactin (YBT) is a siderophore isolated from *Yersinia* spp., for example, *Yersinia enterocolitica*, *Y. pestis*, or *Y. pseudotuberculosis* (Perry et al. 1999). It belongs to the group of siderophores in which five-membered heterocycles are involved in iron coordination (Miller et al. 2006). Also PCH represents this class. The stability constant for ferric–YBT complex is $\log\beta_{110} = 36.6$ (Perry et al. 1999).

Desferriferrithiocin (DFFT) was isolated from *Streptomyces antibioticus* (Naegeli and Zahner 1980). Similarly to PCH, DFFT contains thiazoline ring. In coordination process, the phenolate oxygen atom, nitrogen from thiazoline ring, and oxygen from carboxylic group are involved (Langemann et al. 1996). Nordesferriferrithiocin (NDFT) has one methyl group, localized on chiral carbon, replaced by hydrogen. Both DFFT and NDFT are tridentate ligands; therefore, they form complexes of FeL$_2$ stoichiometry. Stability constants were characterized as $\log\beta_{120} = 31.04$ and 29.09 for ferrithiocin and norferrithiocin, respectively (Langemann et al. 1996).

To successfully deliver iron ion into the cell, the Fe(III)–siderophore complex must be recognized by specific receptor in the outer membrane. Then it is transported to the cytoplasm, where iron ion is released and free siderophore is excreted outside the cell or hydrolyzed. As the mechanism of recognition and transport was more intensively investigated in bacteria than in fungi, here we will focus on bacterial iron uptake via siderophores.

In the first step, siderophores are synthesized under iron starvation conditions (Saha et al. 2013). When iron level in the cell decreases significantly, transcription factor Fur (ferric uptake regulator) dissociates from its complex with DNA, and gene expression associated with siderophores' biosynthesis occurs. In siderophores' excretion to the environment or a host, proteins of MFS (major facilitator superfamily), ABC (ATP-binding cassette) transporters, and efflux pumps are involved, for example, enterobactin from *Escherichia coli* is excreted by EntS protein of MFS, ABC transporters are involved in the secretion of siderophores in *Staphylococcus aureus* and *Mycobacterium tuberculosis*, and PVD from *S. aureus* is released by efflux pump (Furrer et al. 2002; Bleuel et al. 2005; Rodriguez and Smith 2006; Farhana et al. 2008; Grigg et al. 2010; Yeterian et al. 2010). Outside a cell, siderophores bind iron with very high affinity, as demonstrated by stability constants of formed complexes, remaining within the scope of $\log\beta = 20$–50.

The key step of iron assimilation by Gram-negative bacteria (Figure 15.4) is the binding of Fe(III)–siderophore complex with specific receptor.

Receptors present in outer membrane are specific for one of Fe(III)–siderophore complexes. Through the cytoplasmic membrane, one group of siderophores, for example, hydroxamates, are transported by the same transport protein. Receptors do not recognize free siderophores.

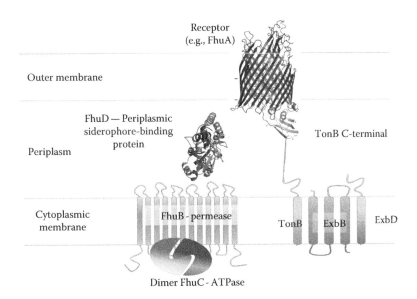

FhuD + FhuB + FhuC = ABC transporter

FIGURE 15.4 Schematic representation of arrangement of proteins involved in transport of ferric complexes of hydroxamate siderophores into the cell Gram-negative bacteria; examples of proteins are given for *Escherichia coli*.

Resolved crystal structures of FhuA (receptor for ferrichrome from *E. coli*), FepA (ferric enterobactin receptor), and FptA (ferric PCH receptor) indicate that the common features of these outer membrane transporters are β-barrel tertiary structure, composed of 22 β-strands, with plug or cork globular domain, which prevents entering the cell until change in conformation occurs after the binding of ferric siderophore complex (Crumbliss and Harrington 2009). Transport through the receptor is an energy-dependent process; therefore, the receptor is connected with TonB protein complex, which consists of two other proteins, that is, ExbB and ExbD, embedded in cytoplasmic membrane (Ferguson et al. 1998). TonB transduces energy required for complex uptake. In the case of Gram-negative bacteria, the transport by receptor is followed by the transfer through periplasm. In this part, one protein, siderophore periplasmic binding protein (SPBP), is able to transport a whole group of siderophores' complexes, for example, hydroxamates or catecholates (Clarke et al. 2000). As Gram-positive bacteria do not possess outer membrane, the ferric siderophore complex outside a cell is bound by siderophore-binding protein (SBP), and the next steps are identical in both Gram-positive and Gram-negative species. Transfer through cytoplasmic membrane is also energy dependent and thus is provided by ABC transporters (Figure 15.4), which consist of permease (e.g., FhuB in *E. coli*) and ATPase (dimer FhuC in *E. coli*) (Neilands 1995; Winkelmann 2002; Chu et al. 2010; Hider and Kong 2010; Schalk and Guillon 2013). SPBP or SBP interacts with permease, probably by forming salt bridges. The process of interaction of transporter with membrane permease is coupled with ATP hydrolysis provided by ATPase (Mademidis et al. 1997; Mademidis and Koster 1998; Hvorup et al. 2007). Through the hydrolysis of ATP, the accumulation contrary to concentration gradient is possible (Higgins and Linton 2001; Hollenstein et al. 2007; Locher 2009). Inside the cytoplasm, ferric siderophore complex may be reduced to ferrous complex (Albrecht-Gary and Crumbliss 1998; Boukhalfa and Crumbliss 2002). The resulting Fe(II)–siderophore complex is much less stable, with stability constant around 20 orders of magnitude lower than for respective ferric complex, and therefore, the hydrolysis may occur, resulting in iron release (Klein and Lewinson 2011). Ferrous ion may be transported to the places of demand or storage. Free siderophore is excreted again outside a cell.

15.3 SIDEROPHORES AND OTHER METALS

Although their primary role is iron acquisition, siderophores are also able to chelate many other metal ions, including those both with biologically relevant and with no known biological function (toxic to living cells). Siderophores have been shown to bind manganese, zinc, copper, nickel, or molybdenum and vanadium, all important for proper functioning of cellular processes, and are present in μM to mM concentrations. They also interact with aluminum, chromium, lead, arsenic, cadmium, or actinides—metal ions with absolutely no biological function and that are toxic to cells. Of importance, even biologically important metals can become deleterious when present in excessive levels, and to accurately control their uptake and homeostasis, microorganisms have evolved a variety of sophisticated regulatory systems; siderophores seem to constitute a part of these mechanisms, participating not only in metal acquisition but also as defense against their toxicity.

15.3.1 METALS WITH A BIOLOGICAL FUNCTION

Zinc is an essential trace element that is found in proteins, enzymes, and some ribosomal proteins and is necessary for every life-form (Hantke 2005). It is the second most abundant transition metal ion in living organisms after iron, and in contrast to it, zinc is redox inactive. Highly flexible coordination number and geometry of zinc complexes depend on ligand size and charge. Ligands that coordinate zinc ions contain usually oxygen, nitrogen, and sulfur as donor atoms. Zinc plays important structural and catalytic function in bacterial proteins—forms part of recognition motif for some transcription factors and is found in certain proteolytic enzymes (Beveridge et al. 1997). It was found that microorganisms incorporate zinc into approximately 4–6% of all proteins. Almost 5% of the *E. coli* proteome contains zinc-binding proteins (Andreini et al. 2006; Kehl-Fie and Skaar 2010). The significance of the zinc acquisition for bacteria is underscored by the fact that upon microbial infection, humans secrete zinc-sequestering proteins to reduce the availability of zinc ions and inhibit the growth of infection (Kehl-Fie and Skaar 2010; Hood and Skaar 2012). On the other hand, high concentration of zinc may be harmful to cells, as zinc has a high affinity to thiols and therefore can block essential reactions in the cell. Taking into account this fact, it is obvious that microorganisms possess mechanisms of maintaining cellular zinc homeostasis to ensure their own survival. Two major types of zinc transporters have been the most widely described: ZnuABC, which is a high-affinity transporter belonging to the cluster C9 family of ABC transporters, and ZupT (low-affinity uptake system) from the ZIP family (Hantke 2005). One can find a few examples, indicating that bacterial siderophores could play a role as zinc chelators, termed zincophores or tsinkophores, which could be involved in the zinc homeostasis strategies (Hood and Skaar 2012).

Physicochemical studies of zinc–siderophore complexes indicate that siderophores are devoted not only to iron ions. PCH binds zinc ions with high affinity. The stability constants is $\log\beta_{120} = 26.0$, while in the case of iron ions, $\log\beta_{120} = 28.8$. PCH has been shown to form bischelate complexes with zinc over a wide range of pH (Brandel et al. 2012). The natural siderophore, coprogen (COP), produced by *Penicillium chrysogenum* and *N. crassa*, was also found to be an effective zinc-binding ligand with $\log\beta_{110} = 11.8$. Complexes formed by siderophore very similar to COP, DFOB, are slightly less stable, with $\log\beta_{110} = 10.36$ (Farkas et al. 1999; Enyedy et al. 2004). A tridentate biochelator, that is, desferrithiocin (DFT) produced by *S. antibioticus* and dihydroaeruginoic acid from *Pseudomonas fluorescens*, has been also shown to chelate zinc ions with $\log\beta_{111} = 9.28$ and 3.8, respectively (Anderegg and Raber 1990). Unfortunately, the transport of these zinc–siderophore complexes has not been demonstrated *in vivo*.

It was found that a small-molecule metal chelator, pyridine-2,6-bis(thiocarboxylic acid) (ptdc), produced by a variety of pseudomonas, has the ability to form soluble complexes with zinc ions, with 1:1 and 1:2 metal-to-ligand stoichiometry (Cortese et al. 2002). The transport of zinc as zinc–ptdc complex can be realized by strains possessing pdtc utilization systems, through outer membrane receptor and inner membrane permease components, but it is much less efficient than for iron–ptdc complex (Leach et al. 2007).

A putative siderophore activity was also ascribed to two molecules, coelichelin and coelibactin, produced by the antibiotic-producing bacterium, *Streptomyces coelicolor*. Two reports have indicated that gene cluster encoding production of coelibactin and its product are indirectly involved in regulation of antibiotic production via its role as a zincophore. Increased expression of this gene cluster proved to result in defect in antibiotic production (Hesketh et al. 2009; Kallifidas et al. 2010; Zhao et al. 2012).

The *Y. pestis* siderophore, YBT, has been developed to serve as a zincophore for zinc acquisition (Bobrov et al. 2014). It has been shown that *Y. pestis* mutants, lacking functional ZnuABC transporter, preserved nearly full virulence, which support the presence of another high-affinity zinc transporter (Desrosiers et al. 2010). After that several studies have been performed to indicate that YBT can be used as a zinc uptake system. Two different mutations in the irp2 gene encoding synthetase for YBT caused growth defects. In addition, growth was restored when complementation with the cloned irp2 gene was used as well as after zinc supplementation to 2.5 μM. Finally, addition of exogenous, purified apo-YBT supported the growth of mutant *Y. pestis*. All of these studies provide evidence that YBT may serve as a zincophore (Bobrov et al. 2014). In addition, the antibiotic micacodin, produced by *Pseudomonas* spp. and *Ralstonia solanacearum*, with a chemical structure similar to YBT, is able to bind zinc ions and has been crystallized as a zinc complex (Nakai et al. 1999; Kreutzer et al. 2011).

Copper is another essential metal ion that, similarly to iron, can cycle between oxidized, Cu(II), and reduced, Cu(I), states. Coordination preferences for Cu(I) and Cu(II) ions are slightly different due to the fact that Cu(I) is a soft acid (in terms of the HSAB [Hard and Soft Acid and Bases] classification) and prefers sulfur-containing ligands, while Cu(II) is a hard acid and favors oxygen and nitrogen ligands. Cu(I) complexes are usually linear or tetrahedral, while Cu(II) complexes are square planar or distorted octahedral (Crichton 2001). Redox potentials of copper-dependent enzymes range between +0.25 and 0.75 V, allowing their utility in a range of biochemical reactions (Crichton 2001). Copper is essential for cellular respiration (cytochrome oxidase), iron transport (ceruloplasmin, hephaestin), superoxide dismutation (superoxide dismutases), pigmentation (tyrosinase, laccase), and other vital processes. On the other hand, redox activity of copper may contribute to its toxicity to microbes (and mammals) through classic copper-catalyzed Fenton chemistry and an increase of ROS or non-Fenton destruction of iron–sulfur clusters in proteins and loss of their function

(Lemire et al. 2013; Chaturvedi and Henderson 2014). The exact mechanisms remain incompletely understood and await elucidation (Djoko et al. 2015). To maintain and control copper homeostasis, bacteria employ copper transport proteins and chaperones, and these systems are relatively well described (Ma et al. 2009). Comparatively, few bacterial import systems have been recognized, with P-type ATPases in *P. aeruginosa* and *Listeria monocytogenes* being two exceptions (Francis and Thomas 1997; Lewinson et al. 2009).

The role of siderophores in the uptake of copper was studied in *P. aeruginosa*, where exposure to elevated copper concentrations induced upregulation of genes involved in the biosynthesis of PVD and downregulation of those involved in the biosynthesis of PCH, pointing to the role of the latter siderophore in the uptake process (Teitzel et al. 2006). PVD is highly iron specific, with a formation constant of $\log\beta_{110}$ ~ 30.8, and binds copper with much lower affinity, $\log\beta_{110}$ ~ 22 (Albrecht-Gary et al. 1994; Chen et al. 1994).

PCH has been shown to form bischelate complexes with both iron and copper, at pH range 6–9, with much lower selectivity between these two metal ions, represented by $\log\beta_{120} = 28.8$ and 25, respectively (Brandel et al. 2012). However, only iron has been efficiently transported by PVD or PCH pathways of *P. aeruginosa* (Braud et al. 2009; Cunrath et al. 2015). These findings support the idea about an additional role for siderophores—extracellular protection of microbial cells against elevated concentrations of metals, enabling their proper growth under such toxic conditions (Braud et al. 2010; Schalk et al. 2011).

Indeed, in contrast to iron withholding used by mammalian hosts as the nutritional immunity tool against pathogens' iron thievery, excess of copper is used by hosts as an antimicrobial weapon. It has been demonstrated that host immune cells, like macrophages, are able to increase copper concentration and expression of copper-transporting machinery in response to bacterial, *Mycobacterium* spp. (Wagner et al. 2005; White et al. 2009), and fungal, *C. neoformans* (Ding et al. 2011), infections, most probably with the purpose to force pathogens to deal with copper excess. Copper resistance systems, including efflux proteins, oxidases, and copper sequestering molecules, have been described in a number of invading pathogens (Hernandez-Montes et al. 2012). Siderophores seem to serve as one of these defensive mechanisms. It has been evidenced that uropathogenic strains of *E. coli* often produce distinct siderophores: (1) prototypical, genetically conserved enterobactin, (2) virulence-associated YBT, and (3) aerobactin (Chaturvedi et al. 2012; Koh and Henderson 2015). Catecholate enterobactin, specialized for iron acquisition, sensitizes bacteria to copper toxicity through the reduction of extracellular Cu(II) to more toxic Cu(I). To prevent this process, phenolate-/thiazoline-based Ybt binds Cu(II) into a stable complex; the role of citrate-hydroxamate aerobactin has to be clarified (Chaturvedi et al. 2012). Of importance, the expression of a cuprous oxidase CueO can also be augmented to alleviate the damaging effects of copper, acting through the oxidation of the enterobactin precursor 2,3-dihydroxybenzoic acid to 2-carboxymuconate and therefore preventing catechol-mediated reduction of Cu(II) (Grass et al. 2004). Moreover, Cu(II)–YBT complex may also perform superoxide dismutase-like activity (via equations 2–3) (Chaturvedi and Henderson 2014), protecting pathogens from the respiratory burst within copper-containing phagosomes, during which superoxide anion is produced by the NADPH oxidase system (Beaman and Beaman 1984). Thus, YBT, regarded previously as a secondary Fe(III)-scavenging siderophore, emerges as a dual-function defense metabolite, capable of protecting pathogens from phagocytic killing via toxic metal binding and its exploitation for catalytic activity.

$$Cu(II) - YBT + O_2 \cdot^- \rightleftharpoons Cu(I) - YBT + O_2 \qquad (15.2)$$

$$Cu(I) - YBT + O_2 \cdot^- + 2H^+ \rightleftharpoons Cu(II) - YBT + H_2O_2 \quad (15.3)$$

Analogous SOD-like behavior was evidenced in methanobactin (Choi et al. 2008), a small copper chelator found in aerobic methane-oxidizing bacteria, or methanotrophs (DiSpirito et al. 1998; Téllez et al. 1998). The physiological function suggested for methanobactin is that of copper binding and transport, thus being chalkophore (from Greek "chalko" for copper) (Graham and Kim 2011). Methanobactin binds Cu(I) with high affinity, through the oxazole nitrogen and thioamide sulfur atoms ($log\beta_{Cu(I)-(Mb-OB3b)} = 20.8$) (El Ghazouani et al. 2011), and is hypothesized to mediate copper acquisition from the environment (Kenney and Rosenzweig 2012). Although methanobactin is able to coordinate Cu(II), it gets quickly reduced to Cu(I) via an unknown mechanism (Choi et al. 2006).

Small amounts of manganese are essential for the proper functioning of most of the organisms, primarily because of its redox properties. The functions of manganese-containing enzymes include, inter alia, process of the photosynthesis, oxygen protection, and nitrogen fixation. Excess of manganese ions may exhibit toxicity (Das et al. 2014); however, the required amount of manganese is indiscernible, and therefore, manganese limitation is mostly hard to determine (Nealson et al. 1988). A great attention is focused on the chemistry of Mn(II) and Mn(IV). It was assumed that all of dissolved manganese is Mn(II), while any solid particles of this metal are insoluble minerals MnO_x, where manganese is in the oxidation state +3 and +4, whereby the vast majority is Mn(IV) (Johnson 2006). However, organic and inorganic chelates may affect the dissolution of Mn(III) and prevent expected disproportionation to Mn(II) and Mn(IV) (Trouwborst et al. 2006). Mn(III) is an intermediary stage during the oxidation of Mn(II) to Mn(IV). This pattern remains in accordance to the known mechanism of electron transfer of the multicopper oxidases, which is an essential part of this process (Parker et al. 2004).

It was reported that, in spite of the preferred binding of iron, siderophores have the ability to bind manganese ions, and in some cases manganese–siderophore complexes exhibit greater stability constants than for iron. Among siderophores described as excellent chelators for manganese, there

are rhizoferrin, protochelin, DFOB, DFOE, and coprogen (Springer and Butler 2016).

Mn(III)-bearing mineral, manganite, was proved to be susceptible to dissolution in the presence of DFOB. It was found that this process can by carried out by two reaction pathways, both reductive and nonreductive. For pH below 6.5, the reductive dissolution promoted by DFOB was a dominant reaction. It can be easily explained by the fact that under such pH conditions, Mn(III)–DFOB complex is unstable, and an internal electron transfer may contribute to the disintegration of the complex, yielding Mn(II) and oxidized siderophore. Also, the unbound Mn(III) from manganite may participate in redox reaction with adsorbed DFOB. In the nonreductive pathway, the dissolution rate depends on the concentration of DFOB (Duckworth and Sposito 2005). The stability constant of Mn(III)–DFOB complex is $log\beta_{110} = 29.9$, which is comparable with $log\beta_{110} = 30.6$ for Fe(III)–DFOB complex. In relation to these values, $log\beta_{110} = 7.7$ for Mn(II)–DFOB complex is relatively small (Springer and Butler 2016). In 1994, Faulkner and colleagues reported that ferrioxamine-type siderophore complexes with Mn(III) ions may be a superoxide dismutase biomimetics (Faulkner et al. 1994).

Two derivatives of PVD, PVD_{PaA} and pyoverdine Gb1 (PVD_{Gb1}), are characterized by higher stability constants for Mn(III) complexes than for Fe(III). For Mn(III)–PVD_{PaA} complex, stability constant is $log\beta_{110} = 35.4$, while for Fe(III)–PVD_{PaA} complex, stability constant is $log\beta_{110} = 30.0$. Similarly, stability constant for PVD_{Gb1} is $log\beta_{110} = 35.3$ for Mn(III) and $log\beta_{110} = 31.06$ for Fe(III) ions (Parker et al. 2004). Although the reason why this situation occurs is not fully understood, the presence of the hydroxycarboxylate donor group seems to accommodate the Jahn–Teller distortion of coordination of Mn(III) ions (Parker et al. 2014). Parker et al. (2002) suggested that Mn(III)–PVD complexes may create reservoirs of Mn(III) in natural conditions.

The carboxylate and hydroxy-carboxylate groups of rhizoferrin are capable to bind Mn(II) and Mn(III) ions with stability constants of $log\beta_{110} = 5.95$ and $log\beta_{110} = 29.8$, respectively (Springer and Butler 2016). This is another example of siderophore that binds Mn(III) ions with greater affinity than iron ions ($log\beta_{110} = 25.3$). It has been suggested that strong interactions of Mn(III) ions with the hydroxyl moiety from hydroxy-carboxylate group are a result of, first, acidity of manganese ions, which is higher than for Fe(III), and, second, capacity to accommodate the Jahn–Teller deformation (Harrington et al. 2012).

The ability of siderophores to form stable complexes with manganese may result in competing of manganese and iron ions for siderophore via ligand exchange, sorption of free siderophores to MnO_x surface, or dissolution of manganese minerals promoted by siderophores (Duckworth et al. 2009). The high affinity of those ligands for Mn(III) ions affect the oxidation state of dissolved manganese, what may disturb the assay of total manganese in studied samples, usually divided into soluble Mn(II) and insoluble precipitate, assumed to be primarily Mn(IV) (Parker et al. 2004). For those reasons, interactions between manganese ions and siderophore molecules can have significant impact on the environment and the process of iron transport.

The process of N_2 fixation follows the mechanism of six-proton and six-electron reduction, requiring the participation of nitrogenases. Those enzymes, generated by diazotrophs, must contain one of the three types of special metal cofactors. First cofactor has only iron center, the other two, in addition to iron, contain molybdenum or vanadium ions (Wichard et al. 2009). The concentration of dissolved molybdenum in water varies between around 100 nM in ocean and less than 30 nM in freshwater, which is why microorganisms have to acquire the ability to release and uptake of molybdenum chelators (Liermann et al. 2005). Bacterium *A. vinelandii* produces siderophores, such as catecholate protochelin, azotochelin and aminochelin, and PVD-type azotobactin, all able to chelate molybdenum and vanadium (Bellenger et al. 2008; Wichard et al. 2009). Secretion of catechol siderophores may be regulated by both molybdate and ferric ions concentrations (Yoneyama et al. 2011). It has been reported that limitation in the concentration of Fe(III) and MoO_4^{2-} leads to an increased production of aminochelin and azotochelin, whereas increased concentrations of molybdenum induce the production of protochelin (Springer and Butler 2016). The decrease in secretion of protochelin and azotochelin is also observed when the concentration of vanadium increases (Bellenger et al. 2008).

Azotochelin binds molybdate less efficiently than ferric ions ($\log\beta_{110} = 35$, $\log\beta_{110} = 53.47$) (Cornish and Page 1998; Duhme et al. 1998); protochelin exhibits similar affinity for Fe and Mo in complexes with 1:1 metal-to-ligand molar ratio (pFe is 27.7 and pMo is 26, where $pM = -\log [M^{z+}]_{free}$ with a total ligand concentration of $[L] = 10^{-5}$ M and a total metal concentration of $[M] = 10^{-6}$ M); however, it was reported that MoO_4^{2-}–protochelin complexes prefer formation of 2:3 MoO_4^{2-}–siderophore complex (pMo is about 16) (Cornish and Page 2000). Protochelin and azotochelin are also capable of binding VO_4^{3-}, which may begin the uptake process of vanadate and incorporation of vanadium into the nitrogenase (Springer and Butler 2016).

Much lower affinity for molybdenum and vanadium than for iron is also observed for azotobactin, which sometimes is referred to as the "true" siderophore of *A. vinelandii*. The secretion of this siderophore seems to occur only under serious limitation of ferric ions. However, the solubility of those oxyanions may promote their binding to siderophore, due to the competition with sparingly soluble FeO_x minerals. Siderophores able to bind molybdenum and vanadium ions may also affect the environment and protect it from the harmful effects of molybdate and vanadate, which can be toxic in too high concentrations (Wichard et al. 2009).

15.3.2 Heavy Metals

Heavy metals, such as cadmium, lead, arsenic, or chromium, are among the major environmental pollutants, particularly in areas with high anthropogenic activity. Heavy metal accumulation in soils is of great concern in agricultural production (Gill 2014), due to the adverse effects on soil organisms, food

safety, and impact on human health, and has contributed to an increasing interest in the development of biological remediation technologies.

Naturally, the effects of metal pollutants as environmental stressors affect also microorganisms, which develop homeostatic mechanisms to avoid any uptake of abiotic elements (Nies 1999, 2003). As microorganisms can modulate heavy metal toxicity by the synthesis and excretion of siderophores, the potential of using siderophores for decontamination purposes has been intensively evaluated. In particular, lots of interest has been focused on a combination of phytoremediation with siderophore-producing bacteria inoculation (Rajkumar et al. 2010). Studies on the processes of interactions between specific siderophores and various metal ions were carried out in order to explain their potential role in bioremediation.

Several siderophores have been shown to bind toxic metal ions. Physicochemical studies on metal–siderophore complexes' stability have been devoted to DFOB, COP, DFT, PVD, and PCH. Stability constants, $\log\beta$, for DFOB complexes with aluminum, gallium, and indium are between 22.18 and 27.56. DFOB complexes with nickel are less stable, with the stability constant of 8.89 (Farkas et al. 1997, 2000; Enyedy et al. 2004). COP forms complexes with very similar stability to DFOB, with $\log\beta$ from 22.51 to 25.8 for aluminum, gallium, and indium and $\log\beta = 11.42$ for nickel (Enyedy et al. 2004). Also DFT can chelate a variety of metal ions, such as manganese, cobalt, and cadmium ($\log\beta_{111} \sim 7.28 - 9.13$); nickel and aluminum form bischelate complexes ($\log\beta_{120}$ of 17.74 and 22.2, respectively) (Anderegg and Raber 1990).

The formation constants for PVD in complex with cadmium, nickel, and manganese are between 8.5 and 17.3 (Chen et al. 1994; Ferret et al. 2015).

In microbiological studies, PVD and PCH have been shown to sequester silver, aluminum, gallium, cadmium, cobalt, chromium, nickel, mercury, manganese, lead, tin, europium, terbium, and thallium from the extracellular medium (Braud et al. 2009a, Braud et al. 2009b). Although microbial uptake systems were able to bind PCH–M and PVD–M complexes, they were very selective at the stage of uptake, allowing efficient transport only for iron complexes. In the case of PCH, cobalt, gallium, and nickel were also transported, but with much lower uptake rates (23–35-fold lower than for iron) (Braud et al. 2009a). PVD transport system also allowed the passage of copper, gallium, manganese, and nickel, with 7- to 42-fold lower uptake rates (Enyedy et al. 2004). None of the other metals was taken up. Moreover, both siderophores were shown to be able to reduce the accumulation of toxic metals and increase metals' tolerance in *P. aeruginosa* (Braud et al. 2010). However, PVD is more involved in heavy metal sequestration outside the bacteria and therefore is more effective in the protection of bacteria against metal toxicity. In agreement, toxic metals stimulated only the production of PVD and not PCH (Braud et al. 2010). Under iron-limited conditions, the presence of 10 μM aluminum, copper, gallium, manganese, and nickel induced PVD production by PCH-deficient PAO6297 strain (Braud et al. 2009b); however, in iron-supplemented medium (100 μM), PVD production was only increased with copper and nickel (290% and 380%, respectively) and to a lesser extent with chromium (134%), suggesting that these metals may activate the signaling cascade regulating PVD production (Braud et al. 2010). The metal concentration was varying from 0 to 100 μM. Lead-enhanced production of PVD and PCH was revealed when $Pb(NO_3)_2$ was used up to 0.5 mM; above this concentration of $Pb(NO_3)_2$, a significant decline in siderophores production was observed (Naik and Dubey 2011).

A phenolate-type alcaligin E (not yet isolated in a pure state), produced by heavy metal-resistant *Alcaligenes eutrophus* CH34 strain, has been also reported to interact with cadmium and affect its bioavailability and toxicity. Addition of this siderophore in the presence of cadmium significantly stimulated the growth of alcaligin E–deficient bacteria (Gilis et al. 1998). The behavior was explained in terms of a decreased bioavailability of cadmium to the bacterium in the presence of alcaligin E; however, a facilitated iron uptake as a growth-stimulating factor was not excluded.

These studies opened up a perspective that siderophores may serve as very efficient tools for heavy metal binding and dissemination in the environment. In several studies, inoculation of plants with siderophore-producing bacteria was applied in order to test the role of siderophores in environmental metal mobilization via phytoextraction, and opposite mechanisms seemed to operate. For example, bioaugmentation of maize on a soil contaminated by chromium and lead has been positively influenced by the presence of *P. aeruginosa*, in which siderophores facilitated metal ions' mobilization and removal from soil (Braud et al. 2009c). In another example, it has been revealed that the presence of *P. aeruginosa* helped reduce cadmium uptake in the mustard and pumpkin plants (Sinha and Mukherjee 2008). Cadmium-induced PVD production was observed, and the strain was able to tolerate up to 8 mM of cadmium ions via intracellular accumulation (Sinha and Mukherjee 2008). In both studies, promotion of the plant growth has been observed (Sinha and Mukherjee 2008; Braud et al. 2009c). Recently, Cornu and coworkers revealed a nice agreement between PVD coordination properties toward copper and cadmium ions and its impact on metal phytoextraction (Cornu et al. 2014). Much higher stability constant of PVD–copper ($logK^{L'Cu}=20.1$) complex than that of PVD–cadmium ($logK^{L'Cd}=8.2$; L′ designates the system for which the ionizable sites of the arginine and succinate moieties have been omitted) complex was in line with enhanced mobility, phytoavailability, and phytoextraction of copper from PVD-supplied calcareous soil by the monocotyledonous (strategy II) and dicotyledonous (strategy I) plants; the fate of cadmium, as well as iron, zinc, nickel, and manganese, was not affected (Naik and Dubey 2011). The effect of PVD was restricted only to the roots; root copper was 2 times higher, while shoot copper was not affected. Similar relationship was evidenced between PVD–cadmium and PVD–nickel complexation ability and metal mobilization, speciation, and phytoavailability in hydroponics (Ferret et al. 2015). Again,

higher stability constant of PVD–nickel ($logK_{L'Ni}=10.9$) complex in comparison to that of PCD–cadmium ($logK_{L'Cd}=8.2$) complex resulted in selective PVD-dependent mobilization of nickel from smectite. PVD decreased plant nickel and cadmium contents and the free ionic fractions of these two metals, in agreement to the free ion activity model. Inoculation with *P. aeruginosa* had a similar effect on nickel phytoavailability to the direct supply of PVD (Ferret et al. 2015).

Preventing the entry of toxic metal ions into the bacteria has been reported for *Streptomyces tendae* F4 producing a variety of hydroxamate siderophores, such as DFOB and DFOE, and coelichelin; the production of the siderophores has been shown to be upregulated by the presence of cadmium ions (Dimkpa et al. 2009). Additionally, the role of siderophores in an increased uptake of cadmium by sunflower plant has been revealed. Using nickel-resistant *Streptomyces acidiscabies* E13, it has been reported that the DFOB, DFOE, and coelichelin siderophores have been able to chelate nickel ions and promote cowpea plant growth under nickel stress. The siderophores were playing a dual role—chelating iron for plant and protecting against nickel toxicity (Dimkpa et al. 2008). A simultaneous production of the same hydroxamate siderophores, that is, DFOB, DFOE, and coelichelin, and auxins—phytohormones that enhance root growth—have been examined in *Streptomyces* spp.; the studies were carried out in the context of metal-induced inhibition of auxins synthesis. The authors concluded that binding of aluminum, cadmium, copper, and nickel by siderophores reduces the metals' ability to inhibit the synthesis of auxins and in effect stimulates the plant growth and their bioremediation potential (Dimkpa et al. 2008).

In the literature, one can find other examples of siderophores that could be associated with a decrease, or a dissemination, of metal toxicity in the environment. Neubauer et al. showed that the hydroxamate siderophore DFOB can drastically affect heavy metal adsorption on clay minerals, what may diminish their toxicity to plants. They used montmorillonite and kaolinite as model systems and showed that, dependent on pH conditions, DFOB can bind copper, zinc, and cadmium better than iron (Neubauer et al. 2000). On the other hand, siderophores (not defined) produced by several strains of arsenic-hypertolerant bacteria, present in ancient gold mine, were proposed as compounds responsible for the extreme arsenic resistance of isolates and, concomitant, As(V) mobilization and release into the environment (Drewniak et al. 2008). Produced in order to facilitate iron uptake from insoluble minerals, siderophores assist the mobilization of As(V) from the solid to the aqueous phase. The toxic effects of arsenic for bacterial cells are neutralized by the activity of the arsenate reductase, and as a result of the reduction process, As(III) is released into the environment. However, the same process, siderophore-induced mobilization of the As(V) in the soil in the process of iron ions' uptake, rendering arsenic more soluble, served as an eco-friendly tool for removing arsenic from the environment (Wang et al. 2011). Inoculation with *Agrobacterium radiobacter* contributed to the increase in the arsenic tolerance of poplar, promotion of the plant growth, increase in the uptake efficiency, and enhancement of arsenic translocation (Wang et al. 2011). In a separate study, siderophores produced by *Pseudomonas azotoformans* (hydroxamate and catecholate type) were applied for washing of contaminated soil and let to the removal of 92.8% of arsenic, both easily available and strongly bound to soil. The use of siderophores as decontamination agents showed no toxic effect on soil microbial population (Nair et al. 2007).

15.3.3 ACTINIDES

To date, there are 441 nuclear power plants operating in 31 countries and another 67 are under construction (International Atomic Energy Agency website). In OECD countries (Organization for Economic Co-operation and Development), the annual production of toxic wastes is maintained at about 300 million tonnes, including about 10,000 m^3 of high-level radioactive waste (World Nuclear Association website), which are typically conditioned under the ground in special repositories (Behrends et al. 2012). This solution may entail the risk from environmental release of radionuclides. In 1993, Riley and Zachara reported that many places with significant radionuclide contamination were located. More than half of them were characterized by the presence of uranium and plutonium ions in groundwater and soil (Riley and Zachara 1993). The inherence of actinide ions in nuclear wastes is a major problem, due to their toxicity, rather easy migration into environment, and long half-lives (Francis 1998).

The alternative to the waste storage is a PUREX process (*p*lutonium *u*ranium *r*edox *ex*traction), used for nuclear fuel reprocessing by separating uranium and Pu(IV) ions. The whole process is based on the extraction of U(VI) and Pu(IV) to diluted tributyl phosphate from the solution obtained by dissolving the waste in nitric acid. Separation of uranium and plutonium can be obtained by the reduction of Pu(IV) to Pu(III) by U(IV) combined with hydrazine, what prevents the reoxidation of Pu(III) (Denniss and Jeapes 1996). After the separation, Pu(IV) and U(VI) ions could be reused as nuclear fuel. The side effect of PUREX process is Np(VI) reduction to extractable Np(IV), contaminating uranium ions, what may reduce the energy value (Taylor and May 1999). This example shows that although PUREX process is successfully used for the reprocessing of spent nuclear fuel, there is still a necessity to enhance this method and to reduce both the cost and the environmental impact (Taylor et al. 2002). It has been shown that simple hydroxamic acids permit simple and selective elimination of neptunium and plutonium ions from the solvent (Taylor et al. 1998, 2002), provided by the presence of hard oxygen donors, which strongly bind to those hard Lewis acids. For this reason, siderophore molecules appear to offer the possibility of forming stable complexes with actinide ions. Actinide–siderophore complex formation may affect not only the ease of nuclear fuel reprocessing but also the bioaccumulation of radionuclides in the environment. DFOB and enterobactin have been described as capable to solubilize solid hydrous plutonium oxide as well as uranium compounds, for example, UO_2 (Brainard et al. 1992).

Plutonium (IV) ions exhibit chemical similarity to iron in respect of the type of hydrolysis products, formation of insoluble hydroxides and oxides, coordination chemistry, and stability constants of their complexes (John et al. 2001), allowing formation of Pu(IV) complexes with molecules intended for iron transport and storage. DFO-type siderophores have been reported to bind Pu(IV) with $log\beta_{110}$ about 30.8 (Jarvis and Hancock 1991) for Pu(IV)–DFOB complex; stability of Pu(IV)–DFOE complex was sufficient to determine its crystal structure by x-ray diffraction methods (Neu et al. 2000). Pu(IV)–DFOB complexes possess the ability to be taken up by bacterium *Microbacterium flavescens*. It has been shown that Pu(IV)–siderophore complexes compete with Fe(III) analogs for receptor proteins, however, the Pu(IV) absorption process if about four times slower than Fe(III) uptake. Furthermore, DFOB complexes with Pu(IV), although recognizable by bacteria, are not able to internalize and remain located in the cell wall of microorganisms (John et al. 2001). This fact is most probably caused by varying coordination number of Fe(III) and Pu(IV). Six-coordinate Fe(III) ion is approximately coplanar with the DFO molecule, while the coordination sphere of Pu(IV) is supplemented by three additional water molecules, what results in location of the metal ion outside the plane of the siderophore. The difference between the two complexes is relatively small; however, it is presumed that the deviation is sufficient to affect the metal ion uptake (John et al. 2001).

The danger associated with the release of uranium comes primarily from its relatively long half-life and high toxicity. It has been reported that exceeding the limit concentration of 20µg/l for uranium leads to a significant mortality of larval fish (Holdway 1992). However, in other studies, considerable uranium toxic effects were confirmed only at the concentrations beyond 300 mg U/kg of soil, which is much higher than the levels determined in most of the areas polluted by industry. It is very probable that in those places other pollutants would be even more important threats, since arsenic and several other radionuclides often occur simultaneously with an increased level of uranium (Sheppard et al. 1992). Uranium in aerobic environments occurs primarily as U(VI) form, while under accordingly reducing conditions U(IV) oxidation state is the predominating species (Behrends et al. 2012). However, it has been evidenced that the majority of uranium that has been released into the environment after the Chernobyl disaster remained in the tetravalent U(IV) form more than a decade after the incident; this proves that the solid fields containing U(IV) may exist in oxygen environment for a long time, most probably due to a slow kinetics of the oxidation reaction. Earlier studies on Pu(IV)–DFOB complex, and the similarity of chemical properties of tetravalent actinides, inspired Frazier et al., who have shown that DFOB affects the mobilization of uranium (Frazier et al. 2005). Later on, DFOB has been shown to form with U(VI) complexes stable in a wide pH range, with $log\beta_{110}$ of about 17.1 (Mullen et al. 2007). In contrast to Pu(IV)–DFOB complexes, U(VI) compounds were not recognized by the uptake proteins of *M. flavescens* and therefore could not be introduced into bacterial cells

(John et al. 2001). Also PVD has been reported to possess the ability to chelate U(VI). PVD produced by *P. fluorescens* has been shown to bind 90.48% and 86.41% of U(VI) from the monazite and (Th–U) concentrate, respectively (Hussien et al. 2013). The most optimal conditions for the binding process were up to pH 6.5; above this pH, the complexation performance decreased successively, most probably due to complex hydrolysis (Hussien et al. 2013).

15.4 CONCLUSIONS

Despite the presence of large amount in Earth's crust, essential metal ions, like iron, are not always easily accessible, especially for microorganisms. During years of evolution, bacteria and fungi have developed the ability to synthesize and excrete siderophores, molecules with very high affinity for iron. Additionally, they have evolved specialized receptors present on the cell surface, by which the complexed iron can be selectively transported into the cell. However, over the years, it became clear that siderophores are not exclusively designed for iron; they can coordinate a range of other metal ions as well as shuttle some selected ones. Although, in general, stability constants for siderophore complexes with metal ions other than iron are inferior to the corresponding iron species, the complexes are efficiently used by the microorganisms for the control and maintenance of metal homeostasis. Using biologically relevant metal ions, like copper, and its siderophore complexes, the additional role of siderophores, that is, the extracellular protection of microbial cells against elevated concentrations of essential metals, was exposed. A decrease of intracellular penetration of metals is realized via extracellular formation of stable metal–siderophore complexes that cannot be recognized by very selective siderophore receptors nor diffused through porins (the usual entrance used by free metals). An analogous approach is used by microorganisms to protect themselves against environmental heavy metal toxicity and has contributed to an increasing interest in their applications in biological remediation technologies. The combination of phytoextraction with siderophore-producing bacteria inoculation has been investigated several times, but contradictory results were revealed. Therefore, understanding the mechanisms of siderophore-mediated transport of metals and the role of siderophores in the bioremediation processes deserves further attention, with specific emphasis on the relation between the thermodynamic and kinetics of siderophore–metal complexation and metal phytoavailability; also the link between siderophores production and heavy metal resistance should be clarified. One cannot forget that siderophore production in response to heavy metal exposure could also have detrimental effects, like metal dissemination in the environment (especially dangerous for radionuclides) or providing a microbial secondary entrance of toxic metals (if not specific enough at the stage of complex–siderophore recognition). Further studies are awaited to clarify also these aspects.

REFERENCES

Abergel, R.J., Warner, J.A., Shuh, D.K., Raymond, K.N., 2006. Enterobactin protonation and iron release: Structural characterization of the salicylate coordination shift in ferric enterobactin1. *Journal of the American Chemical Society*, 128: 8920–8931.

Albrecht-Gary, A.M., Blanc, S., Rochel, N., Ocaktan, A.Z., Abdallah, M.A., 1994. Bacterial iron transport: Coordination properties of pyoverdin PaA, a peptidic siderophore of *Pseudomonas aeruginosa*. *Inorganic Chemistry*, 33: 6391–6402.

Albrecht-Gary, A.M., Crumbliss, A.L. 1998. Iron transport and storage in microorganisms, plants, and animals. In *Metal Ions in Biological Systems*. A. Sigel and H. Sigel (eds.), pp. 239–327. New York: Marcel Dekker.

Anderegg, G., Leplatte, F, Schwarzenbach, G., 1963. Hydroxamatkomplexe.3. Eisen(Iii)-Austausch Zwischen Sideraminen und Komplexonen—Diskussion der Bildungskonstanten der Hydroxamatkomplexe. *Helvetica Chimica Acta*, 46: 1409.

Anderegg, G., Raber, M., 1990. Metal-complex formation of a new siderophore desferrithiocin and of 3 related ligands. *Journal of the Chemical Society, Chemical Communications*, 17: 1194–1196.

Andreini, C., Banci, L., Bertini, I., Rosato, A., 2006. Zinc through the three domains of life. *Journal of Proteome Research*, 5: 3173–3178.

Beaman, L., Beaman, B.L., 1984. The role of oxygen and its derivatives in microbial pathogenesis and host defense. *Annual Review of Microbiology*, 38: 27–48.

Behrends, T., Krawczyk-Baersch, E., Arnold, T., 2012. Implementation of microbial processes in the performance assessment of spent nuclear fuel repositories. *Applied Geochemistry*, 27(2): 453–462.

Bellenger, J.P., Wichard, T., Kustka, A.B., Kraepiel, A.M.L., 2008. Uptake of molybdenum and vanadium by a nitrogen-fixing soil bacterium using siderophores. *Nature Geoscience*, 1: 243–246.

Beveridge, T.J., Hughes, M.N., Lee, H., Leung, K.T., Poole, R.K., Savvaidis, I., Silver, S., Trevors, J.T., 1997. Metal-microbe interactions: Contemporary approaches. *Advances in Microbial Physiology*, 38: 177–243.

Bickel, H., Hall, G.E., Keller-Schierlein, W., Prelog, V., Vischer, E., Wettstein, A., 1960. Stoffwechselprodukte von Actinomyceten. 27. Mitteilung. über die konstitution von Ferrioxamin B. *Helvetica Chimica Acta*, 43: 2129–2138.

Bleuel, C., Große, C., Taudte, N., Scherer, J., Wesenberg, D., Krauß, G.J., Nies, D.H., Grass, G., 2005. TolC is involved in enterobactin efflux across the outer membrane of *Escherichia coli*. *Journal of Bacteriology*, 187: 6701–6707.

Bobrov, A.G., Kirillina, O., Fetherston, J.D., Miller, M.C., Burlison, J.A., Perry, R.D., 2014. The Yersinia pestis siderophore, yersiniabactin, and the ZnuABC system both contribute to zinc acquisition and the development of lethal septicaemic plague in mice. *Molecular Microbiology*, 93: 759–775.

Boelaert, J.R., De Locht, M., Van Cutsem, J., Kerrels, V., Cantinieaux, B., Verdonck, A., Van Landuyt, H.W., Schneider, Y.J., 1993. Mucormycosis during deferoxamine therapy is a siderophore-mediated infection. In vitro and in vivo animal studies. *Journal of Clinical Investigation*, 91: 1979.

Boukhalfa, H., Brickman, T.J., Armstrong, S.K., Crumbliss, A.L., 2000. Kinetics and mechanism of iron (III) dissociation from the dihydroxamate siderophores alcaligin and rhodotorulic acid. *Inorganic Chemistry*, 39: 5591–5602.

Boukhalfa, H., Crumbliss, A.L., 2002. Chemical aspects of sidero-phore mediated iron transport. *Biometals*, 15: 325–339.

Boukhalfa, H., Reilly, S.D., Michalczyk, R., Iyer, S., Neu, M.P., 2006. Iron (III) coordination properties of a pyoverdin sid-erophore produced by *Pseudomonas putida* ATCC 33015. *Inorganic Chemistry*, 45: 5607–5616.

Brainard, J.R., Strietelmeier, B.A., Smith, P.H., Langston-Unkefer, P.J., Barr, M.E., Ryan, R.R., 1992. Actinide binding and solu-bilization by microbial siderophores. *Radiochimica Acta*, 58: 357–364.

Brandel, J., Humbert, N., Elhabiri, M., Schalk, I.J., Mislin, G.L., Albrecht-Gary, A.M., 2012. Pyochelin, a siderophore of *Pseudomonas aeruginosa*: Physicochemical characterization of the iron (III), copper (II) and zinc (II) complexes. *Dalton Transactions*, 41: 2820–2834.

Braud, A., Geoffroy, V., Hoegy, F., Mislin, G.L., Schalk, I.J., 2010. Presence of the siderophores pyoverdine and pyochelin in the extracellular medium reduces toxic metal accumulation in *Pseudomonas aeruginosa* and increases bacterial metal toler-ance. *Environmental Microbiology Reports*, 2: 419–425.

Braud, A., Hannauer, M., Mislin, G.L., Schalk, I.J., 2009a. The *Pseudomonas aeruginosa* Pyochelin-iron uptake pathway and its metal specificity. *Journal of Bacteriology*, 191: 3517–3525.

Braud, A., Hoegy, F., Jezequel, K., Lebeau, T., Schalk, I.J., 2009b. New insights into the metal specificity of the *Pseudomonas aeruginosa* pyoverdine-iron uptake pathway. *Environmental Microbiology*, 11: 1079–1091.

Braud, A., Jézéquel, K., Bazot, S., Lebeau, T., 2009c. Enhanced phytoextraction of an agricultural Cr- and Pb-contaminated soil by bioaugmentation with siderophore-producing bacteria. *Chemosphere*, 74: 280–286.

Braun, V., Killmann, H., 1999. Bacterial solutions to the iron-supply problem. *Trends in Biochemical Sciences*, 24: 104–109.

Budde, A.D., Leong, S.A., 1989. Characterization of siderophores from ustilago-maydis. *Mycopathologia*, 108: 125–133.

Carrano, C.J., Cooper, S.R., Raymond, K.N., 1979. Coordination chemistry of microbial iron transport compounds. 11. Solution equilibria and electrochemistry of ferric rhodotorulate com-plexes. *Journal of the American Chemical Society*, 101: 599–604.

Carrano, C.J., Drechsel, H., Kaiser, D., Jung, G., Matzanke, B., Winkelmann, G., Rochel, N., Albrecht-Gary, A.M., 1996. Coordination chemistry of the carboxylate type siderophore rhizoferrin: The iron (III) complex and its metal analogs. *Inorganic Chemistry*, 35: 6429–6436.

Caza, M. and Kronstad, J.W., 2013. Shared and distinct mecha-nisms of iron acquisition by bacterial and fungal pathogens of humans. *Frontiers in Cellular and Infection Microbiology*, 3: article 80, 1–23.

Chaturvedi, K.S., Henderson, J.P., 2014. Pathogenic adaptations to host-derived antibacterial copper. *Frontiers in Cellular and Infection Microbiology*, 4: article 3, 1–12.

Chaturvedi, K.S., Hung, C.S., Crowley, J.R., Stapleton, A.E., Henderson, J.P., 2012. The siderophore yersiniabactin binds copper to protect pathogens during infection. *Nature Chemical Biology*, 8: 731–736.

Chaturvedi, K.S., Hung, C.S., Giblin, D.E., Urushidani, S., Austin, A.M., Dinauer, M.C., Henderson, J.P., 2013. Cupric yersinia-bactin is a virulence-associated superoxide dismutase mimic. *ACS Chemical Biology*, 9: 551–561.

Chen, Y., Jurkevitch, E., Bar-Ness, E., Hadar, Y., 1994. Stability constants of pseudobactin complexes with transition metals. *Soil Science Society of America Journal*, 58: 390–396.

Choi, D.W., Semrau, J.D., Antholine, W.E., Hartsel, S.C., Anderson, R.C., Carey, J.N., Dreis, A.M. et al., 2008. Oxidase, superox-ide dismutase, and hydrogen peroxide reductase activities of methanobactin from types I and II methanotrophs. *Journal of Inorganic Biochemistry*, 102: 1571–1580.

Choi, D.W., Zea, C.J., Do, Y.S., Semrau, J.D., Antholine, W.E., Hargrove, M.S., Pohl et al., 2006. Spectral, kinetic, and thermodynamic properties of Cu (I) and Cu (II) binding by methanobactin from Methylosinus trichosporium OB3b. *Biochemistry*, 45: 1442–1453.

Chu, B.C., Garcia-Herrero, A., Johanson, T.H., Krewulak, K.D., Lau, C.K., Peacock, R.S., Slavinskaya, Z., Vogel, H.J., 2010. Siderophore uptake in bacteria and the battle for iron with the host; a bird's eye view. *Biometals*, 23: 601–611.

Clarke, T.E., Ku, S.Y., Dougan, D.R., Vogel, H.J., Tari, L.W., 2000. The structure of the ferric siderophore binding protein FhuD complexed with gallichrome. *Nature Structural & Molecular Biology*, 7: 287–291.

Coad, J., Pedley, K., 2014. Iron deficiency and iron deficiency anemia in women. *Scandinavian Journal of Clinical and Laboratory Investigation Supplement*, 244: 82–89; discussion 89.

Cornish, A. S, Page, W.J., 1998. The catecholate siderophores of *Azotobacter vinelandii*: Their affinity for iron and role in oxy-gen stress management. *Microbiology*, 144: 1747–1754.

Cornish, A.S., Page, W.J. 2000. Role of molybdate and other transi-tion metals in the accumulation of protochelin by Azotobacter vinelandii. *Applied and Environmental Microbiology*, 66: 1580–1586.

Cornu, J.Y., Elhabiri, M., Ferret, C., Geoffroy, V.A., Jezequel, K., Leva, Y., Lollier, M., Schalk, I.J., Lebeau, T., 2014. Contrasting effects of pyoverdine on the phytoextraction of Cu and Cd in a calcareous soil. *Chemosphere*, 103: 212–219.

Cortese, M.S., Paszczynski, A., Lewis, T.A., Sebat, J.L., Borek, V., Crawford, R.L., 2002. Metal chelating properties of pyridine-2,6-bis (thiocarboxylic acid) produced by *Pseudomonas* spp. and the biological activities of the formed complexes. *Biometals*, 15: 103–120.

Crichton, R.R., 2001. *Inorganic Biochemistry of Iron Metabolism: From Molecular Mechanism to Clinical Consequences*. New York: Wiley.

Crichton, R.R., Pierre, J.L., 2001. Old iron, young copper: From Mars to Venus. *Biometals*, 14: 99–112.

Crumbliss, A.L., Harrington, J.M., 2009. Iron sequestration by small molecules: Thermodynamic and kinetic studies of natu-ral siderophores and synthetic model compounds. *Advances in Inorganic Chemistry*, 61: 179–250.

Cunrath, O., Gasser, V., Hoegy, F., Reimmann, C., Guillon, L., Schalk, I.J., 2015. A cell biological view of the siderophore pyochelin iron uptake pathway in *Pseudomonas aeruginosa*. *Environmental Microbiology*, 17: 171–185.

Das, A.P., Ghosh, S., Mohanty, S., Sukla, L.B., 2014. Consequences of manganese compounds: A review. *Toxicological & Environmental Chemistry*, 96: 981–997.

Denic, S., Agarwal, M.M., 2007. Nutritional iron deficiency: An evolutionary perspective. *Nutrition*, 23: 603–614.

Denniss, I.S., Jeapes, A.P., 1996. Reprocessing irradiated fuel. In *The Nuclear Fuel Cycle: From Ore to Waste*. P.D. Wilson (eds.). Oxford, U.K.: Oxford Science Publications.

Desrosiers, D.C., Bearden, S.W., Mier, I., Abney, J., Paulley, J.T., Fetherston, J.D., Salazar, J.C., Radolf, J.D., Perry, R.D., 2010. Znu is the predominant zinc importer in *Yersinia pestis* during in vitro growth but is not essential for virulence. *Infection and Immunity*, 78: 5163–5177.

Dimkpa, C., Svatoš, A., Merten, D., Büchel, G., Kothe, E., 2008. Hydroxamate siderophores produced by *Streptomyces acidiscabies* E13 bind nickel and promote growth in cowpea (*Vigna unguiculata* L.) under nickel stress. *Canadian Journal of Microbiology*, 54: 163–172.

Dimkpa, C.O., Merten, D., Svatoš, A., Büchel, G., Kothe, E., 2009. Siderophores mediate reduced and increased uptake of cadmium by Streptomyces tendae F4 and sunflower (*Helianthus annuus*), respectively. *Journal of Applied Microbiology*, 107: 1687–1696.

Dimkpa, C.O., Svatoš, A., Dabrowska, P., Schmidt, A., Boland, W., Kothe, E., 2008. Involvement of siderophores in the reduction of metal-induced inhibition of auxin synthesis in *Streptomyces* spp. *Chemosphere*, 74: 19–25.

Ding, C., Yin, J., Tovar, E.M.M., Fitzpatrick, D.A., Higgins, D.G., Thiele, D.J., 2011. The copper regulon of the human fungal pathogen *Cryptococcus neoformans* H99. *Molecular Microbiology*, 81: 1560–1576.

DiSpirito, A.A., Zahn, J.A., Graham, D.W., Kim, H.J., Larive, C.K., Derrick, T.S., Cox, C.D., Taylor, A., 1998. Copper-binding compounds from Methylosinus trichosporium OB3b. *Journal of Bacteriology*, 180: 3606–3613.

Djoko, K.Y., Cheryl-lynn, Y.O., Walker, M.J., McEwan, A.G., 2015. The role of copper and zinc toxicity in innate immune defense against bacterial pathogens. *Journal of Biological Chemistry*, 290: 18954–18961.

Drago-Serrano, M.E., Parra, S.G., Manjarrez-Hernández, H.A., 2006. EspC, an autotransporter protein secreted by enteropathogenic *Escherichia coli* (EPEC), displays protease activity on human hemoglobin. *FEMS Microbiology Letters*, 265: 35–40.

Drewniak, L., Styczek, A., Majder-Lopatka, M., Sklodowska, A., 2008. Bacteria, hypertolerant to arsenic in the rocks of an ancient gold mine, and their potential role in dissemination of arsenic pollution. *Environmental Pollution*, 156: 1069–1074.

Duckworth, O.W., Bargar, J.R., Sposito, G., 2009. Coupled biogeochemical cycling of iron and manganese as mediated by microbial siderophores. *Biometals*, 22: 605–613.

Duckworth, O.W., Sposito, G., 2005. Siderophore-manganese(III) interactions II. Manganite dissolution promoted by desferrioxamine B. *Environmental Science & Technology*, 39: 6045–6051.

Duhme, A.K., Hider, R.C., Naldrett, M.J., Pau, R.N., 1998.The stability of the molybdenum-azotochelin complex and its effect on siderophore production in *Azotobacter vinelandii*. *Journal of Biological Inorganic Chemistry*, 3: 520–526.

El Ghazouani, A., Baslé, A., Firbank, S.J., Knapp, C.W., Gray, J., Graham, D.W., Dennison, C., 2011. Copper-binding properties and structures of Methanobactins from Methylosinus trichosporium OB3b. *Inorganic Chemistry*, 50: 1378–1391.

Enyedy, E.A., Pocsi, I., Farkas, E., 2004. Complexation and divalent of desferricoprogen with trivalent Fe, Al, Ga, InFe, Ni, Cu, Zn metal ions: Effects of the linking chain structure on the metal binding ability of hydroxamate based siderophores. *Journal of Inorganic Biochemistry*, 98: 1957–1966.

Farhana, A., Kumar, S., Rathore, S.S., Ghosh, P.C., Ehtesham, N.Z., Tyagi, A.K., Hasnain, S.E., 2008. Mechanistic insights into a novel exporter-importer system of mycobacterium tuberculosis unravel its role in trafficking of iron. *PLoS One*, 3: article e2087, 1–16.

Farkas, E., Csóka, H., Micera, G., Dessi, A., 1997. Copper (II), nickel (II), zinc (II), and molybdenum (VI) complexes of desferrioxamine B in aqueous solution. *Journal of Inorganic Biochemistry*, 65: 281–286.

Farkas, E., Enyedy, E.A., Csoka, H., 1999. A comparison between the chelating properties of some dihydroxamic acids, desferrioxamine B and acetohydroxamic acid. *Polyhedron*, 18: 2391–2398.

Farkas, E., Enyedy, E.A., Csoka, H., 2000. Some factors affecting metal ion-monohydroxamate interactions in aqueous solution. *Journal of Inorganic Biochemistry*, 79: 205–211.

Farkas, E., Enyedy, E.A., Fabian, I., 2003. New insight into the oxidation of Fe(II) by desferrioxamine B (DFB): Spectrophotometric and capillary electrophoresis (CE) study. *Inorganic Chemistry Communications*, 6: 131–134.

Farkas, E., Enyedy, É.A., Zékány, L., Deák, G., 2001. Interaction between iron (II) and hydroxamic acids: Oxidation of iron (II) to iron (III) by desferrioxamine B under anaerobic conditions. *Journal of Inorganic Biochemistry*, 83: 107–114.

Faulkner, K.M., Stevens, R.D., Fridovich, I., 1994. Characterization of Mn(III) complexes of linear and cyclic desferrioxamines as mimics of superoxide-dismutase activity. *Archives of Biochemistry and Biophysics*, 310: 341–346.

Ferguson, A.D., Hofmann, E., Coulton, J.W., Diederichs, K., Welte, W., 1998 Siderophore-mediated iron transport: Crystal structure of FhuA with bound lipopolysaccharide. *Science*, 282: 2215–2220.

Ferret, C., Cornu, J.Y., Elhabiri, M., Sterckeman, T., Braud, A., Jezequel, K., Lollier, M., Lebeau, T., Schalk, I.J., Geoffroy, V.A., 2015. Effect of pyoverdine supply on cadmium and nickel complexation and phytoavailability in hydroponics. *Environmental Science and Pollution Research*, 22: 2106–2116.

Francis, A.J., 1998. Biotransformation of uranium and other actinides in radioactive wastes. *Journal of Alloys and Compounds*, 271: 78–84.

Francis, M.S., Thomas, C.J., 1997. The *Listeria monocytogenes* gene ctpA encodes a putative P-type ATPase involved in copper transport. *Molecular & General Genetics*, 253: 484–491.

Frazier, S.W., Kretzschmar, R., Kraemer, S.M., 2005. Bacterial siderophores promote dissolution of UO2 under reducing conditions. *Environmental Science & Technology*, 39: 5709–5715.

Fuchs, R., Budzikiewicz, H., 2001. Structural studies of pyoverdins by mass spectrometry. *Current Organic Chemistry*, 5: 265–288.

Furrer, J.L., Sanders, D.N., Hook-Barnard, I.G., McIntosh, M.A., 2002. Export of the siderophore enterobactin in *Escherichia coli*: Involvement of a 43 kDa membrane exporter. *Molecular Microbiology*, 44: 1225–1234.

Gilis, A., Corbisier, P., Baeyens, W., Taghavi, S., Mergeay, M., Van Der Lelie, D., 1998. Effect of the siderophore alcaligin E on the bioavailability of Cd to *Alcaligenes eutrophus* CH34. *Journal of Industrial Microbiology and Biotechnology*, 20: 61–68.

Gill, M., 2014. Heavy metal stress in plants: A review. *International Journal of Advanced Research*, 2: 1043–1055.

Graham, D.W., Kim, H.J., 2011. Production, isolation, purification, and functional characterization of methanobactins. *Methods in Enzymology: Methods in Methane Metabolism*, 495: 227–245.

Grass, G., Thakali, K., Klebba, P.E., Thieme, D., Müller, A., Wildner, G.F., Rensing, C., 2004. Linkage between catecholate siderophores and the multicopper oxidase CueO in *Escherichia coli*. *Journal of Bacteriology*, 186(17): 5826–5833.

Grigg, J.C., Cooper, J.D., Cheung, J., Heinrichs, D.E., Murphy, M.E., 2010. The *Staphylococcus aureus* siderophore receptor HtsA undergoes localized conformational changes to enclose Staphyloferrin A in an arginine-rich binding pocket. *Journal of Biological Chemistry*, 285: 11162–11171.

Gumienna-Kontecka, E., Pyrkosz-Bulska, M., Szebesczyk, A., Ostrowska, M., 2014. Iron chelating strategies in systemic metal overload, neurodegeneration and cancer. *Current Medicinal Chemistry*, 21: 3741–3767.

Hantke, K., 2005. Bacterial zinc uptake and regulators. *Current Opinion in Microbiology*, 8: 196–202.

Harrington, J.M., Parker, D.L., Bargar, J.R., Jarzecki, A.A., Tebo, B.M., Sposito, G., Duckworth, O.W., 2012. Structural dependence of Mn complexation by siderophores: Donor group dependence on complex stability and reactivity. *Geochimica Et Cosmochimica Acta*, 88: 106–119.

Hernandez-Montes, G., Argueello, J.M., Valderrama, B., 2012. Evolution and diversity of periplasmic proteins involved in copper homeostasis in gamma proteobacteria. *BMC Microbiology*, 12: 249.

Hesketh, A., Kock, H., Mootien, S., Bibb, M., 2009. The role of absC, a novel regulatory gene for secondary metabolism, in zinc-dependent antibiotic production in *Streptomyces coelicolor* A3(2). *Molecular Microbiology*, 74: 1427–1444.

Hider, R.C., Kong, X., 2010. Chemistry and biology of siderophores. *Natural Product Reports*, 27: 637–657.

Higgins, C.F., Linton, K.J., 2001. Structural biology—The xyz of ABC transporters. *Science*, 293: 1782–1784.

Höfte, M., 1992, *Classes of Microbial Siderophores*. New York: Academic Press.

Holdway, D.A., 1992, Uranium toxicity to 2 species of australian tropical fish. *Science of the Total Environment*, 125: 137–158.

Hollenstein, K., Frei, D.C., Locher, K.P., 2007, Structure of an ABC transporter in complex with its binding protein. *Nature*, 446: 213–216.

Hood, M.I., Skaar, E.P., 2012. Nutritional immunity: Transition metals at the pathogen-host interface. *Nature Reviews Microbiology*, 10: 525–537.

Hou, Z., Sunderland, C.J., Nishio, T., Raymond, K.N., 1996, Preorganization of ferric alcaligin, Fe(2)L(3). The first structure of a ferric dihydroxamate siderophore. *Journal of the American Chemical Society*, 118: 5148–5149.

Howard, D.H., 1999. Acquisition, transport, and storage of iron by pathogenic fungi. *Clinical Microbiology Reviews*, 12: 394–404.

Hussien, S.S., Desouky, O.A., Abdel-Haliem, M.E., El-Mougith, A.A., 2013. Uranium (VI) complexation with siderophores-pyoverdine produced by *Pseudomonas fluorescens* SHA 281. *International Journal of Nuclear Energy Science and Engineering*, 3: 95–102.

Hvorup, R.N., Goetz, B.A., Niederer, M., Hollenstein, K., Perozo, E., Locher, K.P., 2007. Asymmetry in the structure of the ABC transporter-binding protein complex BtuCD-BtuF. *Science*, 317: 1387–1390.

International Atomic Energy Agency. Web Site IAEA Power Reactor Information System. Retrieved January 16, 2016, from https://www.iaea.org.

Jarvis, N.V., Hancock, R.D., 1991. Some correlations involving the stability of complexes of transuranium metal-ions and ligands with negatively charged oxygen donors. *Inorganica Chimica Acta*, 182: 229–232.

John, S.G., Ruggiero, C.E., Hersman, L.E., Tung, C.S., Neu, M.P., 2001. Siderophore mediated plutonium accumulation by Microbacterium flavescens (JG-9). *Environmental Science & Technology*, 35: 2942–2948.

Johnson, K.S., 2006. Manganese redox chemistry revisited. *Science*, 313(5795): 1896–1897.

Kallifidas, D., Pascoe, B., Owen, G.A., Strain-Damerell, C.M., Hong, H.J., Paget, M.S., 2010. The zinc-responsive regulator Zur controls expression of the Coelibactin gene cluster in *Streptomyces coelicolor*. *Journal of Bacteriology*, 192: 608–611.

Kehl-Fie, T.E., Skaar, E.P., 2010. Nutritional immunity beyond iron: A role for manganese and zinc. *Current Opinion in Chemical Biology*, 14: 218–224.

Kenney, G.E., Rosenzweig, A.C., 2012. Chemistry and biology of the copper chelator methanobactin. *ACS Chemical Biology*, 7: A-I.

Klein, J.S., Lewinson, O., 2011. Bacterial ATP-driven transporters of transition metals: Physiological roles, mechanisms of action, and roles in bacterial virulence. *Metallomics*, 3: 1098–1108.

Koh, E.-I., Henderson, J.P., 2015. Microbial copper-binding siderophores at the host-pathogen interface. *Journal of Biological Chemistry*, 290: 18967–18974.

Kreutzer, M.F., Kage, H., Gebhardt, P., Wackler, B., Saluz, H.P., Hoffmeister, D., Nett, M., 2011. Biosynthesis of a complex yersiniabactin-like natural product via the mic Locus in Phytopathogen *Ralstonia solanacearum*. *Applied and Environmental Microbiology*, 77: 6117–6124.

Küpper, F.C., Carrano, C.J., Kuhn, J.U., Butler, A., 2006. Photoreactivity of iron(III)—Aerobactin: Photoproduct structure and iron(III) coordination. *Inorganic Chemistry*, 45: 6028–6033.

Langemann, K., Heineke, D., Rupprecht, S., Raymond, K.N., 1996 Nordesferriferrithiocin. Comparative coordination chemistry of a prospective therapeutic iron chelating agent. *Inorganic Chemistry*, 35: 5663–5673.

Leach, L.H., Morris, J.C., Lewis, T.A., 2007. The role of the siderophore pyridine-2,6-bis (thiocarboxylic acid) (PDTC) in zinc utilization by *Pseudomonas putida* DSM 3601. *Biometals*, 20: 717–726.

Lemire, J.A., Harrison, J.J., Turner, R.J., 2013. Antimicrobial activity of metals: Mechanisms, molecular targets and applications. *Nature Reviews Microbiology*, 11: 371–384.

Lewinson, O., Lee, A.T., Rees, D.C., 2009. A P-type ATPase importer that discriminates between essential and toxic transition metals. *Proceedings of the National Academy of Sciences of the United States of America*, 106: 4677–4682.

Liermann, L.J., Guynn, R.L., Anbar, A., Brantley, S.L., 2005 Production of a molybdophore during metal-targeted dissolution of silicates by soil bacteria. *Chemical Geology*, 220: 285–302.

Locher, K.P., 2009. Structure and mechanism of ATP-binding cassette transporters. *Philosophical Transactions of the Royal Society B: Biological Sciences*, 364: 239–245.

Loomis, L.D., Raymond, K.N., 1991. Solution equilibria of enterobactin and metal enterobactin complexes. *Inorganic Chemistry*, 30: 906–911.

López-Millán, A.F., Grusak, M.A., Abadía Bayona, A., Abadía Bayona, J., 2013. Iron deficiency in plants: An insight from proteomic approaches. *Frontiers in Plant Science*, 4: 254.

Ma, Z., Jacobsen, F.E., Giedroc, D.P., 2009. Coordination chemistry of bacterial metal transport and sensing. *Chemical Reviews*, 109: 4644–4681.

Mademidis, A., Killmann, H., Kraas, W., Flechsler, I., Jung, G., Braun, V., 1997. ATP-dependent ferric hydroxamate transport system in *Escherichia coli*: Periplasmic FhuD interacts with a periplasmic and with a transmembrane cytoplasmic region of the integral membrane protein FhuB, as revealed by competitive peptide mapping. *Molecular Microbiology*, 26: 1109–1123.

Mademidis, A., Koster, W., 1998. Transport activity of FhuA, FhuC, FhuD, and FhuB derivatives in a system free of polar effects, and stoichiometry of components involved in ferrichrome uptake. *Molecular and General Genetics*, 258: 156–165.

McDougall, G.J., Hancock, R.D., Boeyens, J.C.A., 1978. Empirical force-field calculations of strain-energy contributions to thermodynamics of complex-formation. 1. Difference in stability between complexes containing 5-membered and 6-membered chelate rings. *Journal of the Chemical Society, Dalton Transactions*, 1438–1444.

Messenger, A.J.M., Barclay, R., 1983. Bacteria, iron and pathogenicity. *Biochemical Education*, 11: 54–64.

Miller, M.C., Parkin, S., Fetherston, J.D., Perry, R.D., DeMoll, E., 2006. Crystal structure of ferric-yersiniabactin, a virulence factor of *Yersinia pestis*. *Journal of Inorganic Biochemistry* 100: 1495–1500.

Mullen, L., Gong, C., Czerwinski, K., 2007. Complexation of uranium(VI) with the siderophore desferrioxamine B. *Journal of Radioanalytical and Nuclear Chemistry*, 273: 683–688.

Naegeli, H.U., Zahner, H., 1980. Metabolites of microorganisms.193. *Ferrithiocin. Helvetica Chimica Acta*, 63: 1400–1406.

Naik, M.M., Dubey, S.K., 2011. Lead-enhanced siderophore production and alteration in cell morphology in a Pb-resistant *Pseudomonas aeruginosa* strain 4EA. *Current Microbiology*, 62: 409–414.

Nair, A., Juwarkar, A.A., Singh, S.K., 2007. Production and characterization of siderophores and its application in arsenic removal from contaminated soil. *Water Air and Soil Pollution*, 180: 199–212.

Nakai, H., Kobayashi, S., Ozaki, M., Hayase, Y., Takeda, R., 1999. Micacocidin A. *Acta Crystallographica Section C: Crystal Structure Communications*, 55: 54–56.

Nealson, K.H., Tebo, B.M., Rosson, R.A., 1988. Occurrence and mechanisms of microbial oxidation of manganese. *Advances in Applied Microbiology*, 33: 279–318.

Neilands, J.B., 1995. Siderophores—Structure and function of microbial iron transport compounds. *Journal of Biological Chemistry*, 270: 26723–26726.

Neu, M.P., Matonic, J.H., Ruggiero, C.E., Scott, B.L., 2000. Structural characterization of a plutonium(IV) siderophore complex: Single-crystal structure of Pu-desferrioxamine E. *Angewandte Chemie-International Edition*, 39: 1442–1444.

Neubauer, U., Nowack, B., Furrer, G., Schulin, R., 2000, Heavy metal sorption on clay minerals affected by the siderophore desferrioxamine B. *Environmental Science & Technology*, 34: 2749–2755.

Nies, D.H., 1999. Microbial heavy-metal resistance. *Applied Microbiology and Biotechnology*, 51: 730–750.

Nies, D.H., 2003. Efflux-mediated heavy metal resistance in prokaryotes. *Fems Microbiology Reviews*, 27: 313–339.

Palanché, T., Blanc, S., Hennard, C., Abdallah, M.A, Albrecht-Gary, A.M., 2004, Bacterial iron transport: Coordination properties of azotobactin, the highly fluorescent siderophore of *Azotobacter vinelandii*. *Inorganic Chemistry*, 43: 1137–1152.

Parker, D.L., Lee, S.W., Geszvain, K., Davis, R.E., Gruffaz, C., Meyer, J.M., Torpey, J.W., Tebo, B.M., 2014, Pyoverdine synthesis by the Mn (II)-oxidizing bacterium Pseudomonas putida GB-1. *Frontiers in Microbiology*, 5: 10–3389.

Parker, D.L., Sposito, G., Tebo, B.M. 2004, Manganese(III) binding to a pyoverdine siderophore produced by a manganese(II)-oxidizing bacterium. *Geochimica et Cosmochimica Acta*, 68: 4809–4820.

Perry, R.D., Balbo, P.B., Jones, H.A., Fetherston, J.D., DeMoll, E., 1999, Yersiniabactin from *Yersinia pestis*: Biochemical characterization of the siderophore and its role in iron transport and regulation. *Microbiology*, 145: 1181–1190.

Philpott, C.C., Leidgens, S., Frey, A.G. 2012, Metabolic remodeling in iron-deficient fungi. *Biochimica Et Biophysica Acta-Molecular Cell Research*, 1823: 1509–1520.

Pierre, J.L., Fontecave, M. 1999, Iron and activated oxygen species in biology: The basic chemistry. Biometals, 12: 195–199.

Pierre, J.L., Fontecave, M., Crichton, R.R. 2002, Chemistry for an essential biological process: The reduction of ferric iron. *Biometals*, 15: 341–346.

Rajkumar, M., Ae, N., Prasad, M.N.V., Freitas, H., 2010, Potential of siderophore-producing bacteria for improving heavy metal phytoextraction. *Trends in Biotechnology*, 28: 142–149.

Raymond, K.N., Telford, J.R. 1995, Siderophore-mediated iron transport in microbes. In *Bioinorganic Chemistry: An Inorganic Perspective of Life*. D.P. Kessissoglou (eds.), pp. 25–37. Kluwer Academic Publisher, Dordrecht, Netherlands.

Riley, R.G., Zachara, J.M. 1993, Chemical contaminants on DOE lands and selection of contaminant mixtures for subsurface science research, U.S. Department of Energy, Office of Energy Research. Washington, USA.

Rivault, F., Schons, V., Liébert, C., Burger, A., Sakr, E., Abdallah, M.A., Schalk, I.J., Mislin, G.L., 2006. Synthesis of functionalized analogs of pyochelin, a siderophore of *Pseudomonas aeruginosa* and *Burkholderia cepacia*. *Tetrahedron*, 62: 2247–2254.

Rodriguez, G.M., Smith, I., 2006. Identification of an ABC transporter required for iron acquisition and virulence in *Mycobacterium tuberculosis*. *Journal of Bacteriology*, 188: 424–430.

Romheld, V., 1987. Different strategies for iron acquisition in higher-plants. *Physiologia Plantarum*, 70: 231–234.

Saha, R., Saha, N., Donofrio, R.S., Bestervelt, L.L., 2013. Microbial siderophores: A mini review. *Journal of Basic Microbiology*, 53: 303–317.

Scarrow, R.C., Riley, P.E., Abu-Dari, K., White, D.L., Raymond, K.N., 1985, Ferric ion sequestering agents.13. Synthesis, structures, and thermodynamics of complexation of cobalt(iii) and iron(iii) tris complexes of several chelating hydroxypyridinones. *Inorganic Chemistry*, 24: 954–967.

Schalk, I.J., Guillon, L., 2013. Fate of ferrisiderophores after import across bacterial outer membranes: Different iron release strategies are observed in the cytoplasm or periplasm depending on the siderophore pathways. *Amino Acids*, 44: 1267–1277.

Schalk, I.J., Hannauer, M., Braud, A., 2011. New roles for bacterial siderophores in metal transport and tolerance. *Environmental Microbiology*, 13: 2844–2854.

Schmidt, H., Günther, C., Weber, M., Spörlein, C., Loscher, S., Böttcher, C., Schobert, R., Clemens, S., 2014. Metabolome analysis of *Arabidopsis thaliana* roots identifies a key metabolic pathway for iron acquisition. *PLoS ONE*, 9: 102444.

Shakoury-Elizeh, M., Protchenko, O., Berger, A., Cox, J., Gable, K., Dunn, T.M., Prinz, W.A., Bard, M., Philpott, C.C., 2010. Metabolic response to iron deficiency in *Saccharomyces cerevisiae. The Journal of Biological Chemistry*, 285: 14823–14833.

Sheppard, S.C., Evenden, W.G., Anderson, A.J., 1992. Multiple assays of uranium toxicity in soil. *Environmental Toxicology and Water Quality*, 7: 275–294.

Sinha, S., Mukherjee, S.K., 2008. Cadmium-induced siderophore production by a high cd-resistant bacterial strain relieved cd toxicity in plants through root colonization. *Current Microbiology*, 56: 55–60.

Solti, Á., Gáspár, L., Mészáros, I., Szigeti, Z., Lévai, L., Sárvári, É., 2008. Impact of iron supply on the kinetics of recovery of photosynthesis in Cd-stressed poplar (*Populus glauca*). *Annals of Botany*, 102: 771–782.

Spasojevic, I., Boukhalfa, H., Stevens, R.D., Crumbliss, A.L., 2001. Aqueous solution speciation of Fe(III) complexes with dihydroxamate siderophores alcaligin and rhodotorulic acid and synthetic analogues using electrospray ionization mass spectrometry. *Inorganic Chemistry*, 40: 49–58.

Springer, S.D., Butler, A., 2016. Microbial ligand coordination: Consideration of biological significance. *Coordination Chemistry Reviews*, 306: 628–635.

Taylor, R.J., Mason, C., Cooke, R., Boxall, C., 2002. The reduction of Pu(IV) by formohydroxamic acid in nitric acid. *Journal of Nuclear Science and Technology Supplement*, 3: 278–281.

Taylor, R.J., May, I., 1999. The reduction of actinide ions by hydroxamic acids. *Czechoslovak Journal of Physics*, 49: 617–621.

Taylor, R.J., May, I., Wallwork, A.L., Denniss, I.S., Hill, N.J., Galkin, B.Y., Zilberman, B.Y., Fedorov, Y.S., 1998a. The applications of formo- and aceto-hydroxamic acids in nuclear fuel reprocessing. *Journal of Alloys and Compounds*, 271: 534–537.

Teitzel, G.M., Geddie, A., De Long, S.K., et al. 2006. Survival and growth in the presence of elevated copper: Transcriptional profiling of copper-stressed Pseudomonas aeruginosa. *Journal of Bacteriology* 188(20): 7242–7256.

Téllez, C.M., Gaus, K.P., Graham, D.W., Arnold, R.G., Guzman, R.Z., 1998. Isolation of copper biochelates from Methylosinus trichosporium OB3b and soluble methane monooxygenase mutants. *Applied and Environmental Microbiology*, 64: 1115–1122.

Trouwborst, R.E., Clement, B.G., Tebo, B.M., Glazer, B.T., Luther, G.W., 2006. Soluble Mn (III) in suboxic zones. *Science*, 313: 1955–1957.

Wagner, D., Maser, J., Lai, B., Cai, Z., Barry, C.E., zu Bentrup, K.H., Russell, D.G., Bermudez, L.E., 2005. Elemental analysis of *Mycobacterium avium*-, *Mycobacterium tuberculosis*-, and *Mycobacterium smegmatis*-containing phagosomes indicates pathogen-induced microenvironments within the host cell's endosomal system. *Journal of Immunology*, 174: 1491–1500.

Wang, Q., Xiong, D., Zhao, P., Yu, X., Tu, B., Wang, G., 2011. Effect of applying an arsenic-resistant and plant growth-promoting rhizobacterium to enhance soil arsenic phytoremediation by *Populus deltoides* LH05-17. *Journal of Applied Microbiology*, 111: 1065–1074.

Wasielewski, E., Atkinson, R.A., Abdallah, M.A., Kieffer, B., 2002. The three-dimensional structure of the gallium complex of azoverdin, a siderophore of *Azomonas macrocytogenes* ATCC 12334, determined by NMR using residual dipolar coupling constants. *Biochemistry*, 41: 12488–12497.

White, C., Lee, J., Kambe, T., Fritsche, K., Petris, M.J., 2009. A role for the ATP7A copper-transporting ATPase in macrophage bactericidal activity. *Journal of Biological Chemistry*, 284: 33949–33956.

Wichard, T., Bellenger, J.P., Morel, F.M., Kraepiel, A.M., 2009. Role of the siderophore azotobactin in the bacterial acquisition of nitrogenase metal cofactors. *Environmental Science & Technology*, 43: 7218–7224.

Winkelmann, G., 2002. Microbial siderophore-mediated transport. *Biochemical Society Transactions*, 30: 691–696.

Wong, G.B., Kappel, M.J., Raymond, K.N., Matzanke, B., Winkelmann, G., 1983. Coordination chemistry of microbial iron transport compounds.24. Characterization of coprogen and ferricrocin, 2 ferric hydroxamate siderophores. *Journal of the American Chemical Society*, 105: 810–815.

World Nuclear Association, n.d. Radioactive waste management. Retrieved October 7, 2016 from http://www.world-nuclear.org/.

Yeterian, E., Martin, L.W., Lamont, I.L., Schalk, I.J., 2010. An efflux pump is required for siderophore recycling by *Pseudomonas aeruginosa*. *Environmental Microbiology Reports*, 2: 412–418.

Yoneyama, F., Yamamoto, M., Hashimoto, W., Murata, K., 2011. *Azotobacter vinelandii* gene clusters for two types of peptidic and catechol siderophores produced in response to molybdenum. *Journal of Applied Microbiology*, 111: 932–938.

Zhao, B., Moody, S.C., Hider, R.C., Lei, L., Kelly, S.L., Waterman, M.R., Lamb, D.C., 2012. Structural analysis of cytochrome P450 105N1 involved in the biosynthesis of the zincophore, coelibactin. *International Journal of Molecular Sciences*, 13: 8500–8513.

16 Microbial Biosorption and Improved/Genetically Modified Biosorbents for Toxic Metal Removal and Thermodynamics

Shubhalakshmi Sengupta, Papita Das, Aniruddha Mukhopadhyay, and Siddhartha Datta

CONTENTS

ABSTRACT

The world in this century is witnessing industrialization along with urbanization at an ever-increasing rate. This has resulted in the generation of large amounts of aqueous effluents with high levels of toxic pollutants, such as heavy metals and ionic dyes. Research on the use of biosorbents started in the late 1980s, and throughout the 1990s, major research in this field gained momentum (Wase and Wase, 2002). The use of biosorbents is an emerging field of research in environmental or bioresource technology processes. The uses of bioadsorbents have major advantages over the conventional processes that include high efficiency, low cost, minimal use of biological or chemical sludges, possibility of metal recovery after absorption, and regeneration of the bioadsorbent. A wide variety of microbial biomasses, both living and dead, have been investigated for biosorption properties. The microorganisms or the virgin biosorbents used usually lack specificity in metal binding that may cause difficulties in metal recovery and recycling. Genetic modification does provide a solution to this problem and results in enhanced properties in the biosorbents. These biosorption studies have been characterized through a variety of models.

16.1 INTRODUCTION

The world in this century has witnessed rapid industrialization along with urbanization at an ever-increasing rate. This has resulted in the generation of large amounts of aqueous effluents with high levels of toxic pollutants, such as heavy metals and ionic dyes. So researchers are engaged in developing novel technologies for the effective remediation of the industrial wastewaters before they are released into various water bodies. Heavy metals when released into the environment pose serious environmental hazards (Park et al., 2010). These heavy metals comprise elements having atomic weights between 63.5 and 200.6 and a specific gravity greater than 5.0. Toxic heavy metals that are a concern for industrial wastewater treatment consist of mainly zinc, copper, nickel, mercury, cadmium, lead, and chromium (Fu and Wang, 2011). In the treatment of these heavy metals, conventional treatment processes like chemical precipitation and coagulation are used, but their effectiveness is reduced when the adsorbates are in low concentration and subsequently become more expensive. Ion-exchange resins and activated carbons (ACs) are also used for treating adsorptive industrial pollutants but have limited economic viability (Park et al., 2010). Membrane filtration and electrochemical treatment technologies are also some of the methods used (Fu and Wang, 2011). Research on the use of biosorbents started in the late 1980s, and throughout the 1990s, major research in this field gained momentum (Wase and Wase, 2002). The use of biosorbents is an emerging field of research in environmental or bioresource technology processes. The uses of bioabsorbents have major advantages over the conventional processes that include high efficiency, low cost, minimal use of biological or chemical sludges, possibility of metal recovery after absorption, and regeneration of the bioabsorbents (Park et al., 2010). In general, biosorption may be simply defined as the "removal of substances from solution by biological material." However, researches are mainly concentrated on the removal of heavy metals and metalloids from wastewater by biological organisms or their biomass. A wide variety of microbial biomasses both living and dead have been investigated for biosorption properties. The studies included mixed organism/biomass systems also. The organisms used in these studies include bacteria, cyanobacteria, algae (including macroalgae), and fungi (filamentous forms, yeasts, fungal fruiting bodies, and lichens) (Gadd, 2008). The microorganisms or the virgin biosorbents used usually lack specificity in metal binding that may cause difficulties in metal recovery and recycling. Genetic modification does provide a solution to this problem and results in enhanced properties in the biosorbents (Vijayaraghavan and Yun, 2008). These biosorption studies have been characterized through a variety of models. These range from simple single-component models like Langmuir and Freundlich to complex multicomponents derived from Langmuir/Freundlich models. These interpretations have been used in comparing different metal–biosorbent systems (Gadd, 2008). This chapter would deal with various genetically modified microbial biosorbent systems used in the removal of toxic metals from wastewater and the thermodynamics adhered to in these systems.

16.2 HEAVY METALS

Industrial developments, especially in mining, fertilizer, tannery, batteries, paper, and pesticide sectors, have led to the discharge of heavy metals in the environment, mainly in the developing countries. These metals are not biodegradable like organic pollutants, and living organisms tend to accumulate the metal ions that are either toxic or carcinogenic. In Table 16.1, some of the toxic heavy metals present in industrial wastewater and their effects on human beings when consumed in excess amount are listed.

16.2.1 TREATMENT METHODS FOR REMOVAL OF HEAVY METALS

There are various conventional and nonconventional methods in use for the removal of toxic heavy metal from the industrial wastewaters. In Table 16.2, a brief description of all the methods used toward this purpose is given.

16.3 BIOSORPTION AND BIOSORBENTS

As a process, biosorption is difficult to define because many mechanisms contribute to the overall process specifically depending on the substance to be sorbed, the biosorbent used, environmental parameters, and the metabolic processes in the case of living organisms. Sorption is a term that is used for both absorption and adsorption. Absorption implies incorporation of a substance from one state into another of a different state, whereas adsorption is the physical adherence or bonding of ions and molecules onto the surface of another molecule. Most solids, including microorganisms, have functional groups like –SH, –OH, and –COOH on their surfaces (Vijayaraghavan and Yun, 2008). These help in the adsorption of metal cations and results in the formation of a competitive complex. Thus, the term "biosorption can describe any system where a sorbate (e.g., an atom, molecule, a molecular ion) interacts with a biosorbent (i.e., a solid surface of a biological matrix) resulting in an accumulation at the sorbate–biosorbent interface, and therefore a reduction in the solution sorbate concentration" (Gadd, 2008).

TABLE 16.1

Heavy Metals and Their Effects

Metals	Effects
Zinc	Stomach cramps, skin irritations, vomiting, nausea, anemia
Copper	Vomiting, cramps, convulsions, death
Nickel	Lung disorder, kidney problems, gastrointestinal distress, pulmonary fibrosis, skin dermatitis, carcinogenic effect
Mercury	Central nervous system damage, impairment of pulmonary and kidney function, chest pain, and dyspnea
Lead	Anemia, insomnia, headache, dizziness, irritability, weakness of muscle, hallucination, renal damage, central nervous system damage, kidney, liver and reproductive system damage, impairment of brain functions
Chromium(VI)	Lung carcinoma

Sources: Oyaro, N. et al., *Int. J. Food Agric. Environ.*, 5, 119, 2007; Paulino, A.T. et al., *J. Colloid Interface Sci.*, 301, 479, 2006; Borba, C.E. et al., *Biochem. Eng. J.*, 30, 184, 2006; Namasivayam, C. and Kadirvelu, K., *Carbon*, 37, 79, 1999; Naseem, R. and Tahir, S.S., *Water Res.*, 35, 3982, 2001; Fu, F. and Wang, Q., *J. Environ. Manage.*, 92, 407, 2011.

TABLE 16.2

Treatment Methods for the Removal of Toxic Metals

Methods	Description
Chemical precipitation	It is the most widely used industrial process. In this process, chemical reacts with heavy metal ions to produce insoluble precipitators. The precipitates are removed by sedimentation or filtration. The precipitation methods most widely used are hydroxide precipitation and sulfide precipitation.
Chemical precipitation combined with other methods	In this approach, another method is employed along with chemical precipitation. For example, sulfide precipitation was followed by nanofiltration to reuse and recover heavy metal ions. It was reported that sulfide precipitation reduced the metal content and nanofiltration yielded solutions that could be directly reused in the plant.
Heavy metal chelating precipitation	This method is used as an alternative to chemical precipitation with many companies using chelating precipitants to precipitate heavy metals from aqueous systems. For example, 1,3-benzenediamidoethanethiol ($BDET^{2-}$) dianion can effectively precipitate mercury in the leachate solution and also other heavy metals from acid mine drainage.
Ion exchange	These processes have been widely used to remove heavy metals from wastewater due to their many advantages, such as high treatment capacity, high removal efficiency, and fast kinetics. In this process, resins are used having the specific ability to exchange its cations with the metals in the wastewater. Among resins, synthetic resins are commonly used as they are effective in removing the heavy metals from the solution.
Membrane filtration	Membrane filtration technologies are also of different types of membranes. These processes show great promise for heavy metal removal for their high efficiency, easy operation, and space saving. The different types of membrane processes used to remove metals from the wastewater are ultrafiltration, reverse osmosis, nanofiltration, and electrodialysis.
Adsorption	Nowadays, adsorption is recognized as an effective method for heavy metal wastewater treatment and is also economically viable. The adsorption process offers flexibility in design and operation and in many cases produce high-quality treated effluents. AC adsorbents are used. Carbon nanotubes (CNTs) are also widely studied for their excellent properties and applications. This new adsorbent (CNT) has been proven to have great potential for removing heavy metal ions such as lead, cadmium, chromium, nickel, and copper from wastewater.
Biosorption	Biosorption of heavy metals from aqueous solutions is a novel technology that is studied extensively by researchers for the removal of metals from wastewater. The major advantages of biosorption are its high effectiveness in reducing the heavy metal ions and the use of inexpensive biosorbents. These processes are particularly suitable to treat dilute heavy metal wastewater. Biosorbents can be derived from three sources: (1) nonliving biomass such as bark, lignin, shrimp, krill, squid, and crab shell, (2) algal biomass, and (3) microbial biomass, for example, bacteria, fungi, and yeast. The microbial bioabsorbents could be genetically modified for achieving effective results.

Source: Fu, F. and Wang, Q., *J. Environ. Manage.*, 92, 407, 2011.

16.3.1 MICROBIAL BIOSORBENTS

Accumulations of metallic elements by microorganisms were first reported in the 1980s. Volesky (1987) reported the accumulation of heavy metals due to metabolic activity of the cells. However, further research resulted in the revelation that dead microbial biomass can passively bind metals through physicochemical mechanisms. Gradually, research on biosorption of metals by biosorbents increased, and it was found that biosorption depends not only on the type of composition of the biomass but also on solution chemistry and physicochemical factors. The various mechanisms involved include one or more mechanisms of ion exchange, complexation, coordination, adsorption, electrostatic interaction, chelation, and microprecipitation. Potent metal biosorbents under the class of bacteria include the genera of *Bacillus*, *Pseudomonas*, *Streptomyces*, *Aspergillus*, and *Penicillium*. These microorganisms are used widely in pharmaceutical and food industry and are therefore attained free or at low cost from the industries (Vijayaraghavan and Yun, 2008). The biosorptive processes using nonliving organisms are more applicable than using live microorganisms that require nutrient supply and complicated bioreactor system along with the maintenance of a healthy microbial population that becomes difficult as a result of toxicity of the pollutants and other unsuitable environmental conditions like temperature and pH of the solution being treated.

Thus, using nonliving organisms renders greater feasibility to large-scale operations (Vijayaraghavan and Yun, 2008). The recovery of the valuable metals is limited to living cells since these may be bound intracellularly. For these reasons, researches have been focused on the use of nonliving biomass as biosorbents (Volesky, 1990). Advantages and disadvantages of using nonliving microbial biomass are listed in Table 16.3.

The challenge faced by researchers is to select the most promising type of biomasses from a large pool of readily available and inexpensive biomaterials (Kratoshvil and Volesky, 1998). In choosing a biomass for large-scale industrial uses, its availability and economic viability are considered first (Vieira and Volesky, 2000; Volesky, 1994; Volesky and Holan, 1995). A broad range of biomass types have been tested for biosorptive processes under various conditions but there is a need to explore novel biomass types having high efficiency and low cost. Table 16.4 summarizes the different types biomasses used as biosorbents.

TABLE 16.3

Advantages and Disadvantages of Using Nonliving Microbial Biomass

Advantages	Disadvantages
1. It is growth independent; the nonliving biomass is not subject to toxicity limitation of cells. Costly nutrients are not required for the growth of cells in feed solutions. Therefore, problems regarding the disposal of surplus nutrients or metabolic products do not arise.	1. Early saturation of the biomass could be a problem, that is when metal interactive sites are occupied, metal desorption is required prior to further use, irrespective of the metal value.
2. The biomass can be procured from the existing fermentation industries, which is essentially a waste after fermentation.	2. The potential for biological process improvement (e.g., through genetic engineering of cells) is often limited because cells are not metabolizing as production of the adsorptive agent occurs during pregrowth. As a result, there is no biological control over characteristics of biosorbent. This becomes true particularly in the case of waste biomass from fermentation units being utilized.
3. This process is not dependent on the physiological constraints prevalent among living microbial cells.	
4. The nonliving biomass acts as an ion exchanger; thus, the process is very rapid and takes less time.	3. There is no potential for biologically altering the metal valency state like those of less soluble forms or even for the degradation of organometallic complexes.
5. Metal loading on biomass is very high, which leads to a very efficient metal uptake.	
6. The cells are nonliving and, therefore, processing conditions are not governed by conditions conducive to the growth of cells. Thus, a wider range of conditions like pH, temperature, and metal concentration are possible.	
7. Aseptic conditions are not required for this process.	
8. Metal can be desorbed readily and then recovered if they are valuable.	
9. The biomass is plentiful, and therefore, the metal-loaded biomass could be incinerated to eliminate further treatments.	

Source: Ahluwalia, S.S. and Goyal, D., *Bioresour. Technol.*, 98, 2243, 2007.

TABLE 16.4

Types of Native Biomass Used as Biosorbents

Category	Examples
Bacteria	Gram-positive bacteria (*Bacillus* sp., *Corynebacterium* sp., etc.), Gram-negative bacteria (*Escherichia* sp., *Pseudomonas* sp., etc.), Cyanobacteria (*Anabaena*, *Synechocystis* sp., etc.)
Fungi	Molds (*Aspergillus* sp., *Rhizopus* sp., etc.), mushrooms (*Agaricus* sp., *Trichaptum* sp., etc.), and yeast (*Saccharomyces* sp., *Candida* sp., etc.)
Algae	Microalgae (*Chlorella* sp., *Chlamydomonas* sp., etc.), macroalgae (green seaweed (*Enteromorpha* sp., *Codium* sp., etc.), brown seaweed (*Sargassum* sp., *Ecklonia* sp., etc.), red seaweed (*Gelidium* sp., *Porphyra* sp., etc.)
Industrial wastes	Fermentation wastes, food beverage wastes, activated sludges, anaerobic sludges, etc.
Agricultural wastes	Fruit and vegetable wastes, rice straws, wheat bran, soybean hull, etc.
Natural residues	Plant residues, sawdust, tree barks, weeds, etc.
Others	Chitosan-driven materials, cellulose-driven materials, etc.

Source: Reproduced from Park, D. et al., *Biotechnol. Bioprocess Eng.*, 15, 86, 2010. With permission.

The biosorbent's capability to remove metal pollutants could be derived through simple chemical and physical methods that yield biosorbent particles. The development of biosorbents from biomasses is given in Figure 16.1.

Some biomass particles have been either immobilized in a synthetic polymer matrix or grafted onto an inorganic support like silica that yields required mechanical properties (Kratochvil and Volesky, 1998). Newer biosorbents could be manipulated for better efficiency and for multiple reuses to render economic viability in comparison with conventional adsorbents like ion-exchange resins or ACs (Vieira and Volesky, 2000).

16.3.1.1 Bacterial Biosorbents

Bacteria is a major group of unicellular living organisms belonging to the prokaryotes. These are present in soil and water and also serve as symbionts of other organisms. Bacteria could be found in a wide variety of shapes and sizes, like cocci (such as *Streptococcus*), rods (such as *Bacillus*), spiral (such as *Rhodospirillum*), and filamentous (such as *Sphaerotilus*). Eubacteria also possess a relatively simple cell structure, which lack cell nuclei. Bacteria have anionic functional groups present in the peptidoglycan, teichoic acids, and teichuronic acids of Gram-positive bacteria. The peptidoglycan, phospholipids, and lipopolysaccharides of

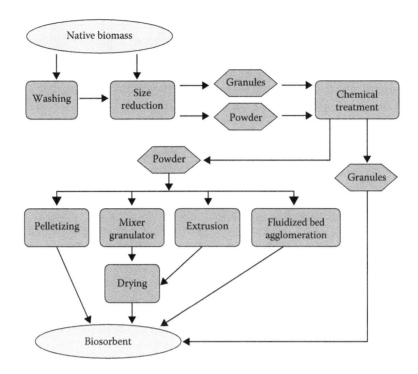

FIGURE 16.1 Schematic diagram for processing different types of native biomass into biosorbents. (Reproduced from Park, D. et al., *Biotechnol. Bioprocess Eng.*, 15, 86, 2010. With permission.)

Gram-negative bacteria are the components primarily responsible for the anionic character and metal-binding capability of the cell wall. Extracellular polysaccharides are also capable of binding metals (McLean et al., 1992).

16.3.1.2 Bacterial Cell Wall and Biosorption Process

The Gram-positive and Gram-negative cell walls of all bacteria are not at all identical. This cell wall structure study is thus very important for analyzing their metal-absorbing capabilities. Among the two general types of bacteria that exist, Gram-positive bacteria (Figure 16.2) comprise a thick peptidoglycan layer (Beveridge, 1981; Dijkstra and Keck, 1996) connected by amino acid bridges. Imbedded inside the Gram-positive cell wall are polyalcohols that are known to be teichoic acids, some of which are lipid linked to form lipoteichoic acids. Since lipoteichoic acids are covalently linked

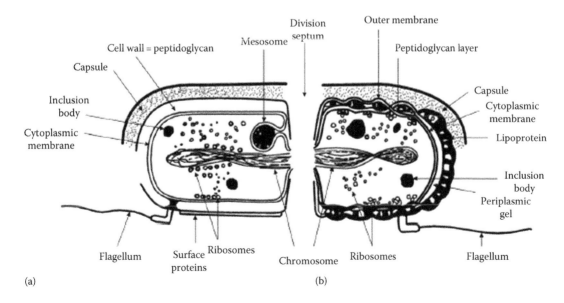

FIGURE 16.2 Structure of (a) Gram-positive and (b) Gram-negative bacteria. (Reproduced from Vijayaraghavan, K. and Yun, Y.S., *Biotechnol. Adv.*, 26, 266, 2008. With permission.)

to lipids within the cytoplasmic membrane, they are responsible for linking peptidoglycan to the cytoplasmic membrane. These cross-linked peptidoglycan molecules form a network that covers the cell like a grid. Teichoic acids also give the Gram-positive cell wall an overall negative charge, due to the presence of phosphodiester bonds between the teichoic acid monomers (Sonnenfeld et al., 1985). In general, 90% of the Gram-positive cell wall comprises peptidoglycan. On the contrary, the cell wall of Gram-negative bacteria (Figure 16.2) is thinner and composed of only 10%–20% peptidoglycan (Beveridge, 1999; Kolenbrander and Ensign, 1968). In addition, the cell wall contains an additional outer membrane (OM) composed of phospholipids and lipopolysaccharides (Sheu and Freese, 1973). The highly charged nature of these lipopolysaccharides confers an overall negative charge on the Gram-negative cell wall.

Sherbert (1978) reported that the anionic functional groups that were present in the peptidoglycan, teichuronic acids, and teichoic acids of Gram-positive bacteria and in the peptidoglycan, lipopolysaccharides, and phospholipids of Gram-negative bacteria were the main components responsible for the metal-binding capability and anionic properties of the cell wall. Metal-binding properties were also present in extracellular polysaccharides (McLean et al., 1992). However, their availability depends on the bacterial species and growth conditions; and they can easily be removed by simple mechanical disruption or chemical washing (Yee and Fein, 2001).

The bacterial cell wall is the first component that come in contact with metal ions/dyes, whereas the solutes can be deposited on the surface or within the cell wall structure (Doyle et al., 1980). Since the mode of solute uptake by dead/inactive cells is extracellular, the chemical functional groups of the cell wall do play vital roles in biosorption. Due to the nature of the cellular components, various functional groups are present on the bacterial cell wall, including carboxyl, phosphonate, amine, and hydroxyl groups (Doyle et al., 1980). As they are negatively charged and are abundantly available, carboxyl groups actively participate in the binding of metal cations. Dye molecules, which exist as dye cations in solutions, are also attracted toward carboxyl and other negatively charged groups. Golab and Breitenbach (1995) reported that the carboxyl groups of the cell wall of peptidoglycan of *Streptomyces pilosus* were responsible for the binding of copper. Also, certain amine groups are very effective at removing metal ions, as it not only chelates cationic metal ions but also adsorbs anionic metal species or dyes via electrostatic interaction or hydrogen bonding. Kang et al. (2007) had observed that amine groups protonated at pH 3 and attracted negatively charged chromate ions via electrostatic interaction. Vijayaraghavan and Yun (2007) gave the confirmation that the amine groups of *Corynebacterium glutamicum* were capable of binding reactive dye anions through electrostatic attraction. In general, increasing the pH increases the overall negative charge on the surface of cells until all the relevant functional groups are deprotonated. This favors the electrochemical attraction and adsorption of cations. Anions would also be expected to interact more strongly with cells with increasing concentration of positive charges, due to the protonation of functional groups at lower pH values. The solution chemistry affects not only the bacterial surface chemistry but also metal/dye speciation. Metal ions in solution undergo hydrolysis as the pH increases. Characterization of bacterial biomass and the biosorption mechanisms could be elucidated using different methods, including potentiometric titrations, Fourier transform infrared spectroscopy, x-ray diffraction, scanning electron microscopy, transmission electron microscopy, and energy-dispersive x-ray microanalysis (Vijayaraghavan and Yun, 2008).

16.3.1.3 Chemically Modified Bacterial Biomass as Biosorbents

The known chemical modification procedures include pretreatment, binding site enhancement, binding site modification, and polymerization. The commonly used chemical pretreatments include acid, alkaline, ethanol, and acetone treatments of the biomass (Vijayaraghavan and Yun, 2007). This employed several chemical agents (mineral acids, NaOH, Na₂CO₃, CaCl₂, and NaCl) for the pretreatments of *C. glutamicum* in the biosorption of Reactive Black 5. The authors identified 0.1 M HNO₃ as being the most suitable for opening new binding sites, which has an enhanced Reactive Black 5 uptake capacity of 1.3 times compared to that of the raw biomass. Care should be taken for the screening methods employed for selecting appropriate chemicals for pretreatment. Sar et al. (1999) observed that the metal (Cu^{2+} and Ni^{2+}) uptake capacity of lyophilized *Pseudomonas aeruginosa* cells was enhanced when pretreated with NaOH, NH₄OH, or toluene, whereas oven heating (80°C), autoclaving, and acid, detergent, and acetone treatments were found to be inhibitory. Even though these chemical pretreatments are almost essential for most biosorbents, especially in industrial wastes, vast improvements were thus observed in their biosorption capacities. Introduction of functional groups onto the biomass surface is done by grafting of long polymer chains onto the biomass surface via direct grafting or by polymerization of a monomer. Research on this aspect is scarce. Deng and Ting (2005a,b,c, 2007) had worked extensively with polyethylenimine, which is composed of large numbers of primary and secondary amine groups. These were also cross-linked with biomass and exhibited very effective biosorption abilities toward metals like chromium(VI), copper, lead, nickel, and arsenic.

Deng and Ting (2005b) had copolymerized acrylic acid on the biomass surface to enhance the carboxyl groups. This had resulted in five- and sevenfold increase in the uptake of copper and cadmium with respect to the pristine biomass. Poly(amic acid), from the reaction of pyromellitic dianhydride and thiourea, which comprises a large number of carboxyl and secondary amine groups in a molecule, was then grafted on the biosorbent surface. This had exhibited 15- and 11-fold increases in the uptakes of toxic metals like cadmium and lead (Yu et al., 2007).

16.3.1.4 Genetically Modified Bacterial Biosorbents

Genetic engineering is an important biotechnological tool that has the potential to improve or redesign microorganisms, where biological metal-sequestering systems will have a higher intrinsic capability as well as specificity and greater resistance to ambient conditions. Virgin biosorbents lack specificity in metal binding. This causes difficulties in recycling and recovery of metal(s). Genetic modification has the potential solution to enhance the selectivity as well as the accumulative properties of the cells. Genetic modification would be effectively feasible especially when the microbial biomass is produced from fermentation processes where genetically engineered microorganisms are used. Nowadays, many kinds of amino acids and nucleic acids are being produced in an industrial scale by using genetically engineered microbial cells. The cell surface adsorption of metal ions has also attracted attention as an alternative approach to bioaccumulation. Cell surface engineering enables the modification of cell surface properties by fusing the various functional proteins/peptides with cell surface anchoring protein (Vijayaraghavan and Yun, 2008). Cell surface engineering has been established in the Gram-negative and Gram-positive bacteria, and surface-engineered cells with useful functions have been constructed for a wide variety of applications (Kuroda and Yueda, 2011). Higher organisms respond to the presence of metals, with the production of cysteine-rich peptides, such as glutathione (Singhal et al., 1997), phytochelatins (PCs), and metallothioneins (MTs) (Mehra and Winge, 1991), which could bind and sequester metal ions in biologically inactive forms. Bacterial cells with overexpression of MTs will result in enhanced metal accumulation. This offers a promising strategy for future development of microbial-based biosorbents. This would lead to remediation of metal contamination (Vijayaraghavan and Yun, 2008).

In addition to the high selectivity and accumulative capacity, uptake by recombinant *Escherichia coli* (expressing the *Neurospora crassa* MT gene within the periplasmic space) was rapid. Greater than 75% Cd uptake occurred in the first 20 min, with maximum uptake achieved in less than 1 h. However, the expression of such cysteine-rich proteins is not devoid of problems, due to the interference with redox pathways in the cytosol. Moreover, the intracellular expression of MTs may prevent the recycling of the biosorbents, as the accumulated metals cannot be easily released suggesting a solution to bypass this transport problem by expressing MTs on the cell surface (Vijayaraghavan and Yun, 2008). Sousa et al. (1996) had demonstrated the possibility of inserting MTs into the permissive site 153 of the LamB sequence. The expression of the hybrid proteins onto the cell surface dramatically increased the whole-cell accumulation of cadmium. The expression of proteins on the surface offers an inexpensive alternative for the preparation of affinity adsorbents. PCs are short, cysteine-rich peptides, with the general structure (γGlu-Cys)nGly ($n = 2$–11) (Zenk, 1996). PCs offer many advantages over MTs, due to their unique structural characteristics, particularly the continuously repeating γGlu–Cys units. PCs have

been found to exhibit higher metal-binding capacity (on a per cysteine basis) than MTs (Mehra and Mulchandani, 1995). The development of organisms overexpressing PCs requires knowledge of mechanisms that are carried out in the synthesis and chain elongation of these peptides. Many biosorbents have been successfully engineered to display metal-binding peptides on their cell surface. A typical example includes creating a repetitive metal-binding motif, consisting of (Glu-Cys) nGly (Bae et al., 2000).

These peptides emulate the structure of PCs. However, they differ in the fact that the peptide bond between the glutamic acid and cysteine is a standard α peptide bond. PC analogs were found to be present on the bacterial surface, which enhanced the accumulation of Cd^{2+} and Hg^{2+} by 12- (Bae et al., 2000) and 20-fold (Bae et al., 2001), respectively. Attempts to create recombinant bacteria with improved metal-binding capacity have so far been restricted to mostly *E. coli*. Since, *E. coli* facilitates genetic engineering experiments and also possesses more surface area per unit of cell mass, this potentially gives higher rates for metal removal from solution. Nevertheless, a Gram-positive bacteria also possesses its merits compared to Gram-negative bacteria: like, translocation through only one membrane is required, and a more rigid cell wall which is less sensitive to shear and potentially more suitable for field applications such as bioadsorption. For example, bioadsorption by generated recombinant *Staphylococcus xylosus* and *Staphylococcus carnosus strains*, with surface exposed chimeric proteins containing polyhistidyl peptides. Owing to their high selectivity, genetically engineered biosorbents may prove very competitive for the separation of toxins and other pollutants from dilute contaminated solutions (Vijayaraghavan and Yun, 2008). In *E. coli*, *Ralstonia eutropha*, *S. xylosus*, *S. carnosus*, and *Saccharomyces cerevisiae*, various metal-binding proteins/peptides such as MTs, PCs, mercury-responsive metalloregulatory protein, hexa-histidine, histidine-rich peptide (GHHPHG; HP), and cysteine-rich peptide (GCGCPCGCG; CP) were displayed on the cell surface for the biosorption of the metal-polluted water (Kuroda and Ueda, 2011). Microbial absorption of metal ions by cell surface display of metal-binding proteins/peptides is given in Table 16.5.

As witnessed in the literature, there have been attempts to develop improved bacterial strains for metal bioremediation. Bacteria possessing different metal-binding complexes on their surfaces have been engineered, and these recombinant cells exhibited enhanced adsorption of heavy metals to varying degrees (Biondo et al., 2012). In the majority of these studies, *E. coli* was used for the cell surface display of different metal-chelating peptide/proteins, even though this bacterium is not suited for metal remediation purposes, given its low level of metal resistance. Metal-resistant bacteria, such as *Cupriavidus metallidurans* or *Pseudomonas putida*, were also used for metal adsorption improvement (Biondo et al., 2012). However, these recombinant strains are dependent on the addition of external inducers for the expression of the metal-chelating protein, raising limitations for scale-up

TABLE 16.5

Microbial Absorption of Metal Ions by Cell Surface Display of Metal-Binding Proteins and Peptides

Displayed Metal-Binding Protein/Peptide	Target Metal Ion	Anchor Protein	Microorganisms
Human metallothionein, yeast metallothionein	Cu, Cd, Zn	LamB	*E. coli*
Short metal-binding peptides	Cu, Cd, Zn	LamB	*E. coli*
Hexa-histidine	Cd	LamB	*E. coli*
Phytochelatin	Cd	Lpp-OmpA	*E. coli*
Screened peptides	Au, Cr	LamB	*E. coli*
Screened peptides	Cd	OmpA	*E. coli*
Screened peptides	Zn	FimH	*E. coli*
Screened peptides	Co, Cr, Mn, Pb	FimH	*E. coli*
Mer R	Hg	Ice nucleation protein	*E. coli*
Phosphate binding proteins (PhoS)	P	Ice nucleation protein	*E. coli, P. putida*
CBD variants	Ni	SPA	*S. carnosus*
Polyhistidyl peptides	Cd, Ni	SPA	*S. xylosus, S. carnosus*
Mouse metallothionein	Cd	IgAβ domain	*E. coli, R. eutropha*

Source: Reproduced from Kuroda, K. and Ueda, M., *Curr. Opin. Biotechnol.*, 22, 427, 2011. With permission.

processes and environmental use (Biondo et al., 2012). The aim to obtain a bacterial strain gathering the characteristics of high metal resistance, genetic stability and metal remediation ability. Thus, they added external inducers for gene expression and antibiotics for the selection of plasmid carrying cells. To this end, the surface display of the synthetic PC EC20sp, with the structure [(Glu-Cys)20Gly],15, was created in the *C. metallidurans* CH34 strain, which resulted in increased capability of the cells to immobilize several metal ions (Biondo et al., 2012). The DNA sequence coding for a synthetic PC with the structure (Glu-Cys)20Gly was devoid of a stop codon and named EC20sp. It was synthesized *in vitro* and cloned in the pHEβ plasmid.

This plasmid encoded the signal sequence (SS), His6 sequence, E-tag sequence, and the β-domain sequence of the *Neisseria gonorrhoeae* IgA protease secretion system (IgAβ). The SS-IgAβ fragment encoded an auto transporter protein that initially crosses the inner membrane (IM) via the Sec pathway. This was followed by the insertion of the C-terminal β-domain in the OM. Thus allowing the translocation and anchoring of the heterologous attached protein on the bacterial cell surface. The His6 coding sequence was then removed from the pHEβ plasmid and replaced by the EC20sp gene that was then inserted in frame with the SS coding sequence, thus resulting in a cell surface display cassette, which encodes the PC EC20sp (4.5 kDa) fused to the E-tag-IgAβ protein (~45 kDa). The E-tag epitope also allowed the immunodetection by a monoclonal antibody (anti-Etag). The new plasmid, named pCM1 (Figure 16.3), carried the SS-EC20sp-E-tag-IgAβ gene fusion under the control of the *E. coli* lac promoter. The pCM1 plasmid was used to transform *E. coli* UT5600, and the transformant cells were found to express EC20sp in the OM. Considering that the pCM1

plasmid was unable to replicate in *C. metallidurans* CH34, in order to ensure expression in this bacterium, the SS-EC20sp-E-tag-IgAβ fragment was then transferred from pCM1 to the pBB-panEGFP plasmid, which derives from the broad-host-range pBBR1MCS plasmid vector (30–40 copies/cell). The pBB-panEGFP plasmid carried the pan promoter, which is a strong promoter for the expression of *C. metallidurans*. The SSEC20sp E-tag-IgAβ fragment was then cloned under pan promoter control, replacing the egfp gene in pBB-panEGFP, and the final plasmid, named pCM2 (Figure 16.4), was then used in the genetic transformation of *C. metallidurans* CH34 cells. To determine the expression and the subcellular localization of the recombinant protein, cells were then harvested and the proteins of the soluble, IM, and OM fractions were analyzed by SDS-PAGE, blotted, and then probed with the anti-E-tag monoclonal antibody. As can be seen in Figure 16.3, a single protein band of the expected size (~50 kDa) was present only in the OM fraction of the *C. metallidurans* CH34/ pCM2 recombinant cells, showing that the PC EC20sp was correctly targeted, presenting no sign of proteolytic degradation (Biondo et al., 2012).

To find out whether the expression and cell surface display of EC20sp increased the ability of *C. metallidurans* CH34 cells in adsorbing metal ions, cells were then grown to mid-log phase and harvested, and the bacterial biomass was inoculated in Milli-Q water containing metal. Given that most of the results in the literature refer to Cd^{2+} adsorption, experiments were initially performed with cadmium. When adsorption was carried out in the 0.10 mM Cd^{2+}, the recombinant cells adsorbed, upon 2 h of incubation, 20.90 nM of Cd^{2+}/mg cell dw (22% more than the control cells) and, upon 24 h incubation, 30.40 nM of Cd^{2+}/mg cell dw (50% more than the control cells). The biosorption experiments were then

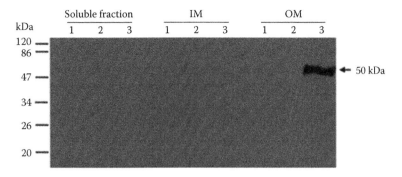

FIGURE 16.3 Western blotting of the EC20sp hybrid protein in *C. metallidurans* strains. 1, *C. metallidurans* CH34 wild type; 2, *C. metallidurans* CH34/pBBR1MCS; and 3, *C. metallidurans* CH34/pCM2. The proteins were fractionated in soluble, IM, and OM fraction. The band corresponding to the EC20sp-Etag-IgAβ protein is indicated by an arrow. (Reproduced from Biondo, R. et al., *Environ. Sci. Technol.*, 46, 8325, 2012. With permission.)

performed with 1 mM Cd^{2+}, Co^{2+}, Cu^{2+}, Mn^{2+}, Ni^{2+}, Pb^{2+}, or Zn^{2+} ions, since this metal concentration is inhibitory to the growth of several bacteria but not to *C. metallidurans* CH34. No significant variation was then found in colony-forming units as was observed after 24 h of incubation with any of the examined metal ions. This showed that cell viability was not affected. Accordingly, the growth kinetics of *C. metallidurans* CH34/pCM2 cells also showed no significant difference with that of control cells, indicating that the genetic modification did not affect normal cell functioning. As can be observed from Figure 16.4, in comparison to the control cells, the recombinant *C. metallidurans* CH34/pCM2 cells showed considerable increase in the bioaccumulation of several metals, particularly when the cells were pregrown in the presence of the metal (induced) (Biondo et al., 2012).

For biosorption purposes, this was a meaningful result, since the presence of the metallic contaminant itself enhanced the bioremediation capability of the cells. The metal adsorption capability of the recombinant strain relative to the control was increased approximately 219%, 210%, 76%, 59%, 45%, and 31% for Zn^{2+}, Pb^{2+}, Cu^{2+}, Cd^{2+}, Ni^{2+}, and Mn^{2+}, respectively (Figure 16.4). The affinity of *C. metallidurans* CH34/pCM2 cells for the different metal ions was also found in the following order: $Pb^{2+} > Zn^{2+} > Cu^{2+} > Cd^{2+} > Ni^{2+} > Mn^{2+} > Co^{2+}$, while for the control cells the metal affinity was $Cu^{2+} > Pb^{2+} > Zn^{2+} > Cd^{2+} > Ni^{2+} > Mn^{2+} > Co^{2+}$, showing that the presence of EC20sp on the cell surface modified the interaction between cell and metal. In order to visualize the metal adsorption onto the cell surface, an experiment was performed using lead. *C. metallidurans* CH34/pCM2 along with the control cells were incubated separately in water containing 1 mM of Pb^{2+}. The bacterial mass was treated and fixed in resin. The electronic micrographs (Figure 16.4) showed black spots on the cell surface, forming structures surrounding the recombinant bacterial cell in higher amount than the wild type, which was an indication of the interaction between Pb^{2+} and the peptides displayed, confirming that the cell surface display of EC20sp increased with the amount of Pb^{2+} bound to the OM of the recombinant cells (Figure 16.4a [A]). In contrast, only a few particles were observed on the surface of wild-type cells (Figure 16.4b [B]).

The results of the present work reinforced the biotechnological potential of PC surface display to increase metal accumulation in *C. metallidurans* CH34. The use of a strong promoter, which does not require external inducers, to express a PC analog with a broad range of metal-binding ability was shown to have increased the bacterial remediation capability of several metals. Using this bioremediation approach, it was also possible to recover the adsorbed metals from the biomass, of particular relevance when precious metals were involved (Biondo et al., 2012).

In another study, a gene coding for a de novo peptide sequence containing a metal-binding motif was chemically synthesized and then expressed in *E. coli* as a fusion with the maltose-binding protein. Bacterial cells expressing this metal-binding peptide fusion demonstrated the enhanced binding of Cd^{2+} and Hg^{2+} compared to bacterial cells lacking the metal-binding peptide. The potential use of this genetically engineered bacteria as biosorbents for the removal of heavy metals from wastewaters is immense (Pazirandeh et al., 1998).

16.3.1.5 Fungal Biosorbents

A fungus is a member of a group of eukaryotic organisms that includes unicellular microorganisms such as yeasts and molds, as well as the multicellular fungi that produce familiar fruiting forms known as mushrooms. Fungus contains chitin in its cell wall, unlike the cell walls of plants, bacteria, and some protists. Similar to animals, fungi are also heterotrophs and acquire their food by absorbing dissolved molecules, typically by secreting digestive enzymes into their environment. Nonliving fungal biomasses have been effectively used as biosorbents. Various novel researches have been conducted in this field. In 1990, a study was reported where copper biosorption by nonliving wood-rotting fungus *Ganoderma lucidum* was conducted, and it was found that protein interaction with metals did not play a significant role in copper(II) uptake (Muraleedharan and Venkobachar, 1990). Waste mycelia of *Aspergillus niger*, *Phanerochaete chrysogenum*, and *Claviceps paspali* from the industrial fermentation plants were utilized as a biosorbent for zinc from the aqueous environment in batch as well as in column modes. In optimized conditions, *A. niger* and *C. paspali* were superior to

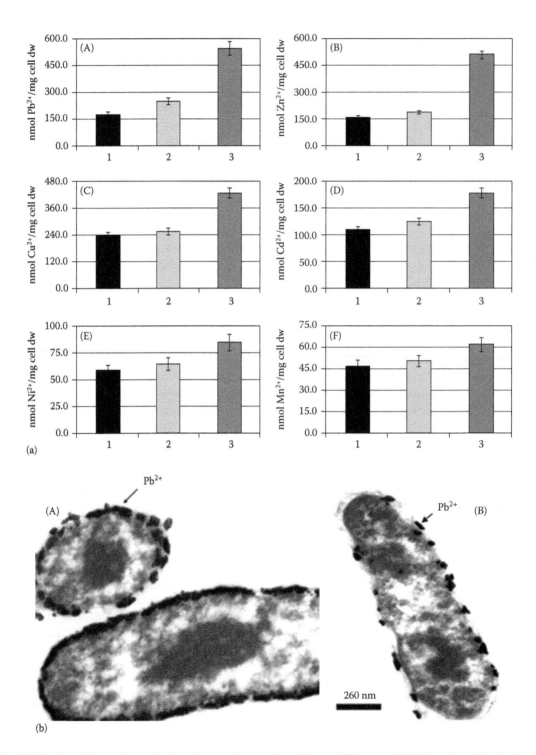

FIGURE 16.4 Metal absorption and transmission electron micrographs of *C. metallidurans*. (a) Metals absorption by *C. metallidurans* CH34 strain. 1, *C. metallidurans* CH34/pBBR1MCS; 2, *C. metallidurans* CH34/pCM2; and 3, *C. metallidurans* CH34/pCM2/induced (cultured in medium containing metal). (A) Pb^{2+}; (B) Zn^{2+}; (C) Cu^{2+}; (D) Cd^{2+}; (E) Ni^{2+}, (F) Mn^{2+}. The values shown are the average of two independent experiments using duplicate samples for each determination ±S.D. (b) Transmission electron micrographs of *C. metallidurans* CH34 cells. (A) *C. metallidurans*/pCM2 cells after 24 h of incubation with 1 mM of lead (Pb^{2+}), and (B) *C. metallidurans* CH34 wild-type cells after 24 h of incubation with 1 mM of lead. (Reproduced from Biondo, R. et al., *Environ. Sci. Technol.*, 46, 8325, 2012. With permission.)

P. chrysogenum (Luef et al., 1991). Biosorption of lead by *P. chrysogenum* biomass was also strongly affected by pH. At pH 4–5, the saturated uptake capacity for lead sorption was found to be higher than that of activated charcoal and that reported for some other organisms (Niu et al., 1993). Dead cells of *S. cerevisiae* were found to remove 40% more uranium or zinc than live cultures, and biosorption rapidly reached up to 60% of the final uptake value within 15 min of contact, and uranium was deposited as fine needlelike crystals inside the cells and also on the outer cell surface. The nonliving waste biomass of *A. niger* was attached to wheat bran and was used as a biosorbent for the removal of zinc and copper from aqueous solutions. The metal uptake was found out as a function of the initial metal concentrations, pH, and biomass loading. Metal uptake was found to decrease in the presence of Co ions, which was dependent on the concentration of metal ions of the two compounds in aqueous solution. Alkali-treated biomass of *A. niger*, referred to as Biosorb, was then found to sequester Cd^{2+}, Cu^{2+}, Zn^{2+}, Ni^{2+}, and Co^{2+} efficiently up to 10% of its weight (w/w), and it exhibited higher metal-binding capacity as compared to *Neurospora*, *Fusarium*, and *Penicillium*. The kinetics of the metal binding by Biosorb indicated that it is a rapid process, and about 70%–80% of the metal is removed from the solution in 5 min followed by a slower rate. Removal of Cr(VI) from aqueous solution was carried out in batch mode using dead biomass of fungal strains *A. niger* NCIM-501, *A. oryzae* NCIM-637, *Rhizopus arrhizus* NCIM-997, and *R. nigricans* NCIM-880. Basic parameters such as pH (2.0–8.0), initial metal ion concentration (100–500 mg/L), contact time (2–24 h), and varying biomass concentration (0.5–3.0 g) were optimized. *R. nigricans* and *R. arrhizus* possessed very good specific uptake of 11 mg Cr(VI)/g of biomass at the pH range of 2.0–7.0. Metal uptake capacity was found to be in the order of *R. nigricans* > *R. arrhizus* > *A. oryzae* > *A. niger*. *Mucor miehei*, a fermentation industry waste, was found to be an effective biosorbent for the removal of hexavalent chromium from leather industry effluent. In comparative studies with ion-exchange resins, *Mucor* biomass also showed biosorption levels corresponding to the strongly acidic commercial resins. Response to pH was similar to the weakly acidic resins in solution, and chromium elution characteristics were found to be similar to both weakly and strongly acidic resins. Nonliving free and immobilized biomass of *R. arrhizus* were also used to study the biosorption of chromium(VI). The rates of the removal were marginally more in free biomass conditions than in immobilized state. Stirred tank reactor studies also indicated higher chromium biosorption at 100 rpm and 1:10 biomass–liquid ratio. The fluidized bed reactor was found to be more efficient in chromium removal than the stirred tank reactor (Ahluwalia and Goyal, 2007).

16.3.1.6 Genetically Modified Fungal Biosorbents

Modified fungal cell surface by genetic engineering is nowadays used for the development of novel biosorbent systems. Metal-binding proteins from living organisms can often not distinguish metal ions that are placed adjacent in the periodic table (Kuroda and Ueda, 2011). Thus, the modification of the metal-binding pocket is necessary. For example, in the modification of the metal-binding pocket, a single amino acid mutation was introduced in the molybdate binding domain of ModE, which was converted to ModE in yeast cells to specifically adsorb tungstate in the presence of molybdate. In cases where known proteins that could bind target metal ions are absent, improved proteins having the potential to bind metal ions adjacent to target ions were developed using cell surface engineering. Their screening among a peptide library could be an efficient strategy to create a protein/peptide that specifically binds a target metal ion. Packed bed column systems combining individual columns for specific adsorption and recovery could also contribute to the fractional recovery of target metal ions (Figure 16.5) (Kuroda and Ueda, 2011).

In a different kind of study, fungal cellulose-binding domain (CBD) has been expressed on bacterial cell surface for metal biosorption. Wernérus et al. (2001) had also studied Ni^2-binding staphylococci that were generated through the surface display of combinatorially engineered variants of a fungal CBD from *Trichoderma reesei* cellulase Cel7A. Novel CBD variants that were generated by combinatorial protein engineering through the randomization of 11 amino acid positions and eight potentially Ni^2-binding CBDs were selected by phage display technology. The new variants were then subsequently introduced genetically into chimeric surface proteins for the surface display on *S. carnosus* cells. The expressed chimeric proteins were also shown to be properly targeted to the cell wall of *S. carnosus* cells, since full-length proteins could be extracted and affinity purified. Surface accessibility for the chimeric proteins was then demonstrated, and furthermore, the engineered CBDs, now devoid of cellulose-binding capacity, were also shown to be functional with regard to metal binding. Since the recombinant staphylococci had gained Ni^2-binding capacity, this biosorbent had various environmental applications (Wenereneus et al., 2001).

16.4 BIOSORPTION: MODELS, ISOTHERM, AND THERMODYNAMICS

The biosorption as a process involves a solid phase (sorbent) and a liquid phase (solvent, normally water) containing the dissolved species that is adsorbed (adsorbate). Quantification of adsorbate–adsorbent interactions is required for the evaluation of potential implementation strategies. In order to compare the pollutant uptake capacities of the different types of biosorbents, this phenomena on could be expressed as batch equilibrium isotherm curves. These could be modeled by mechanistic or empirical equations; the former could explain, represent, and predict the experimental behaviors, while the latter does not reflect the mechanism but does reflect the experimental curves. As shown in Table 16.6, empirical models involving 2, 3, and 4 parameters have been used to fit in batch equilibrium isotherm curves of the biosorbents. Among these, the Langmuir and Freundlich models, which have been most commonly used, have a high rate of success. There are

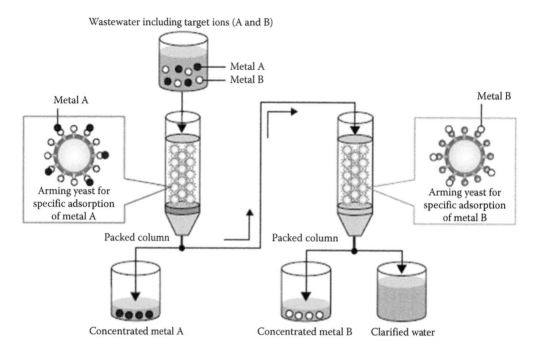

FIGURE 16.5 Packed bed column system for the specific adsorption and recovery of enriched target metal ions by surface-engineered cells. (Reproduced from Kuroda, K. and Ueda, M., *Curr. Opin. Biotechnol.*, 22, 427, 2011. With permission.)

no critical reasons to use the more complex models if the two parameter models can fit the experimental data reasonably well. Since many industrial wastewaters do contain several components to be bound onto the biosorbent, a very judicious use is necessary for the practical applications of effective multicomponent biosorption models. Single-component isotherm models have frequently been extended to the formation of multicomponent ones (Table 16.6). Among these, in many cases, the extended Langmuir equation is found to be fit for the experimental data reasonably well. To characterize the competitive adsorption of biosorbents, a few researchers have successfully employed the ideal adsorbed solution theory using only single-component isotherm parameters (Vijayaraghavan and Yun, 2008).

An increase in the temperature affects not only the diffusion rate of adsorbate molecules from the solution to the adsorbent but also the solubility of the adsorbate molecules (especially in case of dye biosorption). The adsorption characteristics of a material could be expressed in thermodynamic parameters such as ΔG (Gibbs free energy change), ΔH (enthalpy change), and ΔS (entropy change). These parameters can be calculated accurately by using the thermodynamic equilibrium coefficient that are obtained at different temperatures and concentrations using Arrhenius, Gibbs, van't Hoff, Clausius–Clapeyron expressions, and their evaluation could give an insight into the possible mechanisms of adsorption. At constant temperature and pressure, the ΔG value is the fundamental criterion of spontaneity. If this value is found to be negative, then adsorption would take place, indicating the spontaneity of the reaction. Thus, by using the equilibrium

constant obtained from the Langmuir model at each temperature, ΔG can be easily calculated according to the Gibbs expression, while ΔH and ΔS could be determined using the van't Hoff plot. It is to be noted that ΔG could be estimated from the equilibrium adsorption data. It should be under the assumption that the adsorption of a molecule is reversible and that the equilibrium condition has been established in batch system (Crini and Badot, 2008). In addition, it should be noted that, theoretically, the concentration of adsorbate used in the Langmuir isotherm equation must be expressed specifically as its molar concentration. However, in the literature, the volumetric concentration of adsorbate has been commonly used in the Langmuir isotherm equation without any theoretical consideration. Thus, this has eventually led to misapplication of the Langmuir isotherm equation in calculating the ΔG in thermodynamic studies (Park et al., 2010).

16.5 CONCLUSIONS

Biosorption and the use of novel biosorbent are a very important area of research in the field of biotechnology and are rapidly evolving. This field of research is in its developmental stages and requires further improvement in both performance and costs, which can be expected in the future. Fundamental research in this field is required for better understanding before their application industrially. Tools of modern molecular biology will enhance this field through better and efficient genetic engineering of microbial biosorbents for toxic metal removal. Both academia and industry need to work together in order to achieve noteworthy results.

TABLE 16.6
Equilibrium Isotherm Models

System	Expression	Equation Form	Remarks
Single component	Langmuir	$q = \dfrac{q_m b C_e}{1 + b C_e}$	Monolayer sorption
	Freundlich	$q = K C_e^{1/n}$	Simple expression
	Temkin	$q = \dfrac{RT}{b} \ln(a C_e)$	Considering temperature
	Dubinin–Radushkevich	$q = q_D \left[-B_D \left\{ RT \ln \left(1 + \dfrac{1}{C_e} \right) \right\}^2 \right]$	Considering temperature
	Langmuir–Freundlich	$q = \dfrac{q_m b C_e^{1/n}}{1 + b C_e^{1/n}}$	Combination approaches
	Redlich–Peterson	$q = \dfrac{a C_e}{1 + b C_e^n} a$	Freundlich at high concentration
	Sips	$q = \dfrac{a C_e^n}{1 + b C_e^n}$	Complicated
	Radke–Prausnitz	$q = \dfrac{a r C_e^n}{a + r C_e^{n-1}}$	Complicated
	Khan	$q = \dfrac{q_m b C_e}{(1 + b C_e)^n}$	Complicated
	Toth	$q = \dfrac{q_m b C_e}{\left\{ (1 + (b C_e)^{1/n} \right\}^n}$	Complicated
	Brunauer (BET)	$q = \dfrac{q_m B C_e}{(C_1 - C_e)\left\{ 1 + (B - 1)(C_e / C_1) \right\}}$	Multilayer sorption
Multi component	Langmuir	$q_i = \dfrac{q_{mi} b_i C_i}{1 + \sum_{i=1}^{N} (b_i C_i)}$	Competitive
	Langmuir–Freundlich	$q_i = \dfrac{q_{mi} b_i C_i^{1/n_i}}{1 + \sum_{i=1}^{N} \left(b_i C_i^{1/n} \right)}$	Competitive
	Redlich–Peterson	$q_i = \dfrac{a_i C_i}{1 + \sum_{i=1}^{N} \left(b_i C_i^{n_i} \right)}$	Competitive

Source: Reproduced from Park, D. et al., *Biotechnol. Bioprocess Eng.*, 15, 86, 2010. With permission.

REFERENCES

Ahluwalia, S.S., Goyal, D., 2007. Microbial and plant derived biomass for removal of heavy metals from wastewater. *Bioresource Technology*, 98: 2243–2257.

Bae, W., Chen, W., Mulchandani, A., Mehra, R.K., 2000. Enhanced bioaccumulation of heavy metals by bacterial cells displaying synthetic phytochelatins. *Biotechnology Bioengineering*, 70: 518–524.

Bae, W., Mehra, R., Mulchandani, A., Chen, W., 2001. Genetic engineering of *Escherichia coli* for enhanced uptake and bioaccumulation of mercury. *Applied Environmental Microbiolgy*, 67: 5335–5338.

Beveridge, T.J., 1981. Ultrastructure, chemistry and function of the bacterial wall. *International Reviews in Cytology*, 72: 229–317.

Beveridge, T.J., 1999. Structures of Gram-negative cell walls and their derived membrane vesicles. *Journal of Bacteriology*, 181: 4725–4733.

Biondo, R., da Silva, F.A., Vicente, E.J., Souza Sarkis, J.E., Schenberg, A.C.G., 2012. Synthetic phytochelatin surface display in *Cupriavidus metallidurans* CH34 for enhanced metals bioremediation. *Environmental Science & Technology*, 46: 8325–8332.

Borba, C.E., Guirardello, R., Silva, E.A., Veit, M.T., Tavares, C.R.G., 2006. Removal of nickel(II) ions from aqueous solution by biosorption in a fixed bed column experimental and theoretical breakthrough curves. *Biochemical Engineering Journal*, 30: 184–191.

Crini, G., Badot, P.M., 2008. Application of chitosan, a natural amino-polysaccharide, for dye removal by aqueous solutions by adsorption processes using batch studies: A review of recent literature. *Progress in Polymer Science*, 33: 399–447.

Deng, S., Ting Y.P., 2005a. Characterization of PEI-modified biomass and biosorption of Cu(II), Pb(II) and Ni(II). *Water Research*, 39: 2167–2177.

Deng, S., Ting Y.P., 2005b. Fungal biomass with grafted poly-acrylic acid for enhancement of Cu(II) and Cd(II) biosorption. *Langmuir*, 21: 5940–5948.

Deng, S., Ting Y.P., 2005c. Polyethylenimine-modified fungal biomass as a high-capacity biosorbent for Cr(VI) anions: Sorption capacity and uptake mechanisms. *Environmental Science & Technology*, 39: 8490–8496.

Deng, S., Ting Y.P., 2007. Removal of As(V) and As(III) from water with a PEI-modified fungal biomass. *Water Science and Technology*, 55: 177–185.

Doyle, R.J., Matthews, T.H., Streips, U.N., 1980. Chemical basis for selectivity of metal ions by the *Bacillus subtilis* cell wall. *Journal of Bacteriology*, 143: 471–480.

Dijkstra, A., Keck, W., 1996. Peptidoglycan as a barrier to transenvelope transport. *Journal of Bacteriology*, 178: 5555–5562.

Fu, F., Wang, Q., 2011. Removal of heavy metal ions from wastewaters: A review. *Journal of Environmental Management*, 92: 407–418.

Gadd, G.M., 2009. Biosorption: Critical review of scientific rationale, environmental importance and significance for pollution treatment. *Journal of Chemical Technology and Biotechnology*, 84: 13–28.

Golab, Z., Breitenbach, M., 1995. Sites of copper binding in *Streptomyces pilosus*. *Water, Air, & Soil Pollution*, 82: 713–721.

Kang, S.-Y., Lee, J.-U., Kim, K.-W., 2007. Biosorption of Cr(III) and Cr(VI) onto the cell surface of *Pseudomonas aeruginosa*. *Biochemical Engineering Journal*, 36: 54–58.

Kolenbrander, P.E., Ensign, J.C., 1968. Isolation and chemical structure of the peptidoglycan of *Spirillum serpens* cell walls. *Journal of Bacteriology*, 95: 201–210.

Kratochvil, D., Volesky, B., 1998. Advances in the biosorption of heavy metals. *Trends in Biotechnology*, 16: 291–300.

Kuroda, K., Ueda, M., 2011. Molecular design of the microbial cell surface toward the recovery of metal ions. *Current Opinion in Biotechnology*, 22: 427–433.

Luef, E., Prey, T., Kubicek, C.P., 1991. Biosorption of zinc by fungal mycelial waste. *Applied Microbiology and Biotechnology*, 34: 688–692.

McLean, R.J.C., Beauchemin, D., Beveridge, T.J., 1992. Influence of oxidation-state on iron binding by *Bacillus licheniformis* capsule. *Applied Environmental Microbiology*, 55: 3143–3149.

Mehra, R.K., Mulchandani, P., 1995. Glutathione-mediated transfer of Cu(I) into phytochelatins. *Biochemical Journal*, 307: 697–705.

Mehra, R.K., Winge, D.R., 1991. Metal ion resistance in fungi-molecular mechanisms and their regulated expression. *Journal of Cellular Biochemistry*, 45: 30–40.

Muraleedharan, T.R., Venkobachar, C., 1990. Mechanism of biosorption of copper (II) by *Ganoderma lucidum*. *Biotechnology & Bioengineering*, 35: 320–325.

Namasivayam, C., Kadirvelu, K., 1999. Uptake of mercury (II) from wastewater by activated carbon from unwanted agricultural solid by-product: Coirpith. *Carbon* 37: 79–84.

Naseem, R., Tahir, S.S., 2001. Removal of Pb(II) from aqueous solution by using bentonite as an adsorbent. *Water Research*, 35: 3982–3986.

Niu, H., Xu, X.S., Wang, J.H., 1993. Removal of lead from aqueous solutions by penicillin biomass. *Biotechnology & Bioengineering*, 42: 785–787.

Oyaro, N., Juddy, O., Murago, E.N.M., Gitonga, E., 2007. The contents of Pb, Cu, Zn and Cd in meat in Nairobi, Kenya. *International Journal of Food Agricultural Environment*, 5: 119–121.

Park, D., Yun, Y.S., Park, J.M., 2010. The past, present, and future trends of biosorption. *Biotechnology and Bioprocess Engineering*, 15: 86–102.

Paulino, A.T., Minasse, F.A.S., Guilherme, M.R., Reis, A.V., Muniz, E.C., Nozaki, J., 2006. Novel adsorbent based on silkworm chrysalides for removal of heavy metal from wastewaters. *Journal of Colloid and Interface Science*, 301: 479–487.

Pazirandeh, M., Wells, B.M., Ryan, R.L., 1998. Development of bacterium-based heavy metal biosorbents: Enhanced uptake of cadmium and mercury by *Escherichia coli* expressing a metal binding motif. *Applied and Environmental Microbiology*, 64: 4068–4072.

Sar, P., Kazy, S.K., Asthana, R.K., Singh, S.P., 1999. Metal adsorption and desorption by lyophilized *Pseudomonas aeruginosa*. *International Journal of Biodeterioration and Biodegradation*, 44: 101–110.

Sherbert, G.V., 1978. *The Biophysical Characterization of the Cell Surface*. London, U.K.: Academic Press.

Sheu, C.W., Freese, E., 1973. Lipopolysaccharide layer protection of gram negative bacteria against inhibition by long-chain fatty acids. *Journal of Bacteriology*, 115: 869–875.

Singhal, R.K., Andersen, M.E., Meister, A., 1997. Glutathione, a first line of defense against cadmium toxicity. *FASEB Journal*, 1: 220–223.

Sonnenfeld, E.M., Beveridge, T.J., Koch, A.L., Doyle, R.J., 1985. Asymmetric distribution of charge on the cell wall of *Bacillus subtilis*. *Journal of Bacteriology*, 163: 1167–1171.

Sousa, C., Cebolla, A., de Lorenzo, V., 1996. Enhanced metal-loadsorption of bacterial cells displaying poly-His peptides. *Nature Biotechnology*, 14: 1017–1020.

Vieira, R.H.S.F., Volesky, B., 2000. Biosorption: A solution to pollution? *International Microbiology*, 3: 17–24.

Vijayaraghavan, K., Yun, Y.S., 2007. Utilization of fermentation waste (*Corynebacterium glutamicum*) for biosorption of Reactive Black 5 from aqueous solution. *Journal of Hazardous Materials*, 141: 45–52.

Vijayaraghavan, K., Yun, Y.S., 2008. Bacterial biosorbents and biosorption. *Biotechnology Advances*, 26: 266–291.

Volesky, B., 1987. Biosorbents for metal recovery. *TIBTECH*, 5: 96–101.

Volesky, B., 1990. *Biosorption of Heavy Metals*. Boca Raton, FL: CRC Press.

Volesky, B., 1994. Advances in biosorption of metals: Selection of biomass types. *FEMS Microbiological Reviews*, 14: 291–302.

Volesky, B., Holan, Z.R., 1995. Biosorption of heavy metals. *Biotechnological Progress*, 11: 235–250.

Wase, D. J., Wase, J. (eds.), 2002. Biosorbents for metal ions. CRC Press, London, U.K.

Wernérus, H., Lehtiö, J., Teeri, T., Nygren, P.Å., Ståhl, S., 2001. Generation of metal-binding *Staphylococci* through surface display of combinatorially engineered cellulose-binding domains. *Applied and Environmental Microbiology*, 67: 4678–4684.

Yee, N., Fein, J., 2001. Cd adsorption onto bacterial surfaces: A universal adsorption edge? *Geochim Cosmochim Acta*, 65: 2037–2042.

Yu, J., Tong, M., Sun, X., Li, B.A., 2007. Simple method to prepare poly(amic acid)-modified biomass for enhancement of lead and cadmium adsorption. *Biochemical Engineering Journal*, 33: 126–133.

Zenk, M.H., 1996. Heavy metal detoxification in higher plants: A review. *Gene*, 179: 21–30.

17 Biosorption of Metals by Microorganisms in the Bioremediation of Toxic Metals

Sanaz Orandi

CONTENTS

ABSTRACT

Metal-contaminated water/wastewater is one of the most important environmental issues resulting from different industries including mining activities. Conventional water treatment techniques are not cost-effective for mine sites and are inefficient for removing elements at low concentrations. Bioremediation, using live cells, can be an alternative solution for removing metals from mine waters. The selection of resistant microbes is a prerequisite prior to designing a biotreatment system to withstand the harsh conditions of mine waters, usually acidic and containing toxic metals. The extremophile indigenous mining microorganisms, particularly phototrophic eukaryotes (green microalgae), were found to be efficient metal biosorbents. These algal–microbial consortiums, as attached biofilms to drainage substrates, rely on low concentrations of available nitrate and phosphate in mine waters. The microbes are also adapted to the metal-contaminated waters, while they are able to remove and accumulate many metals actively and passively on/in their cells, through various biochemical and biological mechanisms. To immobilize the algal–microbial consortium as biofilm and achieve efficient treatment, a suitable configuration of photobioreactors is required. From the most commonly used types of biofilm reactors for water treatment, rotating biological contactors facilitate biofilm formation and provide a simple operation for potentially sustainable water treatment at mine sites. The application of bioreactors for removing metals from contaminated water and the efficiency of a biological treatment system are discussed in this chapter.

17.1 INTRODUCTION

Industrialization has been presented as the hallmark of civilization. Although industrial activities created important environmental issues such as water contamination, industrial wastewaters, mainly resulting from mining, milling, and surface finishing industries, are contaminated with toxic elements and heavy metals. The polluted waters are drained or discharged into the water bodies that are often the source of irrigation or drinking water for the towns downstream. Since, in many countries, municipal wastewater treatment facilities are not equipped to remove traces of heavy metals, every consumer is exposed to a quantity of pollutants in the consumed water (Gadd, 2010; Gilmour and Riedel, 2009).

Mining industry is one of the major units that releases toxic elements and heavy metals such as As, Cd, Cu, Zn, Pb, and Ni into water bodies and soil. Mining activities increase the concentration of these elements in water compared to their normal concentrations in each area (Lottermoser, 2010). Mining wastewaters, typically known as acid mine drainage (AMD), are acidic and contain the elements Ag, As, Be, Cd, Cr, Cu, Hg, Ni, Pb, Sb, Se, Tl, and Zn that are considered the major pollutants (Sparks, 2005). These elements are not biodegradable and their presence in water affects the ecological balance (Baker and Banfield, 2003).

Biotreatment of AMD through biosorption of metals by microorganisms, in particular indigenous mining microorganisms, is the focus of this chapter. The water treatment processes or techniques that apply to these microorganisms and some related case studies are discussed.

17.2 IMPACT OF MINING ACTIVITIES ON WATER QUALITY

Water is an essential resource for mining activities such as mineral processing, hydrometallurgical extraction, coal washing, and dust suppression. Additionally, water is an unwanted by-product from mine dewatering processes in open pits and underground mining operations. Mines located in wet climates may have to pump more than 100,000 L/min. The water quality in used and unwanted water both are influenced by mining activities (Lottermoser, 2010).

Mining excavation and extraction of ore minerals from a reductive environment expose them to the oxidative environment of surface and accelerate their oxidation (Nganje et al., 2010). The common minor constituents of the Earth's crust are sulfide minerals. Nevertheless, these minerals comprise the major proportion of the rocks in some mine areas, in particular metallic ore deposits (Cu, Fe, Zn, Pb, Au, Ni, U), phosphate ores, coal seams, oil shales, and mineral sands (Brake et al., 2001a,b; Lottermoser, 2010). Sulfide minerals are not stable in an oxidative environment. These minerals react with oxygen and water leading to the release of hydrogen and sulfate ions, which influence the water quality by decreasing the pH. In some cases, the pH of water decreases to 2 (Bhattacharya et al., 2006; Das et al., 2009a). Surface waters that are affected by these influences are referred to as AMD (Kalin et al., 2004; Lottermoser, 2010). Sulfide minerals, typically pyrite and chalcopyrite, contain heavy metals such as Cd, Sn, Pb, Cu, Fe, Hg, Ni, Zn, and Cr in their atomic lattice. The oxidation of the sulfide minerals causes the release of embedded heavy metals (Fe, Cu, Pb, Zn, Cd, Co, Cr, Ni, Hg), metalloids (As, Sb), and other elements (Al, Mn, Si, Ca, Na, K, Mg, Ba) (Bhattacharya et al., 2006; Das et al., 2009a,b; Lottermoser, 2010). Most of these metals and metalloids are dissolved under acidic conditions. Therefore, AMD is an extreme and common example of poor mine water quality and can be a source of severe contamination in surface water and groundwater (Costley and Wallis, 2000; Niyogi et al., 2002).

The excavation of mines, particularly opencut mines, produces a huge amount of overburden and low-grade waste rocks that create massive piles around mine sites. These wastes typically contain sulfide minerals and are one of the major resources that effectively contribute to AMD formation around mine sites (Lottermoser, 2010). An average of 15.5 mL/day was reported as the discharged volume of AMD (pH ~3) from the West Rand Mining Basin in Gauteng Province, South Africa, which highlights the huge volume of released AMD (Hobbs and Cobbing, 2007). AMD formation can also occur through the natural weathering of sulfide-rich and carbonate-poor rocks, along the outcrops or the scree slopes of ore deposit (Lottermoser, 2010).

17.3 IMPORTANCE OF AMD TREATMENT

Discharging AMD to water bodies, rivers, and lakes reduces their water quality. The affected waters are a persistent and potentially severe source of surface water and groundwater contamination around mine sites that can continue for a long time, even after mining closure. Dissolved heavy metals are introduced into the food chain through aquatic habitats or growing vegetables and grains in contaminated soils (Baker and Banfield, 2003; Lottermoser et al., 1999). The accumulation of heavy metals through the trophic chain creates toxic effects and teratogenic changes that threaten the health of humans, plants, and animals (Ahluwalia and Goyal, 2007; Malik, 2004). The extreme toxicity of heavy metals is due to their damaging effects on the nerves, liver, and bones. These elements block functional groups of vital enzymes. As an example, Ni is listed as a possible human carcinogen (group 2B) and causes reproductive problems and birth defects. A range of detrimental effects of heavy metals on fauna and flora are well documented (Gadd, 2010; Malik, 2004).

The contamination of surface water and groundwater by mining effluents, especially in arid areas with scarce water resources, is a serious environmental issue. Additionally, the increased solubility of metals in the acidic waters leads to the loss of considerable amount of precious metals through mining wastewaters. These outcomes and stricter environmental regulations for mine sites add the necessity of treating AMD and removing heavy metals before discharging to the environment.

17.4 CONVENTIONAL TREATMENT TECHNIQUES FOR METAL ION REMOVAL FROM AQUEOUS SOLUTIONS

The conventional established methods for removing metal ions from aqueous solutions mainly include reverse osmosis, electrodialysis, ultrafiltration, ion exchange, chemical precipitation, and phytoremediation. These methods are listed in Table 17.1, with their general description and disadvantages of application. The disadvantages including incomplete metal ion removal, high reagent requirement, and generation of toxic sludge, which require further treatment and additional costs, limit the application of most methods for metal-contaminated waters including AMD (Ahalya et al., 2003; Das et al., 2008; Fu and Wang, 2011; Silverira et al., 2009).

17.5 ESTABLISHED METHODS FOR METAL ION REMOVAL FROM AMD

AMD treatment technologies are site specific, and multiple remediation strategies are commonly required to achieve successful removal of metal ions from AMD (Brown et al., 2002). Currently, the most commonly used methods in mine sites include evaporation, neutralization and precipitation, controlled release, and dilution by natural waters.

Neutralization and chemical precipitation method involve collecting AMD and selecting an appropriate neutralizing reagent. A variety of natural, by-product, or manufactured chemical reagents are used for neutralization such as limestone ($CaCO_3$), caustic lime (CaO), and hydrated lime $Ca(OH)_2$ (Lottermoser, 2010). This method requires continuous addition of reagent to AMD and a mechanical device to mix it with water so as to raise the pH to an alkaline level. An active maintenance and monitoring of the system are also required.

Beneficial chemical reagents such as limestone are low-cost materials with ease of use and form a dense and easily handled sludge. Neutralization has the potential to remove many heavy metals such as Cd, Cu, Fe, Pb, Ni, and Zn from AMD. However, this process leads to the precipitation of metals as hydroxides, which are then typically landfilled, and the precipitation of all elements does not always occur and many metals are dissolved in water within a broad range of pH (Brookins, 1988). Additionally, neutralization is ineffective in treating aqueous solutions where concentrations of the contaminants are low (1–100 mg/L). Slow reaction rates of chemical agents where ions bond with limestone particles forming iron precipitates are also a disadvantage in this method. A rapid mixing unit is required to prevent the coating of the chemicals with reaction products (Perry and Green, 2008).

TABLE 17.1
Established Techniques for Removing Metal Ions from Aqueous Solutions

Technique	Process	Disadvantage
Reverse osmosis	Metal ions are separated by a semipermeable membrane at a pressure greater than osmotic pressure caused by dissolved solids in wastewater.	It is expensive.
Electrodialysis	Metal ions are separated through semipermeable ion-selective membrane using an electrical potential between two electrodes that causes a migration of cations and anions toward respective electrodes.	There is clogging of membrane due to metal hydroxides formation.
Ultrafiltration	Pressure-driven membrane operations use porous membranes for metal ions removal.	There is sludge generation.
Ion exchange	Metal ions from dilute solutions are exchanged with ions held by electrostatic forces on the exchange resin or polymers.	It is costly and partial removal of certain ions.
Chemical precipitation	The metals are precipitated by the addition of coagulants such as alum, lime, iron salts, and other organic polymers.	It produces large amount of sludge containing toxic compounds.
Phytoremediation	This technique uses certain plants to clean up soil, sediment, and contaminated water with metals.	It requires a long time for metal ion removal, and plant regeneration for further biosorption is difficult.

Source: Ahalya, N. et al., *Res. J. Chem. Environ.*, 7, 71, 2003.

Furthermore, the precipitates inhibit the neutralization reactions and cause excessive reagent consumption (Lottermoser, 2010). Active mixing requires high energy and large sedimentation tanks and infrastructure requirements. In addition, any valuable metals present in AMD are lost in the solid-phase sludge (Perry and Green, 2008).

After the neutralization process, the treated AMD is diverted to an open pond system for evaporation, which is not a sustainable remedy and results in huge volumes of laden heavy metal sediments (Aube and Zinck, 2003; Silverira et al., 2009; Zinck, 2006).

AMD treatment has been limited to chemical precipitation in most mine sites, which is not effective for removing pollutant elements present in low concentrations. To remedy this problem, advanced wastewater treatment is required. Biotreatment and/or bioremediation has emerged as an alternative technique and has been successfully used in the municipal wastewater treatment system. This technique depends on the efficiency of microorganisms and a designed system that are discussed in the succeeding text.

17.6 BIOTREATMENT/BIOREMEDIATION OF AMD

Biotreatment or bioremediation has been considered as an alternative, efficient, and cost-effective technique to treat different types of wastewaters including AMD (Ahluwalia and Goyal, 2007; Das et al., 2009a,b; Gadd, 2010; Natarajan, 2008; Prasad, 2007). These biotechnological methods can be defined as any process that uses microorganisms, green plants, or their enzymes to return to the original or uncontaminated nature of the contaminated environment (Gadd, 2004; Malik, 2004). Biotreatment is based on metal-binding capacities of various biological materials and depends on the considerable efficiency of microorganisms. Microalgae, bacteria, and fungi have been proven to be potential metal biosorbents (Gadd, 2010). Biological methods have been developed during the last decades as a polishing stage in wastewater treatment schemes (Ahalya et al., 2003; Ahluwalia and Goyal, 2007; Das et al., 2009a,b; Gadd, 2004, 2010; Kalin et al., 2004; Natarajan, 2008; Prasad, 2007).

Biotreatment can be undertaken by dead or pretreated cells and live cells. Depending on the cell's metabolism, these mechanisms can be divided into two groups: (1) nonmetabolism dependent/passive and (2) metabolism dependent/active (Ahalya et al., 2003).

When biosorbents are exposed to metal ions, ions can be entrapped in the cellular structure and subsequently biosorbed onto the binding sites present in the cellular structure. This method of uptake is independent of the biological metabolic cycle and is known as "biosorption" or "passive uptake," which is the basis of biotreatment using dead/pretreated cells (Fomina and Gadd, 2014; Volesky, 2001).

Using live cells is the basis of the other technique in biotreatment. In this technique, the metal ions can be adsorbed onto the cell walls and also passed across the cell membrane through the metabolic cycle that is absorption. This method, which is referred to as "bioaccumulation" or "active uptake," benefits from the metal uptake by both active and passive modes. This biphasic uptake of metals is carried out by an initial rapid phase of biosorption followed by the slower, metabolism-dependent/active uptake of metals (Malik, 2004).

The main advantages of biotreatment/bioremediation by the metabolically mediated or physicochemical removal potential of microbial biomass over conventional treatment methods include (Ahalya et al., 2003)

- Availability and low cost of biologic materials
- Environmentally friendly concerning energy and material consumption
- Reduced amount of produced sludge for disposal compared to chemical precipitation
- Efficient for very low residual metal concentration (less than mg/L level) compared to other common physicochemical processes
- Can be metal specific
- Possibility of metal recycling and recovery

17.6.1 Biosorption: Passive Systems

During the 1980s and 1990s, a few pilot installations and commercial-scale units were constructed in the United States and Canada based on biosorption process by inactive cells (Tsezos, 2007). In these biotreatment systems, biological sorbent particles (biosorbents) were used that were made by the immobilization of biomass in a matrix such as polyethylene (Bennett et al., 1991), silica, or polyacrylamide gels (Darnal et al., 1986). The results confirmed the applicability of biosorption as the basis for metals sequestering/recovery, particularly from high volumes of dilute complex wastewaters. However, these pilot plants helped the researchers to realize the limitations associated with the application of inactive microbial biomass in an industrial biotreatment system (Tsezos, 2007). The limitations were mainly the cost of formulating the biomass into biosorbent material, the negative effect of solution matrix co-ions on the targeted metals, and the reduced flexibility of the biological material that made recycling and reuse of the biosorbent more difficult (Tsezos et al., 2007). Therefore, developing a continuous biotreatment system based only on biosorptive removal of metals using inactive microbial biomass is not sustainable, efficient, and effective (Ahalya et al., 2003; Malik, 2004; Tsezos et al., 2007). The application and development of a biotreatment system by active microbial cells offer a better solution for removing metal/metalloids because the absorption process is also included and contributes as a parallel mechanism to metabolically mediated uptake mechanisms (Ahalya et al., 2003; Malik, 2004; Tsezos et al., 2007).

17.6.2 Biosorption: Active Systems

The application of live microbial cells for biotreatment has recently received a lot of attention (Gadd, 2010). Application of active and growing microbial cells benefits from (Ahalya et al., 2003; Malik, 2004; Tsezos, 2007)

- Active and passive biosorption and bioaccumulation mechanisms in the cells
- The ability of self-replenishment
- Continuous metabolic uptake of metals after physical adsorption
- Potential for optimization through the development of resistant species
- The ability of the cells for detoxification of the diffused metals into the cells and bond to intracellular proteins or chelating them before being incorporated into vacuoles and other intracellular sites
- Avoiding separate biomass production process, for example, cultivation, harvesting, drying, processing, and storage prior to usage
- The ability of microbes for removing most pollutants via a single-stage process
- Unlimited capacities of live cells for removing dissolved and fine-dispersed metallic elements via immobilization

17.7 BIOSORPTION MECHANISMS: METAL–MICROBIAL CELL INTERACTIONS

Microbial cells are capable of immobilizing and mobilizing metal/metalloid ions through various physicochemical or metabolically dependent mechanisms. These metal–cell interactions can occur simultaneously and play important roles in the biosorption process (Gadd, 2004; Tsezos, 2007). The complex structure of microbial cells contributes to these mechanisms. According to the place where biosorption occurs in microbial cells, they can be further classified as (Ahalya et al., 2003; Gadd, 2010)

1. Extracellular accumulation/precipitation
2. Cell surface sorption and precipitation
3. Intracellular accumulation

The dominant mechanisms of immobilization and mobilization of elements by microbial cells are explained in the following text (Gadd, 2004, 2010).

17.7.1 INTRACELLULAR ACCUMULATION

Intracellular accumulation results from the transport of metals/metalloids across a microbial cell membrane. This kind of biosorption is dependent on the cell's metabolism and only occurs with viable cells. The accumulation process may occur due to the transportation of essential ions such as K^+, Mg^{2+}, and Na^+ during the cell's metabolism and therefore the transportation of other ions including heavy metals with the same charge and ionic radius (Tsezos, 2007).

17.7.2 PHYSICAL ADSORPTION

Physical adsorption is conducted by van der Waals forces. Previous studies showed that the electrostatic interactions

between metal ions in solutions and cell walls contribute to the biosorption of U, Cd, Zn, Cu, and Co in algae, fungi, and yeasts. Examples are biosorption of Cu by the bacterium *Zoogloea ramigera* and the alga *Chiarella vulgaris* (Aksu et al., 1992) and biosorption of Cr by the fungi *Ganoderma lucidum* and *Aspergillus niger* (Ahalya et al., 2003).

17.7.3 ION EXCHANGE

Polysaccharides enclosed in microbial cell walls contain bivalent ions such as Ca^{2+} and Mg^{2+} that can be exchanged with the metal ions such as Co^{2+}, Cu^{2+}, Cd^{2+}, and Zn^{2+} resulting in the adsorption of heavy metals (Brake and Hasiotis, 2010). Muraleedharan and Venkobachar (1990) showed that Cu was adsorbed by fungi *G. lucidum* and *A. niger* due to the ion-exchange mechanism.

17.7.4 COMPLEXATION

Complex formation on microbial cell surface resulting from the interaction between metals and active groups is one of the other biosorption processes that remove metals from solution. Carboxyl groups in the carbohydrates of microbial polysaccharides are responsible for metal complexation. Aksu et al. (1992) stated that the biosorption of Cu by *Calluna vulgaris* and *Z. ramigera* occurred through both adsorption and formation of coordination bonds between metals and active groups of cell wall polysaccharides, amino, and carboxyl groups. Complexation was found to be the only mechanism attributed for the accumulation of Ca, Mg, Cd, Zn, Cu, and Hg by *Pseudomonas syringae* (Ahalya et al., 2003). However, complexation does not always result in metal removal and accumulation in microbial cells. Microorganisms may also produce organic acids such as citric, oxalic, gluconic, fumaric, lactic, and malic acids that can chelate toxic metals, resulting in the formation of metallo-organic molecules (Gadd, 1999; Gadd and Sayer, 2000). These organic acids help in the solubilization of metal compounds and their leaching from their surfaces, which is known as the refurbishing trait of live cells to survive in metal-contaminated waters (Ahalya et al., 2003).

17.7.5 PRECIPITATION

Precipitation takes place by either metabolism-dependent or independent processes. The metal removal from the solution in the former case is often associated with the active defense system of microorganisms. Microbial cells produce some compounds such as phosphate and sulfides in response to the presence of toxic metals that favor the precipitation of metals. Consequently, soluble metals (Me^{2+}) are transformed to insoluble hydroxides, carbonates, phosphates, and sulfides (reactions 1–4 shown in the following text). These reactions may take place simultaneously or sequentially in a biological process. The relative importance of the reactions depends on the microbial culture composition and environmental conditions such as dissolved oxygen and the presence of alternative electron acceptors such as sulfates.

1. $Me^{2+} + 2OH^- \rightarrow Me(OH)_2$
2. $Me^{2+} + HCO_3^- \rightarrow MeCO_3 + H^+$
3. $3Me^{2+} + 2HPO_4^{2-} \rightarrow Me_3(PO_4)_2 + 2H^+$
4. $Me^{2+} + HS^- \rightarrow MeS + H^+$

Bioprecipitation can also result from the ability of microorganisms to alter the pH or alkalinity of their microenvironment. The cells alter their microenvironment due to their normal or induced metabolic activity, resulting in the microprecipitation of metal ions (Gadd, 2004; Tsezos, 2007).

17.7.6 CHELATION

Chelation is conducted by microbial metabolites and siderophores. Fe is an essential element for the growth of microorganisms. To obtain the required Fe, some microbes excrete siderophores, which are low-molecular-weight ligands to aid iron assimilation from precipitates (Gadd, 2001). In fact, this ability of microbes can be useful in biotreatment as the siderophores are also able to bind other metals such as Mg, Mn, Cr(III), and Ga(III) and radionuclides such as Pt(IV) and decrease their bioavailability. The chelated metals are adsorbed to the biomass and/or precipitated (Gadd, 2004, 2010).

17.7.7 LEACHING PROCESS

Microorganisms are able to acidify their environment that results from proton (H$^+$) efflux through the plasma membrane H$^+$-ATPases, charge balance maintenance, and accumulation of respiratory carbon dioxide. Acidification occurs when there is competition between hydrogen and metal ions in a metal–anion complex or in a sorbed form, resulting in the release of free metal cations (Gadd, 1999; Gadd and Sayer, 2000). For example, a strain of *Penicillium simplicissimum* was used to leach Zn from the insoluble ZnO contained in industrial filter dust (Gadd, 2004, 2010).

17.7.8 OXIDATION AND REDUCTION

The immobilization and mobilization of elements are carried out by some bacteria through reduction and oxidation processes, that is named redox transformations. Anaerobic bacteria increase the solubility of elements by reduction such as Fe(III) to Fe(II) and Mn(IV) to Mn(II) (Gadd, 2004; McLean et al., 2002). Although the results of reduction depends on the elemental property and for some elements such as U(VI) and Cr(VI) leads to their immobilization as U(IV) and Cr(III), respectively (Smith and Gadd, 2000).

17.7.9 METHYLATION

Enzymatically, the transformation of different metal/metalloids by methyl groups of microbial cells is called methylation. The solubility of some elements including Hg, As, Se, Sn, Te, and Pb through this process is mediated by a range of bacteria and fungi under aerobic and anaerobic conditions. The formation of methylated metal compounds by these processes differs in their solubility, volatility, and toxicity. For example, methylated species of Se, $(CH_3)_2Se$, and $(CH_3)_2Se_2$ are volatile and are often lost from the substrates (Gadd, 1993, 2010). The bioremediation of contaminated land and water was successfully achieved using microbial methylation of Se at the Kesterson Reservoir, California, reducing the selenium concentrations to acceptable levels via volatilization (Gadd, 2004, 2010).

The efficiency of the biosorption processes mentioned earlier is affected by environmental conditions such as pH and temperature. The presence of cations with similar charge and radius in a multi-ion solution adversely affects the biosorption efficiency. Decreasing pH, temperature, and microbial mass density are of the other preventing parameters (Ahalya et al., 2003).

17.8 BIOSORPTION OF METALS BY MICROORGANISMS FROM AMD

Many studies have reported that specific species of microorganisms have considerable potential for removing heavy metals from different types of contaminated wastewaters, typically at neutral or alkaline pH (Gadd, 2010; Mehta and Gaur, 2005; Munoz and Guieysse, 2006; Romera et al., 2006). However, for the biotreatment of acidic and contaminated AMD waters, selection of resistant and effective species of microorganisms that are able to survive in AMD is a prerequisite step (Boshoff et al., 2004; Russell et al., 2003). Extremophilic indigenous microorganisms in AMD may provide this requirement.

17.9 EXTREMOPHILE INDIGENOUS MICROORGANISMS IN AMD

AMD exerts an environmental pressure on aquatic life by chemical stress from low pH and high concentration of heavy metals and by physical stress from metal oxide deposits (Das et al., 2009a; De la pena and Barreiro, 2009). Additionally, low macronutrient levels such as organic carbon, nitrate, and phosphate in AMD restrict microbial growth (Brake et al., 2001; Lottermoser, 2010). Many studies investigated the impact of AMD on aquatic life (Bray et al., 2008; Smucker and Vis, 2011). These studies show that there are limited biodiversities in AMD, which included extremophilic microalgae, bacteria, fungi, and yeasts (Brake et al., 2001; Johnson and Hallberg, 2003; Malik, 2004; Nordstrom, 2000).

17.9.1 GREEN MICROALGAE IN AMD

The acidophilic species of microalgae, *Klebsormidium, Euglena, Microspora, Mougeotia, Ulothrix, Stigeoclonium, and Chlamydomonas*, are often found in AMD and are able to grow at low pH ~0.05 (Das et al., 2009a,b; Orandi et al., 2009, 2007;

Prasad, 2007). Abundance and distribution of green microalgae, for example, *Klebsormidium* sp. and *Chlamydomonas* sp., were reported as the indicators of AMD and high iron concentration (Novis and Harding, 2007; Valente and Gomes, 2007).

Klebsormidium sp. is an unbranched filamentous microalgae classified in the phylum of Chlorophyta (green algae). This alga is widespread and often abundant in dense mats in streams impacted by AMD (Novis, 2006; Rindi et al., 2012). In addition to filamentous microalgae, unicellular microalgae such as *Chlamydomonas* are also abundant in AMD (Das et al., 2009a,b; Kalin et al., 2004). Microalgae are the only phototrophic form of life in AMD that provide organic carbon for heterotrophic microorganisms such as fungi and bacteria (Amaral Zettler et al., 2002).

17.9.2 FUNGI AND YEASTS IN AMD

Fungi growth occurs over a wide pH range (1–11) and fungi were detected in AMD or acidic industrial wastewater (Brake and Hasiotis, 2010). The resistant strains of fungi that are able to thrive in AMD are reviewed by Das et al. (2009a,b). Fungi such as *Aspergillus*, *Fusarium*, and *Penicillium* were reported mainly from mine sites (Das et al., 2009a). Representatives of yeasts, for example, *Candida*, *Rhodotorula*, and *Trichosporon*, are also usually isolated from streams carrying AMD (Baker et al., 2004). Fungi and yeasts are considered heterotrophic organisms as their metabolism relies solely on carbon fixed by other organisms (Gadd, 2007).

The ecological study of fungal populations in the acidic Tinto River in Southern Spain revealed unexpected levels of microbial richness, as 154 strains of filamentous fungi and 90 strains of yeasts were isolated from this river. The isolated fungal strains belonged to the genus *Penicillium* and *Scytalidium*, *Bahusakala*, *Phoma*, and *Heteroconium*. The isolated strains of yeasts belonged to six genera *Rhodotorula*, *Cryptococcus*, *Tremella*, *Holtermannia*, *Leucosporidium*, and *Mrakia* (Lopez-Archilla et al., 2004). Another investigation on the microbial diversity of a highly acidic runoff (pH ~0.9) from the Richmond Mine at Iron Mountain (northern California) revealed that the majority of isolated microorganisms (68%) belonged to fungi, *Dothideomycetes* and *Eurotiomycetes* sp. (Baker et al., 2004). The presence of *Geotrichum* sp. and *Aspergillus* sp. in Sarcheshmeh copper mine runoff was also reported (Orandi et al., 2007).

17.9.3 BACTERIA IN AMD

Numerous types of bacteria including *Thiobacillus thiooxidans*, *Thiobacillus ferrooxidans*, *Ferrobacillus sulfooxidans*, *Leptospirillum ferrooxidans*, *Thiobacillus concretivorus*, *Thiobacillus thioparus*, *Sulfobacillus thermosulfidooxidans*, and *Metallogenium* were reported commonly from mine sites (Das et al., 2009a,b; Lottermoser, 2010; Novis and Harding, 2007; Palleroni, 2010; Raja and Omine, 2012). Aguilera et al. (2010) reported *Acidithiobacillus ferrooxidans* and *L. ferrooxidans* as the most abundant bacterial species in the AMD

in the Rio Tinto, Spain. Orandi et al. (2007) reported the bacteria, *Pseudomonas* spp. and *Thiobacillus* spp. isolated from acidic (pH ~3) drainages at Sarcheshmeh copper mine, Iran. The optimal pH for the growth of *Ac. ferrooxidans* is 1.5–2.5 (Valdes et al., 2008). Formation of tough and protective endospores allows the organism to tolerate the extreme conditions of AMD. They are mostly reported as an obligate aerobe (Madigan and Martinko, 2005).

Most of these acidophilic/extremophilic bacteria are chemolithotrophic, for example, *T. ferrooxidans* and *L. ferrooxidans*, and are sustained by the energy derived from pyrite oxidation in mine sites (Baker et al., 2004). They need small amounts of nitrogen and phosphorus for their metabolism, and their energy requirements are provided through the oxidation of Fe^{2+}, hydrogen sulfide, thiosulfate, sulfur, and metal sulfides. They also have the ability to transform inorganic carbon into cell building material, which may originate from the atmosphere or from the dissolution of carbonates (Baker and Banfield, 2003; Cabrera et al., 2005; Escobar et al., 2008; Yang et al., 2008).

17.9.4 ROLE OF INDIGENOUS MICROORGANISMS IN AMD

Many investigations have been carried out to understand the role of indigenous microorganisms in AMD resources or AMD-contaminated waters (Brake et al., 2001; Das et al., 2009b; Malkoc and Nuhoglu, 2003; Niyogi et al., 2002). The investigations showed that the extremophilic microbial community is not only adapted to survive under the harsh environment of AMD but also participates significantly in improving the degraded quality of AMD waters (Gadd, 2010; Lottermoser, 2010; Prasad, 2007).

Microorganism cell walls generally contain functional groups including hydroxyl (–OH), phosphoryl (–PO_3O_2), amino (–NH_2), carboxyl (–COOH), and sulfhydryl (–SH), which result in an overall negative charge on the cell surface. Metal ions with positive charges are adsorbed onto the cell surface (Mehta and Gaur, 2005). The cell walls of green microalgae are mainly composed of cellulose that has a fiber-like structure and an amorphous matrix of various polysaccharides (Bayramoglu et al., 2006). The polysaccharides, typically carbohydrates, carry carboxyl groups and are found in the microorganism cell walls. Fungal cell walls contain up to 80%–90% polysaccharides, with proteins, lipids, and polyphosphates. Bacterial cell walls also contain peptidoglycans, which are made of polysaccharides. The negative charge of carboxyl group in these microorganisms contributes in adsorbing metals onto the cell walls (Romera et al., 2006; Wang and Chen, 2009). The potential of bacteria, algae, and fungi for removing heavy metals and accumulating them on the cell surface or into the cytoplasm has been documented extensively (Das et al., 2009b; Gadd, 2010; Mehta and Gaur, 2005).

Recent studies highlight the potential of indigenous AMD microorganisms, especially microalgae, for removing metal/metalloids from AMD (Aguilera et al., 2006a,b, 2007, 2010; Bayramoglu et al., 2006; Munoz and Guieysse, 2006;

Souza-Egipsy et al., 2011; Trzcinska and Skowronska, 2012). A previous study by Skowronska (2003) investigated resistance and accumulation of elevated Zn concentrations for two strains of green algae. One species, referred to as Zn tolerant, was isolated from ditches containing mining water and another species referred to as Zn sensitive was isolated from unpolluted lake water. The Zn-tolerant species was able to accumulate significantly more Zn and Pb than the Zn-sensitive species. Additionally, the Zn-tolerant species was capable of detoxifying the excess of accumulated zinc more efficiently than the Zn-sensitive species. The indigenous microorganisms isolated from AMD can offer efficient biosorbents for AMD treatment. However, there is no adequate information that quantifies the metal removal trends of these microbes over a prolonged treatment period. Understanding the biosorption trend of these microorganisms is required before the exploitation in a biotreatment system can be achieved.

17.10 BIOSORPTION: MOBILIZATION/IMMOBILIZATION OF ELEMENTS

Most metals/metalloids are in soluble and mobile forms under the acidic conditions of AMD. Water treatment processes or techniques are mainly based on the immobilization of these elements to the less mobile species.

Despite a dramatic increase in the published literature relating to the biosorption of metals/metalloids from AMD, this technology is not yet applicable to mine sites (Ahalya et al., 2003; Tsezos, 2007). Many research processes are still at the laboratory scale and cannot be exploited in an industrial context (Fomina and Gadd, 2014; Gadd, 2010). The majority of researches have been conducted to prove the potential of biosorption by employing different microbial strains in batch systems. In these biosorption studies, the removal/immobilization of elements by the microbial cells has been mainly emphasized (Das et al., 2009a; Gadd, 2009). Immobilization takes place through a number of processes that reduce the free metal/metalloid ion species. However, immobilization may promote the solubilization of metals in some circumstances by shifting the equilibrium to release more metal into solution, which is referred to as mobilization or desorption in biotreatment studies (Gadd, 2004, 2010). The efficiency of biotreatment systems is controlled by immobilization and mobilization processes. To date, biosorption studies that demonstrated immobilization have been extensively documented, whereas mobilization/desorption studies have not been addressed adequately (Das et al., 2009). Characteristics of metal removal by microorganisms are not truly determined unless the sorption studies are complemented by desorption studies (Das et al., 2009).

Costley and Wallis (1999) used a laboratory-scale rotating biological contactor (RBC) to investigate the efficiency of an immobilized activated sludge for removing Cu, Zn, and Cd from a contaminated synthetic wastewater. They investigated the effect of three rotational speeds of 3, 15, and 25 rpm, on heavy metal removal within a 7-day continuous period.

The treatment result of >90% was very promising for Cu removal. However, the results for Zn and Cd showed that more desorption was achieved than absorption (Costley and Wallis, 1999). The results and discussion of this research were expanded mainly on the metal removal capacity of the system, whereas desorption trends were not discussed adequately. However, the authors stated that the microbial cells may exhibit resistance mechanisms to tolerate high concentrations of heavy metals as a prevention mechanism for the initial metal uptake or alternatively provide a means of expelling ions from the cells (Costley and Wallis, 2001). The resolubilization of Zn and Cd that sorbed during the adaptation period of inoculum was also attributed to their desorption results (Costley and Wallis, 1999). To design an efficient and applicable system for AMD biotreatment at mine sites, adequate understanding of the microbial performances, in particular by indigenous mine microbes, for immobilization and mobilization of metals/metalloids, is required.

17.11 IMMOBILIZED MICROBIAL BIOFILM FOR AMD TREATMENT

To select metal-resistant species for AMD treatment, instead of depending on the isolation of a single species, a better approach could be achieved by designing a consortium of metal-resistant and biosorptive potential of microorganisms (Gadd, 2010). A multispecies consortium can withstand a wider range of extreme conditions encountered with AMD, such as low pH and toxic elements. A good example of a multispecies consortium can be found in a biofilm community. A biofilm is an assemblage of single or multiple species of bacteria, fungi, and algae that are attached to an abiotic or biotic surface by their secreted extracellular polymeric substances (EPSs) (Ahluwalia and Goyal, 2007; Mehta and Gaur, 2005). The rich EPS content of the biofilms is beneficial for both biosorption of dissolved metals and entrapping dispersed solids (Malik, 2004). Additionally, biofilm formation facilitates the microbial immobilization, which is one of the objectives of biotreatment techniques. Biofilm formation processes, indigenous AMD biofilms, and the importance of biofilm communities for AMD treatment are described in the next sections.

17.12 BIOFILM FORMATION AND ITS SIGNIFICANCE IN WATER TREATMENT

More than 99% of microorganisms existing on Earth are biofilms in almost every environment that is moist with adequate nutrient flow (Singh et al., 2006). Biofilm formation is a natural complex process where clusters of microbial cells attach to a rough surface and stick together by secreted EPS.

A biofilm formation process is depicted in Figure 17.1. The life of a biofilm initiates from planktonic or free-floating cells. In order to form a biofilm, a planktonic cell must first interact with a surface for attachment. The process starts with the attachment of microbial cells (Figure 17.1; Step 1).

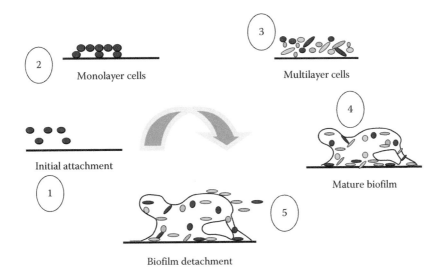

FIGURE 17.1 Cyclic process of biofilm formation. (Adapted from Singh, R. et al., *Trends Microbiol.*, 14, 389, 2006. With permission.)

After initial association with the surface, a planktonic cell can dissociate from the surface and return to the planktonic state or irreversibly be reattached to the surface. Irreversible attachment involves the production of EPS, and a monolayer of cells is produced in this step (Figure 17.1; Step 2). The process continues with cell agglomeration forming a multilayer cell structure (Figure 17.1; Step 3). A mature biofilm is developed (Figure 17.1; Step 4), and finally, the cell detachment step releases some cells that proceed to a new formation cycle (Figure 17.1; Step 5). The thick layer of multicellular aggregates is known as biofilm (Mehta and Gaur, 2005; Qureshi and Blaschek, 2005; Singh et al., 2006).

EPS plays an important role in binding cells to a surface and protecting them from the surrounding environment. EPS is mainly composed of polysaccharides, proteins, nucleic acids, and phospholipids. Carbohydrates and proteins are the prerequirements for biofilm establishment. EPS protects the cells in the biofilm by providing a diffusive barrier to any toxic compounds that could harm the cells. The other role of EPS is to provide a barrier to contain nutrients for cell growth (Qureshi and Blaschek, 2005).

The biofilm structure is composed of microbial cells immobilized in a heterogeneous and very porous matrix with a high amount of water as it comprises 95%–97% of the matrix. Water is bound to the capsules of microbial cells or solvent. Secreted cells products (EPS), cell lysis products, absorbed nutrients, and even particulate material and detritus from the nearby surrounding environment distributed within the interstitial voids form the matrix (Bae et al., 2000). The diffusion function within the matrix is related to the water-binding capacity and mobility of the biofilm (Van Hullebusch et al., 2003).

Water and nutrient diffusion into the interior of a biofilm is highly limited. Once a mature biofilm is developed, water channels facilitate the deeper conveyance of water and nutrients into the biofilm. The architecture of the biofilm develops

in response to shear forces. In low-shear environments, the biofilm forms thick layers as mushroomlike masses. Under high-shear stress environment, the biofilm is flatter or forms long strands (Singh et al., 2006). However, high-shear forces lead to biofilm detachment. Additionally, nutrient starvation is another reason for biofilm detachment (Qureshi and Blaschek, 2005).

Microorganisms that secrete polymers and form biofilms gain high microbial biomass. The high density of this microbial biomass provides and maintains optimal conditions for growth. The optimal conditions include pH, localized solute concentrations, and reductive–oxidative potential in the vicinity of the cells. This is achieved by the unique architecture of the biofilm and controlled circulation of fluids within it. Additionally, high microbial biomass facilitates immobilization of minerals, nutrients, and metals from the surrounding liquid phase (Singh et al., 2006; Van Hullebusch et al., 2003).

Spath et al. (1998) investigated the role of EPS in binding aqueous metal species from wastewater. He evaluated the sorption of Cd and Zn to the EPS and cell components of a biofilm collected from a sequenced batch biofilm reactor. Although EPS has a high density of charged functional groups, 80% of the total Cd content was sorbed to the cellular component of the biofilm. A laboratory study also investigated EPS production and metal adsorption by *Spirulina* sp. A steady increasing amount of Cu removal was observed along with an increased level of EPS production (Das et al., 2009).

17.13 INDIGENOUS ALGAL–MICROBIAL BIOFILM IN AMD

Indigenous AMD microorganisms thrive as extensive biofilms along the acidic drainages, attached to a substrate (Aguilera et al., 2006a,b, 2007, 2010; Levings et al., 2005).

Biofilm development was studied by Aguilera et al. (2006b, 2007, 2010) in an extremely acidic river, Rio Tinto in Spain. The biofilm was mainly composed of the autotrophic species of flagellated or filamentous green microalgae (60%), chemolithotrophic bacteria, and heterotrophic microbes such as bacteria, fungi, amoebae, small flagellates, and ciliates. Coexistence and synergistic relationships among the multispecies of the indigenous AMD biofilm protected them from environmental stresses and enabled them to survive under the extreme conditions (Aguilera et al., 2010; Amaral-Zettler et al., 2002; Das et al., 2009a,b; Levings et al., 2005; Niyogi et al., 2002).

Green microalgae play a significant role among the extremophilic microbes that participate in AMD biofilm. The bright green color of the algal mat denotes the presence of chlorophyll. The evidence of their photosynthesis activity is the presence of oxygen bubbles at the water–air interface (Espana et al., 2007). The photosynthetically produced oxygen enhances the functions of iron-oxidizing bacteria. Microalgae provide organic carbon for heterotrophic microorganisms through photosynthesis and they are the basis of food chains in AMD (Amaral Zettler et al., 2002). These extremophilic microalgae have adapted to grow under low nutrient levels in AMD. Mining wastewaters including AMD are considered to be limiting in NO_3 and PO_4, which are the major nutrient requirements for algae (Dunbabin and Bowmer, 1992). NO_3 is mainly derived from explosive materials, which is used for blasting operations, and phosphorus is resulted from slow weathering and erosion of the surrounding rocks (Lottermoser, 2010; Thomson and Tracey, 2005). The autotrophic property of algae influences AMD waters as they assimilate nitrate and inorganic carbon from the AMD environment. This occurrence raises the alkalinity nature of water, therefore directly decreasing the acidity of water (Das et al., 2009b).

The algal biofilms colonize acidic drainages released from waste piles and tailings at mine sites. The biofilms cover almost 100% of the stream substrate from the discharge point up to several meters along the stream. The development of the concentrated biofilm within this distance is due to the availability of Fe ion. At the release point of AMD, the oxygen content is depleted due to the respiration of heterotrophic bacteria that oxidize pyrite (FeS_2), indicating that Fe is available in AMD in its ionic form. Fe is an essential element for the growth of microalgae (Espana et al., 2007). Additionally, under the low pH (~3) conditions found in AMD, higher organisms (e.g., grazers) are not present, and microalgae can dominate the biofilm.

Microalgae have also the ability to produce EPS, especially under nutrient stress conditions. EPS participates in biofilm formation and metal removal (Sutherland, 2005). Aguilera et al. (2008a,b) analyzed the composition of EPS extracted from 12 biofilms, isolated from the Rio Tinto, Spain. The results showed that the heavy metal content of the biofilm closely resembled the water composition (Das et al., 2009).

Indigenous AMD biofilms dominated with microalgae are efficient biosorbents for the primary biotreatment of AMDs with low pH and high Fe^{2+} ion composition.

17.14 AMD TREATMENT USING BIOFILM REACTORS

Novel biotreatment technologies for industrial wastewater treatment, including AMD, are based on the establishment and maintenance of specific active biofilms on an appropriate solid support medium. The key advantages of using biofilm treatment systems include the following (Tsezos et al., 2007):

1. The natural attachment of biofilms creates high densities of microorganisms per unit volume of reactor that directly improves the treatment performance in small reactor volumes, which is economically advantageous.
2. The immobilized microorganisms in biofilm structures minimize the requirement for sedimentation after the treatment process.
3. The immobilized biofilms survive and withstand concentrated AMD discharges of toxic wastes due to the protective nature of EPS in biofilms.
4. Once a biofilm is established, the microbes are reinforced and protected by the matrix, so the elimination of biofilm is very difficult. Therefore, a sustainable system can be designed that requires minimal maintenance and dosing of nutrients.

The significant advantage of immobilized microorganisms in a biofilm structure has led to the development of biofilm reactors in biotreatment industry for metal-bearing wastewaters such as AMD. Biofilm reactors facilitate an effective contact between biofilms and wastewaters (Ahalya et al., 2003; Tsezos et al., 2007).

During the late 1980s and 1990s, novel biofilm reactor systems were developed and applied (Tsezos et al., 2007). Systems such as small-granule fixed beds, fluidized beds, RBCs, trickling filters, and hybrid-suspended biofilm that support development and maintenance of biofilms were developed. The relatively recent pilot and industrial biofilm reactors used in mining and metallurgical wastewater polishing treatment for sequestering various targeted pollutants including toxic metal/metalloids are moving-bed sand filter reactor (Diels et al., 2003), anaerobic packed bed reactors (Jong and Parry, 2003), and RBCs (Costley and Wallis, 2001). The configuration and operation process method of these reactors are described in the following sections.

17.15 MOVING-BED SAND FILTER REACTORS

In moving-bed sand filters, sand granules act as supporting carrier material for biomass immobilization. The microbial biofilm that forms on sand grains interacts with the soluble targeted metal species in wastewater. The key characteristic of this reactor is that the sand is continuously circulated and washed. The excess of produced biomass with the sequestered metal ions is also continuously removed. Biofilm growth is supported by the addition of selected carbon sources such as acetate. Acetate is metabolized by microbial biomass and increases alkalinity that alters the chemical microenvironment of the biofilm by increasing the pH (Tsezos et al., 2007).

The application of moving-bed sand filters for the development of metal-reducing biofilms was tested by Diels et al. (2003). A moving-bed sand filter with a diameter of 3 m, a filter bed height of 4 m, and a filter bed volume of 12 m³ was used that was filled with quartz sand with a grain-size diameter between 1.2 and 2.0 mm. The applied filtration velocity and feed flow were between 6 and 7 m/h and 20 and 45 m³/h, respectively. The sand circulation speed of 0.5–1.0 cm/min and wash water flow of 3–5 m³/h were used. The contact time to empty the reactor was about 36 min. The inoculum was a consortium composed of resistant, metal-biosorbing/bioprecipitating bacteria including *Pseudomonas mendocina* AS302, *Arthrobacter* spp. BP7/26, and *Ralstonia eutropha* CH34. The used wastewater for treatment contained mainly Co (1.5 mg/L), Ni (0.7 mg/L), Zn (0.2 mg/L), and Cu (0.1 mg/L). Nitrate (3.4 mg/L) and carbon substrate (8 mg/L) were supplied to the system as nutrients to maintain biofilm growth. The reactor was operated for 18 months continuously without loss of activity. Ni, Co, Zn, and Cu were removed between 80% and 100%. Fe removal was 60%–80%, and for other metals, such as Al, Ag, Cr, As, and Se, the removal was up to 80%. The biosludge analysis revealed the concentration of heavy metals up to 10% of the dry weight. The results showed the reliable technology of sand filter reactor for metals' removal from a low-concentration solution by specific resistant microbial strains (Diels et al., 2003). However, in the reported study, few metals were included in the wastewater composition at low concentrations, whereas AMD contains many of these elements at more than 100 mg/L. Additionally, the acidic nature of AMD with high amounts of sulfate that limit bioremoval efficiency was not considered.

17.16 ANAEROBIC BIOFILM REACTORS USING SULFATE-REDUCING BACTERIA

Anaerobic biofilm reactors by the exploitation of sulfate-reducing bacteria (SRB) are the most common biofilm reactors in mining. These reactors are used to produce biogenic hydrogen sulfide under anaerobic conditions. Reactor configurations such as packed or fluidized beds have been tested for pilot or commercial applications (Kaksonen and Puhakka, 2007). The wastewater is fed into the bioreactor where the SRB biomass is grown and metal precipitation occurs simultaneously in the cells' microenvironment. It is also possible to collect the produced H₂S to use subsequently for the precipitation of the metal ions in the wastewater as metal sulfides. This technology can be applied in mining and metallurgical industry for the treatment of AMD and metal-/sulfate-bearing wastewaters (Tsezos et al., 2007).

Jong and Parry (2003) used a bench-scale upflow anaerobic packed bed reactor for removing sulfate and heavy metals by SRB in short-term runs. Contaminated water with Cu, Zn, Ni, Fe, Al, As (5–50 mg/L), Mg (>500 mg/L), and sulfate

(~2500 mg/L) was treated over a 14-day period at 25°C. The reactor was filled with silica sand, and SRB was employed as inoculum. Organic substrate and sulfate were added to the reactor at loading rates of 7.43 and 3.71 kg/day m³, respectively. The results showed that more than 97.5% of the initial concentrations of Cu, Zn, and Ni were removed, while only >77.5% and >82% of As and Fe were removed, respectively. Additionally, >82% reduction in sulfate concentration was recorded. In contrast, the concentration of Mg and Al remained unchanged during the treatment period. The pH of water was increased from ~4.5 to 7.0 due to the SRB activity. Additionally, the removal of sulfate and metals was enhanced in comparison to controls not inoculated with SRB (Jong and Parry, 2003). The results of the aforementioned study highlighted the potential of SRB bacteria for removing various elements. However, some elements such as Mg and Al were not removed. Additionally, SRB bacteria are heterotrophic and nutrient provision is required for their growth (Ahalya et al., 2003).

17.17 RBC

The RBC facilitates the immobilization of microorganisms as a fixed biofilm, attached to the moving support media (Ibrahim et al., 2012; Rodgers and Zhan, 2003). An RBC unit typically consists of a series of closely spaced flat or corrugated discs that are mounted on a horizontal shaft. Generally, the discs are partly submerged (40%) in the solution. The shaft continually rotates by a mechanical motor or a compressed air drive. The media rotation promotes oxygen transfer and maintains the biomass in aerobic conditions. The rotation of the discs creates shear forces that enhance the stripping off of the excess biomass (Cortez et al., 2008; Patwardhan, 2003; Rodgers and Zhan, 2003). In the case of using mesh material for discs, the rotation and shear stress maintain a constant population of microorganism and conserve the media from clogging. Several units can be constructed as series or parallel arrangements to improve the treatment results (Cortez et al., 2008; Ibrahim et al., 2012; Patwardhan, 2003).

17.18 SIGNIFICANT ADVANTAGES OF RBC AND DESIGN PARAMETERS

RBCs are one of the most efficient bioreactors that are commonly used to treat municipal and industrial wastewater (Ibrahim et al., 2012; Mathure and Patwardhan, 2005). Easy construction and expansion, simple process control and monitoring, simple and feasible design and operation, and high interfacial areas are generated by discs that are independent of the rotation speed, resistance to shock and toxic loads, lowland occupancy, low energy consumption, and low cost of operation and maintenance, are of the included advantages of using RBCs. The efficiency of RBCs depends on several design parameters that are listed in Table 17.2 (Cortez et al., 2008; Costely and Wallis, 2000; Ibrahim et al., 2012; Kargi and Eker, 2001; Mathure and Patwardhan, 2005).

TABLE 17.2
RBC Design and Operation Parameters

Factors	Effect	Optimized Parameter
Rotational speed	This is important in biofilm growth, metal removal efficiency, and nutrient and oxygen mass transfer.	1–10 rpm for 1–4 m diameter mounted on shafts around 5–10 m long
Hydraulic flow rates	Increasing flow rates decrease removal efficiency.	The range in full-scale size: 1.292–6.833 dm^3/m^2 h
HRT	Longer contact times improve removal efficiency.	24 h
RBC media	More surface area is favorable (corrugated and cellular mesh), considering costs.	PVC, polycarbonate sheets, HDPE
Staging	Multistaging increases efficiency.	Using baffles in a tank or using a series of tanks
Temperature	It directly affects the microbial activity.	Optimized temperature: between 20°C and 30°C
RBC medium submergence	Medium submersion increases the volume capacity and reduces RBC staging requirement, depending on operation type, microorganisms, and characteristics of wastewater.	40%–60%

From the point of configuration, RBCs provide flat support media for the growth of indigenous AMD algal–microbial biofilms that are used as the biosorbent in this study. The discs' surfaces can resemble AMD substrate surface for the attachment and growth of algal–microbial biofilms, and the rotational speed is adjustable to protect the biofilm from detachment. Additionally, the rotation of discs allows the biofilm to be exposed to a light source, which can be installed on top of the reactor or side area, or exposed to sunlight. The rotational speed is easily adjustable to an optimized speed for maximum metal uptake and biofilm growth. The partial submersion and rotation of discs of the RBC provide enough air around the layer of biofilm that enhance the aerobic microbial performance in biofilm structure. The biofilm in RBCs aids the accumulation of metal/metalloids in a low volume of biosorbent that can be recovered in low volumes of water. In the case of biofilm detachment, it can regrow due to the presence of an initial layer and the resistant nature of the biofilm that provides a viable biosorbent in RBCs.

17.19 RBC APPLICATION FOR REMOVING METAL/METALLOIDS FROM WASTEWATERS

A series of studies were carried out by Costley and Wallis (1999, 2000, 2001) using an RBC for removing Cu, Zn, and Cd from a synthetic wastewater. They used a laboratory-scale RBC, inoculated with the enriched culture of sewage-activated sludge for biofilm development. A nutrient broth was formulated and used to supply carbon while minimizing the possibility for metal complexation. The RBC contained 10 L wastewater, and the discs were submersed to a level of 40%. The rotation speed of disc was adjusted at 10 rpm (revolution/min), and the flow rate was set to 6.9 mL/min. Both parameters were already optimized for achieving maximum biofilm growth and metal removal. The RBC was operated continuously for multiple sorption (84 days) and desorption (48 h)

cycles with a hydraulic retention time (HRT) of 24 h. After the operation period (84 days), the average metal removal of Cu, Zn, and Cd achieved was up to 81.8%, 49.7%, and 30.1%, respectively. The results were relatively constant during the operation (Costley and Wallis, 2001).

In another example, Travieso et al. (2002) used a rotary drum that was covered by 0.5 mm wide polyurethane bands for the immobilization of a microalgal biofilm. The pure culture of microalga *Scenedesmus obliquus* was used as an inoculum that was obtained through serials of dilution and antibiotic treatment. A synthetic wastewater containing 140 mL of municipal sewage and 0.1 g/L of cobalt salt ($CoSO_4 \cdot 7H_2O$) was used in a reactor that gave a final concentration of 3000 µg/L and pH ranging between 8.6 and 8.9. The reactor was operated in a 20-day batch mode at a constant rotational speed of 2 rpm. The authors reported 94.5% cobalt removal after 10 days. In the earlier-mentioned research, algae were successfully immobilized on the rotary system, and the result was promising for Co removal that is a typical toxic element in wastewaters such as AMD.

Kapoor et al. (2004) studied the biological oxidation of Fe^{2+} under acidic conditions by using oxidizing bacteria in a bench-scale RBC. The RBC feed wastewater was composed of $FeSO_4 \cdot 7H_2O$ at 0.08 g/L and $MgSO_4 \cdot 7H_2O$ at 0.04 g/L concentration. $(NH_4)_2SO_4$ and KH_2PO_4 were used at concentrations of 0.08 and 0.1 g/L to supply nutrients. Potable water was used for making up the solution, and concentrated sulfuric acid was added to bring the pH from 1.9 to 2.0. The RBC was able to achieve 50% Fe^{2+} oxidation efficiency after 24 h operation.

The examples described previously used RBCs for removing metals/metalloids from metal-bearing synthetic wastewaters, not particularly AMD. The applications of RBC for AMD treatment have been mostly focused on Fe^{2+} removal, which is one of the major environmental issues related to AMD. For example, Olem and Unz (1977, 1980), who possibly were the pioneers for AMD treatment by RBCs, used a pilot-scale RBC to evaluate ferrous iron oxidation. The actual AMD (pH ~2.7), which was released from a coal mine, was introduced to the RBC. The indigenous iron-oxidizing bacteria developed a biofilm in the RBC that mediated the transformation of Fe^{2+} to the less soluble ferric state, Fe^{3+}. The RBC was operated for 11 months at the optimum operating speed of 10 rpm. The hydraulic loadings of 0.11 and 0.22 m^3/day m^2 resulted in the oxidation of 240 mg/L influent Fe^{2+} to produce effluent Fe^{2+} of 2 and 5 mg/L, respectively. The results indicated the applicability and efficiency of the system to be used for the primary treatment of AMDs with low pH (~3) and high Fe^{2+} ions. However, the applicability of RBC for removing other metals/metalloids was not assessed in the reported study.

The earlier-mentioned studies reported on the application of sand bed filters and anaerobic reactors and RBCs indicate the significant potential of microbial biofilms, mainly from bacteria, developed in the biofilm reactors for removing metal/metalloids from contaminated wastewaters. However, there are significant gaps that limit the exploitation of their results as an applicable biotreatment system on mine sites that include the following:

1. Exploitation of bacterial biofilms requires regular and appropriate nutrient dosing for their growth that adds to costs and maintenance requirements and limits the feasibility of their usage. Costley and Wallis (2001) stated that biotechnological approaches for metal removal from low metal-contaminated voluminous wastewaters can be economically viable only if cheap carbon and nutrient sources can be provided. In the reported study, only the case studies conducted by Olem and Unz (1977, 1980) did not require nutrient supply as the iron-oxidizing bacteria used the released energy from the oxidation process of iron and captured carbon dioxide from air.

2. The majority of biotreatment investigations have reported metal removal efficiency using synthetic AMD or synthetic metal-bearing solutions, composed of only a few specific heavy metals. The multi-ion composition of AMD composed of various metal/metalloids at high or low concentrations, high sulfate concentration, and low pH, which can adversely affect the biotreatment results, was not considered adequately. The application of the biotreatment systems with real AMD resulted in low removal efficiency (Ahalya et al., 2003; Gadd, 2009, 2010).

3. Many biotreatment studies were conducted in batch systems and do not show the metal removal trends in continuous process. Furthermore, the reported results emphasize removal efficiency and do not describe the negative results or desorption stages adequately.

4. The cleansing role of indigenous AMD microorganisms in particular microalgae and fungi has not been taken into account for the biotreatment systems.

17.20 RBC APPLICATION FOR REMOVING METAL/METALLOIDS FROM AMD BY INDIGENOUS MICROORGANISM

A recent study by Orandi et al. (2012) investigated the application of indigenous microbial biofilm in a modified RBC. The microorganisms are obtained from AMD resources at Sarcheshmeh copper mine, Iran. The microbial sample contained mainly filamentous and unicellular green microalgae, *Klebsormidium* sp. and *Chlamydomonas* sp.; bacteria, *Ac. ferrooxidans*, *L. ferrooxidans*, and *Pseudomonas* sp.; and fungi, *Aspergillus* sp. and *Penicillium* sp. In this study, the AMD from which the indigenous microbial consortium was collected was analyzed to quantify its cation and anion (including nutrients PO_4^{-3} and NO_3^{-}) contents. The analysis data were used to synthesize a multi-ion AMD composed of 25 components (cations and anions at concentrations 0.005–100 mg/L), high sulfate (>1000 mg/L), and low pH (~3) (Orandi and Lewis, 2013b). The indigenous microbial assembly was maintained in synthetic AMD (Syn-AMD) *in vitro*. For the biotreatment investigations, a laboratory-scale photo-rotating biological contactor (PRBC) was designed and used to immobilize the microbial consortium as an algal–microbial biofilm (Orandi and Lewis, 2013a). The PRBC was initially operated in batch mode, using Syn-AMD and indigenous microbes as PRBC solution and inoculum, respectively. An algal–microbial biofilm (60 g dry weight) was successfully grown on the discs' surfaces in the PRBC after 12 weeks. The PRBC was then operated at both batch and continuous modes to investigate the efficiency of the system for removing different elements from the Syn-AMD. Batch systems were conducted in 7-day periods under pH 3 and 5. The batch results showed that the algal–microbial biofilm system was able to reduce the concentration of major elements from 10% to 60% at pH 3 in the order of Na > Cu > Ca > Mg > Mn > Ni > Zn, whereas higher results (40%–70%) were recorded for these elements at pH 5 in the order of Cu > Mn > Mg > Ca > Ni > Zn > Na. The removal trend for each element contained maximum and minimum removal values that occurred during the experiment. The removal efficiency of the system for trace elements varied extensively between 3% and 80% under both pH conditions (Orandi and Lewis, 2013a,b).

The efficiency of the system was also evaluated in continuous condition, by introducing Syn-AMD (pH ~3) into the PRBC at the flow rate of 10 mL/min and HRT of 24 h (Orandi et al., 2012). The operation of PRBC within a 28-day period showed similar removal efficiency (10%–60%) compared with the batch operation, for most of the elements. The chemical composition of treated water was examined daily within 28 days, and the results revealed absorption (7 days) and desorption periods occurring alternatively. The increase and decrease of pH by 0.5 and 0.2 were recorded at the same time of absorption and desorption periods, which was attributed to mobilization and immobilization mechanisms occurring in the algal–microbial biofilm.

The system was operated for a further 10 weeks continuously, and the results demonstrated the average weekly removal for major elements from 20% to 50% in the order of Cu > Mg > Ni > Na > Mn > Ca > Zn, whereas for trace elements varied broadly between 10% and 80%. Scanning electron microscopy analysis illustrated the accumulation of heavy metals in/on the biofilm. Biofilm analysis also revealed the presence of different elements of up to 10% of the dried biomass. The results demonstrated the effectiveness and sustainability of indigenous environmentally friendly algal–microbial biofilm to be exploited for removing most of elements from AMD. The results offer a potentially sustainable approach for the primary treatment of AMD at mine sites (Orandi et al., 2012).

17.21 SUMMARY

The principal achievement of this chapter is considering the exploitation of indigenous mining algal–microbial biofilms for the development of an applicable biotreatment system in mine sites to reduce the elemental content of mine water (AMD) or wastewaters. The results of recent studies demonstrated the effectiveness of these environmentally friendly biofilms for removing most elements from AMD. The biosorption results offer a potentially sustainable approach for the primary treatment of AMD at mine sites where pH is ≤3. To improve the removal efficiency, in particular for the elements such as Zn, Mn, Mo, Cr, Pb, Se, Sb, Co, and Al with lower adsorption (~20%), a series of sequential PRBCs must be designed and used. Additionally, sequential reactors could supplement the adsorption results where desorption occurred.

REFERENCES

Aguilera, A., Gomez, F., Lospitao, E., and Amils, R. 2006a. Molecular approach to the characterisation of eukaryotic communities of an extreme acidic environment: Methods for DNA extraction and DGGE analysis. *Systematic and Applied Microbiology*. 29: 593–605.

Aguilera, A., Gonzalez-Toril, E., Souza-Egipsy, V., Amaral-Zettler, L., Zettler, E., and Amils, R. 2010. Phototrophic biofilms from Rio Tinto, an extreme acidic environment, the prokaryotic component. *Microbial Mats*. 14: 469–481.

Aguilera, A., Manrubia, S.C., Gomez, F., Rodriguez, N., and Amils, R. 2006b. Eukaryotic community distribution and their relationship with the water physicochemical parameters in an extreme acidic environment, Rio Tinto (SW, Spain). *Applied and Environmental Microbiology*. 72: 5325–5330.

Aguilera, A., Souza-Egipsy, V., Gomez, F., and Amils, R. 2007. Development and structure of eukaryotic biofilms in an extreme acidic environment Rio Tinto (SW, Spain). *Microbial Ecology*. 53: 294–305.

Aguilera, A., Souza-Egipsy, V., Martin-Uriz, P.S., and Amils, R. 2008a. Extracellular matrix assembly in extreme acidic eukaryotic biofilms and their possible implications in heavy metal adsorption. *Aquatic Toxicology*. 88(4): 257–266.

Aguilera, A., Souza-Egipsy, V., Martin Uriz, P.S., and Amils, R. 2008b. Extraction of extracellular polymeric substances from extreme acidic microbial biofilms. *Applied Microbiology and Biotechnology*. 78(6): 1079–1088.

Ahalya, N., Ramachandra, T., and Kanamadi, R. 2003. Biosorption of heavy metals. *Research Journal of Chemistry and Environment*. 7: 71–79.

Ahluwalia, S.S. and Goyal, D. 2007. Microbial and plant derived biomass for removal of heavy metals from wastewater. *Bioresource Technology*. 98: 2243–2257.

Aksu, Z., Sag, Y., and Kutsal, T. 1992. The biosorption of copper by *C. vulgaris* and *Z. ramigera*. *Environmental Technology*. 13: 579–586.

Amaral Zettler, L.A., Gomez, F., Zettler, E., Keenan, B.G., Amils, R., and Sogin, M.L. 2002. Eukaryotic diversity in Spain's River of Fire. *Nature*. 417: 137.

Aube, B., Zinck, J. 2003. Lime treatment of acid mine drainage in Canada. In *Brazil–Canada Seminar on Mine Rehabilitation*, Florianópolis, Brazil, pp. 23–40.

Bae, J.W., Rhee, S.K., Hyun, S.H., Kim, I.S., and Lee, S.T. 2000. Layered structure of granules in upflow anaerobic sludge blanket reactor gives microbial populations resistance to metal ions. *Biotechnology Letters*. 22: 1935–1940.

Baker, B.J. and Banfield, J.F. 2003. Microbial communities in acid mine drainage. *FEMS Microbiology Ecology*. 44: 139–152.

Baker, B.J., Lutz, M.A., Dawson, S.C., Bond, P.L., and Banfield, J.F. 2004. Metabolically active eukaryotic communities in extremely acidic mine drainage. *Applied and Environmental Microbiology*. 70: 6264–6271.

Bayramoglu, G., Tuzun, I., Celik, G., Yilmaz, M., and Arica, M.Y. 2006. Biosorption of mercury, cadmium and lead ions from aqueous system by microalgae *Chlamydomonas reinhardtii* immobilised in alginate beads. *International Journal of Mineral Processing*. 81: 35–43.

Bennett, P.G., Ferguson, C.R., and Jeffers, T.H. 1991. Biosorption of metal contaminants from acidic mine waters. In *Process Mineralogy XI—Characterization of Metallurgical and Recyclable Products*. (eds.) D.M. Hausen, W. Petruk, R.D. Hagni, and A. Vassiliou, pp. 213–222.

Bhattacharya, J., Islam, M., and Cheong, Y.W. 2006. Microbial growth and action: Implications for passive bioremediation of acid mine drainage. *Mine Water Environment*. 25: 233–240.

Boshoff, G., Duncan, J., and Rose, P.D. 2004. The use of micro-algal biomass as a carbon source for biological sulphate reducing systems. *Water Research*. 38: 2659–2666.

Brake, S., Connors, K., and Romberger, S. 2001a. A river runs through it: Impact of acid mine drainage on the geochemistry of West Little Sugar Creek pre- and post-reclamation at the Green Valley coal mine, Indiana, USA. *Environmental Geology*. 40: 1471–1481.

Brake, S.S., Dannelly, H.K., and Connors, K.A. 2001b. Controls on the nature and distribution of an alga in coal mine-waste environments and its potential impact on water quality. *Environmental Geology*. 40: 458–469.

Brake, S.S. and Hasiotis, S.T. 2010. Eukaryote-dominated biofilms and their significance in acidic environments. *Geomicrobiology Journal*. 27(6–7): 534–558.

Bray, J.P., Broady, P.A., Niyogi, D.K., and Harding, J.S. 2008. Periphyton communities in New Zealand streams impacted by acid mine drainage. *Marine and Freshwater Research*. 59: 1084–1091.

Brown, M., Barley, B., and Wood, H. 2002. *Mine Water Treatment: Technology, Application and Policy*. International Water Association Publishing, London, U.K.

Brookins, D.G. 1988. *Eh-pH Diagrams for Geochemistry*. Springer-Verlag, Berlin, Germany, p. 176.

Cabrera, G., Gomez, J.M., and Cantero, D. 2005. Influence of heavy metals on growth and ferrous sulphate oxidation by *Acidithiobacillus ferrooxidans* in pure and mixed cultures. *Process Biochemistry*. 40: 2683–2687.

Cortez, S., Teixeira, P., Oliveira, R., and Mota, M. 2008. Rotating biological contactors: A review on main factors affecting performance. *Reviews in Environmental Science and Biotechnology*. 7: 155–172.

Costley, S.C. and Wallis, F.M. 1999. Effect of disk rotational speed on heavy metal accumulation by rotating biological contactor (RBC) biofilms. *Letters in Applied Microbiology*. 29: 401–405.

Costley, S.C. and Wallis, F.M. 2000. Effect of flow rate on heavy metal accumulation by rotating biological contactor (RBC) biofilms. *Journal of Industrial Microbiology and Biotechnology*. 24: 244–250.

Costley, S.C. and Wallis, F.M. 2001. Bioremediation of heavy metals in a synthetic wastewater using a rotating biological contactor. *Water Research*. 35: 3715–3723.

Darnal, D.W., Green, B.H., McPherson, R.A., Henzl, M., and Alexander, M.D. 1986. Recovery of heavy metals by immobilized algae. In *Trace Metal Removal from Aqueous. Solution*, (ed.) Thompson, R. The Royal Society of Chemistry, pp. 1–24, 10.

Das, B., Roy, A., Singh, S., and Bhattacharya, J. 2009a. Eukaryotes in acidic mine drainage environments: Potential applications in bioremediation. *Reviews in Environmental Science and Biotechnology*. 8: 257–274.

Das, B.K., Roy, A., Koschorreck, M., Mandal, S.M., Wendt-Potthoff, K., and Bhattacharya, J. 2009b. Occurrence and role of algae and fungi in acid mine drainage environment with special reference to metals and sulfate immobilisation. *Water Research*. 43: 883–894.

Das, N., Vimala, R., and Karthika, P. 2008. Biosorption of heavy metals—An overview. *Indian Journal of Biotechnology*. 7: 159–169.

De la pena, S. and Barreiro, R. 2009. Biomonitoring acidic drainage impact in a complex setting using periphytons. *Environmental Monitoring and Assessment*. 150(1–4): 351–363.

Diels, L., Spaans, P.H., Van Roy, S., Hooyberghs, L., Ryngaert, A., Wouters, H., Walter, E. et al. 2003. Heavy metals removal by sand filters inoculated with metal sorbing and precipitating bacteria. *Hydrometallurgy*. 71: 235–241.

Dunbabin, J.S. and Bowmer, K.H. 1992. Potential use of constructed wetlands for treatment of industrial wastewaters containing metals. *Science of the Total Environment*. 111: 151–168.

Escobar, B., Bustos, K., Morales, G., and Salazar, O. 2008. Rapid and specific detection of *Acidithiobacillus ferrooxidans* and *Leptospirillum ferrooxidans* by PCR. *Hydrometallurgy*. 92: 102.

Espana, J.S., Pastor, E.S., and Pamo, E.L. 2007. Iron terraces in acid mine drainage systems: A discussion about the organic and inorganic factors involved in their formation through observations from the Tintillo acidic river (Riotinto mine, Huelva, Spain). *Geosphere*. 3(3): 133–151.

Fomina, M. and Gadd, G.M. 2014. Biosorption: Current perspectives on concept, definition and application. *Bioresource Technology*. 160: 3–14.

Fu, F. and Wang, Q. 2011. Removal of heavy metal ions from wastewaters: A review. *Journal of Environmental Management*. 92(3): 407–418.

Gadd, G.M. 2004. Microbial influence on metal mobility and application for bioremediation. *Geoderma*. 122: 109–119.

Gadd, G.M. 2007. Geomycology: Biogeochemical transformations of rocks, minerals, metals and radionuclides by fungi, bioweathering and bioremediation. *Mycological Research*. 111(1): 3–49.

Gadd, G.M. and White, C. 1993. Microbial treatment of metal pollution – a working biotechnology. *Trends in Biotechnology* 11: 353–359.

Gadd, G.M. 1999. Fungal production of citric and oxalic acid: importance in metal speciation, physiology and biogeochemical processes. *Advances in Microbial Physiology* 41: 47–92.

Gadd, G.M. 2001. Accumulation and transformation of metals by microorganisms. In *Biotechnology, a Multi-volume ComprehensiveTreatise*, vol. 10, Special Processes. (eds.) H.-J. Rehm, G. Reed, A. Puhler, and P. Stadler. Weinheim: Wiley-VCH, pp. 225–264.

Gadd, G.M. 2009. Biosorption: critical review of scientific rationale, environmental importance and significance for pollution treatment. *Journal of Chemical Technology and Biotechnology* 84: 13–28.

Gadd, G.M. 2010. Metals, minerals and microbes: Geo microbiology and bioremediation. *Microbiology*. 156: 609–643.

Gadd, G.M. and Sayer, J.A. 2000. Fungal transformations of metals and metalloids. In *Environmental Microbe-Metal Interactions*, (ed.) D.R. Lovely. American Society for Microbiology, Washington, DC, pp. 237–256.

Gilmour, C. and Riedel, G. 2009. Biogeochemistry of trace metals and metalloids. In *Encyclopedia of Inland Waters* (ed.) G.E. Likens. Elsevier, Amsterdam, The Netherlands, pp. 7–15.

Hobbs, P.J. and Cobbing, J.E. 2007. Hydrogeological assessment of acid mine drainage impacts in the West Rand Basin, Gauteng Province. Report no. CSIR/NRE/WR/ER/2007/0097/C. CSIR/THRIP. Pretoria, South Africa, p. 109.

Ibrahim H.T., Qiang, H., Al-Rekabi, W.S., and Qiqi, Y. 2012. Improvements in biofilm processes for wastewater treatment. *Pakistan Journal of Nutrition*. 11: 610–636.

Johnson, B.D. and Hallberg, K.B. 2003. The microbiology of acidic mine waters. *Research in Microbiology*. 154: 466–473.

Jong. T. and Parry. D. L. 2003. Removal of sulphate and heavy metals by sulphate reducing bacteria in short-term bench scale upflow anaerobic packed bed reactor runs. Water Research. 37: 3379–3389.

Kaksonen, A.H. and Puhakka, J.A. 2007. Sulfate reduction based bioprocesses for the treatment of acid mine drainage and the recovery of metals. *Engineering Life Sciences*. 7: 541–564.

Kalin, M., Wheeler, W.N., and Meinrath, G. 2004. The removal of uranium from mining wastewater using algal/microbial biomass. *Journal of Environmental Radioactivity*. 78: 151–177.

Kapoor, A., Dinardo, O., and Kuiper, A. 2004. Biological oxidation of ferrous ions under acidic conditions using rotating biological contactor. *Environmental Engineering Science*. 3: 311–318.

Kargi, F. and Eker, S. 2001. Rotating-perforated-tubes biofilm reactor for high-strength wastewater treatment. *Journal of Environmental Engineering*. 127: 959–963.

Levings, C.D., Varela, D.E., Mehlenbacher, N.M., Barry, K.L., Piercey, G.E., Guo, M., and Harrison, P.J. 2005. Effect of an acid mine drainage effluent on phytoplankton biomass and primary production at Britannia Beach Howe Sound, British Columbia. *Marine Pollution Bulletin*. 50: 1585–1594.

Lopez-Archilla, A.I., Gonzalez, A.E., Terron, M.C., and Amils, R. 2004. Ecological study of the fungal populations of the acidic Tinto River in southwestern Spain. *Canadian Journal of Microbiology*. 50: 923–934.

Lottermoser, B.G. 2010. *Mine Wastes: Characterisation, Treatment and Environmental Impacts*. Springer-Verlag, Berlin, Germany.

Lottermoser, B.G., Ashley, P.M., and Lawie, D.C. 1999. Environmental geochemistry of the Gulf Creek copper mine area, north-eastern NSW, Australia. *Environmental Geology*. 39: 61–74.

Madigan, M. and Martinko, J. 2005. *Brock Biology of Microorganisms*. Prentice Hall, Englewood Cliffs, NJ, 992pp.

Malik, A. 2004 Metal bioremediation through growing cells. *Environment International*. 30: 261–278.

Malkoc, E. and Nuhoglu, Y. 2003. The removal of chromium (VI) from synthetic wastewater by *Ulothrix zonata*. *Fresenius Environmental Bulletin*. 12: 376–381.

Mathure, P. and Patwardhan, A. 2005. Comparison of mass transfer efficiency in horizontal rotating packed beds and rotating biological contactors. *Journal of Chemical Technology and Biotechnology*. 80: 413–419.

McLean, J.S., Lee, J.-U. and Beveridge, T.J. 2002. Interactions of bacteria and environmental metals, fine-grained mineral development and bioremediation strategies. In *Interactions Between Soil Particles and Microorganisms*, (eds.) P.M. Huang, J.-M. Bollag, and N. Senesi. New York: Wiley, pp. 228–261.

Mehta, S.K. and Gaur, J.P. 2005. Use of algae for removing heavy metal ions from wastewater: Progress and prospects. *Critical Reviews in Biotechnology*. 25: 113–152.

Munoz, R. and Guieysse, B. 2006. Algal-bacterial processes for the treatment of hazardous contaminants: A review. *Water Research*. 40: 2799–2815.

Muraleedharan, T.R. and Venkobachar, C. 1990. Mechanism of biosorption of copper (II) by *Ganoderma lucidum*. *Biotechnology and Bioengineering*. 35: 320–325.

Natarajan, K.A. 2008. Microbial aspects of acid mine drainage and its bioremediation. *Transactions of Nonferrous Metals Society of China*. 18: 1352–1360.

Nganje, T.N., Adamu, C.I., Ntekim, E.E.U., Ugbaja, A.N., Neji, P., and Nfor, E.N. 2010. Influence of mine drainage on water quality along River Nyaba in Enugu south-Eastern Nigeria. *African Journal of Environmental Science and Technology*. 4: 132–144.

Niyogi, D.K., Lewis, W.M., and McKnight, D.M. 2002. Effects of stress from mine drainage on diversity, biomass, and function of primary producers in mountain streams. *Ecosystem*. 5: 554–567.

Nordstrom, D.K. 2000. Advances in the hydrochemistry and microbiology of acid mine waters. *International Geology Review*. 42: 499–515.

Novis, P.M. 2006. Taxonomy of *Klebsormidium* (*Klebsormidiales*, *Charophyceae*) in New Zealand streams, and the significance of low pH habitats. *Phycologia*. 45: 293–301.

Novis, P.M. and Harding, J.S. 2007. Extreme acidophiles: Freshwater algae associated with acid mine drainage. In *Algae and Cyanobacteria in Extreme Environments*, (ed.) J. Seckbach. Springer, Dordrecht, the Netherlands, pp. 443–463.

Orandi, S. and Lewis, D.M. 2013a. Biosorption of heavy metals in a photo-rotating biological contactor—A batch process study. *Applied Microbiology and Biotechnology*. 97: 5113–5123.

Orandi, S. and Lewis, D.M. 2013b. Synthesising acid mine drainage to maintain and exploit indigenous mining micro-algae and microbial assemblies for biotreatment investigations. *Environmental Science and Pollution Research*. 2: 950–956.

Orandi, S., Lewis, D.M., and Moheimani, N.R. 2012. Biofilm establishment and heavy metal removal capacity of an indigenous mining algal-microbial consortium in a photo-rotating biological contactor. *Journal of Industrial Microbiology and Biotechnology*. 39: 1321–1331.

Orandi S., Yaghubpur, A., Nakhaei M., Mehrabian B., Sahraei H., and Behruz M. 2009. Distribution and role of green algae in acid mine drainage at sarcheshmeh copper mine. *Journal of Geoscience*. 72: 173–180.

Orandi, S., Yaghubpur, A., Sahraei, H., and Behruz, M. 2007. Influence of acid mine drainage on aquatic life at Sarcheshmeh copper mine. Goldschmidt Conference Abstracts A742, Cologne, Germany.

Olem, H. and Unz, R.F. 1977. Acid mine drainage treatment with rotating biological contactors. *Biotechnology and Bioengineering*. 19: 1475–1491.

Olem, H. and Unz, R.F. 1980. Rotating-disc biological treatment of acid mine drainage. *Water Pollution Control Federation*. 52: 257–269.

Palleroni, N.J. 2010. The pseudomonas story. *Environmental Microbiology*. 12(6): 1377–1383.

Patwardhan, A.W. 2003. Rotating biological contactors: A review. *Industrial and Engineering Chemistry Research*. 42: 2035–2051.

Perry, R.H. and Green, D.W. 2008. *Perry's Chemical Engineers' Handbook*. McGraw-Hill, New York, 2400pp.

Prasad, M.N.V. 2007. Aquatic plants for phytotechnology. In *Environmental Bioremediation Technologies*, (eds.) S.N. Singh and R.D. Tripathi. Springer, Berlin, Germany, pp. 259–274.

Qureshi, N. and Blaschek, H.P. 2005. *Butanol Production from Agricultural Biomass*. CRC Press, New York.

Raja, C.E. and Omine, K. 2012. Characterisation of boron resistant and accumulating bacteria *Lysinibacillus fusiformis M1*, *Bacillus cereus M2*, *Bacillus cereus M3*, *Bacillus pumilus M4* isolated from former mining site, Hokkaido, Japan. *Journal of Environmental Science and Health. Part A, Toxic/ Hazardous Substances & Environmental Engineering*. 47: 1341–1349.

Rindi, F., Mikhailyuk, T.I., Sluiman, H.J., Friedl, T., and López-Bautista, J.M. 2012. Phylogenetic relationships in Interfilum and *Klebsormidium* (Klebsormidiophyceae, Streptophyta). *Molecular Phylogenetics and Evolution*. 58: 218–231.

Rodgers, M. and Zhan, X. 2003. Moving-medium biofilm reactors. *Reviews in Environmental Science and Biotechnology*. 2: 213–224.

Romera, E., Gozalez, F., Ballester, A., Blazquez, M.L., and Munoz J.A. 2006. Biosorption with algae: A statistical review. *Critical Reviews in Biotechnology*. 26: 223–235.

Russell, R.A., Holden, P.J., Wilde, K.L., and Neilan, B.A. 2003. Demonstration of the use of *Scenedesmus* and *Carteria* biomass to drive bacterial sulphate reduction by *Desulfovibrio alcoholovorans* isolated from an artificial wetland. *Hydrometallurgy*. 71: 227–234.

Silverira, A.N., Silva, R., and Rubio, J. 2009. Treatment of acid mine drainage (AMD) in South Brazil comparative active processes and water reuse. *International Journal of Mineral Processing*. 93: 103–109.

Singh, R., Paul, D., and Jain, R.K. 2006. Biofilms: Implications in bioremediation. *Trends in Microbiology*. 14: 389–397.

Skowronska, B.P. 2003. Resistance, accumulation and allocation of zinc in two ecotypes of the green alga *Stigeoclonium tenue Kütz* coming from habitats of different heavy metal concentrations. *Aquatic Botany*. 75: 189–198.

Smith, W.L. and Gadd, G.M. 2000. Reduction and precipitation of chromate by mixed culture sulphate-reducing bacterial biofilms. *Journal of Applied Microbiology.* 88: 983–991.

Smucker, N.J. and Vis, M.L. 2011. Acid mine drainage affects the development and function of epilithic biofilms in streams. *Journal of the North American Benthological Society.* 30: 728–738.

Souza-Egipsy, V., Altamirano, M., Amils, R., and Aguilera, A. 2011. Photosynthetic performance of phototrophic biofilms in extreme acidic environments. *Environmental Microbiology.* 13: 2351–2358.

Sparks, D.L. 2005. Toxic metals in the environment: The role of surfaces. *Elements.* 1: 193–196.

Spath, R., Flemming, H.C., and Wuertz, S. 1998. Sorption properties of biofilms. *Water Science & Technology.* 37: 207–210.

Sutherland, I.W. 2005. Microbial exopolysaccharides. In *Polysaccharides: Structural Diversity and Functional Versatility,* (ed.) S. Dumitriu. Marcel Dekker, New York, pp. 431–458.

Thomson, C. and Tracey, D. 2005. Nitrogen and phosphorus cycles. *River Science.* 4: 1–7.

Travieso, L., Pelln, A., Bentez, F., Snchez, E., Borja, R., O'Farrill, N., and Weiland, P. 2002. BIOALGA reactor: Preliminary studies for heavy metals removal. *Biochemical Engineering Journal.* 12: 87–91.

Trzcinska, M. and Skowrońska, B.P. 2012. Differences in Zn and Pb resistance of two ecotypes of the microalga *Eustigmatos* sp. inhabiting metal loaded calamine mine spoils. *Journal of Applied Phycology.* 25: 277–284.

Tsezos, M. 2007. Biological removal of ions: Principles and applications. *Advanced Materials Research.* 20–21: 589–596.

Valdes, J., Pedroso, I., Quatrini, R., Dodson, R.J., Tettelin, H., Blake, R., Eisen J.A., and Holmes, D.S. 2008. *Acidithiobacillus ferrooxidans* metabolism: From genome sequence to industrial applications. *BMC Genomics.* 9: 597.

Valente, T.M. and Gomes, C.L. 2007. The role of two acidophilic algae as ecological indicators of acid mine drainage sites. *Journal of Iberian Geology* 33(2): 283–294.

Van Hullebusch, E.D., Zandvoort, M.H., and Lens, P.N.L. 2003. Metal immobilisation by biofilms: Mechanisms and analytical tools. *Reviews in Environmental Science and Biotechnology.* 2: 9–33.

Volesky, B. 2001. Detoxification of metal-bearing effluents: Biosorption for the next century. *Hydrometallurgy.* 59: 203–216.

Wang, J. and Chen, C. 2009. Biosorbents for heavy metals removal and their future. *Biotechnology Advances.* 27: 195–226.

Yang, Y., Wan, M., Shi, W., Peng, H., Qiu, G., Zhou, J., and Liu, X. 2008. Bacterial diversity and community structure in acid mine drainage from Dabaoshan Mine, China. *Aquatic Microbial Ecology.* 47: 141–151.

Zinck, J.K. 2006. Disposal, reprocessing and reuse options for acidic drainage treatment sludge. In International Conference on Acid Rock Drainage (ICARD), St. Louis, MO, USA, pp 2604–2617.

18 Biosurfactants for the Remediation of Metal Contamination

Catherine N. Mulligan

CONTENTS

ABSTRACT

Biosurfactant applications for the remediation of contaminated soil and water are promising due to their biodegradability, low toxicity, and critical micelle concentration and high effectiveness in enhancing biodegradation and affinity for metals. Various studies have been performed with regard to enhanced bioremediation, flushing and washing of soils, and water contaminated with metals alone or with hydrocarbons. Removal of heavy metals by biosurfactants occurs through solubilization, complexation, and ion exchange. Most biosurfactants investigated, including rhamnolipids, are anionic. Increased yields and new substrates and biosurfactants need to be investigated to increase economic feasibility. Newer applications are related to enhance electrokinetics, phytoremediation, and nanotechnology techniques. Despite the promising applications, field and full-scale remediations are still limited.

18.1 INTRODUCTION

Insufficient methods for waste management, treatment, and disposal have contributed to the contamination of many sites (Yong et al., 2014). Some examples of these liquid and solid wastes include solvents from dry cleaning, lubricating, automotive, hydraulic and fuel oils, organic sludges, and wastewaters. Soil contamination is mainly the result of accidental spills and leaks, leachates from waste materials, and cleaning of equipment containing residues left in used containers and outdated materials. Chemical contaminants may also originate from improperly managed landfills, automobile service establishments, maintenance shops, and photographic film processors. Municipal wastes include pesticides, paints, cleaning, and automotive products (LaGrega et al., 2001).

Metals and heavy metals in the geo-environment result from anthropogenic activities such as management and disposal of wastes in landfills, generation and storage of chemical waste leachates and sludges, extraction of metals in metal processing industries, metal plating works, and even in municipal solid wastes. Some common heavy metals are cadmium (Cd), chromium, (Cr), copper (Cu), iron (Fe), lead (Pb), mercury (Hg), nickel (Ni), and zinc (Zn). Radioactive metals also enter the environment through fallout from nuclear testing, radioactive waste leakage, or accidents such as Chernobyl or Fukushima.

Various *in situ* and *ex situ* remediation techniques exist for the management of the contaminated sites with heavy metals and organic pollutants. *Ex situ* techniques include excavation, contaminant fixation or isolation, incineration or vitrification, and washing and biological treatment processes. *In situ* processes include (1) extraction methods for soluble components; (2) chemical treatments for oxidation or detoxification; (3) stabilization/solidification with cement, lime, or resins; and (4) phytoremediation. Most *in situ* remediation techniques tend to be less expensive and disruptive than *ex situ* ones and reduce worker exposure and environmental disruption. The sustainability of reduced contaminant solubility is required for the long-term success of the treatment.

Biosorption is another approach that has been studied over the last decades for metal remediation. It involves the use of low-cost organic materials to remove heavy metals from contaminated water. Many materials have been used, including microbial biomass, seaweed, molds, yeasts, and other organic waste materials. They are of low cost and high efficiency, with the potential for metal desorption for reuse and material regeneration. Factors such as temperature, pH, and type of material influence sorption capacity (Arjoon et al., 2013).

Soil remediation can be performed with excavation via soil washing or without via *in situ* flushing (Mulligan et al., 2001a). Solubilization of the contaminants is the main mechanism achieved by water alone or with additives. Contaminants such as trichloroethylene (TCE), polycyclic aromatic hydrocarbons (PAHs), and polychlorinated biphenyls (PCBs) are not easily solubilized. Effective bioremediation of organic contaminants leads to complete mineralization of the contaminants. Often, the process is enhanced by the addition of nutrients, electron acceptors, or bacteria (bioaugmentation) (Yong and Mulligan, 2004). Metals and radionuclides are commonly found with the organic contaminants, further complicating the bioremediation process. The inorganic contaminants may be toxic to the bacteria, are biodegradable, but may be converted from one form to another.

Various approaches have been developed to overcome the challenges of mixed contaminants (Sandrin and Maier, 2003). Cyclodextrin is a cyclic oligosaccharide produced from starch by bacteria. Ehsan et al. (2007) showed that ethylenediaminetetraacetic acid (EDTA) (2 mM) and cyclodextrin (10% solution) could enhance the desorption of PCBs and heavy metals from a contaminated soil.

In this chapter, the effectiveness of various biological surface agents for *in situ* flushing and soil washing processes is examined. In the case of mixed contamination, they can be used to enhance bioremediation of the organic contaminants. The main focus is on biologically produced surfactants due to their biodegradability, low toxicity, and effectiveness.

18.2 USE OF SURFACTANTS IN ENVIRONMENTAL TECHNOLOGIES

18.2.1 SOIL FLUSHING

In soil flushing, contaminants in the groundwater are extracted by pumping them up to the soil surface. The main disadvantages are that substantial periods of time may be required and its effectiveness may be limited. Drinking water standards of the recovered water can be achieved by sorption on activated carbon, ion exchange, membrane filtration, and other methods. Solutions can be introduced by soil flooding, sprinklers, leaching fields, and horizontal or vertical drains to enhance the removal rates of the contaminants (Figure 18.1).

Due to the sorption of residuals during flushing, the additives must be nontoxic and biodegradable. Contaminant removal efficiencies are affected by various soil factors including 1 pH, composition, porosity, organic matter and moisture contents, cation exchange capacity, particle size distribution, permeability, and the type of contaminants. Soil flushing is appropriate for highly permeable soils (hydraulic conductivity greater than 1×10^{-3} cm/s). Costs increase as the depth to the groundwater increases. Control of these infiltrating agents and contaminants may be difficult, particularly if the site hydraulic characteristics are not well understood. Emissions of volatile organic compounds (VOCs) should be

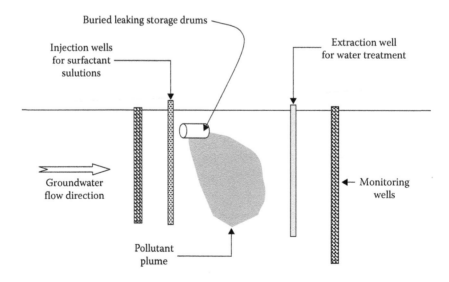

FIGURE 18.1 Schematic of soil-flushing process.

monitored and treated if required. Recycling the additives can improve process costs and reduce material requirements. Metals, VOCs, PCBs, fuel hydrocarbons, and pesticides can be extracted through soil flushing.

In choosing the most appropriate remediation technology, the factors to be considered must include exposure routes, future land use, acceptable risks, regulatory guidelines, level and type of contaminants, site characteristics, and quantity of emissions. Laboratory and field treatability tests are necessary to generate site-specific information. Soil flushing has been demonstrated at numerous Superfund sites with costs in the range of $18–$50 per cubic meter for large easy to small difficult sites.

18.2.2 SOIL WASHING

Soil washing has been used for a variety of soils contaminated with metals, mixed contaminants, and organic contaminants (El-Shafey and Canepa, 2003). Soil washing is a process that uses water to remove contaminants from soil and sediments by physical and/or chemical techniques to the wash solution. It is most appropriate for metals in the form of hydroxides, oxides, and carbonates that are more weakly bound. Metals (with the exception of metal sulfides) can also be removed from the wash solution by precipitation or ion exchange or by electrochemical processes if the levels of organic compounds are not significant.

Figure 18.2 illustrates a soil washing process that includes size separation, washing, rinsing, and other technologies similar to the mineral processing industry. Mixed metal and organic contaminants may require different additives to target the various contaminants sequentially. Removal of the smaller particles that contain higher contamination levels is less costly as it decreases the volumes of soil to be treated. The more contaminated size range is 0.24–2 mm due to the

surface charges and higher surface area of the soil clay particles that attract anionic metal ions. The organic soil fraction binds organic contaminants. Filter presses, conveyer filtration, or centrifugal separation, are used to dewater the soil. Froth flotation by the introduction of air bubbles into slurry may also be used (Venghuis and Werther, 1998). The disposal of the treated fine particles varies depending on the type and level of the contaminants.

Wash water and additives should be recycled, or treated prior to disposal. The most sustainable wastewater management systems use less energy, eliminate, or beneficially reuse biosolids, restore natural nutrient cycles, have much smaller footprints, are more energy efficient, and are designed to eliminate odors and hazardous by-products (Daigger and Crawford, 2005). In addition to the technical aspects of wastewater treatment, selection of a particular technology should be based on the human and environmental activities that surround it. Costs of soil washing are usually in the order of $70–$190 US/m^3 depending on site characteristics (Racer software, Remedial action plan, 2006). Full-scale processes are less common in the United States although they are common in Europe. Feasibility tests should be conducted to determine the conditions for optimal scale-up (chemical type and dosage, contact time, agitation, temperature, and extraction steps to meet regulatory requirements).

18.3 SURFACTANTS

18.3.1 SURFACTANT CHARACTERISTICS

Surfactants are amphiphilic compounds with a hydrophobic portion with little affinity for the bulk medium and a hydrophilic group that is attracted to the bulk medium. The free energy of the system is reduced by replacing the bulk molecules of higher energy at an interface. Surfactants concentrate

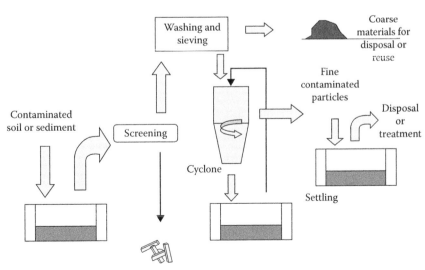

FIGURE 18.2 Schematic of soil or sediment washing of contaminated material.

at interfaces (solid–liquid, liquid–liquid, or vapor–liquid). At an interfacial boundary between two immiscible phases, the hydrophobic portion concentrates at the surface, while the hydrophilic portion is oriented toward the solution. As the surfactant concentration increases, micelles start to form. The minimum concentration is the critical micelle concentration (CMC). The hydrophilic surfaces face outward, while the lipophilic portions face inward. The micelles can enhance the solubility of contaminants such as PAHs that are of low solubility. Surfactants of low CMC are more economic as less surfactant is needed. Applications are based on their abilities to lower surface tensions and increase solubility, detergency, wetting, and foaming capacities (Mulligan and Gibbs, 1993). For example, due to the properties of solubilizing and mobilizing contaminants, they are used for soil washing or flushing.

In general, surfactants are used to save energy, such as the energy required for pumping in pump and treat techniques, to enhance metal removal (Holden, 1989). Cost, charge type, physicochemical behavior, solubility, toxicity, and adsorption behavior are some of the most important selection criteria for surfactants, in addition to their ability to enhance the bioremediation of contaminated land sites (Oberbremer et al., 1990, Samson et al., 1990). Selection of biocompatible surfactants is essential for a more sustainable remediation process.

18.3.2 Biosurfactants

Biosurfactants are produced by bacteria or yeast using various substrates including sugars, oils, wastes (Lin, 1996), and hydrocarbons (Syldatk and Wagner, 1987). Classifications are as glycolipids, lipopeptides, phospholipids, neutral lipids, fatty acids, and polymeric and particulate compounds (Biermann et al., 1987). Most compounds are either anionic or neutral. There are only a few cationic such as those containing amine groups. The hydrophobic part of the molecule is a long-chain fatty acid, hydroxy fatty acid, or α-alkyl-β-hydroxy fatty acid. The hydrophilic portion usually consists of a carbohydrate, amino acid, peptide, carboxylic acid, phosphate, or alcohol. Various microorganisms can produce these compounds with CMCs generally ranging from 1 to 200 mg/L and molecular masses from 500 to 1500 Da (Lang and Wagner, 1987).

Most biosurfactants are growth associated. In this case, they can either use the emulsification of the substrate (extracellular) or facilitate the passage of a hydrophobic substrate through the membrane (cell membrane associated). Biosurfactants, however, can also be produced from carbohydrates, which are very soluble. They can have higher specificity, biocompatibility, and biodegradability than synthetic surfactants (Cooper, 1986). For example, in naphthalene solubilization tests, glycolipids from *Rhodococcus* sp. 413A were 50% less toxic than Tween 80 (Kanga et al., 1997).

In this chapter, most of the emphasis will be placed on three well-studied low-molecular-weight biosurfactants including rhamnolipids, surfactin, and sophorolipids. In each case, environmental applications for metal removal will be

examined for soil (*in situ* and *ex situ*), water, and waste treatment, including enhancing solubilization and biodegradation in the case of mixed contaminants. A summary of selected biosurfactants that will be discussed further is shown in Table 18.1.

18.4 ENVIRONMENTAL APPLICATIONS OF BIOSURFACTANTS

18.4.1 Rhamnolipids

18.4.1.1 Soil Washing of Heavy Metals

A group of biosurfactants that have been studied extensively are the rhamnolipids from *Pseudomonas aeruginosa* (Hitsatsuka et al., 1971, Guerra-Santos et al., 1984). Up to seven homologues have now been identified (Abalos et al., 2001) with surface tensions of 29 mN/m. Two types of rhamnolipids contain either two rhamnoses attached to β-hydroxydecanoic acid (R1) or one rhamnose connected to the identical fatty acid (R1). Other variants, R3 and R4, are also shown (Figure 18.3). *P. aeruginosa* produces rhamnolipids from substrates such as C11 and C12 alkanes, glucose, glycerol, olive oil, succinate, pyruvate, citrate, fructose, and mannitol (Robert et al., 1989). Substrate type, fermentor design, pH, nutrient composition, and temperature used affect the composition and yields (Mulligan and Gibbs, 1993).

Rhamnolipids are anionic in nature with the ability to complex metals from soil and ions such as cadmium, copper, lanthanum, lead, and zinc (Tan et al., 1994; Herman et al., 1995; Ochoa-Loza, 1998). Using ion-exchange resins, Ochoa-Loza et al. (2001) determined stability constants. The lowest

TABLE 18.1
Selected Biosurfactants Used in Environmental Applications

Type of Surfactant	Microorganism
Glycolipids	
Rhamnolipids	*Pseudomonas aeruginosa*
	Pseudomonas sp.
	Serratia rubidaea
Sophorose lipids	*Candida apicola*
	Candida bombicola
	Candida lipolytica
	Candida bogoriensis
Lipopeptides and proteins	
Fengycin	*Bacillus* sp.
Surfactin	*Bacillus subtilis*
	Bacillus pumilus
Viscosin	*Pseudomonas fluorescens*
Glycoprotein	Biosur-PM

Source: Adapted from Mulligan, C.N. and Gibbs, B.F., Factors influencing the economics of biosurfactants, in Kosaric, N. (Ed.), *Biosurfactants, Production, Properties, Applications*, Marcel Dekker, New York, 1993, pp. 329–371.

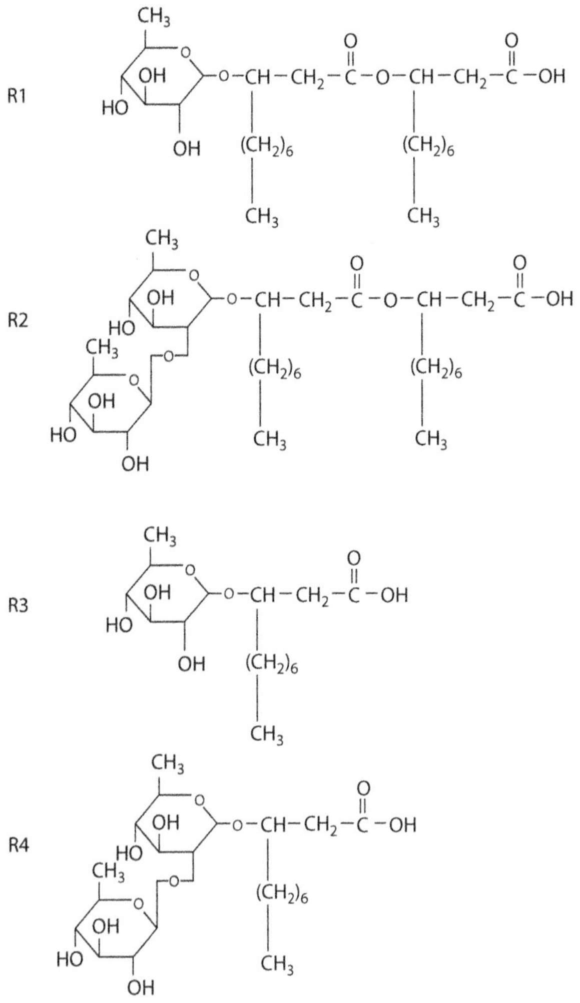

FIGURE 18.3 Structure of rhamnolipids. (Adapted from Mulligan, C.N., *Environ. Pollut.*, 133, 183, 2005.)

to highest affinity of cations for rhamnolipid was $K^+ < Mg^{2+} < Mn^{2+} < Ni^{2+} < Co^{2+} < Ca^{2+} < Hg^{2+} < Fe^{3+} < Zn^{2+} < Cd^{2+} < Pb^{2+} < Cu^{2+} < Al^{3+}$. The affinities were about the same or higher than those of organic acids (acetic, citric, fulvic, and oxalic acids) with metals, indicating rhamnolipid's potential for metal remediation. Ratios of the rhamnolipid to individual metals (molar) were 1.91 for cadmium, 2.31 for copper, 2.37 for lead, 1.58 for zinc, 0.93 for nickel, and 0.84 and 0.57 for magnesium and potassium (common soil cations), respectively.

In the presence of mixed contamination, rhamnolipids were added to soil and sediment for heavy metal removal (Mulligan et al., 1999a,b, 2001b). Although 80% to 100% of cadmium and lead can be removed from artificially contaminated soil, from field samples, the results were more in the range of 20%–80% due to increased bonding of the contaminants over time (Fraser, 2000). Biosurfactant could also be added as a washing agent for excavated soil. Due to the foaming ability of the biosurfactant, metal–biosurfactant complexes can be removed and then the biosurfactant can be recycled through precipitation by reducing the pH to 2.

Neilson et al. (2003) show that a 10 mM solution of rhamnolipids was able to remove about 15% of the lead after 10 washes. Mulligan et al. (1999a, 2001b) showed that lead could be removed from the iron oxide, exchangeable, and carbonate fractions. These low removal levels could be improved if the biosurfactants could be added in multiple cycles (Neilson et al., 2003).

Rhamnolipids have also been added to another media, mining residues, to enhance metal extraction (Dahr Azma and Mulligan, 2004). Batch tests with a 2% rhamnolipid concentration showed that 28% of the copper was extracted. Higher concentrations, although beneficial for copper extraction, could not be used due to its highly viscous nature. Supplementing the 2% rhamnolipid with 1% NaOH enhanced the removal of up to 42%, but removal decreased with increased surfactant concentrations. Sequential extraction studies showed that approximately 70% of the copper was associated with the oxide fraction, 10% with the carbonate, 5% with the organic matter, and 10% with the residual fraction. Washing with 2% biosurfactant (pH 6) for 6 days removed 50% of the carbonate fraction and 40% of the oxide fraction. In summary, rhamnolipids are effective for heavy metal removal and could also be effective for the removal of mixed (hydrocarbon and metal) contaminants.

Other studies evaluated metal removal by biosurfactants. In column tests, Juwarkar et al. (2007) compared the removal of cadmium and lead by a biosurfactant produced by *P. aeruginosa* BS2. Cadmium removal was higher than Pb (92% of Cd and 88% of Pb) by the rhamnolipid (0.1%). The rhamnolipid decreased toxicity and allowed microbial activity (*Azotobacter* and *Rhizobium*) to take place that improved soil quality. Cost-effectiveness, though, was not evaluated.

Asci et al. (2007) demonstrated that rhamnolipid could remove Cd(II) from kaolinite. Various sorption models were compared, and the Kolbe–Corrigan model was found to fit best. The effects of pH sorbed Cd(II) and rhamnolipid concentrations were determined. The conditions for maximal Cd removal of 71.9% were pH 6.8 for an initial Cd concentration of 0.87 mM and a rhamnolipid concentration of 80 mM.

Asci et al. (2008a) compared two soil components, sepiolite and feldspar, for their ability to sorb cadmium. Sepiolite was shown as a superior accumulator of cadmium to feldspar. Thus, desorption of the rhamnolipid from feldspar (96%) was much higher than from sepiolite (10%). Asci et al. (2008b) then examined the removal of zinc from Na feldspar by a rhamnolipid biosurfactant. Optimal pH for removal was found to be 6.8 due to the small vesicles and micelles at a pH > 6.0. Low interfacial tensions facilitated the sorption of the biosurfactant and metal contact. A 25 mM biosurfactant concentration was optimal for 98.8% removal of 2.2 mM of zinc (a 12.2:1 molar ratio).

Biosurfactant sorption can reduce their effectiveness for contaminant removal from the soil. Preliminary tests by Guo and Mulligan (2006) indicated that higher organic and clay contents increased sorption and that the mono rhamnolipid (R1) adsorbed more than the dirhamnolipid form (R2). Further rhamnolipid sorption tests performed by Ochoa-Loza et al. (2007) showed that R1 sorption was dependent on concentration and soil component and followed the order of hematite > kaolinite > MnO$_2$ ~ illite ~ Ca-montmorillonite > gibbsite > humic acid–coated silica for low R1 concentrations. At higher concentrations, the order was illite >> humic acid–coated silica > Ca-montmorillonite > hematite > MnO$_2$ > gibbsite ~ kaolinite. R1 was also found to sorb more strongly than dirhamnolipid (R2) and was more effective for metal removal. This information helps in the evaluation of the feasibility of rhamnolipid treatment and the quantity of rhamnolipid needed.

Kim and Vipulanandan (2006) determined that the removal of lead from contaminated soil (kaolinite) by a biosurfactant could be represented by a linear isotherm. In addition, more than 75% of the lead could be removed from a 100 mg/L contaminated water at a biosurfactant concentration of 10 times the CMC. This represents a biosurfactant to lead ratio for the optimal removal of 100:1. FTIR spectroscopy indicated that the carboxyl group of the biosurfactant was implicated in the removal. The biosurfactant micelle partitioning could also be represented by Langmuir and Freundlich models and that it was more favorable than the synthetic surfactants, sodium dodecyl sulfate (SDS) and Triton X-100.

Batch washes of three different biosurfactants, rhamnolipids, saponin, and mannosylerythritol lipids (MELs), were evaluated to remove heavy metals from a construction site soil containing 890 mg/kg zinc, 260 mg/kg copper, 230 mg/kg total petroleum hydrocarbons, and 170 mg/kg nickel and a lake sediment (4440 mg/kg zinc, 474 mg/kg lead, and 94 mg/kg copper) (Mulligan et al., 2007). Five washings of the soil with saponin (30 g/L) removed the highest amount of zinc (88%) and nickel (76%). Rhamnolipids (pH 6.5) at 2% removed the most copper (46%). Multiple washings of the soil with 4% MELs (pH 5.6) were not very effective (17% of the zinc and nickel and 36% of the copper). The sediment showed similar trends as for the soil. The highest level of zinc (33%) and lead removals (24%) were achieved with 30g/L saponin (pH 5), while the highest copper removal (84%) was achieved with 2% rhamnolipids (pH 6.5). Sequential extraction tests indicated that the removal of zinc was from the oxide fraction zinc and copper from the organic fraction by the biosurfactants.

Slizovsky et al. (2011) compared several surfactants including rhamnolipid for metal removal and reduction of soil ecotoxicology. The amounts of 39, 56, 68, and 43% of Zn, Cu, Pb, and Cd, respectively, were removed by the biosurfactant from an aged field contaminated soil. Reduction in toxicity of the treated soil was demonstrated by reduced bioaccumulation of these metals by two worm species (*Eisenia fetida* and *Lumbricus terrestris*) and increased biomass levels and survival.

The small-angle neutron scattering technique was used to evaluate the size and morphology of rhamnolipid micelles by Dahrazma et al. (2008). At high pH, large aggregates and micelles in the order of 17 Angstroms were found, but in acidic conditions, larger 500–600 Å diameter vesicles were formed. This indicates that there should not be any filtering effect in the soil during flushing through pores that are typically in the

order of 200 nm. Complexation of the micelles with metals did not affect the size of the micelles.

Xia et al. (2009) examined a combination of citric acid and rhamnolipid to improve cost-effectiveness and improve metal removal (mixture of 0.05 mol/L citric acid and 0.05 mol/L rhamnolipid). However, the mixture showed no significant improvement over citric acid alone for a mixture of Pb, Zn, and Cd soil contamination, although there was a slight removal enhancement for Cu.

A study was conducted on the removal of Cr(III) by rhamnolipids from chromium-contaminated kaolinite by Massara et al. (2007). Rhamnolipids had the ability to extract from the kaolinite up to 25% of the more stable form of chromium, Cr(III). The removal of hexavalent chromium by rhamnolipids was also enhanced. Rhamnolipids removed Cr(III) from the kaolinite carbonate and oxide/hydroxide portions. It was also shown for the first time that the rhamnolipids have also the capability of reducing the extracted Cr(VI) to Cr(III) almost completely over 24 days. The rhamnolipids thus could be beneficial for the removal of Cr(VI) and its long-term conversion of Cr(VI) to Cr(III).

This work was continued to evaluate the use of rhamnolipid for the removal and reduction of hexavalent chromium from contaminated soil and water in batch experiments (Ara and Mulligan, 2015). The initial chromium concentration, rhamnolipid concentration, pH, and temperature affected the reduction efficiency. Complete reduction in water of the initial Cr(VI) by rhamnolipid at optimum conditions (pH 6, 2% rhamnolipid concentration, 25°C) occurred at low concentration (10 ppm). For higher initial concentrations (400 ppm), 24 h was required to reduce Cr by 24.4%. In the case of soil, rhamnolipid alone could remove the soluble fraction of the chromium present in the soil. The extraction increased as the initial concentration increased in the soil but decreased slightly when the temperature increased above 30°C. The exchangeable and carbonate fractions comprised 24% and 10% of the total chromium, respectively. The oxide and hydroxide portions bound 44% of the total chromium in the soil. On the other hand, 10% and 12% of the chromium was associated with the organic and residual fractions. Rhamnolipid was able to remove most of the exchangeable (96%) and carbonate (90%) fractions and some of the oxide/hydroxide portion (22%) but not from the other fractions. This information can assist in designing the conditions for soil washing.

Another anionic contaminant, arsenic, was investigated for the treatment of mining residues by rhamnolipids (Wang and Mulligan, 2009a). Only the As(V) form was extracted from the residues at high pH. Significant removal of Cu, Zn, and Pb positively correlated with that of arsenic. The arsenic mobilization mechanism was due either to organic complex formation or metal bridging. The development of mobilization isotherms of arsenic from the residues to predict the mobilization was the focus of Wang and Mulligan (2009b). Easily and moderately extractable arsenic could be removed, but redox or methylation reactions did not occur to any significant effect. The rhamnolipid thus might be potentially useful for the removal of As from mining tailings.

Enhanced electrokinetic treatments were evaluated for dredged harbor sediment contaminated with heavy metals and PAHs. A mixture of a chelating agent (citric acid) and a surfactant was used as additives in the processing fluids (Ammami et al., 2015). Promising results were obtained with solutions of rhamnolipids (0.028%) and a viscosin-like biosurfactant from *Pseudomonas fluorescens* Pfa7B (0.025%). Although the rhamnolipid and the viscosin-like compounds exhibited a higher electrical current than Tween 20, metals and PAHs were removed less from the sediment. Higher concentrations of these biosurfactants should be tested to overcome sorption onto sediment particles.

Barajas-Aceves et al. (2015) investigated the effect of a crude rhamnolipid biosurfactant produced by *Bacillus israelensis* and compost addition to soil and mine tailings that were subjected to phytoremediation. Irrigation with the biosurfactant did not enhance the translocation of heavy metals, Cr, Cu, and Ni by the three plant species. Sheng et al. (2008) indicated that heavy metal bioavailability could be improved by biosurfactants to improve phytoremediation.

18.4.1.2 Soil Flushing

A foam produced by a 0.5% rhamnolipid solution was evaluated for metal removal from a contaminated sandy soil (1710 ppm of Cd and 2010 ppm of Ni) by Mulligan and Wang (2004). After 20 pore volumes, 73.2% of Cd and 68.1% of Ni were removed compared to the biosurfactant liquid solution (61.7% Cd and 51.0% Ni). Triton X-100 foam removed less Cd and Ni (64.7% and 57.3%, respectively) and liquid Triton X-100 that removed 52.8% Cd and 45.2% Ni. Distilled water removed only 18% of the Cd and Ni. The average hydraulic conductivity was 1.5×10^{-4} cm/s for 95% and 2.9×10^{-3} cm/s for 99% that showed that increasing foam quality decreases the hydraulic conductivity. In comparison, the conductivity of water was 0.02 cm/s. This higher viscosity will allow better control of the surfactant mobility during *in situ* use. Further efforts will be required to enable its use at field scale.

A continuous flow configuration was subsequently evaluated by Dahrazma and Mulligan (2007) to simulate a flow through remediation technique. The rhamnolipid solution was pumped within a column constantly through the sediment. The effect of rhamnolipid concentration and flow rate, supplements, and time was investigated. The rhamnolipid heavy metal removal was up to 37% of Cu, 13% of Zn, and 27% of Ni. Addition of 1% NaOH to 0.5% rhamnolipid increased the removal of copper by up to four times compared to 0.5% rhamnolipid alone.

An evaluation of the capability of a rhamnolipid biosurfactant (JBR425) foam for the treatment of contaminated freshwater sediments with elevated levels of Pb, Zn, and Ni and PAH (Alavi and Mulligan, 2011) was performed. The biosurfactant foam was injected in the sediment column with foam quality rhamnolipid between 85% and 99% and stabilities from 15 to 43 min. PAH and metal removals were then

evaluated for sediment samples. Total removal efficiency including mobilization and volatilization for the biosurfactant foam was 56.4% of pyrene, 41.2% of benz(a)anthracene, and 45.9% of chrysene. No volatilization of PAHs was observed. Metal removal with 0.5% rhamnolipid foam (99% quality, pH 10.0) was 53.3% of Ni, 56.8% of Pb, and 55.2% of Zn. A liquid solution 0.5% rhamnolipid solution showed lower levels of metal removal by 11%–17%. Further efforts will be required to optimize the performance of the foam.

A strain of *P. aeruginosa* A11 that produces a biosurfactant and exhibits plant growth promotion and metal resistance was isolated (Singh and Cameotra, 2013). A yield of 4.4 g/L of biosurfactant was obtained from glycerol. Resistance against metals with the exception of Hg and Ni was shown. The strain was able to solubilize phosphorus and produce ammonia, siderophores and HCN. The strain showed potential for the bacterial enhancement of phytoremediation at metal contaminated sites. The biosurfactants were classified as di-Rhl (L-rhamnosyl-L-rhamnosyl-β-hydroxydecanoyl-β-hydroxydecanoate) with single mono-Rhls (L-rhamnosyl-β-hydroxydecanoyl-β-hydroxydecanoate).

Rhamnolipid was studied for its effect on the electrokinetic and rheological behavior on nanozirconia particles and found to adsorb increasingly onto the zirconia as the concentration increased (Biswas and Raichur, 2008). It dispersed zirconia particles at pH 7 as shown by zeta potential measurements, sedimentation, and viscosity tests. It can serve as an eco-friendly flocculation and dispersion product for high solid contents of microparticles.

Nickel oxide (NiO) nanorods (Palanisamy, 2008) were synthesized using a water-in-oil microemulsion technique, with rhamnolipid in a *n*-heptane hydrocarbon phase. The nanorods were found to be approximately 150–250 nm in length and 22 nm in diameter. The pH of the solution had an important effect on the morphology of the nanoparticle. This method with the biosurfactant provides a new approach for the eco-friendly synthesis of nanomaterials.

Fatisson et al. (2010) evaluated various components on the stabilization of carboxymethylcellulose-coated zerononvalent iron nanoparticles (nZVI). Stabilization is important to enhance the transport of the nZVI particles to the zone of contamination. The effects of fulvic acid and rhamnolipid were evaluated as these are naturally found in the groundwater and soil environment. The presence of the rhamnolipid led to the lowest rate of deposition of the particles on silica.

Experiments were conducted to investigate the effect of presence of rhamnolipid on the production and stabilization of iron nanoparticles (Farshidy et al., 2011). In addition, the effect of rhamnolipid on the remediation of chromium (VI) from water using iron nanoparticles was tested. Iron nanoparticles were produced in the presence of different concentrations of rhamnolipid. Then, unmodified nanoparticles were treated with different concentrations of rhamnolipid and carboxymethylcellulose. The TEM electron micrographs indicated that without adding rhamnolipid during the production process, the size of iron nanoparticles is high due to the formation of micron-sized clusters. In the presence of low concentrations of rhamnolipid (90 mg/L), the diameter of the particles was reduced to less than 10 nm due to the coating and stabilization of nanoparticles by the rhamnolipid as confirmed by zeta potential measurements on the modified iron nanoparticles. Furthermore, the effect of the presence of rhamnolipid on reductive remediation of hexavalent chromium, Cr(VI), to trivalent, Cr(III), was investigated. By combining 0.034 g/L chromium (VI) with 0.08 g/L iron nanoparticles and 2% (w/w) of rhamnolipid, the remediation of chromium increased by 123% in 15 h compared to the controls.

18.4.1.3 Treatment of Water

Ultrafiltration is not effective for the removal of heavy metals from water due to their small size. Addition of surfactants in a process called micellar-enhanced ultrafiltration (MEUF) can enhance metal removal due to their binding to large micelles. Rhamnolipid addition (El-Zeftawy and Mulligan, 2011) was studied in MEUF of heavy metals from contaminated waters. The MEUF performance was investigated for copper, zinc, nickel, lead, and cadmium retention by two different membranes with molecular weight cutoffs (MWCOs) of 10,000 and 30,000 Da. Six contaminated wastewaters from metal-refining industries were successfully treated using the two membranes (>99% rejection ratio). Optimization of the influential parameters was done by the response surface methodology approach. The best operating conditions were biosurfactant-to-metal molar ratios of approximately 2:1, a transmembrane pressure of 69 ± 2 kPa, a temperature of 25°C ± 1°C, and pH of 6.9 ± 0.1. The resulting heavy metal concentrations in permeate were all reduced to be in accordance with the federal Canadian regulations. In addition, compared to chemical surfactants, biosurfactants are less toxic, which is advantageous due to some leakage into the permeate.

Further studies were to evaluate the effect of a rhamnolipid biosurfactant on the removal of a mixture of contaminants from aqueous solutions using MEUF (Ridha and Mulligan, 2011). The amount of rhamnolipid to remove different concentrations of ionic copper and benzene separately was determined. The molar ratio (MR) of biosurfactant to contaminant was 1.33 to obtain the 100% rejection of benzene molecules and 6.25 for 100% rejection of copper ions. When copper and benzene were together in the water, the molar ratios improved for benzene from 1.33 to 0.56 but were unchanged for copper. In all cases, rhamnolipid enabled a rejection of 100% for copper and benzene either separately or together, indicating the potential of this approach for wastewater treatment.

Rhamnolipid (JBR425) was applied to reduce Cr(VI) to Cr(III) and then subjected to an MEUF technique by Abbasi-Garravand and Mulligan (2014). A 10,000 Da polysulfone hollow-fiber membrane was used in the MEUF experiments. A reduction of hexavalent chromium of 98.7% (initial concentration of 10 mg/L at pH 6) and a rhamnolipid concentration of 2% (vol/vol) were obtained. Thus, the rhamnolipid proved to be an effective agent for removing hexavalent and trivalent chromium from water.

18.4.2 Surfactin

18.4.2.1 Remediation of Soil

A lipopeptide called surfactin produced by *Bacillus subtilis* (Figure 18.4) consists of seven amino acids attached to the carboxyl and hydroxyl groups of a 14-carbon acid (Kakinuma et al., 1969). Low surfactin concentrations (0.005%) can reduce the surface tension to 27 mN/m, making it a powerful biosurfactant. The primary structure of surfactin, determined many years ago by Kakinuma et al. (1969), is a heptapeptide with a β-hydroxy fatty acid within a lactone ring. Subsequently, the 3D structure was determined by ^{1}H NMR techniques (Bonmatin et al., 1995). Surfactin forms a β-sheet structure resembling a horse saddle in aqueous solutions at the air/water interface (Ishigami et al., 1995). The surface-active properties of the surfactin depend on the orientation of the residues. Mass spectrometry was used to characterize the mixtures of surfactin produced by *B. subtilis* (Hue et al., 2001). It showed that the amino acid composition or length of the acyl chain can vary from 12 to 15 carbons.

The presence of two negative charges on the aspartate and the glutamate residues of surfactin has enabled the binding of the metals, magnesium, manganese, calcium, barium, lithium, and rubidium (Thimon et al., 1992). Subsequently, batch soil washing experiments with surfactin from *B. subtilis* were performed for the removal of heavy metals from contaminated soil and sediments (Mulligan et al., 1999b). The contaminated soil had levels of copper, zinc, and cadmium of 550, 1200, and 2000 mg/kg, respectively. The sediments contained 110 mg/kg of copper and 3300 mg/kg of zinc. A series of five washings of the soil with 0.25% surfactin (1% NaOH) were required to remove 70% of the copper and 22% of the zinc compared to water alone that removed minimal amounts of copper and zinc (less than 1%). The mechanism of metal removal by surfactin was investigated by ultrafiltration, octanol–water partitioning, and the measurement of zeta potential. Surfactin removed the metals by sorption at the

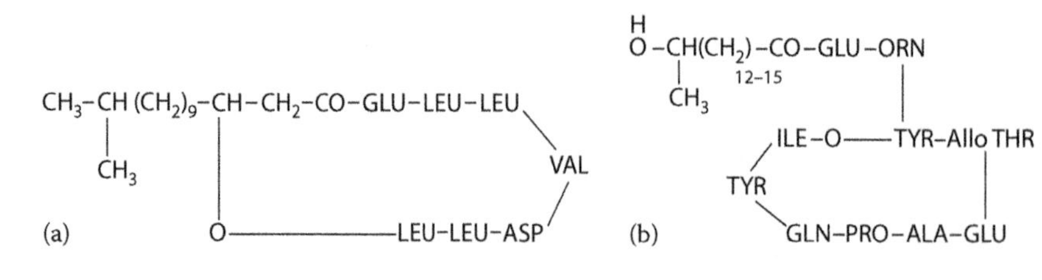

FIGURE 18.4 Structure of (a) surfactin and (b) fengycin. (Adapted from Mulligan, C.N., *Environ. Pollut.*, 133, 183, 2005.)

soil interface, metal complexation, desorption of the metal by reducing soil–water interfacial tension and fluid forces, and finally complexation of the metal by the micelles.

Mercury has been recovered by foam fractionation using surfactin (Chen et al., 2011). Air is bubbled into the solution to separate the metal contaminant. Various parameters were evaluated. A surfactin concentration in the order of 10× CMC, low Hg concentration (2 mg/L), and high pH (8–9) were optimal. Results were superior to the chemical surfactants, SDS and Tween 80.

Other studies showed that surfactin and fungicide were produced by *B. subtilis* A21 (Singh and Cameotra, 2013). Removal of various contaminants was effective (petroleum hydrocarbons (65%), and Cd, Co, Zn, Pb, Ni, Cu (26%–44%)) by the lipopeptides and was evaluated under various conditions. The efficiency was decreased by approximately half of the biosurfactant due to sorption on the soil. Mustard seed germination was improved by the soil washing process, an indication of its environmental friendly nature.

Another study was performed to evaluate the use of foam fractionation and soil flushing to remove heavy metals (Cu, Zn, and Pb) from an industrial contaminated soil (Maity et al., 2013). Two biosurfactants were compared, surfactin and saponin. Saponin was superior to surfactin, while foam fractionation was better than soil flushing. Pb was extracted more efficiently followed by Cu and then Zn. A pH of 4 gave better results, and concentrations of 0.15 g/L were preferential to half the concentration.

18.4.2.2 Remediation of Heavy Metal–Contaminated Water by Surfactin

Using MEUF, Mulligan et al. (1999b) studied the removal of various concentrations of metals from water by various concentrations of surfactin by anultrafiltration membrane with a 50,000 Da MWCO. Cadmium and zinc rejection ratios were close to 100% at basic pH values of 8.3 and 11, while copper rejection ratios were 85% at a slightly acidic pH (6.7). The addition of 0.4% of the oil as a co-contaminant slightly decreased the retention of the metals by the membrane. The ultrafiltration membranes also indicated that metals were associated with the surfactin micelles that remained in the retentate shown in Figure 18.5. The metal-to-surfactin ratio was 1.2:1 that was only slightly higher than the theoretical value of 1 mole metal/1 mole surfactin due to the two charges on the surfactin molecule.

Surfactin was found to be capable of interacting with cesium to form a complex with the surface of the micelles (Taira et al., 2015). A 91% effective separation was achieved with centrifugal ultrafiltration with 3 kDa MWCOs. Thus, it could be a new approach to treat contaminated soil and water.

Another application has been investigated by Reddy et al. (2008). They showed that the synthesis of silver nanoparticles could be stabilized by surfactin. These nanoparticles have unique physical, chemical, and magnetic properties. Various pH and temperature conditions were evaluated. Surfactin, an environmentally friendly additive, was used to stabilize the nanoparticles for two months.

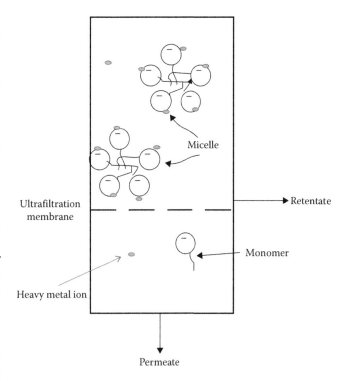

FIGURE 18.5 Removal of heavy metals from aqueous solutions utilizing micellar-enhanced ultrafiltration. (Adapted from Mulligan, C.N. et al., *Environ. Sci. Technol.*, 33, 3812, 1999b.)

18.4.3 SOPHOROLIPIDS

Candida bombicola (formerly known as *Torulopsis bombicola*) is a yeast that produces a sophorolipid biosurfactant (Cooper and Paddock, 1984) that is shown in Figure 18.6. The sophorolipid can be produced in high yields (67 g/L) from soybean oil and glucose (0.35 g/g of substrate). Higher concentrations of sophorolipid of 150 g/L were obtained from the substrates canola oil and lactose (Zhou and Kosaric, 1995). Sophorose lipids at 10 mg/L can lower the surface tension to 33 mN/m and interfacial tensions of *n*-hexadecane and water from 40 to 5 mN/m (Cooper and Paddock, 1984). These properties are stable between pH values of 6–9, various salt concentrations, and temperatures from 20°C to 90°C.

Although sophorolipids are able to release bitumen from tar sands (Cooper and Paddock 1984), few applications have been reported concerning its ability to enhance the remediation of hydrocarbon-contaminated soils. A crude preparation of biosurfactants from *C. bombicola* could solubilize a North Dakota Beulah Zap lignite coal (Polman et al., 1994). Sophorolipids have been studied for *in situ* bioremediation and degradation of hydrocarbons in soils and groundwater (Ducreux et al., 1997; Kang et al., 2010). The sophorolipids can also enhance secondary oil recovery and hydrocarbons from muds and dregs (Baviere et al., 1994; Marchal et al., 1999; Pesce, 2002).

High yields of the sophorolipid make this a potentially useful and economic biosurfactant. However, metal-contaminated soils and sediments have been treated with crude sophorolipids (Mulligan et al., 1999a, 2001b). Schippers et al. (2000)

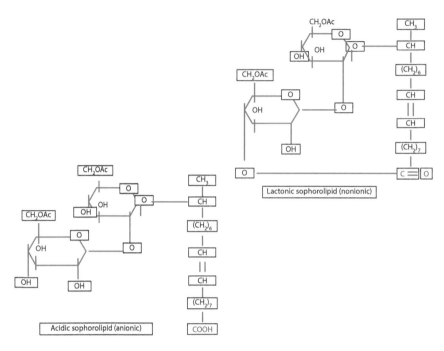

FIGURE 18.6 Structure of sophorolipids. Ac is defined as acetate. (Adapted from Mulligan, C.N., *Environ. Pollut.*, 133, 183, 2005.)

showed that in a 10% soil slurry, phenanthrene concentration was decreased by biodegradation from 80 to 0.5 mg/L with 500 mg/L of the surfactant compared to 2.3 mg/L without surfactant within 36 h. The maximal degradation rate by *Sphingomonas yanoikuyae* was 1.3 mg/L/h with the sophorolipid instead of 0.8 mg/L/h. Tests showed that 232 mg of phenanthrene/g of sophorolipid in water and 80.7 mg/g of sophorolipid in soil were solubilized, 10 times higher than other surfactants such as SDS. In addition, toxicity of the sophorolipid of concentrations up to 1 g was low. Adsorption of the surfactant onto the soil was indicated by the increase in CMC of the sophorolipid in water from 4 to 10 mg/L in the presence of a 10% soil suspension. Therefore, sophorolipids can increase the biodegradation of the phenanthrene through enhanced solubilization. Studies on more types of contaminants and more understanding of the mechanisms will be required.

Arab and Mulligan (2016) evaluated the efficiency of sophorolipids to remove arsenic and heavy metals from mine tailings. Using a 1% sophorolipid (pH 5) solution, 0.7% of the total removed arsenic was removed from the mine tailings samples of the water soluble, 0.7% was from the exchangeable, 0.6% was from the carbonate, 29.9% was from the oxide/hydroxide, 3.0% was from the organic, and 65.1% was from the residual fraction. The results from this study can help in the development of a sustainable solution for the remediation of mine tailings.

18.4.4 OTHER BIOSURFACTANTS FOR ENVIRONMENTAL APPLICATIONS

Efforts have been made to isolate and study other biosurfactants. Hong et al. (2002) examined the removal of zinc and

cadmium, by saponin, a plant-derived biosurfactant from three types of soils (Andosol, Cambisol, and Regosol). The best results were with Regosol with removals of 90%–100% for cadmium and 85%–98% for zinc. Saponin concentrations of 3% were optimal for metal removal within 6 h at a pH of 5.0 to 5.5. Sequential extraction tests indicated that exchangeable and carbonate fractions were removed by the saponin that is similar to surfactin and rhamnolipids tests for zinc performed by Mulligan et al. (2001a,b). Precipitation of the heavy metals from the soil supernatants was achieved by increasing the pH to 10.7. Saponin was recycled for reuse. At pH 10.7, 80% of Cu, 86% of Cd, 90% of Pb, and 91% of Zn were recovered.

Other studies for saponin were performed for removing organic and metal contaminants, phenanthrene and cadmium, from soil (Song et al., 2008) (87.7% and 76.2% for phenanthrene and cadmium, respectively). The mechanisms were phenanthrene removal by solubilization and cadmium complexation by saponin carboxylic groups. Further studies were performed by Chen et al. (2008) for the treatment of copper and nickel by saponin from kaolin. Saponin (2000 mg/L) could remove 85% of the nickel and 83% of the copper (pH 6.5). Desorption of the metals followed the order of EDTA > saponin >> SDS. A three-step washing mechanism consisted of adsorption, then formation of ion pairs with the adsorbed metal, and finally desorption of the metal.

A biosurfactant from marine bacterium was isolated and used for metal removal (Das et al., 2009). A 5× the CMC concentration enabled total removal of lead and cadmium. Metal binding on the micelle surface was indicated by TEM analysis. Precipitation of the metal ions indicated the potential for wastewater treatment.

Yuan et al. (2008) tested the removal of heavy metals by a tea saponin by ion flotation. The biosurfactant acted as both collector and frother. Lead (90%), copper (81%), and cadmium (71%) were adsorbed onto the surfaces of the air bubbles. Complexation via carboxylate groups with the divalent metal ions was determined.

Gusiatin (2014) examined plant biosurfactants, tannic acid, and saponin. Removal efficiencies for both were in the range of 50%–64%. As (V) was removed more effectively than As (III). The mechanism of removal is degradation of the iron-arsenic complex and then complexation of Fe. The reducible and residual fractions were affected.

Mixed contaminants (PCBs and metals Cu and Pb) were desorbed from soil by a combination of saponin and a biodegradable chelating agent (SS-ethylene diamine succinic acid [EDDS]) (Cao et al., 2013). EDTA was previously used, but due to its toxicity it presents a risk to groundwater. The two agents worked in a complementary manner. The EDDS functioned as a chelating agent, while the saponin enhanced the solubilization of the metal–EDDS complex. Up to 45.7%, 85.7%, and 99.8% of the PCBs, Cu, and Pb, respectively, were desorbed by 3000 mg/L of saponin and 10 mM of EDDS. Sorption on the soil of each was reduced by the other.

Various bacterial strains were screened for biosurfactant production from agro-based substrates and the ability to bind heavy metals (Hazra et al., 2011). *P. aeruginosa* AB4 was isolated, and its glycolipid biosurfactant was produced from renewable non-edible seed cakes. Preliminary tests also indicated that Pb and Cd could be chelated by the biosurfactant.

Two biosurfactant-producing strains (*Bacillus cereus* and *B. subtilis*) were obtained from a chromium-contaminated wastewater (Saranya et al., 2015). Palm and coconut oils were used as substrates. High yields of 0.32 g/g palm oil and 0.45 g/g coconut oil were produced by *B. subtilis* and *B. cereus*, respectively. Anionic and cationic lipopeptide biosurfactants were produced. Confirmation of the interaction of Cr(III) with the biosurfactants was confirmed by FT-IR, SEM-EDX, and atomic absorption spectrometry (AAS). An amount of 2.5 g of the biosurfactant could sequester 458 mg Cr(III). The cationic form (Figure 18.7) was more effective than the anionic

biosurfactant. pH 5 is the optimized pH. The amino acids a histidine, phenylalanine, and tyrosine, were protonated and then released protons to become deprotonated. The nitrogen in these amino acids stabilizes the bond with Cr(III) to form a coordinate bond. The anionic biosurfactant contained cysteine, a sulfur-containing amino acid. At pH 6, ionization of cysteine leads to a negative charge. An ionic bond is formed with Cr(III) that is not as strong as the cationic biosurfactant bonding and thus the less effective removal (98% compared to 85%).

A new strain of biosurfactant producing strain of *B. subtilis* was isolated from a Brazilian mangrove (Lima de Franca et al., 2015). Various agricultural wastewaters were evaluated as substrates such as glycerol, sunflower oil, cheese whey, and cashew apple juice. Up to 1.3 g/L could be produced on glycerol. The surface tension was reduced to 30 mN/m by the biosurfactant with a CMC of 25 mg/L. Removal of chromium, copper, and zinc was evaluated and high pH showed favorable conditions for removal from an effluent. Hydrocarbons such as crude oil and motor oil could also be removed from contaminated soil. Against brine shrimp, it was shown to be non toxic.

Hong et al. (1998) have also evaluated a sodium salt of 2-(2-carboxyethyl)-3-decyl maleic anhydride derived from the dehydration of spiculisporic acid produced by *Penicillium spiculisporum*. Membranes with MWCOs of 1000 and 3000 Da were studied for the rejection of cadmium, copper, zinc, and nickel. Highest rates were achieved when the molar biosurfactant and metal ion concentrations were equal.

Wastes from urban food and gardens were evaluated as substrates for emulsifier production. Soluble biobased organic substances isolated from composted biowastes were better than anaerobic digestor biowastes (Vargas et al., 2014). They are a mixture of organic compounds with high molecular weights (67,000–463,000 gmol). They were shown to form stable emulsions that are more environmentally friendly.

Bioremediation of sites contaminated with metals and hydrocarbons can be difficult due to the inhibition by the metals, and thus, various approaches have been developed to overcome the challenges (Sandrin and Maier, 2003). Cyclodextrin is a cyclic oligosaccharide produced from starch by bacteria. Ehsan et al. (2007) showed that EDTA (2 mM) and cyclodextrin (10% solution) could enhance the desorption of PCBs and heavy metals from a contaminated soil (Arjoon et al., 2013). Figure 18.8 shows the mechanism of surfactant interaction with mixed contaminants.

18.5 CONCLUSIONS AND FUTURE DIRECTIONS

Various applications of biosurfactants including remediation are emphasized such as *in situ* or *ex situ* soil remediation projects and water treatment of heavy metals. The high cost of producing the biosurfactants has limited commercial applications. Yields, rates, and recoveries can be improved and also crude preparations and inexpensive or waste substrates can be employed (Mulligan and Gibbs, 1993). Most of the information is available concerning the biosynthesis of rhamnolipids and surfactin, but

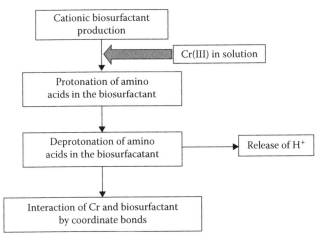

FIGURE 18.7 Removal of Cr(III) by cationic biosurfactants. (Modified from Saranya, P. et al., *RSC Adv.*, 5, 80596, 2015.)

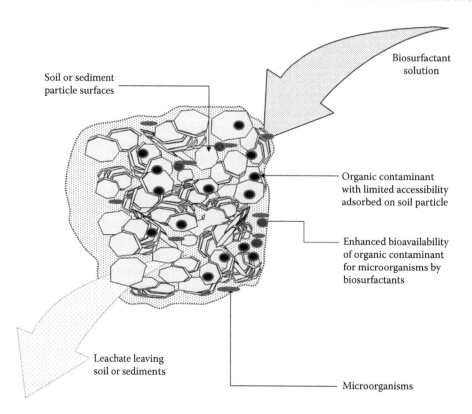

FIGURE 18.8 Enhancement of the bioavailability of pollutants by biosurfactants. (Adapted from Yong, R.N. et al., *Sustainable Practices in Geoenvironmental Engineering*, 2nd edn., CRC Press/Taylor & Francis, Boca Raton, FL, 2014.)

there is still a lack of information regarding secretion of other biosurfactants, metabolic routes, and primary cell metabolism (Peypoux et al., 1999). More research is needed in this area to enhance the applications and economics of the surfactant.

Jeneil Biotech Corp (http://www.jeneilbiotech.com) has produced biosurfactants such as rhamnolipids commercially. AGAE Technologies (http://www.agaetech.com/) also produces rhamnolipids from a unique strain of the bacterium *P. aeruginosa*. The Belgian company Ecover NV (http://www.ecover.com) and the Japanese company Saraya Co., LTD (http://www.saraya.com) are also working in the area of biosurfactants. The French company Soliance (http://www.groupesoliance.com) and the Korean MG Intobio Co. Ltd produce sophorolipid-based products for skin care and cosmetic products. More screening of native or modified sophorolipid producers is required to exploit this molecule for more commercial assets.

A summary of some of the studies involving the use of biosurfactants for biodegradation for mixed contaminants is shown in Table 18.2 and for washing or flushing is shown in Table 18.3. Despite the variety of biosurfactants that have been studied, most involve rhamnolipids. More investigation is needed in the evaluation of other biosurfactants.

Production of biosurfactants could also be *in situ* that would be cost-effective due to lower labor and transport requirements. The process would be ecologically acceptable. Strains of *Pseudomonas* and *Bacillus* have been found as biosurfactant producers at hydrocarbon-contaminated sites (Jennings and Tanner, 2000), and thus, exploiting their presence may be the best strategy. As the bioavailability of the contaminant is an important factor for biodegradation, the role of biosurfactant production could significantly enhance natural attenuation (Yong and Mulligan, 2004), but the fate and transport of the contaminants in the subsurface studies need to be clearly understood. Waste materials as substrates will improve the sustainability of the production process through cost reduction and waste reduction.

Only a few studies have been completed regarding mixed organic and inorganic contamination. Research is needed, particularly in light of the potential for the remediation of metal contamination by the biosurfactants.

Methods to enhance the economics of biosurfactant production will need to be developed further. For example, ultrafiltration (Mulligan and Gibbs, 1990) can be used to concentrate, purify, and then reuse the biosurfactants, thus decreasing costs.

The effects of soil fractions such as hydroxides/oxides and organics on metal desorption by biosurfactants have been evaluated in sediments, mining residues, and soil (Mulligan et al., 1999a,b; Mulligan and Dahr Azma, 2003; Dahr Azma and Mulligan, 2004; Arab and Mulligan 2016). Further methods will need to be developed to enable one to predict and model the efficiency of the washing or flushing processes and enhanced biodegradation with biosurfactants under various hydrological and soil conditions.

Rhamnolipids and other biosurfactants including saponin have shown the potential for the remediation of contaminated soil and water. More biosurfactants need to be isolated and investigated. The biosurfactants enhance biodegradation by

TABLE 18.2
Summary of Biodegradation Studies Involving Biosurfactants

Biosurfactant	Medium	Microorganism	Contaminant	Reference
Rhamnolipid	Soil	*P. aeruginosa* ATCC 9027	Phenanthrene and cadmium	Maslin and Maier (2000)
Rhamnolipid	Soil	*P. aeruginosa* ATCC 9027	Naphthalene and cadmium	Sandrin et al. (2000)
Rhamnolipid	Soil	*Luteibacter* sp.	Petroleum and heavy metals	Zhang et al. (2011)
Crude biosurfactant	Soil	*Bacillus subtilis* ICA 56	Hydrocarbons and heavy metals	Lima de Franca et al. (2015)
Rhamnolipid	Soil	*Pseudomonas aeruginosa* A11	Hg, Ni resistance, promotion of plant growth	Singh and Cameotra (2013)

Source: Adapted from Makkar, R.S. and Rockne, K.J., *Environ. Toxicol. Chem.*, 22, 2280, 2003.
All these studies showed a positive effect of the biosurfactant on biodegradation.

TABLE 18.3
Summary of Soil Washing or Flushing Studies or Translocation with Phytoremediation with Biosurfactants

Biosurfactant	Medium	Contaminant	Reference
Rhamnolipid, Surfactin	Soil, sediments	Pb, Zn, Cu	Mulligan et al. (1999a,b)
Rhamnolipid	Soil	Pb	Neilson et al. (2003)
Rhamnolipid	Mining residues	Cu	Dahrazma and Mulligan (2004)
Lecithin	Kaolinite	Cd, Pb	Juwarkar et al. (2007)
Rhamnolipid	Sepiolite, feldspar	Cd	Asci et al. (2008a)
Rhamnolipid	Feldspar	Zn	Asci et al. (2008b)
Rhamnolipid, surfactin	Kaolinite	Pb	Kim and Vipulanandan (2006)
Rhamnolipid, MEL, saponin	Soil, sediment	Zn, Cu, Pb, Oil	Mulligan et al. (2007)
Rhamnolipid	Water	Cr	Massara et al. (2007)
Rhamnolipid	Soil, water	Cr	Ara and Mulligan (2015)
Rhamnolipid	Water	Cr	Abbasi-Garravand and Mulligan (2014)
Rhamnolipid	Mining residues	As	Wang and Mulligan (2009a,b)
Saponin	Soil	Phenanthrene, Cd	Song et al. (2008)
Saponin	Kaolin	Cu, Ni	Chen et al. (2008)
Saponin	Water	Cu, Cd, Pb	Yuan et al. (2008)
Surfactin	Tannery sludge	Cr	Kilic et al. (2011)
Rhamnolipid	Liquid	Hg	Chen et al. (2011)
Rhamnolipid	Soil	Cd, Ni	Mulligan and Wang (2004)
Rhamnolipid	Sediments	PAH, Pb, Zn, Ni	Alavi and Mulligan (2011)
Rhamnolipid	Water	Cu, Zn, Ni, Pb, Cd	El-Zeftawy and Mulligan (2011)
Rhamnolipid	Water	Cu, benzene	Ridha and Mulligan (2011)
Surfactin, saponin	Soil	Cu, Zn, Pb	Maity et al. (2013)
Saponin	Soil	PCB, Cu, Pb	Cao et al. (2013)
Saponin, tannic acid	Soil	As	Gusiatin (2014)
Surfactin, fengycin	Soil	Petroleum hydrocarbons, Cd, Co, Zn, Pb, Ni, Cu	Singh and Cameotra (2013)
Cationic, anionic lipopeptide biosurfactants	Water	Cr(III)	Saranya et al. (2015)
Surfactin	Water	Cs	Taira et al. (2015)
Sophorolipids	Mining residues	As	Arab and Mulligan (2016)
Rhamnolipid, viscosin	Soil	PAHs, metals	Ammami et al. (2015)
Rhamnolipid with phytoremediation	Soil and mining residues	Cr, Cu, and Ni	Barajas et al. (2015)
Cyclodextrin and EDTA	Soil	PCBs and heavy metals (Cd, Cr, Cu, Mn, Ni, Pb, Zn)	Ehsan et al. (2007)

solubilization and emulsification of the contaminants. Due to their biodegradability and low toxicity, rhamnolipid, biosurfactants, and others are very promising for assisting remediation technologies. Also, there is the potential for *in situ* production, which is advantageous in comparison to synthetic surfactants. New applications for the biosurfactants regarding nanoparticles are developing. Some cases have resulted in the inhibition of the biodegradation, which are not well understood. Modeling of the biodegradation is not reliable, and experimental testing is required. For soil washing applications, solubilization and mobilization are the two main mechanisms as indicated by the effectiveness of the biosurfactants mentioned earlier and below the CMC (Franzetti et al., 2010). Arsenic and chromium have recently been added to the list.

Research on biosurfactant–metal interactions is large and continues to grow. The potential for green, economical remediation technologies based on these interactions is enormous, but there remain several challenges that must be resolved before biosurfactants become viable alternatives for use in remediation technologies. The first challenge is the product cost. Biosurfactants are not currently very competitive with synthetic surfactants. The second challenge is in understanding the chemical properties of individual biosurfactant congeners so that they can be optimized for application in remediation technologies. The ability to manipulate bacteria to produce congeners is appropriate for specific applications. The third challenge is the lack of full-scale applications. Research needs to be performed in pilot demonstration tests in the field before full-scale applications can be performed to demonstrate effectiveness. Although there has been substantial evidence of the benefits of biosurfactants on bioremediation, inhibition has also been observed. Therefore, more research on the complex interactions between the pollutants, biosurfactants, and microorganisms is needed. The research in the field of biosurfactants is advancing rapidly, and it encompasses fields as diverse as medicine, surface science, organic chemistry, molecular biology, etc. Thus, there is a significant increase in biosurfactant applications and thus the need for its increased production. Traditional chemical synthesis methods may increase toxicity and environmental impact. This favors the switch to more sustainable synthesis options. Biosurfactants that are produced from environmentally renewable resources are preferred as they use ambient temperatures, are cost-effective, retain sustainability, and generate minimum waste. Methods for the recovery of biosurfactants must be modified to improve process economics and impact on the environment.

REFERENCES

Abalos, A. Pinazo, A. Infante, M.R. Casals, M. García, F. Manresa, A. 2001. Physicochemical and antimicrobial properties of new rhamnolipids produced by *Pseudomonas aeruginosa* AT10 from soybean oil refinery wastes. *Langmuir* 17: 1367–1371.

Abbasi-Garravand, E., Mulligan, C.N. 2014. Using micellar enhanced ultrafiltration and reduction techniques for removal of Cr (VI) and Cr (III) from water. *Separation and Purification Technology* 132: 505–512.

Alavi, A., Mulligan, C.N. 2011. Remediation of a heavy metal and PAH-contaminated sediment by a rhamnolipid foam. Geo-Environmental Engineering, Takamatsu, Japan, May 21–22, 2011.

Ammami, M.T., Portet-Kol talo, F., Benamar, A., Duclairoir-Poc, C., Wang, H., Le Derf, F. 2015. Applications of biosurfactants and periodic voltage gradient for enhanced electrokinetic remediation of metals and PAHs in dredged marine sediments. *Chemosphere* 125: 1–8.

Ara, I., Mulligan, C.N. 2015. Reduction of chromium in water and soil using a rhamnolipid biosurfactant geotechnical engineering. *Journal of the SEAGS & AGSSEA* 46: 25–31.

Arab, F., Mulligan, C.N. 2016. Efficiency of sophorolipids for arsenic removal from mine tailings. *Environmental Geotechnics.* http://dx.doi.org/10.1680/jenge.15.00016.

Arjoon, A., Olaniran, A.O., Pillay, B. 2013. Co-contaminated of water with chlorinated hydrocarbons and heavy metals: Challenges and current bioremediation strategies. *International Journal of Environmental Science and Technology* 10: 395–412.

Asci, Y., Nurbas, M., Acikel, Y.S. 2007. Sorption of Cd (II) onto kaolin as a soil component and desorption of Cd(II) from kaolin using rhamnolipid biosurfactant. *Journal of Hazardous Materials* B139: 50–56.

Asci, Y., Nurbas, M., Acikel, Y.S. 2008a. A comparative study for the sorption of Cd(II) by K-feldspar and sepiolite as soil components and the recovery of Cd(II) using rhamnolipid biosurfactant. *Journal of Environmental Management* 88: 383–392.

Asci, Y., Nurbas, M., Acikel, Y.S. 2008b. Removal of zinc ions from a soil component Na-feldspar by a rhamnolipid biosurfactant. *Desalination* 233: 361–365.

Barajas-Aceves, M., Camarillo-Ravelo, D., Rodriguez-Vasquez, R. 2015. Mobility and translocation of heavy metals from mine tailings in three plant species after amendment with compost and biosurfactant. *Soil and Sediment Contamination: An International Journal* 24: 224–249.

Baviere, M., Degouy, D., Lecourtier, J. 1994. Process for washing solid particles comprising a sophoroside solution. US Patent, 5,326,407.

Biermann, M., F. Lange, R. Piorr, U. Ploog, H. Rutzen, J. Schindler, R. Schmidt, 1987. Surfactants in consumer products. In: *Theory, Technology and Application,* Falbe, J. (Ed.). Springer-Verlag, Heidelberg, Germany, pp. 86–106.

Biswas, M., Raichur, A.M. 2008. Electrokinetic and rheological properties of nano zirconia in the presence of rhamnolipid biosurfactant. *Journal of American Ceramic Society* 91: 3197–3201.

Bonmatin, J.M., Genest, M., Labbé, H., Grangemard, I., Peypoux, F., Maget-Dana, R., Ptak, M., Michel, G. 1995. Production, isolation and characterization of [Leu4]- and [Ile4] surfactins from *Bacillus subtilis. Letters in Peptide Science* 2: 41–47.

Cao, M., Hu, Y., Sun, Q., Wang, L., Chen, J., Lu, X. 2013. Enhanced desorption of PCB and trace element metals (Pb and Cu) from contaminated soils by saponin and EDSS mixed solution. *Environmental Pollution* 174: 93–99.

Chen, W.-J., Hsiao, L.-C., Chen, K.K.-Y. 2008. Metal desorption from copper (II)/nickel(II)-spiked kaolin as a soil component using plant-derived saponic biosurfactant. *Process Biochemistry* 43: 488–498.

Chen, H.-R., Chen, C.-C., Reddy, A.S., Chen, C.-Y., Li, W. R., Tseng, M.-J., Liu, H.-T., Pan, W., Maity, J.P., Atla, S. B. 2011. Removal of mercury by foam fractionation using surfactin, a biosurfactant. *International Journal of Molecular Sciences* 12: 8245–8258.

Cooper D. 1986. Biosurfactants. *Microbiological Science* 3(5): 145–149.

Cooper, D.G., Paddock, D.A. 1984. Production of a biosurfactant from *Torulopsis bombicola*. *Applied and Environmental Microbiology* 47: 173–176.

Dahr Azma B., Mulligan, C.N. 2004. Extraction of copper from mining residues by rhamnolipids. *Practice Periodical on Hazardous Toxic and Radioactive Waste Management* 8: 166–172.

Dahrazma, B., Mulligan, C.N. 2007. Investigation of the removal of heavy metals from sediments using rhamnolipid in a continuous flow configuration. *Chemosphere* 69: 705–711.

Dahrazma, B., Mulligan, C.N., Nieh, M.P. 2008. Effects of additives on the structure of rhamnolipid (biosurfactant): A small-angle neutron scattering (SANS) study. *Journal of Colloid and Interfacial Science* 319: 590–593.

Daigger, G.T., Crawford, G.V. 2005. Wastewater treatment plant of the future—Decision analysis approach for increased sustainability. In *Second IWA Leading-Edge Conference on Water and Wastewater Treatment Technology*, Water and Environment Management Series, London, U.K.: IWA Publishing, pp. 361–369.

Das, P., Mukherjee, S., Sen, R. 2009. Biosurfactant of marine origin exhibit in heavy metal remediation properties. *Bioresource Technology* 100: 4887–4890.

Ducreux, J., Ballerini, D., Baviere, M., Bocard, C., Monin, N. 1997. Composition containing a surface active compound and glycolipids and decontamination process for a porous medium polluted by hydrocarbons. US Patent 5,654,192.

Ehsan, S., Prasher, S.O., Marshall, W.D. 2007. Simultaneous mobilization of heavy metals and polychlorinated biphenyl (PCB) compounds from soil with cyclodextrin and EDTA in admixture. *Chemosphere* 68: 150–158.

El-Shafey, E.L., Canepa, P. 2003. Remediation of a Cr (VI) contaminated soil: Soil washing followed by Cr (VI) reduction using a sorbent prepared from rice husk. *Journal de Physique IV (Proceedings)* 107: 415–418.

El-Zeftawy, M.A.M., Mulligan, C.N. 2011. Use of rhamnolipid to remove heavy metals from wastewater by micellar-enhanced ultrafiltration (MEUF). *Separation and Purification Technology* 77: 120–127.

Farshidy, M., Mulligan, C.N., Bolton, K. 2011. Stabilization of iron nanoparticles by rhamnolipid for remediation of chromium contaminated water, ICEPR, Ottawa, Ontario, Canada, August 17–19, 2011.

Fatisson, J., Ghoshal, S., Tufenkji, N. 2010. Deposition of carboxymethylcellulose-coated zero-valent iron particles onto silica: Roles of solution chemistry and organic molecules. *Langmuir* 26: 12832–12840.

Franzetti, A., Caredda, P., Ruggeri, C., La Colla, P., Tamburini, E., Papacchini, M., Bestetti, G. 2009. Potential applications of surface active compounds by *Gordonia sp.* BS29 in soil remediation technologies. *Chemosphere* 75: 801–807.

Fraser, L. 2000. Innovations: Lipid later removes metals. *Environmental Health Perspectives* 108: A320.

Guerra-Santos, L.H., Käppeli, O., Fiechter, A. 1984. *Pseudomonas aeruginosa* biosurfactant production in continuous culture with glucose as carbon sources. *Applied and Environmental Microbiology* 48: 301–305.

Gusiatin, Z.M. 2014. Tannic acid and saponin for removing arsenic from brownfield soils: Mobilization, distribution and speciation. *Journal of Environmental Sciences* 26: 855–864.

Guo, Y., Mulligan, C.N. 2006. Combined treatment of styrene-contaminated soil by rhamnolipid washing followed by anaerobic treatment. Chapter 1. In *Hazardous Materials in Soil and Atmosphere*, R.C. Hudson (Ed.). Nova Science Publishers, New York, pp. 1–38.

Hazra, C., Kundu, D., Ghosh, P., Joshi, S., Dandi, N., Chaudhar, A. 2011. Screening and identification of *Pseudomonas aeruginosa* AB4 for improved production, characterization and application of a glycolipid biosurfactant using low-cost agrobased raw materials. *Journal of Chemical Technology and Biotechnology* 86: 185–198.

Herman, D.C., Artiola, J.F., Miller, R.M. 1995. Removal of cadmium, lead and zinc from soil by a rhamnolipid biosurfactant. *Environmental Science and Technology* 29: 2280–2285.

Hitsatsuka, K., Nakahara, T., Sano, N., Yamada, K. 1971. Formation of a rhamnolipid by *Pseudomonas aeruginosa* and its function in hydrocarbon fermentation. *Agricultural and Biological Chemistry* 35: 686–692.

Holden, J. 1989. *How to Select Hazardous Waste Treatment Technologies for Soils and Sludges*. Noyes Data Corp., Park Ridge, NJ.

Hong, J.-J., Yang S.-M., Lee, C.-H., Choi, Y.-K., Kajiuchi, T. 1998. Ultrafiltration of divalent metal cations from aqueous solutions using polycarboxylic acid type biosurfactant. *Journal of Colloid and Interfacial Science* 202: 63–73.

Hong, K.J., Tokunaga, S., Kajiuchi, T. 2002. Evaluation of remediation process with plant-derived biosurfactant for recovery of heavy metals from contaminated soils. *Chemosphere* 49: 379–387.

Hue, N., Serani, L., Laprévote, O. 2001. Structural investigation of cyclic peptidolipids from *Bacillus subtilis* by high energy tandem mass spectrometry. *Rapid Communications in Mass Spectrometry* 15: 203–209.

Ishigami, Y., Osman, M., Nakahara, H., Sano, Y., Ishiguro, R., Matusumoto, M. 1995. Significance of β-sheet formation for micellization and surface adsorption on surfactin. *Colloids and Surfaces B* 4: 341–348.

Jennings, E.M., Tanner, R.S., 2000. Biosurfactant-producing bacteria found in contaminated and uncontaminated soils. *Proceedings of the 2000 Conference on Hazardous Waste Research*, Denver, CO. pp. 299–306.

Juwarkar, A.A., Nair, A., Dubey, K.V., Singh, S.K., Devotta, S. 2007. Biosurfactant technology for remediation of cadmium and lead contaminated soils. *Chemosphere* 68: 1996–2002.

Kakinuma, A., Oachida, A., Shima, T., Sugino, H., Isano, M., Tamura, G., Arima, K. 1969. Confirmation of the structure of surfactin by mass spectrometry. *Agricultural and Biological Chemistry* 33: 1669–1672.

Kang, S.W., Kim, Y.B., Shin, J.D., Kim, E.K. 2010. Enhanced biodegradation of hydrocarbons in soil by microbial biosurfactant, sophorolipid. *Applied Biochemistry and Biotechnology* 160: 780–790.

Kanga, S.H., Bonner, J.S., Page, C.A., Mills, M.A., Autenrieth, R.L. 1997. Solubilization of naphthalene and methyl-substituted naphthalenes from crude oil using biosurfactants. *Environmental Science and Technology* 31: 556–561.

Kiliç, E., Font, J., Puig, R., Colak, S., Celik, D. 2011. Chromium recovery from tannery sludge with saponin and oxidative remediation. *Journal of Hazardous Materials* 185: 456–462.

Kim, J., Vipulanandan, C. 2006. Removal of lead from contaminated water and clay soil using a biosurfactant. *Journal of Environmental Engineering* 132: 777–786.

LaGrega, M.D., Buckingham, P.L., Evans, J.C. 2001. *Hazardous Waste Management*. McGraw Hill, Boston, MA.

Lang, S., Wagner, F. 1987. Structure and properties of biosurfactants. In *Biosurfactants and Biotechnology*, N. Kosaric, W.L. Cairns, and N.C.C. Gray, (Eds.). Marcel Dekker, New York, pp. 21–45.

Lima de Franca, I.W., Lima, A.P., Lemos, J.A.M., Lemos, C.G.F., Melo, V.M.M., de Sant'ana, H.B., Gonçalves, L.R.B. 2015. Production of a biosurfactant by *Bacillus subtilis* ICA 56 aiming bioremediation of impacted soils. *Catalysis Today* 255: 10–15.

Lin, S.C. 1996 Biosurfactants: Recent advances. *Journal of Chemical Technology and Biotechnology* 66(2): 109–120.

Maity, J.P., Huang, Y.M., Hsu, C-M., Wu, C.-I., Chen, C-C., Li, C.-Y., Jean, J.-S., Chang, Y.-F., Chen, C.-Y. 2013. Removal of Cu, Pb, and Zn by foam fractionation and a soil washing process from contaminated industrial soils using soap-berry-derived saponin: A comparative effectiveness assessment. *Chemosphere* 92: 1286–1293.

Makkar, R.S., Rockne, K.J. 2003. Comparison of synthetic surfactants and biosurfactants fin enhancing biodegradation of polycyclic aromatic hydrocarbons. *Environmental and Toxicological Chemistry* 22: 2280–2292.

Marchal, R., Lemal, J., Sulzer, C., Davila, A.M. 1999. Production of sophorolipid acetate acids from oils or esters. US Patent, 5,900,366.

Maslin, P., Maier, R.M. 2000. Rhamnolipid-enhanced mineralization of phenanthrene in organic-metal co-contaminated soils. *Bioremediation Journal* 4: 295–308.

Massara, H., Mulligan, C.N., Hadjinicolaou, J. 2007. Effect of rhamnolipids on chromium contaminated soil. *Soil Sediment Contamination: International Journal* 16: 1–14.

Mulligan, C.N. 2005. Environmental applications for biosurfactants. *Environmental Pollution* 133: 183–198.

Mulligan, C.N., Dahr Azma, B. 2003. Use of selective sequential extraction for the remediation of contaminated sediments, ASTM STP 1442 Contaminated Sediments: Characterization, Evaluation, Mitigation/Restoration and Management Strategy Performance, West Conshocken, PA, pp. 208–223.

Mulligan, C.N., Gibbs, B.F. 1990. Recovery of biosurfactants by ultrafiltration. *Journal of Chemical Technology and Biotechnology* 47: 23–29.

Mulligan, C.N., Gibbs, B.F. 1993. Factors influencing the economics of biosurfactants. In *Biosurfactants, Production, Properties, Applications*, Kosaric, N. (Ed.). Marcel Dekker, New York, pp. 329–371.

Mulligan, C.N., Oghenekevwe, C., Fukue, M., Shimizu, Y. 2007. Biosurfactant enhanced remediation of a mixed contaminated soil and metal contaminated sediment. In *Seventh Geoenvironmental Engineering Seminar, Japan-Korea-France*, Grenoble, France, May 19–24, 2007.

Mulligan, C.N., Yong R.N., Gibbs, B.F. 1999a. On the use of biosurfactants for the removal of heavy metals from oil-contaminated soil. *Environmental Progress* 18: 50–54.

Mulligan, C.N., Yong, R.N., Gibbs, B.F. 1999b. Metal removal from contaminated soil and sediments by the biosurfactant surfactin. *Environmental Science and Technology* 33: 3812–3820.

Mulligan, C.N., Yong, R.N., Gibbs, B.F. 2001a. Surfactant-enhanced remediation of contaminated soil: A review. *Engineering Geology* 60: 371–380.

Mulligan, C.N., Yong R.N., Gibbs, B.F., 2001b. Heavy metal removal from sediments by biosurfactants. *Journal of Hazardous Materials* 85: 111–125.

Mulligan, C. N., Wang, S., 2004. Remediation of a heavy metal contaminated soil by a rhamnolipid foam. In *Proceedings of the Fourth BGA Geoenvironmental Engineering Conference*, Stratford-Upon-Avon, U.K., June 2004.

Neilson, J.W., Artiola, J.F., Maier, R.M., 2003. Characterization of lead removal from contaminated soils by non-toxic soil-washing agents. *Journal of Environmental Quality* 32: 899–908.

Oberbremer, A., Muller-Hurtig, R., Wagner, F. 1990. Effect of the addition of microbialsurfactants on hydrocarbon degradation in a soil population in a stirred reactor. *Applied Microbiology and Biotechnology* 32: 485–489.

Ochoa-Loza, F. 1998. Physico-chemical factors affecting rhamnolipid biosurfactant application for removal of metal contaminants from soil. PhD dissertation, University of Arizona, Tucson, AR.

Ochoa-Loza, F.J., Artiola, J.F., and Maier, R.M. 2001. Stability constants for the complexation of various metals with a rhamnolipid biosurfactant. *Journal of Environmental Quality* 30(2): 479–485.

Ochoa-Loza, F.J., Noordman, W.H., Jannsen, D.B., Brusseau, M.L., Miller, R.M. 2007. Effects of clays, metal oxides, and organic matter on rhamnolipid sorption by soil. *Chemosphere* 66: 1634–1642.

Palanisamy, P. 2008. Biosurfactant mediated synthesis of NiO nanorods. *Materials Letters* 62: 743–746.

Pesce, L. 2002. A biotechnological method for the regeneration of hydrocarbons from dregs and muds, on the base of biosurfactants. World Patent, 02/062495.

Peypoux, F., Bonmatin J.M., Wallach, J. 1999. Recent trends in the biochemistry of surfactin. *Applied Microbiology and Biotechnology* 51: 553–563.

Polman, J.K., Miller, K.S., Stoner, D.L., Brakenridge, C.R. 1994. Solubilization of bituminous and lignite coals by chemically and biologically synthesized surfactants. *Journal of Chemical Technology and Biotechnology* 61: 11–17.

Reddy, A.S., Chen, C.-Y., Baker, S.C., Chen, C-C., Jean, J.-S., Fan, C.-W., Chen, H.-R., Wang, J.-C. 2008. Synthesis of silver nanoparticles using surfactin: A biosurfactant stabilizing agent. *Materials Letters* 63: 1227–1230.

Ridha, Z.A.M., Mulligan, C.N. 2011. Simultaneous removal of benzene and copper from contaminated water using micellar-enhanced ultrafiltration. In *CSCE General Conference*, Ottawa, Ontario, Canada, June 14–17, 2011.

Robert, M., Mercadé, M.E., Bosch, M.P., Parra, J.L., Espiny, M.J., Manresa, M.A., Guinea, J. 1989. Effect of the carbon source on biosurfactant production by *Pseudomonas aeruginosa* 44T1. *Biotechnology Letters* 11: 871–874.

Samson, R., Cseh, T., Hawari, J., Greer, C.W., Zaloum, R. 1990. Biotechnologies appliquées à la restauration de sites contaminés avec d'application d'une technique physico chimique et biologique pour les sols contaminés par des BPC. *Science et Techniques de l'Eau* 23: 15–18.

Sandrin, T.R., Maier, R.M. 2003. Impact of metals on the biodegradation of organic pollutants. *Environmental Health Perspectives* 111: 1093–1101.

Sandrin, T.R., Chech, A.M., and Maier, R.M. 2000. A rhamnolipid biosurfactant reduces cadmium toxicity during naphthalene biodegradation. *Applied and Environmental Microbiology* 6(10): 4585–4588.

Saranya, P., Bhavani, P., Swarnalatha, S., Sekaran, G. 2015. Biosequestration of chromium (III) in an aqueous solution using cationic and anionic biosurfactants produced from two different *Bacillus* sp.—A comparative study. *RSC Advances* 5: 80596–80611.

Schippers, C., Geβner, K., Muller, T., Scheper, T. 2000. Microbial degradation of phenanthrene by addition of a sophorolipid mixture. *Journal of Biotechnology* 83: 189–198.

Sheng, X., He, L., Wang, Q., Ye, K.H., and Jiang, C. 2008. Effects of inoculation of biosurfactant-producing *Bacillus* sp. J119 on plant growth and cadmium uptake in a cadmium-amended soil. *Journal of Hazardous Materials* 155: 17–22.

Singh, A.K., Cameotra, S.S. 2013. Efficiency of lipopeptide biosurfactants in removal of petroleum hydrocarbons and heavy metals from contaminated soil. *Environmental Science and Pollution Research* 20: 7367–7376.

Song, S., Zhu, L., Zhou, W. 2008. Simultaneous removal of phenanthrene and cadmium from contaminated soils by saponin, a plant-derived biosurfactant. *Environmental Pollution* 156: 1368–1370.

Slizovsky, I.B., Klsey, J.W., Hatzinger, P.B. 2011. Surfactant-facilitated remediation of metal-contaminated soils: Efficacy and toxicological consequences to earthworms. *Environmental Toxicology and Chemistry* 30: 112–123.

Syldatk, C., Wagner, F. 1987. Production of biosurfactants. In: *Biosurfactants and Biotechnology*, N. Kosaric, W. L. Cairns, N. C. Gray, (Ed.), Vol. 25, Surfactant Science Series. Marcel Dekker, New York, pp. 89–120.

Taira, T., Yanagisawa, S., Nagano, T., Zhu, Y., Kuroiwa, T., Koumara, N., Kitamoto, D., Imura, T. 2015. Selective encapsulation of cesium ions using the cyclic peptide moiety of surfactin: Highly efficient removal based on an aqueous giant micellar system. *Colloids and Surfaces B: Biointerfaces* 134: 59–64.

Tan, H., Champion, J.T., Artiola, J.F., Brusseau, M.L., Miller, R.M. 1994. Complexation of cadmium by a rhamnolipid biosurfactant. *Environmental Science and Technology* 28: 2402–2406.

Thimon, L., Peypoux, F., Michel, G. 1992. Interactions of surfactin, a biosurfactant from *Bacillus subtilis* with inorganic cations. *Biotechnology Letters* 14: 713–718.

Vargas, A.K.N., Prevot, A.B., Montoneri, E., Le Rous, G.C., Savarino, P., Cavalli, R., Guardani, R., Tabasso, S. 2014. Use of biowaste-derived biosurfactants in production of emulsions for industrial use. *Industrial and Engineering Chemistry Research* 53: 8621–8629.

Venghuis, T., Werther, J. 1998. Flotation as an additional process step for the washing of soils contaminated with heavy metals. In *Proceedings of the Sixth International FZK/TNO Conference, ConSoil' 98*, Edinburgh, U.K., May 1998, Vol. 1, pp. 479–480, London, U.K.: Thomas Telford Publishing.

Wang, S., Mulligan, C.N. 2009a. Rhamnolipid biosurfactant-enhanced soil flushing for the removal of arsenic and heavy metals from mine tailings. *Process Biochemistry* 44: 296–301.

Wang, S., Mulligan, C.N. 2009b. Arsenic mobilization from mine tailings in the presence of a biosurfactant. *Applied Geochemistry* 24: 938–935.

Xia, W.-B., Li, X., Gao, H., Huang, B.-R., Zhang, H.-Z., Liu, Y.-G., Zeng, G.-M., Fan, T. 2009. Influence factors analysis of removing heavy metals from multiple metal-contaminated soils with different extractant. *Journal of Central South University of Technology* 16: 108–111.

Yong, R.N., Mulligan, C.N., 2004. *Natural Attenuation of the Contaminants in Soil*. CRC Press, Boca Raton, FL.

Yong, R.N., Mulligan, C.N., Fukue, M. 2014. *Sustainable Practices in Geoenvironmental Engineering*, 2nd edn. CRC Press/Taylor & Francis, Boca Raton, FL.

Yuan, X.Z., Meng, Y.T., Zwng, G.M., Fang, Y.Y., Shi, J.G. 2008. Evaluation of tea-derived biosurfactant on removing heavy metal ions from dilute wastewater by ion flotation. *Colloid and Surfaces A* 317: 256–261.

Zhang, J., Li, J., Chen, L., Thring, R.W. 2011. Remediation of refinery oily sludge using isolated strain and biosurfactant. In *Water Resource and Environmental Protection (ISWREP), 2011 International Symposium*, Xian, China, May 20–22, 2011, Vol. 3, pp. 1649–1653.

Zhu, L.Z., Zhang, M. 2008. Effect of rhamnolipids on the uptake of PAHs on rye grass. *Environmental Pollution* 156: 46–52.

Zhou, Q.H., Kosaric, N. 1995. Utilization of canola oil and lactose to produce biosurfactant with *Candida bombicola*. *Journal of the American Oil Chemists Society* 72: 67–71.

19 Bacterial Biofilms and Genetic Regulation for Metal Detoxification

Sudhir K. Shukla, Neelam Mangwani, Dugeshwar Karley, and T. Subba Rao

CONTENTS

ABSTRACT

Biofilm-mediated bioremediation is an emerging field and needs to be studied extensively to achieve pragmatic results toward the betterment of the environment. In the past decade, contamination of the environment with hazardous heavy metals has increased at an alarming rate. Although there are many chemical treatment methods for metal detoxification, biological processes are preferred since they are environmentally safe and the bioprocess is techno-economically viable. Among the various methodologies available for heavy metal remediation, the biofilm-based bioremediation technologies are promising. Therefore, it is necessary to understand how bacteria develop heavy metal/toxic metal resistance in a biofilm mode of adaptation in environments contaminated with toxic metals. With the available literature, it is known that metal resistance mechanisms for planktonic microbial cells and biofilm matrix embedded cells are entirely different. Therefore, the genetic basis of these resistance mechanisms needs to be studied extensively for a proper insight into the genetic approaches for metal bioremediation. Many important metal detoxification genes have been discovered; however, owing to the vast majority of the microbial diversity and environmental conditions, a complete understanding of the microbial genetic machinery is required to develop a suitable toxic metal remediation strategy. This chapter describes the physical and genetic bases of heavy metal tolerance and its detoxification bioprocesses by microbial biofilms.

19.1 INTRODUCTION

19.1.1 Microbial Biofilm

The formation of nontransient adherent communities of microorganisms on submerged surfaces, which is known as biofilms, is ubiquitous. More than 90% of the entire aquatic bacteria are found to be associated with interfaces such as the sediment–water interface and surface microlayer of aquatic systems (Cowan et al., 1991; Lappin-Scott and Costerton, 1989). Biofilms constitute a consortium of biotic and abiotic elements like bacteria, cyanobacteria, and algae attached to a substratum by microbially produced extracellular polysaccharide matrix that entraps soluble and particulate matter, immobilizes extracellular enzymes, and acts as a natural sink for nutrients and inorganic elements. The biofilm composition may vary both spatially and temporally pertaining to different waters, and the greatest differences are usually associated with shifts in the relative importance of autotrophic and heterotrophic microorganisms (Rao et al., 1997a; Robb, 1984).

Mutual interactions in biofilms among various microbial species result in community formation, which are habitat specific. The presence of different species of bacteria and algae in successive layers of the biofilm provides clues about the energy and substrate transfer within the biofilm. Water quality constraints affect the biofilm formation because of their role in the regulation of bacterial metabolism (Rao et al., 1997a). Recently, a limited number of investigations have been carried out to understand the mechanisms of biofilm development and the physiological activities of the microorganisms associated with it (Liu et al., 1993). Differences and alterations in the structural or physicochemical properties of the biofilm are determined by the organic and inorganic constituents of the surrounding environment (Keiding and Nielsen, 1997). Some important reports on studies on biofilm characterization in marine and freshwater systems include Characklis and Cooksey (1983), Rao (2003), Rao et al. (1997a,b), and Saravanan et al. (2006). Various physical, chemical, and biological factors govern the formation of biofilms. Among them, nutrient concentrations attain much significance (Rao et al., 1997b).

The adhesion of microorganisms to solid substratum has practical implications in industrial applications particularly in effluent treatment (Allison and Gilbert, 1992; Sjollema and Busscher, 1990). The development of biofilm reactor (Pedersen, 1982) and flow cells (Lawrence and Caldwell, 1987; Sjollema and Busscher, 1990) and studies on microcolony/biofilm development under laboratory environment got augmented (Saravanan et al., 2006). The outcome of the research exertions provided valuable information on the behavior of mono and mixed culture bacterial biofilms at various flow conditions imitating the industrial cooling circuits (Characklis and Marshall, 1990; Sutherland, 1997), ordering characteristics (Rao et al., 1997b), and behavior of bacteria in the hydrodynamic boundary layer (Lawrence and Caldwell, 1987). Costerton et al. (1995) stated that adhesion of microbes to surfaces activates the expression of a σ factor that derepresses a large number of genes so that biofilm cells are phenotypically different from their planktonic counterpart.

19.1.2 Process of Biofilm Development

Generally, biofilms constitute a diverse growth phase of microorganisms that are distinctly different from the planktonic microbiota. The typical structure of biofilm formation on surfaces is outlined into the following phases and schematically described in Figure 19.1a through e, as multiple images and schematics.

19.1.2.1 Phase 1: Reversible Adhesion

This important step in biofilm formation is the approach of microorganisms to the conditioned surface (Figure 19.1a). This process may be active or passive, depending on whether the bacteria are motile or transported by the surrounding aqueous phase. During the initial attachment phase, the physicochemical properties of the bacterial cell surface are important in determining whether the cell may attach or not. The primary forces causing cell adhesion as well as adsorption to any surfaces are polar (hydrogen bonding) forces or hydrophobic interactions and to a smaller degree apolar Van der Waals forces (Busscher and Weerkamp, 1987; Rao et al., 1997a). Bacteria are of colloidal dimensions and possess a net negative charge at pH levels normally encountered in natural habitats. This creates an apparent problem in the adsorption process, as the substratum surfaces in nature either are negatively charged or promptly acquire a negatively charged conditioning film (Marshall, 1980). It is reported that the main body of the bacterium does not make direct contact with the substratum surface (Lawrence and Caldwell, 1987). The degree of their alignment at the solid–liquid interface may be variable and can be elucidated in terms of colloidal chemistry. However, when considering such theories, it is important to remember that the majority of bacteria are not smooth, round particles. Although their (microbial) surfaces are charged with the same sign as that of substratum, still the bacterium may approach a surface, attracted by Van der Waals forces (Allison and Gilbert, 1992; Busscher et al., 1990). Desorption may occur during this stage as a result of the release of reversibly adsorbed cells due to fluid shear forces (Allison and Gilbert, 1992; Characklis and Marshall, 1990).

19.1.2.2 Phase 2: Irreversible Adsorption

This is a very crucial step in biofilm development wherein the irreversible attachment of the microorganism occurs. Though forces of repulsion inhibit the body of the bacterium from making direct contact with the surface, they can still adhere to the surface by producing surface appendages. Since electrostatic repulsion depends on the radius of curvature, bacterial surface appendages, for example, flagella, fimbriae, and exopolysaccharide fibrils, can readily penetrate the energy barrier and move in the primary minimum (Allison and Sutherland, 1987). Cells can hover at this point and take part in a variety of short-range interactions comprising dipole–dipole, ion–dipole, and hydrophobic interactions. Polymeric fibrils form a connection between the bacterium and the surface, thereby, irreversibly reinforcing the association. During this period, the attached cells trigger exopolymeric substance (EPS) secretions that further cement the cells to the surface and stabilize the colony against variabilities in the surrounding macroenvironment (Figure 19.1b).

19.1.2.3 Phase 3: Microbial Biofilm Formation and Maturation

Bacterial colonization of the surface may result in the utilization of the macromolecules almost as fast as they adsorb at the surface, and it depends on the nutritional characteristics of the bacteria found at the surface (Rao, 2003). Biofilm demonstrates significant functional homogeneity; the individual cells

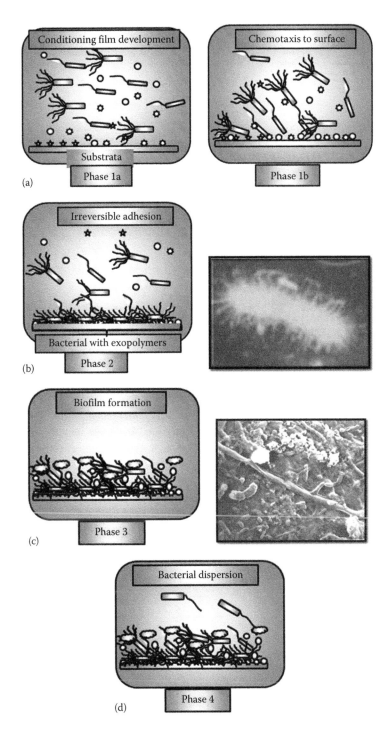

FIGURE 19.1 (a) Initial events in biofilm formation (bacteria and biomolecules interaction). (b) Irreversible phase of biofilm formation/photomicrograph (2750X) showing exopolymer structure of a bacterial cell. (c) Biofilm maturation stage and SEM image (2225X) of a matured freshwater biofilm. (d) Biofilm dispersal phase. *(Continued)*

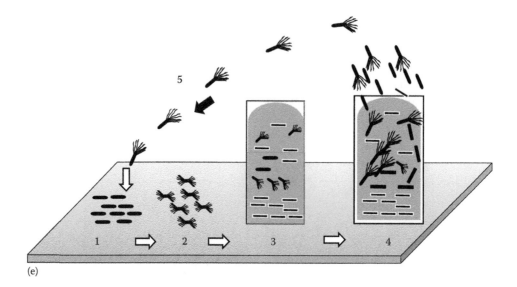

FIGURE 19.1 (*Continued*) (e) Schematic illustration of biofilm formation: 1, bacterial settlement and microcolony formation; 2, irreversible phase of adhesion with exopolymer production; 3, biofilm formation with a consortium of bacteria; 4, mature biofilm and dispersion stage, that is, dispersal phase for resettlement; 5, dispersed gregarious bacteria for settlement on new surface.

and their activities are reliant on the structural integrity of the biofilm *in toto*. Costerton et al. (1995) suggested that biofilms could be considered as a "quasi-tissue" with quantifiable rates of respiration and nutrient uptake. Biofilms resemble a tissue in their physiological cooperativity and protect themselves from variations in the bulk phase conditions by a primitive homeostasis (Costerton et al., 1995). The spatial organization of microbial cells in the biofilm is also subjected to variations. The growth characteristics and metabolic activities of biofilm bring in structural heterogeneity. The physiological congruence detected in biofilms is a typical feature of mixed microbial communities (Costerton et al., 1995). Eventually, with time the biofilm reaches a plateau phase with a certain thickness and metabolic capacity (Figure 19.1c). The biofilm can now provide the basis for succession development, whereby colonization by higher organisms is supported (Characklis and Marshall, 1990). This leads to the process of macrofouling, a common problem described especially in marine systems (Fleming, 2002).

19.1.2.4   Phase 4: Biofilm Detachment and Dispersal

In order to propagate, the biofilm bacteria detach, disperse, and colonize new niches (Figure 19.1d). This requirement is analogous to that of higher organisms such as the mycelial fungi that grow attached to surfaces. At different points in their life cycle, these organisms differentiate and produce spores that are spread to new locations. Bacterial growth in biofilms also brings with it the need to disperse (Costerton et al., 1995). Biofilm accumulation increases surface roughness and also provides shelter from shear forces and increases both the surface area and convective mass transport near the surface. The sessile accretions of microbial cells would produce surface roughness that increases turbulence and mass transport at the colonized surface (Allison and Gilbert, 1992; Bott, 1993; Characklis and Marshall, 1990). The whole

process of biofilm formation in various stages is depicted schematically in Figure 19.1e.

19.1.3   HEAVY METAL TOXICITY AND NEED OF HEAVY METAL REMEDIATION

Metals that have a specific density of more than 5 g/cm³ are termed as heavy metals. Heavy metals have been used by human civilization since the Iron Age. Despite knowing several adverse effects of heavy metals on health exposure to heavy metals, its emission and exposure were not only continued but even increased particularly in less developed countries. Rapid industrialization, population growth, and complete neglect for the environmental concerns have led to global pollution. The release of various pollutants into the environment occurs accidentally or due to anthropogenic activities that ultimately result in soil, water, and air pollution leading to many health hazards. The main threats to human health from heavy metals are associated with exposure to lead, cadmium, mercury and arsenic. These metals have been extensively studied and their effects on human health regularly reviewed by international bodies such as the WHO.

19.1.4   MECHANISMS OF METAL TOXICITY IN MICROORGANISMS

Many metal ions can exert toxicity as they affect multiple biochemical pathways simultaneously. These mechanisms can be divided under five major mechanistic categories.

19.1.4.1   Substitutive Ligand Binding

In this mechanism, a heavy metal ion replaces another metal ion at the binding site of a specific biomolecule (e.g., enzymes), thus altering metal–ligand binding. This replacement of metal ions causes the change in the biological function or complete loss of the function of the target molecule (Nieboer and Fletcher, 1996).

19.1.4.2 Oxidative Damage to DNA, Lipids, and Proteins

Reactive oxygen species (Van der Kooij et al., 1995) are transient and highly reactive molecular species that can damage all biological macromolecules by reduction–oxidation (redox) reactions (Pomposiello and Demple, 2002). The second mechanism by which heavy metals generate toxicity is due to production of reactive oxidative species (ROS) by carrying out redox reactions with cellular biomolecule containing thiol group (R-SH) (Stohs and Bagchi, 1995; Zannoni et al., 2007) particularly with glutathione (Turner et al., 2001). Thiol groups are often present at the active site of carrier molecule and aid in transport mechanism. They are also involved in disulfide bond and tertiary protein structure; thus, destruction of these functional thiol groups on nascent proteins by metal species impairs protein folding or the binding of apoenzymes to cofactors. Thereby, redox reaction involving thiol group is destroyed by heavy metals impairing the normal biological activity of such proteins and in turn the vitality of an organism (Nieboer and Fletcher, 1996). For example,

$$\text{Ex. 1: } 3Ag^+ + 2RSSR + 2HO^{\cdot} \rightarrow 3RSAg + RSO_2H + H^+$$

$$\text{Ex. 2: } 2CuSR + 2Cu^{2+} \rightarrow RSSR + 4Cu^{2+}$$

$$2RSH + O_2 \xrightarrow{Cu} RSSR + H_2O_2$$

Few metal oxyanions such as Se and Te oxyanions (SeO_4^{2-}, SeO_3^{2-}, TeO_4^{2-}, and TeO_3^{2-}) participate in Painter-type reaction with thiols shown as follows, which liberates the toxic ROS superoxide ($O_2^{\cdot-}$) as a by-product of reduction (Kessi and Hanselmann, 2004; Tremaroli et al., 2007). Superoxide ($O_2^{\cdot}$) greatly attributes to extensive damage to all biomolecules that include DNA, lipids, and proteins.

$$4RSH + H_2SeO_3 \rightarrow RS-Se-SR + RSSR + 3H_2O$$

$$6GSH + 3H_2SeO_3 \rightarrow 3\ GS-Se-SG + O_2^{\cdot}$$

19.1.4.3 Fenton-Type Reaction

Fenton reaction is catalyzed by few transition metals, such as Cu, Ni, and Fe, in the presence of H_2O_2 and produce highly reactive oxidative species, such as hydroxyl ($OH^{\cdot}$ and OH^-) and peroxide $HO_2^{\cdot}$ radicals (Geslin et al., 2001; Inaoka et al., 1999; Stohs and Bagchi, 1995).

$$Cu^+ + H_2O_2 \rightarrow HO^{\cdot} + HO^- + Cu^{2+}$$

$$Cu^{2+} + H_2O_2 \rightarrow HOO^{\cdot} + H^+ + Cu^+$$

19.1.4.4 Inhibition of Membrane Transport Processes

Apart from the previously described mechanisms, heavy metals also interfere with membrane transport processes. In this mechanism, the normal functioning of a membrane transporter is interfered by a toxic heavy metal species that outcompetes the binding of any specific essential substrate to a membrane transporter by occupying the binding sites (Foulkes, 2000).

19.1.4.5 Electron Siphoning

The fifth category involves the indirect siphoning of electrons from the respiratory chain by thiol-disulfide oxidoreductases (Borsetti et al., 2007), thereby destroying the proton motive force of the cell membrane (Lohmeier-Vogel et al., 2004). The susceptibility of microorganisms to a particular toxic metal is dependent on multiple factors such as the standard reduction potential (ΔE_0) and electronegativity (χ) of the metal, the solubility product of the metal sulfide complex (pK_{SP}), electron density, and the covalent index (Harrison et al., 2007; Nies, 2003; Workentine et al., 2008).

19.1.5 BACTERIAL RESISTANCE TO TOXIC HEAVY METALS

Since the beginning of life on planet Earth and during the course of evolution, microbes have been exposed to toxic heavy metals that have always been present in abundance (Silver and Phung, 2005). Therefore, it is easily understandable why bacteria do have specific genes for tolerating the toxic ions of the heavy metals. These resistance determinants have been in existence for billions of years. Many of these resistance systems have been found on plasmids, but frequently related systems are subsequently found determined by chromosomal genes in other organisms (Silver, 1996). Essentially all bacteria have genes for toxic metal ion tolerance, and these include Ag^+, AsO^{2-}, AsO_4^{3-}, Cd^{2+}, Co^{2+}, CrO_4^{2-}, Cu^{2+}, Hg^{2+}, Ni^{2+}, Pb^{2+}, TeO_3^2, Tl^+, and Zn^{2+}. The tolerance/resistance systems that are involved include energy-dependent efflux of toxic ions and enzymatic transformations of heavy metals to a lesser toxic forms or specific metal-binding proteins (see Table 19.1). Among these systems, energy-dependent efflux of toxic ions comprises the largest group of resistance systems. The efflux system derives energy either from ATPases or from chemiosmotic ion/proton exchangers (Nies, 2003). For example, Cd^{2+} efflux pumps of bacteria are either inner membrane P-type ATPases or three-polypeptide RND chemiosmotic complexes (involved in metal Resistance/Nodulation of legumes/cell Division). Three-polypeptide complexes comprised a periplasmic-bridging protein, an inner membrane pump, and an outer membrane channel. CZC (Cd^{2+}, Zn^{2+}, and Co^{2+}) chemiosmotic system is one of the best-studied three-polypeptide system, and it is also reported to be involved in the efflux of Ag^+, Cu^+, and Ni^{2+} (Nies, 1995). Enzymatic transformations involve oxidation, reduction, methylation, and demethylation of metals or metal oxyanions to a lesser toxic forms. Metal-binding proteins, as in the case of inorganic mercury, Hg^{2+}, involve a series of metal-binding and membrane transport proteins as well as the enzymes mercuric reductase and organomercurial lyase, which overall convert more toxic to less toxic forms (Silver and Phung, 2005). Table 19.1 presents a current summary of tolerance/resistance systems and their biochemical mechanisms.

Earlier studies on biofilms and planktonic cells have shown that mechanisms of heavy metal tolerance are different between the two modes of growth (Workentine et al., 2008). In other words, biofilms are shown to be more tolerant than that of their planktonic counterparts by reducing metal toxicity in multiple ways, which will be discussed later in this chapter.

TABLE 19.1

A Current Summary of Resistance Systems and Their Biochemical Mechanisms

Metal	Resistance Mechanism	Gene(s) Involved	Protein Function	References
Arsenic	Efflux pumps, enzymatic oxidation	*ars, aso, aar*	Arsenate reductase and transport Arsenite oxidase and transport Respiratory arsenate reductase	Borremans et al. (2001); Messens et al. (2004); Mukhopadhyay et al. (2002)
Mercury	Reduction	*mer*	Mercuric reductase and transport	Brim et al. (2000); Dash et al. (2014)
Cadmium	Cation diffusion facilitator family of single-polypeptide chemiosmotic efflux systems	*Cad, czc*	P-type efflux ATPase, CBA efflux permease	Kermani et al. (2010); Nies (1995)
Chromium	Reduction	*chr*A	Chromate reductase	Ohtake and Silver (1994)
Lead	Intracellular and extracellular binding (accumulation)	*pbr*D	Lead resistance and efflux	Borremans et al. (2001)
Copper	Multicomponent efflux transport, efflux	*cop, pco* *cus*F	Copper resistance and transport	Cooksey (1994); Franke et al. (2003); Rensing and Grass (2003)
	Oxidation	*cue*O	Oxidizes Cu(I) to lesser toxic Cu(II)	Roberts et al. (2003); Singh et al. (2004)
Nickel	Efflux	*ncc, nre, cnr*	CBA efflux permease	Hebbeln and Eitinger (2004); Mergeay et al. (2003)
Cobalt	Efflux	*czc, cnr*	CBA efflux permease	Cavet et al. (2003); Hebbeln and Eitinger (2004)
Silver	Chemiosmotic and efflux pump	*sil*	Silver resistance and binding	Gupta et al. (1999, 2001)
Zinc	Efflux transporter	*Czc* *zup*T	CBA efflux permease ZupT transporter	Blencowe and Morby (2003); Cavet et al. (2003); Grass et al. (2005)

19.1.6 Use of Biofilms in Remediation of Heavy Metals

The promising application of biofilms is in toxic metal remediation (Lloyd and Renshaw, 2005). Bacterial biofilms represent an efficient method for the bioremediation of toxic metals in a process of bioaccumulation or biosorption by the biofilm matrix components. EPS plays a significant role in the structural integrity of the biofilm. Although the production of the EPS by the bacterial cell is an energy-intense process, it supports the bacterial biofilm in many ways such as protection from predation and harsh environment, reducing the penetration/diffusion of antibiotics in the biofilm matrix (Xavier and Foster, 2007). The attachment of bacterial cells and subsequent growth in the form of microbial communities presents special benefits to the sessile cells, while the planktonic cells do not have these advantages. The EPS matrix composed of ionic sugars can effectively bind to a variety of heavy metals. The high biomass of biofilms and their enzyme activities are helpful in the reduction of toxic metal ions and also aid in the biosorption of heavy metals, thus minimizing the environment hazards. Removal of heavy metals by biofilms is chiefly due to the binding capacity of EPS and other cell membrane components (Zhang et al., 2012). The nonaqueous liquid phases of biofilms are helpful in maintaining the static metabolic state and ion mobility inside the matrix. This phase also helps in increasing the bioavailability of the toxic metals and aids in their remediation (Cameotra and Makkar, 2010).

Naturally, biofilms maintain a high biomass density that assists effectively in the mineralization process, since the microniche of the biofilm is optimized (Stoodley et al., 2002). According to Perumbakkam et al. (2006), bacteria utilize organic compounds for their energy and secrete exopolymers to aid in the biofilm formation. As a result, they gain the capability to degrade refractory and slowly degrade complex organic compounds. The exopolymers mainly provide the structural framework for the biofilms and their stability (Allison and Sutherland, 1987; Christensen, 1989; Fletcher et al., 1991). EPS-producing bacteria are an assorted group of bacteria belonging to the genera *Pseudomonas*, *Bacillus*, *Flavobacterium*, and *Aerobacter*. The EPS bacteria are the principal colonizers of metallic surfaces; these biopolymers form a gel-like matrix and give rigidity to the biofilm architecture. These biopolymers apart from aiding in adhesion, also protect the bacteria from bacterivores. The EPS matrix embedded with microbial cells gives a quasi-tissue character to the microbial biofilm and also influences interfacial processes by active water channels at the biofilm–substrate interface. They trap inorganic constituents from the water and also form complexes with some organic compounds, resulting in the decrease of the diffusion phenomena. The EPS-producing bacteria scrub oxygen and create an ideal site for the growth of anaerobic bacteria. Most exopolymers have acidic groups and also contain functional groups that bind metal ions from the surrounding water. Biofilm growth also influences the fate of other compounds in their near vicinity, which is a consequence of the dynamic metabolic activity of the consortium. Horizontal gene transfer (HGT), conjugation, and transformation processes happening in the biofilm make the microorganisms more resilient. This will result in acquiring new metabolic properties and also tolerance to high concentration of toxicants (Roberts and Mullany, 2006). This property of the biofilm aids in the efficient bioremediation of heavy metal.

19.1.7 Biofilm and Metal Detoxification/Tolerance

Metals are essential micronutrients involved in a variety of cellular processes. Not all metals are essential for cellular processes, and some are required in very minute concentrations. To avoid the toxicity of excess or nonessential metals, the cell effluxes them out. Metals in excess hamper enzyme activity and cellular function and damage the cell membrane and DNA. Resistance systems to toxic metal in bacteria are diverse, which occur mainly via efflux pumps, enzymatic detoxification, metallic reduction, exclusion by permeability barriers, sequestration, and metallothionein (Bruins et al., 2000). However, for biofilm, the mechanism of detoxification is highly versatile as compared to planktonic cells. The metabolic gradient, HGT, signaling process, hydrated EPS matrix (enables efficient biosorption of metals to the cell surface/biofilm), etc., are some phenomenon involved in minimizing hazards and modulating the response of cell to a metal (Harrison et al., 2007).

19.1.7.1 Biosorption: Immobilization of Metal by EPS

Exopolymeric compounds and their composition mostly attribute toward physical features of the biofilm. EPS is composed of many vital biomolecules that provide the polyionic charge. The charge is added by functional groups of biomolecules (i.e., carbohydrate, protein, and lipids). Carboxylate (R-COO$^-$), phosphate (R-HPO$_4$.$^-$), sulfhydryl (R-SH), amino (R-NH$_3^+$), and phenolic (R-C$_6$H$_5$OH) groups are some of the charge adding groups in EPS (Zhang et al., 2010). However, EPS is rich in a variety of negatively charged groups at nearly neutral pH conditions (pH 6.5–7.5), which make easy the formation of organometal complex by electrostatic attraction with heavy metals (Beech and Sunner, 2004; Beveridge, 1989). As a result, EPS is one of the first lines of defense to metal resistance/detoxification for cells within biofilm (Harrison et al., 2007). In the environment, biofilm EPS entraps various metallic ions, which can be extracted from the bulk fluid. The EPS composition of a cell and its metal-binding capacity changes with genotype and the surrounding environment. Thus, the detoxification via biosorption varies from organism to organism and the environment. Increase in EPS synthesis is also one of the detoxification mechanisms to avoid metal toxicity, which was reported by Chakraborty and Das (2014). In their study, they observed increase in EPS by *Pseudomonas aeruginosa* biofilm with an increase in Cd^{2+} concentration. Previously, Huang et al. (2013) reported that biosorption and resultant protection from Cd^{2+} in *Bacillus cereus* RC-1 is largely related to EPS rather than intracellular mechanism. Dash et al. (2014) reported that marine bacterium *Bacillus thuringiensis* strain PW-05 possesses *mer* operon and does Hg^{2+} detoxification by volatilization. They observed that although 90% of Hg^{2+} is volatilized by the biofilm, the remaining 10% is bound to the biofilm EPS via interactions with –SH and –COOH groups providing additional protection

to the cells from metal toxicity. Detoxification of Hg, Cd, and Pb by putative entrapment by EPS is also reported in *Bacillus pumilus*, *Bacillus* sp., *P. aeruginosa*, and *Brevibacterium iodinum* (De et al., 2008). EPS plays a vital role in metal tolerance, and monomeric units such as uronic acid and glucuronic acid can enhance the tolerance limit of a cell (Ozturk et al., 2009).

19.1.7.2 Metabolic and Population Heterogeneity

Metabolic changes in response to metal exposure vary significantly between biofilm and planktonic cultures. Biofilms have phenotypic and metabolic heterogeneity due to gradient in nutrients, oxygen, and pH. Many times there is restricted diffusion of them within the community, and as a result, the physiological state of a cell within a biofilm changes with thickness. It has been reported that for every unit thickness of biofilm the physiological state of cells varies differentially (Harrison et al., 2007, 2009). Biofilms have a portion of dead cells, which are one of the major contributors toward pH discontinuities and physiological heterogeneity. Dead cells are reactive biomass, which provide biosorption site as well as protons. These protons compete with metal for cell surface binding site. Therefore, binding of metal to active cell is less, and as a result, biofilms are resistant to metal toxicity. Bacterial cells undergo many metabolic shifts to protect it from oxidative stress induced by metals. In a study by Chenier et al. (2008) on *Pseudomonas fluorescens*, it was observed that oxidative stress induced by Al increases the production of NADPH (an antioxidant) and decreases the production of NADH (a pro-oxidant). In another study, Booth et al. (2011) observed that Cu also induces oxidative stress in *P. fluorescens* planktonic cells, and defense mechanism is similar to Chenier et al. (2008) study. However, in *P. fluorescens* biofilm, these changes were not observed by Chenier et al's. (2008). As an alternative, changes were observed in metabolic pathway linked to EPS synthesis in biofilm providing enhanced protection. From their study, it can be concluded that metal stress in mature biofilm induces protective changes rather than reactive changes to combat metal toxicity. However, for planktonic culture, the response to stress can affect TCA cycle, glycolysis, and pyruvate and nicotinate and niacotinamide metabolism. Even immature biofilms at early or adherent layers stage have higher resistance to toxic compounds as compared to planktonic cells. Surface attachment can trigger physiological changes to multimetal-resistant and/or multimetal-tolerant state in biofilm cells (Davies et al., 2007; Harrison et al., 2005; LaFleur et al., 2006).

19.1.7.3 Signaling Events

Many signaling events in biofilm are responsible for metal susceptibility or tolerance of a cell. One of the auto-inducer-based signaling processes known as quorum sensing (QS) is known to have a conditional role in metal tolerance of biofilm. The cellular signaling defense against oxidative stress protects microbial biofilms from toxic metal species. Increase in the level of detoxifying enzymes indirectly/directly via QS can contribute toward biofilm metal resistance or tolerance. In *P. aeruginosa*, resistance to metal Zn, Co, and Cd is regulated by efflux mechanism by the expression of *czcR* and *czcCBA* efflux proteins. *czcR* is also involved in *P. aeruginosa* QS system. Moreover, Cu toxicity in the bacterium is also regulated by *lasI/R* QS circuit (Dieppois et al., 2012; Thaden et al., 2010). Acidophilic bacterium *A. ferrooxidans* is highly resistance to heavy metal. *A. ferrooxidans* has *afeI/R* QS genes structurally homologous to *luxI/R*. *afeI/R* QS has been reported to regulate Cu resistance in the bacterium (Wenbin et al., 2011). *Acinetobacter junii*, a metal-tolerant bacterium, can tolerate Ni, Hg, Cd, and As. In the bacterium *A. junii*, metal resistance is positively regulated by QS (Sarkar and Chakraborty, 2008).

19.1.7.4 Horizontal Gene Transfer

HGT is a frequent phenomenon between cells in biofilm. Cells in biofilm are very competent in taking extracellular DNA, and as a consequence, the transfer of detoxifying genes is feasible in biofilm community (Hausner and Wuertz, 1999; Singh et al., 2006). Natural transformation is reported to be one of the major reasons behind biofilm resistant to metal, antibiotic, and organics (Nancharaiah et al., 2008; Perumbakkam et al., 2006). Plasmids or eDNA are important structural and functional component of biofilm, stabilized by plasmid-coded mechanisms. Plasmid stability is high in biofilms because of the quiescent nature of cells in the biofilm (Madsen et al., 2012). Metal and antibiotic resistances are often associated in many bacteria (Gómez Calderón et al., 2013). In a study on *Escherichia coli* biofilm, Krol et al. (2011) observed high frequency of gene transfer at the air–liquid interface owing to high densities of donor and recipient cells. In contrast, with decreasing oxygen concentration inside a biofilm, gene transfer events happen at much lower rate. Thus, it can be hypothesized that at surface layer the presence of metal resistance genes protects biofilm cells from toxicity. Besides, secondary mechanisms such as EPS matrix or population heterogeneity can assist cells to combat metal toxicity.

19.1.7.5 Biofilm-Specific Genetic and Metabolic Changes in Response to Toxic Metals

The metabolic constraint and the expression pattern of genes differ when biofilm and planktonic mode of growth are considered. Correspondingly, it is speculated that the expression of genes to combat metal toxicity varies significantly in biofilm. According to Koechler et al. (2015), metal can affect many genetic formats in biofilm. As the primary defense to metal is EPS matrix for biofilm cells, EPS genes prominently respond to the presence of metal. In *P. fluorescens,* Cu stress has been reported to upregulate the expression of EPS genes/protein thereby increasing the synthesis of EPS (Booth et al., 2011). A similar observation has been reported in *Xylella fastidiosa* by Navarrete and Fuente (2014) upon Zn exposure. An increase in detoxification gene expression is another response that can be changed by the presence of toxic metal. In *Desulfovibrio desulfuricans,* the expression of Hg methylase gene and resulting Hg methylation is more rapid as compared to planktonic cells (Lin et al., 2013). In *X. fastidiosa*, copper stress can induce phage-related, pathogenic, virulence, and adaptation genes. However, genes involved in

protein and RNA synthesis get repressed (Muranaka et al., 2012). In bacteria, copper toxicity induces the formation of persister cells by elevating the expression of genes involved in the induction of toxin–antitoxin system within biofilm. Persister cells are metabolically inactive cells, which neither grows nor dies (Harrison et al., 2007) thereby providing resistance to toxic metals. In general, metals such as Ru^{3+}, Cu^{2+}, Ag^+, Ni^{2+}, and Pb^{2+} increase the formation of colony variants. Small colony variant (SCV) can be recovered from biofilm under chemically stressed conditions. SCV formation is one of the important providers for biofilm metal susceptibility/resistance populations. In *P. fluorescens*, genetic rearrangements in *gacA–gacS* loci on exposure to toxins are linked with phenotypic variation within the biofilm (Martínez-Granero et al., 2005).

It is possible to understand multimetal resistance and multimetal tolerance of microbial biofilms when different components of heavy metal tolerance as described earlier are put together under a single multifactorial model (Figure 19.2). Microorganisms within a biofilm behave as a population of cells that display complex and coordinated developmental behavior (Harrison et al., 2007). Multimetal resistance or tolerance is acquired by microbes right from the bacterial adhesion phase to a surface. A number of coordinated events such as persister-cell formation by a fraction of the microbial population and cellular diversification, that is, formation of phenotypic variants within the population during the natural growth of the microorganisms at the surface, contribute to multimetal resistance and tolerance. At the same time, coordinated QS leads to increased production and altered characteristics of biofilm extracellular polysaccharides, which, in turn, affect the diffusion processes and biosorption of toxic metal species. Microbial biofilm populations, therefore, may be able to survive fluxes in toxic metal species because of the diverse cell types that are present in such biofilms. These different cell types might also function to protect each other.

19.1.8 APPLICATION OF DETOXIFICATION PROPERTY OF MICROBIAL BIOFILMS

Human and ecosystem health can be unfavorably affected by all kinds of waste, from its generation to its disposal. Industrialization is necessary for the economy of any nation, but unplanned and mindless exploitation of the natural recourses leads to the degradation of the environment (Hossain and Rao, 2014). Generally, industries have been considered to be a leading consumer of natural resources and energy and a significant contributor of pollutants to the environment. The process of detoxification in some way diminished the chances of polluting the environment. In the modern age, detoxification can be done by various applications depending upon the need and availability. The use of biofilm-mediated bioremediation process is a necessity today for highly contaminating sites, which are results of emissions by mining activities, metal processing, surface treatment, electronic or paint industry etc. These processes contribute to the heavy metal contamination as one of the largest environmental problems. Microorganism can use these metal contaminants for their benefits, such as a

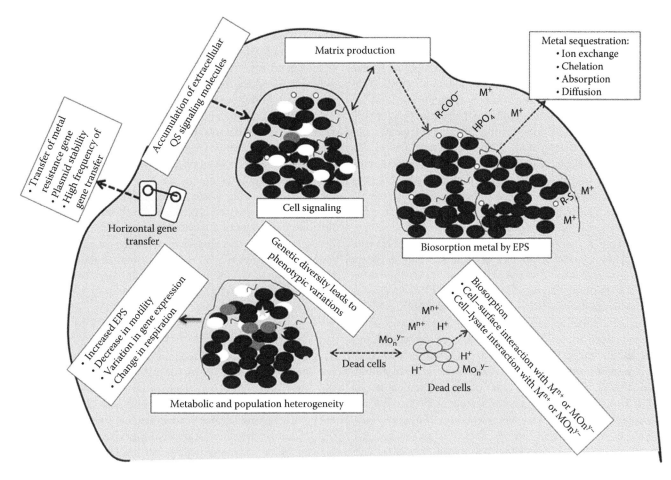

FIGURE 19.2 Mechanism of heavy metal resistance and tolerance in microbial biofilms. Heavy metal resistance and tolerance in microbial biofilms are dependent on a number of factors. In general, the multimetal tolerance of biofilms is a multifactorial property that arises from several interrelated physiological and chemical parameters. EPS, extracellular polymeric substance, and QS, quorum sensing, are the main factors. (Adapted from Harrison, J.J. et al., *Nat. Rev. Microbiol.*, 5, 928, 2007.)

nutrient, a ladder for genetic, and a defense from the surroundings. Cost-effective removal of contamination is not an easy job because of the scale of contaminants. Immobilization of microbes is needed to resolve the issue, which can be solved by naturally formed biofilms. Microbial processes that have grabbed the attention of most researchers are bioremediation, biomining, and bioleaching in this context.

19.1.8.1 Heavy Metal Bioremediation

Bioremediation is a process of degrading environmental contaminants by using biological processes or microbial organisms (Shukla et al., 2014). Huge amount of toxicants and pollutants are produced every year because of human activity. Using chemicals to treat wastewater is very common practice; however, it might produce another set of harmful products (Gallard and von Gunten, 2002). On the other hand, bioremediation is a process of detoxification of the contaminated soil or water using microbial community; therefore, it is free from such side effects and provides pollution-free environment. Thus, bioremediation is a desirous, competent, side effects free practice by which our environment can be cleaned up very efficiently leaving behind a clean, pollution-free environment. During the last few years, bioremediation

is getting popular not only because of its wide application in different area of remediation but also due to its efficiency and cost-effective technology. It is useful to mitigate highly toxic metals, chemicals, effluent, and pollutant. Heavy metal accumulated in water can be harmful or even can cause death to fishes and marine animals. Bioremediation is useful in decontaminating in metal wastes like mercury and cadmium.

Biofilms do have inherent durability and nutrient cycling capability for the process of bioremediation. To survive in toxic environment, microbes form biofilm and the processes are connected for mineral transformation (Gadd, 2010). Metal-binding capacity can result from several physical, chemical, and biological processes, and removal of heavy metals by biofilms is chiefly due to the binding capacity of EPS (Zhang et al., 2010, 2012). EPS is polyionic in nature, and it is able to form organometal complex by electrostatic attraction with heavy metals (Beech and Sunner, 2004). This adds a promising application of biofilm in heavy metal and radionucleotide remediation (Lloyd and Renshaw 2005). Biofilms are effective in the remediation of heavy metals and in the expression of certain genes involved in resistance mechanism. A recent report has showed mercury detoxification by *B. thuringiensis* PW-05 biofilm via mercury reductase (Dash et al., 2014).

White and Gadd (2000) have studied the removal of copper by sulfate-reducing bacteria biofilm. They reported increase in biofilm-associated EPS production in the presence of copper. Costley and Wallis (2001) used bacterial biofilms for the removal of cadmium and zinc from synthetic water. Biofilms can accumulate substantial quantities of metal. Metal binding by biofilm is also influenced by carbohydrate/protein ratio (C/P ratio). Lower C/P ratio facilities binding of metals ions like Ni^{2+} and Pb^{2+} (Jang et al., 2001). Biofilm formation also increases resistance to heavy metals, thus making remediation process efficient (Chien et al., 2013). Biofilm is also effective in radionuclide remediation. The remediation occurs mainly through physicochemical interactions (van Hullebusch et al., 2003). Sarró et al. (2005) reported the ability of bacterial biofilms to remove multiple radionuclides from radioactive wastewaters. They reported abundance of bacteria from the genera *Ralstonia*, *Bacillus*, *Stenotrophomonas*, *Mycobacterium*, *Nocardia*, and *Staphylococcus* involved in biofilm remediation of radionuclide such as ^{60}Co, ^{65}Zn, ^{54}Mn, ^{51}Cr, ^{59}Fe, and ^{95}Nb.

Uranium pollution is one of the common radionuclide contaminations in soil, sediments, and groundwater. Using *D. desulfuricans* G20 biofilms, Beyenal et al. (2004) used metal immobilization for the treatment of hexavalent uranium [U(VI)]. There are two types of bioremediation, *in situ* and *ex situ*: *in situ* treatment is carried out at the site of waste production, and *ex situ* treatment refers to the waste treatment process done away from the site of production. In last several years, *in situ* bioremediation has shown to be a promising technology that facilitates bioremediation with addition of specialized contaminant-degrading microorganisms. Such targeted molecular diagnostics to better engineer and manage the bioremediation process is much more effective in contrast to conventional method of decontamination. These processes and techniques make the options bigger and available to regulatory decision-making and thus affect contaminated site closures. This "diagnostics and therapeutics" approach is much more effective and gaining popularity because it is a site-specific treatment process. A case study shows the effectiveness of this approach, in which the method has been tested and implanted at almost two dozen sites nationwide (Ritalahti et al., 2005). Treatment of contaminated site by bioremediation approach is getting popular and also a useful strategy with respect to established *ex situ* procedure. The process of monitoring remediation at *in situ* processes is more effective (Frische and Höper, 2003).

Bioremediation has many salient features over traditional methods of cleaning waste. Its effectiveness increases when we use this process with combination of other conventional methods. Conventionally used remediation methods like dredging, capping, incineration, or extraction can be assisted with bioremediation that will give smaller number of end products that are nonpolluting and might also result in a lesser environmental footprint.

1.8.1.1 Genetic Engineering in Biofilm-Mediated Heavy Metal Remediation: Future Prospective

The genetic engineering technology allows us to manipulate the bacterial systems as per our requirements. Microbes that thrived in contaminated sites with high concentration of heavy metals are adapted against the toxicity of heavy metals by modulating various genetic mechanisms (Gadd, 2005). These strains have intrinsic ability to survive in polluted sites, and recombinant strains from such bacterium by introduction of a single gene or gene cluster provide an opportunity to create highly robust strain for the metal remediation applications (Brim et al., 2000, 2003). The most common techniques include engineering with a single gene or operon, gene sequence alterations of existing genes, and pathway switching. Table 19.2 lists out few attempts to manipulate bacterial

---

TABLE 19.2
Genetic Engineering in Bacteria to Enhance the Toxic Metal Bioremediation Potential

Bacteria	Genetic Machinery	Genetic Alterations	Applications	References
E. coli JM109	*Plasmid*	Introduction of *mt-1*, *ppk* genes	Efficient Hg accumulation/transformation	Ruiz et al. (2011)
Deinococcus radiodurans	*merA*	Introduction of *merA* gene	Growth in the presence of radiation and Hg, Hg volatilization	Brim et al. (2000)
Bacillus cereus BW-03(pPW-05)	*merA*	Introduction of *merA* gene	Efficient Hg volatilization and biosorption	Dash and Das (2015); Dash et al. (2014)
Cupriavidus metallidurans strain MSR33	*mer* operon	Alteration of *merB*, *merG*, and other *mer* gene	Tolerance to inorganic and organic Hg, along with Cu and C	Rojas et al. (2011)
Pseudomonas aeruginosa	*cad* operon	Mutation of *cad* operon by acridine orange and acriflavine	Increase in Cd^{2+} tolerance to 7 mM	Kermani et al. (2010)
Deinococcus geothermalis	*Plasmid*	Introduction of plasmid pMD66	Reduction of Hg^{2+}, Fe^{3+}, U^{6+}, and Cr^{6+}	Brim et al. (2003)
E. coli	Consortium	Overexpression of *nfs*A	1.5-fold increase in Cr^{6+} reduction	Ackerley et al. (2004)
Pseudomonas putida	*chr* operon	Modification of *chr*R gene	Increased Cr^{6+} reduction by 2.4 times	Gonzalez et al. (2005)
E. coli	*Plasmid*	NiCoT from *Novosphingobium aromaticivorans*, synthetic adherence operon from *E. coli*	Increased Ni^{2+} and Co^{2+} sequestration used for biofiltration	Duprey et al. (2014)

Source: Das, S., Dash, H. R., Chakraborty, J., 2016. Genetic basis and importance of metal resistant genes in bacteria for bioremediation of contaminated environments with toxic metal pollutants. *Applied Microbiology and Biotechnology* 100: 2967–2984.

system to enhance their heavy metal remediation potential. However, such attempts combined with biofilm-mediated remediation also need to be studied.

19.1.8.2 Bioleaching or Biomining

Apart from bioremediation, microbes are also used to isolate metals from their respective ores. Microbial leaching is another application that shows the significance of microbial community over traditional chemical methods like use of cyanide for isolating metals. The solubilization of metal(s) from sulfidic ores or solid wastes into aqueous solutions using living microorganisms is known as biomining or bioleaching. Biomining technologies utilize microorganisms to extract and recover metals from ores and waste concentrates. Bioleaching for the removal of metal from sewage can decrease the overall expenditure of the process by decreasing the sludge volume to be treated and escalating the metal concentration in the leachate. Mehta and Pandey (1999) performed bioleaching of copper, nickel, and cobalt from cobalt converter slag using *Thiobacillus ferrooxidans*, which is a mesophilic bacterium.

Bioleaching is very much dependent on pH; the relationship between the pH of the sludge solution and the efficiency of metal leaching was acquired by Cho et al. (2002) in which quantitative investigation on the effect of pH reduction on leaching of each metal has been performed. Other reports also showed that metal leaching was principally influenced not by sludge solids concentration but by the pH of the sludge solutions (Ryu et al., 2003). Copper printed on circuit board of waste computers is not easy to recover chemically or by means of physical forces. The bioleaching of copper from the printed circuit boards by *A. ferrooxidans* showed a great promise of bioleaching process in electronic waste management.

Bioleaching process has been applied at a commercial scale to extract base metals (e.g., Cu, Co, and Ni) and for processing sulfidic and uranium ores (Schippers et al., 2013). The economical property of microbes makes these processes more efficient and apt. Leaching of low-grade secondary copper sulfide ores containing a high level of impurities such as arsenic and fluoride can be done with bioleaching. It is a well-established technology and an alternative to conventional pyrometallurgical process for copper sulfide treatment (Rodrigues et al., 2016). Scaling up of bioleaching process at industrial scale can make the process further easier and efficient. As already discussed, the use of microbes in electronic waste treatment is very efficient. Recently, Maneesuwannarat et al. (2016) demonstrated the gallium recovery by bioleaching process using microbes. The results showed a potential application of isolated bacteria to leach Ga from semiconductors or electronic wastes. Bhatti (2015) demonstrated the extraction of metals from organic-carbon-rich polymetallic black shale using mixed cultures of acidophilic iron and sulfur-oxidizing microorganisms without any additional nutrient supply. This makes the process cheaper at commercial scale and in that way it has an upper hand on various conventional methods. This study also showed why bioleaching is more effective than chemical leaching. In another example, V, Ni, and Cu were recovered using *A. ferrooxidans* from the oil-fired ash,

which is a major by-product of thermal power plants (Rastegar et al., 2015). Effective removal of Zn, Pb, Mn, Cd, Cu, and As from cafeteria sewer sludge is possible with bioleaching. Yang et al. (2016) showed the use of biosurfactant-producing strain isolated from cafeteria sewer sludge with bioleaching capability of Zn, Pb, Mn, Cd, Cu, and As. In another recent study, the process of bioleaching has used six *Aspergillus* species and showed more than 95% of metal recovery from waste Ni–Cd batteries (Kim et al., 2015). Using improved operation technologies such as immobilization and biofilm-based bioleaching process has great potential and assurance. This is where development of biofilm-based bioreactors would be very helpful in bringing down the total cost and in making downstream process simpler.

1.8.2.1 Biofilms: Role of Interfacial Processes in Bioleaching

Bioleaching process proceeds at the interface of mineral and bacterium (Kinzler et al., 2003; Rohwerder et al., 2003; Sand and Gehrke, 2006). Most of the bacteria taking part in bioleaching grow attached on the surfaces of sulfide ores, as natural biofilms. Various reports on bioleaching bacteria showed that more than 80% of inoculated cells disappeared from the cell suspension within 24 h when a nonlimiting surface of the mineral is provided, and when the inoculum exceeds the available surface area, very few cells remained in the planktonic state (Africa et al., 2013; Bagdigian and Myerson, 1986; DiSpirito et al., 1983; Gehrke et al., 1998; Harneit et al., 2006; Noël et al., 2010). This observation suggested that bioleaching bacteria have strong biofilm-forming capability. The reasons for this finding seem to be connected with electrochemical phenomena and still are not fully elucidated. However, some assumptions with a high probability are discussed. In general, the attachment process of bacteria to a surface is predominantly mediated by the EPS component that surrounds the cells, mainly comprising sugars, lipids, proteins, nucleic acids, etc. Attachment to a surface further stimulates the EPS production considerably up to 100-fold (Noël et al., 2010). Some studies showed that >75 proteins/genes are involved in the case of *A. ferrooxidans* biofilm formation (Bellenberg et al., 2011, 2012; Vera et al., 2013). QS molecules are also found to be involved in the process of surface attachment and biofilm formation on mineral sulfides (Gehrke et al., 2001).

The primary attachment to metal sulfide surfaces occurs mainly by electrostatic interactions between positively charged cells and the negatively charged pyrite surface (Blake et al., 1994; Solari et al., 1992; Vilinska and Rao, 2009) and up to some extent due to the hydrophobic interactions (Gehrke et al., 1998; Sampson et al., 2000). Few biofilms, which grow on elemental sulfur surface, predominantly use hydrophobic interactions for the attachment. Microbial cells grown on elemental sulfur do not adhere to pyrite as their EPS composition is significantly different from those which grow on pyrite. The EPS contains very fewer sugars and uronic acid moieties but has many more fatty acids than the EPS of cells grown on pyrite. The most important difference, however, is the total lack of complexed Fe(III) ions or other positively charged ions (Fuchs et al., 1995). Kinzler et al. (2003)

demonstrated that EPS-containing complexed iron(III) ions are the pertinent agents in the dissolution process of pyrite. Therefore, the composition of the EPS earns considerable attention to standardization of bioleaching processes or to constrain the acid rock drainage (Kinzler et al., 2003).

Interestingly, few recent findings showed that few leaching bacteria, such as *Acidithiobacillus caldus*, do not attach to metal sulfide surfaces; it needs a preformed biofilm of *Leptospirillum ferrooxidans* for attachment (Florian et al., 2011). This finding suggests that in order to enhance/optimize or to inhibit (in acid mine/rock drainage) a bioleaching process by a bioleaching bacterium, it is necessary to understand the ecological niche of a leaching consortium. Furthermore, genetic manipulations can also play a role in optimizing the leaching process, especially when leaching bacterium does not form biofilm, to impart biofilm forming using genetic engineering.

REFERENCES

Ackerley, D., Gonzalez, C., Keyhan, M., Blake, R., Matin, A., 2004. Mechanism of chromate reduction by the *Escherichia coli* protein, NfsA, and the role of different chromate reductases in minimizing oxidative stress during chromate reduction. *Environmental Microbiology* 6: 851–860.

Africa, C. J., van Hille, R. P., Sand, W., Harrison, S. T., 2013. Investigation and in situ visualisation of interfacial interactions of thermophilic microorganisms with metal-sulphides in a simulated heap environment. *Minerals Engineering* 48: 100–107.

Allison, D., Gilbert, P., 1992. Bacterial biofilms. *Science Progress* 1933: 305–321.

Allison, D. G., Sutherland, I. W., 1987. The role of exopolysaccharides in adhesion of freshwater bacteria. *Microbiology* 133: 1319–1327.

Bagdigian, R. M., Myerson, A. S., 1986. The adsorption of *Thiobacillus ferrooxidans* on coal surfaces. *Biotechnology and Bioengineering* 28: 467–479.

Beech, I. B., Sunner, J., 2004. Biocorrosion: Towards understanding interactions between biofilms and metals. *Current Opinion in Biotechnology* 15: 181–186.

Bellenberg, S., Vera, M., Sand, W., 2011. Transcriptomic studies of capsular polysaccharide export systems involved in biofilm formation by *Acidithiobacillus ferrooxidans*. In *Biohydrometallurgy: Biotech Key to Unlock Mineral Resources Value*, Proceedings of the 19th International Biohydrometallurgy Symposium, Central South University Press, Changsha, China Vol. 1, pp. 460–464.

Bellenberg, S., Leon-Morales, C. F., Sand, W., Vera, M., 2012. Visualization of capsular polysaccharide induction in *Acidithiobacillus ferrooxidans*. *Hydrometallurgy* 129: 82–89.

Beveridge, T. J., 1989. Role of cellular design in bacterial metal accumulation and mineralization. *Annual Reviews in Microbiology* 43: 147–171.

Beyenal, H., Sani, R. K., Peyton, B. M., Dohnalkova, A. C., Amonette, J. E., Lewandowski, Z., 2004. Uranium immobilization by sulfate-reducing biofilms. *Environmental Science & Technology* 38: 2067–2074.

Bhatti, T. M., 2015. Bioleaching of organic carbon rich polymetallic black shale. *Hydrometallurgy* 157: 246–255.

Blake, R. C., Shute, E. A., Howard, G. T., 1994. Solubilization of minerals by bacteria: electrophoretic mobility of *Thiobacillus ferrooxidans* in the presence of iron, pyrite, and sulfur. *Applied and Environmental Microbiology* 60: 3349–3357.

Blencowe, D. K., Morby, A. P., 2003. Zn (II) metabolism in prokaryotes. *FEMS Microbiology Reviews* 27: 291–311.

Booth, S. C., Workentine, M. L., Wen, J., Shaykhutdinov, R., Vogel, H. J., Ceri, H., Turner, R. J., Weljie, A. M., 2011. Differences in metabolism between the biofilm and planktonic response to metal stress. *Journal of Proteome Research* 10: 3190–3199.

Borremans, B., Hobman, J., Provoost, A., Brown, N., van Der Lelie, D., 2001. Cloning and functional analysis of thepbr lead resistance determinant of *Ralstonia metallidurans* CH34. *Journal of Bacteriology* 183: 5651–5658.

Borsetti, F., Francia, F., Turner, R. J., Zannoni, D., 2007. The thiol: Disulfide oxidoreductase DsbB mediates the oxidizing effects of the toxic metalloid tellurite (TeO_3^{2-}) on the plasma membrane redox system of the facultative phototroph *Rhodobacter capsulatus*. *Journal of Bacteriology* 189: 851–859.

Bott, T., 1993. Aspects of biofilm formation and destruction. *Corrosion Reviews* 11: 1–24.

Brim, H., Venkateswaran, A., Kostandarithes, H. M., Fredrickson, J. K., Daly, M. J., 2003. Engineering *Deinococcus geothermalis* for bioremediation of high-temperature radioactive waste environments. *Applied and Environmental Microbiology* 69: 4575–4582.

Brim, H., McFarlan, S. C., Fredrickson, J. K., Minton, K. W., Zhai, M., Wackett, L. P., Daly, M. J., 2000. Engineering *Deinococcus radiodurans* for metal remediation in radioactive mixed waste environments. *Nature Biotechnology* 18: 85–90.

Bruins, M. R., Kapil, S., Oehme, F. W., 2000. Microbial resistance to metals in the environment. *Ecotoxicology and Environmental Safety* 45: 198–207.

Busscher, H., Bellon-Fontaine, M. N., Mozes, N., van der-Mei, H., Sjollema, J., Cerf, O., Rouxhet, P., 1990. Deposition of *Leuconostoc mesenteroides* and *Streptococcus thermophilus* to solid substrata in a parallel plate flow cell. *Biofouling* 2: 55–63.

Busscher, H. J., Weerkamp, A. H., 1987. Specific and non-specific interactions in bacterial adhesion to solid substrata. *FEMS Microbiology Reviews* 3: 165–173.

Cameotra, S. S., Makkar, R. S., 2010. Biosurfactant-enhanced bioremediation of hydrophobic pollutants. *Pure and Applied Chemistry* 82: 97–116.

Cavet, J. S., Borrelly, G. P., Robinson, N. J., 2003. Zn, Cu and Co in cyanobacteria: Selective control of metal availability. *FEMS Microbiology Reviews* 27: 165–181.

Chakraborty, J., Das, S., 2014. Characterization and cadmium-resistant gene expression of biofilm-forming marine bacterium *Pseudomonas aeruginosa* JP-11. *Environmental Science and Pollution Research* 21:14188–14201.

Characklis, W., Cooksey, K., 1983. Biofilms and microbial fouling. *Advances in Applied Microbiology* 29: 93–138.

Characklis, W. G., Marshall, K. C., 1990. Biofilms. Wiley, New York.

Chenier, D., Beriault, R., Mailloux, R., Baquie, M., Abramia, G., Lemire, J., Appanna, V., 2008. Involvement of fumarase C and NADH oxidase in metabolic adaptation of *Pseudomonas fluorescens* cells evoked by aluminum and gallium toxicity. *Applied and Environmental Microbiology* 74: 3977–3984.

Chien, C.-C., Lin, B.-C., Wu, C.-H., 2013. Biofilm formation and heavy metal resistance by an environmental *Pseudomonas* sp. *Biochemical Engineering Journal* 78: 132–137.

Cho, K.-S., Ryu, H. W., Lee, I. S., Choi, H.-M., 2002. Effect of solids concentration on bacterial leaching of heavy metals from sewage sludge. *Journal of the Air & Waste Management Association* 52: 237–243.

Christensen, B. E., 1989. The role of extracellular polysaccharides in biofilms. *Journal of Biotechnology* 10: 181–202.

Cooksey, D. A., 1994. Molecular mechanisms of copper resistance and accumulation in bacteria. *FEMS Microbiology Reviews* 14: 381–386.

Costerton, J. W., Lewandowski, Z., Caldwell, D. E., Korber, D. R., Lappin-Scott, H. M., 1995. Microbial biofilms. *Annual Reviews in Microbiology* 49: 711–745.

Costley, S., Wallis, F., 2001. Bioremediation of heavy metals in a synthetic wastewater using a rotating biological contactor. *Water Research* 35: 3715–3723.

Cowan, M. M., Warren, T. M., Fletcher, M., 1991. Mixed-species colonization of solid surfaces in laboratory biofilms. *Biofouling* 3: 23–34.

Das, S., Dash, H. R., Chakraborty, J., 2016. Genetic basis and importance of metal resistant genes in bacteria for bioremediation of contaminated environments with toxic metal pollutants. *Applied Microbiology and Biotechnology* 100: 2967–2984.

Dash, H. R., Das, S., 2015. Bioremediation of inorganic mercury through volatilization and biosorption by transgenic *Bacillus cereus* BW-03(pPW-05). *International Biodeterioration & Biodegradation* 103: 179–185.

Dash, H. R., Mangwani, N., Das, S., 2014. Characterization and potential application in mercury bioremediation of highly mercury-resistant marine bacterium *Bacillus thuringiensis* PW-05. *Environmental Science and Pollution Research* 21: 2642–2653.

Davies, J. A., Harrison, J. J., Marques, L. L., Foglia, G. R., Stremick, C. A., Storey, D. G., Turner, R. J., Olson, M. E., Ceri, H., 2007. The GacS sensor kinase controls phenotypic reversion of small colony variants isolated from biofilms of *Pseudomonas aeruginosa* PA14. *FEMS Microbiology Ecology* 59: 32–46.

De, J., Ramaiah, N., Vardanyan, L., 2008. Detoxification of toxic heavy metals by marine bacteria highly resistant to mercury. *Marine Biotechnology* 10: 471–477.

Dieppois, G., Ducret, V., Caille, O., Perron, K., 2012. The transcriptional regulator CzcR modulates antibiotic resistance and quorum sensing in *Pseudomonas aeruginosa*. *PLoS One* 7: e38148.

DiSpirito, A. A., Dugan, P. R., Tuovinen, O. H., 1983. Sorption of *Thiobacillus ferrooxidans* to particulate material. *Biotechnology and Bioengineering* 25: 1163–1168.

Duprey, A., Chansavang, V., Frémion, F., Gonthier, C., Louis, Y., Lejeune, P., Springer, F., Desjardin, V., Rodrigue, A., Dorel, C., 2014. "NiCo Buster": Engineering *E. coli* for fast and efficient capture of cobalt and nickel. *Journal of Biological Engineering* 8: 1.

Flemming, H.C., 2002. Biofouling in water systems -cases, causes and counter measures. *Applied Microbiology and Biotechnology*. 59:629–640.

Fletcher, M., Lessmann, J. M., Loeb, G. I., 1991. Bacterial surface adhesives and biofilm matrix polymers of marine and freshwater bacteria. *Biofouling* 4: 129–140.

Florian, B., Noël, N., Thyssen, C., Felschau, I., Sand, W., 2011. Some quantitative data on bacterial attachment to pyrite. *Minerals Engineering* 24: 1132–1138.

Foulkes, E., 2000. Transport of toxic heavy metals across cell membranes. *Proceedings of the Society for Experimental Biology and Medicine* 223: 234–240.

Franke, S., Grass, G., Rensing, C., Nies, D. H., 2003. Molecular analysis of the copper-transporting efflux system CusCFBA of *Escherichia coli*. *Journal of Bacteriology* 185: 3804–3812.

Frische, T., Höper, H., 2003. Soil microbial parameters and luminescent bacteria assays as indicators for in situ bioremediation of TNT-contaminated soils. *Chemosphere* 50: 415–427.

Fuchs, T., Huber, H., Teiner, K., Burggraf, S., Stetter, K. O., 1995. *Metallosphaera prunae*, sp. nov., a novel metal-mobilizing, thermoacidophilic archaeum, isolated from a uranium mine in Germany. *Systematic and Applied Microbiology* 18: 560–566.

Gadd, G. M., 2005. Microorganisms in toxic metal-polluted soils. In *Microorganisms in Soils: Roles in Genesis and Functions*. Springer, Berlin, Germany, pp. 325–356.

Gadd, G. M., 2010. Metals, minerals and microbes: geomicrobiology and bioremediation. *Microbiology* 156: 609–643.

Gallard, H., von Gunten, U., 2002. Chlorination of natural organic matter: kinetics of chlorination and of THM formation. *Water Research* 36: 65–74.

Gehrke, T., Telegdi, J., Thierry, D., Sand, W., 1998. Importance of extracellular polymeric substances from *Thiobacillus ferrooxidans* for bioleaching. *Applied and Environmental Microbiology* 64: 2743–2747.

Gehrke, T., Hallmann, R., Kinzler, K., Sand, W., 2001. The EPS of *Acidithiobacillus ferrooxidans*: A model for structure-function relationships of attached bacteria and their physiology. *Water Science and Technology* 43: 159–167.

Geslin, C., Llanos, J., Prieur, D., Jeanthon, C., 2001. The manganese and iron superoxide dismutases protect *Escherichia coli* from heavy metal toxicity. *Research in Microbiology* 152: 901–905.

Gómez Calderón, W. A., Ball Vargas, M. M., Botello Suárez, W. A., Yarzábal Rodríguez, L. A., 2013. Horizontal transfer of heavy metal and antibiotic-resistance markers between indigenous bacteria, colonizing mercury contaminated tailing ponds in southern Venezuela, and human pathogens. *Revista de la Sociedad Venezolana de Microbiología* 33: 110–115.

Gonzalez, C. F., Ackerley, D. F., Lynch, S. V., Matin, A., 2005. ChrR, a soluble quinone reductase of *Pseudomonas putida* that defends against H_2O_2. *Journal of Biological Chemistry* 280: 22590–22595.

Grass, G., Franke, S., Taudte, N., Nies, D. H., Kucharski, L. M., Maguire, M. E., Rensing, C., 2005. The metal permease ZupT from *Escherichia coli* is a transporter with a broad substrate spectrum. *Journal of Bacteriology* 187: 1604–1611.

Gupta, A., Matsui, K., Lo, J.-F., Silver, S., 1999. Molecular basis for resistance to silver cations in Salmonella. *Nature Medicine* 5: 183–188.

Gupta, A., Phung, L. T., Taylor, D. E., Silver, S., 2001. Diversity of silver resistance genes in IncH incompatibility group plasmids. *Microbiology* 147: 3393–3402.

Harneit, K., Göksel, A., Kock, D., Klock, J.-H., Gehrke, T., Sand, W., 2006. Adhesion to metal sulfide surfaces by cells of *Acidithiobacillus ferrooxidans*, *Acidithiobacillus thiooxidans* and *Leptospirillum ferrooxidans*. *Hydrometallurgy* 83: 245–254.

Harrison, J. J., Turner, R. J., Ceri, H., 2005. Persister cells, the biofilm matrix and tolerance to metal cations in biofilm and planktonic *Pseudomonas aeruginosa*. *Environmental Microbiology* 7: 981–994.

Harrison, J. J., Ceri, H., Turner, R. J., 2007. Multimetal resistance and tolerance in microbial biofilms. *Nature Reviews Microbiology* 5: 928–938.

Harrison, J. J., Wade, W. D., Akierman, S., Vacchi-Suzzi, C., Stremick, C. A., Turner, R. J., Ceri, H., 2009. The chromosomal toxin gene yafQ is a determinant of multidrug tolerance for *Escherichia coli* growing in a biofilm. *Antimicrobial Agents and Chemotherapy* 53(6), 2253–2258.

Hausner, M., Wuertz, S., 1999. High rates of conjugation in bacterial biofilms as determined by quantitative in situ analysis. *Applied and Environmental Microbiology* 65: 3710–3713.

Hebbeln, P., Eitinger, T., 2004. Heterologous production and characterization of bacterial nickel/cobalt permeases. *FEMS Microbiology Letters* 230: 129–135.

Hossain, K., Rao, A. R., 2014. Environmental change and it's affect. *European Journal of Sustainable Development* 3: 89.

Huang, F., Dang, Z., Guo, C. L., Lu, G. N., Gu, R. R., Liu, H. J., Zhang, H., 2013. Biosorption of Cd (II) by live and dead cells of *Bacillus cereus* RC-1 isolated from cadmium-contaminated soil. *Colloids and Surfaces B: Biointerfaces* 107: 11–18.

Inaoka, T., Matsumura, Y., Tsuchido, T., 1999. SodA and manganese are essential for resistance to oxidative stress in growing and sporulating cells of *Bacillus subtilis*. *Journal of Bacteriology* 181: 1939–1943.

Jang, A., Kim, S., Kim, S., Lee, S., Kim, I. S., 2001. Effect of heavy metals (Cu, Pb, and Ni) on the compositions of EPS in biofilms. *Water Science and Technology* 43, 41–48.

Keiding, K., Nielsen, P. H., 1997. Desorption of organic macromolecules from activated sludge: Effect of ionic composition. *Water Research* 31: 1665–1672.

Kermani, A. J. N., Ghasemi, M. F., Khosravan, A., Farahmand, A., Shakibaie, M., 2010. Cadmium bioremediation by metalresistant mutated bacteria isolated from active sludge of industrial effluent. *Iranian Journal of Environmental Health Science & Engineering* 7: 279.

Kessi, J., Hanselmann, K. W., 2004. Similarities between the abiotic reduction of selenite with glutathione and the dissimilatory reaction mediated by *Rhodospirillum rubrum* and *Escherichia coli*. *Journal of Biological Chemistry* 279: 50662–50669.

Kim, M.-J., Seo, J.-Y., Choi, Y.-S., Kim, G.-H., 2015. Bioleaching of spent Zn–Mn or Ni–Cd batteries by *Aspergillus* species. *Waste Management* 51: 168–171.

Kinzler, K., Gehrke, T., Telegdi, J., Sand, W., 2003. Bioleaching—A result of interfacial processes caused by extracellular polymeric substances (EPS). *Hydrometallurgy* 71: 83–88.

Koechler, S., Farasin, J., Cleiss-Arnold, J., Arsène-Ploetze, F., 2015. Toxic metal resistance in biofilms: Diversity of microbial responses and their evolution. *Research in Microbiology* 166: 764–773.

Krol, J. E., Nguyen, H. D., Rogers, L. M., Beyenal, H., Krone, S. M., Top, E. M., 2011. Increased transfer of a multidrug resistance plasmid in *Escherichia coli* biofilms at the air-liquid interface. *Applied and Environmental Microbiology* 77: 5079–88.

LaFleur, M. D., Kumamoto, C. A., Lewis, K., 2006. *Candida albicans* biofilms produce antifungal-tolerant persister cells. *Antimicrobial Agents and Chemotherapy* 50: 3839–46.

Lappin-Scott, H. M., Costerton, J. W., 1989. Bacterial biofilms and surface fouling. *Biofouling* 1:323–342.

Lawrence, J. R., Caldwell, D. E., 1987. Behavior of bacterial stream populations within the hydrodynamic boundary layers of surface microenvironments. *Microbial Ecology* 14: 15–27.

Lin, T. Y., Kampalath, R. A., Lin, C. C., Zhang, M., Chavarria, K., Lacson, J., Jay, J. A., 2013. Investigation of mercury methylation pathways in biofilm versus planktonic cultures of *desulfovibrio desulfuricans*. Environmental Science & Technology, 47: 5695–5702.

Liu, D., Lau, Y., Chau, Y., Pacepavicius, G., 1993. Characterization of biofilm development on artificial substratum in natural water. *Water Research* 27: 361–367.

Lloyd, J. R., Renshaw, J. C., 2005. Bioremediation of radioactive waste: radionuclide–microbe interactions in laboratory and field-scale studies. *Current Opinion in Biotechnology* 16: 254–260.

Lohmeier-Vogel, E. M., Ung, S., Turner, R. J., 2004. In vivo 31P nuclear magnetic resonance investigation of tellurite toxicity in *Escherichia coli*. *Applied and Environmental Microbiology* 70: 7342–7347.

Madsen, J. S., Burmølle, M., Hansen, L. H., Sørensen, S. J., 2012. The interconnection between biofilm formation and horizontal gene transfer. *FEMS Immunology & Medical Microbiology* 65: 183–195.

Maneesuwannarat, S., Vangnai, A. S., Yamashita, M., Thiravetyan, P., 2016. Bioleaching of gallium from gallium arsenide by *Cellulosimicrobium funkei* and its application to semiconductor/electronic wastes. *Process Safety and Environmental Protection* 99: 80–87.

Marshall, K. C., 1980. Microorganisms and interfaces. *BioScience* 30: 246–249.

Martínez-Granero, F., Capdevila, S., Sánchez-Contreras, M., Martín, M., Rivilla, R., 2005. Two site-specific recombinases are implicated in phenotypic variation and competitive rhizosphere colonization in *Pseudomonas fluorescens*. *Microbiology* 151: 975–983.

Mehta, K., Pandey, B., 1999. Bio-assisted leaching of copper, nickel and cobalt from copper converter slag. *Materials Transactions, JIM* 40: 214–221.

Mergeay, M., Monchy, S., Vallaeys, T., Auquier, V., Benotmane, A., Bertin, P., Taghavi, S., Dunn, J., van der Lelie, D., Wattiez, R., 2003. *Ralstonia metallidurans*, a bacterium specifically adapted to toxic metals: Towards a catalogue of metal-responsive genes. *FEMS Microbiology Reviews* 27: 385–410.

Messens, J., Van Molle, I., Vanhaesebrouck, P., Van Belle, K., Wahni, K., Martins, J. E. C., Wyns, L., Loris, R., 2004. The structure of a triple mutant of pI258 arsenate reductase from *Staphylococcus aureus* and its 5-thio-2-nitrobenzoic acid adduct. *Acta Crystallographica Section D: Biological Crystallography* 60: 1180–1184.

Mukhopadhyay, R., Rosen, B. P., Phung, L. T., Silver, S., 2002. Microbial arsenic: From geocycles to genes and enzymes. *FEMS Microbiology Reviews* 26: 311–325.

Muranaka, L. S., Takita, M. A., Olivato, J. C., Kishi, L. T., de Souza, A. A., 2012. Global expression profile of biofilm resistance to antimicrobial compounds in the plant-pathogenic bacterium *Xylella fastidiosa* reveals evidence of persister cells. *Journal of Bacteriology* 194: 4561–4569.

Nancharaiah, Y. V., Joshi, H. M., Hausner, M., Venugopalan, V. P., 2008. Bioaugmentation of aerobic microbial granules with *Pseudomonas putida* carrying TOL plasmid. *Chemosphere* 71: 30–35.

Navarrete, F., De La Fuente, L. 2014. Response of *Xylella fastidiosa* to zinc: decreased culturability, increased exopolysaccharide production, and formation of resilient biofilms under flow conditions. *Applied and Environmental Microbiology* 80: 1097–1107.

Nieboer, E., Fletcher, G., 1996. Determinants of reactivity in metal toxicology. In *Toxicology of Metals*. CRC Press, Boca Raton, FL, pp. 113–132.

Nies, D. H., 1995. The cobalt, zinc, and cadmium efflux system CzcABC from *Alcaligenes eutrophus* functions as a cation-proton antiporter in *Escherichia coli*. *Journal of Bacteriology* 177: 2707–2712.

Nies, D. H., 2003. Efflux-mediated heavy metal resistance in prokaryotes. *FEMS Microbiology Reviews* 27: 313–339.

Noël, N., Florian, B., Sand, W., 2010. AFM & EFM study on attachment of acidophilic leaching organisms. *Hydrometallurgy* 104: 370–375.

Ohtake, H., Silver, S., 1994. Bacterial detoxification of toxic chromate. In *Biological Degradation and Bioremediation of Toxic Chemicals*. Chapman and Hall, London, U.K., pp. 403–415.

Ozturk, S., Aslim, B., Suludere, Z., 2009. Evaluation of chromium(VI) removal behaviour by two isolates of *Synechocystis* sp. in terms of exopolysaccharide (EPS) production and monomer composition. *Bioresource Technology* 100: 5588–5593.

Pedersen, K., 1982. Factors regulating microbial biofilm development in a system with slowly flowing seawater. *Applied and Environmental Microbiology* 44: 1196–1204.

Perumbakkam, S., Hess, T. F., Crawford, R. L., 2006. A bioremediation approach using natural transformation in pure-culture and mixed-population biofilms. *Biodegradation* 17: 545–557.

Pomposiello, P. J., Demple, B., 2002. Global adjustment of microbial physiology during free radical stress. *Advances in Microbial Physiology* 46: 319–341.

Rao, T., 2003. Temporal variations in an estuarine biofilm: with emphasis on nitrate reduction. *Estuarine, Coastal and Shelf Science* 58: 67–75.

Rao, T., Kesavamoorthy, R., Babu Rao, C., Nair, K., 1997a. Influence of flow on ordering characteristics of a bacterial biofilm. *Current Science* 73: 69–74.

Rao, T., Rani, P., Venugopalan, V., Nair, K., 1997b. Biofilm formation in a freshwater environment under photic and aphotic conditions. *Biofouling* 11: 265–282.

Rastegar, S. O., Mousavi, S. M., Shojaosadati, S. A., Sarraf Mamoory, R., 2015. Bioleaching of V, Ni, and Cu from residual produced in oil fired furnaces using *Acidithiobacillus ferrooxidans*. *Hydrometallurgy* 157: 50–59.

Rensing, C., Grass, G., 2003. Escherichia coli mechanisms of copper homeostasis in a changing environment. *FEMS Microbiology Reviews* 27: 197–213.

Ritalahti, K. M., Löffler, F. E., Rasch, E. E., Koenigsberg, S. S., 2005. Bioaugmentation for chlorinated ethane detoxification: Bioaugmentation and molecular diagnostics in the bioremediation of chlorinated ethene-contaminated sites. *Industrial Biotechnology* 1: 114–118.

Robb, I., 1984. Stereo-biochemistry and function of polymers. In *Microbial Adhesion and Aggregation*. Springer, Berlin, Germany, pp. 39–49.

Roberts, A. P., Mullany, P., 2006. Genetic basis of horizontal gene transfer among oral bacteria. *Periodontology* 42: 36–46.

Roberts, S. A., Wildner, G. F., Grass, G., Weichsel, A., Ambrus, A., Rensing, C., Montfort, W. R., 2003. A labile regulatory copper ion lies near the T1 copper site in the multicopper oxidase CueO. *Journal of Biological Chemistry* 278: 31958–31963.

Rodrigues, M. L. M., Lopes, K. C. S., Leôncio, H. C., Silva, L. A. M., Leão, V. A., 2016. Bioleaching of fluoride-bearing secondary copper sulphides: Column experiments with *Acidithiobacillus ferrooxidans*. *Chemical Engineering Journal* 284: 1279–1286.

Rohwerder, T. G., Kinzler, K., Sand, W., 2003. Bioleaching review (part A): Progress in bioleaching: fundamentals and mechanisms of bacterial metal sulfide oxidation. *Applied Microbiology Biotechnology* 63: 239–248.

Rojas, L. A., Yáñez, C., González, M., Lobos, S., Smalla, K., Seeger, M., 2011. Characterization of the metabolically modified heavy metal-resistant *Cupriavidus metallidurans* strain MSR33 generated for mercury bioremediation. *PLoS One* 6: e17555.

Ruiz, O. N., Alvarez, D., Gonzalez-Ruiz, G., Torres, C., 2011. Characterization of mercury bioremediation by transgenic bacteria expressing metallothionein and polyphosphate kinase. *BMC Biotechnology* 11: 82.

Ryu, H. W., Moon, H. S., Lee, E. Y., Cho, K. S., Choi, H., 2003. Leaching characteristics of heavy metals from sewage sludge by *Acidithiobacillus thiooxidans* MET. *Journal of Environmental Quality* 32: 751–9.

Sampson, M., Phillips, C., Blake, R., 2000. Influence of the attachment of acidophilic bacteria during the oxidation of mineral sulfides. *Minerals Engineering* 13: 373–389.

Sand, W., Gehrke, T., 2006. Extracellular polymeric substances mediate bioleaching/biocorrosion via interfacial processes involving iron (III) ions and acidophilic bacteria. *Research in Microbiology* 157: 49–56.

Saravanan, P., Nancharaiah, Y. V., Venugopalan, V. P., Rao, T. S., Jayachandran, S., 2006. Biofilm formation by *Pseudoalteromonas ruthenica* and its removal by chlorine. *Biofouling* 22: 371–381.

Sarkar, S., Chakraborty, R., 2008. Quorum sensing in metal tolerance of *Acinetobacter junii* BB1A is associated with biofilm production. *FEMS Microbiology Letters* 282: 160–165.

Sarró, M. I., García, A. M., Moreno, D. A., 2005. Biofilm formation in spent nuclear fuel pools and bioremediation of radioactive water. *International Microbiology* 8: 223–230.

Schippers, A., Hedrich, S., Vasters, J., Drobe, M., Sand, W., Willscher, S., 2013. Biomining: Metal recovery from ores with microorganisms. In *Geobiotechnology I*. Springer, Heidelberg, Germany, pp. 1–47.

Shukla, S. K., Mangwami, N., Rao, T. S., Das, S., 2014. 8 Biofilm-mediated bioremediation of polycyclic aromatic hydrocarbons. In *Microbial Biodegradation and Bioremediation*. Elsevier, Amsterdam, Netherlands, p. 203.

Silver, S., 1996. Bacterial resistances to toxic metal ions: A review. *Gene* 179: 9–19.

Silver, S., Phung, L. T., 2005. A bacterial view of the periodic table: genes and proteins for toxic inorganic ions. *Journal of Industrial Microbiology and Biotechnology* 32: 587–605.

Singh, R., Paul, D., Jain, R. K., 2006. Biofilms: Implications in bioremediation. *Trends in Microbiology* 14: 389–397.

Singh, S. K., Grass, G., Rensing, C., Montfort, W. R., 2004. Cuprous oxidase activity of CueO from *Escherichia coli*. *Journal of Bacteriology* 186: 7815–7817.

Sjollema, J., Busscher, H., 1990. Deposition of polystyrene particles in a parallel plate flow cell. 2. Pair distribution functions between deposited particles. *Colloids and Surfaces* 47: 337–352.

Solari, J. A., Huerta, G., Escobar, B., Vargas, T., Badilla-Ohlbaum, R., Rubio, J., 1992. Interfacial phenomena affecting the adhesion of *Thiobacillus ferrooxidans* to sulphide mineral surface. *Colloids and Surfaces* 69: 159–166.

Stohs, S., Bagchi, D., 1995. Oxidative mechanisms in the toxicity of metal ions. *Free Radical Biology and Medicine* 18: 321–336.

Stoodley, P., Sauer, K., Davies, D., Costerton, J. W., 2002. Biofilms as complex differentiated communities. *Annual Reviews in Microbiology* 56: 187–209.

Sutherland, I. W., 1997. Microbial exopolysaccharides-structural subtleties and their consequences. *Pure and Applied Chemistry* 69: 1911–1918.

Thaden, J. T., Lory, S., Gardner, T. S., 2010. Quorum-sensing regulation of a copper toxicity system in *Pseudomonas aeruginosa*. *Journal of Bacteriology* 192: 2557–68.

Tremaroli, V., Fedi, S., Zannoni, D., 2007. Evidence for a tellurite-dependent generation of reactive oxygen species and absence of a tellurite-mediated adaptive response to oxidative stress in cells of *Pseudomonas pseudoalcaligenes* KF707. *Archives of Microbiology* 187: 127–135.

Turner, R. J., Aharonowitz, Y., Weiner, J. H., Taylor, D. E., 2001. Glutathione is a target in tellurite toxicity and is protected by tellurite resistance determinants in *Escherichia coli*. *Canadian Journal of Microbiology* 47: 33–40.

Van der Kooij, D., Veenendaal, H. R., Baars-Lorist, C., van der Klift, D. W., Drost, Y. C., 1995. Biofilm formation on surfaces of glass and teflon exposed to treated water. *Water Research* 29: 1655–1662.

van Hullebusch, E. D., Zandvoort, M. H., Lens, P. N., 2003. Metal immobilisation by biofilms: Mechanisms and analytical tools. *Reviews in Environmental Science and Biotechnology* 2: 9–33.

Vera, M., Krok, B., Bellenberg, S., Sand, W., Poetsch, A., 2013. Shotgun proteomics study of early biofilm formation process of *Acidithiobacillus ferrooxidans* ATCC 23270 on pyrite. *Proteomics* 13: 1133–1144.

Vilinska, A., Rao, K. H., 2009. Surface thermodynamics and extended DLVO theory of Acidithiobacillus ferrooxidans cells adhesion on pyrite and chalcopyrite. *The Open Colloid Science Journal* 2:1–14.

Wenbin, N., Dejuan, Z., Feifan, L., Lei, Y., Peng, C., Xiaoxuan, Y., Hongyu, L., 2011. Quorum-sensing system in *Acidithiobacillus ferrooxidans* involved in its resistance to Cu(2)(+). *Letters in Applied Microbiology* 53: 84–91.

White, C., Gadd, G., 2000. Copper accumulation by sulfate-reducing bacterial biofilms. *FEMS Microbiology Letters* 183: 313–318.

Workentine, M. L., Harrison, J. J., Stenroos, P. U., Ceri, H., Turner, R. J., 2008. *Pseudomonas fluorescens* view of the periodic table. *Environmental Microbiology* 10: 238–250.

Xavier, J. B., Foster, K. R., 2007. Cooperation and conflict in microbial biofilms. *Proceedings of the National Academy of Sciences of the United States of America* 104: 876–881.

Yang, Z., Zhang, Z., Chai, L., Wang, Y., Liu, Y., Xiao, R., 2016. Bioleaching remediation of heavy metal-contaminated soils using *Burkholderia* sp. Z-90. *Journal of Hazardous Materials* 301: 145–152.

Zannoni, D., Borsetti, F., Harrison, J. J., Turner, R. J., 2007. The bacterial response to the chalcogen metalloids Se and Te. *Advances in Microbial Physiology* 53: 1–312.

Zhang, D., Lee, D.-J., Pan, X., 2012. Fluorescent quenching for biofilm extracellular polymeric substances (EPS) bound with Cu (II). *Journal of the Taiwan Institute of Chemical Engineers* 43: 450–454.

Zhang, D., Pan, X., Mostofa, K. M., Chen, X., Mu, G., Wu, F., Liu, J., Song, W., Yang, J., Liu, Y., 2010. Complexation between Hg (II) and biofilm extracellular polymeric substances: An application of fluorescence spectroscopy. *Journal of Hazardous Materials* 175: 359–365.

20 Geomicrobiology of Arsenic-Contaminated Groundwater of Bengal Delta Plain

Pinaki Sar, Balaram Mohapatra, Soma Ghosh,
Dhiraj Paul, Angana Sarkar, and Sufia K. Kazy

CONTENTS

ABSTRACT

Naturally occurring toxic arsenic (As) in alluvial groundwater systems represents one of the most serious abiotic contaminations with enormous public health concern. Globally, more than 100 million people are affected, while the Bengal Delta Plain (BDP), spread over the large areas of West Bengal (India) and Bangladesh, is the worst affected. Mobilization of As from alluvial sediments rich in Fe/Mn (Al) oxides/hydroxides is caused by complex interactions of hydrogeochemical and microbial processes. The evolutionary stratigraphy, geological settings, mineralogy of aquifer sediments, and biogeochemical factors are found to have extensive implications on the hydrogeochemistry of As in these aquifers. Bacterial community inhabiting As-rich aquifers play an important role in the geochemical transformation of As, mainly by facilitating redox transformation of As species. Bacterial groups inhabiting the western parts of BDP showed high metabolic diversity, particularly with respect to their abilities to utilize broad ranges of carbon sources, electron donors, and acceptors. Culture-independent molecular analysis of bacterial diversity revealed abundance of aerobic/facultative anaerobic, denitrifying, Fe^{2+}/As^{3+}-oxidizing, and As-resistant *Pseudomonas*, *Brevundimonas*, *Microbacterium*, *Acidovorax*, *Acinetobacter*, and *Hydrogenophaga*, anaerobic Fe^{3+}- and SO$_4^{2-}$-reducing *Geobacter* and *Geothrix*, and methanogenic and methylotrophic populations. The revolutionary technology of sequencing, starting from shotgun to single-molecule long-read approach, its impact on our understanding of microbial evolution, and its function in ecophysiology and phylogenomics are discussed. In spite of high toxicity, inhabitant microbes are found to possess catabolic repertoire to transform As, to withstand its toxicity, or to use it as metabolic resource. Based on the current state of knowledge, an overall spectrum of microbiology of As-rich groundwater is presented, highlighting bacteria–As interaction and diversity of bacterial communities within contaminated groundwater of BDP.

20.1 INTRODUCTION

Natural enrichment of groundwater with arsenic (As) concentration exceeding the safe level (10 µg/L) of the World Health Organization (WHO) causes a critical water quality problem in many parts of the world including the South and Southeast Asia, the United States, and countries of the European Union (Fendorf et al., 2010). Groundwater concentration of As shows a very large range (<0.5–5000 µg/L) spreading across more than 70 countries (Sharma et al., 2014). Among the affected areas, the worst mass poisoning has been documented in the alluvial aquifers of Southeast Asia, particularly in the Bengal Delta Plain (BDP) of India and Bangladesh, claiming that more than 50 million people are at risk. Epidemiological studies have shown that chronic As exposure through drinking water and food can cause serious health implications including skin disorders, cardiovascular disease, neurological problems, reproductive disorders, and respiratory effect as well as several types of cancers (Rahman et al., 2015 and references therein).

Extensive work carried out to identify the distribution pattern of As and its underlying geochemical mechanisms established that aqueous As is mainly derived from subsurface solid sediment phases due to a complex interplay of biogeochemical processes between sediment and microbe and water systems. In As-contaminated aquifer sediments, As^{5+} is found to be coprecipitated in or coadsorbed on various Fe- and Mn-rich clastic and authigenic minerals or mineral phases (Fe hydroxides, Fe oxides–coated sand, phyllosilicates, Mn and Al oxides, Mn and Al hydroxides, and authigenic pyrites). Mobilization of solid phase As is caused by a complex interplay of hydrogeobiochemical processes and human interactions with a critical role played by inhabitant microbes (Sarkar et al., 2013 and references therein). In the case of BDP, release of As is believed to be often closely related to release of Mn, but not with Fe, and it is driven by microbial processes that facilitate the reduction of sediment-associated elements (e.g., As, Fe, or Mn) utilizing metabolizable carbon substrates. Among the various mechanisms proposed, release of As due to weathering and/or dissolution of host minerals (Fe/Mn oxides hydroxides) by complex geomicrobiological activities remains the most significant. Biogeochemical activities of indigenous bacteria surviving under As-rich, nutrient-limiting aquifer environment can control the mobility of this metalloid by transformation of host minerals or weathering of rocks. During the past decade, understanding the As–bacteria interaction and geomicrobiology of As-rich aquifers has gained considerable attention to develop sustainable strategies to supply drinking water to the affected populations. To explain subsurface mobilization of As in groundwater, a number of mechanisms including (1) reductive dissolution of As-rich Fe oxyhydroxides, (2) oxidation of As-rich pyrite, and (3) weathering of minerals that contain phosphate, ammonia, iron, etc., have been proposed (Islam et al., 2004; Mailloux et al., 2009; Hery et al., 2010; Sarkar et al., 2013). Each mechanism including both biotic and abiotic components probably plays a role under certain condition(s). Inhabiting microorganisms within aquifer sediments and/or groundwater have been shown to affect the As geochemistry by catalyzing redox transformations and other reactions that affect the mobility of this metalloid in subsurface environment (Oremland and Stolz, 2005). Microorganisms have evolved dynamic mechanisms for facing the toxicity of As in the environment. In this sense, As speciation and mobility are also affected by microbial metabolism (Oremland and

Stolz, 2005). The organisms inhabiting in As-rich habitats and undertaking important biogeochemical reactions are taxonomically diverse and metabolically versatile (Oremland and Stolz, 2005; Slyemi and Bonnefoy, 2012). Some bacteria can reduce As^{5+} to As^{3+} during their anaerobic respiration or as a means of As detoxification, while others oxidize As^{3+} to As^{5+} during their chemolithoautotrophic/heterotrophic metabolism.

Fendorf et al. (2010) have explained the probable principal factors for the occurrence of As in groundwater. Precisely, location and amount of As release to the aquifer from sediment depend on the availability of labile organic carbon (C) and As in the sediment. Recalcitrant organic C will lead to lesser As release than highly reactive forms. Arsenic release is also limited by the presence of sulfate as it forms sparingly soluble As sulfides, a process promoted by organic C. However, previous studies have established the fact that the geological settings and chronological events leading to the formation of the Ganga-Brambhaputra-Meghna (GBM) basin, its mineralogy, and As hydrogeochemistry play important roles in As biogeotransformation and microbial behavior.

In this chapter, a broad overview on geological aspect and geomicrobiology of As-contaminated aquifer, particularly with reference to BDP, is presented. Emphasis is given on (1) geological evolution of As in GBM basin and hydrogeochemistry of As-contaminated aquifers of GBM basin, (2) metabolic abilities of bacteria from As-rich groundwater, (3) bacterial diversity in As-rich groundwater, and (4) the advent of whole genome sequencing of bacterial strains in understanding bacteria–As interaction.

20.2 GEOLOGICAL EVOLUTION OF ARSENIC IN GANGA–BRAHMAPUTRA BASIN

The Ganga–Brahmaputra Delta in the Bengal Basin forms the largest delta in the world channelling suspended solids and enormous amount of dissolved particulates to the Bay of Bengal. Both the rivers have their origins in the Himalayan mountain range of China.

The Ganga Basin, formed in response to the upliftment of the Himalayas due to the collision of the Indian and Asian plates during the Eocene, shows all the major components of a foreland basin system: an orogen (the Himalayas), deformed foreland basin deposits adjacent to the orogen (Siwalik Hills), a depositional basin (Ganga Plain), and peripheral cratonic bulge (Bundelkhand Plateau) (Singh, 1996). Complete establishment of the basin dates back to middle Miocene (Sinha et al., 2005). It received huge quantity of sediments from the Himalayan orogen. During middle Miocene to the mid-Pleistocene, the orogenward part of the Ganga Plain was uplifted and the last major thrust pushed it toward the basin. It is believed that since mid-Pleistocene a considerable amount of craton-derived sediments were contributed to the Ganga foreland basin. The older Pleistocene sediments were subjected to a long period of oxidative weathering due to the lowering of the sea level. Under this period, the sediment acquired the FeOOH coatings needed to adsorb any dissolved As. But much of the Pleistocene deposits have been eroded away from the Siwalik thrust sheets (Singh, 1996). In the early Holocene, vigorous sediment deposition occurred under rapid sea-level rise and high river discharge. The Holocene sediments were derived under a period of low temperatures, which resulted in little chemical weathering and therefore a low degree of FeOOH coatings. During the mid-Holocene to recent times, the discharge and sea-level rise declined, leading to the development of thick, fluvial, and coastal marine deposits.

The Brahmaputra plain, on the other hand, is covered by young alluvial sediments that are deposited from the great sediment load carried by the river and its tributaries. Mineralogy on the south and north side of the river differs significantly. Sediments of the north of Brahmaputra are derived from the young Himalayas, while sediments on the south side originated from the older Assam plateau. The Surma basin (Barak valley) of south Assam forms part of the greater Bengal Basin. Geologically, sediments of Barak valley comprise (1) unconsolidated alluvial deposits of the Holocene age and (2) semiconsolidated deposits of the Surma Group—consisting of Bhuban formation, which forms oldest rocks, overlain by Bokabil subgroup, and Tipam Groups of Mio, the Pliocene age and Holocene Dihing formation. Consistent high concentrations of As ranging from 2 to 18 mg/kg have been found in the Tipam group of rocks when compared to Barail and Surma group of rocks (2–5 mg/kg). Therefore, the high As concentration in the alluvial areas perhaps owes its source to the Tipam rock suite and is probable that As could have been released into the sediments under favorable chemical environment (Thambidurai et al., 2013). A common observation for the GBM basin has been that As contamination is majorly found in the Holocene sedimentary aquifers as compared to that formed in Pleistocene sediments. This fact may be attributed to the environmental conditions during the time of sediment formation (McArthur et al., 2004).

20.2.1 MINERALOGY OF GBM BASIN

Sedimentation is a constant process in a river basin. Characteristics of sediments are indicative of their age. The Pleistocene sediments, deposited as flood plains of the former Ganga–Brahmaputra river systems, are well oxidized and typically reddish brown, have low water content, and contain appreciable amounts of organic material (Parua, 2010). However, sediments carried by Ganga and Brahmaputra differ in their clay content and mineralogy. Clay content of Ganga was found to be enriched with smectite (43%), as compared to the Brahmaputra (3%) where illite (63% vs 41%) and kaolinite (29%–18%) abundance was found. Chlorite (3% vs 1%) is also found to be mostly associated with the Brahmaputra sediments, supplied in physically weathered Himalayan highlands. Brahmaputra, a braided channel, is found to have a coarser bed material. Ganga, a meandering river, has higher mica and carbonate (dolomite) content (Allison et al., 2003). Epidote-to-garnet ratios in sand fractions are also diagnostic of their source, with high (>1) and low (<1) E/G indicating Brahmaputra and Ganga provenances, respectively (Heroy et al., 2003).

20.2.2 HYDROGEOCHEMISTRY OF As

Heavy metal distribution shows pronounced temporal and spatial variations in the GBM basin. Suspended sediments are 5–10 times richer in heavy metal concentration than the bed sediments. The Ganges sediments show more erratic and variable heavy metal distribution than that of the Brahmaputra. Among the other heavy metals, As contamination in the groundwater of GBM delta poses serious threat to mankind, as well as to the ecosystem. The source, the depositional history, and the stratigraphic architecture of GBM delta sediments form the key controllers of groundwater As behavior, which is widespread but shows heterogeneous distribution within the Holocene aquifers (McArthur et al., 2008; Goodbred et al., 2014). Therefore, groundwater extracted from shallow aquifers is usually highly contaminated with As. Thus, the weathering of Himalayan bedrocks (Fe oxyhydroxides) releases As to surface waters that are strongly adsorbed by sediments and oxides under neutral pH and aerobic conditions, which has been reported in many studies (Mok and Wai, 1994; Thornton, 1996; Nickson et al., 2000; Gault et al., 2005; Reza et al., 2010). Arsenic sorbed on the sediments are then codeposited with organic matter in alluvial sediments, which are then subsequently released into the groundwater by dissimilatory As^{5+}-reducing prokaryotes (DARPs) or Fe(III)-reducing bacteria (Ghosh and Sar, 2013). Two hypotheses have been proposed for As contamination in groundwater. Pyrite oxidation hypothesis would result in high sulfate and As concentration near the surface, while the iron oxide reduction hypothesis would predict a buildup of As at greater depths with generally low sulfate concentrations (Kinniburgh and Smedley, 2001). The naturally occurring As minerals are realgar (AsS), orpiment (As_2S), arsenopyrite (FeAsS), arsenolite (As_4O_6), Claudelite (As_2O_3), As pentoxide (As_2O_5), and scorodite ($FeAsO_4 \cdot 2H_2O$). Leaching of geological minerals, mineral precipitation, dissolution of unstable As minerals, adsorption–desorption, chemical transformation, and input from geothermal sources, pesticides, fertilizers, and industrial waste effluents are some of the natural as well as anthropogenic reasons of groundwater As concentration.

Strong reducing conditions in the groundwater aquifers are largely responsible for the mobilization of As. As forms oxyanions and is generally mobile at pH values 6.5–8.5 at low temperatures. Higher values of As^{3+}-to-As^{5+} ratio are sometimes used as redox indicator. Adsorption reactions, precipitations, and coprecipitation generally control As mobility at near-neutral pH. Hydrous ferric oxide (HFO) has shown very high sorption properties for As oxyanions. As^{5+} binds strongly as compared to As^{3+}, which is more mobile in aqueous phase. With increase in pH, even As^{5+} tends to become less strongly sorbed to HFO. The role of organic carbon in As mobilization has been shown by Meharg et al. (2006) who suggested codeposition of As and organic carbon to provide reducing conditions to dissolve iron (III) oxides and release As^{3+} into the pore water (Edmunds et al., 2015). Holocene sediments are characterized by a gray color and contain significant amount of organic matter, while the Pleistocene sediments are characteristically reddish-brown and contain less organic matter (Bhattacharya et al., 2002).

20.3 BACTERIAL INTERACTION WITH ARSENIC

Bacteria play important roles in the geochemical transformation of As (Mukhopadhyay et al., 2002). All free living microorganisms have been found to possess metabolic machinery to deal with As in order to get rid of its toxic effects, while some organisms have evolved further to utilize As compounds and/or their redox transformation in their metabolism (Rosen, 2002). Bacterial uptake of As^{3+} and As^{5+} occurs adventitiously through essential nutrient transport systems such as phosphate (Pst, Pit, or both) and glycerol transporters (aquaglyceroporin/glycerophosphate transporter) (Oremland and Stolz, 2005; Escudero et al., 2013). Despite its severe toxicity, As is readily used by a large number of prokaryotes for cell growth and metabolism. Microbial As transformation includes four fundamental processes: reduction, oxidation, methylation, and demethylation. Bacteria can use As as an electron donor or as a terminal electron acceptor (TEA) during their anaerobic growth respiration. Methylation and/or demethylation of As compounds has also been noted, although less frequently. Figure 20.1a represents several mechanisms of As metabolism in prokaryotes.

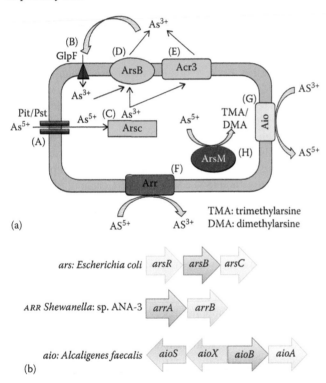

(a)

ars: *Escherichia coli*   arsR arsB arsC

ARR *Shewanella*: sp. ANA-3   arrA arrB

aio: *Alcaligenes faecalis*   aioS aioX aioB aioA
(b)

FIGURE 20.1 (a) Schematic diagram of mechanism of bacteria–As interaction. As^{5+} enters the cell through Pi transporter (A); after entering into a cell, As^{5+} gets reduced to As^{3+} by cytosolic reductase, *arsC* (B); following reduction, As^{3+} gets extruded from cell through efflux pump *arsB* (C). As^{5+} can be used as terminal e^- acceptor during respiration (D). As^{3+} can be oxidized to serve as an e^- donor or a mechanism of resistance via periplasmic aio system (E). As^{3+} enters into cell through aquaglyceroporins channel (F) and is directly extruded out from the cells via another As^{3+} transporter acr3p (G). Furthermore, inorganic As can also be transformed into organic species via methylation (H). (b) Arrangement of As-related gene clusters in different bacterial members.

Since the first documented account of microbially mediated As transformation nearly a century ago (Green, 1919), over 240 strains of bacteria and archaea are known that are capable of transforming As (either oxidize As^{3+} or reduce As^{5+} for energy gain or detoxification) (Amend et al., 2014). Based on 16S ribosomal RNA gene sequences, these strains were found to be affiliated to 11 phyla (*Aquificae, Deinococcus-Thermus, Chloroflexus, Firmicutes, Actinobacteria, Deferribacteres, Chrysiogenetes, Cyanobacteria, Proteobacteria* [including members of the α, β, γ, δ, and ε subgroups], *Bacteroidetes,* and *Crenarchaeota*). As reviewed by Amend et al. (2014), most of the isolates are among the γ-*Proteobacteria* (75), followed by the *Firmicutes* (52), β-*Proteobacteria* (40), and α-*Proteobacteria* (31), which together account for >80% of the total.

Depending upon the metabolic interactions, that is, whether the organisms are using As in their energy metabolisms or only transforming it for detoxification and whether the organisms are heterotrophic or lithotrophic, As transforming microorganisms could be categorized into five groups (Amend et al., 2014): (1) DARPs, (2) As-resistant microorganisms, (3) chemoautotrophic As^{3+} oxidizers (CAOs), (4) heterotrophic As^{3+} oxidizers (HAOs), and (5) phototrophic As^{3+} oxidizers (PAOs). Salient features of each of these groups are summarized in Table 20.1.

20.3.1 As^{5+} Reduction

Bacteria undergo As reduction (As^{5+} to As^{3+}) via two distinct mechanisms: cytosolic and respiratory (Oremland and Slotz, 2005; Escudero et al., 2013) (Figure 20.1a). The first one is widespread in nature providing detoxification of the cells via expression of *ars* operon, found either in chromosomal locations or in plasmids of Gram-negative bacteria belonging to the α-, β-, and γ-*Proteobacteria* as well as in Gram-positive *Firmicutes* and *Actinobacteria* (Escudero et al., 2013). The mechanism conferring As resistance by *ars* operon has been extensively studied in more than 50 genera including *Escherichia, Staphylococcus, Corynebacterium, Ochrobactrum, Pseudomonas, Acinetobacter, Exiguobacterium, Aeromonas, Vibrio, Psychrobacter, Enterobacter, Bacillus, Pantoea,* and *Halanaerobium*. The *ars* genes appear systematically and are cotranscribed by a large variety of genomic configurations, and the core genes of this system include cytosolic As^{5+} reductase *arsC*, membrane-bound As^{3+} efflux pump *arsB*, and transcriptional repressor *arsR* (Escudero et al., 2013) (Figure 20.1b). Although different studies have reported a common origin for the *arsC* genes, based on the source of reducing power, three unrelated groups of ArsC with common biochemical function but different evolutionary relationships have been identified. These three groups are (1) glutaredoxin–glutathione-coupled enzyme associated with plasmids and chromosomes of Gram-negative bacteria as well as in both As^{3+}-oxidizing and (respiratory) As^{5+}-reducing bacteria, (2) the less-studied glutaredoxin-dependent As^{5+} reductase found in yeasts, and (c) thioredoxin-coupled As^{5+} reductase found in Gram-positive as well as Gram-negative *Proteobacteria*. This

TABLE 20.1

Nutritional Behavior, Metabolic Properties, and Taxonomic Affiliation of Diverse Groups of As-Metabolizing Microbes

Groups	C Metabolism	Electron Donor	O_2 Requirement	Terminal Electron Acceptor	Other properties	Taxonomic Affiliation
DARP (dissimilatory As-respiring prokaryotes)	Heterotrophic (use complex organic compounds or a few low mt. organic acids, e.g., acetate, lactate, and pyruvate)	Either the same as C source or for a few reduced S or H_2	Anaerobic	As^{5+} (occasionally Fe, SO_4, etc.)		*Deferribacteres, Chrysiogenetes,* and δ- and ε-*Proteobacteria*
As-resistant microorganisms	Heterotrophic	Organic	Aerobic and anaerobic	O_2, NO_3, SO_4, Fe, etc.	Reduce As^{5+} for detoxification not for energy conservation, presence of *arsC* gene encoding cytosolic As^{5+} reductase	*Firmicutes, Actinobacteria,* α- and γ-*Proteobacteria; Bacteroidetes*
CAO (chemotrophic As^{3+} oxidizer)	Autotrophic	Oxidation of As^{3+} to As^{5+}	Aerobic	O_2, NO_3, Se, ClO_4		α- and β-*Proteobacteria*
HAO (heterotrophic As^{3+} oxidizer)	Heterotrophic	Oxidation of H_2, Fe, S compounds	Aerobic	O_2	Catalyze As^{3+} oxidation for detoxification only (also termed as detoxifying As^{3+} oxidizers DAOs).	*Firmicutes* and α-, β-, and γ-*Proteobacteria; Chloroflexi* and *Cyanobacteria*
PAO (phototrophic As^{3+} oxidizer)	Autotrophic, fix CO_2 using solar energy	As^{3+}	Anaerobic		Grow optimally at > 45°C, pH 9.3.	γ-*Proteobacteria*

divergence may occur mainly due to horizontal gene transfer (HGT) of *arsC* gene among the microorganisms through convergent evolution (Sarkar et al., 2014).

The other type of As^{5+} reductase is dissimilatory or respiratory As^{5+} reductase (*arr*A). This has been described mainly in obligate or facultative anaerobic bacteria affiliated to diverse phylogenetic groups. During this process, bacterial cells gain energy by *breathing* or *respiring* As^{5+} as a TEA (Oremland and Stolz, 2005). Dissimilatory As^{5+} reduction was first observed in *Geospirillum arsenophilus* MIT-13 strain. Subsequently, many other bacteria with respiratory As^{5+} reductase activity have been reported. These include *Desulfotomaculum auripigmentum, Sulfurospirillum barnesii,* and *S. arsenophilum, Bacillus* spp., *Desulfusporosinus* sp., *Wolinella succinogenes, Alkaliphilus metalliredigens, Clostridium* sp., *Shewanella* sp., *Alkaliphilus oremlandii, Halarsenatibacter silvermanii, Desulfohalophilus alkaliarsenatis, Geobacter* sp., and *Anaeromyxobacter* sp. Though the core enzyme ArrAB is highly conserved, the *arr* operon differs from organism to organism with respect to number of genes. *Shewanella* sp. strain ANA has only two genes in the core enzyme (*arrAB*), while *Desulfitobacterium hafniense* has only one gene encoding a putative membrane anchoring peptide as well as a multicomponent regulatory system (*arrABC* or *arrSKRCAB*)

(Malasarn et al., 2008) (Figure 20.1b). Although these two reduction mechanisms (*ars* operon and *arr* operon mediated) are not directly correlated, in some cases, the *arr* and *ars* operons lie in close proximity, within the genome of the organism suggesting an "arsenic metabolism island," which may be in *cis* or *trans* position (Cavalca et al., 2013).

20.3.2 As^{3+} Oxidation

As^{3+} oxidation is widely distributed in metabolically and taxonomically diverse bacteria. Till date more than 50 phylogenetically diverse As^{3+}-oxidizing strains distributed among 25 genera isolated from various contaminated niches (soils, mine tailings, river sediments, groundwater, and geothermal springs) have been described for their ability to oxidize As^{3+} enzymatically (Escudero et al., 2013). Since the first report of As^{3+}-oxidizing bacterium belongs to the genus *Achromobacter*, many As^{3+}-oxidizing bacteria distributed among different genera including *Thermus, Alcaligenes, Achromobacter, Agrobacterium, Herminiimonas, Thiomonas, Ochrobactrum, Pseudomonas, Polaromonas, Bosea, Ancylobacter, Acidovorax,* and *Bordetella* (Paul et al., 2014a; Sar et al., 2015) have been reported. These include both heterotrophic and chemolithoautotrophic As^{3+} oxidizers

(Oremland and Stolz, 2005), wherein they use the energy and reduce power from As^{3+} oxidation during CO_2 fixation and cell growth under both aerobic and anaerobic nitrate-reducing conditions (Oremland and Stolz, 2005). Bacterial As^{3+} oxidase belongs to the dimethyl sulfoxide reductase of the molybdenum family consisting of (1) small subunit with a Rieske [2Fe–2S] cluster and (2) large subunit harboring molybdopterin guanosine dinucleotide at the active site and iron-binding [3Fe–4S] cluster (Oremland and Stolz, 2005) (Figure 20.1b). Arsenite oxidase gene encoding two subunits (formerly known as *aoxA* and *aoxB*) has been recognized and characterized for the first time in the heterotrophic bacteria *Herminimonas arsenicoxydans*. Latter homologues of these genes were identified in a variety of organisms (Escudero et al., 2013). Different nomenclatures were adopted like *aroB* or *asoB* for the small and *aroA* or *asoA* for the large subunit genes in the chemolithoautotrophic As^{3+} oxidizer NT-26 and in *Alcaligenes faecalis* strain NCBI8687. The small and the large subunits of the As^{3+} oxidase have now been termed as *aioB* and *aioA*, respectively. In most As^{3+}-oxidizing bacteria, the synthesis of this enzyme is generally regulated by As^{3+}.

20.3.3 Methylation/Demethylation of As

Methylation of As is considered as a basic detoxification process. The pathway of methylation was proposed for the first time in the fungus *Scopulariopsis brevicaulis* involving a series of steps in which reduction of As^{5+} was followed by oxidative addition of a methyl group. Although many of the enzymes involved in this complex process of As methylation remain unknown, a methyl transferase, ArsM, from *Rhodobacter sphaeroides* conferring resistance to As and generating trimethylarsine has been identified (Sar et al., 2015).

20.4 METABOLIC DIVERSITY OF BACTERIAL GROUPS DETECTED IN BDP

Versatile metabolic abilities of the bacterial communities present in As-rich groundwater of western parts of the BDP (West Bengal) were evaluated vividly using broad arrays of phenotypic and genetic tests (Sarkar et al., 2014; Paul et al., 2015). A large number of bacterial strains isolated from As-rich groundwater were characterized for their As (As^{3+} and As^{5+}) resistance under aerobic and anaerobic conditions, As^{5+} reduction and As^{3+} oxidation activities, and ability to utilize a broad spectrum of carbon compounds (including aromatic and aliphatic hydrocarbons) as electron donor/C source and different inorganic compounds as TEAs (Figure 20.2).

Resistance to As (either As^{3+} or As^{5+} or both) under aerobic to anaerobic condition with considerably high MTC values was found to be an omnipresent phenotype of the bacterial strains. The ability to grow with elevated As^{5+} (≥ 100 mM) or As^{3+} (≥ 10 mM) under anaerobic condition was observed in 30% or 50% strains. Under aerobic condition, nearly 70% strains showed their ability to withstand the same As

species. *Actinobacteria*, *Rhizobium*, and *Rhodococcus* strains showed relatively higher MTC values (≥ 100 mM) for As^{5+} under both conditions. Members of the genera *Pseudomonas*, *Rheinheimera*, *Brevundimonas*, and *Herbaspirillum* although showed higher aerobic As^{5+} resistance, while their inability to withstand the same under anaerobic condition was noted. Aerobic As^{3+} (MTC ≥ 5 mM) resistance was found widely distributed among *Rhizobium*, *Rhodococcus*, *Pseudomonas*, *Bacillus*, *Acinetobacter*, and *Rheinheimera* strains. Under anaerobic condition, comparable As^{3+} resistance was mainly found within *Rhizobium* and *Rhodococcus* strains. Arsenate (As^{5+}) reductase activity was detected in 60% strains, whereas As^{3+} oxidase activity was found to be relatively less abundant. Notably, bacterial members belonging to the genera *Brevundimonas*, *Pseudomonas*, *Rhodococcus*, *Rhizobium*, *Phyllobacterium*, *Acinetobacter*, and *Arthrobacter* were found to have arsenate reductase. Arsenite oxidase activity was present within the strains mostly affiliated to the genera *Rhodococcus* and *Pseudomonas*.

Among all the tested carbon sources, casein was most preferred, utilized by 95% of the all strains, followed by starch, glycerol, glucose, acetate, and pyruvate. Interestingly, members of a few genera like *Bacillus*, *Phyllobacterium*, *Staphylococcus*, and *Stenotrophomonas* showed their lack of ability to use many of the tested C sources (e.g., glucose, glycerol, acetate, pyruvate). Noticeably, more than 80% strains could use long-chain alkanes (e.g., pentadecane, nonadecane) and polyaromatic hydrocarbons (e.g., pyrene, fluorene) as sole C source. Sarkar et al. (2014) have shown that strains affiliated to the genera *Achromobacter*, *Brevundimonas*, and *Rhizobium* were capable of utilizing As^{3+} as the electron donor and grew chemoautotrophically devoid of any "fixed carbon" in the medium. In an earlier study, Sarkar et al. (2013) found that siderophore and phosphatase activities were abundant in 67% and 31% of the isolates obtained from a very high As-contaminating groundwater sample. Both these activities are found to be quite common in strains belonging to unclassified *Rhizobiaceae* and members of the genera *Achromobacter*, *Ochrobactrum*, and *Hydrogenophaga*. Noticeably, unclassified *Rhizobiaceae* bacteria (strains KAs5-23, KAs5-25, KAs3-22, and KAs3-25), *Ochrobactrum* spp. (strains KAs5-19, KAs3-16, KAs3-21, and KAs3-24), and *Achromobacter* spp. (strains KAs5-10, KAs5-12, and KAs3-6) were found to be positive for siderophores, acid phosphatase, and arsenite oxidase or arsenate reductase activities.

With respect to utilization of various inorganic TEAs during anaerobic growth, half of the isolated strains showed their capability to metabolize As^{5+} followed by Se^{6+}, $S_2O_3^{2-}$, and Fe^{3+}. Anaerobic growth using As^{5+} as TEA was found to be relatively frequent within a few genera isolated from specific samples (e.g., *Rhizobium*, *Rhodococcus* strains, and *Bacillus* strains). Ability to use Fe^{3+} as TEA was mainly present in *Rhodococcus*, *Pseudomonas*, *Brevundimonas* *Rhizobium*, *Staphylococcus*, and *Phyllobacterium* strains. In order to ascertain the possible linkage between the taxonomic groups and observed metabolic abilities, Sarkar et al. (2014) used a two-component PCA (Figure 20.3).

The analysis revealed that with respect to the selected metabolic traits, the test bacteria formed a noncoherent cluster branching out from the group of type strains. The data indicated that the repertoire of metabolic versatility of such bacterial strains from the As-rich groundwater may not be related to their taxonomic affiliation. The presence of As homeostasis genes was also observed within the strains. It was found that more than 50% strains possess cytosolic As^{5+} reductase gene *ars*C and As^{3+} transporter gene *ars*B, whereas nearly 10% strains contain As^{3+}-oxidizing gene

*aio*B. Nearly 50% of the isolated As-resistant strains belonging to the genera *Pseudomonas, Rhizobium, Brevundimonas, Rhodococcus*, etc., showed presence of cytosolic As^{5+} reductase (*ars*C) gene. Similarity and conservedness among the sequences of *ars*C gene obtained from *Acinetobacter* BAS123i, *Arthrobacter* CAS410i, *Pseudomonas* CAS907i, *Rhodococcus* CAS922i, and *Staphylococcus* CAS108i showed their highest identity and lineages with As^{5+} reductase of several *Proteobacteria* (*Vibrio* sp., *Escherichia coli*, *Acinetobacter* sp., *Ochrobactrum* sp., *Pseudomonas*, etc.).

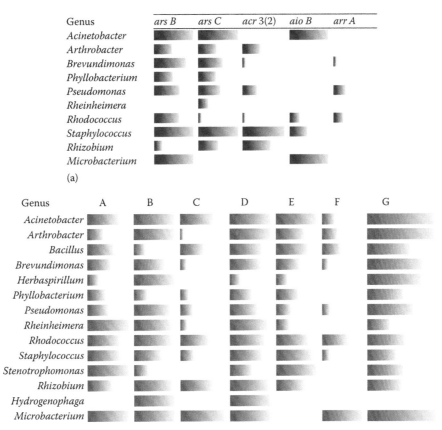

(a)

A, Anaerobic growth; B, As^{3+} aerobic tolerance; C, As^{3+} anaerobic tolerance; D, As^{5+} aerobic tolerance; E, As^{5+} anaerobic tolerance; F, As^{3+} oxidase; G, As^{5+} reductase.

(b)

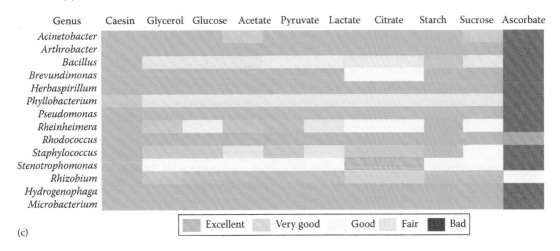

(c)

FIGURE 20.2 (a–d) Metabolic diversity of bacterial strains commonly detected in the western part of BDP. *(Continued)*

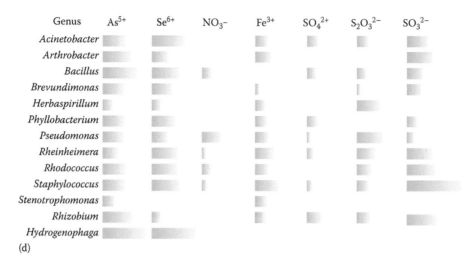

(d)

FIGURE 20.2 (*Continued*)   (a–d) Metabolic diversity of bacterial strains commonly detected in the western part of BDP.

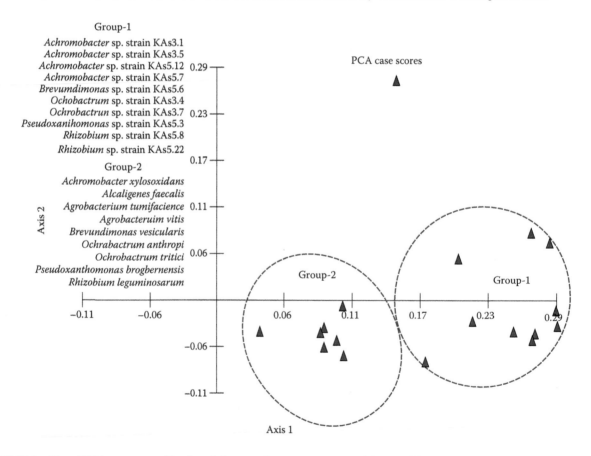

FIGURE 20.3   Plot of PCA scores on utilization of alternate electron acceptors and donors of few commonly occurring strains in As-rich groundwater of West Bengal and related type of strains. (Reproduced from Sarkar, A. et al., *Environ. Sci. Pollut. Res.*, 21, 8645, 2014. With permission.)

Gene coding for arsenate transporter (*ars*B) was also detected in nearly half of the populations studied and was mostly identical to the similar gene possessed by the members affiliated to the genera *Pseudomonas, Brevundimonas, Rhodococcus, Staphylococcus*, etc. Sequences of *ars*B gene obtained from *Brevundimonas* CAS4005i, *Acinetobacter* BAS108i, *Rhodococcus* CAS922i, and *Arthrobacter* CAS411i showed

high relatedness with putative As^{3+} efflux pump protein from *Serratia, Staphylococcus*, etc. Arsenite transporter gene *acr3(2)* was detected in nearly 2/3 of the strains belonging to the genera *Pseudomonas, Staphylococcus, Arthrobacter, Rhizobium*, etc. Sequences of *acr3(2)* gene detected in these strains showed lineages with As^{3+} transporters/efflux pumps reported from genera *Ochrobactrum, Rhizobium*, etc. The *aio*B gene encoding As^{3+} oxidase was present in five bacterial strains affiliated to the genera *Rhodococcus*, *Acinetobacter*, *Staphylococcus*, and *Microbacterium*. Nucleotide sequences of this gene showed their closeness with As^{3+} oxidase of *Bacillus* and *Achromobacter* and with several uncultured β- and α-proteobacterial members. Dissimilatory As^{5+} reductase gene *arr*A was relatively less abundant and could be amplified successfully from only seven strains affiliated to the genera *Rhodococcus*, *Pseudomonas*, and *Brevundimonas*. The sequence obtained from *Rhodococcus* CAS922i revealed its close lineage with arrA of *Shewanella* sp. Noticeably, a number of strains showed simultaneous presence of *arsC*, *arsB*, and *acr3(2)* or *arsC* and *arrA* genes. Possible relation among the presence of all these genes detected within the bacterial strains and their As resistance and transformation phenotypes were studied by Unweighted Pair Group Method with Arithmetic Mean (UPGMA) (Figure 20.4). The presence of *arsC* gene was found to be strongly related to As^{5+} reductase activity, and together with *acr3(2)* gene, a close agreement between As resistance, transformation, and the presence of both these As-related genes was evident.

The discrepancy between 16S rRNA gene-based evolutionary relationship and phylogeny of As homeostasis genes [*aioB*, *arsC*, *arsB*, *acr3(1)*, and *acr3(2)*] was observed while studying the phylogeny of As-related genes (Sarkar et al., 2014; Paul et al., 2015). It was suggested that the As-related genes were most likely acquired by HGT. Guanine-cytosine (GC) content of As homeostasis genes was compared with the characteristic GC% range for reference genomes of closest taxa. The comparison further supports our phylogenetic interpretation with respect to the occurrence of horizontal transfer of *arsC*, *arsB*, and *acr3* genes among the members of the community. For instance, the GC% values calculated for *arsC* genes detected in our strains (affiliated to the genera *Brevundimonas*, *Ochrobactrum*, and *Rhizobium* of α-Proteobacteria, *Achromobacter* of β-Proteobacteria, and *Pseudoxanthomonas* of γ-Proteobacteria) were close to the characteristic GC% (45%–55%) of γ-proteobacterial members *E. coli* or *Vibrio* sp. Similarly, discrepancy in GC content of reference genome average and that of either *arsB* gene from *Rhizobium* spp. strains KAs5-8 and KAs5-22 and *Ochrobactrum* sp. strain KAs3-7 or *acr3(1)* gene from all the strains except *Brevundimonas* sp. strain KAs5-6 also supported the incidence of HGT. Regarding the *acr3(2)* gene, phylogenetic incongruence was observed only in the case of *Achromobacter* spp., which corresponded with the discrepancy among GC content of this gene and the reference genome average for the genus *Achromobacter*. Similarity between *aioB* gene from *Rhizobium* sp. strain KAs5-8 and *Achromobacter* sp. in terms of sequences homology as well as GC content also indicated possible transfer of *aioB* gene from Achromobacter sp. to *Rhizobium* sp. Overall, the metabolic landscape of bacterial communities in As-contaminated groundwater portrayed a number of very relevant properties, including the ability to withstand toxic As even at anaerobic condition, abundance of As^{5+} reductase activity and relevant genes, utilization of As^{5+} as TEA, and use of multiple organic compound including complex aromatic and aliphatic hydrocarbons as sole C sources.

The observed metabolic versatility of indigenous bacterial community seems remarkable. Members of the genera *Acinetobacter*, *Brevundimonas*, *Pseudomonas*, and *Rhizobium* isolated earlier from various As-contaminated groundwater have been shown to utilize multiple electron donors and/or acceptors (Gihring et al., 2001; Ghosh and Sar, 2013; Sarkar et al., 2014). Along with superior As resistance, the presence of As^{5+} reductase activity as detected in the test isolates corroborated earlier reports on bacterial strains isolated from As-rich groundwater of BDP and elsewhere (Liao et al., 2011; Sarkar et al., 2013). The presence and abundance of As-transforming ability within the indigenous bacteria of As-rich groundwater are considered to be an evolutionary outcome to allow aerobic/facultative anaerobic metabolism within the alluvial aquifers (Liao et al., 2011). Abilities of the test isolates to utilize multiple sugar molecules, different hydrocarbon compounds as their energy and/or carbon source, and use of alternate electron acceptor(s) during anaerobic condition clearly indicate that the indigenous organisms are well equipped to survive and flourish under As-rich oligotrophic condition.

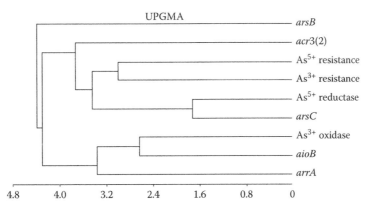

FIGURE 20.4 Correlation among various metabolic properties and As-related genes detected within bacterial strains from western part of BDP. (Courtesy of Paul, D. et al., *PLoS One*, 10, 0118735, 2015.)

20.5 MICROBIAL DIVERSITY, ACTIVITY, AND ROLE IN AS-CONTAMINATED AQUIFERS

Microbial communities within the As-contaminated aquifers of BDP and several other places are explored by analyzing groundwater and sediment samples collected from parts of Bangladesh (Islam et al., 2004; Sutton et al., 2009; Liao et al., 2011; Sultana et al., 2011). A few studies have used samples from West Bengal and Assam (Ghosh and Sar, 2013; Sarkar et al., 2013; Paul et al., 2014b). Characteristics of bacterial populations and/or their potential role in As mobilization within the subsurface environment are mainly investigated by monitoring the population shifts using laboratory-based microcosm experiments and/or culture-based approaches. It is observed that the most abundant populations in different communities are represented by members of *Proteobacteria*, particularly of class β > γ > α-*Proteobacteria* along with *Bacteroidetes*, *Actinobacteria*, *Firmicutes*, etc. Identification of bacterial groups at lower taxonomic level revealed the predominance of a number of genera (viz., *Pseudomonas*, *Rhizobium*, *Methylophilales*, *Burkholderiales* of γ-*Proteobacteria*, α-*Proteobacteria*, and β-*Proteobacteria*) representing the major populations. Abundance of the genera *Thiobacillus*, *Herminimonas*, *Acidovorax*, *Hydrogenophaga*, and *Gallionella* (members of β-*Proteobacteria*); *Pseudomonas*, *Methylophaga*, *Enterobacter*, and *Actinobacteria* (members of γ-*Proteobacteria*); and *Symbiobacterium*, *Planococcus*, and *Bacillus* (members of *Firmicutes*) along with *Planctomyces*, *Acidobacteria*, *Verrucomicrobiales*, *Acitinobacter*, *Sulfuricuruum*, *Thermotogae*, and Fe-reducing *Geobacter* have been noted in other sites (Jiang et al., 2014) (Figure 20.5).

Studies done in the authors' own laboratory have revealed metabolic and taxonomic details of bacterial communities present in contaminated groundwater and their biogeochemical significance. Clone library, DGGE, and culture-based studies indicated diverse, yet near-consistent, community composition represented by the genera *Pseudomonas*, *Flavobacterium*, *Brevundimonas*, *Polaromonas*, *Rhodococcus*, *Methyloversatilis*, and *Methylotenera*. Culture-dependent analysis (using more than 300 bacterial strains) showed abundance of the genera *Arthrobacter*, *Agrobacterium–Rhizobium*, *Achromobacter*, *Acidovorax*, *Ochrobactrum*, *Pseudomonas*, *Brevundimonas*, *Rheinheimera*, *Rhodococcus*, etc. Coupled with high resistance to As (either to As^{3+} or to As^{5+} or both), abundance of As^{5+} reductase activity was detected within the isolated strains. With respect to utilization of different TEAs, 50% of the strains showed their ability to use As^{5+} followed by Se^{6+}, S$_2$O$_3^{2-}$, Fe^{3+}, and HSO$_4^-$. Ability to utilize different carbon sources ranging from C2 to C6 compounds including a few complex sugars as well as long-chain aliphatic and aromatic hydrocarbons was observed. Genes encoding As^{3+} oxidase (*aioA*), As^{5+} reductase (*arsC*), and As^{3+} efflux pump (*arsB* and *acr3*) were detected within the test isolates. The presence of dissimilatory As^{5+} reductase gene (*arr*) was found in relatively fewer strains. Along with diverse metabolic abilities, many bacterial strains showed their ability to use petroleum-derived hydrocarbon as an electron donor and transform As. This could play a critical role in As biogeochemical cycling within aquifer. Potential role of the bacteria in As mobilization and immobilization was studied with different sets of sediment and groundwater microcosms.

Considering the overall ecophysiological and molecular properties of bacterial populations inhabiting As-contaminated groundwater and prevailing geochemical conditions in a subsurface aquifer, a synergistic microbial activity leading toward As mobilization has been envisaged (Sarkar et al., 2013). We have hypothesized that under low-nutrient aquifer condition, autochthonous bacterial members capable of growing chemolithotrophically catalyzed important reactions mediated by siderophore, phosphatase, and As^{5+} reductase activity. Sarkar et al. (2013) reported through their experiments that siderophore and phosphatase produced by bacteria act on insoluble mineral phases allowing their availability/acquisition to/by the bacteria, thereby facilitating the release of mineral-bound As^{5+} to soluble phase. The released As^{5+} is reduced to As^{3+} by *ars* system. The toxic effect of As^{3+} is circumvented by specific efflux pumps [*arsB* and/or *acr3*p] along with As^{3+} oxidase activity. A combined role of bioweathering activity, anaerobic As^{5+} and Fe^{3+} reduction, and As^{5+} detoxification is hypothesized to explain the mechanistic role of indigenous bacteria in releasing As into the groundwater.

20.5.1 Comparing the Microbiology of Ganga and Brahmaputra Basins

The present study aims at enumerating and comparing the dominant bacterial genera isolated from five locations of the Ganga–Brahmaputra–Meghna deltaic region, namely, Barasat, Chakdah, Kolsur, Brahmaputra South (Jorhat), and Brahmaputra North (Arunachal Pradesh). Distribution of dominant bacterial genera found in As-contaminated aquifers of BDP of Indian subcontinent has been shown in Figure 20.6.

20.5.2 Selection of Bacterial Genera

In this study, selection of bacterial genera to study their interrelationship depending on their distribution has been done on the basis of their frequency of occurrence across the As-contaminated groundwater of Barasat, Chakdah, and Kolsur, and West Bengal as well as from north and south of river Brahmaputra. Nine dominant genera were selected, and their abundance has been compared with the same as previously found at other As-contaminated aquifers. Members of α *Proteobacteria*, *Brevundimonas*, and *Rhizobium* dominated in most of the samples, followed by γ- and β-proteobacterial members *Pseudomonas*, *Acidovorax*, *Hydrogenophaga*, *Acinetobacter*, *Rhodococcus*, *Arthrobacter*, and few *Firmicutes* such as *Bacillus*. Among these genera, *Brevundimonas* and *Rhizobium* were previously found in Australia and Italy (Santini et al., 2000; Drewniak et al., 2008; Cavalca et al., 2013); *Acidovorax*, *Hydrogenophaga*, and *Acinetobacter* were reported from China (Duan et al., 2013). *Actinobacteria* and *Firmicutes* were sparsely distributed but are ubiquitous in most of the samples. Figure 20.6 shows percentage abundances (in logarithmic scale) of the

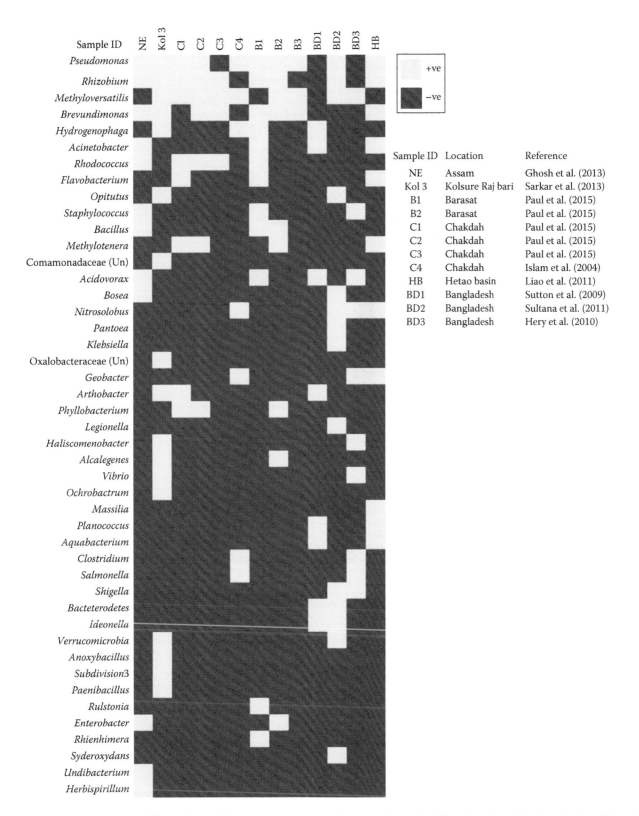

FIGURE 20.5 The presence of diverse bacterial genera across the arsenic-contaminated aquifers throughout Southeast Asia; yellow and red boxes indicate the presence and absence, respectively. Sample locations: NE, North East India; Kol, Kolsure (WB); C1–4, Chakdah (WB); B1-3, Barasat (WB); BD1, 2, 3, Bangladesh samples; HB, Hetao Basin (China).

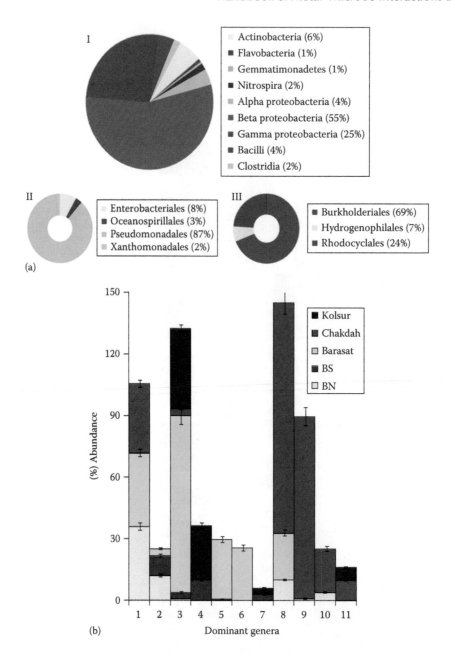

FIGURE 20.6 Distribution of dominant bacterial genera found in As-contaminated aquifers of BDP of the Indian subcontinent (a) I. Percentage abundances of bacterial classes in As-contaminated groundwater, II. Abundance of bacterial orders within the class γ-*Proteobacteria*, III. Abundance of bacterial orders within the class β-*Proteobacteria*. (b) Percentage abundances of dominant bacterial genera: 1. *Brevundimonas*, 2. *Acidovorax*, 3. *Rhizobium* 4. *Acinetobacter*, 5. *Bacillus*, 6. *Phyllobacterium*, 7. *Hydrogenophaga*, 8. *Pseudomonas*, 9. *Rhodococcus*, 10. *Herbaspirillum*, and 11. *Arthrobacter* at five different locations of BDP in India.

bacterial genera at five sampling sites (data collected from Ghosh and Sar, 2013; Paul et al., 2015; Sarkar et al., 2016).

20.5.3 PHYLOGENETIC STUDY

16S rRNA gene sequences retrieved from NCBI and RDP searches from other As-contaminated sites and those isolated from the five sampling sites of GBM basin were subjected to phylogenetic analysis using neighbor-joining tree and corresponding distance matrices. Phylogenetic tree and distance matrices were prepared using Molecular and

Evolutionary Genetic Analysis (MEGA) six following the Jukes Cantor model. Distance matrices were built to ascertain the exact distances (D values) among the sequences of our isolates and similar isolates reported from other sites elsewhere. Figure 20.7 represents phylogenetic relatedness of the dominant classes of bacteria and their genera found in the GBM basin.

The phylogenetic lineage and evolutionary distance among diverse bacterial genera may also have implications on their functional relatedness. *Acinetobacter* KF442760 and KF220447 isolated from Barasat and north of Brahmaputra

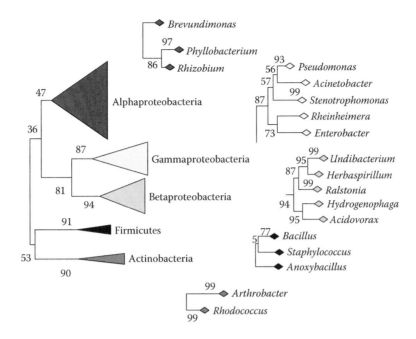

FIGURE 20.7 Neighbor-joining phylogenetic tree showing relatedness among class levels and their reported genera in As-contaminated aquifers of BDP.

lie in the same phylogenetic clade and show a distance value of 0.023 among them. These isolates show close proximity to *Acinetobacter* SeAH-As2w FJ607348 isolated from As-contaminated coastal sediments along the Gwangyang Bay (Chang et al., 2011) with a distance value of 0.062. *Acinetobacter* SeAH-As2w FJ607348 contains *aox* genes and is capable of oxidizing As^{3+} to As^{5+}. *Pseudomonas* isolated from Barasat and Chakdah were highly similar along with few isolated from north of Brahmaputra as they fall within a single clade with negligible D values ranging from 0.005 to 0.040, but a high distance is found for *Pseudomonas* BNA21-65 and BNIIIA2-7 (0.08, 0.1) with the isolates mentioned earlier. These isolates fall in a completely different clade to that of *Pseudomonas* SeAH- As4s, which is found to be close only to *Pseudomonas* BAS232i. *Pseudomonas* SeAH-As4s FJ607351 is found to contain *aox*B. *Pseudomonas* found from other sites (Cai et al., 2009) has been reported to have *aox* cluster (*arsDarsA-aoxA-aoxB*) and the As resistance *ars* gene cluster (*arsC1-arsR-arsC2-acr3-arsH*). Another study (Chang et al., 2010) reports a bacterial isolate identified to be *Pseudomonas stutzeri* strain GIST-BDan2 (EF429003) showing similar oxidation properties with *aoxB* and *aoxR* genes. It is notable that *Pseudomonas* could not be isolated from two sites: Kolsur and south of Brahmaputra.

Members of β-*Proteobacteria* could be isolated from all the five sites. *Acidovorax* from north and south of Brahmaputra are phylogenetically well related to the least distances ranging from 0.002 to 0.01 but show a distance in the range of 0.02 with *Acidovorax* BAS309i (KF442763), isolated from Barasat. All these bacteria are well in compliance with *Acidovorax* GW2 (EF550172), which is reported to be an As^{3+}-oxidizing bacteria (Fan et al., 2008). Few other studies report this group of bacteria to be As^{3+}-oxidizing denitrifying

bacteria in anoxic condition (Sun et al., 2009). Others have shown *Acidovorax* to be a well-known aerobic nitroaromatic-degrading bacterium (Lessner et al., 2003). *Hydrogenophaga* found from Kolsur and Chakdah are phylogenetically very close to each other with very high sequence similarity (D = 0.005–0.007). Although these strains fall in the same clade with *Hydrogenophaga* YED-18 with a bootstrap value of 100, they are diverse as the evolutionary distance found among them (0.025–0.032) is considerable when compared with the same between Kolsur and Chakdah. Strain YED-18 has been seen to oxidize As^{3+}. Another report from Australian gold mines shows *Hydrogenophaga* NT-14 to be capable of arsenite oxidation (Hoven and Santini, 2004). *Hydrogenophaga* has been reported to be "knallgas" bacteria or facultative autotrophs, capable of fixing CO$_2$ while using H$_2$ and O$_2$ as an electron donor and acceptor, respectively (Aragno and Schlegel, 1992). *Rhizobium* and members of *Rhizobiaceae* isolated mostly from Kolsur and few from Barasat and Brahmaputra South are in well compliance with each other as they fall in the same phylogenetic clade with the least distance values around 0.00–0.007, except two. Isolates JX110534 and KF442778 are much distant to others with D values of 0.07–0.08 and 0.02–0.03, respectively. These *Rhizobia* are least related to arsenite-oxidizing chemolithotrophic *Rhizobium*-NT26 (Santini et al., 2000) as they show maximum distance of 0.08, although they root at the same clade with high bootstrap value of 82. *Rhizobium* has been well established to be an As^{3+} oxidizer with *aox* and *arsR* genes in them in various studies. Along with such properties, they have also been found to possess *nod*C gene showing symbiotic activity with leguminous plants (Mandal et al., 2008, Santini et al., 2000, Carrasco et al., 2005). Most of the sequences representing *Brevundimonas* were isolated from north of

Brahmaputra and few from Barasat and Chakdah. All these sequences are highly similar and also share sequence similarity with *Brevundimonas* sp. (EF491966) (Drewniak et al., 2008). Not much has been reported about *Brevundimonas* in the aspect of environmental As, but one among the very few shows hypertolerance of this bacteria to As and occurs in highly As-contaminated sites of Hetao Basin, Mongolia, with high levels of methane and Fe(II) and low SO_4^{2-} and NO_3^- (Li et al., 2013). This observation well corroborates our findings of high As tolerance of this bacteria.

Among the members of class *Actinobacteria* and *Rhodococcus* isolated mainly from Chakdah and Brahmaputra North samples and that reported by Drewniak et al. (2008) are highly similar as they fall in the same clade with bootstrap value of 100 and D values of 0. *Rhodococcus erythropolis* strain OS-1 is reported to be As tolerant. Another study (Bag et al., 2010) reported *Rhodococcus equi* to be an As-tolerant bacteria and proposed its use as a biosorbent from As-contaminated water. Several species of *Rhodococcus* have been reported to hydrolyze hydrocarbons such as dibenzothiophene, dibenzofuran, hexane, methyl-S-triazenes, nitriles, nitrophenol, and polychlorinated biphenyls (Alverez, 2010). On the other hand, *Arthrobacter* isolated from Chakdah and Kolsur are relatively unrelated with considerable distance values of 0.10–0.15. They are also found to be distant from previously reported As-tolerant *Arthrobacter* sp. TS-18 (Cai et al., 2009). *Arthrobacter* is a well-known As^{3+} oxidizer, which can also be used for biosorption purpose (Prasad et al., 2013). Duan et al. (2013) have also reported high oxidation rates of As^{3+} to As^{5+} by an *Arthrobacter* sp. isolated from Datong Basin, China.

Dominant members of *Firmicutes* were represented only by *Bacillus* species. Although few, they might play important roles in As dissemination in aquifers of the GBM delta region. *Bacillus* isolated from Chakdah and *Bacillus arsenicus*–type strain show high-sequence similarity and phylogenetic closeness among themselves. *Bacillus* strain isolated from Brahmaputra South and *Bacillus* AR-9 reported by Liao et al. (2011) depict great variation up to D values till 1.5 with the isolates mentioned earlier. But these two strains are very close to each other with negligible D values. *Bacillus* AR-9 has been reported to possess *arsC* gene conferring As^{5+} resistance. *Bacillus selenitireducens* has been shown to be a potent As^{5+} respirer (Afkar et al., 2003). Other novel species such as *Bacillus indicus* reported by Suresh et al. (2004) from Chakdah, West Bengal, were found to be As resistant.

20.6 INSIGHTS ON AS–BACTERIA INTERACTION FROM WHOLE GENOME SEQUENCING

In the recent past, knowledge on bacterial metabolism of As as well as other heavy metal evolution of resistance genes and their regulation has been the center of attraction for microbiology researchers. Since the microbial potency of tolerating the toxic effects of the heavy metals is deep beneath their genomes, it is exciting to uncover the very amazing genomes of such microbes, which is termed as genomics. In the context of genomics, whole genome sequencing is of high value for understanding how bacteria function, evolve, and interact with each other with their ecosystems, thus providing numerous avenues for newer genomic functions (Mueller et al., 2007). Bacterial genome sequencing work is now nearly two decades old. With reference to microbiology of As, the combination of genome sequencing and computational-driven analysis of sequence data (bioinformatics) has transformed our understanding on the mechanism of As homeostasis and genomic plasticity for better niche colonization (Mueller et al., 2006). The process of sequencing has been remarkable these days so that sequencing projects that used to take years and cost hundreds of thousands of dollars can now be completed in a few days or even in hours. Currently, there are many platforms available in the market launched by various sequencing industries, aimed for different purposes like whole genome sequencing, metagenome sequencing, exome and targeted gene sequencing, and amplicons sequencing, as well as total and meta-RNA sequencing (RNA-Seq). The platforms are commonly known as next-generation sequencing platforms (NGS). Prior to the use of NGS, genome sequencing was burdened high cost, was labor-intensive and time-consuming with Sanger's protocol. Roche 454 sequencing system was the first NGS technology based on pyrosequencing (Margulies et al., 2005), followed by Illumina and Ion Torrent. Illumina was originally developed by Solexa and is based on bridge amplification (Bentley, 2006) and is provided as benchtop versions. Ion Torrent (Life Technologies) uses a different technology called emulsion PCR for template amplification and semiconductor method that uses a change in pH as a detectable signal instead of light. Apart from NGS, third-generation technologies (provided by Pacific Biosciences [PacBio], Oxford Nanopore Technologies) are becoming widely used. PacBio uses a single-molecule real-time sequencing with no template amplification that occurs in a zero mode waive-guide after-bell template library preparation with long sequencing reads of greater than 10Kb (Eid et al., 2009). The details of the sequencing technology platforms, their run time, and necessary details are summarized in Table 20.2.

With an accuracy of 99%, these NGS platforms provide massive amount of data with cheaper cost. The output provided by various NGS technologies is called data (mostly in Mb [million base] or Gb [gigabase]) and in the form of reads (nucleotides with quality scores) as FASTQ files. The data are processed and assembled into contigs of high accuracy following specific protocol called pipelines (Figure 20.8).

Through various software and computational skills, the data are generally analyzed for interpretation. Along with the development of genome sequencing technologies, massive computational advancement in the analysis of genomes has been achieved. There are numerous numbers of online (web-based) as well as system-based tools and software for processing the annotated genomes (Table 20.3). With the advent of these technologies, whole genome sequencing has become accessible, affordable, and emerging as a critical tool for gaining better understanding on microbial genomes and molecular details of metabolic capabilities.

TABLE 20.2

Basic Features of Next-Generation Sequencing Technologies for Whole Genome Sequencing

Platforms Available	Run Time	Read Length	Yield per Run	Throughput
Roche				
454 GS Junior	10–12 h	400–500 bp	~50 Mb	~1
454 FLX Titanium XL+	~23 h	700–1000 bp	~1 Gb	~20
Illumina				
GAIIX	14 days	2 × 150 bp	80–90 Mb	170–190
MiSeq	~39 h	2 × 250 bp	7.5–8.5 Gb	15–17
HiSeq	~40 h	2 × 150 bp	75–85 Gb	160–180
NexSeq	~11–28 h	2 × 75–2 × 150 bp	25–120 Gb	>200
Life Technologies				
Ion PGM system (316 chip V2)	4.9 h	400 bp	600–1000 Mb	12–20
Proton PGM system (PI chip)	2.5–4 h	200 bp	8–10 Gb	~200
Ion S5	2.5–5 h	200/400 bp	15 Gb	>200
Pacific Bioscience				
PacBio RSII	~2.5 h	3–5 kbp (up to >20 kb)	~500 Mb (100×)	~5
Oxford Nanopore technology (MinIon): yet to come in the market.				

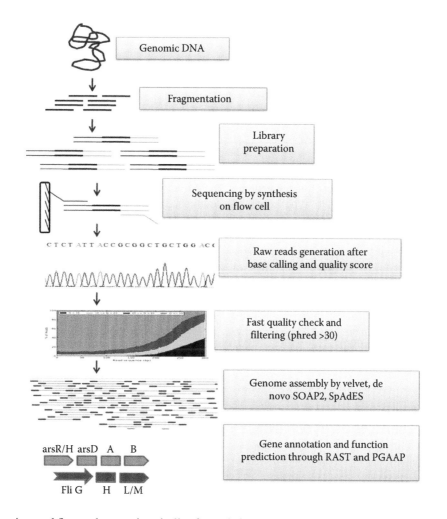

FIGURE 20.8 Sequencing workflow and annotation pipeline for a whole genome sequence analysis of bacterial genomes.

TABLE 20.3

Online Analysis Tools and Software with Their Functions Used for Genome Analysis

Online Analysis Tools and Software	Function
RAST	Automated annotation and function prediction of the genome contigs
AmiGene	Annotation of microbial genes for automatically identifying the most likely coding sequences.
Biology WorkBench	Blast, multiple sequence alignment, randomize sequence
Count codon	Codon usage analysis
Cpgreport	Identify and report CpG islands in nucleotide sequences
FMAP	Sequence alignment with genome
GeneMark	Gene prediction in bacteria, archaea, and metagenomes
Genscan	Predicting the locations and exon–intron structures of genes in genomic sequences from a variety of organisms
make-na	Automatic production of DNA double helix with sequence input
Molecular Toolkit	Programs for analysis and manipulation of nucleic acid and protein sequence; includes dot plot similarity matrix, inverse, complement, and double-stranded conversions
Open reading frame (ORF) finder	ORF finder
Pictogram	Tool to visualize sequence alignments and consensus sequences showing the relative frequencies of the bases at each position
PlasMapper	Generates and annotates plasmid maps using only plasmid DNA sequence as input
Restriction Mapper	Maps sites for restriction enzymes (restriction endonucleases) in DNA sequences
RNA/DNA secondary structure prediction	PairFold predicts the minimum free energy secondary structure formed by two input DNA or RNA molecules
SOAP Short Oligonucleotide Analysis	Program for faster alignment of short oligonucleotides onto reference sequences for next-generation sequencing data analysis
Trait-o-matic	Open-source tool to find and classify phenotypic correlations for variations in whole genomes
Vista	Programs and databases for comparative analysis of genomic sequences
WebACT	Online visualization of comparisons between up to five prokaryotic genome sequences, using Artemis Comparison Tool
WEGO	Web Gene Ontology Annotation Plot
MobilomeFINDER	In silico and experimental discovery of bacterial genomic islands
SyntTax	Linking synteny to prokaryotic taxonomy
AutoGRAPH	Web server for multispecies comparative genomic analysis
CGView Server	Visualization of the circular genome
ResFinder 2.1	Identification of acquired antimicrobial resistance genes

In the prokaryotic domain, a total of 20,584 eubacteria and 907 archaebacteria have been described, out of which genome of 9,966 bacterial (nontype) strains, 3,890 type bacterial strains, and 210 archaebacterial strains have been sequenced (Table 20.4).

Compared to bacterial strains from diverse habitats and metabolic abilities, only 38 whole genomes of As-metabolizing strains (NCBI Taxonomy and genome browser, http://www.ncbi.nlm.nih.gov/Taxonomy/taxonomyhome.html/; RDP hierarchy browser http://www.ncbi.nlm.nih.gov/genome, https://rdp.cme.msu.edu/hierarchy/hb_intro.jsp) have been sequenced till date, the complete genome of the As-oxidizing bacterium *H. arsenicoxydans*

(a heterotroph from industrial wastewater plant contaminated with As and other heavy metals) being the first to be published (Arsène-Ploetze et al., 2010). Till date more than 20 whole genomes of As-oxidizing and As-reducing strains have been publicly available in the database, isolated from diverse environments with diverse taxonomic groups. Gene for As oxidation (*aioA*) and efflux (*ars*), chemotaxis and motility, production of exopolysaccharides to bind As, and DNA repair caused by As^{3+} were the most distinguished features in the genome of *Herminimonas* (Mueller et al., 2006, 2007). A few years later, the genome of the bacterium *Thiomonas* sp. 3, a facultative chemolithoautotroph (which was isolated from acid mine drainage containing high As), was sequenced

TABLE 20.4

Current Status of Prokaryotic Taxonomy and Genome Sequencing (as of 2015)

Domain	Published Names	16S rRNA Gene Sequence (Type Strains)	Genome Sequence (Any Strain)	Genome Sequence (Type Strain)	Genome Sequence (As Metabolizing)
Bacteria	20,584	10,132	9966	3890	38
Archaea	907	403	150	60	NA
Total prokaryote	21,491	10,535	4890	4188	—

(Lin et al., 2012). A complete operon for As detoxification (*arsCABR*) and oxidation (aioBA) was highlighted and comparative genomics of eight different *Thiomonas* strains showed the resolution of genomic islands, which were found to be evolved and influenced by the extreme conditions of the habitat (either heavy metal stress or antibiotics) (Bryan et al., 2009, Cleiss Arnold et al., 2010). The same As genome island was also discovered in the genome of As-oxidizing strain *Achromobacter arsenitoxydans* SY8, which was isolated from As-contaminated soil of a pig farm. Along with it, genes for As resistance (*ars* operon), As^{3+} oxidation (*aio* operon) and phosphate uptake (*pst* operon), and metal transporters (metal resistance) were the most significant findings that came out of the sequencing project. The same research group sequenced the genomes of *Halomonas* sp. strain HAL1 and *Acidovorax* sp. strain NO1, which were both isolated from As-contaminated soil of a gold mine and found similar operons to *A. arsenitoxydans* and *Rhizobium* NT strains (Andres et al., 2013; Kruger et al., 2013). A whole genome sequencing of As (III)-oxidizing strain *A. tumefaciens* strain 5A revealed the expression of *aio* operon and the possible pattern of its regulation by a two-component signal transduction system and by quorum-sensing mechanisms (Hao et al., 2012). Till date, several As-metabolizing strains (*Lysobacter* sp., *Lysinibacillus* sp., *Rhizobium* sp., *Shewanella* sp., *Exiguobacterium* sp., *Bacillus* sp., *Marinobacter* sp, *Chrysiogenes* sp., *Sulfurospirillum* sp., etc.) have been sequenced and unconventional findings have emerged, with newer genetic mechanisms of tackling environmental stress and strategies for reclamation of the contaminants. A comparison of whole genomes also enable us to find gene order; gene presence; regulation events like insertion or deletion, which frequently happens in bacterial genome; and mobile

genetic elements like plasmid, transposons, integrons, insertion sequences, and genomic islands (Figure 20.9), as well as status of the HGT (Binneweise et al., 2006).

With the developments of high throughput DNA sequencing coupled with bioinformatic pipelines for downstream sequence assembly-annotation, genome and metagenome sequencing has moved far away for delineating the elaborative and unconfined taxonomic positions of bacterial species termed "phylogenomics." The recent availability of complete sequences of a number of bacterial genomes has made it possible for the first time to study the genetic and functional relatedness between organisms at a whole-cell level and hence to provide novel insights into the issues related to the identification of species through the 16S rRNA-based system. However, genomic studies to date have mostly been focused on assessing the accuracy of phylogenetic reconstruction, particularly in the light of HGT in all prokaryotic taxa. Since genome sequencing is now affordable and accessible to general microbiology laboratories, it has been suggested that comparison of whole genome sequence data, as a form of digital, in silico DNA–DNA relatedness between several taxa, may be considered as the most reliable method for taxonomic purposes (Konstantinidis and Tiedje, 2005). Average nucleotide identity is another way of representing relatedness between two microbial candidates that represent a mean of identity values between multiple sets of orthologous regions shared by two genomes. The use of genome BLAST distance phylogeny (GBDP) is a distance-type genome relatedness index (Henz et al., 2005) that also counts as a reliable method for assessing species relatedness along with the determination of G% + C% content of the microbial taxa. The details of the tools used for phylogenomics have been presented (Table 20.5).

At present, a combination of whole genome sequencing, proteomic, and transcriptomic studies is providing new and

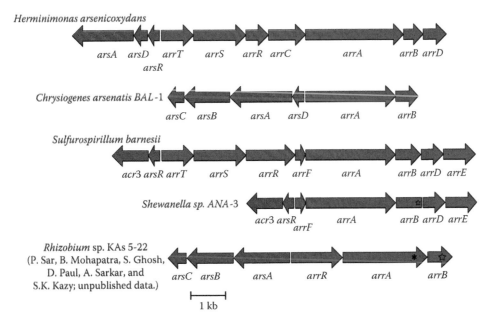

FIGURE 20.9 Arsenic genomic island and gene orientation of annotated whole genome of various As-metabolizing strains. Asterisks indicate the detected putative conserved domain part of the arsenate respiratory reductase subunits. Scale bar 1 kb.

TABLE 20.5

Bioinformatic Tools and Resources for Genome-to-Genome Comparison for Taxonomic Purposes

Genome Relatedness Parameters	Threshold for Species Demarcation	Tools for Genome Analysis (URL)	Description and URL
ANI	95%–96%	JSpecies (http://www.imedea.uib.es/jspecies/) EzGenome (http://www.ezbiocloud.net/ezgenome/ani)	JAVA-based (Richter and Rosselló-Móra, 2009)
GBDP	0.258	Genome-to-genome distance calculator (http://ggdc.dsmz.de/)	Web-based service for the calculation of pairwise GBDP distance using NCBIBLAST, BLAST+, BLAT, BLASTZ
MUMi	Not available	MUMi (http://genome.jouy.inra.fr/mumi/)	Web-based tool for the calculation of MUMi
Nucleotide identity	96.5%	specI (http://vm-lux.embl.de/~mende/specI//)	Web-based server for the analysis of 40 single copy marker genes in bacterial domain

a more detailed information on microbial metabolism and interaction with As, and a new field of As–microbe interaction with omics concept "arsenomics" is proposed (Kruger et al., 2013; Sacheti et al., 2014). Considerable progress has also been made in genomic science to decipher the As stress physiology, but the interplay of gene transcription and expression pattern is the foremost cause of As homeostasis in the contaminated ecosystem. Hence, the use of proteomics and transcriptomics is the most needed tool as a complementary study with the genomics. The proteomic approach has been used to study the response of arsenic in *Pseudomonas, Leptospirillum, Herminiimonas, Klebsiella, Ferroplasma, Rhizobia, Chromobacterium, Caenibacter,* and *Comamonas* (all Gram-negative organisms) and *Staphylococcus* sp. and *Exiguobacterium* sp. (Gram-positive organisms) (Carapito et al., 2006; Baker-Austin et al., 2007; Patel et al., 2007; Zhang et al., 2007; Mandal et al., 2009; Weiss et al., 2009; Li et al., 2010; Ciprandi et al., 2012; Daware et al., 2012; Srivastava et al., 2012; Sacheti et al., 2014). Hence, the marker proteins have been identified during As stress across various organisms, hierarchy (Sacheti et al., 2014). For transcriptomics (transcript expression), the study of As stress for bacterial domain has been a reliable tool for a long time, but the established conventional approach of microarray and dot blot analysis posed many drawbacks. In the near future, the combination of genomics, proteomics, transcriptomics, and metabolomics will provide a detailed and comprehensive study of an organism's behavior in As stress.

REFERENCES

Afkar, E., Lisak, J., Saltikov, C., Basu, P., Oremland, R.S., and Stolz, J.F. 2003. The respiratory arsenate reductase from *Bacillus selenitireducens* strain MLS10. *FEMS Microbiology Letters* 226: 107–112.

Allison, M.A., Khan, S.R., Goodbred, S.L., and Kuehl, S.A. 2003. Stratigraphic evolution of the late Holocene Ganges–Brahmaputra lower delta plain. *Sedimentary Geology* 155: 317–342.

Alvarez, H.M. 2010. *Biology of Rhodococcus* (Vol. 16). Springer Science & Business Media, Heidelberg, Germany.

Amend, J.P., Saltikov, C., Lu, G.S., and Hernandez, J., 2014. Microbial arsenic metabolism and reaction energetics. *Reviews in Mineralogy and Geochemistry* 79(1): 391–433.

Andres, J., Arsène-Ploetze, F., Barbe, V. et al. 2013. Life in an arsenic-containing gold mine: Genome and physiology of the autotrophic arsenite-oxidizing bacterium *Rhizobium* sp. NT-26. *Genome Biology and Evolution* 5: 934–953.

Aragno, M. and Schlegel, H.G. 1992. The Mesophilic Hydrogen-Oxidizing (Knallgas) Bacteria. In *The Prokaryotes. A handbook on the biology of bacteria: Ecophysiology, isolation, identification, applications.* 2nd edition. A. Balows, H.G. Trüper, M. Dworkin, W. Harder, K.H. Schleifer (Eds.), Vol. 1, pp. 344–384. Berlin, Heidelberg, New York: Springer-Verlag.

Arsène-Ploetze, F., S. Koechler, M. Marchal et al. 2010. Structure, function, and evolution of the *Thiomonas* spp. genome. *PLoS Genetics* 6: 1000859.

Bag, P., Bhattacharya, P., and Chowdhury, R. 2010. Bio-detoxification of arsenic laden ground water through a packed bed column of a continuous flow reactor using immobilized cells. *Soil and Sediment Contamination* 19: 455–466.

Baker-Austin, C., Dopson, M., Wexler, M., Sawers, R.G., Stemmler, A., Rosen, B.P., and Bond, P.L. 2007. Extreme arsenic resistance by the acidophilic archaeon *Ferroplasma acidarmanus* Fer1. *Extremophiles* 11: 425–434.

Bentley, D.R. 2006. Whole-genome re-sequencing. *Current Opinion in Genetics and Development* 16: 545–552.

Bhattacharya, P., Jacks, G., Ahmed, K.M., Routh, J., and Khan, A.A. 2002. Arsenic in groundwater of the Bengal Delta Plain aquifers in Bangladesh. *Bulletin of Environmental Contamination and Toxicology* 69: 538–545.

Binnewies, T.T., Motro, Y., Hallin, P.F., Lund, O., Dunn, D., La, T., Hampson, D.J., Bellgard, M., Wassenaar, T.M., and Ussery, D.W. 2006. Ten years of bacterial genome sequencing: Comparative-genomics-based discoveries. *Functional and Integrative Genomics* 6: 165–185.

Bryan, C.G., Marchal, M., Battaglia-Brunet, F., Kugler, V., Lemaitre-Guillier, C., Lièvremont, D., Bertin, P.N., and Arsène-Ploetze, F. 2009. Carbon and arsenic metabolism in *Thiomonas* strains: Differences revealed diverse adaptation processes. *BMC Microbiology* 9: 127.

Cai, L., Liu, G., Rensing, C., and Wang, G. 2009. Genes involved in arsenic transformation and resistance associated with different levels of arsenic-contaminated soils. *BMC Microbiology* 9: 4.

Carapito, C., Muller, D., Turlin, E., Koechler, S., Danchin, A., Van Dorsselaer, A., Leize-Wagner, E., Bertin, P.N., and Lett, M.C. 2006. Identification of genes and proteins involved in the pleiotropic response to arsenic stress in *Caenibacter arsenoxydans*, a metalloresistant beta-proteobacterium with an unsequenced genome. *Biochimie* 88: 595–606.

Cavalca, L., Corsini, A., Zaccheo, P., Andreoni, V., and Muyzer, G. 2013. Microbial transformations of arsenic: Perspectives for biological removal of arsenic from water. *Future Microbiology* 8: 753–768.

Chang, J.S., Lee, J.H., and Kim, I.S. 2011. Bacterial aox genotype from arsenic contaminated mine to adjacent coastal sediment: Evidences for potential biogeochemical arsenic oxidation. *Journal of Hazardous Materials* 193: 233–242.

Chang, J.S., Yoon, I.H., Lee, J.H., Kim, K.R., An, J., and Kim, K.W. 2010. Arsenic detoxification potential of aox genes in arsenite-oxidizing bacteria isolated from natural and constructed wetlands in the Republic of Korea. *Environmental Geochemistry and Health* 32: 95–105.

Ciprandi, A., Baraúna, R.A., Santos, A.V., Gonçalves, E.C., Carepo, M.S.P., Schneider, M.P.C., and Silva, A. 2012. Proteomic response to arsenic stress in *Chromobacterium violaceum*. *Journal of Integrated OMICS* 2: 69–73.

Cleiss-Arnold, J., Koechler, S., Proux, C., Fardeau, M.L., Dillies, M.A., Coppee, J.Y., Arsène-Ploetze, F., and Bertin, P.N. 2010. Temporal transcriptomic response during arsenic stress in *Herminiimonas arsenicoxydans*. *BMC Genomics* 11: 709.

Daware, V., Kesavan, S., Patil, R., Natu, A., Kumar, A., Kulkarni, M., and Gade, W. 2012. Effects of arsenite stress on growth and proteome of *Klebsiella pneumoniae*. *Journal of Biotechnology* 158: 8–16.

Drewniak, L., Styczek, A., Majder-Lopatka, M., and Sklodowska, A. 2008. Bacteria, hypertolerant to arsenic in the rocks of an ancient gold mine, and their potential role in dissemination of arsenic pollution. *Environmental Pollution* 156: 1069–1074.

Duan, M., Wang, Y., Xie, X., Su, C., and Li, J. 2013. Arsenite oxidizing bacterium isolated from high arsenic groundwater aquifers from Datong Basin, Northern China. *Procedia Earth and Planetary Science* 7: 232–235.

Edmunds, W.M., Ahmed, K.M., and Whitehead, P.G. 2015. A review of arsenic and its impacts in groundwater of the Ganges–Brahmaputra-Meghna delta, Bangladesh. *Environmental Science: Processes and Impacts* 17: 1032–1046.

Eid, J., Fehr, A., Gray, J., Luong, K., Lyle, J., Otto, G., Peluso, P., Rank, D., Baybayan, P., Bettman, B., and Bibillo, A. 2009. Real-time DNA sequencing from single polymerase molecules. *Science* 323: 133–138.

Escudero, L.V., Casamayor, E.O., Chong, G., Pedrós-Alió, C., and Demergasso, C. 2013. Distribution of microbial arsenic reduction, oxidation and extrusion genes along a wide range of environmental arsenic concentrations. *PLoS One* 8: 78890.

Fan, H., Su, C., Wang, Y., Yao, J., Zhao, K., and Wang, G. 2008. Sedimentary arsenite-oxidizing and arsenate-reducing bacteria associated with high arsenic groundwater from Shanyin, Northwestern China. *Journal of Applied Microbiology* 105: 529–539.

Fendorf, S., Michael, H.A., and van Geen, A. 2010. Spatial and temporal variations of groundwater arsenic in South and Southeast Asia. *Science* 328: 1123–1127.

Gault, A.G., Islam, F.S., Polya, D.A., Charnock, J.M., Boothman, C., Chatterjee, D., and Lloyd, J.R. 2005. Microcosm depth profiles of arsenic release in a shallow aquifer, West Bengal. *Mineralogical Magazine* 69: 855–863.

Ghosh, S. and Sar, P. 2013. Identification and characterization of metabolic properties of bacterial populations recovered from arsenic contaminated ground water of North East India (Assam). *Water Research* 47: 6992–7005.

Gihring, T.M., Druschel, G.K., McCleskey, R.B., Hamers, R.J., and Banfield, J.F. 2001. Rapid arsenite oxidation by *Thermus aquaticus* and *Thermus thermophilus*: Field and laboratory investigations. *Environmental Science and Technology* 35: 3857–3862.

Goodbred, S.L., Paolo, P.M., Ullah, M.S., Pate, R.D., Khan, S.R., Kuehl, S.A., Singh, S.K., and Rahaman, W. 2014. Piecing together the Ganges-Brahmaputra-Meghna River delta: Use of sediment provenance to reconstruct the history and interaction of multiple fluvial systems during Holocene delta evolution. *Geological Society of America Bulletin* 126: 1495–1510.

Green, H.H. 1919. Isolation and description of a bacterium causing oxidation of arsenite to arsenate in cattle dipping baths. Union of South Africa of Agriculture, Fifth and Sixth Reports of the Director of Veterinary Research, pp. 595–610.

Hao, X., Y. Lin, L. Johnstone et al. 2012. Genome sequence of the arsenite-oxidizing strain *Agrobacterium tumefaciens* 5A. *Journal of Bacteriology* 194: 903.

Henz, S.R., Huson, D.H., Auch, A.F., Nieselt-Struwe, K., and Schuster, S.C. 2005. Whole-genome prokaryotic phylogeny. *Bioinformatics* 21: 2329–2335.

Heroy, D.C., Kuehl, S.A., and Goodbred, S.L. 2003. Mineralogy of the Ganges and Brahmaputra Rivers: Implications for river switching and Late Quaternary climate change. *Sedimentary Geology* 155: 343–359.

Héry, M., Van Dongen, B.E., Gill, F., Mondal, D., Vaughan, D.J., Pancost, R.D., Polya, D.A., and Lloyd, J.R. 2010. Arsenic release and attenuation in low organic carbon aquifer sediments from West Bengal. *Geobiology* 8: 155–168.

Islam, F.S., Gault, A.G., Boothman, C., Polya, D.A., Charnock, J.M., Chatterjee, D., and Lloyd, J.R., 2004. Role of metal-reducing bacteria in arsenic release from Bengal delta sediments. *Nature* 430: 68–71.

Jiang, Z., Li, P., Wang, Y., Li, B., Deng, Y., and Wang, Y. 2014. Vertical distribution of bacterial populations associated with arsenic mobilization in aquifer sediments from the Hetao plain, Inner Mongolia. *Environmental Earth Sciences* 71: 311–318.

Kar, S., Maity, J.P., Jean, J.S., Liu, C.C., Nath, B., Yang, H.J., and Bundschuh, J. 2010. Arsenic-enriched aquifers: Occurrences and mobilization of arsenic in groundwater of Ganges Delta Plain, Barasat, West Bengal, India. *Applied Geochemistry* 25: 1805–1814.

Kinniburgh, D.G. and Smedley, P.L. 2001. Arsenic contamination of groundwater in Bangladesh. Volume 1: Summary. BGS Technical Report WC/00/19, British Geological Survey & Department of Public Health Engineering, p. 15.

Konstantinidis, K.T. and Tiedje, J.M. 2005. Genomic insights that advance the species definition for prokaryotes. *Proceedings of National Academy of Science of the United States of America* 102: 2567–2572.

Kruger, M.C., Bertin, P.N., Heipieper, H.J., and Arsène-Ploetze, F. 2013. Bacterial metabolism of environmental arsenic-mechanisms and biotechnological applications. *Applied Microbiology and Biotechnology* 97: 3827–3841.

Lessner, D.J., Parales, R.E., Narayan, S., and Gibson, D.T. 2003. Expression of the nitroarene dioxygenase genes in *Comamonas* sp. strain JS765 and *Acidovorax* sp. strain JS42 is induced by multiple aromatic compounds. *Journal of Bacteriology* 185: 3895–3904.

Li, B., Lin, J., Mi, S., and Lin, J. 2010. Arsenic resistance operon structure in *Leptospirillum ferriphilum* and proteomic response to arsenic stress. *Bioresource Technology* 101: 9811–9814.

Li, P., Wang, Y., Jiang, Z., Jiang, H., Li, B., Dong, H., and Wang, Y. 2013. Microbial diversity in high arsenic groundwater in Hetao Basin of Inner Mongolia, China. *Geomicrobiology Journal* 30: 897–909.

Li, X., Hu, Y., Gong, J., Lin, Y., Johnstone, L., Rensing, C., and Wang, G. 2012. Genome sequence of the highly efficient arsenite-oxidizing bacterium *Achromobacter arsenitoxydans* SY8. *Journal of Bacteriology* 194: 1243–1244.

Liao, V.H.C., Chu, Y.J., Su, Y.C., Hsiao, S.Y., Wei, C.C., Liu, C.W., Liao, C.M., Shen, W.C., Chang, F.J. 2011. Arsenite-oxidizing and arsenate-reducing bacteria associated with arsenic-rich groundwater in Taiwan. *Journal of Contaminant Hydrology* 123: 20–29.

Lin, Y., Fan, H., Hao, X., Johnstone, L., Hu, Y., Wei, G., Alwathnani, H.A., Wang, G., Rensing, C. 2012. Draft genome sequence of *Halomonas* sp. strain HAL1, a moderately halophilic arsenite-oxidizing bacterium isolated from gold-mine soil. *Journal of Bacteriology* 194: 199–200.

Mailloux, B.J., Alexandrova, E., Keimowitz, A.R., Wovkulich, K., Freyer, G.A., Herron, M., Stolz, J.F., Kenna, T.C., Pichler, T., Polizzotto, M.L., and Dong, H. 2009. Microbial mineral weathering for nutrient acquisition releases arsenic. *Applied and Environmental Microbiology* 75: 2558–2565.

Malasarn, D., Keeffe, J.R., and Newman, D.K. 2008. Characterization of the arsenate respiratory reductase from *Shewanella* sp. strain ANA-3. *Journal of Bacteriology* 190: 135–142.

Mandal, S., Mandal, M., Pati, B., Das, A., and Ghosh, A. 2009. Proteomics view of a *Rhizobium* isolate response to arsenite [As (III)] stress. *Acta Microbiologica et Immunologica Hungarica* 56: 157–167.

Mandal, S.M., Pati, B.R., Das, A.K., and Ghosh, A.K. 2008. Characterization of a symbiotically effective *Rhizobium* resistant to arsenic: Isolated from the root nodules of *Vigna mungo* (L.) Hepper grown in an arsenic-contaminated field. *The Journal of General and Applied Microbiology* 54: 93–99.

Margulies, M., Egholm, M., Altman, W.E., Attiya, S., Bader, J.S., Bemben, L.A., Berka, J., Braverman, M.S., Chen, Y.J., Chen, Z., and Dewell, S.B. 2005. Genome sequencing in microfabricated high-density picolitre reactors. *Nature* 437: 376–380.

McArthur, J.M., Banerjee, D.M., Hudson-Edwards, K.A. et al. 2004. Natural organic matter in sedimentary basins and its relation to arsenic in anoxic ground water: The example of West Bengal and its worldwide implications. *Applied Geochemistry* 19: 1255–1293.

McArthur, J.M., Ravenscroft, P., Banerjee, D.M. et al. 2008. How paleosols influence groundwater flow and arsenic pollution: A model from the Bengal Basin and its worldwide implication. *Water Resources Research* 44: W11411.

Meharg, A.A., Scrimgeour, C., Hossain, S.A., Fuller, K., Cruickshank, K., Williams, P.N., and Kinniburgh, D.G. 2006. Codeposition of organic carbon and arsenic in Bengal Delta aquifers. *Environmental Science and Technology* 40: 4928–4935.

Mok, W.M. and Wai, C.M. 1994. *Mobilization of Arsenic in Contaminated River Waters* (Vol. 26, p. 99). Advances in Environmental Science and Technology, New York.

Mukhopadhyay, R., Rosen, B.P., Phung, L.T., and Silver, S., 2002. Microbial arsenic: From geocycles to genes and enzymes. *FEMS Microbiology Reviews* 26(3): 311–325.

Muller, D., Médigue, C., Koechler, S. et al. 2007. A tale of two oxidation states: bacterial colonization of arsenic-rich environments. *PLoS Genetics* 3: 53.

Muller, D., Simeonova, D.D., Riegel, P., Mangenot, S., Koechler, S., Lièvremont, D., Bertin, P.N., and Lett, M.C. 2006. *Herminiimonas arsenicoxydans* sp. nov., a metalloresistant bacterium. *International Journal of Systematic and Evolutionary Microbiology* 56: 1765–1769.

Nickson, R.T., McArthur, J.M., Ravenscroft, P., Burgess, W.G., and Ahmed, K.M. 2000. Mechanism of arsenic release to groundwater, Bangladesh and West Bengal. *Applied Geochemistry* 15: 403–413.

Oremland, R.S. and Stolz, J.F. 2005. Arsenic, microbes and contaminated aquifers. *Trends in Microbiology* 13: 45–49.

Parua, P.K. 2010. The Ganga geology. In *The Ganga* (pp. 15–21). Springer, Dordrecht, Netherlands.

Patel, P.C., Goulhen, F., Boothman, C., Gault, A.G., Charnock, J.M., Kalia, K., and Lloyd, J.R. 2007. Arsenate detoxification in a *Pseudomonad* hypertolerant to arsenic. *Archives of Microbiology* 187: 171–183.

Paul, D., Kazy, S. K., Gupta, A. K., Pal, T., and Sar, P. 2015. Diversity, metabolic properties and arsenic mobilization potential of indigenous bacteria in arsenic contaminated groundwater of West Bengal, India. *PLoS One* 10: 0118735.

Paul, D., Poddar, S., and Sar, P. 2014a. Characterization of arsenite-oxidizing bacteria isolated from arsenic-contaminated groundwater of West Bengal. *Journal of Environmental Science and Health A* 49: 1481–1492.

Paul, D., Sar, P., Kazy, S.K., and Pal, T. 2014b. Molecular survey on bacterial diversity in arsenic contaminated subsurface sediment in West Bengal, India. *Journal of Environmental Research and Development* 9: 15–23.

Prasad, K.S., Ramanathan, A.L., Paul, J., Subramanian, V., Prasad, R. 2013. Biosorption of arsenite (As^{+3}) and arsenate (As^{+5}) from aqueous solution by *Arthrobacter* sp. biomass. *Environmental Technology* 34: 2701–2708.

Rahman, M.M., Saha, K.C., Mukherjee, S.C., Pati, S., Dutta, R.N., Roy, S., Quamruzzaman, Q., Rahman, M., and Chakraborti, D., 2015. Groundwater arsenic contamination in bengal delta and its health effects. In *Safe and Sustainable Use of Arsenic-Contaminated Aquifers in the Gangetic Plain* (pp. 215–253). Springer International Publishing, Cham, Switzerland.

Reza, A.S., Jean, J.S., Yang, H.J., Lee, M.K., Woodall, B., Liu, C.C., Lee, J.F., and Luo, S.D. 2010. Occurrence of arsenic in core sediments and groundwater in the Chapai-Nawabganj District, northwestern Bangladesh. *Water Research* 44: 2021–2037.

Richter M, Rosselló-Móra R (2009). Shifting the genomic gold standard for the prokaryotic species definition. Proc Natl Acad Sci USA 106(45): 19126–19131.

Rosen, B.P. 2002. Biochemistry of arsenic detoxification. *FEBS Letters* 529: 86–92.

Sacheti, P., Patil, R., Dube, A. et al. 2014. Proteomics of arsenic stress in the gram-positive organism *Exiguobacterium* sp. PS NCIM 5463. *Applied Microbiology and Biotechnology* 98: 6761–6773.

Salmassi, T.M., Walker, J.J., Newman, D.K., Leadbetter, J.R., Pace, N.R., and Hering, J.G. 2006. Community and cultivation analysis of arsenite oxidizing biofilms at Hot Creek. *Environmental Microbiology* 8: 50–59.

Santini, J.M., Sly, L.I., Schnagl, R.D., and Macy, J.M. 2000. A new chemolithoautotrophic arsenite-oxidizing bacterium isolated from a gold mine: Phylogenetic, physiological, and preliminary biochemical studies. *Applied and Environmental Microbiology* 66: 92–97.

Santini, J.M., Sly, L.I., Wen, A., Comrie, D., Wulf-Durand, P.D., and Macy, J.M. 2002. New arsenite-oxidizing bacteria isolated from Australian gold mining environments-phylogenetic relationships. *Geomicrobiology Journal* 19: 67–76.

Sar, P., Paul, D., Sarkar, A., Bharadwaj, R., and Kazy, S.K. 2015. Microbiology of arsenic contaminated groundwater. In *Microbiology for Minerals, Metals, Materials and Environment*, B.D.P. Abhilash and K.A. Natarajan (Eds.), CRC Press, Taylor & Francis Group, Boca Raton, FL.

Sarkar, A., Kazy, S.K., and Sar, P. 2013. Characterization of arsenic resistant bacteria from arsenic rich groundwater of West Bengal, India. *Ecotoxicology* 22: 363–376.

Sarkar, A., Kazy, S.K., and Sar, P. 2014. Studies on arsenic transforming groundwater bacteria and their role in arsenic release from subsurface sediment. *Environmental Science and Pollution Research* 21: 8645–8662.

Sarkar, A., Paul, D., Kazy, S. K., and Sar, P. 2016. Molecular analysis of microbial community in arsenic-rich groundwater of Kolsor, West Bengal. *Journal of Environmental Science and Health, Part A* 51: 229–239.

Sharma, A.K., Tjell, J.C., Sloth, J.J., and Holm, P.E. 2014. Review of arsenic contamination, exposure through water and food and low cost mitigation options for rural areas. *Applied Geochemistry* 41: 11–33.

Singh, I.B. 1996. Geological evolution of Ganga Plain—An overview. *Journal of the Palaeontological Society of India* 41: 99–137.

Sinha, R., Tandon, S.K., Gibling, M.R., Bhattacharjee, P.S., and Dasgupta, A.S. 2005. Late quaternary geology and alluvial stratigraphy of the Ganga basin. *Himalayan Geology* 26: 223–240.

Slyemi, D. and Bonnefoy, V. 2012. How prokaryotes deal with arsenic. *Environmental Microbiology Reports* 4: 571–586.

Srivastava, S., Verma, P.C., Singh, A., Mishra, M., Singh, N., Sharma, N., and Singh, N. 2012. Isolation and characterization of *Staphylococcus* sp. strain NBRIEAG-8 from arsenic contaminated site of West Bengal. *Applied Microbiology and Biotechnology* 95: 1275–1291.

Sultana, M., Härtig, C., Planer-Friedrich, B., Seifert, J., and Schlömann, M. 2011. Bacterial communities in Bangladesh aquifers differing in aqueous arsenic concentration. *Geomicrobiology Journal* 28: 198–211.

Sun, W., Sierra-Alvarez, R., Fernandez, N., Sanz, J.L., Amils, R., Legatzki, A., Maier, R.M., and Field, J.A. 2009. Molecular characterization and in situ quantification of anoxic arsenite-oxidizing denitrifying enrichment cultures. *FEMS Microbiology Ecology* 68: 72–85.

Suresh, K., Prabagaran, S. R., Sengupta, S., and Shivaji, S. 2004. *Bacillus indicus* sp. nov., an arsenic-resistant bacterium isolated from an aquifer in West Bengal, India. *International Journal of Systematic and Evolutionary Microbiology* 54: 1369–1375.

Sutton, N.B., van der Kraan, G.M., van Loosdrecht, M.C., Muyzer, G., Bruining, J., and Schotting, R.J. 2009. Characterization of geochemical constituents and bacterial populations associated with As mobilization in deep and shallow tube wells in Bangladesh. *Water Research* 43: 1720–1730.

Thambidurai, P., Chandrashekhar, A.K., and Chandrasekharam, D. 2013. Geochemical signature of arsenic-contaminated groundwater in Barak Valley (Assam) and surrounding areas, northeastern India. *Procedia Earth and Planetary Science* 7: 834–837.

Thornton, I. 1996. *Sources and Pathways of Arsenic in the Geochemical Environment: Health implications*. Geological Society, London, U.K., Special Publications, Vol. 113, pp. 153–161.

Vanden Hoven, R.N. and Santini, J.M. 2004. Arsenite oxidation by the heterotroph *Hydrogenophaga* sp. str. NT-14: The arsenite oxidase and its physiological electron acceptor. *Biochimica et Biophysica Acta (BBA)-Bioenergetics* 1656: 148–155.

Weiss, S., Carapito, C., Cleiss, J. et al. 2009. Enhanced structural and functional genome elucidation of the arsenite-oxidizing strain *Herminiimonas arsenicoxydans* by proteomics data. *Biochimie* 91: 192–203.

Zhang, Y., Ma, Y.F., Qi, S.W., Meng, B., Chaudhry, M.T., Liu, S.Q., and Liu, S.J., 2007. Responses to arsenate stress by *Comamonas* sp. strain CNB-1 at genetic and proteomic levels. *Microbiology* 153: 3713–3721.

Section IV

Interdisciplinary Approaches for Enhanced Metal Bioremediation

21 Immobilization Techniques for Microbial Bioremediation of Toxic Metals

Niloofar Nasirpour, Seyed Morteza Zamir, and Seyed Abbas Shojaosadati

CONTENTS

ABSTRACT

Accumulation of heavy metals in water or soil due to industrial activities is one of the main sources of environmental pollution. Immobilization of microorganisms is considered as a well-known technique that can enhance the chemical and physical performance of metal uptake from the environment. Immobilization provides more flexibility in the design of reactors compared with conventional suspension systems, along with higher cell density, no washout of cells, and better stability in operation. There are several methods for the immobilization of bacterial, fungal, and algal cells. Biofilm formation on natural or synthetic packing, entrapment of cells inside polymeric matrices, adsorption and covalent binding of cells to a surface, and encapsulation of cells are considered as the main immobilization methods applied in the metal removal processes. The characteristics of matrices have an important effect on the metal uptake process.

21.1 INTRODUCTION

Heavy metal pollution is one of the most concerning environmental problems today, especially in relation to water contamination (Wang and Chen, 2006). Various industries produce wastes containing different heavy metals into the environment, such as mining, fertilizer, pesticide, metallurgy, energy and fuel production, iron and steel, electroosmosis, electrolysis, electroplating, electric appliance manufacturing, metal surface treating, leatherworking, aerospace, and atomic energy installation. Metal deposition brings about serious environmental pollution, threatening human health and ecosystem. Large quantities of metals can be accumulated by different kinds of living microorganisms dependent or independent on metabolism. Both living and dead biomass can be used to remove metal by biosorption process (Wang and Chen, 2009). Unfortunately, there are some problems with the use of freely suspended alive or dead biomass for the uptake of heavy metal ions, including separation of the suspended biomass from the aqueous medium, which is difficult and expensive (Abu Al-Rub et al., 2004), and the possible clogging of pipelines and filters. Moreover, the biomass is not mechanically strong and has a wide size distribution, which can lead to some problems including channeling in column operation (Lu and Wilkins, 1996). Immobilization of the biomass overcomes many of these processing difficulties. Immobilization converts biomass to a particulate form with an appropriate size range (0.5–1.5 mm) and better chemical and physical performance (Texier et al., 2002). The application of immobilization technology to metal removal provides more flexibility in the reactor design when compared with conventional suspension systems. Enhanced metal removal capacity due to higher cell density (Abu Al-Rub et al., 2004) and no washout of cells are the additional advantages of immobilized biomass over freely suspended biomass (Brouers et al., 1989). Immobilization provides cell reuse and eliminates the costly processes of cell recovery and recycling (Shuler and kargi, 2002). In the case of living biomass, immobilization provides protection to cells from metal toxicity (Bozeman et al., 1989). The major disadvantage of cell immobilization is the mass transfer resistance (Khoo and Ting, 2001). However, this can be avoided by careful selection of the immobilization method and the nature of the matrix (Eroglu et al., 2015).

21.2 BIOMASS IMMOBILIZATION TECHNIQUES AND APPLIED MATRICES FOR BIOSORPTION OF METAL IONS

Immobilization of biomass is defined as a technique that results in the restriction of the free movement of biomass (Leenen et al., 1996). Immobilization techniques can be generally classified into two groups of active and passive immobilization. Active immobilization includes entrapment, encapsulation, covalent binding, and adsorption, while biofilm formation is in the category of passive immobilization (Shuler and Kargi, 2002). Figure 21.1 shows the classification and schematic representation of these techniques.

21.2.1 ACTIVE IMMOBILIZATION OF CELLS

Active immobilization is entrapment of cells by physical or chemical forces. The two major methods of active immobilization are entrapment and binding (Shuler and Kargi, 2002). Entrapment within porous matrices is the most widely used method of biomass immobilization for metal removal. Entrapment methods are based on the restriction of cells in the 3D lattice of the gel. The cells are free within their compartments and the pores in the material allow substrates and products diffusion (Mallick, 2002). Most of the entrapment processes have similar protocols, as mixing the biomass suspension with the monomers of the selected polymer, followed by solidification of the resulting biomass/polymer mixture by some physical or chemical processes (Eroglu et al., 2015). The main solidification processes are gelation of polymers, precipitation of polymers, ion-exchange gelation, polycondensation, and polymerization (Shuler and Kargi, 2002). Description of each method is completely reviewed in the literature. Among these methods, ion-exchange gelation is the most commonly used method for obtaining biomass immobilized beads for metal removal. Ion-exchange gelation occurs when a water-soluble polyelectrolyte and a salt solution are mixed together. This material is then formed into droplets by forcing it through a nozzle or orifice to an interacting salt solution in a dropwise manner (Mallick, 2002). Solidification occurs when

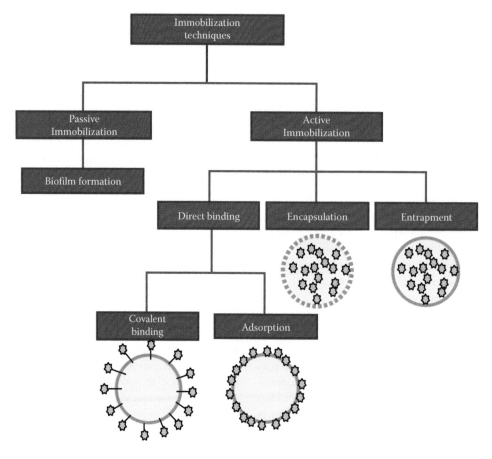

FIGURE 21.1 Classification and schematic representation of immobilization techniques.

the polyelectrolyte and salt solution react together to form a solid gel. The most common example of this kind of gelation is the formation of calcium-alginate gel by mixing sodium-alginate solution with a $CaCl_2$ solution (Eroglu et al., 2015).

Techniques have developed for the large-scale production of biomass-polymer beads. These techniques are based on the forced flow of the gelling material through multiple nozzles. It has also been shown that the addition of vibrational energy onto the bead formation process facilitates the production of monodispersed beads (Hulst et al., 1985). Brandenberger and Widmer (1998) developed a new multinozzle system with 13 nozzles for the immobilization of microorganisms and enzymes. Based on the laminar jet breakup, monodispersed beads of calcium-alginate in size range of 0.2–1 mm are produced under sterile and reproducible conditions (Mallick, 2002).

Encapsulation is another method for the entrapment of various cell types. Microcapsules are spherical, hollow particles bound by semipermeable membranes. Cells are entrapped within the hollow capsule volume. The transport of material in and out of the capsule takes place through the membrane. Microcapsules have several advantages over gel beads. More cells can be entrapped per unit volume of matrix, and intraparticle diffusion limitations are less serious in capsules because of the presence of liquid cell suspension in the intracapsule space (Shuler and Kargi, 2002).

Direct binding of biomass to a support is another immobilization technique. Immobilization of cells on the surfaces of support can be maintained by physical adsorption or covalent binding. Adsorption is a simple, inexpensive method of biomass immobilization. The major advantage of immobilization by adsorption is direct contact between nutrient and matrix. But limited cell loadings and weak binding forces reduce the attractiveness of this method. A ratio of pore to cell diameter of 4–5 is recommended for the immobilization of living cells onto the inner surface of porous support particles. At small pore sizes, diffusion into inner surfaces of pores may be the limiting factor, while the specific surface area may be the limiting factor at large pore sizes (Shuler and Kargi, 2002).

The binding forces between support surfaces and the cell may be different depended on the surface properties of the support material and the type of cells. For positively charged support surfaces (ion-exchange resins, gelatin), the binding force is mostly electrostatic forces. Cells also adhere on negatively charged surfaces by covalent binding or hydrogen bonding. The adsorption of cells on neutral polymer surfaces may be facilitated by chemical bonding, such as Van der Waals forces, and covalent or hydrogen bonding types. Some specific chelating agents can be used to establish stronger cell–surface interactions. Hydrodynamic shear around physically adsorbed cells should be very mild to avoid the separation of cells from support surfaces (Shuler and Kargi, 2002).

21.2.2 Passive Immobilization: Biological Films

Biological films are the multilayer growth of cells on the outer surface of solid supports. The support material can be biologically active or inert. Biofilm formation is common in fermentation processes, such as mold fermentations and wastewater treatment. The interaction among cells and the binding forces between the support and cell may be very complicated (Shuler and Kargi, 2002).

21.2.3 Applied Matrices and Supports for Biomass Immobilization

A suitable matrix for practical use in biosorption must be mechanically and/or chemically stable to resist actual process conditions and successive sorption–desorption cycles. Mass transfer considerations are also of great importance (Jen et al., 1996). Thus, polymer porosity and pore size are critical parameters in matrix selection (Cassidy et al., 1996). The matrix must be porous enough to facilitate the diffusion of the sorbate to the surface of sorbent. Size of the matrix is also a crucial factor to be considered for biosorption process (Mehta and Gaur, 2005). Bead size between 0.7 and 1.5 mm is recommended, corresponding to the size of commercial resins used for removing metal ions (Volesky, 2001). For adsorption mode of immobilization, a good support material should be rigid and chemically inert. It should bind cells strongly and also have high loading capacity (Shuler and Kargi, 2002).

Both natural and synthetic polymers have been used for biomass immobilization via different immobilization techniques. The most common matrices that are used for the immobilization of microorganisms for obtaining a metal removal biosorbent bead are described as follows:

21.2.3.1 Natural Matrices

Matrices having natural origins, mostly extracted from seaweed, algae, etc., attracted great attention for practical use in metal removal. The immobilization procedures for natural gel materials are mild, and cells grow well in these matrices. Furthermore, the effective diffusion coefficients for substrates are close to those in water. These matrices are soluble, biodegradable, and susceptible to abrasion (Leenen et al., 1996). Chitin and algal polysaccharides such as alginate, agar, agarose, and carrageenan are classified as natural polymers (Kolot, 1981).

21.2.3.1.1 Alginate

Alginate is the most common matrix that is used for different bacterial, fungal, and algal strain immobilization for application in metal removal. Alginic acid or the salt of alginic acid—alginate—is the name given to the family of linear anionic polysaccharides including L-guluronic (G) and 1,4-linked–D-mannuronic (M) acid arranged in an irregular, block-wise order along the chain; it is mostly found in the cell walls of brown algae (Andrade et al., 2004; Arica et al., 2004). Extracellular alginate is also produced by certain bacteria such as *Azotobacter vinelandii* and several *Pseudomonas* spp. Different algae produce alginates that differ in chain length, monomer composition, and arrangement. Thus, the properties of alginates are variable (Cassidy et al., 1996). Monovalent ions for different salts of alginic acid are soluble in aqueous medium, while those with divalent or polyvalent metal ions (except Mg^{2+}) are insoluble (Eroglu et al., 2015).

Major advantages of alginate gel are as follows: having hydrophilic properties, being cost-effective and nontoxic, and containing carboxylic groups (de-Bashan and Bashan, 2010). Other additional advantages are their low density and biodegradability that make them highly suitable for many biotechnological applications. In contrast to these advantages, alginate beads have some problems such as structural instability in the presence of high phosphate concentrations or high content of some cations such as K^+ or Mg^{2+} (Kuu and Polack, 1983); it is also reported to be stable in only a narrow pH range of 6–9 (Khoo and Ting, 2001).

21.2.3.1.2 Carrageenan

Carrageenan is produced by red algae, mostly *Chondrus crispus*, *Eucheuma cottonii*, *Gigartina stellata*, and *G. radula*. Three types of carrageenan are produced and labeled by the Greek suffixes ι, λ, and κ. All carrageenans have a backbone of alternating β(1,3)-D-galactose and α(1,4)-D-galactose. The carrageenans differ in the number and sulfonation site on both compounds. Gelation in carrageenans is primarily temperature dependent, and additional reinforcing of the polymer network occurs in the presence of K^+ or Al^{3+}, as they may interact with the sulfate residues on the polymer (Guiseley, 1989). K-Carrageenan, composed mainly of linear chains of alternating 3–0-substituted/3-D-galactopyranose-4-sulfate and units of 4–0-substituted 3,6-anhydro-*c*-D-galactopyranose, is often used in encapsulation processes because of its firmer gelling potential. Carrageenan is usually less expensive than alginate, which can considerably reduce costs in a large-scale environmental application, but most of the heavy metal biosorption studies have used Ca-alginate as an immobilizing matrix (Cassidy et al., 1996).

21.2.3.1.3 Carboxymethylcellulose

Carboxymethylcellulose (CMC) is a water-soluble compound derived from cellulose. Carboxymethylation adds functional groups of carboxylic to the cellulose derivative for cross-linking by trivalent metal ions. CMC can easily convert into hydrogels by cross-linking with trivalent metal ions such as ferric chloride and aluminum chloride. It is preferred more than other matrices because of its several benefits such as biodegradability, hydrophilicity, and the presence of functional carboxylic groups (Bayramoglu et al., 2003).

21.2.3.1.4 Loofah Sponge

Loofah sponge is a porous material obtained from dry fruit of *Luffa cylindrica*. Loofah sponge has been suggested as an immobilization matrix for algal, fungal, and yeast cells, and recently, it has been used in microbial immobilization for metal uptake (Akhtar et al., 2004). It is an effective low-cost and robust carrier for the biomass immobilization for metal ion removal (Iqbal and Edyvean, 2005). Loofah sponge immobilized biomass is a very stable biosorbent over the wide pH range of 1–13 (Iqbal and Edyvean, 2007). The only disadvantage of this matrix is the long biomass immobilization period (Akhtar et al., 2004).

21.2.3.2 Synthetic Matrices

Synthetic matrices have better mechanical properties in comparison with natural ones but mostly involve in lower substrate diffusion. Immobilization conditions for synthetic matrices are less mild resulting in low biomass retention (Leenen et al., 1996). Polyacrylamide, polystyrene, and polyvinyl alcohol (PVA) are synthetic polymers (Kolot, 1981).

21.2.3.2.1 Polyvinyl alcohol (PVA)

The use of PVA as an immobilization matrix was initiated around 25 years ago. Application of PVA-immobilized biomass for heavy metal removal is particularly compared with Ca-alginate-immobilized biomass; advantages of PVA over Ca-alginate are low cost, high durability and chemical stability, and nontoxicity to viable cells. PVA beads are stable in solutions of various sodium salts and urea and over a wide pH range of 1–13. PVA beads are highly elastic, providing high mechanical stability. Large surface area and pore size of PVA beads enable high biomass loading and reduce mass transfer resistance (Khoo and Ting, 2001). Several methods for immobilization using PVA have been reported. These methods include the PVA–boric acid method (Hashimoto and Furukawa, 1986), cross-linking by ultraviolet radiation (Imai et al., 1985), freezing and incubating in liquid methanol (Ariga et al., 1994), and iterative freezing and thawing in liquid paraffin methods (Ariga et al., 1987).

21.2.3.2.2 Polyacrylamide

Polyacrylamide has been the most extensively employed immobilization synthetic material for laboratory research studies (Nakajima et al., 1982; Pons and Fuste, 1993). The advantage of the use of polyacrylamide encapsulation technique is the simple and rapid experimental protocol. The small pore size and the mechanical stability of polyacrylamide-immobilized biomass make it useful for applications in bioreactors because of the minimum release of cells from the matrix (Martins Dos Santos et al., 1997). However, the toxic nature of the monomers and high activity losses of the immobilized living cells (more than 90%) are the major disadvantages of this polymer (Sumino et al., 1992; Tanaka et al., 1991).

21.2.4 Immobilized Bacterial Cells for Metal Removal

Bacteria were used as biosorbents because of their small size and ubiquitousness, their ability to grow under controlled conditions, and their resilience to a wide range of environmental situations (Wang and Chen, 2009). Immobilized forms of bacterial species such as *Bacillus*, *Pseudomonas*, *Streptomyces*, *Escherichia*, and *Desulfovibrio* have been tested for uptake of metals. Texier et al. (2002) have demonstrated the good capability of polyacrylamide-immobilized *Pseudomonas aeruginosa* powder in lanthanide ions removal from aqueous solution in a continuous fixed-bed bioreactor system. Maximum adsorption capacity of the immobilized biomass was 342 μmol g^{-1} (±10%) of lanthanum ions from a 6 mM solution. They reported the order of preferential biosorption as $Eu^{3+} \geq Yb^{3+} > La^{3+}$. Desorption with 0.1 M of EDTA (pH 5) resulted in 96% recovery of lanthanum ions. It was suggested that regenerating and reusing the biomass in three adsorption/desorption cycles were practical. Lanthanum removal was slightly affected by the superficial velocity based on empty column (U_0) in the range of 0.23–0.76 m·h^{-1}. Increasing U_0 values in the range of 0.76–2.29 m·h^{-1} resulted in the decrease of the column sorption capacity. Lopez et al. (2002) immobilized living *Pseudomonas fluorescens* in agar for Ni^{2+} removal in batch studies. Equilibrium established in 5 min and 24 h for free and immobilized biomass, respectively. They reported that maximum nickel uptake occurred at pH 8. Metal ion removal capacity of free and immobilized cells increased when the cell concentration decreased. Agar concentration in immobilized biomass affected the Ni^{2+} accumulation with the optimum agar concentration of 2%. Dipicolinic acid was used as desorbing agent. It was noted about 60% decrease in the rate of metal sorption by *P. fluorescens* cells immobilized in agar bead, as compared with free cells (Lopez et al., 2002). Bang and Pazirandeh (1999) used a recombinant strain of *Escherichia coli* expressing a metal-binding gene encapsulated in alginate, chitosan-alginate and κ-carrageenan for the removal of lower levels of cadmium ions. They reported that chitosan-alginate-immobilized biomass had the highest physical and chemical stability, whereas carrageenan-immobilized biomass had better metal removal ability.

Jezequel and Lebeau (2008) investigated the phyto-available cadmium uptake from soil by two bacteria of *Bacillus* and *Streptomyces* in both immobilized and free form. They used Ca-alginate for the cell entrapment. The two bacteria survived and colonized the soil; however, the immobilization did not improve the cell survival in the bio-augmented soil.

Pires et al. (2011) isolated three bacterial species belonging to the genera *Cupriavidus*, *Sphingobacterium*, and *Alcaligenes* from a contaminated area. They used alginate, pectate, and a synthetic cross-linked polymer as immobilization matrices. They reported that the immobilized form of these strains has great capability in the removal of zinc and cadmium. However, immobilization in the synthetic cross-linked polymer was the most promising one. Immobilized

living cells of *Chryseomonas luteola* TEM 05 were used for Cr^{6+}, Cd^{2+}, and Co^{2+} removal from aqueous solution. Carrageenan and chitosan-coated carrageenan were used as immobilization matrices. Chitosan-coated carrageenan showed higher mechanical strength and thermal stability. In addition, the *C. luteola* TEM 05–immobilized carrageenan-chitosan gel was more efficient for the quick adsorption of ions from aqueous solution than the carrageenan gels without biomass (Baysal et al., 2009). *Pseudomonas maltophilia* cells immobilized in polyacrylamide gel showed a high ability for gold biosorption. The gold adsorbed on the immobilized cells was easily desorbed with 0.1 M thiourea ($SC(NH_2)_2$) solution. The immobilized *P. maltophilia* cells can be used in several biosorption/desorption cycles (Tsuruta, 2004). Piccirillo et al. (2013) used immobilized bacterial strains of *P. fluorescens*, *Microbacterium oxydans*, and *Cupriavidus* sp. on the surface of hydroxyapatite of natural origin, for zinc and cadmium ions' removal from aqueous solution. Immobilized form of the biomass has higher capability in the removal of both ions compared with the free cells. Fayyaz Ahmad et al. (2014) used *Bacillus subtilis* immobilized in Ca-alginate for cadmium ions' removal. They also applied artificial neural network for the prediction of adsorption capacity. The applied model successfully predicted Cd biosorption capacity with R^2 of 0.997. Carpio et al. (2014) obtained a heavy metal–resistant bacteria consortium from a contaminated river; they used this consortium for biofilm formation in a packed bed setup on granular activated carbon supports. They attributed the metal sorption to hydroxyl and carboxyl groups of the microbial consortium by FTIR spectroscopy. The biofilm column retained 45% of the copper mass present in the influent for an initial copper concentration of 15 ppm with the flow rate of 5 mL·min^{-1}. Branco et al. (2016) developed a method for the immobilization of the mutant cells of *Ochrobactrum tritici* As5 on a commercial polymeric net modified by the deposition of polytetrafluoroethylene thin films. They reported that the immobilized biomass exhibited a great capability in arsenite removal. Table 21.1 shows different combination of bacteria and matrices used for metal uptake from aqueous solutions and the capacity of the resulted beads.

TABLE 21.1
Immobilized Bacterial Systems Used for Metal Removal

Microorganism	Matrices	Metal Ion	Capacity (mg·g^{-1})	Reference
Bacillus arsenicus MTCC 4380	Sawdust/MnFe$_2$O$_4$	As (III)	87.6	Podder and Majumdar (2016)
Bacillus arsenicus MTCC 4380	Sawdust/MnFe$_2$O$_4$	As (V)	88.9	Podder and Majumdar (2016)
Bacillus sp.	Tea waste biomass	Cr (VI)	741.4	Gupta and Balomajumder (2015)
Effective microorganism (EM)	Ca-alginate	Cr (III)	0.94	Ting et al. (2013)
Bacillus subtilis	Chitosan	Cu (II)	100.70	Liu et al. (2013)
Bacillus subtilis	Ca-alginate	Cd (II)	251.91	Fayyaz Ahmad et al. (2014)
Pseudomonas aeruginosa	Polyacrylamide	La (III)	55.2	Texier et al. (2002)
Pseudomonas aeruginosa	Polyacrylamide	Eu (III)	44.1	Texier et al. (2002)
Pseudomonas aeruginosa	Polyacrylamide	Yb (III)	56.4	Texier et al. (2002)
Pseudomonas fluorescens	Agar	Ni (II)	37	Lopez et al. (2002)

21.2.5 Immobilized Fungal Cells for Metal Removal

Filamentous fungi and yeasts are often able to bind metallic elements. Large amounts of some waste of mycelia are available for the removal of metals (Kapoor and Viraraghavan, 1995; Wang and Chen, 2006). Both living and dead fungal cells provide a significant ability for absorbing toxic metals (Wang and Chen, 2009). But dead fungal biomass seems to offer several advantages over the living one as it may be obtained cheaply from several industrial sources. Also, it is not subject to metal toxicity or adverse operating conditions and needs no nutrient supply. Moreover, surface-bound metals may be recovered by relatively simple nondestructive treatments (de Rome and Gadd, 1991).

The yeast biomass has been successfully used as biosorbent for the removal of gold, silver, cadmium, cobalt, chromium, copper, nickel, lead, uranium, and zinc from aqueous solution. *Saccharomyces*, *Candida*, and *Pichia* spp. are efficient biosorbents of heavy metal ions (Wang and Chen, 2009). Several studies investigated the ability of immobilized *Saccharomyces cerevisiae* for the removal of toxic metals and radionuclides.

Al Saraj et al. (1999) used a sol-gel matrix for the entrapment of *S. cerevisiae*. They assessed the capability of obtained bio-gel in heavy metal removal. For mercury and cadmium, the metal uptake capacity of the immobilized cells was the same as the free one. However, free yeast cells exhibited better adsorption capacity for lead and copper, while a lower capacity was observed in the cases of cobalt, nickel, and zinc.

De Rome and Gadd (1991) evaluated the biosorption of uranium, strontium, and cesium ions by pelleted mycelium of two fungi species of *Rhizopus arrhizus* and *Penicillium chrysogenum* and immobilized *S. cerevisiae* in both batch and continuous flow systems. The metal uptake mechanism for the pelleted fungal biomass differed from that of the immobilized yeast; the former was metabolism independent, while uptake in the latter was biphasic. Equilibrium time for mycelial pellets was established in about 2 h. Pelleted *R. arrhizus* had the highest uranium uptake of about 90%, and pelleted *P. chrysogenum* had the highest removal for cesium, while strontium removal was approximately similar in all immobilized fungi. Blank Ca-alginate beads have shown weak ability in uranium uptake; thus, beads with the greater ratio of yeast to alginate have higher capacity of uranium or metal removal (de Rome and Gadd, 1991).

Tobin et al. (1993) immobilized *R. arrhizus* on different matrices including alginate, polyacrylamide, epoxy resin, and polyvinyl formal (PVF); they tested the biosorbents for cadmium removal. Among them, Ca-alginate-immobilized biomass has the highest cadmium removal capacity. Blank alginate itself showed high metal uptake capacity that was approximately equivalent to the free cells of *R. arrhizus*. All the epoxy resin–immobilized biomass showed some degrees of cadmium uptake, but the adsorption capacity values were considerably decreased compared with the results obtained by free biomass. They reported that PVF-immobilized biomass as a biosorbent is suitable for industrial applications since it exhibited the combined characteristics of high cell loading, good metal adsorption levels, and mechanical strength.

PVF-immobilized inactive mycelia of *R. arrhizus* were successfully used for uranium removal from aqueous solution. It was shown that the main portion of the metal is adsorbed on biomass and the immobilizing polymer has negligible uranium uptake. Influence of polymer content and particle size on biosorption capacity of the biosorbents was also investigated. Particles with the same diameter, but with lower polymer content, have a thinner outer polymeric layer that results in a higher overall mass transfer coefficient. Thus, the metal uptake capacity of the biosorbent increases as the polymer percentage decreases (Tsezos and Deutschmann, 1990).

Zhou and Kiff (1991) used reticulated polyester foam–immobilized *R. arrhizus* for copper ions' removal from aqueous solution in a packed bed column. They found pH as a critical factor in metal uptake, with the optimum range of 6.7–7. pH influences both cell surface metal-binding sites and metal chemistry in water. The presence of other cations and anions inhibited copper uptake particularly manganese and EDTA ions. Tsekova et al. (2010) determined the ability of *Aspergillus niger* immobilized in two different polymers of PVA and Ca-alginate for removing heavy metal ions of copper, manganese, zinc, nickel, iron, lead, and cadmium from real wastewater in shaking flasks. Total capacities of biosorption were in the following order: free cells (33.3 mg·g^{-1}) < PVA biomass (39.8 mg·g^{-1}) < Ca-alginate biomass (44.6 mg·g^{-1}). Both of the immobilized biosorbents displayed high removal potential for Cd^{2+}, Pb^{2+}, and Fe^{3+}, in comparison to other heavy metal ions from the industrial effluent, while the immobilized biomass in Ca-alginate showed the highest biosorption capacities for Mn^{2+} and Cu^{2+}ions.

Fomitopsis carnea immobilized in PVA beads and Ca-alginate was used for gold uptake. Multipoint BET analysis was used for the determination of the surface area. PVA beads exhibited larger surface area with the value of 48.1 m^2·g^{-1}, while it was 6.25 m^2·g^{-1} for the alginate. The alginate beads have a distinct monotonic peak distribution, with the majority of the pores having a diameter of about 20 Å. In contrast, most of the PVA beads had a pore size between 20 and 50 Å. Both PVA beads and alginate are microporous, but the pore diameters in the PVA beads are larger than those in the alginate beads; thus, lower mass transfer resistance for PVA beads may be achieved. Blank alginate had higher capacity in gold removal in comparison with blank PVA. Equilibrium for the three systems was reached after 76, 120, and 372 h, for freely suspended biomass, PVA-immobilized biomass, and alginate-immobilized biomass, respectively (Khoo and Ting, 2001). Penthkar and Paknikar (1998) obtained 90 min as an equilibrium time for the uptake of gold (initial concentration at 100 mg·dm^{-3}) by *Cladosporium cladosporioides* that was immobilized in a keratinous material of natural origin. Increasing the cell and gold concentration increased the initial rate of sorption and the total removal of gold ions (Khoo and Ting, 2001).

Polysulfone-immobilized heat-inactivated *A. niger* has been successfully applied for cadmium, nickel, copper, and lead removal from aqueous solution in continuous mode. Maximum adsorption capacity for cadmium, nickel, copper, and lead uptake was 3.6, 1.08, 2.89, and 10.05 mg·g^{-1}, respectively (Kapoor and Virarghavan, 1998). Yahaya et al. (2009)

used Ca-alginate-immobilized *Pycnoporus sanguineus* living cells for the removal of copper ions from aqueous solution. In their study, the effect of pH, initial ion concentration, biosorbent loading, and temperature on metal uptake were investigated. The optimum uptake of Cu^{2+} ions was reported at pH 5 with a value of 2.76 $mg \cdot g^{-1}$. Increasing the initial copper concentration resulted in higher metal uptake. Increasing the biosorbent loading from 1 to 6 g resulted in increased removal; however, the additional increase of the biosorbent loading decreased the metal uptake. Higher biosorbent loading could produce a "screening" effect on the cell wall, preserving the binding sites, thus resulting in lower Cu^{2+} uptake (Mashitah et al., 1999; Pons and Fuste., 1993). Increasing the temperature from 303 to 313 K increased the copper uptake from 1.44 to 2.18 $mg \cdot g^{-1}$. Dependence of adsorption on temperature shows that copper biosorption takes place with a chemical reaction. Some thermodynamical parameters such as enthalpy and entropy changes were also indicated. Enthalpy change of 10.16 $kJ \cdot mol^{-1}$ and entropy change of 33.78 $J \cdot mol^{-1} \cdot K^{-1}$ were calculated from the biosorption data. The FTIR analysis showed that OH, NH, CH, CO, COOH, and CN groups were engaged in the biosorption of copper ions onto immobilized cells of *P. sanguineus*. Lu and Wilkins (1996) used heat and caustic-treated *S. cerevisiae* immobilized in alginate for Cu^{2+}, Cd^{2+}, and Zn^{2+} removal, and they compared the ability of the immobilized treated biomass with the free native, immobilized native yeast, and blank Ca-alginate. The order of metal uptake capacity of these biosorbents for all three ions was blank Ca-alginate > immobilized treated biomass > native free biomass > immobilized biomass. Thus, the amount of heavy metal binding to the biomass decreased about 10%–25% by immobilization in the Ca-alginate, when compared to the biosorption by the native yeast. Cross-linking of alginate gel with metal-binding sites and masking of active sites due to the higher density of the biomass are mentioned as the reasons for this decrease in biosorption. The adsorption capacities did not considerably change during the six repeated cycles of adsorption–desorption operations. The affinity of the biosorbents for copper ions was more than cadmium and then for zinc.

Spores of *Phanerochaete chrysosporium* were immobilized into Ca-alginate beads via entrapment for the removal of mercury and cadmium ions from aqueous solution in the concentrations range of 30–500 $mg \cdot L^{-1}$. More than 97% of alginate-fungus beads could be regenerated using 10 mM HCl. The biosorbents were reused in three biosorption–desorption cycles with trivial decrease in biosorption capacity (Kacar et al., 2002). Loofah sponge–immobilized *P. chrysosporium* was successfully used for the removal of lead, copper, and zinc ions from aqueous solution. The metal uptake capacity of immobilized *P. chrysosporium* was higher than that of the free biomass and blank loofah sponge for all the three ions. About 60 min is required for attaining equilibrium for all the metal ions. The rate of metal uptake was independent of initial concentrations. The optimum pH of biosorption for this biosorbent system was found to be 6, while temperature ranging from 10°C to 50°C was not an effective parameter. The maximum metal uptake capacity of immobilized biomass at 100 $mg \cdot L^{-1}$ metal solution was 88.2, 69.5, and 43.4 $mg \cdot g^{-1}$ for Pb^{2+}, Cu^{2+}, and Zn^{2+}, respectively (Iqbal and Edyvean, 2004). In another study, loofah sponge–immobilized *P. chrysosporium* was used for cadmium removal from aqueous solution. The optimum pH for the adsorption was 6. Metal uptake by immobilized and free biomass was 89 and 74 $mg \cdot g^{-1}$, respectively. Thus, immobilization has enhanced the metal uptake capacity. Equilibrium was achieved within 1 h, and biosorption was well defined by the Langmuir isotherm. Regeneration of biomass was done using 50 mM HCl, with up to 99% metal recovery. After 10 biosorption–desorption cycles, negligible loss of capacity was observed (Iqbal and Edyvean, 2005). Also, loofah sponge–immobilized living *P. chrysosporium* was used for the single lead ion removal from aqueous solution. Immobilization has a positive impact on the growth of *P. chrysosporium* as evidenced by the 19.2% increase in biomass production by the immobilized biomass as compared with free biomass over a period of 8 days. Biosorption equilibrium was established in about 1 h. Loofah sponge immobilization has no diffusional limitations as the rate of Pb^{2+} removal by immobilized system was higher than that of the free biomass. Furthermore, the metal uptake capacity for the immobilized biomass was 24.3% higher as compared with free biomass. Regeneration was carried out using 50 mM HCl and reused in seven cycles of biosorption–desorption without any considerable loss in biosorption capacity (Iqbal and Edyvean, 2007).

Bai and Abraham (2003) immobilized *Rhizopus nigricans* in five different polymeric matrices including Ca-alginate, PVA, polyacrylamide, polyisoprene, and polysulfone. They compared the metal uptake capacity of these biosorbents, and the chromium sorption capacity order was free biomass, 119.2 > polysulfone entrapped, 101.5 > polyisoprene immobilized, 98.76 > PVA immobilized, 96.6 > calcium-alginate entrapped, 84.29 > polyacrylamide, 45.56, at 500 $mg \cdot L^{-1}$ concentration of Cr^{6+}. They also compared the mechanical stability and chemical resistance of these different biosorbents, and the order was polysulfone > polyisoprene > PVA > polyacrylamide > calcium-alginate. They used 0.01 N NaOH, $NaHCO_3$, and Na_2CO_3 for the regeneration of bound metal. The successive sorption–desorption studies employing polysulfone-entrapped biomass showed that the biomass beads could be regenerated and reused in more than 25 cycles and the regeneration efficiency was 75%–78%.

Ca-alginate-entrapped *Trametes versicolor* fungus in both dead and live form has adsorbed cadmium ions from aqueous medium significantly. Maximum biosorption capacity for immobilized live and dead fungal mycelia of *T. versicolor* was found as 102.3 mg and 120.6 mg Cd per g of biosorbent. The dead immobilized white-rot fungus *T. versicolor* had a higher adsorption capacity than the living immobilized form. Thus, surface properties of the fungal cells were improved by the application of heat. The amount of cadmium ions adsorbed per unit mass of the biosorbent (i.e., biosorption capacity) increased, when the initial concentration of metal ions increased (Arica et al., 2001). Adsorbed cadmium ions were eluted with 10 mM HCl. Almost 95% of ions was desorbed. The adsorption capacities did not considerably change—only 3%—after three repeated cycles of adsorption–desorption operations (Arica et al., 2001).

Significant accumulation of lead, zinc, and copper ions was reported for *T. versicolor* immobilized in CMC beads via entrapment; CMC beads was prepared by cross-linking via ion exchange and metal coordination mechanism, with trivalent ferric ions. These beads were very stable over the experimental pH range of 3.0–8.0. Metal uptake capacities for living and heat-inactivated immobilized mycelia were compared. Influence of pH in the range of 3–7 was investigated on biosorption. While the biosorption of Cu^{2+}, Pb^{2+}, and Zn^{2+} immobilized by CMC is independent of the temperature in the range of 15°C–45°C for both live and inactivated preparations, it is dependent on pH and maximum metal uptake occurred at pH of 4–6. The amount of Cu^{2+}, Pb^{2+}, and Zn^{2+} ions adsorbed per unit mass of the biosorbent increased, with the increased initial concentration of metal ions. The immobilized inactivated *T. versicolor* had a higher adsorption capacity for all metal ions than the living immobilized one. The equilibrium time for biosorption for tested metal ions was reached within about 1 h. The correlation regression coefficients also showed that the adsorption process could be well defined by both Langmuir and Freundlich equations (Bayramoglu et al., 2003).

Ca-alginate-immobilized live and heat-inactivated *Funalia trogii*—wood-rotting fungus—via entrapment has shown a great potential in mercury, cadmium, and zinc ions removal from aqueous solution. Ca-alginate beads were provided by cross-linking with divalent calcium ions. The biosorption equilibrium time for tested metal ions was about 1 h. Influence of pH on the biosorption capacity was assessed, which resulted to a maximum removal at pH 6. The amount of Hg^{2+}, Cd^{2+}, and Zn^{2+} ions adsorbed per unit mass of the biosorbent increased with an increase in the initial concentration of metal ions in the adsorption medium (Arica et al., 2004). The heat treatment of immobilized white-rot fungi increases the biosorption capacities of metal ions, which can be attributed to a variety of resistance mechanisms including (1) extracellular complexation with metal-binding proteins like phytochelatins and metallothionein that are proteins that contain large amounts of cysteine and can bind heavy metal ions and (2) efficient pumping out of metal ions from the living cell (Arica et al., 2004). The correlation regression coefficients also showed that the adsorption process can be well defined by both Langmuir and Freundlich equations. HCl (10 mM) was used as a desorbing agent. The adsorption capacities did not noticeably change during the three repeated cycles of adsorption–desorption operations. The order of biosorbent affinity on a molar basis was Hg^{2+} > Cd^{2+} > Zn^{2+}.

Mahmoud et al. (2011) immobilized heat-inactivated *Aspergillus ustus* fungus on the silicon dioxide nano-powder matrix. Maximum adsorption capacity of cadmium in the biosorbent was 112 mg·g^{-1}. Similar to the majority of the studies, the metal uptake capacity of the cell immobilized system was higher than that of the free cells. The heat treatment could erode microbial cell surface integrity causing the walls to become leaky with an increase in the passive diffusion of metal ion to the interior part of the cell wall (Churchill et al., 1995). Table 21.2 shows different combination of fungus and matrices used for metal uptake from aqueous solutions and the capacity of the resulted bio-bead.

21.2.6 IMMOBILIZED ALGAL CELLS FOR METAL REMOVAL

Immobilized algal cells have shown a great potential to heavy metal removal. *Chlorella vulgaris* cells immobilized in alginate are a good biosorbent to remove heavy metals (Ilangovan et al., 1998; Lau et al., 1998; Tam et al., 1998; Abdel Hameed, 2002; Tajes-Martinez et al., 2006). Abu Al-Rub et al. (2004) have shown higher nickel uptake by dead *C. vulgaris* cells immobilized in Ca-alginate in comparison with free living cell. They also reported that blank alginate beads resulted in better nickel removal than the free algal cells. The form of algal cells did not significantly affect nickel removal. However, using dead cells removes the possibility of metal toxicity limitations and the need for growth media, and repeated use of cells will be easier. Significant accumulation of cobalt, zinc, and manganese was also recorded for *Chlorella salina* cells immobilized in alginate (Garnham et al., 1992). Fluidized beds of Ca-alginate entrapped cells of *C. vulgaris* were successfully applied to recover gold from a synthetic gold-bearing process solution containing $AuCl_4$, $CuCl_2$, $FeCl_2$, and $ZnCl_2$ (Vieira and Volesky, 2000). Rai and Mallick (1992) and Mallick and Rai (1993) also determined a greater potential of immobilized *C. vulgaris* and *Anabaena doliolum* in accumulating heavy metal ions of copper, nickel, and iron. Accumulation of mercury by free and immobilized *Chlorella emersonii* was determined by Wilkinson et al. (1990). About 90% recovery was achieved after 12 days, and the immobilized cell system was found to uptake more mercury than free cell. Abdel Hameed (2002) reported that the efficiency of immobilized *C. vulgaris* in iron, nickel, and zinc removal was higher than the free cells by 27%, 23%, and 25%, respectively. In another study, *Chlorella sorokiniana* was immobilized on loofah sponge matrix for nickel removal from liquid (Akhtar et al., 2004). The biosorption capacities for the immobilized and bare loofah sponge were found to be 60.38 and 6.1 mg nickel (II).g^{-1}, respectively. The biosorption kinetics were fast with 96% of adsorption within the first 5 min. Biosorption capacity was pH dependent, and maximum biosorption was in the pH range of 4–5. Different solutions including Na_2CO_3, $NaHCO_3$, NH_4Cl, CH_3COOH, H_2SO_4, HCl, and EDTA with 0.1 M concentration were used for the regeneration of the bound metal. Among them, HCl and EDTA with up to 99% recovery have higher recovery efficiencies. The immobilized biomass exhibited robust and stable properties with only slight decrease in the nickel uptake capacity when used in seven biosorption–desorption cycles. Akhtar et al. (2008) also studied the ability of *C. sorokiniana* immobilized on loofah for the removal of Cr(III) from aqueous solution. Maximum metal uptake capacities for the immobilized biomass and free biomass were 69.26 and 58.80 mg Cr(III).g^{-1}, respectively, whereas the amount of Cr(III) ions adsorbed onto blank loofah sponge was 4.97 mg.g^{-1}. Maximum sorption occurred at the solution pH of 4.0. Desorption with 0.1 M HNO_3 resulted in 98% of metal recovery, while other desorbing agents were less effective in the following order: EDTA > H_2SO_4 > CH_3COOH > HCl. The regenerated immobilized biomass could maintain 92.68% of the initial Cr(III) binding

TABLE 21.2
Immobilized Fungal Systems Used for Metal Removal

Microorganism	Matrices	Metal Ion	Capacity (mg·g⁻¹)	Reference
Rhizopus nigricans	Ca-alginate	Cr (VI)	84.3	Bai and Abraham (2003)
Rhizopus nigricans	PVA	Cr (VI)	96.7	Bai and Abraham (2003)
Rhizopus nigricans	Polyacrylamide	Cr (VI)	45.56	Bai and Abraham (2003)
Rhizopus nigricans	Polysulfone	Cr (VI)	101.5	Bai and Abraham (2003)
Rhizopus nigricans	Polyisoprene	Cr (VI)	98.76	Bai and Abraham (2003)
Ustilago maydis spores	Chitosan	Cr (III)	35.88	Sargin et al. (2016)
Ustilago digitariae spores	Chitosan	Cr (III)	49.40	Sargin et al. (2016)
Saccharomyces cerevisiae caustic treated	Ca-alginate	Cu (II)	8.4	Lu and Wilkins (1996)
Saccharomyces cerevisiae	Sepiolite	Cu (II)	4.7	Bag et al. (1999)
Saccharomyces cerevisiae	Sol-gel	Cu (II)	4.45	Al-Saraj et al. (1999)
Aspergillus niger	Ca-alginate	Cu (II)	17	Tsekova et al. (2010)
Aspergillus niger	PVA	Cu (II)	15.6	Tsekova et al. (2010)
Aspergillus niger	Polysulfone	Cu (II)	2.89	Kapoor and Virarghavan (1998)
Phanerochaete chrysosporium	Loofah sponge	Cu (II)	102.8	Iqbal and Edyvean (2004)
Pycnoporus sanguineus	Ca-alginate	Cu (II)	2.76	Yahaya et al. (2009)
Trametes versicolor heat inactivated	CMC	Cu (II)	1.84	Bayramoglu et al. (2003)
Trichoderma asperellum heat inactivated	Ca-alginate	Cu (II)	134.22	Tan and Ting (2012)
Ustilago maydis spores	Chitosan	Cu (II)	66.72	Sargin et al. (2016)
Ustilago digitariae spores	Chitosan	Cu (II)	69.26	Sargin et al. (2016)
Saccharomyces cerevisiae caustic treated	Ca-alginate	Cd (II)	6.20	Lu and Wilkins (1996)
Saccharomyces cerevisiae	Sepiolite	Cd (II)	10.9	Bag et al. (1999)
Aspergillus niger	Ca-alginate	Cd (II)	0.3	Tsekova et al. (2010)
Aspergillus niger	PVA	Cd (II)	0.25	Tsekova et al. (2010)
Aspergillus niger	Polysulfone	Cd (II)	3.6	Kapoor and Virarghavan (1998)
Trametes versicolor dead cell	Ca-alginate	Cd (II)	120.6	Arica et al. (2001)
Phanerochaete chrysosporium	Loofah sponge	Cd (II)	89.0	Iqbal and Edyvean (2005)
Aspergillus ustus heat inactivated	Silicon dioxide nano-powder	Cd (II)	112	Mahmoud et al. (2011)
Funalia trogii heat inactivated	Ca-alginate	Cd (II)	191.6	Arica et al. (2004)
Mucor rouxii	Polysulfone	Cd (II)	3.76	Yan and Viraraghavan (2001)
Ustilago maydis spores	Chitosan	Cd (II)	49.46	Sargin et al. (2016)
Ustilago digitariae spores	Chitosan	Cd (II)	53.96	Sargin et al. (2016)
Saccharomyces cerevisiae	Sol-gel	Pb (II)	41.9	Al-Saraj et al. (1999)
Aspergillus niger	Ca-alginate	Pb (II)	0.15	Tsekova et al. (2010)
Aspergillus niger	PVA	Pb (II)	0.14	Tsekova et al. (2010)
Aspergillus niger	Polysulfone	Pb (II)	10.05	Kapoor and Virarghavan (1998)
Phanerochaete chrysosporium	Loofah sponge	Pb (II)	135.3	Iqbal and Edyvean (2004)
Phanerochaete chrysosporium	Loofah sponge	Pb (II)	136.75	Iqbal and Edyvean (2007)
Trametes versicolor heat inactivated	CMC	Pb (II)	1.11	Bayramoglu et al. (2003)
Mucor rouxii	Polysulfone	Pb (II)	4.06	Yan and Viraraghavan (2001)
Saccharomyces cerevisiae	Sol-gel	Zn (II)	35.3	Al-Saraj et al. (1999)
Saccharomyces cerevisiae	Sepiolite	Zn (II)	8.37	Bag et al. (1999)
Saccharomyces cerevisiae caustic treated	Ca-alginate	Zn (II)	5.75	Lu and Wilkins (1996)
Aspergillus niger	PVA	Zn (II)	2.5	Tsekova et al. (2010)
Aspergillus niger	Ca-alginate	Zn (II)	2.4	Tsekova et al. (2010)
Phanerochaete chrysosporium	Loofah sponge	Zn (II)	50.9	Iqbal and Edyvean (2004)
Trametes versicolor heat inactivated	CMC	Zn (II)	1.67	Bayramoglu et al. (2003)
Mucor rouxii	Polysulfone	Zn (II)	1.36	Yan and Viraraghavan (2001)
Ustilago maydis spores	Chitosan	Zn (II)	30.73	Sargin et al. (2016)
Ustilago digitariae spores	Chitosan	Zn (II)	60.81	Sargin et al. (2016)
Saccharomyces cerevisiae	Sol-gel	Hg (II)	55.76	Al-Saraj et al. (1999)
Saccharomyces cerevisiae	Sol-gel	Co (II)	18.98	Al-Saraj et al. (1999)
Saccharomyces cerevisiae in the presence o/f 50 mM of glucose	Ca-alginate	Sr (II)	16.4	De Rome and Gadd (1991)
Rhizopus arrhizus	Pelleted	Sr (II)	7.7	De Rome and Gadd (1991)

(Continued)

TABLE 21.2 (Continued)
Immobilized Fungal Systems Used for Metal Removal

Microorganism	Matrices	Metal Ion	Capacity (mg·g⁻¹)	Reference
Penicillium chrysogenum	Pelleted	Sr (II)	8.0	De Rome and Gadd (1991)
Saccharomyces cerevisiae in the presence of 50 mM of glucose	Ca-alginate	Cs (I)	23.8	De Rome and Gadd (1991)
Rhizopus arrhizus	Pelleted	Cs (I)	10.9	De Rome and Gadd (1991)
Penicillium chrysogenum	Pelleted	Cs (I)	15.8	De Rome and Gadd (1991)
Saccharomyces cerevisiae in the presence of 50 mM of glucose	Ca-alginate	UO_2^{2+}	93.15	De Rome and Gadd (1991)
Rhizopus arrhizus	Pelleted	UO_2^{2+}	48.6	De Rome and Gadd (1991)
Rhizopus arrhizus	PVA	UO_2^{2+}	200	Tsezos and Deutschmann (1990)
Penicillium chrysogenum	Pelleted	UO_2^{2+}	39.69	De Rome and Gadd (1991)
Aspergillus niger	Polysulfone	Ni (II)	1.08	Kapoor and Virarghavan (1998)
Aspergillus niger	PVA	Ni (II)	1.5	Tsekova et al. (2010)
Aspergillus niger	Ca-alginate	Ni (II)	1.6	Tsekova et al. (2010)
Mucor rouxii	Polysulfone	Ni (II)	0.36	Yan and Viraraghavan (2001)
Ustilago maydis spores	Chitosan	Ni (II)	41.67	Sargin et al. (2016)
Ustilago digitariae spores	Chitosan	Ni (II)	33.46	Sargin et al. (2016)
Aspergillus niger	PVA	Mn (II)	19.3	Tsekova et al. (2010)
Aspergillus niger	Ca-alginate	Mn (II)	22.6	Tsekova et al. (2010)

capacity up to five reuse cycles in continuous-flow fixed-bed columns. Although the majority of studies resulted in higher metal uptake of immobilized biomass in comparison with free biomass, there are limited reports on higher metal uptake by free biomass. For example, immobilization of *Stichococcus bacillaris* algal cell in silica gel decreased the lead biosorption capacity by about 40% in comparison with free biomass (Mahan and Holcombe, 1992). Moreno-Garrido et al. (2005) used the marine microalgae *Tetraselmis chui* for heavy metal removal experiment. Ca-alginate-immobilized algal cells were exposed to 820 µg·L⁻¹ copper and 870 µg·L⁻¹ cadmium during a 24 h period. They reported that practically 20% of cadmium and approximately all copper ions were removed by the immobilized biomass, while blank Ca-alginate removed half of that percentage. *Sphagnum* peat moss immobilized in porous polysulfone has been successfully used to remove zinc, cadmium, and manganese from acid mine drainage waters. Greene and Bedell (1990) demonstrated that

immobilized, nonliving algae are able to bind to some heavy metal ions. Spinti et al. (1995) showed that immobilized biomass efficiently removed heavy metals from wastewater under appropriate conditions. The uptake of copper by inactivated *Sargassum baccularia*—brown seaweed—immobilized into PVA was investigated by Tan et al. (2002). They reported that the stability of the immobilized biomass could lead to the development of an efficient technique for the removal of toxic metals from aqueous solutions. Ulrich et al. (2010) assessed the metal uptake capability of different microalgae and macroalgae powder. They used various sol-gel materials for the immobilization of the algal biomass for the removal of nickel, chromium, copper, and lead from aqueous solution. Among the different biosorbents, *Laminaria saccharina* immobilized in sol-gel had the highest metal uptake capacity of 89.91 µmol·g⁻¹ of biosorbent. Table 21.3 shows different combination of algal cells and matrices used for metal uptake from aqueous solutions.

TABLE 21.3
Immobilized Algal Systems Used for Metal Removal

Microorganism	Matrices	Metal Ion	Capacity (mg·g⁻¹)	Reference
Chlorella vulgaris	K-carrageenan	Cr (III)	—	Travieso et al. (1999)
Chlorella vulgaris	K-carrageenan	Cd (II)	—	Travieso et al. (1999)
Chlorella vulgaris	K-carrageenan	Zn (II)	—	Travieso et al. (1999)
Chlorella vulgaris dead cell	Ca-alginate	Ni (II)	31.3	Abu-Al Rub et al. (2004)
Chlorella sorokiniana	Loofah sponge	Ni (II)	60.38	Akhtar et al. (2004)
Chlorella sorokiniana	Loofah sponge	Cr (III)	69.26	Akhtar et al. (2008)
Chlamydomonas reinhardtii	Ca-alginate	Cd (II)	79.7	Bayramoglu et al. (2006)
Chlamydomonas reinhardtii	Ca-alginate	Pb (II)	308.7	Bayramoglu et al. (2006)
Chlamydomonas reinhardtii	Ca-alginate	Hg (II)	106.6	Bayramoglu et al. (2006)

21.3 BIOLEACHING

Bioleaching is one of the most important types of metal bioremediation in which the metal is oxidized and solubilized by the microbe. The solubilized metal ion can be further concentrated and purified. Few studies have evaluated the capability of the immobilized form of bacteria for bioleaching; these cases are limited only to *Acidithiobacillus ferrooxidans*. Lilova and Karamanev (2005) assessed the direct oxidation of synthetic copper sulfide by a biofilm of the mentioned bacteria. Yujian et al. (2006) applied a new immobilization method for *A. ferrooxidans* entrapment. They used a complex of PVA and Ca-alginate as an immobilization matrix and achieved a maximum oxidation rate of 4.6 g $Fe^{2+} \cdot L^{-1} \cdot h^{-1}$. Giaveno et al. (2008) immobilized *A. ferrooxidans* on chitosan beads and reached the ferric ion productivity of 1.5 $g \cdot L^{-1} \cdot h^{-1}$ in a packed bed bioreactor.

21.4 BIOREACTORS

A bioreactor is a system that can contain and support an organism during a desired process. The bioreactors have defined and controllable environment. The internal environment is usually manipulated and monitored easily. Factors affecting the efficiency of a process and a system have to be defined and optimized. Temperature, pH, mixing, and mass transfer are among affecting factors in the case of metal uptake. There are numerous bioreactors with varying configurations to meet different requirements. Two main categories of bioreactors are used in metal removal, fixed beds and fluidized beds (Gavrilescu, 2004).

21.4.1 PACKED BED BIOREACTORS

Packed bed reactors have been used for immobilized cellular processes more than any other bioreactor configurations. In general, such systems are efficient when maintaining rather long retention times are necessary and external biomass buildup is minimal. Different bioreactors in terms of design and operation have been studied, including horizontal packed bed and multiple column sequences (Mallick, 2002).

A packed bed is simply a piece of pipe, standing on its end and filled with immobilized beads. Solution containing the metal ions flows into one end of the pipe and out the other side. Initially, most of the solute is adsorbed so that the solute concentration is low in the effluent. As adsorption continues, the effluent concentration increases, slowly at first but then abruptly. When this sudden rise—breakthrough—happens, the flow is stopped. The adsorbed metal ion is then eluted by washing the bed with an eluent (Belter et al., 1988). Numerous kinetic models and mathematical correlations have been presented to determine the different sorption rate parameters. It is well recognized that the roughness and topography of a sorbent are important factors that affect the sorption rate parameters. Also, both film resistance and pore diffusion play important roles in the overall transfer of the sorbate. Therefore, different approaches have been suggested

to determine the overall mass transfer in an adsorption system (Gavrilescu, 2004).

Adsorption in a packed bed bioreactor can be expressed by four equations. The first is mass balance on the metal ion concentration in the solution (Belter et al., 1988):

$$\varepsilon \frac{\partial y}{\partial t} = -v \frac{\partial y}{\partial z} + E \frac{\partial^2 y}{\partial z^2} - (1 - \varepsilon) \frac{\partial q}{\partial t} \qquad (21.1)$$

where

 y is the metal ion concentration in the solution

 q is the amount of metal adsorbed per amount of immobilized biomass

 ε is the void fraction in the bed

 $v\ (=H/A)$ is the superficial velocity

 E is a dispersion coefficient

The second term on the right-hand side represents dispersion in the bed. Such dispersion leads to mixing of solute (metal ion) and solvent even in the absence of any adsorption. Dispersion is like diffusion; however, dispersion coefficient in most packed beds is much larger than the diffusion coefficient. As a result, diffusion has been omitted from Equation 21.1.

The second equation is mass balance on the adsorbed metal ion:

$$(1 - \varepsilon) \frac{\partial q}{\partial t} = r \qquad (21.2)$$

The adsorption rate r depends on the mechanism responsible for adsorption that is controlled by mass transfer from the bulk of solution to the surface of the immobilized biomass or by the diffusion within the beads. We make the major assumption that the adsorption rate is linear:

$$r = ka(y - y^*) \qquad (21.3)$$

where

 r represents the rate per bed volume and has the dimension of mass adsorbed per volume per time

 ka is the rate constant

 y^* is the concentration that would exist in solution at equilibrium

The fourth and final key equation is an isotherm:

$$q = k(y^*)^n \qquad (21.4)$$

Therefore, there is a system of nonlinear and coupled equations and so must be solved numerically. There are some approximate methods of characterizing fixed-bed performance. First, the breakthrough can be modeled as a ramp, and second, modeling of these curves with two parameters. If the adsorption isotherm is linear, the equations can be solved in analytical form (Belter et al., 1988).

21.5 SUMMARY

Different matrices and microorganisms including bacteria, yeast, fungi, and algae are used for obtaining an appropriate biosorbent, which could be used in either continuous or batch processes for heavy metal removal. Generally, immobilization improves the chemical and physical stability of the biosorbent and for most of the studied microorganisms immobilization increases metal uptake capacity. Immobilization provides repetitive adsorption/desorption use of the biosorbent system, which is one of the greatest advantage of using immobilization techniques for sequestering metal ions. Biosorption by immobilized microorganisms like free one is extremely pH dependent, while temperature in most of the biosorbent systems is a negligible parameter.

Application of immobilized microorganisms for metal removal is still an open area of research in different fields. Screening for new matrices, isolating, or finding tolerant microorganisms to metal ions are essential prerequisites of designing an efficient biosorbent system that could be used commercially in large-scale processes. Design of automated production systems of immobilized microorganisms and introduction of new bioreactor configurations may enable the extensive utilization of the immobilized systems for metal ion removal, which can guarantee a cleaner environment.

REFERENCES

Abdel Hameed, M.S. 2002. Effect of Immobilization on growth and photosynthesis of the green alga *Chlorella vulgaris* and its efficiency in heavy metals removal. *Bulletin of the Faculty of Science* 31: 233–240.

Abu Al-Rub, F.A., El-Naas, M.H., Benyahia, F., and Ashour, I. 2004. Biosorption of nickel on blank alginate beads, free and immobilized algal cells. *Process Biochemistry* 39: 1767–1773.

Akhtar N., Iqbal, J., and Iqbal, M. 2004. Removal and recovery of nickel (II) from aqueous solution by loofah sponge-immobilized biomass of *Chlorella sorokiniana*: Characterization studies. *Journal of Hazardous Materials* 108: 85–94.

Akhtar, N., Iqbal, M., Zafar, S.I., and Iqbal, J. 2008. Biosorption characteristics of unicellular green alga *Chlorella sorokiniana* immobilized in loofah sponge for removal of Cr (III). *Journal of Environmental Sciences* 20: 231–239.

Al-Saraj, M., Abdel-Latif, M.S., El-Nahal, I., and Baraka, R. 1999. Bioaccumulation of some hazardous metals by sol-gel entrapped microorganisms. *Journal of Non-Crystalline Solids* 248: 137–140.

Andrade, L.R., Salgado, L.T., Farina, M., Pereira, M.S., Mourao, P.A.S., and Amado Filho, G.M. 2004. Ultrastructure of acidic polysaccharides from the cell walls of brown algae. *Journal of Structural Biology* 145: 216–225.

Arica, M.Y., Kacar, Y., and Gence, O. 2001. Entrapment of white-rot fungus *Trametes versicolor* in Ca-alginate beads: Preparation and biosorption kinetic analysis for cadmium removal from an aqueous solution. *Bioresource Technology* 80: 121–129.

Arica, M.Y., Bayramoglu, G., Yilmaz, M., Bektas, S., and Gence, O. 2004. Biosorption of Hg^{2+}, Cd^{2+}, and Zn^{2+} by Ca-alginate and immobilized wood-rooting fungus *Funalia trogii*. *Journal of Hazardous Materials* 109:191–199.

Ariga, O., Takagi, H., Nishizawa, H., and Sano, Y. 1987. Immobilization of microorganisms with PVA hardened by iterative freezing and thawing. *Journal of Fermentation Technology* 65: 651–658.

Ariga, O., Itoh, K., Sano, Y., and Nagura, M. 1994. Encapsulation of biocatalysts with PVA capsules. *Journal of Bioscience and Bioengineering* 78: 74–78.

Bag, H., Lale, M., and Turker, A.R. 1999. Determination of Cu, Zn and Cd in water by FAAS after preconcentration by baker's yeast (*Saccharomyces cerevisiae*) immobilized on sepiolite. *Fresenius Journal of Analytical Chemistry* 363: 224–230.

Bai, R.S. and Abraham, T.E. 2003. Studies on chromium (VI) adsorption–desorption using immobilized fungal biomass. *Bioresource Technology* 87: 17–26.

Bang, S.S. and Pazirandeh, M. 1999. Physical properties and heavy metal uptake of encapsulated *Escherichia coli* expressing a metal binding gene (NCP). *Journal of Microencapsulation* 16: 489–499.

Bayramoglu, G., Bekta, S., and Arıca, M.Y. 2003. Biosorption of heavy metal ions on immobilized white-rot fungus *Trametes versicolor*. *Journal of Hazardous Materials* 101: 285–300.

Bayramoglu, G., Tuzun, I., Celik, G., Yilmaz, M., and Arica, M.Y. 2006. Biosorption of mercury (II), cadmium (II) and lead (II) ions from aqueous system by microalgae *Chlamydomonas reinhardtii* immobilized in alginate beads. *International Journal of Mineral Processing* 81: 35–43.

Baysal, S.H., Onal, S., and Ozdemir, G. 2009. Biosorption of chromium, cadmium, and cobalt from aqueous solution by immobilized living cells of *Chryseomonas luteola* TEM 05. *Preparative Biochemistry and Biotechnology* 39: 419–428.

Belter, P.A., Cussler, E.L., and Hu, W. 1998. *Biseparations: Downstream Processing for Biotechnology*. Wiley, Sterling Heights, MI.

Belter, P.A., Cussler, E.L., and Hu, W. 1988. Biseparations: Downstream processing for biotechnology.

Bozeman, J., Koopman, B., and Bitton, G. 1989. Toxicity testing using immobilized algae. *Aquatic Toxicology* 14: 345–352.

Branco, R., Sousa, T., Piedade, A.P., and Morais, P.V. 2016. Immobilization of *Ochrobactrum tritici* As5 on PTFE thin films for arsenite biofiltration. *Chemosphere* 146: 330–337.

Brandenberger, H. and Widmer, F., 1998. A new multi-nozzle encapsulation/ immobilization system to produce uniform beads of alginate. *Journal of Biotechnology* 63: 73–80.

Brouers, M., Dejong, H., Shi, D.J., and Hall, D.O. 1989. Immobilized cells: An appraisal of the methods and applications of cell immobilization techniques. In: Cresswell R.C., Rees T.A.V., Shah N., *Algal and Cyanobacterial Biotechnology*, pp. 272–293.

Carpio, I.E.M., Machado-Santelli, G., Sakata, S.K., Filho, S.S.F., and Rodrigues, D.F. 2014. Copper removal using a heavy-metal resistant microbial consortium in a fixed-bed reactor. *Water Research* 62: 156–166.

Cassidy, M.B., Lee, H., and Trevors, J.T. 1996. Environmental applications of immobilized microbial cells. *Journal of Industrial Microbiology and Biotechnology* 16: 79–101.

Churchill, S.A., Walters, J.V., and Churchill P.F. 1995. Sorption of heavy metals by pretreated bacterial cell surfaces. *Journal of Environmental Engineering* 121: 706–711.

de-Bashan, L.E. and Bashan, Y. 2010. Immobilized microalgae for removing pollutants: Review of practical aspects. *Bioresource Technology* 101: 1611–1627.

de Rome, L. and Gadd, G.M. 1991. Use of pelleted and immobilized yeast and fungal biomass for heavy metal and radionuclide recovery. *Journal of Industrial Microbiology and Biotechnology* 7: 97–104.

Eroglu, E., Smith, S.M., Raston, C.L. 2015. Application of various immobilization techniques for algal bioprocesses. In: Moheimani, N.R., McHenry, M.P., de Boer, K., Bahri, P., *Biomass and Biofuels from Microalgae: Advances in Engineering and Biology.* Switzerland: Springer international pub, pp. 19–44.

Fayyaz Ahmad, M., Haydar, S., Bhatti, A.A., and Bari, A.J. 2014. Application of artificial neural network for the prediction of biosorption capacity of immobilized *Bacillus subtilis* for the removal of cadmium ions from aqueous solution. *Biochemical Engineering Journal* 84: 83–90.

Garnham, W.G., Codd, G.A., and Gadd, M.G. 1992. Accumulation of Cobalt, Zinc and Manganese by the estuarine green *Chlorella salina* immobilized in alginate microbeads. *Environmental Science and Technology* 26: 1764–1770.

Gavrilescu, M. 2004. Removal of heavy metals from the environment by biosorption. *Engineering in Life Sciences* 4: 219–232.

Giaveno, A., Lavalle, L., Guibal, E., and Donati, E. 2008. Biological ferrous sulfate oxidation by *A. ferrooxidans* immobilized on chitosan beads. *Journal of Microbiological Methods* 72: 227–234.

Greene, B. and Bedell, G.W. 1990. Algal gels or immobilized algae for metal recovery. In: Akatsuka I. *An Introduction to Applied Phycology.* SPB Academic Publishing: The Hague, pp. 137–149.

Guiseley, K.B. 1989. Chemical and physical properties of algal polysaccharides used for cell immobilization. *Enzyme and Microbial Technology* 11: 706–716.

Gupta, A. and Balomajumder, C. 2015. Simultaneous removal of Cr(VI) and phenol from binary solution using *Bacillus* sp. immobilized onto tea waste biomass. *Journal of Water Process Engineering* 6: 1–10.

Hashimoto, S. and Furukawa, K. 1986. Immobilization of activated sludge by PVA–boric acid method. *Biotechnology and Bioengineering* 30: 52–59.

Hulst, A.C., Tramper, J., van Riet, K., and Weesterbeek, J.M.M. 1985. A new technique for the production of immobilized biocatalyst in large quantities. *Biotechnology and Bioengineering* 27: 870–876.

Ilangovan, K., Canizares-Villanueva, R.O., Gonzalez Moreno, S., and Voltolina, D. 1998. Effect of Cadmium and Zinc on Respiration and Photosynthesis in Suspended and Immobilized Cultures of *Chlorella vulgaris* and *Scenedesmus acutus*. *Bulletin of Environmental Contamination and Toxicology* 60: 936–943.

Iqbal, M. and Edyvean, R.G.J. 2004. Biosorption of lead, copper and zinc ions on loofah sponge immobilized biomass of *Phanerochaete chrysosporium*. *Minerals Engineering* 17: 217–223.

Iqbal, M. and Edyvean, R.G.J. 2005. Loofah sponge immobilized fungal biosorbent: A robust system for cadmium and other dissolved metal removal from aqueous solution. *Chemosphere* 61: 510–518.

Iqbal, M. and Edyvean, R.G.J. 2007. Ability of loofah sponge-immobilized fungal biomass to remove lead ions from aqueous solution. *Pakistan Journal of Bio Technology* 39: 231–238.

Imai, K., Shiomi, T., Uchida, K., and Miya, M. 1985. Immobilization of enzyme into polyvinyl alcohol membrane. *Biotechnology and Bioengineering* 28: 1721–1726.

Jen, A.C., Wake, C., and Mikos, A.G. 1996. Review: Hydrogels for cell immobilization. *Biotechnology and Bioengineering* 50: 357–364.

Jezequel, K. and Lebeau, T. 2008. Soil bioaugmentation by free and immobilized bacteria to reduce potentially phytoavailable cadmium. *Bioresource Technology* 99: 690–698.

Kacar, Y., Arpa, C., Tan, S., Denizli, A., Genc, O., and Arica, M.Y. 2002. Biosorption of Hg(II) and Cd(II) from aqueous solutions: Comparison of biosorptive capacity of alginate and immobilized live and heat inactivated *Phanerochaete chrysosporium*. *Process Biochemistry* 37: 601–610.

Kapoor, A. and Viraraghavan, T. 1995. Fungal biosorption—An alternative treatment option for heavy metal bearing wastewaters: A review. *Bioresource Technology* 53: 195–206.

Kapoor, A. and Viraraghavan, T. 1998. Removal of heavy metals from aqueous solutions using immobilized fungal biomass in continuous mode. *Water Research* 32: 1968–1977.

Khoo, K.M. and Ting, Y.P. 2001. Biosorption of gold by immobilized fungal biomass. *Biochemical Engineering Journal* 8: 51–59.

Kolot, F.B. 1981. Microbial carriers-strategy for selection. *Process Biochemistry* 5: 2–9.

Kuu, W.Y. and Polack, J.A. 1983. Improving immobilized biocatalysts by gel phase polymerization. *Biotechnology and Bioengineering* 25: 1995–2006.

Lau, P.S., Tam, N.F.Y., and Wong, Y.S. 1998. Operational optimization of batch wise nutrient removal from wastewater by carrageenan immobilized *Chlorella vulgaris*. *Water Science and Technology* 38: 185–192.

Leenen, E.J., Dos Santos, V.A., Grolle, K.C., Tramper, J., and Wijffels, R. 1996. Characteristics of and selection criteria for support materials for cell immobilization in wastewater treatment. *Water Research* 30: 2985–2996.

Lilova, K. and Karamanev, D. 2005. Direct oxidation of copper sulfide by a biofilm of *Acidithiobacillus ferrooxidans*. *Hydrometallurgy* 80: 147–154.

Liu, Y.G., Ting, L.I.A.O., He, Z.B., Li, T.T., Hui, W.A.N.G., Hu, X.J. et al. 2013. Biosorption of copper (II) from aqueous solution by *Bacillus subtilis* cells immobilized into chitosan beads. *Transactions of Nonferrous Metals Society of China* 23: 1804–1814.

Lopez, A., Lazaro, N., Morales, S., and Marques, A.M. 2002. Nickel biosorption by free and immobilized cells of *Pseudomonas fluorescens* 4F39: A comparative study. *Water, Air, and Soil Pollution* 135: 157–172.

Lu, Y. and Wilkins, E. 1996. Heavy metal removal by caustic-treated yeast immobilized in alginate. *Journal of Hazardous Materials* 49: 165–179.

Mahan, C.A. and Holcombe, J.A. 1992. Immobilization of algae cells on silica gel and their characterization for trace metal pre-concentration. *Analytical Chemistry* 64: 1933–1939.

Mahmoud, M.E., Yakout, A.A., Abdel-Aal, H., and Osman, M.M. 2011. Enhanced biosorptive removal of cadmium from aqueous solutions by silicon dioxide nano-powder, heat inactivated and immobilized *Aspergillus ustus*. *Desalination* 279: 291–297.

Mallick, N. 2002. Biotechnological potential of immobilized algae for wastewater N, P and metal removal. *BioMetals* 15: 377–390.

Mallick, N. and Rai, I.C. 1993. Influence of culture density, pH, organic acids and divalent cations on the removal of nutrients and metals by immobilized *Anabaena doliolum* and *Chlorella vulgaris*. *Journal of Microbiology and Biotechnology* 9: 196–201.

Martins Dos Santos, V.A., Leenen, E.J., Rippoll, M.M., van der Sluis, C., van Vliet, T., Tramper, J., and Wijffels, R.H. 1997. Relevance of rheological properties of gel beads for their mechanical stability in bioreactors. *Biotechnology and Bioengineering* 56: 517–529.

Mashitah, M.D., Zulfadhfy, Z., and Bhatla, S. 1999. Ability of *Pycnoporus sanguineus* to remove copper ions from aqueous solution. *Artificial Cells, Blood Substitutes, and Biotechnology* 27: 429–433.

Mehta, S.K. and Gaur, J.P. 2005. Use of algae for removing heavy metal ions from wastewater: Progress and prospects. *Critical Reviews in Biotechnology* 25: 113–152.

Moreno-Garrido, I., Campana, O., Lubian, L.M., and Blasco, J. 2005. Calcium alginate immobilized marine microalgae: Experiments on growth and short-term heavy metal accumulation. *Marine Pollution Bulletin* 51: 823–829.

Nakajima, A., Horikoshi, T., and Sakaguchi, T. 1982. Recovery of uranium by immobilized microorganisms. *European Journal of Applied Microbiology and Biotechnology* 16: 88–91.

Penthkar, A.V. and Paknikar, K.M. 1998. Recovery of gold from solutions using *Cladosporium cladosporioides* biomass beads. *Journal of Biotechnology* 63: 121–136.

Piccirillo, C., Pereira, S.I.A., Marques, A.P.G.C., Pullar, R.C., Tobaldi, D.M., Pintado, M.E. and Castro, P.M.L. 2013. Bacteria immobilisation on hydroxyapatite surface for heavy metals removal. *Journal of Environmental Management* 121: 87–95.

Pires, C., Marques, P.G.C.A., Guerreiro, A., Magan, N., and Castro, M.L.P. 2011. Removal of heavy metals using different polymer matrixes as support for bacterial immobilization. *Journal of Hazardous Materials* 191: 277–286.

Podder, M.S. and Majumder, C.B. 2016. Fixed-bed column study for As (III) and As (V) removal and recovery by bacterial cells immobilized on Sawdust/MnFe$_2$O$_4$ composite. *Biochemical Engineering Journal* 105: 114–135.

Pons, M.P. and Futse, C.M. 1993. Uranium uptake by immobilized cells *Pseudomonas* strain EPS 5028. *Applied Microbiology and Biotechnology* 39: 661–665.

Rai, L.C. and Mallick, N. 1992. Removal and assessment of toxicity of Cu and Fe to *Anabaena doliolum* and *Chlorella vulgaris* using free and immobilized cells. *World Journal of Microbiology and Biotechnology* 8: 110–114.

Sargın, I., Arslan, G., and Kaya, M. 2016. Microfungal spores (*Ustilago maydis* and *U. digitariae*) immobilized chitosan microcapsules for heavy metal removal. *Carbohydrate Polymers* 138: 201–209.

Shuler, M.L. and Kargi, F. 2002. *Bioprocess Engineering Basic Concepts*, 2nd edn. Prentice Hall, Upper Saddle River, NJ.

Spinti, M., Zhuang, H., and Trujillo, E.M. 1995. Evaluation of immobilized biomass beads for removing heavy metals from wastewaters. *Water Environment Research* 67: 943–952.

Sumino, T., Nakamura, H., Mori, N., Kawaguchi, Y., and Tada, M. 1992. Immobilization of nitrifying bacteria in porous pellets of urethane gel for removal of ammonium nitrogen from wastewater. *Applied Microbiology and Biotechnology* 36: 556–560.

Tajes-Martinez, P., Beceiro-Gonzalez, E., Muniategui-Lorenzo, S., and Prada-Rodriguez, D. 2006. Micro-columns packed with *Chlorella vulgaris* immobilised on silica gel for mercury speciation. *Talanta* 68: 1489–1496.

Tam, N.F.Y., Wong, Y.S., and Simpson, C.G. 1998. Removal of copper by free and immobilized microalga, *Chlorella vulgaris*. In: Tam, N.F.Y., Wong, Y.S., *Wastewater Treatment with Algae*. Springer, Berlin, Germany, pp. 17–36.

Tan, K.F., Chu, K.H., and Hashim, M.A. 2002. Continuous packed bed biosorption of copper immobilized seaweed biomass. *European Journal of Mineral Processing and Environmental Protection* 2: 246–252.

Tan, W. and Ting, A.S.Y. 2012. Efficacy and reusability of alginate-immobilized live and heat-inactivated *Trichoderma asperellum* cells for Cu (II) removal from aqueous solution. *Bioresource Technology* 123: 290–295.

Tanaka, K., Tada, M., Kimata, T., Harada, S., Fujii, Y., Mizuguchi, T., Mori, N., and Emori, H. 1991. Development of new nitrogen removal systems using nitrifying bacteria immobilized in synthetic resin pellets. *Water Science and Technology* 23: 681–690.

Texier, C., Andres, Y., Faur-Brasquet, C., and Le Cloirec, P. 2002. Fixed-bed study for lanthanide (La, Eu, Yb) ions removal from aqueous solutions by immobilized *Pseudomonas aeruginosa*: Experimental data and modelization. *Chemosphere* 47: 333–342.

Ting, A.S.Y., Rahman, N.H.A., Isa, M.I.H.M., and Tan, W.S. 2013. Investigating metal removal potential by effective microorganisms (EM) in alginate-immobilized and free-cell forms. *Bioresource Technology* 147: 636–639.

Tobin, J.M., L'homme, B., and Roux, J.C., 1993. Immobilisation protocols and effects on cadmium uptake by *Rhizopus arrhizus* biosorbents. *Biotechnology Techniques* 7: 739–744.

Travieso, L., Canizares, R.O., Borja, R., Benitez, F., Dominguez, A.R., Dupeyrón y, R., and Valiente, V. 1999. Heavy metal removal by microalgae. *Bulletin of Environmental Contamination and Toxicology* 62: 144–151.

Tsekova, K., Todorova, D., and Ganeva, S. 2010. Removal of heavy metals from industrial wastewater by free and immobilized cells of *Aspergillus niger*. *International Biodeterioration and Biodegradation* 64: 447–451.

Tsezos, M. and Deutschmann, A.A. 1990. An investigation of engineering parameters for the use of immobilized biomass particles in biosorption. *Journal of Chemical Technology and Biotechnology* 48: 29–39

Tsuruta, T. 2004. Biosorption and recycling of gold using various microorganisms. *The Journal of General and Applied Microbiology* 50: 221–228.

Ulrich, S., Sabine, M., Gunter, K., Wolfgang, P., and Horst, B. 2010. Algae-silica hybrid materials for biosorption of heavy metals. *Journal of Water Resource and Protection* 115–122.

Vieira R.H.S.F. and Volesky B. 2000. Biosorption: A solution to pollution? *International Microbiology* 3: 17–24.

Volesky, B. 2001. Detoxification of metal-bearing effluents: Bio-sorption for the next century. *Hydrometallurgy* 59: 203–216.

Wang, J.L. and Chen, C. 2006. Biosorption of heavy metals by *Saccharomyces cerevisiae*: A review. *Biotechnology Advances* 24: 427–451.

Wang, J. and Chen, C. 2009. Biosorbents for heavy metals removal and their future. *Biotechnology Advances* 27: 195–226.

Wilkinson, S.C., Goulding, K.H., and Robinson, P.K. 1990. Mercury removal by immobilized algae in batch culture systems. *Journal of Applied Phycology* 2: 223–230.

Yahaya, Y.A., Don, M.M., and Bhatia, S. 2009. Biosorption of copper (II) onto immobilized cells of *Pycnoporus sanguineus* from aqueous solution: Equilibrium and kinetic studies. *Journal of Hazardous Materials* 161: 189–195.

Yan, G. and Viraraghavan, T. 2001. Heavy metal removal in a biosorption column by immobilized *M. rouxii* biomass. *Bioresource Technology* 78: 243–249.

Yujian, W., Xiaojuan, Y., Hongyu, L., and Wei, T. 2006. Immobilization of *Acidithiobacillus ferrooxidans* with complex of PVA and sodium alginate. *Polymer Degradation and Stability* 91: 2408–2414.

Zhou, J.L. and Kiff, R.J. 1991. The uptake of copper from aqueous solution by immobilized fungal biomass. *Journal of Chemical Technology and Biotechnology* 52: 317–330.

22 Bioreactor for Detoxification of Heavy Metals through Bioleaching Technique

Tiyasha Kanjilal, Pramita Sen, Arijit Nath, and Chiranjib Bhattacharjee

CONTENTS

ABSTRACT

The extent of Bioleaching phenomenon, over the years have been observed, not only for recovery of precious metals but to a widespread other emerging applications. There is requirement for less costly and more environmentally feasible processes. Further advancement is necessary regarding technical and biological features. The latter incorporates increasing the rate of leaching and the acceptance of the potential microbial strains to heavy metals. Microbes properly function only when the waste components permit them to produce nutrients and energy for the growth of more cells. When unfavorable conditions strike, their capacity of degradation is reduced. In such conditions, genetically engineered microorganisms have to be used, although stimulating indigenous microorganisms is preferred.

22.1 INTRODUCTION

Prospective long-term sustainable development requires measures to reduce the reliance on nonrenewable raw materials and the demand for primary resources. Development of metal resources is required with the support of novel technologies; additionally, enhancement of age-old mining techniques can result in the recovery of metal from sources that have less economical attention even today (Bosecker, 1997). Metal-winning processes based on the activity of microorganisms offer a prospect to obtain metals from mineral resources not available in conventional mining (Brierley, 1978). Microorganisms such as bacteria and fungi convert metal components into their easily soluble forms and act as biocatalysts in these leaching processes. Moreover, applying microbiological solubilization processes can likely recover valuable metals from industrial wastes that can be used as secondary raw materials. For any nation, sustainable increase in metal values through utilization of mineral resources is a benchmark of its prudential growth. India's natural mineral assets have been exploited significantly to a larger extent during the past 50 years. The advancement of industrialization together with population growth has increased the demand for metals and is likely to expand in the future. This has resulted in irreversible effects such as degradation of high-grade ores with concurrent generation of solid wastes and effluent-loaded metals. As such, there is a necessity to tackle the problem of preventing pollution and recovery of metal values by a cost-effective process.

Sustainable mineral biotechnology has paved the way in mineral engineering for the advancement of economically feasible processes of metal bioremediation, wastes, and low-grade ore utilization through methods of biochemical leaching, ore upgradation through biobeneficiation, treatment of effluents through bioaccumulation and bioprecipitation, respectively. The innate potentiality of microorganisms belonging to different groups has been efficiently used in all these processes. Worldwide reserves of metal ores are reducing at an alarming rate owing to the increasing demand for metals, though there subsist large stockpiles of low-grade ores still to be mined. However, with the rise in the expense of high energy and capital inputs, recovery of metals from them using conventional techniques is cumbersome. In addition, there are various environmental costs due to the high level of pollution from these technologies. Environmental norms continue to stiffen, mainly toward toxic wastes, so outlays regarding environmental protection will continue to rise. The recovery of metals by the application of microorganisms is now a worldwide promising biotechnological process (Vera et al., 2013). It is one of the revolutionary solutions to the hitch of conventional techniques, such as pyrometallurgy or industrial metallurgy. It offers the promise of vividly decreasing capital costs and reducing environmental pollution. In general, biological processes are mild and often end up in solution acquiescent to treatment than toxic gaseous waste.

Frequently, this technique is mainly employed for metals including copper, cobalt, nickel, zinc, and uranium. These are typically extracted either from insoluble oxides or from sulfides. However, for the recovery of metals like gold and silver from their ores, the utility of leaching bacteria is applied only to digest interfering metal sulfides prior to cyanidation treatment (Ehrlich, 2009). Here, the term *bio-oxidation* is used suitably because the solubilized bioleached metals, in most cases iron and arsenic, are not proposed to be recovered. In general, the term covering both bioleaching and biooxidation techniques is "biomining" (Olson et al., 2003). The solubilization of inorganic pollutants in aqueous solution is performed by exogenous microbial activity, mainly for their capability to dissolve metals and nonmetal targets. Hence, the technique of bioleaching essentially includes a posttreatment clearing or regeneration of the aqueous phase after metal extraction (Simona and Micle, 2011). A principal design of the biotreatment of resource-contaminated leachate is shown in Figure 22.1.

Promising potentialities from developments in the fields of gene biology, omics studies, surface science study (nanobiotechnology), and chemical investigation have added to an advanced understanding of this bioprocess. However, which processes are actually happening at the molecular level at bacterial–mineral interfaces is yet to be known (Vera et al., 2013).

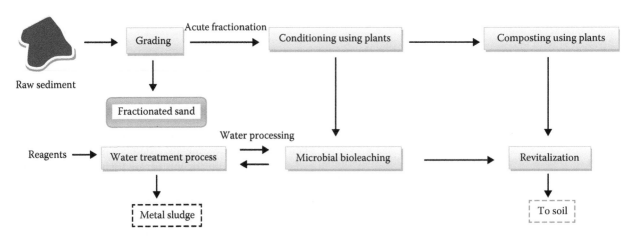

FIGURE 22.1 Schematic showing the design of the biotreatment of resource-contaminated leachate.

22.2 PRINCIPLES OF METAL MICROBE LEACHING

Bioleaching effects of microorganisms on metals are primarily based on three basic principles, namely, complexolysis, acidolysis, and redoxolysis. Bacteria and fungi are able to solubilize metals through the formation of organic or inorganic acids (protons), oxidation and reduction reactions, or excretion of complex agents. Sulfuric acid is by far the main inorganic acid obtained in leaching environments. It is formed by sulfur-oxidizing microorganisms such as thiobacilli. Particularly in the case of bioleaching, thermophiles have shown to have substantial advantages since industrial processes, namely, tank leaching, suffer from expense caused by cooling along with the problem of acid mine/rock drainage, mainly because of mesophilic and psychrophilic leaching bacteria, which are of environmental concern (Johnson and Hallberg, 2005). The fundamental biochemical aspect of leaching reactions has been the focus of intensive research for about a decade (Rohwerder and Sand, 2003). The subsequent discovery that microorganisms had a role in the production of acid mine drainage, the first acidophilic iron- and sulfur-oxidizing bacterium, *Thiobacillus ferrooxidans*, was isolated and described (Temple and Colmer, 1951). The general outlook of the bioleaching principle is depicted in Figure 22.2.

22.2.1 MECHANISM MODEL

In general, the process of bioleaching can be best described as the solubilization of metals from their mineral source by some indigenous microorganisms or by the application of microorganisms to convert elements so that the elements can be extracted from a material when water is passed through it.

A simplified reaction to denote the biological oxidation of a metal sulfide occupied in leaching is

$$BS + 2O_2 \rightarrow BSO_4$$

where B is a bivalent metal.

Formerly, two types of mechanistic models were proposed for metal microbe solubilization, which are discussed next.

22.2.1.1 Direct Mechanistic Model

In this model, microorganisms can oxidize metal sulfides attaining electrons directly from the reduced metals. Cells have to be adsorbed to the metal surface, and a close contact is required. Here, the microorganisms can thoroughly oxidize the sulfide moiety of the sulfide metal and elemental sulfur according to the following equations:

$$S^{2-} + 2O_2 \, (T. \; ferrooxidans) \rightarrow SO_4{}^{2-}$$

$$Fe^{3+} \, (T. \; ferrooxidans) \rightarrow Fe^{3+} + e^-$$

$$S + H_2O + \frac{3}{2}O_2 (Bacteria) \rightarrow H_2SO_4$$

Cells adhere discerningly to metal surfaces occupying the indiscretion of the surface structure. In addition, chemotaxis-occupying genes were also studied in *T. ferrooxidans* and *T. thiooxidans* (Acuna et al., 1992). The direct model was by far confirmed through leaching of iron free of synthetic sulfides as per the following equations:

$$CuS \, (Covelite) + 2O^{2-} \rightarrow CuSO_4$$

$$ZnS \, (Sphalerite) + 2O^{2-} \rightarrow ZnSO_4$$

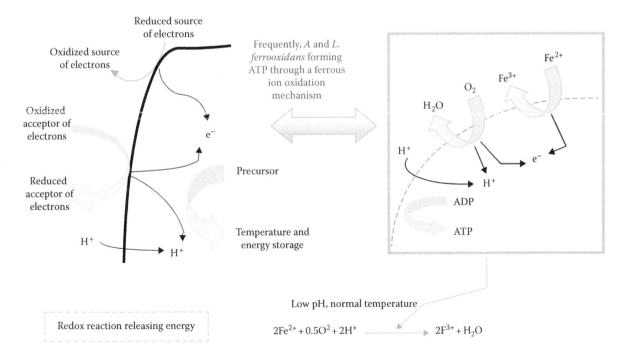

FIGURE 22.2 General outlook of the bioleaching principle.

Since the demonstration of direct electron transfer via enzymes, nanowires, etc., between the metal sulfide and attached cell, a direct mechanism is thought to be nonexistent. Instead, attached cells provide an efficient periplasmic space reaction compartment for indirect leaching with ferrous (III) ions (Sand et al., 2001).

22.2.1.2 Indirect Mechanistic Model

This model suggests a ferric-ferrous cycle of microbial oxidation. Here, iron plays the role of the electron carrier, and no physical contact is required for its oxidation. The following equations describe the indirect mechanistic model for the oxidation of pyrite:

$$4FeSO_4 + O_2 + H_2SO_4 (L. \text{ and } T. ferrooxidans)$$

$$\rightarrow 2Fe_2(SO_4)_3 + 2H_2O$$

$$FeS_2 + Fe_2(SO_4)_3 \text{ (Chemical oxidation)} \rightarrow 3FeSO_4 + 2S$$

$$2S + 3O_2 + H_2O (T. thiooxidans) \rightarrow 2H_2SO_4$$

On combining both mechanisms, a realistic and more suitable model was deduced, which has the following features: cells have to be attached in physical contact to the meal surface, cells form and excrete exopolymers, and these cellular envelopes of exopolymers surrounding the ferric iron compounds form complexes to the residues of glucuronic acid. These are part of the major attack mechanism, along with the notion that thiosulfate is formed as an intermediate during sulfur compounds oxidation; also, sulfur or polythionate granules are formed in the periplasmic space inside the cellular envelope. In the past 4 years, research has been done on the analysis of the redox reactions of aerobic ferrous (II)-oxidizing bacteria such as *Acidithiobacillus ferrooxidans* and *Leptospirillum ferrooxidans* (Blake and Griff, 2012). Although almost all biochemical studies are best known for *A. ferrooxidans*, certain biochemical spectroscopic and omics investigations stated that the ferrous (II)-oxidizing systems of other acidophilic ferrous-oxidizing bacteria are different from that of the redox components used. Figure 22.3 shows a schematic of the bioleaching mechanistic model (Rawlings et al., 1999).

Figure 22.3 shows a format of the model combining the electron transport sequence proposed earlier with concepts evolving from the debate on leaching mechanisms (Sand et al., 1995).

22.2.2 TARGETING THE BIOPROCESS

In order to target the effectiveness of the leaching bioprocess, the efficiency and potentiality of the microorganisms are necessary. Further, the adherence of the microbial strain with the chemical and mineral composition of the metal to be leached is important. Optimal metal extraction yields can be achieved only when the leaching conditions are relative to the optimum growth conditions of the microorganisms. Likewise, there is an indication that the growth of *T. ferrooxidans* and the oxidation of Fe^{2+} are rigidly coupled; also, a direct relationship between Fe^{2+} oxidation and O_2 uptake/CO_2 fixation has been considered. It was also observed that the rate of Fe^{2+} oxidation by *T. ferrooxidans* was relative to the concentration of nitrogen. It is also established that the presence of toxic metals could produce a similar effect. Several studies demonstrate that the growth of *T. ferrooxidans* and its capability to oxidize metal ions are reliant on pH, temperature, and concentrations of Fe^{2+} and Fe^{3+} ions. Additionally, the accessibility of oxygen and CO_2 is the other significant feature that enables the metabolism of bacteria. The metal bioleaching mechanism in acidic environments is influenced by a string of different factors (Table 22.1). The physicochemical and biological factors of the leaching environment vividly influence the rate and efficiency of this bioprocess. Moreover, properties of the solids to be leached are of major significance. For instance, pulp density, pH, and particle size were characterized as prime aspects for pyrite bioleaching by *Sulfolobus acidocaldarius*.

The effect of several parameters, namely, microbial activities, energy source, composition of metals, pulp density, particle size, and temperature, was observed for sphalerite oxidation using *T. ferrooxidans*. The best zinc solubilization was obtained at reduced pulp densities of 50 g/L, low particle sizes, and at a temperature of approximately 35°C. Oxidation of metals using acidophilic organisms can be inhibited by different factors like organic compounds, surface-active agents, solvents, or specific metals. Likewise, in the presence of organic components, namely, yeast extract, pyrite oxidation using *T. ferrooxidans* is inhibited. During the leaching bioprocess, precipitation of metals with mineral contents like jarosites can slow down effective bioleaching. Moreover, amino acids addition in small amounts can effect increased corrosion of pyrites by *T. ferrooxidans*.

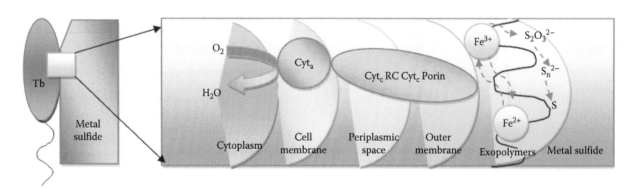

FIGURE 22.3 Schematic bioleaching mechanistic models (Tb, thiobacillus; Cyta,c, cytochrome a and c; RC, rusticyanin).

TABLE 22.1

Features and Conditions Affecting Bacterial Metal Oxidation and Mobilization

Features	Conditions
Physicochemical factors of a bioleaching environment	Adequate temperature
	Adequate pH
	Redox potential
	Water potential
	Content and supply of oxygen
	Content of carbon dioxide
	Mass transfer
	Availability of nutrients
	Concentration of Fe^{3+}
	Source of light
	Source of pressure
	Proper surface tension
	Existence of inhibitors
Biological factors of a bioleaching environment	Diversity of microorganisms
	Population density
	Activities of microbes
	Spatial distribution of microorganisms
	Adherence with metals
	Adaptability with microorganisms
Properties of metal to be leached	Type of metal
	Composition of metal
	Dissemination of metal
	Size of grain
	Surface area
	Porosity
	Hydrophobicity
	Galvanic interactions
	Secondary product formation
Bioprocessing	Mode of leaching
	Pulp density
	Rate of stirring, geometry

22.2.3 METHOD OF METAL EXTRACTION

The metal bioleaching process is effortless and valuable technology for the processing of metal ores and is utilized usually for the recovery of metals like copper and uranium. The efficiency and economics of the bioleaching process depend mainly on the solubilizing activity of microorganisms along with the chemical composition of the metal ore. Therefore, processes are very individualistic and cannot be transferred. In the leaching bioprocess, two major techniques used for the extraction of metals are percolation and agitation leaching. Percolation leaching generally accumulates the percolation of the desired liquid medium for metal ore extraction by means of a static bed, while agitation leaching involves agitation of smaller and smoother elements present in the desired liquid medium for metal ore extraction. Due to the large-scale procedures involved in bacterial leaching, percolation leaching is commercially preferred.

From the commercial point of view, methods including *in situ* heap, vat, and dump bioleaching techniques are frequent. *In situ* processes involve impelling solution and air under pressure into ore bodies that are made permeable by explosive action. The resulting metal-enriched solutions are recovered from the wells drilled below the ore body. The dump-leaching technique requires piled-up uncrushed waste rock. These dumps normally contain about 0.2%–0.7% copper, quite low to recover profitably by conventional processes. Some of these dumps are massive, containing in excess of 50 million tons of waste rock. The heap leaching technique involves the preparation of the ore, primarily size reduction, so as to increase metal–lixiviant interaction and setting of an impermeable base to prevent lixiviant loss and pollution of water bodies. Essentially, both dump and heap leaching techniques utilize the application of lixiviant to the top of the dump or heap surface and the recovery of the metal-laden solution that seeps to the bottom due to gravity. Copper is generally extracted from acid runoff by cementation or solvent extraction or electrowinning (SX/EW). All the processes mentioned here are crucially uncontrolled from a biological and engineering viewpoint. Besides, these processes are quite slow and require long phases to recover a portion of the metal. The vat leaching technique involves mineralization of complex coarse metal material inside a dual-chambered closed tank. Several process-controlling parameters can be incorporated for efficient reactions and improved recovery with the utilization of bioreactors, chiefly considered for ore concentrates and valuable or expensive metals extraction. Large-scale bioleaching operations are carried out in some of the enormous bioprocess reactors in the world. Bioleaching of metals is an intensive process requiring huge technically agitated and aerated processors with volumes of 2000–5000 m^3, and the leaching bioprocess is performed by constructing metal ore heaps with dimensions of several kilometers. The bioleaching technique using bioreactors will be discussed in the subsequent sections of this chapter.

22.3 PERSPECTIVES OF THE BIOLEACHING SYSTEM

To give an outlook on bioleaching, for instance, there are certain mechanistic details on the activity of *A. ferrooxidans* on iron metal:

Two types of contact mechanistic bioprocess have been established using the potential bacteria: a direct contact mechanism where bacteria directly oxidize the metals by biological means without any requirements for Fe^{3+} or Fe^{2+} ions. Another is the indirect contact mechanism where the bacteria oxidize Fe^{2+} ions in the bulk solution to Fe^{3+} ions that leach the mineral ores. Also, in the indirect mechanism, bacteria oxidizes Fe^{2+} ions to Fe^{3+} within the layer of bacteria and exopolymeric materials and the Fe^{3+} ions within this layer leach the minerals.

Among heterotrophic bacteria, members of genera *Thiobacillus* and *Pseudomonas* have been found to be effective in the leaching of nonsulfidic metal ores. Fungi of the genera *Penicillium* and *Aspergillus* have also been applied in the mineral leaching bioprocess.

As reported, microorganisms solubilize metals by the following ways:

- Organic acid formation
- Oxidation or reduction reaction
- Extraction of complexing agents
- Formation of chelates

Research depicts about 55%–60% of nickel and cobalt extraction from Greek lateries when strains of indigenous *Penicillium* sp. and *Aspergillus niger* were used for bioleaching. Swamy et al., (1995) studied the increased stability of *A. niger* in nickel bioleaching, achieving 95% nickel leaching with strains of *A. niger* ultrasound pretreated in 14 days in contrast with 92% nickel leaching in 20 days with untreated *A. niger*. According to reports on the exploitation of filamentous fungus *Penicillium* in the bioleaching of Sukinda lateritic nickel ore under optimum conditions, the fungus could leach a maximum of 12.5% nickel.

22.3.1 Utilizing Microbial Diversity

The metal-solubilizing mechanism performed by ubiquitous microorganisms is found in leaching environments and has been obtained from leachates and acidic mine drainage (Table 22.2). Microbial biodiversity has been seen in extreme environmental conditions, pH as low as 3.6, metal concentrations as high as 200 g/L, and temperature as high as 65°C. Bacteria capable of bioleaching are distributed among the Nitrospirae (*Leptospirillum*), Proteobacteria (*Acidithiobacillus, Acidiphilium, Acidiferrobacter, Ferrovum*), Firmicutes (*Alicyclobacillus, Sulfobacillus*), and Actinobacteria (*Ferrimicrobium, Acidimicrobium, Ferrithrix*). Considering most groups, mesophilic and reasonably thermophilic microorganisms can be identified in almost all of them (Norris et al., 2000). Among the bioleaching organisms, the first researched iron-oxidizing bacterium is *A. ferrooxidans*, formerly *T. ferrooxidans* (Kelly and Wood, 2000). Structural and genetic data explained the diversity of *T. ferrooxidans*, for which 23 strains were classified into seven subgroups based on DNA–DNA hybridization patterns (Harrison, 1982). Recently, 21 *A. ferrooxidans* strains have been identified using molecular techniques like multilocus sequence analysis. This work exposed the existence of four subgroups corresponding to different species (Amouric et al., 2010). In another new study, strain m-1, for a long time classified as *A. ferrooxidans*, has been suggested to belong to the new genus *Acidiferrobacter* among the proteobacterial division and closely related to the alkaliphilic *Ectothiorhodospira* sp. It was shown that this strain is a moderate osmophile (Hallberg et al., 2011).

A. ferrooxidans is also bequeathed with an amazingly broad metabolic facility. This species exists on the oxidation of Fe^{2+} ions and inorganic sulfur compounds and, besides, is able to

TABLE 22.2

Diversified Microbial World Utilized for Bioleaching

Domain	Organism	Nutrition Form	Primary Leaching Agent	Range of pH	Optimum pH	Temperature (°C)
Archaea	*Acidianus brierleyi*	Facultative heterotrophic	Sulfuric acid	Acidophile	1.5	45–75
	Ferroplasma acidiphilum	Facultative heterotrophic	Ferric ions	1.5–2.0	1.7	10–40
	Metallosphaera sedula	Chemolithoautotrophic	Ferric ions, sulfuric acid	Acidophile		Extreme thermophilic
	Sulfolobus acidocaldarius	Chemolithoautotrophic	Ferric ions, sulfuric acid	0.9–5.8	2.0	55–85
	Sulfolobus mirabilis	Mixotrophic	Sulfuric acid	Acidophile		Extreme thermophilic
Bacteria	*Acidiphilium cryptum*	Heterotrophic	Organic acids	2.0–5.0	3.5	Mesophilic
	Acidobacterium capsulatum	Chemoorganotrophic	Organic acids	3.0–6.0		Mesophilic
	Crenothrix sp.	Facultative autotrophic	Ferric ions	5.5–6.0		18–24
	Leptothrix discophora	Facultative autotrophic	Ferric ions	5.5–6.0		10–40
	Thiobacillus albertis	Chemolithoautotrophic	Sulfuric acid	2.0–4.0	3.5	25–30
	Thiobacillus ferrooxidans	Chemolithoautotrophic	Ferric ions, sulfuric acid	1.5–5.0	2.5	28–35
Fungi	*Aspergillus fumigatus*	Heterotrophic	Citrate oxalate			28
	Fusarium sp.	Heterotrophic	Oxalate malate			28
	Penicillium funiculosum	Heterotrophic	Citrate			26
	Trichoderma viride	Heterotrophic	Oxalate			32
Yeast	*Candida lipolytica*	Heterotrophic				30
		Heterotrophic				
	Saccharomyces cerevisiae	Heterotrophic				28
	Torulopsis sp.	Heterotrophic				

oxidize molecular hydrogen, formic acid, and other metal ions (Johnson et al., 2012). Anaerobic growth is likely by sulfur compounds oxidation or Fe^{3+} coupled with hydrogen reduction (Osorio et al., 2013). Various other acidophiles are reported to grow through the reduction of Fe^{3+} using inorganic electron donors in dissimilatory mechanism as in the case of *Acidiphilium* and *Ferroplasma* (Johnson et al., 2012). Diversified strains of Gram-positive organisms in relation to *Acidimicrobium* and *Ferrimicrobium* were isolated from sulfide mine dumps, partially signifying novel species (Schippers et al., 2010).

22.3.2 POTENTIAL FUTURE IMPACTS

As the saying goes, "Much of the future of biomining is likely to be hot" (Rawlings, 2002). Likewise, the application of moderately thermophilic bacteria and archaea is gaining momentous attention for commercial relevance. Billiton Process Research (Dew et al., 2000) accounts benefits of temperate thermophiles for the leaching of nickel concentrates. The pilot-scale test research reveals extreme thermophiles attain efficient bioleaching of primarily copper and nickel sulfide concentrates, giving increased recoveries than accomplished by bioleaching with either mesophilic or mild thermophilic strains. The industrial appliance of microbial leaching bioprocess is related mostly to the chemolithotrophic iron-oxidizing strains of *T. ferrooxidans* and *L. ferrooxidans*. Metal industrial waste products that frequently contain heavy amounts of expensive metals such as in fly ash and slag are present mostly as oxides rather than sulfides. Experiments revealed that *T. thiooxidans* can be utilized for the leaching of the metal oxides in such residues. Certain metals like vanadium, chromium, copper, and zinc can be almost entirely recovered depending on the metal components in the residues. Research on silicate nickel ores revealed that Ni is dissolved by organic acids produced by microbes. Among the effective ones was citric acid by using nickel-tolerant strains of *Penicillium* sp. About 80% of the nickel was extracted depending on the solubilization. Besides the recovery of valuable metals from nonsulfide minerals, heterotrophic microbes can also be exploited for upgrading mineral raw materials by the elimination of impurities. Most of the bacteria dynamic in iron removal are related to *Bacillus* and *Pseudomonas* genera. Among the fungi, *Aspergillus* and *Penicillium* were found to be the most effective ones. Silicate bacteria are known for solubilizing silicon- and silicate-bearing metals. These strains are related to different genera of heterotrophic bacterial species and do not signify a taxonomic unit. Mostly, they belong to the species *Bacillus circulans* and *Bacillus mucilaginous*. Further genetic enhancement of bioleaching microbial strains would give results more rapidly than conventional procedures like screening and adaptation.

22.4 GENETIC BASIS OF THE BIOLEACHING TECHNIQUE

The bioleaching subjects to the microbial-catalyzed method of conversion of insoluble minerals into soluble forms. As a functional biotechnology process universally used, it symbolizes an extremely interesting field of research where omics systems can be utilized in terms of knowledge development, process design, control, and optimization. Within the microbial world, extremophiles, studied for their extreme capabilities in survival and leaching process performance, and, consequently, the works of proteomics and metabolomics will sort the discovery of biomarkers applied in the search for the confirmation of existing or past extra-earthly life (Bonnefoy and Holmes, 2012). Extremophilic microorganisms have been obtained from mining operations (Tyson et al., 2004), and their function in the dynamics and evolution of metals has been widely discussed (Santelli et al., 2009). Also, techniques like bioleaching dumps and heaps can be seen in mines all over the world like Morenci (United States), Radomiro Tomic (Chile), Barberton (South Africa), Talvivaara (Finland), Cerro Verde (Peru), and Dexing and Zijinshan (China), among several other mining locations. As a result, the omics system has a unique globally used biotechnological niche to be impacted not only in terms of knowledge development but also for bio-hydrometallurgical process design, control, and optimization, and there lies the genetical emphasis of leaching bioprocess.

22.4.1 OMICS OF LEACHING

The concurrent work of the complete set of genetic, protein, and metabolic matter of living things will probably lead science to decode at least some of the mysteries of life. Since genomics is not the only accountable omics for life's complexity, more -omes are required to solve this puzzle (Ball, 2013); as such, we must account for the "dark matter" of the biological universe (Varki, 2013). Genomics has definitely made a major impact on our information of bioleaching. Primarily, partial and full-genome sequencing has permitted the accession of biodiversity within leaching environments and accessing molecular-based techniques to examine the temporal population dynamics of different bioleaching processes. Likewise, the diversity of microbial populations in bioleaching operations can be deduced from the isolation of moderate thermophilic heterotrophs of the genus *Sulfobacillus* as the dominant population of the heap operation at the Agnes Gold Mine in Barberton, South Africa (Coram-Uliana et al., 2006), and the presence of chemolithoautotrophic Fe^{2+}-oxidizing *Leptospirilli* in a tailings impoundment at the La Andina copper mine in Chile (Diaby et al., 2007). Future emphasis of genomics, by means of more inclusive, reliable, rapid, and economical next-generation sequencing technologies, will lead to better bioleaching operational design and control (Demergasso et al., 2010). Also, genomics provides the information for in silico renovation of the metabolic potential of formerly unknown essential features of bioleaching, such as the metabolic course for iron and reduced inorganic sulfur compound (RISC) oxidation. Conceptual models on genomics data have revealed the intricate and alternative metabolic pathways for RISC compounds showing common features between bacteria and archaea (Chen et al., 2012), new metabolic characteristics, and environmental adaptation in *A. ferrooxidans* and *Leptospirillum ferriphilum* (Levicán et al., 2008), *Acidithiobacillus caldus* (Mangold et al., 2011), and *Sulfobacillus thermosulfidooxidans* (Justice et al., 2014). Genomics assessments have shown chief differences between different strains of the same species, for example, *Ferroplasma acidarmanus*–related population displaying a mosaic genome structure, where a small number of sequence types prevail, proposing the structure as a common feature of natural archaeal populations (Allen et al., 2007). Also, strains of *A. ferrooxidans* ATCC 23270T and ATCC 53993 have 2397 genes in common, which represent between 78% and 90% of their genomes, where variation in terms of a genomic island provides higher copper resistance in strain ATCC 53993 (Orellana and Jerez, 2011). Recently, a metabolic reconstruction of a simplified bioleaching consortium formed by *L. ferriphilum* and *Ferroplasma acidiphilum* was available (Merino et al., 2014), based on which the existence of a particular species under a specific enriched nutrient and growth condition is predicted. Regardless of the physicochemical nature of most leaching mechanisms depicted so far (Rawlings, 2002), several studies have exhibited that protein-related carriers present in the extracellular polymeric substance layers are able to accumulate colloidal sulfur and enhance the bioleaching of metal sulfides (Bobadilla-Fazzini et al., 2011). Essential sections of the microbial function in bioleaching processes are iron and RISC oxidation, which take place largely in the extracellular space or in the cytoplasmic space. To facilitate a better understanding of the sulfur oxidation metabolism in *A. ferrooxidans* ATCC 23270, a large throughput proteomic research exposed the presence of 131 proteins in the periplasmic fraction of thiosulfate-grown cells. In the pattern of discovery and employment of biomarkers, proteomics data were utilized to conclude the genomic type of natural populations in an acid mine drainage sample. A proteomics-inferred genome typing (PIGT) offered evidence of recombination between these two closely related *Leptospirillum* group II populations (Lo et al., 2007). Later on, PIGT was applied to genotype the dominant *Leptospirillum* group II population in 27 biofilm samples from the Richmond Mine over a 4-year period. Proteomics studies of biofilms have also been worked with laboratory-cultivated biofilms, inoculated with these environmental samples (Belnap et al., 2011). Such research presents examples of how the understanding extracted with genomics and proteomics can be used to comprehend the role of key metabolic pathways in extreme microbial strains, with a massive potential to be applied in industrial operations.

22.4.2 Genetic Transfer Phenomenon

The genetic transfer phenomenon can be utilized to express heterologous genes in the recipient bacteria or to bring back genes that have been customized in more accurate hosts by genetic engineering. The latter activity allows the construction of mutants, specifically defined at the molecular level, which can assist in explaining the physiology of these microbes or improvise some of their detailed metabolic properties.

Such phenomenon, in bioleaching microbes, is a genuine challenge because their life conditions are extreme and unlike

TABLE 22.3

Bioleaching Microbes in Which Genetic Transfer Phenomenon Was Performed

Bacterium	Genetic Transfer Technique
Acidiphilium sp.	Electropermeabilization conjugation
Acidithiobacillus thiooxidans	Conjugation
Acidithiobacillus ferrooxidans	Electropermeabilization
Acidithiobacillus ferrooxidans	Conjugation

"classical" bacterium *Escherichia coli*. Genetic engineering is usually executed in this Eubacterium that is subsequently used as donor cells. While the former strains are acidophiles in moderate or strict form and chemoautolithotrophs in obligatory or facultative form, the latter is neutrophilic and heterotrophic. The primary drawback, as such, lies in analyzing the culture conditions where both the donor and recipient cells have sufficient internal energy for existence. Moreover, bioleaching microbes grow slowly with reduced cell yields, making them complicated to culture. However, genetic transfer was made possible in certain bioleaching microbial strains (Table 22.3).

This has caused the expression of the phosphofructokinase gene (*ptkA*) of *E. coli* in *A. thiooxidans* (Tian et al., 2003). More interestingly, the transposon Tn5 was reported to be able to transpose into the chromosome of *A. ferrooxidans*, throwing light on random transposon insertion mutagenesis (Peng et al., 1994). Finally, conjugation by marker exchange mutagenesis has allowed the construction of the *recA* mutant in *A. ferrooxidans* (Liu et al. 2000), the lone "constructed" mutant illustrated so far in bioleaching microbes (Figure 22.4).

22.5 APPLICATION OF BIOLEACHING IN BIOREMEDIATION

One of the unconstructive disadvantages of industrial globalization is heavy metal contamination in ecosystems. Mining industries, metallurgical manufacturing (electroplating technology, kitchen utensils production, surgical instruments manufacturing, automobile batteries technology), and wastewater from different process industries (pesticides, fungicide, and herbicide process, wood preservatives process, plastic stabilizers process, inorganic fertilizers production) create severe problems in the environment (Su et al., 2014; Wong and Fung, 1997). In Table 22.4, contaminants from different process industries are reported.

Heavy metals are detoxified through the microbial route by different processes, such as *in situ* bioremediation and *ex situ* bioremediation (Vidali, 2001).

22.5.1 In Situ Bioremediation

In situ bioremediation technology is a less costly, more effective technology for metal-contaminated wastewater treatment. The word "*in situ* bioremediation" signifies that there is no need to remove or excavate soil or water in order to accomplish

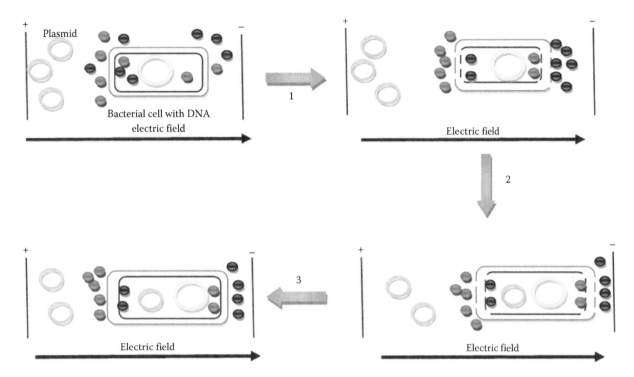

FIGURE 22.4 Illustration of genetic transfer in bioleaching microbes through electrotransformation.

remediation. In this process, naturally occurring consortia are used. Oxygen and nutrients are supplied in the intermediate step of the process. Chemotaxis is important to investigate *in situ* bioremediation processes because microbes with chemotactic capabilities can shift into an area containing pollutants. Although the method is cost-effective, it poses some disadvantages, such as long time for degradation and seasonal

TABLE 22.4

Heavy Metal Contamination by Different Process Industries

Contaminant	Industry
Arsenic	Pesticides, fungicide, and herbicide process, wood preservatives process, smelting, and ore mining technology
Chromium	Tanneries and steel industries
Cadmium	Electroplating technology, paints and pigments process, plastic stabilizers process, and inorganic fertilizers production
Copper	Pesticides, fungicide, and herbicide process, fertilizers production, smelting, and ore mining technology
Nickel	Kitchen utensils production, surgical instruments process, and automobile batteries technology
Mercury	Au–Ag mining technology, coal combustion process, and medical waste
Lead	Aerial emission from combustion of leaded fuel, pesticides, fungicide, and herbicide process, and waste of batteries
Iron	Ore mining technology, steel manufacturing, and electroplating technology

variation of the microbial activity. There are different types of *in situ* bioremediation process practiced in industries, which are described in the subsequent sections.

22.5.1.1 Intrinsic *In Situ* Bioremediation

Intrinsic bioremediation is performed by naturally occurring microbial communities to degrade metal pollutants without taking any engineering steps, except supplying nutrients and oxygen to increase the metabolic activities of microbes and to enhance the process.

22.5.1.2 Engineered *In Situ* Bioremediation

The other type of *in situ* bioremediation, known as engineered *in situ* bioremediation, has been adopted when intrinsic *in situ* bioremediation does not provide a satisfactory result. Engineered *in situ* bioremediation is promoted by the proper control of the physicochemical growth conditions of microorganisms. Generally, oxygen, electron acceptors, and nutrients (carbon, nitrogen, and phosphorus) are provided to promote microbial growth.

22.5.2 *Ex Situ* Bioremediation

In this process, pumping of groundwater or excavation of contaminated soil is adopted to facilitate microbial treatment to detoxify heavy metals. Depending on the state of the contaminant, *ex situ* bioremediation is classified into slurry-phase system (industrial and domestic wastewater) and solid-phase system (leaves and forest waste, agricultural waste, animal manures, domestic waste, industrial waste, sewage sludge, and municipal solid waste).

22.5.2.1 Slurry Phase

Slurry-phase bioremediation is performed in bioreactors, generally operated in a batch mode. It is a stepwise controlled treatment process that involves the excavation of the contaminated soil, mixing it with water, and placing it in a bioreactor. In this process, the separation of rubbles and stones from the contaminated soil is adopted as a pretreatment step. Subsequently, the soil is mixed with a predetermined amount of water to form the slurry. Here, the concentration of water mixed with the solid depends on the absorption of contaminants, physicochemical features of the soil, and the effect of biodegradation by microbes. After completion of this process, the soil is removed and dried up using pressure filters, vacuum filters, or centrifuges. Finally, the resulting fluid is treated for further use.

22.5.2.1.1 Bioreactors

Slurry or aqueous reactors are applied for *ex situ* conduction of contaminated water, sediment, sludge, soil, etc. Contaminated water, sediment, sludge, and soil are pumped up from a contaminated plume to the bioreactor. In a slurry bioreactor, a mechanical agitator is used for mixing purposes to increase the bioremediation rate. In the bioreactor, three phases, for example, solid, liquid, and gas, are generally created. Compared to *in situ* or in solid-phase systems, the extent and rate of biodegradation are greater because the treatment process is more controllable. In this process, pretreatment of contaminated soil is required. Alternatively, in many cases, the contaminant is stripped from the soil via physical extraction or soil washing before it is placed in the bioreactor.

22.5.2.1.2 Bioventing

Bioventing is the most common *in situ* treatment of wastewater, sediment, sludge, soil, etc. In this process, low airflow rate (only provides less amount of oxygen) and nutrients are supplied to contaminated water, sediment, sludge, and soil for stimulating the indigenous bacteria and degradation. In bioventing technique, less amount of oxygen is provided because of minimizing volatilization and release of by-product contaminant gasses to the atmosphere. Even though this know-how is used for simple hydrocarbons degradation, it can be used for the treatment of deep contaminated water.

22.5.2.1.3 Biosparging

In this method, air is injected under pressure at deep point of the water level. This air increases oxygen concentrations in contaminated water and enhances the rate of biological degradation of contaminants by naturally occurring bacteria. This process is considered as a cost-effective and efficient route for wastewater treatment.

22.5.2.1.4 Bioaugmentation

In this process, microorganisms, indigenous or exogenous, are added frequently to the contaminated site. This process is generally used for wasteland treatment. However, this takes a long time to degrade the contaminants, but from the commercial point of view, it is satisfactory.

22.5.2.2 Solid Phase

In the case of solid-phase *ex situ* bioremediation technology, contaminated soil is excavated and placed into piles. Generally, this process is used for the treatment of organic wastes, like leaves and forest waste, agricultural waste, animal manures, domestic waste, industrial waste, sewage sludge, and municipal solid waste. Bacterial growth is stimulated and distributed through and supported by pipe network. Air through the pipes is provided for ventilation and microbial respiration. A solid-phase *ex situ* bioremediation system requires a lot of space. Also, cleanup is required in this process. Land farming, soil biopiles, and composting are included in solid-phase *ex situ* bioremediation processes. Among the different solid-phase *ex situ* bioremediation technologies, soil biopiles are used for petroleum-contaminated soil treatment.

22.5.2.2.1 Land Farming

In the case of *ex situ* bioremediation technique, aerobic degradation of contaminants by microorganisms is considered. Nutrients are added at intermediate steps to promote the growth of microorganisms. Land farming *ex situ* bioremediation technique is a simple one, where contaminated soil is excavated and spread over a prepared bed and periodically tilled until pollutants are degraded.

22.5.2.2.2 Composting

Composting is a well-practiced waste treatment technology where contaminated soil is mixed with a bulking agent, such as leaf, straw, hay, corn cobs, and rice husk, to enable microbial growth and decomposition of toxic metal. Generally, static-bed composting and mechanically agitated composting technology are used for the treatment of contaminated soil and water. To achieve aeration in composting vessel, mechanical agitation is used. In static-bed composting, the contaminated soil mixed with bulking agent is placed in long piles known as window and periodically mixed by a tractor. The typical soil to compost ration is 75:25 (wt:wt), but this ratio changes depending on the soil type, characteristics, and level of contaminants.

From a long prior process, biodetoxification of heavy metals through microbial route has been practiced both at the small scale (laboratory scale) and at the industrial scale (pilot scale). Some laboratory-scale experimental results from the present investigation are reported in Table 22.5 (Kumar et al., 2011; Kulshreshtha et al., 2014).

22.6 EMPLOYING BIOREACTORS IN THE BIOLEACHING PROCESS

The selection and design of a suitable reactor for an industrial bioleaching process are based on the physicochemical and biological features of the system. Adequate attention is paid to the complex nature of the reacting sludge, composed by an aqueous liquid, suspended and attached cells, suspended solids, and air bubbles (Acevedo, 2000). Since large volumes of material need to be processed (approximately 200 tons concentrate per day with a yearly production of 20,000 tons copper) (Gentina

TABLE 22.5
Metal Bioremediation Through Microbial Route

Microorganism	Type of Bioreactor	Removal Efficiency	Reference
Arthrobacter strain D9	Lab-scale reactor (25 Ml working volume), initial cadmium concentration 3–150 µg/mL	22% at 48 h	Roane et al. (2001)
Pseudomonas strain IIa		11% at 48 h	
Bacillus strain H9		36% at 48 h	
Pseudomonas strain H1		36% at 48 h	
Enterobacter cloacae CMCB-Cd1	Lab-scale reactor, initial cadmium concentration 100 mg/L	86% at 24 h	Haq et al. (1999)
Klebsiella spp. CMBL-Cd2		87% at 24 h	
Klebsiella spp. CMBL-Cd3		85% at 24 h	
Aspergillus terreus	Lab-scale reactor (250 mL working volume), initial cadmium concentration 10 mg/L	70% at 13 days	Massaccesi et al. (2002)
Cladosporium cladosporioides		63% at 13 days	
Fusarium oxysporum		63% at 13 days	
Gliocladium roseum		65% at 13 days	
Penicillium spp.		60% at 13 days	
Talaromyces helicus		70% at 13 days	
Trichoderma koningii		64% at 13 days	
Tetraselmis suecica	Lab-scale reactor (2.0 L working volume), initial cadmium concentration 0.6–45 mg/L	59.6% at 6 days	Perez-Rama et al. (2002)
Chlorella spp. NKG16014	Lab-scale reactor (100.0 mL working volume), initial cadmium concentration 5.6 mg/L	48.7% at 14 days	Matsunaga et al. (1999)
Bacillus subtilis	Lab-scale reactor	—	Costa et al. (2001)
Chlorella vulgaris	Lab-scale reactor, initial nickel concentration 10 – 40 mg/L	33%–41% at 24 h	Wong et al. (2000)
Chlorella miniata (WW1)		>99% at 24 h	
Anabaena cylindrica	Lab-scale reactor, initial nickel concentration <20 mg/L	—	Corder and Reeves (1994)
Anabaena flos-aquae		—	
Nostoc spp.		—	
Candida spp. (nonadapted)	Lab-scale reactor (100.0 mL working volume), initial nickel concentration (66–514 mg/L)	29%–57% at 5–15 days	Donmez and Aksu (2001)
Candida spp. (adapted)		44–71 (5–15 days)	
Aspergillus niger	Lab-scale reactor, initial nickel concentration 381 mg/L	98 (4 days)	Magyarosy et al. (2002)
Pseudomonas spp.	Lab-scale reactor	—	Ghozlan et al. (1999)
Escherichia hermannii CNB50	Lab-scale reactor (10.0 mL working volume), initial nickel concentration 100 mg/L	—	Hernandez et al. (1998)
Enterobacter cloacae CNB60			
Neurospora crassa	Lab-scale reactor (10.0 mL working volume), initial cobalt concentration 500 mg/L	90 (24 h)	Karna et al. (1996)
Candida spp. (nonadapted)	Lab-scale reactor (100 mL working volume), initial copper concentration 100–1500 mg/L	22–52 (8 days)	Donmez and Aksu (2001)
Candida spp. (adapted)		5–68 (8–13 days)	
Saccharomyces cerevisiae	Lab-scale reactor (100 mL working volume), initial copper concentration 50.0–700 mg/L	13–74 (4 days)	Donmez and Aksu (1999)
Kluyveromyces marxianus		10–90 (4 days)	
Schizosaccharomyces pombe		11–25 (4 days)	
Candida spp.		13–73 (4 days)	
Aspergillus niger	Lab-scale reactor (100 mL working volume), initial copper concentration 25.0–150 mg/L	19–57 (7 days)	Dursun et al. (2003)
	Lab-scale reactor (100 mL working volume), initial lead concentration 25.0–500 mg/L	13–88 (7 days)	
	Lab-scale reactor (100 mL working volume), initial chromium concentration 25.0–75 mg/L	21–36 (7 days)	
Pseudomonas aeruginosa	Lab-scale reactor (20 mL working volume), initial chromium concentration 69 mg/L	46 (2 days)	Hassen et al. (1998)
Bacillus thuringiensis	Lab-scale reactor (20 mL working volume), initial copper concentration 18 mg/L	34 (2 days)	Hassen et al. (1998)

(Continued)

TABLE 22.5 (*Continued*)

Metal Bioremediation Through Microbial Route

Microorganism	Type of Bioreactor	Removal Efficiency	Reference
Thiobacillus ferrooxidans	Lab-scale reactor (1.5 L working volume), initial copper concentration 1000 mg/L	25 (15 min)	Boyer et al. (1998)
Saccharomyces cerevisiae SN41	Lab-scale reactor (100 mL working volume), initial copper concentration 320 mg/L	90 (7 days)	Brandolini et al. (2002)
Pseudomonas aeruginosa PU21 (Rip64)	Lab-scale reactor (20 mL working volume), initial lead concentration 18 mg/L	80 (2 days)	Chang et al. (1997)
	Lab-scale reactor (20 mL working volume), initial copper concentration 18 mg/L	75 (2 days)	

and Acevedo, 2013), bioleaching is best performed in a continuous mode of operation in which volumetric productivity is high (1.42 g/L metal per day) (Valencia and Acevedo, 2009) and reactor volumes can be kept low. Considering the kinetic characteristics of microbial growth, a continuous stirred-tank reactor appears as the first choice (Gahan et al., 2012). An important consideration in selecting a suitable bioreactor is the autocatalytic nature of microbial growth. During industrial fermentations, nutrients are selected based on their high affinity with the microbial population, while in bioleaching, the mineral species involved are usually recalcitrant to microbial action, implying that the affinity is quite low. Other types of reactors that have been studied for their application in biomining are the percolation column, the Pachuca tank, the airlift column, and some special designs such as rotary reactors (Rossi, 2001).

Currently, bacterial leaching of copper and biooxidation of refractory gold concentrates are well-established industrial processes that are carried out using heaps and tank reactors (Watling, 2015). Heap operation is simple and adequate to handle huge amounts of minerals, but its productivity and yields are limited because of severe difficulties encountered during implementation of an adequate process control (Pradhan et al., 2008). On the other hand, tank reactors can economically handle reasonable amounts of material, but they permit for a close control of the process variables, rendering significantly better performances (Acevedo, 2000). Rawlings (1997) classified the engineering approaches used in biomining into two broad categories, which are discussed next.

22.6.1 Reactors for Irrigation-Based Bioleaching

An irrigation-type method is employed in the irrigation of crushed rock with a leaching solution, followed by the collection and dispensation of the leachate or pregnant liquor solution to recover the target metals, normally by an SX/EW process (Gahan et al., 2012). These processes are categorized as heap bioleaching, dump bioleaching, and *in situ* bioleaching depending on the types of metal resources to be processed. Heap bioleaching deals with freshly mined materials (intermediate grade oxides, secondary sulfides, etc.) deposited in the form of a heap on an impervious support or a synthetically

prepared pad leached with circulation, percolation, and irrigation of leaching medium (Mishra et al., 2005). *In situ* bioleaching is utilized with abandoned and/or underground mines where the ore deposits cannot be extracted by the traditional methods since they are either low grade or in small deposits or both (Pradhan et al., 2008). In dump bioleaching, waste rock, low-grade ore, or concentrator tailings (low-grade oxides and secondary sulfides) are leached at the place of disposal.

22.6.1.1 Heap Bioreactors

The bioleaching of ores is a considerably slow process and is operated by constructing ore heaps with dimensions measured in kilometers. Heap bioreactors are economical to construct and conduct and as a result more suited to the treatment of lower-grade ores. While building a heap, ore (with particle sizes 2–10 mm) (Miller et al., 2003) is piled onto an impermeable base supplied with a leach liquor distribution and collection system (Pradhan et al., 2008; Watling, 2006). Acidic leach solution is percolated through the crushed ore, and the bacteria growing on the mineral surfaces of the heap generate ferric ion and acid that promote mineral dissolution and metal mineralization. As most of the metal leaching bacteria are aerobic and chemolithotrophic in nature, aeration considers supply of both O_2 and CO_2 to the leaching system. Aeration in such processes can be passive, with air being strained into the reactor due to the flow of liquids, or active in which air is blown into the heap through piping installed at the bottom. Active aeration may be conducted using the air distribution networks that include 500 mm headers and 50 mm diameter laterals at 2 m spacing (Andrews, 1997). Air is injected into the heap through a combination of low-pressure high-volume fans or blowers. Bioleaching establishes the oxygen concentration profile in relation with the height of the aerated heap. Underneath the pile, air is more or less forced into the heap, and oxygen is kept in near proximity to saturation, but as the air surges upward through the void spaces, the bacteria catalyzing the oxidation of sulfide consume oxygen, and as a result, the degree of oxygen depletion near the top of the heap increases (Pradhan et al., 2008). A significant oxygen concentration gradient is observed with increase in depth. Hence, increasing the rate of aeration may improve metal leaching. A typical process flow diagram of heap bioleaching is shown in Figure 22.5 (Pradhan et al., 2008).

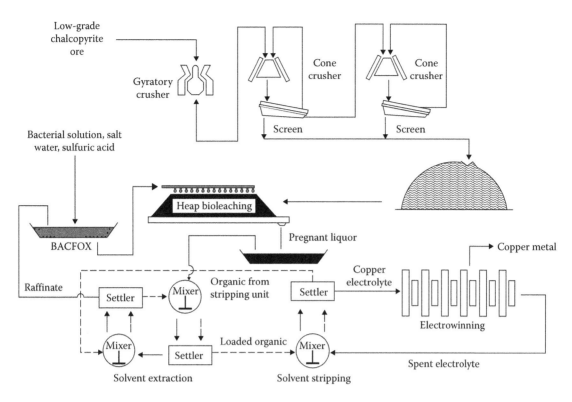

FIGURE 22.5   Process flow diagram of a heap bioleaching process.

The production of heap leach plants can profit from a more specific, quantitative requirement of the maximum plausible irrigation and aeration rates as a purpose of heap stacking height (Pradhan et al., 2008). Conventionally, empirical flooding tests in columns are frequently used in association with experience-based scaling factors to confirm ore permeability. However, methodical geomechanical methods can permit a more quantifiable evaluation of liquor and airflow versus stacking height. A geomechanical experiment rig installed at Mintek is applied to exemplify ores with respect to their flow. Experimental works include "compression tests," where the ore is compressed to simulate rising lift overburden pressures in a cover of ore at the base of the heap. Additionally, "hydrodynamic column tests" permit the measurement of the degree of saturation as a function of percolation rate, which offers an indication of the percolation "regime" where the heap leach is functioning. Heap bioleaching can be applied for treating whole ores or concentrates containing secondary copper sulfides, zinc, nickel, and uranium (Staden et al., 2008).

22.6.1.1.1 Characteristics of Heap Bioreactors
The features of an industrial heap bioreactor are as follows:

1. Nonuniform: Aeration, irrigation, nutrient addition, pH, etc., vary substantially right through the heap.
2. Complicated to control: Inoculation, leaching rate, and temperature are complex to control in the various areas of heap and during several phases of the leaching cycle.

3. Several ecological niches exist; as a result, potentially a large variety of microbes are exploited during bioleaching, and some may have detrimental effects on the reaction.
4. The process of mineral bio-oxidation is slow; biofilms exist and no strong selection is made for faster growth of microbes.

22.6.1.2 Dump Bioleaching Bioreactors
Dump bioleaching of copper has been employed in several countries, like Bulgaria (Groudev and Groudeva, 1993). Leach solutions containing different agents, namely, chemolithotrophic bacteria, dissolved oxygen, sulfuric acid, and iron ions, are pumped to the top of the dumps. The solutions percolate through the dumps and dissolve copper. Dump effluents are sent to precipitation plants where copper is removed by sedimentation with iron (Groudev and Groudeva, 1993; Casas et al., 2004). The barren solutions are supplemented with makeup water and, if needed, with sulfuric acid to adjust the pH to the desired value. The ores being leached using this technique, however, differed markedly from each other with respect to their mineral and chemical composition. The sulfur and copper contents of the dumped ores in Vlaikov Vrah and Assarel were found to be in the range of 2.0–3 and 0.1%–0.30%, respectively (Groudev and Groudeva, 1993). The main copper-bearing mineral in the Vlaikov Vrah ore was chalcopyrite, whereas in the ore of Assarel, the copper was presented mainly as secondary copper sulfides (Groudev and Groudeva, 1993).

22.6.1.3 Flood-Drain Bioreactors

This type of bioreactor seeks to combine high flow rates with low costs while processing intermediate-sized particles (approximately 1/4 in. × 30 mesh) (Andrews and Noah, 1996). The flood-drain bioreactor comprises a lined container with a perforated pipe as a solution distribution system into which an ore is placed to a depth of several meters. The reactor is divided into sections. Intermittently, for 1–2 min, the bacterial culture is pumped from below to fluidize one section of the bed in order to serve two purposes: first, to uniformly wet and inoculate the ore and, second, to destroy any anoxic zones and wash out secondary precipitates or fine particles that might plug the bed (Andrews, 1997). Air (containing O_2, CO_2) is drawn down into the bed as the fluid drains between periods of fluidization. Sequential fluidization and draining of sections, together with controlled flow rate and solution management, result in a high degree of size separation and variation in residence time and allow the washing of the product solids in the countercurrent mode of operation (Walting, 2015).

22.6.1.4 Aerated Trough Reactor

It is an airlift slurry bioreactor, comprising a long, rectangular tank with a V-shaped base provided with a perforated pipe for air sparging (Rossi, 2001). The trough is divided into six sections by solid baffles each having a small hole at the slurry level for flow between sections. This is done following the suggestion that the highest rate could be achieved with little axial mixing (Walting, 2015) and much cell recycle (Rossi, 2001). Each section has an independent aeration system and a drain for high-density solids removal (the pyrite-enriched fraction, subsequently the feed for the bacterial feed to the reactor).

22.6.1.5 Airlift Bioreactor

Airlift reactors have been used frequently for laboratory studies on bacterial leaching of soils, sediments, and sludges (Andrews, 1997) but, less frequently, for bioleaching and metals extraction from ores or concentrates (Li et al., 2011). One of the reasons is that the rate of metal extraction decreases rapidly when the solids loading is greater than 20% (Loi et al., 1994), which, at the time, was one of the key improvements that was being sought in bioreactor design. Chen and Lin (2004) performed bioleaching of heavy metals like copper, manganese, and nickel in an airlift bioreactor. Figure 22.6 shows a schematic of an airlift bioreactor used for this purpose. The bioreactor consists of three parts: a top part, a main column with water jacket, and a bottom part with an air diffuser. The main column is composed of an internal tube of 5 cm internal diameter and 150 cm height, an external tube of 15 cm diameter and 150 cm height, and a total working volume of 25 L. Compressed air was sparged at a rate of 6 L/min from the bottom of the column into the internal tube, which produced a recirculation flow pattern of the sediment slurry in the column. In each bioleaching experiment, the airlift reactor was fed with 25 L sediment

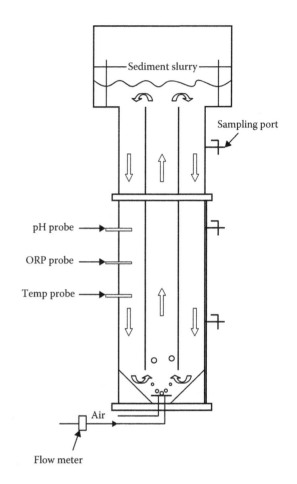

FIGURE 22.6 Schematic diagram of an airlift bioreactor used for bioleaching of heavy metals.

slurry of 1% total solids content, 2.5 L well-acclimatized inoculums, and a predetermined amount of sterilized (tyndallized) elemental sulfur. The rate of metal solubilization from the contaminated sediment was found to be the highest in the case of zinc (96%–99%) followed by nickel (73%–87%) and manganese (62%–68%). The rate of metal solubilization was found to be affected significantly by the rate of acid production. Sethurajan et al. (2012) conducted bioleaching of copper from black shale ore using mixed populations of chemolithotrophic bacteria in a 5 L air uplift bioreactor using a pulp density of 10% w/v. The copper extraction efficiency was obtained as >70% at pH 2, temperature 37°C, and aeration rate 150 L/h for 15 days of continuous operation. Bakhtiari et al. (2008) investigated bioleaching of copper from smelter dust in a series of airlift bioreactors at different pulp densities and obtained an overall copper extraction efficiency of 90%, 89%, and 86% for pulp densities of 2%, 4%, and 7% at mean retention times of 2.7, 4, and 5 days, respectively.

22.6.1.6 Rotating Drums Bioreactor (Biorotor)

This is a novel bioreactor designed for the purpose of bioleaching with an intention to transform biohydrometallurgy from the bench scale to the pilot scale (Walting, 2015). It comprises baffled cylindrical barrel with an aeration inlet (CO_2 and O_2) at one end and an air outlet at the other end. The cylinder sits on rollers and during rotation, the baffles lift solids and then discharge them to fall through the solution, maximizing mixing, eliminating "dead" zones, and minimizing shear stresses on bacterial cells (Rossi, 2001). Tests carried out with a 30% solids suspension, the solids being museum grade pyrite, yielded solubilization rates as high as about 600 g/m^3 h (Walting, 2015). The rotating bioreactor has attracted recent attention with the development of a modified continuous reactor capable of processing 40% solids suspension (Walting, 2015). Such types of bioreactors constitute a double-layered closed cylindrical vessel that has lifters, which are extensively packed and regularly spaced in the inner wall of the vessel. A particular, regular-spaced lifter within the inner wall of the vessel forms a type of collecting tray that collects the processed suspension. The suspension and the air–carbon dioxide mixture are conveyed through a pipe fitted into one front head of the barrel, and a pipe fitted into the opposite head serves as the exhaust outlet. As the barrel revolves on its rollers, the suspension is lifted upward and when the tray reaches the top position, it is discharged as a thin cataracting film with length equal to the height of the cylindrical barrel (Rossi, 2001). Oxygen mass transfer mostly occurs during the cataracting and when the freefalling film plunges into the pool of suspension in the lower part of the barrel. The most interesting feature of this bioreactor is, however, that the oxygen transfer rate actually matches the mass transfer coefficient, that is, the oxygen is made available to the microflora that can attain much higher growth kinetics than those obtained in conventional reactors. However, at high solid concentrations, the demand for oxygen exceeds the maximum possible supply for bioreactors and these bioreactors are not technically feasible for conducting bioleaching (Jin et al., 2010).

22.6.2 Agitation-Based Bioleaching Processes

Agitation-based bioleaching techniques were previously emphasized for the ore concentrate dissolution, but unfortunately, these techniques have very less marketable relevance with bioleaching of only refractory gold ores (Staden et al., 2008). However, base-metal processes have been developed at the pilot and demonstration scales for the bioleaching of copper, nickel, and zinc sulfide concentrates using mesophiles, moderate thermophiles, and thermophiles, and a process for the extraction of cobalt from pyrite concentrate has been commercialized (Walting, 2015). Agitated leaching is conducted in continuously operated, highly aerated, mechanically agitated tank bioreactors (Pradhan et al., 2008). Direct agitated bioleaching is a process in which the bacterial growth and leaching of mineral occur in the same agitated vessel (Staden et al., 2008). A modification of the stirred-tank technology comprises separation of the leaching and metals recovery process from the biological

regeneration of ferric ions; these are considered to be two-stage processes, termed "indirect bioleaching" or "effects separation" processes with which it is possible to optimize the leaching and biological processes independently and thus maximize metals recovery (Staden et al., 2008). Another alteration for the stirred-tank know-how is the function of aerobic reactors followed by anaerobic reactors to cause leaching of acid (air and carbon dioxide spurge) and also bioreduction under anaerobic state (nitrogen and carbon dioxide spurge).

22.6.2.1 Stirred-Tank Leaching Reactors

In these bioreactors, the most important role is played by the impeller, which has to accomplish three major tasks: solids suspension, mixing, and dissolution of the required atmospheric oxygen into the aqueous phase, maximizing the interfacial area between the gaseous and aqueous phases (Rossi, 2001). Initially, and for many years, the Rushton-type turbine was the most widely used impeller for these reactors, but lately the curved blade, axial flow impeller (Rossi, 2001) has been shown to outperform the Rushton turbine as it requires less power for achieving the same performance and induces smaller shear stresses in the suspensions. The floating agitation devices built on the model of Turboxal agitators are also frequently used for the purpose of stirring (Guezennec et al., 2014). Chemical engineering has provided some correlations that help in establishing, as a first approximation, the machine's characteristic parameters also with reference to the different types of agitators (Rossi, 2001). Hence, for the impeller speed necessary to satisfy the just-suspended condition for solid particles in the vessel, Zwietering's criterion is usually adopted, for which the following expression is used (Rossi, 2001; Pangarkar, 2015; Atiemo-Obeng et al., 2004):

$$\left(\frac{ND^2}{\vartheta}\right)^{0.1}\left(\frac{N^2 D}{g}\frac{\rho_1}{\Delta\rho}\right)^{0.45}\left(\frac{D}{d}\right)^{0.2} = k\left(\frac{T}{D}\right)^{\alpha} B^{0.13} \qquad (22.1)$$

where

$\dfrac{ND^2}{\vartheta}$ = Reynolds number

$\dfrac{N^2 D}{g}\dfrac{\rho_1}{\Delta\rho}$ = Froude's number

D/d is the ratio of impeller diameter to particle diameter

N is the rotational frequency of the impeller

D is the impeller diameter (m)

υ is the kinematic viscosity of solution (m^2/s)

ρ_1 is the liquid density (kg/m^3)

g is the acceleration due to gravity (m/s^2)

$\Delta\rho$ is the density difference, gas–liquid (kg/m^3)

d is the particle diameter

k is the constant

T is the tank diameter (m)

B is the distance from impeller midplane to tank bottom (m)

For predicting impeller power, either Mill's correlation (Rossi, 2001) (Equation 22.2) or van't Riet's correlation (Gill et al., 2008) (Equation 22.3) is used:

$$\frac{P_T}{V} = \left(\frac{k_L a}{0.1 - 0.0018\varepsilon} \right)^{1.49} \frac{1}{V_G^{0.046}} \tag{22.2}$$

$$\frac{P}{V} = \frac{(k_L a)^{1.4286}}{0.0001} \frac{1}{V_G^{0.2857}} \tag{22.3}$$

where

P_T is the total power input (kW)

P is the power to produce suspension to height H (kW)

V is the liquid volume plus particle volume below the air–liquid interface (m³)

k_L is the overall liquid phase mass transfer coefficient (m/s)

a is the interfacial area (m²/m³)

ε is the volume fraction of liquid in the suspension, dimensionless

V_G is the air superficial velocity based on tank cross section, gas flow rate divided by the cross-sectional area of the tank (1/m)

For a six-flat-blade Rushton disc impeller, Neale and Pinches (Kadijani et al., 2013) employed the following equations:

For calculating power requirement,

$$k_{La} = 0.0069 \left(\frac{P_G}{V} \right)^{0.52} V_G^{0.24} \tag{22.4}$$

and for BX04 impeller,

$$k_{La} = 0.0084 \left(\frac{P_G}{V} \right)^{0.79} V_G^{0.58} \tag{22.5}$$

where P_G is the agitation shaft power (gassed) (kW).

22.6.3 REACTION KINETICS OF BIOLEACHING

Bioleaching of sulfide minerals comprises three major sub-processes, namely, acid ferric leaching of the sulfide mineral, microbial oxidation of the sulfur moiety, and microbial oxidation of ferrous to ferric form. There are many studies on the kinetics of microbial ferrous iron oxidation in both batch and continuous cultures. Some of the published rate equations for microbial ferrous-ion oxidation are shown in Table 22.6 (Ojumu, 2006).

In the area of industrial bioleaching, most of the rate equations depicted in Table 22.6 have been applied for tank bioleaching since these are engineered processes operated under controlled conditions at which the overall rate of bioleaching is a near-optimal value (Ojumu, 2006). Modeling of heap bioleaching is a complex task considering several phenomena, such as solution and gas mass transport and heat transport, multimineral kinetics, bacterial kinetics, and diffusion effects (Ojumu, 2006). Modeling of these effects remains strictly hypothetical, and none of the adapted rate equations (shown in Table 22.6) have been confirmed to be valid under such extreme conditions.

Furthermore, considering the variable conditions, microorganisms are classified based on the temperature range in which they survive: mesophiles with optimum temperature in the range 35°C–40°C, moderate thermophiles around 50°C, and extreme thermophiles at 65°C. The models proposed by Hinshelwood (Ojumu, 2006) and MacDonald and Clark (1970) are used to show the dependency of bacterial growth on temperature:

Hinshelwood model: $\mu_{max} = K_1 e^{-E_a/RT} - K_2 e^{-E_b/RT}$

Ratkowsky model: $\mu_{max} = b(T - T_{min})\left\{ 1 - e^{c(T - T_{max})} \right\}$

TABLE 22.6
Published Kinetic Models for Bioleaching

Model Equation	Mode of Operation	Reaction Conditions	Values of Parameters	Reference
$\mu = \dfrac{\mu_{max}[Fe^{2+}]}{Y_{SX}K_m + [Fe^{2+}]}$	Batch	$T = 25°C–30°C$, pH = 2–2.3	$\mu_{max} = 0.20$ h⁻¹, $K_m = 1.05$ g/L at 25°C	Ojumu (2006)
$\mu = \dfrac{\mu_{max}[Fe^{2+}]}{K_m + [Fe^{2+}]}$	Continuous	$T = 28°C$, pH = 2.2	$\mu_{max} = 0.161$ h⁻¹, $K_m = 0.215$ g/L	MacDonald and Clark (1970)
$\mu = \dfrac{\mu_{max}([Fe^{2+}] - [Fe^{2+}]_t)}{K_m + ([Fe^{2+}] - [Fe^{2+}]_t)}$	Continuous	$T = 22°C$, $Fe_T = 9–22$ mM	$\mu_{max} = 0.070$ h⁻¹, $K_m = 0.78$ mM	Braddock et al. (1984)
$q_{O_2} = \dfrac{q_{O_2}^{max}([Fe^{2+}] - [Fe^{2+}]_t)}{1 + \dfrac{K_s}{[Fe^{2+}] - [Fe^{2+}]_t} + \dfrac{K_s}{K_i}\dfrac{[Fe^{3+}]}{[Fe^{2+}] - [Fe^{2+}]_t}}$	Continuous	$T = 30°C$, pH = 1.8–1.9, $Fe_T = 0.05–0.36$ mM	$q_{max} =$ mol C/mol h, $K_s/K_i = 0.1$	Boon et al. (1999)
$\mu = \dfrac{\mu_{max}([Fe^{2+}])}{[Fe^{2+}] + K_s(1 + K_i([Fe^{3+}]))}$	Continuous	$T = 35°C$, pH = 1.8, $Fe_T = 0.52–3.29$ g/L	$\mu_{max} = 0.11$ h⁻¹, $K_s = 0.048$ g/L, $K_i = 2.27$ L/g	Liu et al. (1988)

22.6.4 Global Operations of Bioleaching Plants

The various bioleaching techniques discussed in the previous sections have been successfully employed worldwide at the industrial scale. Table 22.7 represents the different bioleaching operations employed by various countries for the extraction of metals from ores.

22.6.5 Case Study

In 2011, Ghoshal, with her coworker, reported on the removal of mercuric ions using the mercury-resistant bacteria *Bacillus cereus* (JUBT1). They isolated the indigenous bacterial strain from the sludge of a local chloralkali industry. In the microbial growth medium, they used sucrose as a carbon source, and growth kinetics of the proposed bacteria with or without mercury ion was reported. They concluded that there is a multiplicative, noncompetitive relationship between sucrose and mercury ions with respect to bacterial growth. The detoxification of mercuric ions (Hg^{2+}) from water was achieved using a biofilm bioreactor (1 m in length and 0.5 m in diameter) as well as an externally associated charcoal filter bed bioreactor (1 m long and 0.25 m diameter). The former having attached growth of Bacillus cereus (JUBT1) on rice husk as packing material of the reactor (Ghoshal et al. 2011). Simulated mercury-contaminated wastewater and air were fed continuously into the packed bed reactor in a downward direction using a peristaltic pump and a compressor, respectively. The performance of the packed bed bioreactor was studied using different inlet concentrations of mercuric ions (Hg^{+2}) (5–30 ppb) and volumetric flow rate (0.00942–0.0176 m^3/h) of the feed solution, etc. Samples collected from the end of the packed bed reactor and from the carbon filter bed were analyzed to determine the percentage of (Hg^{2+}) removal. The optimum parameters, the inlet concentration of Hg^{2+} and the superficial velocity, were found to be 30 ppb and 0.048 m. h^{-1}, respectively. At the optimum condition, the maximum value of mercury removal achieved after the treatment by the biofilm reactor was 94.4%. Furthermore, authors reported that while the packed bed biofilm reactor could ensure the removal efficiency of Hg^{2+} in the range of 91%–94.4%, the subsequent treatment by carbon filter bed increases the efficiency in the range of 95%–96.4% (Ghoshal et al., 2011).

In 2008, Samanta, with her coworker, reported on the bioconversion of hexavalent chromium (Cr^{+6}) to trivalent chromium (Cr^{+3}). Tannery wastewater is generally contaminated by hexavalent chromium (Cr^{+6}) and has been considered for investigation. Two types of microbial consortium, such as *Pseudomonas* sp. (JUBTCr1) and *Bacillus* sp. (JUBTCr3) isolated from tannery wastewater, were used for that study. A 5 L double-jacketed chemostat with a 4 L working volume was used for the biodegradation process (Samanta et al., 2008). Different feed volumetric flow rates, ranging from 118 to 133 mL/h, and different inlet hexavalent chromium concentrations (30–90 mg/dm^3) have been used for the study. It has been previously studied that the Haldan-type substrate inhibition form presumes the effective possibility of bioconversion at several rate of dilution. The investigators reported that the Haldan-type substrate inhibited model can satisfactorily predict the extent of bioconversion (Cr^{+6} --> Cr^{+3}) for different dilution rate. The maximum Cr^{+6} reduction occurred at dilution rate 0.029 1/h for *Pseudomonas* sp. and *Bacillus* sp., 77.21% and 79.22%, respectively (Samanta et al., 2008).

22.7 CONCLUSION

Efforts in scientific research and findings of bioleaching technology have increased over the years globally. New innovative approaches, competitive sources and comparative ideas

TABLE 22.7
Mode of Operation and Capacity of Bioleaching Plants Worldwide

Country	Bioleached Metal	Type of Bioleaching Operation	Capacity	Reference
South Africa (Fairview, Barberton)	Gold from arsenopyrite, pyrite flotation concentrates	Stirred tank	35 tons/day	Gahan et al. (2013)
Brazil (Sao Bento)	Gold from arsenopyrite, pyrite flotation concentrates	Stirred tank	150 tons/day	Gahan et al. (2013)
Uzbekistan (Kokpatas)	Gold	Stirred tank	1,069 tons/day	Gahan et al. (2013)
Chile (Cerro Colorado)	Copper	Heap	115,000 tons/year	Gentina and Acevedo (2013)
Australia	Copper from oxide/sulfide ores	Heap bioleaching	16,000 tons/year	Gahan et al. (2013)
USA (Duval)	Copper from oxide/sulfide ores	Dump bioleaching	2,500 tons/year	Gentina and Acevedo (2013)
Cananea (Mexico)	Copper from oxide/sulfide ores	Dump/*in situ* bioleaching	9,000 tons/year	Gentina and Acevedo (2013)
Australia (Tasmania)	Gold from concentrates	Countercurrent airlift and aerated bioreactors based on Bactech/Mintek technology	120,000 oz/year	Neale et al. (2000)

have evolved in this field. With improved installation and processes of pilot-size bioleaching plants, the present panorama of this bioprocess technology is heartening even in developing nations. Nevertheless, heap leaching is taken as the best option for low grade ores and tailings whereas, tank leaching is preferred for precious and base-metal concentrates. Moreover, utilization of potential extremophilic bacteria and archea is of chief importance for incrementing the rate of leaching bioprocess, recovery of valuable metals along with treatment of recalcitrant ores. The improved findings on unique metabolites of these extremophilic microorganisms with biotechnological characteristics as well as application of certain metabolites as biomarkers to evaluate their metabolic activity within this bioprocess emphasize its future industrial relevance.

REFERENCES

Acevedo, F. 2000. The use of reactors in biomining processes. *Electronic Journal of Biotechnology* 3: 10–11.

Acuña, J., J. Rojas, A. M. Amaro, H. Toledo, and C. A. Jerez. 1992. Chemotaxis of *Leptospirillum ferrooxidans* and other acidophilic chemolithotrophs: Comparison with the *Escherichia coli* chemosensory system. *FEMS Microbiology Letters* 75: 37–42.

Allen, E. E., G. W. Tyson, R. J. Whitaker, J. C. Detter, and P. M. B. J. Richardson. 2007. Genome dynamics in a natural archaeal population. *Proceedings of the National Academy of Sciences of the United States of America* 104: 1883–1888.

Amouric, A., C. Brochier-Armanet, D. B. Johnson, V. Bonnefoy, and K. B. Hallberg. 2010 Phylogenetic and genetic variation among Fe(II)-oxidizing acidithiobacilli supports the view that these comprise multiple species with different ferrous iron oxidation pathways. *Microbiology* 157: 111–122.

Andrews, G. 1997. The optima design of bioleaching process. *Mineral Processing and Extractive Metallurgy Review* 19: 149–155.

Andrews, G. F., and K. S. Noah. 1996. Flood drain bioreactor for coal and mineral processing. *Fuel and Energy Abstracts* 38: 7.

Atiemo-Obeng, V. A., W. R. Penny, and P. Armenante. 2004. *Solid Liquid Mixing in Handbook of Industrial Mixing: Science and Practice*. Hoboken, NJ: Wiley Interscience.

Ball, P. 2013. Celebrate the unknowns. *Nature* 496: 419–420.

Bakhtiari, F., M. Zivdar, H. Atashi, and S. A. S. Bagheri. 2008. Bioleaching of copper from smelter dust in a series of airlift bioreactors. *Hydrometallurgy* 90: 40–45.

Belnap, C. P., C. Pan, V. J. Denef, N. F. Samatova, R. L. Hettich, and J. F Banfield. 2011. Quantitative proteomic analyses of the response of acidophilic microbial communities to different pH conditions. *International Society for Microbial Ecology Journal* 5: 1152–1161.

Blake, R. C. and M. N. Griff. 2012. In situ spectroscopy on intact *Leptospirillum ferrooxidans* reveals that reduced cytochrome 579 is an obligatory intermediate in the aerobic iron respiratory chain. *Frontiers in Microbiology* 3: 136–146.

Bobadilla-Fazzini, R. A., G. Levican, and P. Parada. 2011. *Acidithiobacillus thiooxidans* secretome containing a newly described lipoprotein *Licanantase* enhances chalcopyrite bioleaching rate. *Applied Microbiology and Biotechnology* 89: 771–780.

Boon, M., C. Ras, and J. J. Heijnen. 1999. The ferrous ion oxidation kinetics of *Thiobacillus ferrooxidans* in batch cultures. *Applied Microbiology and Biotechnology* 51: 813–819.

Bonnefoy, V. and D. S. Holmes. 2012. Genomic insights into microbial iron oxidation and iron uptake strategies in extremely acidic environments. *Environmental Microbiology* 14: 1597–1611.

Bosecker, K. 1997. Bioleaching: Metal solubilization by microorganisms. *FEMS Microbiology Reviews* 20: 591–604.

Boyer A., J. P. Magnin, and P. Ozil. 1998. Copper ion removal by *Thiobacillus ferrooxidans* biomass. *Biotechnology Letters* 20: 187–190.

Braddock, J. F., H. V. Luong, and E. J. Brown. 1984. Growth kinetics of *Thiobacillus ferrooxidans* isolated from arsenic mine drainage. *Applied Microbiology and Biotechnology* 48: 48–55.

Brandolini, V., P. Tedeschi, A. Capece, A. Maietti, D. Mazzotta, and G. Salzano. 2002. *Saccharomyces cerevisiae* wine strains differing in copper resistance exhibit different capability to reduce copper content in wine. *World Journal of Microbiology and Biotechnology* 18: 499–503.

Brierley, J. A. 1978. Thermophilic iron-oxidizing bacteria found in copper leaching dumps. *Applied Environmental Microbiology* 36: 523–525.

Casas, J. M., T. Vargas, J. Martinez, and L. Moreno. 1998. Bioleaching model of a copper-sulfide ore bed in heap and dump configurations. *Metallurgical and Materials Transactions B* 29: 899–909.

Chang, J. O., R. Law, and C. C. Chang. 1997. Biosorption of lead, copper and cadmium by biomass of *Pseudomonas aeruginosa* PU21. *Water Research* 31: 1651–1658.

Chen, S. Y., and J. G. Lin. 2004. Bioleaching of heavy metals from contaminated sediment by indigenous sulphur-oxidizing bacteria in an air-lift bioreactor: Effects of sulphur concentration. *Water Research* 38: 3205–3214.

Chen, L., Y. Ren, J. Lin, X. Liu, X. Pang, and J. Lin. 2012. *Acidithiobacillus caldus* sulfur oxidation model based on transcriptome analysis between the wild type and sulfur oxygenase reductase defective mutant. *PLoS ONE* 7: e39470.

Coram-Uliana, N. J., R. P. van Hille, W. J. Kohr, and S. T. L. Harrison. 2006. Development of a method to assay the microbial population in heap bioleaching operations. *Hydrometallurgy* 83: 237–244.

Corder, S. L. and M. Reeves. 1994. Biosorption of nickel in complex aqueous waste streams by cyanobacteria. *Applied Biochemistry and Biotechnology* 46: 847–859.

Costa, A. D., A. Carlos, and F. Pereira. 2001. Bioaccumulation of copper, zinc, cadmium and lead by *Bacillus* sp., *Bacillus cereus*, *Bacillus sphaericus* and *Bacillus subtilis*. *Brazilian Journal of Microbiology* 32: 1–5.

Demergasso, C., F. Galleguillos, P. Soto, M. Serón, and V. Iturriaga. 2010. Microbial succession during a heap bioleaching cycle of low grade copper sulfides. *Hydrometallurgy* 104: 382–390.

Dew, D. W., C. Van Buuren, K. McEwan, and C. Bowker. 2000. Bioleaching of base metal sulphide concentrates: a comparison of high and low temperature bioleaching. *Journal of the South African Institute of Mining and Metallurgy* Nov/Dec: 409–413.

Diaby, N., B. Dold, H. R. Pfeifer, C. Holliger, D. B. Johnson, and K. B. Hallberg. 2007. Microbial communities in a porphyry copper tailings impoundment and their impact on the geochemical dynamics of the mine waste. *Environmental Microbiology* 9: 298–307.

Donmez, G. and Z. Aksu. 1999. The effect of copper(II) ions on growth and bioaccumulation properties of some yeasts. *Process Biochemistry* 35: 135–142.

Donmez, G. and Z. Aksu. 2001. Bioaccumulation of copper(II) and nickel(II) by the non-adapted and adapted growing *Candida* spp. *Water Research* 35: 1425–1434.

Dursun, A. Y., G. Ulsu, Y. Cuci, and Z. Aksu. 2003. Bioaccumulation of copper(II), lead(II) and chromium(VI) by growing *Aspergillus niger*. *Process Biochemistry* 38: 1647–1651.

Ehrlich, H. L. 2009. *Geomicrobiology*. Boca Raton, FL: CRC.

Gahan, C. S., H. Srichandan, D. J. Kim, and A. Akcil. 2012. Biohydrometallurgy and biomineral processing technology: A review on its past, present and future. *Research Journal of Recent Sciences* 1: 85–99.

Gentina, J. C., and F. Acevedo. 2013. Application of bioleaching to copper mining in Chile. *Electronic Journal of Biotechnology* 16: 1–14.

Ghoshal, S., P. Bhattacharya, and R. Chowdhury. 2011. De-mercurization of wastewater by *Bacillus cereus* (JUBT1): Growth kinetics, biofilm reactor study and field emission scanning electron microscopic analysis. *Journal of Hazardous Materials* 194: 355–361.

Ghozlan, H. A., S. A. Sabry, and R. A. Amer. 1999. Bioaccumulation of nickel, cobalt, and cadmium by free and immobilized cells of *Pseudomonas* spp. *Fresenius Environmental Bulletin* 8: 428–435.

Gill, N. K., M. Appleton, F. Baganz, and G. J. Lye. 2008. Quantification of power consumption and oxygen transfer characteristics of a stirred miniature bioreactor for predictive fermentation scale-up. *Biotechnology and Bioengineering* 100: 1144–1155.

Groudev, S. N. and V. I. Groudeva. 1993. Microbial communities in four industrial copper dump leaching operations in Bulgaria. *FEMS Microbiology Reviews* 11: 261–268.

Guezennec, A. G., M. Delclaud, F. Savreux, J. Jacob, and P. D'Hugues. 2014. The use of bioleaching methods for the recovery of metals contained in sulphidic mining wastes. *Hydrometallurgy* 1: 1–7.

Hallberg, K. B., S. Hedrich, and D. B. Johnson. 2011. *Acidiferrobacter thiooxydans*, gen. nov. sp. nov.; an acidophilic, thermo-tolerant, facultatively anaerobic iron- and sulfur-oxidizer of the family Ectothiorhodospiraceae. *Extremophiles* 15: 271–279.

Harrison, A. P. 1982. Genomic and physiological diversity amongst strains of *Thiobacillus ferrooxidans*, and genomic comparison with *Thiobacillus thiooxidans*. *Archives of Microbiology* 131: 68–76.

Hassen, A., N. Saidi, M. Cherif, and A. Boudabous. 1998. Effects of heavy metals on *Pseudomonas aeruginosa* and *Bacillus thuringensis*. *Bioresource Technology* 65: 73–82.

Haq, R., S. K. Zaidi, and A. R. Shakoori. 1999. Cadmium resistant *Enterobacter cloacae* and *Klebsiella* sp. isolated from industrial effluents and their possible role in cadmium detoxification. *World Journal of Microbiology and Biotechnology* 15: 283–290.

Hernandez, A., R. P. Mellado, and J. L. Martinez. 1998. Metal accumulation and vanadium-induced multidrug resistance by environmental isolates of *Escherichia hermannii* and *Enterobacter cloacae*. *Applied Environmental Microbiology* 64: 4317–4320.

Jin, J., G. L. Liu, S. Y. Shi, and W. Cong. 2010. Studies on the performance of a rotating drum bioreactor for bioleaching processes- oxygen transfer, solids distribution and power consumption. *Hydrometallurgy* 103: 30–34.

Johnson, D. B. and K. B. Hallberg. 2005. Acid mine drainage remediation options: A review. *Science of the Total Environment* 338: 3–14.

Johnson, D. B., T. Kanao, and S. Hedrich. 2012. Redox transformations of iron at extremely low pH: Fundamental and applied aspects. *Frontiers in Microbiology* 3: 96.

Justice, N. B., A. Norman, C. T. Brown, A. Singh, B. C. Thomas, J. F. Banfield. 2014. Comparison of environmental and isolate Sulfobacillus genomes reveals diverse carbon, sulfur, nitrogen, and hydrogen metabolisms. *BioMed Central Genomics* 15: 1107.

Kadijani, S. Z., J. Safdari, M. A. Mousavian, and A. Rashidi. 2013. Study of oxygen mass transfer coefficient and oxygen uptake rate in a stirred tank reactor for uranium ore bioleaching. *Annals of Nuclear Energy* 53: 280–287.

Karna, R. R., L. S. Sajani, and P. Maruthi. 1996. Bioaccumulation and biosorption of Co^{2+} by *Neurospora crassa*. *Biotechnology Letters* 18: 1205–1208.

Kelly, D. P. and A. P. Wood. 2000. Reclassification of some species of *Thiobacillus* to the newly designated genera *Acidithiobacillus* gen. nov., Halothiobacillus gen. nov. and *Thermithiobacillus* gen. nov. *International Journal of Systemic and Evolutionary Microbiology* 50: 511–516.

Kulshreshtha, A., R. Agrawal, M. Barar, and S. Saxena. 2014. A review on bioremediation of heavy metals in contaminated water. *IOSR Journal of Environmental Science, Toxicology and Food Technology* 8: 44–50.

Kumar, A., B. S. Bisht, V. D. Joshi, and T. Dhewa. 2011. Review on bioremediation of polluted environment: A management tool. *International Journal of Environmental Sciences* 1: 1079–1093.

Levicán, G., J. A. Ugalde, N. Ehrenfeld, A. Maass, and P. Parada. 2008. Comparative genomic analysis of carbon and nitrogen assimilation mechanisms in three indigenous bioleaching bacteria: Predictions and validations. *BioMed Central Genomics* 9: 581.

Li, D., D. W. Li, and S. J. Zhang. 2011. The study of toxic elements removal and valuable metals recovery from mine tailings in gas-liquid-solid internal circulation bioreactor. *Research Journal of Chemistry and Environment* 15: 990–993.

Liu, M. S., R. M. R. Branion, and D. W. Duncan. 1988. The effects of ferrous iron, dissolved oxygen, and inert solids concentration on the growth of *Thiobacillus ferrooxidans*. *The Canadian Journal of Chemical Engineering* 66: 445–451.

Liu, Z., N. Guiliani, C. Appia-Ayme, F. Borne, J. Ratouchniak, and V. Bonnefoy. 2000. Construction and characterization of a recA mutant of *Thiobacillus ferrooxidans* by marker exchange mutagenesis. *Journal of Bacteriology* 182: 2269–2276.

Lo, I., V. J. Denef, N. C. Verberkmoes, M. B. Shah, D. Goltsman, G. DiBartolo, G. W. Tyson et al. 2007. Strain-resolved community proteomics reveals recombining genomes of acidophilic bacteria. *Nature* 446: 537–541.

Loi, G., A. Mura, P. Troi, and G. Rossi. 1994. Bioreactor performance *vs.* solids concentration in coal bio depyritization. *Fuel Processing Technology* 40: 251–260.

MacDonald, D. G. and R. H. Clark. 1970. The oxidation of aqueous ferrous sulphate by *Thiobacillus ferrooxidans*. *Canadian Journal of Chemical Engineering* 48: 669–676.

Magyarosy, A., R. D. Laidlaw, R. Kilaas, C. Echer, D. S. Clark, and J. D. Keasling. 2002. Nickel accumulation and nickel oxalate precipitation by *Aspergillus niger*. *Applied Microbiology and Biotechnology* 59: 382–388.

Mangold, S., J. Valdés, D. S. Holmes, and M. Dopson. 2011. Sulfur metabolism in the extreme acidophile *Acidithiobacillus caldus*. *Frontiers in Microbiology* 2: 17.

Massaccesi, G., M. C. Romero, M. C. Cazau, and A. M. Bucsinszky. 2002. Cadmium removal capacities of filamentous soil fungi isolated from industrially polluted sediments, in La Plata (Argentina). *World Journal of Microbiology and Biotechnology* 18: 817–820.

Matsunaga, T., H. Takeyama, T. Nakao, and A. Yamazawa. 1999. Screening of microalgae for bioremediation of cadmium polluted seawater. *Journal of Biotechnology* 70: 33–38.

Merino, M.P., B.A. Andrews, and J.A. Asenjo. 2014. Stoichiometric model and flux balance analysis for a mixed culture of *Leptospirillum ferriphilum* and *Ferroplasma acidiphilum*. *Applied Cellular Physiology and Metabolic Engineering* 31(2): 307–315.

Merino, M. P., B. A. Andrews, and J. A. Asenjo. 2010. Stoichiometric model and metabolic flux analysis for *Leptospirillum ferrooxidans*. *Biotechnology and Bioengineering* 107: 696–706.

Miller, J. D., C. L. Lin, C. Roldan, and C. Garcia. 2003. Particle size distribution for copper heap leaching operations as established from 3D mineral exposure analysis by x-ray micro CT. *Hydro Metallurgy of Copper (Book 1)* 4: 83–97.

Mishra, D., D. J. Kim, J. G. Ahn, and Y. H. Rhee. 2005. Bioleaching: A microbial process of metal recovery; a review. *Metals and Materials International* 11: 249–256.

Neale, J. W., A. Pinches, and V. Deeplaul. 2000. Mintek-BacTech's bacterial oxidation technology for refractory gold concentrates: Beaconsfield and beyond. *The Journal of South African Institute of Mining and Metallurgy* 100: 415–422.

Norris, P. R., N. P. Burton, and N. A. M. Foulis. 2000. Acidophiles in bioreactor mineral processing. *Extremophiles* 4: 71–76.

Ojumu, T. V. 2006. A review of rate equations proposed for microbial ferrous-iron oxidation with a view to application to heap bioleaching. *Hydrometallurgy* 83: 21–28.

Olson, G. J., J. A. Brierley, and C. L. Brierley. 2003. Bioleaching review part B: Progress in bioleaching: applications of microbial processes by the minerals industries. *Applied Microbiology and Biotechnology* 63: 249–257.

Orellana, L. H. and C. A. Jerez. 2011. A genomic island provides *Acidithiobacillus ferrooxidans* ATCC 53993 additional copper resistance: A possible competitive advantage. *Applied Microbiology and Biotechnology* 92: 761–767.

Osorio, H., S. Mangold, Y. Denis, I. Nancucheo, M. Esparza, D. B. Johnson, V. Bonnefoy, M. Dopson, and D. S. Holmes. 2013. Anaerobic sulfur metabolism coupled to dissimilatory iron reduction in the extremophile *Acidithiobacillus ferrooxidans*. *Applied Environmental Microbiology* 79: 2172–2181.

Pangarkar, V. G. 2015. *Stirred tank Reactors for Chemical Reactions in Design of Multiphase Reactors*. Hoboken, NJ: John Wiley & Sons.

Peng, J., W. Yan, and X. Bao. 1994. Expression of heterologous arsenic resistance genes in the obligately autotrophic biomining bacterium *Thiobacillus ferrooxidans*. *Applied Environmental Microbiology* 60: 2653–2656.

Perez-Rama, M., J. A. Alonoso, C. H. Lopez, and E. T. Vaamonde. 2002. Cadmium removal by living cells of the marine microalgae *Tetraselmis suecica*. *Bioresource Technology* 84: 265–270.

Pradhan, N., K. C. Nathsarma, K. S. Rao, and L. B. Sukla. 2008. Heap bioleaching of chalcopyrite: A review. *Minerals Engineering* 21: 355–365.

Rawlings, D. E. 1997. *Theory, Microbes and Industrial Processes*. Berlin, Germany: Springer Verlag.

Rawlings, D. E., H. Tributsch, and G. S. Hansford. 1999. Reasons why 'Leptospirillum'-like species rather than *Thiobacillus ferrooxidans* are the dominant iron-oxidizing bacteria in many commercial processes for the biooxidation of pyrite and related ores. *Microbiology* 145: 5–13.

Rawlings, D. E. 2002. Heavy metal mining using microbes. *Annual Review of Microbiology* 56: 65–91.

Roane, T. M., K. L. Josephson, and I. L. Pepper. 2001. Dual-bioaugmentation strategy to enhance remediation of cocontaminated soil. *Applied Environmental Microbiology* 67: 3208–3215.

Rohwerder, T., T. Gehrke, K. Kinzler, and W. Sand. 2003. Bioleaching review part A: progress in bioleaching: Fundamentals and mechanisms of bacterial metal sulfide oxidation. *Applied Microbiology and Biotechnology* 63: 239–248.

Rossi, G. 2001. The design of bioreactors. *Hydrometallurgy* 59: 217–231.

Sand, W., T. Gehrke, R. Hallmann, and A. Schippers. 1995. Sulfur chemistry, biofilm, and the (in)direct attack mechanism—A critical evaluation of bacterial leaching. *Applied Microbiology and Biotechnology* 43: 961–966.

Sand, W., T. Gehrke, P. G. Jozsa, and A. Schippers. 2001. (Bio) chemistry of bacterial leaching—Direct vs. indirect bioleaching. *Hydrometallurgy* 59: 159–175.

Samanta, K., R. Chowdhury, and P. Bhattacharya. 2008. Effect of substrate concentration on the transient dynamics of specific cell growth during bioconversion of Cr^{+6} to Cr^{+3} using polyculture consortia. *Indian Journal of Chemical Technology* 15: 209–215.

Santelli, C. M., V. P. Edgcomb, W. Bach, and K. J. Edwards. 2009. The diversity and abundance of bacteria inhabiting seafloor lavas positively correlate with rock alteration. *Environmental Microbiology* 11: 86–98.

Schippers, A., A. Breuker, A. Blazejak, K. Bosecker, D. Kock, and T. L. Wright. 2010. The biogeochemistry and microbiology of sulfidic mine waste and bioleaching dumps and heaps, and novel Fe(II)-oxidizing bacteria. *Hydrometallurgy* 104: 342–350.

Sethurajan, M., R. Aruliah, O. P. Karthikeyan, and R. Balasubramaniam. 2012. Bioleaching of copper from black shale ore using mesophilic mixed populations in an air uplift bioreactor. *Environmental Engineering and Management Journal* 11: 1839–1848.

Simona, C. C. and V. Micle. 2011. Consideration concerning factors influencing bioleaching processes. *ProEnvironment* 4: 76–79.

Staden, P. J. V., M. Gericke, and P. M. Craven. 2008. Minerals Biotechnology: trends, opportunities and challenges. *Hydrometallurgy* 1: 6–14.

Su, C., L. Jiang, and W. J. Zhang. 2014. A review on heavy metal contamination in the soil worldwide: Situation, impact and remediation techniques. *Environmental Skeptics and Critics* 3: 24–38.

Swamy, K. M., L.B. Sukla, K.L. Narayana, R. N. Kar, and V. V. Panchanadikar. 1995. Use of ultrasound in microbial leaching of nickel from laterites. *Ultrasonics Sonochemistry* 2(1): 5–9.

Temple, K. L. and A. R. Colmer. 1951. The autotrophic oxidation of iron by a new bacterium: *Thiobacillus ferrooxidans*. *Journal of Bacteriology* 62: 605–611.

Tian, K. L., J. Q. Lin, X. M. Liu, Y. Liu, C. K. Zhang, and W. M. Yan. 2003. Conversion of an obligate autotrophic bacteria to heterotrophic growth: expression of a heterogeneous phosphofructokinase gene in the chemolithotroph *Acidithiobacillus thiooxidans*. *Biotechnology Letters* 25: 749–754.

Tyson, G. W., J. Chapman, P. Hugenholtz, E. E. Allen, R. J. Ram, P. M. Richardson, V. V. Solovyev, E. M. Rubin, D. S. Rokhsar, and J. F Banfield. 2004. Community structure and metabolism through reconstruction of microbial genomes from the environment. *Nature* 428: 37–43.

Valencia, P. and F. Avecedo 2009. Are bioleaching rates determined by the available particle surface area concentration?. *World Journal of Microbiology and Biotechnology* 25: 101–106.

Varki, A. 2013. Omics: Account for the "dark matter" of biology. *Nature* 497: 565.

Vera, M., B. Krok, S. Bellenberg, W. Sand, and A. Poetsch. 2013. Shotgun proteomics study of early biofilm formation process of *Acidithiobacillus ferrooxidans* on pyrite. *Proteomics* 13: 133–1144.

Vidali, M. 2001. Bioremediation An overview. *Pure and Applied Chemistry* 73: 1163–1172.

Watling, H. R. 2006. The bioleaching of sulphide minerals with emphasis on copper sulphides—A review. *Hydrometallurgy* 84: 81–108.

Watling, H. R. 2015. Review of biohydrometallurgical metals extraction from polymetallic mineral resources. *Minerals* 5: 1–60.

Wong, J. P. K., Y. S. Wong, and N. F. Y. Tam. 2000. Nickel biosorption by two Chlorella species, *C. Vulgaris* (a commercial species) and *C. Miniata* (a local isolate). *Bioresource Technology* 73(2): 133–137.

Wong, P. K. and K. Y. Fung. 1997. Removal and recovery of nickel ion (Ni^{2+}) from aqueous solution by magnetite-immobilized cells of *Enterobacter* sp. 4–2. *Enzyme and Microbial Technology* 20: 116–121.

23 Computational Approaches for Metal-Binding Site Prediction and Design of Effective Metal Bioremediation Strategies

Ahalyaa Subramanian, Maziya Ibrahim, and Sharmila Anishetty

CONTENTS

ABSTRACT

In this chapter, we have reviewed computational methods to identify metal-binding proteins using sequence and/or structural information. A methodology to generate putative metal-binding sequence motifs is described. Also, a list of available tools and databases for the prediction of metal-binding sites is given. Understanding the genome features of the organisms in metal-rich environments will provide insights into their adaptations, which will help in designing effective metal bioremediation strategies. The use of "omics" and systems biology approaches will aid in revealing interactions within the microbial community and the surrounding environment. A few case studies using these approaches are highlighted here.

23.1 INTRODUCTION

Metal pollution in the environment occurs through emissions from industrial effluents, acid mine drainages, fertilizers, pesticides, sewage sludge, and combustion of fuel. This poses a threat to human life and environment. Heavy metals such as lead, cadmium, mercury, cobalt, copper, nickel, zinc, and arsenic are commonly found in contaminated environments. These metals are important for some biochemical reactions, essential to the growth and development of microorganisms, plants, and animals. However, higher concentrations may exert toxic effects (Kavamura and Esposito, 2010). The use of

microorganisms to break down environmental pollutants into less toxic forms is termed as "bioremediation" (Vidali, 2001).

Microorganisms that thrive in metal-polluted areas have evolved metal resistance mechanisms. Microorganisms that are resistant to high concentrations of metals are called metallophiles (Sinha et al., 2014). The resistance mechanisms used by such microorganisms include (1) efflux of toxic metal out of the cell, (2) enzymatic conversion, (3) intra- or extracellular sequestration, (4) exclusion by permeability barrier, and (5) reduction in sensitivity of cellular targets (Nies, 1999). Uptake of heavy metals also occurs through metabolism-independent biosorption, which happens on the surface of cells, that is, cell wall or cell membranes. Biosorption is possible by both living and nonliving biomass. Glycoproteins present on cell walls have more potential binding sites for Cd^{2+} than phospholipids and lipopolysaccharides (Gourdon et al., 1990). Microorganisms have evolved various metalloregulatory proteins that control intracellular metal ion levels and resistance in microorganisms (Yang et al., 2015).

The rise of -omics (genomics, proteomics, metallomics, metabolomics) in the past decade has resulted in an overwhelming increase in the availability of biological data. This has provided an opportunity to probe and annotate metal resistance systems in microbes. Knowledge about these microbe–metal interactions can be gained using bioinformatics methods. This can aid in effective bioremediation strategies.

In this chapter, we review a few computer-based methods for the identification of metal-binding proteins and illustrate how metal-binding sequence motifs can be designed. We also underscore how computational methods and systems approaches can aid in the design of bioremediation strategies.

23.2   METALLOPROTEOMES

Metalloproteome refers to all the metalloproteins encoded by an organism. The coupling of metallomics and proteomics revealed that the metalloproteomes of many microorganisms remain largely uncharacterized (Cvetkovic et al., 2010). An in silico approach, combining several bioinformatics methods, makes it possible to deduce the metalloproteomes. For example, the content of zinc, nonheme iron, and copper proteins in a representative set of organisms taken from the three domains of life has been predicted based on this approach. The predicted metalloproteins can then be analyzed to identify their function and evolution (Andreini et al., 2009). A subset of these metalloproteins is related to metal resistance systems in metal-tolerant organisms.

Hidden Markov models (HMMs) and support vector machines are machine learning techniques that have also been used for identifying metalloproteins. A computational model based on profile HMMs, in bacterial proteomes with special focus on iron–sulfur proteins designed by Estellon et al. (2014), represents a prototype for the use of linear models to identify metalloprotein superfamilies.

Nucleotide and amino acid sequences of metallophiles can be retrieved from public databases such as the NCBI Genome (Benson et al., 2007) and UniProtKB Consortium (2014). Based on these sequences, putative metalloproteins such as zinc-finger proteins can be predicted by sequence and structural similarity with known metalloproteins or by the presence of specific metal-binding sites and metal-binding domains. Tools such as PSI-BLAST (Altschul et al., 1997) are used for sequence similarity searches. Domain information can be obtained from databases such as InterPro (Mitchell et al., 2015).

Currently, UniProtKB (release 2015_12) has a total of 31,262 proteomes wherein 26,737 belong to bacteria, 3,360 are from viruses, 314 from archaea, and 849 from eukaryotes. A search for proteins under the Gene Ontology (GO) term wherein the molecular function is metal ion binding (GO:0046872) shows a total of 4,715,220 entries.

Computational methods for the prediction of metal-binding sites in proteins require sequence and/or structural information. Due to the paucity of 3D structural information of many proteins, generating a metal-binding motif from sequence data is one of the widely used methods to identify metal-binding sites in a protein.

23.3   METAL UPTAKE SYSTEMS AND METALLOTHIONEINS

In order to acquire metal ions from the extracellular environment, high-affinity active transport systems in the outer membrane or embedded in the plasma or inner membranes are required by the microorganism. Metal transporters facilitate the selective movement of metal ions across the cell membranes. Metal transporters include ATP-binding cassette-type ATPases, P-type ATPases, resistance and nodulation (RND) proteins, cation diffusion facilitator (CDF) proteins, CorA (Co resistance), Ni and Co transporters (NiCoT) proteins, natural resistance associated with macrophage proteins (NRAMPs), and Zrt/Irt-like protein-family transporters (Waldron and Robinson, 2009). RND efflux systems contain an outer membrane protein and a membrane fusion protein for the transport of the substrate, whereas P-type ATPases and CDF proteins are single subunit transporters. Representative protein sequences of these transporters are available in the Transporter Classification Database (TCDB). The NRAMP transporter family (2.A.55) is classified under APC superfamily in TCDB. TCDB lists four families under CDF superfamily, namely, (1) 1.A.52, the Ca^{2+} release-activated Ca^{2+} (CRAC) channel (CRAC-C) family; (2) 2.A.4, the CDF family; (3) 2.A.19, the Ca^{2+}: cation antiporter family; and (4) 2.A.103, the bacterial murein precursor exporter (MPE) family. TCDB can be used as a genome-transporter annotation tool to identify different types of transporters in a particular microorganism of interest (Saeir et al., 2013).

Metallothioneins (MTs) are low-molecular-weight proteins with high cysteine (Cys) and metal content. They bind to a variety of metal ions, preferably to Zn, Cu, and Cd. They contain 15%–30% of Cys residues that are present as CxC and CxxC motifs. Small amino acids such as Gly and Ala are common, whereas bulky aromatic amino acids are rare (Blindauer and Leszczyszyn, 2010). MT genes have been expressed in transgenic bacteria to incorporate high mercury resistance and accumulation that renders the microorganism suitable for bioremediation purposes (Ruiz et al., 2011). Resistance to mercury in bacteria is encoded by the genes of the *mer* operon. merRTPADE operon provides resistance toward inorganic mercury and merRTPAGBDE operon confers resistance to inorganic and organic mercury. *Cupriavidus metallidurans* is a model organism for metal bioremediation that possesses such a *mer* operon. The mechanism of resistance involves an uptake and transport of Hg^{2+} by periplasmic protein MerP and inner membrane protein MerT. MerT is a membrane protein that is thought to act as a broad mercury transporter. MerR is an activator or repressor of the transcription of *mer* genes. Cytosolic mercuric reductase MerA reduces Hg^{2+} to less toxic Hg. MerG reduces the cellular permeability to organomercurial compounds. MerB is an organomercurial lyase that catalyzes the protonolytic cleavage of carbon–mercury bonds (Rojas et al., 2011). Mercury transporters contain two to four transmembrane helical segments. Bioinformatics analyses have found transmembrane segments 1 and 2 to be homologous among the Mer proteins (Mok et al., 2012).

23.4   METAL-BINDING SITES

Sequence motifs or patterns are conserved subsequences characteristic of a protein or gene family. The presence of conserved sequence motifs characteristic of a protein family helps in functional annotation of proteins. Generating metal-binding motifs from protein sequence data will be useful in annotating candidate members of a protein family. PROSITE (Sigrist et al., 2012) is a database of protein families and domains. The database contains patterns and profiles specific for more than a thousand protein families or domains. For example, the database describes a "heavy-metal-associated domain" (PDOC00804) as a conserved domain of about 70 amino acid residues found in proteins that transport or detoxify heavy metals. Proteins such as mercuric reductase (merA) and copper-binding protein copP contain this heavy-metal-associated domain.

23.4.1 Generating Putative Metal-Binding Sequence Motifs

In addition to the sequence motifs available in the databases, one can generate metal-binding motifs for different protein families. Multiple sequence alignment of representative members of a protein family from different species will help in identifying motifs/sequence patterns around conserved regions. Often, there may be more than one conserved subsequence observed through a multiple sequence alignment. In the context of metal-binding proteins, knowledge about the type of ligand/metal that may bind to this site can further help in narrowing down the potential metal-binding motif. For example, it is well known that anionic or acidic residues coordinate the binding of divalent cationic metal ions (Frausto da Silva and Williams, 2001). In this case, conserved stretches that include acidic residues like Asp or Glu may be explored for generating divalent cationic metal-binding motifs. Since the motif is generated using only representative members of a protein family, it should be further validated. One approach is to use ScanProsite (De Castro et al., 2006) tool at the PROSITE database to scan UniProtKB/TrEMBL or PDB (Berman et al., 2000) databases

in order to identify true positives, false positives, and false negatives. True positives in this context are proteins that belong to the said category. For example, if a motif is generated for a cobalt transport protein family, all the proteins of this family recovered from the corresponding databases with this motif as the query are termed *true positives*. Members of this family that are not picked up in this search are termed *false negatives*. False positives are the ones that do not belong to the said family but are being picked up by the motif. However, it should be noted that some of the uncharacterized or hypothetical proteins that are picked up in this search should further be examined for assessing if they are new candidate members of the family. This, in turn, will help in functional annotation of new members of the family. The motif should be refined until the number of true positives is maximized, and the number of false positives and false negatives is minimized. A metric for assessing the reliability of the motif is by calculating the precision, recall, and accuracy. Precision or specificity of the motif is calculated as the ratio of true positive hits to the total of true positives and false positives. Recall or sensitivity is calculated as the ratio of true positive hits to the total number of true positives and false negatives. Accuracy is calculated as the ratio of true positives to the total number of hits. For example, signature pattern for ArsR family (arsenic resistance genes) in the PROSITE database is given as "C-x(2)-D-[LIVM]-x(6)-[ST]-x(4)-S-[HYR]-[HQ]." The precision and recall for this pattern is 100% and 96%, respectively. Literature reports on metal-binding residues from proteins belonging to a family can also be used to validate the motif that is generated for that family. In cases where PDB structures complexed with the metal ligand are available, one can validate the motif by ascertaining if the motif falls in the region binding the metal. Figure 23.1 shows a multiple sequence alignment generated from representative

```
Q58490|CBIN_METJA    ----METKHIILLAIVAIIIALPLIIYAGKGEEEGYFGGSDDQGCEVVEE--LGYKPWFH
Q26234|CBIN_METTH    ----MDKRHILMLLAVIIISVAPLIIYSGHGEDDGYFGGADDSAGDAITE--TGYKPWFQ
Q50799|CBIN_METTM    ----MDKRHTIMLIAVAVIAIAPLVIYSGLGEDQGYFGGAADDSASKAISE--TGYKPWFQ
O68104|CBIN_RHOCA    ---MSSKRTLWLLAGTVALVVVPL-------LMGGEFGGADGQAAELIEATVPGFAPWAD
Q9HPH5|CBIN_HALN1    -MNRWLAAGGILLGA---LVVFSF-------VSAGAWGGADGVAGDTITTINPSYEPWFQ
Q54189|CBIN_STRCO    -MSRNTRINALLLLAVAALAVLPLVLGLGD-HKEEPFAGADAEAETAITEIEPDYWPWFS
O0YYX7|C0YYX7_LACRE  -MKKRTKTNIILAICVILLVLIPFI------FVKGEYSGSDDQGTEQIKKFDPSYKAWAH
O29529|CBIN_ARCFU    -----MKKLLLLLILLIF-----AAK-----VTAEEWAGADEKAEEVIKELKPDYEPWFS
Q8XNY9|CBIN_CLOPE    MKNKRVLTNVILLLLVVFITTIPFFV-----AKNGEFGGSDDQAEEFITQIDENYEPWFS
Q8Z5N3|CBIN_SALTI    -----MKKTLMLLAMVVALVILPFFI-----NHGGEYGGSDGEAESQIQALAPQYKPWFQ
Q05595|CBIN_SALTY    -----MKKTLMLLAMVVALVILPFFI-----NHGGEYGGSDGEAESQIQAIAPQYKPWFQ
                             :                          :.*:*   .  :      : *



Q58490|CBIN_METJA    PIW|EPPSGEIESLLFALQ|AAIGAIIIGYYIGYYNAKRQVAA------------------
Q26234|CBIN_METTH    PIW|EPPSGEIESLLFALQ|AAIGALIIGYVFGYYRGRGESSE------------------
Q50799|CBIN_METTM    PIW|EPPSGEIESLLFALQ|AAIGALIIGYVFGYYRGRGESPE------------------
O68104|CBIN_RHOCA    PLW|EPPSGEVESLFFALQ|AALGAFVVGLVIGRRQGAAKTREQNAPAPRSFPAE-------
Q9HPH5|CBIN_HALN1    SLW|TPPSGEIESLLFSIQ|AAVGGIIIGYYLGRDRPRGQSQDMGSDLP------------
O54189|CBIN_STRCO    PLH|EPPSGEIESALFALQ|AALGAGVLAYYFGLRRGRRQGEERASAASGAAAAPGDAPEGD
C0YYX7|C0YYX7_LACRE  PVW|TPPSGEIESLLFTVQ|GSLGTGIICYFIGAAHGKKKAQQNKTKQTVKQ---------
O29529|CBIN_ARCFU    PIF|EPPSGEIESMLFSLQ|AAIGSLIIGYFLGYYRGLKHARNA----------------
Q8XNY9|CBIN_CLOPE    PLF|EPASGEIESLLFALQ|AAIGAGVIGFGLGYLKGKKKVNDEVNDKHR-----------
Q8Z5N3|CBIN_SALTI    PLY|EPASGEIESLLFTLQ|GSLGAAVIFYILGYCKGKGKQRRDDRA---------------
Q05595|CBIN_SALTY    PLY|EPASGEIESLLFTLQ|GSLGAAVIFYILGYCKGKGKQRRDDRA---------------
                      : |* ***:** :*::*|::*  ::   :*  .   .
```

FIGURE 23.1 Multiple sequence alignment of representative cobalt transporters (CbiN). Conserved motif is shown in a box.

members of a cobalt transport family encoded by "CbiN" gene. The putative cobalt metal-binding motif is shown as a boxed region. The motif is "[ET]P[AP]SGE[IV]ES[LAM] [LF]F[SAT][ILV]Q." This region is predicted to bind to cobalt metal since it is conserved across different species and contains acidic residues. In general, acceptable amino acids for the particular position in a motif are given in square brackets "[]." Curly brackets "{}" represent amino acids that are not acceptable at that given position and "x" is used for a position where any amino acid is accepted. In the multiple sequence alignment, the positions marked with "*" are said to be invariant (i.e., conserved in all the sequences considered in the multiple sequence alignment) while the positions marked with a ":" and a "." are said to be conservation between strong and weak property groups, respectively. The positions within the motif that are not marked with any of these symbols have different types of residues.

This methodology has been used in a study where putative metal-binding motifs were generated for ions of the following metals: arsenic, cadmium, cobalt, magnesium, manganese, mercury, molybdenum, nickel, and zinc. Where possible, these motifs were validated with literature reports, mutant studies, or PDB structures. For example, CzcCBA pump, which is an RND-driven transenvelope exporter, contains three components: CzcA, CzcB, and CzcC. Motifs for CzcA and CzcC have been designed as "D[FI]G[IMVA][IM][LIV]D[AGS]A[IV] V[MIV][VT]E" and "L[DQ]VLDAQ[RN][TE]L," respectively. Similarly, motifs for P-type ATPases have been identified. The conserved Asp and Pro in the motif "[IV]G<u>D</u>G[IV]NDA<u>P</u>[AT] LA" are crucial for catalytic activity (Thilakaraj et al., 2007).

23.4.2 Other Sequence-Based Methods for Metal-Binding Site Prediction

Other sequence-based methods to predict metal-binding sites incorporate "bonding state prediction," a binary classification task where individual residues are predicted as metal binding or not. Current approaches are focused on Cys-, His-, Asp-, and Glu-binding ions (Passerini et al., 2012). A support vector machine prediction system, based on physicochemical properties derived from the sequence of metal-binding proteins, was developed for 10 metal-binding classes (Lin et al., 2006).

23.4.3 Prediction of Metal-Binding Sites Using Protein Structures

The feature of the binding site is that metals bind at the centers of high hydrophobicity contrast, and this function can be used to locate, characterize, and design metal-binding sites (Yamashita et al., 1990). Several computational algorithms have been developed over the years that predict binding sites for specific metals. In the absence of obvious metal-binding sequence motifs, a combination of sequence profile information with low-resolution structural data can help in detecting the metal-binding residues (Sodhi et al., 2004).

The structure-based methods of prediction use the structure information available in PDB. A FEATURE-based framework computes descriptions of protein microenvironments such as features at the atomic, chemical group, residue, and secondary structural level (Wei and Altman, 1997; Halperin et al., 2008). These are used as statistical predictors of metal-binding sites. Successive algorithms along the idea of FEATURE-based method also integrate information from binding motifs and molecular dynamics simulations to improve the accuracy of prediction (Glazer et al., 2008). Based on the metal-binding sites of known metalloproteins, 3D structural motifs with three-residue or four-residue templates can be used for identifying metal-binding sites in the query structures. The use of geometrical parameters enables functional annotation of new structures based on predicted metal-binding sites (Goyal and Mande, 2008). Free energy calculations from empirical force field are also used to predict the metal-binding sites of Mg^{2+}, Ca^{2+}, Zn^{2+}, Mn^{2+}, and Cu^{2+} and their affinities (Schymkowitz et al., 2005).

The analysis of conserved metal-coordinating residues that are commonly found in PDB structures showed that certain residues are preferred to others for binding to certain metals. Also, side-chain-coordinating residues are highly conserved compared to main-chain-coordinating residues. For instance, Cu is predominantly coordinated by side-chain residues of the acidic amino acids Asp and Glu (Kasampalidis et al., 2007).

On the basis of amino acid composition and geometry of the metal-binding site, the frequently occurring amino acid residues at a specific metal-binding site are determined. This is termed as philicity or likeness profiles of the amino acids for various biologically important metals. For example, His, Glu, Asp, and Cys are likely to be present near the cadmium-binding sites (Kumar Kuntal et al., 2010).

Compared to the previously mentioned methods that distinguish the metal-binding sites from the nonmetal-binding sites, newer algorithms such as mFASD are capable of discriminating between different types of metal-binding sites. Along with the microenvironment characteristics, mFASD uses a distance measure based on functional atoms that are in contact with the bound metal (He et al., 2015).

23.4.4 Tools and Databases Related to Metal-Binding Site Prediction and Biodegradation

A list of available tools and databases for the prediction and analysis of metal-binding sites in proteins based on the use of sequence and structural data is given in Table 23.1. Also, databases related to biodegradation pathways and microorganisms are listed in Table 23.1. FINDSITE-metal (Brylinski and Skolnick, 2011) integrates structure/evolution information from threading with machine learning to improve the accuracy of metal-binding site prediction. MetalDetector (Lippi et al., 2008) uses a decision tree approach that combines predictions from two methods, namely, DISULFIND and Metal Ligand Predictor. MetaRouter (Pazos et al., 2005) is a database that links biodegradation information with corresponding protein

TABLE 23.1

List of Tools and Databases for Prediction and Analysis of Metal-Binding Sites in Proteins

S. No.	Resources	Comment	URL	Reference
Structure-Based Tools				
1.	FINDSITE-metal	Threading-based method that detects metal-binding sites in modeled protein structures	http://cssb.biology.gatech.edu/findsite-metal	Brylinski and Skolnick (2011)
2.	MIB: metal ion–binding sites prediction server	Uses fragment transformation method to predict metal ion–binding sites	http://bioinfo.cmu.edu.tw/MIB/	Lu et al. (2012)
3.	MDB: metalloprotein database and browser	Contains information on geometrical parameters of metal-binding sites in protein structures from PDB	http://metallo.scripps.edu	Castagnetto et al. (2002)
4.	Metal PDB	Collects and allows easy access to the knowledge on metal sites in biological macromolecules starting from structural information contained in PDB	http://metalweb.cerm.unifi.it/	Andreini et al. (2013)
5.	CheckMyMetal	Evaluates metal-binding sites in macromolecular structures	http://csgid.org/csgid/metal_sites/	Zheng et al. (2014)
6.	BioMe	A web-based platform for the calculation of various statistical properties of metal-binding sites	http://metals.zesoi.fer.hr/	Tus et al. (2012)
7.	MIPS: metal interactions in protein structures	An open resource for the analysis and visualization of all metals and their interactions with macromolecular structures	http://dicsoft2.physics.iisc.ernet.in/cgi-bin/mips/query.pl	Hemavathi et al. (2009)
Sequence-Based Tools				
8.	SeqCHED	Models the query sequence to PDB template sequences and predicts the putative metal-binding sites	http://ligin.weizmann.ac.il/seqched/	Levy et al. (2009)
9.	MetalDetector	Uses sequence information to predict Cys and histidine metal-binding sites	http://metaldetector.dsi.unifi.it/	Lippi et al. (2008)
Biodegradation-Related Databases				
10.	BacMet	Bioinformatics resource of manually curated and experimentally confirmed antibacterial biocide and metal resistance genes; also contains predicted resistance genes based on sequence similarity	http://bacmet.biomedicine.gu.se/	Pal et al. (2014)
11.	BioSurfDB	Curated relational information system integrating data from a. Metagenomes b. Organisms c. Biodegradation d. Relevant genes metabolic pathways e. Bioremediation experiment results f. Biosurfactant curated list	http://aleph.inesc-id.pt/~biosurfdb/	Oliveira et al. (2015)
12.	UM-BBD	Public resource for microbial biocatalytic reactions and biodegradation pathways	http://eawag-bbd.ethz.ch/	Ellis and Wackett (2012)
13.	MetaRouter	System for maintaining heterogeneous information related to biodegradation	http://pdg.cnb.uam.es/MetaRouter	Pazos et al. (2005)
14.	Bionemo	Manually curated information about proteins and genes directly implicated in the biodegradation metabolism	http://bionemo.bioinfo.cnio.es/	Carbajosa et al. (2009)

and genome data, thus providing a framework to understand the global properties of the bioremediation network.

23.5 MOLECULAR DYNAMICS OF METALLOPROTEINS

Molecular dynamics simulations are useful in understanding the structure and function of biomolecules (Karplus and McCammon, 2002). Quantum mechanics/molecular mechanics approach (QM/MM) is considered ideal for the study of metalloproteins as it accurately describes the nature of the metal coordination sphere. Using QM/MM methods, mechanistic studies on the formation and reduction of high redox intermediates in heme enzymes and antibiotic degradation by Zn-dependent lactamases have been reported (Vidossich and Magistrato, 2014). Nitroarene dioxygenases are members of the naphthalene family of Rieske nonheme iron dioxygenases that are capable of oxidizing the aromatic ring of nitroarene compounds, resulting in the elimination of the nitro group, and thus are targets for bioremediation processes. Molecular dynamics simulations of such nitrobenzene dioxygenases have helped identify key hydrogen bonds that contribute to the stability of the active site and also determine the role of an Asn residue in substrate binding (Pabis et al., 2014).

23.6 SYSTEMS BIOLOGY APPROACHES

It is essential to understand the microbial diversity in metal-rich environments and their metabolic capabilities to design effective bioremediation strategies. Systems biology approaches would aid in this direction. Systems biology is an integrative approach to study biological systems at the molecular, cellular, community, and ecosystem levels through interactions and networks (Chakraborty et al., 2012). This integrative approach utilizes data from literature, public databases, experimental results, phenotype studies, and analyses of the genome, proteome, and metabolome (Ng et al., 2006).

Identification and quantification of the population of microbes present in metal-rich environments can be done using 16S rRNA technologies such as PhyloChip (Schatz et al., 2010). PhyloChip analysis of the chromium-contaminated environment showed an enrichment of sulfate-reducing *Desulfovibrio vulgaris*, nitrate-reducing strain *Pseudomonas stutzeri*, and iron-reducing strain *Geobacter metalliredu-cens* that are capable of Cr(VI) reduction (Faybishenko et al., 2008).

In silico genome-scale models of microorganisms can serve as important tools to improve our understanding of the physiology of microorganisms. Computational systems biology approach revealed cobalt and nickel assimilation systems of methanogens in metal-rich environments (Chellapandi, 2011). In-depth proteome and transcriptome analysis of *Pseudomonas putida* KT2440, a widely used organism for bioremediation, has unraveled the degradation pathway of *n*-butanol and also served to transform the strain from a native *n*-butanol consumer to an engineered *n*-butanol producer

(Vallon et al., 2015). Mathematical modeling of *Bacillus cereus* M6 isolated from heavy crude oil helped describe the mechanisms by which the microbe tolerates and/or resists the effects of toxic metals. These included the use of efflux pumps, intracellular accumulation of metals, and extracellular adsorption in the membrane (Shaw and Dussan, 2015).

Constraint-based reconstruction and analysis methods are proven to be effective for genome-scale modeling of organisms. Genome sequence data and existing experimental information of the organism of interest are used to generate a metabolic network. It also requires information about all known metabolic reactions and the corresponding genes in the target organism. The metabolic network is further represented in a mathematical form. This can be used for the computation of metabolic fluxes through the network that can optimize a specified function of the network, such as maximal growth or production of a particular metabolite (Thiele and Palsson, 2010). Networks can also describe protein–protein interactions and processes like gene regulation and signaling. Model SEED pipeline is an automated tool for generating, optimizing, and analyzing draft genome-scale metabolic models. The metabolic models describe what pathways are present and how these pathways are utilized by the organism (Henry et al., 2010). Other tools used for model building and analysis are COPASI, JSim, CellDesigner, OpenCell, and GENESIS (Ghosh et al., 2011).

In metal-polluted sites, microorganisms exist as a consortium and therefore are capable of interspecies interactions as well as interactions with the surrounding environment. In the bioremediation of uranium-contaminated groundwater, it has been observed that *Geobacter* spp. can rapidly outgrow the *Rhodoferax* spp. in the presence of acetate as the former can fix atmospheric nitrogen and does not rely on ammonium for growth. This complex interplay between the microbes was studied through the integration of genome-based metabolic models. A combination of metabolic models with transport and geochemical models can be used for optimizing uranium bioremediation (Mahadevan et al., 2011).

Ferroplasma acidiphilum and *Leptospirillum ferriphilum* are found together as a mixed culture in acid mine drainage and bioleaching environments. A stoichiometric model and flux balance analysis of this mixed culture enabled understanding of their metabolic capabilities. Knockout simulations of this metabolic model predicted key enzymes for growth, thus providing strategies to optimize productivity in bioleaching processes (Merino et al., 2015).

23.7 CONCLUSION

Microbes have developed different mechanisms to adapt and thrive in metal-polluted sites. With the completion of a large number of genome sequencing projects, it is now possible to utilize computer-based methods to predict metal-binding motifs and annotate metal-binding proteins. Some of these proteins may belong to metal resistance systems. Systems approaches can provide a holistic perspective on complexity of the biological systems by investigating interactions and networks. Iterative approaches wherein integrative genome-scale modeling is coupled with experimental validation for optimizing bioremediation strategies are gaining importance. Knowledge on metal resistance gained through interdisciplinary approaches would aid in the design of effective metal bioremediation strategies.

REFERENCES

Altschul, S.F., Madden, T.L., Schäffer, A.A., Zhang, J., Zhang, Z., Miller, W., and Lipman, D.J. 1997. Gapped BLAST and PSI-BLAST: A new generation of protein database search programs. *Nucleic Acids Research* 25: 3389–3402.

Andreini, C., Bertini, I., and Rosato, A. 2009. Metalloproteomes: A bioinformatic approach. *Accounts of Chemical Research* 42: 1471–1479.

Andreini, C., Cavallaro, G., Lorenzini, S., and Rosato, A. 2013. MetalPDB: A database of metal sites in biological macromolecular structures. *Nucleic Acids Research* 41: D312–D319.

Benson, D.A., Karsch-Mizrachi, I., Lipman, D.J., Ostell, J., and Wheeler, D.L. 2007. GenBank. *Nucleic Acids Research* 35: D21.

Berman, H.M., Westbrook, J., Feng, Z., Gilliland, G., Bhat, T.N., Weissig, H., Shindyalov, I.N., and Bourne, P.E. 2000. The protein data bank. *Nucleic Acids Research* 28: 235–242.

Blindauer, C.A. and Leszczyszyn, O.I. 2010. Metallothioneins: Unparalleled diversity in structures and functions for metal ion homeostasis and more. *Natural Product Reports* 27: 720–741.

Brylinski, M. and Skolnick, J. 2011. FINDSITE-metal: Integrating evolutionary information and machine learning for structure-based metal-binding site prediction at the proteome level. *Proteins: Structure, Function, and Bioinformatics* 79: 735–751.

Carbajosa, G., Trigo, A., Valencia, A., and Cases, I. 2009. Bionemo: Molecular information on biodegradation metabolism. *Nucleic Acids Research* 37: D598–D602.

Castagnetto, J.M., Hennessy, S.W., Roberts, V.A., Getzoff, E.D., Tainer, J.A., and Pique, M.E. 2002. MDB: The metalloprotein database and browser at the Scripps Research Institute. *Nucleic Acids Research* 30: 379–382.

Chakraborty, R., Wu, C.H., and Hazen, T.C. 2012. Systems biology approach to bioremediation. *Current Opinion in Biotechnology* 23: 483–490.

Chellapandi, P. 2011. In silico description of cobalt and nickel assimilation systems in the genomes of methanogens. *Systems and Synthetic Biology* 5: 105–114.

Cvetkovic, A., Menon, A.L., Thorgersen, M.P., Scott, J.W., Poole II, F.L., Jenney Jr., F.E., Lancaster, W.A. et al. 2010. Microbial metalloproteomes are largely uncharacterized. *Nature* 466: 779–782.

De Castro, E., Sigrist, C.J., Gattiker, A., Bulliard, V., Langendijk-Genevaux, P.S., Gasteiger, E., Bairoch, A., and Hulo, N. 2006. ScanProsite: Detection of PROSITE signature matches and ProRule-associated functional and structural residues in proteins. *Nucleic Acids Research* 34: W362–W365.

Ellis, L.B. and Wackett, L.P. 2012. Use of the University of Minnesota Biocatalysis/Biodegradation Database for study of microbial degradation. *Microbial Informatics and Experimentation* 2: 1.

Estellon, J., De Choudens, S.O., Smadja, M., Fontecave, M., and Vandenbrouck, Y. 2014. An integrative computational model for large-scale identification of metalloproteins in microbial genomes: A focus on iron–sulfur cluster proteins. *Metallomics* 6: 1913–1930.

Faybishenko, B., Hazen, T.C., Long, P.E., Brodie, E.L., Conrad, M.E., Hubbard, S.S., Christensen, J.N. et al. 2008. In situ long-term reductive bioimmobilization of Cr (VI) in groundwater using hydrogen release compound. *Environmental Science & Technology* 42: 8478–8485.

Frausto da Silva, J.J.R. and Williams, R.J.P. 2001. *The Biological Chemistry of the Elements: The Inorganic Chemistry of Life.* Oxford, U.K.: Oxford University Press.

Ghosh, S., Matsuoka, Y., Asai, Y., Hsin, K.Y., and Kitano, H. 2011. Software for systems biology: From tools to integrated platforms. *Nature Reviews Genetics* 12: 821–832.

Glazer, D.S., Radmer, R.J., and Altman, R.B. 2008. Combining molecular dynamics and machine learning to improve protein function recognition. In *Pacific Symposium on Biocomputing*, NIH Public Access, p. 332–343.

Gourdon, R., Bhende, S., Rus, E., and Sofer, S.S. 1990. Comparison of cadmium biosorption by gram-positive and gram-negative bacteria from activated sludge. *Biotechnology Letters* 12: 839–842.

Goyal, K. and Mande, S.C. 2008. Exploiting 3D structural templates for detection of metal-binding sites in protein structures. *Proteins: Structure, Function, and Bioinformatics* 70: 1206–1218.

Halperin, I., Glazer, D.S., Wu, S., and Altman, R.B. 2008. The FEATURE framework for protein function annotation: Modeling new functions, improving performance, and extending to novel applications. *BMC Genomics* 9: 1.

He, W., Liang, Z., Teng, M., and Niu, L. 2015. mFASD: A structure-based algorithm for discriminating different types of metal-binding sites. *Bioinformatics* 31: 1938–1944.

Hemavathi, K., Kalaivani, M., Udayakumar, A., Sowmiya, G., Jeyakanthan, J., and Sekar, K. 2009. MIPS: Metal interactions in protein structures. *Journal of Applied Crystallography* 43: 196–199.

Henry, C.S., DeJongh, M., Best, A.A., Frybarger, P.M., Linsay, B., and Stevens, R.L. 2010. High-throughput generation, optimization and analysis of genome-scale metabolic models. *Nature Biotechnology* 28: 977–982.

Karplus, M. and McCammon, J.A. 2002. Molecular dynamics simulations of biomolecules. *Nature Structural & Molecular Biology* 9: 646–652.

Kasampalidis, I.N., Pitas, I., and Lyroudia, K. 2007. Conservation of metal-coordinating residues. *Proteins: Structure, Function, and Bioinformatics* 68: 123–130.

Kavamura, V.N. and Esposito, E. 2010. Biotechnological strategies applied to the decontamination of soils polluted with heavy metals. *Biotechnology Advances* 28: 61–69.

Kumar Kuntal, B., Aparoy, P., and Reddanna, P. 2010. Development of tools and database for analysis of metal binding sites in protein. *Protein and Peptide Letters* 17: 765–773.

Levy, R., Edelman, M., and Sobolev, V. 2009. Prediction of 3D metal binding sites from translated gene sequences based on remote-homology templates. *Proteins: Structure, Function, and Bioinformatics* 76: 365–374.

Lin, H.H., Han, L.Y., Zhang, H.L., Zheng, C.J., Xie, B., Cao, Z.W., and Chen, Y.Z. 2006. Prediction of the functional class of metal-binding proteins from sequence derived physicochemical properties by support vector machine approach. *BMC Bioinformatics* 7: 1.

Lippi, M., Passerini, A., Punta, M., Rost, B., and Frasconi, P. 2008. MetalDetector: A web server for predicting metal-binding sites and disulfide bridges in proteins from sequence. *Bioinformatics* 24: 2094–2095.

Lu, C.H., Lin, Y.F., Lin, J.J., and Yu, C.S. 2012. Prediction of metal ion-binding sites in proteins using the fragment transformation method. *PLoS ONE* 7: e39252.

Mahadevan, R., Palsson, B.Ø., and Lovley, D.R. 2011. In situ to in silico and back: Elucidating the physiology and ecology of *Geobacter* spp. using genome-scale modelling. *Nature Reviews Microbiology* 9: 39–50.

Merino, M.P., Andrews, B.A., and Asenjo, J.A. 2015. Stoichiometric model and flux balance analysis for a mixed culture of *Leptospirillum ferriphilum* and *Ferroplasma acidiphilum*. *Biotechnology Progress* 31: 307–315.

Mitchell, A., Chang, H.Y., Daugherty, L., Fraser, M., Hunter, S., Lopez, R., McAnulla, C. et al. 2015. The InterPro protein families database: The classification resource after 15 years. *Nucleic Acids Research* 43: D213–D221.

Mok, T., Chen, J.S., Shlykov, M.A., and Saier, Jr. M.H. 2012. Bioinformatic analyses of bacterial mercury ion (Hg^{2+}) transporters. *Water, Air, & Soil Pollution* 223: 4443–4457.

Ng, A., Bursteinas, B., Gao, Q., Mollison, E., and Zvelebil, M. 2006. Resources for integrative systems biology: From data through databases to networks and dynamic system models. *Briefings in Bioinformatics* 7: 318–330.

Nies, D.H. 1999. Microbial heavy-metal resistance. *Applied Microbiology and Biotechnology* 51: 730–750.

Oliveira, J.S., Araújo, W., Sales, A.I.L., de Brito Guerra, A., da Silva Araújo, S.C., de Vasconcelos, A.T.R., Agnez-Lima, L.F., and Freitas, A.T. 2015. BioSurfDB: Knowledge and algorithms to support biosurfactants and biodegradation studies. *Database* 2015: bav033.

Pabis, A., Geronimo, I., York, D.M., and Paneth, P. 2014. Molecular dynamics simulation of nitrobenzene dioxygenase using AMBER force field. *Journal of Chemical Theory and Computation* 10: 2246–2254.

Pal, C., Bengtsson-Palme, J., Rensing, C., Kristiansson, E., and Larsson, D.J. 2014. BacMet: Antibacterial biocide and metal resistance genes database. *Nucleic Acids Research* 42: D737–D743.

Passerini, A., Lippi, M., and Frasconi, P. 2012. Predicting metal-binding sites from protein sequence. *IEEE/ACM Transactions on Computational Biology and Bioinformatics* 9: 203–213.

Pazos, F., Guijas, D., Valencia, A., and De Lorenzo, V. 2005. MetaRouter: Bioinformatics for bioremediation. *Nucleic Acids Research* 33: D588–D592.

Rojas, L.A., Yáñez, C., González, M., Lobos, S., Smalla, K., and Seeger, M. 2011. Characterization of the metabolically modified heavy metal-resistant *Cupriavidus metallidurans* strain MSR33 generated for mercury bioremediation. *PLoS ONE* 6: e17555.

Ruiz, O.N., Alvarez, D., Gonzalez-Ruiz, G., and Torres, C. 2011. Characterization of mercury bioremediation by transgenic bacteria expressing metallothionein and polyphosphate kinase. *BMC Biotechnology* 11: 82.

Saier, M.H., Reddy, V.S., Tamang, D.G., and Västermark, Å. 2013. The transporter classification database. *Nucleic Acids Research* 42: D251–D258.

Schatz, M.C., Phillippy, A.M., Gajer, P., DeSantis, T.Z., Andersen, G.L., and Ravel, J. 2010. Integrated microbial survey analysis of prokaryotic communities for the PhyloChip microarray. *Applied and Environmental Microbiology* 76: 5636–5638.

Schymkowitz, J.W., Rousseau, F., Martins, I.C., Ferkinghoff-Borg, J., Stricher, F., and Serrano, L. 2005. Prediction of water and metal binding sites and their affinities by using the Fold-X force field. *Proceedings of the National Academy of Sciences of the United States of America* 102: 10147–10152.

Shaw, D.R. and Dussan, J. 2015. Mathematical modelling of toxic metal uptake and efflux pump in metal-resistant bacterium *Bacillus cereus* isolated from heavy crude oil. *Water, Air, & Soil Pollution* 226: 1–14.

Sigrist, C.J., De Castro, E., Cerutti, L., Cuche, B.A., Hulo, N., Bridge, A., Bougueleret, L., and Xenarios, I. 2012. New and continuing developments at PROSITE. *Nucleic Acids Research* 41: D344–D347.

Sinha, A., Sinha, R., and Khare, S.K. 2014. Heavy metal bioremediation and nanoparticle synthesis by metallophiles. In: Parmar, N. and Singh, A. (Eds.), *Geomicrobiology and Biogeochemistry*, Springer, Berlin, Germany, pp. 101–118.

Sodhi, J.S., Bryson, K., McGuffin, L.J., Ward, J.J., Wernisch, L., and Jones, D.T. 2004. Predicting metal-binding site residues in low-resolution structural models. *Journal of Molecular Biology* 342: 307–320.

Thiele, I. and Palsson, B.Ø. 2010. A protocol for generating a high-quality genome-scale metabolic reconstruction. *Nature Protocols* 5: 93–121.

Thilakaraj, R., Raghunathan, K., Anishetty, S., and Pennathur, G. 2007. In silico identification of putative metal binding motifs. *Bioinformatics* 23: 267–271.

Tus, A., Rakipović, A., Peretin, G., Tomić, S., and Šikić, M. 2012. BioMe: Biologically relevant metals. *Nucleic Acids Research* 40(W1): W352–W357.

UniProt Consortium. 2014. UniProt: A hub for protein information. *Nucleic Acids Research* 43: D204–D212.

Vallon, T., Simon, O., Rendgen-Heugle, B., Frana, S., Mückschel, B., Broicher, A., Siemann-Herzberg, M. et al. 2015. Applying systems biology tools to study n-butanol degradation in *Pseudomonas putida* KT2440. *Engineering in Life Sciences* 15: 760–771.

Vidali, M. 2001. Bioremediation: An overview. *Pure and Applied Chemistry* 73: 1163–1172.

Vidossich, P. and Magistrato, A. 2014. QM/MM molecular dynamics studies of metal binding proteins. *Biomolecules* 4: 616–645.

Waldron, K.J. and Robinson, N.J. 2009. How do bacterial cells ensure that metalloproteins get the correct metal? *Nature Reviews Microbiology* 7: 25–35.

Wei, L. and Altman, R.B. 1997. Recognizing protein binding sites using statistical descriptions of their 3D environments. In *Pacific Symposium on Biocomputing*, pp. 497–508.

Yamashita, M.M., Wesson, L., Eisenman, G., and Eisenberg, D. 1990. Where metal ions bind in proteins. *Proceedings of the National Academy of Sciences of the United States of America* 87: 5648–5652.

Yang, T., Chen, M.L., and Wang, J.H. 2015. Genetic and chemical modification of cells for selective separation and analysis of heavy metals of biological or environmental significance. *Trends in Analytical Chemistry* 66: 90–102.

Zheng, H., Chordia, M.D., Cooper, D.R., Chruszcz, M., Müller, P., Sheldrick, G.M., and Minor, W. 2014. Validation of metal-binding sites in macromolecular structures with the CheckMyMetal web server. *Nature Protocols* 9: 156–170.

24 Heavy Metal Remediation Potential of Metallic Nanomaterials Synthesized by Microbes

Samya Sen and Keka Sarkar

CONTENTS

ABSTRACT

The profound thrust of this chapter is concerned with the removal of heavy metal contaminants, which cause severe human health damage even when present in small concentrations, utilizing surface-active nanoparticles (NPs) with high affinity toward metallic contaminants. Here, the different metal nanoparticles useful for the removal of heavy metal contaminants in a cost-effective manner are discussed. However, the understanding of the toxic effects of synthesized/engineered nanoparticles on the biological system before being introduced into the environment is also discussed. Also, the biogenic synthesis of these metallic nanoparticles has been discussed in detail. This chapter's objective, therefore, is to club together different aspects of nanoremediation, such as biogenic synthesis of nanoparticles, biocompatibility and toxicity, and utility for heavy metal removal resulting in feasible low-cost heavy metal–contaminated water treatment.

24.1 INTRODUCTION

Heavy metal contamination is introduced through industrial activities such as electroplating, textile dyeing, tanneries, leaching from fly ash ponds, and sludge from these industries.

A number of conventional/traditional or energy-driven membrane processes are often utilized to make water comply with drinking water standards. One of the disadvantages of conventional water treatment methods is that they cannot remove soluble salts and some dissolved inorganic and organic substances. Nanotechnology helps to alleviate water contamination by removing various contaminants, including bacteria, viruses, pesticides, and hazardous heavy metals like arsenic, chromium, and nickel.

Nanomaterials having unique physicochemical properties may comprise an alternative approach for the removal of heavy metals released in wastewater from various industries. Recently, a number of researches have been carried out to synthesize nanoparticles from biological origin and stabilize these virgin nanoparticles to reduce agglomeration. The stabilization processes were frequently used to obscure toxicity or to make biocompatible nanoparticles. This chapter describes the following three parts: (1) usage of metal and metal oxide nanoparticles for remediation of toxic substances especially heavy metals from contaminated sites and water, (2) synthesis of nanoparticles mediated by microorganisms, and (3) their toxicity and challenges.

Heavy metals are natural constituents of the environment, but their uncontrolled and anthropogenic uses alter their balance in ecosystem and geochemical cycles. Prolonged exposure and excess accumulation of such heavy metals causes deleterious effect on human health and aquatic biota. Various techniques of thermal, physical, chemical, and biological origin are generally used for remediation, though no readily available treatment was observed that could clean all types of pollutants. Bioremediation is a promising technology for the removal and recovery of heavy metal–polluted habitats, but this process is vulnerable and affected by various environmental factors. Its application is also limited as introduced microbes may become a source of secondary pollution. Thus, to remove toxic contaminants from the environment to a safe level, rapid, efficient, and cost-effective methods are to be developed.

Nanotechnology plays an important role in this regard, having enhanced reactivity, surface area, and adsorption characteristics. Thus, the development of novel nanomaterial with increased affinity, capacity, and selectivity has demonstrated an emerging area of research. Due to the increased specificity toward contaminants, nanomaterials could be more effectively removed, even at low concentrations. Contaminants that were previously impossible to remove could be removed through novel reactions at the nanoscale due to the increased number of surface atoms. Besides, nanotechnology could reduce the steps and energy needed to purify water. The potential of nanotechnology-based methods for water remediation can be categorized into the following functional aspects: (1) nanoparticles as antimicrobial agents used for water disinfection, (2) nanomaterials as catalysts that can degrade pesticides and other organic matter, (3) nanomaterials as sorbents used for sequestration and removal of heavy metals and inorganic contaminants, and

(4) nanomaterials as filtering agents used for the removal of contaminants by filtration.

Biological materials as well as microbes synthesize and accumulate gold, silver, gold–silver alloy, selenium, tellurium, platinum, palladium, silica, titanium, zirconia, quantum dots, magnetite, and uraninite into nanoforms (Narayanan and Sakthivel, 2010). Diverse physiological properties, small size, genetic manipulability, and controlled cultivability make microbial cells ideal producers of nanostructures. Microbial synthesis of nanoparticles is accepted as novel process of microbial fermentation that utilizes low-cost waste materials and extraction techniques of these biogenic, monodispersed nanoparticles. Currently, a lot of interest has grown for the biogenic synthesis of nanoparticles that are synthesized without any toxic chemicals or would not necessarily require extreme conditions for synthesis. Many biological systems have been studied for metal and metal oxide nanoparticle synthesis. Metallic nanoparticles are generally synthesized using toxic chemicals and remain unstable without chemical stabilizers. Further, extreme conditions are needed sometimes for these physical and chemical methods. Thus, synthesis and functionalization of nanoparticles using microorganisms are often accepted as an important alternative to traditional chemical methods (Krumov et al., 2009). The metal NP obtained from bioleaching of waste can find application in the field of medicine, food, and water purification.

Thus, it is possible to converge two diverse fields, that is, biogenic synthesis of nanoparticles and heavy metal bioremediation, thereby providing a cost-effective mode of research for pollution-free environment. This chapter is concerned with the removal of heavy metals utilizing nanoparticles synthesized metabolically, especially from microbes.

24.2 HEAVY METAL ADSORPTION USING METAL AND METAL OXIDE NANOPARTICLES

Nanotechnology is the technique of manipulating materials very close to the nanoscale. When the bulk material has achieved its size in nanoscale, it displays unique and sometimes surprising properties. Thus, it may become stronger, conductive, reactive, and responsive to the various molecules. Metal nanoparticles have potential for application in various commercial sectors, such as fillers, catalysts, semiconductors, cosmetics, microelectronics, pharmaceuticals, drug carriers, energy storage, and antifriction coatings; in the removal of pollutants from wastewater and drinking water; and in medical application such as targeted drug delivery and magnetic resonance imaging (MRI). Nanotechnology has attracted much attention as a potential agent for the removal of toxic contaminants that threaten health risk and the environment.

Conventional approaches for heavy metal removal mainly include chemical precipitation, adsorption, oxidation–reduction, evaporation, ionic exchange, electrochemical treatment, and membrane separation techniques (Flores and

Cabassud, 1999; Rahmani et al., 2010; Mobasherpour et al., 2011). Adsorption is considered as the most attractive technique as it is simple, cost-effective, and feasible to implement (Blanchard et al., 1984; Rahmani et al., 2010). Several inorganic and organic adsorbents including zeolites, clay minerals, fly ash, biosorbents, and activated carbon have been proposed for the removal of metallic content (Abollino et al., 2003; Wingenfelder et al., 2005; Ghorbel-Abid et al., 2009; Afkhami et al., 2011). A large amount of interest has grown for the use of nanoparticles in heavy metal removal (Yantasee et al., 2007; Afkhami et al., 2010; Feng et al., 2012). Nanoparticles have several special properties, such as high surface area and adsorption capacity, unsaturated surfaces, simple operation, and easy synthesis (Navrotsky, 2000; Afkhami et al., 2010, Rahmani et al., 2010). Nanoscale zerovalent iron (Fe^0), iron oxides, TiO_2, SiO_2, and Al_2O_3 are commonly used as adsorbents (Hu et al., 2006; Afkhami and Norooz-Asl, 2009; Sharma and Srivastava, 2009; Chen and Li, 2010; Rahmani et al., 2010; Recillas et al., 2011).

The nanoparticles with higher affinity toward heavy metal contaminants may be embedded in membranes or immobilized on supportive media that can efficiently, inexpensively, and rapidly adsorb contaminants on its surface, which can render polluted water into usable form. Because membrane processes are most suitable for effective water purification, there is a continuous search for novel materials. Keeping that in perspective, various nanomaterials like carbon nanotubes (CNTs) and dendrimers have attracted much attention as cost-effective agents used in water filtration processes.

Bruggen and Vandecastule (2003) have reviewed the use of nanofiltration to remove biological contaminants, organic pollutants, nitrates, and arsenic from surface water and groundwater. Zeolites are crystalline porous materials that comprise silicon (Si), aluminum (Al), and oxygen (Camblor et al., 1998). They have a widespread use in catalysis and separation technique (Margeta et al., 2011). CNTs have attracted considerable attention due to their large surface area and tubular structure, which make CNTs a promising adsorbent material (Balasubramanian and Burghard, 2005). Cerium oxide–supported CNTs were effective sorbent for As(v) and Cr(VI) from drinking water (Peng et al., 2005). The oxidized CNT sheets are promising materials for the preconcentration of metal ions from large sample volumes. Since this article focuses on the use of metallic nanoparticles that can be synthesized by microbes for heavy metal remediation, the biogenic particles should have minimum toxic potential.

Metal or metal oxide nanoparticles are also used broadly to remove heavy metal ions from wastewater. Nanosized silver nanoparticles (Fabrega et al., 2011), ferricoxides (Feng et al., 2012), manganese oxides (Gupta et al., 2011), titanium oxides (Luo et al., 2010), magnesium oxides, copper oxides (Goswami, 2012), and cerium oxides (Cao et al., 2010) provide high surface area and specific affinity toward heavy metals. Further, due to low solubility, the metal oxides provide minimum impact on the environment and thus are widely accepted as sorbents to remove heavy metals.

Hristovski et al. (2007) conducted experiment using 16 nanopowders in packed bed columns for the removal of arsenate. Four nanopowders, TiO_2, Fe_2O_3, ZrO_2, and NiO, were fitted in Freundlich adsorption isotherm parameters. Cao et al. (2012) synthesized titanate nanoflowers having high surface area that can selectively adsorb more toxic heavy metal ion (Cd^{2+}) than less toxic Zn^{2+}, Ni^{2+} ions. The adsorption mechanism fitted well with the Langmuir model, whereas adsorption kinetics followed the pseudo-second-order model. Zhang et al. (2006) synthesized porous anodic alumina membranes filled with $Mg(OH)_2$ nanotube that were used to remove Ni^{2+} ions from wastewater. Further, adsorbed Ni^{2+} ions can be easily removed from the composite membranes by heating, and it becomes ready for the next cycle of adsorption. Nanosized metal oxide acts as an excellent adsorbent for heavy metal removal from wastewater due to their higher surface area with large surface active sites than bulk materials, but their separation from the wastewater becomes difficult. However, Fe_3O_4 nanoparticles are most widely used in wastewater treatment (Hu et al., 2006; Panneerselvam et al., 2011) owing to their magnetic property as they can be separated out using magnet.

24.2.1 Factors Affecting Adsorption

Several factors can affect the adsorption process, such as pH, temperature, metal concentration, adsorbent dosage, and the presence of other ions along with the contaminants. The nanoparticle (NP) size and shape are also important factors (Huang et al., 2012).

24.2.1.1 Effect of pH

Adsorption of heavy metal ions by metal oxide nanoparticles is highly dependent on the pH of the reaction mixture. High acidic conditions correlate with lower adsorption of metal ions. The functional groups in metal oxides are generally covered with hydroxyl groups that differ in forms at different pH (Banerjee and Chen, 2007; Afkhami et al., 2010; Bian et al., 2011; Panneerselvam et al., 2011). Thus, the pH of zero point of charge varies, in which the surface charge of adsorbent becomes neutral. As the pH of solution increases, adsorption increases due to successive deprotonation of hydroxyl group and electrostatic attraction between negative sites on adsorbent and metals. When the pH was decreased, there was competitive adsorption between H^+ ions and strongly competing heavy metals in solution. The binding sites of the nanoadsorbent were dominated by H^+ ions at a low pH value, which led to protonated functional groups, thus reducing adsorption. The number of adsorption sites available for heavy metal ions decreased as the number of protonated metal-binding adsorbent groups increased (Mahdavi et al., 2012).

24.2.1.2 Effect of Temperature

The temperature also influences the adsorption of heavy metals onto the surface of nanoadsorbents. Sharma et al. (2010) demonstrated Cr(VI) removal utilizing silicon-functionalized iron oxide nanocomposites within a range of increasing temperature from 25°C to 45°C in an optimum condition for other parameters. It was observed that with increasing temperature, the adsorption amount increased by 2%. Similar observation was achieved when adsorption capability increased from 150.8 to 186.9 mg/g on increasing the temperature from 2°C to 62°C, suggesting that the adsorption process is endothermic in nature (Luo et al., 2013).

24.2.1.3 Effect of Initial Concentration of Heavy Metal Contamination

This is an important factor when adsorption capacities of different adsorbents are being compared for the adsorption of particular heavy metals. Adegoke et al. (2014) investigated that the percentage adsorption efficiency of hematite was increased with increasing heavy metal concentration, whereas in another experiment (Sharma et al., 2010), the percentage removal was decreased with increasing initial Cr(VI) concentration that may have been due to lack of available active sites per turnover.

24.2.1.4 Effect of Amount of Adsorbent

Removal efficiency also depends on the adsorbent (NP) dosage. An increase in the adsorbent dosage led to an increase in the removal efficiency of Cd^{2+}, Cu^{2+}, Ni^{2+}, and Pb^{2+} (Mahdavi et al., 2012). Due to the increasing number of binding sites with increased amount of adsorbent (Lazaridis and Charalambous, 2005; Chen and Yang, 2006), minimum amount of adsorbent is needed for adsorption process. Jiang et al. (2013) observed that Cr(VI) concentration was effectively reduced when maghemite nanoparticle dosage was increased from 0.1 to 1.5 g/L.

24.2.1.5 Effect of Coexisting Ions

Industrial wastewater contaminated by various anions and cations has a competition for available active sites on an adsorbent. Hu et al. (2005) investigated the influence of commonly coexisting anions present in chrome plating wastewater at pH 2.5, on adsorption of Cr(VI) by phosphonium silane–coated magnetite nanoparticles (Yuan et al., 2009). It was reported that among all the contaminants, phosphate anions were influenced effectively and reduced the adsorption efficiency (Chowdhury and Yanful, 2010; Badruddoza et al., 2011).

24.2.1.6 Adsorption Isotherm

Adsorption of heavy metals on active surface of nanomaterials can be analyzed when equilibrium is established between the concentration of heavy metal adsorbed and the concentration of that metal present in water that becomes constant and the relationship is called an adsorption isotherm. Various mathematical models have been used to describe adsorption characteristics, while Langmuir and Freundlich models are used widely (Ho et al., 2005; Hu et al., 2006; Xie et al., 2009).

The Langmuir model was derived to describe the homogeneous adsorption on the monolayer active surface of an adsorbent, and all the sites are considered as identical and energetically equivalent. Once the active site is occupied, further adsorption does not take place until the removal of adsorbed heavy metal(s) is facilitated (Huang and Chen,

2009). The nonlinear Langmuir isotherm equation can be expressed as

$$q_e = \frac{K_L q_L C_e}{1 + K_L C_e}$$

where

q_e is the amount of sorbent (heavy metals) adsorbed per unit adsorbent surface (mg/g)
C_e is the equilibrium concentration (mg/L) of heavy metals
q_L is the saturation capacity or maximum adsorption capacity of the adsorbent (mg/g)
K_L is the Langmuir isotherm constant (L/g)

The linear Langmuir model can be expressed as

$$\frac{C_e}{q_e} = \frac{C_e}{q_L} + \frac{1}{K_L} q_L$$

A plot C_e/q_e versus C_e would produce a straight line. Maximum adsorption capacity and bond energy can be calculated from the slope and intercept (Wang et al., 2012a). Unlike Langmuir isotherm model, Freundlich isotherm model is used to describe the multilayer adsorption on heterogeneous surfaces of an adsorbent (Huang and Chen, 2009). The energy required for adsorption is not constant, and once heavy metals are attached on the heterogeneous layer of the adsorbent, they will continuously keep binding to the active surfaces. The nonlinear equation is expressed as

$$q_e = K_F \cdot C_e^{1/n}$$

The linear form is expressed as

$$\log q_e = \log K_F + \log (C_e/n)$$

where

q_e is the concentration of sorbent on adsorbent at equilibrium (mg/g)
K_F is the constant that indicates sorption capacity (mg/g)
n is the adsorption constant for Freundlich in L/mg, usually greater than one
C_e is the aqueous concentration of sorbent at equilibrium (mg/L)

In order to explain the nature of adsorption process, it can be stated that if the $1/n$ is below unity, the adsorption is a chemical process, whereas the adsorption is a favorable physical process if the value is above unity (Gimbert et al., 2008). Besides, there are several well-known models that have been used to analyze adsorption studies.

24.2.1.7 Adsorption Kinetic Models

In order to interpret the affinity of adsorbent toward heavy metals in a particular environmental condition (pH, temp), several models are proposed. Solute uptake performance of an adsorbent can be expressed as rate kinetics. It determines the residence time required to complete the adsorption reaction (Qiu et al., 2009).

Pseudo-first-order kinetic model

$$\frac{dq_t}{dt} = K_1(q_e - q_t)$$

where

K$_1$ is the pseudo-first-order rate constant for adsorption (min^{-1})

q$_e$ and q$_t$ are the amount of metal ions adsorbed per gram of adsorbent (mg/g) at equilibrium and at time t (min), respectively

On an initial condition when time t is 0, its linear form observed

$$Ln (q_e - q_t) = \ln q_e - K_1 t$$

The plot ln $(q_e - q_t)$ versus t produces a straight line, and pseudo-first-order rate constant K1 can be calculated from the slope.

Pseudo-second-order kinetic model
The sorption process follows second-order kinetics and represented in the linear fashion

$$\frac{t}{q_t} = \frac{t}{K_2 q_e^2} + \frac{t}{q_e}$$

where K$_2$ (g/mg min) is the pseudo-second-order rate constant of adsorption.

Other kinetic models are also proposed by many researchers but will not be discussed in this chapter.

24.2.2 REMEDIATION BY MAGNETIC IRON OXIDE NANOPARTICLES

Facilitated separation from the bulk sample using magnetic force is the most appreciated property of magnetic nanoparticles (Mohmood et al., 2013). Magnetic iron oxides, such as magnetite (Fe$_3$O$_4$) and maghemite (γ-Fe$_3$O$_4$), have been investigated for environmental applications (Wang and Chen, 2009). In addition to magnetic properties and low toxicity, iron oxide nanoparticles exhibit high surface area. Further, surface chemical modification enhances its capacity for metal uptake during water treatment procedures. Surface of magnetic NPs is modified with inorganic shell or organic molecules. Surface modification with specific functionalities not only stabilizes the NP by protecting from oxidation, it renders NPs to react selectively for ion uptake. Various functional groups, such as carboxylic acid, phosphoric acid, and polymer, could be attached to modify the surface of the nanoparticles (Table 24.1).

Fe$_3$O$_4$ nanoparticles have been utilized for adsorbing Pb(II) ions from aqueous solution. The achieved adsorption equilibrium followed the Langmuir and Freundlich adsorption isotherm models. The thermodynamics of Pb(II) adsorption indicate that the adsorption was spontaneous, endothermic, and physical in nature. The Pb(II) adsorption by Fe$_3$O$_4$ nanoparticles indicates an endothermic reaction as the reaction increases with increasing temperature. The adsorption reaction is specific and not affected by coexisting cations (Nassar, 2010). Maghemite

TABLE 24.1
Surface-Functionalized Magnetic Nanoadsorbents Used for the Removal of Heavy Metals

Nanomaterials	Pollutants	Reference
Ascorbic acid–coated Fe$_3$O$_4$	As (V) and As(III)	Feng et al. (2010)
Cellulose matrix on Fe$_2$O$_3$	As(III) and As(V)	Yu et al. (2013)
β-Cyclodextrin on Fe$_3$O$_4$	As(III) and As(V)	Chalasani and Vasudevan (2012)
6-Hexanediamine on Fe$_3$O$_4$	Cu(II)	Hao et al. (2010)
Silica coated, functional amine group on Fe$_3$O$_4$ using 3-aminopropyl triethoxysilane	Cu(II)	Hui et al. (2013)
Gum Arabic chitosan–coated Fe$_3$O$_4$	Cu	Banerjee and Chen (2007)
α-Ketoglutaric acid on Fe$_3$O$_4$	Cu(II)	Zhou et al. (2009)
Polyacrylic acid–chitosan–coated Fe$_3$O$_4$	Cu(II)	Yan et al. (2012)
Polyvinyl acetate iminodiacetic acid with EDTA on Fe$_3$O$_4$	Cu(II)	Tseng et al. (2009)
Carboxy methyl-β-cyclodextrin grafted on magnetic nanoparticle by carbodiimide method	Cu(II)	Badruddoza et al. (2013)
Magnetic graphene nanocomposite	Cu(VI)	Zhu et al. (2011)
Magnetite polypyrrole composite	Pb(II)	Wang et al. (2012c)
Thiol-functionalized Fe$_3$O$_4$ using dimercaptosuccinic acid (DMSA)	Hg(II), Ag(I), Pb(II), Cd(II), and Tl(I)	Yantasee et al. (2007)
Carboxyl (succinic acid), amine (ethylenediamine), and thiol (2,3-dimercaptosuccinic acid) groups attached to Fe$_3$O$_4$	Cr(III), Co(II), Ni(II), Cu(II), Cd(II), Pb(II), and As(III)	Singh et al. (2011)
Humic acid (HA)-coated Fe$_3$O$_4$	Hg(II), Pb(II), Cd(II), and Cu(II)	Liu et al. (2008)
Carboxymethyl-β-cyclodextrin-modified Fe$_3$O$_4$ by carbodiimide method	Cu(II)	Badruddoza et al. (2011)
Polyethylenimine-modified Fe$_3$O$_4$	Cu(II)	Goon et al. (2010)
Dendrimer-conjugated magnetic nanoparticles	Zn(II)	Chou and Lien (2011)
Amino-functionalized Fe$_3$O$_4$–SiO$_2$ magnetic nanomaterial	Cu(II), Pb(II), and Cd(II)	Wang et al. (2010)

(g-Fe$_2$O$_3$) nanoparticle enables the removal of Pb(II) selectively from electroplating wastewater, where adsorption increases with an increase in the pH (Cheng et al., 2012). MNP-Ca-alginate coated with *Phanerochaete chrysosporium* adsorbed about 90% of Pb(II) on the surface of magnetic nanoparticles (Xu et al., 2012; Madrakian et al., 2013). Water-soluble MNP showed high affinity for Pb(II) and Cr(VI) than water-insoluble MNP (Wang et al., 2012b). Moreover, Iron oxide nanoparticles have been shown to be an effective sorbent for Cr(VI) remediation from wastewater through reduction of Cr(VI) to Cr(III) by the Ferrous ion followed by precipitation of the Cr(III) by increasing the pH of the solution (López-Téllez et al., 2011). Flowerlike and hollow-like nest morphology of Fe$_2$O$_3$ was effective for the removal of As(V) and Cr(V) from water (Cao et al., 2012; Wei et al., 2012). Maghemite nanotubes detoxified Cu(II), Zn(II), and Pb(II) and followed pseudo-second-order equation when adsorption equilibrium was established (Roy and Bhattacharya, 2012). Thiol-functionalized multiwalled CNT–magnetite nanocomposite also proved useful for the removal of Pb and Hg(II). 2-Mercaptobenzothiazole-modified Fe$_3$O$_4$ efficiently removes Hg(II) than modified iron nanoparticles from polluted water (Parham et al., 2012). The mesoporous amine–functionalized Fe$_3$O$_4$ successfully removed lead, copper, and cadmium (Xin et al., 2012). Acrylic acid– and crotonic acid–functionalized Fe$_3$O$_4$, further modified by 3-aminopropyltrioethoxysilane, efficiently detoxified heavy metals such as Cu, Pb, Zn, and Cd(II) from aqueous solutions (Ge et al., 2012). Humic acid–coated magnetite nanoparticles are stable in tap and natural water and are able to remove Hg, Pb(II), Cu(II), and Cd(II) by up to 98% from contaminated natural and tap water (Liu et al., 2008). Surface modification with succinic acid, ethylenediamine, and 2,3-dimercaptosuccinic acid provides the carboxyl, amine, and thiol groups on magnetic nanoparticles. These groups form complexes with heavy metals such as Cr(III), Ni(II), Co(II), Cu(II), Cd(II), Pb(II), and As(III) on the surface of the magnetic particles, and adsorption capacities are increased with increasing adsorbent groups (Singh et al., 2011). Carboxymethyl-β-cyclodextrin-functionalized magnetic nanoadsorbent selectively removes PB(II), Cd(II), and Ni(II) (Badruddoza et al., 2013).

The chemical affinity, capacity, kinetics, and stability of dimercaptosuccinic acid (DMSA)-modified magnetic adsorbents were compared with conventional resin-based sorbents (GT-73), activated carbon, and nanoporous silica in groundwater, river, and seawater and human blood and plasma (Yantasee et al., 2007). Though we are more concerned with biologically synthesized magnetic nanoparticles, it is worth mentioning that the average particle size of chemically synthesized magnetic nanoparticles is 8–35 nm and is useful for the removal of heavy metals such as Ni(II), Cd(II), and Cu(II) from wastewater.

The polymer of aminoethyl methacrylate hydrochloride functionalized with dithiocarbamate on Fe$_3$O$_4$ nanoparticles exhibits high chelating affinity toward Hg(II) than monolayer analogue (Farrukh et al., 2013). MnO$_2$-coated magnetic nanocomposite with flowerlike morphology has shown good adsorption ability of heavy metals due to high surface area (Kim et al., 2013), whereas 100% recovery of heavy metals is possible using Fe$_3$O$_4$–ZnO$_2$ magnetic semiconductor

nanoparticles (Singh et al., 2013). Magnetic nanorods with average diameter of 60 nm were utilized for the removal of heavy metals (Karami, 2013).

Silica-modified magnetic nanoparticles increase their efficiency for removing heavy metals. Cetylpyridinium bromide and 8-hydroxyquinoline formed complex with silica-functionalized magnetic nanoparticles used for the detection of a number of heavy metals such as Cd(II), Co(II), and Ni(II) (Karatapanis et al., 2011). The silica modified magnetic nanoparticles (MNPs) obeyed Langmuir isotherm equation, which is temperature dependent, and can remove Cu(II), Pb(II), and Cd(III) selectively with high efficiency, except Cu(II), in the presence of coexisting chemicals like humic acid, Na, K, and Mg salt (Zhang et al., 2013).

24.2.3 Synthesis of Nanoparticles

A top-down (physical) approach involves thermal decomposition of bulk materials into nanoforms, whereas a bottom-up (chemical and biological) approach involves synthesis of nanoparticles from molecules, for example, chemical reduction, electrochemical synthesis, seeded growth method, and biological entities for the generation of nanoparticles. Different types of chemical, physical, and biological agents are employed to yield nanoparticles of different shapes and sizes. Out of these various methods (Figure 24.1), the green method of nanoparticle synthesis has gained a lot of attention in the last decade due to its utilization of biological agents for the synthesis. This method further diverges into an array of other biosynthetic processes that use bacteria, fungi, algae, actinomycetes, various plants, and yeasts for the synthesis of nanoparticles.

Biological synthesis of nanoparticles or nanobiosynthesis includes the synthesis of nanoparticles by known biological entities. Once produced, these nanoparticles must be properly scrutinized to check their physical characteristics that have a direct impact on their functioning. Figure 24.2 describes the generalized strategy of nanoparticle biosynthesis.

The understanding of all the biological processes and mechanisms involved in these biosynthetic preparations is a vast area of learning. Thus, we will be concentrating on the microbial synthesis of nanoparticles further in this chapter.

24.2.3.1 Microbial Synthesis of Nanoparticles

Though reported far back in 1984 by *Haefeli* that the AG529 strain of *Pseudomonas stutzeri* can produce silver nanoparticles in silver mines, the idea of microbial synthesis of nanoparticles is relatively new, and there has not been extensive research on the topic. But a lot of study has been conducted in the last decade on the synthesis of various metal and nonmetal nanoparticles by microbes like bacteria, fungi, actinomycetes, algae, and even yeast. The various mechanisms employed for the synthesis of nanoparticles by microbes include

- Alteration of solubility and toxicity through reduction or oxidation using various intracellular and extracellular enzymes
- Lack of specific metal transport system

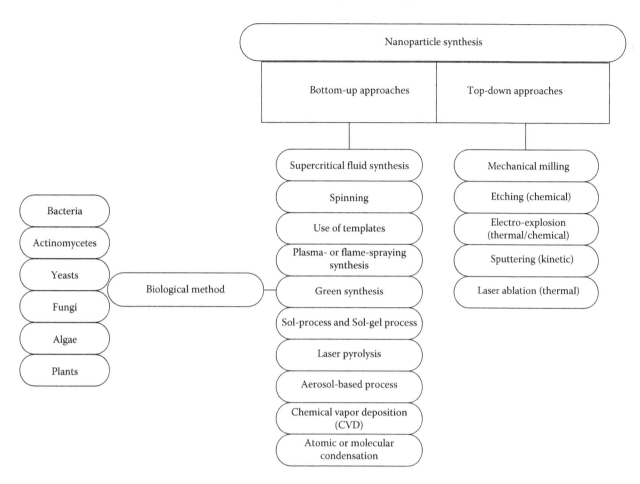

FIGURE 24.1 Some important manufacturing methods used in nanoparticle synthesis. (Reproduced from Iravani, S., *Chemistry*, 13, 2638, 2011. With permission.)

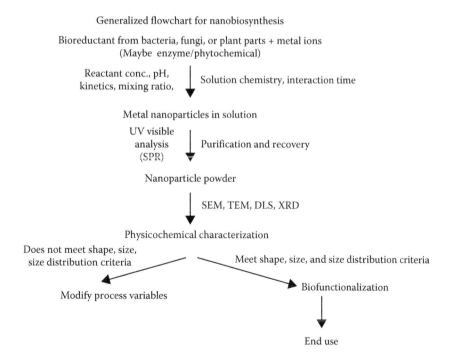

FIGURE 24.2 Flowchart denoting the biosynthesis of nanoparticles. (Reproduced from Prathna, T.C. et al., Biomimetic synthesis of nanoparticles: science, technology and applicability, *Biomimetics Learning from Nature*, A. Mukherjee (Ed.), InTech, Rijek, Croatia, 2010. With permission.)

- Biosorption, extracellular complexation, or precipitation of metals
- Bioaccumulation and efflux systems

24.3 MECHANISMS FOR NANOPARTICLE SYNTHESIS BY MICROORGANISMS: IN A NUTSHELL

The different mechanisms of nanosynthesis by microbes differ from organism to organism, explained as follows.

The cationic metal ions are trapped on the anionic cell walls of the organisms or are transported inside the cell itself. These metal ions are then reduced to metal nanoparticles and either stored inside the cellular periplasmic space or ejected out of the cell where they may be released into the surrounding media or remain attached to the cell surface of the organism. This mechanism has been studied in *Verticillium* sp. where it has been observed that the silver or gold nanoparticles synthesized by the organism remain attached to its mycelia through the negative carboxylate groups of the enzymes present in the mycelia wall that are responsible for attracting the metal ions to the cell surface as well as their transformation into metal nuclei that grow through further reduction and accumulation. In some *Lactobacilli,* it is observed that certain metals tend to interact electrostatically to form metal clusters on the cell wall of the bacteria. Enzymatic transformations of the metal clusters then take place to form nanoclusters. These nanoclusters then pass slowly through the cell wall of the *Lactobacilli* (Nair and Paradeep, 2002).

Some organisms have a different way of performing the bioreduction process. They secrete the reducing enzyme(s) into the surrounding medium. Classic examples of such organisms are *Bacillus licheniformis* (Kalishwaralal et al., 2008) and *Aspergillus niger* that secrete NADH and NADH-dependent enzymes like nitrate reductase (or similar enzymes) into its vicinity that transforms Ag^+ to Ag^0 forming silver nanoparticles.

As already mentioned, bacteria that inhabit habitats with high heavy metal concentration must have a rock solid mechanism for metal homeostasis, failing which they may face the threat of elimination from that specific habitat. Thus, they must have adaptable genetic and proteomic tools to counter the heavy metal toxicity. Thus, it is hypothesized that wide arrays of genetic clusters and associated proteins may have evolved to take care of this predicament of the bacteria tackling the problem in the molecular level.

The synthetic mechanism of another class of nanoparticles has been studied extensively, that is, magnetic nanoparticles and bacterial magnetic particles (BacMPs), which as discussed already are instrumental in bioremediation purposes. Their synthesis takes place in sequential steps and will be discussed later in this chapter.

These aforesaid mechanisms of nanoparticle formation are well understood and represent most of the cases unearthed so far.

24.3.1 NANOPARTICLE BIOSYNTHESIS BY BACTERIA AND ACTINOMYCETES

Bacteria and actinomycetes have existed on this planet for millions of years, during which time they have had to tolerate a range of extreme environments in the fight for existence. It is thus quite obvious that they have had to face and detoxify a variety of toxic entities that have threatened their existence over the ages. A very good example of such toxic agents is heavy metals. Bacteria have learned to detoxify many of those heavy metals primarily through the process of reduction and also by other methods of biotransformation. Thus, many species of bacteria have been reported to live in environments with very high heavy metal concentration. There are many such examples, some of which must be mentioned for the understanding of the phenomenon. Let us note a few of the notable examples of nanoparticle formation by bacteria that include metallic, nonmetallic, metal oxide, magnetic, nonmagnetic, and various other kinds of nanoparticles.

24.3.1.1 Gold Nanoparticles Synthesis

Gold nanoparticles have been in synthesis from the time of the flourishing Roman Civilization. The Romans were instrumental in using them for decoration as well as for curing many diseases. Faraday (some 150 years ago) famously postulated that nanogold may differ from bulk gold as far as physical properties are concerned. And thus scientists all around the world began to experiment with various preparatory methods of gold nanoparticles by methods mentioned previously. Many bacteria have been found useful in synthesizing gold nanoparticle. Nair and Pradeep (2002) showed how *Lactobacillus* can form nanocrystals of gold on its cell surface. The alkali-tolerant actinomycete genus *Rhodococcus* houses members that have the capability to synthesize monodispersed gold nanoparticles from gold complexes when provided with an alkaline environment and slightly high temperatures (Ahmad et al., 2002). Both *Stenotrophomonas maltophilia* and *Rhodopseudomonas capsulata* employ NADH and NADH-dependent reductases as electron carriers to reduce gold ions to Au^0 through electron shuttle enzymatic metal reduction property of the reductases (He et al., 2007; Nangia et al., 2009). Sastry and coworkers have described the extracellular nanogold synthesis by the actinomycete *Thermomonospora* sp. By similar reductases, primarily nitrate reductases secreted into the surrounding environment, facilitating the bioconversion (Mukherjee et al., 2002; Ahmad et al., 2003). Konishi and colleagues have used *Shewanella algae*, which itself is a gram-negative soil and aquatic bacteria to synthesize silver nanoparticles from $AuCl_4^-$ ions. It naturally reduces Fe^{3+} ions by scavenging electrons from lactate or H_2 as electron donor (Konishi et al., 2006). Actinomycetes like *Streptomyces viridogens*, *S. hygroscopicus*, *Thermoactinomycete* sp., *Nocardia farcinica*, and *Thermomonospora* sp. have been reported to produce intracellular as well as extracellular gold nanoparticles (spherical, 5–60 nm) by similar reduction–oxidation mechanisms, as discussed previously (Sastry et al., 2003; Balagurunathan et al., 2011; Oza et al., 2012, Kalabegishvili et al., 2013;

Waghmare et al., 2014). A detailed report regarding the synthesis of biogenic gold nanoparticles has been made.

24.3.1.2 Silver Nanoparticles Synthesis

Silver nanoparticles just like bulk silver have excellent antimicrobial properties and are synthesized to counter a wide range of gram-positive and gram-negative organisms including the much dreaded methicillin-resistant *Staphylococcus aureus*. Lactic acid bacteria have been reported to synthesize silver nanoparticles by gradual increase in pH causing competition between protons and silver ions to bind to the negatively charged cell wall of the lactic acid bacteria (Sintubin et al., 2009). *P. stutzeri* AG259, isolated by Slawson et al. (1992) from a silver mine, has shown the capability of reducing silver nitrate to elemental silver nanoparticles and storing them in their periplasmic space when subjected to supersaturated silver nitrate solutions. *Bacillus subtilis* supernatants have been used to synthesize monodispersed (5–50 nm) silver nanoparticles by Saifuddin and his team (2009). The γ-proteobacterium, *Shewanella oneidensis* MR-1, upon incubation with aqueous silver nitrate solution has been reported to produce monodispersed silver nanospheres in the 2–11 nm size range (Suresh et al., 2011).

Aeromonas sp. SH10 and *Corynebacterium* sp. SH09 have been shown to produce silver NPs from $[Ag(NH_3)_2]^+$ by the addition of NaOH to the reaction mixture. The NaOH provides the OH^- to the system that is bioconverted by the bacteria to produce Ag_2O through reaction with $[Ag(NH_3)_2]^+$ that through further independent processing by the microbes itself forms uniform, monodispersed silver NPs (Fu et al., 2006). Bacteria on different poles of existence have been shown to produce NPs and are extremely interesting indeed. Psychrophilic bacteria *Pseudomonas proteolytica*, *Arthrobacter kerguelensis*, *Arthrobacter gangotriensis*, *Phaeocystis antarctica*, and *Pseudomonas meridiana* have been utilized to biosynthesize silver NPs (Shivaji et al., 2011). Similarly, bacteria of the *Enterobacteriaceae* family, namely, *Klebsiella pneumonia*, *Escherichia coli*, and *Enterobacter cloacae* (Shahverdi et al., 2007), have also been very efficient in producing silver NPs from silver ions. Thus, it is very clear that a wide variety of bacteria have these biotransformation tools inherently in them in case the situation arises where these conversions can save its existence. Silver nanoparticles produced by *Streptomyces* sp. can act against a range of human pathogens like *Pseudomonas aeruginosa*, *S. aureus*, *A. niger*, *K. pneumonia*, *E. coli*, *B. subtilis*, *Enterococcus faecalis*, *Salmonella typhimurium*, and *Candida albicans* (Sadhasivam et al., 2010; Shirley et al., 2010; Prakasham et al., 2012; Subashini and Kannabiram, 2013).

24.3.1.3 Synthesis of Metallic Nanoparticles

We have already discussed how microbes detoxify their habitable environments to survive. In the previous sections, we have discussed the formation of gold and silver NPs by various bacteria and actinomycetes. But they do form a large variety of other metal nanoparticles as well. It is achieved by a certain amount of chemical transformation as well as with the help of energy-driven membrane proteins like ATPases or proton antitransporters.

The genus *Shewanella* has shown us time and again how good they are as far as biological synthesis of NPs is concerned. *S. algae* have been shown to produce platinum nanoparticles from $PtCl_6^{2-}$ at room temperature, neutral pH, and fewer than 60 min, provided that lactate is present as the electron donor. The insoluble platinum nanoparticles of about 5 nm are then stored in the periplasmic space of the bacteria (Konishi et al., 2007). *S. oneidensis* has been reported to reduce soluble palladium (II) into insoluble palladium (0) with lactate, pyruvate, formate, or H_2 as the electron donor (Windt et al., 2005).

There are reports of *S. maltophilia SELTE02*, *Rhodospirillum rubrum*, *E. cloacae*, *E.coli*, *P. stutzeri*, and *Desulfovibrio desulfuricans* with the ability to transform selenite (SeO_3^{2-}) to elemental selenium (Se^0). The NPs formed are of various shapes and sizes (Narayanan and Sakthivel, 2010). Bacteria belonging to *Tetrathiobacter*, *Sulfurospirillum*, *Bacillus*, *Selenihalanaerobacter*, and *Pseudomonas* genera have been reported for their ability to form various kinds of selenium nanoparticles from selenite (Iravani et al., 2014).

Some *Enterobacter* spp. were reported to synthesize mercury nanoparticles, first demonstrated by Sinha and Khare. The nanoparticles were produced intracellularly under low mercury concentration and pH 8 and were uniform (2–5 nm) as well as monodispersed (Sinha and Khare, 2011). An anaerobic, hyperthermophile *Pyrobaculum islandicum* can reduce a multitude of metals like U(VI), Te(VII), Cr(VI), Co(III), and Mn(IV) using H_2 as electron donor. Some notable anaerobic bacteria like *Bacillus selenitireducens* and *Sulfurospirillum barnesii* have shown the ability to reduce tellurite (TeO_3^{2-}) to form elemental tellurium (Te^0) (Baesman et al., 2007). Rajamanickam et al. (2012) have reported the synthesis of zinc nanoparticles *in vitro* by some species of the genus *Streptomyces* that has shown potent antibacterial activity against *S. aureus* and *E. coli*. Usha et al. (2010) and Waghmare et al. (2011) have shown that copper and manganese nanoparticles have antibacterial activity against *E. coli* and *S. aureus*.

24.3.1.4 Synthesis of Magnetic Nanoparticles

As we have already discussed in this chapter, magnetic nanoparticles carry great importance as far as bioremediation is concerned; their synthesis is being studied extensively not only for bioremediation purposes but also for use in targeted drug delivery, MRI, DNA analysis studies, and so on. These nanoparticles or BacMPs are synthesized by a special group of bacteria known as magnetotactic bacteria where NPs are in the form of iron oxides, sulfides, or both (Donaghay and Hanson, 1995; Arakaki et al., 2008). These magnetic nanoparticles are known to carry single magnetic domains (magnetosomes) or magnetites that give them superior magnetic properties (Thornhill et al., 1995). Both the cultured and uncultured magnetotactic bacteria are being studied further. While Thornhill has done extensive work to study the diversity of uncultured magnetotactic bacteria in sediments, Spring has done extensive research for the better understanding of the diversity of these fascinating microbes that show extreme variation in morphology (Spring and Schleifer, 1995; Thornhill et al., 1995). Their synthesis follows a sequence of steps. First, a cytoplasmic invagination forms on the cell surface by an unknown cellular mechanism. This is followed by the arrangement of the vesicles in a linear fashion along with the cytoskeletal filaments. Now, the transmembrane ion transporters and siderophores take over and start transporting ferrous ions into the vesicles. To finish, tightly bound BacMP proteins now trigger the magnetite crystal nucleation and/or regulate the size and shape of the crystals. These BacMP proteins are being studied in detail as these are instrumental in the formation of magnetite from the ferrous ions through the formation of ferrihydrite (Arakaki et al., 2008).

Very few examples of magnetotactic bacteria are known so far that can be cultured and maintained in laboratories. Some culturable magnetotactic bacteria have been isolated from the surface of aquatic sediments like the culturable microaerophilic coccus MC-1. Estuarine salt marshes have been found to harbor a few of these rare culturable bacteria that show the typical morphological characteristics of *Vibrio*. These are MV-1, MV-2, and MV-4. All of these bacteria have been found to belong to the group α-proteobacteria possibly belonging to the Rhodospirillaceae family. They have shown the ability to grow both chemoorganotrophically and chemolithotrophically. Freshwater sediments have also shown to harbor magnetotactic bacteria from the family Magnetospirillaceae. The first member of the family that was successfully isolated was *Magnetospirillum magnetotacticum* strain MS-1. Other isolated members of the family that are being extensively studied are *Magnetospirillum gryphiswaldense* strain MSR-1and *Magnetospirillum magneticum* (Spring and Schleifer, 1995; Thornhill et al., 1995; Arakaki et al, 2008). Bacteria can also produce a variety of other nanoparticles including metal, metal oxide, metal dioxide, and sulfide nanoparticles (Table 24.2).

24.3.2 Nanoparticle Biosynthesis by Fungi and Yeast

Just like bacteria, many fungi, algae, and yeast have been reported to produce various kinds of nanoparticles using different substrates for growth. Sastry and coworkers were successful in synthesizing both gold and silver nanoparticles using the acidophilus fungus *Verticillium* sp. supplemented with $AuCl_4^-$ and Ag^+, respectively (Sastry et al., 2003), where the nanoparticles synthesized were predominantly localized on the external cell surface of the fungi. *Fusarium oxysporum* has proved useful in synthesizing gold nanoparticles as reported by Mukherjee et al. (2002) through bioreduction of gold ions. AgNPs have also been biosynthesized by various other teams using *Verticillium* sp., *F. oxysporum*, *Aspergillus flavus*, and *Aspergillus fumigatus* (Senapati et al., 2004; Bhainsa and D'souza, 2006; Vigneshwaran et al., 2007; Jain et al., 2011). Further, the production of Au–Ag nanoparticles by *F. oxysporum* has been reported, and it is shown how the secreted cofactor NADH plays a vital role in Au–Ag composition (Senapati et al., 2005). Zheng et al. (2010) have studied the production of Au–Ag nanoparticles by yeast extracellularly and have characterized the particles as irregular polygonal (Zheng et al., 2010). Another member of the genus *Fusarium*, namely, *F. semitectum*, has been reported to biosynthesize core–shell Au–Ag nanoparticles (Sawle et al., 2008). Other than these genera, other fungi that can biosynthesize gold nanoparticles are *Neurospora* along with the yeasts *Candida utilis* and *Yarrowia lipolytica* (Agnihotri et al., 2009; Chauhan et al., 2011). *Neurospora crassa* and *Phanerochaete chrysosporium* have been reported to produce both extracellular and intracellular gold nanoparticles (Castro-Longoria et al., 2011; Sanghi et al., 2011). Silver nanoparticles have been biosynthesized extracellularly using the fungal genera *Trichoderma* sp. (Basavaraja et al., 2008; Vahabi et al., 2011), *Penicillium* sp. (Naveen et al., 2010), *Rhizoctonia* sp. (Raudabaugh et al., 2013), *Pleurotus* sp. (Devika et al., 2012), *Aspergillus* sp. (Bhainsa and D'souza, 2006), and *Cladosporium* sp. (Vahabi et al., 2011). Other fungi and yeasts synthesizing different kinds of nanoparticles are listed as follows (Table 24.3).

The production of nanoparticles of various natures by a large variety of fungal species is being investigated all over the world (Fayaz et al., 2009). Hopefully, a lot more names will come up in the next few decades that can really be extremely useful for a variety of physical, chemical, and biological processes. As we come to a close about the biosynthesis of

TABLE 24.2
Production of Metal, Metal Oxide, Metal Dioxide, and Sulfide Nanoparticles by Bacteria

Microorganism	Product	Size (nm)	Shape	Location	Reference
Shewanella oneidensis	Fe_3O_4	40–50	Hexagonal, rhombic, rectangular	Ext. cellular	Perez-Gonzalez et.al. (2010)
Aeromonas hydrophila	ZnO	57.72 (avg)	Spherical, oval	Int. cellular	Jayaseelan et.al. (2012)
Shewanella oneidensis MR-1	Fe_3O_4	30–43	Pseudohexagonal, irregular, rhombohedral	Int. cellular	Bose et al. (2009)
Lactobacillus sp.	$BaTiO_3$	20–80	Tetragonal	Ext. cellular	Jha and Prasad (2010)
Lactobacillus sp.	TiO_2	8–35	Spherical	Ext. cellular	Jha et al. (2009a)
Rhodopseudomonas palustris	CdS	8	Cubic	Int. cellular	Bai and Zhang (2009)
Lactobacillus	CdS	4.9 ± 0.2	Spherical	Int. cellular	Prasad and Jha (2010)
E. coli	CdS	2–5	Wurtzite crystal	Int. cellular	Sweeney et al. (2004)
Desulfobacteraceae	CdS	2–5	Hexagonal lattice	Int. cellular	Labrenz et al. (2000)
Sulfate-reducing bacteria	FeS	2	Spherical	Ent. cellular	Watson et al. (1999)
Klebsiella aerogenes	CdS	20–200	Crystallites	Int. cellular	Holmes et al. (1995)
Propionibacterium jensenii	TiO_2	15–80	Spherical	Int. cellular	Babitha and Korrapati (2013)

TABLE 24.3
Production of Different Nanoparticles by Fungi and Yeasts

Microorganism	Product	Size (nm)	Shape	Location	Reference
Candida utilis	Au	NA	NA	Int. cellular	Gericke and Pinches (2006)
Verticillium luteoalbum	Au	NA	NA	Int. cellular	Gericke and Pinches (2006)
Yarrowia lipolytica	Au	15	Triangles	Ext. cellular	Agnihotri et al. (2009)
Trichoderma viride	Ag	5–40	Spherical	Ext. cellular	Fayaz et al. (2010)
Phanerochaete chrysosporium	Ag	50–200	Pyramidal	Ext. cellular	Vigneshwaran et al. (2006)
Trichoderma viride	Ag	2–4	NA	Ext. cellular	Mohammed et al. (2009)
Saccharomyces cerevisiae	Sb_2O_3	2–10	Spherical	Int. cellular	Jha et al. (2009b)
Fusarium oxysporum	TiO_2	6–13	Spherical	Ext. cellular	Bansal et al. (2005)
Fusarium oxysporum	$BaTiO_3$	4–5	Spherical	Ext. cellular	Bansal et al. (2006)
Fusarium oxysporum	ZrO_2	3–11	Spherical	Ext. cellular	Bansal et al. (2004)
Schizosaccharomyces pombe	CdS	1–1.5	Hexagonal lattice	Int. cellular	Sweeney et al. (2004)
Schizosaccharomyces pombe and Candida glabrata	CdS	2	Hexagonal lattice	Int. cellular	Damero et al. (1989)
Fusarium oxysporum	CdS	5–20	Spherical	Ext. cellular	Ahmad et al. (2002)
Fusarium oxysporum	$PbCO_3$, $CdCO_3$	120–200	Spherical	Ext. cellular	Sanyal et al. (2005)
Fusarium oxysporum	$SrCO_3$	10–50	Needle-like	Ext. cellular	Rautaray et al. (2004)
Yeasts	$Zn_3(PO_4)_2$	10–80 × 80–200	Rectangular	Ext. cellular	Yan et al. (2009)
Fusarium oxysporum	CdSe	9–15	Spherical	Ext. cellular	Kumar et al. (2007)

Note: NA, not applicable.

TABLE 24.4
Production of Different Nanoparticles by Algae, Cyanobacteria, and Others

Microorganism	Product	Size (nm)	Shape	Location	Reference
Plectonema boryanum UTEX 485	Au	10 nm–6 μm	Octahedral	Ext. cellular	Lengke et al. (2006a)
Plectonema boryanum	Au	<10–25	Cubic	Int. cellular	Lengke et al. (2006b)
Rhodobacter sphaeroides	CdS	8	Hexagonal Lattice	Int. cellular	Dameron et al. (1989)
Rhodobacter sphaeroides	ZnS	10.5 ± 0.15	Spherical	Ext. cellular	Bai et al. (2006)
Sulfate-reducing bacteria	FeS	2	Spherical	Ext. cellular	Watson et al. (1999)
Sargassum longifolium	Ag	NA	Spherical to oval	Ext. cellular	Rajeshkumar et al. (2014)
Stoechospermum marginatum	Au	40–85	Spherical	Ext. cellular	Rajathi et al. (2012)
Turbinaria conoides	Au	NA	NA	Ext. cellular	Vijayaraghavan et al. (2011)
Microcoleus sp.	Ag	44–79	Spherical	Ext. cellular	Sudha et al. (2013)
Sargassum muticum	Ag	5–15	Spherical	Ext. cellular	Azizi et al. (2013)

Note: NA, not applicable.

nanoparticles by microorganisms, we must keep in mind that other than bacteria, fungi, and actinomycetes, other organisms are also under investigation for nanoparticle biosynthesis. The following table contains a few such organisms comprising algae, cyanobacteria, and a few other kinds of microorganisms (Table 24.4).

24.4 TOXICITY POTENTIAL OF NANOMATERIALS: NANOTOXICOLOGICAL ASPECTS

Every new technology has its own pros and cons. We have discussed how nanoparticles have brought a revolution in physical, chemical, and biological sciences and how useful they are in bioremediating purposes. But it is also true that they do have some negatives that are worth discussing for the welfare of the environment as well as the living organisms in them. As we clearly understand by now, nanoparticles due to their high surface area to volume ratio are much more reactive than their bulk counterparts and that their reactivity increases significantly and can be toxic to humans as well as the environment as a whole (Lyn, 2009). Unfortunately, the understanding of these toxicological effects is still in its early stages and will require more research.

Nanoparticle application on a living system will follow the same protocol as that of drug administration; hence, doses must be formulated for different nanoparticles even when the application is environmental, as incessant killing of beneficial microbes must be at the minimum during the bioremediation process. This is truly the tricky part. Humans and other animals on exposure to the environment may expose themselves to these nanoparticles via airways, skin, and ingestion. When high doses of a certain nanoparticle are applied to an environmental test area, appearance of exposure-related symptoms will acutely signal the need for the change in the dosage of the NP (Song et al., 2009). The problem is that when the exposure is low and chronic in nature, it may increase the risks of grave degenerative diseases. More research is urgently required in this area. Classical toxicology has always correlated dose with mass of the test substance. But as nanoparticles are composed of fine "particles" (Wittmaack, 2007; Oberdörster., 2010), a more practical way of nanoparticle utility will be the application of a definitive number of nanoparticles and assessing how many of those reach the target cell rather than mass of the NPs. What happens to nanoparticles when they encounter a biological system is thus a huge area of understanding for the nanotoxicologists of this era.

It is worth mentioning that nanoparticles even if synthesized as "bare" particles go through a definite degree of change (Navarro et al., 2008); that is, the heterogeneous nature of the environment promotes weak or strong attachment of definite molecules or macromolecules on its surface. This gives rise to a so-called corona (Cedervall et al., 2007; Lynch et al., 2007). This new entity that is formed determines the properties of the once "bare" nanoparticle. These include the degree or NP aggregation, overall surface charge, and other surface properties (Handy et al., 2008). Studies are ongoing where the effect of different nanoparticles on the environment is being studied, primarily in soil or water ecosystems (Kiser et al., 2010; Quik et al., 2010).

24.4.1 Toxicity Effect on Human Health

As mentioned previously, the small size of the nanoparticles gives rise to a great increase in their surface area to volume ratio further giving rise to their chemical and biological reactivity. On entering human cells, they may increase the production of reactive oxygen species (ROS) in the cell that includes free radicals (Nel et al., 2006). ROS as we know affect the cells negatively when present in an abnormal amount and have the potential to extensively damage the cell and kill it (Gou et al., 2010). Moreover, ROS is a key player in inflammatory responses of the body via upregulation of proinflammatory response genes through the activation of transcription factors NF-κB and AP-1 (Conner and Grisham., 1996; Li et al., 2008). Free radicals have been shown to reduce cellular integrity (Uchino et al., 2002; Li et al., 2003).

24.4.2 Toxic Effect at Cellular Level

Mitochondria, the power house of the cells, are also the target of many NPs like CNTs (Zhu et al., 2006; Tofighy and Mohammadi, 2011) and fullerenes (Foley et al., 2002). TiO_2, CNTs, polystyrene, and silver NPs have shown the ability to induce apoptosis by altering mitochondrial functions (Hussain et al., 2005; Jia et al., 2005; Xia et al., 2006). Another very important target of certain nanoparticles is cellular membranes. These phospholipid bilayer structures encapsulate living cells and also separate various cellular compartments like the mitochondria, nucleus, and other cellular vesicles from the cytosolic component of the cell (Vasir and Labhasetwar, 2008). They maintain cellular homeostasis by selective transport of various ions and molecules.

This transport mechanism is one of the major targets of nanoparticles (Auffan et al., 2008), and its disruption proves fatal for the cell. Due to definite surface charges of NPs, they can physically damage the membranes on impact like by forming membrane nanoholes (Ginzburg and Balijepalli, 2007) and also have the potential to damage them via upregulating production of oxidants that can damage the membrane by oxidation as discussed in point a. Different nanoparticles have different physicochemical identities enabling them to alter the cellular stability and cellular morphology in many cases (Ginzburg and Balijepalli, 2007). Many metallic NPs being positively charged experience electrostatic attraction from cells due to negative charge of the membranes. This causes the nanoparticles to cover the cell surface, interfering with membrane permeability and transport of food and wastes (Seabra and Durán, 2015). Further, research has shown they can definitively alter the functions of membrane-bound structures like mitochondria and lysosomes (Al-Rawi et al., 2011; Greulich et al, 2011) and can breach nuclear membranes (Godbey et al., 1999; Panté, N. and Kann, 2002; Hoshino et al., 2004; Parfenov et al., 2006; Williams et al., 2009).

24.4.3 Effect on Cellular Macromolecules and Proteins

The cell as a working machine relies hugely on the proper functioning of various macromolecules among which proteins are perhaps the most significant. Cell signaling molecules, enzymes, and structural components of the cell (e.g., tubulin) must function properly to maintain cellular viability. Marano and colleagues showed how NPs of the same size and magnitude of a certain signaling molecule interfered with the proper functioning of the signaling system (Marano et al., 2011). It is absolutely necessary for a protein to hold its proper conformation for efficient functioning, and we know how minute changes in conformation of a protein can absolutely destroy its function. Many NPs have been shown to promote aggregation and fibrillation in some peptides, causing the formation of amyloid-like structures (Aili et al., 2008; Wu et al., 2008; Wagner et al., 2010).

Certain NPs have been shown to mimic chaperones to alter proper folding and conformation of proteins (Akiyoshi et al., 1999; Ishii et al., 2003; Takahashi et al., 2010). There have been reports of NPs interacting with many cellular proteins (De and Rotello, 2008; Dawson et al., 2009). Many groups around the globe are investigating further to find other detrimental NP–protein interactions that may take place inside a living cell.

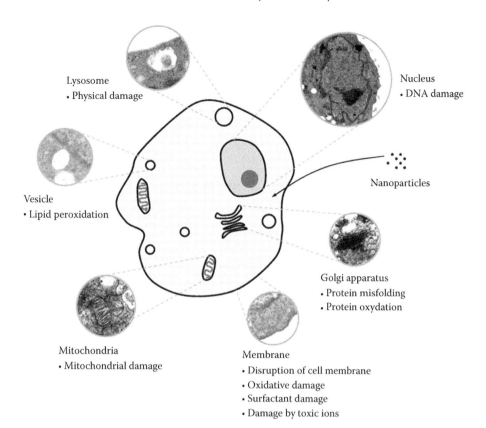

Lysosome
• Physical damage

Nucleus
• DNA damage

Nanoparticles

Vesicle
• Lipid peroxidation

Golgi apparatus
• Protein misfolding
• Protein oxydation

Mitochondria
• Mitochondrial damage

Membrane
• Disruption of cell membrane
• Oxidative damage
• Surfactant damage
• Damage by toxic ions

FIGURE 24.3 Nanoparticle interaction with cells: intracellular targets and nanotoxicological mechanisms. (Reproduced from Elsaesser, A. and Howard, C.V., *Adv. Drug Deliv. Rev.*, 64, 129, 2012. With permission.)

As mentioned previously, NPs have been shown to breach nuclear membranes (Figure 24.3). This is a matter of great concern as the nucleus contains the most important macromolecule in evolution, that is, DNA. Again, some NPs have been shown to increase cellular ROS production that may also affect DNA stability. Genotoxicity can be induced by ROS by dramatically applying oxidative stress on the DNA molecules in the cell (Bhabra et al., 2009; Myllynen, 2009). There have been experiments studying the effect of different NPs on DNA. In a very recent study, Huk and team demonstrated how nanosilver has deleterious effects on DNA. AgNPs were introduced into lymphocytes, and then the lymphocytes were further examined for abnormalities. The cells showed an alarming increase in micronuclei on treatment indicating DNA fragmentation (Huk et al., 2015). It is worth mentionable that the exact mechanism of DNA damage is still not clear, in cases where ROS is not involved. There is evidence though that certain NPs may intercalate into DNA or may interact with DNA molecules electrostatically or by other physical mechanisms (Mehrabi and Wilson, 2007; Xie et al., 2009). Some metal oxides like copper oxide, cobalt oxide, and uraninite have been shown to exert significant stress on DNA molecules (Seabra and Durán, 2015).

Finally, it has to be kept in mind that the nanoparticles on entering the human body tend to accumulate in different organs (Borm et al., 2006), and a lot of research is currently going on in understanding this phenomenon. Apart from

bioaccumulation, the routes of excretion of NPs should also be monitored to understand the dynamics of these particles better. Nanoparticle-mediated frustrated phagocytosis and ROS exposure in the lungs and other organs result in chronic inflammatory responses in the body (Donaldson et al., 2010) that may prove fatal. Some gold nanoparticles have been reported to cross the maternofetal barrier which is of grave concern indeed (Myllynen et al., 2008), and some other studies in mouse have shown how fullerenes have a fatal effect on mouse embryos (Tsuchiya et al., 1996).

24.4.4 Ecological Hazards

The toxicity of nanoparticles in the environment is dependent on the interaction between the exposed nanoparticles and the environmental components. Aggregation of nanoparticles serves as an efficient method for the removal of nanoforms from the environment. Aggregated CNT suspension is added to filter pond water (Zhu et al., 2011), whereas CNT (Hyung et al., 2007) and nZVI (Giasuddin et al., 2007) have been found to remain particulate without aggregation in the presence of humic and fulvic acids.

The toxic effect of nanoparticles in the aquatic system has been reported widely based on indicator aquatic organisms. The acute toxicity of TiO_2 nanoparticles in *Ceriodaphnia dubia*, a dominant freshwater crustacean, in a lake water system exposes the threat of nanoparticles. The exposure of nanoparticles in the air can cause pulmonary, cardiovascular, dermal, and even central nervous system toxicity. The first incidence of nanoparticle toxicity on human health hazard was reported in late 2009 in China where workers suffered permanent lung damage leading toward death when smoke and fumes containing nanoparticles were excreted from a paint factory (Lyn et al., 2009).

Soil serves as the matrix of the earth and provides the support for human existence in the form of agriculture. The effect of nanoparticles is worst toward the plant system. The toxic effect of several metal oxide particles on *Arabidopsis thaliana* is indicative of nanoparticle toxicity in soil system. Phytotoxicity of TiO_2 nanoparticles on seed germination has also been reported on experimentation with fennel (Feizi et al., 2013).

Thus, apart from the technological benefits, concern for safety issue has been raised as an alarm before the scientific community.

24.5 CONCLUSION

Many areas of India have seriously contaminated or damaged ecosystems with uncontrolled anthropogenic use of natural resources, causing tremendous hazards on the environment as well as human health. Removing, reducing, or neutralizing processes of pollution to a safe level within reasonable cost has attracted much attention currently. In this regard, nanotechnology is going to play an important role in improving the quality of natural habitats. It has the potential to supply portable water through the development of low-cost portable filters and purifiers where the accessible water is heavily contaminated. Nanotechnology may be used as a catalyst for the treatment of industrial effluent before being released into the environment. Nanomaterials as sorbents can be used to remove heavy metals and inorganic pollutants.

It is clear from the sections discussed that microbes are an excellent source of nanoparticles. It is true that the bioreduction processes are a lot slower than the more energetic processes like chemical and physical methods of nanoparticle synthesis, and they have been found to be similarly competent in their application compared to the conventionally synthesized NPs. Additionally, their production has been found to be quite inexpensive compared to conventionally synthesized NPs that require a high energy input for the transformation process. Another very striking advantage that the microbial synthesis process has over the conventional synthesis processes is that the microbial synthesis process is far less technical and sophisticated compared to the chemical, physical, and thermal methods.

Bioremediation using microbial synthesized nanoparticles are increasing day by day all around the globe. We have discussed how metal-derived nanoparticles produced by microbes have extreme potential in bioremediation. We have learned elaborately about the use of magnetic nanoparticles in bioremediation. The nanoparticles synthesized are being studied thoroughly so far as their surface properties are concerned, and many modifications are being attempted to apply

them to the environment with minimum nanotoxicity; as we have discussed, the toxicity of a nanoparticle does not depend on the properties of the bare particle itself but on the groups that attach to it when it comes in contact with a heterogeneous environment. Regulation from authorities, proper monitoring, nanoparticle toxicity, and strategies to overcome their toxicity should be undertaken as the key toward a safe nanobiology-driven era for human welfare.

REFERENCES

Abollino, O., Aceto, M., Malandrino, M., Sarzanini, C., and Mentasti, E. 2003. Sorption of heavy metals on Na-montmorillonite effect of pH and organic substances. *Water Research*, 38: 1619–1627.

Adegoke, H.I., AmooAdekola, F., Fatoki, O.S., and Ximba, B.J. 2014. Adsorption of Cr (VI) on synthetic hematite (α-Fe_2O_3) nanoparticles of different morphologies. *Korean Journal of Chemical Engineering*, 31(1): 142–154.

Afkhami, A., Bagheri, H., and Madrakian, T. 2011. Alumina nanoparticles grafted with functional groups as a new adsorbent in efficient removal of formaldehyde from water samples. *Desalination*, 281: 151–158.

Afkhami, A. and Norooz-Asl, R. 2009. Removal, preconcentration and determination of Mo (VI) from water and wastewater samples using maghemite nanoparticles. *Colloids and Surfaces A: Physicochemical and Engineering Aspects*, 346: 52–57.

Afkhami, A., Saber-Tehrani, M., and Bagheri, H. 2010. Simultaneous removal of heavy-metal ions in wastewater samples using nano-alumina modified with 2,4-dinitrophenylhydrazine. *Journal of Hazardous Materials*, 181: 836–844.

Agnihotri, M., Joshi, S., Kumar, A.R., Zinjarde, S., and Kulkarni, S. (2009). Biosynthesis of gold nanoparticles by the tropical marine yeast *Yarrowia lipolytica*, NCIM 3589. *Materials Letters*, 63(15): 1231–1234.

Ahmad, A., Mukherjee, P., and Mandal, D. 2002. Enzyme mediated extracellular synthesis of CdS nanoparticles by the fungus, *Fusarium oxysporum. Journal of the American Chemical Society*, 124: 12108–12109.

Ahmad, A., Senapati, S., Khan, M.I., Kumar, R., Ramani, R., Srinivas, V., and Sastry, M. 2003. Intracellular synthesis of gold nanoparticles by a novel alkalotolerant actinomycete, *Rhodococcus* species. *Nanotechnology*, 14(7): 824.

Aili, D., Enander, K., Rydberg, J., Nesterenko, I., Björefors, F., Baltzer, L., and Liedberg, B. 2008. Folding induced assembly of polypeptide decorated gold nanoparticles. *Journal of the American Chemical Society*, 130: 5780–5788.

Akiyoshi, K., Sasaki, Y., and Sunamoto, J.1999. Molecular chaperone-like activity of hydrogel nanoparticles of hydrophobized pullulan: Thermal stabilization with refolding of carbonic anhydrase B. *Bioconjugate Chemistry*, 10(3): 321–324.

Al-Rawi, M., Diabaté, S., and Weiss, C. 2011. Uptake and intracellular localization of submicron and nano-sized SiO_2 particles in HeLa cells. *Archives of Toxicology*, 85(7): 813–826.

Arakaki, A., Nakazawa, H., Nemoto, M., Mori, T., and Matsunaga, T. 2008. Formation of magnetite by bacteria and its application. *Journal of the Royal Society Interface*, 52: 977–999.

Auffan, M., Achouak, W., Rose, J., Roncato, M.A., Chanéac, C., Waite, D.T., and Bottero, J.Y. 2008. Relation between the redox state of iron-based nanoparticles and their cytotoxicity toward *Escherichia coli. Environmental Science and Technology*, 42(17): 6730–6735.

Azizi, S., Namvar, F., Mahdavi, M., Ahmad, M.B., and Mohamad, R. 2013. Biosynthesis of silver nanoparticles using brown marine macroalga, *Sargassummuticum* aqueous extract. *Materials*, 6: 5942–5950.

Babitha, S. and Korrapati, P.S. 2013. Biosynthesis of titanium dioxide nanoparticles using a probiotic from coal fly ash effluent. *Materials Research Bulletin*, 48: 4738–4742.

Badruddoza, A.Z.M., Shawon, Z.B.Z., Tay, W.J.D., Hidajat, K., Uddin, M.S. 2013. Fe_3O_4/cyclodextrin polymer nanocomposites for selective heavy metals removal from industrial wastewater. *Carbohydrate Polymers*, 91: 322–332.

Badruddoza, A.Z.M., Tay, A.S.H., Tan, P.Y., Hidajat, K., and Uddin, M.S. 2011. Carboxymethyl-β-cyclodextrin conjugated magnetic nanoparticles as nano-adsorbents for removal of copper ions: Synthesis and adsorption studies. *Journal of Hazardous Materials*, 185: 1177–1186.

Baesman, S.M., Bullen, T.D., Dewald, J., Zhang, D., Curran, S., Islam, F.S., Beveridge, T.J., and Oremland, R.S. 2007. Formation of tellurium nanocrystals during anaerobic growth of bacteria that use Te oxyanions as respiratory electron acceptors. *Applied and Environmental Microbiology*, 73: 2135–2143.

Bai, H.J. and Zhang, Z.M. 2009. Microbial synthesis of semiconductor lead sulfide nanoparticles using immobilized *Rhodobacter sphaeroides*. *Materials Letters*, 63: 764–766.

Bai, H.J., Zhang, Z.M., and Gong, J. 2006. Biological synthesis of semiconductor zinc sulfide nanoparticles by immobilized *Rhodobacter sphaeroides*. *Biotechnology Letters*, 28: 1135–1139.

Balagurunathan, R., Radhakrishnan, M., BabuRajendran, R., and Velmurugan, D. 2011. Biosynthesis of gold nanoparticles by actinomycete *Streptomyces viridogens* strain HM10. *Indian Journal of Biochemistry and Biophysics*, 48: 331.

Balasubramanian, K. and Burghard, M. 2005. Chemically functionalized carbon nanotubes. *Small*, 1: 180–192.

Banerjee, S.S. and Chen, D.H. 2007. Fast removal of copper ions by gum arabic modified magnetic nano-adsorbent. *Journal of Hazardous Materials*, 147: 792–799.

Bansal, V., Poddar, P., Ahmad, A., and Sastry, M. 2006. Room-temperature biosynthesis of ferroelectric barium titanate nanoparticles. *Journal of the American Chemical Society*, 128: 11958–11963.

Bansal, V., Rautaray, D., Ahmad, A., and Sastry, M. 2004. Biosynthesis of zirconia nanoparticles using the fungus *Fusarium oxysporum*. *Journal of Materials Chemistry*, 14: 3303–3305.

Bansal, V., Rautaray, D., Bharde, A., Ahire, K., Sanyal, A., Ahmad, A., and Sastry, M. 2005. Fungus-mediated biosynthesis of silica and titanium particles. *Journal of Materials Chemistry*, 15: 2583–2589.

Bansal, V., Sanyal, A., Rautaray, D., Ahmad, A., and Sastry, M., 2005. Bioleaching of sand by the fungus *Fusarium oxysporum* as a means of producing extracellular silica nanoparticles. *Advanced Materials*, 17(7): 889–892.

Basavaraja, S., Balaji, S.D., Lagashetty, A., Rajasab, A.H., and Venkataraman, A. 2008. Extracellular biosynthesis of silver nanoparticles using the fungus *Fusarium semitectum*. *Materials Research Bulletin*, 43(5): 1164–1170.

Bhabra, G., Sood, A., Fisher, B., Cartwright, L., Saunders, M., Evans, W.H., Surprenant, A. et al. 2009. Nanoparticles can cause DNA damage across a cellular barrier. *Nature Nanotechnology*, 4: 876–883.

Bhainsa, K.C. and D'souza, S.F. 2006. Extracellular biosynthesis of silver nanoparticles using the fungus *Aspergillus fumigatus*. *Colloids and Surfaces B: Biointerfaces*, 47: 160–164.

Bian, S.W., Mudunkotuwa, I.A., Rupasinghe, T., and Grassian, V.H. 2011. Aggregation and dissolution of 4 nm ZnO nanoparticles in aqueous environments: Influence of pH, ionic strength, size, and adsorption of humic acid. *Langmuir*, 27: 6059–6068.

Blanchard, G., Maunaye, M., and Martin, G. 1984. Removal of heavy metals from waters by means of natural zeolites. *Water Research*, 18: 1501–1507.

Borm, P.J., Robbins, D., Haubold, S., Kuhlbusch, T., Fissan, H., Donaldson, K., Schins, R. et al. 2006. The potential risks of nanomaterials: A review carried out for ECETOC. *Particle and Fibre Toxicology*, 3: 1.

Bose, S., Hochella, M.F., Gorby, Y.A., Kennedy, D.W., McCready, D.E., Madden, A.S., and Lower, B.H. 2009. Bioreduction of hematite nanoparticles by the dissimilatory iron reducing bacterium *Shewanella oneidensis* MR-1. *Geochimica et Cosmochimica Acta*, 73: 962–976.

Camblor, M.A., Corma, A., and Valencia, S. 1998. Characterization of nanocrystalline zeolite Beta. *Microporous and Mesoporous Materials*, 25(1): 59–74.

Cao, C.Y., Cui, Z.M., Chen, C.Q., Song, W.G., and Cai, W. 2010. Ceria hollow nanospheres produced by a template-free microwave-assisted hydrothermal method for heavy metal ion removal and catalysis. *The Journal of Physical Chemistry C*, 114: 9865–9870.

Cao, C.Y., Qu, J., Yan, W.S., Zhu, J.F., Wu, Z.Y., and Song, W.G., 2012. Low-cost synthesis of flowerlike α-Fe_2O_3 nanostructures for heavy metal ion removal: Adsorption property and mechanism. *Langmuir*, 28: 4573–4579.

Castro-Longoria, E., Vilchis-Nestor, A.R., and Avalos-Borja, M. 2011. Biosynthesis of silver, gold and bimetallic nanoparticles using the filamentous fungus *Neurospora crassa*. *Colloids and Surfaces B: Biointerfaces*, 83: 42–48.

Cedervall, T., Lynch, I., Lindman, S., Berggård, T., Thulin, E., Nilsson, H., Dawson, K.A., and Linse, S. 2007. Understanding the nanoparticle–protein corona using methods to quantify exchange rates and affinities of proteins for nanoparticles. *Proceedings of the National Academy of Sciences of the United States of America*, 104: 2050–2055.

Chalasani, R. and Vasudevan, S. 2012. Cyclodextrin functionalized magnetic iron oxide nanocrystals: A host-carrier for magnetic separation of non-polar molecules and arsenic from aqueous media. *Journal of Materials Chemistry*, 22: 14925–14931.

Chauhan, A., Zubair, S., Tufail, S., Sherwani, A., Sajid, M., Raman, S.C., Azam, A., and Owais, M. 2011. Fungus-mediated biological synthesis of gold nanoparticles: Potential in detection of liver cancer. *International Journal of Nanomedicine*, 6: 2305–2319.

Chen, J.P. and Yang, L., 2006. Study of a heavy metal biosorption onto raw and chemically modified *Sargassum* sp. via spectroscopic and modeling analysis. *Langmuir*, 22: 8906–8914.

Chen, Y.H. and Li, F.A. 2010. Kinetic study on removal of copper (II) using goethite and hematite nano-photocatalysts. *Journal of Colloid and Interface Science*, 347: 277–281.

Cheng, Z., Tan, A.L.K., Tao, Y., Shan, D., Ting, K.E., and Yin, X.J. 2012. Synthesis and characterization of iron oxide nanoparticles and applications in the removal of heavy metals from industrial waste water. *International Journal of Photoenergy*, 2012: 1–5.

Chou, C.M. and Lien, H.L. 2011. Dendrimer-conjugated magnetic nanoparticles for removal of zinc (II) from aqueous solutions. *Journal of Nanoparticle Research*, 13: 2099–2107.

Chowdhury, S.R. and Yanful, E.K. 2010. Arsenic and chromium removal by mixed magnetite–maghemite nanoparticles and the effect of phosphate on removal. *Journal of Environmental Management*, 91: 2238–2247.

Conner, E.M. and Grisham, M.B. 1996. Inflammation, free radicals, and antioxidants. *Nutrition*, 12: 274–277.

Dameron, C.T., Reese, R.N., Mehra, R.K., Kortan, A.R., Carroll, P.J., Steigerwald, M.L., Brus, L.E., and Winge, D.R., 1989. Biosynthesis of cadmium sulphide quantum semiconductor crystallites. *Nature*, 338: 596–597.

Dawson, K.A., Salvati, A., and Lynch, I. 2009. Nanotoxicology: Nanoparticles reconstruct lipids. *Nature Nanotechnology*, 4: 84–85.

De, M., and Rotello, V.M. 2008. Synthetic "chaperones": Nanoparticle-mediated refolding of thermally denatured proteins. *Chemical Communications*, 30: 3504–3506.

Devika, R., Elumalai, S., Manikandan, E., and Eswaramoorthy, D. 2012. Biosynthesis of silver nanoparticles using the fungus *Pleurotus ostreatus* and their antibacterial activity. *Open Access Scientific Reports*, 1: 1–5.

Donaghay, P.L. and Hanson, A.K. 1995. Controlled biomineralization of magnetite (Fe_3O_4) and greigite (Fe_3S_4) in a magnetotactic bacterium. *Applied and Environmental Microbiology*, 61: 3232–3239.

Donaldson, K., Murphy, F.A., Duffin, R., and Poland, C.A. 2010. Asbestos, carbon nanotubes and the pleural mesothelium: A review of the hypothesis regarding the role of long fibre retention in the parietal pleura, inflammation and mesothelioma. *Particle and Fibre Toxicology*, 7: 1.

Elsaesser, A. and Howard, C.V. 2012. Toxicology of nanoparticles. *Advanced Drug Delivery Reviews*, 64: 129–137.

Fabrega, J., Luoma, S.N., Tyler, C.R., Galloway, T.S., and Lead, J.R. 2011. Silver nanoparticles: Behaviour and effects in the aquatic environment. *Environment International*, 37: 517–531.

Farrukh, A., Akram, A., Ghaffar, A., Hanif, S., Hamid, A., Duran, H., and Yameen, B. 2013. Design of polymer-brush-grafted magnetic nanoparticles for highly efficient water remediation. *ACS Applied Materials and Interfaces*, 5: 3784–3793.

Fayaz, A.M., Balaji, K., Girilal, M., Yadav, R., Kalaichelvan, P.T., and Venketesan, R. 2010. Biogenic synthesis of silver nanoparticles and their synergistic effect with antibiotics: A study against gram-positive and gram-negative bacteria. *Nanomedicine: Nanotechnology, Biology and Medicine*, 6(1): 103–109.

Fayaz, A.M., Balaji, K., Kalaichelvan, P.T., and Venkatesan, R. 2009b. Fungal based synthesis of silver nanoparticles—An effect of temperature on the size of particles. *Colloids and Surfaces B: Biointerfaces*, 74(1): 123–126.

Feizi, H., Kamali, M., Jafari, L., and Moghaddam, P.R. 2013. Phytotoxicity and stimulatory impacts of nanosized and bulk titanium dioxide on fennel (*Foeniculum vulgare* Mill). *Chemosphere*, 91(4): 506–511.

Feng, L., Cao, M., Ma, X., Zhu, Y., and Hu, C. 2012. Superparamagnetic high-surface-area Fe_3O_4 nanoparticles as adsorbents for arsenic removal. *Journal of Hazardous Materials*, 217: 439–446.

Feng, Y., Gong, J.L., Zeng, G.M., Niu, Q.Y., Zhang, H.Y., Niu, C.G., Deng, J.H., and Yan, M., 2010. Adsorption of Cd (II) and Zn (II) from aqueous solutions using magnetic hydroxyapatite nanoparticles as adsorbents. *Chemical Engineering Journal*, 162: 487–494.

Flores, V. and Cabassud, C. 1999. A hybrid membrane process for Cu (II) removal from industrial wastewater Comparison with a conventional process system. *Desalination*, 126: 101–108.

Foley, S., Crowley, C., Smaihi, M., Bonfils, C., Erlanger, B.F., Seta, P., and Larroque, C. 2002. Cellular localisation of a water-soluble fullerene derivative. *Biochemical and Biophysical Research Communications*, 294: 116–119.

Fu, J., Ji, J., Fan, D., and Shen, J., 2006. Construction of antibacterial multilayer films containing nanosilver via layer-by-layer assembly of heparin and chitosan-silver ions complex. *Journal of Biomedical Materials Research Part A*, 79(3): 665–674.

Ge, F., Li, M.M., Ye, H., and Zhao, B.X. 2012. Effective removal of heavy metal ions Cd^{2+}, Zn^{2+}, Pb^{2+}, Cu^{2+} from aqueous solution by polymer-modified magnetic nanoparticles. *Journal of Hazardous Materials*, 211: 366–372.

Gericke, M. and Pinches, A. 2006. Biological synthesis of metal nanoparticles. *Hydrometallurgy*, 83: 132–140.

Ghorbel-Abid, I., Jrad, A., Nahdi, K., Trabelsi-Ayadi, M. 2009. Sorption of chromium (III) from aqueous solution using bentonitic clay. *Desalination*, 246: 595–604.

Giasuddin, A.B., Kanel, S.R., and Choi, H., 2007. Adsorption of humic acid onto nanoscale zerovalent iron and its effect on arsenic removal. *Environmental Science and Technology*, 41(6): 2022–2027.

Gimbert, F., Morin-Crini, N., Renault, F., Badot, P.M., and Crini, G. 2008. Adsorption isotherm models for dye removal by cationized starch-based material in a single component system: Error analysis. *Journal of Hazardous Materials*, 157(1): 34–46.

Ginzburg, V.V. and Balijepalli, S. 2007. Modelling the thermodynamics of the interaction of nanoparticles with cell membranes. *Nano Letters*, 7: 3716–3722.

Godbey, W.T., Wu, K.K., and Mikos, A.G. 1999. Tracking the intracellular path of poly (ethylenimine)/DNA complexes for gene delivery. *Proceedings of the National Academy of Sciences of the United States of America*, 96: 5177–5181.

Goon, I.Y., Zhang, C., Lim, M., Gooding, J.J., and Amal, R. 2010. Controlled fabrication of polyethylenimine-functionalized magnetic nanoparticles for the sequestration and quantification of free Cu^{2+}. *Langmuir*, 26: 12247–12252.

Goswami, A., Raul, P.K., and Purkait, M.K. 2012. Arsenic adsorption using copper (II) oxide nanoparticles. *Chemical Engineering Research and Design*, 90: 1387–1396.

Gou, N., Onnis-Hayden, A., and Gu, A.Z. 2010. Mechanistic toxicity assessment of nanomaterials by whole-cell-array stress genes expression analysis. *Environmental Science and Technology*, 44: 5964–5970.

Greulich, C., Diendorf, J., Simon, T., Eggeler, G., Epple, M., and Köller, M. 2011. Uptake and intracellular distribution of silver nanoparticles in human mesenchymal stem cells. *Actabiomaterialia*, 7: 347–354.

Gupta, K., Bhattacharya, S., Chattopadhyay, D., Mukhopadhyay, A., Biswas, H., Dutta, J., Ray, N.R., and Ghosh, U.C. 2011. Ceria associated manganese oxide nanoparticles: Synthesis, characterization and arsenic (V) sorption behavior. *Chemical Engineering Journal*, 172: 219–229.

Haefeli, C., Franklin, C., and Hardy, K. 1984. Plasmid-determined silver resistance in *Pseudomonas stutzeri* isolated from a silver mine. *Journal of Bacteriology*, 158(1): 389–392.

Handy, R.D., Von der Kammer, F., Lead, J.R., Hassellöv, M., Owen, R., and Crane, M. 2008. The ecotoxicology and chemistry of manufactured nanoparticles. *Ecotoxicology*, 17: 287–314.

Hao, Y.M., Man, C., and Hu, Z.B. 2010. Effective removal of Cu (II) ions from aqueous solution by amino-functionalized magnetic nanoparticles. *Journal of Hazardous Materials*, 184: 392–399.

He, S., Guo, Z., Zhang, Y., Zhang, S., Wang, J., and Gu, N. 2007. Biosynthesis of gold nanoparticles using the bacteria *Rhodopseudomonascapsulata*. *Materials Letters*, 61: 3984–3987.

Ho, Y.S., Chiu, W.T., and Wang, C.C. 2005. Regression analysis for the sorption isotherms of basic dyes on sugarcane dust. *Bioresource Technology*, 96: 1285–1291.

Holmes, J.D., Smith, P.R., Evans-Gowing, R., Richardson, D.J., Russell, D.A., and Sodeau, J.R. 1995. Energy-dispersive x-ray analysis of the extracellular cadmium sulfide crystallites of *Klebsiella aerogenes*. *Archives of Microbiology*, 163: 143–147.

Hoshino, A., Fujioka, K., Oku, T., Nakamura, S., Suga, M., Yamaguchi, Y., Suzuki, K., Yasuhara, M., and Yamamoto, K. 2004. Quantum dots targeted to the assigned organelle in living cells. *Microbiology and Immunology*, 48: 985–994.

Hristovski, K., Baumgardner, A., and Westerhoff, P. 2007. Selecting metal oxide nanomaterials for arsenic removal in fixed bed columns: From nanopowders to aggregated nanoparticle media. *Journal of Hazardous Materials*, 147: 265–274.

Hu, J., Chen, G., and Lo, I.M.C. 2005. Removal and recovery of Cr(VI) from wastewater by maghemite nanoparticles. *Water Research*, 39: 4528–4536.

Hu, J., Chen, G., and Lo, I.M. 2006. Selective removal of heavy metals from industrial wastewater using maghemite nanoparticle: Performance and mechanisms. *Journal of Environmental Engineering*, 132: 709–715.

Huang, J., Cao, Y., Liu, Z., Deng, Z., Tang, F., and Wang, W. 2012. Efficient removal of heavy metal ions from water system by titanate nanoflowers. *Chemical Engineering Journal*, 180: 75–80.

Huang, S.H. and Chen, D.H. 2009. Rapid removal of heavy metal cations and anions from aqueous solutions by an amino-functionalized magnetic nano-adsorbent. *Journal of Hazardous Materials*, 163(1): 174–9.

Hui, L.I., Xiao, D.L., Hua, H.E., Rui, L.I.N., and Zuo, P.L. 2013. Adsorption behavior and adsorption mechanism of Cu (II) ions on amino-functionalized magnetic nanoparticles. *Transactions of Nonferrous Metals Society of China*, 23: 2657–2665.

Huk, A., Izak-Nau, E., El Yamani, N., Uggerud, H., Vadset, M., Zasonska, B., Duschl, A., and Dusinska, M. 2015. Impact of nanosilver on various DNA lesions and HPRT gene mutations–effects of charge and surface coating. *Particle and Fibre Toxicology*, 12: 1.

Hussain, S.M., Hess, K.L., Gearhart, J.M., Geiss, K.T., and Schlager, J.J., 2005. In vitro toxicity of nanoparticles in BRL 3A rat liver cells. *Toxicology in Vitro*, 19(7): 975–983.

Hyung, H., Fortner, J.D., Hughes, J.B., and Kim, J.H., 2007. Natural organic matter stabilizes carbon nanotubes in the aqueous phase. *Environmental Science and Technology*, 41(1): 179–184.

Iravani, S. 2011. Green synthesis of metal nanoparticles using plants. *Green Chemistry*, 13: 2638.

Iravani, S., Korbekandi, H., Mirmohammadi, S.V., and Zolfaghari, B. 2014. Synthesis of silver nanoparticles: Chemical, physical and biological methods. *Research in Pharmaceutical Sciences*, 9(6): 385.

Ishii, D., Kinbara, K., Ishida, Y., Ishii, N., Okochi, M., Yohda, M., and Aida, T., 2003. Chaperonin-mediated stabilization and ATP-triggered release of semiconductor nanoparticles. *Nature*, 423: 628–632.

Jain, N., Bhargava, A., Majumdar, S., Tarafdar, J.C., and Panwar, J. 2011. Extracellular biosynthesis and characterization of silver nanoparticles using *Aspergillus flavus* NJP08: A mechanism perspective. *Nanoscale*, 635–641.

Jayaseelan, C., Rahuman, A.A., Kirthi, A.V., Marimuthu, S., Santhoshkumar, T., Bagavan, A., Gaurav, K., Karthik, L., and Rao, K.B. 2012. Novel microbial route to synthesize ZnO nanoparticles using *Aeromonas hydrophila* and their activity

against pathogenic bacteria and fungi. *Spectrochimica Acta Part A: Molecular and Biomolecular Spectroscopy*, 90: 78–84.

Jha, A.K. and Prasad, K. 2010. Ferroelectric BaTiO$_3$ nanoparticles: Biosynthesis and characterization. *Colloids and Surfaces B: Biointerfaces*, 75: 330–334.

Jha, A.K., Prasad, K., and Kulkarni, A.R. 2009a. Synthesis of TiO$_2$ nanoparticles using microorganisms. *Colloids and Surfaces B: Biointerfaces*, 71: 226–229.

Jha, A.K., Prasad, K., and Prasad, K. 2009b. A green low-cost biosynthesis of Sb$_2$O$_3$ nanoparticles. *Biochemical Engineering Journal*, 43: 303–306.

Jia, G., Wang, H., Yan, L., Wang, X., Pei, R., Yan, T., Zhao, Y., and Guo, X. 2005. Cytotoxicity of carbon nanomaterials: Single-wall nanotube, multi-wall nanotube, and fullerene. *Environmental Science and Technology*, 39: 1378–1383.

Jiang, H.M., Yang, T., Wang, Y.H., Lian, H.Z., and Hu, X. 2013. Magnetic solid-phase extraction combined with graphite furnace atomic absorption spectrometry for speciation of Cr (III) and Cr (VI) in environmental waters. *Talanta*, 116: 361–367.

Kalabegishvili, T.L., Murusidze, I.G., Kirkesali, E.I., Rcheulishvili, A.N., Ginturi, E.N., Gelagutashvili, E.S., and Holman, H.Y. 2013. Development of biotechnology for microbial synthesis of gold and silver nanoparticles. *Journal of Life Sciences*, 7(2): 110.

Kalishwaralal, K., Deepak, V., Ramkumarpandian, S., Nellaiah, H., and Sangiliyandi, G. 2008. Extracellular biosynthesis of silver nanoparticles by the culture supernatant of *Bacillus licheniformis*. *Materials Letters*, 62: 4411–4413.

Karami, H. 2013. Heavy metal removal from water by magnetite nanorods. *Chemical Engineering Journal*, 219: 209–216.

Karatapanis, A.E., Fiamegos, Y., and Stalikas, C.D. 2011. Silica-modified magnetic nanoparticles functionalized with cetylpyridinium bromide for the preconcentration of metals after complexation with 8-hydroxyquinoline. *Talanta*, 84: 834–839.

Kim, E.J., Lee, C.S., Chang, Y.Y., and Chang, Y.S. 2013. Hierarchically structured manganese oxide-coated magnetic nanocomposites for the efficient removal of heavy metal ions from aqueous systems. *ACS Applied Materials and Interfaces*, 5: 9628–9634.

Kiser, M.A., Ryu, H., Jang, H., Hristovski, K., and Westerhoff, P. 2010. Biosorption of nanoparticles to heterotrophic wastewater biomass. *Water Research*, 44: 4105–4114.

Konishi, Y., Ohno, K., Saitoh, N., Nomura, T., Nagamine, S., Hishida, H., Takahashi, Y., and Uruga, T. 2007. Bioreductive deposition of platinum nanoparticles on the bacterium *Shewanella algae*. *Journal of Biotechnology*, 128: 648–653.

Konishi, Y., Tsukiyama, T., Ohno, K., Saitoh, N., Nomura, T., and Nagamine, S. 2006. Intracellular recovery of gold by microbial reduction of AuCl$_4^-$ ions using the anaerobic bacterium *Shewanella algae*. *Hydrometallurgy*, 81: 24–29.

Krumov, N., Perner-Nochta, I., Oder, S., Gotcheva, V., Angelov, A., and Posten, C. 2009. Production of inorganic nanoparticles by microorganisms. *Chemical Engineering and Technology*, 32(7): 1026–1035.

Kumar, S.A., Ansary, A.A., Ahmad, A., and Khan, M.I. 2007. Extracellular biosynthesis of CdSe quantum dots by the fungus, *Fusarium oxysporum*. *Journal of Biomedical Nanotechnology*, 3: 190–194.

Labrenz, M., Druschel, G.K., Thomsen-Ebert, T., Gilbert, B., Welch, S.A., Kemner, K.M., Logan, G.A. et al. 2000. Formation of sphalerite (ZnS) deposits in natural biofilms of sulfate-reducing bacteria. *Science*, 290: 1744–1747.

Lazaridis, N.K. and Charalambous, C. 2005. Sorptive removal of trivalent and hexavalent chromium from binary aqueous solutions by composite alginate–goethite beads. *Water Research*, 39: 4385–4396.

Lengke, M.F., Fleet, M.E., and Southam, G. 2006a. Morphology of gold nanoparticles synthesized by filamentous cyanobacteria from gold (I)-thiosulfate and gold (III)-chloride complexes. *Langmuir*, 22(6): 2780–2787.

Lengke, M.F., Ravel, B., Fleet, M.E., Wanger, G., Gordon, R.A., and Southam, G. 2006b. Mechanisms of gold bioaccumulation by filamentous *cyanobacteria* from gold (III)-chloride complex. *Environmental Science and Technology*, 40: 6304–6309.

Li, N., Sioutas, C., Cho, A., Schmitz, D., Misra, C., Sempf, J., Wang, M., Oberley, T., Froines, J., and Nel, A. 2003. Ultrafine particulate pollutants induce oxidative stress and mitochondrial damage. *Environmental Health Perspectives*, 111: 455.

Li, N., Xia, T., and Nel, A.E. 2008. The role of oxidative stress in ambient particulate matter-induced lung diseases and its implications in the toxicity of engineered nanoparticles. *Free Radical Biology and Medicine*, 44: 1689–1699.

Liu, J.F., Zhao, Z.S., and Jiang, G.B. 2008. Coating Fe_3O_4 magnetic nanoparticles with humic acid for high efficient removal of heavy metals in water. *Environmental Science and Technology*, 42: 6949–6954.

López-Téllez, G., Barrera-Díaz, C.E., Balderas-Hernández, P., Roa-Morales, G., and Bilyeu, B. 2011. Removal of hexavalent chromium in aquatic solutions by iron nanoparticles embedded in orange peel pith. *Chemical Engineering Journal*, 173(2): 480–485.

Luo, T., Cui, J., Hu, S., Huang, Y., and Jing, C. 2010. Arsenic removal and recovery from copper smelting wastewater using TiO_2. *Environmental Science and Technology*, 44: 9094–9098.

Luo, Y., Teng, Z., Wang, X., and Wang, Q. 2013. Development of carboxymethyl chitosan hydrogel beads in alcohol-aqueous binary solvent for nutrient delivery applications. *Food Hydrocolloids*, 31(2): 332–339.

Lyn, T.E. 2009. Deaths, lung damage linked to nanoparticles in China, Reuters.

Lynch, I., Cedervall, T., Lundqvist, M., Cabaleiro-Lago, C., Linse, S., and Dawson, K.A. 2007. The nanoparticle–protein complex as a biological entity; a complex fluids and surface science challenge for the 21st century. *Advances in Colloid and Interface Science*, 134: 167–174.

Madrakian, T., Afkhami, A., and Ahmadi, M. 2013. Simple in situ functionalizing magnetite nanoparticles by reactive blue-19 and their application to the effective removal of Pb^{2+} ions from water samples. *Chemosphere*, 90: 542–547.

Mahdavi, S., Jalali, M., and Afkhami, A. 2012. Removal of heavy metals from aqueous solutions using Fe_3O_4, ZnO, and CuO nanoparticles. *Journal of Nanoparticle Research*, 14: 1–18.

Marano, F., Hussain, S., Rodrigues-Lima, F., Baeza-Squiban, A., and Boland, S. 2011. Nanoparticles: Molecular targets and cell signalling. *Archives of Toxicology*, 85: 733–741.

Margeta, K., Vojnovic, B., and Zabukovec Logar, N. 2011. Development of natural zeolites for their use in water-treatment systems. *Recent Patents on Nanotechnology*, 5(2): 89–99.

Mehrabi, M. and Wilson, R. 2007. Intercalating gold nanoparticles as universal labels for DNA detection. *Small*, 3: 1491–1495.

Mobasherpour, I., Salahi, E., and Pazouki, M. 2011. Removal of nickel (II) from aqueous solutions by using nano-crystalline calcium hydroxyapatite. *Journal of Saudi Chemical Society*, 15: 105–112.

Mohmood, I., Lopes, C.B., Lopes, I., Ahmad, I., Duarte, A.C., and Pereira, E. 2013. Nanoscale materials and their use in water contaminants removal—a review. *Environmental Science and Pollution Research*, 20: 1239–1260.

Mukherjee, P., Senapati, S., Mandal, D., Ahmad, A., Khan, M.I., Kumar, R., and Sastry, M. 2002. Extracellular synthesis of gold nanoparticles by the fungus *Fusarium oxysporum*. *ChemBioChem Journal*, 3: 461–463.

Myllynen, P. 2009. Nanotoxicology: Damaging DNA from a distance. *Nature Nanotechnology*, 4: 795–796.

Myllynen, P.K., Loughran, M.J., Howard, C.V., Sormunen, R., Walsh, A.A., and Vähäkangas, K.H. 2008. Kinetics of gold nanoparticles in the human placenta. *Reproductive Toxicology*, 26: 130–137.

Nair, B. and Pradeep, T. 2002. Coalescence of nanoclusters and formation of submicron crystallites assisted by *Lactobacillus* strains. *Crystal Growth and Design*, 2: 293–298.

Nangia, Y., Wangoo, N., Sharma, S., Wu, J.S., Dravid, V., Shekhawat, G.S., and Suri, C.R. 2009. Facile biosynthesis of phosphate capped gold nanoparticles by a bacterial isolate *Stenotrophomonas maltophilia*. *Applied Physics Letters*, 94(23): 233901.

Narayanan, K.B. and Sakthivel, N. 2010. Biological synthesis of metal nanoparticles by microbes. *Advances in Colloid and Interface Science*, 156: 1–13.

Nassar, N.N. 2010. Rapid removal and recovery of Pb (II) from wastewater by magnetic nanoadsorbents. *Journal of Hazardous Materials*, 184: 538–546.

Navarro, E., Baun, A., Behra, R., Hartmann, N.B., Filser, J., Miao, A.J., Quigg, A., Santschi, P.H., and Sigg, L. 2008. Environmental behavior and ecotoxicity of engineered nanoparticles to algae, plants, and fungi. *Ecotoxicology*, 17: 372–386.

Naveen Hemath, K.S., Kumar, G., Karthik, L., and Bhaskara Rao, K.V. 2010. Extracellular biosynthesis of silver nanoparticles using the filamentous fungus *Penicillium* sp. *Archives of Applied Science Research*, 2: 161–167.

Navrotsky, A. 2000. Nanomaterials in the environment, agriculture, and technology (NEAT). *Journal of Nanoparticle Research*, 2: 321–323.

Nel, A., Xia, T., Mädler, L., and Li, N. 2006. Toxic potential of materials at the nanolevel. *Science*, 311: 622–627.

Oberdörster, G. 2010. Safety assessment for nanotechnology and nanomedicine: Concepts of nanotoxicology. *Journal of Internal Medicine*, 267: 89–105.

Oza, G., Pandey, S., Gupta, A., Kesarkar, R., and Sharon, M. 2012. Biosynthetic reduction of gold ions to gold nanoparticles by Nocardiafarcinica. *Journal of Microbiology and Biotechnology*, 511–515.

Panneerselvam, P., Morad, N., and Tan, K.A. 2011. Magnetic nanoparticle (Fe_3O_4) impregnated onto tea waste for the removal of nickel (II) from aqueous solution. *Journal of Hazardous Materials*, 186: 160–168.

Panté, N. and Kann, M. 2002. Nuclear pore complex is able to transport macromolecules with diameters of ~39 nm. *Molecular Biology of the Cell*, 13: 425–434.

Parfenov, A.S., Salnikov, V., Lederer, W.J., and Lukyanenko, V. 2006. Aqueous diffusion pathways as a part of the ventricular cell ultrastructure. *Biophysical Journal*, 90: 1107–1119.

Parham, H., Zargar, B., and Shiralipour, R. 2012. Fast and efficient removal of mercury from water samples using magnetic iron oxide nanoparticles modified with 2-mercaptobenzothiazole. *Journal of Hazardous Materials*, 205: 94–100.

Peng, X., Luan, Z., Ding, J., Di, Z., Li, Y., and Tian, B. 2005. Ceria nanoparticles supported nanotubes for the removal of arsenate from water. *Materials Letters*, 59: 399–403.

Perez-Gonzalez, T., Jimenez-Lopez, C., Neal, A.L., Rull-Perez, F., Rodriguez-Navarro, A., Fernandez-Vivas, A., and Iañez-Pareja, E. 2010. Magnetite biomineralization induced by *Shewanella oneidensis*. *Geochimica et Cosmochimica Acta*, 74(3): 967–979.

Prakasham, R.S., Sudheer Kumar, B., Sudheer Kumar, Y., and Girija Shanker, G. 2012b. Synthesis and characterization of silver nanoparticles from marine *Streptomyces* species. *Journal of Microbiology and Biotechnology*, 22: 614–621.

Prasad, K. and Jha, A.K. 2010. Biosynthesis of CdS nanoparticles: An improved green and rapid procedure. *Journal of Colloidand Interface Science*, 342: 68–72.

Prathna, T.C., Lazar Mathew, N., Chandrasekaran, N., Raichur, A., and Mukherjee, A. 2010. Biomimetic synthesis of nanoparticles: Science, technology and applicability. In *Biomimetics Learning from Nature*, A. Mukherjee (Ed.), InTech, Rijek, Crotia.

Qiu, H., Xue, L., Ji, G., Zhou, G., Huang, X., Qu, Y., and Gao, P. 2009. Enzyme-modified nanoporous gold-based electrochemical biosensors. *Biosensors and Bioelectronics*, 24(10): 3014–3018.

Quik, J.T., Lynch, I., Van Hoecke, K., Miermans, C.J., De Schamphelaere, K.A., Janssen, C.R., Dawson, K.A., Stuart, M.A.C., and Van De Meent, D. 2010. Effect of natural organic matter on cerium dioxide nanoparticles settling in model fresh water. *Chemosphere*, 81: 711–715.

Rahmani, A., Mousavi, H.Z., and Fazli, M. 2010. Effect of nanostructure alumina on adsorption of heavy metals. *Desalination*, 253: 94–100.

Rajamanickam, U., Mylsamy, P., Viswanathan, S., and Muthusamy, P. 2012. Biosynthesis of zinc nanoparticles using actinomycetes for antibacterial food packaging. In *International Conference on Nutrition and Food Sciences*, Singapore, p. 39.

Rajathi, F.A.A., Parthiban, C., Kumar, V.G., and Anantharaman, P. 2012. Biosynthesis of antibacterial gold nanoparticles using brown alga, *Stoechospermum marginatum* (kützing). *Spectrochimica Acta Part A: Molecular and Biomolecular Spectroscopy*, 99, 166–173.

Rajeshkumar, S., Malarkodi, C., Paulkumar, K., Vanaja, M., Gnanajobitha, G., and Annadurai, G. 2014. Algae mediated green fabrication of silver nanoparticles and examination of its antifungal activity against clinical pathogens. *International Journal of Metals*, 11: 1–8.

Raudabaugh, D.B., Tzolov, M.B., Calabrese, J.P., and Overton, B.E. 2013. Synthesis of silver nanoparticles by a bryophilous *Rhizoctonia* species. *Nanomaterials and Nanotechnology*, 3: 3–2.

Rautaray, D., Sanyal, A., Adyanthaya, S.D., Ahmad, A., and Sastry, M. 2004. Biological synthesis of strontium carbonate crystals using the fungus *Fusarium oxysporum*. *Langmuir*, 20: 6827–6833.

Recillas, S., García, A., González, E., Casals, E., Puntes, V., Sánchez, A., and Font, X. 2011. Use of CeO_2, TiO_2 and Fe_3O_4 nanoparticles for the removal of lead from water: Toxicity of nanoparticles and derived compounds. *Desalination*, 277: 213–220.

Roy, A. and Bhattacharya, J. 2012. Removal of Cu (II), Zn (II) and Pb (II) from water using microwave-assisted synthesized maghemite nanotubes. *Chemical Engineering Journal*, 211: 493–500.

Sadhasivam, S., Shanmugam, P., and Yun, K. 2010. Biosynthesis of silver nanoparticles by Streptomyces hygroscopicus and antimicrobial activity against medically important pathogenic microorganisms. *Colloids and Surfaces B: Biointerfaces*, 81(1): 358–362.

Saifuddin, N., Wong, C.W., and Yasumira, A.A. 2009. Rapid biosynthesis of silver nanoparticles using culture supernatant of bacteria with microwave irradiation. *Journal of Chemistry*, 6(1): 61–70.

Sanghi, R., Verma, P., and Puri, S. 2011. Enzymatic formation of gold nanoparticles using *Phanerochaete chrysosporium*. *Advances in Chemical Engineering and Science*, 1(03): 154.

Sastry, M., Ahmad, A., Khan, M.I., and Kumar, R. 2003. Biosynthesis of metal nanoparticles using fungi and actinomycete. *Current Science*, 85(2): 162–170.

Sawle, B.D., Salimath, B., Deshpande, R., Bedre, M.D., Prabhakar, B.K., and Venkataraman, A. 2008. Biosynthesis and stabilization of Au and Au-Ag alloy nanoparticles by fungus *Fusarium seminectum*. Science and technology of advanced materials, 9(3).

Seabra, A.B. and Durán, N. 2015. Nanotoxicology of metal oxide nanoparticles. *Metals*, 5: 934–975.

Senapati, S., Ahmad, A., Khan, M.I., Sastry, M., and Kumar, R. 2005. Extracellular biosynthesis of bimetallic Au–Ag alloy nanoparticles. *Small*, 1: 517–520.

Senapati, S., Mandal, D., Ahmad, A., Khan, M.I., Sastry, M., and Kumar, R. 2004. Fungus mediated synthesis of silver nanoparticles: A novel biological approach. *Indian Journal of Physics*, 78(A): 101–105.

Shahverdi, A.R., Minaeian, S., Shahverdi, H.R., Jamalifar, H., and Nohi, A.A. 2007. Rapid synthesis of silver nanoparticles using culture supernatants of Enterobacteria: A novel biological approach. *Process Biochemistry*, 42(5): 919–923.

Sharma, Y.C. and Srivastava, V. 2009. Separation of Ni (II) ions from aqueous solutions by magnetic nanoparticles. *Journal of Chemical and Engineering Data*, 55: 1441–1442.

Shirley, A.D., Dayanand, A., Sreedhar, B., and Dastager, S.G. 2010. Antimicrobial activity of silver nanoparticles synthesized from novel Streptomyces species. *Digest Journal of Nanomaterials and Biostructures*, 5(2): 447–451.

Shivaji, S., Madhu, S., and Singh, S. 2011. Extracellular synthesis of antibacterial silver nanoparticles using psychrophilic bacteria. *Process Biochemistry*, 46(9): 1800–1807.

Singh, S., Barick, K.C., and Bahadur, D. 2011. Surface engineered magnetic nanoparticles for removal of toxic metal ions and bacterial pathogens. *Journal of Hazardous Materials*, 192: 1539–1547.

Singh, S., Barick, K.C., and Bahadur, D. 2013. Functional oxide nanomaterials and nanocomposites for the removal of heavy metals and dyes. *Nanomaterials and Nanotechnology*, 3: 3–20.

Sinha, A. and Khare, S.K. 2011. Mercury bioaccumulation and simultaneous nanoparticle synthesis by *Enterobacter* sp. cells. *Bioresource Technology*, 102(5): 4281–4284.

Sintubin, L., De Windt, W., Dick, J., Mast, J., van der Ha, D., Verstraete, W., and Boon, N. 2009. Lactic acid bacteria as reducing and capping agent for the fast and efficient production of silver nanoparticles. *Applied Microbiology and Biotechnology*, 84: 741–749.

Slawson, R.M., Van Dyke, M.I., Lee, H., Trevors, J.T. 1992. Germanium and silver resistance, accumulation, and toxicity in microorganisms. *Plasmid*, 27(1): 72–79.

Song, Y., Li, X., and Du, X. 2009. Exposure to nanoparticles is related to pleural effusion, pulmonary fibrosis and granuloma. *European Respiratory Journal*, 34(3): 559–567.

Spring, S. and Schleifer, K.H. 1995. Diversity of magnetotactic bacteria. *Systematic and Applied Microbiology*, 18: 147–153.

Subashini, J. and Kannabiran, K. 2013. Antimicrobial activity of *Streptomyces* sp. VITBT7 and its synthesized silver nanoparticles against medically important fungal and bacterial pathogens. *Der Pharmacia Letter*, 5, 192–200.

Sudha, S.S., Rajamanickam, K., and Rengaramanujam, J. 2013. Microalgae mediated synthesis of silver nanoparticles and their antibacterial activity against pathogenic bacteria. *Indian Journal of Experimental Biology*, 51(5): 393–399.

Suresh, A.K., Doktycz, M.J., Wang, W., Moon, J.W., Gu, B., Meyer, H.M., Hensley, D.K., Allison, D.P., Phelps, T.J., and Pelletier, D.A. 2011. Monodispersed biocompatible silver sulfide nanoparticles: Facile extracellular biosynthesis using the γ-proteobacterium, *Shewanella oneidensis*. *Acta Biomaterialia*, 7(12): 4253–4258.

Sweeney, R.Y., Mao, C., Gao, X., Burt, J.L., Belcher, A.M., Georgiou, G., and Iverson, B.L., 2004. Bacterial biosynthesis of cadmium sulfide nanocrystals. *Chemistry and Biology*, 11(11): 1553–1559.

Takahashi, H., Sawada, S.I., and Akiyoshi, K. 2010. Amphiphilic polysaccharide nanoballs: A new building block for nanogel biomedical engineering and artificial chaperones. *ACS Nano*, 5: 337–345.

Thornhill, R.H., Burgess, J.G., and Matsunaga, T. 1995. PCR for direct detection of indigenous uncultured magnetic cocci in sediment and phylogenetic analysis of amplified 16S ribosomal DNA. *Applied and Environmental Microbiology*, 61: 495–500.

Tofighy, M.A. and Mohammadi, T. 2011. Adsorption of divalent heavy metal ions from water using carbon nanotube sheets. *Journal of Hazardous Materials*, 185(1): 140–147.

Tseng, J.Y., Chang, C.Y., Chang, C.F., Chen, Y.H., Chang, C.C., Ji, D.R., Chiu, C.Y., and Chiang, P.C. 2009. Kinetics and equilibrium of desorption removal of copper from magnetic polymer adsorbent. *Journal of Hazardous Materials*, 171: 370–377.

Tsuchiya, T., Oguri, I., Yamakoshi, Y.N., and Miyata, N. 1996. Novel harmful effects of fullerene on mouse embryos in vitro and in vivo. *FEBS Letters*, 393: 139–145.

Uchino, T., Tokunaga, H., Ando, M., and Utsumi, H. 2002. Quantitative determination of OH radical generation and its cytotoxicity induced by TiO_2–UVA treatment. *Toxicology in Vitro*, 16: 629–635.

Usha, R., Prabu, E., Palaniswamy, M., Venil, C.K., and Rajendran, R. 2010. Synthesis of metal oxide nano particles by *Streptomyces* sp. for development of antimicrobial textiles. *Global Journal of Biotechnology and Biochemistry*, 5: 153–160.

Vahabi, K., Mansoori, G.A., and Karimi, S. 2011. Biosynthesis of silver nanoparticles by fungus *Trichoderma reesei* (a route for large-scale production of AgNPs). *Insciences Journal*, 1: 65–79.

Van der Bruggen, B. and Vandecasteele, C. 2003. Removal of pollutants from surface water and groundwater by nanofiltration: Overview of possible applications in the drinking water industry. *Environmental Pollution*, 122(3): 435–445.

Vasir, J.K. and Labhasetwar, V. 2008. Quantification of the force of nanoparticle-cell membrane interactions and its influence on intracellular trafficking of nanoparticles. *Biomaterials*, 29: 4244–4252.

Vigneshwaran, N., Ashtaputre, N.M., Varadarajan, P.V., Nachane, R.P., Paralikar, K.M., and Balasubramanya, R.H. 2007. Biological synthesis of silver nanoparticles using the fungus *Aspergillus flavus*. *Materials Letters*, 61: 1413–1418.

Vigneshwaran, N., Kathe, A.A., Varadarajan, P.V., Nachane, R.P., and Balasubramanya, R.H. 2006. Biomimetics of silver nanoparticles by white rot fungus, *Phaenerochaete chrysosporium*. *Colloids and Surfaces B: Biointerfaces*, 53: 55–59.

Vijayaraghavan, K., Mahadevan, A., Sathishkumar, M., Pavagadhi, S., and Balasubramanian, R. 2011. Biosynthesis of Au(0) from Au(III) via biosorption and bioreduction using brown marine alga *Turbinaria conoides*. *Chemical Engineering Journal*, 167: 223–227.

Waghmare, S.S., Deshmukh, A.M., Kulkarni, S.W., and Oswaldo, L.A. 2011. Biosynthesis and characterization of manganese and zinc nanoparticles. *Universal Journal of Environmental Research and Technology*, 1: 64–69.

Waghmare, S.S., Deshmukh, A.M., and Sadowski, Z. 2014. Biosynthesis, optimization, purification and characterization of gold nanoparticles. *African Journal of Microbiology Research*, 8: 138–146.

Wagner, S.C., Roskamp, M., Pallerla, M., Araghi, R.R., Schlecht, S., and Koksch, B. 2010. Nanoparticle-induced folding and fibril formation of Coiled-coil-based model peptides. *Small*, 6: 1321–1328.

Wang, J., Zheng, S., Shao, Y., Liu, J., Xu, Z., and Zhu, D. 2010. Amino-functionalized Fe_3O_4 at SiO_2 core–shell magnetic nanomaterial as a novel adsorbent for aqueous heavy metals removal. *Journal of Colloid and Interface Science*, 349: 293–299.

Wang, J., Chen, C. 2009. Biosorbent for heavy metal removal and their future. *Biotechnology Advance*, 27: 195–226.

Wang, L., Li, J., Jiang, Q., and Zhao, L., 2012a. Water-soluble Fe_3O_4 nanoparticles with high solubility for removal of heavy-metal ions from waste water. *Dalton Transactions*, 41: 4544–4551.

Wang, X., Guo, Y., Yang, L., Han, M., Zhao, J., and Cheng, X. 2012b. Nanomaterials as sorbents to remove heavy metal ions in wastewater treatment. *Journal of Environmental and Analytical Toxicology*, 2: 7

Wang, Y., Zou, B., Gao, T., Wu, X., Lou, S., and Zhou, S. 2012c. Synthesis of orange-like Fe_3O_4/PPy composite microspheres and their excellent Cr (VI) ion removal properties. *Journal of Materials Chemistry*, 22: 9034–9040.

Watson, J.H.P., Ellwood, D.C., Soper, A.K., and Charnock, J. 1999. Nanosized strongly-magnetic bacterially-produced iron sulfide materials. *Journal of Magnetism and Magnetic Materials*, 203: 69–72.

Wei, Z., Xing, R., Zhang, X., Liu, S., Yu, H., and Li, P. 2012. Facile template-free fabrication of hollow nestlike α-Fe_2O_3 nanostructures for water treatment. *ACS Applied Materials and Interfaces*, 5: 598–604.

Williams, Y., Sukhanova, A., Nowostawska, M., Davies, A.M., Mitchell, S., Oleinikov, V., Gun'ko, Y., Nabiev, I., Kelleher, D., and Volkov, Y. 2009. Probing cell-type-specific intracellular nanoscale barriers using size-tuned quantum dots. *Small*, 5: 2581–2588.

Windt, W.D., Aelterman, P., and Verstraete, W. 2005. Bioreductive deposition of palladium(0) nanoparticles on *Shewanella oneidensis* with catalytic activity towards reductive dechlorination of polychlorinated biphenyls. *Environmental Microbiology*, 7: 314–325.

Wingenfelder, U., Hansen, C., Furrer, G., and Schulin, R. 2005. Removal of heavy metals from mine waters by natural zeolites. *Environmental Science and Technology*, 39(12): 4606–4613.

Wittmaack, K. 2007. In search of the most relevant parameter for quantifying lung inflammatory response to nanoparticle exposure: Particle number, surface area, or what?. *Environmental Health Perspectives*, 115(2): 187–194.

Wu, W.H., Sun, X., Yu, Y.P., Hu, J., Zhao, L., Liu, Q., Zhao, Y.F., and Li, Y.M. 2008. TiO_2 nanoparticles promote β-amyloid fibrillation in vitro. *Biochemical and Biophysical Research Communications*, 373: 315–318.

Xia, T., Kovochich, M., Brant, J., Hotze, M., Sempf, J., Oberley, T., Sioutas, C., Yeh, J.I., Wiesner, M.R., and Nel, A.E. 2006. Comparison of the abilities of ambient and manufactured nanoparticles to induce cellular toxicity according to an oxidative stress paradigm. *Nano Letters*, 6: 1794–1807.

Xie, W., Wang, L., Zhang, Y., Su, L., Shen, A., Tan, J., and Hu, J. 2009. Nuclear targeted nanoprobe for single living cell detection by surface-enhanced Raman scattering. *Bioconjugate Chemistry*, 20: 768–773.

Xin, X., Wei, Q., Yang, J., Yan, L., Feng, R., Chen, G., Du, B., and Li, H. 2012. Highly efficient removal of heavy metal ions by amine-functionalized mesoporous Fe_3O_4 nanoparticles. *Chemical Engineering Journal*, 184: 132–140.

Xu, P., Zeng, G.M., Huang, D.L., Lai, C., Zhao, M.H., Wei, Z., Li, N.J., Huang, C., and Xie, G.X., 2012. Adsorption of Pb (II) by iron oxide nanoparticles immobilized *Phanerochaete chrysosporium*: Equilibrium, kinetic, thermodynamic and mechanisms analysis. *Chemical Engineering Journal*, 203: 423–431.

Yan, H., Yang, L., Yang, Z., Yang, H., Li, A., and Cheng, R. 2012. Preparation of chitosan/poly (acrylic acid) magnetic composite microspheres and applications in the removal of copper (II) ions from aqueous solutions. *Journal of Hazardous Materials*, 229: 371–380.

Yan, S., He, W., Sun, C., Zhang, X., Zhao, H., Li, Z., Zhou, W., Tian, X., Sun, X., and Han, X., 2009. The biomimetic synthesis of zinc phosphate nanoparticles. *Dyes and Pigments*, 80: 254–258.

Yantasee, W., Warner, C.L., Sangvanich, T., Addleman, R.S., Carter, T.G., Wiacek, R.J., Fryxell, G.E., Timchalk, C., and Warner, M.G. 2007. Removal of heavy metals from aqueous systems with thiol functionalized superparamagnetic nanoparticles. *Environmental Science and Technology*, 41: 5114–5119.

Yantasee, W., Warner, C.L., Sangvanich, T., Addleman, R.S., Carter, T.G., Wiacek, R.J., Fryxell, G.E., Timchalk, C., and Warner, M.G. 2007. Removal of heavy metals from aqueous systems with thiol functionalized superparamagnetic nanoparticles. *Environmental Science and Technology*, 41: 5114–5119.

Yu, X., Tong, S., Ge, M., Zuo, J., Cao, C., and Song, W. 2013. One-step synthesis of magnetic composites of cellulose@iron oxide nanoparticles for arsenic removal. *Journal of Materials Chemistry A*, 1: 959–965.

Yuan, P., Fan, M., Yang, D., He, H., Liu, D., Yuan, A., Zhu, J., and Chen, T. 2009. Montmorillonite-supported magnetite nanoparticles for the removal of hexavalent chromium [Cr (VI)] from aqueous solutions. *Journal of Hazardous Materials*, 166(2): 821–829.

Zhang, S., Cheng, F., Tao, Z., Gao, F., and Chen, J. 2006. Removal of nickel ions from wastewater by $Mg(OH)_2/MgO$ nanostructures embedded in Al_2O_3 membranes. *Journal of Alloys and Compounds*, 426: 281–285.

Zhang, S., Zhang, Y., Liu, J., Xu, Q., Xiao, H., Wang, X., Xu, H., and Zhou, J. 2013. Thiol modified $Fe_3O_4@SiO_2$ as a robust, high effective, and recycling magnetic sorbent for mercury removal. *Chemical Engineering Journal*, 226: 30–38.

Zheng, D., Hu, C., Gan, T., Dang, X., and Hu, S. 2010. Preparation and application of a novel vanillin sensor based on biosynthesis of Au–Ag alloy nanoparticles. *Sensors and Actuators B: Chemical*, 148: 247–252.

Zhou, Y.T., Nie, H.L., Branford-White, C., He, Z.Y., and Zhu, L.M. 2009. Removal of Cu^{2+} from aqueous solution by chitosan-coated magnetic nanoparticles modified with α-ketoglutaric acid. *Journal of Colloid and Interface Science*, 330: 29–37.

Zhu, J., Wei, S., Gu, H., Rapole, S.B., Wang, Q., Luo, Z., Haldolaarachchige, N., Young, D.P., and Guo, Z. 2011. One-pot synthesis of magnetic graphene nanocomposites decorated with core@double-shell nanoparticles for fast chromium removal. *Environmental Science and Technology*, 46: 977–985.

Zhu, Y., Zhao, Q., Li, Y., Cai, X., and Li, W. 2006. The interaction and toxicity of multi-walled carbon nanotubes with *Stylonychiamytilus*. *Journal of Nanoscience and Nanotechnology*, 6: 1357–1364.

25 Fungi-Mediated Biosynthesis of Nanoparticles and Application in Metal Sequestration

Madhurima Bakshi, Shouvik Mahanty, and Punarbasu Chaudhuri

CONTENTS

ABSTRACT

The science of nanotechnology, dealing with particles ranging from 1 to 100 nm, is composed of multidisciplinary approaches involving physics, chemistry, biology, material science, and medicine. Growing research in this field has stressed on different synthesis procedures for nanomaterials and their application potential. Green synthesis of nanoparticles mostly consists of bioreduction processes using plants or microbes. Fungi can act as a good bioreductant to synthesize metal nanoparticles both intracellularly and extracellularly, mostly due to the proteins, organic acids, enzyme hydrogenase, and nitrate-dependent reductase released by them. Some common fungi successfully employed for the biosynthesis of nanoparticles are *Aspergillus* sp., *Cladosporium* sp., *Fusarium* sp., *Trichothecium* sp., *Penicillium* sp., and *Trichoderma* sp. mostly from the phyla Ascomycetes, Basidiomycetes, and Phycomycetes. Exclusive properties of nanoparticles offer us a wide range of applications in fields of medicine to environmental science, including environmental remediation, hazardous waste management, and metal sequestration. Nanomaterials with a high surface-to-volume ratio have been explored to detect and treat pollutants in various environmental matrixes like wastewater, soil, and sediment. Several batch and column experiments incorporating nanomaterials (nanoscale zerovalent iron, iron complexes, nanopolymers, etc.) have been successfully carried out for the sequestration of Ag, As, Cd, Co, Cr, Cu, H$_2$S, Hg, Ni, Pb, Zn, chlorinated pollutants, etc., by means of adsorption, oxidation–reduction, surface complexation, and other mechanisms. However, most of the nanomaterials utilized in sequestration methods have been synthesized chemically, triggering risk to human and ecological health. Fungi-mediated biosynthesized nanomaterials can be a good alternative for this purpose, being environment friendly for both the synthesis and sequestration part.

25.1 INTRODUCTION

The science of nanotechnology is composed of multidisciplinary aspects including physics, chemistry, biology, material science, and medicine dealing with particles of size ranging from 1 to 100 nm (Uskokovic 2008; Rai et al. 2009b; Narayanan and Sakthivel 2010). Nanotechnology also refers to the ability of designing, characterization, production, and application of the structures at nanometer scale (Mansoori 2005).

Properties of matter are significantly different in the nanoscale level. The unique properties of nanoparticles generate predominantly owing to features like large surface atom, large surface energy, quantum confinement, and reduced imperfections (Fendler 1992; Alivisatos 1996; Narayanan and Sakthivel 2010). Small nanoparticles are much more effective than their larger counterparts because of their characteristic surface plasmon resonance (SPR), enhanced Rayleigh scattering, and surface-enhanced Raman scattering in metal nanoparticles, and properties like quantum size effect in semiconductors and supermagnetism in magnetic materials (Wong and Schwaneberg 2003; Ramanaviciusa et al. 2005). The striking characteristics of nanoparticles have been applied in the fields of optoelectronics, catalysis, photoelectrochemistry, medicine, textile, energy saving, environment, magnetic appliances, cosmetics, and many more (Wang and Herron 1991; Hoffman et al. 1992; Schmid 1992; Colvin et al. 1994; Symonds 1995; Sastry et al. 2003; Ramteke et al. 2013).

A growing need of research in this field has been stressed on different synthesis procedures of nanomaterials along with their characterization and application potential. The conventional synthesis procedure of nanoparticles includes physical and chemical methods, which mostly require high pressure, energy, and temperature, and toxic chemicals and solvents, thus restricting their application in sensitive areas like medical fields (Narayanan and Sakthivel 2010; Gnanadesigan et al. 2012). The alternative green synthesis of nanoparticles involves the use of nontoxic chemicals, environmentally benign solvents, and renewable materials (Vigneshwaran 2006), which are able to produce biocompatible nanoparticles (Iravani 2011) in a cost-effective way (Mani et al. 2013). The basic approach behind these methods is to use plants and microbes (both bacteria and fungi) as reducing agents for the formation of noble metal nanoparticles, and the method is thus popularly known as bioreduction or green synthesis (Ahmad et al. 2003; Shankar et al. 2004; Balaji et al. 2009; Kathiresan et al. 2009; Nanda and Saravanan 2009; Nabikhan et al. 2010; Ponarulselvam et al. 2012). A combination of appropriate strain selection and proper optimization of the incubation parameters (pH, temperature, time, concentration of metal ions, and amount of biological material) can overcome the present constraints of biosynthesis of nanomaterials such as time-consuming culture method; lack of control over

size distribution, shape, crystallinity, and monodispersity; and slow rate of production. Particles like gold, silver, and platinum have drawn considerable technological and scientific attention (Mukherjee et al. 2001a; Ahmad et al. 2003; Ahmad et al. 2005; Gericke and Pinches 2006a; Riddin et al. 2006; Gericke and Pinches 2006b; Vigneshwaran et al. 2007; Afreen et al. 2011; Singh et al. 2013.)

Nanoscience is rapidly being utilized in environmental remediation and hazardous waste management vis-à-vis metal sequestration (Li and Zhang 2006). It has been explored to treat various environmental matrixes like polluted water, soil, sediment, or other contaminated environmental sites (Karn et al. 2009; Crane and Scott 2012). The limitation of other conventional sequestration methods involves relatively high time duration, high surface area adsorbents such as carbon black and absorbing polymers, metal specificity, lesser yield, higher cost, and lack of practical implication (Garrett and Prasad 2004; Thayer 2005; McEleney et al. 2006; Macdonald et al. 2007). On the contrary, the rapid, useful, low-cost, and direct approach of this alternative metal sequestration method using nanomaterials for removing catalytic metal ions from polluted aqueous solutions of organic reaction products acquires major significance. The role of nanoparticles in metal sequestration and the brief mechanisms are depicted in Figure 25.1.

25.2 FUNGI-MEDIATED BIOSYNTHESIS OF NANOPARTICLES

The interaction between organic biological resource and inorganic metal salts has catalyzed the process to explore alternatives of physical and chemical methods of nanoparticle synthesis. Green synthesis of nanoparticles involving microorganisms successfully interconnects nanotechnology and microbial biotechnology. Biosynthesis of gold, silver,

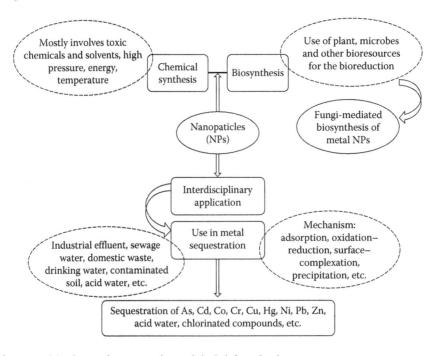

FIGURE 25.1 Role of nanoparticles in metal sequestration and the brief mechanisms.

gold–silver alloy, selenium, tellurium, platinum, palladium, silica, titania, zirconia, quantum dots, magnetite, and uraninite nanoparticles by several plant, bacteria, fungi, yeasts, and viruses have been established following this technology. Proteins, organic acids, and polysaccharides released by these biological resources acted as the key of the bioreduction method leading toward sustainable production of nanoparticles.

25.2.1 Importance of Fungi as Mediator of Green Synthesis

Mycology conjugating with nanotechnology has built up a new approach known as myconanotechnology (Rai et al. 2009a). Fungi play an essential role in producing nanoparticles by reducing the metal to its insoluble complexes such as metal sulfides in colloidal particle form (Mehra and Winge 1991). Compared to the other microorganisms, fungi are more promising in terms of their bioreduction potential. Fungi have a wide range of diversity and can be isolated, cultured, and maintained easily without much sophisticated instrumentation. They can secrete a high amount of proteins and other enzymes, which enhances the process of bioreduction triggering the synthesis of biocompatible nanoparticles. Fungi are also able to hydrolyze metal ion easily. In the bioreduction method, fungal mycelium gets exposed to salts of metal ion. The exclusive mycelial structures of fungi are capable of sustaining varied physicochemical conditions in the bioreduction process (Sastry et al. 2003; Mandal et al. 2006; Gade et al. 2008; Narayanan and Sakthivel 2010). Proteins, organic acids, and polysaccharides released by the fungi are believed to be able to differentiate different crystal shapes and directed to growth into extended spherical crystals. The resultant inorganic nanoparticles synthesized by the catalytic bioreduction process involving fungi can thus lead to a contemporary and viable approach for nanomaterial synthesis. The reducing enzymes secreted from fungi play the vital role in the process. The enzyme hydrogenase in fungi, for example, *Fusarium oxysporum*, *Trichoderma reesei*, and *Trichoderma viridie*, is found in cell suspension grown aerobically or anaerobically in a medium made up of glucose and nitrate salts (Ahmad et al. 2003; Rautio et al. 2006). Nitrate-dependent reductase present in the fungi has been recognized as the main mediator of the bioreduction progression along with a shuttle quinine extracellular process (Ahmad et al. 2003).

Many fungi bearing these properties are capable of reducing Ag(I) or Au(III) to respective metal nanoparticles (Lloyd 2003). It is observed that exposure of an aqueous solution of metal salts to various fungi has produced nanoparticles of sizes even less than 10 nm (Mukherjee et al. 2002).

25.2.2 Different Fungi Strains Used in Green Synthesis of Nanoparticles

In the last two decades, fungi such as *Aspergillus fumigatus*, *Aspergillus niger*, *Coriolus versicolor*, *Colletotrichum* sp., *Cladosporium cladosporioides*, *F. oxysporum*, *Fusarium semitectum*, *Fusarium solani*, *Trichothecium* sp., *Phaenerochaete*

chrysosporium, *Phoma glomerata*, *Penicillium brevicompactum*, *Penicillium fellutanum*, *Trichoderma asperellum*, *Trichoderma viride*, and *Volvariella volvaceae* have been explored for both intracellular and extracellular nanoparticles synthesis. Synthesis of metal nanoparticles (gold and silver) has been reported by many authors on the surface and on the cytoplasmic membrane of fungal mycelium (Pighi et al. 1989; Mukherjee et al. 2001a; Chen et al. 2003; Ahmad et al. 2005; Gericke and Pinches 2006a,b; Vigneshwaran et al. 2007; Singh et al. 2013). The shape of the particles ranges from hexagonal to spherical and size less than 10–25 nm. The extracellular synthesis of nanoparticles employing fungi has captured exclusive attention of researchers because of the easy purification process of the particles as they do not have any cellular attachment and hence are easy to be applied in different fields without much effort. The common phyla of fungi, which have been successfully employed for the green synthesis of nanoparticles, mostly include Ascomycetes, Basidiomycetes, and Phycomycetes.

The different species of fungi from different phyla used to produce biocompatible metal nanoparticles are given in Table 25.1.

25.3 CHARACTERIZATION OF BIOSYNTHESIZED NANOPARTICLES

Nanoparticles conceptualized and designed for several multidisciplinary applications have an extensive range of materials, varied size, diverse surface morphology, unique physicochemical properties, and wide biological activity, which necessitate their detailed characterization study. This includes the study of both their morphological and chemical characteristics as well as their toxicity analysis (Hall et al. 2007) incorporating the methods of their synthesis through inorganic and organic pathways (Utsunomiya and Ewing 2003). The following techniques are used for the general characterization study of nanoparticles.

25.3.1 Scattering Techniques

An appropriate scattering technique or an amalgamation of different techniques like static and dynamic light scattering, small-angle x-ray scattering, wide-angle x-ray diffraction, and small-angle neutron scattering is frequently employed for the estimation of shape, internal structure, interparticle interactions, molecular weight, size distribution, and hydrodynamic radius of the nanoscale particles (Chu and Liu 2000).

25.3.2 Microscopic Technique

Some of the important microscopic techniques used to characterize nanoparticles are scanning electron microscopy (SEM), field emission scanning electron microscopy, transmission electron microscopy, atomic force microscopy (AFM), scanning transmission electron microscopy, energy-dispersive x-ray spectrometry, electron energy loss spectrometry, scanning tunneling microscopy, etc. (Utsunomiya and Ewing 2003).

TABLE 25.1

Fungi Used in the Biosynthesis of Nanoparticles

Fungi	Phylum	Type of NPs	Size (nm)	Location	Morphology	Reference
Aspergillus fumigatus	Ascomycetes	Ag	5–25	Extracellular	Spherical, triangular	Bhainsa and D'Souza (2006)
Agaricus bisporus	Basidiomycetes	Ag	5–50	Extracellular	Spherical	Sujatha et al. (2013)
Aspergillus flavus	Ascomycetes	Ag	8.92	Cell wall	—	Vigneshwaran et al. (2007)
Aspergillus niger	Ascomycetes	Ag	20	Extracellular	Spherical	Gade et al. (2008)
Aspergillus oryzae	Ascomycetes	Fe	10–24.6	Intracellular	—	Tarafdar and Raliya (2013)
Calocybe indica	Basidiomycetes	Ag	5–50	Extracellular	Spherical	Sujatha et al. (2013)
Cladosporium cladosporioides	Ascomycetes	Ag	10–100	Extracellular	Spherical	Balaji et al. (2009)
Colletotrichum sp.	Ascomycetes	Au	20–40	Extracellular	Decahedral and icosahedral	Shankar et al. (2003)
Coriolus versicolor	Basidiomycetes	Ag	25–75	Extracellular	Spherical	Sanghi and Verma (2009)
Fusarium acuminatum	Ascomycetes	Ag	5–40	Extracellular	Spherical	Ingle et al. (2008)
Fusarium oxysporum	Ascomycetes	Ag	5–50	Extracellular	—	Senapati et al. (2004)
F. oxysporum	Ascomycetes	Au–Ag	8–14	Extracellular	—	Senapati et al. (2005)
F. oxysporum	Ascomycetes	Si	5–15	Extracellular	Quasispherical	Bansal et al. (2005)
F. oxysporum	Ascomycetes	Ti	6–13	Extracellular	Spherical	Bansal et al. (2005)
F. oxysporum	Ascomycetes	Zr	3–11	Extracellular	Quasispherical	Bansal et al. (2004)
F. oxysporum	Ascomycetes	Magnetite	20–50	Extracellular	Quasispherical	Bharde et al. (2006)
F. oxysporum	Ascomycetes	CdSe	9–15	Extracellular	Spherical	Kumar et al. (2007)
F. oxysporum	Ascomycetes	BaTi$_3$	4 ± 1	Extracellular	Quasispherical	Bansal et al. (2006)
F. oxysporum	Ascomycetes	Bi$_2$O$_3$	5–8	Extracellular	Quasispherical	Uddin et al. (2008)
F. oxysporum	Ascomycetes	Pt	10–50	Extracellular	Triangle, hexagons, square, rectangles	Riddin et al. (2006)
F. oxysporum	Ascomycetes	SrCO$_3$	—	Extracellular	Needle shaped	Rautaray et al. (2004)
Fusarium semitectum	Ascomycetes	Ag	10–60	Extracellular	Spherical	Basavaraja et al. (2008)
Fusarium solani	Ascomycetes	Ag	16.23	Extracellular	Spherical	Ingle et al. (2009)
F. oxysporum	Ascomycetes	Au	20–40	Extracellular	Spherical, triangular	Mukherjee et al. (2002)
F. oxysporum	Ascomycetes	Au	100	Intracellular	Spherical	Gericke and Pinches (2006a)
Hormoconis resinae	Ascomycetes	Ag	20–80	Extracellular	Spherical with some nanotriangles	Varshney et al. (2009)
Lentinus edodes	Basidiomycetes	Au	5–50	Extracellular	Spherical	Vetchinkina et al. (2014)
Pleurotus platyus	Basidiomycetes	Ag	5–50	Extracellular	Spherical	Sujatha et al. (2013)
Penicillium brevicompactum	Ascomycetes	Ag	58.35 ± 17.88	Extracellular	—	Shaligram et al. (2009)
Penicillium fellutanum	Ascomycetes	Ag	5–25	Extracellular	Spherical	Kathiresan et al. (2009)
Penicillium sp.	Ascomycetes	Ag	25	—	Spherical	Singh et al. (2013)
Phaenerochaete chrysosporium	Basidiomycetes	Ag	5–200	Extracellular	Pyramidal	Vigneshwaran et al. (2006)
Phoma glomerata	Ascomycetes	Ag	60–80	Extracellular	Spherical	Birla et al. (2009)
Pichia jadinii	Ascomycetes	Au	100	Intracellular	Spherical	Gericke and Pinches (2006a)
Pleurots florida	Basidiomycetes	Ag	5–50	Extracellular	Spherical	Sujatha et al. (2013)
Rhizopus stolonifer	Phycomycetes	Ag	5–50	Extracellular	Spherical	Afreen et al. (2011)
Trichoderma viride	Ascomycetes	Ag	5–40	Extracellular	Spherical, rodlike	Fayaz et al. (2010)

(Continued)

TABLE 25.1 (*Continued*)
Fungi Used in the Biosynthesis of Nanoparticles

Fungi	Phylum	Type of NPs	Size (nm)	Location	Morphology	Reference
Trichoderma asperellum	Ascomycetes	Ag	13–18	Extracellular	—	Mukherjee et al. (2008)
T. asperellum	Ascomycetes	Ag	13–18	Extracellular	—	Mukherjee et al. (2008)
Trichothecium sp.	Ascomycetes	Au	5–200	Extracellular	Triangle, hexagonal	Ahmad et al. (2005)
Tricoderma sp.	Ascomycetes	Ag	8–60	Extracellular	Spherical	Devi et al. (2013)
Verticillium dahliae	Ascomycetes	Au	<10	Intracellular	Spherical	Gericke and Pinches (2006a)
Verticillium luteoalbum DSM 63545	Ascomycetes	Au	100	Intracellular	Spherical	Gericke and Pinches (2006a)
Verticillium sp.	Ascomycetes	Ag	2–30	Intracellular	Spherical	Mukherjee et al. (2001a)
Verticillium sp.	Ascomycetes	Au	20 ± 8	Cell wall, cytoplasmic membrane	Spherical, quasi-hexagonal	Mukherjee et al. (2001b)
Verticillium sp.	Ascomycetes	Magnetite	100–400	Extracellular	Cubo-octohedral	Bharde et al. (2006)
Volvariella volvacea	Basidiomycetes	Au, Ag, Au–Ag	20–150	Extracellular	Spherical, hexagonal	Philip (2009)

The additional significant characterization study to determine the shape, stability, and nature of the particles involves the use of UV–Vis spectrophotometer (the characteristic absorption maximum should correspond with the SPR of the specific metal) (Mulvaney 1996), zeta potential (to measure the stability), x-ray diffraction (to study their phase structure and exact material identification), Fourier transform infrared spectroscopy (to find out the binding properties of biosynthesized NPs), thermogravimetric analyzer (to investigate the weight loss of the surface capped NPs), differential scanning calorimeter (to study the thermal behavior of NP), inductively coupled plasma mass spectrometry (for the identification of the different particles based on their elemental distribution and their concentration), and x-ray photoelectron spectroscopy (to determine the elemental composition of the surface and uniformity of elemental composition). Along with these, an inclusive characterization approach for nanomaterials also include physicochemical characterization, sterility assessment, bioavailability and biodistribution (absorption, distribution, metabolism, and excretion), and toxicity assessment, which requires both *in vitro* and *in vivo* studies (Hall et al. 2007).

25.4 MULTIDISCIPLINARY USES OF NANOPARTICLES

Metal nanoparticles are presently getting significant attention because of their various applications in the field of optoelectronics, catalysis, plasmonics, biological sensor, and pharmaceutical applications. Exclusive properties of nanoparticles mostly controlled by their composition, size, and shape offer us a wide range of applications in multidisciplinary fields like medicine to environmental science, thus boosting development of science and technology (Donaldson et al. 2004).

25.4.1 APPLICATION OF NANOPARTICLES IN DIFFERENT FIELDS

For example, single-walled carbon nanotubes (SWNTs) are utilized as probe tips in AFM for imaging. Nanotubes are efficient as a probe in SEM as well because of small diameter of the tubes and stiffness (Hafner et al. 2001). SWNTs bearing biomolecules are for molecular recognition while attached in AFM tips for studying chemical forces between molecules (Hafner et al. 2001). Nanofiber scaffold can be exploited for generating cells of central nervous system and other organs. Experiments on a hamster provided evidence of generation of axonomal tissue by peptides of nanofiber (Ellis-Behnke et al. 2006). Some nanofibers possess antimicrobial properties whose application can cause 90% death of *E. coli*, viruses, and other bacterial species (Koper et al. 2002; Bosi et al. 2003). Nanoparticles of silver and titanium dioxide (less than 100 nm) are used as coating mask in surgical purpose (Li et al. 2006). Nanotubes are the suitable candidate acting as channels for high selective transport of ions and solutes in a specific solution, which are present on both sides of a membrane (Jirage et al. 1997). For example, nanotubes having less than 1 nm diameter can separate small molecule on the basis of molecular size, while nanotubes having a diameter of 20–60 nm can be utilized for separating proteins (Martin and Kohli 2003). Nanoparticles possess the property to penetrate a huge number of organs contributing to its toxicity; however, this particular property can be used in nanomedicine. Nanospheres possessing biodegradable polymers can carry drugs, so when the membrane degrades by cellular mechanism the drugs can be released (Uhrich et al. 1999). These nanospheres degrading in acidic environment such as in tumor cells or any inflammation sites allow site-specific drug delivery. Nanoparticles bearing specific function can be exploited to permeate a cell membrane more effectively than nanoparticles not bearing a specific function (Maïté et al. 2000). Nanosphere possessing silicon labeled with cationic group such as ammonium on the outer surface can bind to DNA, which is a polyanion, through electrostatic interaction, thus delivering DNA into the cells (Kneuer et al. 2000). *Noninvasive* techniques are used in medical technologies. For example, dextrans packed superparamagnetic magnetite particles are utilized in image-enhancement in magnetic resonance imaging (Harisinghani et al. 2003). Intercellular imaging also can be obtained with precision through quantum dots attachment, allowing the observation of intercellular processes. Nanospheres bearing antigen-coated polystyrene can be used as a vaccine carrier targeting the dendritic cells, although the whole theme of this research is still in the developmental stage (Matsusaki et al. 2005). Detecting of nucleic acid sequences, which is unique for viruses and bacteria or specific proteins, and convulsant to a cancer cell, can be achieved by targeting and identifying these novel nucleic acids or proteins (Rosi and Mirkin 2005). Nanoparticles are also effective detection assays. Polymerase chain reaction with molecular fluorophore is an effective way for the detection of nucleic acid, but it also possesses a few drawbacks like complexity, contaminant sensitivity, and cost (Rosi and Mirkin 2005). Enzyme-linked immunosorbent assay is utilized in detection of protein concentration in patients. Nanotubes of different function and recognition capacities can be utilized for the removal of specific molecular solutes or particles (Martin and Kohli 2003).

25.4.2 APPLICATION IN METAL SEQUESTRATION

Nanoparticles play a major role in the field of environmental problems where a specific contaminant within a mixture of materials may affect the environment by causing pollution. Nanoparticles with a very high surface-to-volume ratio can detect not only the pollutants but also act as an acute treatment system by means of their highly active interactions. Several batch and column experiments have been successfully carried out incorporating nanomaterials for metal sequestration by means of adsorption, oxidation–reduction, surface complexation, and other mechanisms. Nanoscale zero valent iron (nZVI), iron complexes, and nanopolymers have been effectively exploited for the sequestration of Ag, As, Cd, Co, Cr, Cu, H_2S, Hg, Ni, Pb, Zn, chlorinated pollutants, etc.

nZVI having a core–shell structure exhibits characteristics of both hydrous iron oxides (i.e., as a sorbent) and metallic iron (i.e., as a reductant) (Li and Zhang 2006). They are capable of removing major water contaminants like arsenite, lead, and copper by inducing oxidation and reduction of As(III) and adsorption, coprecipitation, reduction of Pb(II), and sorption of Cu^{2+} primarily via a redox mechanism that results in the formation of Cu_2O and Cu^0 (Karabelli et al. 2008; Yan et al. 2012a; Zhang et al. 2013) and is used for the treatment of the real acid water system containing mixture of pollutants (Klimkova et al. 2011). Chemically synthesized gold nanoparticles capped with 4-aminothiophenol have been

reported to be used for the sequestration of toxic heavy metals from processed water (Chauhan et al. 2011). Iron nanoparticles of different states have been extensively studied and used for heavy metal remediation. Stable and nonvolatile complexes of iron nanoparticles and sulfide compounds have been successfully synthesized, which were applied to degrade harmful organic pollutants and heavy metals such as PCBs, chlorinated pesticides, mercury, and lead (Li et al. 2007). Superparamagnetic iron oxide (Fe_3O_4) nanoparticles with a surface-functionalized dimercaptosuccinic acid (DMSA) were used as an effective sorbent material for toxic metals such as Hg, Ag, Pb, Cd, and Tl, (which effectively bind to the DMSA ligands) and As (which binds to the iron oxide lattices) (Yantasee et al. 2007), whereas polyrhodanine-coated c-Fe_2O_3 nanoparticles and nZVI were utilized for the removal of heavy metal ions from an aqueous solutions (Song et al. 2011; Li and Zhang 2007). Nanoparticulate akaganeite, goethite, ferrihydrite, schwertmannite, and hematite present in soils, sediments, and mine drainage outflows have high sorption capacities for metal and anionic contaminants such as arsenic, chromium, lead, mercury, and selenium (Waychunas et al. 2005). Kaolinite-supported zerovalent iron nanoparticles (kaolinite–nZVI) synthesized using liquid-phase reduction under ambient atmosphere have been applied to adsorb Ni(II) from an aqueous solution (Wang et al. 2014). Polymer-coated magnetic iron oxide nanoparticles formed using coprecipitation method are capable of removing heavy metal Cr(VI) and dye (alizarin) from the water of different concentrations (Hanif and Shahzad 2014).

Chemically synthesized surface-functionalized polymer nanoparticles have the potential for selective removal of heavy metal ions from water at below ppm levels (Bell et al. 2006). Magnetic biochar composites synthesized using low-cost bioresource, low-cost pine bark waste, and $CoFe_2O_4$ have been successfully applied for the adsorption of Ni(II) from an aqueous solution (Reddy and Lee 2014). Heavy metal adsorption can also be mediated by using chemically produced nitrogen-doped magnetic carbon nanoparticles (Shin et al. 2011). The various metal sequestrations (Ag, As, Cd, Co, Cu, Hg, Zn, Ni, Pb) from several contaminated sites and systems using nanomaterials have been described in Table 25.2.

25.5 CONCLUSION

Extensive use of nanomaterials in metal sequestration and environmental remediation is a promising example of a rapidly developing technology with substantial societal benefits. Growing researches in this field have already overcome many uncertainties about the fundamental features of this technology, thus gradually making it suitable for technological and industrial applications. However, most of the nanomaterials utilized in various sequestration methods have been synthesized chemically using several toxic chemicals and thus increasing the risk to human and ecological health. The development of green synthesis of nanomaterials especially using fungi can be a good alternative for this

TABLE 25.2

Different Nanoparticles Used for Different Metal Sequestration

Metal	Type of NPs	Application/Mechanism	Reference
Acid water	Nanoscale zerovalent iron (nZVI)	nZVI were applied for the treatment of real acid water system containing miscellaneous mixture of pollutants, where the various removal mechanisms occur simultaneously.	Klimkova et al. (2011)
Ag	Nanoscale zerovalent iron (nZVI)	The removal mechanism of Ag(I) is predominantly reduction.	Li and Zhang (2007)
As	Nanosized hydrated ferric oxide (HFO)-loaded polymer sorbents	Arsenate removal from aqueous media.	Zhang et al. (2008)
As	Magnetite	Arsenite is sequestrated via surface adsorption and surface precipitation reactions.	Wang et al. (2008)
As	Nanoscale zerovalent iron (nZVI)	nZVI induces oxidation and reduction of As(III).	Yan et al. (2012b)
As	Magnetite nanoparticles	Removal of As(V) through adsorption and precipitation method.	Wang et al. (2011)
As	Nanoscale zerovalent iron (nZVI)	As(III) species underwent two stages of transformation upon adsorption at the nZVI surface, which involves breaking of As–O bonds at the particle surface, and further reduction and diffusion of arsenic across the thin oxide layer enclosing the nanoparticles, resulting in arsenic forming an intermetallic phase with the Fe(0) core.	Yan et al. (2012a)
As	Nanoscale zerovalent iron (nZVI)	Arsenite oxidation and reduction occurring parallelly at different subdomains of the nanoparticles owing to the particle's core–shell structure.	Yan et al. (2010b)
As	Nanoscale zerovalent iron	Reactions of nZVI with As(III) generated As(0), As(III), and As(V) on the nanoparticle surfaces, indicating both reduction and oxidation of As(III).	Ramos et al. (2009)
Cd	Nanostructured hydrous titanium(IV) oxide	Scavenging of Cd(II) from contaminated water.	Debnath et al. (2011)
Cd	Nano-hydroxyapatite particles	Surface complexation and intraparticle diffusion account for Cd sequestration from water and sediment.	Zhang et al. (2010)
Cd	Nanoscale zerovalent iron (nZVI)	Cd(II) removed via sorption/surface complex formation.	Li and Zhang (2007)
Cd	Nanoscale zerovalent iron (nZVI)	Removal of Cd(II) by adsorption onto nano-zerovalent iron particles.	Boparai et al. (2011)
Cd	Silver nanoparticle	Adsorption of Cd(II) onto Ag nanoparticles used as a water contaminant removal agent.	Alqudami et al. (2012)
Cd	Fe nanoparticle	Adsorption of Cd(II) onto Fe nanoparticles used as a water contaminant removal agent.	Alqudami et al. (2012)
Cd	Titanate ($Na_2Ti_3O_7$) nanotubes	Nanotubes acted as an effective sorbent to remove cadmium ions from water.	Dua et al. (2011)
Cd	Nanoscale zerovalent iron (nZVI)	Cd(II) is sequestrated from polluted water within nZVI by adsorption or surface complex formation with no apparent reduction of Cd(II).	Zhang et al. (2014)
Chlorinated solvents	Nanoscale zerovalent iron (nZVI)	Iron reduces chlorinated compounds by reductive dehalogenation while being oxidized.	O'Carroll et al. (2013)
Co	Nanoscale zerovalent iron supported by kaolinite clay (nZVI-kaol).	Co^{2+} was mainly fixed by the oxyhydroxyl groups of iron nanoparticles.	Üzüm et al. (2009)
Cr	Nanoparticle polyelectrolyte complexes	Entrapped Cr(III) while maintaining colloidal stability in water or gelant.	Cordova et al. (2008)
Cr	Chitosan–Fe^0 nanoparticles	*In situ* reductive removal of Cr(VI) from water by both physical adsorption of Cr(VI) onto the chitosan–Fe^0 surface and subsequent reduction of Cr(VI) to Cr(III).	Li et al. (2009)
Cr	Chitosan–Fe^0 nanoparticles	Physical adsorption of Cr(VI) onto the chitosan–Fe^0 surface and subsequent reduction of Cr(VI) to Cr(III).	Geng et al. (2009a)
Cr	Chitosan-stabilized Fe^0 nanoparticles	Cr(VI) was reduced to Cr(III) and Fe(III) was the only component present on the Fe^0 nanoparticles surface.	Geng et al. (2009b)
Cr	Nanoscale zerovalent iron (nZVI)	*In situ* reductive immobilization of Cr(VI) in water and sandy loam soil.	Xu and Zhao (2007)
Cr	Nanoscale zerovalent iron (nZVI)	Reduction of Cr(VI) from groundwater samples collected from a contaminated site.	Singh et al. (2012)

(Continued)

TABLE 25.2 (*Continued*)
Different Nanoparticles Used for Different Metal Sequestration

Metal	Type of NPs	Application/Mechanism	Reference
Cr	Iron nanoparticles embedded in orange peel pith	Removal of Cr(VI) from aqueous solution using nanobiocomposite.	López-Téllez et al. (2011)
Cr	Nanoscale zerovalent iron supported on pumice	Cr(VI) from wastewater were removed by a rapid physical adsorption in the first 0.5 min and predominantly by reduction.	Liu et al. (2014)
Cr	Nanoscale zerovalent iron (nZVI)	Removal of Cr(VI) from the soil spiked with Cr(VI) via reduction.	Singh et al. (2011)
Cu	Nanoscale zerovalent iron (nZVI)	Cu^{2+} ions were sorbed primarily via a redox mechanism that resulted in the formation of Cu_2O and Cu^0.	Karabelli et al. (2008)
Cu	Magnetite nanoparticles functionalized with polyethylenimine	Selectively adsorb toxic-free cupric ions but not the less toxic EDTA complexed copper.	Goon et al. (2010)
Cu	Nanoscale zerovalent iron (nZVI)	The removal mechanism of Cu(II) is predominantly reduction.	Li and Zhang (2007)
Cu	Nanoscale zerovalent supported by kaolinite clay (nZVI-kaol).	Cu^{2+} ions were fixed by a redox mechanism, leading to the formation of Cu_2O and Cu^0.	Üzüm et al. (2009)
H_2S	Nanoscale zerovalent iron (nZVI)	H_2S was immobilized on the nZVI surface as disulfide (S_2^{2-}) and monosulfide (S^{2-}) species. The retention of hydrogen sulfide occurs via reactions with the oxide shell to form iron sulfide (FeS) and subsequent conversion to iron disulfide (FeS_2).	Yan et al. (2010a)
Hg	Gold nanoparticle supported on alumina	Removal of inorganic Hg from water.	Lisha et al. (2009)
Hg	Magnetic nanosized $(Fe_{3x}Mn_x)_{1-\delta}O_4$ particle	Gaseous elemental mercury capture from Flue Gas.	Yang et al. (2011)
Hg	Nanoscale zerovalent iron (nZVI)	The removal mechanism of Hg(II) is predominantly reduction.	Li and Zhang (2007)
Hg	Nanoscale zerovalent iron (nZVI)	Hg(II) from water was sequestrated via chemical reduction to elemental mercury.	Yan et al. (2010a)
Hg	Nanoscale zerovalent iron supported on pumice	Hg (II) from wastewater was removed by a rapid physical adsorption in the first 0.5 min and predominantly by reduction.	Liu et al. (2014)
Ni	Nanoscale zerovalent iron (nZVI)	Ni(II) is immobilized at the nanoparticle surface by both sorption and reduction.	Li and Zhang (2007)
Ni	Nanoscale zerovalent iron (nZVI)	Ni(II) in water gets removed by sorption and reduction.	Li and Zhang (2006)
Pb	Nanoscale zerovalent iron (nZVI)	Adsorption, coprecipitation, and reduction of Pb(II) by nZVI particles with $Fe(OH)_3$ shell.	Zhang et al. (2013)
Pb	Nano-hydroxyapatite particles	Dissolution–precipitation is the primary immobilization mechanism for Pb sequestration from water and sediment.	Zhang et al. (2010)
Pb	Nanoscale zerovalent iron (nZVI)	Pb(II) is immobilized at the nanoparticle surface by both sorption and reduction.	Li and Zhang (2007)
Pb	Iron nanoparticle	Adsorption of Pb(II) onto Fe nanoparticles used as a water contaminant removal agent.	Alqudami et al. (2012)
Zn	Hematite (Fe_2O_3) nanoparticles	Zn adsorption and the formation of precipitates over a range of Zn(II) surface loadings on hematite nanoparticles.	Ha et al. (2009)
Zn	Nanoscale zerovalent iron (nZVI)	Zn(II) removal mechanism is sorption/surface complex formation.	Li and Zhang (2007)
Zn	Nanoscale zerovalent iron (nZVI)	Zn(II) removal was achieved via sorption to the iron oxide shell followed by zinc hydroxide precipitation.	Yan et al. (2010a)

purpose serving as an environment-friendly approach for both the synthesis and sequestration part. Further research is being carried out to investigate the scope for real large-scale implementation, associated risk evaluation, and sustainability of this technology, which will lead us to a better world for sure.

ACKNOWLEDGMENT

The authors are thankful to the Department of Biotechnology, GoI, for providing financial support (BT/PR9465/NDB/39/360/2013).

REFERENCES

Afreen, V.R. and Ranganath, E., 2011. Synthesis of monodispersed silver nanoparticles by *Rhizopus stolonifer* and its antibacterial activity against MDR strains of *Pseudomonas aeruginosa* from burnt patients. *International Journal of Environmental Sciences* 1: 1582–1592.

Ahmad, A., Mukherjee, P., Senapati, S., Mandal, D., Khan, M.I., Kumar, R., and Sastry, M., 2003. Extracellular biosynthesis of silver nanoparticles using the fungus *Fusariumoxysporum*. *Colloids and Surfaces B: Biointerfaces* 27: 313–318.

Ahmad, A., Senapati, S., Khan, M.I., Kumar, R., and Sastry, M., 2005. Extra-/intracellular biosynthesis of gold nanoparticles by an alkalotolerant fungus, *Trichothecium* sp. *Journal of Biomedical Nanotechnology* 1: 47–53.

Alivisatos, A.P., 1996. Semiconductor clusters, nanocrystals, and quantum dots. *Science* 271: 933–937.

Alqudami, A., Alhemiary, N.A., and Munassar, S., 2012. Removal of Pb(II) and Cd(II) ions from water by Fe and Ag nanoparticles prepared using electro-exploding wire technique. *Environmental Science and Pollution Research* 19: 2832–2841.

Balaji, D.S., Basavaraja, S., Deshpande, R., Mahesh, D.B., Prabhakar, B.K., and Venkataraman, A., 2009. Extracellular biosynthesis of functionalized silver nanoparticles by strains of *Cladosporium cladosporioides* fungus. *Colloids and Surfaces B: Biointerfaces* 68: 88–92.

Bansal, V., Poddar, P., Ahmad, A., and Sastry, M., 2006. Room-temperature biosynthesis of ferroelectric barium titanate nanoparticles. *Journal of American Chemical Society* 128: 11958.

Bansal, V., Rautaray, D., Ahmad, A., and Sastry, M., 2004. Biosynthesis of zirconia nanoparticles using the fungus *Fusarium oxysporum*. *Journal of Materials Chemistry* 14: 3303–3305.

Bansal, V., Rautaray, D., Bharde, A., Ahire, K., Sanyal, A., Ahmad, A., and Sastry, M., 2005. Fungus-mediated biosynthesis of silica and titania particles. *Journal of Materials Chemistry* 15: 2583–2589.

Basavaraja, S., Balaji, S.D., Lagashetty, A., Rajasab, A.H., and Venkataraman, A., 2008. Extracellular biosynthesis of silver nanoparticles using the fungus *Fusarium semitectum*. *Material Research Bulletin* 43: 1164–1170.

Bell, C.A., Smith, S.V., Whittaker, M.R., Whittaker, A.K., Gahan, L.R., and Monteiro, M.J., 2006. Surface-functionalized polymer nanoparticles for selective sequestering of heavy metals. *Advanced Materials* 18: 582–586.

Bhainsa, K.C. and S. F. D'Souza. 2006. Extracellular biosynthesis of silver nanoparticles using the fungus *Aspergillus fumigatus*. *Colloids and Surfaces B: Biointerfaces* 47: 160–164.

Bharde, A., Rautaray, D., Bansal, V., Ahmad, A., Sarkar, I., Yusuf, S.M., Sanyal, M., and Sastry, M., 2006. Extracellular biosynthesis of magnetite using fungi. *Small* 1: 135–141.

Birla, S.S., Tiwari, V.V., Gade, A.K., Ingle, A.P., Yadav, A.P., and Rai, M.K., 2009. Fabrication of silver nanoparticles by *Phoma glomerata* and its combined effect against *Escherichia coli*, *Pseudomonas aeruginosa* and *Staphylococcus aureus*. *Letters in Applied Microbiology* 48: 173–179.

Boparai, H.K., Joseph, M., and O'Carroll D.M., 2011. Kinetics and thermodynamics of cadmium ion removal by adsorption onto nano zerovalent iron particles. *Journal of Hazardous Materials* 186: 458–465.

Bosi, S., da Ros, T., Spalluto, G., and Prato, M. 2003. Fullerene derivatives: An attractive tool for biological applications. *European Journal of Medicinal Chemistry* 38: 913–923.

Chauhan, N., Gupta, S., Singh, N., Singh, S., Islam, S.S., Sood, K.N., and Pasricha, R., 2011. Aligned nanogold assisted one step sensing and removal of heavy metal ions. *Journal of Colloid and Interface Science* 363: 42–50.

Chen, J.C., Lin, Z.H., and Ma, X.X., 2003. Evidence of the production of silver nanoparticles via pretreatment of *Phoma* sp. 3.2883 with silver nitrate. *Letters in Applied Microbiology* 37: 105–108.

Chu, B. and Liu, T. 2000. Characterisation of nanoparticles by scattering techniques. *Journal of Nanoparticle Research* 2: 29–41.

Colvin, V.L., Schlamp, M.C., and Alivisatos, A.P., 1994. Light-emitting diodes made from cadmium selenide nanocrystals and a semiconducting polymer. *Nature* 370: 354–357.

Cordova, M., Cheng, M., Trejo, J., Johnson, S.J., Willhite, G.P., Liang, J.T., and Berkland, C., 2008. Delayed HPAM gelation via transient sequestration of chromium in polyelectrolyte complex nanoparticles. *Macromolecules* 41: 4398–4404.

Crane, R.A. and Scott, T.B., 2012. Nanoscale zero-valent iron: Future prospects for an emerging water treatment technology. *Journal of Hazardous Materials* 211–212: 112–125.

Debnath, S., Nandi, D., and Ghosh, U.C., 2011. Adsorption desorption behavior of cadmium(ii) and copper(ii) on the surface of nanoparticle agglomerates of hydrous titanium(IV) oxide. *Journal of Chemical and Engineering Data* 56: 3021–3028.

Devi, T.P., Kulanthaivel, S., Kamil, D., Borah, J.L., Prabhakaran, N., and Srinivasa, N., 2013. Biosynthesis of silver nanoparticle from *Tricoderma* species. *Indian Journal of Experimental Biology* 51: 543–547.

Donaldson, K., Stone, V., Tran, C.L., Kreyling, W., and Borm, P.J., 2004. Nanotoxicology. *Occupational and Environmental Medicine* 61: 727–728.

Dua, A.J., Suna, D.D., and Leckie, J.O., 2011. Sequestration of cadmium ions using titanate nanotube. *Journal of Hazardous Materials* 187: 401–406.

Ellis-Behnke, R.G., Liang, Y.X., You, S.W., Tay, D.K., Zhang, S., So, K.F., and Schneider, G.E., 2006. Nano neuro knitting: Peptide nanofibers scaffold for brain repair and axon regeneration with functional return of vision. *Proceedings of the National Academy of Sciences of the United States of America* 103: 5054–5059.

Fayaz, A.M., Balaji, K., Girilal, M., Yadav, R., Kalaichelvan, P.T., and Venketesan, R., 2010. Biogenic synthesis of silver nanoparticles and their synergistic effect with antibiotics: A study against gram-positive and gram-negative bacteria. *Nanomedicine: Nanotechnology, Biology and Medicine* 6: 103–109.

Fendler, J.H. 1992. *Membrane Mimetic Chemistry Approach to Advanced Materials*, Springer-Verlag, Berlin, Germany.

Gade, A.K., Bonde, P., Ingle, A.P., Marcato, P.D., Duran, N., and Rai, M.K., 2008. Exploitation of *Aspergillus niger* for synthesis of silver nanoparticles. *Journal of Biobased Material and Bioenergy* 2: 243–247.

Garrett, C.E. and Prasad, K. 2004. The art of meeting palladium specifications in active pharmaceutical ingredients produced by Pd catalyzed reactions. *Advanced Synthesis and Catalysis* 346: 889–900.

Geng, B., Jin, Z., Li, T., and Qi, X., 2009a. Kinetics of hexavalent chromium removal from water by chitosan-Fe^0 nanoparticles. *Chemosphere* 75: 825–830.

Geng, B., Jin, Z., Li, T., and Qi, X., 2009b. Preparation of chitosan-stabilized Fe^0 nanoparticles for removal of hexavalent chromium in water. *Science of the Total Environment* 407: 4994–5000.

Gericke, M. and Pinches, A., 2006a. Biological synthesis of metal nanoparticles. *Hydrometallurgy* 83: 132–140.

Gericke, M. and Pinches, A. 2006b. Microbial production of gold nanoparticles. *Gold Bulletin* 39: 22–28.

Gnanadesigan, M., Anand, M., Ravikumar, S., Maruthupandy, M., Ali, M.S., Vijayakumar, V., and Kumaraguru, A.K., 2012. Antibacterial potential of biosynthesised silver nanoparticles using *Avicennia marina* mangrove plant. *Applied Nanoscience* 2: 143–147.

Goon, I.Y., Zhang, C., Lim, M., Gooding, J.J., and Amal, R., 2010. Controlled fabrication of polyethylenimine-functionalized magnetic nanoparticles for the sequestration and quantification of free Cu^{2+}. *Langmuir* 26: 12247–12252.

Ha, J., Trainor, T.P., Farges, F., and Brown Jr, G.E., 2009. Interaction of aqueous Zn(II) with hematite nanoparticles and microparticles. Part 1. EXAFS study of Zn(II) adsorption and precipitation. *Langmuir* 25: 5574–5585.

Hafner, J.H., Cheung, C.L., Woolley, A.T., and Lieber, C.M., 2001. Structural and functional imaging with carbon nanotube AFM probes. *Progress in Biophysics and Molecular Biology* 77: 73–110.

Hall, J.B., Dobrovolskaia, M.A., Patri, A.K., and McNeil, S.E., 2007. Characterization of nanoparticles for therapeutics. *Nanomedicine* 2: 789–803.

Hanif, S. and Shahzad, A. 2014. Removal of chromium (VI) and dye Alizarin Red S (ARS) using polymer-coated iron oxide (Fe₃O₄) magnetic nanoparticles by co-precipitation method. *Journal of Nanoparticle Research* 16: 2429.

Harisinghani, M.G., Barentsz, J., Hahn, P.F., Deserno, W.M., Tabatabaei, S., van de Kaa, C.H., de la Rosette, J., and Weissleder, R., 2003. Noninvasive detection of clinically occult lymph-node metastases in prostate cancer. *The New England Journal of Medicine* 19: 2491–2499.

Hoffman, A.J., Mills, G., Yee, H., and Hoffmann, M.R., 1992. Q-sized cadmium sulfide: Synthesis, characterization, and efficiency of photoinitiation of polymerization of several vinylic monomers. *The Journal of Physical Chemistry* 96: 5546–5552.

Ingle, A., Gade, A., Pierrat, S., Sonnichsen, C., and Rai, M., 2008. Mycosynthesis of silver nanoparticles using the fungus *Fusarium acuminatum* and its activity against some human pathogenic bacteria. *Current Nanoscience* 4: 141–144.

Ingle, A., Rai, M., Gade, A., and Bawaskar, M., 2009. *Fusarium solani*: A novel biological agent for the extracellular synthesis of silver nanoparticles. *Journal of Nanoparticle Research* 11: 2079–2085.

Iravani, S., 2011. Green synthesis of metal nanoparticles using plants. *Green Chemistry* 13: 2638–2650.

Jirage, K.B., Hulteen, J.C., and Martin, C.R., 1997. Nanotubule based molecular-filtration membranes. *Science* 278: 655–658.

Karabelli, D., Üzüm, C., Shahwan, T., Eroglu, A.E., Scott, T.B., Hallam, K.R., and Lieberwirth, I., 2008. Batch removal of aqueous Cu²⁺ ions using nanoparticles of zero-valent iron: A study of the capacity and mechanism of uptake. *Industrial & Engineering Chemistry Research* 47: 4758–4764.

Karn, B., Kuiken, T., and Otto, M., 2009. Nanotechnology and in situ remediation: A review of the benefits and potential risk. *Environmental Health Perspectives* 117: 1823–1831.

Kathiresan, K., Manivannan, S., Nabeel, M.A., and Dhivya, B., 2009. Studies on silver nanoparticles synthesized by a marine fungus, *Penicillium fellutanum* isolated from coastal mangrove sediment. *Colloids and Surfaces B: Biointerfaces* 71: 133–137.

Klimkova, S., Cernik, M., Lacinova, L., Filip, J., Jancik, D., and Zboril, R., 2011. Zero-valent iron nanoparticles in treatment of acid mine water from in situ uranium leaching. *Chemosphere* 82: 1178–1184.

Kneuer, C., Sameti, M., Bakowsky, U., Schiestel, T., Schirra, H., Schmidt, H., and Lehr, C.M., 2000. A nonviral DNA delivery system based on surface modified silica-nanoparticles can efficiently transfect cells in vitro. *Bioconjugate Chemistry* 11: 926–932.

Koper, O.B., Klabunde, J.S., Marchin, G.L., Klabunde, K.J., Stoimenov, P., and Bohra, L., 2002. Nanoscale powders and formulations with biocidal activity toward spores and vegetative cells of *Bacillus* species, viruses, and toxins. *Current Microbiology* 44: 49–55.

Kumar, S.A., Ansary, A.A., Ahmad, A., and Khan, M.I., 2007. Extracellular biosynthesis of CdSe quantum dots by the fungus, *Fusarium oxysporum*. *Journal of Biomedical Nanotechnology* 3: 190–194.

Li, T., Li, B., Geng, N., Zhang, Z.J., and Qi, X., 2009. Hexavalent chromium removal from water using chitosan-Fe⁰ nanoparticles. *Journal of Physics: Conference Series* 188: 012057.

Li, X., Brown, D.G., and Zhang, W., 2007. Stabilization of biosolids with nanoscale zero-valent iron (nZVI). *Journal of Nanoparticle Research* 9: 233–243.

Li, X. and Zhang, W., 2006. Iron nanoparticles: The core-shell structure and unique properties for Ni(II) sequestration. *Langmuir* 22: 4638–4642.

Li, X. and Zhang, W., 2007. Sequestration of metal cations with zerovalent iron nanoparticless a study with high resolution x-ray photoelectron spectroscopy (HR-XPS). *Journal of Physical Chemistry Letters C* 111: 6939–6946.

Li, Y., Leung, P., Yao, L., Song, Q.W., and Newton, E., 2006. Antimicrobial effect of surgical masks coated with nanoparticles. *Journal of Hospital Infection* 62: 58–63.

Lisha, K.P. and Anshup, T.P., 2009. Towards a practical solution for removing inorganic mercury from drinking water using gold nanoparticles. *Gold Bulletin* 42: 144–152.

Liu, T., Wang, Z., Yan, X., and Zhang, B., 2014. Removal of mercury (II) and chromium (VI) from wastewater using a new and effective composite: Pumice-supported nanoscale zero-valent iron. *Chemical Engineering Journal* 245: 34–40.

Lloyd, J.R., 2003. Microbial reduction of metals and radionuclides. *FEMS Microbiology Reviews* 27: 411–425.

López-Téllez, G., Barrera-Díaz, C.E., Balderas-Hernández, P., Roa-Morales, G., and Bilyeu, B., 2011. Removal of hexavalent chromium in aquatic solutions by iron nanoparticles embedded in orange peel pith. *Chemical Engineering Journal* 173: 480–485.

Macdonald, J.E., Kelly, J.A., and Veinot, J.G., 2007. Iron/iron oxide nanoparticle sequestration of catalytic metal impurities from aqueous media and organic reaction products. *Langmuir* 23: 9543–9545.

Maïté, L., Carlesso, N., Tung, C.H., Tang, X.W., Cory, D., Scadden, D.T., and Weissleder, R., 2000. Tat peptide derivatized magnetic nanoparticles allow in vivo tracking and recovery of progenitor cells. *Nature Biotechnology* 18(4): 410–414.

Mandal, D., Bolander, M.E., Mukhopadhyay, D., Sarkar, G., and Mukherjee, P., 2006. The use of microorganisms for the formation of metal nanoparticles and their application. *Applied Microbiology and Biotechnology* 69: 485–492.

Mani, U., Dhanasingh, S., Arunachalam, R., Paul, E., Shanmugam, P., Rose, C., and Mandal, A.B., 2013. A simple and green method for the synthesis of silver nanoparticles using *Ricinus communis* leaf extract. *Progress in Nanotechnology and Nanomaterials* 2: 21–25.

Mansoori, G.A., 2005. *Molecular-Based Study of Condensed Matter in Small Systems: Principles of Nanotechnology*, World Scientific Publishing Co., Hackensack, NJ.

Martin, C.R. and Kohli, P., 2003. The emerging field of nanotube biotechnology. *Nature Reviews Drug Discovery* 2: 29–37.

Matsusaki, M., Larsson, K., Akagi, T., Lindstedt, M., Akashi, M., and Borrebaeck, C.A., 2005. Nanosphere induced gene expression in human dendritic cells. *Nano Letters* 5: 2168–2173.

McEleney, K., Allen, D.P., Holliday, A.E., and Crudden, C.M., 2006. Functionalized mesoporous silicates for the removal of ruthenium from reaction mixtures. *Organic Letters* 8: 2663–2666.

Mehra, R.K. and Winge, D.R., 1991. Metal ion resistance in fungi: Molecular mechanisms and their regulated expression. *Journal of Cellular Biochemistry* 45: 30–40.

Mukherjee, P., Ahmad, A., Mandal, D., Senapati, S., Sainkar, S.R., Khan, M.I., Parishcha, R. et al., 2001a. Fungus-mediated synthesis of silver nanoparticles and their immobilization in the mycelial matrix: A novel biological approach to nanoparticle synthesis. *Nano Letters* 1: 515–519.

Mukherjee, P., Ahmad, A., Mandal, D., Senapati, S., Sainkar, S.R., Khan, M.I., Ramani, R. et al., 2001b. Bioreduction of AuCl$_{(4)}$$^{(-)}$ ions by the fungus, *Verticillium* sp. and surface trapping of the gold nanoparticles formed. *Angewandte Chemie International Edition* 40: 3585–3588.

Mukherjee, P., Roy, M., Mandal, B.P., Dey, G.K., Mukherjee, P.K., Ghatak, J., Tyagi, A.K., and Kale, S.P., 2008. Green synthesis of highly stabilized nanocrystalline silver particles by a non-pathogenic and agriculturally important fungus *T. asperellum*. *Nanotechnology* 19: 075103–075109.

Mukherjee, P., Senapati, S., Mandal, D., Ahmad, A., Khan, M.I., Kumar, R., and Sastry, M., 2002. Extracellular synthesis of gold nanoparticles by the fungus *Fusarium oxysporum*. *ChemBiochem* 3: 461–463.

Mulvaney, P., 1996. Surface plasmon spectroscopy of nanosized metal particles. *Langmuir* 12: 788–800.

Nabikhan, A., Kandasamy, K., Raj, A., and Alikunhi, N.M., 2010. Synthesis of antimicrobial silver nanoparticles by callus and leaf extracts from saltmarsh plant, *Sesuvium portulacastrum*, L. *Colloids and Surfaces B: Biointerfaces* 79: 488–493.

Nanda, A. and Saravanan, M., 2009. Biosynthesis of silver nanoparticles from *Staphylococcus aureus* and its antimicrobial activity against MRSA and MRSE. *Nanomedicine: Nanotechnology, Biology and Medicine* 5: 452–456.

Narayanan, K.B. and Sakthivel, N. 2010. Biological synthesis of metal nanoparticles by microbes. *Advances in Colloid and Interface Science* 156: 1–13.

O'Carroll, D., Sleep, B., Krol, M., Boparai, H., and Kocur, C., 2013. Nanoscale zero valent iron and bimetallic particles for contaminated site remediation. *Advances in Water Resources* 51: 104–122.

Philip, D., 2009. Biosynthesis of Au, Ag and Au–Ag nanoparticles using edible mushroom extract. *Spectrochimica Acta Part A* 73: 374–381.

Pighi, L., Pumpel, T., and Schinner, F. 1989. Selective accumulation of silver by fungi. *Biotechnology Letters* 11: 275–280.

Ponarulselvam, S., Panneerselvam, C., Murugan, K., Aarthi, N., Kalimuthu, K., and Thangamani, S., 2012. Synthesis of silver nanoparticles using leaves of *Catharanthus roseus* Linn. G. Don and their antiplasmodial activities. *Asian Pacific Journal of Tropical Biomedicine* 2: 574–580.

Rai, M., Yadav, A., Bridge, P., and Gade, A. 2009a. Myconanotechnology: A new and emerging science. In: *Applied Mycology*, Rai, M., Bridge, P. (eds.). CAB International, New York, Vol. 14, pp. 258–267.

Rai, M., Yadav, A., and Rai Gade, A. 2009b. Silver nanoparticles: As a new generation of antimicrobials. *Biotechnology Advances* 27: 76–83.

Ramanaviciusa, A., Kausaite, A., and Ramanaviciene, A., 2005. Biofuel cell based on direct bioelectrocatalysis. *Biosensors and Bioelectronics* 20: 1962–1967.

Ramos, M.A., Yan, W., Li, X.Q., Koel, B.E., and Zhang, W.X., 2009. Simultaneous oxidation and reduction of arsenic by zero-valent iron nanoparticles: Understanding the significance of the core-shell structure. *The Journal of Physical Chemistry Letters C* 113: 14591–14594.

Ramteke, C., Chakraborty, T., Sarangi, B.K., and Pandey, R.A., 2013. Synthesis of silver nanoparticle from the aqueous extract of leaves of *Ocimum sanctum* for enhanced antibacterial activity. *Journal of Chemistry* 1: 1–7.

Rautaray, D., Sanyal, A., Adyanthaya, S.D., Ahmad, A., and Sastry, M., 2004. Biological synthesis of strontium carbonate crystals using the fungus *Fusarium oxysporum*. *Langmuir* 20: 6827–6833.

Rautio, J.J., Smit, B.A., Wiebe, M., Penttilä, M., and Saloheimo, M., 2006 Transcriptional monitoring of steady state and effects of anaerobic phases in chemostat cultures of the filamentous fungus *Trichoderma reesei*. *BMC Genomics* 7: 247.

Reddy, D.H.K. and Lee, S.M., 2014. Magnetic biochar composite: Facile synthesis, characterization, and application for heavy metal removal. *Colloids and Surfaces A: Physicochemical and Engineering Aspects* 454: 96–103.

Riddin, T.L., Gericke, M., and Whiteley, C.G., 2006. Analysis of the inter and extracellular formation of platinum nanoparticles by *Fusarium oxysporum* f. sp *lycopersici* using response surface methodology. *Nanotechnology* 17: 3482–3489.

Rosi, N.L. and Mirkin, C.A., 2005. Nanostructures in biodiagnostics. *Chemical Reviews* 105: 1547–1562.

Sanghi, R. and Verma, P., 2009. Biomimetic synthesis and characterisation of protein capped silver nanoparticles. *Bioresource Technology* 100: 501–504.

Sastry, M., Ahmad, A., Khan, M.I., and Kumar, R., 2003. Biosynthesis of metal nano particles using fungi and actino-mycete. *Current Science* 85: 162–170.

Schmid, G., 1992. Large clusters and colloids. Metals in the embryonic state. *Chemical Reviews* 92: 1709–1727.

Senapati, S., Ahmad, A., Khan, M.I., Sastry, M., and Kumar, R., 2005. Extracellular biosynthesis of bimetallic Au-Ag alloy nanoparticle. *Small* 1: 517.

Senapati, S., Mandal, D., and Ahmad, A., 2004. Fungus mediated synthesis of silver nanoparticles: A novel biological approach. *Indian Journal of Physics* 78: 101–105.

Shaligram, N.S., Bule, M., Bhambure, R., Singhal, R.S., Singh, S.K., Szakacs, G., and Pandey, A., 2009. Biosynthesis of silver nanoparticles using aqueous extract from the compactin producing fungal strain. *ProcBiochem* 44: 939–943.

Shankar, S.S., Ahmad, A., Pasricha, R., and Sastry, M., 2003. Bioreduction of chloroaurate ions by geranium leaves and its endophytic fungus yields gold nanoparticles of different shapes. *Journal of Materials Chemistry* 13: 1822–1826.

Shankar, S.S., Rai, A., Ahmad, A., and Sastry, M., 2004. Rapid synthesis of Au, Ag, and bimetallic Au core–Ag shell nanoparticles using Neem (*Azadirachta indica*) leaf broth. *Journal of Colloid and Interface Science* 275: 496–502.

Shin, K.Y., Hong, J.Y., and Jang, J. 2011. Heavy metal ion adsorption behavior in nitrogen-doped magnetic carbon nanoparticles: Isotherms and kinetic study. *Journal of Hazardous Materials* 190: 36–44.

Singh, D., Rathod, V., Ninganagouda, S., Herimath, J., and Kulkarni, P., 2013. Biosynthesis of silver nanoparticle by endophytic fungi *Pencillium* sp. isolated from *Curcuma longa* (turmeric) and its antibacterial activity against pathogenic gram negative bacteria. *Journal of Pharmacy Research* 7: 448–453.

Singh, R., Misra, V., and Singh, R.P., 2011. Synthesis, characterization and role of zero-valent iron nanoparticle in removal of hexavalent chromium from chromium-spiked soil. *Journal of Nanoparticle Research* 13: 4063–4073.

Singh, R., Misra, V., and Singh, R.P., 2012. Removal of hexavalent chromium from contaminated ground water using zero-valent iron nanoparticles. *Environmental Monitoring and Assessment* 184: 3643–3651.

Song, J., Kong, H., and Jang, J., 2011. Adsorption of heavy metal ions from aqueous solution by polyrhodanine-encapsulated magnetic nanoparticles. *Journal of Colloid and Interface Science* 359: 505–511.

Sujatha, S., Tamilselvi, S., Subha, K., and Panneerselvam, A., 2013. Studies on biosynthesis of silver nanoparticles using mushroom and its antibacterial activities. *International Journal of Current Microbiology and Applied Sciences* 2: 605–614.

Symonds, J.L., 1995. Magnetroelectronics today and tomorrow. *Physics Today* 48: 26–32.

Tarafdar, J.C. and Raliya, R., 2013. Rapid, low-cost, and ecofriendly approach for iron nanoparticle synthesis using *Aspergillus oryzae* TFR9. *Journal of Nanoparticles.* 2013: 4, Article ID 141274, http://dx.doi.org/10.1155/2013/141274.

Thayer, A. and Houston, E.N., 2005. Metal Scavengers and immobilized catalysts may make for cleaner pharmaceutical products. *Chemical & Engineering News* 83: 55–58.

Uddin, I., Adyanthaya, S., Syed, A., Selvaraj, K., Ahmad, A., and Poddar, P., 2008. Structure and microbial synthesis of sub-10 nm Bi_2O_3 nanocrystals. *Journal of Nanoscience and Nanotechnology* 8: 3909–3913.

Uhrich, K.E., Cannizzaro, S.M., Langer, R.S., and Shakesheff, K.M., 1999. Polymeric systems for controlled drug release. *Chemical Reviews* 99: 3181–3198.

Uskokovic, V., 2008. Nanomaterials and nanotechnologies: Approaching the crest of this big wave. *Current Nanoscience* 4: 119–129.

Utsunomiya, S. and Ewing, R.C., 2003. Application of high-angle annular dark field scanning transmission electron microscopy, scanning transmission electron microscopy-energy dispersive X-ray spectrometry, and energy-filtered transmission electron microscopy to the characterization of nanoparticles in the environment. *Environmental Science & Technology* 37: 786–791.

Üzüm, Ç., Shahwan, T., Eroğlu, A.E., Hallam, K.R., Scott, T.B., and Lieberwirth, I., 2009. Synthesis and characterization of kaolinite-supported zero-valent iron nanoparticles and their application for the removal of aqueous Cu^{2+} and Co^{2+} ions. *Applied Clay Science* 43: 172–181.

Varshney, R., Mishra, A.N., Bhadauria, S., and Gaur, M.S., 2009. A novel microbial route to synthesize silver nano particles using fungus *Hormoconis resinae. Digest Journal of Nanomaterials and Biostructures* 4: 349–355.

Vetchinkina, E.P., Loshchinina, E.A., Burov, A.M., Dykman, L.A., and Nikitina, V.E., 2014. Enzymatic formation of gold nanoparticles by submerged culture of the basidiomycete Lentinus edodes. *Journal of Biotechnology* 182–183: 37–45.

Vigneshwaran, N., 2006. A novel one-pot "green" synthesis of stable silver nanoparticles using soluble starch. *Carbohydrate Research* 341: 2012–2018.

Vigneshwaran, N., Ashtaputre, N.M., Varadarajan, P.V., Nachane, R.P., Paralikar, K.M., and Balasubramanya, R.H., 2007. Biological synthesis of silver nanoparticles using the fungus *Aspergillus flavus. Materials Letters* 61: 1413–1418.

Vigneshwaran, N., Kathe, A.A., Varadarajan, P.V., Nachane, R.P., and Balasubramanya, R.H., 2006. Biomimetics of silver nanoparticles by white rot fungus, *Phaenerochaete chrysosporium. Colloids and Surfaces B: Biointerfaces* 53: 55–59.

Wang, J., Liu, G., Zhou, C., Li, T., and Liu, J., 2014. Synthesis, characterization and aging study of kaolinite-supported zero-valent iron nanoparticles and its application for Ni(II) adsorption. *Materials Research Bulletin* 60: 421–432.

Wang, Y. and Herron, N., 1991. Nanometer-sized semiconductor clusters: Materials synthesis, quantum size effects and photophysical properties. *Journal of Physical Chemistry* 95: 525–532.

Wang, Y., Morin, G., Ona-Nguema, G., Juillot, F., Calas, G., and Brown Jr., G.E., 2011. Distinctive arsenic(V) trapping modes by magnetite nanoparticles induced by different sorption processes. *Environmental Science & Technology* 45: 7258–7266.

Wang, Y., Morin, G., Ona-Nguema, G., Menguy, N., Juillot, F., Aubry, E., Guyot, F., Calas, G., and Brown, G.E., 2008. Arsenite sorption at the magnetite–water interface during aqueous precipitation of magnetite: EXAFS evidence for a new arsenite surface complex. *Geochimica Cosmochimica Acta* 72: 2573–2586.

Waychunas, G.A., Kim, C.S., and Banfield, J.F., 2005. Nanoparticulate iron oxide minerals in soils and sediments: Unique properties and contaminant scavenging mechanisms. *Journal of Nanoparticle Research* 7: 409–433.

Wong, T.S. and Schwaneberg, U., 2003. Protein engineering in bioelectrocatalysis. *Current Opinion in Biotechnology* 14: 590–596.

Xu, Y. and Zhao, D., 2007. Reductive immobilization of chromate in water and soil using stabilized iron nanoparticles. *Water Research* 41: 2101–2108.

Yan, W., Herzing, A.A., Kiely, C.J., and Zhang, W., 2010a. Nanoscale zero-valent iron (nZVI): Aspects of the core-shell structure and reactions with inorganic species in water. *Journal of Contaminant Hydrology* 118: 96–104.

Yan, W., Ramos, M.A., Koel, B.E., and Zhang, W.X., 2012a. As(III) sequestration by iron nanoparticles: Study of solid-phase redox transformations with x-ray photoelectron spectroscopy. *The Journal of Physical Chemistry C* 116: 5303–5311.

Yan, W., Ramos, M.A.V., Koelb, B.E., and Zhang, W., 2010b. Multi-tiered distributions of arsenic in iron nanoparticles: Observation of dual redox functionality enabled by a core–shell structure. *Chemical Communication* 46: 6995–6997.

Yan, W., Vasic, R., Frenkel, A.I., and Koel, B.E., 2012b. Intraparticle reduction of arsenite (As(III)) by nanoscale zerovalent iron (nZVI) investigated with in situ x-ray absorption spectroscopy. *Environmental Science & Technology* 46: 7018–7026.

Yang, S., Yan, N., Guo, Y., Wu, D., He, H., Qu, Z., Li, J., Zhou, Q., and Jia, J., 2011. Gaseous elemental mercury capture from flue gas using magnetic nano sized $(Fe_{3-x}Mn_x)_{1-\delta}O_4$. *Environmental Science & Technology* 45: 1540–1546.

Yantasee, W., Warner, C.L., Sangvanich, T., Addleman, R.S., Carter, T.G., Wiacek, R.J., Fryxell, G.E., Timchalk, C., and Warner, M.G., 2007. Removal of heavy metals from aqueous systems with thiol functionalized superparamagnetic nanoparticles. *Environmental Science & Technology* 41: 5114–5119.

Zhang, Q., Pan, B., Zhang, W., Pan, B., Zhang, Q., and Ren, H., 2008. Arsenate removal from aqueous media by nano sized hydrated ferric oxide (HFO)-loaded polymeric sorbents: Effect of HFO loadings. *Industrial & Engineering Chemistry Research* 47: 3957–3962.

Zhang, Y., Li, Y., Dai, C., Zhou, X., and Zhang, W., 2014. Sequestration of Cd(II) with nanoscale zero-valent iron (nZVI): Characterization and test in a two-stage system. *Chemical Engineering Journal* 244: 218–226.

Zhang, Y., Su, Y., Zhou, X., Dai, C., and Keller, A.A., 2013. A new insight on the core–shell structure of zero valent iron nanoparticles and its application for Pb(II) sequestration. *Journal of Hazardous Materials* 263: 685–693.

Zhang, Z., Li, M., Chen, W., Zhu, S., Liu, N., and Zhu, L., 2010. Immobilization of lead and cadmium from aqueous solution and contaminated sediment using nano-hydroxyapatite. *Environmental Pollution* 158: 514–519.

26 Application of Surface-Active Compounds of Microbial Origin to Clean Up Soils Contaminated with Heavy Metals

Verónica Leticia Colin, Marta Alejandra Polti, and María Julia Amoroso

CONTENTS

ABSTRACT

Soil contamination with heavy metals is one of the most serious environmental problems, and more stringent regulations are necessary to confront this scourge. Although some metals perform specific biological functions, many others do not have a known role. Thereby, both prokaryotic and eukaryotic cells possess homeostatic mechanisms to regulate heavy metal concentrations, minimizing the potentials of toxic effects on living cells. The use of microbial surface-active compounds instead of those obtained by chemical synthesis to enhance the recovery of metals and other pollutants from soils is more acceptable due to their natural origin and singular properties that include low toxicity and high biodegradability. In fact, current challenges such as the design of strategies to minimize the production cost of these biomolecules at a large scale are being addressed successfully today. Application strategies of surface-active compounds varies depending on whether the producing microorganism is present in the cleanup process of soil or these biomolecules are added to the soil washing solution without the presence of the producing microorganism. In any case, microbial surface-active compounds tend to form complexes with heavy metals encouraging their effective desorption and removal from soils and sediments. Thereby, a significant reduction of these pollutants to acceptable levels can be achieved through a natural and eco-friendly process without environmental damages.

26.1 INTRODUCTION

Nowadays, heavy metals are considered as one of the priority pollutants, and more stringent regulations are emerging in order to mitigate this serious environmental problem. A variety of remediation technologies are currently available to address these classes of pollutants (Fu and Wang 2011; Yao et al. 2012). However, many of them have several disadvantages such as high cost and the risk of secondary contamination (de Souza et al. 2013). In this context, new technologies for the reduction of heavy metals to acceptable levels but at more manageable costs are continually being evaluated.

Bioremediation is a process that aims at detoxification and degradation of toxic pollutants through microbial assimilation or transformation toward less toxic compounds. The success of this process relies on the availability of microbes and their accessibility to the contaminants. With regard to the recovery of heavy metals from environments, scientific evidences indicate that certain microbial metabolic pathways can be exploited to reduce its concentrations in the diverse systems until acceptable levels (Colin et al. 2012; Polti et al. 2014). On the other hand, some microbial products can be obtained by fermentation processes for their subsequent recovery and direct application in environmental remediation technologies without the presence of producer microorganisms in the process (Karigar and Rao 2011; Colin et al. 2013a,b, 2016). Surface-active compounds (SACs) are among these products, and they are carving a niche for themselves in the market today. The microbial SACs comprise a structurally diverse group of amphipathic molecules produced by bacteria, fungi, and yeasts (Colin et al. 2010, 2016; Luna et al. 2012). These molecules have a combined polar–nonpolar structure, which gives them unique properties that have attracted the interest of scientists and industry. Due to their natural origin, these products are considered biodegradable and less toxic than their counterparts produced by chemical synthesis (Silva et al. 2014; De et al. 2015). Thereby, they have emerged in the last decades as a promising alternative to be used in diverse biotechnological sectors.

In the bioremediation field, microbial SACs can help in the cleanup of diverse pollutants, including heavy metals, through mechanisms that will be considered in detail throughout this chapter.

Despite their potential applications in multiple biotechnological sectors, the real bottleneck for the extensive application of microbial SACs lies in their limited marketing. This is mainly due to economic obstacles to achieve a production at a large scale, which is based on the use of synthetic media. In this context, the finding of new producer strains able to use inexpensive substrates could be a strategic issue to overcome these obstacles (Colin et al. 2014).

Based upon this background, this chapter reviews the main advances on this topic, emphasizing on the potential applications of SACs produced by microorganisms for the recovery of heavy metals from contaminated soils.

26.2 MAIN SOURCE OF CONTAMINATION WITH HEAVY METALS

Heavy metals in the soil have two main sources: natural and anthropogenic pollution. The natural sources included rocks, volcanic eruption, erosion of mineral deposits, evaporation of the oceans, and pedogenic process, among others. Usually, basal soil content is at trace levels (Bradl 2005; Violante et al. 2010). However, several authors have reported elevated baseline concentrations because the parent rock contains a high concentration of heavy metals (Oliveira et al. 2011; Li et al. 2014a).

Anthropogenic sources include agricultural practices, urban and industrial waste, fossil fuel combustion, metal mining, milling, and smelting processes, among other industrial activities (Fernández et al. 2014; Li et al. 2015; Luo et al. 2015; Wei et al. 2015; Wu et al. 2015). Heavy metals are disposed of as gases or liquid effluents, their final destinations usually are soils and sediments, and eventually they are incorporated into the food chain. This is because heavy metals from anthropogenic sources tend to be more mobile, hence bioavailable than pedogenic or lithogenic ones (Wuana and Okieimen 2011).

Agricultural practices contribute to heavy metal pollution through the use of pesticides, fertilizers, and manures. Certain pesticides widely used in agriculture contain metals as impurities or as active ingredients. The most common are copper-containing fungicides, such as copper sulfate or copper oxychloride, and arsenical herbicides, such as monosodium methyl arsenic acid (Paranjape et al. 2014). Moreover, although

mercury-containing pesticides have already been banned, this metal still remains in soils (Edwards 2013; Paranjape et al. 2014). Fertilizers are regularly added to provide adequate N, P, and K for crop growth. Also, Cu and Mn are added in certain cases. The compounds used to supply these elements contain trace amounts of heavy metals (e.g., Cd, F, Hg, and Pb). Also, repeated application of manures produced from animals whose diets contain high concentrations of As, Cu, and Zn can cause considerable accumulation of these metals in the soil in the long run (Wuana and Okieimen 2011).

As a result of urbanization, large quantities of waste, wastewater, and sewage sludge are produced. At present, the land irrigation with municipal and industrial wastewater is common practice. Although the metal concentrations in wastewater generally are below the toxic levels, long-term irrigation results in heavy metal accumulation in the soil (Sherameti and Varma 2015). On the other hand, inadequate disposal of municipal solid waste in sanitary or open landfills produces contaminated leachate affecting ground and groundwater quality. The landfill leachate transferred different pollutants from the waste material to the percolating water (Fernández et al. 2014).

The disposal of electronic waste or e-waste is a growing problem, because of the rapid development of information technology and constant upgrade of electronic products. Due to the large content of heavy metals from e-waste, it is processed to recover these compounds. However, the techniques used are inappropriate and have low recovery, so an important secondary pollution is generated (Wu et al. 2015).

During mining, tailings are directly discharged into natural depressions, including on-site wetlands resulting in elevated concentrations (Wuana and Okieimen 2011). Several studies have reported contamination by heavy metals in neighboring areas to mining activities, which include Cd, Pb, Cu, Zn, Hg, As, and Ni, among others (Khalil et al. 2013; Krishna et al. 2013; Li et al. 2014b).

Airborne sources of metals include stack or duct emissions of air, gas, or vapor streams and emissions from storage areas or waste piles. Stack emissions can be distributed over a wide area by natural air currents until dry and/or wet precipitation mechanisms remove them from the gas stream and eventually deposited on land or sea.

Fossil fuels generally contain heavy metals and its combustion releases them at the atmosphere. For example, combustion of petrol containing tetraethyl contributes substantially to the content of Pb in soils in urban areas (Wuana and Okieimen 2011).

26.3 TOXICITY OF HEAVY METALS ON ENVIRONMENT AND HUMAN HEALTH

Soil heavy metal environmental risk to humans is related to bioavailability. Assimilation pathways include the direct ingestion or contact with contaminated soil, the food chain (soil–plant–human or soil–plant–animal–human), drinking of contaminated groundwater, reduction in food quality (safety and marketability) via phytotoxicity, reduction in land usability for agricultural production causing food insecurity, and land tenure problem (Wuana and Okieimen 2011).

The toxicology of metals is concerned with approximately 80 elements and their compounds, including simple ionic salts to complicated structures, such as organometallic compounds and metallic complexes (Nordberg et al. 2015).

Some heavy metals are essential for organisms, while many of them have no known benefit for human physiology. Among the essential elements are Fe, Zn, Cu, Mn, Se, Cr, Co, and Mo. They exert biochemical and physiological functions in plants and animals, being important constituents of several key enzymes.

The toxicity of heavy metals depends on a number of factors, including the dose, route of exposure, and chemical form, as well as the age, gender, genetics, and nutritional status of exposed individuals.

The metal bioavailability is influenced by physical, chemical, and biological factors, including temperature, phase association, adsorption, speciation, solubility, and biochemical/physiological adaptation (Tchounwou et al. 2012).

On the other hand, for essential metals, there are three concentration ranges: (1) the deficiency range, where biological activity is increased with increasing dose; (2) the normal range, where biological activity is optimal; and (3) the toxicity range, where concentration inhibits normal metabolism. By contrast, nonessential metals are toxic even at trace concentrations (Nordberg et al. 2015).

Toxic effects by heavy metals can be classified into carcinogenic and noncarcinogenic. Specific symptomatology varies according to the metal in question, the total dose absorbed, and whether the exposure was acute or chronic. In general, acute exposures can lead to death without prompt treatment, while chronic exposures produce different types of cancer. Heavy metals affect cellular components such as the cell membrane, mitochondrial, lysosome, endoplasmic reticulum, nuclei, and some enzymes involved in metabolism, detoxification, and damage repair. Metal ions interact with cell components such as DNA and nuclear proteins, causing DNA damage and conformational changes that may lead to cell cycle modulation, carcinogenesis, or apoptosis. Reactive oxygen species production and oxidative stress play a key role in the toxicity and carcinogenicity of several metals. However, each metal has unique characteristics and physicochemical properties that lead to a specific toxicological mechanism of action and consequently specific symptoms. Table 26.1 reviews typical symptoms associated with intoxication with selected heavy metals.

26.4 SURFACE-ACTIVE COMPOUNDS OF MICROBIAL ORIGIN: CLASSIFICATION AND PROPERTIES

Diverse criteria have been applied in order to classify the microbial SACs. For example, based on the mode of action, SACs have been grouped into two categories: *biosurfactants* and *bioemulsifiers*. The terms *biosurfactants* and *bioemulsifiers* have often been used interchangeably to describe metabolites with surface activity since both biomolecules present an amphipathic structure. However, it is important to remark some differences between them. For example, the chemical composition of biosurfactants and bioemulsifiers is different, and this may contribute to their specific roles in nature as well as their potentials in biotechnological applications. Biosurfactants are commonly low-molecular-weight products that include glycolipids, phospholipids, and/or lipopeptides (Shoeb et al. 2013). They are known for their excellent surface activity, which involves lowering the surface and interfacial tension between different phases (liquid–air, liquid–liquid, and liquid–solid). These unique characteristics promote the application of biosurfactants in diverse industrial processes; thereby, they themselves are carving their own niche in the market today. Bioemulsifiers, on the other hand, are high-molecular-weight biopolymers, which can be complex mixtures of heteropolysaccharides, lipopolysaccharides, lipoproteins, and proteins (Perfumo et al. 2009; Sekhon-Randhawa 2014). Bioemulsifiers are less effective in reducing the surface tension. However, these molecules are very efficient in emulsifying two immiscible liquids (water–oil, water–hydrocarbon, etc.) even at low concentrations. For example, Figure 26.4 shows the emulsions formed between a SAC produced by actinobacterium *Streptomyces* sp. MC1 isolated from our region (Tucuman, Argentina) with diverse hydrophobic substrates. When water, an oily compound, and a SAC are mixed, this last rests at the water–oil interface forming systems called macroemulsions or microemulsions. This classification is dependent on their stability and on the droplet's size formed. Macroemulsions have been known for thousands of years and consist of mixtures of two immiscible liquids, one of them being dispersed in the form of droplets with diameter greater than 0.1 µm in the other liquid (Figure 26.1a). Such systems are turbid, milky in color, and thermodynamically unstable, that is, the macroemulsions will be ultimately separated into two original immiscible liquids with time. In contrast, microemulsions have been defined as the clear dispersions (transparent or translucent) (Figure 26.1b) and thermodynamically stable of two immiscible liquids whose dispersed phase consists of small droplets with diameter that ranged 10–100 nm (Tandel et al. 2012). Interestingly, the use of these last systems for the cleaning of contaminated soils has been effectively demonstrated (Zheng and Wong 2010; Pérez et al. 2012).

In general lines, many of the synthetic SACs currently used are toxic and hardly degraded, causing damage to the environment (Masakorala et al. 2011). Therefore, they have increasingly been replaced by biotechnology-based compounds, because they can be produced using natural sources (Olkowska et al. 2014). Biosurfactants have several advantages over chemical surfactants, including lower toxicity, higher biodegradability, effectiveness at extreme temperatures, pH and salinity, biocompatibility, and digestibility values (Zheng et al. 2012; Colin et al. 2013a). These eco-friendly properties have encouraged their application in diverse technological areas, including the bioremediation field such as summarized in Figure 26.2 (Shete et al. 2006).

The replacement of synthetic SACs by those of biological origin requires however the development of new strategies to ensure profitable production processes. This is because that practical applications of microbial surfactants/emulsifiers,

TABLE 26.1

Typical Symptoms Caused by Acute and Chronic Exposition to Selected Heavy Metals

Metal	Acute Exposition	Chronic Exposition
Arsenic	Severe nausea and vomiting, colicky abdominal pain, and profuse diarrhea	Melanosis, hyperkeratosis, and desquamation Anemia and leukopenia Thrombocytopenia
	Psychosis associated with paranoid delusions, hallucinations, and delirium	Multicentric basal cell and squamous cell carcinomas
	Multisystem organ damage (Saha et al. 1999)	Diabetes, hypopigmentation/hyperkeratosis, cancer: lung, bladder, skin, encephalopathy (Saha et al. 1999)
Cadmium	*Pneumonitis*: shortness of breath, lung edema, and destruction of mucous membranes (inhalation)	Proteinuria, kidney stones, glomerular and tubular damage
		Testicular necrosis, estrogen-like effects, affection of steroid hormone synthesis
	Vomiting and diarrhea (oral)	Loss of bone density and mineralization, itai-itai disease (Godt et al. 2006)
Chromium	Vomiting, diarrhea, hemorrhage, and blood loss into the gastrointestinal tract, causing cardiovascular shock; liver and kidney necrosis (oral)	Chronic ulcers of the skin Ulceration and perforation of the nasal septum Rhinitis, bronchospasm, and pneumonia
	Irritative dermatitis	Lung cancer (Costa and Klein 2006; Wilbur et al. 2012)
	Allergic contact dermatitis	
Cobalt	Dermatitis	Auditory and optic neuropathy
	Lethal cardiomyopathy	Goiter and myxedema Asthma (Nordberg et al. 2015)
Copper	Acute gastrointestinal disturbances with vomiting, epigastric burns, and diarrhea	Vineyard sprayer's lung (inhaled); Wilson disease (hepatic and basal ganglia degeneration)
	Metal fume fever	Mucosal and suspected atrophic changes in the mucous membranes of the nose
	Acute hemolysis	Dyspnea, thoracic pain, emphysema, and pulmonary fibrosis Vineyard sprayer's lung (Nordberg et al. 2015)
Lead	Encephalopathy (ataxia, coma, and convulsions)	Problems with attention/concentration/memory, visuospatial and visuomotor skills, and speed of learning and problem-solving ability
	Renal dysfunction with hypertension, hyperuricemia	Low skeletal growth "Saturnine gouty arthritis" Chronic renal failure (Nordberg et al. 2015)
Mercury	Erosive bronchitis and bronchiolitis with interstitial pneumonitis; respiratory insufficiency; tremor or increased excitability (inhalation)	Nausea, metallic taste, gingivostomatitis, tremor, neurasthenia, nephrotic syndrome; hypersensitivity (Pink disease)
	Gastroenteritis (ingestion)	Micromercurialism: asthenic-vegetative syndrome (weakness, fatigue, anorexia, loss of weight, and disturbance of gastrointestinal functions) (Nordberg et al. 2015)
Nickel	Headache, vertigo, nausea, vomiting, nephrotoxic effects, and pneumonia followed by pulmonary fibrosis	Rhinitis, sinusitis, nasal septum perforations, and asthma (Nordberg et al. 2015)
	Contact allergy	
Zinc	Metal fume fever	Copper deficiency: anemia, neurologic degeneration, and osteoporosis (Nordberg et al. 2015)

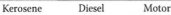

Kerosene   Diesel   Motor oil

(a)

Sunflower oil   Grape oil   Olive oil

(b)

FIGURE 26.1 Emulsions formed between a SAC produced by *Streptomyces* sp. MC1 and diverse hydrophobic substrates: (a) macroemulsions and (b) microemulsions.

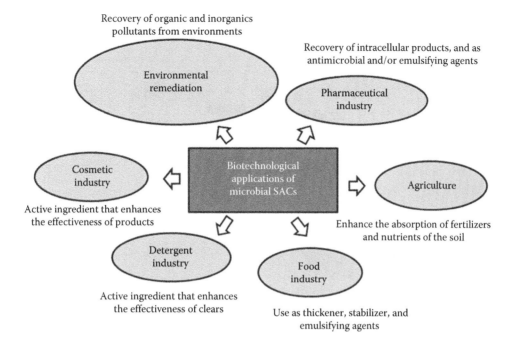

FIGURE 26.2 Main biotechnological applications of the microbial SACs.

particularly in technology of environmental remediation, depends on whether they can be produced economically at an industrial scale. One alternative to overcome these limitations consists of encouraging the SAC production from agro-industrial waste material. While microbial SACs could be economically produced, waste volumes would be also reduced, minimizing their environmental impact (Guerra de Oliveira et al. 2013; Colin et al. 2016).

In the bioremediation field, the capability of SACs and SAC-producing strains to promote the effective pollutant recovery from environments was reported by many authors (Pacwa-Płociniczak et al. 2011; Colin et al. 2013a; Silva et al. 2014). In the next section, available technologies that use microbial SACs to clean up soils contaminated with heavy metals are considered.

26.5 SURFACE-ACTIVE COMPOUNDS APPLIED TO THE RECOVERY OF HEAVY METALS FROM SOIL

During the last decades, there was a major awareness that the soil is not an inexhaustible resource and that, if used improperly, its characteristics can be lost in a short period of time (Nortcliff 2002). Microorganisms can be used in the bioremediation of soils contaminated with heavy metals through diverse processes, some mediated by the SAC production. Bioremediation processes that involve microbial surfactants/emulsifiers can be approached from two perspectives: one that involves the presence of the SAC-producing microorganisms in the contaminated soil and the other where SACs can be produced by microbial fermentation for their subsequent recovery and direct application as a soil washing agent. Both approaches are considered in detail as follows.

26.5.1 Use of Surface-Active Compound–Producing Microorganisms

Successful use of SAC-producing microorganisms to promote the cleanup of soils contaminated with heavy metals is today a methodology in clear expansion. Heavy metals are partially water soluble, and they tend to be absorbed on the soil matrix, which obstruct its recovery. Response of microorganisms facing this situation can be the synthesis and release of SACs to the environment. Once released, the SACs tend to form complexes with the metals through its polar portion. The bonds between SACs and metals are stronger than the metal's bonds with the soil. Therefore, metal–SAC complexes are finally desorbed from the soil matrix due to the lowering of the interfacial tension (Pacwa-Płociniczak et al. 2011). Finally, metals stabilized on the surface of complexes can be taken up and accumulated into the microbial cell itself. Figure 26.3 summarizes this process in four critical steps.

To date, some studies have shown the potential of SAC-producing microorganisms to reduce the heavy metal concentrations in the environments until acceptable levels. Gnanamani et al. (2010), for example, used a biosurfactant produced by the marine bacterium *Bacillus* sp. MTCC 5514 to entrap the trivalent chromium in micelles. These authors proposed that the metal remediation is carried out via two processes: (1) reduction of Cr(VI) to Cr(III) by an extracellular reductase enzyme and (2) entrapment of Cr(III) by the biosurfactants. The first process transforms the toxic state of chromium into less toxic state, while the second step prevents the bacterial cells from the exposure of Cr(III). Both reactions provide tolerance and resistance to the bacterial cells against high chromium concentrations, either in its hexavalent or trivalent state. Other biosurfactant-producing bacteria such as

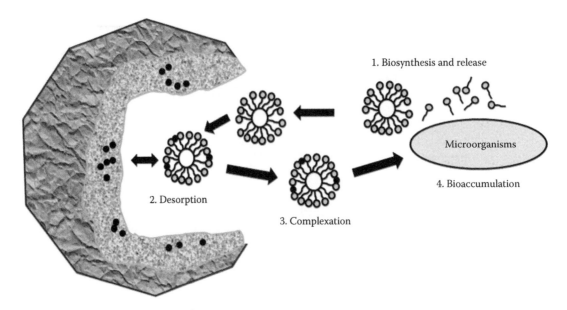

FIGURE 26.3 Flowchart summarizing the desorption process of heavy metals mediated by SAC-producing microorganisms.

Pseudomonas fluorescens G7 and *Bacillus subtilis* TP8 were also assessed for their ability to remove Cd and Zn from contaminated soil (Sarin and Sarin 2010). In this study, it was noted that approximately 19% of Zn and 16.7% of Cd could be removed after incubation with immobilized bacteria during 2 weeks.

Despite this background, the use of SAC-producing microorganisms to clean up contaminated soils has certain limitations because the producer strains should be able to grow and survive in the presence of the pollutants (Pacwa-Płociniczak et al. 2011; Colin et al. 2013a). In fact, little is today known about the SAC production by microorganisms *in situ* since most of the described studies were done at a laboratory scale. Therefore, more efforts could be required to evaluate their prospect of application in technologies of environmental remediation.

26.5.2 Direct Application of Surface-Active Compounds Produced by Microbial Fermentation

Direct use of SACs produced by microbial fermentation is a pragmatic approach where the disadvantages associated with the presence of producer microorganisms in the process are practically eliminated. The recovered SACs from culture supernatants can be added to the washing solutions in order to assist in the solubilization, dispersal, and desorption of heavy metals from soils and sediments. The event sequence that occurs in this case is summarized in Figure 26.4. The same principles described in Figure 26.3 concerning to the transfer of the metal–SAC complexes from soil matrix to the aqueous phase govern these events. However, because the absence of the producer microorganism in the process, it is necessary the continuous addition of new portions of SACs to the washing solution. This could indicate saturation of the binding sites

for the metals on the SACs, since it is assumed that high concentrations of SACs are required to mediate effective metal removal (Colin et al. 2013b).

The "soil washing technologies" that use products of microbial origin are a clear example of a cleanup combined methodology that includes a physicochemical process (washing itself) and a biological process based upon the use of microbial products as washing agents. To date, numerous reports on the use of microbial SACs in soil washing technologies are available. Some studies have been pointed that metal removal effectiveness depends on the metallic species. For example, it was reported on a rhamnolipid surfactant produced by *Pseudomonas aeruginosa* strain BS2, which selectively favors the mobilization of Cd on Pb from soil (Juwarker et al. 2007). Gutierrez et al. (2008) reported also on the differential capacity of an emulsifier exopolymer produced by *Pseudoalteromonas* sp. strain TG12 to desorb metals from marine sediment of a selective form according these are mono-, di-, or trivalent species. More recently, Colin et al. (2013a) used bioemulsifiers produced by the actinobacterium *Amycolatopsis tucumanensis* DSM45259 to wash artificially contaminated soils with Cu(II) and hexavalent Cr(VI). These authors noted that produced bioemulsifiers were not effective in removing Cu(II). However, they were able to mediate Cr(VI) recovery from soil (up to 17%).

On the other hand, it has been reported that the SAC chemical nature as well as their concentration in the washing solution can also affect the performance of the process. Aniszewski et al. (2010), for example, evaluated the bioemulsifier production by *Microbacterium*'s strains isolated from urban mangrove sediments using two carbon sources such as sucrose and glucose. They noted that the performance of bioemulsifiers to remove Cd and Zn from hazardous industrial residues depended on the carbon substrate used for the production. Maximum Cd removal (41%) was reached

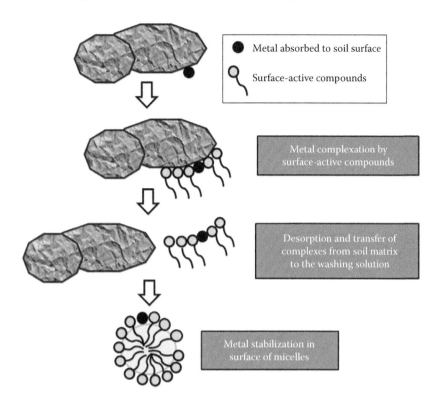

FIGURE 26.4 Flowchart summarizing the desorption and transfers of heavy metals from soil to aqueous phase during the washing process with microbial SACs. (Modified from Pacwa-Płociniczak, M. et al., *Int. J. Mol. Sci.*, 12, 633, 2011.)

during the washing experiments with *Microbacterium* Mc6b bioemulsifier produced from glucose. However, optimal Zn removal (68%) was detected by using *Microbacterium* MC1 emulsifier as washed agent, which was also produced from glucose. Also, Hidayati et al. (2014) researched the effectiveness of four chemically different biosurfactants of microbial origin for heavy metal removal (Pb, Zn, and Cu) from sediments of sludge by washing technologies. They detected the highest removal of Pb (up to 14.04%) by using surfactants produced by *Acinetobacter* sp., while the highest removal for Zn and Cu (up to 6.5% and 2.01%, respectively) was obtained by using the biosurfactant of *Pseudomonas putida* T1(8).

Concerning SAC optimal amount, Wang and Mulligan (2009) reported that high concentrations of rhamnolipid (approximately 100 mg·L^{-1}) are required to enhance the As mobilization by reducing the interfacial tension between the metal and the mine tailings.

Finally, a recent study reported on the high biosurfactant concentrations necessary to achieve an good efficient during the washed of soil samples collected from industrial dumping site which contain high concentrations of heavy metals, among them, Fe, Pb, Ni, Cd, Cu, Co and Zn (Singh and Cameotra 2013). The authors pointed out that SAC concentration in washing solution should be above the micelle critical concentration to achieve an effective metal removal. This is because about 50% of biosurfactant is sorbed to the soil particles, decreasing their effective concentration in washing process.

26.6   SUMMARY

In the last years, an increasing interest has been shown on the use of SACs or those producer microorganisms for the design of remediation strategies of soil contaminated with heavy metals. Application prospect of these natural products in bioremediation strategies has been certainly demonstrated at a small scale. However, many facets remain even to be explored in order to provide valuable progress in this topic, particularly in regard to the high cost of production of microbial SACs.

Currently, our research group is addressing many of these issues to make feasible the real application of these microbial products in soil washing technologies. Although there are many facets to explore, first steps for heavy metal removal from soil at a laboratory scale have already been given.

REFERENCES

Aniszewski, E., Peixoto, R.S., Mota, F.F., Leite, S.G.F., Rosado, A.S. 2010. Bioemulsifier production by *Microbacterium* sp. strains isolated from mangrove from mangrove and their application to remove cadmium and zinc from hazardous industrial residues. *Brazilian Journal of Microbiology* 41: 235–245.

Bradl, H. 2005. *Heavy Metals in the Environment: Origin, Interaction and Remediation.* Academic Press, Amsterdam, the Netherlands.

Colin, V.L., Baigorí, M.D., Pera, L.M. 2010. Bioemulsifier production by *Aspergillus niger* MYA 135: Presumptive role of iron and phosphate on emulsifying ability. *World Journal of Microbiology and Biotechnology* 26: 2291–2295.

Colin, V.L., Castro, M.F., Amoroso, M.J., Villegas, L.B. 2013a. Production of bioemulsifiers by *Amycolatopsis tucumanensis* DSM 45259 and their potential application in remediation technologies for soils contaminated with hexavalent chromium. *Journal of Hazardous Materials* 261: 577–583.

Colin, V.L., Cortes, Á.J., Rodríguez, A., Amoroso, M.J. 2014. Surface-active compounds of microbial origin and their potential application in technologies of environmental remediation. In: *Bioremediation in Latin America: Current Research and Perspectives*, Alvarez, A., Polti, M.A. (eds.), pp. 255–264. Springer, Cham, Switzerland.

Colin, V.L., Juárez Cortes, A.A., Aparicio, J.D., Amoroso, M.J. 2016. Potential application of a bioemulsifier-producing actinobacterium for treatment of vinasse. *Chemosphere* 144: 842–847.

Colin, V.L., Pereira, C.E., Villegas, L.B., Amoroso, M.J., Abate, C.M. 2013b. Production and partial characterization of a bioemulsifier produced by a chromium-resistant actinobacteria. *Chemosphere* 90: 1372–1378.

Colin, V.L., Villegas, L.B., Abate, C.M. 2012. Indigenous microorganisms as potential bioremediators for environments contaminated with heavy metals. *International Biodeterioration and Biodegradation* 69: 28–37.

Costa, M., Klein, C.B. 2006. Toxicity and carcinogenicity of chromium compounds in humans. *Critical Reviews in Toxicology* 36: 155–163.

De, S., Malik, S., Ghosh, A., Saha, R., Saha, B. 2015. A review on natural surfactants. *RSC Advances* 5: 65757–65767.

Edwards, C. 2013. *Environmental Pollution by Pesticides*. Springer Science and Business Media, New York.

Fernández, D.S., Puchulu, M.E., Georgieff, S.M. 2014. Identification and assessment of water pollution as a consequence of a leachate plume migration from a municipal landfill site (Tucumán, Argentina). *Environmental Geochemistry and Health* 36: 489–503.

Fu, F., Wang, Q. 2011. Removal of heavy metal ions from wastewaters: A review. *Journal of Environmental Management* 92: 407–418.

Gnanamani, A., Kavitha, V., Radhakrishnan, N., Rajakumar, G.S., Sekaran, G., Mandal, A.B. 2010. Microbial products (biosurfactant and extracellular chromate reductase) of marine microorganism are the potential agents reduce the oxidative stress induced by toxic heavy metals. *Colloids and Surfaces B* 79: 334–339.

Godt, J., Scheidig, F., Grosse-Siestrup, C., Esche, V., Brandenburg, P., Reich, A., Groneberg, D.A. 2006. The toxicity of cadmium and resulting hazards for human health. *Journal of Occupational Medicine and Toxicology* 1: 22.

Guerra de Oliveira, J., García-Cruz, C.H. 2013. Properties of a biosurfactant produced by *Bacillus pumilus* using vinasse and waste frying oil as alternative carbon sources. *Brazilian Archives of Biology and Technology* 56: 155–160.

Gutierrez, T., Shimmield, T., Haidon, C., Black, K., Green, D.H. 2008. Emulsifying and metal ion binding activity of a glycoprotein exopolymer produced by *Pseudoalteromonas* sp. strain TG12. *Applied and Environmental Microbiology* 15: 4867–4876.

Hidayati, N., Surtiningsih, T., Ni'matuzahroh. 2014. Removal of heavy metals Pb, Zn and Cu from sludge waste of paper industries using biosurfactant. *Journal of Bioremediation and Biodegradation* 5: 255.

Juwarkar, A.A., Nair, A., Dubey, K.V., Singh, S.K., Devotta, S. 2007. Biosurfactant technology for remediation of cadmium and lead contaminated soils. *Chemosphere* 68: 1996–2002.

Karigar, C.S., Rao, S.S. 2011. Role of microbial enzymes in the bioremediation of pollutants: A review. *Enzyme Research* 2011: Article ID 805187.

Khalil, A., Hanich, L., Bannari, A., Zouhri, L., Pourret, O., Hakkou, R. 2013. Assessment of soil contamination around an abandoned mine in a semi-arid environment using geochemistry and geostatistics: Pre-work of geochemical process modeling with numerical models. *Journal of Geochemical Exploration* 125: 117–129.

Krishna, A.K., Mohan, K.R., Murthy, N., Periasamy, V., Bipinkumar, G., Manohar, K., Rao, S.S. 2013. Assessment of heavy metal contamination in soils around chromite mining areas, Nuggihalli, Karnataka, India. *Environmental Earth Sciences* 70: 699–708.

Li, P., Lin, C., Cheng, H., Duan, X., Lei, K. 2015. Contamination and health risks of soil heavy metals around a lead/zinc smelter in southwestern China. *Ecotoxicology and Environmental Safety* 113: 391–399.

Li, Y., Zhang, H., Chen, X., Tu, C., Luo, Y., Christie, P. 2014a. Distribution of heavy metals in soils of the Yellow River Delta: Concentrations in different soil horizons and source identification. *Journal of Soils and Sediments* 14: 1158–1168.

Li, Z., Ma, Z., van der Kuijp, T.J., Yuan, Z., Huang, L. 2014b. A review of soil heavy metal pollution from mines in China: Pollution and health risk assessment. *Science of the Total Environment* 468: 843–853.

Luna, J., Rufino, R., Campos, G., Sarubbo, L. 2012. Properties of the biosurfactant produced by *Candida sphaerica* cultivated in low-cost substrates. *Chemical Engineering Transactions* 27: 67–72.

Luo, Z., Gao, M., Luo, X., Yan, C. 2015. National pattern for heavy metal contamination of topsoil in remote farmland impacted by haze pollution in China. *Atmospheric Research* 170: 34–40.

Masakorala, K., Turner, A., Brown, M.T. 2011. Toxicity of synthetic surfactants to the marine macroalga, *Ulva lactuca*. *Water, Air, & Soil Pollution* 218: 283–291.

Nordberg, G.F., Fowler, B.A., Nordberg, M. 2015. *Handbook on the Toxicology of Metals*, 4th edn. Academic Press, San Diego, CA.

Nortcliff, S. 2002. Standardization of soil quality attributes. *Agriculture, Ecosystems and Environment* 88: 161–168.

Oliveira, S., Pessenda, L.C., Gouveia, S.E., Favaro, D.I. 2011. Heavy metal concentrations in soils from a remote oceanic island, Fernando de Noronha, Brazil. *Anais da Academia Brasileira de Ciências* 83: 1193–1206.

Olkowska, E., Ruman, M., Polkowska, Z. 2014. Occurrence of surface active agents in the environment. *Journal of Analytical Methods in Chemistry* 2014: 769708.

Pacwa-Płociniczak, M., Płaza, G.A., Piotrowska-Seget, Z., Cameotra, S.S. 2011. Environmental applications of biosurfactants: Recent advances. *International Journal of Molecular Science* 12: 633–654.

Paranjape, K., Gowariker, V., Krishnamurthy, V., Gowariker, S. 2014. *The Pesticide Encyclopedia*. Cabi, Wallingford, U.K.

Pérez, V., Sánchez, M., Coronel, V., Pereira, J., Alvarez, R. 2012. Use of microemulsions for cleaning of crude oil-contaminated soils. *Revista Ingenieria* 19: 61–68.

Perfumo, A., Smyth, T.J.P., Marchant, R., Banat, I.M. 2009. Production and roles of biosurfactant and bioemulsifiers in accessing hydrophobic substrates. In: *Microbiology of Hydrocarbons, Oils, Lipids and Derived Compounds*, Timmis Kenneth, N. (ed.), pp. 1502–1512. Springer-Verlag, Berlin, Germany.

Polti, M.A., Aparicio, J.D., Benimeli, C.S., Amoroso, M.J. 2014. Simultaneous bioremediation of Cr(VI) and lindane in soil by Actinobacteria. *International Biodeterioration and Biodegradation* 88: 48–55.

Saha, J.C., Dikshit, A.K., Bandyopadhyay, M., Saha, K.C. 1999. A review of arsenic poisoning and its effects on human health. *Critical Reviews in Environmental Science and Technology* 29: 281–313.

Sarin, C., Sarin, S. 2010. Removal of cadmium and zinc from soil using immobilized cell of biosurfactant producing bacteria. *Environment Asia* 3: 49–53.

Sekhon-Randhawa, K.K. 2014. Biosurfactants produced by genetically manipulated microorganisms: Challenges and opportunities. In: *Biosurfactants*, Kosaric, N., Sukan, F.V. (eds.), pp. 49–67. CRC Press, Boca Raton, FL.

Sherameti, I., Varma, A. 2015. *Heavy Metal Contamination of Soils: Monitoring and Remediation*. Springer, Cham, Switzerland.

Shete, A.M., Wadhawa, G., Banat, I.M., Chopade, B.A. 2006. Mapping of patents on bioemulsifiers and biosurfactants: A review. *Journal of Scientific and Industrial Research* 65: 91–115.

Shoeb, E., Akhlaq, F., Badar, U., Akhter, J., Imtiaz, S. 2013. Classification and industrial applications of biosurfactants. *Natural and Applied Sciences* 4: 243–252.

Silva, R.D.C.F., Almeida, D.G., Rufino, R.D., Luna, J.M., Santos, V.A., Sarubbo, L.A. 2014. Applications of biosurfactants in the petroleum industry and the remediation of oil spills. *International Journal of Molecular Science* 15: 12523–12542.

Singh, A.K., Cameotra, S.S. 2013. Efficiency of lipopeptide biosurfactants in removal of petroleum hydrocarbons and heavy metals from contaminated soil. *Environmental Science and Pollution Research* 20: 7367–7376.

Tandel, H., Raval, K., Nayani, A., Upadhay, M. 2012. Preparation and evaluation of cilnidipine microemulsion. *Journal of Pharmacy and Bioallied Science* 4: 114–115.

Tchounwou, P.B., Yedjou, C.G., Patlolla, A.K., Sutton, D.J. 2012. Heavy metals toxicity and the environment. In *Molecular, Clinical and Environmental Toxicology*, Luch, A. (ed.), pp. 133–164. Springer, Basel, Switzerland.

Violante, A., Cozzolino, V., Perelomov, L., Caporale, A., Pigna, M. 2010. Mobility and bioavailability of heavy metals and metalloids in soil environments. *Journal of Soil Science and Plant Nutrition* 10: 268–292.

Wang, S., Mulligan, C.N. 2009. Arsenic mobilization from mine tailings in the presence of a biosurfactant. *Applied Geochemistry* 24: 928–935.

Wei, X., Gao, B., Wang, P., Zhou, H., Lu, J. 2015. Pollution characteristics and health risk assessment of heavy metals in street dusts from different functional areas in Beijing, China. *Ecotoxicology and Environmental Safety* 112: 186–192.

Wilbur, S., Abadin, H., Fay, M., Yu, D., Tencza, B., Ingerman, L., Klotzbach, J., James, S. 2012. Toxicological profile for chromium. https://www.ncbi.nlm.nih.gov/books/.

Wu, Q., Leung, J.Y.S., Geng, X., Chen, S., Huang, X., Li, H., Huang, Z., Zhu, L., Chen, J., Lu, Y. 2015. Heavy metal contamination of soil and water in the vicinity of an abandoned e-waste recycling site: Implications for dissemination of heavy metals. *Science of the Total Environment* 506–507: 217–225.

Wuana, R.A., Okieimen, F.E. 2011. Heavy metals in contaminated soils: A review of sources, chemistry, risks and best available strategies for remediation. *ISRN Ecology* 2011: 20.

Yao, Z., Li, J., Xie, H., Yu, C. 2012. Review on remediation technologies of soil contaminated by heavy metals. *Procedia Environmental Sciences* 16:722–729.

Zheng, C., Wang, M., Wang, Y., Huang, Z. 2012. Optimization of biosurfactant-mediated oil extraction from oil sludge. *Bioresource Technology* 110: 338–342.

Zheng, G., Wong, J.W.C. 2010. Application of microemulsion to remediate organochlorine pesticides contaminated soils. *Proceedings of the Annual International Conference on Soils, Sediments, Water and Energy* 15: 22–35.

27 Modeling Microbial Energetics and Community Dynamics

Bhavna Arora, Yiwei Cheng, Eric King, Nicholas Bouskill, and Eoin Brodie

CONTENTS

ABSTRACT

Biogeochemical reactions and their rates are largely controlled by microbes. There is thus a growing interest in the inclusion of complex microbial communities, their functioning, and interactions into reactive transport models. Conventionally, these models have included thermodynamic-based approaches that describe the stoichiometry and energetics of microbial reactions. In this chapter, we describe two emergent approaches that can be used to model microbial community dynamics—trait-based approach and dynamic energy allocation model. We explain their practice, limitations, and strategies for improvement.

27.1 INTRODUCTION

A major goal of environmental modeling is to incorporate complex, microbially driven reaction networks in reactive transport models to be able to provide a comprehensive and quantitative understanding of biogeochemical transformations and mass transfers in the subsurface. The application of the traditional microbial energetics approach to environmental remediation problems has been quite successful and led to several new insights (Hunter et al. 1998; Dale et al. 2006; Istok et al. 2010; Hubbard et al. 2014; Riley et al. 2014).

However, there are challenges associated with the traditional thermodynamically based microbial models and there is much that needs to be done (Regnier et al. 2005; Beller et al. 2014; Steefel et al. 2015). First, there is a need to understand the relationship between microbial community composition and ecosystem function as it is shaped by, and in turn shapes, the environment. Second, the phylogenetic and functional diversity of microbial taxa is enormous and distributed through both stochastic and deterministic processes across every biome on Earth (Chesson 2003; Martiny et al. 2006; Chase and Myers 2011). These characteristics of the microbial community have precluded moving beyond the black box model of microbial function toward a systematic approach to represent the diversity in microbial function in biogeochemical and reactive transport models.

Attempts to simplify the diversity of microbial communities into coherent functional units have recently found traction in the concept and application of trait-based approaches (Green et al. 2008; Allison 2012; Martiny et al. 2015). Our definition of microbial trait-based approaches includes models that incorporate the functional diversity of microbial communities through parameterization of key traits that determine their fitness within the environment. Such microbial trait-based models have been applied at both the microscale (Allison 2012; Bouskill et al. 2012; Manzoni et al. 2014) and scaled

445

up to global representation of photoautotrophic (Follows et al. 2007) and diazotroph (Monteiro et al. 2011) distribution (Wieder et al. 2015 and references therein).

Another important aspect of microbial modeling is the ability to understand the consequences of environmental perturbations and how they impact microbial community dynamics. A promising modeling approach that can account for microbial response to environmental conditions is the dynamic energy allocation method. This method deviates from the traditional (fixed energy partitioning) model by dynamically partitioning intracellular energy requirements based on the physical conditions and the availability of relevant substrate and nutrients (Algar and Vallino 2014). This approach is therefore a system-centric view of microbial ecosystem functioning.

There are several approaches that can be used to model ecosystem dynamics within the subsurface environment (Figure 27.1). Song et al. (2014) provide an excellent overview of the different mathematical approaches that are used, ranging from low-resolution supraorganismal modeling unit to high-resolution individual-based modeling. In this chapter, we discuss the traditional thermodynamic-based models and two emergent approaches, functional trait-based models and dynamic energy partitioning models, whose goals are to better represent the linkages between microbial metabolic diversity and ecosystem functioning, and consider the dynamic evolution of microbial ecosystems. We also consider current obstacles to implementation of these microbial population-based models and suggest opportunities for the integration of culture-independent (meta-omic) and culture-dependent experiments to structure and parameterize these models. Finally, we highlight the importance of representing biochemical trade-offs

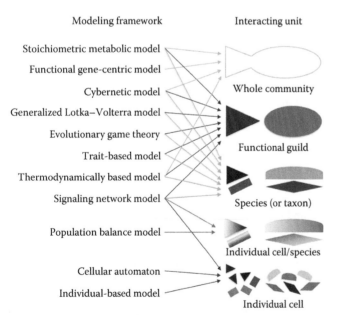

FIGURE 27.1 Mathematical framework of microbial models can depend on the representative interacting unit. This modeling/interacting unit can vary from a single-cell level to the whole community. (Courtesy of Song, H.-S. et al., *Processes*, 2(4), 711, 2014, Figure 2.)

and intracellular resource allocation (dynamic partitioning) to enable a more realistic representation of the ecology and dynamics of microorganisms within diverse ecosystems.

This chapter is organized as follows. Section 27.2 introduces the traditional thermodynamic approach for describing the stoichiometry and energetics of microbial reactions. Section 27.3 provides a description of novel approaches used for modeling microbial community dynamics in reactive transport models. As we move from traditional to modern population-based approaches, we present example applications for each of these. These examples highlight the step-by-step calculations required to represent microbial dynamics in such models. Section 27.4 provides a summary and outlook for the key aspects that we believe could be further developed to improve the predictive value of microbial community models.

27.2 A HISTORICAL VIEW OF MODELING MICROBIAL GROWTH AND SUBSTRATE UTILIZATION

27.2.1 General Kinetic Formulation

In all environments including the subsurface, microorganisms catalyze redox reactions to conserve energy for growth and maintenance. Traditional thermodynamic-based microbial models take into consideration microbial growth, substrate kinetic, and thermodynamic limitations as follows (e.g., Steefel et al. 2015):

$$R_i^j = \mu_i^{ED} \cdot B_i \cdot F_K \cdot F_T \qquad (27.1)$$

where

R (concentration time^{-1}) and μ (time^{-1}) are the reaction rate and maximum growth rate of microbe i, achieved through the utilization of a specific electron donor (*ED*), respectively

B is the concentration of the microbe i

F_K represents the kinetic limitation function ($0 < F_K < 1$)

F_T represents the thermodynamic limitation function ($0 < F_T < 1$) (described in more detail in Section 27.2.1.1)

The kinetic limitation function, F_K, on the microbial population is mathematically described by the general Monod formulation as follows:

$$F_K = \frac{[ED]}{[ED] + K_{ED}} \times \frac{[EA]}{[EA] + K_{EA}} \times \frac{K_{INB}}{[INB] + K_{INB}} \qquad (27.2)$$

where K (concentration) is the half saturation or affinity constant of the electron donor (*ED*)/electron acceptor (*EA*)/inhibitor (*INB*). The first two terms in Equation 27.2 describe the limitation of the electron donor and electron acceptor concentrations on the reaction rate following the Michaelis–Menten (MM) equations. For each MM formulation, if the concentration becomes significantly smaller relative to the half saturation constant, the formulation approaches [concentration]/K. On the other hand, if the concentration becomes significantly greater relative to the half saturation constant, the formulation approaches 1. The last term describes how the concentration of a particular inhibitor limits the reaction rate. In contrast to the MM formulation, for this inhibition term, the higher the concentration of inhibitor, the closer the term approaches 0.

Energy obtained through the redox reaction is used by microorganisms for growth and maintenance. The energetics concepts described by Rittmann and McCarty (2001) provide a framework for quantifying the fraction of energy used for cell synthesis, fs (anabolic), and the fraction of energy used for energy production, fe (catabolic). fs is closely related to cell growth yield per electron equivalent. Using the conceptual framework of Rittmann and McCarty (2001), Roden and Jin (2011) calculated free energies of 38 metabolic pathways with data compiled from 123 experiments. Their work demonstrated a linear relationship between catabolic free energy and growth yield. Values of fs and fe are dependent on the electron donors and acceptors involved in the reactions (Rittmann and McCarty 2001; Roden and Jin 2011). As an example, for a redox reaction between acetate (electron donor) and sulfate (electron acceptor), $fs = 0.08$ (Rittmann and McCarty 2001) and because $fs + fe = 1$, $fe = 0.92$.

The parameters fs and fe are crucial in combining electron donor, electron acceptor, and cell synthesis half equations into one stoichiometric equation. Following Rittmann and McCarty (2001),

$$R = fe \cdot R_a + fs \cdot R_c - R_d \qquad (27.3)$$

where

R is the overall equation
R_a is the electron acceptor half equation
R_d is the electron donor half equation
R_c is the cell synthesis half equation

Continuing with our sulfate reduction example,

$$R_a: 1/8SO_4^{2-} + 9/8H^+ + e^- = 1/8HS^- + 1/2H_2O \qquad (27.4)$$

$$R_d: 1/4HCO_3^- + 9/8H^+ + e^- = 1/8CH_3COO^- + 1/2H_2O \qquad (27.5)$$

$$R_c: 1/4HCO_3^- + 1/20NH_4^+ + 6/5H^+ + e^- \\ = 1/20C_5H_7O_2N + 13/20H_2O \qquad (27.6)$$

Following Equation 27.3, the final balanced reaction relating electron donor, electron acceptor, and microbial cell can be written as

$$R: 0.115SO_4^{2-} + 0.125CH_3COO^- + 0.004NH_4^+ + 0.006H^+ \\ \rightarrow 0.004C_5H_7O_2N + 0.23HCO_3^- + 0.115HS^- + 0.012H_2O \qquad (27.7)$$

A stoichiometric formula of $C_5H_7O_2N$ is commonly used to represent microbial cell biomass. For other stoichiometric formulae, the reader is referred to Chapter 2, Table 2.1 (Rittmann and McCarty 2001). Note that in this example, ammonium is the nitrogen source. Cell synthesis equations can also be set up for other sources of nitrogen (i.e., nitrate, nitrite, or dinitrogen; see Rittmann and McCarty [2001], Chapter 2, Table 2.4).

27.2.1.1 Thermodynamic Limitation on Reaction Rates

The function, F_T (from Equation 27.1), represents a dimensionless thermodynamic potential factor that constrains microbially mediated reaction rates. In the past decade, different F_T formulations have demonstrated the increasing recognition of the need for a more robust method to model microbially mediated reaction rates in natural settings, which are typically of low or intermittent energy availability (Jin and Bethke 2002, 2003, 2005, 2009; LaRowe et al. 2012). At present, there are two common formulations of F_T, which differ in the proxy utilized to represent ΔG_{min}, the minimum amount of energy microbial cells harvest for growth and maintenance (Jin and Bethke 2009; LaRowe et al. 2012). In the Jin and Bethke (2009) formulation, ΔG_{min} is represented by the energetics of ATP synthesis by microorganisms, such that

$$F_T = 1 - e^{\left(\frac{\Delta G_r + m \cdot \Delta G_{ATP}}{\chi \cdot R \cdot T} \right)} \qquad (27.8)$$

where

R is the gas constant (8.314 J mol^{-1} K^{-1})
T is the temperature (K)
ΔG_r is the Gibbs free energy of a reaction per electron transferred (see Equation 27.10)
m is the number of moles of ATP produced per reaction
χ represents the average stoichiometric number for the reaction or the number of times the rate determining step occurs in the overall reaction

ΔG_{ATP} is the Gibbs free energy required to synthesize 1 mol of ATP, which is typically assumed to be ~60 kJ mol^{-1} (Thauer et al. 1977). $m\Delta G_{ATP}$ is therefore the energy threshold required to synthesize ATP. While m is generally not well known, one can determine $m\Delta G_{ATP}$ for many terminal electron accepting processes. For example, iron reducers can grow with $m\Delta G_{ATP}$ as low as −7 kJ mol^{-1}, sulfate reducers with −4 kJ mol^{-1}, and methanogens with −0.5 kJ mol^{-1} (Watson et al. 2003 and references therein). This methodology suffers, however, from requiring multiple adjustable parameters (i.e., m and χ) for each redox reaction.

On the other hand, LaRowe et al. (2012) proposed ΔG_{min} to be represented by the energetics required to maintain a membrane potential:

$$F_T = \frac{1}{e^{\left(\frac{\Delta G_r + F\Delta\Psi}{RT} \right)} + 1} \qquad (27.9)$$

where

F is Faraday's constant (96485.34 C mol^{-1})
$\Delta\Psi$ is the membrane potential

$\Delta\Psi$ is set at 120 mV, an optimal value for ATP production (Dimroth et al. 2003; Kadenbach 2003; Toei et al. 2007). This method has the added benefit of needing only one adjustable parameter. For both formulations, ΔG_r is the Gibbs free energy of the redox reaction per electron transferred (kJ mol e^{-1}) and is calculated as

$$\Delta G_r = -RT \ln\left(\frac{K_{eq}}{Q}\right) \tag{27.10}$$

where

K_{eq} is the reaction equilibrium constant
Q is the reaction quotient of the same reaction

Q is calculated as

$$Q = \prod_i a_i^{v_i} \tag{27.11}$$

where a_i and v_i are the activity coefficient (concentration^{-1}) and stoichiometric coefficient of chemical species i in the redox reaction, respectively. For example, consider the microbial reduction of sulfate with acetate as an electron donor under anaerobic conditions:

$$CH_3COO^- + SO_4^{2-} \leftrightarrow 2HCO_3^- + HS^-$$

For this reaction, Q can be calculated as

$$Q = \frac{a_{HCO_3^-}^2 \cdot a_{HS^-}^1}{a_{SO_4^{2-}}^1 \cdot a_{CH_3COO^-}^1} \tag{27.12}$$

Note that both formulations of F_T are able to demonstrate thermodynamic limitation on microbially mediated reactions. The first method relates the energy threshold to $m\Delta G_{ATP}$, whereas the latter uses $F\Delta\Psi$. Using a membrane potential of 120 mV and assuming $\chi = 1$ (Dale et al. 2006) result in a threshold value of $F\Delta\Psi = \sim11.6$ kJ mol^{-1}. This value falls within the range of $m\Delta G_{ATP}$ values for anaerobic processes.

27.2.1.2 Example Application

We present an example representing acetate biostimulation experiments conducted at the U.S. Department of Energy (DOE) Integrated Field Research Challenge at Rifle Colorado (Anderson et al. 2003; Vrionis et al. 2005; Williams et al. 2011; Yabusaki et al. 2007). The Rifle site is a former uranium ore processing facility that was contaminated with uranium (in the micromolar range) in its shallow aquifer (Yabusaki et al. 2007, 2011). As is commonly known, bioremediation of uranium in subsurface environments can be achieved through the injection of organic carbon. The influx of organic carbon stimulates electron transfer by indigenous bacteria that mediate the reduction of soluble U(VI) to insoluble U(IV) limiting migration in the environment.

Over multiple experiments, acetate was injected into the Rifle subsurface to stimulate Fe(III)-reducing microbes and reduce U(VI) concentrations in the groundwater (Anderson et al. 2003; Vrionis et al. 2005; Williams et al. 2011). In 2010,

in addition to acetate, bicarbonate was also injected to better understand uranium complexation and desorption from sediments (Shiel et al. 2013). These field experiments coupled with reactive transport modeling studies (Fang et al. 2009; Li et al. 2009; Yabusaki et al. 2011; Bao et al. 2014) yielded numerous insights pertaining to bioremediation of uranium in subsurface environments. For example, Li et al. (2009, 2010) demonstrated that biostimulation experiments often result in mineral transformation and biomass accumulation, both of which can potentially change flow paths and the efficacy of bioremediation.

The major microbial reactions included in these reactive transport modeling studies are the reduction of Fe(III), U(VI), and sulfate (Fang et al. 2009; Li et al. 2009; Yabusaki et al. 2011; Bao et al. 2014). Based on the bioenergetics concepts as outlined in Rittmann and McCarty (2001) and described earlier, the stoichiometric reactions are

Sulfate reduction

$$SO_4^{-2} + 1.082CH_3COO^- + 0.052H^+ + 0.035NH_4^+ \\ \rightarrow 0.035C_5H_7O_2N_{(SRB)} + 2HCO_3^- + 0.104H_2O + HS^- \tag{27.13}$$

Fe(III) reduction

$$FeOOH_{(s)} + 1.925H^+ + 0.033NH_4^+ + 0.208CH_3COO^- \rightarrow Fe^{+2} \\ + 0.033C_5H_7O_2N_{(FeRB)} + 0.25HCO_3^- + 1.6H_2O \tag{27.14}$$

U(VI) reduction

$$UO_2^{+2} + 0.067NH_4^+ + 0.417CH_3COO^- + 0.8H_2O \rightarrow UO_{2(s)} \\ + 0.067C_5H_7O_2N_{(FeRB)} + 0.5HCO_3^- + 2.15H^+ \tag{27.15}$$

where

SRB refer to the sulfate-reducing bacteria
FeRB denote the iron-reducing bacteria

Following Equation 27.1, the rates of these microbially mediated kinetic reactions can be obtained as

Sulfate reduction

$$R_{sulfate} = \mu_{SRB}^{acetate} \cdot B_{SRB} \cdot \frac{C^{acetate}}{K_s^{acetate} + C^{acetate}} \cdot \frac{C^{sulfate}}{K_s^{sulfate} + C^{sulfate}} \cdot \\ \times \frac{K_I^{Fe(III)}}{K_I^{Fe(III)} + C^{Fe(III)}} \cdot F_T \tag{27.16}$$

Fe(III) reduction

$$R_{Fe(III)} = \mu_{FeRB}^{Fe(III)} \cdot B_{FeRB} \cdot \frac{C^{Fe(III)}}{K_s^{Fe(III)} + C^{Fe(III)}} \cdot \\ \times \frac{C^{acetate}}{K_s^{acetate} + C^{acetate}} \cdot F_T \tag{27.17}$$

U(VI) reduction

$$R_{U(VI)} = \mu_{FeRB}^{U(VI)} \cdot B_{FeRB} \cdot \frac{C^{U(VI)}}{K_s^{U(VI)} + C^{U(VI)}} \cdot \frac{C^{acetate}}{K_s^{acetate} + C^{acetate}} \cdot F_T$$

(27.18)

1D and 2D reactive transport modeling studies at the Rifle site demonstrated the importance of quantifying biomass growth and mineral transformation/accumulation since they can impact reactions rates during field bioremediation (Li et al. 2009, 2010). Yabusaki et al. (2011) developed a 3D reactive transport model of the Rifle field experiment that showed the important interactions between subsurface hydrology (e.g., falling water table) and biogeochemical reactions governing uranium reduction rates. Recent modeling work at the site includes using dynamic energy consumption rates (see Section 27.3.2.1).

27.3 AN EMERGENT VIEW OF MODELING MICROBIAL COMMUNITY DYNAMICS

27.3.1 Trait-Based Approaches to Representing Microbial Ecosystems

The impetus behind the advancement of microbial trait theory has been derived from the successful application of trait theory to depict and predict plant distribution as a function of the contemporary and anticipated environment (e.g., Shipley et al. 2006; Ackerly and Cornwell 2007; Cornwell and Ackerly 2009). In such cases, traits are defined as morphological, physiological, phenological, or behavioral features of individuals that regulate their fitness within an environment (Violle et al. 2014) and include traits such as leaf area, leaf nitrogen content, and wood density (Díaz et al. 2015). Through this approach, traits can be disaggregated into response traits, which determine the response of key facets of a microbial population (e.g., biomass abundance) to environmental change, and effect traits, which exert strong effects on ecosystem functioning (Lavorel and Garnier 2002). Furthermore, frameworks have recently been advanced for the quantitative scaling of trait theory to ecosystem or global scale (Enquist et al. 2015).

A similar approach can be taken to reduce the complexity of microbial communities to constitutive traits, whereby traits are characteristics (again, morphological, functional, physiological) determining the phenotype, and therefore the fitness, of a given organism under specific environmental conditions (Martiny et al. 2015). These include traits involved in carbon or nutrient cycling, such as cellulose, nitrogenase, or phosphatase expression, traits influencing population viability within heterogeneous environments, including extracellular polysaccharide production or dormancy, and traits that facilitate function under stress, such as osmolyte synthesis (see Martiny et al. 2014, 2015).

27.3.1.1 Example Application

An example of trait-based modeling is the work of Follows et al. (2007), where the authors parameterized an ecosystem of interacting phytoplankton through the stochastic assignment of multiple physiological traits determining light, temperature, and nutrient requirements. Phytoplankton members were determined through a "virtual coin flip" that classified them as either large or small. All large organisms were given higher fixed growth and sinking rates with half saturation constants for nutrient uptake randomly selected from a higher range of values compared to smaller organisms. This work was among the first study to demonstrate the utility of using a self-emerging model of microbial ecosystems to address global-scale questions in ecology and biogeochemistry.

Transitioning from the ocean to a terrestrial setting, Allison (2012) developed a trait-based model linking community composition with physiological and extracellular enzymatic traits to depict community dynamics and interactions between heterotrophic bacteria on the surface of leaf litter. The model explicitly represents the trade-offs involved in production of extracellular enzymes, which, in part, determine emergent bacterial community competitive dynamics dependent on the extent of intracellular resource investment. That the traits influencing enzyme production are shown to have a significant role in rates of decomposition has implications for the next generation of land models steeped in mechanistic fidelity.

27.3.1.2 Representation of Trade-Offs and Allocation Strategies Important for Resolving Diversity

While the complement of traits determines the functional potential of an individual, fitness within a given environment is critically dependent on the understudied role of metabolic or biochemical trade-offs between traits (Johnson et al. 2012; Maharjan et al. 2013). Trade-offs between traits serve to optimize microbial fitness in the environment. For example, Maharajan and coworkers (Maharjan et al. 2013) demonstrated, using engineered *Escherichia coli*, that trade-offs between growth rate and resistance to perturbation determine community diversity under fluctuating conditions. These findings can be explained to some extent by two discrete regulatory components in *E. coli*, the *rpoD* and *rpoS* systems, which direct RNA polymerase to genes involved in housekeeping functions or stress response, respectively (Hengge-Aronis 2002). Because there is a finite concentration of transcriptionally available polymerase, balancing between the requirement to maximize growth rate (by increasing investment into *rpoD*) and bet-hedging against future perturbation (through investment in *rpoS*) can determine the competitiveness of different genotypes within an ecosystem.

Beardmore et al. (2011) demonstrated the importance of trade-offs to maintaining community diversity using a mathematical model to resolve previously reported observations of significant diversity within clonal microbial populations grown in homogeneous environments (Maharjan et al. 2006). The model represented diversity as a function of the number of trade-offs and the mutation rate. Organisms growing with one trade-off (between resource uptake and cellular yield) and at low mutation rates showed a "survival of the fittest" scenario with an ecosystem of low diversity dominated by a high narrow fitness peak inhabited by a couple of individuals. At high-mutation rates, this fitness peak flattens out and is dominated by a larger number of more inefficient genotypes. At intermediate mutation rates, both fit and flat genotypes can be maintained, the range of which widens considerably with additional trade-offs.

Despite the clear importance of coupled trait and trade-off systems, few studies have sought to quantitatively or qualitatively understand their relevance in determining the structure and robustness of complex microbial communities. However, models of microbial physiology have begun to represent the internal allocation of resources within an individual cell that optimize biomass development and response to the external environment (e.g., polysaccharide exudation, osmolyte production), to maintain competitiveness within an ecosystem (e.g., Manzoni et al. 2014; Kooijman 2010). These dynamic energy allocation models show promise by representing a potential physiological response of microbial ecosystems to changing environmental conditions.

27.3.2 Dynamic Energy Allocation

In the previously considered model formulations (Section 27.2.1), a theme emerges in that energy allocated from redox reactions into catabolic and anabolic processes is constant over time. However, the energetics of microbial reactions can be affected by changes in environmental conditions. This variability can result naturally (e.g., through seasonal fluctuations in temperature and precipitation) or can be anthropogenic in nature (e.g., through fertilizer runoff or bioremediation). These dynamics can affect microbial abundances and rates of nutrient cycling, which in turn can result in fluctuations in bioavailable energy. Models describing biogeochemical dynamics, however, do not typically involve explicit linking of reactions to microbial groups, let alone dynamic calculation of energetics and its impact on growth rates. This is in contrast to the fact that when there is more energy provided by a redox reaction, a greater portion can be utilized for growth (Rittmann and McCarty 2001). Section 27.2.1 describes a classical approach for calculating energy allocation from redox reactions into catabolic and anabolic processes (Dale et al. 2006; Rittmann and McCarty 2001). Dynamic energy partitioning takes this idea a step forward by allocating the proportion of energy derived from the metabolism of substrates based upon dynamic environmental conditions.

Before exploring the concept of dynamic energy partitioning, let us first separate this concept from calculations on the thermodynamic feasibility of a reaction. Two methodologies are commonly used (see Section 27.2.1.1) based upon limitations arising from the energetics required for ATP synthesis (Jin and Bethke 2009) and to sustain a membrane potential (LaRowe et al. 2012). These methods, however, only take into account whether or not a redox reaction produces enough energy to be feasible for an organism, but not where the energy is ultimately allocated. González-Cabaleiro et al. (2015) used a simplified kinetic rate expression to investigate microbial growth in wastewater treatment plants by

dynamically allocating energy, for example, between maintenance and anabolic reactions. This allowed for the derivation of dynamic microbial growth yields, which were used to successfully investigate two microbial ecosystems associated with glucose fermentation and nitrogen oxidation/reduction.

Maximum entropy production (MEP) is a concept used in ecosystem models that differs significantly from those previously considered but serves as an alternative method for dynamic energy allocation. Algar and Vallino (2014) utilized this technique to study changes in prevalent nitrogen transformation processes as a consequence of environmental perturbations. As opposed to methods that take an organism-centric view, MEP takes a system perspective that utilizes information on environmental conditions and energy availability (Vallino 2010, 2011; Algar and Vallino 2014). This concept builds upon the theory that systems have evolved to maximize the average rate of entropy production by all organisms over a characteristic timescale. Systems can therefore efficiently extract the most energy over time, for example, over a period of weeks rather than hours. Briefly, this technique can be considered in the context of the examples given in Section 27.2.1.2. Following the formulation given in Algar and Vallino (2014), sulfate, iron, and uranium reduction reactions would be written as follows:

$$a_1CH_2O + b_1SO_4^{-2} + c_1H^+ + \gamma_1NH_4^+ \rightarrow f_{e,1}\mathbb{S}_1 + (1 - f_{e,1}) \\ HCO_3^- + d_1HS^- + e_1H_2O \quad (27.19)$$

$$a_2CH_2O + b_2FeOOH + c_2H^+ + \gamma_2NH_4^+ \rightarrow f_{e,2}\mathbb{S}_2 + (1 - f_{e,2}) \\ HCO_3^- + d_2HS^- + e_2H_2O \quad (27.20)$$

$$a_3CH_2O + b_3UO_2^{+2} + c_3H^+ + \gamma_3NH_4^+ \rightarrow f_{e,3}\mathbb{S}_3 + (1 - f_{e,3}) \\ HCO_3^- + d_3HS^- + e_3H_2O \quad (27.21)$$

where
  CH_2O is the organic matter
  fe is the proportion of energy for anabolic reactions
    $(0 < fe < 1)$
  γ is the stoichiometry of nitrogen in the formulation of biomass
  $\mathbb{S}$ stoichiometries of chemical species a, b, c, d, and e depend on fe

Simulations optimize fe among other parameters (not shown here) for a set of reactions with the goal of maximizing the Gibbs free energy (as a proxy for entropy production) over a certain timescale. As a comparison to the formulation of Rittmann and McCarty (2001), if one were to use $fe_1 = 0.08$, $\gamma_1 = 0.2$, and acetate as the source of organic matter, reaction (27.19) would be equivalent to reaction (27.13).

27.3.2.1 Example Application

The importance of dynamic energy allocation can be seen from the example application introduced in Section 27.2.1.2. A trait-based reaction-transport model, parameterized in part from metagenomic data sets, is being developed to investigate

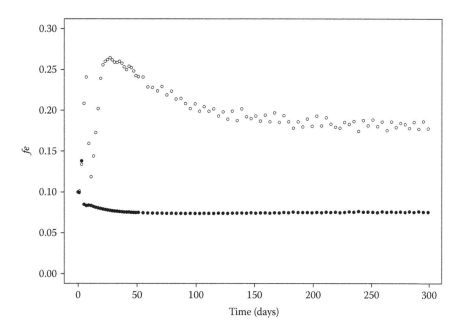

FIGURE 27.2 Fraction of energy (*fe*) going toward anabolic reactions for iron reduction at two spatial locations within the Rifle floodplain (open vs. closed circles).

biogeochemical dynamics at the Rifle, Colorado field site. The model framework incorporates dynamic energy calculations for a set of reactions as they evolve over time, revealing the importance of dynamic versus static calculations of *fe* and *fs*. For example, iron reduction was modeled using *Geobacter* through Equations 27.14 and 27.17. However, Equation 27.14 was allowed to emerge based on dynamic *fe* in this approach as opposed to being fixed at 0.4 as in Li et al. (2009).

This results in spatial variability as can be seen from Figure 27.2 where *fe* differs by a factor of 2–3 between two locations at the Rifle site, which directly correlates to a two- to threefold difference in maximum growth rates. Note that in both locations, F_T is approximately 1, illustrating that variability can exist for energy allocation while thermodynamic constraints are constant.

27.3.3 OBSTACLES TO FURTHER DEVELOPMENT OF POPULATION-BASED MODELS AND OPPORTUNITIES FOR ADVANCEMENT

27.3.3.1 Model Parameterization

Model results can depend on how reaction rates are parameterized, which typically include maximum specific growth rates, half-saturation constants, and inhibition terms. These parameters can be determined using experimental studies by direct estimation or inverse analysis (Bouskill et al. 2012; Fang et al. 2009). For trait-based models, this approach can yield the mean values for given traits and estimate the variance around that mean necessary for understanding the plasticity of a given population. However, few studies consider the interaction between multiple different traits associated with a genetically coherent population. The covariance between multiple traits governs the trajectory of a population across

environmental gradients (Laughlin and Messier 2015) and is important measurements to quantify in future experiments.

In addition, these parameter values may not exist for uncultured organisms or may not be well constrained. For example, parameters describing sulfate reduction can vary over two orders of magnitude (Hunter et al. 1998), making it difficult to choose appropriate values. Taking a step back to successfully model a system, one should also determine the reactions applicable for the study site. For example, Arora et al. (2016) showed the importance of including chemolithoautotrophic pathways based on field genomic and transcriptomic evidence (Jewell et al. 2016), in a reaction network for accurately predicting CO_2 fluxes from a floodplain environment. Therefore, microbial models can benefit from the wealth of recent data being generated from "omics" based analyses.

27.3.3.2 Technological Advancements and the Path Forward

Early attempts at identifying model parameters and fitness traits were derived from single isolates grown in batch culture (Del Re et al. 2000); however, increasingly, the identification of parameters/traits has been made within complex communities using molecular data (Barberán et al. 2012) or via chemical imaging tools linked to real-time biochemistry (Holman et al. 2009). Analysis of this genomic information through cultivation-independent approaches thus provides an indication of the organisms present and their functional roles. Certain parameters can be directly estimated from this type of data, such as optimal growth temperature of a given microbe using amino acid sequences (Zeldovich et al. 2007) and the use of codon usage bias to parameterize minimum generation times that can be converted to growth rates (Vieira-Silva and Rocha 2010). Therefore, one can model, for example, two different sulfate reducers using this approach, each with their own distinct parameterization. Furthermore, these extensive genomic data sets can also reveal key trait linkages within individuals (e.g., the coupling between), which facilitate the representation of biochemical trade-offs. These advances signify that the field of environmental modeling is primed to make significant contributions toward our understanding of microbial community assembly and succession over space and time.

27.4 SUMMARY AND OUTLOOK

Applications of reactive transport models are growing in part due to their predictive capabilities to represent the dynamics and emergent properties of microbial communities. This chapter summarizes the transition from the traditional thermodynamic-based microbial model to a next generation of microbial population-based methods that are aimed at understanding function, dynamics, and emergence in microbial ecosystems. Trait-based models, which represent microbial diversity through a small number of functional traits, are an attractive approach to link community and ecosystem dynamics to environmental change (Allison 2012). Another important approach is the dynamic energy allocation method that accounts for wide ranging environmental gradients to shape the evolution and emergence of community structure.

While the concept of trait-based modeling and dynamic energy allocation are not mutually exclusive, there is a need to integrate these models toward a more comprehensive description of the dynamic nature of microbial communities. An example of such an integration is the work of Tang and Riley (2014). In this study, the authors demonstrated the importance of appropriate representation of key traits within models by focusing on only one microbial trait, carbon use efficiency (CUE). Taking a dynamic energy budget (DEB) approach, they represent the interactions between polymeric soil organic matter, monomeric dissolved organic matter, microbial populations, and mineral surfaces. Microbial enzymes, their target substrates, and microbes themselves can interact with mineral surfaces implying that CUE is not a static variable (as is commonly assumed in models) but a hysteretic, emergent property of the ecosystem. The authors go on to show that under perturbation their DEB approach implies that soil carbon stocks show a weaker, albeit variable, response to rising temperatures, divergent from the response of current earth system models. Therefore, such model integration provides important insights on the future stability of soil carbon stocks under warming.

The careful parameterization of microbial models is crucial in order to minimize equifinality. The importance of accurately parameterizing these sensitive models (e.g., Maggi et al. 2008) calls for a database compiling key functional trait and trade-off measurements across microbial phylogenies. The advancement in "omics" based approaches is primed to make significant contributions in the identification of traits and development of microbial reaction networks for reactive transport models and the path forward will require critical evaluations on how much complexity or genomic information is necessary versus sufficient to parameterize or structure these models.

ACKNOWLEDGMENT

This material is based upon the work supported as part of the Genomes to Watersheds Subsurface Biogeochemical Research Science Focus Area at Lawrence Berkeley National Laboratory and was funded by the U.S. Department of Energy, Office of Science, Office of Biological and Environmental Research under Award Number DE-AC02-05CH11231.

REFERENCES

Ackerly, D.D. and Cornwell, W.K., 2007. A trait-based approach to community assembly: Partitioning of species trait values into within- and among-community components. *Ecology Letters*, 10(2), 135–145.

Algar, C. and Vallino, J., 2014. Predicting microbial nitrate reduction pathways in coastal sediments. *Aquatic Microbial Ecology*, 71(3), 223–238.

Allison, S.D., 2012. A trait-based approach for modelling microbial litter decomposition. *Ecology Letters*, 15(9), 1058–1070.

Anderson, R.T. et al., 2003. Stimulating the in situ activity of *Geobacter* species to remove uranium from the groundwater of a uranium-contaminated aquifer. *Applied and Environmental Microbiology*, 69(10), 5884–5891.

Arora, B. et al., 2016. Influence of hydrological, biogeochemical and temperature transients on subsurface carbon fluxes in a flood plain environment. *Biogeochemistry*, 127(2), 367–396.

Bao, C. et al., 2014. Uranium bioreduction rates across scales: biogeochemical "hot moments" and "hot spots" during a biostimulation experiment at Rifle, Colorado. *Environmental Science and Technology*, 48(17), 10116–10127.

Barberán, A. et al., 2012. Exploration of community traits as ecological markers in microbial metagenomes. *Molecular Ecology*, 21(8), 1909–1917.

Beardmore, R.E. et al., 2011. Metabolic trade-offs and the maintenance of the fittest and the flattest. *Nature*, 472(7343), 342–346.

Beller, H.R. et al., 2014. Divergent aquifer biogeochemical systems converge on similar and unexpected Cr(VI) reduction products. *Environmental Science and Technology*, 48(18), 10699–10706.

Bouskill, N.J. et al., 2012. Trait-based representation of biological nitrification: Model development, testing, and predicted community composition. *Frontiers in Microbiology*, 3, 364.

Chase, J.M. and Myers, J.A., 2011. Disentangling the importance of ecological niches from stochastic processes across scales. *Philosophical Transactions of the Royal Society of London. Series B, Biological Sciences*, 366(1576), 2351–2363.

Chesson, P., 2003. Mechanisms of maintenance of species diversity.

Cornwell, W.K. and Ackerly, D.D., 2009. Community assembly and shifts in plant trait distributions across an environmental gradient in coastal California. *Ecological Monographs*, 79(1), 109–126.

Dale, A.W., Regnier, P., and Van Cappellen, P., 2006. Bioenergetic controls on anaerobic oxidation of methane (AOM) in coastal marine sediments: A theoretical analysis. *American Journal of Science*, 306(4), 246–294.

Del Re, B. et al., 2000. Adhesion, autoaggregation and hydrophobicity of 13 strains of *Bifidobacterium longum*. *Letters in Applied Microbiology*, 31(6), 438–442.

Díaz, S. et al., 2015. The global spectrum of plant form and function. *Nature*, 529(7585), 167–171.

Dimroth, P. et al., 2003. Electrical power fuels rotary ATP synthase. *Structure*, 11(12), 1469–1473.

Enquist, B.J. et al., 2015. Scaling from traits to ecosystems: Developing a general Trait Driver Theory via integrating trait-based and metabolic scaling theories. *Advances in Ecological Research*, 52, 96.

Fang, Y. et al., 2009. Multicomponent reactive transport modeling of uranium bioremediation field experiments. *Geochimica et Cosmochimica Acta*, 73(20), 6029–6051.

Follows, M.J. et al., 2007. Emergent biogeography of microbial communities in a model ocean. *Science*, 315(5820), 1843–1846.

González-Cabaleiro, R. et al., 2015. Microbial catabolic activities are naturally selected by metabolic energy harvest rate. *The ISME Journal*, 9(12), 2630–2641.

Green, J.L., Bohannan, B.J.M., and Whitaker, R.J., 2008. Microbial biogeography: from taxonomy to traits. *Science*, 320(5879), 1039–1043.

Hengge-Aronis, R., 2002. Signal transduction and regulatory mechanisms involved in control of the S (RpoS) subunit of RNA polymerase. *Microbiology and Molecular Biology Reviews*, 66(3), 373–395.

Holman, H.-Y.N. et al., 2009. Real-time molecular monitoring of chemical environment in obligate anaerobes during oxygen adaptive response. *Proceedings of the National Academy of Sciences of the United States of America*, 106(31), 12599–12604.

Hubbard, C.G. et al., 2014. Isotopic insights into microbial sulfur cycling in oil reservoirs. *Frontiers in Microbiology*, 5, 480.

Hunter, K.S., Wang, Y., and Van Cappellen, P., 1998. Kinetic modeling of microbially-driven redox chemistry of subsurface environments: Coupling transport, microbial metabolism and geochemistry. *Journal of Hydrology*, 209(1–4), 53–80.

Istok, J.D. et al., 2010. A thermodynamically-based model for predicting microbial growth and community composition coupled to system geochemistry: Application to uranium bioreduction. *Journal of Contaminant Hydrology*, 112(1–4), 1–14.

Jewell, T.N. et al., 2016. Metatranscriptomic evidence of pervasive and diverse chemolithoautotrophy relevant to C, S, N and Fe cycling in a shallow alluvial aquifer. *ISME*.

Jin, Q. and Bethke, C.M., 2003. A new rate law describing microbial respiration. *Applied and Environmental Microbiology*, 69(4), 2340–2348.

Jin, Q. and Bethke, C.M., 2009. Cellular energy conservation and the rate of microbial sulfate reduction. *Geology*, 37(11), 1027–1030.

Jin, Q. and Bethke, C.M., 2002. Kinetics of electron transfer through the respiratory chain. *Biophysical Journal*, 83(4), 1797–1808.

Jin, Q. and Bethke, C.M., 2005. Predicting the rate of microbial respiration in geochemical environments. *Geochimica et Cosmochimica Acta*, 69(5), 1133–1143.

Johnson, D.R. et al., 2012. Metabolic specialization and the assembly of microbial communities. *The ISME Journal*, 6(11), 1985–1991.

Kadenbach, B., 2003. Intrinsic and extrinsic uncoupling of oxidative phosphorylation. *Biochimica et Biophysica Acta (BBA)—Bioenergetics*, 1604(2), 77–94.

Kooijman, S.A.L.M., 2010. *Dynamic Energy Budget Theory for Metabolic Organisation*. Cambridge, U.K.: Cambridge University Press.

LaRowe, D.E. et al., 2012. Thermodynamic limitations on microbially catalyzed reaction rates. *Geochimica et Cosmochimica Acta*, 90, 96–109.

Laughlin, D.C. and Messier, J., 2015. Fitness of multidimensional phenotypes in dynamic adaptive landscapes. *Trends in Ecology and Evolution*, 30(8), 487–496.

Lavorel, S. and Garnier, E., 2002. Predicting changes in community composition and ecosystem functioning from plant traits: Revisiting the Holy Grail. *Functional Ecology*, 16(5), 545–556.

Li, L. et al., 2010. Effects of physical and geochemical heterogeneities on mineral transformation and biomass accumulation during biostimulation experiments at Rifle, Colorado. *Journal of Contaminant Hydrology*, 112(1–4), 45–63.

Li, L. et al., 2009. Mineral transformation and biomass accumulation associated with uranium bioremediation at Rifle, Colorado. *Environmental Science and Technology*, 43(14), 5429–5435.

Maggi, F. et al., 2008. A mechanistic treatment of the dominant soil nitrogen cycling processes: Model development, testing, and application. *Journal of Geophysical Research: Biogeosciences*, 113, 1–13.

Maharjan, R. et al., 2006. Clonal adaptive radiation in a constant environment. *Science*, 313(5786), 514–517.

Maharjan, R. et al., 2013. The form of a trade-off determines the response to competition. *Ecology Letters*, 16(10), 1267–1276.

Manzoni, S. et al., 2014. A theoretical analysis of microbial ecophysiological and diffusion limitations to carbon cycling in drying soils. *Soil Biology and Biochemistry*, 73, 69–83.

Martiny, J.B.H. et al., 2006. Microbial biogeography: Putting microorganisms on the map. *Nature reviews. Microbiology*, 4(2), 102–112.

Martiny, J.B.H. et al., 2015. Microbiomes in light of traits: A phylogenetic perspective. *Science*, 350(6261), aac9323–aac9323.

Monteiro, F.M., Dutkiewicz, S., and Follows, M.J., 2011. Biogeographical controls on the marine nitrogen fixers. *Global Biogeochemical Cycles*, 25(2), n/a.

Regnier, P. et al., 2005. Incorporating geomicrobial processes in reactive transport models of subsurface environments. In G. Nützmann, P. Viotti, and P. Aagaard, eds. *Reactive Transport in Soil and Groundwater*. Berlin, Germany: Springer, pp. 109–125.

Riley, W.J. et al., 2014. Long residence times of rapidly decomposable soil organic matter: Application of a multi-phase, multi-component, and vertically resolved model (BAMS1) to soil carbon dynamics. *Geoscientific Model Development*, 7(4), 1335–1355.

Rittmann, B.E. and McCarty, P.E., 2001. *Environmental Biotechnology: Principles and Applications*. New York: McGraw-Hill.

Roden, E.E. and Jin, Q., 2011. Thermodynamics of microbial growth coupled to metabolism of glucose, ethanol, short-chain organic acids, and hydrogen. *Applied and Environmental Microbiology*, 77(5), 1907–1909.

Shiel, A.E. et al., 2013. No measurable changes in (238)U/(235)U due to desorption-adsorption of U(VI) from groundwater at the Rifle, Colorado, integrated field research challenge site. *Environmental Science and Technology*, 47(6), 2535–2541.

Shipley, B., Vile, D., and Garnier, E., 2006. From plant traits to plant communities: a statistical mechanistic approach to biodiversity. *Science*, 314(5800), 812–814.

Song, H.-S. et al., 2014. Mathematical modeling of microbial community dynamics: A methodological review. *Processes*, 2(4), 711–752.

Steefel, C.I. et al., 2015. Reactive transport codes for subsurface environmental simulation. *Computational Geosciences*, 19(3), 445–478.

Tang, J. and Riley, W.J., 2014. Weaker soil carbon–climate feedbacks resulting from microbial and abiotic interactions. *Nature Climate Change*, 5(1), 56–60.

Thauer, R.K., Jungermann, K., and Decker, K., 1977. Energy conservation in chemotrophic anaerobic bacteria. *Bacteriological Reviews*, 41(1), 100–180.

Toei, M. et al., 2007. Dodecamer rotor ring defines H+/ATP ratio for ATP synthesis of prokaryotic V-ATPase from Thermus thermophilus. *Proceedings of the National Academy of Sciences of the United States of America*, 104(51), 20256–20261.

Vallino, J.J., 2011. Differences and implications in biogeochemistry from maximizing entropy production locally versus globally. *Earth System Dynamics*, 2(1), 69–85.

Vallino, J.J., 2010. Ecosystem biogeochemistry considered as a distributed metabolic network ordered by maximum entropy production. *Philosophical Transactions of the Royal Society of London. Series B, Biological Sciences*, 365(1545), 1417–1427.

Vieira-Silva, S. and Rocha, E.P.C., 2010. The systemic imprint of growth and its uses in ecological (meta)genomics. *PLoS Genetics*, 6(1), e1000808.

Violle, C. et al., 2014. The emergence and promise of functional biogeography. *Proceedings of the National Academy of Sciences*, 111(38), 13690–13696.

Vrionis, H.A. et al., 2005. Microbiological and geochemical heterogeneity in an in situ uranium bioremediation field site. *Applied and Environmental Microbiology*, 71, 6308–6318.

Watson, I.A. et al., 2003. Modeling kinetic processes controlling hydrogen and acetate concentrations in an aquifer-derived microcosm. *Environmental Science and Technology*, 37(17), 3910–3919.

Wieder, W.R. et al., 2015. Explicitly representing soil microbial processes in Earth system models. *Global Biogeochemical Cycles*, 29(10), 1782–1800.

Williams, K.H. et al., 2011. Acetate availability and its influence on sustainable bioremediation of uranium-contaminated groundwater. *Geomicrobiology Journal*, 28(5–6), 519–539.

Yabusaki, S.B. et al., 2007. Uranium removal from groundwater via in situ biostimulation: Field-scale modeling of transport and biological processes. *Journal of Contaminant Hydrology*, 93(1–4), 216–235.

Yabusaki, S.B. et al., 2011. Variably saturated flow and multicomponent biogeochemical reactive transport modeling of a uranium bioremediation field experiment. *Journal of Contaminant Hydrology*, 126(3–4), 271–290.

Zeldovich, K.B., Berezovsky, I.N., and Shakhnovich, E.I., 2007. Protein and DNA sequence determinants of thermophilic adaptation. *PLoS Computational Biology*, 3(1), e5.

28 Microbial Electroremediation of Metal-Contaminated Wetland Wastewater

Resource Recovery and Application

Tiyasha Kanjilal, Arijit Nath, and Chiranjib Bhattacharjee

CONTENTS

ABSTRACT

The recovery of electricity from waste or wastewater continues to attract several researchers and research options since it offers the potentiality of reducing the overall expense of treatment while decreasing biomass production. From the electric current and power production viewpoint, the innovation of novel components and cell materials is becoming more significant as striking price and efficient performance will immensely expand the utilization of microbial fuel cells (MFCs). Together with the advantage of providing continuous and logistically simple available fuels with high energy density, microbial electrochemical systems or bioelectrochemical systems (BESs) can be operated for portable applications. While the early studies were mainly based on the development of MFCs with bioanodes, the research ideas of BESs are quickly branching due to attractive developments in the study of biocathodes as well as microbial X cells. The important factors for developing BESs to an economical level are the pH components, the increasing ohmic resistance, and the rising overpotentials. For the purpose of wetland wastewater treatment full of landfill leachate, the combination of MFCs along with the newly developed treatment processes seems to be more effective, plausible, and cost-efficient. At present, the materialization of microbial electrosynthesis focuses on an alternative innovation for sustainable generation through bioelectrochemical route. This technique functions by either obtaining from or providing electric current to potential microbial strains so as to stimulate chemical production.

28.1 INTRODUCTION

The demand for inventive and lucrative *in situ* remediation technologies in waste management kindled the effort to employ conduction phenomenon in environment under an electric field to remove chemical species from soil and water. Globally, large amounts of capital and resources are being exhausted for treating trillions of liters of wastewater annually, consuming momentous amounts of energy. Recent innovative strategies for environmental management are centered exclusively on energy-efficient and energy-gaining processes for waste treatment. Approximately 3%–4% of the national electrical energy consumption is used in water and wastewater treatment, with aerobic processes representing some 60% of sewage treatment operation cost and pumping representing 65% of total water treatment operation (McCarty et al., 2011). As a result of fossil fuel exhaustion and energy price increment, it is suggested that the estimated treatment cost will markedly rise in the coming years. Therefore, wastewater remediation vitally needs energy-sustainable progressions to meet hard-hitting discharge guidelines, which would make a predicament for treatment processes (Howe, 2008). Predictably, wastewater remediation processes have mainly stressed on the removal of organic and inorganic pollutants and solid waste removal, though this pattern has steadily moved toward accounting waste as a potential resource with the advancement of environmental technologies (Angenent et al., 2004). The change from the current aerobic to anaerobic process will proceed with energy cost saving, reduction of sludge production, and effective recovery of energy vectors and functional resources (Iranpour et al., 1999). Revival and reuse of useful resources from wastewater treatment will be of superior consideration to reduce carbon footprint with concurrent increase in the sustainability of treatment technologies and essential material recycling. It is reported that the theoretical energy density in wastewater is in the order of 10^7 J/m³ (2.8 kWh/m³), which is five times as much energy consumed to clean the wastewater in the conventional treatment process (Heidrich et al., 2011). Bioprocess engineering definitely comes under environmentally benign management strategies and useful material recovery using a novel interdisciplinary biotechnology, bioelectrochemical system (BES).

28.2 BES

Microbial or bioelectrochemical systems (BESs) are multidimensional systems that can achieve momentous change in wastewater treatment by accounting them as renewable-energy-based repository units (ElMekawy et al., 2014). Biology and electrochemistry are related genetically as Professor A. Ksenzhek pointed out in his paper (Kuzminskiy et al., 2013). The capability of utilizing electrochemistry for explaining "biological" problems is similar in biochemical and electrochemical redox reactions: (1) both the reactions are heterogeneous; (2) proceed in the similar pH range, temperature, and ionic strength; (3) take place in aqueous environments; and (4) include the stage of the substrate orientation (Kuzminskiy et al., 2013). Electrochemical processes play a significant role in nature because the electrochemical changes occur in all the courses of energy transformation in the cells of micro- and macroorganisms including the human body. In humans, the entire cellular surface membranes cover a few hectares, and all cells are filled with electrolyte ion–containing biological fluids, while biological systems do not contain metal electrodes but have individual systems of electron transport chains that perform the function of electron transfer. These are the conventional purposes

of electrochemistry so far. Besides, the practices of photosynthesis and respiration include electrochemical stages as a vital part. In BESs, organic wastes can be employed as an electron donor for microbes. The electrons released from this microbial oxidation process can either be used as energy or be employed to produce useful chemical products (Kelly and He, 2014). Microbial electrochemical process has been widely examined in the effort of developing sustainable wastewater remediation as well as innovative biotechnology platform aroused by reducing power of organics.

28.2.1 PROCESS DESIGNING

From the past decades, BESs have been in function with good examples of microbial X cells (MXCs) and cellular-extract-free biofuel cells (Shroder, 2011). In the MXCs, the "X" stands for the definitive applications for a particular group of the microbial electrochemical systems. Likewise, from the direct current sources, electrons can be captured or ionic migrations can be effectively done using microbial fuel cells (MFCs) or microbial desalination cells (MDCs) (Mehanna et al., 2010), also chemical production microbial chemical cells (MCCs) (Butler et al., 2010), or enhanced with external power input for fuel production, including hydrogen and methane gas microbial electrolysis cells (MECs) (Cheng et al., 2009). Some of the frequent microbial electrochemical systems are listed in Table 28.1. Several different configurations have been created for BESs, though regarding the wastewater and environmental remediation purpose, a frequently applied design is adopted. It comprises a two-chambered bioreactor vessel, including two compartments that are divided by an ion exchange membrane (IEM). Often, the chambers are neatly connected by plain salt bridge (Morris et al., 2009) or combined into one single unit (Erable et al., 2011). Usually, BESs with double-chambered vessel are preferred for environmental treatment research works where decontamination of particular waste materials is generally examined separately in anodic or cathodic chambers for oxidation or reduction reactions, respectively.

For the anodic bioremediation studies, a usual design is a two-chambered BES reactor, for instance, MFCs that utilize unsustainable or important cathode constituents, which are likely noble metals for the purpose of catalyzing reduction reaction of an electron acceptor mainly oxygen or ferricyanide at the cathode (Luo et al., 2009). Generally, proton exchange membrane is also an important component of the MFCs. These designs make it unfeasible to scale up methods for wastewater treatment owing to high material expenses and system frailty in the environment, like contaminated or polluted soils and groundwater. As for the remedial measures, salt bridges or having no proton link may reduce the expense obtained for MFC manufacture (Morris et al., 2009). However, such systems usually produce weak current, depicting their high internal resistance and waste of ample of electrons on heat. Since the effect of distance between the salt bridge ends on the internal resistance is still vague, the infield applicability of MFCs is affected. In the research

TABLE 28.1

Microbial Electrochemical Systems Applied for Wastewater Treatment

Type of Bioelectrochemical Systems	Description	Reference
Microbial fuel cell (MFC)	Two-chambered fuel cell reactor, comprising anode and cathode, where electron transfer is microbial interceded. Basic application includes wastewater treatment centered on electricity production.	Mehanna et al. (2010)
Microbial electrolysis cell (MEC)	Double-chambered fuel cell reactor consisting of cathode and anode. Here, anodic microbes oxidize organic materials with the simultaneous reduction of H^+ to H_2. The cathodic reactions result in CH_4 formation according to the functions of microbes present.	Cheng et al. (2009)
Microbial desalination cell (MDC)	Double-chambered closed fuel cell reactor consists of cathode and anode. Here, oxidation of organic materials occurs through microorganisms producing electrical potential that transfers ion through IEM.	Cao et al. (2009)
Microbial chemical cell (MCC)	Double-chambered closed fuel cell reactor consists of cathode and anode, where microbial-catalyzed electrical reactions result in the production of target chemicals like hydroxides and hydrogen peroxides.	Butler et al. (2010)
External voltage microbial electrochemical system	Reactor system with external voltage supply for the proliferation of microbial reactions. Effective in wastewater treatment, carbon sequestration, and green remediation.	Modin and Wilen (2012)
Open-type microbial electrochemical system	Reactor system with electrode and designs not confined by any boundaries. Due to exclusion of unit boundaries causing unconventional design setup for specific functions. Primarily utilized for *in situ* green remediation and wastewater treatment.	Rabaey and Rozendal (2010)
Sediment fuel cell	Closed or open type of reactor system with anode dipped in organic sediments and cathode kept at surface exposed to air. Primarily utilized in electric generation from the sediment.	Jung et al. (2014)
Microbial electrochemical snorkel	Open fuel cell reactor system comprising single unit; conducting anodic reactions at one end and cathodic reactions at the other end. There is free direct electron transfer from anode to cathode. Application includes biodegradation of organic material in effluents.	Erable et al. (2011)
Biofuel cells or enzymatic fuel cell	Fuel cell reactor with anode and cathode, where electron transfer is microbial enzyme interceded. Application includes waste treatment and organic fuel source utilization.	Falk et al. (2012)

on wastewater treatment, where benthic MFCs are usually applied, the anode is inserted into the water area or marine sediment rich in organic components and the cathode into the polluted water (He et al., 2013). For utilizing the cathode in research, the designs of BESs have gained less significance due to the setting of potential difference between the electrodes by using a potential workstation or power supply (Mu et al., 2009b). As an alternative to bioanode for sustainable generation of electricity necessary to deliver the electrons for the cathodic remediation, solar or wind power could be effectively used.

28.2.2 MICROBIOLOGY AND REMEDIAL DEVELOPMENT

Several studies have reported the application of an anode as the main electron acceptor to introduce biodegradation of organic pollutants in an anode chamber. However, the majority of these works have efficiently used mixed cultures as the primary inoculums for the growth and development of anode-respiring bacterial species, and also characterizing and screening the bacterial communities that can consume certain toxicants are still a potential challenge. On the contrary, certain bacterial strains enriched on the cathode that can deliberately utilize electrons for dissimilatory reactions have been characterized as *Dechloromonas* sp., *Azospira* sp., and *Dechlorospirillum* sp. (Thrash et al., 2007), also *Geobacter* sp. (Gregory and Lovley, 2005), *Trichococcus* sp., and *Pseudomonas* sp.

(Tandukar et al., 2009), and *Desulfitobacterium* sp. (Aulenta et al., 2009). Optimal application of BES in natural remediation will possibly assist in accepting the fact of using anode- or cathode-respiring microbes exchange electrons with the electrode surface. Therefore, the basic anodic processes have been observed in relation to electron transfer technique. The suggested mechanism includes the use of redox components that are present on the outer cellular surface of microbes (i.e., the appendages) for direct electron transfer and the use of redox mediators that are diffusible for mediated electron transfer. Figure 28.1 illustrates the schematics of MFC for wastewater treatment. The majority of the process of direct electron transfer by microorganisms to electrodes is obtained from the reports of dissimilatory metal–reacting bacteria like *Geobacter* sp. and *Shewanella* sp. (Lovley, 2011). Reports have depicted that bioanodes can be converted to biocathodes on reversing the operating variables (Cheng et al., 2010) that show simultaneous roles of electrochemically activated microbes during electron-releasing and electron-accepting reactions. A current study on the gene expression and mutation research of *Geobacter sulfurreducens* showed that processes of electron transfer from potential electrode differ noticeably from the process of electron transfer to electrode (Strycharz et al., 2011). Nevertheless, certain *Geobacter* spp. have been studied to access electrons directly off the surface of electrode; the electron mediators, namely, antraquinone-2,6-disulfonate and methyl viologen, were present in most

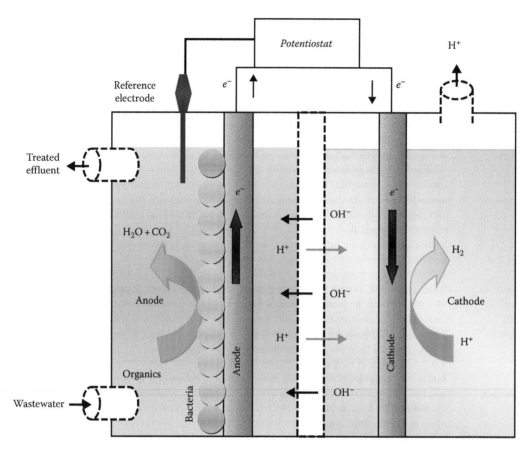

FIGURE 28.1 Schematic illustration of MFC for wastewater treatment.

of the microbial electrochemical remediation works. Apart from antraquinone-2,6-disulfonate and methyl viologen, several other electrochemically active compounds, like quinones (Thrash et al., 2007), phenazines (Rabaey and Verstraete, 2005), and humic substances (Lovley et al., 1999), can be applied in a noncontaminating manner by microbial electron donors or acceptors (Thrash and Coates, 2008). These compounds are either mixed in aqueous solution or attached to the electrode surface; these can be specifically oxidized or reduced by bacteria without being consumed. These compounds usually facilitate the transfer of electrons from potential electrode to the active bacteria and some naturally occurring mediators. However, the supplementation of viable electron mediators may not always be optimizing for wastewater treatment due to the high expense of treatment and extensive procedures of removal from the treated effluent (Figure 28.1).

28.2.3 FEASIBLE OPERATIVE CONDITIONS

In the microbial electrochemical system, the usual metabolic energy (viz., adenosine triphosphate [ATP]) uptake from a particular metabolic process can be expressed by the following equation:

$$\text{ATP} = \sum \left(E_{acceptor}^{0} - E_{donor} \right) \times Q$$

where

E is considered as standard potential of electron acceptor and electron donor

Q is the electric current applied in time t (Huang et al., 2011)

Therefore, optimum electric potential can provide suitable and specific pressure for the growth and differentiation of microorganisms. This specific and independent pressure can cause improved ability of microbes toward electrochemical interaction with electrode and eventual current production. In most degradation studies using BES, the setup potential at electrode generally occupies the position of electron acceptor or donor. The cathode or the anode potential is usually set up in two modes: double-electrode system sets up the working electrode against a counter electrode and three-electrode system sets up the working electrode potential against a reference electrode of constant potential.

The function of microorganisms in the anode is to recover the chemical energy in nonelectrolyte (such as glucose) to an adequate form essential for electrochemical reaction and further convert to electrical energy (Schroder and Harnisch, 2010). Microorganisms then change biochemical energy into ATP by following a series of redox reactions and ultimately transmitting electrons from organic compounds (like glucose) to solid anode. The microbes capable of producing electricity in MFCs are termed as anode-respiring microorganisms since they utilize solid anode as the final electron acceptor (Lee et al., 2008). The rate of growth of such microbial strains depends on the difference between the redox potential of electron donor and the anode potential. Similarly, the lower potential of anode (negative potential) offers a high voltage gradient and as a result maximizes the chance of high current densities in MDCs. Several factors including the kind and concentration of electron donor, electrical properties of MDC, the option of electrode and membranes, and also physiological conditions (like temperature, concentration of substrates, nutrients, agitation, pH), all impose a significant impact on anode potential, which should be taken into account (Borole et al., 2011). Conventionally, the increase in anode potential is observed to yield an increase energy gain for the growth of anode-respiring microbes, though reports predict that a decrease in anode potential is beneficial to the growth of anode-respiring microbial biofilm on the surface of anode (Lee et al., 2008). In one article, comprehensive experiments were conducted to study the effect of anode potential on the developmental pattern of biofilm community on the surface of anode and its simultaneous effect on the electrical performance of MFCs. The experiment made use of activated sludge wastewater as inoculum, acetate was used as an electron donor, and several hydrodynamic factors were standardized in continuous flow reactor. The anode potential of four different MFCs at four different values of 0.15, −0.09, +0.02, and +0.37 V was set using a potentiostat. It was observed that MFC set at the reduced anode potential (−0.09) and revealed a rapid biofilm growth and increased current densities (10.3 A/m²) (Lee et al., 2008). In order to maintain the anode potential at an optimal value that delivers precise growth of anode-respiring microbes in MFCs, a potentiostat can be used. In its absence, the anode potential is analyzed using composition of bacterial communities and physiological conditions like electron donor, temperature, electrochemical loss, and even pH.

28.3 POSSIBLE MECHANISM

Energy production in bacterial metabolism that includes anabolic and catabolic processes is a simultaneous action of fermentation and respiration phenomenon, that is, substrate oxidation and reduction, respectively. Such processes require an electron source that lies in the metabolic flow of bacteria and a strong electron sink to establish the electron transport mechanism. An environment is created to connect the energy given out by bacteria in the form of current density with potential difference created between these two processes (VenkataMohan et al., 2014). The bacteria use the possible substrate, thereby generating the reduced ions (protons and electrons) at the anode. Protons are generated to cathode through electrode interface in solution across ion selective membrane, thereby causing a potential difference between cathode and anode against which electrons move through electric circuit across external load (Pant et al., 2012). The reducing ions produced during BES operation have several applications in the energy production as well as waste treatment areas. For the broader aspect, BES

application can be divided into generation of power, treatment of waste, and recovery of value-added resources. Reducing ions produced from substrate oxidation get metabolized in the occurrence of electron acceptor at the cathode, and thereby power generation occurs. Then again, considering the waste functions as electron donor or electron acceptor, its degradation gets manifested either by anodic oxidation or by cathodic reduction under specific conditions (Pant et al., 2010). Current studies show that during BES operation there is also a reduction of certain substrates such as carbon dioxide as electron acceptors, thereby accelerating its commercial feasibility (Srikanth et al., 2014).

28.3.1 ANODIC ELECTROREMEDIATION

The anode chamber is one of the important factors in the treatment of waste particularly through bacterial metabolism and partly through induced electrochemical oxidation technique. Apart from the electrochemical oxidation, direct and indirect oxidation techniques are also applied for contaminant treatment at anode in BES (VenkataMohan and Srikanth, 2011). Substrate reduction in the anode chamber is generally influenced by oxygen present in the cathode chamber that acts as terminal electron acceptor. Tough electron acceptor conditions at the cathode improve the electron flux in the electric circuit and their release from the bacterial metabolism of wastes. Generally, the contaminants are adsorbed on the surface of anode and get degraded by the electron transfer during direct oxidation, whereas these contaminants will be oxidized by the oxidants produced as a result of electrochemical operation on anode surface during indirect anodic oxidation system. Moreover, the reactions between water and free radicals occurring at the anode surface produce secondary oxidants like nascent oxygen, free chlorine radicals, and hydrogen peroxide, which are applied for organic oxidation purpose (VenkatMohan and Srikanth, 2011). On the contrary, these oxidants can also function as mediators of electron transport between bacteria and surface of anode, thereby helping in their reduction with respective power enhancement. Potential reduction reaction at the cathode influences the substrate's breakdown by inducing oxidation at the anode.

First, the plain organic fraction of pollutant will get oxidized at the anode through bacterial metabolism releasing reducing equivalents, which react with water particles in anode biopotential forming hydroxyl radicals (Israilides et al., 1997). These reactive radical species will be adsorbed onto the anodic sites and facilitate direct anodic oxidation. Water particles and oxygen react with the reactive radical species adsorbed on the electrode, thereby producing secondary oxidants that initiate indirect anodic oxidation process. With the increase in concentration of primary oxidants, the generation of secondary oxidants also speeds up in the electrolyzed solution (Israilides et al., 1997).

The general reactions for oxidant formation are

$$O + e^- \rightarrow O°$$

$$O^\circ + E_C \rightarrow E_C O^\circ$$

$$S + E_C O^\circ \rightarrow S - O^\circ + E_C$$

where

O is the potential oxidant
O° is the oxidant in excited state
E_C is the electrode with active site
S is the substrate used

For the generation of primary oxidants, the following are the reactions:

$$H_2O + E_C + Cl^- \rightarrow E_C \cdot ClOH^- + H^+ + 2e^-$$

$$H_2O + E_C \cdot ClOH^- + Cl^- \rightarrow Cl_2 + E_C + O_2 + 3H^+ + e^-$$

$$C + E_C \cdot ClOH^- \rightarrow E_C + CO + H^+ + Cl^- + e^-$$

For the formation of secondary oxidants, the following are the reactions:

$$H_2O + E_C \cdot ClOH^- + Cl_2 \rightarrow ClO_2 + E_C + 2Cl^- + 3H^+ + e^-$$

$$O_2 + E_C \cdot OH^- \rightarrow E_C + O_3 + H^+ + e^-$$

$$H_2O + E_C \cdot OH^- \rightarrow E_C + H_2O_2 + H^+ + e^-$$

28.3.2 Cathodic Electroremediation

Quite like the anode, the cathode is also occupied in efficient remediation of waste toxicants like azo dyes, nitrobenzene, nitrates, and sulfates. Theoretically, it can be predicted that these toxicants act as terminal electron acceptors at the cathode to shut the electrical circuit in the absence of oxygen, though their purpose as electron acceptor is based upon thermodynamic hierarchy. In contrast with anode, the cathode chamber can be kept under variable microenvironments in order to increase the efficacy of treatment considering the nature of toxicants (Srikanth and VenkataMohan, 2012). Considering the terminal electron acceptor at the cathode, they are of two types: aerobic and anaerobic biocathodes (He and Angenent, 2006). With the aerobic biocathode, the oxidation process results in increased substrate removal. Utilization of H^+ and e^- during aerobic metabolic process will be increased and in turn help in additional removal efficiency of substrate. Many treatment methods undergo simultaneously in the system, causing biochemical reactions resulting in higher toxicant removal. Among the terminal electron acceptor, oxygen initiates the release of hydroxyl ion at the cathode, thereby increasing the oxidation species formation as shown in Figure 28.2. The generation of oxidation species and reactive radicals at cathode biopotential augments the possibility of removal of other toxicants (Aulenta et al., 2010). From the aerobic metabolism, bicarbonates are formed from the reaction of carbon dioxide and water further reacting to form salts, which also function as buffering agent, thereby reducing the chances of redox shifts. The possibility of salt removal through salt splitting mechanism at

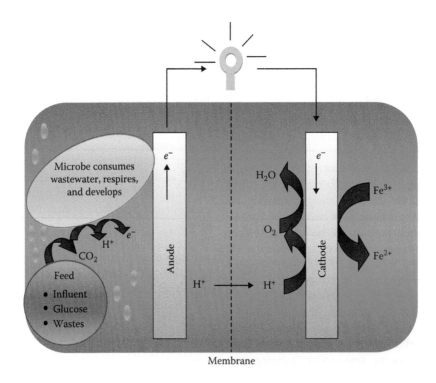

FIGURE 28.2 Schematic diagram of possible microbial electrochemical reactions occurring at anode and cathode during BES operation.

microbial electrochemical system is depicted in the following equations:

$$H_2O + c - a \rightarrow c^+ + a^- + H^+ + OH^-$$

$$H_2O + CO_2 \rightarrow H^+ + HCO_3^-$$

$$HCO_3^- + c^+ + H^+ + OH^- + e^- + O_2 \rightarrow cHCO_3 + H_2O$$

where

c is the cationic type
a is the anionic type

$$c - a + E_C + H_2O \rightarrow E_C \cdot c^+ + a^+ + H^+ + OH^-$$

$$E_C \cdot c^+ + H^+ + OH^- + e^- + O_2 \rightarrow E_C + c - OH + H_2O$$

Cathodic pH maintenance is very important to sustain the bacterial activity. The *in situ* bicarbonate buffering technique produced at the cathode aids to overcome this pH drop, which is by far essential for reduction reaction continuation and microbial metabolism. Physiologically, significant redox reactions in cathode chamber help to increase the metabolic activities of aerobic microbial consortia, thus supporting in efficient substrate removal (Torres, 2014). In general, maintenance of biopotential in such type of reaction will be reduced due to similarity of metabolic functions of microbes and their competition as electron donors and not as electron acceptors. Such a situation will not cause the system to operate the induced oxidation reactions. Alternatively, at the cathode, maintenance of microaerophilic environment has an advantage over the aerobic or anaerobic biocathode operating factors that further help in the treatment of wastes.

28.4 APPLICATION OF BES IN MICROBIAL ELECTROREMEDIATION

The microbial electrochemical system function as treatment unit of wastewater has been expanding currently due to the increased efficiency of waste degradation compared to conventional anaerobic processes (Velvizhi and VenkataMohan, 2011). The fundamentals of microbial electrochemical process depend on the hypothesis that electrochemically activated bacteria can transfer electrons from reduced electron donor to oxidized electron acceptor generating power (Pant et al., 2013). Simultaneous function of bioanode and counter electrode will have increased influence on general wastewater treatment percentage along with the recovery of energy that is tapped.

The prospect of coupling diverse compounds like biological, physical, and chemical components during electrochemical operation initiates numerous reactions like biochemical, bioelectrochemical, electrochemical, and physicochemical reactions, respectively, which are together termed as bioelectrochemical reaction processes. *In situ* occurring biopotential function in increment of the remediation of different toxicants in both anode and cathode chambers. Generation of oxidants and radical species, namely, OH^- and O^-, is advantageous of BES over other expensive systems of treatment (Israilides et al., 1997). At times, wastewater toxicants act as intermediaries in electron transfer. Likewise, elemental sulfur present in waste effluents acts as intermediary at anode and transfers itself to sulfate that is easier to remediate (Dutta et al., 2009). Equally, azo dyes act as intermediaries and decolorize during reduction (Mu et al., 2009a), and estrogenic components get oxidized in BES techniques (Kiran Kumar et al., 2012). The BES technique is also reported for significant reduction of toxicity, color, and TDS from waste solution along with carbon content (VenkataMohan et al., 2014) (Table 28.2). Utilization of BES was also expanded to treat solid waste along with polluted aromatic hydrocarbons under *in situ* conditions (VenkataMohan and Chandrasekhar, 2011). Reports in relation to the pollutant reduction mechanism and electron transfer provide a spectrum of practical viability of such technology for the improved removal of toxic contaminants.

28.5 ISSUES RELATED TO WETLAND WASTEWATER

The formation of substrate is considered as one of the most important biological aspects signifying the treatment efficiency of microbial electroremediation system that affects the recovery of electron. Such system exploits different substrates as electron acceptor and electron donor, along with inorganic and organic compounds, though electron recovery efficiency relies on the oxidation state of electron donor and its relation with microbial strain that can oxidize it. Among the primary substrates, glucose and acetate are commonly used anodic fuels, but other substrates including sucrose, starch, butyrate, dextran, peptone, and ethanol were also considered in microbial electrochemical system due to their power generation properties. Besides the primary substrates, microbial electrochemical system also featured versatility in exploiting a wide range of complex organic pollutants. Such pollutants generated from various origins like industries, agriculture, commercial areas, and residential areas were identified as potential electron donors in the microbial electrochemical system, though pollutants having lower biodegradability such as wetland wastewaters including landfill leachate contamination with high metal and saline concentration will have decreased power generation capacity; still, such pollutants can form a good substrate for any MFC due to its advantage of converting the negatively valued waste products into bioenergy.

28.5.1 TREATMENT OF LOW BIODEGRADABLE WATERS

The microbial electrochemical system can inherently handle highly complex and low biodegradable wastewater such as extremely saline- or metal-contaminated wastewaters as in

TABLE 28.2

Remediation of Dyes, Organic Colorants, and Leachates by Bioelectrochemical System

Types of Dyes or Organic Colorants or Leachates	Types of BES Applied	% of Efficient Treatment	Inferences	Reference
Active brilliant red X-3B azo dye	Air cathode single chamber MFC with glucose as co-substrate.	90% (when initial concentration is 100 mg/L) and 77% (when initial concentration is 150 mg/L).	Generation of electricity in MFCs was unaffected by dye reduction due to competition between anode and azo dye for electron from carbon sources.	Sun et al. (2009)
Acid orange 7 azo dye	Double-chambered BES with oxidation of acetate occurring at anode and dye decolorization at cathode.	78.7% (0.19 mM original dye concentration) and 35% (0.7 mM original dye concentration).	Cathodic decolorization increased on controlling the cathode potential in the range of −350 to −550 MV against standard hydrogen electrode.	Mu et al. (2009)
Methyl orange azo dye	Double-chambered MFC with glucose oxidation at anode by *Klebsiella pneumoniae* and dye decolorization at cathode.	100% (0.05 mM of the original methyl orange solution) in 3 h.	Controlling the redox potentials of pollutant containing cathode is important for the control of power output as well as the degradation rate.	Liu et al. (2009)
Congo red azo dye	Double-chambered MFC with a loop to flow the effluent from anode to cathode. Cell operated initially with glucose (1 g/L) and later added with artificial wastewater containing 100 mg/L of dye.	69.3% (with original glucose concentration of 100 mg/L) and 92.7% (with original glucose concentration of 4000 mg/L).	Recovering electricity during a sequential aerobic–anaerobic azo dye treatment process enhanced COD removal and did not decrease the azo dye removal.	Li et al. (2010)
Landfill leachate (9810 mg/L)	Both the dual-chamber and single-chamber MFC/MDC.	69.54%–98% COD removal depending on initial COD.	Maximum power density of 2060 mW/m^3 for single-chamber MFC.	You et al. (2006)
Landfill leachate (908–3200 mg/L COD)	Three designs, a square (995 mL), circle (934 mL), large state MFC (18.3 L) using graphite as anode and carbon cloth with Pt catalysts as cathode.	BOD, TOC, and ammonium were removed at 50%–72%, 17.55%, and 7%–69%, respectively.	Leachate was used as both the substrate and inoculum, and no additional anaerobic bacteria or nutrient was added. Maximum power density of 669–844 mW/m^3 for circular MFC was obtained.	Jambeck and Damiano (2010)
Landfill leachate (6000 mg/L)	Double-chambered tubular MFC/MDC with carbon veil electrode (360 cm^2)	MFC columns showed low BOD removal (<34% of BOD) as compared to biological aerated filter (66%).	A maximum current density of 0.0004 mA cm was obtained and power density increased linearly with the leachate strength.	Greenman et al. (2009)

the case of wetlands polluted with landfill leachate (Zhang et al., 2008). Leachate is a chief environmental problem affecting the surroundings of landfill in wetland sites. The leachate if not carefully disposed in the landfill might enter groundwater table and pollute the water. Human populations near the landfill using the water for drinking or any other purpose would be in higher risk of health-related issues. The leachate, in particular having higher concentration of metal ions, salts, or organics, is very harmful for human and child life. These are certainly the low biodegradable waste contained in water. The complex form and reduced degradability of the wetland waters contaminated with heavy metal–containing leachate create intricacy in their conversion to reducing equivalents. Moreover, the protons and electrons formed will be accepted by the contaminants in wastewater for further oxidation and generation of lower current densities. These contaminants are recalcitrant to aerobic remediation and can also be used as electron acceptors in microbial electrochemical systems or BES.

28.5.2 Sustainable Additional Microbial Treatment

Application of microbial electrochemical system in the treatment of wetland wastewater including landfill leachate contamination with high metal and saline concentration is considered to be a hopeful technology. This method uses microbial energy from wastewater treatment to drive the ions through IEMs, causing desalination (ElMekawy et al., 2014). This upcoming method can decrease or totally remove the electricity needed for desalination, and it is called microbial desalination cell (MDC). The fundamental feature of the MDC is the microbial strains that are exoelectrogenic, producing electrical potential from the breakdown of organics, which can be used to desalinate water by causing ion transport through IEMs (Kim and Logan, 2013). The wastewater is mainly used as source of organic matter requiring potential gradient development, and therefore, the MDC can attain the goals of energy production, wastewater treatment, and desalination (Kim and Logan, 2013). Another very specialized and recent design included submerged microbial

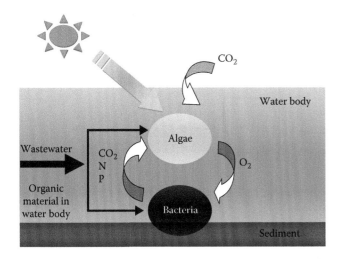

FIGURE 28.3 MFC applying algae for environmental applications.

desalination denitrification cell (SMDDC) to *in situ* eliminate nitrate from groundwater, also producing electric energy together with wastewater treatment (Zhang and Angelidaki, 2013). The SMDDC can be frequently utilized in subsurface environments. When bacteria produced current on the anode, the nitrate and sodium ions were driven to anode and cathode through the anion and cation exchange membrane. The effluent at the anode was driven to the cathode where nitrate was reduced to inorganic nitrogen via autotrophic denitrification. Such a design can remove about 90.5% of nitrate from groundwater in 12 h and also produce 3.4 A/m² of current density. The external nitrification was valuable to generation of current and removal rate of nitrate without affecting the removal efficiency of nitrogen (Zhang and Angelidaki, 2013). The photosynthetic form of MDC was designed and functions applying algae (Figure 28.3) as catalyst in cathode enhancing COD removal and also utilizing the effluent as growth media for acquiring useful biomasses for high-value bioproducts (Kokabian and Gude, 2013). The rise in saline concentrations in anode gives favorable factors for specific microbial types, resulting in the enrichment of the specific microbes and elimination of others that cannot withstand saline conditions (Mehanna et al., 2010). Combination of several techniques of bioprocesses with numerous products can be valuable in upgrading the sustainability of desalination cells using microbes for the treatment of wetland wastewater.

28.6 STUDIES OF BES COMBINING WASTEWATER TREATMENT AND RESOURCE RECOVERY

The primary plus point of microbial electrochemical system or BES is its diversified range of utilizing substrates that include wastes or wastewater and other natural sources. This is cost-effective and sustainable along with its potential to adapt to environment in the form of biofilm further improving its performance through metabolic and genetic engineering. Though this microbial electrochemical system or BES has wide areas

of application, the production of power in the scale-up system would be 120 mW/L, due to its excess capability induced by limitation of mass transport and reaction system kinetics on electrode. Therefore, in order to develop a specific strategy for proper use of low and adequate power generation of BES simultaneously for effective wastewater treatment and resource recovery, the reducing as well as oxidizing potential would be significant. It is required in the current trend to progress in chemical or material production industry and optimized wastewater treatment plants. Currently, several reports suggest the possibility of controlling microbial respiration energy in order to utilize the material recovery of nutrients (ammonia and phosphate) and heavy metal ion in wastewater (Kim et al., 2015) through the upcoming novel approach of BES bioprocess.

28.6.1 APPROACHES FOR BETTER RESOURCE RECOVERY

The conventional organic and nutrient treatment utilizes aerobic techniques that transform nitrogen elements in water into dinitrogen gas. Usually, phosphorus species in components are eliminated biologically or chemically through precipitation in sludge. The traditional oxic or anoxic techniques cause excess energy consumption and additional chemical dose of substrates (acetate and methanol) with heavy cost of sludge disposal. As a result, consumption of energy in nutrient removal treatment techniques is replaced by energetically significant and naturally sustainable systems. In addition, progress of nutrient recovery along with its reusability is very necessary. It has been reported that ammonium ion (NH_4^+) can be driven across cation exchange membrane and mounted up against the concentration gradient in an MFC coupled with generation of electricity (Kim et al., 2008). The study showed implications of microbial electrochemical reactions occurring in MFC that can be used for ammonia elimination or accumulation by IEM together with organic pollutant removal and electricity production (Kim et al., 2008). Also, since heavy metal pollution is of concern and the global trend is moving toward microbial systems (Lloyd et al., 2003), the bacterial reduction and oxidation reactions are applied to change the solubility of heavy metal ions (Pb^{2+}, Fe^{3+}, Mn^{4+}) and to make a feasible separation. A wide number of research have been made on the application of MFC in metal recovery (Tandukar et al., 2009) for chromium, copper, and metal-rich AMD (Heijne et al., 2010). MFCs can also be utilized for the bioremediation of uranium through conversion of radioactive soluble uranium (VI) to insoluble uranium (IV) (Lovley, 1993). The synthesis of biological nanoparticles (Te and Se) has been greatly studied for its extensive biomedical applications like quantum dots (Turner et al., 2012). The genetical modifications of extracellular polymeric substances, biofilm formation, and definite redox reactions have been used efficiently for metal bioremediation and resourcefully in metal nanoparticle generation using the expression of metal-binding protein and appendage (Turner et al., 2012). These research show that microbial respiration or adsorption of metals can be beneficial for separation and selective accumulation of valuable metals with possible improvement of enrichment and biosorption

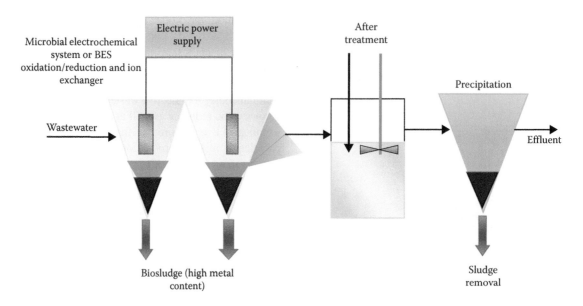

FIGURE 28.4 Metal recovery with hybrid system of BES and metal recovery process.

process. Recent studies implied that chitin-rich deep-sea sediment can power up MFC, producing about 64–76 mW/m^2 (Rezaei et al., 2007), suggesting that microbial-stimulated resource recovery can also be applied to recover metals powered by profuse marine organic matter. Such bacterial respiration of oxidation and reduction of metals can very well convert soluble metal components into less soluble forms, which thereby assist in liquid–solid separation. As a result, electrostatically combined metal ions and bacteria are able to increase the recovery of metals for postprocessing work as described in Figure 28.4. The microbial electrochemical system or BES bioprocess recommends flexibility and applicability for the treatment of metal-loaded multipollutant waste streams (Kim et al., 2015).

28.6.2 ECONOMIC AND ENVIRONMENTAL IMPACT ON THE PROCESSES

Recovery and recycling of valuable components in wastewater have been considered the prime issue in pollution control and environmental sustainability. The generation of bioenergy in the form of biomethane and biohydrogen from wastewater is presently being improved and commercialized. On the other hand, recovery and recycling of valuable components have not been researched upon and utilized, though the concept has been introduced early. Interestingly, these components are profusely distributed in nature, although their economic use is currently hostile due to their heavy cost and analytical complexity of recovery from concentrated sources. The problem of recovery from wastewater is primarily due to ineffective technology and scattered target resources resulting in analytical setbacks and reduced economics. The microbial electrochemical system or BES has been effectively observed to exploit biodegradable organic components in wastewater for figuring separation of valuable resources, utilizing an energy-efficient metal recovery, and recycling. BES technology will

make significant contribution in preventing the depletion of valuable nutrient and materials for human beings (Gilbert, 2012). Such technology will also be involved in academics and industry and national and international organizations through the significant process of sustainable resource management, permitting multidisciplinary advancement of microbiology, biotechnology, electrochemistry, and chemical, environmental, and bioprocess engineering (Kim et al., 2015).

28.7 FACTORS AFFECTING BES AND BOTTLENECKS TO SCALE-UP

The development in the field of microbial electrochemical systems or BES, in terms of material and bioprocess engineering as well as microbial and biotechnological viewpoint, has gone simultaneously. Advancements in design engineering have caused greater power output and higher efficiency along with wide understanding of the factors and methods involved in electron transfer from microbes to the surface of electrode. From the beginning, the primary bottleneck in biofuel cells is the reduced power densities and power generation per unit surface area of electrode due to low transport across cellular membranes (Palmore and Whitesides, 1994). Other major limitations include ohmic voltage losses due to charge transport resistance, overpotential of charge activation from bacteria to electrode, and overpotential of concentrations due to mass transport (Clauwaert et al., 2008). In general, the pH factor, increased ohmic cell resistance, and overpotentials are the prime reasons that prevent the industrial applications of BES (Clauwaert et al., 2008). Another limitation in relevant to BES is related to the upscale feature. Scale-up is always one of the major and difficult barriers, and presently, very few feasible choices for proper and cost-effective increase in scale exist (Logan, 2010). The pilot-scale bioreactor plants require achieving at least similarity of performance as bench-scale reactors, considering the economical production costs and environmentally plausible operations. In order to energize the real-world utilization, MFC units must be operated in stacks, though such stacked configuration is thoroughly susceptible to cell reversal. However, the scale-up from flat sheet to increase in the surface area of packed bed reactor did not form any increment in current due to issues of its distribution and mass transport limitations (Lorenzo et al., 2010). According to a recent report, few of the important challenges related to the scale-up of the microbial electrochemical system or BES include maintenance of low internal resistance while increasing the electrochemically active biomass levels and also reactor design optimization and creating innovative and novel approaches of separating anode from cathode (Fornero et al., 2010). Besides, a frequent limitation, often a common one with BES, is the optimization of reported analyses since experimental parameters are not given or specific comparative measurements of electrical output are not observed (Noll, 2006). Apart from this, numerous designs varying from double chamber to single chamber, mediator based or without mediator, and membrane or without membrane make the comparison cumbersome. In the present scenario, it has been emphasized that power density growth in relation to biocatalyst is improving, and innovative materials including improved electrodes for anode and cathode (Zhang et al., 2009), separators (Harnisch and Schroder, 2009), and new models of MFCs (Wang et al., 2011) are on the verge of striking the global market. The current progress of anode or cathode materials and filling components as 3D electrodes for BES was analyzed (Zhou et al., 2011). It was suggested that different electrodes depicted different features and electromodification proved to be a possible alternative for increasing the performance of BES. Application of electrodes with valuable metal catalyst like Pt and a membrane as separator has been observed as the most expensive component of MFCs (Rozendal et al., 2008). With innovative approaches in designs and development of cost-effective materials, the production expense associated with BES is expected to reduce. According to a recent report, MFC can be efficiently made commercially plausible and environmentally competitive with the conventional anaerobic treatment processes, if its performance exceeds 500 W/m³, since the advantage of microbial electrosynthesis mainly lies in the on-site use of electricity for bioproduction and its independence from the availability of arable land (Rabaey et al., 2011).

28.8 CONCLUSION AND FUTURE OUTLOOK

The recovery of electricity from waste or wastewater continues to attract several researchers and research options since it offers the potentiality of reducing the overall expense of treatment while decreasing biomass production. From the electric current and power production viewpoint, the innovation of novel components and cell materials is becoming more significant as striking price and efficient performance will immensely expand the utilization of MFCs. Together with the advantage of providing continuous and logistically simple available fuels with high energy density, microbial electrochemical systems or BESs can be operated for portable

applications. While the early studies were mainly based on the development of MFCs with bioanodes, the research ideas of BESs are quickly branching due to attractive developments in the study of biocathodes as well as MXCs. The important factors for developing BESs to an economical level are the pH components, the increasing ohmic resistance, and the rising overpotentials. For the purpose of wetland wastewater treatment full of landfill leachate, the combination of MFCs along with the newly developed treatment processes seems to be more effective, plausible, and cost-efficient. At present, the materialization of microbial electrosynthesis focuses on an alternative innovation for sustainable generation through bioelectrochemical route. This technique functions by either obtaining from or providing electric current to potential microbial strains so as to stimulate chemical production.

REFERENCES

Angenent, L.T., K. Karim, M.H. Al-Dahhan, B.A. Wrenn, and R. Domiguez-Espinosa. 2004. Production of bioenergy and biochemicals from industrial and agricultural wastewater. *Trends in Biotechnology* 22: 477–485.

Aulenta, F., A. Canosa, P. Reale, S. Rossetti, S. Panero, and M. Majone. 2009. Microbial reductive dechlorination of trichloroethane to ethane with electrodes serving as electron donors without the external addition of redox mediators. *Biotechnology and Bioengineering* 103: 85–91.

Aulenta, F., V. DiMaio, T. Ferri, and M. Majone. 2010. The humic acid analogue anthraquinone-2,6-disulfonate (AQDS) serves as an electron shuttle in the electricity-driven microbial dechlorination of trichloroethane to cis-dichloroethene. *Bioresource Technology* 101: 9728–9733.

Borole, A.P., G. Reguera, B. Ringeisen, Z.W. Wang, Y. Feng, and B.H. Kim. 2011. Electroactive biofilms: Current status and future research needs. *Energy and Environmental Science* 4: 4813–4834.

Butler, C.S., P. Clauwaert, S.J. Green, W. Verstraete, and R. Nerenberg. 2010. Bioelectrochemical perchlorate reduction in a microbial fuel cell. *Environmental Science and Technology* 44: 4685–4691.

Cao, X.X., X. Huang, P. Liang, K. Xiao, Y.J. Zhou, and X.Y. Zhang. 2009. A new method for water desalination using microbial desalination cells. *Environmental Science and Technology* 43: 7148–7152.

Cheng, K.Y., G. Ho, and R. Cord-Ruwisch. 2010. Anodophilic biofilm catalyzes cathodic oxygen reduction. *Environmental Science and Technology* 44: 518–525.

Cheng, S., D.F. Xing, D.F. Call, and B.E. Logan. 2009. Direct biological conversion of electrical current into methane by electromethanogenesis. *Environmental Science and Technology* 43: 3953–3958.

Clauwaert, P., P. Aelterman, T.H. Pham, L. DeSchamphelaire, M. Carballa, K. Rabaey, and W. Verstraete. 2008. Minimizing losses in bio-electrochemical systems: The road to applications. *Applied Microbiology and Biotechnology* 79: 901–913.

Dutta, P.K., J. Keller, J. Yuan, R.A. Rozendal, and K. Rabaey. 2009. Role of sulfur during acetate oxidation in biological anodes. *Environmental Science and Technology* 43: 3839–3845.

ElMekawy, A., S. Srikanth, K. Vanbroekhoven, H. DeWever, and D. Pant. 2014. Bio-electrocatalytic valorization of dark fermentation effluents by acetate oxidizing bacteria in bioelectrochemical system (BES). *Journal of Power Sources* 262: 183–191.

Erable, B., L. Etcheverry, and A. Bergel. 2011. From microbial fuel cell (MFC) to microbial electrochemical snorkel (MES): Maximizing chemical oxygen demand (COD) removal from wastewater. *Biofueling* 27: 319–326.

Falk, M., Z. Blum, and S. Shleev. 2012. Direct electron transfer based enzymatic fuel cells. *Electrochimica Acta* 82: 191–202.

Fornero, J.J., M. Rosenbaum, and L.T. Angenent. 2010. Electric power generation from municipal, food, and animal wastewaters using microbial fuel cells. *Electroanalysis* 22: 832–843.

Gilbert, N. 2012. African agriculture: Dirt poor. *Nature* 483: 525–527.

Greenman, J., A. Galvez, L. Giusti, and I. Ieropoulos. 2009. Electricity from landfill leachate using microbial fuel cells: Comparison with a biological aerated filter. *Enzyme and Microbial Technology* 44: 112–119.

Gregory, K.B. and D.R. Lovley. 2005. Remediation and recovery of uranium from contaminated subsurface environments with electrodes. *Environmental Science and Technology* 39: 8943–8947.

Harnisch, F. and U. Schroder. 2009. Selectivity versus mobility: Separation of anode and cathode in microbial electrochemical systems. *Chemsuschem* 2: 921–926.

He, Y.R., X. Xiao, W.W. Li, P.J. Cai, S.J. Yuan, and F.F. Yan. 2013. Electricity generation from dissolved organic matter in polluted lake water using a microbial fuel cell (MFC). *Biochemical Engineering Journal* 71: 57–61.

He, Z. and L.T. Angenent. 2006. Application of bacterial biocathodes in microbial fuel cells. *Electroanalysis* 18: 2009–2015.

Heidrich, E.S., T.P. Curtis, and J. Dolfing. 2011. Determination of the internal chemical energy of wastewater. *Environmental Science and Technology* 45: 827–832.

Heijne, A.T., F. Liu, R. Weijden, J. Weijma, C.J. Buisman, and H.V. Hamelers. 2010. Copper recovery combined with electricity production in a microbial fuel cell. *Environmental Science and Technology* 44: 4376–4381.

Howe, A. 2008. *Greenhouse Gas Emissions of Water Supply and Demand Management Options*. Bristol, U.K.: Environmental Agency.

Huang, L., X. Chai, S. Cheng, and G. Chen. 2011. Evaluation of carbon based materials in tubular biocathode microbial fuel cells in terms of hexavalent chromium reduction and electricity generation. *Chemical Engineering Journal* 166: 652–661.

Iranpour, R., M. Stenstrom, G. Tchobanoglous, D. Miller, J. Wright, and M. Vossoughi. 1999. Environmental engineering: Energy value of replacing waste disposal with resource recovery. *Science* 285: 706–711.

Israilides, C.J., A.G. Vlyssides, V.N. Mourafeti, and G. Karvouni. 1997. Olive oil wastewater treatment with the use of an electrolysis system. *Bioresource Technology* 61: 163–170.

Jambeck, J.R. and L. Damiano 2010. Microbial fuel cells in landfill applications. Environment Research Education Foundation (EREF) project report, Raleigh, North Carolina.

Jung, S.P., M.H. Yoon, S.M. Lee, S.E. Oh, H. Kang, and J.K. Yung. 2014. Power generation and anode bacterial community compositions of sediment fuel cells differing in anode materials and carbon sources. *International Journal of Electrochemical Science* 9: 315–326.

Kelly, P.T. and Z. He. 2014. Nutrients removal and recovery in bioelectrochemical systems: A review. *Bioresource Technology* 153: 351–360.

Kim, J.R., Y.E. Song, G. Munussami, C. Kim, and B.H. Jeon. 2015. Recent applications of bioelectrochemical system for useful resource recovery: Retrieval of nutrient and metal from wastewater. *Geosystem Engineering* 18: 173–180.

Kim, J.R., Y. Zuo, J.M. Regan, and B.E. Logan. 2008. Analysis of ammonia loss mechanisms in microbial fuel cells treating animal wastewater. *Biotechnology and Bioengineering* 99: 1120–1127.

Kim, Y. and B.E. Logan. 2013. Microbial desalination cells for energy production and desalination. *Desalination* 308: 122–130.

Kiran Kumar, A., M.V. Reddy, K. Chandrasekhar, S. Srikanth, and S. Venkatamohan. 2012. Endocrine disruptive estrogens role in electron transfer: Bio-electrochemical remediation with microbial mediated electrogenesis. *Bioresource Technology* 104: 547–556.

Kokabian, B. and V.G. Gude. 2013. Photosynthetic microbial desalination cells (PDMCs) for clean energy, water and biomass production. *Environmental Science Processes and Impacts* 15: 2178–2185.

Kuzminsky, Y., K. Shchurska, I. Samarukha, and G. Lagod. 2013. *Bioelectrochemical Hydrogen and Electricity Production*. Lublin, Poland: Lublin University of Technology.

Lee, H.S., P. Parameswaram, A. Kato-Marcus, C.I. Torres, and B.B. Rittmann. 2008. Evaluation of energy conversion efficiencies in microbial cells (MFCs) utilizing fermentable and nonfermentable substrates. *Water Research* 42: 1501–1510.

Lloyd, J.R., D.R. Lovley and L.E. Macaskie. 2003. Biotechnological applications of metal reducing microorganisms. *Advances in Applied Microbiology* 53: 85–128.

Li, J., G.L. Liu, R.D. Zhang, Y. Luo, C.P. Zhang, and M.C. Li. 2010. Electricity generation by two types of microbial fuel cells using nitrobenzene as the anodic or cathodic reactants. *Bioresource Technology* 101: 4013–4020.

Liu, L., F.B. Li, C.H. Feng, and L.X. Zhong. 2009. Microbial fuel cell with an azo dye feeding cathode. *Applied Microbiology and Biotechnology* 85: 175–183.

Logan, B.E. 2010. Scaling up microbial fuel cells and other bioelectrochemical systems. *Applied Microbiology and Biotechnology* 85: 1665–1671.

Lorenzo, M.D., K. Scott, T.P. Curtis, and I.M. Head. 2010. Effect of increasing anode surface area on the performance of a single chamber microbial fuel cell. *Chemical Engineering Journal* 156: 40–48.

Lovley, D.R. 1993. Dissimilatory metal-reduction. *Annual Review of Microbiology* 47: 263–290.

Lovley, D.R. 2011. Reach out and touch someone: Potential impact of DIET (direct interspecies energy transfer) on anaerobic biogeochemistry, bioremediation and bioenergy. *Reviews in Environmental Science and Biotechnology* 10: 101–105.

Lovley, D.R., J.L. Fraga, J.D. Coates, and E.L. Blunt-Harris. 1999. Humics as an electron donor for anaerobic respiration. *Environmental Microbiology* 1: 89–98.

Luo, H., G. Liu, R. Zhang, and S. Jin. 2009. Phenol degradation in microbial fuel cells. *Chemical Engineering Journal* 147: 259–264.

McCarty, P.L., L. Bae, and J. Kim. 2011. Domestic wastewater treatment as a net energy producer-can this be achieved?. *Environmental Science and Technology* 45: 7100–7106.

Mehanna, M., T. Saito, J.L. Yan, M. Hickener, X.X. Cao, and X. Huang. 2010. Using microbial desalination cells to reduce water salinity prior to reverse osmosis. *Energy Environment and Science* 3: 1114–1120.

Modin, O. and B.M. Wilen. 2012. A novel bioelectrochemical BOD sensor operating with voltage input. *Water Research* 46: 6113–6120.

Morris, J.M., S. Jin, B. Crimi, and A. Pruden. 2009. Microbial fuel cell in enhancing anaerobic biodegradation of diesel. *Chemical Engineering Journal* 146: 161–167.

Mu, Y., K. Rabaey, A. Rene, R. Zhigou and J. YuanKeller. 2009a. Decolourization of azo dyes in bioelectrochemical systems. *Environmental Science and Technology* 43: 5137–5143.

Mu, Y., R.A. Rozendal, K. Rabaey, and J. Keller. 2009b. Nitrobenzene removal in bioelectrochemical systems. *Environmental Science and Technology* 43: 8690–8695.

Noll, K. 2006. Microbial fuel cells. In *Fuel Cell Technology*, ed. N. Sammes, pp. 277–296. London, U.K.: Springer.

Palmore, G.T.R. and G.M. Whitesides. 1994. Microbial and enzymatic biofuel cells. In *Enzymatic Conversion of Biomass for Fuels Production*, eds. E. Himmel, J.O. Baker, and R.P. Overend, pp. 271–290. Washington, DC: American Chemical Society.

Pant, D., D. Arslan, G. vanBogaert, Y.A. Gallego, H. DeWever, L. Deis, and K. Vanbroekhoven. 2013. Integrated conversion of food waste diluted with sewage into volatile fatty acids through fermentation and electricity through a fuel cell. *Environmental Technology* 34: 1935–1945.

Pant, D., A. Singh, and G. vanBogaert. 2012. Bioelectrochemical systems (BES) for sustainable energy production and product recovery from organic wastes and industrial wastewater. *RSC Advances* 2: 1248–1263.

Pant, D., G. vanBogaert, and L. Deis. 2010. A review of the substrates used in microbial fuel cells (MFCs) for sustainable energy production. *Bioresource Technology* 101: 1533–1543.

Rabaey, K., P. Girguis, and L.K. Nielsen. 2011. Metabolic and practical considerations on microbial electrosynthesis. *Current Opinion in Biotechnology* 22: 371–377.

Rabaey, K. and R.A. Rozendal. 2010. Microbial electrosynthesis—Revisiting the electrical route for microbial production. *Nature Reviews Microbiology* 8: 706–716.

Rabaey, K. and W. Verstraete. 2005. Microbial fuel cells: Novel biotechnology for energy generation. *Trends in Biotechnology* 23: 291–298.

Rezaei, F., T.L. Richard, R.A. Brennan, and B.E. Logan. 2007. Substrate enhanced microbial fuel cells for improved remote power generation from sediment based systems. *Environmental Science and Technology* 41: 4053–4058.

Rozendal, R.A., H.V.M. Hamelers, and K. Rabaey. 2008. Towards practical implementation of bioelectrochemical wastewater treatment. *Trends in Biotechnology* 26: 450–459.

Schroder, U. and F. Harnisch. 2010. From MFC to MXC: Chemical and biological cathodes and their potential for microbial bioelectrochemical systems. *Chemical Society Reviews* 39: 4433–4448.

Shroder, U. 2011. Discover the possibilities: microbial bioelectrochemical systems and the revival of a 100-year-old discovery. *Journal of Solid State Electrochemistry* 15: 1481–1486.

Strycharz, S.M., R.H. Glaven, M.V. Coppi, S.M. Gannon, L.A. Perpetua, and A. Liu. 2011. Gene expression and deletion analysis of mechanisms for electron transfer from electrodes to *Geobacter sulfurreducens. Bioelectrochemistry* 80: 142–150.

Srikanth, S., M. Maesen, X. Dominguez-Benetton, K. Vanbroekhoven, and D. Pant. 2014. Enzymatic electrosynthesis of formate through CO$_2$ sequestration/reduction in a bioelectrochemical system (BES). *Bioresource Technology* 165: 350–354.

Srikanth, S. and S. Venkatamohan. 2012. Influence of terminal electron acceptor availability to the anodic oxidation on the electrogenic activity of microbial fuel cell (MFC). *Bioresource Technology* 123: 480–487.

Sun, J., Y.Y. Hu, and Y.Q. Cao. 2009. Simultaneous decolorization of azo dye and bioelectricity generation using a microfiltration membrane air cathode single chamber microbial fuel cell. *Bioresource Technology* 100: 3185–3192.

Tandukar, M., S.J. Huber, T. Onodera, and S.G. Pavlostathis. 2009. Biological chromium (VI) reduction in the cathode of a microbial fuel cell. *Environmental Science and Technology* 43: 8159–8165.

Thrash, J.C. and J.D. Coates. 2008. Review: Direct and indirect electrical stimulation of microbial metabolism. *Environmental Science and Technology* 42: 3921–3931.

Thrash, J.C., J.I. vanTrump, K.A. Weber, E. Miller, L.A. Achenbach, and J.D. Coates.2007. Electrochemical stimulation of microbial perchlorate reduction. *Environmental Science and Technology* 41: 1740–1746.

Torres, C.I. 2014. On the importance of identifying, characterizing and predicting fundamental phenomenon towards microbial electrochemistry applications. *Current Opinion in Biotechnology* 27: 107–114.

Turner, R.J., R. Borghese, and D. Zannoni. 2012. Microbial processing of tellurium as a tool in biotechnology. *Biotechnology Advances* 30: 954–963.

Velvizhi, G. and S. VenkataMohan. 2011. Biocatalyst behavior under self induced electrogenic micro- environment in comparison with anaerobic treatment: Evaluation with pharmaceutical wastewater for multi pollutant removal. *Bioresource Technology* 102: 10784–10793.

VenkataMohan, S. and K. Chandrasekhar. 2011. Self-induced biopotential and graphite electron accepting conditions enhances petroleum sludge degradation in bio-electrochemical systems with simultaneous power generation. *Bioresource Technology* 102: 9532–9541.

VenkataMohan, S. and S. Srikanth. 2011. Enhanced wastewater treatment efficiency through microbially catalyzed oxidation and reduction: Synergistic effect of biocathode microenvironment. *Bioresource Technology* 102(22):10210–10220.

VenkataMohan, S., G. Velvizhi, M. LeninBabu, and S. Srikanth. 2014. Microbial catalyzed electrochemical systems: Critical factors and recent advancements. *Renewable and Sustainable Energy Reviews* 165: 355–364.

Wang, H.Y., A. Bernarda, C.Y. Huang, D.J. Lee, and J.S. Chang. 2011. Micro sized microbial fuel cells: A mini review. *Bioresource Technology* 102: 235–243.

You, S.J., Q.L. Zhao, J.N. Jiang, J.N. Zhang, and S.Q. Zhao. 2006. Sustainable approach for leachate treatment: Electricity generation in microbial fuel cell. *Journal of Environmental Science and Health Part A* 41: 2721–2734.

Zhang, F., S.A. Cheng, D. Pant, G. VanBogaert, and B.E. Logan. 2009. Power generation using an activated carbon and metal mesh cathode in a microbial fuel cell. *Electrochemistry Communications* 11: 2177–2179.

Zhang, J.N., Q.L. Zhao, S.J. You, J.Q. Jiang, and N.Q. Ren. 2008. Continuous electricity production from leachate in a novel upflow air-cathode membrane free microbial fuel cell. *Water Science and Technology* 57: 1017–1021.

Zhang, Y. and I. Angelidaki. 2013. A new method for *in situ* nitrate removal from groundwater using submerged microbial desalination-denitrification cell (SMDDC). *Water Research* 47: 1827–1836.

Zhou, M., M. Chi, J. Luo, H. He, and T. Jin.2011. An overview of electrode materials in microbial fuel cells. *Journal of Power Sources* 196: 4427–4435.

Section V

Advances in Metal Bioremediation Research

29 Bioremediation Approach for Handling Multiple Metal Contamination

Deepak Gola, Nitin Chauhan, Anushree Malik, and Ziauddin A. Shaikh

CONTENTS

ABSTRACT

The pollution caused by heavy metals has become a serious problem due to its detrimental impacts on environmental and human health. The current situation demands an eco-friendly, economical, and effective approach for the remediation of varying concentrations of heavy metals from wastewater. Microorganisms such as bacteria, fungi, and algae have gained appreciable attention due to their ability in remediating different heavy metals from wastewater. However, for the sake of simplicity, investigations are concerned with remediation of high concentrations of single metal rather than the low levels of complex mix of different heavy metals, a situation that is encountered in actual contaminated environments. This chapter covers those selected studies that describe simultaneous removal of multiple heavy metals by different microorganisms. This chapter discusses the uptake capacity of different microbes and the effect of interfering metal ions and various factors influencing the uptake rate of heavy metals. It is apparent from the studies that although microbes are efficient biosorbents to uptake heavy metals even at low concentrations, their capacity declines when subjected to binary or ternary mixtures. Very few attempts have been made to understand the response of metabolically active microbes/consortium under multiple metal stress and to harness diverse mechanisms of metal uptake under complex scenarios. More research is needed toward understanding the process mechanisms, selecting the best microbial strain/consortia for multimetal removal, and developing suitable bioreactors to improve the simultaneous uptake of multiple heavy metals from wastewater.

29.1 INTRODUCTION

In the past few decades, the rapid increase in industrialization and urbanization has caused adverse environmental impacts globally (Helios Rybicka, 1996; Olawoyin et al., 2012; Gola et al., 2015; Idris et al., 2015; Mathur et al., 2015). Industries utilize natural resources and produce large amount of wastes responsible for polluting air, water, and soil. Large water bodies act as a sink for dumping tons of industrial waste with or without treatment. The waste that is easily degradable does not pose much threat to environment, but nondegradable wastes are a great threat to environment and health. Heavy metals are one of the nondegradable contaminants present in waste. Although these heavy metals are naturally occurring elements present in the earth's crust, excessive anthropogenic activities such as electroplating, paint, leather, mining, foundries, and smelting industries have contributed to alarming heavy metal contamination. The presence of the heavy metals in the various water bodies around the world is well documented and summarized in Table 29.1.

In the recent past, studies have pointed to the ecological and public health–related issues associated with the presence of heavy metals in the environment (Singh et al. 2010; Khan et al., 2014). Concentration of heavy metals higher than required threshold value is potentially toxic and leads to change in various biochemical as well as physiological processes of living organisms. Since wastes in any water body come from different industrial discharge, they always contain a mixture of different heavy metals (Table 29.1). The combinational effect of multiple heavy metals may differ from a single metal ion's effect. Models such as concentration addition

TABLE 29.1

Concentration of Heavy Metals in River Water in Different Parts of the World

Sites	Cr	Cu	Cd	Pb	Zn	Ni	Reference
Kasardi River, Maharashtra	**26.3**	**41.1**	**26.7**	**30.1**	**14.4**	**10.3**	Lokhande et al. (2011)
Ganga River, West Bengal, India	0.016–0.022	—	0.002–0.007	0.033–0.141	0.075–0.280	—	Paul and Sinha (2013)
Subarnarekha River, India	0–0.0026	**0–0.944**	—	—	0.0013–0.0446	0.0007–0.1238	Giri and Singh (2013)
Pardo River, Brazil	0.0018–0.0025	0.0043–0.0098	0.0000–0.0002	0.0867–0.0047	0.0798–0.0260	0.0068–0.0233	Alves et al. (2014)
Dzindi River, South Africa	**0.030–0.110**	0.03–0.05	—	0.01–0.05	0.05–0.21	—	Edokpayi et al. (2014)
Nile River, Egypt	—	0.0009–0.0043	0.0038–**0.0150**	0.0081–0.3548	0.178–0.2680	—	Badr et al. (2014)
Ismailia Canal, Egypt	—	0.07	**0.45**	0.018	0.015	0.010	Goher et al. (2014)
Cross River (pond), Nigeria	—	—	—	0.455	0.389	0.012	Adamu et al. (2014)
Cross River (stream), Nigeria	—	—	—	0.048	0.066	0.002	Adamu et al. (2014)
Gomti River, India	0.063	—	0.024	0.018	0.067	**0.440**	Gupta et al. (2014)
Kali River, India	0.002–0.087	—	0.001–0.024	0.010–0.340	0.004–0.376	—	Malik and Maurya (2015)
Khambhat Region, Gujarat, India	BDL-0.026	BDL-0.0274	BDL-0.0118	BDL-0.015	0.0030–0.0795	0.0001–0.008	Upadhyaya et al. (2014)
Jayakwadi Dam Water, Maharashtra, India	—	**0.193–0.516**	0.006–0.046	0.226–0.560	0.053–0.310	—	Patil et al. (2014)
Yamuna River, India	**0–0.42**	0.02–0.64	0–0.07	0.03–0.27	0.13–2.22	0.01–0.13	Bhattacharya et al. (2015)
Korotoa River, Bangladesh	0.083–0.073	0.073–0.061	0.011–0.008	0.035–0.027	—	0.039–0.032	Islam et al. (2015)

Note: Bold letter depict concentration above the FAO (Food and Agriculture Organization) permissible limits

and independent action were introduced to study the toxicity of heavy metals in combination (Loewe and Muischnek, 1926; Bliss, 1939), but both of these models assume that no interaction occurred between heavy metals when present in mixture, and therefore, each heavy metal gives rise to an individual effect. Studies have highlighted that the impact of multiple heavy metal on a particular organism may vary from expected sum of individual heavy metal, that is, antagonistic or additive or synergetic effect, respectively (Norwood et al. 2003; Borgmann et al., 2008; Pan et al., 2009). Zhu et al. (2011) observed that binary combinations of Cu(II)–Zn(II) and Cu(II)–Cd(II) had synergistic toxicity on the larvae of *Gobiocypris rarus*. Vosylienė et al. (2003) have observed that the synergistic effect of multiple heavy metal is more dangerous to rainbow trout (*Oncorhynchus mykiss*) than any other effect of multiple heavy metals. In human beings, exposure to heavy metals can cause vomiting, diarrhea, breathing problem (such as asthma, nose irritation, cough, wheezing), allergic reactions, fragile bones, number of cancer (lungs, kidney, skin, prostate, larynx, liver, etc.), and neurological disorders (Jarup, 2003). Therefore, the remediation of heavy metals from water bodies is undoubtedly the need of the hour.

Industries use various physical and chemical methods to reduce the load of heavy metals before its final discharge. Methods such as physical, chemical, or combination of both, that is, physiochemical, have been developed and adopted by these industries (Hashim et al., 2011). Chemical precipitation, ion exchange, reverse osmosis, ultrafiltration, adsorption, and coagulation–flocculation are some of the technologies that are generally used to remove heavy metals from wastewater. But there are problems associated with these methods such as toxic gas intermediate, difficulty in handling, contaminant specificity, high cost, and low effectiveness (Hashim et al., 2011). Hence, the remediation of heavy metals requires appropriate technology and systemic efforts to efficiently remove the unwanted load of heavy metals through environmentally safe ways.

Microorganisms such as bacteria, fungi, and algae play a significant role in the remediation of heavy metal from contaminated soil and wastewater in an eco-friendly way (Malik, 2004; Hashim et al., 2011). Studies have also established the role of microbes in remediating heavy metals from wastewater even at low concentration (Kosińska and Miśkiewicz, 2012). In the past few decades, extensive research on the use of biological techniques as an eco-friendly alternative for removing heavy metals from wastewater and soil has been conducted (Sag et al., 2002; Fang et al., 2012; Mishra and Malik, 2012). Most of the studies have investigated the uptake of individual metals by microorganism. Limited studies consider toxicity, accumulation, and bioremediation of multiple metals from contaminated matrix (Mishra and Malik, 2014a, 2014b). Therefore, the present review focuses exclusively on multiple metal removal using different microbial groups such as bacteria, fungi, and algae followed by description of the working technologies and bioreactor designs for simultaneous removal of multiple metal ions.

29.2 MECHANISM FOR HEAVY METAL UPTAKE BY MICROORGANISM

Biosorption and bioaccumulation are the two mechanisms used by microorganisms during the metal removal process. Active uptake of metals by microorganism that is metabolism dependent is termed *bioaccumulation*, while passive uptake employing dead biomass is termed *biosorption*.

Depending upon the mechanism of uptake, metal may be found in the following locations (Figure 29.1) (Saǧ, 2001; Malik, 2004; Ahemad and Kibret, 2013):

1. *On the surface or biosorption*: A number of surface groups present on the cell wall and cell membrane such as phosphate and carboxyl act as the binding sites for the heavy metal present in the solution (Table 29.2). It can be seen that while some groups are specific for a metal, other may bind several metals

at a given pH. This process does not require energy and is reversible in nature. It involves precipitation, ion exchange, complexation, and physical absorption. This process is more rapid than active uptake.

2. *Inside the cell or bioaccumulation*: Heavy metal from the solution enters the cell via transmembrane proteins present on the cell membrane (Table 29.2). Inside the cell, various metabolic processes detoxify the effect of heavy metal. This process involves energy and is irreversible in nature.

3. *Extracellular accumulation or precipitation*: Heavy metals form complexes with the various anions, polymers, and enzymes released extracellularly by the microorganism such as phosphate, sulfide, oxalic acids, and pigments. Different organisms are known to produce different categories of pigments, which play an important role in metal removal (Table 29.2). Further, under different types of metal exposure, the same organism may produce different pigmentations to deal with the changing metal stress, as depicted in Figure 29.2.

29.3 HEAVY METAL REMOVAL IN MULTIMETAL ION SOLUTION

As mentioned earlier, a large pool of studies investigated metal removal ability of microorganism in the single metal ion solution, and these studies have been extensively reviewed. Only a fraction of literature contributed toward the details of heavy metal removal from the multimetal ion solution. In multimetal ion solution, different metal ions interact with each other and

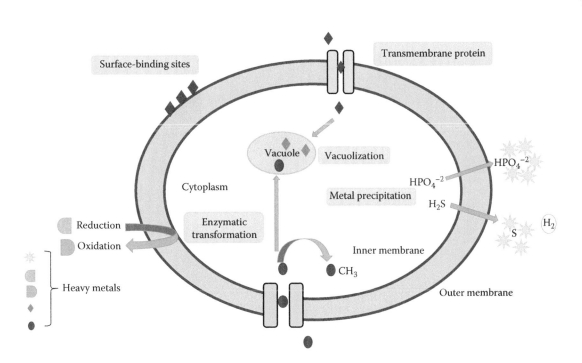

FIGURE 29.1 Different mechanisms for heavy metal removal by microorganism.

TABLE 29.2

Different Pigments, Functional Groups, and Transmembrane Proteins on Microbial Surface That Interact with Heavy Metals

Moieties	Description	Microorganism	Interacting Heavy Metal/s	Reference
Pigments	Melanin	*Gaeumannomyces graminis*	Cu(II)	Caesar-Tonthat et al. (1995)
	Melanin	*Aureobasidium pullulans*	Pb(II)	Suh et al. (1999)
	Cell wall bound pigment	*Oidiodendron maius*	Zn(II)	Martino et al. (2000)
	Anthraquinone type	*Penicillium oxalicum*	Hg(II), Pb(II), and Cd(II)	Svecova et al. (2006)
Functional groups	Phosphoryl and carboxyl	*Penicillium chrysogenum*	Pb	Sarret et al. (1998)
	Phosphoryl	*Penicillium chrysogenum*	Zn	Sarret et al. (1998)
	Amino	*Rhizopus nigricans*	Cr	Bai and Abraham (2002)
	Carboxyl, phosphoryl, and hydroxyl	*Bacillus subtilis*	Cd	Boyanov et al. (2003)
	Hydroxyl and amino	*Botrytis cinerea*	Cd and Cu	Akar et al. (2005)
	Thiol	*Bacillus subtilis*	Hg	Wang and Sun (2013)
Transmembrane proteins	ABC	Bacteria	Mn(II), Zn(II), Ni(II), and Fe(II)	Fath and Kolter (1993)
	RND	Bacteria	Cu(II), Zn(II), Cd(II), and Ni(II)	Saier et al. (1994)
	CDF	Bacteria	Zn(II), Cd(II), Co(II), and Fe(II)	Nies and Silver (1995)
	MIT	Bacteria and fungi	All cations	Paulsen and Saier (1997)
	HoxN	Bacteria	Co(II) and Ni(II)	Nies (1999)

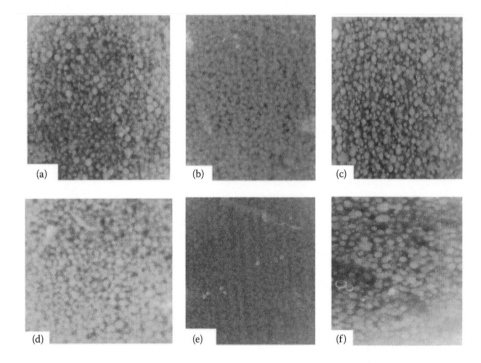

FIGURE 29.2 Different pigment productions by entomopathogenic fungi under different metal exposures. (a) Pb-30 mg L^{-1}; (b) Pb-25 mg L^{-1} + Cr-5 mg L^{-1}; (c) Pb-20 mg L^{-1} + Cr-5 mg L^{-1} + Cu-5 mg L^{-1}; (d) Pb-15 mg L^{-1} + Cr-5 mg L^{-1} + Cu-5 mg L^{-1} + Zn-5 mg L^{-1}; (e) Pb-10 mg L^{-1} + Cr-5 mg L^{-1} + Cu-5 mg L^{-1} + Zn-5 mg L^{-1} + Ni-5 mg L^{-1}; and (f) Pb-5 mg L^{-1} + Cr-5 mg L^{-1} + Cu-5 mg L^{-1} + Zn-5 mg L^{-1} + Ni-5 mg L^{-1} + Cd mg L^{-1}.

organism may respond by producing three types of effects, as follows (Ting and Teo, 1994; Pan et al., 2009):

1. *Synergism*: Toxic effect of two or more metals on microorganism is greater than that of each of the individual effects of heavy metal in the solution.

2. *Antagonistic*: Toxic effect of two or more metals on microorganism is less than that of each of the individual effects of heavy metal in the solution.

3. *Noninteraction*: Toxic effect of two or more metals on microorganism is no more or less than that of each of the individual effects of heavy metal in the solution.

Moreover, interaction between the different metal ions in the solution and with the mixture exposed to multimetal stress depends on the following factors (Sağ, 2001; Ahemad and Kibret, 2013):

1. *Microorganism*: Various types of microorganisms such as bacteria, fungi, and algae are used for the bioremediation of heavy metal from the solution. These microorganisms possess different functional sites and transmembrane proteins, which define the metal uptake capacity of microorganism.

2. *Binding sites*: Functional sites such as carboxyl, amide, thiol, phosphate, and hydroxide are present on the surface of microorganism. The Fourier transform infrared spectroscopy has clearly highlighted the masking/shifting of functional sites when heavy metals bind with functional sites (Akar et al., 2005; Gupta and Rastogi, 2008; Singh et al., 2011).

3. *Metal combination*: Toxicity of multimetal mixture can show synergetic, antagonistic, and noninteraction effect on the microorganism; hence, the combination of different heavy metals in the solution governs the overall removal efficacy of microorganism.

4. *Concentration of interfering metal ion*: Concentration of heavy metal is another factor that decides the metal uptake capacity from the solution. Higher concentration of interfering heavy metals can reduce the uptake capacity of other heavy metals as they occupy the same binding site and block the entry for other heavy metals.

5. *pH of the solution*: The pH of the solution can significantly affect the removal of heavy metals from the solution as low pH can cause precipitation of heavy metals. pH also determines the charge of the function site present on the surface of microorganism (Lesmana et al., 2009).

   a. *Biosorption models*: Biosorption is a simple process in which a metal ion binds specifically to various functional sites present on the surface of dead or living microorganisms. There are various models that describe the uptake of particular heavy metal from the solution using different microorganisms such as algae, bacteria, and fungi. In multimetal solution, these models require various correction and interaction factors. These factors describe the competition between different metal ions present in the solution for the same functional sites on the surface of microorganism. The various isotherm models to describe the metal biosorption data in the single and binary metal ion solution are summarized in Table 29.3. Although both biosorption and bioaccumulation can be used for the remediation of heavy metal ion from the solution, biosorption is the widely used method because of its rapidity and reversibility (Sağ, 2001; Mudhoo et al., 2012).

   b. *Multimetal removal by different microorganism*: To facilitate the understanding and discussion, the studies have been categorized in the subsequent sections based on the type of microorganism employed such as bacteria, fungi, and algae.

29.3.1 HEAVY METAL REMOVAL BY BACTERIA

Bacteria are single-celled organisms. The cell structure of bacteria is simple with no nucleus or membrane-bound organelles. Literature studies have discussed the removal of heavy metal using bacteria (Malik, 2004; Hashim et al., 2011). It was also reported that some bacteria can even grow at significantly high concentrations of heavy metals (Table 29.4). There are various mechanisms like internalization of heavy metals, chelation, and enzymatic transformation to withstand the high concentration of heavy metals (Ahemad and Kibret, 2013). A number of transmembrane proteins such as cop for $Cu(II)$, chr for $Cr(VI)$, and czc for $Cd(II)$ lead to internalization of various heavy metals inside the bacteria (Silver, 1996). Some transmembrane proteins are common for multiple metal ions. For example, Czc model that comprises three protein complexes (CzcA, CzcC, and CzcB) that act as a proton/cation antiporter across membrane for the efflux of $Cd(II)$, $Zn(II)$, and $Co(II)$ in bacterial system. After internalization, various biochemical metabolic processes detoxify the heavy metals.

Several studies have discussed the capability of bacteria to remove heavy metals from synthetic solution or wastewater (Table 29.5). However, these studies are limited to a single metal ion in the solution. The presence of other heavy metal ions in the solution changes the overall scenario of metal ions due to competition. In the binary or ternary metal ion solution, metal ion will compete for the same functional site on the cell surface as well as on transmembrane proteins (which are common for some metal ions) for the influx of heavy metals. Nevertheless, studies discussing the metal removal capacity of the bacteria in binary or ternary metal ion solution are limited. Table 29.8 summarizes the metal removal uptake by bacterial isolate in binary or ternary metal ion solution.

The overall metal removal efficiency of each bacterial cell is different as compared to the other bacteria. This is because each bacterial cell holds its own mechanism to deal with the different metals. In most of the studies, the removal efficiency of particular heavy metal decreases in the presence of other heavy metals (Table 29.6). The overall mechanism of multiple metal ion uptake by bacteria is quite complex. This might be due to the competition for the same binding site on the cell surface of the bacteria, and this competition partially depends on the metal ion characteristic such as ionic radius, atomic weight, covalent index, and concentration and solution characteristic such as pH and temperature (Puranik and Paknikar, 1999; Uslu and Tanyol, 2006). Kim et al. (2015) investigated the biosorption behavior of *Escherichia coli* for $Pd(II)$ and $Pt(II)$ in the binary metal ion solution. A study was performed on two different metal ion concentration ratios of $Pd(II)$ and $Pt(II)$, that is, 1:1 (214:183.5 mg L^{-1}) and 1:2 (218.6:369.7 mg L^{-1}). It was observed that biosorption of

TABLE 29.3

Different Models for Heavy Metal Biosorption under Single- and Multimetal Environments

Models	Single metal ion	Multimetal ion
Langmuir	$Q = \dfrac{Q_{max}\, b_L C_{eq}}{1 + b_L C_1}$	$Q_1 = \dfrac{Q_{max,1}\, b_1 (C_1/\eta_{1,1})}{1 + b_1\left(\dfrac{C_1}{\eta_{1,1}}\right) + b_2 (C_2/\eta_{1,2})}$
		$Q_2 = \dfrac{Q_{max,2}\, b_2 (C_2/\eta_{1,2})}{1 + b_1\left(\dfrac{C_1}{\eta_{1,1}}\right) + b_2 (C_2/\eta_{1,2})}$
Freundlich	$Q = K_f C_{eq}^{1/n_f}$	$Q_1 = K_{f1} C_{f1} (C_{f1} + a_{12} C_{f2})^{((1/n_1)-1)}$
		$Q_2 = K_{f2} C_{f2} (C_{f2} + a_{21} C_{f1})^{((1/n_2)-1)}$
Redlich–Peterson	$Q = \dfrac{K_{RP} C_{eq}}{1 + a_{RP} C_{eq}^{\beta_{RP}}}$	$Q_1 = \dfrac{K_{RP,1}(C_1/\eta_{RP,1})}{1 + \left(a_{RP,1}(C_1/\eta_{RP,1})^{\beta_{RP,1}}\right) + \left(a_{RP,2}(C_2/\eta_{RP,2})^{\beta_{RP,2}}\right)}$
		$Q_2 = \dfrac{K_{RP,2}(C_2/\eta_{RP,2})}{1 + \left(a_{RP,1}(C_1/\eta_{RP,1})^{\beta_{RP,1}}\right) + \left(a_{RP,2}(C_2/\eta_{RP,2})^{\beta_{RP,2}}\right)}$

Source: Compiled from Vijayaraghavan, K. and Joshi, U.M., *Environ. Prog. Sustainable Energy*, 33, 147, 2014.

Q is the metal uptake (mg g^{-1}); C_{eq} is the equilibrium metal ion concentration (mg L^{-1}); Q_{max} is the maximum metal uptake (mg g^{-1}); b_L is the Langmuir equilibrium constant (L mg^1); K_f is the Freundlich constant (mg g^{-1}) (L mg^{-1})$^{1/n}$; n_f is the Freundlich exponent; K_{RP} is the Redlich–Peterson isotherm constant (L g^{-1}); a_{RP} is the Redlich–Peterson isotherm constant (L mg^{-1})$^{1/\beta}$RP; β_{RP} is the Redlich–Peterson model experiment; $Q_{max,1}$, $Q_{max,2}$, b_1, and b_2 are the single-component Langmuir parameter for the first and second metal ion in the solution; $\eta_{1,1}$ and $\eta_{1,2}$ are the interaction factors, for the first and second metal ion in the solution, respectively; a_{12} and a_{21} are the competition factors; $K_{RP,1}$, $K_{RP,2}$, $a_{RP,1}$, $a_{RP,2}$, $\beta_{RP,1}$, and $\beta_{RP,2}$ are the single-component Redlich–Peterson parameters for the first metal and second metal ion, respectively; and $\eta_{RP,1}$ and $\eta_{RP,2}$ are the interaction factors for the first metal ion and second metal ion, respectively.

Pd(II) was more as compared to Pt(II) in both experimental cases by dead biomass of *E. coli*. It was found that *E. coli* contains primary amine group on the cell surface that selectively binds to the Pd(II) ion and attributes to its higher biosorption in the binary metal solution. Ziagova et al. (2007) observed maximum uptake of Cr(VI) by *Staphylococcus xylosus* and *Pseudomonas* sp., with absorption capacity of 143 and 95 mg g^{-1}, respectively. *S. xylosus* displayed maximum uptake at low pH, due to the protonation of peptidoglycans, teichoic and teichuronic acids present on the cell surface, thus favoring Cr(VI) uptake that exists as an oxyanion (HCrO$_4^-$, Cr$_2$O$_7^{-2}$, and CrO$_4^{-2}$). *Pseudomonas* sp. lacks these groups and hence shows low Cr(VI) uptake. The study also highlighted that when Cr(VI) was present as a dominant metal ion, *S. xylosus* prefers Cr(VI) more than Cd(II), due to protonation of cell wall at low pH.

Percentage removal for Ni(II) and Cd(II) by *Brevundimonas vesicularis* decreases by 15.4% and 14% in the case of binary metal ion solution containing equal concentration (100 mg L^{-1}) of each metal ion. The decrease in the percentage removal of heavy metal is due to the competition between Ni(II) and Cd(II) ion for the same binding sites on the cell surface of

B. vesicularis, as suggested by the authors (Singh and Gadi, 2012). Uslu and Tanyol (2006) studied the removal efficiency of *Pseudomonas putida* in a single as well as binary metal ion solution containing Cu(II) and Pb(II). It was observed that both the uptake capacity and biosorption efficiency decrease in binary solution as compared to the individual metal ion in the solution. The uptake capacity (54.17 mg g^{-1}) as well as absorption efficiency (81.8%) of Pb(II) was on the higher side as compared to the Cu(II) for which it was 27.91 mg g^{-1} and 44.6%, respectively, at 60 mg L^{-1} and pH 5.5. In binary metal solution, uptake capacities of Pb(II) and Cu(II) were 30.08 and 18.91 mg g^{-1}, while the absorption efficiency decreased to 61.69% and 30.6%, respectively. The presence of two metal ions led to the competition for the absorption site on the surface, which led to the low uptake of heavy metal by *P. putida* in binary metal solution.

Masood and Malik (2014) observed the biosorption capacity of *Acinetobacter* sp. FM4 in the binary metal ion solution containing various combinations of three metal ions Cr(VI), Cu(II), and Ni(II). The concentration of the first metal ranged from 0 to 400 mg L^{-1}, and the concentration of other metals was varied within the same range. It was observed that

TABLE 29.4

Minimum Inhibitory Concentration of Different Heavy Metals for Different Bacteria under Single Metal Exposure

Bacteria	MIC (mg L^{-1})	Reference
P. aeruginosa	Pb 800, Ni 700, Cd 500, and Cr 400	Raja et al. (2006)
Staphylococcus sp.	Cr 500, Co 150, and Ni 150	Rajbanshi (2009)
Acinetobacter sp.	Cd 150, Cu 200, and Co 180	
Pseudomonas sp.	Cu 300, Co 150, and Cd 120	
Staphylococcus sp.	Ni 150 and Co 150	
Bacillus sp.	Ni 200 and Co 110	
Natronobacterium magadii	Ni 250, Al 600, and Hg 25	Williams et al. (2012)
Natronococcus occultus	Ni 250, Al 600, and Hg 25	
Halobacterium sodomense	Ni 250, Al 600, and Hg 25	
Rhizobium leguminosarum	Zn 70, Co 50, Ni 30, and Cd 10	Abd-Alla et al. (2012)
Bacillus thuringiensis	Cr 1600, Cu 1400, Pb 1500, Ni 1200, and Cd 600	Oves et al. (2013)
Comamonas acidovorans	Cr 600, Hg 200, Pb 500, and Al 500	Rudakiya and Pawar (2013)
Comamonas testosteroni	Cu 200, Cd 100, Cr 300, and Ni 200	Pandit et al. (2013)
Bacillus cereus	Cu 200, Cd 200, Cr 200, and Ni 200	
Exiguobacterium sp.	Cu 200, Cd 200, Cr 200, and Ni 200	
Enterococcus raffinosus	Fe 500, Pb 700, Zn 300, and Cr 400	Issazadeh and Razban (2014)
Enterococcus faecium	Fe 800, Pb 800, Zn 500, and Cr 800	
Enterococcus mundtii	Fe 500, Pb 700, Zn 300, and Cr 200	
Enterococcus gallinarum	Fe 700, Pb 800, Zn 500, and Cr 400	
Enterococcus hirae	Fe 700, Pb 800, Zn 400, and Cr 300	
Enterococcus pseudoavium	Fe 500, Pb 700, Zn 300, and Cr 300	
Enterococcus saccharolyticus	Fe 700, Pb 800, Zn 500, and Cr 500	
Cupriavidus metallidurans CH34	Cu 750 and Cr 100	Fan et al. (2014)
Ochrobactrum intermedium LBr	Cu 300 and Cr 1000	

TABLE 29.5

Uptake Capacity of Bacterial Biomass for Heavy Metals

Bacteria	Heavy Metal	Uptake Capacity (mg g^{-1})	Reference
Pseudomonas putida	Pb	180.41	Uslu and Tanyol (2006)
	Cu	49.50	
Staphylococcus xylosus	Cd	250	Ziagova et al. (2007)
	Cr	143	
Pseudomonas	Cd	278	Ziagova et al. (2007)
	Cr	95	
Bacillus sp. FM1	Cr	64.10	Masood and Malik (2011)
	Cu	78.12	
Rhizobium leguminosarum	Cd	135.3	Abd-Alla et al. (2012)
	Co	167.5	
Bacillus thuringiensis OSM29	Cd	59.17	Oves et al. (2013)
	Cr	71.94	
	Cu	39.84	
	Pb	30.76	
	Ni	43.13	
Acinetobacter sp. FM4	Cr	90.00	Masood and Malik (2014)
	Cu	93.30	
	Ni	66.70	
Cupriavidus metallidurans	Cu	86.78	Fan et al. (2014)
	Cr	47.79	
Escherichia coli	Pd	45.65	Kim et al. (2015)
	Pt	38.87	

TABLE 29.6

Heavy Metal Removal by Bacteria in Multimetal Ion Solution

Bacterial Isolates	Uptake/Removal Capacity in Single Metal Ion Solution	Uptake/Removal Capacity in Multimetal Ion Solution	Remark	Reference
Pseudomonas putida	Pb(II) (54.17 mg g^{-1}) Cu(II) (27.91 mg g^{-1})	Pb(II) + Cu(II) [Pb(II) (30.08 mg g^{-1}) and Cu(II) (18.91 mg g^{-1})]	Absorption of Cu(II) decreases with the increase in concentration of Pb(II)	Uslu and Tanyol (2006)
Staphylococcus xylosus	Cr(VI) (143 mg g^{-1}) Cd(II) (250 mg g^{-1})	Cr*(VI) and Cd(II) [90% Cr(VI) removal] [Cd*(II)(130 mg g^{-1}) and Cr(VI)] (Uptake data of Cr(VI) not available)	Cr(VI) absorbed more dominantly when Cr(VI) is the dominant metal ion in the solution. Cd(II) absorbed more dominantly when Cd(II) is the dominant metal ion in the solution.	Ziagova et al. (2007)
Pseudomonas sp.	Cr(VI) (95 mg g^{-1}) Cd(II) (278 mg g^{-1})	Cd*(II) (150 mg g^{-1}) and Cr(VI) (Uptake data of Cr(VI) not available)	Cd(II) is the dominant metal ion in the solution.	Ziagova et al. (2007)
Escherichia coli WS11	Cd(II) (29.83 mg g^{-1}) Ni(II) (36.03 mg g^{-1})	Cd + Ni [Cd (16.68 mg g^{-1}) + Ni (11.99 mg g^{-1}]	The biosorption of Cd(II) was slightly higher than that of Ni(II) in binary metal solution	Ansari and Malik (2007)
Brevundimonas vesicularis	Ni (52%) Cu (65.4%)	Ni(II) + Cu(II) [Ni (37%) and Cu (51.4%)]	Removal percentage decreases in the presence of binary heavy metal solution as compared to single metal ion solution in each case.	Singh and Gadi (2012)
Bacillus subtilis	Cd (208.04 mg g^{-1}) Hg (382.43 mg g^{-1}) Pb (400.75 mg g^{-1}) Hg (382.43 mg g^{-1})	Cd + Hg (0.5 millimolar L^{-1} of each heavy metal) Pb + Hg (0.5 millimolar L^{-1} of each heavy metal)	The presence of Cd and Hg did not affect the overall biosorption capacity. The presence of Pb did not affect the overall biosorption capacity.	Wang and Sun (2013)
Bacillus megaterium	Pb (60 mg g^{-1}) Cd (40.32 mg g^{-1}) Cr (18.92 mg g^{-1})	Pb*(II) (47.62 mg g^{-1}) + Cr(II) (14.91 mg g^{-1}) Pb*(II) (55.55 mg g^{-1}) + Cd(II) (36.52 mg g^{-1}) Cr*(II) (14.91 mg g^{-1}) + Pb(II) (47.62 mg g^{-1}) Cr*(II) (16.66 mg g^{-1}) + Cd(II) (32.55 mg g^{-1}) Cd*(II) (36.52 mg g^{-1}) + Pb(II) (55.55 mg g^{-1}) Cd*(II) (32.55 mg g^{-1}) + Cr(II) (16.66 mg g^{-1})	Suppression effect of Cr was greater than the suppression effect of Cd on biosorption of Pb(II). Biosorption of Cr(II) ion in the presence of Pb(II) and Cd(II) did not affect significantly. Suppression effect of Cr(VI) was greater than the suppression effect of Pb(II) on biosorption of Cd(II).	Sati and Verma (2014)
Ochrobactrum intermedium LBr	Cr(VI) (10.94 mg g^{-1}) and Cu(II) (75.14 mg g^{-1})	Cr(VI) and Cu(II) [4:1, 2:2 and 1:4 Cu(II):Cr(VI) concentration ratio] (Uptake data not available)	Biosorption of Cu(II) is dominant as compared to the biosorption of Cr(VI) in all cases.	Fan et al. (2014)
Cupriavidus metallidurans CH34	Cr(VI) (47.79 mg g^{-1}) and Cu(II) (86.78 mg g^{-1})	Cr(VI) and Cu(II) [4:1, 2:2 and 1:4 Cu(II):Cr(VI) concentration ratio] (Uptake data not available)	Biosorption of Cu(II) is dominant as compared to the biosorption of Cr(VI) in all cases.	Fan et al. (2014)
Escherichia coli	Pd(II) (38.87 mg g^{-1}) Pt(II) (45.65 mg g^{-1})	Pd(II) (33.1687 mg g^{-1}) and Pt(IV) (7.3287 mg g^{-1})	Pd (II) absorbed more dominantly than Pt(IV)	Kim et al. (2015)
Rhodococcus opacus	Pb (0.45 mmol g^{-1}) Cr (1.40 mmol g^{-1}) Cu (0.50 mmol g^{-1})	Pb(II) and Cu(II) Pb(II), Cu(II), and Cr(III)	Uptake of Pb(II) inhibited dominantly in the presence of Cu(II) alone, as compared to Cu(II) and Cr(III) present simultaneously.	Bueno et al. (2008)

Cu(II) uptake was decreased from 83.80 to 52.0 mg g^{-1} when the concentration of Cr(VI) in binary metal solution increased from 0 to 400 mgL^{-1}. Similar decrease in Cr(VI) uptake was observed from 100.8 to 74.02 mg g^{-1} with the increase in concentration of Cu(II). The overall uptake capacity of *Acinetobacter* sp. for heavy metals depends on the ionic radius of investigated metal ion. Heavy metal removal increases with the increase in the metal ion radius, and the order follows the pattern as Cr(81) > Cu(71) > Ni(63) in the binary metal ion solution. Fan et al. (2014) also observed that the removal of Cu(II) by dried biomass of *Ochrobactrum intermedium* LBr was 0.5–2 times as compared to the removal of Cr(VI) in binary metal solution containing both heavy metals at 100 mg L^{-1} concentration with different Cr(VI):Cu(II) ratios (4:1, 1:1 and 1:4). It was observed that biosorption of Cd(II) was always on the higher side as compared to the Ni(II) by *E. coli* (WS11) in binary metal system even with the increase in initial binary metal concertation from 50 to 400 µg mL^{-1} (Ansari and Malik, 2007). This might be due to the competition between both metal ions for the same biosorption sites. Similarly, higher uptake of Cd(II) in the presence of Ni(II) was observed with biomass of *Citrobacter* (Puranik and Paknikar, 1999). The presence of interfering metal ion (Zn(II), Cd(II), Cu(II), Co(II), and Ni(II)) did not affect the biosorption of Pb(II), whereas the biosorption rate of Cd(II) and Zn(II) was affected significantly in the presence of other metal ions (Cu(II), Pb(II), Ni(II), and Co(II)) by *Citrobacter strain* (MCMB-181) (Puranik and Paknikar, 1999).

In some studies, *qe'/qe* ratio is used to determine dynamics behind the biosorption of heavy metal in binary or ternary metal ion solution, where *qe'* denotes the biosorption capacity in multimetal ion solution and *qe* is the biosorption capacity in single metal ion solution. Three possible outcomes such as >1, =1, and <1 of ratio *qe'/qe* are possible. The value greater than 1 denotes the synergistic effect, while the value less than 1 denotes the antagonistic effect and value equal to 1 indicates no interaction between two or more metal ions in the solution. Sati and Verma (2014) discussed the biosorption ability of *Bacillus megaterium* for Pb(II), Cd(II), and Cr(II) in binary heavy metal solution containing a combination of two metals at a time. The initial metal ion concentration of primary metal was kept constant at 30 mg L^{-1}, and the concentration of secondary metal ion was varied from 10 to 50 mg L^{-1}. It was observed that *qe'/qe* ratio for Pb(II) in the presence of Cr(II) was 0.36, which was increased to 0.65 in the presence of Cd(II). The greater the ratio value of *qe'/qe* ratio, the less will be the suppression effect of secondary metal ion on the primary metal ion. Hence, the biosorption of Pb(II) by *B. megaterium* was affected greatly by Cr(II) as compared to Cd(II) in binary metal ion solution. The biosorption capacity of Pb(II) by *B. megaterium* decreases from 60.00 to 47.62 mg g^{-1} in the presence of 50 mg L^{-1} Cr(II) as compared to 55.55 mg g^{-1} in the presence of 50 mg L^{-1} Cd(II) at pH 7 and 40°C. However, the presence of Pb(II) and Cd(II) in the solution did not affect the biosorption of Cr(II) significantly. The biosorption capacity of Cr(II) decreased from 18.92 to 14.91 mg g^{-1} and 16.66 mg g^{-1} in the presence of Pb(II) and Cd(II), having

0.83 and 0.85 *qe'/qe* ratio, respectively. This indicates that the antagonistic effect of Pb(II) and Cd(II) for Cr(II) is less than that of the case mentioned earlier. Similarly, the biosorption of Cd(II) (40.32 mg g^{-1}) was greatly affected in the presence of Cr(II), that is, 32.55 mg g^{-1}, as compared to that (36.52 mg g^{-1}) in the presence of Pb(II). Thus, biosorption order was Pb(II) > Cr(II), Cd(II) > Cr(II), and Pb(II)~Cd(II) in binary metal ion solution.

As per another study, the biosorption of Cd(II) by UV-mutant *Bacillus subtilis* was significantly affected by the presence of Pb(II) as compared to Hg(II). The biosorption capacity decreased from 220.17 to 206.04 mg g^{-1} and 58.62 mg g^{-1} in the presence of Hg(II) and Pb(II), respectively. The *qe'/qe* ratio also increased in presence of Hg(II) (0.98) as compared to 0.36 in the presence of Pb(II), indicating the greater antagonistic effect of Pb(II), whereas the biosorption of Hg(II) (primary metal) was not affected by the presence of Cd(II) and Pb(II) as secondary metal ion species in the metal solution (with *qe'/qe* ratio approaching to 1 in the presence of Cd(II) and Pb(II), indicating that Hg(II) has no interaction with Cd(II) and Pb(II) in binary metal ion solution). Although chemical properties like coordination number, charge, and electronegativity are similar for Hg(II) and Cd(II), biosorption behavior is different for these metals. Hg(II) preferably binds to thiol group, whereas Cd(II) and Pb(II) bind to carboxyl, phosphoryl, and hydroxyl as mentioned by the authors (Wang and Sun, 2013).

On the other hand, the situation will become more complicated when more than two metal ions are present in the solution, that is, ternary metal ion solution. Very few studies have discussed the removal capacity of bacteria in the ternary metal ion solution. The biosorption ability of the *B. megaterium* was determined in the ternary metal ion solution containing Pb(II), Cd(II), and Cr(II) (Sati and Verma, 2014). The concentration of primary ion was kept constant (30 mg L^{-1}), and the concentration of the secondary ion was varied from 10 to 50 mg L^{-1}. The *qe'/qe* ratio was decreased to 0.32 in the ternary metal ion solution with Pb(II) as the primary metal ion as compared to the 0.36 and 0.65 in binary metal ion solution with Cr(II) and Cd(II), respectively. The uptake capacity decreased from 60 to 43.48 mg g^{-1} in ternary metal ion. Similarly, decrease in *qe'/qe* ratio as well as uptake capacity was observed with other two ternary metal ion mixtures containing Cd(II) and Cr(II) as a primary metal ion. It was clearly indicated by the authors that high absorption of Pb(II) might be due to its largest atomic weight as well as high electronegativity, as compared to Cr(II) and Cd(II). In another ternary metal ion study employing *Bacillus* sp., Cd(II) and Pb(II) did not affect the uptake rate of Hg(II) (primary metal ion), whereas the presence of Cd(II)+Hg(II) and Pb(II)+Hg(II) significantly reduced the uptake rate of Pb(II) and Cd(II), respectively. Uptake capacities of living biomass of *Bacillus* in the single metal ion for Hg(II), Cd(II), and Pb(II) were 397.16 mg g^{-1}, 220.17 mg g^{-1}, and 339 mg g^{-1} and were reduced to 368.33, 57.33, and 277.60 mg g^{-1}, respectively, in ternary metal ion solution (Wang and Sun, 2013). This overall decrease in uptake capacity could be due to the increase in electrostatic repulsion among all the three heavy metal ions that limit the absorption of each other on the cell surface. Hence, the studies indicated that metal with higher atomic radius and electronegativity will strongly attract to the cell surface.

Removal of heavy metal in multimetal ion solution is a complex scenario affected by the factors that depend on the surface properties of bacteria such as functional sites, their charges, pH of solution, number of metal ions, ionic radius of metal ions, and their electronegativity. As mentioned earlier, the presence of interfering metal ions affects the uptake capacity of other metal ions, but the insight on actual mechanisms and their proper functioning is lacking in most of the reports. Therefore, the detailed exploration to elucidate this system needs more systematic efforts and expertise.

29.3.2 Heavy Metal Removal by Fungi

The fungal cell wall is mainly composed of about 90% polysaccharides. The cell wall contains two layers: outer layer is made up of glucan, mannan, galactans, etc., whereas inner layer contains chitin and cellulose. The cell wall of fungi also contains negatively charged phosphate and glucuronic acid groups. The metals in the solution bind to phosphate, glucuronic, chitin, glucan, etc., and lead to the passive uptake of heavy metal by fungi. The presence of chitin increases the metal removal capacity of the fungi (Sağ, 2001). In addition to that, a number of transmembrane proteins are involved in active uptake of heavy metals by fungi. The transmembrane proteins such as Zrt1p, Nramp2, Fet4p, Fet3p, Ftr1p, CorA, ALR1p, and ALR2p lead to the uptake of heavy metal from the surrounding environment (Nies, 1999). These transport membrane proteins alone or in combination with other membrane protein cause accumulation of heavy metal inside the cell. For example, Cu(II) is first reduced to Cu(I) by iron–copper specific reductase (FRE1p, FRE2p, and FRE7p) and Cu(I) is then transported via CTR1p (Nies, 1999).

The advantage of using fungi over bacteria is their ability to tolerate high concentration of heavy metals. Table 29.7 summarizes the MIC values of different heavy metals for different fungi. As compared to bacteria, fungi exhibit MIC on higher side. Table 29.8 summarizes the fungal metal uptake capacity from synthetic solution or wastewater. In the presence of interfering metal ion, uptake of heavy metal gets affected similarly as in bacteria. Different studies indicate different scenarios of heavy metal removal in the presence of multiple heavy metals (Table 29.9).

Sağ et al. (2000b) investigated the biosorption behavior of *Rhizopus arrhizus* for Pb(II), Cu(II), and Zn(II) in the single and binary metal ion solution. The uptake capacity was maximum for Pb(II), followed by Cu(II) and Zn(II) with initial metal ion concentration varied from 5 to 250 mg L^{-1}. To investigate the absorption capacity in the binary metal ion solution, the initial concentrations of the dominant metal in the solution were varied between 20 and 210 mg L^{-1}. The presence of Zn(II) and Cu(II) did not significantly reduce the absorption of Pb(II). But the uptake of Cu(II) and Zn(II) by fungi got reduced significantly in the presence of Pb(II), indicating the antagonistic effect in the binary metal ion solution. The affinity of heavy metal in binary metal ion solution toward fungus was in the following order: Pb(II) > Cu(II) > Zn(II). Tsekova et al. (2007) observed that *Penicillium brevicompactum* shows higher affinity for Co(II) as compared to Cu(II) in the equimolar concentration of both metals in binary metal solution. The maximum metal uptake in single metal ion solution for Cu(II) and Co(II) was 25.32 and 54.64 mg L^{-1}, respectively, whereas it reduced to 17.39 and 30.96 mg L^{-1} for Cu(II) and Co(II) in binary metal ion solution. The results indicate direct competition between the two metal ions for the same binding site on the fungus surface as suggested by the authors. But no evidence was provided by the authors in this regard. The study suggested that Co(II) was more preferable ion as compared to Cu(II) both in single and binary metal ion solutions. Bayramoğlu and Arica (2008) observed that the absorption capacity for Hg(II), Cd(II), and Zn(II) by dead pellet was on higher side as compared to live pellet, in both single and ternary metal ion solutions. The absorption capacities for Hg(II), Cd(II), and Zn(II) by live pellet were 0.58, 0.43, and 0.21 mmol g^{-1} of biosorbent, whereas it increased to 0.81, 0.52, and 0.35 mmol g^{-1}, respectively, for dead pellet. Reduction in absorption capacity by the live biomass may be due to the binding of heavy metal extracellularly with metal-binding proteins (metallothionein and phytochelatins) to reduce the toxicity. These proteins are mainly composed of cysteine and bind heavy metal. The other factor that may lead to the low absorption uptake is efficient efflux of heavy metal from the living pellets, as suggested by the investigators. However, no experimental evidences were provided in this regard. But the total biosorption capacity for each heavy metal was on lower side in the ternary metal ion solution as compared to the single metal ion solution. The biosorption affinity for the heavy metal followed the order Hg(II) > Cd(II) > Zn(II) by both live and dead pellet in single as well as ternary metal ion solution. This affinity may be due to their electronegativity value (2.00 > 1.69 > 1.65) that also follows the same trend. The higher the electronegativity of the element, the higher is the affinity toward biosorbent.

The difference in the biosorption ability for different metals might also be due to the other factors such as electrode potential, ionic charges, and ionic radii. Ahmad et al. (2006) investigated and compared the difference in biosorption capacity of *Aspergillus* sp. and *Penicillium* sp. in single as well as multimetal ion solution at different heavy metal ion concentrations ranging from 2 to 6 mM. The maximum biosorption capacities of *Aspergillus* sp. for Cr(VI), Ni(II), and Cd(II) were 18.05, 25.05, and 19.40 mg g^{-1}, respectively, at 4 mM, which reduced to 11.35, 13.50, and 10.37 mg g^{-1} in multimetal solution. Similar reduction in uptake capacity was observed with *Penicillium* sp. The uptake capacities for Cr(VI), Ni(II), and Cd(II) were reduced to 8.36, 11.78, and 13.80 mg g^{-1} in multimetal ion solution as compared to 19.30, 17.90, and 18.60 mg g^{-1} in single metal ion solution. The overall uptake capacity in the multimetal ion solution was 35.22 and 33.94 by *Aspergillus* sp. and *Penicillium* sp., respectively, which was more than the uptake capacity observed in any single metal ion solution.

TABLE 29.7
Minimum Inhibitory Concentration of Different Heavy Metals for Different Fungi under Single Metal Exposure

Fungus	MIC (mg L^{-1})	Reference
Penicillium sp.	Ni 850, Cd 550, and Cr 600	Ahmad et al. (2006)
Aspergillus sp.	Ni 400, Cd 150, and Cr 300	
Alternaria sp.	Ni 600, Cd 500, and Cr 500	
Rhizopus sp.	Ni 850, Cd 500, and Cr 500	
Aspergillus flavus	Cr 800 and Pb 800	Iram et al. (2012)
Humicola grisea	Cr 400 and Pb 800	
Fusarium sp.	Cr 1,000 and Pb 1,000	
Helminthosporium	Cr 800 and Pb 1,000	
Aspergillus niger	Cr 1,000 and Pb 800	
Aspergillus versicolor	Cr 1,000 and Pb 1,000	
Aspergillus niger	Cr 3,500, Ni 4,500, and Zn 3,000	Kumar et al. (2012)
Aspergillus heteromorphus	Cr 200, Ni 500, and Zn 600	
Aspergillus sydoni	Cr 1,800, Ni 3,500, and Zn 2,000	
Aspergillus flavus	Cr 500, Ni 1,200, and Zn 700	
Aspergillus fumigatus	Cr 400, Ni 700, and Zn 500	
Trichoderma viride	Cr 1,500, Ni 2,500, and Zn 2,000	
Penicillium janthinellum	Cr 1,200, Ni 2,000, and Zn 1,500	
Penicillium fusarium	Cr 300, Ni 700, and Zn 200	
Aspergillus lentulus	Ni 300, Cu 850, Pb 5,000, and Cr(III) 120,000	Mishra (2013)
Fusarium graminearum	Cu 2,000, Cd 100, Pb 4,000, and Zn 3,000	Wolny-Kołaḋka (2014)
Fusarium avenaceum	Cu 8,000, Cd 3,000, Pb 10,000, and Zn 10,000	
Fusarium sporotrichioides	Cu 8,000, Cd 5,000, Pb 10,000, and Zn 10,000	
Fusarium oxysporum	Cu 4,000, Cd 500, Pb 10,000, and Zn 10,000	
Fusarium culmorum	Cu 8,000, Cd 100, Pb 10,000, and Zn 10,000	

TABLE 29.8
Uptake Capacity of Fungal Biomass for Heavy Metals

Fungi	Heavy Metal	Uptake Capacity(mg g^{-1})	Reference
Phanerochaete chrysosporium	Ni, Pb	55.9, 53.6	Çeribasi and Yetis (2001)
Phanerochaete chrysosporium	Cd, Pb, Cu	27.79, 85.86, 28.55	Say (2001)
Aspergillus terreus	Fe, Cr, Ni	164.5, 96.5, 19.6	Dias et al. (2002)
Penicillium canescens	As, Hg, Cd, Pb	26.4, 54.8, 102.7, 213.2	Say et al. (2003)
Funalia trogii	Hg, Cd, Zn	403.2, 191.6, 54	Arica et al. (2004)
Penicillium simplicissimum	Cd, Zn, Pb	52.50, 65.60, 76.90	Fan et al. (2008)
Lentinus edodes	Hg, Cd, Zn	403, 274.3, 57.7	Bayramoğlu and Arica (2008)
Penicillium sp. A1	Pb, Zn	372.54, 123.98	Pan et al. (2009)
Fusarium sp. A19	Cd, Cu, Pb, Zn	22.98, 10.87, 166.11, 475.21	Pan et al. (2009)
Aspergillus niger	Cu, Pb	20.91, 540	Iskandar et al. (2011)
Beauveria bassiana	Cd, Pb	46, 83.33	Ka et al. (2011)
Trametes versicolor	Pb, Cd	208.3, 166.6	Subbaiah et al. (2011)
Aspergillus lentulus	Cu, Cr, Pb	124.50, 331.48, 1120.65	Mishra and Malik (2012)
Auricularia polytricha	Cd, Cu, Pb	63.3, 73.7, 221	Huang et al. (2012)
Brevundimonas vesicularis	Ni, Cu	91.6, 129.4	Singh and Gadi (2012)

In multimetal ion solution, *Aspergillus* sp. showed high affinity to Cr(VI) and Ni(II) as compared to Cd(II). Saĝ et al. (2000) observed simultaneous biosorption of Cr(VI) and Fe(II) using *R. arrhizus*. To determine the biosorption characteristic of Cr(VI) and Fe(III) in binary metal ion solution, the concentration of one ion was varied [25–150 mg L^{-1} for Cr(VI) and 10–250 mg L^{-1} for Fe(III)], while the concentration of other ions was held constant. The Cr(VI) absorption rate decreased to 4.42 from 8.43 mg g^{-1} min^{-1} in the presence of 50 mg L^{-1} Fe(III) at 150 mg L^{-1} initial Cr(VI) concentration.

TABLE 29.9

Heavy Metal Removal by Fungi in Multimetal Ion Solution

Fungus	Uptake/Removal Capacity in Single Metal Ion Solution	Uptake/Removal Capacity in Multimetal Ion Solution	Remark	Reference
Rhizopus arrhizus	—	Zn (II) and Cu(II)	Uptake of Cu(II) and Zn(II) reduced significantly in the presence of Pb(II).	Sağ et al. (2000)
		Cu(II) and Pb(II)	Uptake of Pb(II) did not reduce in the presence of Cu(II) and Zn(II).	
Phanerochaete chrysosporium	Cd (15.2 mg g⁻¹) Pb (12.34 mg g⁻¹)	Cd(II) and Pb(II)	Biosorption of Pb(II) was dominant as compared to Cd(II).	Li et al. (2004)
Botrytis cinerea	Pb(41.60 mg g⁻¹)	Pb(II) (17.87 mg g⁻¹) and Cu(II) Pb(II) (34.85 mg g⁻¹) and Ni(II) Pb(II) (36.83 mg g⁻¹) and Cd(II)	Inhibition effect of Cu(II) was more dominant than Ni(II) and Cd(II) on Pb(II) uptake.	Akar et al. (2005)
Aspergillus flavus	Pb (13.46 mg g⁻¹) Cu (10.82 mg g⁻¹)	Pb(II) and Cu(II) [Pb(II) (6.41 mg g⁻¹) and Cu(II) (3.04 mg g⁻¹)]	Biosorption of Pb(II) was dominant as compared to Cu(II).	Akar and Tunali (2006)
Aspergillus niger	Cr (11.35 mg g⁻¹) Cd (19.4 mg g⁻¹) Ni (25.05 mg g⁻¹)	Cr, Ni(II) and Cd(II) [Cr (11.35 mg g⁻¹), Ni(II) (13.50 mg g⁻¹), and Cd(II) (10.37 mg g⁻¹)]	Biosorption of Cr and Ni was more preferred as compared to Cd.	Ahmad et al. (2006)
Penicillium brevicompactum	Cu (25.32 mg g⁻¹) Co (54.64 mg g⁻¹)	Cu(II) and Co(II) [Cu(II) (17.39 mg g⁻¹) and Co(II) (30.96 mg g⁻¹)]	Co(II) absorbed more dominantly than Cu(II).	Tsekova et al. (2007)
Cunninghamella echinulata	Cu (30%) Cd (50%) Pb (85%)	Cu(II) and Pb(II) [Cu(II) (24%) and Pb(II) (77%)] Cd(II) and Pb(II) [Cd(II) (45%) and Pb(II) (70%)]	Removal efficiency of Pb(II) reduced dominantly in the presence of other metal ion.	Shoaib et al. (2011)
Flammulina velutipes	Cd (88.97%) Pb (70.97%)	Pb(II) and Cd(II) [Pb(II) (70.97%) and Cd(II) (52.15%)]	Biosorption of Pb(II) not affected with the increase in concentration of Cd(II).	Zhang et al. (2012)

Similarly, Fe(III) absorption rate decreased to 1.52 from 3.90 mg g⁻¹ min⁻¹ in the presence of 50 mg L⁻¹ Cr(VI) at 150 mg L⁻¹ initial Fe(III) concentration. Antagonistic effect was observed between the two metal ions when present simultaneously in the solution.

The metal removal efficiency of Cu(II), Cd(II), and Pb(II) by *Aspergillus niger* and *Cunninghamella echinulata* was studied in single, binary, and ternary metal ion solution at 100 mg L⁻¹ initial heavy metal concentration (Shoaib et al., 2011). With *A. niger*, it was observed that the presence of Cu(II) and Cd(II) in binary metal ion solution reduces the metal removal efficiency by 3% and 4% for Cu(II) and Cd(II), respectively. Similarly, decrease in metal removal efficiency by 1% and 14% was observed for Cu(II) and Pb(II) in Cu(II)+Pb(II) metal ion solution. Investigator observed that the presence of Cd(II) and Pb(II) in the solution decreases the removal capacity by 3% and 15%, respectively. Similarly, decrease in metal removal efficiency by *Cunninghamella echinulata* was observed in all the dual metal ion solutions. The removal efficiency further decreases by 5%, 8%, and 23% for Cu(II), Cd(II), and Pb(II) in ternary metal ion solution by *A.*

niger, whereas the removal efficiency decreases by 8.5%, 9%, and 22% for Cu(II), Cd(II), and Pb(II) by *Cunninghamella echinulata* in ternary metal ion solution. The metal removal efficiency of Pb(II) reduced dominantly in the presence of other metal ions. The site for Pb(II) and other metal (Cu(II) and Cd(II)) might be the same that leads to the decrease in metal removal efficiency of the Pb(II).

Cairney et al. (1997) investigated the effect of four toxic metal ions (Pb(II), Cd(II), Zn(II), and Sb³⁻) on the biomass production by ectomycorrhizal fungi. Different combinations of all four metal ions were used in the experiment. The reduction in toxicity of Cd(II) and increment in biomass of *Lactarius deliciosus* occurs when Sb³⁻ was present in medium. The combination of Sb³⁻ + Cd(II) + Zn(II) also leads to an increase in the biomass production as compared to the Cd(II) alone. The combination of Cd(II), Pb(II), and Zn(II) was also less toxic than the individual and paired combination treatment that leads to up to 50% increase in biomass. The interaction between Cd(II) and Zn(II) leads to marginal increase in the biomass production as compared to individual Cd(II) or Zn(II). In the case of *Suillus granulatus*, the addition of Pb(II) and Sb^{3-} to Cd(II)-containing media reduced the toxicity to a such level that biomass production was equal to the control one. This study clearly indicates the synergistic effect of combinational metal ions on the growth of fungi.

Chitin, the main constituent of fungal cell wall, provides suitable binding sites for heavy metals on the cell surface, making fungi a good biosorbent for heavy metal removal. The mechanism for antagonistic and synergistic effects in multimetal ion solution was not fully discussed and understood, although some studies have discussed the effect of electronegativity, that is, metals with higher electronegativity will be absorbed more efficiently (Shoaib et al., 2011). However, for more clarifications on common functional sites and for blockage due to other interfering ions, more research inputs and flair in the subject are required.

29.3.3 Heavy Metal Removal by Algae

Algae are simple plants, ranging from microscopic unicellular to more than hundred feet in length with multicellular form. Algae possess many features such as high absorption capacity, large surface area-to-volume ratio, and expression of various phytochelatin that bind metals, which make them suitable candidate for heavy metal removal from the solution (Kumar and Oommen, 2012; Suresh Kumar et al., 2015). Cell wall component of the algae is negatively charged, and this negatively charged cell wall leads to the high affinity against positively charged heavy metals in solution (Chekroun and Baghour, 2013). Some algae also contain high levels of sulfated polysaccharides as well as alginates in cell wall, which increase the overall heavy metal absorption capacity of the algae (Davis et al., 2003). Table 29.10 summarizes the biosorption capacity of various algae.

Scientists studied the removal capacity of the algae from the single metal ion solution. But studies discussing heavy metal removal in multimetal ion solution are limited. Table 29.11 summarizes the algal heavy metal removal behavior in multimetal ion solution. Mohapatra and Gupta (2005) investigated the biosorption capacity of *Oscillatoria angustissima* and found that absorption capacity decreases in the order Zn(II) > Co(II) > Cu(II) in a single metal ion solution. It was revealed that in some combination of the metal ions mentioned earlier, the absorption capacity was dependent on pH and initial metal ion concentration. The addition of Cu(II) at different concentrations (1 and 2 mM) significantly inhibits the absorption of Zn(II) from 0.31 mmol g^{-1} (at 1.5 mM initial concentration) to 0.14 and 0.07 mmol g^{-1} at pH 4, respectively. The absorption value of Zn(II) was within the same range as observed in the absence of any interfering metal ion, that is, 0.43–0.52 mmol g^{-1} at pH 6, showing that the sorption value of Zn(II) was on lower side at low pH as compared to high pH, which might be due to the H^+ ion interference at low pH for the same binding sites. Moreover, the absorption capacity of Zn(II) dropped with increase in initial concentration of interfering ion at pH 4. But initial concentration of interfering ion at higher pH had no effect on sorption capacity of Zn(II). In the case of Cu(II), the sorption value was independent of pH. Moreover, sorption of Cu(II) remained unaffected up to 2 mM initial concentration of interfering ion (Zn(II) and Co(II)). On the other hand, sorption of Co(II) was independent of pH in the presence of Zn(II), whereas it was dependent on pH as well as initial metal ion concentration in the presence of Cu(II). The absorption of Cu and Pb on *Chlamydomonas reinhardtii* was independent of each other and showed no interaction. This independence is due to different binding sites for Cu(II) and Pb(II). Cu binds mainly to amino group, whereas Pb(II)

TABLE 29.10
Uptake Capacity of Algal Biomass for Heavy Metals

Algae	Heavy Metal	Uptake Capacity (mg g^{-1})	References
Laminaria japonica	Cu(II), Cd(II), Pb(II)	76.26, 124.77, 275.56	Yu et al. (1999)
Laminaria hyperbola	Cu(II), Cd(II), Pb(II)	75.53, 92.17, 279.72	
Ascophyllum nodosum	Cu(II), Cd(II), Pb(II)	75.62, 115.78, 263.14	
L. flavicans	Cu(II), Cd(II), Pb(II)	81.98, 130.39, 300.44	
E. maxima	Cu(II), Cd(II), Pb(II)	77.53, 125.34, 290.08	
Ecklonia radiata	Cu(II), Cd(II), Pb(II)	70.54, 116.91, 261.07	
Durvillaea potatorum	Cu(II), Cd(II), Pb(II)	83.25, 132.64, 321.16	
Lyngbya taylorii	Cd(II), Zn(II), Pb(II)	41.59, 32.04, 304.58	Klimmek et al. (2001)
Padina tetrastromatica	Cd(II), Pb(II)	59.58, 217.35	Wang and Chen (2009)
Cladophora fracta	Cu(II), Zn(II), Cd(II), Hg(II)	2.388, 1.623, 0.240, 0.288	Ji et al. (2011)
Spirulina plantensis	Cd(II), Ni(II)	73.64, 69.04	Çelekli and Bozkurt (2011)
Chlorella sp.	Cu(II), Zn(II)	33.4, 28.5	Wan Maznah et al. (2012)
Cystoseira barbata	Ni(II), Cd(II), Pb(II)	61.93, 124.21, 253.21	Yalçın et al. (2012)
Spirogyra hyalina	Cd(II), Hg(II), Pb(II), As(II), Co(II)	18.18, 35.71, 31.25, 4.80, 12.82	Kumar and Oommen (2012)
Colpomenia sinuosa	Cu(II), Cd(II), Zn(II), Pb(II)	36.61, 1.67, 1.04, 565.99	Cirik et al. (2012)
Chlorella minutissima	Zn(II), Cd(II), Mn(II), Cu(II)	33.71, 35.36, 21.19, 154.17	Yang et al. (2015)

TABLE 29.11

Heavy Metal Removal by Algae in Multimetal Ion Solution

Algae	Uptake/Removal Capacity in Single Metal Ion Solution	Uptake/Removal Capacity in Multimetal Ion Solution	Remark	Reference
Oscillatoria angustissima	Cu (0.27 mmol g^{-1} at pH 4), Cu (0.35 mmol g^{-1} at pH 6)	Cu(II) and Zn(II) [Cu(II) (0.27 mmol g^{-1})] Cu(II) and Co(II) [Cu(II) (0.35 mmol g^{-1})]	Absorption of Cu(II) remains unaffected in the presence of Zn(II) and Co(II) at low concentration.	Mohapatra and Gupta (2005)
Chlorella vulgaris	Cu(II) (57.8 mg g^{-1})	Cu(II) and Pb(II) Cu(II) and Zn(II)	Suppression effect of Pb(II) was greater than the suppression effect of Zn(II) on biosorption of Cu(II).	Al-Rub et al. (2006)
Sargassum sp.	Pb (95%)	Pb(II) and Cu(II) [Pb (78%)] Pb(II) and Cd(II) [Pb (89%)]	Suppression effect of Cu(II) was greater than the suppression effect of Cd(II) on biosorption of Pb(II).	Sheng et al. (2007)
Spirogyra neglecta	Cu (0.396 mmol g^{-1}) Pb (0.297 mmol g^{-1})	Cu(II) and Pb(II) [Cu(II) (0.055 mmol g^{-1}) and Pb(II) (0.213 mmol g^{-1})]	Absorption of Pb(II) was more dominant than biosorption of Cu(II). The presence of Pb(II) decreases the absorption of Cu(II) significantly.	Singh et al. (2007)
Pithophora oedogonia	Cu (0.458 mmol g^{-1}) Pb (0.467 mmol g^{-1})	Cu(II) and Pb(II) [Cu(II) (0.209 mmol g^{-1}) and Pb(II) (0.127 mmol g^{-1})]	Absorption of Cu(II) was more dominant than biosorption of Pb(II).	Kumar et al. (2008)
Laminaria japonica	—	Cu(II) and Ni(II)	Absorption of Cu(II) was more dominant than absorption of Ni(II).	Wang and Li (2009)
Sargassum filipendula	Cd(II) (1.17 mmol g^{-1}) Zn(II) (0.70 mmol g^{-1})	Cd(II) and Zn(II)	The presence of Cd(II) at low concentration did not affect the absorption of Zn(II). The presence of Zn(II) significantly decreases the absorption of Cd(II).	Luna et al. (2010)
Sargassum filipendula	Cu(II) (0.87 mmol g^{-1}) Ni(II) (0.72 mmol g^{-1})	Cu(II) and Ni(II) [Cu(II) (0.75 mmol g^{-1}) and Ni(II) (0.12 mmol g^{-1})]	The presence of Ni(II) did not affect the absorption capacity of Cu(II).	Kleinübing et al. (2011)
Chlamydomonas reinhardtii	—	Pb(II) and Cu(II)	Absorption of Pb(II) was not affected by the presence of Cu(II). Absorption of Cu(II) was not affected by the presence of Pb(II).	Flouty and Estephane (2012)
Cystoseira indicia	Cu(II) (103.09 mg g^{-1}) Co(II) (59.52 mg g^{-1})	Cu(II) and Co(II)	Absorption of Cu(II) increases in the presence of Co(II). Absorption of Co(II) increases in the presence of Cu(II).	Akbari et al. (2015)

binds to carboxyl group (Flouty and Estephane, 2012). Akbari et al. (2015) observed synergistic effect during biosorption of Co(II) and Cu(II) on *Cystoseira indica* when present simultaneously in the solution. It was suggested that a complex is formed with these two ions, and this complex binds to the surface functional group with higher capacity. Biomass of *C. indica* showed more tendency toward the absorption of Cu(II), which might be due to the higher electronegativity of Cu(1.90) as compared to Co(1.88). The authors suggested that binding of Cu(II) to biosorbent changes its chemical characteristics and makes the surface weakly basic in nature. In this situation, Co(II) can interact with the biosorbent and lead to multilayer biosorption.

Ramsenthil and Meyyappan (2010) investigated biosorption of Zn(II) and Cu(II) by immobilized microalgae on silica gel. Antagonistic effect was found to be dominant in binary metal ion solution. Biosorption process was found to be slower in binary metal mixture as compared to single metal ion solution. Moreover, the presence of NO_3^{-} and SO_3^{-} did not affect the biosorption process in both cases. But the presence of Cl^{-} ion in the solution decreased the biosorption process in single and binary metal ion solution. On the other hand, the presence of Ca and Mg in the solution did not affect the sorption process of Cu(II) but decreased the Zn(II) biosorption considerably. So the presence of other ions also affects the sorption process. Sheng et al. (2007) observed biosorption of Cd(II), Cu(II), and Pb(II) in different binary combinations by *Sargassum* sp. and concluded that no significant reduction of the sorption of Cd(II) was seen in the presence of Pb(II) and Cu(II), whereas the sorption of Pb(II) and Cu(II) decreased by

3% and 11% in the presence of Cd(II) with equal concentration of each metal. Similar decline in biosorption of Cr(II) and Mn(II) up to 30% and 62.1%, respectively, was observed in binary metal ion solution as compared to single metal ion solution by *Ulva* sp. (Vijayaraghavan and Joshi 2014).

The multimetal ion solution in the presence of algae produces the same effect as displayed by bacterial and fungal cells. The elements other than heavy metals such as Na$^+$, Ca^{2+}, Cl$^-$, H$^+$ ions can also affect the removal of heavy metals by algae. But studies to determine the chemical/physical factors responsible for heavy metal removal by algae need more comprehensible approach for better understanding.

29.4 BIOREACTOR CONFIGURATION FOR MULTIMETAL REMOVAL

The use of microorganism in a well-defined vessel (bioreactor) with the optimized conditions can provide a good option for heavy metal removal from wastewater. The metal removal efficacy directly depends on parameters like temperature, nutrient availability and its types, stirring rate, type of microorganism, pH of the growth media, dissolve oxygen concentration, rate of aeration, and heavy metal concentration (Mishra, 2013; Mishra and Malik 2013). Bioreactor can be very useful to treat water with low level of heavy metal mixture (Stoll and Duncan, 1997). Anaerobic and aerobic bioreactors are two types of system employed for the treatment of wastes generated from industries and municipal bodies. In general, anaerobic reactors are used to treat highly polluted wastewater with high chemical oxygen demand, that is, industrial wastewater, whereas aerobic reactors are used to treat low strength wastewater. Different configurations of anaerobic and aerobic reactors can be employed for the treatment of wastewater such as membrane bioreactor, continuous stir tank reactor, and sequential reactor (Costley and Wallis, 2001a; Katsou et al., 2011; Kieu et al., 2011; Zeiner et al., 2012).

Kieu et al. (2011) studied the removal of various heavy metals like Cu(II), Zn(II), Cr(II), and Ni(II) in anaerobic semicontinuous stirred tank reactor using a consortium of sulfate-reducing bacteria with over 94%–100% removal efficiency. Hawari and Mulligan (2007) studied the heavy metal removal capacity of flow through system using anaerobic sludge as biosorbent in the presence of calcium. The study revealed that the presence of Cu(II), Cd(II), and Ni(II) decreases the uptake of Pb(II) by 6%–11%. Factors such as hydration effects, hydrolysis effect, and covalent binding of the heavy metal affect the overall uptake capacity of the system. The affinity order of biomass toward heavy metal in flow through system follows the pattern Pb(II) > Cu(II) > Ni(II) > Cd(II). Cibati et al. (2013) reported that biogenic generation of H$_2$S in anaerobic baffled bioreactor increases the precipitation of heavy metals at different pH. Over 23%, 16%, 72%, and 70% recovery was observed for Ni(II), Co(II), Mo, and V, respectively, in this setup. The anaerobic semicontinuous stirred tank reactor gives the removal of Cu(II) and Cr(VI)

(96%–100%) and Ni(II) and Zn(II) (94%–100%) after 12 weeks of operation (Kieu et al., 2014).

Stoll and Duncan (1996) designed a continuous-flow stirred bioreactor system using yeast biomass (*Saccharomyces cerevisiae*) for the removal of Cu(II), Cd(II), Cr(VI), Ni(II), and Zn(II) from effluent. They observed that the effluent concentration of Cu(II), Ni(II), and Cr(VI) after treatment drops significantly and lies within the limits of drinking water. Jackson et al. (2009) evaluated the performance of various bioreactors (two laboratory scales and one on-site) for Al, Ni(II), Fe, Mn(II), and Zn(II) removal from river water. The reactor having microbial consortium (*Pseudomonas* sp., *Sphingomonas* sp., *and Bacillus* sp.) significantly removed Al (85%), Ni (65%), Fe (44%), Mn (57%), and Zn (97%). Oliveira et al. (2007) proved that biological treatment plant can remove the heavy metal such as Hg(II), Cd(II), Zn(II), Cu(II), Pb(II), Cr(VI), and Mn(II) with removal efficiency up to 61.5%, 60%, 44.5%, 44.2%, 39.7%, 16.5%, and 10.4%, respectively. Moreover, sludge generated from the dewatering of the discharge can be used for agricultural purposes. Zeiner et al. (2010) operated horizontal rotating tubular bioreactor and obtained up to 38.1%–95.5% removal of Fe(III), Cr(VI), Ni(II), and Zn(II) from the wastewater. Similarly, metal removal efficiency of mixed microbial culture was studied in horizontal rotating tubular bioreactor and over 87%–93.6%, 99.7%–100%, and 89%–95% removal was obtained for Mn(II), Cr(VI), and Co(II), respectively (Zeiner et al., 2012).

Barros et al. (2006) obtained up to 90% removal of Ni(II) and Cr(VI) in packed bed bioreactor having initial heavy metal concentration of 10 mg L^{-1} each. Costley and Wallis (2001b) observed the removal efficiency of heavy metal in a three-stage biological contactor with 73%, 42%, and 33% removal efficiency for Cu(II), Zn(II), and Cd(II), respectively, from the synthetic wastewater. Kapoor and Viraraghavan (1998) used dead biomass of *A. niger* in packed bed column system and obtained removal up to 38% (Cu), 58% (Pb), and 16% (Zn) from the industrial wastewater containing 0.08 mg L^{-1} (Cu), 0.66 mg L^{-1} (Pb), and 1.78 mg L^{-1} (Zn), respectively. Singh et al. (2013) found that the presence of Fe(III) in the synthetic wastewater increases the metal removal capacity for Cr(VI) (72.9%) and Cd(II) (87.63%) by *Pseudomonas aeruginosa* in batch reactor. Heavy metal removal was observed in stirred tank bioreactor by immobilizing *Polyporus squamosus* on Ca-alginate. The uptake capacity of 42.8 ± 1.52, 153.04 ± 5.78, 23.72 ± 0.7, 31.6 ± 1.29, and 26.52 ± 1.10 was obtained for Cr(VI), Fe(III), Ni(II), Cu(II), and Pb(II), respectively, in untreated wastewater (Chuma 2007). Malakahmad et al. (2011) investigated the heavy metal as well as organic content removal efficiency of a lab-scale sequencing batch reactor for treating wastewater from petrochemical industry. The concentration of Hg(II) (9.03 mg L^{-1}) and Cd(II) (15.52 mg L^{-1}) was spiked in a reactor, and up to 76%–90% and 96%–98% removal was obtained for Hg(II) and Cd(II), respectively. Similarly, up to 70%–80% and 50% removal was obtained in sequencing batch reactor for Cu(II) (10 mg L^{-1}) and Cd(II)

(30 mg L^{-1}), when present individually and in combination (Lim et al., 2002). No synergistic effect on microorganism ability to remove the heavy metal was observed in binary metal system. Sirianuntapiboon and Hongsrisuwan (2007) observed that the presence of Cu(II) in the system decreases the overall efficiency of sequencing batch reactor as compared to Zn(II). The removal capacity of up to 92.62% and 83.77% was obtained for Cu(II) and Zn(II) from synthetic industrial wastewater. Similarly, the investigators observed up to 88.6% and 94.6% removal of Pb(II) and Ni(II) from the synthetic wastewater in sequencing batch bioreactor. Uptake capacities of sludge for Pb(II) and Ni(II) were 840 and 720 mg g^{-1}, respectively.

The use of microbial consortia in a bioreactor was reported to be more efficient than individual microbe for remediating multiple heavy metals from wastewater. Factors such as aeration rate, pH, temperature, metal concentration, nutrient availability, and type of microorganism directly affect the metal removal efficiency of the reactor. There are different models of bioreactor (aerobic, anaerobic, other configurations of bioreactor) that can treat different concentrations of heavy metals from wastewater. It was reported that anaerobic reactor performs better with high-strength wastewater at low pH using sulfate-reducing bacteria. Moreover, in some bioreactors (continuous-flow stirred bioreactor), significant removal was achieved, and after treatment, the quality of water fell within the limits of drinking water. However, most of the studies used small setups (lab-scale bioreactor); hence, research on developing large-scale bioreactors needs keen investigation to generate an environmentally friendly option (bioreactor) for large amount of heavy metal remediation.

29.5 CONCLUSIONS

Bioremediation is an emerging technology that is now perceived as an alternative option to remove heavy metals from low- to high-strength wastewater. Although studies have highlighted the significant achievement in remediating heavy metal contamination, there are a few challenges that need to be addressed, such as proper understanding on mechanism of metal uptake by different microbes, interfering metal ions effect, physical/chemical factors of metal ions, and surface properties of microorganisms. Due to complexity, studies concerning multimetal removal represent a small fraction of a large pool of studies on biological metal removal dominated by single metal studies. Among the multimetal studies, majority of the investigations target biosorption mode that simply reveals interaction between metals and biomass surface groups under single versus multimetal exposure. Negligible studies used live system to study multimetal remediation. Live systems use multiple mechanisms as compared to biosorption where only surface groups play a primary role (Section 29.1). With multiple mechanisms, live microorganisms offer a more robust system to counter the impact of multimetal toxicity. Wider aspects of multiple metal toxicity, stress response, protein expression, and pigmentation by living organism need to

be researched to elaborate the mechanism behind simultaneous removal of multiple metals. In addition, since most of the studies reported were confined to lab/small setups, research to develop setups (high capacity bioreactor) for the treatment of huge volumes of wastewater assumes significance. Interestingly, the bioreactor development studies on multimetal removal often involved microbial consortium and live systems as opposed to the batch studied discussed in Section 29.4. These results indicate strong potential of consortium-based approach to develop field-worthy technologies.

ACKNOWLEDGMENTS

The authors gratefully acknowledge NFBSFARA, Indian Council of Agricultural Research (grant no. NFBSAFARA/WQ-2023/2012-13) for financial support. One of the authors (Nitin Chauhan) thanks University Grants Commission (UGC), Govt. of India, for providing funds through RGNF. The authors are also grateful to the technical staff and students of Applied Microbiology Lab, CRDT, IIT Delhi for their cooperation.

REFERENCES

Abd-Alla, M.H., Morsy, F.M., El-Enany, A.-W.E., and Ohyama, T., 2012. Isolation and characterization of a heavy-metal-resistant isolate of *Rhizobium leguminosarum* bv. viciae potentially applicable for biosorption of Cd^{2+} and Co^{2+}. *International Biodeterioration and Biodegradation* 67: 48–55.

Adamu, C.I., Nganje, T.N., and Edet, A., 2014. Heavy metal contamination and health risk assessment associated with abandoned barite mines in Cross River State, southeastern Nigeria. *Environmental Nanotechnology Monitoring and Management* 3: 10–21.

Ahemad, M. and Kibret, M., 2013. Recent trends in microbial biosorption of heavy metals: A review. *Biochemistry and Molecular Biology* 1: 19–26.

Ahmad, I., Ansari, M.I., and Aqil, F., 2006. Biosorption of Ni, Cr and Cd by metal tolerant *Aspergillus niger* and *Penicillium* sp. using single and multi-metal solution. *Indian Journal of Experimental Biology* 44: 73–76.

Akar, T. and Tunali, S., 2006. Biosorption characteristics of *Aspergillus flavus* biomass for removal of Pb(II) and Cu(II) ions from an aqueous solution. *Bioresource Technology* 97: 1780–1787.

Akar, T., Tunali, S., and Kiran, I., 2005. Botrytis cinerea as a new fungal biosorbent for removal of Pb(II) from aqueous solutions. *Biochemical Engineering Journal* 25: 227–235.

Akbari, M., Hallajisani, A., Keshtkar, A.R., Shahbeig, H., and Ali Ghorbanian, S., 2015. Equilibrium and kinetic study and modeling of Cu(II) and Co(II) synergistic biosorption from Cu(II)-Co(II) single and binary mixtures on brown algae C. indica. *Journal of Environmental Chemical Engineering* 3: 140–149.

Al-Rub, F.A.A., El-Naas, M.H., Ashour, I., and Al-Marzouqi, M., 2006. Biosorption of copper on Chlorella vulgaris from single, binary and ternary metal aqueous solutions. *Process Biochemistry* 41: 457–464.

Alves, R.I.S., Sampaio, C.F., Nadal, M., Schuhmacher, M., Domingo, J.L., and Segura-Muñoz, S.I., 2014. Metal concentrations in surface water and sediments from Pardo River, Brazil: Human health risks. *Environmental Research* 133: 149–155.

Ansari, M.I. and Malik, A., 2007. Biosorption of nickel and cadmium by metal resistant bacterial isolates from agricultural soil irrigated with industrial wastewater. *Bioresource Technology* 98: 3149–3153.

Arica, M.Y., Bayramoglu, G., Yilmaz, M., Bektaş, S., and Genç, O., 2004. Biosorption of Hg^{2+}, Cd^{2+}, and Zn^{2+} by Ca-alginate and immobilized wood-rotting fungus *Funalia trogii*. *Journal of Hazardous Materials* 109: 191–199.

Badr, A.M., Mahana, N.A., and Eissa, A., 2014. Assessment of heavy metal levels in water and their toxicity in some tissues of Nile Tilapia (*Oreochromis niloticus*) in River Nile Basin at Greater Cairo, Egypt. *Global Veterinaria* 13: 432–443.

Bai, R.S. and Abraham, T.E., 2002. Studies on enhancement of Cr(VI) biosorption by chemically modified biomass of *Rhizopus nigricans*. *Water Research* 36: 1224–1236.

Barros, A.J.M., Prasad, S., Leite, V.D., and Souza, A.G., 2006. The process of biosorption of heavy metals in bioreactors loaded with sanitary sewage sludge. *Brazilian Journal of Chemical Engineering* 23: 153–162.

Bayramoğlu, G. and Arica, M.Y., 2008. Removal of heavy mercury(II), cadmium(II) and zinc(II) metal ions by live and heat inactivated *Lentinus edodes* pellets. *Chemical Engineering Journal* 143: 133–140.

Bhattacharya, A., Dey, P., Gola, D., Mishra, A., Malik, A., and Patel, N., 2015. Assessment of Yamuna and associated drains used for irrigation in rural and peri-urban settings of Delhi NCR. *Environmental Monitoring and Assessment* 187: 4146.

Bliss, C.I., 1939. The toxicity of poisons applied jointly. *Annals of Applied Biology* 26: 585–615.

Borgmann, U., Norwood, W.P., and Dixon, D.G., 2008. Modelling bioaccumulation and toxicity of metal mixtures. *Human and Ecological Risk Assessment: An International Journal* 14: 266–289.

Boyanov, M.I., Kelly, S.D., Kemner, K.M., Bunker, B.A., Fein, J.B., and Fowle, D.A., 2003. Adsorption of cadmium to *Bacillus subtilis* bacterial cell walls: A pH-dependent x-ray absorption fine structure spectroscopy study. *Geochimica et Cosmochimica Acta* 67: 3299–3311.

Bueno, B.Y.M., Torem, M.L., Molina, F., and de Mesquita, L.M.S., 2008. Biosorption of lead(II), chromium(III) and copper(II) by *R. opacus*: Equilibrium and kinetic studies. *Minerals Engineering* 21: 65–75.

Caesar-Tonthat, T., Van Ommen Kloeke, F., Geesey, G., and Henson, J., 1995. Melanin production by a filamentous soil fungus in response to copper and localization of copper sulfide by sulfide-silver staining. *Applied Environmental Microbiology* 61: 1968–1975.

Cairney, H.J.W.G., Sanders, F.E., and Meharg, A., 1997. Toxic interactions of metal ions Pb^{2+}, Zn^{2+} and Sb^{3-}) on in vitro biomass production of ectomycorrhizal fungi. *New Phytologist* 137: 551–562.

Çelekli, A. and Bozkurt, H., 2011. Bio-sorption of cadmium and nickel ions using *Spirulina platensis*: Kinetic and equilibrium studies. *Desalination* 275: 141–147.

Çeribasi, I.H. and Yetis, Ü., 2001. Biosorption of Ni(II) and Pb(II) by *Phanerochaete chrysosporium* from a binary metal system—Kinetics. *Water SA* 27: 15–20.

Chekroun, K.B. and Baghour, M., 2013. The role of algae in phytoremediation of heavy metals: A review. *Journal of Materials and Environmental Science* 4: 873–880.

Chuma, P.A., 2007. Biosorption of Cr, Mn, Fe, Ni, Cu and Pb metals from petroleum refinery effluent by calcium alginate immobilized mycelia of *Polyporus squamosus*. *Scientific Research and Essay* 2: 217–221.

Cibati, A., Cheng, K.Y., Morris, C., Ginige, M.P., Sahinkaya, E., Pagnanelli, F., and Kaksonen, A.H., 2013. Selective precipitation of metals from synthetic spent refinery catalyst leach liquor with biogenic H_2S produced in a lactate-fed anaerobic baffled reactor. *Hydrometallurgy* 139: 154–161.

Cirik, Y., Molu Bekci, Z., Buyukates, Y., Ak, İ., and Merdivan, M., 2012. Heavy metals uptake from aqueous solutions using marine algae (*Colpomenia sinuosa*): Kinetics and isotherms. *Chemistry and Ecology* 28: 469–480.

Costley, S.C. and Wallis, F.M., 2001a. Bioremediation of heavy metals in a synthetic wastewater using a rotating biological contactor. *Water Research* 35: 3715–3723.

Costley, S.C. and Wallis, F.M., 2001b. Treatment of heavy metal-polluted wastewaters using the biofilms of a multistage rotating biological contactor. *World Journal of Microbiology and Biotechnology* 17: 71–78.

Rudakiya, D.M. and Pawar, K.S., 2013. Evaluation of remediation in heavy metal tolerance and removal by *Comamonas acidovorans* MTCC 3364\n. *IOSR Journal of Environmental Science, Toxicology and Food Technology* 5: 26–32.

Davis, T.A., Volesky, B., and Mucci, A., 2003. A review of the biochemistry of heavy metal biosorption by brown algae. *Water Research* 37: 4311–4330.

Dias, M.A., Lacerda, I.C.A., Pimentel, P.F., de Castro, H.F., and Rosa, C.A., 2002. Removal of heavy metals by an *Aspergillus terreus* strain immobilized in a polyurethane matrix. *Letters in Applied Microbiology* 34: 46–50.

Edokpayi, J.N., Odiyo, J.O., and Olasoji, S.O., 2014. Assessment of heavy metal contamination of Dzindi River, In Limpopo Province, South Africa contribution originality. *International Journal of Natural Sciences Research* 2: 185–194.

Fan, J., Onal Okyay, T., and Frigi Rodrigues, D., 2014. The synergism of temperature, pH and growth phases on heavy metal biosorption by two environmental isolates. *Journal of Hazardous Materials* 279: 236–243.

Fan, T., Liu, Y., Feng, B., Zeng, G., Yang, C., Zhou, M., Zhou, H., Tan, Z., and Wang, X., 2008. Biosorption of cadmium(II), zinc(II) and lead(II) by *Penicillium simplicissimum*: Isotherms, kinetics and thermodynamics. *Journal of Hazardous Materials* 160: 655–661.

Fang, D., Zhang, R.C., Deng, W.J., and Li, J., 2012. Highly efficient removal of Cu(II), Zn(II), Ni(II) and Fe(II) from electroplating wastewater using sulphide from sulphidogenic bioreactor effluent. *Environmental Technology* 33: 1709–1715.

Fath, M.J. and Kolter, R., 1993. ABC transporters: Bacterial exporters. *Microbiology and Molecular Biology Review* 57: 995–1017.

Flouty, R. and Estephane, G., 2012. Bioaccumulation and biosorption of copper and lead by a unicellular algae *Chlamydomonas reinhardtii* in single and binary metal systems: A comparative study. *Journal of Environmental Management* 111: 106–114.

Giri, S. and Singh, A.K., 2013. Assessment of surface water quality using heavy metal pollution index in Subarnarekha River, India. *Water Quality, Exposure and Health* 5: 173–182.

Goher, M.E., Hassan, A.M., Abdel-Moniem, I.A., Fahmy, A.H., and El-sayed, S.M., 2014. Evaluation of surface water quality and heavy metal indices of Ismailia Canal, Nile River, Egypt. *The Egyptian Journal of Aquatic Research* 40: 225–233.

Gola, D., Namburath, M., and Kumar, R., 2015. Decolourization of the azo dye (direct brilliant blue) by the isolated bacterial strain. *Journal of Basic and Applied Engineering Research* 2: 1462–1465.

Gupta, S.K., Chabukdhara, M., Kumar, P., Singh, J., and Bux, F., 2014. Evaluation of ecological risk of metal contamination in river Gomti, India: A biomonitoring approach. *Ecotoxicology and Environmental Safety* 110: 49–55.

Gupta, V.K. and Rastogi, A., 2008. Equilibrium and kinetic modelling of cadmium(II) biosorption by nonliving algal biomass *Oedogonium* sp. from aqueous phase. *Journal of Hazardous Materials* 153: 759–766.

Hashim, M.A., Mukhopadhyay, S., Sahu, J.N., and Sengupta, B., 2011. Remediation technologies for heavy metal contaminated groundwater. *Journal of Environmental Management* 92: 2355–2388.

Hawari, A.H. and Mulligan, C.N., 2007. Effect of the presence of lead on the biosorption of copper, cadmium and nickel by anaerobic biomass. *Process Biochemistry* 42: 1546–1552.

Helios Rybicka, E., 1996. Impact of mining and metallurgical industries on the environment in Poland. *Applied Geochemistry* 11: 3–9.

Huang, H., Cao, L., Wan, Y., Zhang, R., and Wang, W., 2012. Biosorption behavior and mechanism of heavy metals by the fruiting body of jelly fungus (*Auricularia polytricha*) from aqueous solutions. *Applied Microbiology and Biotechnology* 96: 829–840.

Idris, A.M., Said, T.O., Omran, A.A., and Fawy, K.F., 2015. Combining multivariate analysis and human risk indices for assessing heavy metal contents in muscle tissues of commercially fish from Southern Red Sea, Saudi Arabia. *Environmental Science and Pollution Research* 22: 17012–17021.

Iram, S., Parveen, K., Usman, J., Nasir, K., and Akhtar, N., 2012. Heavy metal tolerance of filamentous fungal strains isolated from soil irrigated with industrial wastewater. *Biologija* 58: 107–116.

Iskandar, N.L., Ain, N., Mohd, I., and Tan, S.G., 2011. Tolerance and biosorption of copper (Cu) and lead (Pb) by filamentous fungi isolated from a freshwater ecosystem. *Journal of Environmental Sciences* 23: 824–830.

Islam, M.S., Ahmed, M.K., Raknuzzaman, M., Habibullah-Al-Mamun, M., and Islam, M.K., 2015. Heavy metal pollution in surface water and sediment: A preliminary assessment of an urban river in a developing country. *Ecological Indicators* 48: 282–291.

Issazadeh, K. and Razban, S., 2014. Isolation and identification of Enterococcus species and determination of their susceptibility patterns against antibiotics and heavy metals in coastal waters of. *International Journal of Advanced Biological and Biomedical Research* 2: 2026–2030.

Jackson, V.A., Paulse, A.N., Bester, A.A., Neethling, J.H., Khan, S., and Khan, W., 2009. Bioremediation of metal contamination in the Plankenburg River, Western Cape, South Africa. *International Biodeterioration and Biodegradation* 63: 559–568.

Jarup, L., 2003. Hazards of heavy metal contamination. *British Medical Bulletin* 68: 167–182.

Ji, L., Xie, S., Feng, J., Li, Y., and Chen, L., 2011. Heavy metal uptake capacities by the common freshwater green alga *Cladophora fracta*. *Journal of Applied Phycology* 24: 979–983.

Ka, H., Sh, H., and Jh, J., 2011. Potential capacity of *Beauveria bassiana* and *Metarhizium anisopliae*. *Journal of General and Applied Microbiology* 57: 347–355.

Kapoor, A. and Viraraghavan, T., 1998. Application of immobilized Aspergillus niger biomass in the removal of heavy metals from an industrial wastewater. *Journal of Environmental Science and Health Part A, Toxic/Hazardous Substances and Environmental Engineering* 33: 1507–1514.

Katsou, E., Malamis, S., and Loizidou, M., 2011. Performance of a membrane bioreactor used for the treatment of wastewater contaminated with heavy metals. *Bioresource Technology* 102: 4325–4332.

Khan, M.U., Malik, R.N., Muhammad, S., Ullah, F., and Qadir, A., 2014. Health risk assessment of consumption of heavy metals in market food crops from sialkot and gujranwala districts, Pakistan. *Human and Ecological Risk Assessment: An International Journal* 21: 327–337.

Kieu, H.T.Q., Horn, H., and Müller, E., 2014. The effect of heavy metals on microbial community structure of a sulfidogenic consortium in anaerobic semi-continuous stirred tank reactors. *Bioprocess and Biosystems Engineering* 37: 451–460.

Kieu, H.T.Q., Müller, E., and Horn, H., 2011. Heavy metal removal in anaerobic semi-continuous stirred tank reactors by a consortium of sulfate-reducing bacteria. *Water Research* 45: 3863–3870.

Kim, S., Song, M.-H., Wei, W., and Yun, Y.-S., 2015. Selective biosorption behavior of *Escherichia coli* biomass toward Pd(II) in Pt(IV)–Pd(II) binary solution. *Journal of Hazardous Materials* 283: 657–662.

Kleinübing, S.J., da Silva, E.A., da Silva, M.G.C., and Guibal, E., 2011. Equilibrium of Cu(II) and Ni(II) biosorption by marine alga *Sargassum filipendula* in a dynamic system: Competitiveness and selectivity. *Bioresource Technology* 102: 4610–4617.

Klimmek, S., Wilke, H.S., Bunke, G., and Buchholz, R., 2001. Comparative analysis of the biosorption of cadmium, lead, nickel, and zinc by algae comparative analysis of the biosorption of cadmium, lead, nickel, and zinc by algae. *Environmental Science and Technology* 35: 4283–4288.

Kosińska, K. and Miśkiewicz, T., 2012. Precipitation of heavy metals from industrial wastewater by *Desulfovibrio desulfuricans*. *Environment Protection Engineering* 38: 51–60.

Kumar, D., Singh, A., and Gaur, J.P., 2008. Mono-component versus binary isotherm models for Cu(II) and Pb(II) sorption from binary metal solution by the green alga *Pithophora oedogonia*. *Bioresource Technology* 99: 8280–8287.

Kumar, J.I.N. and Oommen, C., 2012. Removal of heavy metals by biosorption using freshwater alga *Spirogyra hyalina*. *Journal of Environmental Biology* 33(1): 37–31.

Kumar, R., Bhatia, D., Singh, R., and Bishnoi, N.R., 2012. Metal tolerance and sequestration of Ni(II), Zn(II) and Cr(VI) ions from simulated and electroplating wastewater in batch process: Kinetics and equilibrium study. *International Biodeterioration and Biodegradation* 66: 82–90.

Lesmana, S.O., Febriana, N., Soetaredjo, F.E., Sunarso, J., and Ismadji, S., 2009. Studies on potential applications of biomass for the separation of heavy metals from water and wastewater. *Biochemical Engineering Journal* 44: 19–41.

Li, Q., Wu, S., Liu, G., Liao, X., Deng, X., Sun, D., Hu, Y., and Huang, Y., 2004. Simultaneous biosorption of cadmium(II) and lead(II) ions by pretreated biomass of *Phanerochaete chrysosporium*. *Separation and Purification Technology* 34: 135–142.

Lim, P.E., Ong, S.A., and Seng, C.E., 2002. Simultaneous adsorption and biodegradation processes in sequencing batch reactor (SBR) for treating copper and cadmium-containing wastewater. *Water Research* 36: 667–675.

Loewe, S. and Muischnek, H., 1926. Über Kombinationswirkungen. I Mitteilung: Hilfsmittel der Fragestellung. *Archive Experimentelle Pathologie und Pharmakologie* 114: 313–326.

Lokhande, R.S., U. Singare, P., and Pimple, D.S., 2011. Quantification study of toxic heavy metal pollutants in sediment samples collected from kasardi river flowing along the taloja industrial area of Mumbai, India. *New York Science Journal* 4: 66–71.

Luna, A.S., Costa, A.L.H., da Costa, A.C.A., and Henriques, C.A., 2010. Competitive biosorption of cadmium(II) and zinc(II) ions from binary systems by *Sargassum filipendula*. *Bioresource Technology* 101: 5104–5111.

Malakahmad, A., Hasani, A., Eisakhani, M., and Isa, M.H., 2011. Sequencing Batch Reactor (SBR) for the removal of Hg^{2+} and Cd^{2+} from synthetic petrochemical factory wastewater. *Journal of Hazardous Materials* 191: 118–125.

Malik, A., 2004. Metal bioremediation through growing cells. *Environment International* 30: 261–278.

Malik, D.S. and Maurya, P.K., 2015. Heavy metal concentration in water, sediment, and tissues of fish species (*Heteropneustes fossilis* and *Puntius ticto*) from Kali River, India. *Toxicological and Environmental Chemistry* 1–12.

Martino, E., Turnau, K., Girlanda, M., Bonfante, P., and Perotto, S., 2000. Ericoid mycorrhizal fungi from heavy metal polluted soils: Their identification and growth in the presence of zinc ions. *Mycological Research* 104: 338–344.

Masood, F. and Malik, A., 2011. Biosorption of metal ions from aqueous solution and tannery effluent by *Bacillus* sp. FM1. *Journal of Environmental Science and Health. Part A, Toxic/Hazardous Substances and Environmental Engineering* 46: 1667–1674.

Masood, F. and Malik, A., 2014. Single and multi-component adsorption of metal ions by *Acinetobacter* sp. FM4. *Separation Science and Technology* 6395 (December), 141206225327009.

Mathur, M., Vijayalakshmi, K.S., Gola, D., and Singh, K., 2015. Decolourization of textile dyes by *Aspergillus lentulus*. *Journal of Basic and Applied Engineering Research* 2: 1469–1473.

Mishra, A., 2013. Development of biological system employing microbial consortium for pollutant removal from mixed waste stream (PhD thesis) Indian Institute of Technology New Delhi, India.

Mishra, A. and Malik, A., 2012. Simultaneous bioaccumulation of multiple metals from electroplating effluent using *Aspergillus lentulus*. *Water Research* 46: 4991–4998.

Mishra, A. and Malik, A., 2013. Recent Advances in microbial metal bioaccumulation. *Critical Reviews in Environmental Science and Technology* 43: 1162–1222.

Mishra, A. and Malik, A., 2014a. Novel fungal consortium for bioremediation of metals and dyes from mixed waste stream. *Bioresource Technology* 171: 217–226.

Mishra, A. and Malik, A., 2014b. Metal and dye removal using fungal consortium from mixed waste stream: Optimization and validation. *Ecological Engineering* 69: 226–231.

Mohapatra, H. and Gupta, R., 2005. Concurrent sorption of Zn(II), Cu(II) and Co(II) by Oscillatoria angustissima as a function of pH in binary and ternary metal solutions. *Bioresource Technology* 96: 1387–1398.

Mudhoo, A., Garg, V.K., and Wang, S., 2012. Removal of heavy metals by biosorption. *Environmental Chemistry Letters* 10: 109–117.

Nies, D.H., 1999. Microbial heavy-metal resistance. *Applied Microbiology and Biotechnology* 51: 730–750.

Nies, D.H. and Silver, S., 1995. Ion efflux systems involved in bacterial metal resistances. *Journal of Industrial Microbiology* 14: 186–199.

Norwood, W.P., Borgmann, U., Dixon, D.G., and Wallace, A., 2003. Effects of metal mixtures on aquatic biota: A review of observations and methods. *Human and Ecological Risk Assessment: An International Journal* 9: 795–811.

Olawoyin, R., Oyewole, S.A., and Grayson, R.L., 2012. Potential risk effect from elevated levels of soil heavy metals on human health in the Niger delta. *Ecotoxicology and Environmental Safety* 85: 120–130.

Oliveira, A.D.S., Bocio, A., Trevilato, T.M.B., Takayanagui, A.M.M., Domingo, J.L., and Segura-Muñoz, S.I., 2007. Heavy metals in untreated/treated urban effluent and sludge from a biological wastewater treatment plant. *Environmental Science and Pollution Research International* 14: 483–489.

Oves, M., Khan, M.S., and Zaidi, A., 2013. Biosorption of heavy metals by Bacillus thuringiensis strain OSM29 originating from industrial effluent contaminated north Indian soil. *Saudi Journal of Biological Sciences* 20: 121–129.

Pan, R., Cao, L., and Zhang, R., 2009. Combined effects of Cu, Cd, Pb, and Zn on the growth and uptake of consortium of Cu-resistant *Penicillium* sp. A1 and Cd-resistant *Fusarium* sp. A19. *Journal of Hazardous Materials* 171: 761–766.

Pandit, R.J., Patel, B., Kunjadia, P.D., and Nagee, A., 2013. Isolation, characterization and molecular identification of heavy metal resistant bacteria from industrial effluents, Amala-khadi—Ankleshwar, Gujarat. *International Journal of Environmental Sciences* 3: 1689–1699.

Patil, S., Thakur, V., and Ghorade, I., 2014. Analysis of heavy metals in Jaikwadi dam water, Maharashtra (India). *International Journal of Research in Applied, Natural and Social Sciences* 2: 69–74.

Paul, D. and Sinha, S.N., 2013. Assessment of various heavy metals in surface water of polluted sites in the lower stretch of river Ganga, West Bengal: A study for ecological impact. *Discovery Nature* 6: 8–13.

Paulsen, I.T. and Saier Jr., M.H., 1997. A novel family of ubiquitous heavy metal ion transport proteins. *Journal of Membrane Biology* 156: 99–103.

Puranik, P.R. and Paknikar, K.M., 1999. Biosorption of lead, cadmium, and zinc by *Citrobacter* strain MCM B-181: Characterization studies. *Biotechnology Progress* 15: 228–237.

Raja, C.E., Anbazhagan, K., and Selvam, G.S., 2006. Isolation and characterization of a metal-resistant *Pseudomonas aeruginosa* strain. *World Journal of Microbiology and Biotechnology* 22: 577–585.

Rajbanshi, A., 2009. Study on heavy metal resistant bacteria in guheswori sewage treatment plant. *Our Nature* 6: 52–57.

Ramsenthil, R. and Meyyappan, R., 2010. Single and multi-component biosorption of copper and zinc ions using microalgal resin. *International Journal of Environmental Science and Development* 1: 298–301.

Sağ, Y., 2001. Biosorption of heavy metals by fungal biomass and modeling of fungal biosorption: A review. *Separation and Purification Reviews* 30: 1–48.

Sağ, Y., Akcael, B., and Kutsal, T., 2002. Ternary biosorption equilibria of chromium(VI), copper(II), and cadmium(II) on *Rhizopus arrhizus*. *Separation Science and Technology* 37: 279–309.

Sağ, Y., Ataçoğlu, I., Kutsal, T., Ataçoğlu, I., and Kutsal, T., 2000a. Equilibrium parameters for the single- and multicomponent biosorption of Cr(VI) and Fe(III) ions on R. arrhizus in a packed column. *Hydrometallurgy* 55: 165–179.

Sağ, Y., Kaya, A., and Kutsal, T., 2000b. Lead, copper and zinc biosorption from biocomponent systems modelled by empirical Freundlich isotherm. *Applied Microbiology and Biotechnology* 53: 338–341.

Saier, M.H., Tam, R., Reizer, A., and Reizer, J., 1994. 2 novel families of bacterial-membrane proteins concerned with nodulation, cell-division and transport. *Molecular Microbiology* 11: 841–847.

Sarret, G., Manceau, A., Spadini, L., Roux, J.C., Hazemann, J.L., Soldo, Y., Eybert-Bérard, L., and Menthonnex, J.J., 1998. Structural determination of Zn and Pb binding sites in *Penicillium chrysogenum* cell walls by EXAFS spectroscopy. *Environmental Science and Technology* 32 (11), 1648–1655.

Sati, M. and Verma, M., 2014. Biosorption of heavy metals from single and multimetal solutions by free and immobilized cells of *Bacillus megaterium*. *International Journal of Advanced Research* 2: 923–934.

Say, R., 2001. Biosorption of cadmium(II), lead(II) and copper(II) with the filamentous fungus *Phanerochaete chrysosporium*. *Bioresource Technology* 76: 67–70.

Say, R., Yilmaz, N., and Denizli, A., 2003. Removal of heavy metal ions using the fungus *Penicillium canescens*. *Adsorption Science and Technology* 21: 643–650.

Sheng, P.X., Ting, Y.P., and Chen, J.P., 2007. Biosorption of heavy metal ions (Pb, Cu, and Cd) from aqueous solutions by the marine alga *Sargassum* sp. in single- and multiple-metal systems. *Industrial and Engineering Chemistry Research* 46: 2438–2444.

Shoaib, A., Badar, T., and Aslam, N., 2011. Removal of Pb (II), Cu (II) and Cd (II) from aqueous solution by some fungi and natural adsorbents in single and multiple metal systems. *Pakistan Journal of Botany* 43: 2997–3000.

Silver, S., 1996. Bacterial resistances to toxic metal ions—A review. *Gene* 179: 9–19.

Singh, A., Kumar, D., and Gaur, J.P., 2007. Copper(II) and lead(II) sorption from aqueous solution by non-living Spirogyra neglecta. *Bioresource Technology* 98: 3622–3629.

Singh, A., Sharma, R.K., Agrawal, M., and Marshall, F.M., 2010. Health risk assessment of heavy metals via dietary intake of foodstuffs from the wastewater irrigated site of a dry tropical area of India. *Food and Chemical Toxicology* 48: 611–619.

Singh, N. and Gadi, R., 2012. Bioremediation of Ni(II) and Cu(II) from wastewater by the nonliving biomass of *Brevundimonas vesicularis*. *Journal of Environmental Chemistry and Ecotoxicology* 4: 137–142.

Singh, R., Bishnoi, N.R., Kirrolia, A., and Kumar, R., 2013. Synergism of *Pseudomonas aeruginosa* and Fe0 for treatment of heavy metal contaminated effluents using small scale laboratory reactor. *Bioresource Technology* 127: 49–58.

Singh, R., Kumar, A., Kirrolia, A., Kumar, R., Yadav, N., Bishnoi, N.R., and Lohchab, R.K., 2011. Removal of sulphate, COD and Cr(VI) in simulated and real wastewater by sulphate reducing bacteria enrichment in small bioreactor and FTIR study. *Bioresource Technology* 102: 677–682.

Sirianuntapiboon, S. and Hongsrisuwan, T., 2007. Removal of Zn^{2+} and Cu^{2+} by a sequencing batch reactor (SBR) system. *Bioresource Technology* 98: 808–818.

Stoll, A. and Duncan, J.R., 1997. Implementation of a continuous-flow stirred bioreactor system in the bioremediation of heavy metals from industrial wastewater. *Environmental Pollution* 97: 247–251.

Stoll, A. and Duncan, J.R., 1996. Enhanced heavy metal removal from waste water by viable, glucose pretreated *Saccharomyces cerevisiae* cells. *Biotechnology Letters* 18: 1209–1212.

Subbaiah, M.V., Yuvaraja, G., Vijaya, Y., and Krishnaiah, A., 2011. Equilibrium, kinetic and thermodynamic studies on biosorption of Pb(II) and Cd(II) from aqueous solution by fungus (*Trametes versicolor*) biomass. *Journal of the Taiwan Institute of Chemical Engineers* 42: 965–971.

Suh, J.H., Yun, J.W., and Kim, D.S., 1999. Effect of extracellular polymeric substances (EPS) on Pb^{2+} accumulation by *Aureobasidium pullulans*. *Bioprocess Engineering* 21: 1–4.

Suresh Kumar, K., Dahms, H.-U., Won, E.-J., Lee, J.-S., and Shin, K.-H., 2015. Microalgae—A promising tool for heavy metal remediation. *Ecotoxicology and Environmental Safety* 113: 329–352.

Svecova, L., Spanelova, M., Kubal, M., and Guibal, E., 2006. Cadmium, lead and mercury biosorption on waste fungal biomass issued from fermentation industry. I. Equilibrium studies. *Separation and Purification Technology* 52: 142–153.

Ting, Y.P. and Teo, W.K., 1994. Uptake of cadmium and zinc by yeast: Effects of co-metal ion and physical/chemical treatments. *Bioresource Technology* 50: 113–117.

Tsekova, K., Ianis, M., Dencheva, V., and Ganeva, S., 2007. Biosorption of binary mixtures of Copper and Cobalt by *Penicillium brevicompactum*. *Zeitschrift für Naturforschung C* 62: 261–264.

Upadhyaya, D., Survaiya, M.D., Basha, S., Mandal, S.K., Thorat, R.B., Haldar, S., Goel, S. et al., 2014. Occurrence and distribution of selected heavy metals and boron in groundwater of the Gulf of Khambhat region, Gujarat, India. *Environmental Science and Pollution Research International* 21: 3880–3890.

Uslu, G. and Tanyol, M., 2006. Equilibrium and thermodynamic parameters of single and binary mixture biosorption of lead(II) and copper(II) ions onto *Pseudomonas putida*: Effect of temperature. *Journal of Hazardous Materials* 135: 87–93.

Vijayaraghavan, K. and Joshi, U.M., 2014. Application of *Ulva* sp. biomass for single and binary biosorption of chromium(III) and manganese(II) ions: Equilibrium modeling. *Environmental Progress and Sustainable Energy* 33: 147–153.

Vosyliene, M.Z., Kazlauskiene, N., and Svecevičius, G., 2003. Effect of a heavy metal model mixture on biological parameters of rainbow trout oncorhynchus mykiss. *Environmental Science and Pollution Research* 10: 103–107.

Wan Maznah, W.O., Al-Fawwaz, A.T., and Surif, M., 2012. Biosorption of copper and zinc by immobilised and free algal biomass, and the effects of metal biosorption on the growth and cellular structure of *Chlorella* sp. and *Chlamydomonas* sp. isolated from rivers in Penang, Malaysia. *Journal of Environmental Sciences (China)* 24: 1386–1393.

Wang, J. and Chen, C., 2009. Biosorbents for heavy metals removal and their future. *Biotechnology Advances* 27: 195–226.

Wang, T. and Sun, H., 2013. Biosorption of heavy metals from aqueous solution by UV-mutant *Bacillus subtilis*. *Environmental Science and Pollution Research International* 20: 7450–7463.

Wang, X.S. and Li, Z.Z., 2009. Competitive adsorption of Nickel and Copper ions from aqueous solution using nonliving biomass of the marine brown alga *Laminaria japonica*. *CLEAN—Soil, Air, Water* 37: 663–668.

Williams, G.P., Gnanadesigan, M., and Ravikumar, S., 2012. Biosorption and bio-kinetic studies of halobacterial strains against Ni^{2+}, Al^{3+} and Hg^{2+} metal ions. *Bioresource technology* 107: 526–529.

Wolny-Koł, K.A., 2014. In vitro effects of various xenobiotics on Fusarium spp. strains isolated from cereals. *Journal of Environmental Science and Health. Part. B, Pesticides, Food Contaminants, and Agricultural Wastes* 49: 864–870.

Yalçın, S., Sezer, S., and Apak, R., 2012. Characterization and lead(II), cadmium(II), nickel(II) biosorption of dried marine brown macro algae Cystoseira barbata. *Environmental Science and Pollution Research International* 19: 3118–3125.

Yang, J., Cao, J., Xing, G., and Yuan, H., 2015. Lipid production combined with biosorption and bioaccumulation of cadmium, copper, manganese and zinc by oleaginous microalgae *Chlorella minutissima* UTEX2341. *Bioresource Technology* 175: 537–544.

Yu, Q., Matheickal, J.T., Yin, P., and Kaewsarn, P., 1999. Heavy metal uptake capacities of common marine macro algal biomass. *Water Research* 33: 1534–1537.

Zeiner, M., Rezić, T., and Šantek, B., 2010. Monitoring of Cu, Fe, Ni, and Zn in wastewater during treatment in a horizontal rotating tubular bioreactor. *Water Environment Research* 82: 183–186.

Zeiner, M., Rezić, T., Šantek, B., Rezić, I., Hann, S., and Stingeder, G., 2012. Removal of Cr, Mn, and Co from textile wastewater by horizontal rotating tubular bioreactor. *Environmental Science and Technology* 46: 10690–10696.

Zhang, D., Hea, H., Li, W., Gao, T., and Ma, P., 2012. Biosorption of cadmium(II) and lead(II) from aqueous solutions by fruiting body waste of fungus Flammulina velutipes. *Desalination and Water Treatment* 20: 160–167.

Zhu, B., Wu, Z.-F., Li, J., and Wang, G.-X., 2011. Single and joint action toxicity of heavy metals on early developmental stages of Chinese rare minnow (*Gobiocypris rarus*). *Ecotoxicology and Environmental Safety* 74: 2193–2202.

Ziagova, M., Dimitriadis, G., Aslanidou, D., Papaioannou, X., Litopoulou Tzannetaki, E., and Liakopoulou-Kyriakides, M., 2007. Comparative study of Cd(II) and Cr(VI) biosorption on *Staphylococcus xylosus* and *Pseudomonas* sp. in single and binary mixtures. *Bioresource Technology* 98: 2859–2865.

30 Arsenic Microbiology: From Metabolism to Water Bioremediation

Anna Corsini and Lucia Cavalca

CONTENTS

ABSTRACT

Arsenic affects subsurface and surface waters and is released by various natural processes and anthropogenic actions. Bacteria are responsible for cycling arsenic in the environment, due to metabolic and detoxification reactions. Since diverse health problems arise in humans due to chronic arsenic contamination of waters, new hybrid solutions (arsenic bacteria processing and adsorbing materials) need to be implemented for water decontamination.

30.1 INTRODUCTION

Arsenic (As) contamination in groundwater is distributed all over the world and has been a serious health problem among the people in Southeast, Southwest, and Northeast United States, Mongolia (China), Southwest Taiwan coastal regions, Sonora (Mexico), Pamplonian Plain (Argentina), West Bengal (India), Northern Chile, and Bangladesh (Singh et al. 2015). Natural occurrence of As in groundwater in many parts of the world is widely reported as a consequence of weathering and leaching of geological formations and mine wastes rather than being derived from identifiable point sources of pollution (Bahar et al. 2013). Geogenic materials like organic-rich or black shales, Holocene alluvial sediments with slow flushing rates, gold deposits, volcanogenic sources, and hot springs can significantly enhance the As release in groundwater (Smedley and Kinniburgh 2002). The main mechanisms controlling arsenic release into aquifers are reductive dissolution, alkali desorption, sulfide oxidation, and induced mobilization from parent rocks by geothermal waters (Ravenscroft et al. 2009). Groundwater concentrations of As are reported to be of a very wide range from less than 0.5 to 5000 mg/L covering natural As contamination found in more than 70 countries (Cavalca et al. 2013).

Arsenic is present in surface and groundwater mainly in inorganic form [arsenate, As(V), and arsenite, As(III)]. Although normally groundwater does not contain organic forms of arsenic [monomethylarsonic acid, $CH_3AsO(OH)_2$ (MMA); dimethylarsinic acid, $(CH_3)_2AsOOH$ (DMA); arsenobetaine, $(CH_3)_3As+CH_2COOH$ (AsBet), trimethylarsine, $(CH_3)_3As$ (TMA)], some aquatic environments were found to contain As(V), As(III), as well as AsBet, MMA, and DMA (Mandal and Suzuki 2002; Ronkart et al. 2007). At moderate or high redox potentials, inorganic arsenic can be stabilized as a series of pentavalent As(V) species (Gupta et al. 2012). However, under most reducing (acid and mildly alkaline) conditions and lower redox potential, the trivalent arsenite species, As(III), predominate. As(V) has greater affinity for adsorption to oxyhydroxides and clay minerals, thus resulting in less mobility than As(III). MMA(V) and DMA(V) are both stable in oxidized systems, while DMA(III) and MMA(III) are unstable under oxidizing conditions (Gong et al. 2001).

30.2 ARSENIC TOXICITY

Although arsenic compounds have been used for many centuries as medicinal agents for the treatment of diseases, such as psoriasis, syphilis, rheumatosis, and, more recently, cancer (Dilda and Hogg 2007), it is considered to be one of the most toxic elements on Earth for humans. Based on several epidemiological studies, the International Agency for Research on Cancer (IARC 1987) has classified inorganic arsenic as a known human carcinogen. In 2013, the Environmental Protection Agency ranked arsenic first on Comprehensive Environmental Response, Compensation, and Liability Act (CERCLA) Priority List of Hazardous Substances, based on a combination of its frequency, toxicity, and potential for human exposure (Agency for Toxic Substances and Disease Registry, ATSDR 2013).

Recent findings suggested the following order in terms of acute As toxicity: MMA(III) > As(III) > As(V) > DMA(V) > MMA(V), where the MMA(III) metabolite and As(III) are the most toxic compounds (EFSA 2009). As(III) is generally regarded as more soluble and 60 times more toxic than As(V). As(III) has the ability to bind with sulfhydryl groups of cysteine residues in protein, thereby inactivating them. Conversely, As(V) is a chemical analog of phosphate that can interfere with the normal oxidative phosphorylation (Ordóñez et al. 2005). Long-term exposure to even small concentrations of inorganic arsenic can cause various health effects, such as "arsenicosis" and cancer due to DNA damage.

Contamination of drinking water and consumption of food grown in soils irrigated with As-contaminated water represent the main sources of As intake for humans, causing a life-threatening problem for millions of people in large areas of the world (EFSA 2009). In 2001, the World Health Organization (WHO) set the limit of arsenic in drinking water at 10 μg/L.

Nevertheless, a worldwide population of more than 100 million people are at risk, and out of these more than 45 million people living in developing Asian countries currently are at risk of being exposed to As concentrations that are greater than the national standard of 50 μg/L (Ravenscroft et al. 2009). Therefore, there is a great demand for efficient methods to remove arsenic from drinking water.

30.3 ARSENIC CYCLE IN THE ENVIRONMENT: THE ROLE OF BACTERIA

The fate of As released into the environment is determined by a complex interplay among processes of As mobilization, sequestration, and transformation, most of which are directly or indirectly driven by microbial activity (Table 30.1). Numerous phylogenetically diverse prokaryotes involved in As cycle have been isolated from a variety of terrestrial and aquatic habitats and in a wide range of environmental conditions (Cavalca et al. 2013). Studies on microbial diversity in natural samples to estimate As behavior are widely reported in the literature (Meyer-Dombard et al. 2013; Tomczyk-Żak et al. 2013; Corsini et al. 2014a; Ying et al. 2015). Natural abundance of arsenic in the environment has led many prokaryotes to develop different strategies to utilize arsenic either for metabolic purposes or for detoxification mechanisms (Figure 30.1). Arsenate enters the cell via phosphate transporters and can then interfere with oxidative phosphorylation by replacing phosphate. Entrance of As(III) into cells (at neutral pH) is mediated by the so-called aqua-glyceroporins, membrane channels for water and small nonionic solutes, such as glycerol. The most common phenomenon in many different bacteria is resistance to arsenic based on the presence of an "Ars" detoxification system. More rare is arsenotrophy, which is defined as the oxidation of As(III) or reduction of As(V) as part of respiratory or phototrophic processes (Hoeft et al. 2010).

At present, there are plenty of researches focusing on characterizing the ability of microorganism to directly interact with As (Silver and Phung 2005; Paez-Espino et al. 2009; Tsai et al. 2009; Slyemi and Bonnefoy 2012; Dhuldhaj et al. 2013); however, the biogeochemical cycle of arsenic is also

---

TABLE 30.1
Microbial Processes Influencing As Mobility in the Environment

Processes	Comments
Mobilization	
As(V) reduction	As(III) is more mobile than As(V).
Arsenic methylation	(Gaseous) Me–As is more mobile than inorganic As species.
Immobilization	
As(III) oxidation	As(V) has great affinity for adsorption.
Demethylation	Inorganic As is less mobile than Me–As.
Biomineralization	Formation of As-containing minerals.
Biosorption	Extracellular sequestration.
Bioaccumulation	Intracellular sequestration.
Iron oxidation	Iron oxide formation for As adsorption.

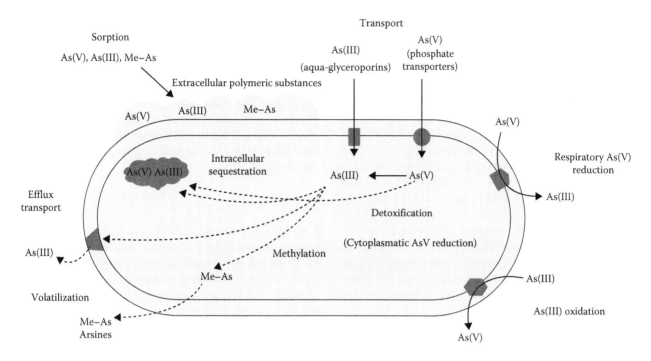

FIGURE 30.1 Metabolic and detoxification reactions of arsenic species by bacteria. As(III), arsenite; As(V), arsenate; Me–As, methylated arsenic.

influenced by many environmental and biotic factors that may critically control arsenic speciation. For instance, iron-reducing bacteria can reduce arsenic-containing iron and aluminum oxides with the release of As(V) in solution, which subsequently can be reduced to the more mobile As(III) by As(V)-reducing bacteria (Paez-Espino et al. 2009). In addition, phosphorous acquisition from arsenic-bearing minerals by *Burkholderia fungorum* was demonstrated as a mechanism of arsenic release (Mailloux et al. 2009). Some microorganisms can also methylate inorganic arsenic or demethylate organic forms (Huang 2014). Moreover, a selenium- and sulfur-mediated pathway for arsenic detoxification has been proposed (Couture et al. 2012), although it remains to be further studied in detail.

30.3.1 As(III) Oxidation

Since the first report of As(III)-oxidizing bacteria by Green (1918), several strains from approximately 21 genera of As(III)-oxidizing prokaryotes have been reported to date. Phylogenetically, they are grouped under α-, β-, and γ-Proteobacteria, *Deinococcus–Thermus*, and *Crenarchaeota* (Cavalca et al. 2013).

For heterotrophic bacteria, the As(III) oxidation process is described as a detoxification mechanism that converts As(III) into the less toxic As(V) form. In contrast to heterotrophic As(III)-oxidizing bacteria, autotrophic As(III) oxidizers utilize As(III) as an electron donor, whereas oxygen is used as the electron acceptor and carbon dioxide as the carbon source. Recently, chemoautotrophic As(III) oxidation has been found to occur via phototrophy (Song et al. 2009)

or anaerobic nitrate-dependent respiration in *Citrobacter* and *Alkalilimnicola* species (Hoeft et al. 2007; Li et al. 2015).

The As(III) oxidation efficiency varies significantly among the As(III)-oxidizing strains depending on their physiological attributes and growth conditions. Salmassi et al. (2002) isolated a heterotrophic As(III)-oxidizing bacterium (*Agrobacterium albertimagni* AOL15) from the surface of aquatic macrophytes collected in a Hot Creek. The strain oxidized 585 µM As(III) within 24 h in mannitol medium. Its maximum oxidation rate (V_{max}) is 3.4 µM, which is reported as the lowest to date. Conversely, the maximum reported oxidation rate (V_{max}) of 1.23×10^{-7} µM/min cell was observed in a resting cell kinetic assay with *Variovorax* sp. strain MM-1 (Bahar et al. 2013a).

Among the γ-*Proteobacteria*, *Stenotrophomonas* sp. strain, MM-7 emerged as being more efficient in As(III) oxidation. It can oxidize 500 µM As(III) within 12 h incubation in minimal salts medium (MSM) with an initial cell density of 1.5×10^{7} cell/mL (Bahar et al. 2012). This strain was isolated from an unpolluted soil (8.8 mg/kg d.w., corresponding to 117 µM) but indicated exceptional hypertolerance to As(III) in culture medium (up to 60 mM). Similarly, Chang et al. (2010) isolated *Pseudomonas stutzeri* strain GIST-BDan2 from wetlands water containing only 1.95 µg/L (corresponding to 0.026 µM), which was shown to oxidize up to 1 mM As(III) within 25–30 h incubation.

In contrast to heterotrophic As(III) oxidizers, chemolithoautotrophs gain energy from As(III) oxidation, and their growth and oxidation rates are lower than those of heterotrophic As(III) oxidizers. Santini et al. (2000) isolated a new chemolithoautotrophic As(III)-oxidizing bacterium

(NT-26) from a gold mine rock that can grow in MSM containing As(III) only as the electron donor with a doubling time of 7.6 h.

As(III) oxidase, the enzyme catalyzing As(III) oxidation, has been characterized in both autotrophic and heterotrophic bacteria. The genes encoding As(III) oxidase show a great degree of divergence, and the sequences of the As(III) oxidase genes found in autotrophic As(III) oxidizers are phylogenetically distinct from those found in heterotrophic As(III) oxidizers (Quéméneur et al. 2008). The gene designation *aox* is also described as *aro* and *aso* in some studies and more complicatedly the large subunit *aoxB* as *aroA* and *asoA*. Likewise, the designation of a small subunit is reported as vice versa. To bring some coherence to the designation of the genes involved in As(III) oxidation, Lett et al. (2012) proposed a new nomenclature, *aio*, for As(III) oxidase and the two genes encoding the large and small subunits of arsenite oxidase designated as *aioA* and *aioB*. The new nomenclature of arsenite oxidase gene has been used in this chapter.

Aio genes have been identified in bacteria isolated from various arsenic-rich environments. Bacteria carrying *aio* belong to alpha-, beta-, and gamma-*Proteobacteria* as well as to *Chloroflexi* and *Deinococcus–Thermus* (Quéméneur et al. 2008; Cai et al. 2009; Tang et al. 2011; Andreoni et al. 2012). Homologues of As(III) oxidase have also been identified in the genomes of the *Crenarchaeota Aeropyrum pernix* and *Sulfolobus tokodaii* (Lebrun et al. 2003).

As(III) oxidase contains two heterologous subunits: a large catalytic subunit (AioA, ~90 kDa) that contains the molybdenum cofactor together with a 3Fe–4S cluster and a small subunit (AioB, ~14 kDa) that contains a Rieske 2Fe–2S cluster (Ellis et al. 2001). The inducible As(III) oxidation system of *Ralstonia* sp. 22 possesses a soluble c554 cytochrome as a second electron acceptor, in addition to the heterodimeric membrane-associated enzyme (Lieutaud et al. 2010). To date, only four species (*Agrobacterium tumefaciens* 5A, *Thiomonas* sp. 3As, *Herminiimonas arsenicoxydans*, and *Ochrobactrum tritici*) have been reported to have a cytochrome C gene cotranscribed with the *aioBA* genes (Kashyap et al. 2006; Branco et al. 2009).

The enzymology of AioA has some features in common with the As(V) respiratory reductase, ArrA. A novel type of As(III) oxidase gene (*arxA*) in the genome of the chemolithotrophic bacterium *Alkalilimnicola ehrlichii* MLHE-1 (Hoeft et al. 2007) showed a higher sequence similarity to *arrA* than to *aioA* (Zargar et al. 2010). ArxA of MLHE-1 is implicated in reversible As(III) oxidation and As(V) reduction *in vitro*. Based on comparative sequence analysis, ArrA and AioA form distinct phylogenetic clades within the dimethyl sulfoxide reductase family of proteins, which probably evolved separately from a common ancestor (Oremland et al. 2009). Recently, an *arx* operon similar to that of MLHE-1 was identified in the genome of the photosynthetic purple sulfur bacterium *Ectothiorhodospira* sp. strain PHS-1, isolated from the hydrothermal waters of the halo-alkaline Mono Lake (Kulp et al. 2008; Zargar et al. 2012). In addition to these, strain ML-SRAO has been isolated from Mono Lake, which

is able to oxidize As(III) anaerobically, while reducing selenite (Fisher and Hollibaugh 2008). The lack of amplification of the As(III) oxidase gene and the positive amplification of the *arrA* gene from strain ML-SRAO is indicative that this ArrA, similar to that of MLHE-1, acts as an oxidoreductase, although further research is necessary to confirm this finding. This new mechanism of As(III) oxidation enables biological oxidation of arsenic in other environments, including other soda lakes, hydrothermal vents, or metal-polluted soils and waters (Cavalca et al. 2013).

30.3.2 As(V) Reduction

Some microorganisms can reduce As(V) to As(III) for detoxification purposes (As(V)-resistant microbes [ARMs]) or use As(V) as an electron acceptor in anaerobic respiration (dissimilatory As(V)-respiring prokaryotes [DARPs]).

While ARMs are widespread in all of the different bacterial phyla, DARPs are found in the *Firmicutes*, γ-, δ-, and ε-*Proteobacteria*, *Aquificae*, *Deferribacteres*, *Chrysiogenetes*, and *Archaea* (Cavalca et al. 2013). Arsenate-detoxifying reducing bacteria were found to play a major role in As mobilization under oxic conditions (Corsini et al. 2010). Differently to As detoxification mechanisms, respiratory As(V) reduction has been shown to be capable of mobilizing solid-associated As(V), including adsorbed and mineral As(V) (Huang, 2014).

Extremophiles from soda lakes have also been recently found to be As(V)-respiring bacteria. *Desulfuribacillus alkaliarsenatis* can reduce As(V) and elemental sulfur completely and thiosulfate incompletely (Sorokin et al. 2012). *D. alkaliarsenatis* was shown to preferentially respire As(V) over sulfate (Blum et al. 2012). Anaerobic bacteria can display As(V) reduction abilities both as As(V)-respiring heterotrophs gaining energy from the oxidation of small organic molecules (Perez-Jimenez et al. 2005) or aromatic compounds (Liu et al. 2004) and as chemolithoautotrophs gaining energy from hydrogen and sulfide (Hoeft et al. 2010). The rate of As(V) reduction is known to be influenced by the binding forms in which As(V) became associated with the mineral phases and coupled strongly with As(V) adsorption and desorption rates. Reductive dissolution of Al-ferrihydrite by *Shewanella* sp. ANA-3 results in the enrichment of Al sites and As(V) reduction accelerates As release due to the low affinity of As(III) on these non-ferric sites (Masue-Slowey et al. 2011).

The resistance mechanisms of ARMs are encoded by the *ars* operon and have been extensively studied. The configuration of the operon is different for different strains (Paez-Espino et al. 2009); the most simple configuration (*arsRBC*) consists of the regulatory protein ArsR, which possesses an As(III)-specific binding site, the As(V) reductase ArsC, and the As(III) efflux pump ArsB. ArsC mediates the reduction of As(V) with glutaredoxin, glutathione, or thioredoxin. This detoxification system requires energy in the form of ATP (Ordónez et al. 2005). ArsC is localized in the cytoplasm, and it reduces As(V) that has entered the cells. Two families of transmembrane efflux pumps are known:

the ArsB and the ACR3 families. The ACR3 type is more widespread in nature, being found in bacteria, animals, and plants, while ArsB is only present in bacteria (Achour et al. 2007). A second operon configuration (*arsRDABC*) contains the additional presence of the ATPase ArsA. A third operon configuration consists of the *ars* genes arranged in two operons, *arsRC* and *arsBH*, that are transcribed in opposite directions. The function of ArsH is not completely clear: it is present in almost all of the Gram-negative bacteria that carry an *ars* operon and it is absent in Gram-positive bacteria. Branco et al. (2008) demonstrated that *arsH* confers to *Ochrobactrum tritici* strain SCII24T the ability to grow at high arsenic concentrations.

In the case of DARPs, the key enzyme is an As(V) reductase, ArrA. The *arr* operon comprises two genes, *arrA* and *arrB*, encoding large and small subunits, respectively (Afkar et al. 2003). A third component, *arrC*, has been retrieved in some organisms (i.e., *Desulfitobacterium hafniense*, *Alkaliphilus metalliredigens*, and *Wolinella succinogenes*). An additional *arrD*, coding for a chaperone, is present in *Alkaliphilus oremlandii*, *Bacillus selenitireducens* MLS10, strain MLMS-1, *Geobacter lovleyi*, *D. hafniense*, and *Halarsenatibacter silvermanii* (Blum et al. 2009). Arr is a heterodimer periplasmic protein composed by two subunits: the large subunit (ArrA), which contains the molybdenum catalytic center, and the small subunit (ArrB), which contains (Fe–S) clusters.

The expression and activity of the respiratory As(V) reductase were assessed for *Shewanella* sp. strain ANA-3 (Malasarn et al. 2008). Arr of strain ANA-3 is expressed at the beginning of the exponential growth phase and the expression persists throughout the stationary phase, when it is released from the cell.

ArrA is a biochemically reversible enzyme (Richey et al. 2009) acting as an oxidase or a reductase depending on its electron potential and the constituents of the electron transfer chain. The reversible ability was also demonstrated for several As(V)-respiring bacteria such as *A. oremlandii* (Fisher and Hollibaugh 2008) and *Shewanella* sp. ANA-3 (Saltikov et al. 2005) in *Marinobacter santoriniensis* (Handley et al. 2009) and in *Thermus* sp. HR13 (Gihring and Banfield 2004).

30.3.3 Arsenic Methylation

While the conversion of As(III) into less toxic As(V) cannot remove As from soils and waters, As(III) methylation and subsequent volatilization are important pathway for As removal. Arsenic methylation involves transformation of the inorganic As to MMA, then to DMA, and finally to the volatile arsines. Arsines have low boiling points and methylated As has low adsorption affinity; thus, they can partition into the atmosphere from aqueous phase (Mestrot et al. 2011). It has been estimated that as much as 2.1×10^4 t As could be lost annually through volatilization from land surfaces to the atmosphere (Srivastava et al. 2011).

A recent report has shown that low redox potentials (i.e., reducing conditions) promote the production and mobilization of methylated As (Frohne et al. 2011). Under reducing conditions, the reductive dissolution of Fe/Mn hydroxides and the reduction of As(V) to As(III) may increase the level of dissolved As in the environment and thereby enhanced microbial methylation of As (Huang 2014).

Bacteria are known to transform aqueous- or solid-associated inorganic As into gaseous arsines and trivalent and pentavalent methyl As species, and the ability to methylate As(III), As(V), or methyl As and arsenic species volatilized seems to be microorganism dependent (Huang et al. 2014; Wang et al. 2014). The majority of works on As volatilization published so far have focused on bacteria in anaerobic ecosystems. Meyer et al. (2007) isolated from an alluvial soil a strain affiliated to the species *Clostridium glycolicum* that showed a versatility in transforming metals and As to volatile compounds. Similar to alluvial soils, rice fields could be the source of a number of bacteria potentially involved in As methylation. Xiao et al. (2016) in a metagenomic study on As metabolism genes in paddy soils gave evidence of the presence As methylation. Several studies suggested that the unique biogeochemical conditions in paddy soils, such as oxic/anoxic alternate due to root oxygen release, organic matter input, and the presence of autotrophic sulfate-reducing bacteria as well as methanoarchaea, can favor As mobilization (Jia et al. 2013; Zhang et al. 2015). As for anaerobic bacteria, almost all methanoarchaea studied so far were able to volatilize a broad spectrum of metalloids, including As (Michalke et al. 2002; Meyer et al. 2008).

In the past, microbial methylation in aquifers was often considered insignificant and the As cycle in these environments was assumed to be limited to transformations between the inorganic forms. This perception was refuted by the results of Maguffin and coauthors (2015) who analyzed the occurrence and significance of As biomethylation in a volcaniclastic aquifer. The authors, using laboratory incubation experiments and an aquifer injection test, showed that microbes produced dimethylarsinate at rates comparable to dimethylarsinate production on surface environments.

Gut microbiota, dominated by anaerobic bacteria and archaea, significantly affect As metabolism, facilitating As volatilization (Lu et al. 2013). In a study using an *in vitro* gastrointestinal model developed to investigate volatilization of As by human intestinal microbiota, Van de Wiele et al. (2010) showed that intestinal bacteria had capacities to volatilize As, increasing the metalloid mobility. More recently, another *in vitro* work to assess soil As metabolism by human gut microbiota confirmed that human gut microbiota actively metabolized As not only changing the speciation of As but also inducing changes from inorganic As to organic As (Yin et al. 2015).

Aerobic bacteria also exhibited the ability of As volatilization. One aerobic bacterium, belonging to the *Flavobacterium–Cytophaga* group, was isolated by Honschopp et al. (1996) from As-contaminated soil. The authors gave evidences that the bacterial growth was enhanced when in the presence of As and of the ability of the bacterium to methylate As. More recently, Miyatake and Hayashi (2009) used cells of *Bacillus cereus* strain R2 to study the relationship between

As methylation activity and the production of DMA. They found that As methylation activity in the bacterium was higher under aerobic conditions than under anaerobic conditions. Similarly, Kuramata and colleagues (2015) isolated one strain of *Streptomyces* sp. from roots of rice grown in As-contaminated paddy soils and demonstrated the ability of the bacterium to produce DMA and MMA when cultured in liquid medium containing As(III).

Although microbial methylation of As(III) has been observed as early as 1935 (Challenger and Higginbottom 1935), the molecular mechanism has only recently been identified in pure cultures of bacteria. It was found that this process is catalyzed by the enzyme As(III) S-adenosylmethionine methyltransferase (ArsM) involving methyl transfer from S-adenosyl-L-methionine (SAM) to As(III) (Qi et al. 2006; Yuan et al. 2008; Qin et al. 2009; Yin et al. 2011; Mestrot et al. 2013). While examining microbial genomes, Qin and coauthors (2006) identified a large number of genes for bacterial and archaeal homologues of arsM that set apart from genes for other homologues because they are each downstream of an *arsR* gene, encoding the archetypal arsenic-responsive transcriptional repressor that controls the expression of ars operons (Paez-Espino et al. 2009), suggesting that these ArsMs evolved to confer arsenic resistance. In the same article, the authors reported that the heterologous expression of *arsM* from *Rhodopseudomonas palustris* conferred As(III) resistance to an arsenic-sensitive strain of *Escherichia coli* and that the As(III) resistance cells in *E. coli* expressing recombinant *arsM* correlated with conversion of As to the methylated species DMA and TMA gas.

Jia et al. (2013) designed specific primers for amplification of *arsM* genes and successfully amplified the target in several soils. They found that bacteria carrying *arsM* gene were phylogenetically diverse: *Actinobacteria*, *Gemmatimonadales*, α-*Proteobacteria*, δ-*Proteobacteria*, β-*Proteobacteria*, *Chloroflexi*, *Firmicutes*, CFB group bacteria, Archaea, and five other unknown groups.

30.3.4 Microbial Intracellular and Extracellular Sequestration

Microorganisms have developed several strategies to cope with toxic compounds, that is, binding toxicants outside the cell, as in the case of metals, or accumulating them inside the cell, as in the case of hydrophobic molecules. Thus, it is plausible that bacteria also have physical interactions with arsenic both intra- and extracellularly. In this section, the potential of microbial intra- and extracellular physical interaction with arsenic is discussed, focusing on biosorption and bioaccumulation.

30.3.4.1 Biosorption

Biosorption by living, nonliving, nongrowing biomass or active groups of some molecules is a nonmetabolism-dependent (passive) metal uptake process. Bacteria can produce an extracellular polymeric substance (EPS) matrix in which they are embedded. The matrix consists of a conglomeration of different types of biopolymers that form the scaffold for the 3D architecture of the biofilm and is responsible for adhesion to surfaces, for cohesion in the biofilm. The EPS protects bacteria against environmental stresses (i.e., desiccation, ultraviolet radiation) and acts as a molecular sieve for cations, anions, apolar molecules, and particles from the water phase (Flemming and Wingender 2010).

During the last two decades, extensive attention has been paid on the retention of cations on the microbial surface (Akcil et al. 2015). Recent growing interest in the biosorption of As species with bacteria has led to a new insight into the interaction between As and the microbial cell surfaces (Huang 2014).

The sorption of As(III), As(V), and monomethylarsonic acid has been evidenced in *Acidithiobacillus ferrooxidans* (Yan et al. 2010; Chandraprabha and Natarajan 2011), *Bacillus subtilis* (Hossain and Anantharaman 2006; Vishnoi et al. 2014), *B. cereus* (Giri et al. 2013), *Halobacterium saccharovorum*, *Halobacterium salinarium*, *Natronobacterium gregoryi* (Williams et al. 2013), *Pseudomonas macerans* (Vishnoi et al. 2014), and *Rhodococcus* sp. WB-12 (Prasad et al. 2011).

Sorption of As to bacterial cell surface was indicated in the form of electrostatic interaction involving hydroxyl, amide, and amino groups by Fourier transform infrared spectroscopy studies (Yan et al. 2010; Prasad et al. 2011; Giri et al. 2013; Vishnoi et al. 2014).

These interactions are pH dependent and are explained by the change behavior of cell surface properties with changing pH of the surrounding medium. In the case of living cells of *P. macerans* and *B. subtilis*, the maximum surface binding of As(III) was recorded at pH 8.0 (Vishnoi et al. 2014), while for *B. cereus* it was at pH 7.5 (Giri et al. 2013). Similarly, other studies using dry biomass of *B. cereus* W2 and *Rhodococcus* sp. WB-12 showed the highest As sorption capacity at pH 7.0 (Miyatake and Hayashi 2011; Prasad et al. 2011).

Thermodynamic works revealed that As sorption is also affected by temperature. It has been observed that sorption abilities of *B. cereus* and *A. ferrooxidans* are greater at higher temperatures (40°C) (Yan et al. 2010; Giri et al. 2013). On the other hand, the As(III) sorption extent of *B. subtilis* and *P. macerans* increased passing from 25°C to 30°C (Vishnoi et al. 2014).

It has been hypothesized that an increase in the sorption of metals with increasing temperature could be correlated with an increase in the number of active surface sites (Meena et al. 2005) or with the formation of an inner-sphere complex between As(III) and the cell surface (Giri et al. 2013); however, further research is needed to clarify these aspects.

Recent studies have shown that As species affect differently the sorption extent of the cells, as in the case of *A. ferrooxidans* that preferably bind MMA rather than As(III) (Yan et al. 2010). Conversely, biosorption study performed by Chandraprabha and Natarajan (2011) with ferrous-grown cells showed another As cell–binding mechanism that favors As(V) due to the strong affinity of ferric ions toward arsenate ions.

30.3.4.2 Bioaccumulation

Different from biosorption, bioaccumulation is a metabolism-dependent (active) uptake process that infers intracellular accumulation and complexation of As inside the cell membranes and cytoplasm instead of at the cell surface (Huang 2014). Although the underlying mechanisms are not yet clear, bioaccumulation of arsenic into the cell would be a result of higher uptake and lower efflux of arsenic.

A number of studies using scanning electron microscopy (SEM), transmission electron microscopy (TEM), and bioaccumulation experiments gave evidence on As bioaccumulation for a variety of As-resistant bacteria in the environment such as *Acinetobacter*, *Bacillus*, *Exiguobacterium*, *Lysinibacillus*, *Pseudomonas*, and *Vogesella* species (Banerjee et al. 2011; Majumder et al. 2013; Xie et al. 2013; Aminur et al. 2014; Pandey and Bhatt 2015). In these studies, the authors reported SEM images demonstrating changes in bacterial cell volume and morphology when bacteria were grown in the presence of arsenic. These changes in morphological structure might be a possible strategy for cells to accumulate metals inside the cells, as suggested by TEM images that evidenced the presence of electron-dense deposits throughout the cytoplasm of all the bacterial isolates.

As confirmation of microscopic images, Aminur et al. (2014) reported that *Lysinibacillus sphaericus* B1-CDA, isolated from a contaminated soil of Bangladesh, was capable of accumulating As as much as 5.0 mg/g of the cells' dry biomass reducing the As concentration in the contaminated medium by 50%. Similarly, Podder and Majumder (2015) revealed that the living cells of *Bacillus arsenicus* MTCC 4380 were proficient in bioaccumulating As(III) and As(V) up to 79.4% and 85.03%, respectively, in the growth medium containing 1000 mg/L arsenic. Comparable results were also reported by Pandey and Bhatt (2015) who found that the bioaccumulation capacity of *Exiguobacterium* sp. and *Bacillus* sp. was >60% for both As species, in the growth medium containing 1500 mg/L As or 100 mg/L of both the forms of arsenic.

30.4 ARSENIC REMOVAL FROM WATERS

Several conventional treatment technologies for As removal from waters, such as flocculation, coagulation, ion exchange, and adsorption onto iron oxide or activated alumina, have been specifically developed for industrial-scale plants (Singh et al. 2015).

It is difficult to compare the costs of various treatment technologies as the efficiency depends on different parameters (i.e., maximum contaminant level, co-occurrence of solutes, quality of the source water, operations and maintenance expenditures, permission requirements, and waste-disposal issues) (Mondal et al. 2006). Furthermore, the majority of these technologies remain too expensive for poor countries. For this reason, research for cost-effective and efficient alternative sorbents (or surface-coated sorbents) is still in progress.

In addition, any effective treatment of arsenic-contaminated water has to remove both As(III) and As(V) forms, but sometimes classical technologies are not efficient enough in the removal of As(III). Therefore, a preoxidation procedure is typically required to oxidize As(III) to As(V) through the addition of chemical reagents, such as ozone, hydrogen peroxide, or manganese oxide (Singh et al. 2015). However, these reagents may cause several problems such as the presence of residuals, formation of by-products, and cost inefficiency.

Biological water treatment methods are considered a suitable approach to overcome these problems, and they have attracted considerable research interest over recent years. Removal of arsenic can be performed by using natural consortia, pure cultures of arsenic-resistant bacteria, or iron- and manganese-oxidizing bacteria that can transform and/or capture arsenic forms indirectly (Cavalca et al. 2013).

30.4.1 BIOLOGICAL As(III) OXIDATION

Several microbial-assisted arsenic removal technologies (Table 30.2) have been attempted from laboratory to pilot scale (Bahar et al. 2013b; Cavalca et al. 2013). These treatment methods incorporate a biological arsenic transformation by microbial biofilms (Bag et al. 2010; Ito et al. 2012) or a biological transformation of arsenic coupled with adsorption by different materials such as goethite (Corsini et al. 2014b), granular-activated carbon (GAC) (Mondal et al. 2008), biogenic iron oxide (Katsoyiannis and Zouboulis 2004), manganese oxide (Katsoyiannis et al. 2004), and activated alumina and charcoal (Mokashi and Paknikar 2002).

Bag et al. (2010) developed a packed bed column of a continuous flow reactor with *Rhodococcus equi* cells immobilized on rice husks. They observed that the As(III) removal efficiency of the cells was of 95% from simulated arsenic-laden water and naturally occurring water with arsenic concentrations ranging from 50 to 100 μg/L.

Similarly, Ito et al. (2012) developed a bioreactor with *Ensifer adhaerens* cells immobilized on polyvinyl alcohol gel droplets to study the As(III) oxidation efficiency of the strain in synthetic groundwater containing 1 mg/L of As(III). The authors demonstrated that As(III) was oxidized to As(V) over the complete time course of the experiment, resulting in a removal efficiency of 90%. As(III) oxidation can be performed not only by pure culture biofilms but also by bacterial consortia.

The formation and activity of an As(III)-oxidizing biofilm developed by the CAsO1 bacterial consortium in a bioreactor using pozzolana (volcanic material) as growth support were investigated by Michel et al. (2007) and Michon et al. (2010). The first research group stated that the biofilm structure of CAsO1 consortium as well as of a pure culture of *Thiomonas arsenivorans* represented a physical barrier decreasing As(III) access to sessile cells and thus to As(III) oxidase activity induction (Michel et al. 2007). The authors suggested further implementation of the process in terms of optimization of operating parameters that can affect biofilm formation. Furthermore, Michon et al. (2010) showed that the

TABLE 30.2

Recently Reported Studies on Different Processes for Biological Arsenic Removal from Aqueous Phases

Removal Process	Water/Medium	Microorganisms/Sorbents	Technology	Main Observations	Reference
Arsenic sorption	Synthetic acid mine drainage (25 mg/L As, As(III):As(V) = 1:1)	*Ralstonia eutropha* MTCC 2487 immobilized on the granular-activated carbon	Upflow column reactor	At the initial stage, As(V) removal was slightly higher than As(III). After 2 days of operations, both forms were equally removed.	Mondal et al. (2008)
	As(III) solution (20 mg/L)	Pottery granules coated with cyst of *Azotobacter* strain SSB81 and Portland cement	Batch experiment	96% of As(III) removed at pH 5.0–6.0.	Gauri et al. (2011)
	Six different concentrations of As(III) and MMA(V) (0.5, 1.0, 1.5, 2.0, 2.5, and 3.0 mg/L)	Suspension of *Acidithiobacillus ferrooxidans* BY-3	Batch experiment	Maximum sorption was achieved within 30′ for As(III) and 40′ for MMA(V). pH strongly influences the biosorption of organic forms.	Lei et al. (2010)
	As(III) solution (20 mg/L)	Suspension of *Rhodococcus* sp. WB-12	Batch experiment	1 g/L of cell biomass removed 77.3 mg/g As(III) with contact time of 30′, 30°C pH 7.0.	Prasad et al. (2011)
	As(III) and As(V) solutions (100 mg/L)	Suspension of *Arthrobacter* sp.	Batch experiment	1 g/L of cell biomass removed 74.91 mg/g As(III) (pH 7.0) and As(V) 81.63 mg/g (pH 4.0) with contact time of 28′, 30°C.	Prasad et al. (2013)
	Groundwater (Mn 0.4 mg/L) spiked with 10–50 μg/L As(III) or As(V)	Fe- and Mn-oxidizing bacteria immobilized on polystyrene beads	Fixed-bed upflow filtration unit	Complete removal of 35 μg/L As(III) and of 42 μg/L As(V). Bacteria accelerate As(III) oxidation and generate reactive Mn oxide surfaces. Presence of phosphates inhibited the overall As removal but not As(III) oxidation.	Katsoyiannis et al. (2004)
As precipitation	Groundwater (Fe(II) 2.8 mg/L) spiked with 20–200 μg/L As(III) or As(V)	*Gallionella ferruginea* and *Leptothrix ochracea*	Fixed-bed upflow filtration unit	Up to 95% removal of As onto biogenic Fe oxides. Under optimized redox conditions, As(III) oxidation is catalyzed by bacteria.	Katsoyiannis and Zouboulis (2004)
	Tap water supplemented with Fe(II) 0.8–1.5 mg/L, Mn(II) 1–1.2 mg/L, and As(III) 100–150 mg/L	Iron-oxidizing bacteria (i.e., *Gallionella*, *Leptothrix*), manganese-oxidizing bacteria (i.e., *Leptothrix*, *Pseudomonas*, *Hyphomicrobium*, *Arthrobacter*), and As(III)-oxidizing bacteria (i.e., *Alcaligenes*, *Pseudomonas*)	Biofilter	Removal efficiencies of 96.2%, 97.7%, and 98.2% for Fe(II), Mn(II), and As(III), respectively.	Yang et al. (2014)
	Mineral medium containing 4.8 mM Fe(II) and 8 mM NO_3^- and various concentrations of As(III) (2.85, 6.85, and 13.65 mM)	Anaerobic nitrate-reducing Fe(II)-oxidizing *Citrobacter freundii* PXL1	Batch experiment	The biogenic amorphous iron oxides coprecipitated with As(III).	Li et al. (2015)
	Mineral medium containing 0.7 g/L Fe(II) and 1.9 g/L As(V)	Thermoacidophilic iron-oxidizing archaeon *Acidianus sulfidivorans*	Batch experiment	The isolate is able to precipitate scorodite $FeAsO_4 \cdot 2H_2O$ at 80°C and pH 1 in the absence of any primary minerals or seed crystals.	Gonzalez-Contreras et al. (2010)

(Continued)

TABLE 30.2 (*Continued*)

Recently Reported Studies on Different Processes for Biological Arsenic Removal from Aqueous Phases

Removal Process	Water/Medium	Microorganisms/Sorbents	Technology	Main Observations	Reference
	Mineral medium containing 1.5 g/L As(V) and 0.14–1.42 g/L Na sulfate	As(V)- and sulfur-respiring *Desulfosporosinus auripigmentum*	Batch experiment	The isolate precipitates arsenic trisulfide (As_2S_3).	Newman et al. (1997)
	Mineral medium containing 1.12 g/L thiosulfate and 0.375 g/L As(V)	*Shewanella* sp. strain HN-41	Batch experiment	The strain is able to produce photoactive realgar (As–S) nanotubes.	Jiang et al. (2009)
	Mining water (13 mg/L As(III))	CAsO1 bacterial consortium and *Thiomonas arsenivorans* strain b6 immobilized on pozzolana	Upflow column reactor	As(III) oxidation of *T. arsenivorans* was nine-fold higher for planktonic cells than for sessile ones and it was induced by As(III). Efficiency of bed reactor in As(III) removal is decreased by the biofilm formation.	Michel et al. (2007)
Biological As(III) oxidation	Simulated and natural laden groundwater (50–100 µg/L As(III))	*Rhodococcus equi* (JUBTAs02) immobilized on rice husk	Packed bed reactor	*R. equi* oxidized As(III). Maximum As(III) removal efficiency was 95%.	Bag et al. (2010)
	Synthetic water (<0.1 mg/L As(III))	CAsO1 bacterial consortium immobilized on pozzolana	Fixed-bed reactor	CAsO1 consortium was able to oxidize small amount of As(III).	Michon et al. (2010)
	Natural groundwater containing 150 µg/L As(III)	As(III) oxidizer *Aliihoeflea* sp. 2WW and goethite	Batch experiment	The two-phase process removed 95% of As(III) in 48 h.	Corsini et al. (2014b)
	Synthetic groundwater (1 mg/L As(III))	*Ensifer adhaerens* strain AOB C-1 immobilized on spherical polyvinyl alcohol gel	Batch experiment and Cylindrical bioreactor	92% As(III) oxidation efficiency in the reactor under continuous flow.	Ito et al. (2012)
Biological As(III) oxidation and As sorption	Tris-HCl buffer (100 mg/L As(III) or As(V))	ULPAs1 strain and sorbent (kutnahorite and chabazite)	Batch experiment	Fast As(III) oxidation by As induced and not induced bacteria in the presence of chabazite. Kutnahorite sorbed As(V), chabazite alone performed As(III) oxidation (0.3 mmol As/g).	Lièvremont et al. (2003)
	Basal salt medium (75 mg/L As(III))	Mixed culture of heterotrophic As(III)-oxidizing and arsenic-tolerant bacteria and sorbent (activated alumina)	Batch experiment	Complete oxidation within 5 days. 98% of As(V) was sorbed by alumina.	Ike et al. (2008)
	Natural water	As(III)-oxidizing bacteria *Mycobacterium lacticum* followed by sorption onto activated alumina/charcoal	Field scale	Five integrated arsenic removal systems with 1000 l/day filtration.	Pal and Paknikar (2012)

As, arsenic; As(III), arsenite; As(V), arsenate; MMA(V), monomethyl arsonate; Fe, iron; Mn, manganese.

CAsO1 consortium was able to oxidize not only in the range of the "mg/L order of magnitude" but also in the concentration scale of the "µg/L order of magnitude."

After the biological oxidation of As(III), it is necessary to remove the produced As(V) by using sorbents. The combined processes of biological oxidation and chemical removal onto different materials have been performed as one- and two-step processes by several authors.

In one of the first applications of this technique, Lièvremont et al. (2003) found that two mineral phases, kutnahorite and chabazite, showed different abilities in adsorbing arsenic after its biological oxidation and in performing abiotic oxidation. The authors also reported that the biological As(III) oxidation was rapid in the presence of chabazite and that As(V) produced was efficiently adsorbed by kutnahorite. A two-phase detoxification process was therefore suggested.

Accordingly, Ike et al. (2008) showed that arsenic removal by activated alumina was greatly enhanced by As(III) oxidation performed by a mixed culture of heterotrophic As(III) oxidizers, suggesting that the two processes must be performed consecutively for the attainment of optimum conditions in each step. More recently, Corsini et al. (2014b) by comparing the As removal performances of one-phase treatment (adding the As(III) oxidizer *Aliihoeflea* sp. 2WW together with goethite) and two-phase treatment (the biological step is performed before the adsorption step) proposed that As removal water treatments must incorporate a biological transformation of As and subsequent adsorption by different materials. Their results evidenced that the two-phase process reduced competition between oxyanions (such as phosphate) and As for sorption sites, confirming previous results of Lièvremont et al. (2003).

Although a number of attempts have been made for the microbial-assisted removal of arsenic in laboratory experiments and pilot-scale reactors, only one technology has been set at field scale, which has operated since 2006. Five integrated arsenic removal systems with 1000 L/day filtration capacities have been installed in arsenic-affected villages in India (Pal and Paknikar 2012). This treatment plant is based on As(III) oxidation by *Mycobacterium lacticum* and subsequent alumina/charcoal adsorption. Even with some limitations, the authors propose this technology as a promising one for arsenic-affected low-income communities.

Further research is needed to implement the results of laboratory experiments conducted so far in order to render the lab- and pilot-scale reactors a potentially effective and cost-efficient technology for the treatment of As(III)-contaminated groundwater.

30.4.2 As Biosorption

In the last years, lab-scale and pilot-scale studies have been conducted with the aim to propose innovative biofiltration technologies for arsenic removal using reactors filled with immobilized bacterial cells capable of arsenic removal.

Recent laboratory studies on As biosorption from aqueous solution by using bacteria as biosorbent evidenced that the process is strongly affected by several factors, that is, pH, contact time, initial arsenic concentration, biomass dose, and temperature (Lei et al. 2010; Prasad et al. 2011, 2013). Lei et al. (2010) evaluated the capability of *A. ferrooxidans* BY-3 to remove As(III) and organic arsenic (MMA(V)) compounds from aqueous solution. The authors evidenced a significant pH effect only on the biosorption of the organic arsenic form. Moreover, the results highlighted that the adsorption process was rapid and maximum sorption capacities were achieved within 30 min for As(III) and 40 min for MMA(V). Similarly, Prasad and colleagues (2011, 2013) studied the biomass sorption characteristics of *Rhodococcus* sp. WB-12 and *Arthrobacter* sp. as a function of biomass dose, contact time, and pH. Their results evidenced that both microorganisms had great potential as biosorbent in As removal from As-contaminated water, and their features were dependent on

the operating conditions. In particular, 1 g/L of *Rhodococcus* sp. WB-12 cell biomass was capable of removing As(III) 77.3 mg/g with the contact time of 30 min at 30°C, pH 7.0; and 1 g/L of *Arthrobacter* sp. cell biomass removed As(III) 74.91 and As(V) 81.63 mg/g with the contact time of 30 min at 28°C (at pH 7.0 and 4.0, respectively).

Taking into account these findings, pilot-scale experiments using bioreactors consisting of immobilized bacterial cells capable of arsenic adsorption have been performed. A novel, cost-effective biocomposite—granules of cement coated with cysts of *Azotobacter*—has been used for arsenic removal from drinking water (Gauri et al. 2011). This biocomposite removed approximately 96% of arsenic, probably due to the presence of polysaccharides and other macromolecules that interact with arsenic. Mondal et al. (2008) utilized the cells of *Ralstonia eutropha* immobilized on a GAC bed in a column reactor to remove arsenic from a synthetic industrial effluent, and they found that, after an initial stage of adaptation and biofilm formation, the cells were able to capture both As(III) and As(V).

30.4.3 As Precipitation

Arsenic can be removed not only by direct sorption but also by coprecipitation on the preformed biogenic iron manganese hydroxides (Cavalca et al. 2013). The removal is achieved by a combination of biological and physicochemical sorption processes, including oxidation and adsorption onto the biogenic iron and manganese oxides (Sahabi et al. 2009).

Katsoyiannis and Zouboulis (2004) investigated the removal of arsenic during biological iron oxidation in a fixed-bed upflow filtration unit containing polystyrene beads. They reported that iron oxides were deposited in the filter medium, along with the iron-oxidizing bacteria *Gallionella ferruginea* and *Leptothrix ochracea*, offering a favorable environment for arsenic to be adsorbed and consequently removed from the aqueous streams. The results indicated that, under the experimental conditions used, As(III) was oxidized by microorganisms that colonized the filter medium, contributing to an overall increase of arsenic removal (up to 95%). The authors demonstrated that this removal was related to a combination of the oxidation of As(III) to As(V) and biosorption of As to the microorganisms.

A biofilter was developed for the simultaneous removal of Fe(II), Mn(II), and As(III) from simulated groundwater, and removal efficiencies of 96.2%, 97.7%, and 98.2% were, respectively, obtained (Yang et al. 2014). Iron-oxidizing bacteria (such as *Gallionella*, *Leptothrix*), manganese-oxidizing bacteria (such as *Leptothrix*, *Pseudomonas*, *Hyphomicrobium*, *Arthrobacter*), and As(III)-oxidizing bacteria (such as *Alcaligenes*, *Pseudomonas*) were found to be dominant in the biofilter.

Nitrate-reducing Fe(II)-oxidizing anaerobic bacteria can be promising microbes to *in situ* treatment of nitrate- and As-contaminated waters. Li et al. (2015) reported that *Citrobacter freundii* PXL1 efficiently removes As(III) from water, associated with Fe(II) oxidation and nitrate reduction. The bacterial cells had little As adsorption capacity and the biogenic amorphous iron oxides play a major role in removing As(III) due to their strong adsorption capacity and their coprecipitation with As(III).

Similar to these respiring bacteria, the thermoacidophilic iron-oxidizing archaeon *Acidianus sulfidivorans* is able to simultaneously crystallize iron (FeII) and arsenate to scorodite ($FeAsO_4·2H_2O$) (Gonzalez-Contreras et al. 2010).

As(V) reducers were thought to increase the element's mobility until the discovery of *Desulfosporosinus auripigmentum*, an As(V)- and sulfur-respiring microorganism that precipitates arsenic trisulfide (As_2S_3), leading to the biogenic formation of auripigment (Newman et al. 1997). More recently, photoactive realgar (As–S) nanotubes have been shown to be produced by *Shewanella* sp. strain HN-41, an anaerobic bacterium that utilizes $S_2O_3{}^{2-}$ as an electron acceptor and lactate as an electron donor and concomitantly reduces As(V) to As(III) for detoxification purposes (Lee et al. 2007).

30.5 FUTURE PERSPECTIVES

Water resource is not considered endless anymore, and its use is becoming more and more conflictual. Human activities are often responsible for reducing the amount of pure and clear water for human purposes, but naturally occurring pollution (i.e., due to rock substrate composition) may even enlarge the impact of pollutants. Besides a rational use of such a scarce resource, it is important to develop innovative depuration processes to remediate waters used for drinking purposes or cultivated land irrigation. The apparent limitations of the conventional treatments for arsenic-contaminated waters have made urgent the development of suitable and environmentally friendly technologies. Biological water treatment could be the most cost-effective and efficient alternative for arsenic removal technology. A number of microbial As transformations and the potential to use bacteria in As removal processes have been reviewed in this chapter.

Microbial oxidation of As(III) into less mobile and toxic As(V) is performed by As(III)-oxidizing bacteria either as detoxification strategy or in cellular metabolism to support growth. Nowadays, the biological oxidation is considered a potential alternative in substituting the chemical oxidation methods used to transform As(III) to As(V). Due to their independence from organic carbon for their metabolism, autotrophic As(III)-oxidizing bacteria may be the best candidates for this process. However, their growing conditions (i.e., low growing rate) can limit their use in a broader scale. Heterotrophic As(III)-oxidizing bacteria may have an advantage of being used in remediation processes, since they can utilize the dissolved organic carbon in contaminated water. Although pure cultures have shown good results in arsenic removal from artificial and natural waters, they might be affected by the fluctuating composition of contaminated waters. To overcome this limitation, field-scale water purification can be achieved by using natural consortia. Bioelectrochemical systems (BES) have been proposed for the bioremediation of different subsurface contaminants and recently also of arsenic. A BES comprising an As(III)-oxidizing bacterial consortium and a polarized graphite electrode serving as direct electron acceptor at laboratory scale was demonstrated to be effective in the conversion of arsenic and suggested as a promising tool for the remediation of anoxic groundwaters (Pous et al. 2015). In a different experiment, the microbial As(III) oxidation activity was coupled to a zerovalent iron electrode to carry out simultaneously arsenic conversion and adsorption (Xue et al. 2013). Once As(III) is oxidized to As(V), a cost-effective and efficient sorbent material is required to completely remove the arsenic from waters. Arsenic adsorption on the biological materials seems a promising technology as it is cheap and renewable.

Besides the joint use of As(III)-oxidizing bacteria and sorbents, As-accumulating bacteria have been used in a number of works in order to elucidate the sustainable role of these bacteria in the As removal process. Bioaccumulation and biosorption of As have been accomplished by bacteria as detoxification mechanisms, which can be considered a tool for bioremediation processes. These mechanisms result in the reduction of As concentration in waters, but essentially As remains in the environment through the biogeochemical cycle. Therefore, more research is needed to find feasible solutions to remove/regenerate the accumulated As in microorganisms, reducing the risk of inappropriate disposal and further pollution of the environment.

In conclusion, there are good prospects for developing and implementing cost-effective and environmentally friendly technologies that offer great opportunities for treating As-contaminated waters. Further investigations on optimization of the process as well as discovering new As(III)-oxidizing bacteria and potential biosorbents are required. The recent advances on As removal researches allow the scientific community to consider the use of bacteria, with and without sorbent materials, a technology of great interest.

ACKNOWLEDGMENT

The work reported in this chapter was conducted with the precious contribution of Fondazione CARIPLO Project 2014-1301.

REFERENCES

Achour, A.R., Bauda, P., Billard, P., 2007. Diversity of arsenite transporter genes from arsenic-resistant soil bacteria. *Research in Microbiology* 158: 128–137.

Afkar, E., Lisak, J., Saltikov, C., Basu, P., Oremland, R.S., Stolz, J.F., 2003. The respiratory arsenate reductase from *Bacillus selenitireducens* strain MLS10. *FEMS Microbiology Letters* 226: 107–112.

Akcil, A., Erust, C., Ozdemiroglu, S., Fonti, V., Beolchini, F., 2015. A review of approaches and techniques used in aquatic contaminated sediments: Metal removal and stabilization by chemical and biotechnological processes. *Journal of Cleaner Production* 86: 24–36.

Aminur, R., N. Nahar, N.N. Nawani, J. Jass, P. Desale, B.P. Kapadnis, K. Hossain, A.K. Saha, S. Ghosh, B. Olsson, A. Mandal. 2014. Isolation and characterization of a *Lysinibacillus* strain B1-CDA showing potential for bioremediation of arsenics from contaminated water, *Journal of Environmental Science and Health*, Part A, 49: 1349–1360.

Andreoni, V., Zanchi, R., Cavalca, L., Corsini, A., Romagnoli, C., Canzi, E., 2012. Arsenite oxidation in *Ancylobacter dichloromethanicus* As3–1b strain: Detection of genes involved in arsenite oxidation and CO_2 fixation. *Current Microbiology* 65: 212–218.

ATSDR (Agency for Toxic Substances and Disease Registry), 2013. Comprehensive environmental response, compensation and liability act (CERCLA) priority list of hazardous substances. http://www.atsdr.cdc.gov/spl/. Accessed on Oct 13, 2016.

Bag, P., Bhattacharya, P., Chowdhury, R., 2010. Bio-detoxification of arsenic laden groundwater through a packed bed column of a continuous flow reactor using immobilized cells. *Soil Sediment Contamination* 19: 455–466.

Bahar, M.M., Megharaj, M., Naidu, R., 2012. Arsenic bioremediation potential of new arsenic oxidizing bacterium *Stenotrophomonas* sp. MM-7 isolated from soil. *Biodegradation* 23: 803–812.

Bahar, M.M., Megharaj, M., Naidu, R. 2013a. Bioremediation of arsenic-contaminated water: Recent advances and future prospects. *Water, Air, and Soil Pollution* 224: 1722.

Bahar, M.M., Megharaj, M., Naidu, R. 2013b. Kinetics of arsenite oxidation of by *Variovorax* sp. MM-1 isolated from a soil containing low arsenic and identification of arsenite oxidase gene. *Journal of Hazardous Materials* 262: 997–1003.

Banerjee S., Datta, S., Chattyopadhyay, D., Sarkar, P. 2011. Arsenic accumulating and transforming bacteria isolated from contaminated soil for potential use in bioremediation. *Journal of Environmental Science and Health, Part A* 46: 1736–1747.

Blum, J.S., Han, S., Lanoil, B., Saltikov, C., Witte, B., Tabita, F.R., Langley, S., Beveridge, T.J., Jahnke, L., Oremland, R.S., 2009. Ecophysiology of "*Halarsenatibacter silvermanii*" strain SLAS-1T, gen. nov., sp. nov., a facultative chemoautotrophic arsenate respirer from salt-saturated Searles Lake, California. *Applied and Environmental Microbiology* 75: 1950–1960.

Blum, J.S., Kulp, T.R., Han, S., Lanoil, B., Saltikov, C.W., Stolz, J.F., Miller, L.G., Oremland, R.S., 2012. *Desulfohalophilus alkaliarsenatis* gen. nov., sp. nov., an extremely halophilic sulfate- and arsenate-respiring bacterium from Searles Lake, California. *Extremophiles* 16: 727–742.

Branco R., Chung, A.P., Morais, P.V., 2008. Sequencing and expression of two arsenic resistance operons with different functions in the highly arsenic-resistant strain *Ochrobactrum tritici* SCII24T. *BMC Microbiology* 8: 95.

Branco, R., Francisco, R., Chung, A.P., Morais, P.V., 2009. Identification of an aox system that requires cytochrome C in the highly arsenic-resistant bacterium *Ochrobactrum tritici* SCII24. *Applied and Environmental Microbiology* 75: 5141–5147.

Cai, L., Rensing, C., Li, X., Wang, G., 2009. Novel gene clusters involved in arsenite oxidation and resistance in two arsenite oxidizers: *Achromobacter* sp. SY8 and *Pseudomonas* sp. TS44. *Applied Microbiology and Biotechnology* 83: 715–725.

Cavalca, L., Corsini, A., Zaccheo, P., Andreoni, V., Muyzer, G., 2013. Microbial transformations of arsenic: Perspectives for biological removal of arsenic from water. *Future Microbiology* 8: 753–768.

Challenger, F., Higginbottom, C., 1935. The production of trimethylarsine by *Penicillium brevicaule* (*Scopulariopsis brevicaulis*). *Biochemical Journal* 29: 1757–1778.

Chandraprabha, M.N., Natarajan, K.A. 2011. Mechanism of arsenic tolerance and bioremoval of arsenic by *Acidithiobacilus ferrooxidans*. *Journal of Biochemical Technology* 3: 257–265.

Chang, J.S., Yoon, I.H., Lee, J.H., Kim, K.R., An, J., Kim, K.W., 2010. Arsenic detoxification potential of aox genes in arsenite-oxidizing bacteria isolated from natural and constructed wetlands in the Republic of Korea. *Environmental Geochemistry and Health* 32: 95–105.

Chris Le, X., 2001. Unstable trivalent arsenic metabolites, monomethylarsonous acid and dimethylarsinous acid. *Journal of Analytical Atomic Spectrometry* 16: 1409–1413.

Corsini, A., Zaccheo, P., Muyzer, G., Andreoni, V., Cavalca, L., 2014a. Arsenic transforming abilities of groundwater bacteria and the combined use of *Aliihoeflea* sp. strain 2WW and goethite in metalloid removal. *Journal of Hazardous Materials* 269: 89–97.

Corsini, A., Cavalca, L., Muyzer, G., Zaccheo, P. 2014b. Effectiveness of various sorbents and biological oxidation in the removal of arsenic species from groundwater. *Environmental Chemistry* 11: 558–565.

Corsini, A., Cavalca, L., Crippa, L., Zaccheo, P., Andreoni, V., 2010. Impact of glucose on microbial community of a soil containing pyrite cinders: Role of bacteria in arsenic mobilization under submerged condition. *Soil Biology and Biochemistry* 42: 699–707.

Couture, R.M., Sekowska, A., Fang, G., Danchin, A., 2012. Linking selenium biogeochemistry to the sulfur-dependent biological detoxification of arsenic. *Environmental Microbiology* 14: 1612–1623.

Dhuldhaj, U.P., Yadav, I.C., Singh, S., Sharma, N.K., 2013. Microbial interactions in the arsenic cycle: Adoptive strategies and applications in environmental management. In D.M. Whitacre (Ed.), *Reviews of Environmental Contamination and Toxicology*. Vol. 224, pp. 1–38. Springer, New York.

Dilda, P.J., Hogg, P.J., 2007. Arsenical-based cancer drugs. *Cancer Treatment Reviews* 33: 542–564.

EFSA, 2009. EFSA panel on contaminants in the food chain, scientific opinion on arsenic in food. *EFSAJ* 7: 1351.

Ellis, P.J., Conrads, T., Hille, R., Kuhn, P., 2001. Crystal structure of the 100 kDa arsenite oxidase from *Alcaligenes faecalis* in two crystal forms at 1.64 A and 2.03 A. *Structure* 9: 25–132.

Fisher, J., Hollibaugh, T.J., 2008. Selenate-dependent anaerobic arsenite oxidation by a bacterium from Mono Lake, California. *Applied and Environmental Microbiology* 74: 2588–2594.

Flemming, H.C., Wingender, J., 2010. The biofilm matrix. *Nature Reviews Microbiology* 8: 623–633.

Frohne, T., Rinklebe, J., Diaz-Bone, R.A., Du Laing, G., 2011. Controlled variation of redox conditions in a floodplain soil: Impact on metal mobilization and biomethylation of arsenic and antimony. *Geoderma* 160: 414–424.

Gauri, S.S., Archanaa, S., Mondal, K.C., Pati, B.R., Mandal, S.M., Dey, S., 2011. Removal of arsenic from aqueous solution using pottery granules coated with cyst of *Azotobacter* and portland cement: Characterization, kinetics and modelling. *Bioresource Technology* 102: 6308–6312.

Gihring, T.M., Banfield, J.F., 2004. Arsenite oxidation and arsenate respiration by a new *Thermus* isolate. *FEMS Microbiology Letters* 204: 335–340.

Giri, A.K., Patel, R.K., Mahapatra, S.S., Mishra, P.C., 2013. Biosorption of arsenic(III) from aqueous solution by living cells of *Bacillus cereus*. *Environmental Science and Pollution Research* 20: 1281–1291.

Gong Z., X. Lu, W.R. Cullen, X.C. Le. 2001. Unstable trivalent arsenic metabolites, monomethylarsonous acid and dimethylarsinous acid. *Journal of Analytical Atomic Spectrometry* 16: 1409–1413.

Gonzalez-Contreras, P., Weijma, J., Weijden, R.V.D., Buisman, C.J., 2010. Biogenic scorodite crystallization by *Acidianus sulfidivorans* for arsenic removal. *Environmental Science and Technology* 44: 675–80.

Green, H.H. 1918. Description of a bacterium that oxidizes arsenite and one which reduces arsenate to arsenite from a cattle-dipping tank. *South African Journal of Science* 14: 465–467.

Gupta, A., Yunus, M., Sankararamakrishnan, N., 2012. Zerovalent iron encapsulated chitosan nanospheres—A novel adsorbent for the removal of total inorganic Arsenic from aqueous systems. *Chemosphere* 86: 150–155.

Handley, K.M., Héry, M., Lloyd, J.R., 2009. *Marinobacter santoriniensis* sp. nov., an arsenate-respiring and arsenite-oxidizing bacterium isolated from hydrothermal sediment. *International Journal of Systematic and Evolutionary Microbiology* 59: 886–892.

Hoeft, S.E., Blum, J.S., Stolz, J.F., Tabita, F.R., Witte, B., King, G.M., Santini, J.M., Oremland, R.S., 2007. *Alkalilimnicola ehrlichii* sp nov., a novel, arsenite oxidizing haloalkalophilic gammaproteobacterium capable of chemoautotrophic or heterotrophic growth with nitrate or oxygen as the electron acceptor. *International Journal of Systematic and Evolutionary Microbiology* 57: 504–5012.

Hoeft, S.E., Kulp, T.R., Han, S., Lanoil, B., Oremland, R.S., 2010. Coupled arsenotrophy in a hot spring photosynthetic biofilm at Mono Lake, California. *Applied and Environmental Microbiology* 76: 4633–4639.

Honschopp, S., Brunken, N., Nehrkorn, A., Breunig, H.J., 1996. Isolation and characterization of a new arsenic methylating bacterium from soil. *Microbiological Research* 151: 37–41.

Hossain, S.M., Anantharaman, N., 2006. Studies on bacterial growth and arsenic(III) biosorption using *Bacillus subtilis*. *Chemical and Biochemical Engineering Quarterly* 20: 209–216

Huang, J.H., 2014. Impact of microorganisms on arsenic biogeochemistry: A review. *Water, Air, and Soil Pollution* 225: 1848–1873.

IARC, 1987. Monographs on the evaluation of the carcinogenic risk to humans: Arsenic and arsenic compounds (Group 1). Supplement 7, International Agency for Research on Cancer, Lyon, France, pp. 100–103.

Ike, M., Miyazaki, T., Yamamoto, N., Sei, K., Soda, S., 2008. Removal of arsenic from groundwater by arsenite-oxidizing bacteria. *Water Science and Technology* 58: 1095– 1100.

Ito, A., Miura, J.I., Ishikawa, N., Umita, T., 2012. Biological oxidation of arsenite in synthetic groundwater using immobilised bacteria. *Water Research* 46: 4825–483.

Jia, Y., Huang, H., Zhong, M., Wang, F.H., Zhang, L.M., Zhu, Y.G., 2013. Microbial arsenic methylation in soil and rice rhizosphere. *Environmental Science and Technology* 47: 3141–3148.

Jiang, S., Lee J-H, Kim, M-G, Myung, N.V., Fredrickson, J.K., Sadowsky, M.J., Hurl, H-G, 2009. Biogenic formation of As-S Nanotubes by diverse *Shewanella* strains. *Applied and Environmental Microbiology* 75: 6896-9.

Kashyap, D.R., Botero, L.M., Franck, W.L., Hassett, D.J., McDermott, T.R., 2006. Complex regulation of arsenite oxidation in *Agrobacterium tumefaciens*. *Journal of Bacteriology* 188: 1081–1088.

Katsoyiannis, I.A., Zouboulis, A.I., Jekel, M., 2004. Kinetics of bacterial As(III) oxidation and subsequent As(V) removal by sorption onto biogenic manganese oxides during groundwater treatment. *Industrial and Engineering Chemistry Research* 43: 486–493.

Katsoyiannis, I.A., Zouboulis, A.I., 2004. Application of biological processes for the removal of arsenic from ground waters. *Water Research* 38: 17–26.

Kulp, T.R., Hoeft, S.E., Asao, M., Madigan, M.T., Hollibaugh, J.T., Fisher, J.C., Stolz, J.F., Culbertson, C.W., Miller, L.G., Oremland, R.S., 2008. Arsenic(III) fuels anoxygenic photosynthesis in hot spring biofilms from Mono Lake, California. *Science* 321: 967–970.

Kuramata, M., Sakakibara, F., Kataoka, R., Abe, T., Asano, M., Baba, K., Takagi, K., Ishikawa, S., 2015. Arsenic biotransformation by *Streptomyces* sp. isolated from rice rhizosphere. *Environmental Microbiology* 17: 1897–1909.

Lebrun, E., Brugna, M., Baymann, F., Muller, D., Lièvremont, D., Lett, M.C., Nitschke, W., 2003. Arsenite oxidase, an ancient bioenergetic enzyme. *Molecular Biology and Evolution* 20: 686–693.

Lee, J.H., Kim, M.G., Yoo, B., Myung, N.V., Maeng, J., Lee, T., Dohnalkova, A.C., Fredrickson, J.K., Sadowsky, M.J., Hur, H.G., 2007. Biogenic formation of photoactive arsenic-sulfide nanotubes by *Shewanella* sp. strain HN-41. *Proceedings of the National Academy of Sciences of the United States of America* 104: 20410–20415.

Lei, Y., H. Yin, S. Zhang, F. Leng, W. Nan, H. Li. 2010. Biosorption of inorganic and organic arsenic from aqueous solution by *Acidithiobacillus ferrooxidans* BY-3. *Journal of Hazardous Materials* 178: 209–217.

Lett, M.C., Muller, D., Lièvremont, D., Silver, S., Santini, J., 2012. Unified nomenclature for genes involved in prokaryotic aerobic arsenite oxidation. *Journal of Bacteriology* 194: 207–208.

Li, B., Pan, X., Zhang, D., Lee, D.J., Al-Misned, F.A., Mortuza, M.G., 2015. Anaerobic nitrate reduction with oxidation of Fe(II) by *Citrobacter freundii* strain PXL1—A potential candidate for simultaneous removal of As and nitrate from groundwater. *Ecological Engineering* 77: 196–201.

Lieutaud, A., Van Lis, R., Duval, S., Capowiez, L., Muller, D., Lebrun, R., Lignon, S. et al., 2010. Arsenite oxidase from *Ralstonia* sp. 22. *Journal of Biological Chemistry* 285: 20433–20441.

Lievremont, D., N'negue, M.A., Behra, P.H., Lett, M.C., 2003. Biological oxidation of arsenite: Batch reactor experiments in presence of kutnahorite and chabazite. *Chemosphere* 51: 419–428.

Liu, A., Garcia-Dominguez, E., Rhine, E.D., Young, L.Y., 2004. A novel arsenate respiring isolate that can utilize aromatic substrates. *FEMS Microbiology Ecology* 48: 323–332.

Lu, K., Cable, P.H., Abo, R.P., Ru, H., Graffam, M.E., Schlieper, K.A., Parry, N.M. et al., 2013. Gut microbiome perturbations induced by bacterial infection affect arsenic biotransformation. *Chemical Research in Toxicology* 26: 1893–1903.

Maguffin, S.C., Kirk, M.F., Daigle, A.R., Hinkle, S.R., Jin, Q., 2015. Substantial contribution of biomethylation to aquifer arsenic cycling. *Nature Geoscience* 8: 290–293,

Mailloux, B.J., Alexandrova, E., Keimowitz, A.R., Wovkulich, K., Freyer, G.A., Herron, M., Stolz, J.F. et al., 2009. Microbial mineral weathering for nutrient acquisition releases arsenic. *Applied and Environmental Microbiology* 75: 2558–2565.

Majumder, A., Ghosh, S., Saha, N., Kole, S.C., Sarkar, S., 2013. Arsenic accumulating bacteria isolated from soil for possible application in bioremediation. *Journal of Environmental Biology* 34: 841–846.

Malasarn, D., Keeffe, J.K., Newman, D.K., 2008. Characterization of the arsenate respiratory reductase from *Shewanella* sp. strain ANA-3. *Journal of Bacteriology* 190: 135–142.

Mandal, B.K., Suzuki, K.T., 2002. Arsenic round the world: A review. *Talanta* 58: 201–235.

Masue-Slowey, Y., Loeppert, R.H., Fendorf, S., 2011. Alteration of ferrihydrite reductive dissolution and transformation by adsorbed As and structural Al: Implications for As retention. *Geochimica et Cosmochimica Acta* 75: 870–886.

Meena, A.K., Mishra, G.K., Rai, P.K., Rajagopal, C., Nagar, P.N., 2005. Removal of heavy metal ions from aqueous solutions using carbon aerogel as an adsorbent. *Journal of Hazardous Materials* 122: 161–170.

Mestrot, A., Merle, J.K., Broglia, A., Feldmann, J., Krupp, E.M., 2011. Atmospheric stability of arsine and methylarsines. *Environmental Science and Technology* 45: 4010–4015.

Mestrot, A., Planer-Friedrich, B., Feldmann, J., 2013. Biovolatilisation: A poorly studied pathway of the arsenic biogeochemical cycle. *Environmental Science: Processes and Impacts* 15: 1639–1651.

Meyer, J., Michalke, K., Kouril, T., Hensel, R., 2008. Volatilisation of metals and metalloids: An inherent feature of methanoarchaea? *Systematic and Applied Microbiology* 31: 81–87.

Meyer, J., Schmidt, A., Michalke, K., Hensel, R., 2007. Volatilization of metals and metalloids by the microbial population of an alluvial soil. *Systematic and Applied Microbiology* 30: 229–238.

Meyer-Dombard, D.R., Amend, J.P., Osburn, M.R., 2013. Microbial diversity and potential for arsenic and iron biogeochemical cycling at an arsenic rich, shallow-sea hydrothermal vent (Tutum Bay, Papua New Guinea). *Chemical Geology* 348: 37–47.

Michalke, K., Meyer, J., Hirner, A.V., Hensel, R., 2002. Biomethylation of bismuth by the methanogen *Methanobacterium formicicum*. *Applied Organometallic Chemistry* 16: 221–227.

Michel, C., Jean, M., Coulon, S., Dictor, M.C., Delorme, F., Morin, D., Garrido, F., 2007. Biofilms of As(III)-oxidising bacteria: Formation and activity studies for bioremediation process development. *Applied Microbiology and Biotechnology* 77: 457–467.

Michon, J., Dagot, C., Deluchat, V., Dictor, M.C., Battaglia-Brunet, F., Baudu, M., 2010. As(III) biological oxidation by CAsO1 consortium in fixed-bed reactors. *Process Biochemistry* 45: 171–178.

Miyatake, M., Hayashi, S., 2009. Characteristics of arsenic removal from aqueous solution by *Bacillus megaterium* strain UM-123. *Journal of Environmental Biology* 9: 123–129.

Miyatake, M., Hayashi, S., 2011. Characteristics of arsenic removal by *Bacillus cereus* strain W2. *Resources Processing* 58: 101–107.

Mokashi, S.A., Paknikar, K.M., 2002. Arsenic(III) oxidizing *Microbacterium lacticum* and its use in the treatment of arsenic contaminated groundwater. *Letters in Applied Microbiology* 34: 258–262.

Mondal, P., Majumder, C.B., Mohanty, B., 2006. Laboratory based approaches for arsenic remediation from contaminated water: Recent developments. *Journal of Hazardous Materials* 137: 464–479.

Mondal, P., Majumder, C.B., Mohanty, B., 2008. Treatment of arsenic contaminated water in a laboratory scale up-flow bio-column reactor. *Journal of Hazardous Materials* 153: 136–145.

Newman, D.K., Beveridge, T.J., Morel, F., 1997. Precipitation of arsenic trisulfide by *Desulfotomaculum auripigmentum*. *Applied and Environmental Microbiology* 63: 2022–2028.

Ordóñez, E., Letek, M., Valbuena, N., Gil, J.A., Mateos, L.M. 2005. Analysis of genes involved in arsenic resistance in *Corynebacterium glutamicum* ATCC 13032. *Applied and Environmental Microbiology* 71: 6206–6215.

Oremland, R.S., Saltikov, C.W., Wolfe-Simon, F., Stolz, J.F., 2009. Arsenic in the evolution of Earth and extraterrestrial ecosystems. *Geomicrobiology Journal* 26: 522–536.

Paez-Espino, D., Tamames, J., de Lorenzo, V., Canovas, D., 2009. Microbial responses to environmental arsenic. *Biometals* 22: 117–130.

Pal, A., Paknikar, K.M. 2012. Bioremediation of arsenic from contaminated water. In T. Satyanarayana et al. (Eds.), *Microorganisms in Environmental Management: Microbes and Environment*. pp. 477–523. Springer, Dordrecht, the Netherlands.

Pandey, N., Bhatt, R., 2015. Arsenic resistance and accumulation by two bacteria isolated from a natural arsenic contaminated site. *Journal of Basic Microbiology* 55: 1275–128.

Perez-Jimenez, J.R., DeFraia, C., Young, L.Y., 2005. Arsenate respiratory reductase gene (arrA) for *Desulfosporosinus* sp. strain Y5. *Biochemical and Biophysical Research Communications* 338: 825–829.

Podder, M.S., Majumder, C.B., 2015. Modelling of optimum conditions for bioaccumulation of As(III) and As (V) by response surface methodology (RSM). *Journal of Environmental Chemical Engineering* 3: 1986–2001.

Pous N., Casentini, B., Rossetti, S., Fazi, S., Puig, S., Aulenta, F., 2015. Anaerobic arsenite oxidation with an electrode serving as the sole electron acceptor: A novel approach to the bioremediation of arsenic-polluted groundwater. *Journal of Hazardous Materials* 283: 617–622.

Prasad, K.S., Srivastava, P., Subramanian, V., Paul, J., 2011. Biosorption of As(III) ion on *Rhodococcus* sp. WB-12: Biomass characterization and kinetic studies. *Separation Science and Technology* 46: 2517–2525.

Prasad, K.S., Ramanathan, A.L., Paul, J., Subramanian, V., Prasad, R., 2013. Biosorption of arsenite (As+3) and arsenate (As+5) from aqueous solution by *Arthrobacter* sp. biomass, *Environmental Technology* 34: 2701–2708.

Qin, J., Rosen, B.P., Zhang, Y., Wang, G., Franke, S., Rensing, C., 2006. Arsenic detoxification and evolution of trimethylarsine gas by a microbial arsenite S-adenosylmethionine methyltransferase. *Proceedings of the National Academy of Sciences of the United States of America* 103: 2075–2080.

Qin, J., Lehr, C.R., Yuan, C., Le, X.C., McDermott, T.R., Rosen, B.P., 2009. Biotransformation of arsenic by a Yellowstone thermoacidophilic eukaryotic alga. *Proceedings of the National Academy of Sciences of the United States of America* 106: 5213–5217.

Quéméneur, M., Heinrich-Salmeron, A., Muller, D., Lièvremont, D., Jauzein, M., Bertin, P.N., Garrido, F., Joulian, C., 2008. Diversity surveys and evolutionary relationships of aoxB genes in aerobic arsenite-oxidizing bacteria. *Applied and Environmental Microbiology* 74: 4567–4573.

Rahman, A., Nahar, N., Nawani, N.N., Jass, J., Desale, P., Kapadnis, B.P., Hossain, K. et al., 2014. Isolation and characterization of a *Lysinibacillus* strain B1-CDA showing potential for bioremediation of arsenics from contaminated water. *Journal of Environmental Science and Health, Part A* 49: 1349–1360.

Ravenscroft, P., Brammer, H., Reichards, K., 2009. *Arsenic Pollution: A Global Synthesis*, Wiley-Blackwell, Chichester, U.K.

Richey, C., Chovanec, P., Hoeft, S.E., Oremland, R.S., Basu, P., Stolz, J.F., 2009. Respiratory arsenate reductase as a bidirectional enzyme. *Biochemical and Biophysical Research Communications* 382: 298–302.

Ronkart, S.N., Laurent, V., Carbonnelle, P., Mabon, N., Copin, A., Barthélemy, J.-P. 2007. Speciation of five arsenic species (arsenite, arsenate, MMAAV, DMAAV and AsBet) in different kind of water by HPLC-ICP-MS. *Chemosphere* 66: 738–745.

Sahabi, D.M., Takeda, M., Suzuki, I., Koizumi, J., 2009. Adsorption and abiotic oxidation of arsenic by aged biofi lter media: Equilibrium and kinetics. *Journal of Hazardous Materials* 168: 1310–1318.

Salmassi, T.M., Venkateswaren, K., Satomi, M., Newman, D.K., Hering, J.G., 2002. Oxidation of arsenite by *Agrobacterium albertimagni* AOL15, sp. nov., isolated from hot creek, California. *Geomicrobiology Journal* 19: 53–66.

Saltikov, C.W., Wildman, R.A. Jr., Newman, D.K., 2005. Expression dynamics of arsenic respiration and detoxification in *Shewanella* sp. strain ANA-3. *Journal of Bacteriology* 187: 7390–7396.

Santini, J.M., Sly, L.I., Schnagl, R.D., Macy, J.M., 2000. A new chemolithoautotrophic arsenite-oxidizing bacterium isolated from a gold mine: Phylogenetic, physiological, and preliminary biochemical studies. *Applied and Environmental Microbiology* 66: 92–97.

Silver, S., Phung, L.T., 2005. Genes and enzymes involved in bacterial oxidation and reduction of inorganic arsenic. *Applied and Environmental Microbiology* 71: 599–608.

Singh, R., Singh, S., Parihar, P., Singh, V.P., Prasad, S.M., 2015. Arsenic contamination, consequences and remediation techniques: A review. *Ecotoxicology and Environmental Safety* 112: 247–270.

Slyemi, D., Bonnefoy, V., 2012. How prokaryotes deal with arsenic. *Environmental Microbiology Reports* 4: 571–586.

Smedley, P., Kinniburgh, D., 2002. A review of the source, behaviour and distribution of arsenic in natural waters. *Applied Geochemistry* 17: 517–568.

Song, B., Chyun, E., Jaffé, P.R., Ward, B.B., 2009. Molecular methods to detect and monitor dissimilatory arsenate respiring bacteria (DARB) in sediments. *FEMS Microbiology Ecology* 68: 108–117.

Sorokin, D.Y., Tourova, T.P., Sukhacheva, M.V., Muyzer, G., 2012. *Desulfuribacillus alkaliarsenatis* gen. nov. sp. nov., a deep-lineage, obligately anaerobic, dissimilatory sulfur and arsenate-reducing, haloalkaliphilic representative of the order Bacillales from soda lakes. *Extremophiles* 16: 597–605.

Srivastava, P.K., Vaish, A., Dwivedi, S., Chakrabarty, D., Singh, N., Tripathi, R.D., 2011. Biological removal of arsenic pollution by soil fungi. *Science of the Total Environment* 409: 2430–2442.

Tang, K.H., Barry, K., Chertkov, O., Dalin, E., Han, C.S., Hauser, L.J., Honchak, B.M. et al., 2011. Complete genome sequence of the filamentous anoxygenic phototrophic bacterium *Chloroflexus aurantiacus*. *BMC Genomics* 12: 334.

Tomczyk-Żak, K., Kaczanowski, S., Drewniak, Ł., Dmoch, Ł., Sklodowska, A., Zielenkiewicz, U., 2013. Bacteria diversity and arsenic mobilization in rock biofilm from an ancient gold and arsenic mine. *Science of the Total Environment* 461: 330–340.

Tsai, S.L., Singh, S., Chen, W., 2009. Arsenic metabolism by microbes in nature and the impact on arsenic remediation. *Current Opinion in Biotechnology* 20: 659–667.

Van de Wiele, T., Gallawa, C.M., Kubachka, K.M., Creed, J.T., Basta, N., Dayton, E.A., Whitacre, S., Du Laing, G., Bradham, K., 2010. Arsenic metabolism by human gut microbiota upon in vitro digestion of contaminated soils. *Environmental Health Perspectives* 118: 1004–1009.

Vishnoi, N., Dixit, S., Singh, D.P., 2014. Surface binding and intracellular uptake of arsenic in bacteria isolated from arsenic contaminated site. *Ecological Engineering* 73: 569–578

Wang, P., Sun, G., Jia, Y., Meharg, A.A., Zhu, Y., 2014. A review on completing arsenic biogeochemical cycle: Microbial volatilization of arsines in environment. *Journal of Environmental Sciences* 26: 371–381.

WHO, 2011. *Guidelines for Drinking-Water Quality*. Vol. 4, pp. 315–318. World Health Organisation, Geneva, Switzerland.

Williams, G.P., Gnanadesigan, M., Ravikumar, S., 2013. Biosorption and bio-kinetic properties of solar saltern Halobacterial strains for managing Zn^{2+}, As^{2+} and Cd^{2+} Metals. *Geomicrobiology Journal* 30: 497–500.

Xiao, K.Q., Li, L.G., Ma, L.P., Zhang, S.Y., Bao, P., Zhang, T., Zhu, Y.G., 2016. Metagenomic analysis revealed highly diverse microbial arsenic metabolism genes in paddy soils with low-arsenic contents. *Environmental Pollution* 211: 1–8.

Xie, Z., Luo, Y., Wang, Y., Xie, X., Su, C., 2013. Arsenic resistance and bioaccumulation of an indigenous bacterium isolated from aquifer sediments of Datong Basin, Northern China. *Geomicrobiology Journal* 30: 549–556.

Xue A., Shen, Z.Z., Zhao B., Zhao, H.Z., 2013. Arsenite removal from aqueous solution by a microbial fuel cell zerovalent iron hybrid process. *Journal of Hazardous Materials* 261: 621–627.

Yan, L., Yin, H.H., Zhang, S., Leng, F.F., Nan, W.B., Li, H.Y. 2010. Biosorption of inorganic and organic arsenic from aqueous solution by *Acidithiobacillus ferrooxidans* BY-3. *Journal of Hazardous Materials* 178: 209–217.

Yang, L., Li, X., Chu, Z., Ren, Y., Zhang, J., 2014. Distribution and genetic diversity of the microorganisms in the biofilter for the simultaneous removal of arsenic, iron and manganese from simulated groundwater. *Bioresource Technology* 156: 384–388.

Yin, N., Zhang, Z., Cai, X., Du, H., Sun, G., Cui, Y., 2015. In vitro method to assess soil arsenic metabolism by human gut microbiota: Arsenic speciation and distribution. *Environmental Science and Technology* 49: 10675–10681.

Yin, X., Chen, J., Qin, J., Sun, G., Rosen, B., Zhu, Y., 2011. Biotransformation and volatilization of arsenic by three photosynthetic cyanobacteria. *Plant Physiology* 156: 1631–1638.

Ying, S.C., Damashek, J., Fendorf, S., Francis, C.A., 2015. Indigenous arsenic(V)-reducing microbial communities in redox-fluctuating near-surface sediments of the Mekong Delta. *Geobiology* 13: 581–587.

Yuan, C.G., Lu, X.F., Qin, J., Rosen, B.P., Le, X.C. 2008. Volatile arsenic species released from *Escherichia coli* expressing the AsIII S-adenosylmethionine methyltransferase gene. *Environmental Science and Technology* 42: 3201–3206.

Zargar, K., Conrad, A., Bernick, D.L., Lowe, T.M., Stolc, V., Hoeft, S., Oremland, R.S., Stolz, J., Saltikov, C.W. 2012. ArxA, a new clade of arsenite oxidase within the DMSO reductase family of molybdenum oxidoreductases. *Environmental Microbiology* 14: 1635–1645.

Zargar, K., Hoeft, S., Oremland, R., Saltikov, C.W., 2010. Identification of a novel arsenite oxidase gene, arxA, in the haloalkaliphilic, arsenite-oxidizing bacterium *Alkalilimnicola ehrlichii* strain MLHE-1. *Journal of Bacteriology* 192: 3755–3762.

Zhang, S.Y., Zhao, F.J., Sun, G.X., Su, J.Q., Yang, X.R., Li, H., Zhu, Y.G., 2015. Diversity and abundance of arsenic biotransformation genes in paddy soils from southern China. *Environmental Science and Technology* 49: 4138–4146.

31 Investigation on Arsenic-Accumulating and Arsenic-Transforming Bacteria for Potential Use in the Bioremediation of Arsenics

Aminur Rahman, Noor Nahar, Neelu N. Nawani, and Abul Mandal

CONTENTS

ABSTRACT

In this chapter, arsenic-accumulating and arsenic-transforming bacterial strains that can be employed as a source for cost-effective and eco-friendly bioremediation of arsenics from contaminated environments have been reviewed. This chapter demonstrates that many naturally occurring bacterial strains like B1-CDA have the potential for reducing arsenic content in contaminated sources to safe levels. Therefore, the socioeconomic impact of this kind of microorganisms is highly significant for those countries, especially in the developing world, where impoverished families and villages are most impacted. Therefore, this discovery should be considered to be the most significant factor in formulating national strategies for effective poverty elimination. Besides human arsenic contamination, these bacterial strains will also benefit livestock and native animal species, and the outcome of these studies is vital not only for people in arsenic-affected areas but also for human populations in other countries that have credible health concerns as a consequence of arsenic-contaminated water and foods.

31.1 INTRODUCTION

As the world undergoes vast human demographic shifts in socioeconomic status, there is an immense opportunity to improve global living standards. However, with these increases come ever higher consumption and pollution rates. This is especially true in the regions of the world where socioeconomic status is quickly rising, industry is vastly expanding, and pollution emanating from such industrial processes is overwhelming the environmental regulatory infrastructure. Arsenic is one of the toxic pollutants that has been reported by the United States Environmental Protection Agency (U.S. EPA, 1998) to be among the five most toxic substances found at contaminated sites (Johnson and Derosa, 1995). Different forms of arsenic induce distinct types of cellular damage, and the forms most prevalent in nature are as inorganic species of arsenic: arsenite (AsIII) and arsenate (AsV).

Arsenic problem in many countries of the world, particularly in South Asia, is mainly arsenic-contaminated groundwater. Contamination of groundwater threatens the health of hundreds of millions of people in these regions. The levels of arsenic (As) in drinking water are so high that the WHO describes arsenic contamination of water supply in South Asian regions as "the largest poisoning of a population in human history." In South Asia, large-scale use of groundwater for irrigation began with the "green revolution," which was initiated almost 30 years ago. Available groundwater irrigation resulted in a significant increase in food production. However, unfortunately, it was shown that this water is highly contaminated with arsenic. The chronic arsenic poisoning by drinking tube-well water has now become a national problem in these regions. The presence of high concentrations of arsenic in crops produced by arsenic-contaminated irrigation water is a credible health concern. Before 2001, it was reported that the arsenic problem was linked only to drinking water, but in 2001, FAO studied its multifarious effects and produced evidences supporting arsenic accumulation in different varieties of cultivated crops, for example, rice and wheat. A few years later, Williams et al. (2006) reported that rice and wheat are the predominant sources of inorganic As in the human food chain in South Asia. Rice and wheat straw may pose another threat for domestic animals as arsenic is readily accumulated in this part of these crops which are used as fodder (Abedin et al., 2002). In South Asia, rice/wheat straw is used for many purposes, among which livestock feed is the major one. The high arsenic concentrations in ingested straw by livestock may have adverse health effects on livestock and further introduce arsenic via arsenic-contaminated meat/milk into the human food chain.

Long-term exposure to arsenic can lead to a variety of skin, neurological, and peripheral vascular disorders, as well as cancers of the skin, bladder, liver, lung, kidney, and colon. In addition, health-related illnesses linked with arsenic exposure include diabetes, ischemic disease, reproductive defects, and impairment of liver function (U.S. EPA, 1998).

One possible way to avoid arsenic contamination in humans and the environment could be bioremediation, the removal of arsenics from the contaminated sources like drinking water and cultivated lands. Bioremediation is a method of removing toxic pollutants using naturally occurring or genetically modified living organisms. When plants are employed for this purpose, the process is called as phytoremediation, whereas when microorganisms are used, it is called as microbial remediation. Phytoremediation is an efficient process where green plants are used to remove, detoxify, or immobilize environmental contaminants from soil, water, or sediments (Greipsson, 2011; Zhu et al., 2012; Hazrat et al., 2013). Attempts for developing a new variety of cultivated crops with elevated features to metabolize arsenic or to prevent uptake of arsenic from contaminated soil have been reported (Nahar et al., 2014). Phytoremediation is not the theme of this chapter, and therefore, it is not discussed further.

Microbial remediation of arsenics, on the other hand, relies on the use of microbes, such as bacterial strains, that can survive and grow in arsenic-contaminated environments, and, at the same time, accumulate arsenics inside the cells and/or convert this toxic pollutant to a nontoxic or less toxic compound. Uptake and accumulation of arsenics inside the cells will result in the reduction or elimination of arsenics from the contaminated environment. Such bacterial strains (usually found in the contaminated sites), once identified and characterized, are normally put through a cutting-edge and yet comparatively much less expensive technique called directed evolution (de Crécy et al., 2007, 2009; Singh et al., 2009). In this method, microorganisms are not modified in a way different from how people have historically selected for and thus directed the evolution of many organisms such as every food crop, as well as domesticated animal, yeast strain (baking), bacterial strain (dairy, alcoholic products, pest control, bioremediation), and fungus (cultured food, pest control, bioremediation). This method simply increases the rate of evolutionary change for many human purpose such as bioremediation of toxic pollutants. Directed evolution simply selects from the isolated strains individuals that are the best at remediating arsenics and allows only those to proliferate. After strains are developed and/or verified, they may be further characterized by using molecular methods such as gene expression analysis and cloning and sequencing of target genes. Chromatography and other methods of analysis like high-performance liquid chromatography and scintillation counter or mass spectroscopy for the determination of the toxic chemicals are utilized to measure water samples in the lab and in the environment. Prior to any actual application, the bacterial strains are extensively exposed to "biosafety tests."

31.2 ARSENIC

Arsenic is a brittle and crystalline gray-appearing chemical element also known as a metalloid having properties of both metals and nonmetals. It is considered to be a heavy metal, and its toxicity shares some features manifested in other heavy metal poisonings (Emsley, 2011). Arsenic is odorless and tasteless, and usually, it exists as part of other chemical compounds (Mandal and Suzuki, 2002). These compounds are divided into two groups: (1) inorganic compounds that combine with oxygen, iron, chlorine, and sulfur and (2) organic compounds that combine with carbon and other atoms. Inorganic arsenic compounds are found in industry, building materials, underground rocks, and arsenic-contaminated groundwater. Inorganic forms of arsenic are more toxic than the organic forms, and they have been linked to various cancerous, neurological, and vascular diseases (Meliker et al., 2007). On the other hand, even though organic arsenic compounds are much less toxic, they are harmful and found in some foods, such as fish and shellfish. Thus far, there is no report on their involvement in the induction of cancerous diseases in humans.

31.3 AVAILABILITY OF ARSENIC

Arsenic is one of the most abundant elements that can be found naturally in rocks, soil, water, air, plants, and animals (Reimer et al., 2010; Emsley, 2011). Minerals that contain arsenics include arsenopyrite (FeAsS), realgar (AsS), orpiment (As_2S_3), and arsenolite (As_2O_3). Although elemental arsenic has several allotropic forms (gray, yellow, and black arsenic), only gray arsenic is ordinarily stable. Elemental arsenic is produced by reducing arsenic trioxide with carbon. This metalloid is also disposed to the environment due to many industrial and agricultural activities, such as fertilizer and pesticide production. Arsenic trioxide, As_2O_3, is the main commercial compound, and as much as 100,000 tons are produced worldwide, with most of it obtained as a byproduct of smelting of copper, lead, cobalt, and gold ores. The quantity of arsenic associated with lead and copper ores may range from 2% to 3%, whereas gold ores may contain up to 11%.

31.4 USE OF ARSENIC

Arsenic is widely used in different fields ranging from agricultural farming to military products. The primary use of arsenic is for the preservation of wood by getting rid of insects, fungi, and bacteria. Chromated copper arsenate–treated wood is used in commercial, industrial, and residential applications, including decking, highway noise barriers, landscaping, playground equipment, retaining walls, signposts, and utility poles. Extremely high-purity arsenic (99.99%) is used to make gallium arsenide or indium arsenide, which is employed in the manufacture of semiconductors. Gallium arsenide is also used in light-emitting diodes and solar cells. Indium arsenide is used to produce infrared devices and lasers (Brooks, 2008). Arsenic is used as a semiconductor agent in solid-state devices such as transistors. Another use of arsenic is as an antifriction additive in ball bearings and for hardening of lead. Also, germanium–arsenide–selenide optical materials are manufactured by using arsenic. Indium–gallium–arsenide was formerly used for shortwave infrared technology.

Recently, most countries banned its use in consumer products due to its high toxicity. Smaller quantities of arsenic are used for the production of agricultural chemicals such as insecticides, herbicides, and algaecides, and it is extensively used as a weed killer. Due to the growing popularity of organic pesticides, the amount of arsenic used for agricultural purposes has drastically decreased. Another use of arsenic is for the production of glass and nonferrous alloys and in electronic industries. Prior to the introduction of DDT in the 1940s, most pesticides were made from inorganic arsenic compounds. Arsenic is also used in leather tanning industries as a preservative.

31.5 ARSENIC IN TREATMENT

In spite of the well-known toxic effects of arsenic, Fowler's solution (Fowler, 1786) containing 1% potassium arsenite (K_3AsO_3) was the most widely used medication for a variety of illnesses for almost 150 years (Gibaud and Gérard, 2010). Donovan's solution (Bailar et al., 1939) containing arsenic triiodide (AsI_3) and de Valagin's solution (Smith, 1919) containing arsenic trichloride ($AsCl_3$) were also recommended to

treat rheumatism, arthritis, malaria, trypanosome infections, tuberculosis, and diabetes. The drug Salvarsan, whose chemical name is arsphenamine, was discovered in 1909 and was the main treatment for syphilis until it was replaced by penicillin in the 1940s. Arsenic trioxide, a recent medicine shown to be effective in treating acute promyelocytic leukemia and APL, received FDA approval in 2001. While the use of arsenic in contemporary medicine has been severely prohibited, it is continually used in treating severe parasitic diseases. Although arsenics have very few uses in modern medicine, arsenic-based drugs have a long history (Nicholis et al., 2009). Prior to the 1980s, most of these drugs were marketed without serious clinical trials and without any knowledge of their mechanism of action. In the 1980s and 1990s, drug regulation organizations considered that the benefits/risk balance was unfavorable and all the pharmaceutical specialties were withdrawn in Europe and in the United States. Nevertheless, the mechanism of action of As_2O_3 in cancer therapy is partly known (Warrell et al., 1998) and a favorable benefits/risk balance may appear for the treatment that may explain the renewed interests in arsenical drugs.

31.6 FOOD CHAIN CONTAMINATION WITH ARSENIC

Humans, considered the topmost consumers in an ecosystem, can be exposed to arsenic contamination from staple food sources and drinking water directly through the plants–humans pathway (Figure 31.1). Rice is the most important staple food for over half of the world's population, especially in South Asia (Fageria, 2007). In Asia as a whole, much of the population consumes arsenic-contaminated rice in every meal (Halder et al., 2012). In many countries, due to irrigation of rice using contaminated water, arsenic poisoning of human foods and the environment is widely spread, exceeding the safety level (e.g., in Bangladesh, the safety level of arsenic in water is 50 µg As/L) and accounts for more than 70% of human caloric intake. Almost 30%–50% of the cultivated lands in Bangladesh and West Bengal state of India are irrigated with As-contaminated groundwater to grow paddy rice (Haldar et al., 2012; Neidhardt et al., 2012). Paddy rice accumulates high levels of arsenite (AsIII) due to irrigation with arsenic contaminated water (Williams et al., 2007). Considerably higher levels of total As, a major fraction of which were in the form of organic As species, were also found in the rice grains grown in the United States (Zavala and Duxbur, 2008; Zavala et al., 2008). Furthermore, rice straw is used as animal fodder in many countries, including the United States, India, China, and Bangladesh (Christopher and Haque, 2012). High As concentrations in straw may have adverse health effects on cattle and may result in an increased As exposure in humans via the "plant–animal–human" pathway (Rahman et al., 2009; as depicted in Figure 31.1).

Therefore, this raises a significant concern on accumulation of As in meat, dairy products, consumable crops, and other vegetables grown in arsenic-affected areas (Arora et al., 2008; Pandey and Pandey, 2009).

FIGURE 31.1 Schematic pathway of arsenic contamination in humans and the environment. Arsenic poisoning takes place directly through consumption of As-contaminated water or foods and indirectly via the meat–milk pathway. (a) Exposure of arsenics to the atmosphere (due to anthropogenic activities) and its precipitation on the surface of the earth, (b) downstream movement of arsenic-contaminating groundwater, (c) shallow tube wells pumping water for irrigation of cultivated crops, (d) uptake and accumulation of arsenic in plants, (e) consumption of arsenic-contaminated foods, (f) human exposure to arsenic, (g, h) consumption of arsenic-contaminated straw or leaves by cattle, (i, j) human consumption of arsenic-contaminated milk and meat, and (k) contaminated drinking water hand tube wells.

31.7 ARSENIC TOXICITY

Arsenic and some of its derivative compounds react with the thiol groups of proteins (Rey et al., 2004). By inhibiting oxidative phosphorylation, arsenic can inactivate most cellular functions (Das et al., 2005; Eblin et al., 2006). Furthermore, arsenite inhibits not only the formation of acetyl-CoA but also the enzyme succinic dehydrogenase. Also, arsenate can replace phosphate in many reactions. Consequently, arsenic and some of its derivative substances are deadly to most biological systems (Hughes, 2002), except for a few bacterial species such as *Bacillus* sp.

31.7.1 Arsenic Toxicity in Humans

The effects of long-term exposure to high amounts of inorganic arsenics (in water and food) include peripheral gastrointestinal symptoms, cardiovascular diseases, diabetes, renal system effects, enlarged liver, destruction of bone marrow and erythrocytes, high blood pressure, neuropathy, and many cancerous diseases (Jensen et al., 2008; Rossman et al., 2008; Rossman and Klein, 2011). Manifestation of arsenic poisoning, pigmentation, and keratosis are quite distinctive as depicted in Figure 31.2 (Cöl et al., 1999). Hyperpigmentation is marked by raindrop-shaped discolored spots, diffused

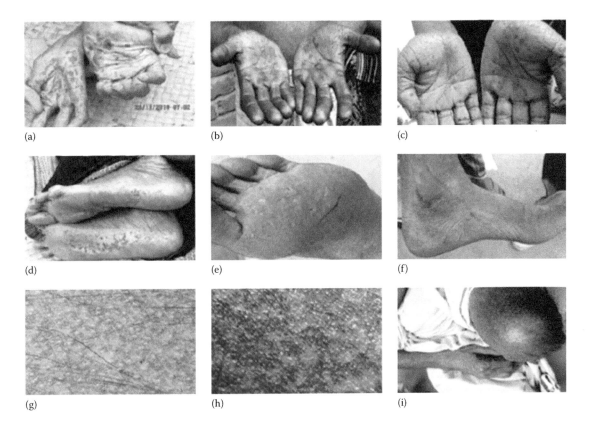

FIGURE 31.2 Effects of long-term exposure of humans to arsenic poisoning through drinking of water and consumption of contaminated foods. The pictures were taken in November 2010 when a research team from the University of Skövde in Sweden visited the arsenic-affected areas, particularly the Chuadanga District in the southwest regions of Bangladesh. These diseases, also called as "arsenicosis," are direct evidence of human suffering resulting from arsenic contamination. (a–c) Hypermelanosis leading to the development of cancer, (d–f) keratosis with hyperpigmentation, (g and h) hyper- and hypomelanosis respectively leading to development of skin cancer, and (i) amputated leg due to gangrene.

dark brown spots, or diffused darkening of the skin on the limbs and trunk. Spotty depigmentation (leucomelanosis) also occurs in arsenicosis. Normally, simple keratosis appears as two-sided thickening of the soles and palms, while nodular keratosis appears as raised keratotic lesions in the palms and soles (Figure 31.2). Moreover, skin lesions pose an important public health problem because advanced forms of keratosis are painful, and consequent disfigurement can lead to social isolation, particularly in villages (Argos et al., 2010). In contrast to cancer caused by arsenic poisoning, which takes decades to develop, skin lesions are generally developed after 5–10 years of exposure. Ingestion of arsenic-contaminated drinking water in adults leads to weakness, conjunctive congestion, hepatomegaly, portal hypertension, lung disease, polyneuropathy, solid edema of the limbs, ischemic heart disease, peripheral vascular disease, hypertension, and anemia (Ahmed et al., 2008; Zhao et al., 2010).

31.8 ARSENIC-ACCUMULATING BACTERIA AND BIOREMEDIATION OF ARSENICS

To date, many conventional remediation methods have been developed for removing arsenic, such as ion exchange, electrochemical treatment, solvent extraction, reverse osmosis,

evaporation, precipitation, and adsorption on activated coal (Balasubramanian et al., 2009; Moussavi and Barikbin, 2010). Nevertheless, most of these methods have some disadvantages, such as inefficient removal rates, especially when contamination levels are low, very high operational and reagent costs as well as requirement of technically skilled manpower for operation and a large area of land (Sharma and Sohn, 2009).

Alternatively, over the past decades, various cost-effective and eco-friendly biological approaches have been considered for the bioremediation of toxic pollutants (Congeevaram et al., 2007; Desale et al., 2014; Rahman et al., 2014, 2015). Often, microorganisms metabolize chemicals to produce carbon dioxide or methane, water, and biomass (Harms et al., 2011; Silar et al., 2011). Fortunately, microorganisms can affect the reactivity and mobility of toxic metals and pollutants. Microorganisms that affect the reactivity and mobility of metals can be used to detoxify some metals and prevent further metal contamination. *Lysinibacillus, Bacillus, Staphylococcus, Pseudomonas, Citrobacter, Klebsiella,* and *Rhodococcus* are bacterial genera commonly used in the bioremediation of toxic metals (Desale et al., 2014; Rahman et al., 2014). These mechanisms consist of bioaugmentation and biostimulation. Bioaugmentation is based on the delivery of nutrients and microbes to the polluted sites to increase

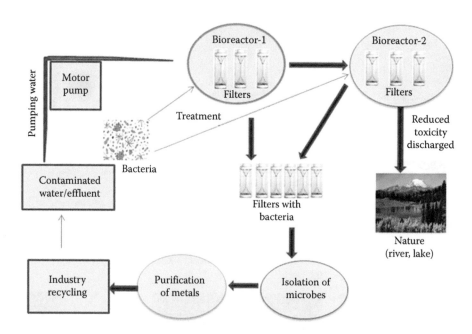

FIGURE 31.3 A schematic diagram of the microbial remediation of arsenics disposed as either effluents or solid wastes from industries.

the capability of the microbial community to degrade certain compounds. On the other hand, biostimulation relies on the supplement of nutrients and enzymes to the microbes already existing in the contaminated sites. Currently, more research is being performed on the use of microbes to metabolize metals. This investment validates the need to research the utilization of microbial processes to clean up contaminated sites. The strategy of arsenic bioremediation proposal has been described in Figure 31.3.

Effluents from industries discharged to drainage or sewage can be pumped up and collected in containers (bioreactors). Alternatively, effluents can be disposed directly to bioreactors. Previously investigated and selected bacterial strain/s by involving directed evolution can be applied to the Bioreactor-1 for the treatment of effluents. After a certain period of exposure, the bacteria will absorb and accumulate arsenic inside the cells, thus reducing the arsenic content in the effluents. Special filters/membranes can be installed in the Bioreactor-1 to capture the bacteria after the treatment period is over. Effluent samples from the bioreactors will be analyzed continuously for monitoring the level of toxicity. This treatment can be repeated several times by using Bioreactors-2, 3, etc., until the arsenic is eliminated or its content in the effluents is reduced to a "safe" level. The "clean" effluent can be discharged into nature. Also, the bacterial cells captured in the filters/membranes can then be used for isolation and purification of arsenics that can be reused for recycling, such as in industries, medical companies, nanotech or biotech industries, and fertilizer factories.

31.9 BIOACCUMULATION OF ARSENIC

In bacteria, arsenic can be stored intercellular, and the process is called bioadsorption (Volesky, 1990). To determine the bioaccumulation of arsenics in the cells, bacteria are grown

for a certain period on arsenic-containing media. Then the cell biomasses are separated from the culture media by centrifugation at 10,000 rpm for 10 min. Cells are digested with suprapure nitric acid according to a ratio of 7.5 mL nitric acid per gram of dry biomass using microwave digestion. The samples are heated until the biomass becomes colorless. The samples are brought to a constant volume prior to the determination of arsenate and arsenite contents. The arsenic content can then be measured by various atomic absorption spectrophotometry (AAS) (Hossain et al., 2003; Rahman et al., 2014) or by using voltammetry with the Computrace 797 VA (Metrohm Ltd., Switzerland) system. Arsenate content is obtained by subtracting arsenite from the total arsenic content. It has been observed in several studies that arsenic bioaccumulation increases in 24 h of exposure of bacteria to the arsenic incorporated in the growth media, but arsenic accumulation decreases after 48 h (Banerjee et al., 2013; Rahman et al., 2014). This could be because of bacterial efflux of arsenic in the media after 48 h. Subsequently, the intracellular accumulation of arsenic increases after 72 h (Banerjee et al., 2013; Rahman et al., 2014). The arsenic transformation ability, however, decreases after 48 h and later increases after 72 h (Banerjee et al., 2013).

In the past decade, a number of microorganisms were reported to resist/tolerate arsenic by utilizing periplasmic biosorption, intracellular bioaccumulation, and/or biotransformation to a less toxic speciation state through direct enzymatic reaction, including *Kocuria* sp. (Banerjee et al., 2011), *Stenotrophomonas*, *Alcaligenes*, *Herminiimonas* sp. (Bahar et al., 2012), *Lactobacillus* sp. (Halttunen et al., 2007), *Rhodococcus*, *Achromobacter*, *Aliihoeflea* sp. (Corsini et al., 2014), *Bacillus* sp. (Mondal et al., 2008), *Lysinibacillus* sp. (Lozano and Dussan, 2013; Rahman et al., 2014), *Pseudomonas* sp. (Kao et al., 2013), and *Brevibacillus* sp. (Banerjee et al., 2013). Most of them have been isolated from soil, industrial

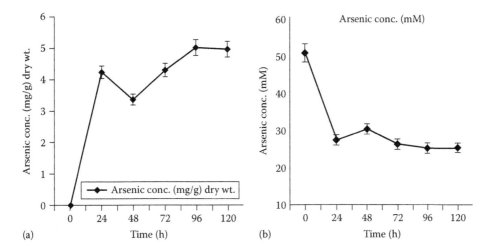

FIGURE 31.4 Estimation of arsenics (a) accumulated inside the bacterial cell biomass and (b) in the cell-free growth medium. (From Rahman, A. et al., *J. Environ. Sci. Health Part A*, 49, 1349, 2014. With permission.)

sewage, evaporation ponds, or discharged water or purchased from culture collection centers. Temperature and pH play major roles in the growth and metal accumulation properties of bacterial strains (Doenmez and Aksu, 2001). Rahman et al. (2014), while working with a soilborne Gram-positive bacterial strain *Lysinibacillus sphaericus* B1-CDA, observed that the highest amount of arsenic accumulation in the cell biomass (5.0 mg/g dry weight) took place after 120 h of exposure of bacteria to 50 mM of arsenate (Figure 31.4). Figure 31.4 also indicates that after 120 h of exposure, the arsenic concentration in the growth medium decreased to 25 mM (50% reduction).

Besides AAS, the time-of-flight secondary ion mass spectroscopy (TOF-SIMS), inductively coupled plasma mass spectroscopy, and inductively coupled plasma atomic emission spectroscopy analyses can be used to confirm the intracellular accumulation of arsenic. By using TOF-SIMS technology, Rahman et al. (2014) demonstrated the presence of arsenate (AsV) and arsenite (AsIII) ions inside the cells of B1-CDA, although the cells were exposed only to arsenate (Figure 31.5). These results suggest that B1-CDA cells when exposed to growth medium containing 50 mM arsenate take up arsenate and transform it to different forms of arsenites inside the cells, as shown in Figure 31.5.

31.10 VERIFICATION OF THE ARSENIC-TRANSFORMING ABILITY

The verification of the arsenic-transforming ability of bacteria can be performed by the $AgNO_3$ method. The silver nitrate test is based on the reaction between $AgNO_3$ and arsenite or arsenate ions. Agar medium can be flooded with a solution of 0.1 M $AgNO_3$, in which a brownish precipitate reveals the presence of arsenate in the medium (arsenite-oxidizing bacteria), while the presence of arsenite can be detected by a bright yellow precipitate (arsenate-reducing bacteria) (Krumova et al., 2008).

31.11 GENES INVOLVED IN ARSENIC BIOREMEDIATION

Bacteria possess many genetic systems for maintaining resistance against toxic metals or maintaining intracellular homeostasis of metal ions. In bacteria, the most well-known genetic mechanisms of metal resistance are the presence of metal-binding proteins and heavy metal efflux systems.

Many genes present in bacteria are involved in specific metal binding, transport, and resistance. Certain bacteria have evolved the necessary genetic mechanisms, allowing them to tolerate arsenic toxicity so that they can grow and survive in an environment where other organisms will not survive at all. The high-level tolerance to As in bacteria is conferred by the arsenical resistance *ars* operon comprising either three (*arsRBC*) (Liao et al., 2011) or five (*arsRD-ABC*) genes arranged in a single transcriptional unit located on plasmids (Owolabi and Rosen, 1990) or chromosomes (Diorio et al., 1995). *ArsB*, an integral membrane protein that pumps arsenite out of the cell, is often associated with an ATPase subunit, *arsA* (Achour et al., 2007). The *arsC* gene encodes the enzyme for arsenate reductase, which is responsible for the biotransformation of arsenate (As(V)) to arsenite (As(III)) prior to efflux. *ArsR* is a transacting repressor involved in the basal regulation of the *ars* operon, while *arsD* is a second repressor controlling the upper levels of expression of *ars* genes (Silver and Phung, 2005). In addition, several other genes act as arsenic-responsive genes like the *arrA* gene for dissimilatory As(V) respiration (Malasarn et al., 2004; Kulp et al., 2007; Song et al., 2009) and the *aoxB* gene for As(III) oxidation (Rhine et al., 2007; Hamamura et al., 2009). Moreover, some studies detected that in spite of clear evidence for the As-transforming activity by microorganisms, no amplicon for arsenite oxidase (*aoxB*) or As(V) respiratory reductase (*arrA*) was attained using the reported polymerase chain reaction primers and protocols (Kulp et al., 2008; Handley et al., 2009).

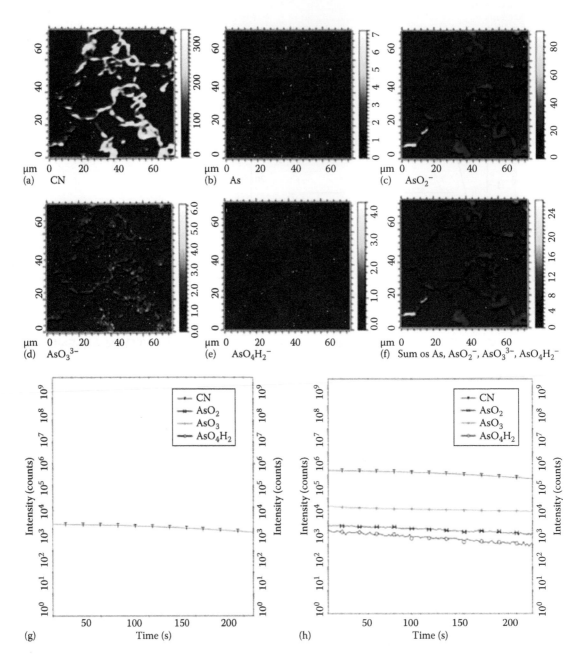

FIGURE 31.5 Analysis of arsenic species inside the bacterial cells by using TOF-SIMS. (a–f) Ion imaging, (g, h) depth profiling, (a) total protein signals, (b) free form of arsenic, (c) meta-arsenite ions (AsO_2^-) signal, (d) orthoarsenite ion AsO_3^{3-} signal, (e) arsenate (AsO_4H_2) ion signal, and (f) sum of different types of arsenics. The scale represents the intensity of ion imaging. (g) Depth profiling of bacterial cells grown on medium without arsenate (control). Blue color represents protein signals, whereas dark brown color stands for background activity of arsenics. (h) Depth profiling of bacterial cells grown on medium containing arsenate. Blue, orange, red, and dark brown colors represent protein signals, meta-arsenite, orthoarsenite, and arsenate, respectively. (From Rahman, A. et al., *J. Environ. Sci. Health Part A*, 49, 1349, 2014. With permission.)

31.12 EFFECT OF ARSENIC ON BACTERIAL MORPHOLOGY

The morphological analysis of bacteria in the presence and absence of metals is carried out by a scanning electron microscope (SEM) with an attached x-ray energy-dispersive system (EDS). To perform the SEM-EDS, bacterial cells are grown in the presence of arsenic and are rinsed to remove unbound arsenic from the cell surface. The cells are fixed and dehydrated in ethanol and subjected to SEM-EDS. The cellular morphology reveals distinct changes in the presence of arsenic, where the cells become elongated and aggregate to support the membrane transport process hampered due to metal stress (Vijayakumar et al., 2011). Nongrowing but live cells when suspended in arsenic-containing water aggregate to form clumps. Biofilm formation or aggregation under stress conditions is a common phenomenon; however, this is unusual for nongrowing but living cells. These changes in morphology might be a possible strategy for cells to accumulate arsenic inside them. Rahman et al. (2014) reported a significant elongation of bacteria when studying the effects of arsenate on the *L. sphaericus*. Also, Desale et al. (2014) reported morphological changes of bacteria when studying the effects of nickel on *Lysinibacillus* sp. Several other researchers have shown bacterial elongation in the presence of toxic heavy metals (Nepple et al., 1999; Banerjee et al., 2011).

REFERENCES

Abedin, M. J., Feldmann, J., and Meharg, A. A. 2002. Uptake kinetics of arsenic species in rice plants. *Plant Physiology* 128: 1120–1128.

Achour, A. R., Bauda, P., and Billard, P. 2007. Diversity of arsenite transporter genes from arsenic-resistant soil bacteria. *Research in Microbiology* 158: 128–137.

Ahmed, K., Akhand, A. A., Hasan, M., Islam, M., and Hasan, A. 2008. Toxicity of arsenic (sodium arsenite) to fresh water spotted snakehead *Channa punctatus* (bloch) on cellular death and DNA content. *Agricultural and Environmental Science* 4: 18–22.

Argos, M., Kalra, T., Rathouz, P. J., Chen, Y., Pierce, B., Parvez, F., Islam, T. et al. 2010. Arsenic exposure from drinking water, and all-cause and chronic-disease mortalities in Bangladesh (HEALS): A prospective cohort study. *The Lancet* 376: 252–258.

Arora, M., Kiran, B., Rani, S., Rani, A., Kaur, B., and Mittal, N. 2008. Heavy metal accumulation in vegetables irrigated with water from different sources. *Food Chemistry* 111: 811–815.

Bahar, M. M., Megharaj, M., and Naidu, R. 2012. Arsenic bioremediation potential of a new arsenite-oxidizing bacterium *Stenotrophomonas* sp. MM-7 isolated from soil. *Biodegradation* 23: 803–812.

Bailar Jr, J. C., Johnson, W. C., and Chenicek, A. G. 1939. Arsenic triiodide. In: Booth, H. S. (ed.) *Inorganic Syntheses*. New York: The McGraw-Hill Book Company, Inc., John Wiley & Sons, Inc., Hoboken, NJ Vol. 1, pp. 103–104. doi: 10.1002/9780470132326.ch36.

Balasubramanian, N., Kojima, T., Ahmed, B. C., and Srinivasakannan, C. 2009. Removal of arsenic from aqueous solution using electrocoagulation. *Journal of Hazardous Materials* 167: 966–969.

Banerjee, S., Datta, S., Chattyopadhyay, D., and Sarkar, P. 2011. Arsenic accumulating and transforming bacteria isolated from contaminated soil for potential use in bioremediation. *Journal of Environmental Science and Health Part A* 46: 1736–1747.

Banerjee, S., Majumdar, J., Samal, A. C., Bhattachariya, P., and Santra, S. C. 2013. Biotransformation and bioaccumulation of arsenic by *Brevibacillus brevis* isolated from arsenic contaminated region of West Bengal. *Journal of Environmental Science, Toxicology and Food Technology* 3: 1–10.

Brooks, W. E. 2008. *Minerals Year Book*. Arsenic (Advance Release). Reston, VA: U.S. Geological Survey.

Christopher, O. A. and Haque, A. M. M. 2012. Arsenic contamination in irrigation water for rice production in Bangladesh: A review. *Trends in Applied Sciences Research* 7: 331–349.

Cöl, M., Cöl, C., Soran, A., Sayli, B. S., and Ozturk, S. 1999. Arsenic-related Bowen's disease, palmar keratosis, and skin cancer. *Environmental Health Perspective* 107: 687–689.

Congeevaram, S., Dhanarani, S., Park, J., Dexilin, M., and Thamaraiselvi, K. 2007. Biosorption of chromium and nickel by heavy metal resistant fungal and bacterial isolates. *Journal of Hazardous Materials* 146: 270–277.

Corsini, A., Zaccheo, P., Muyzer, G., Andreoni, V., and Cavalca, L. 2014. Arsenic transforming abilities of groundwater bacteria and the combined use of *Aliihoeflea* sp. strain 2WW and goethite in metalloid removal. *Journal of Hazardous Materials* 269: 89–97.

Das, S., Santra, A., Lahiri, S., and Guha, M. D. N. 2005. Implications of oxidative stress and hepatic cytokine (TNF-alpha and IL-6) response in the pathogenesis of hepatic collagenesis in chronic arsenic toxicity. *Toxicology Applied Pharmacology* 204: 18–26.

de Crécy, E., Jaronski, S., Lyons, B., Lyouns, T. J., and Keyhani, N. O. 2009. Directed evolution of a filamentous fungus for thermotolerance. *BMC Biotechnology* 9: 74.

de Crécy, E., Metzgar, D., Allen, C., Pénicaud, M., Lyons, B., Hansen, C. J., and de Crécy-Lagard, V. 2007. Development of a novel continuous culture device for experimental evolution of bacterial populations. *Applied Microbiology and Biotechnology* 77: 489–496.

Desale, P., Kashyap, D., Nawani, N., Nahar, N., Rahman, A., Kapadnis, B., and Mandal, A. 2014. Biosorption of nickel by *Lysinibacillus* sp. BA2 native to bauxite mine. *Ecotoxicology and Environmental Safety* 107: 260–268.

Diorio, C., Cai, J., Marmor, J., Shinder, R., and DuBow, M. S. 1995. An *Escherichia coli* chromosomal *ars* operon homolog is functional in arsenic detoxification and is conserved in gram-negative bacteria. *Journal of Bacteriology* 177: 2050–2056.

Doenmez, G. and Aksu, Z. 2001. Bioaccumulation of copper (II) and Nickel (II) by the non-adapted and adapted growing *Candida* sp. *Water Research* 35: 1425–1434.

Eblin, K. E., Bowen, M. E., Cromey, D. W., Bredfeldt, T. G., Mash, E. A., Lau, S. S., and Gandolfi, A. J. 2006. Arsenite and monomethylarsonous acid generate oxidative stress response in human bladder cell culture. *Toxicology and Applied Pharmacology* 217: 7.

Emsley, J. 2011. *Arsenic. Nature's Building Blocks: An A-Z Guide to the Elements*. Oxford, U.K.: Oxford University Press, pp. 47–55.

Fageria, N. K. 2007. Yield physiology of rice. *Journal of Plant Nutrition* 30: 843–879.

Fowler, T. 1786. Medical reports of the effects of arsenic in the cure of agues, remittent fevers, and periodic headaches. London, U.K.: Johnson, pp. 105–107.

Gibaud, S. and Gérard, J. 2010. Arsenic based drugs: From Fowler's solution to modern anticancer chemotherapy. *Topics in Organometallic Chemistry* 32: 1–20.

Greipsson, S. 2011. Phytoremediation. *Nature Education Knowledge* 3(10): 7.

Halder, D., Bhowmick, S., Biswas, A., Mandal, U., Nriagu, J., Guha, M. D. N., Chatterjee, D., and Bhattacharya, P. 2012. Consumption of brown rice: A potential pathway for arsenic exposure in rural Bengal. *Environmental Science and Technology* 46: 4142–4148.

Halttunen, T., Finell, M., and Salminen, S. 2007. Arsenic removal by native and chemically modified lactic acid bacteria. *International Journal of Food Microbiology* 120: 173–178.

Hamamura, N., Macur, R. E., Korf, S., Ackerman, G., Taylor, W. P., Kozubal, M., Reysenbach, A. L., and Inskeep, W. P. 2009. Linking microbial oxidation of arsenic with detection and phylogenetic analysis of arsenite oxidase genes in diverse geothermal environments. *Environmental Microbiology* 11: 421–431.

Handley, K. M., Héry, M., and Lloyd, J. R. 2009. Redox cycling of arsenic by the hydrothermal marine bacterium *Marinobacter santoriniensis*. *Environmental Microbiology* 11: 1601–1611.

Harms, H., Schlosser, D., and Wick, L. Y. 2011. Untapped potential: Exploiting fungi in bioremediation of hazardous chemicals. *Nature Reviews Microbiology* 9: 177–192.

Hazrat, A., Khan, E., and Sajad M. A. 2013. Phytoremediation of heavy metals—Concepts and applications. *Chemosphere* 91: 869–881.

Hossain, M. M., Sattar, M. A., Hashem, M. A., and Islam, M. R. 2003. Arsenic status at different depths in some soil of Bangladesh. *Journal of Biological Sciences* 1: 1116–1119.

Hughes, M. F. 2002. Arsenic toxicity and potential mechanisms of action. *Toxicology Letters* 133: 1.

Jensen, T. J., Novak, P., Eblin, K. E., Gandolfi, J. A., and Futscher, B. W. 2008. Epigenetic remodeling during arsenical-induced malignant transformation. *Carcinogenesis* 29: 1500–1508.

Johnson, B. L. and DeRosa, C. T. 1995. Chemical mixtures released from hazardous waste sites: Implications for health risk assessment. *Toxicology* 105: 145–156.

Kao, A. C., Chu, Y. J., Hsu, F. L., and Liao H. V. C. 2013. Removal of arsenic from groundwater by using a native isolated arsenite-oxidizing bacterium. *Journal of Contaminant Hydrology* 155: 1–8.

Krumova, K., Nikolosova, M., and Groudeva, V. 2008. Isolation and identification of arsenic-transforming bacteria from arsenic contaminated sites in Bulgaria. *Biotechnology and Biotechnological Equipment* 22: 721–728.

Kulp, T. R., Han, S., Saltikov, C. W., Lanoil, B. D., Zargar, K., and Oremland, R. S. 2007. Effects of imposed salinity gradients on dissimilatory arsenate reduction, sulfate reduction, and other microbial processes in sediments from two California soda lakes. *Applied and Environmental Microbiology* 73: 5130–5137.

Kulp, T. R., Hoeft, S. E., Asao, M., Madigan, M. T., Hollibaugh, J. T., Fisher, J. C., Stolz, J. F., Culbertson, C. W., Miller, L. G., and Oremland, R. S. 2008. Arsenic (III) fuels anoxygenic photosynthesis in hot spring biofilms from Mono Lake, California. *Science* 321: 967–970.

Liao, V. H. C., Chu, Y. J., Su, Y. C., Hsiao, S. Y., Wei, C. C., Liu, C. W., Liao, C. M., Shen, W. C., and Chang, F. J. 2011. Arsenite-oxidizing and arsenate-reducing bacteria associated with arsenic-rich groundwater in Taiwan. *Journal of Contamination and Hydrology* 123: 20–29.

Lozano, L. C. and Dussan, J. 2013. Metal tolerance and larvicidal activity of *Lysinibacillus sphaericus*. *World Journal of Microbiology and Biotechnology* 29: 1383–1389.

Malasarn, D., Saltikov, C. W., Campbell, K. M., Santini, J. M., Hering, J. G., and Newman, D. K. 2004. *arrA* is a reliable marker for As(V) respiration. *Science* 306: 455.

Mandal, B. K. and Suzuki, K. T. 2002. Arsenic round the world: A review. *Talanta* 58: 201–235.

Meliker, J. R., Wahl, R. L., Cameron, L. L., and Nriagu, J. O. 2007. Arsenic in drinking water and cerebrovascular disease, diabetes mellitus, and kidney disease in Michigan: A standardized mortality ratio analysis. *Environmental Health* 6: 4.

Mondal, P., Majumder, C. B., and Mohanty, B. 2008. Growth of three bacteria in arsenic solution and their application for arsenic removal from wastewater. *Journal of Basic Microbiology* 48: 521–525.

Moussavi, G. and Barikbin, B. 2010. Biosorption of chromium (VI) from industrial wastewater onto *Pistachio hull* waste biomass. *Chemical Engineering Journal* 162: 893–900.

Nahar, N., Rahman, A., Moś, M., Warzecha, T., Ghosh, S., Hossain, K., Nawani, N. N., and Mandal, A. 2014. In-silico and in-vivo studies of molecular structures and mechanisms of AtPCS1 protein involved in binding arsenite and/or cadmium in plant cells. *Journal of Molecular Modelling* 20: 2104.

Neidhardt, S. N., Tang, X., Guo, H., and Stuben, D. 2012. Impact of irrigation with high arsenic burden groundwater on the soil-plant system: Result from a case study in the Inner Mongolia, China. *Environmental Pollution* 163: 8–13.

Nepple, B. B., Flynn, I., and Bachofen, R. 1999. Morphological changes in phototrophic bacteria induced by metalloid oxyanions. *Microbiological Research* 154: 191–198.

Nicholis, I., Curis, E., Deschamps, P., and Bénazeth, S. 2009. Arsenite medicinal use, metabolism, pharmacokinetics and monitoring in human hair. *Biochimie* 91: 1260–1267.

Owolabi, J. B. and Rosen, B. P. 1990. Differential mRNA stability controls relative gene expression within the plasmid-encoded arsenical resistance operon. *Journal of Bacteriology* 172: 2367–2371.

Pandey, J. and Pandey, U. 2009. Accumulation of heavy metals in dietary vegetables and cultivated soil horizon in organic farming system in relation to atmospheric deposition in a seasonally dry tropical region of India. *Environmental Monitoring and Assessment* 148: 61–74.

Rahman, A., Nahar, N., Nawani, N. N., Jass, J., Desale, P., Kapadnis, B. P., Hossain, K. et al. 2014. Isolation of a *Lysinibacillus* strain B1-CDA showing potentials for arsenic bioremediation. *Journal of Environmental Science and Health Part A* 49: 1349–1360.

Rahman, A., Nahar, N., Nawani, N. N., Jass, J., Hossain, K., Alam, Z. A., Saha, A. K., Ghosh, S., Olsson, B., and Mandal, A. 2015. Bioremediation of hexavalent chromium (VI) by a soil borne bacterium, *Enterobacter cloacae* B2-DHA. *Journal of Environmental Science and Health Part A* 50: 1136–1147.

Rahman, M. M., Owens, G., and Naidu, R. 2009. Arsenic levels in rice grain and assessment of daily dietary intake of arsenic from rice in arsenic-contaminated regions of Bangladesh-implications to groundwater irrigation. *Environmental Geochemistry and Health* 1: 179–187.

Reimer, K. J., Koch, I., and Cullen, W. R. 2010. Organoarsenicals. Distribution and transformation in the environment. *Metal Ions in Life Sciences* 7: 165–229.

Rey, N. A., Howarth, O. W., and Pereira-Maia, E. C. 2004. Equilibrium characterization of the As(III)-cysteine and the As(III)-glutathione systems in aqueous solution. *Journal of Inorganic Biochemistry* 98: 1151.

Rhine, E. D., Ní Chadhain, S. M., Zylstra, G. J., and Young, L. Y. 2007. The arsenite oxidase genes (aroAB) in novel chemoautotrophic arsenite oxidizers. *Biochemistry and Biophysics Research Communications* 354: 662–667.

Rossman, T. G. and Klein, C. B. 2011. Genetic and epigenetic effects of environmental arsenicals. *Metallomics: Integrated Biometal Science* 3: 1135–1141.

Rossman, T. G., Uddin, A. N., and Burns, F. J. 2008. Evidence that arsenite acts as a co-carcinogen in skin cancer. *Toxicology and Applied Pharmacology* 198: 394–404.

Sharma, V. K. and Sohn, M. 2009. Aquatic arsenic: Toxicity, speciation, transformations, and remediation. *Environmental International* 35: 743–759.

Silar, P., Dairou, J., Cocaign, A., Busi, F., Rodrigues-Lima, F., and Dupret, J. M. 2011. Fungi as a promising tool for bioremediation of soils contaminated with aromatic amines, a major class of pollutants. *Nature Reviews Microbiology* 9: 477.

Silver, S. and Phung, L. T. 2005. Genes and enzymes involved in bacterial oxidation and reduction of inorganic arsenic. *Applied and Environmental Microbiology* 71: 599–608.

Singh, A., Kuhad, R. C., and Ward, O. P. 2009. *Advances in Applied Bioremediation*, 1st edn. Berlin, Germany: Springer, p. 2.

Smith, *Industrial & Engineering Chemistry* 11: 109; *Reisener in Ullmann's Encyklopadie der Technischen Chemie.* Munich, Germany: Urban and Schwarzenberg, 1953, Vol. 3 p. 850; Brauer, G. (Ed.) 1963. *Schenk in Handbook of Preparative Inorganic Chemistry*, 2nd edn. New York: Academic Press, Vol. 1, pp. 596, 1919.

Song, B., Chyun, E., Jaffé, P. R., and Ward, B. B. 2009. Molecular methods to detect and monitor dissimilatory arsenate-respiring bacteria (DARB) in sediments. *FEMS Microbiology Ecology* 68: 108–117.

U.S. EPA. 1998. Human health risk assessment protocol for hazardous waste combustion facilities, Vol. 2, peer review draft. Office of Solid waste and Emergency response. EPA-542-R97-004, p. 8. United State.

Vijayakumar, G., Tamilarasan, R., and Dharmendra, M. K. 2011 Removal of Cd^{2+} ions from aqueous solution using live and dead *Bacillus subtilis. England Research Bulletin* 15: 18–24.

Volesky, B. 1990. *Biosorption of Heavy Metals.* Boca Raton, FL: CRC Press.

Warrell, R. P., Soignet, S. L., Gabrilove, J. L., Calleja, E. M., Maslak, P., Scheinberg, D. A., Chen, Y. W., and Pandolfi, P. P. 1998. Initial western study of arsenic trioxide (As$_2$O$_3$) in acute promyelocytic leukemia (APL). *Proceedings of American Society of Clinical Oncology* 17: 6a.

Williams, P. N., Islam, M. R., Adomako, E. E., Raab, A., Hossain, S. A., Zhu, Y. G., and Meharg, A. A. 2006. Increase in rice grain arsenic for regions of Bangladesh irrigating paddies with elevated arsenic in groundwater. *Environmental Science and Technology* 40: 4903–4908.

Williams, P. N., Villada, A., Deacon, C., Raab, A., Figuerola, J., Green, A. J., Feldmann, J., and Meharg, A. A. 2007. Greatly enhanced arsenic shoot assimilation in rice leads to elevated grain levels compared to wheat and barley. *Environmental Science and Technology* 41: 6854–6859.

Zavala, Y. J. and Duxbury, J. M. 2008. Arsenic in rice: I. Estimating normal levels of total arsenic in rice grain. *Environmental Science and Technology* 42: 3856–3860.

Zavala, Y. J., Gerads, R., Gürleyük, H., and Duxbury, J. M. 2008. Arsenic in rice: II. Arsenic speciation in USA grain and implications for human health. *Environmental Science & Technology* 42: 3861–3866.

Zhao F. J., McGrath, S. P., and Meharg, A. 2010. Arsenic as a food chain contaminant, mechanisms of plant uptake and metabolism and mitigation strategies. *Annual Review in Plant Biology* 61: 535–559.

Zhu, Y., Bi D., Yuan, L., and Yin, X. 2012. Phytoremediation of cadmium and copper contaminated soils. In: Yin, X. and Yuan, L. (Eds.) *Phytoremediation and Biofortification: Two Sides of One Coin.* Springer Briefs in Molecular Science. pp. 75–81.

32 Microbial Transformations of Arsenic
From Metabolism to Bioremediation

Odile Bruneel, Marina Héry, Elia Laroche, Ikram Dahmani,
Lidia Fernandez-Rojo, and Corinne Casiot

CONTENTS

ABSTRACT

Due to the toxicity of this metalloid, arsenic contamination currently represents one of the most severe threats for the environment and public health in many countries worldwide with millions of people chronically exposed, mainly in Southeast Asia. Redox transformations strongly impact arsenic environmental fate, since its solubility bioavailability and toxicity depend on its speciation. Microorganisms play a key role in these transformations with the presence in the environment of many microorganisms able to metabolize this toxic element using detoxification or energy conservation reactions based on redox reactions, methylation, or demethylation. This chapter highlights the current knowledge concerning the ecology, physiology, and genomics of these microorganisms and provides an update on the development of processes removing arsenic present in soil and water based on microbial transformations.

32.1 INTRODUCTION

Arsenic is a naturally occurring element with a long history of toxicity (Hettick et al. 2015). This element is ubiquitous in the environment (Cullen and Reimer 1989), with abundance in the earth's crust of around 0.0001% (Yamamura and Amachi 2014). However, local concentrations can vary a lot from one site to another depending on the geological and physicochemical context (Oremland and Stolz 2003). Millions of people worldwide (particularly in Southeast Asia) are exposed to drinking water exceeding the recommended limit by the World Health Organization (WHO) (10 µg/L) (Hossain 2006). The consumption of rice irrigated with As-contaminated water is another important source of human contamination (Williams et al. 2006, Yamamura and Amachi 2014). Human inputs contributed also extensively to environmental arsenic contamination through mining and industrial activities, waste processing, wood preservative, and pesticide and herbicide manufacturing and application (Nordstrom 2002, Belluck et al. 2003). Water resource can also be affected by arsenic contamination associated with generation of acid mine drainage (AMD) originating from mining wastes (Hallberg 2010, Drewniak and Sklodowska 2013). Such phenomenon represents a severe threat for both the environment and public health and precludes the use of water for domestic and agricultural purposes (Johnson and Hallberg 2005, Hudson-Edwards and Santini 2013).

Arsenic occurs with valence states of As(-III) (arsine, AsH$_3$), As(0) (metallic), As(III) (arsenite, H$_3$AsO$_3{}^0$, and H$_2$AsO$_3{}^-$), and As(V) (arsenate, H$_2$AsO$_4{}^-$, and HAsO$_4{}^{2-}$) depending on the environmental conditions (Oremland and Stolz 2003). Arsenate is the predominant form in soil and surface water, while arsenite prevails in reducing conditions in anaerobic groundwater (Cavalca et al. 2013).

Prokaryotes and eukaryotes have developed diverse resistance and metabolic mechanisms to deal with the presence of As in their environment (Yamamura and Amachi 2014). Microbial interactions with arsenic (e.g., based on methylation, demethylation, or redox reactions) modify arsenic speciation and thus its mobility and toxicity. Thus, environmental microorganisms play a key role in arsenic biogeochemical cycling in both aquatic and terrestrial ecosystems. For example, As(III) can be oxidized in the less toxic form As(V), which is also less soluble and more likely to be removed from a liquid phase by precipitation (Oremland and Stolz 2005). These microbial processes represent thus a great potential to develop cost-effective and efficient bioremediation techniques for arsenic removal from contaminated environments. In this review, we will describe the microbial transformations of arsenic at a metabolic, molecular, and genetic level. We will also discuss the environmental impacts and significance of these processes and then their potential for bioremediation applications.

32.2 GENERALITIES: BACKGROUND

Arsenic is a ubiquitous element originating from natural or anthropogenic sources present in the atmosphere, pedosphere, hydrosphere, and biosphere (Mandal and Suzuki 2002, Lievremont et al. 2009). This toxic metalloid, responsible for the contamination of the environment, poses a major risk to human health.

32.2.1 Origins of As

32.2.1.1 Natural Origins

Despite the quantitative uncertainty regarding arsenic fluxes, As is a ubiquitous element, being the 20th most abundant element in the earth's crust, with a background concentration in soils generally less than 15 mg kg^{-1} (Cullen and Reimer 1989, Oremland and Stolz 2003, Yamamura and Amachi 2014).

This metalloid (having properties of metals and nonmetals) is mainly associated with sulfur, iron, and various other metals (especially Au, Ag, Cu, Sb, Ni, Co; Cullen and Reimer 1989, Lievremont et al. 2009) and has been detected in more than 245 different minerals (Mandal and Suzuki 2002, Hoang et al. 2010).

The most important source of As in the environment (hydrosphere, pedosphere, biosphere, and atmosphere) is the release of arsenic originating from As-enriched minerals (Singh et al. 2015). Levels of As in unpolluted surface water and groundwater vary generally from 1 to 10 μg L^{-1}; however, concentration of As strongly depends on lithology (Smedley and Kinniburgh 2002, Ahoulé et al. 2015, Singh et al. 2015). Arsenic concentration varies from 0.15 to 0.45 μg L^{-1} in freshwater systems (rivers and lakes) (Smedley and Kinniburgh 2002, Bissen and Frimmel 2003a,b) and represents usually less than 2 μg L^{-1} in marine waters (Singh et al. 2015). The average concentration in igneous and sedimentary rocks is 2 mg kg^{-1} and from 0.5 to 2.5 mg kg^{-1} in most rocks (Mandal and Suzuki 2002). The U.S. Environmental Protection Agency fixed the permissible limit of arsenic in soil to 24 mg kg^{-1} (Singh et al. 2015). In the atmosphere, As is originating from wind erosion, volcanic emissions, sea spray, forest fires, and the volatilization from biological processes such as biomethylation (Cullen and Reimer 1989). Because arsenic is widely distributed and can be cumulative in living tissue, high concentration can be found in living organisms. In plants, the levels of arsenic range from 0.001 to 5 mg of arsenic per gram of dry matter and vary according to the species and the type of soil involved (Fitz and Wenzel 2002, Lievremont et al. 2009, Alvarez-Ayuso et al. 2016). Domestic animals and humans generally contain less than 0.3 μg g^{-1} on a wet weight basis, and the total As content in human body reaches 3–4 mg and tends to increase with age (Mandal and Suzuki 2002).

32.2.1.2 Anthropogenic Origins

Arsenic enters into the composition of a large number of compound originating from different sectors (mining/smelting precious metal, burning of fossil fuels, power plants, pharmaceuticals, wood treatment, glassmaking industry, electronics industry, chemical weapons, etc.; Mandal and Suzuki 2002). The burning of coal, leading to arsenic volatilization, represents the main anthropogenic source of atmospheric arsenic. Arsenic retention time in the atmosphere was estimated to be less than 10 days before precipitation by rainfall (Cullen and Reimer 1989, Matschullat 2000). In agriculture, sodium arsenite has been extensively used for protecting grapevines from excoriosis, until 2001 in France (Grillet et al. 2004, Lievremont et al. 2009). Arsenic is also used for the preparation of insecticides, pesticides, herbicides, desiccants, and fertilizers (Cullen and Reimer 1989, Mandal and Suzuki 2002), and despite the huge controversy, As is still used in the fabrication of wood preservatives in the United States (Bissen and Frimmel 2003a). Phenylarsonic acid compounds like roxarsone have been widely used as additives to animal feeds to control parasites and promote growth (Jackson et al. 2006).

Banned in the United States in the last few years (USFDA 2014), these compounds are always widely used in chicken feed or swine production in many countries like China (Nachman et al. 2013, Liu et al. 2015). Medicinal use of arsenic dates back from more than 2400 years in Greece and Rome (Miller et al. 2002), and until the discovery of antibiotics, arsenic compounds were widely used in therapeutic treatment for a variety of illnesses such as leukemia, psoriasis, rheumatism, arthritis, asthma, malaria, trypanosomiasis, tuberculosis, and diabetes (Bissen and Frimmel 2003a). Today, arsenic compounds are used in the manufacturing of glass to improve hardness and corrosion resistance and are currently used in the semiconductor industry and in the production of catalysts (Bissen and Frimmel 2003a).

Mine tailings can contain high concentrations of arsenic that may contaminate groundwater and surface waters (Williams 2001, Drewniak and Sklodowska 2013, Hudson-Edwards and Santini 2013). Arsenic-bearing sulfide minerals present in these wastes can be naturally subject to oxidation when brought into contact with air and water leading to the formation of AMD. These waters, characterized by acid pH and high concentrations of iron, sulfate, and numerous toxic elements, constitute one of the main environmental problems arising from mining industries (Johnson and Hallberg 2003, Hallberg 2010, Klein et al. 2013). As a consequence, high concentration of arsenic can be found in Tinto and Odiel Rivers in Spain (reaching up to 800 $\mu g\ L^{-1}$; Sanchez-Rodas et al. 2005). In the Carnoulès mine in France, As concentration reached 10 g L^{-1} of As in subsurface waters draining the tailings stock, and 80–350 mg L^{-1} is present at the spring of the Reigous, a small creek draining the site (Casiot et al. 2003a,b).

32.2.2 Chemical Properties (Speciation/Solubility)

Arsenic occurs in four oxidation states (−3, 0, +3, +5), as various inorganic or organic forms. The two highest oxidation states predominate in the environment, whereas the two lowest are rare. Under oxidizing conditions, As(V) is the prevailing form as oxyanions: $H_2AsO_4^-$ (pH = 2–7) and $HAsO_4^{2-}$ (pH > 7). Arsenite is more present under reducing conditions, with uncharged and charged forms (respectively, H_3AsO_3 at pH = 0–9 and $H_2AsO_3^-$ at pH > 9.2). Arsenate sorbs strongly to a range of solid phases, while arsenite is more mobile (Smedley and Kinniburgh 2002). Furthermore, arsenic toxicity is highly dependent on its speciation, with the following toxicity gradient: dimethylarsenite (DMA(III)) > monomethylarsenite (MMA(III)) > inorganic arsenite > inorganic arsenate > dimethylarsenate (DMA(V)); monomethylarsenate (MMA(V)) (Hirano et al. 2004).

The solubility of arsenic and its resulting bioavailability are also closely related to its speciation (Lievremont et al. 2009). For example, the reduction of arsenate into arsenite results in the mobilization of this element into the environment (Oremland and Stolz 2003, 2005, Sarkar et al. 2014). Arsenate may be sequestered by coprecipitation with ferric iron or sulfur or can be adsorbed by clay, calcite, organic matter, or hydroxides, in particular ferric oxyhydroxides (Morin et al. 2003, Klein et al. 2014). The remediation procedures currently used are generally based on the chemical oxidation of arsenite into arsenate, followed by As(V) immobilization (Lievremont et al. 2009).

32.2.3 Toxicity and Health-Related Problems

Environmental contamination with toxic element like arsenic represents a great concern due to the serious health impact of this metalloid on living organisms. Arsenic was one of the first elements recognized as a cause of cancer (Smith et al. 2002), and the International Agency for Research on Cancer classified it, since 2012, as "carcinogenic to humans" (group 1; Ungureanu et al. 2015). Arsenic is known to exert mutagenic, cytotoxic, and genotoxic effects, and chronic exposure to inorganic As in human causes several diseases from skin lesions (melanosis or hyperkeratosis) to hematological disorders, immunological, neurological, and reproductive alterations, cardiovascular disease, or different types of cancers (liver, kidney, brain, lung, bladder, or stomach; Mandal and Suzuki 2002, Argos et al. 2010, Naujokas et al. 2013, Paul et al. 2013, Agusa et al. 2014, Sun et al. 2014, Mohammed et al. 2015).

The ingestion of contaminated drinking water and food (e.g., rice) represent the main sources of As intake for humans (Nordstrom 2002, Rahman et al. 2014). The WHO established in 1993 10 $\mu g\ L^{-1}$ as the guideline value for arsenic in drinking water (Ungureanu et al. 2015). Currently, around 150 million people worldwide suffer from severe and chronic arsenic poisoning (Chowdhury et al. 2000, Chakraborti et al. 2003, Agusa et al. 2014). The impact of As toxicity is particularly alarming in Bangladesh or India (Mukherjee and Bhattacharya 2001, Rahman et al. 2001, Chakraborti et al. 2004, Bhattacharya et al. 2007). Geogenic As was also retrieved in many countries over the world like Nepal (Tandukar et al. 2006), Vietnam, Cambodia (Berg et al. 2007, Merola et al. 2015), or Pakistan (Nickson et al. 2005). Arsenic is also found in highly geographically dispersed areas like in the United States and Canada or in Latin America such as Mexico, Argentina, Bolivia, Brazil, or Nicaragua (Matschullat 2000, Mandal and Suzuki 2002, Nordstrom 2002, Smedley and Kinniburgh 2002; Bhattacharya et al. 2007, Naujokas et al. 2013).

32.3 MICROBIAL TRANSFORMATIONS

Arsenic is a major contaminant of many ecosystems worldwide and is responsible for serious public health issues. However, life, and microorganisms in particular, has developed various strategies for coping with this toxic element, enabling them to resist and metabolize this chemical (Andres and Bertin 2016). Many microorganisms have the capacity to mediate redox transformations of As via As(III) oxidation and As(V) reduction, and a large variety of As(V)-reducing and As(III)-oxidizing prokaryotes have been isolated from different As-contaminated as well as uncontaminated environments (Oremland and Stolz 2003, 2005,

Yamamura et al. 2009, Heinrich-Salmeron et al. 2011, Jareonmit et al. 2012, Sarkar et al. 2013). Since As(V) reduction and As(III) oxidation directly affect the mobility and the bioavailability of this toxic metalloid, microbial activities play a central role in As cycling and have the ability to promote As removal from contaminated soils and waters (Yamamura and Amachi 2014). Many review papers and book chapters were dedicated to microbial transformation of arsenic within the last few years (Stolz et al. 2006, Bhattacharya et al. 2007, Haferburg and Kothe 2007, Lievremont et al. 2009, Tsai et al. 2009, Stolz et al. 2010, Rahman et al. 2012, Slyemi and Bonnefoy 2012, Bahar et al. 2013, Cavalca et al. 2013, Drewniak and Sklodowska 2013, Hudson-Edwards and Santini 2013, Kruger et al. 2013, Van Lis et al. 2013, Dopson and Holmes 2014, Huang 2014, Rahman and Hassler 2014, Rahman et al. 2014, Yamamura and Amachi 2014, etc.). The following paragraphs review the latest findings related to microbial As metabolism at a metabolic, molecular, and genetic level and highlight the potential use of these processes in bioremediation techniques. A general and schematic representation of microbial transformations of arsenic is presented in Figure 32.1.

32.3.1 Microbial Reduction of As(V)

32.3.1.1 Reduction of Arsenic Based on Detoxification Mechanisms

Arsenic is one of the earliest toxic elements that microorganisms had to deal with in their environment. Because of its structure similarity with phosphate, As(V) can enter the cell through phosphate transporters (Rosen and Liu 2009) (e.g., Pit and Pst in *Escherichia coli* (Rosenberg et al. 1977, Willsky and Malamy 1980) and Pho87p in *Saccharomyces cerevisiae* (Persson et al. 1999)). Inside the cell, As(V) inhibits the oxidative phosphorylation, stopping the main energy-generating system (Oremland and Stolz 2003, Rahman et al. 2014). Because uncharged As(III) mimics glycerol, it can enter the cells through aquaglyceroporins as GlpF in *E. coli* (Meng et al. 2004), GlpF homologue in *Leishmania major* (Gourbal et al. 2004), and FpsIp in yeast (Wysocki et al. 2001). As(III) binds to thiols and vicinal sulfhydryl groups blocking important proteins, such as pyruvate dehydrogenase and 2-oxoglutarate dehydrogenase, and, therefore, the respiration (Oremland and Stolz 2003). The most common microbial detoxification strategy is based on the extrusion of As(III) from the microbial

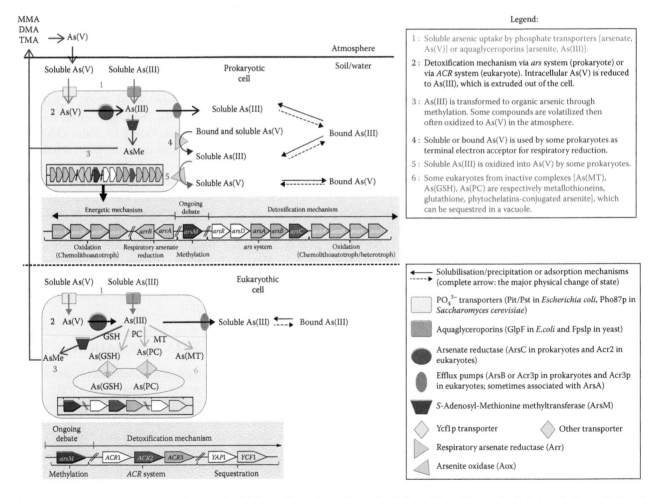

FIGURE 32.1 Schematic representation of microbial transformations of arsenic. (Adapted from Bossy, A., Origines de l'arsenic dans les eaux, sols et sediments du district aurifere de St-Yrieix-la-Perche (Limousin, France): Contribution du lessivage des phases porteuses d'arsenic, These de doctorat. Universite de Tours, Tours, France, 2010; Tsai, S.L. et al., *Curr. Opin. Biotechnol.*, 20, 659, 2009. With permission.)

cell by an efflux pump. Elimination of As(V) requires prior reduction to As(III) by an arsenate reductase, occurring both under aerobic and anaerobic conditions (Macur et al. 2004). Distinct systems described in prokaryotes and eukaryotes display the same detoxification function; they originated from convergent evolution or horizontal gene transfer according to different hypotheses (Bobrowicz et al. 1997, Mukhopadhyay et al. 2002, Cai et al. 2009).

32.3.1.1.1 The Ars System in Prokaryotes

The most common and well-known arsenic resistance system is based on the prokaryotic *ars* operon located on the chromosome or a plasmid. The minimal operon *arsRBC* is found in the pI258 and pSX267 plasmids of *Staphylococcus aureus* (Silver 1998), in the chromosome of *E. coli* (Carlin et al. 1995), or *Pseudomonas fluorescens* (Prithivirajsingh et al. 2001), whereas the R773 plasmid of *E. coli* contains the enlarged operon *arsRDABC* (Silver 1998, Chen et al. 1986). Operon *arsRDABC* may have evolved from the operon *arsRBC*, which originated from a primitive operon *arsRB* (Rosen 1999). Other genomic configurations and genes can be present, that is, *arsM*, *arsH*, *arsO*, and *arsT* encoding, respectively, an S-adenosylmethionine methyltransferase, an unknown protein, a monooxygenase, and a thioredoxin reductase (Wang et al. 2006). However, the *ars* operon always contains three genes encoding a transcriptional regulator (*arsR*, *arsD*), an arsenate reductase (*arsC*), and an efflux system (*arsB*, *arsA*).

ArsR and ArsD are the most studied transcriptional regulators, acting as repressors. ArsR is a transcriptional repressor of the ArsR/SmtB family. In *E. coli*, arsenite induces the transcription of the *ars* operon by separating ArsR from the DNA (Wu and Rosen 1993a). ArsD stops the operon transcription when its own concentration exceeds a certain level. Moreover, ArsD acts as an arsenite chaperone for the ArsAB pump (Wu and Rosen 1993b, Lin et al. 2006).

The arsenate reductase is a small cytoplasmic protein (13–15 kDa) catalyzing the reduction of As(V) to As(III). Three clades of arsenate reductases exist, presenting a low sequence homology (Mukhopadhyay et al. 2002): two different prokaryotic ArsC and one eukaryotic Acr2. In gram-negative bacteria such as *E. coli*, ArsC uses glutathione and glutaredoxin as a source of reducing power (Gladysheva et al. 1994). Thioredoxin-dependent arsenate reductase ArsC is found in gram-positive bacteria (*S. aureus*) and some *Proteobacteria* (Ji and Silver 1992, Butcher et al. 2000).

Arsenite is excreted out of the cell by two major families of efflux pump: ArsB or Acr3p (also called Arr3p). Acr3p is highly prevalent in the environment according to recent studies (Rosen 1999, Achour et al. 2007, Yang et al. 2015). It is widespread in both eukaryotes (yeast, fungi, and some plants) and prokaryotes (bacteria and archaea), whereas ArsB seems more specific of bacteria (Yang et al. 2015). Acr3p is divided into two subfamilies based on their phylogenetic dissimilarities: Acr3(1)p and Acr3(2)p (Cai et al. 2009). In several bacteria (e.g., *E. coli*), ArsB is associated with an ATPase ArsA (Wu et al. 1992). An association ArsY(Acr3p)/ArsA has also recently been described in *Bacillus subtilis* (Zheng et al. 2013).

32.3.1.1.2 Arsenic Detoxification in Eukaryotes

Eukaryotes possess resistance genes *ACR* or *ARR* (Arsenic Compound Resistant) homologue of *ars* genes. In the chromosome XVI of *S. cerevisiae*, *ACR1*, *ACR2*, and *ACR3* genes encode proteins with the same function as those encoded by *arsR*, *arsC*, and *arsB* bacterial genes (Bobrowicz et al. 1997). In *S. cerevisiae*, Acr1 (also called Arr1 and Yap8) is a transcriptional regulator of the Yap (yeast AP-1-like) family that regulates *ACR* genes (Menezes et al. 2004). Acr2 (Arr2) corresponds to the eukaryotic clade of arsenate reductases mentioned earlier, which reduces As(V) to As(III) with electrons donated by the reduced glutathione and glutaredoxin (Mukhopadhyay et al. 2002). Arsenite is then excreted out of the cell by efflux pump Acr3p (Arr3), first identified in *S. cerevisiae* (Bobrowicz et al. 1997).

In addition to efflux system, some eukaryotes use sequestration as resistance mechanism. Indeed, As(III) can bind to intracellular chelating proteins or peptides with thiol ligands, such as phytochelatins (PCs), glutathione (GSH), and metallothioneins (MTs), to form inactive complexes (Cobbett and Goldsbrough 2002). In *S. cerevisiae*, Ycf1p is a vacuolar protein of the ABC transporter superfamily, responsible for the ATP-dependent transport of GSH-conjugated arsenite in the vacuole (Ghosh et al. 1999). The *YCF1* gene is regulated by the transcriptional regulator Yap1. In *Fucus vesiculosus*, arsenite is bound to MTs (Merrifield et al. 2004). In yeast, PCs have been shown to complex arsenite efficiently (Tsai et al. 2009). Moreover, some yeasts use reduced sulfide as a source of reducing power and form a stable high-molecular-weight PC–metal–sulfide complex in the vacuole (Rochette et al. 2000, Mendoza-Cozatl et al. 2006).

32.3.1.2 The Respiratory Arsenate Reduction: Aresenic as Life-Supporting Substrate

In addition to the previously described reduction system (corresponding to a detoxification process), specialized prokaryotes have the capacity to perform the dissimilatory arsenate reduction. Under anoxic conditions, dissimilatory arsenate-reducing prokaryotes (DARP) use As(V) as final electron acceptor for anaerobic respiration (Oremland and Stolz 2003). Arsenate reduction is then coupled to the oxidation of inorganic (e.g., H_2 or S^{2-}) or organic (e.g., acetate, lactate, or aromatic compounds) substrates to produce energy for growth (Stolz et al. 2006). This system, first evidenced in *Sulfurospirillum arsenophilum* and *S. barnesii* (ε-*Proteobacteria*), was then described in other phylogenetic groups: the γ- and δ-*Proteobacteria*, *Firmicutes*, thermophilic *Eubacteria*, and *Crenarchaeota* (Oremland and Stolz 2003).

DARP were identified in a large variety of habitats and are characterized by a wide metabolic diversity (Ahmann et al. 1994, Macy et al. 1996, Switzer Blum et al. 1998, Gihring and Banfield 2001). Indeed, most of them can use a large range of electron donors and acceptors (including selenate, nitrate, nitrite, fumarate, Fe(III), thiosulfate, elemental sulfur, dimethyl sulfoxide (DMSO), and trimethylamine oxide; Stolz et al. 2006). This diversity allows their survival when arsenate concentration is lower in the environment (Lloyd et al. 2011). To date, only one obligate arsenate-reducing prokaryote has been identified (bacterial strain MLMS-1 isolated from Mono Lake; Hoeft et al. 2004).

Respiratory arsenate reductases have been characterized in *Chrysiogenes arsenatis* (Krafft and Macy 1998), *Bacillus selenitireducens* (Afkar et al. 2003), and *Shewanella* sp. ANA-3 (Malasarn et al. 2008). These are heterodimer periplasmic proteins including two subunits: ArrA (87 kDa) and ArrB (29 kDa). The large catalytic subunit ArrA (encoded by *arrA* gene) belongs to the DMSO reductase family containing a molybdenum center and a [4Fe–4S] cluster. The small subunit ArrB (encoded by *arrB* gene) contains up to four [Fe–S] center proteins, which allow electron transfer. In *Shewanella* sp. ANA-3, *arrA* and *arrB* genes are sufficient to perform respiratory arsenate reduction (Saltikov and Newman 2003). Chaperone protein (ArrD) and putative regulatory proteins involved in signal transduction system (ArrR, ArrS, and ArrT) have been evidenced in *Desulfitobacterium hafniense* DCB-2. However, their function remains hypothetical (Yamamura and Amachi 2014).

32.3.1.3 Environmental Impacts and Implication of As(V) Reduction in Bioprocess

As(V) reduction will increase arsenite concentration in the environment. *ars*- or *ACR*-based reduction (detoxification mechanism) occurs both under aerobic and anaerobic conditions and requires soluble As(V) (Oremland and Stolz 2003, Macur et al. 2004). On the other hand, under anoxic conditions only, DARP can reduce soluble and sorbed As(V) (Lloyd et al. 2011), possibly leading to the mobilization of toxic and soluble As(III) in water from arsenic-bearing material. Implication of DARP in the contamination of groundwater by arsenite has been evidenced by several studies (Islam et al. 2004, Héry et al. 2010, 2015, Lloyd et al. 2011). Recently, the genetic potential for As(III) release from contaminated stream bed sediments due to As(V)-respiring microorganisms was evidenced in AMD-impacted waters (Héry et al. 2014, Desoeuvre et al. 2016).

Arsenate, bound to soil particles, is often the major species of arsenic found in telluric environments. Therefore, the use of DARP may represent a way to reduce it to soluble As(III) that can be then removed from the soil. More surprisingly, As(V)-reducing microorganisms can also be used to remove As from the aqueous phase: they reduce As(V) to As(III), which precipitates with sulfide to form insoluble complexes (Newman et al. 1997). This reaction, conducted in bioreactors, has resulted in the decrease of arsenic in water (Chung et al. 2006, Upadhyaya et al. 2010, 2012).

32.3.2 As(III) Oxidation

32.3.2.1 A Detoxification or a Metabolic Process

More than 85 bacteria, archaea, and one algae identified in contrasted environments (soils, mining sediments, wastewaters, hypersaline lakes, etc.) are able to oxidize arsenite to less

toxic arsenate through an arsenite oxidase (Lehr et al. 2007, Bahar et al. 2013). Prokaryotic As(III) oxidizers belong to: the α-, β-, and γ-*Proteobacteria*; *Deinocci* (i.e., *Thermus*); and *Crenarchaeota* (Oremland and Stolz 2003, Bahar et al. 2013). They can be divided into two metabolic groups: the heterotrophic arsenite oxidizers (HAOs) and chemolithoautotrophic arsenite oxidizers (CAOs). Most of the HAOs and CAOs use arsenite oxidation as a detoxification mechanism. Some CAOs are able to couple the oxidation of arsenite to the reduction of oxygen to generate energy for their growth (Santini et al. 2000, Vanden Hoven and Santini 2004, Battaglia-Brunet et al. 2006a).

Furthermore, anaerobic oxidation of arsenite has been evidenced with nitrates used as terminal electron acceptor (Oremland et al. 2002, Rhine et al. 2006). Photosynthetic bacteria can also oxidize As(III) in anaerobic conditions, suggesting that As(III) can be an electron source for photosynthesis (Stolz et al. 2006).

Arsenite oxidases are membrane (*Alcaligenes faecalis* (HAO); Anderson et al. 1992) or periplasmic (*Rhizobium* sp. NT-26 (CAO) (Santini and Vanden Hoven 2004) and *Hydrogenophaga* sp. NT-24 (HAO) (Vanden Hoven and Santini 2004)) proteins belonging to the DMSO reductase family of molybdenum enzymes. They are composed of one large subunit (AsoA for *A. faecalis*; AroA for *Rhizobium* sp. NT-26 and *Hydrogenophaga* sp. NT-24) and one smaller (AsoB for *A. faecalis*; AroB for *Rhizobium* sp. NT-26 and *Hydrogenophaga* sp. NT-24).

The large catalytic subunits contain a molybdenum center and a [4Fe–4S] cluster, similarly to the respiratory arsenate reductase. The small subunits contain a rieske-type site [2Fe–S] carrying a TAT leader sequence, which allows electron transfer (Ellis et al. 2001).

The genes encoding the small and large subunits of the arsenite oxidase have been described in different bacteria by several authors who named it differently (e.g., *aoxA/B* in *Herminiimonas arsenicoxydans* ULPAs1 (Muller et al. 2003) and *Agrobacterium tumefaciens* 5A (Kashyap et al. 2006); *asoB/A* in *A. faecalis* NCIB 8667 (Silver and Phung, 2005); *aroB/A* in *Rhizobium* sp. NT-26 (Santini and Vanden Hoven 2004)). Currently, the abbreviation *aox* is generally accepted. *aoxA* and *aoxB* are present in all the operons described so far, sometimes associated with other genes (Páez-Espino et al. 2009). *aoxC*, *aoxR*, and *aoxC* encode, respectively, a c2-type cytochrome isoform, a NtrC-like protein acting as a regulator of the system, and a periplasmic arsenite-binding receptor. These genes are located on the *aoxRSABC* operon identified in both *A. tumefaciens* (Kashyap et al. 2006) and *Rhizobium* sp. NT-26 (Santini and Vanden Hoven 2004).

One *Thermus* species and *Marinobacter santoriniensis* sp. nov. can do both arsenate dissimilatory reduction and arsenite oxidation, respectively, under anaerobic or aerobic condition (Gihring and Banfield 2001, Handley et al. 2009). A new group of arsenite oxidases Arx, belonging to the DMSO reductase family of molybdopterin enzymes, was shown to have bifunctional activity for arsenite oxidation and arsenate reduction (Richey et al. 2009). This enzyme has been identified in the haloalkaliphilic *Alkalilimnicola ehrlichii* strain MLHE-1 (Oremland et al. 2002, Hoeft et al. 2007, Zargar et al. 2010), a purple sulfur bacterium *Ectothiorhodospira* sp. strain PHS-1 (Kulp et al. 2008, Zargar et al. 2012), and ArxA-like sequences were also detected in Yellowstone National Park hot springs (Zargar et al. 2012). *In vivo*, ArxA only functions as an arsenite oxidase (Richey et al. 2009), although this protein is more closely related to ArrA than to AoxA (Zargar et al. 2012, Van Lis et al. 2013).

32.3.2.2 As(III) Oxidation as a Natural Bioremediation Process

Arsenite-oxidizing prokaryotes oxidize As(III) to less toxic As(V) under anaerobic or aerobic condition depending on the strains. Less soluble As(V) adsorbs more easily than As(III) on iron oxides hydrated, clays, organic matters, and manganese oxides (Mandal and Suzuki 2002). Then, As(III) oxidation generally results in a decrease of the soluble arsenic concentration associated with the immobilization of As(V) into mineral phases. In Carnoulès AMD (Gard, France), a substantial decrease of the arsenic concentration in water is due to the coprecipitation of As(V) with Fe(III) after microbially driven oxidation reactions (Leblanc et al. 2002, Casiot et al. 2003b, Egal et al. 2010, Bertin et al. 2011, Bruneel et al. 2011).

The physicochemical treatments of As-rich waters generally require As(III) oxidation as a pretreatment. However, the use of strong oxidizers (e.g., potassium permanganate or hydrogen peroxide) presents several disadvantages with generally a high cost and high toxicity; thus, the application of arsenite-oxidizing prokaryotes appears relevant in this context (Lievremont et al. 2009).

32.3.3 OTHER BIOTRANSFORMATIONS OF ARSENIC

32.3.3.1 Arsenic Methylation/Demethylation

Poisoning due to arsenical exhalations was suspected since the seventeenth century. In the 1890s, the biomethylation of As(III) to garlic-odored trimethylarsine by a fungus was evidenced by G. Gosio. Since then, a wide range of microorganisms showed the capacity to methylate arsenic (bacteria, archaea, algae, and fungi; Bentley and Chasteen 2002), including photosynthetic organisms (Ye et al. 2012). In 1945, F. Challenger proposed a mechanism for As methylation pathway based on the study of a fungus, *Scopulariopsis brevicaulis*. This complex mechanism seems to be similar in other microorganisms (Bentley and Chasteen 2002, Stolz et al. 2006, Thomas et al. 2007). A series of reduction and methylation–oxidation reactions lead to the formation of the successive methylated arsenical species. Ultimately, arsenic can be volatilized leading to the decrease of As concentration in cells (Challenger 1945, Cullen 2005).

However, because intermediate products as MMA(III) and DMA(III) are highly toxic, arsenic biomethylation as a detoxification process is questionable (Styblo et al. 2000,

Dopp et al. 2010a,b). Moreover, the arsenical substrate [As(V), As(III), and MMA(V)], the methyl donor, and the ending product (mono-, di-, and trimethylarsine) of the methylation depend on the aerobic or anaerobic microorganism (McBride and Wolfe 1971, Shariatpanahi et al. 1983, Michalke et al. 2000). If S-adenosylmethionine (SAM) is the main methyl donor methylcobalamin can be used by the methanogen archaea *Methanobacterium bryantii* and anaerobic bacteria McBride and Wolfe 1971, Bentley and Chasteen 2002).

The S-adenosylmethionine methyltransferase ArsM catalyzes methyl group transfers. In spite of considerable sequence variations, some sequence motifs and cysteine residues are well conserved in this protein (Kagan and Clarke 1994, Thomas et al. 2007, Wang et al. 2015). These conserved regions are probably critical for the transfer of methyl group to As, while the rest of the protein is species specific (Ye et al. 2012). The possible contribution of a methylarsenate reductase in the biomethylation mechanism has been suggested but remains to be elucidated (Qin et al. 2006). *ArsM* homologues have been identified in 125 bacterial and 16 archaeal genomes (Qin et al. 2006). In these microorganisms, *arsM* is adjacent to or part of an *ars* operon and it is often regulated by *arsR* (Ye et al. 2012).

Some microorganisms (i.e., *Alcaligenes*, *Pseudomonas*, and *Mycobacterium* spp.) have the ability to demethylate mono- and dimethyl arsenic compounds (Bentley and Chasteen 2002). *Pseudomonas* can even use DMA as a carbon source (Maki et al. 2004). However, these pathways are still poorly understood.

Another way to remove arsenic in the environment is the biovolatilization, which corresponds to the final step of arsenic biomethylation. Many bacteria, fungi, and algae are able to mediate As biomethylation (Huysmans and Frankenberger 1991, Wang et al. 2014, Roy et al. 2015). Recently, environmental diversity studies of *arsM* gene highlighted an unexpected diversity of microorganisms with the genetic potential for arsenic methylation in terrestrial (Jia et al. 2013, Zhao et al. 2013) and aquatic ecosystems (Desoeuvre et al. 2016). These findings strengthen the idea that arsenic biomethylation may significantly contribute to the global As biogeochemical cycling in terrestrial (Mestrot et al. 2011) and freshwater ecosystems (Héry et al. 2014, Desoeuvre et al. 2016). Arsenic methylation in soils increases with the decrease of redox potential, suggesting that anaerobic microorganisms are more efficient than aerobic microorganisms (Frohne et al. 2011, Wang et al. 2014). Genetically modified microorganisms may improve biomethylating capacity. *arsM* from *Rhodopseudomonas palustris* was expressed with success in *E. coli*, *Sphingomonas desiccabilis*, and *Bacillus idriensis* (Yuan et al. 2008, Liu et al. 2011).

32.3.3.2 Sequestration

Various strategies have been developed to deal with arsenic pollution based on the capacity of sequestration of microorganisms. Some eukaryotes bind arsenite to intracellular chelating proteins or peptides, which are stored in a vacuole

(Cobbett and Goldsbrough 2002). As a consequence, arsenic concentration decreases in the environment. As previously mentioned, sequestration efficiency may be improved in genetically engineered microorganisms by increasing inactive complex production or their transport in the vacuole (Song et al. 2003, Merrifield et al. 2004, Singh et al. 2008, 2010). Some bacteria are also known to hyperaccumulate As like the non-genetically engineered strain XZM002 belonging to the genus *Bacillus* (Xie et al., 2013) or the genetically modified strain of *E. coli*, overexpressing ArsR (Kostal et al. 2004, Huang et al., 2014).

Microbial biofilms can also be used in bioremediation processes as secretion of exopolymers could ultimately immobilize As by passive sequestration (Singh et al. 2006, Marchal et al. 2011, Koechler et al. 2015). These properties are used to develop remediation methods such as biofilm based bioreactors (Chang et al. 2006, Lièvremont et al. 2009).

32.4 APPLICATIONS IN BIOREMEDIATION

Microorganisms interfere with the biogeochemical cycling of arsenic through diverse biotransformations, impacting the speciation and the toxicity of this pollutant and its transfer in the environment (Lievremont et al. 2009). The activity of microorganisms is thus of prime importance to understand the fate of arsenic in terrestrial and aquatic ecosystems. A better understanding of these microbial processes in the global arsenic cycle can lead to promising strategies for bioremediation applications.

32.4.1 BIOREMEDIATION OF SOILS

The methods of treatment available for As-contaminated soils are mainly limited to soil replacement, containment, and solidification/stabilization; thus, alternative methods using microorganisms represent a great potential as novel bioremediation strategy for As removal from soils (Yamamura and Amachi 2014). As As(V) is generally the major species of As found in soils (Bissen and Frimmel 2003a), its reduction in As(III) can promote the removal of arsenic from solid and its release to the aqueous phase. This can be used for the remediation of soils (Yamamura and Amachi 2014). DARP appear thus as interesting agents because they can only reduce aqueous As(V) entered inside their cell and not sorbed As(V). Many studies have been conducted on contaminated sediments; however, the real applicability of this bioremediation strategy at an industrial scale is unknown because most studies conducted to date only focused on biogeochemical aspects (Zobrist et al. 2000, Yamamura et al. 2005, Kudo et al. 2013, Ohtsuka et al. 2013, Yamamura and Amachi 2014). Strains such as *Bacillus selenatarsenatis* SF-1 are able to remove up to 56% of As from industrially contaminated soils (Yamamura et al. 2008). Its application in a slurry bioreactor has been suggested, with a mathematical model to predict the dissolution of As from soil (Soda et al. 2009).

Formation of volatile trimethylarsine by microorganisms expressing *arsM* gene may also represent a relevant bioremediation strategy (Sun et al. 2009, Singh et al. 2011, Chen et al. 2014) as well as microbial intracellular or extracellular sequestration (Huang 2014).

Phytoremediation, an eco-friendly method using plants to extract arsenic and other contaminants from soil via their roots, has also emerged as a promising potential technique for reducing arsenic and heavy metals from contaminated areas (Tripathi et al. 2007, Mirza et al. 2011, Mandal et al. 2012, Hettick et al. 2015). Arsenic can be phytoremediated using distinct mechanisms in soil: phytoextraction–phytoaccumulation, phytostabilization–phytoimmobilization–phytorestoration, phytovolatilization, and phytostimulation (Roy et al. 2015). Several plants were found to be highly resistant to arsenic and some of them have also the capability to hyperaccumulate large amounts of this toxic metalloid in their overground parts (Du et al. 2005, Srivastava et al. 2006, Garcia-Salgado et al. 2012, Vithanage et al. 2012, Hettick et al. 2015). For example, *Pteris vittata* (a Chinese brake fern) is capable to accumulate 1442–7526 mg kg^{-1} of As from contaminated soils and up to 27,000 mg kg^{-1} from hydroponic cultures (Ma et al. 2001, Kalve et al. 2011). Multigenic engineering approach has been also developed to create super-hyperaccumulator plants (Padmavathiamma and Li 2007, Tripathi et al. 2007). The phytostabilization plant focuses on the contrary on the immobilization of As in the rhizosphere but not in the harvestable biomass plant, reducing food chain transfer risk (Madejon et al. 2002, Ma et al. 2011, Roy et al. 2015). In this system, some microbes can affect plant growth by different ways (like bacteria called plant growth–promoting bacteria) and play a key role in the phytoremediation of pollutants as they can actively or passively promote the growth of plants through different mechanisms (like solubilization of phosphate, nitrogen fixation, production of siderophores, phytohormone, or ACC deaminase; Ma et al., 2011). Some microorganisms can also affect directly metal mobilization/immobilization in soil through metal reduction or oxidation, chelator production (siderophores, organic acids or biosurfactants), production of metabolites (glomoline or extracellular polymeric substances), metal biosorption etc. (Ma et al. 2011, Rajkumar et al. 2012, Roy et al. 2015).

32.4.2 BIOREMEDIATION OF WATER

Bioremediation of waters contaminated with arsenic has been an issue in many review papers and book chapters within the last years (Mondal et al. 2006, Wang and Mulligan 2006, Mohan and Pittman 2007, Lievremont et al. 2009, Malik et al. 2009, Tsai et al. 2009, Wang and Zaho 2009, Bundschuh et al. 2010, Litter et al. 2010, Lizama et al. 2011, Pal and Paknikar 2012, Bahar et al. 2013, Cavalca et al. 2013, Kruger et al. 2013, Yamamura and Amachi 2014, Katsoyiannis et al. 2015). It is considered as an alternative to conventional techniques, being cheapest and requiring less maintenance. However, few arsenic removal technologies that use biological processes have been developed to date until the stage of full-scale treatment.

The research studies listed in the following address remediation processes at the scale of continuous flow laboratory treatment reactors, field trials, and a few full-scale treatment experiences. They mainly focus on processes based on bacterial iron and manganese oxidation, with or without concomitant As(III) oxidation. A few studies focus on arsenic trapping during sulfate reduction. Studies dealing with bacterial As methylation, biosorption, intracellular accumulation and biovolatilization by microalgae or fungi, and filtration by plants will not be considered here because they generally did not reach large-scale treatment stage. The potentialities of different materials of biological origin to be exploited for As biosorption in the treatment of arsenic-contaminated waters have been explored by Mohan and Pittman (2007) and will not be included in this chapter.

32.4.2.1 Bioremediation of As-Contaminated Groundwater

32.4.2.1.1 Treatments Based on Biological Iron and/or Manganese Oxidation, and Arsenic Removal

During the last years, the most developed technology for groundwater treatment is biological oxidation of iron or manganese by indigenous bacteria and subsequent retention of As(III) and As(V) onto precipitated Fe-oxyhydroxides or Mn oxides (Katsoyiannis et al. 2002, 2008, 2013, 2015, Katsoyiannis and Zouboulis 2004, 2006). In some cases, this process is accompanied by bacterial oxidation of As(III) into As(V) (Katsoyiannis and Zouboulis 2004, Casiot et al. 2006). In the study of Katsoyiannis and Zouboulis (2004), the bioreactor was filled with polystyrene beads. Indigenous iron-oxidizing bacteria identified as *Gallionella ferruginea* and *Leptothrix ochracea* accumulate in the filtration medium. Both As(III) and As(V) were removed, for concentrations ranging from 50 to 200 μg L^{-1}. The oxidation of As(III) was catalyzed by bacteria. This system was successful in removing arsenic from groundwater containing 60–80 μg L^{-1} As(III) over a 10-month period (Zouboulis and Katsoyiannis, 2005). Similarly, Pokhrel and Viraraghavan (2009) showed that 100 μg L^{-1} arsenic could be removed below 5 μg L^{-1} through sand filters, with an optimum ratio (40:1) of iron and arsenic in the tap water. In the study of Hassan et al. (2009), a biological fixed-bed reactor containing coconut husk and layers of iron matrix, charcoal, and sand filter allowed to decrease arsenic concentration from 500 μg L^{-1} to less than 15 μg L^{-1}. Simultaneous significant iron removal efficiency (over 95%) was also achieved.

Katsoyiannis et al. (2004) used a fixed-bed upflow filtration unit filled with polystyrene beads for the removal of low-level arsenic concentrations (35 and 42 μg L^{-1} for As(III) and As(V), respectively) during Mn removal. Rapid oxidation of As(III) occurred before sorption onto manganese oxide surfaces. The rates of As(III) oxidation yielded an apparent first-order constant of 0.23 min^{-1} (half-life of 3 min). The presence of phosphates at concentrations of around 600 μg L^{-1} had an adverse effect on As(III) removal, reducing the overall removal efficiency by 50%.

In Katsoyiannis et al. (2013), the removal of As(III) and As(V) from groundwater containing 20–250 μg As L^{-1} by biological oxidation of dissolved Fe(II) (2.9 mg L^{-1}) and Mn(II) (0.6 mg L^{-1}) was carried out in a 21 m long pipe reactor, followed by microfiltration that enabled oxyhydroxide particle filtration and removal of adsorbed arsenic. More than 98% iron, 95% manganese, and 90% arsenic removal was achieved, corresponding to final concentrations well below the drinking water regulation limits. In addition, the presence of phosphate at concentrations up to 1.6 mg L^{-1} did not affect the removal of roughly 100 μg L^{-1} As(III). The As(V) removal capacity of this hybrid unit was higher than that achieved by conventional coagulation–filtration with Fe(III).

Yang et al. (2014) analyzed the distribution and genetic diversity of microorganisms along the depth of a biofilter that removed As(III) from 150 to 10 mg L^{-1} during biological iron and manganese oxidation. Results suggested that iron-oxidizing bacteria (*Gallionella*, *Leptothrix*), manganese-oxidizing bacteria (*Leptothrix*, *Pseudomonas*, *Hyphomicrobium*, *Arthrobacter*), and As(III)-oxidizing bacteria (*Alcaligenes*, *Pseudomonas*) are dominant in the biofilter. The removal zone of Fe(II), Mn(II), and As(III) was located at depths 20, 60, and 60 cm, respectively, and the corresponding removal efficiencies were 86%, 84%, and 87%, respectively.

Biological Fe and Mn oxidation with subsequent As removal has been applied in full-scale plants in Greece (Katsoyiannis et al. 2015). Arsenic is present in concentrations between 15 and 20 μg L^{-1}, from which almost 70% is in the form of As(III). A treatment unit was constructed to treat groundwater containing Fe(II), Mn(II), NH$_3$, and phosphate. The applied method is based on the succession of Fe(II) oxidation, followed by the oxidation of Mn(II), ammonium, and As(III) in upflow filters and subsequent coagulation and direct filtration in downflow filters. Besides this biological treatment, coagulation with iron salts is additionally applied to reach arsenic removal efficiency higher than 95% and final arsenic concentrations below 10 μg L^{-1}.

Based on biological Fe and Mn oxidation processes, subsurface groundwater treatment has been developed and applied in the field in Bangladesh (Van Halem et al. 2010). Six groundwater treatment plants have been constructed in West Bengal, India, for supplying potable water to rural communities (Sen Gupta et al. 2006, 2009). Total As in treated water was less than 10 μg L^{-1} and operating costs were low. The arsenic adsorbents were regenerated, thus reducing the volume of wastes by nearly two orders of magnitude (Sarkar et al. 2008). Arsenic leaching during disposal is limited by the containment of As-bearing solids on well-aerated coarse-sand filters.

Sun et al. (2009) demonstrated that nitrate injection in anoxic groundwater contaminated with co-occurring Fe(II) and As(III) might be considered in bioremediation. This process stimulated the microbial oxidation of As(III) to As(V) and Fe(II) to Fe(III) linked to denitrification and the subsequent formation of Fe(III) (hydr)oxide-coated sands with adsorbed As(V). High levels of As(III) (500 μg L^{-1}) were decreased to around 10 μg L^{-1} in the model column used after

250 days. In another study with a bench-scale upflow anaerobic sludge bed bioreactor, Sun et al. (2011) showed that a chlorate (ClO_3^-)-reducing microbial consortium immobilized in the biofilm reduced ClO_3^- to Cl^- and H_2O while oxidizing As(III) to As(V). The culture was dominated by *Dechloromonas* and *Stenotrophomonas* as well as genera within the family *Comamonadaceae*. More than 98% of As(III) was oxidized at volumetric loadings ranging from 0.45 to 1.92 mmol As L^{-1} day^{-1}.

32.4.2.1.2 Treatments Based on Biological Arsenic Oxidation

Several studies have investigated the efficiency of As(III)-oxidizing bacteria immobilized onto a solid support in continuous flow reactors. A culture of *Microbacterium lacticum* able to oxidize As(III) at a high rate was tested for groundwater treatment (Mokashi and Paknikar 2002). The culture was immobilized onto brick pieces for arsenite oxidation; another packed bed activated alumina column was used for arsenate removal, and then packed bed charcoal column was added for cell removal in the washout and UV irradiation for disinfection (Paknikar 2003). Simulated groundwater with 1 mg L^{-1} arsenite supplemented with 0.03% sucrose (to support bacterial growth) was treated at a flow rate of 700 mL min^{-1}. The residual arsenic concentration was 10–25 µg L^{-1}. A treatment plant based on this technology with a 1000 L day^{-1} filtration capacity has been designed and installed in arsenic-affected villages in India (Pal and Paknikar 2012).

A biocolumn reactor with cells of *Ralstonia eutropha* MTCC 2487 immobilized on granular activated carbon has been used for the treatment of synthetic industrial effluent containing 25 mg As L^{-1} (Mondal et al. 2008). After biofilm formation, the biocolumn reactor is capable to reduce arsenic concentration in the effluent water below 0.15 mg L^{-1} using feed flow rate of 1.7–5.1 mL min^{-1} (360–180 min contact time) and periodic backwashing.

In Wan et al. (2010), the As(III) oxidizer *Thiomonas arsenivorans* was inoculated in two upflow fixed-bed reactors filled with sand and zerovalent iron powder. The highest As(III) oxidation rate was 8.36 mg h^{-1} L^{-1} in the first column, and about 45% of total As was removed within 1 h in the second column. Bacterial As(III) oxidation rate by the biofilm present in the column was correlated with the axial length of reactor. A durable removal of total As was realized, and zerovalent iron was not saturated within 33 days.

Continuous experiments were conducted by Ito et al. (2012) using a bioreactor with cells of arsenite-oxidizing bacteria closely related to *Ensifer adhaerens* immobilized on polyvinyl alcohol gel droplets. With an initial As(III) concentration of 1 mg L^{-1} at a hydraulic retention time of 1 h, As(III) oxidation efficiency reached 92%.

Arsenic-resistant bacterial strain, *Rhodococcus equi*, has been used by Bag et al. (2010) for the treatment of synthetic water and natural groundwater containing 50–100 µg As L^{-1} in a continuous packed bed reactor. A maximum value of arsenite removal efficiency of 95% was achieved.

An autotrophic As(III)-oxidizing bacterial consortium CAsO1, containing at least two strains phylogenetically close to *Ralstonia pickettii* and to the genus *Thiomonas*, isolated from a gold-mining site, has been inoculated into a pilot plant consisting of two biological fixed-bed reactors filled with pozzolana (Battaglia-Brunet et al. 2002, Michel et al. 2007, Michon et al. 2010). It showed efficient arsenic oxidation in the concentration scale of µg L^{-1} to mg L^{-1}, with a first-order kinetic constant of 0.04 min^{-1}.

As(III) oxidation by the chemoautotrophic bacterium *T. arsenivorans* strain b6 was investigated in a fixed-film reactor packed with spherical Pyrex glass beads under variable influent As(III) concentrations (500–4000 mg L^{-1}) and hydraulic residence times (0.2–1 day) for a duration of 137 days (Dastidar and Wang 2012). As(III) oxidation efficiency ranged from 48% to 99%.

32.4.2.1.3 Treatments Based on Biological Sulfate Reduction and Arsenic Removal

Biological sulfate reduction has been much less studied than biological As and Fe oxidation for the treatment of high groundwater As. Upadhyaya et al. (2010) used two biologically active carbon reactors in series for the simultaneous removal of nitrate and arsenic from a synthetic groundwater supplemented with acetic acid; they used an inoculum originating from a mixed community of microbes indigenous to groundwater, able to use dissolved oxygen, nitrate, arsenate, and sulfate as the electron acceptors while acetic acid was the electron donor. Biologically produced sulfides effectively removed arsenic from the water, from ~200 µg L^{-1} to below 20 µg L^{-1} likely through the formation of arsenic sulfides, and/or surface precipitation and adsorption on iron sulfides.

A pilot-scale constructed wetland was built by Schwindaman et al. (2014) for the removal of arsenic from simulated Bangladesh groundwater. The system allowed to compare arsenic removal in four series of four reactors: two series were designed to promote the coprecipitation and sorption of arsenic with iron oxyhydroxides under oxidizing conditions, and two other series promoted the precipitation of arsenic sulfide and coprecipitation of arsenic with iron sulfide under reducing conditions. Arsenic removal performance was greater in series with oxidizing conditions and with zerovalent iron amendment. Removal efficiency reached 40%–95%, with rate coefficients 0.13–0.77 day^{-1}.

32.4.2.2 Bioremediation of As-Contaminated Mine Waters

As already seen for groundwaters, there are two main types of biological mechanisms employed to remove arsenic from mine waters. The first mechanism involves the precipitation of Fe-oxyhydroxides catalyzed by Fe-oxidizing bacteria in aerobic conditions and subsequent adsorption or coprecipitation of arsenic (Johnson and Hallberg 2005). The second mechanism exploits the precipitation of arsenic sulfides (such as As_2S_3, As_4S_4) catalyzed by sulfate-reducing bacteria (SRB) in anaerobic conditions (Johnson and Hallberg 2005, Hedrich and Johnson 2014).

32.4.2.2.1 Treatments Based on Iron Oxidation and Arsenic Removal

Mine drainage containing arsenic also contain high concentrations of reduced Fe(II), and they often exhibits a low-pH. Abiotic Fe(II) oxidation is relatively slow at acid pH, but it is greatly accelerated by iron-oxidizing prokaryotes (Bacteria and Archaea), which are naturally present in mine drainage, the most popular being *Acidithiobacillus ferrooxidans*. Therefore, the aerobic biological treatment of As-rich mine drainage involves the biotic oxidation of ferrous iron to ferric iron and the adsorption of As onto ferric iron colloids or its coprecipitation (Johnson and Hallberg 2005). Because As(III) has a poor affinity for mineral surfaces at acid pH, it is not efficiently retained during this process. Therefore, oxidation of As(III) to As(V) effectively contributes to reduce arsenic mobility in AMD. Arsenite-oxidizing bacteria such as *Thiomonas* spp. have been isolated from various mine waters (Bruneel et al. 2003, Coupland et al. 2003, Battaglia-Brunet et al. 2006a), and their possible use in the treatment of As-rich mine waters has been highlighted in several papers (Duquesne et al. 2007, Lievremont et al. 2009). Bioremediation systems developed to remove arsenic from mine drainage by iron oxidation consist principally of aerobic wetlands/ponds and column bioreactors, as described in the following.

32.4.2.2.1.1 Wetlands/Ponds
Aerobic wetlands are open-air systems planted with macrophytes that treat shallow surface flow with a supporting media. Aerobic wetlands are suitable for the removal of Fe, Mn, and Al from mine waters with low metal concentrations and near-neutral pH (Kalin 2004, Johnson and Hallberg 2005). However, there are few studies dealing with As removal in aerobic wetlands.

One of the first experiences was carried out by Buddhawong et al. (2005) who tested different wetland configurations planted with *Juncus effusus* in a 90-day batch experience at an initial pH of 4. The initial arsenic concentration was around 0.5 mg L^{-1}, and during the experiment time, the arsenic removal reached more than 95% in the planted gravel beds. In contrast, in the hydroponic system the arsenic removal was between 25% and 35%, and in the algae ponds (without gravel bed) the arsenic concentration in water remained stable. They concluded that arsenic removal mechanism in the wetland involved coprecipitation with iron (originally coming from the soil) in the oxic zones of the rhizoplane rather than plant uptake or adsorption on the gravel bed.

Another experience combined aerobic basin and anaerobic wetland to treat high As concentrations (50–250 mg L^{-1}) from the AMD of Carnoulès (France; Elbaz-Poulichet et al. 2006). The remediation was carried out *in situ* and combined (1) aerobic ponds (8 m^2) equipped either with ridges or with small cascades, (2) settling ponds at the outflows to recover the precipitated colloidal matter, and (3) anaerobic wetland (70 m^2) filled with compost. In the aerobic ponds, As removal was more efficient in the cascade system (20%) than in the ridge system (11%–15%). The performance of the aerobic ponds appeared to be limited in particular by the raising of the water level and by the formation of a thick bacterial film at the surface of the water, these two phenomena limiting oxygen diffusion. The resulting anoxic conditions in the ponds induced an unexpected release of soluble arsenic from the bottom sediments.

32.4.2.2.1.2 Column-Type Anaerobic Bioreactors
Column-type bioreactors have the advantage of increasing the specific surface area in contact with the contaminated water. This system was used by Battaglia-Brunet et al. (2006b) to study As removal from circumneutral pH waters. The aerobic column bioreactor was filled with pozzolana onto which indigenous bacteria (including As(III) and Fe(II) oxidizers) from an As-contaminated mine drainage were immobilized. Increased As(III) and Fe(II) oxidation rates were obtained in the bioreactor compared to on-site natural attenuation, with an As(III) removal rate of 1900 µg L^{-1} h^{-1}. This bioreactor treatment was applied *in situ*, at the pilot scale, to mine water containing 400–1800 µg L^{-1} of As in a resurgence originating from a small auriferous deposit of arsenopyrite in Lopérec (France). Efficient rates of bacterial activity were obtained within 15 days, and the residual total As concentration was less than 100 µg L^{-1}.

32.4.2.2.2 Treatments Based on Biological Sulfate Reduction and Arsenic Removal

Sulfate reduction mediated by bacteria is an anaerobic process that can naturally occur at depth in AMD (Rowe et al. 2007) where oxygen is depleted. This process has been exploited for the precipitation of metal sulfides from lab to field scale (Huisman et al. 2006, Lewis and van Hille 2006, Kieu et al. 2011). It involves two stages: (1) the production of H$_2$S by SRB that generates alkalinity and (2) the precipitation of metals by the biologically produced H$_2$S. In this second step, metals like Cu or Zn are removed as pure sulfides (Hammack et al. 1993). Such processes have been commercialized in order to produce metal concentrates close to operating metallurgical sites (Scheeren et al. 1993, De vegt et al. 1997).

In the field of mine water remediation, passive and active systems based on sulfate reduction by SRB have been developed. These systems consist of anaerobic wetlands, column-type bioreactors, organic substrate injection, and permeable reactive barriers. However, to our knowledge, for the latter two systems, the arsenic removal efficiency was not monitored in any publication. The other systems are presented in the following.

32.4.2.2.2.1 Anaerobic Wetlands or Combined Compost/Limestone Systems
In passive anaerobic bioremediation system, the water flows through a bed of limestone and biodegradable organic matter, before entering the upper zone of the wetland that can be vegetated. Toxic metals are precipitated together with valuable metals in the compost/limestone

mixture layer. The efficiency of such systems to remove arsenic has been rarely investigated.

One of the first devices set in a real mine was the anaerobic "biofilter" at Wood Cadillac Mine (Abitibi, Canada; Tassé et al. 2003). Previous column laboratory tests with different barks acting as biofilters showed an As removal efficiency often higher than 80% (depending on the bark and on the tested residence time). This experience allowed to develop an *in situ* system consisting of a bed of birch barks where water flowed through, with a residence time of 25 h. This bed promoted the sulfate reduction and precipitation of arsenic, probably in the orpiment form (As_2S_3).

In Trail, Canada, a multistage biological treatment system was installed for the removal of As and other contaminants (Duncan et al. 2004, Mattes et al. 2004, 2011). It collected the leachate from a landfill near an integrated zinc lead smelter. The treatment consisted of two anaerobic bioreactors that worked like subsurface-flow wetlands followed by three ponds planted with different macrophytes. Removal efficiency for arsenic was around 98% in the whole system, with around 74% of arsenic being eliminated in the first anaerobic bioreactor. Mineralogical analysis indicated that arsenic was mainly sequestered by kottigite ($Zn_3(AsO_4)_2 \cdot 8H_2O$; Mattes et al. 2011). There was also some evidence of orpiment formation, but arsenopyrite was not detected.

32.4.2.2.2.2 Column-Type Bioreactors

One of the first experiences to remove arsenic from mine drainage with column-type anaerobic bioreactors was developed by Jong and Parry (2003). They worked with an anaerobic packed bed reactor filled with silica to treat acidic synthetic water. They found that arsenic seemed to be removed by adsorption or coprecipitation with the Cu, Zn, Ni, and Fe metal sulfides, instead of forming orpiment.

Similar results were reported by Altun et al. (2014) who worked with a fixed-bed column bioreactor filled with sand and inoculated with a sludge containing SRB. The feeding solution was adjusted to pH 3.5–5 and included ascorbic acid and ethanol as a carbon and electron source for the bacteria. Arsenic removal efficiency reached 96% with the feeding solution containing 20 mg L^{-1} of As(V) and 200 mg L^{-1} of Fe(II). Arsenic was more easily removed in the presence of iron than without iron, due to the coprecipitation of arsenic with iron sulfide (FeS), pyrite (FeS_2), or the formation of arsenopyrite (FeAsS). Likewise, Luo et al. (2008) obtained As removal efficiencies between 40% and 80% with a packed bed bioreactor filled with small crushed rhyolite rock inoculated with SRB. When they added iron in the form of $FeCl_2$, the As removal yield improved to 95%.

Sahinkaya et al. (2015) used an upflow anaerobic sludge blanket reactor (UASB) to treat synthetic AMD. Unlike the packed bed/fixed-bed reactors, UASB have different operating conditions to optimize the production of biogenic sulfide. In this experience, the reactor bed was composed of biomass granules and the system was fed with ascorbic acid and ethanol, to enrich the SRB. The reactor reached a maximum

As removal rate of 0.41–0.42 mg L^{-1} h^{-1}, with an influent As(V) concentration of 10 mg L^{-1}, in the presence of different metal loadings (Fe, Cu, Ni, and Zn), and low sulfide concentrations (close to 0 mg L^{-1}) in the effluent. In this case, As removal efficiency seemed to be associated with the formation of arsenopyrite or the adsorption of arsenic on metal sulfide precipitates. However, when the AMD contained only arsenic and not the other metals, there was no As removal. The authors hypothesized that this could be due to the high dissolved sulfide concentration that implied the formation of thioarsenite species instead of promoting the precipitation of orpiment at basic pH. Likewise, Altun et al. (2014) found that when there was only As(V) in the reactor, in the presence of high sulfide concentrations (300–600 mg L^{-1}) and alkaline pH, the precipitation of orpiment was delayed, because soluble thioarsenite species limited the precipitation of this mineral (Smieja and Wilkin 2003).

The feasibility of bioprecipitation of arsenic sulfide by SRB tolerant to acidity was shown in fixed-bed reactors filled with pieces of pozzolana continuously fed with solutions containing up to 100 mg L^{-1} As, at pH values between 2 and 5 (Battaglia-Brunet et al. 2009, 2012). During these experiments, the highest As removal rate reached 2.5 mg L^{-1} h^{-1} using As(V) as the initial arsenic form and glycerol as the electron donor, in conditions of limited SRB activity. This study was the first to show that arsenic could be removed with sulfide in acidic water by the precipitation of orpiment.

Cohen and Ozawa (2013) also provided some evidence of orpiment precipitation. They worked with a column bioreactor filled with a mixture of composted livestock manure and hay. The feeding solution was an AMD from Big Five Tunnel (United States) that was adjusted to an influent As(V) concentration of 0.35 and 1.55 mg L^{-1}, in two different testing, a short and a long residence time of 38.3 and 73.2 h, respectively. In both cases, As removal efficiency was higher than 90%. With the support of an adjusted model and the laboratory evidences, the authors hypothesized that arsenic was first reduced to As(III) by microorganisms (abiotic reduction was less than 19%) and then precipitated with the sulfides in the orpiment form.

In a bench-scale biochemical reactor filled with biosolids from Trail (Canada), Jackson et al. (2013) demonstrated that arsenic sulfide precipitation controlled As removal in the bench and in the field scale, whereas the presence of Zn enhanced As removal efficiency from 76% up to 94% although the mechanism involved was not clear.

32.5 CONCLUSION

Arsenic contamination is a worldwide problem, representing a serious risk to human health and ecosystems with more than 150 million people drinking As-contaminated water. The toxicity and bioavailability of this pollutant strongly depend on its speciation. To cope with the presence of this toxic metalloid, microorganisms have developed various interaction mechanisms, including reduction, oxidation, and methylation.

Consequently, it is possible to exploit natural microbial processes to remove As from contaminated environments. This paper reviews the development of bioprocesses for the treatment of arsenic-contaminated soil and waters based on microbial metabolism.

Biological methods available for treating contaminated soils and waters seem promising. Nevertheless, these processes must be tested under field conditions to really assess their efficiency, long-term functioning, economic viability, and robustness.

ACKNOWLEDGMENTS

The authors thank the French National Research Agency ANR for funding through the IngECOST-DMA project (ANR-13-ECOT-0009)

REFERENCES

Achour, A.R., Bauda, P., and Billard, P. 2007. Diversity of arsenite transporter genes from arsenic-resistant soil bacteria. *Research in Microbiology* 158: 128–137.

Afkar, E., Lisak, J., Saltikov, C., Basu, P., Oremland, R.S., and Stolz, J.F. 2003. The respiratory arsenate reductase from *Bacillus selenitireducens* strain MLS10. *FEMS Microbiology Letters* 226: 107–112.

Agusa, T., Trang, P.T.K., Lan, V.M., Anh, D.H., Tanabe, S., Viet, P.H., and Berg, M. 2014. Human exposure to arsenic from drinking water in Vietnam. *Science of the Total Environment* 488: 562–569.

Ahmann, D., Roberts, A.L., Krumholz, L.R., and Morel, F.M. 1994. Microbe grows by reducing arsenic. *Nature* 371: 750.

Ahoulé, D.G., Lalanne, F., Mendret, J., Brosillon, S., and Maïga, A.H. 2015. Arsenic in African waters: A review. *Water, Air, and Soil Pollution* 226: 1–13.

Altun, M., Sahinkaya, E., Durukan, I., Bektas, S., and Komnitsas, K. 2014. Arsenic removal in a sulfidogenic fixed-bed column bioreactor. *Journal of Hazardous Materials* 269: 31–37.

Alvarez-Ayuso, E., Abad-Valle, P., Murciego, A., and Villar-Alonso, P. 2016. Arsenic distribution in soils and rye plants of a cropland located in an abandoned mining area. *Science of the Total Environment* 542: 238–246.

Anderson, G.L., Williams, J., and Hille, R. 1992. The purification and characterization of arsenite oxidase from *Alcaligenes faecalis*, a molybdenum-containing hydroxylase. *The Journal of Biological Chemistry* 267: 23674–23682.

Andres, J. and Bertin, P.N. 2016. The microbial genomics of arsenic. *FEMS Microbiology Reviews* 39: 1–24.

Argos, M., Kalra, T., Rathouz, P.J., Chen, Y., Pierce, B., Parvez, F., Islam T. et al. 2010. Arsenic exposure from drinking water, and all-cause and chronic-disease mortalities in Bangladesh (HEALS): A prospective cohort study. *Lancet* 376: 252–258.

Bag, P., Bhattacharya, P., and Chowdhury, R. 2010. Bio-Detoxification of rrsenic laden ground water through a packed bed column of a continuous flow reactor using immobilized cells. *Soil and Sediment Contamination: An International Journal* 19: 455–466.

Bahar, M.M., Megharaj, M., and Naidu, R. 2013. Bioremediation of arsenic-contaminated water: Recent advances and future prospects. *Water, Air, and Soil Pollution* 224: 1–20.

Battaglia-Brunet, F., Crouzet, C., Morin, D., Joulian, C., Burnol, A., Coulon, S., Morin, D., and Joulian, C. 2012. Precipitation of arsenic sulphide from acidic water in a fixed-film bioreactor. *Water Research* 46: 3923–3933.

Battaglia-Brunet, F., Dictor, M.-C., Garrido, F., Crouzet, C., Morin, D., Dekeyser, K., Clarens, M., and Baranger, P. 2002. An arsenic(III)-oxidizing bacterial population: Selection, characterization, and performance in reactors. *Journal of Applied Microbiology* 93: 656–667.

Battaglia-Brunet, F., Itard, Y., Garrido, F., Delorme, F., Crouzet, C., Greffie, C., and Joulian, C. 2006b. A simple biogeochemical process removing arsenic from a mine drainage water. *Geomicrobiology Journal* 23: 201–211.

Battaglia-Brunet, F., Joulian, C., Garrido, F., Dictor, M.C., Morin, D., Coupland, K., Barrie Johnson, D., Hallberg, K.B., and Baranger, P. 2006a. Oxidation of arsenite by *Thiomonas* strains and characterization of *Thiomonas arsenivorans* sp. nov. *Antonie van Leeuwenhoek* 89: 99–108.

Battaglia-Brunet, F., Morin, D., Coulon, S., and Joulian, C. 2009. Bioprecipitation of arsenic sulphide at low pH. *Advanced Materials Research* 71–73: 581–584.

Bentley, R. and Chasteen, T.G. 2002. Microbial methylation of metalloids: Arsenic, antimony, and bismuth. *Microbiology and Molecular Biology Reviews* 66: 250–271.

Berg, M., Stengel, C., Trang, P.T. K, Viet, P.H., Sampson, M.L., Leng, M., Samreth, S., and Fredericks, D. 2007. Magnitude of arsenic pollution in the Mekong and Red River Deltas Cambodia and Vietnam. *Science of the Total Environment* 372: 413–425.

Bertin, P.N., Heinrich-Salmeron, A., Pelletier, E., Goulhen-Chollet, F., Arsène-Ploetze, F., Gallien, S., Lauga, B. et al. 2011. Metabolic diversity between main microorganisms inside an arsenic-rich ecosystem revealed by meta- and proteo-genomics. *The ISME Journal* 5: 1735–1747.

Bhattacharya, P., Mukherjee, A.B., Bundschuh, J., Zevenhoven, R., and Loeppert, R.H. 2007. *Arsenic in Soil and Groundwater Environment: Biogeochemical Interactions, Health Effects and Remediation*. Trace Metals and Other Contaminants in the Environment (Series Editor Nriagu JO), Elsevier, Amsterdam, the Netherlands.

Bissen, M. and Frimmel, F.H. 2003a. Arsenic, a review. Part I: Occurrence, toxicity, speciation, mobility. *Acta Hydrochimica et Hydrobiologica* 31: 9–18.

Bissen, M. and Frimmel, F.H. 2003b. Arsenic: A review. Part II: Oxidation of arsenic and its removal in water treatment. *Acta Hydrochimica et Hydrobiologica* 31: 97–107.

Belluck, D.A., Benjamin, S.L., Baveye, P., Sampson, J., and Johnson, B. 2003. Widespread arsenic contamination of soils in residential areas and public spaces: An emerging regulatory or medical crisis? *International Journal of Toxicology* 22: 109–128.

Bobrowicz, P., Wysocki, R., Owsianka, G., Goffeau, A., and Ułaszewski, S. 1997. Isolation of three contiguous genes, ACR1, ACR2 and ACR3, involved in resistance to arsenic compounds in the yeast *Saccharomyces cerevisiae*. *Yeast* 13: 819–828.

Bossy, A. 2010. Origines de l'arsenic dans les eaux, sols et sédiments du district aurifère de St-Yrieix-la- Perche (Limousin, France): Contribution du lessivage des phases porteuses d'arsenic. Thèse de doctorat. Université de Tours, Tours, France.

Bruneel, O., Personné, J.C., Casiot, C., Leblanc, M., Elbaz-Poulichet, F., Mahler, B.J., Le Flèche, A., and Grimont, P.A.D. 2003. Mediation of arsenic oxidation by *Thiomonas* sp. in acid-mine drainage (Carnoulès, France). *Journal of Applied Microbiology* 95: 492–499.

Bruneel, O., Volant, A., Gallien, S., Chaumande, B., Casiot, C., Carapito, C., Bardil, A. et al. 2011. Characterization of the active bacterial community involved in natural attenuation processes in arsenic-rich creek sediments. *Microbial Ecology* 61: 793–810.

Buddhawong, S., Kuschk, P., Mattusch, J., Wiessner, A., and Stottmeister, U. 2005. Removal of arsenic and zinc using different laboratory model wetland systems. *Engineering in Life Sciences* 5: 247–252.

Bundschuh, J., Litter, M., Ciminelli, V.S., Morgada, M.E., Cornejo, L., Hoyos, S.G., Hoinkis, J., Alarcón-Herrera, M.T., Armienta, M.A., and Bhattacharya, P. 2010. Emerging mitigation needs and sustainable options for solving the arsenic problems of rural and isolated urban areas in Latin America—A critical analysis. *Water Research* 44: 5828–5845.

Butcher, B.G., Deane, S.M., and Rawlings, D.E. 2000. The chromosomal arsenic resistance genes of *Thiobacillus ferrooxidans* have an unusual arrangement and confer increased arsenic and antimony resistance to *Escherichia coli*. *Applied and Environmental Microbiology* 66: 1826–1833.

Cai, L., Liu, G., Rensing, C., and Wang, G. 2009. Genes involved in arsenic transformation and resistance associated with different levels of arsenic-contaminated soils. *BMC Microbiology* 9: 4.

Carlin, A., Shi, W., Dey, S., and Rosen, B.P. 1995. The ars operon of *Escherichia coli* confers arsenical and antimonial resistance. *Journal of Bacteriology* 177: 981–986.

Casiot, C., Leblanc, M., Bruneel, O., Personné, J.-C., Koffi, K., and Elbaz-Poulichet, F. 2003a. Geochemical processes controlling the formation of As-rich waters within a tailings impoundment. *Aquatic Geochemistry* 9: 273–290.

Casiot, C., Morin, G., Juillot, F., Bruneel, O., Personné, J.C., Leblanc, M., Duquesne, K., Bonnefoy, V., and Elbaz-Poulichet, F. 2003b. Bacterial immobilization and oxidation of arsenic in acid mine drainage (Carnoulès creek, France). *Water Research* 37: 2929–2936.

Casiot, C., Pedron, V., Bruneel, O., Duran, R., Personné, J.C., Grapin, G., and Elbaz-Poulichet, F. 2006. A new bacterial strain mediating As oxidation in the Fe-rich biofilm naturally growing in a groundwater Fe treatment pilot unit. *Chemosphere* 64: 492–496.

Cavalca, L., Corsini, A., Zaccheo, P., Andreoni, V., and Muyzer, G. 2013. Microbial transformations of arsenic: Perspectives for biological removal of arsenic from water. *Future Microbiology* 8: 753–768.

Chakraborti, D., Mukherjee, S.C., Pati, S., Sengupta, M.K., Rahman, M.M., Chowdhury, U.K., Lodh, D., Chanda, C.R., Chakraborti, A.K., and Basu, G.K. 2003. Arsenic groundwater contamination in Middle Ganga Plain, Bihar, India: A future danger? *Environmental Health Perspectives* 111: 1194–1201.

Chakraborti, D., Sengupta, M.K., Rahaman, M.M., Ahamed, S., Chowdhury, U.K., Hossain, M.A., Mukherjee, S.C. et al. 2004. Groundwater arsenic contamination and its health effects in the Ganga–Megna–Brahmaputra Plain. *Journal of Environmental Monitoring* 6: 74–83.

Challenger, F. 1945. Biological Methylation. *Chemical Reviews* 36: 315–361.

Chang W.C., Hsu, G.S., Chiang S.M., and Su, M.C. 2006. Heavy metal removal from aqueous solution by wasted biomass from a combined AS-biofilm process. *Bioresource Technology* 97: 1503–1508.

Chen, C.M., Misra, T.K., Silver, S., and Rosen, B.P. 1986. Nucleotide sequence of the structural genes for an anionpump. The plasmid-encoded arsenical resistance operon. *The Journal of Biological Chemistry* 261: 15030–15038.

Chen, J., Sun, G.X., Wang, X.X., De Lorenzo, V., Rosen, B.P., and Zhu, Y.G. 2014. Volatilization of arsenic from polluted soil by Pseudomonas putida engineered for expression of the arsM arsenic(III) Sadenosine methyltransferase gene. *Environmental Science and Technology* 48: 10337–10344.

Chowdhury, U.K., Biswas, B.K., Chowdhury, T.R., Samanta, G., Mandal, B.K., Basu, G.C., Chanda, C.R. et al. 2000. Groundwater arsenic contamination in Bangladesh and West Bengal, India. *Environmental Health Perspectives* 108: 393.

Chung, J., Xiaohao, L., and Rittmann, B.E. 2006. Bio-reduction of arsenate using a hydrogenbased membrane biofilm reactor. *Chemosphere* 65: 24–34.

Cobbett, C. and Goldsbrough, P. 2002. Phytochelatins and metallothioneins: Roles in heavy metal detoxification and homeostasis. *Annual Review of Plant Biology* 53: 159–182.

Cohen, R.R. and Ozawa, T. 2013. Microbial sulfate reduction and biogeochemistry of arsenic and chromium oxyanions in anaerobic bioreactors. *Water, Air and Soil Pollution* 224: 1–14.

Coupland, K., Battaglia-Brunet, F., Hallberg, K.B., Dictor, M.C., Garrido, F., and Johnson, D.B. 2003. Oxidation of iron, sulfur and arsenic in mine waters and mine wastes: An important role for novel *Thiomonas* spp. In: Tsezos, M., Hatzikioseyian, A., and Remoundaki, E. (Eds.), *15th International Biohydrometallurgy Symposium*. National Technical University of Athens, Zografou, Greece, Athens, Hellas, pp. 639–646.

Cullen, W.R. 2005. The toxicity of trimethylarsine: An urban myth. *Journal of Environmental Monitoring* 7: 11–15.

Cullen, W.R. and Reimer, K.J. 1989. Arsenic speciation in the environment. *Chemical Reviews* 89: 713–764.

Dastidar, A. and Wang, Y. 2012. Modeling arsenite oxidation by chemoautotrophic *Thiomonas arsenivorans* strain b6 in a packed-bed bioreactor. *Science of the Total Environment* 432: 113–121.

De vegt, A.L., Bayer, H.G., and Buisman, C.J. 1997. Biological sulfate removal and metal recovery from mine waters. In *SME Annual Meeting*, Denver, CO.

Desoeuvre, A., Casiot, C., and Héry, M. 2016. Diversity and distribution of arsenic-related genes along a pollution gradient in a river affected by Acid Mine Drainage. *Microbial Ecology* 71: 672–685.

Dopp, E., Kligerman, A.D., and Diaz-Bone, R.A. 2010a. Organoarsenicals. Uptake, metabolism, and toxicity. *Metal Ions in Life Sciences* 7: 231–265.

Dopp, E., von Recklinghausen, U., Diaz-Bone, R., Hirner, A.V., and Rettenmeier, A.W. 2010b. Cellular uptake, subcellular distribution and toxicity of arsenic compounds in methylating and non-methylating cells. *Environmental Research* 110: 435–442.

Dopson, M. and Holmes, D.S. 2014. Metal resistance in acidophilic microorganisms and its significance for biotechnologies. *Applied Microbiology and Biotechnology* 98: 8133–8144.

Drewniak, L. and Sklodowska, A. 2013. Arsenic-transforming microbes and their role in biomining processes. *Environmental Science and Pollution Research* 20: 7728–7739.

Du, W.B., Li, Z.A., Zhou, B., and Peng, S.L. 2005. *Pteris multifida Poir.*, a new arsenic hyperaccumulator: Characteristics and potential. *International Journal of Environment and Pollution* 23: 388–396.

Duquesne, K., Lieutaud, A., Ratouchniak, J., Yarzábal, A., and Bonnefoy, V. 2007. Mechanisms of arsenite elimination by *Thiomonas* sp. isolated from Carnoulès acid mine drainage. *European Journal of Soil Biology* 43: 351–355.

Duncan, W.F.A., Mattes, A.G., Gould, W.D., and Goodazi, F. 2004. Multi-stage biological treatment system for removal of heavy metal contaminants. In *The Fifth International Symposium on Waste Processing and Recycling in Mineral and Metallurgical Industries*, Canadian Institute of Mining, Metallurgy and Petroleum, Hamilton, Ontario, Canada, pp. 469–483.

Egal, M., Casiot, C., Morin, G., Elbaz-Poulichet, F., Cordier, M.A., and Bruneel, O. 2010. An updated insight into the natural attenuation of As concentrations in Reigous Creek (southern France). *Applied Geochemistry* 25: 1949–1957.

Elbaz-Poulichet, F., Bruneel, O., and Casiot, C. 2006. The Carnoulès mine. Generation of as-rich acid mine drainage, natural attenuation processes and solutions for passive in-situ remediation. In *Difpolmine (Diffuse Pollution From Mining Activities)*, Montpellier, France.

Ellis, P.J., Conrads, T., Hille, R., and Kuhn, P. 2001. Crystal structure of the 100 kDa arsenite oxidase from *Alcaligenes faecalis* in two crystal forms at 1.64 A and 2.03 A. *Structure* 9: 125–132.

Fitz, W.J. and Wenzel, W.W. 2002. Arsenic transformations in the soil–rhizosphere–plant system: Fundamentals and potential application to phytoremediation. *Journal of Biotechnology* 99: 259–278.

Frohne, T., Rinklebe, J., Diaz-Bone, R.A., and Du Laing, G. 2011. Controlled variation of redox conditions in a floodplain soil: Impact on metal mobilization and biomethylation of arsenic and antimony. *Geoderma* 160: 414–424.

Garcia-Salgado, S., Garcia-Casillas, D., Quijano-Nieto, M.A., and Bonilla-Simon, M.M. 2012. Arsenic and heavy metal uptake and accumulation in native plant species from soils polluted by mining activities. *Water, Air and Soil Pollution* 223: 559–572.

Ghosh, M., Shen, J., Rosen, B.P., 1999. Pathways of As(III) detoxification in *Saccharomyces cerevisiae*. Proc Natl Acad Sci U S A 96, 5001–5006.

Gihring, T.M. and Banfield, J.F. 2001. Arsenite oxidation and arsenate respiration by a new *Thermus* isolate. *FEMS Microbiology Letters* 204: 335–340.

Gladysheva, T.B., Oden, K.L., and Rosen, B.P. 1994. Properties of the arsenate reductase of plasmid R773. *Biochemistry* 33: 7288–7293.

Gourbal, B., Sonuc, N., Bhattacharjee, H., Legare, D., Sundar, S., Ouellette, M., Rosen, B.P., and Mukhopadhyay, R. 2004. Drug uptake and modulation of drug resistance in *Leishmania* by an aquaglyceroporin. *The Journal of Biological Chemistry* 279: 31010–31017.

Grillet, J.P., Adjemian, A., Bernadac, G., Bernon, J., Brunner, F., and Garnier, R. 2004. Arsenic exposure in the wine growing industry in ten French departments. *International Archives of Occupational and Environmental Health* 77: 130–135.

Haferburg, G. and Kothe, E. 2007. Microbes and metals: Interactions in the environment. *Journal of Basic Microbiology* 47: 453–467.

Hallberg, K.B. 2010. New perspectives in acid mine drainage microbiology. *Hydrometallurgy* 104: 448–453.

Hammack, R.W., Dvorak, D.H., and Edenborn, H.M. 1993. The use of biogenic hydrogen sulfide to selectively recover copper and zinc from severely contaminated mine drainage, In: Torma, A.E., Wey, J.E., Lakshmanan, V.L. (Eds.), *Biohydrometallurgical Technologies*. The Minerals, Metals and Materials Society, Warrendale, PA, pp. 631–639.

Handley, K.M., Héry, M., and Lloyd, J.R. 2009. *Marinobacter santoriniensis* sp. nov., an arsenate-respiring and arsenite-oxidizing bacterium isolated from hydrothermal sediment. *International Journal of Systematics and Evolutionary Microbiology* 59: 886–892.

Hassan, K.M., Fukuhara, T., Hai, F.I. Bari, Q.H., and Islam, K.M.S. 2009. Development of a bio-physicochemical technique for arsenic removal from groundwater. *Desalination* 249: 224–229.

Hedrich, S. and Johnson, D.B. 2014. Remediation and selective recovery of metals from acidic mine waters using novel modular bioreactors. *Environmental Science and Technology* 48: 12206–12212.

Heinrich-Salmeron, A., Cordi, A., Brochier-Armanet, C., Halter, D., Pagnout, C., Abbaszadeh-Fard, E., Montaut, D. et al. 2011. Unsuspected diversity of arsenite-oxidizing bacteria revealed by a widespread distribution of the aoxB gene in prokaryotes. *Applied and Environmental Microbiology* 77: 4685–4692.

Héry, M., Casiot, C., Resongles, E., Gallice, Z., Bruneel, O., Desoeuvre, A., and Delpoux, S. 2014. Release of arsenite, arsenate and methyl-arsenic species from streambed sediment affected by acid mine drainage: A microcosm study. *Environmental Chemistry* 11: 514–524.

Héry, M., Rizoulis, A., Sanguin, H., Cooke, D.A., Pancost, R.D., Polya, D.A., and Lloyd, J.R. 2015. Microbial ecology of arsenic-mobilizing Cambodian sediments: Lithological controls uncovered by stable-isotope probing. *Environmental Microbiology* 17: 1857–1869.

Héry, M., van Dongen, B.E., Gill, F., Mondal, D., Vaughan, D.J., Pancost, R.D., Polya, D.A., and Lloyd, J.R. 2010. Arsenic release and attenuation in low organic carbon aquifer sediments from West Bengal. *Geobiology* 8: 155–168.

Hettick, B.E., Cañas-Carrell, J.E., French, A.D., and Klein, D.M. 2015. Arsenic: A review of the element's toxicity, plant interactions, and potential methods of remediation. *Journal of Agricultural and Food Chemistry* 63: 7097–7107.

Hirano, S., Kobayashi, Y., Cui, X., Kanno, S., Hayakawa, T., and Shraim, A. 2004. The accumulation and toxicity of methylated arsenicals in endothelial cells: Important roles of thiol compounds. *Toxicology and Applied Pharmacology* 198: 458–467.

Hoang, T.H., Ju-Yong, K., Sunbaek, B., and Kyoung-Woong, K. 2010. Source and fate of as in the environment. *Geosystem Engineering* 13: 35–42

Hoeft, S.E., Blum, J.S., Stolz, J.F., Tabita, F.R., Witte, B., King, G.M., Santini, J.M., and Oremland, R.S. 2007. *Alkalilimnicola ehrlichii* sp. nov. a novel, arsenite-oxidizing haloalkaliphilic gammaproteobacterium capable of hemoautotrophic or heterotrophic growth with nitrate or oxygen as the electron acceptor. *International Journal of Systematic and Evolutionary Microbiology* 57: 504–512.

Hoeft, S.E., Kulp, T.R., Stolz, J.F., Hollibaugh, J.T., and Oremland, R.S. 2004. Dissimilatory arsenate reduction with sulfide as electron donor: Experiments with Mono Lake water and isolation of strain MLMS-1, a chemoautotrophic arsenate respirer. *Applied and Environmental Microbiology* 70: 2741–2747.

Hossain, M.F. 2006. Arsenic contamination in Bangladesh: An overview. *Agriculture, Ecosystems and Environment* 113: 1–16.

Huang, J.H. 2014. Impact of microorganisms on arsenic biogeochemistry: A review. *Water, Air and Soil Pollution* 225: 1–25.

Hudson-Edwards, K.A. and Santini, J.M. 2013. Arsenic-microbe-mineral interactions in mining-affected environments. *Minerals* 3: 337–351.

Huisman, J.L., Schouten, G., and Schultz, C. 2006. Biologically produced sulphide for purification of process streams, effluent treatment and recovery of metals in the metal and mining industry. *Hydrometallurgy* 83: 106–113.

Huysmans, D.K. and Frankenberger, W.T. 1991. Evolution of trimethylarsine by a *Penicillium* sp. isolated from agricultural evaporation pond water. *Science of the Total Environment* 105: 13–28.

Islam, F.S., Gault, A.G., Boothman, C., Polya, D.A., Charnock, J.M., Chatterjee, D., and Lloyd, J.R. 2004. Role of metal-reducing bacteria in arsenic release from Bengal delta sediments. *Nature* 430: 68–71.

Ito, A., Miura, J., Ishikawa, N., and Umita, T. 2012. Biological oxidation of arsenite in synthetic groundwater using immobilized bacteria. *Water Research* 46: 4825–4831.

Jackson, B.P., Seaman, J.C., and Bertsch, P.M. 2006. Fate of arsenic compounds in poultry litter upon land application. *Chemosphere* 65: 2028–2034.

Jackson, C.K., Koch, I., and Reimer, K.J. 2013. Mechanisms of dissolved arsenic removal by biochemical reactors: A bench- and field-scale study. *Applied Geochemistry* 29: 174–181.

Jareonmit, P., Mehta, M., Sadowsky, M., and Sajjaphan, K. 2012. Phylogenetic and phenotypic analyses of arsenic-reducing bacteria isolated from an old tin mine area in Thailand. *World Journal of Microbiology and Biotechnology* 28: 2287–2292.

Ji, G. and Silver, S. 1992. Reduction of arsenate to arsenite by the ArsC protein of the arsenic resistance operon of *Staphylococcus aureus* plasmid pI258. *Proceedings of the National Academy Sciences of the United States of America* 89: 9474–9478.

Jia, Y., Huang, H., Zhong, M., Wang, F.H., Zhang, L.M., and Zhu, Y.G. 2013. Microbial arsenic methylation in soil and rice rhizosphere. *Environmental Science and Technology* 47: 3141–3148.

Johnson, D.B. and Hallberg, K.B. 2003. The microbiology of acidic mine waters. *Research in Microbiology* 154: 466–473.

Johnson, D.B. and Hallberg, K.B. 2005. Acid mine drainage remediation options: A review. *Science of the Total Environment* 338: 3–14.

Jong, T. and Parry, D.L. 2003. Removal of sulfate and heavy metals by sulfate reducing bacteria in short-term bench scale upflow anaerobic packed bed reactor runs. *Water Research* 37: 3379–3389.

Kagan, R.M. and Clarke, S. 1994. Widespread occurrence of three sequence motifs in diverse Sadenosylmethionine-dependent methyltransferases suggests a common structure for these enzymes. *Archives of Biochemistry and Biophysics* 310: 417–427.

Kalin, M. 2004. Passive mine water treatment: The correct approach? *Ecological Engineering* 22: 299–304.

Kalve, S., Sarangi, B.K., Pandey, R.A., and Chakrabarti, T. 2011. Arsenic and chromium hyperaccumulation by an ecotype of *Pteris vittata*-prospective for phytoextraction from contaminated water and soil. *Current Science* 100: 888–894.

Kashyap, D.R., Botero, L.M., Franck, W.L., Hassett, D.J., and McDermott, T.R. 2006. Complex regulation of arsenite oxidation in *Agrobacterium tumefaciens*. *Journal of Bacteriology* 188: 1081–1088.

Katsoyiannis, I.A., Mitrakas, M., and Zouboulis, A.I. 2015. Introduction to remediation of arsenic toxicity: Application of biological treatment methods for remediation of arsenic toxicity from groundwaters. In: Chakrabarty, N. (Ed.), *Arsenic Toxicity: Prevention and Treatment*. CRC Press, Taylor & Francis Group, LLC, Boca Raton, FL, pp. 113–131.

Katsoyiannis, I.A., Zikoudib, A., and Huga, S.J. 2008. Arsenic removal from groundwaters containing iron, ammonium, manganese and phosphate: A case study from a treatment unit in northern Greece. *Desalination* 224: 330–339.

Katsoyiannis, I.A. and Zouboulis, A.I. 2004. Application of biological processes for the removal of arsenic from groundwaters. *Water Research* 38: 17–26.

Katsoyiannis, I.A. and Zouboulis, A.I. 2006. Comparative evaluation of conventional and alternative treatment methods for the removal of arsenic from contaminated groundwaters. *Reviews on Environmental Health* 21: 25–41.

Katsoyiannis, I.A., Zouboulis, A.I., Althoff, H.W., and Bartel, H. 2002. As(III) removal from groundwater using fixed bed upflow bioreactors. *Chemosphere* 47: 325–332.

Katsoyiannis, I.A., Zouboulis, A.I, and Jekel, M. 2004. Kinetics of bacterial As(III) oxidation and subsequent As(V) removal by sorption onto biogenic manganese oxides during groundwater treatment. *Industrial & Engineering Chemistry Research* 43: 486–493.

Katsoyiannis, I.A., Zouboulis, A.I., Mitrakas, M., Althoff, H.W., and Bartel, H. 2013. A hybrid system incorporating a pipe reactor and microfiltration for biological iron, manganese and arsenic removal from anaerobic groundwater. *Fresenius Environmental Bulletin* 22: 3848–3853.

Kieu, H.T.Q., Müller, E., and Horn, H. 2011. Heavy metal removal in anaerobic semi-continuous stirred tank reactors by a consortium of sulfate-reducing bacteria. *Water Research* 45: 3863–3870.

Klein, R., Tischler, J.S., Mühling, M., and Schlömann, M. 2014. Bioremediation of mine water. *Advances in Biochemical Engineering/Biotechnology* 144: 109–172.

Koechler, S., Farasin, J., Cleiss-Arnold, J., and Arsène-Ploetze, F. 2015. Toxic metal resistance in biofilms: Diversity of microbial responses and their evolution. *Research in Microbiology* 166: 764–773.

Kostal, J., Yang, R., Wu, C. H., Mulchandani, A., and Chen, W. 2004. Enhanced arsenic accumulation in engineered bacterial cells expressing ArsR. *Applied and Environmental Microbiology* 70: 4582–4587.

Krafft, K. and Macy, J.M. 1998. Purification and characterisation of the respiratory arsenate reductase of *Chrysiogenes arsenatis*. *European Journal of Biochemistry* 255: 647–653.

Kruger, M.C., Bertin, P.N., Heipieper, H.J., and Arsène-Ploetze, F. 2013. Bacterial metabolism of environmental arsenic—Mechanisms and biotechnological applications. *Applied Microbiology and Biotechnology* 97: 3827–3841.

Kudo, K., Yamaguchi, N., Makino, T., Ohtsuka, T., Kimura, K., Dong, D.T., and Amachi, S. 2013. Release of arsenic from soil by a novel dissimilatory arsenate reducing bacterium, *Anaeromyxobacter* sp. strain PSR-1. *Applied and Environmental Microbiology* 79: 4635–4642.

Kulp, T.R., Hoeft, S.E., Asao, M., Madigan, M.T., Hollibaugh, J.T., Fisher, J.C., Stolz, J.F., Culbertson, C.W., Miller, L.G., and Oremland, R.S. 2008. Arsenic(III) fuels anoxygenic photosynthesis in hot spring biofilms from Mono Lake, California. *Science* 321: 967–970.

Leblanc, M., Casiot, C., Elbaz-Poulichet, F., and Personné, J. Ch. 2002. Arsenic removal by oxidising bacteria in a heavily arsenic contaminated acid mine drainage system (Carnoulès, France). Volume special "Mine water hydrogeology and geochemistry." Geological Society, London, U.K., Special Publication No. 198, pp. 267–274.

Lehr, C.R., Kashyap, D.R., and McDermott, T.R. 2007. New Insights into microbial oxidation of antimony and arsenic. Applied and *Environmental Microbiology* 73: 2386–2389.

Lewis, A. and van Hille, R. 2006. An exploration into the sulphide precipitation method and its effect on metal sulphide removal. *Hydrometallurgy* 81: 197–204.

Lievremont, D., Bertin, P.N., and Lett, M.C. 2009. Arsenic in contaminated waters: Biogeochemical cycle, microbial metabolism and biotreatment processes. *Biochimie* 91: 1229–1237.

Lin, Y.F., Walmsley, A.R., and Rosen, B.P. 2006. An arsenic metallochaperone for an arsenic detoxificationpump. *Proceedings of the National Academy and Sciences of the United States of America* 103: 15617–15622.

Litter, M.I., Morgada, M.E., and Bundschuh, J. 2010. Possible treatments for arsenic removal in Latin American waters for human consumption. *Environmental Pollution* 158: 1105–1118.

Liu, S., Zhang, F., Chen, J., and Sun, G. 2011. Arsenic removal from contaminated soil via biovolatilization by genetically engineered bacteria under laboratory conditions. *Journal of Environmental Sciences* 23: 1544–1550.

Liu, X., Zhang, W., Hu, Y., Hu, E., Xie, X., Wang, L., and Cheng, H. 2015. Arsenic pollution of agricultural soils by concentrated animal feeding operations (CAFOs). *Chemosphere* 119: 273–281.

Lizama, K.A., Fletcher, T.D., and Sun, G. 2011. Removal processes for arsenic in constructed wetlands. *Chemosphere* 84: 1032–1043.

Lloyd, J.R., Gault, A.G., Héry, M., and MacRae, J.D. 2011. Microbial transformations of arsenic in the subsurface. In: Oremland, R.S. and Stolz, J.F. (Eds.), *Microbial Metal and Metalloid Metabolism*. American Society of Microbiology, Washington, DC, pp. 77–90.

Luo, Q., Tsukamoto, T.K., Zamzow, K.L., and Miller, G.C. 2008. Arsenic, selenium, and sulfate removal using an ethanol-enhanced sulfate-reducing bioreactor. *Mine Water and the Environment* 27: 100–108.

Ma, L.Q., Komar, K.M., Tu, C., Zhang, W., Cai, Y., and Kennelley, E.D. 2001. A fern that hyperaccumulates arsenic. *Nature* 409: 579.

Ma, Y., Prasad, M.N.V., Rajkumar, M., and Freitas, H. 2011. Plant growth promoting rhizobacteria and endophytes accelerate phytoremediation of metalliferous soils. *Biotechnology Advances* 29: 248–258.

Macur, R.E., Jackson, C.R., Botero, L.M., McDermott, T.R., and Inskeep, W.P. 2004. Bacterial populations associated with the oxidation and reduction of arsenic in an unsaturated soil. *Environmental Science and Technology* 38: 104–111.

Macy, J.M., Nunan, K., Hagen, K.D., Dixon, D.R., Harbour, P.J., Cahill, M., and Sly, L.I. 1996. *Chrysiogenes arsenatis* gen. nov., sp. nov., a new arsenate-respiring bacterium isolated from gold mine wastewater. *International Journal of Systematic and Evolutionary Microbiology* 46: 1153–1157.

Madejon, P., Murillo, J.M., Maranon, T., Cabrera, F., and Lopez, R. 2002. Bioaccumulation of As, Cd, Cu, Fe and Pb in wild grasses affected by the Aznalcollar mine spill (SW Spain). *Science of the Total Environment* 290: 105–120.

Maki, T., Hasegawa, H., Watarai, H., and Ueda, K. 2004. Classification for dimethylarsenate-decomposing bacteria using a restrict fragment length polymorphism analysis of 16S rRNA genes. *Analytical Sciences* 20: 61–68.

Malasarn, D., Keeffe, J.R., and Newman, D.K. 2008. Characterization of the arsenate respiratory reductase from *Shewanella* sp. strain ANA-3. *Journal of Bacteriology* 190: 135–142.

Malik, A.H., Khan, Z.M., Mahmood, Q., Nasreen, S., and Bhatti, Z.A. 2009. Perspectives of low cost arsenic remediation of drinking water in Pakistan and other countries. *Journal of Hazardous Materials* 168: 1–12.

Mandal, A., Purakayastha, T.J., Patra, A.K., and Sanyal, S.K. 2012. Phytoremediation of arsenic contaminated soil by *Pteris vittata* L. II. Effect on arsenic uptake and rice yield. *International Journal of Phytoremediation* 14: 621–628.

Mandal, B.K. and Suzuki, K.T. 2002. Arsenic round the world: A review. *Talanta* 58: 201–235.

Marchal, M., Briandet, R., Koechler, S., Kammerer, B., and Bertin P.N. 2011. Effect of arsenite on swimming motility delays surface colonization in *Herminiimonas arsenicoxydans*. *Microbiology* 6: e23181.

Matschullat, J. 2000. Arsenic in the geosphere—A review. *Science of the Total Environment* 249: 297–312.

Mattes, A., Duncan, W.F.A., and Gould, W.D. 2004. Biological removal of arsenic in a multi-stage engineered wetlands: Treating a suite of heavy metals. In Proceedings of the 28th Annual British Columbia Mines Reclamation Symposium, Cranbrook, BC, June 23.

Mattes, A., Evans, L.J., Douglas Gould, W., Duncan, W.F.A., and Glasauer, S. 2011. The long term operation of a biologically based treatment system that removes As, S and Zn from industrial (smelter operation) landfill seepage. *Applied Geochemistry* 26: 1886–1896.

McBride, B.C. and Wolfe, R.S. 1971. Biosynthesis of dimethylarsine by *Methanobacterium*. *Biochemistry* 10: 4312–4317.

Mendoza-Cozatl, D.G., Rodriguez-Zavala, J.S., Rodriguez-Enriquez, S., Mendoza-Hernandez, G., Briones-Gallardo, R., and Moreno-Sanchez, R. 2006. Phytochelatin–cadmium–sulfide high molecular-mass complexes of *Euglena gracilis*. *FEBS Journal* 273: 5703–5713.

Menezes, R.A., Amaral, C., Delaunay, A., Toledano, M., and Rodrigues-Pousada, C. 2004. Yap8p activation in *Saccharomyces cerevisiae* under arsenic conditions. *FEBS Letters* 566: 141–146.

Meng, Y.-L., Liu, Z., and Rosen, B.P., 2004. As(III) and Sb(III) uptake by GlpF and efflux by ArsB in *Escherichia coli*. *Journal of Biological Chemistry*. 279: 18334–18341.

Merola, R.B., Hien, T.T., Quyen, D.T.T., and Vengosh, A. 2015. Arsenic exposure to drinking water in the Mekong Delta. *Science of the Total Environment* 511: 544–552.

Merrifield, M.E., Ngu, T., and Stillman, M.J. 2004. Arsenic binding to *Fucus vesiculosus* metallothionein. *Biochemical and Biophysical Research Communications* 324: 127–132.

Mestrot, A., Feldmann, J., Krupp, E.M., Hussain, M.S., Roman-Ross, G., and Meharg, A.A. 2011. Field fluxes and speciation of arsines emanating from soils. *Environmental Science and Technology* 45: 1798–1804.

Michalke, K., Wickenheiser, E.B., Mehring, M., Hirner, A.V., and Hensel, R. 2000. Production of volatile derivatives of metal(loid)s by microflora involved in anaerobic digestion of sewage sludge. *Applied and Environmental Microbiology* 66: 2791–2796.

Michel, C., Jean, M., Coulon, S., Dictor, M.C., Delorme, F., Morin, D., and Garrido, F. 2007. Biofilms of As(III)-oxidising bacteria: Formation and activity studies for bioremediation process development. *Applied and Microbiology and Biotechnology* 77: 457–467.

Michon, J., Dagot, C., Deluchat, V., Dictor, M.C., Battaglia-Brunet, F., and Baudu, M. 2010. As(III) biological oxidation by CAsO1 consortium in fixed-bed reactors. *Process Biochemistry* 45: 171–178.

Miller, W.H., Schipper, H.M., Lee, J.S., Singer, J., and Waxman, S. 2002. Mechanisms of action of arsenic trioxide. *Cancer Research* 62: 3893–3903.

Mirza, N., Pervez, A., Mahmood, Q., Shah, M.M., and Shafqat, M.N. 2011. Ecological restoration of arsenic contaminated soil by *Arundo donax* L. *Ecological Engineering* 37: 1949–1956.

Mohammed Abdul, K.S., Jayasinghe, S.S., Chandana, E.P.S., Jayasumana, C., and De Silva, P.M.C.S. 2015. Arsenic and human health effects: A review. *Environmental Toxicology and Pharmacology* 40: 828–846.

Mohan, D. and Pittman Jr., C.U. 2007. Arsenic removal from water/wastewater using adsorbents: A critical review. *Journal of Hazardous Materials* 142: 1–53.

Mondal, P., Majumder, C.B., and Mohanty, B. 2006. Laboratory based approaches for arsenic remediation from contaminated water: Recent developments. *Journal of Hazardous Materials* 137: 464–479.

Mondal, P., Majumder, C.B., and Mohanty, B. 2008. Treatment of arsenic contaminated water in a laboratory scale up-flow bio-column reactor. *Journal of Hazardous Materials* 153: 136–145.

Mokashi, S.A. and Paknikar, K.M. 2002. Arsenic (III) oxidizing *Microbacterium lacticum* and its use in the treatment of arsenic contaminated groundwater. *Letters in Applied Microbiology* 34: 258–262.

Morin, G., Juillot, F., Casiot, C., Bruneel, O., Personné, J.C., Elbaz-Poulichet, F., Leblanc, M., Ildefonse, P., and Calas, G. 2003. Bacterial formation of tooeleite and mixed arsenic(III) or arsenic(V)-Iron(III) gels in the Carnoulès acid mine drainage, France. A XANES, XRD, and SEM study. *Environmental Science and Technology* 37: 1705–1712.

Mukherjee, A.B. and Bhattacharya, P. 2001. Arsenic in groundwater in the Bengal delta plain: Slow poisoning in Bangladesh. *Environmental Reviews* 9: 189–220.

Mukhopadhyay, R., Rosen, B.P., Phung, L.T., and Silver, S. 2002. Microbial arsenic: From geocycles to genes and enzymes. *FEMS Microbiology Reviews* 26: 311–325.

Muller, D., Lievremont, D., Simeonova, D.D., Hubert, J.C., and Lett, M.C. 2003. Arsenite oxidase aox genes from a metal-resistant beta-proteobacterium. *Journal of Bacteriology* 185: 135–141.

Nachman, K.E., Baron, P.A., Raber, G., Francesconi, K.A., Navas-Acien, A., and Love, D.C. 2013. Roxarsone, inorganic arsenic, and other arsenic species in chicken: A U.S.-based market basket sample. *Environmental Health Perspectives* 121: 818–824.

Naujokas, M.F., Anderson, B., Ahsan, H., Aposhian, H.V., Graziano, J., Thompson, C., and Suk, W.A. 2013. The broad scope of health effects from chronic arsenic exposure: Update on a worldwide public health problem. *Environmental Health Perspectives* 121: 295–302.

Newman, D.K., Beveridge, T.J., and Morel, F. 1997. Precipitation of arsenic trisulfide by *Desulfotomaculum auripigmentum*. *Applied and Environmental Microbiology* 63: 2022–2028.

Nickson, R.T., McArthur, J.M., Shrestha, B., Kyaw-Mint, T.O., and Lowry, D. 2005. Arsenic and other drinking water quality issues in Muzaffargarh District, Pakistan. *Applied Geochemistry* 20: 55–68.

Nordstrom, D.K. 2002. Public health. Worldwide occurrences of arsenic in ground water. *Science* 296: 2143–2145.

Ohtsuka, T., Yamaguchi, N., Makino, T., Sakurai, K., Kimura, K., Kudo, K., Homma, E., Dong, D.T., and Amachi, S. 2013 Arsenic dissolution from Japanese paddy soil by a dissimilatory arsenate-reducing bacterium *Geobacter* sp. OR-1. *Environmental Science and Technology* 47: 6263–6271.

Oremland, R.S., Hoeft, S.E., Santini, J.M., Bano, N., Hollibaugh, R.A., and Hollibaugh, J.T. 2002. Anaerobic oxidation of arsenite in Mono Lake water and by a facultative, arsenite oxidizing chemoautotroph, strain MLHE-1. *Applied and Environmental Microbiology* 68: 4795–4802.

Oremland, R.S. and Stolz, J.F. 2003. The ecology of arsenic. *Science* 300: 939–944.

Oremland, R.S. and Stolz, J.F. 2005. Arsenic, microbes and contaminated aquifers. *Trends in Microbiology* 13: 45–49.

Padmavathiamma, P.K. and Li, L.Y. 2007. Phytoremediation technology: Hyperaccumulation metals in plants. *Water Air and Soil Pollution* 184: 105–126.

Páez-Espino, D., Tamames, J., de Lorenzo, V., and Cánovas, D. 2009. Microbial responses to environmental arsenic. *Biometals* 22: 117–130.

Paknikar, K.M. 2003. An integrated process for the removal of groundwater arsenic. *Journal of Applied Hydrology* 17: 8–17.

Pal, A. and Paknikar, K.M. 2012. Bioremediation of arsenic from contaminated water. In: Satyanarayana, T., Johri, B.N., and Prakash, A. (Eds.), *Microorganisms in Environmental Management: Microbes and Environment.* Springer, Dordrecht, the Netherlands, pp. 477–523.

Paul, S., Das, N., Bhattacharjee, P., Banerjee, M., Das, J.K., Sarma, N., Sarkar, A. et al. 2013. Arsenic-induced toxicity and carcinogenicity, a two-wave cross-sectional study in arsenicosis individuals in West Bengal, India. *Journal of Exposure Science and Environmental Epidemiology* 23: 156–162.

Persson, B.L., Petersson, J., Fristedt, U., Weinander, R., Berhe, A., and Pattison, J. 1999. Phosphate permeases of *Saccharomyces cerevisiae*: Structure, function and regulation. *Biochimica et Biophysica Acta* 1422: 255–272.

Pokhrel, D. and Viraraghavan, T. 2009. Biological filtration for removal of arsenic from drinking water. *Journal of Environmental Management* 90: 1956–1961.

Prithivirajsingh, S., Mishra, S.K., and Mahadevan, A. 2001. Detection and analysis of chromosomal arsenic resistance in *Pseudomonas fluorescens* strain MSP3. *Biochemical and Biophysical Research Communications* 280: 1393–1401.

Qin, J., Rosen, B.P., Zhang, Y., Wang, G., Franke, S., and Rensing, C. 2006. Arsenic detoxification and evolution of trimethylarsine gas by a microbial arsenite S-adenosylmethionine methyltransferase. *Proceedings of the National Academy of Sciences of the United States of America* 103: 2075–2080.

Rahman, M.A., Hasegawa, H., and Lim, R.P. 2012. Bioaccumulation, biotransformation and trophic transfer of arsenic in the aquatic food chain. *Environmental Research* 116: 118–135.

Rahman, M.A. and Hassler, C. 2014. Is arsenic biotransformation a detoxification mechanism for microorganisms? *Aquatic Toxicology* 146: 212–219.

Rahman, M.M., Chowdhury, U.K., Mukherjee, S.C., Mondal, B.K., Paul, K., Lodh, D., Biswas, B.K. et al. 2001. Chronic arsenic toxicity in Bangladesh and West Bengal, India—A review and commentary. *Journal of Toxicology: Clinical Toxicology* 39: 683–700.

Rahman, S., Kim, K.H., Saha, S.K., Swaraz, A.M., and Paul, D.K. 2014. Review of remediation techniques for arsenic (As) contamination: A novel approach utilizing bio-organisms. *Journal of Environmental Management* 134: 175–85.

Rajkumar, M., Sandhya, S., Prasad, M.N.V., and Freitas, H. 2012. Perspectives of plant-associated microbes in heavy metal phytoremediation. *Biotechnology Advances* 30: 1562–1574.

Richey, C., Chovanec, P., Hoeft, S.E., Oremland, R.S., Basu, P., and Stolz, J.F. 2009. Respiratory arsenate reductase as a bidirectional enzyme. *Biochemical Biophysical Research Communications* 382: 298–302.

Rhine, E.D., Phelps, C.D., and Young, L.Y. 2006. Anaerobic arsenite oxidation by novel denitrifying isolates. *Environmental Microbiology* 8: 899–908.

Rochette, E.A., Bostick, B.C, Li, G.C., and Fendorf, S. 2000. Kinetics of arsenate reduction by dissolved sulfide. *Environmental Science and Technology* 34: 4714–4720.

Rosen, B.P. 1999. Families of arsenic transporters. *Trends in Microbiology* 7: 207–212.

Rosen, B.P. and Liu, Z. 2009. Transport pathways for arsenic and selenium: A minireview. *Environmental International* 35: 512–515.

Rosenberg, H., Gerdes, R.G., and Chegwidden, K. 1977. Two systems for the uptake of phosphate in *Escherichia coli*. *Journal of Bacteriology* 131: 505–511.

Rowe, O.F., Sánchez-España, J., Hallberg, K.B., and Johnson, D.B. 2007. Microbial communities and geochemical dynamics in an extremely acidic, metal-rich stream at an abandoned sulfide mine (Huelva, Spain) underpinned by two functional primary production systems. *Environmental Microbiology* 9: 1761–1771.

Roy, M., Giri, A.K., Dutta, S., and Mukherjee, P. 2015. Integrated phytobial remediation for sustainable management of arsenic in soil and water. *Environmental International* 75: 180–198.

Sahinkaya, E., Yurtsever, A., Toker, Y., Elcik, H., Cakmaci, M., and Kaksonen, A.H. 2015. Biotreatment of As-containing simulated acid mine drainage using laboratory scale sulfate reducing upflow anaerobic sludge blanket reactor. *Minerals Engineering* 75: 133–139.

Saltikov, C.W. and Newman, D.K. 2003. Genetic identification of a respiratory arsenate reductase. *Proceedings of the National Academy of Sciences of the United States of America* 100: 10983–10988.

Sanchez-Rodas, D., Gomez-Ariza, J.L., Giraldez, I., Velasco, A., and Morales, E. 2005. Arsenic speciation in river and estuarine waters from southwest Spain. *Science of the Total Environment* 345: 207–217.

Santini, J.M., Sly, L.I., Schnagl, R.D., and Macy, J.M. 2000. A new chemolithoautotrophic arsenite-oxidizing bacterium isolated from a gold mine: Phylogenetic, physiological, and preliminary biochemical studies. *Applied and Environmental Microbiology* 66: 92–97.

Santini, J.M. and Vanden Hoven, R.N. 2004. Molybdenum-containing arsenite oxidase of the chemolithoautotrophic arsenite oxidizer NT-26. *Journal of Bacteriology* 186: 1614–1619.

Sarkar, A., Kazy, S.K., and Sar, P. 2013. Characterization of arsenic resistant bacteria from arsenic rich groundwater of West Bengal, India. *Ecotoxicology* 22: 363–376.

Sarkar, A., Kazy, S.K., and Sar, P. 2014. Studies on arsenic transforming groundwater bacteria and their role in arsenic release from subsurface sediment. *Environmental Science and Pollution Research* 21: 8645–8662.

Sarkar, S., Blaney, L.M., Gupta, A., Ghosh, D., and Sengupta, A.K. 2008. Arsenic removal from groundwater and its safe containment in a rural environment: Validation of a sustainable approach. *Environmental Science and Technology* 42: 4268–4273.

Scheeren, P.J.H., Koch, R.O., and Buisman, C.J.N. 1993. Geohydrological containment system and microbial water treatment plant for metal-contaminated groundwater at Budelco. In *International Symposium—World Zinc'93*, Hobart, Tasmania, Australia, pp. 373–384.

Schwindaman, J.P., Castle, J.W., and Rodgers Jr., J.H. 2014. Biogeochemical process-based design and performance of a pilot-scale constructed wetland for arsenic removal from simulated Bangladesh groundwater. *Water, Air and Soil Pollution* 225: 1–11.

Sen Gupta, B. 2006. Subterranean arsenic removal technology for in-situ groundwater arsenic & iron treatment. Kolkata, India. www.insituarsenic.org.

Sen Gupta, B., Chatterjee, S., Rott, U., Kauffman, H., Bandopadhyay, A., DeGroot, W., Nag, N.K., Carbonell-Barrachina, A.A., and Mukherjee, S. 2009. A simple chemical free arsenic removal method for community water supply—A case study from West Bengal, India. *Environmental Pollution* 157: 3351–3353.

Shariatpanahi, M., Anderson, A.C., Abdelghani, A.A., Englande, A.J. 1983. Microbial metabolism of an organic arsenical herbicide, p. 268–277. In: Oxley, T.A. and Barry, S. (Eds.), *Biodeterioration*. John Wiley & Sons, Chichester, U.K., pp. 3268–3274.

Silver, S. 1998. Genes for all metals, a bacterial view of the periodic table. *Journal of Industrial Microbiology and Biotechnology* 20: 1–12.

Silver, S. and Phung, L.T. 2005. Genes and enzymes involved in bacterial oxidation and reduction of inorganic arsenic. *Applied & Environmental Microbiology* 71: 599–608.

Singh, J.S., Abhilash, C., Singh, N.B., Singh, R., and Singh, D. 2011. Genetically engineered bacteria: An emerging tool for environmental remediation and future research perspectives. *Gene* 480: 1–9.

Singh, R., Paul, D., and Jain, R.K. 2006. Biofilms: Implications in bioremediation. *Trends in Microbiology* 14: 389–397.

Singh, R., Singh, S., Parihar, P., Singh, V.P., and Prasad, S.M. 2015. Arsenic contamination, consequences and remediation techniques: A review. *Ecotoxicology and Environmental Safety* 112: 247–270.

Singh, S., Kang, S.H., Lee, W., Mulchandani, A., and Chen, W. 2010. Systematic engineering of phytochelatin synthesis and arsenic transport for enhanced arsenic accumulation in *E. coli*. *Biotechnology and Bioengineering* 105: 780–785.

Singh, S., Lee, W., DaSilva, N.A., Mulchandani, A., and Chen, W. 2008. Enhanced arsenic accumulation by engineered yeast cells expressing *Arabidopsis thaliana* phytochelatin synthase. *Biotechnology and Bioengineering* 99: 333–340.

Slyemi, D. and Bonnefoy, V. 2012. How prokaryotes deal with arsenic. *Environmental Microbiology Reports* 4: 571–586.

Smedley, P.L. and Kinniburgh, D.G. 2002. A review of the source, behaviour and distribution of arsenic in natural waters. *Applied Geochemistry* 17: 517–568.

Smieja, J.A. and Wilkin, R.T. 2003. Preservation of sulfidic waters containing dissolved As(III). *Journal of Environmental Monitoring* 5: 913–916.

Smith, A.H., Lopipero, P.A., Bates, M.N., and Steinmaus, C.M. 2002. Arsenic epidemiology and drinking water standards. *Science* 296: 2145–2146.

Soda, S., Kanzaki, M., Yamamura, S., Kashiwa, M., Fujita, M., and Ike, M. 2009. Slurry bioreactor modeling using a dissimilatory arsenate-reducing bacterium for remediation of arsenic-contaminated soil. *Journal of Bioscience and Bioengineering* 107: 130–137.

Song, W.Y., Sohn, E.J., Martinoia, E., Lee, Y.J., Yang, Y.Y., Jasinski, M., Forestier, C., Hwang, I., and Lee, Y. 2003. Engineering tolerance and accumulation of lead and cadmium in transgenic plants. *Nature Biotechnology* 21: 914–919.

Srivastava, M., Ma, L.Q., and Santos, J.A.G. 2006. Three new arsenic hyperaccumulating ferns. *Science of the Total Environment* 364: 24–31.

Stolz, J.F., Basu, P., and Oremland, R.S. 2010. Microbial arsenic metabolism: New twists on an old poison. *Microbe* 5: 53–59.

Stolz, J.F., Basu, P., Santini, J.M., and Oremland, R.S. 2006. Arsenic and selenium in microbial metabolism. *Annual Review of Microbiology* 60: 107–130.

Styblo, M., Del Razo, L.M., Vega, L., Germolec, D.R., LeCluyse, E.L., Hamilton, G.A., Reed, W., Wang, C., Cullen, W.R., and Thomas, D.J. 2000. Comparative toxicity of trivalent and pentavalent inorganic and methylated arsenicals in rat and human cells. *Archives of Toxicology* 74: 289–299.

Sun, H.J., Rathinasabapathi, B., Wu, B., Luo, J., Pu, L.P., and Ma, L.Q. 2014. Arsenic and selenium toxicity and their interactive effects in humans. *Environment International* 69: 148–158.

Sun, W., Sierra-Alvarez, R., and Field, J.A. 2011. Long term performance of an arsenite-oxidizing-chlorate-reducing microbial consortium in an upflow anaerobic sludge bed (UASB) bioreactor. *Bioresource Technology* 102: 5010–5016.

Sun, W., Sierra-Alvarez, R., Milner, L., Oremland, R., and Field, J.A. 2009. Arsenite and ferrous iron oxidation linked to chemolithotrophic denitrification for the immobilization of arsenic in anoxic environments. *Environmental Science and Technology* 43: 6585–6591.

Switzer Blum, J., Burns Bindi, A., Buzzelli, J., Stolz, J.F., and Oremland, R.S. 1998. *Bacillus arsenicoselenatis*, sp. nov., and *Bacillus selenitireducens*, sp. nov.: Two haloalkaliphiles from Mono Lake, California that respire oxyanions of selenium and arsenic. *Archives in Microbiology* 171: 19–30.

Tandukar, N., Bhattacharya, P., Neku, A., and Mukherjee, A.B. 2006. Extent and severity of arsenic poisoning in Nepal. In: Naidu, R., Smith, E., Owens, G., Bhattacharya, P., and Nadebaum, P. (Eds.), *Managing Arsenic in the Environment: From Soil to Human Health*. CSIRO Publishing, Melbourne, Victoria, Australia, pp. 595–604.

Tassé, N., Isabel, D., and Fontaine, R. 2003. Wood cadillac mine tailings: Designing a biofilter for arsenic control. In *Proceedings of Sudbury*, Sudbury, Ontario, Canada.

Thomas, D.J., Li, J., Waters, S.B., Xing, W., Adair, B.M., Drobna, Z., Devesa, V., and Styblo, M. 2007. Arsenic (+3 oxidation state) methyltransferase and the methylation of arsenicals. *Experimental Biology and Medicine* 232: 3–13.

Tripathi, R.D., Srivastava, S., Mishra, S., Singh, N., Tuli, R., Gupta, D.K., and Maathuis, F.J.M. 2007. Arsenic hazards: Strategies for tolerance and remediation by plants. *Trends in Biotechnology* 25: 158–164.

Tsai, S.L., Singh, S., and Chen, W. 2009. Arsenic metabolism by microbes in nature and the impact on arsenic remediation. *Current Opinion in Biotechnology* 20: 659–667.

Ungureanu, G., Santos, S., Boaventura, R., and Botelho, C. 2015. Arsenic and antimony in water and wastewater: Overview of removal techniques with special reference to latest advances in adsorption. *Journal of Environmental Management* 151: 326–342.

Upadhyaya, G., Clancy, T.M., Brown, J., Hayes, K.F., and Raskin, L. 2012. Optimization of arsenic removal water treatment system through characterization of terminal electron accepting processes. *Environmental Science and Technology* 46: 11702–11709.

Upadhyaya, G., Jackson, J., Clancy, T.M., Hyun, S.P., Brown, J., Hayes, K.F., and Raskin, L. 2010. Simultaneous removal of nitrate and arsenic from drinking water sources utilizing a fixed-bed bioreactor system. *Water Research* 44: 4958–4969.

USFDA (U.S. Food and Drug Administration). 2014. Withdrawal of approval of new animal drug applications for combination drug medicated feeds containing an arsenical drug. *Federal Register* 79: 10974–10976.

Van Halem, D., Olivero, S., de Vet, W.W.J.M., Verberk, J.Q.J.C., Amy, G.L., and van Dijk, J.C. 2010. Subsurface iron and arsenic removal for shallow tube well drinking water supply in rural Bangladesh. *Water Research* 44: 5761–5769.

Van Lis, R., Nitschke, W., Duval, S., and Schoepp-Cothenet, B. 2013. Arsenics as bioenergetic substrates. *Biochimica et Biophysica Acta* 1827: 176–188.

Vanden Hoven, R.N. and Santini, J.M. 2004. Arsenite oxidation by the heterotrophe *Hydrogenophaga* sp. str. NT-14: The arsenite oxidase and its physiological electron acceptor. *Biochimica et Biophysica Acta* 1656: 148–155.

Vithanage, M., Dabrowska, B.B., Mukherjee, B., Sandhi, A., and Bhattacharya, P. 2012. Arsenic uptake by plants and possible phytoremediation applications: A brief overview. *Environmental Chemistry Letters* 10: 217–224.

Wan, J., Klein, J., Simon, S., Joulian, C., Dictor, M.C., Deluchat, V., and Dagot, C. 2010. AsIII oxidation by *Thiomonas arsenivorans* in up-flow fixed-bed reactors coupled to As sequestration onto zero-valent iron-coated sand. *Water Research* 44: 5098–5108.

Wang, L., Chen, S., Xiao, X., Huang, X., You, D., Zhou, X., and Deng, Z. 2006. arsRBOCT arsenic resistance system encoded by linear plasmid pHZ227 in *Streptomyces* sp. strain FR-008. *Applied and Environmental Microbiology* 72: 3738–3742.

Wang, P., Sun, G., Jia, Y., Meharg, A.A., and Zhu, Y. 2014. A review on completing arsenic biogeochemical cycle: Microbial volatilization of arsines in environment. *Journal of Environmental Sciences* 26: 371–381.

Wang, P.P., Bao, P., and Sun, G.X. 2015. Identification and catalytic residues of the arsenite methyltransferase from a sulfate-reducing bacterium, *Clostridium* sp. BXM. *FEMS Microbiology Letters* 362: 1–8.

Wang, S. and Mulligan, C.N. 2006. Natural attenuation processes for remediation of arsenic contaminated soils and groundwater. *Journal of Hazardous Materials* 138: 459–470.

Wang, S. and Zhao, X. 2009. On the potential of biological treatment for arsenic contaminated soils and groundwater. *Journal of Environmental Management* 90: 2367–2376.

Williams, M. 2001. Arsenic in mine waters: An international study. *Environmental Geology* 40: 267–278.

Williams, P.N., Islam, M.R., Adomako, E.E., Raab, A., Hossain, S.A., Zhu, Y.G., Feldmann, J., and Meharg, A.A. 2006. Increase in rice grain arsenic for regions of Bangladesh irrigating paddies with elevated arsenic in groundwaters. *Environmental Sciences and Technology* 40: 4903–4908.

Willsky, G.R. and Malamy, M.H. 1980. Characterization of two genetically separable inorganic phosphate transport systems in *Escherichia coli*. *Journal of Bacteriology* 144: 356–365.

Wu, J. and Rosen, B.P. 1993a. Metalloregulated expression of the ars operon. *The Journal of Biological Chemistry* 268: 52–58.

Wu, J. and Rosen, B.P. 1993b. The arsD gene encodes a second trans-acting regulatory protein of the plasmidencoded arsenical resistance operon. *Molecular Microbiology* 8: 615–623.

Wu, J., Tisa, L.S., and Rosen, B.P. 1992. Membrane topology of the ArsB protein, the membrane subunit of an anion-translocating ATPase. *The Journal of Biological Chemistry* 267: 12570–12576.

Wysocki, R., Chéry, C.C., Wawrzycka, D., Van Hulle, M., Cornelis, R., Thevelein, J.M., and Tamás, M.J. 2001. The glycerol channel Fps1p mediates the uptake of arsenite and antimonite in *Saccharomyces cerevisiae*. *Molecular Microbiology* 40: 1391–1401.

Xie, Z., Luo, Y., Wang, Y., Xie, X., and Su, C. 2013. Arsenic resistance and bioaccumulation of an indigenous bacterium isolated from aquifer sediments of Datong Basin, Northern China. *Geomicrobiology Journal* 30: 549–556.

Yamamura, S. and Amachi, S. 2014. Microbiology of inorganic arsenic: From metabolism to bioremediation. *Journal of Bioscience and Bioengineering* 118: 1–9.

Yamamura, S., Watanabe, M., Kanzaki, M., Soda, S., and Ike, M. 2008. Removal of arsenic from contaminated soils by microbial reduction of arsenate and quinone. *Environmental Science and Technology* 42: 6154–6159.

Yamamura, S., Watanabe, M., Yamamoto, N., Sei, K., and Ike, M. 2009. Potential for microbially mediated redox transformations and mobilization of arsenic in uncontaminated soils. *Chemosphere* 77: 169–174.

Yamamura, S., Yamamoto, N., Ike, M., and Fujita, M. 2005. Arsenic extraction from solid phase using a dissimilatory arsenate-reducing bacterium. *Journal of Bioscience and Bioengineering* 100: 219–222.

Yang, L., Li, X., Chu, Z., Ren, Y., and Zhang, J. 2014. Distribution and genetic diversity of the microorganisms in the biofilter for the simultaneous removal of arsenic, iron and manganese from simulated groundwater. *Bioresource Technology* 156: 384–388.

Yang, Y., Wu, S., Lilley, R.M., and Zhang, R. 2015. The diversity of membrane transporters encoded in bacterial arsenic-resistance operons. *Peer Journal* 3: e943.

Ye, J., Rensing, C., Rosen, B.P., and Zhu, Y.G. 2012. Arsenic biomethylation by photosynthetic organisms. *Trends in Plant Science* 17: 155–162.

Yuan, C., Lu, X., Qin, J., Rosen, B.P., and Le, X.C. 2008. Volatile arsenic species released from *Escherichia coli* expressing the AsIII S-adenosylmethionine methyltransferase gene. *Environmental Science and Technology* 42: 3201–3206.

Zargar, K., Conrad, A., Bernick, D.L., Lowe, T.M., Stolc, V., Hoeft, S., Oremland, R.S., Stolz, J., and Saltikov, C.W. 2012. ArxA, a new clade of arsenite oxidase within the DMSO reductase family of molybdenum oxidoreductases. *Environmental Microbiology* 14: 1635–1645.

Zargar, K., Hoeft, S., Oremland, R., and Saltikov, C.W. 2010. Identification of a novel arsenite oxidase gene, arxA, in the haloalkaliphilic, arsenite-oxidizing bacterium *Alkalilimnicola ehrlichii* strain MLHE-1. *Journal of Bacteriology* 192: 3755–3762.

Zhao, F.J., Harris, E., Yan, J., Ma, J., Wu, L., Liu, W., McGrath, S.P., Zhou, J., and Zhu, Y.G. 2013. Arsenic methylation in soils and its relationship with microbial arsM abundance and diversity, and As speciation in rice. *Environmental Science and Technology* 47: 7147–7154.

Zheng, W., Scifleet, J., Yu, X., Jiang, T., and Zhang, R. 2013. Function of arsATorf7orf8 of *Bacillus* sp. CDB3 in arsenic resistance. *Journal of Environmental Sciences* 25: 1386–1392.

Zobrist, J., Dowdle, P.R., Davis, J.A., and Oremland, R.S. 2000. Mobilization of arsenite by dissimilatory reduction of adsorbed arsenate. *Environmental Science and Technology* 34: 4747–4753.

Zouboulis, A.I. and Katsoyiannis, I.A. 2005. Recent advances in the bioremediation of arsenic contaminated groundwaters. *Environment International* 31: 213–219.

33 Efficacy of Lead-Resistant Microorganisms for Bioremediation of Lead-Contaminated Sites

Raj Mohan B., Jaya Mary Jacob, and Dhilna Damodharan

CONTENTS

ABSTRACT

Lead pollution and its associated aftereffects have existed as a major issue affecting the total quality of the environment. Traditional methods of remediating lead-contaminated sites include a variety of physical, thermal, and chemical treatments, which in general are expensive *ex situ* approaches. In this context, the efficiency of lead-resistant microorganisms for the bioremediation of lead-contaminated soils has emerged as a promising cost-effective technology for practical utilization. In general, the mechanisms of metal resistance in microbes include precipitation of metals such as phosphates, carbonates, and/or sulfides, volatilization via methylation or ethylation, physical exclusion of electronegative components in the membranes and extracellular polymeric substances, energy-dependent metal efflux systems, and intracellular sequestration with low-molecular-weight, cysteine-rich proteins. Such mechanisms have been employed in the macrofungus *Galerina vittiformis* and the marine fungus *Aspergillus terreus* for the effective Pb uptake from soil and for the biosynthesis of PbSe quantum dots, respectively. The metal removal kinetic data for Pb(II) follow pseudo-second-order equation indicating the removal mechanism is a function of both metal ions and nature of microorganism.

33.1 INTRODUCTION

The past few decades have evinced unprecedented human-initiated developmental activities for an improvement in their quality of living. Although the quest toward development at the cost of environment continues, human civilization has begun to realize and respond toward the issues concerning the various pollutants and contaminants in their niche. Although the terminologies "pollution" and "contamination" are used in similar contexts in day-to-day life, "contamination" refers to the presence of a chemical in a given sample with no evidence of harm and "pollution" in cases where the presence of the chemical is known to cause environmental harm (Harrison, 2001). In either of these cases, the afteref-fects are demonstrated in various strata including the soil, the aquatic systems, or the atmosphere.

There are two main classes of pollutants—biodegradable and nonbiodegradable:

1. Biodegradable pollutants:
   a. Can be rendered harmless by natural processes
   b. No permanent harm if adequately dispersed or treated
   c. Examples: paper, wood, and domestic waste
2. Nonbiodegradable pollutants:
   a. Accumulate in the environment; need immediate mitigation measures
   b. May be concentrated in the food chain—long-term harm
   c. Examples: heavy metals and chlorinated components

The soil and groundwater quality data reported by the Central Pollution Control Board in December 2009 have revealed that heavy metals like cadmium, lead, mercury, chromium, cobalt, zinc, nickel, and manganese are the major pollutants and need immediate mitigation measures to bring down their levels in the environment. Of the various heavy metal pollutants, lead contamination has received wide attention. Lead has been used in human civilizations for about 5000 years and is still a major tool in modern technological applications. Its uses and toxic properties at higher concentrations have been recognized since antiquity. Since the early ages of Pb use, evidences of lead poisoning have been found by medical historians, with dramatic effects on the destiny of ancient civilizations (see Box 33.1). The fall of Roman Empire was related to the wide use of lead in paints, water distribution pipes, and wine storage vessels. It was suggested that the declining birthrate, the epidemics of stillbirths and miscarriages, and apparently the increased incidence of psychosis in Rome's ruling class, which may have been at the root of the Empire's dissolution, were a result of exposure to lead in food and wine (Rau et al., 2015).

It has been recently proposed that similar or additional effects of human population exposure to lead may contribute to the decline of current societies through lead-induced impairment of intelligence, increase in delinquent behavior, or psychological changes such as behavioral difficulties (Morgan, 2001). While a few years ago, it would have been possible to apportion the sources that result in Pb contamination, the present scenario of continuous developmental activities in the name of urbanization and globalization has made this a phenomenal task. However, efforts are underway to mitigate the aftereffects of the already existing Pb contaminant load and to adopt environmentally friendly techniques for remediating the contaminated sites.

BOX 33.1   EXAMPLE OF LEAD POISONING: AN ENDEMIC SINCE THE 1880S (NEEDLEMAN, 1991)

In 1887, there was an outbreak of mass lead poisoning in Philadelphia traced to lead chrome yellow dye in cakes and buns. This series included 64 cases and was reported to have developed convulsions leading to death. In 1932, a 7-year-old girl was admitted in coma to the Harriet Lane Home at John Hopkins Hospital, Baltimore. Careful diagnosis uncovered that the persistent lead exposure due to the burning of discarded battery casings for domestic fuel was the causative for the sickness. During the next 2 months, 40 similar cases were identified in the country and the illness was dubbed as the "Depression Disease," since it was common during those years for junk dealers to provide free battery casings to poor families for kitchen fuel. Following the outbreak of lead poisoning, the Baltimore City Health Department halted the distribution of battery casings for fuel. Although several regulations on lead release and emission levels had been declared over the years, like the ban on leaded gasoline, the ubiquitous presence of lead was evident in developing societies. However, India only outlawed leaded petrol in March 2000, and since then, the country has moved a little way toward protecting its citizens from exposure to the metal (Mukhopadhyay, 2009).

33.2   LEAD IN THE ENVIRONMENT: SOURCES AND TOXICOLOGY

33.2.1   Sources

In nature, lead (Pb) is a ubiquitous but biologically nonessential element (Merian et al., 2004). However, during the last 50 years, the use of lead in batteries, bearing metals, cable covering, gasoline additives, explosives, and ammunition, as well as in manufacture of pesticides, antifouling paints, and analytical reagents has caused widespread environmental contamination. The EPA reports that more than 99% of the public's drinking water contains less than 0.05 parts per million of lead, which is the maximum contaminant level for lead (ASTDR, 2007). Of the various sources of Pb contamination, refining facilities, brass foundries, rubber refineries, and steel welding operations are considered to be settings in which exposure to lead is optimal. The Agency for Toxic Substances and Disease Registry states that around 0.5–1.5 million workers are exposed to lead at the workplace. In general, this hazardous material enters the air we breathe, the water we drink, and the plants and animals we consume. The adverse health effects induced by lead exposure are dependent on two important components: dose (how much of a contaminant) and duration (how long there has been contact with the contaminant).

33.2.2 TOXICITY

Chemical properties of lead have been given in Figure 33.1. The toxicity of lead is a consequence of the ability of Pb^{2+} to interfere with several enzymes (Merian, 2004). Because lead causes a large variety of toxic effects, including gastrointestinal, muscular, reproductive, neurological, and behavioral and genetic malfunctions (Johnson, 1998), the fate of lead in the environment is of great concern. In the presence of inorganic compounds (such as Cl^-, CO_3^{2-}, SO_3^{3-}, PO_4^{3-}) and organic ligands (such as humic, folic acids, EDTA), low-soluble lead compounds form. The concentration of dissolved salts, pH, and minerals affects the amounts of lead found in surface water and groundwater. In surface water and groundwater systems, lead can appear in its precipitate form ($PbCO_3$, Pb_2O, $Pb(OH)_2$, $PbSO_4$) and is capable of absorbing into the surface of minerals (Evanko and Dzombak, 1997).

Lead most commonly enters the body through ingestion and then travels to the lungs then swiftly through the bloodstream to other parts of the body. Lead does not change form once it enters the body. Lead in the bloodstream travels to the soft tissues of the body, such as the liver, kidneys, lungs, brain, spleen, muscles, and heart. Over time, lead particles have the ability to move into the bones and teeth (Lalor, 2006). Lead impairs communication between cells and modification of neuronal circuitry (Klaassen and Watkins, 2010). Literature reports that 94% of the total amount of lead in the body is contained in bones and teeth. For children, 73% of lead in their body is stored in their bones (ASTDR, 2007). Lead that enters the bones can remain there for decades. However, over time, lead that is stored in the bones can reenter the bloodstream and organs.

Lead toxicity usually begins with the nervous system in both adults and children. Exposure to lead can cause weakness in fingers, wrists, or ankles. Studies show that long-term exposure to lead at work can lead to decreased performance. High levels of lead can cause damage to the brain and kidney and can cause anemia in both adults and children (ASTDR, 2007).

Lead is considered one of the most frequently encountered heavy metals of environmental concern and is the subject of much remediation research (Rajalakshmi et al., 2011). Severe Pb contamination in soils may cause a variety of environmental problems, including loss of vegetation, groundwater contamination, and Pb toxicity in plants, animals, and humans (Tchounwou, 2012). In addition to industrial and mining activities, elevated soil Pb levels have also occurred due to the disposal of municipal sewage sludge enriched in Pb (Sharma and Dubey, 2005). The remediation of Pb-contaminated soils represents a significant expense to many industries and governmental agencies. Hence, efforts are underway to develop cost-effective and efficient technologies for cleaning up lead-contaminated soils or detoxifying lead.

33.3 BIOREMEDIATION OF PB-CONTAMINATED SOIL

Traditional methods of remediating lead-contaminated sites include a variety of physical, thermal, and chemical treatments (Mulligan et al., 2001). Conventional technologies involve the removal of metals from polluted soils by soil washing (Voglar and Lestan, 2014). These decontamination strategies are *ex situ* approaches and are found to be expensive, resulting in irreversible damage in the soil structure and ecology (Salt, 1995). The common methods to remediate metal-contaminated soil are soil flushing, solidification/stabilization, vitrification, thermal desorption, and encapsulation (Jadia and Fulekar, 2009). However, these methods are applicable for smaller areas of soil sites with high metal contamination. Further, these remediation methods require high-energy input and expensive machinery. Literature review on the *in situ* remediation methods reveals that these methods lead to the destruction of soil structure and decrease their productivity.

The solubilization of metal from soil can be done by chemical leaching. The metal displaced in this way can be recovered via precipitation, adsorption, transformation, and complexation processes. Immobilization of heavy metals through the addition of lime (Krebs et al., 1999), phosphate (Ebbs et al., 1998), and calcium carbonate ($CaCO_3$) (Chen, 2000) has also been practiced as common remediation techniques. These remediation technologies have the advantage of immediately reducing the risk factors arising from metal contamination but may only be considered temporary alternatives because the metals will not be removed completely from the soil environment.

However, in response to the growing needs of environmental issues due to soil Pb contamination, many of these remediation technologies have been used in consortia to treat soil leachate, wastewater, and groundwater contamination (Aboulroos et al., 2007). A particular contaminated site may require a combination of procedures to allow the optimum remediation for the prevailing conditions. Figure 33.2 shows

Chemical properties of lead-"Pb"

- Oxidation state: 0 or +II

- Molecular weight of 207.20 Da

- Bluish-gray colored solid

- Melting point: 327.4°C

- Boiling point of 1740°C

- Most stable form: lead (Pb^{2+}) and lead hydroxy compounds

- Most common form: Pb^{2+} → Produces mononuclear and polynuclear oxides and hydroxides (Evanko and Dzombak, 1997).

FIGURE 33.1 Chemical properties of lead.

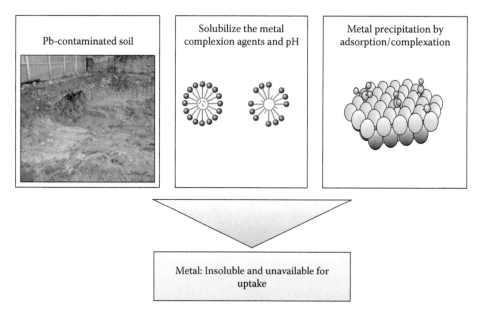

FIGURE 33.2　Sequence of physicochemical methods for the remediation of metal-contaminated soil.

TABLE 33.1

Cost Involved in Various Kinds of Soil Heavy Metal Remediation

Type of Remediation	Cost/Cubic Meter	Time Required (Months)
Excavation and removal	$100–$400	6–9
In situ fixation (including soil amendments)	$90–$200	6–9
Bioremediation	$15–$40	12–18

the sequence of physicochemical methods for the remediation of metal-contaminated soil. Physical, chemical, and biological cleanup technologies may be used in conjunction with one another to reduce the contamination to safer and acceptable levels. However, bioremediation techniques are typically more economical than thermal and physicochemical remediation methods. The comparative costs of bioremediation with other soil remediation approaches like excavation and *in situ* fixation are presented in Table 33.1.

33.3.1　General Concepts in Bioremediation

Recent years have professed the development of bioremediation methods as eco-friendly and cost-effective heavy metal detoxification alternatives to the traditional physicochemical approaches. Bioremediation processes have been classified into two broad categories, according to the place and soil handling/conditioning: *in situ* bioremediation and *ex situ* bioremediation. *In situ* methods are useful to handle large amounts of pollutants and it is an economical technique, whereas the second classes of processes are useful for the remediation of sludge or sediments polluted with high concentration of recalcitrant contaminants. However, *ex situ* bioremediation processes are usually expensive. A summary of various known bioremediation methods is explicated in Table 33.2.

Plant-mediated metal removal from polluted soil (Phytoremediation) is often used for soil remediation. But the plants take significantly longer to grow and to complete the recovery of these metals from their parts (leaves, twigs, etc.) and hence found to be not economically and technologically favorable. Moreover, there may be a possibility of the release of adsorbed/accumulated heavy metals back into the environment. Phytoremediation solely cannot solve these issues of time span, because of its limitations like selectivity of plant, climatic inhibitions, tolerance to heavy metals, and back contamination. Hence, there is a need for a robust methodology that can go hand in hand with other techniques to remediate contaminated sites more quickly, effectively, and economically.

33.3.2　Mechanism of Pb Bioremediation

There are a number of biomaterials that can be used to remove metal from wastewater, such as molds, yeasts, bacteria, and seaweeds. The capability of the microbial strains to survive heavy metal stress in natural environment has aided their use in wastewater treatment where they are directly involved in the decomposition of organic matter in biological processes for wastewater treatment, because often the inhibitory effect of heavy metals is a common phenomenon that occurs in the biological treatment of wastewater and sewage.

TABLE 33.2
Summary of Bioremediation Strategies

Technology	Examples	Benefits	Limitations	Factors to Consider
In situ	Biosparging	Noninvasive	Environmental constraints	Biodegradative property of indigenous microbes
	Bioventing	Relatively passive and natural attenuation	Extended treatment time	Environmental parameters
	Bioaugmentation	Treats soil and water	Monitoring difficulties	Chemical solubility
				Distribution and geography
Ex situ	Land farming	Cost efficient	Space requirements	Same as *in situ*
	Composting	Can be done on site	Extended treatment time	
	Biopiles		Mass transfer problems	
Bioreactors	Slurry reactors	Rapid degradation	Soil excavation required	Same as *in situ*
	Aqueous reactors	Enhances mass transfer	High cost involved	Bioaugmentation
		Effective use of inoculants and surfactants	High operation cost	Toxicity of amendments
				Toxic concentrations of contaminants

Source: Vidali, M. 2001. Bioremediation. An overview, *Pure Appl. Chem.*, 73(7): 1163–1172.

In general, the mechanisms of metal resistance in microbes include (Figure 33.3)

- Precipitation of metals as phosphates, carbonates, and/or sulfides
- Volatilization via methylation or ethylation
- Exclusion of extracellular polymeric substances (EPS) and electroactive species
- Energy-dependent metal efflux systems
- Intracellular sequestration with low-molecular-weight, cysteine-rich proteins

Antibiotic resistance has also been found associated with metal resistance in most of the microorganisms.

The bioaccumulation in macrofungi comprises two distinctive processes: in the first step, metal ions are bound to the surface of the cells—the process is metabolically passive and is identical with biosorption—and in the second step, metal ions are transported to the cellular interior (Figure 33.4). In order to perform this step, the cells must be metabolically active. The entire process of bioaccumulation of metal from the soil to the fungal cells consists of the following steps: (1) transport of metal in the soil near to the mycelial surface, (2) diffusion across the liquid film surrounding the mycelial surface, (3) diffusion of soluble metals in the liquid contained in the pores in the mycelia along the pore walls, (4) sorption and desorption on the external surface of the mycelia or within the mycelial pore surface, and (5) transportation to the cellular interior.

It is documented that the movement of metal ions in the soil and their transport in the mainly depends on various factors like physical and chemical property of metal ions, that is, its molecular mass, ionic radii, and electronegativity. In the organism, the metal mobility also depends on the tolerance mechanism prevailing in the organism and its growth stages. In an earlier study, it was reported that the Pb(II) has the

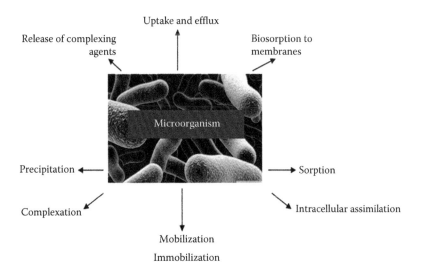

FIGURE 33.3 The microbial metal detoxification strategies.

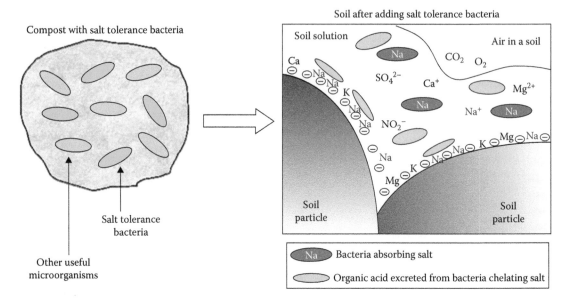

FIGURE 33.4 Salt tolerance mechanism in bacteria.

highest electronegativity compared to other common metal ions like Cu(II) and Zn(II), indicating a higher mobility of Pb(II) ions across the microorganism—an obvious advantage for the utilization of microbes for Pb removal from environmental systems.

33.4 CASE STUDY 1: PB TOLERANCE MECHANISM IN *ASPERGILLUS TERREUS* FOR THE BIOSYNTHESIS OF PBSE QUANTUM DOTS

Mechanisms of metal detoxification by biomolecules proceed as cascade of events, such as induction of proteins such as metallothionein (MT), heat shock protein (HSP), phytochelatins (PC), and ferritin, transferring; or by the activation of antioxidant enzymes such as superoxide dismutase, catalase, glutathione (GSH), and peroxidase. A high turnover of organic acids such as malate, citrate, oxalate, succinate, aconitate, and α-ketoglutarate is another usual stress response demonstrated by microbes. Primarily, the prominent metal complexation processes are the synthesis of PCs and of other metal-chelating peptides (Carpenè et al., 2007). Figure 33.5 shows the scanning electron micrograph of *Aspergillus terreus* biomass in the presence and absence of Pb stress. It can be observed that the induction of Pb stress results in typical protrusions and swelling on the biomass in contrary to the untreated biomass. The characteristic protrusions may have resulted due to the uptake of Pb ions and the resultant cytosolic pressure buildup.

The inherent metal tolerance in the fungi was further utilized for the biosynthesis of PbSe quantum dots (QDs). Initially, the introduction of metal/metalloid precursors activates the cell surface functional groups such as oxalic acid and thiol compounds that reversibly bind the metals on the cell surface as a first line of cellular defense. Oxalate

secretion is well documented in other fungi, and this process has been reported to be stimulated under metal stress (Jarosz-Wilkolazka and Gadd, 2003). The bulk formation of water-insoluble metal-oxalate crystals is undoubtedly an efficient way to prevent toxic metal ions entering fungal cells (Jarosz-Wilkolazka and Gadd, 2003). Also, metal chelation by small molecular mass metabolites, peptides, and proteins is also documented as a crucially important element of almost all metal detoxification processes (Tamas and Martinoia, 2006; Azcón-Aguilar, 2009; Wysocki and Tamás, 2010). The presence of these metal ions in the biological system activates PC synthase enzyme that utilizes GSH from the cells to assemble PCs.

The metal/metalloid ions tend to bind to the thiol groups leading to the formation of low-molecular-weight PC–metal complexes. These complexes are more likely to be transported by ATP-binding cassette membrane transport proteins into a vacuole (Slocik et al., 2004). Subsequently, the selenide ions, in the reaction mixture (produced due to the reducing atmosphere in the growth media) (Prasad and Jha, 2010), also enter the fungal cytosol. Once within the cytosol, the selenide ions complex with the thiol groups of the MTs. Our observations are in concordance with that of Pal and Das (2005), who report that upon exposure to metal ions, fungi synthesize MT and PCs and cellular resistance to heavy metal cytotoxicity due to the binding of metal ions either to MT or to PC.

The metal toxicity is also believed to have induced the ROS, thus activating the enzymes to detoxify ROS, namely, SODs. These enzymes utilize the phenolics as preferential electron donors and initiate a series of redox reactions within the fungus (Figure 33.6). Along with this, a number of simple hydroxy/methoxy derivatives of benzoquinones and toluquinones are elaborated by lower fungi (especially *Penicillium* and *Aspergillus* species) (Goodwin, 1965) that facilitate redox reactions due to its tautomerization (Prasad and Jha, 2010).

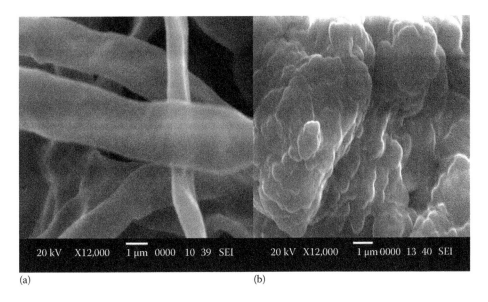

FIGURE 33.5 SEM images of *Aspergillus terreus* (a) untreated and (b) PbSe treated.

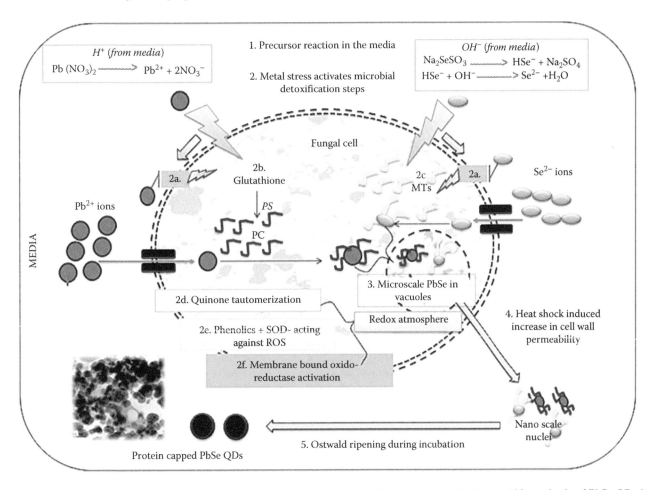

FIGURE 33.6 Schematic representation of the proposed mechanism for the Pb and Se detoxification and biosynthesis of PbSe QDs by *Aspergillus terreus*. (1) Precursors undergo initial redox reactions in the media generating metal/metalloid ions. (2) The metal stress activates the fungal detoxification mechanisms: (a) Surface functional groups; oxalic acids, thiols, etc., get activated, thus reversibly binding the metal ions. (b) Metal stress activates phytochelatin synthase (PS) to convert GSH to PCs, which bind the metal ions and transport them via the ATP-binding cassette. (c) Similarly, MTs bind the metal/metalloid ions; (d) Quinone tautomerization; (e) superoxide dismutase (SOD) activity; and (f) other oxidoreductases create a redox atmosphere to form (3) microscale PbSe and to further initiate transformation to nanoscale PbSe in the vacuoles. (4) Heat shock increases the cell wall permeability to drain out the contents into the media where (5) Ostwald ripening of the nuclei takes place to result in protein capped QD formation.

The induced heat shock is anticipated to increase the permeability of the fungal cells, to transport the metal–peptide complexes to the extracellular environment, wherein the redox atmosphere and involvement of GSHs/MTs assist the process of nanofabrication (Jha and Prasad, 2010) and the process of nanotransformation followed by the Oswald ripening, leading to the fabrication of PbSe QDs (Bao, 2010).

33.5 CASE STUDY 2: PB TOLERANCE MECHANISM IN MACROFUNGUS FOR SOIL BIOREMEDIATION

Bioremediation studies using microorganisms reveal that bacteria, algae, and microfungi are not very efficient in soil heavy metal removal as it is difficult to remove their biomass from the soil after the remediation process. Hence, bioremediation using mushrooms is proven to fetch promising results by the bioaccumulation of pollutants especially the heavy metals in their large biomass and the ease of their separation from the soil.

The effect of Pb on the morphology of *Galerina vittiformis* during the bioaccumulation process was studied through the analysis of the SEM image. It was revealed that the hyphae of *G. vittiformis* were cylindrical, septate, and branched before exposure to the heavy metal. However, a characteristic change in the morphology consisting of curling and formation of hyphal coils in response to Pb(II) stress (at a concentration of 50 mg/kg) has been observed. Canovas et al. (2004) reported that the surface of *Aspergillus* sp. also had rough texture due to protrusions on the hyphae on exposure to 50 mM of heavy metal solution. Such modifications on the surface of fungi indicate the production of intracellular compounds due to heavy metal stress, which results in an increase in pressure within the mycelia, leading to the outward growth of the cell wall structures (Parazkeiwicz et al., 2011). Courbot et al. (2007) have also observed that the impact of metal stresses had led to the production of thiol compounds, especially GSH and MT due to intracellular detoxification of cadmium in the fungus *Paxillus involutus*. According to them, the cell wall protrusions indicate increased formation of intracellular vacuoles that serve as storage compartments for thiol-containing compounds. These compounds are responsible for the binding of metal ions into the intracellular regions and accumulate them in the vacuoles, thereby reducing their toxicity in the cytoplasm and improving tolerance levels.

33.5.1 Description of the Mechanism

Figure 33.7 describes the mechanism of metal uptake in *Galerina vittiformis*. It can be inferred that, initially, the fungal mycelia (roots of the fruiting bodies) adsorbs the metals followed by the passive absorption resulting in the uptake and storage in their periplasmic space. Thereafter, PC and acid production in response to metal stress is activated; acids act

as HSPs that bind to the metal and store them to periplasmic space, and finally to result in the transport and accumulation of metals in vacuole.

FTIR spectra of fruiting body extracts of *Galerina vittiformis* after bioaccumulation studies were analyzed to determine the presence and disappearance of any functional groups involved in metal accumulation mechanism. The FTIR graphs of fruiting body obtained from Pb(II)-laden soil system indicated the presence of stress-related components like oxalic acid, that is, 1658 ± 5 and 1253 ± 5, and thiol group, that is, 2550 ± 5. The role of PCs and MTs as efficient candidates in the Pb detoxification pathway by the macro- and microfungus has also been affirmed by LC–MS analysis.

33.5.1.1 Pb Bioaccumulation Kinetics by *Galerina vittiformis*

In general, kinetics of sorption describes the rate of pollutant uptake on the sorbent, which controls the equilibrium time. The kinetic parameters are helpful for the prediction of sorption rate, which gives important information for designing and modeling the processes. Different types of kinetic models including pseudo-first-order, pseudo-second-order, and intraparticle diffusion have been employed to investigate the mechanism of biosorption and the potential rate-controlling steps such as mass transport and chemical reaction processes.

It is well established that the overall rate of biosorption on a porous solid can be described by the mechanism of three consecutive steps, (1) external mass transport, (2) intraparticle diffusion, and (3) biosorption rate at the interior sites, and it is important to study the rate at which metal ions are removed from soil/aqueous solution in order to apply the principle of bioremediation for industrial uses. The preceding sections elaborate the kinetics of lead sorption and uptake from soil by micro- and macrofungi.

33.5.1.2 Intraparticle Diffusion Kinetics for Pb

Biosorption through pore diffusion has been considered as noteworthy step that affects the overall rate of heavy metal uptake by the microorganism. The most widely applied intraparticle diffusion equation for biosorption system as given by Weber and Morris (1963) is given as follows:

$$q_t = k_{id}\sqrt{t} + C \tag{33.1}$$

where

q_t (mg/kg) is the metal uptake at time t (min)
k_{id} (mg/g min$^{-0.5}$) is the intraparticle diffusion rate constant
C is a constant (mg/kg)

If the rate-limiting step is intraparticle diffusion, a plot of solute uptake against square root of contact time should yield a straight line passing through the origin.

In a study on the Pb bioaccumulation kinetics by *Galerina vittiformis*, it was observed that the intraparticle diffusion was one of rate-controlling steps. Generally, the plot of q_t vs

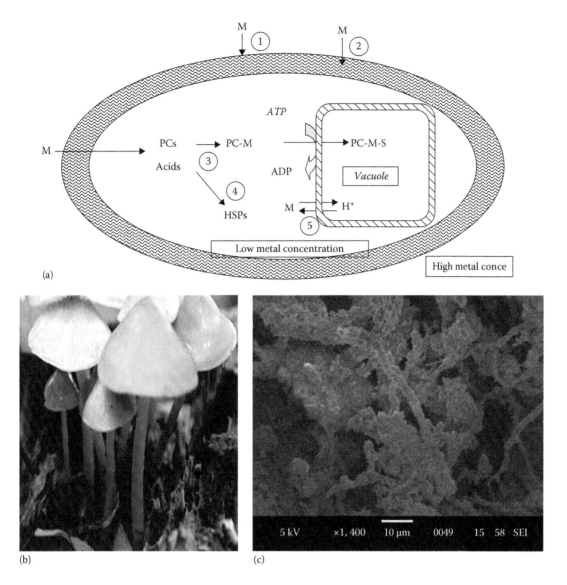

(a)

(b)

(c)

FIGURE 33.7 (a) Schematic representation of the proposed mechanism of metal uptake by *Galerina vittiformis*. (b) The mushroom *Galerina vittiformis*. (c) SEM image of the macrofungal morphology after Pb sorption.

√t passing through the origin indicates that intraparticle diffusion is the sole rate-limiting step in the process. In a study, it was evident that the intercept of the linear fit of the diffusion equation did not pass through the origin for Pb, indicating that the intraparticle diffusion is not the only rate-limiting step for the metal removal process. Hence, surface reaction kinetics, external film mass transfer, or transportation to the cell interior may also simultaneously control the rate of metal uptake by their significant contribution to the overall rate along with intraparticle diffusion. Initial curved portion or multilinear pattern of the q_t vs √t plots is attributed to boundary layer effects, surface reaction, or transportation to the cell interior (Figure 33.8). The value of intercept C was found to be 69.76, which is in turn correlated to be directly proportional to the boundary layer thickness for external film mass transfer.

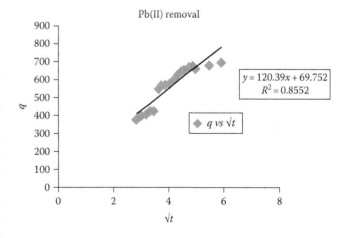

FIGURE 33.8 Intraparticle diffusion kinetic plots for Pb(II) removal.

33.5.1.3 Pseudo-First-Order and Second-Order Kinetic Models

Boyd et al. (1947) proposed a model to describe biosorption kinetics when external film diffusion surrounding the solid surface is the rate-controlling step. The model form proposed by Boyd et al. (1947) is the same as the Lagergren's pseudo-first-order rate equation, indicating the nonfeasibility in differentiating between film diffusion control and pseudo-first-order reaction (Ho et al., 2002). This means that, in the case of Lagergren's pseudo-first-order rate equation, the metal removal rate depends on the diffusion ability of molecule through boundary liquid film and sorption kinetics as a chemical phenomenon (Boyd et al., 1947). Lagergren's pseudo-first-order rate equation is usually preferred in the case of metal uptake by mushrooms as both the phenomena, that is (1) transport by diffusion through boundary liquid film and (2) the chemical phenomena of the binding of metals onto the chelators or organic ligands (secreted by the mycelia) present on the surface, may occur.

The pseudo-first-order rate equation by the Lagergren's is given as

$$\frac{dq_e}{d_t} = k_1(q_e - q_t) \tag{33.2}$$

Integrating Equation 33.2 with the boundary conditions of $q_t = 0$ at $t = 0$ and $q_t = q_t$ at $t = t$ yields

$$\log(q_e - q_t) = \log(q_e) - \frac{k_1 t}{2.303} \tag{33.3}$$

where

- q_e (mg/kg) and q_t (mg/kg) are the equilibrium metal uptake and the metal uptake at any given time t (min), respectively
- k_1 is the rate constant (min^{-1})

The second-order model describes the kinetics of metal uptake involving valence forces through the sharing or exchange of electrons between the surface and the metal as covalent forces and ion exchange (Ho and McKay, 2000).

The pseudo-second-order rate equation by the Lagergren's is given as

$$\frac{dq_t}{dt} = k_2(q_e - q_t)^2 \tag{33.4}$$

Integrating Equation 33.4 with the boundary conditions of $q_t = 0$ at $t = 0$ and $q_t = q_t$ at $t = t$ yields

$$\frac{t}{q_t} = \frac{1}{k_2 q_e^2} + \frac{1}{q_e} t \tag{33.5}$$

where k_2 is the second-order rate constant.

In general, the second-order model better explains the kinetic data for Pb(II) interaction with the microbe when compared to first-order kinetic model. This implies that the

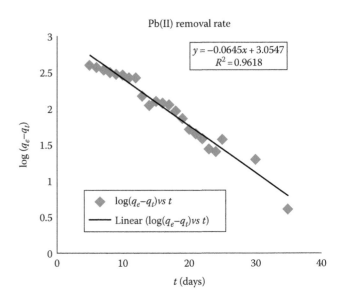

FIGURE 33.9 Pseudo-first-order kinetic plot for Pb(II) removal.

kinetics of the metal uptake by the microbe proceed based on the assumption that reaction between Pb(II) and the metal surface involving valence forces may be one of the prominent rate-limiting steps. Initial sharing or exchange of electrons between the mycelia and the metals further leads to sequence of other steps involved in the bioaccumulation of the metal. Similar results were obtained in the studies of biosorption of heavy metals onto live cells (Figure 33.9).

33.5.1.4 Effect of Chelating Agents on Pb Bioaccumulation Kinetics by Microbes

A chelating agent is a substance whose molecules can form different bonds to a single metal ion. Chelating agents play an important role in metal mobilization especially from soil environment. Chelating agents are organic or inorganic compounds capable of binding metal ions to form complex ringlike structure called "chelates." Chelating agents possess "ligand" binding atoms that form either two covalent linkages or one covalent and one coordinate or two coordinate linkages. ETDA, DTPA, EDDHA, and similar molecules are examples of synthetic chelating agents, while amino acids are examples of natural chelates.

Chelating agents have been used to mitigate metal toxicity to organic-degrading microorganisms; hence, they have been found to play an important role in the metal microbe interaction for bioremediation. For instance, EDTA, a commonly used chelator, has been shown to reduce the toxicity of cadmium to *Chlorella* sp. (Spencer and Nichols, 1983) and copper to bacteria and algae (Brand et al., 1986). However, studies on the effect of chelating agents on Pb uptake by biological agents are scarce. The effect of these agents is generally studied by analyzing the metal content in the biomass.

In this regard, the effect of citric acid and gallic acid in the Pb interaction with *Galerina vittiformis* was studied. It is reported that while 50 mg citric acid per kilogram of soil resulted in a Pb bioaccumulation of 620 mg/kg, gallic acid (10 mmol/kg soil) exhibited comparatively higher Pb bioaccumulation of around 630 mg/kg soil. However, EDTA at 5–10 mmol/kg of soil led to Pb bioaccumulation of around 730 mg/kg soil in *G. vittiformis*. Similar results were reported by various researchers on bioaccumulation studies in bacteria, mycorrhizae, and plants. The toxicity of EDTA to many microorganisms and its limited biodegradability may reduce its suitability for the application of cocontaminated environments. Further, the biodegradation of metal–EDTA complexes may be slow, thus highlighting the utilization of other chelating agents for Pb bioremediation. The detailed literature on chemical chelators and biobased chelators is described in Table 33.3.

33.5.1.5 Effect of Multimetals on Bioaccumulation

Recent studies of bioaccumulation have been majorly focused on single-metal system than multimetal system, even though metal-contaminated soil areas often contain several metal ions dominantly. When more than one metal ion is present, the study becomes more complicated as the interaction of one metal accumulation in the presence of other metal ions may be synergetic, antagonistic, or nonreactive. The effect of interaction of the metals in multimetal soil system containing five metals, namely, Pb(II), Cd(II), Cr(VI), Zn(II), and Cu(II), on removal of metals through bioaccumulation by *Galerina* sp. has been analyzed in soil contaminated with metal ions. In this study, the percentage removal of Pb(II), Cd(II), Cr(VI), Zn(II), and Cu(II) was studied with respect to their concentration in the soil and the pH of the solution.

33.5.1.6 Effect of Soil pH and Multimetals on Pb(II) Removal

Studies indicate the active role of soil pH on Pb(II) removal by microbes. The removal increases with the increase in pH in the acidic range, and a maximum removal of Pb was observed in the range of 100–200 mg/kg of soil and at pH 6.5 in the presence of other metals. The maximum removal of Pb(II) was at a pH of around 6.5, irrespective of concentration of different metals being present. But the maximum percentage removal obtained at pH 6.5 depends on the concentration of the metals present. As the soil pH increases to alkaline range (6.6–8), the removal efficiency decreases drastically for all the studied metal concentrations. Further, the response plots for Pb(II) removal indicated that the removal percentage of Pb(II) increases initially as Cd(II) and Zn(II) concentrations increase, but beyond 150 mg/kg the removal percentage decreases from 93% to nearly 75%. Similar interaction pattern is observed in the case of Pb(II) removal for combinations like Cr:Cd, Cd:Pb, Pb:Zn, Cr:Zn, Cu:Cr, Cr:Pb, Cu:Pb, and Zn:Cu.

It was observed that the percentage of Pb(II) removal by the microbe increases with the increase in concentration of other metals. Once the maximum Pb(II) removal is attained, a decrease in Pb(II) uptake and removal with any further increase in the concentration of the coexisting metals is usually observed. Hence, it can be concluded that these metal ions interact with each other in influencing Pb(II) removal by the microorganism.

Based on experimental observations on Pb(II) removal by microorganisms, it can be inferred that in the presence of lower concentrations of metals, the cells may not reach their maximum bioaccumulation capacity, that is, the vacuoles may still have storage capacity to store the metals in their bound

---

TABLE 33.3
Chelators Affecting Bioaccumulation: A Literature Overview

Sl. No.	Type of Chelators Used	Metals under Study	References
		Biodegradable	
1.	EDDS and MGDA	Pb(II) and Zn(II)	Cao et al. (2007)
	Ethylene diaminedisuccinate (EDDS)	Pb(II)	Mancini and Bruno (2011)
	Citric acid, gallic acid, vanillic acid, oxalic acid	Cd, Pb, Zn, Cu, and Ni	Zhao et al. (2010)
	Nitrilotriacetic acid	Pb(II)	Nascimento (2006)
	IDSA	Zn, Cu, Pb, and Cd	Mancini and Bruno (2011)
	Citric acid	Cd, Cu, Pb, and Zn	Zhao et al. (2010)
			Sinhal et al. (2010)
			Sun et al. (2009)
		Nonbiodegradable	
2.	EDTA	Pb(II), Cr(V)	Dipu et al. (2012); Evangelou et al. (2007)
		Pb(II), Cd(II), and Ni(II)	Chen and Cutright (2001)
		Pb(II), Cu(II), and Cd(II)	Chigbo and Batty (2013); Zhao et al. (2010)
		Hg(II)	Wenger et al. (2005)
		Pb, Cd, Cr,	Blaylock and Huang (2000)

form with the chelators. So at lower metal concentrations, percentage removal increases with the increase in metal concentrations. But at higher metal concentrations in the environment, the total metals present in the environment may exceed the bioaccumulation capacity of the species, and hence, all metal ions may not be transported to the cell. This leads to an increase in residual metal concentration in the environment (decrease in percentage metal removal) with an increase in the initial concentration of the metals. This pattern in Pb(II) removal may also be because of the increase in its lethality on the microorganisms at their higher concentrations and due to inhibition of transporters by the saturated environment of positively charged ions leading to neutralization of membrane charges (Foulkes, 2008).

33.6 CONCLUSION

Lead pollution and its associated aftereffects have existed as a major issue affecting the total quality of the environment. Traditional methods of remediating lead-contaminated sites include a variety of physical, thermal, and chemical treatments, which in general are expensive *ex situ* approaches. In this context, the efficiency of lead-resistant microorganisms for the bioremediation of lead-contaminated soils has emerged as a promising cost-effective technology for practical utilization. Generally, microorganisms precipitate the metals as phosphates, carbonates, or sulfides, volatilization via methylation or ethylation, physical exclusion of electronegative components in the membranes and EPS, energy-dependent metal efflux systems, and intracellular sequestration with low-molecular-weight, cysteine-rich proteins. Such mechanisms have been employed in the macrofungus *Galerina vittiformis* and the marine fungus *Aspergillus terreus* for the effective Pb uptake from soil and for the biosynthesis of PbSe QDs, respectively. The metal removal kinetic data for Pb(II) follow pseudo-second-order equation indicating the removal mechanism being a function of both metal ions and nature of microorganism.

REFERENCES

Aboulroos, S.A., Helal, M.I.D., and Kamel, M.M. 2007. Remediation of Pb and Cd polluted soils using in situ immobilization and phytoextraction techniques. *Soil & Sediment Contamination.* 15(2): 199–215.

ATSDR. 2007. Toxicological profile: Lead. Available at: http://www.atsdr.cdc.gov/toxprofiles/tp.asp?id=96&tid=22. Accessed on November 11, 2015.

Azcón-Aguilar, C. 2009. *Mycorrhizas: Functional Processes and Ecological Impact.* Berlin, Germany: Springer.

Bao, H. 2010. Biosynthesis of biocompatible cadmium telluride quantum dots using yeast cells. *Nano Research* 3: 481–489.

Blaylock, M.J. and Huang, J.W., 2000. Phytoextraction of metals. In: Raskin, I., Ensley, B.D. (Eds.), Phytoremediation of Toxic Metals Using Plants to Clean up the Environment. John Wiley, New York, pp. 53–70.

Boyd, G.E., Adamson, A.W., and Myers, L.S. 1947. The exchange adsorption of ions from aqueous solutions by organic zeolites. ii. kinetics 1. *Journal of the American Chemical Society* 69: 2836–2848.

Brand, L.E., Sunda, W.G., and Guillard, R.R.L. 1986. Reduction of marine phytoplankton reproduction rates by copper and cadmium. *Journal of Experimental Marine Biology and Ecology* 96: 225–250.

Cao, A., Carucci, A., Lai, T., La Colla, P., and Tamburini, E. 2007. Effect of biodegradable chelating agents on heavy metals phytoextraction with *Mirabilis jalapa* and on its associated bacteria. *European Journal of Soil Biology* 43: 200–206.

Carpenè, E., Andreani, G., and Isani, G. 2007. Metallothionein functions and structural characteristics. *Journal of Trace Elements in Medicine and Biology* 21: 35–39.

Cánovas, D., Vooijs, R., Schat, H., and de Lorenzo, V., 2004. The role of thiol species in the hypertolerance of *Aspergillus* sp. P37 to Arsenic. *The Journal of Biological Chemistry* 279, 51234–51240.

Chen, H. 2000. Chemical methods and phytoremediation of soil contaminated with heavy metals. *Chemosphere* 41: 229–234.

Chen, H. and Cutright, T. 2001. EDTA and HEDTA effects on Cd, Cr, and Ni uptake by *Helianthus annuus*. *Chemosphere* 45: 21–28.

Chigbo, C. and Batty, L. 2013. Phytoremediation potential of *B. juncea* in Cu-pyrene co-contaminated soil: Comparing freshly spiked soils with aged soils. *Journal of Environmental Management* 129: 18–24.

Courbot, M., Willems, G., Motte, P., Arvidsson, S., Roosens, N., Saumitou-Laprade, P., and Verbruggen, N. 2007. A major quantitative trait locus for cadmium tolerance in *Arabidopsis halleri* colocalizes with *HMA4*, a gene encoding a heavy metal ATPase. *Plant Physiology* 144: 1052–1065.

Dalby, A. and Grainger, S. 1996. *The Classical Cookbook.* Getty Publications, Los Angeles, CA.

Dipu, S., Anju, A., and Salom, G.T. 2012. Effect of chelating agents in phytoremediation of heavy metals. *Advanced Science, Engineering and Medicine* 2: 364–372.

Ebbs, S.D., Brady, D.J., and Kochian, L.V. 1998. Role of uranium speciation in the uptake and translocation of uranium by plants. *Journal of Experimental Botany* 49: 1183–1190.

Evangelou, M.W., Ebel, M., and Schaeffer, A. 2007. Chelate assisted phytoextraction of heavy metals from soil. Effect, mechanism, toxicity, and fate of chelating agents. *Chemosphere* 68: 989–1003.

Evanko, C.R. and Dzombak, D.A. 1997. Remediation of metals-contaminated soils and groundwater. *Gwrtac Series* 01: 1–61.

Foulkes, E.C. 2008. Transport of toxic heavy metals across cell membranes. *Proceedings of the Society for Experimental Biology and Medicine* 223: 234–240.

Goodwin, T.W. 1965. Chemistry and biochemistry of plant pigments. Available at: http://www.cabdirect.org/abstracts/19651607717.html;jsessionid=2F99C24BA1982062ACF80FF1B45B75A0. Accessed on November 11, 2015.

Harrison, R.M. 2001. *Pollution: Causes, Effects and Control.* Royal Society of Chemistry, London, U.K.

Ho, Y.S., Ng, J.C.Y., and McKay, G. 2000. Kinetics of pollutant sorption by biosorbents: Review. Separation and purification methods. 29(2): 189–232.

Ho, Y.S., Huang, C.T., and Huang, H.W. 2002. Equilibrium sorption isotherm for metal ions on tree fern. Process Biochemistry 37: 1421–1430.

Jadia, C. and Fulekar, M. 2009. Phytoremediation of heavy metals: Recent techniques. *African Journal of Biotechnology* 8(6): 921–928.

Jarosz-Wilkolazka, A. and Gadd, G.M. 2003. Oxalate production by wood-rotting fungi growing in toxic metal-amended medium. *Chemosphere* 52: 541–547.

Jha, A.K. and Prasad, K. 2010. Leaf. *International Journal of Green Nanotechnology: Physics and Chemistry* 1: P110–P117.

Johnson, F.M. 1998. The genetic effects of environmental lead. *Mutation Research/Reviews in Mutation Research* 410(2): 123–140.

Klaassen, C. and Watkins, J.B. III. 2010. *Casarett & Doull's Essentials of Toxicology,* 2nd edn. McGraw Hill Professional, New York.

Krebs, R. et al. 1999. Gravel sludge as an immobilizing agent in soils contaminated by heavy metals: A field study. *Water, Air, and Soil Pollution* 115: 465–479.

Lalor, G. 2006. Acute lead poisoning associated with backyard lead smelting in Jamaica. *West Indian Medical Journal* 55: 394–398.

Mancini, G., Bruno, M., Polettini, A., and Pomi, R. 2011. Chelant-assisted pulse flushing of a field Pb contaminated soil. *Chemistry and Ecology* 27(3): 251–262.

Merian, E. 2004. *Elements and Their Compounds in the Environment.* Weinheim, Germany: Wiley-VCH Verlag GmbH.

Morgan, R.E. 2001. Early lead exposure produces lasting changes in sustained attention, response initiation, and reactivity to errors. *Neurotoxicology and Teratology* 23: 519–531.

Mukhopadhyay, K. 2009. *Air Pollution in India and Its Impact on the Health of Different Income Groups.* Nova Science Publishers, New York.

Mulligan, C.N., Yong, R.N., and Gibbs, B.F. 2001. Remediation technologies for metal-contaminated soils and groundwater: An evaluation. *Engineering Geology* 60: 193–207.

Nascimento, C.W.A., Amarasiriwardena, D., and Xing, B., 2006. Comparison of natural organic acids and synthetic chelates at enhancing phytoextraction of metals from a multi-metal contaminated soil. *Environmental Pollution* 140: 114–123.

Needleman, H.L. 1991. *Human Lead Exposure.* CRC Press, Boca Raton, FL.

Pal, S.K. and Das, T.K. 2005. Biochemical characterization of *N*-methyl *N*'-nitro-*N*-nitrosoguanidine-induced cadmium resistant mutants of *Aspergillus niger. Journal of biosciences* 30: 639–646.

Paraszkiewicz, K., Bernat, P., Naliwajski, M., and Długonski, J. 2011. Lipid peroxidation in the fungus Curvularia lunata exposed to nickel. *Archives of Microbiology* 192(2): 135–141.

Prasad, K. and Jha, A.K. 2010. Biosynthesis of CdS nanoparticles: An improved green and rapid procedure. *Journal of Colloid and Interface Science* 342: 68–72.

Rajalakshmi, K., Haribabu, T.E., and Sudha, P.N. 2011. Toxicokinetic studies of antioxidants of amaranthus tricolour and marigold (*Calendula oficinalis* l) plants exposed to heavy metal lead *International Journal of Plant, Animal and Environmental Sciences,* 1 (2): 105–109.

Rau, T., Urzúa, S., and Reyes, L. 2015. Early exposure to hazardous waste and academic achievement: Evidence from a case of environmental negligence. *Journal of the Association of Environmental and Resource Economists* 2: 527–563.

Salt, D.E. 1995. Mechanisms of cadmium mobility and accumulation in Indian mustard. *Plant Physiology* 109: 1427–1433.

Sharma, P. and Dubey, R.S. 2005. Lead toxicity in plants. *Brazilian Journal of Plant Physiology* 17: 35–52.

Sinhal, V.K., Srivastava, A., and Singh, V.P. 2010. EDTA and citric acid mediated phytoextraction of Zn, Cu, Pb and Cd through marigold (Tagetes erecta). *Journal of Environmental Biology* 31: 255–259.

Slocik, J.M., Knecht, M.R., and Wright, D.W. 2004. Biogenic nanoparticles. *Encyclopedia of Nanoscience and Nanotechnology* 1: 293–308.

Spencer, D.F. and Nichols, L.H. 1983. Free nickel ion inhibits growth of two species of green algae. *Environmental Pollution Series A, Ecological and Biological* 31: 97–104.

Sun, Y.B., Zhou, Q.X., An, J., Liu, W.T., and Geoderma, R.L. 2009. Chelator-enhanced phytoextraction of heavy metals from contaminated soil irrigated by industrial wastewater with the hyperaccumulator plant (Sedum alfredii Hance). *Geoderma* 150: 106–112.

Tamas, M.J. and Martinoia, E. eds., 2006. *Molecular Biology of Metal Homeostasis and Detoxification.* Berlin, Germany: Springer.

Tchounwou, P.B. 2012. Heavy metal toxicity and the environment. *EXS* 101: 133–164.

Vidali, M. 2001. Bioremediation. An overview, *Pure Appl. Chem.,* 73(7): 1163–1172.

Voglar, D. and Lestan, D. 2014. Chelant soil-washing technology for metal-contaminated soil. *Environmental Technology* 35: 1389–1400.

Weber, W.J. and Morris, J.C. 1963. Kinetics of adsorption on carbon from solution. *Journal of the Sanitary Engineering Division* 89: 31–60.

Wenger, K., Tandy, S., and Nowack, B. 2005. Effects of chelating agents on trace metal speciation and uptake. In: Nowack, B., VanBriesen, J. (Eds.), Biogeochemistry of chelating agents, ACS Symposium Series, Vol. 910, Switzerland, pp. 204–224.

Wysocki, R. and Tamás, M.J. 2010. How *Saccharomyces cerevisiae* copes with toxic metals and metalloids. *FEMS Microbiology Reviews* 34: 925–951.

Zhao, Z., Xi, M., Jiang, G., Liu, X., Bai, Z., and Huang, Y. 2010. Effects of IDSA, EDDS and EDTA on heavy metals accumulation in hydroponically grown maize (*Zea mays,* L.). *Journal of Hazardous Materials* 181: 455–459.

34 Cadmium and Lead Tolerance Mechanisms in Bacteria and the Role of Halotolerant and Moderately Halophilic Bacteria in Their Remediation

M. Sowmya and A.A. Mohamed Hatha

CONTENTS

ABSTRACT

In this chapter, the general mechanisms of heavy metal resistance in bacteria and the potential of halophilic bacteria in bioremediation with special reference to cadmium and lead are reviewed. In the present scenario, we need strains with high metal resistance and removal capacity in varying salinities. It is a well-known fact that the constituents of bacterial growth media and also the salt will react with cadmium and lead. The toxicity of a metal to bacteria strongly depends on its bioavailability rather than its concentration. At higher concentrations, lead tends to precipitate more in the media. So when conducting metal reduction experiments, care should be given to constituents of media and the concentration of the metal tested. Though there were a number of studies dealing with the interference of culture medium components and salinity with metals, there is still a long way to go.

34.1 INTRODUCTION

Anthropogenic activities release many toxic elements and compounds in quantities that exceed the carrying capacity of the environment. They impair the self-cleansing mechanisms present in nature and have become a source of pollution. The toxic chemicals released to the environment tend to bioaccumulate in the living organisms. Metals/metalloids are a major class of these compounds that cause undesirable effects in living systems from micro- to macrobiota including human beings. Even micronutrients such as calcium, cobalt, chromium, copper, iron, magnesium, manganese, nickel, potassium, sodium, and zinc are toxic to living systems at higher concentrations (Nies, 1999). Based on the disability-adjusted life year (DALY) estimates, deaths from chemical pollution of soil, water, and air are increasing (GBD, 1990; World's Worst Pollution Problems, 2015). In the "Priority List of Hazardous Substances" prepared by ATSDR (2015), lead and cadmium are in the second position and seventh position, respectively.

The International Agency for Research on Cancer has classified cadmium and cadmium compounds as a Group 1 carcinogen and lead compounds as Group 2 carcinogens.

Globally, 26 million people are at risk of exposure to lead, with an estimated burden of disease of 9 million DALYs. According to the estimates by Pure Earth, 5 million people are at risk of exposure to cadmium globally, with an estimated burden of disease of 250,000 DALYs. As of 2015, the Toxic Site Identification Program has identified over 150 sites around the world where exposure to cadmium threatens the health of the population (World's Worst Pollution Problems, 2015). Many studies from all over the world have reported the toxic effects of lead and cadmium (Haefliger et al., 2009; Checconi et al., 2013; Choi and Han, 2015; Han et al., 2015).

Chemical precipitation, oxidation/reduction, filtration, ion exchange, reverse osmosis, membrane technology, evaporation, and electrochemical treatment are the major physicochemical processes used for the remediation of heavy metals from wastewater. Physical separation methods become ineffective in the separation of metal salts because most of the heavy metal salts are water soluble (Hussein et al., 2004). The chemical methods are also not better when the heavy metal concentration is less than 100 mg/L (Ahluwalia and Goyal, 2007). Bioremediation is an effective and economical alternative when the physicochemical methods fail to remove small concentrations of heavy metals from wastewater.

34.2   BIOREMEDIATION

Bioremediation involves a variety of processes such as biosorption, bioaccumulation, and biotransformation, and it can be done by using various biological means such as plants, algae, fungi, yeast, and bacteria. Because of their small size, short regeneration time, and ability to utilize multiple substrates for growth, bacteria are a favorite choice for the bioremediation of toxic metals.

The industries that manufacture chemicals such as pesticides, pharmaceuticals, and herbicides and oil and gas recovery processes produce wastewaters with salinity fluctuating from time to time. Conventional microbiological treatment processes do not function effectively at these salinity variations because of the inadaptability of microbes to varying salt concentrations. As the extreme halophiles—haloarchaea— that thrive at high salinities cannot withstand low salinity, moderately halophilic or halotolerant bacteria assume significance.

According to Kushner (1978), moderate halophiles are organisms growing optimally between 0.5 and 2.5 M salt. Bacteria able to grow in the absence of salt as well as in the presence of relatively high salt concentrations are designated as halotolerant (or extremely halotolerant if growth extends above 2.5 M). Therefore, the use of moderately halophilic or halotolerant bacteria is a better option (Oren et al., 1992, 1993) for bioremediation in saline environments.

34.3   MECHANISMS OF METAL RESISTANCE

Cadmium and lead toxicity in microorganisms involves thiol binding and protein denaturation, interaction with calcium metabolism and membrane damage, interaction with zinc metabolism, or loss of a protective function, inhibition of enzyme activity, disruption of membrane functions, and oxidative phosphorylation as well as alterations of osmotic balance (Vallee and Ulmer, 1972; Nies, 1999; Bruins et al., 2000; Ahemad, 2012).

There are different mechanisms for bacteria to resist the heavy metals. They can prevent the entry of toxic metals into the cell or accumulate them inside the cell in a form that is not toxic to the cell or expel the metals outside the cell after converting their chemical composition. The general mechanisms of heavy metal resistance in bacteria are summarized in Figure 34.1.

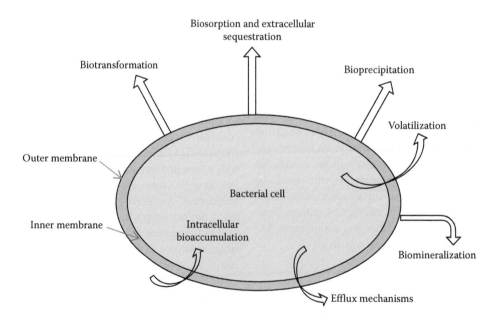

FIGURE 34.1   General mechanisms of heavy metal resistance in bacteria.

34.3.1 Biosorption

Biosorption is the phenomenon of removing heavy metals from aqueous solutions applying inactive or dead biomass (Volesky, 2001; Ilamathi et al., 2014). It is an energy-independent phenomenon. In this process, metal ions can interact on bacterial surface via physical or chemical mechanisms. Aryal and Liakopoulou-Kyriakides (2015) have summarized the effects of different external factors such as pH, contact time, temperature, concentration of biomass and metal ions, and the nature of aqueous environment in the degree of biosorption efficiency of bacterial cells.

34.3.2 Cell Wall and Extracellular Polymeric Substances

The metal homeostasis achieved by preventing the entry of metals inside the cell is known as surface biosorption or extracellular sequestration. In gram-negative bacteria lipopolysaccharides and in gram-positive bacteria, peptidoglycan together with teichoic and teichuronic acids acts as the main binding site for heavy metals. Many studies show that cell envelope is a major barrier for lead. It prevents the entry of lead inside the cell (Massadeh et al., 2005; Kim et al., 2007; Issazadeh et al., 2011; Syed and Chinthala, 2015).

Exopolymers or slime layers are composed of carbohydrates, polysaccharides, and sometimes nucleic acids and fatty acids. Extracellular polymeric substance (EPS) performs a variety of functions such as protection against desiccation, phagocytosis, and parasitism and helps in the formation of biofilm, (Maier et al., 2006). EPSs are generally negatively charged, and consequently, the efficiency of metal–exopolymer binding is pH dependent. Studies on the metal-binding behavior of EPSs revealed their great ability to complex heavy metals through various mechanisms, which include proton exchange and microprecipitation of metals (Comte et al., 2008; Fang et al., 2010). Many reports establish the importance of EPS molecules in metal removal (Arias et al., 2003; Mata et al., 2006; De et al., 2008).

Cadmium and lead are biosorbed differently by different bacteria. The bacterial cell walls are primarily made up of polysaccharides, proteins, and lipids. Polysaccharides have ion-exchange properties. These basic units host a variety of functional groups such as amino, carboxylic, sulfhydryl, phosphate, and thiol groups, which differ in their affinity and specificity for metal binding. Biosorption of heavy metals can be influenced by the composition of EPS, especially by the presence of sulfates and uronic acid (Iyer et al., 2005). The uronic acids and sulfates form gels in the presence of metal ions and thus help in removing them from polluted environments.

The metal-binding capacity of a polymer depends upon the quantity and accessibility of charged groups to the ions (Sutherland, 1994; Arias et al. 2003). *Halomonas maura* strain S-30—a diazotrophic bacterium—produces mauran, with an EPS contributing to the formation of biofilms and binding of some heavy metals. High sulfate and uronic acid contents make mauran an excellent biosorbent for heavy metals especially lead. Native mauran showed high binding capacity to Pb, while deacetylated mauran had significant affinities for both Ag and Pb (Arias et al., 2003). It may be because acetyls bring more electron-donating groups into the vicinity of the binding site, thus allowing the larger Pb ions to bind more strongly (Geddie and Sutherland, 1993, 1994; Arias et al., 2003). The removal of acetyl groups leaves the binding sites more accessible to the cations. However, anionic polysaccharides prefer on the whole to bind cations with large ionic radii (Geddie and Sutherland, 1993).

Mata et al. (2006) demonstrated that EPSs from *Halomonas ventosae* strain Al12T had a significant affinity for lead and that from strains *H. ventosae* Al16 *Halomonas anticariensis* FP35T and FP36 bound copper most efficiently. All the polymers showed a lower affinity for cobalt.

EPS solutions from *H. maura*, *H. ventosae*, and *H. anticariensis* did not produce stable gels in the presence of the metal salts, but they did bind several metals with considerable efficiency. Ion uptake by EPS from all the strains started almost immediately and increased with time.

Enterobacter cloacae P2B isolated from the effluent of lead battery manufacturing company was found to increase its EPS production when exposed to lead nitrate in Tris-buffered minimal medium (Naik et al., 2012). Fourier transform infrared spectroscopy (FTIR) of this EPS revealed carboxyl, hydroxyl, and amide groups along with glucuronic acid. It also showed the presence of several neutral sugars such as rhamnose, arabinose, xylose, mannose, galactose, and glucose contributing to lead-binding hydroxyl groups.

Attenuated total reflectance FTIR spectroscopy suggested that sulfhydryl group (-SH) within the EPS of *Pseudomonas aeruginosa* JP-11 plays a major role in the binding of cadmium (Chakraborty and Das, 2014). A novel EPS called EPS-R produced by a slightly halophilic marine bacterium *Hahella chejuensis* showed specific emulsifying capacity higher than that of commercial polysaccharides and stability over a wide range of temperatures, pH, salinities, etc. (Ko et al., 2000).

34.3.3 Biosurfactants

Biosurfactants being amphiphiles consist of two parts—a polar (hydrophilic) moiety and nonpolar (hydrophobic) group that are produced by microorganisms. A hydrophilic group consists of mono-, oligo-, or polysaccharides, peptides, or proteins, and a hydrophobic moiety usually contains saturated, unsaturated, and hydroxylated fatty acids or fatty alcohols (Lang, 2002). A characteristic feature of biosurfactants is a hydrophilic–lipophilic balance that specifies the portion of hydrophilic and hydrophobic constituents in surface-active substances. Pacwa-Płociniczak et al. (2011) summarized the current status of the environmental applications of biosurfactants. Glycolipid (rhamnolipids), fatty acids, phospholipids and neutral lipids (phosphatidylethanolamine), and lipopeptides (surfactin) are the major groups and classes (in parenthesis) of biosurfactants produced by bacteria that are used for the bioremediation of metals. *Pseudomonas* sp., *Acinetobacter* sp., *Rhodococcus erythropolis*, and *Bacillus subtilis* are the major bacteria reported to produce these surfactants (Jennema et al., 1983; Appanna et al., 1995; Herman et al., 1995).

Biosurfactants remediate metals by forming complexes with them. Metal–anionic biosurfactant bonds are stronger than metal–soil bonding, and eventually, the metal–biosurfactant complex gets desorbed to the soil solution due to the lowering of the interfacial tension. The cationic biosurfactants can replace some of the cationic metal ions by competition, but it is not possible in the case of all negatively charged surfaces (ion exchange). However, metal ions in the soil surfaces can be removed through the biosurfactant micelles.

The polar head groups of micelles can bind metals that mobilize the metals in water (Deziel et al., 1996; Singh and Cameotra, 2004; Juwarkar et al., 2007; Asci et al., 2008).

Juwarkar et al. (2008) reported the use of dirhamnolipid biosurfactant produced by *P. aeruginosa* BS2 for the mobilization of metals from multimetal-contaminated soil. The dirhamnolipid selectively removed heavy metals from soil in the order of Cd = Cr > Pb = Cu > Ni. The use of rhamnolipid together with inorganic compounds (e.g., NaOH) was found to increase the efficiency of the removal of copper and nickel from sediments (Dahrazma and Mulligan, 2007). The addition of OH– group solubilizes the naturally existing organic fraction of the metals and thus increases their availability to the surfactant. Wang and Mulligan (2004) proved that biosurfactant-foam technology is more efficient in the removal of Cd and Ni from sandy soil than the rhamnolipid solution. The rate of heavy metal removal from soil strongly depends on its chemical composition.

The removal of heavy metals by biosurfactant produced by marine bacterium was also reported (Das et al., 2009). The study revealed that tested anionic biosurfactant was able to bind the metal ions and form insoluble precipitate. The efficiency of the removal of Pb and Cd depends on the concentrations of metals and biosurfactants.

34.3.4 Efflux Systems

Efflux systems protect the cell from harmful metals by transporting the metal cations from inside the cell through the cell membrane and thus excluding them. The energy-requiring efflux systems are one of the most effective resistance mechanisms present in bacteria against heavy metals (Bruins et al., 2000).

According to Nies (1995) and Nies and Silver (1989), resistance to cadmium in bacteria is based on cadmium efflux. The efflux of Cd in bacteria is facilitated by P-type ATPases, CBA transporters, and cation-diffusion facilitator (CDF) chemiosmotic transporters. Magnesium and/or manganese uptake systems are also responsible for the uptake of Cd (Tynecka et al., 1981; Laddaga et al., 1985). In gram-negative bacteria, cadmium seems to be detoxified by resistance–nodulation–cell division (RND)-driven systems like Czc, which is mainly a zinc exporter, and Ncc, which is mainly a nickel exporter

(Schmidt and Schlegel, 1994). The product of genes *czcA*, *czcB*, and *czcC* forms the membrane-bound cation efflux protein complex *czcABC* that includes three subunits: CzcA, a cation/proton antiporter (in the cytoplasmic membrane); CzcB, a membrane fusion protein (in the periplasm); and CzcC, an outer membrane protein. They are responsible for the transport of the cations across the cytoplasmic membrane, the periplasmic space, and the outer membrane, thus resulting in the decrease of intracellular cations. In gram-positive bacteria, the primary resistance mechanism is by the cadmium-exporting P-type ATPase, that is, the CadA pump from *Staphylococcus aureus* (Nucifora et al., 1989). This protein was the member of a subfamily of heavy metal P-type ATPases, and all the copper, lead, and zinc transporters found later are related to this protein. Liu et al. (1997) reported that CadA-like proteins mediate cadmium resistance in other gram-positive bacteria like *Bacillus firmus* and *Listeria monocytogenes*.

Eleven distinct *Staphylococcus*, *Micrococcus*, and *Halobacillus* genera carrying the *cadA* gene were reported by Oger et al. (2003). This *cadA* determinant was mostly plasmid borne in the *Staphylococcus* genus, and IS257 sequences, which are known to participate in antibiotic resistance gene dissemination in *S. aureus*, were found to be located near to the *cadA* gene in 16/31 cadmium-resistant *Staphylococcus* strains and one *Micrococcus* strain. This suggests that IS257 has also contributed to the dissemination of the cadA resistance gene among staphylococci. All of the *Staphylococcus* and *Micrococcus* species studied carried the *cadA* gene on plasmids, whereas the *cadA* gene is chromosomal in members of the *Bacillus* (MSA 235) and *Halobacillus* (MSA 234) genera.

Studies on Cd/Zn chemiosmotic efflux transporter (CzcCBA1) in Cd-tolerant *Pseudomonas putida* KT2440 by Gibbons et al. (2011) proved that heavy metal cation efflux mechanisms facilitate shorter lag phases in the presence of metals and the maintenance and expression of tolerance genes carry quantifiable energetic costs and benefits. In *Pseudomonas stutzeri*, numerous genes responsible for resistance to several metals (As, Cu, Hg, Cr, Cd, and Zn) were identified (Li et al., 2012). Transmission electron microscopic (TEM) analysis of two cadmium-resistant bacteria CdRB1 revealed that it actively accumulated cadmium in the cytoplasm (Dabhi et al., 2013). Chakraborty and Das (2014) demonstrated that the cadmium-resistant gene expression increases on Cd stress up to the tolerance level, but optimum pH and salinity are the crucial factors for the proper functioning of cadmium-resistant gene.

Pb^{2+} resistance also involves P-type ATPases. Hynninen et al. (2009) confirmed that lead resistance in *C. metallidurans* is achieved through the cooperation of the Zn/Cd/Pb-translocating ATPase PbrA and the undecaprenyl pyrophosphate phosphatase PbrB. While PbrA nonspecifically exported Pb, Zn, and Cd, a specific increase in lead resistance was observed when PbrA and PbrB were coexpressed. As a model of action for PbrA and PbrB, they proposed a mechanism where Pb is exported from the cytoplasm by PbrA and then sequestered as a phosphate salt with the inorganic phosphate produced by PbrB. Similar operons containing genes for heavy metal–translocating ATPases and phosphatases were found in several different bacterial species, suggesting that lead detoxification through active efflux and sequestration is a common lead resistance mechanism.

The scanning electron microscopy (SEM)-EDX and HR-TEM-EDX techniques revealed that the bacterium *Klebsiella* sp. 3S1 largely adsorbed lead on the entire cell surface and lead accumulated within the cytoplasm also, which indicated the presence of bioaccumulation mechanism (Munoz et al., 2015).

34.3.5 SIDEROPHORES

Siderophores are low-molecular-weight, high-affinity iron (Fe)-chelating organic compounds secreted by various microorganisms. They are excreted by microbes in iron-deficient environments. Siderophore–Fe complex formation is influenced by pH. Siderophores were reported to interact with other metals and in fact there is a competition for siderophore-binding sites between iron and other free protons and metal ions such as Cd^{2+}, Cu^{2+}, Ni^{2+}, Pb^{2+}, and Zn^{2+} (Albrecht-Gary and Crumbliss, 1998; Schalk et al., 2011; Saha et al., 2012) and Mn^{3+}, Co^{3+}, and Al^{3+} and actinides such as Th^{4+}, U^{4+}, and Pu^{4+} (Peterson et al., 2004). Pyoverdine siderophores—defining the characteristic for *Pseudomonas*—were produced by *P. aeruginosa* 4EA and endophytic *P. putida* KNP9. Both were reported to complex Pb (Tripathi et al., 2005; Naik and Dubey, 2011). The pyoverdines produced by the plant growth-promoting strain *P. putida* KNP9 were found to reduce the concentration of Pb in mung bean roots and shoots (Tripathi et al., 2005). The presence of Pb doubled the production of siderophores in *P. aeruginosa* 4EA (Naik and Dubey, 2011). The siderophore pyochelin produced by *P. aeruginosa* PAO1 showed higher affinity toward Pb than pyoverdine (Braud et al., 2010). Braud et al. (2009) reported that though the transporter FpvAI has broad metal specificity at the binding stage, it is highly selective for Fe^{3+} only during the uptake process.

Homann et al. (2009) have isolated a suite of amphiphilic siderophores, that is, loihichelins A–F, from the cultures of the marine heterotrophic Mn(II)-oxidizing bacterium *Halomonas* sp. LOB-5. They suggested that these siderophores may have a role in sequestering Fe(III) released during basaltic rock weathering and in the promotion of Mn(II) and Fe(II) oxidation. Sinha and Mukherjee (2008) reported Cd-induced siderophore production and accumulation of Cd by *P. aeruginosa* KUCd1 in a culture medium containing cadmium.

34.3.6 BIOMINERALIZATION

Biomineralization is the process by which some halophilic bacteria synthesize inorganic solids resulting in organic–inorganic hybrids. The process depends mainly on the organism and factors such as pH and temperature. Many products of biomineralization are capable of binding environmental contaminants and removing them from wastewaters.

The moderately halophilic bacterium *Halomonas halophila* mineralizes calcium carbonate in the calcite polymorph. Rothenstein et al. (2012) studied the biomineralization process in the presence of zinc ions as a toxic model contaminant. The medium used for the experiments contained 3% sea salts. The results showed that *H. halophila* can adapt to zinc-contaminated medium, maintaining the ability for the biomineralization of calcium carbonate. The growth of the bacterium was not much affected by the presence of zinc, and it accumulated zinc ions on its cell surface causing a depletion of zinc in the medium. They also confirmed that Zn ions influence the biomineralization process. In the presence of zinc, the polymorphs monohydrocalcite and vaterite were mineralized, instead of calcite that was synthesized in zinc-free medium. That is, zinc influenced in the modification of the synthesized calcium carbonate polymorph. In addition, the shape of the mineralized inorganic material is changing through the presence of zinc ions. Furthermore, the moderately halophilic bacterium *H. halophila* can be applied for the decontamination of zinc from aqueous solutions.

Carbonate precipitation in a medium containing artificial sea salt of 7.5% (w/v) mediated by moderately halophilic bacteria *Chromohalobacter marismortui* was reported by Rivadeneyra et al. (2010).

34.3.7 BIOSYNTHESIS OF NANOPARTICLES

Nanoparticles (NPs) are defined as materials that are less than 100 nm in size. Biosynthesis of lead and cadmium NPs is reported by various organisms like bacteria, yeast, and fungi and also by various leaf extracts (Ahmad et al., 2002; Kowshik et al., 2002a,b; Prasad et al., in press). The ability of the bacteria to produce NPs is beneficial in bioremediation as well as it provides a less toxic and precise method for large-scale synthesis of these highly beneficial particles. The importance of this emerging field of study is evident from numerous reviews published in the past 5–6 years (Sinha et al., 2009; Arya, 2010; Ramezani et al., 2010; Talebi et al., 2010; Li et al., 2011; Habeeb, 2013; Ingale and Chaudhari, 2013; Hamidi et al., 2014; Iravani, 2014; Malik et al., 2014; Pantidos and Horsfall, 2014; Jacob et al., 2015)

Cysteine desulfhydrase (C-S-lyase)—an intracellular enzyme located in the cytoplasm—was reported to be an important factor in the biosynthesis of metal sulfide NPs (Wang et al., 2001). Bai et al. (2009) found out that the synthesis of CdS NPs in *R. sphaeroides* was mediated by C-S-lyase. According to Kang et al. (2008), NPs synthesized by microorganisms were stabilized by peptides such as phytochelatins, thus preventing aggregation.

Bacillus megaterium synthesized Ag, Pb, and Cd NPs when grown aerobically with the solutions of metal salts. The NPs were accumulated on the surface of the cell wall of the bacteria with particle size in the range of 10–20 nm (Prakash et al., 2010). The biomass of *Bacillus licheniformis* MTCC 9555 when challenged with cadmium chloride and sodium sulfide was found to produce CdS NPs (5.1 ± 0.5 nm) (Tripathi et al., 2014). Shivashankarappa and Sanjay (2015) studied the effect of various ratios of $CdCl_2$ and Na_2S on CdS NP formation by *B. licheniformis*. They also demonstrated the antimicrobial activity of CdS NPs on food-borne bacteria such as *Escherichia coli*, *B. licheniformis*, *P. aeruginosa*, *Bacillus cereus*, and *S. aureus* and fungi such as *Fusarium oxysporum*, *Aspergillus flavus*, and *Penicillium expansum*. The size of CdS varied between 20 and 40 nm. The CdS NP ratio of 4:1, that is, the ratio between $CdCl_2$ and Na_2S at a concentration of 40 mg/ml, showed the highest zone of inhibition in *P. aeruginosa* and *A. flavus* (Shivashankarappa and Sanjay, 2015).

The marine bacterium *Idiomarina* sp. PR58–8 was reported to synthesize silver NPs intracellularly (Seshadri et al., 2012). Holmes et al. (1997) have reported that CdS particles synthesized by *K. pneumoniae* (>5 nm in diameter) on its outer cell wall grow continuously and some reached the size >200 nm, which are known as Q particles. *E. coli* was found to synthesize intracellular cadmium sulfide (CdS) nanocrystals (2–5 nm) when incubated with cadmium chloride and sodium sulfide (Sweeney et al., 2004). Immobilized photosynthetic bacteria *Rhodobacter sphaeroides* produce CdS at room temperature with a single-step process. CdS NPs formed intracellularly were then transported into extracellular solution. The size of sulfide NPs synthesized was found to vary with culture time—2.3–36.8 nm from 36 to 48 h (Bai et al., 2009). They found out that the synthesis of NPs was mediated by cysteine desulfhydrase (C-S-lyase)—an intracellular enzyme located in the cytoplasm.

Desulfotomaculum sp. produced 13 nm diameter PbS NPs (Jun et al., 2007). *Enterobacter* sp. that is able to synthesize lead oxide (PbO) NPs within the periplasmic space and a virulent strain of *B. anthracis* that is able to synthesize lead sulfide (PbS) NPs extracellularly were reported by El-Shanshoury et al. (2012a).

Rapid and low-cost biosynthesis of CdS using the culture supernatants of *E. coli* ATCC 8739, *B. subtilis* ATCC 6633, and *Lactobacillus acidophilus* DSMZ 20079T was reported by El-Shanshoury et al. (2012b). The sizes of the particles were found to vary between 2.5 and 5.5 nm. Seshadri et al. (2011) have reported the intracellular synthesis of stable lead sulfide NPs by the marine yeast *Rhodosporidium diobovatum*. Singha and Nara (2013) synthesized PbS NPs using the supernatants of two bacterial strains (NS2 and NS6) isolated from the heavy metal–rich soil samples of Allahabad and Kanpur. Extracellular PbS NPs were synthesized when supernatant was challenged with lead chloride and calcium sulfate salts. The enzymes involved in metal ion reduction were present only in the supernatant (Singha and Nara, 2013).

34.4 CADMIUM/LEAD RESISTANCE IN SALINE MEDIUM BY HALOTOLERANT/ MODERATELY HALOPHILIC BACTERIA

Hahne and Kroontje (1973) have reported that cadmium and lead [Cd(II) and Pb(II)] react with Cl^- to form a variety of coordination complexes, depending on the concentration of Cl^-. The MCl^+ species of Cd and Pb appear at chloride concentrations above 10^{-3} M (35 ppm), and MCl_2 complexes occur above 10^{-2} M (350 ppm Cl^-). The respective MCl_3^- and MCl_4^{2-} species become important above 10^{-1} M Cl^- (3500 ppm). Babich and Stotzky (1982) investigated the effect of seawater on Cd toxicity to nonmarine fungi and found that the toxicity decreased as the concentration of chloride (Cl^-) or of seawater increased, indicating the lower toxicity of Cd–Cl complexes than that of Cd^{2+}.

Onishi et al. (1984) reported a decrease in the toxicity of cadmium in a moderately halophilic *Pseudomonas* sp. with an increase in NaCl concentration from 1 to 3 M. According to them, the cadmium tolerance of 41 strains of halophilic bacteria differed in their salt requirement and cadmium tolerance. They also showed that Cd toxicity was apparently enhanced by $NaNO_3$ and Na_2SO_4. The toxicity was measured as cell death, and accommodation to Cd required an extended lag time that varied with both the concentration of Cd and that of salts.

Garcia et al. (1987) studied 58 strains of the moderate halophilic bacteria *Vibrio costicola*, including both culture collection strains and freshly isolated strains from solar salterns, regarding their susceptibility to 10 heavy metals in basal medium with 5% and 15% salt. All strains were sensitive to cadmium, copper, silver, zinc, and mercury. Mercury showed the highest toxicity even at the lowest concentration tested. The isolates showed tolerance in the following pattern, Pb > Ni > Cr > As > Co. The majority of strains (96.4%) were multimetal tolerant, with three different metal ion tolerances as the major pattern. But no significant differences were observed for the MIC values for each individual strain when they were tested in media with 5% or 15% salts.

One of the earlier extensive works on metal-tolerant moderately halophilic eubacteria is that by Nieto et al. (1989). They studied the tolerance of moderate halophiles from culture collection and fresh isolates to 10 heavy metal ions. The tested isolates belonged to different taxonomic groups, namely, *Deleya halophila*, *Acinetobacter* sp., *Flavobacterium* sp., and gram-positive cocci (*Marinococcus*, *Sporosarcina*, *Micrococcus*, and *Staphylococcus*). All collection strains were sensitive to silver, mercury, and zinc and tolerant of lead, while they differed in their response to arsenate, cadmium, chromium, and copper. They reported variations in metal resistance pattern among strains irrespective of their taxonomic group. Mercury, silver, and zinc were reported to be the most toxic metals against the tested isolates, while arsenate was the least toxic. Two genera, namely, *Acinetobacter* and *Flavobacterium*, were found to include the most heavy metal–tolerant and heavy metal–sensitive strains, respectively. Lowering the salinity enhanced sensitivity to cadmium and, in some cases, to cobalt and copper. This may be either due to the osmotic changes in the cellular level induced by salinity resulting in a higher availability of toxic ions to bacteria or due to the formation of complex metal species that exert higher toxicities than free metal cations. However, increasing the salinity resulted in only a slight decrease in cadmium, copper, and nickel toxicities. However, zinc showed no reduction in toxicity with an increase in salinity.

In the study conducted by Gaballa et al. (2003), the moderately halophilic bacterium *Staphylococcus* sp. exhibited maximum resistance to nickel (Ni) in halophilic medium. Cd and Cu ions showed the highest toxic effect on the strain when combined in nontoxic levels with Ni. The strain accumulated Ni ions inside the cell. The reduced Ni toxicity in high salt is attributed to the formation of the less toxic forms of the metal and/or the change in the membrane of the cell in such a way that causes higher level of metal tolerance.

Many *Halomonas* species have been reported to harbor plasmids of ~600 and ~70 Mbp, as well as other extrachromosomal elements. *H. elongata* and *H. subglaciescola* (Nieto et al., 1989) harbor ~600 and ~70 kbp plasmids (Argandona et al., 2003). These plasmids could be responsible for some of the adaptive advantages in the genus *Halomonas*, including tolerance to metals.

Amoozegar et al. (2005) isolated 10 moderately halophilic spore-forming bacilli from saline soils in Iran. All isolates were resistant to higher concentrations of arsenate, sodium chromate, and potassium chromate. Maximum and minimum tolerances against oxyanions were seen in selenite and biselenite, respectively. All isolates were susceptible to silver, nickel, zinc, and cobalt, while seven isolates were resistant to lead. Susceptibility to copper and cadmium varied among the isolates. Silver had the maximum toxicity, whereas lead and copper showed minimum toxicity. An increase in salinity from 5% (w/v) to 15% (w/v) enhanced tolerance to toxic oxyanions and metals. The only exception was cobalt, for which the maximum tolerance was seen in 5% (w/v) NaCl-containing medium. The maximum effect of the NaCl concentration on the toxic metal resistance was seen in the Pb-containing medium.

Massadeh et al. (2005) have reported isolation of 10 gram-positive and gram-negative bacterial cultures able to grow at 10% NaCl from water, mud, and soil samples from the shores of the Dead Sea in Suwaymah. Among them, maximum resistance to Pb was shown by two bacterial cultures in nutrient media, and seven bacterial cultures exhibited maximum resistance to Cd. However, the isolates were more resistant to lead than cadmium. The most tolerant gram-negative rod and gram-positive cocci were able to remove lead and cadmium from different concentrations with almost similar efficiency after 2–3 weeks. The accumulation of the absorbed metals was found to be maximum in the protoplast of all the cultures. The accumulation on the cell wall and between the cell wall and the plasma membrane varied for both the metals, and it differed according to strains also.

Bioluminescent bacteria *Vibrio harveyi* and TEMO5 and TEMS1 were very resistant to Pb and As and Pb, respectively, in nutrient medium with 2% NaCl. The strains are very sensitive to Cu, Cd, and Hg. MIC values of Mn, Ni, Zn, Pb, and Cr against *V. harveyi* TEMO5 were different from that of *V. harveyi* TEMS1. This indicated isolate-specific differences in the metal sensitivity (Omeroglu et al., 2007). Amoozegar et al. (2007) reported complete reduction of 0.2 m MCr(VI) after 24 h by halophilic *Nesterenkonia* sp. strain MF2 in the presence of 1 M NaCl.

Kim et al. (2007) in their study on a Pb-resistant *Bacillus* sp. strain CPB4 isolated from heavy metal–contaminated soil

in Korea showed that 90% of Pb was distributed in the cell wall and cell membrane and the rest in the cytoplasm. In Pb uptake cells, electron-dense granules were mainly found on the cell wall and cell membrane. Its metal uptake capacity was in the following order: Pb > Cd > Cu > Ni > Co > Mn > Cr > Zn. In Cd uptake cells, electron-dense granules were mainly found on the cell membrane. However, in Cu-adsorbed cells, electron-dense granules were mainly found on the outside but located very close to the cells. These results show that the CPB4 cell has a variable pattern of biosorption with different heavy metals. The amount of heavy metal uptake was remarkably decreased by reducing the crude protein contents when cells were treated by alkali solutions (Kim et al., 2007).

De et al. (2008) reported that marine bacteria highly resistant to mercury were capable of detoxifying not only Hg but also Cd and Pb. These bacteria identified belong to the species *Alcaligenes faecalis*, *Bacillus pumilus*, *Bacillus* sp., *P. aeruginosa*, and *Brevibacterium iodinium*.

The mechanisms of heavy metal detoxification were through volatilization (for Hg), putative entrapment in the EPS (for Hg, Cd, and Pb) as revealed by the SEM and energy-dispersive x-ray spectroscopy (EDS), and/or precipitation as sulfide (for Pb). These bacteria removed Cd and Pb effectively from a growth medium.

Kawasaki et al. (2008) isolated three strains of genus *Staphylococcus* and one strain of genus *Halobacillus* with high salt resistance and cadmium-absorbing capability from salt-fermented food. Strains showed high removal of cadmium in a medium with salt concentrations of 0%–20%. The isolates could be used for reducing the cadmium in foodstuffs.

Ibrahim et al. (2011) isolated and characterized a novel potent Cr (VI)-reducing alkaliphilic *Amphibacillus* sp. KSUCr3 from hypersaline soda lakes. The strain tolerated very high Cr (VI) concentration in addition to high concentrations of other heavy metals including Pb, Ni, Mo, Co, Mn, Zn, and Cu. Strain KSUCr3 could rapidly reduce 5 mM of Cr (VI) to a nondetectable level over 24 h, and its reduction capacity increased with the increase in NaCl concentration. Khodabakhsh et al. (2011) reported a moderately halophilic bacterium grown in the Ventosa medium with 98% homology with *Salinivibrio costicola* species that is resistant to some toxic metals and has the potency of removing nickel from the contaminated environment.

Periplasmic metal-binding protein termed histidine-rich metal-binding protein (HP)—characterized by high histidine content—was reported from a moderate halophile *Chromohalobacter salexigens* (Yamaguchi et al., 2012). The study on the purified protein showed high affinity toward Ni- and Cu-loaded chelate columns and moderate affinity toward Co and Zn columns. Its structure was found to be stable at 0.2–2.0 M NaCl, and the thermal transition pattern was considerably shifted to higher temperature with increasing salt concentration. The melting temperature was raised to ~20°C at 2.0 M NaCl over the melting temperature at 0.2 M NaCl. HP showed reversible refolding from thermal melting in 0.2–1.15 M NaCl, while it formed irreversible aggregates upon thermal melting at 2 M NaCl.

Solanki and Kothari (2012) isolated *Virgibacillus salaries*, *Staphylococcus epidermidis*, *Bacillus atrophaeus*, *Halomonas shengliensis*, and *Halomonas koreensis* from saline soil of Gujarat. *V. salaries* exhibited better metal tolerance/resistance than other isolates. In certain cases, the stimulatory effect of metal ions on their growth was also observed. According to them, silver had the most toxic effect on all isolates followed by cadmium.

Amoozegar et al. (2012) isolated 24 moderately halophilic bacteria from saline environments of Iran that were used to study their ability to bioremediation of lead and cadmium. Among them, a *Halomonas* strain D showed remarkable ability for the removal of Pb and Cd. The strain D could uptake lead more effectively than cadmium. Biomass showed the best lead removal at 5% NaCl (w/v), while EPS showed maximum removal at 10% NaCl (w/v). For cadmium removal by biomass and EPS, the best results were obtained at 1% NaCl (w/v). In this study, the removal rate was much higher when using the biomass of strain D. The best metal removal results by biomass were obtained at NaCl concentrations near to optimum NaCl concentration for the growth of strain D. Whereas in the case of metal removal by exopolymeric substances, increasing NaCl concentration led to the best Pb removal rate and decreasing NaCl concentration led to the best Cd removal rate.

Multimetal-resistant bacteria belonging to alkaliphilic (*Natronobacterium magadii*, *Natronococcus occultus*, and *Nb. Gregory*) and halophilic (*Halobacterium saccharovorum*, *Hb. Sodomense*, and *Hb. salinarium*) groups were isolated from solar salt pan sediment at the coastal area of Tamil Nadu, India. They were resistant to Ni, Al, Cd, Zn, Hg, and As. The maximum percentage of resistance was identified for Ni, while the minimum percentage of tolerance was identified with Hg. None of the selected isolates showed the presence of plasmids. This indicated that the resistance of the chosen metals was directly controlled by the chromosomal DNA and the resistance activity of the halobacterial species might be related to metal-binding proteins (Williams et al., 2013).

Guo and Mahillon (2013) isolated pGIAK1—a 38 kb plasmid—originating from the obligate alkaliphilic and halotolerant *Bacillaceae* strain JMAK1. The strain was originally isolated from the confined environments of the Antarctic Concordia station. The analysis on the pGIAK1 38,362-bp sequence revealed that in addition to its replication region, this plasmid contains the genetic determinants for cadmium and arsenic resistances, putative methyl transferase, tyrosine recombinase, spore coat protein, and potassium transport protein, as well as several hypothetical proteins.

Halomonas lionensis RHS90T is a halotolerant bacterium highly sensitive to Ag and Cd but grew very well at high concentrations of Cs and was also resistant to Mn. The strain RHS90T contains one or several plasmids >10 kbp that might possibly be involved in metal tolerance (Gaboyer et al., 2014).

Halomonas zincidurans B6T capable of resisting high concentrations of heavy metals in liquid halophilic medium, including Mn, Co, Cu, and Zn, was reported from sediments of the South Atlantic Mid-Ocean Ridge. Its resistance to Zn was much higher when incubated on a marine agar 2216 medium. This *H. zincidurans* B6T was found to encode 31 genes related to heavy metal resistance (Huo et al., 2014).

The genome of *Lunatimonas lonarensis* AK24T shows the presence of heavy metal tolerance genes, including two genes providing resistance against arsenic and 11 genes for cobalt–zinc–cadmium resistance. This gram-negative, pinkish-orange-pigmented, half-moon-shaped bacterium was first isolated from water and sediment samples of Lonar Lake, Buldhana district, Maharashtra, India (Srinivas et al., 2014).

Vela-Cano et al. (2014) isolated eight strains from a sewage sludge compost tea with high resistance to Pb, Zn, Cu, and Cd and one with (*Rhodococcus* sp.) a special tolerance to every heavy metal tested. The strains belonged to *Rhodococcus*, *Virgibacillus*, *Leifsonia*, *Achromobacter*, *Cupriavidus*, and *Oceanobacillus* sp. *Rhodococcus* sp. strain 3 was able to remove different amounts of heavy metals from the culture media with intracellular and surface accumulation of Cu, Zn, Pb, and Cd. The tolerance of the tested strains was observed to be in the order of Cu > Pb > Zn > Cd. *Rhodococcus* sp. strain 3 showed maximum removal of lead (Pb) followed by Cu, Zn, and Cd. TEM analysis of *Rhodococcus* strain 3 grown in the presence of Cu, Cd, Zn, and Pb showed that Cu was mainly deposited on the cell surface structures, while small amounts of Cu were detected inside the cells. In the case of Cd, more metal could be detected inside the cells than outside. The analysis confirmed the bioaccumulation of Pb and Zn because both metals were mainly deposited inside the cells.

Moderately halophilic and halotolerant bacteria showing resistance to cadmium and lead were isolated by Sowmya et al. (2014) from the sediment samples of Vembanad Lake, India. Bacterial strains belonging to different genera such as *Alcaligenes*, *Vibrio*, *Kurthia*, and *Staphylococcus* and members of the family *Enterobacteriaceae* were reported. The isolates showed higher resistance to lead than cadmium at 5%, 10%, and 15% salt concentrations. Selected isolates removed lead more efficiently than cadmium.

Syed and Chinthala (2015) reported the detoxification potential of three *Bacillus* species isolated from solar salterns, namely, *B. licheniformis* NSPA5, *B. cereus* NSPA8, and *B. subtilis* NSPA13 against lead, chromium, and copper, by biosorption in metal biosorption medium (8.1% NaCl). *B. cereus* NSPA8 showed maximum lead biosorption. The biosorption of copper and chromium was relatively low in comparison with lead. Energy-dispersive x-ray system and EDS spectral images gave evidences of the binding of metal ions [Cd, Cu, and Pb ions] on the surface of the cell wall of the bacterial cells.

The works demonstrating the cadmium and lead removal efficiency of halotolerant/moderately halophilic bacteria in a saline medium or the effect of salinity variable on metal reduction are relatively less (Table 34.1).

Nonhalophilic halotolerant bacteria like *Pseudomonas* and *Staphylococcus* are extensively studied for their metal tolerance and tolerance mechanisms. However, even in those studies salinity was not included as a variable. A brief summary of those works is given in Table 34.2.

TABLE 34.1

Metal Resistance of Halotolerant/Moderately Halophilic Bacteria in Saline Medium

Sl. No.	Name of Bacteria	Medium/Salt Percentage (w/v)	Tolerant Heavy Metals	Resistance Mechanism(s)	References
1.	Bacteria and fungi	NB[a]	Zn	—	Babich and Stozky (1978)
2.	Moderately halophilic *Pseudomonas* sp.	SGC[b] 5.8%–17.5%	Cd		Onishi et al. (1984)
3.	*Vibrio costicola* (different strains)	BM[c] + 5% and 15% NaCl	Co < As < Cr < Ni < Pb (Cu, Hg, Ag, Cd, and Zn)[n]	—	Garcia et al. (1987)
4.	*Deleya, Acinetobacter, Flavobacterium,* gram-positive cocci	SW-10[d] broth	As, Cd, Cr, and Cu (Ag, Hg, and Zn)[n]	—	Nieto et al. (1989)
5.	Moderately halophilic *Staphylococcus* sp.	HM[e]	Ni (Cd and Cu)[n]	—	Gaballa et al. (2003)
6.	Moderately halophilic spore-forming bacilli	NB[a] + 5%, 10%, and 15%	Cd < Cu < Pb (Ag, Ni, Zn, and Co)[n]	—	Amoozegar et al. (2005)
7.	Moderately halophilic gram-negative rod and gram-positive cocci	SW-10[d] broth medium	Cd < Pb	Biosorption and bioaccumulation	Massadeh et al. (2005)
8.	*Halomonas maura* S-30		Pb	Binding on EPS—mauran	Arias et al. (2003); Llamas et al. (2006)
9.	*Halomonas ventosae* Al12T *H. ventosae* Al16, *Halomonas anticariensis* FP35T and FP36		Co < Pb Co < Cu	Binding on EPS	Mata et al. (2006)
10.	*Vibrio harveyi* TEMO5 and *Vibrio harveyi* TEMS1	NB[a] + 2%	As < Pb Pb = As (Cu, Cd, and Hg)[n]	—	Omeroglu et al. (2007)
11.	Halophilic *Nesterenkonia* sp. MF2	5.8%	Cr		Amoozegar et al. (2007)
12.	*Bacillus* sp. strain CPB4		Zn < Cr < Mn < Co < Ni < Cu < Cd < Pb	Electron-dense granules on cell surface and cytoplasm	Kim et al. (2007)
13.	*Alcaligenes faecalis, Bacillus pumilus, Bacillus* sp., *Pseudomonas aeruginosa* and *Brevibacterium iodinium*	SWN[f]	Hg Hg, Cd, and Pb Pb	Volatilization Binding on EPS Precipitation as sulfide	De et al. (2008)
14.	*Staphylococcus* sp. *Halobacillus* sp.	0%–20%	Cd	—	Kawasaki et al. (2008)
15.	Alkaliphilic *Amphibacillus* sp. KSUCr3	AM[g] + 0%–25%	Cu = Zn < Co < Cr(VI) = Mo = Pb < Ni = Mn	—	Ibrahim et al. (2011)
16.	*Salinivibrio costicola*	VM[h]	Ni, Cd, Co, Zn, Cu, Pb, Ag	—	Khodabakhsh et al. (2011)
17.	*Chromohalobacter salexigens*		Co and Zn < Ni and Cu	HP indicates Histidine-rich metal binding protein	Yamaguchi et al. (2012)
18.	*Virgibacillus salaries, Staphylococcus epidermidis, Bacillus atrophaeus, Halomonas shengliensis,* and *Halomonas koreensis*		Ag < Cd < Ni	—	Solanki and Kothari (2012)

(Continued)

TABLE 34.1 (*Continued*)

Metal Resistance of Halotolerant/Moderately Halophilic Bacteria in Saline Medium

Sl. No.	Name of Bacteria	Medium/Salt Percentage (w/v)	Tolerant Heavy Metals	Resistance Mechanism(s)	References
19.	*Halomonas* strain D	NA[i] + 10%; SM[j] + 5%	Cd < Pb	Biosorption, bioaccumulation, binding on EPS	Amoozegar et al. (2012)
20.	*Natronobacterium magadii, Natronococcus occultus,* and *Nb. Gregory; Halobacterium saccharovorum, Hb. sodomense,* and *Hb. salinarium*	HA[k]	Hg, Cd, As, Zn, Al, Ni	—	Williams et al. (2013)
21.	*Halomonas lionensis* RHS90T		Mn < Cs (Ag and Cd)[n]	Plasmid encoded	Gaboyer et al. (2014)
22.	*Halomonas zincidurans* strain B6T	HM[e]	Mn, Co, Cu, and Zn	Gene encoded	Huo et al. (2014)
23.	*Rhodococcus, Virgibacillus, Leifsonia, Achromobacter, Cupriavidus,* and *Oceanobacillus* sp.	LPM[l]	Cd < Zn < Pb < Cu	Intracellular and surface accumulation	Vela-Cano et al. (2014)
24.	*Alcaligenes, Vibrio, Kurthia, Staphylococcus,* and members of the family *Enterobacteriaceae*	NB[a] + 5%, 10%, and 15%	Cd < Pb	—	Sowmya et al. (2014)
25.	*Pseudomonas aeruginosa* JP-11	SWN[f] agar	Cd	Binding on EPS and gene encoded	Chakraborty and Das (2014)
26.	*Bacillus licheniformis* NSPA5, *Bacillus cereus* NSPA8, and *Bacillus subtilis* NSPA13	MBM[m]	Cr, Cu < Pb	Binding to cell wall	Syed and Chinthala (2015)

[a] Nutrient broth.
[b] Sehgal and Gibbons complex medium.
[c] Basal medium.
[d] Saline yeast extract medium.
[e] Halophilic medium.
[f] Seawater nutrient agar.
[g] Alkaline medium.
[h] Ventosa medium.
[i] Nutrient agar.
[j] Saline medium.
[k] Halophilic agar.
[l] LPM medium.
[m] Metal biosorption medium.
[n] Metals to which the isolates were sensitive.

34.5 CADMIUM AND LEAD RESISTANCE MECHANISMS

From the earlier discussions, it is clear that both cadmium and lead are externally sequestered on the EPS and cell wall of various bacteria. It is evident in SEM, TEM, and AAS analyses. In the biosynthesis of NPs, many bacteria were found to produce NPs externally. In most studies, we can see that lead tends to bind more on EPS/cell wall than cadmium. Though external and internal precipitation was reported for lead in some bacteria, only few works were reported on halotolerant bacteria. In the case of cadmium, intracellular sequestration and efflux mechanisms mediated by RND, CDF, and P-type ATPase transporters are more important. The main mechanisms of cadmium and lead tolerance are shown in Figures 34.2 and 34.3.

TABLE 34.2

Metal-Resistant Nonhalophilic Bacteria

Sl. No.	Bacteria	Metals Studied	Reference
1.	*Pseudomonas tolaasii* RP23 and *Pseudomonas fluorescens* RS9	Zn and Cd	Amico et al. (2005)
2.	*Pseudomonas aeruginosa* KUCd1	Cd	Sinha and Mukherjee (2008)
3.	Gram-positive bacilli	Fe, Pb, Hg, Zn, Al, Cu, Ni, Cd, and Ag	Roy et al. (2008)
4.	*E. coli* ASU7	Cu, Co, Zn, Ni, Pb, Cd, Cr^{6+}, and Cr^{3+}	Abskharon et al. (2008)
5.	*Pseudomonas aeruginosa*, *P. alcaligenes*, *Methylococcus* sp., and *Desulfotomaculum* sp.	Zn, Cu, Ni, Cd, and Cr	Mishra et al. (2009)
6.	*Proteus vulgaris* (BC1), *Pseudomonas aeruginosa* (BC2 and BC5), and *Acinetobacter radioresistens* (BC3)	As, Ni, Cd, Pb, Hg, and Cr	Raja et al. (2009)
7.	*Pseudomonas aeruginosa* KUCd1	Cd	Sinha and Mukherjee (2009)
8.	Purple nonsulfur bacteria (PNB) NW16 and KMS24	Pb, Cu, Cd, Zn, and Na	Panwichian et al. (2010)
9.	*Pseudomonas aeruginosa* MCCB 102	Cu, Zn, Cd, and Pb	Zolgharnein et al. (2010)
10.	*Aeromonas* sp. and *Pseudomonas* sp.	Cu, Pb	Matyar et al. (2010)
11.	*Bacillus cereus* SIU1	As, Pb, and Cs	Singh et al. (2010)
12.	*Pseudomonas aeruginosa*	Cd	Kermani et al. (2010)
13.	*Pseudomonas* sp.	Pb	Dinu et al. (2011)
14.	*Pseudomonas putida*	Cd	Nanganuru and Korrapati (2012)
15.	*Pseudomonas*, *Aeromonas*, *Bacillus*, *Escherichia*, *Micrococcus*, and *Proteus* species	Fe, Zn, and Pb	Mgbemena et al. (2012)
16.	*Bacillus* sp., *Pseudomonas* sp., *Corynebacterium* sp., *Staphylococcus* sp., and *E. coli*	Pb	Kafilzadeh et al. (2012)
17.	*Pseudomonas* sp., *Klebsiella* sp., *Staphylococcus* sp., *Proteus* sp., and *Bacillus* sp.	Cd and Pb	Nath et al. (2012)
18.	*Bacillus cereus*	Pb	Murthy et al. (2012)
19.	*Actinomyces* sp., *Streptomyces* sp., and *Bacillus* sp.	Cd and Ni	Karakagh et al. (2012)
20.	*Bacillus* sp.	Cd, Cr, Ni, and Co	Samanta et al. (2012)
21.	*Enterobacter cloacae* P2B	Pb	Naik et al. (2012)
22.	*Massilia* sp., *Pseudomonas* sp., *Pseudomonas fulva*, *Bacillus* sp., and *Serratia entomophila*	Cd	Złoch et al. (2013)
23.	*Bacillus cereus*, *Bacillus subtilis* subsp. *subtilis*	Hg, Pb, Ag, Zn, and Cu	Hookoom and Puchooa (2013)
24.	*Pseudomonas putida* and *Bacillus licheniformis*	Mn, Cu, Ni, Co, V, Pb, Ti, Al, and Zn	Kamika and Momba (2013)
25.	*Staphylococcus aureus*	Cd	Shrivastava et al. (2013)
26.	Cadmium-resistant bacterium CdRB1	Cd	Dabhi et al. (2013)
27.	*Bacillus cereus*	Pb	Murthy et al. (2014)
28.	*Stenotrophomonas maltophilia*, *Bacillus halodurans*, *Exiguobacterium homiense*, *Pseudomonas putida*, and *Pseudomonas geniculata*	Cd, Cr, Zn, Mg, Co, and Ni	Prabhu et al. (2014)
29.	*Bacillus*, *Lysinibacillus*, *Micrococcus*, *Stenotrophomonas*, *Bacillus pumilus* Jo2, *Bacillus subtilis* BD18–B23, and *Lysinibacillus fusiformis* B	Cr, Cu, Ni, Co, Cd, Zn, and Pb	Tomova et al. (2014)
30.	*Aeromonas* sp., *Arthrobacter* sp. *Corynebacterium* sp., *Pseudomonas* sp., *Streptococcus* sp.	Pb, Cr, and Cd	Owolabi and Hekeu (2014)
31.	*Bacillus subtilis* KPA	Ag, Hg, Cu, and Cr	Khusro at al. (2014)
32.	*Pseudomonas* sp. PMDZnCd2003, *Serratia* sp. PDMCd2007, and *Serratia* sp. PDMCd0501	Cd and Zn	Nakbanpote et al. (2014)
33.	*Klebsiella pneumoniae*, *Pseudomonas aeruginosa*, *Providencia rettgeri*, *Alcaligenes faecalis*, *Morganella morganii*, and *Pseudomonas putida*	Zn, Hg, and Cd	Yamina et al. (2014)
34.	*Massilia* sp., *Pseudomonas* sp., and *Bacillus* sp.	Cd	Hrynkiewicz et al. (2015)
35.	*Bacillus* sp., *Enterobacter* sp., *Aeromonas* sp., and *Pseudomonas* sp.	Cd	Mathivanan and Rajaram (2014)
36.	*Staphylococcus epidermidis* ATCC 35984	Cd	Wu et al. (2015)
37.	*Klebsiella* sp. 3S1	Pb	Munoz et al. (2015)

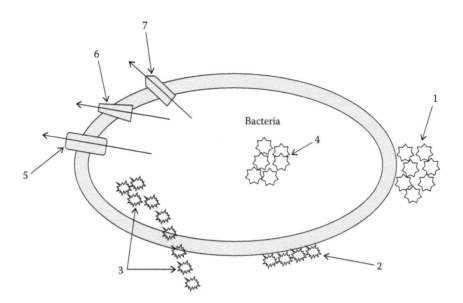

FIGURE 34.2 Cadmium-resistant mechanisms in bacteria. 1, extracellular sequestration on EPS; 2, CdS NPs binding to cell surface; 3, CdS NPs formed intracellulary and coming outside through cell decay; 4, intracellular sequestration/precipitation; 5, efflux mechanisms by RND (Czc, Ncc); 6, efflux mechanisms by CDF (ZntA); 7, efflux mechanisms by P-type ATPase (CadA). (Modified from Jarosławiecka, A. and Piotrowska-Seget, A., *Microbiology* (*United Kingdom*), 160, 12, 2014.)

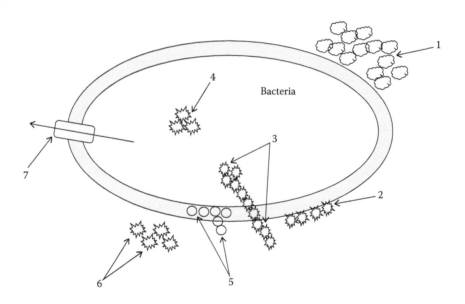

FIGURE 34.3 Lead resistance mechanisms in bacteria. 1, extracellular sequestration on EPS; 2, PbS NPs binding to cell surface; 3, PbS NPs formed intracellulary and coming outside through cell decay; 4, intracellular sequestration/precipitation; 5, periplasmic PbO NPs coming outside through cell decay; 6, extracellular precipitation; 7, efflux mechanisms like PbrA, CadA, and ZntA. (Modified from Naik, M.M. and Dubey, S.K., *Ecotoxicol. Environ. Saf.*, 98, 1, 2013; Jarosławiecka, A. and Piotrowska-Seget, A., *Microbiology* (*United Kingdom*), 160, 12, 2014.)

34.6 SHORTCOMINGS AND FUTURE ASPECTS OF HALOTOLERANT AND MODERATELY HALOPHILIC BACTERIA IN CADMIUM AND LEAD BIOREMEDIATION

In this chapter, the general mechanisms of heavy metal resistance in bacteria and the potential of halophilic bacteria in bioremediation with special reference to cadmium and lead have

been reviewed. In the present scenario, we need strains with high metal resistance and removal capacity in varying salinities. It is a well-known fact that the constituents of bacterial growth media and also the salt will react with cadmium and lead. The toxicity of a metal to the bacteria strongly depends on its bioavailability rather than its concentration. At higher concentrations, lead tends to precipitate more in the media. So when conducting metal reduction experiments, care should be given to the constituents of media and the concentration of the metal tested. Though there were a number of studies dealing with the interference of culture medium components and salinity with metals, there is still a long way to go.

In the case of biosurfactants and NPs, studies on halotolerant bacteria are less compared with other microorganisms. The efficiency of these moderately halophilic and halotolerant microbes has to be exploited in saline conditions for the production of Cd/Pb-binding biosurfactants and Cd/Pb NPs. These organisms have to be more elaborately studied for their metal removal efficiencies.

Biomineralization also presents a prospective area to be studied in detail. It is a very important natural process. So the influence of Cd/Pb on the process and the ability of bacteria to remove them from the environment in the due course have to be explored.

Different aspects have to be studied together. With multimetal resistances, other capabilities of the microbe like COD removal have to be addressed. Emphasis has to be given on the recovery of metals from saline effluents.

REFERENCES

Abskharon, R. N. N., S. H. A. Hassan, G. S. M. F. El-Rab, and A. A. M. Shoreit. 2008. Heavy metal resistant of *E. coli* isolated from wastewater sites in Assiut City, Egypt. *Bull Environ Contam Toxicol* 81: 309–315.

Ahemad, M. 2012. Implications of bacterial resistance against heavy metals in bioremediation: A review. *IIOABJ* 3: 39–46.

Ahluwalia, S. S. and D. Goyal. 2007. Microbial and plant derived biomass for removal of heavy metals from wastewater. *Bioresour Technol* 98: 2243–2257.

Ahmad, A., P. Mukherjee, D. Mandal et al. 2002. Enzyme mediated extracellular synthesis of CdS nanoparticles by the fungus, *Fusarium oxysporum. J Am Chem Soc* 124: 12108–12109.

Albrecht-Gary, A. M. and A. L. Crumbliss. 1998. Coordination chemistry of siderophores: Thermodynamics and kinetics of iron chelation and release. *Met Ions Biol Syst* 35: 239–327.

Amico, E. D., L. Cavalca, and V. Andreoni. 2005. Analysis of rhizobacterial communities in perennial *Graminaceae* from polluted water meadow soil, and screening of metal-resistant, potentially plant growth-promoting bacteria. *FEMS Microbiol Ecol* 52: 153–162.

Amoozegar, M. A., A. Ghasemi, M. R. Razavi, and S. Naddaf. 2007. Evaluation of hexavalent chromium reduction by chromate-resistant moderately halophile, *Nesterenkonia* sp. strain MF2. *Process Biochem* 42: 1475–1479.

Amoozegar, M. A., N. Ghazanfari, and M. Didari. 2012. Lead and cadmium bioremoval by *Halomonas* sp., an exopolysaccharide-producing halophilic bacterium. *P Bio Sci* 2: 1–11.

Amoozegar, M.A., J. Hamedi, M. Dadashipour, and S. Shariatpanahi. 2005. Effect of salinity on the tolerance to toxic metals and oxyanions in native moderately halophilic spore-forming *bacilli. World J Microbiol Biotechnol* 21: 1237–1243.

Appanna, V. D., H. Finn, and M. St. Pierre. 1995. Exocellular phosphatidylethanolamine production and multiple-metal tolerance in *Pseudomonas fluorescens. FEMS Microbiol Lett* 131: 53–56.

Argandona, M., F. Martinez-Checa, I. Llamas, E. Quesada, and A. Moral. 2003. Megaplasmids in gram-negative, moderately halophilic bacteria. *FEMS Microbiol Lett* 227: 81–86.

Arias, S., A. del Moral, M. R. Ferrer, R. Tallon, E. Quesada, and V. Bejar. 2003. Mauran, an exopolysaccharide produced by the halophilic bacterium *Halomonas maura*, with a novel composition and interesting properties for biotechnology. *Extremophiles* 7: 319–326.

Arya, V. 2010. Living systems: Eco-friendly nanofactories. *Dig J Nanomater Biostruct* 5: 9–21.

Aryal, M. and M. Liakopoulou-Kyriakides. 2015. Bioremoval of heavy metals by bacterial biomass. *Environ Monit Assess* 187: 4173.

Asci, Y., M. Nurbas, and Y. S. Acikel. 2008. A comparative study for the sorption of Cd(II) by soils with different clay contents and mineralogy and the recovery of Cd(II) using rhamnolipid biosurfactant. *J Hazard Mater* 154: 663–673.

ATSDR—Priority List of Hazardous Substances. 2015. Summary data for 2015 priority list of hazardous substances. http://www.atsdr.cdc.gov/spl/ (accessed December 20, 2015).

Babich, H. and G. Stozky. 1978. Toxicity of zinc to fungi, bacteria and coliphages: Influence of chloride ions. *Appl Environ Microbiol* 36: 906–914.

Babich, H. and G. Stotzky. 1982. Influence of chloride ions on the toxicity of cadmium to fungi. *Zbl Bakt Hyg I Abt Orig C* 3: 421–426.

Bai, H., Z. Zhang, Y. Guo, and W. Jia. 2009. Biological synthesis of size-controlled cadmium sulfide nanoparticles using immobilized *Rhodobacter sphaeroides. Nanoscale Res Lett* 4: 717–723.

Braud, A., V. Geoffroy, F. Hoegy, G. L. A. Mislin, and I. J. Schalk. 2010. Presence of the siderophores pyoverdine and pyochelin in the extracellular medium reduces toxic metal accumulation in *Pseudomonas aeruginosa* and increases bacterial metal tolerance. *Environ Microbiol Rep* 2: 419–425.

Braud, A., F. Hoegy, K. Jezequel, T. Lebeau, and I. J. Schalk. 2009. New insights into the metal specificity of the *Pseudomonas aeruginosa* pyoverdine–iron uptake pathway. *Environ Microbiol* 11: 1079–1091.

Bruins, M. R., S. Kapil, and F. W. Oehme. 2000. Microbial resistance to metals in the environment. *Ecotoxicol Environ Saf* 45: 198–207.

Chakraborty, J. and S. Das. 2014. Characterization and cadmium-resistant gene expression of biofilm-forming marine bacterium *Pseudomonas aeruginosa* JP-11. *Environ Sci Pollut Res.* 21: 14188–14201.

Checconi, P., R. Sgarbanti, I. Celestino et al. 2013. The environmental pollutant cadmium promotes influenza virus replication in MDCK cells by altering their redox state. *Int J Mol Sci* 14: 4148–4162.

Choi, W. J. and S. H. Han. 2015. Blood cadmium is associated with osteoporosis in obese males but not in non-obese males: The Korea National Health and Nutrition Examination Survey 2008–2011. *Int J Environ Res Public Health* 12: 12144–12157.

Comte, S., G. Guibaud, and M. Baudu. 2008. Biosorption properties of extracellular polymeric substances (EPS) towards Cd, Cu and Pb for different pH values. *J Hazard Mater* 151: 185–193.

Dabhi, B., K. N. Mistry, S. Lal, D. Mehta, and R. Thakar. 2013. Bioaccumulation of cadmium by bacterial strains Cdrb1 isolated from cadmium contaminated soil of Gujarat. *JBB* 4: 72–79.

Dahrazma, B. and C. N. Mulligan. 2007. Investigation of the removal of heavy metals from sediments using rhamnolipid in a continuous flow configuration. *Chemosphere* 69: 705–711.

Das, P., S. Mukherjee, and R. Sen. 2009. Biosurfactant of marine origin exhibiting heavy metal remediation properties. *Bioresour Technol* 100: 4887–4890.

De, J., N. Ramaiah, and L. Vardanyan. 2008. Detoxification of toxic heavy metals by marine bacteria highly resistant to mercury. *Mar Biotechnol* 10: 471–477.

Deziel, E., G. Paquette, R. Villemur, F. Lepine, and J. G. Bisaillon. 1996. Biosurfactant production by a soil *Pseudomonas* strain growing on polycyclic aromatic hydrocarbons. *Appl Environ Microbiol* 62: 1908–1912.

Dinu, L. D., L. Anghel, and S. Jurcoane. 2011. Isolation of heavy metal resistant bacterial strains from the battery manufactured polluted environment. *Rom Biotechnol Lett* 16: 102–106.

El-Shanshoury, A. E. R. R., S. E. Elsilk, P. S. Ateya, and E. M. Ebeid. 2012a. Synthesis of lead nanoparticles by *Enterobacter* sp. and avirulent *Bacillus anthracis* PS2010. *Ann Microbiol* 62: 1803–1810.

El-Shanshoury, A. E. R. R., S. E. Elsilk, and E. M. Ebeid. 2012b. Rapid biosynthesis of cadmium sulfide (CdS) nanoparticles using culture supernatants of *Escherichia coli* ATCC 8739, *Bacillus subtilis* ATCC 6633 and *Lactobacillus acidophilus* DSMZ 20079T. *Afr J Biotechnol* 11: 7957–7965.

Fang, L. C., Q. Y. Huang, X. Wei et al. 2010. Microcalorimetric and potentiometric titration studies on the adsorption of copper by extracellular polymeric substances (EPS), minerals and their composites. *Bioresour Technol* 101: 5774–5779.

Gaballa, A., R. Amer, H. Hussein, H. Moawad, and S. Sabry. 2003. Heavy metals resistance pattern of moderately halophytic bacteria. *Arab J Biotechnol* 6: 267–278.

Gaboyer, F., O. Vandenabeele-Trambouze, J. Cao et al. 2014. Physiological features of *Halomonas lionensis* sp. nov., a novel bacterium isolated from a Mediterranean Sea sediment. *Res Microbiol* 165: 490–500.

Garcia, M. T., J. J. Nieto, A. Ventosa, and F. Ruiz Berraquero. 1987. The susceptibility of the moderate halophile *Vibrio costicola* to heavy metals. *J Appl Bacteriol* 63: 63–66.

Global Burden of Disease (GBD). 1990. http://www.who.int/health info/global_burden_disease/about/en.html (accessed October 01, 2015).

Geddie, J. L. and I. W. Sutherland. 1993. Uptake of metals by bacterial polysaccharides. *J Appl Bacteriol* 74: 467–472.

Geddie, J. L. and I. W. Sutherland. 1994. The effect of acetylation on cation binding by algal and bacterial alginates. *Biotechnol Appl Biochem* 20: 117–129.

Gibbons, S. M., K. Feris, M. A. McGuirl et al. 2011. Use of microcalorimetry to determine the costs and benefits to *Pseudomonas putida* Strain KT2440 of harboring cadmium efflux genes. *Appl Environ Microbiol* 77: 108–113.

Guo, S. and J. Mahillon. 2013. pGIAK1, a Heavy metal resistant plasmid from an obligate alkaliphilic and halotolerant bacterium isolated from the Antarctic Concordia station confined environment. *PLoS One* 8: e72461.

Habeeb, M. K. 2013. Biosynthesis of nanoparticles by microorganisms and their applications. *IJAST* 3(1): 44–51.

Haefliger, P., M. Mathieu-Nolf, S. Lociciro et al. 2009. Mass lead intoxication from informal used lead-acid battery recycling in Dakar, Senegal. *Environ Health Perspect* 117: 1535–1540.

Hahne, C. H. and W. Kroontje. 1973. Significance of pH and chloride concentration on behavior of heavy metal pollutants: Mercury(II), Cadmium(II), Zinc(II), and Lead(II). *J Environ Qual* 2: 444.

Hamidi, M., H. Mollaabbaszadeh, and N. Hajizadeh. 2014. Bioproduction of nanoparticles by microorganisms and their applications. *TLS* 3(Spl Issue 1): 2319–5037.

Han, S. J., K. H. Ha, J. Y. Jeon, H. J. Kim, K. W. Lee, and D. J. Kim. 2015. Impact of cadmium exposure on the association between lipopolysaccharide and metabolic syndrome. *Int J Environ Res Public Health* 12: 11396–11409.

Herman, D. C., J. F. Artiola, and R. M. Miller. 1995. Removal of cadmium, lead and zinc from soil by a rhamnolipid biosurfactant. *Environ Sci Technol* 29: 2280–2285.

Holmes, J. D., D. J. Richardson, S. Saed, R. Evans-Gowing, D. A. Russell, and J. R. Sodeaul. 1997. Cadmium-specific formation of metal sulfide Q-particles' by *Klebsiella pneumoniae*. *Microbiology* 143: 2521–2530.

Homann, V. V., M. Sandy, J. A. Tincu, A. S. Templeton, B. M. Tebo, and A. Butler. 2009. Loihichelins A-F, a suite of amphiphilic siderophores produced by the marine bacterium *Halomonas* LOB-5. *J Nat Prod* 72: 884–888.

Hookoom, M. and D. Puchooa. 2013. Isolation and identification of heavy metals tolerant bacteria from industrial and agricultural areas in Mauritius. *Curr Res Microbiol Biotechnol* 1: 119–123.

Hrynkiewicz, K., M. Zloch, T. Kowalkowski, C. Baum, K. Niedojadlo, and B. Buszewski. 2015. Strain-specific bioaccumulation and intracellular distribution of Cd^{2+} in bacteria isolated from the rhizosphere, ectomycorrhizae, and fruitbodies of ectomycorrhizal fungi. *Environ Sci Pollut Res.* 22: 3055–3067.

Huo, Y. Y., Li, Z. Y., Cheng, H., Wang, C. H. and Xu, X. W. 2014. High quality draft genome sequence of the heavy metal resistant bacterium *Halomonas zincidurans* type strain B6T. *Standards in Genomic Sciences* 9: 30.

Hussein, H., S. Farag, and H. Moawad. 2004. Isolation and characterization of *Pseudomonas* resistant to heavy metals contaminants. *Arab J Biotechnol* 7: 13–22.

Hynninen, A., T. Touzé, L. Pitkänen, D. Mengin-Lecreulx, and M. Virta1. 2009. An efflux transporter PbrA and a phosphatase PbrB cooperate in a lead-resistance mechanism in bacteria. *Mol Microbiol* 74: 384–394.

Ibrahim, A. S. S., M. A. El-Tayeb, Y. B. Elbadawi, and A. A. Al-Salamah. 2011. Isolation and characterization of novel potent Cr(VI) reducing alkaliphilic *Amphibacillus* sp. KSUCr3 from hypersaline soda lakes. *Electron J Biotechnol* 14: 1–14.

Ilamathi, R., G. S. Nirmala, and L. Muruganandam. 2014. Heavy metals biosorption in liquid solid fluidized bed by immobilized consortia in alginate beads. *Int J ChemTech Res* 6: 652–662.

Ingale, A. G. and A. N. Chaudhari. 2013. Biogenic synthesis of nanoparticles and potential applications: An eco-friendly approach. *J Nanomed Nanotechol* 4: 165.

Iravani, S. 2014. Bacteria in nanoparticle synthesis: Current status and future prospects. *Int Sch Res Notices* 2014: Article ID 359316.

Issazadeh, K., M. R. M. K. Pahlaviani, and A. Massiha. 2011. Bioremediation of toxic heavy metals pollutants by *Bacillus* sp. isolated from Guilan Bay sediments, north of Iran. *ICBEE* 18: 67–71.

Iyer, A., K. Mody, and B. Jha. 2005. Biosorption of heavy metals by a marine bacterium. *Mar Pollut Bull* 50: 340–343.

Jacob, J. M., P. N. L. Lens, and R. M. Balakrishnan. 2015. Microbial synthesis of chalcogenide semiconductor nanoparticles: A review. *Microb Biotechnol* 9: 11–21.

Jarosławiecka, A. and Z. Piotrowska-Seget. 2014. Lead resistance in micro-organisms. *Microbiology* (*United Kingdom*) 160: 12–25.

Jennema, G. E., M. J. McInerney, R. M. Knapp et al. 1983. A halotolerant, biosurfactants-producing *Bacillus* species potentially useful for enhanced oil recovery. *Dev Ind Microbiol* 24: 485–492.

Jun, G., Z. Z. Ming, B. H. Juan, and Y. E. Guan. 2007. Microbiological synthesis of nanophase PbS by *Desulfotomaculum* sp. *Sci China Ser E Tech Sci* 50: 302–307.

Juwarkar, A. A., K. V. Dubey, A. Nair, and S. K. Singh. 2008. Bioremediation of multi-metal contaminated soil using biosurfactant—A novel approach. *Indian J Microbiol* 48: 142–146.

Juwarkar, A. A., A. Nair, K. V. Dubey, S. K. Singh, and S. Devotta. 2007. Biosurfactant technology for remediation of cadmium and lead contaminated soils. *Chemosphere* 68: 1996–2002.

Kafilzadeh, F., R. Afrough, H. Johari, and Y. Tahery. 2012. Range determination for resistance/tolerance and growth kinetic of indigenous bacteria isolated from lead contaminated soils near gas stations (Iran). *Euro J Exp Biol* 2: 62–69.

Kamika, I. and M. N. B. Momba. 2013. Assessing the resistance and bioremediation ability of selected bacterial and protozoan species to heavy metals in metal-rich industrial wastewater. *BMC Microbiol* 13: 1–14.

Kang, S. H., K. N. Bozhilov, N. V. Myung, A. Mulchandani, and W. Chen. 2008. Microbial synthesis of CdS nanocrystals in genetically engineered *E. coli*. *Angew Chem Int Ed* 47: 5186–5189.

Karakagh, R. M., M. Chorom, H. Motamedi, Y. K. Kalkhajeh, and S. Oustan. 2012. Biosorption of Cd and Ni by inactivated bacteria isolated from agricultural soil treated with sewage sludge. *Ecohydrol Hydrobiol* 12: 191–198.

Kawasaki, K. I., T. Matsuoka, M. Satomi, M. Ando, Y. Tukamasa, and S. Kawasaki. 2008. Reduction of cadmium in fermented squid gut sauce using cadmium-absorbing bacteria isolated from food sources. *JFAE* 6: 45–49.

Kermani, A. J. N., M. F. Ghasemi, A. Khosravan, A. Farahmand, and M. R. Shakibaie. 2010. Cadmium bioremediation by metal-resistant mutated bacteria isolated from active sludge of industrial effluent. *Iran J Environ Health Sci Eng* 7: 279–286.

Khodabakhsh, F., S. Nazeri, M. A. Amoozegar, and G. Khodakaramian. 2011. Isolation of a moderately halophilic bacterium resistant to some toxic metals from Aran & Bidgol Salt Lake and its phylogenetic characterization by 16S rDNA gene. *Feyz J Kashan Univ Med Sci* 15: 53–60.

Khusro, A., P. J. P. Raj, and S. G. Panicker. 2014. Multiple heavy metals response and antibiotic sensitivity pattern of *Bacillus subtilis* strain KPA. *J Chem Pharm Res* 6: 532–538.

Kim, S. U., Y. H. Cheong, D. C. Seo, J. S. Hur, J. S. Heo, and J. S. Cho. 2007. Characterisation of heavy metal tolerance and biosorption capacity of bacterium strain CPB4 (*Bacillus* sp.). *Water Sci Technol* 55: 105–111.

Ko, S. H., H. S. Lee, S. H. Park, and H. K. Lee. 2000. Optimal conditions for the production of exopolysaccharide by marine microorganism *Hahella chejuensis*. *Biotechnol Bioprocess Eng* 5: 181–185.

Kowshik, M., N. Deshmukh, W. Vogel, J. Urban, S. K. Kulkarni, and K. M. Panikar. 2002a. Microbial synthesis of semiconductor CdS nanoparticles, their characterization and their use in the fabrication of an ideal diode. *Biotechnol Bioeng* 78: 583–588.

Kowshik, M., W. Vogel, J. Urban, S. K. Kulkarni, and K. M. Panikar. 2002b. Microbial synthesis of semiconductor PbS nanocrystallites. *Adv Mater* 14: 815–818.

Kushner, D. J. 1978. Life in high salt and solute concentrations halophilic bacteria. In: *Microbial Life in Extreme Environments*, ed. D. J. Kushner, pp. 317–368. Academic Press, London, U.K.

Laddaga, R. A., R. Bessen, and S. Silver. 1985. Cadmium-resistant mutant of *Bacillus subtilis* 168 with reduced cadmium transport. *J Bacteriol* 162: 1106–1110.

Lang, S. 2002. Biological amphiphiles (microbial biosurfactants). *Curr Opin Colloid Interface Sci* 7: 12–20.

Li, X., J. Gong, Y. Hu et al. 2012. Genome sequence of the moderately halotolerant, arsenite-oxidizing bacterium *Pseudomonas stutzeri* TS44. *J Bacteriol* 194: 1–2.

Li, X., H. Xu, Z. S. Chen, and G. Chen. 2011. Biosynthesis of nanoparticles by microorganisms and their applications. *J Nanomater* 2011: Article ID 270974.

Liu, C. Q., N. Khunajakr, L. G. Chia, Y. M. Deng, P. Charoenchai, and N. W. Dunn. 1997. Genetic analysis of regions involved in replication and cadmium resistance of the plasmid pND302 from *Lactococcus lactis*. *Plasmid* 38: 79–90.

Llamas, I., A. del Moral, F. Martı́nez-Checa, Y. Arco, S. Arias, and E. Quesada. 2006. *Halomonas maura* is a physiologically versatile bacterium of both ecological and biotechnological interest. *Antonie van leeuwenhoek* 89: 395–403.

Maier, R. M., I. L. Pepper, and C. P. Gerba. 2006. *Environmental Microbiology*. Academic Press, San Diego, CA.

Malik, P., R. Shankar, V. Malik, N. Sharma, and T. K. Mukherjee. 2014. Green chemistry based benign routes for nanoparticle synthesis. *J Nanoparticles* 2014: Article ID 302429.

Massadeh, A. M., F. A. Al-Momani, and H. I. Haddad. 2005. Removal of lead and cadmium by halophilic bacteria isolated from the Dead Sea shore, Jordan. *Biol Trace Elem Res* 108: 259–269.

Mata, J. A., V. Bejar, I. Llamas et al. 2006. Exopolysaccharides produced by the recently described halophilic bacteria *Halomonas ventosae* and *Halomonas anticariensis*. *Res Microbiol* 157: 827–835.

Mathivanan, K. and R. Rajaram. 2014. Tolerance and biosorption of cadmium (II) ions by highly cadmium resistant bacteria isolated from industrially polluted estuarine environment. *Indian J Geomarine Sci* 43: 580–588.

Matyar, F., T. Akkan, Y. Ucak, and B. Eraslan. 2010. *Aeromonas* and *Pseudomonas*: Antibiotic and heavy metal resistance species from Iskenderun Bay, Turkey (northeast Mediterranean Sea). *Environ Monit Assess* 167: 309–320.

Mgbemena, I. C., J. C. Nnokwe, L. A. Adjeroh, and N. N. Onyemekar. 2012. Resistance of bacteria isolated from Otamiri River to heavy metals and some selected antibiotics. *Cur Res J Biol Sci* 4: 551–556.

Mishra, R. R., T. K. Dangar, B. Rath, and H. N. Thatoi. 2009. Characterization and evaluation of stress and heavy metal tolerance of some predominant Gram negative halotolerant bacteria from mangrove soils of Bhitarkanika, Orissa, India. *Afr J Biotechnol* 8: 2224–2231.

Munoz, A. J., F. Espínola, M. Moya, and E. Ruiz. 2015. Biosorption of Pb(II) ions by *Klebsiella* sp. 3S1 isolated from a wastewater treatment plant: Kinetics and mechanisms studies. *Biomed Res Int* 2015: Article ID 719060.

Murthy, S., G. Bali, and S. K. Sarangi. 2012. Biosorption of lead by *Bacillus cereus* isolated from industrial effluents. *Br Biotechnol J* 2: 73–84.

Murthy, S., G. Bali, and S. K. Sarangi. 2014. Effect of lead on growth, protein and biosorption capacity of *Bacillus cereus* isolated from industrial effluent. *J Environ Biol* 35: 407–411.

Naik, M. M. and S. K. Dubey. 2011. Lead-enhanced siderophore production and alteration in cell morphology in a Pb-resistant *Pseudomonas aeruginosa* strain 4EA. *Curr Microbiol* 62: 409–414.

Naik, M. M. and S. K. Dubey. 2013. Lead resistant bacteria: Lead resistance mechanisms, their applications in lead bioremediation and biomonitoring. *Ecotoxicol Environ Saf* 98: 1–7.

Naik, M. M., A. Pandey, and S. K. Dubey. 2012. Biological characterization of lead-enhanced exopolysaccharide produced by a lead resistant *Enterobacter cloacae* strain P2B. *Biodegradation* 23: 775–783.

Nakbanpote, W., N. Panitlurtumpai, A. Sangdee, N. Sakulpone, P. Sirisom, and A. Pimthong. 2014. Salt-tolerant and plant growth-promoting bacteria isolated from Zn/Cd contaminated soil: Identification and effect on rice under saline conditions. *J Plant Interact* 9: 379–387.

Nanganuru, Y. H. and N. Korrapati. 2012. Studies on biosorption of cadmium by *Pseudomonas putida*. *IJERA* 2: 2217–2219.

Nath, S., B. Deb, and I. Sharma. 2012. Isolation and characterization of cadmium and lead resistant bacteria. *Glo Adv Res J Microbiol* 1: 194–198.

Nies, D. H. 1995. The cobalt, zinc, and cadmium efflux system CzcABC from *Alcaligenes eutrophus* functions as a cation-proton-antiporter in *Escherichia coli*. *J Bacteriol* 177: 2707–2712.

Nies, D. H. 1999. Microbial heavy-metal resistance. *Appl Microbiol Biotechnol* 51: 730–750.

Nies, D. H. and S. Silver. 1989. Plasmid-determined inducible efflux is responsible for resistance to cadmium, zinc, and cobalt in *Alcaligenes eutrophus*. *J Bacteriol* 171: 896–900.

Nieto, J. J., R. Fernandez-Castillo, M. C. Marquez, A. Ventosa, E. Quesada, and F. Ruiz-Berraquero. 1989. Survey of metal tolerance in moderately halophilic eubacteria. *Appl Environ Microbiol* 55: 2385–2390.

Nucifora, G., L. Chu, T. K. Misra, and S. Silver. 1989. Cadmium resistance from *Staphylococcus aureus* plasmid pI258 *cadA* gene results from a cadmium-efflux ATPase. *Proc Natl Acad Sci USA* 86: 3544–3548.

Oger, C., J. Mahillon, and F. Petit. 2003. Distribution and diversity of a cadmium resistance (cadA) determinant and occurrence of IS257 insertion sequences in *Staphylococcal* bacteria isolated from a contaminated estuary (Seine, France). *FEMS Microbiol Ecol* 43: 173–183.

Omeroglu, E. E., I. Karaboz, A. Sukatar, I. Yasa, and A. Kocyigit. 2007. Determination of heavy metal susceptibilities of *Vibrio harveyi* strains by using 2,3,5-triphenyltetrazolium chloride (TTC). *Rapp Commint Mer Medit* 38: 364.

Onishi, H., T. Kobayashi, N. Morita, and M. Baba. 1984. Effect of salt concentration on the cadmium tolerant *Pseudomonas* sp. *Agric Biol Chem* 48: 2441–2448.

Oren, A., P. Gurevich, M. Azachi, and Y. Henis. 1992. Microbial degradation of pollutants at high salt concentrations. *Biodegradation* 3: 387–398.

Oren, A., P. Gurevich, M. Azachi, and Y. Henis. 1993. Microbial degradation of pollutants at high salt concentrations. In: *Microorganisms to Combat Pollution*, ed. E. Rosenberg, pp. 263–274. Kluwer Academic Publishers, Dordrecht, the Netherlands.

Owolabi, J. B. and M. M. Hekeu. 2014. Heavy metal resistance and antibiotic susceptibility pattern of bacteria isolated from selected polluted soils in Lagos and Ota, Nigeria. *IJBAS-IJENS* 14: 6–12.

Pacwa-Płociniczak, M., G. A. Płaza, Z. Piotrowska-Seget, and S. S. Cameotra. 2011. Environmental applications of biosurfactants: Recent advances. *Int J Mol Sci* 12: 633–654.

Pantidos, N. and L. E. Horsfall. 2014. Biological synthesis of metallic nanoparticles by bacteria, fungi and plants. *J Nanomed Nanotechnol* 5: 233.

Panwichian, S., D. Kantachote, B. Wittayaweerasak, and M. Mallavarapu. 2010. Isolation of purple nonsulfur bacteria for the removal of heavy metals and sodium from contaminated shrimp ponds. *Electron J Biotechnol* 13: 1–12.

Peterson, R. L., H. B. Massicotte, and L. H. Melville. 2004. *Mycorrhizas: Anatomy and Cell Biology*. NRC Research Press, Ottawa, Quebec, Canada.

Prabhu, A., G. Melchias, A. Edward, and P. Kumaravel. 2014. Phylogenetic and functional characterization of halophilic bacterial isolates of upper cretaceous fossils in Ariyalur peninsula. *World J Pharm Pharm Sci* 3: 794–812.

Prakash, A., S. Sharma, N. Ahmad, A. Ghosh, and P. Sinha. 2010. Bacteria mediated extracellular synthesis of metallic nanoparticles. *Int Res J Biotechnol* 1(5): 071–079.

Prasad, K. S, T. Amin, S. Katuva, M. Kumari, and K. Selvaraj. Synthesis of water soluble CdS nanoparticles and study of their DNA damage activity. *Arabian J Chem*. http://dx.doi.org/10.1016/j.arabjc.2014.05.033. In press.

Raja, C. E., G. S. Selvam, and K. Omine. 2009. Isolation, identification and characterization of heavy metal resistant bacteria from sewage. *International Joint Symposium on Geodisaster Prevention and Geoenvironment in Asia*. JS-Fukuoka, Japan.

Ramezani, F., M. Ramezani, and S. Talebi. 2010. Mechanistic aspects of biosynthesis of nanoparticles by several microbes. *Nanocon* 10: 12–14. Olomouc, Czech Republic, EU.

Rivadeneyra, M. A., A. Martin-Algarra, M. Sanchez-Roman, A. Sanchez-Navas, and J. D. Martın-Ramos. 2010. Amorphous Ca phosphate precursors for Ca-carbonate biominerals mediated by *Chromohalobacter marismortui*. *ISME J* 4: 922–932.

Rothenstein, D., J. Baier, T. D. Schreiber, V. Barucha, and J. Bill. 2012. Influence of zinc on the calcium carbonate biomineralization of *Halomonas halophila*. *Aquat Biosyst* 8: 31.

Roy, S., K. Mishra, S. Chowdhury, A. R. Thakur, and S. Raychaudhuri. 2008. Isolation and characterization of novel metal accumulating extracellular protease secreting bacteria from marine coastal region of Digha in West Bengal, India. *Online J Biol Sci* 8: 25–31.

Saha, R., N. Saha, R. S. Donofrio, and L. L. Bestervelt. 2012. Microbial siderophores: A mini review. *J Basic Microbiol* 52: 1–15.

Samanta, A., P. Bera, M. Khatun et al. 2012. An investigation on heavy metal tolerance and antibiotic resistance properties of bacterial strain *Bacillus* sp. isolated from municipal waste. *J Microbiol Biotechnol Res* 2: 178–189.

Schalk, I. J., M. Hannauer, and A. Braud. 2011. New roles for bacterial siderophores in metal transport and tolerance. *Environ Microbiol* 13: 2844–2854.

Schmidt, T. and H. G. Schlegel. 1994. Combined nickel-cobalt-cadmium resistance encoded by the ncc locus of *Alcaligenes xylosoxidans* 31A. *J Bacteriol* 176: 7045–7054.

Seshadri, S., A. Prakash, and M. Kowshik. 2012. Biosynthesis of silver nanoparticles by marine bacterium, *Idiomarina* sp. PR58-8. *Bull Mater Sci* 35: 1201–1205.

Seshadri, S., K. Saranya, and M. Kowshik. 2011. Green synthesis of lead sulfide nanoparticles by the lead resistant marine yeast, *Rhodosporidium diobovatum*. *Biotechnol Prog* 27: 1464–1469.

Shrivastava, A., V. Singh, S. Jadon, and S. Bhadauria. 2013. Heavy metal tolerance of three different bacteria isolated from industrial effluent. *Int J Pharm Bio-Sci* 2: 137–147.

Shivashankarappa, A. and K. R. Sanjay. 2015. Study on biological synthesis of cadmium sulfide nanoparticles by *Bacillus licheniformis* and its antimicrobial properties against food borne pathogens. *Nanosci Nanotechnol Res* 3: 6–15.

Singh, P. and S. S. Cameotra. 2004. Enhancement of metal bioremediation by use of microbial surfactants. *Biochem Biophys Res Commun* 319: 291–297.

Singh, S. K., V. R. Tripathi, R. K. Jain, S. Vikram, and S. K. Garg. 2010. An antibiotic, heavy metal resistant and halotolerant *Bacillus cereus* SIU1 and its thermoalkaline protease. *Microb Cell Fact* 9: 59.

Singha, N. and S. Nara. 2013. Biological synthesis and characterization of lead sulphide nanoparticles using bacterial isolates from heavy metal rich sites. *IJAFST* 4: 16–23.

Sinha, S. and S. K. Mukherjee. 2008. Cadmium–induced siderophore production by a high Cd-resistant bacterial strain relieved Cd toxicity in plants through root colonization. *Curr Microbiol* 56: 55–60.

Sinha, S. and S. K. Mukherjee. 2009. *Pseudomonas aeruginosa* KUCd1, a possible candidate for cadmium bioremediation. *Braz J Microbiol* 40: 655–662.

Sinha, S., I. Pan, P. Chanda, and S. K. Sen. 2009. Nanoparticles fabrication using ambient biological resources. *J Appl Biosci* 19: 1113–1130.

Solanki, P. and V. Kothari. 2012. Metal tolerance in halotolerant bacteria isolated from saline soil of Khambhat. *Res Biotechnol* 3: 1–11.

Sowmya, M., M. P. Rejula, P. G. Rejith, M. Mohan, M. Karuppiah, and A. A. M. Hatha. 2014. Heavy metal tolerant halophilic bacteria from Vembanad Lake as possible source for bioremediation of lead and cadmium. *J Environ Biol* 35: 655–660.

Srinivas T. N. R., S. Aditya, V. Bhumika, and P. A. Kumar. 2014. *Lunatimonas lonarensis* gen. nov., sp. nov., a haloalkaline bacterium of the family *Cyclobacteriaceae* with nitrate reducing activity. *Syst Appl Microbiol* 37: 10–16.

Sutherland, I. W. 1994. Structure-function relationships in microbial exopolysaccharides. *Biotechnol Adv* 12: 393–448.

Sweeney, R. Y., C. Mao, X. Gao et al. 2004. Bacterial biosynthesis of cadmium sulphide nanocrystals. *Chem Biol* 11: 1553–1559.

Syed, S. and P. Chinthala. 2015. Heavy metal detoxification by different *Bacillus* species isolated from solar salterns. *Scientifica* 2015: Article ID 319760.

Talebi, S., F. Ramezani, and M. Ramezani. 2010. Biosynthesis of metal nanoparticles by microorganisms. *Nanocon* 10: 12–14. Olomouc, Czech Republic, EU.

Tomova, I., M. Stoilova–Disheva, and E. Vasileva Tonkova. 2014. Characterization of heavy metals resistant heterotrophic bacteria from soils in the Windmill Islands region, Wilkes Land, East Antarctica. *Pol Polar Res* 35: 593–607.

Tripathi, M., H. P. Munot, Y. Shouche, J. M. Meyer, and R. Goel. 2005. Isolation and functional characterization of siderophore-producing lead and cadmium resistant *Pseudomonas putida* KNP9. *Curr Microbiol* 50: 233–237.

Tripathi, R. M., A. S. Bhadwal, P. Singh, A. Shrivastav, M. P. Singh, and B. R. Shrivastav. 2014. Mechanistic aspects of biogenic synthesis of CdS nanoparticles using *Bacillus licheniformis*. *Adv Nat Sci Nanosci Nanotechnol* 5: 025006.

Tynecka, Z., Z. Gos, and J. Zajac. 1981. Reduced cadmium transport determined by a resistance plasmid in *Staphylococcus aureus*. *J Bacteriol* 147: 305–312.

Vallee, B. L. and D. D. Ulmer. 1972. Biochemical effects of mercury, cadmium, and lead. *Annu Rev Biochem* 41: 91–128.

Vela-Cano, M., A. Castellano-Hinojosa, A. F. Vivas, and M. V. M. Toledo. 2014. Effect of heavy metals on the growth of bacteria isolated from sewage sludge compost tea. *Adv Microbiol* 4: 644–655.

Volesky, B. 2001. Detoxification of metal-bearing effluents: Biosorption for the next century. *Hydrometallurgy* 59: 203–216.

Wang, C. L., D. S. Clark, and J. D. Keasling. 2001. Analysis of an engineered sulfate reduction pathway and cadmium precipitation on the cell surface *Biotechnol Bioeng* 75: 285–291.

Wang, S. and C. N. Mulligan. 2004. Rhamnolipid foam enhanced remediation of cadmium and nickel contaminated soil. *Water Air Soil Pollut* 157: 315–330.

Williams, G. P., M. Gnanadesigan, and S. Ravikumar. 2013. Isolation, identification and metal tolerance of halobacterial strains. *Indian J Mar Sci* 42(3): 402–408.

World's Worst Pollution Problems. 2015. The new top six toxic threats: A priority list for remediation. http://www.greencross. ch/uploads/media/pollution_report_2015_top_six_wwpp.pdf (accessed October 26, 2015).

Wu, X., R. R. Santos, and J. Fink-Gremmels. 2015. Cadmium modulates biofilm formation by *Staphylococcus epidermidis*. *Int J Environ Res Public Health* 12: 2878–2894.

Yamaguchi, R., T. Arakawa, H. Tokunaga, M. Ishibashi, and M. Tokunaga. 2012. Halophilic properties of metal binding protein characterized by high histidine content from *Chromohalobacter salexigens* DSM3043. *Protein J* 31: 175–183.

Yamina, B., B. Tahar, M. Lila, H. Hocine, and F. M. Laure. 2014. Study on cadmium resistant-bacteria isolated from hospital wastewaters. *Adv Biosci Biotechnol* 5: 718–726.

Złoch, M., K. Niedojadło, T. Kowalkowski, and K. Hrynkiewicz. 2013. Analysis of subcellular accumulation of cadmium (Cd) by metal tolerant bacteria using transmission electron microscopy (TEM) with energy-dispersive X-ray microanalysis (EDS). In: *Endophytes for Plant Protection: The State of the Art*, eds. C. Schneider, C. Leifert, and F. Feldmann. Deutsche Phytomedizinische Gesellschaft, Braunschweig, Germany.

Zolgharnein, H., K. Karami, M. M. Assadi, and A. D. Sohrab. 2010. Investigation of heavy metal biosorption on *Pseudomonas aeruginosa* strain MCCB 102 isolated from the Persian Gulf. *Asian J Biotechnol* 2: 99–109.

35 Elucidation of Cadmium Resistance Gene of Cd resistant Bacteria involved in Cd Bioremediation

Tapan Kumar Das

CONTENTS

ABSTRACT

The biochemical modes of cadmium toxicity and its mechanisms for the development of cadmium resistance in bacterial systems have been briefly emphasized in this chapter. In addition to that, the genetic aspects behind cadmium resistance have also been described in detail. Different proteins like metallothioneins showing the binding capacity of cadmium and other metals are reported. The antagonistic role as well as the synergistic effect of different heavy metals on cadmium is presented. Different bacterial genes related to cadmium resistance such as *cadA*, *cadB*, *cadC*, *cadD*, and *cadX* and the sequences of the genes involved are explained in this chapter. The use of cadmium-resistant bacteria as a tool for the bioremediation of cadmium is described along with the mechanism of cadmium bioremediation.

35.1 INTRODUCTION

Global industrialization causes severe pollution in the environment due to emergence of organic and inorganic contaminants from industrial effluents and industrial solid waste, from deposition of airborne industrial waste, or from sewage sludge, mining, fertilizers, agricultural chemicals, etc. In today's world, heavy metal pollution is considered a major environmental problem. Our surroundings are being polluted by different heavy metals due to production and discharge of waste materials containing heavy metal ions, threatening

human health (Lesmana et al., 2009). Cadmium is one among the most toxic metals with devastating effects on the environment. It is introduced into our environment through industrial effluents, sewage wastes, agricultural runoffs, and hospital discharges. In response to cadmium pollution, bacteria isolated from hospital effluents are found to be highly resistant to cadmium as well as various antimicrobial agents. So the use of these cadmium-resistant bacteria is recommended for wastewater bioremediation contaminated with cadmium. Twelve cadmium-resistant bacteria were screened from hospital wastes and preliminarily identified as both Gram-positive and Gram-negative bacteria, and their mechanism of cadmium uptake was studied (Benmalek et al., 2014). Most of the bacteria are Gram negative such as the *Klebsiella pneumoniae* and *Pseudomonas aeruginosa* species.

Cadmium is widely used in different industries and affects people through occupational and environmental exposure. It affects cell proliferation, differentiation, apoptosis, and other cellular activities. Cadmium toxicity modulates metabolic energy production, neural excitability, and brain cholinergic mechanism and causes autoimmune diseases. Moreover, it reduces cellular antioxidative function. It may be treated as carcinogen owing to direct or indirect interaction with adenine and guanine in DNA. It also interferes with the placental progesterone production and induces subchronic nephrotoxicity. Cadmium induces superoxide ion production in the enhancement of lipid oxidation (Stacey and Klaassen, 1981; Amoruso et al., 1982). People around Jinstsu River had been suffering from the itai-itai disease after eating rice grown in the cultivation lands contaminated with cadmium, zinc, and lead.

Microorganisms are of increasing importance in biotechnological processes applied for the removal of heavy metal ions like biosorption and intracellular absorption (Kapoor and Viraraghavan, 1995; Pal and Das, 2005). Different mutants of microorganisms can be used as a tool for partial bioremediation of cadmium. Consequently, several transformation experiments on the *Aspergillus* species have been done to study the homologous and heterologous gene expression. In this aspect, homologous and nonhomologous recombination of *Aspergillus nidulans* has been widely studied (Tillburn et al., 1983). Since cadmium is very toxic to microorganisms and other higher organisms, cadmium-resistant mutants like *Pseudomonas putida* (Denise et al., 1989), *P. aeruginosa* (De Vicente et al., 2011), and *K. pneumoniae* strain CBL-1 (Shamim and Rehman, 2012) have been used as biochemical and genetic tools for understanding the biochemistry, genetics, and molecular physiology of toxicity and detoxification in different microorganisms. Cadmium resistant mutants of *Pseudomonas* isolated by genetic mutation with chemical mutagen like acridine orange, and acriflavine could be utilized for bioremediation of cadmium from cadmium-contaminated waste water (Jabbari Nezhad Kermani et al., 2010). A biofilm forming marine *P. aeruginosa* JP-11 was screened from Odisha coast that showed cadmium resistance up to 1000 ppm (Chakarborty and Das, 2014a). In addition, cadmium-resistant bacterium identified as *Escherichia coli* by the analysis of 16S rRNA was isolated from industrial wastewater used as a tool for the bioremediation of cadmium from cadmium-contaminated aqua environment (Khan et al., 2015). Bacteria isolated from soil sample could grow in a medium containing heavy metal ions and were identified as *Bacillus* species confirmed by 16S rRNA sequencing (Hookoom and Puchooa, 2003). The cadmium-resistant species *K. pneumoniae* and *P. aeruginosa* were isolated from hospital wastewater to understand the mechanism of cadmium resistance in the same bacterial strain (Benmalek et al., 2014). To understand the mechanism of cadmium resistance and genetics in both prokaryotes and eukaryotes, bacterial cadmium-resistant mutants were selected to study the detoxification mechanism of cadmium in the contaminated biological system. An attempt was made to isolate cadmium-resistant mutants of *Aspergillus niger* by genetic mutation with a chemical mutagen, *N*-methyl-*N*′-nitro-*N*-nitrosoguanidine, and to characterize it biochemically. This strain was used to reduce cadmium toxicity of the materials released as industrial waste products in drainage water from industries of electronics, plating, battery, and ammunition manufacture into the rivers (Pal and Das, 2005).

35.2 CHEMISTRY OF THE HEAVY METAL CADMIUM

Elements with atomic numbers greater than 25 are generally considered as heavy metal. Some of them belong to the transitional group of elements. Elements with incompletely filled d orbital or f orbital in the penultimate cell are generally considered as transitional elements. Cadmium (Cd) is such an element whose ground state electronic configuration is $1s^2 2s^2 2p^6 3s^2 3p^6 4s^2 3d^{10} 4p^6 5s^2 4d^{10}$. However, from this electronic configuration, cadmium should not be treated as a transitional element. But from a purely chemical point of view, it is also appropriate to consider this element as transition element because its chemical behavior is, on the whole, quite similar to that of other transition elements. So far as the stereochemistry of cadmium ion is concerned, it is found that Cd^{+2} with its d^{10} configuration shows no stereochemical preferences arising from ligand field stabilization effects. Thus, it displays a variety of coordination numbers and geometries and follows the interplay of electrostatic forces, covalency, and the size factor. Based on the earlier factors, Cd^{+2} more often exists with a coordination number of six and as octahedral or rock salt structure.

35.3 CADMIUM TOXICITY AND EFFECTS

Valberg et al. (1977) estimated the amount of cadmium absorption in the form of intestinal cadmium–thionein complex from the intestinal lumen to evaluate the cadmium toxicity on the intestinal mucosa. Metallothionein (MT) in the proximal intestine could bind cadmium excreted into the lumen with the loss of bids of the outer skin of the epithelium. The equivalent amounts of cadmium administered as $CdCl_2$ or cadmium–thionein entered the mucosa by duodenal insertion pump. Exposure of the mucosa to $CdCl_2$ for 1 h led to minor abnormalities in the form of broadening of villi with significant damage of epithelium and swelling of mitochondria, whereas cadmium–thionein produced extensive necrosis in absorptive cells. So it was suggested that cadmium–thionein seemed to play a contradictory role, giving safeguard to intracellular milieu as well as enhancement of cadmium toxicity when intestinal lumen contains adequate quantity of cadmium–thionein.

The acute and subacute effect of cadmium chloride on homeostasis of calcium and trace metals was discussed by Bornner et al. (1981). The decreased level of plasma calcium along with reduced level of femur concentration in both calcium and zinc was identified after injection of cadmium chloride (1.5 mg Cd^{+2}/kg) in the male rat. It was reported that repeated administration of cadmium chloride (1.5 mg Cd^{+2}/kg) daily, for 28 days, caused a marked hypocalciuria that remained throughout the period of cadmium treatment. There was an increasing excretion of alkaline phosphatase into the urine that resulted in the elevated level of inorganic phosphate in the plasma of these animals that severely damage the kidney in the long run. They concluded that a possible mechanism for this cadmium-induced effect may create an interference of the renal biotransformation of vitamin D and decreased the bioavailability of the essential micro metals on account of the synthesis of MT as well as excessive loss into the urine.

Stacey and Klaassen (1981) enumerated the interaction of different metal ions with cadmium-induced cellular toxicity by incubation of isolated hepatocytes with cadmium ion

(200 or 400 μM) or Cd plus Cr, Mn, Zn, Ni, Pb, Se, or Fe (200–1000 μM) at 37°C. They found the loss of intracellular K⁺ and aspartate aminotransferase (ASAT) from the hepatocytes after cadmium ion administration. The effects on lipid peroxidation were calculated by measuring the concentration of the thiobarbituric acid reactants. They found that the cell injury due to Cd was reduced by Cr, Mn, Zn, Pb, and Fe and lipid peroxidation as the Cd supplied was interfered by Mn, Zn, and Cr. All metals apart from nickel could help in increasing the accumulation of cadmium by hepatocytes, but it was irrelevant to the relationship between decreasing cellular toxicity and inhibition of lipid peroxidation with uptake of Cd. So they came to conclusion that *in vivo* studies of cells protective properties from some metals toxicity can be demonstrated by interaction of metals at the cellular level and protective effects of metals, ordinarily of cadmium induced cellular toxicity are not due to decrease either in Cd uptake or in lipid peroxidation.

Cd-induced cellular toxicity does not occur due to a decrease in either Cd uptake or lipid peroxidation. Moreover, heavy metals like cadmium, lead, and mercury have very deleterious effects on microbial communities at higher concentrations. Increased exposure to such heavy metals leads to the disruption of microbial population by target inhibition of their crucial metabolic processes (Figure 35.1).

Amoruso et al. (1982) demonstrated that cadmium ion induced the enhanced production of superoxide ion in rat and human phagocyte *in vitro*. In the course of their assessment, they observed that the production of superoxide ion was increased by a factor of 2.11 ± 0.25 above control levels in human granulocytes in the concentration of 3.6×10^{-5} M cadmium ion, but the production of ion was gradually reduced with a further increase in cadmium concentration. So they

concluded that the level of cadmium within the range of those occurring during *in vivo* toxicity might provide a clarification of the oxidative role of this metal ion.

Gill et al. (1989) found that two strains of *Drosophila* (v;bw and Austin) produced MT in response to Cd ions in their diet. Although Austin produced more MT than v;bw, under the identical condition, v;bw became more resistant toward cadmium ion than Austin. Hence, they concluded that differences in total MT content did not explain the genetically demonstrable difference in Cd^{+2} resistance between v;bw and Austin and as two MT genes were identified in *Drosophila*.

Ghosh and Bhattacharya (1992) discussed the effect of cadmium on thyroid malfunction. They injected cadmium chloride and mercury chloride in rabbits that resulted in their suffering from severe thyrotoxicosis. It was found that within 24 h of intramuscular administration of cadmium and mercury ions the thyroid peroxidase activity along with tri-iodothyronine (T3) increased significantly over the control, with a remarkable reduction in the thyroxine (T4) level. So the T3/T4 ratio was enhanced compared to the control. It was also indicated that acute heavy metal lethality would induce immediate hyperthyroidism. Production of T3-toxicosis by preferential synthesis of T3 may be possible as suggested, and/or preferential deiodination of T4 to T3 and the measurement of T3 and T4 levels may thus be utilized as a reliable indicator of heavy metal lethality.

Martinez et al. (1993) investigated the relationship between the hypothermic effect induced by lead or cadmium chloride and the cerebral content of metal level. They injected mice with different doses of lead acetate or cadmium chloride at 22°C and 35°C, and rectal temperatures, as well as brain metal levels, were determined. It was found that hypothermia with the rise of lead level in the brain of mice was reflected when mice were given

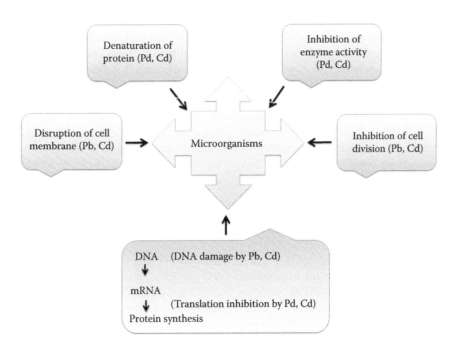

FIGURE 35.1 Toxicity of cadmium and other heavy metals on the microbial cells. (Modified from Ahemad, M., *J. Inst. Integr. Omics Appl. Biotechnol.*, 3, 39, 2012.)

the increased level of lead acetate e.g., >25mg/Kg with their diet at 22°C. This hypothermic effect was gradually reduced with an increase in temperature although at 2 or 5 mg/kg lead acetate at 22°C or 35°C, no significant changes were observed in body temperature after the continuation of excretion, while levels of lead in the brain increased significantly. Treatment with cadmium chloride at 22°C produced a significant fall in body temperature. Rise in brain cadmium levels at doses of 2–4 mg/kg caused a lessening of body temperature that partially repressed at 35°C. Therefore, they suggested that the hypothermic effects elicited by the metals with high molecular weight are interrelated both to the metal level absorbed by the brain and its rate of uptake.

Shibasaki et al. (1993) described the effect of polyaspartic acid (PAA) on cadmium ion–induced nephrotoxicity in rat by producing an animal model of $CdCl_2$ nephrotoxicity in rats and treated them with PAA to prevent renal damage. They injected the $CdCl_2$ in male Sprague–Dawley (SD) rats (190–200 g) for 2 weeks. $CdCl_2$-exposed SD rats showed a remarkable increase in the volume of urine, urinary excretion of N-acetyl-beta-D-glucosaminidase (NAG), alanine aminopeptidase (AAP), and fractional excretion of sodium (FENa) along with a decrease in the percentage of tubular reabsorption of phosphate (%TRP). Out of these indicators of proximal tubular function, AAP and %TRP were found to be more sensitive as compared to NAG or FENa. In the cortex, concentrations of cadmium were three times higher than in the medulla. However, cadmium-treated rat could not be differentiated from PAA-treated rats. Horiguchi et al. (1993) demonstrated that rigorous damage of organs occurred due to acute disclosure to cadmium through infiltration of leukocytes and neutrophils. They tried to investigate Cd-induced interleukin-8 (IL-8) production in human peripheral blood mononuclear cells (PBMC), which is a novel neutrophile chemotactic-activating cytokine. Cd, over a wide range of concentrations, induced human PBMC to produce large amounts of bioactive IL-8, the maximal induction being observed at 10^{-4} M. It was also found that this production was inhibited specifically by a metal chelating agent, ethylenediaminetetraacetic acid. The quantity of IL-8 mRNA increased within 30 min after the addition of Cd and reached a maximal level in 2 h, decreasing thereafter. Cycloheximide, an inhibitor of protein synthesis, failed to inhibit IL-8 mRNA accumulation, indicating that there was no requirement for new protein synthesis on account of induction of IL-8 mRNA. Concomitant with the induction of synthesis of a new protein, within 10 min, intermediates of reactive oxygen were generated by cadmium in human PBMC. They investigated that a radical scavenger, N-acetyl-L-cysteine (NAC), inhibited both IL-8 production and generated reactive oxygen intermediate, for the participation of ROI in the production of IL-8. They also found that a superoxide generating mediator, paraquat, provoked the production of IL-8 in human PBMC and that NAC blocked this paraquat-induced IL-8 production.

Shibasaki et al. (1993) developed a rat model of cadmium (Cd)-induced nephrotoxicity and tried to prevent renal damage by treating the animals with pentoxifylline (PTX). When Sprague–Dawley (SD) rats were given 3.0mg/Kg $CdCl_2$ with diet for 2 weeks, renal proximal tubular was found to damage with the increase in urine volume, urinary excretion of NAG and AAP, and incomplete exposure of sodium (FeNa) accompanied by decrease in %TRP was also observed. The quantity of urine was increased threefold in the $CdCl_2$-treated rats and fivefold in the Cd+PTX-treated rats, to the control treated with saline. No change in the total protein, AAP, and creatinine clearance was reported after PTX administration. The concentration of Cd in the renal cortex was three times higher than that in the renal medulla, whereas the Cd-treated rats could not be differentiated from the Cd+PTX-treated rats on concentration. So, they concluded that PTX is useful for improving the nephrotoxicity of Cd.

Liu et al. (1994) demonstrated that cadmium ion induces nephrotoxicity in rat kidney proximal tubules and LLC-PK1 cells owing to the formation of Cd–MT complex. In intact animals, the rate of absorption of CdMT is more than $CdCl_2$ that results in renal damage. However, it is not clear yet about the mechanism(s) by which renal injury could be produced by CdMT. They used cultured renal proximal tubular cells to study the nephrotoxicity induced by CdMT and $CdCl_2$. Collagenase perfusion, pursued by percoll isopycnic centrifugation, was carried out for the isolation of kidney proximal tubules from rats. ^{14}C-alpha-methylglucose uptake and lactate dehydrogenase outflow were utilized for the induction of nephrotoxicity. Unexpectedly, CdMT was less toxic than $CdCl_2$ to the cultured rat proximal tubule cells and as to the cultured LLC-PK1 cells (a pig kidney proximal tubular cell line). Consistent with these observations on nephrotoxicity, 109CdMT uptake into these cultured renal cells was much less than that of 109 $CdCl_2$. The toxicity and uptake of $CdCl_2$ and CdMT following basolateral and apical exposure were examined with the use of transwell cultures of LLC-PK1 cells. The uptake of both $CdCl_2$ and CdMT from basolateral exposure was higher than that from apical exposure. Again, more 109 $CdCl_2$ was taken up, and more cytotoxicity was observed in the $CdCl_2$ than CdMT-exposed cells. In summary, $CdCl_2$ is highly lethal compared to CdMT to proximal tubules of the cultured rat kidney and LLC-PK1 cells. This is in contradiction to the greater *in vivo* nephrotoxic effects of CdMT than $CdCl_2$. Therefore, cultured renal cells are not considered as a suitable model for understanding the nephrotoxicity of CdMT; transport of CdMT into proximal tubular cells *in vivo* does not appear to be maintained *in vitro*.

Ossola and Tomaro (1995) demonstrated that cadmium chloride ($CdCl_2$) acts as an inducer of heme oxygenase that increases the rat liver chemiluminescence (QLV) after 3 h of administration in "*in vivo*" rat. The activity of heme oxygenase was ongoing to increase after 5 h treatment and gradually increased with time reaching a maximum value around 12–15 h after $CdCl_2$ administration. Such induction appeared before by a decrease of glutathione (GSH) pool in the intrahepatic and a rise in the steady-state concentration of H_2O_2, both effects taking place several hours before induction of heme oxygenase. The activity of antioxidant enzymes, superoxide dismutase, catalase, and GSH peroxidase was found to be significantly reduced after 5 h of $CdCl_2$ injection. Organization of bilirubin as an end product of the catabolic degradation of heme in mammals and the applied α-tocopherol as antioxidant could inhibit the induction of heme oxygenase along with decrease in the formation of hepatic GSH as well as rising the chemiluminescence when the mammalian cells were controlled two hours before the treatment with $CdCl_2$ and the results indicated treatment of mammalian cells with $CdCl_2$ may be correlated with the induction of heme oxygenase and oxidative stress. Bucio et al. (1995) studied the toxic effects of cadmium (Cd) and mercury (Hg) ions using a hepatic human fetal cell line (WRL-68 cells). They treated the cell in three different ways: acute low-dose treatment; chronic treatment; and acute high-dose treatment. Although WRL-68 cells growing in the presence of Cd exhibited the same proliferative curve as control cells, the cells increased their proliferative capacity in the case of Hg. Both metals produced ultrastructural alterations in different degrees in mitochondrial and RER depending on the treatment and concentration of the metal used. Cytotoxicity was assessed by measuring the release of lactate dehydrogenase from the cells. Acute high-dose-treated cells showed the highest value for this parameter and Cd-treated cells presented higher lactate dehydrogenase release than the Hg-treated system. Assay of alanine aminotransferase and ASAT activities could be considered for the measurement of damage of the cells, and release of the highest amount of enzyme occurs when the system is treated with an acute high dose of cadmium. Lipid peroxidation was found to be higher compared to the control cells when cells were treated with high dose of cadmium and mercury. Synthesis of MT was not induced by the treatment of mercury (Hg) as noticed during detections. Even a dramatic induction of MT protein was found when the cells were treated with cadmium. The fetal hepatic cells could be used as a very useful tool for understanding the toxicity mechanism of cadmium and mercury as WRL-68 cells acted differentially in response to Cd and Hg.

Inouhe et al. (1996) studied the possible mechanisms of cadmium resistance in different wild-type yeasts. This resistance (Cd resistance) and possible formation of cadmium-binding complexes were examined in eight different wild-type yeasts. It was found that strain 301N could show higher cadmium-resistant character while *Torulaspora delbrueckii* and *Schizosaccharomyces octosporus* showed partially, but *Saccharomyces cerevisiae* X2180–1B, *Zygosaccharomyces rouxii*, and *Saccharomyces carlsbergensis* were cadmium-sensitive strains. It was found that all the partial cadmium-resistant mutants except the mutants of *S. exiguus* contain variable concentrations of cadmium ions in the cytoplasmic fraction. *S. octosporus* included a Cd-binding complex that contained (gamma EC)nG peptides known as cadystins or chelating-like compounds available from plants, whereas cadmium-binding proteins could be synthesized by *P. farinosa* and *T. delbrueckii* that was found to be similar to cadmium–MT with the composition of amino acids and molecular weight synthesized by

Sa. cerevisiae 301N. These results suggested that cadmium tolerance could be developed in the previously mentioned strains due to the presence of such type of cadmium-chelating proteins available in the cytoplasm. On the other hand, *Sa. exiguus* adsorbed most cadmium in the cell wall fraction, and no Cd-binding complex was found in the cytoplasm indicating the role of the cell wall in Cd tolerance. So, it was concluded that different types of resistance mechanisms play behind the cadmium resistance in different yeasts. Yiin et al. (1999) studied the induced effect of cadmium upon lipid peroxidation in rat testes and simultaneous protection by selenium. It was found that cadmium aided the accumulation of metal ions and lipid peroxidation in the testes of rats. The concentrations of cadmium, copper, zinc, iron, and selenium in the tissue were estimated by an atomic absorption spectrophotometer, and the enhanced level of lipid peroxidation in the testes was measured by spectrophotometric method. It was also noted that these cadmium-induced changes were further accompanied by an increase in iron and copper and a decrease in zinc ion content in the testes. It was found that cadmium-induced alterations in lipid peroxidation were decreased in simultaneous treatment with cadmium and selenium and essential metal levels. So it was suggested that lipid peroxidation associated with cadmium toxicity in the testes and selenium somehow interferes with this effect.

Cols et al. (1999) stated that the beta domain of mouse MT 1 (betaMT) was synthesized in *E. coli* in the presence of copper or cadmium. They used the homogeneous preparations of Cu-betaMT and Cd-betaMT to differentiate the corresponding *in vivo*–conformed metal clusters with the species obtained *in vitro* by metal replacement to a canonical Zn3-betaMT structure. It was observed that betaMT clusters with copper inside the cells were found to be very steady, whereas emerging beta peptide, despite its cadmium-binding ability, was highly unstable whose stoichiometry depended upon cultural conditions. They came to the conclusion that the absence of betaMT protein in *E. coli* having protease crowds cultivated in a medium supplemented with cadmium indicated severe proteolysis of a poorly folded beta peptide, somehow enhanced in the presence of cadmium.

Bilgen et al. (2003) described the role of cholinoceptors in Cd-induced endothelial dysfunction in the male rat. They found that Cd significantly increased the phenylephrine response with a decrease in basal dilator prostanoid release. They compared the changes in the tension of the aortic rings to constrictor and dilator agonists with those of controls. It was found that the depression in the acetylcholine (ACh) response and increase of the receptor-free dilation from calcium ionophore A23187 transpired in cadmium-induced rings, whereas smooth muscle cell response to the NO donor and sodium nitroprusside remained unaltered. It was also noted that cadmium decreased both the highest response to ACh (10^{-5} M) and sensitive component of its pirenzepine. The M1-type cholinoceptor-mediated response to ACh reduced in Cd-exposed rings from 38.40% ± 6.90% (p < 0.001) to 10.30% ± 5.00%. It was further reported that cadmium also reduced the share

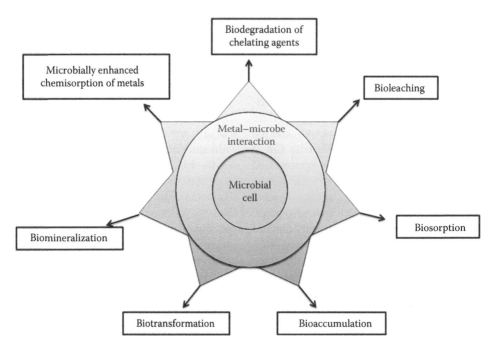

FIGURE 35.2 Heavy metal–microbe interactions and bioremediation of heavy metals. (Modified from Ahemad, M., *J. Inst. Integr. Omics Appl. Biotechnol.*, 3, 39, 2012.)

of indomethacin from 1.64% to 13.92% ± 2.89% (p < 0.01), and one of the most deleterious effects of Cd appeared to be restricted to the M1-dependent ACh response. So it was suggested that Cd produced an endothelial dysfunction by the impairment of the M1-type cholinoceptor-mediated response involved in prostanoid release.

Chaffei et al. (2003) performed the recovery experiment to find out the effects of heavy metals on the growth of plants through a change in chemical and physiological parameters in cadmium prestressed tomato seedlings after the removal of cadmium from growth media. They showed that the revival of growth activity and increment in the content of biomass of stems and leaves were feasible after the removal of Cd from media required for growth. The increase in root biomass with simultaneous enhancement of nitrate content compared to those of leaves and stems was also found after the withdrawal of cadmium from growth media. The activities of the enzymes nitrate reductase, nitrite reductase, and glutamine synthetase increased rapidly after the removal of cadmium. It was also noted that NAD($^+$)-dependent glutamate dehydrogenase (GDH-NAD$^+$) activity increased progressively during the recovery time, but the cognate NADH-dependent glutamate dehydrogenase (GDH-NADH) activity decreased. Finally, they concluded that detoxification and reassimilation of the ammonia that were produced by the stress-induced protein catabolism occurred by the GDH-NADH isoenzyme.

35.4 BIOCHEMICAL MECHANISM INVOLVED IN THE DEVELOPMENT OF CADMIUM RESISTANCE IN BACTERIA

Metals are trapped in extracellular components of bacteria forming complexes that would not be allowed to enter the bacterial cells. Additionally, microbial cell surfaces contain anionic groups like carboxylic, amino, thiol, hydroxyl, and carboxylic groups attracting cationic metal-like cadmium. This could be bound firmly with the respective anionic groups present in cell surfaces/outer membrane components of bacteria like phosphoryl and lipopolysaccharides and prevent entry of metallic ions, that is, cadmium ions inside of the microbial cells. Such types of mechanisms may be found in microbes in aquatic environment helping to adsorb metallic ions and to be utilized in the bioremediation of cadmium or other heavy metals (Figure 35.2).

Four general mechanisms are present in natural microbes for the development of resistance to various heavy metals as described.

35.4.1 EXTRACELLULAR POLYMERS

Extracellular polymers consisting of polysaccharides, carbohydrates, nucleic acids, and fatty acids moieties in their structure help in the binding of metallic ions, such as cadmium, to form polymeric complex extracellularly, which is a common process for the bioremediation of heavy metal ions. Most of the functional groups of extracellular polymers bearing negatively charged groups are influenced by variable pH of the aquatic environmental system that has been considered as the basic principle of detoxification of cadmium and prevention of metal entry into bacterial cells. Extracellular polymers present in *Staphylococcus aureus*, *Micrococcus luteus*, *and Azotobacter* spp. help in cadmium removal (Maier et al., 2009). Chakraborty and Das (2014a) reported the removal of Cd through biomass and extracellular polymeric substances (EPS) secreted by the cells of *P. aeruginosa* JP-11 that removed 58.760% ± 10.62% and 29.544% ± 8.02% of Cd, respectively.

35.4.2 Siderophores

Siderophores, iron-chelating low-molecular-weight organic molecules, are found to be present in bacteria that can accumulate iron in aqua-environmental system containing a very low concentration of iron. Metals chemically similar to iron may be bound to siderophores, minimizing the concentration of metal and reducing the metal toxicity (Roane and Pepper, 2000).

35.4.3 Biosurfactants

Particular kinds of compounds as biosurfactants are excreted by several bacteria outside the cell that help in cadmium removal from toxic environments (Miller, 1995). It is found that these types of biosurfactants are capable of binding cadmium/other metals to make metal complexes, increasing their solubility that is found to be nontoxic to the bacterial cells. Several marine bacteria can reduce the interfacial tension of the medium due to increase supply of the organic substances and it can also entrap heavy metals from the soil (Chakraborty and Das, 2014b). Biosurfactants increase the substrate availability for microorganisms and involve the interaction with the cell surface to increase hydrophobic substrates to associate more easily with the bacterial cells (Chakraborty and Das, 2016).

35.4.4 Metabolic By-Products

Several types of bacteria produce common metabolic by-products that have the capacity of diminishing the metal concentration by reducing the solubility into less soluble salts. For example, *Citrobacter* under aerobic condition could synthesize phosphates enzymatically. These phosphates could reduce the concentration of metal by precipitation resulting in the diminution of metal toxicity toward the bacterial cells. *Desulfovibrio* spp as sulfur oxidation-reduction bacteria could reduce sulfate under anaerobic conditions as a result precipitation of metal ions occurs.

35.4.5 Metallothioneins

Microbial cytosol carries out the synthesis of some ligands like MT to sequester heavy metal, such as cadmium. They are heavy metal sequestering proteins that are expressed by increasing resistance toward heavy metals. Some microbes follow the principle of energy-dependent metal efflux mechanism to eliminate heavy metals from the inside of the cell for protecting the cell from heavy metal toxicity. In the same mechanism, ATPases are involved in pumping out the toxic metal out from microbial cell via this process.

35.5 GENETICS OF CADMIUM-RESISTANT MICROBIAL STRAINS

Flavobacteria and *Bacillus* strains have been considered for the identification of genes related to cadmium resistance that was detected as *cadA* gene. Phylogenetic analysis between 16S rDNA gene suggested that the same gene may be transferred in a lateral way among *Bacillus* and *Flavobacterium* spp. showing the dominating role for promotion to the development of cadmium-resistant phenotypes in soil microorganisms. Additionally, two genes, *cadC* and *cadA*, located on the chromosome have been identified in *Streptococcus thermophilus* that is responsible for acquiring cadmium resistance. It may be assumed that these genes may be present in an operon, and the presence of cadmium may induce their genetic transcription in *Str. thermophilus* for the production of P-type cadmium efflux ATPases. The *cadC* gene encodes a regulatory protein regulating the transcriptional process. The adjacent genes, *cadA* and *cadC*, were first identified in *Sta. aureus* in detail. In addition to this, the homologous genes have also been identified in the bacteria like *Bacillus* sp., *Listeria* sp., *Lactococcus* sp., *Stenotrophomonas maltophilia*, and *Geobacillus stearothermophilus*. It has been found that *cadA* gene of *St. aureus* encodes CadA protein showing the characteristic of cadmium efflux ATPase of the P-type and *cadC* gene encodes cadC protein, a transacting DNA-binding negative regulatory protein that diminishes DNA-binding capacity in the presence of cadmium. A plasmid with *cadA* and *cadC* genes homologous to *St. aureus*and*L. monocytogenes* was also identified in *L. lactis*, but the mechanism of gene regulation is not established yet. The cadmium resistance cassette is residing on the genome of *Str. thermophilus* and *L. lactis*; the strain has been identified and considered as a model for the study of genetic regulation related to cadmium resistance in bacteria.

Staphylococcus, *Pseudomonas*, *Bacillus*, and *Escherichia* species contain the plasmid with *cadA* gene related to cadmium resistance that encodes cadmium-specific ATPase in the respective bacterial system. *Alcaligenes eutrophus* bearing *czc* operon in its chromosome can show the characteristic of cadmium resistance in the same bacterial strain. The *ncc* gene as operator showing cadmium resistance in *Alcaligenes xylosoxidans* was identified that was considered as cadmium resistance operon on the chromosome of the same bacteria (Schmidt and Schlegel, 1994).

Sta. aureus bearing two genetic operons, *cadA* and *cadB*, shows two well-established mechanisms of plasmid-mediated cadmium resistance in the same bacterial strain. First, one is well established and is associated with plasmid p1258, and another one is 3.5 kb operon located on the plasmid containing two genes, *cadA* and *cadC*. The *cadA* gene is capable of encoding 727 amino acids reflecting the same sequence as found in P-class of ATPase. Cadmium is capable of entering *St. aureus* cell through active transport system with the help of an energy-dependent cadmium efflux ATPase encoded by *cadA* gene (Nucifora et al., 1989), whereas *cadC* encodes 122 amino acids building a protein of small molecular weight and functioning as a regulator of transcription of the cadmium operon. The products of *cadA* and *cadC* genes are responsible for the development of cadmium resistance in bacterial strains.

In another way, that is not the well-established mechanism of cadmium resistance, the *cadB* operon related to cadmium resistance is located on a plasmid, pII147. This plasmid bears two genes, *cadB* and *cadX* (showing similar sequence as found in *cadC* protein). It has been assumed that *cadB* is not responsible for the promotion of cation efflux, but it can protect from toxicity by binding cadmium in the membrane (Ji and Silver, 1995). Plasmid pLUG10 isolated from *S. lugdunensis* is involved in showing the third mechanism of cadmium resistance as reported by Chaouni et al. Two genes, *cadB* and *cadX*, are found to be located on the same plasmid where *cadB* encodes cadmium resistance. Additionally, *cadX* has 40% similarity in sequence with *cadC*. Positive regulator of resistance encodes a regulatory locus; as a result, high cadmium resistance may be developed in the bacterial strain (Scott and Churchward, 1995). A unique characteristic of cadmium resistance mechanism is found in chemolithotroph Gram-negative bacteria *Ralstonia metallidurans* containing two "mega plasmid" related multiple heavy metal resistance. This mega plasmid has three *mer* operons, *czc* operon, and *cnr* operon (Nies et al., 1989). The *czc* operon is identified in *R. metallidurans* located in pMOL30. The *cad* operon is present in *Sta. aureus* located in pI258 and also available in *Staphylococcus lugdunensis* located in pLUG10.

The *czc* operon associated with the resistance to Cd^{2+}, Zn^{2+}, and Co^{2+} functions as cationic exchanger (not as ATPase) effluxing cadmium ions out of the cells. Plasmid pMOL30 isolated from *R. metallidurans* was investigated and found to contain three genes, *czcA*, *czcB*, and *czcC*, generating products to form a complex membrane cation efflux pump, and the gene, *czcD*, helps in *czc* operon expression. The *czcA* and *czcB* genes encode hydrophobic membrane proteins, and CzcC is an outer membrane channel protein (Nies et al., 1989; Silver and Phung, 1996). All these are involved in the development of cadmium resistance in the microbe. It has been reported that *Burkholderia* sp. was found to be highly resistant to heavy metals, but the mechanism of cadmium resistance is not clear till date. Mini-Tn5 transposon-inserted mutants of *Burkholderia cenocepacia* clarified cadmium tolerance and found that the same strain loses its cadmium tolerance activity in its absence (Schwager et al., 2012). It has been reported that three out of four genes are involved in affecting outer membrane biogenesis and integrity or DNA repair, and the fourth gene could encode a P1-type ATPase belonging to *cadA* family that is related to showing cadmium resistance. *cadA*-deficient strains were not able to grow in the presence of cadmium but showed resistance to some heavy metal ions like Ni, Cu, and Co ions. It could show the highest activity of *cadA* under the *gfp* promoter at neutral pH in the presence of 30 mM cadmium. Therefore, cadmium resistance genotype is attributed to the presence of several cad operons in diverse bacteria helping in cadmium bioremediation.

35.6 CONCLUSION

Heavy metals have a critical role in the survival of all living cells with the interference of some important biochemical processes. The environment is being polluted ions like cadmium as it has been used abundantly in industries and agriculture. Microorganisms like bacteria can show uptake and efflux mechanisms to adapt them to environments contaminated with heavy metal ions like cadmium. These microorganisms can be considered as a useful tool for the bioremediation of cadmium, and the induced mutants of cadmium-resistant microbes can show high potential in the removal of cadmium from cadmium-enriched medium. Studies of cadmium-resistant genes of bacteria, like *cadA*, *cadB*, and *cadC*, have been identified to correlate the study of genetics in adaptation, tolerance, and resistance to cadmium ions. The genetic character of bacteria can be used for the study of bioremediation of environment contaminated with heavy metal ions and also to open up new avenues related to methodology of decontamination processes. The genetic capacity of bacteria may be utilized as a tool for the bioremediation of heavy metal from systems polluted with heavy metal compounds. Genetic improvement may help to develop the field of existing methodologies for various decontamination processes. In addition, the principle of bioinformatics, like genomics, proteomics, and metabolomics, may be applied to know the genome sequencing of microorganisms and to know the biochemical events and metabolic routes that may provide information of genes related to cadmium microbial resistance. Profiles of cadmium resistance gene (*czcABC*) of *P. aeruginosa* were expressed on cadmium exposure showing an upregulation for resisting cadmium in cadmium-contaminated environment, and downregulation was also found with a variation of pH value and salinity level of the growth media. In this context, the rate of gene expression responsible for Cd resistance increases on Cd stress up to the tolerance level, but definite values of pH and salinity are considered as important factors for the proper functioning of cadmium resistance gene. Different mechanisms are found to involve in bacteria for becoming resistant, tolerant and adapted to cadmium as a result precipitation of metals as phosphates, carbonates, and/or sulfides, volatilization via methylation, physical exclusion of electronegative components in the membranes and EPS, energy-dependent metal efflux systems, and intracellular chelation with cysteine-rich proteins occur. Therefore, bioremediation of cadmium can be achieved with the help of bacteria from cadmium-contaminated soils, sediments, and aquatic environment that can remove cadmium by immobilization, precipitation, and biotransformation from toxic contaminated sites.

REFERENCES

Ahemad, M. 2012. Implications of bacterial resistance against heavy metals in bioremediation: A review. *Journal of Institute of Integrative Omics and Applied Biotechnology (IIOABJ)* 3: 39–46.

Amoruso, M.A., Witz, G., and Goldstein, B.D. 1982. Enhancement of rat and human phagocyte superoxide anion radical production by cadmium in vitro. *Toxicology Letters* 10: 133–138.

Benmalek, Y., Benayad, T., Madour, L., Hacene, H. and Fardeau, M. L. 2014. Study on Cadmium resistant bacteria isolated from hospital waste waters. *Advances in Bioscience and Biotechnology* 5: 718–726.

Bilgen, I., Oner, G., Edremitlioglu, M., Alkan, Z., and Cirrik, S. 2003. Involvement of cholinoceptors in cadmium-induced endothelial dysfunction. *Journal of Basic and Clinical Physiology and Pharmacology* 14: 55–76.

Bonner, F.W., King, L.J., and Parke, D.V. 1981. The acute and subacute effects of cadmium on calcium homeostasis and bone trace metals in the rat. *Journal of Inorganic Biochemistry* 14: 107–114.

Bucio, L., Souza, V., Albores, A., Sierra, A., Chavez, E., Carabez, A., and Gutierrez-Ruiz, M.C. 1995. Cadmium and mercury toxicity in a human fetal hepatic cell line (WRL-68). *Toxicology* 102: 285–299.

Chaffei, C., Gouia, H., Masclaux, C., and Ghorbel, M.H. 2003. Reversibility of the effects of cadmium on the growth and nitrogen metabolism in the tomato (*Lycopersicon esculentum*). *Comptes Rendus Biologies* 326: 401–412.

Chakraborty, J. and Das S. 2014a. Characterization and cadmium-resistant gene expression of biofilm-forming marine bacterium *Pseudomonas aeruginosa*. *Environmental Science and Pollution Research* 21: 14188–14201.

Chakraborty, J. and Das, S. 2014b. Biosurfactant based bioremediation of toxic metals. In: *Microbial Biodegradation and Bioremediation*. Elsevier, Philadelphia, PA, pp. 167e201.

Chakraborty, J. and Das, S. 2016. Characterization of the metabolic pathway and catabolic gene expression in biphenyl degrading marine bacterium *Pseudomonas aeruginosa* JP-11. *Chemosphere* 144: 1706–1714.

Chaoui, A., Mazhaudi, S., Ghorbal, M.H., and Ferlani, E.E. 1997. Cadmium and zinc induction of lipid peroxidation and effects on antioxidant enzyme activities in bean (*Phaseolus vulgaris* L.). *Plant Science* 127: 139–147.

Cols, N., Romero-Isart, N., Bofill, R., Capdevila, M., Gonzalez-Duarte, P., Gonzalez-Duarte, R., and Atrian, S. 1999. In-vivo copper-and cadmium-binding ability of mammalian metallothionein β domain. *Protein Engineering* 12: 265–269.

De Vicente, G., Cloetingh, S., Van Wees, J.D., and Cunha, P.P. 2011. Tectonic classification of Cenozoic Iberian foreland basins. *Tectonophysics* 502: 38–61.

DeNise, S.K., Robinson, G.H., Stott, G.H., and Amstrong, D.V. 1989. Effects of passive immunity on subsequent production in dairy heifers. *Journal of Dairy Science* 72: 552–554

Ghosh, N. and Bhattacharya, S. 1992. Thyrotoxicity of the chlorides of cadmium and mercury in rabbit. *Biomedical and Environmental Science* 5: 236–240.

Gill, H.J., Nida, D.L., Dean, D.A., England, M.W., and Jacobson, K.B. 1989 Resistance of Drosophila to cadmium: Biochemical factors in resistant and sensitive strains. *Toxicology* 56: 315–321.

Hookoom, M.M. and Puvhooa, D. 2003 Isolation and identification of heavy metals tolerant bacteria from industrial and agricultural areas in Mauritius. *Current Research in Microbiology and Biotechnology* 1: 119–123.

Horiguchi, H., Mukaida, M., Okamoto, S., Teranishi, H., Kasuya, M., and Matsushima, K. 1993. Cadmium induces interleukin-8 production in Human peripheral blood mononuclear cells with the concomitant generation of Super oxide radicals. *Lymphokine Caroline Research* 12: 421–428.

Inouhe, M., Sumiyoshi, M., Tohoyama, H., and Joho, M. 1996. Resistance to cadmium ions and formation of a cadmium-binding complex in various wild-type yeasts. *Plant and Cell Physiology* 37: 341–346.

Ji, G.Y. and Silver, S. 1995. Bacterial resistance mechanisms for heavy metals of environmental concern. *Journal of Industrial Microbiology* 14: 61–75.

Kapoor, A. and Viraraghavan, T. 1995. Fungal biosorption-an alternative treatment option for heavy metal bearing wastewaters: A review. *Bioresource Technology* 53: 195–206.

Kermani, A.J.N., Ghasemi, M.F., Khosravan, A., Farahmand, A., and Shakibaie, M.R. 2010. Cadmium bioremediation by metal-resistant mutated bacteria isolated from active sludge of industrial effluent. *Iranian Journal of Environmental Health Science and Engineering* 7: 279–286.

Khan Z., Nisar M.A., Hussain S.Z., Arshad M.N., and Rehman A. 2015. Cadmium resistance mechanism in *Escherichia coli* P4 and its potential use to bioremediate environmental cadmium. *Applied Microbiology and Biotechnology* 99: 10745–10757

Lazzaro, A., Hartmann, M., Blaser, P., Widmer, F., Schulin, R., and Frey, B. 2006. Bacterial community structure and activity in different Cd-treated forest soils. *FEMS Microbiology Ecology* 25: 1993–1999.

Lesmana, S.O., Febriana, N., Soetaredjo, F. E., Sunarso, J., and Ismadji, S. 2009. Studies on potential applications of biomass for the separation of heavy metals from water and wastewater. *Biochemical Engineering Journal* 44: 19–41

Liu, J.I.E., Lu, Y.P., and Klaassen, C.D. 1994. Nephrotoxicity of CdCl2 and Cd-metallothionein in cultured rat kidney proximal tubules and LLC-PK 1 cells. *Toxicology and Applied Pharmacology* 128: 264–270.

Llamas, M.A., Ramos, J.L., and Rodríguez-Herva, J.J. 2000. Mutations in each of the tol genes of *Pseudomonas putida* reveal that they are critical for maintenance of outer membrane stability. *Journal of Bacteriology* 182: 4764–4772.

Maier, R.M., Pepper, I.L., and Gerba, C.P. 2009. *Environmental Microbiology*, Academic Press, San Diego, CA, p. 397.

Martinez, F., Vicente, I., García, F., Peñafiel, R., and Cremades, A. 1993. Effects of different factors in lead-and cadmium-induced hypothermia in mice. *European Journal of Pharmacology: Environmental Toxicology and Pharmacology* 248: 199–204.

Miller, R.M. 1995. Biosurfactant-facilitated remediation of metal-contaminated soils. *Environmental Health Perspectives* 103: 59–62.

Nies, D.H., Nies, A., Chu, L., and Silver, S. 1989. Expression and nucleotide sequence of a plasmid-determined divalent cation efflux system from *Alcaligenes eutrophus*. *Proceedings of the National Academy of Sciences of the United States of America* 86: 7351–7355.

Nucifora, G., Chu, L., Misra, T.K., and Silver, S. 1989. Cadmium resistance from Staphylococcus aureus plasmid pl258 cadA gene results from a cadmium-efflux ATPase. *Proceedings of the National Academy of Sciences of the United States of America* 86: 3544–3548.

Ossola, J.O. and Tomaro, M.L. 1995. Heme oxygenase induction by cadmium chloride: Evidence for oxidative stress involvement. *Toxicology* 104: 141–147.

Pal, S.K. and Das, T.K. 2005. Biochemical characterization of n-methyl N'-nitro-N-nitrosoguanidine-induced cadmium resistant mutants of *Aspergillus niger*. *Journal of Biosciences* 30: 639–646.

Roane, T. M. and Pepper, I. L. 2000. Microorganisms and Metal Pollutants. In Environmental Microbiology. Maier, R.M., Pepper, I.L. and Gerba, C.P. (eds), Academic Press, San Diego, CA, pp. 403–423.

Schmidt, T. and Schlegel, H. G. 1994. Combined nickel-cobalt-cadmium resistance encoded ncc locus of *Alcaligenes xylosoxidans* 3LA. *Journal of Bacteriology* 176(22): 7045–7054.

Schwager, S., Lumjiaktase, P., Stöckli, M., Weisskopf, L., and Eberl, L. 2012. The genetic basis of cadmium resistance of *Burkholderia cenocepacia*. *Environmental Microbiology Reports* 4: 562–568

Scott, J.R. and Churchward, G.G. 1995. Conjugative transformation. *Annual Reviews in Microbiology* 49: 367–397.

Shamim, S. and Rehman, A. 2012. Cadmium resistance and accumulation potential of Klebsiella pneumonia strain CBL-1 isolated from industrial wastewater. *Pakistan Journal of Zoology* 44: 203–208.

Shibasaki, T., Nakano, H., Ohno, I., Ishimoto, F., and Sakai, O. 1993. Effect of polyaspartic acid on CdCl2-induced nephrotoxicity in the rat. *Biological Trace Element Research* 37: 261–267.

Silver, S. and Phung, L.T. 1996. Bacterial heavy metal resistance: New surprises. *Annual Reviews in Microbiology* 50: 753–789.

Stacey, N.H. and Klaassen, C.D. 1981. Interaction of metal ions with cadmium-induced cellular toxicity. *Journal of Toxicology and Environmental Health, Part A Current Issues* 7: 149–158.

Tilburn, J., Scazzocchio, C., Taylor, G.G., Zabicky-Zissman, J.H., Lockington, R.A., and Davies, R.W. 1983. Transformation by integration in *Aspergillus nidulans*. *Gene* 26: 205–221.

Valberg, L.S., Haist, J., Cherian, M.G., Delaquerriere-Richardson, L., and Goyer, R.A. 1977. Cadmium-induced enteropathy: Comparative toxicity of cadmium chloride and cadmium-thionein. *Journal of Toxicology and Environmental Health, Part A Current Issues* 2: 963–975.

Yiin, S.J., Chern, C.L., Sheu, J.Y., and Lin, T.H. 1999. Cadmium induced lipid peroxidation in rat testes and protection by selenium. *Biometals* 12: 353–359.

36 Bioremediation of Copper and Other Heavy Metals Using Microbial Biomass

Rosa Olivia Cañizares-Villanueva, Alfredo de Jesús Martínez-Roldán,
Hugo Virgilio Perales-Vela, Maribel Vázquez-Hernández,
and Orlando Melchy-Antonio

CONTENTS

ABSTRACT

Biosphere pollution by heavy metals has been increasing since the last century. In contrast to other contaminants of anthropogenic origin, heavy metals cannot be degraded biologically; therefore, these need to be recovered from soil or water and then be safely confined. Copper is an abundant heavy metal in soil and water, and it is an essential element for all living organisms as it is involved in cellular metabolism. However, at high concentrations, it is toxic, particularly to photosynthetic organisms, which are the basis of all food chains in nature. This problem caused by heavy metal pollution has led to the development of efficient technologies for their removal. Physicochemical methods provide a quick alternative for the recovery of these contaminants when they are in high concentration; however, they are not effective at low concentrations. The biotechnology bioremediation alternative is the use of higher organisms such as plants or microorganisms such as bacteria, fungi, algae, and microalgae. Metal removal via bioremediation may be accomplished by various processes such as bioextraction, biofiltration, bioleaching, biostabilization, biosorption (biosorption or bioaccumulation/bioabsorption), biooxidation, bioreduction, and biovolatilization (biomethylation), among others. The technologies for the recovery of soil and water contaminated by heavy metalslooks for to reduce the energetic requirement with the aim to lower the operating costs and to ensure longer lifetime of the process, making the processes of bioremediation a viable alternative to physicochemical processes. This chapter is addressed to analyze the problems related to the contamination by heavy metals, and the use of bioremediation as an alternative for their elimination particularly copper, and bioremediation as a biological alternative for their elimination.

36.1 INTRODUCTION

In the last decades, pollution by heavy metals has increased in industrial and urban wastewaters (Nriagu and Pacyna, 1988). The impact on the environment is evident as heavy metals reside in wastewaters that are untreated or partially treated having lakes, rivers, and oceans as their final destination. The pollution in these ecological niches directly impacts the primary producers as they are responsible for introducing heavy metals into the food chain (Rai et al., 1981). Unlike other organic compounds such as herbicides, pesticides, and some petroleum-derived products, heavy metals cannot be biologically degraded and remain in soils and sediments where they are slowly released into the waters (Devars and Sanchez-Moreno, 1999). Therefore, wastewater treatment will be the biggest problem for humanity in the coming years. Because of this, it is necessary to develop new technologies to reduce heavy metal concentration in acceptable biological levels and also they need to be economically affordable.

Heavy metals are a group of chemical elements with density greater than or equal to 5 g cm^{-3} and atomic number greater than 20. These metals are associated with biological toxicity and pollution of ecosystems (Jjemba, 2004). Due to its persistence and toxicity for living organisms, heavy metals are hazardous pollutants and their application in industrial processes has resulted in an increase in its concentration in the ecosystems (Hart et al., 1991).

In the environment, the presence of heavy metals could come from different sources such as anthropogenic or natural mineral deposit. Anthropogenic pollution by heavy metal increased with mine and industrial activities at the end of the nineteenth century and the beginning of the twentieth century. Among the main metals involved in serious incidents at the

beginning of the last century were caused by Pb, Cd, and Hg. At the present and with the arrival of new and modern extraction techniques for processing of minerals and as a consequence of the global demand of certain metals, man mobilizes each time a major number and volume of these metals. Table 36.1 shows the geological natural rate of mobilization, the global production, as well as the anthropogenic discharges of some heavy metals into the marine ecosystem and the heavy metal concentration needed to cause the 50% inhibition (EC$_{50}$) of phytoplankton growth (Páez-Osuna, 1996; Pinto et al., 2003).

As shown in Table 36.1, mining significantly increases the flow of heavy metals compared to the natural flow due to geological weathering.

Mining or metal refinement together with industrial manufacturing of battery alloys, electronic and electrical components, paint preservatives, and pesticides liberate heavy metals such as Cd, Cu, Pb, Hg, and Ni to the air, soil, and water. Table 36.2 shows heavy metal pollution associated with some industries.

TABLE 36.1

Rate of Geological Mobilization, Mine Production, Annual Discharge into Oceans, and Toxicity of Copper (Cu), Cadmium (Cd), Lead (Pb), and Mercury (Hg) on Marine Phytoplankton Environment

Metal	Geological Rate (ton year^{-1})	Production (ton year^{-1})	Download into Oceans (ton year^{-1})	CE$_{50}$ Growth (ng mL^{-1})
Hg	3×10^3	2×10^3	30	>0.4
Cd	—	1×10^4	60	>25
Pb	180×10^3	3.5×10^3	2350	>250
Cu	375×10^3	9×10^6	4500	>10

Sources: Adapted from Páez-Osuna, F., Fuentes de metales en la zona costera marina, in: *Golfo de México, Contaminación e Impacto Ambiental: Diagnóstico y tendencias*, eds. Botello, A.V., Rojas-Galaviz., J.L., Benitez, J.A., and Zárate, L.D., Universidad Autónoma de Campeche, EPOMEX, Campeche, México, 1996, pp. 297–308; Pinto, E. et al., *J. Phycol.*, 39, 1008, 2003.

TABLE 36.2

Heavy Metal Pollution Associated with Industrial Processes

Industry	Cd	Cr	Cu	Fe	Hg	Mn	Pb	Ni	Sn	V	Zn
Cellulose and paper		X	X		X		X	X			X
Petrochemical		X		X	X		X		X	X	X
Chemical	X	X		X	X		X		X		X
Fertilizer	X	X	X	X	X	X	X	X	X	X	X
Oil refinery	X	X	X	X			X	X			X
Steel	X	X	X	X	X		X	X	X	X	X
Nonferrous smelting	X	X	X		X		X				X
Machinery	X	X	X		X				X		
Cement		X								X	X
Textile		X									
Tannery		X									
Thermoelectric		X	X								X

Source: Adapted from Páez-Osuna, F., Fuentes de metales en la zona costera marina, in: *Golfo de México, Contaminación e Impacto Ambiental: Diagnóstico y tendencias*, eds. Botello, A.V., Rojas-Galaviz., J.L., Benitez, J.A., and Zárate, L.D., Universidad Autónoma de Campeche, EPOMEX, Campeche, México, 1996, pp. 297–308.

On the other hand, domestic wastewaters carry great amounts of heavy metals to the aquatic environment; Förstner and Wittmann (1979) reported that such waters on average add 60 mg L^{-1} de Ni, 240 mg L^{-1} Cr, 700 mg L^{-1} Cu, 450 mg L^{-1} Pb, 2600 mg L^{-1} Zn, 10 mg L^{-1} Cd, and 20 mg L^{-1} Ag; this authors explain the enrichment not only by the corrosion of the urban water supply net but also by domestic and industrial drainages not separated. The use of detergents has been reported also to be a source of heavy metal pollution upon containing heavy metals such as Mn, Cr, Zn, Sr, and B, which arrive in the aquatic environments by the nontreated domestic drainages (Páez-Osuna, 1996).

Generally speaking, metals can be classified according to their function and biological effect as (1) essential with a known biological function, (2) toxic, and (3) nonessential and nontoxic with an unknown biological function. It is important to mention that in spite of this classification, in some cases, there are heavy metals that have biological functions, for example, Cu, Co, Fe, and Zn, but in higher concentrations they can be toxic.

The U.S. Environmental Protection Agency (EPA) and Mexican NOM-CRP-001-ECOL/93 have identified heavy metals as dangerous wastes (Cañizares-Villanueva, 2000). In their rating of heavy metals that are toxic to human beings, animals, and plants, they find Hg, Cd, Cr, Pb, Cu, and Al, among others, to be the most dangerous (Maier et al., 2000; Pawlik-Skowronska and Skowronski, 2001).

The exposure or intake of high concentration of heavy metals in human beings and animals produces physiological disorders such as skin rashes, irritation of the respiratory system, and even death. The heavy metals that arrive in water bodies accumulate mainly in sediments. However, in the long term, these could be liberated to the water column intoxicating the aquatic communities; in this regard, it is especially important to know the type of the heavy metal present and its concentration in the environment, because this information will determine the pollution scale and the toxic effects to the organisms (Liu et al., 2015).

In low concentrations (0.1–0.01 mg L^{-1}), copper toxicity is lower in human beings compared to the effects caused to photosynthetic organisms; it affects not only their growth but also their metabolic functions. Heavy metal toxicity in photosynthetic organisms can be summed up in an irreversible increase of their membrane permeability causing not only loss of essential solutes and changes in cellular volume but also reduction of CO_2 fixation and photosynthetic electronic transport. It also inhibits mitochondrial respiration, interrupts nutrient uptake processes, inhibits enzymatic activity by displacement of essential metals, and may even cause death (Reed and Gadd, 1990).

36.2 COPPER: NATURE, TOXICITY, AND DISTRIBUTION

Copper (Cu) is a relatively abundant element in the Earth's crust, which exists at approximately 50 parts per million (ATSDR, 2011). It is a transition metal that occurs in nature in four oxidation states: elemental Cu, solid metal, cuprous ion, and cupric ion. The origin of its use dates back more than 2500 years ago. Traces of Cu-containing tools developed by the Egyptians have been found; there is also evidence that Cu salts were used in the treatment of eye diseases (Koul, 1975).

Cu is distributed across the entire surface of the Earth; it is found in soils, waters, and sediments. Similar to other metals, its presence in the environment is mainly due to geological mobilizations (375 × 10^3 tons year^{-1}), such as volcanic activity, weathering, and erosion of minerals enriched with metals (igneous, metamorphic, and sedimentary rocks) and anthropogenic activities (9 × 10^8 tons year^{-1}) such as industrial mining operations (González-Macías et al., 2006; Pakzad et al., 2014; Srivastava et al., 2015). As occurs with other industrial wastes, Cu wastes are discharged into the environment, causing pollution of groundwater and soils due to oxidation of pyrite (FeS_2), acidification of drainage, and the release and transport of heavy metals (Páez-Osuna, 1996; Pinto et al., 2003; Song et al., 2013). Cu, similar to other heavy metals found in the environment, is found mainly in the form of cations. Its soil and sediment behavior is affected by its physicochemical properties, and its movement in the soil depends on its geochemical form, which affects its solubility and bioavailability. Additionally, Cu cations cause pollution in the bodies of water due to the movement that results from the change in environmental conditions and the buffer capacity of the soil (Bretzel and Calderesi, 2006; Gu et al., 2015).

Some of the major products of the Cu industry are cables, electrical devices, plumbing tools, pipes, and locks. Additionally, some Cu compounds are used in agriculture to treat diseases such as mold fungi. In other industries, Cu is used for water treatment and preserving wood, leather, and fabrics (ATSDR, 2011). The aquaculture industry uses Cu sulfate in treatments against algae and protozoans and Cu oxide is used to manufacture pesticides (Shao et al., 2014).

The worldwide demand for Cu is increasing, and so it is its introduction to the environment, primarily due to anthropogenic processes. China is the largest producer and consumer of refined Cu, but its production is not enough to supply its own demand, so it imports Cu concentrates from Chile, Japan, and some European countries (Song et al., 2013).

Copper contamination in aquatic ecosystems is due to its liberation by different human industrial activities such as the following:

1. Mining. This industry generates liquid and solid wastes with high concentrations of Cu or Cu salts.
2. Manufacturing industries that produce huge amounts of wastewater highly contaminated with Cu.
3. The use of Cu salts as herbicide or algaecide that causes the mobilization of Cu to the soil and afterward to the groundwater affecting different aquatic environments (Fishel, 2005; Ochoa-Herrera et al., 2011).
4. The liberation of trace amounts of Cu by the consumption of fossil fuels and industrial processes such as cement, automobile, and tire manufacturing (Jung, 2001; Romero et al., 2008).

The most employed copper mining process is selective flotation. It produces fine dust that can impact the environment in many different ways depending on weather conditions (Romero et al., 2008). In rainy areas, the dust absorbs high quantities of water causing the dissolution of a portion of the Cu contained in the solid; this liquid can leach into the ground and be mobilized to water bodies (superficial water or groundwater) impacting the ecosystems. In desert areas, wind dispersion occurs because of air currents; this phenomenon causes pollution in bigger areas and higher environmental impact than that caused by hydrologic dispersion. Wind dispersion can pollute both water bodies and soils (Jung, 2001; Romero et al., 2008).

Cu is also present in the wastewaters of different industries such as melting, automobile production, semiconductor production, metallurgy, electroplating, and metal finishing industry. Cu in these wastewaters reaches high concentrations up to 1 g L^{-1}. Instead, in domestic water, the concentrations are too much lower because of an important dilution process (Stanković and Božić, 2009).

Industrial wastewaters contain Cu amounts high enough to use conventional process (chemical) for its removal. There are so many processes, and their employment depends on the characteristics of the wastewater, the concentration of Cu, and the presence or absence of other pollutants (other heavy metals, organic matter, pH, etc.). The most used processes are precipitation, adsorption, ionic exchange, chemical deposition, electrodialysis, membrane filtration, and, recently, photocatalysis (Barakat, 2011). But chemical precipitation is the most popular because of its simplicity and low cost compared to other technologies. Chemical precipitation consists in the generation of copper hydroxides that are insoluble and these can be sedimented using a coagulant (Wang et al., 2004).

But chemical processes are not very useful when wastewaters contain organic matters. In this case, the employment of a biological process as aerobic oxidation is recommended. In this process, a high amount of Cu, present in the effluent, would be removed in the secondary treatment by the sludge (Ochoa-Herrera et al., 2011). The removal is caused by biological activity and ionic exchange that can be very efficient (up to 90%) (Crane et al., 2010; Ochoa-Herrera et al., 2011). But in most of the cases, the remaining Cu is enough to cause an imbalance in aquatic ecosystems (Crane et al., 2010).

It has been even proved that the use of wastewater containing low amounts of Cu for the irrigation of agricultural crops causes a decrease in the final yield, and in the long term, the soil becomes highly contaminated (Cao and Hu, 2000).

36.3 EFFECTS OF CU ON THE ENVIRONMENT

The accumulation of heavy metals in water bodies and the Earth's surface is due mostly to industrial development and urbanization. Heavy metals are transported through the soil, water, and atmosphere. They are considered a risk to the environment and human health because they are resistant to bacterial decomposition, they are persistent, and they are not biodegradable. They accumulate in tissues and alter the physiological processes of organisms such as bacteria, fungi, and algae (Srivastava and Goyal, 2010; Krupskaya and Zvereva, 2014; Ding et al., 2015; Srivastava et al., 2015).

Biomagnification is caused by the tendency of certain metal ions to accumulate in the tissues of both plants (roots, stems, leaves, seeds, and fruits) and animals (bone, liver, kidney, muscle) (Brunner et al., 2008, Macek and Mackova, 2011, Lu et al., 2014, Mahmood et al., 2015). As a result, the consumption of organisms contaminated with metals causes accumulation through the highest trophic levels of the food chain, causing the concentration of metals in tissues of organisms at higher trophic levels to increase up to 10-fold relative to the concentration of the metal in the ecosystem (Das, 2014a; De et al., 2014).

The biogeochemical cycle of Cu in the water column is complex. The factors that determine its environmental fate are related to the nature of the metal and the physicochemical properties of water. Speciation of Cu is different in fresh, salt, and estuarine water bodies, and these differences determine the bioavailability and toxicity of the metal in aquatic organisms. For example, Cu oxide dissociates in water, and the increased prevalence of Cu^{2+} (cupric ion) is responsible for biocidal effects.

Although Cu is essential for various biochemical functions, there is interest in conducting research on high concentrations of this metal. For example, Shao et al. (2014) performed tests to evaluate the toxicity of Cu on two fry species of the genus *Schizothorax* and found that the mortality rates of these species increased with exposure time and concentration of Cu, which presents an LC$_{50}$ of 0.31 and 0.28 mg L^{-1}. Also reported that an 8 h exposure to Cu sulfate (concentration of 3.2 mg L^{-1}) killed all fry specimens of both species; also presented morphological deformity, whitening of the body, and the formation of blue crystals on the body surface, which indicated that the fish were continuously secreting large amounts of mucous to adsorb the Cu and probably neutralize its toxic effect. At a concentration of 1.6 mg L^{-1}, the mortality rate was 100% after 24 h; at concentrations of 0.1 and 0.2 mg L^{-1}, the mortality rate was low after 96 h; and at 0.1 mg L^{-1}, the effects were detected after 264 h. It is important to mention that the criteria of Cu concentrations in drinking water are more stringent, with 1 mg L^{-1} being the value recommended by the EPA; higher concentrations may cause adverse effects on species, as demonstrated by Shao et al. (2014).

36.3.1 SOURCES OF CONTAMINATION

One source of heavy metal contamination are sediments, which release the metals into surface waters, primarily from industrial activities (González-Macías et al., 2006; Wang et al., 2010, Pakzad et al., 2014; Ding et al., 2015). Heavy metals are introduced to water bodies through different chemical processes, such as diffusion, desorption, and hydrodynamics, thus affecting water quality and posing a potential risk to the ecosystem and human health (Pakzad et al., 2014). The concentration of metals such as Cu and Zn in the water is determined by the contribution of the sediment and by other factors, such as flora, fauna, microorganisms, organic matter, and pH (Ding et al., 2015). The concentration of heavy metals in the sediment depends on their depth. For example, Ding et al. (2015) reported that the concentrations of Cu, Zn, Pb, Cd, and Ni decreased with depth, which is attributable to the influence of industrial activities and the discharges from rivers to water bodies. Heavy metal concentration in the sediment is also related to sediment composition. For example, the presence of biotite and muscovite minerals in rocks from water bodies is closely related to the concentrations of Zn and Cu contained in the sediment, while the concentrations of Co and Ni fluctuate similar to the variation of minerals such as amphiboles and intrusive igneous rock fragments, which are formed inside the Earth's crust (Pakzad et al., 2014).

Studies about the presence of Cu in soils are important because plants absorb the compounds that are present in soils. Shi et al. (2013) reported contamination by As, Cu, and Zn in agricultural soils with Cu concentrations up to 52.7 mg kg^{-1} in wheat crops near coal mines. The largest concentrations of Cu and Zn in wheat seeds and grains were found in the zone with the most mining activity. The Cu content in most of the collected grains was up to 277% higher than the permissible limits of 30 mg kg^{-1} Cu established by the FAO (2003).

Areas of land cultivated for the production of renewable energy are another source of accumulation of heavy metals in soils. The metals become concentrated in the tissues of plants and are reemitted into the atmosphere during the process of biomass combustion. However, it is important to note that Cu accumulation in plants does not always depend only on the concentration of metals in the soil but also on other factors that affect the final concentration during the growth period, for example, the atmosphere (Shi et al., 2013). Massas et al. (2009) conducted a study in green areas and open spaces of urbanized lands in Greece and found concentrations of 117, 155, and 440 mg kg^{-1} for Cu, Pb, and Zn, respectively, where the largest concentrations of metals were located in areas with anthropogenic disturbances.

Because chemical fertilizers also contribute to the accumulation of heavy metals in soils, replacing them with biofertilizers is an alternative to reduce metal concentration in soils and agricultural products (Koupaie and Eskicioglu, 2015). Cu oxide is used as an active agent in several pesticides. For example, in agriculture, it is used as a fungicide to protect crops of coffee, cocoa, banana, and citrus, among others. It is also used in antifouling paint for the protection of boats and ships. Cu is an effective biocide that can affect organisms and produce environmental complications.

36.3.2 Cu Effects on Human Health

Heavy metals accumulate in organisms through the trophic chain and biochemical processes such as respiration, absorption, and ingestion, thus generating free radicals and inducing oxidative stress (Saïdi et al., 2013), which is the reason why some metals may cause cancer. Furthermore, their accumulation in organisms causes damage to the nervous, circulatory, endocrine, and immune systems. They are also a risk factor for liver, kidney, and lung complications (Brewer, 2010). The concentration of heavy metals in the human body can

be related to food, as concentrations of metals (Ni, Cr, Cd, Pb, and As) higher than those recommended by the FAO have been reported in meat, fish, milk, vegetables, fruits, and cereals (Islam et al., 2015).

Khan et al. (2015) showed that in soils contaminated with heavy metals, the vegetables that are grown can cause problems to human health. These effects can be reduced by taking some protective measures such as those mentioned by Cherfi et al. (2015), who conducted a study using effluents from urban wastewater treatment plants to irrigate crops of potato, tomato, and cucumber; at the end of the experiment, the contents of heavy metals in the vegetables were analyzed, and the concentrations in the food irrigated with treated water were lower than those found in commercial vegetables. For example, for the cucumber, the concentrations of Zn, Pb, and Cr decreased with the treated water from 2.89 to 0.86, from 2.59 to 0.03, and from 3.72 to 1.72 mg kg^{-1} dry weight, respectively. Importantly, the consumption of these foods had no adverse health effects, but this finding depended on intake. The results mentioned earlier demonstrate the need to treat wastewater containing heavy metals before using it for irrigation or discharging it into water bodies. Some of the methods that can be used are adsorption, precipitation, membrane separation, flotation, or electrochemical treatment; adsorption is the most effective alternative but entails problems with final disposal. Therefore, the use of microbial bioprocesses is suggested for heavy metal adsorption (Srivastava et al., 2015).

Islam et al. (2015) found that the consumption of foods containing high levels of As and Pb is associated with cancer risk. Saïdi et al. (2013) analyzed tuna meal used as a protein source in animal diets and found heavy metals such as Cd, Pb, and Hg and reported that the former had a cytotoxic effect at 30 μM.

It is important to note that the Cu entry point to the bodies of animals, including humans, is through ingestion. Its low concentration in ruminants can be attributed to a diet low in Cu or to the consumption of high concentrations of antagonist elements such as Mo and S, which form thiomolybdates that create insoluble Cu complexes that could lead to malabsorption.

Cu plays an important role in the maturation of proteins that form collagen and elastin. It is involved in the synthesis of neurotransmitters and in Fe oxidation. Cu, similar to Zn, is necessary for the function of some enzymes, acts as an extracellular and intracellular antioxidant, and has a protective effect against obstructive lung disease. It is also necessary in biochemical processes such as electron transport, the dismutation of superoxide, and melanin synthesis (Abakay et al., 2012; Sadeghi et al., 2014).

Although Cu, as is the case for other metals, is indispensable for biochemical functions in plants and animals, the excess of free Cu ions can cause serious damage to organisms and even death (Sahin and Tokgöz, 2010; Durukan et al., 2011; Rahman et al., 2014; Srivastava et al., 2015). For example, the exposure of workers to Cu in Cu and coal mines has been related to respiratory diseases and disabilities. Furthermore, the increase of the total Cu concentration in the body has been related to disorders at the pulmonary, gastrointestinal, and gynecological levels (Abakay et al., 2012; Sadeghi et al., 2014). Constant inputs of high concentrations of Cu and Zn to the body can cause anemia and pancreas, liver, and kidney damage (Shi et al., 2013). Through laboratory analysis and spirometric tests and blood samples, high concentrations of Cu that cause lung diseases have been found in people who work with this metal (Dagli et al., 2010). Therefore, it is important to ensure that the Cu levels in the air of workplaces do not exceed 0.1 mg L^{-1} for Cu fumes and 1.0 mg L^{-1} for Cu dusts (ATSDR, 2011).

36.3.3 Effect of Cu on Algae

Cu is an essential element for the growth of microalgae in saltwater and freshwater, mainly because it is a cofactor of enzymes and electron carriers from photosynthesis and respiration (Perales-Vela et al., 2006, 2007; Lelong et al., 2012). However, concentrations above 2 ppb are toxic to certain species of microalgae (Perales-Vela et al., 2007).

The physiological effects caused by high concentrations of Cu occur similarly in marine and freshwater organisms, inhibiting or modifying many physiological processes. Excess Cu reduces growth, photosynthetic and respiratory activities, pigment synthesis, and other processes (Perales-Vela et al., 2007; Lelong et al., 2012; Jamers et al., 2013). It is also able to inactivate enzymes by oxidizing their hydroxyl groups and generating oxidative stress due to the production of reactive oxygen species (superoxide or hydroxyl radicals), which occurs by Fenton reactions (Johnson et al., 2007; Perales-Vela et al., 2007; Debelius et al., 2009; Jianrong and Qiran, 2009).

At the macro level, the presence of Cu causes changes in the shape and size of cells and colonies. Such is the case of *Scenedesmus incrassatulus*, which, in the absence of this metal, has colonies consisting of a four-cell cenobium, while exposure to 7 ppm Cu causes the disintegration of most of the cenobia and the appearance of individual cells (Peña-Castro et al., 2004). It is also noteworthy that this process is reversible because individual cells reform cenobia when grown in metal-free medium.

Because Cu is an essential element for the growth of microalgae, it usually enters through the membrane at a certain rate; however, high concentrations of Cu increase this rate (Jamers et al., 2013). Subsequently, pathways related to energy metabolism, protein synthesis, and responses to stresses, such as thermal shock, are activated at the transcriptional level (Jamers et al., 2006). In addition to this genetic response, overexposure to Cu modifies some enzymatic processes because certain enzymes are sensitive to this metal; such processes include electron transport (cytochrome oxidase and plastocyanin) and the antioxidant system (Cu/Zn superoxide dismutase) (Masmoudi et al., 2013). After a few hours of exposure to Cu, decreases are observed in the rates of photosynthesis, respiration, and chlorophyll synthesis (Perales-Vela et al., 2007; Jamers et al., 2013). Detoxification pathways are also activated in certain species.

In addition to ion immobilization in organelles such as the nucleus, microalgae are capable of synthesizing peptide molecules with chemical compositions that enable them to immobilize metal ions and reduce their toxicity (Perales-Vela

et al., 2006). These compounds are known as metallothioneins (MTs). They are peptide chains weighing between 2 and 10 kDa, and their production is activated under stress conditions, including the presence of heavy metals such as Cd and Cu (Rai and Gaur, 2001). The production of MTs is directly proportional to the concentration of heavy metals in the environment.

MTs trap metal ions and reduce their toxicity, allowing for their transport through the cell. Metals bound to MT can be transported into a vacuole, chloroplast, or mitochondrion, decreasing the portion of free metal present in the cell (Perales-Vela et al., 2006). The mechanism of this remains uncertain, but Mendoza-Cózatl et al. (2005) have proposed that MTs are produced in the cytosol, bind to the metal, and facilitate its transit to organelles. In their model, the MTs are produced by the organelles, and their binding to free metal ions occurs by a combination of both processes. However, despite the efficiency of the detoxification process by MT, there are limits to the amount of metal that can be trapped, so exposure to concentrations above this threshold will continue to damage cellular metabolism.

The final effect of all these specific conditions is reduced growth, which, in some species, is more pronounced than the effects on metabolic processes such as photosynthesis, respiration, and synthesis of pigments. The strong effect on growth has been attributed to the fact that Cu not only modifies the photosynthetic function but also affects processes related to cellular division (Perales-Vela et al., 2007). Particularly for diatoms, high concentrations of Cu affect silicon uptake, which decreases cell division because a poor supply of silicon makes it impossible to synthesize ornamentations (Martin-Jézéquel et al., 2000).

36.4 BIOREMEDIATION OF COPPER

Cu, similar to other heavy metals, is a persistent contaminant that cannot be degraded as in the case of organic compounds. Currently, remediation of heavy metals is reduced to their confinement or their transformation to less toxic species (Kotrba et al., 2011a; Mahmood et al., 2015). The inability to degrade heavy metals causes metal ions discharged into the ecosystem to be easily accumulated, be moved to long distances, and be biomagnified within food chains (Mudhoo et al., 2012). This is alarming because the accumulation of heavy metals in humans is a cancer risk (Mohammed et al., 2011; Mudhoo et al., 2012; Das et al., 2014b; Dixit et al., 2015; Rahman et al., 2016).

Heavy metals can be mobilized by leaching into agricultural lands and groundwater, causing diffuse pollution problems over large areas and making it difficult to remediate these sites (Inglea and Seabrab, 2014; Rahman et al., 2016). Due to the problem of heavy metal contamination, the development of a wide variety of technologies for removal has been promoted. The efficiency of the techniques depends on the metal to be removed, its concentration, the contaminated site (soil, water, etc.), and the final concentration that needs to be achieved (Kotrba et al., 2011b). In the case of contamination with high concentrations of heavy metals in water or soil, the use of physicochemical techniques is more advisable because they eliminate large amounts of metals in relatively short periods of time. However, when metal concentrations are low

(~2 ppm), the physicochemical removal processes are inefficient and, because they require the addition of large amounts of reagents and energy, are also expensive (Inglea and Seabrab, 2014). Under these conditions, bioremediation processes are most useful. Bioremediation of heavy metals may be carried out using higher organisms such as plants or microorganisms such as bacteria, fungi, algae, and microalgae.

Metal removal through bioremediation may be accomplished by various processes such as bioextraction, biofiltration, bioleaching, biostabilization, biosorption (biosorption or bioaccumulation/bioabsorption), biooxidation, bioreduction, and biovolatilization (biomethylation). The choice of the process depends on the organism used and the contaminant to be removed (Mulligan et al., 2001; Rahman et al., 2016). For phytoremediation, plants can be used for the remediation of both heavy metal–contaminated soils and aquatic ecosystems (rivers, lakes, groundwater, etc.) (Mulligan et al., 2001; Mani and Kumar, 2014; Mahmood et al., 2015; Rahman et al., 2016).

Processes involving microbial biomass (bacteria, fungi, algae, microalgae) are preferred for the removal of metal ions from effluents (Vieira and Volesky, 2000; Perales-Vela et al., 2006; Jácome-Pilco et al., 2009; Zaidi et al., 2011; Chekroun and Baghour, 2013). Other processes using fungi have been proven useful in the bioremediation of soils impacted by heavy metals, mainly through processes of biosorption, bioreduction, and biostabilization (precipitation and solubilization) (Zaidi et al., 2011; Damodaran et al., 2014; Damodaran, 2015).

36.4.1 Bioremediation of Heavy Metals Using Plants

Phytoremediation involves the use of plants for the recovery of a metal from the contaminated phase (soil or water) either by absorption by the plant (bioextraction), by stabilization in the rhizosphere zone, or by transforming it into less toxic species by modifying the oxidation state (Jadia and Fulekar, 2009; Shakoor et al., 2013).

In particular, metal extraction using plants (phytoextraction) consists of the translocation of the metal from the contaminated matrix (soil or water) to the plant tissue through its roots (Shakoor et al., 2013). The purpose of this technology is to extract the metal from the biomass by chemical processes such as incineration. Phytoextraction is widely used for the recovery of both toxic and precious metals (Rahman et al., 2016), as it offers the advantages of being an economic and environmentally friendly process. A variety of plants have shown the ability to accumulate metals in different tissues (Kotrba et al., 2011a; Malik and Biswas, 2012; Ali et al., 2013; Inglea and Seabrab, 2014). Table 36.3 shows some genera of plants considered hyperaccumulators of metal ions.

For phytostabilization, it has been proposed that the mobilization of metal ions in the rhizosphere of the plant is due to the excretion of organic acids (acetic, malic, etc.), chelatin molecules such as ethylenediaminetetraacetic acid (EDTA), nitrilotriacetic acid (NTA), or diethylenetriaminepentaacetic acid. The role of these substances is to solubilize the metal ions, facilitating their entry into the plant tissue through the roots (Shakoor et al., 2013; Anjum et al., 2015).

Remediation of sites contaminated with heavy metals by phytoextraction has several advantages including being an economic and eco-friendly technology that could even use the plants present in the contaminated site, which prevents the introduction of exotic species to the ecosystem. Additionally, energy can be obtained by burning biomass and recovering the extracted metal, resulting in a biomining process (Glass, 2000; Rahman et al., 2016). Some of the disadvantages of using phytoextraction include the time necessary to carry it out because it is estimated that at least 18 months is required to achieve an acceptable efficiency, in addition to the fact that not all plants have a tendency to accumulate metal ions in their tissues (Malik and Biswas, 2012).

Rhizostabilization and phytostabilization occur in the rhizosphere as a result of the action of bacteria present (rhizostabilization) and the action of root exudates of the plant (phytostabilization), which causes the precipitation of the metal, immobilizing it in the soil matrix and preventing its transportation through the food chain or into groundwater (Brunner et al., 2008). Unlike phytoextraction, phytostabilization (the action of roots or the rhizosphere) does not allow for the recovery of the metal but decreases its toxicity and helps to restore contaminated sites.

Another process of metal removal using plants is phytovolatilization, which occurs for certain metals that, upon entering plant tissue, are converted into volatile compounds via chemical reactions and then released into the atmosphere through the stomata (Heaton et al., 1998; Abou, 2011; Wu et al., 2015). The disadvantage of this process is that the metal can reenter the ecosystem by the action of rain; however, it will be as a less toxic species. Particularly for mercury(Hg), this process has the advantage that the Hg released into the atmosphere is elemental Hg, and despite returning to the contaminated site (water or soil) when raining, it will be as a less toxic species than the Hg^{2+} than it was before (Abou, 2011). This technique has also been proven efficient for treating sites contaminated with Se, As, and B, which have volatile forms (Malik and Biswas, 2012; Rahman et al., 2016). Table 36.3 shows some plants in which the phytovolatilization process has been demonstrated.

36.4.2 Bioremediation Using Microbial Biomass

The proposed use of bacterial biomass for metal removal dates from late last century. Beveridge and Fyfe (1985) proposed that the negative charge of the cell membrane of microorganisms could trap positive ions by a charge interaction. The use of microbial biomass (bacteria, yeasts, fungi, algae, and microalgae) for capturing metal ions regardless of whether the biomass is alive or dead is known as biosorption (Cañizares-Villanueva, 2000). Metal biosorption in bacterial biomass occurs by two independent processes, namely, absorption and adsorption. These two processes can be performed simultaneously in living biomass, while dead biomass can perform only adsorption (Cañizares-Villanueva, 2000; Vieira and Volesky, 2000; Gavrilescu, 2004; Hashim and Mukhopadhyay, 2011; Ghosh et al., 2015).

TABLE 36.3

Plant Species Employed to Remove Heavy Metals

Process	Species	Metal	Metal Accumulation
Phytoextraction	*Pandiaka metallorum*	Co, Cu	2.1, 6.3 g kg⁻¹
	Ascolepis metallorum	Co, Cu	1.1, 1.2 g kg⁻¹
	Crotalaria cobalticola	Co	3.0 g kg⁻¹
	Aeollanthus subacaulis	Co, Cu	5.2, 13.7 g kg⁻¹
	Thlaspi careulenscens	Cd	1.8 g kg⁻¹
	Ipomea alpine	Cu	12.3 g kg⁻¹
	Haumaniastrum robertii	Co	10.2 g kg⁻¹
	Thlaspi rotundifolium	Pb	8.2 g kg⁻¹
	Macadamia neurophylla	Mn	51.8 g kg⁻¹
	Psychotria douarrei	Ni	47.5 g kg⁻¹
	Thlaspi careulenscens	Zn	51.6 g kg⁻¹
	Alyssum	Ni	4.7–19.1 g kg⁻¹
	Azolla pinnata	Cd	0.74 g kg⁻¹
	Eleocharis acicularis	Cu, Zn, Cd, As	20.2, 11.2, 0.24, 1.4 g kg⁻¹
	Corrigiola telephiifolia	As	2.1 g kg⁻¹
	Euphorbia cheiradenia	Pb	1.1 g kg⁻¹
	Pteris	As	1.0–8.3 g kg⁻¹
	Pteris vittata	Cr	20.6 g kg⁻¹
	Solanum photeinocarpum	Cd	0.16 g kg⁻¹
	Brassica juncea[b]	Se, B, Fe, Mn, Zn	3.1, 275, 250, 177, 55 mg kg⁻¹
Phytovolatilization	*Typha latifolia*	Se	Released as dimethyl-Se
	Brassica napus, Brassica juncea Czern L., and *Hordeum vulgare*[a]	Se	Removal until about 40%; 50% of the removed is not present in biomass
	Brassica juncea[c]	Se	~2.5 mg kg⁻¹ day⁻¹
	Brassica oleracea var *italica*[c]	Se	~2.25 mg kg⁻¹ day⁻¹
	Brassica oleracea var. *capitata*[c]	Se	~2.25 mg kg⁻¹ day⁻¹
	Brassica oleracea var. *botrytis*[c]	Se	~2.25 mg kg⁻¹ day⁻¹
	Oryza sativa[c]	Se	~1500 mg kg⁻¹ day⁻¹
	Lepidium latifolium, Artemisia douglasiana, Caulanthus sp., *Fragaria vesca, Eucalyptus globulus*[d]	Hg	Most volatilization occurring during the day

Sources: Compiled from Malik, N. and Biswas, A., *Sci. Rev. Chem. Commun.*, 2, 141, 2012; Ali, H. et al., *Chemosphere*, 91, 869, 2013; Kotrba et al., 2011a.

[a] Bañuelos et al. (2000).
[b] Bañuelos (1999).
[c] Zayed et al. (1999).
[d] Leonard et al. (1998).

Some of the main advantages of using microbial biomass as a heavy metal adsorbent are its characteristics of having high surface area per unit of mass, a variety of environments in which they can grow (if using living biomass), their ability to generate ion tolerance, and the selectivity of living cells to absorb certain metal ions, allowing for the recovery of only the metal of interest or precious metals from a mixture (Zabochnicka-Świątek and Krzywonos, 2014; Ghosh et al., 2015).

Table 36.4 compares various aspects of the use of live or dead cells. Choosing one or the other will depend on a particular process because an efficient process will be less selective and vice versa (Vijayaraghavan and Yun, 2008; Zabochnicka-Świątek and Krzywonos, 2014).

In the natural environment, microorganisms are exposed to different chemical compounds and elements that often reach toxic levels, which is the reason why cells have developed various strategies to tolerate these conditions (Cañizares-Villanueva, 2000). When the cells and a contaminated effluent are in contact, the first thing that occurs is that the metal ions begin to interact with components of the cell wall (Vijayaraghavan and Yun, 2008; Yun et al., 2011). After some time, the metal ions enter the cell; however, the cell possesses mechanisms that allow it to maintain a balance between intra- and extracellular metal concentrations in order to prevent levels causing physiological effects (Kuroda and Ueda, 2011).

36.4.2.1 Bioadsorption

The term *bioadsorption* refers to the adsorption of metal ions by ion exchange. This process occurs in both living and dead

TABLE 36.4

Characteristics of Heavy Metals Biosorption with Biomass

Aspect	Dead Cells	Live Cells
pH	Variable. Wide range can be used for process optimization.	Generally high. Metal removal is difficult by the affectation of chemical speciation.
Selectivity	Very low. Any metal ion interacts with electrical charges of a cell wall.	Better than the sorption. Some microorganisms have been reported for the accumulation of one single ion.
Cost	Relatively low. The biomass can stem from industrial process wastes.	High. The control of culture optimal conditions is necessary.
Reusing	Easy. The sorbent regeneration is simple (washing, pH change, etc.).	Low. The metal is accumulated intracellularly; thus, the sorbent regeneration is difficult.
Metal recovery	Simple. The metallic ion elution is easy to carry out.	Difficult. The metal is inside a biomass, so for its recovery it is necessary to employ incineration or another destructive method.

Sources: Cañizares-Villanueva, R., *Revista latinoamericana de microbiología*, 42, 131, 2000; Vijayaraghavan, K. and Yun, Y.S., *Biotechnol. Adv.*, 26, 266, 2008; Kuroda, K. and Ueda, M., *Yeast Biosorption and Recycling of Metal Ions by Cell Surface Engineering Cell Surface Engineering*, 2011; Monteiro, C.M. et al., *Biotechnol. Prog.*, 28, 299, 2012; Zabochnicka-Świątek, M. and Krzywonos, M., *Pol. J. Environ. Stud.*, 23, 551, 2014.

cells (Cañizares-Villanueva, 2000). It is fast, is simply physical, does not depend on energy consumption, and does not generate metabolic reactions for the microorganism (Kuroda and Ueda, 2011). Bioadsorption is a simple interaction between the positive charge of the metal ion and the negative charge of functional groups of the cell wall components, such as carboxyl, phosphate, amino, sulfhydryl, sulfate, ester, carbonyl, and hydroxyl groups (Gavrilescu, 2004; Wang and Chen, 2009; Yun et al., 2011; Ghosh et al., 2015). During this charge interaction, coordinate bonds are generated that permit metal adsorption from the contaminated matrix (Monteiro et al., 2012).

Bioadsorption does not involve metabolic reactions. The amount of active sites (functional groups) in the cell wall is finite and could become saturated, hence limiting the adsorption of metal ions (Vijayaraghavan and Yun, 2008), making the process inefficient, and making it necessary to change the sorbent used or regenerate it by a chemical process (Kotrba et al., 2011b).

One of the great advantages of using microbial biomass as a sorbent is its characteristics of having high surface area per unit mass. However, the microscopic size makes it difficult to separate the biomass from the effluent. To avoid this problem, it has been proposed to use packed columns or the immobilization of cells in various matrices such as alginate, polyvinyl alcohol, polyacrylamide, silica gel, cellulose, glutaraldehyde, polysulfone, κ-carrageenan, polyurethane foam, agar, agarose, and casein (Cañizares-Villanueva et al., 1993, 2001; Cañizares et al., 1994; Ilangovan et al., 1998; Cañizares-Villanueva, 2000; Pellón et al., 2008; de-Bashan and Bashan, 2010). The lifetime and efficiency of the support depend on the type of biomass and the activation process used.

The adsorption of metals with microbial biomass using dead cells is very attractive because, due to the lack of metabolic activity, the process conditions can be changed (temperature, pH, biomass content, initial metal concentration) to obtain optimal conditions to trap the metal ion of interest (Gagrai et al., 2013; Verma and Hasan, 2014).

Once the solvent has been saturated, it is necessary to regenerate it by washing with acidic substances that cause the release of metal ions due to the loss of affinity for the biomass. To perform these washes, acidic solutions (HNO_3, HCl, H_2SO_4, citric acid, lactic acid, acetic acid, etc.), basic solutions (NH_4OH, NaOH), salt ($KHCO_3$), or chelators (EDTA, NTA, etc.) can be used (Gupta et al., 2000; de-Bashan and Bashan, 2010; Dostálek, 2011; Long et al., 2014; Zabochnicka-Świątek and Krzywonos, 2014; Ghosh et al., 2015). However, because the elution of metal ions is carried out by washes with slightly corrosive substances such as those mentioned earlier, the integrity of the biomass decreases with each use-wash cycle, thus decreasing the efficiency of each cycle and also the life of the sorbent (Naja and Volesky, 2011; Long et al., 2014).

To expose the active sites present on the bioadsorbent and increase their numbers, the biomass can be subjected to both physical and chemical treatments (Wang and Chen, 2009; Mudhoo et al., 2012). Physical treatments include biomass lyophilization, heating, freezing, and drying; chemical treatments include washing with detergent and basic, acidic, and saline solutions (NaOH, NH_4OH, Na_2CO_3, $CaCl_2$, NaCl, HNO_3) or other solvents such as ethanol, acetone, or toluene. The biomass can be even subjected to saponification reactions or polysaccharide additions (Figueira et al., 1999; Sar et al., 1999; Gupta et al., 2000; Vijayaraghavan and Yun, 2008; Long et al., 2014).

Different biosorbents based on immobilized biomass are available:

1. Algasorb®, a portable equipment for the removal of heavy metals, with a sorbent made from biomass of the dead microalgae *Chlorella vulgaris* immobilized on silica gel. It is useful for the removal of metal ions and ensures a life span of up to 18 months with regular use (Cañizares-Villanueva, 2000).

2. AMT-Bioclaim, a product based on *Bacillus subtilis* biomass (Dostálek, 2011).

3. BioFix®, a mixture of cyanobacterium (*Spirulina*), yeasts, algae (*Sphagnum* sp.), and a plant (*Lemna* sp.) biomass that is immobilized on polyethylene, polypropylene, or polysulfone with a particle size of up to 2.5 mm. The cyanobacterium is useful in the removal of Al, Cd, Zn, and Mn. A life span of up to 120 adsorption–desorption cycles is guaranteed (Cañizares-Villanueva, 2000; Dostálek, 2011).

4. BV Sorbex Inc. develops systems for biosorption employing different kinds of sorbents and adaptable equipment. Some of the sorbents of this company are prepared using biomass of brown (*Sargassum*, *Ascophyllum*, etc.), green (*Halimeda*), and red (*Chondrus*) algae and a green microalga (*Chlorella*) (BV Sorbex Inc n.d., Dostálek, 2011).

36.4.2.2 Bioabsorption

Bioabsorption is performed by living cells. It includes the incorporation of metal ions into the biomass using transport processes that depend on cellular metabolism and consume energy (Perales-Vela et al., 2006; Chojnacka, 2009; Zabochnicka-Świątek and Krzywonos, 2014). Once the metal is inside the cell, it is immobilized or trapped (Perales-Vela et al., 2006). Immobilization occurs through complex formation or by binding with low-molecular-weight organic acids

such as citrate or malate (Cobbett, 2000; Kaplan, 2013). The trapping involves binding of the metal to macromolecules such as MT, phytochelatins, or polysaccharides, confining them inside an organelle such as a vacuole, chloroplast, or mitochondrion (Mendoza-Cózatl and Moreno-Sánchez, 2005; Mendoza-Cózatl et al., 2005; Perales-Vela et al., 2006; Naja and Volesky, 2011).

Table 36.5 shows the genera of bacteria that have a high potential as biosorbents of heavy metals such as *Bacillus*, *Pseudomonas*, *Streptomyces*, *Micrococcus*, and *Staphylococcus* (Vieira and Volesky, 2000; Dostálek, 2011; Monteiro et al., 2012). The genera *Alcaligenes*, *Pseudomonas*, and *Arthrobacter* can also absorb large quantities of valuable metals such as palladium, gold, and platinum (Tsezos, 2014), and *Bacillus cereus* can absorb up to 90 mg g⁻¹ Ag¹⁺. The genera of fungi with proven ability to remove various metal ions are *Aspergillus*, *Mucor*, *Ganoderma*, *Trichoderma*, *Penicillium*, *Phanerochaete*, *Neurospora*, *Saccharomyces*, *Fusarium*, *Pycnoporus*, and *Rhizopus* (Rajendran et al., 2003; Viraraghavan and Srinivasan, 2011; Monteiro et al., 2012). The latter one stands out because some of its members are able to bioadsorb Cu^{2+}, Th, Am, U, Al, Au, Ag, Pb, Zn, and Cr (Naja and Volesky, 2011; Viraraghavan and Srinivasan, 2011; Tsezos, 2014). Microalgae that are capable of removing heavy metals are mostly

---

TABLE 36.5
Bacterial Species Employed for Heavy Metal Removal

Bacteria	Metal Removed	Results
Micrococcus luteus	Cu, Pb	Specific uptake(mg g⁻¹) 208
Bacillus cereus	Hg	Specific uptake (mg g⁻¹) 104
Bacillus firmus	Cr(VI)	Full reduction from Cr⁺⁶ to Cr⁺³
Stenotrophomonas maltophilia	Cr	90% removed
Pseudochrobactrum saccharolyticum	Cr(VI), Cu(II)	95% removed
Ochrobactrum intermedium	Cr(VI)	High efficiency for dead or live cells
Cupriavidus metallidurans		
Bacillus amyloliquefaciens	Cr(VI)	High production of reductase
Bacillus methylotrophicus	Cr	High reductase activity
Pseudomonas putida	Cu, Pb	Specific uptake (mg g⁻¹) 185.5 y 30.5
Rhodococcus opacus	Cr, Cu, Pb	Removal efficiency of 70%, 52%, and 95%
Lysinibacillus sp.	Ni	Specific uptake (mg g⁻¹) 238
Pseudomonas aeruginosa[a]	U	Efficient adsorption with the employ of live cells stabilized by heat or lyophilization
Escherichia coli[b]	Au	Removal efficiency of 100%
Desulfovibrio desulfuricans[b]	Au	Removal efficiency of 100%
Bacillus licheniformis[c]	Ag	Specific uptake (mg g⁻¹) 73.6
Bacillus subtilis (cellular fragments)[d]	Sr, Fe, Ni, Pb, Cu, Ag, Au, Zn, Hg	Specific uptake (mg g⁻¹) 2.4, 200, 6.3, 4.1, 190, 3.8, 71.4, 45.8, 8.0
Pseudomonas aeruginosa[e]	Ni, Cu	Specific uptake (mg g⁻¹) 265 and 137.6

Source: Modified from Ghosh, A. et al., *J. Environ. Eng.*, C4015003, 2015.

[a] Hu et al. (1996).
[b] Deplanche and Macaskie (2008).
[c] Sun et al. (2013).
[d] Beveridge and Murray (1976).
[e] Sar et al. (1999).

chlorophycean algae and cyanobacteria, especially *Chlorella, Spirulina, Thalassiosira, Phaeodactylum, Scenedesmus, Chlamydomonas, Desmodesmus, Phormidium, Micrasterias, Spirogyra,* and *Tetraselmis* (Cañizares-Villanueva et al., 2000, 2001; Peña-Castro et al., 2004; Perales-Vela et al., 2007; Pellón et al., 2008; Jácome-Pilco et al., 2009; Chekroun and Baghour, 2013). Additionally, some macroalgal genera have been used: *Sphagnum, Sargassum, Ascophyllum, Halimeda, Chondrus, Fucus, Cladophora, Caulerpa, Laminaria,* and *Ulva* (Dostálek, 2011; Monteiro et al., 2012; Chekroun and Baghour, 2013; Tsezos, 2014).

In bacteria, genes have been identified that confer tolerance to certain metals and also the ability to accumulate large quantities of metals. Recently, the ability of *Citrobacter* sp. to accumulate Cu^2, Cd^{2+}, Pb^{2+}, UO_2^{2+}, PuO_2^{2+}, and AmO_2^{2+} has been demonstrated through the formation of phosphate crystals containing the metal (Finlay et al., 1999; Kotrba et al., 2011a). The use of immobilized cells of *Citrobacter* sp. in polyacrylamide has also been studied to allow survival at concentrations of 7 g of Cd^{2+}, 4 g of Pb^{2+}, and 9 g of UO_2^{2+} (Finlay et al., 1999).

Similar to plants, some genera of gram-positive and gram-negative bacteria are capable of changing the redox state of certain metals, either to reduce their toxicity or to remove them from the environment (Barkay et al., 2003), making possible their use to eliminate metals by biotransformation. Specifically, the presence of the Hg reductase enzyme allows bacteria to tolerate high concentrations of Hg, as well as its reduction to Hg^0; this ability is exploited in the process of removing this ion by percolation filters that are inoculated with Hg-resistant bacterial cultures, allowing a good recovery of the metal using mixed cultures (von Canstein et al., 2002; Barkay et al., 2003; Wagner-Döbler, 2003). Fungi have also shown great potential for the removal of heavy metals because they are able to develop different strategies to reduce their toxicity. Some strategies are immobilization by the formation of oxalate salts, many of which are insoluble, and bioabsorption and bioadsorption through the polymers present in the fungal cell wall, especially the amino groups of chitin and chitosan (Gadd, 2000; Zaidi et al., 2011; Tsezos, 2014). This affinity for metal deposition is due to the cell wall being largely composed of polysaccharides (~90%); the inner layer consists of chitin fibers, while the outer layer is predominantly composed of glucan, mannans, galactans, and chitosans (Viraraghavan and Srinivasan, 2011). The carboxyl and hydroxyl phosphate groups present in these polysaccharides interact with the metal ions participating in their retention.

In addition to absorption and adsorption, *Aspergillus clavatus, A. niger, A. oryzae, Trichoderma viride, Penicillium glabrum, P. chrysogenum, P. roqueforti, Cladosporium cladosporioides,* and *Alternaria alternata* are able to carry out biovolatilization of heavy elements such as As and I by mechanisms similar to those that are carried out in plants (Thompson-Eagle et al., 1989; Ban-nai et al., 2006; Urík et al., 2007). Likewise, other genera, such as *Rhizopus, Candida, Phanerochaete,* and *Funalia,* have been used for the removal

by bioadsorption of the heavy metals Cu, Cd, Al, Pb, Hg, Zn, U, and Th; their efficiency has been demonstrated and in many cases is greater when living biomass is used (Zaidi et al., 2011; Tsezos, 2014).

36.4.3 Removal of Heavy Metals Using Microalgae

The metal ion contamination in liquid effluents can be removed by the use of microalgal biomass, which presents some advantages over the use of microbial biomass, such as the presence of negative radicals in their cell wall, a larger surface area per gram of biomass, and the versatility to be used only as a sorbent (dead biomass) or as a bioabsorbent (living biomass) (Cañizares-Villanueva, 2000). The main disadvantage is that microalgae are highly sensitive to certain heavy metals, such as Cu, hindering the use of living cells for removal in effluents with high concentrations of these metals (Perales-Vela et al., 2006, 2007).

Intracellular microalgal strategies for immobilizing metal ions are similar to those displayed by plants. Metallothionein, phytochelatins, or polysaccharides are produced intracellularly in order to reduce their toxicity. Metal ions can also be insolubilized in polyphosphate bodies or can be confined to cellular organelles such as the vacuoles or the nucleus (Mendoza-Cózatl and Moreno-Sánchez, 2005; Mendoza-Cózatl et al., 2005; Perales-Vela et al., 2006; Naja and Volesky, 2011).

Microalgal biomass, similar to bacterial biomass, can be used alive or dead. However, one of the advantages of microalgal biomass is that when used alive, it is not necessary to add an organic carbon source because it is autotrophic (Cañizares-Villanueva, 2000). Nevertheless, its tolerance to certain metal ions, such as Cu, is very low, making it impossible to use living cells in effluents with high concentrations of these ions (Perales-Vela et al., 2007; Jácome-Pilco et al., 2009). To generate a physical barrier against the actual concentration of exposure, different systems of cell immobilization have been developed. In such systems, the cells are embedded in an inert polymer matrix that protects them, making the concentration detected by the cells lower than that in liquid medium (Cañizares et al., 1994; Ilangovan et al., 1998; Cañizares-Villanueva et al., 2001, 1993; de-Bashan and Bashan, 2010). Table 36.6 shows some genera of microalgae that are used to remove metals from contaminated effluents, the condition of the biomass used (living, dead, immobilized), and the removal efficiency or capacity.

The removal of heavy metals using live cells is more efficient if microorganisms that were isolated from highly contaminated sites are used, such as industrial effluents from tanneries, mines, and contaminated waters (von Canstein et al., 2002; Barkay et al., 2003; Wagner-Döbler, 2003; Long et al., 2013; Oves et al., 2013; Das et al., 2014a; Ghosh et al., 2015). This is because chronic exposure to heavy metals generates a type of resistance that causes changes in the cellular metabolic response, allowing the microorganisms to survive at higher metal concentrations (Rehman and Shakoori, 2001;

TABLE 36.6

Removal of Heavy Metals with Employing of Microalgal Biomass

Metal	Species	Condition	Amount of Metal Removal
Au	*Chlorella homosphaera*	Immobilized in alginate	~90%
Cd	*Chlorella vulgaris*[a]	Immobilized in κ-carrageenan	66%
	Scenedesmus acutus[a]	Immobilized in κ-carrageenan	73%
	Chlorella vulgaris	Immobilized in polyurethane foam	57%
	Scenedesmus acutus	Immobilized in polyurethane foam	69%
	Spirulina platensis	Immobilized on silica gel	36.6 mg g^{-1}
	Chlorella	Dead cells	7.7 mg g^{-1}
	Porphyridium	Dead cells	7.5 mg g^{-1}
	Chaetoceros calcitrans	Free live cells	1055.3 mg g^{-1}
	Desmodesmus pleiomorphus	Free live and dead cells	85.3 y 58.6 mg g^{-1}
	Planothidium lanceolatum	Free live cells	275.5 mg g^{-1}
	Scenedesmus abundans	Live cells	574 mg g^{-1}
	Scenedesmus obliquus	Dead cells	7.3–108.5 mg g^{-1}
	Spirulina platensis	Live cells	44.5 y 12.1 mg g^{-1}
Co	*Oscillatoria angustissima*	Dead cells	15.3 mg g^{-1}
	Spirogyra hialina	Dead cells	12.8 mg g^{-1}
	Spirulina spp.	Dead cells	0.01 mg g^{-1}
Cr	*Scenedesmus acutus*[a]	Immobilized in κ-carrageenan	36%
	Chlorella vulgaris[a]	Immobilized in κ-carrageenan	48%
	Anacystis nidulans	Continuous culture	65%–85%
	Chlorella vulgaris	Immobilized in polyurethane foam	34%
	Scenedesmus acutus[a]	Immobilized in polyurethane foam	31%
	Chlorella miniata	Dead cells	14.1–41.1 mg g^{-1}
	Spirulina	Live and dead cells	304 y 167 mg g^{-1}
	Chlamydomonas reinhardtii	Dead cells	18–25.6 mg g^{-1}
Cs	*Chlorella salina*	Immobilized in alginate	~70%
Cu	*Chlorella fusca*	Live cells	3.2 mg g^{-1}
	Anabaena cylindrica	Live cells	12.6 mg g^{-1}
	Ceratium hirundinella	Live and dead cells	2.3 y 5.7 mg g^{-1}
	Chlorella miniata	Dead cells	23.3 mg g^{-1}
	Chlorella pyrenoidosa	Live cells	2.4 mg g^{-1}
	Chlorella sp.	Live cells, dead cells immobilized in alginate	220, 108 y 33.4 mg g^{-1}
	Chlorella vulgaris	Live cells	
	Chlorella vulgaris	Dead cells	1.8–48.1 mg g^{-1}
	Chlorella vulgaris	Immobilized in alginate	63.1 mg g^{-1}
	Scenedesmus obliquss	Live and dead cells	1.8 y 20 mg g^{-1}
	Scenedesmus subspicatus	Dead cells	13.3 mg g^{-1}
	Scenedesmus quadricauda	Dead cells	2.8
Hg	*Calothrix parietina*	Dead cells	19 mg g^{-1}
	Chlamydomonas reinhardtii	Dead cells	72.2 mg g^{-1}
	Chlorella vulgaris	Dead cells	16–18
	Scenedesmus acutus	Dead cells	20 mg g^{-1}
	Scenedesmus subspicatus	Dead cells	9.2 mg g^{-1}
Ni	*Chlorella vulgaris*	Dead cells immobilized in alginate	>90%
	Chlorella miniata	Live and dead cells	1.4 y 20.4 mg g^{-1}
	Chlorella vulgaris	Live cells	0.64–59.3 mg g^{-1}
	Chlorella vulgaris	Dead cells	12.1–42 mg g^{-1}
	Spirulina	Live cells	1378 mg g^{-1}
Pb	*Arthrospira platensis*	Dead cells	102.6 mg g^{-1}
	Chlamydomonas reinhardtii	Dead cells	96.3 mg g^{-1}
	Chlamydomonas reinhardtii	Immobilized in alginate	230–380 mg g^{-1}
	Chlorella vulgaris	Dead cells	17.1–127 mg g^{-1}

(Continued)

TABLE 36.6 (*Continued*)
Removal of Heavy Metals with Employing of Microalgal Biomass

Metal	Species	Condition	Amount of Metal Removal
	Scenedesmus acutus	Dead cells	90 mg g^{-1}
U	*Chlorella regularis*	Live and dead cell suspension	14.3 y 28.3 mg g^{-1}
Zn	*Chlorella vulgaris*[a]	Immobilized in polyurethane foam	78%
	Scenedesmus acutus[a]	Immobilized in polyurethane foam	84%
	Chlorella vulgaris[a]	Immobilized in κ-carrageenan	85%
	Scenedesmus acutus[a]	Immobilized in κ-carrageenan	91%
	Chlorella vulgaris	Dead cells	6.4–43.4 mg g^{-1}
	Planothidium lanceolatum	Live cells	118.6 mg g^{-1}
	Scenedesmus obliquus	Dead cells	6.7–22.3 mg g^{-1}
	Scenedesmus quadricauda	Dead cells	5 mg g^{-1}
	Scenedesmus subspicatus	Live cells	72.1 mg g^{-1}

Sources: Modified from de-Bashan, L.E. and Bashan, Y., *Bioresource Technol.*, 101, 1611, 2010; Suresh Kumar, K. et al., *Ecotoxicol. Environ. Saf.*, 113, 329, 2015.

[a] Travieso et al. (1999).

Rehman et al., 2006; Kalinowska and Pawlik-Skowrońska, 2010). Species tolerant or resistant to various metal ions can be generated in controlled laboratory conditions by successive passes in culture media with gradually increasing metal concentrations (Díaz-Ravina and Baath, 1996; Biagianti-Risbourg et al., 2013).

The two strategies presented earlier allow for the treatment of effluents with high levels of metal, even enabling the accumulation of higher concentrations per unit biomass (Klerks and Weis, 1987; Díaz-Raviña and Baath, 2001; von Canstein et al., 2002; Biagianti-Risbourg et al., 2013).

36.4.3.1 Cu Removal Using Microalgae

Currently, Cu contamination is highly significant in different parts of the world, and it has been estimated that annual discharges are approximately 90,000 tons. Much of this contamination is due to industrial processes or the use of $CuSO_4$ to prevent algal proliferation in lakes and water reservoirs (Perales-Vela et al., 2007). Due to the large number of sources of Cu, the large volumes of discharges, and the accumulation of Cu through food chains, Cu has been classified by different international organizations as a priority pollutant (Ahmed and Häder, 2010).

Although there are physicochemical methods for eliminating this metal from contaminated effluents, the efficacy of such methods decreases at low concentrations (Cañizares-Villanueva, 2000). Previously, it was considered that the levels achieved after physicochemical Cu removal were not toxic to plants; however, recently, it has been demonstrated that this metal is highly toxic even at concentrations of 2 ppm, thus significantly reducing photosynthetic processes (Perales-Vela et al., 2007).

Due to the high toxicity of Cu, removal technologies using microbial biomass have been developed, taking advantage of the high affinity of biomass to Cu ions at even very low concentrations (Cañizares-Villanueva, 2000). Biosorption of the ions of a liquid effluent is a solid–liquid process that depends on operation factors such as agitation and pH and can be carried out in batch or continuous systems, depending on the effluent to be treated and the objectives to be achieved (Dostálek, 2011).

Several different configurations of photobioreactors have been developed to cover specific needs of the microalgae cultivated (Martínez-Roldán and Cañizares-Villanueva, 2015); however, for the removal of Cu from contaminated effluents, the bubble column for live cells (Peña-Castro et al., 2004; Perales-Vela et al., 2007) and the stirred tank or packed column for dead cells (Vieira and Volesky, 2000; Naja and Volesky, 2011) have been the most commonly used settings. Both have achieved high Cu removal efficiency (Table 36.6).

For the treatment of wastewater contaminated with metals, a process is sought that requires little energy, is inexpensive, and has a longer lifetime than current processes. The use of living cells in suspension presents some operational problems, one of them being the separation of the biomass from the liquid because, due to the small cell size, it is necessary to employ solid–liquid separation processes such as centrifugation, filtration, and flotation (Naja and Volesky, 2011). Many attempted solutions are too expensive to be economically feasible (Suresh Kumar et al., 2015).

As a way to reduce the costs of biomass recovery and provide a less aggressive environment for cell metabolism, immobilized systems have been developed (Cañizares et al., 1994; Ilangovan et al., 1998; Cañizares-Villanueva et al., 2001; Dostálek, 2011; Suresh Kumar et al., 2015). These systems facilitate biomass separation, requiring only agitation (usually by bubbling) to precipitate particles and recover them. Furthermore, in systems of immobilized cells, it is possible to manipulate different conditions to optimize the process. Some of the operating conditions that may be varied are the nature of the support (agar, alginate, polyurethane, etc.), particle size (0.7–1.5 mm), and cell density per unit support (Cañizares et al., 1994; Ilangovan et al., 1998;

Cañizares-Villanueva et al., 2001; de-Bashan and Bashan, 2010; Dostálek, 2011; Suresh Kumar et al., 2015).

Despite the high efficiency achieved in Cu removal processes with living cells, the difficulty of keeping them active and the severe physiological effects generated by this metal have driven the trend to use dead cells packed in columns, through which is flowed the Cu-contaminated effluent (Dostálek, 2011). By chemical affinity alone, the metal is bound to the cell wall and removed from the effluent; once the support is saturated, the metal can be recovered by subjecting the support to a chemical washing, thus allowing the reuse of the biomass for several cycles without substitution (Vieira and Volesky, 2000; Naja and Volesky, 2011).

36.5 CONCLUSION

In the last century, population growth, food demand, and scaling of industrial and technological processes have increased the demand for water in the world. This increase in the use of water has resulted in an increase in the volume of wastewater, containing a complex mixture of contaminants generated anthropocentrically. Heavy metals unlike the rest of the contaminants found in the wastewater cannot be degraded by physical or biological methods; for this reason they must be removed from the water. Although physicochemical methods are efficient, they have problems of selectivity, cost, and environmental toxicity; for these reasons, methods for biological wastewater treatment are being studied as an alternative. The bioremediation using higher plants, fungi, bacteria, and algae biomass offers an option to treatment of wastewater, as they are inexpensive, selective, and environmentally friendly.

REFERENCES

Abakay, A. et al. 2012. Relationships between respiratory function disorders and serum copper levels in copper mineworkers. *Biological Trace Element Research* 145: 151–157.

Abou, R.A.E.A.I. 2011. Bioremediation: New approaches and trends, in: Khan, M.S., Zaidi, A., Goel, R., Musarrat, J. (Eds.), *Biomanagement of Metal-Contaminated Soils*. Springer, Utretch, the Netherlands, pp. 65–94.

Ahmed, H., Häder, D. 2010. A fast algal bioassay for assessment of copper toxicity in water using *Euglena gracilis*. *Journal of Applied Phycology* 22: 785–792.

Ali, H., Khan, E., Sajad, M.A. 2013. Phytoremediation of heavy metals—Concepts and applications. *Chemosphere* 91: 869–881.

Anjum, N.A. et al. 2015. Jacks of metal/metalloid chelation trade in plants-an overview. *Frontiers of Plant Science* 6: 192.

ATSDR: Agency for toxic substances and disease registry. 2011. Toxicological profile for copper. U.S. Department of Health and Human Services. Georgia USA. https://www.atsdr.cdc.gov/es/.

Ban-nai, T., Muramatsu, Y., Amachi, S. 2006. Rate of iodine volatilization and accumulation by filamentous fungi through laboratory cultures. *Chemosphere* 65: 2216–2222.

Bañuelos, G. 1999. Factors influencing field phytoremediation of selenium-laden soils, in: Terry, N., Banuelos, G.S. (Eds.), *Phytoremediation of Contaminated Soil and Water*. CRC Press, Boca Raton, FL, pp. 41–60.

Bañuelos, G., Zambrzuski, S., Mackey, B. 2000. Phytoextraction of selenium from soils irrigated with selenium-laden effluent. *Plant Soil* 224: 251–258.

Barakat, M.A. 2011. New trends in removing heavy metals from industrial wastewater. *Arabian Journal of Chemistry* 4: 361–377.

Barkay, T., Miller, S.M., Summers, A.O. 2003. Bacterial mercury resistance from atoms to ecosystems. *FEMS Microbiol Reviews* 27: 355–384.

Beveridge, T., Fyfe, W. 1985. Metal fixation by bacterial cell walls. *Journal of Earth Science* 22: 1893–1898.

Beveridge, T., Murray, R. 1976. Uptake and retention of metals by cell walls of *Bacillus subtilis*. *Journal of Bacteriology* 127: 1502–1518.

Biagianti-Risbourg, S., Paris-Palacios, S., Mouneyrac, C., Amiard-Triquet, C. 2013. Pollution Acclimation, Adaptation, Resistance and Tolerance in Ecotoxicology, in: Férard, J.F., Blaise, C. (Eds.), *Encyclopedia of Aquatic Ecotoxicology*. Springer Netherlands, Dordrecht, pp. 135–146.

Bretzel, F., Calderisi, M. 2006. Metal contamination in urban soils of coastal Tuscany (Italy). *Environmental Monitoring and Assessment* 118: 319–335.

Brewer, G.J. 2010. Copper toxicity in the general population. *Clinical Neurophysiology* 121: 459–460.

Brunner, I., Luster, J., Günthardt-Goerg, M., Frey, B. 2008. Heavy metal accumulation and phytostabilisation potential of tree fine roots in a contaminated soil. *Enviromental Pollution* 152: 559–568.

Cañizares, R.O., Rivas, L., Montes, C., Domínguez, A.R., Travieso, L., Benitez, F. 1994. Aerated swine-wastewater treatment with K-carrageenan-immobilized Spirulina maxima. *Bioresource Technology* 47: 89–91.

Cañizares-Villanueva, R. 2000. Biosorption of heavy metals by microorganisms. *Revista latinoamericana de microbiología* 42: 131–143.

Cañizares-Villanueva, R.O., Dominguez, A.R., Rivas, L., Montes, M.C., Travieso, L., Benitez, F. 1993. Free and immobilized cultures of Spirulina maxima for swine waste treatment. *Biotechnology Letters* 15: 321–326.

Cañizares-Villanueva, R.O., Gonzalez-Moreno, S., Domínguez-Bocanegra, A.R. 2001. Growth, nutrient assimilation and cadmium removal by suspended and immobilized Scenedesmus acutus cultures: Influence of immobilization matrix, in: Chen, F., Jiang, Y. (Eds.), *Algae and Their Biotechnological Potential*. Springer, Utretch, the Netherlands, pp. 147–161.

Cañizares-Villanueva, R.O., Martínez-Jerónimo, F., Espinosa-Chávez, F. 2000. Acute toxicity to Daphnia magna of effluents containing Cd, Zn, and a mixture Cd-Zn, after metal removal by Chlorella vulgaris. *Environmental Toxicology* 15: 160–164.

Cao, Z., Hu, Z. 2000. Copper contamination in paddy soils irrigated with wastewater. *Chemosphere* 41: 3–6.

Chekroun, K.B., Baghour, M. 2013. The role of algae in phytoremediation of heavy metals: A review. *Journal of Materials and Environmental Science* 4: 873–880.

Cherfi, A., Achour, M., Cherfi, M., Otmani, S., Morsli, A. 2015. Health risk assessment of heavy metals through consumption of vegetables irrigated with reclaimed urban wastewater in Algeria. *Process Safety and Environmental Protection* 98: 245–252.

Chojnacka, K. 2009. Biosorption and bioaccumulation-new tools for separation technologies of XXI century, in: Trochimczuk, A.W., Walkowiak, W. (Eds.), *International Symposium on Physico-Chemical Methods of Separation—Ars Separatoria 2009*. Wroclaw, Poland, Vol. 24, pp. 31–35.

Cobbett, C. 2000. Phytochelatins and their roles in heavy metal detoxification. *Plant Physiology* 123: 825–832.

Crane, R.S. et al. 2010. Fate and behaviour of copper and zinc in secondary biological wastewater treatment processes: I. Evaluation of biomass adsorption capacity. *Environmental Technology* 31: 705–723.

Dagli, C.E., Tanrikulu, A.C., Koksal, N., Abakay, A., Gelen, M.E., Demirpolat, G., Tolun, F.I. 2010. Interstitial lung disease in coppersmiths in high serum copper levels. *Biological Trace Element Research* 137: 63–68.

Damodaran, D. 2015. Interaction of heavy metals in multimetal biosorption by *Galerina vittiformis* from soil. *Bioremediation Journal* 19: 56–68.

Damodaran, D., Shetty, K., Mohan, B. 2014. Uptake of certain heavy metals from contaminated soil by mushroom—*Galerina vittiformis*. *Ecotoxicology and Environmental Safety* 104: 414–422.

Das, S. 2014. *Microbial Biodegradation and Bioremediation*, 1st edn. Elsevier. USA. 642pp.

Das, S., Mishra, J., Das, S.K., Pandey, S., Rao, D.S., Chakraborty, A., Sudarshan, M., Das, N., Thatoi, H. 2014a. Investigation on mechanism of Cr(VI) reduction and removal by *Bacillus amyloliquefaciens*, a novel chromate tolerant bacterium isolated from chromite mine soil. *Chemosphere* 96: 112–121.

Das, S., Raj, R., Mangwani, N., Dash, H.R., Chakraborty, J. 2014b. Heavy metals and hydrocarbons: Adverse effects and mechanism of toxicity, in: Das, S. (Ed.), *Microbial Biodegradation and Bioremediation*. Elsevier, USA. pp. 23–54.

De, J., Dash, H.R., Das, S. 2014. Mercury pollution and bioremediation—A case study on biosorption by a mercury-resistant marine bacterium, in: Das, S. (Ed.), *Microbial Biodegradation and Bioremediation*. Elsevier. USA. pp. 137–166.

de-Bashan, L.E., Bashan, Y. 2010. Immobilized microalgae for removing pollutants: Review of practical aspects. *Bioresource Technology* 101: 1611–1627.

Debelius, B., Forja, J., DelValls, Á., Lubián, L. 2009. Toxicity and bioaccumulation of copper and lead in five marine microalgae. *Ecotoxicology and Environmental Safety* 72: 1503–1513.

Deplanche, K., Macaskie, L. 2008. Biorecovery of gold by *Escherichia coli* and *Desulfovibrio desulfuricans*. *Biotechnology and Bioengineering* 99: 1055–1064.

Devars, S., Sánchez-Moreno, R. 1999. Abundancia de los metales pesados en la biosfera, in: Cervantes, C., Moreno-Sánchez, R. (Eds.), *Contaminación ambiental por metales pesados: Impacto en los seres vivos*. AGT Editor, S.A. México.

Díaz-Ravina, M., Baath, E. 1996. Development of metal tolerance in soil bacterial communities exposed to experimentally increased metal levels. *Applied and Environmental Microbiology* 62: 2970–2977.

Díaz-Raviña, M., Baath, E. 2001. Response of soil bacterial communities pre-exposed to different metals and reinoculated in an unpolluted soil. *Soil Biology and Biochemistry* 33: 241–248.

Ding, T., Tian, Y., Liu, J., Hou, J., Guo, Z., Wang, J. 2015. Calculation of the environmental dredging depth for removal of river sediments contaminated by heavy metals. *Environmental Earth Sciences* 74: 4295–4302.

Dixit, R. et al. 2015. Bioremediation of heavy metals from soil and aquatic environment: An overview of principles and criteria of fundamental processes. *Sustainability* 7: 2189–2212.

Dostálek, P. 2011. Immobilized biosorbents for bioreactors and commercial biosorbents commercial biosorbents, in: Kotrba, P., Mackova, M., Macek, T. (Eds.), *Microbial Biosorption of Metals*. Springer, Utretch, the Netherlands, pp. 285–300.

Durukan, I., Arpa Şahin, Ç., Bektaş, S. 2011. Determination of copper traces in water samples by flow injection-flame atomic absorption spectrometry using a novel solidified floating organic drop microextraction method. *Microchemical Journal* 98: 215–219.

Figueira, M., Volesky, B., Mathieu, H. 1999. Instrumental analysis study of iron species biosorption by *Sargassum* biomass. *Environmental Science Technology* 33: 1840–1846.

Finlay, J.A., Allan, V.J., Conner, A., Callow, M.E., Basnakova, G., Macaskie, L.E. 1999. Phosphate release and heavy metal accumulation by biofilm-immobilized and chemically-coupled cells of a *Citrobacter* sp. pre-grown in continuous culture. *Biotechnology Bioengineering* 63: 87–97.

Fishel, F. 2005. *Pesticide Toxicity Profile: Copper-Based Pesticides*. UF/IFAS Extension. University of Florida, USA.

Förstner, U., Wittmann, G.T.W. 1979. *Metal Pollution in the Aquatic Environment*. Springer Verlag, New York, p. 485.

Gadd, G. 2000. Bioremedial potential of microbial mechanisms of metal mobilization and immobilization. *Current Opinion in Biotechnology* 11: 271–279.

Gagrai, M., Das, C., Golder, A. 2013. Reduction of Cr (VI) into Cr (III) by Spirulina dead biomass in aqueous solution: Kinetic studies. *Chemosphere* 93: 1366–1371.

Gavrilescu, M. 2004. Removal of heavy metals from the environment by biosorption. *Engineering in Life Sciences* 4: 219–232.

Ghosh, A., Ghosh-Dastidar, M., Sreekrishnan, T. 2015. Recent advances in bioremediation of heavy metals and metal complex dyes: Review. *Journal of Environmental Engineering* C4015003.

Glass, D. 2000. Economic potential of phytoremediation, in: Raskin, I., Ensley, B. (Eds.), *Phytoremediation of Toxic Metals: Using Plants to Clean up the Environment*. Wiley, New York, pp. 15–31.

González-Macías, C., Schifter, I., Lluch-Cota, D. B., Méndez-Rodríguez, L., Hernández-Vázquez, S. 2006. Distribution, enrichment and accumulation of heavy metals in coastal sediments of Salina Cruz Bay, México. *Environmental Monitoring and Assessment* 118: 211–230.

Gu, Y.-G., Lin, Q., Yu, Z.-L., Wang, X.-N., Ke, C.-L., Ning, J.-J. 2015. Speciation and risk of heavy metals in sediments and human health implications of heavy metals in edible nekton in Beibu Gulf, China: A case study of Qinzhou Bay. *Mpb* 101: 852–859.

Gupta, R., Ahuja, P., Khan, S., Saxena, R.K., Mohapatra, H. 2000. Microbial biosorbents: Meeting challenges of heavy metal pollution in aqueous solutions. *Current Science* 78: 967–973.

Hart, L. E., Cheng, K.M., Whitehead, P.E., Shah, R.M., Lewis, R.J., Ruschkowski, S.R., Blair R.W., Bannett, D.W., Bellward, G.D., Bandiera, S.M. 1991. Effects of dioxin contamination on the growth and development of Great Blue Heron embryos. *Journal of Toxicology and Enviromental Health* 32: 331–344.

Hashim, M., Mukhopadhyay, S. 2011. Remediation technologies for heavy metal contaminated groundwater. *Journal of Environmental Management* 92: 2355–2388.

Heaton, A.C., Rugh, C.L., Wang, N.J., Meagher, R.B. 1998. Phytoremediation of mercury-and methylmercury-polluted soils using genetically engineered plants. *Journal of Soil Contamination* 7: 497–509.

Hu, M.Z., Norman, J.M., Faison, B.D., Reeves, M.E. 1996. Biosorption of uranium by *Pseudomonas aeruginosa* strain CSU: Characterization and comparison studies. *Biotechnology and Bioengineering* 51: 237–247.

Ilangovan, K., Cañizares-Villanueva, R.O., González Moreno, S., Voltolina, D. 1998. Effect of cadmium and zinc on respiration and photosynthesis in suspended and immobilized cultures of *Chlorella vulgaris* and *Scenedesmus acutus*. *Bulletin of Environmental Contamination and Toxicology* 60: 936–943.

Inglea, A., Seabrab, A. 2014. Nanoremediation: A new and emerging technology for the removal of toxic contaminant from environment, in: Das, S. (Ed.), *Microbial Biodegradation and Bioremediation*. Elsevier, New York, pp. 233–250.

Islam, M.S., Ahmed, M.K., Habibullah-Al-Mamun, M., Raknuzzaman, M. 2015. The concentration, source and potential human health risk of heavy metals in the commonly consumed foods in Bangladesh. *Ecotoxicology and Environmental Safety* 122: 462–469.

Jácome-Pilco, C.R., Cristiani-Urbina, E., Flores-Cotera, L.B., Velasco-García, R., Ponce-Noyola, T., Cañizares-Villanueva, R.O. 2009. Continuous Cr(VI) removal by *Scenedesmus incrassatulus* in an airlift photobioreactor. *Bioresource Technology* 100: 2388–2391.

Jadia, C., Fulekar, M.H. 2009. Phytoremediation of heavy metals : Recent techniques. *African Journal of Biotechnology* 8: 921–928.

Jamers, A., Blust, R., De Coen, W., Griffin, J.L., Jones, O.A.H. 2013. Copper toxicity in the microalga *Chlamydomonas reinhardtii*: An integrated approach. *BioMetals* 26: 731–740.

Jamers, A., Van der Ven, K., Moens, L. 2006. Effect of copper exposure on gene expression profiles in *Chlamydomonas reinhardtii* based on microarray analysis. *Aquatic Toxicology* 80: 249–260.

Jianrong, X., Qiran, T. 2009. Early stage toxicity of excess copper to photosystem II of Chlorella pyrenoidosa–OJIP chlorophyll a fluorescence analysis. *Journal of Environmental Sciences* 21: 1569–1574.

Jjemba, P. K. 2004. *Environmental Microbiology: Principles and Applications*. Science Publishers, Inc., Enfield, NH, p. 372.

Johnson, H., Stauber, J., Adams, M., Jolley, D. 2007. Copper and zinc tolerance of two tropical microalgae after copper acclimation. *Environmental Toxicology* 22: 234–244.

Jung, M. 2001. Heavy metal contamination of soils and waters in and around the Imcheon Au–Ag mine, Korea. *Applied Geochemistry* 16: 1369–1375.

Kalinowska, R., Pawlik-Skowrońska, B. 2010. Response of two terrestrial green microalgae (Chlorophyta, Trebouxiophyceae) isolated from Cu-rich and unpolluted soils to copper stress. *Environmental Pollution* 158: 2778–2785.

Kaplan, D. 2013. Absorption and adsorption of heavy metals by microalgae, in: Richmond, A., Hu, Q. (Eds.), *Handbook of Microalgal Culture: Applied Phycology and Biotechnology*, 2nd edn. John Wiley & Sons, Oxford, U.K.

Khan, A., Khan, S., Khan, M. A., Qamar, Z. 2015. The uptake and bioaccumulation of heavy metals by food plants, their effects on plants nutrients, and associated health risk : A review. *Environmental Science and Pollution Research* 22: 13772–13799.

Klerks, P.L., Weis, J.S. 1987. Genetic adaptation to heavy metals in aquatic organisms: A review. *Environmental Pollution* 45: 173–205.

Kotrba, P., Mackova, M., Fišer, J., Macek, T. 2011a. Biosorption and metal removal through living cells, in: Kotrba, P., Mackova, M., Macek, T. (Eds.), *Microbial Biosorption of Metals*. Springer, Utretch, the Netherlands, pp. 197–233.

Kotrba, P., Mackova, M., Macek, T. 2011b. Microbial biosorption of metals, in: Kotrba, P., Mackova, M., Macek, T., *Microbial Biosorption of Metals*. Springer, Utretch, the Netherlands, p. 342.

Kuroda, K., Ueda, M. 2011. Yeast Biosorption and Recycling of Metal Ions by Cell Surface Engineering Cell Surface Engineering, in: Kotrba, P., Mackova, M., Macek, T. (Eds.), *Microbial Biosorption of Metals*. Springer, Utretch, the Netherlands, pp. 235–247.

Koul, R.L. 1975. Copper and health. *Indian Journal of Pediatrics* 42: 87–91.

Koupaie, E.H., Eskicioglu, C. 2015. Health risk assessment of heavy metals through the consumption of food crops fertilized by biosolids : A probabilistic-based analysis. *Journal of Hazardous Materials* 300: 855–865.

Krupskaya, L.T., Zvereva, V.P. 2014. Bioaccumulation of heavy metals with environmental objects and assessment of health risks (the former mining enterprise Khingansky GOK as an example). *Russian Journal of General Chemistry* 84: 2542–2544.

Lelong, A., Jolley, D.F., Soudant, P., Hégaret, H. 2012. Impact of copper exposure on *Pseudo-nitzschia* spp. physiology and domoic acid production. *Aquatic Toxicology* 118–119: 37–47.

Leonard, T.L., Taylor, G.E., Gustin, M.S., Fernandez, G.C.J. 1998. Mercury and plants in contaminated soils: 1. Uptake, partitioning, and emission to the atmosphere. *Environmental Toxicology and Chemistry* 17: 2063–2071.

Liu, L., Zhang, X., Zhong, T. 2015. Pollution and health risk assessment of heavy metals in urban soil in China. *Human and Ecological Risk Assessment* 22: 424–434.

Long, D., Tang, X., Cai, K., Chen, G., Chen, L., Duan, D., Zhu, J., Chen, Y. 2013. Cr(VI) reduction by a potent novel alkaliphilic halotolerant strain *Pseudochrobactrum saccharolyticum* LY10. *Journal of Hazardous Materials* 256–257: 24–32.

Long, Y., Lei, D., Ni, J., Ren, Z., Chen, C., Xu, H. 2014. Packed bed column studies on lead(II) removal from industrial wastewater by modified Agaricus bisporus. *Bioresource Technology* 152: 457–463.

Lu, H., Li, Z., Fu, S., Méndez, A., Gascó, G., Paz-Ferreiro, J. 2014. Can biochar and phytoextractors be jointly used for cadmium remediation? *PLoS One* 9(4): e95218.

Macek, T., Mackova, M. 2011. Potential of biosorption technology, in: Kotrba, P., Mackova, M., Macek, T. (Eds.), *Microbial Biosorption of Metals*. Springer, Utretch, the Netherlands, pp. 7–17.

Mahmood, Q., Mirza, N., Shaheen, S. 2015. Phytoremediation using algae and macrophytes: I, in: Ansari, A.A., Gill, S.S., Gill, R., Lanza, G.R., Newman, L. (Eds.), *Phytoremediation Management of Environmental Contaminants*. Springer International Publishing, Switzerland. Vol. 2, pp. 265–289.

Maier, R.M., Pepper, I.L., Gerba, C.P. 2000. *Environmental Microbiology*. Academic Press, London, U.K., p. 632.

Malik, N., Biswas, A. 2012. Role of higher plants in remediation of metal contaminated sites. *Scientific Reviews and Chemical Communications* 2: 141–146.

Mani, D., Kumar, C. 2014. Biotechnological advances in bioremediation of heavy metals contaminated ecosystems: An overview with special reference to phytoremediation. *International Journal of Environmental Science* 11: 843–872.

Martínez-Roldán, A.J., Cañizares-Villanueva, R.O. 2015. Photobioreactors: Improving the biomass productivity, in: Torres-Bustillos, L. (Ed.), *Microalgae and Other Phototrophic Bacteria: Culture, Processing, Recovery and New Products*. Nova Science Publishers, Hauppauge, NY, pp. 145–170.

Martin-Jézéquel, V., Hildebrand, M., Brzezinski, M.A. 2000. Silicon metabolism in diatoms: Implications for growth. *Journal of Phycology* 36: 821–840.

Masmoudi, S., Nguyen-Deroche, N., Caruso, A., Ayadi, H., Morant-Manceau, A., Tremblin, G., Schoefs, B. 2013. Cadmium, copper, sodium and zinc effects on diatoms: From heaven to hell-A review. *Cryptogam Algol* 34: 185–225.

Massas, I., Ehaliotis, C., Gerontidis, S., Sarris, E. 2009. Elevated heavy metal concentrations in top soils of an Aegean island town (Greece): Total and available forms, origin and distribution. *Environmental Monitoring and Assessment* 151: 105–116.

Mendoza-Cózatl, D., Loza-Tavera, H., Hernández-Navarro, A., Moreno-Sánchez, R. 2005. Sulfur assimilation and glutathione metabolism under cadmium stress in yeast, protists and plants. *FEMS Microbiology Reviews* 29: 653–671.

Mendoza-Cózatl, D.G., Moreno-Sánchez, R. 2005. Cd^{2+} transport and storage in the chloroplast of Euglena gracilis. *Biochimica et Biophysica Acta* 1706: 88–97.

Mohammed, A.S., Kapri, A., Goel, R. 2011. Heavy metal pollution: Source, impact and remedies, in: Khan, M.S., Zaidi, A., Goel, R., Musarrat, J. (Eds.), *Biomanagement of Metal-Contaminated Soils.* Springer, Utretch, The Netherlands, pp. 1–28.

Monteiro, C.M., Castro, P.M., Malcata, F.X. 2012. Metal uptake by microalgae: Underlying mechanisms and practical applications. *Biotechnology Progress* 28: 299–311.

Mudhoo, A., Garg, V.K., Wang, S. 2012. Removal of heavy metals by biosorption. *Environmental Chemistry Letters* 10: 109–117.

Mulligan, C.N., Yong, R.N., Gibbs, B.F. 2001. Remediation technologies for metal-contaminated soils and groundwater: An evaluation. *Engineering Geology* 60: 193–207.

Naja, G., Volesky, B. 2011. The mechanism of metal cation and anion biosorption, in: Kotrba, P., Mackova, M., Macek, T. (Eds.), *Chemistry and Ecology.* Springer, Utretch, the Netherlands, pp. 19–58.

Nriagu, J.O., Pacyna, M.J. 1988. Quantitative assessment of worldwide contamination of air, water and soils by trace metals. *Nature* 33: 134–139.

Ochoa-Herrera, V. et al. 2011. Toxicity of copper(II) ions to microorganisms in biological wastewater treatment systems. *The Science of the Total Environment* 412–413: 380–385.

Oves, M., Khan, M.S., Zaidi, A. 2013. Biosorption of heavy metals by Bacillus thuringiensis strain OSM29 originating from industrial effluent contaminated north Indian soil. *Saudi Journal of Biological Sciences* 20: 121–129.

Páez-Osuna, F. 1996. Fuentes de metales en la zona costera marina, in: Botello, A.V., Rojas, G.J.L., Benitez, J.A., Zárate, L.D. (Eds.), *Golfo de México, Contaminación e Impacto Ambiental: Diagnóstico y tendencias.* Universidad Autónoma de Campeche, EPOMEX, Campeche, México, pp. 297–308.

Pakzad, H.R., Pasandi, M., Soleimani, M., Kamali, M. 2014. Distribution and origin of heavy metals in the sand sediments in a sector of the Oman Sea (the Sistan and Baluchestan province, Iran). *Quaternary International* 345: 138–147.

Pawlik-Skowrońska, B., Skowroński, T. 2001. Freshwater algae, in: Prasad, M.N.V. (Eds.), *Metals in the Environment.* Marcel Dekker, Inc, New York, pp. 59–94.

Pellón, A., Espinosa, M.C., Cañizares-Villanueva, R.O., Frades, J., Chacón, A., Pérez, E., Oña, A., Ramos-Alvariño, C., Mayarí, R., Escobedo, R. 2008. Use of a reactor for the removal of chromium and cadmium with immobilized Scenedesmus obliquus. *Ingeniería hidráulica en México Journal* 23: 139–150.

Peña-Castro, J.M., Martínez-Jerónimo, F., Esparza-García, F., Cañizares-Villanueva, R.O. 2004. Phenotypic plasticity in *Scenedesmus incrassatulus* (Chlorophyceae) in response to heavy metals stress. *Chemosphere* 57: 1629–1636.

Perales-Vela, H.V., González-Moreno, S., Montes-Horcasitas, C., Cañizares-Villanueva, R.O. 2007. Growth, photosynthetic and respiratory responses to sub-lethal copper concentrations in *Scenedesmus incrassatulus* (Chlorophyceae). *Chemosphere* 67: 2274–2281.

Perales-Vela, H.V., Peña-Castro, J.M., Cañizares-Villanueva, R.O. 2006. Heavy metal detoxification in eukaryotic microalgae. *Chemosphere* 64: 1–10.

Pinto, E., Sigaud-kutner, T.C.S., Leitão, M.A.S., Okamoto, O.K., Morse, D., Colepicolo P. 2003. Heavy metal-induced oxidative stress in algae. *Journal of Phycology* 39: 1008–1018.

Rahman, M.A., Rahman, M.M., Reichman, S.M., Lim, R.P., Naidu, R. 2014. Heavy metals in Australian grown and imported rice and vegetables on sale in Australia: Health hazard. *Ecotoxicology and Environmental Safety* 100: 53–60.

Rahman, M.A., Reichman, S.M., De Filippis, L., Sany, S.B.T., Hasegawa, H. 2016. Phytoremediation of toxic metals in soils and wetlands: Concepts and applications, in: Hasegawa, H., Rahman, I.M.M., Rahman, M.A. (Eds.), *Environmental Remediation Technologies for Metal-Contaminated Soils.* Springer, Tokyo, Japan, pp. 161–195.

Rai, L.C., Gaur, J.P. (Eds.). 2001. *Algal Adaptation to Environmental Stresses.* Springer, Berlin, Germany.

Rai, L.C., Gaur, P.J., Kumar, D.H. 1981. Phycology and heavy-metal. *Biological reviews of the Cambridge Philosophical Society* 56: 99–151.

Rajendran, P., Muthukrishnan, J., Gunasekaran, P. 2003. Microbes in heavy metal remediation. *Indian Journal of Experimental Biology* 41: 935–944.

Reed, R.H., Gadd, G.M. 1990. Metal tolerance in eukaryotic and prokaryotic algae, in: Shaw, J. (Ed.), *Heavy Metal Tolerance in Plants: Evolutionary Aspects.* CRC, Boca Raton, FL.

Rehman, A., Shakoori, A.R. 2001. Heavy metal resistance Chlorella spp., isolated from tannery effluents, and their role in remediation of hexavalent chromium in industrial waste water. *Bulletin of Environmental Contamination and Toxicology* 66: 542–547.

Rehman, A., Shakoori, F.R., Shakoori, A.R. 2006. Heavy metal resistant *Distigma proteus* (Euglenophyta) isolated from industrial effluents and its possible role in bioremediation of contaminated wastewaters. *World Journal of Microbiology and Biotechnology* 23: 753–758.

Romero, F.M. et al. 2008. Factores geológicos y climáticos que determinan la peligrosidad y el impacto ambiental de jales mineros. *Revista internacional de contaminación ambiental* 24: 43–54.

Sadeghi, N., Oveisi, M. R., Jannat, B., Hajimahmoodi, M., Behzad, M., Behfar, A., Saadatmand, S. 2014. The relationship between bone health and plasma zinc, copper lead and cadmium concentration in osteoporotic women. *Journal of Environmental Health Science and Engineering* 12: 125.

Sahin, C. A., Tokgöz, I. 2010. A novel solidified floating organic drop microextraction method for preconcentration and determination of copper ions by flow injection flame atomic absorption spectrometry. *Analytica Chimica Acta* 667: 83–87.

Saïdi, S.A., Azaza, M.S., Windmolders, P., van Pelt, J., El-Feki, A. 2013. Cytotoxicity evaluation and antioxidant enzyme expression related to heavy metals found in tuna by-products meal: An in vitro study in human and rat liver cell lines. *Experimental and Toxicologic Pathology : Official Journal of the Gesellschaft Für Toxikologische Pathologie* 65: 1025–1033.

Sar, P., Kazy, S., Asthana, R., Singh, S. 1999. Metal adsorption and desorption by lyophilized *Pseudomonas aeruginosa.* *International Biodeterioration and Biodegradation* 44: 101–110.

Shakoor, M., Ali, S., Farid, M. 2013. Heavy metal pollution, a global problem and its remediation by chemically enhanced phytoremediation: A review. *Journal of Biodiversity and Environmental Sciences* 3: 12–20.

Shao, J., Xie, C., Qin, J., Huo, B., Luo, Y. 2014. Copper acute toxicity tests with Schizothorax o'connori Lloyd and Schizothorax waltoni Regan: A biomonitor of heavy metal pollution in a high altitude area. *Bulletin of Environmental Contamination and Toxicology* 93: 306–310.

Shi, G. L., Lou, L. Q., Zhang, S., Xia, X. W., Cai, Q. S. 2013. Arsenic, copper, and zinc contamination in soil and wheat during coal mining, with assessment of health risks for the inhabitants of Huaibei, China. *Environmental Science and Pollution Research International* 20: 8435–8345.

Song, X., Yang, J., Lu, B., Li, B., Zeng, G. 2013. Identification and assessment of environmental burdens of Chinese copper production from a life cycle perspective. *Frontiers of Environmental Science and Engineering* 8: 580–588.

Srivastava, S., Agrawal, S. B., Mondal, M. K. 2015. A review on progress of heavy metal removal using adsorbents of microbial and plant origin. *Environmental Science and Pollution Research* 22: 15386–15415.

Srivastava, S., Goyal, P. 2010. Novel Biomaterials: Decontamination of Toxic Metals from Wastewater. Springer Science & Business Media, Berlin, p. 190.

Stanković, V., Božić, D. 2009. Heavy metal ions adsorption from mine waters by sawdust. *Chemical Industry and Chemical Engineering Quarterly* 15: 237–249.

Sun, D., Li, X., Zhang, G. 2013. Biosorption of Ag (I) from aqueous solution by *Bacillus licheniformis* strain R08. *Applied Mechanics and Materials* 295: 129–134.

Suresh Kumar, K., Dahms, H.-U., Won, E.-J., Lee, J.-S., Shin, K.-H. 2015. Microalgae—A promising tool for heavy metal remediation. *Ecotoxicology and Environmental Safety* 113: 329–352.

Thompson-Eagle, E.T., Frankenberger, W.T.. J., Karlson, U. 1989. Volatilization of Selenium by *Alternaria alternata*. *Applied and Environmental Microbiology* 55: 1406–1413.

Travieso, L., Cañizares, R.O., Borja, R., Benítez, F., Domínguez, A.R., Dupeyrón, R., Valiente, V. 1999. Heavy metal removal by microalgae. *Bulletin of Environmental Contamination and Toxicology* 62: 144–151.

Tsezos, M. 2014. Biosorption: A mechanistic approach. *Advances in Biochemical Engineering/Biotechnology* 141: 173–209.

Urík, M., Černanský, S., Ševc, J., Šimonovičová, A., Littera, P. 2007. Biovolatilization of arsenic by different fungal strains. *Water Air Soil Pollutants* 186: 337–342.

Verma, D., Hasan, S. 2014. Enhanced biosorptive remediation of hexavalent chromium using chemotailored biomass of a novel soil isolate *Bacillus aryabhattai* ITBHU02: Process variables through artificial neural network linked genetic algorithm. *Industrial and Engineering Chemistry Research* 53: 3669–3681.

Vieira, R.H.S.F., Volesky, B. 2000. Biosorption: A solution to pollution. *International Microbiology* 3: 17–24.

Vijayaraghavan, K., Yun, Y.S. 2008. Bacterial biosorbents and biosorption. *Biotechnology Advances* 26: 266–291.

Viraraghavan, T., Srinivasan, A. 2011. Fungal biosorption and biosorbents, in: Kotrba, P., Mackova, M., Macek, T. (Eds.), *Microbial Biosorption of Metals*. Springer, Utretch, the Netherlands, pp. 143–158.

von Canstein, H., Kelly, S., Li, Y., Wagner-Dobler, I. 2002. Species diversity improves the efficiency of mercury-reducing biofilms under changing environmental conditions. *Applied and Environmental Microbiology* 68: 2829–2837.

Wagner-Döbler, I. 2003. Pilot plant for bioremediation of mercury-containing industrial wastewater. *Applied Microbiology and Biotechnology* 62: 124–133.

Wang, J., Chen, C. 2009. Biosorbents for heavy metals removal and their future. *Biotechnology Advances* 27: 195–226.

Wang, L.K. et al. 2004. Chemical precipitation, in: Wang, L.K., Hung, Y.T., Shammas, N.K. (Eds.), *Physicochemical Treatment Processes*. Humana Press, Clifton, NJ, Vol. 3, pp. 141–198.

Wang, S., Jia, Y., Wang, S., Wang, X., Wang, H., Zhao, Z., Liu, B. 2010. Fractionation of heavy metals in shallow marine sediments from Jinzhou Bay, China. *Journal of Environmental Sciences* 22: 23–31.

Wu, Z., Bañuelos, G., Yin, X., Lin, Z., Terry, N. 2015. Phytoremediation of the metalloid selenium in soil and water, in: Ansari, A.A., Gill, S.S., Gill, R., Lanza, G.R., Newman, L. (Eds.), *Phytoremediation Management of Environmental Contaminants*. Springer International Publishing, London, U.K., Vol. 2, pp. 171–175.

Yun, Y., Vijayaraghavan, K., Won, S. 2011. Bacterial biosorption and biosorbents, in: Kotrba, P., Mackova, M., Macek, T. (Eds.), *Microbial Biosorption of Metals*. Springer, Utretch, the Netherlands, pp. 121–141.

Zabochnicka-Świątek, M., Krzywonos, M. 2014. Potentials of biosorption and bioaccumulation processes for heavy metal removal. *Polish Journal of Environmental Studies* 23: 551–561.

Zaidi, A., Oves, M., Ahmad E., Khan, M.S. 2011. Importance of free living fungi in heavy metal remediation, in: Khan, M.S., Zaidi, A., Goel, R., Musarrat, J. (Eds.), *Biomanagement of Metal Contaminated Soil*. Springer, Utretch, the Netherlands, pp. 479–494.

Zayed, A., Pilon-Smits, E., de Souza, M., Lin, Z.-Q., Terry, N. 1999. Remediation of selenium-polluted soils and waters by phytovolatilization, in: Terry, N., Bañuelos Lewis, G.S. (Eds.), *Phytoremediation of Metal Contaminated Soil and Water*. Lewis Publishers, New York, pp. 61–83.

37 Toxicity of Copper and Remediation Approaches

Fahmi A. Abu Al-Rub, Mohammad M. Fares, and Munther Kandah

CONTENTS

ABSTRACT

Copper is one of the heavy metals that are effluent from many industrial activities, such as plating baths, pulp and paper, fertilizers, mining, petroleum industries, and metal cleaning. Copper concentration in these effluents can reach up to 1000 mg/L. Moreover, copper has been found as a contaminant in some food products including liver, chocolate, mushrooms, shellfish, and nut. The presence of copper even at low concentration can have serious ecological and health hazards as a result of its solubility and mobility. In humans, copper can cause serious problems such as anemia, stomach intestinal distress, and kidney damage. Due to these serious problems, the maximum concentration of copper in industrial effluents and in wastewater has been subjected to stringent standards. Enforcement of these stringent standards was the driving force for active research on copper removal. Different techniques have been formulated for the removal of copper from wastewater. These methods include coagulation, solvent extraction, adsorption, chemical precipitation, reverse osmosis, ion exchange, membrane processes, and biosorption. In this chapter, the environmental impact of copper on humans, animals, and plants is presented. The different conventional remediation methods to remove copper from wastewater are introduced. Bioremediation of copper is the focus of this chapter. The different parameters affecting copper bioremediation, such as temperature, pH, contact time, and copper concentration, are discussed.

37.1 INTRODUCTION

Copper metal is soft, malleable, and ductile with very high electrical and thermal conductivity. It is implemented in various industries and applications. It covers a wide variety of different disciplines including but not limited to telecommunications, architecture, automotive machines, building construction, power generation and transmission, electronic product manufacturing, heating and cooling systems, and consumer and general products. The significance of copper in the industrial sector is justified through its enormous applications, such as for high electrical and thermal conductivity, firm durability, and antimicrobial characteristics. Such features play a central role in the application of copper in a number of sectors, including renewable energy, green buildings, and high efficient sustainable materials.

Copper is a first raw transition element with atomic number 29, symbol Cu, and electron configuration of $[Ar]\ 4s^1 3d^{10}$. Its standard atomic weight is 63.546 as reported by the Commission on Isotopic Abundances and Atomic Weights (CIAAW), International Union of Pure and Applied Chemistry (CIAAW, 2013). In periodic table, it is located in period 4 and group 11, 1B. It is solid, has a reddish-orange metallic luster, and has 29 isotopes. Cu forms just about 69% of the naturally occurring copper with a spin of 3/2 (Audi et al., 2003).

37.2 PHYSICAL AND CHEMICAL PROPERTIES

37.2.1 Physical Properties

Most common oxidation states of copper are +1 and +2 named cuprous (Cu^+) and cupric (Cu^{2+}) ions, respectively. The electronegativity of copper is moderate which equals 1.9 according to Pauling scale; the first and second ionization energies are 745.5 and 1957.9 kJ/mol, respectively. The electron affinity is 118.4 kJ/mol. Empirical atomic radius is 128 pm, and crystal structure is a face-centered cube. Copper has a melting point of 1083°C and a boiling point of 2562°C. Optical, electrical, magnetic, and mechanical parameters of copper are listed in Tables 37.1 and 37.2, respectively.

37.2.2 Chemical Properties

Copper reacts with numerous reactants and forms different compounds. It slowly reacts with atmospheric oxygen and forms copper oxide layer that protects the copper beneath it. Moreover, copper can form many alloys that possess many practical applications; for example, copper–zinc alloy is widely known as brass, copper–tin alloys as bronze, copper–aluminum as aluminum bronze, and copper–nickel as cupronickel used in low-dimensional coins. It is a significant constituent in gold and plays a major role in the jewelry solders industry.

Copper ion salts (Cu^{2+}) have different water solubilities. Although chloride, nitrate, and sulfate salts are highly water soluble, oxide, carbonate, and cyanide salts are water insoluble (Weast, 1969).

Copper ion undergoes precipitation reaction upon treatment with aqueous sodium hydroxide solution to form a light blue shiny copper hydroxide ($Cu(OH)_2$):

$$Cu^{2+} + 2OH^- \rightarrow Cu(OH)_2(s)$$

Since the solubility product (K_{sp}) of $Cu(OH)_2 = 1.6 \times 10^{-19}$ is low, the precipitation of $Cu(OH)_2$ in aqueous solution at pH = 7.0 can start as low as 16 M of copper ion (Cu^{2+}) concentration, which leads to the removal of toxic Cu^{2+} soluble ions from water streams, rivers, lakes, and seas through precipitation process. This is extremely significant for effective microorganisms that suffer from severe damage of their DNA and consequent death and genocide by toxic heavy metals (Zhou et al., 2008).

37.3 SOURCES OF COPPER IN WASTEWATER (INDUSTRY)

Numerous sources of copper contamination in wastewater do exist due to the large implementations of copper metal and copper ions in various applications in domestic uses, commercial businesses, and industrial operations. Unfortunately, many industries use water resources for washing purposes, which ends up with acute and chronic accumulation of copper in sewage water. Moreover, many factories are located at riversides, lakes, or seas, and hence, their copper wastes are thrown in water resources, causing severe damage not only to water validity as drinking water but also to the aquatic marine microorganisms and environments. The copper sources include but not limited to industrial printed circuit board manufacturing, copper polishing, wire drawing, paint manufacturing, electronics plating, copper ores during refinement process, domestic sources, vehicle brake pads, marine antifouling coatings, copper pesticides, copper soldering, copper air emissions, soil erosion, corrosion in copper pipes and cooling water systems, and plumbing materials.

37.3.1 Environmental Impact of Copper on Humans

Copper is a key metal in building strong tissues and producing energy in human cells; it is available in numerous meat and vegetable food sources. Adult daily intake dose of copper for humans is 1.3 mg/day as recommended by the World Health Organization (WHO, 1996). The top 10 world's healthiest foods rich in copper and their Dietary Reference Intake (in daily value (%)) are listed in Table 37.3.

TABLE 37.1
Optical, Electrical, and Magnetic Properties of Copper Metal

Reflectivity	Magnetic Ordering	Thermal Conductivity (W/m/K)	Thermal Expansion (K⁻¹)	Electrical Resistivity (Ω m)	Density (kg/m³)
90%	Diamagnetic	401	16.5×10^{-6}	1.72×10^{-8}	8920

TABLE 37.2
Mechanical Properties of Copper Metal

Young's Modulus (GPa)	Shear Modulus (GPa)	Bulk Modulus (GPa)	Poisson's Ratio	Vickers Hardness (MPa)	Brinell Hardness (MPa)
130	48	140	0.34	369	874

TABLE 37.3

Top Ten World's Healthiest Foods Rich in Copper

Food	Dietary Reference Intake (in Daily Value (%))
Sesame seeds	163
Cashews	98
Soybeans	78
Mushrooms, shiitake	72
Sunflower seeds	70
Tempeh	68
Garbanzo beans	64
Lentils	56
Walnuts	53
Lima beans	49

It is well known that copper is essential to the metabolic processes and organ functionality of human body. It is a critical functional component of several essential enzymes known as cuproenzymes (Prohaska, 2011); it plays an essential role in maintaining the integrity of connective tissues of the heart, blood vessels, and bone formation. Brain function and nervous system are also essentially catalyzed by cuproenzymes (D'Ambrosi and Rossi, 2015; Møller, 2015). Moreover, multicopper oxidase enzymes catalyze ferrous iron (Fe^{2+}) to ferric iron (Fe^{3+}) in human body, which helps in iron metabolism (Vashchenko and MacGillivray, 2013).

A maximum tolerable intake of copper in humans is 10 mg/day, a value based on protection from liver damage as the critical adverse effect (Institute of Medicine, 2001). Above 10 mg/day intake of copper has an adverse accumulation and a toxicological influence on human health. Some features of the adverse effects of high accumulation and acute and chronic exposure of copper on human health can be described as neurological disorder that may lead to amyotrophic lateral sclerosis, Alzheimer and Meknes diseases (Ahuja et al., 2015), neurologic symptoms of Parkinson disease (Skjorringe et al., 2012), mutations in the ATP7B gene that result in Wilson's disease (Loudianos and Gitlin, 2000; Huster, 2010), Indian childhood cirrhosis (Hahn et al., 1995), endemic Tyrolean copper toxicosis (Scheinberg and Sternlieb, 1996), and idiopathic copper toxicosis diseases (Scheinberg and Sternlieb, 1996).

In addition, cancer disease has been strongly correlated with copper accumulation in human body; copper ions can activate proangiogenic factors and enable some nonangiogenic biomolecules such as heparin and the tripeptide glycyl-L-histidyl-L-lysine to become proangiogenic (Saghiri et al., 2015), and trace element copper has been found to promote tumor growth (Brewer, 2001; Lowndes and Harris, 2005). On the contrary, copper can chelate different ligands and form copper complex in human body; for example, the copper complex of salicylaldehyde benzoyl hydrazone derivatives showed increased and potential growth inhibition in several cancer cell lines, when compared with the metal-free salicylaldehyde benzoyl hydrazone derivatives (Johnson et al., 1982; Ainscough et al., 1999).

37.3.2 Environmental Impact of Copper on Animals

Copper metal is toxic to many living organisms due to its ability to generate free radical reactive oxygen species (ROS) and reactive nitrogen species (RNS) (Valko et al., 2007). The ROS and RNS are also products of normal cellular metabolism, and the toxicity of the free radicals is due to its ability to cause potential biological damage termed as oxidative and nitrosative stress, respectively (Kovacic and Jacintho, 2001). Such damage occurs as a result of enzymatic and nonenzymatic antioxidant deficiency and lack of free radical scavengers that leads to overproduction of ROS and RNS. Therefore, not only excess ROS can damage cellular DNA, lipids, or proteins and inhibit their normal function, but it is also responsible for a number of human diseases as well as the aging process (Valko et al., 2007). Copper plays a significant role in the production of ROS, which is strongly correlated with Alzheimer's disease (Butterfield et al., 2002; Dikalov et al., 2004).

Copper accumulation has drastic health consequences on animals as well as on humans; it featured chronic and potential diseases such as inherited copper-associated hepatitis in dogs, which resembles Wilson's disease in humans (Fieten et al., 2012), and Bedlington terrier copper toxicosis, which is caused by a deletion in the COMMD1 gene (Van de Sluis et al., 2002). In sheep, chronic copper toxicity occurs when the sheep's liver (i.e., copper storage) reaches saturation, which results in a large amount of copper release into the bloodstream circulation, causing liver damage, destruction of red blood cells, further tissue damage, and jaundice. Usually, sheep exposed to copper poisoning do not exhibit clinical signs before death (St George-Grambauer and Rac, 1962).

Copper poisoning on marine environment is extremely significant. Many academic and institutional research foundations indicated acute, chronic, and lethal copper poisoning in marine environment and marine creatures; for instance, acute waterborne copper is known to cause toxicity of fish by disrupting sodium homeostasis. Moreover, chronic exposure to elevated concentrations of copper results in copper lethal accumulation in organ tissues of fish (liver, kidney, and gill), which ultimately leads to fish death (Grosell and Wood, 2002).

37.3.3 Environmental Impact of Copper on Plants

Copper poisoning has an acute environmental impact on plants, quite clear symptoms of copper poisoning in plants observed through significant growth reduction. The most common symptom is interveinal foliar chlorosis, which takes place through the formation of cream or white spots or lesions, increasing exposure of copper leads to necrotic leaf tips and margins, and acute copper toxicity leads to wilted leaves, which eventually become necrotic (Taylor and Foy, 1985; Mitchell et al., 1988; Yau et al., 1991). Although copper is an essential micronutrient for plants, its accumulation is poisonous and has an intense effect on root growth and form. It tends to accumulate in the root cells making genotoxic effects and consequent root growth inhibition (Yildiz et al., 2009).

Marine algae, which are also called seaweeds, are a food source for marine animals. It is considered as primary producers in aquatic ecological systems that suffer from chronic toxicity of copper. Usually, copper adsorbed on the surface microalgae sites as described in the following equation:

$$Cu(II)L_{water} + X_{cell} \rightarrow Cu(II)X_{cell} + L_{water} \qquad (37.1)$$

where

L is the ligand

X_{cell} are the sites on the cell membrane (Sunda, 1989)

Copper is adsorbed nonspecifically and rapidly on the surface of carboxylic, sulfhydryl, and phosphate group sites (Crist et al., 1990). However, high concentrations of copper are strongly correlated with phytotoxicity of cells for which it generates an overproduction of ROS and moreover plays a dominant role in the toxicity of algal cells. In addition, heavy metals can directly inhibit cell division and cause chloroplast damage, resulting in growth inhibition, decay, and plant death (Chen et al., 2012).

37.4 REMEDIATION OF COPPER

37.4.1 CONVENTIONAL REMEDIATION OF COPPER

Researchers have used several conventional techniques to remove copper from contaminated wastewaters such as chemical precipitation, coagulation (Amuda et al., 2006), electrochemical treatment (Feng et al., 2003), biological treatment (Sa'idi, 2010), membrane separation (Rivas and Palencia, 2011), solvent extraction (Lin and Juang, 2002), ion exchange (Abdel-Aziz et al., 2013), and adsorption on different materials such as seaweeds (Jalali et al., 2002), alginate (Park et al., 2007), clays, and activated sludge biomass (Gulnaz et al., 2005) and natural materials such as fungi and dead bacteria (Arica et al., 2001). In this chapter, the emphasis is on bioremediation processes.

37.4.2 BIOREMEDIATION

37.4.2.1 What Is Bioremediation?

Bioremediation is an alternative technique for the conventional waste removal methods, in which biological organisms are used to destroy or reduce the concentration of hazardous wastes of contaminated mediums. The definition of bioremediation as per Environmental Protection Act, 1986, is "treatment that uses naturally occurring organisms to breakdown hazardous substances into less toxic or no toxic substances." Bioremediation technique is applied in cleaning up water, soil, sludge, and waste streams. The microorganisms used in bioremediation are called bioremediators. To get the best results from bioremediation, the right microbe should be used in the right place with the right environmental factors. Bioremediation can be done on site and coupled with other physical or chemical treatment methods as shown in Table 37.4 (Boopathy, 2000).

TABLE 37.4

Major Factors Affecting Bioremediation

Factors	Role in Bioremediation
Microbial	Growth until critical biomass is reached
	Mutation and horizontal gene transfer
	Enzyme induction
	Enrichment of the capable microbial populations
	Production of toxic metabolites
Environmental	Depletion of preferential substrates
	Lack of nutrients
	Inhibitory environmental conditions
Substrate	Too low concentration of contaminants
	Chemical structure of contaminants
	Toxicity of contaminants
	Solubility of contaminants
Biological aerobic versus anaerobic process	Oxidation–reduction potential
	Availability of electron acceptors
	Microbial population present in the site
Growth substrate versus cometabolism	Type of contaminants
	Concentration
	Alternate carbon source present
	Microbial interaction (competition, succession, and predation)
Physicochemical bioavailability of pollutants	Equilibrium sorption
	Irreversible sorption
	Incorporation into humic matters
Mass transfer limitations	Oxygen diffusion and solubility
	Diffusion of nutrients
	Solubility/miscibility in/with water

Bioremediation can offer several advantages compared to the conventional methods due to its capability of being applied in polluted areas that are inaccessible without excavation such as groundwater. It also has the ability to introduce the appropriate electron acceptor or donor amendment that can reduce contaminant concentrations after a long time allowing for acclimation. This technique is more efficient and cost-effective than any other technique because it needs no excavation, incineration, regeneration, or pumping energy. In addition to the advantages mentioned earlier, bioremediation can be examined indirectly by measuring the oxidation–reduction potential in soil or groundwater, pH, temperature, oxygen content, electron acceptor/donor concentrations, and concentration of broken-down products. Bioremediation is usually applied in cleaning up waters, soils, and sludge and waste streams.

37.4.2.2 Factors Affecting Bioremediation

Microorganisms may be active in certain environments and not active in others. This means that their efficiency for biodegradation depends on several factors such as moisture content, nutrient availability, dissolved oxygen, temperature, and pH of the contaminated medium.

37.4.2.2.1 Moisture Content

Moisture of contaminated medium is very important and necessary for cell growth and function. The availability of water in the contaminated medium affects the circulation of water and soluble nutrients into and out of microorganisms' cell. On the other hand, excess of water may reduce the amount of available oxygen for aerobic respiration as in the case of saturated soil environment.

37.4.2.2.2 Nutrient Availability

Microbial activity and cell growth need some inorganic nutrients such as nitrogen and phosphorus. These nutrients may increase cell growth rate and decrease the microbial lag phase. An excess of these nutrients may cause microbial inhibition. Copper and other heavy metals act also as nutrients in the active biosorption for specific types of bacteria.

37.4.2.2.3 Temperature

In general, the enzymatic reactions in the cell approximately double for each $10°C$ rise in temperature. This means that temperature affects the rate of biodegradation because it affects the enzymatic reaction rate within microorganisms. Microorganisms as any other creature can resist certain level of temperature. For example, mesophiles are types of bacteria that degrade petroleum hydrocarbons and have an optimum temperature range from $25°C$ to $45°C$, while a thermophilic bacterium that is normally found in hot springs can degrade hydrocarbons in a temperature up to $60°C$.

37.4.2.2.4 pH

Environmental pH range depends on the type of microbe because different microorganisms need specific pH range to survive. Therefore, controlling the pH of the contaminated

environment before and during bioremediation process is very important because the ionization of these functional groups is usually affected by the pH of the contaminated medium and this is also affected by the solubility of the heavy metals. In general, bacterial cell surface is negatively charged due to the presence of different functional groups on the fungal cell wall such as carboxylate and phosphate groups and attracts positive heavy metals such as copper ions. On the other hand, at low pH (acidic medium), the adsorption capacity decreases because of the increase in the positive charge density on the cell surface.

37.4.2.2.5 Heavy Metal Concentration

Usually, the increase in initial heavy metal concentration increases the biosorption capacity because of the large amount of available ions for biosorption either by the adsorption on the cell surface (passive biosorption) or by microorganism degradation (active biosorption).

37.4.2.2.6 Energy Sources

The ability and availability of reduced organic materials to serve as an energy source for the microorganisms are considered as one important factor that may affect the activity of bacteria significantly. For example, higher oxidation states corresponding to lower energy yields provide less energetic incentive for microorganism degradation.

37.4.2.2.7 Contact time

Increasing the contact time between bacteria and copper increases the population of bacteria in the culture that may increase the activity of bacteria significantly because the metabolic uptakes of copper increase with time. Table 37.2 summarizes the major factors affecting bioremediation as explained in Boopathy (2000).

37.5 BIOREMEDIATION OF COPPER

Bioremediation of copper occurs by two methods: using either live microorganisms, which is called active biosorption, or dead microorganisms, which is called passive biosorption. In active biosorption, the removal of copper occurs metabolically, but in passive biosorption, it is adsorbed on the cell surface. In biosorption, both living and dead microbial cells are able to detoxify copper using physicochemical and biological methods (Choi et al., 2009). Activated sludge biosorbent that is composed of both live and dead microbial fractions is a good example because it has carboxylic acid, carboxyl, and amine groups (Gulnaz et al., 2006). Copper can be bioremediated using different types of microorganisms such as copper-resistant bacteria *Stenotrophomonas* sp. PD2 that can be isolated from copper-polluted areas. These isolated bacteria are Gram-negative, coccus-shaped bacteria. Another bacterial product called exopolysaccharide is secreted by a mesophilic bacterium, namely, *Wangia profunda* SM-A87, isolated from deep-sea sediment wastewater (Zhou et al., 2009). These bacteria are also produced by microorganisms originating from soil or wastewater sludge (Li et al., 2008). Different genera of fungi such as *Penicillium*, *Aspergillus*, and *Rhizopus* have been studied extensively as potential microbial agents for copper bioremediation (Volesky and Holan, 1995; Huang and Huang, 1996). Other microorganisms that are used in bioremediation of copper are *Acinetobacter*, *Actinobacter*, *Alcaligenes*, *Arthrobacter*, *Bacillins*, *Beijerinckia*, *Flavobacterium*, *Methylosinus*, *Mycobacterium*, *Myxococcus*, *Nitrosomonas*, *Nocardia*, *Phanerochaete*, *Pseudomonas*, *Serratia*, *Trametes*, and *Xanthobacter* (Singh, 2014).

37.5.1 Mechanisms of Bioremediation of Copper

Bioremediation mechanisms are based on either enhancing the biodegradation processes that occur in nature by enhancing the growth of whatever pollution-eating microbes might already be living at the contaminated site or adding specialized microbes to the pollutants to be degraded or broken down.

Bioremediation of heavy metals involves different mechanisms such as biosorption, biodegradation, biotransformation, metal–microbe interaction, bioaccumulation, and biomineralization. In the biosorption process, microbes adsorb heavy metals at the cell wall due to its high ability to form complex heavy metal by proton exchange and microprecipitation of metals. In biodegradation, organic compounds are broken down, while in biotransformation, inorganic compounds are transferred into environmental friendly compounds by living organisms through reactions that take place as a part of their metabolic processes. In the metal–microbe interaction process, microorganisms dissolve metals and reduce or oxidize them. In this process, microorganisms cause the metal to lose electrons by acting as an oxidizing agent. The removed electrons are accepted by electron acceptors such as nitrate, sulfate, and ferric oxides. In aerobic conditions, the electron acceptor is oxygen, while in anaerobic conditions, microbes oxidize organic pollutants by reducing electron acceptors. In bioaccumulation, both active and passive biosorptions are present in the same process. In the mineralization process, organic pollutants are converted by microorganisms into end products such as CO_2 and H_2O or to metabolic intermediates that are used as primary substrates for cell growth.

In order to get the highest benefit of bioremediation, the right microbe (bacteria or fungi) should be chosen to have the physiological and metabolic abilities to degrade the pollutants. In bioremediation, the pollutant is treated as a nutrient to the bacteria; therefore, there is no need for any regeneration process to recover the pollutant again. This is considered as one big advantage because in the other conventional techniques, pollutants are transferred from one place (wastewater) to another (adsorbent). Bioremediation mechanisms are based on several chemical and physical processes such as adsorption, ion exchange, and covalent bonding in addition to the physiological and metabolic degradation. These processes occur with the adsorptive sites of the microorganisms including carboxyl, hydroxyl, sulfhydryl, amino, and phosphate groups (Frurest and Volesky, 1997).

Copper bioremediation cannot be achieved by using only microorganisms but by using synergism and cometabolism action because microorganisms are able to metabolize and mineralize organic and inorganic pollutants. Microbial cell walls are composed of polysaccharides, lipids, and proteins that provide different negatively charged functional groups such as carboxylate, hydroxyl, amino, and phosphate that bind positively charged heavy metals such as copper ions (Scott and Karanjkar, 1992). In biosorption, microbes require an addition of nutrients for their active uptake of copper, and this increases the chemical or biological oxygen demand in the waste.

37.6 CONCLUSION

Anthropogenic activities and industrial processing play the most dominant role in the accumulation of harmful and poisoning copper in land, sea, and air. The acute and chronic exposure of copper causes potential diseases for humans, animal poisoning, and plant growth inhibition. Therefore, many state and federal restrictions must be put in action to limit and control this nonstop tsunami disaster of copper flux and invasion into our human, animal, and plant environments. This flux, if unstopped and/or controlled, will have growing evil consequences of poisoning, epidemics, and unlimited harm to the planet. Although many conventional treatment methods can be used to remove copper from wastewater, bioremediation provides an alternative technique for the conventional waste removal methods, in which biological organisms are used to reduce the concentration of hazardous wastes of contaminated mediums.

REFERENCES

Abdel-Aziz, M.H., Nirdosh, I., Sedahmed, G.H. 2013. Ion-exchange-assisted electrochemical removal of heavy metals from dilute solutions in a stirred-tank electrochemical reactor: A mass-transfer study. *Industrial & Engineering Chemistry Research* 52: 11655–11662.

Ahuja, A., Dev, K., Tanwar, R.S., Selwal, K.K., Tyagi, P.K. 2015. Copper mediated neurological disorder: Visions into amyotrophic lateral sclerosis, Alzheimer and Menkes disease. *Journal of Trace Elements in Medicine and Biology* 29: 11–23.

Ainscough, E.W., Brodie, A.M., Denny, W.A., Finlay, G.J., Gothe, S.A., Ranford, J.D. 1999. Cytotoxicity of salicylaldehyde benzoylhydrazone analogs and their transition metal complexes: Quantitative structure–activity relationships. *Journal of Inorganic Biochemistry* 77: 125–133.

Amuda, O.S., Amoo, I.A., Ipinmoroti, K.O., Ajayi, O.O. 2006. Coagulation/flocculation process in the removal of trace metals present in industrial wastewater. *Journal of Applied Sciences and Environmental Management* 10: 159–162.

Arıca, M.Y., Kaçar, Y., Genç, Ö. 2001. Entrapment of white-rot fungus *Trametes versicolor* in Ca-alginate beads: Preparation and biosorption kinetic analysis for cadmium removal from an aqueous solution. *Bioresource Technology* 80: 121–129.

Audi, G., Bersillon, O., Blachot, J., Wapstra, A.H. 2003. The NUBASE evaluation of nuclear and decay properties. *Nuclear Physics A* 729: 3–128.

Boopathy, R. 2000. Factors limiting bioremediation technologies. *Bioresource Technology* 74: 63–67.

Brewer, G.J. 2001. Copper control as an antiangiogenic anticancer therapy: Lessons from treating Wilson's disease. *Experimental Biology and Medicine* 226: 665–673.

Butterfield, D.A., Castegna, A., Lauderback, C.M., Drake, J. 2002. Evidence that amyloid beta-peptide-induced lipid peroxidation and its sequelae in Alzheimer's disease brain contribute to neuronal death. *Neurobiology of Aging* 23: 655–664.

Chen, H., Chen, J., Guo, Y., Wen, Y., Liu, J., Liu, W. 2012. Evaluation of the role of the glutathione redox cycle in Cu (II) toxicity to green algae by a chiral perturbation approach. *Aquatic Toxicology* 120: 19–26.

Choi, J., Lee, J.Y., Yang, J.S. 2009. Biosorption of heavy metals and uranium by starfish and *Pseudomonas putida*. *Journal of Hazardous Materials* 161: 157–162.

Crist, R.H., Martin, J.R., Guptill, P.W., Eslinger, J.M., Crist, D.R. 1990. Interaction of metals and protons with algae. 2. Ion exchange in adsorption and metal displacement by protons. *Environmental Science & Technology* 24: 337–342.

D'Ambrosi, N., Rossi, L. 2015. Copper at synapse: Release, binding and modulation of neurotransmission. *Neurochemistry International* 90: 36–45.

Dikalov, S.I., Vitek, M.P., Mason, R.P. 2004. Cupric–amyloid β peptide complex stimulates oxidation of ascorbate and generation of hydroxyl radical. *Free Radical Biology and Medicine* 36: 340–347.

Feng, C., Sugiura, N., Shimada, S., Maekawa, T. 2003. Development of a high performance electrochemical wastewater treatment system. *Journal of Hazardous Materials* 103: 65–78.

Fieten, H., Leegwater, P.A., Watson, A.L., Rothuizen, J. 2012. Canine models of copper toxicosis for understanding mammalian copper metabolism. *Mammalian Genome* 23: 62–75.

Fourest, E., Volesky, B. 1997. Alginate properties and heavy metal biosorption by marine algae. *Applied Biochemistry and Biotechnology* 67: 215–226.

Grosell, M., Wood, C.M. 2002. Copper uptake across rainbow trout gills mechanisms of apical entry. *Journal of Experimental Biology* 205: 1179–1188.

Gulnaz, O., Kaya, A., Dincer, S. 2006. The reuse of dried activated sludge for adsorption of reactive dye. *Journal of Hazardous Materials* 134: 190–196.

Gulnaz, O., Saygideger, S., Kusvuran, E. 2005. Study of Cu (II) biosorption by dried activated sludge: Effect of physico-chemical environment and kinetics study. *Journal of Hazardous Materials* 120: 193–200.

Hahn, S.H., Tanner, M.S., Danke, D.M., Gahl, W.A. 1995. Normal metallothionein synthesis in fibroblasts obtained from children with Indian childhood cirrhosis or copper-associated childhood cirrhosis. *Biochemical and Molecular Medicine* 54: 142–145.

Huang, C., Huang, C.P. 1996. Application of *Aspergillus oryzae* and *Rhizopus oryzae* for Cu(II) removal. *Water Research* 30: 1985–1990.

Huster, D. 2010. Wilson disease. *Best Practice & Research: Clinical Gastroenterology* 24: 531–539.

Institute of Medicine (IMPM). *Dietary Reference Intakes for Vitamin A, Vitamin K, Arsenic, Boron, Chromium, Copper, Iodine, Iron, Manganese, Molybdenum, Nickel, Silicon, Vanadium, and Zinc.* Washington, DC: National Academies Press, 2001.

Jalali, R., Ghafourian, H., Asef, Y., Davarpanah, S.J., Sepehr, S. 2002. Removal and recovery of lead using nonliving biomass of marine algae. *Journal of Hazardous Materials* 92: 253–262.

Johnson, D.K., Murphy, T.B., Rose, N.J., Goodwin, W.H., Pickart, L. 1982. Cytotoxic chelators and chelates 1. Inhibition of DNA synthesis in cultured rodent and human cells by aroyl-hydrazones and by a copper (II) complex of salicylaldehyde benzoyl hydrazone. *Inorganica Chimica Acta* 67: 159–165.

Kovacic, P., Jacintho, J.D. 2001. Mechanisms of carcinogenesis focus on oxidative stress and electron transfer. *Current Medicinal Chemistry* 8: 773–796.

Li, W.W., Zhou, W.Z., Zhang, Y.Z., Wang, J., Zhu, X.B. 2008. Flocculation behavior and mechanism of an exopolysaccharide from the deep-sea psychrophilic bacterium *Pseudoalteromonas* sp. SM9913. *Bioresource Technology* 99: 6893–6899.

Loudianos, G., Gitlin, J.D. 2000. Wilson's disease. *Seminar in Liver Disease* 20: 353–364.

Lowndes, S.A., Harris, A.L. 2005. The role of copper in tumour angiogenesis. *Journal of Mammary Gland Biology and Neoplasia* 10: 299–310.

Mitchell, R.L., Burchett, M.D., Pulkownik, A., McCluskey, L. 1988. Effects of environmentally hazardous chemicals on the emergence and early growth of selected Australian plants. *Plant and Soil* 112: 195–199.

Møller, L.B. 2015. Small amounts of functional ATP7A protein permit mild phenotype. *Journal of Trace Elements in Medicine and Biology* 31: 173–177.

Park, H.G., Kim, T.W., Chae, M.Y., Yoo, I.K. 2007. Activated carbon-containing alginate adsorbent for the simultaneous removal of heavy metals and toxic organics. *Process Biochemistry* 42: 1371–1377.

Prohaska, J.R. 2011. Impact of copper limitation on expression and function of multicopper oxidases (ferroxidases). *Advances in Nutrition: An International Review Journal* 2: 89–95.

Rivas, B.L., Palencia, M. 2011. Removal-concentration of pollutant metal-ions by water-soluble polymers in conjunction with double emulsion systems: A new hybrid method of membrane-based separation. *Separation and Purification Technology* 81: 435–443.

Saghiri, M.A., Asatourian, A., Orangi, J., Sorenson, C.M., Sheibani, N. 2015. Functional role of inorganic trace elements in angiogenesis—Part II: Cr, Si, Zn, Cu, and S. *Critical Reviews in Oncology/Hematology* 96: 143–155.

Sa'idi, M. 2010. Experimental studies on effect of Heavy metals presence in industrial wastewater on biological treatment. *International Journal of Environmental Sciences* 1: 666.

Scheinberg, I.H., Sternlieb, I. 1996. Wilson disease and idiopathic copper toxicosis. *The American Journal of Clinical Nutrition* 63: 842S–845S.

Scott, J.A., Karanjkar, A.M. 1992. Repeated cadmium biosorption by regenerated *Enterobacter aerogenes* biofilm attached to activated carbon. *Biotechnology Letters* 14: 737–740.

Singh, R. 2014. Microorganism as a tool of bioremediation technology for cleaning environment: A review. *Proceedings of the International Academy of Ecology and Environmental Sciences* 4: 1.

Skjørringe, T., Møller, L.B., Moos, T. 2012. Impairment of interrelated iron-and copper homeostatic mechanisms in brain contributes to the pathogenesis of neurodegenerative disorders. *Frontiers in Pharmacology* 3: 169.

St George-Grambauer, T.D., Rac, R. 1962. Hepatogenous chronic copper poisoning in sheep in South Australia due to the consumption of *Echium plantagineum* L. (*Salvation jane*). *Australian Veterinary Journal* 38: 288–293.

Standard Atomic Weights. 2013. Commission on isotopic abundances and atomic weights, http://www.ciaaw.org/atomic-weights.htm. Accessed on September 13, 2015.

Sunda, W.G. 1989. Trace metal interactions with marine phytoplankton. *Biological Oceanography* 6: 411–442.

Tanner, M.S. 1980. Copper and Indian childhood cirrhosis. *Indian Journal of Pediatrics* 47: 467–470.

Taylor, G.J., Foy, C.D. 1985. Differential uptake and toxicity of ionic and chelated copper in *Triticum aestivum*. *Canadian Journal of Botany* 63: 1271–1275.

Valko, M., Leibfritz, D., Moncol, J., Cronin, M.T., Mazur, M., Telser, J. 2007. Free radicals and antioxidants in normal physiological functions and human disease. *The International Journal of Biochemistry & Cell Biology* 39: 44–84.

Van de Sluis, B., Rothuizen, J., Pearson, P.L., van Oost, B.A., Wijmenga, C. 2002. Identification of a new copper metabolism gene by positional cloning in a purebred dog population. *Human Molecular Genetics* 11: 165–173.

Vashchenko, G., MacGillivray, R.T. 2013. Multi-copper oxidases and human iron metabolism. *Nutrients* 5: 2289–2313.

Volesky, B., Holan, Z.R. 1995. Biosorption of heavy metals. *Biotechnology Progress* 11: 235–250.

Weast, R.C. 1969. Handbook of chemistry and physics. *The American Journal of the Medical Sciences* 257: 423.

WHO. 1996. *Trace Elements in Human Nutrition and Health.* Geneva, Switzerland: World Health Organization.

Yau, P.Y., Loh, C.F., Azmil, I.A.R. 1991. Copper toxicity of clove [*Syzygium aromaticum* (L.) Merr. and Perry] seedlings. *Mardi Research Journal* 19: 49–53.

Yildiz, M., Ciğerci, İ.H., Konuk, M., Fidan, A.F., Terzi, H. 2009. Determination of genotoxic effects of copper sulphate and cobalt chloride in *Allium cepa* root cells by chromosome aberration and comet assays. *Chemosphere* 75: 934–938.

Zhou, W., Wang, J., Shen, B., Hou, W., Zhang, Y. 2009. Biosorption of copper (II) and cadmium (II) by a novel exopolysaccharide secreted from deep-sea mesophilic bacterium. *Colloids and Surfaces B: Biointerfaces* 72: 295–302.

38 Suitability of Nickel-Resistant Microbes for Use in Enhanced Nickel Bioremediation

F. Costa, B. Silva, and T. Tavares

CONTENTS

ABSTRACT

This chapter reviews the state of the art on microbial utilization on the rehabilitation of nickel-contaminated soils. A brief description of the main types of soils and of the relevant parameters that may influence their constitution and behavior is presented, as well as a short reference to metal contamination, in particular nickel contamination of soils. The effect of Ni on biota in general and on microbes in particular is discussed, and a general approach to bioremediation technologies is presented. Special attention is paid to microorganisms and their metabolism, and the development of resilience strategies is discussed under the same light. Finally, this chapter focuses on Ni-resistant microbes to be used in bioremediation approaches, with special attention to genetic engineering and to isolation protocols.

38.1 INTRODUCTION

According to several authors (Sousa et al., 2008; Yao et al., 2012), soil is considered a complex, interactive, and dynamic system. It is composed of several layers (Figure 38.1) that differ from site to site in terms of physical–chemical, mineralogical, and biological nature and is strongly influenced by (1) the parent material and its exposition to weathering processes, (2) the climate, and (3) the activities of the living organisms.

Besides acting as an interface between earth, air, and water (E.C., 2006), soil hosts most of the biosphere and acts as a collector filter of organic and inorganic residues, helping in sequestering potential hazardous compounds, thus behaving as a protective buffer toward groundwater resources (Sousa et al., 2008).

In the last decades of the twentieth century, the acknowledgment that soil is not an inexhaustible resource and that its improper use or mismanagement can be translated in the loss of soil characteristics and properties, with restricted opportunities for regeneration, led to an increase in apprehension and concern worldwide. This concern has increased, especially since soil contamination has been considered a primary environmental problem with dire repercussion, the disposal of hazardous residues in the soil has become a practical and inexpensive alternative, and there was a significant increase of soil contamination, especially with heavy metals (Figure 38.2).

According to CCE, 2002, the number of contaminated zones in the European Union may range between 3 hundred thousand and 1.5 million. It is estimated that in England, for example, soil contaminated by sewage irrigation reached about 1×10^{11} m^2 and 1.33×10^9 m^2 by soil stockpiled and solid waste (Yao et al., 2012).

Although heavy metals exist naturally in the soil as *trace* elements (concentrations lower than 1000 mg/kg), the gradual but increasing human-caused interference on their

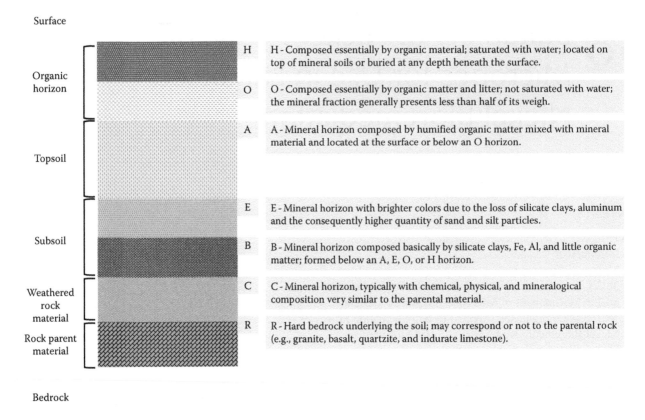

FIGURE 38.1 Soil horizon designations according to the FAO/UNESCO directives. (Data from FAO, World reference base for soil resources, World Soil Resources Reports 84, Food and Agriculture Organization of the United Nations, Rome, Italy, 1998.)

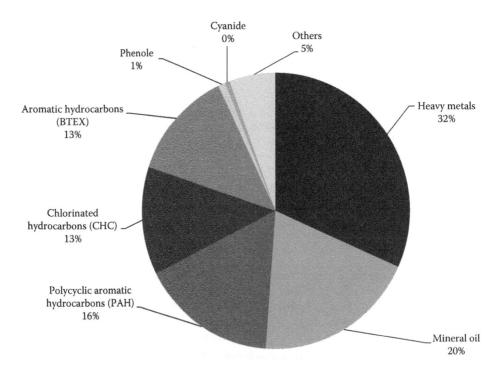

FIGURE 38.2 Overview of the main contaminants at industrial and commercial sites, affecting soil and groundwater in Europe as reported by the European Environment Agency through the EIONET priority data flows on contaminated sites. (Adapted from Panagos, P. et al., *J. Environ. Pub. Health*, 11, 2013, 2013.)

geochemical cycle promotes their accumulation in the soil, posing a serious risk to the human health and to all ecosystems (Pierzynski et al., 2000; Kabata-Pendias, 2001; D'Amore et al., 2005).

Heavy metals become contaminants of the soil because their rates of generation via man-made cycles and subsequent introduction into the environment are higher than the natural ones, and unlike the organic compounds that are oxidized to carbon oxides as a result of microbial activity, most heavy metals do not undergo biological or chemical degradation. Moreover, the concentrations of metals in rejected products are relatively high compared to those in the receiving environment, and the chemical species in which a metal is found in the environmental system may make it more or less bioavailable (Alloway, 1995; Lombi and Gerzabek, 1998). Finally, metals' toxicity is intrinsic to their atomic structure, and they cannot be mineralized or transmuted to an innocuous form (Bonaventura and Johnson, 1997).

Recent studies have shown that not only the characteristics of soil contamination by heavy metals are different in urban and agriculture soil (in type and content, Table 38.1) (Han et al., 2002) but also their sources, transport processes, and fate are distinct (Figure 38.3). Urban soil contamination with metals may result from transportation (leaded gasoline, tire wear debris particles, engine exhaust, weathering street particles), coal combustion, sewage sludge processing, spillage of petrochemicals, atmospheric deposition, electronics applications, electroplating, smelting of metalliferous ores, batteries, petrochemical industries, fine chemistry, jewelry industries, textile mill productions (Ahmady-Asbchin and Jafari, 2013; Merrikhpour and Jalali, 2013), paints and household wastes (Basta et al., 2005; Khan et al., 2008; Zhang et al., 2010). However, in agricultural or rural soils, the main sources of soil contamination are the application of natural and/or artificial fertilizers, animal manures and pesticides in soil, wastewater irrigation, sewage sludge, and smelting minerals (Montagne et al., 2007).

According to several authors (Aydinalp and Marinova, 2003; Bento et al., 2005), the characterization of a contaminated soil would be an advantage for the application of the appropriate remediation techniques, since it would provide knowledge about the source of contamination, elementary chemistry, environmental, and related health effects. Once in the soil, heavy metals may experience a set of different reactions with the soil components (e.g., mineral precipitation and dissolution, ion exchange, adsorption and desorption, aqueous complexation, biological immobilization and mobilization, plant uptake) that are influenced by the chemical form and speciation of the metal and allow the redistribution of the metals into different chemical forms, with different mobility, bioavailability, and toxicity (Shiowatana et al., 2001; Buekers, 2007).

Within the several heavy metals commonly found in contaminated soils, it is possible to highlight nickel. Ni is a natural element in Earth's crust, whose release and subsequent contamination into the soil occurs from natural sources such as wind, dust, manure, biogenic sources, and volcanic eruptions and from artificial sources such as those from anthropogenic activities (combustion of fossil fuels, electroplating, metal plating industries, and nickel mining) (Khodadoust et al., 2004). According to Schroeder et al. (1987), the estimated Ni deposition rates in rural and urban areas ranged from 1 to 50 $mg/m^2/year$ and from 10 to 595 $mg/m^2/year$, respectively.

Once in the environment, most of Ni compounds are strongly adsorbed by sediments or soil particles, where they may become immobilized. In basic soils, Ni adsorption may be irreversible due to the formation of covalent bonds, which in turn limits its mobility and availability. However, in acidic soils, Ni presents higher mobility and is often leached to the adjacent groundwater. Besides the soil pH, the extent of Ni adsorption (and desorption) can be affected by other soil properties such as the nature and amount of clay minerals, groundwater flow, texture, organic matter content, bulk density (ATSDR, 2005), redox potential, cation exchange capacity (CEC), and content of oxides of iron, aluminum, and manganese (Aydinalp and Marinova, 2003).

When released along with industrial discharges, Ni enters the aquatic system, where it can be absorbed by the suspended particles or by the sediments of the water body, thus constituting one of the main removal processes of Ni from the aquatic environment. Once in the aquatic system, Ni adsorption is extensively dependent on partitioning between the soluble and particulate solid phases. Processes such as adsorption, complexation, precipitation, and coprecipitation, which are affected by parameters such as pH, concentration of complexing ions, redox potential, ionic strength of water, and species and metal concentration, affect partitioning and consequently Ni removal from the aquatic systems. The adsorption of Ni on waterborne particulate matter is in competition with adsorption onto dissolved organic matter. This limits the quantity of Ni that can be removed from the water column through the deposition of suspended particles (ATSDR, 2005). Studies

TABLE 38.1
Heavy Metals Content in Urban and Agriculture Soils

Country	Cd	Cr	Cu	Ni	Pb	Zn
Urban Soil (mg/kg)						
France	0.53	42.08	20.06	14.47	43.14	43.14
Iran	1.53	63.79	60.15	37.53	46.59	94.09
Spain	3.76		57.01		1505.45	596.09
Syria		57.00	34.00	39.00	17.00	103.00
Agriculture Soil (mg/kg)						
India	0.82	2.19	1.20	4.34	0.95	28.24
Iran	034	10.36	9.62	11.28	5.17	11.56
Korea	0.12		2.98		5.25	4.78
Slovakia			65.00	29.00	139.00	140.00
Spain	1.42	63.48	107.65	34.75	213.93	427.80
USA	13.5	48.50	48.00	29.00	55.00	88.50

Source: Adapted from Su, C. et al., *Environ. Skeptics Crit.*, 3, 2014.

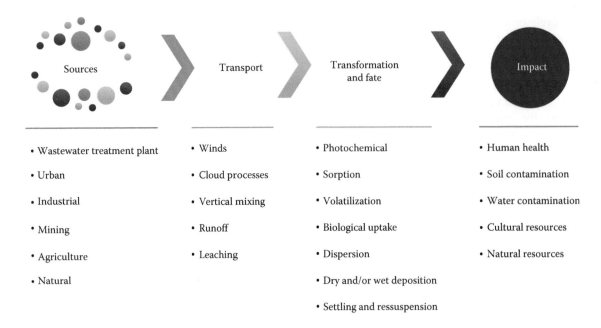

FIGURE 38.3   Sources of soil pollution by heavy metals, their transport, transformation, fate, and impact into the different ecosystems.

conducted by Young et al. (1982) in Lake Onondaga, Syracuse, New York, for a period of 4 months reveal that 36% of the Ni present in Lake Onondaga was lost to sediment and 75% of the Ni added into the lake was soluble and stayed in the lake. According to Evans (1989) and Rai and Zachara (1984), about 75% and 25% of Ni, respectively, in wastewater effluents and runoff, are found strongly adsorbed at inorganic surfaces such as hydrous and oxides of manganese, aluminum, and iron, suggesting that the concentration of Ni in natural water is controlled by adsorption processes.

Ni is released into the air mainly by trash incinerators and by power plants, where it undergoes a set of chemical reactions and settles in the ground. According to Bennett (1984) and Schmidt and Andren (1980), the estimated amount of Ni annually released into the environment by natural sources is about 8.5 million kg, whereas the estimated quantity annually released from artificial sources is about five times higher (Nriagu and Pacyna, 1988). Anthropogenic activities such as burning of residual and fuel oil are accountable for 62% of the total emissions released into the environment (Schmidt and Andren, 1980; Bennett, 1984). According to the Toxic Release Inventory (T.R.I., 2004), the total amount of emission from facilities that produced, processed, or used Ni reached, in 2002 in the United States, a total of 3081 ton and presented the following distribution: 82.2% to the terrestrial system, 6.0% to the air, 2.2% to the aquatic system, and 0.8% to underground injection. It is important to highlight that TRI data do not include all facilities, since only certain types of facilities are obligate to report.

Although Ni may exert an important role in the biology of organisms such as fishes, birds, insects, and animals, through the stimulation and activation of several enzymes (cofactor) (Talwar and Srivastava, 2002) responsible, for example, for the pigmentation and structural organization of ribonucleic

acid (RNA), it can also have a negative impact, especially when the maximum tolerable concentration is exceeded (ATSDR, 2005). Exposure to Ni may cause various types of cancer and acute and chronic health disorders, such as skin dermatitis, extreme weakness, kidney and lung damage, pulmonary fibrosis, nausea, chest pain, diarrhea, renal edema, and cyanosis (Suazo-Madrid et al., 2011).

The development of environment-friendly techniques able to efficiently rehabilitate terrestrial and aquatic systems contaminated with Ni is therefore crucial in order to minimize and reverse the impacts of Ni pollution. In the following sections, special attention will be given to the microbial bioremediation of metals in contaminated environments and to the application of Ni-resistant microorganisms to treat contaminated soils and sediments with this element.

38.2   MICROBIAL BIOREMEDIATION OF METAL-CONTAMINATED ENVIRONMENTS

38.2.1   GENERAL APPROACH TO BIOREMEDIATION TECHNOLOGIES

The exposure to heavy metals can cause serious damages to tissues through several mechanisms, due to oxidative stress. The oxidative stress results from the generation of reactive oxygen species followed by the decrease of antioxidants leading to cell death. Although the conventional remediation techniques have several advantages like easiness of metal recovery and high productivity, they have important drawbacks such as their high cost and production of toxic sludge. With this regard, bioremediation is a promising sustainable process that uses plants, microbes, and their enzymes for the removal of heavy metals in a much more effective and eco-friendly manner. Mostly, bioremediation processes use microbial biomass to degrade xenobiotic and/or toxic compounds contaminating soil or water, but in some cases, it also involves the use of plants applied directly (phytoremediation) or as a colonization support for microorganisms (Gavrilescu, 2004).

Microorganisms need suitable environmental conditions in order to survive and grow. Therefore, the success of a bioremediation process is influenced not only by the presence of the appropriate microorganisms but also by the appropriate environmental conditions. Most of the bioremediation approaches use prokaryotes such as bacteria and archaea, although eukaryotes (fungi and algae) can be also used to transform and degrade pollutants. Microorganisms that live in polluted environments are usually well suited to survival in the presence of the existing pollutants and to the environmental conditions of the site (temperature, pH, and oxidation–reduction potential). These native microorganisms tend to use the nutrients and electron acceptors that are available *in situ*. The bulk of subsurface microbial populations is associated with both microorganisms and dissolved contaminants and the subproducts of their degradation (Tabak et al., 2005).

Metabolic functions allowing growth and reproduction are supported by various chemical reactions performed by microbes. Microorganisms employ such enzyme-catalyzed reactions that are well organized in metabolic pathways for the degradation of a great variety of chemical compounds (Kokate et al., 2011). In heavy metal–polluted habitats, these organisms are known to develop and adopt different detoxifying processes such as biotransformation, biodegradation, biosorption, and bioaccumulation, which can be exploited for bioremediation either *ex situ* or *in situ* (Gadd, 2000; Malik, 2004). The microbial-mediated transformation or degradation of contaminants into nonhazardous or less hazardous substances is known as biotransformation and biodegradation, respectively. Biotransformation is the process whereby the structure of a substance is changed as a result of the action of microorganisms or enzyme preparations derived from biomass. This process results in the loss or alteration of some characteristic properties of the original compound such as solubility, mobility in the environment, or toxicity, with minor or no loss of molecular complexity. In its turn, biodegradation process involves the breakdown of organic compounds, usually performed by microorganisms, into less complex compounds and ultimately to water, carbon dioxide, and oxides or mineral salts of other elements present. Although metals cannot be biodegraded, microorganisms can interact with these contaminants and change their oxidation state through reduction or oxidation processes (Tabak et al., 2005). Due to a change in their oxidation state, heavy metals can be transformed to become either less toxic, less bioavailable, more water soluble (and thus can be removed through leaching), or less water soluble (which allows them to precipitate and become easily removed from the environment) (Garbisu and Alkorta, 2003).

Microbial cells (live or dead) and their products can be very effective bioaccumulators of metals (Gavrilescu, 2004). The intracellular processes include some mechanistic paths for metal accumulation involving chemical linkage, precipitation, and methylation. Metal accumulative bioprocesses are divided into two groups: sorptive uptake by dead biomass that comprises the accumulation of metals without active uptake by a process called biosorption and bioaccumulation by living cells, which is the combination of active and passive uptake modes. Various microbial species have been shown to be relatively efficient in the bioaccumulation of copper (Anand et al., 2006), nickel (Wu et al., 2009; Das et al., 2014), and lead (Yetis et al., 2000), among other metal ions.

38.2.2 MECHANISMS OF MICROBIAL RESISTANCE TO METALS

Microorganisms play an important role in the biogeochemical cycle of heavy metals and also in cleaning up metal-contaminated environments. As mentioned, microorganisms can potentially accumulate metals by either a metabolism-independent, passive uptake or a metabolism-dependent, active uptake. Therefore, the sorption capacity of the cell envelope and the ability for sequestering metals into the cytosol are the two main features of the microbial cells that influence the overall accumulation of metals (Haferburg and Kothe, 2007).

To survive under metal-stressed conditions, microorganisms may develop several types of mechanisms, some of which are used as the basis for potential bioremediation strategies. Those mechanisms may include the physicochemical surface retention of the metal as in biosorption, the ion mobilization through the effect of excreted organic acids or through methylation as in bioleaching, the inverse process as in biomineralization by polymeric complexes or insoluble sulfides, or the intracellular accumulation and enzyme-catalyzed transformation (redox reactions) (Lloyd and Lovley, 2001). The main microbial processes involved in the bioremediation of metals are presented in Figure 38.4.

The mechanisms involved in bacterial metal resistance result from either the active efflux pumping of the toxic metal out of the cell or the enzymatic detoxification (generally redox chemistry) converting a toxic ion into a less toxic or less available metal ion (Mejáre and Bülow, 2001). The reduction of metals can occur through a dissimilatory reduction mechanism, by which bacteria can conserve energy to support growth by coupling the oxidation of organic or inorganic electron donors to the reduction of metal ions. For instance, *Alcaligenes eutrophus* CH34 strain that was found to be resistant to very high levels of Ni appears to possess energy-dependent nickel export systems to maintain low intracellular Ni concentration (Hausinger, 1993).

On the other hand, bacteria may use reduction mechanisms that are not related to respiration processes but are thought to give metal resistance (Juwarkar and Yadav, 2010), for instance, the reduction of Cr(VI) to Cr(III) (Sayel et al., 2012), reduction of Se(VI) to elemental Se (Lloyd et al., 2001), and reduction of Hg(II) to Hg(0) (Brim et al., 2000).

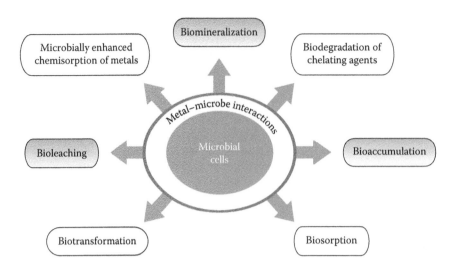

FIGURE 38.4 Metal processing mechanisms of microorganisms. (Adapted from Tabak, H. et al., *Rev. Environ. Sci. Bio/Technol.*, 4, 115, 2005.)

38.2.3 MICROORGANISMS APPLIED TO BIOREMEDIATION PROCESSES

Microorganisms possess certain unique characteristics that make them ideal for bioremediation purposes. The main goal in a bioremediation process is to identify species of microorganisms that are capable of efficient uptake of environmentally and economically important metals. Therefore, screening for microbes with high accumulation capacities and stable resistance characteristics is mandatory for any remediation approach (Haferburg and Kothe, 2007).

The control and optimization of bioremediation processes are dependent upon some other factors, such as the availability of contaminants to the microbial population and environmental parameters (type of matrix, temperature, pH,

the presence of oxygen or other electron acceptors and nutrients). As bioremediation only can be effective where environmental conditions permit microbial growth and activity, its application frequently requires the manipulation of the previously mentioned environmental factors (Chandra, 2015). Microorganisms that are used in bioremediation practices should have acquired metal processing features, as described in Figure 38.5. These mechanisms include the uptake and reflux of contaminating metals, their biosorption, intracellular assimilation, immobilization, complexation and precipitation, and finally their release (Dash and Das, 2012; Das and Dash, 2014). Besides their capability to transform and accumulate an extensive selection of metals, microorganisms suitable for soil bioremediation should possess several other characteristics including the ability to extensively colonize the soil, resist

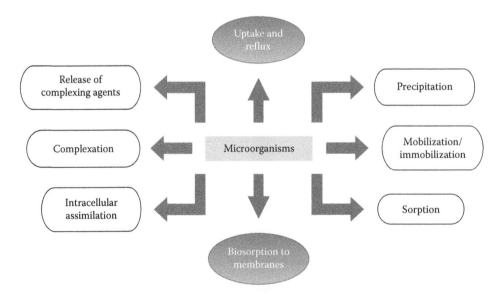

FIGURE 38.5 Metal processing characteristics of bacteria required to use in bioremediation. (Adapted from Dash, H.R. and Das, S., *Int. Biodeterior. Biodegrad.*, 75, 207, 2012.)

at high concentrations of toxic compounds, survive over a long period in limiting conditions, and compete with the other microbes present in the soil matrix (Anastasi et al., 2013).

In many circumstances, accumulated toxic and persistent substances such as heavy metals cannot be removed efficiently just by the natural metabolic diversity of the native microorganisms. Therefore, many studies have searched new possibilities for the cleanup of these contaminants using genetically engineered microorganisms (Kumar et al., 2013). Microorganisms can be genetically modified in order to enhance their intrinsic metal-sequestering capability as well as their resistance to environmental conditions. The production of genetically modified microorganisms is a potential solution for enhancing metal selectivity as well as the accumulating potential of cells (Bae et al., 2000; Krishnaswamy and Wilson, 2000; Deng et al., 2007). Genetic modification is reasonable, particularly when the microbial biomass is produced from fermentation processes where genetically engineered microorganisms are used (Vijayaraghavan and Yun, 2008). These modified microorganisms present a rapid growth rate and resistance to toxicity, features that help to overcome some intrinsic limitations of the bioremediation process. Nevertheless, the impact of the release of such modified organisms into the environment is still unknown.

38.3 NICKEL BIOREMEDIATION BY NICKEL-RESISTANT MICROBES

38.3.1 Biotechnological Approach

Nowadays, environmental protection and rehabilitation tend to involve microorganisms in a successful biotechnological trend as reviewed by Singh et al. (2014). These processes are promoted in controlled operational conditions compatible with ambient temperature and pressure, thus demanding less energy than traditional technology. Biotechnological solutions for the recovery of contaminated systems appear to be cost-effective and fulfill environmental legislation, leading to a very positive image within public opinion. Among distinct approaches to the usage of microorganisms to rehabilitate natural systems, we may find bioremediation, generally defined as the process in which microorganisms are stimulated to grow in contaminated environments in order to degrade or remove hazardous pollutants (Strong and Burgess, 2008). This procedure may rely on indigenous microbial consortia (Agarwal, 1998) or may demand the utilization of manipulated or exogenous microbial communities. These consortia are able to degrade different organic pollutants, by using them as energy or nutrient sources (Fulekar et al., 2009), or eventually to precipitate inorganic pollutants, as in the case of metals. In order to be effective, bioremediation relies on the addition of nutrients and growth factors, on pH and temperature control, and on the presence of electron donors or acceptors (Hess et al., 1997). These and other issues are thoroughly reviewed by Dixit et al. (2015) in an exhaustive assay where heavy metal bioremediation mechanisms are evaluated, focusing inclusively on the action of microbes, of plants, defining the fundamental principles of phytoremediation and on the symbiosis between microbes and some plants. The authors elaborate on the different contamination sources, on bioremediation principles and mechanisms extended to phytoremediation, and on genomics approach and its future prospects. Similar issues were discussed by Kumar et al. (2011), with a more technological emphasis on the *in situ* compared to *ex situ* treatments as well as on the solid phase treatments versus slurry phase treatments.

38.3.2 Genetic Engineering

Since the end of last century, there has been a special effort to manipulate the ability of certain microorganisms to interact with specific pollutants, by genetic engineering procedures. It is now generally accepted that this degrading ability of some microbes is genetically encoded in enzymes present in chromosomal and extrachromosomal DNA. Recombinant DNA techniques such as site-directed mutagenesis, antisense RNA, and PCR are used in a wide variety of microorganisms, selected by their gene diversity and metabolic versatility (Fulekar, 2009). The aim of such procedures is the identification of degradative genes, and their expression in an adequate host through such a vector by adequate promoters, improving the scope and effectiveness of bioremediation technology. Some examples of such bacteria are listed in Table 38.2.

38.3.3 Heavy Metal Resistance

It is known that microbes may develop and present some resistance to metals as reviewed by Hassen et al. (1998). This study refers to toxicological determinations of the resistance and sensitivity of different bacteria toward distinct metals. It is defended that the introduction of such xenobiotics in the environment triggers evident modifications in the microbial communities as well as in their metabolism. Heavy metals in particular do inhibit microbes' behavior by transferring essential ions, blocking functional groups, or even changing the configuration of biologic molecules (Gadd and Griffiths, 1978; Jansen et al., 1994). Nevertheless, some of these elements are essential to life at low concentrations, as they enable the production of enzymes and metalloproteins (Jansen et al., 1994). It is established that the microbial resistance to metals may be developed by two paths: accumulation by association with a protein or blockage at the cell wall and associated transportation systems (Ow, 1993). An evident effort has been made to determine the minimal inhibitory concentrations (MICs), for each metal and specific bacterial strains in distinct matrices, but in spite of some difficulties in simulating real conditions, these determinations proved to be relevant and helpful in the understanding of the relations between metals and bacteria (Hassen et al., 1998). Moreover, the mobility of relevant metal ions is also discussed as there are distinct responses from indigenous biota, depending on the physical state of the contaminated matrix. In fact, the adsorption and the complexation capacity of the metallic compounds are determinant to their toxicity toward the microorganisms under study. Hassen and

TABLE 38.2

Some Genetically Modified Bacteria to Be Applied in Bioremediation Processes

Metal	Genetically Engineered Bacteria	Expressed Gene	References
As	*E. coli* strain	Metalloregulatory protein ArsR	Kostal et al. (2004)
Cd	*E. coli* strain	SpPCS	Seung et al. (2007)
Cr	*Methylococcus capsulatus*	CrR	Hasin et al. (2010)
Cr	*P. putida* strain	Chromate reductase (ChrR)	Ackerley et al. (2004)
Cd, Hg	*Ralstonia eutropha* CH34	merA	Brim et al. (2000); Valls et al. (2000)
	Deinococcus radiodurans		
Hg	*E. coli* strain	Organomercurial lyase	Murtaza et al. (2002)
Hg	*E. coli* JM109	Hg^{2+} transporter	Zhao et al. (2005)
Hg	*Pseudomonas* K-62	Organomercurial lyase	Kiyono and Pan-Hou (2006)
Hg	*Achromobacter* sp. AO22	mer	Ng et al. (2009)
Ni	*P. fluorescens* 4F39	Phytochelatin synthase (PCS)	López et al. (2002); Sriprang et al. (2003)

Source: Adapted from Dixit, R. et al., *Sustainability*, 7, 2189, 2015.

coworkers present a case study with *Pseudomonas aeruginosa* S6 as a bacterial model for environmental applications as it presents high MIC for metals and good antibiotic resistance. Some years later, Bruins et al. (2000) reviewed the resistance mechanisms of microbes to metals and detailed the various possible steps for such a resistance. Metals that may be used as micronutrients participate in redox processes, regulate osmotic pressure, and stabilize enzymatic molecules. Some others are just toxic to life and bond to cellular components by covalent and ionic linkage, damaging their structure. It is stated that this resistance is encoded in genes located in chromosomes, transposons, or plasmids, and microbes may resist to the xenobiotic effect of heavy metals by cellular exclusion, sequestration intra or extra cellules, detoxification by enzymes action, transportation by efflux pumps, or sensitivity reduction of cellular bodies.

38.3.4 Isolation of Resistant Microorganisms

The acknowledgment of the development of resistance mechanisms toward the toxic effect of heavy metals led to several attempts to isolate selected microorganisms from contaminated matrices, in order to use them as detoxifying agents. An earlier publication of Diels et al. (2002) reviewed the possible methods to treat soils and groundwater via bacterial activity. The authors suggested the simultaneous extraction and biomass binding of metal compounds present in contaminated soils, as well as the usage of sulfate-reducing bacteria (SRB) in an upflow anaerobic sludge blanket for the rehabilitation of groundwater. Although SRB just reduces sulfates to sulfides, precipitating metals in the process, in a very low redox potential, this research acknowledged the possible application of specific and resilient bacteria to the resolution of heavy metal contamination. Recently, Kamika and Momba (2013) assessed the resistance and bioremediation ability of protozoan species and bacteria isolated in wastewater concentrated with heavy metals, controlling process parameters such as pH, dissolved oxygen, and chemical oxygen demand. The studied species

were quite resilient to the toxic effect of heavy metals as they have developed resistance mechanisms toward the xenobiotics. The authors focused their attention in some specific species, the protozoan *Peranema* sp. and the bacteria *Pseudomonas putida* and *Bacillus licheniformis*, as being the best candidates for bioremediation procedures. The genetic encoding was also explored, and assays were performed to distinguish the passive biosorption mechanisms from the active bioaccumulation path. Another effort to isolate heavy metal–resistant microbes is presented by Kumar et al. (2015), who were able to collect 14 microbial isolates from agricultural soils irrigated in industrial wastewater. Those isolates proved to be able to bioaccumulate zinc and lead under field conditions. Similar perspective was used by Congeevaram et al. (2007), who compared bacterial and fungal isolates in their biosorptive ability toward Cr and Ni. The species were isolated from electroplating activity–contaminated soils and tested under optimized conditions of temperature and pH. Solid retention time and tolerance threshold were determined for each isolate and each metal in order to design a continuous stirred bioreactor, with the indication of *Micrococcus* sp. and *Aspergillus* sp. as the selected microbes.

38.3.5 Ni-Resistant Microbes

As mentioned earlier, there is a growing concern within the scientific community and society in general related to the increasing discharges of Ni from different industries and anthropogenic activities into the environment, leading to accumulation in food chains. Although Ni is essential to life in general at trace concentrations, it may be stressful to microorganisms and higher organisms when present in higher levels. As for some other metallic elements, Ni-resistant bacterial species have developed resilience strategies to survive the xenobiotic effect, similar to the ones described earlier (Sar et al., 2001).

In fact, there are evident advantages in using biotechnological procedures to remediate and rehabilitate contaminated natural systems. In particular, bacterial biomass is cheaply produced, widely spread in nature, and available to treat relevant volumes of soils, sediments, and leached solutions. Bioremediation with selected bacteria has high kinetics, high selectivity, and favorable isotherms. Nowadays, the isolation and characterization of Ni-resistant indigenous soil bacteria are recognized to be a fundamental condition for effective, competitive, and sustainable environmental rehabilitation process (Grass et al., 2000).

With the aim to develop such a process, initial studies with fungi were presented by Joho et al. (1995) who focused their assays in Ni-resistant mutants of yeasts. The publication forwarded a resistance development mechanism that includes production of extracellular Ni chelating compounds, modification of Mg flux pumps, and Ni sequestration in vacuoles involving H^+-ATPase membrane. The presence of other ions than Ni, of organic matter, pH, and temperature are determinants to the survival of the fungi in Ni environment. At the time, the need of molecular level analysis was acknowledged.

More recently, the stress effect of Ni ions on the growth and entrapment ability of some other microbial species was evaluated in order to optimize a biotechnological tool for bioremediation processes. Wu et al. (2009) considered that stress effect over *Escherichia coli*. It was concluded that Ni ions do affect the growth kinetics of the bacteria and the metal uptake may be optimized with longer adaptation periods of the bacteria to growing concentrations of Ni in the growth broth. The same effect of Ni over *E. coli* was evaluated by Mahalakshmi et al. (2010) and compared to *Saccharomyces cerevisiae*. The eukaryote presents a 10-fold higher MIC to Ni than the prokaryote, but both were evidently affected on their growth rate by the presence of $NiSO_4$. The protein profile of both cultures was notoriously changed with higher expression of the cytosolic protein when the selected microbes were in contact with higher concentrations of Ni. The isolation of Ni-resistant bacteria from wastewater from different industrial sources was achieved by Alboghobeish et al. (2014), who obtained eight distinct species identified by 16SrDNA sequencing. Three Ni-resistant bacteria were selected as the most resilient between the studied ones, based on the maximum tolerable concentration: *Cupriavidus* sp. ATHA3 (8 Mm N^{2+}), *Klebsiella oxytoca* ATHA6 (16 Mm N^{2+}), and *Methylobacterium* sp. ATHA7 (24 Mm N^{2+}). Special attention was paid to *K. oxytoca* that was able to reduce 83 mg/mL of the Ni^{2+} concentration in liquid solution in a 3-day period. The potential of some other bacteria to bioremediate Ni-contaminated matrices is discussed by Das et al. (2014). The isolated strain from Ni-contaminated soil was able to tolerate 7.5–10 mM/mL of Ni and was identified as *Bacillus thuringiensis* KUNi1. The strain showed multimetal resistance, surviving to zinc, copper, cobalt, and cadmium. In vitro culture performed an 82% Ni removal from growing medium. Metabolically inactive cells did not present any metal removal capacity. In addition, the determinant operational parameters were recognized as being temperature, pH, cell density, and the presence of other metals. The authors are convinced of the promising features of such a strain to be used in microbiological bioremediation of Ni-contaminated sites, by *in situ* or *ex situ* procedures.

The effectiveness of microbial bioremediation may be enhanced by the association of bacteria with plants normally used in phytoremediation (Pohontu and Gontariu, 2014). Ni-resistant bacteria (*Bacillus* spp. and *Pseudomonas* spp.) were associated with *Sinapis alba* plants to bioremediate Ni-contaminated soils, in *ex situ* lab-scale assay. Although *Brassica* genus plants show a special competence in uptaking metal ions present in soils through their physiological processes, those ions present low mobility turning harder their biosorption by plant roots. The synergetic association of these plants and rhizosphere Ni-resistant bacteria improved by 30% the phytoremediation ability of such a system, as it was determined by the 60 days follow-up of Ni concentration in the plant roots. The bacterial contamination was performed by inoculation of the bacterial strain within the plant seeds.

All these observations indicate that there is room for optimization of the bioremediation approach to solve Ni-contaminated systems like soils, sediments and groundwater, with a special emphasis on microbes genetically manipulated to entrap relevant amounts of metal ions and on the association of those microorganisms with some other structures. It is the case of permeable barriers where microorganisms may develop a resilient biofilm, preventing contamination of soils or sediments by removing pollutants from percolating contaminated water.

REFERENCES

Ackerley, D.F., Gonzalez, C.F., Keyhan, M., Blake, I.R., and Matin, A. 2004. Mechanism of chromate reduction by the *Escherichia coli* protein, NfsA, and the role of different chromate reductases in minimizing oxidative stress during chromate reduction. *Environmental Microbiology*, 6: 851–860.

Agarwal, S.K. 1998. *Environmental Biotechnology*. New Delhi, India: APH Publishing Corporation.

Ahmady-Asbchin, S. and Jafari, N. 2013. Removal of nickel and zinc from single and binary metal solutions by *Sargassum angustifolium*. *Water Science and Technology*, 68: 1384–1390.

Alboghobeish, H., Tahmourespour, A., and Doudi, M. 2014. The study of Nickel Resistant Bacteria (NiRB) isolated from wastewaters polluted with different industrial sources. *Journal of Environmental Health Science and Engineering*, 12(1): 44–50.

Alloway, B.J. 1995. *Heavy Metals in Soils*. London, U.K.: Blackie Academic and Professional.

Anand, P., Isar, J., Saran, S., and Saxena, R.K. 2006. Bioaccumulation of copper by *Trichoderma viride*. *Bioresource Technology*, 97: 1018–1025.

Anastasi, A., Tigini, V., and Varese, G.C. 2013. The bioremediation potential of different ecophysiological groups of fungi. In *Fungi as Bioremediators*, Goltapeh, E., D. Mohammadi, Y. Rezaee, and A. Varma (eds.), pp. 29–39. Heidelberg, Germany: Springer.

ATSDR, Agency for Toxic Substances and Disease Registry, 2005. Toxicological Profile for Nickel. Atlanta, GA: U.S. Department of Health and Human Services, Public Health Service.

Aydinalp, C. and Marinova, S. 2003. Distribution and forms of heavy metals in some agricultural soils. *Polish Journal of Environmental Studies*, 12: 629–633.

Bae, W., Chen, W., Mulchandani, A., and Mehra, R.K. 2000. Enhanced bioaccumulation of heavy metals by bacterial cells displaying synthetic phytochelatins. *Biotechnology and Bioengineering*, 70: 518–524.

Basta, N.T., Ryan, J.A., and Chaney, R.L. 2005. Trace element chemistry in residual-treated soil: Key concepts and metal bioavailability. *Journal of Environmental Quality*, 34(1): 49–63.

Bennett, B.G. 1984. Environmental nickel pathways in man. In *Nickel in the Human Environment. Proceedings of a Joint Symposium.* IARC Scientific publication no. 53., Lee Chen, S.F. (ed.), pp. 487–495. Lyon, France: International Agency for Research on Cancer.

Bento, F.M., Camargo, F.A.O., Okeke, B.C., and Frankenberger, W.T. 2005. Comparative bioremediation of soils contaminated with diesel oil by natural attenuation, biostimulation and bioaugmentation. *Bioresource Technology*, 96: 1049–1055.

Bonaventura, C. and Johnson, F.M. 1997. Healthy environments for healthy people: Bioremediation today and tomorrow. *Environmental Health Perspectives*, 105: 5–20.

Brim, H., Mcfarlan, S.C., Fredrickson, J.K., Minton, K.W., Zhai, M., Wackett, L.P., and Daly, M.J. 2000. Engineering *Deinococcus radiodurans* for metal remediation in radioactive mixed waste environments. *Nature Biotechnology*, 18: 85–90.

Bruins, M.R., Kapil, S., and Oehme, F.W. 2000. Microbial resistance to metals in the environment. *Ecotoxicology and Environmental Safety*, 45: 198–207.

Buekers, J. 2007. Fixation of cadmium, copper, nickel and zinc in soil: Kinetics, mechanisms and its effect on metal bioavailability, PhD thesis, Katholieke Universiteit Lueven, Lueven, Belgium.

Chandra, R. 2015. *Advances in Biodegradation and Bioremediation of Industrial Waste*. Boca Raton, FL: CRC Press.

Congeevaram, S., Dhanarani, S., Park, J., Dexilin, M., and Thamaraiselvi, K. 2007. Biosorption of chromium and nickel by heavy metal resistant fungal and bacterial isolates. *Journal of Hazardous Materials*, 146: 270–277.

D'amore, J.J., Al-Abed, S.R., Scheckel, K.G., and Ryan, J.A. 2005. Methods for speciation of metals in soils: A review. *Journal of Environmental Quality*, 34(5): 1707–1745.

Das, P., Sinha, S., and Mukherjee, S.K. 2014. Nickel bioremediation potential of *Bacillus thuringiensis* KUNi1 and some environmental factors in nickel removal. *Bioremediation Journal*, 18: 169–177.

Das, S. and Dash, H.R. 2014. 1—Microbial bioremediation: A potential tool for restoration of contaminated areas. In *Microbial Biodegradation and Bioremediation*, Das, S. (ed.), pp. 1–21. Oxford, U.K.: Elsevier.

Dash, H.R. and Das, S. 2012. Bioremediation of mercury and the importance of bacterial mer genes. *International Biodeterioration and Biodegradation*, 75: 207–213.

Deng, X., Yi, X.E., and Liu, G. 2007. Cadmium removal from aqueous solution by gene-modified *Escherichia coli* JM109. *Journal of Hazardous Materials*, 139: 340–344.

Diels, L., Van Der Lelie, N., and Bastiaens, L. 2002. New developments in treatment of heavy metal contaminated soils. *Reviews in Environmental Science and Biotechnology*, 1: 75–82.

Dixit, R., Wasiullah, E., Malaviya, D., Pandiyan, K., Singh, U., Sahu, A., Shukla, R. et al. 2015. Bioremediation of heavy metals from soil and aquatic environment: An overview of principles and criteria of fundamental processes. *Sustainability*, 7: 2189.

E. C. 2006. *European Commission, Thematic Strategy for Soil Protection, COM (2006) 231 Final.* Brussels, Belgium: European Commission.

Evans, L.J. 1989. Chemistry of metal retention by soils. *Environmental Science and Technology*, 23: 1046–1056.

FAO. 1998. World reference base for soil resources, World Soil Resources Reports 84, Food and Agriculture Organization of the United Nations, Rome, Italy.

Fulekar, M.H. 2009. Bioremediation of fenvalerate by *Pseudomonas aeruginosa* in a scale up Bioreactor. *Romanian Biotechnological Letters*, 14: 4900–4905.

Fulekar, M.H., Geetha, M., and Sharma, J. 2009. Bioremediation of Trichlorpyr Butoxyethyl Ester (TBEE) in bioreactor using adapted *Pseudomonas aeruginosa* in scale up process technique. *Biology and Medicine*, 1: 1–6.

Gadd, G.M. 2000. Bioremedial potential of microbial mechanisms of metal mobilization and immobilization. *Current Opinion in Biotechnology*, 11: 271–279.

Gadd, G.M. and Griffiths, A.J. 1978. Microorganisms and heavy metal toxicity. *Microbial Ecology*, 4: 303–317.

Garbisu, C. and Alkorta, I. 2003. Basic concepts on heavy metal soil bioremediation. *The European Journal of Mineral Processing and Environmental Protection*, 3: 58–66.

Gavrilescu, M. 2004. Removal of Heavy Metals from the Environment by Biosorption. *Engineering in Life Sciences*, 4: 219–232.

Grass, G., Große, C., and Nies, D.H. 2000. Regulation of the cnr cobalt and nickel resistance determinant from *Ralstonia* sp. strain CH34. *Journal of Bacteriology*, 182: 1390–1398.

Haferburg, G. and Kothe, E. 2007. Microbes and metals: Interactions in the environment. *Journal of Basic Microbiology*, 47: 453–467.

Han, F., Banin, A., Su, Y., Monts, D., Plodinec, J., Kingery, W., and Triplett, G. 2002. Industrial age anthropogenic inputs of heavy metals into the pedosphere. *Naturwissenschaften*, 89: 497–504.

Hasin, A.A.L., Gurman, S.J., Murphy, L.M., Perry, A., Smith, T.J., and Gardiner, P.H.E. 2010. Remediation of chromium(VI) by a methane-oxidizing bacterium. *Environmental Science and Technology*, 44: 400–405.

Hassen, A., Saidi, N., Cherif, M., and Boudabous, A. 1998. Resistance of environmental bacteria to heavy metals. *Bioresource Technology*, 64: 7–15.

Hausinger, R.P. 1993. Microbial nickel metabolism. In *Biochemistry of Nickel*, Hausinger, R.P. (ed.), pp. 181–201. Springer, New York.

Hess, A., Zarda, B., Hahn, D., Häner, A., Stax, D., Höhener, P., and Zeyer, J. 1997. In situ analysis of denitrifying toluene- and m-xylene-degrading bacteria in a diesel fuel-contaminated laboratory aquifer column. *Applied and Environmental Microbiology*, 63: 2136–2141.

Jansen, E., Michels, M., Van Til, M., and Doelman, P. 1994. Effects of heavy metals in soil on microbial diversity and activity as shown by the sensitivity-resistance index, an ecologically relevant parameter. *Biology and Fertility of Soils*, 17: 177–184.

Joho, M., Inouhe, M., Tohoyama, H., and Murayama, T. 1995. Nickel resistance mechanisms in yeasts and other fungi. *Journal of Industrial Microbiology*, 14: 164–168.

Juwarkar, A.A. and Yadav, S.K. 2010. Bioremediation technology: Recent advances. In *Bioaccumulation and Biotransformation of Heavy Metals*, Fulekar, M.H. (ed.). New Delhi, India: Springer.

Kabata-Pendias, A. 2001. *Trace Elements in Soils and Plants*. Boca Raton, FL: CRC Press.

Kamika, I. and Momba, M.N.B. 2013. Assessing the resistance and bioremediation ability of selected bacterial and protozoan species to heavy metals in metal-rich industrial wastewater. *BMC Microbiology*, 13: 28–41.

Khan, S., Cao, Q., Zheng, Y.M., Huang, Y.Z., and Zhu, Y.G. 2008. Health risks of heavy metals in contaminated soils and food crops irrigated with wastewater in Beijing, China. *Environmental Pollution*, 152: 686–692.

Khodadoust, A.P., Reddy, K.R., and Maturi, K. 2004. Removal of nickel and phenanthrene from Kaolin soil using different extractants. *Environmental Engineering Science*, 21: 691–704.

Kiyono, M. and Pan-Hou, H. 2006. Genetic engineering of bacteria for environmental remediation of mercury. *Journal of Health Science*, 52: 199–204.

Kokate, C., Hurakadle, P.J., and Jalalpure, S.S. 2011. *Textbook of Pharmaceutical Biotechnology*. Haryana, India: Elsevier.

Kostal, J., Yang, R., Wu, C.H., Mulchandani, A., and Chen, W. 2004. Enhanced arsenic accumulation in engineered bacterial cells expressing ArsR. *Applied and Environmental Microbiology*, 70: 4582–4587.

Krishnaswamy, R. and Wilson, D.B. 2000. Construction and characterization of an *Escherichia coli* strain genetically engineered for Ni(II) bioaccumulation. *Applied and Environmental Microbiology*, 66: 5383–5386.

Kumar, A., Bisht, B.S., Joshi, V.D., and Dhewa, T. 2011. Review on bioremediation of polluted environment: A management tool. *International Journal of Environmental Sciences*, 1: 1079–1093.

Kumar, S., Dagar, V., Khasa, Y., and Kuhad, R. 2013. Genetically modified microorganisms (GMOs) for bioremediation. In *Biotechnology for Environmental Management and Resource Recovery*, Kuhad, R.C. and A. Singh (eds.), pp. 191–218. New Delhi, India: Springer.

Kumar, V., Singh, S., Bhadrecha, P., Kaur, P., Bhatia, D., Singla, S., Datta, S. et al. 2015. Bioremediation of heavy metals by employing resistant microbial isolates from agricultural soil irrigated with industrial waste water. *Oriental Journal of Chemistry*, 31: 357–361.

Lloyd, J.R. and Lovley, D.R. 2001. Microbial detoxification of metals and radionuclides. *Current Opinion in Biotechnology*, 12: 248–253.

Lloyd, J.R., Mabbett, A.N., Williams, D.R., and Macaskie, L.E. 2001. Metal reduction by sulphate-reducing bacteria: Physiological diversity and metal specificity. *Hydrometallurgy*, 59: 327–337.

Lombi, E. and Gerzabek, M.H. 1998. Determination of mobile heavy metal fraction in soil: Results of a pot experiment with sewage sludge. *Communications in Soil Science and Plant Analysis*, 29: 2545–2556.

López, A., Lázaro, N., Morales, S., and Marqués, A.M. 2002. Nickel biosorption by free and immobilized cells of *Pseudomonas fluorescens* 4F39: A comparative study. *Water, Air, and Soil Pollution*, 135: 157–172.

Mahalakshmi, T., Ilamathi, M., Siva, R., and Sridharan, T.B. 2010. Effect of nickel stress on *Escherichia coli* and *Saccharomyces cerevisiae*. *Journal of Industrial Pollution Control*, 26: 5–13.

Malik, A. 2004. Metal bioremediation through growing cells. *Environment International*, 30: 261–278.

Mejáre, M. and Bülow, L. 2001. Metal-binding proteins and peptides in bioremediation and phytoremediation of heavy metals. *Trends in Biotechnology*, 19: 67–73.

Merrikhpour, H. and Jalali, M. 2013. Sorption processes of natural Iranian bentonite exchanged with Cd^{2+}, Cu^{2+}, Ni^{2+} and Pb^{2+} cations. *Chemical Engineering Communications*, 200: 1645–1665.

Montagne, D., Cornu, S., Bourennane, H., Baize, D., Ratié, C., and King, D. 2007. Effect of agricultural practices on trace-element distribution in soil. *Communications in Soil Science and Plant Analysis*, 38: 473–491.

Murtaza, I., Dutt, A., and Ali, A. 2002. Biomolecular engineering of *Escherichia coli* organo-mercurial lyase gene and its expression. *Indian Journal of Biotechnology*, 1: 117–120.

Ng, S.P., Davis, B., Palombo, E.A., and Bhave, M. 2009. A Tn5051-like mer-containing transposon identified in a heavy metal tolerant strain *Achromobacter* sp. AO22. *BMC Research Notes*, 2: 38–44.

Nriagu, J.O. and Pacyna, J.M. 1988. Quantitative assessment of worldwide contamination of air, water and soils by trace metals. *Nature*, 333: 134–139.

Ow, D.W. 1993. Phytochelatin-mediated cadmium tolerance in schizosaccharomyces pombe. *In Vitro Cellular and Developmental Biology—Plant*, 29: 213–219.

Panagos, P., Van Liedekerke, M., Yigini, Y., and Montanarella, L. 2013. Contaminated sites in Europe: Review of the current situation based on data collected through a European network. *Journal of Environmental and Public Health*, 2013: 11.

Pierzynski, G.M., Sims, J.T., and Vance, G.F. 2000. *Soils and Environmental Quality*. London, U.K.: CRC Press.

Pohontu, C.M. and Gontariu, I. 2014. Bioremediation of Ni^{2+} polluted soils by plant-associated bacteria in heavy metal phytoremediation processed. *Proceedings of the Seventh International Conference on Environmental and Geological Science and Engineering (EG '14)*, pp. 281–286, Romania.

Rai, D. and Zachara, J.M., 1984. *Chemical Attenuation Rates, Coefficients, and Constants in Leachate Migration*, Volume 1: A Critical Review (EPRI EA-3356). Palo Alto, CA: Electric Power Research Institute.

Sar, P., Kazy, S.K., and Singh, S.P. 2001. Intracellular nickel accumulation by *Pseudomonas aeruginosa* and its chemical nature. *Letters in Applied Microbiology*, 32: 257–261.

Sayel, H., Bahafid, W., Tahri Joutey, N., Derraz, K., Fikri Benbrahim, K., Ibnsouda Koraichi, S., and El Ghachtouli, N. 2012. Cr(VI) reduction by *Enterococcus gallinarum* isolated from tannery waste-contaminated soil. *Annals of Microbiology*, 62: 1269–1277.

Schmidt, J.A. and Andren, A.W. 1980. The atmospheric chemistry of nickel. In *Nickel in the Environment*, Nriagu, J.O. (ed.), pp. 93–135. New York: John Wiley & Sons, Inc.

Schroeder, W.H., Dobson, M., Kane, D.M., and Johnson, N.D. 1987. Toxic trace elements associated with airborne particulate matter: A review. *JAPCA*, 37: 1267–1285.

Seung, H.K., Singh, S., Kim, J.Y., Lee, W., Mulchandani, A., and Chen, W. 2007. Bacteria metabolically engineered for enhanced phytochelatin production and cadmium accumulation. *Applied and Environmental Microbiology*, 73: 6317–6320.

Shiowatana, J., Mclaren, R.G., Chanmekha, N., and Samphao, A. 2001. Fractionation of Arsenic in soil by a continuous-flow sequential extraction method. *Journal of Environmental Quality*, 30(6): 1940–1949.

Singh, R., Singh, P., and Sharma, R. 2014. Microorganism as a tool of bioremediation technology for cleaning environment: A reviewed. *Proceedings of the International Academy of Ecology and Environment*, 4(1): 1–6.

Sousa, A., Pereira, R., Antunes, S.C., Cachada, A., Pereira, E., Duarte, A.C., and Gonçalves, F. 2008. Validation of avoidance assays for the screening assessment of soils under different anthropogenic disturbances. *Ecotoxicology and Environmental Safety*, 71: 661–670.

Sriprang, R., Hayashi, M., Ono, H., Takagi, M., Hirata, K., and Murooka, Y. 2003. Enhanced accumulation of Cd^{2+} by a *Mesorhizobium* sp. transformed with a gene from *Arabidopsis thaliana* coding for phytochelatin synthase. *Applied and Environmental Microbiology*, 69: 1791–1796.

Strong, P.J. and Burgess, J.E. 2008. Treatment methods for wine-related and distillery wastewaters: A review. *Bioremediation Journal,* 12: 70–87.

Su, C., Jiang, L., and Zhang, W. 2014. A review on heavy metal contamination in the soil worldwide: Situation, impact and remediation techniques. *Environmental Skeptics and Critics,* 3(2): 24–38.

Suazo-Madrid, A., Morales-Barrera, L., Aranda-García, E., and Cristiani-Urbina, E. 2011. Nickel(II) biosorption by *Rhodotorula glutinis. Journal of Industrial Microbiology & Biotechnology,* 38: 51–64.

Tabak, H., Lens, P., Van Hullebusch, E., and Dejonghe, W. 2005. Developments in bioremediation of soils and sediments polluted with metals and radionuclides—1. Microbial processes and mechanisms affecting bioremediation of metal contamination and influencing metal toxicity and transport. *Reviews in Environmental Science and Bio/Technology,* 4: 115–156.

Talwar, G.P. and Srivastava, L.M. 2002. *Textbook of Biochemistry and Human Biology.* PHI Learning Pvt. Ltd., New Delhi, India.

T.R.I. 2004. Toxic Release Inventory (TRI explorer): EPA's toxics release inventory data.. Washington, DC: Office of Environmental Information, U.S. Environmental Protection Agency.

Valls, M., Atrian, S., De Lorenzo, V., and Fernández, L.A. 2000. Engineering a mouse metallothionein on the cell surface of *Ralstonia eutropha* CH34 for immobilization of heavy metals in soil. *Nature Biotechnology,* 18: 661–665.

Vijayaraghavan, K. and Yun, Y.-S. 2008. Bacterial biosorbents and biosorption. *Biotechnology Advances,* 26: 266–291.

Wu, L., Yu, J., Sun, X., and Li, B. 2009. The effect of nickel(II) ions on the growth and bioaccumulation properties of *Escherichia coli. Environmental Progress and Sustainable Energy,* 28: 234–239.

Yao, Z., Li, J., Xie, H., and Yu, C. 2012. Review on remediation technologies of soil contaminated by heavy metals. *Procedia Environmental Sciences,* 16: 722–729.

Yetis, U., Dolek, A., Dilek, F.B., and Ozcengiz, G. 2000. The removal of Pb(II) by *Phanerochaete chrysosporium. Water Research,* 34: 4090–4100.

Young, T.C., De Pinto, J.V., and Seger, E.S. 1982. Transport and fate of heavy metals in Onondaga lake, New York. *Bulletin of Environmental Contamination and Toxicology,* 29: 554–561.

Zhang, M.-K., Liu, Z.-Y., and Wang, H. 2010. Use of single extraction methods to predict bioavailability of heavy metals in polluted soils to rice. *Communications in Soil Science and Plant Analysis,* 41: 820–831.

Zhao, X.W., Zhou, M.H., Li, Q.B., Lu, Y.H., He, N., Sun, D.H., and Deng, X. 2005. Simultaneous mercury bioaccumulation and cell propagation by genetically engineered *Escherichia coli. Process Biochemistry,* 40: 1611–1616.

39 Sulfate-Reducing Bacteria
Applications to Metals Bioremediation and Biorecovery

John W. Moreau

CONTENTS

ABSTRACT

The topic of sulfate-reducing bacteria (SRB) in heavy metals bioremediation (or the corollary purpose of metals biorecovery in some environments) has received increased attention over the past couple of decades. Bioremediation, or natural attenuation, by way of harnessing the metabolic activity of SRB to induce the precipitation of contaminant metals as metal sulfides, metal oxides, or elemental metals has become a widely studied and accepted approach to remediating acid-mine drainage and other metalliferous effluents. Yet, a number of factors remain to be understood in better detail for SRB-based bioremediation strategies to be effective. These factors include the nanoparticulate nature of the biogenic precipitates, SRB resistance strategies for metal toxicity and other environmental stressors, community interactions and interdependencies with other microbial species or groups, and the role of organic substrates and natural organic matter in promoting or inhibiting the activity of SRB and influencing metal speciation. This chapter attempts to present a review of seminal and recent studies across these topics and discusses the concept of bacterial sulfate reduction in the context of the evolution and extent of this process within the Earth's biogeochemical sulfur cycle. In this way, the processes of metal precipitation at human and geological time scales are connected through the environmental and ecological significance of this ancient anaerobic respiratory pathway.

39.1 INTRODUCTION

The topic of "sulfate-reducing bacteria (SRB) in bioremediation" has received increasing attention in the past two decades of peer-reviewed scientific literature. Previously, peer-reviewed publications numbering in the tens of papers

per year now range between 100 and 200 annually, and earlier work centered around laboratory experiments has expanded to include more field-based research as well as the application of molecular, cultivation-independent techniques for studying the diversity and activity of SRB in contaminated environments. These advances reflect the increasing interest in, and need for, investigating and harnessing potentially useful natural (i.e., microbially mediated) processes for the neutralization and detoxification of acid- and metal-contaminated waste streams. Such streams unfortunately seem to be in no shortage for the twenty-first century, despite several decades of increased public and corporate awareness, and are produced by a variety of natural and anthropogenic sources: acid-mine and/or acid-rock drainage (AMD/ARD), acid sulfate soil weathering, industrial wastewater production, coal and metallurgical processing effluents, and exposure of naturally or anthropogenically exhumed pyritic sediments and sedimentary rocks.

Detoxification of acidic/metalliferous (waste)waters through natural or engineered bacterial sulfate reduction (BSR) generally involves the precipitation of contaminant heavy metals as metal sulfide minerals, often colloidal or nanoparticulate in particle size (Figure 39.1), following reaction of dissolved metals with aqueous (bi)sulfide (at most natural values of pH), the primary metabolic product of BSR. In some cases, metal contaminants are immobilized and remediated through direct reduction (e.g., Cr^{VI}, U^{VI}, Se^{VI}, Tc^{VII}) or by the production of bicarbonate alkalinity as a neutralizing agent for sulfuric acid–generated protons. These mechanisms by which SRB are harnessed to bioremediation schemes were once thought to be limited mainly by the presence of molecular oxygen, as most SRB are obligate anaerobes. However, more recent studies have shown that oxygen may be less of a problem than previously assumed, as one of the most common (and well-studied) SRB species involved

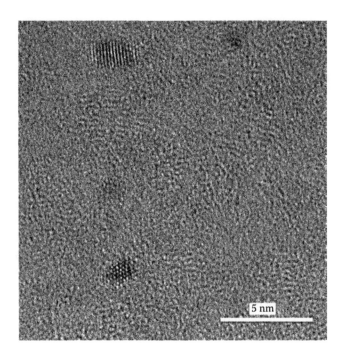

FIGURE 39.1 High-resolution transmission electron microscope image of biogenic zinc sulfide formed by the activity of SRB in an abandoned Pb–Zn mine. (J.W. Moreau; unpublished data.) ZnS nanoparticles can be distinguished from the background SRB biofilm by their diffraction contrast (they appear dark and contain periodic crystal lattice fringes). The ZnS nanoparticle in the lower left center of the image is in near zone axis projection, as can be seen by the regular spacing of ZnS atoms in cubic (A-B-C)-type stacking.

in bioremediation experiments, *Desulfovibrio* spp., displays a remarkable degree of oxygen tolerance (although probably not active growth under oxygen). In contrast, more focus has been placed recently on other potential inhibitors for BSR, particularly the metals to be remediated or the reaction product of BSR itself, bisulfide, the effects of which may require more detailed study if SRB-based bioremediation strategies are to be optimized or improved.

Other factors, such as spatial or temporal variability in the distribution, diversity, and activity of SRB, may also inhibit or promote the efficiency of bioremediation, alternately, and need to be considered on the basis of sufficiently high-resolution spatial and time-series data. The "classical" paradigm of decreasing reduction potential aligned with increasing vertical depth may be insufficient or even entirely inaccurate for describing the actual 3D geometry of redox conditions in mine tailings, mine pit lakes, landfill leachate plumes, or acid sulfate soils/sediments. In this case, the activity of SRB could vary highly, thus undermining the efforts to model BSR predictively across either relevant spatial or temporal scales.

This chapter reviews some recent advances in our understanding of how the metabolism, physiology, phylogenetic diversity, and activity of SRB can be coupled to the bioremediation of acid, heavy metals, and radionuclides. We will examine the topics mentioned earlier, as well as other relevant aspects of SRB in bioremediation, in the context of both the

(roughly) last two decades of research and earlier seminal work involved in establishing the field. Subheadings to be discussed include the following:

- BSR: metabolism and physiology
- The sulfate-reducing prokaryotes (SRP): antiquity, phylogeny, and diversity
- Microbial ecology of SRB: role in biogeochemical sulfur cycling
- Bioremediation using sulfate reducers: impacts on acidity and metal mobility
- Biorecovery of metals: potential for "green" biotechnology?

We will not present here a discussion of hydrocarbon biodegradation/bioremediation by SRB, as this subject encompasses a sufficiently large enough breadth and depth of literature as to warrant its own chapter. A number of excellent past and recent reviews have collated and presented milestones in our accumulating knowledge about microbes and bioremediation (Gadd 2010), particularly the SRP (Pfennig et al. 1981, Postgate 1984, Voordouw et al. 1993, Rabus et al. 2006), including with a focus on metabolism and ecophysiology (Plugge et al. 2011, Rabus et al. 2015), specific genes and enzymes involved in the dissimilatory sulfate (or sulfur) reduction process (Meyer and Kuever 2007, Fauque and Barton 2012, Müller et al. 2015), and comparative genomics of SRP (Muyzer and Stams 2008, Pereira et al. 2011). Specific interactions between SRP and a range of metals have also been surveyed (Barton et al. 2015), as well as the specific application of SRB to the bioremediation of AMD (Kaksonen and Puhakka 2007), treatment of wastewater (Hao et al. 2014), and various other biotechnological applications (Lens et al. 2002).

39.2 BACTERIAL SULFATE REDUCTION: METABOLISM AND PHYSIOLOGY

Microorganisms obtain energy for cell growth and enzyme functionality essentially from the existence of geochemical disequilibria (Trudinger and Bubela 1967, Chyba and Hand 2005). The thermochemical (i.e., abiological) oxidation of organic carbon or molecular hydrogen coupled to the reduction of aqueous sulfate is energetically advantageous, but kinetically challenging at temperatures $\leq 100°C$ (Trudinger et al. 1985, Machel 2001). Dissimilatory SRP evolved somewhere beyond 2–2.4 billion years ago (Shen and Buick 2004) to catalyze enzymatically the eight-electron transfer to transform S^{VI} to S^{II-} at sufficiently high reduction potential (pE ≤ -2 at most circumneutral pH values) and to harvest the free energy released for their catabolic and anabolic needs (Postgate 1984, Odom and Singleton 1993, Hansen 1994, Plugge et al. 2011, Rabus et al. 2015).

The topic of *how* SRB reduce sulfate inevitably leads to detailed discussions of biochemistry that lie beyond the scope of this chapter. It has been known for some time that two key enzymes, *aps* or adenosine 5'-phosphosulfate reductase (Odom et al. 1991, Friedrich 2002, Meyer and Kuever 2007) and *dsr* or dissimilatory sulfite reductase (Wagner et al. 2005, Müller et al. 2015), mediate the transformation of sulfate to (bi)sulfide in two stages, essentially (pre- and postsulfite, SO_3^{2-}). More recently, researchers have identified a key trisulfide protein complex formed through the activity of the DsrC protein as responsible for limiting the rate of BSR (Venceslau et al. 2014, Santos et al. 2015). As might be expected from the required oxidation/reduction potential for BSR to occur, these enzymes require lower redox potentials and the absence of dissolved oxygen, and thus, virtually all SRB are obligate anaerobes. Some studies, however, suggest a remarkable degree of oxygen tolerance for SRB, for example, even just within the highly diverse genus *Desulfovibrio* (Minz et al. 1999, Cypionka 2000). This tolerance is apparently attributed to a reversible blocking of cell division in the presence of molecular oxygen (Fievet et al. 2015). Yet for most SRB species, anoxic conditions are still required for cell growth.

The following equations describe the process of BSR coupled to the oxidation of three common energy substrates: (1) pyruvate, (2) acetate, and (3) molecular hydrogen, arguably the three most common electron sources among a wide range of substrates potentially metabolized by SRB (Hansen 1994), at pH 7:

$$4C_3H_5O_3^- + SO_4^{2-} + 2H^+ \rightarrow 4C_2H_3O_2^- + 4CO_2 + H_2S$$

$$\Delta G^{o\prime} = -360.3\,kJ \tag{39.1}$$

$$C_2H_3O_2^- + SO_4^{2-} + 3H^+ \rightarrow 2CO_2 + H_2S + 2H_2O$$

$$\Delta G^{o\prime} = -57.5\,kJ \tag{39.2}$$

$$4H_2 + SO_4^{2-} + 2H^+ \rightarrow H_2S + 4H_2O \quad \Delta G^{o\prime} = -151.9\,kJ \tag{39.3}$$

Although most SRP oxidize organic compounds like lactate or pyruvate to acetate, or acetate to CO_2, recently one species has been found that can oxidize some organic substrates more complex than acetate completely to CO_2 (Strittmatter et al. 2009). From these and similar redox reactions with sulfate as the terminal electron acceptor, SRB obtain energy for adenosine triphosphate production and organic carbon for cell growth (Postgate 1984, Peck 1993, Rabus et al. 2000). Note that, in these reactions, protons are consumed, while hydrogen sulfide and carbon dioxide (as bicarbonate) are released. Thus, pH and alkalinity will locally increase during BSR, in the absence of titration by acids. Hydrogen sulfide expelled from the SRB cell (and quickly dissociating to bisulfide, HS^-, at circumneutral pH; Millero 1986) can reductively dissolve iron and manganese ox(yhydrox)ides, releasing various commonly adsorbed trace metals such as lead or arsenic (Foster et al. 1998, Morin et al. 1999). However, (bi)sulfide also reacts with a number of dissolved metals, including Fe, Cu, Zn, Pb, Cd, As, and Hg, to precipitate metal sulfides of very low solubilities (log $K_{sp} < -10$) (Sposito 1989, Stumm and Morgan 1996). Thus, in low-temperature anoxic sedimentary environments,

BSR essentially controls the speciation of a wide range of metals (Berner 1967, Boulegue et al. 1982, O'Day et al. 2000). These metals are effectively removed from the hydrosphere for as long as they remain sequestered in metal sulfides. A number of reduced metal oxides (Cr^{III}, U^{IV}) and native metals (Au^0, Se^0, Pd^0) as well as radionuclides (Tc, Np) can also be precipitated inductively by the metabolic activity of SRB.

39.3 THE SULFATE-REDUCING PROKARYOTES: ANTIQUITY, PHYLOGENY, AND DIVERSITY

Two lines of investigation support the antiquity of the evolution of SRP (which include one known Archaean, the genus *Archaeoglobus*). The first comes out of a growing body of molecular geomicrobiological research (Klenk et al. 1997, Heidelberg et al. 2004, Rabus et al. 2004, Wall et al. 2008), including an increasingly large number of fully sequenced SRP genomes (Joint Genome Institute [Genome Portal] n.d.).

Comparative phylogenetic analysis of SRP small subunit ribosomal RNA genes (16S rRNA), a structurally and functionally conservative molecular evolutionary marker for all life (Woese et al 1975), and dissimilatory sulfite reductase (*dsrAB*), the most important of the functional genes coding for the enzymes involved in sulfate respiration (Wagner et al. 1998), reveals an early origin and largely congruent evolution (Klein et al. 2001, Stahl et al. 2002). SRB conferred the ability to respire sulfate to the hyperthermophilic *Archaeoglobus* (Stetter et al. 1987, Stetter 1988) soon after these two domains of life diverged (Klein et al. 2001). This finding suggests that early SRP evolved from thermophilic or hyperthermophilic origins, an interpretation supported by the physiology of the deepest branching SRB, the thermophilic *Thermodesulfovibrio* (Klein et al. 2001). It is interesting to note that both pyruvate and hydrogen could have been available to SRB (Equations 39.1 and 39.3, respectively) from abiotic seafloor hydrothermal processes (Cody et al. 2000).

The second line of evidence for the antiquity of SRB derives from sulfur isotope measurements of ancient sedimentary pyrites that support the interpretation that BSR is at least a 2.5–2.7 Gyr-old process (Ono et al. 2003). It is possible that SRB actually evolved over 2.7 Gyr ago in restricted environments (Ohmoto et al. 1993, Kakegawa and Ohmoto 1999, Shen et al. 2001), given that sulfate concentrations were ≤1% of their modern value (Habicht et al. 2002). However, evidence to support this interpretation has been obscured by Archean atmospheric (abiotic) photochemical processes (Farquhar et al. 2000, Mojzsis et al. 2003). Nonetheless, it is interesting to speculate on what implications the confirmation of an Archean record of BSR would have for our understanding of the origin and evolution of SRB. The strategies that these microbes evolved to capture energy for metabolism reflect the evolution of geochemical disequilibria on the early Earth. Based on our current knowledge, the habitat range, diversity, and antiquity of SRB may still be underestimated (Stahl et al. 2002). The contemporary phylogeny and diversity of SRB, with a focus on environmental and engineered systems and through the application of functional gene-targeted bioinformatics, were most recently reviewed by Thauer et al. (2007) and Pelikan et al. (2015).

39.4 MICROBIAL ECOLOGY OF SULFATE-REDUCING BACTERIA: ROLE IN BIOGEOCHEMICAL SULFUR CYCLING

Dissimilatory sulfate reduction plays a major role in the global biogeochemical sulfur cycle (Brimblecombe and Lein 1989, Ehrlich 1996). Sulfate is the second most abundant anion in modern seawater (~28 mM) and freshwater (~30 μM to 20 mM) (Charlson et al. 1992, Stumm and Morgan 1996, Sparks 2003). Each year, SRB in marine sediments reduce up to ~7 billion metric tons (Gt) of seawater sulfate globally and sequester up to ~1 Gt of pyrite (or ~0.7 Gt of mackinawite, FeS) into the seafloor (Canfield and Raiswell 1999). The flux of FeS formation in coastal and shelf sediments is at least ~1.6 times that of anhydrite formation in seawater circulating through mid-ocean ridges (Charlson et al. 1992). In coastal marine sediments, sulfate is the most abundant terminal electron acceptor, and BSR accounts for 70%–95% of coastal organic matter oxidation (Jørgensen 1982, Mackin and Swider 1989, Blackburn and Blackburn 1993).

Although these environments comprise <10% of the global ocean, they account for ~90% of marine sulfate reduction (Jørgensen 1982). In pelagic sediments, where sulfate is abundant but organic matter may be limited, BSR still oxidizes ~35% of organic carbon. In freshwater sediments (e.g., rice paddies, eutrophic lakes, quiescent streams, flooded soils), between ~30% and 80% of organic carbon can be oxidized via BSR (Lovley and Klug 1983, Capone and Kiene 1988, Achtnich et al. 1995, Holmer and Storkholm 2001).

The ability of SRB to degrade significant quantities of natural organic matter reveals their role as a microbial bridge between the global biogeochemical cycles of sulfur and carbon. This process is also responsible for the so-called "souring" of hydrocarbon deposits under anaerobic conditions (Tanji et al. 2014). More recently, researchers have reexamined the close linkage of BSR to methane oxidation and the adaptability of SRB in the presence of methanogens, concluding that some *Desulfovibrio* display remarkably high enzymatic and metabolic adaptive flexibility that likely sustains them in naturally fluctuating methanogenic environments (Meyer et al. 2013). This metabolic flexibility renders the SRB particularly suitable for the bioremediation of a range of organic contaminants or mixed waste effluents such as AMD/ARD and municipal wastewater. However, as discussed in the following, the nature and abundance of organic matter can also limit the diversity and metabolism of SRB in contaminated systems.

It is worth noting that the waste product of BSR, (bi) sulfide, is also highly reactive for microbial oxidation, and

this process can be mediated both aerobically (coupled to O_2 reduction) and anaerobically (coupled to NO_3^- reduction). Most aqueous biogenic sulfide (~60%–95%), in both marine and continental environments, is reoxidized via abiotic or microbially mediated pathways (Cline and Richards 1969, Jørgensen 1982, Swider and Mackin 1988). This sulfide provides the "fuel" (i.e., electron donors) for diverse sulfide- and sulfur-oxidizing, and ultimately sulfur-, sulfite-, and thiosulfate-disproportionating, microbial populations (Kuhl and Jørgensen 1992, Voordouw et al. 1996, Friedrich 1998). Collectively, these groups of prokaryotes establish the microbial sulfur cycle (Figure 39.2). In marine settings, (re)cycling of sulfide to sulfate is possible because the flux of biogenic sulfide is ≥5 times that of iron from continental weathering and seafloor hydrothermal activity (Canfield and Raiswell 1999). Generally, the extent of BSR exceeds the flux of metal delivery and thus the rate of metal sulfide precipitation, in most low-temperature natural sedimentary environments, with the implication that ample sulfide is bioavailable for sulfide-/sulfur-oxidizing bacteria (SOB). In metal-contaminated systems, however, the ratio of BSR and metals fluxes may be very different from typical sedimentary conditions, with extremely high metal contents being the norm in AMD/ARD streams, for example.

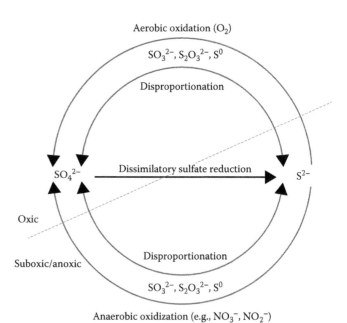

FIGURE 39.2 Schematic representation of the microbial sulfur cycle. The cycle is shown as centered on dissimilatory sulfate reduction. Potential oxidants are shown in parentheses for aerobic and anaerobic sulfide oxidation. Primary reductants for sulfate or intermediate sulfur species (e.g., organic carbon, hydrogen) are not shown. (Modified from Madigan, M.T. et al., *Brock Biology of Microorganisms*, 9th edn., Prentice-Hall, Upper Saddle River, NJ, 2000, p. 687.)

39.5 BIOREMEDIATION USING SULFATE REDUCERS: IMPACTS ON ACIDITY AND METAL MOBILITY

AMD alters natural systems drastically with respect to both sulfate and metal concentrations. The concentration of sulfate in AMD, which by definition originates in continental settings, can be comparable to or enriched relative to seawater, while the concentrations of potentially toxic metals can exceed those found in uncontaminated environments by many orders of magnitude (Nordstrom and Alpers 1999). However, as most known SRB are neutrophilic and obligate anaerobes (Rabus et al. 2000), they cannot take advantage of this sulfate enrichment under such extremes of low pH and high pE. In addition, the small molecular weight organic acids used as electron donors (pyruvate, acetate) are protonated in the low pH milieu outside the cell, but deprotonate when taken into the cell, lowering the internal pH with destructive consequences for key respiratory enzymes (Norris and Ingledew 1992).

High concentrations of potentially toxic metals associated with AMD, especially copper, may also inhibit SRB growth (Sani et al. 2001, Utgikar et al. 2001, Cabrera et al. 2006) by binding to and/or oxidizing enzymes (Dopson et al. 2003). Molybdate is a well-known inhibitor for BSR (Oremland and Capone 1988, Nair et al. 2015), and some studies have evaluated changes in the community structure and dynamics of SRP resulting from heavy metal stress (Hesham et al. 2014). Other studies suggest, however, that certain species of SRB may have developed defense mechanisms for metal toxicity that utilize extracellular metal-binding proteins (Fortin et al. 1994, Bridge et al. 1999). In many of these proteins, the amino acid cysteine, with its functional thiol (HS^-) group, plays an important part in the metal-binding mechanism (Rosen 1996). This is an intriguing and relatively unexplored area of research with regard to SRB and their ability to demonstrate metal resistance.

Even if SRB could tolerate the environmental pressures associated with AMD, the paucity of labile organic matter and fast abiotic oxidation rate of hydrogen in AMD solutions would not favor their growth. It may not be surprising then that most known acid-tolerant SRB are spore-forming genera (*Desulfotomaculum*, *Desulfosporosinus*; Stackebrandt et al. 1997) that may utilize this physiological adaptation to survive until geochemical conditions hospitable for growth return. It is not yet clearly understood whether sporulating SRB can remain metabolically active. However, recent studies, based on sulfur isotope measurements of sedimentary sulfides and pore water sulfate, and cultivation results, suggest that some BSR does occur in and around oxidizing pyrite-rich mine tailings at pH values <3, possibly in less acidic microenvironments (Wielinga et al. 1999, Praharaj and Fortin 2004). The activity of SRB under such low pH conditions remains an area of active research because of the potential role for these organisms in AMD remediation.

In contaminated environments of moderately acidic to circumneutral pH (i.e., 3–8), such as natural waters and sediments receiving diluted and/or neutralized AMD, SRB are known to remain metabolically active (Chang et al. 2001,

Weijma et al. 2002, Suzuki et al. 2003, Kaksonen et al. 2004). As mentioned earlier, BSR itself increases pH, and bisulfide reacts with and reduces the concentrations of many potentially toxic metals in solution (Labrenz et al. 2000). Some studies suggest the potential for harnessing BSR to metals and acid remediation in mine pit lakes (Koschorreck et al. 2010, Kumar et al. 2011, Meier et al. 2012). Studies have demonstrated the efficacy of using SRP to remediate mine drainage in column experiments, also studying the dynamics of SRP populations (Pereyra et al. 2008), or to remediate tannery, dye, and metallurgical wastewaters in similar batch reactor-style experiments (Kosinska and Miskiewicz 2012). Some natural acid-generating environments (acid-rock systems), such as acid sulfate soil systems, can produce copious amounts of acid having the same effect as AMD, and for which the activity of SRB has been studied as a potential tool to which remediation strategies might be harnessed (the East Trinity wetlands, Cairns, Australia; Rio Tinto, Spain) (Sánchez-Andrea et al. 2012, Ling et al. 2015). SRB have also been utilized in metals remediation efforts in landfill leachate plumes (Schmidtova and Baldwin 2011) and tannery wastewater streams. Reductive precipitation of Cr(VI) has been shown to be effective in mixed microbial communities in which SRB played a major role (Brodie et al. 2011, Ahmadi et al. 2015, Joo et al. 2015). Cotreatment of AMD and municipal wastewater is also a current goal of BSR-based bioremediation schemes (Deng et al. 2016). Broader interest exists in the potential for using hyperthermophilic SRP to remediate chalcophilic metals in higher temperature effluents, for example, via the use of *Archaeoglobus fulgidus* (Parey et al. 2013), as well as to couple BSR to metal immobilization with methane generation from anaerobic reactors (Paulo et al. 2015).

The use of SRB to immobilize (directly or indirectly) radionuclides such as U, Tc, or Np has recently gained interest (Michalsen et al. 2007, Junier et al. 2009, Zakharova et al. 2011, Converse et al. 2013, McGuinness et al. 2015, Stylo et al. 2015, Thorpe et al. 2016). Studies have examined the activity of SRB for bioremediation potential via direct enzymatic reduction of U(VI) in U mine tailings (Suzuki et al. 2003, Bondici et al. 2013) and have found good correlation between U immobilization as U(IV) and the activity of SRB (Cardenas et al. 2010). Other lab-based studies, however, have detected no significant correlations between the diversity and structure of iron- or sulfate-reducing bacterial populations and U(VI) concentrations in effluent (Tokunaga et al. 2008). To some extent, the efficacy of U remediation depends on the bicarbonate concentration in the solution (Li et al. 2016) as well as the presence of other potentially toxic metals (Sitte et al. 2015). Interestingly, reduction of U(VI) by SRB may also occur indirectly via mediation by biogenic FeS (i.e., via oxidation of sulfide in FeS coupled to reduction of U[VI]) (Bargar et al. 2013, Veeramani et al. 2013). Close spatial associations among natural organic matter, U(IV) phases (e.g., uraninite, coffinite), iron sulfides (FeS_2), and microbial pseudomorphs can be found in organic-rich sedimentary paleoenvironments and lead to the question of complex redox processes involving one or more intermediate species (phases) (Figure 39.3).

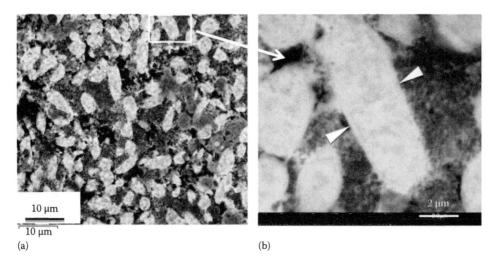

(a)   (b)

FIGURE 39.3 Spatially correlated scanning electron microscope energy-dispersive x-ray (EDX) map (a) and bright-field transmission electron photomicrograph (b) of U-mineralized microbial microfossils (pseudomorphs?) in paleo-lacustrine sediments. Features in the EDX map (a) appear mineralized by coffinite (pink) and distributed in close association with a cluster of framboidal pyrites (green) and silica (purple-gray); a close-up of one such feature (b) shows microbial cell-like structure. (Unpublished data from Cumberland and Moreau.)

These observations have led many researchers to study the use of BSR in constructed wetland bioremediation of AMD and metals (White and Gadd 1996, Gadd 2000, Song et al. 2001, Johnson and Hallberg 2005, Sun et al. 2015). One of the biggest benefits of this approach, after mitigation of acidity and metal toxicity, is the relatively low cost associated with organic substrate amendments suitable for SRB growth. Waste products from the paper and timber industries, along with treated wastewater sludge, have been proposed for this purpose. In some locations, this method of *in situ*, or so-called "passive," bioremediation is already being tried in field-scale experiments (Gammons and Frandsen 2001, Johnson and Hallberg 2005, Nicomrat et al. 2005). One of the biggest risks, obviously, is the potential for unmitigated contamination of wetland environments, with negative consequences for riparian biodiversity and water quality management efforts (Mitsch and Gosselink 2000). A variation of this research direction involves the use of fluidized bed reactors (Sahinkaya et al. 2007), as reviewed recently by Papirio et al. (2013).

Current directions in bioremediation research include the investigation of how natural and constructed wetlands and other aqueous environments receiving AMD or other metalliferous waste streams respond microbially and biogeochemically over extended time periods (Kang et al. 2013) and the impact of these responses with regard to the long-term fate of metals sequestered into biogenic sulfides (Moreau et al. 2013). Some short-term studies have already revealed unexpected challenges, such as the enrichment of acidophilic iron-oxidizing bacteria and/or SOB in *anaerobic* treatment ponds (Johnson and Hallberg 2005). Other studies have noted anomalously high apparent solubility of biogenic metal sulfides and inferred that they are either or both nanoparticulate in size and amorphous or poorly crystalline (Gammons and Frandsen 2001). The implication of this observation is that biogenic metal sulfides can be transported easily by fluid flow,

and aqueous metals can be dispersed from the contaminant source under conditions favorable for oxidative redissolution.

Few studies of long-term (e.g., a decade or more) AMD and/or metal-contaminated natural wetland ecosystems (Figure 39.4) have been conducted, particularly with respect to SRB diversity or activity (Moreau et al. 2010, Pavissich et al. 2010), for comparison to constructed wetland experiments (Ledin and Pedersen 1996). In fact, relatively little is known at all about microbiological processes occurring in natural wetlands and their relevance for bioremediation

FIGURE 39.4 Former site of long-term AMD-contaminated natural wetland. The wetland was a natural tidal marsh that received nearly a century of AMD from buried but oxidizing pyritic mine tailings. This site was remediated in the early to mid-2000s through a combination of active and passive strategies (e.g., removal and disposal of AMD-generating tailings and sediments, and tidal inundation and the promotion of BSR to immobilize dissolved metal contaminants). (Courtesy of J.W. Moreau.)

(Webb et al. 1998). This is especially true for brackish water ecosystems such as coastal salt marshes. The use of constructed wetlands is thus based largely on short-term, limited-contamination experiments or even solely on SRB pure culture studies. A largely unexplored area of research is the natural diversity, distribution, and activity of SRB, in the context of whole microbial community structures, in AMD-contaminated environments. Particular focus should be placed on the identification and cultivation of novel SRB species, especially those that demonstrate some degree of acid and/or metals tolerance (Alazard et al. 2010). A recent study found new species of acid-tolerant SRB that may be useful for remediating heavy metals from AMD/ARD streams (Nancucheo and Johnson 2012, Sánchez-Andrea et al. 2014, 2015). The use of acidophilic sulfur reducers to treat AMD (Florentino et al. 2015) will likely continue to be an applied research goal.

Spatiotemporal interactions with or interplay between SRB and other metals (Cd; Medircio et al. 2007), microbes (e.g., *Geobacter* spp.), and natural organic matter may impact the ability of SRB to immobilize or transform metals into precipitates or other dissolved species (Yabusaki et al. 2011, Barlett et al. 2012, Moreau et al. 2016). The immobilization of SRB on granules or within zeolites to treat AMD has been proposed and studied as a means of controlling or localizing the distribution of active SRB cells (Kim et al. 2015, Zhang and Wang 2016). Interestingly, recent work has suggested that the application of antibiotics may selectively promote the activity of SRP in salt marsh sediments (Córdova-Kreylos and Scow 2007). Different organic substrates will influence the composition and metabolic processing of SRB population in ways that impact geochemical conditions and hence also mineral precipitation (phase, composition, structure) (Kwon et al. 2015). Microbial network or consortium interactions may be important for determining the overall strength of SRB populations (and therefore the efficacy for bioremediation) within a microbial community receiving various organic amendments (Deng et al. 2016). Heterogeneous redox conditions and variable spatial distribution of SRB will likely also cause variability in the volumetric and long-term efficacy of BSR as a method for metals immobilization, particularly under variable (or oscillatory, e.g., tidally influenced) environmental parameters (Dar et al. 2013, Ling et al. 2015). Factorial design analysis and response surface analysis are two approaches being used to optimize SRB-based bioreactor performance in the treatment of AMD against a background of complex variables (Villa-Gomez et al. 2015). For example, Yabusaki et al. (2015) modeled the interplay of a range of factors with the activity of iron-reducing bacteria and SRB involved in a U bioremediation field study. However, certain complexities may be difficult to model or optimize, such as metal-specific transformations that, in turn, depend on a range of other variables. Mercury is one metal for which bioremediation by SRB may be more complicated, as some studies have suggested that Hg(II) may be transformed to methylmercury (MeHg) by SRB rather than immobilized (Vishnivetskaya et al. 2011, Moreau et al. 2016). Researchers have observed interesting and novel lineages of SRB present in mine tailings cocontaminated by MeHg (Winch et al. 2009), and it remains an open question to what degree SRB promote the methylation of mercury or whether MeHg selects for a certain population of SRB (or both). The role of natural organic matter in controlling the chemical form and redox state of metals that could potentially be immobilized or transformed by BSR is another area of active research (Zheng et al. 2011, Moreau et al. 2016). The determination of how and to what extent SRB can cope with various environmental stressors (Zhou et al. 2011) will remain a topic of importance for contemporary bioremediation studies.

Studies have also shown the acute effects of heavy metal toxicity on SRB consortia in treatment bioreactor consortia (Kieu et al 2011), which would obviously have detrimental impacts on any bioremediation effort. To address the toxicity of highly concentrated metals toward SRB, researchers have proposed and studied the cause of SRB with ZVI or Cu–Fe bimetallic particles (to reduce heavy metal concentrations) in AMD treatment (Xin et al. 2008, Bai et al. 2013, Zhou et al. 2013, Ayala-Parra et al. 2016, Zhang et al. 2016). As AMD/ARD (or saline groundwater) streams can also exhibit higher salinities (Santini et al. 2010, Degens 2012), the response of SRB to salt/osmotic stress will also be a key consideration (Zhou et al. 2015). Sulfide itself, although the metabolic waste product of BSR, presents a significant challenge to SRB as a potentially toxic dissolved species (Caffrey and Voordouw 2010). Experimental systems that account for sulfide removal have shown better results for application to bioremediation of metals (Bijmans et al. 2008). Other limitations to applying BSR to metals bioremediation may include biogenic sulfide-induced reductive dissolution of iron oxides that could otherwise serve as sorption sites for As, the conversion of significant quantities of aqueous Hg(II) to MeHg rather than HgS(s) (Bessinger et al. 2012), the proportion of different types of available organic amendments, or the excessive presence of Al and Fe oxides that may encrust and mineralize SRB cells (Lefticariu et al. 2015). However, one study demonstrated that the addition of sulfate to waste streams of As may actually help to immobilize As as an arsenic sulfide mineral (Rodriguez-Freire et al. 2015) rather than via adsorption to Fe-oxides.

The stability of biogenic sulfides and their incorporated metals, with respect to reoxidation/redissolution, will determine the long-term efficacy of BSR-based remediation (Zbinden et al. 2001, Suzuki et al. 2002, Moreau et al. 2004). This factor is controlled by the chemical composition, mineralogy, size, shape, microstructure, aggregation state, and degree of crystallinity of biogenic metal sulfide (or metal oxide or native metal) precipitates. These parameters modulate the solubility, mobility, and residence times of both sulfur and chalcophilic metals in solid versus dissolved phases/species and are under greater or lesser degrees of microbial influence. An increasing number of studies have documented that the earliest-formed biogenic sulfides (or metal oxides such as biogenic U[IV] oxides) are nanoparticulate (Spear et al. 1999, Lloyd et al. 2001, Zbinden et al. 2001, Suzuki et al. 2002, Hockin and Gadd 2003, Moreau et al. 2004, Sharp et al. 2009).

These nanoparticles are easily transported through sediment pores in groundwater (Honeyman 1999) and can be either abiotically or microbially redissolved under more oxidizing conditions (cf. Evangelou and Zhang 1995, Edwards et al. 2000). The hydraulic retention time of biogenic metal sulfide nanoparticles has recently become the focus of some research (Villa-Gomez et al. 2015). The physical and chemical characteristics of biogenic sulfides may partly or wholly reflect the spatiotemporal variability of SRB activity.

A natural corollary of this research is the need for greater study of the diversity of SOB that coexist with SRB in AMD-impacted settings, and by scavenging sulfide, inhibit metal sequestration via biogenic metal sulfide precipitation, or catalyze the redissolution/remobilization of metals from sulfide minerals. In addition to aqueous sulfide, SOB can also oxidize solid sulfides very efficiently via ferrous iron oxidation (either aerobically using oxygen or anaerobically using nitrate or ferric iron, as the terminal electron acceptor; e.g., *Thiobacillus denitrificans*, *Thiomicrospira denitrificans*; Schedel and Truper 1980, Straub et al. 1996, Gevertz et al. 2000, Brofft et al. 2002). The energies available from complete aqueous and solid-phase sulfide oxidation to sulfate, coupled to nitrate reduction at pH ~7, are given in the following (cf. Equations 39.1 through 39.5):

$$HS^- + 4NO_3^- \rightarrow SO_4^{2-} + 4NO_2^- + H^+ \quad \Delta G^{\circ\prime} = -475.82\,kJ \tag{39.4}$$

$$2FeS + 9NO_3^- + 3H_2O \rightarrow 2SO_4^{2-} + 2\alpha - FeOOH + 9NO_2^-$$

$$+ 4H^+ \quad \Delta G^{\circ\prime} = -1072.15\,kJ \tag{39.5}$$

where FeS is the mineral mackinawite ($\Delta G^{\circ}_f = -87.3$ kJ/mole; Lennie et al. 1995) and α-FeOOH is goethite ($\Delta G^{\circ}_f = -488.6$ kJ/mol; Stumm and Morgan 1996). Note that both of these reactions generate protons (i.e., acidity). While SOB are well studied, with regard to AMD generation, under oxic conditions (Fowler and Crundwell 1999, Peccia et al. 2000, Edwards et al. 2003, Severmann et al. 2006), fewer studies have documented the effect(s) on metal sequestration of anaerobic bacterial sulfide oxidation (BSO) under suboxic/anoxic conditions (Schedel and Truper 1980, Beller 2005). The habitat range and physiological functionality of SOB are the areas of recent field-, culturing-, and genome-based research (Beller 2005, Beller et al. 2006). Understanding the interplay between BSR and BSO in metal-contaminated environments, such as treatment wetlands, provides critical contextual information for geochemical/mineralogical studies of the partitioning and speciation of metals (Bostick et al. 2001, Peltier et al. 2003, Moreau et al. 2013).

39.6 BIORECOVERY OF METALS: POTENTIAL FOR "GREEN" BIOTECHNOLOGY?

A corollary to SRB-based bioremediation involves the potential use of these bacteria to concentrate metals, a number of which are economically valuable, into sulfides or native metals (Au, Ni, Pd, Pt) for recovery (e.g., Hutchins et al. 1986, Lloyd et al. 1998, Tabak et al. 2003, Tabak and Govind 2003, Lengke and Southam 2006, Cao et al. 2009, Capeness et al. 2015). This process of "biorecovery" is akin to the well-established practice of "bioleaching" (for reviews, see Bosecker 1997, Krebs et al. 1997) that depends upon the oxidative part of the microbial sulfur cycle. Biorecovery, in various forms, has been gaining interest as a potential means of off-setting costs of (bio)remediating AMD and metals and maximizing the use of nonrenewable resources that are both economically and environmentally expensive to obtain (Wernick and Themelis 1998, Tabak et al. 2003, Lu et al. 2012).

Biorecovery schemes may be aided by close consideration of SRB physiology and the nature of extracellular biomolecules/metabolites apparently produced by at least some SRB species. For example, an association of biogenic zinc sulfides with extracellular proteinaceous or polypeptide molecules has been observed (Moreau et al. 2007) and posited as a potential mechanism by which certain metals may be purified from mixed metal waste streams. Interestingly, a recent study demonstrated that *Desulfovibrio vulgaris* str. DvH did not assimilate Zn as a protein metal cofactor (Lancaster et al. 2014), suggesting that extracellular protein production may be a means by which SRB can complex and remove nanophase ZnS from suspension. Furthermore, Clark et al. (2007) found that *D. vulgaris* str. DvH produced protein filaments to form biofilms, rather than polysaccharides, consistent with the idea that extracellular SRB proteins may interact with dissolved metals. In any case, a recent study noted that extracellular metabolites interact with biogenic ZnS to produce nanoparticles with a higher density of structural defects than incorporated into abiotically formed ZnS (Xu et al. 2016). These defects will influence the solubility of biogenic sulfides, if they are commonly associated with the activity of SRB.

Intriguingly, SRB seem to have been involved in biological "recovery" of economic metals from aqueous solutions for potentially much of the Earth's geologic history, possibly forming some low-temperature sedimentary metal sulfide ore deposits (Figure 39.5) (Schidlowski 1973, Misra 2000, Leach et al. 2010). In fact, a long-standing and commonly held interpretation (Southam and Saunders 2005) holds that BSR contributed significantly to the formation of numerous economic sedimentary metal sulfide ore deposits around the world (Schidlowski 1973, Boast et al. 1981, Kase et al. 1990, Anderson et al. 1998, Fallick et al. 2001, Taylor 2004, Awid-Pascual et al. 2015). If true, then a geochemical model for economic metals biorecovery (possibly as a component of a larger bioremediation scheme) already exists at least in part, if not in similar scale.

The evidence for the biogenic sulfide ore deposits interpretation is grounded mainly in sulfur isotope ratios of the putatively biogenic metal sulfide minerals, with additional petrographic and mineralogical support. Some sulfide ore deposits have characteristics consistent with low-temperature BSR-derived origins: provenance in a marine sedimentary basin; carbonate-hosted, stratiform, and stratabound base-metal sulfide deposition; high organic matter content; widely

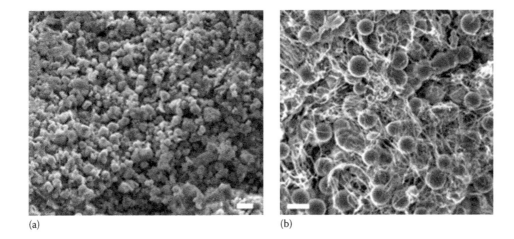

(a) (b)

FIGURE 39.5 Comparison of (a) Middle Tertiary carbonate-hosted ZnS framboids (Courtesy of Trzebionka Mine, Upper Silesia, Poland) to (b) modern SRB biofilm-hosted ZnS spherules (Courtesy of Piquette Mine, Tennyson, WI). For details of the environment in which the biofilm-hosted ZnS was formed. (Modified from Moreau, J.W. et al., *Am. Mineral.*, 89, 950, 2004; Moreau, J.W. et al., *Science*, 316, 1600, 2007.)

varying and negative (with respect to coexisting sulfate) sulfur isotopic compositions; and extremely fine-grained "colliform" (i.e., colloid-like) sulfide textures (Guilbert and Park 1986, Misra 2000, Southam and Saunders 2005). These deposits occur throughout the Phanerozoic and into the Proterozoic, implying that SRB have played a major role in terrestrial sulfur and metals cycling throughout most of geologic history (Schidlowski 1973, Misra 2000, Machel 2001). Indeed, the sulfur isotope composition of mid-Proterozoic (~1.5–1.9 Gyr old) sedimentary sulfides suggests prolonged episodes of complete bacterial reduction of sulfate in the seawater column (Poulton et al. 2004). This interpretation has been offered to explain the regional-scale sedimentary massive sulfide deposition (e.g., McArthur Basin, Northern Territory, Australia) as well as the termination of some banded iron formations (Poulton et al. 2004). Other lines of isotopic evidence (molybdenum in black shales; Arnold et al. 2004) corroborate this explanation. In any case, it is clear that BSR offers a potential means by which some reconcentration and recovery of economic metals can be achieved, preferably on societal time scales, from industrial or natural metalliferous waste streams. An optimal bioremediation system would be one that includes a mechanism by which this process could be used to reduce part of the cost of operation. Indirect metals recovery via biosorption has also been considered as a potential means for turning metalliferous waste streams into resources (de Vargas et al. 2004, Mikheenko et al. 2008).

This review has attempted to examine some of the contemporary issues surrounding the application of BSR, and the use of metabolically flexible and/or environmentally resilient SRB, as well as their metabolic products and interspecies associations in anoxic environments, to the remediation of acidic, metal-rich natural waters or industrial effluents. By no means are the references presented herein exhaustive or even comprehensive in many cases, but rather they could be considered representative of the explosion of interest in many aspects of this subject over the past roughly two decades, and the breadth of topics that perhaps must be considered in order

for bioremediation via BSR to provide a viable solution in metal-contaminated environments worldwide.

REFERENCES

Achtnich, C., Bak, F., Conrad, R. 1995. Competition for electron donors among nitrate reducers, ferric iron reducers, sulfate reducers, and methanogens in anoxic paddy soil. *Biology and Fertility of Soils* 19: 65–72.

Ahmadi, R., Rezaee, A., Anvari, M., Hossini, H., Rastegar, S.O. 2016. Optimization of Cr (VI) removal by sulfate-reducing bacteria using response surface methodology. *Desalination and Water Treatment* 57: 11096–11102.

Alazard, D., Joseph, M., Battaglia-Brunet, F., Cayol, J.L., Ollivier, B. 2010. *Desulfosporosinus acidiphilus* sp. nov.: A moderately acidophilic sulfate-reducing bacterium isolated from acid mining drainage sediments. *Extremophiles* 14: 305–312.

Anderson, I.K., Ashton, J.H., Boyce, A.J., Fallick, A.E. Russell, M.J. 1998. Ore depositional processes in the Navan Zn-Pb deposit, Ireland. *Economic Geology* 93: 535–563.

Arnold, G.L., Anbar, A.D., Barling, J., Lyons, T.W. 2004. Molybdenum isotope evidence for widespread anoxia in mid-Proterozoic oceans. *Science* 304: 87–90.

Awid-Pascual, R., Kamenetsky, V.S., Goemann, K., Allen, N., Noble, T.L., Lottermoser, B.G., Rodemann, T. 2015. The evolution of authigenic Zn–Pb–Fe-bearing phases in the Grieves Siding peat, western Tasmania. *Contributions to Mineralogy and Petrology* 170: 1–16.

Ayala-Parra, P., Sierra-Alvarez, R., Field, J.A. 2016. Treatment of acid rock drainage using a sulfate-reducing bioreactor with zero-valent iron. *Journal of Hazardous Materials* 308: 97–105.

Bai, H., Kang, Y., Quan, H., Han, Y., Sun, J., Feng, Y. 2013. Treatment of acid mine drainage by sulfate reducing bacteria with iron in bench scale runs. *Bioresource Technology* 128: 818–822.

Bargar, J.R., Williams, K.H., Campbell, K.M., Long, P.E., Stubbs, J.E., Suvorova, E.I., Lezama-Pacheco, J.S. et al. 2013. Uranium redox transition pathways in acetate-amended sediments. *Proceedings of the National Academy of Sciences of the United States of America* 110: 4506–4511.

Barlett, M., Zhuang, K., Mahadevan, R., Lovley, D. 2012. Integrative analysis of Geobacter spp. and sulfate-reducing bacteria during uranium bioremediation. *Biogeosciences* 9: 1033–1040.

Barton, L.L., Tomei-Torres, F.A., Xu, H., Zocco, T. 2015. Metabolism of metals and metalloids by the sulfate-reducing bacteria. In Saffarini, D. (Ed.), *Bacteria-Metal Interactions*, pp. 57–83. Springer International Publishing, Switzerland.

Beller, H.R. 2005. Anaerobic nitrate-dependent oxidation of U (IV) oxide minerals by the chemolithoautotrophic bacterium *Thiobacillus denitrificans*. *Applied and Environmental Microbiology* 71: 2170–2174.

Beller, H.R., Chain, P.S.G., Letain, T.E., Chakicherla, A., Larimer, F.W., Richardson, P.M., Coleman, M.A., Wood, A.P., Kelly, D.P. 2006. The Genome sequence of the facultatively anaerobic bacterium *Thiobacillus denitrificans*. *Journal of Bacteriology* 188: 1473–1488.

Berner, R.A. 1967. Thermodynamic stability of sedimentary iron sulfides. *American Journal of Science* 268: 773–785.

Bessinger, B.A., Vlassopoulos, D., Serrano, S., O'Day, P.A. 2012. Reactive transport modeling of subaqueous sediment caps and implications for the long-term fate of arsenic, mercury, and methylmercury. *Aquatic Geochemistry* 18: 297–326.

Bijmans, M.F., Dopson, M., Ennin, F., Lens, P.N., Buisman, C.J. 2008. Effect of sulfide removal on sulfate reduction at pH 5 in a hydrogen fed gas-lift bioreactor. *Journal of Microbiology and Biotechnology* 18: 1809–1818.

Blackburn, T.H., Blackburn, N.D. 1993. Coupling of cycles and global significance of sediment diagenesis. *Marine Geology* 113: 101–110.

Boast, A.M., Coleman, M.L., Halls, C. 1981. Textural and stable isotopic evidence for the genesis of the Tynagh base metal deposit, Ireland. *Economic Geology* 76: 27–55.

Bondici, V.F., Lawrence, J.R., Khan, N.H., Hill, J.E., Yergeau, E., Wolfaardt, G.M., Warner, J., Korber, D.R. 2013. Microbial communities in low permeability, high pH uranium mine tailings: Characterization and potential effects. *Journal of Applied Microbiology* 114: 1671–1686.

Bosecker, K. 1997. Bioleaching: Metal solubilization by microorganisms. *FEMS Microbiology Reviews* 20: 591–604.

Bostick, B.C., Hansel, C.M., La Force, M.J., Fendorf, S. 2001. Seasonal fluctuations in zinc speciation within a contaminated wetland. *Environmental Science and Technology*, 35: 3823–3829.

Boulegue, J., Lord III, C.J., Church, T.M. 1982. Sulfur speciation and associated trace metals (Fe, Cu) in the pore waters of Great Marsh, Delaware. *Geochimica et Cosmochimica Acta* 46: 453–464.

Bridge, T.A.M., White, C., Gadd, G.M. 1999. Extracellular metal-binding activity of the sulphate-reducing bacterium *Desulfococcus multivorans*. *Microbiology* 145: 2987–2995.

Brimblecombe, P., Lein, A.Y., Eds. 1989. *Evolution of the Global Biogeochemical Sulphur Cycle*, p. 241. Wiley & Sons, Chichester, U.K.

Brodie, E.L., Joyner, D.C., Faybishenko, B., Conrad, M.E., Rios-Velazquez, C., Malave, J., Martinez, R. et al. 2011. Microbial community response to addition of polylactate compounds to stimulate hexavalent chromium reduction in groundwater. *Chemosphere* 85: 660–665.

Brofft, J.E., Vaun McArthur, J., Shimkets, L.J. 2002. Recovery of novel bacterial diversity from a forested wetland impacted by reject coal. *Environmental Microbiology* 4: 764–769.

Cabrera, G., Pérez, R., Gomez, J.M., Ábalos, A., Cantero, D. 2006. Toxic effects of dissolved heavy metals on *Desulfovibrio vulgaris* and Desulfovibrio sp. strains. *Journal of Hazardous Materials* 135: 40–46.

Caffrey, S.M., Voordouw, G. 2010. Effect of sulfide on growth physiology and gene expression of *Desulfovibrio vulgaris* Hildenborough. *Antonie van Leeuwenhoek* 97: 11–20.

Canfield, D.E., Raiswell, R. 1999. The evolution of the sulfur cycle. *American Journal of Science* 299: 697–723.

Capeness, M.J., Edmundson, M.C., Horsfall, L.E. 2015. Nickel and platinum group metal nanoparticle production by *Desulfovibrio alaskensis* G20. *New Biotechnology* 32: 727–731.

Cao, J., Zhang, G., Mao, Z., Fang, Z., Yang, C. 2009. Precipitation of valuable metals from bioleaching solution by biogenic sulfides. *Minerals Engineering* 22: 289–295.

Capone, D.G., Kiene, R.P. 1988. Comparison of microbial dynamic in marine and freshwater sediments: Contrasts in anaerobic carbon catabolism. *Limnology and Oceanography* 33: 725–749.

Cardenas, E., Wu, W.M., Leigh, M.B., Carley, J., Carroll, S., Gentry, T., Luo, J. et al. 2010. Significant association between sulfate-reducing bacteria and uranium-reducing microbial communities as revealed by a combined massively parallel sequencing-indicator species approach. *Applied and Environmental Microbiology* 76: 6778–6786.

Chang, Y.J., Peacock, A.D., Long, P.E., Stephen, J.R., McKinley, J.P., MacNaughton, S.J., Anwar Hussain, A.K.M., Saxton, A.M., White, D.C. 2001. Diversity and characterization of sulfate-reducing bacteria in groundwater at a uranium mill tailings site. *Applied and Environmental Microbiology* 67: 3149–3160.

Charlson, R.J., Anderson, T.L., McDuff, R.E. 1992. The sulfur cycle. In Butcher, S.S., Charlson, R.J., Orians, G.H., Wolfe, G.V. (Eds.), *Global Biogeochemical Cycles*, p. 379. Academic Press, London, U.K.

Chyba, C.F., Hand, K.P. 2005. Astrobiology: The study of the living universe. *Annual Review of Astronomy and Astrophysics* 43: 31–74.

Clark, M.E., Edelmann, R.E., Duley, M.L., Wall, J.D., Fields, M.W. 2007. Biofilm formation in *Desulfovibrio vulgaris* Hildenborough is dependent upon protein filaments. *Environmental Microbiology* 9: 2844–2854.

Cline, J.D., Richards, F.A. 1969. Oxygenation of hydrogen sulfide in seawater at constant salinity, temperature, and pH. *Environmental Science and Technology* 3: 838–843.

Cody, G.D., Boctor, N.Z., Filley, T.R., Hazen, R.M., Scott, J.H., Sharma, A., Yoder, Jr., H.S. 2000. Primordial carbonylated iron-sulfur compounds and the synthesis of pyruvate. *Science* 289: 1337–1340.

Converse, B.J., Wu, T., Findlay, R.H., Roden, E.E. 2013. U (VI) reduction in sulfate-reducing subsurface sediments amended with ethanol or acetate. *Applied and Environmental Microbiology* 79: 4173–4177.

Cordova-Kreylos, A.L., Scow, K.M. 2007. Effects of ciprofloxacin on salt marsh sediment microbial communities. *The ISME Journal* 1: 585–595.

Cypionka, H. 2000. Oxygen respiration by *Desulfovibrio* species. *Annual Review of Microbiology* 54: 827–848.

Dar, S.A., Tan, H., Peacock, A.D., Jaffe, P., N'Guessan, L., Williams, K.H., Strycharz-Glaven, S. 2013. Spatial distribution of Geobacteraceae and sulfate reducing bacteria during in situ bioremediation of uranium contaminated groundwater. *Remediation Journal* 23: 31–49.

de Vargas, I., Macaskie, L.E., Guibal, E. 2004. Biosorption of palladium and platinum by sulfate-reducing bacteria. *Journal of Chemical Technology and Biotechnology* 79: 49–56.

Degens, B.P. 2012. Performance of pilot-scale sulfate-reducing bioreactors treating acidic saline water under semi-arid conditions. *Water, Air, & Soil Pollution* 223: 801–818.

Deng, D., Weidhaas, J.L., Lin, L.S. 2016. Kinetics and microbial ecology of batch sulfidogenic bioreactors for co-treatment of municipal wastewater and acid mine drainage. *Journal of Hazardous Materials* 305: 200–208.

Dopson, M., Baker-Austin, C., Ram Koppineedi, P., Bond, P.L. 2003. Growth in sulfidic mineral environments: Metal-resistance mechanisms in acidophile micro-organisms. *Microbiology* 149: 1959–1970.

Edwards, K.J., Bond, P.L., Druschel, G.K., McGuire, M., Hamers, R.J., Banfield, J.F. 2000. Geochemical and biological aspects of sulfide mineral dissolution: Lessons from Iron Mountain, California. *Chemical Geology* 169: 383–397.

Edwards, K.J., McCollum, T.M., Konishi, H., Buseck, P.R. 2003. Seafloor bioalteration of sulfide minerals: Results from in situ incubation studies. *Geochimica et Cosmochimica Acta* 67: 2843–2856.

Ehrlich, H.L. 1996. *Geomicrobiology*, 3rd edn., pp. 508–559. Marcel Dekker, Inc., New York.

Evangelou, V.P., Zhang, Y.L. 1995. A review: Pyrite oxidation mechanisms and acid mine drainage prevention. *CRC Critical Reviews in Environmental Control* 25: 141–199.

Fallick, A.E., Ashton, J.H., Boyce, A.J., Ellam, R.M., Russell, M.J. 2001. Bacteria were responsible for the magnitude of the world-class hydrothermal base metal sulfide orebody at Navan, Ireland. *Economic Geology* 96: 885–890.

Farquhar, J., Bao, H., Thiemens, M. 2000. Atmospheric influence of Earth's earliest sulfur cycle. *Science* 289: 756–758.

Fauque, G.D., Barton, L.L. 2012. Hemoproteins in dissimilatory sulfate- and sulfur-reducing prokaryotes. *Advances in Microbial Physiology* 60: 2.

Fievet, A., Ducret, A., Mignot, T., Valette, O., Robert, L., Pardoux, R., Dolla, A.R., Aubert, C. 2015. Single-cell analysis of growth and cell division of the anaerobe *Desulfovibrio vulgaris* Hildenborough. *Frontiers in Microbiology* 6: 1–11.

Florentino, A.P., Weijma, J., Stams, A.J., Sánchez-Andrea, I. 2015. Sulfur reduction in acid rock drainage environments. *Environmental Science & Technology* 49: 11746–11755.

Fortin, D., Southam, G., Beveridge, T.J. 1994. Nickel sulfide, iron-nickel sulfide and iron sulfide precipitation by a newly isolated Desulfotomaculum species and its relation to nickel resistance. *FEMS Microbiology Ecology* 14: 121–132.

Foster, A.L., Brown, Jr., G.E., Tingle, T.N., Parks, G.A. 1998. Quantitative arsenic speciation in mine tailings using X-ray absorption spectroscopy. *American Mineralogist* 83: 553–568.

Fowler, T.A., Crundwell, F.K. 1999. Leaching of zinc sulfide by Thiobacillus ferrooxidans: Bacterial oxidation of the sulfur product increases the rate of zinc sulfide dissolution at high concentrations of ferrous ions. *Applied and Environmental Microbiology* 65: 5285–5292.

Friedrich, C.G. 1998. Physiology and genetics of sulfur-oxidizing bacteria. *Advances in Microbial Physiology* 39: 235–289.

Friedrich, M.W. 2002. Phylogenetic analysis reveals multiple lateral transfers of adenosine-5′-phosphosulfate reductase genes among sulfate-reducing microorganisms. *Journal of Bacteriology* 184: 278–289.

Gadd, G.M. 2000. Bioremedial potential of microbial mechanisms of metal mobilization and immobilization. *Current Opinion in Biotechnology* 11: 271–279.

Gadd, G.M. 2010. Metals, minerals and microbes: Geomicrobiology and bioremediation. *Microbiology* 156: 609–643.

Gammons, C.H., Frandsen, A.K. 2001. Fate and transport of metals in H2S-rich waters at a treatment wetland. *Geochemical Transactions*, 2, 1 (electronic article).

Gevertz, D., Telang, A.J., Voordouw, G., Jenneman, G.E. 2000. Isolation and characterization of strains CVO and FWKO B, two novel nitrate-reducing sulfide-oxidizing bacteria from oil field brine. *Applied and Environmental Microbiology* 66: 2491–2501.

Guilbert, J.M., Park, Jr., C.F. 1986. *The Geology of Ore Deposits*, pp. 985. W.H. Freeman & Company, New York.

Habicht, K.S., Gade, M., Thamdrupp, B., Berg, P., Canfield, D.E. 2002. Calibration of sulfate levels in the Archean ocean. *Science* 298: 2373–2374.

Hansen, T.A. 1994. Metabolism of sulfate-reducing prokaryotes. *Antonie Van Leeuwenhoek* 66: 165–185.

Hao, T.W., Xiang, P.Y., Mackey, H.R., Chi, K., Lu, H., Chui, H.K., van Loosdrecht, M.C., Chen, G.H. 2014. A review of biological sulfate conversions in wastewater treatment. *Water Research* 65: 1–21.

Heidelberg, J.F., Seshadri, R., Haveman, S.A., Hemme, C.L., Paulsen, I.T., Kolonay, J.F., Eisen, J.A., Ward, N., Methe, B., Brinkac, L.M., Daugherty, S.C. 2004. The genome sequence of the anaerobic, sulfate-reducing bacterium Desulfovibrio vulgaris Hildenborough. *Nature Biotechnology* 22: 554–559.

Hesham, A.E.L., Ahmad, A., Daqiang, C., Khan, S. 2014. PCR-DGGE and real-time PCR dsrB-based study of the impact of heavy metals on the diversity and abundance of sulfate-reducing bacteria. *Biotechnology and Bioprocess Engineering* 19: 703–710.

Heyl, A.V., Agnew, A.F., Lyons, E.J., Behre, Jr., C.H. 1959. The Geology of the Upper Mississippi Valley Zinc-Lead District. United States Geological Survey Professional Paper, p. 305. United States Government Printing Office, Washington, DC.

Hockin, S.L., Gadd G.M. 2003. Linked redox precipitation of sulfur and selenium under anaerobic conditions by sulfate-reducing bacteria. *Applied and Environmental Microbiology* 69: 7063–7072.

Holmer, M., Storkholm, P. 2001. Sulphate reduction and sulphur cycling in lake sediments: A review. *Freshwater Biology* 46: 431–451.

Honeyman, B.D. 1999. Colloidal culprits in contamination. *Nature* 397: 23–24.

Hutchins, S.R., Davidson, M.S., Brierley, J.A., Brierley, C.L. 1986. Microorganisms in reclamation of metals. *Annual Review of Microbiology* 40: 311–336.

Johnson, D.B., Hallberg, K.B. 2005. Biogeochemistry of the compost bioreactor components of a composite acid mine drainage passive remediation system. *Science of the Total Environment* 338: 81–93.

Joint Genome Institute (Genome Portal). n.d. Deltaproteobacteria. http://genome.jgi.doe.gov/deltaproteobacteria/deltaproteobacteria.info.html (Accessed October 12, 2015).

Joo, J., Choi, J.H., Kim, I.H., Kim, Y.K., Oh, B.K. 2015. Effective bioremediation of Cadmium (II), nickel (II), and chromium (VI) in a marine environment by using *Desulfovibrio desulfuricans*. *Biotechnology and Bioprocess Engineering* 20: 937–941.

Jørgensen, B.B. 1982. Mineralization of organic matter in the sea bed—The role of sulphate reduction. *Nature* 296: 643–645.

Junier, P., Frutschi, M., Wigginton, N.S., Schofield, E.J., Bargar, J.R., Bernier-Latmani, R. 2009. Metal reduction by spores of Desulfotomaculum reducens. *Environmental Microbiology* 11: 3007–3017.

Kakegawa, T., Ohmoto, H. 1999. Sulfur isotope evidence for the origin of 3.4 to 3.1 Ga pyrite at the Princeton gold mine, Barberton Greenstone Belt, South Africa. *Precambrian Research* 96: 209–224.

Kaksonen, A.H., Plumb, J.J., Robertson, W.J., Franzmann, P.D., Gibson, J.A.E., Puhakka, J.A. 2004. Culturable diversity and community fatty acid profiling of sulfate-reducing fluidized bed reactors treating acidic metal-containing wastewater. *Geomicrobiology Journal* 21: 469–480.

Kaksonen, A.H., Puhakka, J.A. 2007. Sulfate reduction based bioprocesses for the treatment of acid mine drainage and the recovery of metals. *Engineering in Life Sciences* 7: 541–564.

Kang, S., Van Nostrand, J.D., Gough, H.L., He, Z., Hazen, T.C., Stahl, D.A., Zhou, J. 2013. Functional gene array–based analysis of microbial communities in heavy metals-contaminated lake sediments. *FEMS Microbiology Ecology* 86: 200–214.

Kase, K., Yamamoto, M., Nakamura, T., Mitsuno, C. 1990. Ore mineralogy and sulfur isotope study of the massive sulfide deposit of Filon Norte, Tharsis Mine, Spain. *Mineralium Deposita* 25: 289–296.

Kieu, H.T., Müller, E., Horn, H. 2011. Heavy metal removal in anaerobic semi-continuous stirred tank reactors by a consortium of sulfate-reducing bacteria. *Water Research* 45: 3863–3870.

Kim, I.H., Choi, J.H., Joo, J.O., Kim, Y.K., Choi, J.W., Oh, B.K. 2015. Development of a microbe-zeolite carrier for the effective elimination of heavy metals from seawater. *Journal of Microbiology and Biotechnology* 25: 1542–1546.

Klein, M., Friedrich, M., Roger, A.J., Hugenholtz, P., Fishbain, S., Abicht, H., Blackall, L., Stahl, D.A., Wagner, M. 2001. Multiple lateral transfers of dissimilatory sulfite reductase genes between major lineages of sulfate reducing prokaryotes. *Journal of Bacteriology* 183: 6028–6035.

Klenk, H.P., Clayton, R.A., Tomb, J.F., White, O., Nelson, K.E., Ketchum, K.A., Dodson, R.J., Gwinn, M., Hickey, E.K., Peterson, J.D., Richardson, D.L. 1997. The complete genome sequence of the hyperthermophilic, sulphate-reducing archaeon *Archaeoglobus fulgidus*. *Nature* 390: 364–370.

Kohn, M.J., Riciputi, L.R., Stakes, D., Orange, D.L. 1998. Sulfur isotope variability in biogenic pyrite: Reflections of heterogeneous bacterial colonization? *American Mineralogist* 83: 1454–1468.

Koschorreck, M., Geller, W., Neu, T., Kleinsteuber, S., Kunze, T., Trosiener, A., Wendt-Potthoff, K. 2010. Structure and function of the microbial community in an in situ reactor to treat an acidic mine pit lake. *FEMS Microbiology Ecology* 73: 385–395.

Kosińska, K., Miśkiewicz, T. 2012. Precipitation of heavy metals from industrial wastewater by *Desulfovibrio desulfuricans*. *Environment Protection Engineering* 38: 51–60.

Krebs, W., Brombacher, C., Bosshard, P., Bachofen, R., Brandl, H. 1997. Microbial recovery of metals from solids. *FEMS Microbiology Reviews* 20: 605–617.

Kuhl, M., Jørgensen, B.B. 1992. Microsensor measurements of sulfate reduction and sulfide oxidation ini compact microbial communities of aerobic biofilms. *Applied and Environmental Microbiology* 58: 1164–1174.

Kumar, R.N., McCullough, C.D., Lund, M.A., Newport, M. 2011. Sourcing organic materials for pit lake bioremediation in remote mining regions. *Mine Water and the Environment* 30: 296–301.

Kwon, M.J., O'Loughlin, E.J., Boyanov, M.I., Brulc, J.M., Johnston, E.R., Kemner, K.M., Antonopoulos, D.A. 2015. Impact of organic carbon electron donors on microbial community development under iron-and sulfate-reducing conditions. *PLoS One* 11: e0146689–e0146689.

Labrenz, M., Druschel, G.K., Thomsen-Ebert, T., Gilbert, B., Welch, S.A., Kemner, K.M., Logan, G.A. et al. 2000. Formation of sphalerite (ZnS) deposits in natural biofilms of sulfate-reducing bacteria. *Science* 290: 1744–1747.

Lancaster, W.A., Menon, A.L., Scott, I., Poole, F.L., Vaccaro, B.J., Thorgersen, M.P., Geller, J. et al. 2014. Metallomics of two microorganisms relevant to heavy metal bioremediation reveal fundamental differences in metal assimilation and utilization. *Metallomics* 6: 1004–1013.

Leach, D.L., Bradley, D.C., Huston, D., Pisarevsky, S.A., Taylor, R.D., Gardoll, S.J. 2010. Sediment-hosted lead-zinc deposits in Earth history. *Economic Geology* 105: 593–625.

Ledin, M., Pedersen, K. 1996. The environmental impact of mine wastes—Roles of microorganisms and their significance in treatment of mine wastes. *Earth-Science Reviews* 41: 67–108.

Lefticariu, L., Walters, E.R., Pugh, C.W., Bender, K.S. 2015. Sulfate reducing bioreactor dependence on organic substrates for remediation of coal-generated acid mine drainage: Field experiments. *Applied Geochemistry* 63: 70–82.

Lengke, M., Southam, G. 2006. Bioaccumulation of gold by sulfate-reducing bacteria cultured in the presence of gold (I)-thiosulfate complex. *Geochimica et Cosmochimica Acta* 70: 3646–3661.

Lennie, A.R., England, K.E.R., Vaughan, D.J. 1995. Transformation of synthetic mackinawite to hexagonal pyrrhotite: A kinetic study. *American Mineralogist* 80: 960–967.

Lens, P., Vallerol, M., Esposito, G., Zandvoort, M. 2002. Perspectives of sulfate reducing bioreactors in environmental biotechnology. *Reviews in Environmental Science and Biotechnology* 1: 311–325.

Li, D., Hu, N., Ding, D., Li, S., Li, G., Wang, Y. 2016. An experimental study on the inhibitory effect of high concentration bicarbonate on the reduction of U (VI) in groundwater by functionalized indigenous microbial communities. *Journal of Radioanalytical and Nuclear Chemistry* 307: 1011–1019.

Ling, Y.C., Bush, R., Grice, K., Tulipani, S., Berwick, L., Moreau, J.W. 2015. Distribution of iron-and sulfate-reducing bacteria across a coastal acid sulfate soil (CASS) environment: Implications for passive bioremediation by tidal inundation. *Frontiers in Microbiology* 6: 624.

Lloyd, J.R., Mabbett, A.N., Williams, D.R., Macaskie, L.E. 2001. Metal reduction by sulphate-reducing bacteria: Physiological diversity and metal specificity. *Hydrometallurgy* 59: 327–337.

Lloyd, J.R., Nolting, H.F., Sole, V.A., Bosecker, K., Macaskie, L.E. 1998. Technetium reduction and precipitation by sulfate reducing bacteria. *Geomicrobiology Journal* 15: 45–58.

Lovley, D.R., Klug, M.J. 1983. Sulfate reducers can outcompete methanogens at freshwater sulfate concentrations. *Applied and Environmental Microbiology* 45: 187–192.

Lu, J., Wu, J., Chen, T., Wilson, P.C., Qian, J., Hao, X., Liu, C., Su, Y., Jin, X. 2012. Valuable metal recovery during the bioremediation of acidic mine drainage using sulfate reducing straw bioremediation system. *Water, Air, & Soil Pollution* 223: 3049–3055.

Machel, H.G. 2001. Bacterial and thermochemical sulfate reduction. *Sedimentary Geology* 140: 143–175.

Mackin, J.E., Swider, K.T. 1989. Organic matter decomposition pathways and oxygen consumption in coastal marine sediments. *Journal of Marine Research* 47: 681–716.

Madigan, M.T., Martinko, J.M., Parker, J. 2000. *Brock Biology of Microorganisms*, 9th edn., p. 687. Prentice-Hall, Upper Saddle River, NJ.

McGuinness, L.R., Wilkins, M.J., Williams, K.H., Long, P.E., Kerkhof, L.J. 2015. Identification of bacteria synthesizing ribosomal RNA in response to uranium addition during biostimulation at the Rifle, CO Integrated Field Research site. *PLoS One* 10: e0137270.

Medircio, S.N., Leao, V.A., Teixeira, M.C. 2007. Specific growth rate of sulfate reducing bacteria in the presence of manganese and cadmium. *Journal of Hazardous Materials* 143: 593–596.

Meier, J., Piva, A., Fortin, D. 2012. Enrichment of sulfate-reducing bacteria and resulting mineral formation in media mimicking pore water metal ion concentrations and pH conditions of acidic pit lakes. *FEMS Microbiology Ecology* 79: 69–84.

Meyer, B., Kuehl, J., Deutschbauer, A.M., Price, M.N., Arkin, A.P., Stahl, D.A. 2013. Variation among Desulfovibrio species in electron transfer systems used for syntrophic growth. *Journal of Bacteriology* 195: 990–1004.

Meyer, B., Kuever, J. 2007. Molecular analysis of the distribution and phylogeny of dissimilatory adenosine-5′-phosphosulfate reductase-encoding genes (aprBA) among sulfur-oxidizing prokaryotes. *Microbiology* 153: 3478–3498.

Michalsen, M.M., Peacock, A.D., Spain, A.M., Smithgal, A.N., White, D.C., Sanchez-Rosario, Y., Krumholz, L.R., Istok, J.D. 2007. Changes in microbial community composition and geochemistry during uranium and technetium bioimmobilization. *Applied and Environmental Microbiology* 73: 5885–5896.

Mikheenko, I.P., Rousset, M., Dementin, S., Macaskie, L.E. 2008. Bioaccumulation of palladium by Desulfovibrio fructosivorans wild-type and hydrogenase-deficient strains. *Applied and Environmental Microbiology* 74: 6144–6146.

Millero, F. 1986. The thermodynamics and kinetics of the hydrogen sulfide system in natural waters. *Marine Chemistry* 18: 121–147.

Minz, D., Flax, J.L., Green, S.J., Muyzer, G., Cohen, Y., Wagner, M., Rittmann, B.E., Stahl, D.A. 1999. Diversity of sulfate-reducing bacteria in oxic and anoxic regions of a microbial mat characterized by comparative analysis of dissimilatory sulfite reductase genes. *Applied and Environmental Microbiology* 65: 4666–4671.

Misra, K.C. 2000. *Understanding Mineral Deposits*, pp. 497–539. Kluwer Academic Publishers, Dordrecht, the Netherlands.

Mitsch, W.J., Gosselink, J.G. 2000. The Value of wetlands: Importance of scale and landscape setting. *Ecological Economics* 35: 25–33.

Mojzsis, S.J., Coath, C.D., Greenwood, J.P., McKeegan, K.D., Harrison, T.M. 2003. Mass-independent isotope effects in Archean (2.5 to 3.8 Ga) sedimentary sulfides determined by ion microprobe analysis. *Geochimica et Cosmochimica Acta* 67: 1635–1658.

Moreau, J.W., Fournelle, J.H., Banfield, J.F. 2013. Quantifying heavy metals sequestration by sulfate-reducing bacteria in an acid mine drainage-contaminated natural wetland. *Frontiers in Microbiology* 4: 43.

Moreau, J.W., Gionfriddo, C.M., Krabbenhoft, D.P., Ogorek, J.M., Aiken, G.R. Roden, E.E. 2016. The effect of natural organic matter on mercury methylation by *Desulfobulbus propionicus* 1pr3. *Frontiers in Microbiology*. doi:10.3389/fmicb.2015.01389.

Moreau, J.W., Webb, R.I., Banfield, J.F. 2004. Ultrastructure, aggregation-state, and crystal growth of biogenic nanocrystalline sphalerite and wurtzite. *American Mineralogist* 89: 950–960.

Moreau, J.W., Weber, P.K., Martin, M.C., Gilbert, B., Hutcheon, I.D., Banfield, J.F. 2007. Extracellular proteins limit the dispersal of biogenic nanoparticles. *Science* 316: 1600–1603.

Moreau, J.W., Zierenberg, R.A., Banfield, J.F. 2010. Diversity of dissimilatory sulfite reductase genes (dsrAB) in a salt marsh impacted by long-term acid mine drainage. *Applied and Environmental Microbiology* 76: 4819–4828.

Morin, G., Ostergren, J.D., Juillot, F., Ildefonse, P., Calas, G., Brown, Jr., G.E. 1999. XAFS determination of the chemical form of lead in smelter-contaminated soils and mine tailings: Importance of adsorption processes. *American Mineralogist* 84: 420–434.

Müller, A.L., Kjeldsen, K.U., Rattei, T., Pester, M., Loy, A. 2015. Phylogenetic and environmental diversity of DsrAB-type dissimilatory (bi) sulfite reductases. *The ISME Journal* 9: 1152–1165.

Muyzer, G., Stams, A.J. 2008. The ecology and biotechnology of sulphate-reducing bacteria. *Nature Reviews Microbiology* 6: 441–454.

Nair, R.R., Silveira, C.M., Diniz, M.S., Almeida, M.G., Moura, J.J., Rivas, M.G. 2015. Changes in metabolic pathways of *Desulfovibrio alaskensis* G20 cells induced by molybdate excess. *JBIC Journal of Biological Inorganic Chemistry* 20: 311–322.

Ñancucheo, I., Johnson, D.B. 2012. Selective removal of transition metals from acidic mine waters by novel consortia of acidophilic sulfidogenic bacteria. *Microbial Biotechnology* 5: 34–44.

Newman, D.K., Kennedy, E.K., Coates, J.D., Ahmann, D., Ellis, D.J., Lovley, D.R., Morel, F.M.M. 1997. Dissimilatory arsenate and sulfate reduction in Desulfotomaculum auripigmentum sp. nov. *Archives of Microbiology* 168: 380–388.

Nicomrat, D., Dick, W.A., Tuovinen, O.H. 2006. Assessment of the microbial community in a constructed wetland that receives acid coal mine drainage. *Microbial Ecology* 51: 83–89.

Nordstrom, D.K., Alpers, N. 1999. Geochemistry of acid mine waters. In Plumlee, G., Logsdon, M. (Eds.), *Reviews in Economic Geology*, pp. 133–160. Society of Economic Geologists, Littleton, CO.

Norris, P.R., Ingledew, W.J. 1992. Acidophilic bacteria: Adaptations and applications. In Herbert, R.A. Sharp, R.J. (Eds.), *Molecular Biology and Biotechnology of Extremophiles*, pp. 115–139. Blackie, Glasgow, Scotland.

O'Day, P.A., Carroll, S.A., Randall, S., Martinelli, R.E., Anderson, S.L., Jelinski, J., Knezovich, J.P. 2000. Metal speciation and bioavailability in contaminated estuary sediments, Alameda Naval Air Station, California. *Environmental Science and Technology* 34: 3665–3673.

Odom, J.M., Jessie, K., Knodel, E., Emptage, M. 1991. Immunological cross-reactivities of adenosine-5′-phosphosulfate reductases from sulfate-reducing and sulfide-oxidizing bacteria. *Applied and Environmental Microbiology* 57: 727–733.

Ohmoto, H., Kakegawa, T., Lowe, D.R. 1993. 3.4 Billion-year old biogenic pyrites from Barberton, South Africa: Sulfur isotope evidence. *Science* 262: 555–557.

Ono, S., Eigenbrode, J.L., Pavlov, A.A., Kharecha, P., Rumble III, D., Kasting, J.F., Freeman, K.H. 2003. New insights into Archean sulfur cycle from mass-independent sulfur isotope records from the Hamersley Basin, Australia. *Earth and Planetary Science Letters* 213: 15–30.

Oremland, R.S., Capone, D.G. 1988. Use of "specific" inhibitors in biogeochemistry and microbial ecology. In Marshall, K.C. (Ed.), *Advances in Microbial Ecology*, pp. 285–383. New York, Springer.

Papirio, S., Villa-Gomez, D.K., Esposito, G., Pirozzi, F., Lens, P.N.L. 2013. Acid mine drainage treatment in fluidized-bed bioreactors by sulfate-reducing bacteria: A critical review. *Critical Reviews in Environmental Science and Technology* 43: 2545–2580.

Parey, K., Fritz, G., Ermler, U., Kroneck, P.M. 2013. Conserving energy with sulfate around 100 C–structure and mechanism of key metal enzymes in hyperthermophilic Archaeoglobus fulgidus. *Metallomics* 5: 302–317.

Paulo, L.M., Stams, A.J., Sousa, D.Z. 2015. Methanogens, sulphate and heavy metals: A complex system. *Reviews in Environmental Science and Bio/Technology* 14: 537–553.

Pavissich, J.P., Silva, M., González, B. 2010. Sulfate reduction, molecular diversity, and copper amendment effects in bacterial communities enriched from sediments exposed to copper mining residues. *Environmental Toxicology and Chemistry* 29: 256–264.

Peccia, J., Marchand, E.A., Silverstein, J., Hernandez, M. 2000. Development and application of small subunit rRNA probes for assessment of selected *Thiobacillus* species and members of the genus Acidiphilium. *Applied and Environmental Microbiology* 66: 3065–3072.

Peck, Jr., H.D. 1993. Bioenergetic strategies of the sulfate-reducing bacteria. In Odom, J.M., Singleton, R. Jr. (Eds.), *The Sulfate-Reducing Bacteria: Contemporary Perspectives*, pp. 41–76. Springer-Verlag, New York.

Peevler, J., Fayek, M., Misra, K.C., Riciputi, L.R. 2003. Sulfur isotope microanalysis of sphalerite by SIMS: Constraints on the genesis of Mississippi valley-type mineralization, from the Mascot-Jefferson City district, East Tennessee. *Journal of Geochemical Exploration* 80: 277–296.

Pelikan, C., Herbold, C.W., Hausmann, B., Müller, A.L., Pester, M., Loy, A. 2015. Diversity analysis of sulfite and sulfate reducing microorganisms by multiplex dsrA and dsrB amplicon sequencing using new primers and mock community-optimized bioinformatics. *Environmental Microbiology*. doi: 10.1111/1462-2920.13139.

Peltier, E.F., Webb, S.M., Gaillard, J.F. 2003. Zinc and lead sequestration in an impacted wetland system. *Advanced in Environmental Research* 8: 103–112.

Pereira, I.C., Ramos, A.R., Grein, F., Marques, M.C., Da Silva, S.M., Venceslau, S.S. 2011. A comparative genomic analysis of energy metabolism in sulfate reducing bacteria and archaea. *Frontiers in Microbiology* 2: 88–109.

Pereyra, L.P., Hiibel, S.R., Pruden, A., Reardon, K.F. 2008. Comparison of microbial community composition and activity in sulfate-reducing batch systems remediating mine drainage. *Biotechnology and Bioengineering* 101: 702–713.

Pfennig, N., Widdel, F., Trüper, H.G. 1981. The dissimilatory sulfate-reducing bacteria. In *The Prokaryotes*, pp. 926–940. Springer, Berlin, Germany.

Plugge, C.M., Zhang, W., Scholten, J.C., Stams, A.J. 2011. Metabolic flexibility of sulfate-reducing bacteria. *Frontiers in Microbiology* 2: 81.

Postgate, J.R. 1984. *The Sulphate-Reducing Bacteria*, p. 151. Cambridge University Press, Cambridge, U.K.

Poulton, S.W., Fralick, P.W., Canfield, D.E. 2004. The transition to a sulphidic ocean ~1.84 billion years ago. *Nature* 431: 173–177.

Praharaj, T., Fortin, D. 2004. Indicators of microbial sulfate reduction in acidic sulfide-rich mine tailings. *Geomicrobiology Journal* 21: 457–467.

Rabus, R., Hansen, T., Widdel, F. 2000. Dissimilatory sulfate- and sulfur-reducing prokaryotes. In Dworkin, M., Falkow, S., Rosenberg, E., Schleifer, K.H., Stackebrandt, E. (Eds.), *The Prokaryotes, an Evolving Electronic Resource for the Microbiological Community*, 3rd edn. Springer-Verlag, New York, http://link.springer-ny.com/link/service/books/10125/.

Rabus, R., Ruepp, A., Frickey, T., Rattei, T., Fartmann, B., Stark, M., Bauer, M., Zibat, A., Lombardot, T., Becker, I., Amann, J. 2004. The genome of Desulfotalea psychrophila, a sulfate-reducing bacterium from permanently cold Arctic sediments. *Environmental microbiology* 6: 887–902.

Rabus, R., Hansen, T.A., Widdel, F. 2006. Dissimilatory sulfate-and sulfur-reducing prokaryotes. In *The Prokaryotes*, pp. 659–768. Springer, New York.

Rabus, R., Venceslau, S.S., Wöhlbrand, L., Voordouw, G., Wall, J.D., Pereira, I.A. 2015. Chapter two. A post-genomic view of the ecophysiology, catabolism and biotechnological relevance of sulphate-reducing prokaryotes. *Advances in Microbial Physiology* 66: 55–321.

Rodriguez-Freire, L., Moore, S.E., Sierra Alvarez, R., Root, R.A., Chorover, J., Field, J.A. 2015. Arsenic remediation by formation of arsenic sulfide minerals in a continuous anaerobic bioreactor. *Biotechnology and Bioengineering*.

Rosen, B.P. 1996. Bacterial resistance to heavy metals and metalloids. *Journal of Biological Inorganic Chemistry* 1: 273–277.

Sahinkaya, E., Özkaya, B., Kaksonen, A.H., Puhakka, J.A. 2007. Sulfidogenic fluidized-bed treatment of metal-containing wastewater at low and high temperatures. *Biotechnology and Bioengineering* 96: 1064–1072.

Sánchez-Andrea, I., Knittel, K., Amann, R., Amils, R., Sanz, J. L. 2012. Quantification of Tinto River sediment microbial communities: Importance of sulfate-reducing bacteria and their role in attenuating acid mine drainage. *Applied and Environmental Microbiology* 78: 4638–4645.

Sánchez-Andrea, I., Sanz, J.L., Bijmans, M.F., Stams, A.J. 2014. Sulfate reduction at low pH to remediate acid mine drainage. *Journal of Hazardous Materials* 269: 98–109.

Sánchez-Andrea, I., Stams, A.J., Hedrich, S., Ňancucheo, I., Johnson, D.B. 2015. *Desulfosporosinus acididurans* sp. nov.: An acidophilic sulfate-reducing bacterium isolated from acidic sediments. *Extremophiles* 19: 39–47.

Sani, R.K., Peyton, B.M., Brown, L.T. 2001. Copper-induced inhibition of growth of Desulfovibrio desulfuricans G20: Assessment of its toxicity and correlation with those of zinc and lead. *Applied and Environmental Microbiology* 67: 4765–4772.

Santini, T.C., Degens, B.P., Rate, A.W. 2010. Organic substrates in bioremediation of acidic saline drainage waters by sulfate-reducing bacteria. *Water, Air, & Soil Pollution* 209: 251–268.

Santos, A.A., Venceslau, S.S., Grein, F., Leavitt, W.D., Dahl, C., Johnston, D.T., Pereira, I.A. 2015. A protein trisulfide couples dissimilatory sulfate reduction to energy conservation. *Science* 350: 1541–1545.

Schedel, M., Truper, H.G. 1980. Anaerobic oxidation of thiosulfate and elemental sulfur in *Thiobacillus denitrificans*. *Archives of Microbiology* 124: 205–210.

Schidlowski, M. 1973. Sulfur in the Precambrian Metallogeny. *Geologische Rundschau* 62: 840–863.

Schmidtova, J., Baldwin, S.A. 2011. Correlation of bacterial communities supported by different organic materials with sulfate reduction in metal-rich landfill leachate. *Water Research* 45: 1115–1128.

Severmann, S., Mills, R.A., Palmer, M.R., Telling, J.P., Cragg, B., Parkes, R.J. 2006. The role of prokaryotes in subsurface weathering of hydrothermal sediments: A combined geochemical and microbiological investigation. *Geochimica et Cosmochimica Acta* 70: 1677–1694.

Sharp, J.O., Schofield, E.J., Veeramani, H., Suvorova, E.I., Kennedy, D.W., Marshall, M.J., Mehta, A., Bargar, J.R., Bernier-Latmani, R. 2009. Structural similarities between biogenic uraninites produced by phylogenetically and metabolically diverse bacteria. *Environmental Science & Technology* 43: 8295–8301.

Shen, Y., Buick, R. 2004. The antiquity of microbial sulfate reduction. *Earth-Science Reviews* 64: 243–272.

Shen, Y., Buick, R., Canfield, D.E. 2001. Isotopic evidence for microbial sulphate reduction in the early Archaean era. *Nature* 410: 77–81.

Singleton, Jr., R. 1993. The sulfate-reducing bacteria: An overview. In Odom, J.M., Singleton, R. Jr. (Eds.), *The Sulfate-Reducing Bacteria: Contemporary Perspectives*, pp. 1–40. Springer-Verlag, New York.

Sitte, J., Löffler, S., Burkhardt, E.M., Goldfarb, K.C., Büchel, G., Hazen, T.C., Küsel, K. 2015. Metals other than uranium affected microbial community composition in a historical uranium-mining site. *Environmental Science and Pollution Research* 22: 19326–19341.

Song, Y., Fitch, M., Burken, J., Nass, L., Chilukiri, S., Gale, N., Ross, C. 2001. Lead and zinc removal by laboratory scale constructed wetlands. *Water Environmental Research* 73: 37–44.

Southam, G., Saunders, J.A. 2005. The geomicrobiology of ore deposits. *Economic Geology* 100: 1067–1084.

Sparks, D.L. 2003. *Environmental Soil Chemistry*, 352pp. Academic Press, Amsterdam, the Netherlands.

Spear, J.R., Figueroa, L.A., Honeyman, B.D. 1999. Modeling the removal of uranium U(VI) from aqueous solutions in the presence of sulfate-reducing bacteria. *Environmental Science and Technology* 33: 2667–2675.

Sposito, G. 1989. *The Chemistry of Soils*, 277pp. Oxford University Press, New York.

Stackebrandt, E., Sproer, C., Rainey, F.A., Burghardt, J., Päuker, O., Hippe, H. 1997. Phylogenetic analysis of the genus Desulfotomaculum: Evidence for the misclassification of *Desulfotomaculum guttoideum* and description of *Desulfotomaculum orientis* as *Desulfosporosinus orientis* gen. nov., comb. nov. *International Journal of Systematic Bacteriology* 47: 1134–1139.

Stahl, D.A., Fishbain, S., Klein, M., Baker, B.J., Wagner, M. 2002. Origins and diversification of sulfate-respiring microorganisms. *Antonie van Leeuwenhoek* 81: 189–195.

Stetter, K.O. 1988. *Archaeoglobus fulgidus* gen. nov., sp. nov.: A new taxon of extremely thermophilic archaebacteria. *International Journal of Systematic Bacteriology* 10: 172–173.

Stetter, K.O., Lauerer, G., Thomm, M., Neuner, A. 1987. Isolation of extremely thermophilic sulfate reducers: Evidence for a novel branch of archaebacteria. *Science* 236: 822–824.

Strittmatter, A.W., Liesegang, H., Rabus, R., Decker, I., Amann, J., Andres, S., Henne, A. et al. 2009. Genome sequence of *Desulfobacterium autotrophicum* HRM2, a marine sulfate reducer oxidizing organic carbon completely to carbon dioxide. *Environmental Microbiology* 11: 1038–1055.

Straub, K.L., Benz, M., Schink, B., Widdel, F. 1996. Anaerobic, nitrate-dependent microbial oxidation of ferrous iron. *Applied and Environmental Microbiology* 62: 1458–1460.

Stumm, W., Morgan, J.J. 1996. *Aquatic Chemistry: Chemical Equilibria and Rates in Natural Waters*, Vol. 126, pp. 400–414. Wiley & Sons, New York.

Stylo, M., Neubert, N., Roebbert, Y., Weyer, S., Bernier-Latmani, R. 2015. Mechanism of uranium reduction and immobilization in *Desulfovibrio vulgaris* biofilms. *Environmental Science & Technology* 49: 10553–10561.

Sun, W., Xiao, T., Sun, M., Dong, Y., Ning, Z., Xiao, E., Tang, S., Li, J. 2015. Diversity of the sediment microbial community in the Aha watershed (Southwest China) in response to acid mine drainage pollution gradients. *Applied and Environmental Microbiology* 81: 4874–4884.

Suzuki, Y., Kelly, S.D., Kemner, K.M., Banfield, J.F. 2002. Nanometre-sized products of uranium bioreduction. *Nature* 419: 134.

Suzuki, Y., Kelly, S.D., Kemner, K.M., Banfield, J.F. 2003. Microbial populations stimulated for hexavalent uranium reduction in uranium mine sediment. *Applied and Environmental Microbiology* 69: 1337–1346.

Swider, K.T., Mackin, J.E. 1988. Transformations of sulfur compounds in marsh-flat sediments. *Geochimica et Cosmochimica Acta* 53: 2311–2323.

Tabak, H.H., Govind, R. 2003. Advances in biotreatment of acid mine drainage and biorecovery of metals: 2. Membrane bioreactor system for sulfate reduction. *Biodegradation* 14: 437–452.

Tabak, H.H., Scharp, R., Burckle, J., Kawahara, F.K., Govind, R. 2003. Advances in biotreatment of acid mine drainage and biorecovery of metals: 1. Metals precipitation for recovery and recycle. *Biodegradation* 41: 423–436.

Tanji, Y., Toyama, K., Hasegawa, R., Miyanaga, K. 2014. Biological souring of crude oil under anaerobic conditions. *Biochemical Engineering Journal* 90: 114–120.

Taylor, B.E. 2004. Biogenic and thermogenic sulfate reduction in the Sullivan Pb–Zn–Ag deposit, British Columbia (Canada): Evidence from micro-isotopic analysis of carbonate and sulfide in bedded ores. *Chemical Geology* 204: 215–236.

Thauer, R.K., Stackebrandt, E., Hamilton, W.A. 2007. Energy metabolism and phylogenetic diversity of sulphate-reducing bacteria. In Barton, L.L., Hamilton, W.A. (Eds.), *Sulphate-Reducing Bacteria: Environmental and Engineered Systems*. New York: Cambridge University Press.

Thorpe, C.L., Lloyd, J.R., Law, G.T., Williams, H.A., Atherton, N., Cruickshank, J.H., Morris, K. 2016. Retention of 99mTc at ultra-trace levels in flowing column experiments–insights into bioreduction and biomineralization for remediation at nuclear facilities. *Geomicrobiology Journal*, 33: 199–205.

Tokunaga, T.K., Wan, J., Kim, Y., Daly, R.A., Brodie, E.L., Hazen, T.C., Herman, D., Firestone, M.K. 2008. Influences of organic carbon supply rate on uranium bioreduction in initially oxidizing, contaminated sediment. *Environmental Science & Technology* 42: 8901–8907.

Trudinger, P.A., Bubela, B. 1967. Microorganisms and the natural environment. *Mineralium Deposita* 2: 147–157.

Trudinger, P.A., Chambers, I.A., Smith, J.W. 1985. Low-temperature sulphate reduction: Biological vs. abiological. *Canadian Journal of Earth Science* 22: 1910–1918.

Utgikar, V.P., Chen, B.Y., Chaudhary, N., Tabak, H.H., Haines, J.R., Govind, R. 2001. Acute toxicity of heavy metals to acetate-utilizing mixed cultures of sulfate-reducing bacteria: EC100 and EC50. *Environmental Toxicology and Chemistry* 20: 2662–2669.

Veeramani, H., Scheinost, A.C., Monsegue, N., Qafoku, N.P., Kukkadapu, R., Newville, M., Lanzirotti, A., Pruden, A., Murayama, M., Hochella Jr, M.F. 2013. Abiotic reductive immobilization of U (VI) by biogenic mackinawite. *Environmental Science & Technology* 47: 2361–2369.

Venceslau, S.S., Stockdreher, Y., Dahl, C., Pereira, I.A.C. 2014. The "bacterial heterodisulfide" DsrC is a key protein in dissimilatory sulfur metabolism. *Biochimica et Biophysica Acta (BBA)-Bioenergetics* 1837: 1148–1164.

Villa-Gomez, D.K., Enright, A.M., Rini, E.L., Buttice, A., Kramer, H., Lens, P. 2015. Effect of hydraulic retention time on metal precipitation in sulfate reducing inverse fluidized bed reactors. *Journal of Chemical Technology and Biotechnology* 90: 120–129.

Vishnivetskaya, T.A., Mosher, J.J., Palumbo, A.V., Yang, Z.K., Podar, M., Brown, S.D., Brooks, S.C. et al. 2011. Mercury and other heavy metals influence bacterial community structure in contaminated Tennessee streams. *Applied and Environmental Microbiology* 77: 302–311.

Voordouw, G., Armstrong, S.M., Reimer, M.F., Fouts, B., Telang, A.J., Shen, Y, Gevertz, D. 1996. Characterization of 16S rRNA genes from oil field microbial communities indicates the presence of a variety of sulfate-reducing, fermentative, and sulfide-oxidizing bacteria. *Applied and Environmental Microbiology* 62: 1623–1629.

Voordouw, G., Odom, J. M., Singleton, Jr., R. 1993. *The Sulfate-Reducing Bacteria: Contemporary Perspectives*, p. 89. Brock-Springer Series in Contemporary Biosciences, New York.

Wagner, M., Loy, A., Klein, M., Lee, N., Ramsing, N.B., Stahl, D.A., Friedrich, M.W. 2005. Functional marker genes for identification of sulfate-reducing prokaryotes. *Methods in Enzymology* 397: 469–489.

Wagner, M., Roger, A.J., Flax, J.L. Brusseau, G.A., Stahl, D.A. 1998. Phylogeny of dissimilatory sulfite reductase supports an early origin of sulfate respiration. *Journal of Bacteriology* 180: 2975–2982.

Wall, J.D., Arkin, A.P., Balci, N.C., Rapp-Giles, B. 2008. Genetics and genomics of sulfate respiration in Desulfovibrio. In Dahl, C., Friedrich C. G. (Eds.), *Microbial Sulfur Metabolism*, pp. 1–12. Springer, Berlin, Germany.

Webb, J.S., McGinness, S., Lappin-Scott, H.M. 1998. Metal removal by sulphate-reducing bacteria from natural and constructed wetlands. *Journal of Applied Microbiology* 84: 240–248.

Weijma, J., de Hoop, K., Bosma, W., Dijkman, H. 2002. Biological conversion of anglesite (PbSO4) and lead waste from spent car batteries to galena (PbS). *Biotechnology Progress* 18: 770–775.

Wernick, I.K., Themelis, N.J. 1998. Recycling metals for the environment. *Annual Review of Energy and the Environment* 23: 465–497.

White, C., Gadd, G.M. 1996. Mixed sulphate-reducing bacterial cultures for bioprecipitation of toxic metals: Factorial and response-surface analysis of the effects of dilution rate, sulphate and substrate concentrations. *Microbiology* 142: 2197–2205.

Wielinga, B., Lucy, J.K., Moore, J.N., Seastone, O.F., Gannon, J.E. 1999. Microbiological and geochemical characterization of fluvially deposited sulfidic mine tailings. *Applied and Environmental Microbiology* 65: 1548–1555.

Winch, S., Mills, H.J., Kostka, J.E., Fortin, D., Lean, D.R. 2009. Identification of sulfate-reducing bacteria in methyl mercury-contaminated mine tailings by analysis of SSU rRNA genes. *FEMS Microbiology Ecology* 68: 94–107.

Woese, C.R., Fox, G.E., Zablen, L., Uchida, T., Bonen, L., Pechman, K., Lewis, B.J., Stahl, D. 1975. Conservation of primary structure in 16S ribosomal RNA. *Nature* 254: 83–86.

Yabusaki, S.B., Fang, Y., Williams, K.H., Murray, C.J., Ward, A.L., Dayvault, R.D., Waichler, S.R., Newcomer, D.R., Spane, F.A., Long, P.E. 2011. Variably saturated flow and multicomponent biogeochemical reactive transport modeling of a uranium bioremediation field experiment. *Journal of Contaminant Hydrology* 126: 271–290.

Yabusaki, S.B., Şengör, S.S., Fang, Y. 2015. A uranium bioremediation reactive transport benchmark. *Computational Geosciences* 19: 551–567.

Xin, Y., Yong, K., Duujong, L., Ying, F. 2008. Bioaugmented sulfate reduction using enriched anaerobic microflora in the presence of zero valent iron. *Chemosphere* 73: 1436–1441.

Xu, J., Murayama, M., Roco, C.M., Veeramani, H., Michel, F.M., Rimstidt, J.D., Winkler, C., Hochella, M.F. 2016. Highly-defective nanocrystals of ZnS formed via dissimilatory bacterial sulfate reduction: A comparative study with their abiogenic analogues. *Geochimica et Cosmochimica Acta* 180: 1–14.

Zakharova, E.V., Mikhailina, A.B., Konstantinova, L.I., Proshin, I.M., Luk'yanova, E.A., Nazina, T.N. 2011. Effect of biogeochemical factors on the Np (V) mobility under the conditions of remote zone of deep liquid radioactive waste repositories. *Radiochemistry*, 53: 430–436.

Zbinden, M., Martinez, I., Guyot, F., Cambon-Bonavita, M.A., Gaill, F. 2001. Zinc-iron sulphide mineralization in tubes of hydrothermal vent worms. *European Journal of Mineralogy* 13: 653–658.

Zhang, D., Wang, J., Zhao, J., Cai, Y., Lin, Q. 2016. Comparative study of nickel removal from synthetic wastewater by a sulfate-reducing bacteria filter and a zero valent iron-sulfate-reducing bacteria filter. *Geomicrobiology Journal* 33: 318–324.

Zhang, M., Wang, H. 2016. Preparation of immobilized sulfate reducing bacteria (SRB) granules for effective bioremediation of acid mine drainage and bacterial community analysis. *Minerals Engineering* 92: 63–71.

Zheng, W., Liang, L., Gu, B., 2011. Mercury reduction and oxidation by reduced natural organic matter in anoxic environments. *Environmental Science & Technology* 46: 292–299.

Zhou, A., Hillesland, K.L., He, Z., Schackwitz, W., Tu, Q., Zane, G.M., Ma, Q. et al. 2015. Rapid selective sweep of pre-existing polymorphisms and slow fixation of new mutations in experimental evolution of *Desulfovibrio vulgaris. The ISME Journal* 9: 2360–2372.

Zhou, J., He, Q., Hemme, C.L., Mukhopadhyay, A., Hillesland, K., Zhou, A., He, Z. et al. 2011. How sulphate-reducing microorganisms cope with stress: Lessons from systems biology. *Nature Reviews Microbiology* 9: 452–466.

Zhou, Q., Chen, Y., Yang, M., Li, W., Deng, L. 2013. Enhanced bioremediation of heavy metal from effluent by sulfate-reducing bacteria with copper–iron bimetallic particles support. *Bioresource Technology* 136: 413–417.

40 Microorganism-Mediated Sequestration of Ferric and Ferrous Ion from the Environment

Kurtis Stefan and Domenic Castignetti

CONTENTS

ABSTRACT

As an abundant and versatile element, iron plays an important role in industry and in many ecological systems. Microorganisms are dependent upon iron for electron transport and as the cofactor in numerous enzymes. The sequestration of ferric ion (Fe^{3+}) poses a challenge to microorganisms due to its low solubility in aerobic conditions. Varied systems allow microorganisms to acquire and store iron. Microbes thus invoke mechanisms such as siderophores, heme use, transferrin, lactoferrin, and ferrous ion in order to sequester iron from the environment. This chapter discusses the ways by which microbes sequester, transport, and gain the iron from these sources by both Gram-positive and Gram-negative bacterial species.

40.1 INTRODUCTION

Iron is essential for virtually all microbes. It is a component of heme proteins, including the respiratory protein-electron-carrying cytochromes. Iron is also mandatory for Fe–S proteins, peroxide reduction, amino acid biosynthesis, and oxygen activation (Bagg and Neilands, 1987; Ge and Sun, 2014; Schalk and Guillon, 2013). Iron is a component of nitrogenase and the common form of ribonucleotide reductase (Bagg and Neilands, 1987), which converts ribonucleotides to deoxyribonucleotides, a requirement for DNA biosynthesis. Iron is also a cofactor in enzymes that participate in electron transport, superoxide dismutase, and cellular respiration (Schalk and Guillon, 2013). The need for iron stems from its ability to undergo reversible oxidation/reduction with the corresponding loss or gain of an electron (Emery, 1982).

Iron is plentiful on Earth, being the fourth most common element on the Earth's surface (Andrews et al., 2003), yet its oxidation to ferric ion (Fe^{3+}) results in its precipitation, at neutral (pH 7.0) or alkaline solutions, as ferric oxyhydroxides. These compounds have extremely low solubility constants, 1×10^{-38}, with free Fe^{3+} being present at about 10^{-18} M (Bagg and Neilands, 1987; Braun and Killman 1999; Neilands, 1993, 1995; Schalk and Guillon, 2013). The extremely low concentrations of Fe^{3+} in aerobic neutral or aerobic alkaline environments render it largely unavailable to microbes (Braun and Killman, 1999; Neilands, 1995).

Prior to the genesis of oxygen gas, environments of the ancient planet were anaerobic and reducing; Fe^{2+} predominated (Madigan et al., 2015). As Fe^{2+} has a much greater solubility than Fe^{3+} (Neilands, 1995), iron acquisition by microbes was relatively facile. Oxygenic photosynthesis resulted in significant amounts of Fe^{3+} precipitation (Madison et al., 2015). The changing planet presented a challenge to microbial life: either remain anaerobic and have access to the Fe^{2+} needed for cellular components or adapt to oxygen and the aerobic planet but have far lower access to iron as the insoluble ferric oxyhydroxides would be the predominant form of the metal.

In addition to iron's low solubility in neutral or alkaline aerobic milieus, the element possesses another important characteristic. Iron, by virtue of its ability to transfer a single electron, can chemically generate the reactive oxygen species (ROS) hydroxyl radical (·OH). Superoxide (O_2^-) and hydrogen peroxide (H_2O_2), both by-products of O_2 metabolism or of interactions of O_2 with Fe–S proteins, electron carriers (e.g., cytochromes), flavoproteins, or quinones (Madigan et al., 2015), can react with Fe^{2+} and Fe^{3+}. Through a series of reactions, including the Fenton reaction (#3), hydroxyl radical (·OH) is generated (Bagg and Neilands, 1987; Lipinski, 2011):

1. $Fe^{3+} + O_2^- \rightarrow Fe^{2+} + O_2$
2. $2H^+ + 2O_2^- \rightarrow H_2O_2 + O_2$
3. $Fe^{2+} + H_2O_2 \rightarrow Fe^{3+} + ·OH + {}^-OH$

Microorganisms that use O_2 and reside in aerobic neutral or aerobic alkaline environments have a twofold problem to solve simultaneously: one is to acquire sufficient iron to meet cellular needs and the other is to avoid the toxic effects of the metal brought about by the formation of the hydroxyl radical.

Evolution addressed this dilemma by taking advantage of mechanisms to acquire iron from environments where it is sparingly soluble and to invoke mechanisms to limit the toxicity of the metal. Prokaryotic solutions to these challenges are, respectively, active sequestration-uptake mechanisms, iron storage, and detoxification.

Iron acquisition may be either high or low affinity. Low-affinity mechanisms are not well defined but are generally invoked when iron concentrations are >10 μM (Andrews et al., 2003). The low-affinity uptake systems are believed to be not energy dependent and may use nonbiological mechanisms (nonspecific binding) or uptake systems specific for other metals (Andrews et al., 2003). The distinction between low-affinity and high-affinity iron uptake is not always clear. For example, while ferric citrate is considered a high-affinity uptake system in *Escherichia coli* (Bagg and Neilands, 1987), it is a low-affinity system in the soil bacterium *Mycobacterium smegmatis* (Jones and Niederweis, 2010). Experiments using mutants of *M. smegmatis* indicated that general porins are involved in the low-affinity uptake of ferric ion through the outer membrane of the microbe (Jones and Niederweis, 2010) and thus do not invoke an energy-dependent, high-affinity uptake system.

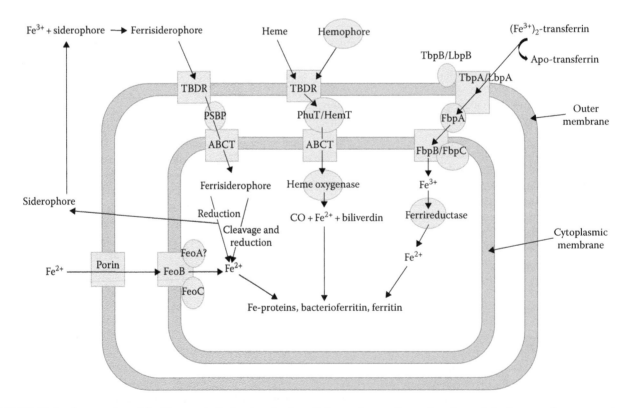

FIGURE 40.1 A general model of iron assimilation in microbes. Refer to the text for explanations of the iron uptake systems shown. *Abbreviations*: PSBPs, periplasmic siderophore–binding proteins; TBDR, TonB-dependent receptor; ABCT, ATP-binding cassette transporter.

High-affinity mechanisms include the uptake of Fe^{2+} by the Feo system (Andrews et al., 2003), the assimilation of Fe^{3+} via the use of siderophores (Greek for "iron bearers"; Neilands 1993), as well as the use of the iron present in heme and hemophore molecules, and the proteins transferrin and lactoferrin (Kelson et al., 2013). Iron storage and detoxification proteins include ferritin, bacterioferritin, and the Dps family, the latter having a role in protecting DNA from oxidative damage (Andrews et al., 2003).

This review will summarize iron acquisition strategies invoked by microbes, but it is not intended to be an all exhaustive review of the literature concerning iron use by microbes. In addition to the current review, the reader is encouraged to examine a variety of articles that examine iron uptake, transport mechanisms, transport proteins, and iron storage and use (Ahmed and Holmström, 2014; Andrews et al., 2003; Braun and Hantke, 2011; Cornelis and Bodilis, 2009; Crosa and Walsh, 2002; Krewulak and Vogel, 2008; Schalk and Guillon, 2013). The focus of this review will be to examine iron acquisition by microbes, primarily bacteria; emphasis will be placed on iron assimilation via siderophores and heme. While other mechanisms of iron sequestration occur in microbes, this chapter will examine in greater detail those iron-sequestration processes most likely to have a major impact on bioremediation.

A general model of iron uptake by microbes is presented in Figure 40.1. As the focus of this chapter is the sequestration of ferric and ferrous ions from the environment, the role of siderophores in iron uptake will be examined first.

40.2 SIDEROPHORES

40.2.1 General Considerations

Siderophores are molecules that bind iron with great affinity, and these Fe^{3+}-chelating molecules are widely distributed throughout the microbial world (Krewulak and Vogel, 2008; Neilands, 1993). Siderophores are synthesized by the microbe and secreted to the external environment where they sequester Fe^{3+}, although some lipophilic siderophores are thought to be tethered to the cell envelope (Kem et al., 2014). When Fe^{3+} is chelated, the siderophore is converted into a ferrisiderophore, which delivers the metal to the microbial cell envelope and is subsequently either stripped of the Fe^{3+} in the envelope or assimilated into the cell as the ferrisiderophore, where the iron is used in cellular processes (Ge and Sun, 2014; Neilands, 1993; Wyckoff et al., 2015). Although both the hydroxamic acids, aspergillic acid and mycobactin, had been previously isolated, the field of siderophore research gained significant momentum when the Neilands laboratory isolated the trihydroxamic acid ferrichrome from *Ustilago sphaerogena* and realized that the compound could bind and deliver Fe^{3+} to the microbe (Neilands, 1952). Ferrichrome has thus come to represent siderophores in general and the hydroxamate [–C(O)–N–OH] siderophores in particular.

A facet of siderophore use is that microbes often assimilate the iron of ferrisiderophores (the Fe^{3+}-siderophore molecule) which they do not make. The siderophore is thus made by another strain or species and referred to as a xenosiderophore (Cornelis and Bodilis, 2009; Cornelis and Dingemans, 2013; Wyckoff et al., 2015). *E. coli* K-12, for instance, synthesizes the siderophores enterobactin and aerobactin, as well as their cognate recognition and uptake systems, yet is able to assimilate Fe^{3+} from these as well as the ferric adducts of coprogen, ferrichrome, ferrioxamine, and rhodotorulic acid (Andrews et al., 2003; Bagg and Neilands, 1987). Species of the fluorescent pseudomonads also contain receptor genes for a variety of ferrisiderophores for which the cognate siderophore is not synthesized by an individual strain (Cornelis, 2010; Cornelis and Bodilis, 2009). *U. sphaerogena* similarly displays the ability to assimilate the Fe^{3+} of a siderophore it does not synthesize (Emery, 1971). The use of exogenous siderophores to sequester Fe^{3+} has two benefits for the assimilating microorganism: one, the metabolic cost of synthesizing the siderophore is not borne by the assimilating organism (Wyckoff et al., 2015), and two, the ability to sequester Fe^{3+} from a wider array of agents helps to ensure that sufficient iron atoms are incorporated into the microbe's biomolecules.

This versatility of ferrisiderophore uptake by a species was at the core of an investigation by D'Onofrio et al. (2010), who noted that a number of marine sediment microorganisms required the presence of a siderophore-excreting bacterium if they were to grow in culture. This observation is in accord with studies noting that a species or strain may contain more siderophore receptors' genes than synthesis genes (Cornelis, 2010; Cornelis and Bodilis, 2009; Moon et al., 2008) and that siderophores are dispersed, at measurable concentrations, in a number of soils (Powell et al., 1980). Of equal interest is the idea that microbes are not limited to invoking a single response to iron deprivation. Martin et al. (2011) noted that *Pseudomonas aeruginosa* from the cystic fibrosis (CF) lung required iron-acquisition mechanisms other than just ferripyoverdine, while Konings et al. (2013) demonstrated that the organism employs siderophore-, heme-, and Fe^{2+}-assimilation systems to sequester the metal from this environment. In a similar manner, Hunter et al. (2013) noted that the source of iron changes in an environment with the passage of time. For organisms such as *P. aeruginosa* infecting the CF lung, Fe^{2+} supplants Fe^{3+} as the dominant source of iron as the infection progresses. Microbes thus respond to changing environments by invoking those uptake systems required to secure adequate amounts of the metal.

Siderophores are diverse in their structures and at least 500 have been characterized (Andrews et al., 2003). Representative structures of a few siderophores are presented in Figure 40.2. Common precursors are lysine, ornithine, citrate, and dihydroxybenzoate; synthesis is often via nonribosomal peptide synthesis (Andrews et al., 2003; Crosa and Walsh, 2002). Ligands that bind the Fe^{3+} include hydroxamates, carboxylates, phenols, catechols, thiazolines, oxazolines, and mixed-function Fe^{3+}-binding moieties (Andrews et al., 2003; D'Onofrio et al., 2010; Wyckoff et al., 2015).

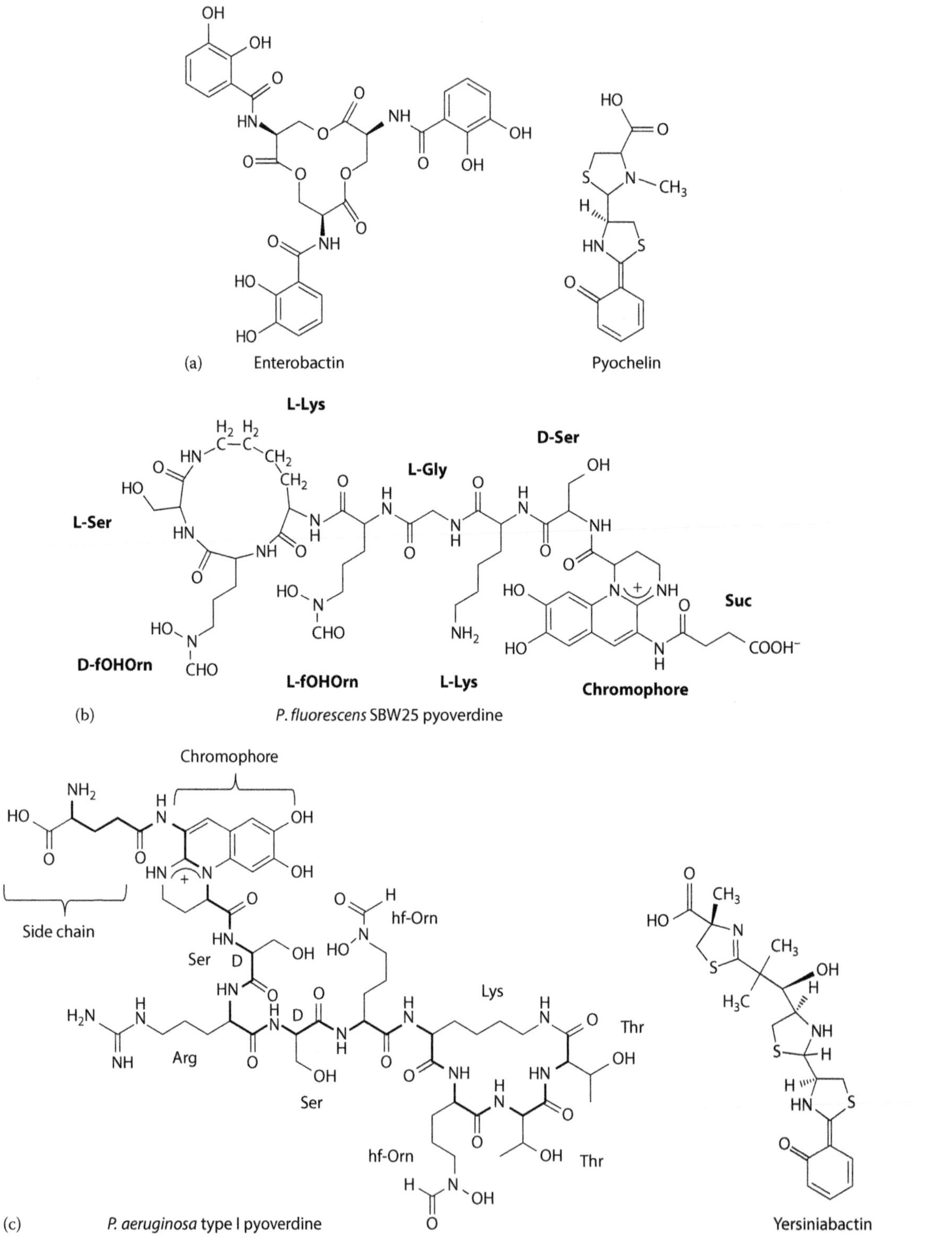

FIGURE 40.2 Structures of the siderophores: (a) enterobactin and pyochelin, (b) pyoverdine from *P. fluorescens* SBW25, and (c) type I pyoverdine of *P. aeruginosa* and yersiniabactin. The structures of enterobactin, pyochelin, and yersiniabactin. The structure of *P. fluorescens* SBW25 pyoverdine, and the structure *of P. aeruginosa* type I pyoverdine. (a: Modified from Adler, C. et al., *PLoS One*, 7(10), e46754, 2012; b: Modified from Moon, C.D. et al., *BMC Microbiol.*, 8, 7, 2008; c: Modified from Cornelis, P. and Dingemans, J., *Front. Cell. Infect. Microbiol.*, 3, 75, 2013.)

40.2.2 Binding

The binding of Fe^{3+} by ferrichrome (and other siderophores) is of such strength that it can mobilize the metal from the poorly soluble Fe hydroxides (Emery, 1982). As with other siderophores, ferrichrome binds Fe^{3+} with great avidity, with an association (formation) constant Ka of 10^{29} (Ka = [CD]/[C][D], where C + D ↔ CD and C = Fe^{3+} and D = the siderophore ligand) (Emery, 1982). The avidity of other siderophores for Fe^{3+} commonly ranges from Ka values of approximately 10^{30} to as high as 10^{52} for ferric enterobactin (Ahmed and Holmström, 2014; Neilands, 1995). Siderophore binding to Fe^{3+} exemplifies the redundancy of using multiple (often 2 or 3) ligands to chelate Fe^{3+}. The additive effect of ligand redundancy results in the truly large formation constants observed (Emery, 1982).

In the case of the representative hydroxamate siderophore ferrichrome, Fe^{3+} binding results in the removal of the 3 protons of the hydroxamate ligands, yielding a neutrally charged molecule. As ferrisiderophores (FeL, where Fe is Fe^{3+} and L is the siderophore) are neutrally charged, movement through the microbial cell envelope, for example, the outer and inner membranes of bacteria, is facilitated. Ferrisiderophore uptake requires specific receptors, and the process is energy dependent (Andrews et al., 2003; Emery, 1982; Krewulak and Vogel, 2008; Wyckoff et al., 2015). Of the four possible isomers of ferrichrome, only the Λ-cis occurs naturally; transport thus requires specific recognition, binding, and uptake by the bacterial envelope proteins (Andrews et al., 2003; Cornelis, 2010; Cornelis and Bodilis, 2009; Cornelis and Dingemans, 2013; Emery, 1982).

Whether the acquisition of Fe^{3+} is from ferrisiderophores, transferrin, lactoferrin, or heme, specific outer membrane receptors to move the ion into the Gram-negative bacterial periplasm are required (Krewulak and Vogel, 2008; Wyckoff et al., 2015). Early models of Fe^{3+} uptake via siderophores (Neilands, 1982) postulated that low-iron stress would invoke the derepression of siderophore synthesis and receptor genes. Siderophore-specific iron-repressible outer membrane proteins, referred to as TonB-dependent (outer membrane) receptors or transporters (Cornelis 2010; Gresock et al., 2015; Liao et al., 2015), are upregulated in the outer membranes of Gram-negative bacteria in response to iron deprivation (Cornelis and Bodilis, 2009; Marugg and Weisbeek, 1991). These receptor transporters were noted to have Mr of 70,000–100,000 (Marugg and Weisbeek, 1991).

Also required are periplasmic-binding proteins (PBPs) and ATP-binding cassette (ABC) inner membrane transporters, which are the associated TonB–ExbB–ExbD proteins that provide the energy necessary to move the Fe^{3+} ligand into the periplasm (Gresock et al., 2015; Krewulak and Vogel 2008; Liao et al., 2015; Wyckoff et al., 2015). The latter proteins supply energy inherent in the proton-motive force (pmf) across the inner membrane to the outer membrane Fe^{3+}-chelating molecule, that is, a ferrisiderophore-, heme-, or hemophore-specific receptor transporter in order that the iron of the ferri-siderophore-, hemin-, or hemophore-mediated uptake systems

is delivered to the periplasmic space (Gresock et al., 2015; Krewulak and Vogel, 2008; Liao et al., 2015; Wyckoff et al., 2015). Gram-positive bacteria, which lack the outer membrane of Gram negatives, utilize cytoplasmic membrane–bound lipophilic-binding proteins and ABC transporters (Hanks et al., 2005; Krewulak and Vogel, 2008).

40.2.3 Uptake

A number of transport processes of the iron of ferrisiderophores have been reported. Included in these mechanisms is what has been termed a shuttle mechanism. In the shuttle procedure, the ferrisiderophore is transported across the cell membrane, its iron removed by reduction, and the siderophore excreted to begin another round of Fe^{3+} assimilation (Ahmed and Holmström, 2014; Emery, 1982). In the "taxicab" mechanism, Fe^{3+} of the ferrisiderophore is transferred across the cell membrane to cellular binding molecules. The hydrolytic mechanism transports the ferrisiderophore into the cell and through a combination of hydrolysis and reduction releases Fe^{3+} from the ferrisiderophore prior to its being reduced to Fe^{2+}, and incorporated into cellular molecules, such as bacterioferritin and ferritin (Ahmed and Holmström, 2014; Wyckoff et al., 2015). A reductive mechanism external to the cell membrane may also occur, such that Fe^{3+} is reduced to Fe^{2+} prior to its being taken into the cell (Ahmed and Holmström, 2014).

The most thoroughly studied assimilation mechanisms are those of bacteria, with Gram-negative transport receiving the greatest attention. Once bound to a siderophore, the first step in the internalization of a ferrisiderophore and the assimilation of its Fe^{3+} begins with the binding of the ferrisiderophore to the Gram-negative bacterium's outer membrane TonB-dependent receptor (also referred to as the TonB-dependent receptor or transporter). These proteins are responsible for the binding of an individual or a few, that is, a class of ferrisiderophore (Cornelis, 2010; Cornelis and Dingemans, 2013; Gresock et al., 2015; Krewulak and Vogel, 2008; Schalk and Guillon, 2013; Wyckoff et al., 2015). Fluorescent pseudomonad species (*P. aeruginosa*, *Pseudomonas syringae*, *Pseudomonas putida/entomophila*, and *Pseudomonas fluorescens*) have a conserved core of TonB-dependent receptor genes as well as others that were most likely acquired via horizontal transfer. Some species cluster their receptor genes in "receptor islands." Other species share receptor genes, and the total number of receptor genes present varies among the different species (Cornelis, 2010; Cornelis and Bodilis, 2009). Of the four strains of *P. aeruginosa* examined, the number of TonB-dependent receptor genes varied from 30 to 37, while a *P. fluorescens* strain (*P. fluorescens* Pf5) contained 45 genes in total (Cornelis and Bodilis, 2009). Further, the pan genome (i.e., all the genes common to a group) of the four *P. aeruginosa* strains examined contained 44 TonB-dependent receptor genes while 26 receptor genes were noted in the core genome, and the *P. putida/entomophila* pan genome contained 53 TonB-dependent receptor genes (Cornelis and Bodilis, 2009). Of interest with the species *P. aeruginosa* is that its major receptor genes for its high-affinity ferrisiderophore (pyoverdine) were most likely acquired via horizontal gene transfer and that a species, but not the genus, contains the core TonB-dependent receptor genes (Cornelis and Bodilis, 2009). The situation with the fluorescent pseudomonads thus serves to note the diversity and multiplicity of ways ferrisiderophores can be assimilated in a species or a group, including receptor genes for siderophores produced by other strains or species, both bacterial and fungal (Cornelis, 2010; Cornelis and Bodilis, 2009). *E. coli* K-12, another model organism examined with respect to ferrisiderophore assimilation and use, similarly invokes a multiplicity of TonB-dependent outer membrane receptors to acquire the iron of ferrisiderophores (Gresock et al., 2015).

Once bound to its cognate TonB-dependent outer membrane receptor, the next step is for the ferrisiderophore to enter the periplasmic space of the Gram-negative cell envelope. The ferrisiderophore-TonB-dependent outer membrane receptor proteins interact with the TonB–ExbB–ExbD proteins to affect the uptake of the ferrisiderophore. The ferrisiderophore-TonB-dependent outer membrane receptors share a structure composed of a 22-antiparallel β-stranded β-barrel that spans the outer membrane (Gresock et al., 2015; Krewulak and Vogel, 2008). Each of these proteins contains a C-terminal periplasmic domain that serves to energize the TonB-gated receptors and an N-terminus cork (also referred to as a plug or hatch) domain that occludes the central opening of the β-barrel. These β-barrel proteins contain 10 short periplasmic loops, a 22-strand β-barrel, and 11 extracellular loops, which are thought to interact with the ferrisiderophore, blocking the opening to noncognate molecules. The binding site is exposed to the solvent on the extracellular face of the TonB-gated receptor (Schalk and Guillon, 2013). Individual receptor proteins vary in their widths and lengths; heights may be as small as 55 angstroms or as large as 70 angstroms with the β-barrel rising above the lipid bilayer of the membrane (Gresock et al., 2015; Krewulak and Vogel, 2008). Binding of a cognate ferrisiderophore is thought to close one or more of the extracellular loops, thus trapping the molecule within the receptor (Krewulak and Vogel, 2008).

The role of the cork domain is to bind the ferrisiderophore, regulate transcription, and interact with the TonB–ExbB–ExbD proteins (Krewulak and Vogel, 2008). The cork domains of these TonB-dependent outer membrane receptors contain a "TonB box" near the N-terminus. The TonB box of the TonB-dependent outer membrane receptors interacts with the TonB–ExbB–ExbD proteins facilitating the transfer of the cytoplasmic membrane–generated pmf energy to move the ferrisiderophore across the outer membrane (Gresock et al., 2015; Krewulak and Vogel, 2008; Sverzhinsky et al., 2015). Recent work with the TonB–ExbD–ExbB proteins of *E. coli* (Gresock et al., 2015) supports the concept that the ExbB proteins have a scaffolding role to the TonB and ExbD proteins, allowing the energization of ExbD units that subsequently associate with TonB dimers to energize the latter and affect the movement of the ferrisiderophore from the outer surface of the TonB-dependent outer membrane receptor into the periplasm of the cell (Gresock et al., 2015). Once spent

of its associated energy, the ExbD units dissociate from their TonB–ExbD dimerization and undergo a subsequent round of reenergization (Gresock et al., 2015). To date, it is not clear whether transport of the ferrisiderophore occurs via a channel opening in the receptor protein or whether partial to complete displacement of the cork domain occurs. Evidence for both models exists (Krewulak and Vogel, 2008).

In addition to its role in energy coupling, the N terminal of the TonB-dependent outer membrane receptors is also involved in signal transduction (Cornelis, 2010; Krewulak and Vogel, 2008). Upon binding of the ferrisiderophore by the receptor protein, signaling is initiated that results in the release of an extracytoplasmic function (ECF) antisigma factor in the cytoplasmic membrane. As a result, the antisigma factor joins the core RNA polymerase (Braun et al., 2003; Cornelis, 2010).

Molecular details have been elucidated in *E. coli* K-12, which serves as the model of such control in Gram-negative cells. With *E. coli*, the binding of dinuclear Fe^{3+} citrate to the outer membrane receptor protein FecA results in its changing structure, initiating the signaling cascade via interacting with the periplasmic C-terminus of the FecR protein. The signal is thus transmitted across the cytoplasmic membrane to the ECF FecI sigma factor, causing the sigma factor to associate with the RNA polymerase core enzyme and initiating the transcription of the ferric citrate (*fecABCDE*) transport genes. In this manner, the availability of ferric citrate in the environment results in the upregulation of its transport genes. The system is subsequently repressed when sufficient iron has been assimilated in the cell and the regulatory Fe^{2+}-Fur protein complex shuts down the transcription of *fecABCDE* (Braun et al., 2003). The end result is efficient control of ferrisiderophore uptake systems and the synthesis of the FecABCDE proteins is prevented when iron is abundant.

Periplasmic siderophore–binding proteins (PSBPs), which belong to cluster 8 of the PBPs, are the next step in the assimilation of Fe^{3+} from most ferrisiderophores by the Gram-negative cell (Krewulak and Vogel, 2008). PSBPs, which function as organic metal-binding and ferrisiderophore-binding complexes, are distinguished as having domains connected via long α-helices (Krewulak and Vogel, 2008). Due to their chemical distinctiveness, each class of siderophores requires a specific PSBP to deliver the molecule to the inner membrane (Braun and Hantke, 2011; Krewulak and Vogel, 2008; Schalk and Guillon, 2013). The role of most of the ferrisiderophore-bound PSBPs is to transfer this ligand to the ABC transporter in the cell membrane, facilitating the delivery of the ferrisiderophore into the cell's cytoplasm. As with the TonB-dependent receptors, the PSBPs bind ferrisiderophores distinctively, with each PSBP responsible for the movement of a particular class of ferrisiderophore. Data suggest that TonB may position a PSBP near the TonB-dependent outer membrane receptor, hence facilitating the movement of the ferrisiderophore to the cognate PSBP (Krewulak and Vogel, 2008).

Not all bacteria, however (*P. aeruginosa* serves as an example), use PSBPs and ABC transporters to assimilate the iron of a siderophore. It is thought that *P. aeruginosa* delivers the Fe^{3+} of ferripyoverdine to the periplasmic space and not directly to a PSBP-associated ABC transporter. A reductive mechanism then releases the iron as Fe^{2+} for assimilation into the cytoplasm (Braun and Hantke, 2011). The deferripyoverdine is then reused as it is shuttled out of the periplasm and into the extracellular medium by an efflux pump. The recycling of pyoverdine by *P. aeruginosa* is energetically favorable as the metabolic cost of synthesizing *siderophores* de novo can be considerable (Braun and Hantke, 2011; Schalk and Guillon, 2013). In addition to this periplasmic mechanism of iron assimilation, reports of pmf-dependent permeases that move PSBP-Fe^{3+} into the cytoplasm exist in the literature. The nature of the interaction, if any, between the pmf and the transport proteins discerned (RhtX from *Sinorhizobium meliloti*, FiuB and FptX from *P. aeruginosa*), however, is yet to be clearly elucidated (Schalk and Guillon, 2013).

Once a ferrisiderophore is bound by a PSBP, the Fe^{3+} of a ferrisiderophore must be moved into the cytoplasm if the microbe is to acquire sufficient amounts of the metal. Grampositive and Gram-negative bacteria accomplish this task via the use of ABC transporter complexes, using the energy of ATP to drive the process (Hanks et al., 2005; Krewulak and Vogel, 2008; Schalk and Guillon, 2013; Wyckoff et al., 2015). Bacterial ABC transporters are commonly composed of five structural domains: two of these domains are responsible for nucleotide (ATP) binding and hydrolysis, thus providing the energy necessary to move the ferrisiderophore into the cell; two transmembrane domains form a membrane channel through which the ferrisiderophore passes; and a fifth domain in the periplasm binds the ligand (ferrisiderophore) (Schalk and Guillon, 2013). Even though a diversity of substrates is accommodated by the ABC transporters, a number of motifs, such as the Walker A and Walker B motifs (which are involved in phosphate and Mg ion binding) and a "signature sequence" that is particular to an ABC transporter family, are shared by the transporters (Krewulak and Vogel, 2008; Schalk and Guillon, 2013).

In order to prevent an errant ATP hydrolysis, evidence suggests that PSBP binding provides the signal to the ATPase component of the ABC transporter. The model then predicts that ATP hydrolysis results in a conformational change of the transmembrane domains such that the ferrisiderophore is delivered to the cytoplasm and ADP and Pi are released from the enzyme (Krewulak and Vogel, 2008; Schalk and Guillon, 2013). Unlike the TonB-dependent receptors, the inner membrane ABC transporters do not require binding specificity of the ferrisiderophores that they transport. Hence, the ferric hydroxamate assimilation system of *E. coli* transports at least four different ferric hydroxamates of different siderophore classes, and the pathogen *Vibrio cholerae* can use two independent ABC transporters to move the same ferrisiderophore into the cell (Schalk and Guillon, 2013).

With Gram-positive bacteria, the lack of an outer membrane allows the ferrisiderophore to bind a cytoplasmic membrane–anchored binding protein that functions in accord with an ABC transporter, the latter which supplies the energy necessary for the ferrisiderophore to cross the cytoplasmic membrane and

into the cytoplasm. The cytoplasmic membrane–anchored binding proteins resemble the PBPs of Gram-negative cells (Krewulak and Vogel, 2008), of which there are numerous examples cited in the NCBI database (https://www.ncbi.nlm.nih.gov/); those PBPs that function in ferrisiderophore uptake across the cytoplasmic membrane belong to PBP cluster 8.

Although differences exist among individual species or genera, *Streptococcus pyogenes* models the uptake of ferrisiderophores in Gram-positive cells. This bacterium contains three ABC transporters with the FtsABCD/SiuABC (ferrichrome transporter system/Streptococcal iron uptake) transporter responsible for ferrichrome use (Ge and Sun, 2014). The lipoprotein FtsB binds ferrichrome and transfers it to FtsCD, the membrane permease responsible for the movement of the ferrisiderophore through the membrane into the cytoplasm. The remaining component of the FtsABCD transporter, FtsA, is the ATPase that provides the energy for ferrichrome assimilation across the membrane and into the cell. Other members of the *Streptococcus* genus have similar ferrisiderophore uptake systems, including FhuD, which binds hydroxamate-type siderophores and shares a good deal of homology to the homologue present in *E. coli* (Ge and Sun, 2014). Interestingly, the ABC transporters of *Streptococcus pneumoniae*, PiuBCDA/Pit1BCDA/FhuDBGC and PiaABCD/Pit2ABCD, were noted to have roles in both siderophore and heme uptake (Ge and Sun, 2014).

40.2.4 Iron Removal

At first glance, the high formation (association) constants (commonly $10^{30}–10^{50}$; see in the preceding text) of Fe^{3+} by siderophores pose the problem of removal of the metal from the siderophore ligand complex. Schalk and Guillon (2013) have summarized the three proposed mechanisms of Fe^{3+} removal from ferrisiderophores: (1) the destruction of at least part of the siderophore structure due to hydrolysis, (2) the use of protons to remove the Fe^{3+} from its siderophore ligands, and (3) the reduction of the Fe^{3+} of the ferrisiderophore, resulting in Fe^{2+} generation. All three mechanisms pose at least potential problems after being invoked by microbes.

The use of H^+ to remove Fe^{3+} requires lower pH values than thought to exist intracellularly in bacteria (Schalk and Guillon, 2013). Hence, the prokaryotic use of this mechanism to remove Fe^{3+} *in vivo* is thus unclear.

Siderophore hydrolysis or modification (acetylation may occur and can lower the affinity of the siderophore for Fe^{3+}) has been noted in a number of bacteria, including *E. coli*, *Salmonella*, and *Bacillus subtilis* (Schalk and Guillon, 2013). In *E. coli*, ferric enterobactin is hydrolyzed at its ester bonds, yielding derivative trimers, dimers, and monomers that have a reduced affinity for Fe^{3+} (Wyckoff et al., 2015). It has long been known (Emery, 1983; Lodge et al., 1982) that microbes use ferrisiderophore reductases to remove Fe^{3+}. Indeed, both ferriphytosiderophore and ferrisiderophore reduction to release Fe^{2+} have been observed by plant NADH nitrate reductase (Castignetti and Smarrelli, 1986; Smarrelli and Castignetti, 1988). *E. coli* employs a flavin adenine dinucleotide reductase to convert the Fe^{3+} attached to the enterobactin hydrolysis products to yield Fe^{2+}. Interestingly, *V. cholera* also synthesizes a catecholate siderophore, but unlike *E. coli*, *V. cholera*'s siderophore (vibriobactin) has no ester bonds to hydrolyze. *V. cholerae* thus employs a reductive mechanism to sequester Fe^{2+} from its siderophore (Wyckoff et al., 2015).

Reduction of the ferrisiderophore allows the deferrisiderophore to be reused and recycled, but the destruction of a siderophore may be an advantage when reduction is either impractical or difficult to accomplish. The destruction of a siderophore is energetically costly; microbes may destroy the siderophore to gain its ferric ion when a sufficient reductant is not available (Emery, 1982).

Reduction of Fe^{3+} can occur either before or after transport in the cytoplasm but may also occur in the periplasm (Schalk and Guillon, 2013; Schroder et al., 2003; Wyckoff et al., 2015). Typical of a number of bacteria, two reductases, one a pyridine nucleotide–disulfide oxidoreductase involved in heme degradation, were noted as having roles in heme and siderophore usage in *Staphylococcus aureus* (Hannauer et al., 2015). As a group, ferrisiderophore reductases are often dependent on flavin cofactors (Emery, 1983; Lodge et al., 1982; Schalk and Guillon, 2013; Schroder et al., 2003), but FhuF is a cytoplasmic 2Fe–2S protein that effectively reduces the Fe^{3+} of ferrioxamine B (Matzanke et al., 2004).

The generation of Fe^{2+} can lead to the Fenton reaction and hydroxyl ion generation (Matzanke et al., 2004; Schalk and Guillon, 2013). The idea of a chaperone-like structure to sequester Fe^{2+} in the cell was thus proposed by Matzanke et al. (2004), and in support of this concept, an oligomeric sugar phosphate compound that effectively bound the Fe^{2+} released from ferrichrome and ferrioxamine E was named ferrochelatin (Matzanke, 1997; Matzanke et al., 2004). As iron is often used to make cellular heme and nonheme iron proteins, the iron realized from siderophores is incorporated into these molecules with excess iron being stored via incorporation into ferritin and bacterioferritin (Schroder et al., 2003).

40.2.5 Regulation

Iron, due to its potential toxicity via the Fenton reaction, requires tight regulation if cellular damage is to be avoided (Ge and Sun, 2014; Noinag et al., 2010; Troxell and Hassan, 2013). The archetypal system, primarily described in *E. coli*, uses the ferric uptake regulator (fur) gene as the transcriptional control master switch. *fur* homologues, with similar cellular functions, exist in numerous other bacteria (Barton et al., 1996; Crosa and Walsh, 2002; Ge and Sun, 2014).

Fur functions as a global regulator of cellular activities such as iron homeostasis and responses to oxidative stress and tricarboxylic acid cycle enzyme synthesis (Becerra et al., 2011). Its role as a transcription factor is often as a negative regulator. Posttranscriptional regulation, via antisense RNA, and positive regulation, however (Chen and Crosa, 1996; Crosa and Walsh, 2002), allow fine tuning of the iron responses by bacteria. Evidence also suggests that Fur may function as an activator via small RNAs, by serving as a *cis* regulatory element that promotes the recruitment of the RNA polymerase holoenzyme or by functioning as an antirepressor blocking or removing DNA transcription repressors (Troxell and Hassan, 2013).

In the classical model, the Fur protein binds Fe^{2+} to repress iron-regulated (homeostasis) genes and control their expression. Under high-iron (Fe^{2+}) cellular conditions, "Fur boxes" are bound by the Fe^{2+}-Fur repressor; the RNA polymerase holoenzyme is thus prevented from binding DNA, thereby repressing genes involved in iron-dependent transport, siderophore biosynthesis, and other iron-associated cellular functions (Moon et al., 2008; Noinaj et al., 2010; Traxell and Hassan, 2013). Upon iron limitation, Fe^{2+} dissociates from Fur; the latter cannot bind DNA and Fur-dependent repression is relieved. Derepression of Fur-controlled monocistronic genes and operons and the upregulation of iron uptake, siderophore biosynthesis, and siderophore transport are thus initiated (Crosa and Walsh, 2002; Ge and Sun 2014; Moon et al., 2008; Noinaj et al., 2010).

Of interest is that the fur gene responds to intracellular signaling whereby oxidative stress regulators (OxyR and SoxRS) promote the synthesis of Fur (Traxell and Hassan, 2013; Zheng et al, 1999). As oxidative stress can comprise Fe–S proteins releasing Fe into the cell, the synthesis of Fur helps to limit this stress by reducing external iron acquisition and indirectly promoting the synthesis of Fe-superoxide dismutase (Traxell and Hassan, 2013). A similar situation exists with reactive nitrogen species (RNS) as Fur regulation is also involved in the RNS response (Mukhopadhyay et al., 2004).

Fur is multifaceted, indirectly controlling several proteins of the tricarboxylic acid (Krebs) cycle and the Fe^{2+}-dependent superoxide dismutase SodB (Noinag et al., 2010; Traxell and Hassan, 2013). The highly conserved small, untranslated RNA *ryhB* is repressed by Fur. *ryhB* base pairs with targeted mRNAs, such as the ones for SodB and the succinate dehydrogenase operon (*sdhCDAB*) translation. It result in the degradation of these mRNAs and limits the expression of their corresponding genes. Thus, when the Fe^{2+} is bound to the Fur repressor, *ryhB* is inactive; SodB and the succinate dehydrogenase proteins are synthesized. Fe^{2+}-Fur therefore indirectly functions as a positive regulator of these genes (Massé and Gottesman, 2002; Noinag et al., 2010; Traxell and Hassan, 2013; Vassinova and Kozyrev, 2000). *ryhB* homologues are present and have virulence functions in a number of Gram-negative pathogens (Traxell and Hassan, 2013).

Fur, however, is not the sole regulatory agent of iron-associated genes. For example, *P. aeruginosa* PAO1 responds to environmental signals concerning the availability of iron. *P. aeruginosa* PAO1's response exemplifies how bacteria adjust to changing environmental conditions, such as may occur in soils or host tissues. In *P. aeruginosa* PAO1, the Fur box controls genes *pvdS* and *fpvI* that encode ECF sigma factors, while the *fpvR* gene encodes an inner membrane antisigma factor. PvdS is an alternative sigma factor; when it joins with the core RNA polymerase, iron-starvation (IS) boxes are bound with the subsequent effect that genes involved in the synthesis of *P. aeruginosa* PAO1's siderophore, pyoverdine, are upregulated and the siderophore is made (Moon et al., 2008).

FpvR and FpvA, the outer membrane ferripyoverdine receptor, contribute to the regulation of PvdS. Binding of ferripyoverdine to FpvA results in signaling that transverses the cytoplasmic membrane resulting in the relieving of FpvR inhibition, and release to the cytoplasm, of PvdS. PvdS then initiates transcription such that pyoverdine and other virulence factors (exotoxin A and the endoprotease PrpL) are synthesized. A similar system exists for the control of FpvA, whereby relief of FpvR inhibition of FpvI results in the upregulation of *fpvA* and synthesis of FpvA (Beare et al., 2003).

The soil bacterium *P. fluorescens* Pf-5 is capable of promoting the growth of cotton seedlings via protection from infection by *Rhizoctonia solani* and *Pythium ultimum* (Lim et al., 2012). This effect is thought due, in part, to the secretion of active secondary metabolites such as pyoluteorin, pyrrolnitrin, hydrogen cyanide, rhizoxin analogues, and 2,4-diacetylphloroglucinol (Lim et al., 2012). In a study to discern the bacterium's response to iron limitation, iron homeostasis genes, such as those for pyoverdine and enantiopyochelin (a second siderophore produced by *P. fluorescens* Pf-5) synthesis, inner membrane transporters, and TonB-dependent receptors, were notably upregulated. Secondary metabolites, such as 2,4-diacetylphloroglucinol, orfamide A, pyrrolnitrin, and the enzyme chitinase, were overexpressed by *P. fluorescens* Pf-5. Downregulated genes, such as those for hydrogen cyanide production and flagellar biosynthesis, were noted under iron limitation. A reduced swarming phenotype was also observed (Lim et al., 2012).

Proteome and transcriptome data indicated that iron assimilation is both transcriptionally and posttranscriptionally regulated in *P. fluorescens* Pf-5 (Lim et al., 2012). Like other important pseudomonads (the plant pathogen *P. syringae* DC3000 and *P. aeruginosa* PAO1), *P. fluorescens* Pf-5 expressed a large number of iron-controlled genes, with the transcription of 180 genes increasing by at least twofold and 121 genes decreasing by at least twofold versus 38 and 86 genes, respectively, increasing or decreasing by twofold when $FeCl_3$ was present. Iron limitation resulted in the upregulation of transport, regulatory function, and binding protein genes, including ECF sigma factor genes with potential roles in iron uptake. Genes for heme acquisition, including *hasA* that codes for a hemophore and *hasI*, which codes for a ECF sigma factor, were notably upregulated as were genes that are part of an alternate heme uptake system (homologues of the *phu* operon) and a putative heme ABC transporter (Lim et al., 2012).

As would be expected when iron was limiting, iron-dependent SodB was downregulated under iron-limiting conditions and the Mn-superoxide dismutase gene was upregulated. Some iron cytochrome genes were downregulated during iron limitation, but others were not significantly iron regulated. The gene *fliA*, which encodes for a sigma factor controlling flagellar synthesis and motility in *P. aeruginosa*, was downregulated, helping to explain why *P. fluorescens* Pf-5 flagellar biosynthesis genes had decreased expression (Lim et al., 2012).

Phenotypic observations positively correlated the amount of iron available with the degree of swarming of the bacterium

(Lim et al., 2012). This result differs from other reports (Déziel et al., 2003; Matilla et al., 2007; Singh et al., 2002; Taguchi et al., 2010) where iron limitation promotes swarming in the closely related species *P. aeruginosa*, *P. putida*, and *P. syringae*. *P. fluorescens* Pf-5 Fe–S cluster binding proteins were not as abundant as expected, and posttranscriptional control was suggested as the reason since their transcriptional level was unaffected (Lim et al., 2012). The conclusion drawn from this elaborate investigation of iron regulation in *P. fluorescens* Pf-5 is that the microbe adjusts its genetic expression and physiology in a manner fine-tuned to environmental conditions. As siderophores are part of the mechanisms that microbes invoke when challenged with environmental challenges such as limiting iron, pollutants, and toxicants, the ability to adjust the expression of these genes serves the needs of bacterium to meet the demands of its environments.

40.2.6 SIDEROPHORES AND BIOREMEDIATION

The role of siderophores in bioremediation has attracted considerable attention. Siderophores are now the focus of investigations as having potential bioremediation capacities. These molecules hold great potential for bioremediation due to their direct engagement in reactions with environmental pollutants or contaminants and their capacity to sequester metals other than iron. In addition, in their roles as suppliers of iron to microbes, siderophores promote microbial metabolism and growth and thus facilitate bioremediation processes.

An example of directly reacting with environmental pollutants was realized when the prime siderophores of the fluorescent pseudomonads, pyoverdine and pyochelin, were examined with organotin, a pollutant introduced into the environments via manufacturing and agricultural practices (Inoue et al., 2003; Sun et al., 2006). Organotin was degraded by pyoverdine from *Pseudomonas chlororaphis*, as well as pyoverdines from other fluorescent *Pseudomonas* species (Inoue et al., 2003). With the pyoverdine from *P. chlororaphis*, a nearly 1:1 ratio of triphenyltin (TPT) degradation was observed with diphenyltin (DPT) and monophenyltin (MPT) being the products. DPT was degraded to MPT, and while the degradation of TPT was highly dependent on deferripyoverdine, DPT degradation was stimulated by the addition of either Cu^{2+} or Sn^{4+}. Conversely, pyochelin-mediated degradation of TPT occurred with either deferri- or ferripyochelin (Sun et al., 2006). The molecule pyridine-2,6-dithiocarboxylic acid is produced by the pseudomonads and chelates a number of transition metals with high affinity (Sebat et al., 2001). Pyridine-2,6-dithiocarboxylic acid also engages in the mineralization of carbon tetrachloride, making it unlikely that the bacterium that secretes this molecule uses it as the prime siderophore for Fe^{3+} sequestration (Lewis et al., 2000, 2001; Sebat et al., 2001).

Similar siderophore-dependent bioremediation/degradation was noted of a siderophore from the brown-rot fungus *Gloeophyllum trabeum*. The siderophore promoted Fe reduction, which is an important step in the bioremediation of pulp and paper wastes (Ahmed and Holmström, 2014). By fostering

Fe reduction, hydrogen peroxide reacted to generate oxygen radicals that promoted the depolymerization of lignocellulose, cellulose, and hemicellulose (Ahmed and Holmström, 2014; Arantes and Milagres, 2007; Wang et al., 2008; Xu and Goodell, 2001). These studies demonstrate that siderophores, which are microbial secondary metabolites (Dimise et al., 2008), can serve as chemical agents that may begin the degradation process of pollutants by reacting with, and converting, such pollutants.

Siderophores may also enhance bioremediation by stimulating the growth of a pollutant-degrading microbe. A *P. aeruginosa*, capable of degrading benzoate, responded normally to elevated iron (54 µM) when grown in a glucose-based medium by limiting the production of its siderophore pyoverdine. In the presence of benzoate, however, the bacterium continued to synthesize pyoverdine in iron concentrations up to 108 µM (Gaonkar et al., 2012). This deviation from the normal mechanism of inhibiting siderophore synthesis solely by iron is not unique to this study. Others (Rashmi et al., 2013) have noted that elevated iron does not necessarily inhibit siderophore synthesis or that other metals may stimulate the production of a siderophore (Rossbach et al., 2000). Metals that stimulate siderophore production (Braud et al., 2010; Diels et al., 1999) include Co, Mo, Mn (Gaonkar and Bhosle, 2013), Pb (Naik and Dubey, 2011), Zn (Gaonkar and Bhosle, 2013; Rossbach et al., 2000), and U (Rashmi et al., 2013), sometimes even when Fe^{3+} is abundant in the medium (Rashmi et al., 2013).

Siderophores are known to have functions outside of their roles as Fe^{3+}-chelating agents. The use of exogenous siderophores (xenosiderophores) has been observed since the early days of siderophore study as it was well known that *E. coli* could acquire the Fe^{3+} from siderophores it did not synthesize (Andrews et al., 2003; Bagg and Neilands 1987; Cornelis and Bodilis 2009; Emery, 1983; Wyckoff et al., 2015). Alternate roles of siderophores include antibiosis compounds, biodegradation catalysts/reactants, and as anti-ROS agents (Adler et al., 2012; Cornelis, 2010). In the latter case, enterobactin (a catecholate) synthesizing wild-type *E. coli* was protected from ROS damage mediated by pyochelin, while enterobactin (catecholate)-deficient mutants were sensitive (Adler et al., 2012). The fungus *Oudemansiella radicata* was protected from the toxic effect of Cd and Pb by the siderophores produced by Cd- and Pb-resistant *Bacillus* sp. and *Pseudomonas* sp. The fungus realized improved growth and ROS damage reduction, manifested in part by a reduced level of superoxide dismutase and peroxidase synthesis, when the fungus was cultured in the presence, as opposed to the absence, of the siderophores (Cao et al., 2012).

Siderophores can chelate metals other than Fe^{3+} with high affinity (Ahmed and Holmström, 2014; Braud et al., 2010; Cornelis, 2010). For example, the hydroxamate siderophore desferrioxamine B has formation (association) constants of 10^{20}–10^{28} M^{-1} for the metals Ga^{3+}, Al^{3+}, and In^{3+}, values close to the constant for Fe^{3+} (10^{30}), while pyoverdine, which has a Fe^{3+} formation constant of 10^{32}, displays formation constants ranging from 10^{17} to 10^{22} for Zn^{2+}, Mn^{2+}, and Cu^{2+} ions (Braud et al., 2010). The main siderophores of the fluorescent pseudomonads, pyochelin and pyoverdine, can chelate numerous metals other than Fe^{3+}, including Ag^+, Al^{3+}, Cd^{2+}, Co^{2+}, Cr^{2+}, Cu^{2+}, Eu^{3+}, Ga^{3+}, Hg^{2+}, Mn^{2+}, Ni^{2+}, Pb^{2+}, Sn^{2+}, Tb^{3+}, Tl^+, and Zn^{2+}. Only the Fe^{3+} adduct of the siderophores, however, is efficiently assimilated and transported into *P. aeruginosa*. Other metals (Co^{2+}, Ga^{3+}, Mn^{2+}, Ni^{2+}) were assimilated, but at markedly (23–35-fold) reduced rates (Braud et al., 2009a,b, 2010). Chelation of nonferrous ions by siderophores may be beneficial to the siderophore-synthesizing microbes as the metals are thus less likely to be assimilated into the cytoplasm where they could otherwise exert toxic effects (Braud et al., 2010).

An interesting mechanism of siderophore-assisted bioremediation stems from the ability of the molecules to chelate metals other than Fe^{3+}. The binding of toxic metals in the periplasm of Gram-negative cells results in their chelation and renders them unable to move into the cell via relatively nonspecific and molecular mass-limited porins (Braud et al., 2010). As lower concentrations of toxic metals exist in the cytoplasm, metabolic pathways and sensitive enzymes are less effected (Braud et al., 2010; Diels et al., 1999). Indeed, siderophore-negative mutants of *P. aeruginosa* PAO1 were more sensitive to a number of metals (Braud et al., 2010; Diels et al., 1999).

As microbes are more adept at resisting the toxic metals with the assistance of siderophores, bioremediation processes, such as the degradation of pollutants and the mobilization of heavy metals, can occur in metal-contaminated environments (Ahmed and Holmström 2014; Diels et al., 1999). Metals such as Fe, Ni, Co, Np, and U were chelated and removed from soils and mine wastes via mobilization (Behrends et al., 2012; Edberg et al., 2010; Rashmi et al., 2013; Wang et al., 2011). Nair et al. (2007) observed that up to 92.8% of the arsenic present in a contaminated soil was removed when the soil was repeatedly (5 times) washed with the siderophore of *Pseudomonas azotoformans*, an effect that was notably better than the removal mediated by EDTA and citric acid. The siderophore made by *P. azotoformans* was of mixed type (catecholate-hydroxamate), chelated up to nine different metals, did not exert any toxic effect on the soil microbes, and demonstrated a notable bioremediation potential for the removal of metals from contaminated soils.

The ability to chelate and deliver Fe^{3+} to bioremediation organisms is associated with the bioremediation of oil spills, such as the Deep Horizon spill in the Gulf of Mexico (Gauglitz et al., 2012). One group (Kem et al., 2014) has shown that such microbes synthesize amphiphilic siderophores that may be of special value to oil-degrading microbes as they have the potential to function as both oil emulsification and iron uptake agents that minimize siderophore diffusion away from the microbe.

Since they are secreted into an environment, siderophores may be considered "public goods." Cordero et al. (2012), studying the dynamics of siderophore use by marine *Vibrionaceae*, noted that "cheaters," that is, cells that do not produce siderophores yet maintain the ability to transport the ferrisiderophore into their cytoplasm, were observed to coevolve with the remainder of the population that was capable of sidero-phore synthesis. Similarly, studying the siderophore dynamics in *P. aeruginosa* from the CF patients' lungs, Andersen et al. (2015) noted that conspecifics lost the ability to make sidero-phores although it was observed that they retained the ability to acquire the iron of the species' siderophore. In this context, it is interesting to note that genome reduction occurs in clones of *P. aeruginosa* from the CF patients' lungs and that the loss of TonB-dependent receptor genes, including those for fer-ripyoverdine uptake, was among the genes culled from the bacterium (Dingemans et al., 2014).

As might be expected, evolution of siderophore-dependent bioremediation processes occurs. Examining *P. aeruginosa* in the context of heavy metal (Cu) bioremediation (O'Brien et al., 2014), the organism, when faced with toxic concentra-tions of Cu, evolved de novo high frequencies of siderophore nonproducers. When cultured separately, nonsiderophore-producing strains fared poorly compared to siderophore-producing strains; the high frequency of *P. aeruginosa* nonproducers suggests that the public goods of siderophore synthesis result in a selective advantage for the "cheats" when in coculture even in the face of toxic concentrations of a metal. Interestingly, O'Brien et al. (2014) concluded that siderophore cheating is possible both within a species and perhaps among different species. The role of iron, other metals, and microbial evolution is thus vibrant and exemplifies not just siderophore protection from toxicants but also the cost/benefit ratio of making the siderophore.

40.3 HEME

40.3.1 HEME

Iron acquisition via heme uptake and its subsequent oxidation is a clinically relevant phenomenon of iron uptake in patho-genic bacteria as well as some soil microbes (Cornelis et al., 2009). Microorganisms contained in human plasma face unique challenges in acquiring heme, as over two-thirds of iron is contained in hemoglobin molecules (Andrews, 1999). With myoglobin and hemoglobin, the greatest sinks of iron in humans, microorganisms must first obtain heme or a heme-bound protein before acquiring free iron. This represents a challenge for pathogens, as iron limitations may inhibit colo-nization (Andrews, 1999).

A variety of mechanisms are present in bacteria to acquire heme and its iron. Iron may be present in different forms, such as heme with Fe^{2+}, ferrous ion, and polyporphyrin or hemin with its iron as Fe^{3+}. An organism may have multi-ple systems of heme acquisition, but differences are found generally between Gram-negative and Gram-positive bacte-rial species. This discussion will present the major systems for heme acquisition described and the degradation of heme within the cell.

Gram-negative bacteria, including *P. aeruginosa* (Konings et al., 2013) and *Bacteroides fragilis* (Rocha and Smith, 2013), use heme as an iron source. Several mechanisms exist for heme uptake in Gram-negative bacteria. The first mechanism consists

of direct heme receptors, transported across the inner mem-brane by ABC transporters (Tong and Guo, 2009). Generally considered members of the hemoglobin outer membrane TonB-linked heme protein family (Marchler-Bauer et al., 2015), this heme acquisition mechanism is modeled by the *phuRSTUV* system in *P. aeruginosa* (Konings et al., 2013) and HemRTUV in *Yersinia enterocolitica* (Wandersman and Stoljijkovic, 2000). Phu proteins collectively assist Gram-negative bacteria in transporting heme across the outer membrane, through the periplasmic space, across the inner membrane, and within the cytoplasm (Konings et al., 2013). The HemR receptor, thought part of the porin superfamily of proteins with features includ-ing ligand-gated channels and ß-stranded barrels (Marchler-Bauer et al., 2015), is the receptor of the heme protein of the hemRTUV operon. Many proteins of this superfamily, includ-ing HemR, are TonB dependent; they are using pmf energy via TonB proteins to drive heme transport.

TonB-dependent processes are inherent to the Phu and Hem systems for heme acquisition and participate in the uptake of ferrisiderophores (Tong and Guo, 2009). The TonB protein uses pmf to transfer energy to the outer membrane receptor via a TonB box, a portion of the receptors N-terminal domain (Tong and Guo, 2009). In this way, TonB and its associated proteins ExbD and ExbB, recently determined to be in a 1:2:7 ratio, respectively, provide the energy required to transport heme into the periplasmic space (Gresock et al., 2015). The TonB–ExbD–ExbB complex, isolated from *E. coli*, is thought to be first formed with ExbD homodimers interacting with TonB homodimers prior to monomeric TonB energetically interact-ing with membrane heme transporters (Gesock et al., 2015).

Pathogenic *P. aeruginosa* is able to use heme as a source for iron when growing in a human host, with hemoglobin serving as a large source of heme, and ultimately iron, within its host (Konings et al., 2013). The ability to transport heme into the cellular environment for *P. aeruginosa* is key to its infective potential, with a significantly lower amount of PhuR mRNA transcript observed in stable compared to postacute patients (Konings et al., 2013). The *P. aeruginosa* PhuR protein is an outer membrane receptor able to bind heme, hemopexin, hemoglobin, myoglobin, or hemoglobin–haptoglobin (Tong and Guo, 2009). Binding of diverse substrates ensures greater likelihood of iron sequestration using this system. In silico work has found the *phuR* gene to be 60%–90% homologous across the *Pseudomonas* genus; PhuR is said to be the univer-sal heme receptor (Cornelis et al., 2009). The analogous HemR transporter in *Y. enterocolitica* also binds diverse substrates including heme and heme transport proteins (Wandersman and Stoljijkovic, 2000). The HemR transporter in *Sodalis glos-sinidius* is used for heme uptake after blood meal in the gut of tsetse flies (Hrusa et al., 2015). Normally symbiotic with the host organism, knockout of HemR in *S. glossinidius* resulted in mutants unable to colonize fly guts (Hrusa et al., 2015).

Using *Y. enterocolitica* as a model, Bracken et al. (1999) determined the importance of two histidine residues in the transportation of heme through the "receptor pore," His128 and His461, as determined by site-directed mutagenesis (1999). They postulated that these histidine residues contribute varying functionality to the heme transport process, as His128 appears to be conserved across the majority of outer membrane TonB-dependent receptors (Bracken et al., 1999). Heme is then transferred to PhuT or HemT using a TonB-dependent process (Hrusa et al., 2015; Tong and Guo, 2009). PhuU, PhuV, and PhuW are required for inner membrane transport and are considered ABC transporters (Tong and Guo, 2009). These proteins are ATP driven, translocating heme into the cytoplasm at the expense of ATP (Tong and Guo, 2009). PhuS chaperones one heme molecule per protein monomer, shuttling it to heme oxygenase (HO) for further degradation (Tong and Guo, 2009).

HO converts the polyporphyrin ring to free iron, biliverdin, and carbon monoxide (Higashimoto et al., 2006; Tong and Guo, 2009). Such oxygenases have been identified in *Neisseria meningitidis* as HemO, in *P. aeruginosa* as PigA, and in *Corynebacterium diphtheriae* as HmuO (Unno et al., 2007; Wilks, 2002). HO is microsomal and, via a multistep process using NADPH-cytochrome P450 reductase (CPR), catalyzes heme degradation to biliverdin IXα, carbon monoxide, and free iron (Higashimoto et al., 2006). Step one is the oxidation of heme to α-hydroxyheme using O_2 and reducing equivalents supplied by CPR. Verdoheme formed in the second step is concomitantly produced along with the hydroxylated α-meso carbon as CO. In the third step, α-verdoheme is converted to a biliverdin–iron chelate. This step requires O_2 and electrons from CPR. In the final step, the biliverdin–iron chelate is reduced with the subsequent release of Fe^{2+} and biliverdin from the enzyme (Higashimoto et al., 2006).

Upon investigating the mechanism of HmuO as a model HO, it was discovered that the iron in heme binds diatomic oxygen with a bent geometry, approximately 110° (Unno et al., 2007). This conformation brings the distant oxygen atom in close contact with the porphyrin ring's α-meso carbon (Unno et al., 2007), promoting a conversion to a hydroperoxide intermediate (Fe^{3+}–OOH) forming α-meso-hydroxyheme (Wilks, 2002). As noted previously, later steps require the reduction of the ferric ion and conversion of hydroxyheme to verdoheme, thought to be the rate-limiting step (Unno et al., 2007), with verdoheme oxidation to biliverdin completing heme conversion and the release of free Fe^{2+} into the cytoplasm (Unno et al., 2007).

The PhuS shuttle protein is thought to be key in regulating heme uptake in the Phu system (O'Neill and Wilks, 2013). In low concentrations of extracellular heme, the equilibrium of PhuS-holo to PhuS-apo is shifted toward the apoprotein (O'Neill and Wilks, 2013). High concentrations of extracellular heme lead to saturation of HemO, driving the PhuS equilibrium toward the holo form and PhuS mRNA is consequently downregulated (O'Neill and Wilks, 2013).

40.3.2 Hemophores

In addition to ABC transport systems for heme, Gram-negative bacteria also utilize hemophores. The heme acquisition system (Has) of *Serratia marcescens* is a model for describing iron acquisition mediated by hemophore secretion in microorganisms (Deniau et al., 2003). In an analogous manner to siderophores (Martin et al., 2011), hemophores are high-affinity molecules that scavenge heme or hemoprotein-bound heme as a source of iron (Deniau et al., 2003). The heme-binding protein HemA is the hemophore component of this system, with a known affinity (formation) constant, Ka, of $5.3 \times 10^{10}\,M^{-1}$ in *Se. marcescens* (Deniau et al., 2003). Using thermodynamic studies with *Se. marcescens* mutants, Deniau et al. (2003) determined that the Tyr75-Fe and internal protein hydrogen bonds are most important in the HasA–heme interaction. HasA–heme is delivered to HasR, the outer membrane component of the Has system. In the bacterium *Se. marcescens*, the HasR receptor binds either the HasA hemophore or free heme (Létoffé et al., 2004). An iron-rich extracellular environment promotes iron acquisition via heme binding to HasR, and overexpression of the TonB complex (TonB–ExbB–ExbD) was required for hemophore uptake in this high-iron environment (Létoffé et al., 2004). As TonB is a limiting factor in hemophore uptake, it is proposed that this complex provides sufficient energy to cause HasR-hemophore dissociation, hence allowing free heme or holo-HasA to bind the receptor (Létoffé et al., 2004).

More recent work proposes that HasR also functions as a free heme sensor in *P. aeruginosa*; in the absence of extracellular heme, anti-σ-factor HasS is associated with HasI σ-factor (Smith and Wilks, 2015). This in turn downregulates Phu and Has heme uptake systems, disallowing their transcription (Smith and Wilks, 2015). When free in the nucleus, HasI also activated the *phuSTUV* system (Smith and Wilks, 2015). Regulation using the HasI σ-factor is analogous to pyoverdine siderophore regulation using σ-factor PvdS, as discussed earlier (Beare et al., 2003). Future work is needed to elucidate the energetics of the HasR receptor and the transfer to it of heme by the HasA hemophore, in addition to exploring the HasR receptor's regulation of heme acquisition systems.

Gram-positive bacteria also use ABC-type transporters for heme acquisition, having been identified in the Group A *Streptococci*, *C. diphtheriae*, and *S. pyogenes* (Tong and Guo, 2009). These bacteria may also contain other systems. *St. aureus*, a pathogenic Gram-positive bacterium, contains the iron-regulated surface determinant systems (Isd), which assists in transporting heme into the cytoplasm (Tong and Guo, 2009). Many components of the Isd system allow Gram-positive bacteria to utilize heme as a source of iron: IsdABCH are receptors anchored to the cell wall. IsdDEF are ABC transporters associated with the cytoplasmic membrane (Fonner et al., 2014). IsdG/I are monooxygenases found in the cytoplasm capable of heme degradation (Torres et al., 2006). Torres et al. (2006) determined that of the cell wall–anchored receptors, IsdB is primarily responsible for the ability of *St. aureus* to bind hemoglobin (2006). The presence of multiple near-iron transporter (NEAT) domains is specific to IsdB and IsdH, both of which are responsible for acquiring heme from hemoglobin and hemoglobin–haptoglobin (Fonner et al., 2014). Broadly, NEAT domains are structurally thought to contain eight ß-strands, an immunoglobulin-like fold and with potential variation in the heme interaction site per protein and organism (Honsa et al., 2014). Because of IsdB's three NEAT domains, the mechanism of heme transfer from hemoglobin to IsdB is of interest. Experiments by Fonner et al. (2014) have determined the importance of *St. aureus'* NEAT domains aromatic motifs in the binding of hemoglobin. Mutation of these aromatic amino acids in IsdH, even when the sequence was made to be a replication of IsdB's aromatic motif, resulted in the loss of hemoglobin-binding ability in the linker NEAT domain (Fonner et al., 2014). The IsdB linker (L) NEAT domain was also observed to be necessary for the high rate of heme transfer from hemoglobin to IsdB, confirmed by shortening the L-NEAT domain that resulted in an inability to successfully transfer heme from hemoglobin (Fonner et al., 2014). Together, the IsD system comprises a significant proportion of heme uptake in pathogenic Gram-positive bacteria, suggesting that NEAT domain targeting could be a viable therapeutic intervention (Honsa et al., 2014).

40.3.3 IRON STORAGE

After heme enters the cytoplasm and is broken down to release Fe^{2+}, microorganisms must store this ion to mediate potential oxidative damage. Many bacteria, both aerobes and anaerobes, contain ferritin proteins able to store free iron within the cytoplasm (Carrondo, 2003). Some bacteria like, *E. coli*, contain both ferritin and bacterioferritin proteins, EcFtnA/EcFtnB and EcBFR, respectively (Bou-Abdallah et al., 2013). Bacterioferritin (Bfr) contains heme, each with 12 intersubunit groups of heme thought to be functional in promoting the release of iron from the solid core (Carrondo, 2003; Yasmin et al., 2011). These proteins store iron as Fe^{3+} inside the core, keeping the intracellular concentration of iron in the range of 10^{-3} to 10^{-5} M (Carrondo, 2003). Structurally, both ferritin and bacterioferritin contain 24 subunits, arranged in dimers, surrounding a hollow cavity capable of fitting 4500 irons (Carrondo, 2003). In the ferroxidase center of EcFtnA, Fe^{2+} oxidation may serve to reduce the formation of hydrogen peroxide and hence free radical induction. Bou-Abdallah et al. (2013) proposed that EcFtnA uses two pathways that account for the ability of the bacterial ferritins to limit hydrogen peroxide production: Fe^{2+} oxidation by O_2 producing H_2O_2, and subsequently "$2Fe^{2+} + H_2O_2 + 2H_2O \rightarrow 2[FeO(OH)]_{CORE} + 4H^+$ or $Fe_A^{2+} Fe_B^{2+} + H_2O_2 \rightarrow (Fe_A^{2+}(OH^-)_2Fe_B^{3+}) + H_2O_2$ on the iron core surface" (Bou-Abdallah et al., 2013). This activity is significant as it prevents damage to the organism via ferrous oxidation by hydrogen peroxide in the Fenton reaction. Bacterial ferritin and bacterioferritin proteins are important for iron ion storage after heme sequestration and also in preventing and suppressing harmful free radicals within the cell.

It is also known that iron–sulfur proteins serve as iron reserves for microorganisms (Romsang et al., 2015). Romsang et al. (2015) determined, by mutational analysis in *P. aeruginosa*, that fprB is involved in storing excess iron in 4Fe–4S clusters, likely due to a decrease in reduced ferredoxin and thereby decreasing possible toxicity of increased free Fe ion (2015). fprB in *E. coli* is thought to have a role in iron–sulfur cluster biogenesis, assisted by the pyridoxal phosphate–dependent desulfurase enzyme Isc and scaffold protein IscU (Yan et al., 2015). The ability to successfully store excess iron mediated by fprB and the *ISC* operon is necessary for survival in conditions of low extracellular iron (Romsang et al., 2015).

These mechanisms for iron storage are necessary after heme uptake to reduce the concentration of cytoplasmic iron because of its reactive nature. Storage in Fe–S complexes, or ferritin-like proteins, ameliorates concerns of intracellular damage due to free Fe ion while still allowing the construction of numerous iron-containing proteins and enzymes.

40.4 TRANSFERRIN AND LACTOFERRIN

Transferrin and lactoferrin are host iron-chelating carrier proteins involved in sequestering the metal, making it unavailable to microbial pathogens (Boradia et al., 2014; Britigan et al., 1993; Rhode and Dyer, 2003). Microbial assault on these molecules may result in iron acquisition via proteolytic cleavage and siderophore-mediated removal, both directly and with assistance from catecholamines, such as norepinephrine, dopamine, and epinephrine. The catecholamines release Fe^{3+} from both transferrin and lactoferrin, making the metal available for chelation and assimilation via siderophores in both Gram-positive and Gram-negative species (Boradia et al., 2014; Britigan et al., 1993; Caza and Kronstad, 2013; Stintzi and Raymond, 2000).

Much attention has been focused on the receptor proteins TbpAB and LbpAB of *Neisseria gonorrhoeae* and *N. meningitidis*, which affect ferric ion uptake from transferrin and lactoferrin, respectively (Caza and Kronstad, 2013; Cornelissen, 2003; Rohde and Dyer, 2003). TbpA, which is TonB dependent, and its lipoprotein coreceptor, TbpB, form a complex of which TbpA is the transmembrane transporter of iron from transferrin. While TbpA, without being associated with TbpB, is capable of iron extraction from transferrin, the extraction of the metal is significantly enhanced by TbpB (Caza and Kronstad, 2013; Cornelissen, 2003; Rohde and Dyer, 2003).

TbpA is responsible for a TonB-dependent conformational change that transports the iron from transferrin and, with assistance from TbpB, releases apotransferrin (Caza and Kronstad, 2013; Cornelissen, 2003). TbpA adjusts the position of its cork domain such that iron is docked within the protein's β-barrel prior to being transported to FbpA. FbpA has a strong homology to, and has been speculated to be, a progenitor of transferrin and is the PBP of this system (Cornelissen, 2003; Rhode and Dyer, 2003; Siburt et al., 2012). It begins the movement of Fe^{3+} across the cytoplasmic membrane with the assistance of FbpB and FbpC. These proteins are thought to have, respectively, the functions of membrane permease and the energy provision (ABC transporter) that result in Fe^{3+} assimilation into the cytoplasm (Caza and Kronstad, 2013). Of interest is that the FbpABC system for transferrin Fe^{3+} uptake by *N. gonorrhoeae* and *N. meningitidis* is analogous to that used by other microbes to assimilate the Fe^{3+} of ferrisiderophores by Gram-negative bacteria (Caza and Kronstad, 2013; Cornelissen 2003). Lactoferrin iron uptake by *N. gonorrhoeae* and *N. meningitidis* is analogous to the TbpAB system, although LbpB is not essential for lactoferrin iron assimilation (Caza and Kronstad, 2013; Rhode and Dyer, 2003).

The study of the pathogen *Mycobacterium tuberculosis* has led to a model whereby the bacterium acquires iron from transferrin by three distinct mechanisms. In the first mechanism, iron withdrawal from transferrin is mediated by the siderophore carboxymycobactin that subsequently transfers the metal to another siderophore, mycobactin, for intracellular delivery. In the second mechanism, carboxymycobactin once again removes the Fe^{3+} of transferrin, but the metal is supplied to the high-affinity transporter IrtAB. In the third, and last, mechanism that is reminiscent of a receptor-mediated transporter, transferrin is internalized into the cell via glyceraldehyde-3-phosphate dehydrogenase and probably other surface proteins. Once inside the cell, ferrireductases remove the iron from the transferrin (Boradia et al., 2014). It is likely that the Fe^{2+} generated is then sequestered into cytoplasmic bacterioferritin (Banerjee et al., 2011).

Work with *St. aureus* assimilation of Fe^{3+} uptake via catecholamines has indicated close mechanistic analogies to the FbpABC system, where the SstABCD proteins serve as a putative ATPase (SstC), cell membrane proteins (SstA and SstB), and a membrane-bound lipoprotein (SstD) (Caza and Kronstad, 2013). Fungi, as exemplified by *Aspergillus fumigatus*, *Candida albicans*, and *Cryptococcus neoformans*, can also acquire iron from transferrin. With *Ca. albicans*, direct contact with transferrin is required in addition to the high-affinity reduction/uptake system composed of the reductase Fre10 and the permease Ftr1 (Caza and Kronstad, 2013).

40.5 FERROUS ION

Fe^{2+} uptake, in general, is less well examined than the assimilation of iron via ferrisiderophores or iron–heme acquisition. Fungi use the Fe^{3+} of ferrisiderophores and heme as well as a reductive iron assimilation mechanism that utilizes a high-affinity iron permease, ferric reductases, and ferroxidases (Bailao et al., 2015). *P. aeruginosa* employs phenazine compounds (phenazine-1-carboxylic acid and pyocyanin) as reductants that reduce the Fe^{3+} of ferriproteins prior to FeoABC system (see in the following text) assimilation of the Fe^{2+} produced. Given that both Fe^{2+} and Fe^{3+} are integral to establishing biofilm in the CF lung (Cornelis and Dingemans, 2013), the production of Fe^{2+} by *P. aeruginosa* has direct clinical relevance.

In the enterobacteria, porins are thought to be the route of Fe^{2+} uptake through the outer membrane, although in the Gram-negative pathogenic bacterium *Francisella tularensis*, a specific protein, FupA, is required for assimilation. In the inner membrane, the ubiquitous Feo ferrous ion uptake system is used to incorporate the metal into the cell's cytoplasm (Pérez and Ramakrishnan, 2014).

The presence of Fe^{2+} is likely under anaerobic or microaerophilic conditions. Organisms, such as enteric bacteria, demonstrate the important role that the Feo has in iron acquisition in such environments (Cartron et al., 2006;

Fetherston et al., 2012; Siburt et al., 2012) and the pathogen *P. aeruginosa* expressed the key Fe^{2+} uptake gene, feoB, in samples taken from the CF lung (Konings et al., 2013). The Feo system is composed of three proteins encoded by the *feoABC* operon (Cartron et al., 2006; Pérez and Ramakrishnan, 2014; Weaver et al., 2013). In *V. cholera* and other bacteria (Cartron et al., 2006; Fetherston et al., 2012; Pérez and Ramakrishnan, 2014; Siburt et al., 2012; Weaver et al., 2013), FeoB is a GTP-dependent inner membrane permease of approximately 84 kDa composed of cytoplasmic N-terminal G-protein domain and the inner membrane integral C-terminal domain (Pérez and Ramakrishnan, 2014; Weaver et al., 2013). The GTPase domain is thought to control the transport of Fe^{2+}, while the C-terminal domain contains eight transmembrane-spanning α-helices thought to form the pore through which Fe^{2+} is assimilated (Weaver et al., 2013). In pathogenic *V. cholerae*, the FeoB protein requires both the FeoA and FeoC proteins if Fe^{2+} is to be assimilated. While it was shown that FeoC interacts with the cytoplasmic domain of FeoB, the exact roles of FeoA and FeoC remain elusive (Weaver et al., 2013). FeoA has been suggested to interact with and stimulate FeoB's GTPase activity, while FeoC may serve as a transcriptional repressor of the *feoABC* locus. As with other iron-acquisition systems, the *feoABC* operon is under Fur-dependent repressible control (Fetherston et al., 2012; Mey et al., 2005; Weaver et al., 2013).

40.6 SUMMARY

The assimilation of iron by microorganisms reflects the group's great metabolic and physiological diversity. This chapter reviews the major mechanisms of iron assimilation by microbes. There are undoubtedly variations and alternatives to the mechanisms of iron assimilation discussed in the current assessment of the topic and the reader is referred to the wealth of iron assimilation literature for further details. Given the tremendous diversity of the group, one expects that yet uncovered mechanisms of iron assimilation await discovery. Indeed, the recent elucidation of a receptor-mediated transport mechanism of assimilation of the iron from transferrin (Boradia et al., 2014) bears witness to such discoveries. It will be of interest to follow the field in the future as details of siderophore, transferrin, lactoferrin, heme, hemophores, and ferrous ion assimilation continue to be elucidated.

REFERENCES

Adler, C., Cobalán, N.S., Seyedsayamdost, M.R., Pomares, M.F., de Cristóbal, R.E., Clardy, J., Kolter, R., and Vincent, P.A. 2012. Catecholate siderophores protect bacteria from pyochelin toxicity, *PLoS One* 7(10): e46754. doi:10.1371/journal.pone.0046754.

Ahmed, E. and Holmström, S.J.M. 2014. Siderophores in environmental research: Roles and applications, *Microbial Biotechnology* 7: 196–208.

Andersen, S.B., Marvig, R.L., Molin, S., Krogh, H., and Griffin, A.S. 2015. Long-term social dynamics drive loss of function in pathogenic bacteria, *Proceedings of the National Academy of Sciences* 112: 10756–10761.

Andrews, N. 1999. Disorders of iron metabolism, *The New England Journal of Medicine* 341: 1986–1995.

Andrews, S.C., Robinson, A.K., and Rodríguez-Quiñones, F. 2003. Bacterial iron homeostasis, *FEMS Microbiology Reviews* 7: 215–237.

Arantes, V. and Milagres, A.M.F. 2007. The effect of a catecholate chelator as a redox agent in Fenton-based reactions on degradation of lignin-model substrates and on COD removal from effluent of an ECF kraft pulp mill, *Journal of Hazardous Materials* 141: 273–279.

Bagg, A. and Neilands, J.B. 1987. Molecular mechanism of regulation of siderophore-mediated iron assimilation, *Microbiology Reviews* 51: 509–518.

Bailao, E.F., de Sousa Lima, P., Silva-Bailao, M.G., Bailao, A.M., da Rocha Fernandes, G., Kosman, D.K., and de Almeida Soares, C.M. 2015. *Paracoccidioides* spp. ferrous and ferric iron assimilation pathways, *Frontiers in Microbiology* 6: 1–12.

Banerjee, S., Farhana, A., Ehtesham, N.Z., and Hasnain, S.E. 2011. Iron acquisition, assimilation and regulation in mycobacteria, *Infection, Genetics and Evolution* 11: 825–838.

Barton, H.A., Johnson, Z., Cox, C.D., Vasil, A.L., and Vasil, M.L. 1996. Ferric uptake regulator mutants of *Pseudomonas aeruginosa* with distinct alterations in the iron-dependent repression of exotoxin A and siderophores in aerobic and microaerobic environments, *Molecular Microbiology* 21: 1001–1017.

Beare, P.A., For, R.J., Martin, L.W., and Lamont, I.L. 2003. Siderophore-mediated cell signalling in *Pseudomonas aeruginosa*: Divergent pathways regulate virulence factor production and siderophore receptor synthesis, *Molecular Microbiology* 47: 195–207.

Becerra, G., Blasco, R., Quesada, A., Merchán, F., and Igeno, M.I. 2011. Role of Fur on cyanide tolerance of *Pseudomonas pseudoalcaligenes* CECT5344, *Biochemical Society Transactions* 39: 1854–1858.

Behrends, T., Krawczyk-Bärsch, E., and Arnold, T. 2012. Implementation of microbial processes in the performance assessment of spent nuclear fuel repositories, *Applied Geochemistry* 27: 453–462.

Boradia, V.M., Malhotra, H., Thakkar, J.S., Tillu, V.A., Vuppala, B., Patil, P., Sheokand, N. et al. 2014. *Mycobacterium tuberculosis* acquires iron by cell-surface sequestration and internalization of human holo-transferrin, *Nature Communications* 5: 1–13. doi: 10.1038/ncomms5730.

Bou-Abdallah, F., Yang, H., Awomolo, A., Cooper, B., Woodhall, M.R., Andrews, S.C., and Chasteen, N.D. 2013. Functionality of the three-site ferroxidase center of *Escherichia coli* bacterial ferritin (EcFtnA), *Biochemistry* 53: 483–495.

Bracken, C.S., Baer, M.T., Abdur-Rashid, A., Helms, W., and Stojiljkovic, I. 1999. Use of heme-protein complexes by the *Yersinia enterocolitica* HemR receptor: Histidine residues are essential for receptor function, *Journal of Bacteriology* 181: 6063–6072.

Braud, A., Geoffroy, V., Hoegy, F., Mislin, G.L., and Schalk, I.J. 2010. Presence of the siderophores pyoverdine and pyochelin in the extracellular medium reduces toxic metal accumulation in *Pseudomonas aeruginosa* and increases bacterial metal tolerance, *Environmental Microbiology Reports* 2: 419–425.

Braud, A., Hannauer, M., Mislin, G.L.A., and Schalk, I.J. 2009a. The *Pseudomonas aeruginosa* pyochelin-uptake pathway and its metal specificity, *Journal of Bacteriology* 191: 3517–3525.

Braud, A., Hoegy, F., Jezequel, K., Lebeau, T., and Schalk, I.J. 2009b. New insights into the metal specificity of the *Pseudomonas aeruginosa* pyoverdine-iron uptake pathway, *Environmental Microbiology* 11: 1079–1091.

Braun, V. and Hantke, K. 2011. Recent insights into iron import by bacteria, *Current Opinion in Chemical Biology* 15: 328–334.

Braun, V. and Killman, H. 1999. Bacterial solutions to the iron-supply problem, *Trends in Biochemical Sciences*, 24: 104–109.

Braun, V., Mahren, S., and Ogierman, M. 2003. Regulation of the FecI-type ECF sigma factor by transmembrane signalling, *Current Opinion in Microbiology* 6: 173–180.

Britigan, B.E., Hayek, M.B., Doebbeling, B.N., and Fick, R.B.J. 1993. Transferrin and lactoferrin undergo proteolytic cleavage in the *Pseudomonas aeruginosa* infected lungs of patients with cystic fibrosis, *Infection and Immunity* 61: 5049–5055.

Cao, Y.R., Zhang, X.Y., Deng, J.Y., Zhao, Q.Q., and Xu, H. 2012. Lead and cadmium-induced oxidative stress impacting mycelial growth of *Oudemansiella radicata* in liquid medium alleviated by microbial siderophores, *World Journal of Microbiology and Biotechnology* 28: 1727–1737.

Carrondo, M.A. 2003. Ferritins, iron uptake and storage from the bacterioferritin viewpoint, *The EMBO Journal* 22: 1959–1968.

Cartron, M.L., Maddocks, S., Gillingham, P., Craven, C.J., and Andrews, C.S. 2006. Feo—Transport of ferrous iron into bacteria, *Biometals* 19: 143–157.

Castignetti, D. and Smarrelli, J.J. 1986. Siderophores, the iron nutrition of plants, and NADH: Nitrate reductase, *FEBS Letters* 209: 147–151.

Caza, M. and Kronstad, J.W. 2013. Shared and distinct mechanisms of iron acquisition by bacterial and fungal pathogens of humans, *Frontiers in Cellular and Infection Microbiology* 3: 80.

Chen, Q. and Crosa, J.H. 1996. Antisense RNA, fur, iron, and the regulation of iron transport genes in *Vibrio anguillarum*, *The Journal of Biological Chemistry* 271: 18885–18891.

Cordero, O.X., Ventouras, L.A., DeLong, E.F., and Polz, M.F. 2012. Public good dynamics drive evolution of iron acquisition strategies in natural bacterioplankton populations, *Proceedings of the National Academy of Sciences of the United States of America* 109: 20059–20064.

Cornelis, P. 2010. Iron uptake and metabolism in pseudomonads, *Applied Microbiology and Biotechnology* 86:1637–1645.

Cornelis, P. and Bodilis, J. 2009. A survey of TonB-dependent receptors in fluorescent pseudomonads, *Environmental Microbiology Reports* 1: 256–262.

Cornelis, P. and Dingemans, J. 2013. *Pseudomonas aeruginosa* adapts its iron uptake strategies in function of the type of infections, *Frontiers in Cellular and Infection Microbiology* 3: 75. doi: 10.3389/fcimb.2013.00075.

Cornelis, P., Matthijs, S., and Oeffelen, L. 2009. Iron uptake regulation in *Pseudomonas aeruginosa*, *Biometals* 22: 15–22.

Cornelissen, C.N. 2003. Transferrin-iron uptake by Gram-negative bacteria, *Frontiers in Bioscience* 8: 836–847.

Crosa, J.H. and Walsh, C.T. 2002. Genetics and assembly line enzymology of siderophore biosynthesis in bacteria, *Microbiology and Molecular Biology Reviews* 66: 223–249.

Deniau, C., Gilli, R., Izadi-Pruneyre, N., Létoffé, S., Delepierre, M., Wandersman, C., Briand, C., and Lecroisey, A. 2003. Thermodynamics of heme binding to the HasA$_{SM}$ hemophore: Effect of mutations at three key residues for heme uptake, *Biochemistry* 42: 10627–10633.

Déziel, E., Lépine, F., Milot, S., and Villemur, R. 2003. rhlA is required for the production of a novel biosurfactant promoting swarming motility in *Pseudomonas aeruginosa*: 3-(3-hydroxyalkanoyloxy) alkanoic acids (HAAs), the precursors of rhamnolipids, *Microbiology* 149: 2005–2013.

Diels, L., De Smet, M., Hooyberghs, L., and Corbisier, P. 1999. Heavy metals bioremediation of soil, *Molecular Biotechnology* 12: 149–158.

Dimise, E.J., Widboom, P.F., and Bruner, S.D. 2008. Structure elucidation and biosynthesis of fuscachelins, peptide siderophores from the moderate thermophile *Thermobifida fusca*, *Proceedings of the National Academy of Sciences of the United States of America* 105: 15311–15316.

Dingemans, J., Ye, L., Hildebrand, F., Tontodonati, F., Craggs, M., Bilocq, F., De Vos, D. et al. 2014. The deletion of TonB-dependent receptor genes is part of the genome reduction process that occurs during adaptation of *Pseudomonas aeruginosa* to the cystic fibrosis lung, *Pathologic Disorders* 71: 26–38.

D'Onofrio, A., Crawford, J.M., Stewart, E.J., Witt, K., Gavrish, E., Epstein, S., Clardy, J., and Lewis, K. 2010. Siderophores from neighboring organisms promote the growth of uncultured bacteria, *Chemistry and Biology* 17: 254–264.

Edberg, F., Kalinowski, B.E., Holström, S.J., and Holm, K. 2010. Mobilization of metals from uranium mine waste: The role of pyoverdines produced by *Pseudomonas fluorescens*, *Geobiology* 8: 278–292.

Emery, T. 1971. Role of ferrichrome as a ferric ionophore in *Ustilago sphaerogena*. *Biochemistry* 10: 1483–1488.

Emery, T. 1982. Iron metabolism in humans and plants, *American Scientist* 70: 626–632.

Emery, T. 1983. Reductive mechanism for fungal iron transport in microbiology. In: ed. D. Schlessinger, *Microbiology*. American Society for Microbiology, Washington, DC, pp. 293–295.

Fetherston, J.D., Mier, I.J., Truszczynska, H., and Perry, R.D. 2012. The Yfe and Feo transporters are involved in microaerobic growth and virulence of *Yersinia pestis* in bubonic plague, *Infection and Immunity* 80: 3880–3891.

Fonner, B.A., Tripet, B.P., Eilers, B.J., Stanisich, J., Sullivan-Springhetti, R.K., Moore, R., Liu, M., Lei, B., and Copié, V. 2014. Solution structure and molecular determinants of hemoglobin binding of the first NEAT domain of IsdB in *Staphylococcus aureus*, *Biochemistry* 53: 3922–2933.

Gaonkar, T. and Bhosle, S. 2013. Effect of metals on a siderophore producing bacterial isolate and its implications on microbial assisted bioremediation of metal contaminated soils, *Chemosphere* 93: 1835–1843.

Gaonkar, T., Nayak, P.K., Garg, S., and Bhosle, S. 2012. Siderophore-producing bacteria from a sand dune ecosystem and the effect of sodium benzoate on siderophore production by a potential isolate, *The Scientific World Journal* 2012 (2012): 857249 (PMC. Web. 10 Oct. 2016).

Gauglitz, J.M., Zhou, H., and Butler, A. 2012. A suite of citrate-derived siderophores from a marine *Vibrio* species isolated following the Deepwater Horizon oil spill, *Journal of Inorganic Biochemistry* 107: 90–95.

Ge, R. and Sun, X. 2014. Iron acquisition and regulation systems in *Streptococcus* species, *Metallomics* 6: 996–1003.

Gresock, M.G., Kastead, K.A., and Postle, K. 2015. From homodimer to heterodimer and back: Elucidating the TonB energy transduction cycle, *Journal of Bacteriology* 197: 3433–3445.

Hanks, T.S., Liu, M., McClure, M.J., and Lei, B. 2005. ABC transporter FtsABCD of *Streptococcus pyogenes* mediates uptake of ferric ferrichrome, *BioMed Central Microbiology* 5: 62.

Hannauer, M., Arifin, A.J., and Heinrichs, D.E. 2015. Involvement of reductases IruO and NtrA in iron acquisition by *Staphylococcus aureus*, *Molecular Microbiology* 96: 1192–1210.

Higashimoto, Y., Sato, H., Sakamoto, H., Takahashi, K., Palmer, G., and Noguchi, M. 2006. The reactions of heme- and verdoheme-heme oxygenase-1 complexes with FMN-depleted NADPH-cytochrome P450 reductase. Electrons required for verdoheme oxidation can be transferred through a pathway not involving FMN, *Journal of Biological Chemistry* 281: 31659–31667.

Honsa, E.S., Maresso, A.W., and Highlander, S.K. 2014. Molecular and evolutionary analysis of NEAr-iron transporter (NEAT) domains, *PLoS One* 9:104794.

Hrusa, G., Farmer, W., Weiss, B.L., Applebaum, T., Roma, J.S., Szeto, L., Aksoy, S., and Runyen-Janecky, L.J. 2015. TonB-dependent heme iron acquisition in the tsetse fly symbiont *Sodalis glossinidius*, *Applied and Environmental Microbiology* 81: 2900–2909.

Hunter, R.C., Asfour, F., Dingemans, J., Osuna, B.L., Samad, T., Malfroot, A., Cornelis, P., and Newman, D.K. 2013. Ferrous iron is a significant component of bioavailable iron in cystic fibrosis airways, *mBio* 4: 00557.

Inoue, H., Takimura, O., Kawaguchi, K., Nitoda, T., Fuse, H., Murakami, K., and Yamaoka, Y. 2003. Tin-carbon cleavage of organotin compounds by pyoverdine from *Pseudomonas chlororaphis*, *Applied and Environmental Microbiology* 69: 878–883.

Jones, C.M. and Niederweis, M. 2010. Role of porins in iron uptake by *Mycobacterium smegmatis*, *Journal of Bacteriology* 192: 6411–6417.

Kelson, A.B., Carnevali, M., and Truong-Le, V. 2013. Gallium-based anti-infectives: Targeting microbial iron-uptake mechanisms, *Current Opinion in Pharmacology* 13: 707–716.

Kem, M.P., Zane, H.K., Springer, S.D., Gauglitz, J.M., and Butler, A. 2014. Amphiphilic siderophore production by oil-associating microbes, *Metallomics* 6: 1150–1155.

Konings, A.F., Martin, L.W., Sharples, K.J., Roddam, L.F., Latham, R., Reid, D.W., and Lamont, I.L. 2013. *Pseudomonas aeruginosa* uses multiple pathways to acquire iron during chronic infection in cystic fibrosis lungs, *Infection and Immunity* 81: 2697–2704.

Krewulak, K.D. and Vogel, H.J. 2008. Structural biology of bacterial iron uptake, *Biochimica et Biophysica Acta* 1778: 1781–1804.

Létoffé, S., Delepelaire, P., and Wandersman, C. 2004. Free and hemophore-bound heme acquisitions through the outer membrane receptor HasR have different requirements for the TonB-ExbB-ExbD complex, *Journal of Bacteriology* 186: 4067–4074.

Lewis, T.A., Cortese, M.S., Sebat, J.L., Green, T.L., Lee, C.H., and Crawford, R.L. 2000. A *Pseudomonas stutzeri* gene cluster encoding the biosynthesis of the CCl4-dechlorination agent pyridine-2,6-bis(thiocarboxylic acid), *Environmental Microbiology* 2: 407–416.

Lewis, T.A., Paszczynski, A.J., Gordon-Wylie, S.W., Jeedigunta, S., Lee, C.-H., and Crawford, R.L. 2001. Carbon tetrachloride dechlorination by the bacterial transition metal chelator pyridine-2, 6-bis(thiocarboxylic acid), *Environmental Science and Technology* 35: 552–559.

Liao, H., Cheng, X., Zhu, D., Wang, M., Jia, R., Chen, S., Chen, X., Biville, F., Liu, M., and Cheng, A. 2015. TonB energy transduction systems of *Riemerella anatipestifer* are required for iron and hemin utilization, *PLoS One* 10: 0127506.

Lim, C.K., Hassan, K.A., Tetu, S.G., Loper, J.E., and Paulsen, I.T. 2012. The effect of iron limitation on the transcriptome and proteome of *Pseudomonas fluorescens* Pf-5, *PLoS One* 7: 39139.

Lipinski, B. 2011. Hydroxyl radical and its scavengers in health and disease, *Oxidative Medicine and Cellular Longevity*, 2011, Article ID 809696, 9 pp. doi:10.1155/2011/809696.

Lodge, J.S., Gaines, C.G., Arceneaux, J.E., and Byers, B.R. 1982. Ferrisiderophore reductase activity in *Agrobacterium tumefaciens*, *Journal of Bacteriology* 149: 771–774.

Madigan, M.T., Martinko, J.M., Bender, K.S., Buckley, D.H., Stahl, D.A., and Brock, T. 2015. *Brock Biology of Microorganisms*, 14th edn. Pearson, Boston, MA., pp. 351–352.

Marchler-Bauer, A., Derbyshire, M.K., Gonzales, N.R., Lu, S., Chitsaz, F., Geer, L.Y., Geer, R.C. et al. 2015. CDD: NCBI's conserved domain database. *Nucleic Acids Research* 43(Database issue): D222–D226.

Martin, L.W., Reid, D.W., Sharples, K.J., and Lamont, I.L. 2011. *Pseudomonas* siderophores in the sputum of patients with cystic fibrosis, *Biometals* 24: 1059–1067.

Marugg, J.D. and Weisbeek, P.J. 1991. Molecular genetics of siderophore biosynthesis in fluorescent pseudomonads. In: ed. G. Winkelmann, *Handbook of Microbial Iron Chelates*. CRC Press, Boca Raton, FL.

Massé, E. and Gottesman, S. 2002. A small RNA regulates the expression of genes involved in iron metabolism in *Escherichia coli*, *Proceedings of the National Academy of Sciences of the United States of America* 99: 4620–4625.

Matilla, M.A., Ramos, J.L., Duque, E., de Dios Alché, J., Espinosa-Urgel, M., and Ramos-González, M.I. 2007. Temperature and pyoverdine-mediated iron acquisition control surface motility of *Pseudomonas putida*, *Environmental Microbiology* 9: 1842–1850.

Matzanke, B.F. 1997. Iron storage in microorganisms. In: eds. G. Winkelmann and C.J. Carrano, *Transition Metals in Microbial Metabolism*. Harwood Academic Publishers, Amsterdam, the Netherlands.

Matzanke, B.F., Anemüller, S., Schünemann, V., Trautwein, A.X., and Hantke, K. 2004. FhuF, part of a siderophore-reductase system, *Biochemistry* 43: 1386–1392.

Mey, A.R., Wyckoff, E.E., Kanukurthy, V., Fisher, C.R., and Payne, S.M. 2005. Iron and fur regulation in *Vibrio cholerae* and the role of fur in virulence, *Infection and Immunity* 73: 8167–8178.

Moon, C.D., Zhang, X., Matthijs, S., Schäfer, M., Budzikiewicz, H., and Rainey, P.B. 2008. Genomic, genetic and structural analysis of pyoverdine-mediated iron acquisition in the plant growth-promoting bacterium *Pseudomonas fluorescens* SBW25, *BMC Microbiology* 8: 7. doi: 10.1186/1471-2180-8-7.

Mukhopadhyay, P., Zheng, M., Bedzyk, L.A., LaRossa, R.A., and Storz, G. 2004. Prominent roles of the NorR and Fur regulators in the *Escherichia coli* transcriptional response to reactive nitrogen species, *Proceedings of the National Academy of Sciences of the United States of America* 101: 745–750.

Naik, M.M. and Dubey, S.K. 2011. Lead-enhanced siderophore production and alteration in cell morphology in a Pb-resistant *Pseudomonas aeruginosa* strain 4EA, *Current Opinion in Microbiology* 62: 409–414.

Nair, A., Juwarkar, A.A., and Singh, S.K. 2007. Production and characterization of siderophores and its application in arsenic removal from contaminated soil, *Water, Air, and Soil Pollution* 180: 199–212.

Neilands, J.B. 1952. A crystalline organo-iron pigment from a Rust Fungus *Ustilago sphaerogena*, *Journal of the American Chemical Society* 74: 4846–4847.

Neilands, J.B. 1982. Microbial envelope proteins related to iron, *Annual Review of Microbiology* 36: 285–309.

Neilands, J.B. 1993. Siderophores, *Archives of Biochemistry and Biophysics* 302: 1–3.

Neilands, J.B. 1995. Siderophores: Structure and function of microbial iron transport compounds, *Journal of Biological Chemistry* 270: 26723–26726.

Noinaj, N., Guillier, M., Barnard, T.J., and Buchanan, S.K. 2010. TonB-dependent transporters: Regulation, structure, and function, *Annual Review of Microbiology* 64: 43–60.

O'Brien, S., Hodgson, D.J., and Buckling, A. 2014. Social evolution of toxic metal bioremediation in *Pseudomonas aeruginosa*, *Proceedings of the Royal Society Biological Sciences* 281: 1787.

O'Neill, M.J. and Wilks, A. 2013. The *P. aeruginosa* heme binding protein PhuS is a heme oxygenase titratable regulator of heme uptake, *ACS Chemical Biology* 8: 1794–1802.

Pérez, N.M. and Ramakrishnan, G. 2014. The reduced genome of the *Francisella tularensis* live vaccine strain (LVSPérez) encodes two iron acquisition systems essential for optimal growth and virulence, *PLoS One* 9: 93448.

Powell, P.E., Cline, G.R., Reid, C.P.P., and Szaniszlo, P.J. 1980. Occurrence of hydroxamate siderophore iron chelators in soils, *Nature Letters* 287: 833–834.

Rashmi, V., ShylajaNaciyar, M., Rajalakshmi, R., D'Souza, S.F., Prabaharan, D., and Uma, L. 2013. Siderophore mediated uranium sequestration by marine cyanobacterium *Synechococcuselongatus* BDU 130911, *Bioresource Technology* 130: 204–210.

Rocha, E.R. and Smith, J.C. 2013. Ferritin-like family proteins in the anaerobe *Bacteroides fragilis*: When an oxygen storm is coming, take your iron to the shelter, *Biometals* 264: 577–591.

Rohde, K.H. and Dyer, D.W. 2003. Mechanisms of iron acquisition by the human pathogens *Neisseria meningitidis* and *Neisseria gonorrhoeae*, *Frontiers in Bioscience* 8: 1186–1218.

Romsang, A., Duang-nkern, J., Wirathorn, W., Vattanaviboon, P., and Mongkolsuk, S. 2015. *Pseudomonas aeruginosa* IscR-regulated ferredoxin NADP(+) reductase gene (fprB) functions in iron-sulfur cluster biogenesis and multiple stress response, *PLoS One*. 2015; 10(7): e0134374.

Rossbach, S., Wilson, T.L., Kukuk, M.L., and Carty, H.A. 2000. Elevated zinc induces siderophore biosynthesis genes and a zntA-like gene in *Pseudomonas fluorescens*, *FEMS Microbiology Reviews* 191: 61–70.

Schalk, I.J. and Guillon, L. 2013. Fate of ferrisiderophores after import across bacterial outer membranes: different iron release strategies are observed in the cytoplasm or periplasm depending on the siderophore pathways, *Amino Acids* 44: 1267–1277.

Schroder, I., Johnson, E., and de Vries, S. 2003. Microbial ferric ion reductases, FEMS Microbiology Reviews 27: 427–447.

Sebat, J.L., Paszczynski, A.J., Cortese, M.S., and Crawford, R.L. 2001. Antimicrobial properties of pyridine-2,6-dithiocarboxylic acid, a metal chelator produced by *Pseudomonas* spp., *Applied and Environmental Microbiology*, 67: 3934–2942.

Siburt, C.J.P., Mietzner, T.A., and Crumbliss, A.L. 2012. FbpA—A bacterial transferrin with more to offer, *Biochimica et Biophysica Acta* 1820: 379–392.

Singh, P.K., Parsek, M.R., Greenberg, E.P., and Welsh, M.J. 2002. A component of innate immunity prevents bacterial biofilm development, *Nature* 417: 552–555.

Smarrelli, J.J. and Castignetti, D. 1988. Iron assimilation in plants: Reduction of a ferriphytosiderophore by NADH: Nitrate reductase from squash, *Planta* 173: 563–566.

Smith, A.D. and Wilks, A. 2015. Differential contributions of the outer membrane receptors PhuR and HasR to heme acquisition in *Pseudomonas aeruginosa*, *The Journal of Biological Chemistry* 290: 7756–7766.

Stintzi, A. and Raymond, K.N. 2000. Amonabactin-mediated iron acquisition from transferrin and lactoferrin by *Aeromonas hydrophila*: Direct measurement of individual microscopic rate constants, *Journal of Biological Inorganic Chemistry* 5: 57–66.

Sun, G.X., Zhou, W.Q., and Zhong, J.J. 2006. Organotin decomposition by pyochelin, secreted by *Pseudomonas aeruginosa* even in an iron-sufficient environment, *Applied and Environmental Microbiology* 72: 6411–6413.

Sverzhinsky, A., Chunga, J.W., Demea, J.C., Fabreb, L., Leveya, K.T., Plesaa, M., Cartera, D.M., Lypaczewskia, P., and Coulton, J.W. 2015. Membrane protein complex ExbB$_4$–ExbD$_1$–TonB$_1$ from *Escherichia coli* demonstrates conformational plasticity, *Journal of Bacteriology* 197: 3433–3445.

Taguchi, F., Suzuki, T., Inagaki, Y., Toyoda, K., Shiraishi, T., and Ichinose, Y. 2010. The siderophore pyoverdine of *Pseudomonas syringae* pv. *tabaci* 6605 is an intrinsic virulence factor in host tobacco infection, *Journal of Bacteriology* 192: 117–126.

Tong, Y. and Guo, M. 2009. Bacterial heme-transport proteins and their heme-coordination modes, *Archives of Biochemistry and Biophysics* 481: 1–15.

Torres, V.J., Pishchany, G., Humayan, M., Schneewind, O., and Skaar, E.P. 2006. *Staphylococcus aureus* IsdB is a hemoglobin receptor required for heme iron utilization, *Journal of Bacteriology* 188: 8421–8429.

Troxell, B. and Hassan, H.M. 2013. Transcriptional regulation by Ferric Uptake Regulator (Fur) in pathogenic bacteria, *Frontiers in Cellular and Infection Microbiology* 3: 59.

Unno, M., Matsui, T., and Ikeda-Saito, M. 2007. Structure and catalytic mechanism of heme oxygenase, *Nature Product Reports* 24: 553–570.

Vassinova, N. and Kozyrev, D. 2000. A method for direct cloning of fur-regulated genes: Identification of seven new fur-regulated loci in *Escherichia coli*, *Microbiology* 146: 3171–3182.

Wandersman, C. and Stojiljkovic, I. 2000. Bacterial heme sources: The role of heme, hemoprotein receptors and hemophores, *Current Opinion in Microbiology* 3: 215–220.

Wang, L., Yan, W., Chen, J., Huang, F., and Gao, P. 2008. Function of the iron-binding chelator produced by *Coriolus versicolor* in lignin biodegradation. *Scientific China Life Sciences* 51: 214–221.

Wang, Q., Xiong, D., Zhao, P., Yu, X., Tu, B., and Wang, G. 2011. Effect of applying an arsenic-resistant and plant growth-promoting rhizobacterium to enhance soil arsenic phytoremediation by *Populus deltoides* LH05-17, *Journal of Applied Microbiology* 111: 1065–1074.

Weaver, E.A., Wyckoff, E.E., Mey, A.R., Morrison, R., and Payne, S.M. 2013. FeoA and FeoC are essential components of the *Vibrio cholerae* ferrous iron uptake system, and FeoC interacts with FeoB, *Journal of Bacteriology* 195: 4826–4835.

Wilks, A. 2002. Heme oxygenase: Evolution, structure, and mechanism, *Antioxidants & Redox Signaling* 4: 603–614.

Wyckoff, E.E., Allred, B.E., Raymond, K.N., and Payne, S.M. 2015. Catechol Siderophore Transport by *Vibrio cholerae*, *Journal of Bacteriology* 197: 2840–2849.

Xu, G. and Goodell, B. 2001. Mechanisms of wood degradation by brown-rot fungi: Chelator-mediated cellulose degradation and binding of iron by cellulose, *Journal of Biotechnology* 87: 43–57.

Yan, R., Adinolfi, S., and Pastore, A. 2015. Ferredoxin, in conjunction with NADPH and ferredoxin-NADP reductase, transfers electrons to the IscS/IscU complex to promote iron–sulfur cluster assembly, *Biochimica et Biophysica Acta* 1854: 1113–1117.

Yasmin, S., Andrews, S.C., Moore, G.R., and Le Brun, N.E. 2011. A new role for heme, facilitating release of iron from the bacterioferritin iron biomineral, *The Journal of Biological Chemistry* 286: 3473–3483.

Zheng, M., Doan, B., Schneider, T.D., and Storz, G. 1999. OxyR and SoxRS regulation of fur, *Journal of Bacteriology* 181: 4639–4643.

41 Zinc Biosorption by Microbial Species for Biotreatment Processes

Griselda Sosa, María Alejandra Daniel, Matías R. Barrionuevo,
Irene C. Lazzarini Behrmann, María Laura Ferreira,
Anita Zalts, Silvana A. Ramírez, and Diana L. Vullo

CONTENTS

ABSTRACT

Metals are ubiquitous in our environment, and humans have been using them for thousands of years from the development of metal mining and smelting techniques, increasing metal distribution. Zinc is a common trace metal in natural waters; and it is also essential for the proper growth and development of humans, animals, and plants. Several treatment techniques have been used and developed over the years to remove dissolved metals from wastewaters. The most conventional technologies are based on chemical precipitation, adsorption, coagulation/flotation, sedimentation, filtration, membrane processes, electrochemical techniques, ion-exchange, and chemical reactions. Although these methods are simple and relatively cheap, most of them present some disadvantages such as a difficult separation of the solid, the need of an adequate disposal of the resulting toxic sludge, and the ineffectiveness for low metal concentration as in industrial wastewaters. The presence of soluble ligands in industrial wastes directly influences $Zn(II)$ availability and its consequent interaction with the sorbent surface. $Zn(II)$ can be efficiently removed by biosorption processes using both mineral and microbiological matrices as sorbents. The implementation of industrial-scale treatments using biosorption techniques will still require a deeper analysis to make them commercially viable, not discarding the combination with other technologies.

41.1 INTRODUCTION

Metals are ubiquitous in our environment, and we, humans, have a long history of thousands of years using them from the development of metal mining and smelting techniques. These activities greatly increased metal distribution throughout the environment. Lee et al. (2008) analyzed Liangzi Lake sediment cores to conclude that in ancient China the concentrations of Cu, Ni, Pb, and Zn increased gradually from about 3000 ± 328 BC, indicating the start of the Bronze Age, although the evidence shows that copper was first used about 7000 years ago. In a similar way, Cooke et al. (2007) have reconstructed a millennium of smelting activity from the Morococha Lake mining region in the central Peruvian Andes using the lake sediment stratigraphy of atmospherically derived metals associated with smelting (Pb, Zn, Cu, Ag, Sb, Bi, and Ti). In this region, the earliest evidence for metallurgy occurs ca. 1000 AD, when smelting appears to have been aimed at copper and copper alloys, because of large increases in Zn and Cu relative to Pb.

It was not until the Industrial Revolution in the eighteenth century that the demand for metals grew exponentially, as did the intensity of metal emissions to the environment. A characteristic of our present technological societies is the development of a broad spectrum of materials in engineering and technological applications. The other side of this development is related to the challenges created by the need of these materials, being resource scarcity and environmental impact two of the most pressing (Harper et al., 2006). The sustained rise in the use of metals over the twentieth century has led to the phenomenon of a considerable shift in metal stocks from the lithosphere to the anthroposphere (Gerst and Graedel, 2008), reflecting not only natural sources affecting the environment but also significant contributions from mankind's industrial activities to metal distribution. Although metals can neither be created nor be destroyed, their chemical forms (e.g., redox state, complexation, dissolution/precipitation, sorption on different surfaces) may be changed, thereby affecting their biological availability and, consequently, their toxicity and environmental impact (Hollenberg, 2010). Metals in the hydrosphere are mostly complexed by ligands or sorbed on surfaces, and only a small fraction is found as free hydrated ions. The physicochemical distribution of a metal among the different chemical species strongly affects its behavior. The presence of ligands and the interactions of metals with solid-phase materials of biological and geological origin modulate metal influence on biological processes.

Since the late 1800s, zinc became a widely used metal and new technological and engineering applications have appeared increasing its consumption in modern living. Zinc is present in the earth's crust mainly as sphalerite (zinc sulfide), although there are other minerals with high zinc content used in Zn production. Although being the 23rd most abundant element, zinc stands fourth among all metals in world production, and its production is only exceeded by iron, aluminum, and copper. U.S. Geological Survey informed that global zinc mine production in 2014 was 13.3 million tons (US Geological Survey, 2015). China (37% share of global production), Australia (11%), and Peru (10%) are the three leading producers of zinc. It is mostly used as coating to protect iron and steel from corrosion: galvanized steel is used extensively in the automotive and construction industries. Other major uses of zinc are the production of zinc alloys such as brass and bronze, chemicals that include zinc oxide used as an activator in the vulcanization process in tire manufacturing industry, and semimanufactures such as zinc sheets for architectural applications (rolled zinc) (USGS Minerals Yearbook, 2013). The nanotechnology field is growing rapidly, and the widespread use of engineered nanomaterials is increasing metal presence in the biosphere. Among metal oxides, zinc-oxide (ZnO) nanoparticles have the third highest volume of global production after TiO_2 and SiO_2 (Piccinno et al., 2012). All these nanoparticles are extensively used in solar panels, cosmetics, paint, paper, plastics, ceramics, and building materials. Kaya et al. (2016) reported that small (10–30 nm) and large (100 nm) ZnO particles have similar toxic effects on *Oreochromis niloticus*, but that smaller ZnO nanoparticles showed a higher toxicity potential to the fish. On the other hand, the dissolution of ZnO nanoparticles often depends upon factors such as interactions of nanoparticles with natural organic matter (NOM). Jiang et al. (2015) reported correlations specifying that aromatic carbon content is a crucial factor responsible for the rate of NOM-promoted dissolution of ZnO nanoparticles. The high surface-to-volume ratio and reactivity of nanomaterials make them highly dynamic in environmental systems (Lowry et al., 2012).

Zinc is a common trace metal in natural waters, and it is also a necessary element for the proper growth and development of humans, animals, and plants; it is the second most common trace metal, after iron, naturally found in the human body. The Annual Average Environmental Quality Standards are expressed as added risk and represent conditions of high bioavailability and so should be protective of sensitive areas. The Scientific Committee on Health and Environmental Risks agrees on the way the AA-EQS for freshwater has been derived (10.9 µg/L bioavailable Zn, as added risk) (SCHER, 2012). The Quality Criteria for Water published by U.S. Environmental Protection Agency in 1976 and 1986 are still current for Zn. A maximum concentration of 5000 µg Zn/L was set taking into account organoleptic effects. Zinc imparts an undesirable astringent taste to water. Tests indicate that 5% of a population could distinguish between zinc-free water and water containing zinc at a level of 4 mg/L (as zinc sulfate). The detection levels for other zinc salts were somewhat higher. Water containing zinc at concentrations in the range 3–5 mg/L also tends to appear opalescent and develops a greasy film when boiled (WHO, 2003). In 1982, Joint FAO/WFO Expert Committee on Food Additives proposed a daily dietary requirement of zinc of 0.3 mg/kg of body weight and a provisional maximum tolerable daily intake of 1.0 mg/kg of body weight (Joint FAO/WHO Expert Committee on Food Additives, 1982). It is recommended that the total concentration of zinc in drinking water should not exceed 5000 µg Zn/L (Environmental Protection Division, 1999). This level appears in many drinking water specifications (see, e.g., Argentine Food Codex, Administración Nacional de Medicamentos, Alimentos y Tecnología Médica, 2007).

Several treatment techniques have been used and developed over the years to remove dissolved metals from wastewaters. The most conventional technologies are based on chemical precipitation, usually using lime, adsorption, coagulation/flotation, sedimentation, filtration, membrane processes, electrochemical techniques, ion exchange, biological processes, and chemical reactions. Although these methods are simple and relatively cheap, most of them present some disadvantages such as a difficult separation of the solid, the need of an adequate disposal of the resulting toxic sludge and the ineffectiveness for low metal concentration (Vullo et al., 2007).

41.2 USE OF METAL SORPTION ON MINERAL MATRICES IN THE TREATMENT OF INDUSTRIAL EFFLUENTS: SECOND RINSING BATHS

Metal protection from corrosion is normally achieved by electroplating. The use of chromium in electroplating and anodizing tanks is among the largest sources of chromium emissions in the United States. The use of zinc is also widespread.

The increasing demand of products and the resulting industrial activities have intensified concerns about environmental pollution issues. In particular, metal finishing industries constitute a highly sensitive sector that produces metal emissions to the environment. Due to increasing regulatory driving forces, many industries face the problem of rising manufacturing costs pursuing "greener" technologies (Baral and Engelken, 2002) or developing better treatment methods.

It is a general practice in Argentina that once the electroplating process is finished, each electroplated piece is rinsed by consecutive immersion in two big pools. Effluents from the second pool, the second rinsing bath (SRB), may be ca. 20 m³/day and typically contain metal ions (e.g., chromium, nickel, cadmium, zinc) and additives. These are basically organic substances such as polyelectrolytes and surfactants, included in the galvanic process for proper coating. Depending on a variety of parameters such as metal content and pH, these wastewaters should be treated if they do not meet the criteria for disposal (Argentina—Decree PEN 999/92).

Additives in these wastewaters can act as ligands toward metals (Barriola et al., 2012). Thus, metals are capable of bonding to form metal complexes of variable strength, keeping metals in solution and affecting the availability when treatments are to be applied.

Adsorption is one of the most effective and economic methods for heavy metal wastewater treatment. This process offers flexibility in design and operation and in many cases good high-quality treated effluents are obtained. Another positive aspect to be considered is that adsorbents may be regenerated by a suitable desorption process when adsorption is reversible.

In a previous paper (Sosa et al., 2016), we have evaluated metal loading and complexing capacity in wastewaters of an electroplating facility. Samples from a SRB were taken during an 8 h workday and analyzed focusing on ligands of moderate strength. Zinc, lead, cadmium, and copper were detected in these SRB samples. The most important changes in metal concentration were related to zinc and lead, which increased up to ca. 90.0 and 1.6 ppm at the end of the working day. Copper and cadmium contents were negligible (Cu < 0.09 ppm and Cd < 0.06 ppm).

The interaction between Zn and ligands was evaluated since zinc was one of the most important components in the SRB. Ligands were detected through an electrochemically monitored titration. The different fractions of the SRB showed two moderate ligand families of constants $K'f_1$ (4.07 ± 0.69) $\cdot 10^6$ and $K'f_2$ (5.56 ± 0.78) $\cdot 10^5$ (pH = 7.5) with Zn. As expected, both ligand families increased their concentration with the use of the SRB from micromolar to millimolar. Considering solely Zn(II) as the metal component, we concluded that important changes were taking place along the use of the SRB. At the beginning of the workday, Zn(II) exceeded ligand concentration and then becomes partially complexed, and by the end of the day, ligand concentration exceeded Zn(II), thus becoming fully complexed. Attention should be paid to Zn(II) and ligand concentrations for a successful treatment prior to disposal.

As a result of our previous studies, we have considered zinc removal from the SRB using a low-cost and easily available adsorbent. We have decided to use pet sanitary pellets (PS), which can be easily obtained in pet shops. PS are basically diatomites. An x-ray analysis of the grinded solid showed different components: smectite, quartz, feldspar, and amorphous particles.

The adsorption isotherm of PS was tested in a batch procedure. A solution of 50 mL of a Zn(II) (Zn(II) content between 2×10^{-6} M and 1×10^{-2} M, pH: 5.5) was added to 55 mg of PS and stirred. Zn(II) concentration in the supernatant was measured after 24 h equilibration using anodic stripping voltammetry (ASV). Results are shown in Figure 41.1.

Different linearization models for the adsorption isotherm have been explored, as shown in Figure 41.2. Several physicochemical models explain the adsorption isotherm behavior in the literature. All these models were developed with inorganic/organic adsorbates, but mainly inorganic surfaces were used as adsorbents. The most common models applied to this isotherm interpretation are Langmuir and Freundlich. The Langmuir model assumes that adsorption process ensues

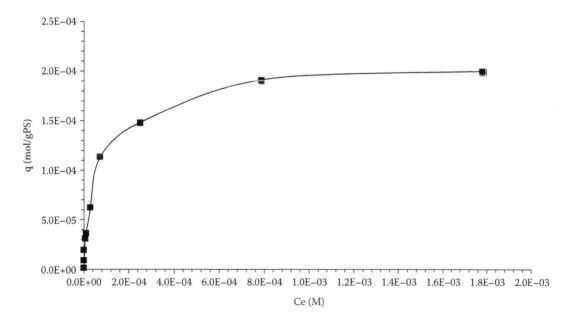

FIGURE 41.1 Zn retention for different Zn(II) concentrations using PS as adsorbent material.

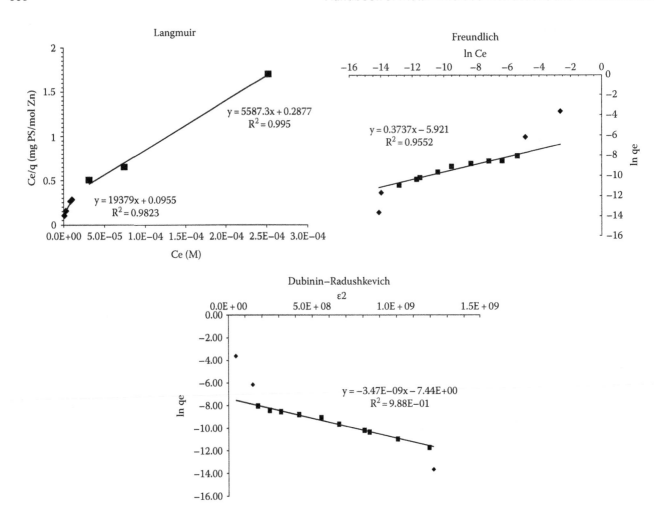

FIGURE 41.2 Fit of experimental results to Langmuir, Freundlich, and Dubinin–Radushkevich isotherms.

in a monolayer, that all adsorption sites are identical, and that no changes in adsorption free energy are observed. According to this, when adsorption follows the Langmuir model, the number of occupied sites, q, is given by

$$q = q_{max} \cdot Ce/(Kd+Ce) \quad or \quad Ce/q = Kd/q_{max} + Ce/q_{max}$$

where

q_{max} is related to the total number of adsorption sites
Ce is Zn(II) final equilibrium concentration in supernatants
Kd is the equilibrium constant for the dissociation of the surface complex

The number of occupied sites, q, was calculated as follows:

$$q = (Ci - Ce) \times \frac{V_t}{m_t}$$

where

Ci is the initial Zn(II) concentration
V_t is the total volume of the biosorption mixture assayed
m_t is the total biosorbent mass as dry weight

When plotting Ce/q versus Ce results in a linear graphic, there is a good agreement to this model and Kd and q_{max} are obtained.

The Freundlich isotherm is represented by

$$\ln q = \frac{1}{n} \cdot \ln Ce + \ln K$$

where

q is the number of Zn(II) mol/g of PS
Ce is Zn(II) concentration measured after equilibration
K is related to the adsorption capacity
n is an indicator of the adsorption intensity

In this empirical model, all sites on the surface are not considered equal. Adsorption becomes progressively more difficult as more and more adsorbate accumulates. It is assumed that once the surface is covered, additional adsorbed species can still be accommodated, leaving no possible prediction of maximal monolayer adsorption. In other words, the Freundlich model does not consider the existence of a monolayer and a maximal adsorption capacity as the Langmuir model does.

The Dubinin–Radushkevich isotherm is represented by

$$\ln q = \ln Q_m - K_d \cdot \varepsilon^2$$

where

$$\varepsilon = R \cdot T \cdot \ln\left(1 + \frac{1}{Ce}\right)$$

In this isotherm, q has the same meaning as in the Freundlich isotherm. Q_m has the same interpretation as K in the Freundlich equation. Considering the parameters obtained from Freundlich or Dubinin–Radushkevich linearizations, Zn(II) normal adsorption on PS is concluded from $1/n = 0.37$, where $n = 2.67$ also suggests a favorable process according to the literature. Both isotherms point to an adsorption capacity around $2 \cdot 10^{-3}$ mol Zn(II)/g PS.

Both equations may seem totally different, but considering that for the whole range of concentrations used in the experiments $1 \ll 1/Ce$, we get to the following approximation for the Dubinin–Radushkevich equation:

$$\ln q = 2 \cdot K_{ad} \cdot (RT)^2 \cdot \ln Ce + \ln Q_m$$

which shows why in the case of our experiments both isotherms are in fair agreement.

At this point, it is common practice to resume all parameters obtained for the three evaluated isotherms. From our point of view, a global analysis would be more helpful. Langmuir's R^2 coefficient is the best of all three (Figure 41.2), but linear behavior is mainly observed in the final portion of the plot. Experimental results are better described by both Freundlich and Dubinin–Radushkevich equations for a wider range of concentrations. Surface heterogeneity is a condition common to both isotherms.

No matter the physicochemical description of the adsorption of Zn(II) onto the PS, it is relevant to determine if the amount of Zn(II) retained is significant and also if this retention is modified by the presence of ligands. Three sets of experiments were done using 55 mg PS and solutions containing Zn(II) and one of the following ligands: HIDA (N-(2-hydroxyethyl) iminodiacetic acid), NTA (nitrilotriacetic acid), and EDTA (ethylenediaminetetraacetic acid). These metal complexing agents are well characterized and were used as model ligands (Martell and Smith, 2010).

Set 1 (S1): PS + Zn(II)
Set 2 (S2): PS + ligand
Set 3 (S3): PS + Zn(II) + ligand

Control experiments with no PS were also done. The solid and the solutions were equilibrated for 48 h (100 rpm). The supernatant was separated by centrifugation, and Zn(II) and ligand concentrations were determined according to the experiment.

For Zn(II) determination in S2 and S3, aliquots of supernatant were conditioned to approximately pH 1 (concentrated HNO_3), transferred to PTFE bottles, and exposed to UV radiation for at least 12 h for organic matter oxidation (Campos et al., 2001). Finally, Zn(II) concentration was determined by ASV. S1 did not require organic matter oxidation.

Ligand concentration in S2 and S3 was determined in an ASV monitored titration. In the case of S3, Zn(II) elimination was required so that ligands were able to interact with the titrant agent. This was performed equilibrating the solution with Chelex 100 for 4 h in an orbital shaker (25°C, 200 rpm). Chelex 100 is a cationic resin that has proved effective

for metal removal including cadmium and zinc (Manouchehri and Bermond, 2006; Leung et al., 2008; Ceretti et al., 2010). After equilibration, the system was filtered through a 0.45 μm pore diameter cellulose membrane and diluted. In the ASV-monitored titration, the peak current (Ip) of the metal used as titrant agent is measured after increasing additions of the metal from a standard solution. After every metal spike, solutions were stirred for equilibration, and N_2 was bubbled to deoxygenate. A titration curve (Ip vs. added Me^{2+} concentration) was plotted. Zn(II) was the titrant metal in this study.

After exposure of PS to the ligands in S2, results show that final concentration was smaller than the initial value (12 mM). NTA in 8.1 mM was determined, representing 32% retention. For HIDA and EDTA, ligand concentration was below detection limit of the procedure (i.e., concentration <1 μM). S2 results are clear evidence that ligands were retained by PS. Titration of the supernatant of S3 containing HIDA as model ligand, before and after treatment with PS, is shown in Figure 41.3. There is no doubt that the curves are different, pointing out a different Zn(II) behavior in each solution. When ligands are absent, Ip should be proportional to Zn(II) solution concentration, and a linear behavior is obtained. This is observed in the linear shape of the "after PS" titration curve confirming that no metal–ligand interactions are preventing Zn(II) to produce the electric signal. The initial shape of the "before PS" curve clearly indicates that a smaller Ip is obtained and that Zn(II) available to produce the signal is smaller than in the final solution for the same amount of added Zn(II). This is a clear indication of the presence of a complexing agent (HIDA) interacting with Zn(II) and preventing the metal ions to produce the signal. The conclusion is that HIDA was retained by PS. Table 41.1 shows a summary of the results obtained for sets S1 and S3.

In S1, ca. 70% Zn(II) retention is obtained. Compared to S3, the presence of dissolved ligands clearly modifies Zn interaction with PS and produces a smaller retention. Another

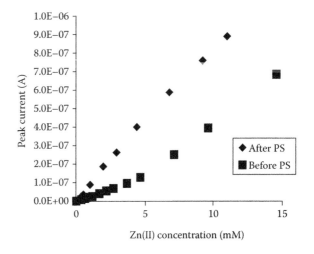

FIGURE 41.3 ASV monitored titration using Zn(II) as titrant agent. Supernatant of S3 containing HIDA [N-(2- hydroxyethyl) iminodiacetic acid)] before exposure to PS (■) and after exposure to PS(◆).

TABLE 41.1

Zn(II) Retention in S1 and S3 in the Presence of Ligands

		S1 (PS[Pet Sanitary Pellets] + Zn(II))	S3 (PS + Zn(II) + Ligand)	
			HIDA (*N*-(2- Hydroxyethyl) Iminodiacetic Acid)	EDTA
Zn(II)	Before PS	10 mM	10 mM	13 mM
	After PS	3.2 mM	8.6 mM	7.3 mM
	Retention %	68.8	14	45
Ligands	Before PS	—	10 mM	10 mM
	After PS	—	<1 nM	<1 nM
	Retention %	—	~100	~100

interesting result is that HIDA and NTA were also retained by the solid, no matter the presence of Zn(II) ions. X-ray Diffraction (XRD) experiments are in progress to explain metal and ligand adsorption on PS. It is clear from these results that the presence of ligands affects metal retention efficiency. Furthermore, from an environmental point of view, the presence of ligands increased solution toxicity by increasing the concentration of the available metal in the solution. The availability of Zn(II) ions is affected, and this fact should be considered in the treatment of this solution.

41.3 METAL–MICROBE INTERACTIONS

41.3.1 Microbes in Biogeochemical Cycles of Transition Metals

Microorganisms play key roles in natural metal cycles, thanks to four main biological processes that can be applied to transition metals in the environment (Tourney and Ngwenya, 2014). The first one represents only physicochemical interactions that promote metal immobilization: cell surface adsorption,

the nonmetabolic uptake of metal ions and other solutes to the cell surfaces. The second one includes the metabolic uptake, with the consequent sequestration of metals within the cell interior. This phenomenon is related to the association of transition metals to several biological activities. The third one considers biomineralization, all mechanisms that encompass the precipitation of minerals within a microbial cell, on the cell surface, or in close association with the cell. This precipitation may be induced or controlled by biological mechanisms. Finally, biotransformation of metal species by changing their oxidation state could be a consequence of electron transfer in metabolic pathways since metals are excellent electron donors and acceptors. Biotransformation would result in either metal precipitation or metal solubilization triggering metal immobilization or metal mobilization, respectively.

Table 41.2 resumes the role of microorganisms in the fate of different transition metals in the environment (adapted from Gadd, 2010). These microbial roles follow the same goal: the survival in the environment. Numerous microbial survival strategies are focused on metabolic requirements, while others are strictly detoxifying mechanisms. Metal

TABLE 41.2

Microbial Roles in Biogeochemical Cycles

Transition Metal	Microbial Roles in Geochemical Cycles
Fe	Bioweathering of Fe-containing minerals in rocks and soils; Fe solubilization by siderophores, organic acids, metabolites, etc.; Fe(III) reduction to Fe(II); Fe(II) oxidation to Fe(III); Fe biomineralization, e.g., oxides, hydroxides, carbonates, sulfides; metal sorption to Fe oxides
Mn	Mn(II) oxidation and immobilization as Mn(IV) oxides; Mn(IV) reduction; indirect Mn(IV)O_2 reduction by metabolites, e.g., oxalate; bioaccumulation of Mn oxides to surfaces and exopolymers; contribution to desert varnish formation; biosorption; accumulation; intracellular precipitation; Mn biomineralization, e.g., oxides, carbonates, sulfides, oxalates; metal sorption to Mn oxides
Cr	Cr(VI) reduction to Cr(III); Cr(III) oxidation; accumulation of Cr oxyanions
Co, Ni, Zn, Cd	Bioweathering of minerals in rocks and soil; biosorption; uptake and accumulation; bioprecipitation, e.g., oxalates, sulfides, phosphates, carbonate; Co(III) reduction
Ag	Reduction of Ag(I) to Ag(0); biosorption; accumulation
Cu	Mobilization from Cu-containing minerals in rocks and soils; CuS formation; biosorption; uptake and accumulation; bioprecipitation, e.g., oxalates
Pb	Biosorption; lead oxalate formation; biomethylation
Au	Reduction of soluble Au species to Au(0); Au mineral dispersion and solubilization
Hg	Hg biomethylation; reduction of Hg(II) to Hg(0); oxidation of Hg(0) to Hg(II); Hg volatilization as Hg(0); degradation of organomercurials; biosorption; accumulation

immobilization promoted by biosorption, bioaccumulation with intracellular complexation or precipitation, and biotransformation with precipitation decrease labile metal concentration in cell surroundings and hence its toxicity. Other processes like biotransformation, efflux mechanisms, influx suppression, impermeability, and release of metal ligands or metabolic products contribute to metal mobilization and bioavailability. Although the presence of efflux, redox mobilization, or methylation mechanisms are beneficial to a particular cell, they enhance metal mobility (Gadd, 2010). The interactions between microorganisms and metals depend on the metal by itself and its speciation in the specific environment.

Within the intracellular processes, an interesting feature is the polyphosphate accumulation in microorganisms, which is also relevant in metal immobilization (Renninger et al., 2004) and biotreatments for phosphate removal in wastewaters (McGrath and Quinn, 2003; Seviour et al., 2003). Polyphosphates in bacteria act as phosphate donors, energy source, and ligands for divalent cations and are involved in global regulatory systems. This is the key of survival for different bacteria, including pathogens under stress conditions (Kulaev and Kulakovskaya, 2000). The enzymes that are involved in this polymer metabolism are polyphosphate kinase (synthesis) and exopolyphosphatase (hydrolysis), which are highly conserved in many bacterial species.

Both biotransformation of metals by changing their oxidation state or precipitation are processes with potential applications in several nanoparticle biosynthesis. The synthesis of the semiconductor ZnS nanoparticules by bacteria is a way to produce new materials under simple and clean conditions (Bai et al., 2006). ZnO nanoparticules are useful as antimicrobial agent and as material for fluorescent bulbs. They are obtained from several bacterial species applying easy and inexpensive mechanisms (Madhumitha et al., 2016). Another interesting biosynthesized nanoparticules are the magnetic nanosized powder with numerous applications in medicine. Moon et al. (2010) described a large-scale production from a Zn-substituted precursor of magnetite by bacterial fermentation. The recovery of metals from wastes as nanoparticles has a high ecological relevance considering the positive environmental implication of metal cycling.

Another example of metal–microbe interactions is *Pseudomonas veronii* 2E, an indigenous bacterium isolated from the polluted Reconquista River basin that showed the ability to retain Cd(II) and Zn(II) mediated by a biosorption process (Vullo et al., 2008; Barrionuevo et al., 2011; Mendez et al., 2011). Ferreira et al. (2013) detected a stimulated siderophore production when grown in an iron-free culture medium or in a supplemented medium with Cd, Cu, Zn, or Ni (0.25 mM). The absorption spectra of these metabolites and their metal complexes are an evidence of metal–siderophore interactions: absorbance peak displacements observed when complexes are formed. Figure 41.4 shows a typical spectrum of the *P. veronii* 2E siderophore in the presence of increasing concentrations of Zn(II). The height of the absorbance peak at 404 nm grows up to 0.1 mM Zn(II), indicating a clear interaction by complexation.

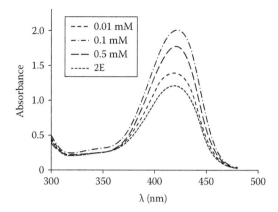

FIGURE 41.4 Spectrum of siderophore extract in the presence of increasing Zn(II) concentrations.

Although bacterial motility is useful for nutrient achievement, it is a fundamental helpful property for microbial interaction with the environment and other bacteria. This environmental interaction includes biofilm development, another survival strategy in nature. Bacterial attachment to surfaces is mostly governed by two different types of motility: swimming and swarming. Swimming is the propagation of cells through liquid or very low viscosity media mediated by flagella and is the most general type of displacement involved on cellular recruit and attachment. Swarming motility is necessary for biofilm expansion as a coordinate surface translocation, depending on multi-flagellation and a strong cell to cell contact. Chemotaxis is the swimming of microorganisms toward or against the influence of a chemical gradient. The positive chemotactic response is a modulation mechanism for swimming because it elicits bacteria to search for nutrients and to transport effectively toward them for growing and surviving as well (Adler et al., 1973; Alon et al., 1998; Alexandre et al., 2000; Boin et al., 2004; Kato et al., 2008). Not only positive chemotaxis is a common bacterial behavior: when a toxic compound is present in the environment, bacteria can detect it and swim away. This last phenomenon is known as negative chemotaxis. Bacterial motility is closely associated with bioremediation since inorganic and organic compounds are commonly present in polluted aquatic environments as a result of industrial discharges. Regarding metals—valid also for metalloids like As—positive or negative chemotaxis may influence in their immobilization or mobilization.

Pseudomonas aeruginosa PA01 is a reference strain with great versatility in adaptive responses to changes of environmental conditions. Siderophore biosynthesis and their regulation mechanisms as well as other characteristics on biofilm development and motility are well described in the literature (Braud et al., 2009, 2010; Kamatkar and Shrout, 2011). *P. aeruginosa* PA01 behavior in terms of swimming, swarming, and chemotaxis in the presence of Zn(II) was deeply studied. Swimming and swarming experiments were performed in PYG broth plus 3 g/L agar-agar and Swarming Agar (SWA) (meat extract 3 g/L, meat peptone 5 g/L, casein peptone 3 g/L, glucose 5 g/L, agar-agar 4 g/L), respectively, both supplemented with 0.01–10 mM Zn(II). Chemotactic response was evaluated with a modified chemical in plug assay using PYGc (casein peptone 10 g/L, yeast extract 5 g/L, glucose 20 g/L, and agarose-gel point [4%] ≤ 35°C 5 g/ L) as attractant plus 0.01–10 mM Zn(II), testing in parallel bacterial activity against metal as control (Barrionuevo and Vullo, 2012). Strains were grown in PYG broth at 32°C, 150 rpm up to an optical density$_{600nm}$ of 0.3–0.5. Then, cells were harvested by centrifugation at 1157 g (10 min). The supernatant was discarded, and the pellet was resuspended in 20 mL 150 mM NaCl. This washing operation was carried out twice, and finally, cells were resuspended in 2.5 mL HEPES buffer (N-[2-hydroxyethyl] piperazine-N'-[2-ethanesulfonic acid], pK_a 7.5, 10 mM, pH 7). This suspension was mixed with 22.5 mL of an agarose solution to give a 0.035 g/L agarose final concentration.

After a week at 32°C, *P. aeruginosa* PA01 swimming motility was only affected by Zn(II) concentrations higher than 1 mM. At 2, 5, and 10 mM Zn(II), only a delay in aerotaxis was observed evidencing a decrease in motility rates

(Figure 41.5). On the other side, swarming motility was gradually altered up to inhibition while increasing Zn(II) concentration (Figure 41.6) during the 48 h of incubation at 32°C.

P. aeruginosa PA01 positive chemotaxis toward nutrients clearly became negative in the presence of Zn(II) when metal concentrations surpassed 2 mM, indicating the ability to swim away when a toxic is present (Figure 41.7), evidenced by a cell density ring around the central plug. When Zn(II) was in the central plug without any attractant compound, no response from the microorganisms was observed.

According to Barrionuevo and Vullo (2012), Zn(II) did not alter the swimming motility of *P. veronii* 2E (Figure 41.8), and after a 48 h incubation at 32°C in the presence of 1 mM Zn(II), swarming motility was reduced but not totally inhibited (Figure 41.9). Hence, there is a good reason to believe that this bacterium would have a persistent motility behavior in environments containing Zn(II). Interestingly, negative chemotaxis was observed when exposing *P. veronii* 2E cells

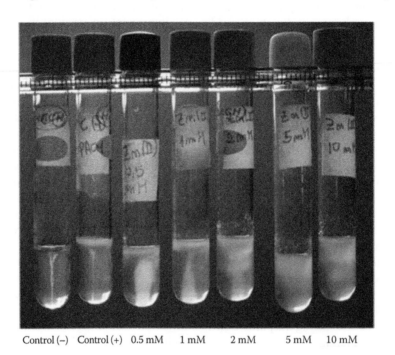

Control (–)   Control (+)   0.5 mM   1 mM   2 mM   5 mM   10 mM

FIGURE 41.5 Swimming motility of *Pseudomonas aeruginosa* PA01 in the presence of Zn(II).

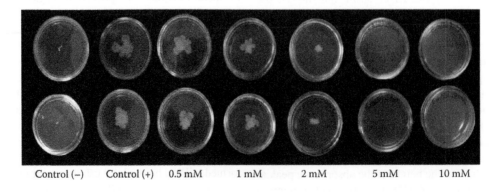

Control (–)   Control (+)   0.5 mM   1 mM   2 mM   5 mM   10 mM

FIGURE 41.6 Swarming motility of *Pseudomonas aeruginosa* PA01 in the presence of 0.5–10 mM Zn(II).

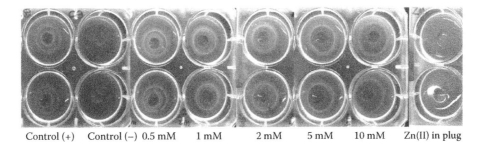

Control (+) Control (–) 0.5 mM 1 mM 2 mM 5 mM 10 mM Zn(II) in plug

FIGURE 41.7 *Pseudomonas aeruginosa* PA01 negative chemotaxis toward attractants in the presence of 0.5–10 mM Zn(II).

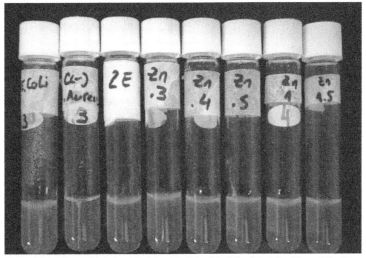

Control (+) Control (–) 0 mM 0.3 mM 0.4 mM 0.5 mM 1.0 mM 1.5 mM

FIGURE 41.8 Swimming motility of *Pseudomonas veronii* 2E with 0–1.5 mM Zn(II).

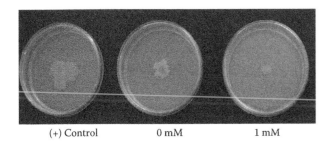

(+) Control 0 mM 1 mM

FIGURE 41.9 Swarming motility of *Pseudomonas veronii* 2E with 1 mM Zn(II).

to Zn(II) in chemical in plug assays, while a positive chemotaxis was denoted with the attractant present despite Zn(II) (Figure 41.10). In summary, positive chemotaxis directly influences cell attachment to any substrate and biofilm development. The biofilm stability will potentiate immobilization events in the case of metals (metalloids as well), respectively. In addition, it is likely to find biofilm establishment in polluted environments despite the negative chemotactic response toward pollutants observed *in vitro*.

0.3 mM 0.5 mM 1 mM

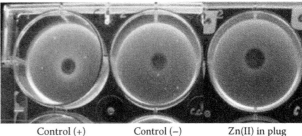

Control (+) Control (–) Zn(II) in plug

FIGURE 41.10 *Pseudomonas veronii* 2E positive chemotaxis toward nutrients in the presence of 0.3–1 mM Zn(II) as cell concentration halo surrounding central plug and cell density ring of negative chemotaxis toward Zn(II) in central plug.

41.3.2 Metal Biosorption

Regarding the chemical structure of bacterial envelopes and matrix exopolymers, carboxylate, hydroxyl, thiol, phosphate, and amine and sulfate groups are exposed to cellular microenvironments and able to interact, for example, with metal ions. The binding capacity of these functional groups allows bacterial surfaces to be considered as a conjunction of ligands, which can decrease the amount of bioavailable metal and consequently reduce the toxic effects on cells. This is the basis of the mechanism that contributes to biosorption, a term that refers to a process of metal uptake by biological matrices. A wide spectrum of definitions of biosorption can be found in bibliography, referring to active or passive phenomena or both of them, depending on the sorbing material and the sorbate nature (Fomina and Gadd, 2014). Some of these definitions contemplate bioaccumulation processes, but in this chapter only physicochemical interactions with biological surfaces will be considered.

41.3.2.1 Bacterial Components and Zn Biosorption

Cd(II) and Zn(II) retention mediated by *P. veronii* 2E through a biosorption process was proved through classical adsorption isotherms and complemented with zeta potential experiments (Vullo et al., 2008; Mendez et al., 2011). *P. veronii* 2E behaved as an efficient multi-ligand particle biosorbing 76.8% Cd(II) in 5 h at pH 7.5 and 32°C. Being an efficient biosorbent, *P. veronii* 2E is still a living organism.

The pH regulation and the units for the concentration expression are crucial for a correct interpretation of the obtained results. When interactions between metal and ligands are object of study, a clear dependence on pH should be expected since there is a great variability in the availability of both metal and functional groups. In addition, the sorbate–sorbent complex is related to the equilibrium of a number of ions with a number of adsorption sites, so molar concentration is the way to describe accurately the process as many equilibrium constants are involved.

Apart from Langmuir and Freundlich, other models were applied when Zn(II) sorption was studied on biological

sorbents (Limousin et al., 2007; Dada et al., 2012). By ignoring the extremely low and large value of concentrations, the Temkin model assumes that heat of adsorption of all molecules in the layer would decrease linearly with coverage and Dubinin–Radushkevich isotherm is generally applied to express the adsorption mechanism with a Gaussian energy distribution onto a heterogeneous surface. The latter model has often successfully fitted high solute activities.

But are all the physicochemical models applicable to cells as adsorbates? What do they really mean? The information extracted from them remains unclear in terms of usefulness. Biological particles introduce complexity in a model since unknown mechanisms are involved in the processes that are usually disregarded. As an example, the isotherm performed in a V_t of 10 mL with 3 g/L *P. veronii* 2E and 0–0.0005 M Zn(II) at 32°C and pH = 7.5 (10 mM HEPES buffer) is shown in Figure 41.11. The curve shape shows a saturation at high Zn(II) concentrations compatible with a Langmuir isotherm, but when the linearization is carried out, adjusting with three models as resumed in Figure 41.12, interesting conclusions can be extracted from these results.

The R^2 obtained for the three curves are close enough to promote a misunderstanding of the approach that could be done. Langmuir says that there is a saturated system; Freundlich does not consider the existence of a monolayer and a maximal adsorption capacity, and Dubinin–Radushkevich associates the curve with the presence of an adsorbate active diffusion through the adsorbent microporosity. To conclude, the analysis and modelling of such multiple mechanisms may be difficult, adding that the models were developed from gas–solid systems. When the aim is focused only on finding an efficient biosorbent to design a new and proper biotreatment process, it may not be necessary to have a deep understanding of what mechanisms are operative.

In addition, the use of these equations for the prediction of metal biosorption behavior under changing pH, ionic strength, and initial metal concentration is impossible.

From a chemical point of view, the macromolecules of the cell surface contain carboxylate, phosphate, and amino

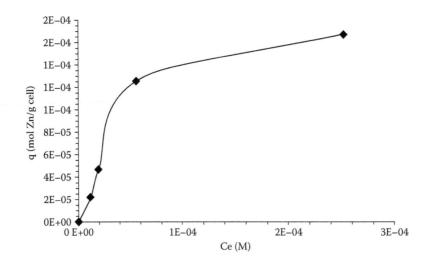

FIGURE 41.11 Zn(II) adsorption isotherm by *Pseudomonas veronii* 2E.

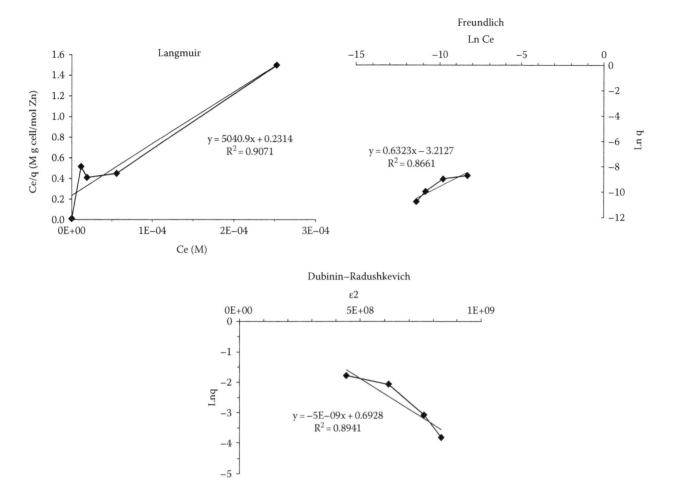

FIGURE 41.12 *Pseudomonas veronii* 2E biosorption isotherm linearizations according to the Freundlich, Langmuir, and Dubinin–Radushkevich models.

functional groups. Ionization of these groups depends on pH conferring electrostatic charge to the cell periphery. Electrostatic charge affects polarity and hydrophilicity, which are fundamental properties for cell functions (Hayashi et al., 2003; Tsuneda et al., 2004; Eboigbodin et al., 2006). The surface charge is also responsible for electrokinetic phenomena, such as electrophoresis. It is assumed that the liquid adhering to the particles surface (in this case the bacteria) and the mobile liquid are separated by a shear plane. The electrokinetic charge is the charge on the shear plane, and the electric potential at the plane is defined as the zeta potential (ζ) that reflects the potential difference between the plane of shear and the bulk phase, being able to be estimated from electrophoretic mobility measurements by the Smoluchowski equation:

$$\zeta = \frac{(\eta \cdot \mu)}{(\varepsilon_0 \cdot \varepsilon)}$$

where
  η is the viscosity of the medium
  μ is the electrophoretic mobility
  ε_0 is the permittivity of vacuum
  ε is the dielectric constant of the medium

A Zn(II)–cell surface concrete interaction could be evidenced with *P. veronii* 2E by measuring zeta potential in the presence or absence of Zn(II) and with buffered pH or not. Electrophoretic mobility is an intensive property that indicates the particles' ability to move when an electrical field is applied. It strictly depends on the particle's charge and the environment's ionic strength. Establishing experimental conditions for the determination of bacteria electrophoretic mobility represents a challenge since ionic strength, pH regulation, and cell concentration need to be carefully established. The ionic strength had to be fixed with 10 mM KCl in order to buffer any possible change in the total ionic concentration due to the different concentrations of metal salts used in this work. The same study demonstrated the pH regulation: buffered solutions were stringently required since bacterial surface charges rely on pH. Cell concentration should be taken into account before testing metal–cell interactions; it must be clearly over the detection limits of the instrument (zetameter), so as to reach reliable μ and consequently ζ measurements.

Figure 41.13 shows ζ determination of the 3 g/L suspension of *P. veronii* 2E in the presence of increasing Zn(II) concentrations under controlled pH (10 mM HEPES), uncontrolled pH, and two different ionic strength fixed with KCl.

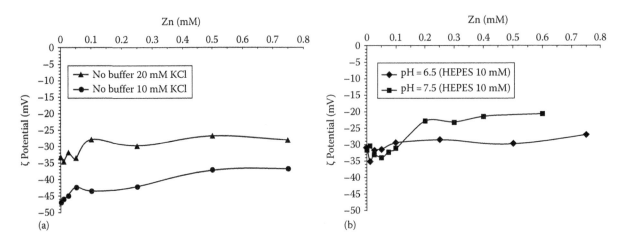

FIGURE 41.13 Zeta potential (ζ) of *Pseudomonas veronii* 2E in the presence of 0–0.75 mM Zn(II). (a) 10 and 20 mM KCl and uncontrolled pH. (b) 10 mM KCl and 10 mM HEPES buffer at two pH constant values.

A high variation to more negative values of ζ was observed when ionic strength decreased from 20 to 10 mM KCl with or without Zn(II). However, a slight increasing tendency of ζ was detected with higher Zn(II) concentrations, specially with 10 mM KCl, poor evidence of a decrease in negative surface charges by Zn(II) complexation (Figure 41.13a). Interestingly, under controlled pH (Figure 41.13b), an initial deep slope is noticed to more positive ζ values when Zn(II) concentration was increased at a constant pH 7.5, while at pH 6.5 practically ζ remained invariable. At pH 6.5, the availability of functional groups on cell surface was clearly low to interact with Zn(II). At pH 7.5, more negatively charged groups were available and ready to behave as ligands, so the interaction with the cation led to a more positive ζ measurements. It is important to highlight that Zn(II) at the highest concentration assayed in this experiment was completely available at both pHs, since no aquo complexes are reported from its speciation diagram.

41.3.2.2 Biofilm Reactors for Zn Biotreatments

Biofilms are structures consisting primarily of an extracellular matrix (EPS), composed by polysaccharides, proteins, and nucleic acids, in which the bacterial cells are embedded. When biosorption is conducted by biofilm-associated microorganisms, it constitutes a more efficient alternative than when mediated by planktonic microorganisms as experiments detailed earlier. Cells in the biofilm have passed through different levels of adaptation that allow their survival (especially during stress periods), and they have the additional protection generated by the EPS matrix that may allow them to tolerate and retain greater concentrations of pollutants, in these

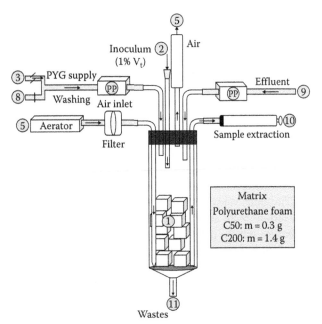

FIGURE 41.14 Lab-scale biofilm bioreactors (C50 and C200) for Zn biosorption experiments.

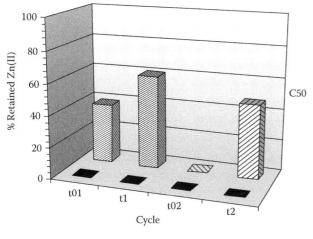

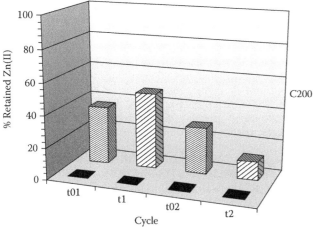

FIGURE 41.15 Zn retention in C50 and C200 bioreactors during cycles 1 and 2: t01 and t1 correspond to the data obtained for the first cycle after filling the reactor and 24 h, respectively. t02 and t2 are the reactor performance for the second cycle. Black squares represent the 0% values of the cell free experiments.

case metals. EPS is commonly found in activated sludge, and its complexing capacity can be determined under different experimental conditions. The metal-binding constants for EPS extracted from activated sludges varied over a wide range: for Cd(II), $-3.7 < \log K < 2.3$, and for Pb(II), $-1.4 < \log K < 4.1$ (Guibaud et al., 2006, 2008, 2009). EPS is a biofilm component that allows bacterial tolerance and boosts metal removal from aquatic environments by promoting metal immobilization. Biofilm-based reactors are usually used for treating large volumes of dilute aqueous solutions such as industrial and municipal wastewaters. The main biofilm reactors are the classical designs: packed bed reactor, fluidized bed reactor, trickled bed reactor, air lift reactor, and upflow anaerobic sludge blanket applied to a wide range of different biotreatments.

In a laboratory scale, two batch bioreactors operated in a packed bed mode with a work volume of 50 mL (C50) and 200 mL (C200) each were designed as shown in Figure 41.14. Both reactors were filled with polyurethane foam cubes as matrix for cellular immobilization. In the first step, *P. veronii* 2E biofilm was developed, obtaining from 0.12 to 0.18 g biomass/g matrix (dry weight) in reactors after two to four weeks with

periodic culture medium renewal, 32°C, constant aeration, and homogenization. The second step corresponded to the biosorption process: metal retention by immobilized *P. veronii* 2E was determined in 0.5 mM Zn(II) aqueous solutions (pH = 7.5, 10 mM HEPES buffer) after 24 and 48 h cycles. Total metal concentration was analyzed in each fraction after treatment by ASV. The results are exposed in Figure 41.15. A total retention of 60% Zn was achieved in the first cycle with the C50 reactor, maintaining the biosorptive capacity to more than a 40% during the second cycle. The matrix did not retain Zn(II) as revealed by cell-free control experiments. The C200 bioreactor sorbed only a 35%, attributed to the different homogenization and biofilm: Zn(II) contact, aspect that needs improvement. Metal retention results are still being evaluated for the possibility of scaling up this biological treatment taking into account the initial design problems detected in these performances.

41.4 CONCLUDING REMARKS

As final conclusions, it can be remarked that Zn(II) present in industrial wastewaters can be efficiently removed by sorption processes using both mineral and biological matrices as sorbents. According to our results, the presence of soluble ligands directly influences on Zn(II) availability and consequent interaction with the mineral sorbent surface. Industrial wastes contain different components that can behave as ligands. This fact should not be disregarded when a successful Zn(II) removal is the aim of the designed treatment.

The scaling up has been the main concern to implement biotreatments based on biological sorbents. In this work, this problem was observed just at a laboratory level, dealing with small volume changes. In agreement to Fomina and Gadd (2014), biosorption is a promissory technology with serious obstacles when applied to industrial processes. The implementation of industrial-scale treatments using biosorption techniques will still require a deeper analysis to make them commercially viable, not discarding the combination with other technologies.

ACKNOWLEDGMENTS

This work was supported by the Universidad Nacional de General Sarmiento and CONICET. We are grateful to Miss Leticia Rossi for the English language revision.

REFERENCES

Adler, J., G.L. Hazelbauer, M.M. Dahl. 1973. Chemotaxis toward sugars in *Escherichia coli*. *Journal of Bacteriology* 115: 824–847.

Administración Nacional de Medicamentos, Alimentos y Tecnología Médica. 2007. Código Alimentario Argentino. http://www.anmat.gov.ar/alimentos/codigoa/CAPITULO_XII.pdf (Accessed October 6, 2016).

Alexandre, G., S.E. Greer, I.B. Zhulin. 2000. Energy taxis is the dominant behavior in *Azospirillum brasilense*. *Journal of Bacteriology* 182: 6042–6048.

Alon, U., L. Camarena, M.G. Surette, B.A. y Arcas, Y. Liu, S. Leibler, J.B. Stock. 1998. Response regulator output in bacterial chemotaxis. *The EMBO Journal* 17: 4238–4248.

Bai, H.J., Z.M. Zhang, J. Gong. 2006. Biological synthesis of semiconductor zinc sulphide nanoparticles by immobilized *Rhodobacter sphaeroides*. *Biotechnology Letters* 28: 1135–1139.

Baral, A., R.D. Engelken. 2002. Chromium-based regulations and greening in metal finishing industries in the USA. *Environmental Science & Policy* 5: 121–133.

Barriola, A., M. Ostra, C. Ubide. 2012. Gas chromatography with flame ionization detection for determination of additives in an electrolytic Zn bath. *Journal of Chromatography A* 1256: 246–252.

Barrionuevo, M., M.A. Daniel, L. Garavaglia et al. 2011. Knowing more about metals, microbes and environment interactions: How to improve wastewater biotreatments. In *Bioremediation: Biotechnology, Engineering and Environmental Management*, ed. A. Mason, pp. 383–404. New York: Nova Science Publishers, Inc.

Barrionuevo, M.R., D.L. Vullo. 2012. Bacterial swimming, swarming and chemotactic response to heavy metal presence: Which could be the influence on wastewater biotreatment efficiency? *World Journal of Microbiology and Biotechnology* 28: 2813–2825.

Boin, M.A., M.J. Austin, C.C. Hase. 2004. Chemotaxis in *Vibrio cholerae*. *FEMS Microbiology Letters* 239: 1–8.

Braud, A., V. Geoffroy, F. Hoegy, G.L. Mislin, I.J. Schalk. 2010. Presence of the siderophores pyoverdine and pyochelin in the extracellular medium reduces toxic metal accumulation in *Pseudomonas aeruginosa* and increases bacterial metal tolerance. *Environmental Microbiology Reports* 2: 419–425.

Braud, A., F. Hoegy, K. Jezequel, T. Lebeau, I.J. Schalk. 2009. New insights into the metal specificity of the Pseudomonas aeruginosa pyoverdine–iron uptake pathway. *Environmental Microbiology* 11: 1079–1091.

Campos, M.L.A.M., L.C. Mello, D.R. Zanette, M.M.S. Sierra, A. Bendo. 2001. Construção e otimização de um reator de baixo custo para a fotodegradação da matéria orgânica em águas naturais e sua aplicação no estudo da especiação do cobre por voltametria. *Química Nova* 24: 257–261.

Ceretti, H., D. Vullo, A. Zalts, S. Ramírez. 2010. Effect of bacterial growth in the complexing capacity of a culture medium supplemented with cadmium (II). *World Journal of Microbiology and Biotechnology* 26: 847–853.

Cooke, C.A., M.B. Abbott, A.P. Wolfe, J.L. Kittleson. 2007. A millennium of metallurgy recorded by lake sediments from Morococha, Peruvian Andes. *Environmental Science & Technology* 41: 3469–3474.

Dada, A.O., A.P. Olalekan, A.M. Olatunya, O. Dada. 2012. Langmuir, Freundlich, Temkin and Dubinin–Radushkevich isotherms studies of equilibrium sorption of Zn^{2+} unto phosphoric acid modified rice husk. *Journal of Applied Chemistry* 3: 38–45.

Eboigbodin, K.E., J.R.A. Newton, A.F. Routh, C.A. Biggs. 2006. Bacterial quorum sensing and cell surface electrokinetic properties. *Applied Microbiology and Biotechnology* 73: 669–675.

Environmental Protection Division. 1999. Water quality. Ambient water quality guidelines for Zinc. Overview report. http://www.env.gov.bc.ca/wat/wq/BCguidelines/zinc/zinc.html.

Ferreira, M.L., I.C. Lazzarini Behrmann, H.M. Ceretti, A. Zalts, D.L. Vullo, S.A. Ramirez. 2013. Towards understanding the influence of cadmium on microbial behavior. In *Cadmium: Characteristics, Sources of Exposure, Health and Environmental Effects*, eds. M. Hasanuzzaman, M. Fujita, pp. 151–170. New York: Nova Science Publishers, Inc.

Fomina, M., G.M. Gadd. 2014. Biosorption: Current perspectives on concept, definition and application. *Bioresource Technology* 160: 3–14.

Gadd, G.M. 2010. Metals, minerals and microbes: Geomicrobiology and bioremediation. *Microbiology* 156: 609–643.

Gerst, M.D., T.E. Graedel. 2008. In-use stocks of metals: Status and implications. *Environmental Science & Technology* 42: 7038–7045.

Guibaud, G., F. Bordas, A. Saaid, P. D'abzac, E. Van Hullebusch. 2008. Effect of pH on cadmium and lead binding by extracellular polymeric substances (EPS) extracted from environmental bacterial strains. *Colloids and Surfaces B: Biointerfaces* 63: 48–54.

Guibaud, G., E. Van Hullebusch, F. Bordas. 2006. Lead and cadmium biosorption by extracellular polymeric substances (EPS) extracted from activated sludges: pH-sorption edge tests and mathematical equilibrium modelling. *Chemosphere* 64: 1955–1962.

Guibaud, G., E. Van Hullebusch, F. Bordas, P. d'Abzac, E. Joussein. 2009. Sorption of Cd (II) and Pb (II) by exopolymeric substances (EPS) extracted from activated sludges and pure bacterial strains: Modeling of the metal/ligand ratio effect and role of the mineral fraction. *Bioresource Technology* 100: 2959–2968.

Harper, E.M., M. Bertram, T.E. Graedel. 2006. The contemporary Latin America and the Caribbean zinc cycle: One year stocks and flows. *Resources, Conservation and Recycling*, 47: 82–100.

Hayashi, H., H. Seiki, S. Tsuneda, A. Hirata, H. Sasaki. 2003. Influence of growth phase on bacterial cell electrokinetic characteristics examined by soft particle electrophoresis theory. *Journal of Colloid and Interface Science* 264: 565–568.

Hollenberg, P.F. ed. 2010. Introduction: Mechanisms of metal toxicity special issue. *Chemical Research in Toxicology* 23: 292–293.

Jiang, C., G.R. Aiken, H. Hsu-Kim. 2015. Effects of natural organic matter properties on the dissolution kinetics of zinc oxide nanoparticles. *Environmental Science & Technology* 49: 11476–11484.

Joint FAO/WHO Expert Committee on Food Additives. 1982. Evaluation of certain food additives and contaminants (WHO Food Additives Series, No. 17). Cambridge, U.K.: Cambridge University Press.

Kamatkar, N.G., J.D. Shrout. 2011. Surface hardness impairment of quorum sensing and swarming for *Pseudomonas aeruginosa*. *PLoS One* 6: 20888.

Kato, J., H.E. Kim, N. Takiguchi, A. Kuroda, H. Ohtake. 2008. Pseudomonas aeruginosa as a model microorganism for investigation of chemotactic behaviors in ecosystem. *Journal of Bioscience and Bioengineering* 106: 1–7.

Kaya, H., F. Aydın, M. Gürkan, S. Yılmaz, M. Ates, V. Demir, Z. Arslan. 2016. A comparative toxicity study between small and large size zinc oxide nanoparticles in tilapia (*Oreochromis niloticus*): Organ pathologies, osmoregulatory responses and immunological parameters. *Chemosphere* 144: 571–582.

Kulaev, I., T. Kulakovskaya. 2000. Polyphosphate and phosphate pump. *Annual Reviews in Microbiology* 54: 709–734.

Lee, C.S., S.H. Qi, G. Zhang, C.L. Luo, L.Y. Zhao, X.D. Li. 2008. Seven thousand years of records on the mining and utilization of metals from lake sediments in central China. *Environmental Science & Technology* 42: 4732–4738.

Leung, K.M., R.W. Furness, J. Svavarsson, T.C. Lau, R.S. Wu. 2008. Field validation, in Scotland and Iceland, of the artificial mussel for monitoring trace metals in temperate seas. *Marine Pollution Bulletin* 57: 790–800.

Limousin, G., J.P. Gaudet, L. Charlet, S. Szenknect, V. Barthes, M. Krimissa. 2007. Sorption isotherms: A review on physical bases, modeling and measurement. *Applied Geochemistry* 22: 249–275.

Lowry, G.V., K.B. Gregory, S.C. Apte, J.R. Lead. 2012. Transformations of nanomaterials in the environment. *Environmental Science & Technology* 46: 6893–6899.

Madhumitha, G., G. Elango, S.M. Roopan. 2016. Biotechnological aspects of ZnO nanoparticles: Overview on synthesis and its applications. *Applied Microbiology and Biotechnology* 100: 571–581.

Manouchehri, N., A. Bermond. 2006. Study of trace metal partitioning between soil–EDTA extracts and Chelex-100 resin. *Analytica Chimica Acta* 557: 337–343.

Martell, E., R.M. Smith. 2010. NIST Standard Reference Database 46. NIST critically selected stability constants of metal complexes: Version 8.0. http://www.nist.gov/srd/nist46.cfm (Accessed December 10, 2015).

McGrath, J.W., J.P. Quinn. 2003. Microbial phosphate removal and polyphosphate production from wastewaters. *Advances in Applied Microbiology* 52: 75–100.

Méndez, N., S.A. Ramírez, H.M. Ceretti, A. Zalts, R. Candal, D.L. Vullo. 2011. *Pseudomonas veronii* 2E surface interactions with Zn (II) and Cd (II). *Global Journal of Environmental Science and Management* 1: 3.

Moon, J.W., C.J. Rawn, A.J. Rondinone, L.J. Love, Y. Roh, S.M. Everett, R.J. Lauf, T.J. Phelps. 2010. Large-scale production of magnetic nanoparticles using bacterial fermentation. *Journal of Industrial Microbiology & Biotechnology* 37: 1023–1031.

Piccinno, F., F. Gottschalk, S. Seeger, B. Nowack. 2012. Industrial production quantities and uses of ten engineered nanomaterials in Europe and the world. *Journal of Nanoparticle Research* 14: 1–11.

Renninger, N., R. Knopp, H. Nitsche, D.S. Clark, J.D. Keasling. 2004. Uranyl precipitation by Pseudomonas aeruginosa via controlled polyphosphate metabolism. *Applied and Environmental Microbiology* 70: 7404–7412.

SCHER (Scientific Committee on Health and Environmental Risks). 2012. Opinion on chemical and the water framework directive: Draft environmental quality standards—Zinc. http://ec.europa.eu/health/scientific_committees/environmental_risks/docs/scher_o_157.pdf.

Seviour, R.J., T. Mino, M. Onuki. 2003. The microbiology of biological phosphorus removal in activated sludge systems. *FEMS Microbiology Reviews* 27: 99–127.

Sosa, G.L., A. Zalts, S.A. Ramírez. 2016. Complexing capacity of electroplating rinsing baths-a twist to the resolution of two ligand families of similar strength. *Journal of Analytical Science and Technology* 7: 1.

Tourney, J., B.T. Ngwenya. 2014. The role of bacterial extracellular polymeric substances in geomicrobiology. *Chemical Geology* 386: 115–132.

Tsuneda, S., H. Aikawa, H. Hayashi, A. Hirata. 2004. Significance of cell electrokinetic properties determined by soft-particle analysis in bacterial adhesion onto a solid surface. *Journal of Colloid and Interface Science* 279: 410–417.

U. S. Environmental Protection Agency. National Recommended Water Quality Criteria. 2016. http://www2.epa.gov/wqc/national-recommended-water-quality-criteria. Accessed November 18, 2016.

U. S. Geological Survey: Minerals Yearbook. 2013. Zinc (advance release). http:// minerals. usgs.gov /minerals/pubs/commodity/zinc/ (Accessed October 22, 2015).

U. S. Geological Survey: Mineral Commodity Summaries. 2015. http://minerals.usgs.gov/minerals/pubs/mcs/2015/mcs2015.pdf (Accessed October 22, 2015).

Vullo, D.L., H.M. Ceretti, M.A. Daniel, S.A. Ramírez, A. Zalts. 2008. Cadmium, zinc and copper biosorption mediated by Pseudomonas veronii 2E. *Bioresource Technology* 99: 5574–5581.

Vullo, D.L., H.M. Ceretti, S.A. Ramírez, A. Zalts. 2007. Heavy metals and microorganisms in the environment: Taking advantage of reciprocal interactions for the development of a wastewater treatment process. In *Progress in Environmental Microbiology*, ed. M. Kim, pp. 111–149. New York: Nova Science Publishers, Inc.

World Health Organization. 2003. Zinc in Drinking-water, Background document for development of WHO Guidelines for Drinking-water Quality. http://www.who.int/water_sanitation_health/dwq/chemicals/zinc.pdf.

42 Hexavalent Chromium (VI) Removal by *Aspergillus niger*

*Ismael Acosta-Rodríguez, Juan F. Cárdenas-González,
María de Guadalupe Moctezuma-Zárate, Adriana Rodríguez Pérez, and
Víctor M. Martínez-Juárez*

CONTENTS

ABSTRACT

We isolated an *Aspergillus niger* strain from polluted air, near to the Faculty of Chemical Science, UASLP, in San Luis Potosi, S.L.P. Mexico, which grows about 50% relative to control at 2 g/L of Cr(VI) (79 μg of dry weight without metal) in Lee's minimal medium, probably is resistant to metal, and also efficiently removes 1 g/100 mL of Cr(VI) after 180 min of incubation and removes metal in 6 days in solution at 100% and 70% in contaminated earth and water samples, respectively. This strain showed an efficient capacity of reduction (100%) of 50 mg/L Cr(VI) in the growth medium after 7 days of incubation at 28°C, pH 5.3 and 6.2, 100 rpm, and with an inoculum of 8.3 mg of dry weight. The Cr(VI) reduction potential of the resting cells was increased by cell permeabilization. The optimum temperature and pH of chromate reductase activity of the cell-free extract were found to be 37°C and 7.0, respectively, and the activity was enhanced in the presence of 0.1 mM NADH and other electron donors. 1 mMol of metal ions such as Cu^{2+}, Na^+, Hg^{2+}, Mg^{2+}, Fe^{3+}, Ca^{2+}, and Cd^{2+} and respiratory inhibitors resulted in a decrease of the activity. Finally, these results suggest the potential applicability of *A. niger* for the remediation of Cr(VI) from polluted soils and waters.

42.1 INTRODUCTION

Effluents from tanneries are a major source of contamination with Cr(VI) for water bodies and soils. This metal is used in tanning leather and skins, as well as in steel alloys, electroplating, textile dyeing, and biocide in water cooling systems in nuclear plants, which invariably results in metal discharges to the environment with their consequences (Ahemad, 2014). Chromium has nine oxidation states from −2 to +6. However, according to the stability in the environment, only hexa- and trivalent species are important (Tejada Tovar et al., 2015). Cr(VI) is mainly found as chromate (CrO_4^{2-}), which becomes acidic pH at chromate protonated ($HCrO_4^-$), dichromate ($Cr_2O_7^{2-}$), and finally acid ($H_2Cr_2O_7$) (Vitti et al., 2014). These anions of Cr(VI) are very soluble salts and their sorption on surfaces of clays and oxyhydroxides is very low (Tejada Tovar et al., 2015), which represent a source of impairment for the bodies of water and soil and a potential risk to human beings. Cr(VI) is a chemical oxidant species that is toxic, mutagenic, carcinogenic, and/or teratogenic (Thatoi et al., 2014). In contrast, Cr(III) forms compounds with less solubility, such as hydroxides, oxides, or sulfates, so their mobility in the environment is very limited. In addition, the toxicity of Cr(III) is 100 times lower than that of Cr(VI), and the mutagenic effect is 1000 times lower (Gutiérrez-Corona et al., 2010). The main techniques to repair or remove Cr(VI) from wastewater include chemical reduction and precipitation, activated carbon adsorption, ion exchange, and reverse osmosis. Currently, the most widely used process is the reduction of species Cr(VI) to form insoluble compounds of Cr(III), using a reducing agent and a core, usually lime milk, the substance to form $Cr(OH)_3$, which is a compound of very low solubility (Srivastava et al., 2015). However, these methods have disadvantages such as high cost, low efficiency, generation of toxic waste or other, requiring controlled disposal, complicating the operation and process control (Shakya et al., 2015), and adsorption and biosorption currently used.

Adsorption is the preferential accumulation of a substance on the surface of a normally porous solid. The substance that is adsorbed is called adsorbate and may be an ion or molecule. Furthermore, the solid on which the adsorption occurs is known as an adsorbent (Basu et al., 2015), and biosorption is a phenomenon widely studied in the bioremediation of sites impacted by pollution. The study of microorganisms and bioadsorbent material is constantly growing with the use of microbial consortia, or macromolecules formed by microorganisms' (polymers) mixed systems, which would increase yield's uptake of certain specified metals or mixtures thereof (Ahemad, 2014; Tejada Tovar

et al., 2015). There are many reports of the use of microbial biomass for the removal of Cr(VI) from industrial wastewater and/or contaminated water (Bahafid et al., 2013; Nguema et al., 2014a; Thatoi et al., 2014), and some have analyzed the capture and accumulation of Cr(VI) and Cr(III) by the different filamentous fungi (Ahemad, 2014; Coreño-Alonso et al., 2014; Vitti et al., 2014; Acosta-Rodríguez et al., 2015; Cárdenas-González et al., 2015; Sallau et al., 2015), with highly satisfactory results. This chapter reports the removal of Cr(VI) in an aqueous solution by a strain of *Aspergillus niger*, which is highly resistant to this metal.

42.2 MATERIALS AND METHODS

42.2.1 SCREENING OF THE MICROORGANISM SHOWING THE RESISTANCE TO CHROMIUM(VI) AND CHROMATE TEST

A fungal strain was isolated from the polluted air near the Faculty of Chemical Science, UASLP (San Luis Potosí, México), and this was used for the screening. The strain was grown on a Petri dish containing modified Lee's minimal medium (LMM) (with 0.25% KH_2PO_4, 0.20% $MgSO_4$, 0.50% $(NH_4)_2SO_4$, 0.50% NaCl, 0.25% glucose, and 2% agar) supplemented with 500 mg/L of K_2CrO_4; the pH of the medium was adjusted and maintained at 5.3 with 100 mmol/L of citrate-phosphate buffer. The plates were incubated at 28°C for 7 days. The strain was identified based on its macroscopic characteristics and macroscopic and microscopic observations (López-Martínez et al., 2012). Fungal cultures grown in thioglycolate broth were used as primary inoculums. Chromate-resistant tests of the isolated strain, filamentous fungus *A. niger*, were performed on liquid LMM containing the appropriate nutritional requirements and different concentrations of Cr(VI) (as potassium chromate), and the dry weight was determined.

42.2.2 BIOSORPTION TESTS BY USING DRY CELLS

The fungus was grown at 28°C in a stirred and aerated liquid media containing thioglycolate broth at a concentration of 8 g/L (p/v). After 5 days of incubation, the cells were recovered by centrifugation (3000 rpm, 10 min) and washed twice in the same conditions with deionized water, and subsequently, it was dried (80°C, 24 h) in an oven. Solutions of Cr(VI) for analysis were prepared by diluting 71.86 mg/L of stock metal solution. The concentration range of chromium(VI) solutions was 50–1000 mg/L. The pH of each solution was adjusted to the required value by adding 1 M H_2SO_4 solution before mixing with the microorganism. The biosorption of the metal by fungal dry cells was determined at different concentrations (50–1000 mg/L) of 100 mL Cr(VI) solution, with 1 g of fungal biomass, at 120 rpm, and the sample was filtered. The filtrate containing the residual concentration of Cr(VI) was determined spectrophotometrically. For the determination of the rate of metal biosorption, the solution of Cr(VI) was used at concentrations of 200, 400, 600, 800, and 1000 mg/L. The supernatant was analyzed for residual Cr(VI) at different times after a contact period. For the determination of the effects of pH and temperature, seven solutions (pH 1, 2, and 3) and temperatures (28°C, 40°C, 50°C, and 60°C) were respectively used. Moreover, biosorption to the contaminated earth and water was examined. Four Erlenmeyer glass flasks which contain 5 g of fungal biomass and 20 g of contaminated earth and 20 mL of water (297 mg Cr(VI)/g earth or 155 mg Cr(VI)/L water) from tannery (Celaya, Guanajuato, Mexico) were completed to 100 mL with trideionized water, and they were incubated during 7 days, stirred at 120 rpm, and filtered in Whatman filter paper No. 1. The concentration of Cr(VI) of the filtrate is analyzed with 1,5-diphenylcarbazide (Greenberg et al., 1992).

42.2.3 REDUCTION OF CR(VI) BY LIVING CELLS

Reduction efficiency of Cr(VI) by living, resting, and permeabilized cells was examined. To examine the living cells, cultures in 100 mL of LMM were inoculated with 5×10^5 spores/mL (28°C, 48 h), the cells were centrifuged (2000 rpm, at 4°C, 10 min) and washed twice with sterile trideionized water, and the pellet was resuspended in 3 mL of the same solution and was transferred at a fresh LMM (100 mL with 50 mg/L Cr(VI)). At different times: 1 mL aliquots were removed and centrifuged (5000 rpm, 10 min), and the concentration of Cr(VI) or total Cr in the supernatant was determined.

Reduction efficiency of Cr(VI) was examined by the resting cells in 100 mL thioglycolate broth (pH 7.0), 5×10^5 spores/mL of *A. niger* was inoculated and incubated for 5 days and was harvested (3000×g at 4°C); cell pellets obtained were washed by centrifugation twice with 100 mM of potassium phosphate buffer (pH 7.0) and resuspended in the same buffer. The suspended cell pellets were added in 2–10 mg/100 mL of Cr(VI) solution, mixed for 30 min, and incubated at 30°C for 8 h. Heat-killed culture pellets (2 mL), which were treated at 100°C for 10 min, were used as control. After the incubation, the tubes were centrifuged, and 100 μL aliquots were withdrawn from each sample to estimate the remaining Cr(VI).

Reduction efficiency of Cr(VI) was also examined by the permeable fungal cells. The culture of *A. niger* was grown for 5 days, harvested, and washed with potassium phosphate buffer (pH 7.0) as described previously. The suspended culture pellets were treated with 0.2% (w/v) sodium dodecyl sulfate, 0.2% Tween 80 (v/v), 0.2% Triton X-100 (v/v), and 0.2% toluene (v/v), by vortexing for 30 min to achieve cell permeabilization. Permeabilized cell suspensions (0.5 mL) were then added with 2–10 mg/100 mL of Cr(VI) as final concentrations and incubated for 8 h at 30°C.

42.2.4 ACTIVITY OF CHROMATE REDUCTASE

Cell-free extracts (CFEs) of *A. niger* were prepared by modifying the previous protocols. The CFEs precipitated were resuspended in 10 mL of 100 mM potassium phosphate buffer (pH 7.0, 5% v/v) and were placed into an ice bath and disrupted using a Mini Bead Beater (Dentsply) with 15 cycles of 60 s for each one. The homogenate obtained was then centrifuged at 3000×g for 10 min at 4°C. The pellet was resuspended in

100 mM of potassium phosphate buffer (pH 7.0), and this is the CFE.

Enzymatic chromate reduction was estimated as described previously using a standard curve of Cr(VI) 0–30 mM. The assay was as follows: the reaction system (1.0 mL) was made up of varying Cr(VI) final concentrations (5–30 mM) in 700 μL of 100 mM of potassium phosphate buffer (pH 7.0) added with 250 μL aliquots of CFE for chromate reduction and 50 μL of NADH 1.0 mM. The system volume of 1.0 mL was kept constant for all experiments. Chromate reductase activity was measured at 37°C at different pH values using several buffers (100 mM of phosphate citrate, pH 5.0; 50 mM of phosphate, pH 6.0–8.0; and 50 mM of Tris–HCl, pH 8.0–9.0). Another important parameter to measure chromate reductase activity is the temperature, which was studied between 20°C and 60°C, at optimum pH. The CFE samples were also treated with several metal ions to a final concentration of 1 mM at optimal pH and temperature; Na^+, Ca^{2+}, Cu^{2+}, Hg^{2+}, Mg^{2+}, Cd^{2+}, and Fe^{3+} were tested by using 10 mM solutions of Na_2SO_4, $CaCl_2$, $CuCl_2$, $HgCl_2$, $MgCl_2$, $CdCl_2$, and $FeCl_3$. The electron donors tested were NADH, glucose, sodium acetate, formic acid, citrate, cystine, lactic acid, and ascorbic acid in a final concentration of 1 mM, and the inhibitors were EDTA, KCN, NaN_3, and β-mercaptoethanol at the same concentration. For chromate reductase activity, one unit was defined as an enzyme that reduces 1 mM of Cr(VI)/min/37°C, and the specific activity was defined as a unit chromate reductase activity/min/mg protein in the CFE. Protein concentrations were determined by the Lowry method (Lowry et al., 1951).

42.2.5 DETERMINATION OF HEXAVALENT, TRIVALENT, AND TOTAL AMOUNT OF CHROMIUM

Hexavalent and trivalent chromium were quantified employing diphenylcarbazide (Greenberg et al., 1992) and chromazurol S (Pantaler and Pulyaeva, 1985), respectively; the total amount of chromium was determined by electrothermal atomic absorption spectroscopy (Greenberg et al., 1992). Three dependent experiments were carried out and the mean value was shown.

42.3 RESULTS AND DISCUSSION

42.3.1 ISOLATION AND IDENTIFICATION OF A FUNGAL STRAIN TOLERANT TO CR(VI)

The microorganism was grown on the LMM agar plates containing 500 mg/L of K_2CrO_4, and the largest colony of fungi was isolated. Colonies isolated grew rapidly within 3–5 days. Colonies consist of a compact white or yellow basal felt covered by a dense layer of dark brown to black conidial heads. Conidial heads were large (up to 3×15–20 mm in diameter), globose, dark brown, and becoming radiate, and in old cultures, they split in loose columns. He has macrosiphonate and septate (2–4 μm) mycelium and reproductive hyphae of 4–8 μm. Its aspergillar heads of 100–200 μm diameter, with long conidiophores coenocytic and vesicles subspherical

50–100 μm, with two series of sterigma (large and small) at an angle of 360°, often septate metulae. Conidia are globose to subglobose (3.5–5.0 μm in diameter), dark brown to black and rough-walled, RG-1 organism (Figure 42.1) (Larone, 2002).

The cells of the isolated strain grew on LMM supplemented with 2 g/L of Cr(VI); about 50% of growth relative to control (0.079 mg of dry weight without metal) was obtained (Figure 42.2) and, therefore, probably is resistant to the metal. Different microorganisms that are Cr(VI) resistant have been isolated from different contaminated sites: Acosta-Rodríguez et al. (2015) reported for a strain of *Penicillium* sp. IA-01, about 50% of growth relative to control (85 mg of dry weight without metal), and Cárdenas-González and Acosta-Rodríguez (2010) observed for a strain of *Paecilomyces* sp., a growth of 183 mg for the control, and 16 mg at a concentration of 2000 mg/L of the metal. In addition, reported resistance is available to Cr(VI) to the bacterium *Corynebacterium hoagii* (176 mg/L) (Vitti et al., 2003), the yeast *Candida* sp.

(640 mg/L) (Guillén-Jiménez et al., 2008), and the filamentous fungus *Trichoderma inhatum* (680 mg/L) (Morales-Barrera and Cristiani-Urbina, 2008). Chromate tolerance has been described in the mutants of stocked culture and in native isolates of contaminated sites, as in this work; in several cases, both yeast and filamentous fungus showed that tolerance to Cr(VI) is due to transport of sulfate disturbance that leads to reduced incorporation of chromate (Cervantes et al., 2001). In other cases, phenotypes of hypersensitivity to Cr(VI) are produced as a result of alteration of the vacuolar ATPase and vacuolar structures (Gharieb and Gadd, 1998) or alteration of proteins that protect the oxidative effect of Cr(VI) as the alkyl hydroperoxide reductase (Nguyen-Nhu and Knoops, 2002) or Cu–Zn superoxide dismutase and methionine sulfoxide peptide reductase (Sumner et al., 2005). However, the mechanism of tolerance in this strain of *A. niger* fungus is not investigated. Thus, we precisely examined the characteristics of this strain to estimate the mechanism in the following experiments.

(a)

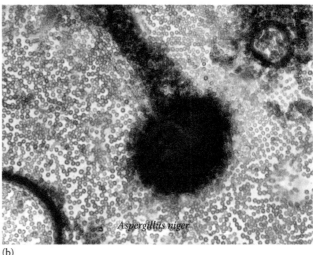

(b)

FIGURE 42.1   (a) Macroscopic and (b) microscopic morphology of the fungus *Aspergillus niger*.

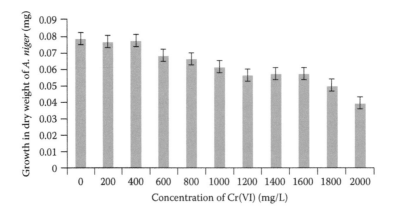

FIGURE 42.2   Growth in dry weight of *A. niger* with different concentrations of Cr(VI), 1×10^5 spores/mL, 28°C, 7 days of incubation, and 100 rpm.

42.3.2 ABSORPTION OF CR(VI) BY THE DRY CELLS OF *ASPERGILLUS NIGER*

42.3.2.1 Effect of the Incubation Time and pH

Figure 42.3 shows the effect of incubation time and pH on the biosorption of 50 mg/L of Cr(VI) by the biomass of *A. niger*. It was found that a higher removal, which is proportional to the biosorption, occurs at 30 min and at a pH of 1.0. Acosta-Rodríguez et al. (2015) reported an incubation time of 150 min, when working with biomass from *Penicillium* sp. IA-01; 48 h at pH 1.0 by fruiting bodies of the jelly fungus *Auricularia polytricha* (Zheng et al., 2014); 24 h for *Cyberlindnera fabianii*, *Wickerhamomyces anomalus*, and *Candida tropicalis*, at pH range between 2 and 4 for the three species (Bahafid et al., 2013); and 5 days at pH 2.0 for *A. niger* with 10 g of biomass (Park et al., 2005). Permeability and porosity of the cell wall can affect the incubation time of each microorganism, giving greater or lesser exposure of the functional groups in the cell wall of the biomass analyzed (Park et al., 2005).

At acid pH used in these experiments, Cr(VI) has negative charge and the removal by protonation of the biosorbent surface is favored, which induces a strong attraction to these anions, increasing biosorption and, therefore, the removal of the solution. However, if the pH increases, the concentration of OH⁻ ions increases and the positive positions are reduced in the adsorbent surface, releasing the ions of Cr(VI) to the solution. This explains why at higher pH, removal of Cr(VI) decreases (Bai and Abraham, 2001), though not always desorption of anions of Cr(VI) is observed with an increase of pH, so biosorption is not the only mechanism occurring, but the reduction of Cr(VI) by organic matter also occurs, due to the high oxidation potential of these species, forming Cr(III), which is an insoluble species in basic medium. It has been reported that when Cr(VI) is removed from contaminated sites by indigenous strains of filamentous fungi, most of the metal is reduced to Cr(III) (Coreño-Alonso et al., 2014; Vitti et al., 2014). While sorption is favored at low concentrations of Cr(VI), precipitation is favored at relatively high concentrations.

42.3.2.2 Effect of the Temperature

Temperature is a critical parameter in the removal of Cr(VI) (Figure 42.4); at higher temperatures, we observe greater removal: at 60°C, 100% of the metal is removed in 13 min and 30 min at 28°C. These results are similar to those reported for *Penicillium* sp. IA-01 (Acosta-Rodríguez et al., 2015), *Paecilomyces* sp. (Cárdenas-González and Acosta-Rodríguez, 2010), and *A. niger* (Park et al., 2005), but these are different for *Rhizopus arrhizus* (Kratochvil and Volesky, 1998). As temperature goes up the rate of removal of Cr(VI) increase the contact time required for complete removal of the metal decrease, so increase the redox reaction rate increase (Agarwal et al., 2006).

42.3.2.3 Effect of the Initial Concentration of Chromium(VI) at 28°C and 60°C

The results obtained in the removal of the metal at 60°C did not depend on the concentration of Cr(VI). After 30 min, 100% of the metal is removed at concentrations of 400–1000 mg/L;

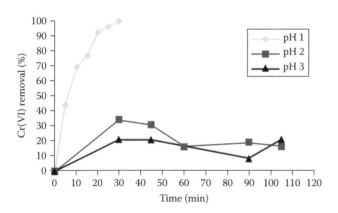

FIGURE 42.3 Effect of incubation time and pH on Cr(VI) removal by *A. niger* 50 mg/L Cr(VI), 100 rpm, 28°C, and 1.0 g of fungal biomass.

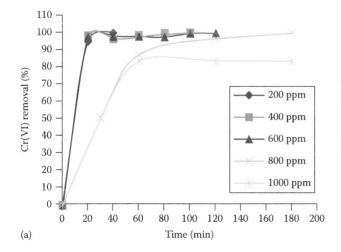

(a)

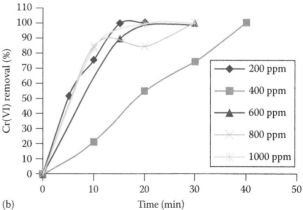

(b)

FIGURE 42.4 Effect of initial metal concentration on chromium(VI) removal by *A. niger*. (a) 28°C, (b) 60°C, pH 1.0, 100 rpm, and 1 g of fungal biomass.

with 200 mg/L, the removal was at 20 min. At 28°C, the concentration influences the removal of the metal; at lower concentration of the same, removal is faster (at 40 min, 200 ppm, and at 180 min, 800 and 1000 ppm, are removed, respectively) (Figure 42.4a and b), which may be due to sorption at low concentrations (which is a very fast process) happens, but at higher concentrations, possibly when positive positions were saturated, precipitation occurs (which is a slower process) (Leyva-Ramos, 2010). For some fungal biomass, it is indicated that the removal of metal increases in direct proportion to the increase of concentration of Cr(VI) in the solution (Morales-Barrera and Cristiani-Urbina, 2008), and others report a lower removal, for example, 25 and 200 mg/L at pH 2.0 using mycelial pellets of *Penicillium simplicissimum* impregnated with powdered biochar (Xu et al., 2015) and 50–100 mg/L with *A. niger* var *tubingensis* strain Ed8 (Coreño-Alonso et al., 2014), while others indicate that there is no influence from the metal concentration (Acosta-Rodríguez et al., 2015) and from the three yeast strains isolated from sediments in Morocco (Bahafid et al., 2013). This could be due to the increased number of those competing for the functional groups of the surface of the biomass ions (Tejada-Tovar et al., 2015).

42.3.2.4 Effect of the Initial Concentration of Biosorbent

From the evaluation of the removal of 50 mg/L of Cr(VI) with different concentrations of biomass, it was found that the higher the concentration of the latter, the removal of Cr(VI) is greater and faster. One gram of biomass with the maximum removal was observed at 30 min, whereas 5 g removal time was 3 min (Figure 42.5). These observations can be explained as the amount of added bioadsorbent determines the number of sites available for biosorption load of chrome anions or any metal contaminant (Kratochvil and Volesky, 1998). Similar results have been reported for *Penicillium* sp. IA-01 (Acosta-Rodríguez et al., 2015), *Paecilomyces* sp. (Cárdenas-González

and Acosta-Rodríguez, 2010), and mycelial pellets of *P. simplicissimum* impregnated with powdered biochar (Xu et al., 2015), and these results are different from those reported for biomass wastes from the mandarin (gabassa), with an optimal concentration of biomass of 100 mg/L (Zubair et al., 2008).

42.3.2.5 Removal of Cr(VI) in Industrial Wastes with Fungal Biomass

In order to analyze the possible use and the ability of *A. niger* biomass to remove chromium (VI) from sediments and effluents, a removal assay was mounted in an aqueous solution in the presence of 5 g biomass, with nonsterile earth contaminated with 297 mg of Cr(VI)g/L and 100 mL of contaminated water with 400 mg of Cr(VI), resuspending the land in trideionized water at 28°C and stirring at 100 rpm. It was observed that at 7 days of incubation, 70.1% of Cr(VI) present in the water contaminated with Cr(VI) was removed, while 62% of Cr(VI) from the ground (Figure 42.6) was removed. The metal removal capability by the biomass of *A. niger* is equal to or greater than the other biomasses that have been studied, *Candida maltose* RR1 (Ramirez-Ramirez et al., 2004) and *Penicillium* IA-01 (Acosta-Rodriguez et al., 2015). In particular, it was more efficient than the other biomasses because the experiments in this study were carried out at acidic pH. Some studies of the removal of Cr(VI) were performed at neutral pH, with an initial concentration of Cr(VI) of 500 mg/L (Fukuda et al., 2008), whereas with the mycelium of *A. niger*, the removal was 8.9 mg/L/g dry weight of mycelium after 7 days of incubation (Park et al., 2005). Furthermore, after incubating the cell extracts of the fungus *A. niger* with contaminated nonsterile earth containing 50 Cr(VI) (50 mg/L) and resuspending in LMM, at pH 4.0, it was observed that after 8 days of incubation, the concentration of Cr(VI) was completely removed (Acosta-Rodríguez et al., 2015), and Xiao et al. (2014) reported a removal between 65.3% and 89.9% for chromium-resistant bacteria.

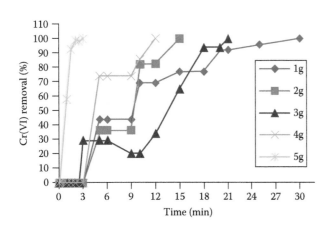

FIGURE 42.5 Effect of biomass concentration on chromium(VI) removal by *A. niger* 50 mg/L Cr(VI), 28°C, pH 1.0, and 100 rpm.

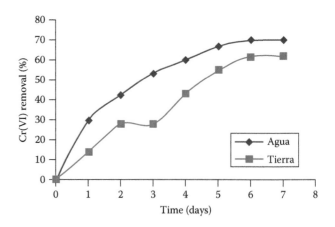

FIGURE 42.6 Removal of Cr(VI) from industrial wastes incubated with 5 g of fungal biomass, 100 rpm, 28°C, 20 g, and 100 mL of contaminated soil and water (297 mg Cr(VI)/g earth and 400 mg Cr(VI)/L), respectively.

42.3.3 Removal of Cr(VI) by the Living Cells of *Aspergillus niger*

42.3.3.1 Effect of the Incubation Time and pH

Figure 42.7 represent the effect of different pH (4.0, 5.3, and 6.2, maintained at phosphate–citrate buffer 100 mM/L) on the removal of Cr(VI). The speed and rate of metal removal increase as much as pH goes up. The maximum removal was observed at pH 5.3 and 6.2 (100% after 6 days incubation at 28°C and 100 rpm). Acosta-Rodríguez et al. (2015) reported a 77% of removal at pH of 5.5, and 4 days at 28°C, using *Penicillium* sp. IA-01, Coreño-Alonso et al. (2014) reported a 95% of removal at pH of 5.3 and 24 h with *A. niger* var *tubingensis* strain Ed8, and Kumari et al. (2015) reported 100% of removal at pH 6.0 and 24 h of incubation with *Pseudomonas stutzeri*. In contrast to our observations, Cárdenas-González and Acosta-Rodríguez (2010) reported a 96% of removal after 7 days of incubation, at 28°C, pH 4.0, 100 rpm, and an inoculum of 38 mg of dry weight of *Paecilomyces* sp. The maximum capacities of adsorption by both living yeasts were found at pH 4.0 for *C. fabianii* HE650139 and *W. anomalus* HE648168; at pH 3.0 for *C. tropicalis* HE650140, with a percentage removal of 100%, by all living microorganisms (Bahafid et al., 2013); and at pH 7.0 by *Enterobacter* sp. DU17 isolated from the tannery waste dump site (Rahman and Singh, 2014) and *Bacillus amyloliquefaciens*, a novel chromate-tolerant bacterium isolated from chromite mine soil (Das et al., 2014). The decreasing of pH causes protonation of the adsorbent surface by attracting ions of Cr(VI) in the solution, so it increases the acidity of the solution, and the biosorption is favored for some microorganisms. As much as the pH increases, the concentration of OH$^-$ ions increases too, favoring the presence of Cr^{6+} ion valence. This stimulates changes in the biosorbent and prevents metal biosorption (Gadd, 1981). The biosorption of the metal according to the reports mentioned earlier is favored for *A. niger* at not very acidic pH and, in some cases, into a pH range from 4 to 7. *A. niger* removal was favored with

increasing pH: varying from 4.5 to 6.2. This is the range of optimal fungal growth.

42.3.3.2 Effect of the Initial Concentration of the Inoculum

Furthermore, the effect of the concentration of cell biomass for the removal capacity of Cr(VI) in the solution (Figure 42.8) was analyzed. At the concentrations tested (8.3, 11.6, and 17.1 mg dry weight), the removal capacity was similar (98.6%, 98.5%, and 97.1%, respectively). Against to our observations, most reports in the literature indicate that the higher amount of biomass increases the percentage of removal: Acosta Rodríguez et al. (2015) for *Penicillium* sp. IA-01, Cárdenas-González and Acosta-Rodríguez (2010) with *Paecilomyces* sp., Cárdenas-González et al. (2015) with *Purpureocillium lilacinum*, Srivastava and Thakur (2006) with *Aspergillus* sp., and Morales-Barrera and Cristiani-Urbina (2008) with *Trichoderma inhamatum*; so the greater the amount of the inoculum, the more binding sites for complexing metal (e.g., ions and HCrO^{4-} Cr$_2$O$_7^{2-}$) (Cervantes et al., 2001).

42.3.3.3 Effect of the Initial Concentration of Chromium(VI)

In this study, it was observed that metal removal is more efficient to low concentration thereof, showing that the removal was 100%, 38.1%, 18.6%, and 17.1% for 50, 100, 150, and 200 mg/L, respectively (Figure 42.9), which may be due to the increased amount of ions competing for free functional groups on the surface of the biomass of *A. niger*. These observations are consistent with most literature reports: Acosta-Rodríguez et al. (2015) for *Penicillium* sp. IA-01, Cárdenas-González and Acosta-Rodríguez (2010) with *Paecilomyces* sp., Cárdenas-González et al. (2015) with *P. lilacinum*, Srivastava and Thakur (2006) with *Aspergillus* sp., Morales-Barrera and Cristiani-Urbina (2008) for *T. inhamatum*, and Gutiérrez-Corona et al. (2010) with *Aspergillus* sp. Ed8 and *Penicillium* sp. H13.

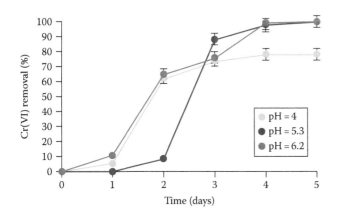

FIGURE 42.7 Effect of pH on chromium(VI) removal by biomass of *A. niger* 50 mg/L Cr(VI), 100 rpm, and 28°C.

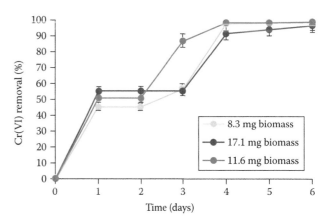

FIGURE 42.8 The effect of cell concentration on the removal of Cr(VI), 50 mg/L Cr(VI), 100 rpm, 28°C, and pH 6.2.

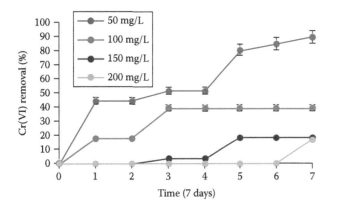

FIGURE 42.9 The effect of the concentration of Cr(VI) in solution on the removal of the metal, 100 rpm, 28°C, and pH 6.2.

42.3.3.4 Removal Capacity of Cr(VI) with Different Carbon Sources

To determine the efficiency of the fungus in the removal of Cr(VI) using different carbon sources such as fermentable (glucose, sucrose, citrate), nonfermentable (succinate), oxidized (glycerol), and commercial (unrefined sugar, brown sugar), the same culture conditions (50 mg/L of Cr(VI), 6 days, 28°C, 100 rpm). The results are shown in Figure 42.10a and b. We found out that the reduction of Cr(VI) is higher when the medium contains fermentable carbon sources (100% glucose, 97.2% sucrose, 93.35% citrate), and removal is high with oxidizable carbon source (89.9% glycerol), unlike the nonfermentable (41.80% succinate), while with the other sources of commercial and economic carbon as unrefined (86.50%) and brown sugar (100%), the removal of chromium(VI) is very similar, observing a complete removal with brown sugar at 4 days of incubation. If we incubate the fungal biomass without a carbon source, there are no changes in the initial Cr(VI) concentration during the experiment (data not shown), suggesting that

a carbon source is required to decrease Cr(VI) concentration in the growth medium. Our view points are consistent with those reported by Cárdenas-González and Acosta-Rodríguez (2010) with the fungus *Paecilomyces* sp., in the presence of glucose, and other sources of commercial and economic carbon as unrefined and brown sugar, or glycerol; the removal of Cr(VI) is very similar in 7 days of incubation (99.17%, 100%, 94.2%, and 99%, respectively). Also, these results are very similar to those reported by Acosta-Rodríguez et al. (2015) for *Penicillium* sp. IA-01, Cárdenas-González et al. (2015) with *P. lilacinum*, Acevedo-Aguilar et al. (2006) using chromate-resistant strains of filamentous fungi indigenous, Prasenjit and Sumathi (2005) for *Aspergillus foetidus*, and Coreño-Alonso (2009), using the strain Ed8 of *A. niger* chromium resistant, all with glucose as the carbon source, but are different from the observations of Srivastava and Thakur (2006), with *Aspergillus* sp., who observed that sodium acetate was the carbon source that induced a greater removal of Cr(VI).

42.3.3.5 Fungal Strain Capacity to Decrease the Initial Concentration of Cr(VI) and Production of Cr(III) in the Culture Medium

As the mobility and toxicity of Cr(VI) depend on its oxidation state, its redox reactions are important to its fate in the environment and health risk. It has been suggested that the reduction in contaminated environments results from a complex interaction between biotic and abiotic processes (Gutiérrez-Corona et al., 2010). Furthermore, Cr(VI) is reduced to Cr(III) in soil for redox reactions with aqueous inorganic species, electron transfer in mineral surfaces, no reaction with humic organic substances (carbohydrates and proteins), or reduction substances in humic soil (Palmer and Wittbrodt, 1991). In addition, microbial activity may contribute to the reduction in these polluted environments, through the release of ferrous

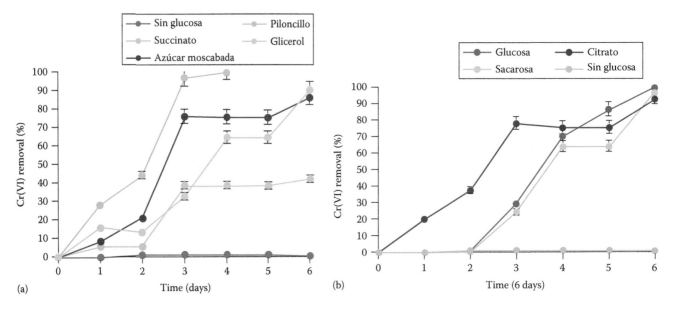

(a)   (b)

FIGURE 42.10 (a, b) The effect of different carbon sources on the capability of *A. niger* to decrease Cr(VI) levels in the growth medium, 100 rpm, 28°C, and pH 6.2.

ion, sulfides, or organic reactive intermediates (Buerge and Hug, 1998).

The ability of the fungus to reduce Cr(VI) with the production of Cr(III) in the culture medium was analyzed. Table 42.1 shows that after 7 days of incubation, *A. niger* decreases fully the concentration of the metal, with the concomitant production of Cr(III), without significant changes in the levels of total Cr in the culture medium without inoculum. This indicates that the fungus is able to reduce Cr(VI) to Cr(III) in medium supplemented with chromate. It describes two mechanisms by which the chromate can be reduced to an oxidation state, which is less toxic: one that involves enzymatic reactions. We do not know if the fungal strain expresses enzymes that reduce Cr(VI), so more studies to understand the effects of ions involved in that reducing activity are required, although capacity reductions have been reported in the literature (Gutiérrez-Corona et al., 2010). The second mechanism is the biosorption (Ahemad, 2014). The fungal cell wall has different functional groups capable of forming coordination complexes with metals, which can facilitate the removal thereof in solution. In addition, our observations are similar to most reports in the literature (Morales-Barrera and Cristiani-Urbina, 2008; Rodríguez et al., 2008; Cárdenas-González and Acosta-Rodríguez, 2010; Acosta-Rodríguez et al., 2013; Das et al., 2014; Nguema et al., 2014; Xu et al., 2015).

TABLE 42.1

Fungal Strain Capacity to Decrease the Initial Concentration of Cr(VI) and Production of Cr(III) in the Culture Medium

	Day 1	Day 7
Cr(VI) (mg/L)[a]	50	0
Cr(III)[b]	0	49.6
Total Cr[c]	50	49.8

[a] Diphenylcarbazide.
[b] Chromazurol S.
[c] Atomic absorption spectrophotometry.

42.3.3.6 Analysis of the Possible Use of the Strain of *Aspergillus niger* to the Removal of Cr(VI) in Nonsterile Earth and Water Contaminated with the Same Metal

Bioremediation study was also carried out, which was inoculated in 5×10^5 spores/mL and 100 mL of LMM (pH 5.3), incubating at 28°C for 48 h at 100 rpm; then 20 g of nonsterile earth and 10 mL of water, contaminated with 50 mg Cr(VI)/g of earth, and 50 mg/L of water, obtained from a factory from the city of Celaya, Guanajuato, Mexico, were added and at different times aliquots of 5 mL were taken and centrifuged at 5000 rpm for 10 min. The supernatant was used to determine the concentration of Cr(VI), observing that after 6 days of incubation, the removal of the metal in the solution at 100% and 70% in earth and water samples, respectively (Figure 42.11), unchanged significantly in the total Cr content. In an experiment conducted in the absence of the fungal strain, the concentration of Cr(VI) of the samples decreases by about 18% in earth and 7% in water (data not shown), which may be caused by native microflora and reducers present in contaminated samples or components. The capacity removal of Cr(VI) by the fungus is equal or better than that from other strains reported such as *Paecilomyces* sp. (Cardenas-González and Acosta-Rodríguez, 2010), different natural biomasses (Acosta-Rodríguez et al., 2013), chromate-resistant strains of filamentous fungi indigenous (Acevedo-Aguilar et al., 2006), *As. foetidus* (Prasenjit and Sumathi, 2005), the strain Ed8 of *A. niger* chromium resistant (Gutiérrez-Corona et al., 2010), and *C. maltose* RR1 (Ramírez-Ramírez et al., 2004). Perhaps, this strain was higher than the other reported, due to its ability to efficiently reduce Cr(VI) under acidic conditions. Most studies of reduction of Cr(VI) by microorganisms have been performed at neutral pH (Ramírez-Ramírez et al., 2004; Acevedo-Aguilar et al., 2006). Reports in the use of microorganisms for bioremediation studies on chromate-contaminated soils are few, as the use of unidentified bacteria native to the contaminated site, finding the maximum reduction of Cr(VI), occurred with the use of 15 mg of biomass bacteria/g of soil (wet weight), with 50 mg of molasses/g soil as a carbon source, and 5.6 mg completely reduce Cr(VI)/g of soil

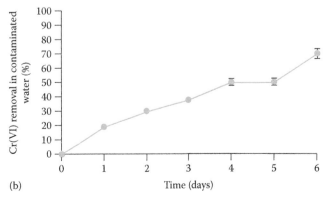

FIGURE 42.11 Bioremediation of Cr(VI) from contaminated earth and water (50 mg Cr(VI)/g earth, and/or 50 mg/L of water), 28°C, pH 6.2, 100 rpm (a) earth, and (b) water.

in 20 days (Jeyasingh and Philip, 2005). In a literature report of Cr(VI)-reducing bacteria, unidentified native to a contaminated site, here the reduction of 50 mg/L of Cr(VI) by bacteria was around 80%, with 10 g/L peptone as a source of electrons and hydraulic retention time of 8 h. The Cr(III) obtained was removed using a column with the fungus *Ganoderma lucidum*, being the specific adsorption capacity of Cr(III) by the fungus column of 576 mg/g (Krishna and Philip, 2005), a water-phase bioremediation assay to explore the possible usefulness of strain of *Paecilomyces* sp.; for eliminating Cr(VI) from industrial wastes, the mycelium biomass was incubated with nonsterilized contaminated earth containing 50 mg Cr(VI)/g of earth, suspended in LMM, at pH 4.0. Eight days of incubation were enough for that the *Paecilomyces* sp. biomass could completely decreased the Cr(VI) concentration of earth sample, without showing major changes in total Cr content, during the experiments (Cárdenas-González et al., 2010), and incubating 100 kg of contaminated earth (345 mg Cr(VI)/g earth), with 20 kg of natural biomass of *Mammea americana*, were incubated in a greenhouse at 28°C, and after 10 weeks of incubation, the natural biomass removal 83% of the metal from contaminated soil (Acosta-Rodríguez et al., 2013).

42.3.4 Adsorption and Reduction by Resting and Permeable Cells

We also estimated the ratio of absorption and/or reduction to adsorption, as we found that the fungi *A. niger* have these abilities as well as adsorption from the results in Section 42.3.3. The resting cells and permeabilized cells were used for the examination, and the heat-killed cells were used to examine the amount of adsorption. The removal was calculated as the value of Cr(VI) in resting cells to the value from total value minus the value of Cr(VI) in heat-killed cells (0% of removal). First, the removal of the metal by resting cells was examined. The cell pellets of *A. niger*, which were cultured in 100 mL thioglycolate broth, were incubated in 100 mM potassium phosphate buffer (pH 7.0) for 8 h at 37°C. The resting cells of the fungus were expedient in reducing 0–10 mg/100 mL Cr(VI) concentrations in 8 h as shown in Figure 42.12. The fungus removal was carried out between 61% and 81% (2–10 mg/100/mL) of the

metal, and these results resemble those reported by the fungi *Penicillium* sp. IA-01 (Acosta-Rodríguez et al., 2015), *P. lilacinum* (Cárdenas-González et al., 2015), *A. niger* and *Aspergillus parasiticus* (Shugaba et al., 2012), *Fusarium solani* (Sen and Ghosh Dastidar, 2011), and *Paecilomyces lilacinus* (Sharma and Adholeya, 2011) and the bacteria *Pseudomonas* sp. G1DM21, isolated from Cr(VI)-contaminated industrial landfill (Desai et al., 2008). Structural properties of the biosorbent including the cellular support and other several factors are known to affect the biosorption rate (Bayramo et al., 2003; Leyva-Ramos, 2010).

It was observed that the cell permeabilization increased the Cr(VI) reduction by the resting cells, as the permeabilized cells with Triton X-100 could reduce 66%, toluene 63%, Sodium Dodecyl Sulphate 60%, and Tween 80 47% (Figure 42.13) of 30 mM Cr(VI) within 8 h, suggesting an efficient intracellular mechanism of chromate reduction. The Cr(VI) reductase activity in the CFE of the cells grown in the absence of Cr(VI) was 27.5 µM/min/mg protein. These results indicate that the Cr(VI) reductase was associated with the CFE. Fungal, yeast, and bacteria chromate reductases have been localized in the CFE of *Penicillium* sp. IA-01 (Acosta-Rodríguez et al., 2015), *A. niger* and *A. parasiticus* (Shugaba et al., 2012), *Pichia jadinii* M9, *Pichia anomala* M10 (Martorell et al., 2012), *Pichia* sp. (Fernández et al., 2012), *Brevibacterium* sp., *Stenotrophomonas* sp. (Shimei et al., 2015), *Bacillus* sp. (Campos et al., 1995), and *Arthrobacter* sp. SUK 1201 (Dey and Paul, 2013). The cytosolic fraction of *Candida maltosa* (Ramírez-Ramírez et al., 2004), *Pichia* sp. (Fernández et al., 2013), and *Pannonibacter phragmitetus* (Xu et al., 2012) and membrane fraction of *P. lilacinum* (Cárdenas-González et al., 2015), *Pseudomonas* sp. G1DM21 (Desai et al., 2012), *Bacillus megaterium* TK W3 (Cheung et al., 2006), *Enterobacter cloacae* (Wang et al., 1980), and *Bacillus cereus* Pf-1 were isolated from sewage treatment plant (Nguema et al., 2014). The results by resting and permeable cells suggest as follows: as shown in Figure 42.13, 61%–81% of Cr(VI) could be removed by resting cells, and the result suggests that the absorption of Cr(VI) occurs without energy of carbon sources or ATP. Additionally, as shown in Figure 42.13, the ratios of the removal of Cr(VI) in case of the pretreatment by Triton X-100, toluene, and SDS of glucose

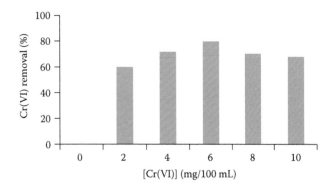

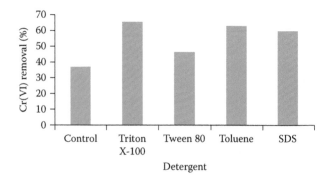

FIGURE 42.12 Resting cell assays for Cr(VI) reduction by *A. niger* performed at initial concentrations of 0–10 mg/100 mL of Cr(VI), pH 7.0, and 37°C in 8 h.

FIGURE 42.13 Permeabilized cell assays for Cr(VI) reduction by *A. niger* performed at initial concentrations of 28 mM of Cr(VI), pH 7.0, and 37°C. 8 h.

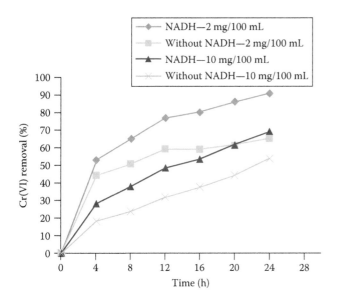

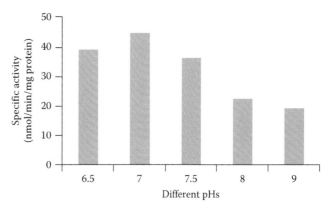

FIGURE 42.14 Cr(VI) reduction by cell-free extracts as a function of time determined with 2 and 10 mg/100 mL Cr(VI) as initial concentration at pH 7.0, 37°C.

FIGURE 42.15 Effect of pH on Cr(VI) reductase activity in cell-free extracts of *A. niger*, as a function of time determined with different buffers (pH 6.5–9.0) with an initial concentration of 5.6 mM of Cr(VI), at 37°C, 1 h.

were 1.8 times higher. Therefore, the transport through the cell membrane is the rate-limiting steps.

The Cr(VI) reduction by the CFEs as a function of time is shown in Figure 42.14, wherein this could reduce 91.3% and 69.3% with 2 and 10 mg/100 mL of the metal, respectively, in the presence of NADH at 24 h, suggesting an effective enzymatic mechanism of Cr(VI) reduction in the CFEs of the fungus. The dependence of Cr(VI) reductases on NADH has been discovered in several previous reports as in the case of *Pseudomonas* sp. (Park et al., 2000), *Escherichia coli* ATCC 33456 (Bae et al., 2005), and *Bacillus* sp. ES 29 (Camargo et al., 2003). As studied in *E. coli*, it has been suggested that the intracellular Cr(VI) reduction using NADH as an electron donor results in the formation of soluble Cr(III)-NAD$^+$ complexes (Puzon et al., 2002), whereas it has been deduced that the soluble chromate reductase of *Pseudomonas ambigua* G-1 catalyzes three-electron reduction of Cr(VI) to Cr(III) consuming three molecules of NADH (Suzuki et al., 1992).

42.3.5 EFFECT OF DIFFERENT *IN VITRO* CONDITIONS ON THE CHROMATE REDUCTASE

42.3.5.1 Effect of pH

The result of permeable cells (Figure 42.15) suggests that *A. niger* has the enzymatic ability of reduction to Cr(VI). Thus, we investigated the reduction of Cr(VI) by *A. niger*. The activity of chromate reductase is examined in the CFE. The function of the chromate reductase of this fungus was characterized in different *in vitro* conditions. In determining the optimal pH for the chromate reductase activity, we used the following buffers at different pH ranges: potassium phosphate, citrate phosphate, and Tris–HCl; and we found the maximum enzymatic activity to be at an optimum pH of 7.0, with potassium phosphate

buffer, as depicted later. The obtained results resemble those reported for a hexavalent chromate reductase activity present in the CFEs of *Penicillium* sp. IA01 (Acosta-Rodríguez et al., 2015), a mixed membrane fraction of *P. lilacinum* (Cárdenas-González et al., 2015), and *B. amyloliquefaciens*, a novel chromate-tolerant bacterium isolated from chromite mine soil (Das et al., 2014), with pH between 7.0 and 7.4 for the bacteria *Pseudomonas* sp. G1DM21 (Desai et al., 2008), between 6.5 and 7.5 for *E. coli* CFE (Bae et al., 2005), and in a range of 5.0–8.0 for *Bacillus* sp. (Elangovan et al., 2006), *A. niger* and *A. parasiticus* (Shugaba et al., 2012), *P. jadinii* M9 (Martorell et al., 2012), and *Arthrobacter* sp. SUK 1201 (Dey and Paul, 2013). But these are different for the reports of Sallau et al. (2015): pH of 5.5 for chromate reductase of *A. niger*, pH of 8.0 for the chromate reductase of a novel *Ochrobactrum* sp. strain Cr-B4 (Hora and Shetty, 2015), pH of 7.0 by dried biomass obtained from a chromium-resistant bacterium *B. cereus* Pf-1 (Nguema et al., 2014), and pH of 2.0 for mycelial pellets of *P. simplicissimum* impregnated with powdered biochar (Xu et al., 2015). But if the pH increases, the concentration of OH$^-$ ions also increases, inducing changes in the adsorbent surface and preventing the biosorption of Cr(VI) negatively charged ions, thereby decreasing the adsorption of metal to high pH values (Gadd, 1981).

42.3.5.2 Effect of Temperature

The optimal temperature for the Cr(VI) reductase activity was 37°C, but the reductase activity was altered significantly at 20°C (25.63% of inhibition); but when the assays were performed at 50°C temperature, the reductase activity showed 7.83% of inhibition (Figure 42.16). For *Penicillium* sp. IA-01, the optimal temperature for the Cr(VI) reductase activity too was 37°C, with 39% and 14.2% of inhibition at 20°C and 50°C, respectively (Acosta-Rodríguez et al., 2015). For *P. jadinii* M9, incubation at 55°C produced a reduction in the activity of 55% (Martorell et al., 2012). For *Brevibacterium* sp. and *Stenotrophomonas* sp., the optimal temperature was 35°C (Shimei et al., 2015), with 66% of inhibition at 20°C; for *Bacillus* sp. DGV19, the temperature was 40°C, with 83% of

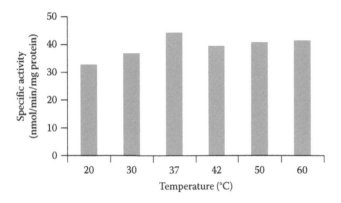

FIGURE 42.16 Effect of temperature on Cr(VI) reductase activity in cell-free extracts of *A. niger* with initial concentrations of 28 mM Cr(VI) at pH 7.0, 1 h.

inhibition at 20°C (Manikandan et al., 2015); for *B. cereus* Pf-1, the maximum Cr(VI) reduction activity was established at 30°C, and this activity decreased at both lower (20°C) and higher (45°C) temperatures (Nguema et al., 2014); in *Ochrobactrum* sp., strain Cr-4, the maximum activity was at 40°C, with 24% of inhibition at 20°C (Hora and Shetty, 2015); and for the activity in CFE of *Arthrobacter* sp. SUK 1201, it was 32°C (Dey and Paul, 2013). On the contrary, *Pseudomonas putida* CFE, and chromate reductase of *A. niger*, proved to be more resistant, keeping its stability up to 50°C (Park et al., 2000; Sallau et al., 2015).

42.3.5.3 Effect of Different Cr(VI) Concentrations

The concentrations analyzed were in a range between 2 and 10 mg/100 mL, exhibiting a higher chromate reductase activity in 2 mg/100 mL, in which the activity decreases with the increase of the chromium initial concentration as seen in Figure 42.17, which also helps us to determine the tolerance of the fungus with the metal. This resembles the report for CFE of *Penicillium* sp. IA-01 (Acosta-Rodríguez et al., 2015), a mixed membrane fraction of *P. lilacinum* (Cárdenas-González et al., 2015), but is different to those for the bacterium *Ochrobactrum* sp., strain Cr-4, in which

the specific activity of chromate reductase was found to increase with the increase in initial Cr(VI) concentration up to 350 µm (Hora and Shetty, 2015). The higher chromate reductase activity not only depends on the active sites but also depends on metal bindings and the initial concentration (Sukumar, 2010).

42.3.5.4 Effect of Different Metal Cations

The effect of different metal cations on the chromate reductase activity of *A. niger* is shown in Figure 42.18. All the metal ions showed inhibition for Cr(VI) reductase activity of the CFE of 5.83% with Cu^{2+} and 61.5% with Na^+, and these results are like those reported for a mixed membrane fraction of *P. lilacinum* (Cárdenas-González et al., 2015) and for CFEs from *Penicillium* IA-01 (Acosta-Rodríguez et al., 2015). They are different to reported for the bacterium *Ochrobactrum* sp., strain Cr-4, in which all the metal ions except iron and copper were found to inhibit the enzyme activity, Cu^{2+}, an ion that significantly stimulated chromate activity by 30.16% (Hora and Shetty, 2015). The role of Cu^{2+} in the stimulation of chromate reductase could be related to its main function as a protective agent for electron transport, as a single-electron redox center, and as a shuttle for electron between protein subunits (Sau et al., 2010). In *Stenotrophomonas* sp., the presence of Cu^{2+}, Fe^{3+}, and Zn^{2+} caused the reduction of Cr(VI) to be enhanced to 160%, 133%, and 147%, respectively, compared to no addition of metal ion by growing cells. In contrast, the presence of Ag^+ and Co^{2+} causes inhibition reduction of Cr(VI) in about 34% and 68%, respectively, compared with no addition of other metal, and when metal ions are present as Ba^{2+}, Mg^{2+}, and Al^{3+}, they do not affect the reduction of Cr(VI) (Shimei et al., 2015). In *Brevibacterium* sp., the presence of Cu^{2+} and Zn^{2+} caused the reduction of Cr(VI) to be increased to 192% and 158%, respectively, compared to no addition of metal ion by growing cells. The presence of the metal Ag^+ and Al^{3+} causes inhibition approximately 34% and 86%, respectively, compared with no addition of metal ion, and other metal ions that are present such as Ba^{2+}, Co^{2+}, Fe^{3+}, and Mg^{2+} affect some reduction of Cr(VI) by growing cells (Shimei et al., 2015).

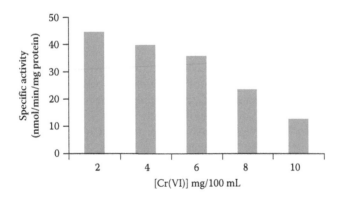

FIGURE 42.17 Effect of different concentration of Cr(VI) on the crude chromate reductase activity in cell-free extract of *A. niger*, 1 h of incubation, pH 7.0, 100 rpm, and 37°C.

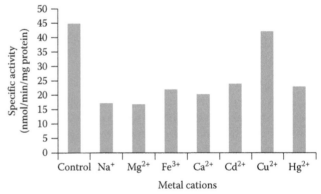

FIGURE 42.18 Effect of different metal cations on Cr(VI) reductase activity in cell-free extracts of *A. niger* at pH 7.0 and 37°C, 1 h of incubation.

42.3.5.5 Effect of Different Electron Donors

The reductase activity increased on supplementation in the reaction mixtures with electron donors. All the electron donors analyzed increased the activity, and the most efficient were ascorbic acid, NADH, glucose, and citrate by 4.7, 4.1, 2.7, and 2.6 times, respectively (Figure 42.19). These results are like those reported for the fungi *P. lilacinum* (Cárdenas-González et al., 2015) and *Penicillium* IA-01 (Acosta-Rodríguez et al., 2015), with citric acid, NADH, and glucose. For the yeasts *P. jadinii* M9 and *P. anomala* chromate reductase with NADH (Martorell et al., 2012), the specific activity of Cr(VI) reduction in the CFE of *B. cereus* Pf-1 showed an increase with the addition of 0.1 mM NADH; an addition of 0.1 mM NADH in the reaction mixture containing CFE also stimulated the reduction of Cr(VI). Additionally, glucose, a possible electron donor, during the reduction of Cr(VI), increased Cr(VI) reduction. Citrate increased Cr(VI) reduction to a significant level. However, succinate had no significant effect on the reduction of Cr(VI) by CFE (Nguema et al., 2014). In the previous reports of *Bacillus* sp., glucose has been reported to act as an electron donor and has been demonstrated to increase Cr(VI) reduction (Liu et al., 2006), and formate-dependent Cr(VI) reductases have been reported in *Shewanella putrefaciens* MR-1 (Myers et al., 2000).

42.3.5.6 Effect of Different Respiratory Inhibitors

Respiratory inhibitors like azide (1 mM), EDTA (1 mM), cyanide (1 mM), and 2-ME caused inhibitions of 63.71%, 50.57%, 25.64%, and 4.05%, respectively (Figure 42.20), in the Cr(VI) reductase activity; these results agree with those obtained in previous studies (Dey and Paul, 2013; Acosta-Rodríguez et al., 2015; Cárdenas-González et al., 2015; Hora and Shetty, 2015), and it has been observed that 2-ME and azide inhibited partially purified chromate reductase of *Ochrobactrum* sp., strain Cr-B4 (Hora and Shetty, 2015), and EDTA inhibited the chromate reductase activity from endophytic *Bacillus* sp. DGV19 of *Albizzia lebbeck* (L.) (Manikandan et al., 2015). On the other hand, urea and thiourea are well-known protein-denaturizing agents; both inhibited Cr(VI) reduction by 85%

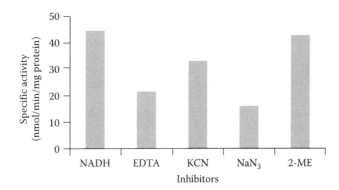

FIGURE 42.20 Effect of different inhibitors on Cr(VI) reductase activity in cell-free extracts of *A. niger* at pH 7.0 and 37°C, 1 h.

and 95%, respectively, in cytoplasmic and cell membrane fractions of chromate-reducing bacterium *B. cereus* Pf-1, isolated from sewage treatment plant (Nguema et al., 2014). Respiratory inhibitors act on de novo protein synthesis or affect the respiratory chain intermediates responsible for Cr(VI) reduction, wherein Cr(VI) serves as a terminal electron acceptor (Park et al., 2000). As shown in Figures 42.15 and 42.16, the optimal pH and temperature of chromate reductase in *A. niger* were pH 7 and 37°C, and the results were supported by the results of living cells. Therefore, the reduction mainly occurred to remove Cr(VI) and to show resistance to high concentration of Cr(VI), whereas the addition of electron donors caused the decrease of the activity, and therefore, these compounds may be the inhibitor of the enzyme.

REFERENCES

Acevedo-Aguilar, F., Espino-Saldaña, A., León-Rodríguez, I., Rivera-Cano, M. E., Ávila-Rodríguez, M., Wrobel, K., Lappe, P., Ulloa, M., and Gutierrez-Corona, J. F. 2006. Hexavalent chromium removal *in vitro* and from industrial wastes, using chromate-resistant strains of filamentous fungi indigenous to contaminated wastes. *Canadian Journal of Microbiology* 52: 809–815.

Acosta-Rodríguez, I., Arévalo-Rangel, D., Cárdenas-González, J. F., Moctezuma-Zárate, M. G., and Martínez-Juárez, V. M. 2015. Hexavalent chromium removal by *Penicillium* sp. IA-01. In: *Bioremediation of Wastewater and Polluted Soil*. Shiomi, N., ed. InTech, Rijeka, Croatia, Chapter 8, pp. 166–192.

Acosta-Rodríguez, I., Cárdenas-González, J. F., Moctezuma-Zárate, M. G., and Martínez Juárez, V. M. 2013. Removal of hexavalent chromium from solutions and contaminated sites by different natural biomasses. In: *Applied Bioremediation—Active and Passive Approaches*. Patil, Y. B. and Rao, P., eds. InTech, Rijeka, Croatia, Chapter 9, pp. 207–224.

Agarwal, G. S., Kumar, H., and Chaudari, S. 2006. Biosorption of aqueous chromium(VI) by *Tamarindus indica* seeds. *Bioresource Technology* 97: 949–956.

Ahemad, M. 2014. Bacterial mechanisms for Cr(VI) resistance and reduction: An overview and recent advances. *Folia Microbiologica* 59: 321–332.

Bae, W. C., Lee, H. K., Choe, Y. C., Jahng, D. K., Lee, S. H., and Kim, S. J. 2005. Purification and characterization of NADPH-dependent Cr(VI) reductase from *Escherichia coli* ATCC 33456. *Journal of Microbiology* 43: 21–27.

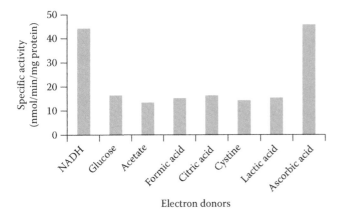

FIGURE 42.19 Effect of different electron donors on Cr(VI) reductase activity in cell-free extracts of *A. niger* at pH 7.0 and 37°C, 1 h.

Bahafid, W., Joutey, N. T., Sayel, H., Iraqui-Houssaini, M., and El Ghachtouli, N. 2013. Chromium adsorption by three yeast strains isolated from sediments in Morocco. *Geomicrobiology Journal* 30: 422–429.

Bai, R. S. and Abraham, T. E. 2001. Biosorption of chromium(VI) from aqueous solution by *Rhizopus nigricans*. *Bioresource Technology* 79: 73–81.

Basu, A., Panda, S. S., and Dhal, N. K. 2015. Potential microbial diversity in chromium mining areas: A review. *Bulletin of Environment, Pharmacology and Life Sciences* 4: 158–169.

Bayramo, G., Bektas, S., and Arica, M. Y. 2003. Biosorption of heavy metal ions on immobilized white-rot fungus *Trametes versicolor*. *Journal of Hazardous Materials* 101: 285–300.

Buerge, I. J. and Hug, S. J. 1998. Influence of organic ligands on chromium(VI) reduction by iron(II). *Environmental Science and Technology* 32: 2092–2099.

Camargo, F. A. O., Okeke, B. C., Bento, F. M., and Frankenberger, W. T. 2003. *In vitro* reduction of hexavalent chromium by a cell-free-extract of *Bacillus* sp. ES 29 stimulated by Cu^{2+}. *Applied Microbiology and Biotechnology* 62: 569–573.

Campos, J., Martinez-Pacheco, M., and Cervantes, C. 1995. Hexavalent-chromium reduction by a chromate-resistant *Bacillus* sp., strain. *Antonie van Leeuwenhoek* 68: 203–208.

Cárdenas-González, J. F. and Acosta-Rodríguez, I. 2010. Hexavalent chromium removal by a *Paecilomyces* sp. fungal strain isolated from environment. *Bioinorganic Chemistry and Applications* 2010: 1–6.

Cárdenas-González, J. F., Terán Figueroa, Y., and Acosta Rodríguez, I. 2015. Hexavalent chromate reductase activity present in the mixed membrane fraction of *Purpureocillium lilacinum*. *Joküll Journal* 65: 99–120.

Cervantes, C., Campos García, J., Devars, S., Gutiérrez Corona, F., Loza Tavera, H., Torres Guzmán, J. C., and Moreno Sánchez, R. 2001. Interactions of chromium with microorganisms and plants. *FEMS Microbiological Review* 25: 335–347.

Cheung, K. H., Lai, H. Y., and Gu, J. D. 2006. Membrane-associated hexavalent chromium reductase of *Bacillus megaterium* TKW3 with induced expression. *Journal of Microbiology and Biotechnology* 16: 855–862.

Coreño-Alonso, A. 2009. *Caracterización del sistema de reducción de Cr (VI) de la cepa Ed8 de Aspergillus niger resistente a cromato*. Tesis de doctorado (Biología), Posgrado en Biología Experimental, Universidad de Guanajuato, Guanajuato, Gto, Mexico.

Coreño-Alonso, A., Solé, A., Diestra, E., Esteve, I., Gutiérrez-Corona, J. F., Reyna-López, G. E., Fernández, F. J., and Tomasini, A. 2014. Cr(VI) Mechanisms of interaction of chromium with *Aspergillus niger* var. *tubingensis* strain Ed8. *Bioresource Technology* 158: 188–192.

Das, S., Mishra, J., Das, S. K., Sony Pandey, S., Rao, D. S., Chakraborty, A., Sudarshan, M., Das, N., and Thatoi, H. 2014. Investigation on mechanism of Cr(VI) reduction and removal by *Bacillus amyloliquefaciens*, a novel chromate tolerant bacterium isolated from chromite mine soil. *Chemosphere* 96: 112–121.

Desai, C., Jain, K., and Madamwar, D. 2008. Hexavalent chromate reductase activity in cytosolic fractions of *Pseudomonas* sp. G1DM21 isolated from Cr(VI) contaminated industrial landfill. *Process Biochemistry* 43(7): 713–721.

Dey, S. and Paul, A. K. 2013. Evaluation of *in vitro* reduction of hexavalent chromium by cell-free extract of *Arthrobacter* sp. SUK 1201. *British Microbiology Research Journal* 3: 325–338.

Elangovan, R., Abhipsa, S., Rohit, B., Ligy, P., and Chandraraj, K. 2006. Reduction of Cr(VI) by a *Bacillus* sp. *Biotechnology Letters* 28: 247–252.

Fernández, P. M., Martorell, M., Farina, J. I., and Figueroa, L. I. 2012. Removal efficiency of Cr^{6+} by indigenous *Pichia* sp. isolated from textile factory effluent. *The Scientific World Journal* 2012: 708–213.

Fukuda, T., Ishino, Y., Ogawa, A., Tsutsumi, K., and Morita, H. 2008. Cr(VI) reduction from contaminated soils by *Aspergillus* sp. N2 and *Penicillium* sp. N3 isolated from chromium deposits. *Journal of General and Applied Microbiology* 54: 295–303.

Gadd, G. M. 1981. Accumulation of metals by microorganisms and algae. In: *Biotechnology: A Comprehensive Treatise*, Vol. 6B. Rhem, H. J. and Reed, G., eds. VCH, Weinheim, Germany, 1981, pp. 401–433.

Gharieb, M. M. and Gadd, G. M. 1998. Evidence for the involvement of vacuolar activity in metal(loid) tolerance: Vacuolar-lacking and defective mutants of *Saccharomyces cerevisiae* display higher sensitivity to chromate, tellurite and selenite. *BioMetals* 11: 101–106.

Greenberg, A. E., Clesceri, L. S., and Eaton, A. D. 1992. *Standard Methods for the Examination of Water and Wastewater*, 18th edn. American Public Health Association, Washington, DC, 1992, pp. 3.58–3.60.

Guillén-Jiménez, F. M., Morales-Barrera, L., Morales-Jiménez, J., Hernández-Rodríguez, H., and Cristiani-Urbina, E. 2008. Modulation of tolerance to Cr(VI) and Cr(VI) reduction by sulfate ion in a *Candida* yeast strain isolated from tannery wastewater. *Journal of Industrial Microbiology and Biotechnology* 35: 1277–1287.

Gutiérrez-Corona, J. F., Espino Saldaña, A. E., Coreño Alonso, A., Acevedo Aguilar, F. J., Reyna López, G. E., Fernández, F. J., Tomasini, A., Wrobel, K., and Wrobel, K. 2010. Mecanismos de interacción con cromo y aplicaciones biotecnológicas en hongos. *Revista Latinoamericana de Biotecnología Ambiental y algal* 1: 47–63.

Hora, A. and Shetty, V. K. 2015. Partial purification and characterization of chromate reductase of a novel *Ochrobactrum* sp. strain Cr-B4. *Preparative Biochemistry and Biotechnology* 45(8): 769–784.

Jeyasingh, J. and Philip, L. 2005. Bioremediation of chromium contaminated soil: Optimization of operating parameters under laboratory conditions. *Journal of Hazardous Materials* 118: 113–120.

Kratochvil, D. and Volesky, B. 1998. Advances in the biosorption of heavy metals. *Reviews Tibtech* 16: 291–300.

Krishna, K. R. and Philip, L. 2005. Bioremediation of Cr(VI) in contaminated soils. *Journal of Hazardous Materials* 121: 109–117.

Kumari, D., Pan, X., Zhang, D., Zhao, Ch., Al-Misned, F. A., and Golam, M. 2015. Bioreduction of hexavalent chromium from soil column leachate by *Pseudomonas stutzeri*. *Bioremediation Journal* 19: 249–258.

Larone, D. H. 2002. *Medically Important Fungi: A Guide to Identification*, 4th edn. ASM Press, Washington, DC, pp. 366–368.

Leyva-Ramos, R. 2010. *Fundamentos de adsorción en sistemas liquido-sólido*. Situación del arsénico en la Región Ibérica e Iberoamericana. Posibles acciones articuladas e integradas para el abatimiento del arsénico en zonas aisladas. CYTED, Argentina, Chapter 3, p. 43.

Liu, Y. G., Xu, W. H., Zeng, G. M., Li, X. and Gao, H. 2006. Cr(VI) reduction by *Bacillus* sp., isolated from chromium landfill. *Process Biochemistry* 41: 1981–1986.

López-Martínez, R., Méndez Tovar, L. J., Hernández Hernández, F., and Castañon Olivares, L. R. 2012. *Hongos contaminantes comunes en el Laboratorio. En Micología Médica*, 3ª edn. Procedimientos para el diagnóstico de Laboratorio, Trillas, México, 2012, pp. 135–142.

Lowry, O. H., Rosebrough, N. J., Farr, A. L., and Randall, R. J. 1951. Protein measurement with the Folin phenol reagent. *The Journal of Biological Chemistry* 193: 265–275.

Manikandan, M., Gopal, J. G., Kumaran, R. S., Kannan, V., and Chun, S. 2016. Purification and characterization of a highly active chromate reductase from endophytic *Bacillus* sp. DGV19 of *Albizzia lebbeck* (L.) Benth. actively involved in phytoremediation of tannery effluent-contaminated sites. *Preparative Biochemistry and Biotechnology* 46(2): 192–199.

Martorell, M. M., Fernández, P. M., Farina, I., and Figueroa, L. I. 2012. Cr(VI) reduction by cell-free extracts of *Pichia jadini* and *Pichia anomala* isolated from textile-dye factory effluents. *International Biodeterioration & Biodegradation* 71: 80–85.

Myers, C. R., Carstens, B. P., Antholine, W. E., and Myers J. M. 2000. Chromium(VI) reductase activity is associated with the cytoplasmic membrane of anaerobically grown *Shewanella putrefaciens* MR-1. *Journal of Applied Microbiology* 88: 98–106.

Morales-Barrera, L. and Cristiani-Urbina, E. 2008. Hexavalent chromium removal by a *Trichoderma inhamatum* fungal strain isolated from tannery effluent. *Water, Air, and Soil Pollution* 187: 327–336.

Nguema, P. F., Luo, Z., and Lian, J. 2014a. The biosorption of Cr(VI) ions by dried biomass obtained from a chromium-resistant bacterium. *Frontiers in Chemical Science of England* 8: 454–464.

Nguema, P. F., Luo, Z., and Lian, J. 2014b. Enzymatic chromium(VI) reduction by cytoplasmic and cell membrane fractions of chromate-reducing bacterium isolated from sewage treatment plant. *International Journal of Biology* 6: 64–76.

Nguyen-Nhu, N. T. and Knoops, B. 2002. Alkyl hydroperoxide reductase 1 protects *Saccharomyces cerevisiae* against metal ion toxicity and glutathione depletion. *Toxicology Letter* 135: 219–228.

Palmer, C. H. and Wittbrodt, P. P. 1991. Processes affecting the remediation of chromium-contaminated sites. *Environmental Health Perspective* 92: 25–40.

Pantaler, R. P. and Pulyaeva, I. V. 1985. A spectrophotometric study of complexation between chromium and chromazurol S. *Journal of Analytical Chemistry* (Moscow) 40: 1634–1639.

Park, C., Keyhan, M., Wielinga, B., Fendorf, S., and Matin, A. 2000. Purification to homogeneity and characterization of a novel *Pseudomonas putida* chromate reductase. *Applied and Environmental Microbiology* 66: 1788–1795.

Park, D., Yun, Y. S., Jo, J. H., and Park, J. M. 2005. Mechanism of hexavalent chromium removal by dead fungal biomass of *Aspergillus niger*. *Water Research* 39: 533–540.

Prasenjit, B. and Sumathi, S. 2005. Uptake of chromium by *Aspergillus foetidus*. *Journal of Material Cycles and Waste Management* 7: 88–92.

Puzon, G. J., Peterson, J. M., Roberts, A. G., Kramer, D. M., and Xun, L. A, 2002. A bacterial flavin reductase system reduces chromate to a soluble chromium(III)-NAD+ complex. *Biochemical, Biophysical and Research Communications* 294: 78–81.

Rahman, Z. and Singh, V. P. 2014. Cr(VI) reduction by *Enterobacter* sp. DU17 isolated from the tannery waste dump site and characterization of the bacterium and the Cr(VI) reductase. *International Biodeterioration & Biodegradation* 91: 97–103.

Ramírez-Ramírez, R., Calvo-Méndez, C., Avila-Rodríguez, M., Lappe, P., Ulloa, M., Vázquez-Juárez, R., and Gutiérrez-Corona, J. F. 2004. Cr(VI) reduction in a chromate-resistant strain of *Candida maltose* isolated from the leather industry. *Antonie van Leeuwenhoek* 85: 63–68.

Rodríguez, M. E., Miranda, R. C., Olivas, R., and Sosa, C. A. 2008. Efectos de las condiciones de operación sobre la biosorción de Pb++, Cd++ y Cr+++, en solución por *Saccharomyces cerevisiae residual*. *Información Tecnológica* 19: 47–55.

Sallau, A. B., Inuwa, H. M., Ibrahim, S., and Nok, A. J. 2015. Isolation and properties of chromate reductase from *Aspergillus niger*. *International Journal of Modern Cellular and Molecular Biology* 3: 10–21.

Sau, G. B., Chaterje, S., and Mukherjee, S. K. 2010. Chromate reduction by cell free extract of *Bacillus firmus* KUCrI. *Polish Journal of Microbiology* 59: 185–190.

Sen, M. and Ghosh Dastidar, M. 2011. Biosorption of Cr(VI) by resting cells of *Fusarium solani*. *Iran Journal of Environmental Health Science Technology* 8: 153–158.

Shakya, M., Sharma, P., Meryem, S., Mahmood, Q., and Kumar, A. 2016. Heavy metal removal from industrial wastewater using fungi: Uptake mechanism and biochemical aspects. *Journal of Environmental Engineering* 142(9): C6015001–C6015001–18.

Sharma, S. and Adholeya, A. 2011. Detoxification and accumulation of chromium from tannery effluent and spent chrome effluent by *Paecilomyces lilacinus* fungi. *International Biodeterioration and Biodegradation* 65: 309–317.

Shimei, G., Shichao, G., Maohong, Z., and Xinjiao, D. 2015. Bioremediation of hexavalent chromate using permeabilized *Brevibacterium* sp. and *Stenotrophomonas* sp. cells. *Journal of Environmental Management* 157: 54–59.

Shugaba, A., Buba, F., Kolo, B. G., Nok, A. J., Ameh, D. A., and Lori, J. A. 2012. Uptake and reduction of hexavalent chromium by *Aspergillus niger* and *Aspergillus parasiticus*. *Journal of Petroleum and Environmental Biotechnology* 3: 8.

Srivastava, S., Agrawal, S. B., and Mondal, M. K. 2015. A review on progress of heavy metal removal using adsorbents of microbial and plant origin. *Environmental Science Pollution Research*. 22(20): 15386–15415.

Srivastava, S. and Thakur, I. S. 2006. Isolation and process parameter optimization of *Aspergillus* sp. for removal of chromium from tannery effluent. *Bioresource Technology* 97: 1167–1173.

Sukumar, M. 2010. Reduction of hexavalent chromium by *Rhizopus Oryzae*. *African Journal of Environmental Science and Technology* 4: 412–418.

Sumner, E. R., Shanmuganathan, A., Sideri, T. C., Willetts, S. A., Houghton, E. J., and Avery, S. 2005. Oxidative protein damage causes chromium toxicity in yeast. *Microbiology* 151: 1939–1948.

Suzuki, T., Miyata, N., Horitsu, H., Kawai, K., Takamizawa, K., and Tai, Y. 1992. NAD(P)H-dependent chromium(VI) reductase of *Pseudomonas ambigua* G-1: A Cr(V) intermediate is formed during the reduction of Cr(VI) to Cr(III). *Journal of Bacteriology* 174: 5340–5345.

Tejada-Tovar, C., Villabona-Ortiz, Á., and Garcés-Jaraba, L. 2015. Adsorción de metales pesados en aguas residuales usando materiales de origen biológico. *Tecno Lógicas* 18: 109–123.

Thatoi, H., Das, S., Mishra, J., Rath, B. P., and Das, N. 2014. Bacterial chromate reductase, a potential enzyme for bioremediation of hexavalent chromium: A review. *Journal of Environmental Management* 146: 383–399.

Vitti, C., Marchi, E., Decorosi, D., and Giovannetti, L. 2014. Molecular mechanisms of Cr(VI) resistance in bacteria and fungi. *FEMS Microbiological Review* 38: 633–659.

Vitti, C., Pace, A., and Giovannetti, L. 2003. Characterization of Cr(VI) resistant bacteria isolated from, chromium-contaminated soil by tannery activity. *Current Microbiology* 46: 1–5.

Wang, P. C., Mori, T., Toda, K., and Ohtake, H. 1980. Membrane associated chromate reductase activity from *Enterobacter cloacae*. *Journal of Bacteriology* 172: 1670–1672.

Xiao, W., Yang, X., He, Z., and Li, T. 2014. Chromium-resistant bacteria promote the reduction of hexavalent chromium in soils. *Journal of Environmental Quality* 43: 507–516.

Xu, A., Jian, H., Liu, Y., Zeng, G., Gu, Y., and Tan, X. 2015. Removal of chromium(VI) from aqueous solution using mycelial pellets of *Penicillium simplicissimum* impregnated with powdered biochar. *Bioremediation Journal* 19: 259–268.

Xu, L., Luo, M., Jiang, C., Wei, X., Kong, P., Lang, X., Zhao, J., Yang, L., and Liu, H. 2012. *In vitro* reduction of hexavalent chromium by cytoplasmic fractions of *Pannonibacter phragmitetus* LSSE-09 under aerobic and anaerobic conditions. *Applied Biochemistry and Biotechnology* 166: 933–941.

Zheng, S., Huang, H., Zhang, R., and Cao, L. 2014. Removal of Cr(VI) from aqueous solutions by fruiting bodies of the jelly fungus (*Auricularia polytricha*). *Applied Microbiology and Biotechnology* 98: 8729–8736.

Zubair, A., Bhatti, H. N., Hanif, M. A., and Shafqat, F. 2008. Kinetic and equilibrium modeling for Cr(III) and Cr(VI) removal from aqueous solutions by *Citrus reticulate* waste biomass. *Water Air Soil Pollution* 191: 305–318.

43 Microbial Remediation of Chromium-Contaminated Wastes

Lai Peng, Yiwen Liu, Jing Sun, and Bing-Jie Ni

CONTENTS

ABSTRACT

Cr(VI) and Cr(III) are the two most ubiquitous forms of chromium. The remediation of chromium contaminant is primarily dependent on transformation from highly toxic and mobile Cr(VI) to insoluble Cr(III) with less toxicity. Biological Cr(VI) removal is not only a promising method but also a very complicated process. This chapter introduces various remediation processes, highlights important Cr(VI)-reducing microorganisms, and reveals underlying mechanisms for bacterial Cr(VI) resistance and reduction. The assessment of key factors limiting and influencing bacterial growth and chromium removal along with reviews on mathematical modeling tended to facilitate further application of microbial Cr(VI) removal in the bench scale, pilot scale, and *in situ* studies under a wide array of environmental conditions.

43.1 INTRODUCTION

Chromium contamination is ubiquitous in soil and water environments, which can be caused by a variety of industrial activities such as electroplating, wood preservation, and leather tanning (Cheung and Gu, 2007). Among the varying valence states from −II to +VI, hexavalent chromium (Cr(VI)) and its trivalent counterpart (Cr(III)) are the dominant species (Thatoi et al., 2014). Cr(VI) prevails in natural aquifer, whereas Cr(III) is dominant in municipal wastewater rich in organics (Cheung and Gu, 2007). Cr(VI) is highly soluble and mobile and of acute toxicity, mutagenicity, and carcinogenicity in natural environment (Somasundaram et al., 2009). In contrast, Cr(III) is nearly insoluble at neutral pH and possesses less toxicity (Somasundaram et al., 2009). As a result, the effective detoxification of Cr(VI) relies on the reduction of chromate to a less mobile and toxic trivalent form.

Several technologies of Cr(VI) remediation in soil and water environment have been developed in the past decades, including ion exchange, reverse osmosis, reduction precipitation, surface adsorption, electrolysis, and membrane technologies (Kikuchi and Tanaka, 2012). However, these physicochemical methods are not suitable for large-scale application due to the fact that the setup, operation, and maintenance are economically expensive (Nourbakhsh et al., 1994). Other drawbacks include environmental hazards caused by toxic sludge generation that requires subsequent treatment, incomplete metal removal, energy requirement, and chemical consumption (Dhal et al., 2013). Hence, microbial Cr(VI) removal has received increasing attentions due to minimum chemical requirements and operating costs (Srivastava and Majumder, 2008).

Cr(VI) can be removed through enzymatic reduction, biosorption, and bioaccumulation. It has been demonstrated that a wide array of bacteria, such as chromate-reducing bacteria (CRB), iron-reducing bacteria (IRB), sulfate-reducing bacteria (SRB), and sulfur-reducing bacteria, are able to reduce Cr(VI) under aerobic and anaerobic conditions in previous studies (Fude et al., 1994; Wang and Shen, 1995; Smith and Gadd, 2000; Wielinga et al., 2001; Somasundaram et al., 2009; Sahinkaya et al., 2012). In addition, fungi (Sanghi et al., 2009), yeast (Chen and Wang, 2007), and algae (Bankar et al., 2009) also possess the capacity for Cr(VI) reduction. Many factors have significant impacts on biological Cr(VI) removal, including pH, temperature, electron donor, oxyanions, Cr(VI) concentration, interfering metal ions, etc. A better understanding of the interactions between various microbial species and its affecting factors will facilitate the application of *in situ* bioremediation in contaminated natural environment and make the treatment more cost-effective, sustainable, and environmentally friendly.

This chapter aims to summarize research outcomes concerning microbial Cr(VI) removal. The chromium toxicity and its biotransformation in natural environment have been discussed. A variety of microorganisms, capable for Cr(VI) bioremediation, have been investigated. The mechanism for microbial Cr(VI) removal under aerobic and anaerobic conditions has been clarified. The effect of environmental conditions (e.g., pH, temperature, interfering metal ions, cell density, etc.) has been examined. Finally, this chapter provides the state-of-the-art research performed on mathematical modeling for description and prediction of biological Cr(VI) contaminant remediation.

43.2 CHROMIUM TOXICITY ON MICROBIAL ORGANISMS

Although chromium in trace quantities is critical for living organisms, an elevated level causes hazards in environment and human health due to the toxicity (Sharma et al., 1995). Large abundance of chromium is present in aquifer, which constitutes 0.1–0.3 mg/kg of the earth's crust (Molokwane et al., 2008). Among the varying valence states from −II to

+VI, Cr(VI) and Cr(III) are the dominant species (Thatoi et al., 2014). Cr(VI) is of the highest toxicity, known to destroy the living cells by its strong oxidation potential (Kotaś and Stasicka, 2000). Moreover, Cr(VI) is of acute teratogenicity, mutagenicity, and carcinogenicity, which has been recognized as one of the priority chemicals that pose greatest threat to human beings by U.S. Environmental Protection Agency (USEPA) with 50–100 μg/L set as maximum allowable level in drinking water (Marsh and McInerney, 2001; Cheung and Gu, 2007). The Occupational Safety and Health Administration (OSHA) has set limits of 52 μg/L for Cr(VI) compounds, 500 μg/L for water-soluble Cr(III) compounds, and 1000 μg/L for metallic chromium(0) and insoluble chromium compounds for 8 h work shifts and 40 h work weeks (Das and Singh, 2011).

The toxicity of Cr(VI) is associated with its ability to cause oxidative stress and membrane damage in eukaryotes and prokaryotes due to inhibition of the electron transport chain and loss of membrane integrity (Hüser et al., 1998; Reynolds et al., 2009). Concomitant with entering the cell via sulfate transporter pathway, Cr(VI) is reduced to short-lived intermediates including free radicals, Cr(V)/Cr(IV), and Cr(III) as the end product during various enzymatic and nonenzymatic processes (Costa, 2003; Cheung and Gu, 2007). A reactive oxygen species (ROS) produced during the one-electron redox cycle of Cr(V) and Cr(VI) combines DNA–protein complexes within the cell, while Cr(IV) hinders physiological functions of the cell by binding to cellular materials (Cervantes et al., 2001; Cheung and Gu, 2007). As for the human beings, several traumata are caused by exposure to Cr(VI). Stomach irritation or ulceration can be caused by short-term exposure at levels above maximum concentration, while long-term exposure results in damage of kidney, liver, circulation, and nerve tissue (Kotaś and Stasicka, 2000). In addition, Cr(VI) may increase the risk of respiratory cancer, allergic contact dermatitis, and skin ulcerations (Nethercott et al., 1994; Williams, 1997; Narayani and Shetty, 2013). Cr(VI) can also elicit hazards in environment. For example, the accumulated Cr(VI) would reduce microbial population, affect microbial community, and thus alter the whole soil structure (Shi et al., 2002; Zhou et al., 2002). In contrast, Cr(III) is less toxic and mobile in environment due to its impermeability to living cell membranes and the formation of insoluble hydroxide/oxides at pH above 5.5 (He et al., 2009). Therefore, the transformation from Cr(VI) to Cr(III) serves as an alternative to reduce chromium hazards.

43.3 TRANSFORMATION OF CHROMIUM IN SOIL AND WATER

As chromium is produced by numerous industries including electroplating, wood preservation, and leather tanning, its compounds are often released in soil, ground, and surface water, near the industrial sites. To eliminate the hazard of chromium, a better understanding of its transforming process in natural environment is of great importance.

43.3.1 THE CHROMIUM CYCLE

Figure 43.1 illustrates the chromium cycle in natural environment. Cr(VI) and Cr(III) are present with opposite characteristics in chemical and physical aspects. Cr(VI) is most mobile and toxic form of chromium. Cr(VI), as an anion, is primarily present as chromate (CrO_4^{2-}) at pH above 6.4, but in the form of bichromate ($HCrO_4^-$) at pH below 6.4 (Dhal et al., 2013). Cr(III) is relatively immobile owing to the strong affinity for negatively charged ions and colloids and the resulting insoluble compounds such as $Cr(OH)_3$ at pH 4–8 (Fendorf, 1995). Oxidized manganese, organic matter, and reduced inorganic species are ubiquitous in soil and water in partial equilibrium with atmospheric system (Stollenwerk and Grove, 1985). The reduction of Cr(VI) oxyanions to Cr(III) by organic matter or reduced inorganic species and the oxidation of Cr(III) to Cr(VI) by manganese oxides occur at the same time (Dhal et al., 2013). The oxidation process is favored by high pH conditions, whereas the reduction process is enhanced at low pH levels.

43.3.2 CHROMIUM IN SOIL

The pH, redox potential, oxygen concentration, and the presence and concentration of reducers determine the fate of chromium in soil. Cr(VI) in soil can react with non-humic organic substances (i.e., carbohydrates and proteins) or soil humic substances with Cr(III) being the end product (Dhal et al., 2013). Cr(VI), in mobile forms as $HCrO_4^-$ and CrO_4^{2-}, can be reduced by different inorganic species such as Fe(II) or S^{2-} (Bartlett and James, 1988). Acidifying conditions may either increase the Cr(VI) reduction by organic matter directly or release Fe(II) from minerals in the soil for reaction with aqueous Cr(VI) species (Eary and Rai, 1991; Arnfalk et al., 1996). Cr(III) oxidation to Cr(VI) can take place under

certain circumstances, causing serious environmental hazards (Bartlett, 1991; Milacic and Stupar, 1995; Chuan and Liu, 1996; Dong et al., 2010). This oxidation process is limited by oxygen availability, the amount of hydroquinone-reduced manganese in the soil, the concentration of water-soluble chromium, pH, initial available surface area, and the ionic strength (James, 2001; Apte et al., 2005; Dong et al., 2010). However, even with favorable pH conditions and the presence of manganese oxides, Cr(III) would not be oxidized to Cr(VI) due to the unavailability of mobile Cr(III) (Apte et al., 2005).

43.3.3 CHROMIUM IN WATER

Cr(VI) and Cr(III) are two stable chromium forms in water environment. Cr(VI) is present as chromate (CrO_4^{2-}) and dichromate ($Cr_2O_7^{2-}$) in a pH-influenced equilibrium: $2CrO_4^{2-} + 2H^+ \rightarrow Cr_2O_7^{2-} + H_2O$ (Vaiopoulou and Gikas, 2012). At lower pH levels (1–6), Cr(VI) is principally in the form of $HCrO_4^-$ with coexistence of $Cr_2O_7^{2-}$, $Cr_3O_{10}^{2-}$, and $Cr_4O_{13}^{2-}$. From pH 6 to pH 8, Cr(VI) coexists as $Cr_2O_7^{2-}$ and CrO_4^{2-}. At pH levels of above 8, CrO_4^{2-} is the primary chromium form. The various forms of Cr(VI) are visible as different colors, varying from yellow (CrO_4^{2-}) to orange ($Cr_2O_7^{2-}$) (Vaiopoulou and Gikas, 2012). As for Cr(III), it is hydrated to form free ion complexes (e.g., $Cr(H_2O)_6^{3+}$) and subsequently follows a stepwise hydrolysis to generate mononuclear hydroxyl complexes ($CrOH^{2+}$, $Cr(OH)_2^+$, $Cr(OH)_4^-$). Cr(III) is able to react with water to produce varying forms of chromium depending on the pH: $[Cr(OH)_6]^{3-}$ at extremely acidic conditions, $Cr(OH)_3$ at neutral pH, and $[Cr(H_2O)_6]^{3+}$ at extremely alkaline conditions. Among them, the insoluble form of $Cr(OH)_3$ can be converted to stable Cr(III) oxide (Cr_2O_3) by heating (Vaiopoulou and Gikas, 2012).

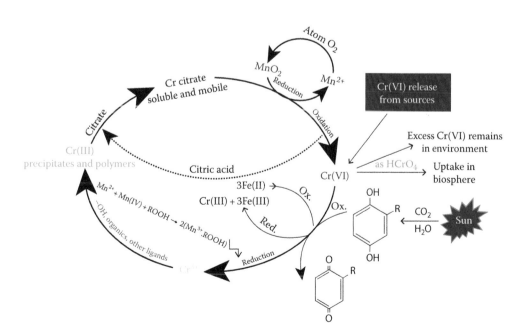

FIGURE 43.1 The chromium cycle in natural environment. (From Dhal, B. et al., *J. Hazard. Mater.*, 250–251, 272, 2013. With permission.)

43.4 MICROBIAL CHROMIUM REMOVAL

43.4.1 BACTERIAL RESISTANCE TO CHROMIUM

Many microbes develop strategies to detoxify chromium via reduction to chromium form with lesser toxicity and mobility through enzyme-mediated processes or catalyzed by the metabolic end products (i.e., HS^- and $Fe(II)$) or indirectly making complexes with metabolites (Hwang et al., 2002; Pei et al., 2009; Soni et al., 2013). Both Cr(VI)-resistant and nonresistant strains have been isolated from natural environment as well as Cr(VI)-contaminated sites (Wang and Shen, 1995; Schmieman et al., 1998). The main difference of the two groups of microorganisms lies in that the growth of the latter is inhibited as Cr(VI) increases above a certain level (Bopp and Ehrlich, 1988). Among the microorganisms with the capacity to reduce Cr(VI), the Gram-positive bacteria are reported to tolerate the toxicity of Cr(VI), while the Gram-negative bacteria are completely inhibited at high Cr(VI) concentration (Coleman, 1988). It has been reported that the Cr(VI)-contaminated ecosystem yields adapted bacteria for Cr(VI) toxicity (Das et al., 2013). However, it should be noted that not all bacteria with Cr(VI) tolerance are able to reduce Cr(VI). In other words, the resistance toward Cr(VI) and the ability for Cr(VI) reduction are not necessarily linked to each other, but rather two independent properties (Bopp and Ehrlich, 1988; Silver, 1997). As such, a full understanding of the resistant mechanism is extraordinarily significant for efficient biological removal of chromium.

In general, six bacteria-resistant mechanisms are summarized and depicted in Figure 43.2, which are reduced uptake of Cr(VI), extracellular Cr(VI) reduction, ROS detoxifying enzymes/intracellular Cr(VI) reduction, DNA repair enzymes, efflux of Cr(VI) from cell, and ROS scavenging (Thatoi et al., 2014).

Reduced uptake of Cr(VI) (Figure 43.2a): The bacterial cell with the aid of nonspecific anionic (SO_4^{2-} or PO_4^{3-}) carriers conveys the CrO_4^{2-} ion through cell membrane via sulfate transporter pathway due to the fact that the CrO_4^{2-} ion structurally resembles SO_4^{2-} (Qi et al., 2000). Ramírez-Díaz et al. (2008) observed that the sulfate transporter pathway was mutated to decrease the uptake of CrO_4^{2-} in some microorganisms growing in metal-contaminated systems. In summary, mechanism A is about bacterial protective reflection associated with decreased CrO_4^{2-} transportation through sulfate transporter pathway.

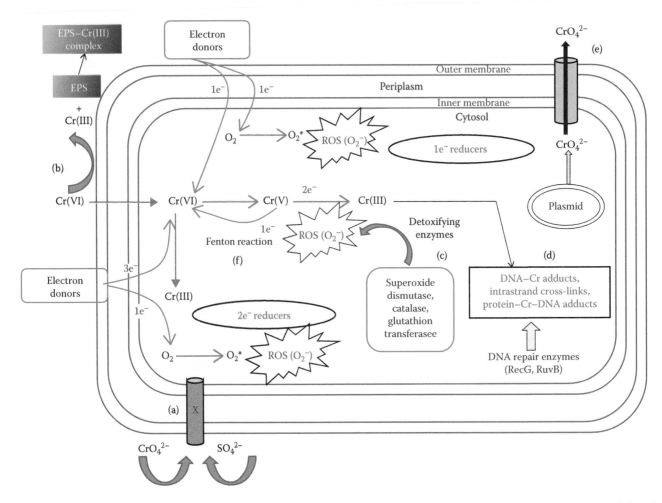

FIGURE 43.2 Bacterial Cr(VI) resistance mechanisms. (a) Reduced uptake of Cr(VI). (b) Extracellular Cr(VI) reduction. (c) ROS detoxifying enzymes/intracellular Cr(VI) reduction. (d) DNA repair enzymes. (e) Efflux of Cr(VI) from cell. (f) ROS scavenging. (From Thatoi, H. et al., *J. Environ. Manage.*, 146, 383, 2014. With permission.)

Extracellular Cr(VI) reduction (Figure 43.2b): Some bacteria have adsorptive properties associated with reactive functional groups, such as sulfhydryl groups, amine, carboxyl, phosphate, and hydroxyl on the membrane surface of the cell (Thatoi et al., 2014). The reduced Cr(III) from Cr(VI) has the potential to bind to these functional groups on cell surface and thus facilitate the removal process (Ngwenya and Chirwa, 2011). Mechanism B involves the extracellular reduction of Cr(VI), which diminishes the Cr(VI) concentration inside the cell.

ROS-detoxifying enzymes/intracellular Cr(VI) reduction (Figure 43.2c): Concomitant with entering the cell via sulfate transporter pathway, Cr(VI) is reduced to short-lived intermediates including free radicals, Cr(V)/Cr(IV), and Cr(III) as the end product during various enzymatic and nonenzymatic processes (Costa, 2003; Cheung and Gu, 2007). ROS produced during the one-electron redox cycle of Cr(V) and Cr(VI) combines DNA–protein complexes within the cell, while Cr(IV) hinders physiological functions of the cell by binding to cellular materials (Cervantes et al., 2001; Cheung and Gu, 2007). Additional Cr(VI) resistance carried out by bacterial proteins is triggered by the oxidative stress, which results from ROS (Ramírez-Díaz et al., 2008). However, the bacterial enzymes (i.e., superoxide dismutase [SOD], glutathione transferase, etc.) help to release this oxidative stress (Ackerley et al., 2004). Mechanism C is associated with the ROS-detoxifying enzymes in response to chromium stress, which takes place inside the cell.

DNA repair enzymes (Figure 43.2d): As mentioned earlier, the reduction of Cr(VI) to Cr(III) during various enzymatic and nonenzymatic processes generates ROS, which have deleterious effect on the activity of proteins and DNA (Thatoi et al., 2014). DNA repair enzymes, induced by the DNA damage, protect the bacterial cell from chromium toxicity. In *Pseudomonas aeruginosa*, the entry of Cr(VI) into bacterial cell causes increase of DNA helicases such as RecG and RuvB, components of the recombinational DNA repair system (Miranda et al., 2005). The SOS response enzymes in *Escherichia coli* serve as a bacterial shield toward the oxidative stress (Llagostera et al., 1986). Zhitkovich (2011) found that the short-lived intermediates including free radicals, Cr(V)/Cr(IV), and Cr(III) generated Cr-DNA adducts and further triggered mutations and chromosomal breaks. Mechanism D is a reflection of bacterial resistance to Cr(VI) in DNA levels.

Efflux of Cr(VI) from cell (Figure 43.2e): It has been demonstrated in several bacteria that the Cr(VI) efflux mechanism involves ChrA protein, a chromate ion transporter CHR superfamily and a hydrophobic membrane protein, encoded by plasmids pUM505 of *P. aeruginosa* and pMOL28 from *Cupriavidus metallidurans* (Cervantes et al., 1990; Nies et al., 1990; Ramírez-Díaz et al., 2008). The proton motive force (PMF), as the driving power, forces ChrA protein to pump the Cr(VI) from cytoplasm or periplasm to outside (Thatoi et al., 2014). Ramírez-Díaz et al. (2008) reported that the efflux of Cr(VI) kept the toxic ions from accumulating intracellularly. Mechanism E is an effective and widespread Cr(VI) resistance mechanism in bacteria, mediated by transporters encoded by specific plasmid-borne genes.

ROS scavenging (Figure 43.2f): After entering the cell, Cr(V), an unstable toxic intermediate, is produced by Cr(VI) reduction with NAD(P)H or organic matters being the electron donor. Cr(V) is further reduced to a relatively stable form (Cr(III)), during which a pair of electrons is transported to chromate reductases (Thatoi et al., 2014). Simultaneously, a small fraction of Cr(V) is converted back to Cr(VI), resulting in a one-electron redox cycle and generating ROS by a Fenton-like reaction. In particular, the reduction from molecular oxygen to O_2· radicals produces hydrogen peroxide (H_2O_2) via dismutation. Subsequently, the reaction between Cr(VI) and H_2O_2 forms ·OH radicals via a Fenton-like reaction (Thatoi et al., 2014). Mechanism F possesses some similarities with the Fenton reaction, during which the ·OH generation from the oxidation of Fe(II) with H_2O_2 is favored, to a large degree, by the formation of Fe(II) complexes that have vacant sites for H_2O_2 coordination (Thatoi et al., 2014).

43.4.2 Microbial Reduction Processes

Physical and chemical processes such as reduction–precipitation, ion exchange, and reverse osmosis are widely used for chromium treatment. However, the costs of design, setup, and operation are prohibitively high for large-scale treatment (Beleza et al., 2001). From bioremediation point of view, the biological reduction of Cr(VI) to Cr(III) is particularly important, which could be an alternative method for treating Cr(VI) contaminants. Cr(VI)-reducing microorganisms have been extensively studied with a variety of bacterial strains being isolated, including *E. coli*, *Pseudomonas*, *Bacillus*, *Enterobacter*, *Deinococcus*, *Shewanella*, *Agrobacterium*, *Escherichia*, *Thermus*, and other species (Thatoi et al., 2014).

43.4.2.1 Aerobic Reduction of Cr(VI)

Pseudomonas reduction of Cr(VI) has been extensively studied. The reduced uptake of Cr(VI) (mechanism A in Figure 43.2) and/or efflux of Cr(VI) from the cell (mechanism E in Figure 43.2) accounts for the resistance to Cr(VI) in *Alcaligenes eutrophus* CH34 and *P. aeruginosa* (Alvarez et al., 1999; Valls et al., 2000; Aguilera et al., 2004; Vaneechoutte et al., 2004). Besides these two *Pseudomonas* species, a close relative of *Pseudomonas synxantha* was found to be able to reduce Cr(VI) (Gopalan and Veeramani, 1994). And bacteria from other genera, such as *Bacillus* spp., *E. coli* ATCC 33456, and *Shewanella alga* BrY-MT, were also reported to have high Cr(VI) reduction potential (Shen and Wang, 1994a; Guha et al., 2001; Camargo et al., 2003; Cheung and Gu, 2007). The electron donors such as bagasse extract, fructose, glucose, and sucrose enhanced the Cr(VI) reduction by *Bacillus* sp. and *Staphylococcus capitis* with glucose offering the highest Cr(VI) removal efficiency (Kathiravan et al., 2011). Aerobic reduction of Cr(VI) by microorganisms is mostly catalyzed by reductases soluble in the cytosol. It was reported that the soluble Cr(VI) reductase was purified in *Pseudomonas* (i.e., *Pseudomonas putida* PRS2000, *P. putida*

MK1, and *Pseudomonas ambigua* G-1) (Ishibashi et al., 1990; Suzuki et al., 1992; Park et al., 2000). On the basis of the amino acid sequences of the N-terminal and internal amino acid segments of the pure enzyme, Park et al. (2002) further identified the ChrR-coding gene, *chrR*, from the genomic sequence of *P. putida* MK1 and an YieF-coding gene, *yieF*, with a high homology to *chrR* on the *E. coli*. Cheung et al. (2006) found a membrane-associated Cr(VI) reductase from the proteome of *Bacillus megaterium* TKW3 detected on a 2D electrophoresis gel.

The mechanism for aerobic Cr(VI) reduction is depicted in Figure 43.3. Under aerobic conditions, Cr(VI) is reduced to short-lived forms as Cr(V)/Cr(IV) and finally stable form as Cr(III) with NAD(P)H and electron from endogenous reserve being the electron donors during enzyme-mediated or spontaneous processes (Cheung and Gu, 2007). As mentioned earlier, ChrR, YieF, and a membrane-associated reductase may be involved in Cr(VI) reduction. However, detailed Cr(VI) reduction pathways are substantially different from one another. As for ChrR, it initially reduces Cr(VI) to Cr(V) through a one-electron shuttle and subsequently to Cr(III) via a two-electron transportation. In contrast, four electrons are translocated during Cr(VI) reduction to Cr(III), catalyzed by YieF Cr(VI) reductase. Three of the electrons are transported to Cr(VI) to generate Cr(III), while one of them passes to O_2. The membrane-associated reductase has been demonstrated on *B. megaterium* TKW3 using NADH as the electron donor (Park et al., 2002). In general, the YieF-catalyzed

Cr(VI) reduction is more effective than the ChrR-mediated one due to lower levels of ROS produced during the former process (Cheung and Gu, 2007). More effort is still in need to be dedicated to the characterization of Cr(VI) reduction by membrane-associated reductase.

43.4.2.2 Anaerobic Reduction of Cr(VI)

A variety of anaerobes are able to reduce Cr(VI) including *P. Ambigua*, *P. aeruginosa*, *Pseudomonas fluorescens*, *Pseudomonas chromatophila*, *Pseudomonas dechromaticans*, *Aeromonas*, *Achromobacter eurydice*, *Bacillus subtilis*, *Bacillus cereus*, *Desulfovibrio vulgaris*, *Desulfovibrio desulfuricans*, *E. coli*, *Enterobacter cloacae*, *Micrococcus roseus*, etc. (Lovley, 1993). The reduction of Cr(VI) by *E. cloacae* HO1, another Cr(VI)-reducing facultative anaerobe, was inhibited by oxygen and other metals due to their toxicity (Kato and Ohtake, 1991; Cheung and Gu, 2007). SRB were extensively studied for Cr(VI) reduction. Some species of SRB involve a c-type cytochrome-catalyzed Cr(VI) reduction, such as *D. vulgaris* and *Desulfomicrobium norvegicum* (Lovley, 1995; Chardin et al., 2002). The sulfide produced by SRB can also chemically reduce Cr(VI) to Cr(III). However, the release of hydrogen sulfide causes secondary environmental pollution and accumulated sulfide exerts an inhibitory effect on the SRB activity (O'Flaherty et al., 1998). IRB are able to oxidize organic matters with Fe(III) as an electron acceptor (Somasundaram et al., 2011). The generated Fe(II) is a stronger reactant for Cr(VI) reduction than sulfide from

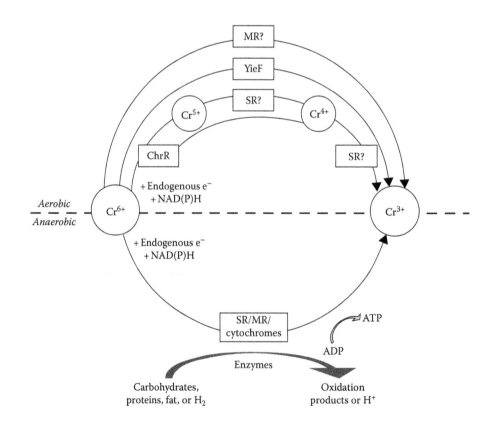

FIGURE 43.3 Possible mechanisms of enzymatic Cr(VI) reduction under aerobic and anaerobic conditions. (From Cheung, K.H. and Gu, J.D., *Int. Biodeterior. Biodegrad.*, 59, 8, 2007. With permission.)

SRB (Khérici-Bousnoubra et al., 2009). Increasing evidence showed that IRB were also capable of microbial Cr(VI) reduction with hydrogen or organic carbon being electron donors (Lovley et al., 1993; Parmar et al., 2000). Simultaneous biological reduction of Fe(III) to Fe(II) and Cr(VI) to Cr(III) by *S. alga* strain BrY (IRB) was observed by Parmar et al. (2000). Wielinga et al. (2001) demonstrated the microbial and chemical reduction of Cr(VI) by *S. alga* strain BrY and Fe(II), respectively, in a stirred-flow reactor. Some dual chemical–microbial processes of anaerobic Cr(VI) reduction have been assessed for their efficiency (Fujie et al., 1996; Tucker et al., 1998).

Both soluble and membrane-associated reductases are involved in Cr(VI) reduction by anaerobes (Cheung and Gu, 2007). In comparison to Cr(VI) reductases in aerobes, the Cr(VI) reduction by anaerobes is linked to their electron transfer systems catalyzing the electron shuttle along the respiratory chains (Cheung and Gu, 2007). Cr(VI) serves as a terminal electron acceptor in the respiratory chain in the absence of oxygen, and the Cr(VI)-reducing microorganisms can utilize various organic compounds such as amino acids, carbohydrates, and fatty acids as wells as NAD(P)H and endogenous electron reserves as electron donors for chromium reduction (Wang and Shen, 1995). Wang and Shen (1995) also reported that the reduction of Cr(VI) by anaerobes offered no energy for microbial growth, while Tebo and Obraztsova (1998) identified an SRB isolate capable of using energy produced by anaerobic Cr(VI) reduction for biomass growth.

43.4.2.3 Fungi for Cr(VI) Reduction

The microbial reduction of Cr(VI) involves some species of fungi. Chromium-resistant fungi such as *Paecilomyces lilacinus*, *Hypocrea tawa*, *Aspergillus* sp. N2, and *Penicillium* sp. N3 were isolated from soil or water environment (Barrera-Díaz et al., 2012). *Paecilomyces lilacinus* was able to reduce Cr(VI) from tannery effluent in a broad array of pH levels (5.5–8.0) (Sharma and Adholeya, 2011). Morales-Barrera et al. (2008) found that *H. tawa* could completely reduce Cr(VI) at very high initial level (4.13 mM). Under the same testing conditions (neutral pH solution containing 50 ppm Cr(VI)), *Aspergillus* sp. N2 possessed a stronger Cr(VI)-reducing capacity than *Penicillium* sp. N3, both of which combined enzyme-mediated reduction and biosorption to mycelia (Fukuda et al., 2008).

Several factors, such as light, varying types of carbon sources and biomaterials, have been reported to impact Cr(VI) removal by fungi (Barrera-Díaz et al., 2012). Lin et al. (2009) observed an 85% increase of Cr(VI)-reducing rate by *Neurospora crassa* in the presence of light comparing to that without light with the same experimental conditions applied. The light-induced enhancement of Cr(VI) reduction by fungi is possibly related to activated biomass upon light absorption. The alteration of carbon source from sodium acetate to sucrose in an aerobic batch reactor inoculated from industrial sludge resulted in a 1.3–2.1-fold increase of Cr(VI) reduction and shift of dominant microbial population from bacterial

species of *Acinetobacter lwoffii*, *Defluvibacter lusatiensis*, *Pseudoxanthomonas japonensis*, *Mesorhizium chacoense*, and *Flavobacterium suncheonense* to fungal strains (*Trichoderma viride* and *Pichia jadinii*) (Tekerlekopoulou et al., 2010). Biomaterials derived from fungi provide surface-involved pathway for Cr(VI) scavenging. Hsu et al. (2010) revealed that 53.8%–59.5% of Cr(VI) reduction was contributed by the dissolved organic compounds (peptides, polysaccharides, and glycoproteins) from *N. crassa*.

43.4.2.4 Yeast and Algae for Cr(VI) Reduction

Some chromate-resistant species of yeast and algae, which are able to reduce toxic Cr(VI) into insoluble Cr(III), are seen as promising candidates for alleviating Cr(VI) contamination. *Cyberlindnera jadinii* M9 and *Wickerhamomyces anomalus* M10 were isolated and shown with the capacity for complete Cr(VI) removal during 48 h of cultivation (Fernández et al., 2013). The selected yeast *Saccharomyces cerevisiae* served as biosorbent to adsorb Cr(VI) in aqueous solution, and the maximum biosorption capacity was determined by the Langmuir isotherm model (Chen and Wang, 2007). Moreover, marine algae can also serve as electron donors for redox reactions, which take place on the surface of the biomass, involving the Cr(VI) reduction to Cr(III) and further binding of Cr(III) to the negatively charged groups (Dittert et al., 2014). de Souza et al. (2016) reported that the brown macroalgae *Sargassum cymosum* was able to reduce Cr(VI) to Cr(III) via biomass oxidation, which was highly dependent on pH level, followed by Cr(III) sequestration from aqueous solutions. Three millimoles of Cr(VI) was reduced by 1 g of biomass. And Cr(III) removal was directly linked to weak acidic carboxylic groups at the biomass surface.

43.4.3 Bioelectrochemical Reduction of Cr(VI)

The bioelectrochemical system (BES) has been enthusiastically pursued for energy recovery and valuable bioproduct production along with contaminant removal (Nancharaiah et al., 2015). Recently, BES provides a novel platform for the removal of metal ions in aqueous solutions from metallurgical wastes, process streams, and wastewaters. The BES systems comprise a cathode, an anode, and a separator membrane in dual/single chamber. The electrons generated by microbial reactions (i.e., oxidation of organic matters) at the anode are transferred to the cathode, where metal ion reduction occurs. The reduced metals are either precipitated/deposited on the cathode or remain soluble in the solution (Nancharaiah et al., 2015).

The reductive precipitation of Cr(VI) has been successfully demonstrated in microbial fuel cells (MFCs). Eastmond et al. (2008) revealed that chromium toxicity affected the metabolism and viability of bacterial cell. Cr(VI) serves to be an efficient terminal electron acceptor due to the fact that the redox potential of half-cell Cr(VI) reduction is +1.33 V (versus the Standard Hydrogen Electrode [SHE]), which is much higher than O_2 (Nancharaiah et al., 2015). Wang et al. (2008) observed a higher Cr(VI) removal efficiency at lower pH levels owing to the pH dependency of the Cr(VI) reduction reaction. Simultaneous removal of Cr(VI) and vanadium (V) at the cathode was reported in a dual-chamber MFC with acetate being the electron donor (Zhang et al., 2012). Li et al. (2009) found that modified cathode with rutile (natural form of TiO$_2$) coating in an MFC system led to increased Cr(VI) reduction due to light-induced photocatalysis at the cathode. The efficiency of Cr(VI) was improved using Cr(VI)-reducing biofilms at the cathode (biocathodes), which received electrons from the poised cathode and catalyzed the reduction of Cr(VI) to Cr(III) (Xafenias et al., 2013). Increased Cr(VI) reduction was also observed at the cathode chamber of MFC inoculated with a denitrifying mixed culture (Tandukar et al., 2009). A simple acclimatization method for the Cr(VI) reduction at a biocathode by first enriching an exoelectrogenic biofilm on anode, followed by direct inversion of the anode to function as the biocathode, significantly enhanced the Cr(VI) reduction efficiency in MFC system owing to the higher microbial density and less resistive Cr(III) precipitates on the cathode comparing to common biocathode acclimatization method (Wu et al., 2015). Cr(VI) removal in BES system is a complex process and may involve adsorption, migration of metal ions from the cathode to the anode chamber, and chemical precipitation.

43.4.4 BIOSORPTION OF CR(VI)

Biosorption of chromium from aqueous solutions is a newly developed and promising technique. Biomass materials such as cells of bacteria, fungi, and microalgae, plant tissues, seaweeds, and low-cost agricultural wastes are able to bind metal cations and anions via physical adsorption (van der Waals forces, ion-exchange reaction electrostatic interactions, surface precipitation, complexation, etc.) (Kikuchi and Tanaka, 2012). The metal ion biosorption is mainly attributed to the surface functional groups on microbial cell walls including amino, carboxyl, hydroxyl, phosphate, sulfonate, etc. (Davis et al., 2003; Ahluwalia and Goyal, 2007). In addition, extracellular polymeric substances (EPS) from microorganisms (i.e., glucoprotein, polysaccharides, lipopolysaccharides, soluble peptides, etc.) with anionic functional groups have the ability to adsorb metal cations (Wang and Chen, 2006; Kikuchi and Tanaka, 2012). The metal biosorption may be affected by biomass species and favored by pretreatment processes such as alkali treatment or crushing, leading to an increased availability of surface area and increased metal-binding sites (i.e., intracellular components) by destroying cell wall and membrane (López Errasquín and Vázquez, 2003; Wang and Chen, 2006).

Cr(VI) reduction by protonated biomass of *Ecklonia* sp. and fungal biomass of *Aspergillus niger* under low pH (2.0) have been demonstrated (Park et al., 2005a,b). Metal biosorption is a complicated process involving surface adsorption, adsorption–complexation, complexation, chemisorption, microprecipitation, ion exchange, and heavy metal hydroxide condensation onto surface of biomass (Owlad et al., 2009). Cr(VI) may also be reduced to Cr(III)

during biosorption via two pathways: (1) direct reduction of Cr(VI) by functional groups of the biomass with lower redox potential and (2) binding of anionic Cr(VI) ion to the positively charged surface functional groups on the biomass such as carboxyl groups, protonated amine, etc., followed by reduction of Cr(VI) to Cr(III) by the adjacent electron donor groups (Nancharaiah et al., 2015). The reduced Cr(III) ion may bind to the adjacent groups by complexation on surface, or released into liquid phase owing to electronic repulsion against the positively charged surface functional groups (Park et al., 2005a,b). Chatterjee et al. (2011) observed a maximum biosorption of Cr(VI) at the initial Cr(VI) concentration of 30 mg/L and pH of 8.0 with carboxyl and amino groups on the bacterial surface binding chromium ions.

Other microbial Cr(VI) removing processes include bioaccumulation and constructed wetland. Bioaccumulation is an integrated Cr(VI) uptake process, where Cr(VI) ions first bind to the cell walls or EPS through physicochemical adsorption or biosorption, subsequently translocated across the membrane and is finally combined with intracellular compounds (Kikuchi and Tanaka, 2012). The Cr(VI) removal in constructed wetlands takes place at the root area, involving complex interactions between plants, microorganisms, and contaminant as well as combination of biological and physicochemical processes such as sedimentation, binding to porous media, precipitation as insoluble forms, and plant uptake (Dotro et al., 2012; Sultana et al., 2014).

43.5 FACTORS AFFECTING BIOLOGICAL CR(VI) REMOVAL

A wide array of factors are influencing microbial Cr(VI) removal, including pH, temperature, Cr(VI) concentration, electron donor, oxyanions, heavy metals, and others as depicted in Figure 43.4.

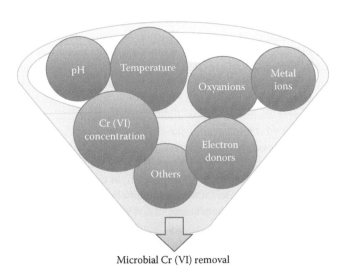

FIGURE 43.4 The affecting factors for microbial Cr(VI) removal.

43.5.1 pH

As mentioned earlier, varying pH levels may alter the chemical speciation of chromium in aqueous environment. As for Cr(VI), dichromate ($Cr_2O_7^{2-}$), hydrochromate ($HCrO_4^-$), and chromate (CrO_4^{2-}) are the dominant chromium forms depending on the pH level. Cr(III) is in the form of hydrated trivalent chromium ($Cr(H_2O)_6^{3+}$) and chromium hydroxide complexes, $Cr(OH)(H_2O)_5^{2+}$ or $Cr(OH)_2(H_2O)^{4+}$ (Narayani and Shetty, 2013). Cr(VI) reduction is a proton consumption process, as shown in Equations 43.1 and 43.2 under aerobic and anaerobic conditions, respectively, with cultures of *E. coli* ATCC 33456 (Chirwa and Wang, 2000):

$$C_6H_6O + 9\frac{1}{3}CrO_4^{2-} + 40\frac{2}{3}H^+ \rightarrow 9\frac{1}{3}Cr^{3+}$$

$$+ 6HCO_3^- + 20\frac{1}{3}H_2O \qquad (43.1)$$

$$C_6H_6O + 2CrO_4^{2-} + 8H^+ \rightarrow 2Cr^{3+} + 2HCO_3^-$$

$$+ 3H_2O + HOOC-(CH_2)_2-COOH \qquad (43.2)$$

Therefore, from the stoichiometric point of view, Cr(VI) production is favored under low pH conditions. pH can also affect the activity of Cr(VI)-reducing microorganisms. A variety of studies have demonstrated that pH plays an important role in the microbial Cr(VI) reduction. The optimum pH range for Cr(VI) reduction by *Bacillus* sp. and *P. fluorescens* is between 7.0 and 8.0 and an extremely low or high pH would result in inhibited Cr(VI) reduction as well as bacterial growth (Parameswari et al., 2009). Similar observations have been obtained in Cr(VI) reduction by *Enterobacter* strain (optimum pH: 6.5–8.5) (Wang et al., 1990), by *E. coli* ATCC 33456 (optimum pH around 7.0) (Shen and Wang, 1994b), by *Agrobacterium radiobacter* EPS-916 (optimum pH: 7.0–7.5) (Bhide et al., 1996), and by *Streptomyces* sp. MS-2 (optimum pH around 7.0) (Mabrouk, 2008). *Leucobacter* sp. reduced Cr(VI) using both NADH and NADPH as the electron donors and exhibited the optimal activity at pH 5.5 (Sarangi and Krishnan, 2015). The different pH levels for Cr(VI) reduction by various cultures indicate that pH control is of great significance for Cr(VI)-contaminant remediation. The effect of pH on microbial Cr(VI) reduction and on microbial growth may not be consistent. Zhang and Li (2011) observed optimum growth of *Serratia* sp. but lowest Cr(VI) removal efficiency at the same pH levels of 5.0. Okeke (2008) reported that the growth of *Exiguobacterium* GS1 increased upon the increase of pH in the range from 6.0 to 8.0, while the removal rate of Cr(VI) showed no significant changes. The effect mechanism of pH on microbial reduction of Cr(VI) is possibly linked to the altered enzyme's activity or protein's conformation caused by pH variations (Narayani and Shetty, 2013).

43.5.2 Electron Donors

Cr(VI)-reducing microorganisms are able to use a wide array of electron donors for Cr(VI) reduction. Acetate, glucose, glycerol, glycine, and propionate were efficiently utilized by *E. coli* ATCC 33456 (Shen and Wang, 1994b), while acetate, lactate, formate, and pyruvate favored Cr(VI) removal by *Pannonibacter phragmitetus* (Xu et al., 2011). Besides, sulfur and organic matter (lactate) may serve as electron donors for simultaneous removal of Cr(VI) and nitrate in a denitrifying biofilm reactor under mixotrophic conditions (Sahinkaya et al., 2011). However, different types of electron donors could result in different Cr(VI) reduction rates. Microbes tend to utilize amino acids, carbohydrates, and fatty acids as electron donors (Wang and Shen, 1995). Lactate and glycerol could significantly enhance Cr(VI) reduction by *Pseudomonas* sp. (McLean and Beveridge, 2001) and by *Vibrio fischeri* (Fulladosa et al., 2006), respectively. Both Komori et al. (1989) and Wani et al. (2007) reported that glucose was a more efficient electron donor for microbial Cr(VI) reduction than acetate. Similarly, optimum Cr(VI) reduction was observed with glucose being the electron donor in *Bacillus* sp. (Kathiravan et al., 2010), *Bacillus coagulans* (Philip et al., 1998), and *E. coli* (Bae et al., 2000). It was hypothesized that glucose was involved in tricarboxylic acid (TCA) cycle via direct transformation to pyruvate through glycolysis, while other electron donors such as sucrose and lactose were disaccharide molecules, which were first converted into monosaccharides and then catabolized. In contrast, the mixture of acetate, ethanol, malate, succinate, glycerol, and amino acid served as a more effective electron donor than the mixture of glucose, citrate, pyruvate, and lactate by *E. cloacae* HO1 (Ohtake et al., 1990). In two sulfate-reducing fluidized bed reactors, the removal efficiency of Cr(VI) with ethanol as the electron donor was higher than that with glucose as the electron donor (Cirik et al., 2013). In addition, microbial Cr(VI) reduction could take place even in the absence of external electron donors. Viamajala et al. (2007) reported that *Cellulomonas* sp. ES6 and *Acinetobacter haemolyticus* could utilize endogenous electron donor to reduce Cr(VI). A low-cost energy source for microbial Cr(VI) removal is of great significance to further improve the possibility of application of this bioprocess. In study by Michailides et al. (2015), molasses, a by-product of sugar processing, was used as the carbon source for Cr(VI) reduction by indigenous microorganisms from industrial sludge.

43.5.3 Temperature

Since the temperature can exert a significant effect on bacterial growth and function of the enzymes, including reductases for Cr(VI) reduction, many studies have been conducted to investigate the temperature impact on microbial Cr(VI) reduction. In essence, a variety of optimum temperatures for Cr(VI) reduction by various microorganisms were reported: 28°C for *Streptomyces griseus* (Laxman and More, 2002), 30°C for *E. cloacae* (Wang et al., 1989) and *Bacillus* sp. (Narayani and Shetty, 2013), and 37°C for *Serratia* sp. *Losi* (Zhang and Li, 2011). As for low temperature, the transportation system of the cell membrane is hindered due to a decrease of fluidity, resulting in a decreasing concentration of intracellular growth substrates (Narayani and Shetty, 2013). On the other hand, the high temperature causes irreversible denaturation of the protein, leading to damaging membrane structure, functional loss of Cr(VI) reductases, and alteration of ribosome conformation. However, Horton et al. (2006) demonstrated that microbial reduction of Cr(VI) by *Arthrobacter aurescens* at 10°C was feasible and microbial reduction of Cr(VI) was observed at 65°C using a thermophilic bacterium (Narayani and Shetty, 2013) and at 100°C using a hyperthermophilic archaea (Kashefi et al., 2002). Temperature was also shown to have a significant effect on the fractionation of chromium isotopes during Cr(VI) reduction by *Bacillus* sp. under aerobic condition (Xu et al., 2015).

43.5.4   CR(VI) CONCENTRATION

As Cr(VI) concentration is below inhibitory limit, the Cr(VI) reduction increases with the increase of initial Cr(VI) concentration (Shen and Wang, 1994b). This observation is attributed to sequential occurrences. First, the increased Cr(VI) concentration leads to increased metal ions. Subsequently, the corresponding increase of collision rate of metal ions onto the active sites on the cell surface causes increased probability of active collisions. As Cr(VI) takes up more enzyme active sites, the enzyme-mediated reaction rates increase with the increase of active collisions. As a result, the activity of Cr(VI) reductases keeps arising till the Cr(VI) ion saturates the enzyme (Narayani and Shetty, 2013). High initial Cr(VI) (above the inhibitory levels) negatively affects both microbial Cr(VI) removal and bacterial growth, accompanied by extended growth phase and morphological alteration owing to Cr(VI) toxicity and stress (Ackerley et al., 2006; Chourey et al., 2006; Desai et al., 2008; Sau et al., 2008). Wang and Shen (1997) reported that as the initial Cr(VI) concentration exceeded a certain limiting level, the decreasing reduction rate upon further increase of metal ion concentration was due to lower biomass growth (irreversible inactivation of the batch culture) and enzyme production rate. Mary Mangaiyarkarasi et al. (2011) investigated microbial Cr(VI) reduction by *Bacillus* sp. with varying initial Cr(VI) concentrations at 50, 100, and 200 mg/L and found Cr(VI) of 100 mg/L yielded the highest reduction rate. Complete Cr(VI) removal by *Bacillus* sp. was obtained at Cr(VI) levels of 10, 20, and 40 mg/L, while residual Cr(VI) was observed at initial concentration of 80 mg/L (Liu et al., 2006). The removal efficiency of COD, sulfate, color, and Cr(VI) in ethanol-fed fluidized bed reactors treating textile wastewater decreased substantially with further increase of Cr(VI) in the influent from 45 mg/L (Cirik et al., 2013). Cr(VI) concentration exceeding the toxicity limit would completely inhibit bacterial growth and Cr(VI) reduction. Completely inhibited Cr(VI) reduction was observed at initial Cr(VI) concentration of 25 mg/L for *P. fluorescens* LB300 (Apel, 1991). Even for the same cultures, the inhibitory concentration of Cr(VI) may be different. The Cr(VI) inhibitory impact was minimized (inhibitory constant value K_I of 148.5 mg/L) in an attached-growth reactor, compared to a suspended-growth reactor, where inhibitory constant value was 8.219 mg/L (Tekerlekopoulou et al., 2013).

43.5.5   INTERFERING METAL IONS

It is essential to assess the combined effect of Cr(VI) together with other metal ions on microorganism as they usually coexist in natural environment, and metallic industry effluent contains multiple metal contaminants (Narayani and Shetty, 2013). It has been found that some Cr(VI)-reducing microbes are able to remove multiple toxic or nontoxic metal ions at the same time (Lovley, 1995). However, the effect of metal ions on microbial Cr(VI) reduction is complex and dependent on their concentrations and species of microorganisms. 30% of Cr(VI) and 65% of Pb in mixed metal solutions were reduced by *P. aeruginosa* BC15 within 48 h (Raja et al., 2006). A pilot-scale anaerobic bioreactor with high levels of microbial sulfate reduction was known to be capable of enzyme-mediated reduction of As(V) to As(III) and chemical reduction of Cr(VI) to Cr(III) by sulfide, followed by precipitation of arsenic sulfides and chromium hydroxide from a metal- and acid-contaminated waste stream (Cohen and Ozawa, 2013). Wielinga et al. (2001) found simultaneous reduction of Fe(III) to Fe(II) and Cr(VI) to Cr(III) by *S. alga* strain BrY. Nevertheless, the presence of other metal ions may enhance or hinder Cr(VI) reduction. Cd and Ni promoted Cr(VI) reduction initially and exerted no significant effect afterward using *Streptomyces thermocarboxydus* (Desjardin et al., 2003). Bae et al. (2000) reported that the presence of Mg induced Cr(VI) reduction by *E. coli* ATCC 33456, whereas the presence of Hg resulted in complete inhibition. The Cr(VI) reduction by *Brevibacterium* strain was enhanced by the presence of Zn as 200 mg/L $ZnSO_4$ and inhibited by Co as 50 mg/L $CoCl_2$ (Faisal and Hasnain, 2004). In contrast, several studies have opposite observations. Sau et al. (2008) found that in the presence of Cd and Zn, Cr(VI) reduction rate by *Bacillus firmus* dropped remarkably. Similarly, 16%–33% decrease in Cr(VI) reduction was observed in *E. coli* in presence of 200 mg/L Zn but without significant inhibitory effect at lower concentration (Shen and Wang, 1994a). Ohtake et al. (1990) investigated Cr(VI) reduction by *E. cloacae* HO1 and found a complete inhibition at Zn level of 32 mg/L. Other metal ions that could exert inhibitory effects on microbial Cr(VI) reduction included As, Cu, Pb, and Ag (Ishibashi et al., 1990; McLean and Beveridge, 2001; Sultan and Hasnain, 2007; Singh et al., 2013). However, both Sultan and Hasnain (2007) and Xu et al. (2009) found that Cu enhanced Cr(VI) reduction by *Ochrobactrum intermedium* SDCr-5 and *Pseudomonas aeruginosa*, respectively. Among the tested metal ions (Hg(II), Fe(II), Ba(II), Li(I), Ag(I), Mn(II), etc.), only Cu(II) had a stimulating effect on the simultaneous reduction of the dye (100 mg/L) and Cr(VI) (2 mg/L) by *P. putida* KI in a mineral salts medium (Mahmood et al., 2015). It is critical to consider the presence of other heavy metal ions in natural environment when assessing Cr(VI) bioreduction and designing the bioreactors for the purpose, as the reduction rate is altered by their presence (Hora and Shetty, 2014).

43.5.6 Oxyanions

Since oxyanions such as SO_4^{2-}, SO_3^{2-}, PO_4^{3-}, and NO_3^- compete for electrons with Cr(VI), they may play important roles in microbial Cr(VI) reduction. Under aerobic conditions, it has been reported that the addition of SO_4^{2-} and NO_3^- (from 40 up to 1000 mg/L) exerts no significant effect on Cr(VI) reduction by *B. coagulans* (Philip et al., 1998), *Bacillus* sp. (Wang and Xiao, 1995; Liu et al., 2006), and *B. subtilis* (Garbisu et al., 1998). In addition, 10 mM of SO_4^{2-}, SO_3^{2-}, PO_4^{3-}, and NO_3^- have no influence on Cr(VI) reduction by *A. haemolyticus* (Zakaria et al., 2007). Cr(VI) possesses stronger affinity for electrons than these oxyanions under aerobic conditions. In comparison, under anaerobic conditions, Cr(VI) reduction would be inhibited by the presence of some of the oxyanions. SO_4^{2-} at higher level (above 0.5 mM) exerted an inhibitory effect on Cr(VI) reduction by *Comamonas testosteroni*, while the inhibition was mitigated at lower concentration (0.1 mM) (Cooke et al., 1995). Cr(VI) reduction rate by *E. coli* dropped in the presence of SO_4^{2-} (8000 mg/L) comparing to that without SO_4^{2-}; however, NO_3^- (8000 mg/L) did not hinder the Cr(VI) reduction (Shen and Wang, 1994b). 25–100 µM $ZnSO_4$ and 5000 µM $NaNO_3$ inhibited Cr(VI) reduction by *E. cloacae* HO1 under anaerobic conditions, while 50 µM Na_2SO_3 elicited no inhibitory effect on Cr(VI) reduction (Komori et al., 1989). This observation is in line with the findings by Xu et al. (2011) using *P. phragmitetus*.

43.5.7 Other Factors

Biomass concentration: It is known that microbial removal of Cr(VI) is positively correlated to the biomass concentration under both aerobic and anaerobic conditions (He et al., 2011; Ibrahim et al., 2013). Inoculum size is also shown to affect microbial Cr(VI) removal (Pal and Paul, 2004; Sultan and Hasnain, 2007). As mentioned earlier, bioaccumulation of Cr(VI) involves physical adsorption or ion exchange at the cell surface, followed by ion transportation into bacterial cell, which are directly linked to the total amount of biomass.

Oxidation and reduction potential: Variation in oxidation and reduction potential conditions can be caused by varying biochemical reactions and metabolites produced in medium during bacterial growth. The redox potential can indirectly affect microbial Cr(VI) reduction through influencing metal bioavailability. Dmitrenko et al. (2006) revealed that microbes tend to utilize electron acceptors with higher standard redox potential in the medium. Both aerobic and anaerobic Cr(VI) reduction by microbes relied on the oxidation–reduction potential of the environment (LLovera et al., 1993). Contradictory observations were obtained showing no effect of redox potential on Cr(VI) reduction (Shen and Wang, 1994a; Wang and Xiao, 1995).

Inhibitory compounds: Salt ions, which are widespread in industrial wastewaters, have toxic impacts on the activity of microbes (Stasinakis et al., 2003; Narayani and Shetty, 2013). Thus, the bioreduction of Cr(VI) may be affected indirectly. The microorganisms show different resistance to high salinity in the wastewater. Many Cr(VI)-resistant bacteria are reported to be highly salt tolerant. Biological Cr(VI) reduction decreased with the increase of salt concentration (Cetin et al., 2008). Maximum reduction rate of Cr(VI) was obtained at 0%–1.5% salt concentration for *Bacillus* sp. KSUCr5 (Ibrahim et al., 2013) and at 2% salt concentration for *Burkholderia cepacia* (Wani et al., 2007). Above 3% salt concentration, the Cr(VI) reduction by *Bacillus* sp. was significantly impeded (Okeke et al., 2008). High salinity causes cell lysis. On one hand, cell lysis decreases the cell density. On the other hand, it exposes the membrane enzyme for Cr(VI) reduction and erases the diffusion barriers. Metabolic inhibitors may coexist with Cr(VI) in plating industries, including azide, cyanide, rotenone, antimycin, succinate, etc. Cr(VI) reduction by varying bacterial species was reported to be largely inhibited by these metabolic inhibitors (Garbisu et al., 1998; Wani et al., 2007; Zakaria et al., 2007; Al Hasin et al., 2010; Alam and Ahmad, 2012). However, Wani et al. (2007) found that 2,4-dinitrophenol (DNP) enhanced the removal efficiency of Cr(VI) possibly due to the accelerated respiratory chain–linked electron transportation.

43.6 MODELING OF MICROBIAL CHROMIUM REMOVAL

Mathematical modeling not only gives insights into underlying mechanisms of microbial Cr(VI) removal but also serves as a powerful tool to assist the design and operation of pilot-scale or full-scale reactors for the treatment of Cr(VI) contaminants.

Several kinetic models have been developed for the description of Cr(VI) reduction in different bioreactors. Gagrai et al. (2013) reported a second-order kinetic model for the reduction of Cr(VI) into Cr(III) by *Spirulina* sp., and higher overall reduction rate constant, kt, was observed at low pH (0.5), high temperature (45°C), and low concentration of background anions (Cl⁻, SO_4^{2-} and NO_3^-) (<0.235 mM). A mechanistic mathematical model was proposed for Cr(VI) removal inside porous aquifer media columns (Mtimunye and Chirwa, 2014). The model results showed that Cr(VI) threshold inhibition concentration was 50 mg/L and post-barrier infusion of biomass into the clean aquifer downstream of the barrier could be limited by depletion of the substrates within the barrier. Lin et al. (2015) proposed a mathematical model to describe the Cr(VI) reduction with acetate biodegradation by *E. coli* in a fixed biofilm reactor, and the removal efficiency for both Cr(VI) and acetate was over 90% under the simulated condition.

Since multiple contaminants may coexist in natural environment or contaminated industrial sites, technologies aiming at simultaneous removal have been experimentally and mathematically evaluated. Simultaneous reduction of nitrate and Cr(VI) in a biofilm reactor was described by a 1D biofilm model under autotrophic, heterotrophic, and mixotrophic denitrifying conditions with sulfur and COD as the electron donors (Peng et al., 2016). The model successfully captured substrate conversion profiles (nitrate, chromate, methanol, and sulfate dynamics) and predicted the complex interactions between the microbes including sulfur-oxidizing bacteria, denitrifiers, and chromate reducers during this process. Hydraulic retention time was shown to play an important role in affecting the microbial distribution and system performance. Contreras et al. (2011) developed a mathematical model for the biological reduction of Cr(VI), carbon, and nitrogen, comprising nongrowing aerobic heterotrophic cells, growing aerobic heterotrophic cells with chromate reductase activity, and growing aerobic heterotrophic cells that have lost the chromate reductase activity. Peng et al. (2015) further developed a mathematical model describing simultaneous reduction of Fe(III) and Cr(VI) in an IRB-based stirred-flow reactor, gave useful insight into the chemical and biological Cr(VI) reduction pathways, and assessed the effect of several parameters (hydraulic retention time and electron donor concentrations) on the contribution of each of the two pathways to Cr(VI) removal.

Biosorption is another effective and widespread method for Cr(VI) remediation. The biosorption process can be described by Langmuir isotherm (Kikuchi and Tanaka, 2012) as

$$\frac{r_{eq}}{r_{max}} = \frac{bC_{eq}}{1+C_{eq}} \qquad (43.3)$$

where

r_{eq} is the amount of metal adsorbed on biosorbent at equilibrium state

r_{max} is the maximum adsorption capacity

b is an affinity constant, related to the affinity of adsorbent for the metal

C_{eq} is the concentration of metal in solution at equilibrium state

The selected yeast *S. cerevisiae* served as biosorbent to adsorb Cr(VI) in aqueous solution, and the maximum biosorption capacity was determined by the Langmuir isotherm model (Equation 43.3) (Chen and Wang, 2007). Besides Langmuir model, Freundlich, D-R, and Temkin isotherm models could also be used for the process of Cr(VI) biosorption (Srivastava et al., 2015).

Simultaneous processes may take place during Cr(VI) removal, including sorption/reduction of Cr(VI) as well as production and binding of reduced Cr(III). A model including all these physical, chemical, and biological processes and the effect of the presence of Cu(II) in these processes has been developed and validated by Liu et al. (2016). To describe Cr(VI) removal by *S. alga Simidu* (BrY-MT) ATCC 55627, Guha (2004) established a model considering dual-Monod

kinetics with Cr(VI) as the electron acceptor and lactate as the electron donor; first-order kinetic adsorption of Cr(III) and lactate; advective–dispersive transport of Cr(VI), Cr(III), lactate, and protein; and conversion of solid-phase β-MnO$_2$ to solid-phase MnOOH due to oxidation of Cr(III). Mohamed and Hatfield (2005) reported similar kinetics involving nonlinear physical transportation, sorption, and microbial reduction processes of heavy metal transformation. The model for the description of Cr(VI) reduction when fully developed will be used in the assessment of proposed *in situ* biological treatment before implementation in full-scale trials.

ACKNOWLEDGMENTS

This work was supported by the Natural Science Foundation of China (51578391). Dr. Bing-Jie Ni acknowledges the support of ARC Discovery Early Career Researcher Award (DE130100451).

REFERENCES

Ackerley, D. F., Barak, Y., Lynch, S. V., Curtin, J., and Matin, A. 2006. Effect of chromate stress on *Escherichia coli* K-12. *Journal of Bacteriology* 188: 3371–3381.

Ackerley, D. F., Gonzalez, C. F., Keyhan, M., Blake, R., and Matin, A. 2004. Mechanism of chromate reduction by the *Escherichia coli* protein, NfsA, and the role of different chromate reductases in minimizing oxidative stress during chromate reduction. *Environmental Microbiology* 6: 851–860.

Aguilera, S., Aguilar, M. E., Chávez, M. P., López-Meza, J. E., Pedraza-Reyes, M., Campos-García, J., and Cervantes, C. 2004. Essential residues in the chromate transporter ChrA of *Pseudomonas aeruginosa*. *FEMS Microbiology Letters* 232: 107–112.

Ahluwalia, S. S. and Goyal, D. 2007. Microbial and plant derived biomass for removal of heavy metals from wastewater. *Bioresource Technology* 98: 2243–2257.

Alam, M. Z. and Ahmad, S. 2012. Toxic chromate reduction by resistant and sensitive bacteria isolated from tannery effluent contaminated soil. *Annals of Microbiology* 62: 113–121.

Al Hasin, A., Gurman, S. J., Murphy, L. M., Perry, A., Smith, T. J., and Gardiner, P. H. 2010. Remediation of chromium(VI) by a methane-oxidizing bacterium. *Environmental Science & Technology* 44: 400–405.

Alvarez, A. H., Moreno-Sánchez, R., and Cervantes, C. 1999. Chromate efflux by means of the ChrA chromate resistance protein from *Pseudomonas aeruginosa*. *Journal of Bacteriology* 181: 7398–7400.

Apel, W. 1991. Bioremediation of hexavalent chromium by bacterial reduction. In: Smith, R. W. and Misra, M. (Eds.), *Mineral Bioprocessing*, TMS, Baltimore, MD. pp. 376–387.

Apte, A. D., Verma, S., Tare, V., and Bose, P. 2005. Oxidation of Cr(III) in tannery sludge to Cr(VI): Field observations and theoretical assessment. *Journal of Hazardous Materials* 121: 215–222.

Arnfalk, P., Wasay, S. A., and Tokunaga, S. 1996. A comparative study of Cd, Cr(III), Cr(VI), Hg, and Pb uptake by minerals and soil materials. *Water, Air, and Soil Pollution* 87: 131–148.

Bae, W. C., Kang, T. G., Kang, I. K., Won, Y. J., and Jeong, B. C. 2000. Reduction of hexavalent chromium by *Escherichia coli* ATCC 33456 in batch and continuous cultures. *The Journal of Microbiology* 38: 36–39.

Bankar, A. V., Kumar, A. R., and Zinjarde, S. S. 2009. Removal of chromium (VI) ions from aqueous solution by adsorption onto two marine isolates of *Yarrowia lipolytica*. *Journal of Hazardous Materials* 170: 487–494.

Barrera-Díaz, C. E., Lugo-Lugo, V., and Bilyeu, B. 2012. A review of chemical, electrochemical and biological methods for aqueous Cr(VI) reduction. *Journal of Hazardous Materials* 223–224: 1–12.

Bartlett, R. J. 1991. Chromium cycling in soils and water: Links, gaps, and methods. *Environmental Health Perspectives* 92: 17–24.

Bartlett, R. J. and James, B. R. 1988. Mobility and bioavailability of chromium in soils. *Chromium in the Natural and Human Environments* 20: 571.

Beleza, V. M., Boaventura, R. A., and Almeida, M. F. 2001. Kinetics of chromium removal from spent tanning liquors using acetylene production sludge. *Environmental Science & Technology* 35: 4379–4383.

Bhide, J. V., Dhakephalkar, P. K., and Paknikar, K. M. 1996. Microbiological process for the removal of Cr(VI) from chromate-bearing cooling tower effluent. *Biotechnology Letters* 18: 667–672.

Bopp, L. H. and Ehrlich, H. L. 1988. Chromate resistance and reduction in *Pseudomonas fluorescens* strain LB300. *Archives of Microbiology* 150: 426–431.

Camargo, F. A., Okeke, B. C., Bento, F. M., and Frankenberger, W. T. 2003. In vitro reduction of hexavalent chromium by a cell-free extract of *Bacillus* sp. ES 29 stimulated by Cu^{2+}. *Applied Microbiology and Biotechnology* 62: 569–573.

Cervantes, C., Campos-García, J., Devars, S., Gutiérrez-Corona, F., Loza-Tavera, H., Torres-Guzmán, J. C., and Moreno-Sánchez, R. 2001. Interactions of chromium with microorganisms and plants. *FEMS Microbiology Reviews* 25: 335–347.

Cervantes, C., Ohtake, H., Chu, L., Misra, T. K., and Silver, S. 1990. Cloning, nucleotide sequence, and expression of the chromate resistance determinant of *Pseudomonas aeruginosa* plasmid pUM505. *Journal of Bacteriology* 172: 287–291.

Cetin, D., Dönmez, S., and Dönmez, G. 2008. The treatment of textile wastewater including chromium(VI) and reactive dye by sulfate-reducing bacterial enrichment. *Journal of Environmental Management* 88: 76–82.

Chardin, B., Dolla, A., Chaspoul, F., Fardeau, M. L., Gallice, P., and Bruschi, M. 2002. Bioremediation of chromate: Thermodynamic analysis of the effects of Cr(VI) on sulfate-reducing bacteria. *Applied Microbiology and Biotechnology* 60: 352–360.

Chatterjee, S., Ghosh, I., and Mukherjea, K. K. 2011. Uptake and removal of toxic Cr(VI) by *Pseudomonas aeruginosa*: Physico-chemical and biological evaluation. *Current Science (Bangalore)* 101: 645–652.

Chen, C. and Wang, J. 2007. Influence of metal ionic characteristics on their biosorption capacity by *Saccharomyces cerevisiae*. *Applied Microbiology and Biotechnology* 74: 911–917.

Cheung, K. H. and Gu, J. D. 2007. Mechanism of hexavalent chromium detoxification by microorganisms and bioremediation application potential: A review. *International Biodeterioration & Biodegradation* 59: 8–15.

Cheung, K. H., Lai, H. Y., and GU, J. D. 2006. Membrane-associated hexavalent chromium reductase of *Bacillus megaterium TKW3* with induced expression. *Journal of Microbiology and Biotechnology* 16: 855–862.

Chirwa, E. N. and Wang, Y. T. 2000. Simultaneous chromium(VI) reduction and phenol degradation in an anaerobic consortium of bacteria. *Water Research* 34: 2376–2384.

Chourey, K., Thompson, M. R., Morrell-Falvey, J., Verberkmoes, N. C., Brown, S. D., Shah, M., Zhou, J., Doktycz, M., Hettich, R. L., and Thompson, D. K. 2006. Global molecular and morphological effects of 24-hour chromium(VI) exposure on *Shewanella oneidensis MR-1*. *Applied and Environmental Microbiology* 72: 6331–6344.

Chuan, M. C. and Liu, J. C. 1996. Release behavior of chromium from tannery sludge. *Water Research* 30: 932–938.

Cirik, K., Dursun, N., Sahinkaya, E., and Cinar, O. 2013. Effect of electron donor source on the treatment of Cr(VI)-containing textile wastewater using sulfate-reducing fluidized bed reactors (FBRs). *Bioresource Technology* 133: 414–420.

Cohen, R. R. and Ozawa, T. 2013. Microbial sulfate reduction and biogeochemistry of arsenic and chromium oxyanions in anaerobic bioreactors. *Water, Air, and Soil Pollution* 224: 1–14.

Coleman, R. N. 1988. Chromium toxicity: Effects on microorganisms with special reference to the soil matrix. In: Nriagu, J. O. and Nieboer, E. (Eds.), *Chromium in Natural and Human Environments*, Wiley-Interscience, pp. 335–350.

Contreras, E. M., Orozco, A. M., and Zaritzky, N. E. 2011. Biological Cr(VI) removal coupled with biomass growth, biomass decay, and multiple substrate limitation. *Water Research* 45: 3034–3046.

Cooke, V. M., Hughes, M. N., and Poole, R. K. 1995. Reduction of chromate by bacteria isolated from the cooling water of an electricity generating station. *Journal of Industrial Microbiology* 14: 323–328.

Costa, M. 2003. Potential hazards of hexavalent chromate in our drinking water. *Toxicology and Applied Pharmacology* 188: 1–5.

Das, A. P. and Singh, S. 2011. Occupational health assessment of chromite toxicity among Indian miners. *Indian Journal of Occupational and Environmental Medicine* 15: 6–13.

Das, S., Ram, S. S., Sahu, H. K., Rao, D. S., Chakraborty, A., Sudarshan, M., and Thatoi, H. N. 2013. A study on soil physico-chemical, microbial and metal content in Sukinda chromite mine of Odisha, India. *Environmental Earth Sciences* 69: 2487–2497.

Davis, T. A., Volesky, B., and Mucci, A. 2003. A review of the biochemistry of heavy metal biosorption by brown algae. *Water Research* 37: 4311–4330.

Desai, C., Jain, K., and Madamwar, D. 2008. Hexavalent chromate reductase activity in cytosolic fractions of *Pseudomonas* sp. G1DM21 isolated from Cr(VI) contaminated industrial landfill. *Process Biochemistry* 43: 713–721.

Desjardin, V., Bayard, R., Lejeune, P., and Gourdon, R. 2003. Utilisation of supernatants of pure cultures of *Streptomyces thermocarboxydus* NH50 to reduce chromium toxicity and mobility in contaminated soils. *Water, Air and Soil Pollution: Focus* 3: 153–160.

de Souza, F. B., de Lima Brandão, H., Hackbarth, F. V., de Souza, A. A. U., Boaventura, R. A., de Souza, S. M. G. U., and Vilar, V. J. 2016. Marine macro-alga *Sargassum cymosum* as electron donor for hexavalent chromium reduction to trivalent state in aqueous solutions. *Chemical Engineering Journal* 283: 903–910.

Dhal, B., Thatoi, H. N., Das, N. N., and Pandey, B. D. 2013. Chemical and microbial remediation of hexavalent chromium from contaminated soil and mining/metallurgical solid waste: A review. *Journal of Hazardous Materials* 250–251: 272–291.

Dittert, I. M., de Lima Brandão, H., Pina, F., da Silva, E. A., de Souza, S. M. G. U., de Souza, A. A. U., Botelho, C. M., Boaventura, R. A., and Vilar, V. J. 2014. Integrated reduction/oxidation reactions and sorption processes for Cr(VI) removal from aqueous solutions using *Laminaria digitata* macroalgae. *Chemical Engineering Journal* 237: 443–454.

Dmitrenko, G. N., Konovalova, V. V., and Ereshko, T. V. 2006. The successive reduction of Cr(VI) and NO$_3^-$ or Mn(IV) ions present in the cultivation medium of denitrifying bacteria. *Mikrobiologiia* 75: 160–164.

Dong, C. X., Dai, R. N., and Xiong, J. J. 2010. Kinetic characteristics of Cr(III) oxidation by delta-MnO$_2$. *Zhongguo Ke Xue Yuan Huan Jing Ke Xue Wei Yuan Hui* 31: 1395–1401.

Dotro, G., Castro, S., Tujchneider, O., Piovano, N., Paris, M., Faggi, A., Palazolo, P., Larsen, D., and Fitch, M. 2012. Performance of pilot-scale constructed wetlands for secondary treatment of chromium-bearing tannery wastewaters. *Journal of Hazardous Materials* 239–240: 142–151.

Eary, L. E. and Rai, D. 1991. Chromate reduction by subsurface soils under acidic conditions. *Soil Science Society of America Journal* 55: 676–683.

Eastmond, D. A., Macgregor, J. T., and Slesinski, R. S. 2008. Trivalent chromium: Assessing the genotoxic risk of an essential trace element and widely used human and animal nutritional supplement. *Critical Reviews in Toxicology* 38: 173–190.

Faisal, M. and Hasnain, S. 2004. Comparative study of Cr(VI) uptake and reduction in industrial effluent by *Ochrobactrum intermedium* and *Brevibacterium* sp. *Biotechnology Letters* 26: 1623–1628.

Fendorf, S. E. 1995. Surface reactions of chromium in soils and waters. *Geoderma* 67: 55–71.

Fernández, P. M., Cabral, M. E., Delgado, O. D., Fariña, J. I., and Figueroa, L. I. C. 2013. Textile-dye polluted waters as a source for selecting chromate-reducing yeasts through Cr(VI)-enriched microcosms. *International Biodeterioration & Biodegradation* 79: 28–35.

Fude, L., Harris, B., Urrutia, M. M., and Beveridge, T. J. 1994. Reduction of Cr(VI) by a consortium of sulfate-reducing bacteria (SRB III). *Applied and Environmental Microbiology* 60: 1525–1531.

Fujie, K., Hu, H. Y., Huang, X., Tanaka, Y., Urano, K., and Ohtake, H. 1996. Optimal operation of bioreactor system developed for the treatment of chromate wastewater using *Enterobacter cloacae* HO-1. *Water Science and Technology* 34: 173–182.

Fukuda, T., Ishino, Y., Ogawa, A., Tsutsumi, K., and Morita, H. 2008. Cr(VI) reduction from contaminated soils by *Aspergillus* sp. N2 and *Penicillium* sp. N3 isolated from chromium deposits. *The Journal of General and Applied Microbiology* 54: 295–303.

Fulladosa, E., Desjardin, V., Murat, J. C., Gourdon, R., and Villaescusa, I. 2006. Cr(VI) reduction into Cr(III) as a mechanism to explain the low sensitivity of *Vibrio fischeri* bioassay to detect chromium pollution. *Chemosphere* 65: 644–650.

Gagrai, M. K., Das, C., and Golder, A. K. 2013. Reduction of Cr(VI) into Cr(III) by Spirulina dead biomass in aqueous solution: Kinetic studies. *Chemosphere* 93: 1366–1371.

Garbisu, C., Alkorta, I., Llama, M. J., and Serra, J. L. 1998. Aerobic chromate reduction by *Bacillus subtilis*. *Biodegradation* 9: 133–141.

Gopalan, R. and Veeramani, H. 1994. Studies on microbial chromate reduction by *Pseudomonas* sp. in aerobic continuous suspended growth cultures. *Biotechnology and Bioengineering* 43: 471–476.

Guha, H. 2004. Biogeochemical influence on transport of chromium in manganese sediments: Experimental and modeling approaches. *Journal of Contaminant Hydrology* 70: 1–36.

Guha, H., Jayachandran, K., and Maurrasse, F. 2001. Kinetics of chromium(VI) reduction by a type strain *Shewanella alga* under different growth conditions. *Environmental Pollution (Barking, Essex: 1987)* 115: 209–218.

He, M., Li, X., Liu, H., Miller, S. J., Wang, G., and Rensing, C. 2011. Characterization and genomic analysis of a highly chromate resistant and reducing bacterial strain *Lysinibacillus fusiformis* ZC1. *Journal of Hazardous Materials* 185: 682–688.

He, Z., Gao, F., Sha, T., Hu, Y., and He, C. 2009. Isolation and characterization of a Cr(VI)-reduction *Ochrobactrum* sp. strain CSCr-3 from chromium landfill. *Journal of Hazardous Materials* 163: 869–873.

Hora, A. and Shetty K. V. 2014. Inhibitory and stimulating effect of single and multi-metal ions on hexavalent chromium reduction by *Acinetobacter* sp. Cr-B2. *World Journal of Microbiology & Biotechnology* 30: 3211–3219.

Horton, R. N., Apel, W. A., Thompson, V. S., and Sheridan, P. P. 2006. Low temperature reduction of hexavalent chromium by a microbial enrichment consortium and a novel strain of *Arthrobacter aurescens*. *BMC Microbiology* 6: 5.

Hsu, L. C., Wang, S. L., Lin, Y. C., Wang, M. K., Chiang, P. N., Liu, J. C., Kuan, W. H., Chen, C. C., and Tzou, Y. M. 2010. Cr(VI) removal on fungal biomass of *Neurospora crassa*: The importance of dissolved organic carbons derived from the biomass to Cr(VI) reduction. *Environmental Science & Technology* 44: 6202–6208.

Hüser, J., Rechenmacher, C. E., and Blatter, L. A. 1998. Imaging the permeability pore transition in single mitochondria. *Biophysical Journal* 74: 2129–2137.

Hwang, I., Batchelor, B., Schlautman, M. A., and Wang, R. 2002. Effects of ferrous iron and molecular oxygen on chromium(VI) redox kinetics in the presence of aquifer solids. *Journal of Hazardous Materials* 92: 143–159.

Ibrahim, A. S., El-Tayeb, M. A., Elbadawi, Y. B., and Al-Salamah, A. A. 2013. Bioreduction of Cr(VI) by potent novel chromate resistant alkaliphilic *Bacillus* sp. strain KSUCr5 isolated from hypersaline Soda lakes. *African Journal of Biotechnology* 10: 7207–7218.

Ishibashi, Y., Cervantes, C., and Silver, S. 1990. Chromium reduction in *Pseudomonas putida*. *Applied and Environmental Microbiology* 56: 2268–2270.

James, B. R. 2001. Remediation-by-reduction strategies for chromate-contaminated soils. *Environmental Geochemistry and Health* 23: 175–179.

Kashefi, K., Holmes, D. E., Reysenbach, A. L., and Lovley, D. R. 2002. Use of Fe(III) as an electron acceptor to recover previously uncultured hyperthermophiles: Isolation and characterization of *Geothermobacterium ferrireducens* gen. nov., sp. nov. *Applied and Environmental Microbiology* 68: 1735–1742.

Kathiravan, M. N., Karthick, R., Muthu, N., Muthukumar, K., and Velan, M. 2010. Sonoassisted microbial reduction of chromium. *Applied Biochemistry and Biotechnology* 160: 2000–2013.

Kathiravan, M. N., Karthick, R., and Muthukumar, K. 2011. *Ex situ* bioremediation of Cr(VI) contaminated soil by *Bacillus* sp.: Batch and continuous studies. *Chemical Engineering Journal* 169: 107–115.

Kato, J. and Ohtake, H. 1991. Effects of heavy metal cations on chromate reduction by *Enterobacter cloacae* strain HO1. *The Journal of General and Applied Microbiology* 37: 519–522.

Khérici-Bousnoubra, H., Khérici, N., Derradji, E. F., Rousset, C., and Caruba, R. 2009. Behaviour of chromium VI in a multilayer aquifer in the industrial zone of Annaba, Algeria. *Environmental Geology* 57: 1619–1624.

Kikuchi, T. and Tanaka, S. 2012. Biological removal and recovery of toxic heavy metals in water environment. *Critical Reviews in Environmental Science and Technology* 42: 1007–1057.

Komori, K., Wang, P. C., Toda, K., and Ohtake, H. 1989. Factors affecting chromate reduction in *Enterobacter cloacae* strain HO1. *Applied Microbiology and Biotechnology* 31: 567–570.

Kotaś, J. and Stasicka, Z. 2000. Chromium occurrence in the environment and methods of its speciation. *Environmental Pollution (Barking, Essex: 1987)* 107: 263–283.

Laxman, R. S. and More, S. 2002. Reduction of hexavalent chromium by *Streptomyces griseus*. *Minerals Engineering* 15: 831–837.

Li, Y., Lu, A., Ding, H., Jin, S., Yan, Y., Wang, C., Zen, C., and Wang, X. 2009. Cr(VI) reduction at rutile-catalyzed cathode in microbial fuel cells. *Electrochemistry Communications* 11: 1496–1499.

Lin, Y. C., Wang, S. L., Shen, W. C., Huang, P. M., Chiang, P. N., Liu, J. C., Chen, C. C., and Tzou, Y. M. 2009. Photo-enhancement of Cr(VI) reduction by fungal biomass of *Neurospora crassa*. *Applied Catalysis B: Environmental* 92: 294–300.

Lin, Y. H., Tu, Y. C., and Chen, G. L. 2015. Kinetics of chromium(VI) reduction with acetate biodegradation by *Escherichia coli* 33456 in a fixed biofilm reactor. *Environmental Engineering Science* 32: 761–772.

Liu, C., Fiol, N., Villaescusa, I., and Poch, J. 2016. New approach in modeling Cr(VI) sorption onto biomass from metal binary mixtures solutions. *The Science of the Total Environment* 541: 101–108.

Liu, Y. G., Xu, W. H., Zeng, G. M., Li, X., and Gao, H. 2006. Cr(VI) reduction by *Bacillus* sp. isolated from chromium landfill. *Process Biochemistry* 41: 1981–1986.

Llagostera, M., Garrido, S., Guerrero, R., and Barbé, J. 1986. Induction of SOS genes of *Escherichia coli* by chromium compounds. *Environmental Mutagenesis* 8: 571–577.

LLovera, S., Bonet, R., Simon-Pujol, M. D., and Congregado, F. 1993. Effect of culture medium ions on chromate reduction by resting cells of *Agrobacterium radiobacter*. *Applied Microbiology and Biotechnology* 39: 424–426.

Lovley, D. R. 1993. Dissimilatory metal reduction. *Annual Review in Microbiology* 47: 263–290.

Lovley, D. R. 1995. Bioremediation of organic and metal contaminants with dissimilatory metal reduction. *Journal of Industrial Microbiology* 14: 85–93.

Lovley, D. R., Giovannoni, S. J., White, D. C., Champine, J. E., Phillips, E. J. P., Gorby, Y. A., and Goodwin, S. 1993. *Geobacter metallireducens* gen. nov. sp. nov., a microorganism capable of coupling the complete oxidation of organic compounds to the reduction of iron and other metals. *Archives of Microbiology* 159: 336–344.

López Errasquín, E. and Vázquez, C. 2003. Tolerance and uptake of heavy metals by *Trichoderma atroviride* isolated from sludge. *Chemosphere* 50: 137–143.

Mabrouk, M. E. 2008. Statistical optimization of medium components for chromate reduction by halophilic *Streptomyces* sp. MS-2. *African Journal of Microbiology Research* 2: 103–109.

Mahmood, S., Khalid, A., Arshad, M., and Ahmad, R. 2015. Effect of trace metals and electron shuttle on simultaneous reduction of reactive black-5 azo dye and hexavalent chromium in liquid medium by *Pseudomonas* sp. *Chemosphere* 138: 895–900.

Marsh, T. L. and McInerney, M. J. 2001. Relationship of hydrogen bioavailability to chromate reduction in aquifer sediments. *Applied and Environmental Microbiology* 67: 1517–1521.

Mary Mangaiyarkarasi, M. S., Vincent, S., Janarthanan, S., Subba Rao, T., and Tata, B. V. 2011. Bioreduction of Cr(VI) by alkaliphilic *Bacillus subtilis* and interaction of the membrane groups. *Saudi Journal of Biological Sciences* 18: 157–167.

McLean, J. and Beveridge, T. J. 2001. Chromate reduction by a pseudomonad isolated from a site contaminated with chromated copper arsenate. *Applied and Environmental Microbiology* 67: 1076–1084.

Michailides, M. K., Tekerlekopoulou, A. G., Akratos, C. S., Coles, S., Pavlou, S., and Vayenas, D. V. 2015. Molasses as an efficient low-cost carbon source for biological Cr(VI) removal. *Journal of Hazardous Materials* 281: 95–105.

Milacic, R. and Stupar, J. 1995. Fractionation and oxidation of chromium in tannery waste- and sewage sludge-amended soils. *Environmental Science & Technology* 29: 506–514.

Miranda, A. T., González, M. V., González, G., Vargas, E., Campos-García, J., and Cervantes, C. 2005. Involvement of DNA helicases in chromate resistance by *Pseudomonas aeruginosa* PAO1. *Mutation Research* 578: 202–209.

Mohamed, M. M. and Hatfield, K. 2005. Modeling microbial-mediated reduction in batch reactors. *Chemosphere* 59: 1207–1217.

Molokwane, P. E., Meli, K. C., and Nkhalambayausi-Chirwa, E. M. 2008. Chromium (VI) reduction in activated sludge bacteria exposed to high chromium loading: Brits culture (South Africa). *Water Research* 42: 4538–4548.

Morales-Barrera, L., de Maria Guillen-Jimenez, F., Ortiz-Moreno, A., Villegas-Garrido, T. L., Sandoval-Cabrera, A., Hernández-Rodríguez, C. H., and Cristiani-Urbina, E. 2008. Isolation, identification and characterization of a *Hypocrea tawa* strain with high Cr(VI) reduction potential. *Biochemical Engineering Journal* 40: 284–292.

Mtimunye, P. J. and Chirwa, E. 2014. Finite difference simulation of biological chromium(VI) reduction in aquifer media columns. *Water SA* 40: 359–368.

Nancharaiah, Y. V., Venkata Mohan, S., and Lens, P. N. 2015. Metals removal and recovery in bioelectrochemical systems: A review. *Bioresource Technology* 195: 102–114.

Narayani, M. and Shetty, K. V. 2013. Chromium-resistant bacteria and their environmental condition for hexavalent chromium removal: A review. *Critical Reviews in Environmental Science and Technology* 43: 955–1009.

Nethercott, J., Paustenbach, D., Adams, R., Fowler, J., Marks, J., Morton, C., Taylor, J., Horowitz, S., and Finley, B. 1994. A study of chromium induced allergic contact dermatitis with 54 volunteers: Implications for environmental risk assessment. *Occupational and Environmental Medicine* 51: 371–380.

Ngwenya, N. and Chirwa, E. M. 2011. Biological removal of cationic fission products from nuclear wastewater. *Water Science and Technology: A Journal of the International Association on Water Pollution Research* 63: 124–128.

Nies, A., Nies, D. H., and Silver, S. 1990. Nucleotide sequence and expression of a plasmid-encoded chromate resistance determinant from *Alcaligenes eutrophus*. *The Journal of Biological Chemistry* 265: 5648–5653.

Nourbakhsh, M., Sag, Y., Özer, D., Aksu, Z., Kutsal, T., and Caglar, A. 1994. A comparative study of various biosorbents for removal of chromium(VI) ions from industrial waste waters. *Process Biochemistry* 29: 1–5.

O'Flaherty, V., Mahony, T., O'Kennedy, R., and Colleran, E. 1998. Effect of pH on growth kinetics and sulphide toxicity thresholds of a range of methanogenic, syntrophic and sulphate-reducing bacteria. *Process Biochemistry* 33: 555–569.

Ohtake, H., Fujii, E., and Toda, K. 1990. A survey of effective electron donors for reduction of toxic hexavalent chromium by *Enterobacter cloacae (strain HO1)*. *The Journal of General and Applied Microbiology* 36: 203–208.

Ohtake, H., Komori, K., Cervantes, C., and Toda, K. 1990. Chromate-resistance in a chromate-reducing strain of *Enterobacter cloacae*. *FEMS Microbiology Letters* 55: 85–88.

Okeke, B. C. 2008. Bioremoval of hexavalent chromium from water by a salt tolerant bacterium, *Exiguobacterium* sp. GS1. *Journal of Industrial Microbiology & Biotechnology* 35: 1571–1579.

Okeke, B. C., Laymon, J., Crenshaw, S., and Oji, C. 2008. Environmental and kinetic parameters for Cr(VI) bioreduction by a bacterial monoculture purified from Cr(VI)-resistant consortium. *Biological Trace Element Research* 123: 229–241.

Owlad, M., Aroua, M. K., Daud, W. A. W., and Baroutian, S. 2009. Removal of hexavalent chromium-contaminated water and wastewater: A review. *Water, Air, and Soil Pollution* 200: 59–77.

Pal, A. and Paul, A. K. 2004. Aerobic chromate reduction by chromium-resistant bacteria isolated from serpentine soil. *Microbiological Research* 159: 347–354.

Parameswari, E., Lakshmanan, A., and Thilagavathi, T. 2009. Chromate resistance and reduction by bacterial isolates. *Australian Journal of Basic and Applied Sciences* 3: 1363–1368.

Park, C. H., Gonzalez, D., Ackerley, D., Keyhan, M., Matin, A. 2002. Molecular engineering of soluble bacterial proteins with chromate reductase activity. In: Pellei, M., Porta, A., and Hinchee, R. E. (Eds.), *Remediation and Beneficial Reuse of Contaminated Sediments*, Vol. 3. Battelle Press, Columbus, OH.

Park, C. H., Keyhan, M., Wielinga, B., Fendorf, S., and Matin, A. 2000. Purification to homogeneity and characterization of a novel *Pseudomonas putida* chromate reductase. *Applied and Environmental Microbiology* 66: 1788–1795.

Park, D., Yun, Y. S., Jo, J. H., and Park, J. M. 2005a. Mechanism of hexavalent chromium removal by dead fungal biomass of *Aspergillus niger*. *Water Research* 39: 533–540.

Park, D., Yun, Y. S., and Park, J. M. 2005b. Studies on hexavalent chromium biosorption by chemically-treated biomass of *Ecklonia* sp. *Chemosphere* 60: 1356–1364.

Parmar, N., Warren, L. A., Roden, E. E., and Ferris, F. G. 2000. Solid phase capture of strontium by the iron reducing bacteria *Shewanella alga* strain BrY. *Chemical Geology* 169: 281–288.

Pei, Q. H., Shahir, S., Raj, A. S. S., Zakaria, Z. A., and Ahmad, W. A. 2009. Chromium(VI) resistance and removal by *Acinetobacter haemolyticus*. *World Journal of Microbiology and Biotechnology* 25: 1085–1093.

Peng, L., Liu, Y., Gao, S. H., Chen, X., and Ni, B. J. 2016. Evaluating simultaneous chromate and nitrate reduction during microbial denitrification processes. *Water Research* 89: 1–8.

Peng, L., Liu, Y., Gao, S. H., Dai, X., and Ni, B. J. 2015. Assessing chromate reduction by dissimilatory iron reducing bacteria using mathematical modeling. *Chemosphere* 139: 334–339.

Philip, L., Iyengar, L., and Venkobachar, C. 1998. Cr(VI) reduction by *Bacillus coagulans* isolated from contaminated soils. *Journal of Environmental Engineering* 124: 1165–1170.

Qi, W., Reiter, R. J., Tan, D. X., Garcia, J. J., Manchester, L. C., Karbownik, M., and Calvo, J. R. 2000. Chromium(III)-induced 8-hydroxydeoxyguanosine in DNA and its reduction by antioxidants: Comparative effects of melatonin, ascorbate, and vitamin E. *Environmental Health Perspectives* 108: 399–402.

Rai, D., Eary, L. E., and Zachara, J. M. 1989. Environmental chemistry of chromium. *Science of the Total Environment* 86: 15–23.

Raja, C. E., Anbazhagan, K., and Selvam, G. S. 2006. Isolation and characterization of a metal-resistant *Pseudomonas aeruginosa* strain. *World Journal of Microbiology and Biotechnology* 22: 577–585.

Ramírez-Díaz, M. I., Díaz-Pérez, C., Vargas, E., Riveros-Rosas, H., Campos-García, J., and Cervantes, C. 2008. Mechanisms of bacterial resistance to chromium compounds. *Biometals* 21: 321–332.

Reynolds, M. F., Peterson-Roth, E. C., Bespalov, I. A., Johnston, T., Gurel, V. M., Menard, H. L., and Zhitkovich, A. 2009. Rapid DNA double-strand breaks resulting from processing of Cr-DNA cross-links by both MutS dimers. *Cancer Research* 69: 1071–1079.

Sahinkaya, E., Dursun, N., Kilic, A., Demirel, S., Uyanik, S., and Cinar, O. 2011. Simultaneous heterotrophic and sulfur-oxidizing autotrophic denitrification process for drinking water treatment: Control of sulfate production. *Water Research* 45: 6661–6667.

Sahinkaya, E., Kilic, A., Altun, M., Komnitsas, K., and Lens, P. N. 2012. Hexavalent chromium reduction in a sulfur reducing packed-bed bioreactor. *Journal of Hazardous Materials* 219: 253–259.

Sanghi, R., Sankararamakrishnan, N., and Dave, B. C. 2009. Fungal bioremediation of chromates: Conformational changes of biomass during sequestration, binding, and reduction of hexavalent chromium ions. *Journal of Hazardous Materials* 169: 1074–1080.

Sarangi, A. and Krishnan, C. 2015. Detoxification of hexavalent chromium by *Leucobacter* sp. uses a reductase with specificity for dihydrolipoamide. *Journal of Basic Microbiology* 55: 1–9.

Sau, G., Chatterjee, S., Sinha, S., and Mukherjee, S. K. 2008. Isolation and characterization of a Cr(VI) reducing *Bacillus firmus* strain from industrial effluents. *Polish Journal of Microbiology* 57: 327–332.

Schmieman, E. A., Yonge, D. R., Rege, M. A., Petersen, J. N., Turick, C. E., Johnstone, D. L., and Apel, W. A. 1998. Comparative kinetics of bacterial reduction of chromium. *Journal of Environmental Engineering* 124: 449–455.

Sharma, D. C., Chatterjee, C., and Sharma, C. P. 1995. Chromium accumulation and its effects on wheat (*Triticum aestivum* L. cv. HD 2204) metabolism. *Plant Science* 111: 145–151.

Sharma, S. and Adholeya, A. 2011. Detoxification and accumulation of chromium from tannery effluent and spent chrome effluent by *Paecilomyces lilacinus* fungi. *International Biodeterioration & Biodegradation* 65: 309–317.

Shen, H. and Wang, Y. T. 1994a. Biological reduction of chromium by *E. coli*. *Journal of Environmental Engineering* 120: 560–572.

Shen, H. and Wang, Y. T. 1994b. Modeling hexavalent chromium reduction in *Escherichia coli* 33456. *Biotechnology and Bioengineering* 43: 293–300.

Shi, W., Becker, J., Bischoff, M., Turco, R. F., and Konopka, A. E. 2002. Association of microbial community composition and activity with lead, chromium, and hydrocarbon contamination. *Applied and Environmental Microbiology* 68: 3859–3866.

Silver, S. 1997. The bacterial view of the periodic table; specific functions for all elements. *Reviews in Mineralogy and Geochemistry* 35: 345–360.

Singh, R., Bishnoi, N. R., and Kirrolia, A. 2013. Evaluation of *Pseudomonas aeruginosa* an innovative bioremediation tool in multi metals ions from simulated system using multi response methodology. *Bioresource Technology* 138: 222–234.

Smith, W. L. and Gadd, G. M. 2000. Reduction and precipitation of chromate by mixed culture sulphate-reducing bacterial biofilms. *Journal of Applied Microbiology* 88: 983–991.

Somasundaram, V., Philip, L., and Bhallamudi, S. M. 2009. Experimental and mathematical modeling studies on Cr(VI) reduction by CRB, SRB and IRB, individually and in combination. *Journal of Hazardous Materials* 172: 606–617.

Somasundaram, V., Philip, L., and Bhallamudi, S. M. 2011. Laboratory scale column studies on transport and biotransformation of Cr(VI) through porous media in presence of CRB, SRB and IRB. *Chemical Engineering Journal* 171: 572–581.

Soni, S. K., Singh, R., Awasthi, A., Singh, M., and Kalra, A. 2013. In vitro Cr(VI) reduction by cell-free extracts of chromate-reducing bacteria isolated from tannery effluent irrigated soil. *Environmental Science and Pollution Research International* 20: 1661–1674.

Srivastava, N. K. and Majumder, C. B. 2008. Novel biofiltration methods for the treatment of heavy metals from industrial wastewater. *Journal of Hazardous Materials* 151: 1–8.

Srivastava, S., Agrawal, S. B., and Mondal, M. K. 2015. Biosorption isotherms and kinetics on removal of Cr(VI) using native and chemically modified *Lagerstroemia speciosa bark*. *Ecological Engineering* 85: 56–66.

Stasinakis, A. S., Thomaidis, N. S., Mamais, D., Papanikolaou, E. C., Tsakon, A., and Lekkas, T. D. 2003. Effects of chromium (VI) addition on the activated sludge process. *Water Research* 37: 2140–2148.

Stollenwerk, K. G. and Grove, D. B. 1985. Adsorption and desorption of hexavalent chromium in an alluvial aquifer near Telluride, Colorado. *Journal of Environmental Quality* 14: 150–155.

Sultan, S. and Hasnain, S. 2007. Reduction of toxic hexavalent chromium by *Ochrobactrum intermedium strain SDCr*-5 stimulated by heavy metals. *Bioresource Technology* 98: 340–344.

Sultana, M. Y., Akratos, C. S., Pavlou, S., and Vayenas, D. V. 2014. Chromium removal in constructed wetlands: A review. *International Biodeterioration & Biodegradation* 96: 181–190.

Suzuki, T., Miyata, N., Horitsu, H., Kawai, K., Takamizawa, K., Tai, Y., and Okazaki, M. 1992. NAD(P)H-dependent chromium(VI) reductase of *Pseudomonas ambigua G-1*: A Cr(V) intermediate is formed during the reduction of Cr(VI) to Cr(III). *Journal of Bacteriology* 174: 5340–5345.

Tandukar, M., Huber, S. J., Onodera, T., and Pavlostathis, S. G. 2009. Biological chromium(VI) reduction in the cathode of a microbial fuel cell. *Environmental Science & Technology* 43: 8159–8165.

Tebo, B. M. and Obraztsova, A. Y. 1998. Sulfate-reducing bacterium grows with Cr(VI), U(VI), Mn(IV), and Fe(III) as electron acceptors. *FEMS Microbiology Letters* 162: 193–199.

Tekerlekopoulou, A. G., Tsiamis, G., Dermou, E., Siozios, S., Bourtzis, K., and Vayenas, D. V. 2010. The effect of carbon source on microbial community structure and Cr(VI) reduction rate. *Biotechnology and Bioengineering* 107: 478–487.

Tekerlekopoulou, A. G., Tsiflikiotou, M., Akritidou, L., Viennas, A., Tsiamis, G., Pavlou, S., Bourtzis, K., and Vayenas, D. V. 2013. Modelling of biological Cr(VI) removal in draw-fill reactors using microorganisms in suspended and attached growth systems. *Water Research* 47: 623–636.

Thatoi, H., Das, S., Mishra, J., Rath, B. P., and Das, N. 2014. Bacterial chromate reductase, a potential enzyme for bioremediation of hexavalent chromium: A review. *Journal of Environmental Management* 146: 383–399.

Tucker, M. D., Barton, L. L., and Thomson, B. M. 1998. Reduction of Cr, Mo, Se and U by *Desulfovibrio desulfuricans* immobilized in polyacrylamide gels. *Journal of Industrial Microbiology & Biotechnology* 20: 13–19.

Vaiopoulou, E. and Gikas, P. 2012. Effects of chromium on activated sludge and on the performance of wastewater treatment plants: A review. *Water Research* 46: 549–570.

Valls, M., Atrian, S., de Lorenzo, V., and Fernández, L. A. 2000. Engineering a mouse metallothionein on the cell surface of *Ralstonia eutropha CH34* for immobilization of heavy metals in soil. *Nature Biotechnology* 18: 661–665.

Vaneechoutte, M., Kämpfer, P., De Baere, T., Falsen, E., and Verschraegen, G. 2004. *Wautersia* gen. nov., a novel genus accommodating the phylogenetic lineage including *Ralstonia eutropha* and related species, and proposal of *Ralstonia (Pseudomonas) syzygii* comb. nov. *International Journal of Systematic and Evolutionary Microbiology* 54: 317–327.

Viamajala, S., Smith, W. A., Sani, R. K., Apel, W. A., Petersen, J. N., Neal, A. L., Roberto, F. F., Newby, D. T., and Peyton, B. M. 2007. Isolation and characterization of Cr(VI) reducing *Cellulomonas* spp. from subsurface soils: Implications for long-term chromate reduction. *Bioresource Technology* 98: 612–622.

Wang, G., Huang, L., and Zhang, Y. 2008. Cathodic reduction of hexavalent chromium (Cr(VI)) coupled with electricity generation in microbial fuel cells. *Biotechnology Letters* 30: 1959–1966.

Wang, J. and Chen, C. 2006. Biosorption of heavy metals by *Saccharomyces cerevisiae*: A review. *Biotechnology Advances* 24: 427–451.

Wang, P. C., Mori, T., Komori, K., Sasatsu, M., Toda, K., and Ohtake, H. 1989. Isolation and characterization of an *Enterobacter cloacae* strain that reduces hexavalent chromium under anaerobic conditions. *Applied and Environmental Microbiology* 55: 1665–1669.

Wang, P. C., Mori, T., Toda, K., and Ohtake, H. 1990. Membrane-associated chromate reductase activity from *Enterobacter cloacae*. *Journal of Bacteriology* 172: 1670–1672.

Wang, Y. T. and Shen, H. 1995. Bacterial reduction of hexavalent chromium. *Journal of Industrial Microbiology* 14: 159–163.

Wang, Y. T. and Shen, H. 1997. Modelling Cr(VI) reduction by pure bacterial cultures. *Water Research* 31: 727–732.

Wang, Y. T. and Xiao, C. 1995. Factors affecting hexavalent chromium reduction in pure cultures of bacteria. *Water Research* 29: 2467–2474.

Wani, R., Kodam, K. M., Gawai, K. R., and Dhakephalkar, P. K. 2007. Chromate reduction by *Burkholderia cepacia* MCMB-821, isolated from the pristine habitat of alkaline crater lake. *Applied Microbiology and Biotechnology* 75: 627–632.

Wielinga, B., Mizuba, M. M., Hansel, C. M., and Fendorf, S. 2001. Iron promoted reduction of chromate by dissimilatory iron-reducing bacteria. *Environmental Science & Technology* 35: 522–527.

Williams, N. 1997. Occupational skin ulceration in chrome platers. *Occupational Medicine (Oxford, England)* 47: 309–310.

Wu, X., Zhu, X., Song, T., Zhang, L., Jia, H., and Wei, P. 2015. Effect of acclimatization on hexavalent chromium reduction in a biocathode microbial fuel cell. *Bioresource Technology* 180: 185–191.

Xafenias, N., Zhang, Y., and Banks, C. 2013. Enhanced performance of hexavalent chromium reducing cathodes in the presence of *Shewanella oneidensis MR-1* and lactate. *Environmental Science & Technology* 47: 4512–4520.

Xu, F., Ma, T., Zhou, L., Hu, Z., and Shi, L. 2015. Chromium isotopic fractionation during Cr(VI) reduction by *Bacillus* sp. under aerobic conditions. *Chemosphere* 130: 46–51.

Xu, L., Luo, M., Li, W., Wei, X., Xie, K., Liu, L., Jiang, C., and Liu, H. 2011. Reduction of hexavalent chromium by *Pannonibacter phragmitetus* LSSE-09 stimulated with external electron donors under alkaline conditions. *Journal of Hazardous Materials* 185: 1169–1176.

Xu, W. H., Liu, Y. G., Zeng, G. M., Xin, L. I., Song, H. X., and Peng, Q. Q. 2009. Characterization of Cr(VI) resistance and reduction by *Pseudomonas aeruginosa*. *Transactions of Nonferrous Metals Society of China* 19: 1336–1341.

Zakaria, Z. A., Zakaria, Z., Surif, S., and Ahmad, W. A. 2007. Hexavalent chromium reduction by *Acinetobacter haemolyticus* isolated from heavy-metal contaminated wastewater. *Journal of Hazardous Materials* 146: 30–38.

Zhang, B., Feng, C., Ni, J., Zhang, J., and Huang, W. 2012. Simultaneous reduction of vanadium(V) and chromium(VI) with enhanced energy recovery based on microbial fuel cell technology. *Journal of Power Sources* 204: 34–39.

Zhang, K. and Li, F. 2011. Isolation and characterization of a chromium-resistant bacterium *Serratia* sp. Cr-10 from a chromate-contaminated site. *Applied Microbiology and Biotechnology* 90: 1163–1169.

Zhitkovich, A. 2011. Chromium in drinking water: Sources, metabolism, and cancer risks. *Chemical Research in Toxicology* 24: 1617–1629.

Zhou, J., Xia, B., Treves, D. S., Wu, L. Y., Marsh, T. L., O'Neill, R. V., Palumbo, A. V., and Tiedje, J. M. 2002. Spatial and resource factors influencing high microbial diversity in soil. *Applied and Environmental Microbiology* 68: 326–334.

44 Bacteria- and Algae-Mediated Remediation of Chromium

Gouri Chaudhuri, Rajdeep Roy, N. Chandrasekaran, and Amitava Mukherjee

CONTENTS

ABSTRACT

There is an increased release of chromium into the environment because of constant industrial and various anthropogenic processes. Chromium is a well-known highly toxic metal for living organisms. Hexavalent and trivalent chromium both are considered as toxic heavy metals, but hexavalent chromium is considered a priority pollutant. Some of the primary techniques that remove heavy metals in the environment result from biological activities. Among these, the successful techniques are biosorption, bioaccumulation, and bioreduction. This chapter is written in order to give an introduction to the readers about cleaning up chromium metal wastes with the help of bacteria and algae.

44.1 INTRODUCTION

A large number of toxic metal ions are being released to the environment by industrial and mining processes (Uslu and Tanyol, 2006). There are certain chemical elements whose specific gravity is at least five times water's specific gravity. These elements are considered as "heavy metals," for example, cadmium, 8.65; chromium, 7.18–7.20 (20°C); cobalt, 8.9 (20°C); and lead, 11.34 (Lide, 1992). Heavy metal constitutes

a significant hazard for the human health and the ecosystem (Boopathy, 2000). Moreover, it usually forms compounds that can be toxic, carcinogenic, or mutagenic even at very low concentrations (Ruiz-Manriquez et al., 1998). There are immense applications of heavy metals in several industries due to their beneficial technological attributes. Heavy metal contamination is at an invariable increase, owing to an increase in anthropogenic activities like mining. There is an adverse impact on the environment due to discharge of unprocessed effluents containing heavy metals from many industries. Industrial wastewaters containing heavy metals induce metal pollution to soil, water, and terrestrial ecosystem. Since heavy metals are nonbiodegradable, a specific problem related to these heavy metals is their accumulation in the food chain and living tissues (Gavrilescu, 2004; Deng et al., 2006).

Thirty-five metals have been listed as metals of major concern because of their occupational or residential exposure (Glanze, 1996), out of which lead (Pb), copper (Cu), chromium (Cr), arsenic (As), bismuth (Bi), cadmium (Cd), cerium (Ce), cobalt (Co), gallium (Ga), antimony (An), iron (Fe), gold (Au), manganese (Mn), nickel (Ni), mercury (Hg), platinum (Pt), silver (Ag), tellurium (Te), thallium (Tl), vanadium (V), tin (Sn), uranium (U), and zinc (Zn) have been listed as considerable toxic metals, while others do not have substantial toxic properties (Hogan, 2010).

Contamination of heavy metal is one of the most notable environmental problems worldwide, and therefore, it is of global concern (Murphy et al., 2008). It is essential to ensure the removal of heavy metals to protect human health and environment.

44.1.1 CHROMIUM AND ITS TOXIC EFFECTS

Chromium is largely used in many industrial processes such as electroplating, wood preserving, pigmenting, textile dyeing, pulp and paper industries, tannery, and wood preservation. Besides this, the chromium manufacturing industry also produces a large quantity of solid and liquid waste containing hexavalent chromium. The wastewater discharged from these processes carries a high amount of chromium metal, which causes hazardous effects on living organisms (Zayed and Terry, 2003). Chromium is widely available in the environment and is present in many natural as well as anthropogenic sources. It is described as the 21st most abundant element in the earth's crust (Ali et al., 2011).

In nature, chromium exists in several forms ranging from −2 to +6, of which trivalent and hexavalent chromium have stable oxidation states. It has been reported that the valence state plays an important role in determining the toxicokinetics of a given chromium compound (ATSDR). Between these hexavalent and trivalent chromium, hexavalent chromium is more carcinogenic than that of trivalent as well as causes contamination of surface water, groundwater, and soil (Cervantes et al., 2001). Since Cr (VI) is highly mobile, it diffuses through the epidermis easily and is then promptly reduced to Cr (III) with the help of gastric fluids and extracellular and intracellular low-molecular-weight proteins and molecules. This Cr (III) further interacts with protein nucleotides, DNA, and nuclear enzymes and eventually contributes to the mutagenic and carcinogenic activity of Cr (VI) (Thomas and Gropper, 1996).

Though Cr (III) is considered as an essential trace element, if present at high doses, it has low acute and chronic toxicity to humans. The inability of Cr (III) to perforate cell membranes limits the possibility of carcinogenic activity.

Therefore, it is required to remove chromium from the environment in order to keep it safe. Chromium has been identified as one of the greatest threat to humans by the U.S. Environmental Protection Agency (USEPA). USEPA has set the allowable level of total chromium in drinking water as 0.05 mg/L (Baral and Engelken, 2002).

44.1.2 CONVENTIONAL METHODS FOR THE REMOVAL OF CHROMIUM

There are various conventional techniques to remove metals from industrial wastewaters, for example, oxidation–reduction, filtration, chemical precipitation, electrochemical techniques, and various sophisticated separation procedures using membranes. Few of them are discussed as follows.

44.1.2.1 Ion Exchange

Ion exchange is the most widely used method for removing chromium from wastewater. Trivalent or hexavalent chromium can be separated from wastewater by using this method where trivalent chromium is removed by using cation-exchange resins and hexavalent chromium is removed by anion resins. But compared to reduction and lime precipitation, the wastewater treatment by chromium reduction and ion exchange would not be economical (Davis and Cornwell, 2008). Moreover, ion exchange is in general costly and less specific for Cr (VI) removal in the presence of other ions (Elangovan and Philip, 2009).

44.1.2.2 Solvent Extraction

It is another flexible and effective technique in order to extract, separate, enrich, and recover the desired metal. There is no report for the recovery of this metal in a commercial scale. Kumar et al. (1998) stated that only limited information is available on the extraction and recovery of Cr (III) and Cr (VI) from the engineered effluents.

44.1.2.3 Sulfide Precipitation

Precipitation of Cr (VI) is possible in the form of its sulfide by using $FeSO_4$ and NaHS. It has been noticed that sulfide precipitation gives lower metal concentration than that of hydroxide precipitation. Moreover, metal sulfides get settled faster. This method may yield toxic sludge requiring proper disposal (Elangovan and Philip, 2009).

44.1.2.4 Reduction–Precipitation Method

Ferrous sulfate is used frequently for the reduction of Cr (VI). Besides this, sulfur dioxide, sodium metabisulfite, barium sulfide, iron, and zinc are also used as reductants. Cr (III) can be precipitated as its hydroxide using alkali treatment. The primary obstruction of this method is that it consists of two major steps. The first step involves reduction or lower pH and the second includes precipitation at higher pH using lime or caustic soda. The by-products of this process are large quantities of chemical sludge (Elangovan and Philip, 2009).

44.1.3 ADVANTAGES OF BIOREMEDIATION

The traditional techniques used for remediation have been to excavate polluted soil and transfer it to a landfill or to cap and contain the contaminated areas of a site. The methods have some disadvantages. The first technique moves the contamination elsewhere and may create significant risks in digging, handling, and transporting hazardous material. Furthermore, it is very difficult and increasingly expensive to get new landfills in order to dispose of the material. The second method is only an interim solution since the contamination remains on-site, requiring continuous monitoring and maintenance of the isolation barriers long into the future along with all the associated costs and potential liability. Therefore, these processes are not cost-effective and most of the cases not eco-friendly.

A better approach than these conventional techniques is the complete destruction of the pollutants if possible, or at least to transform them into innocuous substances. There are few methods that are in use such as high-temperature incineration and several types of chemical decomposition (e.g., UV oxidation, base-catalyzed dechlorination). Though they are very efficient at reducing the levels of a range of contaminants, they have various drawbacks, mainly their technological complexity, the cost for small-scale application, and the lack of public acceptance, especially for incineration, which may increase the exposure to contaminants for both the workers at the site and nearby residents (Vidali, 2001).

On the other hand, bioremediation is an alternative that offers the possibility to destroy or render several harmless contaminants using natural biological process. It uses relatively low-cost, low-technology techniques, which in general have a high public acceptance. It can also be carried out on-site. Thus, bioremediation has the potential to clean up a polluted environment since it involves the use of microbes to detoxify and degrade environmental contaminants (Gadd, 2000; Malik, 2004; Radhika et al., 2006; Farhadian et al., 2008). Bioremediation, being *in situ* treatment, is a safe and economic alternative to conventional physicochemical strategies. In order to clean up contaminated soil and water, biological agents especially microorganisms, for example, yeast, fungi, or bacteria, are used for bioremediation (Strong and Burgess, 2008). It is an emerging technology. In this chapter, discussion has been carried out as part of the development of an integrated process for the bioremediation of chromium using bacteria and algae.

44.2 MICROORGANISMS AND METALS

Exposure of microorganisms to high concentrations of toxic metals had been observed for a long time. Thus, these microorganisms have evolved few indigenous mechanisms of metal resistance and detoxification in order to avoid the hazards caused by metals in the environment. It has been noticed that when the concentration of metal ion increases to a level intolerable to the microorganisms, they tend to adapt, resulting in a change in morphology or cell alignment (Padmini et al., 2009). Microorganisms play a noteworthy role in transformation of carbon, nitrogen, sulfur, and iron. These transformation cycles do operate on different time scales and at distances ranging from the microbial to the global.

44.3 REMEDIATION OF CHROMIUM BY BACTERIA

Application of microbes for remediation is preferred because of their low cost, eco-friendliness, and fewer side effects. It has been noticed that an important role in biogeochemical cycling is played by the microbial community. Many microbes are capable of changing the physical characteristics of metals and metalloids, which include the oxidation–reduction reactions that alter the nutrient's physical and chemical characteristics (Falkowski et al., 2008). It has also been reported that bacteria

indigenous to mine water system can undergo cell elongation and chain formation during their adaptation to high Cr concentration (Samuel et al., 2012).

44.3.1 CHROMIUM REMOVAL BY BIOSORPTION USING BACTERIA

Biosorption is a procedure in which a variety of active groups of cell constituents such as acetamido group of chitin, structural polysaccharide of fungi, amine (amino and peptidoglycosides), sulphahydral and carboxyl groups in protein, and phosphodiester take part in biosorption. It is considered as an efficient technique to remove metals from contaminated water because it is highly efficient for absorbing metals, environment friendly, possess a wide operational range of pH and temperature, and can easily recycle sorbent (Singh et al., 2012).

Samuel et al. carried out studies on the adsorptive removal of Cr (VI) by alginate beads containing Cr (VI)-adapted bacteria such as *Acinetobacter junii*, *Escherichia coli*, and *Bacillus subtilis* in batch and continuous packed bed column reactors. The investigation showed that under optimized conditions (pH 3.0; contact time, 180 min; 30°C; initial Cr (VI) concentration of 100 mg/L), the adsorption capacity was 65.86 mg/g in the batch study. Furthermore, an adsorbent dosage of 1 g/L, a flow rate of 5 mL/min, a bed height of 20 cm, an initial Cr (VI) concentration of 300 mg/L, and a capacity of 657 mg/g were observed for the continuous column assay. The Langmuir isotherm and pseudo-second-order kinetics were followed for the batch sorption. From five sorption/desorption cycles, 100%, 99.63%, 95.31%, 80.7%, and 74.22%, regenerations were noted, respectively. The adsorption studies for Cr (VI) using spiked groundwater, freshwater, and domestic wastewater in a packed bed reactor elucidated Cr (VI) removals of 64.8%, 55.08%, and 56.86%, respectively. Fourier transform infrared (FTIR) and energy-dispersive x-ray spectroscopy were used in order to confirm the Cr (VI) sorption on immobilized bacteria (Samuel et al., 2013).

44.3.2 CHROMIUM REMOVAL BY BIOACCUMULATION USING BACTERIA

Bioaccumulation is a procedure wherein metabolically dependent living cells accumulate the contaminants. Preetha and Viruthagiri (2007) demonstrated bioaccumulation capacity of *Rhizopus arrhizus* wherein it was observed that *R. arrhizus* can accumulate chromium(VI), copper(II), and nickel(II) ions.

Sundar et al. (2010) demonstrated Cr (III) bioremoval capacities of *B. subtilis* VITSCCr0, indigenous and adapted bacterial strains from Palar river basin, Vellore, Tamil Nadu, India. Palar River Basin of Vellore district, Tamil Nadu, India, is mushroomed with many tanneries, which are the major cause of Cr (III) pollution. They isolated 68 strains from the Palar River Basin. Out of 68 strains, 45 strains were Cr (III) tolerant, 52 strains were Cr (VI) tolerant, and 29 bacterial strains were tolerant to both the chromium valencies. The tolerance of the strain *B. subtilis* (VS-1) to Cr (III) was found to be 1500 mg/L, and its Cr (III) bioremoval capacity was 64%, whereas the strain *Bacillus cereus* (RS-1) showed tolerance to Cr(VI) up to 3000 mg/L and removal capacity was 70%. The study showed that exopolysaccharide (EPS) production by the bacteria was increased as the concentration of Cr (III) and Cr (VI) increased. It was observed that EPS productions were 62.33 and 54.32 mg/L for VS-1 and RS-1, respectively. Adaptation process was performed by repeatedly growing the bacteria in the presence of a known concentration of Cr (III). Due to adaptation of *B. subtilis* (VITSCCr01) with higher Cr (III) concentration, that is, 2500 mg/L, the bioremoval capacity was improved to 85%. Similarly, for the adapted *B. cereus* strain (VITSCCr02) with higher Cr (VI) concentration, that is, 4000 mg/L, the removal capacity was 88%. Adaptation caused the increase in EPS production also, for example, 75.62 mg/L and 65.48 mg/L for *B. subtilis* (VITSCCr01) and *B. cereus* strain (VITSCCr02), respectively. The presence of polysaccharides in the Cr (III)-treated bacteria was confirmed by the FTIR spectral studies. Accumulation of chromium was observed by SEM-EDX. A biofilm was also prepared using the adapted strains, and the experimental data showed 98% of total chromium removal from the tannery effluent. The study demonstrated that for *in situ* removal of Cr (III) especially in the polluted tannery environment, *B. subtilis* (VITSCCr01) and *B. cereus* (VITSCCr02) can be used as a tool (Sundar et al., 2010).

44.3.3 CHROMIUM REMOVAL BY BIOREDUCTION USING BACTERIA

Reduction of extremely harmful and mobile Cr (VI) by bacterial strains is believed to be a feasible alternative to reduce Cr (VI) contamination, in soils and water bodies, exhaling from the overburden dumps of chromite ores and mine drainage. A variety of bacteria have been reported to reduce Cr (VI) to Cr (III) such as *Achromobacter* sp., *Synechocystis* sp., *Chroococcus* sp., *Microcystis* sp., *Methylobacterium* sp., *Pseudomonas* sp., *Stenotrophomonas* sp., and *B. subtilis* (Ma et al., 2007; Ozturk and Aslim, 2008; Ozturk et al., 2008; Morel et al., 2009; Sundar et al., 2011). The following are three major mechanisms involved in Cr (VI) reduction (Ramírez-Díaz et al., 2008):

1. Chemical reduction of Cr (VI) by cellular compounds such as vitamins, organic acids, amino acids, nucleotides, sugars, and glutathione.
2. Aerobic Cr (VI) reduction, which is related to soluble proteins requiring NADPH as an electron donor.
3. Anaerobic Cr (VI) reduction by membrane-associated reductases of which some can utilize H_2 as an electron donor. Sometimes, Cr (VI) itself acts as an electron acceptor in the electron transport chain during the reduction of Cr (VI) (Cheung and Gu, 2007).

A number of prior studies reported that reduction of Cr (VI) to a less toxic Cr (III) using various anaerobic and aerobic bacteria, though the amount of Cr (VI) reduction and the mechanisms involved would vary and depend on various factors including the source from which the species are isolated (Pei et al., 2009).

There are reports on different genera of *Bacillus* where they exhibit *in vitro* Cr (VI) reduction (Desai et al., 2008; Molokwane et al., 2008). Dhal et al. reported the isolation of Cr (VI)-resistant bacterial strains (CSB-4 that was identified as *Bacillus* sp.) from an Indian chromite mine soil and their potential use in the reduction of hexavalent chromium. This species was found to be tolerant to as high as 2000 mg/L Cr (VI) concentration. The bacterial strain (CSB-4) was capable of reducing Cr (VI) to Cr (III) in different growth media. The study revealed that under the optimized conditions (pH ~ 7.0; Cr(VI), 100 mg/L; temperature, 35°C; stirring speed, 100 ppm), CSB-4 reduced more than 90% of Cr (VI) in 144 h (Dhal et al., 2010). *Acinetobacter* sp. has also been reported as an efficient agent in both Cr (VI) reduction and oil degradation (Zakaria et al., 2007; Mishra et al., 2010).

Singh et al. have reported the efficient removal of chromium (VI), COD, and sulfate from an aqueous solution as well as real effluent (CETP) in a small-scale bioreactor using sulfate-reducing bacteria consortium. They also studied the effect of different hydraulic retention times, initial metal concentrations, various carbon sources, and temperatures on the removal of chromium (VI), COD, and sulfate. Experimental data showed 96% of chromium (VI) removal at an initial concentration of 50 mg/L using lactates as a carbon source. The absence of NADH-dependent chromate reductase proves the presence of anaerobic consortium. Consortium medium with a strong negative oxidation reduction potential indicated the reducing activity. FTIR spectrum showed the existence of the sulfate ions. This indicated the use of sulfate during the growth phase of sulfate-reducing bacteria (Singh et al., 2011).

A recent study was carried out by Samuel et al. where reduction of chromium was observed by three indigenous isolates from chromate-mining sites' water at Sukinda Valley, Orissa, India. The isolates were *B. subtilis* VITSUKMW1, *A. junii* VITSUKMW2, and *E. coli* VITSUKMW3. In order to increase the Cr (VI) reduction rate and the growth rate of bacteria, adaptation was performed for each isolate serially at 10, 20, 30, 50, 80, and 100 mg/L Cr (VI). Enhancement in Cr (VI) bioreduction rate was observed through adaptation and consortia development. A high tolerance was observed for the isolates at 500–1000 mg/L of Cr (VI). For unadapted isolates, initial Cr (VI) concentrations were 5, 10, and 20 mg/L, and for adapted isolates and consortia 5, 10, 20, 50, and 100 mg/L. The reduction rates for the isolate *B. subtilis* VITSUKMW1 were 0.199, 0.218, and 0.463 mg/L/h for 5, 10, and 20 mg/L of initial Cr (VI) concentration, respectively. *A. junii* VITSUKMW2 showed reduction rate as 0.477, 0.317, and 0.286 mg/L/h for 5, 10, and 20 mg/L of initial Cr (VI) concentration, respectively. Similarly, reduction rates for unadapted *E. coli* VITSUKMW3 were 0.25, 0.2, and 0.333 mg/L/h for 5, 10, and 20 mg/L of initial Cr (VI) concentration, respectively. An enhancement in the reduction rate from 0.199–0.477 to 0.5–1.16 mg/L/h at 5–20 mg/L of initial

Cr (VI) concentration was achieved by the isolates, which got adapted (Samuel et al., 2012).

It was observed that there was a correlation between the reduction rate and the growth rate of the bacteria. The experimental data showed that the specific growth rates of unadapted isolates were in the range of 0.032–0.104 h^{-1} at an initial Cr (VI) range of 5–20 mg/L at 30°C, with a pH of 7.5. It was reported that at 20 mg/L initial Cr (VI) concentration, compared to A. *Junii* VITSUKMW2 and E. *coli* VITSUKMW3, B. *subtilis* VITSUKMW1 had a higher reduction rate of 0.463 mg/L/h. The reduction rate of the unadapted isolates was found to be similar to few previously reported data of Cr (VI) reduction by *Bacillus* sp. (Elangovan et al., 2006; Wani et al., 2007).

Compared to the unadapted isolates, an improved microbial growth was observed for the adapted isolates. An increase in the Cr (VI) reduction rate was observed for the adapted isolates and the highest was 2.364 mg/L/h by B. *subtilis* VITSUKMW1 at 100 mg/L initial Cr (VI) concentration, which was substantially more, compared to the earlier reports (Caravelli and Zaritzky, 2009; Zahoor and Rehman, 2009). Similarly, the reduction rate observed for A. *Junii* VITSUKMW2 at 100 mg/L of initial Cr (VI) concentration was 1.5 mg/L/h, which was higher in comparison with that of the earlier reports (Srivastava et al., 2007; Pei et al., 2009). The reduction rate was found to be increased with increase in Cr (VI) concentration. Similarly, E. *coli* VITSUKMW3 showed a high reduction rate of 1.85 mg/L/h at 100 mg/L of initial Cr (VI) concentration, which is importantly higher than that of the reduction rate by E. *coli* of Cr (VI) in earlier reports (Abskharon et al., 2009).

The binary and tertiary consortium of adapted bacterial isolates was prepared in order to enhance the reduction of Cr (VI). The study revealed notable reduction rate while binary consortia were used. For the binary consortium of A. *junii* VITSUKMW2 + E. *coli* VITSUKMW3, it was observed that the reduction rates were 1.23, 1.49, 2.76, 5.23, and 6.38 mg/L/h, when the initial concentrations of Cr (VI) were 5, 10, 20, 50, and 100 mg/L, respectively. Similarly, for the binary consortium of B. *subtilis* VITSUKMW1 + A. *junii* VITSUKMW2, it was observed that the reduction rates were 2.12, 3.8, 4.73, 11.01, and 22.87 mg/L/h, while the initial concentrations of Cr (VI) were 5, 10, 20, 50, and 100 mg/L, respectively. Another observation for the binary consortium of B. *subtilis* VITSUKMW1 + E. *coli* VITSUKMW3 showed the reduction rates 3.33, 4.5, 7.33, 15.99, and 23.17 mg/L/h when the initial concentrations of Cr (VI) were 5, 10, 20, 50, and 100 mg/L, respectively. On the other hand, an increase in growth rate and Cr (VI) reduction rate of 0.86–2.6813 mg/L/h at the initial Cr (VI) concentration of 5–100 mg/L was observed for the ternary consortium of adapted isolates (B. *subtilis* VITSUKMW1 + A. *junii* VITSUKMW2 + E. *coli* VITSUKMW3). The study showed that the rate of Cr (VI) reduction by the ternary consortium at 100 mg/L of initial Cr (VI) concentration was 6 times more than the highest reduction rate of the unadapted isolates at 20 mg/L of initial Cr (VI) concentration.

44.4 REMEDIATION OF CHROMIUM BY ALGAE

Among the biological materials, algae are also used to remediate chromium (VI) or other heavy metals because of their easy availability and fast growth rate in both freshwater and marine environment. They have relatively high surface area and high binding affinity, which make them potential biosorbents for the removal of chromium (VI) or other heavy metals from the environment. The cell structure of an algae mainly contains cellulose and glycoproteins, which further consist of various functional groups such as amino, carboxyl, sulfate, and hydroxyl. These functional groups may play an important role in bioremediation process (Sari and Tuzen, 2008; Jayakumar et al., 2014).

A vast study of more than 20 algae species for chromium (VI) detoxification has been reported by many researchers. Among them, three groups of algae such as red, green, and brown algae are predominant. Brown algae specifically *Sargassum* sp., *Fucus* sp., and *Ascophyllum* sp. are considered to have higher bioremoval capability of chromium (VI) (Sen and Dastidar, 2010). The possible mechanism of chromium (VI) removal from an aqueous solution by nonliving biomass involves the following steps: first, the negatively charged chromium (VI) ions bind with the positively charged groups on the surface of biomass; second, reduction of chromium (VI) to chromium (III) occurs by adjacent electron-donor groups; and finally, an electronic repulsion occurs between the positively charged groups and chromium (III) ions, which lead to release of chromium (III) ions into the aqueous solution (Park et al., 2004, 2005).

Few literatures state about the algal resistance to chromium, but the specific mechanisms of tolerance are still unrevealed. Chromium resistance in freshwater algae *Scenedesmus acutus* has been reported (Corradi et al., 1995b). The algal species was treated with chromate, which caused to prevent its asexual reproduction, but the chromium-tolerant progeny was developed through sexual reproduction. Decrease in cellular metal uptake is not the actual reason behind the algal tolerance to chromium (VI) (Gorbi et al., 1996). In the presence of chromium (VI), S. *acutus* formed aggregates in culture medium. When the aggregates broke up, biflagellated cells were released, and thus, the algal cells were able to survive under stress condition (Corradi et al., 1995a). It is not necessary that the chromium-tolerant algal strain would show equal resistance capability against other heavy metals too. It was evaluated that chromium-tolerant S. *acutus* showed more sensitivity to zinc compared to copper (Abd-El-Monem et al., 1998). Plastid-deprived gamete induction was found to be favored in the presence of chromium (VI) (Corradi et al., 1995a). The involvement of chromium with plastidial metabolism has been explained in three algae species such as *Euglena gracilis*, S. *acutus*, and *Chlorella pyrenoidosa* (Fasulo et al., 1983; Wong and Chang, 1991; Corradi and Gorbi, 1993). Thus, algae could be used as a suitable biological system for the removal of chromium (VI) from the environment, considering their chromium tolerance capability (Figure 44.1).

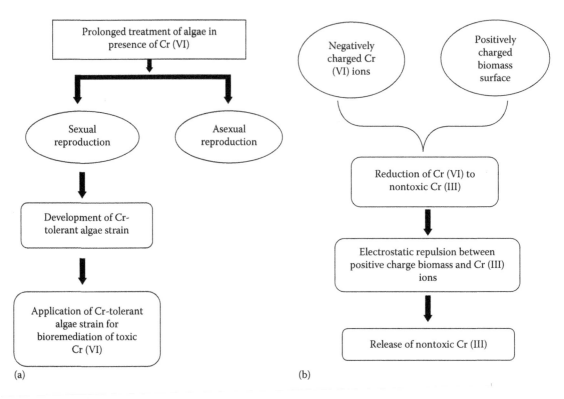

FIGURE 44.1 Schematic illustration of bioremediation of chromium (VI) by algae. (a) Development of chromium-tolerant algal strain for bioremediation purpose. (b) Mechanism of Cr (VI) adsorption and reduction by Cr-tolerant algae.

44.4.1 Chromium Removal by Biosorption Using Algae

Biosorption of hexavalent chromium using chemically modified *Spirulina platensis* algal biomass has been reported (Hegde et al., 2015). The algal biomass was chemically modified by acid treatment, which altered the functional groups present on the cell membrane of biomass. HCl treatment of algal biomass showed the highest sorption capacity for hexavalent chromium at pH 3. The biosorption efficiency of HCl-treated biomass calculated by the Langmuir isotherm model was found to be 5 mg/g for 50 mg/L of initial chromium (VI) concentration at 15 h of contact time.

S. platensis extract (SPE) beads were found to have a chromium (VI) adsorption capacity of 41.12 mg/g at pH 2 with an initial chromium (VI) concentration of 100 mg/L and 1/L of adsorbent dose. The temperature, contact time, and agitation rate for the maximum Cr (VI) removal by SPE beads were found to be 25°C, 24 h, and 180 rpm, respectively. The behavior of chromium (VI) on the SPE beads followed the Brunauer–Emmett–Teller (BET) isotherm (Kwak et al., 2015).

A study conducted by Jayakumar et al. (2015) demonstrated that chemically modified (0.1 M HCl–treated) marine brown algae *Sargassum myriocystum* biomass showed a maximum sorption capacity of 66.66 mg/g at pH 5.2, a sorbent dose of 2.017 g/L, a contact time of 108 min, and an agitation speed of 120 rpm with an initial chromium (VI) concentration of 250 mg/L obtained by the Langmuir and

Toth isotherm models. Hexavalent chromium biosorption capacity of dried biomass of marine green algae *Halimeda gracilis* has also been evaluated by Jayakumar et al. (2014). The algal biomass was chemically modified by treating with 0.1M HCl. The biosorption capacity was found to be maximum, that is, 55.5 mg/g, calculated through the Langmuir isotherm model, at pH 4.9 with 50 mg/L of initial chromium (VI) concentration. The optimum sorbent dose, agitation speed, and contact time were reported as 2.2 g/L, 136 rpm, and 47 min, respectively.

Dried biomass of brown macroalgae species *Sargassum thunbergii* showed the maximum adsorption capacity of 1.855 mmol/g for 2 mmol/L of initial chromium (VI) concentration at pH 2, agitation speed 1500 r/min and contact time 10 min. It was observed that as the pH value increased from 2 to 9, the adsorption rate was gradually decreased. The suitable biomass concentration for maximum biosorption was found to be 1.2 g/L as beyond this concentration up to 4 g/L of biomass dose, the chromium (VI) adsorption capacity per unit biomass weight was decreased up to 0.43 mmol/g (Wang et al., 2014). Chromium (VI) removal capacity of chemically modified (0.1M HCl treated) freshwater algae *Rhizoclonium hookeri* biomass was found to be 67.3 mg/g at pH 2, biomass dose 1 g/L, contact time 45 min, agitation speed 150 rpm with an initial chromium concentration of 1000 mg/L (Kayalvizhi et al., 2015). An increase in dose quantity leads to a decrease of adsorption rate. It could be due to the formation of aggregates of the biomass at higher concentrations, which could decrease the effective surface area for biosorption process

(Mata et al., 2008; Sarı and Tuzen, 2008; Jayakumar et al., 2014; Wang et al., 2014; Kayalvizhi et al., 2015).

Temperature plays an important role in the biosorption process. The specific uptake of chromium (VI) was enhanced to 48.4 mg/g by the algal biomass of *S. myriocystum* as the temperature increased from 20°C to 35°C, but the specific uptake decreased to 46.9 mg/g upon further increase of temperature from 35°C to 40°C (Jayakumar et al., 2015). Similarly, there was an increase in specific uptake of chromium (VI) by the dried biomass of *H. gracilis* when the temperature increased from 293.15 to 308.15 K. But a decrease in specific uptake of chromium (VI) was noted to be 42.8 mg/g when the temperature was further increased from 308.15 to 313.15 K (Jayakumar et al., 2014). Changes in texture and deterioration of absorbent materials at very high temperature could be the possible reason for the reduction in specific uptake of chromium (VI) by algal biomass (Volesky, 2003). But Kayalvizhi et al. (2015) reported that biosorption capacity could be increased with an increase of temperature as pores in the biomass get enlarge, resulting in increased surface available for the sorption process, diffusion, and perforation of chromium (VI) ions within the pores of algal biomass. The highest temperature (45°C) chosen over the range of 25°C, 35°C, and 45°C was found to be optimum for maximum biosorption of chromium (VI) by *S. thunbergii* (Wang et al., 2014).

FTIR spectral analysis confirmed that the modification of functional groups of carbohydrates, proteins, and lipids present in the cell membrane was found to be involved in the biosorption process (Hegde et al., 2015). The possible functional groups such as amido, hydroxyl, C=O, C–O, C–H$_2$, and –COOH could play a major role in the biosorption process (Loukidou et al., 2004; Popuri et al., 2007; Jayakumar et al., 2014; Wang et al., 2014; Kayalvizhi et al., 2015). The additional functional group SO$_3^2$ took part in adsorption with *H. gracilis* (Jayakumar et al., 2014).

44.4.2 Chromium Removal by Bioaccumulation Using Algae

Bioaccumulation consists of a combination of two stages such as passive and active. Passive stage occurs rapidly lasting for few hours. The passive stage is also known as biosorption, wherein the binding of metal ions takes place to the surface of the algal cell wall takes place. The active stage is a relatively slower stage than passive. It lasts for more than 10 h. In this stage, cellular energy is consumed for active transport of metal ions into the cell interior. Thus, the active stage is known as specific bioaccumulation or adequate bioaccumulation (Mehta and Gaur, 2005). Apart from transportation of metal ions into the cell interior, adequate bioaccumulation needs cellular energy for simultaneous release of metal-binding sites on the cell surface, synthesis of metal-binding proteins, and cellular growth. By the end of the bioaccumulation process, all the metal ions are removed from the surface of the cells. The passive biosorption process can take place by both living and nonliving cells, while the active bioaccumulation can take place only by living cells (Chojnacka and

Wojciechowski, 2007). The time consumed during the bioaccumulation process depends on metal ion concentration in the solution as well as on the cell surface. The chemical nature of heavy metals also plays a major role in the bioaccumulation process. Algae are more efficient biological agents in heavy metal ion binding than bacteria or fungi because the process carrying out algae is associated with their metabolic activity and photosynthesis (Sunda, 1989; Barron, 1995; Khoshmanesh et al., 1997; Inthorn, 2001, Perales-Vela et al., 2006).

Chromium (III)-tolerant microalgae *Dictyosphaerium chlorelloides* strain was developed by treating the wild-type *D. chlorelloides* cells (approximately 109 cells) with 112.6 μM chromium chloride (CrCl$_3$·6H$_2$O) (Pereira et al., 2013). Toxicity assay revealed the median growth inhibitory concentration (IC$_{50}$) to be 1883.5 and 53.5 μM of chromium chloride for chromium (III)-tolerant and chromium (III)-nontolerant wild-type *D. chlorelloides* strains, respectively. Biosorption and bioaccumulation of chromium (III) by chromium (III)-tolerant *D. chlorelloides* strain was investigated. An increased vacuolization and thick cell wall were observed through a transmission electron microscopic analysis of the chromium (III)-tolerant strain in comparison with the nontolerant wild-type strain. Distributions of considerable amounts of chromium were quantified through energy-dispersive x-ray analysis in cell wall as well as in cytoplasm, chloroplast, and vacuoles. Observed were 25–150 nm diameter high-density precipitates in cell walls, inside cytoplasms, chloroplast, and vacuoles of a chromium (III)-tolerant strain, while these precipitates were not found inside wild-type algal strains. *D. chlorelloides* cells accumulated chromium (III) on their cell surface through bioadsorption process, while through bioaccumulation process, chromium was transported inside the cell. *D. chlorelloides* cells were found to bioaccumulate chromium around 0.56% of its cell volume. Electrostatic force is considered as the main driving force of the bioadsorption process, whereas the bioaccumulation process is truly controlled by enzymes of the cells. It has been proposed that heavy metals accumulated by a living cell either bind to intracellular components or just precipitate. The activity of heavy metals may affect the cellular macromolecules, enzymes, and metal cofactors (Gadd, 1988). Detoxification of heavy metals occurs by the bioaccumulation process or by binding of heavy metals with proteins in both prokaryotic and eukaryotic cells (Zhang and Majidi, 1994).

Chromium (VI) accumulation by *Scenedesmus incrassatulus* cells in an airlift photobioreactor was evaluated (Jácome-Pilco et al., 2009). On the first day of experiment, maximum chromium (VI) uptake by algal cells was found to be 4.4 mg/g dry biomass. After 14 days, a steady-state condition was achieved where chromium (VI) bioaccumulation in the cells decreased to 1.7 mg/g dry biomass at pH 8.9, temperature 25°C, and air flow rate 997 mL/min with an initial chromium concentration of 1 mg/L. Cervantes et al. and Pinto et al. reported that chromate can diffuse through cellular membrane and generate large amounts of reactive oxygen species (ROS) due to its high oxidation capability. The ROS then reacts with photosynthetic pigments in algal cells and induces morphological changes (Cervantes et al., 2001; Pinto et al., 2003). Thus, photosynthetic pigments such as chlorophyll a, chlorophyll b, and carotenoids of *S. incrassatulus* were decreased by 34.1%, 35.4%, and 27.8%, respectively, at steady-state condition in comparison with control that lacked chromium (VI) feed. The advantage of morphological changes in algal cells due to the oxidation state of chromium indicates different responses between sensitive and tolerant algal strains/species, which could be useful to develop a specific algal biosensor for industrial application (Pereira et al., 2013).

44.4.3 CHROMIUM REMOVAL BY BIOREDUCTION USING ALGAE

Algae are the natural electron donors for oxidation/reduction reactions, which take place on the surface of the cells. An example of one such reaction is reduction of hexavalent chromium to its trivalent form (Dittert et al., 2014). Marine macroalgae *Sargassum cymosum* was used as an electron donor for the reduction of Cr(VI) to Cr(III) from an aqueous solution (de Souza et al., 2016). The reduction process takes place when algal biomass is being oxidized. Cr (VI) reduction mainly depends on pH of the solution as the reduction mechanism is carried out by the involvement of protons. Nearly, complete reduction of Cr (VI) with 0.6 mM/L initial concentration was observed in less than 20 min at pH ≤ 2, algal biomass 2 g/L, temperature 25°C, and agitation speed 190 rpm. pH values between 2.5 and 3 showed removal of more than 80% of total chromium from the solution after 72 h, which confirmed that the biomass of *S. cymosum* served as an electron donor for the reduction of Cr (VI) to Cr (III). At the same time, the algae biomass also acts as a natural cation exchanger for the sequestration of Cr (III) from a solution. It is quite easier for the reduction of hexavalent chromium to its trivalent state when in contact with biological agents (e.g., algae) at an acidic pH of solution because Cr (VI) shows higher redox potential above +1.3 V (Yang and Chen, 2008; Chen et al., 2010; Park et al., 2011). Biomass dose also plays an important role in Cr (VI) removal. Dose-dependent increase in Cr(VI) reduction by *S. cymosum* was noted at pH 2 and temperature 25°C with an initial Cr(VI) concentration of 0.6 mM/ L. Cr (VI) removal rate was found to be doubled when the biomass dose increased from 0.5 to 1 g/L because there was sufficient availability of electrons for reduction reaction. Total chromium removal by *S. cymosum* was found to be 34%, 52%, 71%, and 83% for increased biomass dose of 0.5, 1, 2, and 4 g/L, respectively. FTIR and potentiometric titration techniques revealed that carboxylic groups present on the biomass surface of *S. cymosum* were mainly involved in the Cr (III) removal process.

Adsorption-coupled reduction mechanism of hexavalent chromium removal by immobilized freshwater microalgal consortium has been reported (Samuel et al., 2015). Four freshwater microalgal species such as *Oocystis* sp., *Nostoc* sp., *Syncoccus* sp., and *Desmococcus* sp. were used to develop

consortium for Cr (VI) adsorption as well as reduction through batch and column reactor studies. The adsorption capacity increased to 83.5 and 579.2 mg/g in batch and column reactors, respectively, with an initial Cr (VI) concentration of 300 mg/L at pH 3, temperature 30°C, biomass dose 1 g/L, and contact time 180 min. Adsorption-mediated surface reduction of hexavalent chromium to trivalent form by microalgal consortium was confirmed by FTIR, EDAX, and EPR analysis. The possible mechanism of Cr (VI) reduction could be demonstrated as follows:

1. $R–CHO + 2HCrO_4^- (sorbent) + 2H^+ (aqueous) \rightarrow R–COOH (sorbent) + Cr_2O_3 (sorbent) + 2H_2O (aqueous)$
2. $C_6H_5OH + 2HCrO_4^- (sorbent) + H^+ (aqueous) \rightarrow C_6H_5O^- (sorbent) + Cr_2O_3 (sorbent) + 2H_2O (aqueous)$

44.5 SUMMARY

Applications of chromium in industries are increasing day by day. Thus, release of toxic hexavalent chromium into the environment could cause environmental pollution, which further leads to health risks. It is necessary to remediate toxic hexavalent chromium from the environment. Various conventional methods such as ion exchange, solvent extraction, and sulfate precipitation are used to remediate chromium (VI), but all these methods are costly and may generate chemical by-products that could act as pollutants. Thus, bioremediation is an appropriate option for the removal of chromium to avoid the side effects of conventional methods. Among various biological agents, microbes such as bacteria and algae show diverse interaction with chromium in the environment. They follow three different modes for detoxification and removal of chromium such as biosorption, bioaccumulation, and bioreduction. Their quick adaptive nature under stressed conditions makes them resistant to various heavy metals. Apart from these, fast growth rate, easy availability, and low maintenance cost make them suitable biological agents for the bioremediation of chromium.

REFERENCES

Abd-El-Monem, H.M., Corradi, M.G., Gorbi, G. 1998. Toxicity of copper and zinc to two strains of Scenedesmus acutus having different sensitivity to chromium. *Environmental and Experimental Botany* 40: 59–66.
Abskharon, R.N.N., Gad El-Rab, S.M.F., Hassan, S.H.A., Shoreit, A.A.M. 2009. Reduction of toxic hexavalent chromium by *E. coli. Journal of Biotechnology Biochemistry* 4: 98–103.
Ali, S., Bai, P., Zeng, F., Cai, S., Shamsi, I.H., Qiu, B., Wu, F., Zhang, G. 2011. The ecotoxicological and interactive effects of chromium and aluminum on growth, oxidative damage and antioxidant enzymes on two barley genotypes differing in Al tolerance. *Environmental and Experimental Botany* 70: 185–191.
ATSDR (Agency for toxic substances and disease registry). 1999. Prepared by Clement International Corporation. Under Contract to 205: 88-0608, Atlanta, GA.

Baral, A., Engelken, R.D. 2002. Chromium-based regulations and greening in metal finishing industries in the USA. *Environmental Science & Policy* 5: 121–133.

Barron, M.G. 1995. Bioaccumulation and bioconcentration in aquatic organisms. *Handbook of Ecotoxicology*, Lewis Publisher, Boca Raton, FL, pp. 652–666.

Boopathy, R. 2000. Factors limiting bioremediation technologies. *Bioresource Technology* 74: 63–67.

Caravelli, A.H., Zaritzky, N.E. 2009. About the performance of *Sphaerotilus natans* to reduce hexavalent chromium in batch and continuous reactors. *Journal of Hazardous Materials* 168: 1346–1358.

Cervantes, C., Campos-García, J., Devars, S., Gutiérrez-Corona, F., Loza-Tavera, H., Torres-Guzmán, J.C., Moreno-Sánchez, R. 2001. Interactions of chromium with microorganisms and plants. *FEMS Microbiology Reviews* 25: 335–347.

Chen, S., Yue, Q., Gao, B., Xu, X. 2010. Equilibrium and kinetic adsorption study of the adsorptive removal of Cr (VI) using modified wheat residue. *Journal of Colloid and Interface Science* 349: 256–264.

Cheung, K.H., Gu, J.D. 2007. Mechanism of hexavalent chromium detoxification by microorganisms and bioremediation application potential: A review. *International Biodeterioration & Biodegradation* 59: 8–15.

Chojnacka, K., Wojciechowski, P.M. 2007. Bioaccumulation of Cr (III) ions by blue green-alga Spirulina sp. Part II. Mathematical modeling. *American Journal of Agricultural and Biological Science* 2: 291–298.

Corradi, M., Gorbi, G. 1993. Chromium toxicity on two linked trophic levels II. Morphophysiological effects on *Scenedesmus acutus*. *Ecotoxicology and Environmental Safety* 25: 72–78.

Corradi, M.G., Gorbi, G., Bassi, M. 1995a. Hexavalent chromium induces gametogenesis in the freshwater alga *Scenedesmus acutus*. *Ecotoxicology and Environmental Safety* 30: 106–110.

Corradi, M.G., Gorbi, G, Ricci, A., Torelli, A., Bassi, M. 1995b. Chromium-induced sexual reproduction gives rise to a Cr-tolerant progeny in *Scenedesmus acutus*. *Ecotoxicology and Environmental Safety* 32: 12–18.

Davis, M., Cornwell, D. 2008. *Introduction to Environmental Engineering*, McGraw-Hill Companies, Dubuque, IA.

De Souza, F.B., De Lima Brandão, H., Hackbarth, F.V., de Souza, A.A.U., Boaventura, R.A., de Souza, S.M.G.U., Vilar, V.J. 2016. Marine macro-alga *Sargassum cymosum* as electron donor for hexavalent chromium reduction to trivalent state in aqueous solutions. *Chemical Engineering Journal* 283: 903–910.

Deng, L., Su, Y., Su, H., Wang, X., Zhu, X. 2006. Biosorption of copper (II) and lead (II) from aqueous solutions by nonliving green algae *Cladophora fascicularis*: Equilibrium, kinetics and environmental effects. *Adsorption* 12: 267–277.

Desai, C., Jain, K., Madamwar, D. 2008. Evaluation of in vitro Cr (VI) reduction potential in cytosolic extracts of three indigenous *Bacillus* sp. isolated from Cr (VI) polluted industrial landfill. *Bioresource Technology* 99: 6059–6069.

Dhal, B., Thatoi, H., Das, N., Pandey, B.D. 2010. Reduction of hexavalent chromium by *Bacillus* sp. isolated from chromite mine soils and characterization of reduced product. *Journal of Chemical Technology and Biotechnology* 85: 1471–1479.

Dittert, I.M., de Lima Brandão, H., Pina, F., da Silva, E.A., de Souza, S.M.G.U., de Souza, A.A.U., Botelho, C.M., Boaventura, R.A., Vilar, V.J. 2014. Integrated reduction/oxidation reactions and sorption processes for Cr (VI) removal from aqueous solutions using Laminaria digitata macro-algae. *Chemical Engineering Journal* 237: 443–454.

Elangovan, R., Abhipsa, S., Rohit, B., Ligy, P., Chandraraj, K. 2006. Reduction of Cr (VI) by a *Bacillus* sp. *Biotechnology Letters* 28: 247–252.

Elangovan, R., Philip, L. 2009. Performance evaluation of various bioreactors for the removal of Cr (VI) and organic matter from industrial effluent. *Biochemical Engineering Journal* 44: 174–186.

Falkowski, P.G., Fenchel, T., Delong, E.F. 2008. The microbial engines that drive Earth's biogeochemical cycles. *Science* 320: 1034–1039.

Farhadian, M., Vachelard, C., Duchez, D., Larroche, C. 2008. In situ bioremediation of monoaromatic pollutants in groundwater: A review. *Bioresource Technology* 99: 5296–5308.

Fasulo, M.P., Bassi, M., Donini, A. 1983. Cytotoxic effects of hexavalent chromium in *Euglena gracilis*. II. Physiological and ultrastructural studies. *Protoplasma* 114: 35–43.

Gadd, G.M. 1988. Accumulation of metals by microorganisms and algae. *Biotechnology* 6: 401–433.

Gadd, G.M. 2000. Bioremedial potential of microbial mechanisms of metal mobilization and immobilization. *Current Opinion in Biotechnology* 11: 271–279.

Gavrilescu, M. 2004. Removal of heavy metals from the environment by biosorption. *Chemistry* 28: 30.

Glanze, W. 1996. *Mosby Medical Encyclopedia*, Revised Edition, St. Louis, MO: C.V. Mosby, p. 318.

Gorbi, G., Corradi, M.G., Torelli, A., Bassi, M. 1996. Comparison between a Normal and a Cr-Tolerant Strain of *Scenedesmus acutus* as a Food Source to *Daphnia magna*. *Ecotoxicology and Environmental Safety* 35: 109–111.

Hegde, S.M., Babu, R.L., Vijayalakshmi, E., Patil, R.H., Naveen Kumar, M., Kiran Kumar, K.M., Nagesh, R., Kavya, K., Sharma, S.C. 2015. Biosorption of hexavalent chromium from aqueous solution using chemically modified *Spirulina platensis* algal biomass: An ecofriendly approach. *Desalination and Water Treatment* 57: 1–10.

Hogan, C.M. 2010. *Heavy Metal: Encyclopedia of Earth*, National Council for Science and the Environment, Washington, DC.

Inthorn, D. 2001. *Removal of Heavy Metal by Using Microalgae: Photosynthetic Microorganisms in Environmental Biotechnology*, Springer, Hong Kong, pp. 111–135.

Jácome-Pilco, C.R., Cristiani-Urbina, E., Flores-Cotera, L.B., Velasco-García, R., Ponce-Noyola, T., Cañizares-Villanueva, R.O. 2009. Continuous Cr (VI) removal by *Scenedesmus incrassatulus* in an airlift photobioreactor. *Bioresource Technology* 100: 2388–2391.

Jayakumar, R., Rajasimman, M., Karthikeyan, C. 2014. Sorption of hexavalent chromium from aqueous solution using marine green algae *Halimeda gracilis*: Optimization, equilibrium, kinetic, thermodynamic and desorption studies. *Journal of Environmental Chemical Engineering* 2: 1261–1274.

Jayakumar, R., Rajasimman, M., Karthikeyan, C. 2015. Sorption and desorption of hexavalent chromium using a novel brown marine algae *Sargassum myriocystum*. *Korean Journal of Chemical Engineering* 32: 2031–2046.

Kayalvizhi, K., Vijayaraghavan, K., Velan, M. 2015. Biosorption of Cr (VI) using a novel microalga *Rhizoclonium hookeri*: Equilibrium, kinetics and thermodynamic studies. *Desalination and Water Treatment* 56: 194–203.

Khoshmanesh, A., Lawson, F., Prince, I.G. 1997. Cell surface area as a major parameter in the uptake of cadmium by unicellular green microalgae. *The Chemical Engineering Journal and the Biochemical Engineering Journal* 65: 13–19.

Kumar, V., Agrawal, A., Pandey, B. 1998. Processing of spent tanning and chrome plating solutions for chromium recovery in Bandopadhyay, A., Goswami, N.G., and Rao, P.R. (eds.), *Environmental and Waste Management*. NML: Jamshedpur, India, pp. 174–184.

Kwak, H.W., Kim, M.K., Lee, J.Y., Yun, H., Kim, M.H., Park, Y.H., Lee, K.H. 2015. Preparation of bead-type biosorbent from water-soluble *Spirulina platensis* extracts for chromium (VI) removal. *Algal Research* 7: 92–99.

Lide, D.R. 1992. 1993. *CRC Handbook of Chemistry and Physics*, CRC Press, Boca Raton, FL.

Loukidou, M.X., Zouboulis, A.I., Karapantsios, T.D., Matis, K.A. 2004. Equilibrium and kinetic modeling of chromium (VI) biosorption by *Aeromonas caviae*. *Colloids and Surfaces A: Physicochemical and Engineering Aspects* 242: 93–104.

Ma, Z., Zhu, W., Long, H., Chai, L., Wang, Q. 2007. Chromate reduction by resting cells of *Achromobacter* sp. Ch-1 under aerobic conditions. *Process Biochemistry* 42: 1028–1032.

Malik, A. 2004. Metal bioremediation through growing cells. *Environment International* 30: 261–278.

Mata, Y.N., Blazquez, M.L., Ballester, A., Gonzalez, F., Munoz, J.A. 2008. Characterization of the biosorption of cadmium, lead and copper with the brown alga *Fucus vesiculosus*. *Journal of Hazardous Materials* 158: 316–323.

Mehta, S., Gaur, J. 2005. Use of algae for removing heavy metal ions from wastewater: Progress and prospects. *Critical Reviews in Biotechnology* 25: 113–152.

Mishra, V., Samantaray, D.P., Dash, S.K., Mishra, B.B., Swain, R.K. 2010. Study on hexavalent chromium reduction by chromium resistant bacterial isolates of Sukinda mining area. *Our Nature* 8: 63–71.

Molokwane, P.E., Meli, K.C., Nkhalambayausi-Chirwa, E.M. 2008. Chromium (VI) reduction in activated sludge bacteria exposed to high chromium loading: Brits culture (South Africa). *Water Research* 42: 4538–4548.

Morel, M.A., Ubalde, M.C., Olivera-Bravo, S., Callejas, C., Gill, P.R., Castro-Sowinski, S. 2009. Cellular and biochemical response to Cr (VI) in *Stenotrophomonas* sp. *FEMS Microbiology Letters* 291: 162–168.

Murphy, V., Hughes, H., McLoughlin, P. 2008. Comparative study of chromium biosorption by red, green and brown seaweed biomass. *Chemosphere* 70: 1128–1134.

Ozturk, S., Aslim, B. 2008. Relationship between chromium (VI) resistance and extracellular polymeric substances (EPS) concentration by some cyanobacterial isolates. *Environmental Science and Pollution Research* 15: 478–480.

Ozturk, S., Aslim, B. and Ugur, A. 2008. Chromium (VI) resistance and extracellular polysaccharide (EPS) synthesis by *Pseudomonas*, *Stenotrophomonas* and *Methylobacterium* strains. *ISIJ International* 48: 1654–1658.

Padmini, E., Rani, M.U., Geetha, B.V. 2009. Studies on antioxidant status in *Mugil cephalus* in response to heavy metal pollution at Ennore estuary. *Environmental Monitoring and Assessment* 155: 215–225.

Park, D., Yun, Y.S., Lee, D.S., Park, J.M. 2011. Optimum condition for the removal of Cr (VI) or total Cr using dried leaves of *Pinus densiflora*. *Desalination* 271: 309–314.

Park, D., Yun, Y.S., Park, J.M. 2004. Reduction of hexavalent chromium with the brown seaweed Ecklonia biomass. *Environmental Science & Technology* 38: 4860–4864.

Park, D., Yun, Y.S., Park, J.M. 2005. Studies on hexavalent chromium biosorption by chemically-treated biomass of *Ecklonia* sp. *Chemosphere* 60: 1356–1364.

Pei, Q.H., Shahir, S., Raj, A.S., Zakaria, Z.A., Ahmad, W.A. 2009. Chromium (VI) resistance and removal by *Acinetobacter haemolyticus*. *World Journal of Microbiology and Biotechnology* 25: 1085–1093.

Pei, Q.H., Shahir, S., Tao, L., Ahmad, W.A. 2014. Determination of chromium (VI) reduction by *Acinetobacter haemolyticus* using X-ray absorption fine structure spectroscopy. *Malaysian Journal of Fundamental and Applied Sciences* 4: 415–422.

Perales-Vela, H.V., Pena-Castro, J.M., Canizares-Villanueva, R.O. 2006. Heavy metal detoxification in eukaryotic microalgae. *Chemosphere* 64: 1–10.

Pereira, M., Bartolomé, M.C., Sánchez-Fortún, S. 2013. Bioadsorption and bioaccumulation of chromium trivalent in Cr (III)-tolerant microalgae: A mechanisms for chromium resistance. *Chemosphere* 93: 1057–1063.

Pinto, E., Sigaud-kutner, T., Leitao, M.A., Okamoto, O.K., Morse, D., Colepicolo, P. 2003. Heavy metal–induced oxidative stress in algae. *Journal of Phycology* 39: 1008–1018.

Popuri, S.R., Kalyani, S., Kachireddy, S.R., Krishnaiah, A. 2007. Biosorption of hexavalent chromium from aqueous solution by using prawn pond algae (*Sphaeroplea*). *Indian Journal of Chemistry Section A* 46: 284.

Preetha, B., Viruthagiri, T. 2007. Bioaccumulation of chromium (VI), copper (II) and nickel (II) ions by growing *Rhizopus arrhizus*. *Biochemical Engineering Journal* 34: 131–135.

Radhika, V., Subramanian, S., Natarajan, K.A. 2006. Bioremediation of zinc using *Desulfotomaculum nigrificans*: Bioprecipitation and characterization studies. *Water Research* 40: 3628–3636.

Ramírez-Díaz, M.I., Díaz-Pérez, C., Vargas, E., Riveros-Rosas, H., Campos-García, J., Cervantes, C. 2008. Mechanisms of bacterial resistance to chromium compounds. *Biometals* 21: 321–332.

Ruiz-Manriquez, A., Magana, P.I., Lopez, V., Guzman, R. 1998. Biosorption of Cu by Thiobacillus ferrooxidans. *Bioprocess Engineering* 18: 113–118.

Samuel, J., Paul, M.L., Kumari, J., Ravikumar, K.V.G., Natarajan, C., Mukherjee, A. 2015. Removal of Cr (VI) by immobilized consortium of freshwater microalgae in batch and continuous system. *Asian Journal of Chemistry* 27: 2161.

Samuel, J., Paul, M.L., Pulimi, M., Nirmala, M.J., Chandrasekaran, N., Mukherjee, A. 2012. Hexavalent chromium bioremoval through adaptation and consortia development from Sukinda chromite mine isolates. *Industrial & Engineering Chemistry Research* 51: 3740–3749.

Samuel, J., Pulimi, M., Paul, M.L., Maurya, A., Chandrasekaran, N., Mukherjee, A. 2013. Batch and continuous flow studies of adsorptive removal of Cr (VI) by adapted bacterial consortia immobilized in alginate beads. *Bioresource Technology* 128: 423–430.

Sari, A., Tuzen, M. 2008. Biosorption of total chromium from aqueous solution by red algae (*Ceramium virgatum*): Equilibrium, kinetic and thermodynamic studies. *Journal of Hazardous Materials* 160: 349–355.

Sen, M., Dastidar, M.G. 2010. Chromium removal using various biosorbents. *Iranian Journal of Environmental Health Science & Engineering* 7: 189.

Singh, R., Kumar, A., Kirrolia, A., Kumar, R., Yadav, N., Bishnoi, N.R., Lohchab, R.K. 2011. Removal of sulphate, COD and Cr (VI) in simulated and real wastewater by sulphate reducing bacteria enrichment in small bioreactor and FTIR study. *Bioresource Technology* 102: 677–682.

Singh, S.K., Bansal, A., Jha, M.K., Dey, A. 2012. An integrated approach to remove Cr (VI) using immobilized *Chlorella minutissima* grown in nutrient rich sewage wastewater. *Bioresource Technology* 104: 257–265.

Srivastava, S., Ahmad, A.H., Thakur, I.S. 2007. Removal of chromium and pentachlorophenol from tannery effluents. *Bioresource Technology* 98: 1128–1132.

Strong, P.J., Burgess, J.E. 2008. Treatment methods for wine-related and distillery wastewaters: A review. *Bioremediation Journal* 12: 70–87.

Sunda, W.G. 1989. Trace metal interactions with marine phytoplankton. *Biological Oceanography* 6: 411–442.

Sundar, K., Mukherjee, A., Sadiq, M., Chandrasekaran, N. 2011. Cr (III) bioremoval capacities of indigenous and adapted bacterial strains from Palar river basin. *Journal of Hazardous Materials* 187: 553–561.

Sundar, K., Vidya, R., Mukherjee, A., Chandrasekaran, N. 2010. High chromium tolerant bacterial strains from Palar River basin: Impact of tannery pollution. *Research Journal of Environmental and Earth Sciences* 2: 112–117.

Thomas, V.L., Gropper, S.S. 1996. Effect of chromium nicotinic acid supplementation on selected cardiovascular disease risk factors. *Biological Trace Element Research* 55: 297–305.

Uslu, G., Tanyol, M. 2006. Equilibrium and thermodynamic parameters of single and binary mixture biosorption of lead (II) and copper (II) ions onto *Pseudomonas putida*: Effect of temperature. *Journal of Hazardous Materials* 135: 87–93.

Vidali, M. 2001. Bioremediation: Aan overview. *Pure and Applied Chemistry* 73: 1163–1172.

Volesky, B. 2003. *Sorption and biosorption*. BV Sorbex, Inc./McGill University, Montreal, Quebec, Canada, 316pp.

Wang, Y., Li, Y., Zhao, F.J. 2014. Bisorption of chromium (VI) from aqueous solutions by *Sargassum thunbergii* Kuntze. *Biotechnology & Biotechnological Equipment* 28: 259–265.

Wani, P.A., Khan, M.S., Zaidi, A. 2007. Chromium reduction, plant growth–promoting potentials, and metal solubilizatrion by *Bacillus* sp. isolated from Alluvial Soil. *Current Microbiology* 54: 237–243.

Wong, P.K., Chang, L. 1991. Effects of copper, chromium and nickel on growth, photosynthesis and chlorophyll a synthesis of *Chlorella pyrenoidosa* 251. *Environmental Pollution* 72: 127–139.

Yang, L., Chen, J.P. 2008. Biosorption of hexavalent chromium onto raw and chemically modified *Sargassum* sp. *Bioresource Technology* 99: 297–307.

Zahoor, A., Rehman, A. 2009. Isolation of Cr (VI) reducing bacteria from industrial effluents and their potential use in bioremediation of chromium containing wastewater. *Journal of Environmental Sciences* 21: 814–820.

Zakaria, Z.A., Zakaria, Z., Surif, S., Ahmad, W.A. 2007. Hexavalent chromium reduction by Acinetobacter haemolyticus isolated from heavy-metal contaminated wastewater. *Journal of Hazardous Materials* 146: 30–38.

Zayed, A.M., Terry, N. 2003. Chromium in the environment: Factors affecting biological remediation. *Plant and Soil* 249: 139–156.

Zhang, W., Majidi, V. 1994. Monitoring the cellular response of Stichococcus bacillaris to exposure of several different metals using in vivo 31P NMR and other spectroscopic techniques. *Environmental Science & Technology* 28: 1577–1581.

45 Bacterial Interactions with Chromium and Strategies for the Remediation of Hexavalent Chromium Pollution

Kunal Jain, Chirayu Desai, and Datta Madamwar

CONTENTS

ABSTRACT

The rapidly increasing industrialization has mobilized chromium into the pristine environment. This has resulted in an increase in the concentration of chromium severalfold than usually found in nature. Given its potential mutagenic and carcinogenic effects in humans, hexavalent chromium has been prioritized as one of the major inorganic environmental pollutants by environmental agencies of different countries. In the environment, chromium is prevalent in several oxidation states; however, the two most ecologically stable forms are the Cr(III)–trivalent chromium and Cr(VI)–hexavalent chromium. The toxicity potential of Cr(VI) is considered higher than Cr(III) due to its high solubility and mobility in the aqueous environment. Many indigenous bacterial species have been identified at chromium-contaminated sites. These bacterial strains have evolved different strategies to evade chromium toxicity and also have the ability to detoxify or transform the toxic hexavalent chromium. This ability of certain bacterial species for reductive immobilization of Cr(VI) to innocuous Cr(III) via soluble or cytoplasmic enzymes or via other mechanisms offers a great advantage in the biological treatment of hexavalent chromium. The bacterial mechanisms to evade chromium toxicity generally involve decreased uptake or exclusion of Cr(VI) compounds through the cell membranes, upregulation of genes associated with oxidative stress response, or biosorption. Given the successful results of several studies using microorganisms to mitigate chromium pollution, eco-friendly bioremediation approaches are nowadays considered to be pragmatic over physicochemical treatment processes. This chapter focuses on the bacterial interactions with chromium and bacterial detoxification strategies in treating hexavalent chromium pollution.

45.1 INTRODUCTION

Chromium, the seventh most abundant element comprising about 0.037% of the earth's crust (0.1–0.3 mg/g), is found in the soil, rocks, volcanic dust, and gases (McGrath and Smith, 1990; Ahemad, 2014; Joutey et al., 2015). According to the periodic table, chromium is a redox active 3d transition metal, which belongs to the group VI-B along with molybdenum and tungsten. It was first discovered as an element in the Siberian red crystalline mineral ore, crocoite ($PbCrO_4$), in the year 1797 by the French chemist Louis Nicolas Vauquelin. It is a steel-gray, hard-crystalline, lustrous, odorless, and tasteless metal (Mishra and Bharagava, 2016). In nature, chromium is found in the form of chromium compounds as ores of chromite [(Fe,Mg)O(Cr, Al, Fe)$_2$O$_3$]. The largest deposits are found in Kazakhstan and South Africa, while substantial producers are the Philippines, Southern Zimbabwe, and Turkey, and the total world reserve is estimated at >480,000 metric tons of chromite ore (Papp, 2012; Dhal et al., 2013).

Chromium generally exists in different oxidation states from −2 to +6, but the most commonly found states are 0, +2, +3, and +6. Elemental chromium is not naturally found in the earth's crust and considered as biologically inert (Mishra and Bharagava, 2016). But its two most ecologically stable states are the hexavalent Cr(VI) and trivalent Cr(III) forms (Bartlett, 1991; Smith et al., 2002; Zayed and Terry, 2003). The physicochemical properties and biological reactivity of both states vary considerably in terms of its solubility, mobility, bioavailability, and toxicity. Trivalent chromium (Cr(III)) has greater affinity toward organic compounds and forms insoluble complexes that precipitates in the form of hydroxides, oxides, and sulfates (Nickens et al., 2010). The emission of hexavalent chromates is mainly due to anthropogenic activities and mere 0.001% is reported to be produced by natural processes; moreover, Cr(VI) is highly soluble that makes it exceedingly persistent in the environment (Ahemad, 2014). The soil organic matter can reduce Cr(VI) in Cr(III), but the reducing capacity of soil decreases at higher concentration of Cr(VI), which resulted in the accumulation of Cr(VI) in the soil and thereby in enhances its biological availability (Cheung and Gu, 2007; Nickens et al., 2010). Hexavalent chromium–Cr(VI) is the most stable form as well as strong oxidizing agent and in solutions it exists as hydrochromate ($HCrO_4^-$), chromate (CrO_4^{2-}), and dichromate ($Cr_2O_7^{2-}$) ions. At basic and neutral pH, chromate forms predominate; at lower pH (6.0–6.2), the hydrochromate increases (Mishra and Bharagava, 2016).

Upon the release of chromium-containing compounds into the open environment, Cr(VI) becomes a potential groundwater contaminant and therefore enters the food chains. The U.S. Environmental Protection Agency has identified Cr(VI) as 1 of the 17 chemicals posing the greatest threat to humans (Marsh and McInerney, 2001).

45.2 SOURCE OF CHROMIUM POLLUTION

Industrial manufacturing activities are the major sources of pollution causing an estimated annual discharge of greater than 170,000 tons of chromium into the environment (Kamaludeen et al., 2003). Chromium is used in chemical, leather, refractory brick, wood preservation, dyes and pigments, textiles, paints, inks, fungicides, rubber, ceramics, metallurgical aircrafts, gas, chrome plating, stainless steel welding, mine tailing, pulp and paper production, corrosion control, petrochemicals and nuclear reactor vessels, negative and film making, pyrotechnics, electronics, and cement-producing plants

sectors (Langard, 1980; Losi and Frankenberger, 1994; Baral et al., 2006; Blade at al., 2007; Alam and Ahmad, 2012; Lu et al., 2013).

45.2.1 METALLURGY INDUSTRIES

Chromium metal is widely used in metallurgical operations in the steel industries in alloyed steels and in nonferrous alloys as an anticorrosive agent (Mishra and Bharagava, 2016). In metallurgical processes, chromium is usually alloyed with iron, nickel, and cobalt to enhance their resistance to corrosion and oxidation. It is used in electroplating industries to oxidize metal surfaces for providing smooth, shiny, and clean finishes.

45.2.2 TANNERY INDUSTRY

Tannery industries are among the major anthropogenic sources causing significant chromium pollution. During the tanning process, salts of chromium, such as Cr(III) sulfate ($Cr(H_2O)_5(OH)SO_4$), are widely used as tanning agents for converting hides and skins into a leather (Mishra and Bharagava, 2016). The effluent and wastewater generated during the process of leather making are rich source of heavy metals, organic matter, tannins, phenolics, and chromium (Mwinyihija, 2012).

45.2.3 REFRACTORY INDUSTRY

Refractory materials are those that retain their stability and durability at high temperatures and chromite is refractory in nature. In the refractory industries, chromium is used in cement and lime kilns, oven bricks, and other chemical processes. Chromium is used along with magnesia, the "chrome–mag" refractories. They have high resistance power and hence are used in industries using nonferrous metallurgy, followed by cement and limekilns. Chromium oxide (Cr_2O_3) has a melting temperature of 22°C–75°C, which is used along with other refractories in the productions of special refractory bricks. During the manufacturing of special refractory bricks and their life cycles, significant concentrations of Cr(III) are being oxidized to Cr(VI) on the surface and within the bulk of the refractory to form alkali chromates and calcium chromate (Mishra and Bharagava, 2016).

45.2.4 NATURAL SOURCES

One of the most important sources of chromium contamination in the environment is natural geological rock outcroppings. In nature, chromium exists in concentrated forms as an ore. Chromium composition and its concentration depend on the nature of the rock, atmospheric conditions, and weathering process (Mishra and Bharagava, 2016). It is also observed that many sedimentary rocks and shales including limestone and sandstones have very high chromium concentrations (Cannon et al., 1978). Volcanic eruptions have reported to emit high levels of chromium along with other metals and gases.

45.3 CHROMIUM TOXICITY

As mentioned earlier, chromium exists in different oxidation states and its toxicity is dependent on its oxidation states. Different oxidation states vary in their chemical characteristics and cause different effects within living organisms. Two distinct observations pronounce two different explanations for Cr(III). Conventionally, Cr(III) is considered as essential micronutrient for humans and animals (Viti et al., 2014). Di Bona et al. (2011) observed that rats on a diet with low Cr(III) showed no adverse effect to glucose metabolism, insulin sensitivity, or body composition, compared to rats with a diet having sufficient dose of Cr(III). On the other hand, it was also observed that a high dose (supranutritional level) of Cr(III) improves the insulin sensitivity in rats.

Cr(III) is relatively insoluble as compared to Cr(VI) that is highly soluble in aqueous medium; thus, it has 10–100 times less toxicity effect than Cr(VI). Due to its structural similarity to SO_4^{2-} (sulfate), CrO_4^{2-} crosses the cell membrane via sulfate transport system (Ksheminska et al., 2005). After crossing the cell membrane, under normal physiological conditions, Cr(VI) reacts spontaneously with intracellular reductants like NAD(P)(H), $FADH_2$ pentoses, cysteines, antioxidants such as glutathione or ascorbate, one-electron reducers, glutathione reductases to generate unstable short-lived intermediates Cr(V)/Cr(IV), free radicals, and the end product Cr(III) (Ahemad, 2014; Joutey et al., 2015).

Under aerobic conditions, the unstable Cr(V) upon the availability of electron undergoes one-electron redox cycling to regenerate Cr(VI) by transferring an electron to oxygen (Joutey et al., 2015). During the reduction process of Cr(VI) to Cr(III), reactive oxygen species (ROS), single oxygen (O) and superoxide (O^{2-}) hydroxyl (OH) and hydrogen peroxide (H_2O_2) radicals are generated which easily binds with DNA–protein complexes (Cheng et al., 2009; McNeill and McLean, 2012). Meanwhile, the intermediates cause oxidative damage to proteins and DNA lesions together with DNA–DNA cross-links, Cr–DNA adducts, and DNA–protein cross-links (Cervantes and Campos-García, 2007; Cheung and Gu, 2007; Nickens et al., 2010). However, the genotoxicity of chromium does not solely depend on the action of ROS. Intracellular Cr(III)/Cr(IV) ions interact with negatively charged phosphate groups on a DNA molecule and thereby influence DNA replication and transcription, leading to mutagenesis (Cervantes et al., 2001). Likewise, Cr(III) reacts with thiol and carboxyl groups of several enzymes affecting the enzyme activity and altering the enzyme structures (Cervantes et al., 2001).

45.4 ACCUMULATION OF CHROMIUM IN BIOMASS

As described earlier, the cell membranes of organisms are impermeable to chromium. Hexavalent chromium–Cr(VI) in solution exists as tetrahedral CrO_4^{2-} that is analogous to physiological anions of SO_4^{2-} and PO_4^{2-}. Hexavalent chromium enters the living cells through nonselective anion channel or through sulfate transporters, where competition exists between hexavalent chromium and sulfate. Therefore, by supplementing sulfate, the chromate toxicity may be decreased. In a study by Pereira et al. (2008), a decrease in sulfate assimilation and sulfur metabolite pools was observed in the presence of chromium ions, indicating that the cells suffer sulfur starvation. Once inside the cells hexavalent chromium is reduced to trivalent form. The trivalent chromium outside the cell membrane forms complexes in aqueous medium with biologically important ligand, and these complexes may enter in the cells (Ksheminska et al., 2005).

Different studies have reported about the use of living and dead microbial biomass for removing chromium from polluted aquifers through biosorption and bioaccumulation (Ksheminska et al., 2005; Srivastava and Thakur, 2006; Anjana et al., 2007; Mungasavalli et al., 2007). A recent study using transmission electron microscopy and energy-dispersive x-ray spectroscopy detected the bioaccumulation of chromium within *Pseudochrobactrum asaccharolyticum* LY6 cells (Long et al., 2013). The bioaccumulation and biosorption methods in chromium removal have inherent advantages and limitations. Applying dead biomass solves limitations associated with metal toxicity and maintenance of cell metabolic activity. Furthermore, the adsorbed metal can be collected, and the resultant biomass can be reutilized for sorption or accumulation.

Nevertheless, this method is limited by the fact that reactions cannot occur in dried cells. Therefore, by applying living biomass, metal can be removed as microbes grow, and this avoids storing and reutilization of dried biomass. Even this approach has limited application for the removal of high concentration of metal that may be inhibitory toward the growth of biomass. Therefore, microorganisms that are preadapted and can resist greater concentrations of Cr(VI) should be used with this approach (Holda et al., 2011).

45.5 BIOSORPTION OF CHROMIUM

The bacterial chromium biosorption mechanisms can be broadly classified into four categories:

1. *Anionic adsorption to cationic functional groups*: In this process, negatively charged chromium ions (chromate (CrO_4^{2-})/dichromate ($Cr_2O_7^{2-}$)) bind to positively charged functional groups on the surface of biosorbents through electrostatic attraction. At low pH, hexavalent chromium adsorption increases, and at high pH, Cr(VI) adsorption decreases. Garg et al. (2013) revealed that functional groups like carbonyl and amide of bacterial cells might be involved in adsorbing reduced Cr(III) on the surface of *Pseudomonas putida*.
2. *Adsorption-coupled reduction*: In this mechanism, hexavalent chromium is reduced to trivalent chromium in the presence of acid and adsorbed to the microbial biomass (Sanghi et al., 2009).

3. *Anionic and cationic adsorption*: A fraction of hexavalent chromium is reduced to trivalent chromium and both forms are adsorbed to the biosorbents (Joutey et al., 2015).

4. *Reduction and anionic adsorption*: In this process, a portion of hexavalent chromium is reduced to trivalent form by a biosorbent while the former is adsorbed to biomass and the trivalent chromium remains in the solution (Joutey et al., 2015).

The biomass or cells of bacteria, fungi, algae, seaweed, industrial by-products, and agricultural wastes are widely used as sorbent materials for the removal of metal pollutants. Among these different sorption methods, biosorbents are generally the cheapest and environment-friendly alternatives for removing chromium from the polluted environment.

45.6 BACTERIAL CHROMIUM REDUCTION

Biological reduction of Cr(VI) to Cr(III) was first reported by Romanenko and Korenkov (1977) from an uncharacterized *Pseudomonas* sp., under anaerobic conditions. Since then, several bacteria belonging to different genera, having both aerobic and anaerobic Cr(VI)-reducing ability, have been isolated from diverse environments (Opperman and van Heerden, 2008; He et al., 2009; Alam and Ahmad 2012; Batool et al., 2012; Ge et al., 2012; Narayani and Shetty, 2012; Nguema and Luo, 2012; Shi et al., 2012). The bacteria that catalyze the conversion of Cr(VI) to Cr(III), either aerobically or anaerobically or under both conditions, are ubiquitous in soil, sediments, and water bodies both in chromium-contaminated and chromium-noncontaminated environments (Cervantes et al., 2001; Cervantes and Campos-García, 2007).

Bacterial reduction of Cr(VI) is specifically important in the ecological and biological context, as this process transforms toxic and mobile Cr(VI) compounds into comparatively nontoxic and immobile reduced form Cr(III) (Daulton et al., 2007; Soni et al., 2013). The enzymes responsible for reducing Cr(VI) are known as chromate reductase; however, their biological function is not to reduce Cr(VI), as it is an auxiliary substrate for these enzymes (Ishibashi et al., 1990; Cervantes and Campos-García, 2007). Their primary roles are ferredoxin-NADP+ reductase, glutathione reductase, nitroreductases, iron reductases, and lipoyl reductase (Kwak et al., 2003; Mazoch et al., 2004; Cervantes and Campos-García 2007; Opperman and van Heerden, 2008). Under aerobic conditions, they utilize NADH and endogenous cell reserves; under anaerobic conditions, they utilize chromium in electron transport system containing cytochromes (Zhu et al., 2008; Dey and Paul, 2013).

The chromium reduction within bacterial cells usually occurs under three different conditions—(1) aerobic conditions through cytosolic soluble reductases, (2) anaerobic conditions, and (3) membrane-associated chromate reductases (Cheung and Gu, 2007; Ramírez-Díaz et al., 2008; Chirwa and Molokwane, 2011; Ngwenya and Chirwa, 2011):

1. *Aerobic Cr(VI) reduction*: The soluble NAD(P)H-dependent extracellular chromium reductase under aerobic environment reduces Cr(VI) to Cr(III) and is removed upon reaction with functional groups present on the cell surface. *Escherichia coli* ATCC 33456, *P. ambigua* G-1, *P. putida* PRS2000, and *Desulfovibrio vulgaris* were reported to produce soluble Cr(VI) reductases, which utilizes a range of electron donors (Ishibashi et al., 1990; Suzuki et al., 1992; Shen and Wang, 1993; Lovley and Phillips, 1994; Chen and Hao, 1998). The chromium reduction under aerobic conditions is an energy-requiring step and is a highly regulated mechanism; therefore, these reductases are produced constitutively in the cells. The extracellular Cr(VI) reduction is advantageous for the bacterial cell since the process does not involve transport of chromium ions into the cells and subsequent expulsion out of the cells; it protects the cell from chromium-induced toxicity and DNA damage (Ahemad, 2014).

2. *Anaerobic Cr(VI) reduction*: Under anaerobic conditions, the cell components like glutathione, flavoproteins, carbohydrates, amino acids, nucleotides, vitamins, organic acids, NAD(P)H, and hemeproteins donate electrons to reduce Cr(VI), where in the process chromium ions serve as terminal electron acceptors.

3. *Membrane-associated chromate reductases*: In anaerobic conditions, certain membrane-associated enzymes act as chromate reductases to reduce Cr(VI) that occasionally require glucose or H_2 as electron donors (Ibrahim et al., 2012). As reported for certain bacterial strains, anaerobic membrane-associated reductases utilize Cr(VI) as an electron acceptor in the electron transport chain (Wang, 2000; QuiIntana et al., 2001).

45.7 GENES AND PROTEINS INVOLVED IN HEXAVALENT CHROMIUM RESISTANCE AND REDUCTION

Resistance to hexavalent chromium has been found in all three domains of life: *Archaea*, *Bacteria*, and *Eukarya* (Nies et al., 1998; Flores-Alvarez et al., 2012). The bacterial genes associated with the resistance belong to chromate ion transport superfamily (Ramirez-Diaz et al., 2008). Bacterial genetic elements responsible for the reduction of Cr(VI) to Cr(III) have been found to be mostly of chromosomal origin and only few have been associated with plasmids (Cervantes et al., 2001). Chromium reduction mechanism can be broadly classified into two categories based on the availability of oxygen: (1) Cr(VI) reduction into Cr(III) in the presence of oxygen, which is catalyzed by soluble chromate reductase, mostly requires NADH or NADPH as cofactors (Park et al., 2000), and (2) in the absence of oxygen (anaerobic conditions), certain bacteria utilize Cr(VI) as the terminal electron acceptor in the electron transport chain (Tebo and Obraztova, 1998). Besides, chromium can also be reduced by nonspecific reactions through redox intermediates of organic compounds like sugars, nucleotides, amino acids, organic acids, amino acids, or glutathione (Myers et al., 2000; Robins et al., 2013). Ohtake et al. (1990) first reported the characterization of chromate reductase from *Enterobacter cloacae* HO1, a membrane-bound enzyme, which transfers electron from NADH-dependent cytochromes to hexavalent chromium. Chromate reductases generally exhibit NADH:flavin-dependent oxidoreductase activity and also catalyze the reduction of chromium as an alternate substrate (Viti et al., 2014).

The chromate reductase ChrR, a soluble flavin mononucleotide-binding enzyme from *P. putida*, has been well characterized (Park et al., 2000). The ChrR has a broad substrate specificity that requires NADH as cofactor in reduction catalysis of quinines, uranium(VI), and Cr(VI) (Barak et al., 2006). The ChrR usually catalyzes one- or two-electron transfers to Cr(VI) along with the formation of unstable intermediates Cr(V) and Cr(IV) to form the final end product Cr(III). A fraction of unstable intermediate Cr(V) is spontaneously reoxidized and in the process generating ROS, while major portion is reduced to Cr(III) generating very less amount of harmful radicals. In a different study on purified ChrR from *P. putida*, Gonzalez et al. (2005) observed that the enzyme also possesses quinine reductase activity in addition to chromium reduction, where quinols produced after reduction of quinine confer tolerance toward ROS. The results suggested that ROS generated during chromium reduction by ChrR are neutralized by quinols produced by the quinine reductase activity of the same enzyme, thus providing protection against ROS (Ackerley et al., 2004; Cheung and Gu 2007; Ramirez-Diaz et al., 2008). A different mechanism where Cr(VI) acts as terminal electron acceptor in the absence of oxygen has been comparatively understudied (Richter et al., 2012). However, the global transcriptomic analysis of *Shewanella oneidensis* MR-1 where Cr(VI) (100 µM) provided as sole source of electron acceptor revealed the upregulation of four genes: MtrA, MtrB, MtrC, and OmcA (Bencheikh-Latmani et al., 2005). The expression of these genes is involved in the dissimilatory extracellular reduction of U(VI), Tc(VII), and ferric iron [Fe(III)] (hydr)oxides (Belchik et al., 2011). The MtrC and OmcA expressions have been reported as the terminal reductases in Cr(VI) detoxification mechanism of *S. oneidensis* MR-1 (Belchik et al., 2011).

In addition, the reduction of hexavalent chromium has been also observed through biosorption. In a nonmetabolic chromium reduction, Cr(VI) was found to get adsorbed to bacterial surfaces under nutrient-limiting conditions and subsequently reduced to Cr(III), due to oxidation of organic molecules within the cell wall acting as a probable source of electron donor (Fein et al., 2002). Nancharaiah et al. (2010)

observed that granular bacterial biofilms under aerobic conditions did not reduce hexavalent chromium under nutrient-poor conditions, but the biofilms efficiently reduced Cr(VI) in the presence of acetate provided in the minimal medium.

In different bacteria (*P. aeruginosa*, *Cupriavidus metallidurans*, *Shewanella* sp. ANA3), ChrA protein has been characterized in detail and several putative ChrA homologs have been identified (Cervantes et al., 1990; Nies et al., 1990; Aguilar-Barajas et al., 2008). ChrA gene provides chromium protection in submillimolar range unlike other heavy metal resistance cascade. The chrA orthologs display resistance capacities in the range of 0.35–200 mM Cr(VI) concentrations, which suggest that mere presence of chrA gene is not sufficient for Cr(VI) resistance (Juhnke et al., 2002; Henne et al., 2009; Monsieurs et al., 2011). It was also observed that a strong activation of Chr efflux pump may lead to sulfate extrusion, which results in sulfur starvation conditions not suitable for growth (Branco et al., 2008). Juhnke et al. (2002), on the contrary to the previous observations, found that in bacteria chrA genes are found on plasmid and chromosomal DNA or both, which are generally organized in operons along with other chr genes. In *Ochrobactrum tritici* 5bv11, chr genes are found on transposable operon organized in cluster as chrBACF, while in *C. metallidurans*, heavy metal resistance genes are found on a 34 kb region of pMOL28 plasmid, and they are organized in gene cluster as chrFECAB in a chromosomal operon chr2 (chrB2, chrA2, chrF2 genes) (Branco et al., 2008).

The genes flanking chrA may not be directly related to the chromium functions but may have different regulatory functions. ChrB has a specific role in *O. tritici* 5bv11, where it acts as a chromium-sensitive regulator of the chr operon, while in *C. metallidurans* it is activated by Cr(III), Cr(VI), or sulfate ions (Peitzsch et al., 1998; Juhnke et al., 2002; Branco et al., 2008). The strain of *C. metallidurans* AE126 lacking chrB1 genes exhibits tolerance to chromium as compared to wild type (Juhnke et al., 2002). The reduction of Cr(VI) was less when chrC and chrI were deleted from the same strain, while the activity of chrC and chrF was not found to affect the chromium resistance in *O. tritici* 5bv11. Likewise, in *C. metallidurans* chrC protein exhibited similar properties to the superoxide dismutase enzymes with active manganese or iron centers (Juhnke et al., 2002; Branco et al., 2008). Table 45.1 summarizes selected bacterial interactions and their applicability for the remediation of chromium-contaminated soils/effluents or wastes.

45.8 BIOREMEDIATION STRATEGIES FOR CHROMIUM-CONTAINING SOLID WASTES

Given an increase in awareness for environmental restoration and strict regulatory norms, industries are now working for more robust bioremediation technology for environmental cleanup of heavy metals. Different methodologies are continually being developed for treating the chromium pollution originating from various sources.

TABLE 45.1

Bacterial Interactions with Chromium Ions and the Applicability in Environment Contaminated with Chromium Compounds

Bacterial sp.	Type of Interaction	Application	Reference
Haloalkaliphilic bacterium	Cr(VI) reduction	Remediation of Cr(VI) leachate resulting from chromite ore processing residue (COPR)	Watts et al. (2015)
Pseudomonas aeruginosa CRM100	Cr(VI) reduction	Remediation of wastewater treatment plant containing chromium under anaerobic conditions	Salamanca et al. (2013)
Ochrobactrum intermedium Rb-2	Cr(VI) reduction	Remediation of Tannery effluent containing chromium under aerobic conditions	Batool et al. (2012)
Cellulomonas spp.		Remediation of subsurface soil contaminated with chromium under aerobic conditions	Viamajala et al. (2007)
Serratia marcescens	Cr(VI) reduction	Remediation of Tannery effluent containing chromium under aerobic conditions	Mondaca et al. (2002)
Pseudomonas sp.	Cr(VI) reduction	Remediation of soil contaminated with chromium under chromium conditions	Desai et al. (2008a)
Bacillus spp.	Cr(VI) reduction	Remediation of soil contaminated with chromium under aerobic conditions	Desai et al. (2008b)
Pseudochrobactrum asaccharolyticum LY6	Bioaccumulation	Accumulation of Cr(VI) within the cells from aqueous solutions	Long et al. (2013)
Thermus scotoductus strain SA-01	Cr(VI) reduction	Remediation of chromium from aqueous solutions	Opperman and van Heerden (2007)
Alcaligenes sp. DMA, *Bacillus* sp. DMB, *Stenotrophomonas* sp. DMS, *Enterococcus* sp. DME	Cr (VI) reduction	Remediation of soil contaminated with chromium under aerobic conditions	Desai et al. (2009)
Pseudomonas putida	Biosorption	Remediation of chromium from contaminated sites	Garg et al. (2013)
Bacillus coagulans	Biosorption	Remediation of aqueous solution containing Cr(VI)	Srinath et al. (2003)

45.9 REMEDIATION METHODS FOR CHROMIUM CONTAMINATION FROM MINING WASTES

Chromium-containing mining waste has been successfully treated using indigenous bacterial species in different studies. Dhal et al. (2010a,b) isolated chromium-resistant *Bacillus* sp. from chromite mine soil; the bacterium could tolerate 2000 mg/L of hexavalent chromium. The bacterium showed an impressive rate of reduction in the order of 3.22×10^{-2} to 6.5×10^{-3} per hour in the presence of 10–500 mg/L hexavalent chromium concentration. In another study at Sukinda chromite mine, Samuel et al. (2012) developed the microbial consortia with chromium reduction capacity ranging from 0.199–0.477 to 0.5–1.16 mg/L/h capable of reduction of 5–20 mg/L chromium concentrations. Likewise, a *Bacillus* sp. reportedly isolated from chromite overburden/soil (500 mg/kg) exhibited an impressive ability to reduce nearly 98% of initial chromium in 10 h.

45.10 REMEDIATION OF CHROMIUM CONTAMINATION FROM CHROMIUM ORE PROCESSING RESIDUES

Microorganisms in the presence of HCl have reduced chromium within ore processing residues. Several strains of bacteria such as *Bacillus* sp., *Micrococcus* sp., and *Rhodococcus*

sp. reduce Fe(II) containing chromium ore processing residues, along with reductants such as gastric fluid, prior leaching with water followed by H_2SO_4 treatment and reduction with cow manure (Turick and Apel, 1997; Higgins et al., 1998; Stewart et al., 2010; Whittleston et al., 2011a,b; Yu et al., 2012). Methods like *in situ* and *ex situ* physicochemical and microbiological, sulfate-reducing bacteria, organic matter, and $FeSO_4$ treatment have also been established (Bewley et al., 2001). Additionally, sucrose, starch, and wheat flour were also used for reducing chromium from ore processing residues (Wang et al., 2007a).

45.11 REMEDIATION METHODS FOR CHROMIUM WASTE OF TANNING INDUSTRIES

The Cr(VI) reduction from the tannery wastes has been extensively studied. The methods employed were direct microbial reduction as well as in presence of molasses or using native microbes of cow manure. Other methods have employed indigenous bacteria from sediments of tannery effluent accumulated site or using indigenous iron/sulfur oxidizing bacteria for bioleaching chromates (Jayasingh and Philip, 2005; Sethunathan et al., 2005; Zhou et al., 2006; Wang et al., 2007b; Sundar et al., 2010; Kathiravan et al., 2011). Biosurfactants produced by microbes and plants have been useful in bioremediation applications. Stable chromium was extracted from contaminated kaolinite using rhamno-lipids, and this treatment can aid in long-term conversion of Cr(VI) to Cr(III) (Massara et al., 2007). Biosurfactants of plant origin such as aescin and saponin in aqueous solutions were used for the remediation of different heavy metals from contaminated soil, incinerator fly ash, and kaolin (Hong et al., 1998, 2002; Chen et al., 2008). Nearly one-fourth of chromium was extracted using saponin extracted 24% from the tannery sludge (Kilic et al., 2011).

45.12 REMEDIATION METHODS FOR CHROMIUM-CONTAMINATED ELECTROPLATING SITES

Electroplating wastes containing chromium have been treated using bioleaching methods. Bayat and Sari (2010) showed 34% bioleaching of the chromium and other heavy metals (Zn, Cu, Ni, Pb, Cd) from dewatered electroplating sludge containing heavy metals with no sulfide or sulfate compounds. In another study, A. ferrooxidans was used for bioleaching of 49% chromium along with other heavy metals from continuous stirred tank reactor along with the biooxidized Fe(III) (Dhal et al., 2013). In another study, leaching of chromium at a concentration of 218 mg/kg with mineral salt solution in glucose was observed in electroplating industry wastewater; likewise, 64.5% Cr(VI) was reduced by indigenous bacterial species (Turick et al., 1998).

45.13 IMPLICATIONS

Given the rate of urbanization and increase in the anthropogenic activities, the flow of natural materials has been altered that has resulted in irrevocable loss of pristine environment. It is inevitable and most timely to exploit the microbe-mediated bioremediation technologies to detoxify the environment contaminated with toxic heavy metals including chromium, since the conventional technologies are inherited with indigenous drawbacks with limited success. In this chapter, bacterial interactions with chromium and strategies for its removal from different types of contaminated sources have been reviewed. The chromium detoxification mechanisms vary depending on the strain involved, and it involves efflux of chromium compounds, enzymatic reduction of Cr(VI) to Cr(III), biosorption, bioaccumulations, intracellular or extracellular precipitation, etc. Bacterial chromium removal includes three stages: (1) binding of chromium ions to the bacterial cell surfaces, (2) translocation of chromium ions into the cell, and (3) reduction of Cr(VI) to Cr(III). The uptake of chromium ions by a bacterial cell is a biphasic process. The initial step is known as biosorption, a metabolic energy-independent process followed by bioaccumulations, which is slower and dependent on cell metabolic activity. Once inside the cell, Cr(VI) ions are reduced to Cr(III) directly by either cytosolic soluble chromate reductases or membrane-bound chromate reductases or indirectly through metabolic reduction via different metabolic reductants (Joutey et al., 2015).

Further prospective in chromium remediation is developing the novel trait in chromium-reducing bacteria by genetic engineering and/or random mutation to enhance their detoxifying efficiency (Ahemad, 2014).

ACKNOWLEDGMENTS

The authors are highly grateful to the Department of Biotechnology, Ministry of Science and Technology, New Delhi, for financial support.

REFERENCES

Ackerley, D. F., Gonzalez, C. F., Keyhan, M., Blake, R., Matin, A. 2004. Mechanism of chromate reduction by the *Escherichia coli* protein, NfsA, and the role of different chromate reductases in minimizing oxidative stress during chromate reduction. *Environmental Microbiology* 6: 851–860.

Aguilar-Barajas, E., Paluscio, E., Cervantes, C., Rensing, C. 2008. Expression of chromate resistance genes from *Shewanella* sp. strain ANA-3 in *Escherichia coli*. *FEMS Microbiology Letters* 285: 97–100.

Ahemad, M. 2014. Bacterial mechanisms for Cr(VI) resistance and reduction: An overview and recent advances. *Folia Microbiologica* 59: 321–332.

Alam, M. Z., Ahmad, S. 2012. Toxic chromate reduction by resistant and sensitive bacteria isolated from tannery effluent contaminated soil. *Annals of Microbiology* 62: 113–121.

Anjana, K., Kaushik, A., Kiran, B., Nisha, R. 2007. Biosorption of Cr(VI) by immobilized biomass of two indigenous strains of cyanobacteria isolated from metal contaminated soil. *Journal of Hazardous Materials* 148: 383–386.

Barak, Y., Ackerley, D. F., Dodge, C. J., Banwari, L., Alex, C., Francis, A. J., Matin, A. 2006. Analysis of novel soluble chromate and uranyl reductases and generation of an improved enzyme by directed evolution. *Applied and Environmental Microbiology* 72: 7074–7082.

Baral, A., Engelken, R., Stephens, W., Farris, J., Hannigan, R. 2006. Evaluation of aquatic toxicities of chromium and chromium containing effluents in reference to chromium electroplating industries. *Archives of Environmental Contamination and Toxicology* 50: 496–502.

Bartlett, R. J. 1991. Chromium cycling in soils and water—Links, gaps, and methods. *Environmental Health Perspectives* 92: 17–24.

Batool, R., Yrjälä, K., Hasnain, S. 2012. Hexavalent chromium reduction by bacteria from tannery effluent. *Journal of Microbiology and Biotechnology* 22: 547–554.

Bayat, B., Sari, B. 2010. Comparative evaluation of microbial and chemical leaching processes for heavy metal removal from dewatered metal plating sludge. *Journal of Hazardous Materials* 174: 763–769.

Belchik, S. M., Kennedy, D. W., Dohnalkova, A. C., Wang, Y. M., Sevinc, P. C., Wu, H., Lin, Y. H., Lu, H. P., Fredrickson, J. K., Shi, L. 2011. Extracellular reduction of hexavalent chromium by cytochromes MtrC and OmcA of *Shewanella oneidensis* MR-1. *Applied and Environmental Microbiology* 77: 4035–4041.

Bencheikh-Latmani, R., Williams, S. M., Haucke, L., Criddle, C. S., Wu, L. Y., Zhou, J. Z., Tebo, B. M. 2005. Global transcriptional profiling of *Shewanella oneidensis* MR-1 during Cr(VI) and U(VI) reduction. *Applied and Environmental Microbiology* 71: 7453–7460.

Bewley, R. J. F., Jeffries, R., Watson, S., Granger, D. 2001. An overview of chromium contamination issues in the South-east of Glasgow and the potential for remediation. *Environmental Geochemistry and Health* 23: 267–271.

Blade, L. M., Yencken, M. S., Wallace, M. E., Catalano, J. D., Khan, A., Topmiller, J. L, Shulman, S. A., Martinez, A., Crouch, K. G., Bennett, J. S. 2007. Hexavalent chromium exposures and exposure-control technologies in American enterprise: Result of NIOSH field research study. *Journal of Occupational and Environmental Hygiene* 4: 596–618.

Branco, R., Chung, A. P., Johnston, T., Gurel, V., Morais, P., Zhitkovich, A. 2008. The chromate-inducible chrBACF operon from the transposable element TnOtChr confers resistance to chromium(VI) and superoxide. *Journal of Bacteriology* 190: 6996–7003.

Cannon, H. L., Connally, G. G., Epstein, J. B., Parker, J. G., Wixson, G. 1978. Rocks: Geological sources of most trace elements. In: Report to the workshop at south seas plantation Captiva, Island, FL, US. *Environmental Geochemistry and Health* 3: 17–31.

Cervantes, C., Campos-García, J. 2007. Reduction and efflux of chromate by bacteria. In: Nies, D., Silver, S. (eds.), *Molecular Microbiology of Heavy Metals*. Springer-Verlag, Berlin, Germany, pp. 407–419.

Cervantes, C., Campos-Gracia, J., Devras, S., Gutierrez-Corona, F., Loza-Tavera, H., Torres-Guzman, J. C., Moreno-Sanchez, R. 2001. Interactions of chromium with microorganisms and plants. *FEMS Microbiology Reviews* 25: 335–347.

Cervantes, C., Ohtake, H., Chu, L., Misra, T. K., Silver, S. 1990. Cloning, nucleotide sequence, and expression of the chromate resistance determinant of *Pseudomonas aeruginosa* plasmid pUM505. *Journal of Bacteriology* 172: 287–291.

Chen, J., Hao, O. 1998. Microbial chromium(VI) reduction. *Critical Reviews in Environmental Science and Technology* 28: 219–251.

Chen, W. J., Hsia, L. C., Chen, K. K. Y. 2008. Metal desorption from copper (II) (nickel (II)- spiked kaolin as a soil component using plant-derived saponin biosurfactant. *Process Biochemistry* 43: 488–498.

Cheng, Y., Xie, Y., Zheng, J., Wu, Z., Chen, Z., Ma, X., Li, B., Lin, Z. 2009. Identification and characterization of the chromium(VI) responding protein from a newly isolated *Ochrobactrum anthropi* CTS-325. *Journal of Environmental Science* 21: 1673–1678.

Cheung, K., Gu, J. D. 2007. Mechanism of hexavalent chromium detoxification by microorganisms and bioremediation application potential: A review. *International Biodeterioration and Biodegradation* 59: 8–15.

Cheung, K. H., Lai, H. Y., Gu, J. D. 2006. Membrane-associated hexavalent chromium reductase of *Bacillus megaterium* TKW3 with induced expression. *Journal of Microbiology and Biotechnology* 16: 855–862.

Chirwa, E. M. N., Molokwane, P. E. 2011. Biological Cr(VI) reduction: Microbial diversity, kinetics and biotechnological solutions to pollution. In: Sofo, D. A. (ed.), *Biodiversity*. InTech Croatia, EU. Available from: http:// www.intechopen.com/books/biodiversity/biological-cr-vireduction-microbialdiversity-kinetics-and-biotechnologicalsolutions-to-pollution.

Daulton, T. L., Little, B. J., Jones-Meehan, J., Blom, D. A., Allard, L. F. 2007. Microbial reduction of chromium from the hexavalent to divalent state. *Geochimica et Cosmochimica Acta* 71: 556–565.

Desai, C., Jain, K., Madamwar, D. 2008a. Evaluation of *In vitro* Cr(VI) reduction potential in cytosolic extracts of three indigenous *Bacillus* sp. isolated from Cr (VI) polluted industrial landfill. *Bioresource Technology* 99: 6059–6069.

Desai, C., Jain, K., Madamwar, D., 2008b. Hexavalent chromate reductase activity in cytosolic fractions of *Pseudomonas* sp. G1DM21 isolated from Cr(VI) contaminated industrial landfill. *Process Biochemistry* 43: 713–721.

Desai, C., Jain, K., Madamwar, D. 2009. Efficacy of bacterial consoertium-AIE2 for contemporaneous Cr(VI) and azo dye bioremediation in batch and continuous bioreactor systems, monitoring steady-state bacterial dynamics using qPCR assays. *Biodegradation* 20: 813–826.

Dey, S., Paul, A. K. 2013. Hexavalent chromium reduction by aerobic heterotrophic bacteria indigenous to chromite mine overburden. *Brazilian Journal of Microbiology* 44: 307–315.

Dhal, B., Das, N. N., Thatoi, H. N., Pandey, B. D. December 15–17, 2010a. Kinetic aspects of Cr(VI) bioremediation of water and soil contaminated from chromite mine overburden dumps. In: Singh, R., Das, A., Banerjee, P. K., Bhattacharyya, K. K., Goswami, N. G. (eds.), *Proceedings of the XI International Seminar on Mineral Processing Technology*. NML, Jamshedpur, India, pp. 1007–1014.

Dhal, B., Thatoi, H. N., Das, N. N., Pandey B. D. 2010b. Reduction of hexavalent chromium by *Bacillus* sp. isolated from chromite mine soils and characterization of reduced product. *Journal of Chemical Technology and Biotechnology* 85: 1471–1479.

Dhal, B., Thatoi, H. N., Das, N. N., Pandey, B. D. 2013. Chemical and microbial remediation of hexavalent chromium from contaminated soil and mining/metallurgical solid waste: A review. *Journal of Hazardous Materials* 250–251: 272–291.

Di Bona, K. R., Love, S., Rhodes, N. R. et al. 2011. Chromium is not an essential trace element for mammals: Effects of a 'low-chromium' diet. *Journal of Biological Inorganic Chemistry* 16: 381–390.

Fein, J. B., Fowle, D. A., Cahill, J., Kemner, K., Boyanov, M., Bunker, B. 2002. Nonmetabolic reduction of Cr(VI) by bacterial surfaces under nutrient-absent conditions. *Geomicrobiology Journal* 19: 369–382.

Flores-Alvarez, L. J., Corrales-Escobosa, A. R., Cortes-Penagos, C., Martinez-Pacheco, M., Wrobel-Zasada, K., Wrobel-Kaczmarczyk, K., Cervantes, C., Gutierrez-Corona, F. 2012. The *Neurospora crassa* chr-1 gene is up-regulated by chromate and its encoded CHR-1 protein causes chromate sensitivity and chromium accumulation. *Current Genetics* 58: 281–290.

Garg, S. K., Tripathi, M., Singh, S. K., Singh, A. 2013. Pentachlorophenol dechlorination and simultaneous Cr^{6+} reduction by *Pseudomonas putida* SKG-1 MTCC (10510): Characterization of PCP dechlorination products, bacterial structure and functional groups. *Environmental Science and Pollution Research* 20: 2288–2304.

Ge, S., Zhou, M., Dong, X., Lu, Y., Ge, S. 2012. Distinct and effective biotransformation of hexavalent chromium by a novel isolate under aerobic growth followed by facultative anaerobic incubation. *Applied Microbiology Biotechnology* 97(5): 2131–2137.

Gonzalez, C. F., Ackerley, D. F., Lynch, S. V., Matin, A. 2005. ChrR, a soluble quinone reductase of *Pseudomonas putida* that defends against H$_2$O$_2$. *Journal of Biological Chemistry* 280: 22590–22595.

He, Z., Gao, F., Sha, T., Hu, Y., He, C. 2009. Isolation and characterization of a Cr(VI)-reduction *Ochrobactrum* sp. strain CSCr-3 from chromium landfill. *Journal of Hazardous Materials* 163: 869–873.

Henne, K. L., Turse, J. E., Nicora, C. D. et al. 2009. Global proteomic analysis of the chromate response in *Arthrobacter* sp. strain FB24. *Journal of Proteome Research* 8: 1704–1716.

Higgins, T. E., Halloran, A. R., Dobbins, M. E., Pittignano A. J. 1998. In situ reduction of hexavalent chromium in alkaline soils enriched with chromite ore processing residue. *Journal of the Air & Waste Management Association* 48: 1100–1106.

Hołda, A., Kisielowska, E., Nieboda, T. 2011. Bioaccumulation of Cr(VI) ions from aqueous solutions by *Aspergillus niger*. *Polish Journal of Environmental Studies* 20: 345–349.

Hong, K. J., Choi, Y. K., Tokunaga, S., Ishigami, Y., Kajiuchi, T. 1998. Removal of cadmium and lead from soil using aescin as a biosurfactant. *Journal of Surfactants and Detergents* 1: 247–250.

Hong, K. J., Tokunaga, S., Kajiuchi, T. 2002. Evaluation of remediation process with plant-derived biosurfactant for recovery of heavy metals from contaminated soils. *Chemosphere* 49: 379–387.

Ibrahim, A. S. S., Elbadawi, Y. B., El-Tayeb, M. A., Al-Salamah, A. A. 2012. Hexavalent chromium reduction by novel chromate resistant alkaliphilic *Bacillus* sp. strain KSUCr9a. *African Journal of Biotechnology* 11: 3832–3841.

Ishibashi, Y., Cervantes, C., Silver, S. 1990. Chromium reduction in *Pseudomonas putida*. *Applied and Environmental Microbiology* 56: 2268–2270.

Jayasingh, J., Philip, L. 2005. Bioremediation of chromium contaminated soil: Optimization of operating parameters under laboratory condition. *Journal of Hazardous Materials* 118: 113–120.

Juhnke, S., Peitzsch, N., Heubener, N., Grobe, C., Nies, D. H. 2002. New genes involved in chromate resistance in *Ralstonia metallidurans* strain CH34. *Archives of Microbiology* 179: 15–25.

Joutey, N. T., Sayel, H., Bahafid, W., El Ghachtouli N. 2015. Mechanisms of hexavalent chromium resistance and removal by microorganisms. *Reviews of Environmental Contamination and Toxicology* 233: 45–69.

Kamaludeen, S. P. B., Arunkumar, K. R., Avudainayagam, S., Ramasamy, K. 2003. Bioremediation of chromium contaminated environments. *Indian Journal of Experimental Biology* 41: 972–985.

Kathiravan, M. N., Karthick, R., Muthukumar, K. 2011. Ex situ bioremediation of Cr(VI) contaminated soil by Bacillus sp.: Batch and continuous studies. *Chemical Engineering Journal* 169: 107–115.

Kilic, E., Font, J., Puig, R., Colak, S. Celik, D. 2011. Chromium recovery from tannery sludge with saponin and oxidative remediation. *Journal of Hazardous Materials* 185: 456–462.

Kimbrough, D. E., Cohen, Y., Winer, A. M., Creelman, L., Mabuni, C. 1999. A critical assessment of chromium in the environment. *Critical Reviews in Environmental Science and Technology* 29: 1–46.

Ksheminska, H., Fedorovych, D., Babyak, L., Yanovych, D., Kaszycki, P., Koloczek, H. 2005. Chromium (III) and (VI) tolerance and bioaccumulation in yeast: A survey of cellular chromium content in selected strains of representative genera. *Process Biochemistry* 40: 1565–1572.

Kwak, Y. H., Lee, D. S., Kim, H. B. 2003. *Vibrio harveyi* nitroreductase is also a chromate reductase. *Applied and Environmental Microbiology* 69: 4390–4395.

Langard, S. 1980. A survey of respiratory symptoms and lung function in ferrochromium and ferrosilicon workers. *International Archives of Occupational and Environmental* 46: 1–9.

Long, D., Tang, X., Cai, K., Chen, G., Shen, C., Shi, J., Chen, L., Chen, Y. 2013. Cr(VI) resistance and removal by indigenous bacteria isolated from chromium contaminated soil. *Journal of Microbiology and Biotechnology* 23: 1123–32.

Losi, M. E., Frankenberger, W. T. 1994. Chromium-resistant microorganisms isolated from evaporation ponds of a metal processing plant. *Water Air and Soil Pollution* 74: 405–413.

Lovley, D. R., Phillips, E. J. P. 1994. Reduction of chromate by *Desulfovibrio vulgaris* and its c3 cytochrome. *Applied and Environmental Microbiology* 60: 726–728.

Lu, Z., Ouyang, X., Zhang, W., Lu, X. 2013. Isolation of Cr(VI) resistant bacteria and exploration of Cr(VI) removal mechanism of strain n-9. *Journal of Applied Mechanics* 295: 74–77.

Marsh, T. L., McInerney, M. J. 2001. Relationship of hydrogen bioavailability to chromate reduction in aquifer sediments. *Applied Environmental Microbiology* 67: 1517–1521.

Massara, H., Mulligan, C., Hadjinicolaou, N. J. 2007. Effect of rhamnolipids on chromium-contaminated kaolinite. *Soil and Sediment Contamination* 16: 1–14.

Mazoch, J., Tesařík, R., Sedláček, V., Kučera, I., Turánek, J. 2004. Isolation and biochemical characterization of two soluble iron(III) reductases from *Paracoccus denitrificans*. *European Journal of Biochemistry* 271: 553–562.

McGrath, S. P., Smith, S. 1990. Chromium and nickel. In: Alloway, B. J. (ed.), *Heavy Metals in Soils*. John Wiley & Sons, New York, pp. 125–147.

McNeill, L., McLean, J. 2012. State of the science of hexavalent chromium in drinking water research foundation. USA.

Mishra, S., Bharagava, R. N. 2016. Toxic and genotoxic effects of hexavalent chromium in environment and its bioremediation strategies. *Journal of Environmental Science and Health Part C Environmental Carcinogenesis and Ecotoxicology Reviews* 34(1): 1–32.

Mondaca, M. A., Campos, V., Moraga R., Zaror, C. A. 2002. Chromate reduction in *Serratia marcescens* isolated from tannery effluent and potential application for bioremediation of chromate pollution. *The Scientific World Journal* 2: 972–977.

Monsieurs, P., Moors, H., Van Houdt, R., Janssen, P. J., Janssen, A., Coninx, I., Mergeay, M., Leys, N. 2011. Heavy metal resistance in *Cupriavidus metallidurans* CH34 is governed by an intricate transcriptional network. *Biometals* 24: 1133–1151.

Mungasavalli, D. P., Viraraghavan, T., Chung, Y. J. 2007. Biosorption of chromium from aqueous solutions by pretreated *Aspergillus niger*: Batch and column studies. *Colloid Surface* 301: 214–223.

Mwinyihija, M. 2012. Pollution control and remediation of the tanning effluent. *The Open Environmental Pollution and Toxicology Journal* 3: 55–64.

Myers, C. R., Carstens, B. P., Antholine, W. E., Myers, J. M. 2000. Chromium(VI) reductase activity is associated with the cytoplasmic membrane of anaerobically grown *Shewanella putrefaciens* MR-1. *Journal of Applied Microbiology* 88: 98–106.

Nancharaiah, Y. V., Dodge, C., Venugopalan, V. P., Narasimhan, S. V., Francis, A. J. 2010. Immobilization of Cr(VI) and its reduction to Cr(III) phosphate by granular biofilms comprising a mixture of microbes. *Applied and Environmental Microbiology* 76: 2433–2438.

Narayani, M., Shetty, K.V. 2012. Characteristics of a novel Acinetobacter sp. and its kinetics in hexavalent chromium bioreduction. *Journal of Microbiology and Biotechnology* 22: 690–698.

Nath, K., Singh, S., Sharma, Y. K. 2009. Phytotoxic effects of chromium and tannery effluent on growth and metabolism of *Phaseolus mungo roxb*. *Journal of Environmental Biology* 30: 227–234.

Nguema, P. F., Luo, Z. 2012. Aerobic chromium(VI) reduction by chromium-resistant bacteria isolated from activated sludge. *Annals of Microbiology* 62: 41–47.

Ngwenya, N., Chirwa, E. M. N. 2011. Biological removal of cationic fission products from nuclear wastewater. *Water Science & Technology* 63: 124–128.

Nickens, K. P., Patierno, S. R., Ceryak, S. 2010. Chromium genotoxicity: A double-edged sword. *Chemico-Biological Interactions* 188: 276–288.

Nies, A., Nies, D. H., Silver, S. 1990. Nucleotide sequence and expression of plasmid-encoded chromate resistance determinant from *Alcaligenes eutrophus*. *Journal of Biological Chemistry* 265: 5648–5653.

Nies, D. H., Koch, S., Wachi, S., Peitzsch, N., Saier, M. H. 1998. CHR, a novel family of prokaryotic proton motive force-driven transporters probably containing chromate/sulfate antiporters. *Journal of Bacteriology* 180: 5799–5802.

Ohtake, H., Fujii, E., Toda, K. 1990. Reduction of toxic chromate in an industrial effluent by use of a chromate-reducing strain of *Enterobacter cloacae*. *Environmental Technology* 11: 663–668.

Opperman, D. J., van Heerden E. 2007. Aerobic Cr(VI) reduction by *Thermus scotoductus* strain SA-01. 103: 1907–1913.

Opperman, D. J., van Heerden, E. 2008. A membrane-associated protein with Cr(VI)-reducing activity from *Thermus scotoductus* SA-01. *FEMS Microbiology Letters* 280: 210–218.

Papp J. P. 2012. Chromium. In Salazar, K., McNutt K. M. (Eds.), *Mineral Commodity Summaries*. U. S. Geological Survey. pp. 42–43.

Park, C. H., Keyhan, M., Wielinga, B., Fendorf, S., Matin, A. 2000. Purification to homogeneity and characterization of a novel Pseudomonas putida chromate reductase. *Applied and Environmental Microbiology* 66: 1788–1795.

Peitzsch, N., Eberz, G., Nies, D.H. 1998. Alcaligenes eutrophus as a bacterial chromate sensor. *Applied and Environmental Microbiology* 64: 453–458.

Pereira, Y., Lagniel, G., Godat, E., Cornu, P. B., Junot, C., Labarre, J. 2008. Chromate causes sulfur starvation in yeast. *Toxicological Science* 106: 400–412.

QuiIntana, M., Curutchet, G., Donati, E. 2001. Factors affecting chromium(VI) reduction by *Thiobacillus ferrooxidans*. *Biochemical Engineering Journal* 9: 11–15.

Ramírez-Díaz, M. I., Diaz-Perez, C., Vargas, E., Riveros-Rosas, H., Campos-Garcia, J., Cervantes, C. 2008. Mechanisms of bacterial resistance to chromium compounds. *Biometals* 21: 321–332.

Robins, K. J., Hooks, D. O., Rehm, B. H. A., Ackerley, D. F. 2013. *Escherichia coli* NemA is an efficient chromate reductase that can be biologically immobilized to provide a cell free system for remediation of hexavalent chromium. *PLoS One* 8: 59–200.

Romanenko, V. I., Korenkov, V. N. 1977. A pure culture of bacteria utilizing chromate and dichromate as hydrogen acceptors in growth under anaerobic conditions. *Mikrobiologiya* 46: 414–417.

Richter, K., Schicklberger, M., Gescher, J. 2012. Dissimilatory reduction of extracellular electron acceptors in anaerobic respiration. *Applied and Environmental Microbiology* 78: 913–921.

Salamanca, D., Strunk, I. N., Engesser, K. H. 2013. Chromate reduction in anaerobic systems by bacterial strain *Pseudomonas aeruginosa* CRM100. *Chemie Ingenieur Technik* 85: 1575–1580.

Samuel, J., Paul, M. L., Pulimi, M., Nirmala, M. J., Chandrasekaran, N., Mukherjee, A. 2012. Hexavalent chromium bioremoval through adaptation and consortia development from Sukinda chromite mine isolates. *Industrial and Engineering Chemistry Research* 51: 3740–3749.

Sanghi, R., Sankararamakrishnana, N., Daveb, B. C. 2009. Fungal bioremediation of chromates: Conformational changes of biomass during sequestration, binding, and reduction of hexavalent chromium ions. *Journal of Hazardous Materials* 169: 1074–1080.

Sethunathan, N., Megharaj, M., Smith, L., Kamaludeen, S. P. B., Avudainayagam, S. R., Naidu, R. 2005. Microbial role in the failure of natural attenuation of chromium (VI) in long-term tannery waste contaminated soil. *Agriculture, Ecosystems & Environment* 105: 657–661.

Shen, H., Wang, Y. T. 1993. Characterization of enzymatic reduction of hexavalent chromium by *Escherichia coli* ATCC 33456. *Applied and Environmental Microbiology* 59: 3771–3777.

Shi, Y., Chai, L., Yang, Z., Jing, Q., Chen, R., Chen, Y. 2012. Identification and hexavalent chromium reduction characteristics of *Pannonibacter phragmitetus*. *Bioprocess and Biosystems Engineering* 35: 843–850.

Smith, W. A., Apel, W. A., Petersen, J. N., Peyton, B. M. 2002. Effect of carbon and energy source on bacterial chromate reduction. *Journal of Bioremediation & Biodegradation* 6: 205–215.

Soni, S. K., Singh, R., Awasthi, A., Singh, M., Kalra, A. 2013. In vitro Cr(VI) reduction by cell-free extracts of chromate-reducing bacteria isolated from tannery effluent irrigated soil. *Environmental Science and Pollution Research* 20: 1661–1674.

Srinath, T., Garg, S. K., Ramteke, P.W. 2003. Biosorption and elution of chromium from immobilized *Bacillus coagulans* biomass. *Indian Journal of Experimental Biology* 4: 986–990.

Srivastava, S., Thakur, I. S. 2006. Evaluation of bioremediation and detoxification potentiality of *Aspergillus niger* for removal of hexavalent chromium in soil microcosm. *Soil Biology Biochemistry* 38: 1904–1911.

Stewart, D. I., Burke, I. T., Hughes-Berry, D. V., Whittleston, R. A. 2010. Microbially mediated chromate reduction in soil contaminated by highly alkaline leachate from chromium containing waste. *Ecological Engineering* 36: 211–221.

Sundar, K., Vidya, R., Mukherjee, A., Chandrasekaran, N. 2010. High chromium tolerant bacterial strains from Palar River basin: Impact of tannery pollution. *Research Journal of Environmental and Earth Sciences* 2: 112–117.

Suzuki, T., Miyata, N., Horitsu, H., Kawai, K., Takamizawa, K., Tai, Y., Okazaki, M. 1992. NAD(P)H-dependent chromium(VI) reductase of *Pseudomonas ambigua* G-1: A Cr(V) intermediate is formed during the reduction of Cr(VI) to Cr(III). *Journal of Bacteriology* 174: 5340–5345.

Tebo, B. M., Obraztova, A. Y. 1998. Sulfate-reducing bacterium grows with Cr(VI), U(VI), Mn(IV), and Fe(III) as electron acceptors. *FEMS Microbiology Letters* 162: 193–198.

Turick, C. E., Apel, W. A. 1997. A bio-processing strategy that allows for the selection of Cr(VI) reducing bacteria from soil. *Journal of Industrial Microbiology and Biotechnology* 18: 247–250.

Turick, C. E., Graves, C., Apel, W. A. 1998. Bioremediation potential of Cr(VI)- contaminated soil using indigenous microorganisms. *Bioremediation Journal* 2: 1–6.

Viamajala, S., Smith, W. A., Sani, R. K., Apel, W. A., Petersen, J. N., Neal, A. L., Roberto, F. F., Newby, D. T., Peyton, B. M. 2007. Isolation and characterization of Cr(VI) reducing *Cellulomonas* spp. from subsurface soils: Implications for long-term chromate reduction. *Bioresource Technology* 98: 612–622.

Viti, C., Marchi, E., Decorosi, F., Giovannetti, L. 2014. Molecular mechanisms of Cr(VI) resistance in bacteria and fungi. *FEMS Microbiology Reviews* 38: 633–659.

Wang, T., He, M., Pan, Q. 2007a. A new method for the treatment of chromite ore processing residues. *Journal of Hazardous Materials* 149: 440–444.

Wang, Y. S., Pan, Z. Y., Lang, J. M., Xu, J. M., Zheng, Y.G. 2007b. Bioleaching of chromium from tannery sludge by indigenous *Acidithiobacillus thiooxidans*. *Journal of Hazardous Materials* 147: 319–324.

Wang, Y. T. 2000. Microbial reduction of chromate. In: Lovley, D.R. (ed.), *Environmental Microbe-Metal Interactions*. American Society for Microbiology Press, Washington DC.

Watts, M. P., Khijniak, T. V., Boothman, C., Lloyd J. T. 2015. Treatment of Alkaline Cr(VI) contaminated leachate with an alkaliphilic metal-reducing bacterium. *Applied and Environmental Microbiology* 81: 5511–5518.

Whittleston, R. A., Stewart, D. I., Mortimer, R. J. G., Ashley, D. J., Burke, I. T. 2011b. Effect of microbially induced anoxia on Cr(VI) mobility at a site contaminated with hyperalkaline residue from chromite ore processing. *Geomicrobiology Journal* 28: 68–82.

Whittleston, R. A., Stewart, D. I., Mortimer, R. J. G., Tilt, Z. C., Brown, A. P., Gerakid, K., Burke, I. T. 2011a. Chromate reduction in Fe(II)-containing soil affected by hyperalkaline leachate from chromite ore processing residue. *Journal of Hazardous Materials* 19: 15–23.

Yu, S., Du, J., Luo, T., Huang, Y., Jing, C. 2012. Evaluation of chromium bio-accessibility in chromite ore processing residue using in vitro gastrointestinal method. *Journal of Hazardous Materials* 209–210: 250–255.

Zayed, A. M., Terry, N. 2003. Chromium in the environment: Factors affecting biological remediation. *Plant Soil* 249: 139–156.

Zhou, S., Zhou, L., Wang, S., Fang, Di. 2006. Removal of Cr from tannery sludge by bioleaching method. *Journal of Environmental Sciences* 18: 885–890.

Zhu, W., Chai, L., Ma, Z., Wang, Y., Xiao, H., Zhao, K. 2008. Anaerobic reduction of hexavalent chromium by bacterial cells of *Achromobacter* sp. strain Ch1. *Microbiological Research* 163: 616–623.

46 Uranium Sequestration by Marine Cyanobacteria and its Metabolites

An Insight into Metabolism-Dependent and Metabolism-Independent Interactions

Vijayaraghavan Rashmi, Mohandass ShylajaNaciyar,
Dharmar Prabaharan, and Lakshmanan Uma

CONTENTS

ABSTRACT

Uranium is a radionuclide, which can be used as an abundant source of nuclear energy. It is widespread in many rocks, groundwater, and ocean. However, like other metals, it is seldom sufficiently concentrated to be economically recoverable. Physicochemical methods such as precipitation, oxidation/reduction, solid/liquid separation, membrane technology, and ion exchange though are effective; these processes not only are expensive but also lead to secondary pollution. On the other hand, using microbial biomass to harness uranium will be of an effective and economic alternative. The present report delineates the marine cyanobacteria–radionuclide interaction with a systematic approach through both metabolism-dependent and metabolism-independent mechanisms. This report strengthens the concept that a combination of several mechanisms, such as adsorption, ion exchange, complexation, coordination, or each functioning independently, can contribute to the overall metal sequestration. There is growing recognition that marine cyanobacteria can play a critical role in many bioremediation processes especially in heavy metal sequestration. This biotechnological process can be economical and eco-friendly not leading to secondary pollution, which could be a promising technology in larger-scale applications. The important future perspectives are required for the successful application of biosorption technology.

46.1 INTRODUCTION

Uranium (U) has been called nature's gift to hold up economic development and occupies an imperative position in nuclear energy. This naturally occurring radionuclide has found its way into the environment through activities associated with the nuclear industry (Kalin et al., 2004). Environmentalists have played a valuable role in accomplishing our requirement by uncompromising the environment being contaminated with nuclear spent fuel and radioactive wastes. Hence, to meet the existing nuclear demand and also to protect our environment from contamination in an economic way for sustainable energy has turned into a challenge for universal scientists.

Several techniques have been employed for the treatment of heavy metal–bearing industrial effluents, which usually come under two broad divisions: abiotic and biotic methods. Conventional or abiotic methods include precipitation, adsorption, ion exchange, membrane, and electrochemical technologies. *Hitherto*, several reports have disclosed the negative impact of these conventional techniques (Atkinson et al., 1998; Crini, 2006), which can be summarized as expensive, not environment friendly, and usually dependent on the concentration of the waste. Therefore, the search for efficient, eco-friendly, and cost-effective remedies for wastewater treatment has been initiated, some of which are in the process of commercialization. Biotic methods generally include biological methods for the treatment of effluents and solid wastes. In modern era, applying biotechnology for metal sequestration has been paid much attention and gradually becomes a hot topic in the field of biomining because of its prospective application.

Biosorption involves a number of processes and however it is complex and dependent on the nature of metal ion that interacts with the cell wall. The mechanisms responsible for the pollutant uptake differ according to the biomass type, cell physiology, and physicochemical influence of the environment such as pH, temperature, and metal concentration (Vijayaraghavan and Yun, 2008).

The biosorption of radionuclide with cultures may be (1) it possesses fascinating mechanisms for the interaction of metal ions by either metabolically dependent processes or metabolically independent processes; (2) compared to the conventional processes, biosorption process acquires less time and energy; (3) it facilitates the development of continuous treatment process; and (4) biosorption can be generated in large volume ratio (Volesky, 1994).

Any metal recovery process in natural environment is mediated primarily by two groups of microorganisms, namely, bacteria and fungi. Potent metal biosorbents under the class of bacteria include genera of *Bacillus* (Nakajima and Tsuruta, 2004), *Pseudomonas* (Chang et al., 1997; Uslu and Tanyol, 2006), and *Streptomyces* (Mameri et al., 1999; Selatnia et al., 2004). Important fungal biosorbents include *Aspergillus* (Kapoor and Viraraghavan, 1997; Jianlong et al., 2001; Binupriya et al., 2007), *Rhizopus* (Bai and Abraham, 2002; Park et al., 2005), and *Penicillium* (Niu et al., 1993; Tan and Cheng, 2003). In addition to the effective uranium sequestration by these heterotrophs, several algal biomasses have also been reported for its effective uranium sequestration, for example, *Chlorella* (Choudhary and Sar, 2011), brown algae, and red algae (Kalin et al., 2004). The drawbacks behind its usage are that it requires high nutrient inputs and in turn does not generate large surface/volume ratio.

Hence, biologists have turned their attention in exploring potential cost-effective biosorbent for metal sequestration to be used in a large scale. Among photosynthetic organisms, cyanobacteria seem to be the best candidate for bioremediation in view of their relatively easy photoautotrophic growth and trophic independence for carbon and nitrogen (Priya et al., 2011).

Moreover, cyanobacteria occupy a credential position in metal sequestration process. Many heavy metals such as copper, cadmium, and chromium are reported to be sequestered by cyanobacteria (Karna et al., 1999). Marine organisms are more advantageous because of its extreme adaptability and larger surface/volume ratio of unicellular cyanobacteria, which also suits larger application. Further, they have the ability to perform multiple tasks to sequester radionuclides through a package of extracellular metal-binding phenomenon that involves ionic interaction, precipitation, complexation, coordination, etc. (Rashmi et al., 2013). Natural occurrence of these potential organisms on seacoasts and estuarine waters makes them an economical source of biosorbent material.

The National Facility for Marine Cyanobacteria, sponsored by the Department of Biotechnology, Govt. of India, Bharathidasan University, Tiruchirappalli, India, has the largest marine cyanobacterial germplasm representing different geographical regions. Though it is known that cyanobacteria can sequester metals, with uranium the works are very limited. The prime goal of this chapter is to list out the mode of phenomenon involved in uranium sequestration by marine cyanobacteria.

46.2 SOLUBILITY AND STABILITY OF URANIUM FOR BIOSORPTION STUDIES

Seawater is the largest reservoir of dissolved uranium, and it contains a highly uniform value of 3.3 µg L^{-1}. The concentration of uranium in groundwater is usually in the range of 0.1–50 µg L^{-1}. In groundwaters, weathering of uranium-bearing rocks and minerals is the source of dissolved uranium (Ivanovich and Harmon, 1982).

The physicochemical properties of uranium are of continuing interest because it is a major constituent of highly radioactive spent nuclear fuel, which must be disposed safely. Two oxidation states of uranium are generally considered geochemically, when predicting its fate and transport in the natural environment. (1) Hexavalent uranium, U (VI), is highly soluble in water, while tetravalent uranium, U (IV), is sparingly soluble and easily precipitates to form the mineral uraninite (Kalin et al., 2004).

Uranium in the +VI oxidation state is relatively soluble and can be detected in almost any natural water. The high solubility of uranium (VI) was also reported by Bruno et al. (1986) with the concentration of above 4×10^{-5} mol^{-1} at pH 5.5–10, and the solubility did not vary with the varying pH range. This report was in concurrent with the present data (Rashmi. V thesis, 2014), as the maximum stability was attained in the pH range of 4–8 for the tested concentration ranging from 0.5 to 1 mM (Table 46.1). The solubility of uranium tested in artificial seawater nutrients (ASN) III medium, a marine synthetic growth medium for marine cyanobacteria (Rippka et al., 1979), further lights up the finding of Bruno (1989); under reducing conditions, the release of uranium from an underground repository into the sea is controlled by the solubility of uraninite, UO_2, where uranyl ion UO_2^{2+} behaves like divalent metal ions.

TABLE 46.1

Solubility and Stability of Uranium (Uranyl Acetate) with Respect to Varied pH and Light Intensities

Uranium Concentration (mM)	Solubility		Stability to pH (3–10)	Light (µmol Photons $m^2 s^{-1}$)		
	Water	ASN III Medium		100	200	Field
0.1	Soluble	Soluble	Stable (3–6)	Precipitates	Precipitates	Precipitates
0.5	Soluble	Soluble	Stable (3–6)	Stable	Stable	Stable
1	Soluble	Soluble	Stable (3–6)	Stable	Stable	Stable
5	Soluble	Soluble	Stable (3–6)	Stable	Stable	Stable
10	Soluble	Soluble	Stable (3–6)	Stable	Stable	Stable
50	Soluble	Soluble	Above pH 7 resulted a yellow precipitate	Stable	Stable	Stable
100	Soluble	Soluble	Above pH 7 results in precipitation	Stable	Stable	Stable

46.3 GROWTH PARAMETERS OF MARINE CYANOBACTERIA UNDER URANIUM INFLUENCE

In the aspect of using microbes as live biosorbent, at the concentrations above the threshold level, any metal is shown to inhibit growth, extend the lag phase of cultures, and interfere with several cellular processes including photosynthesis, respiration, enzyme activity, pigment synthesis, and cell division (Suzuki and Banfield, 2004).

Uranium being a radionuclide may have an antagonistic effect on marine cyanobacteria. In the present study, ASN III, a simple inorganic marine synthetic medium, was used to study uranium tolerance for marine cyanobacteria. The pH of the medium used is 7.2, and the toxicity for the tested cyanobacteria ranged between 0.1 and 10 mM U concentration (Table 46.2). Growth in terms of chlorophyll *a* and dry weight in all studied concentrations (0.1, 0.5, 1, 5, 10 mM and control) was determined. The effect of the different concentrations of U on the growth status of selected marine cyanobacteria significantly varied. The growth components were determined on 7th day as its end point growth (Figure 46.1).

TABLE 46.2

Tolerance of Marine Cyanobacteria to the Various Uranium Concentrations

Marine Cyanobacteria	Uranium Concentration (mM)					
	0.1	0.5	1	5	10	100
S. elongates BDU130911	+++	++	++	+	–	–
P. valderianum BDU140081	+++	++	++	+	–	–
O. boryana BDU 141071	+++	++	++	+	–	–
N. calcicola BDU 40302	+++	++	++	–	–	–

+++, Luxurious growth in uranium-supplemented concentration.

++, Tolerance coupled with growth.

+, Tolerance alone exhibited in particular uranium concentration.

–, Organism exhibited no growth and tolerance.

All the selected four strains showed luxurious growth in ASN III control medium. When the growth of the four selected marine cyanobacteria was estimated in uranium-supplemented condition, invariably all the tested strains exhibited tolerance coupled with minimal growth up to 1 mM U concentration. Beyond 1 mM, the tested strains showed only tolerance rather than growth in uranium-supplemented media conditions (Figure 46.1a and b).

Based on the growth components and the tolerance to uranium, unicellular marine cyanobacterium Synechococcus elongatus BDU 130911 was rated as highly tolerant. Among filamentous forms, *Oscillatoria boryana* BDU141071 and *Phormidium valderianum* BDU140081 were rated as potent tolerant, whereas the filamentous heterocystous form *Nostoc calcicola* BDU40302 was rated as least tolerant.

From the tolerance studies, it was identified that marine cyanobacteria could exhibit its tolerance to uranium and were rated in the order *O. boryana* BDU141071> *P. valderianum* BDU140081 >*S. elongatus* BDU130911 >*N. calcicola* BDU40302.

The physicochemical form of metal (speciation) is a critical factor in controlling metal bioavailability and toxicity in natural waters. Uranium in the form of uranyl acetate did not affect the growth of selected marine cyanobacteria at 1mM uranium concentration.

Suzuki and Banfield (2004), reported that, uranium affects several cellular processes in bacteria (*Deinococcus radiodurans* and *Arthrobacter* sp.) isolated from uranium-contaminated site, and this damage of the cell membrane at high metal concentration leads to uncontrolled efflux/influx of electrolytes or even other vital ions, which may be responsible for inhibition of growth. In the current study, only higher concentrations of U showed growth inhibition substantiated by chlorophyll *a* and dry weight (Figure 46.1a and b). Acharya et al. (2009) also showed no effect on the growth or cell morphology of the tested marine cyanobacteria *S. elongatus* BDU75042 up to 10 µM in the laboratory experiments up to 15–20 days. Uranium at 100 µM concentration showed about 25% inhibition that was concurrent with the present

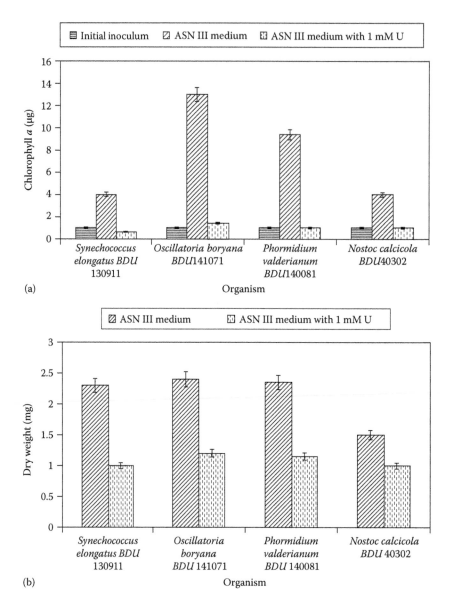

FIGURE 46.1 Relative growth of selected marine cyanobacteria in (i) ASN III medium and (ii) ASN III medium supplemented with 1 mM uranium, measured in terms of (a) chlorophyll *a* and (b) dry weight.

studies. Studies of Acharya et al. (2012) had shown highest tolerance of *S. elongatus* BDU75042 and *Anabaena torulosa* to 100 μM.

A dose-dependent growth inhibition has also been demonstrated with other microbes, and microalgae that have been sensitive to uranium were *Hyalella azteca* (Borgmann et al., 2005). In the work of Li et al. (2004), uranium tolerance is noted with *Microcystis aeruginosa*. With reference to *Chlorella* sp., uranium toxicity was high at pH 5.7 compared to pH 6.5 due to the greater proportion of UO$_2$ and UO$_2$(OH), that is, variation in uranium species (Franklin et al., 2000). So far, there are only few reports that are available to evaluate the influence of toxicity of uranium and the modifying factors during tolerance studies for marine cyanobacteria.

46.4 MARINE CYANOBACTERIAL BIOSORBENTS IN URANIUM BIOSORPTION

Radionuclides such as uranium and thorium are of particular concern due to their high toxicity and long half-lives; thus, they are considered severe ecological and public health hazards. Suitable methods are needed for removing radionuclides and heavy metals from contaminated soils. The ideal process for treating contaminated sites would selectively remove the contaminants of concern in a readily recoverable form, without excessive soil or water body destruction. Dissolved U is found in most natural waters at very low concentrations but is of some concern when concentrations increase to levels above 1 mg dm^3. U^{238}, the most common radionuclide, has a half-life of 4.5 million years, which makes environmental cleanup an important one (Meinrath et al., 1996).

Currently, removal of U from waste streams may be tackled in four ways:

1. Direct chemical methods
2. Electrochemical treatments
3. Ion exchange and biosorption methods
4. Sequestration by growing plant, algal, and microbial cells

Remediation of heavy metals from industries by chemical and electrochemical treatment is expensive and has been an important issue in abating environmental regulations. It was observed by several workers, mostly microbiologists, that microbial cells had the ability to concentrate metals, in their cellular mass, which existed in dilute concentrations in their aqueous environment. As a result, the idea of the use of biomaterial for the uptake of heavy metals has been extensively studied for the last two decades (Volesky, 2007). Generally, biosorption though is very effective depends on the type of biomass (i.e), live, dead biomass, bioabsorbs heavy metals from solutions even at lower concentrations.

46.4.1 Advantages of Using Biomass over Chemical Treatment

- Use of naturally abundant renewable biomaterials that can be economically produced
- Ability to treat large volumes of wastewater due to rapid kinetics
- High selectivity in terms of removal and recovery of specific heavy metals
- Ability to handle multiple heavy metals and mixed wastes
- Operation over a wide range of physiochemical conditions, including temperature, pH, and presence of other ions (including Ca^{2+} and Mg^{2+})
- Relatively low capital investment and low operational cost
- Improved recovery of heavy metals from the biomass
- Minimize the volume of hazardous waste produced

Biosorption and sequestration by living plants provide an environmentally and economically favorable method for removing U from natural waters. Although the ability of living microorganisms to take up metals from aqueous solution was investigated as early as eighteenth and nineteenth centuries, it is only during the last three decades that living or nonliving microorganisms have been used as adsorbents for the removal and recovery of materials from aqueous solutions. However, the environmental implementation of these biomaterials for uranium sequestration is still under limelight.

Cyanobacterial cells have been shown to concentrate several metals within the cells. Metal ions cross the cyanobacterial membranes with the help of channels termed as porins by active or passive mechanisms. Intracellularly, the metal ion sequestration is facilitated by either polyphosphate bodies or small, cysteine-rich proteins, called metallothioneins (Pettersson et al., 1988; Daniels et al., 1998). Live cyanobacterial cells have been shown to concentrate metal ions such as Pb, Sr, Mn, Al, 160 Zn, Cu, Cd, and Hg in the intracellular polyphosphates. Live marine, unicellular cyanobacterium, *S. elongatus* strain BDU/75042, was found to sequester 90%–98% uranium by the strain from simulated seawater containing 13 nM uranyl carbonate at pH 7.8, resulting in a loading of 0.7–4.2 mg U g^{-1} and from aqueous systems at pH 7.8 of 72% (53.5 mg U g^{-1} dry weight) where uranium deposits as uranyl carbonate hydrate with the cell biomass (Acharya et al., 2012, 2013). The filamentous, heterocystous, diazotrophic cyanobacterium *A. torulosa* was found to sequester 48% uranium within 30 min, resulting in a loading of 56 mg U g^{-1} of dry wt, and revealed the involvement of acid soluble polyphosphates in uranium accumulation (Acharya et al., 2012).

Adsorptive pollutants such as metals and dyes not only can be removed by living microorganisms but can also be removed by dead biological material. Feasibility studies for large-scale applications have demonstrated that biosorptive processes using dead biomass are more applicable than using living microorganisms, since the latter require nutrient supply and complicated bioreactor systems. Biosorption process with nonliving biomass is analogous to an ion-exchange process. In addition, maintenance of a healthy microbial population is difficult due to the toxicity of the pollutants and unsuitable environmental factors such as temperature and pH. Recovery of valuable metals is also limited in living cells since these may bound intracellularly. For these reasons, attention has been focused on the use of nonliving biomass as biosorbents (Park et al., 2010).

Cyanobacterial dead biomass harbors functional groups such as carboxyl, phosphoryl, hydroxyl, and amine, which bind metal ions to form metal–ligand surface complexes (Phoenix et al., 2002; Yee et al., 2004). These organisms bind heavy metals on the cell wall found outside its cell wall restricting the metal transport to the cell interior (De Philippis et al., 1998). Dead biomass has several advantages over live biomass and was listed as (1) feasibility to large-scale applications, (2) no demand for nutrient supply, (3) durability in maintaining healthy population, and (4) high recovery of extracted metals. However, many attributes of dead biomass remain unexploited in an industrial context.

It is suggested that the carboxyl groups represent the most important sink for metal ions at near-neutral pH (Yee et al., 2004). Acharya et al. (2009) have documented that from uranyl carbonate solution, uranium sequestration by marine cyanobacterium at pH (7.8) was predominantly associated with carboxyl and amide groups. Also in the previous report, uranyl-binding efficiency of the heat-killed or the nonviable *Synechococcus* cells was also evaluated explaining the extracellular localization of uranium.

Premuzic et al. (1991) established that chemical and structural characteristics of the cell membranes vary with species and should therefore influence their capacity of metal uptake by different microorganisms. The metal uptake process is dependent on the chemistry of the metal ions, specific surface properties of the biomass, cell physiology, and physicochemical influence of the environment, such as pH, temperature, and metal concentration. The metal–microbe interaction can be facilitated by several processes, which include diffusion across the cell membrane, adsorption to cell wall–bound extracellular polysaccharides (EPSs), precipitation, complexation, and oxidation–reduction reactions (Gadd and White, 1993). The bioremediation of heavy metals using microorganisms has received a great deal of attention, not only as a scientific novelty but also for its potential application in the industry (Uslu and Tanyol, 2006).

According to Akar et al. (2013), an initial rapid metal adsorption rate was due to the presence of instantaneous surface adsorption sites. When these sites became gradually roofed, the rate of adsorption attains a plateau. The size of the biosorbent also plays a vital role in biosorption. Park et al. (2010) well illuminate that larger surface area of the biosorbent is not suitable for the biosorbent process due to its low mechanical strength and clogging of the column. In contrary, small surface area biosorbent is more suitable for biosorption, which has higher ionic and mechanical strength, favors biosorption, and results in a shorter equilibration time. Simultaneously, a particle for biosorption should be sufficiently resilient to withstand the applicable pressures and extreme conditions applied during regeneration cycles (Volesky, 2003). Therefore, preliminary experiments are mandatory to decide the suitable size of a biosorbent. Findings of Acharya et al. (2012) revealed maximum uranium sequestration efficiency with *S.elongatus* BDU75042, an unicellular form, whose cell size is 3.5 μm Acharya et al. (2012) where the maximum biosorbent efficiency was observed with *S. elongatus* BDU75042, a unicellular form, which have a cell size of 3.5 μm, and showed its potentiality in uranium sequestration process. Similar studies were also attempted with various dead biomasses of *Saccharomyces cerevisiae* (Selatnia et al., 2004), algae (Kalin et al., 2004), and *Pseudomonas putida* (Chen et al., 2007) revealing the advantages of smaller size biomass in uranium sequestration.

46.4.2 Surface Modification of Biomass for Uranium Sequestration

The surface modeling of biomass could be improved when metal ions were classified according to the valence of metal ions (divalent ions) or hard/soft ions. It was found that different metal ionic characteristics played various roles in metal biosorption for several classes of metal ions (Chen et al., 2007). Metal ions were basically classified as class A (hard) and class B (soft) according to Nieboer and Richardson (1980), but with some variation. Uranium falls at class A metal ion

that generally showed high biosorptive capacity by various biomasses to form stable complex with the metal-binding donor atom in ligands in the order O > N > S (Nieboer and Richardson, 1980).

Normally, physical modification is very simple and inexpensive but is generally less effective than chemical modification. Among various chemical modification methods, chemical pretreatment (washing) has been preferred due to its simplicity and efficiency. Vast improvements in the biosorptive capacity of a biosorbent can be obtained through enhancement or modification of its functional groups (Ngah and Hanafiah, 2008). Amine, carboxyl, hydroxyl, sulfonate, thiol, and phosphonate groups are known binding sites for metals or dyes; thus, these groups can be newly formed, or their amounts increased, to enhance biosorptive capacity.

The cell surfaces of cyanobacteria consist of polysaccharides, proteins, and lipids, thereby offering several functional groups capable of binding to heavy metal ions (Acharya et al., 2012). Metal adsorption by biomass is supposed to occur through interactions with these functional groups that make up the cell wall (Chen et al., 2007). Marine cyanobacterial alkali–treated dead biomass of *Phormidium valderianum* BDU20041 was known to adsorb heavy metals, namely, nickel, copper, zinc, and cobalt (Karna et al., 1999). To maximize the efficacy of the dead biomass of marine cyanobacteria, it is important to spot the functional groups responsible for metal binding. According to Dotto et al. (2012) at acidic pH, cyanobacterial biomass surface is positively charged, and at alkali pH, the surface is negative charged. All these results given earlier suggested that uranium adsorption to the cyanobacterial biomass could be due to a high electrostatic attraction, which envisaged to be a selective process. Effective uranium biosorption was demonstrated in yeast cells *Rhodotorula glutinis* (Bai et al., 2010), through surface modifications of biomass (esterification of carboxyl groups and methylation of amine groups). Rashmi (2014), unpublished has shown that carboxyl group is more likely to have played a role in uranium sequestration as proved by the surface modification of biomass. *Hitherto*, no specific reports were there to support marine cyanobacterial surface modifications for selective uranium recovery.

46.4.3 Biosorption Isotherm

The metal biosorption process involves a solid phase (sorbent) and a liquid phase (solvent, normally water) containing the dissolved species to be adsorbed (adsorbate). Quantification of adsorbate–adsorbent interactions is fundamental for the evaluation of biosorption strategies (Volesky, 2003). To exactly determine the metal uptake capacities of different types of biosorbents, adsorption phenomena can be expressed as equilibrium isotherm curves that can be modeled by mechanistic or empirical equations.

Langmuir isotherm models have been most commonly used, with a high rate of success (Bhat et al., 2008). There are no critical reasons to use more complex models if this parameter model can fit the experimental data reasonably well.

The linearized form of Langmuir isotherm can be represented as follows:

$$\frac{C_e}{q_e} = \left[\left(\frac{1}{q_{max}}\right)\left(\frac{1}{b}\right)\right] + \frac{C_e}{q_{max}}$$

where

q_{max} is the maximum metal uptake (mg g^{-1})

b is the ratio of adsorption/desorption rates related to energy of adsorption

q_e is the equilibrium metal uptake capacity (mg g^{-1})

C_e is the residual uranium concentration in the solution (mg L^{-1})

According to this model, metal binding takes place at specific sites at the outer surface of the biomass, and once a sorbate occupies a binding site, no further biosorption occurs at these sites, suggesting a monolayer adsorption. This isotherm is a mean to interpret hyperbolic adsorption data (Bhat et al., 2008). It is basically the same equation used in Michaelis–Menten enzyme kinetics and describes the adsorption of metal ions to a finite number of ligand sites in a single layer on the cell surface (Bhat et al., 2008).

According to Volesky (2003), Langmuir equation is found to fit the experimental data reasonably well, as it follows single-component isotherm model extended for the multiple-component ones.

46.5 ROLES OF MARINE CYANOBACTERIAL METABOLITES TO COMPLEX WITH URANIUM

Microorganisms can mobilize radionuclide through either autotrophic or heterotrophic organisms, and their several microbial functional groups. Immobilization is one of processes, where radionuclide can get adsorbed to cell components via intracellular sequestration or precipitation. The functional moieties are oxalates, sulfides and phosphates (Boswell et al. 2001).

Cyanobacterial metabolites are accomplished with qualities to react with toxic metals, thereby altering its property and reducing the toxicity. These mechanisms modify the metal speciation, leading to decreased or increased mobility of metals. Such mechanisms include extracellular sequestration, intracellular localization, organic or inorganic precipitation, active/passive transport, and synthesis of the metal-binding proteins such as metallothioneins. Some of these mechanisms harbored by cyanobacteria for the detoxification of metal contaminants have been discussed by Suzuki and Banfield (2004).

Metallothioneins are low-molecular-weight cysteine-rich proteins that bind metal ions in metal thiolate clusters. Their synthesis has been shown to increase in response to increased concentrations of metals like cadmium, copper, and zinc

and forms complexes with metals (Blindauer et al., 2002). SmtA protein from *Synechococcus* sp. PCC 7942 is the only fully characterized prokaryotic metallothionein (Turner and Robinson, 1995). A Zn metallothionein-like sequence has also been reported in *Anabaena* sp. PCC 7120 and *Synechocystis* sp. 6803 (Blindauer et al., 2002). However, *hitherto* marine cyanobacterial metallothioneins have not been reported to bind or detoxify uranium.

46.5.1 EXTRACELLULAR POLYSACCHARIDES

Marine cyanobacteria can interact with heavy metals by both direct and indirect means. Direct interaction processes involve biosorption, whereas indirect processes include the formation of complexes with bioligands secreted by them (Moll et al., 2008). The curiosity of researchers in exploiting high-molecular-weight polysaccharides in biosorption has been greatly augmented. These biopolymers often show advantages over the polysaccharides currently in use, which are mostly extracted from plants or marine macroalgae (Sutherland, 2001). Of late, the use of exopolysaccharide-secreting microorganisms, such as bacteria, cyanobacteria, and algae, in metal biosorption is receiving a lot of attention. This has been primarily due to the reason that exopolysaccharides of these microorganisms are anionic and hence show the enormous ability of sorbing a variety of metal ions, such as Cu(II), Cd(II), Pb(II), and Cr(VI). Cyanobacterial mats, representing a kind of naturally immobilized system of cyanobacteria, may be a good metal biosorbent because they have a great allotment of exopolysaccharide-rich matrix (Pereira et al., 2009).

Cyanobacteria can be included among the potential sources of new polymers, and several species have been characterized by the presence of thick capsules surrounding the cells and by the ability to release polysaccharide material into the culture medium (De Philippis and Vincenzini, 2003). A large number of cyanobacteria are characterized by the presence of mucilagenous outermost investments (Parikh and Madamwar, 2006).

Bioprospecting cyanobacteria for the production of valuable polysaccharides has increased tremendously (Bhatia et al., 2010). These biomolecules have traditionally been used in the food, cosmetic, painting, and pharmaceutical industries as emulsifiers, stabilizers, or thickening agents (Pereira et al., 2009). According to De Philippis and Vincenzini (2003), organisms bind the heavy metals on cell wall or EPSs found outside its cell wall restricting the metal transport to the cell interior. The carboxyl group of cell surface of *Calothrix* sp. KC97, has been identified as the functional group that binds to copper, cadmium and lead (Phoenix et al., 2002). It is also suggested that the carboxyl groups represent the most important sink for metal ions at near-neutral pH (Yee et al., 2004).

Suzuki and Banfield (2004) have proposed that the radionuclides present in aquatic environment are accumulated by the marine microorganisms through direct adsorption from the water, and this property is independent of cellular metabolism. The cell surfaces of cyanobacteria have EPSs that contribute to metal adsorption. Cell surface–associated mucilaginous sheaths of *Gloeothece magna* have been shown to bind cadmium and magnesium effectively (Mohamed, 2001). EPS of *Anabaena cylindrica* gets complexed with copper, zinc, and iron.

Aqueous biopolymer solutions containing polysaccharides of *Nostoc muscorum* coated/immobilized onto a uranium-contaminated steel coupons showed a removal of >80% of the uranium (VI) from such coupons. The biopolymer–radionuclide complex is foreseen to reduce the work load in uranium decontamination activities. The overall metal sorptive capacity of such biopolymer of *Nostoc* was found to be up to 0.2 g U g^{-1} biomass (Davison, 2001).

Most of the cyanobacterial EPSs are composed of at least one uronic acid, several sugars in combination with protein molecules (Moreno et al., 1998). The major part of the EPS consisted of high-molecular-weight polysaccharides that had an anionic character due to uronic acids. De Philippis et al. (2007) found 13% uronic acids in the EPS of *Nostoc* sp. PCC 6720 and 26.7% in the EPS of *Nostoc* sp. PCC 7107. Uronic acids contribute to the anionic nature of the EPS, conferring a negative charge and a "sticky" behavior to the overall macromolecule (Decho, 1990; Sutherland, 1994; De Philippis and Vincenzini, 2003; Nichols et al., 2005). The anionic charge is an important characteristic for the affinity of these EPSs toward cations, notably dyes and metal ions. The presence of sulfate in the EPS contributes a unique feature among prokaryotes that share the similarity with archaea and eukaryotes (Micheletti et al., 2008).

Most of the bound uranium was found to be associated with the EPSs, suggesting its interaction with the surface-active ligands (Acharya et al., 2009), namely, carboxyl, amide, and hydroxyl groups. The uranyl-binding efficiency of the marine cyanobacteria EPS is supposed to be caused by the aggregation of double helical structural conformation of polysaccharides upon alkaline treatment due to ionic bridges between anionic and cationic bond formation (De Philips et al., 2007). EPSs are recognized as valuable biopolymers based on their unique properties to function as an effective biosorbent.

46.5.2 SIDEROPHORES

Biosequestration or bioreduction serves a promising, alternative, and economically viable strategy for metals such as uranium that is in very low concentrations (Cecal et al., 2012). There are many physiological mechanisms that were suggested for biosequestration of uranium by microbes listed as follows: metallic complex formation with metabolites; precipitation of toxic metals in extracellular and cellular compartments such as cell walls, phosphate-rich granules, lipid bodies, vacuoles, and nucleus; and metal binding in specific organic–metallic compounds such as metalloproteins and phytochelatins (Lombardi and Vieira, 2000). Many environmental stresses are overcome by organism through specific defense mechanism.

Marine cyanobacteria exhibit the ability to sequester radionuclides through various mechanisms, namely, direct enzymatic reactions and/or indirect chemical reductions, namely physical adsorption, complexation, ion exchange, and precipitation (Acharya et al., 2012).

Of these, siderophores constitute a major class of naturally occurring chelators that include hydroxamate, catecholate, and carboxylic acid functional groups secreted by microorganisms in various habitats, which bind to iron and mediate its transport to the cell. This metal chaperone, though specific for iron, also bind effectively with a variety of environmentally and biologically important metals, such as arsenic, magnesium, manganese, chromium and various radionuclides.

The biogeochemistry of hexavalent uranium in the ocean is dominated by soluble complexes formed with high-affinity organic ligands that are believed to be microbial siderophores or similar biogenic chelating agents. Siderophores are bioligands, known extensively to bind with Fe(III). Few bacterial (Frazier et al., 2005) and fungal (Renshaw et al., 2002) siderophores are known to bind with uranium and other radionuclides. The exploitation of microbes for biological treatment of radionuclides will largely be determined by the ability of microorganisms to sequester radionuclides and survive under radiation stress. Though a number of organisms are believed to be involved in bioremediation process through siderophores (Frazier et al., 2005), cyanobacteria, the unique oxygen-evolving photosynthetic prokaryotes, have an edge over all of them in view of their tropic independence for nitrogen and carbon.

The selected marine cyanobacteria, namely, *Synchococcus elongatus* BDU130911, *Oscillatoria boryana* BDU141071, *Phormidium valderianum* BDU140081, and *Nostoc calcicola* BDU40302, were estimated for their siderophore production in various media conditions that include [Fe(−)], [Fe(−)U(+)], and [Fe(+)U(+)], where ASN III complete medium served as control.

Among the organisms tested, S.elongatus BDU130911 in [Fe(−)U(+)] condition produced a maximum siderophore (μgmg^{-1} dry weight) of 56 ± 0.22 followed by O.boryana BDU141071 (24 ± 0.45), N. calcicola BDU40302 (22 ± 0.25), and P. valderianum BDU140081 (14 ± 0.82). Siderophore production by *S.elongatus* BDU130911 in [Fe(+)U(+)] was less 11 μg when compared to [Fe(−)U(+)], though *S.elongatus* BDU130911 ranked higher. Further, siderophore production in [Fe(+)U(+)] condition clearly substantiates that uranium had induced siderophore production. Here, siderophore production analyzed by chrome azurol S (CAS) assay confirms iron–siderophore complexation. The overall capacity of marine cyanobacteria to produce siderophore was rated as follows: *S. elongatus* BDU130911 > *O. boryana* BDU141071> *N. calcicola* BDU40302 > *P. valderianum* BDU140081 (Figure 46.2).

Interestingly, it was observed that the behavior of the organism to the growth and siderophore production varied greatly, where in terms of growth *O. boryana* BDU141071 rated higher

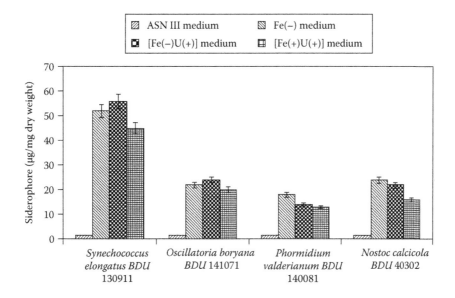

FIGURE 46.2 Siderophore production of marine cyanobacteria by CAS assay (Fe complexation) in (i) ASN III medium, (ii) [Fe(−)], (iii) [Fe(−)U(+)], and (iv) [Fe(+)U(+)] media conditions. (From Rashmi, V. et al., *Bioresour. Technol.*, 130, 204, 2013. With permission.)

(Figure 46.1a and b) and in terms of siderophore production *S. elongatus* BDU130911 rated the best (Figure 46.2).

Variation in siderophore production can be both phenotypic and genotypic and also species specific (Harrison et al., 2008), which coincides with our findings that all the chosen four organisms produced siderophore in varied quantities, though analogous media and incubation conditions were given.

Despite the fact, the organisms produced siderophore in uranium-supplemented media and exhibited their inability to grow as that of normal ASN III Fe(+) media clearly points out the probable binding of uranium to siderophore, thus inhibiting the transport of Fe^{3+}, the essential micronutrient. According to Kraepiel et al. (2009), the uptake of siderophore complex is not only limited to Fe^{3+} ion; however, the cells also take up other metal complexes, only when the metal is required for growth. The presence of uranium in the Fe-supplemented environment probably has made Fe unavailable. According to Kalin et al. (2004), algae and other aquatic plants are known to actively pump metals across their cell membranes. This process was energy driven that was directly linked with growth, as most of the metals taken up were essential nutrients. The problem arises when the external metal concentrations far exceed cell requirements. While U is not a required nutrient, it may mimic a metal that is required. Whatever the reason, U (VI) is taken up by some algal cells (*Scenedesmus, Spirogyra, Chlorella*). Once inside the cell, U is concentrated into vacuoles and sometimes precipitated through uranium reductases.

Siderophore production estimated by CAS assay showed interesting results with varying culture conditions. Experimental studies substantiate that siderophore production was maximum in its stationary phase and inversely proportional to the growth. The maximum siderophore production was observed on fifth day in [Fe(−)U(+)] medium and was 56 ± 0.22 μg mg⁻¹ dry weight. Surprisingly less siderophore production was observed in [Fe(−)] as the trace amount of iron might be present (Figure 46.2) in the media. This beyond doubt makes us to conclude that the condition for siderophore production by marine cyanobacterium *S.elongatus* BDU130911 is influenced only by the presence of uranium in the medium.

In addition, siderophore production in [Fe(+)U(+)] could be well explained by the fact that (Figure 46.2) uranium in the medium could perhaps act as a competitive agent making Fe^{3+} ion unavailable to the organism due to its similar chemistry (John et al., 2001). As reported by John et al. (2001) in aqueous state, UO_2^{2+} and Fe^{3+} share similar chemistry. Since a high chemical similarity between Fe and U exists, the siderophore produced by *S. elongatus* BDU130911 could not efficiently differentiate during complexation. Uranium sequestration ability of siderophore produced in *milieu* by *S. elongatus* BDU130911 grown in [Fe(−)U(+)] was tested for residual concentration of uranium, colorimetrically by the dibenzoylmethane method (Yoe et al., 1953).

Significantly, the residual concentration of uranium in *milieu* decreased up to 50% with the increased siderophore concentration (Figure 46.3). This evidently signifies that the siderophore produced by *S. elongatus* BDU130911 had complexed with uranium. The complexation of >95% uranium with formohydroxamic acids, demonstrated by Nunez and Vandegrift (2001), through a chemical process clearly substantiates uranium–siderophore complexation. Marine cyanobacteria *S. elongatus* BDU130911 produced a maximum siderophore of 56 ± 0.22 μg mg⁻¹ dry weight, and significantly, the residual concentration of uranium in *milieu* decreased up to 50% with the increased siderophore concentration (Rashmi et al., 2013). This report evidently signifies that the siderophore produced by *S. elongatus* BDU130911 had complexed with uranium.

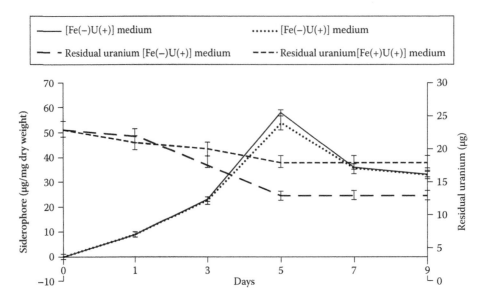

FIGURE 46.3 Time course siderophore production in *S. elongatus* BDU130911 (i) ASN III medium, (ii) [Fe(−)], (iii) [Fe(−)U(+)], and (iv) [Fe(+)U(+)] media conditions. (From Rashmi, V. et al., *Bioresour. Technol.*, 130, 204, 2013. With permission.)

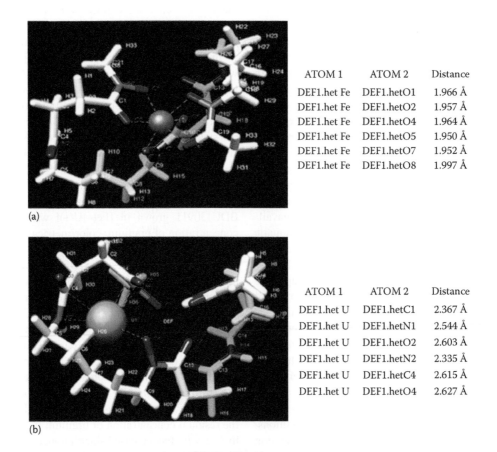

ATOM 1	ATOM 2	Distance
DEF1.het Fe	DEF1.hetO1	1.966 Å
DEF1.het Fe	DEF1.hetO2	1.957 Å
DEF1.het Fe	DEF1.hetO4	1.964 Å
DEF1.het Fe	DEF1.hetO5	1.950 Å
DEF1.het Fe	DEF1.hetO7	1.952 Å
DEF1.het Fe	DEF1.hetO8	1.997 Å

(a)

ATOM 1	ATOM 2	Distance
DEF1.het U	DEF1.hetC1	2.367 Å
DEF1.het U	DEF1.hetN1	2.544 Å
DEF1.het U	DEF1.hetO2	2.603 Å
DEF1.het U	DEF1.hetN2	2.335 Å
DEF1.het U	DEF1.hetC4	2.615 Å
DEF1.het U	DEF1.hetO4	2.627 Å

(b)

FIGURE 46.4 Stereo view of (a) Fe(III)-binding site in DFO-B. DFO-B is shown in ball-and-stick representation with Fe(III) (orange) in the center. The atomic distance between Fe(III) and oxygen atom is shown in blue lines, (b) U(IV)-binding site in DFO-B. DFO-B is shown in ball-and-stick representation with U(IV) (blue) in the center. The atomic distance between U(IV) and oxygen atom is shown in blue lines. (From Rashmi, V. et al., *Bioresour. Technol.*, 130, 204, 2013. With permission.)

Although heuristic, these results suggest that the biogeochemical role of cyanobacterial siderophore may extend beyond their traditional function as iron-complexing ligands. Understanding the coordination chemistry and stability of aqueous metal–siderophore complexes is essential in elucidating the factors that control their biological sequestration by microbes. The donor atoms for uranium binding are predominantly oxygen, which are derived from hydroxamate and catecholate groups (Sayyed and Chincholkar, 2006). In silico docking of Fe and uranium shares similar bonding pattern, which further adds up the light for uranium–siderophore complexation (Figure 46.4a and b). The coordination atoms involved in complexing Fe^{3+} and UO_2^{2+} to the DFO-B were identified as O, C, and N, respectively. Interaction studies revealed that six oxygen atoms (O-1, O-2, O-4, O-5, O-7, and O-8) of hydroxamic acid in DFO-B are involved in Fe (III) complexation (Figure 46.4a) with the distance of less than 2 Å. The deprotonation of hydroxamate groups confers high-affinity binding with iron. The complexation of U(IV) with DFO-B configuration also resulted in metal–ligand hexadentate coordination (C-1, N-1, O-2, O-4 position of hydroxamate groups and N2, C4 of amine group) equivalent to that of DFO-B with Fe (III) complex, where the distance remained less than 3 Å between the U(IV) and DFO-B complex (Figure 46.4b). Till date, no study was evidently reported for the complexation pattern of uranium with the donor atoms. The report of Rashmi et al. (2013) had clearly figured out the donor atoms involved in uranium complexation. These reports further corroborate hydroxamate-type siderophore production in *S. elongatus* BDU130911 and its role in uranium complexation.

So far, reports on cyanobacterial siderophore production have been evidenced only under metal-depleted condition (Gademann et al., 2009), not on metal supplementation like metallothionein (Cavet et al., 2003). This hypothesis has been proved by the findings of Rashmi et al. (2013), where the production and complexation of marine cyanobacterial siderophore with uranium and radionuclide have given a new lead in the biosorption phenomenon.

46.6 ANALYTICAL INSTRUMENTS FOR URANIUM BIOSEQUESTRATION

The information on the active sites involved in the metal binding is obtained through infrared absorption spectroscopy or Fourier transform infrared spectroscopy (FTIR), transmission electron microscopy (TEM), energy-dispersive x-ray spectroscopy, x-ray diffraction (XRD) analysis, thermal gravimetric analysis (TGA), and differential scanning calorimetry (Varma et al. 2004; Wang and Chen, 2006).

TEM analysis of *Arthrobacter*, *Desulfovibrio*, and *Sphingomonas* cells, which had accumulated uranium, showed needlelike fibrils in the cytoplasm (Merroun et al., 2006). In view of Kazy et al. (2009), studies on microbial accumulation of actinide elements including uranium and other heavy metals have indicated a selective sequestration of accumulated metals in different parts of the microbial cell. In many cases

of enzymatic metal sequestration (including U) by dissimilatory iron-reducing bacteria (*Geobacter*, *Desulfovibrio*, etc.) or by metal-resistant bacteria (*Citrobacter*, *Pseudomonas*, and *Escherichia coli*) and even in metabolism-independent biosorption process, the sequestered metal was found to be localized in and around cell boundary (outer membrane, LPS, EPS, periplasm). Kazy et al. (2009) reported that uranium was found to localize in the cell wall and periplasm as black deposits that were analyzed using TEM studies.

Metal exclusion by the cell wall, membrane, or cell envelope results in structural alteration of the cells. Copper-stressed cells of *Synechocystis* sp. PCC 6803 developed thickened calyx around the cell wall that was found to be responsible for binding copper ions on the cell surface evidenced using TEM analysis (Gardea Torresdey et al., 2004).

Previous studies on uranyl interactions identified asymmetric stretching frequencies of uranyl ion, UO_2^{2+}, between 900 and 950 cm^{-1} (Merroun et al., 2003; Schmeide et al., 2003). Ulrich et al. (2006) have elucidated uranyl ions in the spectral samples at 903 cm^{-1}, in addition, to the asymmetric stretching of UO_2^{2+}-carbonato or –hydroxo complexes which are known to excite bands at substantially higher wavenumbers (>920 cm^{-1}) which is reported by Ulrich et al. (2006). According to Cecal et al. (2012), ion exchange is the principal mechanism for uranium sorption, suggesting a high electrostatic attraction between the negatively charged, anionic biomass surface and uranium, a cationic metal.

According to Kazy et al. (2009), the crystalline uranium formation indicates possible complexation of such metals with the cellular functional groups, namely, carboxyl or phosphates, facilitating metal nucleation and metal precipitation in a crystalline state. FTIR spectroscopic analysis indicates the involvement of functional groups in uranium binding. Anionic functional groups present in the cell membrane and wall materials may act as the primary metal-binding sites creating negative surface charge conducive to cation binding (Merroun et al., 2003). Implication of these groups in the complexation of uranium in Gram-negative bacteria *Acidithiobacillus ferrooxidans* has been previously confirmed using different spectroscopic and microscopic techniques (Merroun et al., 2005). The initial binding of metal ions to biomass reactive sites may serve as nucleation sites for further precipitation of metals forming large metal deposits (Volesky, 2003). The formation of crystalline uranium by different bacterial strains has been observed by Beazley et al. (2007).

The energy-dispersive x-ray fluorescence (EDXRF) analysis is based on the fact that the x-rays emitted from an ionized atom have energies that are characteristic of the element involved. The x-ray intensity is proportional to both the elemental concentration and the strength of the ionizing source. For the qualitative and quantitative determination of uranium present in the biomass, EDXRF analysis is appreciated. Photon ionization, which is achieved using either an x-ray tube or radioisotope, is most applicable to the nondestructive assay of nuclear material (Tertian and Claisse, 1982). The elemental selectivity for uranium showed clear identifiable peaks

TABLE 46.3

Proposed Analytical Techniques Used in Identification of Active Sites Involved in Uranium Biosorption for Marine Cyanobacteria

Analytical Techniques	Remarks
UV-Vis spectrophotometer	Uranium concentrations in aqueous phase were measured in terms of color intensity.
TEM	Visual confirmation of inner morphology of biomass, especially cells and the presence of uranium.
EDXRF	Element analysis and chemical characterization of uranium bound on the biosorbent.
XRD analysis	Crystallographic structure and chemical composition of uranium bound on the biosorbent.
FTIR	Determined active sites of the biosorbent.
TGA	Characterized thermal stability of the biosorbent (dead biomass).

of UL x-rays, that is, ULα1 and ULβ1, detected within a range of 13.5–17 keV.

All of these analytical techniques provide the type of information that they yield for understanding and evaluating the biosorption phenomenon of interest (Table 46.3). However, different techniques can often provide distinctive, but complementary, information on the biosorption of a target pollutant; thus, combining different techniques can optimize the exploration of biosorption mechanisms.

46.7 IMMOBILIZATION OF MARINE CYANOBACTERIAL BIOMASS FOR URANIUM RECOVERY

Suspended cultivation of microalgae has been employed to remove various nutrients and inorganic chemicals. However, some difficulties limit the practical application of suspended microalgae that include monospecificity, and good operation conditions are hard to be separated before discharge. The use of immobilized microalgae for the removal of nutrients from wastewater shows the potential to solve the given problems (Lee et al., 1996). Immobilized cells have several advantages as they provide simple and cost-effective treatment, and the durability of the cells remains constant for longer periods (Patnaik et al., 2001). The most commonly used matrices for microalgae are agar, alginate gels, polyurethane, and polyvinyl foams. A disadvantage of agar and alginate is their low mechanical stability for long-term use in bioreactor; moreover, calcium alginate gels are disrupted by phosphate ions. Polyvinyl and polyurethane foams offer better mechanical properties and are neutral to most commonly used ions (Boominathan, 2005).

Immobilization of cyanobacterial biomass allows easier metal recovery along with the regeneration of biomass, without compromising the natural binding capacity of biomass for

the metal. Free cells or cell suspensions have generally low mechanical strength and smaller particle size. High pressure is required to generate suitable flow rates for metal binding leading to disintegration of free cells. These problems can be appropriately addressed by the use of immobilized biomass (Mallick and Rai, 1994). Among the several methods of immobilization of cyanobacteria reported in the literature, entrapment of cyanobacterial cells in natural or synthetic polymers is the most popular method in use (Garbisu et al., 1993; Prakasham and Ramakrishna 1998). Cyanobacteria possess high metal absorption capacity that has encouraged its application for detoxification of effluents and has an edge over conventional wastewater treatment facilities (Darnall et al., 1986).

Various techniques have been used for the immobilization of biomass and, in general, can be divided into three categories: cell immobilization on inert supports, entrapment within a polymeric matrix, and cross-linking. Microbial biomass can be immobilized on various inert supports at specific stages of its cultivation. If living cells are directly used for a biosorption process, these can be easily immobilized within the process, *in situ*, and other pollutants may be removed simultaneously (Park et al., 2010).

Immobilized cyanobacteria show better potential in metal removal than their free-living counterpart. Immobilized *Anabaena doliolum* showed an increased uptake of Cu and Fe, nearly 45% and 23%, respectively, higher than that of the free-living cells (Rai and Mallick, 1992). *Synechococcus* sp. PCC 7942 biomass immobilized in silica successfully bound copper, lead, nickel, and cadmium under flow-through conditions at pH 5. More than 98% of the adsorbed metals could be recovered when treated with 0.1N HCl, providing a recyclable system for adsorption of these metal ions (Gardea Torresdey et al., 1998).

There is no adequate report on the usage of immobilized cyanobacterial biomass for uranium recovery from aqueous solutions above pH > 7 under flow-through conditions. It was shown previously that the adsorption of uranium by immobilized cells of *Streptomyces* and *Chlorella* was unaffected by pH values between 4 and 9 compared to free cells (Nakajima et al., 1982). Acharya et al. (2009) evaluated the potential of the marine unicellular cyanobacterium, *S. elongatus*, in immobilized form for uranium recovery under flow-through conditions where uranium binding in *S. elongatus* is predominantly a surface phenomenon. The use of immobilized *S. elongatus* cells demonstrates their potential for uranium recovery/remediation from aqueous solutions allowing regeneration of the biomass for multiple sorption–desorption cycles above pH 7.

Similar observations have been noticed with the following immobilized cyanobacteria, namely, *A. doliolum* (Mallick and Rai, 1994) and *Synechococcus* sp. PCC 7942 (Gardea-Torresdey et al., 1998). According to Gardea-Torresdey et al. (1998), more than 98% of the adsorbed metals could be recovered when treated with 0.1 N HCl, providing a recyclable system for adsorption of these metal ions.

This was substantiated by Park et al. (2010) that one of the important industrial applications of biosorption is recovery of loaded pollutants (especially valuable metals) from the biosorbent and simultaneous regeneration of the biosorbent for reuse. In fact, the usefulness of a specific biomass as a biosorbent depends not only on its biosorptive capacity but also on the ease of its regeneration and reuse. However, most researchers have tended to focus only on the biosorptive capacity of biosorbent tested, without consideration of the regeneration required for industrial applications in particular for important radionuclides considering its cost and reuse.

Biosorption of uranium is still at the laboratory-scale study, in spite of unquestionable progress. As a first step to the industrialization, the present study proposes the possible phenomenon (Figure 46.5) and standardized parameters (Table 46.4) to exploit marine cyanobacteria for large-scale uranium sequestration. However, much work in this area is necessary to demonstrate its possibilities to be implemented in industrial scale.

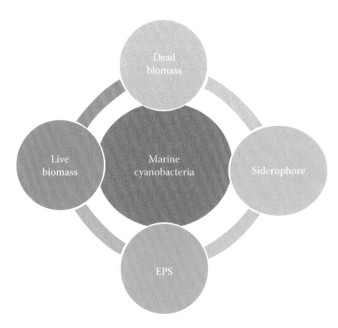

FIGURE 46.5 Schematic illustration of uranium sequestration phenomenon (both metabolism dependent and metabolism independent) by marine cyanobacteria.

46.8 SUMMARY

The present study for the first time has investigated both metabolism-dependent and metabolism-independent phenomena for uranium sequestration in marine cyanobacteria. Natural occurrence of this organism in the earth's oceans combined with its ability to adsorb anionic uranyl complexes, rapid kinetics, high uranyl loading capacity, and tolerance to high salinity makes this group as a promising candidate for uranium recovery from the marine environments. Utilizing these versatile natural bioadsorbents satisfies all the requisites for the development of an easy, inexpensive, green, zero waste process for the recovery of uranium that will open up a new arena in cyanobacterial bioremediation.

Conversely, it is no small feat to replace well-established conventional techniques that are in process. Therefore, through continued research, especially on pilot- and full-scale biosorption process, the situation is likely to change in the near future, with which biosorption technology is becoming more beneficial and attractive than currently used technologies. Although attempts are very few in using marine cyanobacteria for uranium sequestration, the present study gives a lead for commercialization considering its oxygenic, rapid growth, eco-friendly natural, low-cost material for future perspective.

ACKNOWLEDGMENTS

The authors thank the Department of Atomic Energy (Govt. of India) Grant No: (2007/37/29/BRNS/1906) and the Department of Biotechnology, Govt. of India (Grant No: BT/IS/MAIN/1/98), for financial support.

TABLE 46.4
Proposed Biomatrix for Uranium Sequestration by Marine Cyanobacteria

Parameters	Marine Cyanobacterial Biomass	
	Live Biomass	Dead Biomass
pH	7.2	4.5–7.5
Solubility of uranium	Soluble in water and ASN III medium	Soluble in water
Temperature	25°C ± 2°C	25°C ± 2°C
Rotation per minute	120	120
Uranium concentration	0.1–5 mM	0.1–10 mM
Contact time	24 h	24 h
Saturation time	30 min	1 h
Tolerance limit	1 mM	Not applicable
Adsorption isotherm	Monolayer	Monolayer
Functional groups	$COOH^-$, OH^-, NH^+	$COOH^-$, OH^-
Stability	Less stable	Stable
Siderophore type	Hydroxamate type	Not applicable
Regeneration	Less effective	Effective

REFERENCES

Acharya, C., Chandwadkar, P., Apte, S. K. 2012. Interaction of uranium with a filamentous, heterocystous, nitrogen-fixing cyanobacterium, *Anabaena torulosa*. *Bioresource Technology* 116: 290–294.

Acharya, C., Chandwadkar, P., Joseph, D., Apte, S. K. 2013. Uranium (VI) recovery from saline environment by a marine unicellular cyanobacterium, *Synechococcus elongatus*. *Journal of Radio analytical and Nuclear Chemistry* 295: 845–850.

Acharya, C., Joseph, D., Apte, S. K. 2009. Uranium sequestration by a marine cyanobacterium, *Synechococcus elongatus* strain BDU/75042. *Bioresource Technology* 100: 2176–2181.

Akar, T., Ozkara, E., Celik, S., Turkyilmaz, S., Akar, S. T. 2013. Chemical modification of a plant origin biomass using cationic surfactant ABDAC and the biosorptive decolorization of RR45 containing solutions. *Colloids and Surfaces B: Biointerfaces* 101: 307–314.

Atkinson, B. W., Bux, F., Kasan, H. C. 1998. Considerations for application of biosorption technology to remediate metal-contaminated industrial effluents. *Water SA* 24: 129–135.

Bai, J., Yao, H., Fan, F., Lin, M., Zhang, L., Ding, H., Lei, F. et al. 2010. Biosorption of uranium by chemically modified *Rhodotorula glutinis. Journal of Environmental Radioactivity* 101: 969–973.

Bai, R. S., Abraham, T. E. 2002. Studies on enhancement of Cr (VI) biosorption by chemically modified biomass of *Rhizopus nigricans. Water Research* 36: 1224–1236.

Beazley, M. J., Martinez, R. J., Sobecky, P. A., Webb, S. M., Taillefert, M. 2007. Uranium biomineralization as a result of bacterial phosphatase activity: Insights from bacterial isolates from a contaminated subsurface. *Environmental Science & Technology* 41: 5701–5707.

Bhat, S. V., Melo, J. S., Chaugule, B. B., D'souza, S. F. 2008. Biosorption characteristics of uranium (VI) from aqueous medium onto *Catenella repens*, a red alga. *Journal of Hazardous Materials* 158: 628–635.

Bhatia, S., Namdeo, A., Nanda, S. 2010. Factors effecting the gelling and emulsifying properties of a natural polymer. *Systematic Reviews in Pharmacy* 1: 86.

Binupriya, A. R., Sathishkumar, M., Kavitha, D., Swaminathan, K., Yun, S. E. 2007. Aerated and rotated mode decolorization of a textile dye solution by native and modified mycelial biomass of *Trametes versicolor. Journal of Chemical Technology and Biotechnology* 82: 350–359.

Blindauer, C. A., Harrison, M. D., Robinson, A. K., Parkinson, J. A., Bowness, P. W., Sadler, P. J. Robinson, N. J. 2002. Multiple bacteria encode metallothioneins and SmtA-like zinc fingers. *Molecular Microbiology* 45: 1421–1432.

Boominathan, M. 2005. Bioremediation studies on dairy effluent using cyanobacteria, PhD thesis, Bharathidasan University, Tamil Nadu, India, pp. 1–92.

Borgmann, U., Couillard, Y., Doyle, P., Dixon, D. G. 2005. Toxicity of sixty-three metals and metalloids to *Hyalella azteca* at two levels of water hardness. *Environmental Toxicology and Chemistry* 24: 641–652.

Boswell, C. D., Dick, R. E., Eccles, H., Macaskie, L. E. 2001. Phosphate uptake and release by *Acinetobacter johnsonii* in continuous culture and coupling of phosphate release to heavy metal accumulation. *Journal of Industrial Microbiology and Biotechnology* 26: 333–340.

Bruno, J. 1989. A reinterpretation of the solubility product of solid uranium (IV) dioxide. *Acta Chemica Scandinavica* 43: 99–100.

Bruno, J., Casas, I., Lagerman, B. and Munoz, M. 1986. The determination of the solubility of amorphous UO_2 (Sn and the mononuclear hydrolysis constants of uranium (IV) at 25°C. In *MRS Proceedings*, vol. 84, p. 153. Cambridge University Press, Cambridge, U.K.

Cavet, J. S., Borrelly, G. P., Robinson, N. J. 2003. Zn, Cu and Co in cyanobacteria: Selective control of metal availability. *FEMS Microbiology Reviews* 27: 165–181.

Cecal, A., Humelnicu, D., Rudic, V., Cepoi, L., Ganju, D., Cojocari, A. 2012. Uptake of uranyl ions from uranium ores and sludges by means of *Spirulina platensis, Porphyridium cruentum* and *Nostok linckia* alga. *Bioresource Technology* 118: 19–23.

Chang, J. S., Law, R., Chang, C. C. 1997. Biosorption of lead, copper and cadmium by biomass of *Pseudomonas aeruginosa* PU21. *Water Research* 31: 1651–1658.

Chen, X., Shi, J., Chen, Y., Xu, X., Chen, L., Wang, H., Hu, T. 2007. Determination of copper binding in *Pseudomonas putida* CZ1 by chemical modifications and x-ray absorption spectroscopy. *Applied Microbiology and Biotechnology* 74: 881–889.

Choudhary, S., Sar, P. 2011. Identification and characterization of uranium accumulation potential of a uranium mine isolated *Pseudomonas* strain. *World Journal of Microbiology and Biotechnology* 27: 1795–1801.

Crini, G. 2006. Non-conventional low-cost adsorbents for dye removal: A review. *Bioresource Technology* 97: 1061–1085.

Daniels, M. J., Turner-Cavet, J. S., Selkirk, R., Sun, H., Parkinson, J. A., Sadler, P. J., Robinson, N. J. 1998. Coordination of Zn^{2+} (and Cd^{2+}) by Prokaryotic Metallothionein involvement of his-imidazole. *Journal of Biological Chemistry* 273: 22957–22961.

Darnall, D. W., Greene, B., Henzl, M. T., Hosea, J. M., McPherson, R. A., Sneddon, J., Alexander, M. D. 1986. Selective recovery of gold and other metal ions from an algal biomass. *Environmental Science & Technology* 20: 206–208.

Davison B. H. 2001. 'Green' biopolymers for improved decontamination of metals from surfaces: Sorptive characterization and coating properties. Annual report to be submitted to DOE Program Managers for posting on web page. No. R01-111120. ORNL Oak Ridge National Laboratory (US), Oak Ridge, TN.

De Philippis, R., Margheri, M. C., Materassi, R., Vincenzini, M. 1998. Potential of unicellular cyanobacteria from saline environments as exopolysaccharide producers. *Applied and Environmental Microbiology* 64: 1130–1132.

De Philippis, R., Paperi, R., Sili, C. 2007. Heavy metal sorption by released polysaccharides and whole cultures of two exo-polysaccharide-producing cyanobacteria. *Biodegradation* 18: 181–187.

De Philippis, R., Vincenzini, M. 2003. Outermost polysaccharidic investments of cyanobacteria: Nature, significance and possible applications. *Recent Research Developments in Microbiology* 7: 13–22.

Decho, A. W. 1990. Microbial exopolymer secretions in ocean environments: Their role in food webs and marine processes. *Oceanography and Marine Biology* 28: 73–153.

Dotto, G. L., Esquerdo, V. M., Vieira, M. L. G., Pinto, L. A. A. 2012. Optimization and kinetic analysis of food dyes biosorption by *Spirulina platensis. Colloids and Surfaces B: Biointerfaces* 91: 234–241.

Franklin, N. M., Stauber, J. L., Markich, S. J., Lim, R. P. 2000. pH-dependent toxicity of copper and uranium to a tropical freshwater alga (*Chlorella* sp.). *Aquatic Toxicology* 48: 275–289.

Frazier, S. W., Kretzschmar, R., Kraemer, S. M. 2005. Bacterial siderophores promote dissolution of UO_2 under reducing conditions. *Environmental Science & Technology* 39: 5709–5715.

Gadd, G. M., White, C. 1993. Microbial treatment of metal pollution-a working biotechnology? *Trends in Biotechnology* 11: 353–359.

Gademann, K., Kobylinska, J., Wach, J. Y., Woods, T. M. 2009. Surface modifications based on the cyanobacterial sidero-phore anachelin: From structure to functional biomaterials design. *Biometals* 22: 595–604.

Garbisu, C., Hall, D. O., Serra, J. L. 1993. Removal of phosphate by foam-immobilized *Phormidium laminosum* in batch and continuous-flow bioreactors. *Journal of Chemical Technology and Biotechnology* 57: 181–189.

Gardea-Torresdey, J. L., Gonzalez, J. H., Tiemann, K. J., Rodriguez, O., Gamez, G. 1998. Phytofiltration of hazardous cadmium, chromium, lead and zinc ions by biomass of *Medicago sativa* (Alfalfa). *Journal of Hazardous Materials* 57: 29–39.

Gardea-Torresdey, J. L., Peralta-Videa, J. R., Montes, M., De la Rosa, G., Corral-Diaz, B. 2004. Bioaccumulation of cadmium, chromium and copper by *Convolvulus arvensis* L.: Impact on plant growth and uptake of nutritional elements. *Bioresource Technology* 92: 229–235.

Harrison, F., Paul, J., Massey, R. C., Buckling, A. 2008. Interspecific competition and siderophores-mediated cooperation in *Pseudomonas aeruginosa*. *ISME Journal* 2: 49–55.

Ivanovich, M., Harmon, R. S. 1982. *Uranium Series Disequilibrium: Applications to Environmental Problems*. Oxford University Press, Oxford, U.K.

Jianlong, W., Xinmin Z., Decai, D., Ding, Z. 2001. Bioadsorption of lead (II) from aqueous solution by fungal biomass of *Aspergillus niger*. *Journal of Biotechnology* 87: 273–277.

John, S. G., Ruggiero, C. E., Hersman, L. E., Tung, C. S., Neu, M. P. 2001. Siderophore mediated plutonium accumulation by *Microbacterium flavescens* (JG-9). *Environmental Science & Technology* 35: 2942–2948.

Kalin, M., Wheeler, W. N., Meinrath, G. 2004. The removal of uranium from mining waste water using algal/microbial biomass. *Journal of Environmental Radioactivity* 78: 151–177.

Kapoor, A., Viraraghavan, T. 1997. Heavy metal biosorption sites in *Aspergillus niger*. *Bioresource Technology* 61: 22–227.

Karna, R. R., Uma, L., Subramanian, G., Mohan, P. M. 1999. Biosorption of toxic metal ions by alkali-extracted biomass of a marine cyanobacterium, *Phormidium valderianum* BDU 30501. *World Journal of Microbiology and Biotechnology* 15: 729–732.

Kazy, S. K., D'Souza, S. F., Sar, P. 2009. Uranium and thorium sequestration by a *Pseudomonas* sp.: Mechanism and chemical characterization. *Journal of Hazardous Materials* 163: 65–72.

Kraepiel, A. M. L., Bellenger, J. P., Wichard, T., Morel, F. M. M. 2009. Multiple roles of siderophores in free-living nitrogen-fixing bacteria. *Biometals* 22: 573–581

Lee, C. M., Lu, C., Yin, Y. H., Chen, P. C. 1996. Treatment of nitrogenous wastewaters by immobilized cyanobacteria in an air-lift-fluidized photo-bioreactor. *Progress in Biotechnology* 11: 556–562.

Li, P. F., Mao, Z. Y., Rao, X. J., Wang, X. M., Min, M. Z., Qiu, L. W., Liu, Z. L. 2004. Biosorption of uranium by lake-harvested biomass from a cyanobacterium bloom. *Bioresource Technology* 94: 193–195.

Lombardi, A. T., Vieira, A. A. H. 2000. Copper complexation by Cyanophyta and Chlorophyta exudates. *Phycologia* 39: 118–125.

Lueng, W. C., Chua, H., Lo, W. 2001. Biosorption of heavy metals by bacteria isolated from activated sludge. In *Twenty-Second Symposium on Biotechnology for Fuels and Chemicals*, Humana Press, Gatlinburg TN, pp. 171–184.

Mallick, N., Rai, L. C.1994. Removal of inorganic ions from wastewaters by immobilized microalgae. *World Journal of Microbiology and Biotechnology* 10: 439–443.

Mameri, N., Boudries, N., Addour, L., Belhocine, D., Lounici, H., Grib, H., Pauss, A. 1999. Batch zinc biosorption by a bacterial nonliving *Streptomyces rimosus* biomass. *Water Research* 33: 1347–1354.

Meinrath, G., Kato, Y., Kimura, T., Yoshida, Z. 1996. Solid-aqueous phase equilibria of uranium (VI) under ambient conditions. *Radiochimica Acta* 75: 159–168.

Merroun, M., Nedelkova, M., Rossberg, A., Hennig, C., Selenska-Pobell, S. 2006. Interaction mechanisms of bacterial strains isolated from extreme habitats with uranium. *Radiochimica Acta* 94: 723–729.

Merroun, M. L., Geipel, G., Nicolai, R., Heise, K. H., Selenska-Pobell, S. 2003. Complexation of uranium (VI) by three eco-types of *Acidithiobacillus ferrooxidans* studied using time-resolved laser-induced fluorescence spectroscopy and infrared spectroscopy. *Biometals* 16: 331–339.

Merroun, M. L., Raff, J., Rossberg, A., Hennig, C., Reich, T., Selenska-Pobell, S. 2005. Complexation of uranium by cells and S-layer sheets of *Bacillus sphaericus* JG-A12. *Applied and Environmental Microbiology* 71: 5532–5543.

Micheletti, E., Pereira, S., Mannelli, F., Moradas-Ferreira, P., Tamagnini, P., Philippis, R. De. 2008. Sheathless mutant of the cyanobacterium *Gloeothece sp* strain PCC 6909 with increased capacity to remove copper ions from aqueous solutions. *Applied and Environmental Microbiology* 74: 2797–2804.

Mohamed, Z. A. 2001. Removal of cadmium and manganese by a non-toxic strain of the freshwater cyanobacterium *Gloeothece magna*. *Water Research* 35: 4405–4409.

Moll, H., Johnsson, A., Schafer, M., Pedersen, K., Budzikiewicz, H., Bernhard, G. 2008. Curium (III) complexation with pyoverdins secreted by a groundwater strain of *Pseudomonas fluorescens*. *Biometals* 21: 219–228.

Moreno, J., Vargas, M. A., Olivares, H., Rivas, J., Guerrero, M. G. 1998. Exopolysaccharide production by the cyanobacterium *Anabaena* sp. ATCC 33047 in batch and continuous culture. *Journal of Biotechnology* 60: 175–182.

Nakajima, A., Horikoshi, T., Sakaguchi, T. 1982. Recovery of uranium by immobilized microorganisms. *European Journal of Applied Microbiology and Biotechnology* 16: 88–91.

Nakajima, A., Tsuruta, T. 2004. Competitive biosorption of thorium and uranium by *Micrococcus luteus*. *Journal of Radio Analytical and Nuclear Chemistry* 260: 13–18.

Ngah, W. S., Hanafiah, M. A. K. M. 2008. Removal of heavy metal ions from wastewater by chemically modified plant wastes as adsorbents: A review. *Bioresource Technology* 99: 3935–3948.

Nichols, M. C. A., Guezennec, J., Bowman, J. P. 2005. Bacterial exopolysaccharides from extreme marine environments with special consideration of the southern ocean, sea ice, and deep-sea hydrothermal vents: A review. *Marine Biotechnology* 7: 253–271.

Nieboer, E., Richardson, D. H. 1980. The replacement of the non-descript term 'heavy metals' by a biologically and chemically significant classification of metal ions. *Environmental Pollution Series B: Chemical and Physical* 1: 3–26.

Niu, H., Xu, X. S., Wang, J. H. 1993. Removal of lead from aqueous solutions by *Penicillium* biomass. *Biotechnology and Bioengineering* 42: 785–87.

Nunez, L., Vandegrift, G. F. 2001. *Evaluation of Hydroxamic Acid in Uranium Extraction Process: Literature Review*. Argonne National Laboratory, Argonne, IL.

Parikh, A., Madamwar, D. 2006. Partial characterization of extracellular polysaccharides from cyanobacteria. *Bioresource Technology* 97: 1822–1827.

Park, D., Yun, Y. S., Park, J. M. 2005. Use of dead fungal biomass for the detoxification of hexavalent chromium: Screening and kinetics. *Process Biochemistry* 40: 2559–65.

Park, D., Yun, Y. S., Park, J. M. 2010. The past, present, and future trends of biosorption. *Biotechnology and Bioprocess Engineering* 15: 86–102.

Patnaik, S., Sarkar, R., Mitra, A. 2001. Alginate immobilization of *Spirulina platensis* for wastewater treatment. *Indian Journal of Experimental Biology* 39: 824–826.

Pereira, S., Zille, A., Micheletti, E., Moradas-Ferreira, P., Philippis, R. De., Tamagnini, P. 2009. Complexity of cyanobacterial exopolysaccharides: Composition, structures, inducing factors and putative genes involved in their biosynthesis and assembly. *FEMS Microbiology Reviews* 33: 917–941.

Pettersson A., Hallbom, L., Bergmann, B. 1988 Aluminium effects on uptake and metabolism of phosphorus by the cyanobacterium *Anabaena cylindrica*. *Plant Physiology* 86: 112–116.

Phoenix, V. R., Martinez, R. E., Konhauser, K. O., Ferris, F. G. 2002. Characterization and implications of the cell surface reactivity of *Calothrix* sp. strain KC97. *Applied and Environmental Microbiology* 68: 4827–4834.

Prakasham, R., Ramakrishna, S. V. 1998. The role of cyanobacteria in effluent treatment. *Journal of Scientific and Industrial Research* 57: 258–265.

Premuzic, E. T., M. Lin, H. L. Zhu, Gremme, A. M. 1991. Selectivity in metal uptake by stationary phase microbial populations. *Archives of Environmental Contamination and Toxicology* 20: 234–240.

Priya, B., Uma, L., Ahamed, A. K., Subramanian, G., Prabaharan, D. 2011. Ability to use the diazo dye, C.I.Acid Black 1 as a nitrogen source by the marine cyanobacterium *Oscillatoria curviceps* BDU 92191. *Bioresource Technology* 102: 7218–223.

Rai, L. C., Mallick, N. 1992 Removal and assessment of toxicity of Cu and Fe to *Anabaena doliolum* and *Chlorella vulgaris* using free and immobilized cells. *World Journal of Microbiology and Biotechnology* 8: 110–114.

Rashmi, V. 2014. Marine Cyanobacteria—A potential candidate for uranium sequestration, PhD thesis, Bharathidasan University, Tamil Nadu, India, pp. 1–175.

Rashmi, V., ShylajaNaciyar, M., Rajalakshmi, R., D'Souza, S. F., Prabaharan, D., Uma, L. 2013. Siderophore mediated uranium sequestration by marine cyanobacterium *Synechococcus elongatus* BDU 130911. *Bioresource Technology* 130: 204–210.

Renshaw, J. C., Robson, G. D., Trinci, A. P., Wiebe, M. G., Livens, F. R., Collison, D., Taylor, R. J. 2002. Fungal siderophores: Structures, functions and applications. *Mycological Research* 106: 1123–1142.

Rippka, R., Deruelles, J., Waterbury, J. B., Herdman, M., Stainer, R. Y. 1979. Generic assignments, strain histories and properties of pure cultures of cyanobacteria. *Microbiology* 111: 1–61.

Sayyed, R. Z., Chincholkar, S. B. 2006. Purification of siderophores of *Alcaligenes faecalis* on Amberlite XAD. *Bioresource Technology* 97: 1026–1029.

Schmeide, K., Sachs, S., Bubner, M., Reich, T., Heise, K. H., Bernhard, G. 2003. Interaction of uranium (VI) with various modified and unmodified natural and synthetic humic substances studied by EXAFS and FTIR spectroscopy. *Inorganica Chimica Acta* 351: 133–140.

Selatnia, A., Bakhti, M. Z., Madani, A., Kertous, L., Mansouri, Y. 2004. Biosorption of Cd^{2+} from aqueous solution by a NaOH-treated bacterial dead *Streptomyces rimosus* biomass. *Hydrometallurgy* 75: 11–24.

Sutherland, I. W. 1994. Structure–function relationship in microbial exopolysaccharides. *Biotechnology Advances* 12: 393–448.

Sutherland, I. W. 2001. Microbial polysaccharides from Gram negative bacteria. *International Dairy Journal* 11: 663–674.

Suzuki, Y., Banfield, J. F. 2004. Resistance to, and accumulation of, uranium by bacteria from a uranium-contaminated site. *Geomicrobiology Journal* 21: 113–121.

Tan, T., Cheng, P. 2003. Biosorption of metal ions with *Penicillium chrysogenum*. *Applied Biochemistry and Biotechnology* 104: 119–128.

Tertian, R., Claisse, F. 1982. *Principles of Quantitative X-Ray Fluorescence Analysis*. Heyden & Son, Inc., Philadelphia, PA.

Turner, J. S., Robinson, N. J. 1995 Cyanobacterial metallothioneins: Biochemistry and molecular genetics. *Journal of Industrial Microbiology* 14: 119–125.

Ulrich, K. U., Rossberg, A., Foerstendorf, H., Zänker, H., Scheinost, A. C. 2006. Molecular characterization of uranium (VI) sorption complexes on iron (III)-rich acid mine water colloids. *Geochimica et Cosmochimica Acta* 70: 5469–5487.

Uslu, G., Tanyol, M. 2006. Equilibrium and thermodynamic parameters of single and binary mixture biosorption of lead(II) and copper(II) ions onto *Pseudomonas putida*: Effect of temperature. *Journal of Hazardous Materials* 135: 87–93.

Varma, A. J., Deshpande, S. V., Kennedy, J. F. 2004. Metal complexation by chitosan and its derivatives: A review. *Carbohydrate Polymers* 55: 77–93.

Vijayaraghavan, K., Yun, Y. S. 2008. Bacterial biosorbents and biosorption. *Biotechnology Advances* 26: 266–291.

Volesky, B. 1994. Advances in biosorption of metals: Selection of biomass types. *FEMS Microbiology Reviews* 14: 291–302.

Volesky, B. 2003. Biosorption process simulation tools. *Hydrometallurgy* 71: 179–190.

Volesky, B. 2007. Biosorption and me. *Water Research* 41: 4017–4029.

Wang, J., Chen, C. 2006. Biosorption of heavy metals by *Saccharomyces cerevisiae*: A review. *Biotechnology Advances* 24: 427–451.

Yee, N., Benning, L. G., Phoenix, V. R., Ferris, F. G. 2004. Characterization of metal-cyanobacteria sorption reactions: A combined macroscopic and infrared spectroscopic investigation. *Environmental Science & Technology* 38: 775–782.

Yoe, J. H., Will III, F., Black, R. A. 1953. Colorimetric determination of uranium with dibenzoylmethane. *Analytical Chemistry* 25: 1200–1204.

47 Bioremediation
A Promising Technology for Radionuclides and Explosive Compounds

T. Subba Rao, Rajesh Kumar, H. Seshadri, and Francis Vincent

CONTENTS

ABSTRACT

Anthropogenic activities in the context of industrialization are producing a huge amount of pollutants, toxicants, and obnoxious compounds, resulting in the abuse of our ecosystem. These toxicants are turning our environment into a sink at an alarming rate, and if not controlled, they can cause eutrophication, desertification, water pollution, health effects, etc.; this, in turn, makes human existence on planet Earth difficult. To control the impact of these toxicants, techno-economically viable processes that are environmental friendly have to be adopted. Physicochemical methods that are being used to control the pollutant toxicity are inherently problematic and costly and result in an extremely concentrated sludge that is difficult to dispose. Among the various green technologies that are being developed, bioremediation is a cheap, efficient, side effect–free process by which our ecosystem can be decontaminated, leaving behind a clean environment. This chapter comprises the following topics: a brief introduction to bioremediation, some types of bioprocesses, and details of granular and phototrophic biomass. Toxic waste details (1) radioactive waste, (2) explosive wastes, and their treatment using various types of biomass, namely, (1) bacteria, (2) algae, (3) fungi, and (4) phototrophic granules. This chapter is mainly tailored for the bioremediation of radioactive waste, explosives, and its by-products. Finally, it is concluded with a note on efficient management of toxic wastes.

47.1 INTRODUCTION

Bioremediation has acquired international priority to clean up major waste types, which are mostly organic solvents, poly- and halogenated aromatic hydrocarbons, pesticides, and munition wastes (Adams, 2015). During the past decade, there is an interest in biotechnological processes for the treatment of toxicants such as radioactive waste and the nitrogenous explosive compounds. Despite the high potential for this technology, the use of bioremediation is limited by an incomplete understanding of the metabolomics (Villas Boas and Bruheim, 2007). Bioremediation can be defined as a unique process that will accelerate the natural biodegradation of a toxicant by optimization of environmental conditions and is an ecologically and economically viable method. Bioremediation comes from two words: *bios* means life and *remediate* means to work out a solution. The degradation of toxic waste from the environment using microorganisms is referred to as bioremediation. Microorganisms like bacteria, fungi, and algae take part in bioremediation, and there are several reviews and journal publications in support (Vidali, 2001; Sasek and Caithami, 2005; Perumbakkam et al., 2006; Singh et al., 2006; Villas and Bruheim, 2007; Abdulsalan, 2011; Dadrasina and Agamuthu, 2013; Adams et al., 2014, 2015). Bioremediation techniques are currently being used to degrade highly toxic metals, chemicals, effluents, and pollutants. The toxic metals from tanneries produce poisonous oxides, which can cause lung cancer, asthma, paralysis, brain damage, memory loss, etc. If the toxic metals are accumulated into water bodies, they cause the death of fishes and marine animals. Algal blooms are generally formed once the water body becomes eutrophic that have a significant effect; they block sunlight and reduce dissolved oxygen levels in the water, resulting in the death of fishes and other aquatic fauna. Algal blooms in drinking water sources make the water unfit for drinking as the algal growth gives color and bad odor to the water. Oil spills also cause similar effects making the water unfit for biota growth, causing the death of organisms. A large number of marine lives were lost due to these oil spills, thus causing a disturbance in food chains and the functioning of the ecosystem. Fertilizers and agrochemicals are commonly added to such polluted water bodies to increase the metabolic productivity of the microbiota that degrades the pollutant (Al-Sulaiman et al., 2011; Adams et al., 2014, 2015).

According to the site of treatment, there are two types of bioremediations *in situ* and *ex situ*. In the case of *in situ* treatment, it is at the site of production, while *ex situ* treatment is away from the site of production. Some authors (Kumar et al., 2011; Orji et al., 2012; Hamzah et al., 2013) have described the types of bioremediation mentioned previously. However, it is necessary to determine what exactly is done *in situ and ex situ* and use the same process to describe the types of bioremediation. To augment the bioremediation process, microorganisms are genetically modified to scale up the bioprocess, for example, bacteria like *Deinococcus radiodurans* was successfully used to demonstrate the absorption of mercury and in the degradation of aromatic hydrocarbons like toluene (Brim et al., 2006). Bioremediation can easily be observed by keeping a check on the redox potential, pH, temperature, oxygen content, residues from catabolic activity, etc. Among the various bioprocesses, the common methods are bioventing, bio-absorption, bio-augmentation, cometabolism, mycoremediation, and phytoremediation. Based on the type of toxicant/pollutant, the bioremediation process may be natural or intrinsic bioremediation.

47.2 TYPES OF BIOREMEDIATION

47.2.1 Cometabolism

Cometabolism is gaining prominence in the recently emergent bioremediation processes. Cometabolism is referred to as the transformation of a hazardous waste to environmentally benign products indirectly as a consequence of the metabolism of another chemical that the bacterium uses as a source of carbon and energy. An aspect of cometabolism that is of concern is the production of partially oxidized end products from cometabolism that may not be readily degraded by the indigenous microbes because of the small adaptation time. Cometabolism is an aerobic process in which chlorinated ethenes are degraded as a result of fortuitous biochemical interactions that yield no benefit to the bacteria. Alvarez-Cohen and McCarty (1991) reported that tetrachloroethylene (TCE) is degraded under aerobic conditions by methanotrophic bacteria in soil enriched with methane and oxygen. Further studies revealed that the methane monooxygenase enzyme was responsible for catalyzing the oxidation of TCE. This oxidation reaction is called cometabolism because the reaction uses metabolic enzymes but does not contribute any energy in return. Cometabolism tends to be an unsustainable process under stagnant conditions because of substrate competition and enzyme inhibition and inactivation (Ely et al., 1995, 1997). Competitive inhibition occurs when enzymes cometabolize chlorinated solvents to the exclusion of natural substrates, ultimately depleting the bacteria of energy molecules (Chang and Alvarez-Cohen, 1995). The TCE oxidation byproducts such as TCE epoxide may result in the inactivation of the oxygenase activity caused by damage to the enzymes (Ely et al., 1995, 1997). Thus, the process has its limitations when the enhanced conditions of bioremediation are in the process (Kim et al., 2002).

47.2.2 Phytoremediation

Natural plants and transgenic plants accrue some toxicants in their basal parts; this process is termed as phytoremediation (Zhang et al., 2009). In this method, plants are used to remove contaminants. The plants can draw the contaminants into their structures and hold them, thus effectively removing them from soil or water. Commonly, plants cannot readily remediate all heavy metals and toxicants. Phytoextraction is also referred to as phytoaccumulation or phytomining. The process involves the removal of contaminants (metals, radionuclides, and certain organic compounds) from the environment by direct uptake into the plant tissue. The process involves the planting of one or more plant species that are hyper-accumulators of the contaminants of concern. Phytoremediation may be applied wherever the soil or static water bodies are polluted or are suffering from chronic pollution. Phytoremediation has been used to remediate heavy metals (Cd, Co, Pb, Cu, Ni, Se, Zn), radionuclides (Cs, Sr, U), chlorinated solvents (TCE, PCE), petroleum hydrocarbons (BTEX), and nutrients (nitrate, ammonium, phosphate).

There are six primary ways that plants are used for waste treatment and site remediation, which can be termed as mechanism of phytoremediation: (1) stimulation of microorganism bioactivity in the root zone/rhizosphere, (2) precipitation of enzymes from root and stem tissues that can bind aromatic pollutants and degrade organic compounds, (3) adsorption/sequestration of organics in the plant roots (i.e., phytostabilization), (4) hyper-accumulation of heavy metal and radioactive compounds in the plant tissues for soil and water (can be termed as phytoextraction or rhizofiltration), and (5) phytovolatilization.

Plants can stimulate microbial (bacteria and fungi) bioactivity in the root zone/rhizosphere by the excretion of bio-enhancing compounds. Typically, there are few 100 times higher microbial population in the rhizosphere compared to the bulk soil. The plant-excreted root exudates provide a carbon and nitrogen source for soil bacteria. The excreted compounds commonly include amino acids, carbohydrates, polysaccharides, flavonoids, and phenols. Specifically, it has been shown that flavonoids can support the growth of PCB-degrading bacteria. Also, the phenols excreted by some plants such as asparagus plants can stimulate PCB-degrading bacteria and inhibit other microbes. Enhanced herbicide degradation has also been found in the root exudates.

There has been some promising research into plant-based remediation of explosive-contaminated soils and groundwater (Kumar et al., 2014). Poplar trees have been shown to rapidly uptake TNT from hydroponic solutions. TNT within the plant was transformed to metabolites such as amino-2,6-dinitrotoluenes and 2,4,diamino-6-nitrotoluene in the roots and stems. There was no significant production of CO_2 from the TNT Thompson et al. (1998). When the TNT was present in the soil, it was less bioavailable for uptake due to sorption competition (Thompson and Schnoor, 1997). The behavior of the poplars in the presence of RDX was significantly different, with much slower uptake from water, 60% translocation to the leaves within 48 h of initial uptake, and no significant transformation of the RDX within the plant. Treatment of TNT and RDX in wetland systems was also investigated, wherein it was showed that the enzyme nitroreductase can rapidly break down TNT (Oh et al., 2001).

In addition to precipitation and binding of aromatic compounds, other intracellular plant enzymes can degrade and transform organic compounds. For example, the nitroreductase enzyme in plants rapidly degrades explosives and other nitro-aromatics. In laboratory studies, it was shown that hairy root cultures can degrade TNT, and also, some common aquatic plants degrade TNT. Another plant enzyme, a dehalogenase, has been used for the degradation of halogenated hydrocarbons. Also, poplar trees (*Populus* is a genus of deciduous flowering plants belonging to the family Salicaceae, usually characterized by the columnar manner of growth of its branches) have been demonstrated to have the ability to dealkylate and hydrolyze atrazine, and the metabolites are incorporated into plant roots and leaves (Hughes et al., 1996).

Certain types of plants can hyper-accumulate heavy metals and radioactivity (100× to 1000× compared to the biomass of conventional plants), and then plant biomass are harvested and the toxic elements concentrated and reclaimed or disposed of. Growing, harvesting, and processing plant biomass is an inexpensive process, compared to digesting, ashing, smelting, and kiln processing. Some plants may uptake more than one contaminant for multicomponent cleanup and concentrate in different parts of the plants, namely, roots, leaves, and stalk. Toxic metal concentrations in plant saps can range from 1,800 to 51,600 mg/kg for Cd, Cu, Co, Pb, Ni, and Zn. Phytotoxicity may limit the maximum amount of contaminant that can be accumulated. Soil pH manipulation can augment uptake rates.

Though phytoremediation technology is still in research and development phases, various applications have shown potential for success. This has helped to increase interest and research to develop phytoremediation into a commercially viable industry. There are some key technical hurdles that must be addressed like identifying more species that have remediation capabilities; optimizing phytoremediation processes, such as appropriate plant selection and agronomic practices; understanding more about how plants uptake, translocate, and metabolize contaminants; identifying genes responsible for uptake and degradation; decreasing the length of time needed for phytoremediation process; and devising appropriate methods for contaminated biomass disposal, particularly for heavy metals and radionuclides. Phytoremediation offers many potential advantages over conventional technologies and will have public acceptance because of lower cost.

47.2.3 Mycoremediation

Bioremediation process in which fungi are used to remediate the toxic pollutants from the environment is termed *mycoremediation*. Fungi feed on natural decomposers of dead organic matter as they feed on them; hence, they are called saprophytes. Fungal hyphae secrete acids and enzymes that decompose lignin and cellulose. Microfiltration also uses fungal mycelium for the removal of toxic wastes. Fungi aids in the degradation of oil, aromatic compounds, hydrocarbons, and petroleum products, which pollute soil and water (Sasek and Caithami, 2005).

The use of fungi in decomposing *in situ* intractable, persistent, and extremely toxic pollutants, including TNT (2, 4, 6-trinitrotoluene), is well documented (Ponting, 2001; Sasek and Caithami, 2005; Rhodes, 2012). The diesel oil–contaminated soil can be treated by inoculating with oyster mushroom (*Pleurotus ostreatus*) and mycelia. Mycelia converted 95% of the polyaromatic hydrocarbons to nontoxic compounds within 30 days. It seems that the naturally present community of microbes acts in concert with the fungi to decompose the contaminants, finally resulting in full mineralization.

Wood-degrading fungi are extremely effective in decomposing toxic aromatic pollutants from petroleum and also chlorine containing resilient pesticides (Rhodes, 2012, 2013). The prospects of using fungi, principally white-rot fungi, for cleaning contaminated soil are surveyed. White-rot fungi are useful in degrading a wide range of organic compounds since they release extracellular lignin-degrading enzymes. The enzymes include lignin peroxidases and hydrogen peroxide–producing enzymes. The degradation processes can be augmented by adding carbon sources such as sawdust and straw at the polluted sites.

Mycofiltration is a procedure that utilizes fungal mycelia to filter and remove toxic materials and microorganisms from water. Edible mushrooms might be grown for mycoremediation, and the prospects of whether they would be safe to eat after treatment has to be studied (Kulshreshtha et al., 2014). However, this depends on the nature of the pollutant; if they are heavy/toxic metals, it will be a problem (if they are absorbed and concentrated into the mushroom), while hydrocarbon contaminants might be decomposed or mineralized completely without any toxicity. In either case, the benefit is that the land that is contaminated and unfit for agriculture purpose can be cleaned and used for cultivating crops (Ponting, 2001; Sasek and Caithami, 2005; Rhodes, 2012).

47.2.4 Biosorption

Biosorption is a passive and active process wherein the absorption of heavy (toxic) metals like mercury, cadmium, lead, and chromium is done through various types of microbial biomass properties.

The exponential industrial growth worldwide and the increase in abuse of our environment by human population have led to the accumulation of toxic metals. Toxic metal contamination has become a very serious environmental concern in the last few decades as a result of its insusceptibility. Surface and groundwater supplies are prioritized as a major source of inorganic pollutants (toxic/heavy metal ions). Even though the inorganic pollutants are present in dilute and undetectable quantities, their mobility and persistence in natural water systems are a concern. The detection of toxic metals in their elemental state implies that they are not subject to further biodegradative processes.

Physical techniques such as ion exchange, activated carbon sorption and membrane technology, and conventional chemical precipitation/neutralization processes are inherently problematic for the remediation of heavy metal ions. Chemical methods are costly, and the active agent cannot be recovered for use in subsequent treatment cycles. The end product is a highly concentrated metal ion–bearing sludge that is difficult to dewater or dispose of. Although there are many technologies to treat toxic metal waste, biosorption process has shown some promise to decontaminate the toxic metal pollution (Bux and Kasan, 1996). Microbial biosorption is one of the green technologies for the bioremediation of toxic metals. The process can be defined as a technique that can be used for the elimination of pollutants from

waters, particularly those that are not biodegradable such as metals and dyes. A variety of biomaterials are known to bind these pollutants, including bacteria, fungi, algae, and industrial and agricultural wastes. Undoubtedly, the biosorption process is an impending technique for heavy metal decontamination (De et al., 2008).

Biosorption (sorption of metal ions from solutions by live or dried biomass) offers an alternative to the remediation of industrial effluents as well as the recovery of metals contained in other media. Biosorbents are prepared from naturally abundant and waste biomass. Due to the high uptake capability and economical source of the raw material, biosorption is a progression toward a perspective technique. The mechanism by which microorganisms take up metals is relatively unclear, but it has been demonstrated that both living and nonliving biomass may be utilized in biosorption processes, as they often exhibit a marked tolerance toward metals and other hostile conditions. One of the major advantages of the process is the treatment of bulk volumes of effluents with low concentrations of contaminants (Gavrilescu, 2004; Hetzer et al., 2006). Bio-absorption/biosorption can be defined as a passive and active process wherein the toxic metals like mercury, cadmium, lead, and chromium are absorbed by microbial biomass through various types of interactions/reactions (Hetzer et al., 2006; Huang et al., 2010). Figure 47.1 gives a general schematic illustration of a microbial cell interaction in various ways to detoxify, complex, reduce, or oxidize the toxic metal ions. Surface charge is the principal driving force for metal ion biosorption on to the biomass. The higher the biomass electronegativity, the greater the adsorption of toxic metal cations is. A positive relationship exists between the electronegativity of the activated sludge surfaces and their metal biosorption capacity (Bux and Kasan, 1994). Measurement of the biosorbent net charge indicates the potential of a specific biomass type for application in a treatment process. Isothermal studies in conjunction with charge evaluations will reveal the actual capacity of the biosorbent to remove the dissolved toxic metal in an effluent.

Biosorption reaction rates are rapid, usually in the order of seconds and minutes. It was shown that the majority of metal ions are removed from the effluent sample within the first 15 min of the reaction. High rapid biosorption rates are typical for the adsorption of a dissolved toxicant onto a solid substratum. The observed biosorption patterns are biphasic and are similar in most cases, that is, an initial rapid metabolism-independent phase followed by a slower second phase wherein the equilibrium will be attained in a couple of minutes. Higher initial metal ion concentration affects the reaction rates significantly since the kinetics is rapid (Atkinson et al., 1998).

47.2.5 Bioventing

Bioventing is an *in situ* remediation technology that utilizes indigenous microorganisms for degrading organic constituents adsorbed to the soil particles in the unsaturated zone. In the bioventing process, the activity of the indigenous bacteria

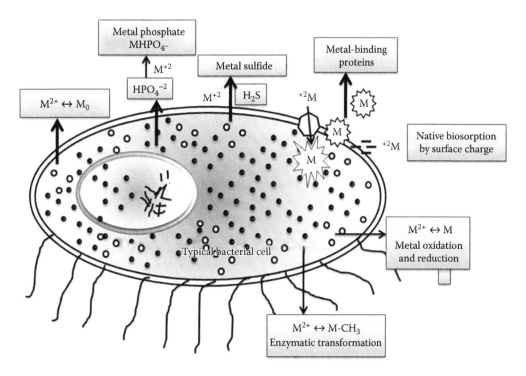

FIGURE 47.1 A schematic illustration of microbe–metal interaction.

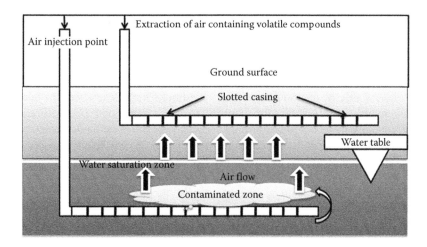

FIGURE 47.2 Schematic illustration of bioventing process.

is enhanced by inducing airflow (or oxygen) inside the unsaturated zone (using injection wells or extraction) and, when required, is augmented with the addition of nutrients.

In the bioventing process, all the chemicals that are aerobically biodegradable can be treated (Figure 47.2). Bioventing has proven to be very successful in remediating petroleum products that include gasoline, jet fuels, kerosene, and diesel fuel. Bioventing is often used at sites with medium range molecular weight petroleum products (i.e., diesel fuel and jet fuel), because lighter products (i.e., gasoline) tend to volatilize readily and can be removed more rapidly by means of soil vapor extraction (SVE). Heavier or high-molecular-weight products (e.g., lubricating oils) take a longer time to biodegrade than the lighter products.

On-site bioremediation is one of the most attractive options for reducing or eliminating toxicants in soil and water. The technology is simple, cost-effective, and has minimal site disturbance. The process aims at reducing the toxic effects to a safe and acceptable limit. Bioventing has noteworthy remediation connections with distinct goals. Air sparging forces the compressed air into the saturated soil that will enhance the biodegradation. The process reduces volatilization and maximizes biodegradation. The advantage is that by minimizing the volatilization the process eliminates the treatment of obnoxious gasses. Bioventing can also be called as a stepped-up version of natural attenuation that happens without intervention wherein the native bacterial population degrades the wastes. Oxygen is the limiting factor since it indirectly limits the microbial degradation process. When extraction wells are used for bioventing, the process is very similar to SVE. However, while SVE removes constituents chiefly through volatilization, a bioventing system promotes the biodegradation of constituents and reduces volatilization (generally by using lower airflow rates than SVE). In practice, some degree of volatilization and biodegradation ensues when either SVE or bioventing is used. It may be noted that the environment is maintained in such a manner so as to promote the growth of the indigenous microorganisms that further the biodegradation of the contaminants (Leahy and Erickson, 1995a). Bioventing eliminates the need for the excavation of the site, therefore eliminating the necessity to shore up streets or transport polluted soil to remote places for processing or disposal. Thus, one can remediate an area with little or no influence to the geological makeup, no recompaction, or restoration of the location is required.

47.2.6 Bio-Augmentation and Bio-Stimulation

Bio-augmentation is the addition of microorganisms to the soil, whereas bio-stimulation is the modification, that is, addition, reduction, or genetic engineering, of the microbes to degrade pollutants.

There are two basic forms of bioremediation presently being practiced: the microbiological approach (bio-augmentation) or the microbial ecological approach (bio-stimulation). On the contrary, the two processes can also be defined as bio-stimulation (addition of limiting nutrients to support microbial growth) and bio-augmentation (addition of living cells capable of degradation). The bio-augmentation approach involves addition of highly concentrated and specialized populations of definite microbes into a polluted site to enhance the rate of contaminant biodegradation in the affected soil or water. On the other hand, in bio-stimulation approach, emphasis is placed on identifying and adjusting certain physical and chemical factors (such as soil temperature, pH, moisture, and nutrient content) that may be impeding the rate of degradation of the pollutant by the indigenous microorganism in the affected site (Abdulsalam and Omale, 2009).

47.2.6.1 Bio-Augmentation

Bio-augmentation, or the addition of oil-degrading microorganisms to supplement the indigenous populations, has been projected as an alternate strategy for the bioremediation of oil contaminated (Head et al., 2006). The rationale for this methodology is that indigenous microbial populations may not be capable of degrading the extensive range of potential substrates present in complex mixtures such as petroleum or that they may be in a stressed state as a result of exposure to the oil spill (Seo et al., 2009). Other conditions under which bio-augmentation may be considered are when the indigenous hydrocarbon-degrading population is low and when the need to speed up decontamination is the primary factor Al-Sulaimani et al. (2011). In this process, microbial seeding may reduce the lag period to start the bioremediation process.

For successful application of this approach in the field, the seed microorganisms must be able to degrade most of the hydrocarbon constituents and should maintain the genetic stability and cell viability during storage/preservation. The microbes should also survive in foreign and hostile environments, effectively compete with indigenous microorganisms, and move through the pores of the sediment to degrade the contaminants. Different microbial species have different enzymatic abilities and preferences for the degradation of hydrocarbon compounds. Some microorganisms degrade linear, branched, or cyclic compounds. Others prefer mono- or polynuclear aromatics and others degrade both alkanes and aromatic hydrocarbons (Thomassin-Lacroix et al., 2002).

Successful bio-augmentation treatments depend on the use of the right inoculum consisting of a variety of microbial strains or microbial consortia that have been well adapted to the site to be decontaminated. The process will be successful only when the microbial efficiency depends on the ability to compete with indigenous microorganisms, predators, and various abiotic factors. Factors affecting proliferation of microorganisms used for bio-augmentation include the chemical structure and concentration of the pollutants, the availability of the contaminant to the microorganisms, the nature and the size of the microbial population, and the surrounding physical environment, which have to be considered when screening for microorganisms Abdulsalam et al. (2011).

47.2.6.2 Bio-Stimulation

Bio-stimulation can be defined as a type of natural remediation that can improve pollutant degradation by optimizing conditions such as aeration, the addition of nutrients, and pH and temperature control. Bio-stimulation can be considered as an appropriate remediation technique for petroleum pollutant removal in soil and requires the evaluation of both the intrinsic degradation capacities of the autochthonous microbiota and the environmental parameters involved in the kinetics of the *in situ* process.

The process was tested for hydrocarbon biodegradation in soil, that is, limited by many factors, including nutrients, pH, temperature, moisture, oxygen, soil properties, and the concentration of the contaminant (Al-Sulaimani, 2011). Bio-stimulation involves the modification of the environment to stimulate existing bacteria capable of bioremediation. This can be done by the addition of various forms of limiting nutrients and electron acceptors, such as phosphorus, nitrogen, oxygen, or carbon (e.g., in the form of molasses), which are otherwise available in low quantities to constrain the microbial activity. According to Dadrasina and Agamuthu (2013), the addition of nutrients, oxygen, or other electron donors and acceptors to the contaminated site should increase the population or activity of naturally occurring microorganisms available for remediation.

The primary advantage of bio-stimulation is that the remediation process will be undertaken by the native microorganisms that are well suited to the subsurface environment and are well distributed spatially. The addition of nutrients will promote the growth of heterotrophic microorganisms that are not innate degraders of the pollutant, thereby creating a competition between the resident microbiota (Adams et al., 2014).

47.3 MICROBIAL BIOFILMS IN BIOREMEDIATION

The successful application of a bioprocess relies upon a detailed understanding of microbial interaction with the compound to be treated. Bacterial biofilms represent an efficient method for bioremediation of organic compounds and toxic metal and waste detoxification. Microbial biofilms are omnipresent in almost all kind of environments (Garrett et al., 2008). The bacterial cells in the biofilms are enclosed in an extracellular polymeric matrix (Beer and Stoodley, 2006). Even though the production of the extracellular matrix is an energy-intensive process, it renders many advantages to the bacterial cells, such as protection from predation, protection from the harsh environment, and limiting the diffusion of antibiotics in the immediate milieu (Xavier and Foster, 2005). Thus, microbial cells in a biofilm are better adapted to the environmental stress (Karthikeyan et al., 2001). The aggregation of bacteria cells into assorted communities also confers special benefits that are not enjoyed by planktonic bacteria. In nature, biofilms support a high biomass density that facilitates the mineralization process more effectively since the conditions in the biofilm are optimized (Stoodley et al., 2002). Bacteria secrete exopolymers that aid in the formation of biofilms on most surfaces and utilize organic compounds for their energy. Consequently, they gain the capability to degrade recalcitrant and slowly degrading compounds (Perumbakkam et al., 2006). Biofilms also influence the fate of other compounds in their vicinity, which is a consequence of the dynamic metabolic machinery of the consortium. At the same time, horizontal gene transfer and recombination processes like conjugation and transformation are additional processes in the biofilm that make them more resilient to a high concentration of toxicants (Roberts and Mullany, 2006). This property aids in the biodegradation of the xenobiotics. Microbial cells in biofilm consortia are highly tolerant to many toxic chemicals and environmental stress. Due to these properties, bacterial biofilms are specifically used in numerous studies, such as bioremediation of chlorophenols (Puhakka et al., 1995; Carvalho et al., 2001), dinitrotoluene (DNT), atrazine (Perumbakkam et al., 2006), azo dyes (Sudha et al., 2014), and naphthalene (Shimada et al., 2012) and hydrocarbon biodegradation from contaminated beach sediments (Atlas and Bragg, 2009). Monoculture, bacterial biofilms from the genus, like *Pseudomonas*, *Rhodococcus*, *Alcaligenes*, *Sphingomonas*, and *Methylosinus*, were reported to degrade a variety of organic pollutants efficiently (Singh et al., 2006; Shimada et al., 2012). Another promising application of biofilms is in heavy metal and radionuclide remediation (Lloyd and Renshaw, 2005). The extracellular polymeric substance (EPS) matrix is composed of ionic sugars that can efficiently bind to a variety of heavy metals. EPS composition, high

density of microbial population, and enzyme activities are helpful in the detoxification and biosorption of heavy metals and radionuclides, thereby minimizing the hazards to the environment. Removal of heavy metals by biofilms is chiefly due to the binding capacity of EPS and other cellular components (Zhang et al., 2010, 2012). Other than water, a biofilm is also composed of nonaqueous liquid phase, which is helpful in maintaining static and chemical mobility inside biofilm. This phase also helps in increasing the bioavailability of the pollutant and aids in its dissolution (Cameotra and Makkar, 2010).

Bioremediation of various recalcitrant compounds is an extensively studied area. The ability of organisms to multiply and propagate themselves offers an in-exhaustible catalyst for carrying out various bioremediation processes. Biological processes are economical, efficient, and sustainable alternates to chemical processes (Gavrilescu and Chisti, 2005; Pandey et al., 2002). Among the various biological processes, biofilms have an edge over their planktonic counterparts, as biofilms offer protection and can easily be separated from bulk (Singh et al., 2006). Biofilms are surface-attached entities, bound to living or nonliving surfaces. They are complex heterogeneous consortia of microbes, generally embedded in a polymeric matrix, called extracellular polymeric substances (EPS), consisting of proteins, carbohydrates, and nucleic acids (Sutherland, 2001). Biofilms have long been demonstrated to be useful in several bioremediation processes. They have been used in several types of bioreactors for applications ranging from the removal of nutrients like nitrates, ammonia, sulfates, and phosphates to degradation of several organics like phenol and its derivatives, naphthalene, chlorinated aromatics, herbicides, and toluene to metal reduction and removal and degradation of complex wastes like molasses, whey, and winery (Artham and Doble, 2008; Andreottola et al., 2009; Cao et al., 2009). Biofilm provides ideal environment for metabolic cooperation among the members, thereby increasing the chances of survival in adverse condition (Davey and O'Toole, 2000). The most extensively used application of substrate-bound biofilms is in domestic wastewater treatment in the form of trickling filters (Daigger and Boltz, 2011). Figure 47.3 gives a general view of biofilm formation, maturation, and dispersal.

Biofilm-mediated bioremediation can be termed as a sustainable technology that can be used in various remediation processes. Biofilms also provide an optimal environment for cell–cell interaction, intercellular exchange of genetic material, communication signal, metabolite diffusion, etc. (Mangwani et al., 2012). Biofilm-mediated bioremediation is also a better methodology for the treatment of contaminated sites and industrial wastewaters. Quorum sensing of microbial cells aids in the development of resilient biofilms that can be used in the development of engineered biofilms Atkinson et al. (2006). Such tailored biofilms have the potential to improve *in situ* bioremediation with enhanced kinetics. However, the involvement of quorum sensing in bioremediation has not been well documented, but its role in stress response has been reported in many bacteria (Parsek

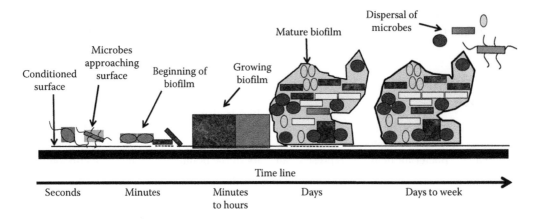

FIGURE 47.3 Schematic of biofilm formation, maturation, and dispersal.

and Greenberg, 2005; De Kievet, 2009; Kang and Park, 2010). Thus, the xenobiotic stress may also induce quorum sensing response in biofilm community that can be of enormous importance and aid in the development of efficient biotechnological processes.

47.3.1 Substratum-Less Biofilms in Bioremediation

Microbes possess the ability to aggregate leading to the formation of flocs (Bossier and Verstraete, 1996). The biological surface area offered by these microbial aggregates is independent of the surface area of the reactor system. The high density of these aggregates provides much higher surface-area-to-volume ratio as compared to surface-bound biofilms, resulting in more efficient bioprocesses. The phenomenon of aggregation of biomass has very profound implications in the area of environmental biotechnology (Maximova and Dahl, 2006). The utilization of microbial flocs is one of the earliest modern era bioremediation applications of microbes in the field of domestic wastewater treatment, generally referred to as activated sludge process (ASP). The ASP is designed in such a way that not only organics are degraded but nutrients like ammonia and phosphates and trapped gases are also removed from the waste (Sakai et al., 1997, Cronje et al., 2002, Liu, 2003). Though the ASP has been used extensively in domestic and industrial water treatment, it has several drawbacks. The ASP requires large area (high footprint) for construction. The poor settling behavior of flocs leads to slow processing of wastewater and hence the possibility of excess sludge discharge, leading to very low biomass retention in the aerated tank that cannot be ruled out.

The shortcomings of the anaerobic sludge have been overcome in microbial aerobic granulation. Aerobic granulation has several advantages over both ASP and anaerobic granulation technology. The foremost advantage of microbial aerobic granules (MAGs) is the requirement of much shorter start-up time (Liu and Tay, 2004). Moreover, MAGs are suitable for removing both organic wastes and nutrients like nitrogen and phosphate (Gao et al., 2011). Microbial aerobic granulation

technology combines the best of both ASP and anaerobic granulation. Apart from the ASP, carrier-less/substratum-less biofilms have also been reported in anaerobic digesters used for wastewater treatment. The aggregation of microbes in anaerobic reactor results in compact biomass called microbial granules, which have far superior settling behavior than activated sludge (Pol et al., 2004). Though anaerobic granular reactors have been extensively used in wastewater treatment throughout the world (Xu et al., 2002), they have their own drawbacks as they require relatively high operational temperature and are not suitable for the removal of nutrients like ammonium nitrogen and phosphate.

Unlike anaerobic granulation and ASP, aerobic granulation is a recently observed phenomenon. It was first reported by Mishima and Nakamura (1991) in an aerobic upflow sludge blanket (AUSB) reactor with a separate oxygenation system for wastewater using pure oxygen and a stirring system. Shin et al. (1992) studied the effect of shear on microbial granulation in AUSB with two different stirring velocities and concluded that COD removal was higher at higher mixing velocities, while bulking due to filamentous growth was seen at lower mixing. Morgenroth et al. (1997) demonstrated for the first time the formation of granules in column-type aerobic sequencing batch reactor (SBR) by gradually reducing cycle time and hydraulic retention time. Beun et al. (1999) and Peng et al. (1999) demonstrated the formation of aerobic granules in bubbled column SBRs. Beun et al. (1999) studied aerobic granulation with varying retention time and COD input, while Peng et al. (1999) demonstrated the efficacy of COD removal at different aeration levels. Beun et al. (2000) proposed a mechanism of granulation, which suggested that initially biomass pellet formation took place due to interweaving filamentous microbes (fungi), which slowly became granular due to shear force. These granules were then colonized by other microbes, leading to bigger size, followed by disintegration due to lack of oxygen in the center of the granule. The resulting particles were still large enough to settle in the reactor and were dominated by bacteria. Beun et al. (2000) established the formation of granules in sequencing batch airlift reactor

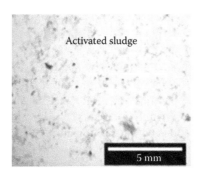

• Right reactor configuration
• Right seed
• Right substrate at right rate
• Right operation strategy
• Right shear force
• Right settling time
• Right hydraulic retention time
• Right feast famine ratio
• Right DO levels

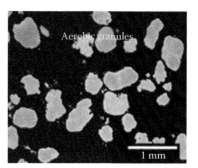

FIGURE 47.4 The process of formation of aerobic granules.

and demonstrated denitrification and nitrification processes, along with utilization of organics in a single reactor. There was no need for any optimization of process conditions for each activity. Thereafter, most of the studies on aerobic granules have been carried out in bubbled column SBRs. The SBRs are single tank reactors, where all the processes of a treatment plant like ASP are carried out in the same tank in a sequential manner.

The phenomenon of aerobic granulation has been studied extensively in the recent past for the evaluation of various process parameters and for the bioremediation of different xenobiotic and recalcitrant chemicals. The starting material for cultivation of aerobic granules has generally been activated sludge. Tay et al. (2001) studied the effect of feed on granulation. The microscopic structure of aerobic granules revealed that the granules fed with glucose were having more filamentous bacteria as compared to acetate-fed granules. Further, they found that the periodic starvation was essential for compact granular biomass formation, which was also corroborated by Beun et al. (2002) and McSwain et al. (2004). The starvation time should be carefully chosen as high starvation time affects the practical application due to lower net efficiency and low starvation time that leads to the instability of sludge in the long run (Liu and Tay, 2007). However, Liu et al. (2007) found that the starvation was not essential for granule formation, but it positively affected the stability of granules.

Size of granules is a parameter that determines the physical properties of granules as well as their biological composition. The increase in the size leads to anoxic zone inside the aerobic granules, which results in the death of cells and subsequent problems in operation of SBR. The ideal operating range for aerobic granule has been suggested to be 1–3 mm diameter (Toh et al. 2003). The specific COD removal rate as well as specific biomass growth rate decreases with an increase in size of granules (Liu et al. 2005). Though aerobic granules are generally cultivated at high aeration levels, strict anaerobes have also been observed in the granules. This has been attributed to oxygen diffusion limitation to the interior of granules. Such a condition can lead to disintegration of granules. Therefore, getting the right size of granule is of paramount importance for the successful application of the technology (Tay et al., 2002). The overall process of formation of granules can be summarized as given in Figure 47.4.

Particulate biofilms are considered a reasonably mature technology in the area of wastewater treatment and have been applied in several field-level applications in various reactor configurations (Nicolella et al., 2000). The environmental biotechnological application of aerobic granules was first demonstrated by Jiang et al. (2004) for the degradation of phenol in the wastewater. De Bruin (2004) explored the possibility of using granular sludge reactors in place of conventional wastewater treatment plant. The authors concluded that a granular sludge plant was more promising from engineering as well as economic perspective, as they would occupy only 25% of the area that a conventional plant occupied. But the granular sludge reactor is sensitive to rain weather and needs further improvements for simultaneous removal of N, P, and COD. Van der Star et al. (2007) successfully demonstrated full-scale ammonium oxidation reactor using granules that were scaled up directly from lab-scale bioreactor.

Oxidation processes like nitrification are likely to occur only in the top layers of granules within 300 μm as availability of oxygen is near zero after this depth (Jang et al., 2003). Ammonia removal has been demonstrated in autotrophic aerobic granular SBRs, where ammonia forms the only source of energy (Matsumoto et al. 2010). The processes of denitrification and biological removal of phosphorus by aerobic granules have also been demonstrated (Jeon et al., 2003). Yang et al. (2003) for the first time demonstrated that it was possible to remove COD as well as nitrogen from the wastewater in a single aerobic granular sludge reactor. Both nitrification and denitrification processes occur in the same reactor. The N/COD ratio determines the population dynamics, but both the populations coexist in aerobic granules. There have been a few more studies on simultaneous removal of both nitrogen and COD (de Kreuk et al., 2005). Nancharaiah and Venugopalan (2011) reported denitrification of concentrated nitrate waste using aerobic granules developed in acetate-fed SBR. The denitrification under anoxic conditions in batch experiments could be achieved in less than 40 h for concentration up to 900 mg L^{-1} NO_3^--N, but higher concentration of 2250 NO_3^--N took 120 h. Higher nitrate load can be denitrified if DO levels in the reactors are maintained at levels lower than 5 mg L^{-1}, resulting in small aerobic region in granule and larger anoxic core (Wichern et al., 2008).

The microbes in biofilms actively transform metals and radionuclides (Barkay and Schaefer, 2001). Aerobic granules have also been demonstrated to be very useful as absorbents of several metals like cadmium, copper, and zinc (Liu et al., 2002, 2003a,b), cerium (Zhang et al., 2005), chromium (Nancharaiah et al., 2010; Sun et al., 2010), and uranium (Nancharaiah et al., 2006).

Aerobic granules can also remove microparticulate material of biotic as well as abiotic origin from wastewater. Ivanov et al. (2004) suggested that the removal was mainly due to porous structure of granular surface and not due to protozoan grazing. However, studies by Schwarzenbeck et al. (2004a) confirmed that protozoans play an important role in the removal of particulates. Nonetheless, this aspect should be considered for full-scale application as actual wastewaters contain particulate matter.

Formation of aerobic granules has been achieved in industrial wastewaters like dairy wastewater (Arrojo et al., 2004; Schwarzenbeck et al., 2005) and malting wastewater (Schwarzenbeck et al., 2004b). Aerobic granules have been used for degradation of several other wastes and organics like tert-butyl alcohol (Tay et al., 2005), chelants in pure as well complexed form (Nancharaiah et al., 2006), pyridine (Adav et al., 2007), petrochemical wastewater (Zhang et al., 2011a), acrylic waste (Wang et al., 2010), methylene blue (Ma et al., 2011a), pigmented textile wastewater (Muda et al., 2011), and swine slurry (Figueroa et al., 2011).

47.3.2　Phototrophic Biofilms

The biofilm systems described in previous sections are predominantly heterotrophic bacterial systems, which also include some heterotrophic eukaryotes, such as protozoans and fungi. Most of the natural microbial systems though include autotrophic organisms that form the ultimate energy source for the heterotrophs. In illuminated conditions, the phototrophic organisms dominate the biofilms systems. Phototrophic biofilms are also called photobiofilms or photosynthetic biofilms or microbial mats. The lotic phototrophic biofilms in rivers and water streams generally consist of triple consortia of bacteria, cyanobacteria, and algae (Cuzman et al. 2010). The photosynthetic organisms fix carbon, and this photosynthetic activity fuels the entire biofilm community, including the heterotrophic fraction, which is generally the pioneering community of a phototrophic biofilms (Canfield and Des Marais, 1993; Roeselers et al., 2007).

Phototrophic biofilms have been studied in different contexts and using a variety of natural and laboratory based systems. Typical pond-type reactors systems are cheaper and scalable but are not as efficient as lab-scale photobioreactors and have large footprint. Some of the important studies on different photobioreactor systems are discussed here. Xu et al. (2002) described a simple airlift reactor for efficient cultivation of algae. To maximize the economic advantage of photobioreactors, Ogbonna et al. (1999) designed a scalable stirred-tank photobioreactor illuminated with solar light using optical fibers, light tracker, and intensity sensor. During the absence of solar light, the reactor is illuminated by artificial lights. Mirón et al. (2000) demonstrated the growth of algae in bubble column and airlift reactors.

Phototrophic biofilms (e.g., microbial mats) are made of very diverse communities of microbes and perform very important functions in nature, including primary production, nutrient recycling, organic matter decomposition, pollutant detoxification, and biogeochemical cycling (Battin et al., 2003; Strauss, 2005; Zippel and Neu, 2005; Tlili et al., 2011). They have high potential to be employed in bioremediation processes (Schumacher et al. 2003). Though cyanobacteria are oxygenic phototrophs, cyanobacterial mats may provide opportunity for the growth of anaerobic bacteria, due to thick nature of the mats. Sulfate reduction, an anaerobic process, has been reported in cyanobacterial mats (Frund and Cohen, 1992). Phototrophic biofilms have been used in bioremediation processes such as oil degradation (Cohen, 2002) and removal of toxins (e.g., microcystins) (Babica et al., 2005).

Munoz and Guieysse (2006) described various advantages of using microalgae along with bacteria in bioremediation. The authors argued that using algae not only enhanced the removal of nutrients, organics, and metals from the waste but also added to the availability of oxygen for the degradation of volatile organics, which may escape due to mechanical aeration in normal bioreactors. Algal biomass itself is a very useful product as it can be used as animal food source, for high-value biomolecules, fertilizers, and biofuels. The treatment of wastewater is generally carried out using bacteria-based systems. But this results in two main greenhouse gases, namely, CO_2 and CH_4. Methane is generally burnt resulting to CO_2 release to the atmosphere. Dahalan et al. (2015) suggested introduction of photosynthetic bacteria in aerobic granular sludge to reduce the load of greenhouse gases from wastewater treatment plant. As an alternative strategy, the waste treatment facility could be associated with an industry that produced CO_2 or flue gas. In their recent review, Subashchandrabose et al. (2011) discussed the biotechnological potential of typical phototrophic biofilms containing the consortia of cyanobacteria, microalgae, and bacteria. The authors argued that not only the triple consortia form a self-sustaining unit where phototrophs help heterotrophs sustain and vice versa but also the detoxification efficiencies would be better due to versatility of the microorganisms. The authors discussed the status of studies demonstrating usefulness of the triple consortia in processes like degradation of organics and removal of metal pollutants and nutrients from water. They argued that unlike genetic engineering, consortia engineering could be a simpler approach, which was more likely to achieve success for desired objective of bioremediation of toxins. Further, the biomass generated could be utilized in various ways.

Similarly, Christenson and Sims (2011) discussed the possibilities of combining algal production with wastewater treatment as a mutually beneficial technology. The authors also delved on the problems associated with downstream processing of algal biomass, such as issues of biomass separation from bulk, emphasizing on the usefulness of flocculation and biofilm systems. Sharma et al. (2011) emphasized the biotechnological potential of cyanobacteria, because of the large spectrum of metabolites produced by them, with very simple nutritional requirements like plants and possibility of high yield cultivation. The authors commented on the need for more research in the area of metabolic engineering and efficient photobioreactors, before cyanobacteria could become resource for the future.

47.3.3 PHOTOTROPHIC GRANULES

Aerobic granulation, as suggested by many authors, requires significant research in terms of understanding the mechanism of granulation and the role of various parameters, improving physical and functional integrity of granules, microbial diversity, genetic modification of microbes, and reduction of reactor start-up time. Similarly, it is generally agreed that, for making significant strides in the area of phototrophic biofilms, it is essential to look for algal strains that are tolerant to toxic waste, to develop photobioreactors that are efficient and scalable but with small footprint and utilize genetic engineering to increase the reactor efficacy and photosynthesis efficiency.

Moreover, for increasing the efficacy of the granular reactor systems, it is essential to try different biomass sources (other than activated sludge) as seed materials, which can tender a variety of useful characteristics to the sludge, apart from reducing the reactor start-up time. Further, the process of MAG development involves addition of cost-intensive carbon/energy source. The availability of reduced organic material (energy source), which can be easily utilized by bacteria involved in the bioremediation, is one of the limiting factors and reasons for the limited success of bioremediation (Boopathy, 2000). Any step toward minimizing this cost can potentially help in enhancing the scope of this technology for various industrial applications. One of the means to achieve this would be to incorporate photosynthetic elements into the granular microbial sludge. Though bioremediation processes are more effective and economical than physical and chemical methods, they have not been used to their fullest potential. Only a fraction of total available microbial diversity (culturable, mostly heterotrophic) has been used in these processes (Paul et al., 2005). Using cyanobacteria and algae in mainstream wastewater treatment can provide much wider metabolic diversity and therefore more efficient treatment, as it has been proved that higher phototrophic component in biofilms results in higher enzymatic activity (Romaní and Sabater, 2000).

Phototrophic biofilm biotechnology has potential applications in the development of clean energy systems, wastewater treatment, bioremediation, fish feed production, and soil fertilization. However, phototrophic biofilms come with their own set of problems. Industrial-scale application of microbial mats has been limited, with a few exceptions like aeration ponds and trickling filters. One of the major limitations is the requirement of large illuminated area for cultivation of the phototrophs such as microalgae and cyanobacteria (Roeselers et al., 2008).

Granular phototrophic biofilm development in photobioreactors can overcome the limitation. Compact photobioreactors can be designed with substantially reduced footprint, and the granular biomass can be continuously kept in suspension such that the organisms are seldom light limited. Granules are easily separated from the bulk, and phototrophic microbes support the heterotrophic microbes in the biofilm, thereby making a self-sustaining minimal ecosystem. Recently, Kumar and Venugopalan (2015) demonstrated the reconstruction of phototrophic biofilms in the form of granules in bubbled column photobioreactors, thereby mitigating limitations of both aerobic granular technology and phototrophic biofilms. The authors also demonstrated that phototrophic granular sludge could successfully be used for the treatment of toxic industrial waste.

47.4 RADIOACTIVE WASTE

Nuclear energy is considered to be safe from the point of view of climate change, air quality, and energy resource availability compared to the energy from conventional sources like coal and petroleum. Generation of waste is inevitable for any industrial activity, and nuclear industry is no exception. The difference in nuclear waste compared to other industrial and explosive waste is that the former is associated with radioactivity that decays with time and is likely to continue to the subsequent generation as well. Nuclear industry for power generation depends on harnessing of fission energy where the fissile isotopes such as ^{235}U, ^{239}Pu, and ^{233}U split to produce a wide variety of fission products (i.e., lighter nuclei from fission) and activation products (nuclei of fuel and surrounding structural materials becoming radioactive after absorbing neutrons). While a majority of the nuclear fission products are short-lived, a small amount of long-lived fission products and actinides are also produced during the nuclear fuel cycle operations.

Careful handling of nuclear waste and proper nuclear waste management program are essential to avoid radiation exposure to human beings and also to avoid contamination of air, groundwater, etc. The main phases of the nuclear fuel cycle that produce radioactive waste include uranium mining and milling, fuel fabrication, nuclear reactor operation, and nuclear fuel reprocessing (Figure 47.5). As a waste management philosophy, thrust is given to waste volume minimization at every stage including design, operation, and maintenance. All types of radioactive wastes are collected and transferred to treatment and conditioning facilities for the purposes of volume reduction, removal of radioactive constituents, and stabilization of their chemical and physical forms (Raj et al., 2006). The treated wastes are contained appropriately and placed in suitable matrices that ensure proper isolation of the waste from the biosphere. Radioactive wastes can arise in different forms, namely, solid, liquid, and

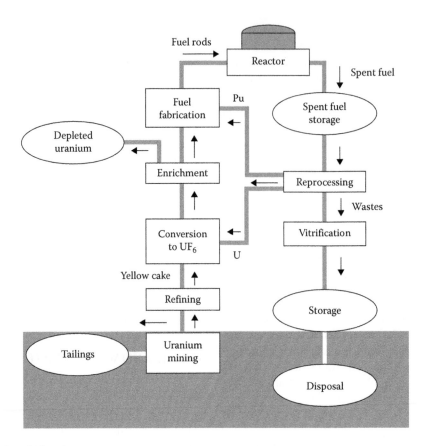

FIGURE 47.5 The nuclear fuel cycle.

TABLE 47.1
Radionuclides Generated during Different Fuel Cycle Operations

Group	Fission Product	Half-Life	Hazards
0	^{85}Kr	10.76 years	Released in elemental state and needs special attention.
I	^{137}Cs, ^{134}Cs	^{137}Cs (30 years)	Cs is a biological hazard due to its long half-life.
		^{134}Cs (2.06 years)	
II	^{89}Sr, ^{90}Sr, ^{140}Ba	^{89}Sr (50.5 day)	Sr is a bone seeker and toxic hazard. ^{140}La, daughter product of ^{90}Ba, is
		^{90}Sr (28.5 years)	a hard γ emitter.
		^{140}Ba (12.8 days)	
III	^{90}Y, ^{91}Y, ^{140}La	^{140}La (40.2 h)	Radiological hazard
	^{141}Ce, ^{144}Ce, ^{144}Pr,	^{141}Ce (32.51 days)	Stable radionuclides such as ^{149}Sm, ^{151}Eu, ^{155}Gd, ^{157}Gd, and ^{164}Dy
	^{143}Pr, and ^{147}Nd	^{144}Ce (248.8 days)	produced pose neutron poisoning problems in reactor.
		^{144}Pr (17.3 months)	
		^{147}Nd (10.89 days)	
		^{90}Y (64.1 days)	
		^{91}Y (58.5 days)	
IV	^{95}Zr	^{95}Zr (64 days)	Forms complexes with the degradation products of tri-n-butyl phosphate
			(TBP) and reduces the efficiency of the extraction processes.
V	^{95}Nb	^{95}Nb (35 days)	Enters the organic phase during solvent extraction and poses problems.
VI	Mo	^{101}Mo (14.6 min)	Radiological hazard with local overheating in storage tanks.
VII	^{99}Tc, ^{131}I, ^{129}I	^{99}Tc (2.41×10^5 years)	Radiological hazard.
		^{131}I (8.1 days)	
		^{129}I (1.57×10^7 years)	
VIII	^{103}Ru, ^{106}Ru	^{103}Ru (39.35 days)	Radiological hazard and complex chemistry of ruthenium.
		^{106}Ru (1 years)	

gas, with a variety of physical and chemical/radiochemical characteristics (IAEA, 2009).

During uranium mining, milling, refining, and fuel fabrication processes, highly active radionuclides are not generated. However, small amounts of uranium in tailings as well as radon gas from leach tank vents and ore piles are generally observed. Nuclear reactor and other back end fuel cycle operations generate bulk of the waste that requires special treatment before they are discharged into the environment. Radioactivity is generally contained in reactor system, and many barriers such as fuel clad, reactor system, and auxiliary buildings are in place to prevent the release of radioactivity into the atmosphere. However, noble gases such as Kr and Xe, alkali metals such as cesium and Rb, alkaline earth metals such as strontium and barium, and halogens such as bromine and iodine are released during the reactor operation. Also, the coolant and structural materials absorb neutrons and get activated. Coolant activation products are generally short-lived except ^3H and ^{14}C. The activation products formed due to the interaction of neutron with structural materials have considerable half-lives and enter into the coolant systems through corrosion and erosion from many parts (e.g., Zr, Co, Mn, and Ni). The processes pertaining to the back end of the fuel cycle including nuclear fuel processing and nuclear waste management involves handling of many fission products in a variety of forms (Table 47.1).

47.4.1 Classification of Radioactive Waste

Considering the safety as well as the process perspective, classification of radioactive waste is extremely essential. Classification of waste into different categories is also important in their segregation, choice of appropriate treatment processes, devising storage, and disposal options. Classification principles may be based on the origin of waste, its physicochemical nature or type of radiation, and half-life of the different radionuclides. Accordingly, radioactive waste streams are commonly classified as exempt waste, low-level waste, intermediate-level waste, and high-level waste (HLW). Figure 47.6 provides a conceptual illustration of nuclear waste classification. Exempt wastes have too low levels of radioactivity and can be disposed of to the environment without chemical treatment. Low- and intermediate-level wastes (LILWs) are further categorized as short-lived and long-lived wastes. Because of their short half-life (<30 years half-life), radiological hazards of these wastes get significantly reduced over a few hundred years by radioactive decay processes. The HLW warrants high levels of isolation from the biosphere as they generate significant amounts of heat and radioactivity, and this is applicable for both short-lived and long-lived HLWs. In India, the categories of different wastes are defined by the Atomic Energy Regulatory Board (AERB) and are given in Table 47.2, and the current IAEA classification of radioactive wastes is given in Table 47.3.

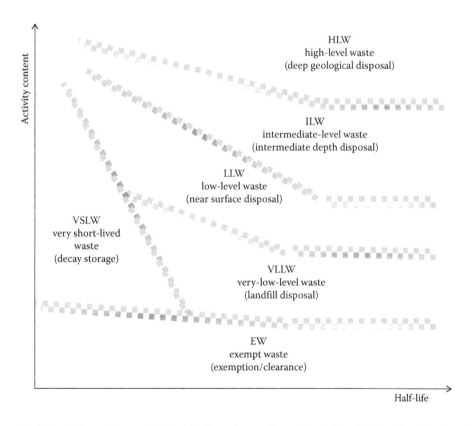

FIGURE 47.6 Conceptual illustration of the waste classification scheme. (Reproduced from IAEA, *Classification of radioactive waste: Safety guide*, IAEA Safety Standard Series No. GSG-1, STI/PUB/1419, International Atomic Energy Agency, Vienna, Austria, 2009, pp. 1–68. With permission.)

TABLE 47.2

IAEA Classification of Radioactive Wastes

Waste Classes		Typical Characteristics	Disposal Options
1.	Exempt waste	Activity below clearance levels that are based on an annual dose to public citizens of less than 0.01 mSv	No radiological restrictions
2.	Low- and intermediate-level waste (LILW)	Activity levels above clearance levels and thermal power below about 2 kW/m^3	Disposal in controlled sites after treatment
2.1	Short-lived waste (LILW-SL)	Restricted long-lived radionuclide concentration, such as alpha-emitting radionuclides to 4000 Bq/g in individual waste packages and to an overall average of 400 Bq/g per waste package	Near surface or geological disposal facility
2.2	Long-lived waste (LILW-LL)	Long-lived radionuclide concentration exceeding limitation for short-lived waste	Geological disposal facility
3.	High-level waste (HLW)	Thermal power above about 2 kW/m^3 and long-lived radionuclide concentration exceeding limitation for short-lived waste	Geological disposal facility

TABLE 47.3

Categorization of Nuclear Waste

Category	Solid Surface Dose (mGy/h)	Liquid Activity (Bq/m^3)	Gaseous Activity (Bq/m^3)
I	<2	<3.7 × 10^4	<3.7
II	2–20	3.7 × 10^4 to 3.7 × 10^7	3.7 to 3.7 × 10^4
III	>20	3.7 × 10^7 to 3.7 × 10^9	>3.7 × 10^4
IV	Alpha bearing	3.7 × 10^9 to 3.7 × 10^{14}	—
V	—	>3.7 × 10^{14}	—

47.4.2 RADIOACTIVE WASTE MANAGEMENT

The primary objective of radioactive waste management is protection of human health, environment, and future generation. Various radioactive waste management practices followed in India are presented in Figure 47.7. In India, the AERB is mandated to review and authorize from the safety angle the siting, the design, the construction, and the operation of the waste management plants and also frame safety guidelines from time to time.

Keeping in view the large variety of radioactive wastes being generated, the treatment processes used are also diverse and need rigorous research and development (IAEA, 2009).

47.4.3 GASEOUS WASTE

Gaseous wastes released from different parts of the nuclear facilities may be in the form of gases, vapors, and particulate suspensions, and the retention of radionuclides is achieved as a result of certain physical, chemical, or physicochemical properties. In order to control and minimize discharge of activity through air route in conformity with the principle of ALARA, the nuclear installations should be equipped with an elaborate off-gas cleaning system and released to the atmosphere through tall stacks. These off-gas systems are designed depending on the specific activity, radioactivity levels and type, particulate density, concentration of volatile isotopes, etc. For the retention of gaseous radionuclides and particulates, methods such as iodine sorption system, particulate filter system, noble gas delay system are being used and high-efficiency particulate air filters are normally used for effective removal of particulate matter. From the processes involved in the back end of the fuel cycle, that is, nuclear fuel reprocessing, radionuclides such as ^{3}H, ^{14}C, ^{85}K, and ^{129}I and aerosols containing fission products are released that need special attention.

47.4.4 SOLID WASTE

Solid wastes arising from different nuclear fuel cycle operations cover a wider range of materials, sizes, shapes, and degrees of contamination. Solid radioactive wastes are categorized depending on the radiation field, concentration, and type of radioactivity. They are further segregated as compressible or noncompressible and combustible or noncombustible. Nearly 60% of solid radioactive wastes produced in nuclear facilities like nuclear power plants, processing plants, or nuclear research centers can be classified as combustible waste and 40% as noncombustible or bulky waste. The objective of solid waste treatment is to reduce as much as possible the waste volumes to be stored or disposed of and to concentrate and immobilize as much as possible the radioactivity contained in the waste. Volume reduction by compaction aims in increasing the overall density of the waste material. Presses and compacting devices are generally employed as a treatment technique in waste management. Incineration of combustible waste offers a high-volume reduction factor of the order of 50 to 80, and for the purpose, incinerators are generally used (Wattal, 2013).

47.4.5 LIQUID WASTE

The overall philosophy for the safe management of radioactive liquid waste relies on the concepts of (1) delay and decay, (2) dilute and disperse, and (3) concentrate and contain. The LILW streams require treatment to reduce their activity to a level at which they are allowed to be discharged as per the regulations. The treatment processes for the low- and intermediate-level liquid wastes include filtration, chemical treatment, ion exchange, steam evaporation, solar evaporation, and membrane processes. Conditioning of low- and

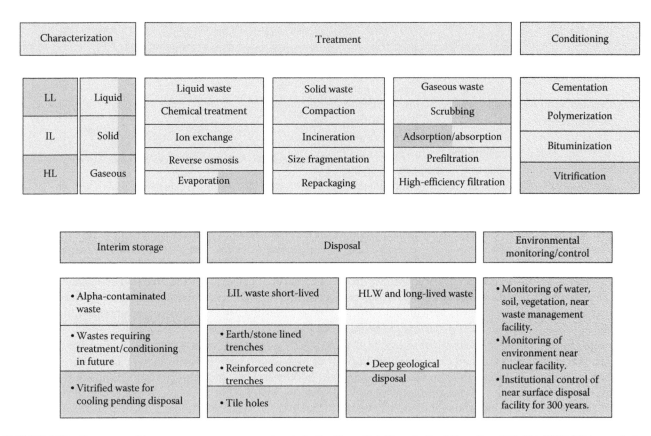

FIGURE 47.7 Summary of the waste management practices in India. (From Raj, K. et al., *Nucl. Eng. Des.*, 236, 914, 2006. With permission.)

intermediate-level liquid waste is based on the compatibility of the matrix with waste, chemical and mechanical durability of the solidified product, cost of processing, throughput, and disposal options. Cementation and polymerization processes are adopted in India for the management of this type of wastes (INCAS, 2007). Liquid HLW generated during reprocessing of spent nuclear fuels is concentrated by evaporation and stored in stainless steel tanks that need sufficient cooling and surveillance. The management strategies of HLW include

1. Immobilization of high-level liquid waste into vitrified borosilicate glasses
2. Interim retrievable storage of the conditioned waste under passive cooling and surveillance
3. Disposal in deep geological formations

47.4.6 Biological Methods for Radioactive Waste Treatment

Bioremediation is a promising technology for reducing the toxicity and amount of radioactive waste that uses bacteria and other live cells or enzymes to clean and reduce the volume of waste. Bioremediation techniques can generally be divided into two categories: *in situ* and *ex situ*. The former involves treating the contaminated area on-site while the latter involves removing it to an alternative location for treatment. The organisms that are used metabolize anaerobically, that is, in nonoxygen conditions, by chemically reacting with

metal ions in solution, reducing them to lower oxidation states (lower ionic charge). For some elements, this reduced ion is insoluble, so it precipitates out of solution as a solid. For example, soluble U^{6+} is reduced to insoluble U^{4+} that precipitates as solid uranium dioxide, UO_2. These precipitates can be recovered easily, then isolated, and contained. The emerging concept is by employing a live cell "bioreactor," wherein some elements such as uranium (U), plutonium (Pu), and technetium (Tc) from contaminated waters can be treated with bio-organisms.

Phytoremediation can be defined as the efficient use of plants to remove, detoxify, or immobilize environmental contaminants in a growth matrix (soil, water, or sediments) through the natural biological, chemical, or physical activities and processes of the plants. Phytoremediation involves growing plants in a contaminated matrix, for a required growth period, to remove contaminants from the matrix or facilitate the immobilization or degradation of the pollutants. The plants can be subsequently harvested, processed, and disposed. Phytoremediation is applicable to a broad range of contaminants including polycyclic aromatic hydrocarbons, surfactants, explosives, heavy metals, and radionuclides. For example, out of the many green plants that have been screened to check the uptake efficiency of radionuclides, it was found that *Catharanthus roseus* has the high potential for radionuclides. Beet (*Beta vulgaris*), quinoa (*Chenopodium kali*), water hyacinth (*Elichornia crassipes*), and *Amaranthus retroflexus* were found to take up ^{137}Cs to a significant extent. The use of economic plants such as biofuel crops for the utilization and remediation of the contaminated sites would be a better choice, as these plants not only remediate contaminated soil but also can produce valuable biomass, which is also a useful commodity. Phytoremediation would play a vital tool in sustainable management of contaminated soil.

Sunflowers have been used successfully in rhizofiltration applications. Successful rhizofiltration techniques require identification of species of plants that have the ability to process large quantities of water and sequester certain contaminants in plant biomass. A hydroponically grown special strain of sunflower on rafts has removed radionuclides from water. The sunflower rhizofiltration system can successfully reduce uranium, strontium, and cesium levels in water to below cleanup standards. The Indian mustard plant (*Brassica juncea*) can extract both heavy metals and radionuclides from the soil. The Indian mustard plant is a high-biomass crop that can help control selenium levels in the agricultural effluent. Also, some of the harvested mustard plants can be blended with hay and fed to animals in areas where selenium deficiency is a problem. The mustard plant can accumulate heavy metals such as lead, chromium, cadmium, nickel, and zinc. The approach requires addition of a chelating agent to the soil for solubilizing the metal lead and allows it to move from the roots into the shoots. Indian mustard plant was also tested to remove radionuclides such as cesium-137 and strontium-90 at Chernobyl site.

Biosorbents and dead biomasses are also promising candidates for the removal/speciation of heavy metal toxic ions from aqueous solutions as well as removal of radionuclides from nuclear waste streams (Tiwari et al., 2006). The selectivity of these biosorbents depends upon their thermal and radiation stability. Rice husk and bark samples of *Mangifera indica* and *Azadirachta indica* trees were explored for the uptake of Sr (85 and 89) from aqueous radioactive streams.

47.5 EXPLOSIVE WASTE

Explosives can be defined as chemical compounds that detonate at high rates, thus producing large amounts of hot gases. Exothermic oxidation–reduction reactions provide the energy released during detonation. Generally used explosives are organic compounds rich in nitro group that can self-oxidize to smaller gaseous molecules (N_2, H_2, O_2, and CO_2). These compounds are broadly divided into two categories:

1. Polynitro aromatic compound that include 2,4,6-trinitrotoluene (TNT), 1,3,5-trinitrobenzene (TNB), dinitrotoluenes (2,4-DNT, 2,6-DNT), dinitrobenzene(DNB), N-methyl-N,2,4,6-tetranitroaniline (tetryl), and 2,4,6-trinitrophenol (picric acid)
2. Polynitro amines RDX (royal demolition explosive) named as 1,3,5-trinitro-1,3,5-triazacyclohexane and HMX (high melting explosive), 1,3,5,7-tetranitro-1,3,5,7-tetraazacyclooctane.

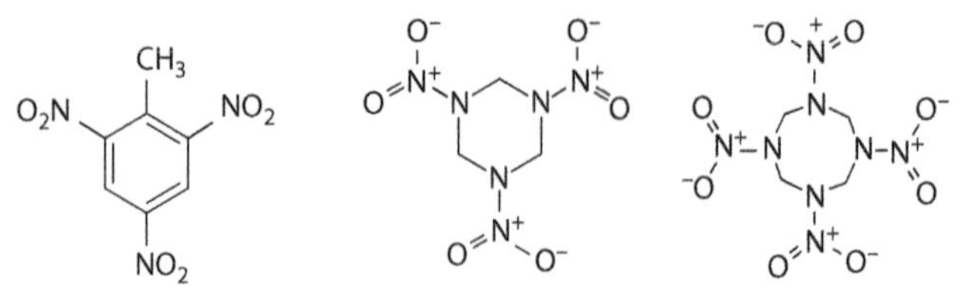

FIGURE 47.8 Molecular structure of commonly used explosives.

Among these explosives, TNT, RDX, and HMX (Figure 47.8) are the most widely used in military and industries. Generally, it is very difficult to degrade these explosives, because of the nitro groups that are strong electron-withdrawing functional groups and hence are resilient to chemical or biological degradation (Podlipná et al., 2015). Due to their chemical similarities, the explosives mentioned earlier have similarity in their production process, their environmental impact, and their disposal and degradation.

TNT contamination at manufacturing sites was found to be as high as 100 ppm in water (Joshi et al., 2005). Exposure to 12,000 ppm leads to health problems including liver damage and anemia. TNT is toxic and is mutagenic at low concentrations and hence its decontamination becomes necessary (Zhang et al. 2011b). DNTs are precursors to TNT, and hence, TNT-contaminated sites are also contaminated with DNTs. Both these DNTs are included in U.S. EPA list as priority pollutants and are known mutagens and suspected carcinogens. Permissible levels for 2,4- and 2,6-DNT for discharge streams are 0.32 and 0.55 mg L^{-1}, respectively (Paca et al., 2009). HMX is one of the nitro explosives produced by nitration of hexamine using ammonium nitrate and acetic anhydride, and hence, the wastewater bears high concentration of nitrate and acetate (Kanekar et al., 2009). The traditional methods to dispose explosives and by-products mostly involved burning; the following are some ways:

1. *Open burning*: In this method, the ammunition is ignited by using liquid fuel or other means. It is carried out in open grounds at remote areas, mostly at military training grounds. A slight variant of this method is open-pit burning wherein the material is burned in a pit usually made of concrete.
2. *Open detonation*: In this method, the ammunition is set off with the help of another explosive charge thus resulting in complete burning of the explosive as compared to open burning; on the other hand, this method also results in more dispersion of waste generated including heavy metal dust compared to open burning method. Both methods being performed in open air have serious implications due to the release of the waste products (solid residues, dust, and combustion gases) directly to the environment, without any control or treatment.
3. *Closed detonation*: This is a modification of open burning method where the ammunition is placed in

a closed chamber and burned, thus giving us control over the emissions into the environment. But appropriate downsizing of explosive is required depending on the type of explosive and the size of the chamber. An added advantage of this method is that this facility can be transported.

4. *Rotary kiln incineration*: This method involves a thick-walled steel cylinder with refractory lining, rotating horizontally with a slight inclination of about 3°–5° to the ground. Solid explosive wastes are fed in through the raised end, and as it slowly passes through the kiln, it gets burned and the residues fallout from the lower end where it gets collected, thus offering control over the residues and emissions.

5. *Fluidized bed combustion*: It is a bed of sand particles (silica) through which hot air is passed that in turn carry the sand thus resulting in the fluid nature of the bed. The slurry or liquid waste is injected as fine droplets to ensure mixing and aeration that will lead to complete decomposition.

47.5.1 Bioremediation of Explosive Waste

Moshe et al. (2009) have reported that TNT can be biodegraded both aerobically and anaerobically. In either case, the products formed are the reduced amino derivatives 4-amino-2,6-DNT and 2-amino-4,6-DNT. Bacteria that degrade 2,4-DNT will yield very small biomass, because several intermediate compounds of its catabolism are known uncouplers of respiration and oxidative phosphorylation (Hudcova et al., 2011). The genus *Rhodococcus* (*Actinomycetales* species) is mainly limited to degradation of 4-nitrophenol, 2,4-dinitrophenol and 2,4,6-trinitrophenol (Martínková et al., 2009; Foster et al., 2014).

RDX is recalcitrant to microbial biodegradation and has a relatively high solubility in water, and the toxic RDX from explosive manufacturing activities reach the groundwater and surface water environments easily since RDX is not absorbed strongly to the top layers of soil. Recent reports have shown the anaerobic biodegradation of RDX, although a number of aerobic RDX-degrading bacteria have been isolated that use RDX as a sole nitrogen source. Aerobic RDX degradation has several advantages over anaerobic biodegradation, due to its faster degradation rate and lower toxicity of the intermediate compounds (Khan, 2015). RDX-degrading bacteria are classified within the phyla *Firmicutes*, *Actinobacteria*, *Proteobacteria*, and *Fusobacteria* species. Although microbial isolates could degrade explosives in laboratory conditions, the same may not be true in the case of treatment at actual contaminated site. This is mainly because the laboratory-grown bacteria may not survive or carry out degradation in field conditions. The biodegradation of RDX occurs under both aerobic and anaerobic conditions, but the anaerobic process is significantly faster. Aerobically, it has been found that the microbial cleavage of one of the N-NO$_2$ bonds yields unstable intermediates, followed by rapid cleavage of the triazine ring. Anaerobically, sequential reduction of the nitro groups resulting in the production of mono-, di-, and trinitroso derivatives MNX, DNX, and TNX, respectively, is the most frequently observed in this process. The nitroso derivatives may be further transformed to produce the unstable hydroxylamine derivatives, leading to ring cleavage. A two-step, mixed immobilized bioreactors where anaerobic phase was followed by aerobic bacteria filter in an upflow submerged assembly was used for the degradation of TNT. Alcohol when added to the culture broth was found to increase the degradation specially in the anaerobic stage, that is, the reduction process. The effluents from the first stage were found to contain only 0.3 ppm of the added 80 ppm of TNT in the effluent, but it contained reduced metabolite of TNT. In the aerobic phase, both reduced metabolites and TNT were not detected, which shows the complete mineralization of the TNT. The ethanol addition to the process was fermented to acetate and hydrogen that served the reduction of TNT to amino derivatives. Microbial community of the bioreactor after characterization showed the presence of *Pseudomonas* sp., *Flavobacterium* sp., and *Chryseobacterium* sp. (Gumuscu and Takinay, 2013). In another study, aerobic biodegradation of 2,4-DNT and 2,6-DNT in continuous packed bed bioreactors was reported by Paca (2009). A novel strain of *Achromobacter spanius* was found to degrade 100 mg TNT within 20 h under aerobic conditions (Gumuscu, 2013). Biodegradation of 2,4-dinitrotoluene by *Pseudomonas fluorescens* sp., isolated from a nitrobenzene-contaminated soil under growth limited conditions. The culture was enriched with 2,4-DNT as the sole source of carbon and energy. Results indicated that the isolate grew optimally at pH 7.5 in the mineral salts medium supplemented with 2,4-DNT with 40 ppm. The study indicated the potential application of the isolate for the remediation of 2,4-DNT-contaminated groundwater (Huang, 2012).

REFERENCES

Abdulsalam, S., Bugaje, I. M., Adefila, S. S., and Ibrahim, S. 2011. Comparison of biostimulation and bioaugmentation for remediation of soil contaminated with spent motor oil. *International Journal of Environmental Science & Technology* 8: 187–194.

Abdulsalam, S. and Omale, A. B. (2009). Comparison of biostimulation and bioaugmentation techniques for the remediation of used motor oil contaminated soil. *Brazilian Archives of Biology and Technology* 52: 747–754.

Adams, G. O., Fufeyin, P. T., Okoro, S. E., and Ehinomen, I. 2015. Bioremediation, biostimulation and bioaugmentation: A review. *International Journal of Environmental Bioremediation & Biodegradation* 3: 28–39.

Adams, G. O., Tawari-Fufeyin, P., and Igelenyah, E. 2014. Bioremediation of spent oil contaminated soils using poultry litter. *Research Journal in Engineering and Applied Sciences* 3: 124–130.

Adav, S. S., Lee, D. J., and Ren, N. Q. 2007. Biodegradation of pyridine using aerobic granules in the presence of phenol. *Water Research* 41: 2903–2910.

Al-Sulaimani, H., Joshi, S., Al-Wahaibi, Y., Al-Bahry, S., Elshafie, A., and Al-Bemani, A. 2011. Microbial biotechnology for enhancing oil recovery: Current developments and future prospects. *Biotechnology, Bioinformatics and Bioengineering* 1: 147–158.

Alvarez-Cohen, L. and McCarty, P. L. 1991. A cometabolic biotransformation model for halogenated aliphatic compounds exhibiting product toxicity. *Environmental Science & Technology* 25: 1381–1387.

Andreottola, G., Foladori, P., and Ziglio, G. 2009. Biological treatment of winery wastewater: An overview. *Water Science & Technology*, 60.

Arrojo, B., Mosquera-Corral, A., Garrido, J. M., and Méndez, R. 2004. Aerobic granulation with industrial wastewater in sequencing batch reactors. *Water Research* 38: 3389–3399.

Artham, T. and Doble, M. 2008. Biodegradation of aliphatic and aromatic polycarbonates. *Macromolecular Bioscience* 8: 14–24.

Atkinson B. W., Bux F. and Kasan H. C. 1998. Waste activated sludge remediation of metal-plating effluents. *Water SA* 24: 355–359.

Atkinson, S., Chang, C. Y., Sockett, R. E., Cámara, M., and Williams, P. 2006. Quorum sensing in *Yersinia enterocolitica* controls swimming and swarming motility. *Journal of Bacteriology* 188: 1451–1461.

Atlas, R. and Bragg, J. 2009. Bioremediation of marine oil spills: When and when not–the Exxon Valdez experience. *Microbial Biotechnology* 2: 213–221.

Babica, P., Blaha, L., and Marsalek, B. 2005. Removal of microcystins by phototrophic biofilms. A microcosm study. *Environmental Science and Pollution Research* 12: 369–374.

Barkay, T. and Schaefer, J. 2001. Metal and radionuclide bioremediation: Issues, considerations and potentials. *Current Opinion in Microbiology* 4: 318–323.

Battin, T. J., Kaplan, L. A., Newbold, J. D., Cheng, X., and Hansen, C. 2003. Effects of current velocity on the nascent architecture of stream microbial biofilms. *Applied and Environmental Microbiology* 69: 5443–5452.

Beer, D. and Stoodley, P. (2006). Microbial biofilms. In: Dworkin, M., Falkow, S., Rosenberg, E., Schleifer, K.H., and Stackebrandt, E. (eds.) *The Prokaryotes: Symbiotic Associations, Biotechnology, Applied Microbiology.* Kiel, Germany: Springer, pp. 904–938.

Beun, J. J., Hendriks, A., Van Loosdrecht, M. C. M., Morgenroth, E., Wilderer, P. A., and Heijnen, J. J. 1999. Aerobic granulation in a sequencing batch reactor. *Water Research* 33: 2283–2290.

Beun, J. J., Van Loosdrecht, M. C. M., and Heijnen, J. J. 2000. Aerobic granulation. *Water Science and Technology* 41(4–5): 41–48.

Beun, J. J., Van Loosdrecht, M. C. M., and Heijnen, J. J. 2002. Aerobic granulation in a sequencing batch airlift reactor. *Water Research* 36: 702–712.

Boopathy, R. 2000. Factors limiting bioremediation technologies. *Bioresource Technology* 74: 63–67.

Bossier, P. and Verstraete, W. 1996. Triggers for microbial aggregation in activated sludge. *Applied Microbiology and Biotechnology* 45: 1–6.

Brim, H., Osborne, J. P., Kostandarithes, H. M., Fredrickson, J. K., Wackett, L.P., and Daly, M. J. 2006. Deinococcus radiodurans engineered for complete toluene degradation facilitates Cr(VI) reduction. *Microbiology* 152: 2469–2477.

Bux, F. and Kasan, H. C. 1994. Comparison of selected methods for relative assessment of surface charge on waste sludge biomass. *Water SA* 20: 73–76.

Bux, F. and Kasan, H. C. 1996. Assessment of wastewater treatment sludge as metal biosorbents. *Resource and Environmental Biotechnology* 1: 163–177.

Cameotra, S. S. and Makkar, R. S. 2010. Biosurfactant-enhanced bioremediation of hydrophobic pollutants. *Pure and Applied Chemistry* 82: 97–116.

Canfield, D. E. and Des Marais, D. J. 1993. Biogeochemical cycles of carbon, sulfur, and free oxygen in a microbial mat. *Geochimica et Cosmochimica Acta* 57: 3971–3984.

Cao, B., Nagarajan, K., and Loh, K. C. 2009. Biodegradation of aromatic compounds: Current status and opportunities for biomolecular approaches. *Applied Microbiology and Biotechnology* 85: 207–228.

Carvalho, M., Vasconcelos, I., Bull, A., and Castro, P. 2001. A GAC biofilm reactor for the continuous degradation of 4-chlorophenol: Treatment efficiency and microbial analysis. *Applied Microbiology and Biotechnology* 57: 419–426.

Chang, H.-l. and Alvarez-Cohen, L. 1995. Model for the cometabolic biodegradation of chlorinated organics. *Environmental Science and Technology* 29: 2357–2367.

Christenson, L. and Sims, R. 2011. Production and harvesting of microalgae for wastewater treatment, biofuels, and bioproducts. *Biotechnology Advances* 29: 686–702.

Cohen, Y. 2002. Bioremediation of oil by marine microbial mats. *International Microbiology* 5: 189–193.

Cronje, G. L., Beeharry, A. O., Wentzel, M. C., and Ekama, G. A. 2002. Active biomass in activated sludge mixed liquor. *Water Research* 36: 439–444.

Cuzman, O. A., Ventura, S., Sili, C., Mascalchi, C., Turchetti, T., D'Acqui, L. P., and Tiano, P. 2010. Biodiversity of phototrophic biofilms dwelling on monumental fountains. *Microbial Ecology* 60: 81–95.

Dadrasnia, A. and Agamuthu, P. 2013. Potential of biowastes to remediate diesel fuel contaminated soil. *Global Network Environmental Science and Technology* 15: 474–484.

Dahalan, F. A., Najib, M. Z. M., Salim, M. R., and Ujang, Z. 2015. Characteristics of developed granules containing phototrophic aerobic bacteria for minimizing carbon dioxide emission. *International Biodeterioration & Biodegradation* 102: 15–23.

Daigger, G. T. and Boltz, J. P. 2011. Trickling filter and trickling filter-suspended growth process design and operation: A state-of-the-art review. *Water Environment Research* 835: 388–404.

Davey, M. E. and O'toole, G. A. 2000. Microbial biofilms: From ecology to molecular genetics. *Microbiology and Molecular Biology Reviews* 64: 847–867.

De, J., Ramaiah, N., and Vardanyan, L. 2008. Detoxification of toxic heavy metals by marine bacteria highly resistant to mercury. *Marine Biotechnology* 10: 471–477.

De Bruin, L. M. M., De Kreuk, M. K., Van der Roest, H. F. R., Uijterlinde, C., and Van Loosdrecht, M. C. M. 2004. Aerobic granular sludge technology: An alternative to activated sludge. *Water Science & Technology* 49: 1–7.

De Kievit, T. R. 2009. Quorum sensing in *Pseudomonas aeruginosa* biofilms. *Environmental Microbiology* 11: 279–288.

De Kreuk, M. K., Heijnen, J. J., and Van Loosdrecht, M. C. M. 2005. Simultaneous COD, nitrogen, and phosphate removal by aerobic granular sludge. *Biotechnology and Bioengineering* 90: 761–769.

Ely, R. L., Hyman, M. R., Arp, D. J., Guenther, R. B., and Williamson, K. J. 1995. A cometabolic kinetics model incorporating enzyme inhibition, inactivation, and recovery: II. Trichloroethylene degradation experiments. *Biotechnology and Bioengineering* 46: 232–245.

Ely, R. L., Williamson, K. J., Hyman, M. R., and Arp, D. J. 1997. Cometabolism of chlorinated solvents by nitrifying bacteria: Kinetics, substrate interactions, toxicity effects, and bacterial response. *Biotechnology and Bioengineering* 54: 520–534.

Figueroa, M., Val del Rio, A., Campos, J. L., Mosquera-Corral, A., and Mendez, R. 2011. Treatment of high loaded swine slurry in an aerobic granular reactor. *Water Science and Technology* 63: 1808.

Foster, A., Barnes, N., Speight, R., and Keane, M. A. 2014. The repertoire of nitrogen assimilation in Rhodococcus: Catalysis, pathways and relevance in biotechnology and bioremediation. *Journal of Chemical Technology and Biotechnology* 89: 787–802.

Fründ, C. and Cohen, Y. 1992. Diurnal cycles of sulfate reduction under oxic conditions in cyanobacterial mats. *Applied and Environmental Microbiology* 58: 70–77.

Gao, D., Liu, L., Liang, H., and Wu, W. M. 2011. Aerobic granular sludge: Characterization, mechanism of granulation and application to wastewater treatment. *Critical Reviews in Biotechnology* 31: 137–152.

Garrett, T. R., Bhakoo, M., and Zhang, Z. 2008. Bacterial adhesion and biofilms on surfaces. *Progress in Natural Science* 18: 1049–1056.

Gavrilescu, M. 2004. Removal of heavy metals from the environment by biosorption. *Engineering in Life Sciences* 4: 219–232.

Gavrilescu, M. and Chisti, Y. 2005. Biotechnology—A sustainable alternative for chemical industry. *Biotechnology Advances* 23: 471–499.

Gumuscu, B. and Tekinay, T. 2013. Effective biodegradation of 2, 4, 6-trinitrotoluene using a novel bacterial strain isolated from TNT-contaminated soil. *International Biodeterioration & Biodegradation* 85: 35–41.

Hamzah, A., Phan, C. W., Abu Bakar, N. F., and Wong, K. K. 2013. Biodegradation of crude oil by constructed bacterial consortia and the constituent single bacteria isolated from Malaysia. *Bioremediation Journal* 17: 1–10.

Head, I. M., Jones, D. M., and Röling, W. F. 2006. Marine microorganisms make a meal of oil. *Nature Reviews Microbiology* 4: 173–182.

Hetzer, A., Daughney, C. J., and Morgan, H. W. 2006. Cadmium ion biosorption by the thermophilic bacteria *Geobacillus stearothermophilus* and *G. thermocatenulatus*. *Applied and Environmental Microbiology* 72: 4020–4027.

Huang, A., Teplitski, M., Rathinasabapathi, B., and Ma, L. 2010. Characterization of arsenic-resistant bacteria from the rhizosphere of arsenic hyperaccumulator *Pteris vittata*. *Canadian Journal of Microbiology* 56: 236–246.

Huang, Z. H., Pu, X., Zhang, H., Yuan, F., and Liu, Z. L. 2012. Biodegradation of 2, 4-dinitrotoluene by *Pseudomonas fluorescens* sp. isolated from a nitrobenzene contaminated soil. *Advanced Materials Research Journal* 610: 1233–1236.

Hudcova, T., Halecky, M., Kozliak, E., Stiborova, M., and Paca, J. 2011. Aerobic degradation of 2, 4-dinitrotoluene by individual bacterial strains and defined mixed population in submerged cultures. *Journal of Hazardous Materials* 192: 605–613.

Hughes, J. B., Shanks, J., Vanderford, M., Lauritzen, J., and Bhadra, R. 1996. Transformation of TNT by aquatic plants and plant tissue cultures. *Environmental Science & Technology* 31: 266–271.

IAEA. 2009. *Classification of radioactive waste: Safety guide.* IAEA Safety Standard Series No. GSG-1. STI/PUB/1419. Vienna, Austria; International Atomic Energy Agency, pp. 1–68.

IANCAS Bulletin. 2007. *Nuclear Waste Management.* Mumbai, India: BARC.

Ivanov, V., Tay, J. H., Tay, S. L., and Jiang, H. L. 2004. Removal of micro-particles by microbial granules used for aerobic wastewater treatment. *Water Science and Technology* 50: 147–154.

Jang, A., Yoon, Y. H., Kim, I. S., Kim, K. S., and Bishop, P. L. 2003. Characterization and evaluation of aerobic granules in sequencing batch reactor. *Journal of Biotechnology* 105: 71–82.

Jeon, C. O., Lee, D. S., and Park, J. M. 2003. Microbial communities in activated sludge performing enhanced biological phosphorus removal in a sequencing batch reactor. *Water Research* 37: 2195–2205.

Jiang, H. L., Tay, J. H., Maszenan, A. M., and Tay, S. T. L. 2004. Bacterial diversity and function of aerobic granules engineered in a sequencing batch reactor for phenol degradation. *Applied and Environmental Microbiology* 70: 6767–6775.

Joshi, S., Ghole, V., and Litake, G., 2005. TNT biotransformation potential of the clinical isolate of *Salmonella typhimurium*-potential ecological implications. *Indian Journal of Occupational and Environmental Medicine* 9: 29.

Kanekar, S. P., Kanekar, P. P., Sarnaik, S. S., Gujrathi, N. P., Shede, P. N., Kedargol, M. R., and Reardon, K. F. 2009. Bioremediation of nitroexplosive wastewater by an yeast isolate Pichia sydowiorum MCM Y-3 in fixed film bioreactor. *Journal of Industrial Microbiology & Biotechnology* 36: 253–260.

Kang, Y. S. and Park, W. 2010. Contribution of quorum-sensing system to hexadecane degradation and biofilm formation in Acinetobacter sp. strain DR1. *Journal of Applied Microbiology* 109: 1650–1659.

Karthikeyan, S., Korber, D. R. Wolfaardt, G. M., and Caldwell, D. E. 2001. Adaptation of bacterial communities to environmental transitions from labile to refractory substrates. *International Microbiology* 4: 73–80.

Khan, M. I., Yang, J., Yoo, B., and Park, J. 2015. Improved RDX detoxification with starch addition using a novel nitrogen-fixing aerobic microbial consortium from soil contaminated with explosives. *Journal of Hazardous Materials* 287: 243–251.

Kim, Y., Arp, D. J., and Semprini, L. 2002. Kinetic and inhibition studies for the aerobic cometabolism of 1, 1, 1-trichloroethane, 1, 1-dichloroethylene, and 1, 1-dichloroethane by a butane-grown mixed culture. *Biotechnology and Bioengineering* 80: 498–508.

Kulshreshtha, S., Mathur, N., and Bhatnagar, P. 2014. Mushroom as a product and their role in mycoremediation. *AMB Express* 4: 29.

Kumar, R. and Venugopalan, V. P. 2015. Development of self-sustaining phototrophic granular biomass for bioremediation applications. *Current Science* 108: 1653–1661.

Kumar, S. S., Celin, M. V., Bishnoi, N. R., and Malvan, S. K. 2014. Phytoremediation of HMX contaminated soil through *Jatropha curcas*. *International Journal of Recent Scientific Research* 5: 1444–1450.

Kumar, A. Bisht, B. S., Joshi, V. D. and Dhewa, T. (2011). Review on Bioremediation of Polluted Environment: A management tool. *International Journal of Environmental Sciences*, 1, 1079–1093.

Leahy, M. C. and Erickson, G. P. 1995a. Bioventing reduces cleanup costs. *Hydrocarbon Processing* 74: 63–654.

Leahy, M. C. and Erickson, G. P. 1995b. Bioventing reduces cleanup costs. Hydrocarbon surface charge on waste sludge biomass. *Water SA* 20: 73–76.

Liu, Y. 2003. Chemically reduced excess sludge production in the activated sludge process. *Chemosphere* 50: 1–7.

Liu, Y. Q., Liu, Y., and Tay, J. H. 2005. Relationship between size and mass transfer resistance in aerobic granules. *Letters in Applied Microbiology* 40: 312–315.

Liu, Y. and Tay, J. H. 2004. State of the art of biogranulation technology for wastewater treatment. *Biotechnology Advances* 22: 533–563.

Liu, Y. Q. and Tay, J. H. 2007. Characteristics and stability of aerobic granules cultivated with different starvation time. *Applied Microbiology and Biotechnology* 75: 205–210.

Liu, Y. Q., Wu, W. W., Tay, J. H., and Wang, J. L. 2007. Starvation is not a prerequisite for the formation of aerobic granules. *Applied Microbiology and Biotechnology* 76: 211–216.

Liu, Y., Xu, H., Yang, S. F., and Tay, J. H. 2003a. A general model for biosorption of Cd $^{2+}$,Cu $^{2+}$ and Zn $^{2+}$ by aerobic granules. *Journal of Biotechnology* 102: 233–239.

Liu, Y., Yang, S. F., Tan, S. F., Lin, Y. M., and Tay, J. H. 2002. Aerobic granules: A novel zinc biosorbent. *Letters in Applied Microbiology* 35: 548–551.

Liu, Y., Yang, S. F., Xu, H., Woon, K. H., Lin, Y. M., and Tay, J. H. 2003b. Biosorption kinetics of cadmium (II) on aerobic granular sludge. *Process Biochemistry* 38: 997–1001.

Lloyd, J. R. and Renshaw, J. C. 2005. Bioremediation of radioactive waste: Radionuclide–microbe interactions in laboratory and field-scale studies. *Current Opinion in Biotechnology* 16: 254–260.

Luo, Y. R., Kang, S. G., Kim, S. J., Kim, M. R., Li, N., Lee, J. H., and Kwon, K. K. 2012. Genome sequence of benzo (a) pyrene-degrading bacterium *Novosphingobium pentaromativorans* US6–1. *Journal of Bacteriology* 194: 907–907.

Ma, D. Y., Wang, X. H., Song, C., Wang, S. G., Fan, M. H., and Li, X. M. 2011. Aerobic granulation for methylene blue biodegradation in a sequencing batch reactor. *Desalination* 276: 233–238.

Mangwani, N., Dash, H. R., Chauhan, A., and Das, S. 2012. Bacterial quorum sensing: Functional features and potential applications in biotechnology. *Journal of Molecular Microbiology and Biotechnology* 22: 215–227.

Martínková, L., Uhnáková, B., Pátek, M., Nešvera, J., and Křen, V. 2009. Biodegradation potential of the genus *Rhodococcus*. *Environment International* 35: 162–177.

Matsumoto, S., Katoku, M., Saeki, G., Terada, A., Aoi, Y., Tsuneda, S., and Van Loosdrecht, M. 2010. Microbial community structure in autotrophic nitrifying granules characterized by experimental and simulation analyses. *Environmental Microbiology* 12: 192–206.

Maximova, N. and Dahl, O. 2006. Environmental implications of aggregation phenomena: Current understanding. *Current Opinion in Colloid & Interface Science* 11: 246–266.

McSwain, B. S., Irvine, R. L., and Wilderer, P. A. 2004. The effect of intermittent feeding on aerobic granule structure. *Water Science & Technology* 49: 19–25.

Mirón, A. S., Garcia Camacho, F., Contreras Gomez, A., Grima, E. M., and Chisti, Y. 2000. Bubble-column and airlift photobioreactors for algal culture. *AIChE Journal* 46: 1872–1886.

Mishima, K. and Nakamura, M. 1991. Self-immobilization of aerobic activated sludge—A pilot study of the aerobic upflow sludge blanket process in municipal sewage treatment. *Water Science and Technology* 23: 981–990.

Moran, B. N. and Hickey, W. J. 1997. Trichloroethylene biodegradation by mesophilic and psychrophilic ammonia oxidizers and methanotrophs in groundwater microcosms. *Applied and Environmental Microbiology* 63: 3866–3871.

Morgenroth, E., Sherden, T., Van Loosdrecht, M. C. M., Heijnen, J. J., and Wilderer, P. A. 1997. Aerobic granular sludge in a sequencing batch reactor. *Water Research* 31: 3191–3194.

Moshe, S. S. B., Ronen, Z., Dahan, O., Weisbrod, N., Groisman, L., Adar, E., and Nativ, R. 2009. Sequential biodegradation of TNT, RDX and HMX in a mixture. *Environmental Pollution* 157: 2231–2238.

Muda, K., Aris, A., Salim, M. R., Ibrahim, Z., van Loosdrecht, M. C., Ahmad, A., and Nawahwi, M. Z. 2011. The effect of hydraulic retention time on granular sludge biomass in treating textile wastewater. *Water Research* 45: 4711–4721.

Munoz, R. and Guieysse, B. 2006. Algal–bacterial processes for the treatment of hazardous contaminants: A review. *Water Research* 40: 2799–2815.

Nancharaiah, Y. V., Dodge, C., Venugopalan, V. P., Narasimhan, S. V., and Francis, A. J. 2010. Immobilization of Cr (VI) and its reduction to Cr (III) phosphate by granular biofilms comprising a mixture of microbes. *Applied and Environmental Microbiology* 76: 2433–2438.

Nancharaiah, Y. V., Joshi, H. M., Mohan, T. V. K., Venugopalan, V. P., and Narasimhan, S. V. 2006. Aerobic granular biomass: A novel biomaterial for efficient uranium removal. *Current Science* 9: 503–509.

Nancharaiah, Y. V., Schwarzenbeck, N., Mohan, T. V. K., Narasimhan, S. V., Wilderer, P. A., and Venugopalan, V. P. 2006. Biodegradation of nitrilotriacetic acid (NTA) and ferric–NTA complex by aerobic microbial granules. *Water Research* 40: 1539–1546.

Nancharaiah, Y. V. and Venugopalan, V. P. 2011. Denitrification of synthetic concentrated nitrate wastes by aerobic granular sludge under anoxic conditions. *Chemosphere* 85: 683–688.

Nicolella, C., Van Loosdrecht, M. C. M., and Heijnen, J. J. 2000. Wastewater treatment with particulate biofilm reactors. *Journal of Biotechnology* 80: 1–33.

Ogbonna, J. C., Soejima, T., and Tanaka, H. 1999. An integrated solar and artificial light system for internal illumination of photobioreactors. *Journal of Biotechnology* 70: 289–297.

Oh, B. T., Sarath, G., and Shea, P. J. 2001. TNT nitroreductase from a Pseudomonas aeruginosa strain isolated from TNT-contaminated soil. *Soil Biology and Biochemistry* 33: 875–881.

Orji, F. A., Ibiene, A. A., and Dike, E. N. 2012. Laboratory scale bioremediation of petroleum hydrocarbon—Polluted mangrove swamps in the Niger Delta using cow dung. Malaysian *Journal of Microbiology* 8: 219–228.

Paca, J., Halecky, M., Barta, J., and Bajpai, R. 2009. Aerobic biodegradation of 2, 4-DNT and 2, 6-DNT: Performance characteristics and biofilm composition changes in continuous packed-bed bioreactors. *Journal of Hazardous Materials* 163: 848–854.

Pandey, G. and Jain, R. K. 2002. Bacterial chemotaxis toward environmental pollutants: Role in bioremediation. *Applied and Environmental Microbiology* 6(12): 5789–5795.

Parales, R. E. and Haddock, J. D. 2004. Biocatalytic degradation of pollutants. *Current Opinion in Biotechnology* 15: 374–379.

Parsek, M. R. and Greenberg, E. P. 2005. Socio-microbiology: The connections between quorum sensing and biofilms. *Trends in Microbiology* 13: 27–33.

Paul, D., Pandey, G., Pandey, J., and Jain, R. K. 2005. Accessing microbial diversity for bioremediation and environmental restoration. *Trends in Biotechnology* 23: 135–142.

Peng, D. C., N. Bernet, J. P. Delgenes and Bernet, N., Delgenes, J. P., and Moletta, R. 1999. Aerobic granular sludge—A case report. *Water Research* 33: 890–893.

Perumbakkam, S., Hess, T.F., and Crawford, R. L. 2006. A bioremediation approach using natural transformation in pure-culture and mixed-population biofilms. *Biodegradation* 17: 545–557.

Podlipná, R., Pospíšilová, B., and Vaněk, T. 2015. Biodegradation of 2, 4-dinitrotoluene by different plant species. *Ecotoxicology and Environmental Safety* 112: 54–59.

Pointing, S. 2001. Feasibility of bioremediation by white-rot fungi. *Applied Microbiology and Biotechnology* 57: 20–33.

Pol, L. H., de Castro Lopes, S. I., Lettinga, G., and Lens, P. N. L. 2004. Anaerobic sludge granulation. *Water Research* 38: 1376–1389.

Puhakka, J. A., Herwig, R. P., Koro, P. M., Wolfe, G. V., and Ferguson, J. F. 1995. Biodegradation of chlorophenols by mixed and pure cultures from a fluidized-bed reactor. *Applied Microbiology and Biotechnology* 42: 951–957.

Raj, K., Prasad, K. K., and Bansal, N. K. 2006. Radioactive waste management practices in India. *Nuclear Engineering and Design* 236: 914–930.

Rhodes, C. J. 2012. Feeding and healing the world: Through regenerative agriculture and permaculture. *Science Progress* 95: 345–446.

Rhodes, C. J. 2013. Applications of bioremediation and phytoremediation. *Science Progress* 96: 417–427.

Roberts, A. P. and Mullany, P. 2006. Genetic basis of horizontal gene transfer among oral bacteria. *Periodontology* 42: 36–46.

Roeselers, G., Van Loosdrecht, M. C. M., and Muyzer, G. 2007. Heterotrophic pioneers facilitate phototrophic biofilm development. *Microbial Ecology* 54: 578–585.

Roeselers, G., Van Loosdrecht, M. C. M., and Muyzer, G. 2008. Phototrophic biofilms and their potential applications. *Journal of Applied Phycology* 20: 227–235.

Romani, A. M. and Sabater, S. 2000. Influence of algal biomass on extracellular enzyme activity in river biofilms. *Microbial Ecology* 40: 16–24.

Sakai, Y., Fukase, T., Yasui, H., and Shibata, M. 1997. An activated sludge process without excess sludge production. *Water Science and Technology* 36: 163–170.

Sasek, V. and Cajthami, T. 2005. Myco-remediation: Current state and perspectives. *International Journal of Medicinal Mushrooms* 7: 360–361.

Schumacher, G., Blume, T., and Sekoulov, I. 2003. Bacteria reduction and nutrient removal in small wastewater treatment plants by an algal biofilm. *Water Science and Technology* 47: 195–202.

Schwarzenbeck, N., Borges, J. M., and Wilderer, P. A. 2005. Treatment of dairy effluents in an aerobic granular sludge sequencing batch reactor. *Applied Microbiology and Biotechnology* 66: 711–718.

Schwarzenbeck, N., Erley, R., Mc Swain, B. S., Wilderer, P. A., and Irvine, R. L. 2004a. Treatment of malting wastewater in a granular sludge sequencing batch reactor (SBR). *Acta Hydrochimica et Hydrobiologica* 32: 16–24.

Schwarzenbeck, N., Erley, R., and Wilderer, P. A. 2004b. Aerobic granular sludge in an SBR-system treating wastewater rich in particulate matter. *Water Science and Technology* 49: 41–46.

Seo, J.S., Keum, Y.S., and Li, Q. X. 2009. Bacterial degradation of aromatic compounds. *International Journal of Environmental Research and Public Health* 6: 278–309.

Sharma, N. K., Tiwari, S. P., Tripathi, K., and Rai, A. K. 2011. Sustainability and cyanobacteria (blue-green algae): Facts and challenges. *Journal of Applied Phycology* 23: 1059–1081.

Shimada, K., Itoh, Y., Washio, K., and Morikawa, M. 2012. Efficacy of forming biofilms by naphthalene degrading *Pseudomonas stutzeri* T102 toward bioremediation technology and its molecular mechanisms. *Chemosphere* 87: 226–233.

Shin, H. S., Lim, K. H., and Park, H. S. 1992. Effect of shear-stress on granulation in oxygen aerobic upflow sludge bed reactors. *Water Science and Technology* 26: 601–605.

Singh, R., Paul, D., and Jain, R. K. 2006. Biofilms: Implications in bioremediation. *Trends in Microbiology* 14: 389–397.

Stoodley, P., Sauer, K., Davies, D., and Costerton, J. 2002. Biofilms as complex differentiated communities. *Annual Reviews in Microbiology* 56: 187–209.

Strauss, E. A. 2005. Microbiology of lotic aggregates and biofilms. In Lehr, J. H. and Keeley, J. (Eds.), *Water Encyclopedia: Surface and Agricultural Water.* New York: John Wiley & Sons, Inc., pp. 305–309.

Subashchandrabose, S. R., Ramakrishnan, B., Megharaj, M., Venkateswarlu, K., and Naidu, R. 2011. Consortia of cyanobacteria/microalgae and bacteria. *Biotechnology Advances* 29: 896–907.

Sudha, M., Saranya, A., Selvakumar, and G. Sivakumar, N. 2014. Microbial degradation of azo dyes: A review. *International Journal of Current Microbiology and Applied Sciences* 3: 670–690.

Sun, X. F., Ma, Y., Liu, X. W., Wang, S. G., Gao, B. Y., and Li, X. M. 2010. Sorption and detoxification of chromium (VI) by aerobic granules functionalized with polyethylenimine. *Water Research* 44: 2517–2524.

Sutherland, I. W. 2001. The biofilm matrix—An immobilized but dynamic microbial environment. *Trends in Microbiology* 9: 222–227.

Tay, J. H., Liu, Q. S., and Liu, Y. 2001. Microscopic observation of aerobic granulation in sequential aerobic sludge blanket reactor. *Journal of Applied Microbiology* 91: 168–175.

Tay, S. L., Ivanov, V., Yi, S., Zhuang, W. Q., and Tay, J. H. 2002. Presence of anaerobic bacteroides in aerobically grown microbial granules. *Microbial Ecology* 44: 278–285.

Tay, S. T. L., Zhuang, W. Q., and Tay, J. H. 2005. Start-up, microbial community analysis and formation of aerobic granules in a tert-butyl alcohol degrading sequencing batch reactor. *Environmental Science & Technology* 39: 5774–5780.

Thomassin-Lacroix, E., Eriksson, M., Reimer, K., and Mohn, W. 2002. Biostimulation and bioaugmentation for on-site treatment of weathered diesel fuel in Arctic soil. *Applied Microbiology and Biotechnology* 59: 551–556.

Thompson, P. L., Ramer, L. A., and Schnoor, J. L. 1998. Uptake and transformation of TNT by hybrid poplar trees. *Environmental Science & Technology* 32: 975–980.

Tiwari, D., Prasad, S. K., Yang, J. K., Choi, B. J., and Lee, S. M. (2006). Inorganic and bio-materials in the removal/speciation of radiocesium and radiostrontium: An overview. *Environmental Engineering Research* 11: 106–125.

Tlili, A., Maréchal, M., Bérard, A., Volat, B., and Montuelle, B. 2011. Enhanced co-tolerance and co-sensitivity from long-term metal exposures of heterotrophic and autotrophic components of fluvial biofilms. *Science of the Total Environment* 409: 4335–4343.

Toh, S., Tay, J., Moy, B., Ivanov, V., and Tay, S. 2003. Size-effect on the physical characteristics of the aerobic granule in a SBR. *Applied Microbiology and Biotechnology* 60: 687–695.

Van der Star, W. R., Abma, W. R., Blommers, D., Mulder, J. W., Tokutomi, T., Strous, M., and van Loosdrecht, M. C. 2007. Startup of reactors for anoxic ammonium oxidation: Experiences from the first full-scale anammox reactor in Rotterdam. *Water Research* 41: 4149–4163.

Vidali, M. 2001. Bioremediation: An overview. *Pure and Applied Chemistry* 73: 1163–1172.

Villas Boas, S., Bruheim, P. 2007. The potential of metabolomics tools in bioremediation studies. *OMICS A Journal of Integrative Biology*: 11: 305–313.

Wang, S., Zhang, S., Zhou, W., and Zhu, J. 2010. Acrylic wastewater treatment using SBR technology with aerobic granules. *International Journal of Environment and Waste Management* 7: 80–89.

Wattal, P. K. 2013. Indian program on radioactive waste management. *Sadhana* 38: 849–857.

Wichern, M., Lübken, M., and Horn, H. 2008. Optimizing sequencing batch reactor (SBR) reactor operation for treatment of dairy wastewater with aerobic granular sludge. *Water Science & Technology* 58: 6.

Xavier, J. B. and Foster, K. R. 2007. Cooperation and conflict in microbial biofilms. *Proceedings of the National Academy of Sciences* 104: 876–881.

Xavier, K. B. and Bassler, B. L. 2005. Interference with AI-2-mediated bacterial cell–cell communication. *Nature* 437: 750–753.

Xu, Z., Baicheng, Z., Yiping, Z., Zhaoling, C., Wei, C., and Fan, O. 2002. A simple and low-cost airlift photobioreactor for microalgal mass culture. *Biotechnology Letters* 24: 1767–1771.

Yang, S. F., Tay, J. H., and Liu, Y. 2003. A novel granular sludge sequencing batch reactor for removal of organic and nitrogen from wastewater. *Journal of Biotechnology* 106: 77–86.

Zhang, D., Pan, X., Mostofa, K. M., Chen, X., Mu, G., Wu, F., and Fu, Q. (2010). Complexation between Hg (II) and biofilm extracellular polymeric substances: An application of fluorescence spectroscopy. *Journal of Hazardous Materials* 175: 359–365.

Zhang, H., Dang, Z., Zheng, L. C., and Yi, X. Y. 2009. Remediation of soil co-contaminated with pyrene and cadmium by growing maize (*Zea mays* L.). *International Journal of Environmental Science & Technology* 6: 249–258.

Zhang, H., He, Y., Jiang, T., and Yang, F. 2011a. Research on characteristics of aerobic granules treating petrochemical wastewater by acclimation and co-metabolism methods. *Desalination* 279: 69–74.

Zhang, L. L., Feng, X. X., Xu, F., Xu, S., and Cai, W. M. 2005. Biosorption of rare earth metal ion on aerobic granules. *Journal of Environmental Science and Health* 40: 857–867.

Zhang, M., Zhao, Q., and Ye, Z. 2011b. Organic pollutants removal from 2, 4, 6-trinitrotoluene (TNT) red water using low cost activated coke. *Journal of Environmental Sciences* 23: 1962–1969.

Zhang, W., Chen, L., and Liu, D. 2012. Characterization of a marine-isolated mercury-resistant *Pseudomonas putida* strain SP1 and its potential application in marine mercury reduction. *Applied Microbiology and Biotechnology* 93: 1305–1314.

Zhu, J. and Mekalanos, J. J. 2003. Quorum sensing-dependent biofilms enhance colonization in *Vibrio cholerae*. *Developmental Cell* 5: 647–656.

Zippel, B. and Neu, T. R. 2005. Growth and structure of phototrophic biofilms under controlled light conditions. *Water Science and Technology* 52: 203–209.

48 Magnetically Responsive Microbial Cells for Metal Ions Removal and Detection

Ivo Safarik, Kristyna Pospiskova, Eva Baldikova, and Mirka Safarikova

CONTENTS

ABSTRACT

Prokaryotic and eukaryotic cells can interact with various types of magnetic nanoparticles and microparticles. Due to the presence of magnetic particles on the cell surface, in protoplasm or in intracellular organelles, magnetically modified cells can be rapidly, easily, and selectively separated from desired environments by means of magnetic separators. The cell surface contains a wide variety of functional groups that can be efficiently employed in adsorption processes. Magnetically responsive cells are very interesting and easily obtainable biomaterials that are highly usable for the removal and detection of metal ions. Such types of biosorbents can be very promising in environmental technologies in the near future.

48.1 INTRODUCTION

Human activity, as well as natural geochemical processes, is accompanied by the pollution of water, soil, and atmosphere by huge amounts of inorganic and organic pollutants. Toxic metals are extensively employed in electronics, machine construction, high-tech technologies, and many products used in everyday life, such as batteries and accumulators. As a result, toxic metals enter water resources and cause harmful pollution. The main sources of contamination include mining wastes, landfill leaches, municipal wastewater, urban runoff, and industrial wastewaters, particularly from the electroplating, electronic, and metal-finishing industries. Metal concentrations in many aquatic environments are substantially higher than official limits. Heavy metal ions may cause serious health problems in human beings; in water sources, they can pose a threat to a variety of fish and invertebrates. Large acute doses can lead to harmful, even fatal effects. Because of these reasons, heavy metal ions have to be removed to very low levels from water resources (Sharma, 2015).

Various treatment methods for the removal of target metal ions from industrial wastewater have been developed. A number of traditional treatment techniques including ion exchange, chemical precipitation, electrochemical precipitation, adsorption, membrane filtration, reverse osmosis, phytoremediation, photocatalysis, or complexation have been described and used. However, the necessity to reduce the amount of heavy metal ions in wastewater streams has led to an increasing interest in alternative procedures (Sharma, 2015).

48.2 BIOSORPTION

During the last several decades, new processes for the removal of both organic and inorganic pollutants have been developed and tested. A huge amount of procedures is based on the use of materials of biological origin, acting as adsorbents (biosorbents). In general, biosorption has been defined as the property of certain biomaterials to sequester metal ions or other molecules from aqueous solutions. During the removal of metal ions, different processes including transport through the cell membrane, biosorption on cell walls, entrapment in the extracellular capsule, and oxidation/reduction reactions have been observed in nonliving and living microorganisms. Selectivity in removing the desired metal ions is an added advantage of bio-based separation techniques. These techniques have been proved to be some of the most economical and eco-friendly procedures for the removal of metal ions. Biosorption of metal ions has thus become an intensively studied field of research in environmental science and technology.

Diverse types of microalgae, bacteria, fungi, and yeasts have been used as efficient biosorbents of metal ions. In addition to cultivated microbial cells, also low-cost waste biomass can be efficiently used. The metal uptake capacity of various biological materials has usually been evaluated during batch adsorption experiments; equilibrium adsorption isotherms of different types are created and, if possible, maximum adsorption capacities are calculated. The effect of various reaction conditions, including contact time, pH, temperature, biomass loading, ionic strength, etc., has also been studied extensively, as well as detailed adsorption kinetic studies. Both living and dead microbial biomass can be used as a biosorbent; the optimum form depends on the specific application.

Various mechanisms have been suggested for metal ions removal from model solutions and wastewater using microbial cells. Metal ions can be biosorbed onto the binding sites present in the cellular structure; this process is known as passive uptake. Metal ions can also pass inside the cell across the cell membrane through the cell metabolic cycle; this mode of metal uptake is known as active uptake. The metal uptake by both active and passive modes can be described as bioaccumulation (Sharma, 2015).

In order to simplify manipulation with biosorbents in difficult-to-handle environments, microbial cells are magnetically modified to get magnetically responsive derivatives, which can be easily separated using an appropriate magnetic separator.

48.3 PREPARATION OF MAGNETICALLY RESPONSIVE MICROBIAL CELLS

An absolute majority of prokaryotic and eukaryotic cells exhibit diamagnetic properties. Nevertheless, there is an extraordinary bacterial group called magnetotactic bacteria (MTB) that can synthesize intracellular biogenic magnetic nanoparticles (based either on magnetite (Fe_3O_4) or on greigite (Fe_3S_4)) that enable their magnetic separation and movement; it is generally assumed that magnetosomes (organelles containing individual magnetic nanoparticles covered with a lipid bilayer) are involved in magnetoreception (Schüler, 2007).

Many procedures enabling conversion of diamagnetic cells into their magnetically responsive derivatives have been described recently (Safarik et al., 2014, 2015a). Magnetic modification of cells is based on their interaction with an appropriate magnetic label, usually nanoparticles and microparticles of magnetite (Fe_3O_4) or maghemite (γ-Fe_2O_3); in some cases, ferrite particles (Lee et al., 2004) or chromium dioxide particles (Widjojoatmodjo et al., 1993) have also been used. Alternatively, the modification can be performed by binding paramagnetic cations on acid groups on the cell surface (Zborowski et al., 1992) or by the binding of magnetoferritin (Zborowski et al., 1996) on the cell surface. In general, microbial cells can especially be modified by the nonspecific attachment of magnetic iron oxide nanoparticles (e.g., by the magnetic fluid treatment) (Safarikova et al., 2009), by binding of magnetic microparticles on the cell surface (Pospiskova et al., 2013), by specific interactions with immunomagnetic nanoparticles and microparticles (Safarik and Safarikova, 1999), by covalent immobilization on magnetic carriers (Safarik et al., 2015b), by cross-linking of the cells or isolated cell walls with a bifunctional reagent in the presence of magnetic particles (Patzak et al., 1997), or by entrapment (together with magnetic particles) into biocompatible polymers (Safarik et al., 2008). The individual modification procedures enabling preparation of magnetically responsive microbial biosorbents will be described in more detail in further text.

Magnetic modification of microbial cells can be efficiently performed using appropriate magnetic fluid (ferrofluid).

In the simplest way, perchloric acid–stabilized magnetic fluid was mixed with baker's or brewer's yeast cells washed with and suspended in acetate buffer, pH 4.6, or in glycine–HCl buffer, pH 2.2; alternatively, tetramethylammonium hydroxide–stabilized magnetic fluid was used for baker's yeast cells modification in 0.1 M glycine–NaOH buffer, pH 10.6. After a short time period, magnetic particles precipitated on the cell surface (Figure 48.1) (Safarikova et al., 2009). The modified cells can be heated in a boiling water bath to kill the cells; a stable adsorbent for the removal of selected organic and inorganic xenobiotics can be prepared (Safarik et al., 2002). Magnetic modification of dried *Kluyveromyces marxianus* (fodder yeast; Figure 48.2) and *Chlorella vulgaris* cells required thorough washing with 0.1 M acetic acid to remove a substantial portion of soluble macromolecules that otherwise caused spontaneous precipitation of magnetic fluid; after washing and suspending the cells in acetic acid solution, the addition of perchloric acid–stabilized magnetic fluid resulted in the formation of magnetically modified yeast and algae cells (Safarik et al., 2007, 2008).

A simple procedure for the magnetic modification of yeast and algae cells based on the use of microwave-synthesized magnetic iron oxide nanoparticles and microparticles has been developed recently. The particles are usually prepared from ferrous sulfate heptahydrate and sodium or potassium hydroxide; after their mixing and precipitation of mixed iron hydroxides, the suspension underwent the microwave treatment in a regular kitchen microwave oven (700 W, 2450 MHz) and nanoparticles and microparticles of magnetic iron oxides formed (Zheng et al., 2010; Pospiskova et al., 2013). Mixing of magnetic particles with algae cells (*Chlorella vulgaris*) and yeast cells (*Saccharomyces cerevisiae*) suspensions caused cell flocculation and magnetically responsive cell aggregates (usually ca. 100–300 μm in diameter) were formed (Figure 48.3) (Pospiskova et al., 2013; Prochazkova et al., 2013). Alternatively, commercial magnetite microparticles were used to capture bacterial cells (Sze et al., 1996; Wong and Fung, 1997) and fungal biomass (Wainwright et al., 1990). Another adsorbent for capturing bacterial cells was prepared from granular activated carbon after its modification with Mn ferrite (Podder and Majumder, 2016).

In general, microbial cells are covalently bound on magnetic carriers by means of an appropriate coupling agent (e.g., aminosilane, carbodiimide, glutaraldehyde) to introduce a specific group on the carrier surface, which subsequently can interact with reactive groups on the cell surface. In a typical example, magnetic chitosan particles activated by glutaraldehyde have been utilized for *Saccharomyces cerevisiae* immobilization (Figure 48.4) (Safarik et al., 2015b). Alternatively, magnetic cellulose microparticles were activated with periodic acid and subsequently utilized for the immobilization of yeast cells (Ivanova et al., 2011). Silanized magnetite nanoparticles, activated by (3)-aminopropyltriethoxysilane followed by glutaraldehyde treatment, were covalently bound to cells of *Bacillus circulans* (Safarikova et al., 2007). An iron-based ammonia synthesis catalyst covered by a stable film of amino groups containing epoxy resin was used for *Saccharomyces*

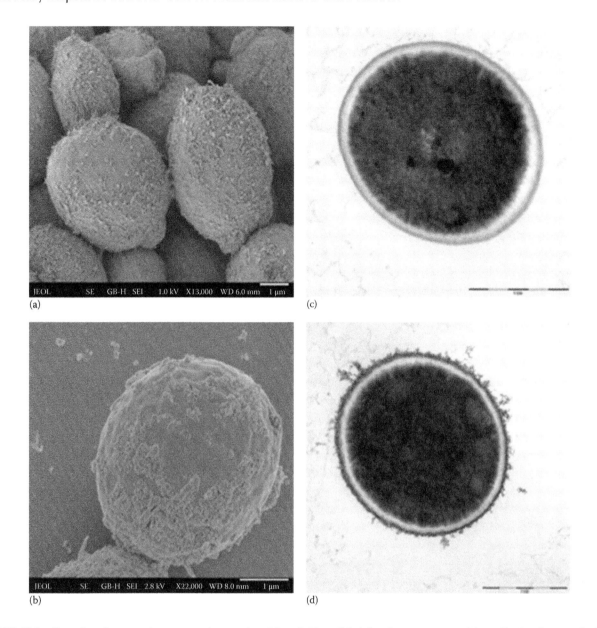

FIGURE 48.1 Scanning electron microscopy micrographs of ferrofluid modified *Saccharomyces cerevisiae* cells showing attached magnetic nanoparticles and their aggregates on the cell surface (a, b; bars: 1 μm). Transmission electron microscopy micrographs of native *S. cerevisiae* cells (c; bar: 1 μm) and ferrofluid modified cells with attached magnetic iron oxide nanoparticles on the cell wall (d; bar: 1 μm). (Reproduced from Safarikova, M. et al., *Food Res. Int.*, 42, 521, 2009. With permission.)

cerevisiae cells immobilization after glutaraldehyde activation (Ivanova et al., 1996).

Microbial cell walls contain free amino and/or carboxyl groups, which can easily be cross-linked by bifunctional or multifunctional reagents such as glutaraldehyde or toluene diisocyanate. The cells are usually cross-linked in the presence of an inert protein like gelatine, albumin, raw hen egg white, and collagen. If magnetic particles are used throughout the cross-linking process, magnetic cells or cell walls' derivatives can be prepared (Patzak et al., 1997).

Microbial cells can be entrapped in natural or biocompatible synthetic carriers (gels) that can be formed by various mechanisms, namely, polymerization (e.g., polyacrylamide, polymethacrylate), cross-linking (e.g., proteins), polycondensation (polyurethane, epoxy resins), thermal gelation (e.g., gelatine, agar, agarose), ionotropic gelation (e.g., alginate, chitosan), and precipitation (cellulose, cellulose triacetate). The gel is formed in the presence of the cells and appropriate magnetic materials. There are various methods available to obtain particles (beads) containing entrapped cells and magnetic particles (Safarik et al., 2014).

Magnetically responsive alginate beads containing entrapped *S. cerevisiae* cells and magnetite microparticles were prepared; larger beads (2–3 mm in diameter) were formed by dropping the mixture into a calcium chloride solution, while microbeads (50–100 μm) were prepared using the water-in-oil emulsification process (Figure 48.5). The presence of magnetic material had no negative effect

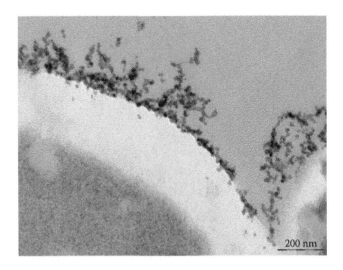

FIGURE 48.2 Transmission electron microscopy picture of magnetically modified dried fodder yeast (*Kluyveromyces marxianus*) cells (bar—200 nm). (Reproduced from Safarik, I. et al., *Enzyme Microb. Technol.*, 40, 1551, 2007. With permission.)

on the activity of cells (Safarik et al., 2008). Bacterial alkaliphilic cells *Amphibacillus* sp. KSUCr3 were immobilized in silica-coated magnetic alginate gel beads and applied for detoxification of hexavalent chromate. In comparison with the cells immobilized into a nonmagnetic matrix, the magnetic beads with cells showed approximately 16% higher reduction activity. Coating of magnetic alginate beads with a dense silica layer (using sol-gel procedure; the silica layer was deposited by addition of ammonia and tetraethyl orthosilicate to dispersed beads) improved the physical and mechanical properties and thermal stability of immobilized cells (Ibrahim et al., 2013).

Bacterial cells of *Pseudomonas delafieldii* R-8 were immobilized in magnetic polyvinyl alcohol beads and utilized for biodesulfurization. The suspension of cells in phosphate buffer was mixed with aqueous solution of PVA and oleic acid stabilized magnetic fluid and subsequently dropped into liquid nitrogen for quick freezing. The formed beads with immobilized cells underwent thawing by slow increase of the temperature under vacuum (Guobin et al., 2005).

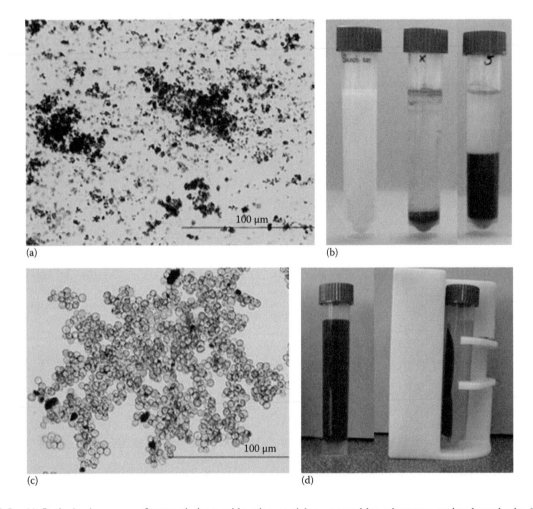

(a)

(b)

(c)

(d)

FIGURE 48.3 (a) Optical microscopy of magnetic iron oxide microparticles prepared by microwave-assisted synthesis; (b) process of magnetic modification of yeast cells (left tube, *Saccharomyces cerevisiae* cells suspension; middle tube, sedimented iron oxide microparticles for magnetic modification; right tube, sedimented magnetically modified yeast cells); (c) optical microscopy of *S. cerevisiae* cells modified by iron oxide microparticles; (d) magnetic separation of magnetically modified yeast cells. (Reproduced from Pospiskova, K. et al., *Lett. Appl. Microbiol.*, 56, 456, 2013. With permission.)

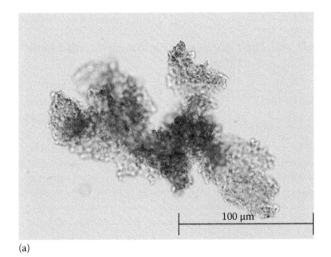

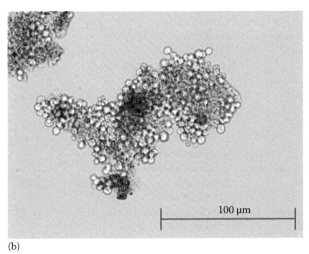

FIGURE 48.4 (a) Optical microscopy images of microwave-synthesized magnetic chitosan microparticles and (b) magnetic chitosan microparticles with immobilized *Saccharomyces cerevisiae* cells. (Reproduced from Safarik, I. et al., *Yeast*, 32, 239, 2015b. With permission.)

48.4 USE OF MAGNETICALLY RESPONSIVE MICROBIAL CELLS FOR METAL IONS REMOVAL AND DETECTION

Various magnetically modified prokaryotic and eukaryotic cells, as well as selected strains of MTB, have been tested as possible biosorbents for a range of metal ions (Tables 48.1 and 48.2). Copper, nickel, cadmium, lead, chromium, mercury, iron, gold, silver, manganese, uranium, arsenic, strontium, plutonium, and cerium ions have served as the tested adsorbates. Most experiments have been performed with model metal ions solutions in a laboratory scale; when the biosorption potential using real industrial wastewater was tested, the efficiency was usually substantially lower.

Yeast biomass represents an important and promising material for metal ions biosorption. Yeast cells of the genus *Saccharomyces* are nonpathogenic, are easily available, and enable simple manipulation. *Saccharomyces cerevisiae* cells (both baker's and brewer's yeasts) can be magnetically modified by contact with perchloric acid–stabilized magnetic fluid (Safarikova et al., 2009; Uzun et al., 2011) or other magnetic nanoparticles (Gorobets et al., 2013; Pospiskova et al., 2013). Alternatively, yeast cells can be magnetically modified with magnetic chitosan particles (Saifuddin and Dinara, 2012; Safarik et al., 2015b), bound to magnetic activated carbon (Abdel-Fattah et al., 2014; Mahmoud et al., 2015) or entrapped in magnetic biopolymer beads (Safarik et al., 2008). Green technologies, especially microwave irradiation, have also been employed during magnetic carriers' synthesis and cells' immobilization (Saifuddin and Dinara, 2012; Pospiskova et al., 2013; Safarik et al., 2015b). In some cases, the yeast cell wall was chemically treated with ethylenediaminetetraacetic dianhydride; this treatment resulted in a high number of carboxyl and amino groups on the cell surface (Xu et al., 2011; Zhang et al., 2011). In order to have stabilized product enabling work for a long period of time, dead yeast cells are

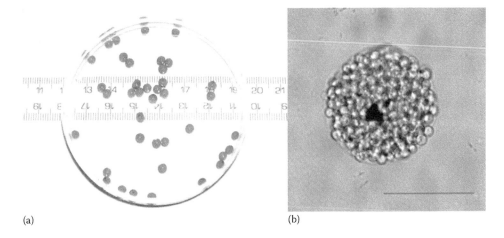

FIGURE 48.5 Magnetically responsive alginate beads containing entrapped *Saccharomyces cerevisiae* cells and magnetite microparticles. (a) Millimeter-sized beads and (b) microbeads. The scale bar corresponds to 50 μm. (Reproduced from Safarik, I. et al., *J. Agric. Food Chem.*, 56, 7925, 2008. With permission.)

TABLE 48.1

Examples of Magnetically Responsive Eukaryotic Microbial Cells and Their Applications as Magnetic Adsorbents for Metal Ions Removal

Microorganism	Magnetic Modification	Adsorbed Ion(s)	Matrix	Other Details	Reference
Baker's yeast	Cells adsorption on activated carbon/nano-Fe_3O_4 composite	Cr(VI) Cr(III)	Aqueous solutions	Removal of chromium species from various real water samples studied	Abdel-Fattah et al. (2014)
Baker's yeast	Cells adsorption on activated carbon/nano-Fe_3O_4 composite	Hg(II)	Model and real aqueous samples	Adsorption follows Langmuir model	Mahmoud et al. (2015)
Baker's yeast biomass	Cross-linked, EDTA-dianhydride modified cells with bound nano-Fe_3O_4	Ca(II) Cd(II) Pb(II)	Model aqueous solutions	Biosorbent with plenty of carboxyl and amino groups introduced by the EDTA molecules	Xu et al. (2011)
Baker's yeast biomass	Cross-linked, EDTA-dianhydride modified cells with bound nano-Fe_3O_4	Pd(II) Cd(II)	Model aqueous solutions	Biosorbent regeneration with 0.1 M HCl	Zhang et al. (2011)
Fungal mycelium (waste)	Composite from waste fungal mycelium, chitosan, and Fe_3O_4 nanoparticles	Cu(II)	Model aqueous solutions	Adsorption equilibrium data fit Langmuir isotherm equation; maximum adsorption capacity 46.25 mg/g	Ren et al. (2008b)
Fungal mycelium (waste)	Composite from waste fungal mycelium, chitosan, and Fe_3O_4 nanoparticles	Cu(II)	Model aqueous solutions	Adsorption equilibrium data fit Langmuir isotherm equation; maximum adsorption capacity 71.36 mg/g	Ren et al. (2008a)
Kluyveromyces marxianus	Treatment with water-based magnetic fluid stabilized with perchloric acid	Sr(II)	Model aqueous solutions	Adsorption equilibrium data fit Langmuir isotherm equation; maximum adsorption capacity 140.8 mg/g	Ji et al. (2010)
Rhizopus cohnii	Immobilization of mycelium and magnetic particles in the matrix of sodium alginate and polyvinyl alcohol	Cr(VI)	Model aqueous solutions	Adsorption equilibrium data fit Langmuir isotherm equation; maximum adsorption capacity 5.79 mg/g	Li et al. (2008)
Rhodotorula glutinis	Treatment with water-based magnetic fluid stabilized with perchloric acid	$UO_2(NO_3)_2$	Model aqueous solutions	Adsorption equilibrium data fit Langmuir isotherm equation; maximum adsorption capacity ca 190 mg/g	Bai et al. (2012)
Saccharomyces cerevisiae	Modification with iron nanoparticles (nZVI)	Ni(II)	Model aqueous solutions	Adsorption equilibrium data fit Langmuir isotherm equation; maximum adsorption capacity 54.23 mg/g	Guler and Ersan (2016)
Saccharomyces cerevisiae	Cells modification with magnetite nanoparticles	Fe(II)	Model aqueous solutions	Adsorption equilibrium data fit Langmuir isotherm equation; maximum adsorption capacity 230 mg/g	Gorobets et al. (2013)
Saccharomyces cerevisiae	Cells bound to cross-linked chitosan-magnetic nanoparticles beads	U(VI)	Model aqueous solutions	Adsorption equilibrium data fit Langmuir isotherm equation; maximum adsorption capacity 72.4 mg/g	Saifuddin and Dinara (2012)
Saccharomyces cerevisiae	Cells immobilized on the surface of chitosan-coated magnetic nanoparticles	Cu(II)	Model aqueous solutions	Adsorption equilibrium data fit Langmuir isotherm equation; maximum adsorption capacity 144.9 mg/g	Peng et al. (2010)
Saccharomyces cerevisiae subsp. *uvarum*	Treatment with water-based magnetic fluid stabilized with perchloric acid	Cu(II)	Model aqueous solutions	Adsorption equilibrium data fit Langmuir isotherm equation; maximum adsorption capacity 1.2 mmol/g	Uzun et al. (2011)
Saccharomyces cerevisiae subsp. *uvarum*	Treatment with water-based magnetic fluid stabilized with perchloric acid	Hg(II)	Model aqueous solutions and artificial wastewater	Adsorption equilibrium data fit Langmuir isotherm equation; maximum adsorption capacity 114.6 mg/g	Yavuz et al. (2006)
Saccharomyces cerevisiae cell walls	Cell walls with covalently and noncovalently bound magnetite	Cu(II) Cd(II) Ag(I)	Model aqueous solutions	Maximum adsorption capacities 50 (Cu), 25 (Cd), and 45 (Ag) μmol/g	Patzak et al. (1997)
Yarrowia lipolytica	Cells modified with phyto-inspired Fe^0/Fe_3O_4 nanoparticles	Cr(VI)	Model aqueous solutions	Adsorption equilibrium data fit Langmuir isotherm equation; maximum adsorption capacity 156.3 mg/g	Rao et al. (2013)
Yeast	Magnetic chitosan/yeast composite	Ce(III)	Aqueous solutions	Adsorption equilibrium data fit Langmuir isotherm equation; maximum adsorption capacity 73.53 mg/g	Ou et al. (2013)
Yeast biomass	Ethylenediamine-modified yeast biomass coated with magnetic chitosan microparticles	Pb(II)	Model aqueous solutions	Adsorption equilibrium data fit Langmuir isotherm equation; maximum adsorption capacity 121.26 mg/g	Li et al. (2013)

TABLE 48.2

Examples of Magnetically Responsive Prokaryotic Microbial Cells and Their Applications as Magnetic Adsorbents for Metal Ions Removal or for Bioreduction

Microorganism	Magnetic Modification	Adsorbed Ion(s)	Matrix	Other Details	Reference
Amphibacillus sp.	Cells immobilized in silica-coated magnetic alginate beads	Cr(VI)	Model aqueous solution	Bioreduction of Cr(VI) studied; magnetically immobilized cells showed 16% higher Cr(VI) reduction activity than nonmagnetic immobilized cells	Ibrahim et al. (2013)
Corynebacterium glutamicum	Cells adsorption on granular activated carbon/$MnFe_2O_4$ composite	As(III) As(V)	Model aqueous solution	Biosorption of both As(III) and As(V) was spontaneous and exothermic under studied conditions	Podder and Majumder (2016)
Desulfovibrio magneticus	Magnetotactic bacterium	Cd(II)	Model aqueous solution	Electron-dense particles present on cell surface when cultivated in the presence of cadmium ions	Arakaki et al. (2002)
Enterobacter sp.	Cells adsorption on commercial magnetite	Ni(II)	Model aqueous solution	Optimum adsorption at pH 9; elution of adsorbed Ni(II) with diluted citric acid	Wong and Fung (1997)
Escherichia coli	Cells functionalization by magnetic nanobeads	Cd(II) Hg(II)	Model aqueous solutions	Electrochemical biosensor for heavy metal detection constructed	Souiri et al. (2009)
Geobacillus galactosidasius	Cells adsorption on magnetite prepared from Fe^{2+} and Fe^{3+} ions	Pb(II) Cd(II)	Tap and mineral water, food extracts	Preconcentrations of Pb and Cd ions by solid-phase extraction before ICP-OES	Özdemir et al. (2016)
Magnetospirillum gryphiswaldense	Magnetotactic bacterium	Au(III)	Model aqueous solution	Reduction of gold ions to gold nanoparticles	Cai et al. (2011)
Magnetotactic bacterium	—	Pu	Extract from Ravenglass Estuary sediments, UK	Orientation magnetic separation used for cells collection	Bahaj et al. (1998)
Magnetotactic bacterium	—	Au(III) Cu(II)	Model aqueous solutions	Optimum adsorption at pH 1–5.5 for Au(III) and 2.0–4.5 for Cu(II)	Song et al. (2007)
Magnetotactic spirillum	—	Fe(III)	Model aqueous solutions	Orientation magnetic separation used for cells collection	Bahaj et al. (1994)
Pedomicrobium manganicum	Cells adsorption on magnetite particles prepared from crude ore	Mn(II)	Model aqueous solutions	Fluidized bed of magnetized biomass used for manganese removal from water	Sly et al. (1995)
Pseudomonas putida	Cells adsorption on commercial magnetite	Cu(II)	Electroplating effluent	Batch and continuous adsorption in a tank reactor; maximum adsorption capacity 14 and 13.4 mg/g, resp.	Sze et al. (1996)
Pseudomonas putida	Cells adsorption on commercial magnetite	Cu(II)	Industrial waste effluent	Pretreatment of cells by 0.6 M HCl greatly enhances the adsorption capacity	Wang et al. (2000)
Pseudomonas putida	Cells adsorption on commercial magnetite	Cu(II)	Synthetic wastewater	Pretreatment of cells by dilute HCl enhances the adsorption capacity	Chua et al. (1998)
Pseudomonas putida	Cells adsorption on commercial magnetite	Cu(II)	Cu(II)-bearing wastewater	Cu(II) accumulated on the surface of the cell walls; efficient desorption by the acidic treatment	Lei et al. (2000)
Stenotrophomonas sp.	Magnetotactic bacterium	Au(III)	Model aqueous solutions	Reduction of Au(III) to Au(0) by the reductants on the biomass; thiourea used for Au desorption	Song et al. (2008)

preferred. Fodder yeast cells (*Kluyveromyces marxianus*) are usually used in the dried (powdered) form enabling both simple magnetic modification and preparation of a low-cost adsorbent; treatment with magnetic fluid was successfully employed (Safarik et al., 2007).

Magnetically modified baker's, brewer's, and fodder yeast cells were tested as efficient adsorbents of various metal ions, including chromium, mercury, cadmium, lead, nickel, iron, uranium, strontium, and silver ones. The adsorption equilibrium data could usually be well fitted to the Langmuir isotherm. The yeast biomass containing adsorbed metal ions could be easily regenerated by diluted inorganic acids with high efficiency (Yavuz et al., 2006; Ji et al., 2010; Zhang et al., 2011).

Magnetically modified *Rhodotorula glutinis* was used to adsorb uranium from aqueous solutions. The presence of competing cations showed only little effect on uranium sorption. The sorption process was endothermic and spontaneous, implying that it becomes favorable at higher temperature (Bai et al., 2012).

The industrially important yeast *Yarrowia lipolytica* was used for Cr(VI) removal. Due to the cells modification by the phyto-inspired Fe^0/Fe_3O_4 nanocomposite, it was concluded that the yeast cells were efficiently involved in the biosorption of Cr(VI) that was followed by the reduction of Cr(VI) to Cr(III) due to the presence of Fe^{2+} reduction sites in the nanocomposite; the regeneration was caused by electron transfer from Fe^0 particles to Fe^{3+} of magnetite (Rao et al., 2013).

In addition to yeast cells, magnetically modified waste fungal mycelium and *Rhizopus cohnii* mycelium were also tested as adsorbents for copper and chromium ions removal (Li et al., 2008; Ren et al., 2008a,b).

Magnetically modified bacterial cells (especially of the genus *Pseudomonas*) were intensively studied as possible adsorbents for heavy metal ions. *Pseudomonas putida* strain isolated from heavy metal ions contaminated samples exhibited high affinity for Cu(II) ions; pretreatment of the cells with diluted hydrochloric acid (0.6 M) led to the increase in the adsorption capacity. Copper ions were mainly accumulated on the cell surface. EDTA solution (0.1 M) or 0.6 M HCl efficiently removed the adsorbed copper ions from the adsorbent (Sze et al. 1996; Chua et al., 1998; Lei et al., 2000; Wang et al., 2000).

Cr(VI)-reducing *Amphibacillus* cells were immobilized by entrapment in agar, agarose, alginate, or gelatine gels in the presence of Fe_3O_4 nanoparticles; alginate was selected as the best immobilization matrix. In comparison with nonmagnetic immobilized cells, the magnetically immobilized cells exhibited approximately 16% higher Cr(VI) reduction activity. To improve their physical and mechanical properties, the magnetic alginate beads were successfully coated with a dense silica layer using sol-gel chemistry (Ibrahim et al., 2013).

Cells of *Enterobacter* sp. immobilized on magnetite particles could remove large amounts of nickel ions from aqueous solutions. The optimal conditions to remove Ni(II) ions were at an alkaline pH. Complete recovery of adsorbed ions from immobilized cells was achieved by washing with diluted citric acid (Wong and Fung, 1997).

Magnetically modified bacterial cells can also be used for analytical purposes. *Geobacillus galactosidasius* cells modified with maghemite nanoparticles were used for the preconcentrations of Pb and Cd ions by solid-phase extraction before inductively coupled plasma-optical emission spectrometry measurement. Linear calibration curves were constructed in the concentration ranges of 1.0–60 ng/mL for both cations. Maximum adsorption capacities were 47.8 mg/g for Pb and 52.9 mg/g for Cd (Özdemir et al., 2016).

Several strains of MTB have also been tested as potential adsorbents of heavy metal ions and radionuclides.

The solution contaminated with plutonium interacted with MTB, and subsequently, orientation magnetic separation was used for cells collection (Bahaj et al., 1998). *Desulfovibrio magneticus* was used in the recovery of cadmium from water solutions (Arakaki et al., 2002). Cells of *Magnetospirillum gryphiswaldense* reduced gold ions to gold nanoparticles that became attached to the bacterial surface (Cai et al., 2011); a similar process was observed in the case of *Stenotrophomonas* sp. (Song et al., 2008).

48.5 SUMMARY

The potential of biosorbents applicable for the removal of heavy metals from aquatic environments has been widely investigated in the last three decades, motivated by the ever-increasing level of water pollution. The sorbents of natural origin are very useful especially for larger-scale applications where good sorption properties combined with lower price (in comparison with other sorbents) are required. Among them, prokaryotic and eukaryotic microbial cells form an intensively studied group of biosorbents enabling the efficient adsorption of heavy metals; sorption processes may also be accompanied with bioremediation. Microbial cells can be modified in order to prepare biosorbents with additional properties; for example, facilitated separation after their magnetic modification is very beneficial and can be one of the future trends in modern technologies improvement. In this chapter, preparation of magnetically responsive microorganisms and their applications for the removal of metal ions have been reviewed. Based on the presented examples of magnetically modified yeast and bacterial cells or fungal mycelia, these types of biosorbents are very promising in environmental technologies. However, their practical applications are not widespread, so further intensive research in this area could be expected in the future.

REFERENCES

Abdel-Fattah, T.M., Mahmoud, M.E., Osmam, M.M., Ahmed, S.B. 2014. Magnetically active biosorbent for chromium species removal from aqueous media. *Journal of Environmental Science and Health, Part A* 49: 1064–1076.

Arakaki, A., Takeyama, H., Tanaka, T., Matsunaga, T. 2002. Cadmium recovery by a sulfate-reducing magnetotactic bacterium, *Desulfovibrio magneticus* RS-1, using magnetic separation. *Applied Biochemistry and Biotechnology* 98: 833–840.

Bahaj, A.S., Croudace, I.W., James, P.A.B., Moeschler, F.D., Warwick, P.E. 1998. Continuous radionuclide recovery from wastewater using magnetotactic bacteria. *Journal of Magnetism and Magnetic Materials* 184: 241–244.

Bahaj, A.S., James, P.A.B., Croudace, I.W. 1994. Metal uptake and separation using magnetotactic bacteria.. *IEEE Transactions on Magnetics* 30: 4707–4709.

Bai, J., Wu, X., Fan, F., Tian, W., Yin, X., Zhao, L., Fan, F., Li, Z., Tian, L., Qin, Z. et al. 2012. Biosorption of uranium by magnetically modified *Rhodotorula glutinis*. *Enzyme and Microbial Technology* 51: 382–387.

Cai, F., Li, J., Sun, J., Ji, Y. 2011. Biosynthesis of gold nanoparticles by biosorption using *Magnetospirillum gryphiswaldense* MSR-1. *Chemical Engineering Journal* 175: 70–75.

Gorobets, S.V., Karpenko, Y.V., Kovalev, O.V., Olishevsky, V.V. 2013. Application of magnetically labeled cells *S. cerevisiae* as biosorbents at treatment plants. *Research Bulletin NTUU 'KPI'* 3: 42–47.

Guler, U.A., Ersan, M. 2016. *S. cerevisiae* cells modified with nZVI: A novel magnetic biosorbent for nickel removal from aqueous solutions. *Desalination and Water Treatment* 57: 7196–7208.

Guobin, S., Jianmin, X., Chen, G., Huizhou, L., Jiayong, C. 2005. Biodesulfurization using *Pseudomonas delafieldii* in magnetic polyvinyl alcohol beads. *Letters in Applied Microbiology* 40: 30–36.

Chua, H., Wong, P.K., Yu, P.H.F., Li, X.Z. 1998. The removal and recovery of copper (II) ions from wastewater by magnetite immobilized cells of *Pseudomonas putida* 5-X. *Water Science and Technology* 38: 315–322.

Ibrahim, A.S.S., Al-Salamah, A.A., El-Toni, A.M., El-Tayeb, M.A., Elbadawi, Y.B., Antranikian, G. 2013. Detoxification of hexavalent chromate by *Amphibacillus* sp KSUCr3 cells immobilised in silica-coated magnetic alginate beads. *Biotechnology and Bioprocess Engineering* 18: 1238–1249.

Ivanova, V., Hristov, J., Dobreva, E., Al-Hassan, Z., Penchev, I. 1996. Performance of a magnetically stabilized bed reactor with immobilized yeast cells. *Applied Biochemistry and Biotechnology* 59: 187–198.

Ivanova, V., Petrova, P., Hristov, J. 2011. Application in the ethanol fermentation of immobilized yeast cells in matrix of alginate/magnetic nanoparticles, on chitosan-magnetite microparticles and cellulose-coated magnetic nanoparticles. *International Review of Chemical Engineering* 3: 289–299.

Ji, Y.-Q., Hu, Y.-T., Tian, Q., Shao, X.-Z., Li, J., Safarikova, M., Safarik, I. 2010. Biosorption of strontium ions by magnetically modified yeast cells. *Separation Science and Technology* 45: 1499–1504.

Lee, D.Y., Oh, Y.I., Kim, D.H., Kim, K.M., Kim, K.N., Lee, Y.K. 2004. Synthesis and performance of magnetic composite comprising barium ferrite and biopolymer. *IEEE Transactions on Magnetics* 40: 2961–2963.

Lei, W., Chua, H., Lo, W.H., Yu, P.H.F., Zhao, Y.G., Wong, P.K. 2000. A novel magnetite-immobilized cell process for heavy metal removal from industrial effluent. *Applied Biochemistry and Biotechnology* 84–86: 1113–1126.

Li, H., Li, Z., Liu, T., Xiao, X., Peng, Z., Deng, L. 2008. A novel technology for biosorption and recovery hexavalent chromium in wastewater by bio-functional magnetic beads. *Bioresource Technology* 99: 6271–6279.

Li, T.-T., Liu, Y.-G., Peng, Q.-Q., Hu, X.-J., Liao, T., Wang, H., Lu, M. 2013. Removal of lead(II) from aqueous solution with ethylenediamine-modified yeast biomass coated with magnetic chitosan microparticles: Kinetic and equilibrium modeling. *Chemical Engineering Journal* 214: 189–197.

Mahmoud, M.E., Ahmed, S.B., Osman, M.M., Abdel-Fattah, T.M. 2015. A novel composite of nanomagnetite-immobilized-baker's yeast on the surface of activated carbon for magnetic solid phase extraction of Hg(II). *Fuel* 139: 614–621.

Ou, H., Bian, W., Weng, X., Huang, W., Zhang, Y. 2013. Adsorption of Ce(III) by magnetic chitosan/yeast composites from aqueous solution: Kinetic and equilibrium studies. In *Energy Engineering and Environmental Engineering*, Pts 1 and 2, ed. T. Sun, pp. 391–394. Zurich, Switzerland: Trans Tech Publications.

Özdemir, S., Kılınç, E., Okumuş, V., Poli, A., Nicolaus, B., Romano, I. 2016. Thermophilic *Geobacillus galactosidasius* sp. nov. loaded γ-Fe_2O_3 magnetic nanoparticle for the preconcentrations of Pb and Cd. *Bioresource Technology* 201: 269–275.

Patzak, M., Dostalek, P., Fogarty, R.V., Safarik, I., and Tobin, J.M. 1997. Development of magnetic biosorbents for metal uptake. *Biotechnology Techniques* 11: 483–487.

Peng, Q., Liu, Y., Zeng, G., Xu, W., Yang, C., Zhang, J. 2010. Biosorption of copper(II) by immobilizing *Saccharomyces cerevisiae* on the surface of chitosan-coated magnetic nanoparticles from aqueous solution. *Journal of Hazardous Materials* 177: 676–682.

Podder, M.S., Majumder, C.B. 2016. Application of granular activated carbon/$MnFe_2O_4$ composite immobilized on *C. glutamicum* MTCC 2745 to remove As(III) and As(V): Kinetic, mechanistic and thermodynamic studies. *Spectrochimica Acta, Part A* 153: 298–314.

Pospiskova, K., Prochazkova, G., Safarik, I. 2013. One-step magnetic modification of yeast cells by microwave-synthesized iron oxide microparticles. *Letters in Applied Microbiology* 56: 456–461.

Prochazkova, G., Safarik, I., Branyik, T. 2013. Harvesting microalgae with microwave synthesized magnetic microparticles. *Bioresource Technology* 130: 472–477.

Rao, A., Bankar, A., Kumar, A.R., Gosavi, S., Zinjarde, S. 2013. Removal of hexavalent chromium ions by *Yarrowia lipolytica* cells modified with phyto-inspired Fe^0/Fe_3O_4 nanoparticles. *Journal of Contaminant Hydrology* 146: 63–73.

Ren, Y., Zhang, M., Zhao, D. 2008a. Synthesis and properties of magnetic Cu(II) ion imprinted composite adsorbent for selective removal of copper. *Desalination* 228: 135–149.

Ren, Y.M., Wei, X.Z., Zhang, M.L. 2008b. Adsorption character for removal Cu(II) by magnetic Cu(II) ion imprinted composite adsorbent. *Journal of Hazardous Materials* 158: 14–22.

Safarik, I., Maderova, Z., Pospiskova, K., Baldikova, E., Horska, K., Safarikova, M. 2015a. Magnetically responsive yeast cells: Methods of preparation and applications. *Yeast* 32: 227–237.

Safarik, I., Maderova, Z., Pospiskova, K., Horska, K., Safarikova, M. 2014. Magnetic decoration and labeling of prokaryotic and eukaryotic cells. In *Cell Surface Engineering: Fabrication of Functional Nanoshells*, eds. R.F. Fakhrullin, I. Choi, and Y.M. Lvov, pp. 185–215. Cambridge, U.K.: The Royal Society of Chemistry.

Safarik, I., Pospiskova, K., Maderova, Z., Baldikova, E., Horska, K., and Safarikova, M. 2015b. Microwave-synthesized magnetic chitosan microparticles for the immobilization of yeast cells. *Yeast* 32: 239–243.

Safarik, I., Ptackova, L., Safarikova, M. 2002. Adsorption of dyes on magnetically labeled baker's yeast cells. *European Cells and Materials* 3(Suppl. 2): 52–55.

Safarik, I., Rego, L.F.T., Borovska, M., Mosiniewicz-Szablewska, E., Weyda, F., Safarikova, M. 2007. New magnetically responsive yeast-based biosorbent for the efficient removal of water-soluble dyes. *Enzyme and Microbial Technology* 40: 1551–1556.

Safarik, I., Sabatkova, Z., Safarikova, M. 2008. Hydrogen peroxide removal with magnetically responsive *Saccharomyces cerevisiae* cells. *Journal of Agricultural and Food Chemistry* 56: 7925–7928.

Safarik, I., Safarikova, M. 1999. Use of magnetic techniques for the isolation of cells. *Journal of Chromatography B* 722: 33–53.

Safarikova, M., Atanasova, N., Ivanova, V., Weyda, F., Tonkova, A. 2007. Cyclodextrin glucanotransferase synthesis by Semicontinuous cultivation of magnetic biocatalysts from cells of *Bacillus circulans* ATCC 21783. *Process Biochemistry* 42: 1454–1459.

Safarikova, M., Maderova, Z., Safarik, I. 2009. Ferrofluid modified *Saccharomyces cerevisiae* cells for biocatalysis. *Food Research International* 42: 521–524.

Safarikova, M., Pona, B.M.R., Mosiniewicz-Szablewska, E., Weyda, F., Safarik, I. 2008. Dye adsorption on magnetically modified *Chlorella vulgaris* cells. *Fresenius Environmental Bulletin* 17: 486–492.

Saifuddin, N., Dinara, S. 2012. Immobilization of *Saccharomyces cerevisiae* onto cross-linked chitosan coated with magnetic nanoparticles for adsorption of uranium (VI) ions. *Advances in Natural and Applied Sciences* 6: 249–267.

Schüler, D., ed. 2007. *Magnetoreception and Magnetosomes in Bacteria*. Berlin, Heidelberg: Springer.

Sharma, S.K., ed. 2015. *Heavy Metals in Water: Presence, Removal and Safety*. Cambridge, U.K.: The Royal Society of Chemistry.

Sly, L.I., Arunpairojana, V., Dixon, D.R. 1995. Method for removing manganese from water. USA Patent 5,443,729.

Song, H.P., Li, X.G., Sun, J.S., Xu, S.M., Han, X. 2008. Application of a magnetotactic bacterium, *Stenotrophomonas* sp to the removal of Au(III) from contaminated wastewater with a magnetic separator. *Chemosphere* 72: 616–621.

Song, H.P., Li, X.G., Sun, J.S., Yin, X.H., Wang, Y.H., Wu, Z.H. 2007. Biosorption equilibrium and kinetics of Au(III) and Cu(II) on magnetotactic bacteria. *Chinese Journal of Chemical Engineering* 15: 847–854.

Souiri, M., Gammoudi, I., Ouada, H.B., Mora, L., Jouenne, T., Jaffrezic-Renault, N., Dejous, C., Othmane, A., Duncan, A.C. 2009. *Escherichia coli*-functionalized magnetic nanobeads as an ultrasensitive biosensor for heavy metals. *Procedia Chemistry* 1: 1027–1030.

Sze, K.F., Lu, Y.J., Wong, P.K. 1996. Removal and recovery of copper ion (Cu^{2+}) from electroplating effluent by a bioreactor containing magnetite-immobilized cells of *Pseudomonas putida* 5X. *Resources, Conservation and Recycling* 18: 175–193.

Uzun, L., Saglam, N., Safarikova, M., Safarik, I., Denizli, A. 2011. Copper biosorption on magnetically modified yeast cells under magnetic field. *Separation Science and Technology* 46: 1045–1051.

Wainwright, M., Singleton, I., Edyvean, R.G.J. 1990. Magnetite adsorption as a means of making fungal biomass susceptible to a magnetic field. *Biorecovery* 2: 37–53.

Wang, L., Chua, H., Wong, P.K., Lo, W.H., Yu, P.H.F., Zhao, Y.G. 2000. An optimal magnetite immobilized *Pseudomonas putida* 5-x cellsystem for Cu^{2+} removal from industrial waste effluent. *Water Science and Technology* 41: 241–246.

Widjojoatmodjo, M.N., Fluit, A.C., Torensma, R., Verhoef, J. 1993. Comparison of immunomagnetic beads coated with protein A, protein G, or goat anti-mouse immunoglobulins. *Journal of Immunological Methods* 165: 11–19.

Wong, P.K., Fung, K.Y. 1997. Removal and recovery of nickel ion (Ni^{2+}) from aqueous solution by magnetite-immobilized cells of *Enterobacter* sp 4–2. *Enzyme and Microbial Technology* 20: 116–121.

Xu, M., Zhang, Y., Zhang, Z., Shen, Y., Zhao, M., Pan, G. 2011. Study on the adsorption of Ca^{2+}, Cd^{2+} and Pb^{2+} by magnetic Fe$_3$O$_4$ yeast treated with EDTA dianhydride. *Chemical Engineering Journal* 168: 737–745.

Yavuz, H., Denizli, A., Gungunes, H., Safarikova, M., Safarik, I. 2006. Biosorption of mercury on magnetically modified yeast cells. *Separation and Purification Technology* 52: 253–260.

Zborowski, M., Fuh, C.B., Green, R., Baldwin, N.J., Reddy, S., Douglas, T., Mann, S., Chalmers, J.J. 1996. Immunomagnetic isolation of magnetoferritin-labeled cells in a modified ferrograph. *Cytometry* 24: 251–259.

Zborowski, M., Malchesky, P.S., Jan, T.F., Hall, G.S. 1992. Quantitative separation of bacteria in saline solution using lanthanide Er(III) and a magnetic field. *Journal of General Microbiology* 138: 63–68.

Zhang, Y., Zhu, J., Zhang, L., Zhang, Z., Xu, M., Zhao, M. 2011. Synthesis of EDTAD-modified magnetic baker's yeast biomass for Pb^{2+} and Cd^{2+} adsorption. *Desalination* 278: 42–49.

Zheng, B.Z., Zhang, M.H., Xiao, D., Jin, Y., Choi, M.M.F. 2010. Fast microwave synthesis of Fe$_3$O$_4$ and Fe$_3$O$_4$/Ag magnetic nanoparticles using Fe^{2+} as precursor. *Inorganic Materials* 46: 1106–1111.

49 Detoxification of Mercury
Bioremediation to Chemical Degradation

*Mainak Banerjee, Ramesh Karri, Karthick Muthuvel,
Ashish Chalana, and Gouriprasanna Roy*

CONTENTS

ABSTRACT

Mercury and mercury-related compounds, especially methylmercury species, are highly toxic to humans and animals due to their high affinity toward thiol or selenol residues found in cellular systems. Due to increasing industrialization as well as urbanization all over the world and the absence of proper chemical waste management system along with the lack of awareness about the health hazards, particularly in the developing countries, people are continuously exposed to various toxic heavy metals, including mercury. In fact, extremely high levels of mercury contamination in air, soils, vegetables, and even milk has been reported in the mining areas, particularly coal mining areas, in China, India, and other parts of the world. As a result, high level of mercury was found in blood, hair, and nail samples collected from people in the mining regions and neighboring areas. Thus, the detoxification of mercury and mercury-related products is of critical importance. In this chapter, we have mainly focused on currently available detoxification processes of mercury and mercury-related products by many microorganisms and chemical methods.

49.1 INTRODUCTION

Mercury and mercury-related compounds, especially methylmercury species ($MeHg^+$, MeHg hereafter), are highly toxic to humans and animals due to their high affinity toward thiol or selenol residues found in cellular systems (Clarkson, 1997; Clarkson and Magos, 2006; Mutter et al., 2007; Branco et al., 2011, 2012; Farina et al., 2011). Due to the structural similarity with methionine, L-cysteine-conjugated methylmercury (MeHgCys) could easily cross the cellular membranes including the blood–brain barrier and the blood–placental barrier, causing irreversible damage to the nervous system especially in early stages of development (Aschner and Aschner, 1990; Carvalho et al., 2008; Yin et al., 2008). Owing to the relatively high stability of the CH_3–Hg bond under physiological conditions, MeHg has the unique ability to biomagnify and bioaccumulate in the aquatic food chain, and therefore, dietary consumption of these toxic species (MeHgCl, MeHgCys, or MeHgOH) poses a serious health threat to humans and other animals (Clarkson et al., 2003). For instance, the first epidemic of severe methylmercury poisoning known as "Minamata disease" was reported in 1956 in Minamata City, Japan, caused by the consumption of large amounts of fish and shellfish heavily contaminated with mercury and methylmercury discharged from a chemical plant (Irukayama, 1969; Rabenstein, 1978; Uchino et al., 1995; Clarkson and Magos, 2006). According to the survey, fishing communities that commonly depend on fish-based diet are mostly exposed to toxic levels of MeHg (Clarkson et al., 2003). Absorption of MeHg in the gastrointestinal (GI) tract is significantly higher (around 90%–95%) as compared to the GI absorption of inorganic salts of mercury Hg^{2+} from food (Nielsen and Andersen, 1992; Farina et al., 2011). However, a little information is available about the demethylation of MeHg and subsequent reduction of Hg^{2+} to less toxic Hg^0. On the other hand, phenylmercury compound ($PhHg^+$, PhHg hereafter) has widely been used as preservatives in various medicines, as antiseptic agents in skin creams, in diaper washes, and as a spermicide. Overexposure of this compound as antiseptic agents was also reported in several places all over the world. The exposure of infants to PhHg has been known to cause acrodynia, which is characterized by pink discoloration of the hands and feet (Clarkson and Magos, 2006).

49.2 MERCURY CYCLE

Though mercury is toxic to humans and animals, causing damage to various organs including the brain, kidneys, and immune system, it is a naturally occurring element and is found in the earth's crust at concentrations ranging from 21 (lower crust) to 56 (upper crust) ppb throughout the world (Barkay et al., 2003). Mercury has always been also present in the environment as stable minerals such as cinnabar, an ore mined to produce mercury, metacinnabar, and tiemannite (UNEP, 2013). They are in fact insoluble materials of either mercury sulfide (HgS) (10^{-6} g/100 mL water for cinnabar) or mercury selenide (HgSe). These minerals, however, are not present in an isolated manner in the earth's crust, and, moreover, mercury is also present as an impurity in many other economically valuable minerals such as in coal, nonferrous metals, and fossil fuels. During combustion at high temperatures (above 1100°C), these stable minerals are vaporized, leading to the formation of many volatile and soluble inorganic and organic mercury compounds. Due to increasing industrialization as well as urbanization all over the world and the absence of proper chemical waste management system along with the lack of awareness about the health hazards, particularly in the developing countries, people are continuously exposed to various toxic heavy metals including mercury. Human activities, especially combustion

of coal in large scale in industries, mining of other metals, refining, manufacturing, waste incineration, and chlor-alkali production, have increased the mobilization of mercury into the environment in the last few decades, raising the amounts of mercury in the atmosphere, soils, freshwaters, and oceans. Apart from discharges from various industries (coal-using power plants, pulp, mining, etc.) mentioned earlier, municipal and medical wastes, fluorescent light fixtures, batteries, dental restorations, and electrodes used in chlor-alkali process are also contributing in large scale in this anthropogenic emission process. Mercury exists within the environment in various chemical forms such as elemental mercury vapor (Hg^0), inorganic mercury salts (Hg^{2+}), and organic mercury ($MeHg^+$, $EtHg^+$, $PhHg^+$) (Clarkson, 2002). Anthropogenic emissions emit both elemental Hg with a long life in the atmosphere and reactive gaseous and particulate mercury, which are short-lived in the air and deposited near the emission source. Mercury emitted into oceans or the atmosphere can travel long distances without being degraded due to its high residence time, and hence, there is a global reservoir of mercury circulating worldwide at any particular time. In the last several decades (the postindustrial era), due to growing industrialization as well as urbanization, the anthropogenic emission of mercury has increased manyfolds compared to the preindustrial era (Li et al., 2009; Wangberg et al., 2010; De et al., 2014). Total anthropogenic emissions of mercury to the atmosphere in 2010 are estimated at 1960 tonnes. Apart from anthropogenic contribution to mercury pollution in the environment, natural sources such as volcanoes and geothermal activities, evasion from surficial soils, water bodies, natural mercury deposits, and volatilization from the river and ocean may be included in this process. Forest fires and burning of fossil fuels such as coal and petroleum also contribute to the natural sources of mercury pollution in the environment. Current anthropogenic sources are responsible for about 30% of annual emissions of mercury to air. Another 10% comes from natural sources, and the rest (60%) is from reemissions of previously released mercury that has built up over decades and centuries in surface soils and oceans (Figure 49.1) (UNEP, 2013). In fact, current demand

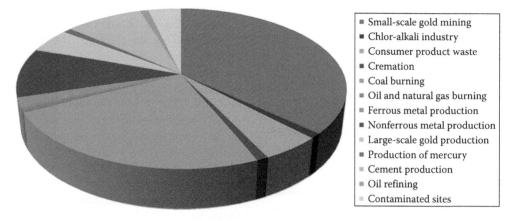

FIGURE 49.1 Relative contributions to estimated emissions to air from anthropogenic sources in 2010. (From UNEP, Global Mercury Assessment 2013: Sources, Emissions, Releases and Environmental Transport, UNEP Chemicals Branch, Geneva, Switzerland, 2013.)

for mercury all over the world is generally met by the supply of mercury recovered from stockpiles and industrial by-products rather than from mercury mining.

Mercury is released to the aquatic environments through anthropogenic process reaching hundreds of tonnes to the amount of mercury cycling in the environment (Figure 49.2), and the annual global input of mercury to the reservoir is estimated to be 4900 tons (Selin, 2009; De et al., 2014). Biotic and abiotic processes facilitate mercury cycling from water and soils to the atmosphere and back to the surface by both wet and dry deposition. Once mercury and mercury-related compounds enter into an aquatic system, it can be converted from one form to another inside the ecosystem as shown in Figure 49.2. Natural processes in aquatic systems convert less toxic elemental and inorganic mercury into much more toxic methylmercury. Methylmercury concentrates and accumulates in the food web, leading to high concentrations in some species of seafood and fish that many people eat, which presents the greatest health risk to humans and wildlife. Global climate change may have profound effects on many aspects of the movement and chemical transformations of mercury in the environment. For instance, warmer temperatures may increase the rates of bacterial activity, hence possibly leading to faster production of methylmercury from inorganic mercury (UNEP, 2013).

High level of mercury contamination in air, soils, crops such as in wheat grain and vegetables, and even milk has been reported in the mining areas, particularly coal mining areas, in China, India, and other parts of the world. As a result of this contamination, high level of mercury was found in blood, hair, and nail samples collected from people in the mining regions and neighboring areas. Thus, the detoxification of mercury and mercury-related products is of critical importance.

49.3 BIODEGRADATION OF MERCURY COMPOUNDS

It is well known that because they are lipophilic in nature as compared to inorganic mercury (Hg^{2+}), organomercurials including methylmercuric (MeHg) and ethylmercuric ($EtHg^+$, EtHg hereafter) compounds are highly toxic to humans and animals. Many biological methods have been known to be employed for the detoxification of organomercurials. Many microorganisms in aerobic sediments including corrinoid-producing anaerobes and sulfate-reducing bacteria (SRB) can methylate Hg^{2+} to produce toxic $MeHg^+$. In case of SRB-mediated methylation process, it has also been reported that the soluble and neutral mercury sulfide acts as a substrate in the microbial MeHg synthesis (Barkay et al., 2003). Besides, several other microorganisms have

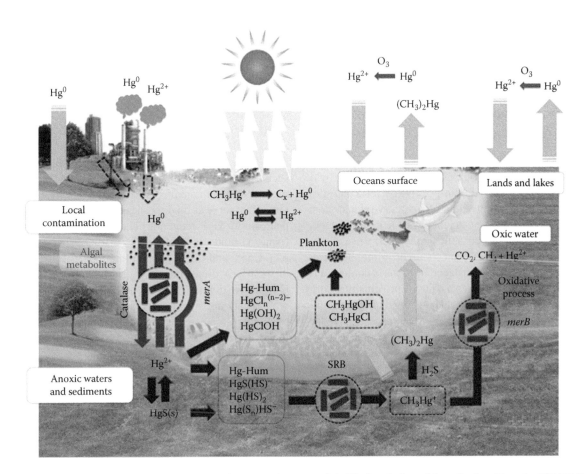

FIGURE 49.2 The biogeochemical cycle of mercury in the environment. (Modified and adapted from Barkay, T. et al., *FEMS Microbiol. Rev.*, 27, 355, 2003; Morel, F.M.M. et al., *Ann. Rev. Ecol. Syst.*, 29, 543, 1998.)

evolved mercury resistance mechanisms to deal with both inorganic and organic mercury compounds (Walsh et al., 1988; Moore et al., 1990; Fitzgerald et al., 2007) either by cleaving the Hg–C bond of toxic MeHg and transforming it into a less toxic inorganic mercury, Hg^{2+}, or by transforming toxic organomercurials into less toxic insoluble mercury sulfide (HgS) (Pan-Hou and Imura, 1981; Baldi et al., 1993; Marvin-Diapsquale et al., 2000; Barkay et al., 2003; Essa et al., 2005; Glendinning et al., 2005; Fitzgerald et al., 2007). Bacteria carry out chemical transformations of mercury by methylation, demethylation, oxidation, and reduction and develop resistance determined by the extra chromosomal genetic material, including plasmids (Silver and Misra, 1984). Various types of mercury resistance mechanisms have been reported in bacteria (Summers and Lewis, 1973), which include (1) enzymatic reduction of Hg^{2+} to Hg^0, (2) enzymatic cleavage of C–Hg bond in various organomercurials including methylmercury followed by the reduction of Hg^{2+} to elemental Hg^0, (3) enzymatic cleavage of C–Hg bond in organomercurials followed by conversion to less toxic insoluble mercuric sulfide compound, (4) sequestration of methylmercury, (5) methylation of mercury, and (6) reduced uptake of mercuric ions (Barkay et al., 2003; De et al., 2014). Enzymatic cleavage of C–Hg bond followed by the reduction of Hg^{2+} to Hg^0 and the enzymatic cleavage of C–Hg bond in organomercurial followed by the reduction of Hg^{2+} to Hg^0 or conversion to less toxic insoluble mercuric sulfide compounds will be discussed here in detail.

49.3.1 ROLE OF MERCURY RESISTANCE ENZYME MERCURIC ION REDUCTASE (MERA) AND ORGANOMERCURIAL LYASE (MERB) ON THE DEGRADATION OF MERCURY COMPOUNDS

49.3.1.1 Role of MerA

Many mercury-resistant bacteria have been found in a wide range of Gram-negative and Gram-positive bacteria that show tolerance to high concentration of mercury-related compounds, both organic (RHg^+) and inorganic (e.g., $HgCl_2$) mercury compounds, due to their ability to convert these soluble toxic mercury compounds to innocuous or immobile forms. A key example is mercury resistance in bacteria conferred by the *mer* operon, which encodes a suite of proteins that carry out the cleavage of otherwise inert Hg–C bond in various toxic organomercurials (RHg^+) followed by the transport and reduction of Hg^{2+} to transform this toxic ion into less toxic, elemental Hg^0. Mercury-resistant determinants vary in number and identity of genes involved (De et al., 2014) and are encoded by the *mer* operon located on chromosomes (Wang et al., 1987, 1989; Inoue et al., 1989, 1991; Iohara et al., 2001), plasmids (Summers and Silver, 1978; Brown et al., 1986; Griffin et al., 1987; Radstrom et al., 1994; Osborn et al., 1997), transposons (Misra et al., 1984; Kholodiet al., 1993; Hobman et al., 1994; Liebert et al., 1997, 1999), and integrons (Liebertet al., 1999). Recent studies suggest the ubiquitous nature of this *mer*

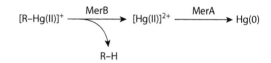

SCHEME 49.1 Detoxification of organomercury compounds in nature by the combined action of two enzymes, namely, (i) bacterial organomercurial lyase (MerB) and (ii) mercuric ion reductase (MerA). (From Parks, J.M. et al., *J. Am. Chem. Soc.*, 131, 13278, 2009.)

operon as it has been found in many different species among bacterial communities (Lian et al., 2014). Work on the mechanism of detoxification of MeHg led to the identification of the two key enzymes found in bacteria in the aquatic sediment, organomercurial lyase (MerB), which catalyzes the protolytic cleavage of the otherwise inert C–Hg bond, and mercuric ion reductase (MerA), which reduces Hg^{2+} to elemental mercury, Hg^0, which is volatile and simply diffuses out of the bacteria, as shown in Scheme 49.1 (Begley et al., 1986a,b; Omichinski, 2007; Strasdeit, 2008).

The most widely studied eubacterial mercury resistance is the reduction of reactive cationic species of mercury (Hg^{2+}) to the volatile and inert form of elemental mercury (Hg^0) by MerA. Although MerA was purified in the early 1970s from both *Pseudomonas* K62 and *Escherichia coli.* significant mechanistic studies (Furukawa and Tonomura and Tonomura, 1972; Izaki et al., 1974; Tezuka and Tonomura, 1976; Schottel, 1978; Fox and Walsh, 1982, Fox et al., 1983; Rinderle et al., 1983; Sahlman and Lindskog, 1983; Williams, 1992) were not initiated until after the discovery of a flavin adenine dinucleotide (FAD) cofactor in the enzyme. MerA is a cytosolic flavin disulfide oxidoreductase that uses NAD(P)H as a source of electrons (reductant). The active form of MerA is a homodimer (~120 kDa) in which the two active sites from each monomer are located within the catalytic core domain present at the interface of the dimer. To compete with other endogenous thiols such as glutathione (GSH) for its Hg^{2+} substrate, each chain of the MerA dimer contains three pairs of strictly conserved cysteines that serve as binding sites for Hg^{2+} transfer and provide a platform for Hg^{2+} reduction. The N-terminal domain of MerA (NmerA) binds and delivers Hg^{2+} to the C-terminal cysteine pair, C558′ and C559′, of the other monomer near the surface of the MerA dimer interface (Lian et al., 2014). The flexible C-terminal domain changes its conformation after binding with dicationic mercury (Hg^{2+}) atom through C558′/C559′ cysteine pair present at the active site to move the complex (C558′–S–Hg^{2+}–S–C559′) from the surface to the protein interior where Hg^{2+} is transferred to the active site cysteine pair, C136 and C141, located adjacent to the isoalloxazine ring of the FAD cofactor. The other reducing substrate dihydronicotinamide adenine dinucleotide phosphate (NADPH), located on the opposite face of FAD, converts FAD to two-electron reduced $FADH^-$ by transferring hydride to FAD and oxidized to $NADP^+$ itself. Subsequently, $FADH^-$ reduces Hg^{2+} present in the C141–S–Hg^{2+}–S–C136 complex to yield the final product, the elemental mercury, Hg^0 (Lian et al., 2014).

49.3.1.2 Role of Merb

Tonomura and coworkers first described the cell-free, substrate-inducible activity of a mercury-resistant soil *Pseudomonas*, strain K62 (Furukawa and Tonomura, 1971; 1972; Tezuka and Tonomura, 1976; Hines et al., 2000), which degraded various organomercurials (PhHg, EtHg, and MeHg), and later they purified and characterized two enzymes (splitting enzyme 1 and splitting enzyme 2) that catalyzed the splitting of carbon–mercury linkage of arylmercury compounds that were found in extracts of mercury-resistant *Pseudomonas* K-62. Silver and coworkers also reported that the penicillinase plasmids of *Staphylococcus aureus*, containing genes that are responsible for the resistance to both organomercurial phenylmercury acetate (PhHgOAc) and inorganic mercury (Hg^{2+}), have shown ability to degrade various organomercurials (Schottel et al., 1974; Weiss et al., 1977). Based on the studies on every bacterium, resistance to inorganic mercury and to PhHgOAc was due to the production of two different enzymes that are responsible for degrading mercury compounds. One is a soluble enzyme that cleaves the C–Hg bond in the RHgX (R=Me, Ph; X=Cl, OAc), releasing a protonated organic moiety (such as methane from MeHgCl or benzene from PhHgOAc) and the Hg^{2+} cation that was then reduced to Hg^0 by another enzyme, mercuric reductase (MerA) (Schottel, 1978; Jackson and Summers, 1982). In Gram-negative *Enterobacteriaceae*, the *merB* gene typically occurs in plasmid-carried *mer* operons immediately 3′ of the *merA* gene (Liebert et al., 1997). In *Pseudomonas* species, merB occurs in a variety of operon structures (Clark et al., 1977; Kiyono et al., 1995; Reniero et al., 1995), usually associated with transposons on plasmids, whereas in Gram-positive bacteria, *merB* is also associated with transposons but may be chromosomal in *Bacillus* (Wang et al., 1989; Huang et al., 1999), *Clostridium*, and *Streptomyces* (Sedlmeier and Altenbuchner, 1992; Ravel et al., 2000) and on plasmids in *Staphylococcus* (Ladagga et al., 1987). Crystal structure of bacterial organomercurial lyase MerB (PDB ID: 3FN8) revealed that three cysteine residues, namely, Cys-96, Cys-159, and Cys-160, are present at the active site of the enzyme in which two highly conserved cysteine residues (Cys-96 and Cys-159) are essential for catalytic activity, as shown in Figure 49.3 (Benison et al., 2004; Lafrance-Vanasse et al., 2009). Cys-96 and Cys-159 play crucial role in binding the substrate MeHg, helping in cleaving carbon–mercury bond, and subsequently in the regulation process of mercury ion release from the active site of enzyme MerB to another enzyme MerA where Hg^{2+} will get reduced to Hg^0 as discussed earlier.

The active site of the enzyme buried within the hydrophobic core eventually helps to inhibit the release of the toxic product ionic mercury (Hg^{2+}) after protolytic cleavage of the carbon–mercury bond, which could otherwise result in extensive cellular damage. Another cysteine residue, the Cys160, located on a flexible loop with catalytic Cys159, may not be essential for the cleavage of carbon–mercury bond but plays an important role in the removal of the Hg^{2+} product after protonolysis from the inside of the active site pocket to

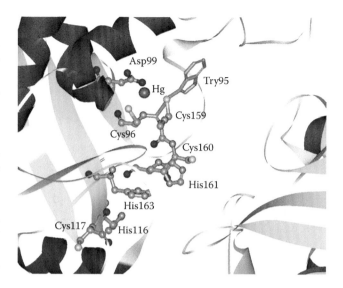

FIGURE 49.3 Stereo view of the active site of the Hg-MerB complex (PDB ID 3FN8).

the outside, assisting in accessing Hg^{2+} by the cysteine pair of mercuric ion reductase, MerA. Moreover, the active site structure of Hg^{2+}-product-bound MerB revealed a trigonal coordination of Hg^{2+} by Cys96, Cys159, and a water molecule, showing that two or more protein sulfur ligands are required to activate carbon–mercury bonds for protonolysis. On the other hand, based on the Asp99Ala mutation analysis, it has been shown that the residue Asp99 present in the active site of the enzyme is essential for its catalytic activity as evidenced by the x-ray structure of Hg^{2+}-product-bound MerB, which reveals a weak coordination of Hg^{2+} with an aspartic acid carboxylate (Figure 49.3) (Lafrance-Vanasse et al., 2009; Wahba et al., 2016).

The mechanism by which MerB achieves its catalytic activity and transfers ion mercury to MerA is still not clear. As part of the mechanism of MerB, deprotonation of the cysteine residue (Cys-96 and/or Cys-159) takes place first followed by nucleophilic attack at positively charged center of mercury of RHgX to form a higher-order structure about the Hg^{2+} that makes the carbon–mercury bond more amenable to protonolysis. Contradictory results, however, have been reported in the literature by two independent research groups on nucleophilic attack by cysteine residue. Benison et al. proposed that Cys96 would initially attack the organomercurial compound after deprotonation of the cysteine residue by Asp99, as shown in Scheme 49.2 (Benison et al., 2004; Lafrance-Vanasse et al., 2009). On the other hand, based on the several biochemical experiments, Pitts et al. proposed the mechanism for the cleavage of C–Hg bond, where Cys159 first reacts more readily with both the substrate and the Hg^{2+} product, as shown in Scheme 49.3 (Pitts and Summers, 2002). Moreover, it has been proposed that Cys159 remains coordinated with Hg throughout the reaction course, both during the cleavage of C–Hg bond of organomercurial compound by MerB and in the process of removal of product

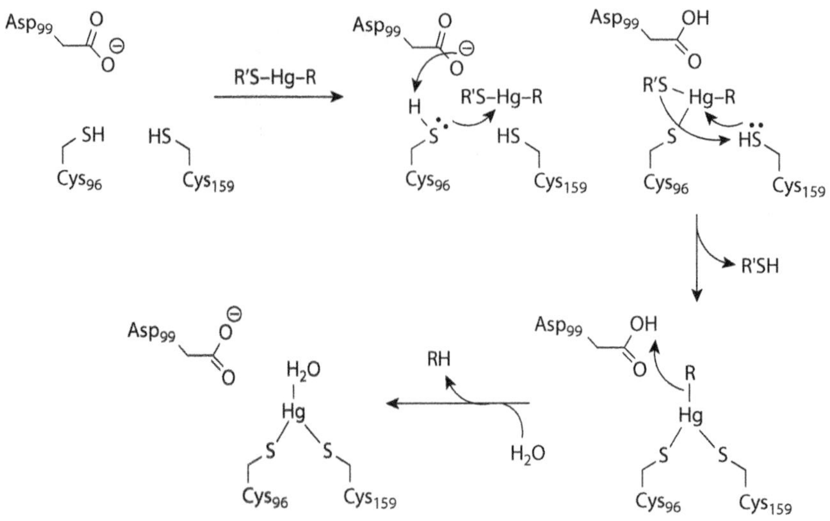

SCHEME 49.2 Showing the cleavage of the C–Hg bond and the formation of HgMerB complex. Here, the initial attack of organomercurial compound takes place by Cys96. (Adapted from Lafrance-Vanasse, J. et al., *J. Biol. Chem.*, 284, 938, 2009.)

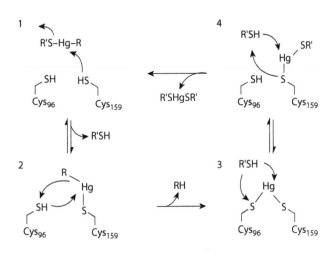

SCHEME 49.3 Another proposed mechanism for the cleavage of C–Hg bond. In this mechanism, it has been shown that the organomercurial compound was initially attacked by Cys159. (Adapted from Pitts, K.E. and Summers, A.O., *Biochemistry*, 41, 10287, 2002.)

Hg^{2+} to MerA. Due to the close proximity of Trp95 to the Hg^{2+} product chelated by Cys96 and Cys159 in Hg^{2+}-bound MerB (HgMerB), recently Trp95 has been used as a fluorescence probe for mechanistic studies of Hg^{2+} binding and release from the active site of the enzyme (Baoyu Hong et al., 2010). Linear decrease of tryptophan fluorescence intensity was observed upon addition of one equivalent $HgBr_2$ to the native enzyme MerB that can be reversed upon addition of excess N-terminal domain of MerA (NmerA), but not by the addition of the mercury-bound NmerA (HgNmerA) complex. In addition, no change in the fluorescence intensity was observed upon addition of NmerA to unligand MerB, indicating that the fluorescence change is specifically associated with the binding/removal of Hg^{2+} from the active site. This clearly indicates the involvement of Trp95 in this process and

during the release of ionic Hg^{2+} from MerB or the reaction of GSH with HgNmerB (Baoyu Hong et al., 2010).

The discharge of heavy metals from the industrial waste materials into the environment is of major concern. The use of microorganisms for heavy metal removal from wastes offers a potential alternative to conventional methods that may be expensive or may produce hazardous by-products in the process. Various types of microbe–metal interactions have been reviewed and applied to waste decontamination. For example, in the case of mercury, its remediation via volatilization as Hg^0 by mercury-resistant bacteria was reported. The reduction of the highly reactive cationic form of mercury, Hg^{2+}, to a relatively inert form of monatomic volatile mercury vapor of Hg^0 is catalyzed by an enzyme called mercuric ion reductase (MerA) found in a wide range of Gram-negative and Gram-positive bacteria. However, this still requires more study, especially on the large scale, for the recovery of the metallic Hg^0 to avoid its escape into the atmosphere.

49.3.2 Role of Sulfate-Reducing Bacteria on the Degradation of Mercury Compounds

Several SRB play a crucial role in metal sulfide immobilization in the environment via dissimilatory reduction of sulfate (SO_4^{2-}) ion in the anaerobic condition in the lake and sea sediments. They are used as an alternative approach for the treatment of wastewaters. In the anoxic environment, SRB such as *Desulfovibrio desulfuricans* play a crucial role in detoxifying MeHg species by forming insoluble mercury sulfide (HgS) via dimethylmercury sulfide, $(MeHg)_2S$ (Pan-Hou and Imura, 1981; Baldi et al., 1993; Marvin-Diapsquale et al., 2000; Barkay et al., 2003; Essa et al., 2005; Glendinning et al., 2005; Fitzgerald et al., 2007). The methylmercury resistance in *D. desulfuricans* is mainly due to the transformation of methylmercury to insoluble $(MeHg)_2S$ that further reacts with hydrogen sulfide (H_2S) produced during metabolism to give dimethylmercury (Me$_2$Hg), methane, and harmless and insoluble metacinnabar (HgS). Demethylation of methylmercury in aquatic environment generally occurs via microbial processes or photolysis. In anoxic sediment, cleavage of the Hg–C bond (demethylation process) of methylmercury is known to occur in the presence of H$_2$S to form white-colored precipitate of (MeHg)$_2$S as an intermediate that gradually decomposes into a black precipitate of β-HgS and (CH$_3$)$_2$Hg (Rowland et al., 1977; Craig and Bartlett, 1978). On the other hand, H$_2$S produced during the anaerobic metabolism of mercury-resistant *Clostridium cochlearium* reacts with mercuric salt and HgCl$_2$ and is converted into insoluble HgS (Pan-Hou and Imura, 1981). Recently, an alternative method of mercury detoxification is reported from a culture of *Klebsiella pneumoniae* grown aerobically (Essa et al., 2006). Dimethyl disulfide (Me$_2$S) was found as a significant component in this culture grown aerobically, and it is believed that Me$_2$S helps to form HgS precipitate when reacted with organomercurials. Several mercury-tolerant Gram-positive bacteria are capable of producing HgS precipitate without the formation of any H$_2$S or having any mercuric reductase activity. Glendinning et al. reported the isolation of thermophilic *Bacillus* sp. (Scholz et al., 1987) and *Ureibacillus* sp. (Fortina et al., 2001) and their ability to immobilize mercury salt by forming crystalline HgS via routes other than by direct production of gaseous H$_2$S (Glendinning et al., 2005). Formation of another volatile organosulfur compound Me$_2$S could be the possible reason and possibly a new route for mercury removal.

49.4 METHYLMERCURY TOXICITY AND TREATMENT

Mercury in any form, inorganic or organic, is highly toxic to human. Being a soft acid, it generally binds with sulfhydryl and selenohydryl groups present in protein in cellular function and alters the tertiary and quaternary structure of proteins. Mercury is also believed to interfere with DNA transcription and protein synthesis in various tissues in humans including the developing brain, with destruction of endoplasmic reticulum and disappearance of ribosomes. Although mercury can potentially affect many tissues including the liver and kidney, it has unique ability to reach high levels in the central nervous system (CNS), leading to neurotoxic effects (Farina et al., 2007). Studies have shown that endogenous thiols and thiol drugs play a significant role in the toxicology of mercury. For instance, many metals exhibit a strong affinity for reduced sulfhydryl groups in GSH, forming metal–GSH complexes with high thermodynamic stability. GSH levels in various

tissues are an important determinant of tissue methylmercury deposition. Evidence suggests that GSH, which has a concentration in rat bile of 1–4 mM, accounts for over 90% of the reduced sulfhydryl groups in bile, plays a very important role in the biliary secretion of several metals, including mercury, copper, silver, zinc, and chromium (Ballatori and Clarkson, 1983, 1984, 1985; Ballatori and Dutczak, 1994). With increasing hepatic GSH level, it has been shown to increase the biliary secretion of methylmercury. On the other hand, the agents that deplete the hepatic content of GSH inhibit the biliary secretion of methylmercury and simultaneously decrease the GSH content of bile (Ballatori and Clarkson, 1985). British anti-lewisite (BAL; 2,3-dimercapto-1-pro-panol) had been used widely since 1951 as a chelating agent for heavy metals such as mercury, lead, and arsenic (Figure 49.4) (Peters and Stocken, 1945; Aposhian et al., 1995). However, due to its severe side effects and low therapeutic index, BAL has been discontinued in chelation therapy (Aposhian, 1982). Currently, two other analogs of BAL such as sodium 2,3-dimercapto-propanesulfate (DMPS) and meso-2,3-dimercaptosuccinic acid are widely used as drugs for the treatment of mercury (Hg) poisoning (Friedheim et al., 1954; Petrunkin, 1956; Aposhian and Aposhian, 1990). DMPS has high affinity for mercury, and due to its superior safety, it has been widely used in Germany for the past 50 years and is available over the counter in that country and also available for compounding in the United States (Bernhoft, 2012). However, evidences suggest that these two clinically used drugs are also not very effective as chelators and fail to remove Hg from the brain. On the other hand, α-lipoic acid, another chelating agent, has been shown recently that it can access all tissues of the body, including the CNS (Carvalho et al., 2011).

49.5 DETOXIFICATION OF MERCURY IN ANIMAL

The detoxification of organic or inorganic form of Hg occurs in natural systems, and the formation of HgSe has been observed in different tissues of marine mammals and wild animals and in various organs of Hg mine workers (Koeman et al., 1973; Kosta et al., 1975; Magos et al., 1984; Cikrt and Benko, 1989; Drasch et al., 1996; Nigro and Leonzio, 1996; Ng et al., 2001; Khan and Wang, 2009a; Korbas et al., 2010; Nakazawa et al., 2011). Marine mammals such as striped dolphin, arctic whales, and ringed seals in the Canadian Arctic accumulate mercury (Hg) in their tissues at high concentration because of their high position in the marine food chain and their long life span (P.-S. Ng et al., 2001). Some

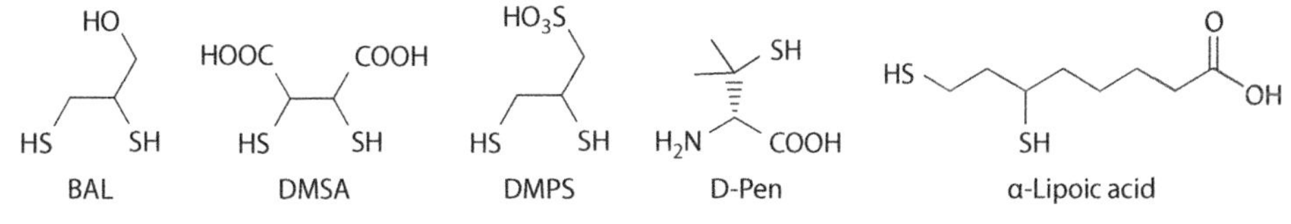

FIGURE 49.4 Chemical structure of the chelating agents employed in the treatment of mercury poisoning.

marine mammals are known to contain high amount of Hg in the liver (more than 10 mg/g dry weight). However, in spite of the high concentration, no indication of Hg intoxication has been reported for marine mammals (Korbas et al., 2010). This clearly indicates that they seem to have an ability to tolerate high concentrations of mercury. Insoluble teimannite (HgSe) granules were identified in various tissues, particularly in the liver, of those marine mammals. The presence of HgSe in the brains of humans exposed to MeHg was detected. Insoluble HgE (E=S, Se) particles are considered to be much less toxic than soluble MeHg species such as MeHgCys. Thus, efficient detoxification of organomercurials including methylmercury species by forming insoluble HgE particles at physiologically relevant conditions may offer a novel approach to treat the patients suffering from methylmercury poisoning.

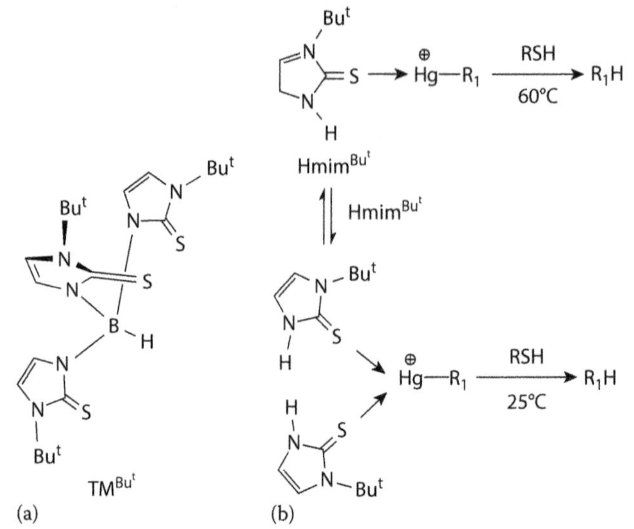

49.6 DEGRADATION OF ORGANOMERCURIALS BY CHEMICAL METHOD

49.6.1 By Functional Mimic of MerB

In an effort to cleave otherwise inert Hg–C bond in various organomercurials such as methylmercury and ethylmercury and convert them into less toxic inorganic Hg^{2+} salts, recently, Melnick et al. have described a facile room temperature protolytic cleavage of Hg–C bond in mercury-alkyl compounds by using tripod thiol compound, tris (2-mercapto-1-t-butyl-imidazolyl) hydroborato ligand [Tm^{But}], which emulates the structure and function of the mercury resistance bacterial organomercurial lyase MerB (Melnick and Parkin, 2007).

Interestingly, in their paper, they have shown that the coordination number of Hg in methylmercury plays a crucial role on the reactivity of the cleavage of C–Hg bond in methylmercury. Multiple thiol coordination to Hg leads to facile cleavage of C–Hg bond in methylmercury as compared to single thiol coordination, as shown in Figure 49.5. The $Hmim^{But}$ ligand that emulates the κ^1-coordination mode (i.e., coordination via a single sulfur atom) of the ligand [Tm^{But}] is not able to cleave C–Hg bond in methylmercury under comparable conditions in the presence of external monothiol such as PhSH, which requires high temperature (60°C) to cleave inert C–Hg bond in methylmercury. However, upon addition of one equivalent excess $Hmim^{But}$ to the mixture of {[$Hmim^{But}$]HgEt}[BF_4] and PhSH, elimination of C_2H_6 occurs rapidly at room temperature, indicating the importance of multiple coordination on the demethylation of methylmercury (Melnick et al., 2009, 2010).

49.6.2 By Imidazole-Based Thione and Selone

The N-methylimidazole-based thione (1) and its selenium analog (2) having a 2-hydroxyethyl substituent have shown remarkable effect in detoxification of various organomercurials including methylmercury by producing insoluble HgS and HgSe nanoparticles as the end products at normal physiological temperature (37°C) (Figure 49.6) (Banerjee et al., 2015).

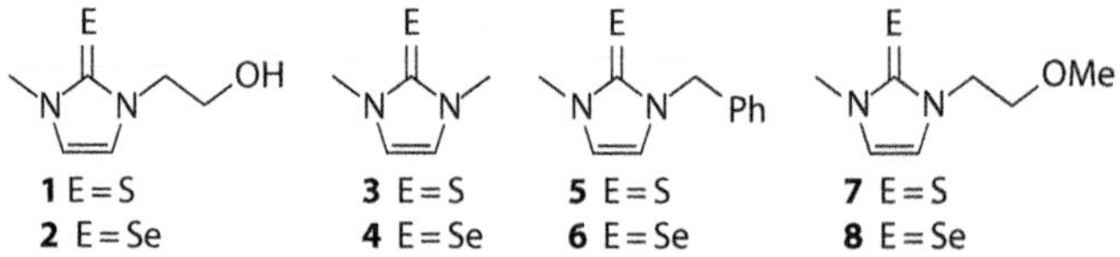

FIGURE 49.5 Chemical structure of tripod thiol ligand (a). Demethylation of methylmercury by thione, $Hmim^{But}$, in the presence of different concentrations of $Hmim^{But}$ (b). (From Melnick, J.G. and Parkin, G., Science, 317, 225, 2007.)

FIGURE 49.6 Chemical structures of some imidazole-based thiones and selones.

On the other hand, compounds 3–8 that lack the N–CH$_2$CH$_2$OH group failed to produce the corresponding HgS and HgSe nanoparticles under identical reaction conditions. When thione 1 (15 mM) was treated with one equivalent of ArHgOH in water/acetonitrile (1:1), the formation of 1:1 mercury-conjugated complex 9 was obtained (Scheme 49.4). Interestingly, after a few hours, a white precipitate was observed in the reaction mentioned earlier that gradually transformed to a black precipitate of HgS. In addition to that, the gradual formation of ketone 11 (m/z of 143.0822) at an LC/MS retention time of 1.8 minutes was observed (Scheme 49.4). The formation of both ketone 11 and HgS nanoparticles was increased with the progress of the reaction. In addition, the formation of (Ar)$_2$Hg was also observed by LC/MS (Figure 49.4).

The FT-IR spectrum of ketone 11 shows an intense peak at 1652 cm^{-1} due to carbonyl stretching $v_{(C=O)}$. The lower value of $v_{(C=O)}$ in ketone 11 clearly suggests that the C=O double bond is involved in conjugation with the lone pairs of adjacent N atoms in the imidazole ring. Density Functional Theory (DFT) calculation results are in good agreement with the experimental results. The C–O bond length (C–O: 1.226 Å; bond order 1.56) in ketone 11 is longer than that observed in cyclopent-3-enone (C–O: 1.205 Å; bond order 1.85), which does not have any nitrogen atom in the five-membered ring (Figure 49.6). On the other hand, the grayish-black precipitate

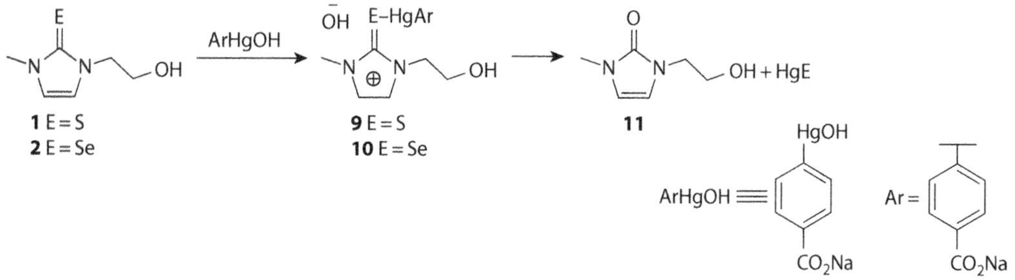

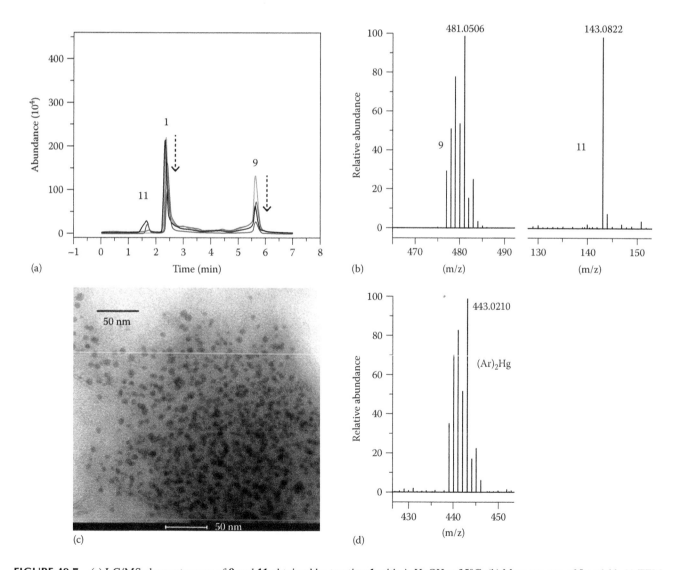

SCHEME 49.4 Formation of HgE (E=S, Se) nanoparticles in the reactions of ArHgOH with **1** and **2**, respectively.

of HgS was found to be insoluble in water and any common organic solvent and was characterized thoroughly by various techniques. TEM images showed that the HgS nanoparticles were spherical and somewhat monodisperse with an average size of ca. 7 nm (Figures 49.7 and 49.8).

On the other hand, selones are expected to be more reactive toward organomercurials due to the soft and nucleophilic

nature of selenium. When selone **2** (15 mM) was treated with one equivalent of ArHgOH in water/acetonitrile (1:1) mixture, at 37°C, it resulted in the formation of 1:1 mercury-conjugated complex **10** that was detected at a retention time of 6.2 min. However, unlike thione **1**, in this case, the black precipitate of HgSe appeared within a few minutes, which was isolated and characterized thoroughly by various techniques. In addition,

FIGURE 49.7 (a) LC/MS chromatogram of **9** and **11** obtained by treating **1** with ArHgOH at 35°C. (b) Mass spectra of **9** and **11**. (c) TEM image of HgS NPs obtained by treating **1** with ArHgOH. (d) Mass spectra of Ar₂Hg.

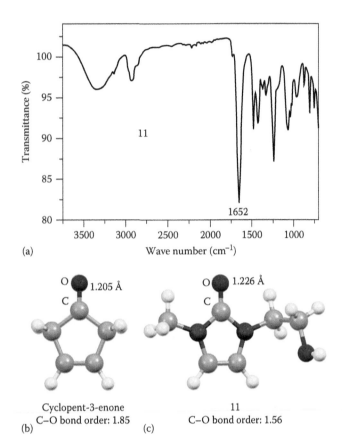

(a)

(b) Cyclopent-3-enone
C–O bond order: 1.85

(c) 11
C–O bond order: 1.56

O 1.205 Å

O 1.226 Å

FIGURE 49.8 Optimized geometries of cyclopent-3-enone and ketone **11**. The structures were optimized at the B3LYP level of theory using 6–311G(d) basis set [OPT in # b3lyp/6-311+G(d,p); NBO in # B3LYP/6-311++G(2d,p)].

similar to thione **1**, ketone **11** and (Ar)$_2$Hg were identified as side products along with spherical-shaped HgSe nanoparticles with an average size of ca. 6 nm. Surprisingly, compounds **3–8** that do not contain any N-CH$_2$CH$_2$OH group in the imidazole ring failed to produce the corresponding HgS and HgSe nanoparticles when reacted with ArHgOH under identical reaction conditions at 37°C. When compound **3** was treated with one equivalent of ArHgOH, the immediate formation of 1:1 mercury-conjugated complex was reported. In this reaction, however, the formation of insoluble black precipitate of HgS was not observed under identical reaction condition even after 15 days of stirring. Similarly, the reactions of **4** and **5** (in 15 nM) with one equivalent of ArHgOH did not result in the formation of the corresponding nanoparticles at 37°C or less temperature for almost 7 days. Interestingly, compounds **7** and **8** with N-CH$_2$CH$_2$OMe group imidazole ring, where the –OH group is replaced with –OMe group, failed to produce any HgS or HgSe nanoparticles upon treatment with ArHgOH or RHgCl (R=Ph, Et) in the presence or absence of KOH under identical reaction conditions. The high reactivity of compounds **1** and **2** is attributed to the ability of –OH group of N-CH$_2$CH$_2$OH moiety in these two compounds to participate in facile desulfurization and deselenization process (Figure 49.9).

Based on the experimental observations, Banerjee et al. 2015 reported a novel pathway on the degradation of various organomercurial compounds by compounds **1** and **2**. In general, compounds **1** and **2** first react with ArHgOH to produce the corresponding 1:1 mercury-conjugated complexes **9** and **10**, which in turn further react with another molecule of ArHgOH to produce ketone **11** and the corresponding nanoparticle, as shown in Scheme 49.5. In fact, it has been observed that the addition of one more equivalent of ArHgOH, ([ArHgOH]/[**2**] ratio of 2:1) drove step 2 in the

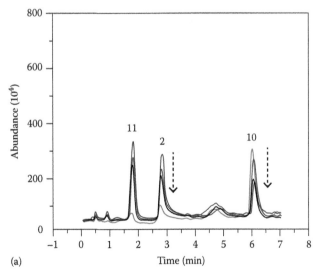

(a)

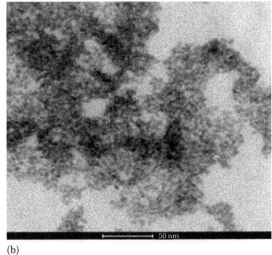

(b)

FIGURE 49.9 (a) LC/MS chromatogram of compound **2** and ketone **11** obtained by treating compound **2** with ArHgOH (1:1). (b) TEM image of HgSe obtained in the reaction of **2** with ArHgOH.

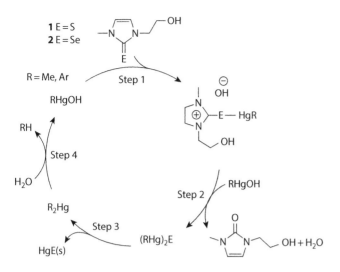

SCHEME 49.5 The possible pathway for the detoxification of various toxic organomercurials by compounds **1** and **2**.

forward direction and consequently the formation of ketone **11** was significantly increased over the time with progress of the reaction.

On the other hand, the formation of white-colored precipitate of $(MeHg)_2S$ was reported in the reaction of MeHgCl with H_2S that gradually decomposed into a black precipitate of β-HgS and $(CH_3)_2$ Hg (Rowland et al., 1977; Craig and Bartlett, 1978; Baldi et al., 1993). Similarly, the formation of selenium analog of $(MeHg)_2S$, $(MeHg)_2Se$, was observed in the blood of rabbit when MeHgCl and sodium selenite were added to the blood (Naganuma and Imura, 1980). Thus, it is likely that the 1:1 mercury-conjugated complexes decompose further into a ketone **11** and the corresponding unstable intermediates $(RHg)_2E$, which subsequently decompose to the corresponding insoluble HgE nanoparticles and R_2Hg, as shown in step 3, Scheme 49.5. Finally, the dialkyl/diaryl mercury, R_2Hg, formed in step 3 further reacts with water to yield RHgOH and RH and thereby complete the cycle (Khan and Wang, 2009b, 2010; Asaduzzaman and Schreckenbach, 2011). Overall, in every cycle, two molecules of RHgOH react with one molecule of **1** or **2** to form one molecule of insoluble HgS(s) or HgSe(s) nanoparticle, which precipitates out from the cycle (Banerjee et al., 2015).

49.7 CONCLUSION

Mercury is a global threat to human and environmental health. Due to increasing human activities, industrialization as well as urbanization, the anthropogenic emission of mercury in the environment has increased manyfold in the last several decades. Organomercurials are highly toxic to humans both at high levels through poisonings in the environment and at lower levels through consumption of seafood. Thus, the efficient detoxification of various organomercurials must be adapted quickly to deal this current situation, by either chemical or biological remediation.

REFERENCES

Aposhian, H. V. 1982. Biological chelation: 2,3-dimercaptopropanesulfonic acid and meso-dimercaptosuccinic acid. *Advances in Enzyme Regulation* 20: 301–319.

Aposhian, H. V. and M. M. Aposhian. 1990. Meso-2,3-dimercaptosuccinic acid: Chemical, pharmacological and toxicological properties of an orally effective metal chelating agent. *Annual Review of Pharmacology and Toxicology* 30: 279–306.

Aposhian, H. V., R. M. Maiorino, D. Gonzalez-Ramirez, M. Zuniga-Charles, Z. Xu, K. M. Hurlbut, P. Junco-Munoz, R. C. Dart, and M. M. Aposhian. 1995. Mobilization of heavy metals by newer, therapeutically useful chelating agents. *Toxicology* 97: 23–38.

Asaduzzaman, A. M. and G. Schreckenbach. 2011. Degradation mechanism of methyl mercury selenoamino acid complexes: A computational study. *Inorganic Chemistry* 50: 2366–2372.

Aschner, M. and J. L. Aschner. 1990. Mercury neurotoxicity: Mechanisms of blood-brain barrier transport. *Neuroscience & Biobehavioral Reviews* 14: 169–176.

Baldi, F., M. Pepi, and M. Filippelli. 1993. Methylmercury resistance in *Desulfovibrio desulfuricans* strains in relation to methylmercury degradation. *Applied and Environmental Microbiology* 59: 2479–2485.

Ballatori, N. and T. W. Clarkson. 1983. Biliary transport of glutathione and methylmercury. *American Journal of Physiology* 244: 435–441.

Ballatori, N. and T. W. Clarkson. 1984. Dependence of biliary secretion of inorganic mercury on the biliary transport of glutathione. *Biochemical Pharmacology* 33: 1093–1098.

Ballatori, N. and T. W. Clarkson. 1985. Biliary secretion of glutathione and of glutathione-metal complexes. *Toxicological Sciences* 5: 816–831.

Ballatori, N. and W. J. Dutczak. 1994. Identification and characterization of high and low affinity transport systems for reduced glutathione in liver cell Canalicular membranes. *Journal of Biological Chemistry* 269: 19731–19737.

Banerjee, M., R. Karri, K. S. Rawat, K. Muthuvel, B. Pathak, and G. Roy. 2015. Chemical detoxification of organomercurials. *Angewandte Chemie* 127: 9455–9459.

Baoyu Hong, B., R. Nauss, I. M. Harwood, and S. M. Miller. 2010. Direct measurement of mercury (II) removal from organomercurial lyase (MerB) by tryptophan fluorescence: NmerA domain of coevolved γ-proteobacterial mercuric ion reductase (MerA) is more efficient than MerA catalytic core or glutathione. *Biochemistry* 49: 8187–8196.

Barkay, T., S. M. Miller, and A. O. Summers. 2003. Bacterial mercury resistance from atoms to ecosystems. *FEMS Microbiological Review* 27: 355–384.

Begley, T., A. Walts, and C. Walsh.1986a. Bacterial organomercurial lyase: Overproduction, isolation, and characterization. *Biochemistry* 25: 7186–7192.

Begley, T., A. Walts, and C. Walsh.1986b. Mechanistic studies of a protonolytic organomercurial cleaving enzyme: Bacterial organomercurial lyase. *Biochemistry* 25: 7192–7200.

Benison, G. C., D. L. Paola, J. E. Shokes, N. J. Cosper, R. A. Scott, P. Legault, and J. G. Omichinski. 2004. A stable mercury-containing complex of the organomercurial lyase MerB: Catalysis, product release, and direct transfer to MerA. *Biochemistry* 43: 8333–8345.

Bernhoft, R. A. 2012. Mercury toxicity and treatment: A review of the literature. *Journal of Environmental and Public Health* 2012: 10 pp.

Branco, V., J. Canario, A. Holmgren, and C. Carvalho. 2011. Inhibition of the thioredoxin system in the brain and liver of zebra-seabreams exposed to waterborne methylmercury. *Toxicology and Applied Pharmacology* 251: 95–103.

Branco, V., J. Canario, A. Holmgren, and C. Carvalho. 2012. Mercury and selenium interaction *in vivo*: Effects on thioredoxin reductase and glutathione peroxidase. *Free Radical Biology & Medicine* 52: 781–793.

Brown, N. L., T. Misra, J. N. Winnie, A. Schmidt, and M. Seiff. 1986. The nucleotide sequence of the mercuric resistance operons of plasmid R100 and transposon Tn501: Further evidence for *mer* genes, which enhance the activity of the mercuric ion detoxification system. *Molecular Genetics and Genomics* 202: 143–151.

Carvalho, C. M. L., E. H. Chew, S. I. Hashemy, J. Lu, and A. Holmgren. 2008. Inhibition of the human thioredoxin system a molecular mechanism of mercury toxicity. *The Journal of Biological Chemistry* 283: 11913–11923.

Carvalho, C. M. L., J. Lu, X. Zhang, E. S. J. Arnér, and A. Holmgren. 2011. Effects of selenite and chelating agents on mammalian thioredoxin reductase inhibited by mercury: Implications for treatment of mercury poisoning. *The FASEB Journal* 25: 370–381.

Cikrt, M. and V. Benko. 1989. Mercury- selenium interaction: Distribution and excretion of ^{203}Hg^{2+} in rats after simultaneous administration of selenite or selenite. *Toxicology Letters* 48: 159–164.

Clark, D. L., A. A. Weiss, and S. Silver. 1977. Mercury and organomercurial resistances determined by plasmids in *Pseudomonas*. *Journal of Bacteriology* 132: 186–196.

Clarkson, T. W. 1997. The toxicology of mercury. *Critical Reviews in Clinical Laboratory Sciences* 34: 369–403.

Clarkson, T. W. 2002. The three modern faces of mercury. *Environmental Health Perspectives Supplement* 110: 11–23.

Clarkson, T. W. and L. Magos. 2006. The toxicology of mercury and its chemical compounds. *Critical Reviews in Toxicology* 36: 609–662.

Clarkson, T, W., L. Magos, and G. Myers. 2003. The toxicology of mercury-current exposures and clinical manifestations. *The New England Journal of Medicine* 349: 1731–1737.

Craig, P. J. and P. D. Bartlett. 1978. The role of hydrogen sulphide in environmental transport of mercury. *Nature* 275: 635–637.

De, J., H. R. Dash, and S. Das. 2014. *Microbial Biodegradation and Bioremediation: A Potential Tool for Restoration of Contaminated Areas*, Chapter 6. London, U.K.: Elsevier Science Publishing Co Inc., pp. 137–166.

Diapsquale, M., J. Agee, C. McGowan, R. S. Oremland, M. Thomas, D. Krabbenhoft, and C. C. Gilmour. 2000. *Environmental Science and Technology* 34: 4904–4917.

Drasch, G., E. Wanghoefer, G. Roider, and S. Strobach. 1996. Correlation of mercury and selenium in the human kidney. *Journal of Trace Elements in Medicine and Biology* 10: 251–254.

Essa, A. M. M., M. Creamer, N. L. Brown, and L. E. Macaskie. 2006. A new approach to the remediation of heavy metal liquid wastes via off-gases produced by *Klebsiella pneumoniae* M426. *Biotechnology and Bioengineering* 95: 576–583.

Essa, A. M. M., L. E. Macaskie, and N. L. Brown. 2005. A new method for mercury removal. *Biotechnology Letters* 27: 1649–1655.

Farina, M., J. B. T. Rocha, and M. Aschner. 2011. Mechanisms of methylmercury-induced neurotoxicity: Evidence from experimental studies. *Life Sciences* 89: 555–563.

Farina. M., M. Aschner, T. Syversen, D. O. Souza, and J. B. T. Rocha. 2007. Involvement of glutamate and reactive oxygen species in methylmercury neurotoxicity. *Brazilian Journal of Medical and Biological Research* 40: 285–291.

Fitzgerald, W. F., C. H. Lamborg, and C. R. Hammerschmidt. 2007. Marine biogeochemical cycling of mercury. *Chemical Reviews* 107: 641–662.

Fortina, M. G., P. Rudiger, P. Schumann, D. Mora, C. Parini, L. P. Manachini, and E. Stackebrandt. 2001. *Ureibacillus gen.* nov., a new genus to accommodate *Bacillus thermosphaericus* (Andersson et al., 1995), emendation of *Ureibacillus thermos-phaericus* and description of *Ureibacillus terrenus* sp., nov. *International Journal of Systematic and Evolutionary Microbiology* 51: 447–455.

Fox, B. and C. T. Walsh. 1982. Mercuric reductase. Mercuric reductase. Purification and characterization of a transposon-encoded flavoprotein containing an oxidation-reduction-active disulphide. *The Journal of Biological Chemistry* 257: 2498–2503.

Fox, B. and C. T. Walsh. 1983. Mercuric reductase: Homology to glutathione reductase and lipoamide dehydrogenase. Iodoacetamide alkylation and sequence of the active site peptide. *Biochemistry* 22: 4082–4088.

Friedheim, E., J. R. Dasilva, and A. V. Martins. 1954. Treatment of *Schistosomiasis mansoni* with antimony a-dimercaptopotassium succinate (TWSb). *The American Journal of Tropical Medicine and Hygiene* 3: 714–727.

Furukawa, K. and K. Tonomura. 1971. Enzyme system involved in the decomposition of phenyl mercuric acetate by mercury-resistant *Pseudomonas*. *Agricultural and Biological Chemistry* 35: 604–610.

Furukawa, K. and K. Tonomura. 1972. Metallic mercury releasing enzyme in mercury-resistant Pseudomonas. *Agricultural and Biological Chemistry* 36: 217–226.

Glendinning, K. J., L. E. Macaskie, and N. L. Brown. 2005. Mercury tolerance of thermophilic *Bacillus* sp. and *Ureibacillus* sp. *Biotechnology Letters* 27: 1657–1662.

Griffin, H. G., T. J. Foster, S. Silver, and T. K Mishra. 1987. Cloning and DNA sequence of mercuric reductase and organomercurial resistance determinants of plasmids pDU 1358. *Proceedings of the National Academy of Sciences USA* 84: 3112–3116.

Haytham M. W., L. Lauriane, M. Stevenson, M. Ahmed, C. Laurent, J. L. Vanasse, K. J. Wilkinson, S. Jurgen, D. E. Wilcox, and J. G. Omichinski. 2016. Structural and biochemical characterization of a copper-binding mutant of the organomercurial lyase MerB: Insight into the key role of the active site aspartic acid in Hg–carbon bond cleavage and metal binding specificity. *Biochemistry* 55: 1070–1081.

Hines, M. E., M. Horva, J. Faganeli, J. C. Bonzongo, T. Barkay, E. B. Major, K. J. Scott, E. A. Bailey, J. J Warwick, and W. B. Lyons. 2000. Mercury biogeochemistry in the Idrija river, Slovenia, from above the mine into the Gulf of Trieste. *Environmental Research* 83: 129–139.

Hobman, J., G. Kholodii, V. Nikiforov, D. A. Ritchie, P. Strike, and O. Yurieva. 1994. The sequence of the *mer* operon of p327/419, 330 and 05. *Gene* 277: 73–78.

Huang, C-C., M. Narita, T. Yamaglata, Y. Itoh, and G. Endo. 1999. Structure analysis of a class II transposon encoding the mercury resistance of the Gram-positive bacterium *Bacillus megaterium* MB1, a strain isolated from Minamata Bay, Japan. *Gene* 239: 361–366.

Inoue, C., K. Sugawara, T. Shiratori, and T. Kusano. 1991. The merR regulatory gene in *Thiobacillus ferrooxidans* is spaced apart from the mer structural genes. *Molecular Microbiology* 5: 2707–2718.

Inoue, C., K. Sugawara, T. Shiratori, T. Kusano, and Y. Kitagawa. 1989. Nucleotide sequence of the Thiobacillus ferrooxidans chromosomal gene encoding mercuric reductase. *Gene* 84: 47–54.

Iohara, K., R. Iiyama, K. Nakamura, S. Silver, M. Sakai, M. Takeshita et al. 2001. The *mer* operon of a mercury-resistant *Pseudoalteromonas haloplanktis* strain isolated from Minamata Bay. *Applied Microbiology and Biotechnology* 56: 736–741.

Irukayama, K. 1969. The course of Minamata disease and its problems' (in Japanese). *Koushueise.* 33: 70–76.

Izaki, K., Y. Tashiro, and T. Funaba. 1974. Mechanism of mercuric chloride resistance in microorganisms. 3. Purification and properties of a mercuric ion reducing enzyme from Escherichia coli bearing R factor. *The Journal of Biochemistry (Tokyo)* 75: 591–599.

Jackson, W. J. and A. O. Summers. 1982. Polypeptides encoded by the mer operon. *Journal of Bacteriology* 149: 479–487.

Khan, M. A. K. and F. Wang. 2009a. Reversible dissolution of glutathione-mediated $HgSe_x S_{1-x}$ nanoparticles and possible significance in Hg–Se antagonism. *Chemical Research in Toxicology* 22: 1827–1832.

Khan, M. A. K. and F. Wang. 2009b. Mercury-selenium compounds and their toxicological significance: Toward a molecular understanding of the mercury-selenium antagonism. *Environmental Toxicology and Chemistry* 28: 1567–1577.

Khan, M. A. K. and F. Wang. 2010. Chemical demethylation of methylmercury by selenoamino acids. *Chemical Research in Toxicology* 23: 1202–1206.

Kholodi, G. Y., O. V. Yurieva, O. L. Lomovskaya, Z. M. Gorlenko, S. Z. Mindlin, and V. G. Nikiforov. 1993. Tn5053, a mercury resistance transposon with integron ends. *Journal of Molecular Biology* 230: 1103–1107.

Kiyono, M., T. Omura, H. Fujimori, and H. Pan-Hou. 1995. Organo-mercurial resistance determinants in *Pseudomonas* K-62 are present on two plasmids. *Archives of Microbiology* 163: 242–247.

Koeman, J. H., W. H. M. Peeters, C. H. M. Koudstaal-Hol, P. S. Tijoe, and J. J. M. de Goeij. 1973. Mercury-selenium correlations in marine mammals. *Nature* 245: 385–386.

Korbas, M., J. L. O. Donoghue, G. E. Watson, I. J. Pickering, S. P. Singh, G. J. Myers, T. W. Clarkson, and G. N. George. 2010. The chemical nature of mercury in human brain following poisoning or environmental exposure. *ACS Chemical Neuroscience* 1: 810–818.

Kosta, L., A. R. Byrne, and V. Zelenko. 1975. Correlation between selenium and mercury in man following exposure to inorganic mercury. *Nature* 254: 238–239.

Ladagga, R. A., L. Chu, T. K. Misra, and S. Silver. 1987. Nucleotide sequence and expression of the mercurial-resistance operon from *Staphylococcus aureus* plasmid pI258. *Proceedings of the National Academy of Sciences USA* 84: 5106–5110.

Lafrance-Vanasse, J., M. Lefebvre, P. D. Lello, J. Sygusch, and J. Omichinski. 2009. Crystal structures of the organomercurial lyase merb in its free and mercury-bound forms insights into the mechanism of methylmercury degradation. *Journal of Biological Chemistry* 284: 938–944.

Li, P., X. B. Feng, G. L. Qiu, L. H. Shang, and Z. G. Li. 2009. Mercury pollution in Asia: A review of the contaminated sites. *Journal of Hazardous Materials* 168: 591–601.

Lian, P., H-B Guo, D. Riccardi, A. Dong, J. M. Parks, Q. Xu, E. F. Pai, S. M. Miller, D-Q Wei, J. C. Smith, and H. Guo. 2014. X-ray Structure of a Hg^{2+} complex of mercuric reductase (MerA) and quantum mechanical/molecular mechanical study of Hg^{2+} transfer between the C-terminal and buried catalytic site cysteine Pairs. *Biochemistry* 53: 7211–7222.

Liebert, C. A., R. M. Hall, and A. O. Summers. 1999. Transposon Tn21, flagship of the floating genome. *Microbiology and Molecular Biology Reviews* 63: 507–522.

Liebert, C. A., J. Wireman, T. Smith, and A. O. Summers. 1997. Phylogeny of mercury resistance (mer) operons from Gram-negative bacteria isolated from the fecal flora of primates. *Applied and Environmental Microbiology* 63: 1066–1076.

Magos, L., T. W. Clarkson, and A. R. Hudson. 1984. Differences in the effects of selenite and biological selenium on the chemical form and distribution of mercury after the simultaneous administration of $HgCl_2$ and selenium to rats. *Journal of Pharmacology and Experimental Therapeutics* 228: 478–483.

Marvin-Diapsquale, M. C., J. Agee, C. McGowan, R. S. Oremland, M. Thomas, D. Krabbenhoft, and C. C. Gilmour. 2000. Methylmercury degradation pathways: A comparison among three mercury-impacted ecosystems. *Environmental Science and Technology* 34: 4908–4917.

Melnick, J. G. and G. Parkin 2007. Cleaving mercury-alkyl bonds: A functional model for mercury detoxification by MerB. *Science* 317: 225–227.

Melnick, J. G., K. Yurkerwich, and G. Parkin. 2009. Synthesis, structure, and reactivity of two-coordinate mercury alkyl compounds with sulfur ligands: Relevance to mercury detoxification. *Inorganic Chemistry* 48: 6763–6772.

Melnick, J. G., K. Yurkerwich, and G. Parkin. 2010. On the chalcogenophilicity of mercury: Evidence for a strong Hg–Se bond in [TmBut] HgSePh and its relevance to the toxicity of mercury. *Journal of the American Chemical Society* 132: 647–655.

Misra, T. K., N. L. Brown, D. C. Fritzinger, R. D. Pridmore, W. M. Barnes, L. Haberstroh et al. 1984. The mercuric-ion resistance operon of plasmid R100 and transposon Tn501: The beginning of the operon including the regulatory region and the first two structural genes. *Proceedings of the National Academy of Sciences USA* 81: 5975–5979.

Moore, M. J., M. D. Distefano, L. D. Zydowsky, R. T. Cummings, and C. T. Walsh. 1990. Organomercurial lyase and mercuric ion reductase: Nature's mercury detoxification catalysts. *Account of Chemical Research* 23: 301–308.

Morel, F. M. M., A. M. L. Kraepiel, and M. Amyot. 1998. The chemical cycle and bioaccumulation of mercury. *Annual Review of Ecology and Systematics* 29: 543–566.

Mutter, J., J. Naumann, and C. Guethlin. 2007. Comments on the article "The Toxicology of Mercury and Its Chemical Compounds" by Clarkson and Magos (2006). *Critical Reviews in Toxicology* 37: 537–549.

Naganuma, A. and N. Imura. 1980. Bis(methylmercuric) selenide as a reaction product from methylmercury and selenite in rabbit blood. *Research Communications in Chemical Pathology and Pharmacology* 27: 163–173.

Nakazawa, E., T. Ikemoto, A. Hokura, Y. Terada, T. Kunito, S. Tanabe, and I. Nakai. 2011. The presence of mercury selenide in various tissues of the striped dolphin: Evidence from μ-XRF-XRD and XAFS analyses. *Metallomics* 3: 719–725.

Ng, P-S., K. Matsumoto, S. Yamazaki, T. Kogure, T. Tagai, and H. Nagasawa. 2001. Striped dolphin detoxificates mercury as insoluble Hg(S, Se) in the liver. *Proceedings of the Japan Academy* 77: 178–183.

Nielsen, J. B. and O. Andersen. 1992. The toxicokinetics of mercury in mice offspring after maternal exposure to methylmercury–effect of selenomethionine. *Toxicology* 74: 233–241.

Nigro, M. and C. Leonzio. 1996. Intracellular storage of mercury and selenium in different marine vertebrates. *Marine Ecology Progress Series* 135: 137–143.

Omichinski, J. G. 2007. Toward methylmercury bioremediation. *Science* 317: 205–206.

Osborn, A. M., K. D. Bruce, P. Strike, and D. A. Ritchie. 1997. Distribution, diversity and evolution of the bacterial mercury resistance (mer) operon. *FEMS Microbiology Reviews* 19: 239–262.

Pan-Hou, H. S. K., and N. Imura. 1981. Role of hydrogen sulfide in mercury resistance determined by plasmid of *Clostridium cochlearium* T-2. *Archives of Microbiology* 129: 49–52.

Parks, J, M., H. Guo, C. Momany, L. Liang, S. M. Miller, A. O. Summers, and J. C. Smith. 2009. Mechanism of Hg–C protonolysis in the organomercurial lyase MerB. *Journal of the American Chemical Society* 131: 13278–13285.

Peters, R. A. and L. A. Stocken. 1945. British anti-lewisite (BAL). *Nature* 156: 616–619.

Petrunkin, V. E. 1956. Synthesis and properties of mercapto derivatives of alkylsulfonic acids. *Ukrainskii Khimicheskii Zhurnal* 22: 603–607.

Pitts, K, E. and A. O. Summers. 2002. The roles of thiols in the bacterial organomercurial lyase (MerB). *Biochemistry* 41: 10287–10296.

Rabenstein, D. L. 1978. The aqueous solution chemistry of methylmercury and its complexes. *Account of Chemical Research* 11: 100–107.

Radstrom, P., O. Skold, G. Swedberg, J. Flensburg, P. H. Roy, and L. Sundstrom. 1994. Transposon Tn5090 of the plasmid R751, which carries integron, is related to Tn7, Mu, and the retroelements. *Journal of Bacteriology* 176: 3257–3268.

Ravel, J., J. DiRuggiero, F. T. Robb, and R. T. Hill. 2000. Cloning and sequence analysis of the mercury resistance operon of *Streptomyces* sp. strain CHR28 reveals a novel putative second regulatory gene. *Journal of Bacteriology* 182: 2345–2349.

Reniero, D., E. Galli, and P. Barbieri. 1995. Cloning and comparison of mercury- and organomercurial-resistance determinants from a *Pseudomonas stutzeri* plasmid. *Gene* 166: 77–82.

Rinderle, S. J., J. E. Booth, and J. W. Williams. 1983. Mercuric reductase from R-plasmid NR1: Characterization and mechanistic study. *Biochemistry* 22: 869–876.

Rowland, I. R., M. J. Davies, and P. Grasso. 1977. Volatilisation of methylmercuric chloride by hydrogen sulphide. *Nature* 265: 718–719.

Sahlman, L. and S. Lindskog. 1983. A stopped-flow study of the reaction between mercuric reductase and NADPH. *Biochemical and Biophysical Research Communications* 117: 231–237.

Schottel, J., A. Mandal, D. Clark, S. Silver, and R. W. Hedges. 1974. Volatilisation of mercury and organomercurials determined by inducible R-factor systems in enteric bacteria. *Nature* 251: 335–337.

Schottel, J. L. 1978. The mercuric and organomercurial detoxifying enzymes from a plasmid-bearing strain of Escherichia coli. *The Journal of Biological Chemistry* 253: 4341–4349.

Scholz, T., W. Demharter, R. Hensell, and O. Kandler. 1987. Bacillus pallidus sp. nov., a new thermophilic species from sewage. *Systematic and Applied Microbiology* 9: 91–96.

Sedlmeier, R. and J. Altenbuchner. 1992. Cloning and DNA sequence analysis of the mercury resistance genes of *Streptomyces lividans. Molecular Gene and Genetics* 236: 76–85.

Selin, N. E. 2009. Global biogeochemical cycling of mercury: A review. *Annual Review of Environment and Resources* 34: 43–63.

Silver, S. and T. K. Misra. 1984. Bacterial transformations of and resistances to heavy metals. *Basic Life Science* 28: 23–46.

Strasdeit, H. 2008. Mercury–alkyl bond cleavage based on organomercury lyase. *Angewandte Chemie International Edition* 47: 828–830.

Summers, A. O. and E. Lewis. 1973. Volatilization of mercuric chloride by mercury resistant plasmid bearing strain of *Escherichia coli, Staphylococcus aureus* and *Pseudomonas aeruginosa. Journal of Bacteriology* 113: 1070–1072.

Summers, A. O. and S. Silver. 1978. Microbial transformations of metals. *Annual Review of Microbiology* 32: 637–672.

Tezuka, T. and K. Tonomura. 1976. Purification and properties of an enzyme catalyzing the splitting of carbon-mercury linkages from mercury-resistant *Pseudomonas* K-62 strain. I. Splitting enzyme 1. *The Journal of Biochemistry (Tokyo)* 80: 79–87.

Uchino, M., Y. Tanaka, Y. Ando, T. Yonehara, A. Hara, I. Mishima, T. Okajima, and M. J. Ando. 1995. Neurologic features of chronic minamata disease (organic mercury poisoning) and incidence of complications with aging. *Journal of Environmental Science and Health Part B* 30: 699–715.

UNEP. 2013. Global Mercury Assessment 2013: Sources, Emissions, Releases and Environmental Transport. UNEP Chemicals Branch, Geneva, Switzerland.

Wahba, H. M., L. Lecoq, M. Stevenson, A. Mansour, L. Cappadocia, J. Lafrance-Vanasse, K. J. Wilkinson, J. Sygusch, D. E. Wilcox and J. G. Omichinski. 2016. Structural and biochemical characterization of a copper-binding mutant of the organomercurial lyase MerB: Insight into the key role of the active site aspartic acid in Hg-carbon bond cleavage and metal binding specificity. *Biochemistry* 55: 1070–1081.

Walsh, C. T., M. D. Distefano, M. J. Moore, L. M. Shewchuk, and G. L. Verdine. 1988. Molecular basis of bacterial resistance to organomercurial and inorganic mercuric salts. *The FASEB Journal* 2: 124–130.

Wang, Y., I. Mahler, H. S. Levinson, and H. O. Halvorson. 1987. Cloning and expression in Escherichia coli of chromosomal mercury resistance genes from a *Bacillus* sp. *Journal of Bacteriology* 16: 94848–94851.

Wang, Y., M. Moore, H. S. Levinson, S. Silver, C. Walsh, and I. Mahler. 1989. Nucleotide sequence of a chromosomal mercury resistance determinant from a *Bacillus* sp. with broad-spectrum mercury resistance. *Journal of Bacteriology* 171: 83–92.

Wangberg, I., J. Moldanova, and J. Munthe. 2010. Mercury cycling in the environment effects of climate change. Report by Swedish Environmental Research Institute. Report for EUROLIMPAC, Stockholm, Sweden.

Weiss, A. A., S. D. Murphy, and S. Silver. 1977. Mercury and organomercurial resistances determined by plasmids in *Staphylococcus aureus. Journal of Bacteriology* 132: 197–208.

Williams, C. H. Jr. 1992. Lipoamide dehydrogenase, glutathione reductase, thioredoxin reductase, and mercuric ion reductase– A family of flavoenzyme transhydrogenases. In: *Chemistry and Biochemistry of Flavoenzymes*, Vol. 3 (Muller, F., Ed.), pp. 121–211. CRC Press, Boca Raton, FL.

Yin, Z., H. Jiang, T. Syversen, J. B. Rocha, and M. Farina. 2008. The methylmercury-L-cysteine conjugate is a substrate for the L-type large neutral amino acid transporter. *Journal of Neurochemistry* 107: 1083–1090.

50 Assessment of the Diversity and Bioremediation Potential of Mercury-Resistant Marine Bacteria in the Bay of Bengal, Odisha, India

Hirak Ranjan Dash and Surajit Das

CONTENTS

ABSTRACT

Marine environment is the most dynamic among the existing environmental conditions. The increase in mercury pollution in the Bay of Bengal, Odisha, India, has substantially increased the number of mercury-resistant marine bacterial (MRMB) population as observed in the study sites. Representative 25 isolates with distinguished colony morphology, 5 from each study sites, were characterized for their phenotypic and genotypic characteristics. The minimum inhibitory concentration (MIC) of the isolates was found to be in the range of 12.5–50 ppm. All the isolates were not only found to tolerate higher amounts of mercury but also showed resistance to other toxic metals such as Cd, Zn, Pb, As, and many antibiotics. Phenotypic mercury resistance in isolated MRMB revealed the mechanism of the resistance phenomenon. 36% and 28% of the screened isolates were found to resist mercury by *mer* operon–mediated volatilization and mercury biosorption mechanisms, respectively. However, in many isolates, the mercury-resistant genetic determinant could not be identified, and the novel search of which may provide many advantageous approaches to deal with the globally rising mercury pollution through bioremediation. In addition to that, volatilization of mercury by the isolates was found to be more effective than biosorption in terms of mercury removal by the isolates.

50.1 INTRODUCTION

Microorganisms are known for their high level of adaptation to noxious environmental conditions. They have evolved a variety of mechanisms that enable them to live in the presence of toxic concentrations of metals (Silver and Phung, 2005). These mechanisms may include efflux of toxic metals that enter cells by essential metal transporters, intra- or extracellular precipitation of metals, or enzymatic transformations of the metals to decrease their toxicity in the environment (Nies, 1999). Concerning the public and environmental health as well as the impact on the environment, a detailed understanding on the resistant bacteria in the contaminated environment is of utmost importance. Ubiquitous presence and diversity of the resistant bacteria in environmental conditions imply the evolution of mercury resistance genotype prior to the spread of metal contaminants by anthropogenic activities (Chatziefthimiou et al., 2007). Environments with geological processes resulting in the formation of metal-rich deposits and evolution of microorganisms in the presence of such toxic metals are taking place since the evolution of life on earth.

The characteristic feature of marine environment is unique in nature, and the microorganisms inhabiting in it possess the enormous potential to overcome the continuously changing pattern of pH, temperature, salinity, sea temperature, and other variable parameters (Dash et al., 2013). There are many reports on the utilization of marine bacteria for the bioremediation of toxic substances (Chikere et al., 2012; Naik et al., 2012). Mercury is a highly toxic metal; and its concentration is increasing at an alarming rate in the environment due to various anthropogenic activities such as discharges from chloralkali industries, coal and metal mining, gold mining, and cement production. However, mercury-resistant bacteria possessing the characteristic feature of transforming both organic and inorganic forms of mercury to a less-toxic elemental form are mostly by *mer* operon–mediated mechanism located on either transposons, plasmids, or bacterial chromosomes (Barkay et al., 2003).

Many organic and/or inorganic mercury-resistant bacteria have been previously reported belonging to genera *Escherichia*, *Bacillus*, *Pseudomonas*, *Staphylococcus*, and *Brevibacterium* from different contaminated environmental conditions (Mindlin et al., 2005; De and Ramaiah, 2007; Poulain et al., 2007; Mirzaei et al., 2008; Bafana et al., 2010; Deng and Wang, 2012). In this regard, mercury is a toxic metal that is difficult to remove from the environment, as it is degraded chemically or biologically and hence becomes indestructible. However, physicochemical remediation techniques of polluted sites including incineration of soils to chemical precipitation and ion-exchange technologies used widely so far are costly and environmentally damaging (Gadd, 2010). Biological processes based on metal-resistant microorganisms have received huge attention as an alternative remediation process. Currently, the biological processes involving mercury removal from the environment includes the utilization of *mer* operon of bacterial system for the reduction of Hg^{2+} to Hg^0. In another approach, live or dead bacterial biomass has also been used for biosorption applications, which consist of passive immobilization of metals by the biomass and can rely on different physicochemical mechanisms, such as adsorption, surface complexation, ion exchange, or surface precipitation (Le Cloirec and Andrès, 2005).

In this regard, microbial population inhabiting in the marine environment are of varied nature. Due to their strain-specific character, some of them exhibit higher tolerance to the toxic metals and some of them tolerate at a lower level.

Hence, it is highly important to distinguish between the potent isolates from the nonpotent ones. This relies on various screening techniques such as determination of MIC, colony morphology, biochemical characters, and others. Marine environment is the relatively untapped environment for microbial study, and many important discoveries have taken place in the past utilizing marine microorganisms. Hence, proper characterization of the mercury-resistant marine bacteria (MRMB) may lead to the discovery of novel bioremediation approaches for the bioremediation of mercury. In this regard, the objective was to screen the potent MRMB isolates and to characterize them both phenotypically and genotypically.

50.2 MATERIALS AND METHODS

50.2.1 STUDY SITES, SAMPLING, ISOLATION, PURE CULTURES, AND SCREENING OF POTENT ISOLATES

Water and sediment samples were collected from five study sites along the Bay of Bengal, Odisha coast, namely, Bhitarkanika National Park (20.7500°N, 87.0000°E), Paradeep (20.3200°N, 86.6200°E), Chilika Lake (19.7167°N, 85.3167°E), Rushikulya (19.07°–20.19°N, 84.01°E–85.0°E), and Gopalpur (19.2700°N, 84.9200°E). MRMB were isolated by plating the samples on seawater nutrient agar (SWNA) plates supplemented with 10 ppm of mercury as $HgCl_2$. The individual colonies were streaked repeatedly on the same plate to obtain pure cultures. To screen the marine bacterial populations for higher resistant bacteria, agar well diffusion technique was conducted (Devillers et al., 1989).

50.2.2 DETERMINATION OF MIC AND BIOCHEMICAL CHARACTERIZATION

The MIC of the isolates was determined following microbroth dilution technique (CLSI, 2006). Various biochemical tests and carbohydrate utilization tests were performed using Rapid Biochemical Identification kit [KB003 Hi25®] (HiMedia, India). The result was interpreted as per the manufacturer's Result Interpretation Chart.

50.2.3 TOLERANCE TO TOXIC METALS AND ANTIBIOTICS

Several other toxic metal tolerance patterns of the isolates were determined using the same protocol as MIC of mercury for four other toxic metals: cadmium ($CdCl_2$), zinc ($ZnSO_4 \cdot H_2O$), lead ($PbNO_3$), and arsenic ($Na_2HAsO_4 \cdot 7H_2O$). Antibiotic susceptibility test was performed using disc diffusion technique (Bauer et al., 1966). The sensitivity of all the MRMB to amoxicillin (AM), vancomycin (VA), ampicillin (AMP), chloramphenicol (C), tetracycline (TE), norfloxacin (NX), ciprofloxacin (CF), azithromycin (AZM), neomycin (N), cefotaxime (CTX), and clavulanic acid (AC) was examined. The isolates were determined to be susceptible or resistant to the antibiotics based on the sensitivity chart supplied by the manufacturer (HiMedia, India).

50.2.4 Mechanism of Mercury Resistance, Amplification of *merA* Gene, and Localization of *mer* Operon

The isolates were analyzed for their resistance mechanism to mercury. Volatilization of mercury by *mer* operon and biosorption of mercury were analyzed during this study. A conserved region of *merA* gene was amplified using the primer sets of F1 merA-'TCGTGATGTTCGACCGCT3' and F2 merA-5'TAXTXXXΓXXΓTTTXXAAT3' (Sotero-Martins et al., 2008). In order to confirm the presence of functional *mer* operon in the isolates, volatilization of mercury was studied using simplified x-ray film assay (Nakamura and Nakahara, 1988). H$_2$S assay was performed to determine the biosorption of mercury by the isolated bacterial cell mass. In order to locate the molecular address of *mer* operon in the MRMB isolates, PCR amplification of *merA* gene was performed independently using genomic DNA and plasmids as templates.

50.2.5 Volatilization and Biosorption Potential of the Isolates

The isolates were inoculated in Sea Water Nutrient (SWN) medium containing sublethal concentration of Hg as HgCl$_2$ and grown at 30°C for 48 h. Both cell pellet and supernatant were collected by centrifugation, and the cell pellet was resuspended in lysis buffer [100 mM NaH$_2$PO$_4$, 10 mM Tris–Cl, 8 M urea (pH 8.0)] for 1 h and centrifuged at 13,000 rpm for 10 min to collect cell-associated mercury. The level of mercury in both supernatant and cells was determined by cold vapor atomic absorption spectrophotometer (Perkin-Elmer AAnalyst™ 200). A standard solution of HgCl$_2$ (Merck, India) was used in this experiment.

50.2.6 Expression Pattern of *merA* Gene by RT-PCR Analysis

Real-time polymerase chain reaction (RT-PCR) was carried out following Zhang et al. (2012) to study the expression pattern of *merA* gene under the same conditions in the isolates. All the isolates were grown in Luria-Bertani (LB) broth supplemented with 10 ppm of mercury as HgCl$_2$ for 48 h. RNA was extracted from cells using RNA purification kit (Fermentas, USA). cDNA was prepared from the extracted RNA using random hexamer primer using RevertAid First Strand cDNA synthesis kit (Thermo Scientific, USA). The synthesized cDNA was used as the template to carry out RT-PCR in an Eppendorf real-time detection system by using SYBR® Green JumpStart™ Taq ReadyMix™ (Sigma-Aldrich, USA). The primers used for this study were merAF (5'GAGATCTAAAGCACGCTAAGGC3') and merAR (5'GGAATCTTGACTGTGATCGGG3'). 16S rRNA was used as an appropriate internal control for this experiment. Dissociation analysis was performed at the end of each PCR run to confirm that only PCR product was amplified and detected. The comparative threshold cycle method (2-ΔΔCT)

was used to analyze the relative mRNA expression. As 16S rRNA gene is regarded as the housekeeping gene in bacteria, its expression level remains constant throughout the experimental conditions. Thus, in order to measure the correct level of expression of *merA* gene, the expression level of 16S rRNA gene was normalized by using Mastercycler ep realplex (Eppendorf, Germany).

50.2.7 Amplification of 16S rRNA Gene, Sequencing, and Phylogeny of the Isolates

The isolated MRMB were identified by the sequencing of 16S rRNA gene, and the obtained sequences were submitted to NCBI GenBank to obtain the accession number. Multiple alignments of sequences were performed with the ClustalX 1.83 (Thompson et al., 1997). A phylogenetic tree was constructed using the neighbor-joining DNA distance algorithm (Saitou and Nei, 1987) using Mega 6. The resultant tree topologies were evaluated by bootstrap analysis (Felsenstein, 1985) of neighbor-joining data sets based on 1000 resamplings.

50.3 RESULT AND DISCUSSION

50.3.1 Potent Isolates, MIC, and Biochemical Characteristics

Based on the lowest zone of inhibition by agar well diffusion technique, a total of 25 isolates, 5 from each study sites, were selected for further study. When the isolates were subjected to determination of MIC, all of them showed a higher level of tolerance toward mercury in the form of HgCl$_2$. The highest tolerance limit was observed by three isolates, namely, 50 ppm in PW-05, GW-702, and BW-03. However, the range of MIC varied between 12.5 and 50 ppm. Isolates PW-216, BW-201T, CS-605, CW-302, RW-401, and GW-605 could tolerate 25 ppm of mercury, whereas the rest of the isolates could resist 12.5 ppm of the same (Figure 50.1). All the 25 selected isolates showed distinct biochemical characteristics. Most of them could utilize various carbon sources and were positive for various biochemical tests (Table 50.1). Out of the 25 bacterial isolates, 20 were found to be Gram positive in nature.

Agar well diffusion technique is the most useful technique to assess the antimicrobial potential of any substances such as antibiotics or toxic metals (Kumar and Rao, 2012). The higher the zone of inhibition, the lesser is the tolerance of bacteria toward those substances. Thus, this technique was appropriate to screen a huge number of bacterial strains for the level of mercury tolerance. In general practice, when a constant concentration of mercury as mercury chloride was used for the screening of higher mercury tolerating bacterial population, the zone of inhibition was found to be crucial. The higher the resistance limit in the isolate, the lesser was the trend of zone of inhibition in the petri plate. The maximum zone of inhibition was observed in the range of 15 mm, whereas 12 isolates did not produce any zone of inhibition.

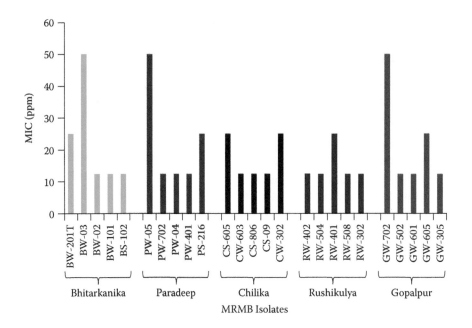

FIGURE 50.1   Minimum inhibitory concentration (MIC) of the screened isolates for mercury as HgCl$_2$ as determined by micro-broth dilution technique.

There are many reports suggesting the higher tolerance limit of mercury by the resistant isolates, that is, in the rage of 2–200 ppm (Moghbeli et al., 2011; Møller et al., 2011; Bhattacharyya et al., 2013). However, most of the marine isolates were considered to be resistant to mercury if they can resist an amount of mercury, that is, equal to or above 10 ppm (Ramaiah and De, 2003). Micro-broth dilution technique is the most useful technique for the determination of MIC of metals or antibiotics. This technique has many advantages over the conventional practices as it consumes lesser time for completion as well as it proves to be cost-effective. In this regard, the MIC level of all the 25 isolates was found to be in the range of 12.5–50 ppm. Many other works also proposed the same range of mercury resistance in the marine isolates (Vetriani et al., 2005; François et al., 2012; Monachese et al., 2012).

Many advancements have taken place regarding the characterization of the bacterial isolates; still biochemical characterization of the bacterial entities provides important information. This is due to the fact that certain closely related bacterial species cannot be distinguished by genotypic characterization. In addition, certain bacterial species possess unique biochemical characteristics such as optochin susceptibility by *Streptococcus pneumoniae* that distinguishes it from other *Streptococcus* spp. (Pikis et al., 2001). Thus, the biochemical analysis of bacteria still persists for their proper analysis and characterization. In most of the cases, the biochemical test results corroborated with the identification of the bacterial isolates by 16S rRNA gene sequencing as described later. Biochemical test provided a greater insight to the utilization of diverse sugars and other nitrogen sources besides their physiological characteristics (Zourari et al., 1992; Brandelli et al., 2010).

50.3.2   TOLERANCE TO TOXIC METALS AND ANTIBIOTICS

All the isolates showed a significant amount of tolerance toward the toxic metals tested during this study. Out of all the isolates, PW-05 was found to resist the higher toxic metal concentration compared to the other isolates. Similarly, based upon the resistance/susceptible/intermediate pattern, the antibiogram of the isolates were determined. Out of all the isolates tested, GW-702 and GW-502 were found to resist the maximum number of antibiotics, whereas CW-603, GS-305, and RW-401 were susceptible to a maximum number of antibiotics tested. The detailed results of the toxic metal tolerance and antibiotics susceptibility of the isolates have been provided in Table 50.2.

There are many reports suggesting the coexistence of metal- and antibiotic-resistant genes along the bacterial genome (Groves et al., 1975; Davies and Davies, 2010; Knapp et al., 2011; Chait et al., 2012; Mathiyazhagan and Natarajan, 2012). As the isolates used in this study have been screened for metal resistance characteristics, the presence of antibiotic resistance in them de novo cannot be ignored. Out of the 25 isolates characterized, 16 of them were resistant to amoxicillin, 13 showed vancomycin resistance, 10 were resistant for ampicillin, 2 for ciprofloxacin, 4 were for tetracycline, none for norfloxacin, 4 for azithromycin, 2 for neomycin, 10 for cotrimoxazole, and 10 showed resistance toward acidoclav. The isolate PW-702 was resistant to most number of antibiotics, that is, 9 out of 11 antibiotics tested. In contrary to that, four isolates (CW-603, RW-302, GW-601, and GS-305) did not show any resistance pattern to any of the antibiotics tested. Antibiotic resistance may develop in environmental strains of bacteria due to the antibiotic stress in the environment. However, antibiotic resistance in marine

TABLE 50.1
Biochemical Characteristics of the 25 Selected Mercury-Resistant Marine Bacterial Isolates from Study Sites along the Bay of Bengal, Odisha, India

Sl. No.	Isolates	Identification	Gram Nature	Morphology	Biochemical Results	
					Positive Results	Negative Results
1.	BS-102	*Bacillus cereus*	+ve	Rods	Catalase, glucose, starch hydrolysis, nitrate reduction, motility, gelatin, casein, VP, maltose, trehalose, citrate	Indole, lactose
2.	BW-03	*B. cereus*	+ve	Rods	Catalase, glucose, starch hydrolysis, nitrate reduction, motility, gelatin, casein, VP, maltose, trehalose, citrate	Indole, lactose
3.	BW-201T	*Vibrio hepatarius*	−ve	Slightly curved cells	Indole, oxidase, catalase, nitrate, glucose, mannitol, sucrose, maltose, fructose, trehalose, glycerol	H_2S, urease, lysine decarboxylase, ornithine, galactosidase, rhamnose, melibiose, rhamnose
4.	BW-02	*Bacillus infantis*	+ve	Rods	Glycogen, inulin, mannitol, salicin, trehalose, starch	Glycerol
5.	BW-101	*Bacillus licheniformis*	+ve	Rods	Arginine, dihydrolase, esculin hydrolysis, beta-galactosidase, phenylalanine, cellobiose, fructose, galactose, glycerol, salicin, starch, glycogen, lactose, mannose, maltose, ribose, sorbitol, sucrose, trehalose	Lysine, ornithine, tryptophan, urea, fructose, arabinose, inulin, raffinose, rhamnose
6.	PW-05	*Bacillus thuringiensis*	+ve	Rods	Catalase, glucose, starch hydrolysis, nitrate reduction, motility, gelatin, casein, VP, maltose, trehalose, citrate	Indole, lactose
7.	PW-401	*Exiguobacterium* sp.	+ve	Rods	Starch, casein, glucose, ribose, maltose, mannitol, glycogen, gluconate	Gelatin, nitrate, sucrose, sorbitol, salicin, starch
8.	PW-216	*Pseudomonas* sp.	−ve	Rods	Oxidase, catalase, tryptophan deaminase, arginine, lysine, ornithine, urea hydrolysis, VP, citrate	β-Galactosidase, gelatin hydrolysis, indole production, H_2S production, arabinose, arabitol, fructose, glucose, galactose, lactose, mannitol, mannose, rhamnose, ribose
9.	PW-702	*Bacillus* sp.	+ve	Rods	Catalase, glucose, starch hydrolysis, nitrate reduction, motility, gelatin, casein, VP, maltose, trehalose, citrate	Lysine, ornithine, tryptophan, urea, fructose, arabinose, inulin, raffinose, rhamnose
10.	PW-04	*B. licheniformis*	+ve	Rods	Arginine, dihydrolase, esculin hydrolysis, beta-galactosidase, phenylalanine, cellobiose, fructose, galactose, glycerol, salicin, starch, glycogen, lactose, mannose, maltose, ribose, sorbitol, sucrose, trehalose	Lysine, ornithine, tryptophan, urea, fructose, arabinose, inulin, raffinose, rhamnose
11.	CS-806	*B. cereus*	+ve	Rods	Catalase, glucose, starch hydrolysis, nitrate reduction, motility, gelatin, casein, VP, maltose, trehalose, citrate	Indole, lactose
12.	CS-605	*Bacillus* sp.	+ve	Rods	Catalase, glucose, starch hydrolysis, nitrate reduction, motility, gelatin, casein, VP, maltose, trehalose, citrate	Lysine, ornithine, tryptophan, urea, fructose, arabinose, inulin, raffinose, rhamnose
13.	CS-09	*Escherichia coli*	−ve	Rods	Catalase, nitrate reduction, acid production from arabinose, mannitol, mannose, trehalose	Oxidase, H_2S production, urease, VP, gelatinase, citrate utilization, inositol
14.	CW-302	*Bacillus subtilis*	+ve	Rods	Galactosidase, esculin hydrolysis, catalase, VP, starch hydrolysis, citrate, nitrate, casein, glucose, mannose, glycerol, glycogen, salicin, cellobiose, fructose, maltose, raffinose, ribose, sucrose, trehalose, mannitol	Tyrosine, phenylalanine, lysine, ornithine, adonitol, arabinose, arabitol, fructose, rhamnose, sorbose, xylose

(Continued)

TABLE 50.1 (Continued)

Biochemical Characteristics of the 25 Selected Mercury-Resistant Marine Bacterial Isolates from Study Sites along the Bay of Bengal, Odisha, India

Sl. No.	Isolates	Identification	Gram Nature	Morphology	Biochemical Results Positive Results	Negative Results
15.	CW-603	*Exiguobacterium* sp.	+ve	Rods	Glucose, sucrose, maltose, lactose, fructose, mannose, mannitol, galactose, arabinose, VP, citrate, H_2S production, starch hydrolysis, casein hydrolysis, gelatin hydrolysis	Indole, MR, nitrate reduction, catalase
16.	RW-402	*B. cereus*	+ve	Rods	Catalase, glucose, starch hydrolysis, nitrate reduction, motility, gelatin, casein, VP, maltose, trehalose, citrate	Indole, lactose
17.	RW-401	*B. subtilis*	+ve	Rods	Galactosidase, esculin hydrolysis, catalase, VP, starch hydrolysis, citrate, nitrate, casein, glucose, mannose, glycerol, glycogen, salicin, cellobiose, fructose, maltose, raffinose, ribose, sucrose, trehalose, mannitol	Tyrosine, phenylalanine, lysine, ornithine, adonitol, arabinose, arabitol, fructose, rhamnose, sorbose, xylose

TABLE 50.2

Toxic Metal Tolerance Pattern and Antibiogram of the Mercury-Resistant Bacterial Isolates

Isolates	Toxic Metals (in ppm) Cd	Zn	Pb	As	Antibiogram[a]
V. hepatarius BW-201T	458.3	1614	2500	232.4	AM^R VA^R AMP^R C^S TE^I NX^S CF^I AZM^I N^I CTX^S AC^R
B. cereus BW-03	458.3	1614	2500	464.8	AM^R VA^R AMP^S C^S TE^S NX^S CF^I AZM^I N^S CTX^S AC^S
B infantis BW-02	458.3	1614	1250	464.8	AM^R VA^R AMP^S C^S TE^S NX^S CF^R AZM^S N^S CTX^S AC^R
B. licheniformis BW-101	458.3	1614	1250	464.8	AM^R VA^R AMP^S C^S TE^I NX^S CF^R AZM^I N^I CTX^I AC^R
B. cereus BS-102	160.8	807.35	625	232.4	AM^S VA^R AMP^R C^I TE^S NX^S CF^R AZM^I N^I CTX^R AC^R
B. thuringiensis PW-05	160.8	807.35	1250	232.4	AM^R VA^S AMP^S C^S TE^S NX^S CF^R AZM^I N^I CTX^R AC^R
Bacillus sp. PW-702	160.8	807.35	625	232.4	AM^R VA^R AMP^R C^R TE^R NX^I CF^R AZM^R N^I CTX^R AC^R
B. licheniformis PW-04	458.3	807.35	2500	464.8	AM^R VA^R AMP^R C^S TE^I NX^S CF^R AZM^I N^I CTX^I AC^R
Exiguobacterium sp. PW-401	20.1	201.83	156.25	29.05	AM^R VA^S AMP^I C^S TE^S NX^S CF^R AZM^S N^I CTX^S AC^S
Pseudomonas stutzeri PW-216	20.1	201.83	312.5	29.05	AM^S VA^R AMP^S C^S TE^S NX^S CF^R AZM^S N^S CTX^S AC^S
Bacillus sp. CS-605	458.3	1614	2500	464.8	AM^R VA^R AMP^I C^S TE^I NX^S CF^R AZM^R N^I CTX^I AC^R
Exiguobacterium sp. CW-603	40.2	201.83	156.25	29.05	AM^S VA^S AMP^S C^S TE^S NX^S CF^I AZM^S N^S CTX^I AC^S
B. cereus CS-806	20.1	201.83	312.5	29.05	AM^R VA^S AMP^R C^S TE^S NX^S CF^R AZM^I N^R CTX^I AC^S
E. coli CS-09	160.8	403.67	625	232.4	AM^R VA^S AMP^R C^S TE^R NX^S CF^R AZM^I N^I CTX^R AC^R
B. subtilis CW-302	20.1	201.83	156.25	29.05	AM^I VA^S AMP^S C^R TE^S NX^S CF^R AZM^S N^S CTX^R AC^I
B. cereus RW-402	20.1	201.83	156.25	29.05	AM^R VA^R AMP^R C^S TE^I NX^S CF^I AZM^S N^I CTX^R AC^S
B. cereus RW-504	20.1	201.83	312.5	29.05	AM^R VA^S AMP^R C^S TE^S NX^S CF^R AZM^R N^S CTX^I AC^S
B. subtilis RW-401	40.2	201.83	156.25	29.05	AM^S VA^S AMP^S C^I TE^I NX^S CF^R AZM^S N^S CTX^R AC^I
Alcaligenes faecalis RW-508	20.1	201.83	312.5	29.05	AM^I VA^S AMP^R C^S TE^I NX^S CF^R AZM^R N^I CTX^R AC^S
Staphylococcus aureus RW-302	160.8	807.35	1250	232.4	AM^S VA^S AMP^S C^S TE^I NX^S CF^S AZM^S N^S CTX^I AC^I
Bacillus safensis GW-702	458.3	1614	2500	464.8	AM^R VA^R AMP^R C^I TE^R NX^S CF^R AZM^I N^I CTX^R AC^I
Pseudomonas aeruginosa GW-502	20.1	201.83	312.5	29.05	AM^R VA^R AMP^R C^I TE^R NX^S CF^R AZM^I N^R CTX^R AC^R
Exiguobacterium arabatum GW-601	160.8	807.35	625	232.4	AM^S VA^S AMP^I C^S TE^S NX^S CF^S AZM^S N^I CTX^I AC^S
Bacillus pumilus GW-605	160.8	807.35	1250	29.05	AM^R VA^R AMP^I C^S TE^S NX^S CF^R AZM^I N^S CTX^I AC^I
B. pumilus GS-305	20.1	201.83	156.25	29.05	AM^S VA^S AMP^S C^S TE^S NX^S CF^I AZM^S N^S CTX^S AC^I

[a] Resistance/intermediate/sensitive toward antibiotics was detected on the basis of zone of inhibition and by comparing with the chart provided by the manufacturer.

bacteria may prove to be problematic if the strains are pathogenic in nature (Grimes et al., 1986; Jayanth et al., 2002; Altug et al., 2012).

50.3.3 Mechanism of Mercury Resistance, Amplification of *merA* Gene, and Localization of *mer* Operon

The presence of *merA* gene and subsequently *mer* operon was confirmed in nine isolates, namely, PW-401, PW-702, PW-05, GW-502, BS-102, BW-201T, CS-09, PW-216, and RW-401. A strain of *Escherichia coli* DH5A harboring *mer* operon was kept as positive control in this experiment. The isolates showed a clear, distinct banding pattern at around 431 bp on agarose gel electrophoresis (Figure 50.2a). The band position of positive control was also found to be at the same region on the gel. This confirms the *mer*-mediated mercury resistance in the marine isolates mentioned earlier. Mere presence of *merA* gene and subsequently *mer* operon does not solve the purpose of mercury bioremediation. Rather, the *mer* operon should be functional to carry out the volatilization of mercury. The functionality of *mer* operon was confirmed by volatilization on x-ray film. The foggy areas developed on the film on corresponding wells containing the organisms confirm the volatilization of mercury and the functional nature of *mer* operon. Out of nine positive isolates harboring *mer* operon, eight of them, namely, RW-401, PW-401, PW-702, PW-05, GW-502, BW-201T, CS-09, and PW-216, showed positive volatilization by this assay (Figure 50.2b).

Plasmid and genomic DNA were used as the targets for the localization of the address of *mer* operon in the MRMB isolates. Out of the nine *mer* operon harboring MRMB isolates, three of them showed the presence of *mer* operon in their genomic DNA (BW-201T, BS-102, and PW-401), whereas in six isolates it was found to be present in the plasmid (PW-702, PW-05, GW-502, CS-09, PW-216, and RW-401). The harboring of plasmid-mediated *mer* operon was also confirmed in the isolates by plasmid curing.

There exist many mechanisms of mercury resistance in bacteria, which includes *mer* operon–mediated volatilization, mercury biosorption, bioaccumulation inside bacterial cell, or sequestration of mercury (Barkay et al., 2003). Out of all the possible resistance mechanisms, *mer* operon–mediated volatilization has been studied extensively. The major component of *mer* operon is *merA* gene, which is present in both broad-spectrum and narrow-spectrum operons (Schelert et al., 2004). It is responsible for the conversion of Hg^{2+} to Hg^0, which is the prerequisite step of mercury volatilization from both inorganic and organic mercury. *merA* gene amplification and volatilization of Hg^{2+} from the supplemented medium confirm the presence of a functional *mer* operon in the bacterial isolate (Leonhauser et al., 2007). PCR amplification confirms the presence of *merA* gene in 9 isolates out of 25 screened isolates.

Thus, a mere 36% of the isolates possess *mer* operon–mediated mechanism for mercury resistance, whereas the rest possess different mechanisms of mercury resistance in them. To a surprise, one of the *mer* positive isolates, that is, BS-102, did not show volatilization of mercury, confirming the presence of nonfunctional or incomplete *mer* operon in it. Many other studies have also confirmed the non-*mer*-mediated mercury resistance in bacteria (Mathema et al., 2011; Boyd and Barkay, 2012). However, the non-*mer*-mediated mercury resistance phenomenon is prevalent in marine bacterial isolates isolated from the Bay of Bengal, Odisha, India. In the search of the alternative mechanism of mercury resistance in the isolates, mercury bioaccumulation test was conducted by H_2S assay. The study confirms the significant accumulation of mercury in seven of the *mer* negative isolates.

50.3.4 Volatilization and Biosorption Potential of the Isolates

The percentage of mercury removal potential of the isolates in terms of either volatilization or biosorption has been determined by atomic absorption spectrophotometer (AAS).

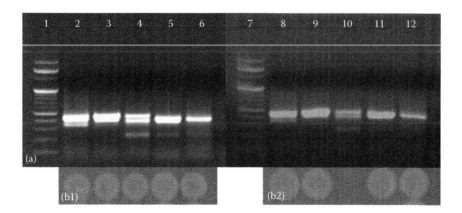

FIGURE 50.2 (a) Gel photograph of amplified *merA* gene after separation by agarose gel electrophoresis and visualization under UV in Gel Documentation System (Bio-Rad, USA) and (b1 and b2) Volatilization of inorganic mercury by *mer* operon–harboring mercury-resistant marine bacterial (MRMB) isolates on x-ray film. 1: 100 bp ladder, 2: BS-102, 3: BW-201T, 4: PW-05, 5: PW-702, 6: PW-216, 7: 100 bp ladder, 8: CS-09, 9: RW-401, 10: GW-502, 11: PW-401, 12: Positive control *E. coli* DU1040.

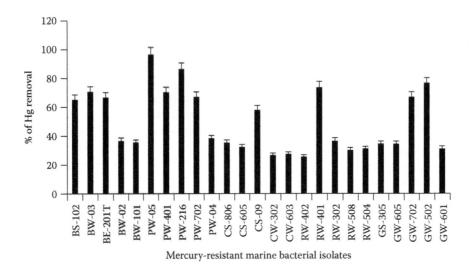

FIGURE 50.3   Comparative account of the mercury removal potential of the marine bacterial isolates.

Out of all the isolates, PW-05 emerged to be the most potential isolate among the volatilizing strains, whereas BW-03 was found to remove the highest percentage of mercury from the supplemented medium under *in vitro* conditions among bioaccumulating isolates (Figure 50.3). A wide variation in mercury removal efficiency was observed by the isolates, which varied from 96.67% (PW-05) to 30.4% (RW-508) in a strain-specific manner.

Mercury removal potential of the isolates showed a noteworthy pattern as most of the mercury-volatilizing isolates outclass the biosorbing isolates. The most potent isolate of mercury-volatilizing strain PW-05 was found to remove almost 98% of the supplemented mercury, whereas the highest removal by mercury biosorbing strain was found to be 70.99%. There exists a limited study on the biosorption of mercury by the bacterial isolates for the removal of mercury (Rehman et al., 2008). However, the biosorption potential of the marine isolates as found in the current study was at a much lower magnitude as compared to the isolates from industrial effluents. Many isolates of *Pseudomonas* spp. have been reported to remove 65%–75% of mercury by biosorption

from the medium supplement (Rehman et al., 2008). Thus, marine bacteria can be considered to remove mercury more efficiently by volatilization than mercury bioabsorption.

50.3.5   EXPRESSION PATTERN OF *merA* GENE BY qRT-PCR ANALYSIS

Under the same laboratory conditions and 10 ppm of initial mercury supplement, the *mer* operon–harboring MRMB showed a wide range of expression pattern of *merA* gene in them (Figure 50.4). The highest level of *merA* expression was observed in the case of PW-05 followed by PW-216, RW-401, PW-401, PW-702, BW-201T, BS-102, and CS-09.

50.3.6   PHYLOGENY OF MRMB ISOLATES BY 16S RRNA GENE SEQUENCING

A consensus result was obtained between the biochemical test results and the molecular identification of the isolates by 16S rRNA gene sequencing. As 16S rRNA gene sequence is conserved among bacterial species, sequencing of this gene

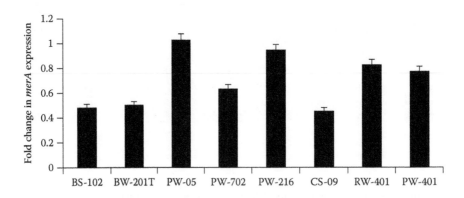

FIGURE 50.4   Expression pattern of *merA* gene of *mer* operon–harboring MRMB isolates grown on seawater nutrient agar medium supplemented with 10 ppm of mercury as $HgCl_2$.

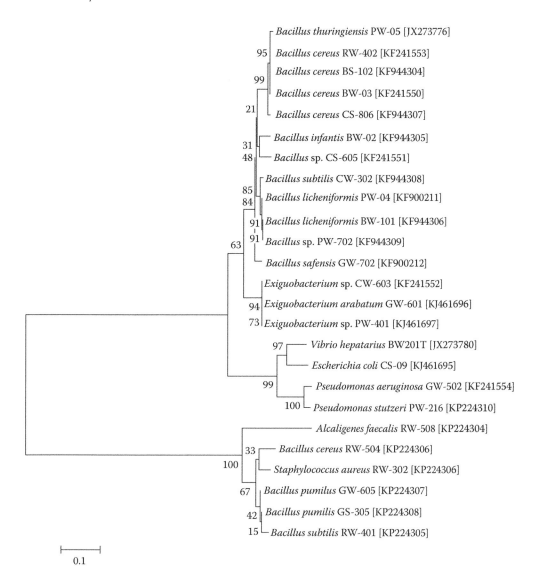

FIGURE 50.5 Phylogenetic relationship among the MRMB isolates from the study sites along the Bay of Bengal, Odisha, India.

is widely used for the identification of the bacterial species. be the most suitable target for bacterial identification practice. The phylogenetic relationship among the MRMB isolates was analyzed and interpreted in the form of phylogenetic tree (Figure 50.5) to reveal the genetic relatedness among them.

Phylogeny from 16S rRNA gene sequences revealed the abundance of Firmicutes along Odisha Coast of the Bay of Bengal as potential MRMB. The present study also determined the presence of diverse lineages and the presence of almost all common genera present in the collected samples. As the most diverse and even community, the marine samples had the highest numbers of *Bacillus* sp. Many mercury-resistant *Bacillus* spp. have also been reported from marine environments along the globe earlier (Nakamura and Silver, 1994; Gupta et al., 1999; Narita et al., 2003; Ramaiah and De, 2003; Dash and Das, 2014). Phylogeny from 16S rRNA sequence reveals the constitution of distinct clusters for different group of bacteria with Gram-positive and Gram-negative isolates forming their own groups, which include the isolates

from different as well as the same study sites. Similar results have also been observed by many previous works (Anukam et al., 2005; Case et al., 2007; Kim et al., 2011; Tripathi et al., 2013), which concludes the utility and usability of 16S rRNA gene sequences as a marker of phylogeny analysis of MRMB.

50.4 CONCLUSION

Marine environment, the most dynamic environmental condition, has witnessed pollution from varieties of toxic chemicals, including mercury. An increase in MRMB populations along the Bay of Bengal, Odisha, India corroborates with the increased level of mercury pollution in this region. All the MRMB isolates showed a higher level of mercury resistance along with resistance to other toxic metals and antibiotics. Though the mechanism of resistance to mercury was found to be by volatilization or biosorption in few isolates; however, the resistance mechanism could not be defined in many isolates, suggesting the presence of some novel mechanisms of mercury resistance in MRMB. The mercury removal efficiency of the isolates revealed that mercury volatilization is more effective than mercury volatilization, suggesting their application in mercury-contaminated environments.

REFERENCES

Altug, G., Gurun, S., Cardak, M., Ciftci, P. S., Kalkan, S. (2012) The occurrence of pathogenic bacteria in some ships ballast water incoming from various marine regions to the Sea of Marmara, Turkey. *Mar. Environ. Res.* **81**, 35–42.

Anukam, K. C., Osazuwa, E. O., Ahonkhai, I., Reid, G. (2005) 16S rRNA gene sequence and phylogenetic tree of *lactobacillus* species from the vagina of healthy Nigerian women. *Afr. J. Biotechnol.* **4**, 1222–1227.

Bafana, A., Krishnamurthi, K., Patil, M., Chakrabarti, T. (2010) Heavy metal resistance in *Arthrobacter ramosus* strain G2 isolated from mercuric salt-contaminated soil. *J. Hazard. Mater.* **177**, 481–486.

Barkay, T., Miller, S. M., Summers, A. O. (2003) Bacterial mercury resistance from atoms to ecosystems. *FEMS Microbiol. Rev.* **27**, 355–384.

Bauer, A., Kirby, W., Sherris, J., Turck, M. (1966) Antibiotic susceptibility testing by a standardized single disk method. *Am. J. Clin. Pathol.* **45**, 493–496.

Bhattacharyya, S., Basu, S., Chaudhuri, P., Santra, S. C. (2013) Assessment of mercury detoxification potentiality of isolated *Streptococcus* sp. MTCC 9724 under different environmental conditions. *Environ. Ecol. Res.* **1**, 62–72.

Boyd, E. S., Barkay, T. (2012) The mercury resistance operon: From an origin in a geothermal environment to an efficient detoxification machine. *Front. Microbiol.* **3**, 1–13.

Brandelli, A., Daroit, D. J., Riffel, A. (2010) Biochemical features of microbial keratinases and their production and applications. *Appl. Microbiol. Biotechnol.* **85**, 1735–1750.

Case, R. J., Boucher, Y., Dahllöf, I., Holmström, C., Doolittle, W. F., Kjelleberg, S. (2007) Use of 16S rRNA and *rpoB* genes as molecular markers for microbial ecology studies. *Appl. Environ. Microbiol.* **73**, 278–288.

Chait, R., Vetsigian, K., Kishony, R. (2012) What counters antibiotic resistance in nature? *Nat. Chem. Biol.* **8**, 2–5.

Chatziefthimiou, A. D., Crespo-Medina, M., Wang, Y., Vetriani, C., Barkay, T. (2007) The isolation and initial characterization of mercury resistant chemolithotrophic thermophilic bacteria from mercury rich geothermal springs. *Extremophiles* **11**, 469–479.

Chikere, C. B., Chikere, B. O., Okpokwasili, G. C. (2012) Bioreactor-based bioremediation of hydrocarbon-polluted Niger Delta marine sediment. *Nigeria 3 Biotech.* **2**, 53–66.

CLSI. (2006) *Clinical and Laboratory Standard Institute Methods for Dilution Antimicrobial Susceptibility Tests for Bacteria that Grow Aerobically*, 7th edn., Approved Standard M7-A7, CLSI, Wayne, PA.

Dash, H. R., Das, S. (2014) Assessment of mercury pollution through mercury resistant marine bacteria in Bhitarkanika mangrove ecosystem, Odisha, India. *Ind. J. Geo-Mar. Sci.* **43**, 1103–1115.

Dash, H. R., Mangwani, N., Chakraborty, J., Kumari, S., Das, S. (2013) Marine bacteria: Potential candidates for enhanced bioremediation. *Appl. Microbiol. Biotechnol.* **97**, 561–571.

Davies, J., Davies, D. (2010) Origins and evolution of antibiotic resistance. *Microbiol. Mol. Biol. Rev.* **74**, 417–433.

De, J., Ramaiah, N. (2007) Characterization of marine bacteria highly resistant to mercury exhibiting multiple resistances to toxic chemicals. *Ecol. Ind.* **7**, 511–520.

Deng, X., Wang, P. (2012) Isolation of marine bacteria highly resistant to mercury and their bioaccumulation process. *Bioresour. Technol.* **121**, 342–347.

Devillers, J., Steiman, R., Seigle-Murandi, F. (1989) The usefulness of the agar-well diffusion method for assessing chemical toxicity to bacteria and fungi. *Chemosphere* **19**, 1693–1700.

Felsenstein, J. (1985) Confidence limits on phylogenies: An approach using the bootstrap. *Evolution* **39**, 783–791.

François, F., Lombard, C., Guigner, J. M., Soreau, P., Brian-Jaisson, F., Martino, G., Vandervennet, M. et al. (2012) Isolation and characterization of environmental bacteria capable of extracellular biosorption of mercury. *Appl. Environ. Microbiol.* **78**, 1097–1106.

Gadd, G. M. (2010) Metals, minerals and microbes: Geomicrobiology and bioremediation. *Microbiology* **156**, 609–643.

Grimes, D. J., Atwell, R. W., Brayton, P. R., Palmer, L. M., Rollins, D. M., Roszak, D. B., Singleton, F. L., Tamplin, M. L., Colwell, R. R. (1986) The fate of enteric pathogenic bacteria in estuarine and marine environments. *Microbiol. Sci.* **3**, 324–329.

Groves, D. J., Short, H., Thewaini, A. J., Young, F. E. (1975) Epidemiology of antibiotic and heavy-metal resistance in bacteria: Resistance patterns in Staphylococci isolated from populations in Iraq exposed and not exposed to heavy metals or antibiotics. *Antimicrob. Agent Chemother.* **7**, 622–628.

Gupta, A., Phung, L. T., Chakravarty, L., Silver, S. (1999) Mercury resistance in *Bacillus cereus* RC607: Transcriptional organization and two new open reading frames. *J. Bacteriol.* **181**, 7080–7086.

Jayanth, K., Jeyasekaran, G., Shakila, R. J. (2002) Isolation of marine bacteria, antagonistic to human pathogens. *Ind. J. Mar. Sci.* **31**, 39–44.

Kim, M., Morrison, M., Yu, Z. (2011) Evaluation of different partial 16S rRNA gene sequence regions for phylogenetic analysis of microbiomes. *J. Microbiol. Met.* **84**, 81–87.

Knapp, C. W., McCluskey, S. M., Singh, B. K., Campbell, C. D., Hudson, G., Graham, D. W. (2011) Antibiotic resistance gene abundances correlate with metal and geochemical conditions in archived Scottish soils. *PLoS One.* **6**(11): e27300. DOI: 10.1371/journal.pone.0027300.

Kumar, S. R. S., Rao, K. V. B. (2012) In vitro antagonistic and hemolytic activity of marine Actinobacteria. *Int. J. Adv. Life Sci.* **3**, 7–12.

Le Cloirec, P., Andrès, Y. (2005) Bioremediation of heavy metals using microorganisms, pp. 97–140. In: Fingerman, M., Nagabhushanam, R. (Eds.), *Bioremediation of Aquatic and Terrestrial Ecosystems*, Science Publishers, Inc., Enfield, NH.

Leonhauser, J., Wang, W., Deckwer, W.D., Wagner-Dobler, I. (2007) Functioning of the mercury resistance operon at extremely high Hg(II) loads in a chemostat: A proteome analysis. *J. Biotechnol.* **132**, 469–480.

Mathema, V. B., Thakuri, B. C., Sillanpää, M. (2011) Bacterial *mer* operon-mediated detoxification of mercurial compounds: A short review. *Arch. Microbiol.* **193**, 837–844.

Mathiyazhagan, N., Natarajan, D. (2012) Metal and antibiotic tolerance potentiality of *Acidithiobacillus* spp and *Pseudomonas* spp from waste dumps of bauxite and magnesite mines. *Arch. Appl. Sci. Res.* **4**, 616–622.

Mindlin, S., Minakhin, L., Petrova, M., Kholodii, G., Minakhina, S., Gorlenko, Z., Nikiforov, V. (2005) Present-day mercury resistance transposons are common in bacteria preserved in permafrost grounds since the Upper Pleistocene. *Res. Microbiol.* **156**, 994–1004.

Mirzaei, N., Kafilzadeh, F., Kargar, M. (2008) Isolation and identification of mercury resistant bacteria from Kor River, Iran. *J. Biol. Sci.* **8**, 935–939.

Moghbeli, M., Shakeri, F., Hashemi-Moghaddam, H. (2011) Separation of mercury resistant bacteria from wastewater of milk, detergent and ceramic industry. *J. Chem. Health Risks* **1**, 19–22.

Møller, A. K., Barkay, T., Al-Soud, W. A., Sørensen, S. J., Skov, H., Kroer, N. (2011) Diversity and characterization of mercury-resistant bacteria in snow, freshwater and sea-ice brine from the High Arctic. *FEMS Microbiol. Ecol.* **75**, 390–401.

Monachese, M., Burton, J. P., Reid, G. (2012) Bioremediation and tolerance of humans to heavy metals through microbial processes: A potential role for probiotics? *Appl. Environ. Microbiol.* **78**, 6397–6404.

Naik, M. M., Pandey, A., Dubey, S. K. (2012) Bioremediation of metals mediated by marine bacteria. In: Satyanarayana, T., Prakash, A., Johri, B. N. (Eds.), *Bacteria in Environmental Management and Biotechnology*, Springer Verlag, New York.

Nakamura, K., Nakahara, H. (1988) Simplified x-ray film method for detection of bacterial volatilization of mercury chloride by *Escherichia coli. Appl. Environ. Microbiol.* **54**, 2871–2873.

Nakamura, K., Silver, S. (1994) Molecular analysis of mercury-resistant *Bacillus* isolates from sediment of Minamata Bay, Japan. *Appl. Environ. Microbiol.* **60**, 4596–4599.

Narita, M., Chiba, K., Nishizawa, H., Ishii, H., Huang, C. C., Kawabata, Z., Silver, S., Endo, G. (2003) Diversity of mercury resistance determinants among *Bacillus* strains isolated from sediment of Minamata Bay. *FEMS Microbiol. Lett.* **223**, 73–82.

Nies, D. H. (1999) Microbial heavy-metal resistance. *Appl. Microbiol. Biotechnol.* **51**, 730–750.

Pikis, A., Campos, J. M., Rodriguez, W. J., Keith, J. M. (2001) Optochin resistance in *Streptococcus pneumoniae*: Mechanism, significance, and clinical implications. *J. Infect. Dis.* **184**, 582–590.

Poulain, A. J., Chadhain, S. M. N., Ariya, P. A., Amyot, M., Garcia, E., Campbell, P. G. C., Zylstra, G. J., Barkay, T. (2007) Potential for mercury reduction by microbes in the high arctic. *Appl. Environ. Microbiol.* **73**, 2230–2238.

Ramaiah, N., De, J. (2003) Unusual rise in mercury-resistant bacteria in coastal environs. *Microb. Ecol.* 45, 444–454.

Rehman, A., Ali, A., Shakoori, A. R. (2008) Biosorption of mercury by bacteria, isolated from industrial effluents: Potential use in bioremediation of wastewater. *Pakistan J. Zool.* **40**, 115–122.

Saitou, N., Nei, M. (1987) The neighbour-joining method: A new method for reconstructing phylogenetic trees. *Mol. Biol. Evol.* **4**, 406–425.

Schelert, J., Dixit, V., Hoang, V., Simbahan, J., Drozda, M., Blum, P. (2004) Occurrence and characterization of mercury resistance in the hyperthermophilic archaeon *Sulfolobus solfataricus* by use of gene disruption. *J. Bacteriol.* **186**, 427–437.

Silver, S., Phung, T. (2005) A bacterial view of the periodic table: Genes and proteins for toxic inorganic ions. *J. Ind. Microbiol. Biotechnol.* **32**, 587–605.

Sotero-Martins, A., Jesus, M. S. D., Lacerda, M., Moreira, J. C., Filgueiras, A. L. L., Barrocas, P. R. G. (2008) A conservative region of the mercuric reductase gene (*merA*) as a molecular marker of bacterial mercury resistance. *Braz. J. Microbiol.* **39**, 307–310.

Thompson, J. D., Gibson, T. J., Plewniak, F., Jeanmougin, F., Higgins, D. G. (1997) CLUSTAL X windows interface: Flexible strategies for multiple sequence alignment aided by quality analysis tools. *Nucleic Acids Res.* **25**, 4876–4882.

Tripathi, P., Banerjee, G., Gupta, M. K., Saxena, S., Ramteke, P. W. (2013) Assessment of phylogenetic affiliation using 16S rRNA gene sequence analysis for *Pseudomonas aeruginosa* in patients of lower respiratory tract infection. *Ind. J. Med. Res.* **138**, 557–559.

Vetriani, C., Chew, Y. S., Miller, S. M., Yagi, J., Coombs, J., Lutz, R. A., Barkay, T. (2005) Mercury adaptation among bacteria from a deep-sea hydrothermal vent. *Appl. Environ. Microbiol.* **71**, 220–226.

Zhang, W., Chen, L., Liu, D. (2012) Characterization of a marine-isolated mercury-resistant *Pseudomonas putida* strain SP1 and its potential application in marine mercury reduction. *Appl. Microbiol. Biotechnol.* **93**, 1305–1314.

Zourari, A., Accolasa, J. P., Desmazeaud, M. J. (1992) Metabolism and biochemical characteristics of yogurt bacteria. A review. *Dairy Sci. Technol.* **72**, 1–34.

Index